CLINICAL FLOW CYTOMETRY
PRINCIPLES AND APPLICATION

CLINICAL FLOW CYTOMETRY

PRINCIPLES AND APPLICATION

EDITED BY

KENNETH D. BAUER, Ph.D.
Genentech, Inc.
South San Francisco, California

Associate Professor
Department of Pathology
Northwestern University Medical School
Chicago, Illinois

RICARDO E. DUQUE, M.D.
Clinical Assistant Professor
Department of Pathology
The University of Alabama at Birmingham
Department of Pathology
Carraway Methodist Medical Center
Norwood Clinic Inc.
Birmingham, Alabama

T. VINCENT SHANKEY, Ph.D.
Assistant Professor
Departments of Urology and Pathology
Loyola University Medical Center
Maywood, Illinois

WILLIAMS & WILKINS
BALTIMORE · HONG KONG · LONDON · MUNICH
PHILADELPHIA · SYDNEY · TOKYO

Editor: Charles W. Mitchell
Project Manager: Kathleen Courtney Millet
Copy Editors: Dominique Van de Stadt, Judith F. Minkove
Designer: Dan Pfisterer
Illustration Planner: Wayne Hubbel

Copyright © 1993
Williams & Wilkins
428 East Preston Street
Baltimore, Maryland 21202, USA

All rights reserved. This book is protected by copyright. No part of this book may be reproduced in any form or by any means, including photocopying, or utilized by any information storage and retrieval system without written permission from the copyright owner.

Printed in the United States of America

Library of Congress Cataloging-in-Publication Data

Clincical flow cytometry : principles and application / edited by
  Kennth D. Bauer, Ricardo E. Duque, T. Vincent Shankey.
    p. cm.
  Includes bibliographical references and index.
  ISBN 0-683-00480-8
    1. Flow cytometry—Diagnostic use. I. Bauer, Kenneth D.
II. Duque, Ricardo E. III. Shankey, T. Vincent.
  [DNLM: 1. Flow Cytometry.]
RB43.C57 1993
616.07'582—dc20
DNLM/DLC
for Library of Congress   92-13746
                                                                       CIP

92 93 94 95 96
1 2 3 4 5 6 7 8 9 10

To Students of Cytometry

Preface

Flow cytometry is emerging as an important adjunctive technology in the delivery of contemporary medicine. Applications span from straightforward enumeration of cell populations with high accuracy, to providing data needed to define disease processes. For example, this technology is used to establish the precise number of peripheral blood CD4+ T-lymphocytes to monitor the disease progression of AIDS, to determine the prognosis in patients with ductal breast carcinoma, or to classify lymphomas.

Flow cytometrists come from varied backgrounds, and operate with technology that requires some familiarity with a variety of disciplines. Individuals who utilize flow cytometry should understand elements of physics, cell biology, chemistry, and medicine. The need for a diverse background is reflected in *Clinical Flow Cytometry: Principles and Applications*. A working knowledge of these areas enhances the successful application of flow cytometry.

Although flow cytometry has grown considerably in its clinical application, more widespread use is limited by several factors. These include the lack of uniformly accepted guidelines for quality control and clinical interpretation. Important efforts are currently underway to provide guidelines for different applications (immunophenotyping, DNA content analysis, etc.), under the auspices of several agencies. The implementation of these guidelines should improve the consistency of results from different laboratories, providing a stronger rationale for the clinical utilization of flow cytometry.

Prospective studies with well-defined patient populations in terms of diagnosis, stage, and treatment will clarify current areas of controversy. The recent efforts of cooperative groups to include flow cytometry studies as part of randomized clinical trials represent a significant advance. These investigations should provide a more rational basis for the appropriate integration of flow cytometric data in clinical decision making.

Our goals are to stimulate the intellect of the reader interested in flow cytometry, to provide some insight into the current and emerging clinical applications, and to provide a critical appraisal of its clinical utility. The underlying assumption is that this effort is a "snapshot." Clearly, advances in cytometry, and in the multiple disciplines that impact on its application, will continue to make our task more demanding.

KENNETH D. BAUER
RICARDO E. DUQUE
T. VINCENT SHANKEY

Acknowledgments

ASSOCIATE EDITORS

Kenneth A. Ault, M.D., Maine Cytometry Research Institute, Portland, Maine
Raul C. Braylan, M.D., University of Florida College of Medicine, Gainesville, Florida
Zbigniew Darzynkiewicz, M.D., Ph.D., The Cancer Research Institute, New York Medical College, Valhalla, New York
David W. Hedley, M.D., Princess Margaret Hospital, Toronto, Ontario, Canada
Douglas E. Merkel, M.D., Northwestern University Medical School, Evanston Hospital, Evanston, Illinois
Peter S. Rabinovitch, M.D., Ph.D., University of Washington School of Medicine, Seattle, Washington

To ensure the overall high scientific quality of the information contained in this book, each chapter has been reviewed by at least one individual knowledgeable in the appropriate areas covered by that chapter. The editors gratefully acknowledge these individuals who have reviewed chapters for this book.

Michael Andreeff, M.D., Ph.D., M.D. Anderson Cancer Center, Houston, Texas
Bruce H. Davis, M.D., Harris Methodist Health Systems, Fort Worth, Texas
Phillip N. Dean, Ph.D., Lawrence Livermore National Laboratory, Livermore California
Thomas M. Ellis, Ph.D., Loyola University Medical Center, Maywood, Illinois
Len Erickson, Ph.D., Loyola University Medical Center, Maywood, Illinois
Jawad Fareed, Ph.D., Loyola University Medical Center, Maywood, Illinois
Charles Goolsby, Ph.D., Lakeside VA Hospital, Chicago, Illinois
Alice Gottleib, M.D., Ph.D., Rockefeller University, New York, New York
Chester Herman, M.D., Ph.D., Emery University Medical Center, Atlanta, Georgia
Robert A. Hoffman, Ph.D., Becton Dickinson Immunocytometry Systems, San Jose, California
James Jacobberger, Ph.D., Case Western Reserve University, Cleveland, Ohio
James M. Kozlowski, M.D., Northwestern University Medical School, Chicago, Illinois
Alan L. Landay, Ph.D., Rush Presbyterian-St. Luke's Medical Center, Chicago, Illinois
Lewis L. Lanier, Ph.D., DNAX Research Institue, Palo Alto, California
Dorothy E. Lewis, Ph.D., Baylor College of Medicine, Houston, Texas
Katharine A. Muirhead, Ph.D., Zynaxis Cell Science Inc., Malvern, Pennsylvania
Maria Pallavacini, Ph.D., University of California/San Francisco, San Francisco, California
Stephen C. Peiper, M.D., James Graham Brown Cancer Center, Louiseville, Kentucky
Michael B. Prystowsky, M.D., Ph.D., University of Pennsylvania, Philadelphia, Pennsylvania
Diether J. Recktenwald, Ph.D., Becton Dickinson Immunocytometry Systems, San Jose, California
Daniel H. Ryan, M.D., University of Rochester Medical Center, Rochester, New York
Larry Sklar, Ph.D., Los Alamos National Laboratory, Los Alamos, New Mexico
John Sleasman, M.D., University of Florida College of Medicine, Gainesville, Florida

Tapas Das Gupta, M.D., Ph.D., University of Illinois College of Medicine, Chicago, Illinois

Funds provided by the following corporations were used to offset some of the costs incurred in the organization and production of this book. The editors gratefully acknowledge the assistance provided.

Becton Dickinson Immunocytometry Systems, San Jose, California
Southern Biotechnology, Inc., Birmingham, Alabama
Coulter Cytometry, Hialeah, Florida
Ortho Diagnostics, Inc., Rahway, New Jersey
Cytometry Associates, Brentwood, Tennessee

Contributors

KENNETH A. AULT, M.D.
Research Director
Maine Medical Center Research Institute
South Portland, Maine

C. BRUCE BAGWELL, M.D., Ph.D.
Maine Medical Center Research Institute
South Portland, Maine
Verity Software House, Inc.
Tompsham, Maine

PETER H. BARTELS, Ph.D.
Professor of Optical Sciences
Optical Sciences Center
University of Arizona
Tucson, Arizona

KENNETH D. BAUER, Ph.D.
Department of Cell Analysis
Genentech, Inc.
South San Francisco, California
Associate Professor of Pathology
Department of Pathology
Northwestern University Medical School
Chicago, Illinois

ENRIQUE BECKMANN, M.D., Ph.D.
Director, Microbiology and Diagnostic Immunology
Department of Pathology
Humana Hospital—Michael Reese
Chicago, Illinois

RAUL C. BRAYLAN, M.D.
Professor of Pathology
Chief of Hematopathology
Department of Pathology and Laboratory Medicine
University of Florida College of Medicine
Gainesville, Florida

WAYNE O. CARTER, D.V.M.
Department of Veterinary Physiology and Pharmacology
Purdue University Cytometry Laboratories
Purdue University School of Veterinary Medicine
West Lafayette, Indiana

CHARLES V. CLEVENGER, M.D., Ph.D.
Assistant Professor
Division of Anatomic Pathology
Department of Pathology and Laboratory Medicine
University of Pennsylvania School of Medicine
Philadelphia, Pennsylvania

CEES J. CORNELISSE, Ph.D.
Department of Pathology
University of Leiden
Leiden, The Netherlands

JOHN D. CRISSMAN, M.D.
Professor and Chairman
Department of Pathology
Wayne State University School of Medicine
Harper Hospital
Detroit, Michigan

ZBIGNIEW DARZYNKIEWICZ, M.D., Ph.D.
Professor of Medicine
The Cancer Research Institute
New York Medical College
Valhalla, New York

BRUCE H. DAVIS, M.D.
Director of Hematopathology and Analytical Cytometry
Harris Methodist Health System
Fort Worth, Texas

PETER DEVILEE, Ph.D.
Department of Pathology
University of Leiden
Leiden, The Netherlands

RICARDO E. DUQUE, M.D.
Clinical Assistant Professor of Pathology
Department of Pathology
The University of Alabama at Birmingham
Department of Pathology
Carraway Methodist Medical Center
Norwood Clinic Inc.
Birmingham, Alabama

THOMAS M. ELLIS, Ph.D.
Associate Professor
Department of Medicine
Section of Hematology and Oncology
Director, FACS Core Laboratory
Loyola University Medical Center
Maywood, Illinois

JOHN F. ENSLEY, M.D., F.A.C.P.
Department of Internal Medicine
Division of Hematology-Oncology
Wayne State University
Detroit, Michigan

LAUREN A. ERNST, Ph.D.
Director of Reagents
Biological Detection Systems, Inc.
Pittsburgh, Pennsylvania

MICHAEL L. FRIEDLANDER, Ph.D., F.R.A.C.P.
Associate Professor of Medicine
Director, Department of Medical Oncology
Institute of Oncology
University of New South Wales
Prince Henry/Prince of Wales Hospitals
Sydney, Randwick, Australia

MACK J. FULWYLER, Ph.D.
Cell Biotech, Inc.
Santa Ana, California

SETH P. HARLOW, M.D.
Laboratory of Flow Cytometry
Roswell Park Cancer Institute
Buffalo, New York

DAVID W. HEDLEY, M.D.
Associate Professor of Medicine and Medical Biophysics
Departments of Medicine and Pathology
Princess Margaret Hospital
Toronto, Ontario, Canada

CHARLES L. HITCHCOCK, M.D., Ph.D.
Sarasota Pathology
Sarasota, Florida

ROBERT A. HOFFMAN, Ph.D.
Associate Scientific Director
Flow Cytometry Systems
Research Department
Becton Dickinson Immunocytometry Systems
San Jose, California

CARL H. JUNE, M.D.
Immune Cell Biology Program
Naval Medical Research Institute
Bethesda, Maryland

TERRANCE J. KAVANAGH, Ph.D.
Research Assistant Professor
Departments of Medicine and Environmental Health
University of Washington School of Medicine
Seattle, Washington

JAMES M. KOZLOWSKI, M.D., F.A.C.S.
Associate Professor of Urology, Surgery and Tumor Cell Biology
Director, Genitourinary Oncology Program
Northwestern University Medical School
Chicago, Illinois

AWTAR KRISHAN, Ph.D.
Division of Experimental Therapeutics
Department of Oncology
University of Miami Medical School
Miami, Florida

ALAN L. LANDAY, Ph.D.
Associate Professor
Department of Immunology/Microbiology, Pathology and Medicine
Director, Clinical Immunology
Director, Flow Cytometry
Rush Presbyterian-St. Luke's Medical Center
Chicago, Illinois

LEWIS L. LANIER, Ph.D.
DNAX Research Institute of Molecular and Cellular Biology
Department of Immunology
Palo Alto, California

DOROTHY E. LEWIS, Ph.D.
Associate Professor, Microbiology and Immunology
Director, Flow Cytometry Core Facility
Baylor College of Medicine
Houston, Texas

GERSHON Y. LOCKER, M.D.
Clinical Associate Professor of Medicine
Evanston Hospital
Evanston, Illinois

A. THOMAS LOOK, M.D.
Department of Hematology/Oncology
St. Jude Children's Research Hospital
Department of Pediatrics
University of Tennessee Center for the Health Sciences
Memphis, Tennessee

JOHN R. LURAIN, M.D.
John and Ruth Brewer Professor of Gynecology and Cancer Research
Head, Section of Gynecologic Oncology
Department of Obstetrics and Gynecology
Northwestern University Medical School
Chicago, Illinois

ROSEMARY MAZANET, M.D., Ph.D.
Instructor in Medicine
Harvard Medical School
Clinical Associate
Dana-Farber Cancer Institute
Boston, Massachusetts

ALVIN W. MARTIN, M.D.
Henry Vogt Cancer Research Institute of the James Graham Brown Cancer Center
University of Louisville
Louisville, Kentucky

THOMAS M. McHUGH, D.Sc.
Department of Laboratory Medicine
University of California at San Francisco
San Francisco, California

DOUGLAS E. MERKEL, M.D.
Assistant Professor of Medicine
Northwestern University Medical School
Evanston Hospital
Evanston, Illinois

KATHARINE A. MUIRHEAD, Ph.D.
Vice President, Research
Director, In Vivo Technologies
Zynaxis Cell Science Inc.
Malvern, Pennsylvania

STEPHEN C. PEIPER, M.D.
Henry Vogt Cancer Research Institute of the James Graham Brown Cancer Center
University of Louisville
Louisville, Kentucky

PETER S. RABINOVITCH, M.D., Ph.D.
Professor
Department of Pathology
University of Washington School of Medicine
Seattle, Washington

DIETHER J. RECKTENWALD, Ph.D.
Research Department
Becton Dickinson Immunocytometry Systems
San Jose, California

J. PAUL ROBINSON, Ph.D.
Associate Professor
Department of Veterinary Physiology & Pharmacology
Purdue University Cytometry Laboratories
Purdue University School of Veterinary Medicine
West Lafayette, Indiana

DANIEL H. RYAN, M.D.
Associate Professor
Department of Pathology and Laboratory Medicine
University of Rochester Medical Center
Rochester, New York

GARY C. SALZMAN, Ph.D.
Cell Growth, Damage and Repair Group
Life Sciences Division
Los Alamos National Laboratory
Los Alamos, New Mexico

ANTONIETA SAUERTEIG, M.Sc.
Division of Experimental Therapeutics
Department of Radiation Oncology
University of Miami Medical School
Miami, Florida

JUAN C. SCORNIK, M.D.
Professor of Pathology
Department of Pathology
University of Florida College of Medicine
Gainesville, Florida

T. VINCENT SHANKEY, Ph.D.
Assistant Professor
Departments of Urology and Pathology
Loyola University Medical Center
Maywood, Illinois

DAVID N. SHAPIRO, M.D.
Department of Hematology/Oncology
St. Jude Children's Research Hospital
Department of Pediatrics
University of Tennessee Center for the Health Sciences
Memphis, Tennessee

VINCENT TH. H.B.M. SMIT, M.D., Ph.D.
Department of Pathology
University of Leiden
Leiden, The Netherlands

LISA STAIANO-COICO, Ph.D.
Associate Professor
Department of Surgery
Cornell University Medical College
New York, New York

CARLETON C. STEWART, Ph.D.
Director, Flow Cytometry Laboratory
Roswell Park Cancer Institute
Buffalo, New York

JOHN L. SULLIVAN, M.D.
Professor of Pediatrics
Director, Division of Immunology/Rheumatology
Department of Pediatrics
Program of Molecular Medicine
University of Massachusetts Medical Center
Worcester, Massachusetts

EARL A. TIMM, JR.
Laboratory of Flow Cytometry
Roswell Park Cancer Institute
Buffalo, New York

SAMUEL G. TAYLOR IV, M.D.
Professor of Medicine
Division of Medical Oncology
Rush Presbyterian-St. Luke's Medical Center
Chicago, Illinois

DANIEL W. VISSCHER, M.D.
Assistant Professor of Pathology
Department of Pathology
Wayne State University School of Medicine
Harper Hospital
Detroit, Michigan

ROBERT F. VOGT, JR., Ph.D.
Division of Environmental Health Laboratory Sciences
Centers for Disease Control
Atlanta, Georgia

ALAN S. WAGGONER, Ph.D.
Professor of Biological Sciences
Department of Biological Sciences
Center for Light Microscope Imaging and Biotechnology
Carnegie Mellon University
Pittsburgh, Pennsylvania

DAVID S. WEINBERG, M.D., Ph.D.
Medical Director, Hematology Laboratory
Assistant Professor of Pathology
Brigham and Women's Hospital
Harvard Medical School
Boston, Massachusetts

JAMES C.S. WOOD, Ph.D.
Coulter Cytometry
Epics Division of Coulter Corporation
Miami Lakes, Florida

Contents

Preface .... *vii*
Acknowledgments .... *iv*
Contributors .... *xi*

PART I.
PRINCIPLES OF CLINICAL FLOW CYTOMETRY

SECTION A. THEORETICAL ASPECTS

1. DNA Content as a Genetic Marker of Cancer Cells 3
   CEES J. CORNELISSE, PETER DEVILEE, VINCENT TH. H.B.M. SMIT

2. The Cell Cycle: Application of Flow Cytometry in Studies of Cell Reproduction 13
   ZBIGNIEW DARZYNKIEWICZ

3. Theoretical Aspects of Flow Cytometry Data Analysis 41
   C. BRUCE BAGWELL

4. Flow Cytometry Reveals the Complexity and Diversity of the Human Immune System 63
   LEWIS L. LANIER

SECTION B. TECHNICAL ASPECTS

5. Clinical Flow Cytometry Instrumentation 71
   JAMES C.S. WOOD

6. Technical Considerations for Dissociation of Fresh and Archival Tumors 93
   CHARLES L. HITCHCOCK, JOHN F. ENSLEY

7. Fluorescence Reagents for Flow Cytometry 111
   ALAN S. WAGGONER, LAUREN A. ERNST

8. Practical Considerations for DNA Content and Cell Cycle Analysis 117
   PETER S. RABINOVITCH

9. Cytochemistry I: Cell Surface Immunofluorescence 143
   DOROTHY E. LEWIS

10. Cytochemistry II: Immunofluorescence Measurement of Intracellular Antigens 157
    CHARLES V. CLEVENGER, T. VINCENT SHANKEY

11. Quality Control for Clinical Flow Cytometry 177
    KATHARINE A. MUIRHEAD

PART II.
CLINICAL APPLICATION

SECTION C. APPLICATIONS IN CLINICAL ONCOLOGY

12. Lymphomas 203
    RAUL C. BRAYLAN

13. Acute Leukemias 235
    RICARDO E. DUQUE

14. Breast Cancer 247
    DAVID W. HEDLEY
    Clinical Commentary 260
    DOUGLAS E. MERKEL

15. Gynecological Cancer 263
    MICHAEL L. FRIEDLANDER
    Clinical Commentary 269
    JOHN R. LURAIN

16. Urological Cancers 271
    T. VINCENT SHANKEY
    Clinical Commentary 301
    JAMES M. KOZLOWSKI

17. Colorectal Neoplasia 307
    KENNETH D. BAUER
    Clinical Commentary 316
    DOUGLAS E. MERKEL, GERSHON Y. LOCKER

18. Upper Aerodigestive and Lower Respiratory Tract Tumors 319
    DANIEL W. VISSCHER, JOHN D. CRISSMAN
    Clinical Commentary 328
    SAMUEL G. TAYLOR, IV

19. Pediatric Solid Tumors 331
    DAVID N. SHAPIRO, A. THOMAS LOOK

20. Soft-Tissue Sarcomas 343
    ENRIQUE BECKMANN
    Clinical Commentary 356
    ROSEMARY MAZANET

21. Relative Applicability of Image Analysis and Flow Cytometry in Clinical Medicine 359
    DAVID S. WEINBERG

SECTION D. APPLICATIONS IN CLINICAL HEMATOLOGY AND IMMUNOLOGY

22. Flow Cytometric Analysis of Red Blood Cells 373
    BRUCE H. DAVIS

23. Flow Cytometric Analysis of Platelets 387
    KENNETH A. AULT

24. Flow Cytometric Analysis of Granulocytes   405
    J. PAUL ROBINSON, WAYNE O. CARTER

25. Role of Flow Cytometric Evaluation of Congenital and Acquired Immunodeficiencies 435
    ALAN A. LANDAY, JOHN L. SULLIVAN

26. Role of Flow Cytometry in Clinical Transplantation 449
    JUAN C. SCORNIK

SECTION E. EMERGING CLINICAL AND RESEARCH APPLICATIONS

27. Flow Cytometric Monitoring of Cellular Resistance to Cancer Chemotherapy 459
    AWTAR KRISHAN, ANTONIETA SAURTEIG

28. Cell-Associated Receptor Quantitation 469
    ROBERT A. HOFFMAN, DIETHER J. RECKTENWALD, ROBERT F. VOGT, JR.

29. Detection of Minimal Residual Disease by Flow Cytometry 479
    DANIEL H. RYAN

30. Flow Cytometric Assays of Cell-Mediated Cytotoxicity 497
    THOMAS M. ELLIS

31. Measurements of Cell Physiology: Ionized Calcium, pH, and Glutathione 505
    PETER S. RABINOVITCH, CARL H. JUNE, TERRANCE J. KAVANAGH

32. Microsphere-Based Fluorescence Immunoassays Using Flow Cytometry Instrumentation 535
    THOMAS M. McHUGH, MACK J. FULWYLER

33. Flow Cytometry in Skin Disorders 545
    LISA STAIANO-COICO

34. Molecular Biology and Flow Cytometry I: Analytical Considerations 557
    SETH P. HARLOW, EARL J. TIMM, JR., DOROTHY E. LEWIS, CARLETON C. STEWART

35. Molecular Biology and Flow Cytometry II: Clinical Potential 573
    STEPHEN C. PEIPER, ALVIN W. MARTIN

36. Expert Systems for Cytometry Data Analysis 585
    GARY C. SALZMAN, PETER H. BARTELS

Appendix: Workshop-Defined Cluster Groups for Monoclonal Antibodies to Human Leukocytes ... 603

Index 609

PART I.

Principles of Clinical Flow Cytometry

SECTION A. THEORETICAL ASPECTS

1

DNA Content as a Genetic Marker of Cancer Cells

CEES J. CORNELISSE, PETER DEVILEE and VINCENT TH. H.B.M. SMIT

INTRODUCTION

Cancer is a disease in which cells escape the growth control mechanisms needed to maintain the normal structure and function of tissues and organs of multicellular organisms. This leads to uncontrolled cell proliferation and invasion into surrounding tissues and, subsequently, to dissemination into distant organs by a process called metastasis. With time, tumors tend to become more malignant, a process known as tumor progression (1, 2). This phenomenon implies that cancer may evolve stepwise through a number of stages, beginning as a relatively benign growth and ending up as a highly malignant tumor. The clinical and biological events of tumor progression are thought to represent the result of the sequential selection of variant subpopulations within an expanding, genetically unstable neoplastic clone (clonal evolution) (3).

DNA-content changes occur frequently in solid tumors and are the most global reflection of the chromosomal and subchromosomal genetic changes that play a key role in tumor development and progression. DNA-content changes in hematological malignancies are, in general, less pronounced than in solid tumors, which parallels the less extensive cytogenetic aberrations in these neoplasms. Flow cytometry is now widely applied to analyze DNA-content distributions of clinical and experimental tumors in order to relate this to clinical and biological parameters of tumor behavior. The basis for DNA cytometry has been laid in the 1930's by Torbjörn Caspersson (4). He pioneered the measurement of DNA and RNA in single cells with microspectrophotometers that he himself built and made eminent contributions both to the field of analytical cytology and to cytogenetics.

In this chapter, we will give an overview of the current knowledge about the genetic mechanisms involved in cancer development, with particular emphasis on solid tumors, and discuss their possible relationship with DNA-ploidy changes.

GENETIC FACTORS IN TUMORIGENESIS
Cancer is a Genetic Disease of Somatic Cells

The concept that cancer is a genetic disease of somatic cells has now gained wide support from studies on the molecular biology and molecular (cyto-)genetics of cancer cells. Tumorigenesis is considered to be the end-result of multiple genetic accidents in somatic cells, ultimately resulting in autonomous cell growth and metastasis (3, 5, 6). Evidence for the somatic mutation theory is presented by the frequent occurrence of chromosomal aberrations in cancer cells (7). In 1914, more than 40 years before the correct number of chromosomes in man was established, Boveri speculated that such chromosomal abnormalities might bear a casual relationship to neoplasia (8). The hypothesis was proven to be correct with the advent of chromosome analysis techniques, the classical example being the discovery of the Philadelphia chromosome (Ph^1) as a specific genetic abnormality in chronic myeloid leukemia (CML), (9, 10). This so-called Ph^1 chromosome is actually the product of a reciprocal translocation between the long arms of chromosomes 9 and 22 (Fig. 1.1).

In contrast to hematological malignancies, solid tumors often show complex karyotypes and, therefore, the identification of tumor-specific chromosomal changes is much more difficult. This notwithstanding, cytogenetic analysis of solid tumors has been helpful in locating genes that are associated with carcinogenesis (11, 12). The identification and description of these genes (i.e., protooncogenes and tumor suppressor genes) in the human genome and, in particular, their alterations in human tumors provide yet further proof that cancer is a genetic disease.

DNA aberrations in cancer cells can be studied at different levels of genomic organization. First, the total amount of DNA in cancer cells can rapidly be evaluated by either flow or image cytometric techniques. Second, if cells enter the mitotic phase of the cell cycle, the DNA is organized into chromosomes that can be studied by different cytogenetic strategies. Finally, molecular-genetic analysis enables us to investigate tumor cells at the level of individual genes, even at the level of single base pairs.

Genetic Analysis at the Level of the Chromosome

Two major steps forward in the field of solid-tumor cytogenetics have been the development of chromosome banding and improved cytogenetic culture techniques (13, 14). Before the advent of banding techniques, chromosomes could only be classified according to their length, their centromere position, and the presence of secondary constrictions. The specificity of the banding patterns, which provide a kind of "bar code," permits the identification of every individual chromosome. Among the many types of banding

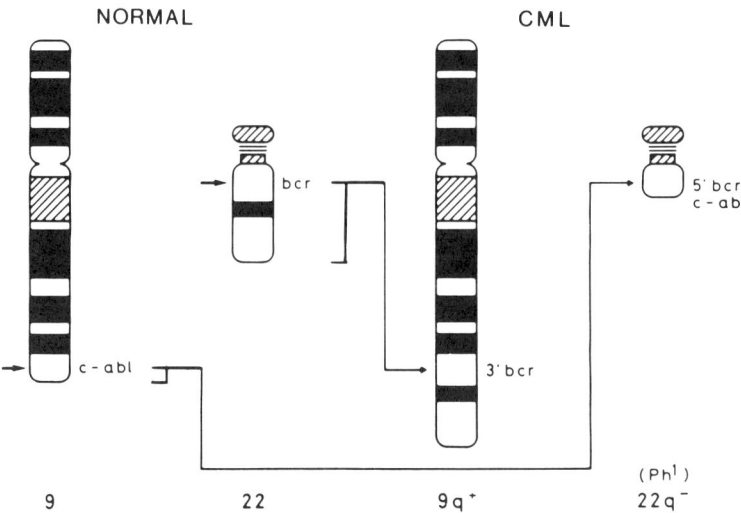

Figure 1.1. A schematic representation of the translocation between chromosomes 9 and 22 resulting in the 9q+ chromosome and the Philadelphia chromosome (22q−).

techniques that have been described, Giemsa trypsin (G)-banding, reverse (R)-banding, Q-banding and C-banding have been the most commonly used in solid-tumor cytogenetics to date (15). In cancer cells, a deviation of the normal diploid chromosome set (n = 46) is often found. In principle, these cytogenetic changes concern the ploidy (number) or the morphology (structure) of chromosomes and/or a combination of both aberrations. Losses (monosomies) or gains (trisomies and polysomies) belong to the first category, while rearrangements within one chromosome (deletions, duplications, inversions) or rearrangements between two or more chromosomes (translocations) belong to the latter. In addition, certain chromosomal regions may become amplified and may show up as homogeneously-stained regions (HSR's) on a chromosome arm or as extrachromosomal "double minutes."

Since the classification of DNA-ploidy abnormalities of tumor cell populations is derived from the cytogenetic nomenclature, it is relevant to briefly summarize the main rules for the cytogenetic classification of tumor cell populations (16, 17). In short, cell populations are considered to be of clonal origin (derived from a single progenitor) when a number of cells have the same or closely related abnormal chromosome complements. A tumor cell population may be characterized further by its modal chromosome number, whereas the stemline indicates the most frequent chromosome constitution. Depending on the modal chromosome number, cell populations may be classified cytogenetically as (near)-diploid, triploid, (hypo)-tetraploid etc.

The already mentioned complexity of karyotypes of solid tumors often makes it difficult to decide between primary changes, which are casually related to tumorigenesis, and secondary changes, which are related to tumor progression. The specificity of chromosome changes in carcinogenesis is illustrated by the consistent involvement of only 71 of the 329 chromosomal bands of the human genome in primary, neoplasia-associated rearrangements (18). Moreover, many of the cellular oncogenes that have been localized to specific chromosomes, localize with these neoplasia-associated breakpoints (18). Comparative chromosome analysis of tumors from unrelated patients has helped to pinpoint important cytogenetic alterations in a number of solid-tumor types (Table 1.1). Although carcinomas account for the greatest proportion of malignant disease, they represent only 15% of the karyotypic data (16).

Several factors may account for this discrepancy. First, it is difficult to obtain good quality chromosome preparations due to technical problems associated with the cytogenetic analysis of solid tumors, including low mitotic indices, overgrowth by normal fibroblasts present in the tumor tissues, in-vitro selection, and probably artificial induction of cytogenetic variants due to culturing (19). Second, the karyotypes of tumor cells frequently show high modal numbers with many bizarre marker chromosomes, which possibly reflects karyotypic evolution, but which, at the same time, obscures primary specific events. Moreover, the limited resolution of the banding techniques and the lack of molecular genetic information (morphology-oriented) has further frustrated progress in solid-tumor cytogenetics.

The recent development of nonradioactive in situ hybridization techniques (ISH) presents exciting prospects in all areas of cytogenetics, including cancer cytogenetics (20, 21). These techniques are based on the use of haptenized, cloned, chromosome-specific DNA sequences as probes for the in situ detection of the cognate target sequences in metaphase chromosomes or interphase nuclei. The implementation of ISH in karyotype analysis can be of great help in improving the classification of chromosomes from solid-tumor cells. In particular, the use of DNA probes that recognize parts of or entire chromosomes, e.g., cosmid-probes and chromosome-specific libraries in combination with multiple-hybridization procedures, will soon be incorporated into solid-tumor cyto-

Table 1.1
Recurrent chromosomal changes in human solid tumors

Tumors	Chromosome Changes
Benign	
Meningioma	-22, 22q
Pleomorphic adenoma	t(3;8)(p21;q12), t(9;12)(p13$-$22; q13$-$15)
Lipoma	t(12;?)(q14;?)
Colonic adenomas	1, 7, $+8$, 12q$^-$13, (structural and numerical)
Carcinomas	
Bladder	i(5p), $+7$, $-9/9q^-$, 11p$^-$, 17p$^-$
Prostate	del(7)(q22), del(10)(q24)
Small-cell lung cancer	del(3)(p14p23)
Colon	$+7$, $+8$, $+12$, $-17/17p^-$, -18
Kidney	del(3)p11$-$p21)
Breast	t(1q), -8, -13, t/del (16q)
Ovary	1 (structural changes), t(6;14)(q21; q24),6q$^-$, $-X$
Sarcomas	
Liposarcoma (myxoid)	t(12;16)(q13;p11)
Synovial sarcoma	t(X;18)(p11.2;ql1.2)
(Alveolar) rhabdomyosarcoma	t(2;13)(q37;q14)
Embryonal and other	
Testicular seminoma, ovarian Dysgerminoma, (germ-cell tumors)	i(12p)
Retinoblastoma	del(13)(q14)/-13, i(6p)
Wilms tumor	del(11)(p13)
Neuroblastoma	del(1)(p31$-$32)
Melanoma	t/del(1)(p12$-$p22), t(1;19)(q12; q13), t/del(6q)/i(6p), $+7$
Mesothelioma	del(3)(p13$-$p23)
Ewings' sarcoma/peripheral neuroepithelioma	t(11;22)(q24;q12)

genetic studies (22, 23). In our opinion, a revolutionary contribution of this technique in this field is to be expected, due to the possibility to combine specific molecular and topological information.

An additional advantage of in situ hybridization techniques is the possibility to evaluate the signals in interphase nuclei ('interphase cytogenetics') (21, 24, 25). However, the accuracy and reliability of signal interpretation in interphase tumor nuclei are restricted by the size and kind of the target DNA. Centromere–specific and region–specific DNA probes are far more suitable for interphase cytogenetics than chromosome–specific libraries. The latter class of probes most often gives a blurred staining pattern of chromosomal domains whose signals may be difficult to interpret. Therefore, in solid tumors, interphase cytogenetics is performed best in situations for which specific karyotypic and/or molecular genetic changes are known. Once these specific genetic alterations have been identified, interphase cytogenetics is very appropriate to increase our insight of the (cyto-)genetic heterogeneity of the tumor cell populations for the particular chromosomal aberration. Even without knowledge about the specific aberrations, interphase cytogenetics is still a more sensitive technique for detecting aneuploidy in solid tumors than DNA flow cytometry, although the latter is much faster (25).

Genetic Analysis at the Gene Level
ONCOGENES

At present, two classes of genes are known to be involved in the development and progression of cancer: viz., protooncogenes and tumor suppressor genes. Protooncogenes are normal cellular genes involved in positive growth control. Mutations can turn protooncogenes into activated oncogenes that drive the growth of tumor cells. At present, the molecular biology of some 70 protooncogenes has been elucidated to some degree. Their gene products are localized in different cell compartments and function as growth factors, growth factor receptors, components of the intracellular signal transduction system, cell cycle control proteins, and transcription factors (26–28). In general, the changes that activate the cellular protooncogenes to become true oncogenes result from the unregulated increase of the protein or gene product or unscheduled expression of the protein or gene product or from mutations or deletions in the gene's regulatory domains.

Activation of rat sarcoma genes (ras) is found in human cancers more often than activation of any other oncogene and ras is one of the most investigated family of protooncogenes. To date, three ras genes, N-ras, H-ras and K-ras, have been identified in the human genome (29). They all encode 21 kD GTP binding proteins with GTPase activity (30) that are localized to the inner cell membranes. These are similar to the various G-proteins that interact with hormone and neurotransmitter receptors to transduce information via second messengers, such as cyclic AMP (31). The ras protooncogene becomes activated by point mutations in one of the critical codons (12, 13, 59, or 61) of the gene (31). The resulting amino acid alteration of p21 leads to a prolongation of the activated state of the ras-GTP complex. Screening human tumors for the presence of a mutated ras gene has revealed that the incidence of ras gene mutations varies widely between different tumor types (31). The incidence of ras mutations probably indicates the relative importance of this genetic event in the carcinogenesis of these tumor types.

A second mechanism by which a protooncogene can be turned into an oncogene is transposition by chromosomal translocation. The already mentioned specific chromosomal aberration in CML, viz., Philadelphia chromosome, results from such a translocation event. In the past decade, it has been demonstrated that the breakpoint on chromosome 9 is within the c-abl oncogene and that the breakpoint on chromosome 22 is within a small region of a gene on chromosome 22 called "breakpoint cluster region" (bcr) (Fig. 1.1). This translocation produces a hybrid gene (5' bcr segment and truncated c-abl gene lacking the first exon), encoding an aberrant protein (p210) with increased tyrosine kinase activity (32, 33).

Finally, gene amplification is an important mechanism that can lead to oncogene overexpression. Amplification of the N-myc oncogene is frequently observed in grade III and IV neuroblastomas and may be associated with the poor prognosis of these tumors (34). Amplification of the HER-2/c-erB/neu oncogene is found in approximately 20% to 30% of the breast and ovarian carcinomas (35, 36).

TUMOR SUPPRESSOR GENES

Tumor suppressor genes form a more recently discovered class of cancer genes that appear to play a role particularly in the development of solid tumors. Functional inactivation of wild-type tumor suppressor genes by mutation releases the negative growth control exerted by these genes. The classical example of a tumor suppressor gene is the Rb1 gene involved in sporadic and hereditary retinoblastoma, a malignancy of early childhood. Hereditary retinoblastoma presents at a younger age than the sporadic form and is more often bilateral and multifocal. Based on the difference in incidence, Knudson postulated a recessive mechanism for the development of retinoblastoma requiring two genetic events or "hits" (37). The molecular-genetic basis of this mechanism in retinoblastoma has now been unravelled and appears to involve the functional inactivation of both copies of a gene in the q14 band of chromosome 13. The first "hit" is a germline (hereditary tumors) or somatic mutation (sporadic tumors) inactivating one of the two alleles of the gene. The second "hit" is a somatic event that inactivates the remaining wildtype allele. The high probability that such a second event will occur in individuals carrying a germ-line mutation accounts for the earlier onset of hereditary disease. Functional inactivation of the remaining wildtype allele appears to occur predominantly by elimination of the chromosome region carrying the gene. Chromosomal mechanisms like nondisjunction, mitotic recombination, or gene conversion may account for the elimination of the wildtype allele and its replacement with a duplicated copy of the mutant gene. Since this usually will also include DNA sequences flanking the tumor suppressor gene, tumors often show loss of heterozygosity (LOH) for polymorphic DNA markers mapping to this region. LOH can be detected by comparing restriction fragment length polymorphisms of tumor DNA with that of the patients constitutive DNA (extracted from normal cells, e.g., peripheral blood leukocytes) on Southern blots (Fig. 1.2). The probability of an independent loss of the second copy of the gene is estimated to be a factor of 100–1000× lower (38). It has been possible to revert the transformed phenotype of retinoblastoma cells by transfering the wildtype gene, which gives a formal proof for the essential role of the mutated Rb gene in retinoblastoma tumorigenesis (38).

Tumor suppressor genes are now implicated in variety of solid tumors. Evidence for this has been obtained largely by LOH studies. The detection of nonrandom LOH at certain chromosome regions strongly implicates these as putative tumor suppressor gene loci. At present, several tumor suppressor genes have been cloned and characterized (Table 1.2). The p53 tumor suppressor gene is now emerging as a major cancer gene because of its frequent and widespread involvement in human malignancies. In

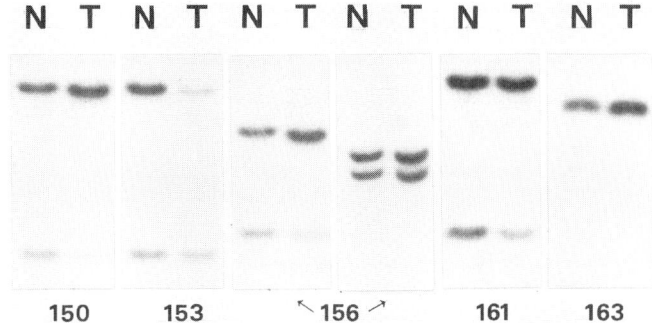

Figure 1.2. Southern blot analysis of matching normal (lymphocyte) DNA (*N*) and tumor DNA (*T*) of five breast cancer patients. Patient identification numbers are below each pair of lanes. The probe p144-D6 (D17S34), which was assigned to the short arm of chromosome 17, was informative in four patients and showed loss of heterozygosity (LOH) in all cases (150, 153, 156, 161). Tumor 153 (right lane) shows loss of the upper allele, whereas the lower allele is lost in tumors 150, 156 and 161. Due to the presence of contaminating normal cells a faint band remains visible in 153, 156 and 161. In one patient (163), the D17S34 probe was not informative. A second probe CRI-L427 (D21S112), which was assigned to 21q, was used in one patient (156/right panel) and this probe did not show LOH at this locus.

Table 1.2
Chromosomal loci associated with cancer

Chromosomal Loci	Associated Tumor Suppressor Gene	Type of Cancer
1p		melanoma, neuroblastoma, medullary thyroid carcinoma, pheochromocytoma, MEN-2, ductal breast carcinoma
1q		breast carcinoma
3p		small-cell lung carcinoma, renal cell carcinoma, cervical carcinoma, von Hippel-Lindau disease
5q	DP2.5	familial adenomatous polyposis
5q	MCC	colorectal cancer
6q		melanoma, breast cancer
9q		bladder carcinoma
10q		astrocytoma, MEN-2
11p	WT1	Wilms tumor, breast carcinoma, rhabdomyosarcoma, hepatoblastoma, bladder carcinoma
11q		MEN-1
13q	RB1	retinoblastoma, osteosarcoma, SCLC, ductal breast carcinoma, colon carcinoma, bladder carcinoma, stomach carcinoma
17p	p53	SCLC, colorectal carcinoma, breast carcinoma osteosarcoma, ovarian carcinoma, bladder cancer
17q	NF1	neurofibromatosis (NF-1)
17q		breast carcinoma
18q	DCC	colorectal carcinoma
22q		NF-2, meningioma, acoustic neuroma, pheochromocytoma

contrast to the Rb1 gene, the mechanism of this gene in oncogenesis is more complex than pure loss of function by deletion and/or mutation. Mutant p53 protein can functionally inactivate the wildtype gene product by forming mixed complexes. Evidence is accumulating that certain point mutations can fix the p53 protein into a conformation state

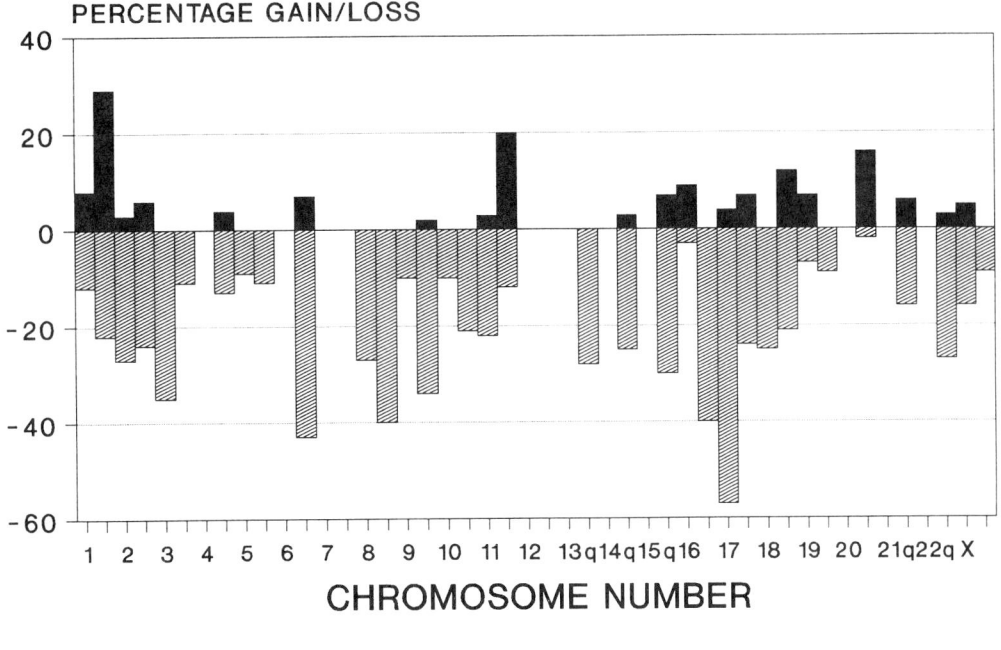

Figure 1.3. Allelic imbalance in 86 breast cancers. Frequency of allelic *gain* (solid bars) and allelic *loss* (hatched bars) as observed on individual chromosome arms in 86 breast tumors. Note that the long arm of chromosomes 1 and 6 and the short arm of chromosome 17 are most frequently involved.

that actually has a growth-promoting effect (40). Mutant p53 protein has a prolonged lifetime and can be demonstrated by immunocytochemical techniques in about 70% of the common cancer types (41). Certainly, as the list of loci showing nonrandom LOH is still growing, we can expect that more tumor suppressor genes will be identified and cloned within the next few years.

Accumulation of Genetic Events in Solid Tumorigenesis

Molecular-genetic studies by Vogelstein et al., clearly indicate that the adenoma-carcinoma sequence in colorectal carcinoma is characterized by an accumulation of distinct genetic events, including oncogene activation and tumor suppressor gene inactivation (42). At least four different tumor suppressor genes and the K-ras protooncogene have been implicated in colorectal carcinogenesis (44).

Similarly the accumulation of multiple genetic lesions appears to be needed for the development of several other solid tumor types, such as breast cancer, for example. Evidence for this can be derived from LOH studies showing loss of chromosomal material at multiple loci within one tumor (43). To get a better estimate of the extent of allelic loss it is necessary to screen all chromosome arms with at least one polymorphic DNA marker. In this way, a molecular counterpart of the karyotype is obtained for which Vogelstein et al. introduced the term "allelotype" (44). Allelotypes have been published for several tumor types, including breast carcinoma, hepatocellular carcinoma, renal cell carcinoma, ovarian carcinoma, and malignant astrocytoma (45–49). Figure 1.3 shows the allelotype of a series of 86 primary breast carcinomas obtained in our laboratory (50). Apart from chromosomes 17p, 1q, and 6q showing allelic imbalance in more than 50% of the informative cases, another 10 chromosome arms showed allelic imbalance in 30% to 40% of the informative cases. The loci at 17p, 13q, and 18q harbor known tumor suppressor genes like p53, Rb1, and the DCC (deleted in colorectal cancer) gene. Point mutations of the p53 gene in breast cancer have been reported (51, 52). Although the expression of Rb1 has been found to be altered in some breast tumors, (53) further studies are needed to establish both Rb1 and DCC as the actual targets for LOH in breast cancer.

It is remarkable that a number of loci appear to be involved in quite different types of solid tumors (Table II). Whether this indicates that the same tumor suppressor genes are also involved in these different types of cancer awaits futher molecular-genetic analysis. Also, it is not clear whether all allelic imbalances involve putative tumor suppressor gene loci or whether, in some cases, they may reflect increased genomic instability of the tumor cell population. As a quantitative parameter for the multiplicity of allelic losses within one tumor, Vogelstein et al. have introduced the term "fractional allelic loss (FAL)" (44), which represents the fraction of evaluable chromosome arms showing allelic loss in an individual tumor. Vogelstein and his coworkers found high fractional allelic loss to be associated with poor prognosis (44). An association between high fractional allelic loss and clinico-pathological features of tumor aggressiveness was also reported for malignant astrocytoma, renal cell carcinoma, and ovarian carcinoma (47–49). For breast cancer such as association has not been convincingly demonstrated as yet (56).

RELATION BETWEEN MOLECULAR-GENETIC CHANGES AND TUMOR PLOIDY

Model for the Genetic Evolution of Solid Tumors

In spite of the widespread occurrence of DNA-content changes (DNA aneuploidy) in solid tumors and its reported association with clinico-pathological signs of tumor progression in a number of tumor types, little is known about the forces that drive the evolution of discrete aneuploid tumor stemlines. It seems likely that the process of aneuploidy evolution is accompanied or generated by a relative increase in genomic instability. However, the presence of discrete DNA-aneuploid stemlines in tumors indicates that the neoplastic clones eventually regain genetic stability to some extent.

Which forces drive the genetic evolution of solid tumors? This question has been addressed recently by Shackney et al., who developed a computer model to simulate the time course of changes that accompany the spontaneous neoplastic transformation of mouse fibroblasts in vitro (55). The model includes processes such as cell growth, cell division, tetraploidization, chromosome loss, and the development of (growth-promoting) structural chromosomal aberrations that have been represented using Monte Carlo simulation techniques.

Certainly, tetraploidization is not the only mechanism by which cells can increase their chromosome copy number. Mitotic nondisjunction is a second common mechanism giving rise to aneuploid cells that has been considered in the model, although not explicitly. Moreover, this seems to be a likely pathway for the development of hyper- and hypodiploid stemlines. A recent study of Giaretti and Santi clearly invokes an early role of abnormal mitotic processes, generating two daughter cells with unequal DNA content, in the ploidy evolution pathway of colorectal cancer (61). They observed a remarkably higher frequency of near-diploid clones in adenomas with mild-to-moderate dysplasia (70%) as compared with that in carcinomas with moderate-to-poor differentiation (12%). Frequency distributions of DNA indices of large series of solid tumors such as breast cancer, colorectal cancer, ovarian cancer, bladder cancer, etc., typically show a bimodal pattern, with clustering both in the (hyper-)diploid and in the (hypo-)tetraploid range, but with a minimum clustering in the triploid range (56, 59, 60–62). Papillary thyroid carcinomas are frequently (near-)diploid, indicating that tetraploidization does not play an important role in the genetic evolution of this tumor type (63). Interestingly, feline and canine mammary carcinomas only infrequently show hypotetraploid stemlines, which reveals a species difference in the genetic evolution of tumors of the same histogenetic origin (64, 65).

Whereas in the computer simulations of Shackney et al., tetraploidized cell populations could gradually drift to a near-diploid mode by continuous chromosome loss, this does not seem to occur frequently in vivo. The paucity of DNA indices in the triploid range argues against such an extensive loss of chromosomes after tetraploidization. Apparently the range above which tetraploidized cells can segregate chromosomes without endangering cell viability is more limited than is predicted by the model. The balance between the genomic instability of an expanding tumor-cell clone (genetic divergence) and the selection forces acting on this population of cells (genetic convergence) will largely determine whether progressive ploidy evolution leading to discrete aneuploid stemlines will occur (66).

Genomic Instability and Development of Aneuploidy

Although genomic instability has been attributed a central role in the process of clonal evolution, its causes are poorly understood (66, 67). Apart from a possible destabilizing effect caused by increased chromosome number, several other mechanisms for genetic destabilization have been suggested. Hypomethylation as well as mutations in specific genes that are components of and regulate the mitotic machinery have been implicated (68). Recently, it has been suggested that the genomic stability of somatic cells is not simply a default position but maybe the result of active checks and balances (69).

In view of the increased genomic instability of tumor cells in general, it is remarkable that DNA-aneuploid stemlines can remain stable over many years. Evidence for this has been obtained by comparing the DNA-ploidy status of the primary tumors with their metastases (70). Although this observation indicates a limited genetic evolution in metastases, it should be emphasized that a discrete aneuploid-tumor cell population with a low coefficient of variation by DNA flow cytometry does not imply a 100% cytogenetic homogeneity and/or stability. Aside from the presence of low-frequency subpopulations with DNA contents scattered around the main G_0/G_1 peak or in the polyploid range, cells within this population can exhibit considerable karyotypic variability below the detection limit of DNA flow cytometry.

A corollary of the stochastic nature of the process of karyotype evolution is the low probability that two independent tumors in the same patient will have a parallel ploidy evolution, with chromosomal loss, gain, and structural rearrangements that ultimately will result in near-identical aneuploid DNA index. These assumptions imply that DNA flow cytometry can be used to probe the inter-relatedness of synchronously occurring or metachronous tumors thought to represent separate primary tumors by DNA flow cytometry (62, 71).

Oncogene Activation and Aneuploidy

Activated protooncogenes would be likely candidates for the dosage-dependent, growth-promoting, structural chromosome abnormalities required in the Shackney model for the evolution of aneuploid stemlines. Indeed, in experimental tumorigenesis models a distinct relation has been found between numerical chromosomes aberrations and overrepresentation of the mutant ras gene (72, 74). An increased dosage of the mutant gene was associated with aneuploidy, and the mechanism by which the dosage of the mutant gene could increase implicated two sequential nondisjunction events of the chromosome harboring the activated oncogene.

Thus far, relatively few studies have been devoted to the relationship between oncogene activation and ploidy status in human tumors. For breast cancer, a strong correlation between HER-2/neu overexpression and DNA content measured by image cytometry has been reported (75, 76). All of the carcinomas with HER-2/neu amplifications were near-tetraploid. In ovarian cancer, elevation of the c-myc oncoprotein was reported to precede the development of aneuploidy in a study by Watson et al. (77). However, no correlation between gene amplification of either c-myc or int-2 and HER-2/neu with DNA ploidy status was found by one group (78).

Studies combining DNA flow cytometry with ras mutation analysis have been performed on seminomas, bladder carcinomas, colorectal tumors, and pancreatic carcinomas (79–82). Whereas ras mutations have been found in both diploid and aneuploid colorectal cancers and pancreatic cancers, in bladder cancers and seminomas, ras mutations are associated only with aneuploidy. Thus, it appears that the time at which ras mutations occur in the genetic evolution of a tumor cell population depends on tumor type and may either precede, coincide with, or perhaps follow aneuploidization. At present, there is insufficient evidence to attribute gene-dosage effects by activated oncogenes a pivotal role in the development of aneuploidy in human malignancies.

Tumor Suppression Gene Inactivation and Aneuploidy

The relationship between tumor suppressor gene inactivation and tumor ploidy has been studied by only a few groups. LOH at 17p and 18q was found in a series of 50 colorectal carcinomas with aneuploid tumors having more frequent allelic loss on chromosome 17p and also a greater mean fractional allelic loss (83). Previously, Monpezat et al. reported LOH at 17p and 18q in a small number of aneuploid colorectal cancers (84). In another study on colorectal cancer (85), LOH at chromosome 17p and amplification of the Rb1 gene were found to be associated with DNA aneuploidy. Recently, we have compared tumor ploidy with fractional allelic imbalance (FAI) in a series of 86 breast carcinomas (54). Mean FAI was significantly higher in flow cytometrically aneuploid tumors than in diploid tumors and it was highest in the hypotetraploid subset of aneuploid tumors (Table 1.3). This observation suggests that tetraploidization followed by chromosome segregation is an important route in the development of allelic imbalances at multiple loci. As was the

Table 1.3
Mean fractional allelic imbalance (FAI) and DNA-ploidy status in 86 breast carcinomas

DNA-Index		Mean Fractional Allelic Imbalance
Near-diploid	(0.90<DI<1.10)	0.17 ± 0.15
Hyperdiploid	(1.11<DI<1.40)	0.23 ± 0.11
Hypotetraploid	(1.41<DI<1.80)	0.37 ± 0.11
Near-tetraploid	(1.81<DI<2.20)	0.20 ± 0.14
Hypertetraploid	(DI>2.21)	0.28 ± 0.15

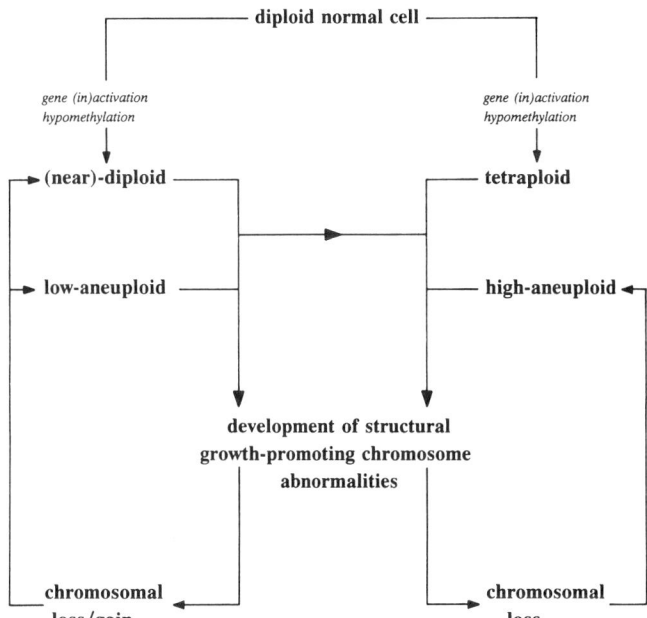

Figure 1.4. Schematic representation of a model for the genetic evolution of human solid tumors (modified from Shackney et al.). Destabilization of normal diploid cells by, i.e., (epi-) genetic factors may result in (near)-diploid or in tetraploid cells. Both routes account for subsequent random development of structural and numeric chromosome abnormalities (losses and gains of chromosomes) that may result in (in-) activation of growth promoting and suppressing genes. Cells containing these abnormalities might be expected to overgrow and emerge as the dominant line. The transition toward a malignant phenotype remains unclear.

case with colorectal cancer, we found that LOH at 17p was significantly more frequent in DNA-aneuploid than in DNA-diploid tumors; the same was found for LOH at 3p. Also, Chen et al. report an association between LOH at 17p and aneuploidy as well as a high S-phase fraction in breast cancer (86).

In ovarian carcinoma, a significant correlation between p53 overexpression (indicating the presence of a mutant gene) and aneuploidy was found in patients with advanced-stage (III/IV) disease (87). Thus, these preliminary data suggest a relationship between DNA-aneuploidy and chromosomal changes associated with tumor suppressor gene inactivation.

CONCLUSIONS

In this chapter we have given an overview of the genetic mechanisms invoked in tumor development and progression in order to appreciate better the significance of DNA ploidy aberrations. Because DNA aneuploidy is more pronounced in solid tumors than in hematopoietic malignancies, emphasis has been put on changes observed in the latter. Molecular-genetic studies strongly support a multi-stage model of tumorigenesis comprising the mutational activation or inactivation of genes with positive or negative growth control functions, respectively. An intriguing question is how these specific genetic events relate to the extensive karyotypic and DNA ploidy changes often seen in solid tumors. Although structural chromosome changes, such as translocations, may lead to activation of protooncogenes, due to the rather limited role of oncogene activation in solid tumorigenesis, this is unlikely to be the only explanation. For the same reason, the significance of tetraploidization in doubling the doses of genes with growth-promoting mutations (i.e., oncogenes) may be insufficient to explain the frequency of this event in solid tumor ploidy evolution.

The recent data indicating that, presumably, multiple-tumor suppressor genes have to be inactivated by a two-step mechanism in most of the common solid tumor types seem to fit better with the widespread occurrence of aneuploidy in these malignancies. Some of the chromosomal mechanisms involved in the elimination of the remaining wild-type allele (nondisjunction, mitotic recombination) might be facilitated by tetraploidization, which doubles the normal chromosome complement. This, for instance, could accelerate tumor progression by eliminating additional genes with negative growth control function. The still limited data on the correlation between allele losses and DNA ploidy support this view. Thus, the extent of DNA ploidy changes may somehow be related to the inactivation of multiple-tumor suppressor genes during tumor development and progression and this may provide the molecular basis for the prognostic effect of DNA aneuploidy. A schematic representation of a modification of the model proposed by Shackney that includes the inactivation of tumor suppressor genes is presented in Figure 1.4. With the rapid progress in the identification of tumor suppressor genes, we expect that, within the next five years, our insight in the genetic evolution of solid tumors, including that of aneuploidy development, will increase dramatically.

REFERENCES

1. Foulds L. Mammary tumors in hybrid mice: growth and progression of spontaneous tumors. Br J Cancer 1949;3:345–375.
2. Nowell PC. The clonal evolution of tumor cell populations. Acquired genetic lability permits stepwise selection of variant sublines and underlies tumor progression. Science 1976;194:23–28.
3. Cairns J. The origin of human cancers. Nature 1981;289:353–357.
4. Caspersson T. Über den chemischen Aufbau der Strukturen des Zellkernes. Skand Arch Physiol 1936;73[Suppl 8]:1-151.
5. Weinberg RA. Oncogenes, antioncogenes, and the molecular basis of multistep carcinogenesis. Cancer Res 1989;49:3713–3721.
6. Bishop JM. Molecular themes in oncogenesis. Cell 1991;64:235–258.
7. Heim S, Mitelman F. Chromosomal abnormalities in specific disorders: solid tumors. In: Cancer cytogenetics, edited by Heim S, Mitelman F. New York: Alan R. Liss Inc 1987, p. 227.
8. Boveri T. Zur Frage der Enstehung maligner Tumoren, Jena: Verlag von Gustav Fisher, 1914.
9. Nowell P, Hungerford DA. A minute chromosome in human granulocytic leukemia. Science 1960; 132:1497–1499.
10. Rowley JD. A new consistent chromosomal abnormality in chronic myelogenous leukemia identified by quinacrine fluorescence. Nature 1973; 243:290–294.
11. Stanbridge EJ, Nowell PC. Origins of human cancer revisited. Cell 1990;63:867–874.
12. Solomon A, Barrow J, Goddard AD. Chromosome aberrations and cancer. Science 1991;254:1153–1160.
13. Caspersson T, Farber S, Folley G, et al. Chemical differentiation along metaphase chromosomes. Exp Cell Res 1968;49:219–222.
14. Yunis JJ. Mid-prophase human chromosomes. The attainment of 2000 bands. Hum Genet 1981; 56:293–298.
15. Sandberg AA. The chromosomes in human cancer and leukemia. 2nd ed. Amsterdam: Elsevier Science Publ., 1990.
16. Heim S, Mitelman F. Cancer Cytogenetics. New York. Alan R. Liss Inc, 1987.
17. Hiddeman W, Schumann J, Andreeff M, et al. Convention on nomenclature for DNA-cytometry. Cytometry 1984;5:445–446.
18. Mitelman F, Heim S. Consistent involvement of only 71 of the 329 chromosomal bands of the human genome in primary neoplasia-associated rearrangements. Cancer Res 1988;48:7115–7119.
19. Teyssier JR. The chromosomal analysis of human solid tumors; A triple challenge. Cancer Genet Cytogenet 1989;37:103–125.
20. Raap AK, Nederlof PM, Dirks RW, Wiegant JCAG, van der Ploeg M. Use of haptenized nucleic acid probes in fluorescent in situ hybridization. In: Harris N, Williams DG eds. In situ hybridization: application to developmental biology and medicine. Great Britain: Cambridge University press 1990, p. 33.
21. Cremer T, Lichter P, Borden J, Ward DC, Manuelides L. Detection of chromosome aberrations in metaphase and interphase tumor cells by in situ hybridization using chromosome-specific library probes. Hum Genet 1988;80:235–246.
22. Pinkel D, Landegent J, Collins C, et al. Fluorescence in situ hybridization with human chromosome-specific libraries: detection of trisomy 21 and translocations of chromosome 4. Proc Natl Acad Sci USA 1988;85:9138–9142.
23. Smit VTHBM, Wessels JW, Molevanger P, et al. Improved interpretation of complex chromosomal rearrangements by combined GTG banding and in situ suppression hybridization using chromosome-specific libraries and cosmid probes. Genes Chromosomes Cancer 1991;3:239–248.
24. Hopman AHN, Ramaekers FCS, Raap AK, et al. In situ hybridization as a tool to study numerical chromosome in solid bladder tumors. Histochem 1988;89:307–316.
25. Devilee P, Thierry RF, Kievits T, et al. Detection of chromosome aneuploidy in interphase nuclei from human breast tumors using chromosome-specific repetitive DNA probes. Cancer Res 1989;48:5825–5830.
26. Hunter T. Cooperation between oncogenes. Cell 1991;64:249–270.
27. Cantley LG, Auger KR, Carpenter C, et al. Oncogenes and signal transduction. Cell 1991;64:281–302.
28. Aaronson S. Growth factors and cancer. Science 1991;254:1146–1153.
29. Barbacid M. Ras genes. Annu Rev Biochem 1987;56:779–827.
30. McCormick F. ras GTPase activating protein. Signal transmittor and signal terminator. Cell 1989;56:5–8.
31. Bos JL. Ras oncogenes in human cancer: A review. Cancer Res 1989; 49:4682–4689.
32. De Klein AS. Oncogene activation by chromosomal rearrangements in chronic myeloid leukemia. Mutat Res 1987;186:161–172.

33. Ben-Neriah Y, Daley GQ, Mes-Masson AM, Witte ON, Baltimore D. The chronic myelogenous leukemia-specific p210 protein is the product of the bcr/abl hybrid gene. Science 1986;233:212–214.
34. Seeger RC, Brodeur GM, Sather H, et al. Association of multiple copies of the n-myc oncogene with rapid progression of neuroblastomas. N Engl J Med 1985;313:1111–1116.
35. Slamon DJ, Clark GM, Wong SG, Levin WJJ, Ullrich A, McGuire WL. Human breast cancer: correlation of relapse and survival with amplification of the HER-2/neu oncogene. Science 1987;235:177–235.
36. Van de Vijver MJ, Peterse JL, Mooi JW, et al. Oncogene activations in human breast cancer. Cancer Cells 1990;7:385–391.
37. Knudson AG. Retinoblastoma: a prototypic hereditary neoplasm. Semin Oncol 1978;5:57–68.
38. Weinberg RA. Tumor suppressor genes. Science 1991;25:1138–1145.
39. Bernards R, Schackleford GM, Gerber MR, et al. Structure and expression of the murine retinoblastoma gene and characterization of its encoded protein. Proc Natl Acad Sci USA 1989;87:6474–6478.
40. Milner J, Medcalf EA. Cotranslation of activated mutant p53 with wild type drives the wild type p53 protein into the mutant conformation. Cell 1991;65:765–774.
41. Hall PA, Ray A, Lemoine NR, Midgley CA, Krausz T, Lane DP. p53 immunostaining as a marker of malignant disease in diagnostic cytopathology. Lancet 1991;338:513.
42. Vogelstein B, Fearon ER, Hamilton SR, et al. Genetic alterations during colorectal-tumor development. N Engl J Med 1988;319:525–532.
43. Devilee P, van den Broek M, Kuipers-Dijkshoorn NJ, et al. At least four different chromosomal regions are involved in loss of heterozygosity in human breast carcinoma. Genomics 1989;4:554–560.
44. Vogelstein B, Fearon ER, Kern SE, et al. Allelotype of colorectal carcinomas. Science 1989;244:207–211.
45. Sato T, Tanigami A, Yamakawa K, et al. Allelotype of breast cancer: accumulative allele losses promote tumor progression in primary breast cancer. Cancer Res 1991;51:4707–4711.
46. Fujimori M, Tokino T, Hino O, et al. Allelotype study of primary hepatocellular carcinoma. Cancer Res 1991;51:89–93.
47. Morita M, Ishikawa J, Tsutsumi M, et al. Allelotype of renal cell carcinoma. Cancer Res 1991;51:820–823.
48. Ehlén T, Dubeau L. Loss of heterozygosity on chromosomal segments 3p, 6q, and 11p in human ovarian carcinomas. Oncogene 1990;5:219–223.
49. Fults D, Pedone CA, Thomas GA, White R. Allelotype of human malignant astrocytoma. Cancer Res 1990;50:5784–5789.
50. Devilee P, van Vliet M, van Sloun P, et al. Allelotype of human breast carcinoma: a second major site for loss of heterozygosity is on chromosome 6q. Oncogene 1991;6:1705–1711.
51. Nigro JM, Baker JS, Preisinger AC, et al. Mutations in the p53 gene occur in diverse human tumor types. Nature 1989;342:705–708.
52. Prosser J, Thompson AM, Cranston G, Evans HJ. Evidence that p53 behaves as a tumor suppressor gene in sporadic breast tumors. Oncogene 1990;5:1573–1579.
53. Varley JM, Armour J, Swallow JE, Jeffreys AJ, Ponder BAJ, Walker RA. The retinoblastoma gene is frequently altered leading to loss of expression in primary breast tumors. Oncogene 1989;4:725–729.
54. Cornelisse CJ, Kuipers-Dijkshoorn NJ, van Vliet M, Hermans J, Devilee P. Fractional allelic imbalance in human breast cancer increases with tetraploidization and chromosome loss. Int J Cancer (in press)
55. Shackney SE, Smith CA, Miller BW et al. Model for the genetic evolution of human solid tumors. Cancer Res 1989;49:3344–3354.
56. Wijkstrom H, Granberg-Ohman I, Tribukait B. Chromosomal and DNA patterns in transitional cell bladder cancer. Cancer 1984;53:1718–1723.
57. Oosterhuis JW, Castedo SMMJ, de Jong B, et al. Ploidy of primary germ cell tumors of the testis. Pathogenetic and clinical relevance. Lab Invest 1989;60:14–21.
58. Oosterhuis JW, Castedo SMMJ, de Jong B. Cytogenetics, ploidy and differentiation of human testicular, ovarian and extragonadal germ cell tumours. Cancer Surv 1990;9:321–332.
59. Beerman H, Kluin PhM, Hermans J, van de Velde CJH, Cornelisse CJ. Prognostic significance of DNA ploidy in a series of 690 primary breast cancer patients. Int J Cancer 1990;45:34–39.
60. Ewers SB, Langstrom E, Baldetorp B, Killander D. Flow cytometric DNA analysis in primary breast carcinomas and clinicopathological correlations. Cytometry 1984;5:408–419.
61. Giaretti W, Santi L. Tumor progression by DNA flow cytometry in human colorectal cancer. Int J Cancer 1990;45:597–603.
62. Smit VTHBM, Fleuren GJ, van Houwelingen JC, Zegveld ST, Kuipers-Dijkshoorn NJ, Cornelisse CJ. Flow cytometric DNA-ploidy analysis of synchronously occurring multiple female genital tract malignancies. Cancer 1990;66:1843–1849.
63. Schelfhout LJDM, Cornelisse CJ, Goslings BM, et al. Frequency and degree of aneuploidy in begnign and malignant thyroid neoplasms. Int J Cancer 1990;45:16–20.
64. Rutteman GR, Cornelisse CJ, Dijkshoorn NJ, Poortman J, Misdorp W. Flow cytometric analysis of DNA ploidy in canine mammary tumors. Cancer Res 1988;48:3411-3417.
65. Minke JMHM, Cornelisse CJ, Stolwijk JAM, Kuipers-Dijkshoorn NJ, Rutteman GR, Misdorp W. Flow cytometric DNA ploidy analysis of feline mammary tumors. Cancer Res 1990;50:4003–4007.
66. Heim S, Mandahl N, Mitelman F. Genetic convergence and divergence in tumor progression. Cancer Res 1988;48:5911-5916.
67. Volpe JPG. Genetic instability of cancer. Cancer Genet Cytogenet 1988;34:125–134.
68. Holliday R. Chromosome error propagation and cancer. TIG 1989;5:42–45.
69. Pardue ML. Dynamic instability of chromosomes and genomes. Cell 1991;66:427–431.
70. Auer G, Fallenius A, Erhardt K, Sundelin B. Progression of mammary adenocarcinomas as reflected by nuclear DNA content. Cytometry 1984;5:420–425.
71. Schwartz D, Banner BF, Roseman DL, Coon JS. Multiple "primary" colon carcinomas: a retrospective flow cytometric study. Cancer 1986;58:2082–2088.
72. Bremner G, Balmain A. Genetic changes in skin tumor progression: correlation between presence of a mutant ras gene and loss of heterozygosity on mouse chromosome 7. Cell 1990;61:407–417.
73. Bianchi AB, Aldaz CM, Conti CJ. Nonrandom duplication of the chromosome bearing a mutated Ha-ras-1 allele in mouse skin tumors. Proc Natl Acad Sci USA 1990;87:6902–6906.
74. Klein G. The role of gene dosage and genetic transposition in carcinogenesis. Nature 1981;294:313–316.
75. Bacus SS, Ruby SG, Weinber DS, Chin D, Ortiz R, Bacus JW. HER-2/Neu oncogene expression and proliferation in breast cancers. Am J Pathol 1990a;137:103–111.
76. Bacus SS, Bacus JW, Slamon DJ, Press MF.HER-2/Neu oncogene expression and DNA ploidy analysis in breast cancer. Arch Pathol Lab Med 1990b;114:164–169.
77. Watson JV, Curling OM, Munn CF, Hudson CF. Oncogene expression in ovarian cancer: A pilot study of c-myc oncoprotein in serous papillary ovarian cancer. Gynecol Oncol 1987;28:137–150.
78. Sasano K, Garrett CT, Wilkinson DS, Silverberg S, Comerford J, Hyde J. Protooncogene amplification and DNA ploidy in human ovarian neoplasms. Hum Pathol 1990;2:382–391.
79. Mulder MP, Keijzer W, Verkerk A, et al. Activated ras genes in human seminomas: evidence for tumor heterogeneity. Oncogene 1989;4:1345–1351.
80. Burmer GC, Loeb LA. Mutations in the KRAS2 oncogene during progressive stages of human colon carcinoma. Proc Natl Acad Sci USA 1989b;86:2403–2407.
81. Czerniak B, Deitch D, Simmons H, Etkind P, Herz F, Koss LG. Ha-ras gene codon 12 mutation and DNA ploidy in urinary bladder carcinoma. Br J Cancer 1990;62:762–763.

82. Smit VTHBM, DNA ploidy analysis and K-ras oncogene activation in pancreatic adenocarcinomas. [Thesis]. Leiden, the Netherlands: The University of Leiden, 1991;103–114.
83. Offerhaus GA, de Feyter EP, Cornelisse CJ, et al. The relationship of DNA-aneuploidy to molecular genetic alterations in colorectal carcinoma. Gastroenterology 1992;102:1612–1619.
84. Monpezat JPh, Delattre O, Bernard A, et al. Loss of chromosome 18 and on the short arm of chromosome 17 in polyploid colorectal carcinomas. Int J Cancer 1988;41:404–408.
85. Meling GI, Lothe RA, Borresen AL, et al. Genetic alterations within the retinoblastoma locus in colorectal carcinomas. Relation to DNA ploidy pattern studied by flow cytometric analysis. Br J Cancer 1991; 64:475–480.
86. Chen LC, Neubauer A, Kurisu W, et al. Loss of heterozygosity on the short arm of chromosome 17 is associated with high proliferative capacity and DNA aneuploidy in primary human breast cancer. Proc Natl Acad Sci USA 1991;88:3847–3851.
87. Marks JR, Davidoff AM, Kerns BJ, et al. Overexpression and mutation of p53 in epithelial ovarian cancer. Cancer Res 1991;51:2979–2984.

2

The Cell Cycle: Application of Flow Cytometry in Studies of Cell Reproduction

ZBIGNIEW DARZYNKIEWICZ

INTRODUCTION

Cells reproduce by doubling their constituents, followed by division. The sum of cell activities that is essential for their reproduction is generally defined as the cell cycle. These activities can be subdivided into two categories: those related to the chromosome cycle and those related to the cytoplasmic cycle. The landmarks of the chromosome cycle are DNA replication and mitosis. Changes in nuclear chromatin structure—associated with preparation for DNA replication—DNA replication itself, and preparation for mitosis, are also considered to be part of the chromosome cycle. Cell growth, i.e., the set of activities resulting in the doubling in quantity of most cell components, and the actual act of cell division, cytokinesis, constitute the framework of the cytoplasmic cycle.

Progress in biochemistry and molecular biology was of paramount importance in the development of our understanding of mechanisms related to cell reproduction. However, our present knowledge in this area also benefited from three specific techniques designed to analyze the individual cells, either alone or in conjunction with the tools of biochemistry and molecular biology. These techniques are autoradiography, time-lapse cinematography, and flow cytometry.

Autoradiography was responsible for establishing the concept of the cell cycle, as we now know it, consisting of the four major phases; G_1, S, G_2 and M. Four decades ago, while studying incorporation of a DNA precursor by autoradiography, Howard and Pelc (1) observed that DNA synthesis was discontinuous and occupied a discrete portion of the cell life (S phase). Mitotic division occurred after a certain period of time following DNA replication. A distinct time interval between mitosis (M) and DNA replication was also apparent. The cell cycle, or the cell generation time, defined as the interval between the midpoint of mitosis and the midpoint of the subsequent mitosis of the daughter cells, was thus subdivided into four consecutive phases, G_1, S, G_2 and M. The G_1 and G_2 phases represented "gaps" ("G") between mitosis and the onset of DNA replication, and between the end of DNA replication and the initiation of mitosis, respectively. This framework of the cell cycle proposed by Howard and Pelc survived, basically unchanged, until now. Analysis of incorporation of precursors into DNA, RNA, or protein by autoradiography yielded most of the data fundamental to our current understanding of the cell cycle. Kinetics of cell progression through the cell cycle could be studied in great detail after the development of the more advanced autoradiographic approaches, such as measurements of the fraction of labeled mitosis ("FLM" curves) or sequential cell labeling with ^{3}H- and ^{14}C-TdR. Autoradiography also provided the means to recognize the presence of noncycling, quiescent, (often called G_0) cells in cell populations (2) and, conversely, to enumerate the proportion of cells in the reproductive pool in tumors, the tumor growth fraction (3). The literature directly related to autoradiographic techniques as applied to research on the cell cycle is the subject of several reviews (4–6).

Compared with autoradiography, the applications of time-lapse cinematography in research of the cell cycle were much more limited. This technique, however, can be credited for providing the data that made it possible to comprehend the heterogeneity of generation times in cell populations and to probe the mechanisms generating this heterogeneity (7, 8).

The history of flow cytometry is relatively short, yet this methodology is already in widespread use, worldwide, especially in studies of the cell cycle (reviews 9–11). Rapidity and accuracy of the cell measurements, combined with the development of a wide spectrum of markers to different cell constituents, contributed to the popularity of this methodology in general. The extensive use of flow cytometry in studies on the cell cycle, in particular, is the consequence of several additional factors. First, there is widespread interest in the scientific community on the subject of cell reproduction. The interest spans various biomedical disciplines and the methodology is used both in basic research on cell reproduction and as a practical tool to monitor cell proliferation. The second factor is the availability, early on in the development of the methodology, of a multitude of the DNA-specific dyes applicable to cell cycle analysis. Still another factor contributing to the wide use of flow cytometry in cell cycle studies is the nature of the correlated, multiparameter measurements (recorded in the list mode fashion) that this method can offer. Namely, the simple act of measurement of cellular DNA content reveals the position of the cell in the cell cycle: The correlated measurement of the second or third constituent provides information about the relationship of this constituent to the position of a cell in the cycle. Therefore, much information pertaining to biology of the cell cycle can be obtained without having to use cumbersome methods of cell synchronization. The wide use of flow cytometry also stems from the fact that this methodology provides a means to study the properties of large cell populations rather than concentrating on either a few cells, as could be obtained by autoradiography, or on average values, such as obtained by biochemical measurements of synchronized populations. In the latter case, the presence of rare cells or minor subpopulations escapes detection.

Reviews of the literature on cell reproduction can be found in several monographs and review articles published recently (12–17). No effort has been made in the present chapter, therefore, to extensively cover this subject or to discuss the bulk of the literature that may be found in these sources. This chapter provides, however, some basic information about changes in the cell during progression through the cell cycle, which may be relevant to researchers willing to apply flow cytometry in their studies. Particular attention is given to changes in nuclear chromatin, because mechanisms regulating the cell cycle progression, which have started to unravel only very recently, appear to operate within the framework of the chromatin cycle. The subject of chromatin changes has been generally neglected in most of the earlier reviews. This chapter also describes kinetic properties of cell populations and review applications of flow cytometry in the study of the cell cycle. The aim of this article is also to describe various strategies by which flow cytometry can be applied to analyze cell reproduction, both in situations when the cell growth is unperturbed as well as when the cell cycle is modulated by various drugs and treatments.

CHROMATIN CHANGES DURING THE CELL CYCLE

The history of investigations of chromatin changes has two distinct phases. The first phase, initiated with the discovery of the nucleosomal subunit structure of chromatin (18, reviews 19, 20), was characterized by extensive studies on chemical modifications of individual histones and nonhistone nuclear proteins in relation to the cell cycle. Biochemical methods, combined with various approaches of cell synchronization, were primarily used in these studies. The second phase is very recent. The modern methods of molecular biology, combined with genetic analysis of the rapidly dividing yeast cells made it possible to characterize molecular mechanisms associated with their reproduction. The enzymes involved in modifications, (especially phosphorylation) of nuclear proteins have been identified and their role in the cell cycle revealed. The mechanisms described in yeasts and higher eukaryots appeared to be strikingly similar; hence, yeast cells are now providing a useful and widely used model to study cell reproduction in general. However, although significant progress in understanding the mechanisms of cell reproduction has been made recently, many questions related to the regulation of the cell cycle still remain unanswered.

Nucleosome Structure; Histone Acetylation

The gross chromatin structure undergoes significant changes during the cell cycle. The most dramatic change occurs when the chromatin condenses prior to mitosis as well as when it decondenses in postmitotic cells. Despite such marked changes in nuclear morphology, the basic chromatin subunit, the nucleosome, remains unaltered during these transitions, at least when probed by classical tests such as sensitivity to digestion by micrococcal nuclease or appearance under the electron microscope (21). Thus, the subunit structure, originally observed in interphase chromatin, is conserved in mitotic chromosomes, and chromatin condensation during the cell cycle does not involve structural changes at the nucleosomal level. During S phase, however, at the sites of DNA replication, a stepwise assembly of nucleosomes takes place, first by association of the newly replicated DNA with histones H3 and H4, and then by addition of H2A and H2B (22).

Although the gross structure of the nucleosome remains unaltered, more subtle changes, in the form of chemical modification of the core particle histones, occur during the cell cycle or during cell transition from quiescence to the cycle. Acetylation of histones was studied extensively in a variety of cell types, both in vivo and in vitro (reviews, 20, 23). It was postulated that acetylation, by neutralizing positive charges mostly on the NH_2-terminal regions of histones of the core particle, decreases electrostatic interactions between histones and DNA and thus "loosens" the structure of nucleosomes (24). However, the acetylation of histones has a rather minor effect on their binding to DNA, as measured by dissociation of the DNA-histone complexes (25) or thermal denaturation of DNA in chromatin (26). This suggests that, at least at low levels of acetylation, protein-protein, rather than protein-DNA interactions, are weakened.

Acetylation of histones H3 and H4 characterizes nucleosomes in transcriptionally active chromatin (20, 23). A much larger proportion of histones in the cell are acetylated and deacetylated, however, than would appear to be necessary if acetylation were involved only in activating specific genes or even specific regions of the genome (27). Therefore, it was proposed that acetylation may have a two–fold function. Low level acetylation, which affects a large portion of chromatin, may be necessary for the transition of the heterochromatin to euchromatin. More extensive acetylation may weaken DNA-protein interactions and such change, affecting only a small portion of chromatin, may be involved in activation of individual genes (27). Indeed, the extent of acetylation of core histones remains generally in reverse proportion to the extent of condensed chromatin. Thus, for example, in deer mouse (*Perymyscus*) cells, the amount of unacetylated histones during interphase is proportional to the amount of constitutively condensed chromatin (28). In synchronized CHO cells, the highest acetylation of histone H4 is seen during interphase; during prophase and metaphase, i.e., at the time of maximal chromatin condensation, acetylation of histone H4 is minimal. Postmitotic decondensation of chromatin is preceded by a rapid increase in the proportion of acetylated histone H4 molecules, already observed in telophase (29). In *Physarum polycephalum,* the proportion of highly acetylated histone H4 is maximal in mid-S phase and minimal in prophase (30).

During S phase, new histones are synthesized and, immediately following synthesis, acetylated (31). Nucleosomes associated with freshly replicated DNA are more sensitive to nuclease digestion and dissociation in media of increasing ionic strength (32). Thus, the nascent chromatin, nearest to the actual site of DNA replication, differs in structure from mature chromatin and undergoes a "maturation" process as the replication advances. Two phases of the maturation have been identified: the brief (5 min) phase is characterized by diffuse, irregular spacing of nucleosomes on new DNA immediately after replication. The second phase (lasting from 5 to 30 min after replication) is characterized by the normal repeat length; the nucleosomes, however, are "immature" and undergo increased sliding during nuclease digestion, in comparison with bulk chromatin (33). As mentioned, the maturation involves the stepwise deposition of H3 and H4, followed by H2A and H2B histones (22).

Acetylation specific to S phase involves all histones; histones H3 and H4 are mono- and di-acetylated. Following S phase, there is acetylation of histones H3 and H4 during G_2 phase, which involves the addition of up to four acetyl groups; this modification, however, appears to correlate with transcription (34).

In addition to its role during transcription and DNA replication, histone acetylation is also essential during the process of histone replacement by protamines in the course of spermiogenesis. Extensive acetylation of all of the core histones takes place in a brief period during spermatid maturation (35). It is believed that such a high level of acetylation is needed to dissociate histones from DNA in order to make DNA accessible and to form complexes with protamines. The latter substitute for histones in providing counterions for DNA and are essential for packaging DNA in sperm chromatin.

The association between histone H3 and H4 acetylation and cell progression through the cell cycle is apparent in the case of histone hyperacetylation during cell growth in the presence of *n*-butyrate. Accumulation of multiacetylated forms of histones H3 and H4 due to inhibition of histone deacetylase by *n*-butyrate (36) is paralleled by cell arrest in G_1 (37, 38). However, although following removal of *n*-butyrate the pattern of histone acetylation returns to normal rather quickly (39), there is a considerable delay before cells resume progression through the cell cycle. Furthermore, *n*-butyrate induces other modifications of chromatin proteins, such as appearance of the histone H1°, which is characteristic of noncycling cells (40, 41). No clear evidence thus exists that there is a cause-effect relationship between histone hyperacetylation such as that induced by *n*-butyrate and cell progression through the cell cycle. The role of histone acetylation or deacetylation, essential for cell cycle progression, appears thus to be demonstrated only in the case of cells replicating DNA.

Histone Methylation

Histones H3 and H4 are methylated on ϵ-amino groups (reviews, 20, 23). This modification is not extensive; mono- or dimethyl–H3 and H4 molecules generally predominate. The evidence is controversial as to whether, and to what extent, histones other than H3 and H4 are methylated. The methylation takes place in the cell nucleus and, in contrast to acetylation, is rather stable (turnover of methyl groups is low). Methylation of H3 and H4 occurs after DNA replication, predominantly in G_2 and mitosis (42). The rates and mechanisms of methylation appear to be different for H3 and H4 (43).

The role of histone methylation in the cell cycle or chromatin structure is unclear. The stability of this modification suggests that it may function as a marker of particular groups of nucleosomes, e.g., for their recognition by regulatory proteins. The fact that methylation takes place in G_2 or M may simply reflect the slow rate of this modification, which occurs some time after histone synthesis and even after their assembly in chromatin. Thus, the timing of the methylation does not necessarily mean that this modification is essential for G_2 or M progression.

Histone Poly(ADP)ribosylation

Modification of proteins by poly(ADP)ribosylation involves enzymatic transfer of the ADP-ribose subunits from nicotinamide adenine dinucleotide (NAD+) to the acceptor protein (20, 44). The poly(ADP)ribose chains are often branched. This modification is common to a variety of proteins. Most studies indicate that histone H1 is the most extensively poly(ADP)ribosylated histone; histone H2A is also significantly modified (45). Poly(ADP)polymerase, the enzyme that catalyzes the reaction, is bound to the chromatin. Maximal poly(ADP)ribosylation occurs during S phase, with two distinct peaks, one in mid-S and the other at the time of the S-G_2 transition (46, 47). The enzyme, however, undergoes translocation from the nucleus to the cytoplasm and its activity in these compartments varies significantly at different phases of the cell cycle. This complicates the interpretation of the experiments in which this modification was studied in vitro. Its maximal nuclear activity was observed during G_2 (47).

The biological role of histone poly(ADP)ribosylation was the subject of controversy (reviewed in 20), and it is still not entirely resolved. Because this modification is extensive during S phase and increases after damage to DNA (48) it is assumed that poly(ADP)ribosylation may provide recognition points in chromatin for DNA ligase, both during DNA replication and DNA repair. Poly(ADP)ribosylation of chromatin produces relaxation of the native "zigzag" appearance of the nucleosomal fiber and precludes condensation of this fiber in media of high ionic strength (49, 50). Thus, poly(ADP)ribosylation of histone H1 may result in relaxation (decondensation) of chromatin fibers, without detachment of H1 from DNA. This relaxation may increase the accessibility of DNA to enzymes involved in DNA replication, repair, or transcription. There is no agreement in the literature, however, as to whether a correlation exists between the degree of histone poly(ADP)ribosylation and the transcriptional activity of chromatin.

Histone Ubiquitination

Modification of histones by isopeptide branching was detected in the form of covalent addition to ubiquitin to histone H2A (51). Ubiquitin is a highly conserved protein containing 74 amino acids. The product of ubiquitination of H2A, initially found in nucleoli of regenerating liver, was defined as protein A24. About 10% of H2A appears to be modified in such a way. A minor fraction of H2B also appears to be ubiquitinated (52).

The ubiquitin residues on H2A and H2B are in rapid turnover and in equilibrium with the pool of free ubiquitin, both in cycling and quiescent cells (52). The enzymatic cleavage of ubiquitin from histones is rapid and occurs at rather constant rates throughout the cell cycle; the rate of ubiquination is high in interphase (maximal early in G_1) and drops during mitosis (52–54). An association between cell cycle progression and histone ubiquitination was reported by Finley et al., (55) who observed that a temperature-sensitive mutant cell line, defective in one of the enzymes involved in attachment of ubiquitin, also shows a defect in phosphorylation of histone H1 and cannot enter mitosis; both defects are removed after reversion of the mutant.

The role of histone ubiquitination is also a subject of controversy. In general, ubiquination serves as a marker targeting proteins for proteolytic degradation. This, however, does not seem to be the case with histones; their turnover is much slower than that of the ubiquitin moiety (52). The ubiquination of H2A may play a role in the regulation of transcription (56). Rapid turnover of this peptide on H2A and H2B, its high evolutionary stability, additional chemical modifications such as poly(ADP)ribosylation of the uniquitinated histones (49), and its loss during mitosis, all suggest that this modification may play an essential role in continuous modulation of the nucleosomal structure and in packing chromatin into metaphase chromosomes.

Histone Phosphorylation

HISTONE KINASES

Among all changes in chromatin that are associated with the cell cycle progression, phosphorylation of histones, and phosphorylation of H1 in particular, are of major importance. Phosphorylation of histone H1 is essential for the cyclic condensation of chromatin during the mitotic cycle (57). Several kinases, which are able to phosphorylate histones in the in vitro assays, have been identified (reviews, 20, 58). They can be subdivided into two categories: those that are dependent on cyclic nucleotide monophosphates for activation and those that are independent. Among the kinases involved in phosphorylation of histone H1 in mammalian cells, two enzymes, which have been characterized in considerable detail, play a major role. In the early studies these kinases were classified as kinase A, and kinase G. Kinase A, designated also as HK-I (H), which is cyclic AMP (cAMP) dependent, was shown to phosphorylate a single serine at position 37 of the rabbit histone H1. Such modification, triggered by hormones, was reported to alter the DNA transcription pattern and induce synthesis of new proteins (59). Phosphorylation at serine-37 may facilitate the displacement of histone H1 from DNA sections at the points of DNA entrance and exit from the nucleosome core (23, 60); the displaced histone H1 may then be substituted by one of the High Mobility Group (HMG) proteins (HMG1 or HMG2). Phosphorylation by kinase A affects only about 2% of the total histone H1 molecules. Modification to this extent is consistent with the portion of genome which is involved in structural rearrangement during changes in transcription, for example, in response to hormones.

Although the cAMP-dependent H1 kinase does not appear to be directly involved in the chromatin cycle, its activity was reported to be associated with cell proliferation (61). This data, however, may be subject to a variety of interpretations, as discussed before (15).

Kinase G ("growth associated kinase") activity is directly related to chromatin changes that occur during the cell cycle. This enzyme phosphorylates several threonines and serines of histone H1 (reviews 20, 58). Early studies of Hardie et al., (62) and Matthews (63) suggested that there are two enzymes (kinase G1 and kinase G2) with different substrate specificities phosphorylating different sites of H1. Mitchelson et al. (64) observed that an increase in the activity of this enzyme during mitosis is a result of activation of the pre-existing enzyme rather than its de novo synthesis, suggesting that the enzyme acquires an activator prior to mitosis. More recent studies indicate that the growth associated histone H1 kinase in eukaryots is a homologue of the 34-kDa serine-threonine protein kinase (65), a product of the *cdc2* gene of the fission yeast *Saccharomyces pombe* (p34^{cdc2}) (reviews 16, 66–69). Homologues of p34^{cdc2} were recently identified in budding yeast *Saccharomyces cerevisiae* (p34^{cdc2}8) and in a variety of animal (70–75) and plant (76) species. The most recent studies indicate that this 34-kDa histone H1 kinase is one of the key molecules involved in the regulation of the cell cycle. This enzyme and the mechanisms involved in its activation are highly conserved during evolution, and are very similar in organisms ranging from yeast to man (Fig. 2.1).

The content of the 34-kDa histone H1 kinase is relatively constant throughout the cell cycle and its turnover is low.

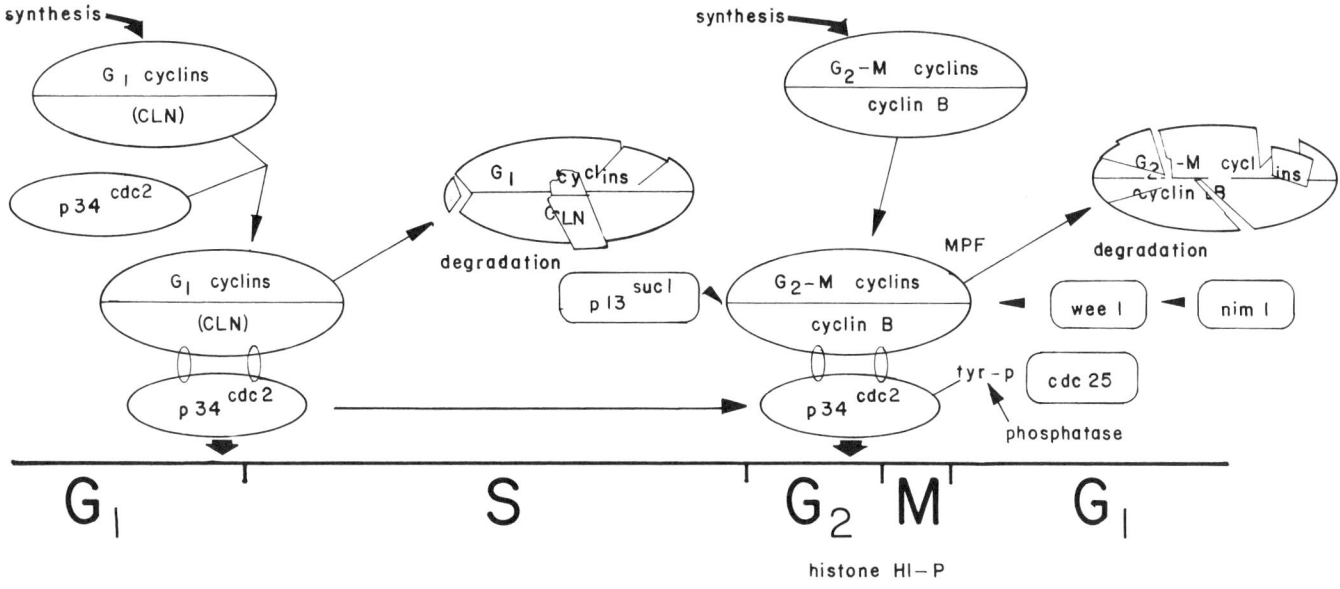

Figure 2.1. Schematic representation of the role of p34^{cdc2} in the regulation of the cell cycle. Histone H1 kinase, "growth-associated kinase," or "kinase-G" are different names given to the same protein, which was described in mammalian cells and later found to be the homologue of the 34 kDa serine-threonine protein kinase, a product of the cdc2 gene of S. cerevisie. The level of p34^{cdc2} is invariant during the cycle and this protein alone, (uncomplexed), has no kinase activity. Late in G$_1$, a special class of proteins, cyclins, are synthesized; the G$_1$ cyclins associate with p34^{cdc2} forming a complex in which p34^{cdc2} gains kinase activity. These G$_1$ cyclins so far have been identified only in S. pombe (products of CLN genes), but their presence in mammalian cells is suspected. Activation of the kinase activity of the p34^{cdc2}/G$_1$ cyclin complex triggers phosphorylation of histone H1 and perhaps other substrates, not yet identified in vivo, that are essential for cell entrance to S phase. During progression through S the complex dissociates and G$_1$ cyclins are degraded. During G$_2$ and M phases, p34^{cdc2} associates with another type of cyclin proteins that are synthesized primarily in G$_2$. Completion of S phase appears to initiate synthesis of a set of cyclin proteins. These G$_2$-M cyclins with a molecular weight of 54-60 kDa, first identified in sea urchins and clams, then in S. cerevisiae (product of cdc13) and in mammalian cells, are class B cyclins, differing from class A cyclins (the latter are synthesized in S and their function is still unclear) by their amino acid sequences. This complex was originally defined as the Maturation-Promoting Factor (MPF); the factor was observed to induce meiotic maturation of G$_2$ arrested oocytes. The protein kinase activity of this heterodimer is triggered when the phosphotyrosine, located at the N-terminus of p34^{cdc2} is dephosphorylated. The dephosphorylation, recently found to be mediated by the phosphatase encoded by the cdc25 gene of S. cerevisiae, is the rate-limiting step in activation of the kinase activity of MPF. Completion of S appears to be necessary for triggering dephosphorylation. Phosphorylation of histone H1 (inducing chromatin condensation) and nuclear lamins (triggering disassembly of the nuclear envelope) has been shown to be the result of p34^{cdc2} protein kinase activity; these events are a prerequisite for cell entrance to mitosis. Many other proteins, whose phosphorylation is known to occur at that time (see the text), are also likely to be the substrates of MPF; their in vivo phosphorylation by this complex, however, has not yet been demonstrated.

The kinase activity of the MPF abruptly falls at the metaphase/anaphase transition, coincident with degradation of the G$_2$-M cyclins. This step is essential for decondensation of chromatin and cell transition to G$_1$. The activity of p34^{cdc2}/G$_2$-M cyclin complex is regulated by several other proteins. Thus, p13, the product of the suc1 gene of S. cerevisiae binds directly to p34^{cdc2} and blocks the process of tyrosine dephosphorylation, thereby preventing activation of the enzyme and entrance of cells to M. Once the kinase is activated, however, p 14^{suc1} has no effect on its activity. Another inhibitory activity is exercised, in a dose dependent fashion, by the produce of wee1 gene of S. cerevisiae, which itself is also a protein kinase. The activity of the wee1 gene product, in turn, is regulated by still another kinase, the product of the nim1 gene. The nim1 product, thus, is an indirect activator of P34^{cdc2} kinase, acting via suppression of the wee1 product's inhibitory effects. Other proteins, including several protooncogenes, are suspected to exercise additional modulating effects on the activity of the p34^{cdc2} complex.

Activity of this enzyme, however, is regulated by its cyclic phosphorylation/dephosphorylation and association with cyclins. Cyclins are proteins characterized by periodic accumulation and degradation during the cell cycle. Cyclins were initially identified in the cleaving egg of marine invertebrates by their intensive synthesis during interphase and rapid degradation during mitosis (77). The G$_1$-specific cyclins (CLN1, CLN2, CLN3 cyclins), were identified in *Saccharomyces cerevisiae;* these proteins are synthesized in G$_1$ and their accumulation peaks late in G$_1$ (77, 80). Association of the G$_1$-cyclins with the 34-kDa histone H1 kinase activates the enzyme; the activation triggers cell entrance to S phase. In yeasts, this activation triggers progression of cells through the START point of the cell cycle and results in initiation of DNA replication (66).

The START point represents a step in the cell cycle during which cell growth in size and the chromosome cycle are coordinated: To pass this point the cells have to acquire a size (rRNA or protein content), and correlates its growth with activation of p34^{cdc2} kinase, is not well understood. The equivalent of the START in other eukaryots is the transition point between G$_{1A}$ and G$_{1B}$ discriminated by cellular rRNA content (21); cells prior to this point have a subthreshold rRNA content and do not enter S phase (see below).

It should be stressed, however, that, although based on the evidence of the "restriction point" in G_1 (74, 75) the presence of G_1 cyclins in mammalian cells is expected, at the time of writing of this article (Jan. 1991) no such proteins were identified in cells other than *Saccharomyces cerevisiae*. Cyclin A, which is expressed maximally in late S and early G_2 phase does not appear to be the G_1-type cyclin; it may, however, be associated with cell progression through the S phase (76).

During G_2 phase, association of $p34^{cdc2}$ histone H1 kinase with G_2 cyclins (B type cyclins) activates this kinase, triggers chromatin condensation, and is essential for cell entrance to mitosis (72, 81, 84, 85). The complex of $p34^{cdc2}$ and cyclin B, termed the "maturation promoting factor" (MPF), was initially isolated from vertebrate eggs and shown to induce meiotic maturation of G_2 arrested oocytes in the absence of the de novo protein synthesis (86, 87). It was postulated by Pines and Hunter (72) that the completion of S phase may signal induction of transcription from the G_2 cyclin gene and also stabilize the mRNA of that cyclin. The rather constant time, which is expected for cyclin to reach a threshold value under these conditions, may explain the almost invariant length of G_2 observed in each cell cycle. The G_2 cyclin, isolated from HeLa cells is a 62-kDa phosphoprotein. Phosporylation of this protein or $p34^{cdc2}$ kinase or both, regulates association between these molecules and activation of the kinase. However, because dephosphorylation of $p34^{cdc2}$ correlates with an increase in kinase activity, (88) it is possible that both this event and phosphorylation of the cyclin within the complex (89) are required to fully activate the histone H1 kinase. Recent studies indicate that dephosphorylation of the tyrosine residue of $p34^{cdc2}$ kinase by the protein encoded by the cdc25 gene, which has phosphatase activity, appears to be the primary event triggering chromosome condensation and entrance of the cell to mitosis (90).

In addition to the effect on phosphorylation of histone H1, the complex of $p34^{cdc2}$ and cyclin B is involved in phosphorylation of the single 67-kDa nuclear lamin; this event triggers disassembly of the nuclear envelope (91). This enzymatic complex therefore is not only responsible for chromatin condensation, but also for restructuring the constituents of the nuclear envelope required for entrance to mitosis. It is likely that many more targets of phosphorylation by the $p34^{cdc2}$-cyclin B complex will soon be identified and that it will be shown that their phosphorylation/dephosphorylation is a prerequisite for cell entrance to and progression through mitosis.

Cyclin B is rapidly degraded after completion of metaphase. This results in the inactivation of $p34^{cdc2}$ kinase, the step required for the cell to complete mitosis and enter G_1. Preventing degradation of the cyclin results in the prolongation of mitosis (85).

Clearly, the $p34^{cdc2}$ protein kinase (and its homologues, found in all species of eukaryots studied so far) plays the key role in regulation of the cell cycle. The recent data of Broek et al., (92) indicate that the cell cycle can be reprogrammed by altering the state of $p34^{cdc2}$. When fission yeast mutants defective in $p34^{cdc2}$ in G_2 have their $p34^{cdc2}$ function restored, they "forget" their position in the cycle and initiate an extra round of DNA replication without entering mitosis. Thus, completion of mitosis is not required for a cell to enter the next S phase, as long as $p34^{cdc2}$ is programming the next step which, in this case, is entrance to S phase. This fact led these authors to postulate that "the cell cycle can be considered a $p34^{cdc2}$ cycle". The $p34^{cdc2}$ cycle thus consists of cyclic conversions of this kinase, as a result of its posttranslational modification and association with the respective cyclins, to "M form" (during G_2) and to "S-form" (during G_1). Thus, within the framework of the $p34^{cdc2}$ cycle, the G_1 and G_2 periods are interchangeable, being determined primarily by the status of $p34^{cdc2}$, which can overrule the past history of the cell cycle (92). Other details related to the role of $p34^{cdc2}$ kinase in the cell cycle are shown schematically in Figure 2.1 and briefly summarized in the legend.

Sites of Phosphorylation. The progression of cells through the cell cycle appears to be associated with the ordered, phase-specific, phosphorylation and dephosphorylation of various sites, primarily on histone H1 and also on histones H3 and H2A. Systematic studies on histone phosphorylation were carried out on two cell systems. The first system consisted of the slime mold *Physarum polycephalum,* which can be easily synchronized in the cycle. Phosphorylation of histone H1 during G_1 in this organism is low and there is either no, or only a single site phosphorylated by the $p34^{cdc2}$ kinase. The rate of phosphorylation rises during S phase and two sites on H1 are then phosphorylated. Finally, a rapid rise in the enzyme activity, observed in mid G_2, coincides in time with the maximal number of phosphorylated sites on H1 (63). The second cell system thoroughly investigated with respect to histone phosphorylation was that of CHO or HeLa cells, synchronized in the cycle by a variety of methods. Progression of CHO cells through G_1 involves phosphorylation of a single serine in the C-terminal region of the molecule (93). During S phase two additional serines in the same region are phosphorylated. During G_2, just prior to mitosis, two additional sites (threonine and serine) in the N-terminal region, as well as one threonine in the C-terminal portion, undergo phosphorylation. This extensive phosphorylation prior to mitosis has been defined as superphosphorylation. The transition from anaphase to interphase is accompanied by a rapid dephosphorylation of H1 back to the form containing 0-3 phosphates per molecule.

Additional complexities in the pattern of histone modifications during the cell cycle were described by Ajiro et al. (94, 95) in studies on HeLA S-3 cells. These authors were able to discriminate two subtypes of histone H1, H1A and H1B. The cationic C-terminal of H1A is shorter and more

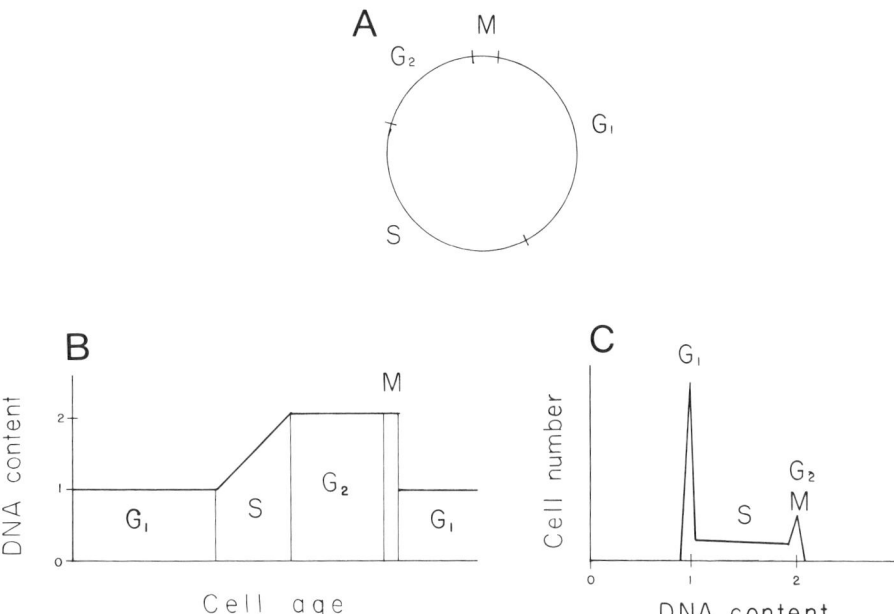

Figure 2.2. Relationship between the changes in DNA content in individual cells progressing through the cell cycle and the frequency distribution of DNA content of the cell population. DNA replication is discontinuous and occupies a discrete interval in the cell lifetime (S phase: **A**), during which cellular DNA doubles in content (**B**). By measuring DNA content of individual cells in the population (**C**), therefore, the probability of finding cells that have not yet replicated DNA (G_1 phase), cells replicating DNA (S), and those that already have replicated DNA (G_2, M), is represented by the frequency distribution histograms and is correlated with the respective duration of these phases. For cells growing exponentially and asynchronously, a correction for the exponential character of the cell age distribution (see Fig. 2.8) has to be introduced to relate the DNA content frequency distributions to the phase durations.

hydrophobic than that of H1B, indicating that the two subtypes may differ in their interaction with DNA and proteins. During G_1, while most H1A molecules contain 0-1 phosphates, H1B molecules are more heterogenous and contain 0-3 phosphates. When cells progress through S phase there is an additional phosphorylation of the two distinct sites in the C-terminal region of each subtype. The authors explain the phosphorylations during S phase as related to the progression of the replication fork (95). During mitosis, H1A and H1B are additionally phosphorylated at 3-4 sites. The threonyl residue at position 17 was identified as one of the sites phosphorylated in H1B; this residue is not present in the H1A subtype. Other sites phosphorylated during mitosis were located in the C-terminal region of both the H1A and H1B subtypes.

Phosphorylation of histone H3 during the cell cycle involves only serine at position 10, which is modified during mitosis (96). Because this serine is highly conserved and appears to be involved in interactions between H3 and H4, (which are responsible for maintenance of the core particle structure), its modification during mitosis suggests that the nucleosomal core particle structure may be modified during the G_2 to M transition.

Phosphorylation of histone H2A does not show any specific pattern that can be traced to cell progression through a particular phase of the cycle. The constitutive character of H2A phosphorylation has been explained as related to maintenance of the heterochromatin (97). Conversely, the association of the phosphorylated H2A with the DNase hypersensitive regions of chromatin suggested that this modification is associated with transcriptionally active chromatin.

Compared with the wealth of information on protein kinases associated with phosphorylation of histones and other proteins during the cell cycle, relatively little is known about phosphatases that may be involved in the specific dephosphorylation of these molecules. Although phosphatase activities were reported in numerous studies dealing with histone phosphorylation, no systematic studies had been carried out to identify these enzymes and analyze their specificity and cell cycle phase activities. It is only most recently that the role phosphatases in the regulation of cell cycle progression was observed. As mentioned, the product of cdc25 has phosphatase activity and it dephosphorylates the phosphotyrosine residue of the p34^{cdc2} kinase, resulting in activation of this enzyme. Two types of protein phosphatases (type 1 and 2A), involved in regulation of the cell cycle were described in fission yeast (98, 99), and a new gene product was identified in these cells (sds22+), which by itself does not appear to have the phosphatase activity, but which enhances the activity of one of the type 1 phosphatases (100).

Histone Modification in Noncycling Cells

A striking feature of the chromatin of noncycling cells is the increase in quantity of a distinct protein, histone H1° (101). This protein is similar to histone H1 and is regarded as its subtype (37, 40). While small amounts of H1° can be detected in cycling cells (37), its level markedly rises as the rate of cell proliferation decreases (37, 40, 102).

Histone H1° is phosphorylated during the cell cycle and the pattern of its phosphorylation resembles that of histone H1 (40). Thus, phosphorylation of H1° is completely inhibited when cells are arrested in G_1. During G_2, up to 35% of H1° molecules are phosphorylated at one or two sites, and during mitosis, all H1° molecules are phosphorylated at four sites (37). Also like histone H1, H1° consist of two subtypes, (H1°a and H1°b) differing in primary structure. Both fractions are phosphorylated at similar rates during the cell cycle progression (37, 40).

In avian erythrocytes, an equivalent of histone H1° is histone H5; H5 replaces histone H1 during the final stages of differentiation of these nucleated erythrocytes (103).

Although there are significant differences in primary structure and conformation of all three H1 histones, they all bind to linker DNA at its entrance and exit from the core particle and are all involved in the condensation of chromatin into the 30-nm fiber (20). This supranucleosomal organization correlates with the resting state of chromatin. There may be differences, however, in the degree of reversibility of this condensation depending on the particular H1 subtype. As was shown recently by Sun et al. (41), an inducible transfection of rat sarcoma cells with the histone H5 gene results in reversible inhibition of DNA replication and cell arrest in G_1. These results clearly indicate that substitution of histone H1 by H5 precludes cell progression through the cell cycle.

Nonhistone Nuclear Proteins

DNA transcription and replication, as well as the maintenance of chromatin in a dynamic state during the cell cycle, cell differentiation, DNA repair, etc., all require the presence of innumerable proteins. Increasing attention is being focused on these proteins because among them are factors regulating the most essential functions of the cell. In live cells, these proteins are localized in the nucleus and they are generally classified as "nonhistone nuclear (chromosomal) proteins." Close to 1500 distinct proteins can be reproducibly detected by gel electrophoresis of nuclear samples (104).

During nuclear isolation, a procedure needed to study these proteins, an uncontrolled exchange of the proteins between the nucleus and cytoplasm takes place. Numerous proteins, which are loosely bound in the nucleus, are generally lost. Thus, in practice, the term "nonhistone nuclear proteins" defines the proteins that remain associated with chromatin following nuclear isolation. Depending on the method of nuclear isolation (presence or absence of detergent, ionic composition of the medium, number of washings) and cell type, marked differences in the content of nuclear proteins are observed. An excellent review of this subject is given by van Holde (20). Isolated cell nuclei are being used in flow cytometry with increasing frequency, not only for determinations of DNA but also for the measurement of "nuclear protein content" or for detection, by immunochemical methods, of the particular nuclear proteins. In light of the point noted above (i.e., uncontrolled loss of proteins), interpretation of this data, especially in quantitative terms, should be done with caution.

The subject of the nonhistone proteins is covered by several reviews (20, 105–107). The present chapter concentrates on the group of proteins that, so far, are the best characterized and appear to be associated both with DNA transcription and cell reproduction, namely, the high mobility group (HMG) proteins. Individual proteins, the presence or content of which correlates with cell proliferation or quiescence, and that therefore can be used as markers in flow cytometry, will also be briefly reviewed further in this chapter.

HIGH MOBILITY GROUP PROTEINS

HMG proteins received their name due to the mobility of individual proteins in polyacrylamide gels (20, 105, 108). Mammalian cells contain four main HMG types: HMG 1, HMG 2, HMG 14 and HMG 17. These proteins are associated with nucleosomes: HMG 1 and 2 bind to linker DNA while HMG 14 and 17 are associated with the core particle DNA. Binding of the latter confers an increase in sensitivity of DNA in chromatin to DNAse I and therefore is considered as facilitating DNA transcription (109, 110). HMG 1 and 2 also interact with different histone H1 subtypes. There is a controversy regarding the quantity of HMG protein per nucleus; most studies indicate, however, that these proteins associate with a relatively small fraction (0.1 to 3%) of nucleosomes (review, 20).

A correlation is apparent between levels of HMG 2 (but not of HMG 1) protein and cell proliferative activity (111). It has been proposed that HMG 2 and histone H1°, which show a reciprocal relationship when their content is compared in proliferating and nonproliferating cells, may have alternative functions in nucleosomal linker regions: HMG 2 may be responsible for maintenance of chromatin in a state ready for replication, whereas H1° precludes replication (111). Other studies suggest, however, that a low content of HMG 2 correlates more with cell differentiation than with quiescence (112).

Studies on synchronized HeLa cells indicate that the cellular content of HMG 1 and 2 increases during mid-S phase, while that of HMG 14 and 17 increases in mid G_1 (113). There are also large differences in the extent of phosphorylation of HMG 14 and 17 throughout the cell cycle. Thus, a sevenfold increase in phosphorylation of HMG 14 is observed in G_2, compared to G_1, and a twofold increase of HMG 17 phosphorylation is observed in early S, as compared to G_1 or G_2.

In addition to HMG 1, 2, 14, and 17 nonhistone proteins, there is also an HMG I family of proteins of which two members, HMG I and HMG Y, are the best characterized. HMG I proteins bind in the minor groove of AT rich stretches of DNA, and are expressed at elevated levels in proliferating or neoplastic cells; their expression is presumed to be associated with rapid cell division or maintenance of the undifferentiated state (114). Recent studies indicate that during mitosis p34^{cdc2} protein kinase phosphorylates a threonine residue at the N-terminal end of HMG I (114). As a result of this phosphorylation HMG I binding to DNA is markedly decreased. Thus, HMG I proteins, like histone H1, may play an important role in the regulation of the chromosome cycle.

CELL GROWTH, RNA METABOLISM

Protein Synthesis

All cell constituents double in amount during the cell cycle. Most proteins are synthesized continuously, at progressively increasing rates, during interphase. A decline in the rate of synthesis to near stop is seen at mitosis and is presumed to be caused by the dissociation of polyribosomes (115). Following mitosis, when individual ribosomes reassociate to form polyribosomes, reprogramming of the translation pattern takes place.

As mentioned, the timing and rates of synthesis of most individual proteins vary little during the cell cycle (104, 116). Some proteins, however, are synthesized discontinuously. The most notable example are the core particle histones. Early studies indicated that their synthesis was limited only to the S phase and was coupled with DNA replication: The DNA/histone ratio was considered to be constant throughout the cycle (review, 117). More recent studies brought evidence, however, that whereas the majority of these proteins are synthesized during S, about 10% of histones ("basal histones") are made in G_1 and G_2 (118, 119). The proportions of different histones and histone variants synthesized during S and in other phases of the cycle vary substantially. Also, the pattern of "basal" histone variants is different in cycling cells in G_1 than in noncycling, G_0 cells (120). Little is known about the possible functional differences between "basal" and S-phase synthesized histones.

There is another group of proteins or peptides detected in cells only at specific periods within the cell cycle, or in cells closely related to cell proliferation. This group can be divided into three categories. To the first belong proteins suspected of having regulatory function during the cell cycle. An example, discussed earlier, is the G_1 and G_2 cyclins, which associate with p34^{cdc2} and activate cell entrance to S or M. The second category consists of proteins directly involved in cell proliferation, such as enzymes of DNA replication. The third category is cell surface proteins or glycoproteins that exhibit cell cycle (phase) dependency and generally serve as receptors to various growth factors (e.g., interleukin 2 or transferrin receptors). The list of proteins showing variation in the cycle is growing rapidly. Their detection and characterization are not only important to unravel the enigma of regulation of cell proliferation, but also have immediate practical value. Namely, antibodies against these proteins can serve as markers of the cells' proliferative potential for characterization of tumor cell populations and can be of clinically prognostic value. The most important proteins in this category will be listed further in the chapter in the discussion of markers of cell proliferation.

RNA METABOLISM, COORDINATION OF CELL GROWTH, AND CYCLE PROGRESSION

To ensure that cell size remains constant from generation to generation, cell growth has to be coordinated with division. Although the mechanism of coordination has not yet been revealed, extensive evidence obtained from studies of various cell systems, ranging from yeasts to mammalian cells, indicates that twice during the cell cycle the cell "senses" its size and couples growth with progression through the chromosome cycle. Under conditions of unperturbed growth, only cells that attain a certain size in G_1 enter S phase (66, 83). In yeasts, this is equivalent to the START point discussed earlier. Cell commitment to enter S phase is thus being made based on passing a size threshold. The second point is during G_2: Cell entrance to mitosis is associated with cells achieving another growth threshold (e.g., 121). Nutritional shortage very often blocks cell reproduction by limiting cell size and preventing entrance to S phase. It should be stressed, however, that this coordination does not necessarily indicate the cause-effect relationship inasmuch as there are several examples showing that cells whose size is below the threshold level, are able to initiate DNA replication (review, 122). Cell size, however, appears to be more critical for cells to enter mitosis. The mechanism by which cells sense their size is not known and is the subject of much speculation.

The consequence of the coupling of cell growth and proliferation is that the markers of cell growth are often also the markers of cell proliferation. This certainly holds true with respect to cellular RNA content. Quiescent cells are generally characterized by low RNA content and markedly diminished RNA synthesis rate, compared to their cycling counterparts (reviews, 15, 123). This is common even in the plant kingdom (124). Activation of quiescent cells is paralleled by the increase in rate of RNA synthesis and the total number of ribosomes per cell. For example, mitogenic stimulation of lymphocytes results in a tenfold rise in number of ribosomes (125) or rRNA content (126). Cell proliferation is also closely associated with the number and size of nucleoli, the organelles which are the site of synthesis and initial processing of rRNA (127–128). Recent studies indicate that there is also a close relationship between the content of nucleolar proteins (e.g., such as nuclear antigen detected by means of Ki-67 antibodies or nucleolar p120 antigen) and cell proliferation (129–132). The association between cellular RNA metabolism and cell proliferation is the subject of a recent review (123).

Given the above, it is not surprising that the RNA content of tumor cells is a good prognostic marker in many malignancies (e.g., 133, 134, reviewed in 123). Clearly, the rate of tumor growth and its sensitivity to drugs, most of which show cell cycle specificity, is expected to correlate with cell kinetics and thus, indirectly, with RNA content. In growing tumors, suppression of cell proliferation under adverse growth conditions (e.g., due to increased tumor mass, worsening nutrient, and oxygen supply) is secondary, being the consequence of the initial suppression of cell growth. Decline in RNA content precedes and thus, in a way, predicts a slowdown in cell proliferation. Cellular or nuclear RNA content (approximately 85% of total cellular RNA is rRNA; almost all nuclear RNA is nucleolar, rRNA or pre-rRNA) is therefore perhaps an even more sensitive parameter of cell proliferation than the actual quantitation of S-phase cells in situations when the transition from cycling to quiescent state is being investigated. Any measure of the activity of the nucleolus is expected to be an equally predictive marker of malignancy as is rRNA content (132).

ANALYSIS OF THE CELL CYCLE BY FLOW CYTOMETRY

Univariate Cellular DNA Content Distributions

Cellular DNA Content as a Marker of the Cell Position in the Cycle. Due to the fact that during S phase the cell doubles its DNA content and that following mitosis the daughter cells inherit half the DNA content of the mother cell, the cells' position in the cycle can be estimated based on the measurement of their DNA content (Fig. 2.2). Cells in G_1 have one unit of DNA, cells in G_2 two units, and cells in S phase have DNA values between one and two units. Mitotic cells, up to the moment of cytokinesis, also have two units of DNA. The frequency histograms showing DNA content values of the cells of a given cell population also represent the distribution of cells with respect to their positions in the cell cycle (G_1, S or $G_2 + M$ phases).

The simple act of measuring cellular DNA content, therefore, provides information on the cell cycle distribution of the population and, with several limitations (as will be discussed further), gives insight into the kinetic properties of the population analyzed. Numerous fluorochromes can be used to stain cellular DNA and a multitude of methods applicable to flow cytometry have been developed to measure DNA content. Thus far, this univariate DNA content analysis is the most common approach in cell cycle studies. Because the duration of S, G_2 and M phases is relatively constant, while the duration of G_1 varies and cells generally enter quiescence with a DNA content typical of that of G_1 cells (G_0, G_{1Q} cells), the proportion of $S + G_2 + M$ cells is often considered to represent the proliferative potential of that population.

All cells in G_1 have a uniform DNA content, as do cells in G_2 phase. Under ideal conditions of DNA staining and measurement, the fluorescence intensities of all G_1 and G_2 cells are expected to be uniform and, after the digitization of the analog electronic signal from the photomultiplier (representing their fluorescence intensity), to have uniform numerical values. This, however, is never the case; the DNA fluorescence values of the G_1 and G_2 cell populations are represented by peaks of various widths on frequency histograms. The coefficient of variation of the mean value of DNA-associated fluorescence of the G_1 cell population (CV_{G1}) is a measure of the width of these peaks.

The CV_{G1} is a very important parameter in flow cytometry. Its value varies due to biological and technical reasons. The biological reason for the increased CV_{G1} is the genuine DNA content variability. Such variability is observed, for example, when the populations consist of cells having minor variations in DNA content, e.g., due to an uneven degree of gene amplification, as in the case of the presence of the additional minute chromosomes, genome instability after X-irradiation, excessive amounts of mitochondrial DNA, etc. A much more common cause for the increasing values of CV_{G1} is technical, resulting from the inaccuracy of DNA measurements. Myriad reasons could account for the latter; their source may originate from the problems associated either with tissue preparation, cytochemistry of DNA staining, or fluorescence measurement. These issues are the subjects of several reviews (135–137) (see also Chapters 6, 8, and 10).

As mentioned, inaccuracies in the measurements of DNA lead to a widening of the G_1 and G_2 peaks; these peaks are generally Gaussian. Therefore, the measured DNA values of S-phase cells overlap with the values of G_1 and $G_2 + M$ cells on the histograms. This overlap creates a problem in distinction between G_1 versus S versus $G_2 + M$ cells. Statistical methods and computer algorithms have been developed to tackle this problem, and there are several different approaches to correct for "widening" of the distributions (reviews: 138–140). These methods of deconvolution of the frequency histograms are based on certain common assumptions pertaining to the character of the change of the distributions as a consequence of inaccuracy of DNA measurements. Generally, if the accuracy of DNA measurement is not much compromised (e.g., when CV_{G1} does not exceed 5%), and when the measurements are done on cells growing exponentially and asynchronously, these methods yield accurate and reproducible results. Problems arise, however, when either the cycle progression of populations being studied is significantly perturbed, and/or when the CV_{G1} is high. In these situations, none of the methods can guarantee correctness of the results.

Limitations of the Univariate DNA Content Analysis. Several limitations are inherent to all the methods evaluating the cell cycle status of cell populations based on a single parameter DNA measurement.

One obvious limitation is the inability to discriminate between G_2 and M cells. Since both have identical DNA content, they are represented by a single peak on the DNA frequency distribution histograms. Actually, because the chromatin of M cells is highly condensed, the accessibility of DNA to many DNA fluorochromes, and thus DNA stainability, is often lower in M than that in G_2 cells (141). Therefore, on the DNA frequency histograms, mitotic cells may overlap the position of S phase rather than G_2 cells. In the case of measurements employing isolated nuclei rather than whole cells, mitotic cells, lacking a nuclear membrane, may be lost altogether from the analysis (142).

Another limitation is the inability to discriminate between quiescent G_0 (G_{1Q}) and G_1 cells. In many cell systems, cells enter quiescence with a G_1 DNA content and, based on DNA content alone, such cells cannot be distinguished from the G_1 (cycling) cells.

The DNA frequency histograms do not provide any information on cell kinetics. Two cell populations that have doubling times, eight and 80 hrs respectively, for example, may be represented by identical DNA frequency histograms if the duration of individual phases are in the same proportion to the length of their whole cycles. Furthermore, under certain conditions, the cells become "frozen" in the cycle and their progression is halted in all phases simultaneously (143). Such a population also cannot be distinguished from the cycling cells by the univariate analysis of cellular DNA content. In all in vitro studies, therefore, it is essential that when the DNA content alone serves as the cell cycle marker, the cell proliferation rates ("growth curves", "doubling times") should be estimated in parallel (see further).

Still another drawback of the univariate analysis is the inability to discriminate different cell subpopulations in the sample. One example is the mixture of diploid tumor cells and host stromal and/or infiltrating normal cells; such a mixture is present in the biopsied or resected tumor samples. The presence of normal cells in these samples precludes estimation of the cell cycle distribution of the diploid tumor cell population. Even when the tumor is aneuploid, the overlap of DNA values of the diploid normal cells progressing through S and G_2+M phases with DNA values of tumor cells complicates the cell cycle analysis of the latter.

DNA-Specific Fluorochromes: Cellular Fluorescence Intensity May Not be Proportional to DNA Content. An assumption generally made is that the fluorescence intensity of individual cells counterstained with dyes specific for DNA is proportional to the DNA content of these cells. Extensive evidence exists, however, that only a fraction of DNA is accessible to the most commonly used fluorochromes and that this fraction may vary depending on chromatin structure (144). Caution, therefore, should be exercised in interpreting the fluorescence intensity values of cells stained with these dyes as absolute markers of their DNA content and thus, in interpreting their position in the cell cycle. This is of particular relevance when cells differing significantly in chromatin structure are compared.

Among the variety of dyes, the binding of DAPI to DNA in situ is the least influenced by nuclear proteins (144). DAPI, therefore, appears to be the most suitable DNA fluorochrome, staining this polymer stoichiometrically. At the other extreme is 7-amino actinomycin D, whose binding is markedly affected by differences in chromatin structure (144, 145). The methods of cell preparation that make use of treatments with acids and/or proteolytic enzymes, such as pepsin (137, 146), result in the release of many nuclear proteins, which otherwise restrict DNA accessibility, and make DNA staining with several dyes more stoichiometric. Conversely, fixation with formaldehyde or glutaraldehyde stabilizes DNA-protein and protein-protein complexes and impedes subsequent DNA stainability. DNA content measurements of samples which were prefixed with these aldehydes, as is the case for nuclei isolated from archival paraffin blocks (147), is therefore very difficult because of the uncertainty of the effect of fixation. Factors such as the concentration of aldehydes in the fixative, duration and temperature of fixation, chromatin structure, and variation in the nuclei isolation protocols, all affect DNA stainability.

Multiparameter Studies
STRATEGIES AND APPLICATIONS

The subject of multiparameter analysis of the cell cycle by flow cytometry has been recently extensively reviewed (148). Only the essential points will be discussed in the present Chapter. Multiparameter analysis makes it possible to resolve some of the limitations of the univariate DNA content analysis, discussed above. In the multiparameter approach, more than one feature of the same cell is measured and, generally, the measurements are recorded in list mode fashion. It is possible therefore to attribute these features to a particular cell and thus to obtain their correlated measurements on a cell by cell basis. When one of the parameters measured is cellular DNA content, the analysis of the correlated data provides information on the quantity of a second constituent relative to the cell's position in the cell cycle. Multiparameter flow cytometry is, therefore, the method of choice to study the relationship between cell growth (e.g., expressed RNA or protein content) or differentiation (measured by the content of a differentiation-specific product) and cell progression through the cell cycle. The method can also estimate the time of appearance and the content of a particular, specific constituent, such as oncogene product, at a given point of the cell cycle. The list of the available antibodies that can be used in the latter application is rapidly growing.

Three strategies may be applied in adapting multiparameter flow cytometry to studies of the cell cycle. The first is the "snapshot," or static analysis of the cell population. This is the most common approach, which, in a single measurement, reveals the relationship between the cell's metabolic state and the cell cycle phase. Based on this analysis, additional subcompartments can be distinguished within the traditional phases of the cell cycle. Cells in these subcompartments, having different metabolic features, are also characterized by different kinetic properties (143, 148, 149). Therefore, particular metabolic features, such as cellular RNA or nuclear RNA content, expression of a proliferation-associated protein, etc., in conjunction with the cellular DNA content, are markers predictive of cell kinetics. Thus, although no actual cell kinetic is measured, the kinetic potential of the cell can be predicted based on its metabolic (general biochemical properties of the cell) and/or molecular (expression of individual proteins) fingerprinting (131, 150–154).

The second strategy represents a combination of multiparameter flow cytometry with kinetic measurements of the cell population. In this approach the static observations, which do not contain any kinetic information per se, can be correlated with the rate of cell progression through the cell cycle. One such experimental design involves the use of synchronized cells. The progression of synchronized populations (e.g., cells released from G_1 or G_2 arrest) can be measured sequentially in time such that the rates of cell entrance to S, progression through S, and entrance to G_2 can be estimated (155). Another design is based on the principle of stathmokinesis. This design has been introduced to measure the rates at which cells traverse through several points of the cell cycle simultaneously (156) and has been found to be extremely useful in studies of drug effects on the cell cycle (157).

The third strategy in the application of multiparameter flow cytometry for cell cycle analysis is based on measurements of DNA replication, using the assays involving cytochemical (158–161) or immunochemical (162, 163) detection of incorporation of the thymidine analog 5-bromodeoxyuridine (BrdUrd). The cells that replicate DNA in the presence of BrdUrd can be recognized, making it possible to estimate their rate of entrance into S or to discriminate cycling from noncycling cells (reviews, 136, 157). This strategy is gaining wide application both in basic research and clinically and will be discussed later in this chapter.

GENERAL METABOLIC FEATURES OF CELLS IN THE CELL CYCLE

Cell Size. Cell size can be estimated in flow by physical methods such as Coulter volume (165), forward light scatter (166), axial light loss (167) or fluorescence-pulse duration (158) measurements. Because, as discussed, progression of cells through the cell cycle is coupled with their growth in size, cells in G_2 are generally larger than in G_1 and an increase in cell size is observed during progression through S. The heterogeneity of cell sizes, however, is so large that it is often impossible to discriminate G_2 from G_1 cells, for example, by differences in their size values. On the other hand, the differences in size between quiescent (G_0) and cycling cells is frequently of such magnitude that the size parameter is used to distinguish these cell types. Most commonly, this is done on lymphocytes; their stimulation by mitogens, which represents a transition from the resting (G_0) to the cycling state and is accompanied by cell enlargement, has been measured by Coulter volume (169), axial light loss (170) or light scatter (171).

RNA and Protein Content. Correlated measurement of cellular RNA and DNA can be accomplished using either the metachromatic dye acridine orange (126, 172), or a combination of Hoechst 33342 and pyronin Y (173, 174). To be interpreted in quantitative terms, both methods require stringent conditions of dye concentration and dye/nucleic acids ratio per sample (172, 174). DNA and RNA content of 3T3 cells growing exponentially and also cells entering quiescence is shown in Fig. 2.3. The G_1 population of the cycling cells (panel a) has a characteristically broad RNA distribution. It is quite evident that, prior to entrance to S, individual cells accumulate a threshold amount of RNA; cells with RNA content below this value (marked by a solid line in Fig. 2.3) have been classified as residing in the G_{1A} compartment (81). The threshold, which reflects the minimal quantity of ribosomes (rRNA) per cell needed to enter S phase, is an equivalent of the START point of the yeast cell cycle (discussed earlier in this chapter), i.e., the point where the cells equalize in size prior to initiation of DNA replication. The characteristic kinetic feature of G_{1A} cells is the exponential-like ("stochastic") distribution of their residence times in this compartment. This is in contrast to late G_1 (G_{1B}) cells, whose transit times through G_{1B} have a rather Gaussian distribution (81). As is also evident in Fig. 2.3a, the progression of cells through S is associated with a further increase in their RNA content.

When cells are cultured at low serum concentration, a marked suppression in the number of cells progressing through S and G_2+M is apparent (Fig. 2.3 b,c). Concomitantly, a loss of cellular RNA content is observed. At first, most cells are arrested in G_1 and are characterized by low RNA values typical of G_{1A} cells. With time, the loss of RNA is more severe and the RNA content of these quiescent cells drops below that of G_{1A} cells. These cells are viable and, when trypsinized and grown in medium with a high serum concentration, (10%) resume progression and enter S phase after a delay of approximately 16 hours (81). Thus, low cellular RNA content is a marker discriminating quiescent from cycling cells. Such deeply quiescent G_1 cells can be distinguished as a separate category (G_{1Q} cells) having distinct metabolic (very low RNA content) and kinetic (long delay before entrance into S phase) properties (123, 143, 149).

Peripheral blood lymphocytes, noncycling and having minimal RNA content, are a classic example of G_{1Q} cells (149). As discussed earlier in the chapter, their mitogenic stimulation triggers synthesis and accumulation of rRNA and, subsequently, these cells enter S phase. Changes in RNA and DNA content during lymphocyte mitogenic stimulation are illustrated in Fig. 2.4. Several compartments, representing different phases of cell growth and progression through the mitotic cycle, can be distinguished by analyzing changes in the bivariate RNA and DNA content distribution of these cells in the course of their stimulation (123, 143, 149; see legend to Fig. 2.4). Multiparameter RNA/DNA analysis of lymphocytes by flow cytometry is now a widely used test to assay mitogenic stimulation of these cells, having many advantages over the traditional tritium labelled thymidine incorporation assay (148).

The cell cycle-related changes in cellular RNA content discussed above, should be distinguished from the tissue-

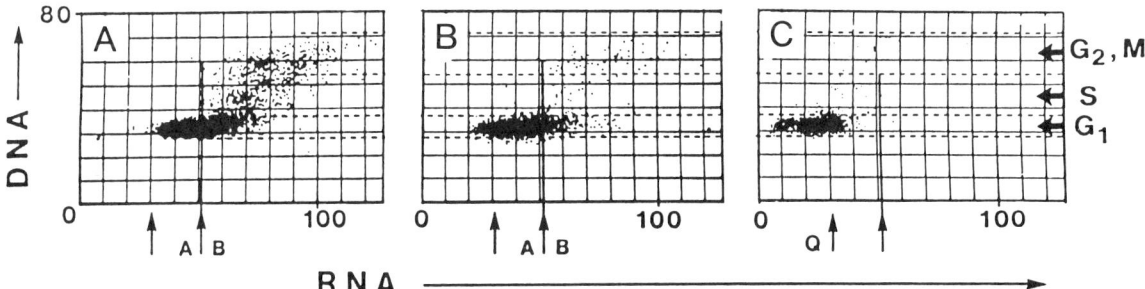

Figure 2.3. DNA versus RNA distributions of 3T3 cells during exponential growth (**A**), and quiescence induced by: (**B**) growth for three days in low (0.5%) serum concentration and (**C**) growth for seven days in 0.5% serum. A stepwise loss of cellular RNA content accompanies cell entrance to the quiescent state, which makes it possible to identify three compartments of the G_1 phase, namely G_{q1} (quiescent compartment, characterized by minimal RNA content); G_{1A} (early G_1, "prethreshold" compartment); and G_{1B} (prereplicative phase). (Reprinted with permission from Darzynkiewicz Z, Traganos F. Multiparameter flow cytometry in studies of the cell cycle. In: Melamed Mr, Lindmo L, Mendelsohn ML, eds. Flow Cytometry and Sorting. Second edition. New York: Wiley-Liss, 1990: 469–501.)

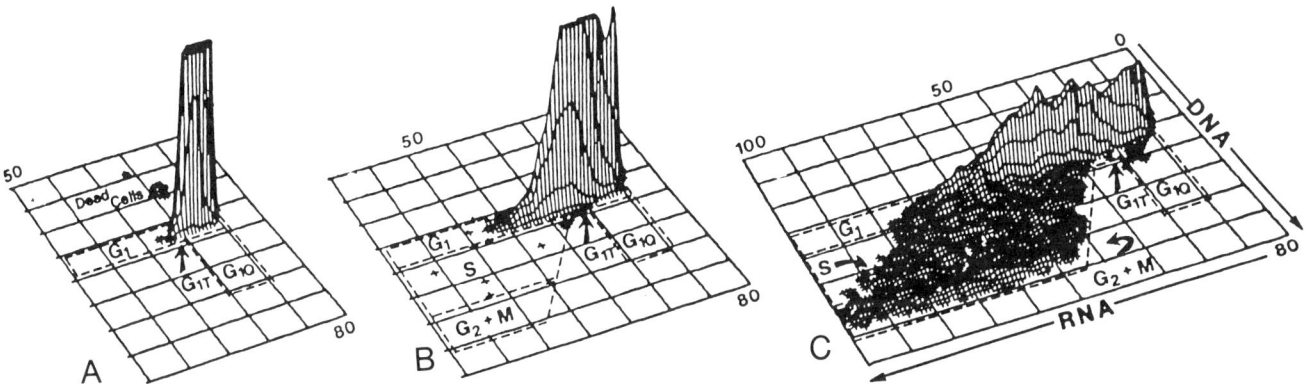

Figure 2.4. Progression of lymphocytes through the cell cycle after their stimulation by phytohemagglutinin (PHA). DNA versus RNA distribution of human lymphocytes incubated for (**A**) 0, (**B**) 1 and (**C**) 3 days with PHA. Stimulated lymphocytes are distinguished from quiescent cells (G_{1Q}) by their increased RNA content and can be identified first as cells in transition (G_{1T}), and then as G_{1A} and G_{1B} cells in one-day-old cultures. In three-day-old cultures there are cells in G_1, S and $G_2 + M$ phase. Identification of S and $G_2 + M$ cells is based on both increased RNA and DNA content and is confirmed by comparison with synchronized cultures. (Reprinted with permission from Darzynkiewicz Z, Traganos F. Multiparameter flow cytometry in studies of the cell cycle. In: Melamed MR, Lindmo L, Mendelsohn ML, eds. Flow Cytometry and Sorting. Second edition. New York: Wiley-Liss, 1990: 469–501.)

specific variation in RNA content. For example, significant differences in RNA content are observed between blast cells of various leukemia types (123, 175), between B and T lymphocytes (176), etc. This cell (tissue) type-specific variation in cellular RNA content is a marker of cell phenotype and has no apparent relation to cell proliferation properties.

Several methods have been designed to simultaneously measure cellular protein and DNA content (177, 178). The cellular protein content is generally correlated with the content of rRNA and with cell size (141). The patterns of the bivariate protein/DNA distributions are thus similar to those of RNA/DNA distributions. Cell heterogeneity, however, is higher when an analysis is based on protein as compared to RNA content (179).

Multiparameter RNA/DNA or protein/DNA analysis is the method of choice to detect the unbalanced cell growth and to quantitatively estimate, in relation to cell position in the cell cycle, the degree of the unbalance (180). Growth unbalance often occurs as a result of cell arrest in the mitotic cycle and is also a characteristic feature of cells exposed to a variety of antitumor drugs that either preferentially suppress DNA replication, transcription, or translation (180). Because the severity of the unbalance is predictive of cell death, its estimate on biopsied tumor samples during treatment, for example, may be an important parameter in clinical oncology. The simultaneous, correlated measurement of cellular DNA, RNA, and protein, which allows one to measure the changes in RNA/protein ratio during the cell cycle (141), is especially suited to study unbalanced growth (181).

Progression of cells through the cell cycle as well as their transition from quiescence to the proliferative state is associated with changes in other biochemical or metabolic parameters that can be measured by flow cytometry, such as mitochondrial transmembrane potential, pH, lysosomal activity, calcium content (influx), intracellular content of thiols, etc. A review of these parameters is not within the scope of this chapter, but the relevant methodologies have been recently presented in a single volume (11) and many of these parameters are discussed in other articles (9, 10, 182, 183).

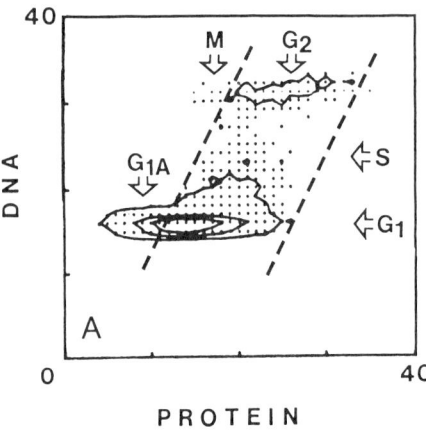 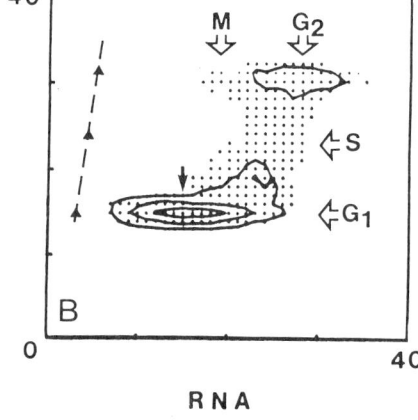

Figure 2.5. Correlated measurements of *DNA* and *protein* (**A**) and *DNA* and *RNA* (**B**) content of nuclei isolated from exponentially growing CHO cells. Nuclei of early G_1 cells contain very low protein and RNA content. Because the histone: DNA ratio is relatively constant throughout the cycle, the changes in protein content in nuclei falling outside the marked outlines (G_{1A}, M cells) represent a deficit in nonhistone proteins in chromatin of these cells, compared to cells in S phase.

MEASUREMENTS OF ISOLATED NUCLEI; CHROMATIN STRUCTURE

As discussed earlier in this chapter, significant changes in chromatin structure parallel cell progression through the cell cycle. Large changes in nuclear chromatin also occur during exit from quiescence and cell entrance to the cycle or in the course of differentiation. Analysis of nuclear constituents with respect to their content, modifications, and conformation, therefore, provides clues as to the functional state of the cell, especially regarding the regulatory mechanisms associated with its proliferative or differentiation status.

Using appropriate probes, several features of nuclear chromatin can be conveniently studied by flow cytometry. Isolated nuclei and permeabilized or prefixed cells can be measured under conditions in which nuclear constituents are in the native state with their spatial structure and the intermolecular interactions preserved (review, 145). This is in contrast to the traditional, biochemical methods of chromatin analysis, which rely on solubilization of chromatin and extraction of the components and destroys these interactions.

Tumor diagnosis and classification are often based on analysis of nuclear morphology and chromatin structure by conventional shape; the pattern ("texture") and degree of chromatin condensation; and the number, prominence, and morphology of nucleoli. All are critical parameters for tumor evaluation. Thus, by offering an objective and quantitative criteria for the evaluation of nuclei, flow cytometry may be of great value for tumor diagnosis and prognosis. Furthermore, the isolation of nuclei from solid tumors by using detergents is more convenient than the enzymatic or mechanical dissociation of tumors and the isolation of whole intact cells. For this practical reason, flow cytometric analysis of isolated nuclei is often preferred over studies of intact cells.

Nuclear size can be evaluated from the intensity of the forward light scatter signal or fluorescence pulse duration. Because isolated nuclei are permeable to ions, their size estimate by electric impedance ("Coulter volume") is not reliable.

Measurements of the nuclear protein content reveal interesting changes during the cell cycle (Fig. 2.5). First, it is evident that when the cell membrane is lysed to isolate the nuclei, metaphase chromosomes remain attached to each other and, despite the lack of a nuclear envelope, the amount of DNA and protein associated with these chromosome clusters can be estimated. Although these structures contain a full 4C DNA content, they can be distinguished from G_2 cells due to their lower protein content or low light scatter properties (184, 185). Thus, the amount of protein associated with chromosomes in mitosis is lower than that present in nuclei of G_2 cells. Postmitotic cells have a minimal content of nuclear proteins. A threshold protein content characterizes the G_1 cell population; cells with a subthreshold nuclear protein content (G_{1A}) do not enter S phase. Assuming that the histone/DNA ratio is constant throughout the cell cycle, the protein/DNA pattern, as shown in Fig. 2.5, can be interpreted as indicating that only those cells that accumulated the critical amount of nonhistone protein can progress to S phase. This data indicates that there is a loss of nonhistone proteins from chromatin in mitosis and that these proteins reaccumulate in nuclei prior to cell entrance to S phase. Quiescent cells have an even lower nuclear protein content than do early G_1 cells (G_{1A}) in cycling cell populations (185).

Changes in nuclear RNA content during the cell cycle are also illustrated in Fig. 2.5. Generally, nuclear RNA content correlates well with whole cell RNA and, based on either total cellular- or nuclear-RNA content, it is possible to discriminate quiescent (G_{1Q}) cells from the cycling G_1 cells as well as to recognize early G_1 (G_{1A}) from late G_1 (G_{1B}) cells (145). Nuclear RNA is localized mainly in nucleoli; relatively little RNA is in the nucleoplasm. It is well established that nucleolar morphology and staining properties correlate with cell proliferation, cell cycle progression, and neoplasia (122, 123). Also, changes in nucleoli (e.g., "nucleolar seg-

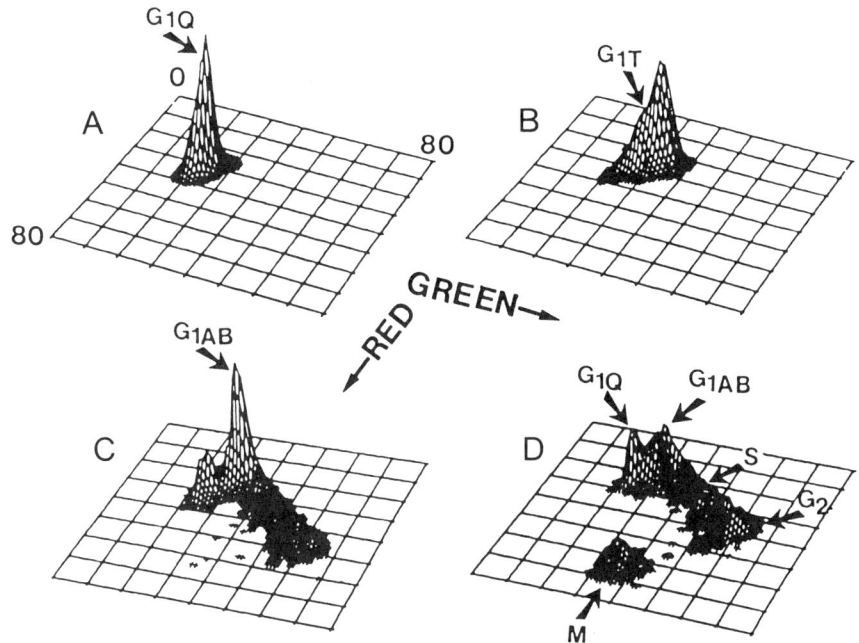

Figure 2.6. Bivariate distribuion of green and red fluorescence intensities of unstimulated and PHA-stimulated human lymphocytes following partial denaturation of DNA in situ by acid and cell staining with acridine orange (AO). The metachromatic fluorochrome AO differentially stains nondenatured (green) and denatured (red) sections of DNA (143, 148). Unstimulated lymphocytes (**A**) form a uniform population. A short time after PHA stimulation (18h; **B**) cells undergoing transition to the proliferative phase (G_{1T}) are characterized by increased green and decreased red fluorescence, which indicates that their DNA is more resistant to denaturation. During maximal stimulation after three days with PHA (**C**), in addition to G_{1Q}, G_{1A}, and G_{1B} populations there are also cells in S, G_2 and M. Treatment with Colcemid for 8hr (**D**) causes accumulation of cells in mitosis and depletion of cells from the postmitotic G_1 peak. DNA in mitotic cells, which shows maximal metachromatic (red) stainability with AO, is the most sensitive to denaturation. The degree of DNA denaturation can also be expressed as the so called α_t index, which represents the ratio of red to total (red plus green) fluorescence and is proportional to the extent of DNA denaturation in situ (145). Reproduced with permission from Darzynkiewicz Z, Traganos F. Multiparameter flow cytometry in studies of the cell cycle. In: Melamed MR, Lindmo L, Mendelsohn ML, eds. Flow Cytometry and Sorting. Second edition. New York: Wiley-Liss, 1990: 469–501.)

regation'') are often among the earliest effects in cells exposed to various antitumor drugs, especially intercalators (122). Because these morphological changes correlate with alterations in nuclear RNA content, the latter may be one of the most sensitive parameters of cell metabolism and drug sensitivity measured by flow cytometry.

DNA IN SITU SENSITIVITY TO DENATURATION

The stability of the double helical DNA structure is influenced by the ionic environment and, in the nucleus, also by interactions of DNA with histones and nonhistone proteins (review, 145). The strength and extent of these interactions may be evaluated by analyzing the profiles of DNA denaturation (''melting curves''). Modifications of histones that weaken their interactions with DNA (phosphorylation, acetylation), decrease the stability of DNA in chromatin (145). Thus, analysis of DNA sensitivity to denaturation provides information on the structure of nuclear chromatin in situ in intact cells. The flow cytometric assay of DNA denaturation is based on the use of the metachromatic fluorochrome acridine orange (AO), which can differentially stain double stranded versus denatured DNA (186). Either isolated nuclei or permeabilized cells are incubated with RNase and then heated or treated with acid to partially denature DNA. Subsequent staining with AO reveals the extent of denatured DNA, which stains metachromatically red, whereas double stranded DNA in reaction with AO gives green fluorescence (Fig. 2.6). Under appropriate staining and measurement conditions, the ratio of red-to-total (red plus green) fluorescence intensities (α_t) represents the fraction of denatured DNA, while the total fluorescence is proportional to total DNA content (Fig. 2.6).

Sensitivity of DNA to heat- or acid-induced denaturation is highly correlated with the degree of chromatin condensation. The most sensitive to denaturation is DNA in metaphase chromatin. The DNA in condensed chromatin of quiescent cells (G_{1Q}) is also highly sensitive to denaturation. The least sensitive is DNA in interphase cells' nuclei (Fig. 2.6). Among the latter, there are differences in DNA sensitivity as well: Late G_1 (G_{1B}) and early S cells have the most resistant DNA, whereas DNA in early G_1 (G_{1A}) or G_2 cells is more sensitive to denaturation. This pattern of DNA sensitivity to denaturation throughout the cell cycle resembles the pattern of varying content of nonhistone proteins in the nucleus, as discussed above (Fig. 2.5). Namely, the cells that have the highest protein-to-DNA ratio (G_{1B}, early S) have the DNA most resistant to denaturation. Conversely, M and G_{1Q} cells have minimal protein content and their DNA is the most sensitive. Thus, based on differences in DNA denaturability, it is possible to distinguish cells which have the same content of DNA but differ in the degree of chromatin condensation. Namely, G_0 (G_{1Q}) cells can be discriminated from G_{1A}, and the latter from G_{1B} cells. Likewise, G_2 cells can be distinguished from mitotic cells (186).

The technique of acid-induced DNA denaturation has recently been applied in pathology to characterize human colon tumors (187), and breast (188) and bladder cancers (189). A correlation was observed, in these studies, between staging of the malignant disease and other parameters relevant to tumor prognosis and chromatin structure of tumor cells probed by this methodology.

DETECTION OF INDIVIDUAL PROTEINS ASSOCIATED WITH CELL PROLIFERATION

Not long ago, the predominant view was that, with the exception of histones, the rates of synthesis of individual proteins in the cell were invariant during the cycle (116, 190). The last decade, however, brought innumerable reports describing changes in the rate of transcription of particular genes, variation in rates of translation or stability of mRNAs, and altered expression (accumulation) of the gene products in relation to cell progression through the cell cycle. As mentioned earlier in this chapter, the constituents that vary during the cell cycle can be subdivided into several categories. One category consists of those that have regulatory function during the cell cycle. Among them are many oncogene, or anti-oncogene products functioning as growth-factor receptors, signal transduction messengers, protein kinases, activators, etc. Another group includes proteins or peptides that are constituents of the machinery of cell replication or that are directly related to proliferation but have no apparent regulatory function. Examples of proteins in this group are enzymes and cofactors of DNA replication, nucleosome core histones, microtubule-associated proteins, and other structural proteins.

Antibodies against many of these proteins are available. Therefore, detection of these proteins by immunocytochemical methods and, with certain limitations, their quantitative estimation in the cell, is possible. Methodologies have been developed, involving optimal cell fixation protocols, permeabilization of the membrane, and other approaches, to make these intracellular antigens accessible to antibodies without a loss of their epitopes' native conformation (154). Their detection, in conjunction with measurements of DNA content and other markers of cell metabolism, by multiparameter flow cytometry, has already provided a wealth of information on the biology of the cell cycle, and has become a recognized tool in pathology to characterize the proliferative status of many tumor types (131, 150, 154, 191–193).

The scope of this article does not allow a thorough review of all the proliferation-associated cellular constituents, or a presentation of the range of applications of their antibodies as markers of cell proliferation. The list of these proteins is very long and is continually growing. The subject was recently reviewed, both with respect to all cellular constituents that change during the cell cycle (148), and in relation to these components that are localized specifically in nuclear chromatin (145). Therefore, only the few antigens that have been the most widely studied so far, are mentioned in the present review (see also Chapter 10).

The Ki-67 antibody developed by Gerdes et al. (131, 151, 193) is the most widely recognized marker of proliferating cells, and is being used in numerous laboratories worldwide. The antigen detected by this antibody is localized primarily in nucleoli and is present only in proliferating cells; its content increases during S and G_2 phase and the antigen appears to be degraded after mitosis. The nature of the protein detected by this antibody is unknown, but it is likely that the antigen is associated with the nuclear matrix (194). It has now become commonplace to estimate the fraction of proliferating cells in tumors based on the percentage of cells labelled with Ki-67 antibody. However, considering that among cycling cells the antigen's expression is transiently minimal during G_1, (prior to their entrance to S), not all Ki-67 negative cells are noncycling. It is inaccurate, therefore, to identify the percentage of Ki-67 positive cells as the "growth fraction" (195) as is commonly practiced.

The Proliferating Cell Nuclear Antigen (PCNA), originally described by Tan and his colleagues (196) as well as cyclin, described by Celis et al., (197), have been later found to be the same 36 kDa protein, an auxiliary factor of DNA polymerase (198, 199). Its expression is elevated in cycling cells, primarily during S phase (196–200). Two fractions of this protein can be distinguished in the nucleus: one, detergent-extractable, localized diffusely in nucleoplasm; and another, bound tightly to nuclear structures resembling replicon clusters in S cells (200). It should be stressed that the term "cyclin" given to this protein bears no relation to cyclin proteins that associate with $p34^{cdc2}$ protein kinase discussed earlier in the chapter.

Expression of the interchromatin p105 antigen, described by Clevenger et al., (150, 201) is also correlated with cell proliferation. A manifold increase in content of this protein parallels cell transition from G_0 to G_1 and its maximal expression occurs in mitosis (Fig. 2.7.) The p105 antigen is stable and can be detected in fixed and paraffin-embedded nuclei, which makes it possible to study its content in archival material (202).

Another nuclear protein showing cell proliferation dependency is the p34 antigen, also described by Clevenger et al., (203). This protein is expressed in cycling cells, but its level is invariant during the cell cycle. This feature and its molecular weight resemble that of $p34^{cdc2}$ protein kinase, but no evidence exists that these proteins are identical.

The enzymes associated with DNA replication, such as DNA polymerase (204), dihydrofolate reductase (205), DNA topoisomerase II (206, 207), ribonucleotide reductase (208), or thymidine kinase (205), all are expressed at higher

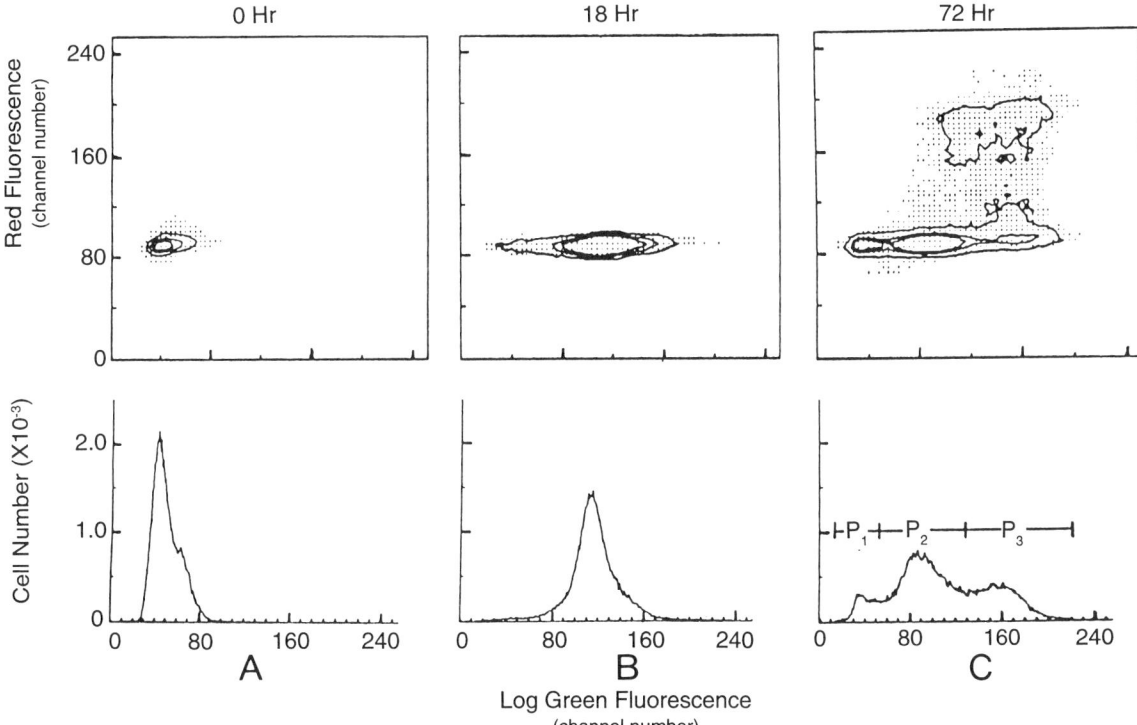

Figure 2.7. Bivariate distribution of DNA content and the expression of proliferation-associated p105 nuclear antigen. Nonstimulated (**A**) and Pokeweed mitrogen-stimulated human lymphocytes, 18(**B**) and 72h (**C**) after stimulation, stained with antibody against the proliferation associated p105 nuclear antigen (201) and with propidium iodide. DNA content is represented on a linear scale, immunofluorescence on log scales; each log scale spans three decades. A low level of p105 is seen in the case of noncycling, quiescent cells (**A**). Expression of p105 is increased soon after stimulation of lymphocytes, prior to their entrance to S (**B**). Three populations of lymphocytes, differing in degree of expression of p105, are evident 72h after stimulation (**C**). (Reprinted with permission from Clevenger CV, Epstein AL, Bauer KD. Modulation of the nuclear antigen p105 as a function of cell-cycle progression. J. Cell Physiol 1987; 130:336–343.)

levels in cycling cells. Thus, antibodies against these enzymes can serve as markers of proliferating cell populations.

Statin, a 57 kDa protein is expressed in nuclei of quiescent or senescent cells but is absent in cycling cells (209). Antibodies against this protein can be used to identify noncycling cells.

Some limitations of the immunocytochemical methods should be kept in mind when applying these antibodies for flow cytometric analysis. First, to detect intracellular antigens, cells have to be fixed and/or permeabilized to allow access of the primary and secondary antibodies. However, this procedure does not always ensure full retention of the antigen in its native conformation, such that its epitope remains unchanged. Fixation, therefore, should be customized for each particular antigen. Isolation of unfixed nuclei for the "washless" procedure for antigen detection (210) on the other hand, may result in uncontrolled loss of various nuclear constituents, including the sought after antigen. Another limitation involves the possible variation in accessibility of the antigen to the antibody. The binding of large molecules, such as antibodies, is certainly suppressed if the antigen is interacting with other constituents and is sequestered, as are most proteins in situ. Furthermore, allosteric changes of the epitope (e.g., due to chemical modifications of the antigen, such as phosphorylation, acetylation, etc.) may result in changes in the binding affinity of the antibody. Caution, therefore, should be exercised in interpretation of the data, especially in quantitative terms, resulting from the immunocytochemical procedures.

Analysis of Cell Kinetics

KINETIC INFORMATION OF THE UNIVARIATE DNA FREQUENCY HISTOGRAMS

As mentioned, the individual "snapshot" of the distribution of cells in particular phases of the cycle, regardless of whether it is based on a single- or multi-parameter measurement, provides no actual information about the kinetics of cell progression through the cycle. A combination of such measurements, however, with the information on the cell doubling time, (the latter easily obtainable by cell counts in vitro) may provide (when several assumptions are made) adequate information to estimate the duration of individual phases of the cell cycle. Namely, when no significant cell death occurs, cells grow exponentially and asynchronously and all cells divide (growth fraction = 1.0), the fractions of cells in G_1, S, G_2 are M are proportional to the duration of these phases, respectively, with a correction for exponential character of the cell age distribution in the cycle (4, 136, 211). The sum of the duration of these phases can be esti-

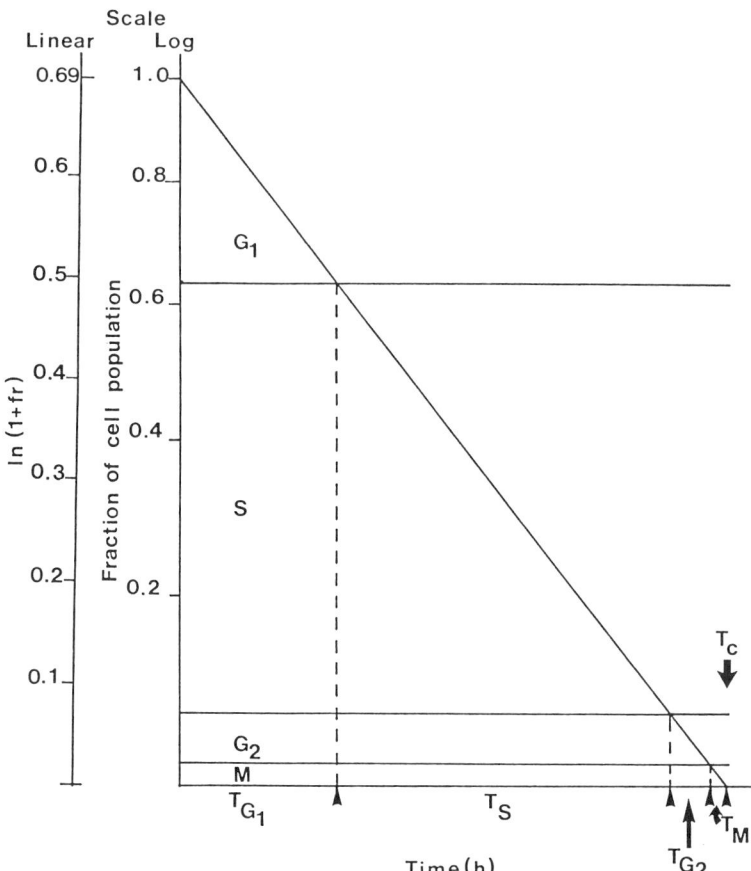

Figure 2.8. A graphic method for the estimation of the duration of the cell cycle phases based on the data obtained from the DNA content frequency histograms (modified from Okada S. Simple graphic method of computing the parameters of the life cycle of cultured mammalian cells in the exponential growth phase. Cell Biol 1967; 34:915–916.) Under the assumption that cells grow exponentially and asynchronously, and that all cells divide (growth fraction = 1.0), the cell doubling time in the cultures corresponds to the cell cycle time (T). Duration of individual phases of the cell cycle can then be estimated by plotting T, (*time, h*) on the horizontal axis (linear scale) and the fractions of cells in particular phases of the cell cycle (estimated from the frequency histograms), on the vertical axis [log scale, or as $\ln(1 + fr)$, where fr is the fraction of cells in a given phase]. The abscissa values, at the points of intersection of the line connecting T_c, and the top of the cell fraction scale (fraction = 1.0), with the ordinate values representing fractions of cells in particular phases of the cell cycle, provide estimates of the duration of these phases. It should be stressed that the average duration of the particular phases, as estimated by this or other methods (e.g., 4, 136), neglect the stochastic component in the cell cycle kinetics (see further), which is of special importance in the estimation of cell residence times in G_1 or G_0 (G_{1Q}).

mated from the cell generation (cycle time), which, under the same assumptions, and with some approximation, is equal to the cell doubling time. Mathematical formulas used to calculate the duration of individual phases of the cell cycle, based on the proportions of cells in these phases and the cell age distributions (in the steady state and exponentially growing populations), are presented elsewhere (4, 136, 211). As a practical matter, a graphic procedure can be very useful. Originally devised by Okada (212), this procedure can be easily adapted to compute the cell phases' durations from the DNA frequency histograms (Fig. 2.8).

CELL SYNCHRONIZATION IN THE CYCLE

Another approach to estimate the rates of cell progression through the cell cycle involves the time-sequential measurements of the cell cycle distributions, following their release after the initial synchronization. The synchronization can be induced by variety of methods, categorized as chemical, nutritional or mechanical (reviews, 148, 213, 214). Among chemical agents, the most common are inhibitors of DNA replication (e.g., high thymidine concentration, hydroxyurea, aphidicolin, methotrexate) or mitotic spindle poisons (vinca alkaloids, colchicine). Unfortunately, most of these agents, even when not immediately cytotoxic under the conditions used for cell synchronization, induce undesirable effects on cell metabolism that may perturb their subsequent progression through the cycle. For example, treatment with mitotic inhibitors often leads to formation of polyploid cells due to endoreduplication (215), whereas the S phase blockers cause an extensive degree of growth unbalance (180). Furthermore, the blocking effects of these agents are not always totally reversible and a lag is often observed following their removal before cells reinitiate progression through the cell cycle at normal rates.

Nutritional synchronization methods are based on the removal of the essential growth factors from the medium. Generally, the methods work quite well for normal, nontransformed cells that arrest in the G_{1Q} phase of the cycle. The most common approach involves cell growth in medium lacking isoleucine (216).

Several different approaches can be classified as mechanical synchronization. Among them, the methods based on cell separation due to differences in cell size (centrifugal elutriation) or density (density gradient centrifugation) are most often used, primarily to obtain enriched cell populations of a particular phase for biochemical studies. The mechanical dislodging of cells in mitosis (the mitotic shake-off procedure) offers the most convenient approach to obtain highly synchronous cell populations with minimal disruption of their metabolism (217). The mitotic shake off, however, can be applied only to relatively few cell types which grow attached to glass or plastic and detach during mitosis.

THE STATHMOKINETIC METHOD

The stathmokinetic approach is based on arresting the cell progression at a certain phase of the cell cycle and analyzing the rate of cell entrance to that phase. In the classic metaphase arrest technique, exponentially growing, asynchronous cell cultures, are treated with an agent that blocks cells in mitosis; the slope of the plot representing the cumulative increase in the number of mitotic cells provides an estimate of the rate of cell entrance into mitosis (218). Mathematical analysis and interpretation of the stathmokinetic data are the subjects of several reviews (157, 218–220).

The combination of flow cytometry with stathmokinesis offers the possibility of rapid and accurate estimation of rates of cell cycle progression through several points of the cell cycle simultaneously (156, 157, 220, 221). Especially useful in this respect is the application of multiparameter analysis of the cell cycle, which allows us to identify cells in various phases of the cell cycle, including mitosis (156, 157, 220). This multiparameter approach makes it possible to estimate the rate of cell entrance to mitosis, the duration of G_2, the rate of progression through different portions of S and through late G_1, and the rate of cell exit from the early portion of G_1 (G_{1A}) in a single experiment (156). Any perturbation of the cell cycle, as that induced by antitumor drugs, for example, can be easily recognized and quantitatively expressed. This method has already been extensively used in studies of the action mechanism of a variety of antitumor agents (155, 222, 223).

INCORPORATION OF BrdUrd

When incorporated into DNA during DNA replication, the thymidine analog BrdUrd can be detected cytochemically (158) or immunochemically (162). Cytochemical detection of BrdUrd is based on the phenomenon of fluorescence quenching of such fluorochromes as Hoechst 33258 (158, 160, 161) or acridine orange (159) in the complexes with DNA in which thymidine is substituted by BrdUrd. Thus, the cells that replicated DNA when exposed to this precursor can be identified as having lowered fluorescence intensity compared to cells that have not synthesized DNA.

Because Hoechst 33258 and ethidium have different binding sites, it is possible to stain cellular DNA simultaneously with both these dyes (Fig. 2.9). In contrast to Hoechst 33258, the fluorescence intensity of DNA stained with ethidium is not affected by the incorporated BrdUrd. Dual staining with Hoechst 33258 and ethidium, therefore, provides information on both the cell cycle position and DNA replication (160). This principle of DNA staining is the basis for the technique for measuring BrdUrd incorporation, originally designed by Bohmer (160) and modified and extensively used by Kubbies, Rabinovitch, and their collaborators (224–227). When cell measurements are made sequentially at different times after their exposure to BrdUrd, it is possible to estimate the duration of the first and even second cell cycle, the rate of exit from G_1, and other parameters of the cycle (224).

The second cytochemical method, developed by Crissman et al. (161), is based on dual cell staining with Hoechst 33258 and mithramycin and on subtraction of the blue fluorescence of Hoechst (quenched by BrdUrd) from the green-yellow fluorescence of mithramycin (unchanged by BrdUrd). The subtraction is done electronically at the stage of processing analog signals of the respective fluorescence intensities. The S-phase cells that incorporated BrdUrd can be easily distinguished from cells that did not incorporate the precursor. The method appears to be sensitive enough to detect a short duration (10 min) pulse of BrdUrd and can be applied to study cell kinetics.

Still another method of cytochemical detection of BrdUrd incorporation is based on quenching of the green fluorescence of acridine orange bound to DNA by intercalation (159). This method combines the differential staining of DNA and RNA (which provides for the discrimination of quiescent cells due to their low RNA content) with the ability to recognize cells that replicated DNA.

Development of antibodies against BrdUrd made it possible to detect cells replicating DNA immunocytochemically (162). This approach, combined with staining of cellular DNA using propidium, provided a very sensitive technique to simultaneously estimate the cell cycle position and detect cells that replicated DNA (163). This methodology, which already has gained wide application in basic and clinical studies on the cell cycle, has been recently thoroughly reviewed (135).

Because the immunochemical method is highly sensitive in detecting the incorporated BrdUrd, it is possible to follow the cohort of BrdUrd-labeled cells during subsequent cell cycles (136, see Fig. 2.10). The method, similar to the earlier techniques based on incorporation of tritiated thymidine (5, 6, 211), can therefore be used to estimate a variety of parameters of the cell cycle. In comparison with these earlier methods, or even with the newer methods utilizing the image

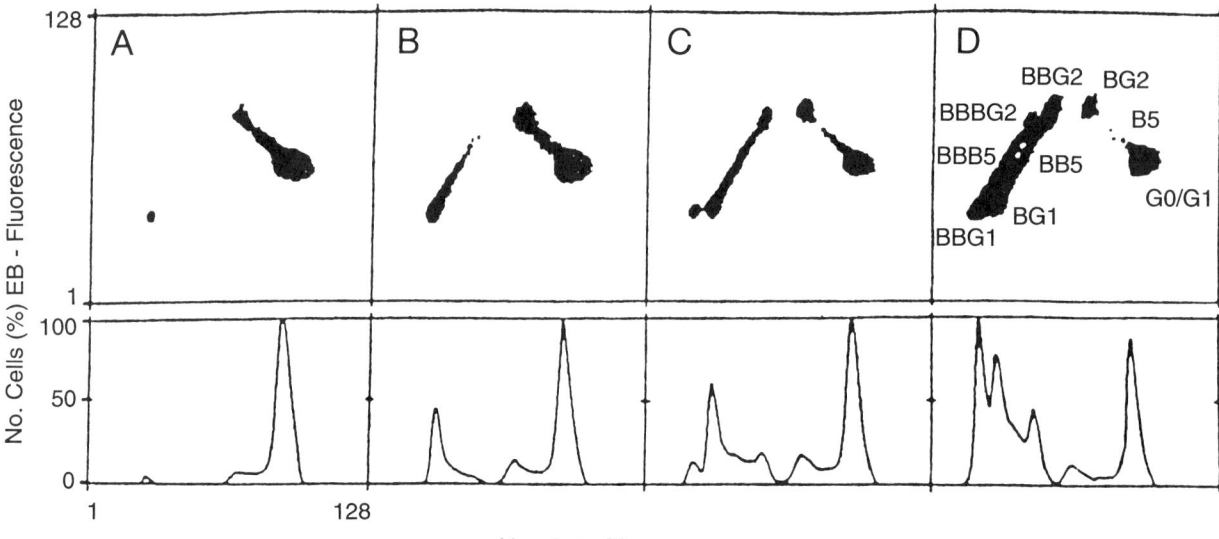

Figure 2.9. Demonstration of cell cycle progression by the bivariate analysis of *Hoechst* 33258 and ethidium *fluorescence*, following cell growth in the presence of BrdUrd.

Hoechst 33258 fluorescence (abscissa, *top and bottom panels*) and ethidium fluorescence (*ordinate, top panels*) of peripheral blood lymphocytes analyzed at (**A**)36h, (**B**)46h, (**C**)60h and (**D**)72h after growth in the presence of PHA and BrdUrd. The bottom panels show histograms of the Hoechst fluorescence. Nomenclature of the three cell cycles displayed is indicated in (**D**), where "*B*" prefixes indicate the number of rounds of replication in the presence of BrdUrd. (Reprinted with permission from Robinovitch PS. Regulation of human fibroblast growth rate by both noncycling cell fraction and transition probability is shown by growth in 5-bromodeoxyuridine followed by Hoechst 33258 flow cytometry. Proc Natl Acad Sci USA 1983;80: 2951–2955).

analysis principle for detection of BrdUrd incorporation and estimation of DNA content, the flow cytometric measurement of these parameters is much more rapid, accurate, and convenient.

PROPERTIES OF CELL POPULATIONS. STOCHASTIC ELEMENTS IN CELL KINETICS

The generation times of individual cells are highly variable, especially in vivo. This kinetic heterogeneity complicates any chemotherapeutic approach to the elimination of tumor cells, especially when treatment is based on cell synchronization and the application of cytotoxic drugs having cell cycle phase specificity. During the past two decades, numerous attempts have been made to explain the biological causes of this variability as well as to develop mathematical models that can accommodate the experimental data demonstrating such heterogeneity (7, 8, 228).

A characteristic feature of intercellular kinetic variability is the presence of an esponential-like component that is evident, for example, in the frequency distributions of intermitotic times of sibling cells also exhibiting an exponential-like distribution (7). Probabilistic models of the cell cycle have been advanced to explain this type of distribution of intermitotic times (7, 229). According to some of these models, the signal for cell proliferation is generated at random, with a constant "transition probability" characteristic for the cell type but modulated by environmental factors. Within the framework of such models, the cell cycle has been subdivided into "probabilistic" and "deterministic" compartments, respectively (7, 8). Because the G_1 phase generally shows maximum variability, the probabilistic compartment (A) was considered to be part of G_1.

Multiparameter analysis of cycling cells by flow cytometry reveals several subcompartments of the cell cycle; the cells in these compartments exhibit different metabolic (rRNA content, degree of chromatin condensation) and kinetic properties (143, 148, 149). From two of these compartments, namely from G_{1Q} and G_{1A}, cells exit with exponential-like kinetics (141, 230). The G_{1Q} compartment is similar to that generally denoted as G_0: Cells in G_{1Q} are temporarily withdrawn from the cycle and are characterized by very low metabolic activity. These dormant cells have minimal RNA content; low mitochondrial mass and transmembrane potential; and very condensed chromatin. An example of G_{1Q} cells are nonstimulated peripheral blood lymphocytes. G_{1A} cells are present in exponentially growing populations. They are postmitotic, early G_1 cells, characterized by low RNA content, low nonhistone nuclear protein content, and rather condensed chromatin (Figs. 2.4, 2.5). Clearly, cell transit through these compartments shows typical probabilistic, stochastic-like characteristics (83, 148, 179, 231). Other phases of the cell cycle are much more uniform and do not show evidence of such type of variability (179).

Despite the apparent "probabilistic" character, the exponential-like component of cell kinetics most likely has a metabolic foundation (228). There is extensive literature suggesting that cell size, overall protein, or RNA content, may be causally associated with cell kinetics (232–234). How-

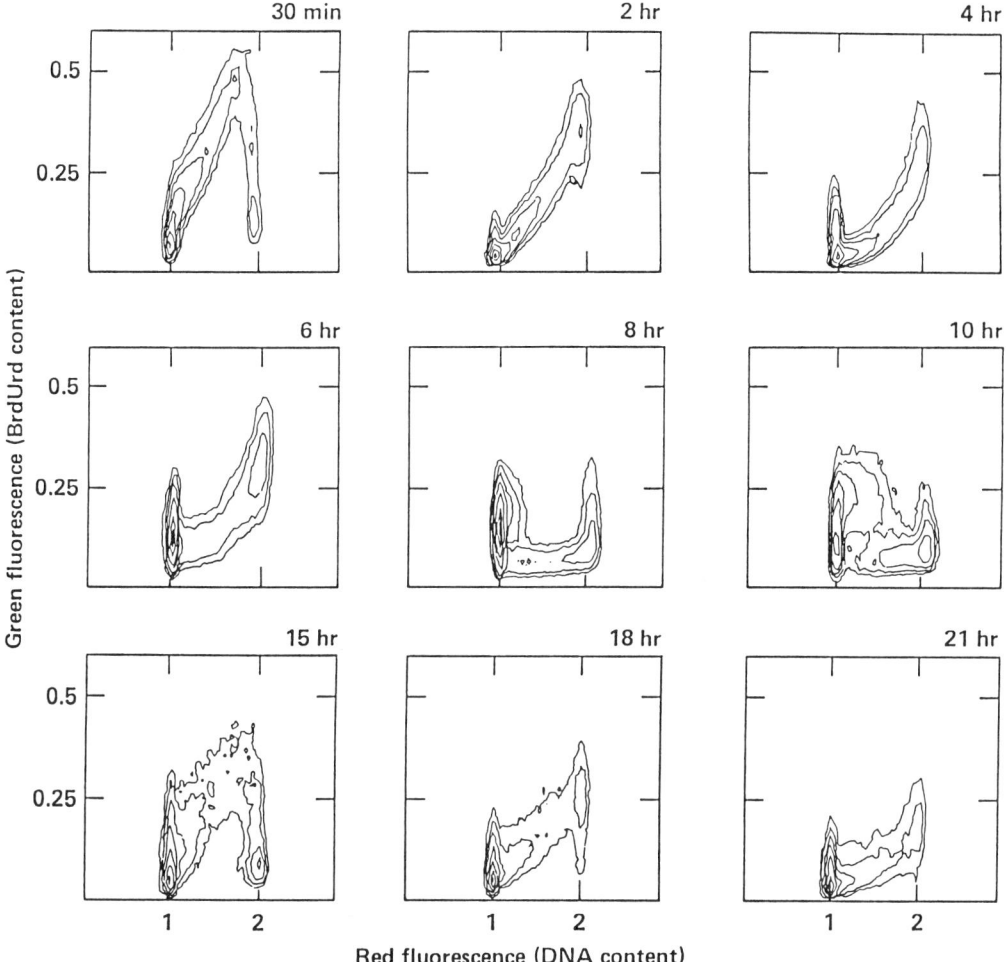

Figure 2.10. Serial bivariate distributions representing fluorescence of CHO cells labelled in vitro with BrdUrd for 10 min and stained with BrdUrd antibody and propidium (136). Cells were analyzed at different times after the pulse of BrdUrd, as indicated above each panel. Kinetic parameters can be estimated from the analysis of progression of the cohort of cells labeled with BrdUrd through the subsequent cycles (Reprinted with permission from Gray WJ, Dolbeare F, Pallavicini MG. Quantitative cell-cycle analysis. In: Melamed MR, Lindmo L, Mendelsohn ML, eds. Flow Cytometry and Sorting. Second edition. New York: Wiley-Liss, 1990: 445–467).

ever, populations of cycling cells are very uniform with respect to cell size, RNA, or protein content and their distributions are generally Gaussian (179). It is unlikely, therefore, that these constituents are causally related to the stochastic component of cell kinetics. On the other hand, cell populations are extremely heterogenous with respect to the distribution of most types of growth factor receptors, oncogene products, or proliferation associated antigens (230). For example, cell distributions of the ras protein, PCNA, or nuclear antigen p105, all require use of logarithmic coordinates (153, 201, 202, 235). This would suggest that a major cause of the kinetic heterogeneity is the high intercellular variability in the content of the individual, proliferation associated, regulatory proteins. Thus, cell exit from G_{1Q} or G_{1A}, may be triggered only after saturation of a threshold number of receptors by growth factors. This saturation, which can be achieved only in the case of cells that already expressed large number of these receptors, may then provide an adequate transduction signal to subsequently activate the key molecules involved in cell cycle progression, such as the combination of the respective cyclins and $p34^{cdc2}$ protein kinase.

It is likely that mechanisms generating metabolic and kinetic heterogeneity have an evolutionary origin, that came about in order to increase survivability of cell populations under adverse growth conditions (e.g., exposure to radiation or to cell-cycle-specific environmental toxins). Extended residence in the prereplicative compartments could allow cells to repair the damage induced by these factors prior to DNA replication or mitosis. Otherwise, if not repaired prior to mitosis, the damage (e.g., double stranded DNA breaks caused by radiation), could be lethal. Cell populations that developed an ability to provide such mechanisms would have an obvious advantage over populations that proliferate rapidly and uniformly. The same mechanism may operate (apart from gene amplification) in providing resistance of tumor cell populations to antitumor drugs. Indeed, the heterogeneity that is manifested as the presence of the stochastic-like components, both in metabolic and kinetic cell properties, appears to ensure optimal protection of cell populations under adverse conditions (236).

Cell Cycle Markers: A New Framework for the Classification of Tumors

Tumor diagnosis and classification still often rely on histological examination of tissue by the staining methods developed nearly a century ago. Unfortunately, the histology is of limited prognostic value. On the other hand, tumor grading, based primarily on the degree of disease advancement, is more predictive but it relates descriptively to the late symptoms rather than to the biological cause of the malignancy. It is a common observation that the clinical outcome of histologically identical tumors is different. Thus, there is an undisputable need for additional prognostic markers, intrinsic to individual tumors, that can allow one to customize therapy for each patient in order to obtain maximal therapeutic effects at the lowest toxicity.

Regardless of its etiology or histological origin, cancer is a disease characterized by abnormal cell proliferation. Malfunction of the regulatory mechanisms of the cell cycle is a common pathway of cancer, and most antitumor drugs are designed to suppress cell proliferation. There is great diversity among these drugs regarding their mechanism of action or cell cycle phase(s) specificity (review 237). Unfortunately, many tumors escape treatment and classical histopathology is unable to predict individual tumor response. It is expected that markers of cell proliferation that better correlate with tumor growth rate will be predictive of tumor sensitivity to cytostatic drugs.

The markers of cell proliferation that can be of clinical interest can be subdivided into three categories. The first group consists of the histone H1 p34^{cdc2} protein kinase and the key molecules involved in the cyclic activation/deactivation of this enzyme complex, as presented in Fig. 2.1. These molecules are the major regulatory factors of the cell cycle; therefore, the malfunction of any one of them alters the essential regulatory mechanisms and cell kinetics. It can be expected that a large portion of most types of tumors will be characterized by some defect in this common regulatory pathway. The defect may either be a primary cause of the tumor or be secondary to the earlier change, e.g., in the expression of an oncogene that interacts with this main cell cycle regulating complex. The markers in this group can also be predictive of tumor sensitivity to a particular type of cytostatic drug.

A variety of the oncogenes which interact with the p34^{cdc2} protein kinase or with one of the key molecules of the p34^{cdc2} complex can be included in the second group of markers. These are predominantly molecules that belong to the signal transduction pathway. They are often kinases or phosphatases that can modify individual proteins of the p34^{cdc2} complex or substrates modified by the p34^{cdc2} kinase. Generally, they can be subdivided into activators or suppressors of cell proliferation. The scope of this article does not allow review of this large group of proteins. As is already known, oncogene defects are often the primary lesions leading to tumor development. Because of their large number and diversity, one can expect that relatively few tumors will have the same type of defect as one of these molecules.

The actual parameters of cell kinetics can be grouped into the third category of markers. This includes the proportion of cells in S phase, the rate of cell progression through the cycle, duration of the cycle, growth fraction, etc. Because, as discussed earlier in this chapter, cell cycle progression and cell growth are coupled, the parameters of cell growth, such as cellular rRNA content or nucleolar activity, belong to this category as well. This third group of markers has been the most extensively explored so far, and there are innumerable papers which emphasize their prognostic value (see the recent review 238).

It now becomes apparent that flow cytometry and associated techniques, rather than simply diagnosing cancer, can be used more successfully to provide information on cell constituents that allow one to evaluate the tumor cells' metabolic status, growth, and kinetic potential or heterogeneity. These features can be of prognostic value and can bear clues with respect to tumor sensitivity toward particular treatment regimens. It is thus expected that a new classification of tumors will evolve based on combined cytogenetic, molecular, biochemical, kinetic, and intercellular heterogeneity profiles of the cell population. The proliferation markers will be essential in this classification. This new classification will complement, or perhaps even replace the traditional one, which is based on qualitative staining methods and microscopy. The new classification will be predictive of tumor malignancy (growth rate, metastatic potential) sensitivity to various drugs and, thus, more useful in clinical applications. Treatment can then be individually customized to the profile of the tumor. Flow cytometry clearly will be of paramount importance for the development of such a classification.

ACKNOWLEDGMENTS

I thank Drs. Frank Traganos and Janusz Skierski and Ms. Irene Logsdon for their help in the preparation of the manuscript.

Supported by NCI Grants R37 CA23296 and RO1 28704, as well as the Carl Inserra Fund.

REFERENCES

1. Howard A, Pelc SR. Nuclear incorporation of 32P as demonstrated on autoradiographs. Exp Cell Res 1951:178–187.
2. Lajtha LG. On the concept of the cell cycle. J Cell Comp Physiol 1963;62(Suppl 1):142–145.
3. Mendelsohn ML. Autoradiographic analysis of cell proliferation in spontaneous breast cancer of C3H mouse. III. The growth fraction. J Natl Cancer Inst 1962;28:1015–1029.
4. Ahern WA, Camplejohn PC, Wright NA. An introduction to cell population kinetics. London: Edward Arnold Ltd., 1977.
5. Simpson-Herren L. Autoradiographic techniques for measurement of the labeling index. In: Gray JW, Darzynkiewicz Z, eds. Techniques in cell cycle analysis. Clifton, N.J.: Humana Press, 1987:1–29.

6. Shackney SE, Ritch PS. Percent labelled mitosis curve. In: Gray JW, Darzynkiewicz Z, eds. Techniques in cell cycle analysis. Clifton, NJ: Humana Press, 1987:31–45.
7. Smith JA, Martin L. Do cells cycle? Proc Natl Acad Sci USA 1973;70:1263–1267.
8. Shields R. Further evidence for a random transition in the cell cycle. Nature 1978;272:755–758.
9. Shapiro HM. Practical flow cytometry. Second edition. New York: Alan R Liss, 1988.
10. Melamed MR, Lindmo T, Mendelsohn ML, eds. Flow cytometry and sorting. Second edition. New York: Wiley-Liss, 1990.
11. Darzynkiewicz Z, Crissman HA, eds. Flow cytometry. Meth Cell Biol Vol 33. San Diego: Academic Press, 1990.
12. Baserga R. The biology of cell reproduction. Cambridge, MA.: Harvard University Press, 1985.
13. Brachet J. Molecular cytology. Vol 1, The cell cycle. San Diego: Academic Press, 1985.
14. Prescott DM. Cell reproduction. Int Rev Cytol 1987;100:93–128.
15. Darzynkiewicz Z. Cell growth and division cycle. Int Encycl Pharmacol Ther 1986;121:1–43.
16. Beach D, Basilico C, Newport J, eds. Cell cycle controls in eukarytes. Cold Spring Harbor Laboratory, 1988.
17. Cross F, Roberts J, Weintraub H. Simple and complex cell cycles. Annu Rev Cell Biol 1989;5:341–395.
18. Worcell A, Benyajati C. Higher order coiling of DNA in chromatin. Cell 1977;12:83–100.
19. Igo-Kemenes T, Horz W, Achan HG. Chromatin. Annu Rev Biochem 1982;1:89–121.
20. van Holde KE. Chromatin. New York: Springer Verlag, 1989.
21. Compton JL, Hancock R, Oudet P, Chambon P. Biochemical and electron microscopic evidence that the subunit structure of Chinese-hamster ovary interphase chromatin is conserved in mitotic chromosomes. Eur J Biochem 1976;70:555–568.
22. Smith S, Stillman B. Stepwise assembly of chromatin during DNA replication in vitro. EMBO J 1991;10:971–980.
23. Allfrey VG. Molecular aspects of the regulation of eukaryotic transcription: Nucleosomal proteins and their postsynthetic modifications in the coltrol of DNA conformation and template function. In: Goldstein L, Prescott D, eds. Cell biology, a comprehensive treaty Vol 3, New York: Academic Press, 1980:374–437.
24. Allfrey VG. Post-synthetic modifications of histone structure: A mechanism for the control of chromosome structure by modulation of histone-DNA interaction. In: Li HJ, Eckhardt RA, eds. Chromatin and chromosome structure, New York: Academic Press, 1977:167–191.
25. Reczek PR, Weissman D, Huvos PE, Fasman GD. Sodium butyrate induced structural changes in HeLa cell chromatin. Biochemistry 1982;21:993–1002.
26. Simpson RT. Structure of chromatin containing excessively acetylated H3 and H4. Cell, 1978;13:691–699.
27. Moore M, Jackson V, Sealy L, Chalkley R. Comparative studies of highly metabolically active histone acetylation. Biochim Biophys Acta 1979;561:248–260.
28. Halleck MS, Gurley LR. Histone acetylation and heterochromatin of cultured Perymyscus cells. Exp Cell Res 1981;132:201–213.
29. D'Anna JA, Tobey R, Barnham SS, Gurley LR. A reduction in the degree of H4 acetylation during mitosis in Chinese hamster cells. Biochem Biophys Res Commun 1977;77:187–194.
30. Chahal SS, Matthews HR, Bradbury EM. Acetylation of histone H4 and its role in chromatin structure and function. Nature 1980;287:76–79.
31. Ruiz-Carrillo A, Wangh LJ, Allfrey VG. Processing of newly synthesized histone molecules. Science 1975;190:117–128.
32. DePamphilis ML, Wassarman PM. Replication of eukaryotic chromosomes: A close-up of the replication fork. Annu Rev Biochem 1980;49:627–666.
33. Smith PA, Jackson V, Chalkley R. Two-stage maturation process for newly replicated chromatin. Biochemistry 1984;23:1576–1581.
34. Waterborg JH, Matthews HR. Patterns of histone H4 acetylation in Physarum polycephalum. H2A and H2B acetylation is functionally distinct from H3 and H4 acetylation. Eur J Biochem 1984;142:329–335.
35. Christensen ME, Rattner JB, Dixon GH. Hyperacetylation of histone H4 promotes chromatin decondensation prior to histone replacement by protamines during spermatogenesis of rainbow trout. Nucleic Acid Res 1984;12:4575–4592.
36. Riggs MG, Whittaker RG, Newman JR, Ingram VM. N-butyrate causes histone modification in HeLa and Friend erythroleukemia cells. Nature 1977;268:462–464.
37. D'Anna JA, Tobey RA, Burley LR. Concentration-dependent effects of sodium butyrate in Chinese hamster cells: Cell cycle progression, inner-histone acetylation, histone H1 dephosphorylation, and induction of an H1-like protein. Biochemistry 1980;19:2656–2671.
38. Darzynkiewicz Z, Traganos F, Xue S-B, Melamed M. Effect of n-butyrate on cell cycle progression and in situ chromatin structure. Exp Cell Res 1981;136:279–293.
39. Vidali G, Boffa LC, Bradbury EM, Allfrey VG. Butyrate suppression of histone deacetylation leads to accumulation of multiacetylated forms of histones H3 and H4 and increased DNase I sensitivity of the associated DNA sequences. Proc Natl Acad Sci USA 1978;75:2239–2243.
40. D'Anna JA, Gurley LR, Tobey RA. Syntheses and modulation in the chromatin contents of histones H1° and H1 during G_1 and S phases in Chinese hamster cells. Biochemistry 1982;21:3991–4001.
41. Sun J-M, Wiaderkiewicz R, Ruiz-Carrillo A. Histone H5 in the control of DNA synthesis and cell proliferation. Science 1989;245:68–71.
42. Borun TW, Pearson D, Paik WK. Studies on histone methylation during the HeLa S3 cell cycle. J Biol Chem 1972;247:4288–4298.
43. Thomas G, Lange HW, Hempel K. Kinetics of histone methylation in vivo and its relation to the cell cycle in Ehrlich ascites tumor cells. Eur J Biochem 1975;51:609–615.
44. Ord MG, Stocken LA. Adenosine diphosphate ribosylated histones. Biochem J 1977;161:583–592.
45. Adamietz P, Rudolph A. ADP-ribosylation of nuclear proteins in vivo. Identification of histone H2B as a major acceptor for monoo- and poly(ADP)ribose in dimethyl sulfate-treated hepatoma SH 7974 cells. J Biol Chem 1984;259:6841–6846.
46. Kidwell WR, Mage MG. Changes in poly(adenosine diphosphateribose) polymerase in synchronous HeLa cells. Biochemistry 1976;15:1213–1217.
47. Tanuma S, Kanai Y. Poly(ADP-ribosyl)ation of chromosomal proteins in the HeLa S3 cell cycle. J Biol Chem 1982;257:6565–6570.
48. Berger NA, Kaichi AS, Steward PG, Klevecz RR, Forrest GL, Gross SD. Synthesis of poly(adenosine diphosphate-ribose) in synchronized Chinese hamster cells. Exp Cell Res 1978;117:127–135.
49. Mandel P, Okazaki H, Niedergang C. Poly(adenosine diphosphate ribose). Prog Nucl Acid Res Mol biol 1982;27:1–51.
50. Poirier GG, deMurcia G, Jongstra-Bilen J, Niedergang C, Mandel P. Poly(ADP)ribosylation of polynuleosomes causes relaxation of chromatin structure. Proc Natl Acad Sci USA 1982;79:3423–3427.
51. Goldknopf IL, Busch H. Isopeptide linkage between nonhistone and histone 2A polypeptides of chromosomal conjugate-protein A 24. Proc Natl Acad Sci USA 1977;74:864–868.
52. Wu RS, Kohn KW, Bonner WB. Metabolism of ubiquitinated histones. J Biol Chem 1981;256:5916–5920.
53. Matsui SI, Seon BK, Sandberg AH. Disappearance of structural protein A24 in mitosis. Implications for molecular basis of chromatin condensation. Proc Natl Acad Sci USA 1979;76:6386–6390.
54. Goldknopf IL, Sudhaker S, Rosenbaum I. Busch H. Timing of ubiquitin synthesis and conjugation into protein A24 during the HeLa cell cycle. Biochem Biophys Res Commun 1980;95:1253–1260.
55. Finley D, Ciechanower A, Varshavsky A. Thermolability of ubiquitin-activating enzyme from mammalian cell cycle mutant ts85. Cell 1984;37:43–55.

56. Goldknopf IL, French MF, Daskal Y, Busch H. A reciprocal relationship between contents of free ubiquitin and protein A24, its conjugate with histone 2A in chromatin obtained by the DNase II, Mg2 procedure. Biochem Biophys Res Commun 1978;84:786–793.
57. Bradbury EM, Inglis RJ, Matthews HR. Control of cell division by very lysine-rich histone (F1) phosphorylation. Nature 1974;247:257–261.
58. Matthews HE, Hubner VD. Nuclear protein kinases. Mol Cell Biochem 1985;59:81–89.
59. Langan TA, Rall SC, Cole RD. Variation in primary structure at a phosphorylation site to lysine-rich histones. J Biol Chem 1971;245:1942–1944.
60. Fasy TM, Inone A, Johnson EM, Allfrey VG. Phosphorylation of H1 and H5 histones by cyclic AMP-dependent protein kinase reduces DNA binding. Biochim Biophys Acta 1979;564:322–334.
61. Laks MS, Harrison JJ, Schwoch G, Jungmann RA. Modification of nuclear protein kinase activity and phosphorylation of of histone H1 subspecies during prereplication phase of rat liver regeneration. J Biol Chem 1981;256:8775–8785.
62. Hardie DG, Matthews HR, Bradbury EM. Cell cycle dependence of two nuclear histone kinase enzyme activities. Eur J Biochem 1976;66:37–42.
63. Matthews HR. Phosphorylation of H1 and chromosome condensation. In: Bradbury EM, Javaherian K, eds. The organization and expression of of the eukaryotic genome. New York: Academic Press, 1977:67–80.
64. Mitchelson K, Chambers T, Bradbury EM, Matthews HR. Activation of histone kinase in G_2 phase of the cell cycle in *Physarum polysephalum*. FEBS Letters 1978;92:339–342.
65. Langan TA, Gautier J, Lohka M, Hollingsworth R, Moreno S, Nurse P, Maller J, Sclafani RA. Mammalian growth-associated H1 kinase: a homolog of cdc2/CDC28 protein kinases controlling mitotic entry in yeast and frog cells. Mol Cell Biol 1989;9:3860–3868.
66. Nurse P. Universal control mechanism regulating onset of M phase. Nature 1990;344:503–505.
67. Hayles J, Nurse P. A review of mitosis in fission yeast *Schizosaccharomyces pombe*. Exp Cell Res 1989;184:273–286.
68. Murray AW, Kirschner MW. Dominos and clocks: the union of two views on the cell cycle. Science 1989;246:614–621.
69. Dunphy EG, Newport JW. Unraveling of mitotic control mechanisms. Cell 1988; 55:925–928.
70. D'Urso G, Marracino RL, Marshak DL, Roberts JM. Cell cycle control of DNA replication by a homologue from human cells of the p34^{cdc2} protein kinase. Science 258;1990:786–791.
71. Furukawa Y, Piwnica-Worms H, Ernst TJ, Kanakura Y, Griffin JD. CDC2 gene expression at the G_1 to S transition in human T lymphocytes. Science 1990;250:805–808.
72. Pines J, Hunter T, Isolation of a human cyclin cDNA: evidence for cyclin mRNA and protein regulation in the cell cycle and for interaction with p34^{cdc2}. Cell 1989;58:833–846.
73. Draetta G, Piwnica-Worms H, Morrison D, Druker B, Roberts T, Beach D. Human cdc2 protein kinase is a major cell cycle regulated tyrosine kinase substrate. Nature 1988;336:738–744.
74. Pardee AB. A restriction point for control of normal animal cell proliferation. Proc Natl Acad Sci USA 1974;71:1286–1290.
75. Campisi J, Medrano EE, Morreo G, Pardee AB. Restriction point control of cell growth by a labile protein: evidence for increased stability in transformed cells. Proc Natl Acad Sci USA 1982;79:436–440.
76. Draetta G, Luca F, Westendorf J, Brizuela L. Ruderman J, Beach D. cdc2 protein kinase is complexed with both cyclin A and B: evidence for proteolytic inactivation of MPF. Cell 1989;56:829–838.
77. Gautier J, Matsukawa T, Nurse P, Maller J. Dephosphorylation and activation of Xenopus p34 protein kinase during the cell cycle. Nature 1990;339:626–629.
78. Feiler HS, Jacobs TW. Cell division in higher plants: a cdc2 gene, its p 34-kDa product, and histone H1 kinase activity in pea. Proc Natl Acad Sci USA 1990;87:5397–5401.
79. Standart N, Minshull J, Pines J, Hunt T. Cyclin synthesis, modification and destruction during meiotic maturation of the starfish oocyte. Dev Biol 1987; 124:248–254.
80. Wittenberg C, Sugimoto K, Reed SI. G_1-specific cyclins of S. cerevisiae: cell cycle periodicity, regulation by mating pheromone, and association with the p34^{cdc28} protein kinase. Cell 1990;62:225–237.
81. Murray AW, Kirschner MW. Cyclin synthesis drives the early embryonic cell cycle. Nature 1989;339:275–280.
82. Solomon MJ, Glotzer M, Lee TH, Philippe M, Kirschner MW. Cyclin activation of p34^{cdc2}. Cell 1990;63:1013–1024.
83. Darzynkiewicz Z, Sharpless T, Staiano-Coico L, Melamed MR. Subcompartments of the G_1 phase of cell cycle detected by flow cytometry. Proc Natl Acad Sci USA 1980;77:6696–6700.
84. Minshull J, Blow JJ, Hunt T. Translation of cyclin mRNA is necessary for extracts of activated Xenopus eggs to enter mitosis. Cell 1989;56:947–956.
85. Murray AW, Solomon MJ, Kirschner MW. The role of cyclin synthesis and degradation in the control of maturation promoting factor activity. Nature 1989;339:380–386.
86. Masui Y, Markert CL. Cytoplasmic control of nuclear behavior during mitotic maturation of frog oocytes. J Exp Zool 1971;177:349–356.
87. Smith LD, Eckert RE. The interaction of steroids with *rana pipiens* oocytes in the induction of maturation. Dev Biol 1971;25:233–247.
88. Dunphy WL, Newport JW. Fission yeast p13 blocks mitotic activation and tyrosine dephosphorylation of Xenopus cdc2 protein kinase. Cell 1989;58:181–191.
89. Meijer L, Arion D, Golsteyn R, Pines J, Brizuela L, Hunt T, Beach D. Cyclin is a component of the sea urchin egg M-phase specific histone H1 kinase. EMBO J. 1989;8:2275–2282.
90. Kumagai A, Dunphy WG. The cdc25 protein controls tyrosine dephosphorylation of the cdc2 protein in a cell free system. Cell 1991; 64:903–914.
91. Dessev G, Iovcheva-Dessev C, Bischoff JR, Beach D, Goldman R. A complex containing p34^{cdc2} and cyclin B phosphorylates the nuclear lamin and dissassembles nuclei of clam oocytes *in vitro*. J Cell Biol 1991;112:523–533.
92. Broek D, Bartlett R, Crawford K, Nurse P. Involvement of p34^{cdc2} in establishing the dependency of S phase on mitosis. Nature 1991;349:388–393.
93. Gurley LR, Walter RA, Tobey RA. Sequential phosphorylation of histone subfractions in the Chinese hamster cells. J Cell Biol 1975;250:3936–3944.
94. Ajiro K, Borun TW, Cohen LH. Phosphorylation states of different histone 1 subtypes and their relationship to chromatin function during HeLa S-3 cell cycle. Biochemistry 1981;20:1445–1454.
95. Ajiro K, Borun T, Shulman SD, McFadden GM, Cohen LH. Comparison of the structure of human histones 1A and 1B and their intramolecular phosphorylation sites during the HeLa S-3 cell cycle. Biochemistry 1981;20:1454–1464.
96. Paulson JR, Taylor SS. Phosphorylation of histones 1 and 3 and nonhistone high mobility group 14 by an endogeneous kinase in HeLa metaphase chromosomes. J Biol Chem 1982;257:6064–6072.
97. Gurley LR, Walters RA, Barham SS, Deaven LL. Heterochromatin and histone phosphorylation. Exp Cell Res 1978;111:373–383.
98. Booher R, Beach D. Involvement of a type 1 phosphatase encoded by bws1+ in fission yeast mitotic control. Cell 1989;57:1009–1016.
99. Kinoshita N, Ohkura H, Yanagida M. Distinct, essential roles of type 1 and 2S protein phosphases in the control of fission yeast cell division cycles. Cell 1990;63:405–415.
100. Okhura H, Yanagida M. S. *pombe* gene *sds22+* essential for a midmitotic transition encodes a leucine-rich repeat protein that positively modulated protein phosphatase-1. Cell 1991;64:149-157.

101. Panyim S, Chalkley R. A new histone found only in mammalian tissues with little cell division. Biochem Biophys Res Commun 1969;37:1042–1043.
102. Chabanas S, Lawrence JJ, Humbert J, Eisen H. Cell cycle regulation of histone H1° in CHO cells: a flow cytofluorimetric study after double staining of the cells. EMBO J 1983;2:833–837.
103. Cary PD, Hines ML, Bradbury ME, Smith BJ, Johns EW. Conformational studies of histone H1° in comparison with histones H1 and H5. Eur J Biochem 1981;120:371–377.
104. Rabilloud T, Pennetier JL, Hibner U, Vincens P, Tarroux P, Rougeon F. Stage transitions in B-lymphocyte differentiation correlate with limited variations in nuclear proteins. Proc Natl Acad Sci USA 1991;88:1830–1834.
105. Johns EW ed. The HMB chromosomal proteins. New York Academic Press, 1982.
106. Studzinski GP. Oncogenes, growth and the cell cycle: an overview. Cell Tissue Kinet 1989;22:405–424.
107. Denhardt DT, Edwards DR, Parfett CLJ. Gene expression during the mammalian cell cycle. Biochim Biophys Acta 1986;865:83–125.
108. Goodwin GH, Walker GM, Johns EW. The high mobility group (HMG) nonhistone chromosomal proteins. In: Busch H ed. The cell nucleus, Vol VI: Chromatin, Part C. New York: Academic Press, 1978:181–291.
109. Weisbrod S, Weintraub H. Isolation of subclass of nuclear proteins responsible for conferring a Nase I-sensitivity structure on globin chromatin. Proc Natl Acad Sci USA 1979;76:630–634.
110. Weisbrod S. Active chromatin. Nature 1982; 297:289–295.
111. Seyedin SM, Kistler WS. Levels of chromosomal protein high mobility group 32 parallel the proliferative activity of testis, skeletal muscle and other organs. J Biol Chem 1979;254:11264–11271.
112. Seyedin SM, Pehrson JR, Cole DR. Loss of chromosomal high mobility group proteins HMG1 and HMG2 when mouse neuroblastome and Friend erythroleukemis cells become committed to differentiation. Proc Natl Acad Sci USA 1981;78:2988–2992.
113. Bhorjee JS. Differential phosphorylation of nuclear nonhistone high mobility group proteins HMG14 and HMG17 during the cell cycle. Proc Natl Acad Sci USA 1981;78:6944–6948.
114. Reeves R, Langan TA, Nissen MS. Phosphorylation of the DNA binding domain of nonhistone high mobility group I protein by cdc2 kinase: Reduction of binding affinity. Proc Natl Acad Sci USA 1991;88:1671–1675.
115. Fan H, Penman S. Regulation of protein synthesis in mammalian cells. J Mol Biol 1970;50:655–670.
116. Milcarek C, Zahn K. The synthesis of ninety proteins including actine throughout the Hela cell cycle. J Cell Biol 1978;79:833–838.
117. Isenberg I. Histones. Annu Rev Biochem 1979;48:159–191.
118. Wu RS, Bonner WM. Separation of basal histone synthesis from S-phase histone synthesis in dividing cells. Cell 1981;27:321–330.
119. Waithe WI, renaud J, Nadeau P, Pallotta D. Histone synthesis by lymphocytes in G_0 and G_1. Biochemistry 1983;22:1778–1783.
120. Wu RS, Tsai S, Bonner WB. Patterns of histone variant synthesis can distinguish G_0 from G_1 cells. Cell 1982;31:367–374.
121. Ronning OW, Lindmo T, Peterson EO, Seglen PO. The role of protein accumulation in the cell cycle control of human NH1K 3025 cells. J Cell Physiol 1981;109:411–418.
122. Baserga R. Growth in size and cell DNA replication. Exp Cell Res 1984;151:1–4.
123. Darzynkiewicz Z. Cellular RNA content, a feature correlated with cell kinetics and tumor prognosis. Leukemia 1988;2:777–787.
124. Bergounioux C, Perennes C, Brown SC, Gadal P. Nuclear RNA quantification in protoplast cell-cycle phases. Cytometry 1988;9:84–87.
125. Ringdahl MH, Cooper HL. Sequential changes in ribosomal activity during the activation and cessation of growth in lymphocytes stimulated by concanavalin A. J Cell Physiol 1978;97:253–264.
126. Darzynkiewicz Z, Traganos F, Sharpless T, Melamed MR. Lymphocyte stimulation: a rapid, multiparameter analysis. Proc Natl Acad Sci USA 1976;73:2881–2884.
127. Busch H, Smetana K. The nucleolus. New York: Academic Press, 1970.
128. Ghosh S. The nucleolus. Int Rev Cytol, Suppl. 17. Cytology and Cell Physiology. 1987:573–598.
129. Ochs R, Lischwe M, O'Leary P, Busch H. Localization of nucleolar phosphoproteins B23 and C23 during mitosis. Exp Cell Res 1983;146:139–149.
130. Spector DC, Ochs RL, Busch H. Silver staining, immunofluorescence and immunoelectron microscopic localization of nucleolar phosphoproteins B23 and C23. Chromosoma 90: 139–148.
131. Gerdes J, Lemke H, Baisch H, Wacker H-H, Schwab U, Stein H. Cell cycle analysis of cell proliferation-associated human nuclear antigen defined by the mononuclear antibody Ki-67. J Immunol 1984;133:1710–1715.
132. Freeman JW, McGrath P, Bondada V, Selliah N, Ownby H, Maloney T, Busch RK, Busch H. Prognostic significance of proliferation associated nucleolar antigen p120 in human breast carcinoma. Cancer Res. 1991;51:1973–1978.
133. Andreeff M, Assing G, Cirricione C. Prognostic value of DNA/RNA flow cytometry in myeloblastic and lymphoblastic leukemia in adults: RNA content and S-phase predict remission duration and survival in multivariate analysis. Ann NY Acad Sci 1986;468:387–406.
134. Barlogie B, Alexanian R, Dixon D, Smith L, Smallwood L, Delasalle K. Prognostic implications of tumor cell DNA and RNA content in multiple myeloma. Blood 1985;66:338–341.
135. Horan PPK, Muirhead KA, Slezak SE. Standards and controls in flow cytometry. In: Melamed MR, Lindmo L, Mendelsohn ML, eds. Flow cytometry and sorting. Second edition. New York: Wiley-Liss, 1990: 397–414.
136. Gray WJ, Dolbeare F, Pallavicini MG. Quantitative cell-cycle analysis. In: Melamed MR, Lindmo L, Mendelsohn ML, eds. Flow cytometry and sorting. Second edition. New York: Wiley-Liss, 1990: 445–467.
137. Hedley DW. Flow cytometry using paraffin-embedded tissue: Five years on. Cytometry 1989;10:229–241.
138. Dean PN. Data processing. In: Melamed MR, Lindmo L, Mendelsohn ML, eds. Flow cytometry and sorting. Second edition. New York: Wiley-Liss, 1990: 415–444.
139. Dean PN. Data analysis in cell kinetics research. In: Gray JW, Darzynkiewicz Z, eds. Techniques in cell cycle analysis. Clifton, New Jersey: Humana Press. 1987:207–253.
140. Bagwell CB, Hudson JL, Irwin GL II. Non parametric flow cytometry analysis. J Histochem Cytochem 1979;27:293–297.
141. Crissman HA, Darzynkiewicz Z, Tobey RA, Steinkamp JA. Correlated measurements of DNA, RNA and protein in individual cells by flow cytometry. Science 228:1321–1324.
142. Darzynkiewicz Z, Traganos F, Melamed MR. Detergent treatment as an alternative to cell fixation for flow cytometry. J Histochem Cytochem 1981;29:329–330.
143. Darzynkiewicz Z, Traganos F, Melamed MR. New cell cycle compartments identified by multiparameter flow cytometry. Cytometry 1980;1:98–108.
144. Darzynkiewicz Z, Traganos F, Kapuscinski J, Staiano-Coico L, Melamed MR. accessibility of DNA in situ to various fluorochromes: Relationship to chromatin changes during erythroid differentiation of Friend leukemia cells. Cytometry 1984;5:355–363.
145. Darzynkiewicz Z. Probing nuclear chromatin by flow cytometry. In: Melamed MR, Lindmo L, Mendelsohn ML, eds. Flow Cytometry and Sorting. Second edition. New York: Wiley-Liss, 1990:315–340.
146. Vindelov L, Christensen IJ. An integrated set of methods for routine flow cytometric DNA analysis. Meth Cell Biol 1990;33:127-138.
147. Hedley DW. DNA analysis from paraffin-embedded blocks. Meth Cell Biol 1990;33:139–148.
148. Darzynkiewicz Z, Traganos F. Multiparameter flow cytometry in studies of the cell cycle. In: Melamed MR, Lindmo L, Mendelsohn ML, eds. Flow cytometry and sorting. Second edition. New York: Wiley-Liss, 1990:469–501.

149. Darzynkiewicz Z. Metabolic and kinetic compartments of the cell cycle distinguished by multiparameter flow cytometry. In: Skehan P, Friedman SJ, eds. Growth, cancer and the cell cycle. Clifton, NJ: Humana Press, 1984:249–280.
150. Clevenger CV, Epstein AL, Bauer KD. Quantitative analysis of nuclear antigen in interphase and mitotic cells. Cytometry 1987;8:280–296.
151. Baish H, Gerdes J. Simultaneous staining of exponentially growing *versus* plateau cells with the proliferation associated antibody kIi-67 an propidium iodide. Analysis by flow cytometry. Cell Tissue Kinet 1987;20:387–391.
152. Czerniak B, Darzynkiewicz Z, Staiano-Coico L, Herz F, Koss LG. Expression of Ca antigen in relation to cell cycle of cultured human tumor cells. Cancer Res 1984;44:4342–4346.
153. Kurki P, Vanderlaan M, Dolbeare F, Gray J, Tan EM. Expression of proliferating cell nuclear antigen (PCNA/cyclin) during the cell cycle. Exp Cell Res 1986;166:209–219.
154. Bauer KD. Analysis of proliferation associated antigens. Methods Cell Biol 1990;33:235–248.
155. Traganos F, Kimmel M, Bueti C, Darzynkiewicz Z. Effects of inhibition of RNA or protein synthesis on CHO cell cycle progression. J Cell Physiol 1987;133:277–287.
156. Darzynkiewicz Z, Traganos F, Xue S-B, Staiano-Coico L, Melamed MR. Rapid analysis of drug effects on the cell cycle. Cytometry 1981;1:279–286.
157. Darzynkiewicz Z, Traganos F, Kimmel M. Assay of cell cycle kinetics by multivariate flow cytometry using the principle of stathmokinesis. In: Gray JW, Darzynkiewicz Z, eds. Techniques for cell cycle analysis. Clifton, NJ: Humana Press, 1987:291–332.
158. Latt SA. Detection of DNA synthesis in interphase nuclei by fluorescence microscopy. J. Cell Biol. 1974;62:546–550.
159. Darzynkiewicz Z, Andreeff M, Traganos F, Sharpless T, Melamed MR. Discrimination of cycling and noncycling lymphocytes by BUdR-suppressed acridine orange fluorescence in a flow cytometric system. Exp Cell Res 1978;115:31–35.
160. Bohmer RM, Elwart J. Cell cycle analysis by combining 5 bromodeoxyuridine/33258 Hoechst technique with DNA-specific ethidium bromide staining. Cytometry 1981;2:31–34.
161. Crissman HA, Steinkamp JA. A new method for rapid and sensitive detection of bromodeoxyuridine in DNA replicating cells. Exp Cell Res 1987;173:256–261.
162. Gratzner HG. Monoclonal antibody to 5-bromo- and 5- iododeoxyuridine. A new reagent for detection of DNA replication. Science 1982;218:474–475.
163. Dolbeare F, Gratzner HG, Pallavicini MG, Gray JW. Flow cytometric measurement of total DNA content and incorporated bromodeoxyuridine. Proc Natl Acad Sci USA 1983;80:5573–5577.
164. Crissman HA, Steinkamp JA. Cytochemical techniques for multivariate analysis of DNA and other cellular constituents. In: Melamed MR, Lindmo L, Mendelsohn ML, eds. Flow Cytometry and Sorting. Second edition. New York: Wiley-Liss, 1990:227–247.
165. Coulter WH. High speed automatic blood cell counter and cell size analyzer. Proc Natl Electron Conf 1956;12:1034–1042.
166. Mullaney PF, Van Dilla MA, Coulter JR, Dean PN. Cell sizing. A light scattering photometer for rapid volume determination. Rev Sci Instr 1969;40:1029–1032.
167. Steinkamp JA. A differential amplifier circuit for reducing noise in axial light loss measurements. Cytometry 1983;4:83–87.
168. Sharpless TK, Melamed MR. Estimation of cell size from pulse shape in flow cytometry. J Histochem Cytochem 1976;24:257–265.
169. Steen HB, Lindmo T. The effect of colchicine and colcemid on the mitogen-induced blastogenesis of lymphocytes. Eur J Immunol. 1978;8:667–671.
170. Monroe JG, Cambier JC. Level of mla expression on mitogen stimulated murine B lymphocytes is dependent on position in cell cycle. J Immunol 1983;130:626–631.
171. Doukas JG, Ruckdedeschel JC, Mardiney MR. Quantitative and qualitative analysis of human lymphocyte proliferation to specific antigen *in vitro* by use of helium neon laser, J Immunol Meth 1977;15:229–238.
172. Darzynkiewicz Z. Differential staining of DNA and RNA in intact cells and isolated nuclei with acridine orange. Meth Cell Biol 1990;33:285–298.
173. Shapiro HM. Flow cytometric estimation of DNA and RNA content in intact cells stained with Hoechst 33342 and pyronin Y. Cytometry 1981;2:143–150.
174. Darzynkiewicz Z., Kapuscinski J, Traganos F, Crissman HA. Aplication of pyronin Y (G) in cytochemistry of nucleic acids. Cytometry 1987;8:138–145.
175. Andreeff M, Darzynkiewicz Z, Sharpless TK, Clarkson B, Melamed MR. Discrimination of human leukemia subtypes by flow cytometric analysis of cellular DNA and RNA. Blood 1980;55:282–293.
176. Andreeff M, Beck JD, Darzynkiewicz Z, Traganos F, Gupta S, Melamed MR, Good RA. RNA content in human lymphocyte subpopulations. Proc Natl Acad Sci USA 1978;75:1938–1942.
177. Stöhr M, Vogt-Schaden M, Knobloch M, Vogel R, Futterman G. Evaluation of eight fluorochrome combinations for simultaneous DNA-protein flow analyses. Stain Technol 1978;53:205–212.
178. Crissman HA, Van Egmond JV, Hodrinet RS, Pennings A, Haanen C. Simplified method for DNA and protein staining of human hematopoietic cell samples. Cytometry 1981;2:59–62.
179. Darzynkiewicz Z, Crissman HA, Traganos F, Steinkamp JA. Cell heterogeneity during the cell cycle. J Cell Physiol 1982;113:465–474.
180. Traganos F, Darzynkiewicz Z, Melamed MR. The ratio of RNA to total nucleic acid content as a quantitative measure of unbalanced cell growth. Cytometry 1982;21:212–218.
181. Crissman HA, Darzynkiewicz Z, Tobey RA, Steinkamp JA. Normal and perturbed CHO cells: correlation of DNA, RNA and protein by flow cytometry. J Cell Biol 1985;101:141–147.
182. Karen DF, ed. Flow cytometry in clinical diagnosis. Chicago: ASCP Press, 1989.
183. Yen A, ed. Flow cytometry: Advanced research and clinical applications, Volume I and II. Boca Raton, FL: CRC Press, 1989.
184. Roti Roti JL, Higashikubo R, Blair CC, Uygur N. Cell-cycle position and nuclear protein content. Cytometry 1982;3:91–96.
185. Pollack A, Moulis H, Block NL, Irvin III GL. Quantitation of cell kinetic responses using simultaneous flow cytometric measurements of NA and nuclear protein. Cytometry 1984;5:473–481.
186. Darzynkiewicz Z. Acid-induced denaturation of DNA *in situ* as a probe of chromatin structure. Meth Cell Biol 1990;33:337–352.
187. Kunicka J, Darzynkiewicz Z, Melamed MR. DNA *in situ* sensitivity to denaturation: a new parameter for flow cytometry of normal colonic epithelium and colon carcinoma. Cancer Res 1987;47:3942–3947.
188. Kunicka JE, Olszewski W, Rosen P, Kimmel M, Melamed MR, Darzynkiewicz Z. DNA *in situ* sensitivity to denaturation as a marker of human breast tumors. Cancer Res 1989;49:6347–6351.
189. Bretton PR, Darzynkiewicz Z, Henry E, Kimmel M, Fair WR, Melamed MR. DNA *in situ* sensitivity to denaturation in bladder cancer and its correlation with tumor stage. Cancer Res 1990;50:7912–7914.
190. Elliott SG, McLoughlin C. Rate of macromolecular synthesis through the cell cycle of the yeast *Saccharomyces cerevisiae*. Proc Natl Acad Sci USA 1978;75:4384–4388.
191. Engelhard HH, Butler AB, Bauer KD. Quantification of the c-myc oncoprotein in human glioblastoma cells and tumor tissue. J Neurosurg 1989;71:224–232.
192. Engelhard HH, III, Krupka JL, Bauer KD. Simultaneous quantification of c-yc oncoprotein, total cellular protein, and DNA content using multiparameter flow cytometry. Cytometry 1991;12:68–76.
193. Van Bockstaele DR, Lan J, Snoeck H-W, Korthout ML, De Bock RF, Peetermans ME. Aberrant Ki-67 expression in normal bone marrow revealed by multiparameter flow cytometric analysis. Cytometry 1991;12:50–63.

194. Verheijen R, Kuijpers JHJ, Van Driel R, Beck JLM, van Dierendock JH, Brackenhoff GJ, Ramaekers FCS. Ki-67 detects a nuclear matrix-associated proliferation-related antigen. II. Localization in mitotic cells and association with chromosomes. J Cell Science 1989;92:531–540.
195. Lopez F, Belloc F, Lacombe F, Dumain P, Reiffers J, Bernard P, Boisseau MR. Modalities of synthesis of Ki67 antigen during the stimulation of lymphocytes. Cytometry 1991;12:42–49.
196. Takasaki Y, Fishwild D, Tan EM. Characterization of proliferating cell nuclear antigen recognized by autoantibodies in lupus sera. J Exp Med 1984;159:981–992.
197. Celis JE, Bravo R, Larsen PM. Fey SJ. Cycin: a nuclear antigen whose level correlates directly with the proliferative state of normal as well as transformed cells. Leukemia Res 1984;8:143–157.
198. Tan EM, Ogata K, Takasaki Y. PCNA/ cyclin: a lupus antigen connected with DNA replication. J Rheumatol 1987;14(suppl 13):89–96.
199. Bravo R, Frank R, Blundell PA, Macdonald-Bravo H. Cyclin/PCNA is the auxillary protein of DNA polymerase δ. Nature (Lond.) 1987; 326:515–517.
200. Bravo R, Macdonald-Bravo H. Existence of two populations of cyclin/proliferating cell nuclear antigen during the cell cycle: Association with DNA replication sites. J Cell Biol 1986;105:1549–1554.
201. Clevenger CV, Epstein AL, Bauer KD. Modulation of the nuclear antigen p105 as a function of cell-cycle progression. J Cell Physiol 1987; 130:343–343.
202. Bauer KD, Clevenger CV, Endov RK, Murad T, Epstein AL, Scarpelli DG. Simultaneous nuclear antigen and DNA content quantitation using paraffin-embedded colonic tissue and multiparameter flow cytometry. Cancer Res 1986;46:2428–2434.
203. Clevenger CV, Bauer KD, Epstein AL. A method for simultaneous nuclear immunofluorescence and DNA content quantitation using monoclonal antibodies and flow cytometry. Cytometry 1985;6:206–216.
204. Nakamura H, Morita T, Masaki S, Yashida S. Intracellular localization and metabolism of DNA polymerase in human cells visualized with monoclonal antibody. Exp Cell Res 1984;151:123–133.
205. Lin HT, Gibson CW, Hirschorn RR, Sittling S, Baserga R, Mercer WE. Expression of thymidine kinase and dihydrofolate reductase genes in mammalian ts mutants of the cell cycle. J Biol Chem 1985; 260:3269–3274.
206. Heck MM, Hittelman WN, Earnshaw WC. Differential expression of DNA topoisomerases I and II during the eukaryotic cell cycle. Proc Natl Acad Sci USA 1988;85:1086–1090.
207. Sullivan DM, Latham MD, Ross WE. Proliferation-dependent topoisomerase II content as a determinant of antineoplastic drug action in human, mouse and Chinese hamster ovary cells. Cancer Res 1987;47:3973–3979.
208. Mann GJ, Musgrove EA, Fox RM, Thelander L. Ribonucleotide reductase M1 subunit in cellular proliferarion, quiescence and differentiation. Cancer Res 1988;48:5151–5156.
209. Wang E. Statin, a nonproliferation-specific protein, is associated with the nuclear envelope and is heterogenously distributed in cells leaving quiescent state. J Cell Physiol 1989;140:418–426.
210. Larsen JK. Washless double staining of a nuclear antigen (Ki-67 or bromodeoxyuridine) and DNA in unfixed nuclei. Meth Cell Biol 1990;33:227–234.
211. Cleaver JE. Thymidine metabolism and cell kinetics. Amsterdam: North-Holand Publ Co., 1967.
212. Okada S. Simple graphic method of computing the parameters of the life cycle of cultured mammalian cells in the exponential growth phase. J Cell Biol 1967;34:915–916.
213. Ashihara T, Baserga R. Cell Synchronization. Meth Enzymol 1989; 58:248–262.
214. Grdina DJ, Meistrich ML, Meyn RE, Johnson TS, White RA. Cell synchrony techniques: a comparison of methods. In: Gray JW, Darzynkiewicz Z, eds. Techniques in cell cycle analysis. Clifton, New Jersey: Humana Press, 1987:367–402.
215. Rizzoni M, Palitti F. Regulatory mechanisms of cell division. I. Colchicine-induced endoreduplication. Exp Cell Res 1973;77:450–458.
216. Ley KD, Tobey RA. Regulation of initiation of DNA synthesis in Chinese hamster cells. II. Induction of DNA synthesis and cell division by isoleucine and glutamine in G_1-arrested cells in suspension culture. J Cell Biol 1970;47:453–459.
217. Petersen DF, Anderson EC, Tobey RA. Mitotic cells as a source of synchronized cultures. In: Prescott DM, ed. Methods in Cell Physiology. New York: Academic Press, 1968:347–370.
218. Puck TT, Steffen J. Life cycle analysis of mammalian cells. I. A method for localizing metabolic events within the life cycle and its application to the action of Colcemid and sublethal doses of X-irradiation. Biophys J 1963;3:379-397.
219. Wright NA, Appleton DR. The metaphase arrest technique. A critical review. Cell Tissue Kinet 1980;13:643–663.
220. Traganos F, Kimmel M. The stathmokinetic experiment: a single-parameter and multiparameter flow cytometric analysis. Meth Cell Biol 1990;33:249–270.
221. Dosik GM, Barlogie B, White AR, Gohde W, Drewinko B. A rapid automated stathmokinetic method for determination of *in vitro* cell cycle transit times. Cell Tissue Kinet 1981;14:121–134.
222. Darzynkiewicz Z, Williamson B, Carswell EA, Old LJ. The cell cycle specific effects of tumor necrosis factor. Cancer Res 1984;44:83–90.
223. Del Bino G, Skierski JS, Darzynkiewicz Z. Diverse effects of camptothecin, an inhibitor of topoisomerase I on the cell cycle of lymphocytic (L1210, MOLT-4) and myelogenous (HL-60, KG1) leukemic cells. Cancer Res 1990;50:5746–5750.
224. Kubbies M, Rabinovitch P. Flow cytometric analysis of factors which influence the BrdUrd-Hoechst quenching effect in cultured human fibroblasts and lymphocytes. Cytometry 1983;3:276–281.
225. Poot M, Kubbies M, Hoehn H, Grossmann A, Chen Y, Rabinovitch PS. Cell cycle analysis using continuous bromodeoxyuridine labeling and Hoechst 33258-ethidium bromide bivariate flow cytometry. Meth Cell Biol 1990;33:185–198.
226. Rabinovitch PS. Regulation of human fibroblast growth rate by both noncycling cell fraction and transition probability is shown by growth in 5-bromodeoxyuridine followed by Hoechst 33258 flow cytometry. Proc Natl Acad Sci USA 1983;80:2951–2955.
227. Rabinovitch PS, Kubbies M, Chen YC, Schindler D, Hoehn H. BrdU-Hoechst flow cytometry: a unique tool for quantitative cell cycle analysis. Exp Cell Res 1988;174:309–318.
228. Castor LN. A G_1 rate model accounts for cell cycle kinetics attributed to "transition probability." Nature 1980;287:76–79.
229. Brooks RF. Continuous protein synthesis is required to maintain the probability of entry into S phase, Cell 1977;12:3111–3117.
230. Darzynkiewicz Z, Traganos F, Kimmel M, Myc A. Kinetic and metabolic heterogeneity of cells in the cell cycle. In: Tautu P, ed. Stochastic modelling in biology. World Scientific Publishing Co. 1991 (in press).
231. Kimmel M, Traganos F, Darzynkiewicz Z. Do all daughter cells enter the "indeterminate" ("A") state of the cell cycle? Analysis of stathmokinetic experiments on L1210 cells. Cytometry 1983;4:191–196.
232. Johnston GC, Pringle JR, Hartwell LH. Coordination of cell growth with cell division in the yeast *Saccharomyces cerevisiae*. Exp Cell Res 1977;105:79–85.
233. Tyson JJ. Size control of Cell division. J Theor Biol 1987;126:381–391.
234. Murray LE, Singer RA, Fenwick RG Jr, Johnson GC. The G_1 interval in the mammalian cell cycle: dual control by mass accumulation and stage-specific activities. Cell Prolif 1991;24:215–228.
235. Zu Y-L, Tsubai F, Namba Y, Hanaoka, Simultaneous measurement of transferrin receptor and DNA content of human Il2 dependent T cells by flow cytometry. Cell Struct Function 1988;13:13–32.
236. Murphy JS, Landsberger FR, Kikuchi T, Tamm I. Occurence of cell division is not exponentially distributed: differences in the generation

times of sister cells can be derived from the theory of survival of populations. Proc Natl Acad Sci USA 1984;81:2374–2383.

237. Bhuyan BK, Groppi VE. Cell cycle specific inhibitors. Pharmac Ther 1989;42:307–348.

238. Koss LG, Czerniak B, Herz F, Wersto RP. Flow cytometric measurements of DNA and other cell components in human tumors: a critical appraisal. Hum Pathol 1989;20:528–548.

3

Theoretical Aspects of Flow Cytometry Data Analysis

C. BRUCE BAGWELL

The cells have been harvested and stained and the cellular preparation has been run through the flow cytometer with all the proper quality control procedures. All the histograms and/or listmode data files have been safely stored on the computer's hard disk and we are finally ready to ask the universal question, "What does it all mean?"

Conceptually, our initial dilemma is a data reduction problem. A typical one-parameter flow cytometric histogram is in reality a sequence of 256 to 1024 numbers. A two-correlated-parameter histogram is generally a sequence ranging between 4,096 and 16,384 numbers. The size of a typical listmode file is in the range of 100,000 to 250,000 bytes. The sheer quantity of numbers representing a single flow cytometric sample is overwhelming without some drastic simplifying data reduction.

The most common method of reducing data into a manageable size is to graph it. In a graph, the numbers generally transform to a few peaks (one parameter) or to clusters of points (two correlated parameters) that usually can be qualitatively interpreted. This qualitative level of data interpretation takes us a long way in understanding the biology of the sample as represented by our measurements of specific fluorescent probes and light scatter emissions. For many applications, no further analysis is necessary. The relative positions and magnitude of the peaks or clusters of points is enough to convey the answer to the question being asked of the sample.

This chapter focuses on the next logical step in the interpretation of the data patterns presented to us in graphical form: the quantification of the objects implicitly defined in the data into descriptive numbers. This process, frequently referred to as parametric analysis, involves computer algorithms that are designed for specific applications. Although hundreds of such algorithms have been proposed, only a few are in general use today. It is the intent of this chapter to describe these methods theoretically and practically. For those of you not interested in mathematical details, skip the sections entitled mathematics. The major applications that will be described are: 1) DNA Cell Cycle Analysis, 2) Immunofluorescence Analysis, and 3) Histogram Comparison Methods.

Not all aspects of flow cytometry data analysis are covered in this chapter. For further information on bivariate, multiparameter, and clustering analysis techniques, Dean's chapter on Data Processing in *Flow Cytometry and Sorting* (1) is recommended. Further details on practical aspects of data analysis are also presented elsewhere in this book (see Chapter 8).

DNA DATA ANALYSIS

Introduction

Since 1976, when Dean and Jett published their landmark paper on DNA analysis (2), there have been hundreds of methods that have been either published or proposed for accurately estimating S phase. This chapter concentrates on those methods that have survived the test of time and are in current use today rather than giving an accurate historical perspective on the entire evolution of this discipline.

Flow cytometry DNA analysis' popularity is, in large part, due to Hedley's early efforts to analyze DNA histograms from paraffin-embedded material (3). Some of the reasons for this popularity are: 1) It is possible to complete large retrospective studies in a relatively short period of time with paraffin-embedded material; 2) some of these studies are showing the prognostic capability of DNA S-phase and ploidy; and 3) the histograms from paraffin-embedded material generally have relatively poor CV's and high debris contamination necessitating computer analysis rather than simple integration methods.

In one way or another, all proposed DNA analysis methods involve the Gaussian function. What is a Gaussian anyway? The formula for a Gaussian located at position μ with area A and standard deviation of σ is shown below:

$$y(x) = \frac{A}{\sqrt{2\pi}\,\sigma} e^{\frac{-(x-\mu)^2}{2\sigma^2}} \qquad [3.1]$$

On first inspection, this function looks ominous at best. Let's attempt to understand the Gaussian from an intuitive point of view. We know that since a Gaussian has a bell shape, it should approach the X-axis asymptotically on both sides. A simple function that has this asymptotic characteristic is the exponential shown in Figure 3.1A.

$$y(x) = k_1\, e^{-k_2 x}$$

The k_1 parameter of this function controls where it intersects the Y-axis and the k_2 parameter, the rate constant, controls the rate at which it approaches the X-axis. However, the exponential is not the function we are looking for since it

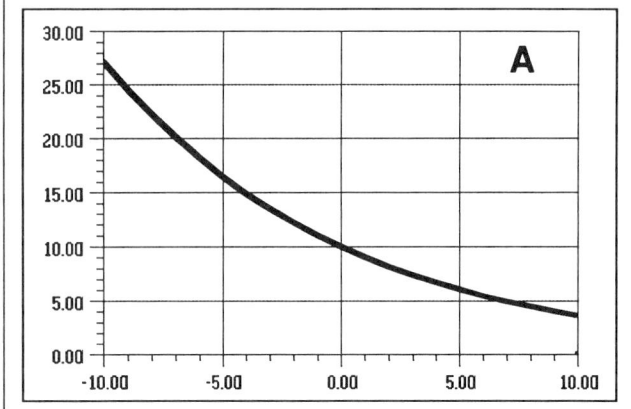
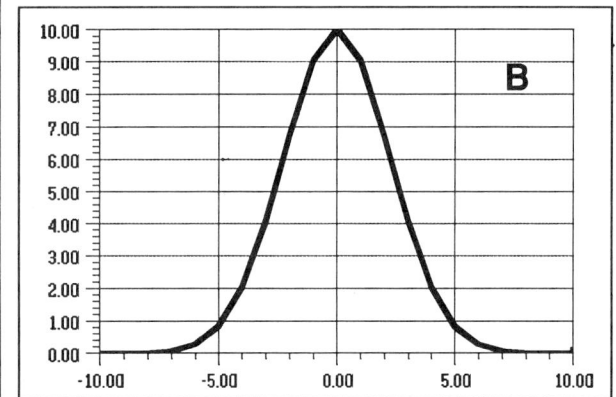
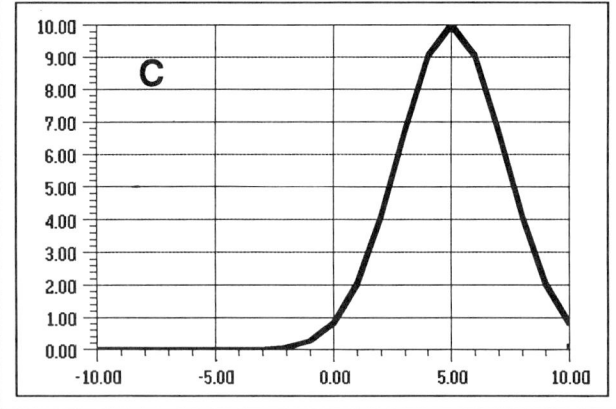

Figure 3.1. Intuitive derivation of a Gaussian. **A** represents an exponential asymptotically approaching the X-axis from one side. **B** demonstrates the bilateral symmetry of the function when the independent variable is squared. **C** represents the general formula for a Gaussian with the addition of a positional degree of freedom.

asymptotically approaches the X-axis from only one side. We want the function to approach the X-axis from both sides. We can give the function this bilateral symmetry by simply squaring the independent variable, x, in the exponent (Fig. 3.1B).

$$y(x) = k_1 e^{-k_2 x^2}$$

Now, negative x values have the same effect on the function as positive values, thus giving it symmetry about the x = 0 axis. To give the function a positional degree of freedom, we subtract a constant, k_3, from the x value before squaring it (Fig. 3.1C). The larger this constant is, the farther to the right the function will be placed. Thus, k_1 controls the height, k_2 controls the width, and k_3 controls the position of our intuitively derived function.

Our final function

$$k_1 e^{-k_2(x-k_3)^2} \qquad [3.2]$$

is simply a general form of the Gaussian function. It can be shown that k_2 is related to the standard deviation, σ, by the equation (4)

$$k_2 = \frac{1}{2\sigma^2}$$

and the height of the Gaussian, k_1, is related to its area by the equation

$$k_1 = \frac{A}{\sqrt{2\pi}\,\sigma}.$$

Substituting these values back into the general function (Eq. 3.2) and letting $k_3 = \mu$, we end up with the Gaussian defined in Equation 3.1. The point of this exercise is to show that the Gaussian is actually a very intuitive function. It has many uses in the field of flow cytometry, but none is as predominant as that of describing and analyzing DNA histograms.

DNA Histogram Theory

What is a DNA histogram? We can deduce from the name that a DNA histogram represents the number of events on the Y-axis and DNA content on the X-axis. But what is a DNA histogram in a theoretical sense? There are three interwoven relationships that describe the theoretical DNA histogram (2, 5). The first relation, DNA versus cell age, is biological, the last two, number of cells versus cell age and signal broadening, are more statistical in nature.

DNA Versus Cell Age

The relationship between DNA content and cell age is depicted in a highly schematized view of the cell cycle (Fig. 3.2). G_0 cells have 2C (C stands for complement) amount of DNA and are either not in the cycle or have a relatively long cell cycle time. Cells initially enter the cycle at the G_1 (G stands for gap) stage and have 2C amount of DNA. During this phase, the cell creates the proteins it needs for DNA synthesis and as soon as the DNA synthetic machinery is turned on, the cell is considered to be in S phase. The cell increases its DNA content during S phase until it has duplicated the genome with 4C amount of DNA and enters the second gap phase called G_2. During G_2 phase the cell produces the proteins necessary for mitosis and, once completed, enters M phase or mitosis phase. When the cells complete the cycle at the end of telophase, the daughter cells

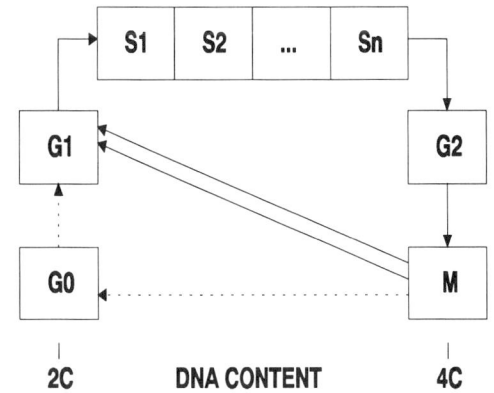

Figure 3.2. Cell cycle compartments and DNA content.

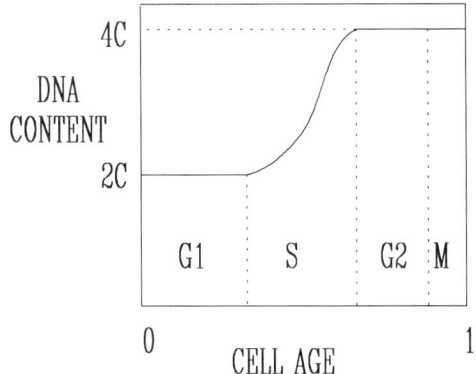

Figure 3.3. DNA content versus cell age.

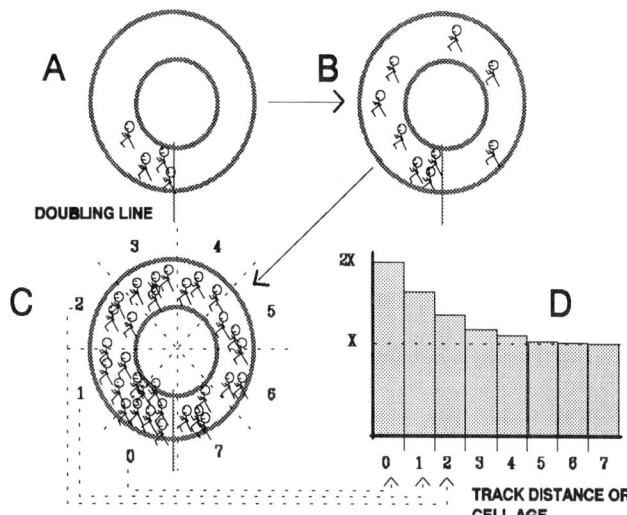

Figure 3.4. Number of cells versus cell age: a running track paradigm.

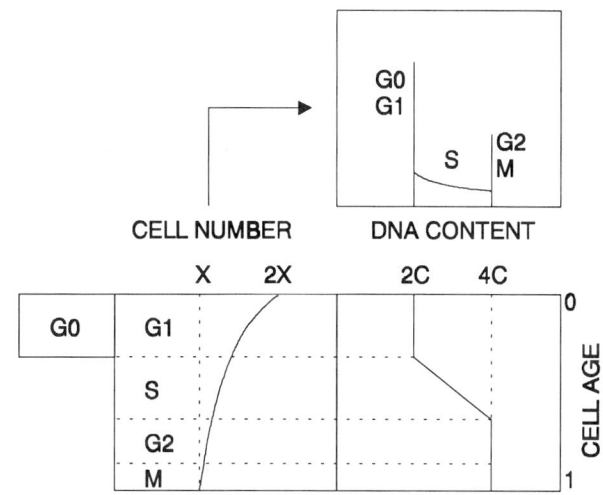

Figure 3.5. Derivation of an ideal DNA histogram.

can re-enter the cell cycle at G_1, creating an exponentially expanding population.

A cell is considered zero cell cycle units old at the point of cytokinesis in telophase and one cell cycle unit old just prior to cytokinesis. The step function between cell age and DNA content is shown in Figure 3.3. The shape of the curve during the S-phase period is a function of the rate of change in DNA synthesis rate during S phase. Relative DNA synthesis rate calculations will be covered later in this chapter.

Number of Cells Versus Cell Age

The next relationship implicit in a DNA histogram is the theoretical relation between the age of a cell and the numbers of cells in an exponentially growing population. We will develop this theory with a simple running track paradigm and then later relate it to the cell cycle.

Imagine that we have a running track and that across one end of the track is a line called the doubling line. We have a few runners poised to begin a special race. When a runner crosses the doubling line, another runner enters the track at that point. Assume that each runner runs at his own speed. Let's now begin the race.

Figure 3.4A and B depict the initial situation with a few runners. If we let the race proceed for a while, more and more runners begin accumulating on the track. Before it gets too crowded on the track, we terminate the race. Figure 3.4C shows the final distribution of runners with the track divided into eight equal segments. As you would expect, the density of runners is highest just after the doubling line. In fact, the density of runners just after the doubling line is approximately twice the density just before the doubling line. The density of runners decreases as we move around the track.

The track represents the cell cycle, the runners represent cells, and the position on the track represents cell age. Thus, the relationship between cell age and number of cells is a decreasing function beginning at 2x and ending at x. The actual relationship can be shown to be a decreasing exponential (Fig. 3.4D) (5).

The Ideal DNA Histogram

Notice that the first relation is between DNA and cell age and the second relationship is between number of cells and cell age. We have two parameters related to the same param-

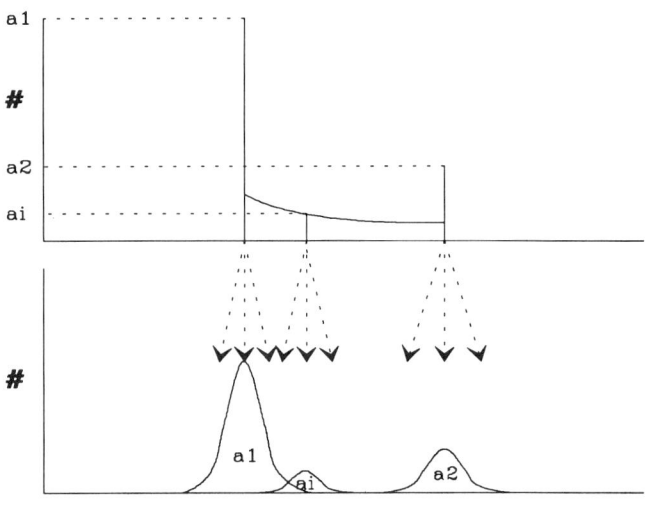

Figure 3.6. Signal broadening.

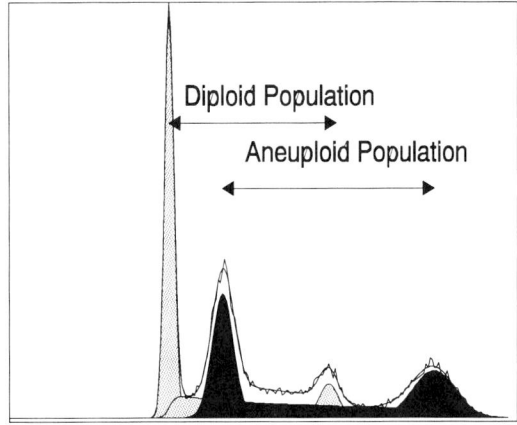

Figure 3.7. DNA aneuploidy.

eter, cell age. We can eliminate the common parameter of cell age and solve for number of cells versus DNA which, by definition, is a DNA histogram. Figure 3.5 demonstrates how this is accomplished.

The area of the number versus cell age curve during the G_1 part of the cycle has an invariant 2C amount of DNA and thus can be represented as a spike on the ideal DNA histogram. Note that those cells in G_0 are simply added to the height of this spike. In a similar manner, the area of the curve over the G_2 interval is represented as a spike at the 4C position on the ideal DNA histogram. Finally, as we increase the DNA content between 2C and 4C, the number of cells decreases as an exponential. Thus, an ideal histogram is represented as two spikes at 2C and 4C flanking a decreasing exponential.

Signal Broadening

If the probes that measured cellular DNA content had a perfect stochiometric relationship with the number of DNA nucleotides and we could measure them exactly, then the mea-

Table 3.1

Parameter	Description
1	Diploid G_0G_1 position
2	Diploid G_0G_1 standard deviation
3	Diploid G_0G_1 area
4	S-phase area (typically 2–3 parameters)
7	Diploid G_2M position
8	Diploid G_2M standard deviation
9	Diploid G_2M area
10	Aneuploid G_0G_1 position
11	Aneuploid G_0G_1 standard deviation
12	Aneuploid G_0G_1 area
13	Aneuploid S-phase (typically 2–3 parameters)
16	Aneuploid G_2M position
17	Aneuploid G_2M standard deviation
18	Aneuploid G_2M area
19	Debris compensation
20	Aggregate compensation (typically 1–3 parameters)

sured DNA histogram would look very similar to the ideal DNA histogram, since normal cell-to-cell DNA content variation is quite small. Unfortunately, the dyes that are used to measure DNA do not exactly bind with the DNA and the means of measuring these dyes have inherent errors as well. The end result of this imprecise measurement is that each point of the ideal histogram is broadened due to the uncertainty of the measured parameter (Fig. 3.6). This broadening process is the final fundamental relationship implicit in the DNA histogram.

DNA Aneuploidy

To complicate matters a bit more, some cells have an aberrant amount of DNA for their cycle components. If these cells are from solid tumors, there are generally normal stromal cells superimposed on the aberrant cycle (Fig. 3.7). Thus, a DNA histogram can be a complex set of peaks and continuous distributions. How can one take this puzzle apart and figure out all the cell fractions that make up a DNA histogram?

Nonlinear Least-Squares Analysis and Models

The method that is almost universally used to analyze DNA histograms is called nonlinear least-squares analysis. The user selects a mathematical model that describes the underlying biology of the sample and asks the computer to find the best values for the model parameters that describe the observed histogram. The models are usually quite complex, cannot be transformed to a simple polynomial function, and thus are called nonlinear models. Typical DNA aneuploid models can have around 20 parameters (Table 3.1).

The nonlinear least-squares algorithm that is most often used is the Marquardt method. A good description of this method as well as the computer source code can be found in Bevington (6). The S-phase portion of the model is the most variable and is often the distinguishing characteristic between many of the DNA models used today. Since the S-phase distribution is a continuous distribution with generally

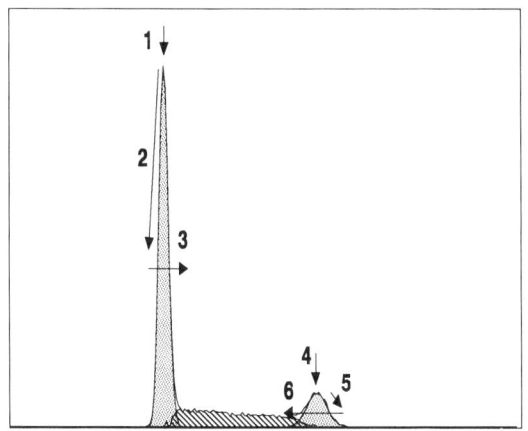

Figure 3.8. Outside-in DNA analysis.

few abrupt changes, it is to be expected that many different mathematical distributions seem to fit it quite well. It is probably safe to say that no one model is the absolute best under all circumstances. Some S-phase model components, such as multiple trapezoids, are good at fitting perturbed DNA histograms, while more conservative model components, such as the single broadened rectangle or trapezoid, are better for solid tumor DNA histograms.

Simple Analysis Methods

Initially, DNA modeling was restricted to large mainframe computers. Simplified approaches to the solution of deriving cell cycle parameters from DNA histograms were created to run on relatively slow microprocessor-based computer systems. These simplified methods provide approximate solutions at best and should never be used if a nonlinear least-squares approach is available. Today, for the most part, simple analysis methods are employed to obtain first estimates to the more rigorous and accurate nonlinear least-squares methods.

There are two fundamentally different simple analysis strategies for estimating S phase. The first method, outside–in, attacks the problem of S-phase determination from the outside of the G_0G_1 and G_2M Gaussians and works toward the S-phase region for the final solution. The second approach, inside-out, works from the middle of S-phase and extrapolates out to the G_0G_1 and G_2M modal channels.

Outside-In or Peak Reflection Method

Figure 3.8 schematizes how the outside-in method works. The computer scans the histogram over a user-defined region looking for the highest peak. Once it finds the peak, it searches for the steepest part of the curve (step 1), moves down to either 1/2 or 1/3 its height, and fits the defined segment of the histogram with a Gaussian function (step 2). The computer subtracts the Gaussian from the histogram (step 3) and proceeds to look for the second G_2M peak at approximately twice the G_0G_1 modal channel. It repeats the fitting process for the G_2M peak (steps 4, 5, 6), culminating in the subtraction of the G_2M Gaussian from the observed histogram. The histogram residue is added up between the G_0G_1 and G_2M modal channels and is used to approximate S phase.

The advantages of the outside-in method are that it is fast and not very sensitive to perturbations in the middle of S phase; however, the method is highly affected by G_0G_1 and G_2M peak skewing and has a bias towards underestimating true S phase (Fig. 3.9).

Inside–Out Methods

Figure 3.10 schematizes how the inside-out method works. The computer begins much the same way as the outside-in method. It first finds the modal channel of the G_0G_1 peak in a user-defined region. It estimates the standard deviation of the peak and then finds the second G_2M peak and estimates its standard deviation. The computer uses these standard deviations to define a fitting window in the interior of S-phase. The left boundary of the S–phase window is generally set to G_0G_1 mode $+3.5 * G_0G_1$ standard deviation, and the right boundary of the S-phase window is to set to G_2M mode $-3.5 * G_2M$ standard deviation. The number of standard deviations is variable in most programs, ranging between 2.5 and 3.5. If the standard deviations are so large that no window is formed, the procedure aborts.

There are three variants of the inside-out procedure. The first fits the S-phase window with a rectangle and extrapolates it to the G_0G_1 and G_2M modal channels known as the Baisch or box method (Fig. 3.10A) (7). The second method fits the S-phase window with a trapezoid and extrapolates it in the same way (Fig. 3.10B) and the third variant uses a polynomial for the extrapolation and is known as the SFit Method (Fig. 3.10C) (8).

The Inside–out method is as fast as the outside–in method but is not as sensitive to G_0G_1 and G_2M skewing. It's more accurate than the outside-in method for well-behaved DNA histograms (Fig. 3.9). Its disadvantage is that it is very sensitive to extra peaks in S phase. This disadvantage is a severe one for solid tumors since many times they have aneuploid DNA peaks in the middle of S phase. Because of the above reason, outside-in methods are generally used for first estimates for complex solid tumor DNA histograms rather than inside-out methods.

The Modeling Process Applied to DNA Histograms

Figure 3.11 summarizes the modeling process. The first and most difficult step is to define the correct mathematical model that is appropriate for the data to be analyzed. As mentioned above, these models can be quite complex and many times there are numerous decisions to be made about the interpretation of the data. A model definition is a list of mathematically derived components that are bound together by a series of dependencies. An example of a dependency is that the standard deviation of the G_2M component should theoretically be twice the standard deviation of the G_0G_1

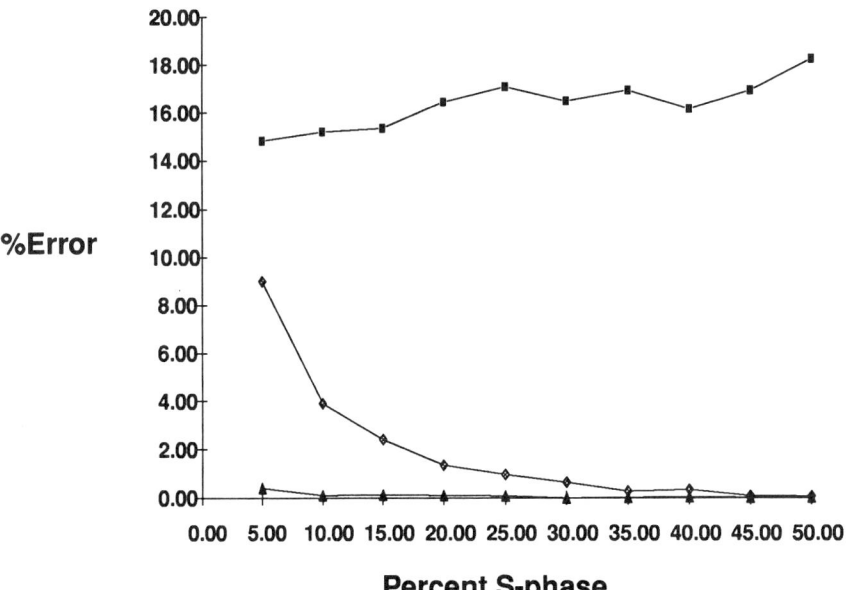

Figure 3.9. Accuracy of Simple DNA Analyses. %Error of S-phase estimate ($100 * \frac{\text{actual} - \text{calculated}}{\text{actual}}$) versus generated *percent S-phase* (%CV = 3, %G_0G_1 = 40, %S = 40, %G_2M = 20) for outside-in (squares), inside-out (polynomial, diamonds), and a five trapezoid S-phase model (triangles).

component. Another less obvious example is that the beginning of S-phase component is dependent on the mean of the G_0G_1 model component by a factor of one. There are usually six or more dependencies that bind together the model components for typical DNA models, much the same as the component pieces of an erector set are bound together with nuts and screws.

The other major part of the modeling system is the nonlinear least-squares algorithm that attempts to find the best set of parameters that control the shape, position, and area of all the model components, such that when added together the model matches, as best as possible, the observed data. Figure 3.12 depicts the important aspects of this process. Imagine that the G_0G_1 position is parameter 1 and the G_0G_1 standard deviation is parameter 2. Suppose we took all possible combinations of these two parameters and asked how well the model fit the data. We generally quantify the "goodness" of fit with a chi-square value where higher chi-squares represent poor fits and lower chi-squares better fits.

$$X^2 = \sum_{i=1}^{u} \frac{(\text{observed}(x) - \text{calculated}(x))^2}{\text{observed}(x)}$$

When we take all the combinations of these two parameters and plot the chi-square value at each point, we end up with a three-dimensional surface similar to the one shown in Figure 3.12. Note that there is generally a single large valley on this surface where values of the parameters minimize the chi-square or maximize the goodness of fit between the model and the observed data. The sole objective of the nonlinear least-squares algorithm is to find this lowest portion of the chi-square surface. Don't be distressed to realize that this example demonstrates this process for only a two-parameter model when most DNA histogram models have considerably more parameters than two. The computer algorithms in use today work well with a large number of parameters in which the only drawback to increasing the number of parameters is that it requires more computations, analysis time, and computer memory to find an optimal solution.

The most common nonlinear least-squares method used today is the Marquardt Compromise method (6, 9). The exact way the method works is outside the purview of this chapter, but it is important to know that the method is actually two different strategies tied together with a variable called λ. When λ is greater than one, the method behaves in a gradient search manner. When λ is less than 0.01, it behaves in a linearization manner.

The gradient search method is the easiest to visualize. The computer calculates the gradient (slope) of a plane defined in multiparameter space and jumps some distance along its steepest descent. The direction it takes is much the same as the direction a drop of water would take along a more solid version of the chi-square valley. The gradient search method always converges toward a solution, but when the chi-square valley is no longer steep the gradient search's rate of convergence slows down.

The other method, called linearization, works in an entirely different but, as you will see, complementary manner. The linearization method also samples its current location on the chi-square surface but instead of calculating a planar surface and traveling along its steepest vector, it attempts to

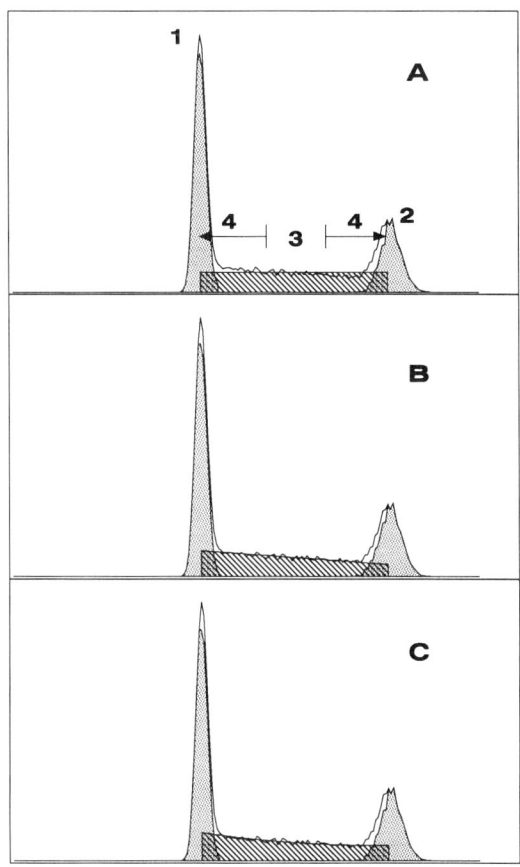

Figure 3.10. Inside-out DNA Analysis. **A** demonstrates the sequence of analysis steps for rectangle extrapolation. **B**, trapezoid extrapolation. **C**, polynomial extrapolation.

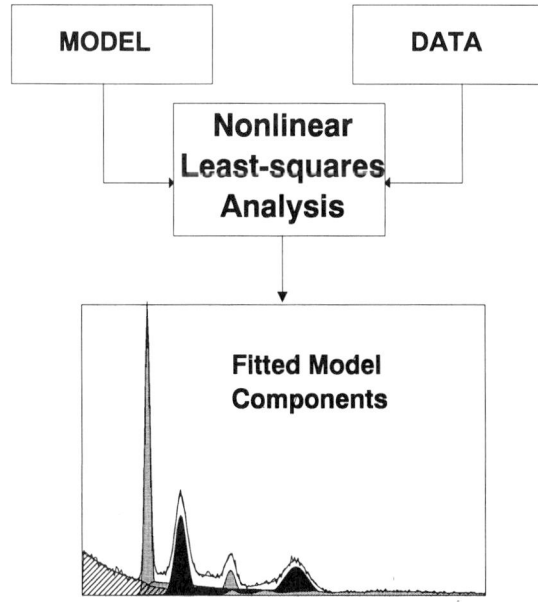

Figure 3.11. Modeling process.

estimate the entire surface by extrapolating its sampled surface in all dimensions and then jumps to that extrapolated

surface's lowest point. The advantage of this method is that its rate of convergence is quite fast, but its disadvantage is that if the local surface that is sampled is quite bumpy due to statistical noise, it may jump to a distant location on the chi-square surface and be further away from the solution than when it started. This sad state of affairs is called non-convergence.

The genius of the Marquardt Compromise is that it combines these two methods with a single parameter called λ and obtains the best convergent characteristics from both. Generally, with the Marquardt method, λ is initially set to 0.001 making the algorithm predominantly behave as a linearization method. If all goes well and the system converges toward a solution, λ is divided by 10 and the process is repeated. If, however, the system does not converge, which is easily detectable by an increasing chi-square value, λ is multiplied by 10. If it still does not converge, λ is multiplied by 10 again and, eventually, the system will behave as a steepest gradient method and begin to converge again.

The starting λ value and the factor by which λ is either multiplied or divided is rather arbitrary and historically based. The normal values published by Marquardt may indeed not be optimal for complex DNA histogram models.

Let us now inspect Figure 3.12 more closely. To begin the modeling process, it is necessary for the computer to somehow be positioned on the chi-square surface near the major valley. Initial estimates of all the nonlinear parameters are necessary to begin the nonlinear least-squares process. In fact, the accuracy of these estimates is the second most important aspect of the modeling process. The first is the initial selection of the model. If an estimate is inaccurate, it is possible for the computer to begin the nonlinear least-squares process near a smaller valley (see Fig. 3.12), eventually converging to a false solution. Thus, accuracy of the estimates is quite important in this entire modeling process. The methods generally used to obtain estimates are the simple analysis strategies covered earlier. The outside-in method is recommended because it is the most robust for complex histograms.

Once the computer obtains accurate estimates it begins to iterate toward the lowest portion of the nearest chi–square valley. The computer knows when to stop this iterative process by a few termination criteria. The algorithm usually terminates when either the number of maximum allowable iterations has been reached (5–15) or when the chi-square changes by less than some fraction (0.01–0.001). The final parameters are then evaluated and translated into a form understandable to the user. It is fairly easy to compute the probable errors made in these estimates and, if they are available, it is recommended that they be included in the reported results.

Most of the proposed DNA models are a composite of well known mathematical functions. Before looking at these components in detail, it is important for us to cover the process of broadening or, in the parlance of a DNA histogram modeler, the process of convolution. We refer again to Fig-

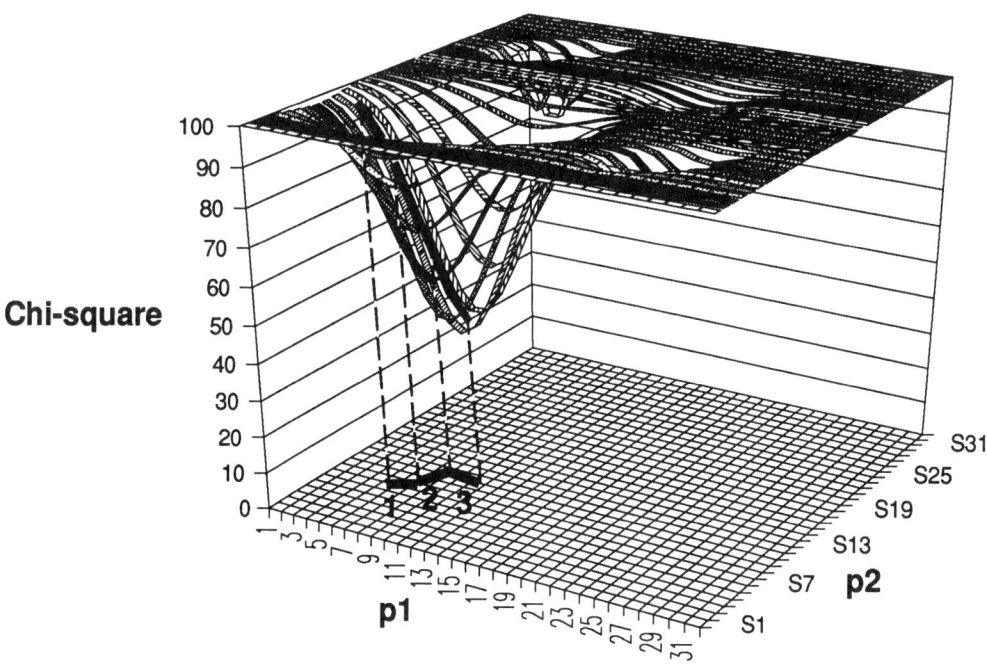

Figure 3.12. Chi-square fitting surface.

ure 3.6, which showed that, because it is impossible to exactly measure the DNA content of a cell as it traverses the laser beam, we broadened each point of the ideal DNA histogram with a Gaussian. There are two common approaches to broadening model components. One involves numerous iterations and the other involves solving a special integral called the convolution formula. We will introduce these methods by means of a simple example, the broadening of a rectangle of height r and domain $x_1 \leq x \leq x_2$. The left-hand edge of the rectangle, x_1, will be broadened with a Gaussian of standard deviation σ_1 and the right-hand edge of the rectangle, x_2, will be broadened with a Gaussian with a standard deviation of σ_2. The intermediate points along the rectangle are broadened with Gaussians with linearly increasing standard deviations.

The iterative method breaks the rectangle into n segments. Usually, the segments are of equal width, but that restriction is not necessary. A description of the process is simplified by creating two x axes. The first, τ, represents the unbroadened function's independent variable and the second, x, represents the broadened function's independent variable. The area of the ith segment is $r\Delta\tau$ where r is the height of the rectangle. There are two ways to distribute the unbroadened segments onto x. We could take each segment and distribute it over x with a Gaussian function with an area equal to the segment area or we could calculate the contribution of all the segments into one channel, x. Although the former method is normally used because it is slightly more efficient, we will discuss the latter because it prepares us for the later convolution formulae. The equation that expresses the contribution of each rectangle segment into a single channel, x, is shown below.

$$R(x) = \frac{r}{\sqrt{2\pi}} \sum_{i=1}^{n} \frac{1}{\sigma_i} e^{-\frac{(x-\tau_i)^2}{2\sigma_i^2}} \Delta\tau_i \qquad [3.3]$$

where,
$R(x)$ = broadened rectangle evaluated at channel x,
r = height of the rectangle,
σ_i = standard deviation at midpoint of segment i,
τ_i = midpoint of the ith segment of the rectangle,
$\Delta\tau_i$ = width of the ith segment of the rectangle, and
n = number of segments.

The iterative process is simple in concept and, for some functions, it is the only way to go; however, a more elegant and efficient solution to the problem can be obtained by using convolution theory. The convolution of a rectangle with a Gaussian has a formula similar to Equation 3.3, except that it is defined over the real domain rather than the integer domain of x:

$$R(x) = \frac{r}{\sqrt{2\pi}} \int_{\tau=x_1}^{x_2} \frac{1}{\sigma(\tau)} e^{-\frac{(x-\tau)^2}{2\sigma(\tau)^2}} d\tau \qquad [3.4]$$

If $\sigma(\tau)$ were assumed to be a constant, σ, over the range of the rectangle, then the integral expressed in Equation 3.4 can be solved exactly (5) as

$$R(x) = \frac{r\left(\text{erf}\left(\frac{x_2-x}{\sqrt{2}\sigma}\right) - \text{erf}\left(\frac{x_1-x}{\sqrt{2}\sigma}\right)\right)}{2} \qquad [3.5]$$

where,
r = height of the rectangle,

x_1, x_2 = beginning and ending rectangle x values,
σ = broadening standard deviation, and

erf(z) = the error function (10, $\frac{2}{\sqrt{\pi}} \int_0^z e^{-t^2} dt$).

If σ is allowed to vary linearly with x from x_1 to x_2, a very accurate approximation of the rectangle-Gaussian convolution (Unpublished results) can be derived as

$$R(x) \simeq r \, \text{derf}(x_1, x_2, \sigma_1, \sigma_2; x) \quad [3.6]$$

where

$$\text{derf}(x_1, x_2, \sigma_1, \sigma_2; x) = \frac{\text{erf}\left(\frac{x_2 - x}{\sqrt{2\sigma_2}}\right) - \text{erf}\left(\frac{x_1 - x}{\sqrt{2\sigma_1}}\right)}{2}. \quad [3.7]$$

The difference error function, derf, can also be used to broaden trapezoids and polynomials.

Model Components

We are now in a position to introduce most all the players in the DNA modeling game. We will give the iterative and, if possible, convolution formulae a graph of the components' typical distribution in a DNA model and provide a brief description of the use of each function.

GAUSSIAN

Standard form:
$$G(x) = \frac{A}{\sqrt{2\pi}\,\sigma} e^{\frac{-(x-\mu)^2}{2\sigma^2}} \Delta x \quad [3.8]$$

Exact form:
$$G(x) = \frac{A}{2\sigma} \left(\text{erf}\left(\frac{x + 0.5 - \mu}{\sqrt{2\sigma}}\right) - \text{erf}\left(\frac{x - 0.5 - \mu}{\sqrt{2\sigma}}\right) \right) \quad [3.9]$$

where,
A = area,
σ = standard deviation,
μ = mean, and
erf = error function.

Although the standard form of the Gaussian (Eq. 3.8) is frequently used in DNA analysis packages, the exact form is preferred especially for Gaussians with low standard deviations. The Gaussian (Fig. 3.13) has many uses in DNA histogram analysis. It is used as a model component to represent G_0G_1 cells, G_2M cells, beads, and aggregates. A series of Gaussians has been proposed to represent S phase (11). Also, a Gaussian can be combined with a polynomial to model perturbed S-phase distributions (12). As mentioned earlier, the Gaussian is also used to broaden most of the rest of the continous functions described below.

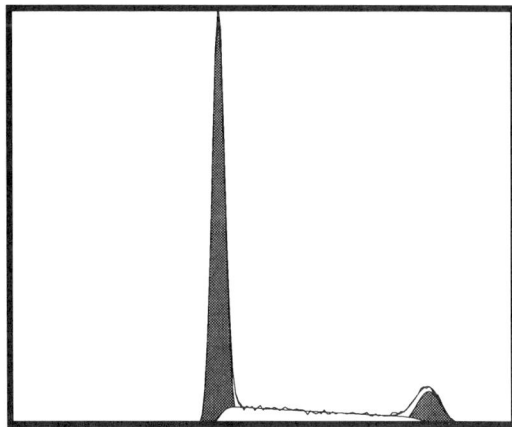

Figure 3.13. Gaussian model components (shaded).

BROADENED RECTANGLE

Iterative form:
$$R(x) = \frac{r}{\sqrt{2\pi}} \sum_{i=1}^{n} \frac{1}{\sigma_i} e^{-\frac{(x-\tau_i)^2}{2\sigma_i^2}} \Delta \tau_i \quad [3.10]$$

Convolution form:
$$R(x) \simeq r \, \text{derf}(x_1, x_2, \sigma_1, \sigma_2; x) \quad [3.11]$$

Multiple rectangle compartments:
$$R(x) \simeq \sum_{i=1}^{m} r_i \, \text{derf}(x_{1i}, x_{2i}, \sigma_{1i}, \sigma_{2i}; x) \quad [3.12]$$

where,
r = height of the unbroadened rectangle,
x_1, x_2 = beginning and ending x values of the rectangle,
σ_i = standard deviation at midpoint of segment i,
σ_1, σ_2 = standard deviations at x_1 and x_2, respectively,
τ_i = midpoint of the ith segment of the rectangle,
$\Delta \tau_i$ = width of the ith segment of the rectangle,
n = number of segments,
m = number of rectangles, and
derf = difference error function (Eq. 3.7).

The broadened rectangle (Fig 3.14) (5) is the most conservative S-phase model component. The word "conservative" is used in the sense that small perturbations in the shape of S phase will have a relatively small impact on the final S-phase area estimation. For DNA histograms generated from solid tumors, this may be the model component of choice for S-phase estimation. More than one broadened rectangle can be employed to model complex S-phase shapes.

One of the problems in using multiple broadened rectangles is the determination of the optimal number of compartments and spacing for particular observed DNA histograms. The optimal spacing can be shown to occur using exponentially increasing compartment sizes (5) and the optimal number can be derived using information theory (5). However, there is such a tolerance for deviations from the optimal

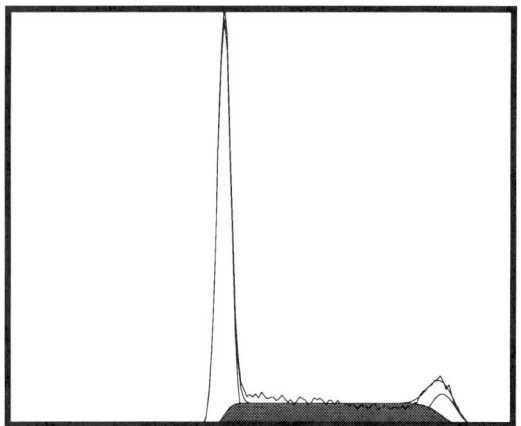

Figure 3.14. Broadened rectangle model component (shaded).

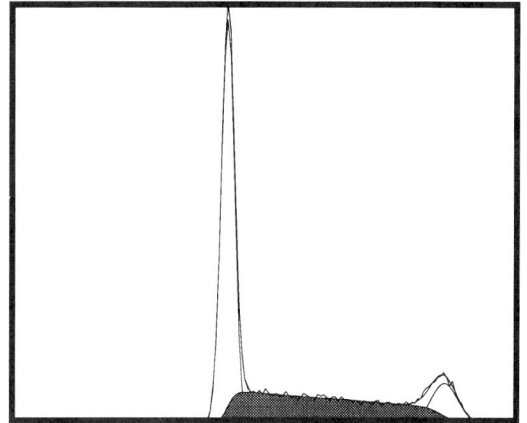

Figure 3.15. Broadened trapezoid model component (shaded).

number that the automatic determination of compartment number is generally not done in commercially available DNA histogram analysis programs. If multiple compartments are used, the most popular number is three since S phase conveniently splits into early, middle, and late compartments.

BROADENED TRAPEZOID

Iterative form:

$$T(x) = \frac{1}{\sqrt{2\pi}} \sum_{i=1}^{n} \frac{\left(\frac{t_2 - t_1}{x_2 - x_1} (\tau_i - x_2) + t_2 \right)}{\sigma_i} e^{-\frac{(x - \tau_i)^2}{2\sigma_i^2}} \Delta\tau_i \quad [3.13]$$

Convolution form:

$$T(x) \simeq \left(\frac{t_2 - t_1}{x_2 - x_1} (x - x_2) + t_2 \right)^* \text{derf}(x_1, x_2, \sigma_1, \sigma_2; x) \quad [3.14]$$

Multiple trapezoid component form:

$$T(x) \simeq \sum_{i=1}^{m} \left(\frac{t_{2i} - t_{1i}}{x_{2i} - x_{1i}} (x - x_{2i}) + t_{2i} \right)^*$$

$$\text{derf}(x_{1i}, x_{2i}, \sigma_{1i}, \sigma_{2i}; x) \quad [3.15]$$

where,
* = domain restricted to $x_1 \leq x \leq x_2$,
t_1, t_2 = beginning and ending heights of the unbroadened trapezoid,
x_1, x_2 = beginning and ending x values of the trapezoid,
σ_i = standard deviation at midpoint of segment i,
σ_1, σ_2 = standard deviations at x_1 and x_2, respectively,
$\Delta\tau_i$ = width of the ith segment of the trapezoid,
n = number of segments,
m = number of trapezoids, and
derf = difference error function (Eq. 3.7).

The broadened trapezoid (Fig. 3.15) (13, 14) is also a relatively conservative model component for S phase. If there is a need for multiple compartments, broadened trapezoids are probably the method of choice. Unlike broadened rectangles, multiple trapezoids are continuous with respect to the Y-axis ($t_{2i} = t_{1i+1}$) and thus are a more realistic representation of a DNA histogram's S phase.

A single broadened trapezoid works well for solid tumors and is currently the most popular of all the methods. The discussion on the optimal number and spacing of rectangle compartments also holds for trapezoid components.

BROADENED POLYNOMIAL

Iterative form:

$$P(x) = \frac{1}{\sqrt{2\pi}} \sum_{i=1}^{n}$$

$$\frac{\left(p_1(\tau_i - x_1)(\tau_i - x_2) + p_2 \frac{x_2 - \tau_i}{x_2 - x_1} + p_3 \frac{\tau_i - x_1}{x_2 - x_1} \right)}{\sigma_i}$$

$$e^{-\frac{(x - \tau_i)^2}{2\sigma_i^2}} \Delta\tau_i \quad [3.16]$$

Convolution form:

$$P(x) \simeq$$

$$\left(p_1(x - x_1)(x - x_2) + p_2 \frac{x_2 - x}{x_2 - x_1} + p_3 \frac{x - x_1}{x_2 - x_1} \right)^*$$

$$\text{derf}(x_1, x_2, \sigma_1, \sigma_2; x) \quad [3.17]$$

where,
* = domain restricted to $x_1 \leq x \leq x_2$,
p_1, p_2, p_3 = parabola concavity, height at x_1, and x_2, respectively,
x_1, x_2 = beginning and ending x values of the parabola,
σ_i = standard deviation at midpoint of segment i,
σ_1, σ_2 = standard deviations at x_1 and x_2, respectively,
$\Delta\tau_i$ = width of the ith segment of the rectangle,
n = number of segments, and

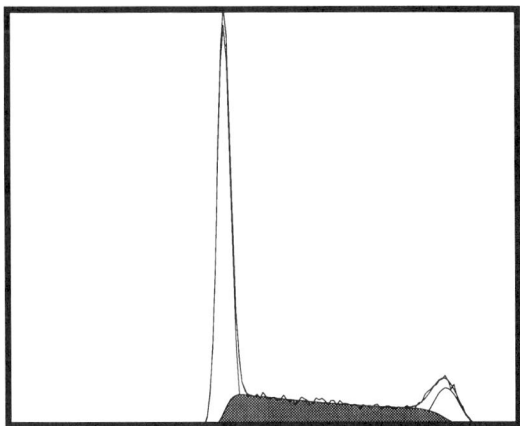

Figure 3.16. Broadened polynomial model component (shaded). (Reprinted with permission from Dean PN, Jett JH. Mathematical analysis of DNA distributions derived from flow microfluorometry. J Cell Biol 1974;60:523–527.)

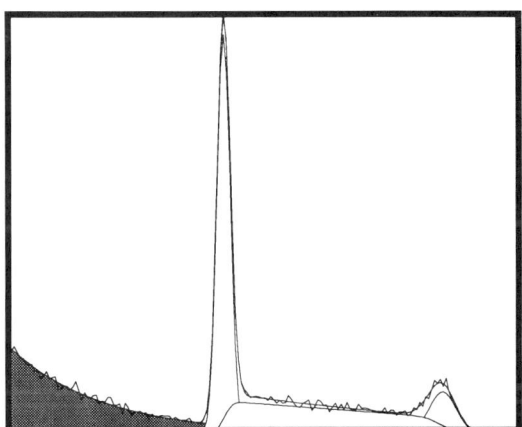

Figure 3.17. Exponential model component (shaded).

erf = error function.

A slightly more complicated form of the 2nd degree polynomial is used in the above equations than is usually presented ($ax^2 + bx + c$). The above formulation allows the computer algorithm to check automatically and correct for the illegal condition of having one or more model component heights, p_2 and p_3, less than zero. Also, the meaning of p_1, p_2 and p_3 is clearer when defined as above.

The broadened polynomial (Fig. 3.16) (2) is the S-phase model component used in the original Dean and Jett DNA model. Since the broadened polynomial can take on numerous shapes because of its three degrees of freedom, it is still the method of choice for well-behaved DNA histograms, such as most tissue culture cell lines. However, the increased flexibility of the model component can present a problem for solid tumor DNA analyses. If the G_0G_1 peak is slightly skewed or if there is significant debris contamination, this model component can give spurious results and thus is not recommended for routine solid tumor DNA analyses.

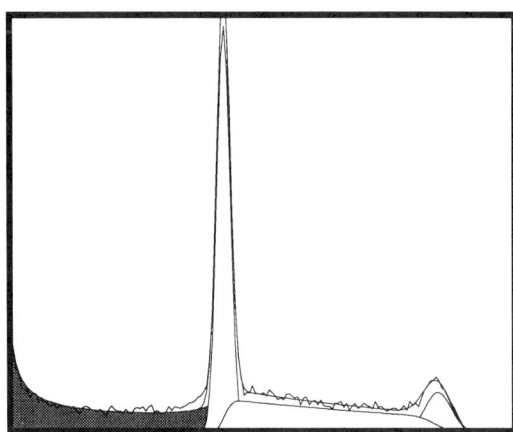

Figure 3.18. Single-cut model component (shaded).

EXPONENTIAL

$$E(x) = k_1 e^{-k_2 x} \qquad [3.18]$$

where,
  k_1 = exponential amplitude and
  k_2 = rate constant.

Historically, the exponential (Fig. 3.17) has been used to compensate for debris contamination into S phase (15). Its routine use is rapidly declining because now there are much better methods to compensate for debris, such as the Single-cut and Multi-cut distributions. Since the exponential has the theoretical shape of an ideal DNA histogram S phase, assuming the rate of DNA synthesis is constant, it can also be used as an S-phase model component.

SINGLE-CUT

$$S(x) \simeq a \sum_{j=x+1}^{n} \sqrt[3]{j}\, Y_j P_s(j,x) \qquad [3.19]$$

where,

$$P_s(j,x) \simeq \frac{2}{\pi j \sqrt{\left(\dfrac{x}{j}\right)\left(1 - \dfrac{x}{j}\right)}},$$

a = amplitude parameter,
Y_j = observed events in channel j, and
$P_s(j,x)$ = the probability of a single cut nuclei from channel j falling into channel x.

The Single–cut (Fig. 3.18) and Multi-cut model components (16–19) are fundamentally different from those discussed thus far. The Gaussian, broadened rectangle, broadened trapezoid, broadened polynomial, and exponential all have an intrinsic shape defined by a characteristic function. The Single-cut, Multi-cut, and aggregate distributions derive their shape from the observed histogram and thus are termed histogram-dependent model components.

The Single-cut distribution is the theoretical probability distribution of pieces formed by a single random cut through

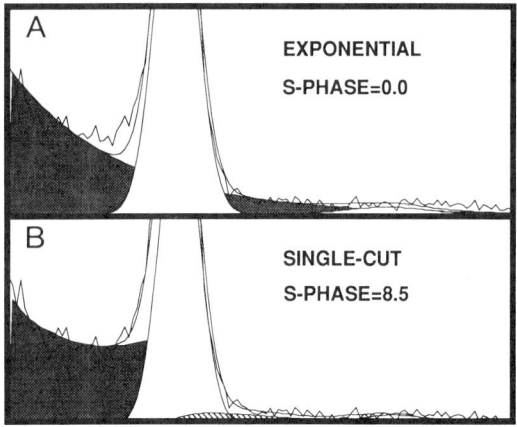

Figure 3.19. Typical paraffin-embedded DNA histogram modeled with an exponential (**A**) and with a single-cut (**B**) debris model component.

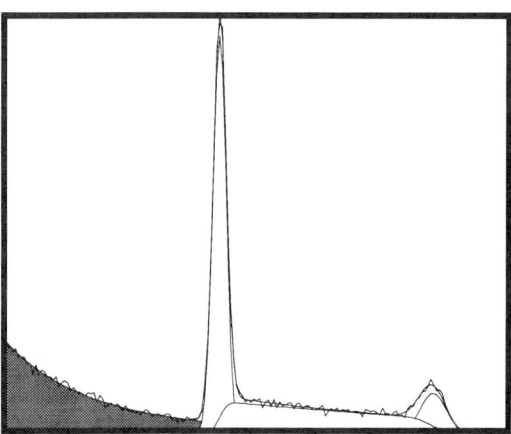

Figure 3.20. Multiple-cut model component (shaded). (Reprinted with permission from Bagwell CB, et al. DNA histogram debris theory and compensetic Cytometry 1991;12:107–118; and Weaver D, et al. Improved flow cytometric determination of cell cycle activity (S-phase fraction from paraffin–embedded tissues. AJCP 1990;94:576–584.)

ellipsoids of arbitrary eccentricity. The distribution is assumed to resemble the distribution of nuclear fragments generated by a knife sectioning paraffin-embedded material. The theoretical distribution is a relatively flat curve that ramps up quickly at its end-points. Figure 3.19 shows a typical paraffin-embedded derived DNA histogram modeled with an exponential and a Single-cut distribution. The exponential has a strong tendency to overfit the S-phase region because of a mismatch of its shape and that of the debris, whereas the Single-cut distribution has a much more intuitively correct fit. The Single-cut distribution has been shown to give better correlation between DNA S-phases from matched frozen and paraffin-embedded samples than the exponential (16, 17). It is recommended to use the Single-cut distribution for DNA histograms derived from paraffin-embedded material. It is further recommended not to attempt to gate out the debris with either light scatter gates or high DNA fluorescence thresholds since that will disturb the underlying distribution and make accurate compensation impossible.

MULTI-CUT

$$M(x) \simeq a\, e^{-kx} \sum_{j=x+1}^{n} Y_j \qquad [3.20]$$

where,

a = amplitude parameter, and
Y_j = observed number of events at channel j.

The Multi-cut distribution (Fig. 3.20) (16, 17) is the theoretical distribution of a large number of random cuts through ellipsoids of arbitrary eccentricity. The distribution is intended to represent the nuclei fragmentation that is an inevitable consequence of current preparative techniques. When the distribution is applied to each channel in an observed histogram, the end result is the Multi-cut model component. We recommend using the Multi-cut distribution for frozen or fresh DNA preparations if there is observable debris.

AGGREGATES

Originally, estimates of the number of doublets contaminating the observed 4C peak were performed by implementing some empirically derived transformation linearizing the height of G_0G_1 singlets, doublets, and triplets (5). A more accurate method was recently proposed that assumes that, for the most part, aggregation is a steady state phenomenon where the rate of aggregate formation equals the rate of disaggregation (20). The theory further assumes that aggregation is driven by particle concentration and exposed surface area. There are currently two fundamentally different forms of this new aggregation theory. The first, which we will call the discrete form, predicts the number of doublets based on the observed number of events at 2C, 4C, and 6C. The second, called the continuous form, predicts all interactions of particles in a DNA sample. The continuous aggregate theory describes a histogram-dependent model component for nonlinear least-squares analysis. The derivation of the equations is included in this section because of the current absence of a reference that explains their origin and assumptions.

DISCRETE AGGREGATE THEORY

The process of aggregation can be posed in a similar manner to substrate/product steady state equations.

$$[G_1] + [G_1] \underset{}{\overset{p}{\rightleftharpoons}} [D] \qquad [3.21]$$

If it is assumed that the above reaction is driven by the exposed surface area of the reactants, G_1 cells, and k_1 is proportional to surface area (20), then Equation 3.21 can be rewritten as

$$k_1[G_1] + k_1[G_1] \underset{}{\overset{p}{\rightleftharpoons}} [D]\,. \qquad [3.22]$$

Applying the mass action law to Equation 3.22 yields the formulation of the equilibrium constant, p.

$$p = \frac{[D]}{k_1^2[G_1]^2}.$$

If we let [2C] represent the number of observed events at 2C, then the equation that describes the composition of [2C] is given by

$$[2C] = [G_1] - 2[D] \text{ or } [2C] = [G_1] - 2pk_1^2[G_1]^2. \quad [3.23]$$

The theory can be extended to include triplets. The full set of equations are:

$$[2C] = [G_1] - 2pk_1^2[G_1]^2 - pk_1k_2[G_1][G_2] - p^2k_1^2k_{11}[G_1]^3 \quad [3.24]$$

$$[4C] = [G_2] + pk_1^2[G_1]^2 - pk_1k_2[G_1][G_2] - p^2k_1^2k_{11}[G_1]^3 \quad [3.25]$$

$$[6C] = pk_1k_2[G_1][G_2] + p^2k_1^2k_{11}[G_1]^3 \quad [3.26]$$

where,
$[G_1]$, $[G_2]$ = G_0G_1 and G_2M concentrations, respectively,
p = aggregation equilibrium constant,
k_1 = G_1 surface area constant (=1),
k_2 = G_2 surface area constant (=$2^{2/3}$), and
k_{11} = G_1 doublet surface area constant (=1.5).

The solution for p, $[G_1]$, and $[G_2]$ is possible by an iterative approach. It is interesting to note that the [6C] population, often inappropriately called triplets, is actually a mixture of G_1-G_2 doublets and G_1 triplets. By making a few simplifying assumptions, this discrete model of aggregation can be extended to the more complete continuous case.

CONTINUOUS AGGREGATE THEORY

The continuous aggregate model component is another example of a histogram-dependent model component. Accurately estimating aggregates involves writing simple equilibrium equations for the formation of each class of aggregates with constants that are either related to surface area (20, 21) or to number of possible aggregate combinations (22). The attractive aspect of these model components is that they may explain and correct for the observation that DNA histograms derived from paraffin-embedded material seem to have consistently higher S-phases than fresh or frozen samples, presumably due to an interaction of debris with G_0G_1 populations (16, 17).

MATHEMATICS

The surface area variant of this theory is described here because of its close relationship with the discrete form. Let p represent the equilibrium constant of the formation of an aggregate from two particles. If we assume that channel number to the 2/3's power is proportional to surface area, then the equation that represents the formation of a doublet at channel i from a particle at channel j aggregating with a particle at channel i-j is

$$j^{2/3}[Y_j] + (i-j)^{2/3}[Y_{i-j}] \underset{}{\overset{p}{\rightleftarrows}} [D_j] \quad [3.27]$$

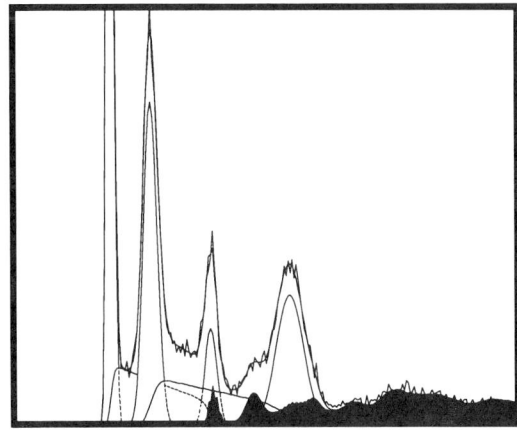

Figure 3.21. Aggregate model component (shaded).

Notice the similarity of this equation to the corresponding discrete aggregate equilibrium equation shown in Equation 3.22. The total number of doublets, triplets, and quadruplets found at channel i are given by:

$$\text{Doublets}(i) = p \sum_{j=1}^{i/2} Y_j \, Y_{i-j} \, (j(i-j))^{2/3}$$

$$\text{Triplets}(i) = p \sum_{j=1}^{i/2} Y_j \, \text{Doublets}(i-j) \, (j(i-j))^{2/3}$$

$$\text{Quadruplets}(i) = p \sum_{j=1}^{i/2} Y_j \, \text{Triplets}(i-j) \, (j(i-j))^{2/3} +$$

$$p \sum_{j=1}^{i/2} \text{Doublets}(j) \, \text{Doublets}(i-j) \, (j(i-j))^{2/3}$$

$$A(i) = \text{Doublets}(i) + \text{Triplets}(i) + \text{Quadruplets}(i) \quad [3.28]$$

The theory can be extended to include quadruplets and higher aggregate forms, but, at least with the surface area variant of the theory, the benefits are marginal.

DISCUSSION

The aggregate model component can have a complex shape, especially with DNA aneuploid histograms (Fig. 3.21). The aggregation model component accounts for all particle interactions. Compensating for debris interacting with G_0G_1 populations has a tendency to lower S-phase estimates with high debris contamination. The aggregate model component works best when the DNA histogram includes the higher order aggregates in the area of interest.

Software aggregation compensation is the newest and thus the least tested of all the DNA model components described thus far. It is included in detail in this chapter because of the current lack of a published reference and its potential in removing unwanted aggregates from a sample. The major limitation of the technique is its assumption that all combinations of particle adhesion can be calculated by a computed surface area and a single aggregation equilibrium constant, p. The real situation is undoubtedly more complex. It remains to be tested how this method compares with pulse processing methods and their inherent source of errors.

The major advantage of the method is that it works with the original integral pulse derived parameters and can be applied long after sample acquisition. As the theory and implementation of software aggregate compensation evolves, data can be reanalyzed and reinterpreted. Another major advantage of this technique is it does not require subjective gating decisions as does the pulse-processing method. The final advantage is that it is possible to calculate the theoretical unaggregated histogram from the analysis results. It is not enough to simply eliminate aggregates from cell cycle calculations. In order to obtain accurate cell cycle estimates, it is necessary to reconstitute the original unaggregated form of the histogram and then apply the methods described earlier to obtain accurate S-phase estimates.

HARDWARE AGGREGATE COMPENSATION

Aggregates can be substantially reduced by peak versus integral and pulse-width versus integral gating. The pulse-processing circuitry takes advantage of the fact that aggregates are generally oblong structures tending to axially align in laminar flow conditions. There are, however, some important caveats to be made about this technique. First, doublet discrimination assumes that aggregates are axially aligned, which, in general, is not true. Significant numbers of aggregates are presented at random orientations to the laser beam. It is not uncommon to see a continuum of events between the hypothetical doublet location on the peak versus integral plot and the unaggregated G_2M position. Thus, it is important to realize that this type of gating is not 100% effective and may only be removing 50%–70% of the doublets. The second problem is that some cells have oblong shapes and are thus mistakenly gated out by this technique. The third problem centers around the positioning of the diagonal gate. Any time software programs allow users to make subjective decisions on gate boundaries, human biases become an unknown factor in the precision of all subsequent analyses. The last problem with hardware doublet discrimination is that once it is performed and the listmode data are saved, the above software methods should not be used, since the underlying distribution has been changed in an unknown way.

Model Construction

Construction of DNA models is accomplished by picking the appropriate model components described earlier that are consistent with the biologic interpretation of the DNA sample. It is important to realize that no one model can fit all DNA histograms. For example, if a sample were derived from paraffin-embedded material with observable debris and also was characterized as a DNA diploid cell type, an appropriate DNA model might be described mathematically as follows:

$$\text{Model}(x) = S(x) + G^1(x) + T(x) + G^2(x) + A(x)$$

where,
$S(x)$ = single cut debris (see Eq. 3.19),
$G^1(x)$ = G_0G_1 Gaussian (see Eq. 3.9),
$T(x)$ = single S–phase broadened trapezoid (see Eq. 3.14),
$G^2(x)$ = G_2M Gaussian (see Eq. 3.9), and
$A(x)$ = aggregates (see Eq. 3.28).

There is a little more complexity to the creation of models than just selecting the appropriate model components. Additional relationships between some of the model component variables need to be defined. For example, the standard deviation of the G_2M Gaussian is often made dependent on the standard deviation of the G_0G_1 Gaussian by a factor of two, and the beginning and ending of the S–phase broadened trapezoid is a factor of one times the G_0G_1 Gaussian and G_2M Gaussian means, respectively. There are generally six or more defined model component dependencies that add rigidity and form to the DNA model.

Reduced Chi-square

As mentioned earlier, the major objective of the nonlinear least-squares algorithm is to adjust all the parameters that comprise a model to find the set that minimizes the chi-square surface. The final chi-square value can be used to evaluate the quality of the fit; however, there is a better method available called the reduced chi-square (6). The problem in using the chi-square as an indicator of fit quality is its dependence on the number of data points. A 256-channel histogram generally yields a lower chi-square value than a corresponding 1024-channel histogram. The reduced chi-square normalizes the chi-square by degrees of freedom:

$$X_r^2 = \frac{X^2}{n-m}$$

where,
n = number of data points and
m = number of model parameters.

The reduced chi-square should be approximately equal to one if the model fits the data appropriately. If it is much greater than one (>4.0), then probably some aspect of the data has not been fit by the model. If the reduced chi-square is much lower than one (<0.7), then the data has been overdetermined by the model. A good example of overfitting is using greater than 10 rectangle or trapezoid compartments in S phase.

Modeling Rules of the Road

Precision and accuracy are the important goals to DNA modeling. Estimates derived from the analysis are only useful if they can be related to a study that demonstrates their benefit. The following set of suggestions should be helpful in achieving these goals.

Overlapping Distributions. Never analyze data with two completely overlapping model components of similar shape. A good example of this problem occurs when attempting to analyze DNA tetraploid histograms with a model that has both the diploid G_2M and the aneuploid G_0G_1 Gaussians defined. Since the Gaussians overlap, the computer has no information to estimate the areas of each Gaussian. Under these conditions, most least-squares algorithms will either fail or give inaccurate results. The solution to this dilemma is to deactivate the diploid G_2M model component.

Another example of overlapping distributions is the potential overlap of DNA diploid S phase with DNA aneuploid S phase in DNA aneuploid histograms. Simulation studies with typical %CV's indicate that if an aneuploid tumor has DNA index of less than 1.3, only one S phase should be modeled to avoid erroneous estimates.

Number of Counts. If there is enough sample and the flow cytometer is stable, the more counts accumulated, the better. Strive for 500–1000 counts in the G_0G_1 peaks. Typical count ranges for 256-channel histograms vary between 10,000 and 20,000 counts. With 1024 histograms, the count ranges are generally 4 times higher.

Be Conservative. With most of the cell cycle packages available today, the user has the capacity and often the inclination to create complex models that overdetermine the cell cycle information implicit in a DNA histogram, frequently yielding worthless calculated parameters. The DNA modeler should choose model components that are well-behaved for the type of data being analyzed. A typical transgression of this conservative approach is to use numerous S-phase compartments (>10) in the multiple rectangle or trapezoid model components. If too many compartments are chosen, the model begins to fit statistical noise in S phase, with often disastrous effects on accuracy.

Use Published Models. Each model has its own biases in S-phase estimates; therefore, if low to high S-phase prognostic cutoffs are to be used, make sure you use the same model as described in the literature or create your own database of S-phase values. Also, if you publish a paper using a particular cell cycle model, it is important to adequately describe the model so others can reproduce the work.

Relative Rates of DNA Synthesis

In areas of high DNA synthesis rate, there is a relative paucity of cells. The same kind of phenomenon can be seen on the highway every day during city rush hour. The chance of finding a car in a particular area of road is inversely related to the average car's speed in the immediate vicinity; thus as the velocity of traffic is allowed to increase, the average distance between cars also increases.

If the DNA sample is derived from cells that are homogeneous, exponentially growing and asynchronous, the relative rate of DNA synthesis across S phase can be calculated (5, 23). Relative DNA synthesis rate is defined as the change in DNA content over the change in cell age. Since DNA histograms have no absolute time information, the time units are in cell age units ranging from zero to one cell cycle.

If S phase is analyzed with more than one Gaussian, rectangle, or trapezoid, it is possible to calculate a relative rate of DNA synthesis for each S-phase compartment. The first step in calculating relative rates of DNA synthesis is to determine the relative transit times for each of the S-phase compartments. This calculation makes extensive use of the number versus cell age relation depicted in Figures 3.4 and 3.5, which can be written mathematically as

$$y = 2Xe^{-\ln 2 \, t}$$

where when $t = 1$ cell cycle time, $y = x$.

The above equation can be reformulated in terms of cell fractions.

$$F(y) = 2\ln 2 \, e^{-\ln 2 \, t}$$

To calculate the G_1 relative transit time, T_{G1}, the above equation is integrated from 0 to T_{G1} and solved for T_{G1}.

$$T_{G1} = \frac{\ln\left(1 - \frac{F(G_1)}{2}\right)}{-\ln 2}$$

where,

$F(G_1)$ = fraction of cells in G_1 phase.

The relative transit time for the first S-phase compartment, T_{S1}, can be calculated in a similar manner.

$$T_{S1} = \frac{\ln\left(1 - \frac{F(G_1)}{2} - \frac{F(S_1)}{2}\right) - T_{G1}}{-\ln 2}$$

The above formula can be extended to solve for any S-phase relative transit time.

$$T_{Si} = \frac{\ln\left(1 - \frac{F(S_i)}{\left(2 - F(G_1) - \sum_{j=1}^{i-1} F(S_j)\right)}\right)}{-\ln 2}$$

Once the S-phase relative transit times are known, the relative DNA synthesis rate is calculated by dividing the change in DNA content over the domain of the S-phase compartment (normally the width of the compartment) by the estimated relative transit time.

If a single parabola or higher order polynomial is used as an S-phase model component, the relative rate of DNA synthesis can be calculated using the formula (23)

$$f(D) = \frac{2M(1 - 0.5F(G_1) - N(D))}{\alpha n(D)}$$

where,

$f(D)$ = relative rate of DNA synthesis at DNA content D,

α = ln2,
F(G₁) = fraction of cells in G₁,
N(D) = integral DNA distribution, and
n(D) = differential DNA distribution.

The above formula is quite general and can be used to approximate the relative rates of DNA synthesis for an observed DNA histogram once the fraction of cells in G₁ is known.

Relative rates of DNA synthesis calculations have been generally restricted to tissue culture systems because of the severity of the homogeneous, exponentially growing, and asynchronous assumptions. Also, since only one S-phase compartment is generally used for solid tumor DNA histograms, it is not normally possible to generate informative DNA rate of synthesis calculations.

Future of DNA Analysis

The great challenge of DNA analysis today can be found in the area of quality control and standardization. Simply put, any laboratory in the country should obtain the same ploidy and S-phase fraction with identical specimens. Several interlaboratory studies (24, 25) have demonstrated that we are still far from achieving this goal. Complete acceptance of DNA analysis as an important clinical test will only come when we have either solved the problem of standardization for all laboratories or when we restrict the test to a few laboratories that have the necessary quality-control procedures for ensuring reproducibility of test results.

Standardization of DNA modeling can best be accomplished by making the available cell cycle analysis programs more automatic in range positioning and model selection. The **Modeling Rules of the Road** section of this chapter should also minimize some of the common pitfalls in DNA modeling.

The obvious extension of single-parameter DNA analysis is to couple the DNA measurement with other information-rich parameters such as RNA, cytokeratins, or interesting nuclear and cytoplasmic proteins. From a data analysis point of view, it is much more probable to expect that these additional parameters will be used as gating parameters to yield single-parameter DNA histograms that will then be subjected to the type of analysis covered earlier in this chapter, rather than as modeling parameters necessitating the extension of the previously described model components to two or more dimensions. Exceptions to the above statement can be found in modeling two-parameter chromosome distributions (1); however, for the most part, two- or more parameter modeling will not gain wide acceptance in the near future because of our inability to adequately describe the other parameters mathematically. As you will see in the next chapter segment, data analysis of immunofluorescence histograms is mostly relegated to separating negatives from positives rather than to explaining subtleties in the shape of the curves.

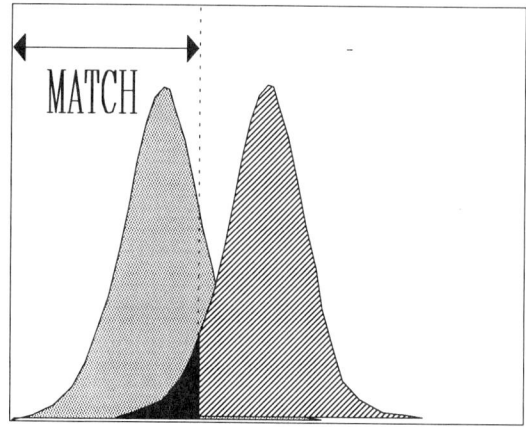

Figure 3.22. Normalized subtraction.

IMMUNOFLUORESCENCE ANALYSIS

Introduction

Mainly because of the AIDS crisis, the analysis of immunofluorescence spectra, such as those derived from αCD4 and αCD8 monoclonals, is of prime importance to most flow cytometry laboratories. In many ways, this type of analysis usually presents little analysis difficulties and can be done by simple integration methods. However, when there is overlap between cell populations expressing different levels of immunofluorescent staining, the problem of accurately quantifying these populations with computer algorithms becomes more complicated. The purpose of this section is to review some of the methods that are currently being used to analyze this more difficult type of immunofluorescence spectra.

The main difficulty in analyzing immunofluorescence spectra is that the characteristic shape of the separate populations' immunofluorescence spectra is not easily posed in mathematical terms, as are those modeling functions implemented in DNA analysis. As a result, clever algorithms, analogous to the inside-out and outside-in DNA algorithms, are used to yield approximate solutions. The two most widely used methods, normalized subtraction and accumulative subtraction, are described below.

Normalized Subtraction

The normalized subtraction method (26, 27) uses a control histogram, normally an isotype control, as well as the test histogram to quantify the percent positives in a population. The method normalizes the height of the control histogram such that the area in a user definable region, the match range, is equal to the corresponding area in the test histogram. The algorithm then subtracts the normalized control histogram from the test histogram. The residue histogram is then assumed to represent the positive population. Figure 3.22 graphically summarizes the method.

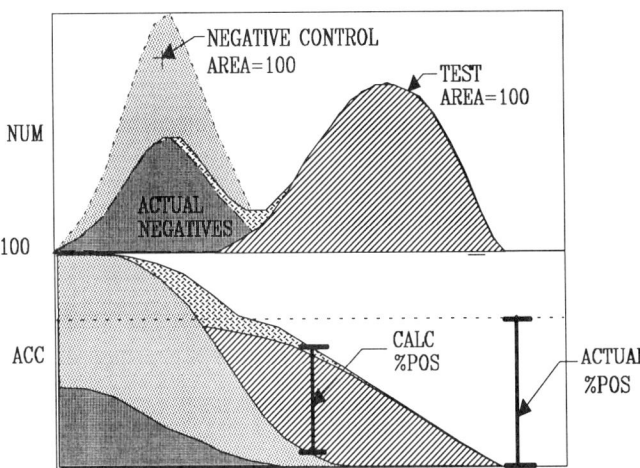

Figure 3.23. Cumulative subtraction.

Mathematics

Let c_i and t_i be the ith channels of the control and test histogram, respectively. The control normalizing factor, a, is given by

$$a = \frac{\sum_{i=m_l}^{m_u} t_i}{\sum_{i=m_l}^{m_u} c_i}$$

where,
m_l and m_u = user-definable lower and upper boundaries of an area called the "match" region.

The estimate for the ith channel of the positive histogram, p_i, is

$$p_i = t_i - ac_i$$

The percent-positive estimate, %pos, is given by

$$\%pos = 100 \frac{\sum_{i=l}^{u} p_i}{\sum_{i=1}^{n} t_i}$$

where,
l,u = the lower and upper boundaries for the positive distribution and
n = the total number of channels in the histogram.

Discussion

The normalized subtraction method is more accurate in estimating percent-positive cells than are integration methods for weakly positive immunofluorescence histograms. It also has the advantage of being able to calculate some statistics on the resultant positive histogram. The major source of error for the technique is the assumption that the control histogram has the same shape and position as does the negative population hidden in the test histogram. Deviation from this assumption can dramatically affect the accuracy of the technique. The method also has an unfortunate reliance on user input for the match region and the positive region. Anytime an algorithm has user-defined regions there is the possibility of human bias entering into the estimates. For interlaboratory and even intralaboratory consistency, it is generally better to use more automated algorithms. The next method, cumulative subtraction, has that important feature.

Cumulative Subtraction

The cumulative subtraction method (28) uses a completely different strategy than the normalized subtraction method to estimate percent positives. However, like the normalized subtraction method, it assumes the control histogram represents the shape and position of the negatives in an immunofluorescence histogram. The algorithm moves a boundary over the domain of all channels to determine the integral percent positives for the test and control histograms. The procedure builds a reverse cumulative histogram (100% at the origin) for both the test and control histograms (Fig. 3.23). The routine then finds the maximum difference between these two curves and assumes that this value is a good estimator of the actual percent positives.

Mathematics

Let c_i and t_i be the ith channels of the control and test histogram, respectively. Two reverse cumulative histograms, C_i and T_i, are formed using the equations:

$$C_i = 100 \frac{\sum_{j=i}^{n-1} c_j}{\sum_{j=0}^{n-1} c_j}$$

$$T_i = 100 \frac{\sum_{j=i}^{n-1} t_j}{\sum_{j=0}^{n-1} t_j}$$

Let $D_i = |C_i - T_i|$. The percent-positive estimate is the maximum D_i value for $0 \leq i \leq n-1$.

Discussion

The main advantage of the cumulative subtraction method is that it requires no user-defined ranges and can be completely automated. The procedure is certainly more accurate than integration methods for overlapping immunofluorescence spectra. Like the normalized subtraction method, the major source of error is a deviation from the assumption that the control histogram represents the shape and position of negatives in a test histogram.

HISTOGRAM COMPARISON METHODS

Introduction

You have two sets of histograms. One set was derived from cells treated with a compound that you have been studying and the other set of histograms is derived from untreated control cells. You want to know if the compound has any effect on the observed histogram parameter. What do you do? If you are like most of us, you will take one representative graph of a histogram from the treated group and one from the untreated group, hold them up to a light source, and look for obvious differences between them.

The two methods described in this section attempt to improve on the above direct visual method. The first method, D-value Comparison, is normally used when there is one histogram in each of the two groups. The second method, Average Histogram Comparison, is generally used when there are numerous histograms in each group.

Historically, cytometrists grouped these methods together as "Nonparametric Analyses" because they did not generate intrinsically meaningful parameters, such as the DNA and immunofluorescence methods described earlier. However, because of confusion with the statistical definitions of nonparametric methods, this term is less frequently used today.

D-value Comparison

In 1977, it was proposed by Young that Komogorov-Smirnov (KS) D-values could be used to quantify the differences between two flow cytometric histograms (29). The first major clinical use of this technique was for the detection of a "clonal excess" of circulating B cells expressing either κ or λ cell surface immunoglobulin light chains in patients with lymphoproliferative disease (30). This application of KS D-value is still the most popular use of the technique.

D-values are the maximum difference between two cumulative histograms. A cumulative histogram is formed by integrating the observed histogram from channel 0 to i, where i varies between 0 and n−1. Figure 3.24 demonstrates the process.

MATHEMATICS

Let $k(i)$ and $l(i)$ represent the ith channel accumulations in two histograms. The two histograms are normalized to the same area of 100 and then corresponding cumulative histograms are formed.

$$K(i) = \frac{100 \sum_{j=0}^{i} k(j)}{\sum_{j=0}^{n} k(j)}$$

$$L(i) = \frac{100 \sum_{j=0}^{i} l(j)}{\sum_{j=0}^{n} l(j)}$$

The D-value typically used in flow cytometry is

$$\text{D-value} = \text{Maximum} |K(i) - L(i)| \text{ for } 0 \leq i \leq n$$

DISCUSSION

The KS D-value can theoretically be transformed into a probability value if the computed D-value is normalized by the number of degrees of freedom (31). The number of degrees of freedom was originally assumed to be related directly to the number of channels in the histograms (29). The probability values never became popular because they generally demonstrated significant differences between histograms that were not known to be different. This problem is thought to be related to the inability to estimate the actual "effective" number of degrees of freedom since adjacent channels have some unknown correlation.

Average Histogram Comparison

The average histogram comparison method, originally published under the name of "Nonparametric Analysis" (32), is used to examine two groups of histograms for regions of possible significant differences. The histograms are normally preprocessed with a positional remapping algorithm to accurately align histograms if some internal standard is available. The algorithm examines the data, channel by channel, and calculates a Student t-test p value for each histogram based on the difference between the means, their standard deviations, and the number of degrees of freedom. The p values are normally displayed in the log domain directly underneath a co-plot of the mean histograms and their respective standard deviations. An example of such a plot is shown in Figure 3.25.

MATHEMATICS

If the histogram is derived from a linear amplifier, then the remapping of histogram Y to Z is given by

$$Z(x) = Y(kx)$$

where,

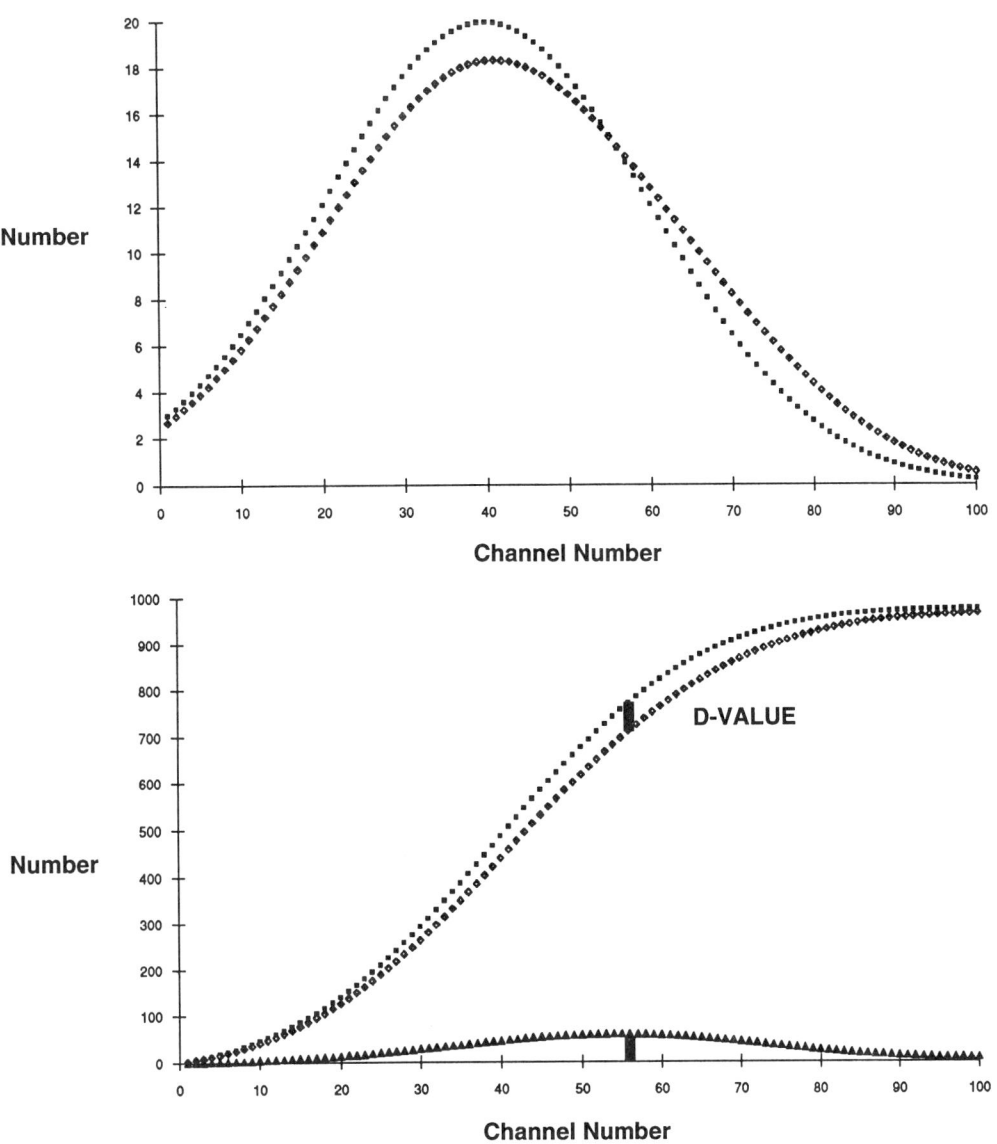

Figure 3.24. D-value comparison.

k = remapping constant normally derived from the current position of an internal standard divided by its desired position.

If kx is not an integer value, then interpolation between adjacent channel accumulations is necessary

$$Z(x) = Y(x_b) + x_f (Y(x_b+1) - Y(x_b))$$

where,
x_b = base integer of k x and
x_f = channel fraction above x_b.

The mean and standard deviation histograms are given by

$$M_i(x) = \frac{\sum_{j=1}^{n_i} Z_{ij}(x)}{n_i} \text{ and } S_i(x) = \sqrt{\frac{\sum_{j=1}^{n_i}(Z_{ij}(x) - M_i(x))^2}{n_i - 1}}$$

where
$M_i(x)$ = group i mean value at x,
n_i = number of histograms in group i, and
$S_i(x)$ = group i standard deviation value at x.

A one- or two-tailed t-test is generally performed for each channel given the means, standard deviations, and degrees of freedom (33).

DISCUSSION

Average histogram comparison is a valuable tool to visualize possible significant areas between groups of histograms.

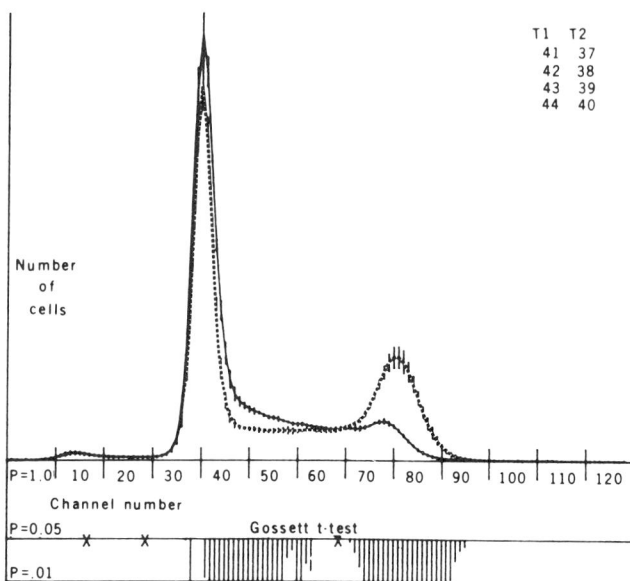

Figure 3.25. Average histogram comparison.

Once these areas are identified, a suitable parametric test is employed to determine possible significance.

SUMMARY

Whether you are a seasoned DNA modeler with a hearty appetite for convolution theory, or a clinician who would rather be tortured than be subjected to the tedium of mathematics, or somewhere in between, you should attempt to understand the limitations of the data analysis technology. Many times the possibilities of a technology are emphasized rather than the limitations, but, if it is to become a useful medical test, we must use it in a consistent and appropriate manner.

The major limitations of DNA modeling lie primarily in choosing the appropriate model that reflects the underlying biology of the sample and in obtaining good estimates for the nonlinear least-squares algorithm. The former problem is best solved through experience, while the latter is overcome by just being careful and insuring reasonably accurate simple analysis estimates.

Be conservative. If debris overwhelms the sample, the CV is high, or there are numerous aggregates, don't be heroic and try to force a solution with complex models. Let common sense prevail; ask for another sample or find the problem in the sample preparation. If you do not feel comfortable with the analysis of the sample, share the data with some respected laboratory and obtain a second opinion. Resist the temptation to fit subtle shoulders in DNA histograms for they probably represent differences in nuclear stainability rather than subtle DNA content aberrations. If the reduced chi-square is much higher than two to three, find out what is failing in the model.

The major limitation in the described immunofluorescence analysis techniques of normalized subtraction and cumulative subtraction is the assumption that the control histogram represents the shape and position of the negatives in the test histogram. Deviations from this assumption can have a profound effect on the accuracy of the method. Don't hesitate to use integration as a quality-control check for these methods.

The use of D values to compare two different histograms is exceedingly sensitive to slight shifts in histogram position. To use this test effectively for an application such as the κ-λ clonal excess test, antibodies must be carefully chosen to yield superimposable histograms with normal samples and the instrumentation must be stable for each acquisition.

Average histogram comparison is a powerful technique to find slight differences in groups of histograms. Observing differences with this method does not demonstrate significance; it merely points out areas in the histogram that may be amenable to some other type of analysis, yielding results that can then be subjected to standard statistical methods.

REFERENCES

1. Dean PN. Data processing. In: Melamed MR, ed. Flow Cytometry and Sorting. New York: Wiley-Liss, 1990:415–444.
2. Dean PN, Jett JH. Mathematical analysis of DNA distributions derived from flow microfluorometry. J Cell Biol 1974;60:523–527.
3. Hedley DW, Freidlander ML, Taylor IW, Rugg CA, Musgrove EA. Methods for analysis of cellular DNA content of paraffin embedded pathologic material using flow cytometry. J Histochem Cytochem 1983;31:1333–1335.
4. Young HD. Statistical Treatment of Experimental Data. New York: McGraw-Hill Book Company, Inc. 1962:73.
5. Bagwell CB. Theory and application of DNA histogram analysis [Dissertation]. Florida: University of Miami. ERIC/EDRS ED 299 130, 1979.
6. Bevington PR. Data Reduction and Error Analysis for the Physical Sciences. New York: McGraw-Hill Book Company, 1969.
7. Baisch H, Gohde W, Linden WA. Analysis of PCP-data to determine the fraction of cells in the various phases of the cell cycle. Radiat Environ Biophys 1975;12:31–39.
8. Dean PN. A simplified method of DNA distribution analysis. Cell Tissue Kinet 1980;13:299.
9. Marquardt DW. An algorithm for least-squares estimation of nonlinear parameters. J Soc Ind Appl Math 1963;11:431–441.
10. Abramowitz M. Handbook of Mathematical Functions. Dover Publications, Inc. NY. 1972:299.
11. Fried J. Method for the quantitative evaluation of data from flow microfluorometry. Comp Biomed Res 1976;9:263.
12. Fox MH. A model for the computer analysis of synchronous DNA distributions obtained by flow cytometry. Cytometry 1980;1:71.
13. Dean PN. Personal communication (1981).
14. Bagwell CB. PARA2 Program, EASY2 Software, Coulter Electronics, Inc., 1981.
15. Haag D, Feichter G, Goerttler K, Kaufmann M. Influence of systematic errors on the evaluation of S-phase portions from DNA distributions of solid tumors as shown for 328 breast carcinomas. Cytometry 1987;8:377–385.
16. Bagwell CB, Mayo SW, Whetstone SD, et al. DNA histogram debris theory and compensation. Cytometry 1991;12:107–118.
17. Weaver D, Bagwell CB, Hitchcox SA, et al. Improved flow cytometric determination of cell cycle activity (S-phase fraction) from paraffin-embedded tissue. AJCP 1990;94:576–584.
18. Bagwell CB. ModFit Program, Verity Software House, Inc. Topsham, Maine, 1989.
19. Rabinovitch PS. Multicycle Program. Phoenix Flow Systems, San Diego, CA, 1988.

20. Sato S. Doublet correction of DNA histogram. Abstract 521A. XIV International Meeting of the Society for Analytical Cytology, Asheville, North Carolina. Cytometry Suppl 4, 1990.
21. Bagwell CB. Aggregate model theory and compensation, in prep., 1991.
22. Rabinovitch PS. Aggregation compensation for DNA histograms. Presentation. XIV International Meeting of the Society for Analytical Cytology, Asheville, North Carolina, 1990.
23. Dean PN, Anderson E. The rate of DNA synthesis during S-phase by mammalian cells in vitro. In: Haenen CAM, Hillen HFP, and Wessels JMC, eds. Pulse Cytophotometry. European Press Medikon, Ghent, 1975, p. 77.
24. Wheeless LL, Coon JS, Cox C, et al. Measurement variability in DNA flow cytometry of replicate samples. Cytometry 1989;10:731–738.
25. Hitchcock C and the "Flow Cytometry Inter-Laboratory Study Group." Inter-laboratory analysis of DNA by flow cytometry. Abstract 493B. XIV International Meeting of the Society for Analytical Cytology, Asheville, North Carolina. Cytometry Suppl 4, 1990.
26. Hoffman R. Simple analysis of immunofluorescence histograms. Abstract 61. VIII Conference on Analytical Cytology and Cytometry, Wentworth-by-the-Sea, New Hampshire: May 19-25. Cytometry 1981; 2:104.
27. Bagwell CB. IMMUNO Program, EASY2 Software, Coulter Electronics, Inc., 1980.
28. Overton RW. Modified histogram subtraction technique for analysis of flow cytometry data. Cytometry 1988;9:619–626.
29. Young IT. Proof without prejudice: Use of the Kogomorov-Smirnov test for the analysis of histograms from flow systems and other sources. J Histochem Cytochem 1977;25:935–941.
30. Ault KA. Detection of small numbers of monoclonal B lymphocytes in the blood of patients with lymphoma. N Engl J Med 1979;300:1401–1405.
31. Zaar JH. Biostatistical Analysis. NJ: Prentice-Hall, Inc. 1984:53–58.
32. Bagwell CB, Hudson JL, Irvin GL, III. Nonparametric flow cytometry analysis. J Histochem Cytochem 1979;27:293.

4

Flow Cytometry Reveals the Complexity and Diversity of the Human Immune System

LEWIS L. LANIER

INTRODUCTION

The immune system is responsible for the recognition and elimination of pathogenic bacteria and viruses. Highly specialized cell types have evolved to provide these critical functions and ensure both immediate and long-term defense against infection. While morphology and histology provided the initial characterization of the body's cell types, these techniques failed to reveal the considerable diversity of cells comprising the immune system. Based on the vast array of functions performed by the immune system, it was evident that the relatively "homogeneous" population of small, round cells called lymphocytes had to include different cell types responsible for these specialized tasks. The availability of inbred, congenic strains of mice permitted identification of a limited number of polymorphic membrane proteins that allowed discrimination of subsets of lymphocytes with distinct functions (1–3). However, these polyclonal serological reagents were available in limited amounts and were often contaminated with unwanted specificities. Early attempts to identify subsets of human lymphocytes were based on techniques such as the fortuitous binding of human T-cells, but not B-cells, to sheep erythrocytes (4), a method to enumerate T-lymphocytes that has persisted in clinical immunology labs until only very recently. While subsets of human T-cells comparable to murine counterparts were initially described using rabbit antisera (5), this approach was generally unreliable.

HYBRIDOMA TECHNOLOGY AS A CATALYST OF FLOW CYTOMETRY

Two major technological advancements have provided the key to unveiling the complexity and diversity of the immune system. First, in 1975, Kohler and Milstein (6) published a simple method to produce limitless quantities of monoclonal antibodies (mAb). Given the vast diversity of the immunoglobulin repertoire (estimated at $>10^9$), it was now possible to generate serological reagents against essentially any immunogenic substance. Almost immediately, this technique was used to generate specific reagents to analyze the cell types comprising the immune system of mice, rats, and man (7–11). The availability of these reliable and specific antibodies prompted the development of analytical techniques to exploit these new reagents. While very sensitive RIA and ELISA methods were available for detecting antigens recognized by mAb, these techniques were inadequate to distinguish diversity within tissues containing many different cell types or to quantitate heterogeneous levels of antigen expression within a population. Fortunately, advances in the nascent field of flow cytometry were underway to complement the development of hybridomas.

DEVELOPMENT OF MULTI-COLOR IMMUNOFLUORESCENCE AND MULTIPARAMETER FLOW CYTOMETRY

The objective of flow cytometry is to measure multiple properties of single cells at a rapid rate to permit detailed quantitative and qualitative analysis. Initially, flow cytometers were limited to one or two parameters, generally one for light-scattering measurements and another for fluorescence detection. Using these early instruments equipped with argon-ion lasers, immunologists began analysis of the immune system typically by one-color immunofluorescence using fluorescein-isothiocyanate (FITC) conjugated antisera or mAb (7, 12–14). However, it soon became evident that many of the mAb reacted with overlapping subsets of cells. There were attempts to estimate the proportion of cells reactive with two different antibodies by staining cells with a mixture of two antibodies and comparing the percentage of reactive cells with the percentage of cells staining with each individual antibody (15) or, alternatively, by complement depletion of cells reactive with one mAb followed by immunofluorescent staining with a second mAb (16). However, these approaches certainly were not satisfying and failed to provide information concerning the amounts of different antigens expressed on the "double-positive" population. The complexity of the immune system revealed by the use of mAb against lymphocyte surface antigens demanded development of new fluorescent dyes, techniques to conjugate fluorochromes to antibodies, and sensitive multiparameter flow cytometers.

The initial problem was to identify dyes that could be conjugated to antibodies without inactivating their binding capacity and to select fluorochromes with distinct emission spectra. The first two-color immunofluorescence systems for flow cytometry employed fluorescein and rhodamine–conjugated antibodies (17). However, considerable spectral overlap of these fluorochromes resulted in less than optimal fluorescence detection. This problem was overcome by the synthesis of a sulfonylchloride derivative of rhodamine, designated Texas Red (18). This dye has been used successfully in combination with FITC for very sensitive two–color immunofluorescence, but requires both an argon ion laser (488 nm) and a dye laser (~600 nm) for excitation of FITC and Texas Red, respectively (18–20).

A key breakthrough in flow cytometry was the development of phycobiliprotein dyes for immunofluorescence applications (21). Phycobiliproteins are components of the photosynthetic apparatus of red algae and blue–green bacteria. These natural fluorochromes are water–soluble, fluorescent at neutral pH, readily conjugated to mAb, and have very high quantum yields. One of these fluorochromes, phycoerythrin (PE), is excited by 488 nm light and can be used in conjunction with FITC to provide a very sensitive and reliable detection system for two-color immunofluorescence. Moreover, FITC, PE, and Texas Red together provided the first practical system for three-color immunofluorescence analysis of the mouse (22) and human (23) immune systems. Allophycocyanin (APC), a phycobiliprotein that can be excited at the same wavelength as Texas Red, permits four-color immunofluorescence using the simultaneous combination of FITC, PE, Texas Red, and APC conjugated reagents (24). More recently, three-color immunofluorescence systems employing a single argon-ion laser (488 nm) have been devised to eliminate necessity for a dual laser cytometry. For use in conjunction with FITC and PE–conjugated reagents, reagents have been conjugated to "tandem" fluorochromes (25). For example, Texas Red can be covalently coupled to PE. When PE is exited by 488 nm light, its emission is transferred to Texas Red for subsequent emission of light above 600 nm. However, this strategy necessitates efficient energy transfer between the dye molecules. An alternative fluorochrome for use with FITC and PE is a dinoflagellate protein, peridinin-chlorophyll-a-protein (26). This fluorochrome can be directly coupled to mAb, is excited with 488 nm light, and emits at >600 nm. Many other fluorochrome combinations for immunofluorescence are available in addition to those mentioned above. The present "world record" for multi-color analysis using 488 nm excitation is undoubtedly the six-color immunofluorescence system described by Recktenwald and colleagues (26).

Thus, unlike the situation a decade ago, investigation of the immune system is rarely constrained by inadequate reagents or instrumentation for flow cytometry. Perhaps the major advantage of flow cytometry is that this technology permits the separation of cells based upon any of the measured criteria. Using the present generation of sorters, a population of interest can be identified on combinations of any of the light scattering or fluorescence parameters and viable cells can be isolated to high purify (typically >95%) for analysis of functional or genetic properties.

CELL LINEAGES OF THE IMMUNE SYSTEM

mAb reactive with leukocyte-cell surface antigens have been instrumental in the study of hematopoiesis, classification of the cell types of the immune system, and analysis of the dynamic processes accompanying an active immune response. Several international workshops have been devoted to determining the specificity of the vast collection of mAb generated against human leukocyte differentiation antigens. Furthermore, these workshops have provided a universal nomenclature for membrane antigens that have been designated "CD" antigens. Seventy-eight CD antigens are currently catalogued, but this number might easily double at the next conference. Proceedings of these workshops have been published and provide an excellent resource for more detailed information on the structure, function, and distribution of these antigens (27).

The immune system is composed of leukocytes, a diverse collection of cell types that arise from hematopoietic stem cells in the bone marrow. Hematopoietic stem cells can be identified in bone marrow by the presence of a membrane antigen designated CD34 (28). Although present in the bone marrow at low frequency (<0.5%), these cells generate all lineages of mature leukocytes (29, 30). Progenitors for myeloid, erythroid, and lymphoid cells, as well as platelets, differentiate from these stem cells when they are provided with the appropriate cytokines and environmental conditions. Myeloid progenitor cells subsequently develop into mature monocyte/macrophage lineage cells and granulocytes, a population that includes neutrophils, basophils, and eosinophils. In vitro culture systems have been devised to study the development of monocyte, granulocyte, and erythroid progenitor cells and commitment to these lineages can be influenced by the presence of a variety of "colony-stimulating factors" or cytokines (31). However, this has not been achieved for human lymphoid progenitor cells as yet. A more detailed investigation of the population of cells bearing CD34 antigen has revealed considerable heterogeneity within this population (32, 33). $CD34^+$ hematopoietic progenitor cells are not confined to the bone marrow, but can also be found in peripheral blood (34, 35) and thymus (36). It is anticipated that within the $CD34^+$ population, it will be possible to identify progenitor cells already committed to different lineages, as reflected by coexpression of other membrane antigens. Prior studies in murine model systems have used multicolor immunofluorescence and cell sorting to identify totipotent as well as lineage-specific hematopoietic progenitor cell populations (37). There is considerable interest in the hematopoietic stem cell population for applications in bone marrow transplantation and gene reconstitution therapy and for a more comprehensive understanding of hematopoiesis.

Since lymphocytes are relatively homogeneous on the basis of morphology, delineation of the predominant lineages and subsets of lymphocytes has required mAb and flow cytometry. Lymphocytes comprise three distinct lineages of cells: T-cells, B-cells, and natural-killer (NK) cells.

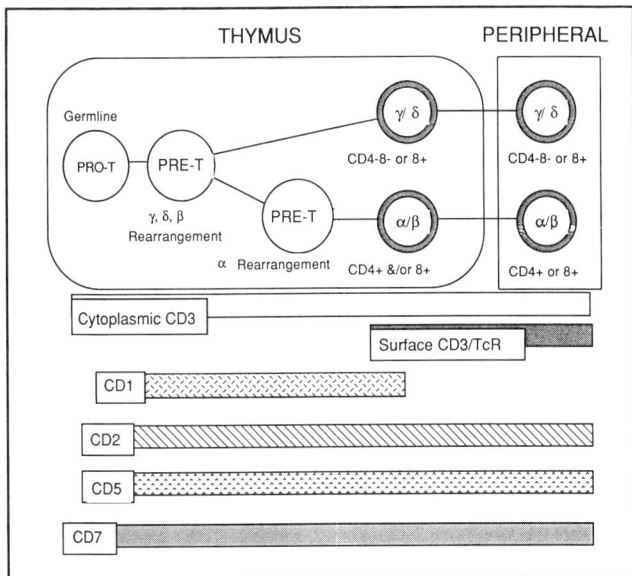

Figure 4.1. Hypothetical scheme of T-cell development.

B-cells arise from bone-marrow progenitor cells and are defined as lymphocytes that productively rearrange and express immunoglobulin (Ig) genes. Ig genes are composed of variable (V), joining (J) and constant (C) segments that are rearranged during B-cell ontogeny (38). Rearrangement of these elements, as well as subsequent somatic mutation of Ig V region segments, results in sufficient structural diversity to specifically recognize $>10^9$ antigens. Igs are tetramolecular proteins consisting of two disulfide-bonded heavy chains, each of which is disulfide-linked to a light chain. Two isoforms of light chains (κ and λ) and nine isoforms of heavy chains ($\mu, \delta, \gamma1, \gamma2, \gamma3, \gamma4, \epsilon, \alpha1, \alpha2$) exist. An integral membrane–anchored form of Ig is expressed on the surface of B-cells and, following activation, B-cells produce a secreted form of Ig that is released into the surrounding tissues and serum. Thus while B-cells may have several functions in the immune system, their predominant role is to provide humoral immunity as a consequence of Ig secretion. While Ig expression is undoubtedly the most definitive characteristic of a B-cell, using this as a criteria to enumerate B-cells is complicated by the fact that Ig (particularly IgM and IgG) is present in high concentrations in the serum. Many non-B-cells express cell surface receptors for the Fc portion of Ig and bind serum Ig "non-specifically". As a consequence, many "Ig$^+$" cells are not B-cells, but have passively acquired these proteins from the serum. Other B-cell-associated membrane antigens have proven more reliable for the identification and enumeration of B-lineage cells. CD19 is essentially exclusively on B-cells and is expressed from the earliest stages of pre-B-cell development throughout the B-cell differentiation pathway, with the exception that it is down-regulated on plasma cells. While B-lymphocytes express a variety of other membrane antigens (i.e., HLA-DR, -DP, -DQ, CD10, CD38, CD20) that are useful in tracking B-cell differentiation, these antigens are not restricted to B-cells, but are present on other cell types.

T-lymphocytes also arise from bone marrow progenitors and most require a unique anatomical location, the thymus, for early development (Fig. 4.1). T-cells are responsible for antigen-specific cell-mediated immunity. Upon antigen-specific activation, T-cells produce a variety of cytokines and growth factors that regulate B-lymphocytes and other hematopoietic cells and acquire the ability to mediate cytotoxicity against virus infected target cells. Four T-cell antigen receptor genes (TcR), designated TcR-α,-β,-γ, and -δ encode proteins that are assembled as either TcR-$\alpha\beta$ or TcR-$\gamma\delta$ heterodimers. Like Ig, these TcR genes are encoded by V, J, and C gene segments that rearrange during T-cell development. Individual T-cells express either $\alpha\beta$-TcR or $\gamma\delta$-TcR (but never both receptors) on the cell surface. In human blood and lymphoid tissues, $\alpha\beta$-TcR bearing cells comprise ~90–99% of total T-cells, whereas $\gamma\delta$-TcR T-cells comprise ~1–10%. The rationale for two types of TcR is presently unknown. The $\alpha\beta$-TcR recognize a vast array of specific peptide antigens that are associated with either HLA class I or class II proteins on the surface of an "antigen-presenting cell" and are the predominant mediators of most T-cell-mediated immune responses. While the antigenic repertoire of $\gamma\delta$-TcR T-cells is less well defined, it is likely that these cells also react with peptide antigens on major histocompatibility complex (MHC) molecules. The $\alpha\beta$-TcR and $\gamma\delta$-TcR are never expressed on the surface of a T-cell alone, but require association with the invariant CD3 polypeptides for membrane expression (39). Four different CD3 genes have been identified, CD3-γ,-δ,-ϵ, and -ζ and these proteins are thought to be involved in T-cell activation subsequent to antigen binding to the TcR heterodimer (40). Most of the anti-CD3 mAbs available (i.e., anti-Leu 4, OKT3) react specifically with the CD3ϵ subunit (41). Expression of cell-surface CD3ϵ is strictly restricted to expression on T-cells and, thus, is the most reliable reagent to quantitate mature T-lymphocytes. Unlike the situation with Ig, T-cells do not secrete TcR into the serum and soluble TcR heterodimers likely have no appreciable affinity for antigen.

NK-cells comprise a third lineage of lymphocytes that arise from bone marrow progenitor cells, but, unlike T-cells, do not require the thymus for development. NK-cells have the capacity to kill virus-infected cells and tumor cells, but, unlike cytotoxic T-lymphocytes, do not require the presence of MHC class I or II antigens to recognize their targets. In addition to mediating cytotoxic function, activated NK-cells produce regulatory cytokines and growth factors, including interferon-γ, granulocyte/monocyte-colony-stimulating factor, tumor-necrosis factor-α, and others (42). The receptors on the membranes of NK-cells that are responsible for recognition of tumors or virus-infected targets are presently unknown. NK-cells do not rearrange Ig or TcR genes, and unlike T-cells do not

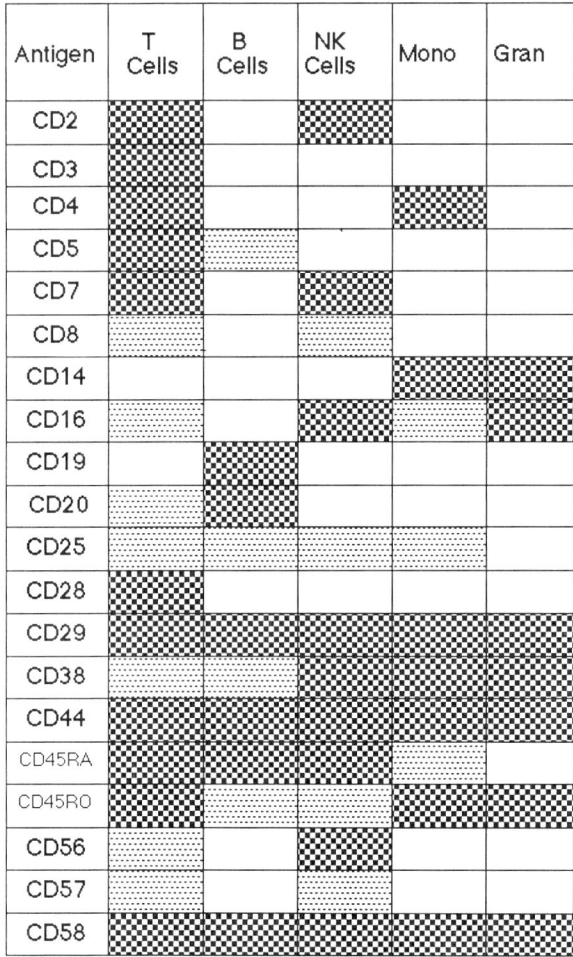

Figure 4.2. Lineage infidelity of differentiation antigen expression distribution of several frequently analyzed cell-surface-differentiation antigens are summarized. An open box indicates that the antigen has not been detected on the particular cell type (yet); checker board indicates that more than 50% of cells within this lineage express detectable amounts of the antigen; dotted lines indicate that less than 50% of cells within this lineage express detectable amounts of the antigen (with today's technology).

LINEAGE "INFIDELITY" OF LEUKOCYTE DIFFERENTIATION ANTIGENS

The antigens present on the surface of leukocytes are not present for the convenience of immunologists to categorize cell lineages and subsets. These structures are carbohydrates, lipids, and proteins that perform essential cellular functions. Many of these functions are required by several different cell types of the immune system; so it should not be surprising that these antigens may be expressed by different cell lineages. Often, an antigen is assumed to be specific for the favorite cell type of the investigator who has discovered this new structure. For example, had CD4 first been observed on monocytes, it would certainly have been considered a monocyte–specific marker, "also present on some T-cells," with the subsequent extrapolation that T-cells must be members of the monocyte/macrophage lineage. Likewise, CD8 might well be considered an "NK-cell subset" marker, also present on a minor subset of T-lymphocytes. It is important not to be mislead by nomenclature or successful marketing. Of the scores of membrane antigens that have been identified on leukocytes, very few are restricted to a single-lineage or even hematopoietic cells. A few examples of this are summarized in Figure 4.2. When analyzing expression of an antigen, it is important to be observant, and open–minded. Real discovery occurs when something that is not supposed to be there is found.

SUBSETS OF SUBSETS OF SUBSETS

Considerable heterogeneity in antigenic phenotype exists within the T, B, and NK populations. Equipped with an extensive panel of mAb labeled with appropriate fluorochromes and a flow cytometer capable of multiparameter detection, it is possible to identify an essentially infinite number of "subsets." Given a sufficient number of antibodies and detectors, the number of "subsets" will equal the number of cells analyzed. This may sound facetious; however, each individual cell is unique if enough parameters are measured. The major point is that one must avoid the temptation to design experiments whose only goal is to determine "how many subsets exist within this population or lineage." While this approach can produce prodigious amounts of flow cytometric data, this strategy is somewhat like acquiring a set of Chinese boxes; opening each box reveals only a smaller box. Identifying subsets is only useful in the context of understanding relationships between different cells with regard to differentiation stage, activation state, or correlation with a particular function of interest. Given this caveat, the ability of flow cytometry to identify both qualitative and quantitative heterogeneity in antigenic phenotype within lymphocyte populations has provided valuable insight into the function and composition of the immune system.

Antigenic heterogeneity exists within the B, T, and NK-cell lineages. T-lymphocytes have been most estensively studied with respect to differential antigen expression. The "major" T-cell subsets are identified by expression of the

express CD3ε on the cell surface. Human NK-cells can be identified by the phenotype surface CD3ε$^-$ and CD16$^+$ and/or CD56$^+$ (43). CD16 is an Fc receptor for IgG expressed on NK-cells, macrophages, neutrophils, and some T-cells. CD56 is the neural cell adhesion molecule (N-CAM) that is expressed on neural tissues, NK-cells, and some T-cells (44). Since a subset of T-lymphocytes can express either CD16 or CD56 (43, 45), the only definitive method to identify NK-cells within the lymphocyte population is by two-color immunofluorescence using a combination of anti-CD3ε and anti-CD56 and/or anti-CD16 mAb (43). Unfortunately, there is no single mAb that defines a structure that is restricted to NK-cells. NK-cells are probably involved in the early defense against certain viral pathogens and permit time for the generation of virus-specific cytotoxic T-lymphocytes during an immune response (46, 47).

CD4 and CD8 antigens. The most immature T-lymphocytes in the thymus begin with the phenotype CD3-,4-,8-, acquire expression of CD4, CD8, and then CD3 (CD3-,4+,8+ and CD3+,4+,8+), and upon subsequent differentiation evolve into mature CD3+,4+,8- and CD3+,4-8+ T-cells (48). Most thymocytes (~70–80%) are either CD3-,4+,8+ or CD3+,4+,8+, whereas these cell types are relatively rare in peripheral tissues (usually <1% of T-cells). By contrast, in peripheral tissues, CD3+,4+,8- and CD3+,4-,8+ represent about 70% and 25% of total CD3+ T-cells, respectively. The remaining 5% of the peripheral T-cells are CD3+,4-,8-, most of which express γδ-TcR (49). CD3+,4-,8+ and CD3+,4+,8- T-cells predominantly express αβ-TcR; however, minor subsets of αβ-TcR+ T-cells are CD3+,4-,8- and a minor proportion of γδ-TcR+ T-cells are CD3+,4-,8+ and CD3+,4+,8- (50). Initially, the CD3+,4-,8+ cells and CD3+,4+,8- cells were designated "cytotoxic/suppressor" T-cells and "helper/inducer" T-cells, respectively, since expression of these phenotypes usually correlated with these functional activities. Cytotoxic T-cells kill target cells that present specific peptide antigens on appropriate cell surface HLA class I molecules. Helper T-cells secrete cytokines that augment Ig synthesis by B-cells after interaction with specific peptide antigens present on appropriate cell surface HLA class II molecules of "antigen presenting cells." CD4 is a specific receptor for HLA class II molecules and CD8 is a receptor for HLA class I molecules, explaining the correlation between these functions and expression of CD4 and CD8. It should be appreciated that there are exceptions to these general correlations. In particular, cytotoxic T-cells with the phenotype CD3+,4+,8- have been identified in many situations, usually in the context of CTL directed against HLA class II alloantigens (51).

Within the CD4 and CD8 T-cells subsets in peripheral blood and thymus there is further diversity that can be resolved by multicolor immunofluorescence (23, 48). A major goal of many laboratories is to correlate this antigenic heterogeneity with unique functions. For example, it is has been observed that most "helper" T-cell function is within the CD3+,4+, Leu 8- (52) or CD3+,4+,29$^{bright+}$ (53) fraction. Recently, it has been proposed that antigenic heterogeneity within the T-cell population may relate to whether the T-cell is "naive" or "virgin" with respect to interaction with a specific antigen or to whether it has "memory" as a consequence of prior antigenic exposure and activation (54, 55). The "naive" and "memory" T-cell populations were initially proposed based on differential expression of CD45 isoforms (a membrane phosphatase) and surface density of CD29 (β1 integrin). "Naive" T-cells express low levels of CD29 and the CD45RA isoform, whereas "memory" T-cells have higher levels of CD29 and the CD45RO isoform. Additionally, it appears that there is coordinate up–regulation of several genes in "memory" T-cell populations since these cells possess high cell-surface density of several differentiation antigens, including CD2, CD11a, CD44, and CD58 compared to the "naive" T-cell population (56). It remains controversial whether the "naive/memory" classification actually reflects a differentiation process within a cell lineage or, alternatively, simply on activation status.

Although less extensively studied than in the T-cell population, phenotypic heterogeneity also exists within the B- and NK-lymphocyte populations. The best defined subsets of B-cells have been delineated based on expression of CD5, an antigen once considered T-cell specific. The minor subset of B-cells that are CD5+ have been shown to be the predominant source of auto-antibody production in mouse and man (57). Heterogeneity within the NK-cell population is demonstrated by analysis of a variety of cell surface antigens, including CD2, CD8, CD25 (p55 IL-2 receptor), CD44, CD57, and Leu 8 (58). The rationale and functional consequences for most of this heterogeneity is not as yet understood. However, there are two examples where antigenic phenotype has been correlated with function for NK-cells. In peripheral blood, the majority (~90%) of NK-cells express CD16, an Fc receptor for IgG, whereas ~10% lack CD16 (58). CD16+ NK-cells possess the capacity to mediate antibody-dependent cellular cytotoxicity (ADCC), whereas CD16- NK-cells lack ADCC function. Additionally, the CD16- NK population is unique in that it is the only cell type found in normal peripheral blood that constitutively expresses detectable high affinity IL-2 receptors composed of both p75 and p55 (CD25) subunits (59, 60). The functional consequence of this finding is that CD16- NK cells preferentially activate and proliferate in response to extremely low concentrations of IL-2 that might be expected to occur in physiological immune reactions in vivo.

RELATIONSHIP OF ANTIGENIC PHENOTYPE AND CELLULAR FUNCTION

A predominant focus of flow cytometric analysis in immunology has been to correlate antigenic phenotype with discrete immune functions. While this approach has been extremely helpful in understanding diversity within the immune system, it must be stressed that in most cases the findings are strictly **correlations** and not absolutes. There is often a tendency to extrapolate from these functional correlations demonstrated in very limited in vitro experimental systems to in vivo situations. For example, many investigators will observe an increase in CD8 T-lymphocytes in blood and conclude that there is an increase in "suppressor" or "cytotoxic" T-cells, implying a parallel augmentation in these functions. Likewise, it is concluded that a patient expressing a lower frequency of NK-cells or CD4+ T-cells must have lower "NK activity" or "helper" function, respectively. Such conclusions are completely inappropriate and likely wrong. In order to conclude that a biological function has been affected, it is necessary to directly measure that function. One important consideration is that function is usually a reflection of cellular activation status, rather than the number or frequency of cells. For example, it is possible to activate cytotoxic function or induce cytokine secretion at

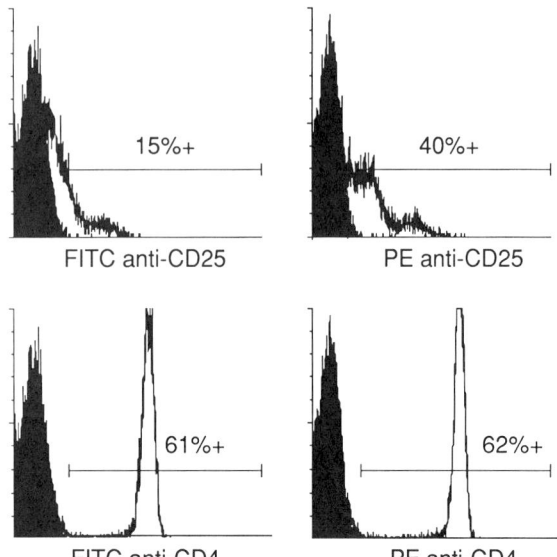

Figure 4.3. Influence of fluorochrome efficiency on flow cytometric results. Peripheral blood T-lymphocytes were stained with FITC-conjugated Ig control, PE-conjugated Ig control, FITC-conjugated anti-CD4 (Leu 3a), PE-conjugated anti-CD4, FITC-conjugated anti-CD25 (p55 IL-2 receptor), or PE-conjugated anti-CD25. Samples were analyzed using a FACScan flow cytometer. Data are presented as histograms with relative fluorescence on the x–axis (4 decade log scale) and number of cells on the y–axis. Histograms from control Ig-stained samples are superimposed on histograms from cells stained with anti-CD4 or anti-CD25 mAb.

a time significantly before any cellular proliferation has occurred. Therefore, in these situations, the number or frequency of cells fails to reflect a substantial change in lymphocyte function. Moreover, it must also be appreciated that lymphocyte functions (i.e., cytokine production, Ig synthesis, cytotoxicity, etc.) are very complex processes involving receptor-mediated signalling, regulation of gene transcription, de novo protein synthesis, and cellular proliferation. Perturbation of any of these intracellular steps may prevent the function. This point is particularly important in the interpretation of antigenic phenotype data obtained from patients affected with disease or malignancy. Often, the number and frequency of the lymphocyte subsets are normal; however, when the functional capacity of the cells is directly measured it is noted that these activities are impaired. In summary, while correlations between antigenic phenotype and cellular function have provided invaluable insight into the role of certain cell types in the immune system, **antigenic phenotype alone is insufficient to draw conclusions concerning the functional capacity of a cell population.** There are no shortcuts; the function of interest must be directly measured.

DYNAMICS OF THE IMMUNE SYSTEM

The immune system is designed to rapidly respond to insult or injury. This process can be monitored both at a single-cell level and at a population level, using flow cytometry. The traditional approach has been to use multicolor immunofluorescence and flow cytometry to detect "activation antigens" on the membrane of leukocytes that have previously been stimulated in vivo or in vitro. It is very encouraging to note that new methodologies are being developed to directly measure cellular activation using flow cytometry. Many of these techniques permit measurement of the very earliest events accompanying leukocyte activation. For example, flow-cytometric methods have been developed to measure the kinetics of oxidative product formation subsequent to interaction between neutrophils and a bacterial stimulus (61). With respect to lymphocytes, flow cytometry has been used to estimate membrane potential (62) and to detect T-cell activation as reflected by increases in intracellular Ca^{++} levels (63, 64) (see also Chapter 31). Alterations in intracellular Ca^{++} concentration result from phosphatidylinositol-mediated signalling in T-cells stimulated through the T-cell antigen receptor or CD2 molecules (63, 65). These changes are readily monitored using the fluorescent Ca^{++} indicators indo-1 or fluo-3 and flow cytometry (64, 66–68).

EXISTENTIAL FLOW CYTOMETRY: THE MEANING OF POSITIVE AND NEGATIVE

Results from flow-cytometric analysis involving antigenic phenotype are most frequently presented as "percentage-positive" cells within a population. Investigators rarely consider the fact that whether a cell is classified as "positive" or "negative" is critically dependent upon the sensitivity of the detection system. In absolute terms, negative implies that a cell has zero copies of a given antigen expressed on the cell surface while a positive cell could conceivably express one molecule. Obviously, this is the extreme case, but it emphasizes the problem in assigning such distinctions. The influence of the fluorochrome used for detection of a cell-surface antigen is demonstrated in Figure 4.3. In this example, a preparation of human peripheral blood T-cells from a "normal" donor were stained with a mAb directed against the p55 low affinity subunit of the IL-2 receptor (CD25) conjugated to either FITC or PE. Using the FITC reagent for detection, the results would be reported as 15%-positive cells. However, when the same cell preparation was stained with the same mAb conjugated to PE, the data indicate that 40% of the T-cells express the antigen. This discrepancy is explained by the higher quantum yield of PE compared to FITC. This dramatic difference is most evident when detecting antigen levels that are at the threshold of detection limits for a particular instrument or fluorochrome. For comparison, the same T-cell preparation stained with FITC and PE-conjugated mAb against a relatively abundant surface antigen, CD4, gives identical percent-positive values, although the fluorescence intensity of the PE-stained cells is obviously much higher (Fig. 4.3). A similar situation exists when flow cytometers with different optical sensitivities are used to evaluate the same fluorochrome-stained sample. For years, it was reported that CD25 was not present on resting peripheral blood T-lymphocytes. However, with the development of more sensitive instrumentation, it has recently been appreciated that a substantial proportion of normal peripheral blood T-cells do in fact express low levels of CD25 antigen (69). As both instrumentation and fluorochromes are improved, it is certain that many of today's "negative" cells will magically become "positive." The general conclusion is that negative should be viewed as "I can't detect it" rather than as "it isn't expressed." When enumerating cells expressing antigens that are present at the lower limits of the system's sensitivity, it might be prudent to report values as "greater than" the observed percentage, leaving open the possibility that more positive cells exist.

FURTHER DIRECTIONS

The remarkable technical advancements in flow-cytometric analysis and cell sorting during the past decade have provided tools that have substantially enhanced our understanding of the human immune system. It is anticipated that the flow cytometer will be increasingly used to monitor the dynamic intracellular events accompanying leukocyte activation and to detect and quantitate cellular components in the cytoplasm and nucleus as well as on the cell membrane. As in the computer field, flow cytometry has progressed from the toy of a few hobbyists to an essential tool of many scientific disciplines. The hardware is available; the challenge now is to design good experiments.

ACKNOWLEDGMENTS

DNAX Research Institute of Molecular and Cellular Biology is supported by the Schering-Plough Corporation. I thank Drs. Anne Jackson and Leon Terstappen for helpful discussions, and Becton Dickinson Immunocytometry Systems for generously providing antibodies.

REFERENCES

1. Reif AE, Allen JMV. Mouse thymic isoantigens. Nature 1966;209: 521–523.
2. Boyse EA, Miyazawa M, Aoki T, Old LJ. Ly-A and Ly-B: two systems of lymphocyte isoantigens in the mouse. Proc R Soc Lond (Biol) 1968;170:175–193.
3. Kisielow P, Hirst JA, Shiku H, Beverley PCL, Hoffmann MK, Boyse EA, Oettgen HF. Ly antigens as markers for functionally distinct subpopulations of thymus-derived lymphocytes of the mouse. Nature 1975; 253:219–220.
4. Jondal M, Holm G, Wigzell H. Surface markers on human T and B lymphocytes. I. A large population of lymphocytes forming nonimmune rosettes with sheep red blood cells. J Exp Med 1972;136:207.
5. Reinherz EL, Schlossman SF. Con A-inducible suppression of MLC: evidence for mediation by the TH_2+ T cell subset in man. J Immunol 1979;122:1335–1341.
6. Kohler G, Milstein C. Continuous cultures of fused cells secreting antibody of predefined specificity. Nature 1975;256:495–497.
7. Reinherz EL, Kung PC, Goldstein G, Schlossman SF. A monoclonal antibody with selective reactivity with functionally mature human thymocytes and all peripheral human T cells. J Immunol 1979;123: 1312–1317.
8. Williams AF, Galfre G, Milstein C. Analysis of cell surface by xenogeneic myeloma hybrid antibodies: Differentiation antigens of rat lymphocytes. Cell 1977;12:663.
9. Springer T, Galfre G, Secher DS, Milstein CA. Monoclonal xenogeneic antibodies to murine cell surface antigens: Identification of novel leukocyte differentiation antigens. Eur J Immunol 1978;8:539.
10. Oi VT, Jones PP, Goding JW, Herzenberg LA, Herzenberg LA. Properties of monoclonal antibodies to mouse Ig allotypes, H-2, and Ia antigens. Curr Top Microbiol Immunol 1978;81:115.
11. McMichael AJ, Pilch JR, Galfre G, Mason DY, Fabre JW, Milstein C. A human thymocyte antigen defined by a hybrid myeloma monoclonal antibody. Eur J Immunol 1979;9:205–210.
12. Loken MR, Herzenberg LA. Analysis of cell populations with a fluorescence-activated cell sorter. Ann NY Acad Sci 1975;254:163–171.
13. Mathieson BJ, Sharrow SO, Campbell PS, Asofsky R. An Lyt differentiated thymocyte detected by flow microfluorometry. Nature 1979;277: 478–480.
14. Ledbetter JA, Herzenberg LA. Xenogeneic monoclonal antibodies to mouse lymphoid differentiation antigens. Immunol Rev 1979;47:63–90.
15. Ledbetter JA, Evans RL, Lipinski M, Cunningham-Rundles C, Good RA, Herzenberg LA. Evolutionary conservation of surface molecules that distinguish T lymphocyte helper/inducer and cytotoxic/suppressor subpopulations in mouse and man. J Exp Med 1981;153:310–323.
16. Reinherz EL, Schlossman SF. The differentiation and function of human T lymphocytes. Cell 1980;19:821–827.
17. Loken MR, Parks DR, Herzenberg LA. Two-color immunofluorescence using a fluorescence-activated cell sorter (FACS). J Histochem Cytochem 1977;25:899.
18. Titus JA, Haugland R, Sharrow SO, Segal DM. Texas red: a hydrophilic, red-emitting fluorophore for use with fluorescein in dual parameter flow microfluorometric and fluorescence microscopic studies. J Immunol Methods 1982;50:193–204.
19. Parks DR, Hardy RR, Herzenberg LA. Dual immunofluorescence—new frontiers in cell analysis and sorting. Immunol Today 1983;4:145–150.
20. Lanier LL, Engleman EG, Gatenby P, Babcock GF, Warner NL, Herzenberg LA. Correlation of functional properties of human lymphoid cell subsets and surface marker phenotypes using multiparameter analysis and flow cytometry. Immunol Rev 1983;74:143–160.
21. Oi VT, Glazer AN, Stryer L. Fluorescent phycobiliprotein conjugates for analyses of cells and molecules. J Cell Biol 1982;93:981–986.
22. Hardy RR, Hayakawa K, Parks DR, Herzenberg LA. Demonstration of B-cell maturation in X-linked immunodeficient mice by simultaneous three-colour immunofluorescence. Nature 1983;306:270.
23. Lanier LL, Loken MR. Human lymphocyte subpopulations identified by using three-color immunofluorescence and flow cytometry analysis: Correlation of Leu-2, Leu-3, Leu-7, Leu-8, and Leu-11 cell surface antigen expression. J Immunol 1984;132:151–156.
24. Hardy RR, Hayakawa K, Parks DR, Herzenberg LA. Murine B cell differentiation lineages. J Exp Med 1984;159:1169–1188.
25. Glazer AN, Stryer L. Fluorescent tandem phycobiliprotein conjugates: Emission wavelength shifting by energy transfer. Biophys 1983;43: 383–386.
26. Recktenwald D, Prezelin B, Chen CH, Kimura J. Biological pigments as fluorescent labels for cytometry. In: Salzman GC, ed. New Technologies in Cytometry and Molecular Biology SPIE, 1990;1206:106–111.
27. Knapp W, Dorken B, Gilks WR, Rieber EP, Schmidt RE, Stein H, Dr. von dem Borne AEG. Leukocyte Typing IV: White Cell Differentiation Antigens. Oxford University Press, Oxford, UK, 1990; pp. 1–1182.
28. Civin CI, Strauss LC, Brovall C, Fackler C, Schwartz JF, Shaper JH. Antigenic analysis of hematopoiesis III. A hematopoietic progenitor cell surface antigen defined by a monoclonal antibody raised against KG-1a cells. J Immunol 1984;133:157–165.
29. Andrews RG, Singer JW, Bernstein ID. Human hematopoietic precursors in long-term culture: Single CD34+ cells that lack detectable T cell, B cell, and myeloid cell antigens produce multiple colony-forming cells when cultured with marrow stromal cells. J Exp Med 1990;172: 355–358.

30. Ema H, Suda T, Miura Y, Nakauchi H. Colony formation of clone-sorted human hematopoietic progenitors. Blood 1990;75:1941–1946.
31. Clark SC, Kamen R. The human hematopoietic colony-stimulating factors. Science 1987;1229–1230.
32. Lansdorp PM, Sutherland HJ, Eaves CJ. Selective expression of CD45 isoforms on functional subpopulations of CD34+ hematopoietic cells from human bone marrow. J Exp Med 1990;172:363–366.
33. Lewinsohn DM, Nagler A, Ginzton N, Greenberg P, Butcher EC. Hematopoietic progenitor cell expression of the H-CAM (CD44) homing-associated adhesion molecule. Blood 1990;75:589–595.
34. Nagler A, Greenberg PL, Lanier LL, Phillips JH. The effect of recombinant interleukin 2-activated natural killer cells on autologous peripheral blood hematopoietic progenitors. J Exp Med 1988;168:47–54.
35. Kessinger A, Armitage JO, Landmark JD, Smith DM, Weisenburger DD. Autologous peripheral hematopoietic stem cell transplantation restores hematopoietic function following marrow ablative therapy. Blood 1988;71:723–727.
36. Picker LJ, Terstappen LWMM, Rott LS, Streeter PR, Stein H, Butcher EC. Differential expression of homing-associated adhesion molecules by T cell subsets in man. J Immunol 1990;145:3247–3255.
37. Spangrude GJ, Heimfeld S, Weissman IL. Purification and characterization of mouse hematopoietic stem cells. Science 1988;241:58–62.
38. Blackwell TK, Alt FW. Mechanism and developmental program of immunoglobulin gene rearrangement in mammals. Annu Rev Genetics 1989;23:605–636.
39. Ohsahi PS, Mak TW, van den Elsen P, Yanagi Y, Yoshikai Y, Calman AF, Terhorst C, Stobo JD, Weiss A. Reconstitution of an active surface T3/T-cell antigen receptor by DNA transfer. Nature 1985;316:606–609.
40. Ashwell JD, Klausner RD. Genetic and mutational analysis of the T cell antigen receptor. Annu Rev Immunol 1990;8:139–167.
41. Transy C, Moingeon PE, Marshall B, Stebbins C, Reinherz EL. Most anti-human CD3 monoclonal antibodies are directed to the CD3 ε subunit. Eur J Immunol 1989;19:947–950.
42. Trinchieri G. Biology of natural killer cells. Adv Immunol 1989;47:187–376.
43. Lanier LL, Le AM, Civin CI, Loken MR, Phillips JH. The relationship of CD16 (Leu-11) and Leu-19 (NKH-1) antigen expression on human peripheral blood NK cells and cytotoxic T lymphocytes. J Immunol 1986;136:4480–4486.
44. Lanier LL, Testi R, Bindl J, Phillips JH. Identity of Leu 19 (CD56) leukocyte differentiation antigen and neural cell adhesion molecule (N-CAM). J Exp Med 1989;169:2233–2238.
45. Lanier LL, Kipps TJ, Phillips JH. Functional properties of a unique subset of cytotoxic CD3+ T lymphocytes that express Fc receptors for IgG (CD16/Leu-11 antigen). J Exp Med 1985;162:2089–2106.
46. Biron CA, Byron KS, Sullivan JL. Severe herpesvirus infections in an adolescent without natural killer cells. N Engl J Med 1989;320:1731–1735.
47. Bukowski JF, Warner JF, Dennert G, Welsh RM. Adoptive transfer studies demonstrating the antiviral effect of natural killer cells in vivo. J Exp Med 1985;161:40–52.
48. Lanier LL, Allison JP, Phillips JH. Correlation of cell surface antigen expression on human thymocytes by multi-color flow cytometric analysis: Implications for differentiation. J Immunol 1986;137:2501–2507.
49. Lanier LL, Weiss A. Presence of Ti (WT31) negative T lymphocytes in normal blood and thymus. Nature 1986;324:268–270.
50. Groh V, Porcelli S, Fabbi M, Lanier LL, Picker LJ, Anderson T, Warnke RA, Bhan AK, Strominger JL, Brenner MB. Human lymphocytes bearing T cell receptor γ/δ are phenotypically diverse and evenly distributed throughout the lymphoid system. J Exp Med 1989;169:1277–1294.
51. Krensky AM, Reiss CS, Mier JW, Strominger JL, Burakoff SJ. Long-term human cytolytic T-cell lines allospecific for HLA-DR6 antigen are OKT4+. Proc Natl Acad Sci USA 1982;79:2365–2369.
52. Gatenby PA, Kansas GS, Xian CY, Evans RL, Engleman EG. Dissection of immunoregulatory subpopulations of T lymphocytes within the helper and suppressor sublineages in man. J Immunol 1982;129:1997–2000.
53. Morimoto C, Letvin NL, Boyd AW, Gagan M, Brown HM, Kornacki MM, Schlossman SF. The isolation and characterization of the human helper inducer T cell subset. J Immunol 1985;134:3762–3769.
54. Sanders ME, Makgoba MW, Shaw S. Human naive and memory T cells: reinterpretation of helper-inducer and suppressor-inducer subsets. Immunol Today 1988;9:195–198.
55. Beverley PCL. Is T-cell memory maintained by crossreactive stimulation? Immunol Today 1990;11:203–205.
56. Sanders ME, Makgoba MW, Sharrow SO, Stephany D, Springer TA, Young HA, Shaw S. Human memory T lymphocytes express increased levels of three cell adhesion molecules (LFA-3, CD2, and LFA-1) and three other molecules (UCHL1, CDw29, and Pgp-1) and have enhanced IFN-γ production. J Immunol 1988;140:1401–1407.
57. Hayakawa K, Hardy RR. Normal, autoimmune, and malignant CD5+ B cells: The Ly-1 B lineage. Annu Rev Immunol 1988;6:197–218.
58. Nagler A, Lanier LL, Cwirla S, Phillips JH. Comparative studies of human FcRIII-positive and negative NK cells. J Immunol 1989;143:3183–3191.
59. Nagler A, Lanier LL, Phillips JH. Constitutive expression of high affinity interleukin-2 receptors on human NK cell subsets. J Exp Med 1990;171:1527–1533.
60. Caligiuri MA, Zmuidzinas A, Manley TJ, Levine H, Smith KA, Ritz J. Functional consequences of interleukin 2 receptor expression on resting human lymphocytes. Identification of a novel natural killer cell subset with high affinity receptors. J Exp Med 1990;171:1509–1526.
61. Bass DA, Parce JW, Dechatelet LR, Szejda P, Seeds MC, Thomas M. Flow cytometric studies of oxidative product formation by neutrophils: a graded response to membrane stimulation. J Immunol 1983;130:1910–1917.
62. Shapiro HM, Natale PJ, Kamentsky LA. Estimation of membrane potentials of individual lymphocytes by flow cytometry. Proc Natl Acad Sci USA 1979;76:5728–5730.
63. Alcover A, Weiss MJ, Daley JF, Reinherz EL. The T11 glycoprotein is functionally linked to a calcium channel in precursor and nature T-lineage cells. Proc Natl Acad Sci USA 1986;83:2614–2618.
64. Rabinovitch PS, June CH, Grossmann A, Ledbetter JA. Heterogeneity among T cells in intracellular free calcium responses after mitogen stimulation with PHA or anti-CD3: Simultaneous use of Indo-1 and immunofluorescence with flow cytometry. J Immunol 1986;137:952–961.
65. Weiss A, Imboden J, Shoback D, Stobo J. Role of T3 surface molecule in human T-cell activation: T3-dependent activation results in an increase in cytoplasmic free calcium. Proc Natl Acad Sci USA 1984;81:4169–4173.
66. Kao JPY, Harootunian AT, Tsien RY. Photochemically generated cytosolic calcium pulses and their detection by Fluo-3. J Biol Chem 1989;264:8179–8184.
67. Grynkiewicz G, Poenie M, Tsien RY. A new generation of Ca2+ indicators with greatly improved fluorescence properties. J Biol Chem 1985;260:3440–3450.
68. Rijkers GT, Justement LB, Griffioen AW, Cambier JC. Improved method for measuring intracellular Ca++ with Fluo-3. Cytometry 1990;11:923–927.
69. Jackson AL, Matsumoto H, Janszen M, Maino V, Blidy A, Shye S. Restricted expression of p55 interleukin 2 receptor (CD25) on normal T cells. Clin Immunol Immunopath 1990;54:126–133.

SECTION B. TECHNICAL ASPECTS

5

Clinical Flow Cytometry Instrumentation

JAMES C. S. WOOD

INTRODUCTION

As the number of flow-cytometry-based clinical tests escalates, it becomes increasingly more important that clinicians and clinical researchers understand the fundamental theory behind the design and operation of a flow cytometer. The goal of this chapter is to provide the clinical user with the knowledge needed to design clinical experiments and protocols and interpret the results. Toward this end, this chapter will review the operation of the major components of a clinical flow cytometer and their interaction with one another.

FLOW SYSTEM

In its simplest form, a flow cytometer is the combination of a fluidic-based sample delivery system with a spectrofluorometer and a light scatter photometer. Starting with a sample consisting of a suspension of cells (or other biological particles), the fluidic sample delivery system provides a convenient and efficient means of presenting individually the cells in the sample to a measurement station. The cell suspension is carried by the sample delivery system to the flow cell where it is injected through a small-bore injection needle into a larger-diameter, rapidly-flowing sheath stream. As the sheath and sample stream enter the flow cell, they undergo a rapid acceleration. The increased acceleration of the sample stream relative to the sheath stream hydrodynamically focuses the injected cell suspension into a core stream having the approximate diameter of one cell. The sample cells are constrained to flow single file past a narrowly-focused excitation light beam that is used to probe the cell properties of interest. A typical system is shown in Figure 5.1.

As the cells are transported past the focused excitation light beam, each cell scatters light and may emit fluorescent light, depending on whether the cell is labeled with a fluorescent dye and/or is autofluorescent. Scattered light is measured in both the forward and perpendicular directions relative to the incident beam. The fluorescent emissions of the cell are measured in the perpendicular direction by photosensitive detectors. Optical filters are used to separate the spectral components of the fluorescent emissions. Measurements of the light-scatter and fluorescent-emissions intensities are used to characterize each cell as it is processed.

Flow Cytometer Fluidics System

The fluidics system of a clinical flow cytometer consists of the sheath fluid, the sample suspension, and the necessary pneumatic controls to deliver both of these fluids to the flow cell at an appropriate rate. The goal is to measure cells on an individual basis. Thus, it is important to start with a sample suspension consisting of single cells. The presence of doublets and larger aggregates in the sample suspension may irreparably corrupt the data set. The relative delivery rates of the sample and sheath fluids to the flow cell directly influence the alignment of the cells for individual and sequential interrogation by the excitation light beam. The hydrodynamic alignment of the sample cells into a narrow core stream within the flow cell is a major factor contributing to the accuracy and precision of the data collected by a flow cytometer. In most instances, it is desirable for the sample core stream to have a cross–sectional diameter approximately equal to the size of the cells being measured.

Since the delivery rate of sheath fluid is either fixed or very rarely adjusted, the rate of sample fluid delivery into the flow cell is most commonly used to control the sample core stream diameter. The sample delivery rate (R in (ml/sec) controls the core stream diameter (D in μ) according to the following relationship

$$D = \sqrt{\frac{R}{V \cdot 3.92 \times 10^{-7}}}$$

where V is the sheath fluid velocity in meters per second (1). For syringe-based sample delivery systems, the sample delivery rate is set directly by adjusting the rate of delivery by the syringe. For pressure-driven systems, the sample delivery rate can be set using a sample of known concentration and adjusting the differential pressure between sheath and sample fluids to achieve the appropriate sample count rate. Some flow cytometer operators use the sample delivery rate to control the sample count rate without regard to sample concentration. Except for minor changes, this is to be avoided. Since the sample delivery rate determines the core stream diameter, it should not be changed to control the sample count rate; rather, the sample concentration should be changed to achieve the desired sample count rate. As a rule of thumb, once the sample delivery rate (or differential pressure) has been set for optimal core stream alignment, it

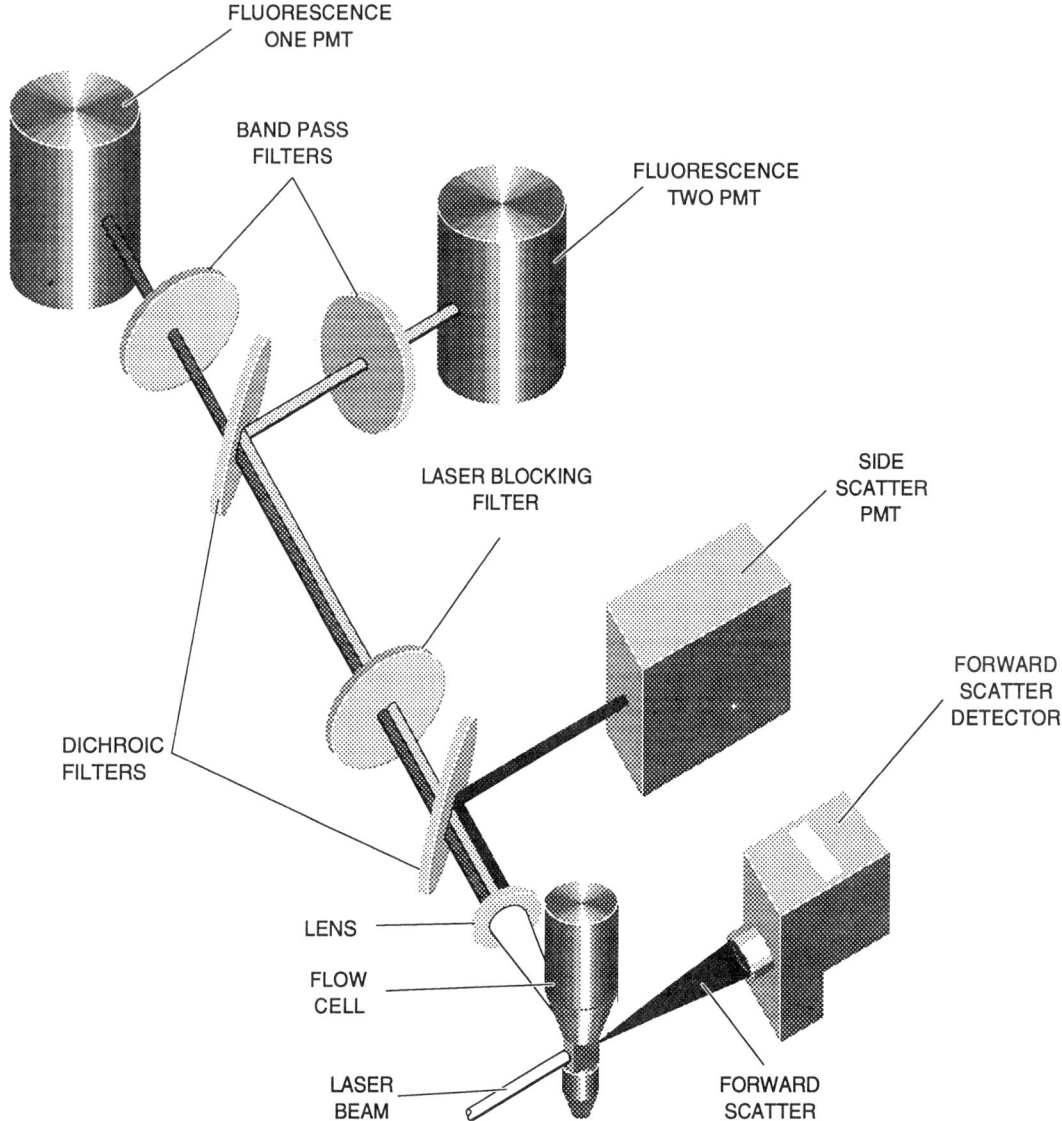

Figure 5.1. Schematic diagram of a flow cytometer showing the flow cell, excitation light beam, light collection optics, filters, and photodetectors (Courtesy of Coulter Corporation).

should not be readjusted unless the flow cell and/or the sheath delivery rate have been changed. The degree to which one should adhere to this rule is dependent on the measurement accuracy and precision dictated by the application. For example, the quantitation of cellular DNA content requires the maintenance of optimal core stream alignment, but for lymphocyte phenotyping the core stream alignment may be relaxed because the biological variation is large and the data is typically collected on a logarithmic scale.

Flow Cell

The flow cell has been aptly called by some the heart of a flow cytometer (2). Within the flow cell, cells are hydrodynamically aligned by the fluidics system and presented one by one to the focused excitation light beam which stimulates the emission of fluorescence and scattered light. A flow cell consists of three sections (Fig. 5.2). In the first part, the sample suspension is injected into a stream of sheath fluid. The injected sample suspension develops into what is called a sample core stream and follows the streamlines of the sheath fluid. The second part of the flow cell is a narrow orifice through which the sheath fluid and sample core stream pass. As both fluid streams pass through the narrow orifice they undergo a rapid acceleration. The rapid acceleration hydrodynamically focuses the sample core stream to a very narrow cross–section of approximately the size of the cells being measured. In the third part of the flow cell, the hydrodynamically aligned cells are presented one-by-one to a focused excitation light source.

The stability of the hydrodynamic focusing within the flow cell is dependent on the design and alignment of the first two components of the flow cell. The manufacturers of flow cytometers have taken great care in the design of these

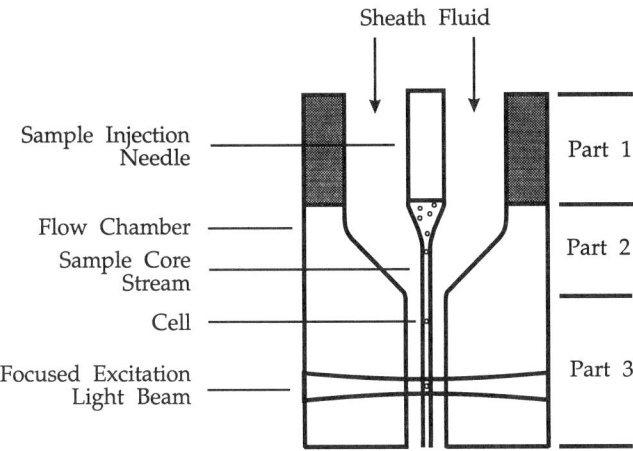

Figure 5.2. Schematic diagram showing the major parts of a flow cell.

parts of the flow cell to assure the highest precision, accuracy, and reliability of the instrument for a wide range of cell samples. However, some cell samples may require additional control over the alignment of the cells with respect to the excitation light source.

For cells whose measurement may be sensitive to cell orientation, it may be advantageous to modify the shape of the tip of the sample injection needle to force a preferred orientation on the cells. An asymmetrically shaped tip on the sample insertion needle can effect a significant improvement in orientation of epithelial cells, chicken erythrocytes, and sperm cells (3, 4, 5). The result is a more precise and accurate measurement.

The optical sensitivity of a flow cell is dependent on the design of the third part of a flow cell. Namely, how the cells are illuminated, how the emitted light is collected and transferred to the collection optics, and the amount of excitation light scattered into the collection optics. There are two basic designs for the sensing portion of the flow cell. The particles are illuminated either within a chamber (closed flow) or just as they exit the flow cell (sense–in–air). The chamber-based flow cell design uses quartz capillaries with circular, square, or rectangular interior cross–sections typically on the order of 100 to 250 μm. The excitation light source intersects the sample core stream within the quartz capillary and the emissions from within the capillary are measured. The sense–in–air flow cell design, used primarily for sorting, forces the sheath and sample core stream out through an orifice with an inside diameter of typically 50 to 100 μm to form a jet. Larger orifices have been used for special applications involving the sorting of large particles (6, 7). The excitation light source is focused to intersect the sample core stream a few hundred μ away from the jetting orifice.

The renewed interest in the use of low-power excitation light sources, i.e., air–cooled lasers and arc lamps, has shifted the attention away from the use of classic sense–in–air flow cells. Sense–in–air flow cells diffract a significant portion of the incoming light and offer few opportunities for easily optimizing the light collection efficiency. Thus, chambered or capillary flow cells have been the subject of significant design advances.

Quartz capillary flow cells can be designed to minimize diffracted light and improve optical sensitivity. Minimizing the amount of light that interacts with the inside walls of the flow cell reduces the diffracted light intensity. This is achieved by ensuring that the width of the focused excitation light beam is smaller than the internal dimension of the quartz flow cell perpendicular to the excitation beam. For example, if the width of the focused excitation light beam is 125 μm then there will be a negligible amount of diffracted light generated by a flow cell with an internal channel cross-section of 200 to 250 μm^2. The sensitivity of a quartz flow cell can be improved substantially by coupling a lens directly to the flow cell. This lens acts as the first lens of an immersion objective. The sensitivity can be further improved by attaching a mirror of appropriate design to the side of the flow cell opposing the above-mentioned first lens. The resultant improvements in light collection efficiency in currently available flow cytometers over a simple quartz flow cell with standard collection optics with a numerical aperture (N.A.) of approximately .6 has been advertised as being 4 (N.A. = 1.2) to 5.4 (N.A. = 1.4) fold. The development of this technology has been instrumental in making it possible to use low-powered, air-cooled lasers as excitation light sources in clinical flow cytometer analyzers. Furthermore, this technology has been applied to the design of flow cells used for sorting to provide the same improvements in sensitivity and signal-to-noise, thereby permitting the successful use of air-cooled lasers on flow cytometer cell sorters.

EXCITATION LIGHT SOURCE

The purpose of the excitation light source is to provide light at a specified wavelength and intensity to excite the fluorophores attached to a cell to a detectable level and/or to produce a detectable level of scattered light. Traditionally, lasers have been the light source of choice because they produce low divergence, single wavelength light of very high intensity for a number of specific wavelengths. Other light sources have been explored and successfully used in a few clinical flow cytometers. The most notable nonlaser light source is the mercury arc lamp, which has been used extensively for its strong UV emissions (8, 9, 10). Nevertheless, with the recent improvements in stability and reliability of low-cost, air-cooled gas and solid-state lasers, the small air-cooled lasers currently are the most common choice for light sources in clinical flow cytometers.

Arc Lamps

Arc lamps are characterized by their emission spectra, spot size, and power. Arc lamps operate by passing an electric arc between two very closely spaced electrodes. The arc ionizes the gas in the lamp to generate UV, visible, and infrared light. The spot size is determined by the distance between

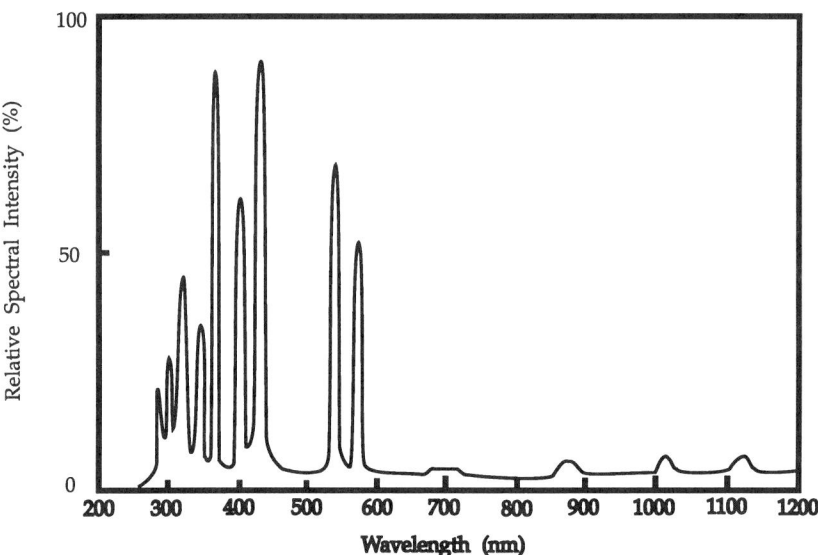

Figure 5.3. Typical relative spectral intensity distribution for a mercury arc lamp (Courtesy of The Ealing Corporation).

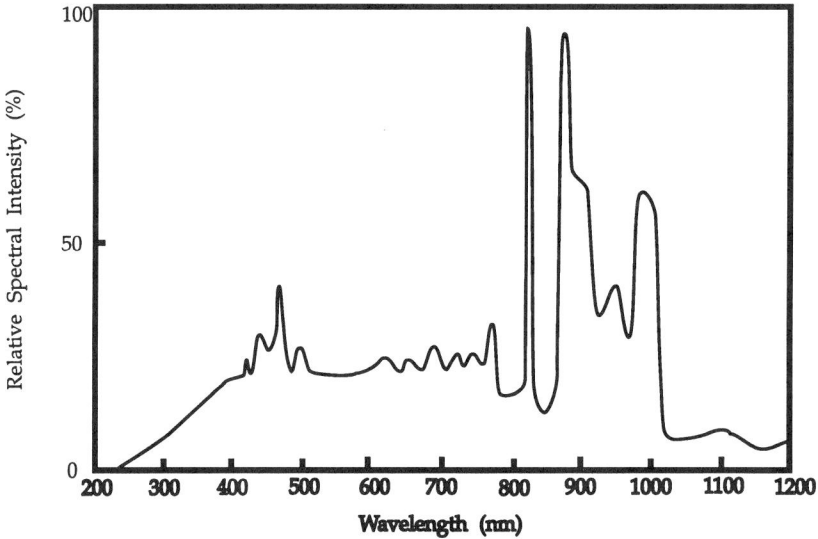

Figure 5.4. Typical relative spectral intensity distribution for a xenon arc lamp (Courtesy of the Ealing Corporation).

the electrodes and the lamp power is set by the current and voltage used to generate the electric arc. The resulting emission spectrum is controlled by the type of gas(es) in the lamp. The two most widely used arc lamps in flow cytometry are the mercury and xenon arc lamps.

The most commonly used arc lamp in flow cytometry is the 100W mercury arc lamp. Its small arc (250μm × 250μm) produces the highest brightness (radiance) of most arc lamps, especially when the amount of radiance available from its strong spectral lines is considered. The mercury arc lamp produces several spectral lines centered at 365, 405, 435, 546, and 578 nm which are superimposed on a weaker background continuum (Fig.5.3). The mercury arc lamp has a particularly strong UV emission (330–380 nm) that has an effective power of approximately 10 mW. This has made the mercury arc lamp particularly attractive in epi-illuminated flow cytometers for applications that require UV excitation, i.e., DNA quantitation (8, 10, 11).

In contrast to mercury arc lamps, the xenon arc lamps are typically not as bright, but they have an emission spectrum with a higher background continuum in the region typically used in flow cytometry (Fig. 5.4). Particularly in the region between 450 and 540 nm, the output of the xenon lamps is higher than the comparable mercury arc lamps (8).

The useful lifetime of an arc lamp is typically only 200 to 400 hr. The light output of a mercury arc lamp decreases by approximately 2% each 8-hr day so that after 200 hours (25 days) the lamp outputs only 60% of its original light output. The stability of the arc also diminishes over the lifetime of the arc lamp as the electrodes deteriorate. This is normally not as serious a problem as the fall off in light output since electronic feedback circuitry can be used to adjust the lamp

Table 5.1
Table of Excitation Wavelengths Available from Lasers Commonly Used in Flow Cytometry

Excitation Wavelength	HeNe	HeCd	Argon	Krypton
325 nm		X		
350–356 nm				X
351–364 nm			X	
406–415 nm				X
441 nm		X		
454 nm			X	
457 nm			X	
466 nm			X	
472 nm			X	
476 nm				X
477 nm			X	
482 nm				X
488 nm			X	
497 nm			X	
502 nm			X	
514 nm			X	
521 nm				X
529 nm			X	
531 nm				X
543 nm	X			
568 nm				X
594 nm	X			
611 nm	X			
633 nm	X			
647 nm				X
676 nm				X
752 nm				X
799 nm				X

current to compensate for small fluctuations in lamp output (8).

Laser

For most users of clinical flow cytometers, the relevant characteristics of a laser are its wavelength, power, and beam intensity profile. In contrast to incandescent and arc lamps, which emit over a broad range of wavelengths, most lasers have a finite set of wavelengths at which they can operate (Table 5.1). The most notable exceptions are dye lasers, which are tunable and cover a broad range of wavelengths. The fluorescent dye in the dye lasers is excited by an external light source, which, for a flow cytometer, is typically a gas laser. The output power available from the dye laser varies with wavelength because the dye laser operates with different efficiencies at each wavelength. For any given application, the output wavelength and power of the laser(s) must be matched to the excitation spectrum, quantum efficiency of the dye(s), and particle–bound dye concentration. The key is to excite the dye molecules bound to the particle so that they all emit enough photons to make an accurate and precise measurement of the amount of dye bound to the particle. The most common approach is to maximize the number of photons emitted by a dye molecule by choosing a laser power high enough to achieve a strong enough emission but not high enough to destroy (i.e., photobleach) the dye molecules. Alternatively, to minimize the effect of inhomogeneities in particle illumination at the expense of sensitivity, some research applications (i.e., chromosome scanning) use high-power lasers to photobleach intentionally all the particle–bound dye molecules to be certain that every dye molecule was excited (12, 13). However, to maximize sensitivity (i.e., the number of photons emitted per dye molecule), it is advantageous to choose either a wavelength closer to the excitation maximum of the dye or a dye with a higher quantum efficiency to reduce the laser power requirements. Maximizing the amount of dye bound to the particle, up to the concentration at which self-quenching becomes significant, can partially compensate for low dye quantum efficiency and excitation at less than optimal wavelength, thereby reducing laser power requirements.

In addition to power and wavelength, lasers may be operated in a number of modes that are characterized by their spatial distribution of light irradiance within the beam cross-section. The preferred beam cross-section has a Gaussian (normal distribution) intensity profile and is referred to as TEM_{00}. This produces the cleanest beam by minimizing the contribution of the laser to the signal coefficient of variation. Also, it is the only beam profile whose shape is preserved as it passes through the beam shaping optics and illuminates the particles in the core stream (11). The presence of other modes will distort the pulse shape and adversely affect the signal coefficient of variation.

For clinical flow cytometers, the interest in low power, air-cooled, compact lasers has been on the rise. Argon, helium-neon (HeNe), and helium-cadmium (HeCd) air-cooled gas lasers are currently being used. Solid-state lasers, such as diode lasers and frequency-doubled neodymium-YAG lasers are being considered. Of the gas lasers, the most commonly used are the low-power argon lasers, which have an output power of approximately 15 mW at a fixed wavelength of 488 nm. These lasers are used to excite the wide range of probes used in flow cytometry. Fluorescein, phycoerythrin, phycoerythrin-based conjugates, propidium iodide, and fluo–3 can all be successfully excited by low-power argon lasers. Higher power air–cooled and water–cooled argon lasers are available for applications that require greater beam intensity and selectable output wavelengths (visible and UV).

The HeNe lasers have generated interest because of their low cost, small size, high power efficiency, stability and reliability. The most widely used HeNe laser has a fixed output wavelength of 633 nm with available output powers of 10 mW or less. These lasers have been successfully used to excite cyanine dyes and allophycocyanin. There is significant interest in finding other dyes that excite at 633 nm. Other HeNe lasers are available with fixed output wavelengths of 543 nm, 594 nm, and 611 nm but their output powers are limited to less than a couple of milliwatts. This limits their utility to measuring only bright-staining particles.

Until very recently, the only air–cooled laser that could emit UV light was the HeCd laser. HeCd lasers are available with fixed output wavelengths of 325 nm or 441 nm. Typical output powers range from 10 mW to 40 mW. The 325-nm-wavelength HeCd laser can be used to excite the DNA probes DAPI and Hoescht 33342, and the calcium probe Indo–1 (10). The 441-nm-wavelength HeCd laser can be used to excite the DNA probes mithramycin and chromomycin A3 (10). Despite the advantages of HeCd lasers, there are some significant disadvantages that require closer consideration. For example, the beam intensity profile of the high-power, multimode HeCd lasers is not very clean, and the use of the HeCd laser for DNA quantitation is limited by the optical noise of the laser beam.

Alternatives to the HeCd laser are some recently announced low-power argon-ion lasers capable of UV operation. These lasers use a water–cooled design, but they can be used with a closed-circulation system in which the water is cooled by a radiator located next to the flow cytometer. It is too early to tell whether these lasers will supplant the HeCd laser in the future.

The appeal of the low cost and reliability of low-power, air–cooled lasers is significant, but there are some limitations imposed on the operation of a flow cytometer when low-power lasers are used. The most significant limitation is that fewer photons are available per unit time to excite the fluorochromes, which means that fewer photons are emitted by the labeled cell as it passes by the laser beam. Unlike higher powered water–cooled lasers, fewer photons can be wasted. To compensate for the lower power and maintain maximum sensitivity, flow cytometers with low-power air–cooled lasers must (a) use low-loss excitation optics, (b) usually run at a slower sheath fluid (i.e., particle or cell) velocity to increase the number of photons collected from a cell, (c) use high-efficiency collection optics, and (d) use high-quality interference filters. There are some penalties associated with the trade-off of lowering the sheath fluid velocity to maximize sensitivity. Most notable is a restriction on the sample throughput if the core stream diameter is to be maintained for optimal alignment at a diameter of approximately one cell. As noted before, this would directly impact applications requiring high measurement precision, such as the quantitation of cellular DNA content. In contrast, immunophenotyping applications would be more favorably affected since the improvement in sensitivity would be significant and the increase in measurement variability at high sample throughput rates would be better tolerated. Clinical flow cytometers have been designed with these factors in mind and, usually, a number of compromises have been made to accommodate the current spectrum of clinical applications.

Future developments in clinical flow cytometers will most likely include, or possibly use exclusively, diode lasers. Diode lasers are being developed in the laser industry. They have outputs that span from the UV to the infrared. Currently, most of these lasers are still in the research laboratories and most of those that are available are too expensive for inclusion in a clinical flow cytometer. The only inexpensive diode lasers are those that emit at 670 nm, 780 nm, and further into the infrared; these require the development of new dyes and reagent systems to be used effectively. Nevertheless, the appeal of diode lasers is significant due to their small size, low power requirements, and high efficiencies.

Beam-Shaping Optics

The purpose of beam-shaping optics is to provide uniform illumination and resolution of each cell as it passes by the focused excitation light beam. Typically, the focused excitation light beam has an elliptical cross–section and is formed by two lenses. One of the lenses is a cylindrical or anamorphic lens. This lens focuses the light beam in one direction only. It is usually a short focal length lens used to control the dimension of the excitation beam spot cross–section along the axis of flow. The size of the beam cross–section (d in μm) along the axis of flow at a distance from the lens equal to the lens' focal length is determined by the laser beam diameter; D in mm; the wavelength of the laser light, λ in μm; and the focal length of the lens, f in mm; using the following formula:

$$d = \frac{1.27 \cdot \lambda \cdot f}{D}$$

The second lens may be either a cylindrical or spherical lens, usually of a longer focal length, and is used to control the dimension of the excitation beam spot perpendicular to the axis of flow. This beam spot dimension is determined by the same equation. Thus, the elliptical beam cross–section is oriented with the short dimension along the axis of flow and the long dimension perpendicular to the axis of flow.

The shape of the excitation light beam strongly influences the information content of data collected with a flow cytometer. Both the vertical (along the axis of flow) and horizontal (perpendicular to the axis of flow) dimensions of the excitation beam need to be considered when setting up an experiment and interpreting the data collected.

The horizontal dimension of the excitation light beam influences the measurement precision, as determined by the coefficient of variation (CV), and the amount of light illuminating the particle. These two parameters are mutually exclusive since increases in precision can lead to reductions in particle illuminance.

The precision is maximized when the beam is widened horizontally. Because of slight fluctuations in the horizontal positioning of the sample core stream, it is important that the intensity of the excitation light beam not change appreciably along over the range over which the sample core stream might wander. Widening the beam increases the distance between the two points on the Gaussian beam profile, where the beam intensity drops to only a small percentage of the maximum intensity. The effect of widening the beam is shown in Figure 5.5. Widening the beam has the corresponding negative effect of decreasing the peak intensity of the excitation light beam at the sample core stream. Thus, if the

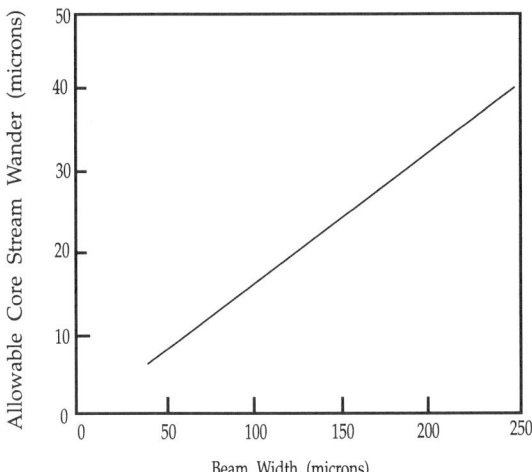

Figure 5.5. Graph showing, for a range of beam sizes, the span over which the beam intensity changes less than 5%. For a 10-μm particle to stay within this span, the distance that the particle can wander is 10 μm less than the noted span.

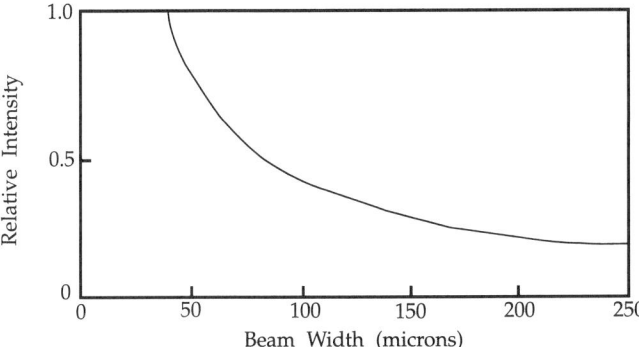

Figure 5.6. Graph showing the relative amount of light available to illuminate a 10-μm particle for a range of beam sizes.

beam is widened excessively it would reduce the radiance of fluorescent emissions, which also would adversely affect precision.

If achieving maximum sensitivity is important, then the excitation light beam must be shortened horizontally. This maximizes the sensitivity by increasing the light that illuminates the sample core stream. The effect of narrowing the beam on sensitivity is shown in Figure 5.6. There is a limit as to how narrowly the excitation light beam can be shortened. The beam must obviously be wider than the width of the biggest particle to be analyzed, but how much wider is determined by the desired precision and sensitivity that one wants to achieve for an experiment.

Applications that require high precision, such as cellular DNA quantitation, benefit from a horizontally wide beam. If sensitivity is more important, as it is for immunophenotyping or platelet enumeration, for example, a horizontally narrow beam is the better choice.

The vertical dimension of the excitation light beam controls the ability of the flow cytometer to resolve two closely spaced particles. Particles can be in close proximity either from clumping in the sample or from the sample running so fast that the probability of two or more particles being spatially unresolvable is significant. Thus it would seem that the beam should be as narrow as possible vertically in order to resolve doublets and larger cell aggregates.

There are practical limitations on the vertical narrowness of the excitation beam. Cytometer precision is adversely effected by making the beam too narrow. Concomitant with the narrowing of the excitation beam is a decrease in the depth of focus. The result is an increased sensitivity to slight movements of the sample core stream. The peak pulse data collected with a narrow beam may not accurately represent the total scattered or fluorescent light emitted by the particle. Collection of integral pulse data under similar conditions has no such limitations beyond those introduced by the reduction in depth of focus.

FORWARD-ANGLE LIGHT SCATTER SENSOR

The forward-angle light scatter (FS) sensor detects the light scattered by a cell in the forward direction near the axis of the incident beam. The signal is approximately proportional to cell size. FS is particularly useful for discrimination between cells and debris and different cell types. For example, FS is the most commonly used signal to trigger the flow cytometer's data acquisition system and has been used to distinguish between live and dead cells. Cell discrimination is affected by the cell's size, index of refraction, and absorptive properties. The degree to which the index of refraction and absorptive properties influence the light-scatter measurement is dependent on the collection angles and on the average index of refraction of the cell. Thus, for certain FS collection angles and cell preparations (i.e., unfixed cells), the index of refraction and absorptive properties can have a significant influence. The excitation wavelength is also an important parameter, since the discrimination at one wavelength may not be as good as that at another (14). However, the use of two wavelengths (351 and 488 nm) simultaneously can provide improved resolution of cell subpopulations (15). Consequently, even though FS has found its place in numerous applications, it is a complex and incompletely understood phenomenon and one should be cautious about interpreting differences in FS intensities purely in terms of size changes.

Optics and Detectors

The FS sensor is placed beside the flow cell opposite the entry point of the excitation light beam. The FS sensor consists of a photodetector, a structure to block the on-axis unscattered laser light, and, usually, some optical elements. A solid-state device called a photodiode is the most commonly used photodetector to measure scattered light. A structure called a beam dump is required in order to keep the high-intensity laser beam from hitting the photodiode and making it blind to the scattered light. Typically, the optics, if any, may include fiber optics, and/or a simple lens system that serves to transfer the light-scatter signals to the photodiode. A unique optical system is used in the EPICS flow cytometers which places the photodiodes at the Fourier plane of the collection lens. This configuration makes the positioning of the signal on the photodiodes relatively insensitive to the position of the light-scatter source (11).

Collection Angle

The collection angle of a forward-angle light-scatter sensor determines the sensitivity of the light-scatter measurement to diffraction and refraction. For higher index of refraction particles (i.e., fixed, fixed and stained cells) diffraction is primarily size-dependent. However, for lower index of refraction particles (i.e., unfixed cells) diffraction is dependent on both size and index of refraction. Typically, the forward-angle light-scatter sensor collects light between a lower limit of .5° to 1° and an upper limit of 10° to 20°. Diffraction predominates at lower angles (.5° to 2°) and refraction becomes more significant at the wider angles (16, 17). Consequently, the output of the forward-angle light-scatter sensor is determined by both the cell's size and index of refraction. At first, this might be considered to be an unnecessary complication, but it has been found that discrimination between cells is improved by including some index of refraction information. For example, the discrimination between live and dead cells is improved if light scattered between the angles of .5° and 12.5° is collected (18). Decreasing the angle of collection reduces the amount of discrimination (18). Obviously, live and dead cells are about the same size but differ significantly in their refractive index.

Depending on the needs of the application, it is possible to control the sensitivities of the light-scatter sensor to cell size and index of refraction by changing the angle of collection. The lower limit is typically controlled by the beam dump or obscuration bar. If either are made smaller, then more diffraction information would be collected from the very narrow angles. Conversely, if they were made larger, it would be possible to reduce significantly the amount of diffraction information included in the measurement. The upper angle of collection can be controlled with an iris or annulus. Decreasing the upper limit would serve to reduce the refractive component from the light-scatter measurement. Unfortunately, since size and index of refraction potentially influences measurements at all angles, other than the above guidelines there is little that can be done to predict with any accuracy an optimal set of angles for an application, except for experimentation.

FLUORESCENCE AND SIDE-SCATTER SENSORS

Located perpendicular to the excitation light beam are the collection optics for the fluorescence and side-scatter sensors. This location was chosen to minimize the amount of excitation light that is scattered into the fluorescence photodetectors. It was discovered that the light scattered at the laser wavelength by the cell in this direction contains information about the cell's internal structure and has proved to be useful in a number of applications (19). For example, in conjunction with the forward-angle light-scatter signal, it has been used to discriminate between the three major subpopulations of white cells, i.e., lymphocytes, monocytes, and granulocytes (20). However, it should be noted that size and index of refraction also affect this signal. Variations in the side-scatter signal may be a reflection of size variability and/or differences in internal cellular structure.

Light Collection Optics

The most important features of the fluorescence light collection optics are their light-collection efficiency and ability to minimize the amount of background light that reaches the photodetectors. Light-collection efficiency is determined by the numerical aperture (NA) and is proportional to the square of the NA; thus, a 20% improvement in N.A. translates to a 44% improvement in light-collection efficiency. Manufacturers are emphasizing the use of low-power, air–cooled lasers in their clinical flow cytometer products and this has necessitated the use of high-N.A. light collection optics systems. As discussed in the section on flow cell design, the advertised numerical apertures range from 1.2 to 1.4. All manufacturers claim to be able to detect approximately 1000 or fewer fluorescein molecules on a cell.

There are two approaches to the design of the flow cytometer's light collection optics for fluorescence and wide-angle light-scatter detection. One approach uses an imaging radiometer design that transfers the collected light to the photodetectors as a real, magnified image of the cell as it passes through the excitation light source. The other approach uses a collimating radiometer design that transfers the collected light to the photodetectors without forming any intermediary images. Both of these designs use apertures or field stops to minimize the amount of background light that reaches the photodetectors. For single-laser applications, the collimating radiometer design is advantageous because it is not constrained by the same depth-of-focus limitations as the imaging radiometer. However, for multilaser applications, in which the beams can be separated by a large distance (i.e., 250 µm), the ability of the imaging radiometer to resolve spatially between two light sources could be advantageous. The collimating radiometer design requires optical filters and temporal electronic gating to separate the two light emissions, but there is no limitation on how far the laser beams need to be separated as long as they do not interfere with one another (11).

Filters

The light gathered by the light collection optics is a combination of all the fluorescence emissions (specific binding, nonspecific binding, autofluorescence), Raman scattering, and scattered laser light. Optical filters are utilized in a flow cytometer to separate the light collected into usable spectral components for the quantitation of the respective fluorescence or scatter signals. Careful selection and proper application of optical filters and the use of proper biological and staining controls are crucial to the successful utilization of a clinical flow cytometer for an application.

The two types of optical filters used in flow cytometry are absorption and interference filters. Absorption filters are usually made of colored glass. As their name implies, absorption filters work by absorbing the light of unwanted wavelengths and passing the light with the desired wavelengths. Depending on the density of the absorbing material in the glass and on the thickness of the filter, absorption filters provide excellent exclusion of the undesired light wavelengths over the complete range of wavelengths encountered in a flow cytometer and high transmittance of the desired light wavelengths. Unfortunately, because absorption filters absorb light they usually fluoresce. This is a particularly pronounced problem when the filters are used to absorb light at the shorter wavelengths, i.e., blue to UV. Conversely, it is a lesser problem with light at the longer wavelengths, i.e., orange to red. In any case, absorption filters must be used with care and should not be the first filter in line between the excitation light source and the detector(s).

Interference filters are the type most often used in a clinical flow cytometer. Instead of absorbing the unwanted light, interference filters attenuate or reflect the unwanted light. This is accomplished by depositing a specific series of metals and dielectrics onto a substrate. The sequence of deposition is designed to attenuate the unwanted light by destructive interference and reflect the rejected light. This type of construction permits interference filters to be designed with sharp cut–on and/or cut–off wavelengths. This is particularly important when it is necessary to resolve between closely spaced fluorescence emissions. Interference filters can be designed to serve almost any type of filter need. There are five types of interference filters, the bandpass, notch, longpass, shortpass, and dichroic-interference filters. The bandpass filters are designed to pass a specific range of light wavelengths centered on a specified wavelength. A notch filter is just the opposite of a bandpass filter in that it blocks light over a specified range of light wavelengths. Longpass filters are designed to pass light above and shortpass filters are designed to pass light below a specified wavelength. In reality both long- and shortpass filters are very wide-band bandpass or notch filters and, as such, must be specified not to turn on or off again within the relevant range of UV, visible, and infrared light wavelengths. For example, a longpass filter designed to pass red wavelengths may transmit in the UV as well or, conversely, a UV short-pass filter may transmit some of the longer red wavelengths. Dichroic filters are designed to reflect a specific range of light wavelengths and to pass all other wavelengths. Dichroic filters can be designed to pass long wavelengths or short wavelengths. To illustrate the use of the above-mentioned filter types, a common filter configuration for detecting right-angle light scatter and three colors of fluorescence emissions is shown in Figure 5.5. Because the light is not absorbed by interference filters, this type of filter is typically not prone to fluoresce. The instances in which interference filters have fluoresced are usually due to the improper selection of the materials used in the protective layer needed to preserve the filter coatings.

Fluorescence and Side-Scatter Photodetectors

In contrast to the light scattered in the forward direction, the fluorescence emissions of dye-labeled cells are many orders of magnitude lower in intensity. This necessitates the use of a high-sensitivity, low-noise photodetector, such as the photomultiplier tube (PMT), to detect the fluorescence emissions. In flow cytometry, the detection of fluorescence emissions has been done exclusively with PMTs.

Photomultiplier tubes are characterized by their gain and spectral sensitivities. PMTs operate by producing photoelectrons released by the interaction of photons with the photocathode. As the photoelectrons are accelerated through a voltage potential onto the surface of the first dynode, they cause the release of more electrons through secondary emission (Fig. 5.7). These secondary electrons stimulate the release of more electrons at each subsequent dynode. The resultant cascade of electrons gives rise to an electrical pulse the amplitude of which is proportional to the number of photoelectrons initially released. The gain of a PMT is determined by the number and design of the dynodes as well as the high voltage applied across the dynodes. The higher the high voltage—up to the tube's maximum limit—applied across the dynodes, the higher the gain (Fig. 5.8). Additionally, there is a minimum voltage that must be applied across the dynodes. Below the minimum voltage, the pulse output of the PMT is not linearly related to the number of photons impinging on the photocathode. Each flow cytometer manufacturer recommends an appropriate PMT operating voltage range. This voltage range must be observed when reducing the PMT high voltage to analyze brightly-stained cells, i.e., cells stained with a DNA stain or bright microspheres. It may be necessary to attenuate the fluorescence signal with a neutral density filter rather than reducing the PMT high voltage below an acceptable level. Conversely, increasing the high voltage above the recommended maximum when analyzing weakly fluorescent cells can lead to a breakdown or continuous discharge of electrons and will be reflected as a high-noise signal.

The spectral sensitivity of a PMT is determined by the photocathode material. The multialkali photocathode is the one most commonly found in PMTs used in flow cytometers. A typical PMT with a multialkali photocathode has a useful spectral response range of approximately 300 nm to 750 nm (Fig. 5.9). This range spans the emissions of the most common dyes used in a clinical flow cytometer. However, there is a strong interest in the development of dyes that can be excited with a HeNe laser at 633 nm (or diode lasers that emit at 670 nm or 780 nm) and fluoresce in the near infrared. Photomultiplier tubes exist that can detect these longer wavelengths, but they are not currently found in

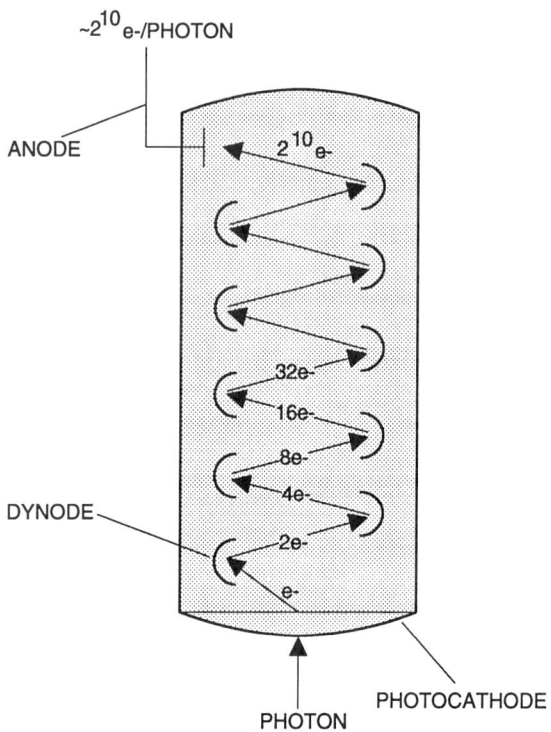

Figure 5.7. Schematic representation of a photomultiplier tube. The amplification process is illustrated by the doubling of the electrons at each successive dynode (Courtesy of Coulter Corporation).

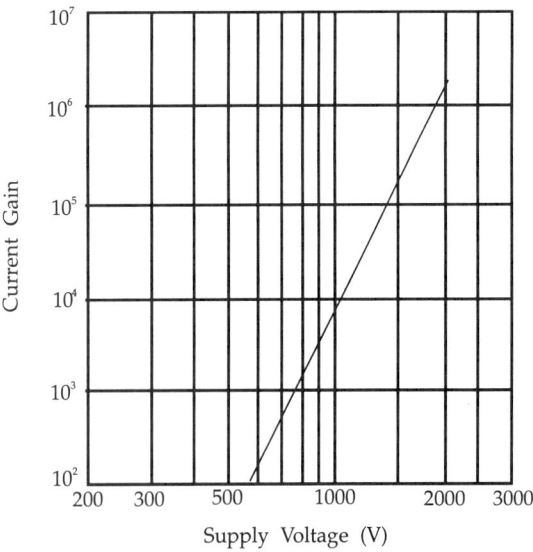

Figure 5.8. Graph illustrating the dependence of the photomultiplier tube (Hamamatsu R1923) current amplification on the voltage applied to the photomultiplier tube (Courtesy of Hamamatsu Corporation).

commercial flow cytometers. Nevertheless, some of these PMT's could be fitted into existing flow cytometers. However, as applications develop around longer wavelength dyes, flow cytometers with appropriate detectors will become more widely available.

Though not as bright a signal as the forward light-scatter signal, the side-scatter signal is still much brighter than a

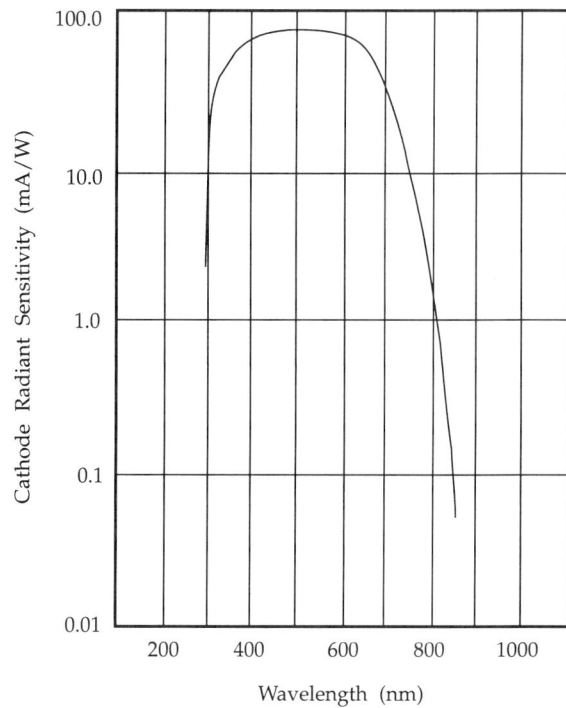

Figure 5.9. Graph illustrating the cathode radiant sensitivity for a photomultiplier tube (Hamamatsu R1923) with a multialkali photocathode (Courtesy of Hamamatsu Corporation).

fluorescence signal. Traditionally, a PMT has been used to detect side scatter but photodiodes have been found to be just as effective as the PMT for detecting wide-angle light scatter.

Solid-state detectors, though useful for side-scatter, have not been used for the detection of fluorescence signals. At the moment, low gain and poor signal-to-noise ratio make them unsuitable as fluorescence detectors. There is much interest in trying to develop a solid-state equivalent of a PMT, but, as of yet, such devices are only research curiosities. It is expected that in the not-too-distant future, solid-state detectors will replace the PMTs used in flow cytometers.

SIGNAL PROCESSING ELECTRONICS

Types of Pulse Signals Generated By a Flow Cytometer

The output of the photodetectors is proportional to the intensity of the light falling upon it. The photodetector outputs are converted to a voltage signal, amplified, and processed by a series of amplifiers. The two types of electronic signals produced by a flow cytometer's photodetector amplifiers are peak pulses and integral pulses. Examples of these pulse types are shown in Figure 5.10. Although these pulse types have distinctly different characteristics, they have been considered interchangeable for many applications. In some instances this will be true. Nevertheless, there still may be advantages to choosing one pulse type over the other.

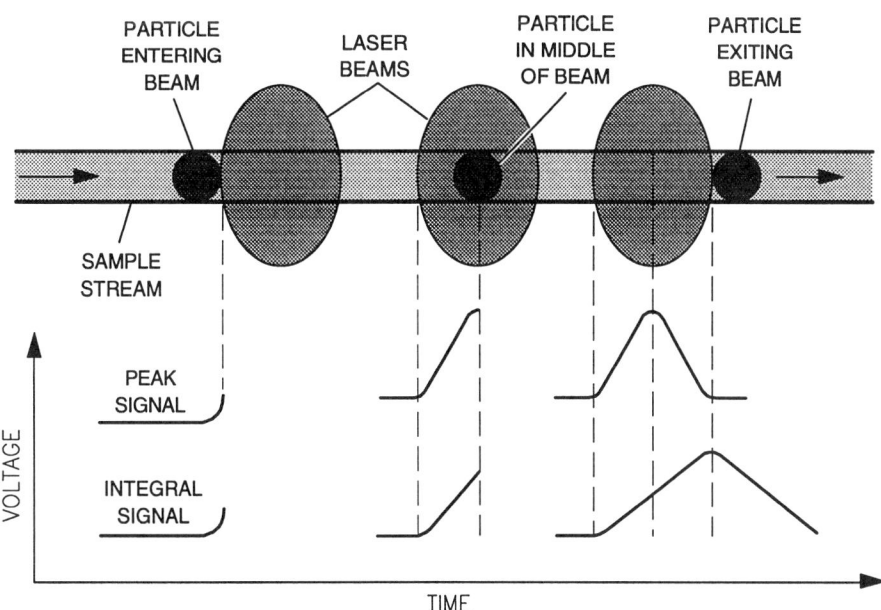

Figure 5.10. Comparison of peak and integral pulses. The peak pulse represents the intensity of the light emitted by a cell as it passes through the focused excitation light beam. The integral pulse represents the total amount of light emitted by a cell, in other words, the area under the peak pulse (Courtesy of Coulter Corporation).

There are well-defined instances where the peak and integral pulses are not equivalent. In these cases, it is not possible to interpret the resulting data without first understanding the impact of the instrument configuration.

In general, it is important not to assume equivalence of pulse types and to make a deliberate decision to select the correct pulse type for the particular application. By understanding the differences between the two pulse types, it is possible to make an informed decision of which pulse type to select to gather the desired data and/or to exploit these differences and derive more information about the particles being analyzed.

PEAK PULSE

As implied by its name, a peak pulse starts at a baseline value as the cell or particle enters the focused excitation light beam, rises to reach a maximum value as the cell or particle reaches the center of the focused excitation light beam, and then returns to baseline as the cell or particle exists the focused excitation light beam. Thus a peak pulse represents the light intensity measured when a particle traverses a sharply focused excitation light beam and corresponds to what one obtains from the output of a photodetector, i.e., photomultiplier tube.

The meaning of the peak pulse height (amplitude) is dependent on whether the dimension of the excitation light beam in the direction of flow is greater or less than the particle diameter. If the width of the excitation light beam in the direction of the particle travel is greater than the particle diameter, then the height of the peak pulse is proportional to the maximum signal that is potentially available from the particle and/or the total amount of the fluorescent marker used to identify a particle component. However, if the width of the excitation light beam in the direction of the particle travel is less than the particle diameter, then the height of the peak pulse is not representative of the maximum emission obtained from the particle. Rather, it is representative only of that part of the particle that, when illuminated, produces the greatest signal.

An illustration of these two cases is made by comparing the peak pulses obtained from a uniformly, fluorescently-labeled 10-μm wide excitation light beam of equal power, and a similarly uniformly labeled 1-μm particle, i.e., a bacterium, traversing the same beams (Fig. 5.11). The fluorescent label used is assumed to have a short fluorescence lifetime. In the first case, the peak pulse obtained when the 10-μm particle goes through the 15-μm beam would be about two-thirds the height of that obtained from the 10-μm particle going through the 5-μm beam. The reason being that the 5-μm excitation beam illuminates approximately half of the 10-μm particle at any one time, but at three times the light intensity delivered with the 15-μm beam. In the second case, the peak pulse obtained when the 1-μm particle passes through the 15-μm excitation light beam would be approximately one-third of that obtained from the 1-μm particle traversing the 5-μm excitation beam. In this latter case, the 1-μm particle is smaller than either excitation light beam and, thus, is fully illuminated by both. The 5-μm excitation light beam concentrates the same excitation light power into one-third the space of the 15-μm excitation light beam; thus, the intensity of the excitation light beam where it intercepts the particle is three times greater for the 5-μm beam width than for the 15-μm beam width.

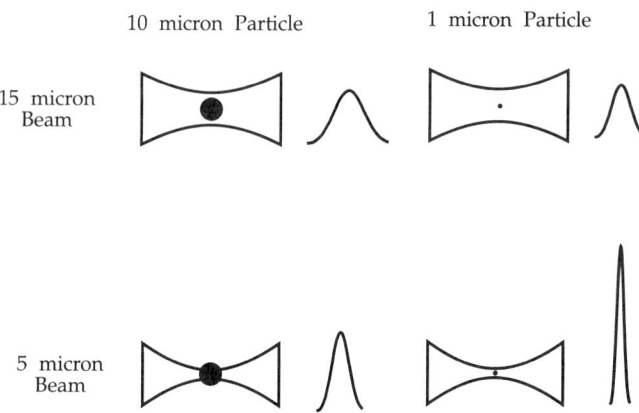

Figure 5.11. Comparison of pulse amplitudes obtained from a 10-μm particle and a 1-μm particle passing through 15-μm- and 5-μm-wide excitation light beams.

A frequently overlooked characteristic of a peak pulse is that the height of the peak pulse is not sensitive to changes in the speed with which a particle traverses the excitation light beam. This assumes that (a) the fluorescent dyes, if used, have short lifetimes, and (b) the electronic amplifiers of the flow cytometer have been designed with a bandwidth (frequency response) wide enough to respond similarly to narrow and wide pulses as well as to rapid sequences of pulses.

INTEGRAL PULSE

An integral pulse is derived from a peak pulse by integrating the peak pulse over time. The height (amplitude) of the integral pulse is proportional to the area (integral) under the corresponding peak pulse. Because all of the photons emitted by the particle contribute to the final integral pulse measurement, the height of the integral pulse is a more precise representation of the total amount of light emitted by a particle, which is proportional to the particle dye content and is independent of the excitation beam dimension or cross–section along the axis of flow.

Based on the theory of counting (Poisson) statistics, the precision of the integral pulse is expected to be greater than that of the peak pulse. The relative error of a counting measurement diminishes as more events (photons) are included in the measurement. Thus, an integral pulse exhibits a lower statistical variation than the corresponding peak pulse. This is particularly evident for particles that emit only small amounts of light, such as might be encountered when immunophenotyping cells that have only a small number of fluorescently-labeled epitopes on their outer membrane. Integral pulses are preferred over peak pulses for these low light measurements.

In contrast to the peak pulse measurement, the integral pulse measurement is insensitive to beam size and profile. As discussed earlier, the peak pulse measurement is only proportional to the total or maximum light emission when the cross-section of the excitation light beam is larger than the particle. For clinical samples, there can be a large heterogeneity in cell sizes, ranging from a few microns to as high as a few tens of microns. A dependence of peak pulse height on the ratio of the cell size to excitation beam cross–section proves to be a source of noise that manifests itself by a broadening or mixing of populations, which can mask the underlying phenomena being studied. This problem does not exist for data collected with integral pulses. The integral pulse is a cleaner measurement of the intensity of light emitted by a particle.

It would seem from the above discussion that an integral pulse should be the preferred pulse form for data collection for all situations, but some consideration needs to be given to the significant limitations that arise from the way current clinical flow cytometers generate integral pulses. Depending on the type of integration circuitry used in the clinical flow cytometer, the most significant limitation can be the sample throughput.

Presently, the electronic pulse integrators used in the clinical flow cytometers of the major manufacturers are passive resistor-capacitor (RC) networks. In a passive RC network circuit, a capacitor is rapidly charged by the peak pulse to a voltage proportional to the area of the peak pulse. After the peak pulse passes, the capacitor is slowly discharged by a resistor. This type of circuit produces an integral pulse that rises to a maximum in a time equal to the width of the peak pulse, but returns to a zero value in a time equal to several peak pulse widths. The resulting wide integral pulse limits the number of particles that can be analyzed in a given time. The problem that arises when one attempts to analyze too many particles per unit time is the occurrence of pulse coincidence or pulse pile up. For example, if the integral pulse width was 10 microseconds, the maximum throughput at which the coincidence rate would be less than 1% would be approximately 1,000 particles per second. Likewise, if the pulse width was 100 microseconds, the maximum throughput would be approximately 100 particles per second. As a result of this problem, in the worst cases, artifactual data may be generated if no pulse coincidence detection circuitry is employed and, in the best case, a significant amount of particle data may be ignored.

A more subtle limitation of passive RC network integrators is that the value of the integral pulse is dependent on the peak pulse width. Thus, if the peak pulse width varies due to changes in particle velocity, trajectory through the excitation beam, or orientation, the resulting integral pulse height will vary accordingly. Most clinical flow cytometers, if well maintained, can provide stable, constant velocity flow through the excitation beam, thereby minimizing this limitation. If unexpected results are obtained, it is important to have instrument controls to be certain that the unexpected results are not due to flow-related variations within the flow cytometer.

Amplification

A modern clinical flow cytometer utilizes both linear and logarithmic amplifier systems for pulse amplification. Linear amplification systems were the first to be used in flow cytometers. This sufficed for most applications (i.e., DNA quantitation) studied at that time. With the subsequent growth in interest in immunology research applications the need arose for the capability to view the distributions of very dimly and very brightly stained cells on the same histogram. Logarithmic amplification systems were designed to fulfill this requirement. The employment of both amplification systems in clinical flow cytometers provides the greatest flexibility for the end user.

LINEAR AMPLIFICATION

Linear amplification systems produce an output directly proportional to the input signal. The output is equal to the input signal times an amplification factor. The only exception to this rule is when the input signal times the amplification factor exceeds the maximum amplifier output, at which point the output is held at the maximum output level. The resultant amplified pulses have a flat-topped appearance. This is known as amplifier saturation and any information about the original amplitude of the input signal is lost. Reducing the gain of the amplifiers will decrease the incidence of amplifier saturation.

Linear amplification lends itself to applications requiring accurate signal quantitation and measurement of relative signal intensities. When linearly amplified signals are displayed on a histogram, the channel resolution is the same across the histogram. Each channel represents the same increment in signal value. Thus, a histogram peak centered at channel 200 represents a signal intensity twice that of a histogram peak centered at channel 100. This straightforward linear relationship between channel positions simplifies the mathematics needed to analyze peaks for mean position, standard deviation, coefficient of variation, and relative intensities.

The resolution of linear amplification is determined by the number of channels in the histogram used to represent the signal intensities, i.e., histogram resolution. Linear amplification is most appropriate for applications that require a moderate dynamic range. Common applications that fit into this category include the quantitation by fluorescence of DNA, RNA, and protein content as well as light-scatter quantitation of cell size and granularity.

LOGARITHMIC AMPLIFICATION

Logarithmic amplification systems produce an output that is proportional to the logarithm of the amplitude of the input pulse over a specified range. For a modern clinical flow cytometer, logarithmic amplifiers are typically designed to work over input pulse amplitude ranges that span three (1000:1) or four (10,000:1) decades. This wide dynamic range enables the experimenter to view, in a single histogram, cell populations that have widely varying characteristics; i.e., cell populations representing prolific producers (i.e., positive cells) and nonproducers (i.e., negative cells) of a cell-surface protein. However, there is obviously a concomitant loss of ability to resolve populations with similar intensities.

A logarithmic amplifier system compresses a wide input range into a manageable range for the data acquisition system by applying the same mathematical principles used in semi–log graphs. An ideal logarithmic amplifier system has an output value that is proportional to the logarithm of the input value. In actual practice, the transfer function is modified due to constraints of the amplifier design. For example, all input pulses with amplitudes between zero and the minimum value have the same output value equivalent to channel zero. Similarly, all input values equal to and greater than the maximum allowable input produce the same maximum output channel value. Since it is not practical to design logarithmic amplifiers that provide a perfect logarithmic transformation at the extremes of the designated input range, it is important that the amplifiers be carefully calibrated to give the widest usable dynamic range. If not done carefully, then a nonlogarithmic response would be noticed in a significant portion of the lower first decade and the upper last decade. Miscalibration would lead to misinterpretation of the data.

When selecting the dynamic range (three or four decades) of the logarithmic amplifier system, the resolution of the histogram will be best utilized if the lowest dynamic range that will serve the application is chosen. To illustrate this for the case of 1024-channel histograms, a three-decade logarithmic amplifier will divide a histogram into three regions (decades) of 341 channels each, whereas a four-decade logarithmic amplifier will divide a histogram into four regions (decades) of 256 channels. The additional resolution achievable with three-decade logarithmic amplifiers may allow the identification of more structure in the intensity distribution. Other operational parameters, i.e., sensitivity, will not be affected.

There are a number of applications in hematology and immunology that can utilize the advantages of the wide dynamic range of logarithmic amplifier systems. By far the most common application to utilize logarithmic amplifier systems is immunophenotyping. Both three- and four-decade logarithmic amplifiers are used for immunophenotyping of clinical samples, with three-decade logarithmic amplifiers usually sufficient for many of the more common clinical applications. Logarithmic amplifier systems have also been applied to light-scatter signals to better differentiate by size, refractive index, and granularity between the different subpopulations within blood, i.e., the identification of platelets (21).

Pulse Processing

In addition to the light-scattering and fluorescence-intensity information, the electronic pulses generated by the fluorescence and light-scatter sensors contain information about the particle size and the distribution of the emitted light. Pulse processing circuitry can extract information from the shape and width of the pulse. Information about the particle size, the distribution of emitted light, and whether the particle was a singlet or a multiplet can be extracted.

Undesirable components that affect the accuracy and precision of the measurement of the desired signal also can be components of a signal pulse. Optical noise sources, such as background fluorescence, scattered light, and spectral overlap of the emissions of fluorescent markers co-staining the particle add to the measured light intensity. In a clinical flow cytometer, the contributions of these undesirable components are partially removed by the pulse processing circuitry.

BASELINE RESTORATION

Baseline restoration is the process by which the effects of background light are reduced by subtracting the average background light level from the pulse amplitude. Significant amounts of undesired light can reach the photosensors despite efforts to minimize this interfering signal. Examples of background light include scattered light from the excitation source that leaks through the optical filters, Raman scattering, and fluorescence from residual dye in the core stream. Baseline restoration circuitry monitors background light levels between cells and attempts to compensate for the average background light. The resultant corrected pulse is then passed on to the pulse amplifiers for further processing.

The baseline restorer works best when the correction for background light is small. When measuring light intensity levels, the measurement error is proportional to the square root of the light intensity. Thus the greater the background light intensity the greater the error associated with the corrected signal. This is particularly serious when measuring low-level light signals. It is possible for the magnitude of the errors to exceed the signal level if the background light levels are not kept to a minimum. In those instances, it is important to reduce the background light levels to a minimum by using appropriate optical blocking filters.

ELECTRONIC FLUORESCENCE SPECTRAL COMPENSATION

Despite attempts to choose optical filters in order to limit the light allowed to reach each photomultiplier tube to the emissions of a single fluorescent dye, it is not always possible to achieve this goal. Because of the broad emission spectra of the commonly used fluorescent dyes, the light reaching a photomultiplier tube may consist of not only the emissions from the intended fluorescent dye but, in addition, of the emissions of other fluorescent dyes. Within the wavelength acceptance range for the specific optical filter, the emission spectra of the other fluorescent dyes overlap the emission spectrum of the intended fluorescence dye. The classic example of this kind of spectral overlap is the overlap between fluorescein and phycoerythrin (Fig. 5.12).

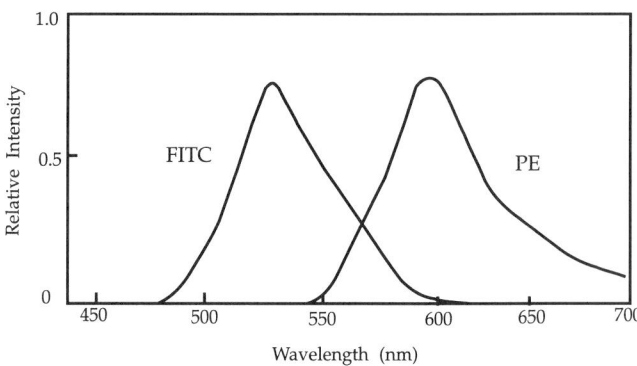

Figure 5.12. Illustration of the spectral overlap between the emission spectra of fluorescein (*FITC*) and phycoerythrin (*PE*).

To correct for spectral overlap, clinical flow cytometers are equipped with an electronic means of subtracting the contaminating components from the measured signal to yield a better measure of the emission intensity of the intended fluorescent dye. Electronic fluorescence spectral compensation is accomplished with an analog subtraction circuit that subtracts from the measured signal a fraction of the other contaminating signals as measured by their respective photomultipliers (22). For example, the amount of fluorescein emission detected by the photomultiplier assigned to monitor the phycoerythrin emission is proportional to the signal detected by the photomultiplier used to detect the fluorescein emission. The amount of compensation required is reported as a percentage of the contaminating signal's measured intensity. Continuing with the above example, the compensation needed to remove acceptably the fluorescein contribution to the light measured by the phycoerythrin photomultiplier tube may be reported as 15% of the signal measured by the fluorescein photomultiplier tube.

Most experiments using two or more markers require electronic fluorescence spectral compensation. The range of applications includes immunophenotyping experiments with two or more markers, measurement of DNA content with nuclear and/or cell-surface markers, and cell function with nuclear and/or cell-surface markers. The success of any particular experiment will be dependent on the proper choice of filters and the use of appropriate spectral compensation controls.

DOUBLET DISCRIMINATION

The major premise of flow cytometry is the measurement of cellular properties on a single-cell basis. Corruption of the measurement process by the inadvertent inclusion of particle doublets and, less commonly, particle multiplets in the data considered to consist only of single particles are often overlooked. The effect of doublets on DNA analysis has been investigated in some detail but little has been done for immunology-related analyses (23, 24). Although the frequency of their occurrence may be low, the presence of doublet contamination in flow data becomes more serious when the populations of interest are rare and the possibility of confusion with a doublet even is high.

Doublets and multiplets can be categorized into two types. By far, the most common type of doublet is a result of sample preparation. Doublets may exist in a sample preparation from inadequate tissue dissociation or from single cells adhering to one another to form aggregates. The only way to minimize the frequency of this kind of doubtlet is by carefully fine-tuning the preparation procedure. It is imperative that the experimenter visually inspect the sample by microscopic examination to ensure that the percentage of doublets and aggregates is adequately small. It is important that that fraction be kept to a minimum (i.e., < 5%) because it is not possible to remove all doublets electronically with doublet discrimination. The second type of doublet is the result of two single particles being so close to each other in the sample core stream that the flow cytometer sees them as a single event. The frequency of this type of doublet is a function of the sample throughput rate (sample concentration and delivery rate), sheath stream velocity, excitation beam width along the axis of flow, and amplifier speed in the detection and data acquisition systems. For a given sample throughput rate and sheath stream velocity, a flow cytometer with a wide excitation beam and/or wide pulse–width would have a greater incidence of this kind of doublet.

There is typically no specific doublet detection hardware in a clinical flow cytometer; rather, there are three graphical approaches to doublet discrimination currently in general use to detect doublets arising from both of the above-mentioned causes. In each method, the integral pulse data representing the cellular DNA content is used in conjunction with another parameter. The secondary parameters are (a) peak pulse amplitude (height), (b) peak pulse width and (c) forward-angle light scatter (23, 24). Of these methods, the first two are the most popular. With rare exceptions, the use of light scatter is precluded by the existence of wide population heterogeneity in which doublets of one sub-population can overlay singlets of another sub-population in the light-scatter distribution.

The first method using integral (area) pulse vs. peak (height) pulse data is based on the inherent information content differences between peak and integral pulses. The amplitude of the integral pulse is a measure of the total amount of dye contained in a cell or cell doublet and is independent of the size of the cell or cell doublet. In contrast, the amplitude of the peak pulse is proportional to the amount of dye within the beam at any time. Thus, for a focused excitation spot that is approximately the size of a cell, the peak pulse is more closely proportional to the amount of dye in each cell regardless of whether the cell is a singlet event or part of a doublet event with one cell following the other. Thus, a cell doublet appears as twice the integral pulse amplitude of a cell singlet, but the peak pulse amplitude of a cell doublet is approximately the same as that of a cell singlet. A typical example of peak vs. integral pulse plot is shown in Figure 5.13(a). It can be seen that not all cell doublets that represent the population in the lower right of the histogram have peak pulse amplitudes equal to those of singlet events. This is due to the doublets not having their long axis perfectly aligned with the axis of flow.

The second method uses peak pulse width in conjunction with integral (area) pulse amplitude to identify doublets. In this method the doublets are identified by their wider peak pulse width. It is intuitively obvious that a doublet particle would have to produce a wider pulse width since two particles are going through the excitation beam typically one after the other. However, the pulse width of a cell doublet is not usually twice the pulse width of a cell singlet. The width of a peak pulse is proportional to the particle size plus the excitation beam width along the axis of flow. Thus, if the excitation beam width is approximately the size of a cell, a cell doublet will have a peak pulse only 50% wider than a cell singlet. An example of a histogram of integral pulse amplitude vs. peak pulse width is shown in Figure 5.13(b).

The successful detection of doublets by either of these two methods depends on the doublet particles having their long axis aligned along the axis of flow. Doublets so aligned are measured as having a wider pulse width than a singlet particle, an integral pulse height twice that of a singlet particle, and a peak pulse height approximately the same as that of a singlet particle. Unfortunately, for particle doublets and larger aggregates this precise condition is seldom exactly met. The degree to which this condition is met is determined by the hydrodynamic focusing, which is a function of the fluidic constants of the flow cytometer system. In practice, most particle doublets are aligned on axis or slightly off axis. Infrequently, their long axis is aligned closer to the axis that is perpendicular to the central axis of flow. Consequently, particle doublets can have pulse widths ranging from significantly larger than a singlet particle with a singlet peak pulse height to a singlet particle pulse width with twice the singlet peak pulse height. Thus, in practice, a percentage of particle doublets will escape detection. These graphical approaches to doublet correction serve only as a backup to catch a large fraction of the few remaining doublets in a sample. In short, there is no substitute for proper sample preparation and microscopic verification that the incidence of doublets is minimal. Furthermore, the graphical correction for doublets is complicated when analyzing heterogeneous populations. In those instances, the identification of the singlet events is very subjective, resulting in the variable success of doublet correction.

RATIO

The absolute fluorescence intensity measured in an experiment is modulated sometimes by not only the phenomena under study but by other parameters, such as differences in dye adsorption, cell volume, and cell surface area. The result of the added modulation is an increased variability in the resultant data that may make the data uninterpretable. A classic example in spectroscopy is the measurement of the absorption of a solute dissolved in a solvent. Light absorption

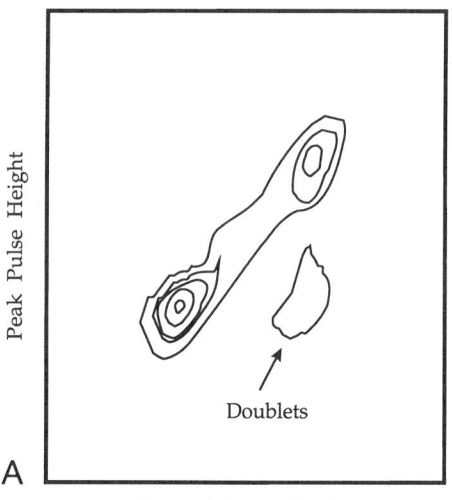
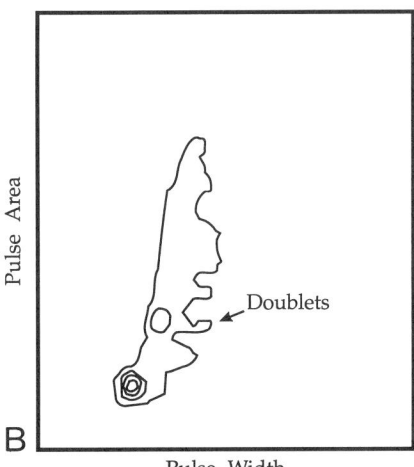

Figure 5.13. Examples of typical two-parameter tissue culture cell DNA histograms used for doublet discrimination by the two most common graphical methods. **A** is used by the peak pulse amplitude vs. integral pulse amplitude method, and **B** is used by the pulse width vs. integral pulse amplitude method. Both methods work well for samples with low numbers of doublets. However, for samples with heterogeneous populations and/or large numbers of doublets, the distinction between the desired singlet events and doublet events will not be as dramatic. For these more complicated cases, it may not be possible to discriminate doublet events without losing some singlet event data.

is a function of both the solute and solvent. Accurate measurement of the absorption of the solute alone depends on measuring the absorption of the solute plus solvent with normalization against the absorption of the solvent, as in a dual-beam spectrophotometer.

In a flow cytometer, ratio circuitry provides a means of normalizing one parameter against another. An analog divider circuit is used to generate the ratio of the two inputs. The inputs are scaled to adjust the desired output range of the analog divider. The output of the analog divider is digitized by the data acquisition system. A notable example of an application that requires the use of the ratio circuitry is the measure of calcium metabolism by Indo-1. The changes in intracellular calcium levels are measured by measuring the ratio of the fluorescent signal intensities from the calcium-bound to calcium-free dye. The use of the ratio measurement makes this calcium measurement largely insensitive to cell-to-cell variations in cellular dye content (11, 25).

DATA ACQUISITION AND STORAGE

Data Acquisition

The typical flow cytometer data acquisition system takes the signals from the outputs of the photosensors and converts the analog signals to digital values for subsequent storage and analysis. The acquisition system consists of an event discriminator, an analog-to-digital converter (ADC), and an interface to the data storage device.

EVENT DISCRIMINATION

The event discriminator is used to determine when the signals from the photodetectors belong to a valid cell event rather than to a piece of small debris or electronic noise. Discrimination serves two purposes. The first is to avoid including nonrelevant data in the data files. The second is to reduce the burden on the data acquisition system. Burdening the data acquisition system with an excessive amount of irrelevant data degrades the effective data acquisition system throughput rate for the events that should be included in the data set of interest.

Usually the event discriminator monitors only one signal line for which signal levels above a certain voltage (discriminator level) have been determined to be a reliable indicator of a cellular event. The most common signal line monitored is the forward-angle light-scatter signal because the amplitude of the forward-angle light-scatter signal is related to particle size. Debris in a sample is typically much smaller than the cells of interest; thus, the event discriminator can be set not to trigger the data acquisition system on a noise event. In instances where size is not a good discriminator, i.e., bacteria cells or chromosome preparations, a signal from a fluorescence detector monitoring DNA, RNA, or protein content is more appropriately used as the source for the event discriminator. Once a valid cellular event is detected, the peak amplitudes of all the peak and integral signals being collected are detected and held so that the analog-to-digital conversion circuitry can process the signals.

ANALOG-TO-DIGITAL CONVERSION

An analog-to-digital converter changes the analog signals being held to digital values so that they can be processed and stored by a digital computer. The performance of an ADC is characterized by its speed of conversion, channel width uniformity (differential linearity), and accuracy. Currently employed ADC's are faster, more linear, and more accurate than those used originally in flow cytometers. Historically, the faster the ADC, the less linear and accurate was the resultant conversion. However, fast and slow are relative terms and what is a slow ADC by today's standards was fast by yesterday's standards.

The type of ADC that is currently used in most commercial flow cytometers utilizes a successive approximation technology to digitize the data. These ADC's have conversion times between 1 and 10 microseconds. Since most instruments use ADC's that convert to values having between 12 and 16 bits, of which only the most significant 10 bits are used, differential linearity is very good. The number of significant bits used determines the maximum histogram resolution. The maximum number of channels is equal to 2^n, where n is the number of significant bits used. Thus, if 10 bits are used, the maximum histogram resolution is 1024 channels.

ADC conversion time is the major contributor to the deadtime of the data acquisition system. Throughput of the data acquisition system is inversely related to the deadtime. Referring to the previous discussion about the event discriminator, the advantage of discriminating against small non-cell particles becomes more apparent. Small debris might be so numerous that it is more probable that the data acquisition system would be processing debris signals and ignoring the relevant signals. The processing of the debris signals consumes a major portion of the data acquisition throughput and effectively reduces the rate at which the cells of interest can be processed.

After the data values are converted they are collected in data buffers and passed to the data storage system. The buffers are designed to be large enough so that the instantaneous rate of data collection does not exceed the rate of data storage.

TIME

As important as the measurement and correlation of the optical signals is the inclusion of time as a parameter. Firstly, with time as a parameter it is possible to do kinetic experiments on a flow cytometer. By adding a time value to each cell data event, the measurement of dye uptake or the response to a stimulus can be followed.

Another use of time is as a quality control parameter. When collecting data with a flow cytometer it is possible that cell staining may change over time or that the fluidics system may drift or change transiently. Typically, the most common cause of the latter are obstructions in the flow cell. The effects of these changes manifest themselves in the data as changes in the measured intensities of the optical signals. If a time value is included with each cell data set, it is possible to track these fluctuations and possibly exclude from the data set those measurements adversely affected by changes in staining or in the fluidics system.

DATA STORAGE

Once converted to digital values, the data is ready for display, analysis, and storage. For display and analysis, the data is usually in a histogram form. For storage, the data may be stored in a histogram format and/or stored directly in a mass storage device as sequential correlated data points. Since the amount of data collected by the data acquisition system can be very large and contain data from nonrelevant particles, flow cytometer data acquisition systems provide the capability to select with gates only the data relevant to the experiment. Gating may be used also to reduce a multidimensional data set to a more manageable dimensionality. This reduced data set can then be stored.

GATING

Flow cytometers are provided with the capability of gating to manage the large data sets that are routinely generated. Gates are used to define populations of cells based on each cell's intensity or value on any number of measured parameters. It is then possible to analyze the gated cell population separately from the rest of the cells in the sample or to store it as a separate data set. The three types of gates available on flow cytometers are the rectilinear, elliptical, and amorphous gates.

Rectilinear gates are a series of lower and upper limits (channel numbers) for one or more parameters. Only cells or particles that have an intensity falling between those limits will be further analyzed by the system. Rectilinear gates are the simplest to set up. Originally defined for use with one-parameter histograms, rectilinear gating has been expanded to two-parameter histograms. In two-parameter histograms, rectilinear gating represents a rectangular region (box); thus it is applicable primarily for data aligned with the histogram axes.

Elliptical gating in a two-parameter histogram is useful to define populations that do not align with the histogram axes and whose shape can be described by a bivariate Gaussian distribution (11).

Amorphous gates provide the greatest versatility. Amorphous gates can be used to define a population of any shape and accommodate situations in which there is some overlap between populations. A common way to implement an amorphous gate in a flow cytometer is in the form of a digital bitmap. A bitmap is a two-parameter array of 1's and 0's set up as a look-up table. The computer uses the inputted boundaries of the amorphous gate to generate a bitmap that, for example, may use 1's to indicate the interior of the amorphous gate. The computer can rapidly determine if an event falls inside or outside the gate by checking if the array element to which the paired intensity or measurement values point is a 1 or a 0. Hence, the reason why amorphous gates are sometimes referred to as bitmap gates.

Gating can also be used to reduce a multidimensional data set into more manageable one- and two-parameter histograms. Examples of this type of gating application are EPICS PRISM and Becton-Dickinson Paint-a-Gate. Both of these tools have found applications in immunophenotyping when two or more labeled antibodies are used and allow the user to visualize the coexpression of multiple surface antigens. The EPICS PRISM provides a one-parameter histogram output showing and quantitating all the possibilities of surface antigen expression. Becton Dickinson Paint–a–Gate provides a series of one- and two-parameter histograms in which the various combinations of surface antigen expression are represented in the histograms with different colors.

HISTOGRAM DATA

To display and analyze the results of an experiment, data is processed and reduced to one or more histograms. The most commonly used are one- or two-parameter histograms. A one-parameter histogram is the frequency distribution of one of the collected parameters. It is usually displayed as the number of counts accumulated for each intensity value (channel number) of that parameter. A two-parameter histogram is a bivariate or two-dimensional map of the frequency distribution of two of the collected parameters. Since two-parameter histograms show the correlation between two cell parameters they are also referred to as correlated data sets. Two-parameter histograms are displayed as dot plots, color or intensity plots, contour plots, and isometric or surface plots. Each of these four plotting techniques differ only in how they represent the number of events in each channel of the histogram.

LIST MODE DATA

The list mode data format is an alternative way to store the results of an experiment. In contrast to a histogram, the data is correlated for each cell event and is stored sequentially in a data list. For example, if four parameters were collected for each cell, the list mode data file would consist of the values of the four parameters for the first cell followed by those of the second cell, followed by those of the third cell, etc. The advantage of storing data in a list mode format is that the data can be gated, put in histogram format, and analyzed multiple times without any loss of data. In essence, it simulates running the same sample over and over again in the flow cytometer. The main limitations are the inability to change any gains or instrument settings used in the collection of the data and the impossibility to compensate for poor instrument set up.

The disadvantage of the list mode data format is the large data storage requirement. The size of the data file is dependent on the number of cells and collected parameters included in the data set. For example, if six parameters are being collected for each cell and each parameter requires two bytes of storage then the storage requirement for each cell would be 12 bytes, so a list mode file for 100,000 events would require 1,200,000 bytes or 1.2 megabytes of storage space. Even for a moderate number of cells, the size of a list mode file can exceed the typical histogram file size. The archiving of large numbers of list mode data files requires large magnetic tape drives or optical disk drives. Additionally, flow cytometer data acquisition systems may need to be connected to networks to adequately manage the very large data sets generated by a single experiment.

FLOW CYTOMETRY STANDARD FILE FORMAT

In an effort to foster the free flow of information between flow cytometry laboratories, a standard file format for flow cytometry data was proposed in 1984 and later revised in 1990 (26, 27). This standard file format has been named the flow cytometry standard (FCS) file format. Since its introduction, instrument manufacturers and software developers have adopted the FCS file format.

The FCS file format provides a universally recognized means of including all relevant information about the experiment within the data file. The information includes a description of the sample, instrument type, instrument settings, data collected, and results of any analyses. The current revision of the FCS file format, FCS2.0, requires that each file consist of at least four sections, the HEADER, TEXT, DATA, and ANALYSIS sections. They may be repeated to allow for multiple data sets within a single file (27).

The HEADER section of the data set is always at the beginning of the data set. It is a fixed-format section beginning with the identification of the version of the FCS format that is being used, FCS1.0 or FCS2.0, followed by the byte offsets from the beginning of the file to the beginnings and ends of the TEXT, DATA, and ANALYSIS sections. These are written to the file in a text format, i.e., ASCII format (27).

The TEXT section consists of a series of key words and values, written in a text format, that serve to describe the instrument and experimental setup. In order to define the start and stop of each key word and value pair, a delimiter is defined as the first character in the text section. For example, if the delimiter is the "/" character and the keyword is $DATE, then the entry is the text section might look like "$DATE/01-JAN-1991/." Six key words ($DATATYPE, $PAR, $MODE, $PnB, $BYTEORD, $NEXTDATA) that describe the format of the data format in the DATA section are required to be in the TEXT section. Definitions of these and other commonly used key words are shown in Table 5.2 (27).

The DATA section contains the numerical values for the data set. The data are written in the format defined in the TEXT section. The data can consist of multiple single-parameter histograms, a single two-parameter histogram, or a list mode data set.

The ANALYSIS section contains the results of any relevant analyses done on the data set. The results are written in a text format. Although, there are no defined key words for this section, it is to be structured in the same format as the TEXT section. Any required key words are left to the user to define (27).

Table 5.2
Summary of the Required and Commonly Used Key Words for the TEXT Section of a FCS2.0 Formatted File[a]

Keyword	Description
$BYTEORD	The order of data bytes in DATA section. (Required)
$DATATYPE	Numeric format of data in the DATA section. (Required)
$MODE	Specifies whether the data is histogram or list mode data. (Required)
$NEXTDATA	Byte offset to the next data set header. (Required)
$PAR	Number of parameters stored in data set. (Required)
$PnB	Number of bytes required for each data value of parameter n. (Required)
$PnR	The range of parameter n. Channels in a histogram or total number of events. (Required)
$DATE	Date file created.
$EXP	Name of experimenter.
$PROJ	Name of project.
$SRC	Description of source of sample.
$FIL	Name of data file.
$CYT	Name of cytometer used.
$PnN	Name of parameter n.
$TOT	Total number of events included in data set.
$CELLS	Description of cells or sample.

[a]Data File Standards Committee of the Society for Analytical Cytology. Data file standard for flow cytometry. Cytometry 1990;11:323–332.

SORTING

A direct extension of the flow cytometer's ability to measure cellular parameters on an individual cell basis is the capability to sort individual cells based on the flow cytometer's measurements. Most flow cytometers, commercial and research, use flow cells with a jetting orifice to generate droplets for sorting. It should be noted, however, that there have been some sorters designed around fluid switches (28).

Flow cytometer cell sorters that sort droplets use a jetting orifice with a diameter that ranges from 50 μm to 100 μm to form a stream that contains the sample cells aligned in single file by hydrodynamic focusing, as discussed above. Depending on the design of the flow cell, the cells are measured either in an enclosed quartz chamber, just before exiting through the jetting orifice, or in the stream itself, just as the cells exit from the jetting orifice. With a piezoelectric crystal device acoustically coupled to the flow cell, the stream is driven with high-frequency (20 to 40 kHz) vibrations. This frequency is known as the droplet break-off frequency. Due to the vibration, the stream starts to break into droplets a short distance after the stream and cells exit the jetting orifice. The location, with respect to the exit of the jetting orifice and the stability of the droplet break-off point, is determined by the frequency and power applied to the piezoelectric crystal.

The droplet break-off point is the location where a charge is applied to a droplet containing the cell of interest. Since the droplet break-off point is a fixed point in space, the time required for a cell to travel between the measurement station and the droplet break–off point is fixed as well and is known as the droplet delay.

After a cell is measured and the information is digitized, the information is made available to the sort control logic to make sorting decisions. The selection of the cells to be sorted is performed in a manner analogous to gating. After delaying for a period equal to the droplet delay, a charge is applied to the droplet at the droplet break-off point. The polarity of the charge depends on whether the cell is to be sorted left, right, or not at all. In the latter case no charge is applied. Retaining its charge after breaking off, the droplet passes by charged deflection plates that deflect the droplet left or right into the receiving vessels.

The rate at which cells can be sorted is determined by the droplet break-off frequency and the number of droplets sorted with each cell. The sort rate is limited by the rate of coincidence, i.e., the rate at which two or more cells occupy the droplet(s) to be sorted. For a given sample throughput rate, the coincidence decreases as the droplet break-off frequency increases and the coincidence increases as more droplets are sorted with each cell. (Each cell is in a droplet but some of the preceding and following droplets will be sorted with the droplet containing the cell thus, guaranteeing the capture of cell.)

This is just a brief introduction to cell sorting. For a more in-depth discussion of the physics of sorting and the limitations of sorting the reader is directed to Chapter 8 in *Flow Cytometry and Sorting* (28).

OBJECTIVE MEASUREMENTS OF FLOW CYTOMETER PERFORMANCE

The performance of a flow cytometer can be summarized in terms of the instrument's precision, accuracy, sensitivity, and throughput. It is important to understand and know what these parameters are and how to determine whether a particular flow cytometer can be appropriately used for an application. (For further details concerning clinical issues regarding instrumentation performance, see Chapter 11)

Precision

The precision of a flow cytometer is monitored by determining the coefficient of variation (CV) of measurements made on a sample of nearly identical particles. The CV is mathematically defined as the standard deviation (σ) divided by the mean (μ). As to what is an acceptable CV, the requirements of the application under investigation need to be considered. Typically, the instrument CV must be less than the CV due to biological variability. In any case, one should strive for a minimum CV value to obtain the highest precision. This is not difficult to attain since most flow cytometers can achieve very low CV values.

A number of ways have evolved over the years to calculate CV, but the two most common methods are called the full-peak and the half-peak methods. The full-peak method starts by having the user define graphically, with cursors, the left and right limits of the peak to be measured. A mean is first calculated and then the standard deviation is calculated.

The CV is then calculated by dividing the standard deviation by the mean. The half peak method starts in the same way and proceeds to calculate a mean, but instead of using the standard deviation it measures the width of the peak at half maximum height. The full–width–at–half–maximum (FWHM) is converted to the standard deviation by dividing the FWHM by 2.35. The CV is then calculated as before. By using the FWHM measure, the half-peak method is most accurate when the peak is clean and symmetrical since it ignores the shape of the peak below the half maximum point. Thus, it is possible to obtain low CV values with a peak that is significantly skewed at the baseline. It is important to visually check CV values if they seem unexpectedly low.

Monitoring instrument precision with the forward-angle light scatter signal offers many advantages over other signals. In particular, it is a very bright signal and one that is very sensitive to sample core stream instabilities. An example of a suitable test particle for this measurement is a polystyrene microsphere.

An alternative to the forward-angle light-scatter signal to monitor very brightly- and uniformly-stained particles is a fluorescence-detection PMT. Examples of particles that are or can be stained very brightly and uniformly are fluorescent polystyrene microspheres and cell nuclei, such as those of calf thymocytes. The signal from the particles must be very bright so that the photon counting statistics do not contribute significantly to the CV measurement.

When multiple optical measurements are about to be made it may be necessary to monitor the CV on all detectors. Minimizing the CV on all detectors simultaneously will assure the proper alignment of the flow cytometer.

Accuracy

The accuracy of a flow cytometer instrument measurement is dependent on nearly every component of the flow cytometer instrument. Among them are the fluidic system, the photodetectors, and all of the electronics up to and including the data acquisition system. If the instrument precision is high and calibrated control particles are available, the accuracy of a flow cytometer can be checked by monitoring the linearity and offsets.

For linearly amplified signals, the use of calibrated control particles (either microspheres or biological particles) with intensities distributed over the whole histogram range can facilitate the generation of a plot of control values vs. histogram channel number. If the flow cytometer system is linear, a linear correlation should be evident. For example, if a $1\times$ particle is in channel 100, the $2\times$ particle should be in channel 200, the $4\times$ in channel 400, etc. Additionally, if the intercept on the control-value axis is non-zero, there is an offset that needs to be corrected.

For logarithmically amplified signals, it is important to be sure that the logarithmic transfer function is the same over the whole histogram. This is the equivalent of linearity for logarithmic amplifiers. To check the transfer function, one uses two populations of fluorescent microspheres with close but resolvable intensities. While monitoring the signals from the microspheres with a fluorescence detection PMT, the high voltage on that PMT can be changed to move the pair of microsphere peaks up and down the histogram and record the distance in channels between the two peaks (29, 30). Care should be taken to not lower the PMT high voltage below the point where the PMT becomes nonlinear or to raise the high voltage above the maximum allowed high voltage. If the transfer function is truly logarithmic, the distance between the peaks should not change. Deviations from the ideal transfer function compromise the utility of the data for making quantitative intensity measurements and comparisons. Ideally, one would like to use logarithmic amplifiers that show a true logarithmic transfer function over the entire histogram. Unfortunately, most real-life logarithmic amplifiers show some deviations at the extreme points of both the lower and upper regions of the histogram. Thus, only if significant deviations are observed toward the middle of the histogram should one become concerned and recalibrate the logarithmic amplifiers.

Sensitivity

The sensitivity of a flow cytometer is typically defined as the minimum number of fluorescent molecules, i.e., fluorescein, that the flow cytometer needs to detect in order to distinguish the labeled particles from the nonlabeled particles or electronic noise and optical noise. This is measured by using a few very dimly stained calibrated samples of particles and one sample of an unstained particle. The stained samples are used to calibrate the histogram channels in fluorescent equivalent units (FEU). The peak channel position of the unstained sample is converted to FEU's and reported as the sensitivity (31). An example of this type of analysis is shown in Figure 5.14. Most commercial flow cytometers have reported sensitivities of 1000 FEU's or less.

The above definition of sensitivity does not follow exactly the usual definition of sensitivity. Rather, it is closer to the definition of the minimum available signal-to-noise ratio than to sensitivity. This test for sensitivity is affected by both light-collection efficiency and the amount of background light excluded from the PMT's. In fact, it is possible to get a better sensitivity value by choosing suitable filters and reducing the amount of background light that reaches the PMT's.

When considering what the sensitivity value means for a particular application, it is important to remember that some cells exhibit a large amount of autofluorescence. Typically, the intensity of the autofluorescence for paraformaldehyde-fixed lymphocytes is approximately 4000 FEU's (31). Consequently, a reported sensitivity of 500 FEU's using microspheres may not be realized for lymphocytes. Most flow cytometers would not be able to distinguish between 4000 FEU's and 4500 FEU's. Therefore, optimization of an instrument for the detection of microspheres without due con-

SIGNAL TO NOISE DETERMINATION

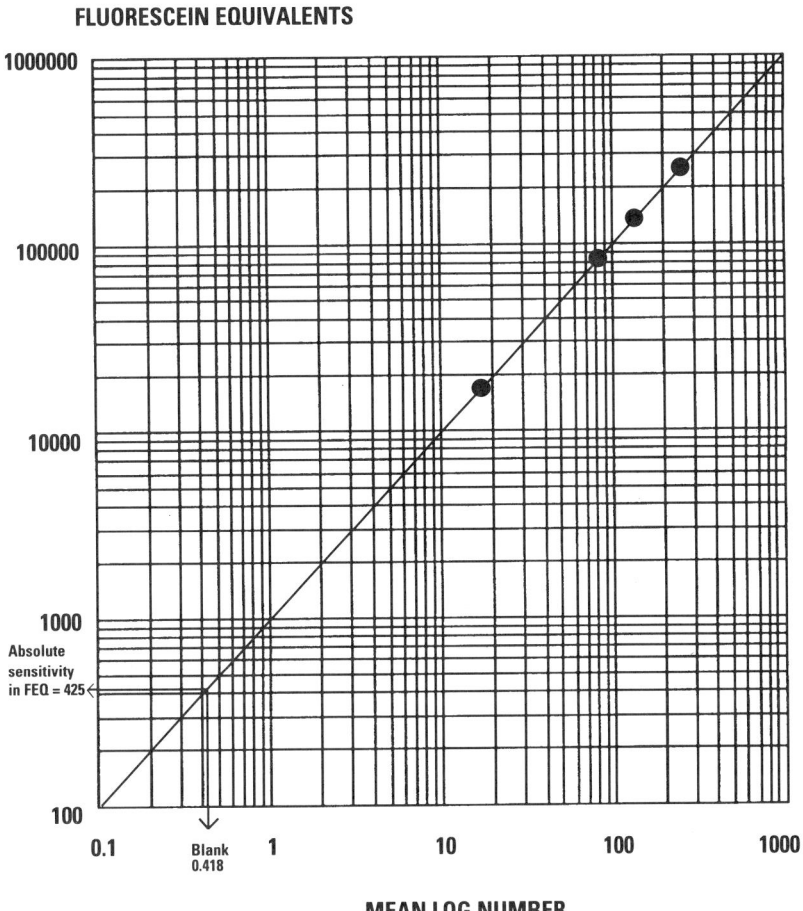

Figure 5.14. An example of the plot required to determine a flow cytometer's sensitivity in FEU's (Courtesy of Coulter Corporation).

sideration of the actual intended biological applications may be a futile exercise, since the limitations are generally biologically related.

Throughput

The throughput of a flow cytometer is an important characteristic that is usually not adequately addressed. This is particularly true when it comes to understanding what factors go into determining the maximum throughput and what are the consequences of exceeding the maximum rate. The primary factors that determine the maximum throughput rate are the pulse width and the allowable miss rate for the data acquisition system. The maximum throughput is simply the minimum of the two rates as determined by the above two factors.

The pulse width is used to determine the maximum rate that cells can pass by the excitation light beam and still have a negligible number of coincidence events. The maximum rate is calculated from the theory of Poisson statistics. If we require that the coincidence rate be 0.5% or less and the pulse width is 10 microseconds, the maximum throughput is approximately 1000 cells per sec. For small coincidence rates, the maximum throughput is approximately inversely proportional to the pulse width. Thus, decreasing the pulse width by a factor of 5 increases the maximum throughput by a factor of 5. Conversely, increasing the pulse width by a factor of 5 decreases the maximum throughput by a factor of 5. The consequence of exceeding the maximum throughput rate is an increase in the number of cells spaced close enough in the core stream to be considered as single cells when they pass through the excitation light beam. As mentioned in the section on doublet correction, there are ways to exclude some of these events from the data set; however, it is better not to have doublets occur since a percentage are always missed. The presence of doublets adversely affects both DNA and immunophenotyping applications.

Determining the throughput rate from the allowable miss rate by the data acquisition system is a problem similar to the above. Again, Poisson statistics are used to determine the number of doublet events that can occur during the time when the data acquisition is blind to further events passing by; i.e., the dead time. In contrast to the above problem, it is not a matter of getting two particles confused as one, rather, the second particle is not seen. For example, if a data acquisition system is capable of synchronously accepting 40,000 events per second, the actual throughput at which only 5% of the cells would be missed would be 4000 events. In this case, for low miss rates, the throughput associated with a given miss rate is proportional to the synchronous acquisition rate of the data acquisition system. The consequence of exceeding this throughput rate is that more cells would be missed, thereby requiring the processing of a larger sample.

It can be seen that for a typical low-power, air-cooled, laser-based flow cytometer with a pulse width of 10 microseconds and a data acquisition system capable of accepting 40,000 events per sec the maximum throughput for the system is 1000 cells per sec and is determined by the pulse width. In contrast, a larger cell sorter with a pulse width of 2 microseconds and a similar data acquisition system would be limited to a maximum throughput rate of 4000 cells per sec, as limited by the data acquisition requirements. Thus, a consequence of using low-power air-cooled lasers and associated design requirements is a reduction in system throughput.

CONCLUSIONS

It can be seen that a large number of factors enter into the design of a flow cytometry experiment. Flow cytometers have been designed with specific applications in mind and we have seen that it is important to understand what impact the design of the flow cytometer may have on experiments to explore new applications. When properly understood, the flow cytometer is a powerful tool that will continue to find itself as the basis of new clinical applications. Furthermore, the continued development and incorporation of new technology will serve to solidify the place of flow cytometry in the clinical laboratory.

REFERENCES

1. Pao RHF. Fluid Dynamics. Columbus, Ohio: Charles E. Merrill Books, Inc., 1967:290.
2. Pinkel D, Stovel R. Flow Chambers and Sample Handling. In: Van Dilla MA, Dean PN, Laerum OD, Melamed MR, ed. Flow Cytometry: Instrumentation and Data Analysis. New York: Academic Press, 1985: 78–129.
3. Kay DB, Wheeless LL. Experimental findings on gynecologic cell orientation and dynamics for three flow nozzle geometries. J Histochem Cytochem 1977;25:870–874.
4. Fulwyler MJ. Hydrodynamic orientation of cells. J Histochem Cytochem 1977;25:881–883.
5. Pinkel D, Lake S, Gledhill BL, Van Dilla MA, Stephenson D, Watchmaker G. High resolution DNA content measurements of mammalian sperm. Cytometry 1982;3:1–9.
6. Harkins KR, Galbraith DW. Factors governing the flow cytometric analysis and sorting of large biological particles. Cytometry 1987;8:60–70.
7. Freyer JP, Wilder ME, Jett JH. Viable sorting of intact multicellular spheroids by flow cytometry. Cytometry 1987;8:427–436.
8. Steen HB. Characteristics of flow cytometers. In: Melamed MR, Lindmo T, Mendelsohn, eds. Flow Cytometry and Sorting. 2nd ed. New York: Wiley-Liss. 1990:11–26.
9. Peters D. A comparison of mercury arc lamp and laser illumination for flow cytometers. J Histochem Cytochem 1979;27:241–245.
10. Shapiro HM. Practical flow cytometry. 2nd ed. New York: Alan R. Liss, Inc., 1988.
11. Wood JCS, Horton AF, Byrne JD, Pedroso RI, Bisnow M, Auer RE. Dual laser excitation flow cytometry: the state of the art. In: Yen A, ed. Flow Cytometry: Advanced Research and Clinical Applications. Boca Raton, FL: CRC Press, Inc., 1989:5–62.
12. Bartholdi MF, Sinclair DC, Cram LS. Chromosome analysis by high illumination flow cytometry. Cytometry 1983;3:395–401.
13. Mathies RA, Stryer L. Single molecule detection: a feasibility study using phycoerythrin. In: Taylor DL, Wagonner AS, Lanni F, Murphy RM, Birge RR, eds. Applications of fluorescence in the biomedical sciences. New York: Alan R. Liss, 1986.
14. Loken MR, Houck DW. Light scattered at two wavelengths can discriminate viable lymphoid cell populations on a fluorescence activated cell sorter. J Histochem Cytochem 1981;29:609–615.
15. Otten GR, Loken MR. Two color light scattering identifies physical differences between lymphocyte populations. Cytometry 1982;3:182–187.
16. Mullaney PF, Van Dilla M, Coulter JR, Dean PN. Cell sizing: a light scattering photometer for rapid volume determination. Rev Sci Instr 1969;40:1029–1032.
17. Leary JF. Laser light scattering detection and characterization of virally infected and transformed mammalian cells [Dissertation]. University Park, PA: The Pennsylvania State University, 1977. 149 p.
18. Loken MR, Herzenberg LA. Analysis of cell populations with a fluorescence activated cell sorter. Ann NY Acad Sci 1975;254:163–171.
19. Salzman GC, Singham SB, Johnston RG, Bohren CF. Light scattering and cytometry. In: Melamed MR, Lindmo T, Mendelsohn, eds. Flow Cytometry and Sorting. 2nd ed. New York: Wiley-Liss, 1990:81–108.
20. Salzman GC, Crowell JM, Martin JC, et al. Cell classification by laser light scattering: identification and separation of unstained leukocytes. Acta Cytol 1975;19:374–377.
21. Sims PJ, Faioni EM, Wiedmer T, Shattil SJ. Complement proteins C5b-9 cause release of membrane vesicles from the platelet surface that are enriched in the membrane receptor for coagulation factor Va and express prothrombinase activity. J Biol Chem 1988;263:18205–18212.
22. Loken MR, Parks DR, Herzenberg LA. Two-color immunofluorescence using a fluorescence-activated cell sorter. J Histochem Cytochem 1977;25:899–907.
23. Gohde W, Schumann J, Fruh J. Coincidence eliminating device in pulse cytophotometry. In: Göhde W, Schumann J, Büchner T, eds. Pulse Cytophotometry II. Ghent: European Press Medikon, 1976:79–85.
24. Sharpless T, Traganos F, Darzynkewicz, Melamed MR. Flow cytofluorimetry: Discrimination between single cells and cell aggregates by direct size measurements. Acta Cytol 1975;19:577–581.
25. Rabinovitch PS, June CH. Measurement of intracellular ionized calcium and membrane potential. In: Melamed MR, Lindmo T, Mendelsohn, eds. Flow Cytometry and Sorting. 2nd ed. New York: Wiley-Liss, 1990:651–668.
26. Murphy RF, Chused TM. A proposal for a flow cytometric data file standard. Cytometry 1984;5:553–555.
27. Data File Standards Committee of the Society for Analytical Cytology. Data file standard for flow cytometry. Cytometry 1990;11:323–332.
28. Lindmo T, Peters DC, Sweet RG. Flow sorters for biological cells. In: Melamed MR, Lindmo T, Mendelsohn, eds. Flow Cytometry and Sorting. 2nd ed. New York: Wiley-Liss, 1990:145–169.
29. Schmid I, Schmid P, Giorgi JV. Conversion of logarithmic channel numbers into relative linear fluorescence intensity. Cytometry 1988;9: 533–538.
30. Parks DR, Bigos M, Moore WA. Logarithmic amplifier transfer evaluation and procedures for logamp optimization and data correction. Cytometry 1988;(Suppl 2):27.
31. Flow Cytometry Standards Corporation. Monograph: fluorescent microbead standards. 1988.

6

Technical Considerations for Dissociation of Fresh and Archival Tumors

CHARLES L. HITCHCOCK and JOHN F. ENSLEY

INTRODUCTION

Flow cytometric analyses of solid tumors require a suspension of single cells or nuclei. These are derived by dissociating samples of fresh tumor or sections of fixed, paraffin-embedded tumor. Discordant results have been reported from flow cytometry studies of such samples (1–4). This problem prevails even when the same methodology is used and arises from an inability to understand or control technical variables. The fact that dissociation methods that are optimal for one tissue may not be optimal for another or even for the same tissue in a different species precludes the establishment of universal procedures. Thus, the end point to be measured and not convenience, past experience, or availability of reagents, should be the major factor determining which tumor dissociation method to use. Table 6.1 depicts the basic criteria that a method for dissociating a solid tumor should meet (5). These criteria are applicable for flow cytometry studies as well as for analysis of cell kinetics and clonogenicity.

In the absence of a universal dissociation method, one must be keenly aware of the variables that affect sampling and subsequent dissociation of tumors, as well as the criteria needed to evaluate these technical differences. The goal of this chapter is to discuss the variables encountered with dissociation of both fresh and fixed, paraffin-embedded tissues, and their effect on flow cytometric analysis, particularly DNA analysis.

SAMPLING

Obtaining a representative sample is the first step in processing a solid tumor for subsequent study. There are several questions to ask when evaluating a sampling method. First, does it yield an amount of material that is sufficient to represent the in vivo characteristics of the tumor? Can the technique be used for both in vivo and in situ sampling? Can several sites be sampled? And finally, what level of expertise is needed for the sampling technique?

The following points have to be considered in order to ensure that the first question is affirmatively answered: (a) Sampling specific areas of larger lesions helps to reduce dilution of the neoplastic cells with stromal and inflammatory elements within the tumor (6); (b) single-site sampling techniques, such as scraping the cut surface of a resected tumor (7–10), should be avoided whenever possible because those techniques may miss subpopulations of tumor cells and cause excessive damage to those obtained (7); (c) sampling multiple sites minimizes the impact of heterogenous tumor cell populations. (This is especially important for sarcomas (11) and carcinomas of the ovary (12), breast (13–15), colon (7, 16), esophagus (17), bladder (18), lung (19), and kidney (20).)

Biopsy

The clinical situation may call for a lesion to be biopsied rather than excised. This is exemplified by the use of endoscopic biopsies to sample malignant and premalignant lesions of the gastrointestinal tract (7, 21–23). Biopsies performed for diagnostic purposes may only sample a limited part of the tumor; therefore, it is important to compare the results of the biopsy and the resected tumor whenever possible (24). As with resected lesions, biopsy specimens should be kept moist at 4°C and sampled as quickly as possible after removal. Such handling slows autolysis and degradation of cellular proteins (25, 26). Areas of grossly nonviable tumor, necrosis, and fat should be removed prior to further sampling and processing. This minimizes the debris and nonviable tumor that may obscure detection of subpopulations of tumor cells as well as impair biochemical and clonogenic assays.

Fine-Needle Aspiration

Fine-needle aspiration (FNA) is a versatile technique for sampling lesions as well as a gentle mechanical dissociation method that encourages extraction and dispersal of single and cohesive plugs of dysplastic and tumor cells (27, 28).

Table 6.1
Criteria for Evaluating Dissociation Techniques

Optimal cell yields per gram of tissue
No preferential selection of cell subpopulations
Retention of desired behavioral, morphologic, molecular, and biochemical phenotypes
Minimal debris and aggregation of cells or nuclei

The size, location, and nature of the lesion directly influences the amount of diagnostic material in a FNA sample. In addition, the number of passes made, the experience and skill of the person performing the aspiration, and the cytopathologist's skill also influence sample adequacy. On-site cytological assessment of the aspirate is an essential guide for determining sample adequacy. Vindelov (29) recommended that a hemocytometer be used to quantitate cell yield and that aspiration continue until 10^6 cells in 200 µl are obtained. Sufficient material is obtainable from lymphoid lesions (30) and testicular aspirates (31, 32); however, it may be more difficult to obtain sufficient amounts from denser lesions. Joensuu et al. (33) obtained as few as 600 cells from thyroid aspirates and concluded that, although DNA ploidy analysis can be made on this low number of cells, a minimum of 3000 cells is needed for an accurate assessment of the cell cycle. In a study of over 300 breast FNAs from women at high risk for breast carcinoma, Martino et al. (28) reported a sufficient cell yield from up to 97% of breast masses for flow cytometric DNA-content analysis.

FNA sampling combined with flow cytometry is useful in a number of clinical settings. Immunophenotyping of FNA samples is reported to improve the diagnostic accuracy of autoimmune endocrinopathies and lymphadenopathies (34–37). DNA analysis of testicular aspirates is useful for analysis of spermatogenesis (30, 31) and can identify males at increased risk for developing testicular carcinoma (38). FNA is being used with increased frequency for in vivo and in situ sampling of solid tumors. The needle's small bore promotes global sampling of a lesion. This minimizes sampling error and helps to characterize the in-situ heterogeneity of tumors (11, 14, 16). Flow cytometric analysis of FNA samples is a diagnostic adjunct for studying breast lesions (27, 28, 39–42) as well as a tool for monitoring treatment response in women with breast cancer (43). In addition, slide preparations from FNA samples can be used for either immunohistochemistry or for static cytofluorometric analysis. The retention of cellular architecture by FNA permits cytologic and histopathologic correlation with flow cytometric or image analysis. This result has led to reported increases in diagnostic sensitivity and specificity for thyroid tumors (33) and palpable metastatic lesions (44). Not all solid lesions are amenable to this combined approach. Fuhr (45) concluded that flow cytometric analysis of FNA samples is an excellent screening technique for lung and breast tumors, but is only of marginal use for liver tumors. Also, the presence of aneuploid cell populations in histologically benign lesions precludes its use in screening thyroid and pancreatic lesions.

DISSOCIATION IN FRESH-TISSUE SAMPLES

Tumor dissociation involves the disruption of intercellular cohesion and the extracellular matrix. Four types of techniques can be used alone or in combination to accomplish this. These include: (a) chemical, (b) mechanical manipulation, (c) enzymatic digestion, and (d) enucleation techniques. The advantages and disadvantages of each have been critically reviewed (46–50) since first described by Rous and Jones in 1916 (51). Each method damages cells; the extent of the damage varies with the tissue. Ideally, a dissociation protocol should be individualized for the tumor of interest and evaluated relative to cell yields and the end point to be measured. This may not be practical in the clinical setting.

Chemical Dissociation

Chemical dissociation, or chelation, of solid tumors is designed to disrupt or sequester the Ca^{+2} and Mg^{+2} ions needed to maintain the intercellular matrix and cell surface integrity (52). Commonly used chelating agents include ethylene-diaminoacetate (EDTA), ethyleneglycol (2-aminoethylether)-N-N¹-tetraacetic acid (EGTA), tetraphenylboron (TBP) or citrate ion (53, 54).

Chelation by itself does not adequately dissociate many types of tissue (55). Solid-tumor dissociation by chelating agents is characterized by low-cell yields, good retention of morphology, and poor clonogenicity due to altered biochemical activity (5). Thus, these agents are best used in conjunction with mechanical or enzymatic dissociation procedures.

Mechanical Dissociation

Mechanical dissociation uses shearing forces as a quick and simple means to break intercellular bonds and to release cells from the extracellular matrix. This process may involve either one or a combination of the following techniques: repeated mincing with scissors or sharp blades, homogenization, scraping a surface, filtration through a nylon or steel mesh, vortexing, repeated aspiration through pipettes or small-gauge needles. These methods can be used easily to obtain a single-cell suspension from lymph nodes, spleen, and thymus. This contrasts with the variable results from the mechanical dissociation of solid tumors.

Enzymatic Dissociation

Enzymes are commonly used to dissociate tumors for studies requiring a high yield of viable cells with near-normal morphology and function (56–58). To meet these needs, enzymes must hydrolyze the various proteins, glycoproteins, lipids, and glycolipids that form the extracellular matrix as well as disrupt intercellular cohesion. Enzymes that can be used for this fall into one of three categories: nonspecific proteases, proteases specific for elastic and collagenous fibers, and hydrolytic enzymes specific for mucopolysaccharides. The use of crude enzymes, whose composition is often undefined (59), has led to confusion in the literature as to which enzyme is best used to dissociate a particular tumor. This problem is best exemplified by trypsin and collagenase; they are the two proteolytic enzymes commonly used for tumor dissociation.

In its pure form, trypsin is a serine protease that hydrolyses peptide bonds involving carboxyl groups of arginine and lysine. Yet this form is ineffective in dissociating tissue due to little selectivity for extracellular matrix proteins. Similarly, purified collagenase is usually inefficient for dissociating tissue due to incomplete hydrolysis of collagenous peptides and inactivity toward other proteins in the extracellular matrix. The dissociative properties of commercial preparations of these two enzymes is due to variable concentrations of contaminating protease, nuclease, lipases, and polysaccharidases. Lot-to-lot variation in the activities and toxicity of contaminants has given rise to conflicting results for both of these enzymes. Other enzymes used to dissociate tumor include pronase, chymotrypsin, pepsin, papain, elastase, and hyaluronidase. DNase is commonly used to hydrolyze the gel-like DNA-protein complexes that arise from damaged and dead cells and often entrap viable tumor cells. The activity of these enzymes is also influenced by lot-to-lot variations in contaminating enzymes.

Enzymatic dissociation may have a deleterious effect on tumor cells (47, 59). This is minimized by monitoring such technical variables as the type and concentration of enzyme used, cell concentration or wet tissue weight, the medium and its pH, and the duration and temperature of digestion. In addition, these factors should be optimized to maximize cell yield while ensuring retention of the parameters being measured. This means that an investigator should evaluate different tumor dissociation protocols in order to obtain a suspension of cells that conforms to the end point being measured.

Cell Enucleation

Enucleation of tumor cells is a quick and simple means to dissociate a tumor and is commonly used in clinical flow cytometry laboratories performing DNA analyses (60, 61). Typically, initial mechanical disruption of the tumor is performed prior to enucleation. Cell lysis and enucleation can be accomplished using hypotonic or hypertonic solutions, nonionic detergents, or proteolytic enzymes alone or in combination. The resulting nuclear suspensions yield DNA histograms characterized by low coefficients of variation (CV) (under 4% in our experience) and minimal debris. These same suspensions are suitable for flow cytometric analysis of nuclear antigens (62), oncogene products (63), and BrdU staining (64). Enucleation precludes analysis of surface and cytoplasmic antigens and obviates the use of these antigens as markers to assess the loss of specific subpopulations of tumor cells.

The first enucleation-DNA-staining solution reported for flow cytometry used a hypotonic citrate-propidium-iodide (PI) solution designed to eliminate the need for fixation and RNase treatment of the released nuclei (65). This technique resulted in nonspecific PI staining of nucleoli, retention of cytoplasmic elements, nonlysis of some cell types, and poor staining stability beyond six hours. Improved osmotic lysis and DNA staining have been reported using a hypotonic Tris-MgCl$_2$-PI solution plus RNase (66).

Detergent lysis of cells with a nonionic detergent in combination with hypotonic (67, 68), hypertonic (69), or isotonic solutions (60, 70), or with a proteolytic enzyme (61) provides a reproducible means to enucleate unfixed or fixed cells. Easy to use, one-step enucleation and DNA-staining techniques have been developed by combining a detergent lysis solution with a DNA fluorochrome (60, 67, 68, 70). These are easy-to-use solutions in which a fresh tissue sample is minced and the resulting nuclei removed for analysis. Their use minimizes or eliminates the need for centrifugation, a major cause of cell clumping (71), and has found widespread clinical use (72, 73).

Vindelov (61) designed a detergent/enzyme enucleation and DNA-staining procedure, applicable to most fresh solid tumors, that has found widespread acceptance in the clinical setting (29). The use of a stock solution containing the nonionic detergent NP-40 to lyse cells, spermine tetrahydrochloride to stabilize unfixed nuclei, and citrate to chelate divalent cations is central to this procedure. Initial trypsinization digests cytoplasmic fragments that can contribute to variation in staining intensity. Enzymatic digestion is inhibited using a trypsin inhibitor followed by RNase, which digests double-stranded RNA. DNA staining is accomplished with PI. Careful adherence to the protocol is needed to obtain the best results (29). Low levels of spermine result in increased debris and high CVs for DNA histograms; an increased spermine concentration or too little trypsin results in clumping of nuclei.

Evaluating Dissociation Methods

The evaluation of a tumor dissociation protocol must take into consideration the criteria in Table 6.1, the type of tumor, and the relevant technical variables. The efficiency of tumor dissociation to yield an adequate number of cells for analysis varies with the tumor as well as with the dissociation protocol used (9, 50, 57, 74–78). When compared to other dissociation techniques, enzymatic dissociation is generally considered to result in increased cell yields; the significance of this difference depends on the tumor. Enzymatic dissociation of squamous cell (74, 76) and colon carcinomas (79) results in a significant increase in cell yields (Table 6.2). This difference is not significant in cases of carcinomas of the breast, ovary, and lung (50, 57), melanomas, (50, 57) and some sarcomas (57, 77). Varying such technical factors as the type of enzyme, its concentration, and duration of digestion will also alter cell yields (77, 79).

Recovery of specific tumor cell subpopulations may vary with the tumor dissociation protocol. When using DNA histograms as the end point being measured, a significant difference in the ratio of DNA-aneuploid to DNA-diploid cells is obtainable with different tumor dissociation protocols (Figs. 6.1–6.3). Compared to mechanical dissociation, enzymatic dissociation can yield a decrease in the relative proportion of DNA-aneuploid cells obtained from different tumors (50, 74–77, 79–81). This can be attributed to an increased release of inflammatory and stromal cells, which dilutes the DNA-aneuploid cell population or selectively destroys them

Table 6.2
Cell Yields from Dissociation of Fresh Solid Tumors

Tumor	N	Mechanical Dissociation (mean ± S.E)	Enzymatic Dissociation (mean ± S.E)	P-Value
SCCHN[a]				
Total cell yield	77	14.1 ± 2.6	55.5 ± 6.1	0.0001
% Cell viability	77	46.2 ± 3.3	91.0 ± 1.3	0.0001
CV	76	4.20 ± 0.16	4.18 ± 0.18	0.692
Colon Carcinoma				
Total cell yield	25	20.6 ± 4.1	37.5 ± 6.2	0.01
% Cell viability	25	49.7 ± 4.8	87.7 ± 2.5	0.01
CV	25	3.81 ± 0.20	4.81 ± 0.41	0.01

[a]SCCHN—squamous cell carcinoma of the head and neck; Total cell yield in 10^6/gr tissue; CV—G0/G1 peak coefficient of variation; % Cell viability by dye exclusion (81).

(81). This dilution effect may be considerable, such that a DNA-aneuploid population would go undetected (Figs. 6.1 and 6.2) (50). The effect of tumor type on the yield of subpopulations of tumor cell is noted in results reported by Ensley and coworkers (74, 76, 77, 81). They found that unlike colon carcinoma (Fig. 6.2) and sarcomas (Fig. 6.3) mechanical dissociation of squamous-cell carcinomas of the head and neck (SCCHN) frequently resulted in an apparent loss of DNA-aneuploid cell subpopulations when compared to a cell suspension obtained by enzymatic dissociation of the same tumor. In over 80% of the tumors studied, either a reduction or loss of DNA-aneuploid cells occurred (Fig. 6.1). Changing the conditions of enzymatic dissociation can also have a significant impact on the recovery of tumor cell subpopulations. This is exemplified by the results with colon carcinoma (Fig. 6.4) and by the studies of Bijman et al. (82) with SCCHN.

The choice of a tumor dissociation protocol is dependent on the retention of the cellular phenotype under study. Mechanical and chemical dissociation protocols often yield low cell viability, altered biochemical functions, and increased debris and cell clumping (9, 50, 57, 75). Thus, enzymatic dissociation is the method of choice for studies of cell kinetics and clonogenicity. The impact that a dissociation protocol will have on these two parameters will vary with the tumor and with the conditions of enzymatic digestion. Costa et al. (50) reported that, regardless of the dissociation protocol used, cell proliferation, as measured by incorporation of ^{3}H-thymidine (TLI), was lower in dissociated melanomas and carcinomas of the ovary and breast than in in situ tumor cells. The extent of this difference varied with the tumor. In contrast, using a murine KHT tumor model, Pallavicini et al. (83) reported no significant difference in the TLI of tumor cells in sections or in suspension after tumor dissociation with either trypsin or neutral protease. McDivitt et al. (9) reported an excellent correlation (r = 0.847) between TLI and the S-phase fraction of collagenase-dissociated breast tumors. This same level of concordance was obtained by two different estimations of the TLI of the same specimens. These results demonstrate the importance of controlling the conditions of enzymatic dissociation so as to preclude a preferential loss of proliferating cells. Clonogenic assays require viable cells to retain their clonogenic efficiency. Variability in this efficiency is attributable to the type of enzyme used, its concentration, the duration of digestion, and its inherent cellular properties (78).

Few studies have evaluated enucleation techniques. Ensley and colleagues, using a murine model of SCCHN, compared results from three enucleation methods (29, 71, 84) with those obtained with enzymatic and mechanical dissociation of the same tumor. In contrast to enzymatic and mechanical dissociation, they found that tumor enucleation resulted in a 90% loss of nuclei, a preferential loss of tumor cell subpopulations, and artefactual pseudo-hyperdiploid DNA histograms. Not all enucleation methods give the same results. Using paired specimens, Risberg et al. (85) compared Thornthwaite's detergent lysis method (71) with Vindelov's detergent/enzyme enucleation technique (61) and noted marked differences in every DNA histogram parameter measured. Vindelov's technique was found to increase the release and detection of DNA aneuploid nuclei, while decreasing: G0/G1 peak CVs, DNA indices (DI), S-phase fraction (SPF), and percent G2/M. Decreased SPF and G2/M percentages were attributed to decreased debris and aggregate formation following detergent/enzyme enucleation.

Evaluating Methods for Handling Dissociated Cells

FIXATION

Flow cytometric results can vary with the fixative used; thus, it is recommended that fixation techniques, along with dissociation techniques, be evaluated. Commonly used fixatives include: formalin, paraformaldehyde, ethanol, glutaraldehyde, methanol, and 20% acetic acid. Varying the fixative will vary staining intensity (86, 87) as well as the quality of the results. Our experience with ethanol fixation indicates that 50% pure (Gold Shield) ethanol yields up to 100% cell recovery after 60 min compared to less than 40% when samples are fixed with 50% denatured ethanol. We have also noted that fixation with denatured ethanol can produce broad CVs and pseudo-DNA-aneuploid peaks (Fig. 6.5). Because of these variations in DNA fluorochrome binding stoichiometry, it is treacherous to combine data that results from different fixation techniques (49, 74).

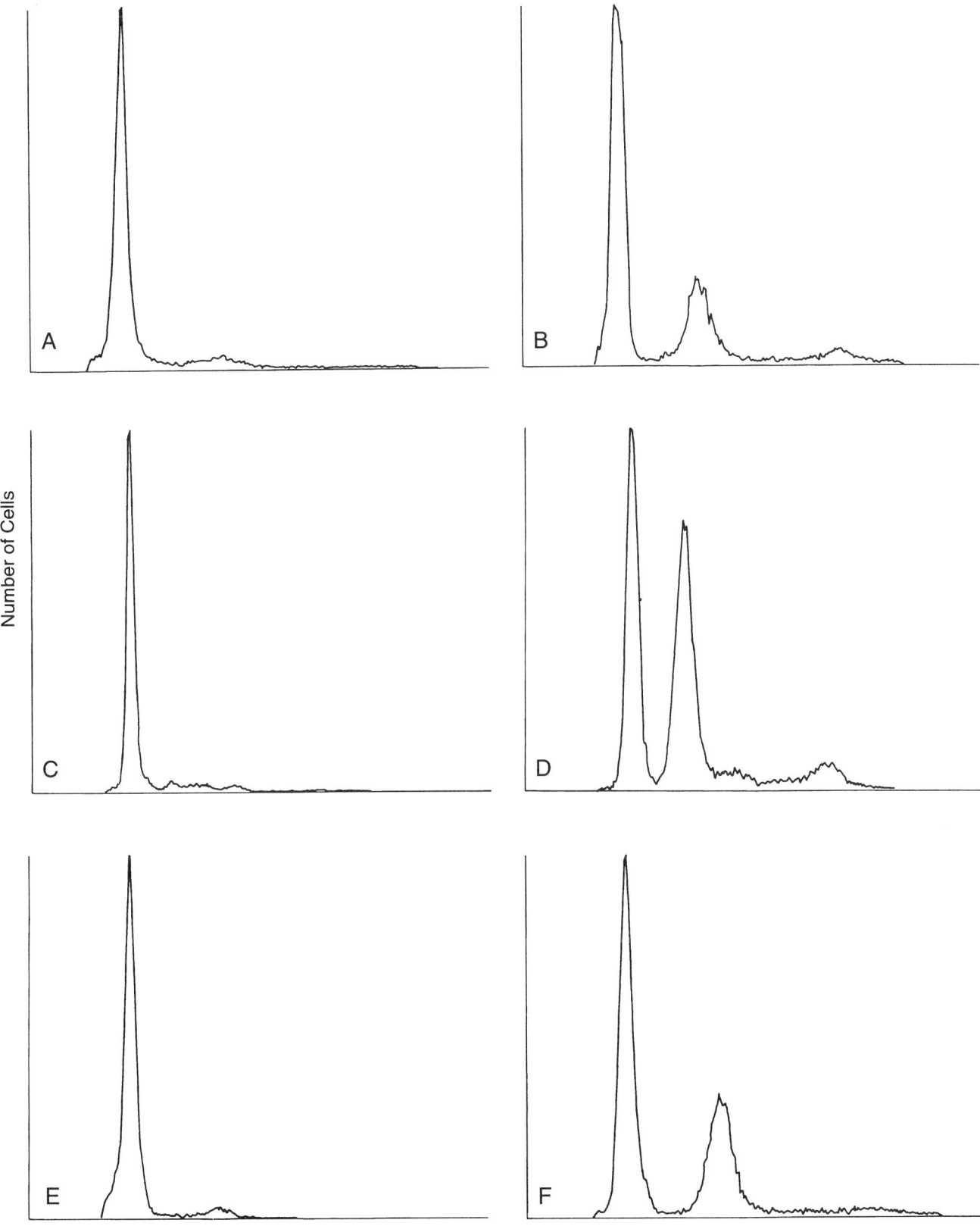

Figure 6.1. DNA histograms from three different squamous cell carcinomas of the head and neck dissociated by mechanical (**A, C,** and **E**) or enzymatic (**B, D,** and **F**) means (G_0/G_1 Peaks: **A** – 51, **B** – 52 & 99; **C** – 56, **D** – 52, 99; **E** – 50, **F** – 55 & 110) (81).

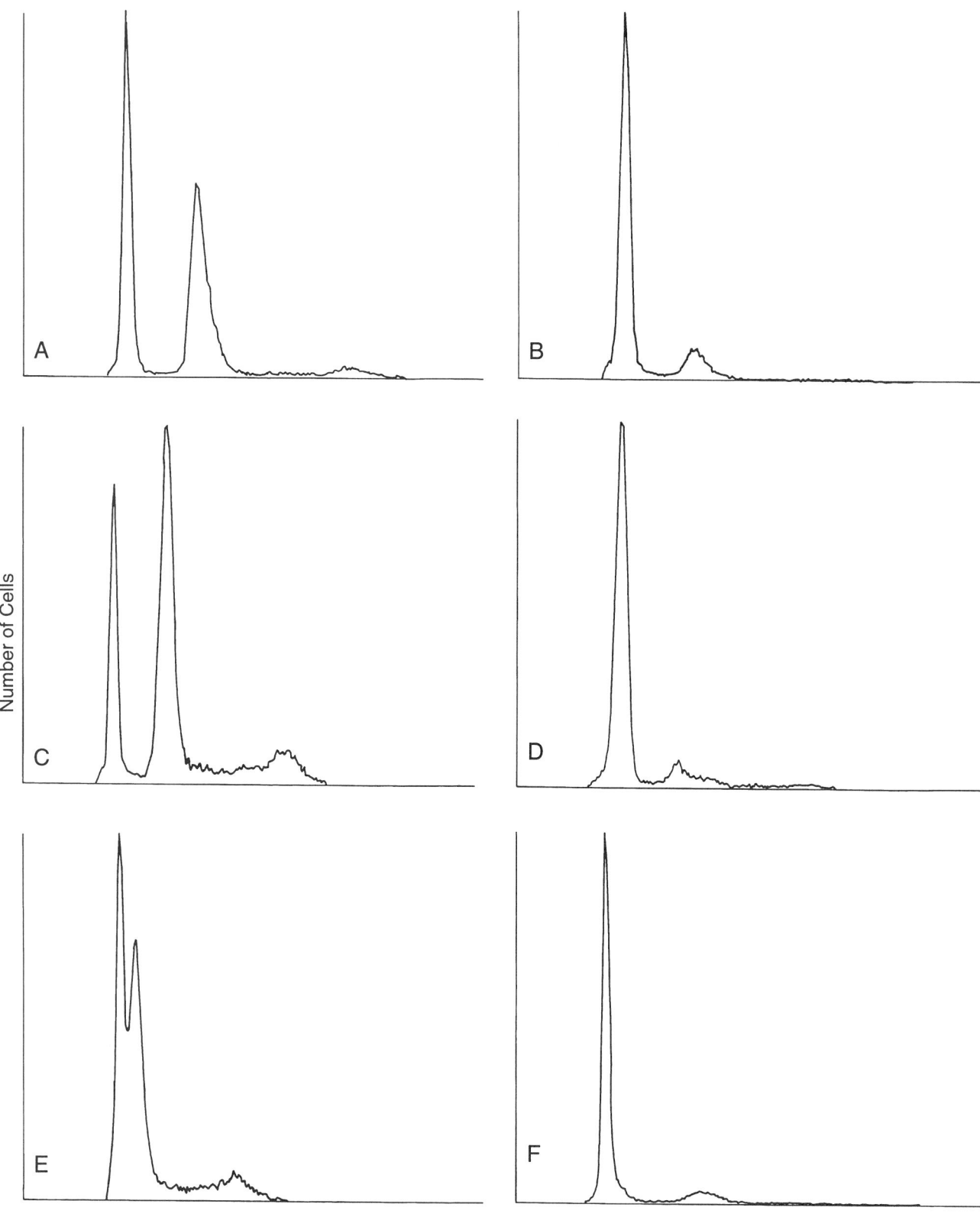

Figure 6.2. DNA histograms from three different colon carcinomas dissociated by mechanical (**A, C,** and **E**) or enzymatic (**B, D,** and **F**) means (G_0/G_1 Peaks: **A** — 59 & 100, **B** — 60; **C** — 50 & 81, **D** — 59; **E** — 56 & 64, **F** — 58) (81).

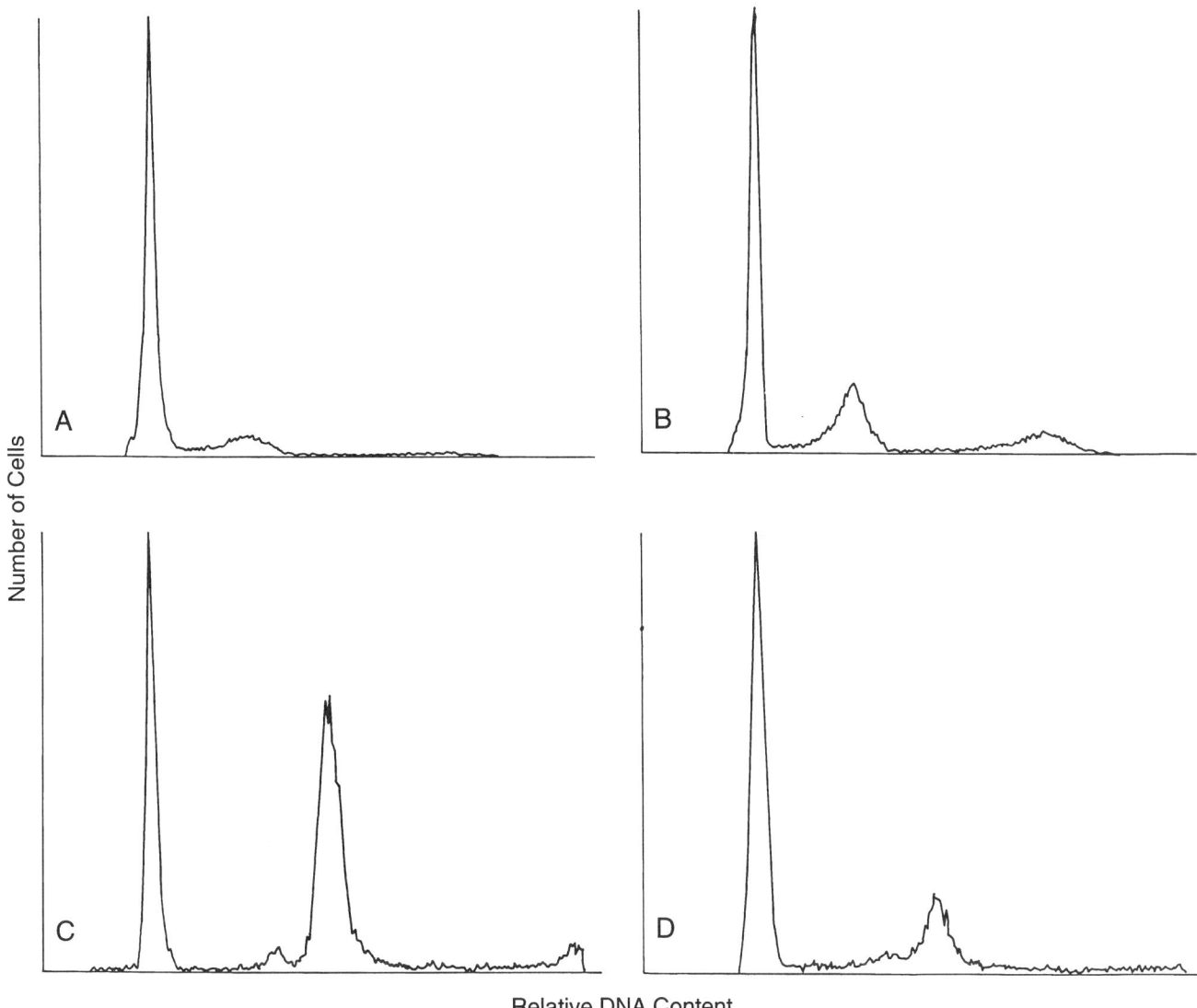

Figure 6.3. DNA histograms comparing mechanical (**A**) and enzymatic (**B**) dissociation of a human soft tissue sarcoma as well as mechanical (**C**) and enzymatic (**D**) dissociation of a human osteosarcoma.

Fixation can also lead to clumping, lysis, and loss of cells. Petersen (88) reported that ethanol fixation of colorectal carcinoma cells resulted in a cell recovery of only 45% after mincing and concluded that loss of DNA-aneuploid cells occurred after fixation. This can be minimized by vortexing the cell suspension at the same time that the fixative is being added in a dropwise manner.

STORAGE

Various cryopreservative solutions have been developed that permit the prolonged storage of samples and dissociated cells prior to processing. Storage of FNA samples for up to five years at −80°C was reported (89) using Vindelov's solution (250 mM sucrose, 40 mM citrate, and 50 ml of DMSO per liter of water). Stone et al. (90) found no significant differences in cell loss or DNA staining in samples stored for up to one month at −70°C in CMRL 1066 media containing 20% fetal calf serum, 1.765 gm NaCl, and 14.5 ml DMSO. Alanen et al. (91) reported improved cell yields and CVs after FNAs were fixed and stored in a solution containing a 1:1 mixture of citric-acid–buffered saline (CABS 85.3 gm sucrose, 11.8 gm trisodium citrate, 50 ml DMSO per liter of water, pH 7.6.) and 99% ethanol. Cells may also be stored for varying lengths of time in ethanol (74) or enucleation media (71) without significant alteration in flow cytometric results.

DEFINING DNA-ANEUPLOID CELLS

One can evaluate protocols for tumor dissociation and subsequent sample handling in the context of their ability to provide a reference cell population for determining the position of the DNA-diploid peak position. An external reference cell can be used. It is recommended that this be added to the suspension of dissociated tumor cells prior to further pro-

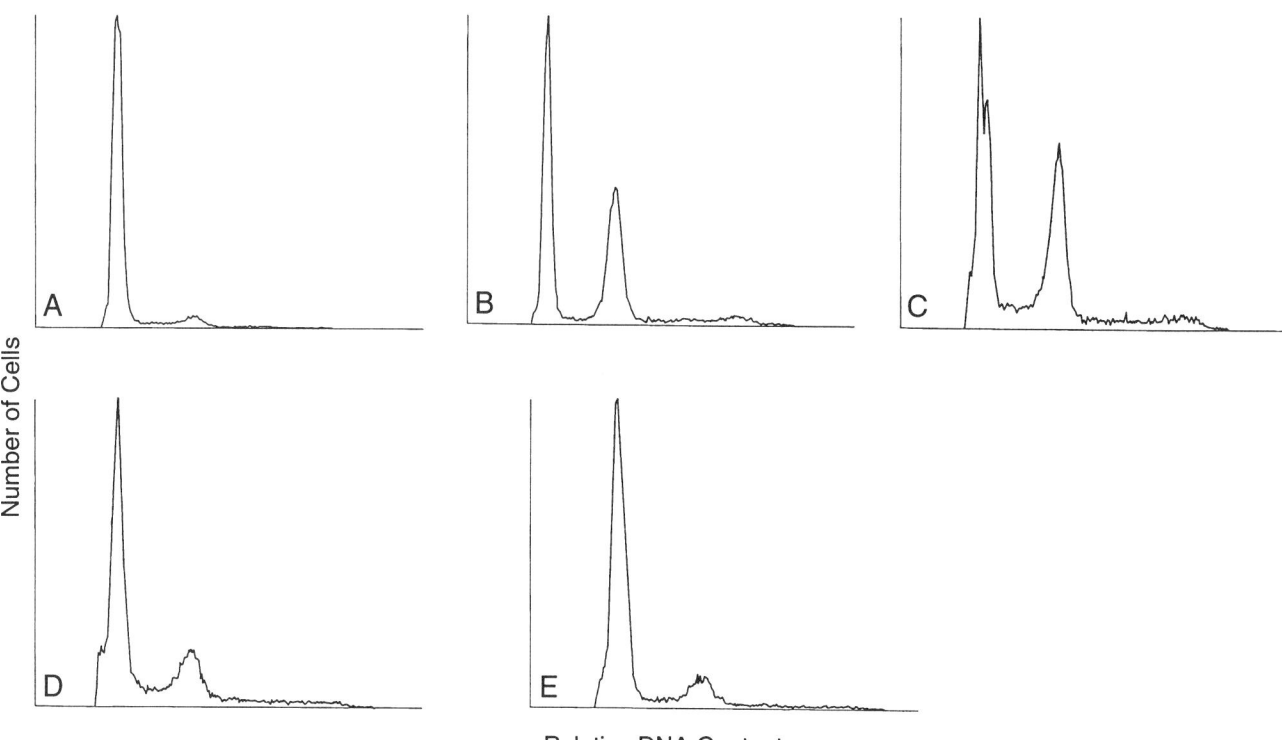

Figure 6.4. DNA histograms from a representative sequential time incubation study of human colon carcinoma with 0.1 mg/ml collagenase II; normal colonic mucosa (**A**) (G_0/G_1 Peak − 55), mechanically dissociated (**B**) (G_0/G_1 Peaks − 55 & 100), 30-min incubation (**C**) (G_0/G_1 Peaks − 55 & 100), 60-min incubation (**D**) (G_0/G_1 Peaks − 55 & 100), and 120-min incubation (**E**) (G_0/G_1 Peaks − 55 & 105) (81).

cessing and staining (92); however, normal DNA-diploid cells in the sample are the best reference population. Although the use of these cells minimizes the impact of processing-induced artifacts, DNA staining intensity of normal human diploid cells may vary after fixation (86) or enucleation (92). This problem is overcome by the use of intact whole cells for concurrent analysis of DNA and cell surface or cytoplasmic antigens. Park et al. (83) combined propidium iodide (PI) and anti-CD45 staining to identify the DNA-diploid lymphoid elements in pleural effusions and lymph nodes containing tumor cells. The use of mechanical techniques insured that the CD45 antigens were not altered during dissociation. Introducd by Feitz et al. (94), the use of labeled antibodies to cytokeratins used together with PI increases the sensitivity of detection of DNA-aneuploid cells in cells dissociated from a solid tumor (79, 95). These dual-staining protocols permit one to negatively or positively select the epithelial component in the tumor and limit cell cycle analysis only to the tumor cells.

DISSOCIATION OF SECTIONS OF FIXED, PARAFFIN-EMBEDDED TISSUE

Since its introduction in 1983, there has been a plethora of articles written using the technique of Hedley and coworkers (96) for dissociating nuclei from fixed, paraffin-embedded tissue. This approach has both prognostic and diagnostic implications. The use of stored paraffin blocks facilitates retrospective studies determining the prognostic significance of DNA content and cell proliferation in well-defined patient groups with known clinical follow-up (96–98). It is a valuable tool in accessing the clinical significance of these parameters in rare and uncommon lesions (99–103). Its diagnostic applicability is seen in the case of products of conception where specific villous and fetal structures must be sampled to determine the presence or absence of a DNA-triploid cell population (104). This ability to sample specific histopathologic regions was used by Bauer et al. (105) to demonstrate that the expression of the p105 nuclear protein differs among closely associated tumor cells. Care should be taken to avoid the routine use of this type of archival material as samples from individual patients. This may not be avoidable in cases where the sample has to be submitted in toto for histopathologic examination or when the pathologist does not realize the need for flow cytometry at the time of sampling.

One must weigh the disadvantages and advantages when deciding whether or not to undertake these types of studies. Flow cytometric studies of nuclei dissociated from sections of fixed, paraffin-embedded tissue are characterized as having wide peak CVs and increased debris and as lacking an adequate internal standard for DNA-ploidy determination. Each of these disadvantages has an impact on results.

DNA histograms from the dissociated nuclei of archival material are often associated with asymmetric and wide peak CVs when compared to fresh samples from the same tumor (Table 6.3). One source for this is the fixation-induced

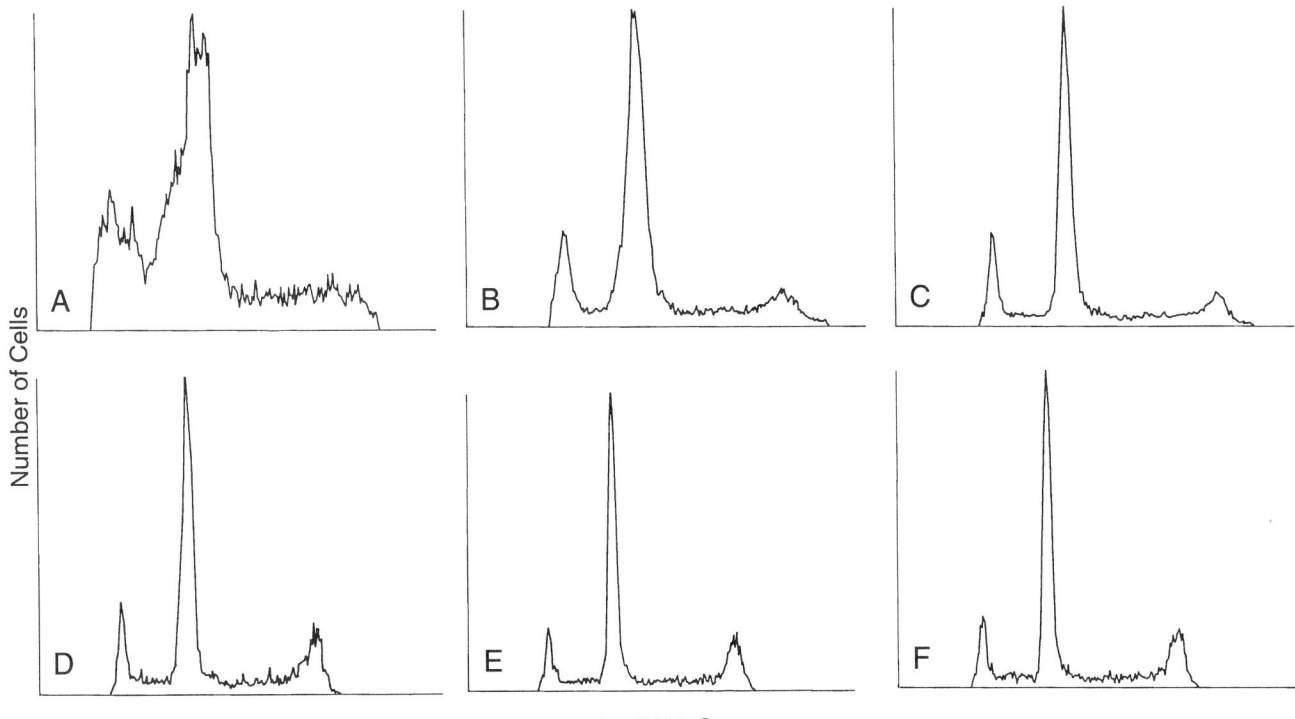

Figure 6.5. DNA histograms of ethanol-fixed, enzymatically dissociated tumor cells from three squamous-cell carcinomas of the head and neck. Upper panel—dissociated tumor cells fixed in 50% denatured ethanol (CVs: **A** – 6.14, **B** – 6.15, **C** – 6.06). Lower panel—the same dissociated tumor cells fixed in 50% Gold Shield ethanol (CVs: **D** – 2.14, **E** – 2.01, **F** – 2.00).

crosslinking of nuclear proteins, which alters the binding of DNA fluorochromes. This is further altered after enzymatic dissociation of the sections. Debris is another source. Together, they can reduce one's ability to detect a tumor cell subpopulation with near-diploid DNA content (111).

Debris arises from necrotic tissue as well as from fragments of cut nuclei and stroma. For paraffin-block sections, the amount of debris is further influenced by fixation, tissue cellularity, section thickness, and enzymatic digestion. Debris may obscure detection of DNA-aneuploid cell populations present at low frequencies. Furthermore, increased debris induces cell/nucleus aggregation, which may contribute to artefactual DNA-aneuploid or DNA-tetraploid peaks. The fact that debris forms a continuum that underlies the DNA distribution affects the accuracy of cell-cycle analysis in an otherwise interpretable DNA histogram. Computer algorithms have been developed to subtract debris (112) and aggregates (113) from DNA histograms so as to minimize their impact on cell-cycle analysis. Although these subtraction algorithms reduce the impact of debris and aggregates, the S-phase fraction (SPF) is generally higher in archival samples than in fresh samples (Table 6.3) (113, 114).

An external or internal DNA-diploid reference cell is often needed to verify the presence and location of DNA-aneuploid cell populations in a DNA histogram. The effects of fixation and enzymatic digestion on DNA staining preclude the use of a reliable external standard and necessitates the use of an internal standard (97, 108). A variety of DNA-diploid reference cells have been used. Arber et al. (115) demonstrated that the position of the G0/G1 peak varied among tissues as well as among the same type of tissues from different patients. This lack of reproducible DNA staining negates the use of blocks of nonmalignant tissue from another patient. Optimally, one should use the nonmalignant cell population found in a paraffin block containing both malignant and nonmalignant tissue as an internal standard. Both tissues can be separated and analyzed independently as well as admixed.

Studies comparing DNA histograms derived from fresh and fixed, paraffin-embedded samples from the same tumor provide valuable information to evaluate the usefulness of

Table 6.3
DNA Distributions from Fresh and Archival Material

| Tumor Type | Mean CV | | Concordance | | Ref # |
	Fresh	Archival	DI (r)[a]	%S (r)	
Bladder	3.7	7.3	0.957	—	106
Breast	3.5	5.1	0.985	0.886	107
Colon	3.6	5.0	0.988	0.588	108
Lymphoma[b]	3.8	4.0	>0.95	0.940	109
	4.4	4.5			
Various[c]	3.7	5.2	>0.95	—	110
SCCHN[d]	3.8	4.4			

[a] r = coefficient of correlation.
[b] Lymphomas stained with DAPI.
[c] Lymphomas stained with PI.
[d] SCCHN—squamous cell carcinoma of the head and neck.

this technique. Differences in tumor DNA histograms between fixed, paraffin-embedded, and unfixed samples were attributed to: tumor heterogeneity (e.g., sampling differences) (116, 117), and/or to fixation (118), tissue dispersal technique (e.g., enzymatic digestion) (75), and DNA-binding fluorochromes (119). In general, the percentage of DNA-aneuploid nuclei in archival samples was lower than that in fresh tissue (55, 114), but the detection of DNA-aneuploid cell populations was similar in both samples (Table 6.3). Concordance in SPF values (Table 6.3) has been associated with the DNA ploidy of the tumor cells (109), and the method used for debris subtraction (114). Studies (1, 3, 4) comparing DNA staining of paraffin sections from the same formalin-fixed tissue suggest that differences in DNA histograms indicate that technical variables are a significant source of inter- and intralaboratory differences.

Hedley's original technique (96), calling for 30-μm-thick sections to be deparaffinized and rehydrated, enzymatically dissociated, and to have their nuclei stained with DAPI, has been extensively modified, and extended to include immunological staining of nuclear antigens (63, 105). Several studies (3, 4, 79, 97, 108, 111, 119–125) have investigated the impact on DNA histograms of changes in such processing variables as fixation; section sampling and thickness; deparaffinization and rehydration protocols; enzymatic digestion conditions; and DNA staining. The remainder of this section discusses points to consider when modifying the steps in this methodology.

Initial Tissue Handling

The use of sections from fixed, paraffin-embedded solid tumors for flow cytometric analysis requires careful control of initial processing techniques (126). This begins with attention to the tissue itself and to its sampling. The effect that fixation has on DNA staining varies among tissues. Kallioniemi (127) reported that the histograms of breast tumors were superior to those generated from thyroid, pancreas, and colon tumors. Similarly, McIntire et al. (128) reported lower CVs from lymph node controls than from prostatic carcinomas.

Tissue sampling is often overlooked. Necrotic and hemorrhagic tissue should be trimmed prior to fixation in order to reduce their contribution to debris and limit the interpretation of DNA histograms. The processing of multiple samples ensures the detection of heterogenous populations of DNA-aneuploid cells (19, 129–131). Kallioniemi (127) reported that intratumor heterogeneity for DI occurred in 13% of breast tumors and in 15% of ovarian cancers. Similar heterogeneity was reported in the SPF of breast and ovarian tumors.

Fixation

Tissue fixation is a critical factor in determining the quality of flow cytometric results. It significantly affects the intensity of DNA (25, 86) and nuclear-protein staining (63, 121). The impact of fixation on these parameters varies with the fixative and the size and density of the tissue.

Different fixatives yield markedly different flow cytometric results (Fig. 6.6) (97, 132). In general, fixation with Bouin's or Zenker's fixatives destroys antigen expression and results in poor to uninterpretable DNA histograms; whereas acceptable results are commonly obtained from tissue fixed in neutral-buffered formalin (NBF). DNA histograms of varying quality are obtained from blocks fixed in B5, ethanol, or Omnifix (Omni) (121, 132, 133). DNA staining intensity is highest with NBF-fixed tissue and lowest with Bouin's and Zenker's fixatives (129, 130, 133). Alanen et al. (25) reported that the DI was consistently larger, but not significantly so, in NBF-fixed samples compared to fresh or ethanol-fixed samples from the same tumor. Fixation also induces wide G0/G1 peak CVs that may preclude detection of tumor cell populations with differences in DNA content less than 15% (DI ≤ 1.15) (134). Of further concern is the report that NBF fixation can result in a preferential loss of G2/M cells (25).

Tissue autolysis, not necessarily evident by light microscopy, must be minimized. This process can yield large CVs, false DNA-aneuploid peaks, increased debris, and artefactual antigen expression (25). Autolysis can continue to occur even during fixation. This happens because fixation tends to lag behind the tissue penetration of the fixative. The rate of autolysis varies with the tissue and contributes to the "intrinsic properties of the tissue" (79). To minimize this problem, tissue samples should be squares of less than 2 cm × 2 cm and less than 4 mm thick. Even thinner tissue sections should be taken from fibrotic or sclerotic samples. These samples should be placed in fixative at room temperature as soon as possible after removal from the patient. Tissue can be kept for up to 24 hr in NBF without a significant increase in CVs (79).

Handling the Block

Careful sampling of paraffin blocks is desirable to ensure that both tumor and normal cells are obtained for analysis. The sampling of specific areas of interest begins with a careful review of a stained thin section. This helps to identify the specific areas to sample as well as to eliminate. A thin section should also be taken at the end of the sampling to determine specimen adequacy. Specific areas can be bored out using a skin-punch biopsy device (135) or a bore mounted onto the objective holder of the microscope (136, 137). By scoring the block face one is able to outline and separate the wanted areas from the unwanted ones during sectioning. Specific regions can also be sampled by mounting a rehydrated section on a slide and using a scalpel to cut away unwanted tissue (124, 131). Inadequate sampling may result in reduced cellularity and cause an inaccurate assessment of a tumor's relative DNA content (138, 139). Selected sampling minimizes unwanted blood and necrotic tissue and it improves the detection of the cells of interest. Sampling two or

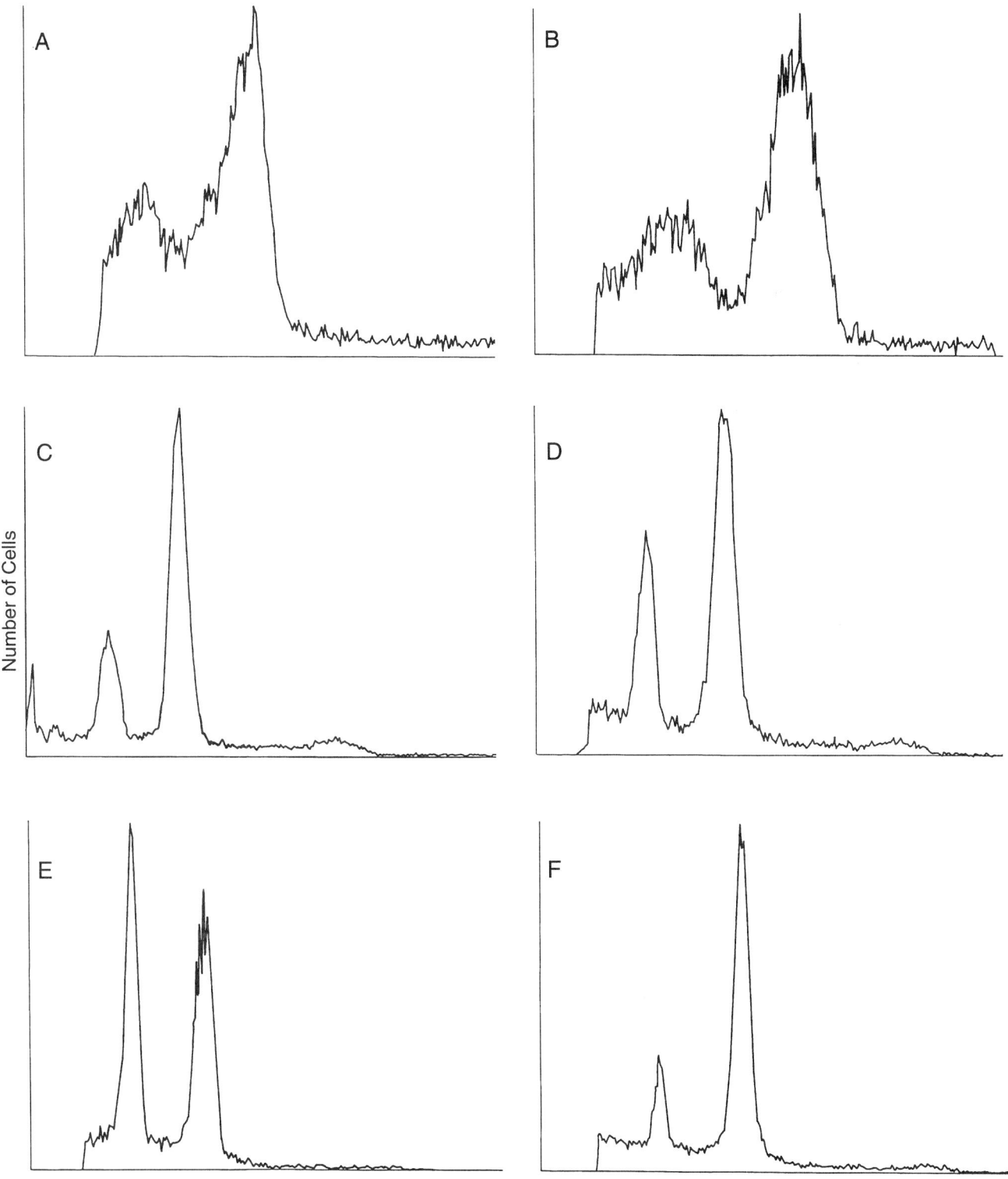

Figure 6.6. Effects of different fixatives on DNA histograms from archival material; (**A**) Bouin's fixative, (**B**) Omni Fix, (**C**) Carnoy's fixative, (**D**) B5, (**E**) ethanol (ETOH), and (**F**) neutral-buffered formalin (NBF).

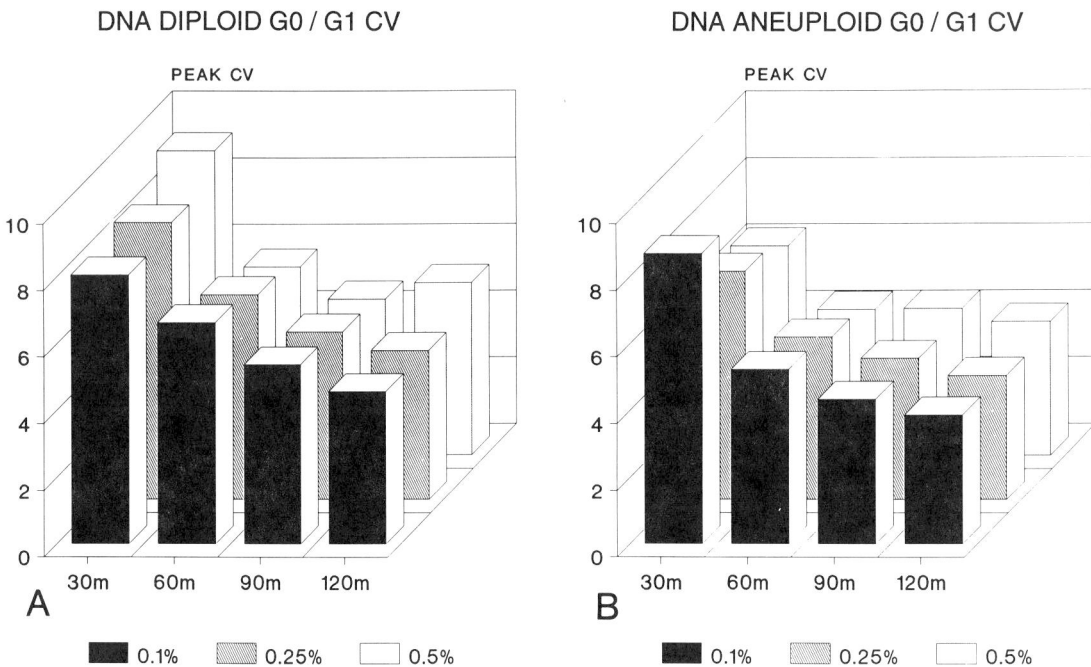

Figure 6.7. Effects of pepsin concentration and duration of digestion on the coefficient of variation (CV) of (**A**) DNA-diploid and (**B**) DNA-aneuploid G0/G1 peaks.

more blocks increases detection of tumor cell populations (15, 19, 130). Such multiple sampling has led to better agreement with results obtained from fresh tumor samples (117). As noted earlier, the normal components in a tumor block need to be carefully sampled so as to provide the necessary internal DNA diploid standard.

Sectioning of paraffin blocks may produce debris by fragmenting nuclei. The amount of debris is inversely related to the thickness of the section and the size of the nuclei (120, 125, 140). Using sections 50 μm or greater in thickness significantly reduces debris (120). The number of sections to be taken is dictated by the cellularity of the specimen. In most cases, two sections yield ample material for subsequent processing.

Deparaffinization and Rehydration

Deparaffinization and rehydration of the paraffin sections is accomplished using standard histopathological techniques. Sections are first deparaffinized in two changes of xylene or such xylene substitutes as Histoclear (National Diagnostics, Somerville, NJ) or Histo-DE (Fisher Scientific). Sections are then rehydrated to water or a saline solution through a series of ethanol solutions of decreasing concentration. Most protocols call for sections to be kept in each solution for 10 to 30 min. Anecdotal evidence indicates that if sections are kept in xylenes for as long as 60 min the CVs are lowered. Automation of this process reduces the need for repeated centrifugation and aspiration. Sections can be placed in nylon mesh bags or wrapped in tissue paper and inserted into embedding cassettes that are then transferred to the different solutions by hand or by running a tissue processing machine in reverse (79, 135, 124, 141, 142). Because variables in this processing method have not been carefully examined, we recommend that the time and volume of each solution be adjusted to insure thorough clearing and rehydration of the tissue sections.

Enzymatic Digestion

Hedley's original protocol (96) called for rehydrated sections to be digested with 0.5% pepsin in acidified saline (pH 1.5) for 30 min. This dissociation step has been extensively modified, often without examining the impact of these modifications on results. Reported pepsin concentrations range from 0.05% to 1.0%. DNA staining intensity and peak CVs can differ substantially over this twenty-fold variation (121, 143, 144) (Figs. 6.7 and 6.8). Antigenic epitopes on nuclear proteins are best retained using low concentrations of pepsin (62, 63, 105). Increasing pepsin digestion time tends to decrease CVs while increasing staining intensity (Figs. 6.7 and 6.8) overall cell yields, and the release of DNA-aneuploid cells (79, 143, 144).

Other enzymes can be used to dissociate and enucleate sections of rehydrated tissue. Overnight digestion in trypsin has been reported to yield DNA histograms with less debris, lower CVs, and more reproducible SPF calculations than DNA histograms produced using pepsinized sections (107, 108, 129). However, Kallioniemi (127) concluded that trypsinization is not optimal for all tumors. Although 90% of sections from breast and ovarian carcinomas yielded interpretable DNA histograms, fewer than 80% of thyroid tumors and colonic polyps yielded adequate DNA histograms.

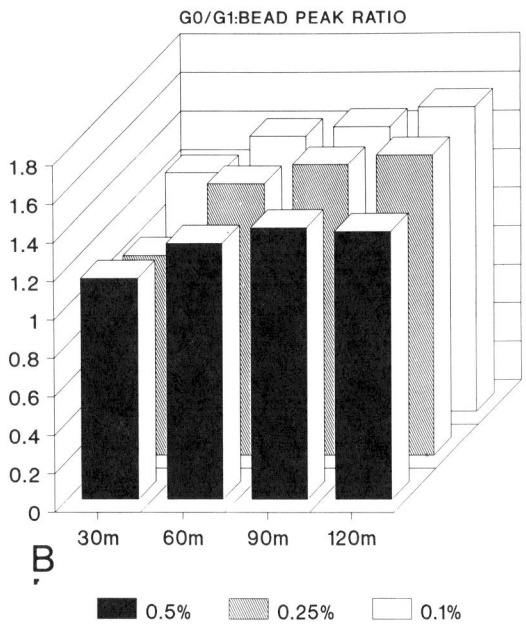

Figure 6.8. Effects of pepsin concentration and duration of digestion on the ratio of **(A)** DNA-diploid and **(B)** DNA-aneuploid G0/G1 peaks to the peak of a known standard (DNA check beads).

Other proteolytic enzymes show promise as well. Wittwer and Bauer (145) used proteinase K and detergent to dissociate sections and found increased staining intensity and decreased CVs as compared to pepsin-treated sections. Yet, the extent of improvement varied with the tissue. Pronase has been used to prepare nuclei for DNA scanning cytophotometry, where it is reported to improve the quality of cellular detail for image analysis (139, 146). Unlike proteinase K, pronase digestion leaves more intact cytoplasm than pepsin digestion (146). Initial flow cytometric studies comparing pronase and pepsin digestion have reported improved histogram quality but poorer nuclear yields (106, 141).

Shearing increases the release of nuclei during dissociation. The nature of the tumor determines which method to use. Mincing is often needed with fibrotic or sclerotic samples. This can be done prior to or during digestion, but it is time consuming. Repeated vortexing is a simple technique applicable to tumors with high cellularity and loosely bound cells. Yields of nuclei from sclerotic specimens and specimens with decreased cellularity are increased by repeated aspiration through a Pasteur pipette, a 23- to 27-gauge needle, or two Luer-locked syringes. Care must be taken that shearing is not so vigorous as to produce increased amounts of debris.

Sample Storage

Samples stored after rehydration, enzymatic digestion, or DNA staining make batching easier. Storing sections or nuclei at 4°C in ethanol, distilled water, or saline may well improve the quality of DNA histograms and decrease intersample staining variation (122, 147). Prolonged freezing is also possible without deleterious effects (148). Overnight storage, at 4°C, of the stained nuclei may improve the quality of the DNA histograms by allowing the dye molecules to achieve complete equilibrium or access to DNA. Stained nuclei can be kept at 4°C for several days without staining degradation. This can lead to increased numbers of aggregates and clumps of nuclei that can be disrupted by repeated aspiration or by sonication (149).

CONCLUSION

The goal of flow cytometric analysis of solid tumors is to provide the clinician with additional diagnostic and prognostic information. The future of flow cytometric analysis of solid tumors lies in the optimization and standardization of technical variables for tissue processing. Such a critical assessment of methodologies is needed if this technology is to be transferred from the research bench to the clinical laboratory. As we have stated, there is no single optimum or standardized technique for processing fresh or fixed, paraffin-embedded solid-tumor samples for flow cytometry. A better understanding of the problems inherent in these techniques is a start.

ACKNOWLEDGMENTS

The opinions or assertions contained herein are the private views of the authors and are not to be construed as official or as reflecting the views of the Department of the Army, Department of the Air Force, or the Department of Defense.

The authors would like to take this opportunity to thank Dr. Joe Griffin, Annette Geisel, and Karen Scott at the AFIP and Mark Zalupski, M.D., Haline Pietraszkiewicz, and Zosia Maciorowski at Wayne State University for their continued efforts to make this work possible. Dr. Hitchcock's work was supported in part by grants from the American Registry of Pathology. Dr. Ensley's work was supported by NIH grants CA 40498 and RFA 90-CA-03; Comprehensive Cancer Center Support Grant, NIH 2P30CA22453; Department of Internal Medicine Fund for Medical Research and Education; Harper Hospital Medical Staff Trust Fund for Medical Research and Education; and the Ben Kasle Flow Cytometry Laboratory, Wayne State University, Detroit, Michigan.

REFERENCES

1. Coon JS, Deitch AD, de Vere White RW, et al. Interinstitutional variability in DNA flow cytometric analysis of tumors. The National Cancer Institute's Flow Cytometry Network experience. Cancer 1988;61:126–130.
2. Wheeless LL, Coon JS, Cox C, et al. Measurement variability in DBA flow cytometry of replicate samples. Cytometry 1989;10:731–738.
3. Hitchcock CL (for "The Flow Cytometry Inter-Laboratory Study Group"). Inter-laboratory analysis of DNA by flow cytometry [Abstract]. Cytometry 1990;(suppl 4):82.
4. Hitchcock CL (for "The Flow Cytometry Inter-Laboratory Study Group"). Variability in flow cytometric results using identical archival samples [Abstract]. Cytometry 1991;(suppl 5):46.
5. Pallavicini M. Solid tissue dispersal for cytokinetic analyses. In: J Gray, Z Darzynkiewicz ed. Techniques in cell cycle analysis. New York: Humana Press, 1986:139–162.
6. Scott NA, Grande JP, Weiland LH, Pemberton JH, Beart RW Jr, Lieber MM. Flow cytometric DNA patterns from colorectal cancers. How reproducible are they? Mayo Clin Proc 1987;62:331–337.
7. Wersto RP, Liblit RL, Deitch D, Koss LG. Variability in DNA measurements in multiple tumor samples of human colonic carcinoma. Cancer 1991;67:106–115.
8. Cornelisse CJ, van de Velde CJH, Caspers RJC, Moolenaar AJ, Hermans J. DNA ploidy and survival in breast cancer patients. Cytometry 1987;8:225–234.
9. McDivitt RW, Stone KR, Meyer JS. A method for dissociation of viable human breast cancer cells that produces flow cytometric kinetic information similar to that obtained by thymidine labeling. Cancer Res 1984;44:2628–2633.
10. van den Ingh HF, Bara J, Cornelisse CJ, Nap M. Aneuploidy and expression of gastric-associated mucus antigens M1 and CEA in colorectal adenomas. Am J Clin Pathol 1987;87:174–179.
11. Xiang JH, Spanier SS, Benson NA, Braylan RC. Flow cytometric analysis of DNA in bone and soft-tissue tumors using nuclear suspensions. Cancer 1987;59:1951–1958.
12. Listinsky CM, Bonfiglio TA, Leary J. Variable ploidy of ovarian clear cell carcinomas. Implications for adequacy of tissue sampling. Anal Quant Cytol Histol 1988;10:21–27.
13. Askensten UG, von Rosen AR, Nilsson RS, Auer GU. Intratumoral variations in DNA distribution patterns in mammary adenocarcinoma. Cytometry 1989;10:326–333.
14. Mullen P, Miller WR. Variations associated with the DNA analysis of multiple fine needle aspirates obtained from breast cancer patients. Br J Cancer 1989;59:688–691.
15. Beerman H, Smit VTHBM, Kluin PM, Bonsing BA, Hermans J, Cornelisse CJ. Flow cytometric analysis of DNA stemline heterogeneity in primary and metastatic breast cancer. Cytometry 1991;12:147–154.
16. Petersen SE, Lorentzen M, Bichel P. A mosaic subpopulation structure of human colorectal carcinomas demonstrated by flow cytometry. In: Laerum OD, Lundmo T, Thorud E, ed. Flow cytometry IV. Oslo: Universitetsforlaget, 1980:412–416.
17. Rabinovitch PS, Reid BJ, Haggitt RC, Norwood TH, Rubin CE. Progression to cancer in Barrett's esophagitis is associated with genomic instability. Lab Invest 1988;60:65–71.
18. Jacobsen A, Mommsen S, Olsen S. Characterization of ploidy level in bladder tumors and selected site specimens by flow cytometry. Cytometry 1983;4:170–173.
19. Carey FA, Lamb D, Bird CC. Intraturmoral heterogeneity of DNA content in lung cancer. Cancer 1990;65:2266–2269.
20. Ljungberg B, Stenling R, Roos G. DNA content in renal cell carcinoma with reference to tumor heterogeneity. Cancer 1985;56:503–508.
21. Reid BJ, Haggitt RG, Rubin CE, Rabinovitch PS. Barrett's esophagitis, correlation between flow cytometry and histology in detection of patients at risk for adenocarcinoma. Gastroenterology 1987;93:1–11.
22. Weiss H, Gutz HJ, Schroter J, Wildner GP. DNA distribution pattern in chronic gastritis. I. DNA ploidy and cell cycle distribution. Scand J Gastroenterol 1989;24:643–648.
23. Lofberg R, Caspersson T, Tribukait B, Ost A. Comparative DNA analyses in longstanding ulcerative colitis with aneuploidy. Gut 1989;30:1731–1736.
24. Ensley J, Alonso M, Maciorowshi Z, Pietraszkiewicz H. Comparisons of DNA content parameters in paired pre-treatment biopsies (PTB) and surgical resections (SR) of squamous cell carcinomas of the head and neck (SCCHN) [Abstract]. Proc AACR 1990;31:24.
25. Alanen KA, Joensuu H, Klemi PJ. Autolysis is a potential source of false aneuploid peaks in flow cytometric DNA histograms. Cytometry 1989;10:417–425.
26. Katz RL, Patel S, Fritsche HA Jr, et al. Comparison of immunocytochemical and biochemical assays for estrogen receptor in fine needle aspirates and histologic sections from breast cancer. Breast Cancer Res Treat 1990;15:191–203.
27. Greenebaum E, Koss LG, Sherman AB, Elequin F. Comparison of needle aspiration and solid biopsy techniques in the flow cytometric study of DNA distributions of surgically resected tumors. Am J Clin Path 1984;82:559–564.
28. Martino S, Ensley JF, Weaver D, et al. Cellular DNA content characteristics of needle aspirates from patients at high risk for developing breast cancer [Abstract]. Proc AAACR 1989;30:A1018.
29. Vindelov LL, Christensen IJ. A review of techniques and results obtained in one laboratory by an integrated system of methods designed for routine clinical flow cytometric DNA analysis. Cytometry 1990;11:753–770.
30. Dunphy CH, Katz RL, Fanning CV, Dalton WT Jr. Leukemic lymphadenopathy: diagnosis by fine needle aspiration. Hematol Pathol 1989;3:35–44.
31. Kaufman DG, Nagler HM. Aspiration flow cytometry of the testes in the evaluation of spermatogenesis in the infertile male. Fertil Steril 1987;48:287–291.
32. Skoog SJ, Evans CP, Hayward IJ, Griffin JL, Hitchcock CL. Flow cytometry of fine needle aspirations of the S-D rat testis: defining normal maturation and the effects of multiple biopsies. J Urol 1991;146:620–623.
33. Joensuu H, Klemi P, Eerola E. Diagnostic value of flow cytometric DNA determination combined with fine needle aspiration biopsy in thyroid tumors. Anal Quant Cytol Histol 1987;9:328–333.
34. Farsetti A, Pontecorvi A, Antonozzi I, Andreoli M, Gaetano C. Cytofluorometric analysis of lymphocyte subsets in thyroid aspirates from patients with autonomously functioning nodules. Clin Endocrinol (Oxf) 1990;32:729–738.
35. Hanson CA, Schnitzer B. Flow cytometric analysis of cytologic specimens in hematologic disease. J Clin Lab Anal 1989;3:207.
36. Katz RL, Gritsman A, Cabanillas F, et al. Fine needle aspiration cytology of peripheral T-cell lymphoma. A cytologic, immunologic, and cytometric study. Am J Clin Pathol 1990;91:120–131.

37. Johnson A, Akerman M, Cavallin-Stahl E. Flow cytometric detection of B-clonal excess in fine needle aspirates for enhanced diagnostic accuracy in non-Hodgkin's lymphoma. Histopathology 1987;11:581–590.
38. Nagler HM, Kaufman DG, O'Toole KM, Sawczuk IS. Carcinoma in situ of the testes: diagnosis by aspiration flow cytometry. J Urol 1990;143:359–361.
39. Shabot MM, Goldberg IM, Schick P, et al. Aspiration cytology is superior to tru-cut needle biopsy in establishing the diagnosis of clinically suspicious breast masses. Ann Surg 1982;196:122–126.
40. Lykkesfeldt AE, Balslev I, Christensen IJ, et al. DNA ploidy and S-phase fraction in primary breast carcinomas in relation to prognostic factors and survival for premenopausal patients at high risk for recurrent disease. Acta Oncol 1988;27:749–756.
41. Palmer JO, McDivitt RW, Stone KR, Rudloff MA, Gonzalez JG. Flow cytometric analysis of breast needle aspirates. Cancer 1988;62:2387–2391.
42. Fallenius AG, Askensten UG, Skoog LK, Auer GU. The reliability of microspectrophotometric and flow cytometric nuclear DNA measurements in adenocarcinoma of the breast. Cytometry 1987;8:260–266.
43. Briffod M, Spryratos F, Tubiana-Hulin M, et al. Sequential cytopunctures during preoperative chemotherapy for primary breast carcinoma, morphologic changes, initial tumor ploidy, and tumor regression. Cancer 1989;63:631–637.
44. Joensuu H, Klemi P, Eerola E. Flow cytometric DNA analysis combined with fine needle aspiration biopsy in the diagnosis of palpable metastases. Anal Quant Cytol Histol 1988;10:256–260.
45. Fuhr JE. Flow cytometry in clinical oncology. Diagn Clin Testing 1990;28:24–29.
46. Waymouth C. Methods for obtaining cells in suspension from animal tissues. In: Pretlow II TG, Pretlow TP ed. Cell separation: methods and applications vol. 1. New York: Academic Press, 1988:1–30.
47. Pretlow TG II, Pretlow TP. Evaluation of data, problems and general approach. In: Pretlow II TG, Pretlow TP ed. Cell separation: methods and applications vol. 1. New York: Academic Press, 1988:31–40.
48. Brattain M. Tissue disaggregation. In: Melamed MR, Mullaney P, Mendelsohn ML, ed. Flow cytometry and sorting. New York, Wiley & Sons, 1979:193–203.
49. Ensley JF, Maciorowski Z, Hassan M, et al. The potential and pitfalls of solid tumor flow cytometry with respect to squamous cell cancers of the head and neck. In: Wolf GT, Carey T, ed. Head and neck oncology research. Proceeding of the second international research conference on head and neck cancer. Amsterdam: Kugler, 1988:213–224.
50. Costa A, Silvestrini R, Del Bino G, Motta R. Implications of disaggregation procedures on biological representation of human solid tumors. Cell Tissue Kinet 1987;20:171–180.
51. Rous P, Jones FS. A method for obtaining suspensions of living cells from the fixed tissues and for the plating of individual cells. J Exp Med 1916;23:549–555.
52. Berwick L, Corman DR. Some chemical factors in cellular adhesion and stickiness. Cancer Res 1962;22:982–986.
53. Oldbring J, Hellsten S, Lindholm K, Mikulowski P, Tribukait B. Flow DNA analysis in the characterization of carcinoma of the renal pelvis and ureter. Cancer 1989;64:2141–2145.
54. Koss LG, Wolley RC, Schreiber K, Mendecki J. Flow-microfluorometric analysis of nuclei isolated from various normal and malignant human epithelial tissue. A preliminary report. J Histochem Cytochem 1977;25:565–572.
55. Koss LG, Czerniak B, Wersto RP. Flow cytometric measurements of DNA and other cell components in human tumors: a critical appraisal. Hum Pathol 1989;20:528–548.
56. Engelholm SA, Spang-Thomsen M, Brunner N, Nohr I, Vindelov LL. Dissagregation of human solid tumors by combined mechanical and enzymatic methods. Br J Cancer 1985;51:93–98.
57. Slocum HK, Pavelic ZP, Kanter PM, Nowak HJ, Rustum YM. The soft agar clonogenicity and characterization of cells obtained from human solid tumors by mechanical and enzymatic means. Cancer Chemother Pharmacol 1981;6:219–225.
58. Slocum HK, Pavelic ZP, Rustum YM, et al. Characterization of cells obtained by mechanical and enzymatic means from human melanoma, sarcoma and lung tumors. Cancer Res 1981;41:1428–1434.
59. Bashor MM. Dispersion and disruption of tissues. Methods Enzymol 1979;58:119–131.
60. Thornthwaite JT, Sugarbaker EV, Temple WJ. Preparation of tissues for DNA flow cytometric analysis. Cytometry 1980;1:229–237.
61. Vindelov LL, Christensen IJ, Nissen NI. A detergent-trypsin method for the preparation of nuclei for flow cytometric DNA analysis. Cytometry 1983;3:323–327.
62. Clevenger CV, Epstein AL, Bauer KD. Quantitative analysis of a nuclear antigen in interphase and mitotic cells. Cytometry 1987;8:280–286.
63. Lincoln ST, Bauer KD. Limitations in the measurement of c-myc oncoprotein and other nuclear antigens by flow cytometry. Cytometry 1989;10:456–462.
64. Ohyama S, Yonenura Y, Miyazaki I. Prognostic value of S-phase fraction and DNA ploidy studied with in vivo administartion of bromodeoxyuridine on human gastric cancers. Cancer 1190;65:116–121.
65. Krishan A. Rapid flow cytofluorometric analysis of mammalian cell cycle by propidium iodide staining. J Cell Biol 1975;66:188–193.
66. Deitch AD, Law H, DeVere White R. A stable propidium iodide staining procedure for flow cytometry. J Histochem Cytochem 1982;30:967–972.
67. Fried J, Perez AG, Clarkson BD. A rapid hypotonic method for flow cytofluorometry of monolayer cell cultures: some pitfalls in staining and DNA analysis. J Histochem Cytochem 1978;26:921–933.
68. Petersen SE. Accuracy and reliability of flow cytometric DNA analysis using a simple, one-step ethidium bromide staining protocol. Cytometry 1986;7:301–306.
69. Vindelov LL. Flow microfluorometric analysis of nuclear DNA in cells from solid tumors and cell suspensions. A new method for rapid isolation and staining of nuclei. Virchows Arch [B] 1977;24:227–242.
70. Taylor IW. A rapid single step staining technique for DNA analysis by flow microfluorimetry. J Histochem Cytochem 1980;28:1021–1024.
71. Thornthwaite JT, Thomas RA, Russo J, et al. A review of DNA flow cytometric preparatory and analytical methods. In: Proceedings of the workshop on immunocytochemistry in tumor diagnosis. Detroit, 1984:380–398.
72. Tribukait B, Gustafson H, Esposti P-L. The significance of ploidy and proliferation in the clinical and biological evaluation of bladder tumors: A study of 100 untreated cases. Br J Urology 1982;54:130–135.
73. Volm M, Drings P, Mattern J, Sonka J, Vogt-Moykopf I, Wayss K. Prognostic significance of DNA patterns and resistance-predictive tests in non-small cell lung carcinoma. Cancer 1985;56:1396–1403.
74. Ensley JF, Maciorowski Z, Pietraszkiewicz H, et al. Solid tumor preparation for flow cytometry using a standard murine model. Cytometry 1987;8:479–487.
75. Chassevent A, Daver A, Bertrand G, et al. Comparative flow DNA analysis of different cell suspensions in breast carcinoma. Cytometry 1984;5:263–267.
76. Ensley JF, Maciorowski Z, Pietraszkiewicz H, et al. Solid tumor preparation for clinical application of flow cytometry. Cytometry 1987;8:488–493.
77. Zalupski M, Ensley J, Ryan J, et al. Comparative dissociation techniques and flow cytometry in soft tissue and osteogenic neoplasms [Abstract]. Proc AACR 1989;30:233.
78. Leith JT, Faulkner LE, Bliven SF, et al. Disaggregation studies of xenograft solid tumors grown from pure or admixed clonal subpopulations from a heterogenous human colon adenocarcinoma. Invasion Metastasis 1985;5:371–335.
79. Crissman JD, Zarbo RJ, Niebylski CD, Corbett T, Weaver D. Flow cytometric DNA analysis of colon adenocarcinomas: a comparative study of preparatory techniques. Mod Pathol 1988;1:198–204.

80. Ljung B-M, Mayhall B, Lottich C, et al. Cell dissociation techniques in human breast cancer—variations in tumor cell viability and DNA ploidy. Breast Cancer Res Treat 1989;13:153–159.
81. Ensley J, Maciorowski Z, Hassan M, Pietraszkiewcz H, Sakr W, Heilbrun L. The loss of DNA aneuploid cells during tumor dissociation in human colon and head and neck cancers analyzed by flow cytometry. Cytometry (In press).
82. Bijman JTh, Wagener DJTh, van Renned H, Wessels JMC, van den Broek P. Flow cytometric evaluation of cell dispersion from human head and neck tumors. Cytometry 1985;6:334–341.
83. Pallavicini MG, Folstad LJ, Dunbar C. Solid KHT tumor dispersal for flow cytometric cell kinetic analysis. Cytometry 1981;2:54–58.
84. Meyer JS, Micko S, Craver JL, McDivitt RW. DNA flow cytometry of breast carcinomas after acetic acid fixation. Cell Tissue Kinet 1984;17:185–197.
85. Risberg B, Stal O, Eriksson L-L, Hussein A. DNA flow cytometry on breast carcinomas: comparison of a detergent and enzyme-detergent preparation method. Anal Cell Pathol 1990;2:287–295.
86. Becker RL, Mikel UV. Interrelation of formalin fixation, chromatin compactness and DNA values as measured by flow cytometry and image cytometry. Anal Quant Cytol Histol 1990;12:333–341.
87. Holtfreter HB, Cohen N. Fixation-associated quantitative variations of DNA fluorescence observed in flow cytometric analysis of hemopoietic cells from adult diploid frogs. Cytometry 1990;11:676–685.
88. Petersen SE. Flow cytometry of human colorectal tumors: nuclear isolation by detergent technique. Cytometry 1985;6:452–460.
89. Vindelov LL, Christensen IJ, Keiding N, Spang-Thomsen M, Nissen NI. Long-term storage of samples for flow cytometric DNA analysis. Cytometry 1982;3:317–322.
90. Stone KR, Craig RB, Palmer JO, Rivkin SE, McDivitt RW: Short-term cryopreservation of human breast carcinoma cells for flow cytometry. Cytometry 1985;6:357–361.
91. Alanen KA, Klemi PJ, Taimela S, Joensuu S. A simple preservative for flow cytometric DNA analysis. Cytometry 1988;9:86–89.
92. Iversen OE, Laerum OD. Trout and salmon erythrocytes and human leukocytes as internal standards for ploidy control in flow cytometry. Cytometry 1987;8:190–196.
93. Park CH, Lee SH, Stephens RL, Smith TK, Park MH. Flow cytometry DNA analysis on tumor cell subpopulation of human tumor specimens by exclusion of lymphohemopoietic cells. J Histochem Cytochem 1988;36:705–709.
94. Feitz WFJ, Beck HLM, Smeets AWGB, et al. Tissue-specific markers in flow cytometry of urological cancers: cytokeratins in bladder carcinoma. Int J Cancer 1985;36:349–356.
95. Visscher DW, Zarbo RJ, Jacobsen G, et al. Multiparametric deoxyribonucleic acid and cell cycle analysis of breast carcinomas by flow cytometry. Clinicopathologic correlations. Lab Invest 1990;62:370–378.
96. Hedley DW, Friedlander ML, Taylor IW, Rugg CA, Musgrove EA. Method for analysis of cellular DNA content of paraffin-embedded pathological material using flow cytometry. J Histochem Cytochem 1983;31:1333–1335.
97. Hedley DW: Flow cytometry using paraffin-embedded tissue: five years on. Cytometry 1989;10:229–241.
98. Merkel DE, McGuire WL. Ploidy, proliferative activity and Prognosis. DNA flow cytometry of solid tumors. Cancer 1990;65:1194–1205.
99. Hitchcock CL, Norris HJ, Khalifa MA, Wargotz ES. Flow cytometric analysis of granulosa tumors. Cancer 1989;64:2127–2132.
100. Hitchcock CL, Norris HJ. Flow cytometric analysis of endometrial stromal sarcomas. Am J Clin Path 1992;97:267–271.
101. Seidman JD, Berman JJ, Hitchcock CL, et al. DNA analysis of cardiac myxomas: flow cytometry and image analysis. Human Pathol 1991;22:495–500.
102. Wenig B, Hitchcock C, Ellis G, Gnepp D. Metastasizing mixed tumor of salivary glands (MMTSG): a clinicopathologic and flow cytometric analysis. [Abstract] Lab Invest 1991;64:66A.
103. Wenig B, Hitchcock C, Griffin J, Heffner D. Olfactory esthesioneuroblastoma: support for a grading system based on flow cytometric and histopathologic correlation. [Abstract] Lab Invest 1991;64:66A.
104. Hitchcock CL, Conran RM, Griffin J: Hydatidiform moles and the use of flow cytometry in their diagnosis, In: Garvin AJ, O'Leary TJ, Bernstein J, Rosenberg HS, ed. Perspectives in pediatric pathology 1991;15:117–141.
105. Bauer KD, Clevenger CV, Endow RK, Murad T, Epstein AL, Scarpelli DG. Simultaneous nuclear antigen and DNA content quantitation using paraffin-embedded colonic tissue and multiparameter flow cytometry. Cancer Res 1986;46:2428–2434.
106. Jacobsen AB, Fossa SD, Thorud E, Lunde S, Melvik JE, Pettersen EO. DNA flow cytometric values in bladder carcinoma biopsies obtained from fresh and paraffin-embedded material. APMIS 1988;96:25–29.
107. Wingren S, Hatschek T, Stal O, Boeryd B, Nordenskjold B. Comparison of static and flow cytometry for estimation of DNA index and S-phase fraction in fresh and paraffin-embedded breast cancer tissue. Acta Oncol 1988;6:793–797.
108. Schutte B, Reynders MMJ, Bosman FT, Blijham GH. Flow cytometric determination of DNA ploidy level in nuclei isolated from paraffin-embedded tissue. Cytometry 1985;6:26–30.
109. Camplejohn RS, Macartney JC, Morris RW. Measurement of S-phase fractions in lymphoid tissues comparing fresh versus paraffin-embedded tissue and 4',6'-diamidino-2 phenylindole dihydrochloride versus propidium iodide. Cytometry 1989;10:410–416.
110. Klemi P, Joensuu H. Comparison of DNA ploidy in routine fine needle aspiration biopsy samples and paraffin-embedded tissue samples. Analyt Quant Cytol Histol 1988;10:195–199.
111. Hedley DW, Friedlander ML, Taylor IW. Application of DNA flow cytometry to paraffin-embedded archival material for the study of aneuploidy and its clinical significance. Cytometry 1985;6:327–333.
112. Bagwell CB, Mayo SW, Whetstone SD et al. DNA histogram debris theory and compensation. Cytometry 1991;12:107–118.
113. Rabinovitch PS. Numerical compensation for the effects of cell clumping on DNA content histograms [Abstract]. Cytometry 1990; (Suppl 4):27.
114. Weaver DL, Bagwell CB, Hitchcox SA, et al. Improved flow cytometric determination of proliferative activity (S-phase fraction) from paraffin-embedded tissue. Am J Clin Pathol 1990;94:576–584.
115. Arber DA, Cook PD, Moser LK, Speights VO. Variation in reference cells for DNA analysis of paraffin-embedded tissue. Am J Clin Pathol (In press).
116. Owainati AAR, Robins RA, Hinton C, et al. Tumor aneuploidy, prognostic parameters and survival in primary breast cancer. Br J Cancer 1987;55:449–454.
117. Frierson HF Jr. Flow cytometric analysis of ploidy in solid neoplasms: comparison of fresh tissues with formalin-fixed paraffin-embedded specimens. Hum Pathol 1988;19:290–294.
118. Pelstring RJ, Hurtubise PE, Swerdlow SH. Flow-cytometric DNA analysis of hematopoietic and lymphoid proliferations: a comparison of fresh, formalin-fixed and B5-fixed tissues. Hum Pathol 1990;21:551–558.
119. McIntire TL, Goldey SH, Benson NA, Braylan RC. Flow cytometric analysis of DNA in cells obtained from deparaffinized formalin-fixed lymphoid tissues. Cytometry 1987;8:474–478.
120. Price J, Herman CJ. Reproducibility of FCM DNA content from replicate paraffin block samples. Cytometry 1990;11:845–847.
121. Herbert DJ, Nishiyama RH, Bagwell CB, Munson ME, Hithcox SA, Lovett EJ III. Effects of several commonly used fixatives on DNA and total nuclear protein analysis by flow cytometry. Am J Clin Pathol 1989;91:535–541.
122. McLemore DD, El Naggar A, Stephens LC, Jardine JH. Modified methodology to improve flow cytometric DNA histograms from paraffin-embedded material. Stain Technology 1990;65:279–291.

123. Oud PS, Hanselaar TGJM, Reubsaet-Veldhuizen JAM, et al. Extraction of nuclei from selected regions in paraffin-embedded tissue. Cytometry 1986;7:595–600.
124. Sickle-Santanello BJ, Farrar WB, DeCenzo JF, et al. Technical and statistical improvements for flow cytometric DNA analysis of paraffin-embedded tissue. Cytometry 1988;9:594–599.
125. Stephenson RA, Gay H, Fair WR, Melamed MR. Effect of section thickness on quality of flow cytometric DNA content determinations in paraffin-embedded tissues. Cytometry 1986;7:41–44.
126. Hedley DW, Friedlander ML, Taylor IW, Rugg CA, Musgrove EA. DNA flow cytometry of paraffin-embedded tissue. Cytometry 1984;5:660.
127. Kallioniemi O-P. Comparison of fresh and paraffin-embedded tissue as starting material for DNA flow cytometry and evaluation of intratumor heterogeneity. Cytometry 1988;9:164–169.
128. McIntire TL, Murphy WM, Coon JS, et al. The prognostic value of DNA ploidy combined with histologic substaging for incidental carcinoma of the prostate gland. Am J Clin Pathol 1988;89:370–373.
129. Emdin SO, Stenling R, Roos G. Prognostic value of DNA content in colorectal carcinoma. A flow cytometric study with some methodologic aspects. Cancer 1987;60:1282–1287.
130. Quirke P, Dixon M, Clayden AD, et al. Prognostic significance of DNA aneuploidy and cell proliferation in rectal adenocarcinomas. J Pathol 1987;151:285–291.
131. Murad T, Bauer K, Scarpelli DG. Histopathologic and flow cytometric analysis of adenomatous colonic polyps. Arch Pathol Lab Med 1989;113:1003–1008.
132. Esteban JM, Sheibani K, Owens M, Joyce J, Bailey A, Battifora H. Effects of various fixatives and fixation conditions on DNA ploidy analysis. A need for strict internal DNA standards. Am J Clin Pathol 1991;95:460–466.
133. Hostetter AL, Hrafnkelsson J, Wingren SOW, Enestrom S, Nordenskjold B. A comparative study of DNA cytometry methods for benign and malignant thyroid tissue. Am J Clin Pathol 1988;89:760–763.
134. van den Ingh HF, Griffien G, Cornelisse CJ. Letter to the editor: DNA aneuploidy in colorectal adenomas. Br J Cancer 1987;55:351.
135. Lundberg S, Carstensen J, Rundquist I. DNA flow cytometry and histopathological grading of paraffin-embedded prostate biopsy specimens in a survival study. Cancer Res 1987;47:1973–1977.
136. Mesker WE, Eysackers MJ, Ouwerkerk-van Velzen MCM, van Driel-Kulker MJ, Ploen JS. Discrepancies in ploidy determination due to sampling errors. Anal Cytol Pathol 1989;1:87–95.
137. van Driel-Kulker AMJ, Eysackers MJ, Dessing MTM, Ploem JS. A simple method to select specific tumor areas in paraffin blocks for cytometry using incident fluorescence microscopy. Cytometry 1986;7:601–604.
138. Kute TE, Gregory B, Galleshaw J, Hopkins M, Buss D, Case D. How reproducible are flow cytometry data from paraffin-embedded blocks? Cytometry 1988;9:494–498.
139. Mesker WE, Eysackers MJ, Ouwerkerk-van Velzen MCM, van Driel-Kulker MJ, Ploen JS. Discrepancies in ploidy determination due to sampling errors. Anal Cytol Pathol 1989;1:87–95.
140. Camplejohn RS. Comments on "Effects of section thickness on quality of flow cytometric DNA content determination in paraffin-embedded tissues". Cytometry 1986;7:612–615.
141. Amberson JB, Wersto RP, Agarwal V, Suhrland M, Koss LG. Preparation of paraffin-embedded tissue for flow and image cytometric analysis: an improved and more efficient procedure. [Abstract] Cytometry 1988;(Suppl 2):34.
142. Babiak J, Poppema S. Automated procedure for dewaxing and rehydration of paraffin-embedded tissue sections for DNA flow cytometric analysis of breast tumors. Am J Clin Pathol 1991;96:64–69.
143. Hitchcock CL, Scott K. Optimization of techniques for flow cytometric analysis of DNA from archival material [Abstract]. Cytometry 1990;(suppl 4):102.
144. Hitchcock C, Scott K. Flow cytometric analysis of archival tissue: The use of a model tumor system to examine the effects of technical parameters [Abstract]. Lab Invest 1991;64:122A.
145. Wittwer CT, Bauer KD. Use of proteinase K, SDS, and heat to decrease light scatter variation and peak width in DNA content analysis of formalin-fixed, paraffin-embedded tissue [Abstract]. Cytometry 1990;(Suppl)4:80–81.
146. van Driel-Kulker AMJ, Mesker WE, van Velzen I, Tanke HJ, Feichtinger J, Ploem JS. Preparation of Monolayer smears from paraffin-embedded tissue for image cytometry. Cytometry 1985;6:268–272.
147. Bowlby LS, DeBault LE, Abraham SR. Flow cytometric analysis of parathyroid glands. Relationship between nuclear DNA and pathologic classifications. Am J Pathol 1987;128:338–344.
148. Morkve O. Long-term storage of nuclear suspensions prepared from paraffin-embedded material for flow cytometric DNA analysis. Anal Cellular Pathol 1990;2:327–331.
149. Gonchoroff NJ, Ryan JJ, Kimlinger, et al. Effect of sonication on paraffin-embedded tissue preparation for DNA flow cytometry. Cytometry 1990;11:642–646.

7

Fluorescent Reagents for Flow Cytometry

ALAN S. WAGGONER and LAUREN A. ERNST

INTRODUCTION

Fluorescence detection is the basis of the power and versatility of flow cytometry. While light-scatter and cell-volume measurements provide useful parameters for identifying major cell classes in heterogeneous populations, fluorescent probes provide the ability to subclassify cells, determine their DNA content, and analyze physiological properties of living cells.

There are three fundamental reasons why fluorescence is so useful in cytometry:

First, fluorescence derives part of its power from its sensitivity in measuring the fluorophore content of cells. The fluorescence signal from each cell can be collected in a few microseconds as the cells pass through the laser beam. Because the dye molecules remain in the excited state only a few nanoseconds before fluorescence occurs, each dye molecule can be excited as many as 10,000 times, if the beam is sufficiently bright. As few as 1000 fluorophores can be detected on a single cell in a flow cytometer provided the cellular autofluorescence and background signals are not too large.

Second, fluorescent probes that emit different colors of light can be used simultaneously to give multiparameter data for each cell. When combined with the light-scatter signals, these multiple signals provide a powerful tool for measuring several subpopulations of cells in a single sample, as well as for correlating combinations of markers and functions in individual cells. The determination of population ratios, such as T-helper/T-suppressor ratios, is generally more accurate when each cell type is detected simultaneously (1). Also, a DNA content indicator used with other probes can correlate the extent of cellular proliferation with cell type.

Third, fluorescent probes can also provide a different kind of sensitivity. Certain types of fluorophores have fluorescence properties (excitation spectra, fluorescence spectra, or fluorescence efficiencies (quantum yields)) that are sensitive to the molecular microenvironment of the fluorophore. When properly designed, these probes can be used as physiological indicators of the function of individual living cells. The commonly used physiological indicator probes, such as the calcium indicators fura2 and indo1, are discussed in the chapter on Flow Cytometric Measurements of Cell Physiology (Chapter 31) by Rabinovitch et al. Several other reviews of the use of these probes are available (2, 3, 4, 5).

The remainder of this paper will be devoted to the discussion of fluorescent probes that are covalent labeling reagents that are generally insensitive to their environment. These can be used to tag antibodies, DNA probes, and other probes for detecting and quantifying specific antigens or nucleic acid sequences in cells. We will not cover probes used for quantitation of the total DNA or RNA content of cells. While these probes are widely used for cell cycle analyis, most interact noncovalently with DNA and RNA. They are discussed in Chapters 2 and 8 of this volume.

PROPERTIES OF FLUORESCENT REAGENTS FOR COVALENT LABELING

A wide selection of low-molecular-weight (<1 kd) fluorescent compounds can be covalently attached to proteins and other biomolecules. Some of these reagents, such as the reagents containing isothiocyanate (-ITC), sulfonyl halide, or active ester (hydroxysuccinimidyl, -NHS), groups react readily with unprotonated primary alkyl amine groups, while reagents containing maleimide (-MAL) or iodoacetamido (-IA) groups are more selective for labeling sulfhydryl groups (6). Conjugation of fluorophore with biomolecule competes with the reaction with water or other buffer components (perhaps TRIS or 2-mercaptoethanol) and is primarily controlled by the pH of the reaction.

Fluorescein and rhodamine reagents have been used for many years, but Bodipy, cyanine dye derivatives, several coumarins, and a number of other fluorophores have been developed as labeling reagents during the past few years. The spectral data contained in Table 7.1, suggest that most of these labeling reagents can form highly fluorescent antibodies. To compare the effectiveness of different reagents, an estimate of the expected fluorescence ("brightness") of a labeled antibody can be defined as the numerical product of the extinction coefficient of the fluorophore at the excitation wavelength (ϵ_{exc}), the quantum yield (Φ) of the fluorophore, and the number of fluorophores bound to the antibody (N);

$$B = \epsilon_{exc} \cdot \Phi \cdot N.$$

Surprisingly, the Φ of the fluorochromes in water or alcohol solvents frequently does not reflect the average Φ of fluorophores bound to the antibody. Figure 7.1 illustrates the tendency of the average Φ of many fluorophores to sharply decrease as the number of fluorophores conju-

Table 7.1
A Selection of Low-Molecular-Weight Fluorescent Labeling Reagents

Probe[a]	Absorption Maximum[b]	Extinction Max.[c]	Emission Maximum[b]	Quantum Yield	Measurement Conditions	References
Fluorescein-ITC, DCT, IA, MAL[d]	490	67	520	0.71	pH7, PBS	[10], W[e], MP[f]
TRITC-amines	554	85	573	0.28	pH7, PBS	[1], MP
XRITC-amines	582	79	601	0.26	pH7, pBS	W
Texas Red®-amines	596	85	620	0.51	pH7, PBS	[2], W, MP
Lissamine rhodamine sulfonamide	570		590	med		MP
CY3.12-OSu-amine	556	130	574	0.05	pH7, PBS	[8]
CY5.12-OSu-amine	650	200	674	0.13	pH7, PBS	[8]
CY7.12-OSu-amine	755	200	784		pH7, PBS	[8]
CY3.18-OSu-antibody	554	130	568	0.14[f]	pH7, PBS	[9]
CY5.18-OSu-antibody	652	200	672	0.18[f]	pH7, PBS	[9]
CY7.18-OSu-antibody	755	200	778	0.02[f]	pH7, PBS	[9]
Cascade Blue®	378, 399	26	423		water	MP
NBD-amine	478	24.6	520–550	0.36/0.21	EtOH/MeOH	[3], [4], [5]
Dansyl-NH-CH3	340	3.4	578	0.068	water	MP, [6]
	(pH 7.4, 0.1M Tris)		539	0.5	ethanol	
			508	0.41	chloroform	
Coumarin-phalloidin	387		470		water	[10]
Phycoerythrin-R	480–565	1960	578	0.68	pH7, PBS	[7]
Allophycocyanine	650	700	660	0.68	pH7, PBS	[7]

[a]Abbreviations: ITC = Isothiocyanato-; DCT = Dichlorotrizinyl-; IA = Iodoacetamido-; MAL = Maleimido-; OSu = Succinimidyl active ester
[b]Measured in nanometers
[c]Multiply value listed by 1000 to get liters/mol. cm.
[d]Covalently bound to amino group (ITC, DCT) or sulfhydryl group (IA, MAL)
[e]W. Waggoner laboratory determination
[f]MP, Molecular Probes, Inc. catalog or personal communication

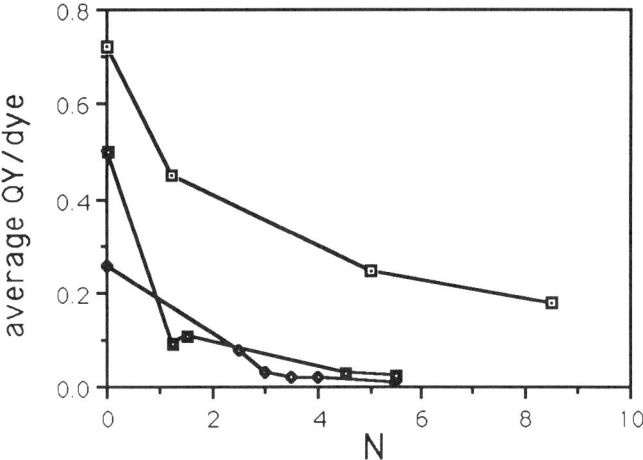

Figure 7.1. Average quantum yields of fluorochromes bound to IgG at different dye/protein ratios (**N**). Open squares: fluorescein; filled diamonds; X-rhodamine; filled squares: Texas Red® [(Used with permission from Mujumdar et al. Cyanine dye labeling reagents: sulfoindocyanine succinimidyl esters. (submitted) 1992.)]

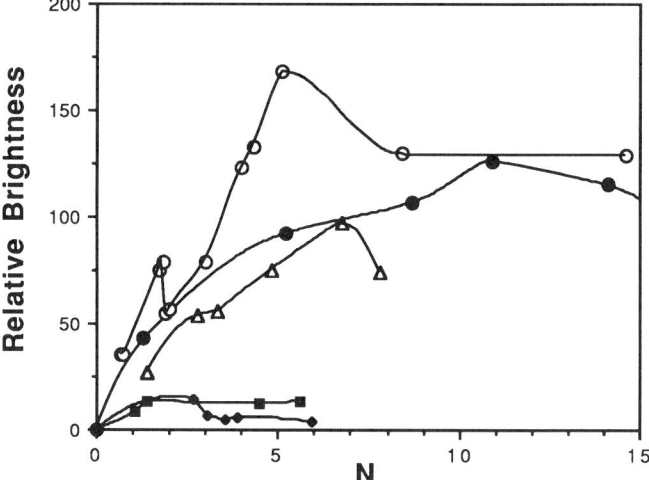

Figure 7.2. Brightness of antibodies labeled with different fluorochromes at different dye/protein ratios (**N**). Filled circles: FITC; open circles: Cyanine5.18; open triangles: Cyanine3.18; filled squares: Texas Red; filled diamonds: X-RITC [(Used with permission from Mujumdar et al. Cyanine dye labeling reagents: sulfoindocyanine succinimidyl esters. (submitted) 1992.)].

gated to the antibody increases. This inverse relationship between the average quantum yield per fluorophore and the dye/protein (d/p) ratio suggests that the brightness of labeled antibodies should reach a maximum. As shown in Figure 7.2, the brightness of sheep IgG tagged with several fluorophores reached a broad maximum or plateau. Optimal labeling would occur with d/p ratios at the lower end of the plateau to minimize changes in the antibody-binding activity.

DEPENDENCE OF ANTIBODY FLUORESCENCE ON THE DEGREE OF LABELING

Quenching of fluorescence has been attributed to contact interactions between dye molecules bound at adjacent sites on the protein molecule (7). In support of this mechanism, our laboratory has obtained absorption spectra of both monomers and dimers of a cyanine dye in concentrated aqueous solu-

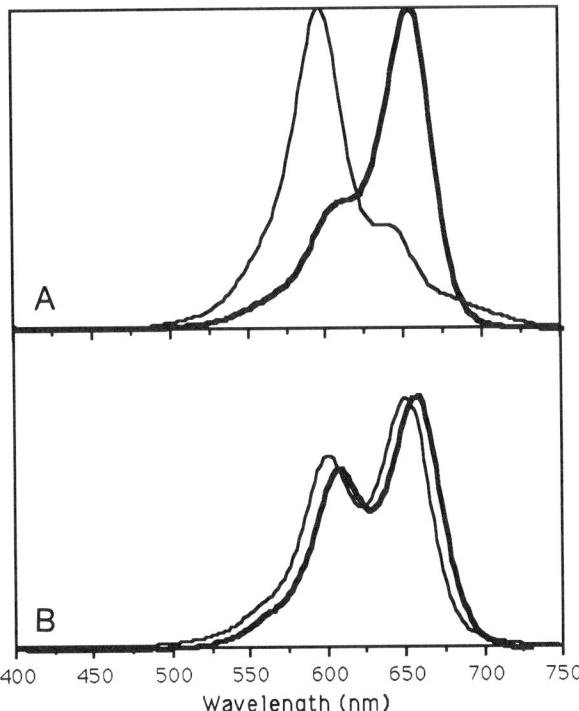

Figure 7.3. Monomer (right-hand peak) and dimer (left-hand peak) absorption spectra of Cyanine5.12 (**A**). Absorption spectrum of IgG antibody over-labeled with Cyanine5.12 (*heavy line*) and absorption spectrum of 1mM Cyanine5.12 in water (*thin line*) (**B**). The wavelength differences arise from the polarity of the microenvironment of the dyes. The concentration of the cyanine reagent in the labeling reaction mixture was about 1 mM.

tion (Fig. 7.3A). The monomeric dyes have a high quantum yield but the dimeric forms are nonfluorescent. A dye with a very similar structure was activated and bound to IgG at different d/p ratios. At higher d/p ratios, where the probability of dye contact increases, the dimer peak becomes a significant part of the absorption spectrum (Fig. 7.3B). As expected, the fluorescence excitation spectrum of the heavily labeled antibody has the shape of the monomer absorption spectrum, thus proving that the antibody–bound dimers do not contribute to the antibody fluorescence.

It is likely that other quenching mechanisms exist. Fluorescein fluorescence is quenched by binding to antifluorescein antibodies. Also, as shown in Figure 7.1, the average Φ of FITC decreases sharply even at d/p levels where no dimers are apparent. Fluorophores covalently linked to proteins exhibit the range of fluorescence efficiencies. At high d/p ratios, it is more probable that one of the fluorophores will be bound to a site that has a reduced quantum yield and will act as an energy sink. Neighboring fluorophores on the protein molecule would be quenched by migration of the excitation energy to the "sink" fluorophore, resulting in a reduction of the total fluorescence of the antibody. The rate of energy migration from one protein-bound fluorophore to another is proportional to the d/p ratio. The decrease in average Φ at higher d/p ratios can arise from a larger number of energy sinks as well as from increased energy migration to these quenching centers.

Clearly, Figure 7.1 indicates that some labeling reagents tend to form aggregates on the antibody surface more readily than others. The rhodamine dyes are a particular nuisance in this regard. These reagents also cause denaturation and precipitation of labeled antibodies at modest d/p ratios. We (and others) have attributed this tendency to the planar, hydrophobic structure of the rhodamines, which, we believe, favors pre-formation of rhodamine dimers in the reaction buffer before either dye molecule has formed a covalent link with the protein.

While the planarity of the aromatic system of a fluorophore cannot be decreased without reducing its fluorescence, the hydrophobic nature of the fluorophore can be modified. Based on this principle, and on the assumption that enhanced water solubility and repulsion between the charges would reduce dye-dye interactions on the protein surface, our laboratory has designed cyanine dye-based labeling reagents that have charged groups attached to the planar ring structure. This strategy was successful as judged by the tendency of cyanines, such as Cyanine3.18 and Cyanine5.18, to remain much more fluorescent than rhodamines as the d/p ratio increases. In fact, commercially sold rhodamine-labeled antibodies often carry only 1 or 2 bound fluorophores, whereas the optimal d/p for cyanine reagents is in the 4-7 range. Indeed, these cyanines can yield a d/p ratio greater than 20 while retaining significant fluorescence and inducing no protein precipitation. We find that for d/p ratios greater than 2, Texas Red (Molecular Probes) and XRITC-labeled antibodies tend to drop out of the solution during the first few days following conjugation.

FLUORESCENT LABELS FOR MULTI-COLOR IMMUNOFLUORESCENCE

Adding multicolor fluorescence to light-scattering parameters greatly enhances the power of flow cytometry by allowing the correlation of the levels of multiple fluorescent probes on each cell analyzed. For example, fluorescein-labeled antibodies and propidium Iodide (PI) can be used with 488-nm argon-laser excitation to quantify both the antigen level and the DNA content of cells (Fig. 7.4). This capability has attracted many new investigators to the field.

For multicolor detection, the fluorescence spectrum can be divided into detection windows or channels, perhaps 50 nm wide for convenience. With 488-nm excitation, typical detection windows are "green" (500 to 550 nm), "orange" (550 to 600 nm), "scarlet" (600 to 650 nm), and "deep red" (650 to 700 nm). Because the emission of most fluorophores tails-off slowly at longer wavelengths, fluorescence usually spills over into more than one detection window (Fig. 7.4). For any type of quantitative analysis, the spillover of signal into the second channel must be removed by compensation. Determination of the required level of compensation may be difficult because of the variable autofluo-

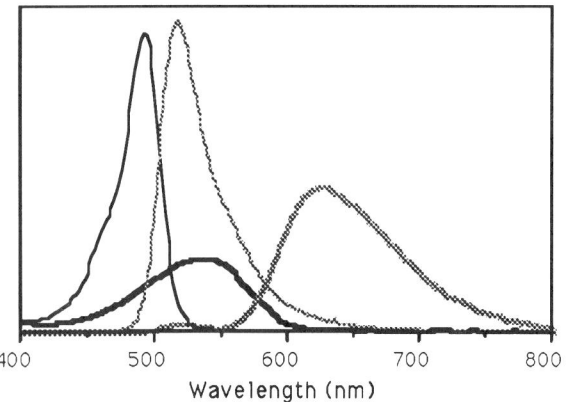

Figure 7.4. Absorption (*black*) and fluorescence (*gray*) spectra of FITC-conjugated antibody (*thin line*) and propidium iodide complexed with DNA (*heavy line*). Notice that both fluorescein and propidium can be excited with a single argon ion laser emitting at 488 nm. The fluorescein signal can be detected with a narrow band interference filter centered near 530 nm, and the propidium signal can be detected with a filter that transmits light at wavelengths beyond 560 nm. The exact region of transmission of the propidium filter determines the extent to which the fluorescein signal spills into the propidium detection channel.

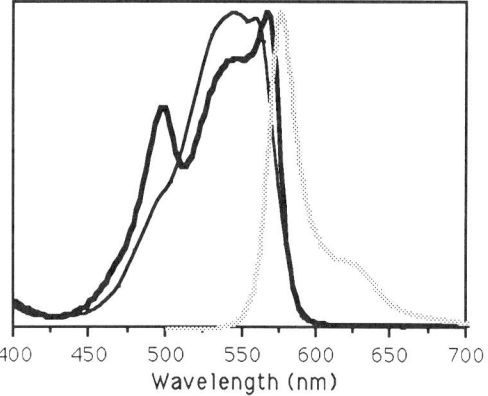

Figure 7.5. Absorption (*heavy black*) and emission (*gray*)spectra of R-phycoerythrin with the absorption spectrum of B-phycoerthrin (*thin black*).

rescence signals from different cells and cell types. Measurement of immunofluorescence intensities on a log scale causes additional problems in setting the proper compensation level (for further details, see chapters 5, 8, 9, and 11).

Based on reliability, available fluorophores, and laser power, the most common light source for flow cytometry is the 488-nm line from an argon-ion laser. For practical purposes, fluorescein derivatives remain the reagents of choice for obtaining green-fluorescing (525 nm) antibodies with excitation at 488 nm. Bodipy has been proposed by Molecular Probes, Inc. (5) as a fluorescein alternative because its narrower emission band has less spillover into the orange detection channel. Thus, less signal compensation is required in two-color analyses using this fluorophore. At this point, few bodipy-labeled primary antibodies are commercially available and reactive bodipy is not commercially available for users to label their own antibodies. Molecular Probes is developing other bodipy analogs that fluoresce at longer wavelengths.

The demonstration by Oi, Stryer and Glazer (9) that protein components of the photosynthetic apparatus, phycobiliproteins, can be used as antibody labels was an important development for flow cytometry. Figure 7.5 shows the absorption spectra of R-Phycoerythrin (RPE) and B-phycoerythrin (BPE). The fluorescence spectra of both compounds are almost identical. The presence of an additional chromophore that absorbs at 490 nm and transfers the energy makes RPE more useful for flow cytometry, even though BPE has a higher quantum yield. With RPE- and fluorescein-tagged antibodies, it is straightforward to quantify simultaneously two different fluorescent antibodies using 488-nm argon-ion laser excitation. Furthermore, many investigators already have flow cytometers with this laser available. RPE is still the fluorophore of choice for obtaining orange-fluorescing (575 nm) antibodies with 488 nm excitation and many RPE-labeled antibodies are available from commercial sources.

Occasionally, it is important to consider a low-molecular-weight-labeling reagent as an alternative to PE. First, it is somewhat difficult for inexperienced biologists to conjugate directly primary antibodies with PE. Second, because PE is approximately 1.5 times the size of an antibody molecule, there may be intracellular targets that exclude PE-labeled markers. Cy3-reactive dye (Biological Detection Systems, Inc.) is a low-molecular-weight, orange-fluorescing (575 nm) label that is optimally excited in the 530-550-nm range but that has sufficient (~20% of maximum) absorption at 488 nm to produce a good signal with antigens expressed at moderate levels. While not as bright as PE, Cy3 reactive dye is small and easy to conjugate to antibodies. Antibodies are optimally labeled with approximately six Cy3 molecules, whereas the average PE-to-antibody ratio is less than 1.

Until recently, three-color flow cytometry could only be accomplished using an instrument with two or more lasers. While there are excellent commercial flow cytometers that have multiple lasers, these systems are more expensive and often more difficult to maintain. For simpler instruments that use an argon laser, several labeling reagents have become available that can be excited with the 488-nm line but that fluoresce in a band at wavelengths longer than PE. One of the reagents is formed by covalently binding Texas Red™ to PE (Duochrome (Becton Dickinson), Red 613 (Life Technologies, Inc.)). Excitation energy from the PE, after excitation at 488 nm, is transferred to the bound Texas Red molecules, which emit scarlet light (620 nm). This reagent has been applied to three-color studies, but it is difficult to use because the energy transfer to Texas Red varies with the success of the Texas Red-PE conjugation process and often is not 100% efficient. Also, the Texas Red emission spills into the PE channel and the PE emission spills into the Texas Red channel. Therefore, the instrument operator must compensate the signals in both of these channels.

Another three-color reagent, a phycobiliprotein that emits at >640 nm (PerCP (Becton Deckinson)), is available. A new three-color reagent (CyChrome (Pharmingen) and Tri-Color (CalTag Laboratories)) has been developed that uses Cy5-reactive dye (Biological Detection Systems, Inc.) instead of Texas Red as the acceptor of energy transfer from the excited PE. Since Cy5 emits further to the red than Texas Red, difficulties in compensating the PE signal are largely eliminated. The Cy5 dye is a very efficient energy absorber and it is easy to conjugate with proteins. Investigators and manufacturers are currently perfecting the production and use of this new reagent, which could make three-color analyses routine with simple, single-laser flow cytometers.

For investigators with multilaser flow cytometers, other multicolor fluorescence measurements are possible. For example, in addition to the two-color PE/fluorescein combination that can be obtained with the argon 488 line, there are red emitting fluorophores that can be excited with krypton lasers (568 nm, 647 nm) and inexpensive air-cooled HeNe lasers (633 nm). Texas Red and XRITC are the commonly used fluorophores excited at 568 nm, and Cy5 is efficiently excited at either 647 nm or 633 nm. The Coulter Elite flow cytometer includes a HeNe laser and users have found that Cy5-labeled materials give excellent signals with this instrument.

In the blue region of the spectrum, there are few satisfactory labeling reagents. Amino-methyl-coumarin-acetic acid (AMCA), with excitation and emission maxima at 354 nm and 442 nm, respectively, is available on some commercial antibodies. Molecular Probes, Inc. sells second antibodies labeled with Cascade Blue (Molecular Probes), which can be excited at 377 and 398 nm and emit at 422 nm. Both of these fluorophores have rather modest extinction coefficients (< 30,000 liter/mole-cm). The laser lines available for excitation of blue-fluorescing dyes are 350-365 nm and 458 from the more powerful water-cooled argon-ion lasers and the 325 and 410 nm lines from the air-cooled HeCd laser. Additional probe development work is needed to produce brighter blue-fluorescing labeling reagents that can optimally be excited at the available laser wavelengths. The brightness of these fluorophores is especially important in this wavelength region since UV and blue wavelengths can produce strong cellular autofluorescence that can bury weak and even moderate signals from tagged antibodies.

FUTURE PROSPECTS

What does the future hold for fluorescent labeling reagents? A number of advances will take place. First, more colors of fluorochromes will be developed that can be excited with the present lasers and the new lasers and laser diodes that will appear. These new colors will be used to reduce the complexity of the analysis of heterogeneous cell systems by helping investigators to eliminate (gate out) irrelevant cells or to select (gate on) relevant subpopulations. As multiparameter analysis software continues to improve, the use of additional fluorescence wavelength windows will be justified. The availability of a wide selection of tagged primary antibodies for direct immunofluorescence will be needed to fully utilize the power of multicolor analyses.

Second, there are classes of investigators who need to detect very few antigens on cells or to obtain large signals from positive cells so that they can be detected unambiguously in the presence of negative cells, such as in rare-cell detection. These experiments require very bright fluorescent labeling reagents. Presently, PE is the brightest label that is frequently used, although some investigators have bound antibodies to highly fluorescent sub-micron polymer particles and liposomes (see chapter 32). The possibility of labeling antibodies with enzymes that convert soluble, non-fluorescent substrates into fluorescent precipitates is intriguing. This method could give large signal amplification, but may be more difficult to quantify (8). While these large-sized labeling reagents may be useful to detect surface antigens, the labels are sometimes too large to readily penetrate into intracellular regions. Therefore, continued efforts to develop small but very bright fluorescent labels are important, particularly in view of the rapidly evolving interest in quantifying intracellular antigens and DNA in situ hybridization probes by flow cytometry.

ACKNOWLEDGMENTS

This work was sponsored by NIH NS-19353.

REFERENCES

1. Dauber JH, Wagner M, Brunsvold S, Paradis IL, Ernst LA, Waggoner AS. Flow cytometric analysis of lymphocyte phenotypes in bronchoalveolar lavage fluid. Comparison of a two color technique with standard immunoperoxidase assay. Amer Rev Resp Dis Molec Biol (in press).
2. Tsien RY, Waggoner AS. Fluorescent probes and photochemistry for confocal microscopy. In: Pauley J, ed. Confocal Microscopy Handbook. Plenum Publishing Co., 1990;169–178.
3. Waggoner AS. Fluorescent probles for cytometry. In: Melamed MR, Lindmo T, Mendelsohn ML, eds. Flow Cytometry and Sorting. 2nd ed. New York: Alan R. Liss, Inc. 1990;209–225.
4. Darzynkiewicz Z, Crissman HA. Flow cytometry. In: Methods in Cell Biology, Vol 33. San Diego: Academic Press, 1990.
5. Haugland R. Bioprobes. Eugene, Oregon: Molecular Probes, Inc., 1991.
6. Means GE, Feeney RE. Chemical Modification of Proteins. San Francisco: Holden-Day, 1971.
7. Mujumdar RB, Ernst LA, Mujumdar SR, Waggoner AS. Cyanine dye labeling reagents containing isothiocyanate groups. Cytometry 1989; 10:11–19.
8. Haugland R. Targeting enzymes in living cells using new fluorogenic substrates. Presentation, International Society of Analytical Cytology, Bergen, Norway. Cytometry Supplement 1991;5:27.
9. Oi V, Glazer AN, Stryer L. Fluorescent phycobiliprotein conjugates for analysis of cells and molecules. J Cell Biol 1982;93:981–986.
10. Haugland RP. Covalent fluorescent probes. In: Steiner RF, ed. Excited States of Biopolymers. New York: Plenum Press, 1983:29–58.
11. Titus JA, Haugland RP, Sharrow SO, Segal DM. Texas Red, a hydrophilic red-emitting fluorophore for use with fluorescein in dual parameter microfluorometry and fluorescence microscopic studies. J Immunol Meth 1982;50:193.
12. Kenner RA, Aboderin AA. A new fluorescent probe for protein and nucleoprotein conformation. Binding of 7-)p-methoxybenzylamino)-4-

nitrobenzoxadiazole to bovine trypsinogen and bacterial ribosomes. J Biochem 1971;10:4433–4440.
13. Allen G, Lowe G. Investigation of the active site of papain with fluorescent probes. J Biochem 1973;133:679–686.
14. Bratcher SC. J Biochem 1979;183:255–268.
15. Chen RF. Dansyl labeled proteins: determination of extinction coefficient and number of bound residues with radioactive dansyl chloride. J Anal Biochem 1968;25:412–416.
16. Southwick PL, Ernst LA, Tauriello EW, et al. Cyanine dye labeling reagents. Carboxymethylindocyanine succinimidyl esters. Cytometry 1990;11:418–430.
17. Mujumdar RB, Ernst LA, Mujumdar SR, et al. Cyanine dye labeling reagents: sulfoindocyanine succinimidyl esters. Bioconjugate Chem (submitted).
18. Small JV, Zobeley S, Rinnerthalen G, Faulstich H. Coumarinphalloidin: a new actin probe permitting triple immunofluorescence microscopy of the cytoskeleton. J Cell Sci 1988;89:21–24.

8

Practical Considerations for DNA Content and Cell Cycle Analysis[1]

PETER S. RABINOVITCH

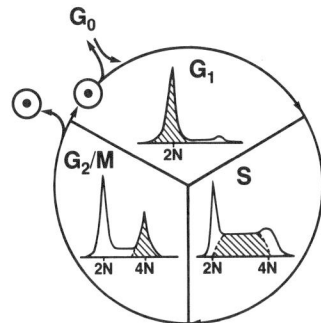

Figure 8.1. A schematic of the cell cycle illustrating flow cytometric G_1, S, and G_2 components.

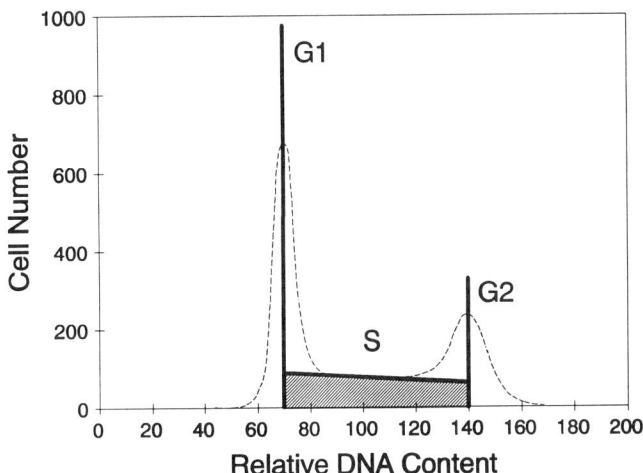

Figure 8.2. The difference between a histogram from a "perfect" flow cytometer with no errors in measurement (*heavy lines*) and the Gaussian broadening of the histogram that is encountered in all real analyses (*dashed lines*). The heavy vertical lines labeled G_1 and G_2 represent the theoretical, exactly reproducible DNA fluorescence seen in G_1 and G_2 cells. The shaded region represents S-phase cells with intermediate DNA contents. The breadth of the broadened histogram peaks will depend upon the experimental CV.

INTRODUCTION

One of the earliest applications of flow cytometry was the measurement of DNA content in cells. The striking feature of this analysis was that, for the first time, the DNA content of individual cells could be identified without laborious and time-consuming metabolic labeling. This analysis is based upon the principle that DNA-specific dyes should stain the cellular DNA in a stoichiometric manner (i.e., the amount of stain is directly proportional to the amount of DNA within the cell). Using such a DNA-specific dye, flow cytometry can rapidly and accurately measure the dye fluorescence in many individual cells, resulting in a histogram of DNA contents of the entire population. The purpose of this chapter is to discuss the practical considerations involved in obtaining and interpreting this data. Important areas of discussion include the departure from stoichiometry in DNA staining, artifacts introduced by cell or tissue processing, and optimal techniques for sample preparation and analysis by flow cytometry. Finally, the interpretation of the DNA content histogram is often not straightforward and guidelines for interpretation, as well as intrinsic limitations in the analysis, are discussed below.

CELL CYCLE FUNDAMENTALS

A characteristic pattern is produced when DNA-specific fluorescence is measured by flow cytometry, reflecting the cell cycle phases within the cell population. When cells with G_1 DNA content are stained, a "narrow" distribution of fluorescent intensities is obtained (Figs. 8.1 and 8.2). Since all of these G_1 cells have exactly the same DNA content, exactly the same fluorescence should be detected, in theory, from every G_1 cell and only a single channel of the histogram would be filled. This would be the case if flow cytometers were perfect and if binding of the DNA-specific dye was entirely uniform. In practice, however, a variety of instrumental errors, as well as biological variability in DNA-dye binding, occur. Consequently, the measured fluorescence from G_2 cells produces, at best, a Gaussian (normally distributed) peak, which is characteristic of theoretical variation in measurement (Fig. 8.2). At worst, the distribution may be

---

[1]Portions of this chapter are derived from the manual to the program MULTICYCLE™, by permission of Phoenix Flow Systems, San Diego, CA

skewed or even bimodal. Greater variation in measurement results in broader DNA-content peaks. The term coefficient of variation (CV) is used to describe the width of the peak, assuming that it is normally distributed (CV = 100 × S.D. / mean of the peak). Similarly, G_2 and mitotic cells, which biologically have twice the normal G_1 DNA content, theoretically produce a Gaussian peak at twice the mean position of the G_1 peak (Fig. 8.1). In practice, the G_2/G_1 ratio is usually less than 2.0, due to the greater condensation in DNA-protein (chromatin) packing in G_2 and mitotic cells compared with G_1 cells. As described subsequently, in such circumstances the DNA–specific dyes typically have slightly impaired accessibility to their DNA binding sites and, thus, a G_2/G_1 ratio below 2.0 (often 1.96–1.98) is more common.

In a theoretically perfect flow cytometer, S-phase cells would be observed in the histogram starting just above the position occupied by all the G_1 cells, and some of the S-phase cells would be found in each channel extending up to just below the position of all the G_2 cells (Fig. 8.2). This is due to the fact that, as cells begin to synthesize DNA in S phase, their DNA content is initially just barely above the G_1 content; their DNA content progressively increases until the completion of S phase with cells having a G_2 DNA content. Actual histograms are not so simple, because the same factors that broaden the G_1 and G_2 peaks, also broaden the S-phase distribution, with the result that early S-phase cells overlap with G_1 cells, and late S-phase cells overlap with G_2 cells. Chapter 3 describes models that can be applied to attempt to extract the proportions of G_1-, S- and G_2-phase cells from such a histogram.

DIPLOID AND ANEUPLOID DNA CONTENTS

As described above, G_1 cells have, in almost all cases, the same DNA content and the same chromosomal complement (humans have two each of 23 chromosomes). This is referred to by cytogeneticists as the **diploid** DNA content, and the designation 2N is used to describe this value (where N refers to a single complement of chromosomes, the **haploid** DNA content). The designation 2C can be used to describe a DNA content equivalent to diploid cells. However, for maximum clarity, the designation DNA Index (DI) 1.0, is usually used to describe the diploid DNA content measured by flow cytometry (1). DNA contents other than 1.0 are not necessarily abnormal. S- and G_2-phase cells have the DNA contents described above, gametes have haploid DNA contents, and a few cells in the body are **tetraploid** (46 chromosomes, DI 2.0). Multinucleated cells are another exception and are commonly found in certain tissue and cell types. All of these DNA contents are together referred to as **euploid** values; all have chromosomes in intact sets and each chromosome is itself an unaltered subunit. Any other DNA content has either an abnormal set of chromosomes or at least one abnormally constructed chromosome and is referred to as **aneuploid** (literally, not euploid). Because whole-cell or whole-nucleus DNA flow cytometry does not measure or examine chromosomes, the terms diploid and aneuploid are terms best used by cytogeneticists. Flow cytometry cannot in fact tell whether a cell that has a DNA index of 1.0 has a normal chromosomal constitution and it should be properly referred to as "indistinguishable from diploid." Sometimes the shorthand designation DNA diploid is applied to imply that the DNA content is apparently diploid as examined by DNA flow cytometry. Similarly, a cell with a DNA index of 2.0 could be a G_2 cell, a tetraploid cell, or an aneuploid cell that has abnormal chromosomes that happen to give rise to a DNA content that is indistinguishable from tetraploid.

Flow cytometry can detect aneuploidy when a population of cells with a DNA content that is not a multiple of DNA index 1.0 is observed, which implies that either the number or the composition of chromosomes has been altered (with the exception of the rare congenital defect of triploidy, in which cells have three sets of normal chromosomes, and haploid gametes). Since total cellular DNA content alone is measured by flow cytometry, it is more correct to refer to it as "DNA aneuploid" to distinguish it from cytogenetic analysis. Aneuploid cell populations are almost always, but not exclusively, associated with malignant tissues. Exceptions that must be noted are some benign tumors (for example, endocrine adenomas) and some premalignant epithelial cells (for example, colonic adenomas or dysplastic epithelium in ulcerative colitis).

The degree of chromosomal abnormality that is detectable as DNA aneuploidy depends upon the resolution and quality of the DNA analysis. Typically, the normal and abnormal DNA content peaks must differ by 5% to 20% in DNA content in order to be distinguishable as separate populations. The distinction of near–tetraploid DNA aneuploidy from DNA tetraploidy (indistinguishable from tetraploid) is perhaps even more challenging than the detection of near-diploid DNA aneuploidy. Both of these subjects are discussed in detail subsequently.

When a malignancy is found to be DNA-aneuploid, the analysis almost invariably shows a component of DNA-diploid cells as well. The latter consist of lymphocytes, endothelial cells, fibroblasts, and other stromal elements that are always present in tissue to a greater or lesser degree. Both malignant and stromal cells have a subset of proliferating cells (the latter usually much smaller than the former), so a DNA-content histogram of an aneuploid tumor usually has two overlapping cell cycles, requiring greater sophistication in the method of cell cycle analysis.

PREPARATION OF CELL SUSPENSIONS FOR DNA ANALYSIS

Fresh Tissue

Mechanical, mechanical-enzymatic, or mechanical-detergent preparation techniques may be employed to produce a dispersed solution of stained cells or nuclei for flow cytometry. A detailed discussion of these techniques is provided in Chapter 6. In brief, methods that yield nuclear suspensions are the most rapid, are technically less demanding, and avoid the use of fixatives that may induce cell aggregation or staining artifacts. Either hypotonic lysis (2) (most commonly used for cells already in suspension) or nonionic detergents are used to disrupt the cell membrane and release nuclei. Detergent may be used with (3) or without (4) proteolytic enzymes (usually trypsin). Dissociation of tissues is aided by mechanical disruption or trituration (syringing) during tissue mincing.

The primary disadvantage of nuclear suspensions is that cell surface and cytoplasmic markers are lost. This severely limits the methods available for use of a second parameter to differentiate among cell types (for example, cytokeratin to identify epithelial cells). The preparation of intact single-cell suspensions is challenging, however, because many of the epithelial cell types of interest have strong intercellular attachments. Methods used to prepare whole-cell suspensions are mechanical (5, 6), or mixed mechanical-enzymatic (6–8). A component of free nuclei or cells that have been partially stripped of cytoplasm is often obtained. This can be a complication in the use of cytoplasmic and surface markers: If cells with little residual cytoplasm are an appreciable portion of the total, interpretation of multivariate data must take into account "false-negative" cells due to this effect (9).

In evaluating a particular protocol, particular attention should be given to these aspects: 1) a low CV and little debris is desired for reasons elaborated upon subsequently; 2) representative sampling and an adequate cell/nuclei yield are necessary. A low cell yield presents the problem that a large sample size is needed to derive adequate cell numbers; more importantly, low yield increases the possibility of nonrepresentative sampling of cellular subsets (aneuploid populations and cell cycle subpopulations).

Paraffin-embedded Tissue

Most preparative techniques for obtaining nuclei from paraffin-embedded tissue are based upon the method originally proposed by Hedley (10). A variety of variations of the original technique have been described (11) (see Chapter 6). Two consistent points have been made in the literature examining this technique: a) Inadequate protease digestion may result in the nonrepresentative release of nuclei from the tissue and, in some cases, the detection of a malignant population may be compromised by falsely low proportions of these cells present in suspension. Some authors suggest lengthening the time in protease. Since prolonged digestion may degrade DNA (broadening CVs, increasing debris, or even preferentially destroying tumor cells (12), an alternative strategy is to harvest the released nuclei (supernatant in a 1g sedimentation) after routine enzyme treatment and then to subject the remaining pellet to a repeated round of digestion and harvest. With careful attention, enzymatic digestion from embedded tissue may result in cell yields as high as for fresh tissue. Adequacy of digestion and trituration can often be gauged by the amount of nondigested tissue remaining and the extent of aggregation in the harvested sample. Underdigestion can lead to nonrepresentative cell release and potential failure to identify abnormal cell populations. Additionally, inadequate enzyme digestion can lower fluorescence staining intensities and can result in wider CVs. This reflects the fact that cross-linking, which occurs during fixation, inhibits DNA dye binding accessibility (see below), but that proteolysis can restore such accessibility. b) Care should be taken to ensure complete dewaxing before enzyme treatment. Residual paraffin and incomplete hydration impair both proteolysis and DNA staining and result in decreased cell yield and wider CVs.

DNA STAINING OF CELLS

Dyes and Dye Classes

A small number of DNA-binding dyes is ordinarily used in flow cytometry for ploidy and cell cycle analysis. Propidium iodide (PI) and ethidium bromide (EtBr), the most common, are related red-fluorescing compounds that are easily excited by the 488-nm line from an argon-ion laser. Both dyes intercalate between base pairs in double-stranded DNA or RNA; hence, RNase treatment after fixation or permeabilization is required in order to impart DNA specificity. The ultraviolet-excited dyes 4,6-diamidino-2-phenylindole (DAPI) and Hoechst (HO) 33258 and 33342 have high specificity for DNA, selectively binding to A-T-rich regions. Because of its rapid staining, ability to yield low CVs, and relatively low dependence on state of chromatin conformation, DAPI is widely regarded as the dye of choice when UV excitation is available. For analysis of viable cells, however, the cell-permeant dye Hoechst 33342 is used almost exclusively (13). Acridine Orange (AO) is used principally for DNA/RNA staining or for differentiating single vs. double-stranded DNA (see Chapter 2). 7-amino-actinomycin D (7AAD) is a bulky intercalating dye that, unless dependence of staining upon DNA conformation is specifically desired, is useful principally because it is 488-nm excited and emits above 600 nm, allowing its use simultaneously with FITC and phycoerythrin stains (14). The DNA-binding antibiotics mithramycin (MI), chromomycin, and related dyes are usually used only in special applications.

Staining Considerations and Artifacts

A central assumption in the usual applications of DNA flow cytometry is that DNA staining is stoichiometric, that is, that DNA binding, and resultant fluorescence, is proportional to DNA content. It is known, however, that only a portion of total DNA is ordinarily accessible to DNA-binding fluorochromes (15, 16). This fraction is highest with DAPI and lowest with 7AAD. More importantly, accessibility to dyes is known to be dependent upon the state of chromatin condensation being influenced, for example, by cell type, cell cycle state, cell differentiation, and cell viability (17, Chapter 2). Illustrations of differences in DNA staining of different cell types have been made by comparing lymphocytes to

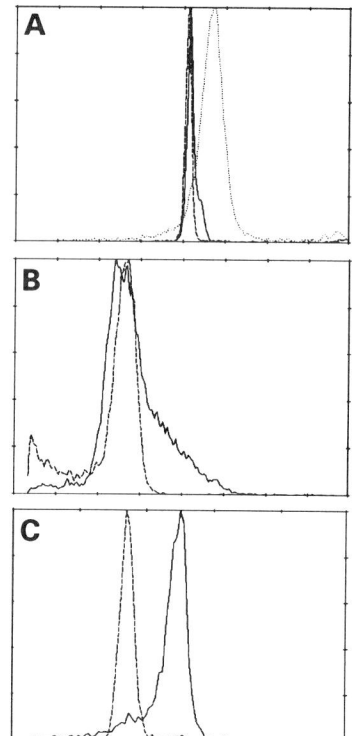
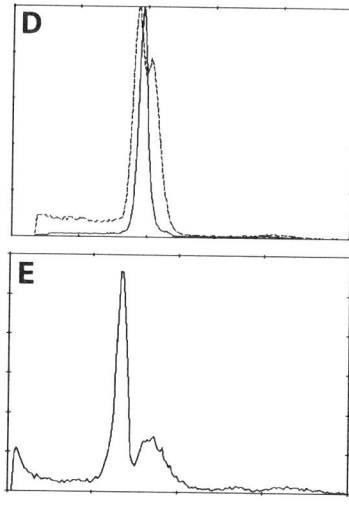

Figure 8.3. Nonstoichiometric DNA staining. In **A**, lymphocytes (*dashed line*) and granulocytes (*solid line*) were isolated from human peripheral blood and mixed 1:1:1 with cultured human diploid fibroblasts (*dotted line*). The mixed cells were fixed with 70% EtOH and stained with PI in phosphate-buffered saline with 0.1% RNAse. Fluorescence from each cell type was identified by gating on forward vs. 90° light scatter. Granulocyte staining appears 1.9% more intense and fibroblast staining appears 26% more intense than lymphocytes. In **B**, lymphocytes and granulocytes were fixed with 1% paraformaldehyde for 2 hr, then washed and stained as above. Note that in comparison to **A**, cell staining intensity is reduced and CVs are broader, especially in the case of granulocytes. **C** shows EtOH-fixed lymphocytes and granulocytes, as in **A**, but stained with 7AAD. Granulocyte staining is considerably more intense than that of lymphocytes (X-axis scale amplified over that of **A** and **B**). **D** (courtesy of Dr. Dennis Ahnen) shows pseudoaneuploidy in PI-stained cells extracted from paraffin-embedded colon mucosa, analyzed with partial (*dotted line*) vs. complete (*solid line*) neutralization of the acid pH used in pepsin digestion. **E** shows pseudoaneuploidy from histologically normal paraffin-embedded pancreas analyzed with DAPI staining.

fibroblasts using AO, PI, and EtBr (18) (Fig. 8.3A); lymphocytes to granulocytes and monocytes using 7AAD (19) (Figs. 8.3A and 8.3C); leukocytes to a variety of benign tissue types using EBMI (20); nucleated red blood cells to leukocytes using PI, HO, and MI (21); and within different types of amniotic fluid cells using DAPI (22). Differences have also been documented within erythroid cells at different stages of differentiation using DAPI, AO, and 7AAD (15) and by examining spermatocyte differentiation with AO and DAPI (23, 24). Nonstoichiometric dye binding is also observed in dead or dying cells, apoptotic cells, or cells with DNA damage (25–29). Finally, differences in chromatin condensation and DNA staining have been observed between cells in different phases of the cell cycle, especially G_1 vs. G_0 and M phases, using AO, PI, EtBr, 7AAD, and MI (14, 28, 30–32).

To further complicate the relationship of dye binding to DNA content, a variety of treatments commonly employed in flow cytometric DNA analysis are known to affect dye accessibility. These include aldehyde fixatives (Fig. 8.3B) (30, 33) and proteases and acid treatment (Fig. 8.3D) (34, Chapter 2). The latter treatments remove nucleoproteins; the removal of histones, in particular, has been shown to allow enhanced dye accessibility and binding (34). It is also possible that dye-staining protocols that purposely remove histones may reduce the occurrence of artifactual staining patterns in nonformalin-fixed tissues, although this has received little attention to date.

Finally, optimum staining and analysis is only achieved when the DNA/Dye ratio is within a favorable range. When the ratio is large, staining conditions are far from saturation and differences between cells in DNA-dye binding affinity will produce larger differences in staining intensity than when the dye is closer to saturating conditions. If the DNA dye concentration is too high, then the fluorescence from nonbound dye will be large and this will reduce the signal–to–noise ratio, increasing the CV. For PI, commonly recommended optimum staining conditions are 50ug/ml dye and a maximum of a few million cells/ml.

The occurrence of nonstoichiometric DNA binding has a great impact upon the routine practice of DNA flow cytometry. This is partly due to the possibility of finding false aneuploidy. Variability in DNA staining also imposes a practical lower limit on CVs obtainable from analyses of tissues containing mixed cell types or may result in skewed DNA-content distributions. Inflammatory cells admixed with epithelial cells, for example, can produce a large rightward tail in the DNA-diploid peak (35). Since such differences in staining can be unrelated to actual DNA content, this also imposes limits on the reliability of detection of near-diploid DNA aneuploidy. For this reason, small DNA staining differences or differences that do not result in bimodally separated DNA-content peaks are not considered reliable evidence of DNA aneuploidy by conventional criteria (1). Also, since cells chosen as a "standard" (for example, lymphocytes) may differ in DNA staining from other diploid cell types (as noted above), small differences in staining intensity relative to "standards" cannot necessarily be interpreted as evidence of DNA aneuploidy. These issues will be treated in depth in subsequent sections of this Chapter.

SAMPLE PRESERVATION AND AUTOLYSIS

DNA is subject to degradation in dead and dying cells. Frequently, but not always, DNA alterations may be accompanied by histologic evidence of necrosis, karyolysis (nuclear disruption), or karyorrhexis (chromatin condensation and clumping). Cellular degeneration and autolysis almost always begins to occur as soon as cells are removed from the body. The rate of autolysis will vary between tissues; the most rapid autolysis occurs within hours in tissues that naturally contain degradative enzymes, such as pancreas. Disruption of the nucleus leads to fragmentation that is visible as debris in the histogram. Debris has an adverse effect upon the accuracy of cell cycle analysis, as detailed below. More subtle autolytic changes result in altered chromatin structure. These changes can lead to altered stoichiometry of dye binding to DNA. Alanen et al. (36) have shown that the increased staining of DNA that accompanies autolysis can result in skewed G_1 distributions, shoulders on the right side of the G_1 peak, and even distinctly bimodal or separate peaks, up to an apparent DI of 1.3. Furthermore, the authors suggest that, because their intercellular connections are weakened, autolytic cells may be preferentially released in mechanically isolated cell suspensions. The presence of a peak that meets conventional criteria for diagnosis of DNA aneuploidy (1) in benign autolyzed tissue seriously complicates the interpretation of DNA histograms. Thus, specimens should be refrigerated immediately upon acquisition, and preparation and analysis of specimens for flow cytometry should not be delayed. In cases in which storage of unfixed samples is required, they should be frozen immediately upon acquisition (some authors suggest buffered medium containing 10% DMSO). Because refrigeration and time of transportation of unfixed specimens between institutions may not be reliable, it may be desirable to ship frozen tissue. In samples in which peaks with DI less than 1.3 are seen, the possibility of autolysis should be investigated by the evaluation of histology as well as transport and processing history.

Similar to the above observations, the analysis of archival blocks of normal tissues has shown false aneuploid peaks with DI 1.15 to 1.4 (37). These were seen in 4.7% of cases examined and were found in the pancreas (Fig. 8.3E), spleen, liver, thyroid, and lymph nodes (in decreasing order). The mechanism for their production may be the same as that for autolysis of fresh tissues, as noted above. Histologic examination, however, did not predict which of the specimens gave rise to bimodal histograms. The possibility that bimodality results from differential effects of fixation on a subpopulation of cells also cannot be excluded.

RUNNING THE SAMPLE ON THE FLOW CYTOMETER

The following are practical considerations in acquiring the DNA content histogram:

TUNING/CV

Following the manufacturer's suggestions, use lymphocytes, nucleated red blood cells, etc., for calibration. CVs from such standards should be <3.0. CV's from fresh tissue are generally <4.0, and should rarely be >6.0. The results obtained with paraffin-derived samples are more heterogeneous. At least some samples with CV's that approach those of fresh samples should be seen, although the average may be several percent higher.

SKEWED PEAKS

In addition to CV, criteria for tuning the instrument should include a symmetric Gaussian (normal, bell-shaped) distribution for the peak. Skewed or tailed (nonsymmetrical) distributions impair the reliability of cell cycle analysis. Occasionally, a tissue sample may produce an abnormal peak shape even when the tuning standard is satisfactory. Check staining conditions first, such as cell and dye concentrations, RNase presence (for PI, EB), and adequate staining time. Also check other tissue samples, if available.

If technical factors are eliminated, the probability increases that the abnormal peak shape is sample-specific. This can be due to true DNA-content heterogeneity (i.e., DNA aneuploidy or multiple DNA-aneuploid populations) or to uncorrectable DNA staining variability in cells with the same DNA content. The latter is more frequently encountered in formalin-fixed tissue and results from variations in formalin fixation within different regions of a specimen or from different degrees of chromatin crosslinking occurring in different cell types.

NUMBER OF CELLS

As described above, there is an optimum concentration range for cells or nuclei in order that the DNA/dye ratio be favorable. Below this cell concentration, the dye is nearer to saturating, which may be advantageous. However, the sample takes longer to analyze and fewer total cells are available for analysis. There is no single criterion for the minimum number of cells required for an accurate DNA-content analysis. As a rule, 10^5 cells represents the minimum desirable number in a histogram; however, ploidy information and, in some cases, accurate cell cycle data can be obtained with fewer cells. More cells should be acquired when debris signals are a large proportion of the total events analyzed in order to maintain the desired number of cell in the actual cell cycle distribution. The object of aquiring larger numbers of cells is to reduce statistical fluctuations in the histogram. This "noise" is most visible in areas of fewer cell numbers, for example, S phase. The least-squares curve fitting models with simple S-phase shapes tend to be less sensitive to such statistical noise than other methods. Smoothing of the histogram is generally not necessary or advisable. For a particular laboratory protocol and instrument, the best test of adequate cell numbers is empirical, performed by acquiring two or more histograms from the same sample, even if each has low cell numbers. Conduct a separate analysis for each, compare numerical results, and establish the standard deviation of each estimated parameter. The inconsistencies will be extremely small for large sample numbers, but will increase as cell numbers decline. The range of cell numbers resulting in acceptable reproducibility can be ascertained for categories of histograms having high and low S-phase fractions, and large and small proportions of aneuploid cells.

TISSUE SAMPLING

Has the Tumor Been Analyzed? A principal weakness of flow cytometry is that information relating to tissue organization has been lost by the time of the analysis. In the clinical setting, it is important to insure that the tumor has been analyzed through gross and histologic examination of the tissue submitted for flow cytometry. For samples analyzed in the unfixed state, the gross evaluation is best confirmed by histological examination of an adjacent section of tissue. If histological examination of this adjacent portion reveals that the tissue submitted for flow cytometry was not representative of the tumor (or contained too small a proportion of malignant cells or too great an inflammatory infiltrate) then the analysis can be repeated from the best of the paraffin-embedded tissue.

For samples submitted in paraffin block(s), histological examination of a thin section can reveal the location of the cells of interest (i.e., malignant). Some blocks may contain diffuse malignancy and, thus, section(s) from the entire block can be processed for flow cytometry. There is an increasing trend, however, for diagnostic and surgical procedures to be performed at earlier stages of neoplasia, that are often associated with only focal malignancy. In this setting, it becomes important to select carefully only the subportion(s) of the block that contain the most abundant malignant cells. This is important because cell cycle analysis is less accurate when samples include admixtures of large proportions of benign cells (see challenge no. 4 in cell cycle analysis, below). This can be an especially severe problem in the presence of an abundant lymphocytic infiltrate. The human eye usually underestimates considerably the numbers of the smaller lymphocytes in proportion to the larger (but often numerically fewer) neoplastic cells. Careful comparison and alignment of the stained thin section to the remaining paraffin block may be necessary to select regions of the block that correspond to the desired areas of the thin section. Care should be taken that the thin section examined is a "recent" section, since the architecture of the tissue can change as the specimen is serially sectioned. In difficult cases, assistance may be provided by applying a DNA stain directly to a fresh surface of the paraffin block. Under examination by fluorescence microscopy, cellular and nuclear architecture within the block may be observed directly (38, 39). Once the desired region of the block is identified, one of three different procedures can be followed: a) the block can be sectioned at 50 um, (prevent the sections from curling by exercising care and/or using adhesive tape, then use a scalpel to cut subregions of each thick section either before or after rehydration) (40); b) the desired region(s) can be marked with cuts made directly into the paraffin block with a scalpel (these will remain apparent in subsequent thick sections); or regions of tissue can be cut or punched out from the block and reembedded for thick sectioning.

Finally, to eliminate any doubt that tumor cells were present in the tissue and that they were extracted for flow cytometry, the cell or nuclear suspension giving rise to a DNA-"diploid" histogram can be directly examined for the presence of tumor cells by a cytologist using fluorescence microscopy.

The Number of Tissue Samples Required. It is now clear that tumors (especially large ones) may have DNA-content heterogeneity manifested either by multiploidy (two or more ploidies) or by aneuploidy and non-aneuploiody in different regions on the same tumor. The literature varies considerably, however, as to the frequency of these findings. In breast cancer, Beerman et al. (41) reported as many as 61% multiploid tumors. This seems high in view of other reports, including a larger study by Meyer and Wittliff (42) who found 26% heterogeneity, with heterogeneity being more common in large tumors. Mixed regions of DNA aneuploidy and absence of aneuploidy in one tumor have been reported in 18% of cases (43) and in only 5% of cases (44).

Other tumor types in which DNA content heterogeneity has been described include lung (45) and colon cancers (46). Extreme examples of multiploidy and heterogeneity in location are presented by premalignant tissues of the gastrointestinal tract, Barrett's esophagus, and Ulcerative colitis, in which 13-14 different aneuploid populations have been reported (47, 48).

The best evidence to date indicates that multiple flow cytometric sampling of tumors may be needed to detect most accurately aneuploidy or multiploidy, especially in large tumors or tissue samples. At the same time, however, it should be noted that most of the literature establishing the clinical utility of DNA analyses has been performed with single-tumor samples for flow cytometry.

THE RANGE OF DATA IN THE HISTOGRAM

An important guideline is: **Don't throw away data**. Valuable information can be lost at this stage and cannot be recovered once the sample is run. The left-hand portion of the histogram contains most of the information relating to the extent and shape of the debris distribution. Ignoring this information is one of the most common inadequacies in the practice of histogram analysis. Each laboratory should establish a procedure to compensate for the effect of the debris—this should not, however, include the cosmetic approach of merely setting a lower discriminator or a gate to exclude the debris from the DNA histogram. The debris that is observed in histogram channels below the G_1 peak is only a symptom of the problem of larger debris fragments, aggregates of debris, and cut or sliced nuclei (discussed below). These debris signals can also be present in channels that underlie the S-phase and G_2 populations. Ignoring this debris can often result in falsely elevated S-phase estimates. Setting a lower limit of data acquisition at a channel that corresponds to DI 0.1 is usually satisfactory. There is usually no need to collect debris channels so low that the debris height exceeds the G_1 peak height(s).

Similarly, do not ignore the hypertetraploid region. Data above the G_2 of the population with highest ploidy contains valuable information relating to the degree of aggregation, and hypertetraploid peaks may not be detected if these "high" data channels are discarded or are accumulated in the last "overflow" channel. As a general rule, observe channels at least 50% above the highest G_2 peak (DNA index = 3 for a diploid sample, higher if hyperdiploid aneuploidy is present). Observation of the position of triplets (DNA index = 3 for diploid triplets) allows the approximation of the extent of aggregation at the time of sample analysis. Use an expanded vertical histogram scale for best visibility. As a rough rule, the triplets should not be in excess of 10% of the peak height of the G_2 population. Use of software aggregation modeling, as discussed below, can quantitate the extent of aggregation and, to a substantial extent, compensate for its effects, but this can only be done if adequate hypertetraploid data is contained within the histogram. This may require that the G_1 peak be placed to the far left on the DNA-content histogram (channel 40 of a 256-channel histogram when near-tetraploid aneuploidy is present, for example). Implications of the use of lower histogram channels are discussed below.

DIGITIZING DATA: HISTOGRAM LINEARITY AND NUMBER OF CHANNELS

No matter how much care is lavished on sample preparation and instrument tuning, when the fluorescence signals are converted to a digital histogram, inaccuracies can be introduced by the analog-to-digital conversion process itself. A common problem is a lack of linearity in the analog-to-digital conversion. This means that differences between two analog signals are not faithfully converted into the same relative differences in digital histogram channel values. Departures from linearity can produce nonstandard G_2/G_1 ratios, altered DNA indices, and potential difficulty in modeling aggregation using computer software. One common source of nonlinearity is an incorrect setting in the analog-to-digital converter for the channel in which a signal of zero magnitude is placed. Although it seems intuitively obvious that a signal of zero size should be placed in channel zero, there is often a small offset that is introduced in the analog circuitry and a corresponding means to adjust this. Problems with nonlinearity from this and other sources are most commonly manifest in the lowest and highest ends of the histogram; this can have the greatest effect on the use of a DNA-content standard in the lower channels or on the evaluation of G_2 peaks and aggregates in the higher channels.

It should not be assumed that the manufacturer of your instrument has insured good linearity, even if the flow cytometer is new. The user should request proof of testing of linearity of new instruments and, for machines already in use, the manufacturer should have guidelines for calibrating linearity. Detailed methods are also available in the literature for assessing linearity (49). If nonlinearity of the histogram is found, the analog-to-digital converters should be adjusted or replaced. Software compensation for the effects of nonlinearity is theoretically possible, but not routinely available. At the very least, the presence of nonlinearity should be recognized so that variant G_2/G_1 ratios and DNA index estimates are not misinterpreted.

The number of channels into which the DNA-content histogram is digitized can have an important effect upon the data. A wide spectrum of histogram "resolutions" is in common use, ranging from 64 to 1024 channels or "bins." Too few channels may not provide the resolution needed to preserve the accuracy of the original analog signal (i.e., channel 10 contains analog values from 9.5 to 10.5, numbers that vary by 10%). Low channel numbers may be used in bivariate cytograms; when one axis is DNA content, the number of channels of DNA resolution may commonly be as low as 64 or 100. When high channel numbers (512 or greater) are used, almost no accuracy in DNA content is lost. But, unless large numbers of cells are analyzed, statistical fluctuations between the values of histogram channels will be greater than if fewer channels are utilized. This produces a less satisfactory appearance and, depending upon the type of histogram analysis, can produce greater uncertainties in data analysis. Histograms with low vs. high channel numbers usually have equal susceptibility to the effects of nonlinearity, since most modern analog-to-digital converters perform the initial conversion at high resolution (greater than 1024 channels) and then decrease the resolution actually displayed.

In practical terms, how low a channel can a DNA peak be placed in and not lose accuracy of the CV and DNA content ratio? Figure 8.4 shows the result of analysis of histo-

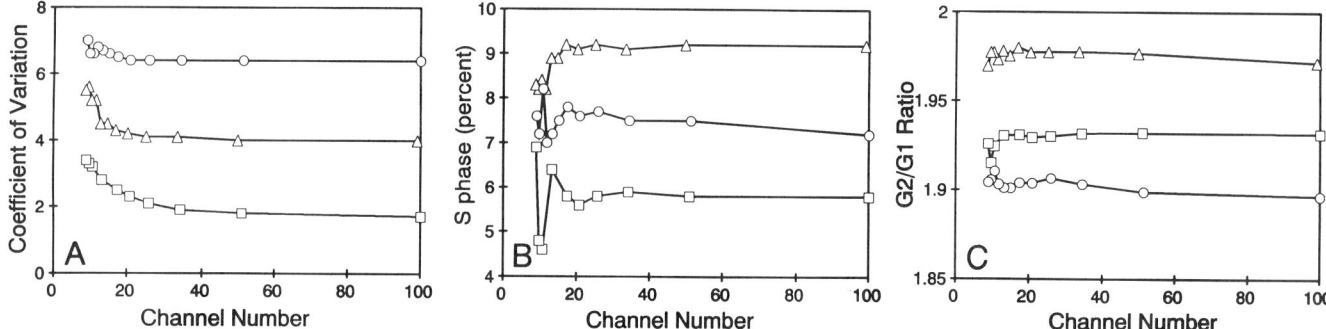

Figure 8.4. The effect of lower numbers of channels or "bins" in the X-axis of a DNA-diploid cell cycle histogram with a 256-channel resolution, with the G_1 peak placed at channel 100. The same histograms were reanalyzed after the X-axis range was reduced by compressing two or more channels of data into a single channel, effectively reducing the number of channels in the histogram. The mean channel position of the G_1 peak for each of these compressed histograms is shown on the abscissa vs. the CV of the G_1 peak (**A**), the percent S-phase estimate (**B**) and the G_2/G_1 ratio (**C**). The three different histograms analyzed had an initial G_1 peak CV of approximately 2 (squares), 4 (triangles), and 6 (circles). The CV 2 population was from a fresh specimen, the CV 4 and 6 samples were extracted from paraffin. Cell cycle analysis was performed using nonlinear least squares fitting to the model of Dean and Jet using a zero order (rectangular) S-phase polynomial; curve-fitting methods such as this are among the least sensitive to the effects of lower numbers of channels. Debris from paraffin-extracted cells was analyzed with the sliced-nucleus model.

grams with progressively lower x-axis ranges, resulting in G_1 peaks located in progressively lower channel numbers. It can be seen that CV rises appreciably when the mean of the peak is below channel 25 (Fig. 8.4A). S-phase estimates become erratic below this value as well (Fig. 8.4B). G_2/G_1 ratio estimates vary by only a few percent when the G_1 mean is below channel 25 (Fig. 8.4C), as do calculations of DI (not shown); if more pronounced changes are encountered, they are almost certainly due to nonlinearity. As discussed above, it may be necessary in some analyses to place the lowest G_1 peak in the lower 15% of the total channels (and DNA-content standards even lower). Sufficient resolution is usually achieved with 256 channels. 100-128-channel histograms sometimes have insufficient resolution and 64-channel histograms commonly have insufficient resolution.

DNA-CONTENT STANDARDS: WHEN AND WHAT TO RUN

For nonformalin-fixed cells, a DNA-content standard should be run whenever it is unclear which peak corresponds to the diploid DNA content. Keep in mind that it is very unusual for a tissue with aneuploid cells present not to also have at least a small component of diploid cells (i.e., stromal cells, endothelial cells, etc.). It is also very unusual for aneuploid DNA contents to be present with DNA indices below 0.8. Because of these factors, for histograms with aneuploid DI above 1.5 it generally is clear that the peak present in the expected diploid channel number is in fact the DNA-diploid peak without having to add a standard to the sample itself. If there is any uncertainty, after storing the original histogram, a known DNA-content standard may be added to the sample in amounts that are as close as possible to concentrations of the original cells. If human lymphocytes are used as a standard, the "diploid" peak position should be elevated in magnitude. As described above, however, differences in DNA staining in different cell types can be observed due to differences in chromatin condensation and dye binding. Thus, small differences in DNA fluorescence between an unknown sample and a cell "standard" should be interpreted with great caution. There also are very reasonable arguments that any DNA-content standard of known size can be used and that a confidence interval of ratios of diploid-to-standard DNA contents can be established in the laboratory for each tissue type and patient sex (women have 1.8% more DNA per cell than men). If the standard has a DNA content that is much less than that of diploid human cells (for example, chicken or trout RBCs), then the standard will appear as a distinct peak in the histogram and will not overlap with the diploid human cells. Software analysis of this peak will provide a ratio to "diploid" for evaluation. Chicken erythrocytes have the advantage that their DNA content is low enough that doublets do not overlap the cell cycle of human cells.

Use of a standard in the analysis of nuclei from paraffin blocks is unlikely to be feasible. Variability in fixation and DNA-dye accessibility prohibits any reproducibility in the position of the standard peak. Use caution even if attempting to use diploid regions of tissue from the same case or block; there is substantial variability in DNA stainability from block to block and even within different portions of the same block (50). The criteria that distinctly bimodal peaks be present to diagnose DNA aneuploidy is especially pertinent. Many laboratories interpret histograms with two or more G_1 DNA peaks by making the assumption (true in at least the large majority of cases) that the DNA aneuploid population(s) are hyperdiploid, rather than hypodiploid. If utilized, this assumption (and appropriate caveats) should be stated.

Some authors have suggested that chicken and trout RBCs be run simultaneously (51). In principle, this allows compensation for instrumental offsets in the X-axis zero position (zero offset). This approach has been used to measure DNA-content differences of 0.5% to 1% between samples

(51). However, DNA-content differences within the same range also have been measured using only one DNA content standard (52). The practical advantage of two standards for DNA-content analysis of tissues may be limited by the following arguments:

1) In most cases, the accuracy of determining DNA-content peaks is more limited by variation in staining of a particular specimen than by instrumental error in determining the diploid:standard ratio. Therefore, any increased accuracy of the standard ratio is probably much less than the variation in sample staining.

2) Since the diagnosis of DNA aneuploidy must be supported by demonstration of distinct DNA-diploid and aneuploid peaks (see the subsequent section, "detection of near-diploid aneuploidy"), the purpose of the standard is only to identify which peak is DNA-diploid. For CVs in the ranges usually obtained, the minimum DNA-content differences that need to be resolved are 5% to 10% (see below). The resolution needed for this is within the capability of a single standard.

3) If the DNA-diploid peak is kept in the same approximate X-axis position when running samples, then the diploid:standard ratio will be reproducible using one standard, even if there is instrumental histogram nonlinearity. A case for the use of two standards may be made if a large uncorrected zero offset in the analog-to-digital conversion is present and the diploid peak is to be run in a changing position on the histogram axis.

CELL CYCLE ANALYSIS OF DNA-CONTENT HISTOGRAMS

Methods for cell cycle analysis from DNA-content histograms range from simple graphical approaches to more complex curve-fitting and deconvolution methods, as described in detail in Chapter 3. All of the simpler methods are based upon the observation that the G_1- and G_2-phase fractions may be extrapolated from the examination of the portions of the histogram where the G_1 or G_2 phases have less overlap with S phase. These methods also assume that the G_1 and G_2 peaks are perfectly symmetrical (DNA staining variability in tissues does not always provide this) and that the midpoint (mean) of each peak is known with precision. Because of the overlap of G_1 and G_2 peaks with the S phase, the mean of these peaks is not always at their maximal height (mode); the situation is even more complex and ambiguous if a second overlapping cell cycle is also present. Consequently, these simpler methods are the most prone to errors resulting from noise, histogram artifacts, and nonsymmetrical peak shapes.

Curve-fitting methods are based upon the prediction that the cell cycle histogram is produced by the Gaussian broadening of the theoretically perfect distribution (Fig. 8.2). The underlying distribution can be recovered or "deconvoluted" by fitting the G_1 and G_2 peaks as Gaussian curves and the S-phase distribution as a Gaussian-broadened distribution. An advantage of such methods is that they can be directly extended, in principle, to analysis of two or even three overlapping cell cycles; the overlapping components are mathematically deconvoluted into their individual parts.

Note that the calculation of the CV can vary depending upon the method used to define the S.D. of a peak. Statistical formulae are based upon a region chosen in the vicinity of a peak and are dependent upon how wide a region is chosen. Curve-fitting to a Gaussian distribution as part of a cell cycle model is less subject to this variability. Furthermore, real data generated in the laboratory, especially from clinical samples, often departs from the theoretical ideal, which limits the accuracy of the simpler techniques and presents challenges to even the most sophisticated methodologies. The most common artifact seen in tissues (especially in the clinical laboratory, where preservation may not be perfect or the cells may have been fixed and embedded in paraffin) is that the "Gaussian" G_1 and G_2 peaks are skewed or broadened at their bases. An additional advantage of more sophisticated curve-fitting methods is that they tend to be less dependent upon the initial or "starting parameters" used to begin the fitting process. Such parameters include initial estimates of peak means and CVs as well as the limits of the region of the histogram included in the fit. When these starting parameters are less important, interoperator variability in producing results decreases (53).

An important aspect of the analysis of histograms that are imperfect (for example, broad CVs or nongaussian peak shapes) or complex (for example, multiple overlapping peaks and cell cycles, appreciable aggregates) is the ability to reduce the model's complexity when needed. For example, some models assume that a skew or broad base in G_0 or G_1 peaks is part of the S phase, which can lead to an overestimation of the true S phase. This applies to the multiple Gaussian technique (54) and the Dean and Jett algorithm (55) when used with a second order polynomial S phase (see Chapter 3). A more conservative approach may be more accurate in situations where there is not high confidence in the quality of the histogram (in general, confidence may be high when individual peaks are narrow and well resolved; multiple peaks, if present, are distinct and minimally overlapping; debris and aggregation is minimal, and CVs are below ~6% (unfortunately these are very approximate rules and it is difficult to apply them uniformly)). The fitting can be made more conservative by simplifying the model used. This reduces its ability to fit the more subtle details of a histogram, but it also reduces the possibility of incorrect fitting of the data. One such simplification is using the Dean and Jett algorithm with a zero, or perhaps first order S phase polynomial (zero order is a "broadened rectangle"), instead of the more flexible, but error-prone second-order curved polynomial. In addition, if the skew or shoulder on a G_1 peak is large, then, unless this feature can be explicitly added to the cell cycle model, it may be best to exclude this portion of the DNA histogram from the fitting procedure.

The model can be further simplified, if necessary, by making one or more of several assumptions that constrain the fitting model, but, at the same time, force the result to be biologically reasonable. An example is that the CVs of the G_2 and G_1 peaks may be constrained to be equal (they are almost always very close anyway). Other examples are that the CVs of DNA-diploid and aneuploid peaks can be constrained to be equivalent or that G_2/G_1 ratios can be constrained to have a user-supplied value (for example, 1.97 or whatever is expected in the laboratory in question). The use of these constraints is discussed in a subsequent section.

The presence of a DNA-aneuploid cell cycle adds additional degrees of complexity to cell cycle analysis. In most cases, the S-phase compartments overlap and, for near-diploid or near-tetraploid DNA aneuploidies, even the G_1 and/or G_2 peaks may be overlapping. Fortunately, it is a feature of the method of deconvolution by curve-fitting that a greater complexity of the histogram is still amenable to the same approach. Fitting of two cell cycles is accomplished by adding two additional peaks and the intervening S phase to the model; the fitting process assigns the appropriate magnitudes to each component. Three cell cycles, seen in multiploid tumors (i.e., diploid and two DNA-aneuploid peaks) can be similarly fit. In many cases, this approach yields a straightforward cell cycle analysis. As the fitting model becomes more complex with larger numbers of fitted parameters, however, the degree of freedom in the fitting model increases and there is a greater chance that the model fits data artifacts, rather than a real biological process. When artifacts are present (non-Gaussian G_1 or G_2 peaks), too few cells are collected, debris is a significant portion of an S-phase region, or aggregation is appreciable, ambiguity in the fitting of the data can result. This ambiguity can be reduced by employing some or all of the constraints in fitting mentioned above. Especially careful attention to fitting of the background debris and aggregation also is required in order to make cell cycle analysis of complex histograms more reliable, as discussed subsequently.

PRACTICAL CHALLENGES AND PROBLEMS IN CELL CYCLE ANALYSIS:

It is often an easier process to perform a ploidy and cell cycle analysis than it is to assess the accuracy and reliability of the estimates obtained. As described above, sample processing (dissociation technique, staining protocol, resultant debris, etc.) and instrumental variables (for example, CV) may contribute to variability in results. Interlaboratory studies that have compared histograms derived from replicate cells or tissues have repeatedly illustrated the great variation that may be encountered (56–60). In addition, choices that the operator must make, both during the fitting process and in the interpretation of the fitted histogram, can lead to substantial variability in results; variation in histogram analysis may be seen both between different software programs analyzing the same data and between laboratories using the same software (57). Some of this variability may potentially be reduced by use of fully automated "semi-intelligent" peak detection and cell cycle analysis software (61, 62) Succeeding paragraphs will address several important factors in histogram analysis. Attention to these considerations should reduce the variation and inaccuracy in the interpretation of DNA histograms:

1. Near–diploid DNA contents—When can they be reliably detected?
2. DNA-diploid and aneuploid populations are both present: What S-phase measurement is relevant?
3. The aneuploid cells comprise only a small proportion of the total number of cells.
4. Fitting results do not make biological sense.
5. G_2 and near-tetraploid populations overlap.
6. Debris is present—How is it fit and how does it affect the accuracy of cell cycle estimates?
7. Aggregation is present—Aggregates overlie G_2 and/or S phases. How does this affect cell cycle estimates?
8. How accurate is a particular cell cycle analysis? Confidence estimates for cell cycle parameters.

Detection of Near-Diploid Aneuploid Populations

Distinguishing two populations with different ploidy becomes increasingly difficult as they become closer in DNA content. In the context of the clinical laboratory, there is a consensus that near-diploid DNA contents can not be reliably diagnosed unless the histogram is bimodal, i.e., there is a depression or trough between two separate peaks. This is the criterion proposed by the convention on nomenclature for DNA cytometry in 1984 (1) and presently required by the College of American Pathologists. This criterion is conservative; in theory, aneuploid peaks overlapping with diploid peaks might be detected by the use of external DNA-content standards, such as lymphocytes or nucleated red blood cells. Relative to the fluorescent standard, a shift in the position of the diploid/near-diploid composite peak away from the expected position of a diploid peak can be taken as evidence of DNA aneuploidy. The problem, however, is that, as described in the section "DNA staining and artifacts" above, a variety of cell and tissue-specific factors can affect DNA staining and produce a shift in "diploid" peak position. Using an external reference standard to establish a range of positions or CVs to define as "diploid" results in appreciably more diagnoses of DNA aneuploidy than using the convention noted above (63, 64). Overdiagnosis of near–diploid aneuploidy may thus occur. In the past, differing criteria of diagnosis of DNA aneuploidy utilized by different investigators complicated the interpretation and comparison of published data.

How close to diploid a population may be depends upon both the CV of the analysis and the relative proportions of diploid and near-diploid cells. Figure 8.5 illustrates this effect: Note that bimodality can be detected in a DI = 1.1 population when the CV is 6 (Fig. 8.5A), but that when the CV rises to 8%, the aneuploidy cannot be discerned as a bimodal curve (Fig. 8.5C). When the CV drops to 4%, the D.I

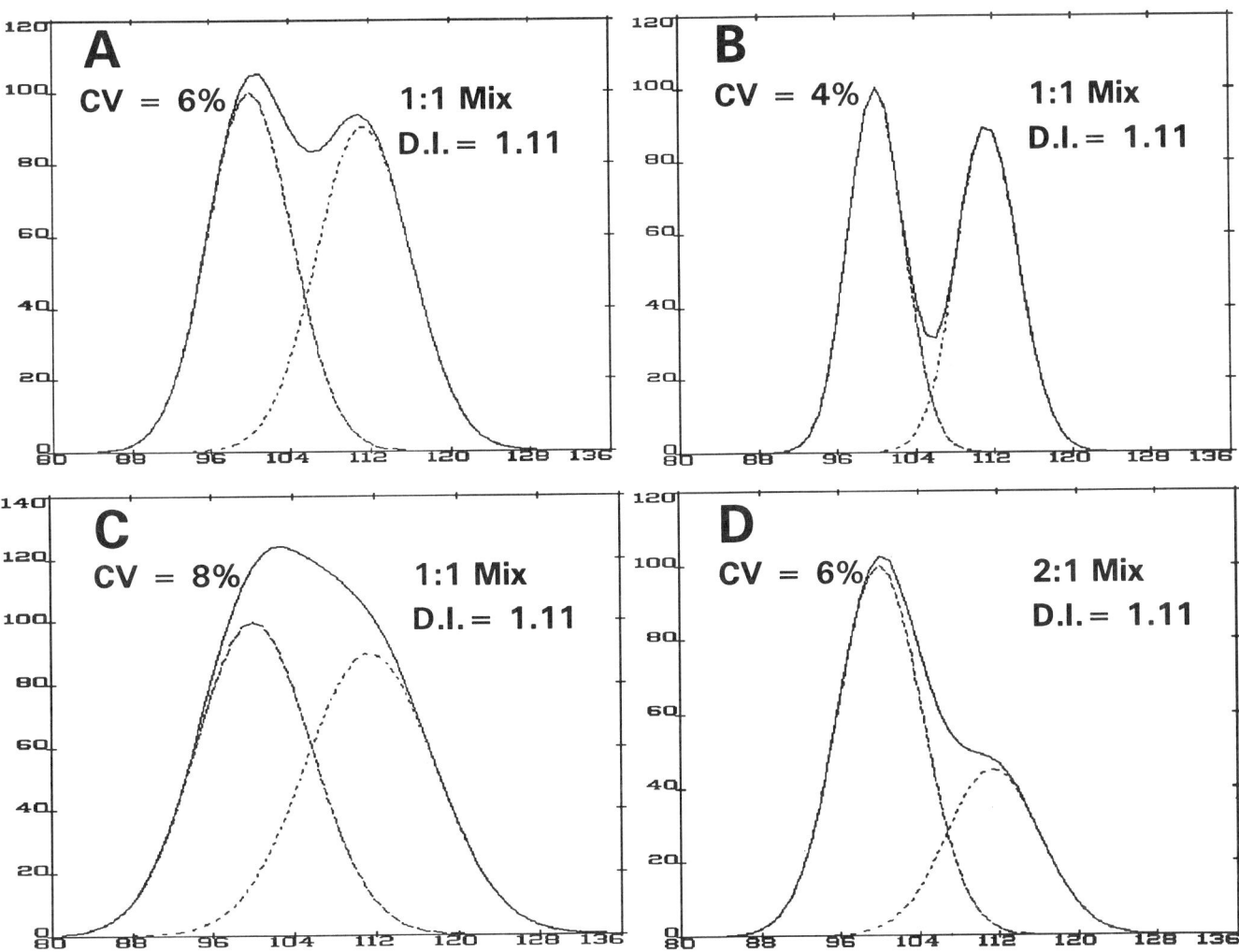

Figure 8.5. Demonstration of the effect of CV and proportion of aneuploid cells on discrimination of an aneuploid population. When the diploid and aneuploid cells are in a 1:1 ratio and the DI is 1.1, then bimodality is barely present at a CV of 6 (**A**), is quite easily apparent with a CV of 4 (**B**), but is not present with a CV of 8 (**C**). When the aneuploid cells are only 33% of the total, bimodality is not seen with a CV of 6 (**D**).

1.1 population becomes much better separated (Fig. 8.5B). Detection is optimal when the diploid and near-diploid cells are in equal proportions, as seen in Figure 8.5A-C. When the proportions of DNA-diploid and aneuploid cells are unequal, the ploidy separation is not as accurate (Fig. 8.5D) and the CV must be lower or peaks must be even further apart to create bimodality.

Using detection of bimodality by a 10% dip between DNA-diploid and aneuploid curves as a criterion, the relationship between the minimum detectable DNA Index and CV and the proportion of aneuploid cells is shown in Figure 8.6. Note that when the diploid:aneuploid peak proportions are 1:1, bimodality is produced only when the percent increase in DNA content of the aneuploid cells is greater than approximately twice the CV. For example, for equal numbers of diploid and aneuploid G_1 cells, and a CV of 4.0, bimodality will result only if the ploidy is greater than about DI 1.08 (an 8% increase) or less than 0.92. It should also be obvious from Figure 8.6 that, if the CV is excessively wide, a near-diploid DNA aneuploidy may escape detection. The values shown in Figure 8.6 are minimum estimates, as problems with real data, such as non Gaussian-shaped peaks or low numbers of cells in the histogram, will adversely affect peak discrimination.

Note also that, when fitting near-diploid DNA aneuploidies, best results are often obtained using a software option to constrain the CVs of peaks to be equal. Choose such an option if the fit without such constraint produces CVs for DNA-diploid and aneuploid peaks that are very dissimilar.

Finally, it should be noted that, in some clinical situations, it may not be important to differentiate DNA-diploid from near-diploid cells. For example, in breast cancer, it has been reported that the most useful ploidy classification is made by grouping cells with DI less than 1.2 together with DNA-diploid cells (65).

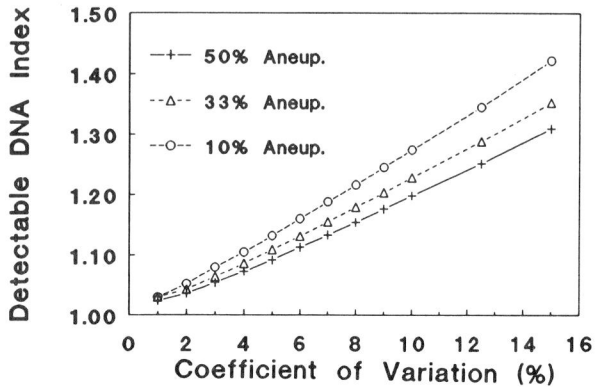

Figure 8.6. The minimum detactable *DI* as a function of *CV* and proportion of *aneuploid* cells. The criteria for detectability is presence of bimodality, with a 10% dip between diploid and aneuploid peaks. The *dashed lines* show diminished capacity to detect aneuploidy when the proportion of aneuploid cells is reduced from 1:1 (+) to 2:1 (triangles), and to 9:1 (circles). Almost exactly the same relationship holds when the proportion of diploid cells is reduced from 1:1 to 1:2 or to 1:9.

DNA-Diploid and Aneuploid Populations are Both Present: What S-Phase Measurement is Relevant?

S PHASES OF TWO POPULATIONS EXTENSIVELY OVERLAP

More sophisticated cell cycle analysis approaches do not require that S phases be nonoverlapping in order to give independent estimates of each. It is simply required that at least some part of the S- and G_2-phase distributions are nonoverlapping so that the relative components of each can be evaluated. For near-diploid aneuploidies, this is often not the case. Exactly how close the two populations can be, and still be independently evaluated, depends mainly upon the CV of the analysis. If there is no region of the histogram in which the two cell cycles are largely nonoverlapping, then **the individual S phases cannot be reliably established and only the estimate of the combined S average phase of the two populations should be reported**.

BOTH DNA-DIPLOID AND ANEUPLOID S PHASES CAN BE CALCULATED

When the ploidy value and quality of the histogram is such that S phases of the DNA-diploid and aneuploid peaks do not completely overlap, then individual S-phase estimates may often be derived. When DNA-aneuploid cells are present, these represent the malignant cells (except in the less common case of those multiploid tumors that have malignant DNA-diploid cells as well). Since the proliferative behavior of the malignant cells is expected to be most relevant to the biological aggressiveness of a tumor, it also would be expected that the aneuploid S phase, when available, would be the clinically relevant parameter. Unfortunately, this ques-

Table 8.1
The Prognostic Value of S-Phase Estimates Obtained by Different Methods in Node-Negative Breast Cancer[a]

	Method	P value[b]
1	Diploid S for diploid tumors, Aneuploid S for aneuploid tumors, average S phase for near-diploid tumors. Aneuploid peaks less than 15% of total cells ignored. Classical exponential debris.	0.004
2	As above but with sliced nuclei debris model	0.0005
3	Average of diploid and aneuploid S phase utilized	.029
4	diploid S phase only utilized	0.13
5	As in (2), but S phase of aneuploid populations of 3% to 15% also utilized	0.014

[a]Modified from Kallioniemi O-P, Visakorpi T, Holli K, Isola JJ, Rabinovitch PS. Automated peak detection and cell cycle analysis of flow cytometric histograms. Submitted for publication, 1992.
[b]Difference (Wilcoxon-Breslow analysis) between five-year survival with above- or below-median S phase.

tion has not been well addressed in the flow cytometry literature and published reports may use the average S phase of both DNA-diploid and aneuploid cells or the diploid S phase (the origin of the S-phase calculation is rarely specified). The study of node-negative breast cancer summarized in Table 8.1 demonstrates clearly that utilizing the DNA-aneuploid S phase, when available (method 2) provides greatly superior prognostic value compared to the use of the average of the DNA-diploid and aneuploid S phase (method 3). In contrast, if only the S phase of the DNA-diploid cells is utilized (method 4), little prognostic information is obtained.

The Aneuploid Cells Comprise Only a Small Proportion of the Total Cells

When a population of cells is a minor fraction of the total cells, the accuracy of single-parameter cell cycle estimates of the minor population is considerably diminished. A common clinical setting in which this may occur is in the measurement of DNA-aneuploid cell parameters in the presence of an abundant lymphocytic infiltrate (66). There are several reasons underlying this complication. Firstly, overlap of cells from the major population into the S- or G_2-phase regions of the minor population impairs the accuracy of estimation of the S and G_2 phases of the rarer cells (the accuracy of deconvolution by curve-fitting is lessened for a population that constitutes a small proportion of total cells). Secondly, even when the two populations do not overlap extensively, debris and aggregates from the more abundant population may still overlap the cell cycle of the rarer population. Both of these conditions are cases of reduced "signal-to-noise," i.e., the signals (S and G_2) become buried within larger amounts of "noise" (overlap, debris, and aggregates). The limitations on accuracy of fitting minor populations will vary from laboratory to laboratory (depending on CV, extent of debris and aggregates, etc.). Cusick has suggested that this situation arises when the aneuploid cell fraction is less than 10% (67). As a general rule, however, we suggest that, when one population becomes less than 15% of the total, the accuracy of its S phase (and perhaps G_2) should be carefully evaluated. Only if debris and aggregation are minimal (or at least well-fit, see below) and the quality of the analysis is otherwise excellent (including adequate cell numbers and low CV) can an accurate S and G_2 phase be estimated when the DNA-aneuploid fraction is below 15%. The clinical relevance of this principle is illustrated in Table 8.1. The inclusion of S-phase measurements from DNA-aneuploid populations comprising less than 15% of total cells (method 5) substantially degraded the strength of the association of the S-phase calculation with clinical outcome.

The analysis of overlapping and/or infrequent populations can be assisted by the use of additional measurement parameters. The use of scattered light parameters has been recommended as useful (68, 69). More promising is the use of cell-differentiation markers, such as cytokeratin staining of epithelial cells (see Chapters 16 and 33).

Non-Biological Results: Aberrant G_2/G_1 Ratios

With certain types of histograms, the software program may yield a G_2/G_1 ratio that is outside the expected range of acceptable values (i.e., below 1.9 and above 2.0 for the author's laboratory). This may be seen, for example, when a near-tetraploid aneuploid population is present. In such cases, constrain the G_2/G_1 ratio to a specified value (one that you usually encounter in other such samples) using the appropriate software option for that purpose. Refit the histogram using a model with two cell cycles to test the fit, on the assumption that the near tetraploid peak is an aneuploid population and not a G_2 peak (see the related topic "Is a DNA-index 2.0 peak a G_2 or a tetraploid population?" below).

Non-Biological Results: CVs of Peaks are Dissimilar

Frequently, CVs of DNA-aneuploid populations are greater than those of diploid cells, probably a reflection of underlying chromosomal instability and variation. This is thus not really a problem but a legitimate observation. It is also possible, especially from paraffin-embedded samples, for the DNA-aneuploid peak to have a lower CV than the DNA-diploid peak, probably due to the mixture of diploid stromal cell types that exhibit variability in DNA staining.

The CV of G_2 cells may occasionally be much larger or smaller than that of the G_1 cells, especially if there is not a visible G_2 peak rising above the S phase. In such a circumstance, use the appropriate software option to constrain the CV of the G_2 to be equal to the CV of the G_1. In unusual cases where the CV of a peak seems to be aberrantly fitted (usually due to overlap with another peak), use a software option that forces the CVs of such peaks to be the same.

Is a DNA Index 2.0 Peak a G_2 or a Tetraploid Population?

Unfortunately, there is no way that any single-parameter cell cycle analysis can provide an exact answer to this question. There are, however, several guidelines:

1) If the DI 2.0 peak is large, it is probably not G_2, but tetraploid or near-tetraploid. What is meant by "large" is open to interpretation. Some laboratories view this as being in excess of 15% of the total G_1 + S + G_2/tetraploid. Some define it as 15% larger than the S-phase value (to account of a larger fraction of "true" G_2 cells in actively proliferating tissue).
2) Is the DNA index of the peak really within the range expected for G_2 cells? Careful recording of the actual G_2/G_1 ratios for a given cell or tumor type will help to establish a confidence interval for true G_2 cells.
3) Is there evidence of S and G_2 phases for the DI 2.0 population? Fitting of the DI 2.0 cells as the G_1 of a second cell cycle should be attempted (usually with a constraint of the G_2/G_1 ratio to a reasonable value in order to help to place the diploid G_2 in the right place). If there are appreciable S and G_2 cycling components to the DI 2.0 cells, then there is strong evidence for a G_1 near-tetraploid population. Note that the aggregation modeling technique described subsequently is often very useful in demonstrating that the G_2 of the near-tetraploid population is greater than can be accounted for by the effects of aggregation alone.

Perhaps more relevant than whether a DI 2.0 peak is G_2 or tetraploid, is the question of whether an increase in this fraction is a meaningful clinical indicator. Well designed studies can address this issue. However, this has not been frequently addressed in the literature. A notable exception is the study of prostatic adenocarcinoma (see Chapter 16) in which elevation in the DI 2.0 population outside the range of nonmalignant controls is a significant prognostic finding (70–72). Even in these studies, different benign histologic categories of tissue were used as the "control" group, and the control range has varied from laboratory to laboratory. The threshold for establishing abnormality is usually set at 2 S.D. outside the normal mean; this threshold has been reported to vary from 7% to 13% in prostate. The presence of aggregates is a large potential problem in obtaining accurate G_2 estimates; using the software correction algorithm discussed subsequently, the normal G_2 range in prostate appears to be below 5% (unpublished observations; see Figure 8.14). Thus, each laboratory may need to examine its own range of nonmalignant control values. A second reported example is the study of Barrett's esophagus; here, DI 2.0 populations in excess of 6% are reported to be predictors of future histologic progression toward cancer (73). In these studies, the DI 2.0 threshold value that had the greatest predictive value was established by statistical procedures (receiver–operator curves) that optimized sensitivity and specificity (74).

Overlap of G_2 Peaks and a Near-Tetraploid G_1 Peak

When a tetraploid or near-tetraploid population is present (see above), it will, by definition, overlap with the G_2 of the "diploid" population. In such a case, how much of the near-tetraploid peak is diploid G_2 and how much is near-tetraploid G_1? If these two populations are very closely overlapping, then there is no way to answer this question with certainty. Cell cycle analysis software can, however, make a reasonable assumption that the G_2 of the diploid (or lower ploidy)

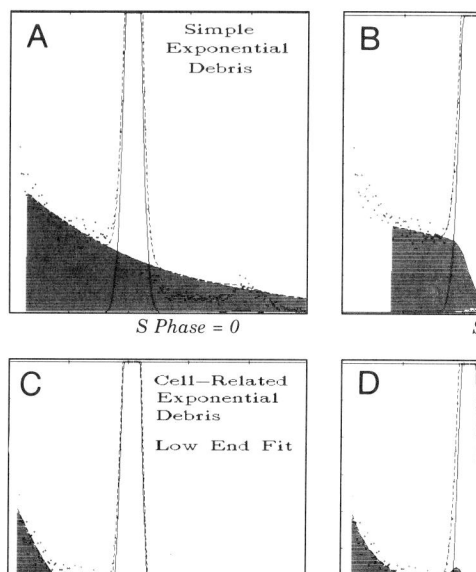

Figure 8.7. Fitting of a histogram derived from paraffin-embedded diploid cells using a simple exponential background debris curve (**A**), a model that assumes that exponential debris is produced by each cellular component of the histogram (**B**) and (**C**), and the sliced-nuclei model (**D**). **B** shows fitting of the background in a region closer to the G_1 peak, **C** shows fitting of a more leftward region of the histogram. **A** and **D** show fitting to a wide region of the debris left of the G_1 peak. Cell cycle analyses in this and other figures were performed using the "Multicycle" software written by the author (Phoenix Flow Systems, San Diego, CA).

population is proportional to the S-phase fraction of the same population; this will prevent the software from giving aberrantly high G_2-phase estimates.

Fitting of Background "Debris" and Effects of Nucleus Sectioning

Almost all cell or nuclear suspensions analyzed by DNA-content flow cytometry contain some damaged or fragmented nuclei, resulting in an additional component of the histogram, usually most visible to the left of the diploid G_1. In samples derived from fresh tissues or cell culture, most of these "debris" signals are at the far left side of the histogram and fall to baseline rapidly. In the best case, the debris signal is small and insignificant in the region of the histogram occupied by the cell cycle. Unfortunately, this is often not the case, and it becomes very important to include modeling of the debris curve in the computer analysis in order to subtract the effects of the underlying debris from the cell cycle fitting. In the past, the classical assumption in debris fitting was that the rapidly declining background debris curve could be fit by an exponential function (e^{-kx}). There are two primary reasons why a simple exponential curve does not usually provide an accurate fit:

1) The shape of most debris curves is not actually exponential; it is more common to observe a component that rapidly declines with increasing DNA content and then a portion that declines more slowly or plateaus. This more slowly declining portion has a much greater effect upon the cell cycle fitting than is otherwise predicted from an exponential curve.

2) Debris is a result of degradation, fragmentation, or actual cutting of nuclei and, as such, it extends only leftward (to smaller DNA contents) from each DNA-content position. This implies that the shape of the debris curve is dependent upon where the peaks in the DNA histogram are and that the shape of the debris cannot be fit independently of the cell histogram. Since the S phase is the lowest and broadest cell cycle compartment in the histogram, S-phase calculations are most affected by the presence and shape of the debris distribution.

Figure 8.7 illustrates these two points. Figure 8.7A shows fitting of a simple exponential curve to the debris region left of the G_1 peak. This model does not take into consideration the fact that much of the debris results from fragments of nuclei with the G_1 DNA content; thus, the curve predicts too much background over the S- and G_2-phase positions. Cell cycle analysis with this debris model yields a zero S. A more sophisticated model of exponential debris assumes that each DNA-content position is associated with the production of exponential debris that extends leftward from that position. Application of this histogram-specific exponential model (also termed multi-cut in Chapter 3) is shown in figures 8.7B and 8.7C, which illustrate that the background debris curve drops rapidly from the left side of the G_1 peak to the right side of the G_1 peak. Figure 8.7B shows fitting of the debris curve in the region closer to the G_1 peak, while Figure 8.7C shows fitting of the region at the lower end of the histogram; different debris curves are generated

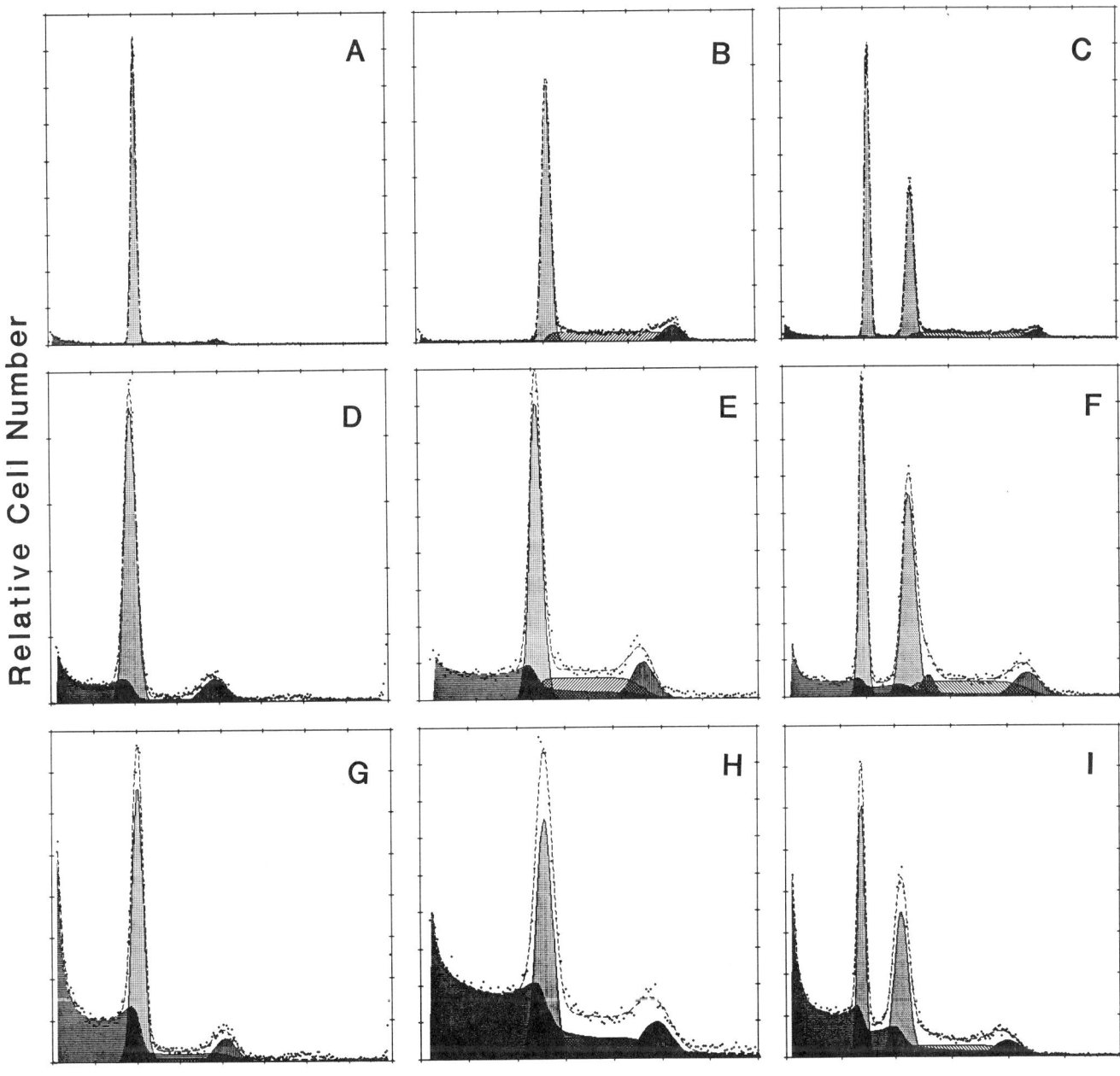

Figure 8.8. Sliced-nucleus debris modeling in cell cycle analysis of lymphocytes (**A, D, G**), Hela cells (**B, E, H**) and mixtures of these cells (**C, F, I**). Analyses were performed on fresh cells (**A, B, C**), paraffin-embedded cells sectioned at 50 μ (**D, E, H**), and paraffin-embedded cells sectioned at 20 μ (**G, H, I**). The debris component of the fitted model is shown by the horizontally hatched portion, and S phase is diagonally hatched.

for each region chosen for the fit. The S-phase estimates also differ (5.5% for Fig. 8.7B vs. 9.3% for Fig. 8.7C). Despite this variability, this model yields better results than the simple exponential curve.

The best fit to this histogram is obtained by combining the histogram-specific exponential model with a model that accounts for the production of debris by slicing nuclei during sectioning from the paraffin block; this model (hereafter referred to as the sliced-nucleus model, which is a combination of single–cut and multi–cut in the terminology of Chapter 3) fits all portions of the debris curve (Fig. 8.7D). In a systematic study of paraffin-embedded tissue, Kallioniemi et al. (53) have shown that the sliced-nucleus model is relatively insensitive to the end points chosen for the fitting region. The utility of this model, especially in analysis of paraffin-derived nuclei, is described below.

The analysis of DNA histograms from cells preserved in paraffin blocks has become an increasingly important part of

Table 8.2
S-phase Estimates (±S.D.) without and with (in Parentheses) Sliced Nuclei Correction (MultiCycle model, n = 3)

	Hela	Lymphocyte	Mixed Hela	Lymphocyte
Fresh	26.2 ± .4	5.5 ± .3	27.5 ± .8	8.7 ± .1
	(25.7 ± .5)	(5.2 ± .3)	(27.7 ± 1.0)	(5.6 ± .4)
50 μm paraffin	29.6 ± 2.0	7.1 ± 1.1	33.7 ± 5.0	28.9 ± 2.8
	(26.6 ± 1.4)	(5.4 ± .7)	(28 ± .3)	(6.1 ± 2.5)
20 μm paraffin	33.8 ± 1.0	12.8 ± 3.8	43.1 ± 6.8	41.6 ± 5.8
	(26.9 ± .8)	(6.6 ± 1.1)	(31.8 ± 3.8)	(18.3 ± 2.6)

DNA flow cytometry. Not only is it possible to conduct retrospective research on such material, thereby establishing relationships of flow cytometry results to long-term patient follow-up, but, in many cases, fresh tissue is not available and the analysis of material extracted from paraffin becomes very important in the clinical setting. In order to derive useful cell cycle information care must be exercised in the isolation of nuclei and in the computer modeling of the cell cycle analysis. As part of the process of extraction of nuclei from paraffin blocks, sections are usually cut with a microtome at a thickness near 50 μm; the sectioning of nuclei is an unavoidable consequence. These nuclear fragments can have a substantial effect upon S-phase calculations, but, as described in Chapter 3, mathematical modeling of the production of sliced nuclei as part of the cell cycle analysis can help to correct for this effect.

To illustrate the practical utility of the sliced or cut nuclei algorithm, Figure 8.8 shows the analysis of DNA content histograms from growing human lymphocytes and Hela cells (derived from an adenocarcinoma) and mixtures of these cell types. The cells were analyzed both fresh and after embedding and extraction from paraffin. Figure 8.8 illustrates that the debris portion of the histograms increases progressively from fresh, to 50 μm-, to 20 μm-section thicknesses. The shape of the debris curve to the left of the G_1 peak in paraffin-derived samples contains a broad plateau, as predicted from the model of random sectioning of nuclei. The ability of the computer model to closely fit this shape is evident in Figures 8.8D through 8.8I. Table 8.2 shows a comparison of S-phase estimates with and without fitting of the background debris.

In the case of fresh tissue, it is only apparent when the Y-axis is magnified that the shape of the debris curve has a small flat-concave component (not shown). For the fresh cells, the effects of the sliced nuclei debris modeling is small, except for the estimate of the lymphocyte S phase in the sample mixed with malignant epithelial Hela cells; in this case, the cut Hela nuclei overlap the lymphocyte S phase, giving rise to a 3% overestimation of S phase without sliced nuclei debris modeling, and a satisfactory estimate with the model. Because many cell or nuclear extraction methods for unfixed tissues utilize cutting, mincing, or forcing through mesh, some component of flat debris will usually be found. It is therefore recommended that sliced nucleus modeling be utilized in histogram analysis, even if this shape can only be visualized when the Y-axis scale is expanded.

For paraffin-derived lymphocytes and Hela cells (unmixed), there is an overestimation of S phase, which increases progressively as the section thickness decreases; this is almost completely corrected by debris modeling. In Figure 8.8, the partitioning of the histogram region between G_1 and G_2 into both S-phase and cut-nuclei components can be seen in D, E, G, and H.

Much more dramatic effects of nuclear slicing are seen in histograms in which there are two cycling populations with different DNA contents. When lymphocytes and Hela cells are mixed, many of the sliced Hela nuclei overlap the lymphocyte cell cycle distribution and result in an artificial elevation of the lymphocyte S phase. This is readily visible in 50 μm sections (Figure 8.8F) and even more in 20 μm sections (Figure 8.8I). Cell cycle fitting using the sliced nucleus model closely fits the raw data and at both 50 μm and 20 μm section thicknesses the model produces S-phase estimates that are closer to that of the fresh cells (Table 8.2), although correction of this effect in 20 μm sections is only partial. Table 8.2 also shows that the standard deviation of S-phase estimates is generally smaller when debris modeling is applied than when it is not. An additional consequence of the correction for sliced nuclei is that a slight broadening of the left side of the G_1 peak is accounted for by the model; at 50 μm section thicknesses the CV of the Hela G_1 peak averaged 5.3 without the sliced nucleus model, and 4.7 with the model.

The impact of sliced-nucleus debris modeling on the clinical utility of cell cycle estimates needs to be established from clinical databases. For example, Kallioniemi et al. (53) have reported that for prostate cancer and node-negative stage I-II breast cancer, the relative risk (RR) of death for high S-phase tumors were both 3.1 times that for low S-phase tumors when analyses were made without background subtraction; the prognostic distinction improved to a RR of 5.3 for prostate cancer and 4.5 for breast cancer when analyses were made with sliced-nucleus modeling. In the same vein, Table 8.1 illustrates that the statistical significance of the predictive value of S-phase estimates in node-negative breast cancer is greatly improved by the use of sliced-nucleus modeling (method 2), compared to the classical simple exponential debris model (method 1).

HOW DOES THE EXTENT OF DEBRIS AFFECT THE RELIABILITY OF CELL CYCLE PARAMETER ESTIMATES?

Even when the best models of the debris component are used, the accuracy of cell cycle estimates declines as the proportion of debris in the histogram increases. Unfortunately, it is impossible to give an absolute rule for the magnitude of this effect, since the success of the computer modeling of debris varies from sample to sample. Several guidelines may be of use, however:

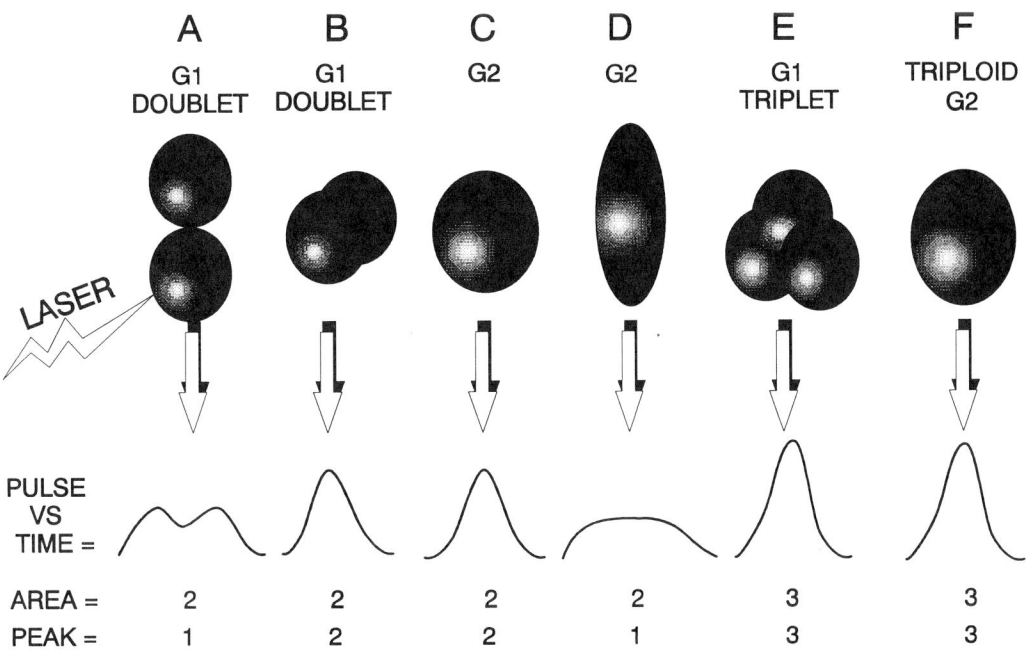

Figure 8.9. Pulse-shape differences between singlets, doublets, and triplets of cells, and the dependence upon orientation with respect to the laser beam. **A**, G_1 doublet aligned parallel to the flow stream. **B**, G_1 doublet aligned perpendicular to the flow. **C**, a spherical G_2 singlet. **D**, an elongated oblate G_2 singlet. **E**, A triplet of G_1 cells. **F**, a spherical triploid G_2 cell.

1. Greater amounts of debris reduce the accuracy of S- and G_2-phase estimates. The S phase is the most sensitive to error. The total proportion of debris in the histogram is one factor and, as an approximate rule, it is unlikely that high accuracy in cell cycle estimates will be achieved when this proportion is more than 40% (measured from DI 0.1 and above). It is important to look carefully at the proportion of S- and G_2-phase regions of the histogram that are predicted to be due to debris, not S or G_2. If half the S-phase "height" is in fact debris, the S-phase estimate is likely to be approximate. When the debris is an even greater proportion of the data (for example, in a diploid S-phase region in a paraffin sample with hyper-diploid aneuploidy), accuracy of S- and G_2-phase estimates can be expected to be poor.
2. Carefully examine the actual data points compared to the computer fit. Discrepancies are indications that the debris shape differs from that predicted by the model, and errors in accounting for the overlap of debris and S and G_2 phases are more likely to occur. The chi-square of the fit will be higher in such cases; since many factors can affect the chi-square, however, it is better to evaluate the data and fit by direct examination of the histogram.
3. Sliced-nucleus debris is more reliably accounted for than exponential debris. Histograms that mainly contain debris of the "flat-concave" sliced-nuclei shape will produce more accurate cell cycle estimates than will histograms in which there is appreciable "exponential" tilt to the left of the DNA-diploid G_1. This is because the sliced-nucleus effect is consistent and reproducible and it requires only one unknown variable to be fit into the computer model. Exponential shapes are much more variable; they are produced by a larger variety of mechanisms (cell death and degeneration, pulverization or multiple cutting of nuclei, etc.), and this shape requires that two variables in the fitting model be estimated. In general, laboratories should use a protocol that results in as little of the exponential component of debris as possible when performing DNA analysis from paraffin-embedded specimens.

Fitting and Correction for the Effects of Cell or Nuclear Aggregation

In the ideal flow cytometric analysis, a cell or nuclear suspension is free of aggregates or clumps and the considerations of cell cycle and debris are sufficient to fit the data. In the majority of "real" histograms, however, careful inspection will reveal evidence of cell aggregation. "Doublets" of G_1 cells will overly the G_2 peak (and will be overlooked), diploid triplets will be seen at DI 3.0, quadruplets at DI 4.0, etc. Not only will the G_1 cells aggregate, but S and G_2 and nuclear fragments (debris) also will aggregate with G_1 cells and with each other. The effects of aggregation are more complex when a sample contains DNA-aneuploid as well as DNA-diploid cells, as aggregates of diploid and aneuploid cells will occur as well as diploid-diploid and aneuploid-aneuploid aggregates. The sources of aggregation can be various. Even if disaggregation is initially complete, some preparative procedures for flow cytometry, such as those that employ ethanol or other solvent fixation or any procedure that uses centrifugation, may reintroduce aggregates.

The conventional approach to the management of aggregation and its effects has centered on attempts to distinguish aggregates by the altered pulse-shape that they may produce when illuminated by a focused laser beam. Subsequent analysis of the DNA histogram is gated on the pulse shape distribution. This approach suffers several notable limitations: Firstly, the pulse shape of aggregates may not be different

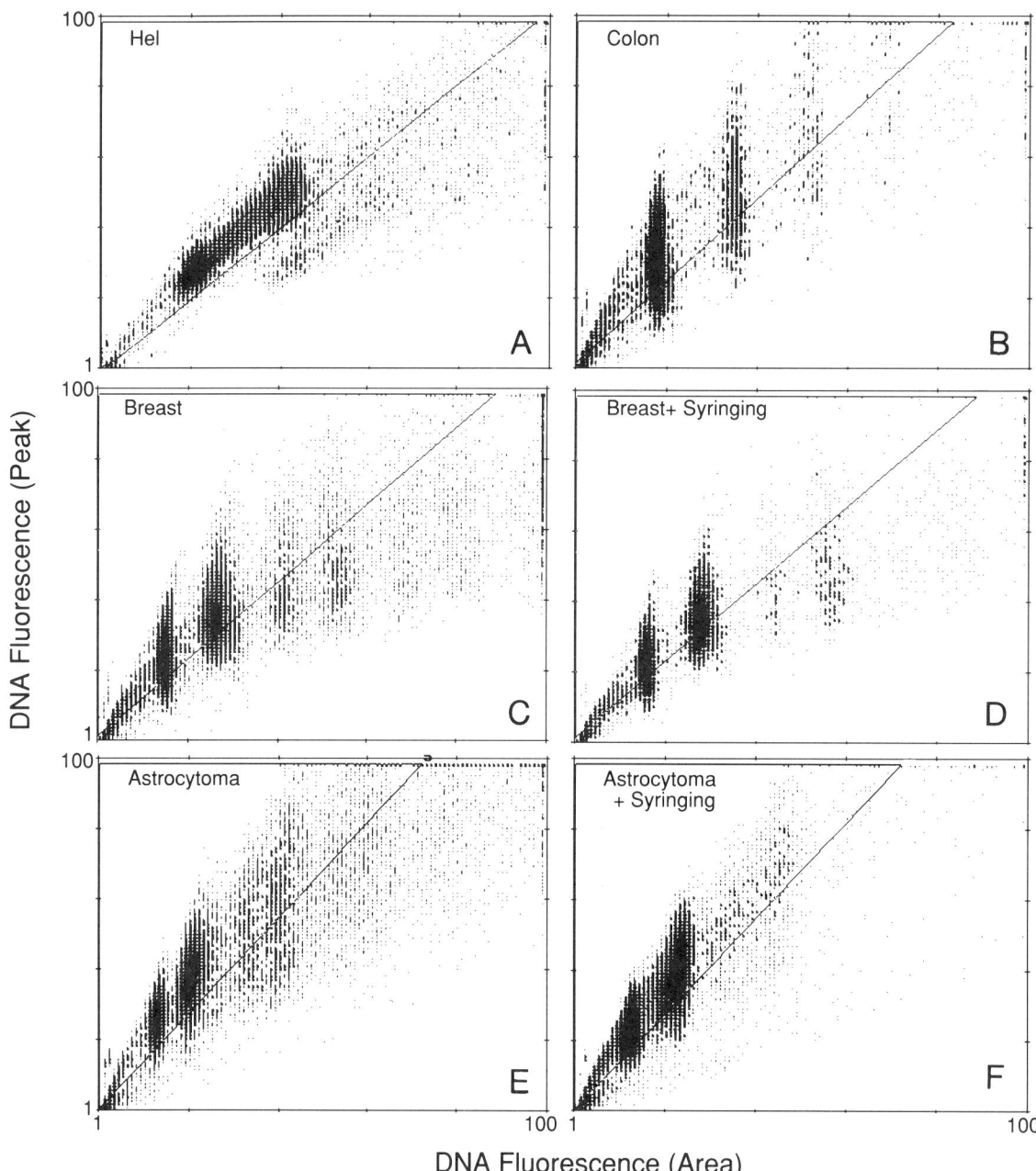

Figure 8.10. Pulse-shape "doublet discrimination" applied to HEL cells (**A**); a round granulocytic cell line; colonic mucosal nuclei (**B**); nuclei derived from a breast adenocarcinoma, before (**C**) and after trituration (**D**); and nuclei from a high-grade astrocytoma, before (**E**) and after (**F**) trituration. In each case, the region above the diagonal line was used for gating to attempt to remove aggregates. Cells and nuclei were stained with DAPI. Analyses were performed on an Ortho Cytofluorograf, an instrument with a narrow (approximately 5μm) vertical dimension to the illuminating laser beam. Further analysis of this data is shown in Figures 8.12 and 8.13.

from that of single cells or nuclei. Spherical cells or nuclei will appear different from a doublet of two such particles as long as they pass through the laser beam in single file: a G_1 doublet produces a fluorescence intensity profile that is longer than that from a single G_2 (Fig. 8.9A vs. 8.9C). If the doublet passes through the laser beam with one cell behind the other (Fig. 8.9B), however, the fluorescence profile cannot be distinguished from that of the G_2 cell.

Secondly, many cells or nuclei derived from solid tissues (especially epithelial cells) are themselves oblong or at least het-erogeneous in shape. If an oblong G_2 cell passes through the laser beam, it cannot be easily distinguished from a G_1 doublet on the basis of peak or width vs. area (Fig. 8.9A vs. 8.9D).

Finally, aggregates of more than two particles may not have a longer axis, and may not be distinguishable from a single large cell: for example, a G_1 triplet (DI>=3.0) may not be distinguishable from a triploid G_2 (DI 3.0) (Fig. 8.9E vs. 8.9F).

Figure 8.10 shows the application of "doublet" discrimination on the basis of pulse peak vs. pulse area analysis for several cell types. Figure 8.10A shows whole fixed HEL cells, a hematopoetic cell line with a roughly spherical shape. As with most peak/area analyses, a diagonal line is drawn, with the assumption that aggregates will fall below the line (i.e., their pulse peak value will be lower than nonaggregates for a given pulse area). For HEL cells, Figure 8.10A shows that a large population of doublets falls below the line, although some particles with DNA content above the G_2 value are above the line and could be undiscriminated aggregates. Figure 8.9B shows a similar analysis for nuclei derived from normal human colon mucosa. Many of these are epithelial nuclei and are oblong in shape, while some are from stromal cells, including lymphocytes, which are more spherical. The distribution of G_1 and G_2 cells on the plot of peak vs. area is very variable in the peak value—the expected result for a mixture of round and oblong cells. It is very difficult to see where on this plot the diagonal should be placed in order to exclude aggregates; in essence, many of the single epithelial nuclei have the pulse shape of round cell doublets, and doublets of epithelial nuclei may not be formed end-to-end and, thus, difficult to differentiate from singlets by pulse shape.

Figure 8.10C shows a somewhat more intermediate pattern from a DNA-aneuploid adenocarcinoma of the breast. A diagonal line is shown that appears to result in most of the G_1 triplets and aggregates with DNA content greater than the aneuploid G_2 being below the line and, thus excluded from the gated analysis. Figure 8.10D shows the same cells after trituration by syringing 18 times through a 26-gauge needle. Appreciable aggregation still remains, mostly below the line, as in Figure 8.10C. However, some aggregates appear to remain above the line. Figure 8.10E and 8.10F show nuclei derived from a DNA-aneuploid astrocytoma before and after syringing, respectively. As for the breast cancer, the diagonal line cannot be placed in a position that excludes all aggregates (without excluding most or all of the G_1 nuclei).

Past attempts to detect aggregates using software have been made by adding an extra peak to the cell cycle model to fit the triplet peak position or by predicting triplets on the basis of the frequency of doublets (75). This approach will not, however, fit the much more complicated patterns of aggregation that result from diploid and aneuploid G_1, S, and G_2 aggregations. In addition, it is desirable to compensate for the effects of aggregates that can not be easily fit as separate peaks because they overlie other cell cycle components.

In order to allow a more general and flexible software algorithm to compensate for the effects of aggregation, a computer model can be applied that allows a generalized approach to the fitting of aggregation in DNA histograms (76). The basis of this model is the assumption that cell aggregation is, or can appear to be, a probabilistic event. It is assumed in this model that any two cells or nuclei have a certain probability to aggregate with each other. On the assumption that this probability is roughly the same for all cells, the distribution of doublets, triplets, quadruplets, etc., follows certain rules and, in fact, the net "aggregate histo-

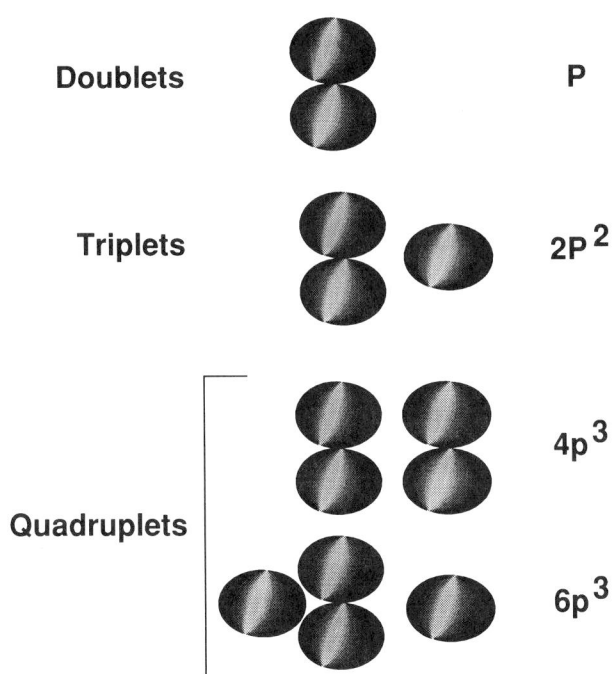

Figure 8.11. A theory of aggregation modeling made by assigning probabilities of aggregate formation to each of the classes of aggregates (*doublets*, *triplets*, and *quadruplets*). *Doublets* form with a probability "p." *Triplets* form by association of a doublet with a singlet; the singlet can "attach to" either of the two cells in the *doublet*, with a net probability of $2p^2$. *Quadruplets* can form in two ways: two *doublets* can aggregate with each other with a probability of $4p^3$ (there are four ways the two doublets can attach to each other, or 4p times p^2), or a *triplet* can combine with a singlet with a probability of $6p^3$ (there are three ways to combine the triplet with the singlet, or 3p times $2p^2$). The constants 2, 4, and 6 are derived here in the simplest fashion; it is possible to calculate these on alternative bases, changing the final aggregate distributions slightly. The computer applies this model to the histogram by finding all possible combinations of aggregation of one cell or nucleus with another. The doublet distribution, D(i), for example, may be mathematically derived from the cell distribution without aggregation. Y(i), by the formula:

$$D(i) = p \cdot \sum_{j=1}^{i} \sum_{k=1}^{i} Y(j) \cdot Y(k)$$

(for all $j+k=i$). Triplets and quadruplets are calculated similarly, and the net distribution of all aggregates is the sum of these distributions. There is only one unknown in the above equations, the value of the probability of aggregation "p," thus, this adds minimal complexity to the least-squares fit.

gram" has a characteristic shape that is predicted on the basis of random probabilities of aggregate formation.

Figure 8.11 illustrates the assumptions made in this model. Software implementing this model can use the least-squares fitting technique to determine the aggregation probability ("p" in Figure 8.11) that gives the best fit to the data. An example of this fitting is shown in Figure 8.12, using the histogram derived from the ungated DNA area analysis of the astrocytoma presented in Figure 8.10E. Note in Figure 8.12B that the events to the right of the aneuploid G_1 are fit as part of the aggregate "background," and that the shape of

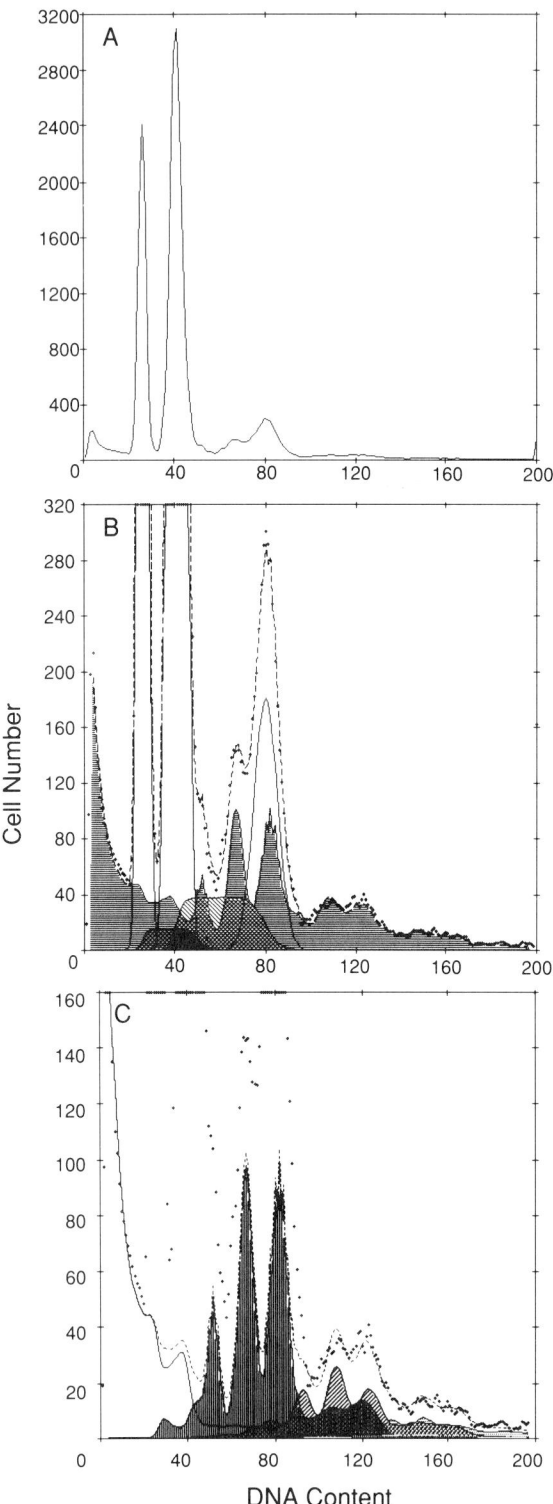

Figure 8.12. Application of the aggregation model to the astrocytoma shown in Figure 8.10C (without gating). **A** shows the raw DNA content histogram. **B** (10× scale) shows the total background fitting (horizontal hatching), including debris and aggregates. Diploid and aneuploid S phases are shown by diagonal hatching, and Gaussian G_1 and G_2 peaks are shown by solid lines. The total fit is indicated by the *dashed line*. **C** (20× scale) shows the individual components of the background fit: sliced-nucleus debris (*solid line* at left), doublets (*vertical hatching*) triplets (*diagonal hatching*) and quadruplets (*strippling*). The total background fit is indicated by a *dashed line*.

this aggregate distribution is correctly modeled. There are several aggregate peaks over the region of the DNA-diploid and aneuploid cell cycles, and there are additional aggregation events (not only peaks) in the regions overlying S and G_2 phases. The net result is an excellent fit to the large numbers of peaks in the data (some being due exclusively to aggregation) and both the S- and G_2-phase fractions resulting from fitting with this model are lower than those without aggregation modeling.

Figure 8.12C shows the components that make up the debris and aggregate distribution shown in Figure 8.12B. At the left of the histogram is the debris predicted by the sliced nucleus model (*solid line*); this curve declines progressively to the right, as seen previously in Figures 8.7 and 8.8. The doublet distribution (shown with *vertical stripes*) is seen to be very complex in shape, reflecting the fact that all histogram components (diploid G_1, S, and G_2; aneuploid G_1, S, and G_2) are predicted to aggregate with each other. This distribution is so complex that to try to model the aggregation peak-by-peak would be impractical. The triplet distribution is shaded with *diagonal stripes*; it has a higher DNA content overall than the doublets, but there is extensive overlap. Similarly, the quadruplet distribution (*stippled*) is higher in DNA content, but overlaps the triplet distribution to a large extent.

Comparison of the effects on cell cycle analysis of a) aggregation modeling, b) pulse processing and gating, and c) trituration by syringing is shown in Figure 8.13. Nuclei from an adenocarcinoma of the breast (Figures 8.10C, 8.10D, and 8.13A) and a high grade astrocytoma (Figures 8.10E, 8.10F, and 8.13B) were subjected to either 0, 4, or 18 passages through a 26-gauge needle. In addition, whole cells were isolated from the astrocytoma. Each sample was analyzed as both a gated pulse-shape "doublet-detected" (above the diagonal in Figure 8.10) histogram and as an ungated DNA histogram. Each resulting histogram was analyzed with and without the aggregation modeling described above.

For the adenocarcinoma of the breast (Fig. 8.13A), the S phase of the DNA-aneuploid cells without trituration was 12.6%. For gating using the region shown in Figure 8.10C, the S phase was 11.2%. In contrast, the aggregation software model applied to the ungated data reduced the S-phase estimate to zero. The aggregate model calculated that 16.8% of the "cells" in the histogram were aggregates and manual enumeration of aggregates by microscopy (two independent observers) was in close agreement. Because some of the aggregates were removed in the gated histogram, when the aggregate model was applied to it the S-phase estimate was not reduced as much as in the ungated histogram. With progressive trituration, aggregation was reduced (software and manual estimates remaining in agreement) and S-phase estimates without aggregation modeling declined also. The estimate with aggregation modeling remained at zero. It seems probable that, if further disaggregation of nuclei had been possible, the S-phase estimate without aggregate modeling would have declined further, perhaps to near zero.

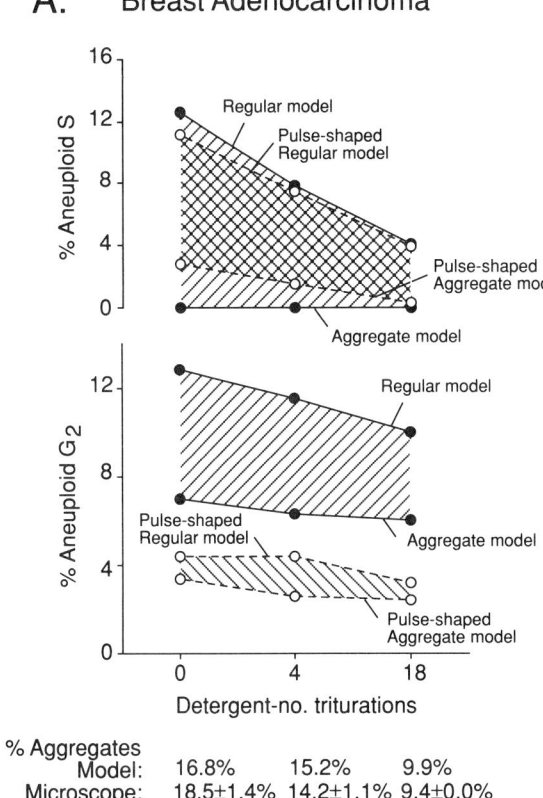
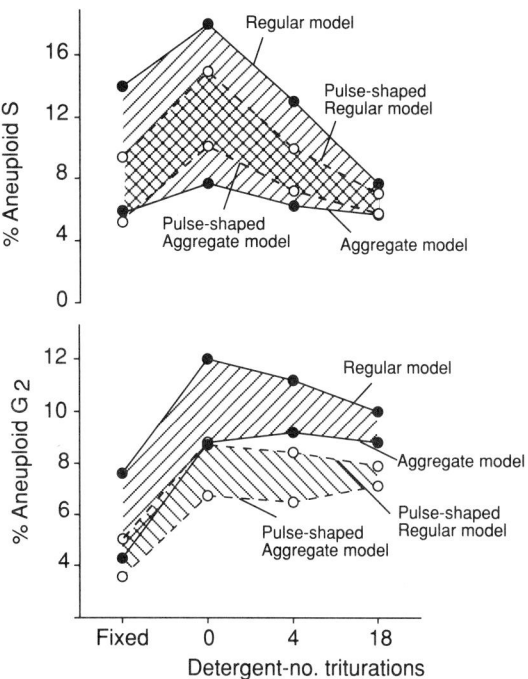

Figure 8.13. S-phase and G_2-phase estimates of the aneuploid cell component of an adenocarcinoma of the breast (**A**) and a high-grade astrocytoma (**B**) using cell cycle fitting with and without software aggregation modeling, and with and without gating on the basis of *pulse-shape* (hardware "doublet discrimination"). The number of triturations (syringing through a 26-gauge needle) is shown on the bottom axis. The *percentage aggregates* estimated to be present in the histogram by the Multicycle software model, and the *percentage aggregates* manually estimated by two observers using microscopy are shown at the bottom. Whole cells were isolated by digestion in collagenase with teasing and mechanical agitation, and were then fixed in ethanol. Nuclei were isolated by mincing in isotonic tris buffer with NP-40 detergent (Rabinovitch, 1981).

The effect on G_2-phase estimates shown in Figure 8.13 indicates that pulse-shape gating removes substantial amounts of events in the aneuploid G_2 position. The aggregate model applied to the ungated histograms shows a reduction also, but not to the same extent. A plausible interpretation of these results is that gating removes not only some aggregates, but also some legitimate G_2 events. Resetting the gating region to remove fewer G_2 cells would result in the elimination of even fewer aggregates over the S phase.

A very similar result was obtained with cells from the astrocytoma (Figure 8.13B). Trituration was more successful in removing aggregates in this example. The higher estimate of aggregation from microscopic examination may have been due to the presence of cells that visually appeared adjacent but that did not remain aggregated within the flow cytometer. Once again, the software aggregation modeling resulted in an S-phase estimate for the aneuploid cells that was almost independent of the degree of aggregation and was similar for fixed and unfixed cells. The regular S-phase estimate was progressively reduced with trituration; the rate of this decline suggests the possibility that, had mechanical disaggregation been complete, the regular model estimate would have equaled the aggregate model estimate. G_2 estimates using the regular model declined with the extent of trituration, while G_2 estimates with aggregation modeling were almost unchanging. Once again, G_2 estimates from pulse-shape gating were slightly lower. The fixed whole-cell preparation appeared to have fewer G_2 cells by all estimates, possibly because the release of G_2 cells by enzymatic digestion was less complete than it was by detergent isolation.

In summary, Figure 8.13 demonstrates that, in samples that contain aggregation, the software aggregation model produces S-phase estimates that are closer to the values seen in triturated, disaggregated samples. Pulse-shape gating appears to be much less effective for these cell types. Application of the aggregation model to pulse-shape–gated histograms is probably inappropriate, because the gating disturbs the random aggregation relationships upon which the model relies.

Finally, it should be noted that microscopic enumeration of aggregation requires careful discrimination between merely adjacent vs. adherent cells and that some of the discrepancies between microscopic enumeration and the software estimate could be due to difficulties with the former

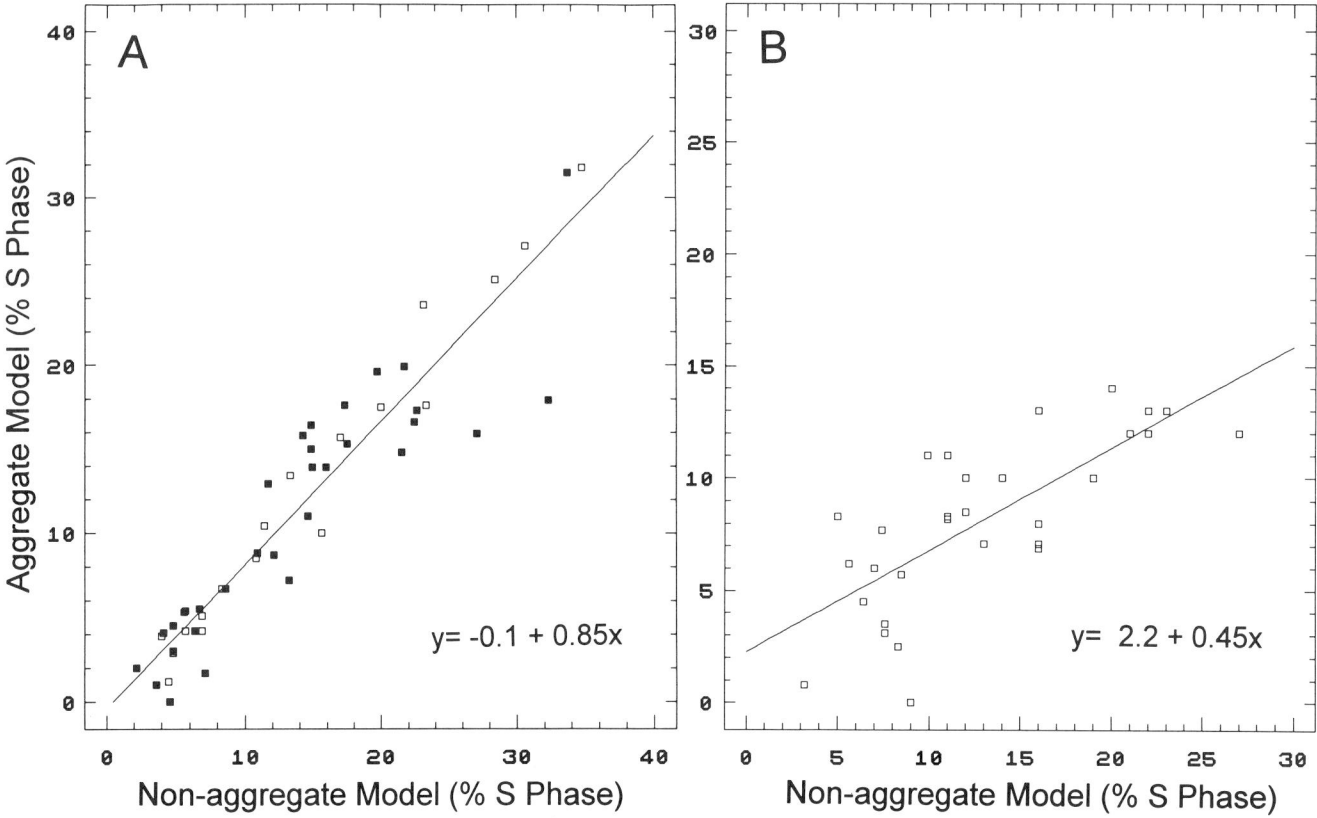

Figure 8.14. **A** shows a plot of the S-phase values of 56 aneuploid breast cancers derived from cell cycle fitting using the sliced-nucleus model without (*abscissa*) or with the addition of aggregation modeling (*ordinate*) to the Multicycle software program (sliced-nucleus modeling). Filled squares are analyses derived from paraffin. **B** (courtesy of Dr. T. Vincent Shankey) shows analysis of the S-phase estimates of DNA-tetraploid populations in 30 cases of prostate cancer with tetraploid fractions of 25% of cells or greater. The estimated S-phase percentage in DNA-tetraploid prostate tumors was reduced by approximately half by software modeling of aggregation.

rather than the latter. If there is a need to quantify aggregation (even if there is no attempt to compensate for its effects), then the software algorithm may, in fact, be more consistent.

Practical Use of Aggregation Estimates

As described above, in the analysis of heterogeneously shaped cells, it may be more accurate to use the software modeling of aggregates than to use hardware approaches. The model is especially appropriate when there is obvious aggregation, but it seems likely that the aggregation model could be applied to less obvious cases as well. An important feature of the aggregation model is that it also assumes that debris as well as whole cells, can aggregate. Thus, when debris is high, the aggregation model may indicate that there is a substantial effect on S and G_2 **even when distinct peaks of aggregation (i.e., DI = 3) do not appear to be significant.** Analyses can be performed with and without aggregation modeling. Appreciable differences will indicate that aggregation may be present and that regular cell cycle estimates should be interpreted with caution in these cases. If retrituration is not successful and the estimates with and without aggregation modeling still vary, then at the very least, these estimates serve to indicate a range of uncertainty, as described in the section below.

The practical impact of the aggregation model on clinical samples is suggested by an analysis of breast and prostate cancer specimens. In DNA-diploid breast tumors fewer aggregates overlay the S phase (those that do are primarily aggregates of debris and G_1 cells). The reduction in %S-phase estimates in these cancers averaged only 0.89, with a maximum difference in %S-phase estimates of 3.4. Aneuploid S-phase estimates, on the other hand, were reduced an average of 2.5, with a maximum difference of 14.5. (Fig. 8.14A). Note that there were a number of examples of S-phase estimates reduced from the high range (i.e., 13%) to the intermediate range (i.e., 7%), or from the intermediate range (i.e., 7%) to the low range (i.e., 2%). In prostate cancer, the detection of aggregates is especially important. As elevations in G_2/tetraploid fractions appear to be linked to adverse prognosis (Chapter 16) S-phase estimates may also have prognostic importance in prostate cancer (Chapter 16). Figure 8.14B shows that, in the analysis of histograms from paraffin-embedded prostate cancers, S-phase estimates are greatly diminished when aggregation is taken into account. Future application of this model to data for which long-term clinical outcome is available should establish whether this model

Table 8.3
Intramodel and Intermodel Confidence Estimation for the Histogram Shown in Figure 8.8F[a]

Model[1]	Diploid S	Diploid G2	Aneuploid S	Aneuploid G2	χ_v^2
1. Sliced-nucleus, zero-order S phase	8.1	9.3	28.6	13.1	1.4
"Intramodel" 95% confidence interval from above	0–16.8	6.8–11.7	23.6–33.2	11.9–14.3	
2. #1 plus G_2/G_1 ratio constraint	9.7	6.3	27.5	13	1.5
3. #2 plus aggregation model	10.1	6.2	26.0	12.1	1.2
4. #1 plus all CVs constrained to be equal	8.1	15.4	31.7	12.7	1.6
5. #1 plus first-order S-phase polynomial	5.3	9.6	27.3	1.4	1.4
"intermodel" confidence interval	5.3–10.1	6.2–15.4	26.0–31.7	12.1–14.4	

[a] Performed using the MultiCycle program.

does indeed improve the prognostic significance of S- and/or G_2/tetraploid-phase estimates.

Confidence Estimation in Cell Cycle Analysis

After performing a cell cycle analysis, how the resulting parameters are used often hinges upon an assessment of the accuracy of the cell cycle estimates. As the preceding seven "practical challenges" indicate, accurate histogram analysis usually involves considerable user-evaluation and examination of the fitting results. This is because addressing these "challenges" involves application of the user's experience and judgment. Unfortunately, this means that the current technology alone cannot provide a definitive answer to the question of the accuracy of cell cycle estimates. Nevertheless, computerized histogram analysis can provide several items that can be used to assist the human interpretation of confidence estimation.

The chi-square statistic

$$\chi^2 = \sum \frac{(yfit_i - ydata_i)^2}{\sigma_i^2}$$

where σ_i^2 is the uncertainty of each data point $ydata_i$) or the reduced chi-square

$$\chi_v^2 = \frac{\chi^2}{\text{degrees of freedeom}}$$

are measures of the deviation of the fitting function from the data, i.e., measures of the goodness of fit. The χ^2 statistic may thus be useful to indicate the extent to which the fitting model matches the histogram distribution, lower values of the χ^2 being better. The problem with presuming that the fit with the lowest χ^2 is the best is that the fitting model may be incorrect for that histogram or that it may be performing an excellent fit to artifacts in the data (for example, peaks with skews or shoulders). Thus, human judgment of the fit is essential, although the χ^2 may be used to aid this judgment. In addition, the user should also be aware that the χ^2 statistic also is affected by the number of cells acquired in the histogram and by the relative proportion of cell-cycle-vs.-background distribution represented in the histogram.

The nonlinear least-squares method of fitting can provide another indicator of the confidence level of each cell cycle parameter. The nonlinear least-squares technique optimizes the fit by adjusting each fitting parameter to minimize the value of χ^2 (see Chapter 3). The rate of change in χ^2 as the parameter is adjusted is an indicator of the uncertainty in the parameter (this is estimated using an "error matrix," (77). "Intramodel" error estimates based on this method may be useful to indicate a range of confidence in cell cycle measurements. For example, Table 8.3 indicates that the diploid S-phase estimate resulting from the fitting shown in Figure 8.8F is unreliable (due to the large component of sliced-nucleus debris that overlaps the diploid S), while the aneuploid S-phase estimate has a narrow confidence interval.

The intramodel error estimates still do not address the question of whether the model used is appropriate for the histogram. The extent to which the cell cycle parameters vary depending upon the model chosen is another indicator of the range of uncertainty in the estimates. This "intermodel" error estimation is illustrated in Table 8.3. In this example, the intramodel and intermodel techniques yield similar indications of reliability and, in the case of G_2 estimates, they agree very closely. In other histograms, the two methods may yield different confidence estimates.

THE CLINICAL INTERPRETATION OF DNA PLOIDY AND CELL CYCLE DATA

The clinical application of flow cytometry to particular organ systems is discussed in depth in other Chapters of this text. In interpreting the relevant literature, it is worth noting that there are several issues that have not yet been clearly resolved.

Near-Diploid and Near-Tetraploid DNA Aneuploidy

This chapter has attempted to address some of the difficulties encountered in interpreting histograms in these two categories. Because identification of near-diploid and near-tetraploid DNA aneuploidy can be affected by interpretation and by instrumental factors such as CV, diagnosis of these categories has varied in published literature. Moreover, in the past, the criteria used in published reports has frequently not been described. Thus, it is difficult in many clinical situations to assess what the prognostic implication of a near-diploid or near-tetraploid DNA aneuploidy is. This uncertainty is underscored by the presence in the literature of a few reports that have examined these categories more carefully, finding, for example, that near-diploid tumors are either more similar to DNA-diploid tumors (65) or more similar to tumors with higher DNA indices (78), and that near-tetraploid tumors may have a more favorable prognosis than other ploidy values (78). If either of these categories prove to have special significance, it will be very important to address the criteria and cutoff values that should be used to designate these conditions.

Prognostic Categories Based on S Phase

Considerable attention has been devoted in this chapter to describing methods and approaches that should increase the accuracy and reproducibility of S- and G_2-phase measurements derived from DNA flow cytometry. Success in this regard, however, still leaves the question of how these parameters should be interpreted in the clinical setting. The literature summarized in other chapters of this text provides evidence that the prognosis of certain tumor types is correlated with flow cytometric cell cycle estimates. Most often, clinical survival or relapse-free interval is compared for patients with above- or below-mean or median S phase, although some authors have determined the S-phase cutoff that yields the maximum intergroup difference in clinical outcome (79). The latter appears to be the soundest statistical approach. Although a single cutoff point yields a binary result—a good or a bad prognosis, for example—this division does not make good biological or statistical sense. Using a binary categorization, cell cycle estimates that differ by insignificant amounts may be placed in opposite prognostic categories (S-phase estimates of 6.9% and 7.1% with a cutoff of 7%, for example). Recognizing that the flow cytometric parameters contain a significant error margin, a growing number of authorities recommend that three prognostic groups be established. In this method, a range centered on the binary cutoff value is designated to be an intermediate prognostic category, and a stronger association with a poorer or a better prognosis is obtained for the low and high ranges on either side of the intermediate category.

Most often, S phase is the parameter evaluated in the literature, although occasionally $S+G_2$ is utilized. The vast majority of studies do not report whether S or $S+G_2$ yields the best correlation with clinical outcome; this question needs to be more uniformly addressed (the reproducibility of the $S+G_2$ measurement may depend upon the success of eliminating aggregates from the G_2 calculation).

Whatever parameter and cutoff value is cited in the relevant literature, a significant problem remains in making use of this data in the clinical laboratory. This is because the values of S and/or G_2 derived by one laboratory cannot be readily compared to those of another laboratory, as several interlaboratory comparisons have described (57, 58). Some of the reasons for this have been previously described and have to do with sample preparation, analysis, and histogram interpretation. Even the same histograms fit using different commercial software programs can produce different values for estimated S phase (Table 8.4). If cutoff values are estab-

Table 8.4
Effect of Analysis Programs on the Determination of the Clinical Significance of S-Phase Fraction for Patients with Merkel Cell Carcinoma (n = 98)[a]

Total % S	ModFit	MultiCycle
Mean	13.85	11.71
Mean + 1 S.D.	19.88	17.30
Survival difference for S above vs. below Mean + 1 S.D.	p=0.001	p<0.001

[a]Data from C.L. Hitchcock et. al., ms. submitted.

lished relative to the mean and S.D. of the S-phase distributions established with the software used in Table 8.4, however, each of the different estimates yield valid prognostic information (Table 8.4).

As of yet (1992), reference standards for cell cycle measurements and a validated protocol to utilize them are not available. The only method available for an individual laboratory to "calibrate" its cell cycle estimates to the published literature is to assemble a local database for which the value of median S phase (or other reference parameter) can be established. Even this laborious method is subject to the caveat that the patient population examined locally may differ from that investigated in a published study.

Hopefully, greater recognition of the importance of the above issues will lead to more detailed reporting of criteria and methods in the literature and a future consensus on clinical procedures and calibration.

CONCLUSION

It is obvious from the above that considerable attention to detail is required in order to optimize the precision of cell cycle analyses by univariate DNA-content analysis. Still, this approach will continue to be important because of the simplicity of both sample preparation and flow cytometric analysis. Careful consideration of the aspects of analysis and interpretation detailed in this chapter may help to improve the overall reliability of ploidy and cell cycle measurements, allowing greater confidence in its application over a wider range of clinical settings.

In the immediate future, improved accuracy will be derived from bivariate analyses, where one parameter is DNA content and the other is an immunofluorescent probe to distinguish cell type or tumor marker. The second parameter can be used to derive a gated DNA distribution. This chapter applies equally well to these analyses and, thus should be relevant to the increasing number of multiparameter studies in the future.

REFERENCES

1. Hiddeman W, Schumann J, Andreeff M, et al. Convention on Nomenclature for DNA Cytometry. Cytometry 1984;5:445–6.
2. Krishan A. Rapid flow autofluorometric analysis of mammolian cell cycle by propidium iodide staining. J Cell Biol 1975;66:188–193.

3. Vindelov LL. Flow microfluorimetric analysis of nuclear DNA in cells from solid tumors and cell suspensions. A new method for rapid isolation and staining of nuclei. Virchows Arch [B] 1977;24:227.
4. Thornthwaighte JT, Sugarbaker EV, Temple WJ. Preparation of tissues for DNA flow cytometric analysis. Cytometry 1980;1:229–237.
5. Crissman JD, Zarbo RJ, Niebylski CD, Corbett T, Weaver D. Flow cytometric DNA analysis of colon adenocarcinomas: a comparative study of preparative techniques. Mod Pathol 1988;1:198–204.
6. Cerra R, Zarbo RJ, Crissman JD. Dissociation of cells from solid tumors. Methods Cell Biol 1990;33:1–12.
7. Ensley JF, Maciorowski Z, Pietraszkiewicz H, et al. Solid tumor prepatation for flow cytometry using a standard murine model. Cytometry 1987;8:479.
8. Engelholm SA, Spang-Thomsen M, Brunner N, et al. Disaggregation of human solid tumors by combined mechanical and enzymatic methods. Br J Cancer 1985;51:93.
9. Visscher DW, Wykes S, Zarbo RJ, Crissmann JD. Multiparameter Evaluation of flow Cytometric synthesis phase fraction determination in dual-labelled breast carcinomas. Anal Quant Cytol Histol 1991;13:246–252.
10. Hedley DW, Friedlander ML, Taylor IW, Rugg CA, Musgrove EA. Method for analysis of cellular DNA content of paraffin-embedded pathological material using flow cytometry. J Histochem Cytochem 1983;31:1333–1335.
11. Hedley DW. Flow cytometry using paraffin-embedded tissue: five years on. Cytometry 1989;10:229–241.
12. Smeets AWGB, Pauwels RPE, Beck HLM, et al. Comparison of tissue disaggregation techniques of transitional bladder carcinomas by flow cytometry and chromosomal analysis. Cytometry 1987;8:14–19.
13. Arndt-Jovin DJ, Jovin TM. Analysis and sorting of living cells according to deoxyribonucleic acid content. J Histochem Cytochem 1977;25:585–589.
14. Rabinovitch PS, Torres RM, Engel D. Simultaneous cell cycle analysis and two-color surface immunofluorescence using 7-amino-actinomycin D and single laser excitation: applications to study of cell activation and the cell cycle of murine LY-1 B cells. J Immunol 1986;136:2769–2775.
15. Darzynkiewicz Z, Traganos F, Kapuscinski J, Staiano-Cuico L, Melamed MR. Accessibility of DNA in situ to various fluorochromes: relationship to chromatin changes during erythroid differentiation of Friend Leukemia cells. Cytometry 1984;5:355–363.
16. Bertuzzi A, D'Agnano I, Gandolfi A, Graziano A, Starace G, Ubezio P. Study of propidium iodido binding to DNA in intact cells by flow cytometry. Cell Biophys 1990;17:257–267.
17. Kubbies M. Flow cytometric histogram analysis: non-stoichiometric fluorochrome binding and pseudo-aneuploidy. J Patho 1992, [in press].
18. Wolley RC, Herz F, Koss LG. Caution on the use of lymphocytes as standards in the flow cytometric analysis of cultured cells. Cytometry 1982;2:370–373.
19. Stokke T, Steen HB. Distinction of Leukocyte classes based on chromatin-structure-dependent DNA-binding of 7-Aminoactinomycin D. Cytometry 1987;8:576–583.
20. Iverson OE, Laerum OD. Trout and Salmon Erythrocytes and human leukocytes as internal standards for ploidy control in flow cytometry. Cytometry 1987;8:190–196.
21. Holtfreter HB, Cohen N. Fixation-associated quantitative Variations of DNA fluorescence observed in flow cytometric analysis of hemopoietic cells from adult diploid frogs. Cytometry 1990;11:676–685.
22. Koch H, Bettecken T, Kubbies M, Salk D, Smith JW, Rabinovitch PS. Flow cytometric analysis of small DNA content differences in heterogeneous populations: human amniotic fluid cells. Cytometry 1984;5:118–123.
23. Evenson D, Darzynkiewicz Z, Jost L, Ballachey B. Changes in accessibility of DNA to various fluorochromes during spermatogenesis. Cytometry 1986;7:45–53.
24. Otto PJ, Hettwer H. Flow cytometric discrimination of human semen cells. Cell Mol Biol 1990;36:225–232.
25. Kubbies M. Flow cytometric recognition of clastogen induced chromatin damage in G_0/G_1 lymphocytes by non-stoichiometric Hoechst fluorochrome binding. Cytometry 1990;11:386–394.
26. Stokke T, Holte H, Erikstein B, Davies CL, Funderud S, Steen HB. Simultaneous assessment of chromatin structure, DNA content, and antigen expression by dual wavelength excitation flow cytometry. Cytometry 1991;12:172–178
27. Telford WG, King LE, Fraker PJ. Evaluation of glucocorticoid-induced DNA fragmentation in mouse thymocytes by flow cytometry. Cell Prolif 1991;24:447–459
28. Roti Roto JL, Wright WD, Higashikubo R, Dethlefsen LA. DNase I sensitivity of nuclear DNA measured by flow cytometry. Cytometry 1985;6:101–108.
29. Nicoletti I, Migliorati G, Pagliacci MC, Grignani F, Riccardi C. A rapid and simple method for measuring thymocyte apoptosis by propidium iodide staining and flow cytometry. J Immunol Methods 1991; 139:271–279.
30. Larsen JK, Munch-Peterson B, Christiansen J, Jorgensen J. Flow cytometric discrimination of mitotic cells: resolution of M, as well as G_1 S and G_2 phase nuclei with mithramycin, propidium iodide and ethidium bromide after fixation with formaldehyde. Cytometry 1986;7:54–63.
31. Bruno S, Crissman HA, Bauer KD, Darzynkiewicz Z. Changes in cell nuclei during S phase: progressive chromatin condensation and altered expression of the proliferation-associated nuclear proteins Ki-67, cyclin (PCNA), p105, and p34. Exp Cell Res 1991;196:99–106.
32. Darzynkiewicz Z, Traganos F, Sharpless TK, Melamed MR. Cell cycle related changes in nuclear chromatin of stimulated lymphocytes as measured by flow cytometry. Cancer Res 1977;37:4635–4640.
33. Becker RL Jr, Mikel UV. Interrelation of formalin fixation, chromatin compactness and DNA values as measured by flow and image cytometry. Anal Quant Cytol Histol 1990;12:333–341.
34. Darzynkiewicz Z. Acid-induced denaturation of DNA in situ as a probe of chromatin structure. Methods Cell Biol 1990;33:337–352
35. Klein FA, White KH. Flow Cytometry deoxyribonucleic acid determinations and cytology of bladder washings: practical experience. J Urology 1988;139:275–278.
36. Alanen Ka, Joensuu H, Klein PJ. Autolysis is a potential source of false aneuploid peaks in flow cytometric DNA histograms. Cytometry 1989;10:417–425.
37. Joensuu H, Alanen K, Klemi P, Aine R. evidence for false aneuploid peaks in flow cytometry analysis of paraffin-embedded tissue. Cytometry 1990;11:431–437.
38. Van Driel-Kulker AMJ, Eyesackers MJ, Dessing MTM, Ploem JS. A simple method to select specific tumor areas in paraffin blocks for cytometry using incident fluorescence microscopy. Cytometry 1986;7:601–604.
39. Mesker WE, Eysakers MJ, Ouwerkerk-Van Velzen MC, Discrepancies in ploidy determination due to specimen sampling errors. Anal Cell Path 1989;1:87–95.
40. Oud PS, Hanselaar TGJM, Reubsaet-Veldhuizen JAM, et al. Extraction of nuclei from selected regions in paraffin-embedded tissue. Cytometry 1986;7:595.
41. Beerman H, Smit VT, Kluin PM, Bonsing BA, Hermans J, Cornelisse CJ. Flow cytometric analysis of DNA stemline heterogeneity in primary and metastatic breast cancer. Cytometry 1991;12(2):147–154.
42. Meyer JS, Wittliff JL. Regional heterogeneity in breast carcinoma: thymidine labelling index, steroid hormone receptors, DNA ploidy. Int J Cancer 1991;47:213–220.
43. Fuhr JE, Frye A, Kattine AA, Van Meter S. Flow cytometric determination of breast tumor heterogeneity. Cancer 1991;67(5):1401–1405.
44. Askensten UG, von Rosen AK, Nilsson RS, Auer GU. Intratumoral variations in DNA distribution patterns in mammary adenocarcinomas. Cytometry 1989;10:326–333.
45. Isobe H, Miyamoto H, Inoue K, et al. Flow cytometric DNA content analysis in primary lung cancer: comparison of results from fresh and paraffin-embedded specimens. J Surg Oncol 1990;43:36–39.

46. Lindmark G, Glimelius B, Påhlman L, Enblad P. Heterogeneity in ploidy and S-phase fraction in colorectal adenocarcinomas. Int J Colorectal Dis 1991;6:115–120.
47. Rabinovitch PS, Reid BJ, Haggitt RC, Norwood TH, Rubin, CE. Progression to cancer in Barett's esophagus is associated with genomic instability. Lab Invest 1988;60:65–71.
48. Levine DS, Rabinovitch PS, Haggitt RC, et al. Distribution of aneuploid cell populations in ulcerative colitis with dysplasia or cancer. Gastroenterology 1991;101:1198–1210.
49. Bagwell CB, Baker D, Whetstone S, et al. A simple and rapid method for determining the linearity of a flow cytometer system. Cytometry 1989;10:689–694.
50. Price J, Herman CJ. Reproducibility of FCM DNA content from replicate paraffin block samples. Cytometry 1990;11:845.
51. Vindelov II, Christenson IJ, Nissen NI. Standardization if high-resolution flow cytometric DNA analyses by simultaneous use of chicken and trout red blood cells as internal reference standards. Cytometry 1983;3:328–331.
52. Rabinovitch PS, O'Brien KO, Simpson M, Callis JB, Hoehn H. Flow Cytogenetics: II. High-resolution ploidy measurements in human fibroblast cultures. Cytogenet Cell Genet 1981;29:65–76.
53. Kallioniemi O-P, Visakorpi T, Holli K, Heikkinen A, Isola J, Koivula T. Improved prognostic impact of S phase values from paraffin-embedded breast and prostate carcinomas after correcting for nuclear slicing. Cytometry 1991;12:413–421.
54. Fried J. Method for the quantitative evaluation of data from flow microfluorometry. Comp Biomed Res 1976;9:263–276.
55. Dean P, Jett J. Mathematical analysis of DNA distributions derived from flow microfluorimetry. J Cell Biol 1974;60:523.
56. Wheeless LL, Coon JS, Cox C, et al. Precision of DNA flow cytometry in inter-institutional analyses. Cytometry 1991;12:405–412.
57. Hitchcock CL. Variability in flow cytometric results using identical archival samples. Cytometry Suppl 1991;5:46.
58. Kallioniemi O-P, Joensuu H, Klemi P, Koivula T. Inter-laboratory comparison of DNA flow cytometric results from paraffin-embedded breast carcinomas. Breast Cancer Res Treat 1990;17:59–61.
59. Wheeless LL, Coon JS, Cox C, et al. Measurement variability in DNA flow cytometry of replicate samples. Cytometry 1989;10;731–738.
60. Coon JS, Deitch AD, de Vere White RW, et al. Check samples for laboratory self-assessment in DNA flow cytometry. The National Cancer Institute's Flow Cytometry Network experience. Cancer 1989;63:1592–1599.
61. Rabinovitch P, Kallioniemi O-P. Automated peak detection and cell cycle analysis of DNA histograms. Cytometry Suppl 5 1991;138.
62. Kallioneimi O-P, Visakorpi T, Rabinovitch P. Evaluation of a software program for automated ploidy and S-phase analysis from DNA histograms. Cytometry Suppl 1991;5137.
63. Wersto RP, Liblit RA, Koss LG. Flow cytometric DNA analysis of human solid tumors: a review of the interpretation of DNA histograms. Hum Pathol 1991;22:1085–1098.
64. Heiden T, Strang P, Stendahl U, Tribukait B. The reproduceability of flow cytometric analysis in human tumors. Methodological aspects. Anticancer Res 1990;10:49–54.
65. Toikkanen S, Joensuu H, Klemi P. Nuclear DNA content as a prognostic factor in T1-2N0 breast cancer. Am J Clin Pathol 1990 Apr;93(4):471–479.
66. Eckhardt R, Feichter GE, Goerttler K. Influence of the elimination of lymphocytes from tumor cell suspensions on the calculation of S phase fractions by flow cytometry. Anal Quant Cytol Histol 1989;11:384–390.
67. Cusick EL, Milton JI, Ewen SWB. The resolution of Aneuploid DNA stemlines by flow cytometry: limitations by the coefficient of variation and the percentage of aneuploid nuclei. Anal Cell Pathol 1990;2:139–148.
68. Banner BF, Chacho MS, Roseman, DL et al. Multiparameter flow cytometric analysis of colon polyps. Am J Clin Pathol 1987;87:313–318.
69. Coon JS, Weinstein RS: Nuclear light scatter as a "second parameter" in flow cytometry of archival tumor specimens. Cytometry Suppl 1988;2:36.
70. Nativ O, Winkler HZ, Raz Y, et al. Stage C prostate adenocarcinoma: flow cytometric nuclear DNA ploidy analysis. Mayo Clin Proc 1989;64:911–919.
71. Jones EC, McNeal J. Bruchovsky N. DNA content in prostatic adenocarcinoma. A flcw cytometric study of the predictive value of aneuploidy for tumor volume, percentage gleason grade 4 and 5, and lymph node metastases. Cancer 1990;66:752–757.
72. Stephenson RA, James BC, Gay H, Fair WR, Whitmore WF, Melamed, MR. Flow cytometry of prostate cancer: relationship of DNA content to survival. Cancer Res 1987;47:2504–2509.
73. Reid BJ, Blount PL, Rubin CE, Levine DS, Haggitt RC, Rabinovitch PS. Flow cytometric and histologic progression to malignancy in-Barett's esophagus: prospective endoscopic surveillance of a cohort. Gastroenterology 1992;102:1212–1219.
74. Reid BJ, Haggitt RC, Dean PJ, Blount PL, Rabinovitch PS. Proliferative abnormalities and histological indicators of neoplastic risk in Barrett's Esophagus. Gastroenterology 1990;98:A111.
75. Beck HP. Evaluation of flow cytometric data of human tumors. Correction procedures for background and cell aggregations. Cell Tissue Kinet 1980;13:173–181.
76. Rabinovitch PS. Numerical compensation for the effects of cell clumping on DNA content histograms. Cytometry Suppl 1991;4:27.
77. Bevington PR. Data reduction and error analysis for the physical sciences. New York: McGraw-Hill, 1969:153–160.
78. Kallioniemi O-P, Blanco G, Alavaikko M, et al. Improving the prognostic value of DNA flow cytometry in breast cancer by combining DNA index and S phase fraction. A proposed classification of DNA histograms in breast cancer. Cancer 1988;62:2183–190.
79. Clark, GM, Dressler LG, Owens, MA, Pounds G, Oldaker P, McGuire WL. Prediction of relapse or survival in patients with node-negative breast cancer by DNA flow cytometry. N Engl J Med 1989;320:627–633.

9

Cytochemistry I: Cell Surface Immunofluorescence

DOROTHY E. LEWIS

GENERAL PRINCIPLES OF ANTIGEN–ANTIBODY PRIMARY AND SECONDARY INTERACTIONS

Introduction

The binding of fluorescently-labeled antibodies to cell-surface determinants was originally described by Coons, in 1941, and has played an important role in the study of the immune system (1). This antigen–antibody interaction is primary in nature and, hence, predates the radioimmunoassay (RIA) and the enzyme-linked immunoadsorbent assay (ELISA) as the first primary antibody technique. Because of the original use of fluorescein isothiocyanate (FITC), which can be conveniently coupled to most proteins with good quantum yield (0.5–0.7 at pH 8), immunofluorescence has been used to characterize cell-surface antigen expression in both single cells and in tissues.

Antibodies in native form can be divalent (IgG) or multivalent (IgM, IgA). The valency of the antibody is determined by the number of binding sites per antibody molecule. As a practical point, even though IgG antibodies have divalent binding capacity and IgM antibodies have a pentameric binding capacity, it is unlikely that all the sites are bound in any given primary antigen–antibody interaction (2).

The relative affinity of each antibody for a given antigen is important because the sensitivity of detection is enhanced with high-affinity antibodies. Researchers, therefore, choose antibodies with the greatest affinity and specificity toward the antigen under study. With the advent of flow-cytometric methods to measure cellular immunofluorescence, several important constraints were placed on the technology of fluorescent antigen–antibody reactions. The first constraint is that the specimens must be in single-cell suspensions in order to be examined in a flow cytometer. In addition, the cells must be of reasonable size and shape or modifications to the instrument are necessary. As with other immunological methods, the reagents used must be pretitered and there is an optimal concentration of cells required.

Structure of Antigen–Antibody Complexes

The structure of the antibody–antigen complex on the surface of the stained cells is assumed to be similar to that of antigen–antibody binding in solution. Thus, the antigen binding site on the antibody is a three-dimensional structure composed of variable regions from both the heavy and light chains. These hypervariable regions are known as complementary determining regions (CDR's). There are six CDR's in an IgG molecule, three on the light chain and three on the heavy chain. The size of the combining site can be small, as with the binding of peptides or haptens, but recent evidence suggests that as much as a 500–750 Å^2 area can be involved with contact of the antigen involving all six CDR's (3). Another important consideration is that the moiety detected by the antibody on the surface of the cell might represent several different amino acid sequences of that molecule that are not necessarily contiguous. The antigen–antigen complex is entirely dependent on noncovalent interactions, such as hydrogen bonds, van der Waals forces, coulombic interactions, and hydrophobic bonds. The variety of determinants that antibodies can recognize is large. For example, with the defined human cell-surface antigens on leukocytes known as CD antigens, the molecules detected have been globular glycoproteins, carbohydrate moieties, or protein enzymes (4).

Affinity and Avidity

Affinity is a measure of the strength of binding between antigen and antibody. The interaction between antigen and antibody is reversible and usually on the order of 10^5 M^{-1} to above 10^{12} M^{-1}. The average affinity of an antibody population is about 10^9 M^{-1}. The time to reach equilibrium depends primarily on the rate of diffusion; however, higher affinity antibodies will take less time to reach equilibrium. In most cases, equilibrium is reached within minutes. In the binding of antibody to cells, the affinity ranges from 10^6 M^{-1}, which would represent a weak signal, to 10^8 M^{-1}, which would be a strong signal on the flow cytometer (3). The binding is usually bivalent and, if indirect immunofluorescence is done, there is likely to be multivalent binding.

Avidity is a measure of the overall stability of the complex and, if the multivalent complexes are formed, the better the avidity. Thus, polyclonal sera, because of multiple specificities, form more stable antigen–antibody complexes than do monoclonal antibodies (mAbs). Indeed, because of this limitation of mAbs, in some cases, a mixture of mAbs reactive to different determinants on the same antigen is used to increase the avidity of the complexes formed.

GENERAL METHODS

Cell Preparation

Lymphocytes from the peripheral blood or suspension cells grown in culture are usually easy to prepare. Greater detail about their preparation will be given in the routine sample section. It is important to be as gentle as possible with the cells, especially if they have been cultured, if there are dead cells in the mixture, or if the cells are tumor cells. In my experience, if the cells have been cultured with a stimulus like a mitogen, the background, both in terms of autofluorescence and other nonspecific binding, can increase. An increased number of dead or fragile cells can also be a problem in cultured cells. In addition, when staining for cell-surface antigens, most tumor cells do better when handled at room temperature than in the cold. Because tumor cells are bigger and more dense, they need less centrifugal force and time to pellet. After the cells have been washed from either the blood or the tissue culture medium, the cells should be counted either by automated counter or by hemocytometer. Viability can be determined by Trypan blue exclusion or the uptake of fluorescein diacetate. The latter is more sensitive since it relies on the cell to use esterases to metabolically produce a fluorescent product, fluorescein (5). Other methods, described by Terstappen et al. and Riedy et al, have the advantage of detection of dead cells after the cellular suspension has been fixed (6, 7). The first reagent is called LDS-751 (6), is excited by 488, and emits in the far red, >670 nm; the other is ethidium monoazide which is also excited by 488 and is detected above 650 nm.

Staining and Fixation Methods

Most cell staining can be done in phosphate buffered saline (PBS) or Hanks balanced salt solution (HBSS) pH 7.2–7.4. A low concentration (1%) of protein is recommended. I prefer HBSS with Hepes as a buffering system. Some investigators have reported interference of phycoerythrin (PE) fluorescence caused by Phenol Red as a pH indicator in the medium, but I have never seen this problem. However, there is interference when using biotin/avidin staining methodology if the medium used to stain or wash contains a high concentration of biotin (i.e., RPMI 1640).

Staining of separated cells can be accomplished by resuspension of the cells at a concentration between $5 \times 10^6 - 2 \times 10^7$/ml in PBS (50-100 μl/test) with 1–3% bovine serum albumin (BSA). As few as 2×10^5 cells/test can be stained with directly conjugated antibodies as long as the procedure is gentle so that reasonable recovery is assured. To minimize capping, staining should be done at 4°C with cold buffers. Some investigators add sodium azide at a final concentration of 0.02%.

The preferred method of fixation is 0.05–1.0% paraformaldehyde in pH 7.2 PBS (0.2–1.0 ml vol) for 30 min. For fixation of cell surface molecules the fixative should be prepared fresh every two-to-three weeks and excessive heat should be avoided when dissolving the paraformaldehyde (8). The cells can be left in this solution if they are to be run within a few days. If they are to be stored longer, they should be centrifuged and resuspended in an equivalent volume of PBS. This procedure is necessary to avoid a significant increase in autofluorescence with storage. This fixation procedure is sufficient to destroy human immunodeficiency virus (HIV) in clinical specimens (9). Such specimens can be stored for months with little loss of fluorescence and only minor changes in the light scatter characteristics. The samples are stored in covered boxes to prevent fluorescence deterioration.

Remember not to use glutaraldehyde as a fixative as the background fluorescence is greatly increased.

Reagent Concentration

Whether you prepare your own antibody reagents or buy commercially prepared ones, it is important to titer the reagents to achieve saturation binding conditions. Thus, if you know the amount of reagent to use for an ELISA or other immunological test, it is no guarantee that the same amount will be adequate for cellular immunofluorescence. In general, considerably more antibody is required for this application. From past experience, saturation of cellular binding sites is generally achieved around 1.0 μg antibody/10^6 cells. Commercially prepared antibodies are usually pretitered; however, with some formulations, it is possible to save money because considerably more antibody is in each vial than is necessary for saturation conditions. Lyophilized preparations of commercial antibodies have long shelf lives; however, significant problems with aggregate formation can occur after reconstitution. If Ig aggregates are formed, the preparation can be ultracentrifuged for 30 min at 100,000xg and the top 70% to 80% of the solution can be used for cell surface staining. The disadvantages of this method are the loss of titer of the reagent and the cost. After ultracentrifugation, it is necessary to retiter the reagents. Titration is accomplished by dilution of the antibody in a suitable diluent, such as PBS or HBSS with a protein source, usually 1% BSA or fetal calf serum (FCS). It is critical that **all** reagents be titered. Some believe that only the primary antibody in indirect immunofluorescence applications need to be titered. if the second step is not titered as well, there is no way to know whether an optimal signal-to-noise ratio has been reached. In addition, a general rule of thumb is to include a protein source such as human AB+ plasma, heat-inactivated FCS, or BSA while staining, both to enhance cellular viability and to reduce nonspecific binding.

Nonspecific Binding Caused by Fc Interactions

By far the biggest problem encountered in cell surface staining is nonspecific binding of antibody to Fc receptors that occur on a variety of cell types. Three types of Fc receptors for IgG have been characterized (10). FcγRI is a 70 kD glycoprotein that appears on monocytes and macrophages and can be induced on neutrophils by γ interferon (IFN-γ). It

COLOR PLATE I

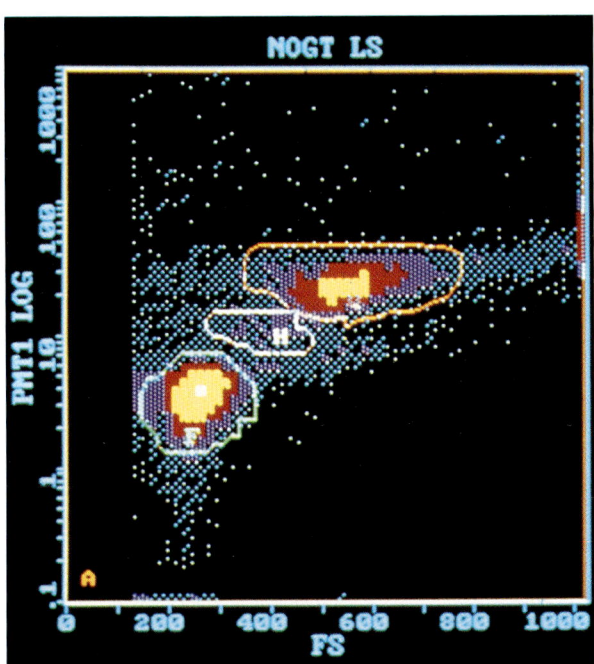

Figure 9.3. Two-parameter forward scatter (FS) versus log-side scatter (SS) of white blood cells prepared by the Q prep™ method. Three populations are detected. The population with the least FS and SS on the bottom left is the lymphocyte population (49%). The middle population is the monocyte population (3%) and the top right population is the granulocyte population (34%). This pattern indicates a relative lymphocytosis, perhaps because the cells come from an individual with asymptomatic HIV infection.

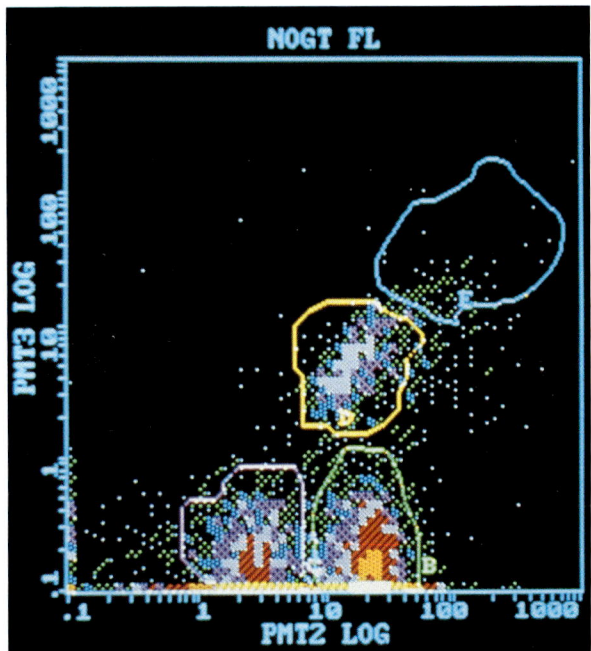

Figure 9.4. Two-parameter FITC CD45 (X axis) and PE CD14 (Y axis) histogram showing three predominant populations of white blood cells. The "bitmap" with the least CD45 fluorescence and no CD14 expression represents the granulocytes (32%). The "bitmap" with $CD45^{bright}$ expression and no CD14 expression represents lymphocytes (59%), and the "bitmap" above it with $CD45^{dim}$ $CD14^+$ cells are the monocytes (5%).

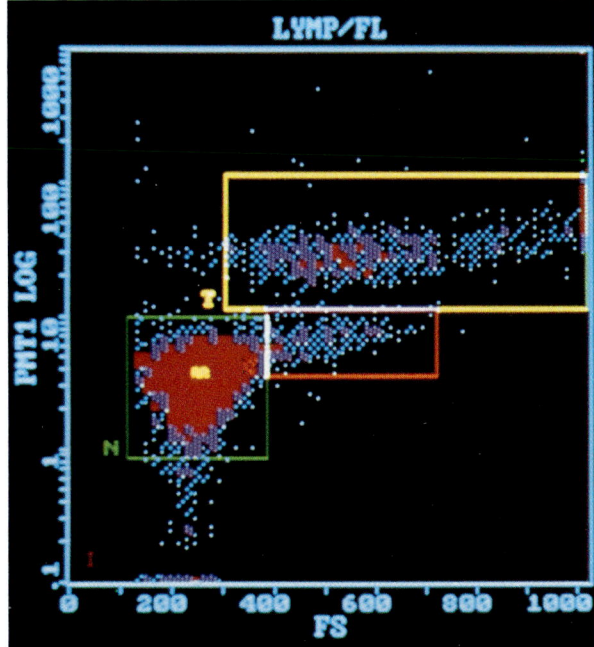

Figure 9.5. Two-parameter FS versus LSS histogram of cells gated on the $CD45^{bright}$ $CD14^-$ fluorescence. The bottom left region, which are lymphocytes, contains 85% of the population. However, 3% of the cells are monocytes and 10% are granulocytes by light scatter characteristics. This indicates that the CD4 CD14 fluorescence characteristics are *not* sufficient to establish lymphocyte region gates.

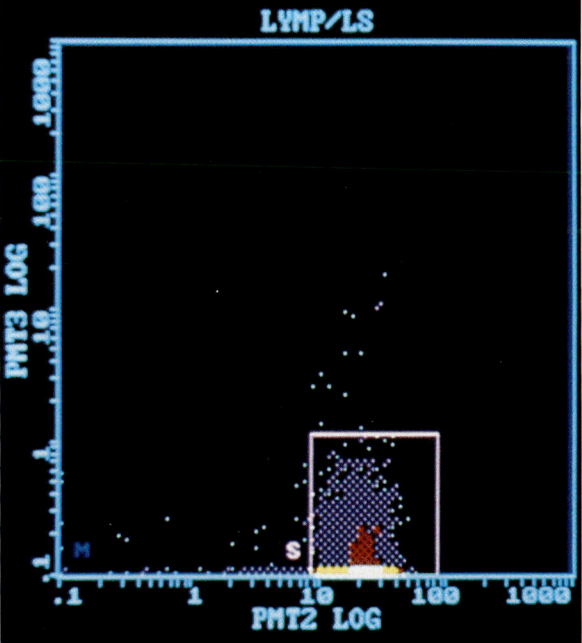

Figure 9.6. Two-parameter FITC CD45 PE CD14 histograms of lymphocytes gated first using CD45 CD14 fluorescence characteristics followed by FS versus LSS characteristics. This backgating procedure results in 98% of the cells in the lymphocyte region, with 1% in the granulocyte region and 1% in the monocyte region.

binds to monovalent IgG at an affinity of about 10^{-9} M for human IgG1, IgG3 and mouse IgG2a, IgG3. FcγRII (CDw32) is a 40 kD glycoprotein that exists in two forms with different cytoplasmic domains. This Fc receptor has low affinity for monovalent human IgG and may be specific for immune complexes and opsonized particles. These Fc receptors are found on virtually all hematopoietic cells including monocytes, macrophages, eosinophils, neutrophils, platelets, and β-cells and are not induced by treatment with IFN-γ. FcγRIII (CD16) is a 55–70 kD glycoprotein with low affinity for monomeric Ig and thus primarily binds IgG immune complexes and polymers. It appears on natural-killer- (NK) cells, neutrophils, eosinophils, and much less on macrophages. Granulocytes express a different form than do NK-cells because of anchoring characteristics and a structural polymorphism.

Nonspecific binding of antibody to Fc receptors on cells can be overcome in some cases by the use of F(ab)₂ reagents that are prepared by enzymatic digestion and column chromatography. Pepsin cleaves Ig molecules on the carboxyl-terminal side of the disulfide bonds that hold the two heavy chains together. Human, rabbit, and mouse IgG are relatively resistant to cleavage at secondary sites in the molecule; however, other mouse Ig isotypes are less resistant (11). Therefore, each antibody to be cleaved may need different conditions. In fact, elastase, instead of pepsin, is recommended for the cleavage of mouse IgG2b, which has been shown to be very sensitive to pepsin cleavage at secondary sites. Fragments can then be purified by removal of intact antibodies and Fc fragments on Protein A. Alternatively, the fragments can be purified on an ion–exchange column followed by gel filtration. Unfortunately, directly conjugated commercial antibodies are not produced in this manner nor are they likely to be in the future. It is possible to purchase secondary reagents which are F(ab)₂ fluorochrome conjugated; but, in my experience, the signal-to-noise ratio is reduced substantially with such reagents. That is, the background is reduced, but there is also disproportionate reduction in the positive signal. An alternative is to use antibodies which are specific for the Fc region of Ig. The idea is that, when the primary antibody has bound via the F(ab)₂ region, the Fc region necessarily is displayed from the surface of the cell. Thus, such a secondary antibody reagent specific for the Fc region would not react with Fc bound primary antibody. This strategy might not work if the secondary antibody itself bound via Fc. In that case, purchase of an anti-Fc F(ab)₂ fluorochrome-labeled reagent should resolve the problem, although there may be a significant signal-to-noise reduction. For some cell types, it is possible to block the Fc region with unconjugated Ig (10 μg/ml) of the same species/isotype as the specific antibody to be used for the test. Another alternative is to use a fluorochrome-conjugated non–antibody reagent as the secondary molecule. Thus, Protein A or G binds to certain antibodies dependent on their class (3). Protein A, a constituent of the cell wall of staphylococci, has four potential binding sites for antibody, but only two are used at one time. The primary binding site is in the Fc domain, but there are great differences in affinity for Protein A between species, i.e. human and rabbit Ig bind Protein A very well, mouse is in the intermediate range, but sheep, goat and rat antibodies do not bind well. Because Protein A shows differential binding for individual antibodies, it is important to note the isotype of the antibody. Thus, all subclasses of human Ig except IgG₃ bind well to Protein A, whereas with mouse Ig there is an order from best to worst, G2a>G2b>G3>G1. Protein G has recently offered an alternative for antibodies which do not bind well to Protein A. With the exception of chicken antibody which only binds weakly to Protein G, all other Igs of different species bind well to Protein G and all subclasses of human and mouse Ig bind well. An important caveat is that Protein G can also bind to albumin so that, if a protein source is used in the medium for wash, it should be free of albumin.

To reduce background problems and to increase signal relative to the background, many investigators use biotin–avidin immunofluorescence systems. The primary antibody is usually easily biotinylated and the secondary reagent is a fluorochrome-conjugated avidin or streptavidin which serves as both an enzyme for biotin and as the secondary amplifying molecule. Streptavidin is recommended because it has a more favorable pI. The affinity of biotin–avidin complexes is on the order of 10^{-14} M and hence, is essentially irreversible (3). This methodology can provide four-to-ten fold amplification over conventional secondary antibody reactions.

Problems with Polyclonal Antibodies

Because polyclonal sera contain large amounts of irrelevant antibodies of unknown specificity, they present special problems when used as either primary or secondary reagents. Titration of the antibody against control sera (usually a prebleed serum from the same animal) can sometimes produce sufficient signal while the background is reduced to nil. Sometimes it is necessary to preadsorb the sera with acetone powders of the same species that the antibody reacts with to reduce nonspecific binding. Alternatively, immunoaffinity purification of the sera can make the reagent monospecific and reduce nonspecific background binding.

Steric Hindrance and Quenching

Steric hindrance is caused by one reagent that inhibits the binding or interferes with the detection of another. A specific example is illustrative. A few years ago, my colleagues and I examined B-cell subpopulations in patients with Juvenile Rheumatoid Arthritis and Kawasaki's disease (12, 13). In particular, we wished to examine B-subpopulations using anti-CD20 and anti-CD21, since it had been reported initially by Anderson et al. that a functional difference was observed between double-positive cells and cells expressing only CD20 (14). CD20 is a Type III (intracytoplasmic COOH and NH₂ terminus) membrane-embedded, non-glycosylated phosphoprotein with a molecular mass of 37

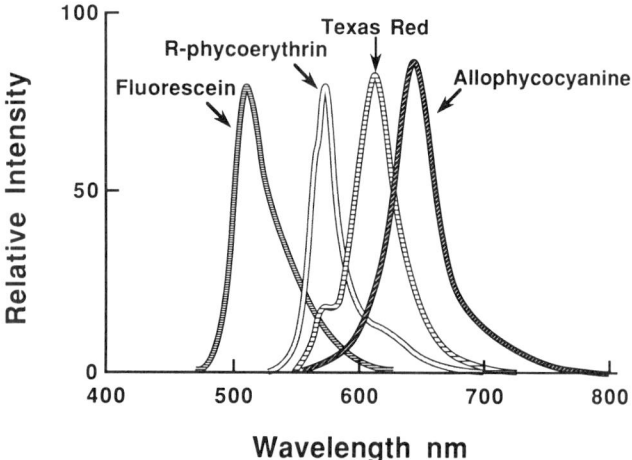

Figure 9.1. Comparison of the fluorescent emission characteristics of commonly used fluorochromes, fluorescein isothiocyanate (FITC), R-phycoerythrin (PE), Texas Red (TR), and allophycocyanine (APC).

kD. It is thought to be involved in ion transport. CD21 (CR2) is a Type I (NH_2 terminus extracytoplasmic) membrane glycoprotein and a member of a gene family of complement regulatory proteins. It is thought to be a receptor for Epstein Barr Virus. Although single-color staining with the anti-CD20 and anti-CD21 antibodies in separate tubes worked well, we could not detect dual-labeled cells stained simultaneously, regardless of which fluorochrome was used on the antibody or whether the order of staining was reversed. The commercial reagent was subsequently found to be different from the originally described antibody in that it was an IgM antibody. The change in isotype from IgG2a to IgM did not affect the ability to recognize CD21; rather, it resulted in steric hindrance when incubated with an antibody that recognized another B-cell-specific antigen, CD20. Thus, for dual staining applications, it is important to be aware of the isotypes of both antibodies. If one reagent is an IgM antibody, it could cause steric interference with another antibody. Another example of steric interference due to the type of fluorochrome that the antibody was coupled with was recently described in our quantitative fluorescence studies (15). FITC was the original molecule conjugated to antibodies for use in immunofluorescence applications. In the last 10 years, several new dyes, including PE and Texas Red (TR), have become commonplace (16). PE is a phycobiliprotein of more than 1×10^6 kD, which has the enviable property of excitation by the same wavelength that excites FITC (488). Therefore, FITC and PE have been used extensively in dual-labeling studies. TR is a rhodamine derivative and is excited by 595 nm, which can be achieved by a dye laser using rhodamine 6G; it emits around 630 nm. Allophycocyanine (APC) is another dye, chemically related to PE, that is excited, as is TR, but emits above 650 nm and, therefore, can be used in quadruple-color immunophenotyping (17) (for further details, see Chapter 7). Figure 9.1 shows the relative emission characteristics of these fluorochromes. In our quantitative fluorescence studies the **same** antibody was conjugated with different fluorochromes, i.e., FITC and PE, and gave significantly different values for the numbers of molecules on cells in quantitative binding studies. The PE values were consistently lower. We suggest that direct PE-conjugation of some antibodies may actually cause steric interference by reducing the ability of the antibody to bind to a given antigen. Thus, even though the phycobiliproteins have large extinction coefficients, and thus absorb light efficiently, conjugation of PE to an antibody may not always be ideal for use as a direct reagent. On the other hand, as an indirect agent, steric interference is not as likely to occur and the advantages of PE are multiple.

It is also possible for two antibodies on the same cell surface to quench the fluorescence of the other signal. This could result if the cellular determinants are in close proximity on the cell surface. Quenching occurs when molecules of dye become close to each other and energy is transferred without the release of photons. The problem of photobleaching, which is the irreversible loss of fluorescence signal caused by extended periods of strong radiation, is more important in fluorescence microscopy where the time of illumination is longer. In that application, an antifading medium such as 2.5% 1,4 diazo biocyclo [2.2.2] octane (DABCO) can be used (2).

Differences in Detection by Flow Cytometry and Microscopy

If the results you obtain on the cell sorter are unexpected, the most appropriate action is to verify them using the microscope. This is important because morphology can be verified and nonspecific staining can be localized. In addition, the flow cytometer can sometimes give different information than does visual inspection. If your eye sees a difference between the control and test specimens and the flow cytometer does not, the most likely explanation is that it stems from the nature of the measurements that flow cytometers make. The flow cytometer either measures a peak fluorescence (the highest level signal of the cell) or an integrated fluorescence (the sum of all the fluorescence across the cell) (see Chapter 5 for further details). The eye is much better at measuring punctate, discrete, peak fluorescence than the cytometer. This is because the peak signal generated after the cytometer encounters the cell represents a single amount of light, and not several points of light per cell, which the eye discerns under a microscope. On the other hand, with integrated measurement of fluorescence in the flow cytometer, a diffuse level of fluorescence over the entire cell could equal the amount of light coming from several discrete entities seen by the eye. This actually happened in a group of experiments that examined the cell cycle of B-cells and the expression of cell-surface actin (18). Thus, by fluorescence microscopy, we could correlate expression of actin to S phase by visualization of punctate areas on the surface of the cells; however, the flow cytometer data showed no differential level of ex-

Table 9.1
Troubleshooting Methods to Optimize Immunofluorescence Cell Surface Staining

1. To Prevent Nonspecific Binding of Antibodies to Cells with Fc Receptors:
   a. Preincubate with antibody of the same species as test antibody (10 μg/ml).
   b. Centrifuge polyclonal sera at 100,000xg for 30 min to reduce Ig complexes.
   c. Use F(ab)$_2$ reagents for direct immunofluorescence.
   d. Use anti-Fc or non-antibody (protein A, streptavidin) reagents for indirect immunofluorescence.
2. To Prevent Other Nonspecific Background Problems:
   a. Use protein source (1-3% BSA or FCS) in wash medium for cell preparation.
   b. Titer antibodies to provide optimal signal to noise.
   c. Alter staining conditions: time of incubation, increase number or duration of washes.
   d. If polyclonal AS used, adsorb with appropriate acetone powders or affinity–purify antibodies.
   e. Use compensation circuit.

pression of cell-surface actin according to cell-cycle stage. This was because the integrated but diffuse background fluorescence was equal to the integrated but punctate fluorescence. As a remedy to this problem, it might be possible to use a slit scanning type system to decrease the beam size or a time of flight ratio to discern the punctate type of immunofluorescence on the flow cytometer.

Proper Controls

For troubleshooting immunofluorescence analysis of cell surface antigens, it is important to use proper controls. This includes the use of a tube referred to as "auto" that undergoes the same treatment as the rest of the tubes. It is designed to determine the true background. This is important because the next control is the second step control or isotype control. In this tube, fluorochrome-conjugated antibody of the same isotype or a non-specific primary antibody is used, followed by the fluorochrome-conjugated secondary reagent. Only by comparison of the autofluorescence control with the second control can you determine whether the background binding occurs via Fc or other factors or whether the cellular autofluorescence is the culprit. Both controls are especially important in staining new cell types or using new reagents. A positive control is also essential to monitor staining techniques and instrument performance. As with other scientific techniques, if the positive and negative controls are incorrect, the rest of the test is uninterpretable. Another important factor in troubleshooting cell surface immunofluorescence is to determine the viability of the cells, especially if the cells are cultured or if they are fragile cells. High fluorescence backgrounds are often observed with cultured cells. In my experience, ficoll separation of the suspension cells can greatly relieve this problem. In some cases, rough treatment of the cells (i.e., excessive washing or too high a centrifugation speed) can make fragile cells die during the staining procedure and cause reduction in the cell number or high background. Remember, the cells are living entities and do not respond well to rough treatment. A summary of troubleshooting methods is given in Table 9.1.

Direct versus Indirect Immunofluorescence

Direct immunofluorescence indicates that the primary antibody used for cell surface immunofluorescence is labeled with a fluorochrome. As such, this method requires a single incubation and wash and the cells are ready for analysis. Most dual- and triple-color immunofluorescence assays are performed using directly conjugated reagents. The background tends to be less with this method; however, the sensitivity also may be reduced. Indirect methods involve application of the primary antibody, incubation, and washing, followed by addition of a fluorochrome-conjugated antibody reactive to the first antibody or another secondary reagent, such as avidin, streptavidin, and Protein A or G. This necessarily creates an amplification of signal. In the case of secondary antibody and Protein A or G it is at least twofold; in the case of tetravalent, streptavidin, or avidin, it is usually four-to-ten fold. Because these methods require more steps, it takes more time and several additional controls are needed. Nonspecific antibodies of the same isotype should be used as the primary reagent followed by the second step reagent, whether it is FITC, Protein A, or PE-labeled streptavidin. This control is necessary to check for background binding caused by the secondary reagent. Again, if an autofluorescence control is used, it is possible to distinguish the true cellular autofluorescence from any nonspecific binding caused by the secondary fluorochrome labeled reagent.

Number of Cells to Analyze

The optimal number of cells to analyze is usually between 2,000 and 20,000. As few as 1000 could be analyzed if the population frequency is greater than 10%. For routine screening of potential mAbs, we analyze 2000 cells; for routine peripheral blood lymphocytes, 5000 is sufficient; but, for analysis of suspensions containing 1-5% of the positive cell type, up to 100,000 cells may be necessary to achieve statistical significance.

Problems Encountered during Flow Cytometric Analysis

The most difficult specimens to analyze are cultured cells and cells that tend to clump and reaggregate. The latter can cause havoc when sorting. Filtration through nylon mesh sometimes is sufficient to prevent clumping. In addition, a small amount of DNase (19) can eliminate problems because sometimes the clumping is caused by dying cells. Avoidance of fibers in the flow cytometer is recommended as they can cause clogs and then no samples can be run.

One of the most frequently encountered problems of the novice is the occasional plug. If the samples have been going smoothly and all of a sudden the data are uninterpretable, the best advice is to rerun a previous sample that gave reasonable results and compare the data. If the previous sample looks the same, it is likely that the problem is with the new specimen and **not** the instrument. If the rerun specimen does not behave as the previous sample did, then the problem is likely to be the instrument.

Summary of Troubleshooting Methods

There are several techniques that have been discussed to reduce background fluorescence. These include the prevention of nonspecific binding by same-species blocking reagents or by using F(ab)$_2$ fragments or other, secondary reagents. In addition, removal of dead cells and debris from cultured cells can reduce the nonspecific background fluorescence. Reduction in autofluorescence from cells can be accomplished in some cases by changing the type of medium or using log-phase cells. Alberti et al. showed that confluent L cells had greater autofluorescence than did cells prior to confluence (20). In the same paper, an electronic way of reducing autofluorescence was described. Autofluorescence of living cells occurs across a great range of excitation wavelengths (and especially below 500 mm) and is likely to be due to the fluorescence of flavins (21). Autofluorescence, therefore, is greater with UV excitation and is the least with excitation in the far red. In the paper by Alberti et al. (20), orange/red autofluorescence was subtracted from the FITC fluorescence signal using the compensation circuitry available on current commercial flow cytometers. This technique is especially useful when the background is significant and expression of the test marker is minimal or in the analysis of rare event populations. In addition, with some tumor cells or with alveolar macrophages that have substantial autofluorescence, the use of TR or APC that are excited by a dye laser tuned to 595 nM might be useful to optimize the signal and reduce the background (16).

ANALYSIS OF ROUTINE SAMPLES

Preparation of Cells

The method of choice for peripheral blood analysis of cell surface molecules as done in clinical flow laboratories is the whole blood lysis technique (22). Peripheral blood can be collected in either ethylenediaminetetraacetic acid (EDTA), ammonium citrate dextrose (ACD), or Heparin, aliquoted (usually 100 μl/tube) and stained with appropriate mAbs for 10-20 min at room temperature. An instrument called a Q prep (Coulter Corporation, Hialeah, FL), which lyses and fixes the red blood cells in a single step, can be used. Alternatively, the red blood cells can be lysed using one of several techniques (23) and the sample can be centrifuged and resuspended in fixative to reduce biohazard concerns. The results with either the Q prep method or the lyse/wash method are similar. However, different preparations of red blood cell lysing solutions can affect the light scatter properties of the white blood cells. For example, the diethylene glycol method significantly shrinks the white blood cells whereas the ammonium chloride or formic acid methods do not. It is, therefore, important to be aware of the methodology used to prepare the samples so that the gains can be appropriately adjusted.

For some applications, functional studies are performed and it is therefore necessary to prepare isolated mononuclear cells. Lymphocytes and monocytes are banded on ficoll gradients. The cells are washed twice in PBS, counted to determine yield, and the viability is determined. The caveat with the ficoll separation methodology is that significant subpopulation losses, especially in the CD8$^+$ subset, can occur (24). Because the separation of mononuclear cells is more technically demanding, there is usually more laboratory–to–laboratory variation using ficoll enrichment of lymphocytes. Therefore, if an accurate determination of the types of lymphocytes in the peripheral blood is to be made, the whole blood method is recommended. After ficoll separation, it may be necessary to again verify the proportion of mononuclear subsets obtained. Platelets also can serve as a contaminate of mononuclear cells and cause annoying problems with nonspecific binding. Also, there are gradients available to enrich for NK-cells and other gradients to enrich for granulocytes (25, 25a). The preparation of adherent cells for cell surface immunofluorescence can pose special problems. In some cases, the trypsin treatment necessary to remove the adherent cells from the plastic (0.25% plus 0.09% EDTA) can result in preparations that are unacceptable for flow cytometry because of excessive clumping. In others, the cell surface antigen may be destroyed by trypsin treatment. We recently examined endothelial cells for the expression of various glycosphingolipids before and after treatment with IFN-γ. Because trypsin treatment of the cells proved unacceptable, we used 5.0 mM EDTA and gentle scraping to remove the cells from the plastic. This method worked well and allowed us to determine that, although IFN-γ upregulates class II expression on endothelial cells, it actually decreases the expression of the glycosphingolipids (26).

As discussed previously, every cell sample type needs three controls, an autofluorescence control, an isotype or second step control, and a positive control. This is true for routine whole blood and for separated cells. Too often novices assume that one control can serve for a number of cell types or samples from different individuals. This **is not** acceptable. Appropriate controls should be run for every cell type.

Whole Blood Cellular Immunofluorescence

To determine the types of lymphocytes in the peripheral blood, an appropriate size/morphology gate is required to examine the fluorescence characteristics of the lymphocytes. A two-parameter histogram (Fig. 9.2, left) is acquired using integrated forward-angle light scatter (FS) versus integrated linear- or log-amplified 90° LS (right-angle light scatter or side scatter [LSS]). At the dimensions of the laser light and the cells that we routinely analyze, FS is roughly equivalent to cell size. Thus, lymphocytes scatter less light than mono-

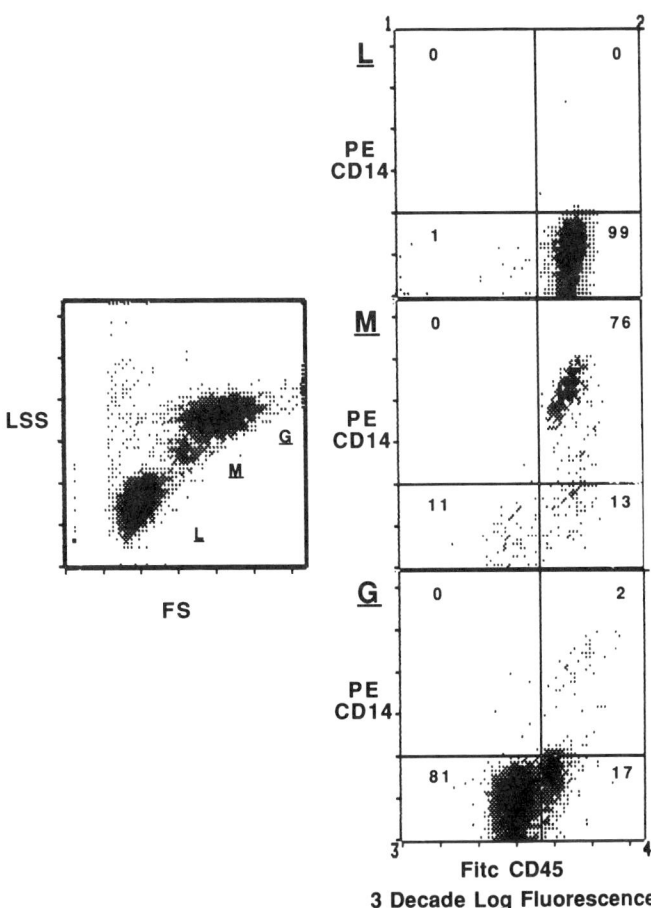

Figure 9.2. Forward scatter (FS) versus log side scatter (LSS) of a whole blood cell preparation after lysis of the red blood cells. Three populations of cells are delineated as shown in the left-hand figure. In order to obtain fluorescence histograms from these three populations, "bitmaps" are drawn encompassing the cells in the populations (designated: L, lymphocytes; M, monocytes; G, granulocytes). The fluorescence histograms of FITC CD45 and PE CD14 generated after FS versus LSS gating are shown on the right-hand series of histograms. The top right panel is CD45 CD14 expression on the cells designated L in the FS versus LSS histograms. Note the high level of CD45-staining (98%) with very few CD45+dim cells (1%) or CD14+ cells (1%) This indicates that there are very few monocytes or granulocytes within the L "bitmap" gate. The middle panel shows the CD45 CD14-staining of the middle FS versus LSS population designated M. Note the dual expression of both CD45 and CD14 on the majority of cells (85%). As can be seen, there is some granulocyte (11%) and lymphocyte (13%) contamination in the gate. The bottom panel is CD45 CD14 expression on cells in the FS versus LSS gate designated G. Note the 10-fold lower expression of CD45 on the majority of the cells (81%) and the contamination by 2% of the CD45+ CD14+ cells. In this individual, there was an abundance of eosinophils, which have about half as much CD45 expression as do lymphocytes, but about three times more CD45 than do other granulocytes.

cytes, and granulocytes can be separated using the FS and SS parameters, it is possible to place electronic gates around the cells in the lymphocyte region to determine the fluorescence properties. These electronic gates are delineators of the characteristics of the cells to be analyzed. Thus, cells that scatter more or less light can be distinguished using gates which can be drawn by rectilinear cursors or be produced by cursors that are multidimensional. Therefore, these "bitmap gates" can be irregular in shape, allowing for more realistic selection of clusters of cells for further analysis. The ability to place selective gates around populations of interest is crucial in the analysis of cell surface antigens because debris, dead cells, and unwanted cells can be excluded on the basis of light scatter or fluorescence properties.

The choice of linear SS versus log SS for gating white blood cell subpopulations depends on the preparation technique and the instrument. Some investigators suggest that the log SS acquisition results in more lymphocytes included in the "bitmap" gate. A three-part differential of the percentage of lymphocytes, monocytes, and granulocytes, can be obtained on a flow cytometer and, at this stage, should be used as a cross-check for automated or manually determined differential counts. The normal distribution in human peripheral blood is about 65% neutrophils, 25% lymphocytes, and 5-10% monocytes with rare eosinophils or basophils. In the future, an accurate leukocyte differential is likely to be obtained on the cytometer. In order to determine whether the cells drawn in the gate are indeed lymphocytes, a positive control is used that consists of a FITC-labeled anti-CD45 antibody plus a PE-labeled anti-CD14. CD45 (T200) is an antigen found on all leukocytes at various expression levels. It is a single-chain glycoprotein that has four isoforms of differing molecular weight and plays a role in signal transduction. Lymphocytes and monocytes express high levels of CD45, whereas the amount of CD45 on granulocytes is about 10-fold less. Interestingly, the amount of CD45 on eosinophils is about half that found on lymphocytes but three times more than on granulocytes. Thus, differential expression of CD45 can be used to distinguish between white blood cell populations. CD14 is a phosphatidylinositol–linked single-chain glycoprotein of molecular weight 55 kD and is found primarily on monocytes. Thus, with these two reagents (CD45 FITC and CD14 PE), we can determine whether the cells defined by the FS versus log SS gate are indeed lymphocytes. An example is shown in Figure 9.2, right panels. Lymphocytes are characterized by CD45[bright] staining with no expression of CD14. Monocytes express both CD45 and CD14 and granulocytes express about 10-fold lower levels of CD45 than do lymphocytes but do not express CD14. The recommended guidelines for immunophenotyping (27) suggest that, if the gate contains less than 85% lymphocytes as determined by CD45[bright] CD14− expression, the gate should be redrawn, the sample rerun, or the blood reacquired. Most of the time >95% of the events in the lymphocyte gate meet the CD45[bright] CD14− criterion (28). Further analysis of the cell types can be done by a process of backgating as dis-

cytes because they are smaller. On the other hand, although dead cells appear larger under the microscope, they actually scatter less light in the flow cytometer. The FS measurement is coupled with SS, which is a reflection of the granularity of the cells. Thus, granulocytes scatter more 90° light than do monocytes or lymphocytes. Because lymphocytes, mono-

cussed by Loken et al. (29) and shown in the color plates elsewhere in the book. Figure 9.3[1] shows an FS versus log-SS distribution from a whole blood preparation which demonstrates the three primary populations of white blood cells in the peripheral blood. These include lymphocytes, monocytes, and granulocytes.

The backgating procedure is based on using the CD45^{bright} expression of the lymphocyte population to establish a gate around that population as shown in Figure 9.4[1] In the second step, a FS versus log-SS histogram gated on CD45^{bright} expression is acquired as shown in Figure 9.5[1], which demonstrates the accuracy of the initial gating on the CD45^{bright} CD14$^-$ population. In this example, 85% of the CD45^{bright} population appears to fall within the lymphocyte FS versus LSS region. There was, however, 10% granulocyte and 3% monocyte contamination. The sample then is rerun with both FS versus log SS **and** fluorescence gates to determine the amount of contamination of other cells. This procedure usually results in >98% lymphocytes as shown in Figure 9.6.[1] Calculation of the cells in both FS versus log-SS and fluorescence gates can give the percentage of other cells or debris. As demonstrated by Loken et al. (29), the events that are not lymphocytes may include monocytes, debris, degranulated granulocytes, or basophils. The backgating procedure allows more accurate determination of the cells analyzed. Unfortunately, such gating is not available on some commercial flow cytometers. For most purposes, definition of FS versus log-SS properties, followed by the determination of the fluorescence of CD45 and CD14 on the white blood cell populations is adequate. Care must be taken, however, to be sure that the lymphocyte FS versus log-SS gates are not too exclusive because NK-cells and activated T cells might be excluded and therefore underrepresented. A comparison of the use of log versus linear SS is useful when setting optimal gates because acquisition of SS under linear amplification spreads the white blood cells across a linear range with the lymphocytes necessarily in the first decade of amplification. On the other hand, log amplification compresses the major white blood cell populations into a fairly narrow region, but in the middle of the three- or four-decade log amplifier. It is thus easier to draw a nonselective gate around the lymphocytes because of this compression. Another possible source of variation in the analysis of whole blood immunofluorescence is that the forward scatter detection methods between FACS and EPICS instruments are different and, thus, give different patterns of white blood cells in the three regions (see Chapter 5).

Compensation/Subtraction

One usually correctable problem with multicolor fluorescence as utilized in routine immunophenotyping is the necessity to control spectral spillover. This is done using compensation circuitry. The problem in routine peripheral blood analysis is that fluorescein has a broad spectral emission which enters the photomultiplier tube (PMT) that measures the PE fluorescence (see Fig. 9.1 for emission spectra of commonly used fluorochromes). The most important thing to remember is that the more intense the FITC emission, the greater the requirement for compensation. Hence, a "one-size-fits-all" standardization method is not really possible. An example of the concept that increasing FITC fluorescence requires more compensation is shown in Figure 9.7. The FITC-labeled mixture from Flow Cytometry Standards Corporation (FCSC, Research Triangle, NC) was used because it contains a series of five beads with increasing amounts of fluorescein fluorescence. As can be seen in the figure, increasing levels of compensation are required for the brightest beads. In the routine analysis of lymphocyte subsets, most of the intensity levels of FITC fluorescence of the recommended peripheral blood phenotyping panel are relatively comparable and compensation levels do not have to be reset. An example of correct compensation in the analysis of FITC- and PE-labeled cultured lymphocytes is shown in Figure 9.8. In this example, CD8$^+$ cells were activated with a mitogen, phytohemagglutinin (PHA) for three days, and then stained for the expression of CD8 and CD38. CD8 consists of two chains (α and β)-encoded by two different genes and expressed as a disulfide linked α-homodimer or an $\alpha\beta$-heterodimer. The molecular mass is 32 kD. T-cells expressing CD8 recognize antigen in the context of class I molecules; they are predominantly cytotoxic and suppressor cells. CD38 is a Type II (intracytoplasmic NH$_2$ terminus) integral membrane protein of 45 kD. It is an activation antigen on T-cells and is found on thymocytes and plasma cells. As can be seen, correct levels of compensation are crucial in determining the number of CD8$^+$ cells which actually express CD38.

Compensation problems are further escalated in triple- or quadruple-color immunofluorescence. It is important to have a set of controls for each combination of mAbs, particularly for the FITC/PE combination since the dichroic filters for triple-color fluorescence are somewhat different than for dual-color fluorescence. In addition, the PE/TR dual combination is necessary to measure spillover of the PE emission. The choice of mAb combinations is crucial in multiple-color immunofluorescence because the excitation and emission of TR with a standard-dye-laser system or a single-laser system is less than optimal. Thus, the biotin-labeled mAb to be used with TR streptavidin should not be directed toward an antigen with low expression. A streptavidin-PE/TR tandem conjugate is available from several sources (Southern Biotechnology, Birmingham, AL and GIBCO BRL, Grand Island, NY) for triple-color immunofluorescence using single wavelength (488 nm) excitation on clinical flow cytometers (30). The tandem conjugate of PE/TR relies on energy transfer to excite the TR molecule. Thus, 488 nm light excites PE, which emits light of a longer wavelength >570, which excites the TR conjugated together with the PE on the same avidin molecule. The system is not perfect because there is

---

[1]See Color Plate I between pages 144 and 145.

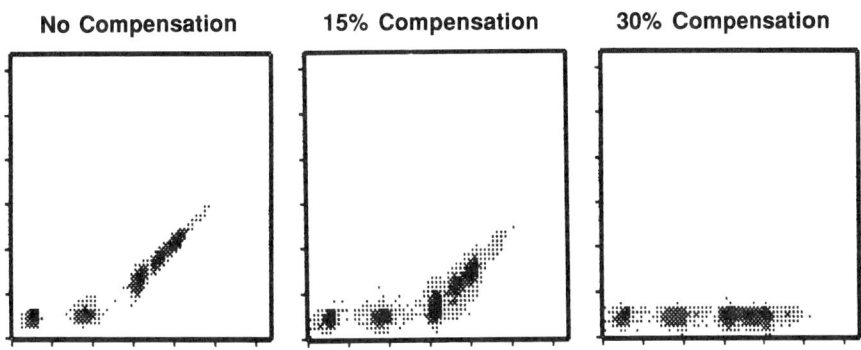

Fitc Fluorescence of 5 Bead Mixture
(3 Decade Log)

Figure 9.7. Increased levels of FITC-fluorescence require more compensation. The five-bead fluorescein-labeled mixture from FCSC was used in a 3-D histogram with FITC on the X-axis and the red PMT and amplifier on the Y-axis. This illustrates the spectral spillover of FITC fluorescence into the red PMT. With no compensation, the three brightest beads demonstrate spillover of light into the red channels of the Y axis. With 15% compensation, the middle bead is fully compensated; however, 30% compensation is required to fully correct the spectral spillover caused by the brightest of the five beads.

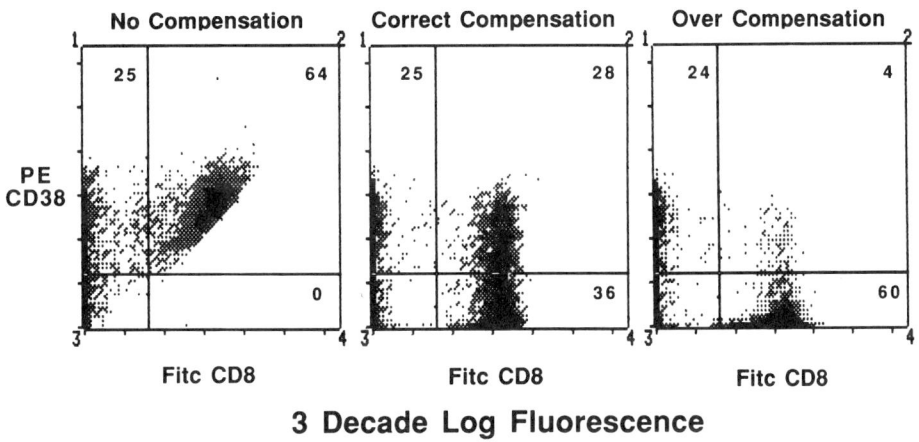

3 Decade Log Fluorescence

Figure 9.8. Activated human peripheral blood cells stained with anti-CD8 and anti-CD38 to detect activated CD8+ cells. In the first panel with no compensation, there is extensive spectral spillover of FITC fluorescence into the red PMT. In the middle panel, correct compensation is applied, and in the last panel, overcompensation is observed. Note that the percentage of single-positive CD8+ cells and double-positive CD8+ CD38+ cells are vastly different between compensation levels.

Table 9.2
Monoclonal Antibodies for Identification of Lymphocyte Subsets in Human Peripheral Blood

Tube No.	Reagent(s) Added (FITC/PE)	Purpose
1	None	To reveal autofluorescence
2	Isotype IgG_1, IgG_2	To detect nonspecific Ig binding
3	CD45 CD14	To verify appropriate lymphocyte gates
4	CD3 CD8	To distinguish T-cells bearing CD8 antigens from non-CD3+ CD8+ cells
5	CD3 CD4	To distinguish T-cells bearing CD4 antigen from dim CD4+ monocytes
6	CD3/CD16 CD56	To distinguish between NK-cells and subsets of T-cells
7	CD19 CD2	To identify B-, T-, and NK-cells

significant emission from the PE/TR complex caused by PE emission that spills into the TR PMT. Significant compensation is thus required to reduce PE emission from the tandem conjugate in the TR channel. In so doing, the sensitivity of both PE and TR detection is limited. If a longer wavelength filter is used to reduce the spillover PE emission, the TR detection is greatly reduced. Thus, the antibody-PE/TR tandem combination should be chosen to react with a cell type that stains brightly for that antigen. In addition, the source of the PE/TR conjugate may be important. It is best to examine several different products and choose the one that gives the best signal to noise for your application.

Choice of Reagents and Recommended Guidelines

The recommended mAb panel in Table 9.2 is designed to account for all the lymphocytes in the peripheral blood. The panel includes the two negative controls (auto, isotype) and the positive gating control (CD45, CD14). The T-cell

Table 9.3
Recommended Guidelines for Human Lymphocyte Phenotyping

1. Anticoagulant: EDTA, heparin, or ACD.
2. Store sample at room temperature and process within 24 hr.
3. Analyze whole blood—if ficoll separation necessary, use care to avoid loss of $CD8^+$ cells.
4. Fix samples for biohazard reduction.
5. Standardize flow cytometer daily, including fluorescence compensation.
6. Use mAb panel that detects dual-labeled populations.
7. Include positive and negative controls for staining.
8. Establish lymphocyte gate based on FS versus SS and $CD45^{bright}$ $CD14^-$ fluorescence.
9. For analysis, compare test samples to negative controls.
10. Perform internal consistency checks as discussed in Table 9.4.

Table 9.4
Internal Consistency Checks for Lymphocyte Phenotyping

1. The sum of the percentages of: T-cells + B-cells + NK-cells = 100%
   CD3 + CD19 + CD16 = 100%
2. The sum of the percentages of: T- NK-cells + B-cells = 100%
   CD2 + CD19 = 100%
3. The sum of the percentages of: T-cells + NK-cells = $CD2^+$ cells
   CD3 + CD16 = CD2
4. The sum of the percentages of CD4 + CD8 = CD3 with approximately 10% variation. Some individuals have large numbers of $CD3^+$ $CD4^-$ $CD8^-$ cells.

marker CD3 is coupled with CD8 in order to distinguish $CD8^+$ cells that are T-cells from those that are not. CD3 consists of five invariable chains (25–28, 20, 16 kD) that are associated with the T-cell receptor for antigen and are involved in signal transduction. Most mAbs to CD3 react to the ε chain. CD3 is also used with CD4 to determine whether all the $CD4^+$ cells are T-cells. CD4 is a 59 kD single chain transmembrane glycoprotein and is found on T-cells that recognize antigen in the context of class II molecules. The function of $CD4^+$ cells is primarily to help other T-cells and B-cells via production of cytokines. CD4 is also a primary receptor for HIV. CD3 is coupled with two NK-cell markers because NK-cells express both CD16 (FcγIII) and CD56 whereas some T-cells express CD56. CD56 is a 140 kD isoform of the adhesion molecule N-CAM that is expressed on NK-cells and a subpopulation of $CD8^+$ T-cells. The last tube in the panel is used to determine the numbers of B-cells using CD19. CD19 is a 90 kD Type I integral membrane glycoprotein that contains two Ig-like domains and is found on B-cells. For detection of all lymphocytes in the peripheral blood, CD19 can be coupled with the anti-CD2 marker, which is the 50 kD single chain transmembrane glycoprotein that recognizes the sheep erythrocyte receptor found on both T- and NK-cells.

The recommended guidelines for immunophenotyping are given in Table 9.3 (27). Briefly, the samples should be collected in one of several anticoagulants and stored at room temperature until stained. The consensus from several studies (27) is that EDTA, heparin, or ACD can be used if the blood is to be analyzed within 24 hrs. If a white count and differential are made from the same collection tube, EDTA is the recommended anticoagulate. In addition, ACD or heparin is better than EDTA if the blood is to be analyzed 24 hrs after collection. This is primarily due to increased granulocyte fragility and degranulation, which results in significant contamination of the lymphocyte gating region. Several groups also have suggested that storage of whole blood on ice before whole blood immunophenotyping results in changes in the SS pattern regardless of the anticoagulate used. Significant reduction in the number of $CD4^+$ cells recovered after ficoll separation occurs in blood samples stored at room temperature longer than four days or after incubation at 4°C overnight. If the sample is to be stored two-to-four days before ficoll separation, dilution of the blood in medium 1:1 is recommended. To avoid the problems associated with ficoll separation, the routine immunophenotyping procedure should be a whole blood method. The samples should be fixed in paraformaldehyde and dual-labeled samples should be tested. The flow cytometer should be standardized, positive and negative controls run, and the test samples compared to control samples for analysis. The panel chosen serves as a means to check for internal consistency because the results of key reagents can be summed to equal 100% (Table 9.4). Lymphocytes consist of predominantly three cell types, T-cells, B-cells, and NK-cells. Because the reagents CD3, CD19, and CD16 react with these cell types, the sum of the percentage-positive of these three reagents should equal 100%. Likewise, the sum of the percentage of $CD4^+$ cells + $CD8^+$ cells ≃ CD3 and the sum of the percentage of $CD3^+$ cells + $CD16^+$ cells should equal the percentage of $CD2^+$ cells. In addition, a within-experiment check of monocyte contamination can be done by the addition of FITC- and PE-labeled anti-CD14. In this case, the monocyte contamination is revealed as a diagonal double-positive bright population.

Special Problems

A summary of special problems encountered in cell surface immunofluorescence is given in Table 9.5. These include background or clumping problems with cultured or activated cells that can be remedied by the removal of dead cells, DNase treatment, or filtration. Adherent cells can pose special problems because they often display high autofluorescence and Fc binding. Fc receptors can be blocked, as discussed, and dead cells should be removed. Trypsin treatment can sometimes cause reduction in cell surface expression, which could be remedied by providing time for reexpression or by alternative adherence removal methods. If specimens have a low cell count, more blood or gentler preparation methods may be required. Pediatric specimens routinely have higher numbers of cells and, therefore, less blood could be used for staining or alternatively more antibody is required for each test. In some cases, the red blood cells fail to

CHAPTER 9 : CELL SURFACE IMMUNOFLUORESCENCE

Table 9.5
Special Problems Associated with Cell-Surface Immunofluorescence Analysis

Type of Problem	Solution
1. Cultured, activated cells	
a. High background	a. Remove dead cells, check fixative
b. Clumping	b. Filter cells, DNAse treatment
2. Adherent cells	
a. High background	a. Block Fc receptors with unconjugated Ig, remove dead cells
b. Trypsin treatment affects cell surface expression	b. Allow time for reexpression, use alternative method to release cells, i.e., EDTA or cold Ca^+Mg^+-free medium plus gentle scraping
3. Special specimens	
a. Low cell count (HIV-infected, anemia, etc.)	a. Use more blood, be gentler to cells
b. High cell count (pediatric specimens)	b. Use less blood or more antibody for each test
c. Failure to lyse red blood cells	c. Lipid interference, tube not at room temperature, increased reticulocytes
4. Other sources of artifacts in cellular immunofluorescence	
Problems	Cause
a. Increased granulocyte fragility	a. AZT, increased age or decreased viability of specimens
b. Increased autofluorescence	b. Some antibiotics and chemotherapy
c. Decreased lymphocyte or CD4 levels	c. Nicotine, corticosteroids, strenuous exercise
d. Variable lymphocyte counts	d. Diurnal variation

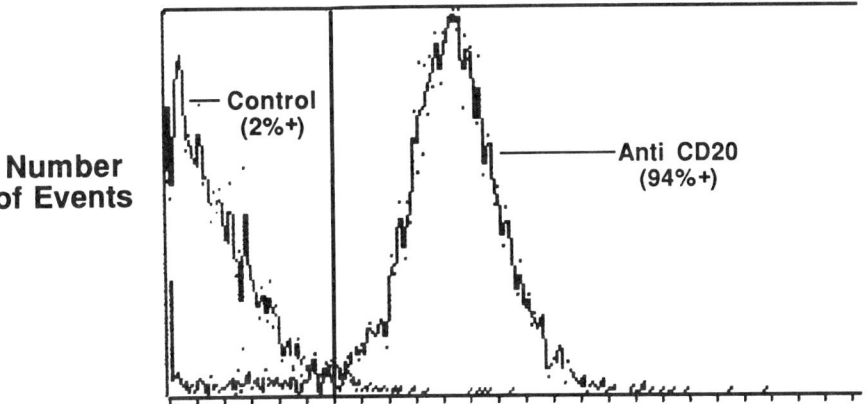

Figure 9.9. Immunofluorescence analysis of a single-parameter histogram of human purified B-cells stained with anti-CD20 compared with the isotype control. The cursor was placed so that <2% of the events in the isotype control were past channel 60. For the CD20-staining, >94% of the cells were CD20+ at the same cursor setting.

lyse. This can be caused by incorrect storage or staining temperature, increased lipids in the serum, or an increased number of reticulocytes. In order to eliminate this problem, it is possible to use fluorochrome-conjugated antiglycophorin, which reacts only with erythrocytes. The cells can then be electronically removed from consideration using a fluorescence regating procedure. In addition, other sources of cell surface staining artifacts can be caused by drugs, storage time, diurnal variation, or strenuous exercise (27).

Analysis of Data

Immunofluorescence is usually analyzed by inspection if there is no real overlap between positive and negative cells. Thus, a cursor is placed where the negative control has no more than 2% of the events found above that cursor. The percentage of positive cells is then determined using the same cursor. The background is not routinely subtracted. Figure 9.9 shows an example of single-color immunofluorescence analysis of purified human B-cells stained with FITC-labeled anti-CD20 and compared to an isotype control. The cursor was placed in the control at channel 60 where <2% of the events were past that channel. In the overlaid histogram of anti-CD20 staining, the cursor at channel 60 shows that 94% of the cells are positive. If there is significant overlap between populations, there are several methods available to compare control and test histograms that are discussed elsewhere (see Chapter 3). In Figure 9.10 an example of the analysis of dual-color staining is given. **A** is the appropriate isotype control where the cursors would be initially set. **B** is a three-dimensional plot of anti-CD3 CD8-staining of human peripheral lymphocytes. The Y cursor is placed at the original isotype position and is also reset at a point higher to enu-

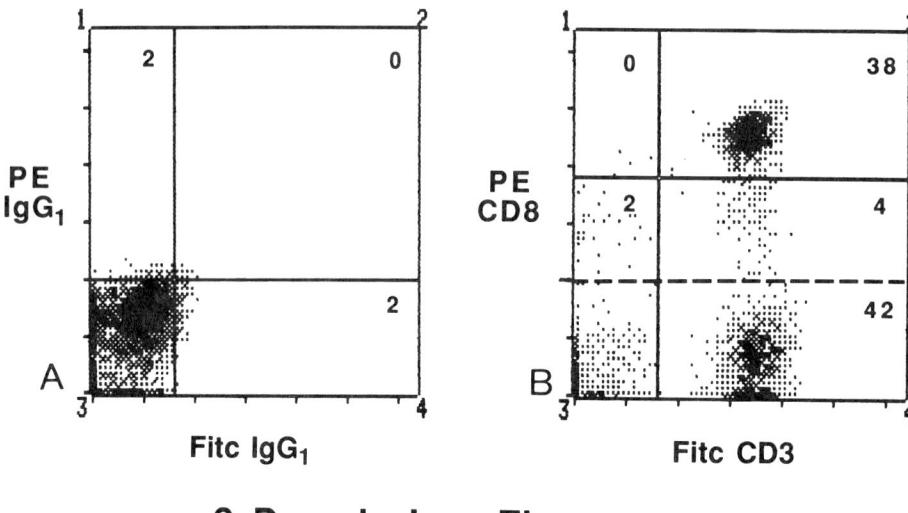

Figure 9.10. Immunofluorescence analysis of dual-color (*FITC* and *PE*) staining of human peripheral lymphocytes. In **A** are the *FITC*- and *PE*-labeled isotype controls and in **B** are *FITC* anti-CD3-plus *PE*-labeled anti-CD8. The percentage of positive cells in each region is given on the figure. The dotted line is drawn to enumerate the CD8^{+dim} cells some of which are CD3$^+$ (4%) and some are CD3$^-$ (2%). The CD3$^-$ cells are likely to be NK-cells.

merate the percentage of CD8$^+$ bright cells that are also CD3$^+$ (38%). In addition to information about dual bearing CD3$^+$ CD8$^+$ cells, the histogram also provides the percentage of CD3$^-$ CD8$^+$ dim cells (2%), which are likely to be NK-cells, and the percentage of CD8$^+$ CD8$^-$ cells (42%), which are likely to be CD4$^+$-cells. Indeed, comparison of the CD3$^+$ CD8$^-$ cell value with the results from CD3$^+$ CD4$^+$ test in the panel should be reasonably close.

In some cases, it is important not to be so rigid about setting the cursor for analysis because subtle and important information can be missed. Thus, a population of dim CD20 cells that were shown to be T-cells was missed by inappropriate placement of the cursor (31). Low-level CD4 intensity could result in the undercounting of CD4$^+$ cells if common sense is not used when placing the cursor. In the example in Figure 9.11, the staining of CD56 on NK-lymphocytes is extremely dim; however, there is a clear cluster of positive cells shown in the figure which are CD2$^+$ CD56$^+$. If the cursor is placed as indicated by the isotype control, approximately 3% of the positive cells would be missed. In this case, it is better to use one's judgement and visual inspection rather than to refuse to alter the cursors.

FUTURE DIRECTIONS FOR CELL SURFACE IMMUNOFLUORESCENCE

Fluorescence Quantitation

The future of cell-surface immunofluorescence measurements by flow cytometry lies in further efforts to standardize the methodology and in ways to quantitate cell surface immunofluorescence. In the clinical laboratory today, much information is discarded because we do not have adequate methodology or the time to determine quantitative levels of fluorescence. Considerations for quantitation of immunofluorescence are discussed in depth in Chapter 29; however, some comment here is justified.

The problem in establishing quantitative immunofluorescence methods lies in developing a way to correlate intensity levels from one instrument to the next. This is because there is no absolute standard, there are only relative ones. Thus, the beads that have been used as calibration standards are determined to contain a mean number of fluorescein equivalents by spectrophotometric comparison with free fluorescein in solution (32). The problem is that the fluorescein placed on the beads is not the same as the fluorescein-conjugated antibody binding to cells. Thus, we either need a reasonable normalization factor to go from soluble fluorescein equivalents, to antibody binding, to cells, or we need a surrogate cell that reacts with antibody in the same way that cells do. This may be an impossible goal to meet; however, there are now beads that have mouse anti-Ig on them that do bind mouse antibody in a somewhat similar fashion (Simply Cellular Beads [FCSC]). Further experiments are necessary to determine the limitations and constraints of this surrogate system. At this point, a relative system would be better than no system at all.

Use of Flow Cytometer for Absolute Counts and for Differential Counts

At present, everyone recognizes the problems with the hematologic measurements made to determine absolute CD4 counts (33). The importance of such counts has increased because of the use of CD4 cellular numbers as a surrogate marker for HIV progression and as an indicator of treatment necessity. The problems with the absolute CD4 counts lie in the natural variability of the white blood count and the differential as well as the inaccuracy of both methodologies

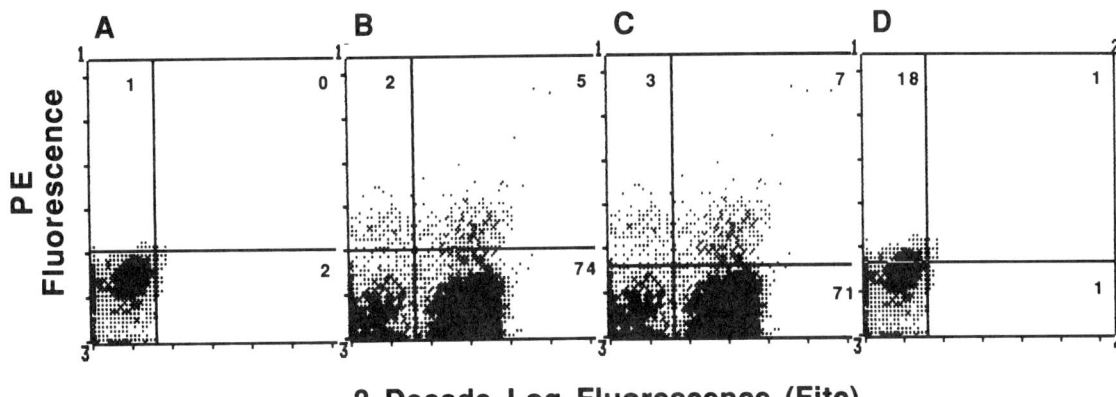

Figure 9.11. Example of too stringent immunofluorescence gates determined by the isotype control (**A**). Cursors were set on the isotype control so that <2% of the events were in quadrants *1, 2,* and *4*. **B** is a 3-D histogram of peripheral blood lymphocytes stained with FITC anti-CD2 and PE anti-CD56, with the cursors placed as in the control. Because the expression of CD56 is weak, the distinction between positive and negative cells is not great; however, there *is* a clear distinction (quadrant *2* in **B** and **C**). The placement of the cursor at the inflection point is much more realistic, as shown in **C**. This is in spite of the fact that the placement of the isotype control cursors in **D** is incorrect according to guidelines.

(33, 34). The white blood count and differential were never meant to be used as part of a tripartite calculation to establish an absolute CD4 number. They were intended only to provide a ballpark determination of the white blood count and the types of white blood cells. Because of the recognition that these measurements have severe limitations as well as the fact that flow cytometric analysis has great reproducibility and less variation between laboratories, there are ongoing efforts to establish a routine white blood count and differential on the flow cytometer. It is anticipated that, in the future, the problems can be overcome and the measurement will be made routine.

It is also anticipated that further automation of the whole blood phenotyping methodology will increase accuracy, reduce the possibility for variation, and reduce potential biohazard problems.

In addition to the use of the flow cytometer to make more accurate measurements of white blood cells or the types of white blood cells, it may also be possible to characterize the important populations in fewer tubes than currently used. Thus, as originally described by Horan et al., it might be possible to use antibodies of different specificity and intensity levels to detect peripheral blood cell populations (35).

Other Areas for Progress

Progress in the methodology to detect multiple populations in very few tubes might be aided by the development and widespread availability of new dyes with high quantum efficiency that are excited by air-cooled argon or helium-neon lasers that have low initial costs and little maintenance. The cyanine dyes appear promising. Another area of progress might be in the use of axial light loss (ALL) to better discriminate between leukocyte populations. It has been reported by Stewart et al. that the combination of anti-CD45 and ALL resulted in the much improved resolution of the major white blood cell populations (30). The use of ALL in existing flow cytometers would require instrument reconfiguration and hence cooperation with the manufacturers. With respect to this kind of interplay in the future, it is hoped that commercial concerns, which are responsible for antibody, instrument, and other reagent development will work together toward the common goal of obtaining quality flow cytometric answers. The techniques and reagents developed in the future should be available for all instruments and not limited by proprietary concerns. Such jockeying in the past has led to confusion and misunderstanding in the scientific community and does a disservice to the public that we serve.

ACKNOWLEDGMENTS

The author would like to thank Ms. Wendy Schober for many helpful suggestions and comments, Ms. Eleanor Chapman for expert secretarial assistance, and Coulter Corporation for use of the ELITE workstation to generate the color photographs.

REFERENCES

1. Coons AH, Creech HJ, Jones RN. Immunological properties of an antibody containing a fluorescent group. Proc Soc Exp Biol Med 1941;47:200–202.
2a. Harlow E, Lane D, eds. Antibody-Antigen Interactions. *Antibodies: A Laboratory Manual*. Cold Spring Harbor, NY: Cold Spring Harbor Laboratory, 1988:23–35.
2b. Harlow E, Lane D, eds. Cell Staining: Binding Antibodies to Cells in Suspension. *Antibodies: A Laboratory Manual*. Cold Spring Harbor, NY: Cold Spring Harbor Laboratory, 1988:394–399.
3. Rees AR. The antibody combining site: Retrospect and prospect. Immunol Today 1987;8:44–45.
4. Knapp W, Dörken B, Gilks WR, Rieber EP, Schmidt RE, Stein H, von dem Borne AEGKr, eds. Appendix A: CD Guide. *Leukocyte Typing IV: White Cell Differentiation Antigens*. New York: Oxford University Press, 1989:1074–1093.
5. Rotman B, Papermaster BW. Membrane properties of living mammalian cells as studied by enzymatic hydrolysis of fluorogenic esters. Proc Natl Acad Sci USA 1966;55:134–141.

6. Terstappen LWMM, Shah VO, Conrad MP, Recktenwald D, Loken MR. Discriminating between damaged and intact cells in fixed flow cytometric samples. Cytometry 1988;9:477–484.
7. Riedy MC, Muirhead KA, Jensen CP, Stewart CC. The use of a photolabeling technique to identify nonviable cells in fixed homologous or heterologous cell populations. Cytometry 1990; in press.
8. Lanier LL, Warner NL. Paraformaldehyde fixation of hematopoietic cells for quantitative flow cytometry (FACS) analysis. J Immunol Methods 1981;47:25–30.
9. Cory JM, Rapp F, Ohlsson-Wilhelm BM. Effects of cellular fixatives on human immunodeficiency virus production. Cytometry 1990;11: 647–651.
10. Fanger MW, Shen L, Graziano RF, Guyre PM. Cytotoxicity mediated by human Fc receptors for IgG. Immunol Today 1989;10:92–99.
11. Parham P. Preparation and purification of active fragments from mouse monoclonal antibodies. In: Weir DM, ed. Cellular Immunology, Vol 1, Chapter 14. 4th ed. California: Blackwell Scientific Publications, 1986.
12. Barron K, DeCunto C, Montalvo J, Orson F, Lewis D. Abnormalities of immunoregulation in Kawaski Syndrome. J Rheumatol 1988;15: 1243–1249.
13. Barron KS, DeCunto CL, Montalvo JF, Orson FM, Lewis DE. Abnormalities of immunoregulation in Juvenile Rheumatoid Arthritis. J Rheumatol 1989;16:940–948.
14. Anderson KC, Boyd AW, Nadler LM, et al. Isolation and functional analysis of human B cell populations. I. Characterization of the B1$^+$B2$^+$ and B1$^+$B2$^-$ subsets. J Immunol 1985;134:820–827.
15. Schober W, Hughes B, Thompson J, Smith W, Lewis D. Standardization of quantitative fluorescence methods in flow cytometry [Abstract]. Fifth Annual Meeting, Clinical Applications of Cytometry, 1990:11.
16. Waggoner AS. Fluorescent probes for cytometry. In: Melamed MR, Lindmo T, Mendelsohn ML, eds. Flow Cytometry and Sorting. 2nd Ed. New York: Wiley-Liss, Inc., 1990:209–225.
17. Schnizlein-Bick CT, Magier MR, Jones RB, Fife KH, Katz BP, Walker EB. Differences among mononuclear cell subpopulations in HIV seropositive or seronegative homosexual and heterosexual men as determined by four-color flow cytometry. J Acquir Immune Defic Syndr 1990;3:747–756.
18. Bach MA, Lewis DE, McClure JE, Parikh N, Rosenblatt HM, Shearer WT. Monoclonal anti-actin antibody recognizes a surface molecule on normal and transformed human B lymphocytes: Expression varies with phase of cell cycle. Cell Immunol 1986;98:364–374.
19. Muirhead KA, Kloszewski ED, Antell LA, Griswold DE. Identification of live cells for flow cytometric analysis of lymphoid subset proliferation in low viability populations. J Immunol Methods 1985;77: 77–86.
20. Alberti S, Parks DR, Herzenberg LA. A single laser method for subtraction of cell autofluorescence in flow cytometry. Cytometry 1987; 8:114–119.
21. Benson R, Meyer RA, Zaruba M, McKhann G. Cellular autofluorescence. Is it due to Flavins? J Histochem Cytochem 1979;27:44–48.
22. Landay L, Muirhead KA. Procedural guidelines for performing immunophenotyping by flow cytometry. Clin Immunol Immunopathol 1989;52:48–60.
23. Muirhead KA, Wallace PK, Schmitt TC, Rescatore RL, Ranco JA, Horan PK. Methodological considerations for implementation of lymphocyte subset analysis in a clinical reference laboratory. In: Andreeff M, ed. Clinical Cytometry. New York: The New York Academy of Sciences. Ann NY Acad Sci, 1986;468:113–127.
24. Renzi P, Ginns LC. Analysis of T cell subsets in normal adults comparison of whole blood lysis technique to Ficoll-Hypaque separation by flow cytometry. J Immunol Methods 1987;98:53–56.
25. Maderazo EG, Ward PA. Leukocyte chemotaxis. In: Rose NR, Friedman H, Fahey JL, eds. Manual of Clinical Laboratory Immunology. Celluar components. 3rd ed. Washington, DC: 1986:290–294.
25a. Herberman RB. Natural killer cell activity and antibody-dependent cell-mediated cytotoxicity. Cellular components. 3rd ed. Washington, DC: 1986:308–314.
26. Gillard BK, Jones MA, Turner AA, Lewis DE, Marcus DM. Interferon-γ alters expression of endothelial cell-surface glycosphingolipids. Arch Biochem Biophys 1990;279:122–129.
27. National Committee for Clinical Laboratory Standards. In: Clinical Applications of Flow Cytometry: Quality Assurance and Immunophenotyping of Peripheral Blood Lymphocytes. NCCLS Document H42-P, Vol. 9(13);765–849.
28. Sucic M, Kolevska T, Kopjar B, et al. Accuracy of routine flow-cytometric bitmap selection for three leukocyte populations. Cytometry 1989;10:442–447.
29. Loken MR, Brosnan JM, Bach BA, Ault, KA. Establishing optimal lymphocyte gates for immunophenotyping by flow cytometry. Cytometry 1990;11:453–459.
30. Stewart CC, Stewart SJ, Habbersett RC. Resolving leukocytes using axial light loss. Cytometry 1989;10:426–432.
31. Landay A, Ohlsson-Wilhelm B, Giorgi JV. Application of flow cytometry to the study of HIV infection. AIDS 1990;4:479–497.
32. Vogt Jr RF, Cross GD, Henderson LO, Phillips DL. Model system evaluating fluorescein-labeled microbeads as internal standards to calibrate fluorescence intensity on flow cytometers. Cytometry 1989;10: 294–302.
33. Koepke JA, Landay AL. Precision and accuracy of absolute lymphocyte counts. Clin Immunol Immunopathol 1989;52:19–27.
34. Taylor JMG, Fahey JL, Detels R, Giorgi JV. CD4 percentage, CD4 number, and CD4:CD8 ratio in HIV infection: Which to choose and how to use. J Acquir Immune Defic Syndr 1989;2:114–124.
35. Horan PK, Slezak SE, Poste G. Improved flow cytometric analysis of leukocyte subsets: Simultaneous identification of five cell subsets using two-color immunofluorescence. Proc Natl Acad Sci USA 1986;83: 8361–8365.

10

Cytochemistry II: Immunofluorescence Measurement of Intracellular Antigens

CHARLES V. CLEVENGER and T. VINCENT SHANKEY

INTRODUCTION

The quantitation of intracellular antigens by flow cytometry (FCM) is dependent upon fixation and permeabilization techniques. Since most antigens quantified by FCM are labeled with immunofluorescence (IF) techniques, the cellular membrane and intracellular sols must be sufficiently permeabilized to permit the passage of these probes. Intracellular antigens can be quantitated by IF technique in unfixed, permeabilized cells (1, 2). These methods, however, can extract soluble cellular constituents and these preparations undergo rapid degeneration. To prevent this, cellular fixation is frequently employed to stabilize intracellular components before or during cellular permeabilization. Since the FCM quantitation of an intracellular antigen may be simultaneously performed with the quantitation of other cellular properties (e.g., cell morphology (forward- and side-light scatter, Coulter volume), other intracellular antigens, and DNA content), a fixative protocol must permit measurement of these other parameters while introducing minimal artifact (3). A useful fixation/permeabilization for intracellular antigen quantitation, therefore, should meet the criteria listed in Table 10.1.

FIXATION AND PERMEABILIZATION

A variety of fixation and permeabilizing agents have been used or have potential use in the FCM quantitation of intracellular antigens, as listed in Table 10.2. Fixatives differ in their ability to induce the coagulation (e.g., denaturation) and/or cross-linking of cellular proteins that lead to the stabilization of protein within cells (reviewed in 4, 5). Unfortunately, coagulation and cross-linking can lead to the destruction of the epitope(s) of interest. It should be stressed that **the optimal fixative/permeabilization technique for a given antigen must be empirically determined**. A brief discussion of the history, mechanism, and use of some of the fixatives and permeabilizing agents follows. It should be remembered that factors such as pH, tonicity, and buffer composition, as well as fixative concentration, duration, and temperature can all significantly affect the preservation of an antigen (4).

Alcohols

Used before formaldehyde, in the 1800's, as a general tissue fixative, methanol and ethanol are classic examples of coagulant fixatives. A coagulant fixative is defined by Hayat (4) as, a "reagent that coagulates proteins (suspended in fluid) into an opaque mixture of granular or reticular solids." Typically diluted with water to a final concentration of 50% to 70%, alcohols denature proteins by displacing water and by disrupting the hydrogen bonds that are requisite for tertiary protein structure (4, 5). While alcohols denature and fix most proteins, they do not appear to fix nucleoproteins (4), although this has been debated by some (6). Alcoholic fixation is hypotonic and, combined with its coagulant effect on proteins, can markedly alter cellular morphology. Alcoholic

---

Table 10.1
Criteria for a Fixation/Permeabilization Technique Useful in Intracellular and DNA Content Quantitation

1. Preservation of immunofluorescence staining patterns and cell morphology.
2. High resolution of DNA content (low CV's).
3. Minimal induction of artifact, i.e., cell clumping and autofluorescence.
4. Ease/speed of cell fixation/permeabilization.

Table 10.2
Fixation and Permeabilization Techniques for FCM Intracellular Antigen Quantitation

1. Pure Coagulant
   A. Ethanol (70%); Methanol (50%) (18, 19, 55, 111)
2. Cross-linking with permeabilization
   A. Paraformaldehyde (0.5% to 4.0%), in conjunction with:
      1. Permeabilizing agents (see below)
      2. Alcohols (26, 27)
3. Coagulant and Cross-linking
   A. Bouin's (10)
   B. B5 (23)
   C. Paraformaldehyde with alcohols or acetone (22, 79, 110)
4. Permeabilizing Agents (with or without prior fixation)
   1. Triton X-100 (3, 20, 30, 112)
   2. Tween-20 (12)
   3. NP-40 (21)
   4. Saponin (113, 114)
   5. Lysolecithin (2)

fixatives act as their own cell membrane permeabilizing agents, extracting significant quantities of phospholipid (7).

Formaldehyde

First synthesized by Butlerov in 1859, the practical uses of formaldehyde were not recognized until 1894 with the seminal work of Ferdinand Blum (for an excellent review of the properties of formaldehyde, and the work of Butlerov and Blum see (8)). Although Blum's initial work explored the antiseptic properties of formaldehyde, he serendipitously recognized its tissue-hardening and fixative properties, resulting in the widespread use of this agent in pathology laboratories. Formaldehyde is a cross-linking fixative, forming reversible methylene bridges between amino, imino, sulfhydryl, and hydroxyl groups within proteins (4). Formaldehyde can react with the amino groups of nucleic acids (9) and catalyzes the cross-linking of DNA to nucleoproteins (6). Formaldehyde also reacts to a lesser extent with lipids and carbohydrates, but, unlike alcoholic fixatives, does not act as a permeabilizing agent. Commercially produced as a gas, saturation of water with formaldehyde produces a 40% solution (weight/volume) alternatively known as formaldehyde or formalin. In solution, the vast majority of formaldehyde is hydrated into the relatively inert methylene glycol. The rapid penetration, but slow fixation, of tissues and cells is the result of the equilibrium between formaldehyde and its hydrate. Although both are rapid penetrators, only formaldehyde is reactive. Thus, conditions favoring the dissociation of methalene glycol (low pH, high concentrations, elevated temperatures) favor cross-linking. Formaldehyde spontaneously deteriorates in solution via a Cannizzaro reaction ($2HOCH_2OH \rightarrow CH_3OH + HCOOH + H_2O$), producing formic acid (5). Oxidation of formaldehyde due to atmospheric exposure, excess heat and light, and manufacturing processes also results in the presence of formic acid. Although formic acid can be neutralized in part by buffering solutions (i.e., neutral-buffered formaldehyde), its harsh protein-denaturing properties could have detrimental effects on antigenic structures. Thus, the preferred solution of formaldehyde is prepared from its solid polymer, paraformaldehyde [$HO(CH_2O)_{6-100}H$]. EM grade paraformaldehyde is solubilized by heat and neutral buffer solutions and is free of formic acid. It is also free of methanol which is found as a stabilizing agent in unbuffered solutions of formaldehyde. Highly pure preparations of paraformaldehyde can be obtained from electron microscopy supply houses (i.e., Polysciences, Electron Microscopy Services, etc.).

Other Fixatives

The cross-linking fixative of choice for electron microscopy is the divalent aldehyde, glutaraldehyde. Its cross-links with proteins and nucleic acids are largely irreversible and morphologic tissue preservation is excellent (4). Glutaraldehyde, as demonstrated later in this chapter, however, is not used in flow cytometry because cell suspensions fixed with this agent yield broad DNA content coefficients of variation (CV's) and high levels of autofluorescence. Furthermore, the extensive cross-linking induced by glutaraldehyde can disrupt some epitopes. Other commonly used fixatives, such as Bouin's (a combination of formaldehyde, picric acid, and glacial acetic acid) or B5 (formaldehyde and mercuric chloride) are combinations of cross-linking and coagulative fixatives (5). They are more toxic and expensive than formaldehyde. Although they are extensively used for their enhancement of light microscopic morphology, relatively little is known regarding their preservation of cellular constituents. Bouin's solution, however, has been used in the preservation of peptide hormones for immunocytochemistry and flow cytometry (10).

Permeabilizing Agents

Numerous agents have been employed to permeabilize cell membranes and the cytoplasmic and nuclear sols. Although permeabilizing agents are disparate in composition, they all disrupt/solubilize phospholipid to some extent. Permeabilizing agents facilitate the intracellular entry of antibody probes. Commonly used permeabilizing agents include Triton X-100, Tween-20, NP-40, saponin, lysolecithin, digitonin, and the alcohols. Unfortunately, little is known regarding the ability of these agents to extract cellular components other than phospholipids (i.e., proteins, DNA, and RNA). Working with unfixed nuclear matrix preparations, Berezney and Coffey demonstrated that treatment of isolated nuclei with 1% Triton X-100 extracts 91% of the phospholipid, 5% of the RNA, 6% of the protein, and essentially 0% of the DNA (11). The extraction of cellular components is also affected by the extent of fixation. Increased intracellular cross-linking induced by increased effective concentrations of paraformaldehyde improved the retention of cellular components at the EM level after permeabilization (12). Improved retention of cellular constituents, however, comes at the expense of overall cell permeability to protein, DNA, and RNA probes. The degree of extraction of these components can alter other cellular parameters, such as light scatter. Thus, while cells fixed with paraformaldehyde alone demonstrate preserved light-scatter distributions (13, 14), the use of a permeabilizing agent with paraformaldehyde can dramatically alter light-scatter distributions. Milder permeabilizing agents, i.e., lysolecithin, may minimize this effect and be preferred when preservation of light scatter is important, such as in the study of lymphocytes. Therefore, the selection of a permeabilizing agent and the conditions under which it is used must be empirically determined for the antigen and cell type under study.

Effect of Fixation Protocols on Antigen Detection

Fixation methods can have profound effects on a given antigen. These effects can differ significantly even for antigens localized within the same cellular subcompartment and may

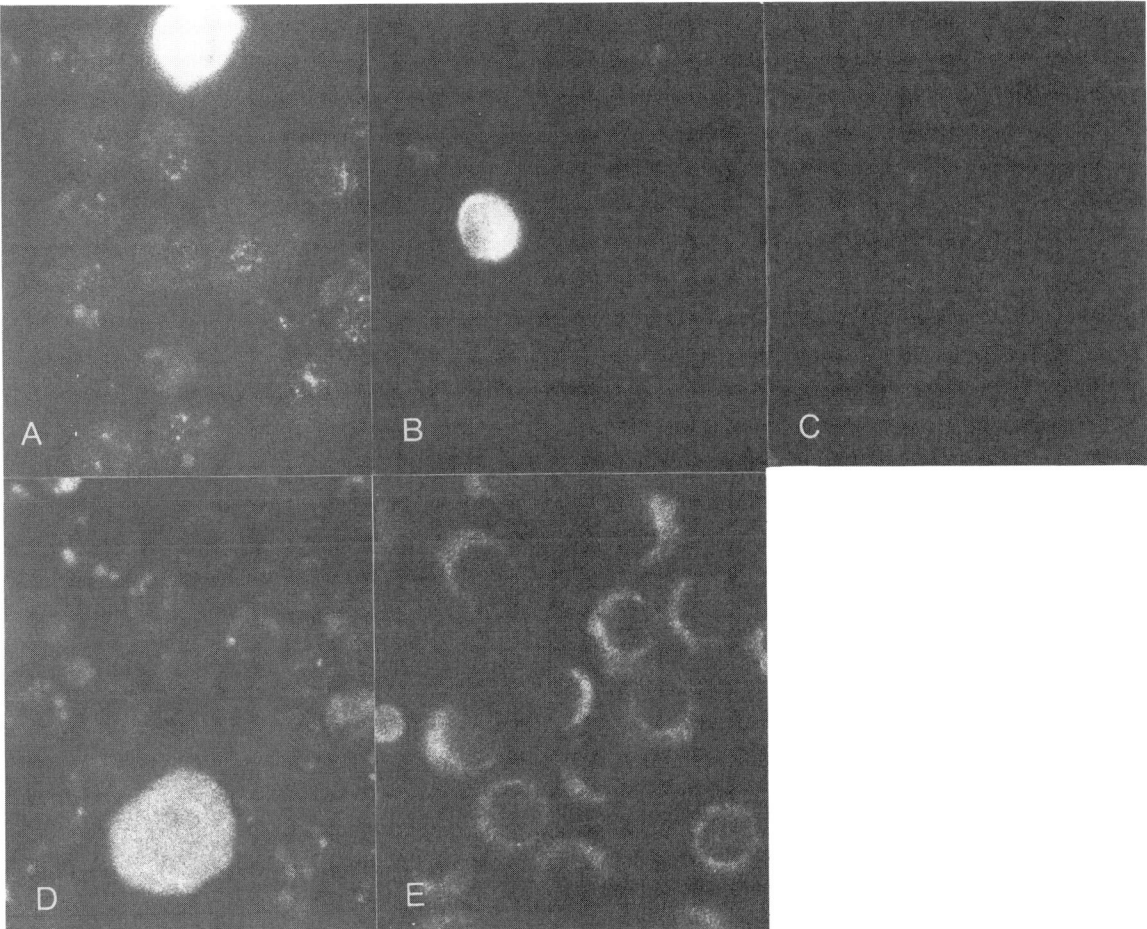

Figure 10.1. Interleukin-2 stimulated cloned murine T-cells labeled with anti-p105 indirect immunofluorescence after various fixation protocols. **A**: 1% Paraformaldehyde/0.1% Triton X-100 (4°C); **B**: 70% Ethanol (4°C); **C**: 50% Methanol (−20°C); **D**: Bouin's solution (20°C); **E**: Acetone (−20°C). All fixations were performed for 10 min. Magnification ×1000.

reflect a given antigen's (in)solubility, accessibility, and susceptibility to denaturation. The effect of various fixative protocols on the cell-cycle-associated intranuclear antigen p105 and prolactin in interleukin-2-stimulated cloned T-cells is shown in Figures 10.1 and 10.2. The intranuclear antigen p105 is expressed biphasically during the cell cycle, with cells in the G_0 phase of the cell cycle containing one arbitrary unit of p105, while cells with G_1, S, and G_2 contain 10 arbitrary units (15). Mitotic cells have the highest levels of p105, containing approximately 100 arbitrary units of p105 (16). The localization, quantity, and cell-cycle expression of p105 is dramatically affected by the fixative protocol utilized. Fixation with 1% paraformaldehyde, followed by permeabilization with 0.1% Triton X-100 is optimal for this antigen. A discrete pattern of nuclear speckling is seen in interphase cells labeled with anti-p105 immunofluorescence. This speckling is due to the association of the p105 antigen with interchromatin granules within the nucleus. The bright cell seen within the photomicrograph (Fig. 10.1) represents a mitotic cell, where p105 antigen can be found diffusely throughout the mitotic cytoplasm. Fixation with ethanol and Bouin's solution denature, mask, and/or extract the p105 antigen from interphase cells, demonstrating positive immunofluorescence only within mitotic cells. Methanol fixation, on the other hand, eliminates p105 immunofluorescence from mitotic cells, preserving p105 immunofluorescence within interphase cells, albeit at reduced levels. Fixation with acetone appears to "strip" the antigen into the cytoplasm of the cell, and obliterates quantitative differences in p105 levels between interphase and mitotic cells.

The localization and quantitation of intranuclear prolactin within T-cells (17) is also affected by fixation (Fig. 10.2). Again, fixation with paraformaldehyde/Triton X-100 is optimal. Alcoholic fixation leads to a partial redistribution of antigen into the cytoplasm; furthermore, ethanol fixation causes an apparant intranuclear "clump" of prolactin. Bouin's solution and acetone appear suboptimal in the preservation of this antigen.

Thus, a variety of fixatives need to be examined when initiating immunofluorescence studies on a previously uncharacterized antigen. Reliance on a single fixative can provide misleading results regarding the distribution within a

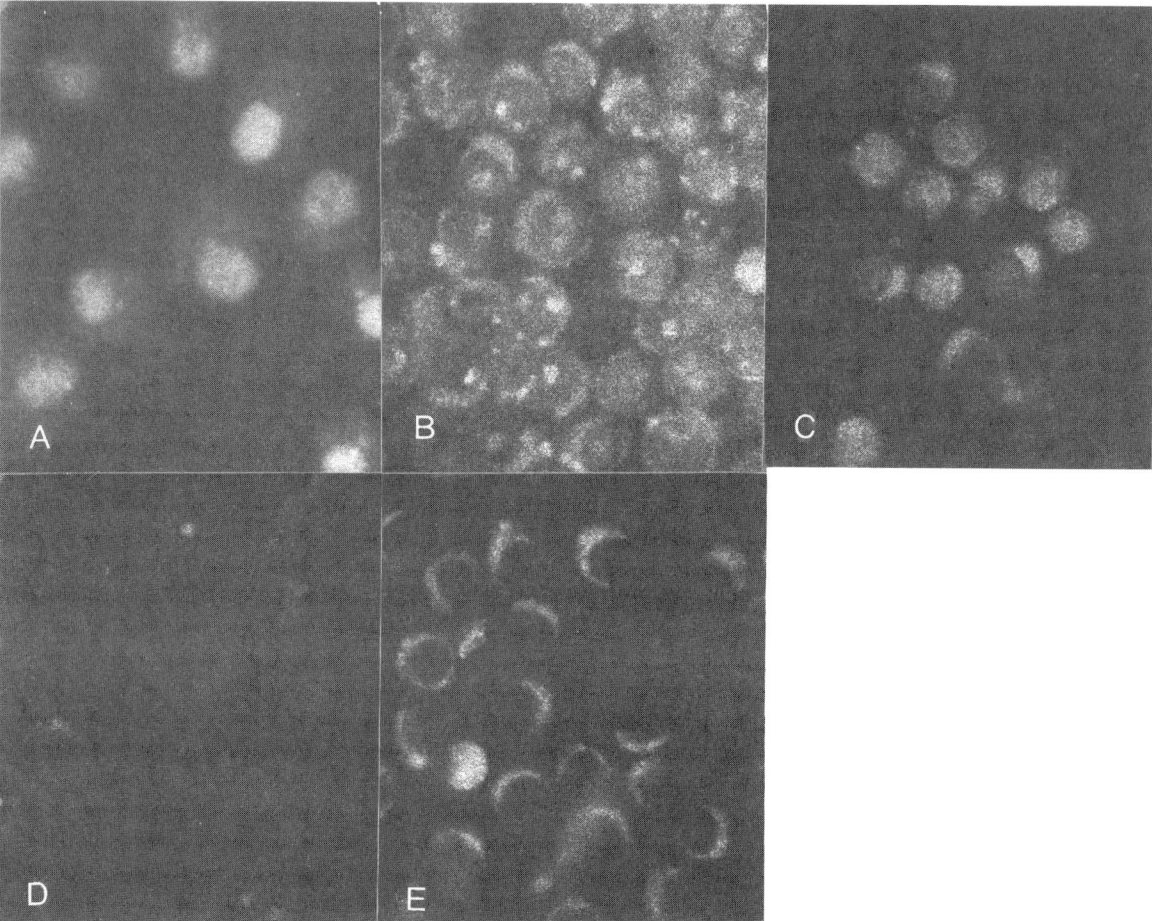

Figure 10.2. Interleukin-2 stimulated cloned murine T-cells labeled with antiprolactin indirect immunofluorescence after various fixation protocols. **A**: 1% Paraformaldehyde/0.1% Triton X-100 (4°C); **B**: 70% Ethanol (4°C); **C**: 50% Methanol (−20°C); **D**: Bouin's solution (20°C); **E**: Acetone (−20°C): All fixations were performed for 10 min. Magnification ×1000.

cell or across the cell cycle. Originally, proliferating cell nuclear antigen (PCNA, also previously known as cyclin) was believed to be localized entirely in the nucleus of methanol-fixed cells in S phase (18, 19). Subsequent studies using other fixative techniques, as well as alternative quantitation methods, revealed that PCNA could also be identified in the nucleus and cytoplasm of cells within the G_1 and G_2 phases of the cell cycle (20, 21). The localization of PCNA may be both fixation- and antibody-(IgM or IgG isotype) dependent (22), although appropriate fixation is likely to result in predominantly nuclear localization with minimal cytoplasmic staining. Similar fixation artifacts have been observed with the c-myc antigen (23). Therefore, the optimal fixative technique for IF should be carefully confirmed by empirical observation and subsequently confirmed by an independent technique as described later in this chapter.

Effect of Fixation Protocols on DNA Content Analysis

Frequently, FCM quantitation of intracellular antigens is performed simultaneously with DNA content analysis. This permits the correlation of antigen levels with cell cycle progression and is particularly useful in characterizing cell-cycle associated antigens. Although fluorescein-conjugated probes, in conjunction with propidium iodide, are most frequently used for this purpose, any combination of probes (i.e., FITC/DAPI, FITC/7-AAD, etc.) with good spectral separation of fluorescence emission is acceptable.

Fixation/permeabilization protocols also affect the quality of the DNA content histogram (24). Optimal fixation for an antigen, however, is not necessarily optimal fixation for DNA content analysis. A balance needs to be struck between cross-linking, if such a fixative is used, and permeabilization. Cross-linking fixatives rapidly couple histone and nonhistone proteins to DNA (6, 9). If used in excessive quantities, they can hinder the interaction of DNA-binding/intercalating dyes. Furthermore, if permeabilization is inadequate, the appropriate intranuclear DNA:dye stoichiometry may not be achieved. These principles, in part, are demonstrated in Figure 10.3. Fixation with paraformaldehyde followed by permeabilization provides excellent DNA histograms with low coefficients of

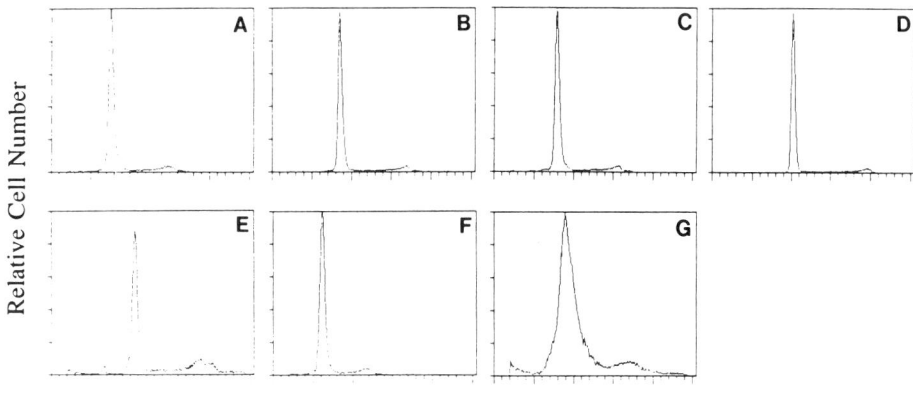

Figure 10.3. Effect of various fixation protocols on the quality of DNA content histograms. Following fixation, the cells were stained with propidium iodide after RNAse treatment, as previously described. All primary fixations were for 10 min at 4°C. **A**: 1% Paraformaldehyde (75/3.5%/<5%); **B**: 1% Paraformaldehyde/0.1% Triton X-100 (86/2.4%/<5%); **C**: 1% Paraformaldehyde/50% Methanol (78/2.8%/<5%); **D**: 50% Methanol (102/2.2%/<5%); **E**: 70% Ethanol (106/3%/15%); **F**: Bouin's solution (62/5%/5%); **G**: 1% Glutaraldehyde (90/>8%/5%). Values in parentheses are as follows: (mean G1 channel/%G1 coefficient of variation/% cells in clumps).

variation (CV), minimal clumping, and adequate mean fluorescence intensities. As seen in Figure 10.3A, fixation with paraformaldehyde alone significantly increases the CV and decreases the fluorescence intensity, demonstrating the effect of inadequate permeabilization. In comparison, the mean DNA fluorescence intensity of alcohol-fixed preparations increased. This is most likely due to the extraction of nuclear protein during alcohol fixation, which increases the accessibility of DNA to the propidium iodide. The hypotonic nature of alcohol fixation alters other cellular parameters as well. The light-scattering properties of alcohol-fixed cells are dramatically changed, with forward light scatter increasing and side light scatter decreasing (data not shown). In addition, ethanol fixation is particularly prone to inducing cellular clumping, as seen in Figure 10.3E. These negative effects of alcoholic fixation can be partially avoided if the alcohol is first prechilled to 70°C. This may initially snap-freeze the cell and provide for more optimal fixation, reducing the effect of hypotonic shock and extraction of the cell (personal communication, K.D. Bauer). It may also aid in the preservation of antigens sensitive to the effects of fixation, such as the cell-proliferation-associated Ki-67 antigen (personal observation). In comparison, a number of fixatives used for tissue preservation are not suitable for DNA content analysis. Paraffin-embedded tissues fixed in Bouin's solution give poor quality histograms with broad CV's (25). Also, a comparison of several different fixatives (24) demonstrated that B-5, Zenker's, and Bouin's fixatives also produce significantly more debris than formalin fixation.

Paraformaldehyde/Permeabilization Protocols

Fixation with paraformaldehyde coupled with a permeabilizing agent is perhaps the most flexible and generally applicable technique for the quantitation of intracellular antigens. It is not without its artifact, however, if carelessly used. A useful starting point in using paraformaldehyde is to fix a cellular suspension (1-3 $\times$ 10^6 cells/ml phosphate buffered saline-PBS) with 1% paraformaldehyde in PBS for 10 min at 4°C. As has been shown in detailed studies (3, 12, 16, 27), decreased levels of fixation can lead to excessive cellular extraction, while increased levels of fixation can lead to excessive cross-linking, hindering antibody and DNA content dye permeability. Depending on the localization, solubility, and "robust" nature of the antigen of interest, the concentration, and/or duration of fixation may need to be altered (i.e., from 0.5% to 2.0% paraformaldehyde for 5-100 min). After washing, cells need to be permeabilized prior to antibody staining. This can be accomplished with a 3 min wash in 0.1% Triton X-100 in PBS at 4°C. Other permeabilization agents (see Table 10.2) should be examined as significant differences exist between permeabilizing agents in their ability to extract cellular constituents and/or expose antigenic sites. This has been most elegantly demonstrated in a paper documenting the multiparameter quantitation of SV40 T antigen after fixation with either paraformaldehyde/Triton X-100 or paraformaldehyde/menthanol (27). Although paraformaldehyde cells permeabilized with methanol demonstrated higher levels of T antigen immunofluorescence, cells fixed with Triton X-100 demonstrated decreased debris, less cell aggregation, lower DNA content CV's, and improved reproducibility of T antigen measurement. On the basis of this data, the authors prostulate that permeabilization with methanol after paraformaldehyde fixation randomizes molecular structure, exposing hidden T antigen epitopes, while Triton X-100 permeabilization preserves higher order protein associations with the cell. Thus, the choice of fixation/permeabilization agents is not a given, but should be based instead on the cellular model systems used and the questions that need to be answered.

ANTIBODIES AND INTRACELLULAR ANTIGENS

Penetration of antibodies into the interior of fixed cells requires the cytoplasmic membrane (and nuclear membrane for nuclear antigens) to be modified. As indicated above, alcohol fixation techniques dissolve part of the lipid components of the membrane, providing pores for the entry and exit of antibody molecules. For formalin fixed cells, the membrane must be partially dissolved following fixation to allow antibodies access to the cell interior. While it is possible, in part, to predict the general nature of the lesions generated by different types of permeabilizing agent, the physical parameters of the resulting membrane pores have not been extensively studied. The number, chemical nature, diameter, and size distribution are poorly understood for different combinations of fixation and permeabilization conditions.

Different combinations of fixative and permeabilization conditions have different effects, and, as stated previously, conditions must be optimized for different cell types, target antigens, and antibodies used. Potential solutions to diffusion-related limitations include increasing the number or size (distribution) of membrane pores. However, precise control of membrane premeabilization is difficult. In addition, increased erosion of the membrane increases the likelihood that internal constituents can diffuse out of the cell (or nucleus).

Diffusion coefficients of IgG antibodies are significantly higher than either IgM or IgA antibody molecules (28). Other physical characteristics, including intrinsic viscosity, partial specific volume, and molecular weight all affect the rate of passive transfer of different antibody isotypes through permeabilized cell membranes. IgG molecules with a maximum dimension of roughly 140 Å, (29) will diffuse through smaller pores at a higher rate than either IgA (dimer maximum dimension of 300 Å) or IgM (28) antibodies. Thus, theoretically, antibodies of the IgG isotype should penetrate faster or require lower concentrations than larger antibody isotypes. Many mouse IgM monoclonal antibodies are ~7S molecules, composed of two heavy and two light chains (180 kD molecular weight). These IgM molecules would act more like IgG antibodies in their membrane diffusion characteristics.

The rate of passive transfer is also proportional to antibody concentration and temperature. Since temperature is measured in degrees Kelvin for thermodynamic processes, the difference between 4°C (277°K) vs 37°C (310°K) would increase the diffusion rate by 10%. Based on passive transport properties alone, incubation at 37°C would not significantly improve the rate of transport of antibody molecules through cell membranes. Temperature has an additional effect on the final equilibrium between antibody combining sites and epitopes. For interactions where the final equilibrium is driven by an increase in enthalpy ($\Delta H°$), the increase in free energy will be greater at 4°C than at higher temperatures. Stabilization of antibody-ligand complexes for these types of interactions are favored at lower temperatures.

Diffusion limitations predict that larger antibodies (19S IgM) will also diffuse out of cells at a lower rate and thus are more likely to show increased background. Temperature influences that rate of passive diffusion, with an equivalent effect on antibody molecules diffusing into and out of the cell. Diffusion of IgM antibodies into the nucleus of intact, fixed cells requires diffusion through two membranes. This could require longer times to reach equilibrium and a lower rate of diffusion of unbound IgM out of the cell.

Factors that affect the equilibrium of antibody binding to antigen include the affinity and avidity of the antibodies used. For monoclonal antibodies, all molecules will have the same binding characteristics and thus should bind to a homogeneous antigen in an identical fashion. However, fixation (particularly by cross-linking agents like formalin) may artifactually cause structural changes in some target epitopes inside a cell. Polyclonal antibodies represent a collection of many different antibodies with a range of affinities for different epitopes on the target antigen. The highest affinity antibodies will tend to preferentially bind target epitopes, and a polyclonal antibody solution could require more time than a monoclonal antibody solution to reach final equilibrium. In all cases, saturation of the target antigen(s) must be determined empirically.

Cell and Antibody Controls

Critical concepts for the detection of intracellular antigens include evidence that the positive signal is significantly higher than an appropriate background or control level, and that the measured positive signal intensity reflects the quantity of antigen within that cell. A number of factors influence the quantitation of specific antigens inside a cell or nucleus. Instrumentation and theoretical aspects of fluorescence quantitation are considered in other chapters (see chapters 5, 7 and 28). Here, we discuss appropriate controls and quantitative assessment of positive immunofluorescence measurements.

ANTIGEN SATURATION

For studies involving quantitative measurements of antigen expression, it is necessary to establish that the signal measured by the flow cytometer reflects the intracellular antigen level. As indicated below, it is important to determine by independent means that antigen expression is not artifactually changed during the steps of cell isolation and fixation. For studies involving quantitation of antigen expression (i.e., proliferation-related antigens) it is important to saturate all the antigenic sites available in each cell with antibodies (and with secondary antibodies for indirect stains). In this way, the measured signal from each cell will be a reflection of the actual antigen content. Simple titration experiments are necessary for direct immunofluorescence to ensure saturation of the antigenic epitopes (and maximal fluorescence signal). For indirect immunofluorescence, either a matrix of different combinations of primary and secondary antibodies or multiple experiments to assure saturation of target antigen, with primary then secondary antibody, are necessary.

ISOTYPE CONTROLS

Applications requiring accurate quantitative measurements of intracellular antigens may require different types of controls to assure the quality of the data. The type of control will depend on the specific application. As with the measurements of cell surface antigens, it is critical that all flow cytometric measurements of intracellular antigens include appropriate negative or background staining controls. Fixed cells have different levels of background fluorescence (depending on the type of cell and the fixation/permeabilization techniques used) that will be increased by nonspecifically-bound fluorochrome in the cell. The amount of nonspecific binding of fluorochrome-protein conjugates is related to the protein concentration. Thus, it is important to match the control protein concentration with the protein concentration of the antibodies to intracellular antigens. Indirect immunofluorescence requires an irrelevant antibody (i.e., it would not bind to the cell via its antibody combining sites) or pooled immunoglobulin. The control and test samples should be treated identically, including all wash steps and incubation with the secondary antibody or avidin-labelled fluorochrome. Considering the importance of concentration-matched controls, it is unfortunate that commercial sources of antibodies (both unlabeled and fluorochrome-labeled) do not always indicate the final protein concentration of immunoglobulin on all of their products.

The level of nonspecific binding to both the cell surface and to the inside of fixed cells in influenced by the isotype of the immunoglobulin. Many cells have cell surface receptors that bind different immunoglobulin isotypes via their Fc region (see Chapter 9). The intracellular trapping of immunoglobulins is influenced by the types or amounts of carbohydrate and the size of the immunoglobulin molecule. For these reasons, it is necessary to use an immunoglobulin control of the same isotype as the test antibody. This is important for IgM antibodies, since IgM consistently shows higher nonspecific intracellular staining in many types of fixed cells than any IgG isotype, with both direct and indirect immunofluorescence techniques.

Antibodies to native or mutated oncogene proteins that are produced by immunization with defined peptides are now available. The use of antibodies to specific peptides is a useful technique to determine antibody binding specificity and nonspecific (background) binding. The level of fluorescence can be compared for cells incubated with antibodies alone and for cells incubated with antibodies plus peptide. Incubation of the appropriate peptide (at a sufficiently high concentration) plus antibodies should block the specific antibody binding to the target antigen. This approach has been used to measure the level of oncogene proteins in lymphoid cells (30, 31).

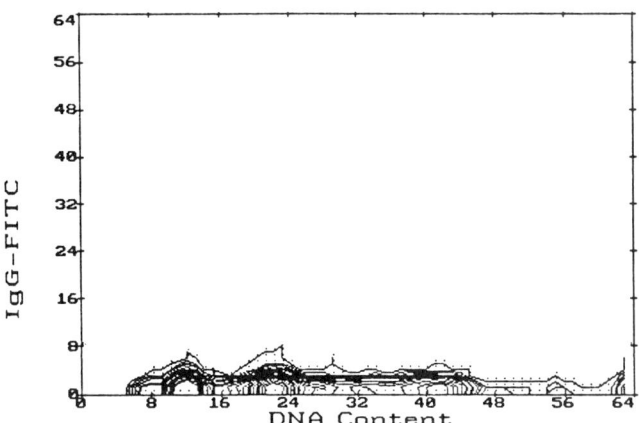

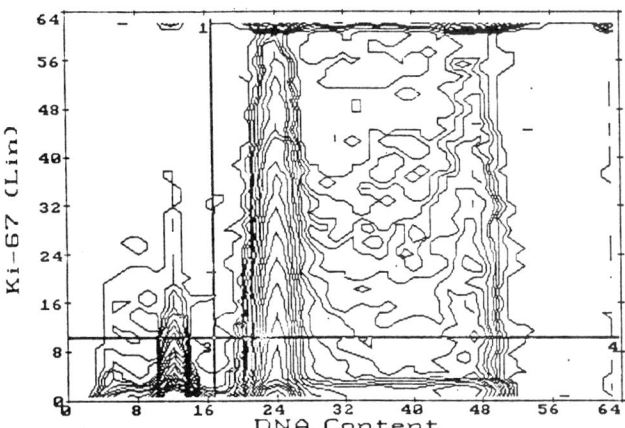

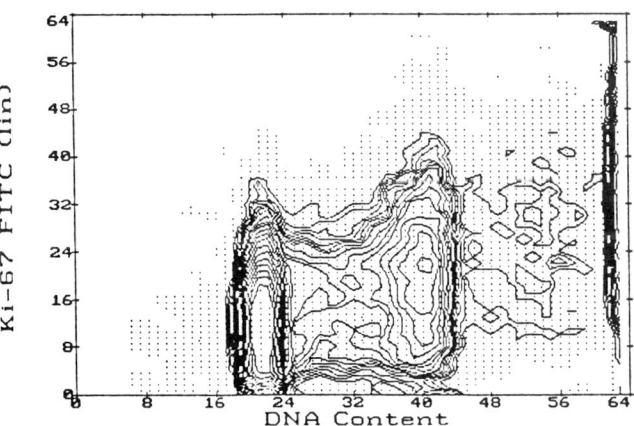

Figure 10.4. Demonstration of the effect of fixation/permeabilization of T24 cells labeled with antibodies (Ki-67, DAKO) to proliferation-related antigen(s). Results of two-parameter analysis (DNA content versus Ki-67 content) demonstrate that only ~50 of these cells express Ki-67 (middle panel). Viability of the cells before fixation was >95%. DNA content analysis showed ~25% S-phase cells. Appropriately fixed and permeabilized T24 cells (bottom panel) show >95% labeling with Ki-67.

An alternative technique to peptide inhibition has been reported (32) using the absorption of the anti-ras antibodies with recombinant ras protein. The fluorescence intensity of ras-positive cells (mitogen-stimulated lymphocytes) was compared by flow cytometry before and after absorption of anti-ras antibodies. This approach offers an interesting alternative when specific antigenic peptide sequences have not been mapped. However, it does present a potential background signal from soluble immune complexes that can bind specifically or nonspecifically.

POSITIVE CONTROLS

As with cell surface immunofluorescence, it is necessary to establish that antibodies to intracellular antigens give an appropriate positive signal. Since the fixation and permeabilization steps typically taken before antibody incubation can introduce potential artifacts, it is important to include cells that are known to express the antigen of interest.

An illustration of this point is provided in Figure 10.4. T24 cells grown in tissue culture were fixed with 0.5% paraformaldehyde and permeabilized (with 0.1% Triton) before indirect immunofluorescence labeling with the proliferation-related antibody Ki-67. As shown here, only half of the cells are labeled with Ki-67 (middle panel). Since these cells showed a typical cell cycle distribution by DNA content analysis (~25% of the cells in S phase), the high percentage of T24 cells not labeled with Ki-67 (particularly in S and G_2/M phases as determined by DNA content) is evidence that either permeabilization was incorrectly performed or less-than-saturating amounts of antibody were used. Probable solutions to this problem include incubation with a higher concentration of antibody, changing the fixation technique, or changing the triton permeabilization time.

A useful control is the inclusion of both positive and negative cells (for the intracellular antigen of interest). As illustrated in Figure 10.5, a mixture of cytokeratin-positive (T24) and negative (peripheral blood lymphocytes) cells, simultaneously fixed and incubated with antibodies to cytokeratin (Cam 5.2), demonstrates that the antibody is reacting appropriately with both cell populations. Here, appropriate isotype controls are also used. The simultaneous measurement of such biological controls provides the best negative control for intracellular antigen expression.

VALIDATION OF ANTIGEN EXPRESSION

Incomplete or inappropriate fixation can change the localization of antigen within the cell, cause loss of antigen by diffusion through cytoplasmic pores, or denature the antigenic epitope, causing loss of antigenicity. Change in antigen localization is not a significant problem provided that the antigen does not leave the cell and the flow cytometer used is not sensitive to fluorescence localization (peak versus integral signals; see Chapter 5). A good fluorescence microscope is essential to a flow cytometry laboratory as it provides quick verification of the quality of the immunofluorescence staining and cell fixation.

For previously unstudied antigens or antibodies, it is essential to validate the staining pattern seen and the specificity

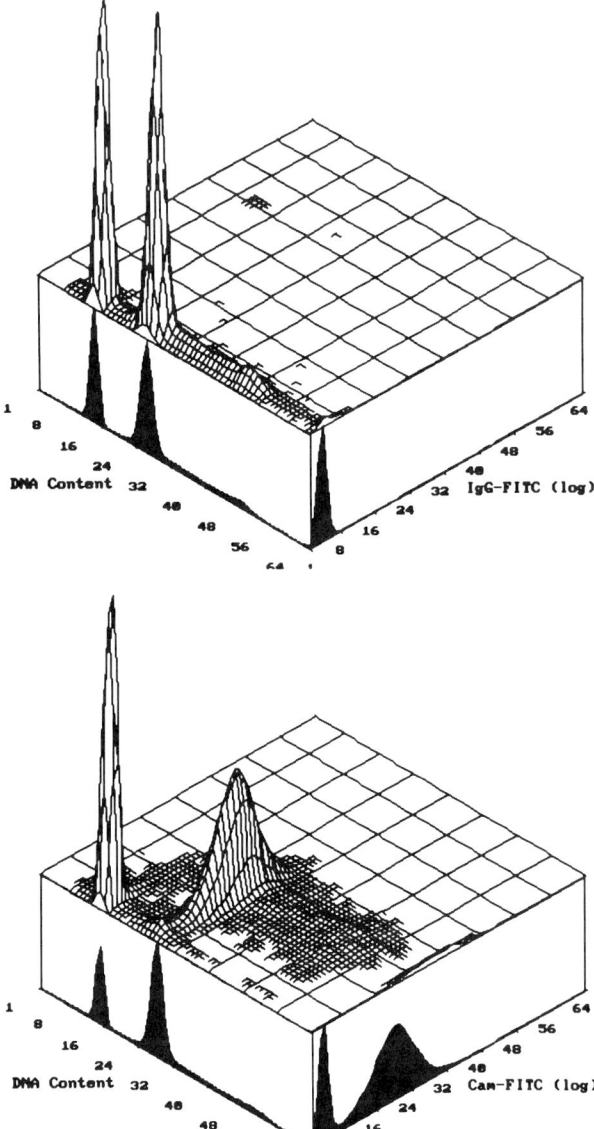

Figure 10.5. Flow cytometric analysis of a mixture of peripheral blood lymphocytes (all G_0 cells) and T24 cells (all G_1, S, or G_2/M), simultaneously fixed (0.5% PF), permeabilized (0.1% TX-100) and labeled with propridium iodide and FITC-conjugated IgG control (top panel), or with propidium iodide plus monoclonal antibodies to cytokeratins (Cam 5.2, BD)(bottom panel). Results shown here demonstrate that all lymphocytes are negative for expression of cytokeratin, while all T24 cells are positive.

of the antibody used. Validity of staining involves appropriate use of negative controls (see above), correlation with other assays when possible, and the use of multiple fixation techniques (34, 35). For antigens with specific cellular localization, it is possible to validate the fixation techniques used with electron microscopy (EM). Using antibodies labeled with electron-dense materials (i.e., colloidal gold), EM studies allow selection of fixation techniques that provide appropriate antigen expression (15). With cell-cycle-related antigens (PCNA, Ki-67, etc.) it is possible to select appropriate negative and positive cells or cells in different cell cycle compartments (using mitotic shake-off, stathmokinetic techniques, flow sorting cells by DNA content, etc.) to validate appropriate expression using fluorescence microscopy as well as optical, or electron microscopy (15, 22, 34, 35).

INDEPENDENT VERIFICATION OF ANTIGEN EXPRESSION

Perhaps the most extreme case requiring independent verification of changes in antigen expression is the measurement of proteins in nuclei recovered from paraffin-embedded tissues. The technique developed by Hedley and coworkers (36) includes the exposure of tissue to the relatively nonspecific protease, pepsin, at a pH of 1.5. This has the potential to denature and digest many different kinds of proteins. In addition, acidic treatment alone denatures proteins and removes many histone proteins from the DNA; these seem to be lost during subsequent processing of nuclei for flow cytometry.

In a careful study of antigen loss, Lincoln and Bauer (37) used EM and flow cytometry to study changes in two nuclear proteins, c-myc and p105, in a model cell system. These studies demonstrated that even brief treatment with pepsin (0.1% for 5 min) causes a 60% reduction in c-myc immunofluorescence. In contrast, little if any p105 immunoreactivity was lost under these same conditions. Related studies from this same laboratory (38) included an important comparison of fresh and paraffin-embedded tissue from the same colonic adenocarcinoma tumor demonstrating no differences in p105 expression. In contrast, a significant loss of immunoreactive c-myc protein in the paraffin-embedded tissues was measured by both flow cytometry and immunogold staining EM (37).

Additional techniques to confirm flow cytometric measurements of intracellular antigens include immunoblotting and Western blot analysis. When flow cytometry shows heterogeneity of antigen expression, positive (and negative) cells can be sorted and subsequently analyzed for antigen expression by dot (immuno) blots or by Western blot analysis (39). In both techniques it is important to analyze equal amounts of total protein for both positive and negative cell populations. This will demonstrate either that the levels measured by flow cytometry are due to artifacts (differences in target antigen accessibility, epitope sequestration, etc.) or that they actually reflect differences in antigen expression in different cells (39).

ANTIBODY SPECIFICITY

A large number of monoclonal and polyclonal antibodies to cellular antigens are now available from commercial sources. While the unique specificity of monoclonal antibodies is widely appreciated, the potential for unexpected crossreactions with unrelated antigens is not. Independent verification of antibody specificity is accomplished by SDS-Polyacrylamide gel electrophoresis followed by Western blot analysis (15, 35, 39). Detection of appropriate molecular weight bands with labeled antibody provides direct verification of the cellular protein(s) recognized. A more rapid, though less rigorous technique, involves immunoblot screening of appropriate positive and negative cell lysates.

INDIRECT VERSUS DIRECT IMMUNOFLUORESCENCE TECHNIQUES

As indicated in the summary of applications later in this chapter, the majority of studies reported have utilized indirect immunofluorescence techniques for flow cytometry. Indirect immunofluorescence requires two independent incubation and washing steps: the first to bind primary antibody, the second to bind reporter groups (i.e., fluorescent dye) to the primary antibody. The need for multiple centrifugation and washing steps increases cell clumping and cell loss. The increased background signal caused by nonspecific binding by the additional (second) antibody is an additional disadvantage. A more important limitation of the indirect technique is the technical difficulties in measuring two or more independent antigens. For the measurement of multiple antigens using indirect immunofluorescence, the primary antibodies must be from different species (or have unique antigenic characteristics, such as different heavy or light chain isotypes) and the secondary antibodies (i.e., FITC-labeled goat anti-mouse Ig and TRITC-labeled burro anti-rat Ig) must not crossreact with each other (heterophile antibody reactions) or react with the inappropriate primary antibody (i.e., goat anti-mouse Ig binding to rat monoclonal antibody). For these types of experiments, it is critical to run multiple single-color controls to establish the appropriate specificity of each signal. Although species-specific secondary antibodies are now commercially available, it is advisable to perform preliminary experiments to establish antibody specificity in a flow cytometric assay.

One approach to circumvent problems caused by multiple indirect antibody systems is the use of biotin-labeled primary antibodies and avidin or streptavidin dye conjugates. This approach has been successfully applied to flow cytometric measurement of intracellular antigens simultaneously labeled with direct (dye) conjugates of monoclonal antibodies to lymphocyte subpopulations (40). For applications using biotin-labeled primary antibodies, it has been reported that the addition of free biotin ($\sim 10^{-5}$ M) increases the fluorescence yield of avidin and streptavidin fluorochrome (R.P. Haugland, manuscript in preparation).

Direct antibody conjugates offer a number of advantages over indirect immunofluorescence techniques. If the antibodies do not compete for related or crossreactive epitopes, two or more fluorochrome-labeled antibodies can be incubated at a single step. The major disadvantages of this approach are the lower signal amplification provided by all single-antibody immunoassay systems and the small number of antibodies to intracellular antigens now available commercially as direct dye conjugates. An extensive search of commercial sources revealed seven different antibodies to intracellular

Table 10.3
Some Commercially Available Monoclonal Antibodies to Intracellular Antigens Currently Available as Direct Conjugates

Antibody	Specificity	Dye/Conjugates	Source
Anti-Ig	Ig Heavy or Light Chain	FITC	Many
		Biotin	Many
		(X)RITC	Many
CAM 5.2	Cytokeratins 8, 18, 19	FITC	Becton Dickinson
CK-1	Cytokeratins 6, 18	FITC	DAKO
Ki-67	Nucleolar Antigen(s)	FITC, PE	DAKO
PCNA	Subunit of DNA Polymerase	FITC	(IgG) Boehringer
		FITC	(IgG) Coulter
		FITC	(IgG) DAKO
TdT	Terminal deoxytransferase	FITC	Coulter, Supertech.
Phosphotyrosine	Phosphorylated Tyrosine Residues	FITC	Boehringer

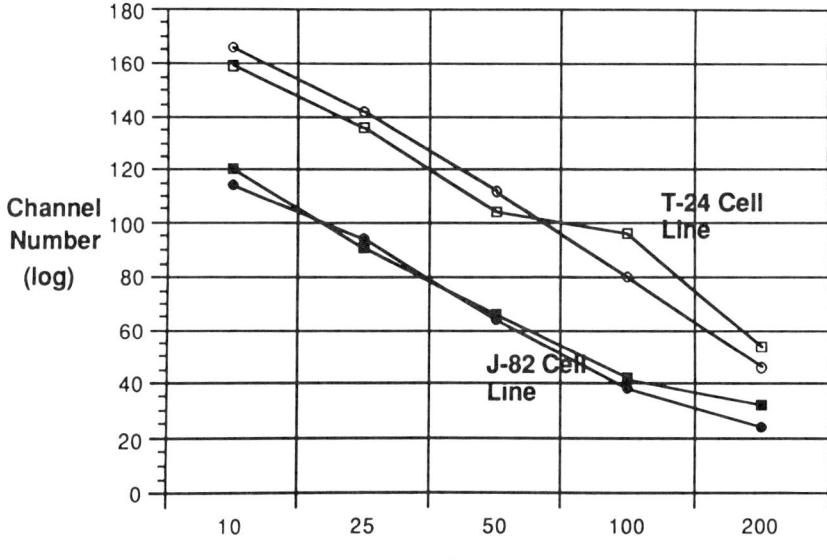

Figure 10.6. Titration of FITC-labeled anti-cytokeratin antibodies (Cam 5.2, BD) using T24 and J82 human bladder cancer cell lines. Flow cytometric analysis results compare the peak (log) green fluorescence measured at each antibody dilution, before and after washing each cell preparation.

antigens currently available as directly-labeled dye conjugates (not including antibodies to immunoglobulin heavy and light chains). As shown in Table 10.3, most of these antibodies are available as FITC conjugates at this time. It is hoped that as flow cytometric applications increase, more direct conjugates of antibodies to intracellular antigens will become commercially available.

One advantage of direct antibody conjugates is that they can provide a "no wash" technique after the addition of antibody. Using human tumor cell lines (J82, T24, HT29) fixed in either cold ethanol (50%) or 1% paraformaldehyde (permeabilized with 0.1% Triton X-100), we have determined the relative fluorescence intensity of FITC-positive cells. A similar fluorescence intensity was measured before and after washing cells to remove unbound antibody. Typical results for ethanol-fixed cells washed and labeled with FITC-conjugated anti-cytokeratin antibodies (CAM 5.2, Becton Dickinson) are shown in Figure 10.6. Aliquots of 2 × 10^6 fixed cells were incubated with different concentrations of antibody and analyzed before or after washing. These results indicate that for antigens present in high intracellular concentration, such as intermediate filaments, most of the antibody measured in the sensing volume of the flow cytometer used (Coulter Epics) is bound to the antigen. Similar results may not be seen for other antigens or for other flow cytometers. The narrow beam optics used here (5 × 160 μm) generate a small sensing region, where a significant part of the total volume is taken up by the cell. Cytometers with large beam spots could produce a condition where a significant part of the total fluorochrome is unbound. Under this condition, it would be difficult to detect differences between cells with high or low concentrations of bound antigen. We have successfully applied this "no-wash" technique with FITC-conjugated anti-cytokeratin antibodies in flow cytometric analysis of urine and bladder wash specimens to transitional cell cancers. This technique demonstrates equal sensitivity to indirect antibody techniques and offers lower cell loss, less cell clumping, and significantly less sample preparation time (T.V. Shankey, results submitted for publication).

Similar "washless" techniques have been reported for cell-surface (41, 42) and nuclear antigens (43, 44). The later studies, using FITC-labeled anti-Ki-67 antibodies, have demonstrated a sensitivity to antibody and cell concentration similar to that seen in our studies (Fig. 10.6) with FITC-labeled antibodies to cytokeratin. These studies suggest that saturation of intracellular antigens may be difficult to achieve with these conjugates and that higher antibody concentrations may be necessary for these "no-wash" techniques.

DYES FOR MEASUREMENT OF INTRACELLULAR ANTIGENS

Most applications for flow cytometry reported thus far have utilized FITC-conjugated antibodies (or FITC-Avidin with biotinylated primary antibodies) for intracellular immunofluorescence. FITC is readily available conjugated to a wide variety of secondary antibodies. For applications with simultaneous DNA content analysis using PI, FITC provides a good separation between the emission signals of both dyes. With the appropriate filters, little spectral (color) compensation is required. It is possible to minimize the compensation for the green (FITC) signal in the red (PI) photomultiplier by using a higher wavelength, multi-cavity long-pass interference filter (i.e., 630 or 640 nm) in front of the red photomultiplier (K. Bauer, unpublished). Although only the higher end of the PI emission is used under these conditions (PI peak emission 610-620 nm), this usually does not cause a problem with large signals generated by PI-stained human diploid cells.

The major disadvantage of FITC is the high background frequently seen, caused by nonspecific binding of dye and cellular autofluorescence. Excitation of fixed cells with low-wavelength light (i.e., 488 nm) can induce autofluorescence by endogenous compounds (aromatic compounds containing flavines). Most of the compounds contributing to autofluorescence demonstrate short Stoke's shifts and fluoresce near their low excitation wavelengths. The hydrophobic character of anthracene derivatives such as FITC leads to nonspecific binding to many hydrophobic cellular components. Often this background is difficult to reduce with simple aqueous buffer systems. One solution is to include low concentrations (0.01% to −0.1%) of detergents or lipophilic agents such as Triton X-100, NP-40, or Tween-20 in all wash buffers. As stated previously, the use of these agents must be optimized in carefully executed control experiments. Treatment of many alcohol-fixed cells with even low concentrations of Triton greatly accelerates the rate of degeneration, and these cells are generally unsuitable for flow cytometric analysis within 24 hr of detergent treatment.

A partial listing of some of the fluorochromes that are currently applied or are potentially useful for flow cytometric analysis of intracellular antigens is given in Table 10.4.

Rhodamine derivatives (i.e., TRITC, XRITC) offer the advantages of extending the number of different independent antigens that can be detected simultaneously. However, their simultaneous use with FITC, PI, and other dyes excited at 488 nm requires a second excitation light source. An additional disadvantage is the increased hydrophobic nature and greater background signal.

Other useful dyes include the UV-excited dyes AMCA (Polysciences, Jackson ImmunoResearch) and Cascade Blue (Molecular Probes), that have similar spectral properties. Both are relatively small dyes that readily penetrate into the interior of fixed, permeabilized cells as antibody dye conjugates or avidin conjugates. Both provide a blue emission signal with high fluorescence yield. AMCA and Cascade Blue have been used for indirect immunofluorescence flow cytometry for the simultaneous measurement of intracellular antigens and DNA content. These studies have taken advantage of the 620 nm emission of propidium iodide bound to DNA with UV excitation (K.D. Bauer, unpublished; T.V. Shankey, unpublished). With a two-laser flow cytometer, this approach allows the additional measurement of FITC conjugates plus additional dyes with a second light source.

The molecular dimensions of Phycoerythrin and other phycobiliproteins may limit their applications to cell surface immunofluorescence. However, it should be noted that the molecular dimensions of monomeric Phycoerythrin (MW ~ 250 kD) are of the same order as intact IgG antibody molecules. The major problem with the application of phycobiliproteins to the measurement of intracellular antigens is their high nonspecific binding inside fixed cells. Experiments are currently underway to determine if this can be reduced without altering levels of antigen expression.

APPLICATIONS OF INTRACELLULAR ANTIGEN MEASUREMENT

In this section, we provide an overview of the types of applications that have been reported thus far using flow cytometry to analyze intracellular antigens. These applications are summarized in Table 10.5. The purpose here is to present the different types of cytoplasmic and nuclear antigens that have been successfully measured and the different types of fixation and cell permeabilization employed. For applications involving clinical specimens or measurement of intracellular antigens of clinical relevance, appropriate references are provided.

INTERMEDIATE FILAMENT PROTEINS

An early report of applications for immunofluorescence measurements of intracellular antigens by flow cytometry used cells permeabilized with lysolecithin without prior fixation (1). Using a carefully determined lysolecithin concentration, these authors reported successful measurements of either cytoplasmic immunoglobulin u chains or cytokeratins, using polyclonal primary antibodies and indirect immunofluorescence techniques. As indicated in Table 10.5, different fixa-

Table 10.4
Fluorochrome Potentially Suitable for Flow Cytometric Analysis of Intracellular Antigens

DYE	Excitation-Emission[a]	Advantages	Disadvantages
FITC	488-525	2-color with PI, Ab Commercially available	Nonspecific binding, Autofluorescence
TRITC	530-580	3 or more colors possible	Need Krypton or other second light source
XRITC	568-620	Same as TRITC	Same as TRITC
AMCA[b]	350-450	Excellent separation from PI (UV)	Need UV light source
Cascade Blue[c]	350-420	Same as AMCA	Same as AMCA
Ultralite[d]	350 to 400-660 to 740[e] (also 650 to 720 nm excitations)	Possibility of 3 colors with one UV source	Emission at wavelength above sensitivity of most PMT's
Cy-3 to 7[f]	488-multiple	Multiple colors with 488 laser	Not evaluated for intracellular Ag
Vita Blue[g]	630-670	3 or more colors possible 2 lasers	pH dependence of absorption and emission Not tested for intracellular Ag
Phycoerythrin	488-575	Possibility of 3 colors with 488 laser	Need large membrane holes. High background
Tandem Conjugates			
(PE-RITC)[h]	488-580	Possibility of 3 colors with one laser	PE Emission signal from conjugate
Red 613[i]	488-613	3 Colors with single laser	Emission overlaps PI. Not tested for intracellular Ag

[a]Suitable wavelength for peak excitation-wavelength for peak emission
[b]7-Amino-4-Methylcumarine-3-acetic acid
[c]Trademark of Molecular Probes
[d]Trademark of Ultra Diagnostics
[e]Different dyes available with different peak emission wavelengths
[f]Available from Jackson ImmunoResearch
[g]Lee, et al., (115)
[h]Tandem conjugate of PE and RITC (Dual Chrome), trademark of Becton Dickinson
[i]Trademark of Gibco BRL

tion techniques have been used for studies of intermediate filament expression by flow cytometry (45–49). In general, intermediate filament protein epitopes are robust and are not denatured by most commonly used fixatives for immunohistochemistries (including acetone and acetic acid/alcohol). One reported exception to this is the epitope recognized by the RGE 53 antibody to cytokeratin 18 (50), which does not react with formalin-fixed, paraffin-embedded tissues. Work in our lab (T.V. Shankey) has demonstrated that the epitope is not affected in cell lines following fixation with formalin or paraformaldehyde, and suggests that the antigenic epitope may be lost as a consequence of processing for paraffin-embedding. Since intermediate filament proteins are naturally cross-linked within the cell and have anchorage or attachment points at the cytoplasmic and nuclear membranes, loss of these proteins by diffusion following permeabilization is highly unlikely.

Flow cytometric analyses of intermediate filament proteins are among the earliest clinical applications for simultaneous measurements of DNA content and intracellular (or cell surface) antigens (45, 46). Clinical applications using flow cytometry and antibodies to cytokeratins were pioneered by Ramaekers and coworkers, and were shown to be useful in the detection of DNA-aneuploid tumors from the bladder (51), kidneys (47), and endometrium (52), and squamous tumors of the head and neck (53). Changes in the pattern of cytokeratin expression of transitional cell cancers of the bladder (50) prompted Hijazi and coworkers (48) to study the expression of cytokeratin 18 as a potential marker for bladder tumor aggressiveness using a flow cytometric analysis of DNA and cytokeratin content. A change in the pattern of cytokeratin expression has also been studied in breast cancer, using simultaneous DNA content and cytokeratin analysis by flow cytometry (54). This study also demonstrated the increased sensitivity in detecting DNA-aneuploid tumors using cytokeratin-specific antibodies.

An additional advantage of using antibodies to cytokeratins in the analysis of tumors has been demonstrated by Visscher and coworkers (55). These authors studied the effect of gating DNA content analysis of breast tumor samples on tumor S phase using cytokeratin-positive cells. Improved correlation with other histopathologic and clinical markers was apparent when the cytokeratin-positive population was used. These studies raise the important question of the overall accuracy of single-parameter tumor S-phase measurements, particularly for DNA-diploid (or near-diploid) tumors where a significant portion of the diploid events are normal or reactive (infiltrating lymphocytes, granulocytes, etc.) cells. The impact of this gating technique for tumor S phase as a predictor of patient survival has not been tested as yet.

CYTOPLASMIC ONCOGENES AND OTHER PROTEINS

As indicated in Table 10.5, a number of different cytoplasmic proteins have been successfully studied using multiparameter flow cytometry. These include the analysis of protooncogene proteins ras (32, 56, 57) and Her-2/neu (Erbb) (32, 58), which are found on the inner aspect of the cytoplasmic membrane; a number of cytoplasmic enzymes (12, 59–61); and the HIV-associated p24 antigen (62). The report of ribonucleotide reductase (12) is highly instructive, as the authors determined that this cytoplasmic antigen was lost af-

Table 10.5
Applications of Immunofluorescence Measurements for Intracellular Antigens Using Flow Cytometry

Cytoplasmic Markers	Protein	Fixation/Perm[a]	Reference
Intermediate Filaments	Cytokeratins	No fix/lysolec	(1)
	Cytokeratins	EtOH	(45)
	Cytokeratins	EtOH	(46)
	Cytokeratin 18	EtOH	(48)
	Vimentin	EtOH	(47)
	GFAP	MeOH	(49)
Cytoplasmic Oncogene Proteins	ras	MeOH	(56)
	ras	PF/TX-100	(57)
	ras	PF-Saponin	(32)
	neu	PF-Saponin	(32)
	neu	EtOH	(58)
Other Cytoplasmic Proteins	Ribonucleotide reductase	PF/TX-100/Tween 20	(12)
	Thymidylate synthase	Formaldehyde or PF	(59)
	Myeloperoxidase	PF/lysolec	(60)
	Ornithine decarboxylase	PF/TX-100	(61)
	HIV p24		(62)
	Ig u chains	Acetic Acid/EtOH	(68)
	Ig	No fix/lysolec	(1)
	Ig	Acetic acid/alcohol	(69)
	Ig	Several	(70)
	Ig u chains	PF/Tween-20	(71)
Nuclear Proteins	PCNA	PF/MeOH	(21)
	PCNA	Several	(22)
	Ki-67	Acetone	(74)
	Ki-67	Formald/NP-40/Citrate	(77)
	Ki-67	Period-lysine-PF	(78)
	Ki-67	EtOH	(79)
	p53	EtOH	(101)
	p53	MeOH	(102)
	SV40 T Ag	MeOH	(26)
	SV40 T Ag	Several	(27)
	c-myc	PF/lysolec	(60)
	c-myc	PF-TX-100	(105)
	c-myb	PF/TX-100	(106)
	c-fos	PF/TX-100	(106)
	p105	PF/TX-100	(16)
	TdT	Acetone/formald.	(110)
	TdT	PF/TX-100	(111)

[a]Abbreviations: lysolec- lysolecithin
No fix- no fixation
EtOH- ethanol fixation
MeOH- methanol fixation
PF- paraformaldehyde
TX-100- Triton X-100
Period- periodate

ter fixation with lower paraformaldehyde concentrations and that only after fixation with 4% paraformaldehyde at 37°C was the antigen detected by immunofluorescence.

Mutations in the ras gene family (Ha-ras, Ki-ras, and N-ras) have been reported in a variety of cancers, including leukemias, lymphomas, and bladder, colon, and prostate cancers. The normal gene product is a GTP-binding protein and the role of the mutated p21 oncoprotein in tumorigenesis and/or metastasis is unclear. Early studies of ras protein expression using multiparameter flow cytometry suggested that ras protein levels might be a marker for G_1 transition (56, 57). Other studies have suggested that increased ras expression may be a marker for cellular differentiation (63, 64). Clinical studies of ras protein expression using flow cytometry have demonstrated increased ras protein levels in DNA-aneuploid multiple myeloma, compared to normal bone marrow or most DNA-diploid multiple myelomas (65).

The protein product of the Her-2/neu oncogene is amplified in some breast tumor cell lines and in less than half of the breast cancers studied. The impact of amplified neu oncoprotein on patient survival remains controversial. In a limited study of breast cancer samples (57), the use of antibodies to neu increased the sensitivity of the flow cytometric DNA content analysis in detecting DNA-aneuploid tumors.

Clinical applications of flow cytometry have been greatly impacted by the increasing appearance of HIV infections. While most of these studies have focused on cell surface molecules, studies of the expression of HIV-related proteins using flow cytometry have been reported (62). These studies demonstrated that significant numbers of HIV-infected (p24 positive) lymphocytes are present in peripheral circulation and that individuals with high percentages of p24-positive cells may progress more rapidly. An additional aspect of this study is the implication that monitoring the percentage of p24-positive/CD4-positive cells may provide a sensitive indicator for individual responsiveness to antiviral therapeutics such as zidovudine (66).

A number of the cytoplasmic proteins listed in Table 10.5 are expressed in a proliferation-related manner and have been used to study cell cycle progression (12, 59). Levels of thymidilate synthetase increase in cycling cells, compared to their G_0 counterparts, peaking in mid S phase; Ribonucleotide reductase increases roughly five-fold during G_1-S transition (67). The use of these markers in the analysis of clinical materials may be limited by the increasing reliance on archival paraffin-embedded tissues. As indicated below, this pattern has focused most of the clinical studies on those nuclear proliferation-associated markers that are applicable to this type of study.

Cytoplasmic immunoglobulin u chains are the earliest reported intracellular antigen analyzed by flow cytometry (68). As shown in Table 10.5, a number of different fixation techniques have been employed, including acetic acid/alcohol (68, 69), methanol (70) and paraformaldehyde (71). In one study (70), different fixation protocols were used and methanol fixation was shown to provide the highest fluorescence intensity for antigen staining with the lowest background. In addition, this study demonstrated that the percent-positive cells by flow cytometry was similar to a microscopic assay for plasmacytes and agreed with the results of a plaque-forming cell bioassay. A flow cytometric assay was used to study cytoplasmic u chain expression in pre-B-cell ALL (69), with the conclusion that the results correlated well with the results obtained by fluorescence microscopy.

NUCLEAR PROTEINS

As indicated in Table 10.5, a number of nuclear proteins have been studied using flow cytometry. These include the protooncogenes myc, myb, and fos; the product of the suppressor gene known as p53; and a number of other proliferation-associated nuclear proteins.

Changes in cell proliferation are a hallmark of many cancers where aggressiveness (or survival) frequently correlated with tumor proliferation. For clinical applications of flow cytometry, nuclear proteins provide a source of information regarding the biological potential of the tumor (see Chapter 2, and (67)). Most of the archival tumor material in pathology departments in the U.S. is formalin-fixed and paraffin-embedded, and the extraction of these tissues by the technique developed by Hedley and coworkers (36) destroys most of the cytoplasm. The remaining nucleus can provide information that is useful in determining disease course. As previously noted, however, it is necessary to perform critical experiments to determine that different nuclear antigens are not significantly altered during fixation, embedding, and the steps used in tissue digestion (37, 38).

Since the initial report describing the monoclonal antibody Ki-67 (74), this has become one of the most frequently studied proliferation-associated proteins in both cell biology and in clinical applications. The target antigen is highly conserved phylogenetically (as are many proliferation-associated antigens) and, although the gene has been localized to chromosome 10 in humans (73), the protein antigen has not yet been isolated. While most studies have indicated that expression is limited to cells in G_1, S, G_2, and reaches maximal levels in M (74), some studies have suggested that aberrant patterns of expression are demonstrated in drug-treated cells (75) and in normal bone marrow (76). The later study should be reviewed with some care, as the author's data show considerable background staining with Ig controls and the low levels of proliferation calculated by single-parameter DNA content (S phase plus G_2M) and BrdUr labelling could be within experimental error of the percent Ki-67-positive cells (values for mean S phase determined by DNA content are not statistically different from mean percent Ki-67-positive cells).

A major limitation to more widespread application of anti-Ki-67 antibodies in clinical studies is the loss of reactivity in paraffin-embedded tissues. As demonstrated in Table 10.5, the antigen is detectable following a variety of different fixation/permeabilization techniques (74–79), including fixation with formalin or paraformaldehyde. For studies using fresh or frozen tissues, the major advantages of alcohol or (low concentration) paraformaldehyde fixation is the lower CV for DNA content measurements and less debris than acetone or acetic acid/alcohol fixatives (3, 27).

Ki-67 antibodies have been used to measure cell proliferation in a variety of human cancers, including leukemias (80, 81), lymphomas (82–84), brain tumors (85, 86), breast cancer (87–91), and lung cancer (92). A number of these studies have not utilized quantitative techniques (flow or image analysis) to determine percent of tumor cells in the cell cycle and have focused on qualitative endpoints. Since clinical studies are limited to fresh or frozen material, large studies of the impact of Ki-67 positivity on patient survival or disease course are generally limited to prospective studies.

PCNA is another nuclear antigen that is closely associated with cell proliferation in normal and transformed cells (93). The protein is a well characterized 36-Kd polypeptide that has been identified as an auxiliary protein of DNA polymerase δ (94). PCNA is part of the replication complex involved in the synthesis and repair of DNA (95). Unlike the antigen recognized by Ki-67, PCNA maintains its antigenicity in paraffin-embedded tissues (96), although detailed studies to measure possible changes during fixation or tissue processing for the Hedley technique have not yet been reported.

The pattern of cell-cycle-related expression of PCNA is controversial. Microscopy studies of antibody-labeled cells have demonstrated cytoplasmic as well as nuclear localization patterns (22) and have sometimes demonstrated diffuse nuclear staining patterns not consistent with replicon clusters (97). Flow cytometric analyses of cell lines have suggested that different patterns of cell cycle expression are seen with different anti-PCNA antibodies (98). While these differences in the results from different studies are not well resolved,

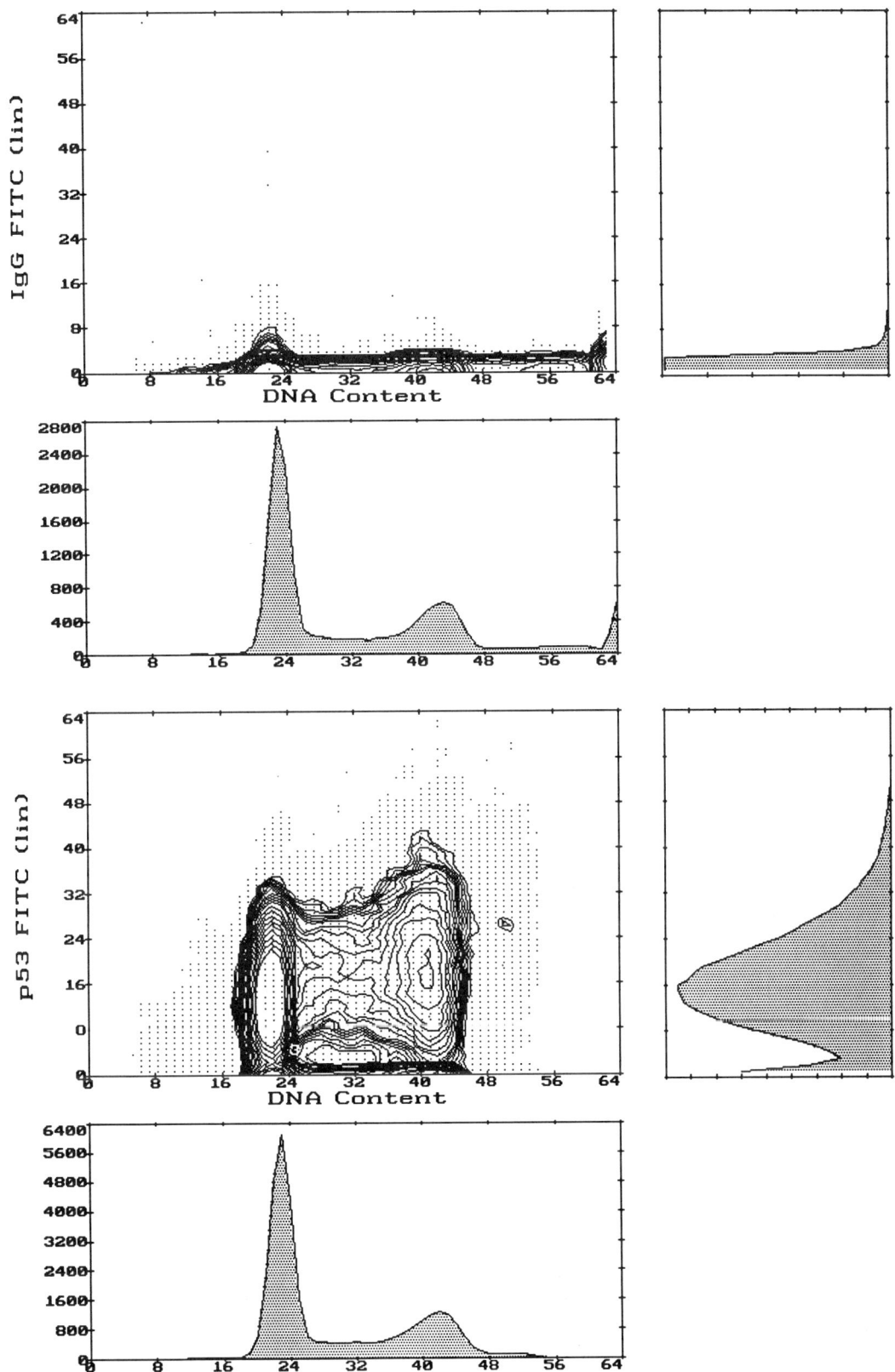

Figure 10.7. Simultaneous DNA content and p53 protein measurement by flow cytometry. SV 40 T Ag positive fibroblasts were fixed with paraformadehyde/lysolecithin (1% PF/10µg lysolecithin/ml) at 4°C for 15 min, washed, and treated with methanol (10 min at −20°C). Cells were incubated with saturating amounts of antibody to wild-type p53 (G59-12, PharMingen)*(bottom panel)*, or with an identical concentration of isotype matched IgG (top panel), followed by FITC-labeled Goat anti-mouse Ig, RNA'se, and PI. Using the Ig control to determine background staining, 92% of the p53 stained cells are positive by this criteria (includes 4.5% aggregated cells in channels 60-64 for DNA Content).

changes in localization patterns (22, 97) are probably due in part to fixation artifacts (as discussed earlier in this chapter). Given optimal fixation conditions (22) proliferating cell lines consistently show a predominant increase in PCNA immunofluorescence in G_1 with minimal further increase through mid S phase. These results are generally in agreement with studies comparing PCNA staining with ^{3}H-thymidine incorporation (99) and indicate that PCNA expression precedes reactivity with Ki-67 (22).

The p53 tumor suppressor gene has received considerable attention and the product of this gene is believed to act as a negative regulator of cell proliferation (reviewed in 100). The gene has been located on the short arm of chromosome 17 in humans and mutational "hot spots" are seen within this locus in cell lines, xenografts, and primary tumors, including colon, lung, and bladder tumors (100). Expression of the p53 protein has been measured in cell lines using flow cytometry (101, 102). In these studies, cells were fixed using cold alcohol. In our lab (T.V. Shenkey), we have investigated the pattern of expression of p53 in tumor cell lines and an SV-40 T antigen expressing fibroblast cell line (provided by Dr. C. Goolsby). As shown in Figure 10.7, transformed fibroblasts fixed with paraformaldehyde and lysolecithin and permeabilized with cold ($-20°C$) methanol show a pattern of expression similar to that seen with Ki-67. Under these fixation conditions, the immunofluorescence is completely localized to the nucleus in these cells.

Expression of p53 protein as measured by flow cytometry has been studied in tumor samples from human lung (103) and colon (104) cancers. The study of lung tumors (103) involved archival formalin-fixed, paraffin-embedded tissues prepared for flow cytometry using the Hedley technique (36), and included important controls to establish that the level of p53 immunoreactivity was not significantly affected in the fixation or nuclear isolation procedures. This study demonstrates that patterns of p53 expression (using antibodies that recognize wild-type p53) are different in different tumors. The impact this may have on disease course or survival has not been reported.

As indicated in Table 10.5, expression of the nuclear oncogene proteins c-myc, c-fos, and c-myb have been measured by flow cytometry using cells fixed with paraformaldehyde (105, 106). These proteins are expressed in a cell-cycle-specific fashion (67), with all three showing maximal levels in G_0-G_1 transition. Using multiparameter flow cytometry to measure intracellular oncogene protein in conjunction with cell surface receptors, Civin and coworkers (106) have studied levels of c-myc, c-fos, and c-myb in different hematopoietic cells from human bone marrow. Levels of c-myc protein have also been studied using simultaneous DNA content and oncoprotein measurements by flow cytometry in archival paraffin-embedded tissues from testicular (107) and cervical (108) cancers. Considering the demonstration of the sensitivity of this protein to proteolytic digestion during the Hedley technique (36), the results of these studies on archival clinical tumors must be interpreted conservatively.

TdT (terminal deoxynucleotidyl transferase) is a template-independent nuclear DNA replication enzyme that attaches deoxynucleotides to the 3' terminus of DNA and may play an important role in Immunoglobulin and T-cell receptor gene rearrangements (109). Flow cytometry studies of the expression of this nuclear antigen (110, 111) have used acetone followed by formalin or paraformaldehyde to fix bone marrow cells from patients with leukemia. This flow cytometric assay allows the analysis of large numbers of cells and provides a sensitive method to detect minimal residual disease in patients following therapy (111).

CONCLUSION

Flow-cytometry-based measurements of intracellular antigens demonstrate the potential for this methodology for applications in both cell biology and clinical medicine. The majority of applications thus far have involved research investigations into the expression of intracellular antigens. However, an increasing use of intracellular antigen measurements in flow cytometric analysis of routine clinical samples is being reported. It is quite likely that multiparameter measurements of clinical specimens using multiple probes to intracellular antigens will have a significant impact on future clinical applications of this technology. Further progress in this area will depend on the development of new probes and new fluorochromes, along with a better understanding of the mechanisms and effects of fixation and permeabilization techniques.

REFERENCES

1. Schroff RW, Bucana CD, Klein RA, Farrell MM, Morgan AC. Detection of intracytoplasmic antigens by flow cytometry. J Immunol Methods 1984;70:167–177.
2. Young HA, Klein RA, Shih TY, Morgan AC, Schroff RW. Detection of the intracellular ras p21 onogen product by flow cytometry Anal Biochem 1986;156:67–71.
3. Clevenger CV, Bauer KD, Epstein AL. A method for simultaneous nuclear immunofluorescence and DNA content quantitation using monoclonal antibodies and flow cytometry. Cytometry 1985;6:208–214.
4. Hayat MA. Fixation. In: Principles and Techniques of Electron Microscopy, Vol 1. New York: Van Nostrand Rheinhold, 1970; pp 5–107.
5. Kiernan JA. Fixation. In: Histological and Histochemical Methods: Theory and Practice. Pergamon Press, Oxford. 1981; pp 8–20.
6. Fraschini A, Pellicciari C, Biggiogera M, Manfredi-Romanini MG. The effect of different fixatives on chromatin: cytochemical and ultrastructural approaches. Histochem J 1981;13:763–779.
7. Dallam RD. Determination of protein and lipid lost during osmic acid fixation of tissues and cellular particulates. J Histochem Cytochem 1957;5:178–181.
8. Fox CH, Johnson FB, Whiting J, Roller PP. Formaldehyde fixation. J Histochem Cytochem 1985;8:845–853.
9. Brutlag D, Schlehuber C, Bonner J. Properties of formaldehyde-treated nucleohistone. Biochemistry 1969;8:3214–3218.
10. Hatfield JM, Hymer WC. Flow cytometric immunofluorescence of rat anterior pituitary cells. Cytometry 1985;6:137–142.

11. Berezney R, Coffey DS. Nuclear matrix: Isolation and characterization of a framework structure from rat liver nuclei. J Cell Biol 1977; 73:616–637.
12. Mann GJ, Dyne M, Musgrove EA. Immunofluorescent Quantification of ribonucleotide reductase M1 subunit and correlation with DNA content by flow cytometry. Cytometry 1987;8:509–517.
13. Lanier LL, Warner NL. Paraformaldehyde fixation of hematopoietic cells for quantitative flow cytometry (FACS) analysis. J Immunol Meth 1981;47:25–30.
14. Lal RB, Edison LS, Chused TM. Fixation and long-term storage of human lymphocytes for surface marker analysis by flow cytometry. Cytometry 1988;9:213–219.
15. Clevenger CV, Epstein AL, Bauer KD. Modulation of the nuclear antigen p105 in lymphocytes as a function of cell cycle progression. J Cell Physiol 1987;130:336–343.
16. Clevenger CV, Epstein AL, Bauer KD. Quantitative analysis of a nuclear antigen in interphase and mitotic cells. Cytometry 1987;8:280–286.
17. Clevenger CV, Russell DH, Shipman P, Prystowsky MB. Regulation of IL2 dependent T cell proliferation by prolactin. Proc Natl Acad Sci USA 1990;87:6460–6464.
18. Celis JE, Celis A. Cell cycle-dependent variations in the distribution of the nuclear protein cyclin proliferating cell nuclear antigen in cultured cells: Subdivision of S phase. Proc Natl Acad Sci USA 1985;82: 3262–3266.
19. Celis JE, Madsen P, Nielsen S, Celis A. Nuclear patterns of cyclin (PCNA) antigen distribution subdivide S-phase in cultured cells. Leukemia Res 1986;10:237–249.
20. Shipman PM, Sabath DE, Fisher AH, et al. Cyclin mRNA and protein expression in recombinant interleukin 2-stimulated cloned murine T-lymphocytes. J Cell Biochem 1988;38:189–198.
21. Kurki P, Ogata K, Tan EM. Monoclonal antibodies to proliferating cell nuclear antigen/cyclin as probes for proliferating cells by immunofluorescence microscopy and flow cytometry. J Immunol Methods 1988;109:49–59.
22. van Dierendonck JH, Wijsman JH, Keijzer R, et al. Cell-cycle-related staining patterns of anti-proliferating cell nuclear antigen monoclonal antibodies. Am J Pathol 1991;138:1165–1172.
23. Loke, S-L, Neckers LM, Schwab G, and Jaffe ES. C-myc-protein in normal tissue: effects of fixation on its apparent subcellular distribution. Am J Pathol 1988;131:29–37.
24. Herbert DJ, Nishiyama RH, Bagwell CB, et al. Effects of several commonly used fixatives on DNA and total nuclear protein analysis by flow cytometry. Amer J Clin Pathol 1989;91:535–541.
25. Hedley, DW. Flow cytometry using paraffin-embedded tissue: Five years on. Cytometry 1989;10:229–241.
26. Jaccobberger JW, Fogelman D, Lehman JM. Analysis of intracellular antigens by flow cytometry. Cytometry 1986;7:356–364.
27. Schimenti KJ, and Jacobberger JW. Fixation of mammalian cells for flow cytometric evaluation of DNA content and nuclear immunofluorescence. Cytometry 1992;13:48–59.
28. Metzger H. Structure and function of gM Macroglobulins. Adv in Immunol 1970;12:57–116.
29. Dorrington KJ, Mihaesco C. The subunit structure of human µM-globulins. Immunochem 1970;7:651–660.
30. Kastan MB, Slamon DJ, Civin CI. Expression of protooncogene c-myb in normal human hematopoietic cells. Blood 1989;73:1444–1451.
31. Kastan MB, Stone KD, Civin CI. Nuclear oncoprotein expression as a function of lineage, differentiation stage, and proliferative status of normal human hematopoietic cells. Blood 1989;74:1517–1524.
32. Rabin H, Trimpe KL, Hamer PJ, et al. Expression of ras and neu oncogene proteins as determined by monoclonal antibodies. Cancer Cells 1989;7:157–160.
33. Shapiro H. *Practical Flow Cytometry*. Alan Liss, New York 1988; p 24.
34. Childs GV. The use of multiple methods to validate immunocytochemical stains. J Histochem Cytochem 1983;31:168–176.
35. Bauer KD. Analysis of proliferation-associated antigens. Methods Cell Biol 1990;33:235–247.
36. Hedley DW, Friedlander ML, Taylor IW, et al. Method for analysis of paraffin-embedded pathological material using flow cytometry. J Histochem Cytochem 1983;31:1333–5.
37. Lincoln ST and Bauer KD. Limitations in the measurement of c-myc oncoprotein and other nuclear antigens by flow cytometry. Cytometry 1989;10:456–462.
38. Bauer KD, Clevenger CV, Endow RK, et al. Simultaneous nuclear antigen and DNA content quantitation using paraffin-embedded colonic tissue and multiparameter flow cytometry. Cancer Res 1986;46: 2428–2434.
39. Bauer KD. Application of multiperameter DNA content and Immunofluorescence analysis in the investigation of human neoplasia. Pathol Immunopathol Res 1988;7:371–380.
40. Hayden GE, Walker KZ, Miller JFAP, et al. Simultaneous cytometric analysis for the expression of cytoplasmic and surface antigens in activated T cells. Cytometry 1988;9:44–51.
41. Hoffman RA and Hansen WP. Immunofluorescent analysis of blood cells by flow cytometry. Int J Immunopharmacol 1981;3:249–254.
42. Caldwell CW, Taylor HM. A rapid, no-wash technique for immunophenotypic analysis by flow cytometry. Am J Clin Pathol 1986;86: 600–607.
43. Larsen JK. Washless double staining of a nuclear antigen (Ki-67 or bromodeoxyuridine) and DNA in unfixed nuclei. Methods Cell Biol 1990;33:227–234.
44. Larsen JK, Christensen IJ, Christiansen J, Mortensen BT. Washless double staining of unfixed nuclei for flow cytometric analysis of DNA and nuclear antigen (Ki-67 or bromodeoxyuridine) Cytometry 1991; 12:429–437.
45. Ramaekers FCS, Beck H, Vooijs GP, Herman CJ. Flow cytometric analysis of mixed cell populations using intermediate filament antibodies. Exp Cell Res 1984;153:249–53.
46. Huffman JL, Garen-Chesa P, Gay H, et al. Flow cytometric identification of human bladder cells using a cytokeratin monoclonal antibody. Ann NY Acad Sci 1986;468:302–315.
47. Feitz WF, Karthaus HFM, Beck HLM, et al. Tissue specific markers in flow cytometry of urological cancers: (II) cytokeratin and vimentin in renal cell tumors. Int J Cancer 1986;37:201–207.
48. Hijazi A, Devonec M, Bouvier R, and Revillard J-P. Flow cytometry study of cytokeratin 18 expression according to tumor grade and deoxyribonucleic acid content in human bladder tumors. J Urol 1989;141: 522–526.
49. Ito M, Nagashima T, Hoshino T. Quantitation and distribution analysis of glial fibrillary acidic protein in human glioma cells in culture. J Neuropharmacol Exp Neurol 1989;48:560–567.
50. Ramaekers FCS, Huysmans A, Moesker O, et al. Cytokeratin expression during neoplastic progression of human transitional cell carcinomas as detected by a monoclonal and a polyclonal antibody. Lab Invest 1985;52:353–361.
51. Feitz WJF, Beck HLM, Smeets AWGB, et al. Tissue specific markers in flow cytometry of urological cancers: cytokeratins in transitional bladder carcinoma. Int J Cancer 1985;36:349–365.
52. Oud PS, Henderik JBJ, Beck HLM, et al. Flow cytometric analysis and sorting of human endometrial cells after immunocytochemical labeling for cytokeratin using a monoclonal antibody. Cytometry 1985; 6:159–164.
53. Bijman JTh, Wagener DJTh, Wessels JMC, et al. Cell size DNA and cytokeratin analysis of human head and neck tumors by flow cytometry. Cytometry 1986;7:76–81.
54. Ferrero M, Spyratos F, Le Doussal V, et al. Flow cytometric analysis of DNA content and keratins using CK7, CK8, CK18, CK19, and KL1 monoclonal antibodies in benign and malignant human breast tumors. Cytometry 1990;11:716–724.

55. Visscher DW, Zarbo RT, Jacobsen G, et al. Multiparameter deoxyribonucleic acid and cell cycle analysis of breast carcinomas by flow cytometry. Clinopathologic correlations. Lab Invest 1990;62:370–378.
56. Andreeff M, Slater D, Bressler J, and Furth ME. Cellular ras oncogenic expression and cell cycle measured by flow cytometry in hematopoietic cell lines. Blood 1986;67:676–681.
57. Czerniak B, Herz F, Wersto RP, Koss LG. Expression of Ha-ras oncogene p21 protein in relation to the cell cycle of cultured human tumor cells. Am J Pathol 1987;126:411–416.
58. Kelsten ML, Berger MS, Maguire HC, et al. Analysis of c-erbB-2 protein expression in conjunction with DNA content using multiparameter flow cytometry. Cytometry 1990;11:522–532.
59. Shibui S, Hoshino T, Iwasaki K, et al. Cell cycle phase dependent emergence of thymidylate synthase studied by monoclonal antibody (M-TS-4). Cell Tissue Kinet 1989;22:259–268.
60. Dent GA, Leglise MC, Pryzwansky KB, Ross DW. Simultaneous paired analysis by flow cytometry of surface markers, cytoplasmic antigen or oncogene expression with DNA content. Cytometry 1989;10:192–198.
61. Robertson FM, Gilmour SK, Beavis AJ, et al. Flow cytometric determination of ornithine decarboxylase activity in epidermal cell populations. Cytometry 1990;11:832–836.
62. Cory JM, Olhsson-Wilhelm BM, Broch EJ, et al. Detection of human immunodeficiency virus-infected lymphoid cells at low frequency by flow cytometry. J Immunol Methods 1987;105:71–78.
63. Studzinski GP and Brelvi ZS. Increased expression of oncogene c-Ha-ras during granulocytic differentiation of HL60 cells. Lab Invest 1987;56:499–504.
64. Czerniak B, Herz F, Wersto RP, and Koss LG. Modification of Ha-ras oncogene p21 expression and cell cycle progression in human colonic cancer cell line HT-29. Cancer Res 1987;47:2826–2830.
65. Tsuchiya H, Epstein J, Selvanayagam P, et al. Correlated flow cytometric analysis of H-ras p21 and nuclear DNA in multiple myeloma. Blood 1988;72:796–800.
66. Landay A, Ohlsson-Wilhelm B, Giorgi JV. Application of flow cytometry to the study of HIV infection. AIDS 1990;4:479–497.
67. Darzynkiewicz Z, Traganos F. Multiparameter flow cytometry in studies of the cell cycle. In: Flow Cytometry and Sorting, 2nd. Ed. Melamed MMR, Lindmo T, and Mendelsohn ML, eds. Wiley-Liss, New York. 1990;p 469–501.
68. Chused T, Moutsopoulos H, Sharrow S, Hansen C. Evidence of a primary B-lymphocyte abnormality in NZB mice. Dev Immunol 1978;3:363–370.
69. Zipf TF, Bryant LD, Koskowich GN, et al. Enumeration of cytoplasmic u immunoglobulin positive acute lymphoblastic leukemia cells by flow cytometry: Comparison with fluorescence microscopy. Cytometry 1984;5:610–613.
70. Levitt D and King M. Methanol fixation permits flow cytometric analysis of immunofluorescent stained intracellular antigens. J Immunol Methods 1987;96:233–237.
71. Schmid I, Iuttenbogaart CH, Giotgi JV. A gentle fixation and permeabilization method for combined cell surface and intracellular staining with improved precision in DNA quantification. Cytometry 1991;12:279–285.
72. Gerdes J, Schwab U, Lemke H, Stein H. Production of a mouse monoclonal antibody reactive with a human nuclear antigen associated with cell proliferation. Int J Cancer 1983;31:13–20.
73. Schonk DM, Kuijpers HJH, van Drunen E, et al. Assignment of the gene(s) involved in the expression of the proliferation-related Ki-67 antigen to human chromosome 10. Hum Genet 1989;83:297–299.
74. Gerdes J, Lemke H, Baisch H, et al. Cell cycle analysis of a cell proliferation-associated human nuclear antigen defined by the monoclonal antibody Ki-67. J Immunol Meth 1984;133:1710–1715.
75. Lopez F, Belloc F, Lacombe F, et al. Modalities of synthesis of Ki67 antigen during the stimulation of lymphocytes. Cytometry 1991;12:42–49.
76. Van Bockstaele DR, Lan J, Snoeck H-W, et al. Aberrant Ki-67 expression in normal bone marrow revealed by multiparameter flow cytometric analysis. Cytometry 1991;12:50–63.
77. Palutke M, Kukuruga D, Tabaczka P. A flow cytometric method for measuring lymphocyte proliferation directly from tissue culture plates using Ki-67 and propidium iodide. J Immunol Methods 1987;105:97–105.
78. Drach J, Gattringer C, Glassl H, et al. Simultaneous flow cytometric analysis of surface markers and nuclear Ki-67 antigen in leukemia and lymphoma. Cytometry 1989;10:743–749.
79. Landberg G, Tan E, Roos G. Flow cytometric multiparameter analysis of proliferating cell nuclear antigen/cyclin and Ki-67 antigen: a new view of the cell cycle. Exp Cell Res 1990;187:111–118.
80. Neckers LM, Funkhouser WK, Trepel JB, et al. Significant non-S-phase DNA synthesis visualized by flow cytometry in activated and in malignant human lymphoid cells. Exp Cell Res 1985;156:429–438.
81. Falini B, Canino S, Sacchi S, et al. Immunocytochemical evaluation of the percentage of proliferating cells in pathological bone marrow and peripheral blood samples with the Ki-67 and anti-bromo-deoxyuridine monoclonal antibodies. Br J Haematol 1988;69:311–320.
82. Gerdes J, Van Baalen J, Pileri S, et al. Tumor cell growth fraction in Hodgkin's disease. Am J Pathol 1987;128:390–393.
83. Schrape S, Jones DB, Wright DH. A comparison of three methods for the determination of the growth fraction in non-Hodgkin's lymphoma. Br J Cancer 1987;55:283–286.
84. Cibull ML, Heryet A, Gatter KC, Mason DY. The utility of Ki-67 immunostaining, nucleolar organizer region counting, and morphology in the assessment of follicular lymphomas. J Pathol 1989;158:189–193.
85. Giangaspero F, Doglioni C, Rivano MT, et al. Growth fraction in human brain tumors defined by monoclonal antibody Ki-67. Acta Neuropathol (Berl) 1987;74:179–182.
86. Nishizaki T, Orita T, Furutani Y, et al. Flow-cytometric DNA analysis and immunohistochemical measurement of Ki-67 and BudR labeling indices in human brain tumors. J Neurosurg 1989;70:379–384.
87. Gerdes J, Lelle RJ, Pickartz H, et al. Growth fractions in breast cancers determined in situ with monoclonal antibody Ki-67. J Clin Pathol 1986;39:977–980.
88. Barnard N, Hall PA, Lemoine NR, Kadar N. Proliferative index in breast carcinoma and its relationship to clinical and pathological variables. J Pathol 1987;152:287–295.
89. Charpin C, Andrac L, Vacheret H, et al. Multiparametric evaluation (SAMBA) of growth fraction (monoclonal Ki-67) in breast carcinoma tissue sections. Cancer Res 1988;48:4368–4373.
90. Walker RA, Camplejohn RS. Comparison of monoclonal antibody Ki-67 reactivity with grade and DNA flow cytometry of breast carcinomas. Br J Cancer 1988;57:281–283.
91. Isola JJ, Helin HJ, Hell MJ, Kallioniemi O-P. Evaluation of cell proliferation in breast carcinoma. Comparison of Ki-67 immunohistochemical study, DNA flow cytometric analysis, and mitotic count. Cancer 1990;65:1180–1184.
92. Gatter KC, Dunnill MS, Gerdes J, et al. New approach to assessing lung tumors in man. J Clin Pathol 1986;39:590–593.
93. Celis JE, Bravo R, Larsen PM, Fey SJ. Cyclin: A nuclear protein whose level correlates directly with the proliferative state of normal as well as transformed cells. Leukemia Res 1984;8:143–157.
94. Prehlich G, Tan CK, Kostura M, et al. Functional identity of proliferating cell nuclear antigen and a DNA polymerase-d auxiliary protein. Nature 1987;326:517–520.
95. Nishida C, Reinhard P, Linn S. DNA repair synthesis in human fibroblasts requires DNA polymerase d. J Biol Chem 1988;263:501–510.
96. Garcia RL, Coltrera MD, Gown AM. Analysis of proliferative grade using anti-PCNA/Cyclin monoclonal antibodies in fixed, embedded tissues. Am J Pathol 1989;134:733–739.
97. Bravo R, MacDonald-Bravo H. Existence of two populations of cyclin/proliferating cell nuclear antigen during the cell cycle: Association with DNA replication sites. J Biol Chem 1987;105:1549–1554.

98. Kurki R, Ogata K, Tan EM. Monoclonal antibodies to proliferating cell nuclear antigen (PCNA)/cyclin as probes for proliferating cells by immunofluorescence microscopy and flow cytometry. J Immunol Methods 1988;109:49–58.
99. Garland P, Degraef C. Cyclin/PCNA immunostaining as an alternative to tritiated thymidine pulse labelling for marking S phase cells in paraffin sections from animal and human tissues. Cell Tissue Kinet 1989;22:383–392.
100. Levine AJ, Momand J, Finlay CA. The p53 tumour suppressor gene. Nature 1991;351:453–456.
101. Darzynkiewicz Z, Staiano-Coico L, Kunicka JE, Deleo AB, and Old LJ. p53 content in relation to cell growth and proliferation in murine L1210 leukemia and normal lymphocytes. Leukemia Res 1986;10:1383–1389.
102. Laffin J, Fogleman D, and Lehman JM. Correlation of DNA content, p53, T antigen and V antigen in Simian Virus 40 infected human diploid cells. Cytometry 1989;10:205–213.
103. Morkve O, Laerum OD. Flow cytometric measurement of p53 protein expression and DNA content in paraffin-embedded tissue from bronchial carcinomas. Cytometry 1991;12:438–444.
104. Remvikos Y, Laurent-Puig P, Salmon RJ, et al. Simultaneous monitoring of p53 protein and DNA content of colorectal adenocarcinomas by flow cytometry. Int J Cancer 1990;45:450–456.
105. Engelhard HH, Krupka JL, Bauer KD. Simultaneous quantification of c-myc oncoprotein, total cellular protein, and DNA content using multiparameter flow cytometry. Cytometry 1991;12:68–76.
106. Kastan MB, Stone KD, Civin CI. Nuclear oncoprotein expression as a function of lineage, differentiation stage, and proliferative status of normal human hematopoietic cells. Blood 1989;74:1517–1524.
107. Watson JV, Stewart J, Evan GI, et al. The clinical significance of flow cytometric c-myc oncoprotein quantitation in testicular cancer. Br J Cancer 1986;53:331–337.
108. Hendy-Ibbs P, Cox H, Evan GI, Watson JV. Flow cytometric quantitation of DNA and c-myc oncoprotein in archival biopsies of uterine cervix neoplasia. Br J Cancer 1987;55:275–282.
109. Alt F, Baltimore D. Joining of immunoglobulin heavy chain gene segments; implications from a chromosome with evidence of theee D-J_H fusions. Proc Natl Acad Sci USA 1982;79:4118–4121.
110. Slaper-Cortenbach ICM, Admirall LG, Kerr JM, van Leeuwen EF, von dem Borne AGK, Tetteroo PAT. Flow cytometric detection of terminal deoxynucleotidyl transferase and other intracellular antigens in combination with membrane antigens in acute lymphatic leukemias, Blood 1988;72:1639–1644.
111. Gore SD, Kastan MB, Goodman SN, Civin CI. Detection of minimal residual T cell acute lymphoblastic leukemia by flow cytometry. J Immunol Methods 1990;132:275–286.
112. Rosette CD, DeTeresa PS, and Pallavicini MG. Simultaneous for cytometric detection of cellular c-myc protein, incorporated bromodeoxyuridine, and DNA. Cytometry 1990;11:547–551.
113. Cooper JA, Loftus DJ, Frieden C, Bryan J, Elson EL. Localization and mobility of gelsolin in cells. J Cell Biol 1988;106:1229–1240.
114. Wikstrom AC, Bakke O, Okret S, Bronnegard M, Gustafsson J. Intracellular lozalization of the glucocorticoid receptor: Evidence for cytoplasmic and nuclear localization. Endocrinol 1987;120:1232–1242.
115. Lee LG, Berry GM, and Chen C-H. Vita blue: a new 633-excitable fluorescent dye for cell analysis. Cytometry 1989;10:151–164.

11

Quality Control for Clinical Flow Cytometry

KATHARINE A. MUIRHEAD

OVERVIEW

Quality control (QC) is integral to the overall laboratory goal of quality assurance, i.e., assuring that final test results are accurate and reproducible. Methods that identify and minimize sources of variation in the test system are therefore essential. Variability may arise during sample handling and preparation procedures prior to measurement, the measurement procedure itself, or subsequent data analysis and processing (Table 11.1). Quality control for flow cytometry is similar to that for other clinical laboratory methodologies in that it requires validation of sample preparation procedures, reagent performance, and instrument performance. Flow cytometry shares with automated hematology the problem of finding stable control materials suitable for cellular analysis. However, for flow cytometry, the problem is aggravated by the fact that typical flow cytometers use a larger number of measurements (i.e., one or two types of light scattering and/or resistance, two or three colors of fluorescence) to resolve cell subpopulations than do typical automated hematology instruments (i.e., resistance and/or capacitance, one type of light scattering and/or absorbance). Further, as is evident from the applications described in this volume, flow cytometry relies heavily on measurements of cellular fluorescence. Since there are no quantitative standards for measurement of fluorescence, flow cytometric measurements are made in terms of relative fluorescence intensity, the units of which may vary significantly from laboratory to laboratory.

The goal of flow cytometric measurements is to reproducibly make accurate and quantitative distinctions among individual cells having different biologies and/or pathologies. In its broadest sense, the purpose of quality control for flow cytometry is to provide valid comparisons of clinically relevant parameters under a number of circumstances:

1) from cell to cell within a sample,
2) from sample to sample within the laboratory,
3) from day to day for a given patient,
4) from laboratory to laboratory and/or patient group to patient group.

Methods for quality control of specimen preparation and reagent performance are tightly coupled to the particular biology or pathology under study, and many of these specific issues are addressed in other chapters. Therefore, this chapter will begin with quality control problems and methods relating to assessment of instrument performance, outline some general issues that must be considered in establishing "biological quality control" for immunofluorescence–based or DNA–based applications, and conclude with the author's biases about keys to successful quality control.

INSTRUMENT QUALITY CONTROL

Since there are at present no standards that can be used to check the absolute accuracy of flow cytometric measurements, the major goal of instrument quality control is first to optimize and then to monitor instrument performance, allowing identification of trends that indicate that preventive maintenance or repair is needed. A related goal is to assure that measured intensities accurately reflect amount(s) of probe originally present on or in the cell. For some tests it may also be desirable to establish that intensity measurements are the same from day to day, i.e., that objects with the same intensity will consistently be classified into the same channel. A brief description of the measurement process will illustrate some of the variables that can alter instrument performance; references 1 and 2 provide further detail.

The Measurement Process

Let us assume for the moment that a specimen has been prepared by methods that maintain representative frequencies of relevant cell types and stained with reagents that bind specifically to the biological component(s) of interest (a large assumption!) The prepared cells, when introduced into the flow cytometer, interact with an exciting light beam, scatter-

Table 11.1
QC Issues for Clinical Flow Cytometry

Step in Testing Process	Potential Problems
Specimen collection	adequate? representative?
Transportation and storage	viability? selective losses? autofluorescence?
Sample preparation	selective losses? aggregation? contaminating cells?
Staining	saturation? linearity? equilibrium? signal amplification? cross reactivity? blocking or interference? viability? washing? fixation?
Measurement	electronic noise? optical noise (scatter, color overlap)? standardization? calibration? gating?
Data analysis	percent-positive? intensity? unimodal? bimodal? multimodal?
Reporting	normal ranges? lab-to-lab variation?

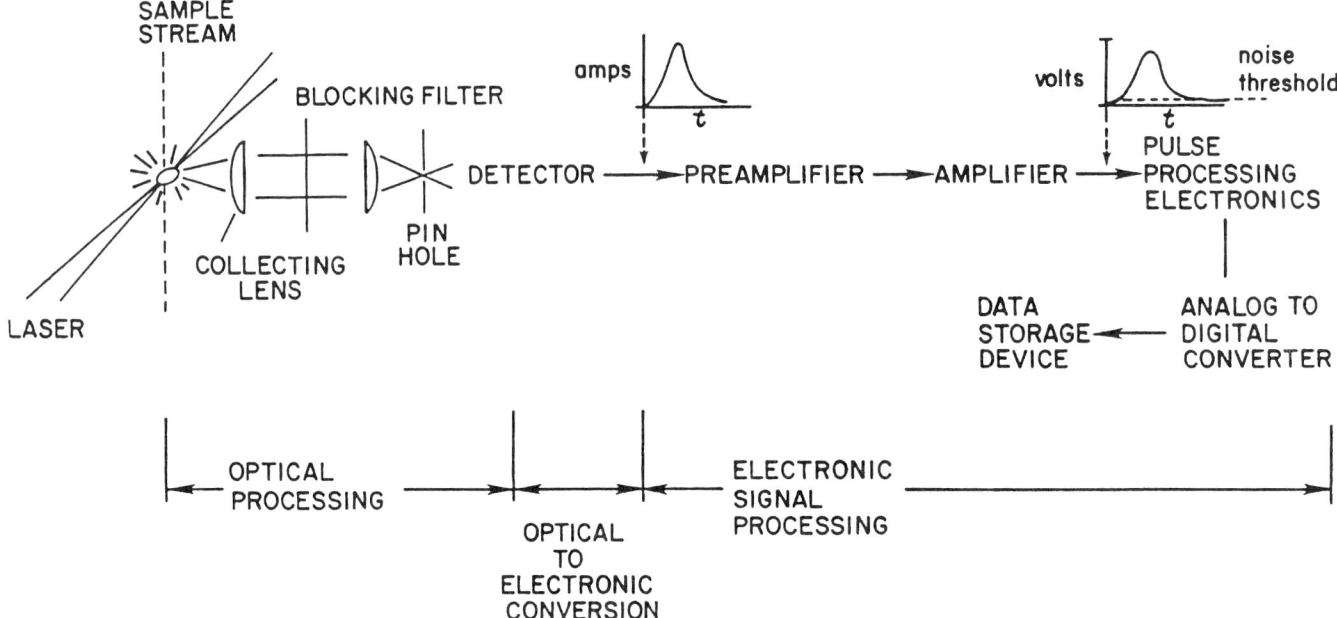

Figure 11.1. Different types of signal processing employed during the flow cytometry measurement process.

ing light in accordance with their intrinsic shape, size, and granularity. Fluorescence of characteristic colors is emitted by fluorochromes naturally present within the cell ("autofluorescence") and by reagent fluorochromes added during the staining process.

As indicated schematically in Figure 11.1, three basic steps are involved in converting the scattered or emitted light signals emanating from each cell into processed data. Step 1 is optical processing. Light originating from the cell is collected using lenses, pinholes, and other optical elements. The signal intensity is a function of the size, shape, and intensity of the exciting beam through which the cell passes, the rate at which the cell traverses the beam, and the efficiency of the collecting optics. Collected light is then separated via color-selective filters into components of different wavelengths that are routed to separate detectors, enabling the flow cytometer to "see" different biological properties of the cell as indicated by different colors (wavelengths) of light associated with specific probes. The goal of optical processing is thus to allow the output of a given detector to be interpreted in terms of a particular biological property.

Step 2 is optical to electronic conversion. An electrical signal proportional to the amount of light received is generated by the photodetector, providing a signal suitable for further electronic evaluation and computer storage. Step 3 is electronic signal processing. The electrical signal produced by the photodetector is amplified, its magnitude (and, in some cases, its shape) is evaluated, and its magnitude may be corrected based on the magnitude of a second signal arising from the same cell. Finally, the signal is converted from continuous (analog) voltage values into discrete (digital) values, classified into 1 of 256 or 1024 categories (channels) based on its magnitude, and recorded by the computer. (For further details regarding instrumentation and signal processing, see Chapter 5)

The final result of the process described in Figure 11.1 may be either or both of the following types of data:

i) a data file that, on a cell-by-cell basis, lists intensity classifications for signals received at each detector for every cell analyzed (a "list mode" file), and/or

ii) a data file recording intensity classifications of signals received at selected detectors for selected cells (a single-parameter "histogram" or a dual-parameter display such as a dot plot or contour plot).

Obviously, many factors ranging from filter selection to photodetector performance characteristics to electronic signal processing can affect instrument performance and, therefore, the values recorded during sample analysis.

Instrument performance assessment is typically carried out for one of two reasons: 1) to select appropriate conditions for sample analysis when a new instrument or test is introduced into the laboratory (instrument setup); 2) to reproduce specific analysis conditions and/or monitor reproducibility (instrument monitoring). With these goals in mind, let us consider some methods for evaluating several different aspects of instrument performance.

Optical System QC

Optimization of the optical components of the flow cytometer includes optical alignment and filter selection.

OPTICAL ALIGNMENT: OPTIMIZATION

Whether carried out by the user or the service representative, optimization of optical alignment involves positioning the cytometer's optical components (laser, laser focusing lenses,

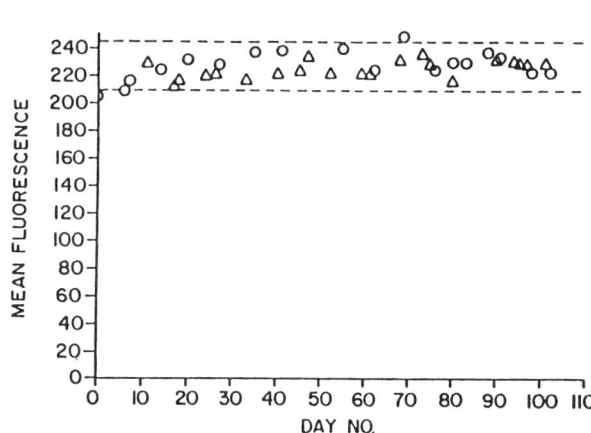
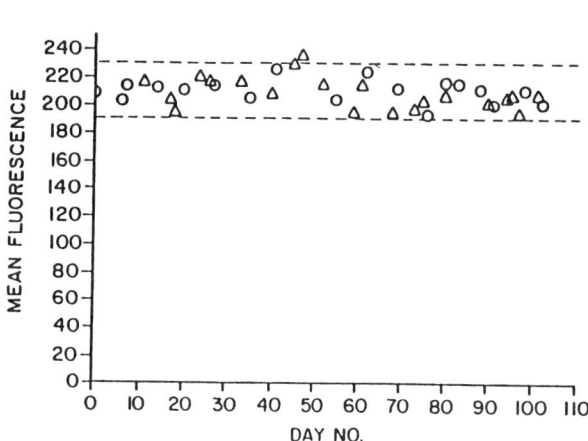
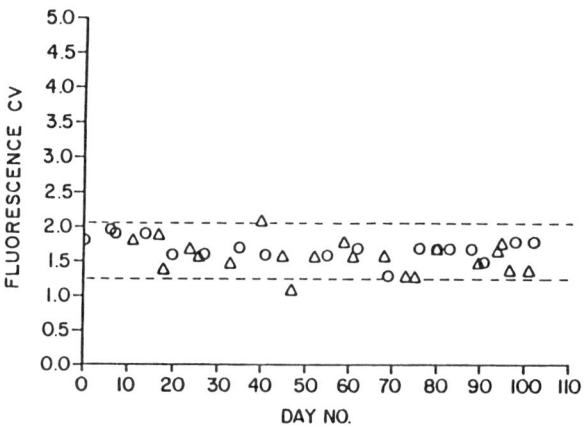
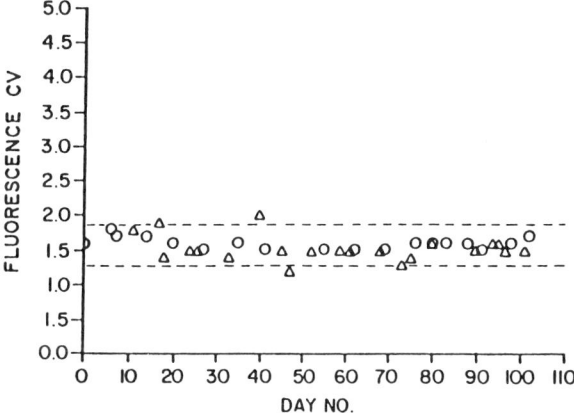

Figure 11.2. Instrument performance monitoring using constant instrument settings. Optical alignment was optimized daily using constant laser power, optical filters, PMT high voltage, and gain settings. Mean intensities and CVs were recorded for plastic microspheres containing a stable broad spectrum dye (Coulter Immuno-check). Different symbols indicate different instrument operators; dashed lines indicate acceptance range (mean ± 2 SD) established during instrument setup phase.

cell stream, collecting lenses, and photodetectors) to give maximum signal intensity (mean channel number) and minimum variability (standard deviation or coefficient of variation) for all parameters that are to be measured on test samples. Homogeneous plastic particles (microspheres) containing fluorescent dyes that emit across a broad range of wavelengths are frequently used to optimize optical alignment. Their uniformity makes it easy to visually determine whether a given adjustment has resulted in increased channel number and/or decreased peak width, and their broad spectrum allows simultaneous evaluation of all filter windows used. For instruments with fixed optical systems, the manufacturer can provide expected mean intensities and CVs for alignment particles that can be used under specific operating conditions (laser power, detector high voltage and gain, etc.) to establish that instrument performance is optimized. For instruments in which filters and other optical components are user–adjustable, optimizing alignment means identifying conditions that simultaneously maximize intensity and minimize CV for all signals of interest.

OPTICAL ALIGNMENT: MONITORING

Once satisfactory optical performance has been established in the instrument or test setup phase, materials to be used for daily monitoring should be selected and acceptance ranges established for each parameter to be monitored. Any of a variety of plastic or cell–based particles may be useful (Table 11.2). Ideally, the material used should give stable values (preferably over months to years) and should be similar in size to test cells so that it is comparably handled by the fluidic and optical systems.

Two different approaches to daily monitoring of optical performance are in common use. Method 1, illustrated in Figure 11.2, using plastic particles, monitors the reproducibility of particle mean intensities and CVs under fixed in-

Table 11.2
Commercially Available Control, Calibration, and Standardization Materials

MANUFACTURER:	Becton-Dickinson Immunocytometry Systems, 2350 Qume Dr., San Jose, CA, 95131-1812; 1-800-223-8226; in CA 800-821-9796
PRODUCT NAME:	Calibrite spheres
Properties:	fluorescein-coated, phycoerythrin-coated, and blank 5 μm beads; intensities similar to immunofluorescence
PRODUCT NAME:	DNA QC kit
Properties:	2 μm fluorescent microspheres, unstained fixed CEN (chick erythrocyte nuclei, and unstained fixed CTN (calf thymocyte nuclei), PI stain, and QC/ troubleshooting instructions
MANUFACTURER:	Coulter Cytometry; P.O. Box 4486, Hialeah, FL 33014; 1-800-327-6531 (for special requests, contact Dr. Jorge Quintana)
PRODUCT NAME:	DNA-check alignment spheres
Properties:	10 μm beads contain broad spectrum dye (uv-orange excitation); relatively bright; 1–2% CVs for light scatter and fluorescence
PRODUCT NAME:	Immuno-check alignment spheres
Properties:	10 μm beads contain broad spectrum dye (uv-orange excitation); relatively bright; CVs somewhat broader than DNA-check
PRODUCT NAME:	Immuno-brite intensity standardization spheres
Properties:	10 μm beads contain broad spectrum dye (uv-orange excitation); 1 blank bead and 4 beads of graded intensity levels; intensity range similar to lymphocyte immunofluorescence
PRODUCT NAME:	Standard-brite intensity standardization spheres
Properties:	10 μm beads contain broad spectrum dye (uv-orange excitation); intensity similar to mid-level Immuno-brites and lymphocyte immunofluorescence
PRODUCT NAME:	Cyto-Trol Control Cells
Properties:	lyophilized preparation of pooled human lymphocytes; light-scatter and antibody-staining characteristics similar to lymphocytes in freshly prepared mononuclear or whole blood preparations; cannot be subjected to lysing agents
MANUFACTURER:	Duke Scientific; 445 Sherman Ave., Palo Alto, CA 94306; 1-415-328-2400
PRODUCT NAME:	Fluorescent polystyrene spheres, Dynospheres
Properties:	wide range of particle sizes (non fluorescent) and several sizes of fluorescent particles containing Rhodamine B or fluorescein, low or high intensity (Dynospheres); CVs 2–3% (diameter basis)
MANUFACTURER:	Flow Cytometry Standards Corporation; P.O. Box 12621, Research Triangle Park, N.C. 27709; 1-919-967-9345, Telex 579447 (for special requests, contact Dr. Abe Schwartz)
PRODUCT NAME:	Fluorescent reference standards
Properties:	2–10 μm beads; Hoechst 33342, fluorescein, propidium iodide, Texas red, phycoerythrin, chlorophyll, Indo-1, Fura-2, acridine orange, allophycocyanine, oxazine, dansyl choride, "autofluorescent" and "blank"; contact Abe Schwartz for other possibilities
PRODUCT NAME:	Quantum kits
Properties:	4.4–9.0 μm beads with surface coated fluorochromes; 3,000 – 2,000,000 fluorescein and/or 10,000 – 50,000 phycoerythrin equivalents per particle, plus "blank" or autofluorescent particles
PRODUCT NAME:	QuickCal system
Properties:	microbead mix with multilevel FITC and/or PE intensities and calibration software for use with DOS or HP 3.0 list mode files
PRODUCT NAME:	Fluorescence Compensation kits
Properties:	blank, fluorescein-only, phycoerythrin-only, and fluorescein+phycoerythrin-coated particles with light scatter similar to cells; multilevel kit for simultaneous calibration and compensation
PRODUCT NAME:	QC3
Properties:	fluorescein+phycoerythrin coated beads; intensity similar to immunofluorescence; for monitoring reproducibility of fluorescence and light-scatter settings using a single particle
PRODUCT NAME:	"Simply Cellular"
Properties:	8–10 μm beads coated with goat anti-mouse serum; defined number of antibody binding sites per bead, usable with mouse Ig labeled with any chromophore to compare fluorochrome/protein ratios, lot-to-lot variation, etc.
PRODUCT NAME:	"Simply Cellular T3"
Properties:	8–10 μm beads coated with OKT3; for determination of serum antibodies against OKT3
PRODUCT NAME:	"FluoroTrol" control particles (in development)
Properties:	fixed preparation of thymocyte nuclei covalently labeled with FITC; provided as mixture of unstained, low and high intensity (similar to immunofluorescence)
MANUFACTURER:	Molecular Probes, Inc., 4849 Pitchford Ave., Eugene, OR 97402; 1-503-344-3007; FAX 1-503-344-3007 telex 858721 (Molecular) [Good general source of fluorescent stains, probes and substrates; catalog has an excellent compilation of excitation/emission maxima, and literature references on applications of various fluorescent probes.]
PRODUCT NAME:	Fluospheres latex microspheres
Properties:	0.2–5 μm beads; blue, yellow-green, orange, or red fluorescence
MANUFACTURER:	Pandex Laboratories; 909 Orchard Rd., Mundelein, IL 60060; 1-312-949-6700
PRODUCT NAME:	Fluoricon particles
Properties:	1.7–2.2 μm beads; wide color range (uv – long red excitations), range of intensities (contact Jeff Wang for detailed information)
MANUFACTURER:	Polysciences, Inc., 400 Valley Rd., Warrington, PA 18976; 1-800-523-2575
PRODUCT NAME:	Fluoresbrites
Properties:	0.5–6.0 μm beads (other sizes available on request); yellow/green (ex. 458nm, em. 540nm) and red (ex. 590nm, em. 657nm); CVs vary with size, tighter for smaller ones (2 μm diameter)
PRODUCT NAME:	Fluorescent intensity kit
Properties:	6 micron yellow/green Fluorebrites with 1×, 1/5×, 1/10×, 1/20×, 1/50× and 1/100× dye intensity
PRODUCT NAME:	Fluorescent Color Range kit
Properties:	1.75 μm beads; excitation at 273nm, 365nm, 458nm, 530nm, 641nm, 763nm
MANUFACTURER:	Riese Enterprises, P.O. Box 9523, San Jose, CA; 1-800-345-2267
PRODUCT NAME:	Bio-Sure Flow Cytometry Control
Properties:	glutaraldehyde-fixed chick red blood cells, autofluorescence similar to immunofluorescence, characteristic bimodal light scatter distribution
MANUFACTURER:	Sigma Diagnostics, P.O. Box 14508, St. Louis, MO 63178; 1-800-325-0250; FAX 1-800-325-5052
PRODUCT NAME:	Accuscan standard

Properties:	fluorescein + phycoerythrin-coated beads; intensity similar to immunofluorescence; for monitoring reproducibility of fluorescence and light-scatter settings using a single particle
PRODUCT NAME:	Accuscan Color compensation kit
Properties:	blank, fluorescein-only, phycoerythrin-only, and fluorescein + phycoerythrin-coated particles
PRODUCT NAME:	Accuscan cellular microspheres
Properties:	8–10 μm beads coated with goat anti-mouse serum; defined number of antibody binding sites per bead, usable with mouse Ig labeled with any chromophore to compare fluorochrome/protein ratios, lot-to-lot variation, etc.
PRODUCT NAME:	Accuscan certified blank
Properties:	unlabeled beads with low autofluorescence; for monitoring instrument sensitivity
MANUFACTURER:	Streck Laboratories, Inc., 14306 Industrial Road, Omaha, NE 68144; 1-800-228-6090; in Nebraska, 402-333-1982.
PRODUCT NAME:	Reticulocyte counting kit
Properties:	contains Retic-Chex (stabilized human red blood cells), new methylene blue, and expected values for manual and FACS assays

strument conditions (i.e., specified laser power, filters, PMT high voltages, and gains). Changes in measured particle intensity give a direct indication of how much instrument performance varies from day to day. Method 2, illustrated in Figure 11.3, using cellular particles, monitors the reproducibility of instrument settings (typically PMT high voltages) required to achieve specified mean channel numbers for the alignment particles. Since mean channel number is exponentially affected by PMT high voltage, it is more difficult to estimate the exact magnitude of performance shifts when using this method. However, values outside the established acceptance range indicate altered instrument performance and the need to reevaluate optical alignment and/or other instrument parameters that may affect the mean channel value.

More important than the particular method chosen for monitoring optical alignment is that some method be used consistently and with clear definition of the range of values defining acceptable instrument performance. The range of variation expected can be defined by running the selected "alignment particle" under conditions of optimal alignment a total of at least 20 times over a minimum of at least five separate days, collecting data for all parameters that will be used to analyze test specimens. Acceptance ranges for mean channel numbers and CVs (method 1) or for PMT settings required to achieve specified mean channel values (method 2) are then established based on the observed range of variation of these parameters (typically mean ± 2 standard deviations;) (3). Acceptance ranges for new lots of particles are determined running them in parallel with a previously characterized lot on an optimized instrument.

FILTER SELECTION AND MONITORING

Optimal filter selection is a function of the combination of fluorochromes being used (which determines the degree of color overlap), the particular biology under study (which determines the relative amounts of each fluorochrome present per cell), and the instrument being used (which determines whether filter configuration is fixed or user–selectable). The number of permutations is obviously legion, but some general rules can be given. For weak signals, use the minimum number of filters required to reduce scattered light or overlapping colors to acceptable levels. No filter is 100% efficient and wavelength selective filters (i.e., dichroic filters) typically give less efficient light transmission than bandpass filters. Where filter configuration is adjustable, select filters to reflect shorter wavelengths and transmit longer ones, since dichroic filters are typically more efficient when used in this fashion. Where maximally efficient light collection is critical to resolution of dimly labeled cells, rerunning the same specimen with different filter combinations may be helpful in selecting an optimal combination.

Filters used in flow cytometers are commonly of two types: a) color filters, containing dyes that absorb light of specified wavelengths and transmit all other wavelengths, and b) interference filters, containing reflective layers spaced to allow transmission of specified wavelengths and reflection of others. Interference filters positioned at a 45° angle to allow collection of both transmitted and reflected wavelengths are referred to as dichroic filters. Color filters are typically quite stable as long as excessive heating or intense light exposure are avoided. Interference filters are more stable to intense light exposure and hence are often used as laser-blocking filters. However, mechanical shock, scratching of the reflective coating, or extremes of heat and humidity may affect their transmission characteristics. Filters are also sometimes mislabeled. Therefore, if at all possible, transmission spectra should be obtained for all filters at the time the instrument is received and monitored at approximately six-month intervals to verify that performance is acceptable. Filter transmission spectra are readily obtained by placing the filter in a scanning spectrophotometer, perpendicular to the light beam and with the reflective coating facing the beam, and recording percent transmission as a function of wavelength.

Signal Processing QC

As mentioned previously, electronic signals arising from a photodetector may be processed in a variety of ways prior to final conversion to a channel number. Different methods of signal processing have different effects on channel number recorded and, thus, on the relative intensity inferred for a particular cell type. Inaccurate signal processing can lead to variability and/or inaccuracy of test results. For example, nonlinear amplification can lead to invalid DNA index determinations; improper color compensation can result in immunofluorescence false-positives or false-negatives. Therefore, verifying acceptable instrument performance requires moni-

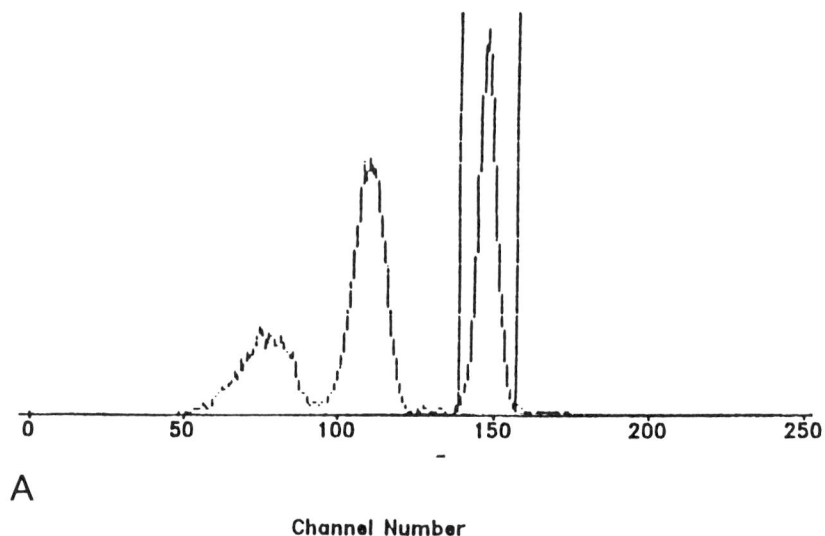

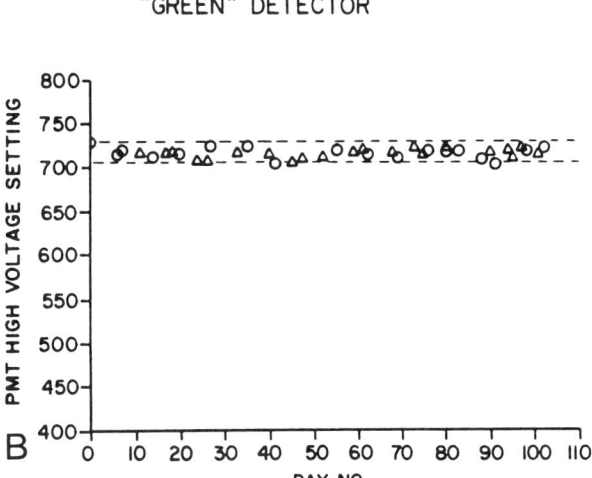
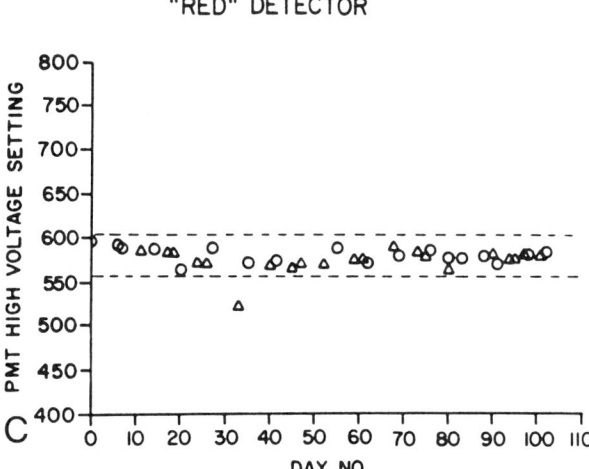

Figure 11.3. Instrument performance monitoring using constant intensity settings. Optical alignment was optimized daily using constant laser power and optical filters. PMT high voltage required to obtain constant mean (±1 channel) for fluorescein–labeled thymocyte nuclei was recorded; no color compensation was used. Different symbols indicate different instrument operators; dashed lines indicate acceptance range (mean + 2 SD) established during instrument setup phase. Data were recorded over the same time period as shown in Figure 11.2. **A**, Log–amplified "green" fluorescence intensity distribution of Fluorotrol-GF thymocyte nuclei (4; provided by Dr. Robert Hoffman) whose peak position was held constant by adjusting PMT high voltage settings is indicated by cursors. **B**, High voltage settings required for "green" PMT (530 ± 15 nm filter). **C**, High voltage settings required for "red" PMT (570 nm long–pass filter). Note out-of-range value observed on day 34. Microsphere values for same day were within established acceptance ranges (Fig. 11.2), indicating that optical alignment was not at fault. Thymocyte nuclei had taken up propidium iodide remaining in system from a previous run, causing an increased red signal and a decreased high voltage setting required to obtain the prescribed mean value.

toring accuracy and adequacy of signal processing. Methods for calibrating intensity measurements are also helpful in comparing data from longitudinal studies and/or from different laboratories.

LOGARITHMIC VS. LINEAR AMPLIFICATION

Figure 11.4 illustrates the effect of logarithmic vs. linear signal amplification on data obtained from a mixture of plastic microspheres of differing intensities. Linear amplification assigns events to a channel number directly proportional to the original signal voltage and each channel represents a constant increment in voltage. The dynamic range of signals that can be displayed is equivalent to the number of channels available. DNA fluorescence measurements, which have a relatively limited biological dynamic range (typically four-to-eight fold), are usually made using linear amplification. Logarithmic amplification, on the other hand, assigns events

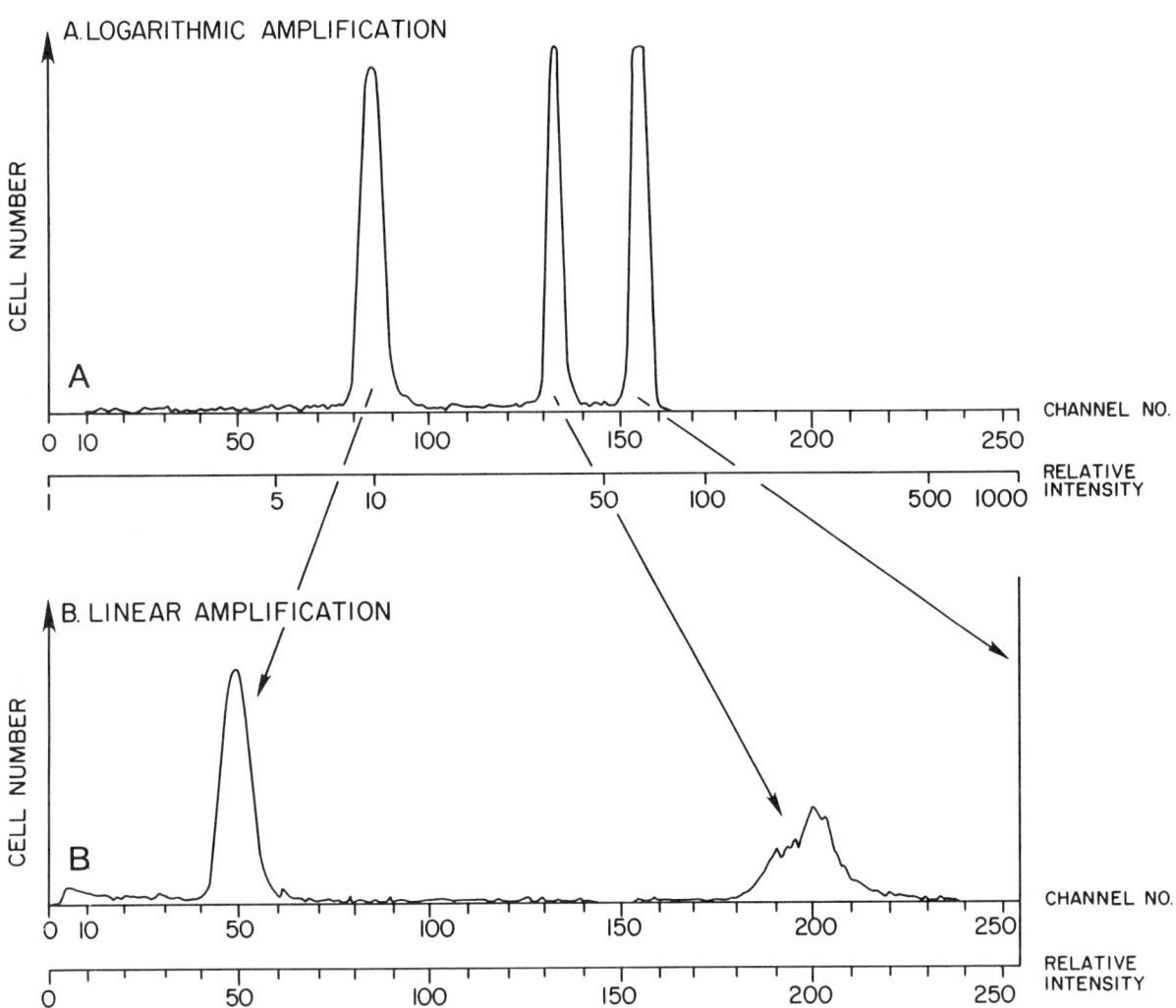

Figure 11.4. Linear vs. logarithmic signal processing. Fluorescence histogram obtained from mixture of microspheres of relative intensity 1 : 3.7 : 7.4 analyzed using a logarithmic amplifier with three-decade dynamic range (**A**) or a linear amplifier with 256-fold dynamic range (**B**). Note that relative signal intensities of 1-10, which occupy only the first tenth (10 channels) of the linear histogram, occupy the first third (85 channels) of the log histogram. Conversely, relative signal intensities of 128–256, which occupy the last half (128 channels) of the linear histogram, are compressed into ≈25 channels in the three-decade log histogram. On a linear scale, resolution of small differences in intensity is best accomplished by displaying the signals of interest at relatively high channel number. On a log scale, however, subtle heterogeneities in intensity distribution (for example, the shoulder on microsphere peak two in **B**) may be lost when they are displayed at high channel numbers.

to a channel number proportional to the log of the original signal voltage and successive channels represent exponential increments in signal voltage. Thus the dynamic range of signals that can be displayed using a logarithmic amplifier (typically 10^3–10^4) is greater than the number of channels (typically 256). Immunofluorescence measurements, which often cover a wide biological dynamic range (10^3–10^4 fold differences in intensity), are typically made using logarithmic amplification.

Amplifier performance characteristics should be assessed at three different times: as part of instrument setup, whenever amplifiers are replaced, and at regular intervals during normal operation (i.e., every three-to-six months). Several methods for determining amplifier and/or detector response curves are summarized below, each of which may be used to assess performance of either linear or logarithmic amplifiers. When data is collected using linear amplification, the ratio of mean channel numbers is used to estimate the relative intensity of two different populations (i.e., to determine DNA Index of tumor vs. normal cells). When data is collected using logarithmic amplification, the difference between modal channels (channel numbers with the most events) is used to estimate relative intensities (5).

One way to assess amplifier performance is to repeatedly run some reference particle, keeping the detector high voltage constant, but inserting blocking filters of varying percent transmission in the light path to produce a known change in signal intensity reaching the photodetector (6). If amplifier performance is linear, the mean channel number observed should be a linear function of filter transmittance; if amplifier performance is logarithmic, the modal channel number observed should be a linear function of filter optical density (or, equivalently, a logarithmic function of filter transmittance). An alternative is to run a calibrated mixture of particles with known relative fluorescence intensities. Linear correlation between observed mean channel values and assigned intensity values (or between observed modal channel number and log of the assigned intensity) indicates proper amplifier performance (7).

Both of the above methods maintain constant detector high voltage in order to assess amplifier response curves independently of detector response curve. However, it may also be desirable to characterize the system response (combined detector–amplifier curve). This may be done using a sample containing two populations with resolved but closely spaced peaks (8). The two populations are moved from low to high channel values by varying detector high voltage. If the amplifier is operating linearly and if the photodetector is operating within its linear response range (i.e., in a range where number of electrons out is linearly proportional to number of photons in) the ratio of mean channel values for the two peaks should remain constant across the scale. If the amplifier is operating logarithmically, the difference between modal values for the two peaks should remain constant across the scale. Finally, where not already provided by the manufacturer, cross-calibration between linear and logarithmic scales can be carried out by simultaneously accumulating both types of data (Fig. 11.3 and Table 1 of Ref. 9).

CALIBRATION AND STANDARDIZATION OF INTENSITY SCALES

Many of the factors that can influence intensity measurements in a flow cytometer are of no biological interest (i.e., laser power, rate of travel through the excitation region, optical collection efficiency, filter characteristics, detector efficiency, fluorochrome:protein ratio; see Chapter 5). Therefore, flow cytometric intensity comparisons are typically made using relative units (channel numbers) rather than absolute ones (number of photons, fluorochromes, DNA base pairs, or antigens per cell). For some applications, it is sufficient to simply adjust instrument conditions so that all populations of interest are on scale and adequately resolved without worrying about whether the intensity scale is positioned reproducibly from day to day. This is the case where reference populations are found within the same sample, i.e., for determination of DNA Index (ratio of fluorescence intensity of the test G_0G_1 population to that of a normal diploid population in the same specimen) or percentage of cells labeled with a specific antibody.

However, for other purposes it may be preferable to use a consistent intensity scale, i.e., to adjust instrument conditions to reproduce the position of a stable reference material. For example, it has been suggested that level of expression of specific cell surface antigens may be useful indices of maturation, differentiation, disease state, and/or responsiveness to therapy (10–12). While general consensus on the clinical utility of such variations has not yet been reached, use of a consistent intensity scale from day to day within the laboratory is one way to speed the process of gathering the necessary information. A consistent intensity scale can also be very helpful in troubleshooting sample preparation methods or reagent performance (Fig. 11.5) as well as in monitoring instrument performance (Fig. 11.3). Any reference material with stable intensity(ies) comparable to those found in test samples can be used to establish a consistent intensity scale, regardless of whether the spectral properties exactly match those of the stained specimen.

As discussed in Chapter 28 and Vogt et al. (13), conversion of relative fluorescence to biologically or clinically meaningful units requires significant effort. Current clinical applications of flow cytometry tend to utilize intensities as diagnostic or prognostic information only when intrasample comparisons are possible (DNA Index, relative CD5 expression on normal vs. neoplastic B-cells, etc.). This may appear surprising in comparison with other clinical laboratory methods, where test results are reported in standard units in all laboratories. However, it is the natural consequence of the fact that there are currently no relevant standards for determination of absolute fluorescence intensities. Looking to the future (last section of this book), reliable use of tests based on quantification of antigen expression or analyte levels will require calibration of intensities in units that are not only reproducible from day to day within a given laboratory but among different laboratories using different instruments. For this purpose, reference materials with spectral properties identical to those of the fluorochrome(s) used for cell analysis and **with stable intensities** are required to insure that variations in filter and detector responses do not bias the results and that stable results are obtained in longitudinal studies (13). Increasing commercial availability of materials with appropriate spectral characteristics (Table 11.2) is an encouraging step toward making the full quantitative capabilities of flow cytometry applicable in a clinical setting. The problem of verifying stability of intensity remains a knotty one, again due to the lack of absolute fluorescence intensity standards to serve as a reference point.

COLOR COMPENSATION: OPTIMIZATION

Use of multiple probes, each bearing a different fluorochrome, to simultaneously evaluate multiple characteristics of each cell is a common strategy in flow cytometric tests. Such an approach allows identification of cell types not uniquely characterized by a single marker. For example, CD3 antigen is expressed on both T_{helper} and $T_{suppressor/cytotoxic}$ lymphocytes and CD8 antigen is expressed on both $T_{suppressor/cytotoxic}$ and NK lymphocytes, but coexpression of CD3 and CD8 antigens uniquely identifies the $T_{suppressor/cytotoxic}$ subset. Multicolor analysis also allows discrimination of contaminating cell types that may result in confusing or inaccurate test results. For example enumeration of T_{helper} lymphocytes as $CD3^+CD4^+$ cells is more accurate than enumeration of all

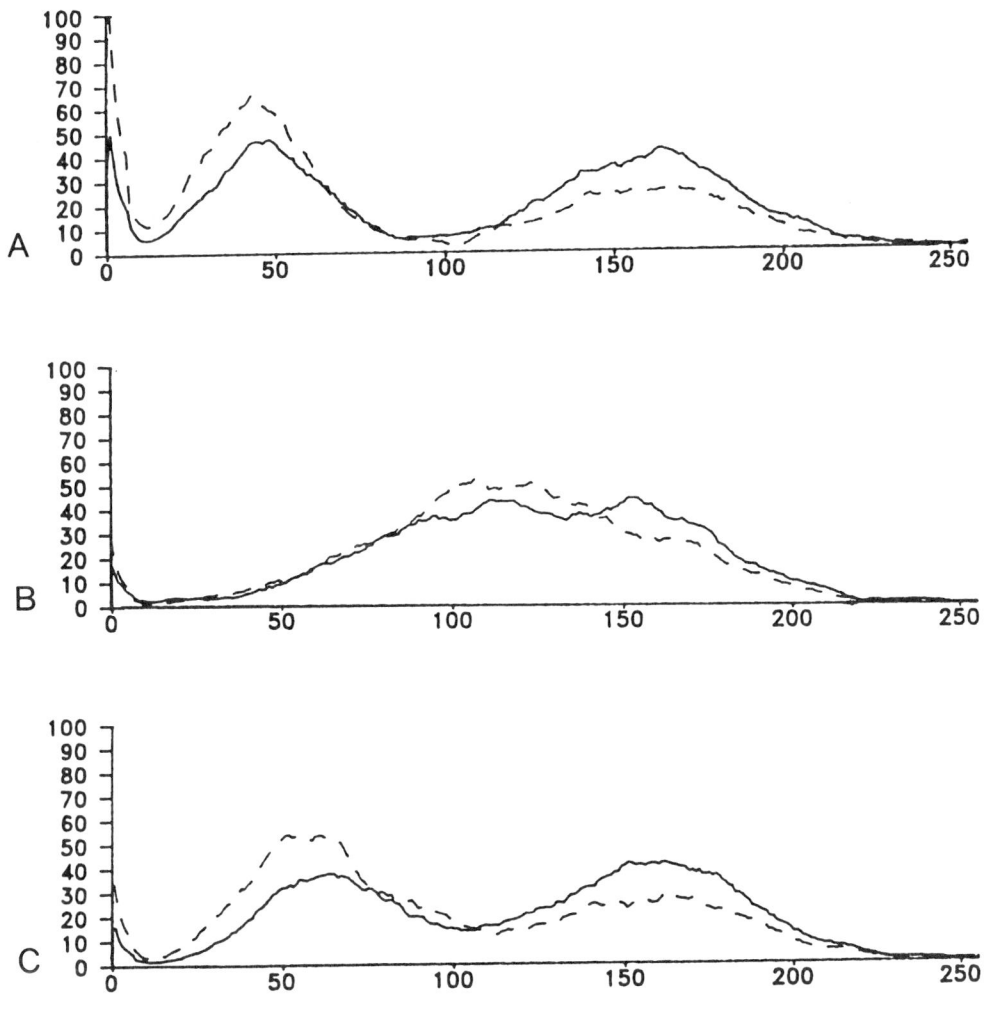

Channel Number

Figure 11.5. Utility of constant intensity scale for troubleshooting sample preparation effects. After optical alignment, PMT high voltages were adjusted to obtain constant mean intensities (±1 channel, compensation set to 0) for Calibrite FITC–labeled and PE–labeled particles. Mononuclear cells were prepared by density gradient separation from blood samples held for 0-48 hr, labelled with rabbit polyclonal sera reactive with human κ (solid line) or λ (dashed line) immunoglobulin light chains, and counterstained with PE–labeled CD20 and FITC–labeled goat anti–rabbit serum (F(ab')$_2$ fraction). B–cell light chain (FITC) distributions were collected by gating on PE-positive cells. **A**, In freshly prepared samples, excellent resolution between light chain positive and negative populations was observed. **B**, Resolution between positive and negative populations was lost after 48 hours in transport medium prepared by admixing equal parts of heparin-anticoagulated whole blood with RPMI culture medium. Since the intensity scale was reproduced from day to day, comparison of the upper and middle panels indicated that the loss of resolution was not due to decreased light chain staining of the positive population but to increased autofluorescence or nonspecific staining of the negative population. **C**, Selection of alternate holding conditions (ACD anticoagulated blood) significantly improved resolution between positive and negative populations at 48 hr (also at 72 hr, data not shown).

CD4$^+$ cells because the requirement for coexpression of CD3 discriminates against inclusion of monocytes, which are also CD4$^+$ but do not bear the CD3 antigen. Similarly, binding of cytokeratin antibodies can be used to distinguish diploid carcinoma cells from infiltrating stromal or immune cells that are also diploid, enabling more accurate assessment of the proliferative status of the tumor cells. Clearly, this approach can be extended to as many probes (i.e., fluorochromes) as one has detectors to monitor, but the following examples will use only two for the sake of clarity.

Using ideal fluorochromes and perfect filters, cells bearing only marker 1 (color 1) would give a signal at detector 1 but not at detector 2, and cells bearing only marker 2 (color 2) would give a signal at detector 2 but not at detector 1. Unfortunately, real fluorochromes emit a relatively broad range of wavelengths, and fluorochromes that can be excited using the same wavelength typically also have overlapping emissions, as exemplified by fluorescein (FITC) and phycoerythrin (PE) in Figure 11.6. If we define "orange" as wavelengths from 560–590 nm, both FITC and PE emit in the

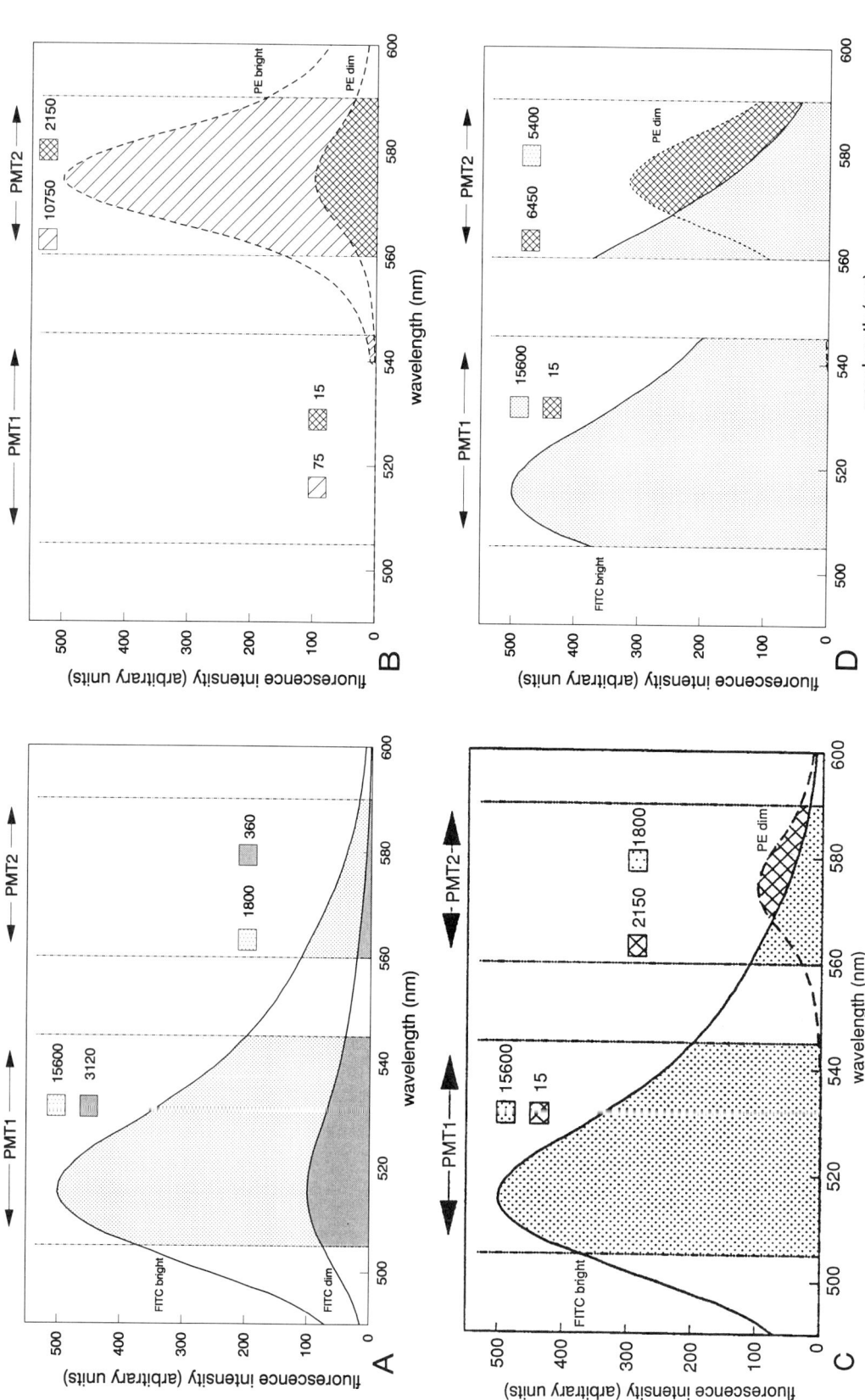

Figure 11.6. Effect of filter selection and signal amplification on electronic color compensation factors. Each curve represents the spectrum of emissions that would be obtained from the assemblage of FITC-labeled and/or PE-labeled antibodies present on a single cell after single- or dual-color labeling. PMT1 ("FITC" detector) represents the range of wavelengths that would be collected using a 525±20 nm bandpass filter. PMT2 ("PE" detector) represents the range of wavelengths that would be collected using a 575±15 nm bandpass filter. Values given for shaded regions represent the relative signal seen by the indicated PMT (i.e., relative area under the emission curve falling within the filter window). **A,** Upper curve represents spectrum of emissions from a brightly FITC-labeled cell; relative signal seen at PMT2 (1800 units) is 11.5% of that seen at PMT1 (15600 units). Lower curve represents spectrum of FITC-labeled cell 1/5 as bright as the upper curve; relative signal seen at PMT2 (360 units) is 11.5% of that seen at PMT1 (3120 units). Correction (compensation) for overlap of FITC into the PE detector would therefore be to subtract 11.5% of the signal observed at PMT2. **B,** Emission spectra from brightly PE-labeled cell (upper curve) and PE-labeled cell only 1/5 as bright (lower curve). Appropriate correction for overlap of RE into the FITC detector would be to subtract 0.7% of the signal observed at PMT2 from the total signal observed at PMT1. **C,** Emission spectra from a dual-labeled cell brightly labeled with FITC (left-most curve) and dimly labeled with PE (right-most curve). Total signal seen at PMT2 is comprised of 2150 units from PE and 1800 units from FITC; correction by subtraction of 11.5% of the PMT1 signal (0.115 × 15615 total units = 1800 units) gives the correct result of 2150 units at PMT2 due to PE. **D,** Same spectra as in panel C, but with heights in filter window two scaled to represent increasing PMT2 amplification by a factor of three. Note that the ratio of FITC signals at PMT1 and PMT2 is no longer 11.5%, i.e., the compensation factor must be changed. The compensation factor would also have to be changed if the width of filter windows one or two were altered.

orange, although PE has a greater proportion of its total signal in the orange than does FITC. We would like to make PMT1 the "FITC" (marker 1) detector and PMT2 the "PE" (marker 2) detector. However, with the filter choices shown in Figure 11.6, there will always be some "false PE" signal from any FITC present on the cell and a much smaller but non-zero amount of "false FITC" signal from any PE present. This problem is exacerbated by the presence of bright and dim cells of each type (Fig. 11.6A and B) and by the fact that some cells can bear both labels (Fig. 11.6C).

Clearly, two conflicting needs must be met in setting up the instrument for two-color analysis. Collecting the maximum signal from each fluorochrome corresponds to selecting a filter window wide enough to span the whole area under the emission curve for that fluorochrome. However, collecting the minimum "false-positive" signal from the second fluorochrome corresponds to selecting a filter window that does not include any region of spectral overlap, which would lead to collecting only a small fraction of the total signal. This conflict can be partly resolved by a process called color compensation, which in most commercial flow cytometers involves a form of signal processing referred to as electronic subtraction. The basic principles of electronic subtraction are illustrated schematically in Figure 11.6. As can be seen by comparing the curves for FITC dim and bright cells in Figure 11.6A, the magnitude of the "false PE" signal is directly proportional to the amount of true FITC signal. More importantly, the proportionality factor is the same for both bright and dim cells. In the example shown, the compensation factor would be 11.5% for FITC overlap into the PE detector and 0.7% for PE overlap into the FITC detector. To estimate how much of the signal at PMT2 was truly due to PE emission and not to overlapping FITC emission, 11.5% of the PMT1 signal would be subtracted from the total signal (FITC+PE) observed at PMT2. Where a cell was labeled only with FITC, this subtraction would result in a value of 0 at PMT2, which is precisely the correct result when no PE is present. However, electronic compensation is not a panacea. Consider the case of a dual-positive cell which is brightly FITC-labeled but dimly PE-labeled. The PMT2 signal consists of a small true PE contribution plus a large "false PE" signal (Figure 11.6C), and subtraction of the large correction factor results in decreased accuracy for the corrected PE signal. For those who are interested, a mathematical description of electronic subtraction can be found in Loken et al. (14).

How are compensation settings (percent subtraction) selected in real life? They will depend on several factors: 1) the fluorochrome combination used, 2) the exact range of wavelengths collected by each filter, and 3) the relative signal amplification occurring at each detector. Therefore, compensation factors must be selected empirically—by choosing instrument settings under which cells labeled only with FITC give no greater signal than unstained cells at the PE detector and vice versa. The most appropriate material to use in establishing compensation settings is cells stained with mutually exclusive markers bearing the fluorochromes of interest.

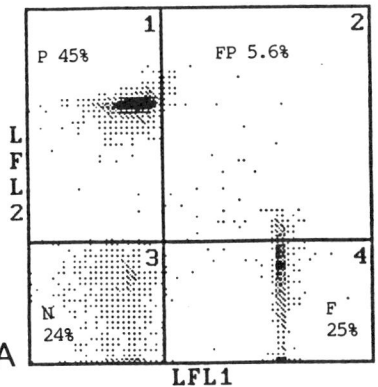

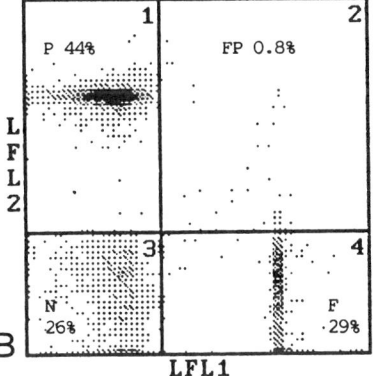

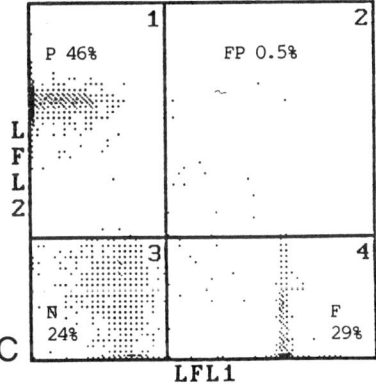

Figure 11.7. Effect of under- and overcompensation on two-color analysis of mutually exclusive markers. Microspheres labeled with no fluorochrome (*N*), FITC only (*F*) or PE only (*P*) (Flow Cytometry Standards Corp.) were analyzed at several different compensation settings. *LFL1* = all fluorescence collected using 525±20 nm bandpass filter; *LFL2* = all fluorescence collected using 575±15 nm bandpass filter. **A**, Slight undercompensation—average *LFL2* signal of FITC beads greater than average *LFL2* signal of unlabeled beads; average *LFL1* signal of PE beads slightly greater than average *LFL1* signal of unlabeled beads. **B**, Correct compensation—average *LFL2* signal of FITC beads same as average *LFL2* signal of unlabeled beads; average *LFL1* signal of PE beads same as average *LFL1* signal of unlabeled beads. **C**, Slight overcompensation—average *LFL2* signal of FITC beads less than average *LFL2* signal of unlabeled beads; average *LFL1* signal of PE beads less than averge *LFL1* signal of unlabeled beads.

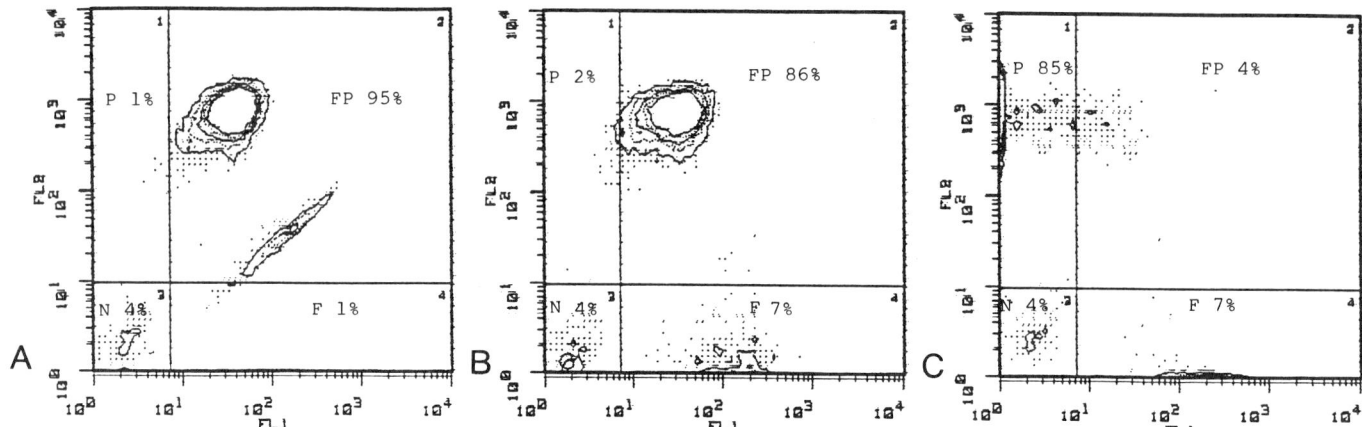

Figure 11.8. Effect of under- and overcompensation on two-color analysis of coexpressed markers. Specimen from patient with B-CLL was labeled with FITC-CD5 (*FL1*) and PE-CD20 (*FL2*) and analyzed at several different compensation settings for frequency of normal T-cells (*F*; CD5^{bright}), normal B-cells (*P*; CD20^{bright}), neoplastic B-cells (*FP*; CD5^{dim}, CD20^{bright}, and antibody–negative cells (*N*). Percent CD20 positive (quadrant 1), CD5/CD20 dual positives (quadrant 2), negative (quadrant 3), and CD5 positive (quadrant 4) events are indicated by numbers in each region. [Data courtesy of Dr. Michael Borowitz.] **A**, Undercompensation—use of perpendicular analysis thresholds shown results in inclusion of normal T-cells as false-positives in region 2 (dual-positive region). Note strong diagonal nature of the normal T-cell cluster, characteristically seen when two parameters are strongly correlated (here because both *FL1* and *FL2* signals reflect FITC intensity). Unless a sample known to contain no FITC-positive cells has been previously analyzed and found to give no significant events in quadrant 2, it is not possible to say whether the CLL population is truly double positive. **B**, Correct compensation—normal T-cells have average *FL2* similar to that of antibody negatives and are correctly counted as single positives in quadrant 4; CLL-cells are dual positive. Note, however, that this can only be said with confidence if a sample known to contain no FITC-positive cells has been previously analyzed and found to give no significant events in quadrant 2. **C**, Overcompensation—use of perpendicular analysis thresholds shown results in analysis of most CLL cells as CD5 negative. Normal T-cells are barely visible along X-axis and have apparent *FL2* intensity significantly less than that of antibody–negative cells; CLL cells have apparent *FL1* intensity significantly less than that of antibody-negative cells.

If normal peripheral lymphocytes are used as the reference material for establishing compensation settings, the following samples might be analyzed:

Tube No.	Reagent(s)
1	FITC-CD8
2	PE-CD4 (or PE-CD8)
3	mix equal volumes of tubes 1 and 2

Ideally, the markers used should be those giving relatively high fluorescence intensities, since undercompensation is more readily visualized with bright than dim signals. Note, however, that it is also possible to overcompensate (electronically overcorrect). Overcompensation will not affect detection of singly-labeled cells, but may cause underestimation of dual-labeled cells, particularly if they are dimly positive.

Figures 11.7 and 11.8 illustrate the effects of under- and overcompensation on two different types of samples. In Figure 11.7 FITC- and PE– labeled microspheres with intensities similar to those that would be found using the samples listed above were analyzed. On a dual parameter display, undercompensation and resulting overlap of FITC–labeled beads into the LFL2 positive (and dual positive) region is obvious (Fig. 11.7A). Analyzing only single-positive events, it is difficult to see the effects of overcompensation (Fig. 11.7C) as opposed to proper compensation (Fig. 11.7B) since the frequencies obtained for single-positive events are not altered. Figure 11.8 illustrates the effect of over- vs. undercompensation on a leukemic specimen containing true double positives. If dual-labeled events differ in FITC intensity from single-labeled events, it may still be possible to distinguish between them even when there is slight over- or undercompensation. However, if single- and dual-positive events have similar FITC intensities, it becomes difficult to correctly discriminate.

Given that undercompensation can result in false-positive events and overcompensation in false-negative events, what is the best strategy for identifying ''correct'' compensation levels? As illustrated above, use of bright, mutually exclusive markers bearing appropriate fluorochromes allows ready visualization of undercompensation, since no true double positives are expected. To avoid false-negatives due to overcompensation, the lowest compensation settings giving satisfactory results for percent double positives with mutually exclusive markers should be selected. This is clearly a case where although a little is good, more is not necessarily better.

Because degree of compensation (percent subtraction) required depends on filter spectral characteristics and level of signal amplification, compensation settings should be reevaluated whenever filters are replaced or when instrument parameters that affect fluorescence amplification are altered (i.e., photomultiplier high voltage and/or gain settings; see Figure 11.6D). What happens if high voltage is changed daily to maintain a constant position for a reference material, as shown in Figure 11.3? If compensation is not set to zero before high voltage is adjusted, things can become very confusing since high voltage and compensation setting affect measured intensity in opposing fashions. If compensation is set to zero and the high voltage is then adjusted to bring the reference material to its preset position, overall system amplification is being held constant and, therefore, the same compensation settings may be used from day to day. In this case, compensation settings need only be reevaluated when filters are replaced. Note, however, that in order to avoid chasing a moving target it is crucial to select a reference material with stable intensity.

COLOR COMPENSATION: MONITORING

Once appropriate color compensation settings have been identified, a material to be used for daily monitoring of instrument performance should be selected and acceptance ranges established. One method is to use freshly prepared specimens and to repeat, on a daily basis, the procedure described above for selecting appropriate compensation settings. Another common method is to use particles stably labeled with relevant fluorochromes as a point of reference. If such particles are to be used in daily compensation monitoring, they should be run a minimum of 20 times on at least five separate days to establish acceptance ranges. After instrument conditions are established to be correct using freshly prepared specimens, mean "green" and "red" fluorescence intensities should be recorded for both labeled and unlabeled particles. If the spectral properties of labeled particles perfectly match those of labeled cells, the "FITC" intensity of PE-labeled and unlabeled particles should be equal and the "PE" intensity of FITC-labeled and unlabeled particles should also be equal. If the spectral match is not perfect, the "PE" intensity of FITC-labeled particles may not precisely match that of the unlabeled particles. However, they can still be used for daily monitoring as long as expected values are established under color compensation settings verified to be appropriate for labeled cells. A variety of fluorochrome-labeled particles useful for compensation monitoring are commercially available (Table 11.2). In some cases, particles come with manufacturer-supplied acceptance values that have been established by going through a process similar to that just described (i.e., Calibrite particles used on a fixed-filter FACScan instrument). Where manufacturer-supplied values are not available or not appropriate (i.e., where Calibrites are to be used on an instrument other than one for which acceptance values have been established or where a different particle is being run on a FACScan), the laboratory must establish its own acceptance ranges.

Instrument Performance Monitoring: Strategy and Record Keeping

Given the plethora of variables which can affect instrument performance, what strategies can be adopted to assure cost-effective use of QC time and materials as well as accurate test results? The first step is to identify reference material(s) sensitive to those instrument parameters most critical to accurate test results. In general, the more similar the reference material to test specimens in terms of size, spectral distribution, and signal intensities, the more accurately it will reflect small shifts in instrument performance that may adversely affect test results. It is encouraging that commercial manufacturers are finally beginning to offer a range of relatively stable materials potentially useful in performance monitoring and even more encouraging that the list is growing rapidly. In some cases, it may be possible to admix different types of commercially available materials to maximize the amount of information obtained in a single QC run (7).

In laboratories where only a limited range of instrument conditions is used, it may be possible to use a single reference material to validate instrument performance under the same conditions used for specimen analysis. Rerunning specimens from a previous day may also be helpful in verifying reproducibility of instrument performance; however, this requires validation of the assumption that holding or fixation does not alter the test result. For laboratories that run a wide variety of applications requiring different instrument conditions, it may be better to approach daily monitoring in two steps: a) establish that instrument performance is reproducible under some standard set of conditions using a general reference material (as in Fig. 11.2), and b) establish that it is acceptable under conditions relevant to the specific test to be performed using a more test-specific reference material (as in Fig. 11.3).

However, even the most appropriate reference materials are no better than the manner in which they are applied. Therefore, two further items are critical to reliable instrument performance: good record-keeping and consistent use of established acceptance ranges. One essential is a complete and easily accessed QC record (for examples, see the proposed guidelines of the National Committee for Clinical Laboratory Standards) (15). Since trends are much more readily identified by looking at a plot than at a column of numbers, graphs to which data are added on a daily basis (Figs. 11.2 and 11.3) are recommended in addition to a numerical log of relevant parameters. Note that even parameters that are not user adjustable often provide useful information about trends in instrument performance and are worth recording (i.e., current required to obtain a given laser power, time required to meet manufacturer's specifications when autoalignment and/or auto-compensation routines are used). The second item crucial to good QC is a set of well-defined acceptance ranges for instrument performance parameters and adherence to them. Sooner or later, statistical variation will give rise to out-of-range values and criteria for which events to ignore and which to investigate should be applied to flow cytometers in the same fashion as to other clinical instruments.

The bottom line is that a well-monitored and maintained flow cytometer should be capable of giving very consistent performance. Selection of appropriate reference materials and their consistent use in instrument performance monitoring are thus indispensable tools for minimizing down time and maximizing accuracy of results.

BIOLOGICAL QUALITY CONTROL

Poorly handled or improperly prepared specimens will give inaccurate test results regardless of how good instrument performance may be. Therefore, quality control also requires validating the assumption that representative frequencies of relevant cell types have been maintained during specimen preparation and that reagents have been used that are capable of binding specifically to the cellular component(s) of interest. This is typically done at two levels: the test system and the individual specimen. Quality control for the test system requires analysis of positive and negative control specimens prepared at the same time using the same reagents and protocols as test specimens. The positive control verifies that the test method identifies cell type(s) of interest when they are present; the negative control verifies that it gives acceptably low false-positive rates when they are not. Once acceptable system performance has been verified using control specimens, quality control for individual specimens involves monitoring properties specific to the biology of the test specimen (i.e., CV and level of debris present for DNA analysis) as well as quality of light-scatter gating (scatter & subset frequencies both refer to immunofluorescence analysis) and internal consistency of subset frequencies for immunofluorescence analysis).

Biological QC concerns can be somewhat arbitrarily subdivided into four categories: patient-related, specimen- and preparation-related, reagent-related, and data-collection-analysis related. Patient-related variables include the clinical question being addressed, potentially interfering medications, quantity of specimen available, expected frequency of cell type of interest, etc. Specimen-related variables include specimen type (blood, bone marrow, tissue, body fluid), holding conditions (anticoagulant and/or holding medium, temperature, fixative), and specimen age. Altered viability and/or selective losses of specific cell types may arise as patient– and specimen–related factors interact with preparation methods (mechanical vs. enzymatic tissue disaggregation, density gradient separation vs. erythrocyte lysis, etc.) These, in turn, affect reagent-related issues, such as concentration required for reagent excess, level of nonspecific staining, and selection of reagents for multiparameter analysis. And all of the above impact on methods for gating during data collection and on selection of data analysis methods. Clearly, while verification of reliable and accurate instrument performance is a necessary starting point, it is certainly not sufficient to guarantee overall performance of the assay system.

Immunofluorescence QC

Reagent and preparation issues for immunofluorescence analysis of cell surface or intracellular antigens are discussed in Chapters 9 and 10, respectively, while reagent selection and sample preparation problems posed by particular pathologies are addressed in Chapters 6, 7, 25, and 26. Therefore, discussion here will be limited to three other QC issues common to many, if not all, immunofluorescence-based applications: selection and validation of light scatter gating parameters, discrimination against nonviable cells, and selection of appropriate analysis regions and/or methods.

A common application of immunofluorescence in clinical flow cytometry is peripheral blood lymphocyte phenotyping. Whole blood specimens prepared by lysis of erythrocytes contain a mixture of leukocyte types; therefore, some type of gating must be applied in order to identify the subpopulation of leukocytes (lymphocytes) whose immunofluorescence is of interest. The "simple" case of defining and validating lymphocyte light scatter gates is a good illustration of the more general problems of quality control for gating. Traditionally, lymphocytes have been identified by the fact that they have lower forward-angle and right-angle light scatter than most other leukocytes and appear to form a discrete cluster in a dual-parameter light-scatter display (Fig. 11.9A). However, even under ideal conditions there is some overlap between cell types (i.e., between large lymphocytes and small monocytes, between larger dead cells and smaller live ones, between small lymphocytes and red cells or platelet aggregates). In addition, the distinction between leukocyte clusters is affected by instrumental, methodological, and biological factors including forward-scatter detector geometry (scattering angle subtended, stream-in-air vs. enclosed flow cell); type of lysing agent used; heterogeneity of light scatter properties among lymphocyte subsets; altered frequency of lymphocyte subsets having light scatter overlapping with small monocytes (i.e., NK cells); and altered cell viability or activation state due to clinical condition, sample age, or mishandling (16–20).

An alternate definition of lymphocytes relies on the observation that normal lymphocytes exhibit higher staining intensities with CD45 antibodies than do other leukocytes and that erythrocytes, platelets, and debris do not stain with such antibodies (16, 20). CD45 antibodies give overlapping intensity distributions for lymphocytes and monocytes, and two–color staining with CD14 antibodies is typically used to completely resolve lymphocytes from monocytes (Fig. 11.9B). A method known as "backgating" combines the use of scatter and immunofluorescence properties to select a final light scatter gate. Some instruments have specialized software for automated "backgating," but the following process can be carried out on almost all multiparameter flow cytometers. $CD45^{bright}CD14^-$ cells with appropriate light scatter are defined as lymphocytes: a dual-parameter light scatter distribution gated on all $CD45^{bright}CD14^-$ cells is collected, and a light scatter gate is established which includes as many of these cells as possible (Fig. 11.9C). However, a lymphocyte light-scatter gate set broadly enough to include most $CD45^{bright}CD14^-$ cells with low–to–moderate light scatter may also include a significant proportion of contaminating cell types (Fig. 11.9D).

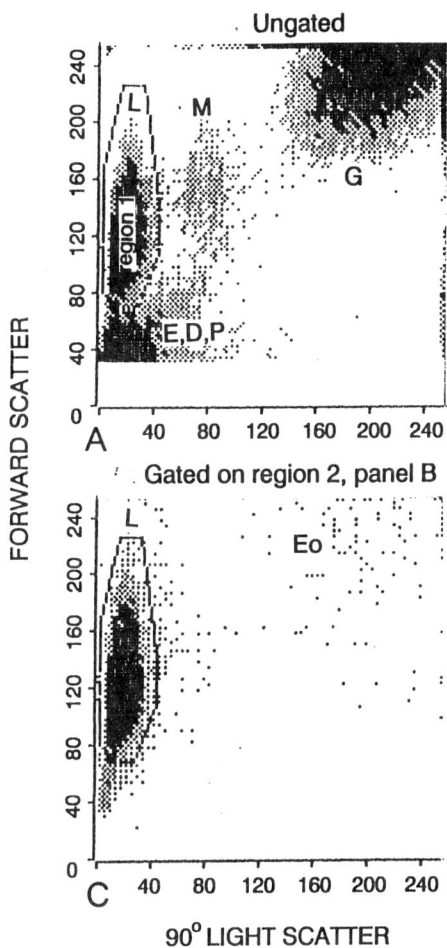
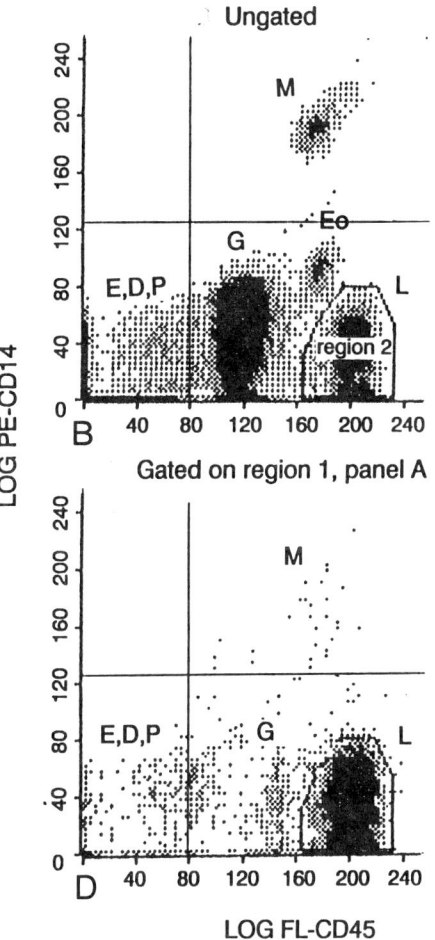

Figure 11.9. Comparison of light-scatter and immunofluorescence methods for identification of lymphocytes. Whole blood was stained with FITC-CD45 and PE-CD14, lysed, washed, and fixed overnight in buffered 2% formaldehyde. **L, M, G, E, D,** and **P** indicate regions believed to be primarily lymphocytes, monocytes, granulocytes, erythrocyte ghosts, debris, and platelets/platelet aggregates, respectively. Ungated light-scatter (**A**) or immunofluorescence (**B**) distributions were collected and used to define gating regions believed to be primarily lymphocytes. Gated data were then accumulated (**C** = light scatter distribution of events in *region 2*, **D** = immunofluorescence distributon of events in *region 1*). [Data courtesy of Dr. B. Ohlsson-Wilhelm.] **A**, An ungated light-scatter distribution was collected and used to define a gating region believed to be primarily lymphocytes (*region 1*, low–moderate light scatter). **B**, Simultaneously with collection of the data shown in **A**, an ungated immunofluorescence distribution was collected and used to define a gating region believed to be primarily lymphocytes (*region 2*, CD45^{bright}CD14$^-$). Events to the left of the vertical threshold are counted a nonleukocytes; events above the horizontal threshold are counted as monocytes. **C**, *Light-scatter* distribution of events in *region 2*; overlay of *region 1* allows comparison with visually selected lymphocyte light scatter region. Note that some events judged to be lymphocytes on the basis of immunofluorescence characteristics were excluded from the visually selected light scatter gate (*region 2*). **D**, Immunofluorescence distribution of events in *region 1*; overlay of *region 2* allows estimation of contaminating cell types (here 0.3% monocytes, 3% non-leukocytes, 5% basophilc + granulocytes). For systems with autogating software, comparison of **A** and **C** aids in optimization of light scatter gates to include a many lymphocytes as possible, while attempting to limit contaminating cells. Comparison of the number of events in *region 2* for **B** and **D** allows estimation of the number of CD45^{bright}CD14$^-$ cells excluded from analysis when *region 1* is used as the light-scatter gate (here 6%). The goal in optimizing lymphocyte gates is to maximize the number of lymphocytes included while minimizing the number of contaminating cells (see text for further discussion on the compromises involved).

Even with the use of CD45+14 reagents, selection of optimal lymphocyte light-scatter gates may not be easy. Resolution of CD45^{bright}CD14$^-$ lymphocytes from CD45^{dim}CD14$^-$ granulocytes is affected by biological and methodological factors, including instrument optical configuration (particularly scatter angle collected for forward scatter) and relative frequency of specific cell types (i.e., the CD45^{bright}CD14$^-$ cluster may be difficult to identify in lymphopenic specimens or in specimens from AZT-treated patients; basophils exhibit light scatter similar to lymphocytes but are CD45^{dim}). Usually, a compromise must be made between including all cells believed to be lymphocytes and excluding all contaminating cell types. Typical compromises are to:

1) Set more conservative light-scatter gates (which then will not include all lymphocytes) to limit the number of contaminating cells, and accept the associated risk of excluding pertinent cells from analysis, or

2) set more generous light-scatter gates and attempt to correct for the proportion of contaminating cells when immunophenotypes are reported (i.e., if 20% CD4+ cells were observed when 85% of light scatter events are lymphocytes and if all positive events were believed to be lymphocytes, a corrected value of 20%/85% = 23.5% CD4+ lymphocytes would be reported). Unfortunately, since non–lymphocytes may constitute either false-positives (i.e., CD4+ monocytes) or false-negatives (i.e., CD4− erythrocytes) depending on the antigen of interest, simply correcting the denominator for contaminating cells may still not give an accurate result.

What about cases for which neither of the above compromises seems appropriate? The concept of using immunofluorescence gating to define cells of interest can again be helpful. For example, by using three-color immunofluorescence, and because the CD3 antigen is expressed only on T-lymphocytes, CD4+ T-cells can be clearly distinguished by setting gates that include all CD3+-cells and exclude CD14+-cells (21). A "cocktail" of immunofluorescent reagents may be used to simultaneously identify cell type(s) of interest and potential contaminating cell types in a single tube, maximizing both the speed and accuracy of subset determinations (22). In leukemic specimens, use of restrictive light scatter gates similar to those used for normal peripheral lymphocytes may well discriminate against the neoplastic population; similarly, CD45 intensities are often different for neoplastic vs. normal lymphocytes (10, 12). Gating on CD19/20 positive cells during analysis of surface immunoglobulin light chains allows ready exclusion of monocytes and improved sensitivity to low frequency monoclonal B-cell populations, without requiring restrictive light scatter gates that risk excluding neoplastic cells different in size from normal lymphocytes (23, Chapter 13). Again, more important than the particular gating method chosen is that it be a) appropriate to the clinical question at hand, b) validated using appropriate positive and negative control specimens, and c) used consistently and with a clear definition of acceptance criteria.

A second problem commonly encountered in immunofluorescence QC (though not always recognized) is that of discriminating against nonviable cells, which represent a significant source of potential false-positives due to their tendency to nonspecifically bind antibodies (see Chapter 9). This problem is further complicated by the desire to fix specimens prior to flow cytometric analysis in order to inactivate possible infectious agents (15, 24). As with light-scatter gating, there is no ideal method of dead-cell gating applicable to all types of specimens. The most commonly used flow cytometric definitions of "dead" are a) decreased forward light scatter intensity (25), b) the ability to exclude charged dye molecules such a ethidium bromide or propidium iodide (25, 26), and c) the inability to enzymatically cleave a nonfluorescent substrate (fluorescein diacetate or BCECF-AM) and retain the charged fluorescent product (27).

Method 1 (light scatter) works well for samples with relatively homogeneous cell sizes but much less reliably for those that are heterogeneous (25) and is further limited by the fact that light scatter resolution typically decreases after fixation. Methods 2 and 3 (dye exclusion or retention) define "live" cells as those having intact membranes impermeable to charged fluorescent dyes; therefore, neither method can be used in fixed samples where all cell membranes become permeabilized. Altered uptake of the dye LDS751 has been described as an indicator of dead cells in fixed specimens, but its mechanism of action and what is being defined as "dead" are unknown; using this method as many as 60% of peripheral leukocytes are defined as "damaged" after ammonium chloride lysis and 1% paraformaldehyde fixation (28). A recently described variant of Method 2 applicable to analysis of fixed cells utilizes labeling with ethidium monoazide (EMA) after preparation and staining but prior to fixation (29). Ethidium monoazide is a photoactivatable analog of ethidium bromide, which becomes covalently crosslinked to DNA after illumination with visible light and is therefore retained even after permeabilization of cell membranes during the fixation process. Because intensity of ethidium monoazide staining is lower than that of equilibrium staining with ethidium bromide or propidium iodide, ability to resolve dead (EMA+) cells from live phycoerythrin positive cells must be verified if EMA is to be added to all samples. An alternative is to use an EMA stained autofluorescence control to identify where dead cells fall in the light scatter distribution and to set light scatter gates to minimize the number of dead cells included (30), although this also introduces some risk of systematic bias since lymphocyte subsets are not homogeneously distributed throughout the light scatter distribution (16). Although ethidium monoazide is excluded by live peripheral leukocytes, low levels of uptake by some activated or actively proliferating cell types have been noted (J. Auger, personal communication). While this typically does not interfere with discrimination between live and dead cells, it may complicate discrimination between live immunofluorescence positive and negative cells. Ability of live cells to exclude EMA should therefore be verified when adapting this method to different specimen types (31).

What level of dead cell contamination is acceptable for immunofluorescence analysis? The only general answer that can be given is "a level that does not affect the reliability or accuracy of the reported result." Empirical comparison of results obtained from specimens analyzed with and without live cell gating (exclusion of dead cells) is probably the most practical way to answer this question. For example, in T-cell subset analysis of lysed whole blood preparations, we found that results from ungated analysis did not being to diverge from the range found for replicate samples analyzed using live cell gating until >15% dead cell events were included in the ungated analysis (32). Therefore, acceptance criteria were set so that any specimen in which light scatter gates could not be set to include fewer than 15% dead cells was flagged. In general, acceptance criteria will depend on the type of clinical specimen under analysis, the frequency and immunofluorescence staining intensity of the target popula-

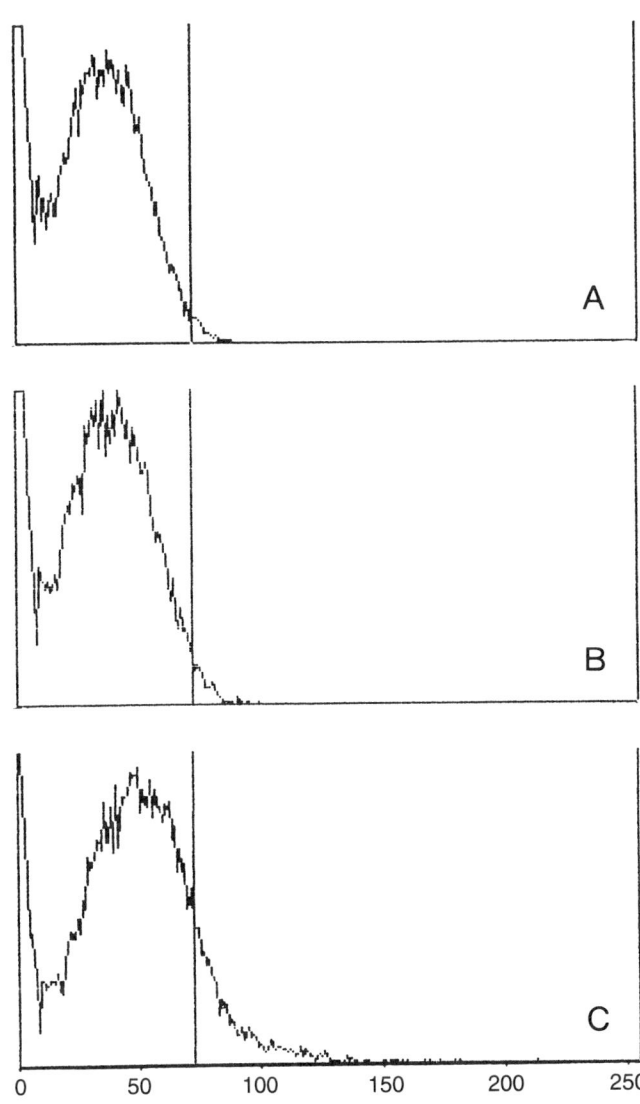

Figure 11.10. Variability in nonspecific binding estimated by different isotype controls. Whole blood from a single individual was incubated with the indicated reagents, lysed, washed, and analyzed, using light-scatter gating to accumulate lymphocyte fluorescence distributions. IgG_{2a} controls were used at equivalent concentrations based on information provided by the manufacturers. Effective fluorescence per antibody was the same for both isotype controls as estimated using Simply Cellular goat-anti-mouse beads (Flow Cytometry Standards Corp.) according to the manufacturer's instructions. **A**, Autofluorescence control (cells treated with buffer only); 99% of events below channel 73 threshold indicated by cursors. **B**, IgG_{2a} isotype control from manufacturer A: 98% of events below channel 73, 99% of events below channel 78. **C**, IgG_{2a} isotype control from manufacturer B: 88% of events below channel 73, 99% of events below channel 120.

tion, and the accuracy required for a clinically useful result. For example, where the frequency and/or staining intensity of true-positives are low, the levels of contaminating false-positive dead cells must necessarily be more stringently controlled. As with light scatter gating, it is important to validate the dead-cell gating method selected by analysis of both positive and negative controls and to set appropriate and consistently applied viability acceptance criteria.

Selection of appropriate analysis regions is also a significant source of intra- and inter-laboratory variablity in immunofluorescence analysis. Where the parameter to be reported is "percent-positive," a negative control sample is used to set an intensity threshold above which cells are regarded as specifically labeled. Since the negative control specimen is believed not to contain any true-positive cells, the threshold value is selected so that an acceptably low frequency of false-positive events (most typically <2%) fall above that threshold. The underlying assumption is that the types and levels of nonspecific staining for the negative control sample closely mimic those for the negative cells in the sample stained with test antibody (see Chapter 9 for further discussion of this and other immunofluorescence analysis methods). What is frequently forgotten is that subtype or isotype controls represent, at best, **estimates** of the nonspecific interactions of the test antibody. It is not uncommon to find that different control antibodies of exactly the same isotype give different threshold values even when they are of similar concentration and fluorochrome:protein ratio (Fig. 11.10). It is also frequently true that different individuals show different levels of nonspecific staining with a given control antibody, as judged by increase above the autofluorescence of a totally unstained sample (33). Rigid adherence to a threshold level set using an isotype control, although easy to set as a consistent criterion for use by different operators within a laboratory, appears to create more problems than it solves when the isotype control gives an intensity distribution significantly different than that observed for the negative region using the test antibody. A recent European interlaboratory sendout of stained fixed specimens found that different methods of selecting analysis regions was one of the major sources of variability in the test results reported, and that this was most pronounced in cases where the control reagent exhibited a significant level of nonspecific staining and where the test antibody exhibited relatively low intensity staining (34).

A related immunofluorescence analysis issue that deserves thought when a new test is being introduced is whether "percent-positive" is really the appropriate test statistic to report out. In cases where the biology or pathobiology gives rise to a clearly bimodal immunofluorescence distribution, percent-positive may indeed be the appropriate statistic, despite the fact that exactly where to set the analysis threshold is not always a simple decision (see above). However, in cases where a unimodal or skewed immunofluorescence distribution is present, it may be both more informative and more reproducible to describe the data in terms of shifts in intensity between samples stained with control and test antibodies. For tests where the underlying distribution is expected to be unimodal (analysis of antiplatelet or antigranulocyte antibodies, estimation of anti-HLA activity in transplant crossmatching), describing the data in terms of intensity shifts seems clearly preferable. For immunophenotyping of leukemic cells or other abnormal specimens, where the pathobiology may lead to unimodal, bimodal, or combination histograms, it is preferable to report the data out in a fashion that provides maximum information for correlation with other clinical data (12, 35). For example, reporting out a leukemic specimen as having 30% CD13 positive cells is ambiguous: this value could come either from a bimodal histogram in which 30% of the cells analyzed exhibited a fluorescence intensity 100X background or from a unimodal histogram in which 100% of the cells displayed a shifted fluorescence intensity but only 30% of them were brighter than the preponderance of cells in the negative control. The importance of using both negative (i.e., normal) and positive (i.e., abnormal) control specimens in selection of appropriate analysis methods and in training of technologists in consistent use of analysis regions cannot be overemphasized.

DNA Analysis QC

General issues of sample preparation, reagents, and staining methods, as well as data analysis techniques for nucleic acid analysis are discussed in Chapters 6, 8, and 10. Relevance of DNA aneuploidy and/or DNA cell cycle distribution, as well as sample preparation and reagent selection issues for specific pathologies are discussed in Section C. Discussion here will therefore be restricted to QC issues common to most DNA–based applications.

While issues of gating are at least as important for DNA analysis as for immunofluorescence, the selection of the most appropriate gating parameter(s) is very dependent on the tissue and disease process being evaluated. Due to the greater heterogeneity of both normal and abnormal cell types found in solid tumors, light scattering properties are perhaps more helpful discriminators in hematologic malignancies (12). However, right-angle light scatter has been described as a helpful gating parameter in some types of solid tumors, perhaps representing the flow cytometric analog of cytologic "nuclear irregularity" (36; C.B. Bagwell, personal communication). Immunofluorescence also serves as a helpful gating parameter. Cytoplasmic antigens (immunoglobulins, cytokeratins, etc.), nuclear antigens (steroid hormone receptors, oncogenes, proliferation-associated antigens), and surface antigens (immunoglobulins, T- or B-cell markers, P-glycoprotein) can be used either for positive identification of neoplastic cells or for discrimination against admixed normal cells. Both positive control (known abnormal) and negative control (known normal) specimens are required to validate that gating parameter(s) are correctly identifying abnormal cells that are then to be analyzed for DNA aneuploidy or cell cycle distribution.

Use of positive and negative control specimens is also crucial for the optimization and daily monitoring of staining conditions and instrument performance. Fluorescent dyes used as DNA stains vary in the extent to which they reflect DNA content (pg of DNA per cell) vs. DNA "stainability," since chromatin condensation and accessibility of DNA binding sites varies with tissue type, differentiation state, and degree of quiescence (37) (see Chapter 2). Staining conditions that produce optimal DNA accessibility may not be those that provide optimal retention of antigenic markers used for gating and vice versa. Preparation of nuclei by hypotonic lysis minimizes cytoplasmic staining but disallows use of surface or cytoplasmic markers and may destroy some nuclear antigens; fixatives that crosslink help preserve antigenic structures but impede the slight DNA "stretching" required for optimal staining by intercalating dyes, etc. (38). Control specimens allow assessment of the range of conditions (specimen holding conditions, cell concentration, stain concentration, length and type of enzyme digestion, sheath and sample flow rates, etc.) under which the desired discrimination between normal and abnormal cells is maintained and provide validation that those conditions have been met each day.

Relevant normal and abnormal specimens are of course the ideal control materials, but both may be difficult to obtain, particularly in sufficient quantity to carry out optimization studies. A commonly used alternative normal control is peripheral blood "lymphocytes" (usually mononuclear cells) from a healthy individual, although it is somewhat arguable how well they approximate either quiescent stromal cells or activated immune cells found in tumor samples. In some cases, it is possible to construct control materials that resemble abnormal specimens in at least certain critical respects, and this may be helpful in minimizing the amount of patient material required for abnormal controls. For example, Figure 11.11 illustrates the use of peripheral blood mononuclear cells ("normal" control) and an admixture of the same cells with a cultured colon carcinoma cell line (LoVo, "abnormal" control) to troubleshoot sample preparation effects on DNA Index determination. It also illustrates the effect of enzymatic digestion on "stainability" of DNA with propidium iodide, as well as the difficulties of using an admixed "internal standard" to verify the location of the DNA-diploid population in a heterogeneous sample. When both the internal standard (normal mononuclear cells) and the abnormal (LoVo) cells are held and prepared identically, the DNA Index obtained for frozen enzyme-digested "specimen" is identical to that obtained for fresh "specimen". However, when they are not treated in the same fashion (i.e., when one cell type is frozen and made more accessible to enzyme prior to digestion and staining but the other is not), the resulting DNA index is altered significantly. A similar effect was observed when the abnormal specimen used was a frozen breast tumor: If freshly prepared (unfrozen) mononuclear cells were admixed with tumor prior to sample preparation by mincing and enzyme digestion, the mononuclear cells did not coincide with what were believed to be the diploid cells in the specimen; if frozen mononuclear cells were used, the two peaks were coincident. Extending this principle to analysis of archival specimens suggests that it will be difficult to find a control material that mimics the effects of fixation, paraffin embedding, rehydration, and en-

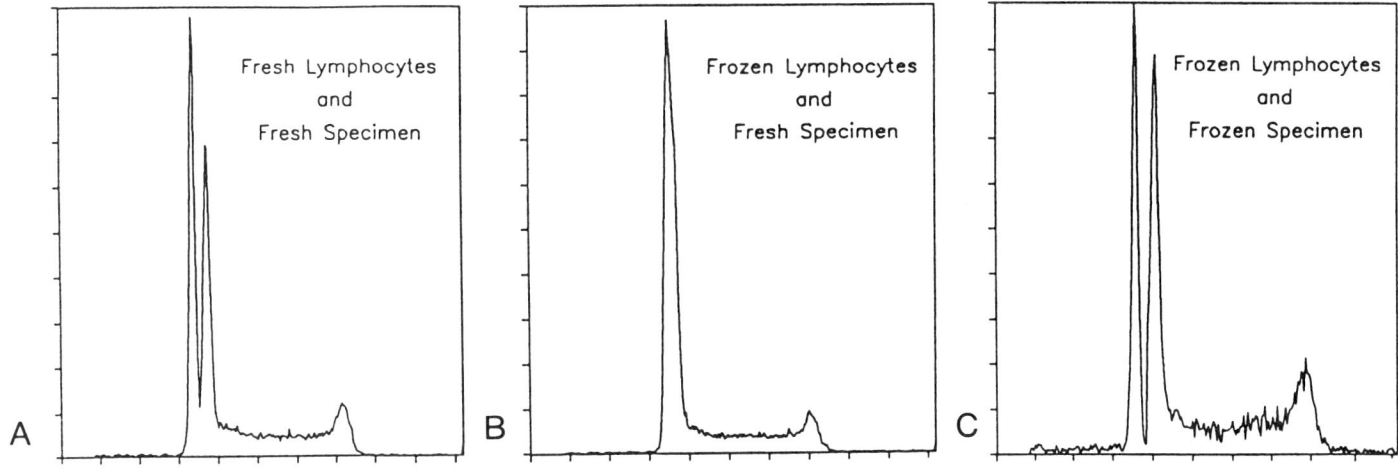

Figure 11.11. Use of composite control material to evaluate preparation effects on DNA Index determination. Peripheral blood mononuclear cells were prepared by density gradient separation and admixed with a "specimen" consisting of cultured colon carcinoma cells (LoVo, DNA Index 1.13±0.02). The mixture was then carried through a collagenase digestion protocol being investigated for preparation of frozen breast biopsies. Data was accumulated on a 256-channel scale using linear amplification. (Data courtesy of Paul K. Wallace.) **A**, freshly-prepared mononuclear cells and LoVo were admixed prior to collagenase digestion and staining with hypotonic propidium iodide/RNAase. **B**, mononuclear cells were frozen and thawed prior to admixture with freshly cultured LoVo, following which the mixture was digested with collagenase and stained. Increased permeability to collagenase after freezing and thawing caused increased DNA stainability for mononuclear cells and loss of resolution between lymph and LoVo peaks. **C**, freshly-prepared mononuclear cells and LoVo were separately frozen and thawed prior to admixture, collagenase digestion, and staining. Increased DNA stainability was seen for both cell types but DNA Index was identical to that obtained for a mixture of fresh cells (upper panel).

zymatic digestion of cross-links. At present, the only alternatives are either to assume that the lowest intensity peak is the DNA diploid population (quick but risky) or to microdissect and prepare in parallel normal and abnormal regions from the same paraffin block (safer, but tedious and not always possible). Thus, a practical control specimen useful in identifying the expected location of DNA-diploid cells in an archival specimen remains a major need in DNA quality control.

The type of control material described in Figure 11.11 can also be helpful in identifying optimal instrument conditions for specimen analysis. Flow rates and optical alignment conditions that produce minimum CVs for microspheres are not necessarily the same as those that produce minimum CVs for cell preparations, particularly when the size and shape of cells differ substantially from those of alignment spheres. In addition, microsphere CVs are insensitive to instrument factors that can perturb DNA-staining equilibrium. Figure 11.12 illustrates the effect of slight turbulence during sample startup on G_1 CV and ability to resolve small differences in DNA Index using the lymph+LoVo "abnormal control." It also illustrates the value of another quality control parameter that is available on an increasing number of instruments but is much underutilized—time. Despite our best efforts to minimize aggregation and consequent flow disturbances, they do occur. Collection of list mode data in which time is included as a parameter allows retrospective review and excision of artifactual data arising from formation and resolution of flow disturbances during a sample run.

Finally, the utility of artificial control materials in identifying and monitoring optimal preparation and analysis conditions is a direct function of how well they imitate the critical properties of patient specimens. The lymph+LoVo mixture described above does not imitate the level of debris found in either mechanically or enzymatically disaggregated solid tumor specimens. And it would be a poor control for monitoring adequacy staining and analysis in a protocol using cytoplasmic immunoglobulin staining in conjunction with DNA staining, although a different cell line with more appropriate antigenic characteristics could certainly be selected if desired. Again, it is encouraging to see the beginning of commercial availability of biological control materials in this area (Table 11.2). Consensus on how frequently it is necessary to prepare and run normal and abnormal patient specimens is even less well established for DNA analysis than for immunofluorescence analysis. However, clear definition of what is being controlled for, what material(s) are appropriate for each purpose, and the range representing acceptable performance are as crucial to successful QC for DNA analysis as they are for any other laboratory test.

SUMMARY: KEYS TO SUCCESSFUL QUALITY CONTROL

The most fundamental commandment for successful quality control is "know your own system." This will require understanding the ways in which things can go wrong, either biologically or instrumentally, and identifying materials and methods for verifying that system performance is acceptable

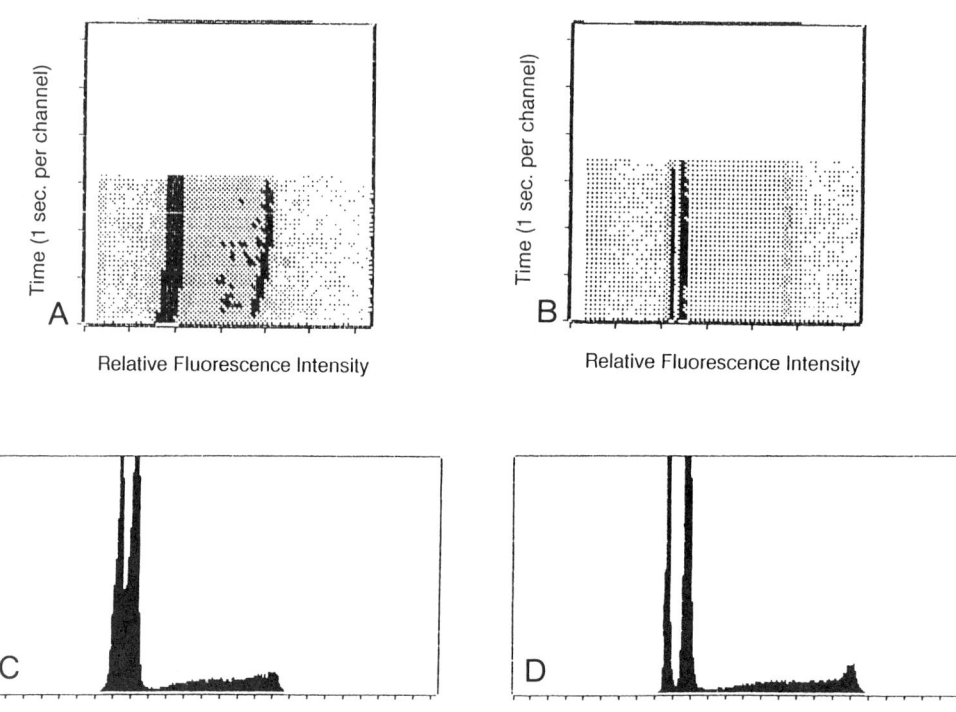

Relative Fluorescence Intensity

Figure 11.12. Use of time as a quality control parameter during data accumulation. A sample prepared as described in Figure 10.11A was analyzed by accumulating fluorescence intensity (linear amplification) and time (1 second per channel) on linear 256-channel scales. **A**, fluorescence vs. time data acquired during the first 2 min after introduction of sample into the instrument. **B**, single-parameter DNA fluorescence histogram for data taken during the first 2 min after introduction of sample into the instrument. **C**, fluorescence vs. time data for the same sample during the third and fourth minutes after sample introduction (sample was allowed to continue running and data acquisition restarted at end of the time shown at upper left). **D**, single-parameter DNA fluorescence histogram for data taken during time period shown at upper right.

on a daily basis. It is particularly valuable if QC materials can be selected to allow distinction among specific ways in which performance may be degraded (Table 11.3). For example, poor resolution of positive and negative immunofluorescence may arise from suboptimal instrument performance (i.e., inadequate fluorescence sensitivity), poor specimen quality (i.e., significant autofluorescence background, high proportion of dead cells), poor reagent quality (i.e., high levels of nonspecific binding, bleaching of fluorochrome), excessive background due to free antibody in "no-wash" procedures, or a combination of the above. A material such as beads containing zero vs. low levels of appropriate fluorochrome can be used to test whether instrument conditions (flow rate, laser power, filter(s), detector high voltages, etc.) are adequate to resolve labeled from unlabeled beads. If blank beads are resolvable from dim ones but unstained cells are not resolvable from stained ones, inadequate instrument performance can clearly be absolved from blame and attention should be focused on specimen or reagent problems. If a positive control specimen gives adequate resolution but the patient specimen does not, reagent performance problems can be eliminated from consideration—assuming, of course, that the positive control specimen has been selected to have an epitope density similar to test specimens so that degradation of reagent performance will have similar effects on both types of samples. If no positive control was run, it is, of course, impossible to say whether the poor result in the test specimen was a reagent problem or not!

The most time- and cost-effective quality control scheme is clearly one in which the maximum amount of information about total system performance is obtained from the minimum number of control materials/specimens. What does this say about strategy in selection and use of such materials? Reference materials used for standardization or calibration of instrument performance should be as sensitive as possible to instrument parameters being optimized for a particular assay. In general, this means selection of materials as similar as possible in size and intensity to the specimen being assayed. Similarity in size helps assure that hydrodynamics and light scatter detection are optimal and reproducible. Similarity in intensity helps assure that fluorescence detection (alignment, filters, and detector sensitivity) is optimal and reproducible. Biological control materials should likewise be as similar as possible to test specimens, so that they exhibit similar responses to factors affecting viability, staining, gating, and/or the interaction of those factors with instrumental variables. Any flow cytometer has certain built-in assumptions and compromises in both hardware and software that affect the

Table 11.3
Use of QC Materials for Troubleshooting

Problem	Possible Cause	Observed Effect on QC Material(s)
Poor resolution of positives from negatives in immunofluorescence measurements	Poor optical alignment, inefficient signal collection	Alignment particles out of limits (mean channel values low and/or high voltage settings high); CVs may be high
	Filter deterioration or detector noise causing increased background	"Blank" sample (beads or unstained cells) out of limits, high for one parameter but not for others; reference particles or reagents that normally give good resolution fail to do so
	Inappropriate color compensation settings (particularly overcompensation)	Correlated values observed for intensities of color 1 (i.e., FITC) and color 2 (i.e., PE) for particles labeled with only one color; easiest to see when multiple-intensity levels of fluorochrome are present
	High background due to carryover of fluorochromes (i.e., propidium iodide) that an label fixed cells	"Blank" bead sample OK, "blank" fixed cell sample out of limits, high for one color but not others
	High levels of nonspecific staining due to dead cells, inappropriate antibody titer, insufficient washing, etc. (see Chapter 9)	Resolution of "dim" from "blank" particles OK, but "negative" cells in sample give significantly higher intensities than unstained cells
	Loss of specificity, decreased titer, or loss of fluorochrome from test antibody	Positive-control specimen expected to show reactivity with reagent fails to do so; instrument and nonspecific reagent controls OK
	Increased autofluorescence of negatives and/or loss of specific fluorescence due to fixation	Instrument and reagent controls OK; fixed samples show decreased positive intensity or increased negative intensity compared with fresh when run on standardized intensity scale
	Level of antigen on positive cells too low to give complete separation from autofluorescence/nonspecific fluorescence of negatives using signal amplification method employed	Instrument, negative reagent controls and positive reagent controls OK; try indirect staining or alternate fluorochromes that minimize contribution from autofluorescence
Poor resolution of DNA aneuploid populations	Poor optical alignment, uneven illumination	Alignment particle CVs out of limits high for all parameters (light scatter may be more sensitive than fluorescence) but minimally affected by sample input rate
	Poor hydrodynamic focusing, uneven illumination	Alignment particle CVs out of limits high and increase with run time; mean fluctuates with time
	Nonlinear instrument response (amplifier and/or PMT)	Nonlinear correlation between mean channel values and assigned relative intensity values for calibrated particles *or* ratio between two closely spaced particles is not constant across the scale
	Non–equilibrium staining conditions	Instrument controls OK and give stable values over time of run but cell intensities drift over time
	Nonlinear biological response due to high levels of non-DNA-related staining (i.e., incomplete RNAase digestion), partial degradation of test sample, etc.	Instrument controls OK; biological controls, expected to give a single narrow peak (i.e., normal lymphocytes), give broadened and/or skewed peaks when processed in parallel with test sample

results obtained. A manufacturer can provide materials that determine whether a given instrument is operating according to established specifications, but if those specifications are inappropriate for the type of specimen being analyzed, the resulting data may nonetheless be inaccurate.

Achieving the goal of accurate and reproducible flow cytometric test results often requires determining which of several different factors are critical for the test at hand. First is sensitivity: How small a signal can accurately be distinguished from background "noise"? Note that "noise" may be instrumental (i.e., spurious optical or electronic signals) or biological (i.e., autofluorescence, nonspecific binding of reagents). Second is resolution: How reliably can cells belonging to different populations be distinguished from each other? This will be a function of biological heterogeneity within populations of interest as well as variability associated with the measurement process. Third is standardization: How reproducibly do samples with the same properties give rise to the same measured values, even if those values cannot be assigned absolute units? Fourth is calibration: In what units are measured values or differences expressed and how reproducible are those units from laboratory to laboratory? Note that in answering any of these questions, the experimental unit for purposes of reproducibility is the sample, not the cell. There is a tendency to believe that because we can rapidly analyze thousands to millions of cells by flow cytometry that the answer is more reliable than that obtained by counting hundreds microscopically. This would be true if the only source of error were statistical counting error. However, the variability introduced by sampling, preparation, staining and measurement is generally much greater than counting error. Good quality control therefore requires: a) identification of the step(s) responsible for greatest variability, b) optimization of the system to control that variability, and c) establishment of acceptance criteria that reflect the sources of variability of greatest impact on the final test result.

Fundamentally, of course, quality control for flow cytometry is much like quality control for any other clinical test: The goal is to start with an adequate and representative specimen and to assure that as little bias as possible is introduced during the preparation, staining, measurement, and analysis processes. However, quality control for flow cytometry differs from other more established technologies in that consensus methods are still being identified and commercially available materials are still being developed. Results of recent interlaboratory surveys suggest a good news/bad news situation. Immunophenotyping performance surveys find generally good agreement on values for percent-positive cells using markers that are well resolved; they find much greater disagreement on tests where markers are less well resolved (34) or where quantitative fluorescence intensities must be reported (39). Similarly, DNA analysis performance surveys find generally good agreement on DNA Index determinations for well resolved peaks but much poorer agreement on cell cycle distribution parameters (40). It is precisely through comparative surveys using well-designed test and control specimens that some of the systematic biases introduced by procedural differences between laboratories can be identified and resolved. However, until the consensus process is considerably further along, responsibility will remain with the individual laboratory to assure its own consistency and accuracy. In meeting this goal, there is no substitute for the combination of carefully selected reference materials and controls, established acceptance criteria, and the thoughtful use of both by an observant technologist.

ACKNOWLEDGMENTS

The author gratefully acknowledges many stimulating discussions with fellow faculty members and participants in the Flow Cytometry Applications courses over the years. Particular thanks are due to Anne Hurley, Bruce Bagwell, Betsy Ohlsson-Wilhelm, Paul Wallace, Carleton Stewart, Abe Schwartz, Greg Stelzer, and Wayne Green for thoughtful commentary on various issues covered in this chapter.

REFERENCES

1. Steen HB. Characteristics of flow cytometers. In: Melamed MR, Lindmo T, Mendelsohn ML, eds. Flow cytometry and sorting. 2nd ed. New York: Wiley-Liss, 1990:11–25.
2. Hiebert RD. Electronics and signal processing. In: Melamed MR, Lindmo T, Mendelsohn ML, eds. Flow cytometry and sorting. 2nd ed. New York: Wiley-Liss, 1990:127–144.
3. National Committee for Clinical Laboratory Standards. Internal quality control testing: principles and definitions; approved guideline. Villanova, PA, 1991: NCCLS Document C24-A.
4. Brown MC, Hoffman RA, Kirchanski S. Controls for flow cytometry in hematology and immunology. Ann. N.Y. Acad. Sci. 1986;468:93–103.
5. Schmid I, Schmid P, Giorgi J. Conversion of logarithmic channel numbers into relative linear fluorescence intensity. Cytometry 1988;9:533–538.
6. Muirhead KA, Schmitt TC, Muirhead AR. Determination of linear fluorescence intensities from flow cytometric data accumulated with logarithmic amplifiers. Cytometry 1983;3:251–256.
7. Vogt RF Jr., Cross GD, Henderson OL, Phillips DL. Model system evaluating fluorescein-labeled microbeads as internal standards to calibrate fluorescence intensity on flow cytometers. Cytometry 1989;10:294–302.
8. Bagwell CB, Baker D, Whetstone S, et al. Simple and rapid method for determining the linearity of a flow cytometer amplification system. Cytometry 1989;10: 689–694.
9. Horan PK, Muirhead KA, Slezak SE. Standards and controls in flow cytometry. In: Melamed MR, Lindmo T, Mendelsohn ML, eds. Flow cytometry and sorting. 2nd ed. New York: Wiley-Liss, 1990:397–414.
10. Caldwell CW, Patterson WP, Hakami N. Alterations of HLE1 (T200) fluorescence intensity on acute lymphoblastic leukemia cells may relate to therapeutic outcome. Leuk Res 1987;11:103–106.
11. Hurwitz CA, Loken MR, Graham ML, Karp JE, Borowitz MJ, Pullen DJ, Civin CI. Asynchronous antigen expression in B lineage acute lymphoblastic leukemia Blood 1988;72:299–307.
12. Duque RE, Braylan R. Applications of flow cytometry in diagnostic hematopathology. In: Coon JS, Weinstein RS, eds. Techniques in Diagnostic Pathology: Diagnostic Flow Cytometry. Baltimore: Williams & Wilkins, 1991:89.
13. Vogt RF, Marti GE, Schwartz A. Quantitative calibration of fluorescence intensities for clinical and research applications of immunophenotyping by flow cytometry. In: Tyrer HW, ed. Critical Issues in Biotechnology and Bioengineering, Vol. 1. Norcross, NJ: Ablex Publishing Co., (in press).
14. Loken MR, Parks DR, Herzenberg LA. Two-color immunofluorescence using a fluorescence-activated cell sorter. J Histochem Cytochem 1977;25:899–907.
15. National Committee for Clinical Laboratory Standards. Clinical applications of flow cytometry: quality assurance and immunophenotyping of peripheral blood lymphocytes; proposed guideline. Villanova, PA; 1989: NCCLS Document H42–P.
16. Loken MR, Brosnan JM, Bach BA, Ault KA. Establishing optimal lymphocyte gates for immunophenotyping by flow cytometry. Cytometry 1990;11:453–459.
17. Salzman GC, Crowell JM, Martin JC, et al. Cell classification by laser light scattering: identification and separation of unstained leucocytes. Acta Cytol 1975;19:374–386.
18. Hoffman RA, Kung PC, Hansen WP, Goldstein G. Simple and rapid measurement of human T lymphocytes and their subclasses in peripheral blood. Proc Nat Acad Sci USA 1980;77:4914–4917.
19. Stewart CC, Stewart SJ, Habersett RC. Resolving leukocytes using axial light loss. Cytometry 1989;10:426–432.
20. Jackson AL, Warner NL. Preparation and analysis by flow cytometry of peripheral blood leukocytes. In: Rose NR, Friedman H, Fahey JL, eds. Manual of Clinical Laboratory Immunology. Washington, DC: American Microbiol Assoc, 1986:226–235.
21. Mandy FF, Bergeron M, Izaguirre CA. Application Tools for clinical flow cytometry: Gating Parameters for immunophenotyping. Clin Immunol. Newsletter 1992;12:25–32.
22. Liu C-M, Muirhead KA, George SP, Landay AL. Flow cytometric monitoring of human immunodeficiency virus-infected patients: simultaneous enumeration of five lymphocyte subsets. American J Clin Pathol 1989;92:721–728.
23. Letwin BW, Wallace PK, Muirhead KA. An improved clonal excess assay using flow cytometry and B-cell gating. Blood 1990;75:1178–1185.
24. Cory JM, Ohlsson-Wilhelm BM, Eyster ME, Rapp F. Effects of cell fixatives used for flow cytometry on the infectivity of human immunodeficiency virus-1. Cytometry 1990;11:647–651.
25. Horan PK, Loken MR. A practical guide for the use of flow systems. In: VanDilla MA, Dean PN, Laerum OD, Melamed MR, eds. Flow Cytometry: Instrumentation and Data Analysis. New York: Academic Press, 1985:260–280.
26. Sasaki DT, Dumas SE, Engleman EG. Discrimination of viable and non-viable cells using propidium iodide in two color immunofluorescence. Cytometry 1987;8:413–420.

27. Dive C, Cox H, Watson JV, Workman P. Polar fluorescein derivatives ad improved substrate probes for flow cytoenzymological assay of cellular esterases. Mol and Cel Probes 1988;2:131–145.
28. Terstappen LWMM, Shah VO, Conrad MP, Recktenwald D, Loken MR. Discriminating between damaged and intact cells in fixed flow cytometric samples. Cytometry 1988;9:477–484.
29. Riedy MC, Muirhead KA, Jensen CP, Stewart CC. Use of a photolabeling technique to identify nonviable cells in fixed homologous or heterologous cell populations. Cytometry 1991;12:133–139.
30. Stewart CC. Clinical applications of flow cytometry: immunologic methods for measuring cell membrane and cytoplasmic antigens. Cancer, in press, 1992; Cancer 69(Suppl. 6)1543–1552.
31. Jensen BJ, Muirhead KA. Flow Cytometry Methods. In: Howard, G., ed. Methods in Non-Radioactive Detection. New York: Elsevier, in press, 1992.
32. Muirhead KA, Wallace PK, Schmitt TC, Frescatore RL, Franco JA, Horan PK. Methodological considerations for implementation of lymphocyte subset analysis in a clinical reference laboratory. Ann NY Acad Sci 1986;468:113–127.
33. Ohlsson-Wilhelm BM, Cory JM, Kessler HA, Eyster ME, Rapp F, Landay A. Circulating human immunodeficiency virus (HIV) p24 antigen-positive lymphocytes: a flow cytometric measure of HIV infection. J Infect Diseases 1990;162:1018–1024.
34. Martini E, D'Hautcourt J-L, Brando B, Lawry J, O'Connor J-E, Sansonetty F. First European quality control of cellular phenotyping by flow cytometry - 1989. Paris, 1990: Editions Frison-Roche.
35. Duque RE, Everett ET, Iturraspe J. Flow cytometric analysis of acute leukemias. Clinical Immunol Newsletter 1990;10:43–50.
36. Benson MC, McDougal DC, Coffey DS. The application of perpendicular and forward light scatter to assess nuclear and cellular morphology. Cytometry 1984;5:515–522.
37. Darzynkiewicz Z, Traganos F, Kapuscinski J, Staiano-Coico L, Melamed MR. Accessibility of DNA in situ to various fluorochromes: relationship to chromatin changes during erythroid differentiation of Friend leukemia cells. Cytometry 1984;5:355–363.
38. Schmid I, Uittenbogaart CH, Giorgi JV. A gentler fixation and permeabilizaiton method for combined cell surface and intracellular staining with improved precision in DNA quantification. Cytometry 1991;12:279–285.
39. Paxton H, Kidd P, Landay A, et al. Results of the flow cytometry ACTG quality control program: analysis and findings. Clin Immunol Immunopathol 1989;52:68–84.
40. Coon JS, Deitch AD, de Vere White RW, et al. Interinstitutional variability in DNA flow cytometric analysis of tumors: the National Cancer Institute's flow cytometry network experience. Cancer 1988;61:126–130.

PART II.

Clinical Application

SECTION C. APPLICATIONS IN CLINICAL ONCOLOGY

12

Lymphomas

RAUL C. BRAYLAN

INTRODUCTION

Lymphomas are tumors of the immune system presenting primarily in the lymph nodes, spleen, bone marrow, mucosa-associated lymphoid tissues and, less frequently, in the skin or solid viscera. Lymphoma cells may occasionally also be detected in large numbers in the circulation.

The basic method used in the diagnosis and classification of lymphomas is still microscopic observation. Over the years, a multiplicity of histological schemes have been utilized for the classification of these tumors. Unfortunately, these morphological approaches are not universally accepted and applied and have been repeatedly shown to lack acceptable reproducibility. One should realize that the complex biology of neoplasms can only be partially appreciated by microscopy. As we gain more knowledge and apply more objective, quantitative, and biologically relevant diagnostic tools, characterization of the lymphomas based purely on hematoxylin-eosin-stained tissue sections is proving to be increasingly unsatisfactory and unreliable.

Among the emerging diagnostic technologies applied to lymphoid pathology, flow cytometry (FCM) is rapidly gaining popularity and is becoming an essential adjunct to more popular and established diagnostic methods. Flow cytometry has been used in the analysis of human lymphomas since the late 1970's. The earlier studies dealt mainly with the measurement of nucleic acids and, in particular, DNA. Later, as monoclonal antibodies became available, numerous investigators who used the fluorescence microscope for the analysis of surface antigens in lymphoid cells turned to the flow cytometer for their measurements.

With few exceptions, laboratories using flow cytometry in the study of lymphomas segregated into two groups: those primarily interested in the measurement of nucleic acid content and those involved mainly in the analysis of surface antigen expression. As it will become apparent in this chapter, both cellular DNA content and antigen measurements provide useful information on lymphomas and the two should be integrated for the proper assessment of these tumors.

In the following sections we will describe methodological aspects of sample preparation and analysis as applied specifically to lymphoproliferative processes. We will then discuss data presentation and interpretation, significance of results, correlations with histological diagnosis, and, finally, we will address practical considerations with regard to the application of this technology to the detection and classification of lymphomas.

SAMPLE PREPARATION

When considering applications for flow cytometric analysis in the study of lymphoid neoplasms, it must be kept in mind that flow cytometry is only one of many technologies that are useful in the characterization of these diseases. Thus, the initial handling of clinical specimens should include various preparation and preservation procedures that ensure that the appropriate morphological, immunological, and genetic studies can be performed. For solid tissues, the following is a list of the various technical steps recommended for the complete characterization of lymphoid neoplasms.

a. Touch imprints for cytological observation
b. Formalin fixation for routine histology and (limited) immunohistochemistry
c. Snap freezing for immunohistochemistry and molecular genetics
d. Ethanol fixation for molecular genetics
e. Glutaraldehyde fixation for electron microscopy
f. Cell suspension preparation for:
    Immunophenotyping
    Ploidy and cell cycle fraction calculations
    Cytogenetics
    Molecular genetics
    Cytospin preparations for cytological observations and immunocytochemistry

Lymphoid cells from blood, bone marrow, and other fluids are already in suspension. Thus only the last step (f) above, applies to these specimens.

For best results in immunophenotypical and DNA analysis, samples should be obtained fresh. Unfortunately, formalin-fixed tissues cannot be used for a complete flow cytometric analysis. Although DNA content analysis may be performed with nuclei extracted from tissues fixed in formalin and embedded in paraffin, fixation modifies cellular antigens and prevents intact cell dispersion.

The size of the tissue sample required for an adequate viable cell yield depends on a variety of factors, such as the amount of fibrous tissue or necrosis in the sample and the cellularity of the tumor. The cell yield and viability are better in low-grade tumors than in rapidly proliferating lymphomas. In general, a sample of at least 0.1 cubic cm is sufficient, but adequate cell yields may even be obtained from smaller specimens and sometimes from material from needle aspirates (1).

Cells can be isolated from bone marrow aspirates or biopsies. It is important to realize that a variable amount of peripheral blood usually contaminates bone marrow aspirates. This contamination, which is seldom appreciated, may significantly alter the interpretation of immunophenotype and DNA analysis in bone marrow aspirates. Cells obtained from bone marrow biopsies, on the other hand, are more representative of the marrow compartment. Simple methods for mechanical disaggregation of marrow core biopsies have been described (2).

In contrast with other solid tumors, lymphoma cells in lymphoid tissues are not firmly attached to each other or to stroma, allowing relatively easy dispersion into single-cell suspensions by simple mechanical means. Studies performed on the same tissues comparing immunophenotypical analysis by flow cytometry and by frozen section immunohistology showed a good correlation of results, suggesting minimal selective cell loss (3, 4). However, it must be kept in mind that low recovery of neoplastic cells may occur when tumor cells do not disaggregate by mechanical dispersion or when they are fragile and are preferentially lost during preparation of the suspension. This may be more often the case in large-cell lymphomas, even when enzymatic digestion is used to isolate cells (5). Also, poor tumor representation may occur when there is focal involvement or in certain lymphomas, such as Hodgkin's disease, the so-called T-rich B-cell lymphoma, or other tumors in which normal cells outnumber neoplastic cells. In these cases, the malignant nature of the process may be undetected. For this reason, a portion of the final cell suspension is routinely used to make cytocentrifuge preparations or smears for microscopic observation. This useful morphological control is then compared with tissue imprints or sections of the original tissue to assess selective loss, a practice that minimizes sampling errors.

The analysis of very small samples may be facilitated by micromethodologies (6) and new fluorochromes that permit the combination of three fluorescent reagents in a single container for the labeling of small numbers of cells.

A frequent concern is sample storage and transportation. Immunophenotypical analysis by flow cytometry requires that samples be processed soon after they are obtained. Since this practice is not always possible when laboratories are distant from the surgical or procedural site, specimens often need to be stored temporarily and/or transported to specialized centers. We recommend that samples be kept and transported at 4°C, wrapped in a gauze soaked in isotonic saline or cell culture medium. In our experience, cell viability may deteriorate with time but this varies greatly depending on the type of tissue and the grade of tumor. When different pieces from the same lymphoma had to be analyzed over periods of two to three days by our laboratory, we never observed modifications in antigenic expression that would change the interpretation of results. However, controlled studies on cell surface antigen expression in unfixed lymphoma tissues during prolonged storage have not been adequately performed. An alternative procedure is to store frozen cells for future analysis.

SAMPLE STAINING

If red cells, granuloytes, or nonviable cells are present in the suspension, they can be removed by centrifugation through ficoll-hypaque solution. However, for surface antigen analysis this step may not be necessary if red cells are lysed and granulocytes and dead cells, which are permeable to fluorescence dyes such as propidium iodide, are eliminated from the analysis by appropriate gating on dual-light-scatter or fluorescence signals, respectively. DNA content analysis of nonviable cells is feasible, in our experience, if the loss of viability is recent and the DNA remains intact. Blood and bone marrow samples containing abundant granulocytes may show artifactual near-diploid DNA-aneuploid peaks, particularly if the cells are ethanol-fixed prior to staining.

Immunostaining

Cell surface antigen staining is performed by direct or indirect immunofluorescence methods and a variety of quality antibodies already labeled with fluorescein isothiocyanate (FITC) or phycoerythrin (PE) are available commercially. Both single-color and dual-color immunofluorescence may be used with these reagents, the latter being particularly useful in the analysis of coexpression of antigens.

For the diagnosis and characterization of lymphomas, antibodies are usually assembled in "panels." The composition of these panels has not been standardized and varies widely. The design of the panels depends on the philosophy of the professional interpreting the results, the questions to be answered, the clinical and other laboratory information, the number of cells in the sample, and other practical matters, such as the number of samples handled by the laboratory and cost concerns.

There are basically two approaches to this issue: One is to apply initially a sufficiently extensive assortment of reagents to allow a complete characterization of the neoplasia, regardless of any other consideration. The other approach is to first perform a limited analysis with a minimum of antibodies aimed at answering simple or directed questions. If necessary, one then returns to the sample for additional analysis with antibodies chosen on the basis of the initial results. Although the latter strategy is probably less expensive and less sample-consuming, it requires the initial input of a professional in the selection of antibodies and may need time-consuming sequential sample staining and analysis. Also, the use of many antibodies is always prudent since neoplastic lymphocytes often display an aberrant expression of surface antigens that may not be detectable if only a few reagents are used.

To lower the cost of the use of multiple antibodies, cells may be stained and analyzed in the wells of microtiter plates instead of conventional tubes. This strategy helps reduce the

Table 12.1
Antibodies Useful in the Diagnosis and Classification of Lymphomas

	Dual-color antibodies (PE[a]/FITC[b])	
Antigens	*CD19/KAPPA*	*CD3/CD7*
	CD19/LAMBDA	*CD5/CD20*
	CD20/KAPPA	CD8/CD4
	CD20/LAMBDA	
Controls[c]	MoPE-IgG$_1$/MoFITC-IgG$_2$	MoFITC-IgGp$_1$/MoPE-IgG$_2$

	Single-color antibodies					
Antigens	CD45	CD19	CD20	CD10	HLADR	
	CD3	CD2	CD7	CD5	CD4	CD8
	CD14	CD71	IgG	IgM	IgA	IgD
Controls[c]	MoIgG$_1$	MoIgG$_2$	GtIgG[d]			

[a]PE: phycoerythrin
[b]FITC: fluorescein isothiocyanate
[c]Mo: mouse
[d]Gt: goat

volume of reagents and cells and eliminates the handling of individual tubes, which lessens technician time and error. Furthermore, using commercially available robotic devices, cells in microtiter plates can be automatically and sequentially resuspended, aspirated, and injected into the fluidics line of the flow cytometer, allowing for a virtual "hands-off" operation during cell analysis.

The standard panel of antibodies used in our laboratory is shown in Table 12.1 and includes monoclonal and polyclonal reagents with specificities against a wide range of lymphocyte and activation/proliferation-related antigens. CD-45 is widely used in the differential diagnosis between lymphoma and metastatic tumors (7). CD19, and CD20 are broadly reactive anti-B-cell antibodies. In particular, the level of expression of CD20 is valuable because it is associated with different tumor types (8). Antibodies against immunoglobulin light and heavy chains are also very helpful in detecting B-cell lymphomas. The pattern of heavy chain expression is associated with histological grade in non-Hodgkin's lylmphomas (9). CD3, CD2, CD7, and CD5 detect most T-cell lymphomas and CD8, CD4, and HLA-DR also help to recognize these tumors. CD14 may detect the extremely infrequent lymphomas of macrophage origin and serves as a good marker for monocytes in blood and bone marrow specimens. CD71 is useful since the lack of expression of transferrin receptors is associated with low cellular proliferation.

Combinations of antibodies labeled with different fluorochromes are helpful in studying coexpression of antigens. For example, one of the best approaches to the analysis of B-cell clonality consists of staining simultaneously B-cell antigens and individual light chains (e.g., PE-CD19 or PE-CD20 plus either FITC-anti-κ or FITC-anti-λ). In this manner, the analysis of immunoglobulin expression can be restricted to B-cells only, enhancing the sensitivity of detection of monoclonal B-cells (see below). Other useful diagnostic combinations include CD5/CD20 for small lymphocytic lymphoma, CD3/CD7 for peripheral and cutaneous T-cell lymphomas, and CD4/CD8 for thymic (lymphoblastic) lymphomas.

In addition to the antibodies listed above, we also use other reagents for special cases. For example, the interleukin-2 receptor (CD25) is expressed in a variety of T- and B-cell neoplasms, Hodgkin's disease, and reactive hyperplasias (10, 11), but is specially valuable in hairy-cell leukemia and adult T-cell lymphoproliferative disorders (see below). Indicated in italics in Table 12.1 are the antibodies that we consider to be the minimum requirement for immunophenotypical analysis of lymphomas. This "minimum" panel includes the combination of a pan-B antibody (CD19) with polyclonal antibodies against either κ or λ immunoglobulin light chains, which will detect most B-cell lymphomas. In addition, the CD3/CD7 combination should recognize many T-cell tumors. CD5/CD20 complements the above and is also useful in detecting small lymphocytic lymphomas.

Intracellular Antigens

Cytoplasmic or nuclear antigens in cells in suspension can be rendered accessible to antibodies by fixation and/or permeabilization of the cell membrane. In this manner, important cellular elements such as immunoglobulins (12–14), enzymes (15) or oncogene products (16, 17) can be detected. The success of this technology usually depends on the nature of the antigen and the degree of "leakage" of antigen or antigen-antibody product following membrane permeabilization. (For further technical details, see Chapter 10, this volume.)

DNA Staining

Propidium iodide (PI) is one the most commonly used dyes for DNA staining in the clinical laboratory. Other dyes are more specific for DNA but they may require UV light sources or may not be as efficient as PI when excited with argon-ion lasers, the most common light source in clinical instruments. Acridine orange (AO) is a useful dye that permits the simultaneous staining of DNA and RNA (18). AO has been extensively applied to the study of leukemias and lymphomas, but the routine application of this dye has been limited to few laboratories because its use requires experience. In our hands, PI staining followed by RNase treatment provides excellent results. Moreover, the fluorescence emission of PI can be conveniently separated from FITC fluorescence, allowing for simultaneous staining of DNA and surface or intracellular antigens.

Normal lymphocytes or other diploid standards such as chicken or trout erythrocytes may be used, preferably mixed with samples, as internal controls for instrument performance and ploidy determination. In our laboratory we prepare two aliquotes of cells from each sample: One contains an internal diploid standard and is used primarily for analysis of ploidy, since the presence of the diploid cells may interfere with cell cycle calculations; the other aliquot contains no internal standard and is used for the analysis of larger number of cells for adequate cell-cycle-fraction measurements.

DNA may also be analyzed in nuclei extracted from paraffin blocks used in histological preparations. Although this analysis provides data that can be used for retrospective clinical correlations, the quality of the results is usually poorer than those obtained with fresh samples (19, 20). Interestingly, in parallel studies, results obtained from paraffin-embedded tissues showed a higher S fraction than those from fresh samples (20, 21), perhaps due to aggregates of intact nuclei and nuclear fragments (22). Most significantly, cell surface antigens are destroyed during the processing and, thus, no direct correlation between DNA content and immunophenotype is possible.

Multiple Staining

Two or three surface or intracellular antigens can be labeled simultaneously. Also, simultaneous staining of DNA and antigens is possible. For simultaneous analysis of DNA and surface antigens, viable cells are first stained with FITC-labeled antibodies as indicated above. The cells are then fixed in alcohols (23) or paraformaldehyde (24, 25), exposed to RNase and stained with PI. Alternatively, detergent treatment may be used without fixation following surface antigen staining (26).

For simultaneous staining of intracellular antigens and DNA, cells are first fixed in alcohol or paraformaldehyde, treated with detergent, and then exposed to PI for DNA analysis or to antibodies for intracellular antigen labeling (16, 27). Very often, internal antigens are stained in conjunction with membrane antigens (28) or DNA (29, 30).

The method of recovering nuclei from paraffin blocks can also be used for simultaneous analysis of DNA and intranuclear antigens (31).

SAMPLE ANALYSIS
Cellular Antigens

For surface antigen analysis, we usually analyze 10^4 cells per antibody. Nonviable cells are identified on the basis of their forward-light-scatter characteristics or (preferably) PI incorporation, and are excluded from the analysis.

In general, it is advisable that the flow cytometric data be stored in list mode. Although this approach requires larger amounts of storage media, it facilitates multiple interactive data retrieval and gated analysis, which may be very valuable in the interpretation of heterogeneous or minimally involved samples.

"Live" gating is used when one wishes to analyze only a certain subpopulation of cells within the total sample and is performed to avoid occupying computer memory and storage media with unnecessary data. Using this approach, for example, data from granulocytes and nonviable cells may be excluded. In our laboratory, live gating of B-cells stained with either anti-κ or anti-λ antibodies is frequently performed to increase the sensitivity of detection of B-cell clonal expansion (see below). A similar approach is applied to any cell subpopulation that may be poorly represented in the sample.

DNA

DNA analysis requires stable and properly calibrated and aligned instrumentation in order to obtain acceptable linearity and coefficient of variation (CV) of the measurements. With modern instruments, CV of 3% or less for diploid standards should be easily obtainable. While the number of cells analyzed is not critical for ploidy analysis, cell-cycle-fraction calculations require a sufficient number of cells for adequate statistics. We routinely analyzed 3×10^4 cells per sample (at a concentration of 10^6/ml). The flow rate is maintained at approximately 200 cells/sec and the data are acquired in list mode and ungated. Samples containing diploid standards are analyzed separately. The same principles apply to samples stained simultaneously for antigens and DNA. In the latter situaiton, one should be careful about proper controls, optical filtration, and electronic color compensation.

EXPRESSION OF RESULTS
Cellular Antigens

In most laboratories, the analysis of surface antigens is performed on cells selected (gated) first on light-scatter properties. Once the cells of interest are chosen, their immunofluorescence is analyzed on single-parameter histograms and results are expressed as "percent-positive" cells per each antibody tested. Percent-positive represents the fraction of cells with fluorescence greater than that of the control sample, which consists of cells exposed to an irrelevant immunoglobulin. This approach is based on a practice initiated many years ago with fluorescence microscopy. It is applicable to discrete cell populations with relatively bright fluorescence and is particularly useful in lymphocyte subset analysis, where a change in the number of cells in each subset is clinically important. In samples suspected to harbor lymphoma, however, information on the number of neoplastic cells present in the sample is not usually relevant. The most pertinent consideration is to determine whether or not neoplastic cells are present and what type of lymphoma they represent. Unfortunately, this information may not always be reflected in the percentage of cells expressing antigens. Several factors contribute to the inadequacy of reporting results of immunofluorescence in lymphoma as percent-positive cells. First, the initial gating strategies based on evaluation of the light-scatter properties of the cells may be difficult and subjective since neoplastic lymphocytes are frequently heterogeneous in size and lymphoma samples usually contain high numbers of non-neoplastic cells. Second, in contrast to non-neoplastic lymphocytes, lymphoma cells often vary in the quantity of antigen expression (32) and dim immunofluorescence for certain antigens is common (8, 33) Third, this approach falsely assumes that, within a tumor, neoplastic cells can be either positive or negative for a given antigen, an attribute seldom observed in lymphoma.

These issues have prompted our laboratory to use a "visual" approach to flow cytometric data analysis of lymphomas and rely less on percentages (see below). We believe that the complex data generated by the flow cytometer in tumor samples are best expressed in graphical form, from which one can obtain excellent information on cell heterogeneity and antibody binding. Results may then be reported descriptively as text, graphically, or, when appropriate, numerically.

Light-scatter (LS) analysis is extremely helpful in the characterization of lymphoid neoplasms. LS signals are of particular value in the analysis of blood and marrow samples or in the analysis of any specimen containing abundant blood (34). Side-LS signals permit the elimination of granulocytes from the analysis and forward-LS signals are extensively used in the assessment of relative cell size. As in many other tumors, neoplastic transformation of lymphocytes leads to changes in cell size (35). In general, tumors composed of large cells tend to be more aggressive than those consisting of small cells. This, of course, is most likely related to the fact that proliferating cells and aneuploid cells are larger than diploid quiescent cells (36–38). In any event, this physical change of cells is used extensively in morphological classifications since there appears to be a relationship between the cell size (or nuclear size) and the clinical behavior of the tumors (39).

In order to determine the relationship between cell size and antigenic expression, we display results as correlated data of forward-light-scatter and immunofluorescence signals for each of the antibodies used (Fig. 12.1). The number of cells may be indicated as dot plots, contour maps or isometric (3D) displays. Antigen expression is simply determined visually by comparing patterns produced by cells exposed to the various antibodies to those obtained with irrelevant, isotype–matched immunoglobulin controls. Within a sample, cells with similar physical and antigenic properties produce relatively discrete data groups or clusters. Depending on the type and composition of the lymphoma and the particular antibody, these clusters display certain graphical distributions such as location, size, shape, and density. Similar types of lymphoma cells display similar forward-scatter patterns and tend to bind the same antibodies in a comparable manner. Thus, the characteristics of these clusters obtained with an appropriate panel of antibodies tend to be analogous and distinctive for each of the major types of lymphomas and serve as "fingerprints" that help one to recognize these tumors without needing precise numerical data.

Of course, this analysis does not preclude the enumeration of various lymphocyte subtypes, either neoplastic or normal. However, precise counting of tumor cells is only performed when there is a justifiable need to determine the level of involvement of the sample (e.g., blood and bone marrow biopsies) These numerical results are best derived from calculations made on multiparametric data. The two-color antibody combination listed in Table 12.1 is not only useful in the analysis of antibody binding patterns, but can also be applied to a precise quantitation of neoplastic cells in the sample. The quantitation can be further refined by restricting (gating) the analysis by light-scatter signals, particularly if lymphoma cells are larger than the normal lymphocytes that are almost always present in the samples and serve as excellent internal controls.

The strategies mentioned above may not apply to intracellular antigens. Antibody penetration requires fixation or other cell permeabilization procedures that modify the cell membrane and may result in changes in light-scatter properties and a decrease or loss of cell surface antigens. With the exception of terminal transferase, cytoplasmic immunoglobulins, and proliferation–related proteins, very few intracellular antigens are presently used in the clinical laboratory for the diagnosis or classification of lymphoma. In most laboratories, these cellular components are still being analyzed by immunocytochemical methods.

SPECIAL ANALYSIS: CLONAL B-CELL EXPANSION

Most lymphomas are composed of neoplastic B-lymphocytes. In each tumor, the neoplastic cells arise from a single B-cell at a stage of post immunoglobulin gene rearrangement. Since at this stage B-lymphocytes are capable of producing only one of the two possible types of immunoglobulin light chain (κ or λ) in a B-cell lymphoma, all neoplastic cells express either κ or λ chains (40) In contrast, in benign hyperplastic lymphoid tissue, the reacting normal B-cells, having arisen from multiple ancestors, consist of a mixture of κ- and λ-bearing cells. In normal or reactive conditions, this mixture is composed of an approximate 1.5:1 ratio of κ-:λ-bearing B-cells. The presence of monoclonal B-cells in a sample produces a change in this ratio that can be advantageously used for diagnostic purposes. This change is the single most useful information in the detection of B-cell lymphoma.

Several approaches have been used to analyze the κ and λ immunoglobulin distributions in normal and neoplastic conditions. Most studies have been based on the comparison of single-parameter immunofluorescence curves obtained with specific anti-κ and anti-λ antibodies. In 1979, three different laboratories reported abnormalities in the distribution of κ- and λ-bearing circulating lymphocytes in patients with non-Hodgkin's lymphomas who had normal lymphocyte counts and morphology. Using flow cytometry, Ligler et al. (41) studied the blood of eight patients with non-Hodgkin's lymphoma and found that seven patients had an increase in brightly-stained cells bearing the same light chain type found in the involved lymph node. Similar changes of the κ-to-λ ratios were not observed in 41 controls without non-Hodgkin's lymphoma. Garrett et al. (42) used fluorescence microscopy to study the expression of immunoglobulin light chain in the blood lymphocytes of 63 patients with non-Hodgkin's lymphoma, 20 normal individuals, and 21 pa-

Figure 12.1. Lymph node with benign reactive hyperplasia. The contour plots indicate forward LS values on the vertical axis (*FSC*) and immunofluorescence values on the horizontal axis. Cells with less fluorescence than channel 10 correspond to negative cells, as judged by the corresponding immunoglobulin control (not shown). The charts demonstrate a mixture of B (*CD19, CD20*) and T (*CD3, CD2, CD5, CD7*) lymphocytes that are predominantly small, although larger B-cells are also noted, most likely representing germinal center cells. CD20 is expressed more intensely than *CD19*, and *CD10* (a follicular marker) is partially expressed. The *CD3-, CD2-, CD5-,* and CD7-positive cell clusters are nearly equal in size, indicating no apparent loss of T-cell antigens. The clusters of *CD4* and *CD8* cells do not to appear overtly abnormal. *HLA-DR* expression shows a distribution that represents a mixture of B-cells and activated T-cells without a clear separation between positive and negative cells. The intensely *HLA-DR*-expressing cells that are located above channel 130 on forward LS appear to be very large cells. These signals do not represent large cells but rather correspond to cell aggregates, which are commonly formed when specific antibodies are added to cells expressing high levels of the corresponding antigen. *CD71* and CD25 show the characteristic distributions of a reactive lymph node. IgM and IgD are strongly expressed. IgG expression is variable and frequently represents non-specifically cell bound serum IgG. The two parallel histograms in the bottom center chart represent κ and λ immunoglobulin light change distributions for B-cells only (*CD19+* gated cells). The vertical axes in this chart indicate relative number of cells. All normal mature B-cells express immunoglobulins (either κ or λ but not both). Thus, the normal expressions of κ and λ immunoglobulins in B-cells follow typical bimodal distributions, each with a negative and a positive population. approximately 60% of B-cells express κ and 40% of B-cells express λ. DNA content analysis (last chart histogram) demonstrates a diploid distribution and a low S fraction.

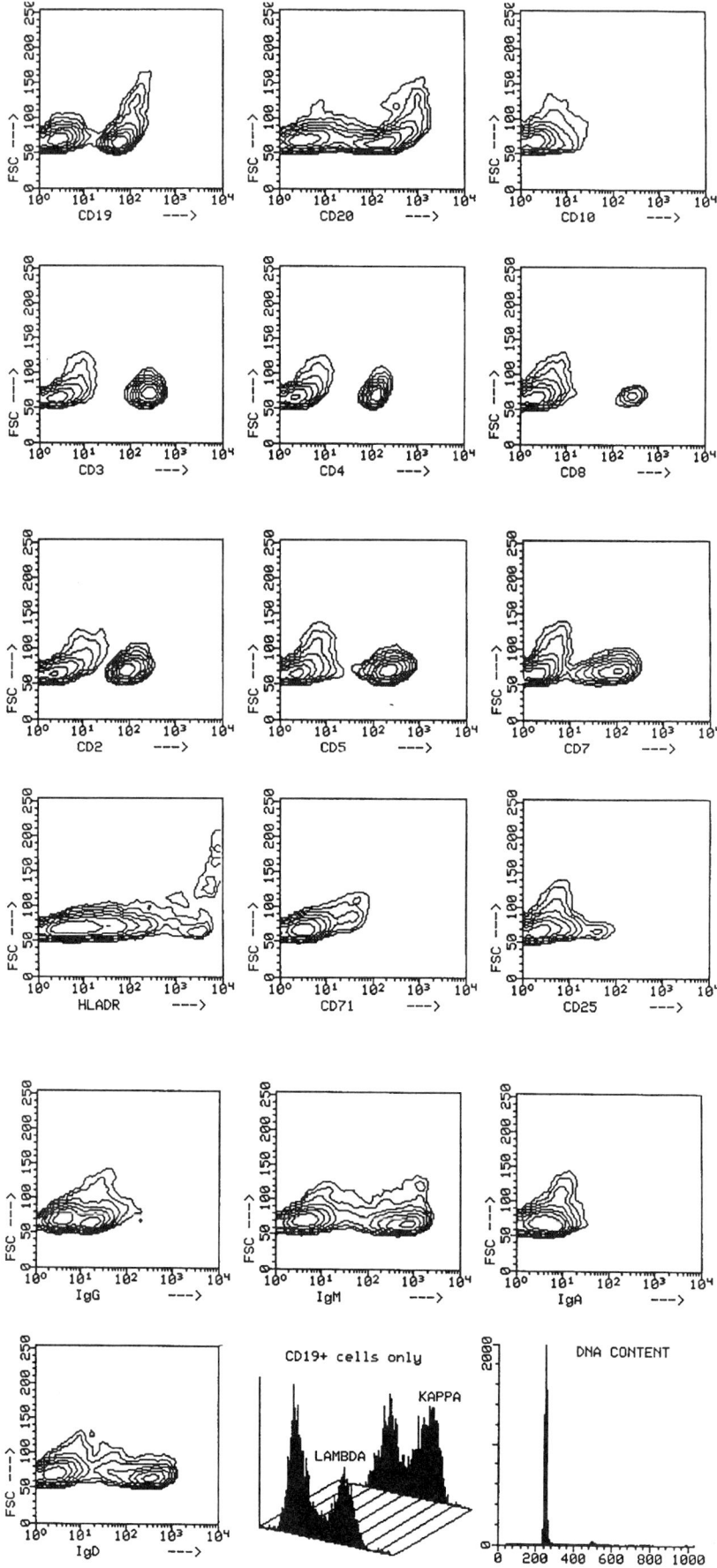

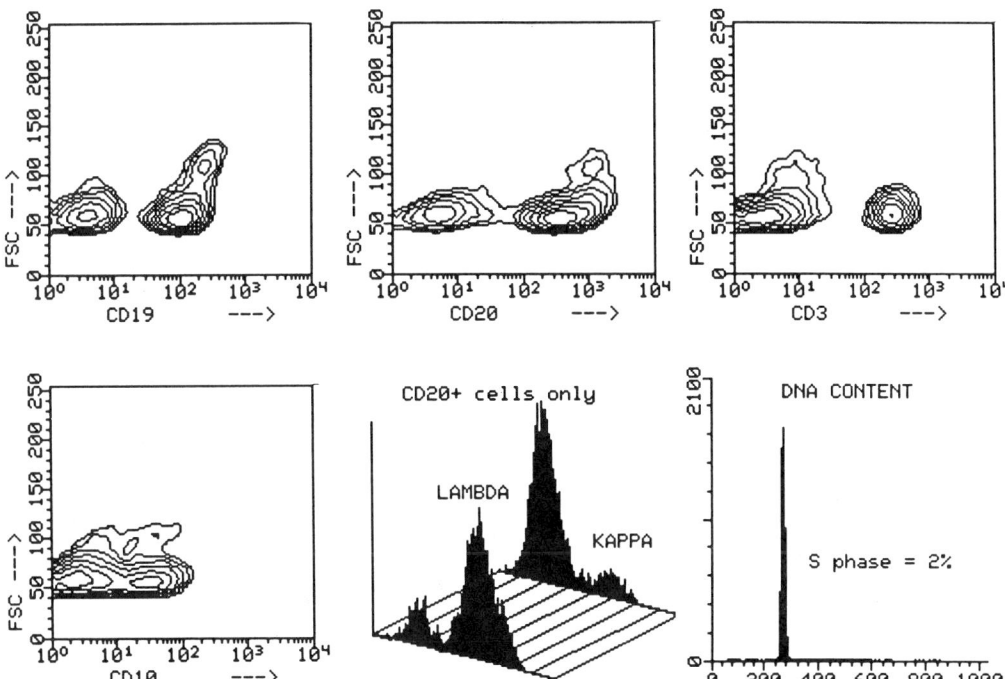

Figure 12.2. Low-grade follicular lymphoma. Correlated analysis of cell size (*FSC*) and immunofluorescence demonstrates a distinct population of predominantly small B-cells expressing *CD19* and *CD20* (left and center top). *CD10*, undetectable or present only in small amounts on normal B-cells, is often well expressed in lymphomas of follicular origin (bottom left). Non-neoplastic *CD3*-expressing T-cells are all small (top right). Analysis of κ and λ chain expression of the B-cells only shows a clear predominance of λ-expressing cells. This light chain restriction is indicative of a monoclonal B-cell expansion. The DNA content distribution demonstrates no overt DNA aneuploidy, and the S fraction is low and consistent with a low-grade tumor, even if the normal T-cells are accounted for.

tients with various diseases and nonhematologic malignancies. In this study, half of the patients with non-Hodgkin's lymphoma showed κ-to-λ ratios outside the range of the controls and the abnormal ratios were more frequently observed in indolent than in aggressive lymphomas. Ault (43) designed a novel flow cytometry approach to study κ and λ expression distributions in blood lymphocytes that detected 10% or less monoclonal B-cells and showed a 30% to 40% incidence of monoclonal B-cells in patients with non-Hodgkin's lymphoma.

These initial reports were subsequently followed by larger studies confirming the earlier observations. Using their "clonal excess" analysis, Ligler et al. (44) observed abnormal κ-to-λ ratios in the blood of 11 of 20 patients with lymphocytic lymphoma that could be analyzed by their technique. By mixing experiments, these authors claimed a sensitivity of 0.1% in the detection of monoclonal B-cells, although they acknowledged difficulties in their assay produced by cytophilic immunoglobulin bound to monocytes. Brudler et al. (45), applying the method of Ligler et al, established a sensitivity of 0.5% detection of monoclonal B-cells and found an overall 36% incidence of abnormal κ-to-λ ratios in the blood of 28 patients with non-Hodgkin's lymphoma. Using various degrees of sophistication in the statistical analysis of the curves, a high sensitivity in the detection of B-cell clonal expansion or "excess" has been claimed (46–48).

Despite the above observations, unacceptable rates of false-positive (49) results have been reported with these techniques. Many of these problems may be due to the fact that the measurements were based on single-color distribution analysis of κ and λ immunofluorescence in the presence of cells with cytophilic immunoglobulin bound to their surface (44). Depending on the tissue, these cells may be activated T-cells, monocytes, macrophages, or myeloid elements that can not be easily removed by density gradiants. Also, other than recognizing an abnormality, many of these measurements do not provide precise information on the number of neoplastic cells present in the sample or are not always capable of determining which light chain is expressed by the neoplastic cells. Furthermore, they will fail to detect abnormalities in cases of B-cell lymphomas that do not express surface immunoglobulins (50).

More recently, procedures for multiparametric measurement of κ- and λ-bearing B-cells have been developed. They are based on the analysis of cells stained simultaneously with both a pan-B-cell and an anti-immunoglobulin light chain antibody (either κ or λ) (5, 51–53). In this manner, B-cells can be analyzed for their expression of immunoglobulin independently from other nonrelevant cells.

We have been using this approach in a variety of tissues and fluids for several years and find it more acceptable than the originally published, single-parameter immunofluorescence clonal excess assays. We first display correlated data

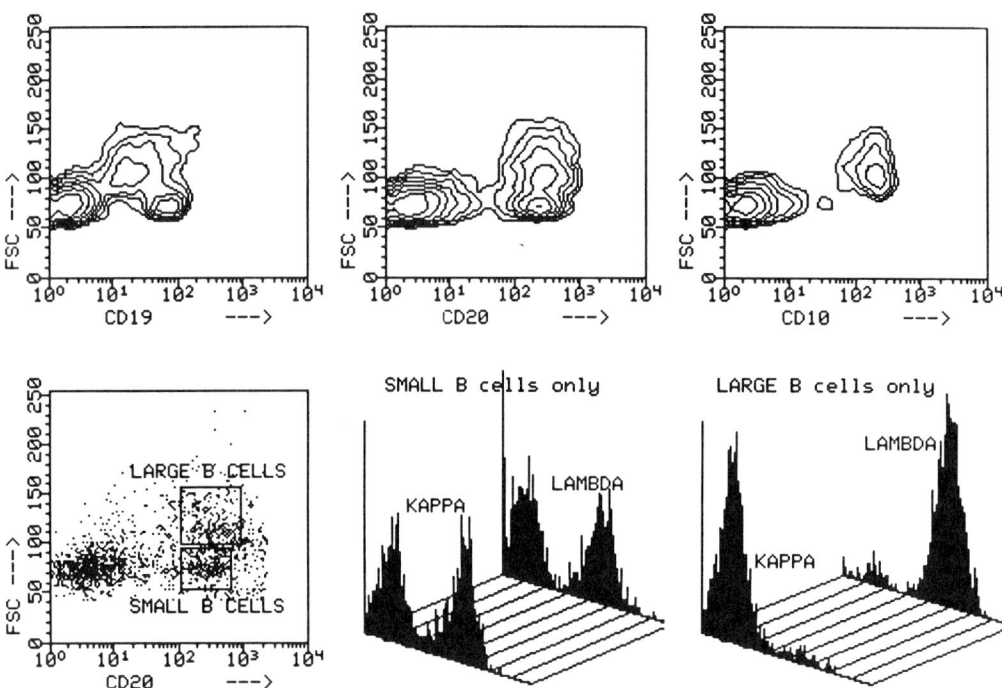

Figure 12.3. Lymph node involved by a large cell lymphoma, follicular type. A population of large B-cells expressing CD19, CD20, and CD10 is observed (top charts). CD19 is expressed more faintly on the large than on the smaller B-lymphocytes. While the small B-lymphocytes represent non-neoplastic cells, as indicated by the normal bimodal distributions of κ and λ Immunoglobulins (bottom center chart), virtually all large B-cells express only λ immunoglobulin.

of forward light scatter and immunofluorescence from pan-B antibody (CD19 or CD20). Then a gate is placed around the B-cells and fluorescence histograms of κ and λ immunoglobulin distributions for B-cells only are produced. If B-cells are non-neoplastic, they are polyclonal and both κ and λ distributions will be bimodal, indicating the presence of a mixture of κ- and λ-bearing B-cells (Fig. 12.1). If a neoplastic B-cell clone is present, its cells will only express either κ or λ chains, resulting in a disproportionate increase in B-cells expressing one and a corresponding decrease in cells expressing the other light chain (Fig. 12.1). Further selection of subpopulations of B-cells according to cell size (50, 54) (Fig. 12.3) or intensity of expression of highly informative antigens, such as CD20 (8) (Fig. 12.4), improves the resolution and increases the sensitivity of detection of monoclonality. The κ and λ histograms can be compared by a variety of statistical methods. As indicated above, the normal ratio of κ-to-λ-bearing B-cells obtained by simple integration of the histogram data, is approximately 1.5:1. In our laboratory, values in normal blood (n=87) ranged from 0.75 to 2.46, these figures may be applied to solid lymphoid tissues as well (Braylan RC, unpublished). Sophisticated statistical analysis of the data may be useful when assessing minimal tumor involvement. However, in the vast majority of B-cell lymphomas the changes are so obvious that a simple visual inspection of appropriately gated data is sufficient to determine clonal expansion. This assay is easy to perform, provides information on the light chain expressed by the neoplastic cells, is informative of the number of neoplastic cells present in the sample, and is useful even in the detection of B-cell lymphomas that express low-level or no surface immunoglobulins.

Unfortunately, no similar assay for clonality in T-cell malignancies is possible. Monoclonal antibodies against V region framework determinants of the T-lymphocyte β receptor gene may serve as "almost" clonotypic reagents in these lymphomas (55). However, this approach may be of only limited value (56).

DNA Content

There is no standard procedure for data analysis of DNA content by flow cytometry. Different laboratories have their own sets of rules and these may not be necessarily comparable. In our laboratory, DNA content analysis is performed in cells fixed with ethanol, which preserves better the light-scatter characteristics of the cells than of isolated nuclei. The fixation allows us to relate ploidy and cell-cycle phase measurements to relative cell size. Thus, our initial approach is to examine bivariate dot plots of forward LS and PI fluorescence signals (Fig. 12.5). The bivariate plots are also helpful in obtaining a general idea of the amount and size of debris and cell aggregates and in detecting small aneuploid populations that may not be obvious in single-parameter DNA histograms. Ethanol often induces cell clumping, so it is advisable to reduce or eliminate cell aggregates by mathematical procedures or by discrimination based on the correlated analysis of signal-pulse width and area (integral).

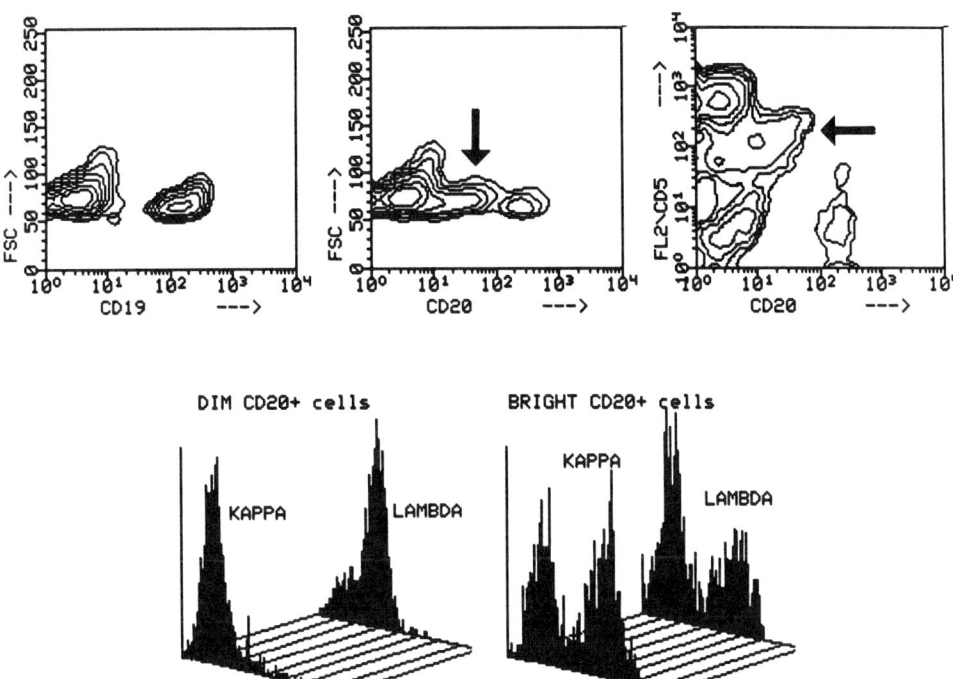

Figure 12.4. Blood sample partially involved by B-chronic lymphocytic leukemia (CLL). The top left and center charts demonstrate the expression of two pan-B-cell antigens, *CD19* and *CD20*. The top right chart shows a correlated analysis of *CD20* and *CD5*, an antigen expressed on T-cells, a small fraction of normal B-cells, and on B-CLL. Only a single cluster of *CD19*-expressing lymphocytes is observed. In contrast, the B-cells identified by *CD20* form two groups with unequal *CD20* expression. Poorly-expressing *CD20* cells (*arrow*) correspond to CLL cells coexpressing *CD20* and *CD5* (top right chart). Note that *CD5* expression on CLL cells is less intense than on normal T-cells. CLL cells express (also dimly) a single immunoglobulin light chain (bottom left chart). In contrast, cells expressing normal levels of *CD20* show a normal κ and λ distribution (bottom right chart).

As indicated above, cell suspensions from lymphomas often contain a large component of non-neoplastic cells. In some cases, abundant normal cells may obscure the presence of aneuploid neoplastic cells in the DNA histogram, leading to the false assumption that the lymphoma is diploid and producing artificially low proliferative fractions. Knowledge of the immunophenotypic results and representative cytological controls is extremely helpful in determining the presence and approximate frequency of neoplastic cells in the cell suspension analyzed for DNA content.

PLOIDY

A presently accepted convention (57) specifies that a tumor should be designated as aneuploid by DNA quantitation when a discrete abnormal peak(s) that is separable from that of diploid cells in the DNA histogram is observed. The DNA index (DI) is the ratio between the mean (or modal) fluorescence of the neoplastic G_1 peak and that of a diploid control. The present convention, however, does not take into consideration that the resolution of two-cell populations with different DNA content depends not only on the true DI but also on the CV of the measurements and the percentage of aneuploid cells in the sample (58, 59). Thus, the DI values for near-diploid tumors may not be very accurate (60). One should also be aware of certain artifacts that produce pseudo aneuploidy (61). Spurious extra-peaks are often observed in DNA histograms of ethanol-fixed, PI-stained whole blood or bone marrow cells due to the presence of granulocytes (Braylan RC, unpublished).

DNA-aneuploid cells can be characterized immunologically by dual-color analysis of DNA and antigens. This approach helps resolve near-diploid DNA-aneuploid populations (38) and is extremely valuable when there is only a small number of malignant cells in the sample (31, 62) (Fig. 12.6). It also allows the detection of lymphomas with coexisting diploid and aneuploid components, an abnormality that may not be appreciated by single-color DNA analysis (38, 62).

CELL CYCLE FRACTIONS

There has been great variability in the procedures for calculating the fraction of cells in the various phases of the cell cycle using flow cytometry. Also, some authors utilize the fraction of S-phase cells for their studies while others measure the proliferative index (S plus G_2 and mitosis). In a DNA histogram with a single diploid distribution, almost always these calculations have not taken into consideration contaminating normal diploid cells. In histograms displaying separable diploid and aneuploid populations, the cell cycle fractions are often calculated for the aneuploid cells with the exclusion of the diploid peak, but other studies report results for both combined. Moreover, many investigators have ex-

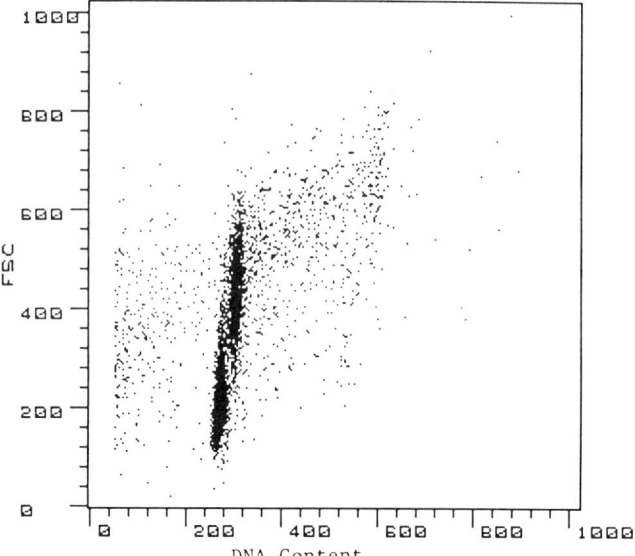

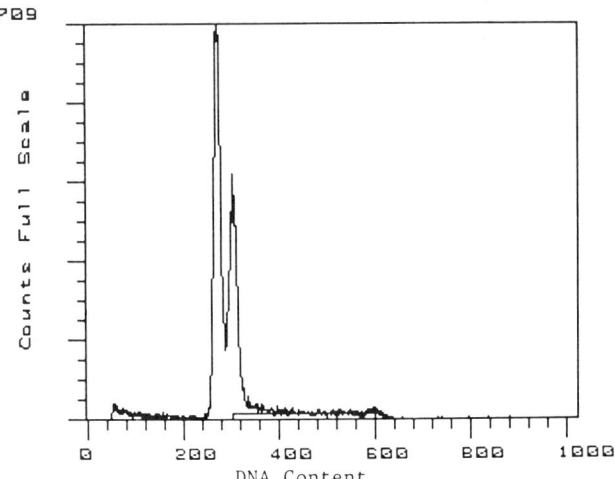

Figure 12.5. Correlated, multiparameter analysis provides more information than single-parameter analysis of DNA content. This figure represents DNA content analysis of a large cell lymphoma. Cells have been fixed in ethanol, treated with RNase, and stained with propidium iodide. Doublets have been eliminated on the basis of signal properties. The top chart shows correlated analysis of *forward LS* and *DNA content* distributions. The bottom chart illustrates the same data, but only the DNA content is displayed. Using a diploid control (not shown) it was established that the left peak represent diploid cells. The data in the top chart clearly illustrate that (neoplastic) hyperdiploid cells are larger than (non-neoplastic) diploid cells, and that the majority of the S-phase cells are associated with the larger aneuploid cells. This information can not be extracted from the DNA histogram seen in the bottom chart.

cluded altogether from analysis the histograms exhibiting overlapping diploid and aneuploid peaks or multiple aneuploid peaks. The computing methods have also varied. Some investigators have used simple manual integration of the different regions of the DNA histograms and others have applied diverse mathematical algorithms of varied complexity.

Since lymphomas are often diploid or near-diploid; a precise calculation of tumor cell cycle phases is often unattainable by examining solely DNA measurements due to contamination with normal cells. However, careful observation of the data and correlation of the results with the phenotypical analysis, or the combined analysis of DNA and tumor-associated antigens should, in most instances, permit a reasonably good estimation of the proliferating fraction of neoplastic cells. The only instance in which we do not attempt to evaluate cell cycle phases by flow cytometry is in cases of poor tumor representation in the sample (less than 10% of the cells). In these instances, other methodologies (such as immunocytochemical detection of cycle-dependent proteins) should be more accurate.

For histograms with low-to-intermediate S fractions (less than 15%), we use the relatively simple approach of Baisch et al. (63). This ''rectangle'' method for S-phase calculation is suitable to most cases of lymphoma and can be easily applied even to data generated by manual integration of the various regions of the histogram. For histograms with very high proliferative fractions, other approaches such as Dean's S-fit method (64) may be used. In cases of complex histograms it may be possible to apply interactive manual measurements (65). Several commercial firms offer excellent programs that are very flexible and can fit even complex data. However, despite their sophistication, these programs require professional judgment and some understanding of the biology of these tumors. The goodness of fit of the model to the data helps to determine the appropriateness of the results.

It has already been mentioned that, if unaccounted for, the presence of normal cells in diploid or near-diploid tumors may significantly affect the calculations. In these cases, the number of normal cells can be separately estimated by their antigenic and light-scatter properties. Except for bone marrow cells, one can assume normal cells are mostly nondividing elements that do not contribute significantly to the cycling pool. They can then be subtracted from the G_0-G_1 peak and the calculations corrected accordingly. Alternatively, an immunological marker for either the neoplastic or the non-neoplastic cells can be used in conjunction with DNA staining. In this manner, the cells of interest can be identified and a relatively uncontaminated tumor cell cycle fraction analysis can be calculated (38) (Fig. 12.7).

The same assumption can be applied to DNA histograms with separable diploid and aneuploid populations. Using only the DNA data, noncycling diploid cells are simply eliminated from the analysis that is only applied to aneuploid cells. As mentioned earlier, these and other complex histograms may be analyzable by a variety of mathematical models that fit the data and help deconvolute the different cell cycle fractions.

SIGNIFICANCE OF THE RESULTS

Cellular Antigens

Lymphoma cells conserve many antigens that are present in normal lymphocytes. Thus, the analysis of surface and intracellular antigens is helpful in the recognition of these tu-

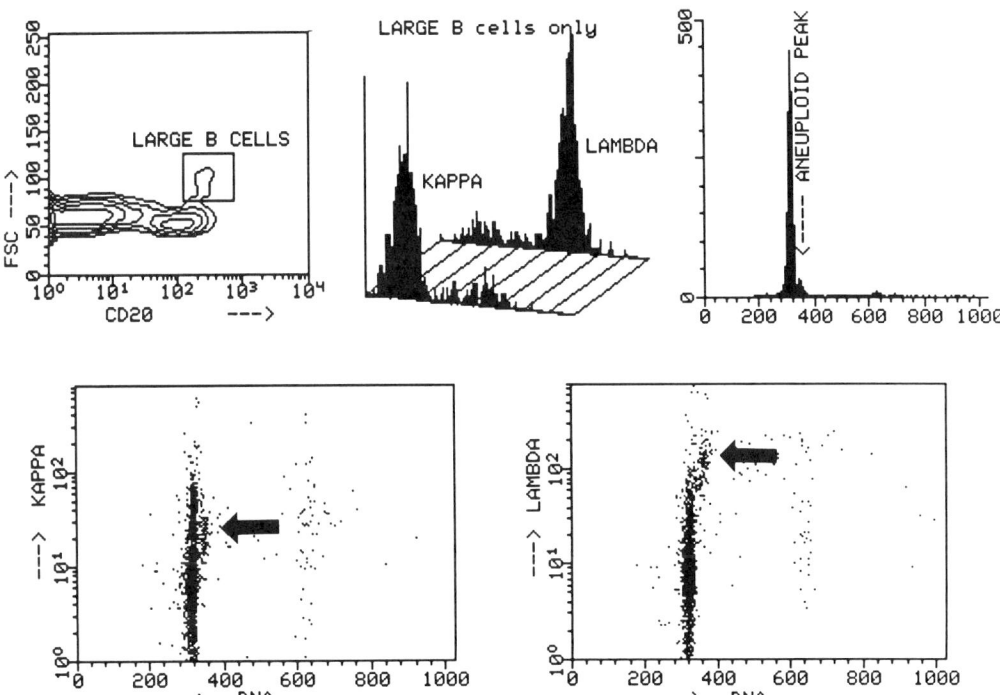

Figure 12.6. Lymph node minimally involved by aneuploid *large B-cell* lymphoma. Top left chart shows the majority of B-cells to be of relatively small size. Some B-cells, however, are larger, as judged by their light-scatter characteristics. Whereas the smaller B-cells demonstrate normal κ and λ distributions (not shown), most of the larger B-cells express λ but no κ Ig (top center). The DNA histograms demonstrate a discrete hyperdiploid population (top right) that expresses λ immunoglobulins only (*arrows*) and corresponds to the large cells seen in the left top chart. Thus, the combined analysis of light scatter, immunoglobulin light chain and *CD20* expression, and DNA distributions permitted the detection and characterization of a poorly represented neoplastic lymphoid population. The scarcity of neoplastic cells in this sample, however, precludes obtaining cell cycle kinetic characteristics for the lymphoma cells.

mors, provides information on cell lineage, and may also be useful in some therapeutic applications. On the other hand, the analysis of lineage-specific or differentiation antigens is only of limited value in determining the grade of the lymphoma.

LINEAGE IDENTIFICATION AND STAGE DIFFERENTIATION

Immunophenotypical analysis is the most helpful technique for differentiating lymphomas from nonhemopoietic neoplasms. Although occasionally absent (66), CD45 expression is an excellent marker for this purpose. Furthermore, specific antibodies against lymphoid antigens allow the identification of tumors derived from B-cells, thymic T-cells, peripheral (post-thymic) T-cells, and macrophages (true histiocytic lymphomas).

Many authors have attempted to assign every lymphoma to a particular developmental step in the T- and B-cell ontogeny, speculating that neoplastic lymphocytes recapitulate normal stages of B- and T-cell differentiation. This is a stimulating idea that has been applied primarily to leukemias. However, in lymphomas there is very little evidence to substantiate the same assumption and such a supposition has no practical relevance in the diagnosis or management of these tumors. Despite the superficial similarities, lymphoma cells often express antigens in a manner that is clearly different from that of any potential normal counterpart and there is not always adequate proof that the target elements in these tumors are present at low frequency in normal tissues. Neoplasms of other origins display markedly altered morphologic or other phenotypes that have no resemblance to those of putative normal counterparts. Similarly, the unusual antigenic profiles observed in lymphoma cells are most likely the consequence of genomic changes resulting from oncogenic events rather than the reflection of phenotypes present at different stages of lymphoid differentiation. Thus, in these tumors, it would be prudent to avoid rigid immunophenotypical classification based solely on cell resemblance to normal counterparts.

LYMPHOMA DETECTION

Clonal Expansion. We mentioned earlier the relatively high incidence of abnormal κ/λ ratios found in the blood of patients with non-Hodgkin's lymphomas. Following Ault's initial findings (43), his approach was used to demonstrate monoclonal B-lymphocytes in the blood of all 12 patients with Waldenström's macroglobulinemia (67). The observations also suggested that serial determinations of these cells reflected the clinical activity of the disease. His clonal excess analysis was later applied to 60 suspensions of various tissues suspected to harbor non-Hodgkin's lymphoma (50).

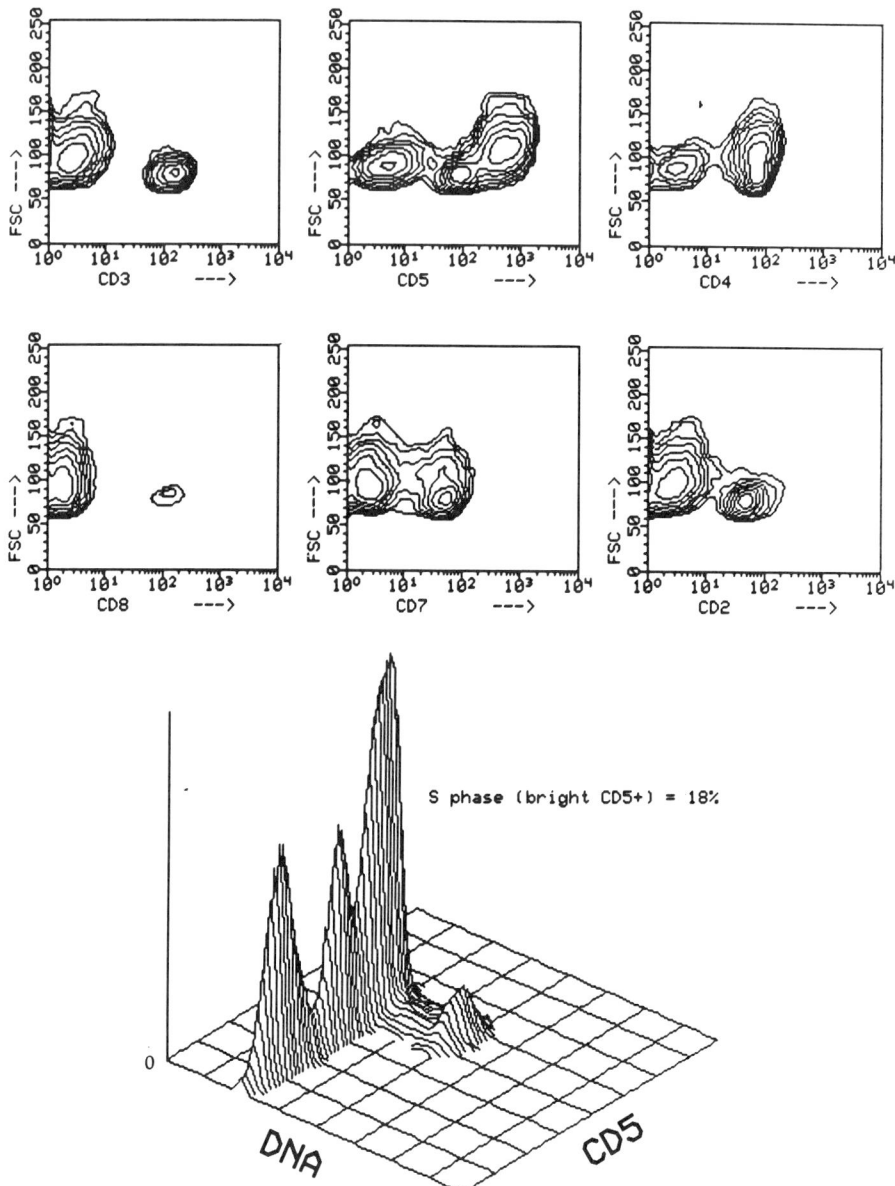

Figure 12.7. Peripheral T-cell lymphoma. Pan-T antibodies CD3 and CD2 label a predominantly small cell population. CD8-positive cells are also small. A cell population of relatively larger size representing lymphoma cells is clearly defined by CD5 (top center). It also expresses CD4 and, only partially, CD7. This is a severely aberrant T-cell phenotype showing an intense CD5 and partial or total absence of other T-cell markers. In this case, cell kinetic analysis of the tumor population could be achieved by dual staining of surface antigens and DNA. The most informative antigen in this case is CD5 and the results of this analysis are shown in the bottom chart. The population of bright CD5 cells corresponding to the larger cells in the top center chart shows a higher associated S fraction than the other populations of CD5-positive and CD5-negative cells.

B-cell clonal excess was found in 24 of 25 B-cell lymphomas but in none of the remaining cases, which included reactive hyperplasia, T-cell lymphoma, and Hodgkin's disease. When restricted to cells gated on size (54), on B-cell surface antigens (5), or on both, the analysis of κ and λ light-chain-bearing cells is highly diagnostic of B-cell lymphoma, even when neoplastic involvement is minimal (Figs. 12.6 and 12.8).

In a study of 91 patients with non-Hodgkin's lymphoma, Smith et al. (68) observed a very high frequency (78%) of circulating monoclonal B-cells, independently of histological subgroup. Other studies (69) demonstrate a much lower, and perhaps more realistic (70) frequency of monoclonal circulating B-cells in patients with non-Hodgkin's lymphoma. Circulating monoclonal B-lymphocytes have been detected in the blood of patients with autoimmune disease (71–73). Also, similar abnormalities have been reported in lymph nodes that were histologically normal or reactive (49). Whether this represents a true phenomenon or an artifact of single-parameter clonal excess analysis remains to be elucidated.

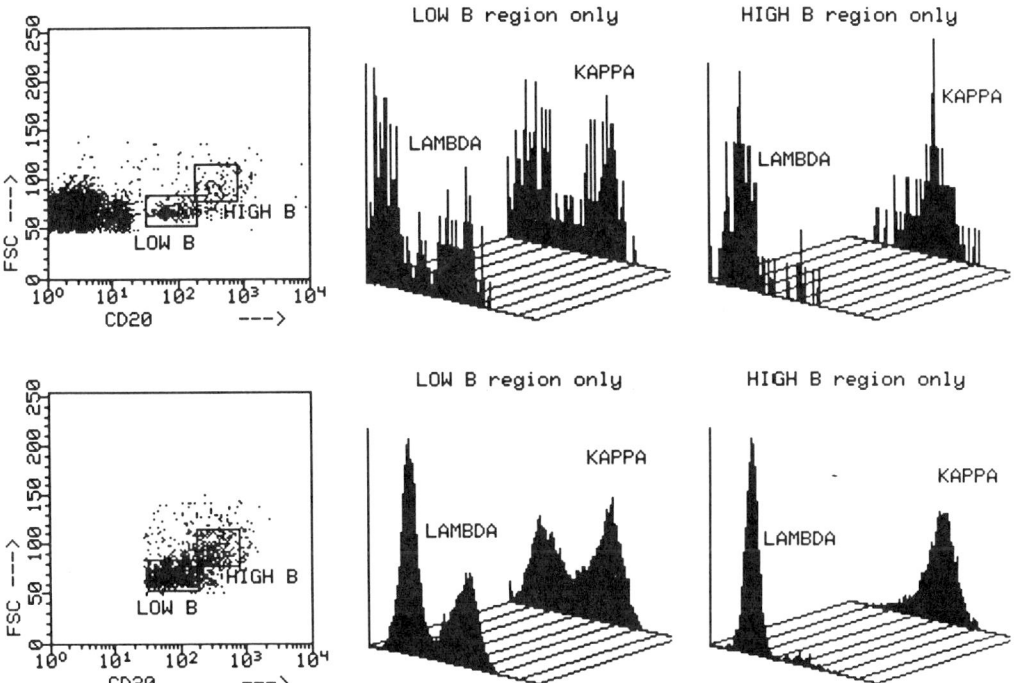

Figure 12.8. Analysis of pleural fluid containing very low numbers of monoclonal B-cells. The cells in the fluid contained predominantly lymphocytes and some mesothelial cells. Most lymphocytes were T-cells and no apparent abnormality was observed when κ and λ immunoglobulin distributions were analyzed in the whole population (not shown). However, when κ- and λ-bearing cells were analyzed on CD20-expressing cells only, there appeared to be an excess of κ over λ cells among larger B-cells with intense expression of CD20 (high B) (above charts), although these histograms appear ragged due to the low number of cells present in the regions studied. To confirm these findings with better distributions, a repeat cell analysis was performed restricting the data acquisition (live gating) only to B-cells (bottom charts). Whereas smaller cells that express CD20 less intensely, display normal light chain distributions, the slightly larger B-cells express more CD20 represent a monoclonal population. The number of neoplastic cells detected by this procedure represents approximately 15 of the cells in the sample. (This patient subsequently underwent a mesentery lymph node biopsy that showed a follicular, IgG κ-expressing lymphoma.)

In contrast to B-cell lymphomas, in which the mutually exclusively expression of immunoglobulin light chains serves as a convenient means to demonstrate clonal expansion, there is no simple immunological procedure to detect monoclonality in T-cell lymphomas at present. The immunological recognition of T-cell lymphomas is mainly based on the detection of abnormal phenotypes (see below).

Aberrant Expression. The aberrant expression of antigens that is often observed in neoplastic lymphoid cells can be used advantageously for diagnostic purposes. Numerous studies in the immunohistochemical literature describe alterations of expression of lymphocyte-associated antigens in lymphomas and excellent reviews have been written on this subject (74). Aberrant antigen expression is mainly exploited in the detection of T-cell lymphomas, which often display phenotypes not observed in normal mature T-cells. Thus, the presence of an early thymic antigenic profile in the lymph nodes or blood cells is a strong indicator for (lymphoblastic) T-cell lymphoma (Fig. 12.9). The absence or weak expression of one or more T-cell antigens in otherwise "mature" T-cells strongly suggests a "peripheral" T-cell lymphoma (75, 76). Less known, however, is the quantitative variation in antigenic density that often occurs in lymphoid tumors (8, 77). These changes as well as the abnormal coexpression of antigens may be difficult to appreciate in immunohistological preparations but are readily detected by flow cytometry (Fig. 12.7).

Therapeutic Applications. Highly sensitive detection of lymphomas cells is not only useful diagnostically, but may also be important therapeutically. For example, the detection of low levels of marrow involvement should be useful in purging autograft marrows and may be significant in determining residual disease or early relapse in patients with aggressive lymphomas who might undergo autologous bone marrow transplant. Also, flow cytometry has been used in monitoring in vivo cellular changes following therapeutic use of antibodies directed against idiotypic and leukocyte-associated determinants (78).

DNA Content

PLOIDY

The abnormal gain or loss of chromosomes frequently observed in solid tumors results in quantitative changes of DNA content that may be detectable by flow cytometry and are designated as "flow aneuploidy." Although aneuploidy has been reported in lymphoid hyperplasias in homosexual men (79), for practical purposes, this abnormality of highly suggestive of neoplastic transformation.

The frequency of aneuploidy reported in the literature varies widely. Overt aneuploidy is present in approximately 10% to 20% of low-grade and 50% to 60% of aggressive non-Hodgkin's lymphomas and is most frequent in B-cell lymphomas of intermediate grade (80). In a limited number of studies, aneuploidy detected by flow cytometry correlated well with cytogenetics findings, showing a low false-positive rate (81). As expected, cytogenetically pseudo-diploid or near-diploid clonal abnormalities are not detectable by flow cytometry (82).

Virtually all studies show that the incidence of overt aneuploidy in indolent lymphomas is lower than in higher grade tumors (38, 83–85) and in intermediate-grade (60, 88) than in high-grade tumors, such as lymphoblastic or Burkitt's lymphomas, which are generally diploid or near-diploid (83, 88). This distribution explains why, when considered in general, no correlation is found between aneuploidy and prognosis in non-Hodgkin's lymphoma (89, 90).

Most aneuploid peaks are seen in the hyperdiploid or in the peri-tetraploid regions (38, 60, 84, 90–92). Hypodiploidy or multiple aneuploid peaks in the same tumor are infrequent (60, 84) but coexisting diploid and aneuploid populations may be seen more often (38). Shackney et al. (60) found no difference in the incidence of aneuploidy at diagnosis and at recurrence. In intermediate and high-grade lymphomas, Srigley et al. (84) observed a higher incidence of aneuploidy in histologically transformed than in de novo tumors with similar histology, and Cowan et al. mentioned frequent changes in ploidy level in recurrent tumors (93).

CELL CYCLE KINETICS

Knowledge of cell cycle fractions in lymphomas is extremely useful in initial tumor grading and in the assessment of conversion of low-grade to aggressive lymphomas. This is supported by the observation that the fraction of proliferative cells in lymphomas correlates very well with major histological grades and is one of the best indicators of tumor growth rate. Thus, histologically low-grade tumors contain low numbers of cycling cells, while in intermediate and high-grade tumors the number of cells in S phase, G_2, and mitosis is higher.

It is remarkable that correlations between proliferative fractions and histological grade have been shown by virtually all laboratories, regardless of the morphological classification applied and despite the variability in methodologies used for cell cycle phase analysis, including sample preparation, cell staining and analysis, and data analysis. Early in vivo studies of lymphomas showed a relationship between the fraction of cells in DNA synthesis, growth fraction, and tumor doubling time (review in (94)). Furthermore, numerous in vitro studies utilizing tritiated thymidine uptake (94–100), DNA quantitation by flow cytometry (35, 60, 83, 84, 88, 92, 96, 101–105) and, more recently, immunohistological and flow cytometric analysis of cell-cycle dependent nuclear proteins, such as Ki-67 (27, 106–108), repeatedly and consistently demonstrated a correlation between proliferation fraction and histological grade in these tumors. It is important to recognize, however, that while low-grade tumors almost always show low proliferative fractions, higher grades demonstrate a wide scatter of values. This is also true even after correcting for normal cell dilution effect (38). The relationship between DNA synthetic fraction and histological expression is even demonstrated within immunologically homogeneous lymphomas (60, 109), although no overt differences can be been in S fractions between T- and B-cell tumors in general (60, 86).

CORRELATIONS WITH HISTOLOGY

On the basis of morphological appearance, the lymphomas have been traditionally classified into two groups: Hodgkin's disease, and the non-Hodgkin's lymphomas. Most investigators using flow cytometry have been concerned with the non-Hodgkin's lymphomas, a highly variable group of neoplasms, and the majority of the studies have used histological parameters as gold standards to establish meaningful correlations. This is understandable, since it is difficult to demonstrate the importance of new biological data without a point of reference. However, it is clear that conventional histopathology for the diagnosis and classification of non-Hodgkin's lymphoma suffers from serious deficiencies. There is still a lack of common nomenclature and, except for broad categories, all morphological systems proposed have shown poor reproducibility and are prone to observer error (110–119).

For this reason, in this chapter, we will depart from the traditional approach of describing phenotypical or other biological features of lymphomas by following strict histological and cytological categories, and will limit our discussion to the major groups, only mentioning subtypes lymphomas when there is reasonable certainty that they represent reproducible, recognizable, and meaningful entities. We will also describe findings in benign hyperplastic adenopathies that occasionally may be confused with lymphomas. Table 12.2 summarizes the flow cytometry findings in the major histologically defined lymphoid processes.

Benign Lymphoid Hyperplasias

Lymphoid tissues can enlarge as a result of an exogenous or an endogenous antigenic stimulus. Similarly, lymphocytes may accumulate as a consequence of antigenic stimulation in sites not normally occupied by lymphoid tissues. The cellular response to these stimuli may be intense and result in clinically worrisome masses that often present diagnostic dilemmas for the clinician.

The immunophenotype of the benign lymphoid hyperplasias shows a mixture of B- and T-cells in highly variable ratios (120) (Fig. 12.1). As stated above, within the B-cell

Table 12.2
Flow Cytometric Findings in Lymphadenopathy

Histologic Diagnosis	Immunologic Phenotype	Ploidy	S-Phase	Cell Size
Non-neoplastic	Pan-B-antigens, including Sig, are nearly equally expressed. K:L ratio ≈ 1.5:1 Pan-T antigens are nearly equally expressed. CD4 and CD8 are expressed on different cells and CD4+ plus CD8+ cells ≈ mature T-(CD3+) cells. Very few cells express thymic antigens.	Diploid.	Usually <10%	Predominantly small
Hodgkin's	As in non-neoplastic nodes. CD4 predominates in many cases.	Usually diploid. Rarely aneuploid.	Low	Predominantly small
Small lymphocytic lymphoma	Increased number of B-cells. CD20 expression is faint. Sig expression is usually faint and restricted to a single type. Neoplastic B-cells express CD5 (often fainter than on normal T-cells).	Diploid.	Low	Predominantly small
Follicular lymphoma	B-cells are usually increased although T-cells may be abundant. Expression of CD 20 is intense. Expression of Sig is intense and restricted to a single type. CD10 is expressed in approximately 75% of cases.	Usually diploid or near-diploid. Occasionally hyperdiploid or tetraploid.	Low	Variable
Diffuse lymphoma, large cell, B-cell type	B-cells may be increased but normal T-cells are usually abundant and may obscure the presence of neoplastic cells. Sig expression is restricted to a single type. Aberrant B-cell antigen and/or Sig expression may be observed. Transferrin receptor is highly expressed.	Often aneuploid (usually hyperdiploid or tetraploid).	Intermediate	Large
Diffuse lymphoma, mixed or large cell, (peripheral) T-cell type	T-cells are increased. T-cell antigen expression is often aberrant (abnormal CD5, absence of CD7). HLA DR is usually expressed. CD25 is often expressed in (HTLV1+) ATL.	Aneuploidy is less frequent than in B-cell types.	Intermediate	Variable
Lymphoblastic lymphoma	Usually early thymic phenotype (often coexpression of CD4, CD8, CD2, CD5, CD7, and CD1, and lack of CD3 and HLA DR). Sometimes early B-cell phenotype (coexpression of CD19, CD10, and HLA DR and lack of CD30 and Sig). TdT is expressed.	Usually diploid. Occasionally tetraploid.	Intermediate to high	Medium
Small noncleaved lymphoma (Burkitt's)	B-cells are increased and few T-cells are present. Sig expression is intense and restricted to a single type. CD10 is usually expressed. Transferrin receptor is highly expressed.	Diploid or slightly hyperdiploid.	Always very high	Medium

population, the κ-to-λ ratio is approximately 1.5 (range 0.8–2.3). CD4/CD8 ratios are higher than in peripheral blood values and activation-related antigen expression in T-cells is frequent in hyperplasias (120). Monocyte-macrophage cells are seldom detectable in significant amounts in suspensions prepared from hyperplastic lymph nodes.

In correlating flow cytometric data with histological findings, it must be kept in mind that sampling error or selective loss of neoplastic cells may occur during solid tissue preparation. Thus, an absence of abnormalities indicated by flow cytometry does not necessarily point to a benign condition nor does it exclude the existence of lymphoma.

Hodgkin's Disease

Hodgkin's disease accounts for approximately 40% of the lymphomas. It is characterized by the presence of morphologically abnormal cells, called Reed-Sternberg and variants, and an associated large number of normal-appearing cells, such as lymphocytes, histiocytes, fibroblasts, granulocytes, and plasma cells.

The origin of the malignant cell in Hodgkin's disease is still undefined (reviewed in (121)). Numerous reports suggest the possibility of a lymphocytic origin (122), but other studies, both in biopsy material as well as in culture lines, support an origin in the antigen-presenting, interdigitating reticulum cell. It must be emphasized that it is extremely difficult to study the neoplastic cells of Hodgkin's disease. In most instances, too few of these cells can be obtained in suspension from tissues affected by the tumors. This is due in part to the fact that normal cells outnumber the malignant elements in these tissues, but also probably to other factors, such as high fragility of the neoplastic cells and/or their tight attachment to stroma. Complicating the problem of the cell origin in this disorder is the lack of an adequate experimental model and the extreme difficulty in obtaining reliable and representative culture lines. Furthermore, this issue is confounded by evidence that Hodgkin's disease may not be a single entity (123, 124), which would explain some of the inconsistencies in the interpretation of results.

The technical difficulties encountered with Hodgkin's disease in suspending sufficient numbers of neoplastic cells has precluded their analysis by flow cytometry methods and explain why most studies of this disease have been done with immunohistological methods. Even if sophisticated strategies for enriching Reed Sternberg cells are applied, an adequate analysis of these cells would be extremely difficult because it is likely that numerous lymphocytes will be firmly attached to their surface (125).

Flow cytometry studies performed on paraffin-embedded materials showed a very low incidence of DNA aneuploidy in Hodgkin's disease (126), confirming earlier results obtained in fresh biopsies (83). The infrequent occurrence of DNA aneuploidy in measurements performed in suspensions of Hodgkin's disease very likely reflects the low frequency of malignant cells in the preparations or perhaps the fact that these cells contain different (abnormal) amounts of DNA within individual tumors. Early studies using Feulgen-stained cytological preparations demonstrated that Reed-Sternberg cells contain increased DNA above the diploid and often beyond the tetraploid range (127, 128). Interestingly, flow cytometry studies using a combined staining of nucleoli (a prominent feature in the malignant cells of Hodgkin's disease) and DNA in paraffin-embedded material showed distinct aneuploid populations in all 15 cases of Hodgkin's disease studied, and two to four aneuploid populations were found in each case (31).

Earlier studies using autoradiography suggested that Reed-Sternberg cells were not capable of DNA synthesis (128). However, these results may have been due to in vitro cell manipulation since in tissue sections most neoplastic-looking elements in Hodgkin's disease are cycling (97, 129).

Normal lymphocyte populations extend from Hodgkin's disease lesions, on the other hand, have been extensively characterized (130, 131). Most studies have found a predominance of T-cells with a relative preponderance of T-helper inducer phenotype. However, there is an extraordinary variation in results and B-cell-rich Hodgkin's disease is not infrequent. In our own experience, and as reported in the literature (132), there are no quantitative or qualitative differences in lymphocyte distribution or expression of lymphoid surface antigens between Hodgkin's disease a benign reactive lymph nodes.

The Non-Hodgkin's Lymphomas

The non-Hodgkin's lymphomas include all lymphomas without the features of Hodgkin's disease. Virtually all are monoclonal lymphoid neoplasms that display extremely heterogeneous morphologic, antigenic, and kinetic phenotypes as well as a highly variable clinical expression.

The working formulation classification divides non-Hodgkin's lymphoma into three groups: low, intermediate, and high grade, these groups are futher arranged into numerous subtypes (39). Most oncologists treat the indolent low-grade lymphomas conservatively. On the other hand, both intermediate and high-grade lymphomas behave aggressively and are treated with curative intent.

As the name implies, the indolent or low grade non-Hodgkin's lymphomas have a low proliferative capacity. However, these tumors are often disseminated at presentation, have a tendency to recur after successful initial therapy, and are rarely cured. Despite these ominous characteristics, most patients with low-grade lymphomas can enjoy a relatively long survival. Aggressive non-Hodgkin's lymphomas, on the other hand, are rapidly progressive tumors that can cause substantial damage or death if left untreated or if treated inadequately. In contrast to those with low-grade lymphomas, patients with aggressive non-Hodgkin's lymphomas may respond with prolonged remissions and even be cured if the tumors are recognized early and are appropriately treated.

The pathogenesis of growth of these two groups of lymphomas is likely to be different. Obviously, in both instances, cell production exceeds cell death. However, while low-grade tumors most likely develop as a consequence of cell accumulation and prolonged cell survival (133, 134), the growth of aggressive lymphomas, despite considerable cell death, is mainly due to an increased number of cells undergoing DNA synthesis and division (99).

It is important to keep in mind, however, that some lymphoid tumors cannot readily be assigned to either of the two major groups and that, in a certain percentage of cases, low-grade malignancies can "transform" into aggressive lymphomas during the course of the disease.

INDOLENT (LOW-GRADE) LYMPHOMAS AND CHRONIC LYMPHOPROLIFERATIVE DISORDERS

These are the most common lymphoid neoplasms of middle-aged and older populations. The vast majority are B-cell monoclonal tumors that are slowly progressive and frequently involve multiple lymphoid sites. Some types tend to permeate nonlymphoid tissues including marrow and blood, in which case they are designated as chronic leukemias or chronic lymphoproliferative disorders.

Small Lymphocytic Lymphoma (SL) and Chronic Lymphocytic Leukemia (CLL). These are chronic neoplasms composed of clonal, predominantly small, mature-appearing lymphocytes. The two terms are applied to essentially the same disease. SL is the histological label used by pathologists when lymphoid tissues are involved, whereas CLL is applied to a condition in which there is overt involvement of blood and bone marrow. Although SL and CLL may differ in the expression of certain adhesion molecules, which may be responsible for their different anatomical distribution (135), the cells in both SL and CLL are otherwise phenotypically identical. They express B-cell antigens such as CD19 and CD20, the latter usually faintly (8, 33, 136) (Fig. 12.4), and only a single immunoglobulin light chain is identifiable. However, cell surface immunoglobulins (sIg) may not be detectable since they are also often poorly expressed (8, 33, 54, 137). IgM and IgD are the most common sIg present. CD5, present only on T-cells and a small fraction of normal B-lymphocytes, is present on SL and CLL cells (138–140). Quantitatively, however, CD5 is less well expressed on SL and CLL cells than on normal T-cells (8, 77, 136) (Fig. 12.4). CD10 is rarely expressed in these neoplasms.

The term Waldenström's macroglobulinemia is applied to patient with SL or CLL who suffer from high content of monoclonal serum IgM. In these cases, numerous plasma cells and lymphoplasmacytoid cells are often present in the tissues involved. These cells are recognizable morphologically and, although the expression of sIg is highly variable, they contain easily detectable cytoplasmic immunoglobulin of a single light chain type. Extra care should be taken in the analysis of Ig light chain expression in tumors from patients with high content of serum monoclonal Ig. In these cases, cytophilic Ig may interfere with the interpretation of results (see clonal B-cell expansion above).

Plasmacytoid cells may be frequent in some SL. They may express interleukin-2 (CD25) and other activation antigens such as transferrin (CD71) receptors (141) or 4F2 (142). B-cell-specific antigens, CD5 and HLA-DR may be weak or absent and even CD45 may be lacking. CD10 is usually not detectable in these cells (143, 144).

Kinetically, SL and CLL neoplasms show a very low synthetic phase and the DNA content analysis demonstrates no overt aneuploidy (83). If transformation to a more aggressive form is present (e.g., Richter's syndrome), numerous larger cells and evidence of higher proliferation are detectable. In these cases, the intensity of expression of sIg and other antigens may also change.

Miscellaneous Lymphoproliferative Disorders. Prolymphocytic leukemia (PLL) is a morphologically defined variant of CLL that may represent a transformation of this disease (145). PLL cells have been described as having abundant cytoplasm and prominent single nucleoli; the condition is associated with high lymphocyte counts, splenomegaly, and minimal lymphadenopathy (146) and the prognosis of these patients, although extremely variable, may be worse than that of patients with classical CLL. The cells of PLL are larger than those of common CLL and mature B-cell markers, such as CD20 and FMC7, are well expressed. Also, a higher density of membrane immunoglobulins can be demonstrated in most cases (145). Although variable, the proliferative fraction in PLL may be higher than in CLL (147).

T-CLL has also been described (148). This lymphoproliferative disorder has not been well defined and probably encompasses a variety of biologically diverse diseases including the so-called large granular lymphocytosis and circulating forms of cutaneous T-cell lymphomas (149, 150).

In general, chronic neoplastic T-cell lymphoproliferative disorders consist of post-thymic T-cells expressing either CD4 or CD8 antigens. Often, these neoplastic cells are recognized on the basis of their abnormal T-cell antigen density which can be detected using two-color analysis with a combination of either CD5 or CD7 and CD3 antibodies. In these bivariate plots, most neoplastic T-cells form clusters that are different from normal T-cells.

Circulating neoplastic B-cells may be seen in patients with follicular and intermediate differentiation lymphoma (see next section). Cytologically, these cells show nuclear irregularities and the phenotype is identical to that of the cells in the involved lymphoid tissues.

Follicular Lymphomas. These tumors tend to grow in nodular aggregates mimicking lymphoid follicles. The neoplastic cells are invariably B-cells, but there is considerable variation in tumor cell size among different tumors. This variation is the basis for the subclassification of these lymphomas into small-, mixed-, and large-cell types. Variation in cell size may relate to the proliferative capacity of the cells or to their DNA content, since large follicular lymphoma cells are often aneuploid (151). Regardless of their significance, large cells may have a proliferative advantage or an increased resistance to therapy, since the large cell variants of these lymphomas often become "diffuse" in growth pattern and may be associated with a worse prognosis than the small and mixed types.

Follicular lymphoma cells express CD20 intensely and there is usually little difficulty in demonstrating single immunoglobulin light chains on their surface (152, 153) (Fig. 12.2). The immunoglobulin heavy chain isotype is highly variable. A small proportion of these tumors, however, despite bearing mature B-cell markers, lack immunoglobulin light chain expression (74, 154). B-cells deficient of Sig may be sporadically observed in reactive lymphoid processes. We have occasionally seen this situation in cases with marked reactive germinal center hyperplasia (e.g., HIV infection). This finding may be related to the observation that normal germinal center cells lack detectable light chain expression (74). In general, aberrant pan-B-cell antigen expression in follicular lymphoma is much less common than in aggressive B-cell lymphomas.

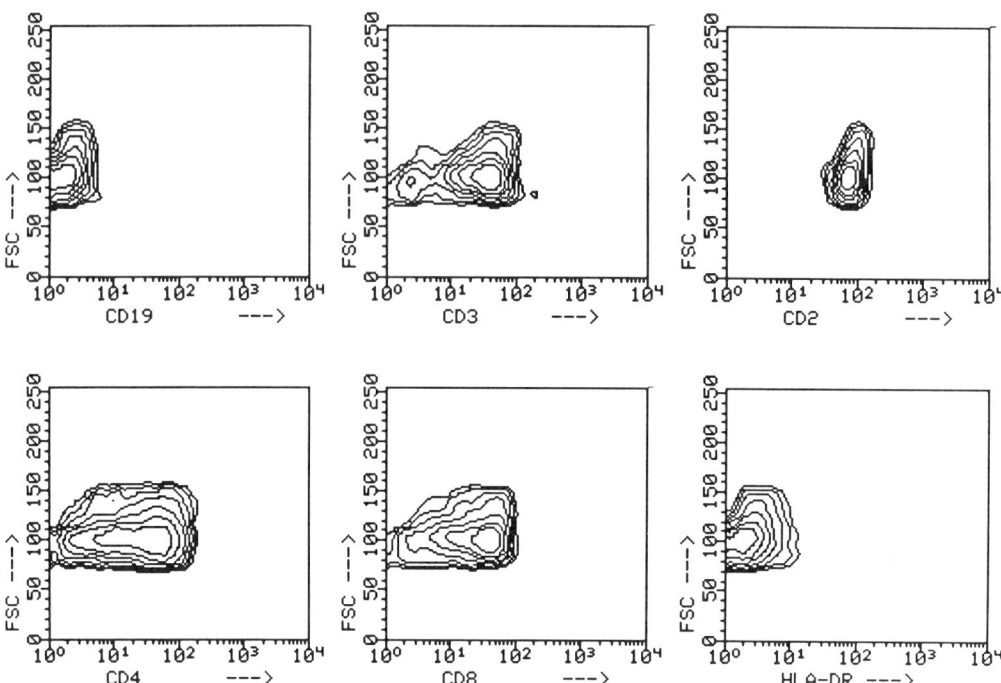

Figure 12.9. T-lymphoblastic lymphoma. Cells from these lymphomas often resemble thymic cells. They lack B-cell antigens and express T-cell antigens (top chart). Both *CD4* and *CD8* are expressed and, as in T-cell acute lymphoblastic leukemias, *HLA-DR* is not usually expressed (bottom charts). Lymphoblastic lymphomas, although resembling thymic cells, may also demonstrate severe aberrancies in antigen expression.

CD10 is detectable in approximately 75% to 80% of cases of follicular lymphomas (153). Activation antigens such as 4F2 are generally expressed poorly (155) and transferrin receptor is more often observed in these tumors than in other low-grade lymphomas (156). In typical follicular lymphomas CD5 expression is observed only exceptionally.

Small non-neoplastic T-cells with a normal distribution of T-cell subtypes are always present in follicular lymphomas and their numbers may vary greatly among tumors (Fig. 12.2). The majority of follicular lymphomas show a relatively low proliferative capacity and this is easily demonstrated by DNA content analysis (38, 83, 84, 88). Many of these tumors are DNA-aneuploid with near-diploid DNA content. These can either demonstrate G_0-G_1 coefficient of variation or identifiable near-diploid DNA-aneuploid peaks when sensitive flow cytometry methods are applied (38). These changes reflect the high frequency of small gains in chromosomal material (80). However, a low percentage of follicular lymphomas may show overt DNA-aneuploid peaks (38, 88).

Predominantly small cell lymphomas that do not readily match the typical description of SL or follicular lymphoma or may have features of both, are often called low-grade tumors of "intermediate differentiation" (157, 158). The so-called mantle zone lymphoma is regarded as a variant of lymphoma of intermediate differentiation (159). These intermediate differentiation lymphomas are of B-cell origin and frequently express CD5 but not CD10 antigens (138, 160, 161).

AGGRESSIVE LYMPHOMAS

These are heterogeneous tumors that vary greatly in morphologic and antigenic expression. Genotypically, they are also diverse. The most common type in this group is the diffuse large-cell lymphoma. The majority (approximately 80% to 90%) display B-cell markers and the remaining are of T-cell phenotype. CD10 is demonstrable in approximately 10% of the B-cell cases (162). Phenotypic aberrancies, such as loss of normal pan-B antigens or surface Ig (74, 154), or expression of Ig heavy chain without detectable Ig light chain, may be seen (Braylan RC, unpublished). Nevertheless, the frequency of aberrant phenotypes in B-cell lymphomas is lower than in T-cell neoplasms. Loss of pan-B-cell antigens such as CD19, CD20, or CD22 is more likely to be observed in lymphomas with higher S fractions (86).

Diffuse large-cell lymphomas of B-cell phenotype are frequently DNA-aneuploid (38, 88) and the proliferative fractions are extremely variable but higher than that of the low-grade lymphomas (38, 83, 84, 88).

Diffuse small cleaved lymphoma is a morphological entity that may represent a variety of biologically different tumors. Some observers have indicated that such tumors are mostly B-cell in origin and lack CD10 (162). However, it would be extremely difficult to expect any meaningful correlations between immunological data and an ill-defined morphological tumor type. The diffuse mixed (small and large) cell lymphomas are immunologically heterogeneous although they often display mature (post-thymic) T-cell markers (163).

Aggressive lymphomas demonstrate higher RNA index than low-grade lymphomas (84). Immunoblastic lymphomas (IL) are a morphological variant of large-cell lymphomas that exhibit the highest RNA index (84). However, there are major uncertainties in categorizing this group of tumors precisely. In the working formulation classification, immunoblastic lymphomas are classified as high-grade lymphomas and, unfortunately, separated from other types of diffuse large-cell lymphomas that are included in the intermediate-grade group. This separation seems artificial since IL may not be reproducibly identified and no prognostic differences between IL and non-IL diffuse large-cell lymphomas have been observed even in large series (119, 164).

The majority of T-cell malignancies are aggressive. Immunophenotypically, the cells of some T-lymphomas resemble thymic lymphocytes, whereas other tumors display a "peripheral" T-cell (post-thymic) antigenic profile. Diffuse large T-cell lymphomas show typically a peripheral T-cell phenotype, with malignant cells often expressing pan-T-cell markers aberrantly and CD4 or (less often) CD8 (165–167). As indicated above, the loss of one or more pan-T-cell antigens is commonly seen in these tumors (74, 75, 165, 166), an observation that has been confirmed by gene rearrangement analysis (168). The one pan-T-antigen most often absent in peripheral T-cell malignancies is CD7, although changes in the intensity of other pan-T antigens are also frequently seen (Fig. 12.7). HLA-DR, an antigen normally expressed on monocytes, B-cells, and activated T-cells is generally present in peripheral T-cell tumors. Some of the aggressive T-cell lymphomas are associated with HTLV-1 infection (see below). Lennert's lymphoma is a lymphoid process characterized by the presence of atypical lymphoid cells of various sizes accompanied by numerous epithelioid cells grouped in ill-defined clusters or sheets. There is good evidence that these lymphomas are composed of malignant T-cells with helper/inducer phenotype (62, 169, 170).

Transferrin receptor (CD71) is expressed on activated cells and is readily detectable in proliferative cells of aggressive lymphomas (86, 171, 172). Other antigens, such as the 4F2, that are expressed on activated or proliferative cells, are also better expressed in aggressive lymphomas (142, 155).

As in the majority of lymphomas, large-cell aggressive types contain normal T-cells. The number of T-cells in these tumors is highly variable, but in some B-cell lymphomas they may be so numerous that the tumors have been labeled as T-cell-rich B-cell lymphomas (173).

Ki-1-expressing large-cell lymphomas (174, 175) are tumors composed of bizarre atypical large cells which, in addition to Ki-1, frequently express T-cell and activation antigens. B-cell antigens have also been detected in some of these tumors, suggesting biological heterogeneity. Recent studies describe a possible association between a specific breakpoint (q35) in chromosome 5 and Ki-1 lymphomas (176). Ki-1 antibodies (CD30) also frequently label Reed-Sternberg and other abnormal cells in Hodgkin's disease.

True histiocyte (macrophage)-derived tumors mimicking lymphomas are exceedingly rare. The term malignant histiocytosis has been applied to a disease characterized by the presence of cells cytologically reminiscent of tissue macrophages (histiocytes) in lymph nodes, spleen, liver, and bone marrow. Modern technologies have shown that this disease represents a viral-mediated hemophagocytic syndrome or, in most instances, tumors of lymphoid origin (177, 178). Interestingly, the reciprocal translocation involving chromosomes 2 and 5 frequently observed in Ki-1 lymphomas (179) has been observed in some cases of malignant histiocytosis (180), suggesting an overlap between these two "entities."

Lymphoblastic Lymphomas (LL). LL are defined morphologically. Many of these tumors are simply lymphoid tissue involvement by acute lymphoid leukemia (ALL). Thus, cytologically, the neoplastic cells are indistinguishable from L1 or L2 ALL blasts and different lymphoid immunophenotypes have been identified (181, 182). LL affect predominantly children and young adults, although cases in older patients are not unusual (183).

The most frequent immunophenotype of LL is that of thymic cells (184) expressing CD1, TdT, CD7, and other pan-T-cell antigens (including sometimes CD3), and coexpressing CD4 and CD8, while HLA-DR and B-cell markers are not detected (Fig. 12.9). Cytoplasmic CD3 may be present in the absence of membrane CD3 (185–187). Less often, LL shows a phenotype of precursor B ALL (CD10+, CD19+, TdT+, HLDR+, SIg−) (188) and, occasionally, mature B-cell (189). Variation in antigen expression and aberrancies within the B- or T-cell lineages are not infrequent. The high frequency of T-cell phenotype in LL probably reflects the greater tendency of T-ALL blasts to invade lymph nodes and nonhemopoietic organs. Despite the morphologic similarities, some studies showed phenotypic differences between T-ALL and T-LL (184). Also, a recent report suggests that in T-ALL the T-cell γ-δ receptor (TcR) is more frequently utilized than the α-β TcR, and that the opposite occurs in the majority of T-LL (190).

Most LL show no detectable DNA aneuploidy although, occasionally, we have seen DNA indexes near 2.0. The proliferative fractions are high but variable (35, 102, 191).

Burkitt's Lymphomas. Burkitt's lymphoma is a high-grade B-cell tumor that is morphologically designated as a small non-cleaved cell type. It has been known from early studies that these tumors are highly proliferative (192, 193). Flow cytometry later confirmed that Burkitt's lymphomas are one of the most proliferative neoplasms, with S fractions usually exceeding 20% (35) (Fig. 12.10). Pan-B antigens are well expressed and monotypic sIg (often IgM with λ light chain) is easily detectable (194). Surface antigen aberrancies are infrequent and CD10 is commonly present (162, 194) (Fig. 12.10).

A group of "non-Burkitt," small, noncleaved cell lymphomas has been defined histologically. However, even experienced pathologists have difficulties reproducing such an observation (195). In a study by Garcia et al. (194) no valua-

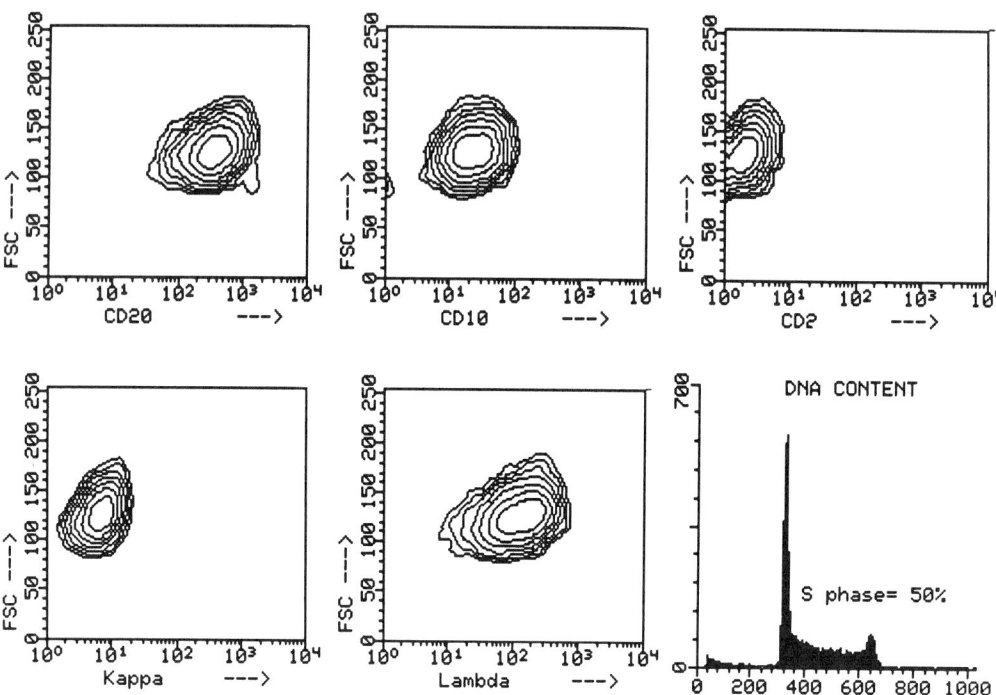

Figure 12.10. Burkitt's lymphoma (small noncleaved cell lymphoma). The cells of these tumors are B-cells that very frequently express CD10 (upper charts). The neoplastic B-cells are monoclonal, as shown by the different expression of κ and λ light chains (bottom left and center charts). As often seen in these tumors, very few normal cells are present. DNA content analysis shows a very large number of proliferating cells, a characteristic feature of these lymphomas (bottom right chart).

ble markers to identify the non-Burkitt's group were found, an observation that is hardly surprising since this group cannot always be recognized with certainty (196). As expected, at least in children, there are no clinical differences between Burkitt's and non-Burkitt's groups (197).

Burkitt's and Burkitt's-type lymphomas do not generally show overt DNA aneuploidy (38, 83, 88).

Aggressive B-cell lymphomas with morphological features of Burkitt's or large-cell lymphomas are seen with increased frequency in patients with congenital or acquired immunodeficiency, or immunosuppression (198, 199). These lymphomas are of rapid onset and often affect extranodal sites, such as the nervous system or gastrointestinal tract.

Special Types of Lymphomas and Lymphoproliferative Diseases

CUTANEOUS T-CELL LYMPHOMAS

These are skin lymphomas that display mature T-cell (postthymic) immunophenotypes (200–202). The term *mycosis fungoides* refers to a condition in which the malignant T-cells form defined skin lesions. Sezary syndrome is a disease characterized by malignant T-cells involving the skin in a diffuse manner, causing erythroderma and desquamation. In the latter, circulating cells with their characteristic cerebriform nuclei are usually recognized by morphological methods. In both conditions, malignant cells can often be detected in lymphoid organs and other viscera by sensitive methods. In this regard, loss of pan-T antigens (203) may be extremely helpful in identifying the malignant cells against a background of normal CD4-expressing cells. Also, the aberrant expression of B-associated antigens, such as CD24, may be of diagnostic value in these disorders (204).

DNA analysis in cutaneous T-cell lymphomas may show slightly abnormal (skewed) G_0-G_1 peaks, even in early stages (205). However, these changes may be seen also in benign cutaneous conditions. Discrete DNA-aneuploid peaks are confined to malignant lesions (206). Distinct DNA-aneuploidy is observed in approximately 40% of cases (207) and is associated with advanced disease, frequent relapse, and death (206, 207). As in other lymphomas, DNA-aneuploid tumors contain predominantly large cells (207).

ADULT T-CELL LYMPHOMAS/LEUKEMIAS

This is a T-cell neoplasia associated with the human T-cell lymphotropic (HTLV-1) retrovirus. The tumors are relatively frequent in Japan but are rare in the USA where they may be seen in the southeastern states. Many patients come from the Caribbean region and are often young black adults (208). The phenotype of the neoplastic cells is that of mature helper T-cells and characteristically express interleukin-2 receptors. The morphologic expressions of these tumors are quite variable although often they are classified as diffuse mixed or large cell types. In general, these lymphomas are aggressive and are associated with a short survival. However, chronic and "smoldering" forms have been described (209). As expected the chronic forms show a significantly lower proliferative fraction than the acute forms (210).

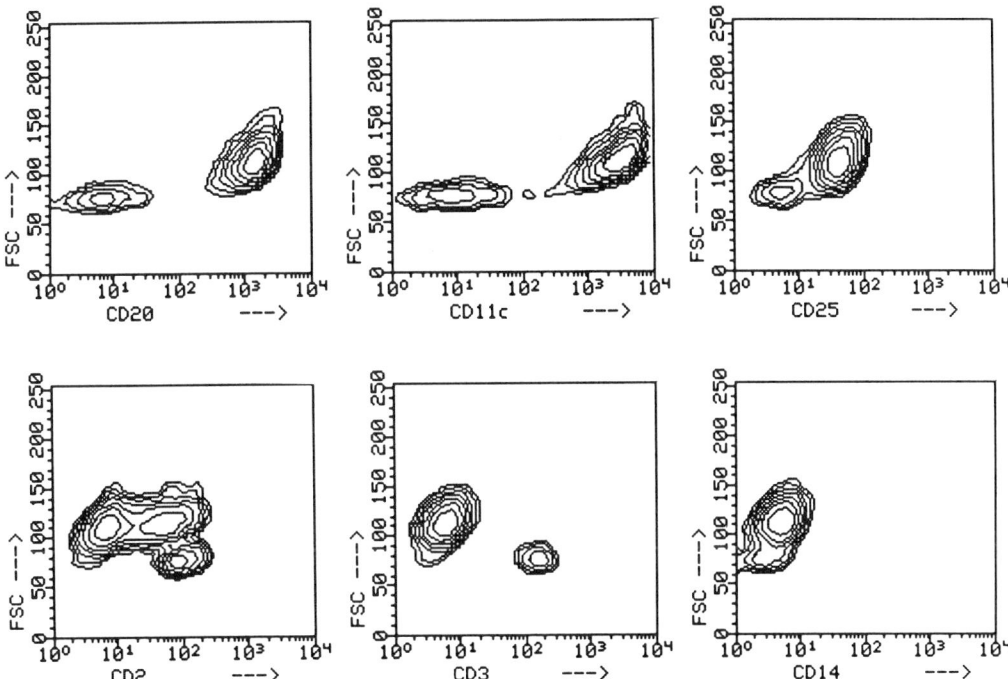

Figure 12.11. Hairy cell leukemia in blood. Two distinct populations differing in cell size are observed. The larger cells express *CD20* and *CD11c* (both intensely), and (*CD25*), the interleukin-2 receptor (top charts). Interestingly, a subpopulation of the large cells also expresses the T-cell-associated sheep erythrocyte receptor (*CD2*) (bottom left), an observation previously noted in some hairy cell leukemias (218). Most small cells are (normal) T-cells (bottom center). Monocytes are not detectable (bottom right).

HAIRY-CELL LEUKEMIA (HCL)

HCL is an infrequent but relatively well-defined chronic disease characterized by the presence of morphologically distinct leukocytes primarily in the spleen and bone marrow. These cells are often undetectable in the blood, although sometimes patients present with an overt leukemic picture. The diagnosis of HCL is of significance since the disease generally responds to some new therapeutic agents, such as α-interferon, with sustained and prolonged remissions.

HCL cells have been extensively characterized immunologically (211–215). Although some features of HCL cells are reminiscent of monocytes (216), rearrangement of the immunoglobulin genes are detectable in these cells. Also, single sIg light chain expression is the norm. HCL of T-cell origin or with T-cell features have been described but are exceedingly rare (217, 218) (Fig. 12.11). A human T-cell lymphotropic virus (HTLV II) has been isolated from some of these T-cell HCL (219). Some authors have suggested that HCL cells are related to preplasma cells. This has been supported by in vitro studies; the occasional case with Ig production; the rare instances of osteolytic lesions; and the partial, shared antigenicity with plasma cells. Clinically and biologically, however, the vast majority of strictly diagnosed HCL cases bear no resemblance to plasma-cell neoplasms. B-cell lymphoproliferative disorders with intermediate features between typical HCL and PLL and resistant to α-interferon therapy have been described (220).

No absolutely specific antibodies recognizing HCL exist. However, for diagnostic purposes, one can take advantage of the unique combined morphologic and antigenic features of HCL cells. For example, CD11c, a 150-kd adhesion receptor present on HCL cells (221), is also detectable in a fraction of CLL (222, 223) and may be seen in SL (141) and other lymphomas (224). However, HCL cells can be recognized with high specificity when CD11c is used in conjunction with CD20 (215) and forward-light-scatter analysis (Fig. 12.11). In contrast to CLL or SL, HCL cells express intense CD20 (Braylan RC, unpublished), their size is much larger than lymphocytes (216), and the small lymphocytes present in the sample are mostly normal T-cells (Fig. 12.11). Also helpful is CD5, which is not present in HCL except for anecdotal cases (225). FMC7, an antibody against determinants of some B-cells that are present in PLL and HCL but only rarely in CLL, may provide additional information.

HCL cells may occasionally show minimal DNA content abnormalities, but the proliferative compartments are invariably low (216, 226).

PLASMA-CELL TUMORS

Plasma-cell neoplasms involve most frequently the bone marrow and produce the clinical syndrome of multiple myeloma. Sometimes, however, neoplastic cells accumulate primarily in lymph nodes or other extraosseous sites. The phenotype of plasma-cell tumors has been extensively studied mainly in bone marrow cells of patients with myeloma (227–229). Typical plasma cells contain cytoplasmic immunoglobulin and express CD38 and 4F2 antigens, but generally lack sIg, HLA-DR, and B-cell antigens. CD10 may be expressed on myeloma plasma cells and this phenotype has been associated with a more aggressive behavior and poorer survival (230). However, the value of CD10 or other antigens for predicting survival has been recently questioned (229). The natural killer-associated antigen CD56 is also frequently expressed in myeloma plasma cells (231, 232). Antibodies reacting with plasma cells (PCA-1, PCA-2, PC-1) have been described (227, 233). Some of these antibodies are not absolutely specific for plasma cells and myeloma cells often fail to express them. Also, like normal plasma cells (234), myeloma cells may express myeloid or myelomonocytic antigens (235, 236).

In general, there is extreme variability in antigenic expression in myeloma, both among different tumors and within the same tumors (229). In our experience, the most valuable criterion for the diagnosis of malignant plasma-cell tumor is the detection cells that are morphologically consistent with plasma cells and contain single Ig light chain in the cytoplasm. Cytoplasm Ig can be detected by microscopic methods, but flow cytometric assays have also been developed (237–239).

In patients with myeloma, the presence of a high concentration of monoclonal serum immunoglobulin may indirectly affect the results of cell analysis since it may be difficult to eliminate nonspecifically bound serum immunoglulin from the cell surfaces. This phenomenon is readily apparent in monocytes, but it also occurs in other cells, creating the false impression that a monoclonal B-cell population is present. In these cases, the artifact is minimized by the simultaneous analysis of sIg and other antigens to identify the exact nature of the cells. In this regard, it is important to emphasize that the presence of monoclonal sIg on cells other than B-cells is highly suggestive of multiple myeloma or other monoclonal gammopathies.

Often, myeloma cells are DNA-aneuploid (237, 240, 241), which also helps in the diagnosis. Combined cytoplasmic Ig and DNA quantitation is an extremely valuable approach, allowing the characterization of DNA-aneuploid peaks and also the analysis of myeloma cell kinetics (242, 243). Interestingly, the level of p21 ras oncoprotein was found to be increased in aneuploid myeloma cells (244). The proliferative pool in patients with myeloma seems quite heterogenous, from early B-cell precursors to terminal plasma cells (243).

T-γ AND LARGE GRANULAR LYMPHOPROLIFERATIVE DISORDERS

These are chronic, indolent diseases characterized by cytopenias, splenomegaly without lymphadenopathy, autoimmune disorders, and an increased number in blood and bone marrow of lymphocytes exhibiting ample cytoplasm containing distinct azurophillic granules (245–249).

Early phenotypic studies of these lymphoctes suggested that they were natural killer cells. However, only a minority of these cases are CD3-negative CD16-positive. In most patients, the expanded lymphocytic population is CD3-positive cells and clonal T-cell gene rearrangements have been detected (250).

ANGIOIMMUNOBLASTIC LYMPHADENOPATHY (AIL)

Almost 20 years ago, several studies called attention to a distinct change in lymph nodes that was associated with a clinical condition characterized by constitutional symptoms, generalized adenopathy, hepatosplenomegaly, hypergammaglobulinemia, and autoimmune phenomena (251, 252). Although believed to be a severely abnormal immune response, the exact nature of the condition has remained unclear (review in (253)). Many of these patients develop malignant lymphoma and, frequently, the lesions demonstrate cytogenetic abnormalities and clonal gene rearrangements (254, 255). These observations suggest that AIL is potentially neoplastic. However, its clinical course, although often poor, is extremely variable and controversy still exists as to the appropriate therapy. The cells in the lymph nodes often express predominantly T-cell antigens, but no antigen loss has been observed. Most of the cell proliferation occurs in the T-cell compartment (256).

PROGNOSTIC SIGNIFICANCE

Studies attempting to determine the prognostic significance of certain biological properties of lymphoma cells often suffer from methodological deficiencies or from uncontrollable variables that are inherent to clinical research. These deficiencies are accentuated when results from different centers are compared. Thus, different therapeutic strategies, histological schemes, staging procedures, sample sources, factors that may vary with each laboratory, or the physician's decisions may influence prognosis. With this caveat in mind, it is worth mentioning some important observations.

Lymphoma Antigens

Since the vast majority of low-grade tumors are of B-cell origin, most efforts directed at determining if cell derivation is an independent prognostic variable have been concentrated on the aggressive lymphomas. This issue is still controversial (163, 257–259), although it appears that T-cell lymphomas are associated with a higher relapse rate and worse survival than their B-cell counterparts (260, 261). In particular, large-cell lymphomas of "peripheral" T-cell phenotype are probably more aggressive than other lymphomas of similar histology and may require different treatment strategies (262). Absence of HLA-DR and other immunophenotypical aberrancies in aggressive B-cell lymphomas have been associated with decreased survival (263, 264).

In a prospective study of a large number of patients with aggressive lymphomas treated in a uniform manner, no significant differences in prognostic variables such as age, stage, symptoms, tumor size, or performance status were found between B- and T-cell lymphoma patients (265).

Also, no significant differences in complete remission rate or overall survival were found between B- and T-cell lymphoma patients when the extent of the disease was relatively limited. However, for stage IV disease, the prognosis was significantly worse for patients with T-cell lymphoma.

Cell-activation antigens such as 4F2 may be used to discriminate between indolent and aggressive lymphomas and to predict prognosis in low-grade tumors (155, 266). Patients with CLL expressing only SIgM show a significantly worse survival than those either expressing IgD only, or coexpressing SIgM and SIgD (267).

Although most follicular lymphomas follow a chronic indolent course, a few become rapidly resistant to therapy or transform into aggressive forms (e.g., diffuse large-cell lymphoma). At present, no specific immunophenotypical finding has been found that would serve as a predictor of such a clinical behavior.

The significance of normal T-cells infiltrating follicular lymphomas or other B-cell lymphomas is not well known. In follicular lymphomas in general, T-cells do not seem to have prognostic significance or play a role in progression or transformation (268). However, in a relatively small series, patients who had spontaneous regression of these tumors had significantly more T-helper cells in the lymphomas than patients with progressive disease (269). Also, low number of host CD4 cells appear to correlate with worse survival in diffuse small B-cell lymphomas (270). It has been reported that, in untreated B-CLL, circulating T-cells are higher than in healthy controls, and that the total T- and CD8 cells are significantly increased in advanced and progressive stages of the disease (271). A recent study of patients with B-cell diffuse large-cell lymphoma demonstrated a reduced relapse-free survival (but not overall survival) for patients whose tumors contained a low percentage of CD8 lymphocytes (272).

The role that oncogene translocations play in uncontrolled cell growth and programmed cell death is being intensively examined at the molecular level. The expression of some oncogenes such as c-myc and bcl-2, has been analyzed in non-Hodgkin's lymphomas (16, 273). Although unregulated oncogene products should be of significant biological importance in these tumors, further studies will be necessary to determine their full clinical impact.

P-glycoprotein is a protein associated with multidrug resistance. It appears to function as a cellular efflux pump and is probably involved in detoxification. P-glycoprotein is infrequently expressed in untreated lymphomas but commonly detected in treated cases resistant to chemotherapy (274).

The expression of lymphocyte homing receptors (LHR), which mediate lymphocyte binding to high endothelial venules, is believed to play a role in the control of normal lymphocyte circulation (275). In diffuse large-cell lymphomas, an association may be seen between a strong expression of LHR antigens and generalized lymph node involvement (beyond stage II) (276, 277). Also, a better prognosis has been observed in patients whose lymphomas exhibited no or low levels of HR (277, 278).

CIRCULATING LYMPHOMA CELLS

Using morphological methods, approximately 10% of patients with follicular lymphomas demonstrate circulating neoplastic cells. As indicated before, the frequency is much higher when sensitive detecting techniques are used. Using their clonal excess analysis, Smith et al. (68) demonstrated a correlation between the presence of circulating monoclonal B-cells and disease stage. In patients with prolonged remission, those with follicular lymphomas often had persisting monoclonal populations, whereas those with diffuse lymphomas did not, suggesting a potential prognostic value for this analysis. The clinical significance of circulating malignant cells in low grade tumors is not the same as in aggressive lymphoma, where this finding usually represents a terminal event (279).

DNA Content

Numerous investigators have studied correlations between tumor ploidy or proliferation fraction, and patient response to therapy or survival. One should emphasize that there is extreme variation among laboratories in sample preparation and staining, instrument type and performance, criteria for definition of aneuploidy, cell fraction calculations, doublet and debris subtraction, and exclusion of cases for analysis (Review in (280)). This variability, when added to the heterogeneity of clinical parameters used in many of these studies, can account for some of the discrepancies reported. However, certain observations are consistently reproduced and are important to emphasize here.

ABNORMAL DNA CONTENT (DNA ANEUPLOIDY)

Ploidy changes in lymphomas have more diagnostic than prognostic value. Correlations between survival and aneuploidy remain controversial (85, 92) and there is no convincing evidence that aneuploidy, as applied to lymphomas in general, has any prognostic significance.

The clinical implications of aneuploidy are also debatable even for relatively homogeneous lymphoma groups, such as the aggressive types, with some authors claiming a survival advantage for aneuploid tumors (21) and others finding the opposite (90, 151), or no correlation (91, 93). Also, the clinical significance of distinct aneuploidy in low-grade lymphomas is not well known. Christensson et al. found that aneuploidy is associated with poorer treatment response and shorter remission duration (102). On the other hand, Macartney et al., in a study of 83 cases of follicular lymphoma, found no difference in survival between DNA-diploid and aneuploid cases (281).

S PHASE AND PROLIFERATIVE FRACTION

In view of the relationship between proliferative fractions and histological grade, it is not surprising that many authors observed excellent correlations between cell cycle data and patient response to therapy or survival (84, 95, 103–105, 191, 266, 282). More interesting is the fact that this correlation was observed even within histologically defined (21, 85, 91, 100, 263, 270, 281, 283–285) or immunologically homogeneous tumors (286). Christensson et al. (102) found that the proliferative fraction not only correlates very well with histological grade a with relapse rate, but may even be a better predictor of survival than histology.

Andreeff et al. (105) used AO stain to measure DNA and RNA content in 96 B-cell non-Hodgkin's lymphomas. These measurements were correlated with a number of other variables, including thymidine labeling index, cell volume, histological classification, and patient survival. Of all the variables investigated, RNA index, S phase, and DNA had prognostic significance. Interestingly, RNA index seemed prognostically more important than any other single analysis. When analyzed with identical statistical methods, S phase and RNA measurements together were superior in predicting survival than four popular morphologic classifications.

Multivariate analysis of DNA and RNA content using AO staining in a large series of heterogeneously treated follicular and diffuse large-cell lymphomas showed a significant survival advantage for patients having tumors with low proliferative fractions, even when the analysis was restricted to the diffuse large-cell category. In this study also, tumors with an ''intermediate'' RNA index were associated with the most favorable prognosis (151).

S-phase cells also appear to have prognostic significance in B-CLL (267). Interestingly, lower S fractions have been shown in transformed follicular lymphomas than in the de novo diffuse forms (35). Also, in low-grade lymphomas that recurred, the S-fraction of the initial biopsy was higher in tumors that transformed than in those that did not (287).

Very few studies dispute the general correlation between the size of the proliferative compartments of the lymphoma and patient survival. Morgan et al. (92), failed to show such a correlation. However, their conclusion was based on a series of patients diagnosed between 1963 and 1967. In their study, no details on therapy were provided, the quality of the analysis appeared suboptimal (mean CV of diploid tumors = 6.4% and abundant debris in their Figure 1), and the proliferative fractions were not calculated in aneuploid tumors (26% of all cases and 38% of high grade lymphomas). Cavalli et al. (89) showed statistically significant better patient survival for low S-phase than high S-phase tumors at 6- to-7 years after diagnosis. However, beyond that period the differences decreased. This finding is to be expected since low grade lymphomas progress slowly but are seldom cured.

Assessment of the proliferative capacity of a lymphoma is of paramount importance in patient management. Thus, rapid determination of cell cycle fractions by flow cytometry may be very useful in designing therapy for individual patients.

With regard to myelomas, several studies also suggest that the proliferative properties of the tumor cells may be of importance in prognosis (241, 288, 289).

PRACTICAL CONSIDERATIONS

Flow cytometric analysis of antigens and DNA content is only one of many new diagnostic technologies available to the modern pathologist handling a potential lymphoma. When is it appropriate to use these techniques? Most investigators would agree that current immunological and genetic methods are valuable adjuncts in the diagnosis of lymphoproliferative diseases. Immunophenotypical analysis by flow cytometry matches the morphologic diagnosis and determines the neoplastic cell lineage in more than 85% of cases of non-Hodgkin's lymphoma (132). A recent study suggests that immunophenotypic analysis helps resolve significant diagnostic problems in approximately half of the cases of lymphoid proliferations considered to be difficult to interpret by routine morphology (290). Certainly, these figures can vary greatly with the pathologist's experience in the field and with his or her degree of confidence on his or her own microscopic abilities. Also, decisions may be influenced by an informed and demanding clinician who is well aware of the shortcomings of conventional histology or wishes to obtain more precise biological information that may help the management of the patient.

Many laboratories perform immunohistochemical, flow cytometric or genetic analyses of lymphoid tissues only when lesions are difficult to interpret by conventional histology. However, even in histologically ''typical cases'' it is valuable to obtain biological information that can help to confirm the diagnosis and may also add prognostic significance to lymphomas of similar histological appearance. Furthermore, immunological or genetic information may be relevant in tumor recurrences. Presently, we are repeatedly confronted with possible lymphoma relapses in small samples from spinal fluid, needle aspirates, or fiberscope biopsies. Simple morphologic examination in these cases often lacks the sensitivity needed for tumor detection. On the other hand, previous information on tumor markers, the presence of an aneuploid peak, or a particular gene rearrangement in the original tumor, may be extremely useful in the interpretation of results. Also, there is increasing interest in treating lymphoma patients with bone marrow transplant procedures that often require knowledge of minimal tumor involvement in the marrow or may be subjected to in vitro purging of neoplastic cells.

For these reasons, we recommend that, whenever possible, even if ancillary analyses are not performed at the time of biopsy, fresh tissue or cells be stored frozen for possible subsequent diagnostic utilization.

THE FUTURE

Despite the extensive data accumulated over the last 15 years, flow cytometry of lymphomas and lymphoproliferative disorders is still a relatively young technique in the clinical laboratory. The major reasons for the slow acceptance of the flow cytometer, of course, have been its cost and complexity. Both of these have been substantially reduced recently. However, pathologists still rely extensively on tissue architecture for their interpretations of lymphoma and, despite the excellent results obtained with cells in fluids, blood, or marrow, as well as with materials from needle aspirates (1), they still feel uneasy about data obtained from tissues dispersed into single-cell suspensions.

Given the variability in recognizing and classifying lymphomas and the lack of a common language in using the histological method, it is exceedingly difficult to evaluate any new diagnostic technology by comparing it with histological schemes. The validity of a new technology will have to await proof of clinical merit independently of morphological bias. Before this is accomplished, the new method will have to demonstrate at least reproducibility and biological pertinence. Highly specific reagents, such as monoclonal antibodies that label relevant cellular constituents, have already proven to have acceptable reproducibility by immunohistology (291). These and other reagents that are capable of recognizing products that reflect meaningful genetic events, in conjunction with the quantitative abilities of flow cytometry, can be expected to be very valuable in the study of lymphomas.

Flow cytometry analysis offers distinctive advantages over other diagnostic methodologies. The use of automated instruments facilitate the complex task of multiple antibody labeling and analysis (6). The ability to analyze multiple cell properties and constituents simultaneously, and to determine the amount per cell of these constituents is unique and provides an unparalleled insight into the biology of lymphomas, which are characteristically composed of heterogeneous cell populations (292). Furthermore, subjectivity is minimized by the quantitative nature of the measurements. The data may be easily and permanently stored, retrieved, copied, and conveniently transmitted and interpreted from remote locations. Eventually, even data interpretation may be handled by computers.

Another exceptional attribute of flow cytometry analysis that is rarely exploited in the clinical laboratory is sensitivity. As shown above and also in the literature, flow cytometry analysis in B-cell lymphomas may be as sensitive as Southern analysis (52, 293). Although this capability has not been rigorously tested in T-cell lymphomas, high sensitivity of detection may also be attainable by taking advantage of the aberrant phenotype frequently observed in these tumors.

The abnormalities in genetic make up of lymphoid tumors is rapidly being characterized in detail. This information should eventually translate into knowledge of quantitative or qualitative changes in the products coded by the abnormal genes, and the generation of specific antibodies that recognize these changes. Also, it is likely that reagents targeted on therapeutically significant cell components such as the multi-drug resistance P-glycoprotein (274), will be more frequently utilized.

In the future, we should expect advances in hardware and software that will help to improve cell and data analysis. The complexity and cost of the instruments will continue to decrease and automation in cell staining and analysis will be routine. Better disaggregation procedures may reduce sampling errors in solid tissues and new fluorochromes that allow multiple and simultaneous labeling should minimize sample size requirements. As mentioned before, the large data provided by the flow cytometer can no longer be easily handled visually or manually and translated into simple percentages. The complex plots generated by the binding of the many reagents used routinely in the characterization of lymphoid tumors and their clinical significance will ultimately have to be interpreted by computer. Undoubtedly, these biological and technological advances will lead to a better understanding of lymphoid neoplasia and to an improvement in disease detection and patient management.

ACKNOWLEDGMENT

The invaluable assistance of Jose Iturraspe and the Hematopathology medical technologists of Shands Hospital is greatly appreciated.

REFERENCES

1. Sneige N, Dekmezian RH, Katz RL, et al. Morphologic and immunocytochemical evaluation of 220 fine needle aspirates of malignant lymphoma and lymphoid hyperplasia. Acta Cytol 1990;34:311–322.
2. Pihan GA, Woda BA. Immunophenotypic analysis of cells isolated from bone marrow biopsies in patients with failed bone marrow aspiration ('dry tap'). Am J Clin Pathol 1990;93:545–548.
3. Witzig TE, Banks PM, Stenson MJ, et al. Rapid immunotyping of B-cell non-Hodgkin's lymphomas by flow cytometry. A comparison with the standard frozen-section method. Am J Clin Pathol 1990;94:280–286.
4. Garcia CF, Weiss LM, Lowder J, et al. Quantitation and estimation of lymphocyte subsets in tissue sections. Comparison with flow cytometry. Am J Clin Pathol 1987;87:470–477.
5. Segal GH, Edinger MG, Owen M, et al. Concomitant delineation of surface Ig, B-cell differentiation antigens, and HLADR on lymphoid proliferations using three-color immunocytometry. Cytometry 1991;12:350–359.
6. Braylan RC, Benson NA. Flow cytometric analysis of lymphomas. Arch Pathol Lab Med 1989;113:627–633.
7. Warnke RA, Gatter KC, Falini B, et al. Diagnosis of human lymphoma with monoclonal antileukocyte antibodies. N Engl J Med 1983;309:1275–1281.
8. Almasri NM, Duque RE, Iturraspe J, Everett ET, Braylan RC. Reduced expression of CD20 antigen as a characteristic marker for chronic lymphocytic leukemia. Am J Hematol 1992;(in Press)
9. Lindemalm C, Christensson B, Biberfeld G, et al. Prognostic significance of immunoglobulin isotype expression in B-cell non-Hodgkin's lymphoma. Med Oncol Tumor Pharmacother 1988;5:243–248.
10. Erber WN, Mason DY. Expression of the interleukin-2 receptor (Tac antigen/CD25) in hematologic neoplasms. Am J Clin Pathol 1988;89:645–648.
11. Sheibani K, Winberg CD, van-de-Velde S, Blayney DW, Rappaport H. Distribution of lymphocytes with interleukin-2 receptors (TAC antigens) in reactive lymphoproliferative processes, Hodgkin's disease, and non-Hodgkin's lymphomas. An immunohistologic study of 300 cases. Am J Pathol 1987;127:27–37.
12. Schroff RW, Bucana CD, Klein RA, Farrell MM, Morgan AC, Jr. Detection of intracytoplasmic antigens by flow cytometry. J Immunol Methods 1984;70:167–177.
13. Bardales RH, Al Katib AM, Carrato A, Koziner B. Detection of intracytoplasmic immunoglobulin by flow cytometry in B-cell malignancies. J Histochem Cytochem 1989;37:83–89.

14. Loftin KC, Reuben JM, Hersh EM, Sujansky D. Cytoplasmic IgM in leukemic B cells by flow cytometry. Leuk Res 1985;9:1379–1387.
15. Hirata M, Okamoto Y. Enumeration of terminal deoxynucleotidyl transferase positive cells in leukemia/lymphoma by flow cytometry. Leuk Res 1987;11:509–518.
16. Holte H, Stokke T, Smeland E, et al. Levels of myc protein, as analyzed by flow cytometry, correlate with cell growth potential in malignant B-cell lymphomas. Int J Cancer 1989;43:164–170.
17. Andreeff M, Slater DE, Bressler J, Furth ME. Cellular ras oncogene expression and cell cycle measured by flow cytometry in hematopoietic cell lines. Blood 1986;67:676–681.
18. Andreeff M, Beck JD, Darzynkiewicz Z, et al. RNA content in human lymphocyte subpopulations. Proc Natl Acad Sci U S A 1978;75:1938–1942.
19. Pelstring RJ, Hurtubise PE, Swerdlow SH. Flow-cytometric DNA analysis of hematopoietic and lymphoid proliferations: a comparison of fresh, formalin-fixed and B5-fixed tissues. Hum Pathol 1990;21:551–558.
20. McIntire TL, Goldey SH, Benson NA, Braylan RC. Flow cytometric analysis of DNA in cells obtained from deparaffinized formalin-fixed lymphoid tissues. Cytometry 1987;8:474–478.
21. Wooldridge TN, Grierson HL, Weisenburger DD, et al. Association of DNA content and proliferative activity with clinical outcome in patients with diffuse mixed cell and large cell non-Hodgkin's lymphoma. Cancer Res 1988;48:6608–6613.
22. Camplejohn RS, Macartney JC. Comparison of DNA flow cytometry from fresh and paraffin embedded samples of non-Hodgkin's lymphoma. J Clin Pathol 1985;38:1096–1099.
23. Braylan RC, Benson NA, Nourse V, Kruth HS. Correlated analysis of cellular DNA, membrane antigens and light scatter of human lymphoid cells. Cytometry 1982;2:337–343.
24. Lakhanpal S, Gonchoroff NJ, Katzmann JA, Handwerger BS. A flow cytofluorometric double staining technique for simultaneous determination of human mononuclear cell surface phenotype and cell cycle phase. J Immunol Methods 1987;96:35–40.
25. Hallden G, Andersson U, Hed J, Johansson SG. A new membrane permeabilization method for the detection of intracellular antigens by flow cytometry. J Immunol Methods 1989;124:103–109.
26. Rigg KM, Shenton BK, Murray IA, Givan AL, Taylor RM, Lennard TW. A flow cytometric technique for simultaneous analysis of human mononuclear cell surface antigens and DNA. J Immunol Methods 1989;123:177–184.
27. Drach J, Gattringer C, Glassl H, Schwarting R, Stein H, Huber H. Simultaneous flow cytometric analysis of surface markers and nuclear Ki-67 antigen in leukemia and lymphoma. Cytometry 1989;10:743–749.
28. Slaper Cortenbach IC, Admiraal LG, Kerr JM, van Leeuwen EF, von dem Borne AE, Tetteroo PA. Flow-cytometric detection of terminal deoxynucleotidyl transferase and other intracellular antigens in combination with membrane antigens in acute lymphatic leukemias. Blood 1988;72:1639–1644.
29. Clevenger CV, Bauer KD, Epstein AL. A method for simultaneous nuclear immunofluorescence and DNA content quantitation using monoclonal antibodies and flow cytometry. Cytometry 1985;6:208–214.
30. Almasri NM, Iturraspe JA, Benson NA, Chen MG, Braylan RC. Flow cytometric analysis of terminal deoxynucleotidyl transferase. Am J Clin Pathol 1991;95:376–380.
31. Anastasi J, Bauer KD, Variakojis D. DNA aneuploidy in Hodgkin's disease. A multiparameter flow-cytometric analysis with cytologic correlation. Am J Pathol 1987;128:573–582.
32. Ratech H. HLA-DR expression in B-cell non-Hodgkin's malignant lymphomas: a multiparameter flow cytometry study. Hum Pathol 1990;21:1275–1282.
33. Cossman J, Neckers LM, Hsu S, Longo D, Jaffe ES. Low-grade lymphomas. Expression of developmentally regulated B-cell antigens. Am J Pathol 1984;115:117–124.
34. Braylan RC, Benson NA, Benson BA, Nourse VA. Analysis of neoplastic leukocyte surface antigens in unfractionated blood. Ann N Y Acad Sci 1986;468:160–170.
35. Braylan RC, Fowlkes BJ, Jaffe ES, Sanders SK, Berard CW, Herman CJ. Cell volumes and DNA distributions of normal and neoplastic human lymphoid cells. Cancer 1978;41:201–209.
36. Shackney SE, Skramstad KS, Cunningham RE, Dugas DJ, Lincoln TL, Lukes RJ. Dual parameter flow cytometry studies in human lymphomas. J Clin Invest 1980;66:1281–1294.
37. Weinberg DS. The role of cell cycle activity in the generation of morphologic heterogeneity in non-Hodgkin's lymphoma. Am J Pathol 1989;135:759–770.
38. Braylan RC, Benson NA, Nourse VA. Cellular DNA of human neoplastic B-cells measured by flow cytometry. Cancer Res 1984;44:5010–5016.
39. National Cancer Institute sponsored study of classifications of non-Hodgkin's lymphomas: summary and description of a working formulation for clinical usage. The Non-Hodgkin's Lymphoma Pathologic Classification Project. Cancer 1982;49:2112–2135.
40. Levy R, Warnke R, Dorfman RF, Haimovich J. The monoclonality of human B-cell lymphomas. J Exp Med 1977;145:1014–1028.
41. Ligler FS, Vitetta ES, Smith RG, et al. An immunologic approach for the detection of tumor cells in the peripheral blood of patients with malignant lymphoma: implications for the diagnosis of minimal disease. J Immunol 1979;123:1123–1126.
42. Garrett JV, Scarffe JH, Newton RK. Abnormal peripheral blood lymphocytes and bone marrow infiltration in non-Hodgkin's lymphoma. Br J Haematol 1979;42:41–50.
43. Ault KA. Detection of small numbers of monoclonal B lymphocytes in the blood of patients with lymphoma. N Engl J Med 1979;300:1401–1405.
44. Ligler FS, Smith RG, Kettman JR, et al. Detection of tumor cells in the peripheral blood of nonleukemic patients with B-cell lymphoma: analysis of "clonal excess." Blood 1980;55:792–801.
45. Brudler O, Han LT, Barcos M, et al. Determination of clonal excess in non-Hodgkin's lymphoma: clinical significance. Prog Clin Biol Res 1983;133:197–202.
46. Ratech H, Litwin S. Surface immunoglobulin light chain restriction in B-cell non-Hodgkin's malignant lymphomas. Am J Clin Pathol 1989;91:583–586.
47. Wong GY, Gebhard D, Mittleman A, Hancu M, Koziner B. Analysis of cell surface light chain immunoglobulin expression by flow cytometry in normal controls: a new mathematical approach. J Histochem Cytochem 1985;33:119–126.
48. Nakano M, Kuge S, Kuwabara LS, et al. The basic study on kappa-lambda imaging by delta-curve for the detection of a monoclonal B-cell population in the peripheral blood. Blood 1988;72:1461–1466.
49. Liendo C, Danieu L, Al Katib A, Koziner B. Phenotypic analysis by flow cytometry of surface immunoglobulin light chains and B and T cell antigens in lymph nodes involved with non-Hodgkin's lymlphoma. Am J Med 1985;79:445–454.
50. Weinberg DS, Pinkus GS, Ault KA. Cytofluorometric detection of B cell clonal excess: a new approach to the diagnosis of B cell lymphoma. Blood 1984;63:1080–1087.
51. Braylan RC, Benson NA, Goldey SH, Nourse VA. Precise quantitation by flow cytometry of light chain-bearing B-lymphocytes in peripheral blood [abstract]. Blood 1985;66:185a.
52. Letwin BW, Wallace PK, Muirhead KA, Hensler GL, Kashatus WH, Horan PK. An improved clonal excess assay using flow cytometry and B-cell gating. Blood 1990;75:1178–1185.
53. Leemhuis T, Srour E, Hanks S, Smith BR, Jansen J. Detection of infiltration of bone marrow by B-cell lymphoma using two-color fluorescence clonal-excess analysis [Abstract]. Blood 1987;70:217a.
54. de Martini RM, Turner RR, Boone DC, Lukes RJ, Parker JW. Lymphocyte immunophenotyping of B-cell lymphomas: a flow cytometric analysis of neoplastic and nonneoplastic cells in 271 cases. Clin Immunol Immunopathol 1988;49:365–379.

55. Charley M, McCoy JP, Deng JS, Jegasothy B. Anti-V region antibodies as "almost clonotypic" reagents for the study of cutaneous T cell lymphomas and leukemias. J Invest Dermatol 1990;95:614–617.
56. O'Grady J, Krajewski AS, Ramage EF. Demonstration of clonality in T-cell lymphoma using an anti-T-cell receptor variable region antibody panel. Histopathology 1990;17:553–556.
57. Hiddemann W, Schumann J, Andreeff M, et al. Convention on nomenclature for DNA cytometry. Committee on Nomenclature, Society for Analytical Cytology. Cancer Genet Cytogenet 1984;13:181–183.
58. Braylan RC, Benson NA. DNA content of solid tumors - Practical considerations. In: Nishiya I, Cram LS, Gray JW, eds. Flow cytometry and image analysis for clinical applications. Amsterdam: Elsevier Science, 1991:27–36.
59. Cusick EL, Milton JI, Ewen SW. The resolution of aneuploid DNA stem lines by flow cytometry: limitations imposed by the coefficient of variation and the percentage of aneuploid nuclei. Anal Cell Pathol 1990;2:139–148.
60. Shackney SE, Levine AM, Fisher RI, et al. The biology of tumor growth in the non-Hodgkin's lymphomas. A dual parameter flow cytometry study of 220 cases. J Clin Invest 1984;73:1201–1214.
61. Cunningham RE, Skramstad KS, Newburger AE, Shackney SE. Artifacts associated with mithramycin fluorescence in the clinical detection and quantitation of aneuploidy by flow cytometry. J Histochem Cytochem 1982;30:317–322.
62. Stonesifer KJ, Benson NA, Ryden SE, Pawliger DF, Braylan RC. The malignant cells in a Lennert's lymphoma are T lymphocytes with a mature helper surface phenotype. A multiparameter flow cytometric analysis. Blood 1986;68:426–429.
63. Baisch H, Gohde W, Linden WA. Analysis of PCP-data to determine the fraction of cells in the various phases of cell cycle. Radiat Environ Biophys 1975;12:31–39.
64. Dean PN. A simplified method of DNA distribution analysis. Cell Tissue Kinet 1980;13:299–308.
65. Ritch PS, Shackney SE, Schuette WH, Talbot TL, Smith CA. A practical graphical method for estimating the fraction of cells in S in DNA histograms from clinical tumor samples containing aneuploid cell populations. Cytometry 1983;4:66–74.
66. Van Eyken P, De Wolf Peeters C, Van den Oord J, Tricot G, Desmet V. Expression of leukocyte common antigen in lymphoblastic lymphoma and small noncleaved undifferentiated non-Burkitt's lymphoma: an immunohistochemical study. J Pathol 1987;151:257–261.
67. Smith BR, Robert NJ, Ault Ka. In Waldenstrom's macroglobulinemia the quantity of detectable circulating monoclonal B lymphocytes correlates with clinical course. Blood 1983;61:911–914.
68. Smith BR, Weinberg DS, Robert NJ, et al. Circulating monoclonal B lymphocytes in non-Hodgkin's lymphoma. N Engl J Med 1984;311:1476–1481.
69. Johnson A, Cavallin Stahl E, Akerman M. Flow cytometric light chain analysis of peripheral blood lymphocytes in patients with non-Hodgkin's lymphoma. Br J Cancer 1985;52:159–165.
70. Horning SJ, Galili N, Cleary M, Sklar J. Detection of non-Hodgkin's lymphoma in the peripheral blood by analysis of antigen receptor gene rearrangements: results of a prospective study. Blood 1990;75:1139–1145.
71. van der Harst D, de Jong D, Limpens J, et al. Clonal B-cell populations in patients with idiopathic thrombocytopenic purpura. Blood 1990;76:2321–2326.
72. Fox DA, Smith BR. Evidence for oligoclonal B cell expansion in the peripheral blood of patients with rheumatoid arthritis. Ann Rheum Dis 1986;45:991–995.
73. Jasani B. Immunohistologically definable light chain restriction in autoimmune disease. J Pathol 1988;154:1–5.
74. Picker LJ, Weiss LM, Medeiros LJ, Wood GS, Warnke RA. Immunophenotypic criteria for the diagnosis of non-Hodgkin's lymphoma. Am J Pathol 1987;128:181–201.
75. Hastrup N, Ralfkiaer E, Pallesen G. Aberrant phenotypes in peripheral T cell lymphomas. J Clin Pathol 1989;42:398–402.
76. Sun T, Ngu M, Henshall J, et al. Marker discrepancy as a diagnostic criterion for lymphoid neoplasms. Diagn Clin Immunol 1988;5:393–399.
77. Wormsley SB, Collins ML, Royston I. Comparative density of the human T-cell antigen T65 on normal peripheral blood T cells and chronic lymphocytic leukemia cells. Blood 1981;57:657–662.
78. Miller RA, Maloney D, Levy R. Monoclonal antibodies in the treatment of human leukemias and lymphomas: applications of flow cytometry. Ann N Y Acad Sci 1984;428:49–56.
79. Srigley JR, Butler JJ, Osborne BM, Guarda L, Barlogie B. Nucleic acid cytometry of homosexual-associated lymphoproliferative disease. Am J Pathol 1986;123:563–569.
80. Levine EG, Arthur DC, Frizzera G, Peterson BA, Hurd DD, Bloomfield CD. There are differences in cytogenetic abnormalities among histologic subtypes of the non-Hodgkin's lymphomas. Blood 1985;66:1414–1422.
81. Cabanillas F, Trujillo JM, Barlogie B, et al. Chromosomal abnormalities in lymphoma and their correlations with nucleic acid flow cytometry. Cancer Genet Cytogenet 1986;21:99–106.
82. Lakkala T, Laasonen A, Franssila KO, Teerenhovi L, Knuutila S. Comparison of DNA and karyotype aneuploidy in malignant lymphomas. Am J Clin Pathol 1990;94:600–605.
83. Diamond LW, Nathwani BN, Rappaport H. Flow cytometry in the diagnosis and classification of malignant lymphoma and leukemia. Cancer 1982;50:1122–1135.
84. Srigley J, Barlogie B, Butler JJ, et al. Heterogeneity of non-Hodgkin's lymphoma probed by nucleic acid cytometry. Blood 1985;65:1090–1096.
85. Rehn S, Glimelius B, Strang P, Sundstrom C, Tribukait B. Prognostic significance of flow cytometry studies in B-cell non-Hodgkin lymphoma. Hematol Oncol 1990;8:1–12.
86. Wain SL, Braylan RC, Borowitz MJ. Correlation of monoclonal antibody phenotyping and cellular DNA content in non-Hodgkin's lymphoma. The Southeastern Cancer Study Group experience. Cancer 1987;60:2403–2411.
87. O'Brien CJ, Holgate C, Quirke P, et al. Correlation of morphology, immunophenotype, and flow cytometry with remission induction and survival in high grade non-Hodgkin's lymphoma. J Pathol 1989;158:31–39.
88. Christensson B, Tribukait B, Linder IL, Ullman B, Biberfeld P. Cell proliferation and DNA content in non-Hodgkin's lymphoma. Flow cytometry in relation to lymphoma classification. Cancer 1986;58:1295–1304.
89. Cavalli C, Danova M, Gobbi PG, et al. Ploidy and proliferative activity measurement by flow cytometry in non-Hodgkin's lymphomas. Do speculative aspects prevail over clinical ones? Eur J Cancer Clin Oncol 1989;25:1755-1763.
90. Lehtinen T, Aine R, Lehtinen M, et al. Flow cytometric DNA analysis of 199 histologically favourable or unfavourable non-Hodgkin lymphomas. J Pathol 1989;157:27–36.
91. Young GA, Hedley DW, Rugg CA, Iland HJ. The prognostic significance of proliferative activity in poor histology non-Hodgkin's lymphoma: a flow cytometry study using archival material. Eur J Cancer Clin Oncol 1987;23:1497–1504.
92. Morgan DR, Williamson JM, Quirke P, et al. DNA content and prognosis of non-Hodgkin's lymphoma. Br J Cancer 1986;54:643–649.
93. Cowan RA. Harris M, Jones M, Crowther D. DNA content in high and intermediate grade non-Hodgkin's lymphoma-prognostic significance and clinicopathological correlations. Br J Cancer 1989;60:904–910.
94. Hansen H, Koziner B, Clarkson B. Marker and kinetic studies in the non-Hodgkin's lymphomas. Am J Med 1981;71:107–123.
95. Braylan RC, Diamond LW, Powell ML, Harty Golder B. Percentage of cells in the S phase of the cell cycle in human lymphoma determined by flow cytometry. Cytometry 1980;1:171–174.

96. Juneja SK, Cooper IA, Hodgson GS, et al. DNA ploidy patterns and cytokinetics of non-Hodgkin's lymphoma. J Clin Pathol 1986;39:987–992.
97. Meyer JS, Higa E. S-phase fractions of cells in lymph nodes and malignant lymphomas. Arch Pathol Lab Med 1979;103:93–97.
98. Costa A, Bonadonna G, Villa E, Valagussa P, Silvestrini R. Labeling index as a prognostic marker in non-Hodgkin's lymphomas. J Natl Cancer Inst 1981;66:1–5.
99. Lang W, Kienzle S, Diehl V. Proliferation kinetics of malignant non-Hodgkin's lymphomas related to histopathology of lymph node biopsies. Virchows Arch Pathol [A] 1980;389:397–407.
100. Kvaloy S, Marton PF, Kaalhus O, Hoie J, Foss-Abrahamsen A, Godal T. 3H-thymidine uptake in B cell lymphomas—relationship to treatment response and survival. Scand J Haematol 1985;34:429–435.
101. Diamond LW, Braylan RC. Flow analysis of DNA content and cell size in non-Hodgkin's lymphoma. Cancer Res 1980;40:703–712.
102. Christensson B, Lindemalm C, Johansson B, Mellstedt H, Tribukait B, Biberfeld P. Flow cytometric DNA analysis: a prognostic tool in non-Hodgkin's lymphoma. Leuk Res 1989;13:307–314.
103. Lenner P, Roos G, Johansson H, Lindh J, Dige U. Non-Hodgkin lymphoma. Multivariate analysis of prognostic factors including fraction of S-phase cells. Acta Oncol 1987;26:179–183.
104. Scarffe JH, Crowther D. The pre-treatment proliferative activity of non-Hodgkin's lymphoma cells. Eur J Cancer 1981;17:99–108.
105. Andreeff M, Hansen H, Cirrincione C, Filippa D, Thaler H. Prognostic value of DNA/RNA flow cytometry of B-cell non-Hodgkin's lymphoma: development of laboratory model and correlation with four taxonomic systems. Ann N Y Acad Sci 1986;468:368–386.
106. Gerdes J, Schwab U, Lemke H, Stein H. Production of a mouse monoclonal antibody reactive with a human nuclear anigen associated with cell proliferation. Int J Cancer 1983;31:13–20.
107. Schwartz BR, Pinkus G, Bacus S, Toder M, Weinberg DS. Cell proliferation in non-Hodgkin's lymphomas. Digital image analysis of Ki-67 antibody staining. Am J Pathol 1989;134:327–336.
108. Weiss LM, Strickler JG, Medeiros LJ, Gerdes J, Stein H, Warnke RA. Proliferative rates of non-Hodgkin's lymphomas as assessed by Ki-67 antibody. Hum Pathol 1987;18:1155–1159.
109. Egerter DA, Said JW, Epling S, Lee S. DNA content of T-cell lymphomas. A flow-cytometric analysis. Am J Pathol 1988;130:326–334.
110. Kim H, Zelman RJ, Fox MA, et al. Pathology Panel for Lymphoma Clinical Studies: a comprehensive analysis of cases accumulated since its inception. J Natl Cancer Inst 1982;68:43–67.
111. Metter GE, Nathwani BN, Burke JS, et al. Morphological subclassification of follicular lymphoma: variability of diagnoses among hematopathologists, a collaborative study between the Repository Center and Pathology Panel for Lymphoma Clinical Studies. J Clin Oncol 1985;3:25–38.
112. Classification of non-Hodgkin's lymphomas. Reproducibility of major classification systems. NCI non-Hodgkin's Classification Project Writing Committee. Cancer 1985;55:91–95.
113. Jones SE, Butler JJ, Byrne GE, Jr., Coltman CA, Jr., Moon TE. Histopathologic review of lymphoma cases from the Southwest Oncology Group. Cancer 1977;39:1071–1076.
114. Velez Garcia E, Durant J, Gams R, Bartolucci A. Results of a uniform histopathologic review system of lymphoma cases. A ten-year study from the Southeastern Cancer Study Group. Cancer 1983;52:675–679.
115. De-Wolf-Peeters C, Caillou B, Diebold J, et al. Reproducibility and prognostic value of different non-Hodgkin's lymphoma classifications: study based on the clinicopathologic relations found in the EORTC trial (20751). Eur J Cancer Clin Oncol 1985;21:579–584.
116. Ezdinli EZ, Costello W, Wasser LP, et al. Eastern Cooperative Oncology Group experience with the Rappaport classification of non-Hodgkin's lymphomas. Cancer 1979;43:544–550.
117. Bird CC, Lauder I, Kellett HS, et al. Yorkshire Regional Lymphoma Histopathology panel: analysis of five years' experience. J Pathol 1984;143:249–258.
118. Wolf BC, Gilchrist KW, Mann RB, Neiman RS. Evaluation of pathology review of malignant lymphomas and Hogkin's disease in cooperative clinical trials. The Eastern Cooperative Oncology Group experience. Cancer 1988;62:1301–1305.
119. Dick F, VanLier S, Banks P, et al. Use of the working formulation for non-Hodgkin's lymphoma in epidemiologic studies: agreement between reported diagnoses and a panel of experienced pathologists. J Natl Cancer Inst 1987;78:1137–1144.
120. Self SE, Burdash NM, Ponzio AD, Lavia MF. Lymphocyte subsets in lymph node hyperplasias and B cell neoplasms as determined by fluoresceinated antibodies and flow cytometry. Ann N Y Acad Sci 1986;468:195–210.
121. Diehl V, von-Kalle C, Fonatsch C, Tesch H, Juecker M, Schaadt M. The cell of origin in Hodgkin's disease. Semin Oncol 1990;17:660–672.
122. Falini B, Stein H, Pileri S, et al. Expression of lymphoid-associated antigens on Hodgkin's and Reed-Sternberg cells of Hodgkin's disease. An immunocytochemical study on lymph node cytospins using monoclonal antibodies. Histopathology 1987;11:1229–1242.
123. Regula DPJ, Hoppe RT, Weiss LM. Nodular and diffuse types of lymphocyte predominance Hodgkin's disease. N Engl J Med 1988;318:214–219.
124. Neiman RS, Rosen PJ, Lukes RJ. Lymphocyte-depletion Hodgkin's disease. A clinicopathological entity. N Engl J Med 1973;288:751–755.
125. Sitar G, Brusamolino E, Bernasconi C, Ascari E. Isolation of Reed-Sternberg cells from lymph nodes of Hodgkin's disease patients. Blood 1989;73:222–229.
126. Morgan KG, Quirke P, O'Brien CJ, Bird CC. Hodgkin's disease: a flow cytometric study. J Clin Pathol 1988;41:365–369.
127. Petrakis NL, Bostick WL, Wiegel BV. The deoxyribonucleic acid (DNA) content of Sternberg-Reed cells of Hodgkin's disease. J Natl Cancer Inst 1959;22:551–554.
128. Peckham MJ, Cooper EH. Proliferation characteristics of the various classes of cells in Hodgkin's disease. Cancer 1969;24:135–146.
129. Gerdes J, Van-Baarlen J, Pileri S, Schwarting R, van-Unnik JA, Stein H. Tumor cell growth fraction in Hodgkin's disease. Am J Pathol 1987;128:390–393.
130. Borowitz MJ, Croker BP, Metzgar RS. Immunohistochemical analysis of the distribution of lymphocyte subpopulations in Hodgkin's disease. Cancer Treat Rep 1982;66:667–674.
131. Knowles DM, Halper JP, Jakobiec FA. T-lymphocyte subpopulations in B-cell-derived non-Hodgkin's lymphomas and Hodgkin's disease. Cancer 1984;54:644–651.
132. Little JV, Foucar K, Horvath A, Crago S. Flow cytometric analysis of lymphoma and lymphoma-like disorders. Semin Diagn Pathol 1989;6:37–54.
133. Themi H, Trepel F, Schick P, Kaboth W, Begemann H. Kinetics of lymphocytes in chronic lymphocytic leukemia: studies using continuous 3H-thymidine infusion in two patients. Blood 1973;42:623–636.
134. Nunez G, London L, Hockenbery D, Alexander M, McKearn JP, Korsmeyer SJ. Deregulated Bcl-2 gene expression selectively prolongs survival of growth factor-deprived hemopoietic cell lines. J Immunol 1990;144:3602–3610.
135. Inghirami G, Wieczorek R, Zhu BY, Silber R, Dalla-Favera R, Knowles DM. Differential expression of LFA-1 molecules in non-Hodgkin's lymphoma and lymphoid leukemia. Blood 1988;72:1431–1434.
136. Marti GE, Faguet G, Bertin P, et al. CD20 and CD5 expression in B-chronic lymphocytic leukemia (B-CLL). Ann N Y Acad Sci 1992;(in Press)
137. Braylan RC, Jaffe ES, Burbach JW, Frank MM, Johnson RE, Berard CW. Similarities of surface characteristics of neoplastic well-differentiated lymphocytes from solid tissues and from peripheral blood. Cancer Res 1976;36:1619–1625.
138. Spier CM, Grogan TM, Fielder K, Richter L, Rangel C. Immunophenotypes in "well-differentiated" lymphoproliferative disorders, with

emphasis on small lymphocytic lymphoma. Hum Pathol 1986;17: 1126–1136.
139. Royston I, Majda JA, Baird SM, Meserve BL, Griffiths JC. Human T cell antigens defined by monoclonal antibodies: the 65,000-dalton antigen of T cells (T65) is also found on chronic lymphocytic leukemia cells bearing surface immunoglobulin. J Immunol 1980;125:725–731.
140. Burns BF, Warnke RA, Doggett RS, Rouse RV. Expression of a T-cell antigen (Leu-1) by B-cell lymphomas. Am J Pathol 1983;113: 165–171.
141. Medeiros LJ, Strickler JG, Picker LJ, Gelb AB, Weiss LM, Warnke RA. "Well-differentiated" lymphocytic neoplasms. Immunologic findings correlated with clinical presentation and morphologic features. Am J Pathol 1987;129:525–535.
142. Lewandrowski KB, Medeiros LJ, Harris NL. Expression of the activation antigen, 4F2, by non-Hodgkin's lymphomas of B-cell phenotype. Cancer 1990;66:1158–1164.
143. Anderson KC, Bates MP, Slaughenhoupt BL, Pinkus GS, Schlossman SF, Nadler LM. Expression of human B cell-associated antigens on leukemias and lymphomas: a model of human B cell differentiation. Blood 1984;63:1424–1433.
144. Gale RP, Foon KA. Biology of chronic lymphocytic leukemia. Semin Hematol 1987;24:209–229.
145. Stark AN, Limbert HJ, Roberts BE, Jones RA, Scott CS. Prolymphocytoid transformation of CLL: a clinical and immunological study of 22 cases. Leuk Res 1986;10:1225–1232.
146. Galton DA, Goldman JM, Wiltshaw E, Catovsky D, Henry K, Goldenberg GJ. Prolymphocytic leukaemia. Br J Haematol 1974;27:7–23.
147. Scott CS, Ramsden W, Limbert HJ, Master PS, Roberts BE. Membrane transferrin receptor (TfR) and nuclear proliferation-associated Ki-67 expression in hemopoietic malignancies. Leukemia 1988;2: 438–442.
148. Brouet JC, Sasportes M, Flandrin G, Preud'Homme JL, Seligmann M. Chronic lymphocytic leukaemia of T-cell origin. Immunological and clinical evaluation in eleven patients. Lancet 1975;2:890–893.
149. Huhn D, Thiel E, Rodt H, Schlimok G, Theml H, Rieber P. Subtypes of T-cell chronic lymphatic leukemia. Cancer 1983;51:1434–1447.
150. Catovsky D, Linch DC, Beverley PC. T cell disorders in haematological diseases. Clin Haematol 1982;11:661–695.
151. McLaughlin P, Osborne BM, Johnston D, et al. Nucleic acid flow cytometry in large cell lymphoma. Cancer Res 1988;48:6614–6619.
152. Stein H, Lennert K, Feller AC, Mason DY. Immunohistological analysis of human lymphoma: correlation of histological and immunological categories. Adv Cancer Res 1984;42:67-147.
153. Borowitz MJ, Bousvaros A, Brynes RK, et al. Monoclonal antibody phenotyping of B-cell non-Hodgkin's lymphomas. The Southeastern Cancer Study Group experience. Am J Pathol 1985;121:514–521.
154. Gregg EO, Al-Saffar N, Jones DB, Wright DH, Stevenson FK, Smith JL. Immunoglobulin negative follicle centre cell lymphoma. Br J Cancer 1984;50:735–744.
155. Salter DM, Krajewski AS, Sheehan T, Turner G, Cuthbert RJ, McLean A. Prognostic significance of activation and differentiation antigen expression in B-cell non-Hodgkin's lymphoma. J Pathol 1989; 159:211–220.
156. Medeiros LJ, Picker LJ, Horning SJ, Warnke RA. Transferrin receptor expression by non-Hodgkin's lymphomas. Correlation with morphologic grade and survival. Cancer 1988;61:1844–1851.
157. Mann RB, Jaffe ES, Berard CW. Malignant lymphomas—a conceptual understanding of morphologic diversity. A review. Am J Pathol 1979;94:105–191.
158. Weisenburger DD, Nathwani BN, Diamond LW, Winberg CD, Rappaport H. Malignant lymphoma, intermediate lymphocytic type: a clinicopathologic study of 42 cases. Cancer 1981;48:1415–1425.
159. Weisenburger DD, Kim H, Rappaport H. Mantle-zone lymphoma: a follicular variant of intermediate lymphocytic lymphoma. Cancer 1982;49:1429–1438.
160. Strickler JG, Medeiros LJ, Copenhaver CM, Weiss LM, Warnke RA. Intermediate lymphocytic lymphoma: an immunophenotypic study with comparison to small lymphocytic lymphoma and diffuse small cleaved cell lymphoma. Hum Pathol 1988;19:550–554.
161. Bookman MA, Lardelli P, Jaffee ES, Duffey PL, Longo DL. Lymphocytic lymphoma of intermediate differentiation: morphologic, immunophenotypic, and prognostic factors. J Natl Cancer Inst 1990;82: 742–748.
162. Ritz J, Nadler LM, Bhan AK, Notis-McConarty J, Pesando JM, Schlossman SF. Expression of common acute lymphoblastic leukemia antigen (CALLA) by lymphomas of B-cell and T-cell lineage. Blood 1981;58:648–652.
163. Cossman J, Jaffee ES, Fisher RI. Immunologic phenotypes of diffuse, aggressive, non-Hodgkin's lymphomas. Correlation with clinical features. Cancer 1984;54:1310–1317.
164. Simon R, Durrleman S, Hoppe RT, et al. The Non-Hodgkin Lymphoma Pathologic Classification Project. Long-term follow-up of 1153 patients with non-Hodgkin lymphomas. Ann Intern Med 1988; 109:939–945.
165. Weiss LM, Crabtree GS, Rouse RV, Warnke RA. Morphologic and immunologic characterization of 50 peripheral T-cell lymphomas. Am J Pathol 1985;118:316–324.
166. Borowitz MJ, Reichert TA, Brynes RK, et al. The phenotypic diversity of peripheral T-cell lymphomas: the Southeastern Cancer Study Group experience. Hum Pathol 1986;17:567-574.
167. Weisenburger DD, Linder J, Armitage JO. Peripheral T-cell lymphoma: a clinicopathologic study of 42 cases. Hematol Oncol 1987;5: 175–187.
168. Henni T, Gaulard P, Divine M, et al. Comparison of genetic probe with immunophenotype analysis in lymphoproliferative disorders: a study of 87 cases. Blood 1988;72:1937–1943.
169. Feller AC, Griesser GH, Mak TW, Lennert K. Lymphoepithelioid lymphoma (Lennert's lymphoma) is a monoclonal proliferation of helper/inducer T cells. Blood 1986;68:663–667.
170. Spier CM, Lippman SM, Miller TP, Grogan TM. Lennert's lymphoma. A clinicopathologic study with emphasis on phenotype and its relationship to survival. Cancer 1988;61:517–524.
171. Habeshaw JA, Lister TA, Stansfeld AG, Greaves MF. Correlation of transferrin receptor expression with histological class and outcome in non-Hodgkin lymphoma. Lancet 1983;1:498–501.
172. Das Gupta A, Shah VI. Correlation of transferrin receptor expression with histologic grade and immunophenotype in chronic lymphocytic leukemia and non-Hodgkin's lymphoma. Hematol Pathol 1990;4:37–41.
173. Ramsay AD, Smith WJ, Isaacson PG. T-cell-rich B-cell lymphoma. Am J Surg Pathol 1988;12:433–443.
174. Stein H, Mason DY, Gerdes J, et al. The expression of the Hodgkin's disease associated antigen Ki-1 in reactive and neoplastic lymphoid tissue: evidence that Reed-Sternberg cells and histiocytic malignancies are derived from activated lymphoid cells. Blood 1985;66:848–858.
175. Agnarsson BA, Kadin ME. Ki-1 positive large cell lymphoma. A morphologic and immunologic study of 19 cases. Am J Surg Pathol 1988;12:264–274.
176. Rimokh R, Magaud JP, Berger F, et al. A translocation involving a specific breakpoint (q35) on chromosome 5 is characteristic of anaplastic large cell lymphoma ('Ki-1 lymphoma'). Br J Haematol 1989; 71:31–36.
177. Wilson MS, Weiss LM, Gatter KC, Mason DY, Dorfman RF, Warnke RA. Malignant histiocytosis. A reassessment of cases previously reported in 1975 based on paraffin section immunophenotyping studies. Cancer 1990;66:530–536.
178. Delsol G, Al-Saati T, Gater KC, et al. Coexpression of epithelial membrane antigen (EMA), Ki-1, and interleukin-2 receptor by anaplastic large cell lymphomas. Diagnostic value in so-called malignant histiocytosis. Am J Pathol 1988;130:59–70.
179. Bitter MA, Franklin WA, Larson RA, et al. Morphology in Ki-1(CD30)-positive non-Hodgkin's lymphoma is correlated with clinical features and the presence of a unique chromosomal abnormality, t(2; 5)(p23;q35). Am J Surg Pathol 1990;14:305–316.

180. Benz-Lemoine E, Brizard A, Huret JL, et al. Malignant histiocytosis: a specific T(2;5)(p23;q35) translocation: Review of the literature. Blood 1988;72:1045–1047.
181. Weiss LM, Bindle JM, Picozzi VJ, Link MP, Warnke RA. Lymphoblastic lymphoma: an immunophenotype study of 26 cases with comparison to T cell acute lymphoblastic leukemia. Blood 1986;67:474–478.
182. Grogan T, Spier C, Wirt DP, et al. Immunologic complexity of lymphoblastic lymphoma. Diagn Immunol 1986;4:81–88.
183. Nathwani BN, Diamond LW, Winberg CD, et al. Lymphoblastic lymphoma: a clinicopathologic study of 95 patients. Cancer 1981;48:2347–2357.
184. Bernard A, Boumsell L, Reinherz EL, et al. Cell surface characterization of malignant T cells from lymphoblastic lymphoma using monoclonal antibodies: evidence for phenotypic differences between malignant T cells for patients with acute lymphoblastic leukemia and lymphoblastic lymphoma. Blood 1981;57:1105–1110.
185. Link MP, Stewart SJ, Warnke RA, Levy R. Discordance between surface and cytoplasmic expression of the Leu-4 (T3) antigen in thymocytes and in blast cells from childhood T lymphoblastic malignancies. J Clin Invest 1985;76:248–253.
186. van-Dongen JJ, Krissansen GW, Wolvers-Tettero IL, et al. Cytoplasmic expression of the CD3 antigen as a diagnostic marker for immature T-cell malignancies. Blood 1988;71:603–612.
187. Campana D, Thompson JS, Amlot P, Brown S, Janossy G. The cytoplasmic expression of CD3 antigens in normal and malignant cells of the T lymphoid lineage. J Immunol 1987;138:648–655.
188. Cossman J, Chused TM, Fisher RI, Magrath I, Bollum F, Jaffee ES. Diversity of immunological phenotypes of lymphoblastic lymphoma. Cancer Res 1983;43:4486–4490.
189. Stroup R, Sheibani K, Misset JL, Szekely AM, Tremblay G, Rappaport H. Surface immunoglobulin-positive lymphoblastic lymphoma. A report of three cases. Cancer 1990;65:2559–2563.
190. Gouttefangeas C, Bensussan A, Boumsell L. Study of the CD3-associated T-cell receptors reveals further differences between T-cell acute lymphoblastic lymphoma and leukemia. Blood 1990;75:931–934.
191. Hall PA, Richards MA, Gregory WM, dArdenne AJ, Lister TA, Stansfeld AG. The prognostic value of Ki67 immunostaining in non-Hodgkin's lymphoma. J Pathol 1988;154:223–235.
192. Cooper EH, Frank GL, Wright DH. Cell proliferation in Burkitt tumours. Eur J Cancer 1966;2:377–384.
193. Iversen OH, Iversen U, Ziegler JL, Bluming AZ. Cell kinetics in Burkitt lymphoma. Europ J Cancer 1974;10:155–163.
194. Garcia CF, Weiss LM, Warnke RA. Small noncleaved cell lymphoma: an immunophenotypic study of 18 cases and comparison with large cell lymphoma. Hum Pathol 1986;17:454–461.
195. Wilson JF, Kjeldsberg CR, Sposto R, et al. The pathology of non-Hodgkin's lymphoma of childhood: II. Reproducibility and relevance of the histologic classification of "undifferentiated" lymphomas (Burkitt's versus non-Burkitt's). Hum Pathol 1987;18:1008–1014.
196. Grogan TM, Warnke RA, Kaplan HS. A comparative study of Burkitt's and non-Burkitt's "undifferentiated" malignant lymphoma: immunologic, cytochemical, ultrastructural, cytologic, histopathologic, clinical and cell culture features. Cancer 1982;49:1817–1828.
197. Hutchison RE, Murphy SB, Fairclough DL, et al. Diffuse small noncleaved cell lymphoma in children, Burkitt's versus non-Burkitt's types. Results from the Pediatric Oncology Group and St. Jude Children's Research Hospital. Cancer 1989;64:23–28.
198. Swinnen LJ, Costanzo-Nordin MR, Fisher SG, et al. Increased incidence of lymphoproliferative disorder after immunosuppression with the monoclonal antibody OKT3 in cardiac-transplant recipients [see comments]. N Engl J Med 1990;323:1723–1728.
199. Biemer JJ. Malignant lymphomas associated with immunodeficiency states. Ann Clin Lab Sci 1990;20:175–191.
200. Haynes BF, Bunn P, Mann D, et al. Cell surface differentiation antigens of the malignant T cell in Sezary syndrome and mycosis fungoides. J Clin Invest 1981;67:523–530.
201. Boumsell L, Bernard A, Reinherz EL, et al. Surface antigens on malignant Sezary and T-CLL cells correspond to those of mature T cells. Blood 1981;57:526–530.
202. Haynes BF, Hensley LL, Jagasothy BV. Phenotypic characterization of skin-infiltrating T cells in cutaneous T-cell lymphoma: comparison with benign cutaneous T-cell infiltrates. Blood 1982;60:463–473.
203. Nasu K, Said J, Vonderheid E, Olerud J, Sako D, Kadin M. Immunopathology of cutaneous T-cell lymphomas. Am J Pathol 1985;119:436–447.
204. Pirruccello SJ, Lang MS. Differential expression of CD24-related epitopes in mycosis fungoides/Sezary syndrome: a potential marker for circulating Sezary cells. Blood 1990;76:2343–2347.
205. Wantzin GL, Larsen JK, Christensen IJ, Ralfkiaer E, Thomsen K. Early diagnosis of cutaneous T-cell lymphoma by DNA flow cytometry on skin biopsies. Cancer 1984;54:1348–1352.
206. Ralfkiaer E, Larsen JK, Christensen IJ, Thomsen K, Wantzin GL. DNA analysis by flow cytometry in cutaneous T-cell lymphomas. Br J Dermatol 1989;120:597–605.
207. Bunn PAJ, Whang-Peng J, Carney DN, Schlam ML, Knutsen T, Gazdar AF. DNA content analysis by flow cytometry and cytogenetic analysis in mycosis fungoides and Sezary syndrome. J Clin Invest 1980;65:1440–1448.
208. Broder S, Bunn PAJ, Jaffe ES, et al. NIH conference. T-cell lymphoproliferative syndrome associated with human T-cell leukemia/lymphoma virus. Ann Intern Med 1984;100:543–557.
209. Yancey WB,Jr, Dolson LH, Oblon D, et al. HTLV-1-associated adult T-cell leukemia/lymphoma presenting with nodular synovial masses. Am J Med 1990;89:676–683.
210. Yamada Y, Murata K, Kamihira S, et al. Prognostic significance of the proportion of Ki-67-positive cells in adult T-cell leukemia. Cancer 1991;67:2605–2609.
211. Jansen J, LeBien TW, Kersey JH. The phenotype of the neoplastic cells of hairy cell leukemia studied with monoclonal antibodies. Blood 1982;59:609–614.
212. Jansen J, Schuit HR, Meijer CJ, van-Nieuwkoop JA, Hijmans W. Cell markers in hairy cell leukemia studied in cells from 51 patients. Blood 1982;59:52–60.
213. Naeim F. Hairy cell leukemia: characteristics of the neoplastic cells. Hum Pathol 1988;19:375–388.
214. Anderson KC, Boyd AW, Fisher DC, Leslie D, Schlossman SF, Nadler LM. Hairy cell leukemia: a tumor of pre-plasma cells. Blood 1985;65:620–629.
215. Kristensen JS, Ellegaard J, Hokland P. A two-color flow cytometry assay for detection of hairy cells using monoclonal antibodies. Blood 1987;70:1063–1068.
216. Braylan RC, Jaffe ES, Triche TJ, et al. Structural and functional properties of the "hairy" cells of leukemic reticuloendotheliosis. Cancer 1978;41:210–227.
217. Hernandez D, Cruz C, Carnot J, Dorticos E, Espinosa E. Hairy cell leukaemia of T-cell origin. Br J Haematol 1978;40:504–506.
218. Cawley JC, Burns GF, Nash TA, Higgy KE, Child JA, Roberts BE. Hairy-cell leukemia with T-cell features. Blood 1978;51:61–69.
219. Kalyanaraman VS, Sarngadharan MG, Robert Guroff M, Miyoshi I, Golde D, Gallo RC. A new subtype of human T-cell leukemia virus (HTLC-II) associated with a T-cell variant of hairy cell leukemia. Science 1982;218:571–573.
220. Sainati L, Latutes E, Mulligan S, et al. A variant form of hairy cell leukemia resistant to alpha-interferon: clinical and phenotypic characteristics of 17 patients. Blood 1990;76:157–162.
221. Schwarting R, Stein H, Wang CY. The monoclonal antibodies alpha S-HCL 1 (alpha Leu-14) and alpha S-HCL 3 (alpha Leu-M5) allow the diagnosis of hairy cell leukemia. Blood 1985;65:974–983.
222. Hanson CA, Gribbin TE, Schnitzer B, Schlegelmilch JA, Mitchell BS, Stoolman LM. CD11c (LEU-M5) expression characterizes a B-cell chronic lymphoproliferative disorder with features of both chronic lymphocytic leukemia and hairy cell leukemia. Blood 1990;76:2360–2367.

223. Wormsley SB, Baird SM, Gadol N, Rai KR, Sobol RE. Characteristics of CD11c+CD5+ chronic B-cell leukemias and the identification of novel peripheral blood B-cell subsets with chronic lymphoid leukemia immunophenotypes. Blood 1990;76:123–130.
224. Vardiman JW, Gilewski TA, Ratain MJ, Bitter MA, Bradlow BA, Golomb HM. Evaluation of Leu-M5 (CD11c) in hairy cell leukemia by the alkaline phosphatase anti-alkaline phosphatase technique. Am J Clin Pathol 1988;90:250–256.
225. Heimann PS, Vardiman JW, Stock W, Platanias LC, Golomb HM. CD5+, CD11c+, CD20+ hairy cell leukemia [Letter]. Blood 1991;77:1617–1619.
226. Braylan RC, Diamond LW. Flow analysis of hairy cell leukemia. Leuk Res 1980;4:177–183.
227. Anderson KC, Park EK, Bates MP, et al. Antigens on human plasma cells identified by monoclonal antibodies. J Immunol 1983;130:1132–1138.
228. Jackson N, Ling NR, Ball J, Bromidge E, Nathan PD, Franklin IM. An analysis of myeloma plasma cell phenotype using antibodies defined at the IIIrd International Workshop on Human Leucocyte Differentiation Antigens. Clin Exp Immunol 1988;72:351–356.
229. San Miguel JF, Gonzalez M, Gascon A, et al. Immunophenotypic heterogeneity of multiple myeloma: influence on the biology and clinical course of the disease. Castellano-Leones (Spain) Cooperative Group for the Study of Monoclonal Gammopathies. Br J Haematol 1991;77:185–190.
230. Durie BG, Grogan TM. CALLA-positive myeloma: an aggressive subtype with poor survival. Blood 1985;66:229–232.
231. Van-Camp B, Durie BG, Spier C, et al. Plasma cells in multiple myeloma express a natural killer cell-associated antigen: CD56 (NKH-1; Leu-19). Blood 1990;76:377–382.
232. Drach J, Gattringer C, Huber H. Expression of the neural cell adhesion molecule (CD56) by human myeloma cells. Clin Exp Immunol 1991;83:418–422.
233. Anderson KC, Bates MP, Slaughenhoupt B, Schlossman SF, Nadler LM. A monoclonal antibody with reactivity restricted to normal and neoplastic plasma cells. J Immunol 1984;132:3172–3179.
234. Terstappen LW, Johnsen S, Segers Nolten IM, Loken MR. Identification and characterization of plasma cells in normal human bone marrow by high-resolution flow cytometry. Blood 1990;76:1739–1747.
235. Stewart AK, Freedman J, Garvey MB. Acute leukemia evolving from multiple myeloma and co-expressing myeloid and plasma cell antigens. Am J Hematol 1990;34:210–214.
236. Grogan TM, Durie BG, Spier CM, Richter L, Vela E. Myelomonocytic antigen positive multiple myeloma. Blood 1989;73:763–769.
237. Barlogie B, Latreille J, Alexanian R, et al. Quantitative cytology in myeloma research. Clin Haematol 1982;11:19–46.
238. Zeile G. Intracytoplasmic immunofluorescence in multiple myeloma. Cytometry 1980;1:37–41.
239. Barlogie B, Alexanian R, Pershouse M, Smallwood L, Smith L, Cytoplasmic immunoglobulin content in multiple myeloma. J Clin Invest 1985;76:765–769.
240. Bunn PAJ, Krasnow S, Makuch RW, Schlam ML, Schechter GP. Flow cytometric analysis of DNA content of bone marrow cells in patients with plasma cell myeloma: clinical implications. Blood 1982;59:528–535.
241. Latreille J, Barlogie B, Johnston D, Drewinko B, Alexanian R. Ploidy and proliferative characteristics in monoclonal gammopathies. Blood 1982;59:43–51.
242. Chan CS, Wormsley SB, Peter JB, Schechter GP. Dual parameter analysis of myeloma cells by flow cytometry. DNA content of cells containing monotypic cytoplasmic immunoglobulin. Am J Clin Pathol 1989;91:12–17.
243. Chan CS, Wormsley SB, Pierce LE, Peter JB, Schechter GP. B-cell surface phenotypes of proliferating myeloma cells: target antigens for immunotherapy. Am J Hematol 1990;33:101–109.

244. Tsuchiya H, Epstein J, Selvanayagam P, et al. Correlated flow cytometric analysis of H-ras p21 and nuclear DNA in multiple myeloma. Blood 1988;72:796–800.
245. McKenna RW, Parkin J, Kersey JH, Gajl-Peczalska KJ, Peterson L, Brunning RD. Chronic lymphoproliferative disorder with unusual clinical, morphologic, ultrastructural and membrane surface marker characteristics. Am J Med 1977;62:588–596.
246. McKenna RW, Arthur DC, Gajl-Peczalska KJ, Flynn P, Brunning RD. Granulated T cell lymphocytosis with neutropenia: malignant or benign chronic lymphoproliferative disorder? Blood 1985;66:259–266.
247. Reynolds CW, Foon KA. T gamma-lymphoproliferative disease and related disorders in humans and experimental animals: a review of the clinical, cellular, and functional characteristics. Blood 1984;64:1146–1158.
248. Pandolfi F, Loughran TP,Jr, Starkebaum G, et al. Clinical course and prognosis of the lymphoproliferative disease of granular lymphocytes. A multicenter study. Cancer 1990;65:341–348.
249. Loughran TP,Jr, Starkebaum G. Large granular lymphocyte leukemia. Report of 38 cases and review of the literature. Medicine 1987;66:397–405.
250. Aisenberg AC, Krontiris TG, Mak TW, Wilkes BM. Rearrangement of the gene for the beta chain of the T-cell receptor in T-cell chronic lymphocytic leukemia and related disorders. N Engl J Med 1985;313:529–533.
251. Frizzera G, Moran EM, Rappaport H. Angio-immunoblastic lymphadenopathy with dysproteinaemia. Lancet 1974;1:1070–1073.
252. Lukes RJ, Tindle BH. Immunoblastic lymphadenopathy. A hyperimmune entity resembling Hodgkin's disease. N Engl J Med 1975;292:1–8.
253. Steinberg AD, Seldin MF, Jaffe ES, et al. NIH conference. Angioimmunoblastic lymphadenopathy with dysproteinemia. Ann Intern Med 1988;108:575–584.
254. Kaneko Y, Larson RA, Variakojis D, Haren JM, Rowley JD. Nonrandom chromosome abnormalities in angioimmunoblastic lymphadenopathy. Blood 1982;60:877–887.
255. Weiss LM, Strickler JG, Dorfman RF, Horning SJ, Warnke RA, Sklar J. Clonal T-cell populations in angioimmunoblastic lymphadenopathy and angioimmunoblastic lymphadenopathy-like lymphoma. Am J Pathol 1986;122:392–397.
256. Namikawa R, Suchi T, Ueda R, et al. Phenotyping of proliferating lymphocytes in angioimmunoblastic lymphadenopathy and related lesions by the double immunoenzymatic staining technique. Am J Pathol 1987;127:279–287.
257. Rudders RA, Ahl ETJ, DeLellis RA. Surface marker and histopathologic correlation with long-term survival in advanced large-cell non-Hodgkin's lymphoma. Cancer 1981;47:1329–1335.
258. Horning SJ, Doggett RS, Warnke RA, Dorfman RF, Cox RS, Levy R. Clinical relevance of immunologic phenotype in diffuse large cell lymphoma. Blood 1984;63:1209–1215.
259. Freedman AS, Boyd AW, Anderson KC, et al. Immunologic heterogeneity of diffuse large cell lymphoma. Blood 1985;65:630–637.
260. Coiffier B, Brousse N, Peuchmaur M, et al. Peripheral T-cell lymphomas have a worse prognosis than B-cell lymphomas: a prospective study of 361 immunophenotyped patients treated with the LNH-84 regimen. The GELA (Groupe d'Etude des Lymphomes Agressives). Ann Oncol 1990;1:45–50.
261. Brown DC, Heryet A, Gatter KC, Mason DY. The prognostic significance of immunophenotype in high-grade non-Hodgkin's lymphoma. Histopathology 1989;14:621–627.
262. Grogan TM, Fielder K, Rangel C, et al. Peripheral T-cell lymphoma: aggressive disease with heterogeneous immunotypes. Am J Clin Pathol 1985;83:279–288.
263. Slymen DJ, Miller TP, Lippman SM, et al. Immunobiologic factors predictive of clinical outcome in diffuse large-cell lymphoma. J Clin Oncol 1990;8:986–993.

264. Spier CM, Grogan TM, Lippman SM, Slymen DJ, Rybski JA, Miller TP. The aberrancy of immunophenotype and immunoglobulin status as indicators of prognosis in B cell diffuse large cell lymphoma. Am J Pathol 1988;133:118–126.
265. Armitage JO, Vose JM, Linder J, et al. Clinical significance of immunophenotype in diffuse aggressive non-Hodgkin's lymphoma. J Clin Oncol 1989;7:1783–1790.
266. Holte H, de Lange Davies C, Beiske K, et al. Ki67 and 4F2 antigen expression as well as DNA synthesis predict survival at relapse/tumour progression in low-grade B-cell lymphoma. Int J Cancer 1989; 44:975–980.
267. Kimby E, Mellstedt H, Nilsson B, et al. Blood lymphocyte characteristics as predictors of prognosis in chronic lymphocytic leukemia of B-cell type. Hematol Oncol 1988;6:47–55.
268. Swerdlow SH, Habeshaw JA, Murray LJ, Stansfeld AG. T cells in follicular centroblastic/centrocytic (cleaved follicular centre cell) lymphomas at diagnosis, relapse and transformation. Hematol Oncol 1989;7:355–363.
269. Strickler JG, Copenhaver CM, Rojas VA, Horning SJ, Warnke RA. Comparison of "host cell infiltrates" in patients with follicular lymphoma with and without spontaneous regression. Am J Clin Pathol 1988;90:257–261.
270. Medeiros LJ, Picker LJ, Gelb AB, et al. Numbers of host "helper" T cells and proliferating cells predict survival in diffuse small-cell lymphomas. J Clin Oncol 1989;7:1009–1017.
271. Kimby E, Mellstedt H, Nilsson B, Bjorkholm M, Holm G. T lymphocyte subpopulations in chronic lymphocytic leukemia of B cell type in relation to immunoglobulin isotype(s) on the leukemic clone and to clinical features. Eur J Haematol 1987;38:261–267.
272. Lippman SM, Spier CM, Miller TP, Slymen DJ, Rybski JA, Grogan TM. Tumor-infiltrating T-lymphocytes in B-cell diffuse large cell lymphoma related to disease course. Mod Pathol 1990;3:361–367.
273. Zutter M, Hockenbery D, Silverman GA, Korsmeyer SJ. Immunolocalization of the Bcl-2 protein within hematopoietic neoplasms. Blood 1991;78:1062–1068.
274. Miller TP, Grogan TM, Dalton WS, Spier CM, Scheper RJ, Salmon SE. P-glycoprotein expression in malignant lymphoma and reversal of clinical drug resistance with chemotherapy plus high-dose verapamil. J Clin Oncol 1991;9:17–24.
275. Berg EL, Goldstein LA, Jutila MA, et al. Homing receptors and vascular addressins: cell adhesion molecules that direct lymphocyte traffic. Immunol Rev 1989;108:5–18.
276. Picker LJ, Medeiros LJ, Weiss LM, Warnke RA, Butcher EC. Expression of lymphocyte homing receptor antigen in non-Hodgkin's lymphoma. Am J Pathol 1988;130:496–504.
277. Horst E, Meijer CJ, Radaszkiewicz T, Ossekoppele GJ, Van Krieken JH, Pals ST. Adhesion molecules in the prognosis of diffuse large-cell lymphoma: expression of a lymphocyte homing receptor (CD44), LFA-1 (CD11a/18), and ICAM-1 (CD54). Leukemia 1990;4:595–599.
278. Jalkanen S, Joensuu H, Klemi P. Prognostic value of lymphocyte homing receptor and S phase fraction in non-Hodgkin's lymphoma. Blood 1990;75:1549–1556.
279. Come SE, Jaffe ES, Andersen JC, et al. Non-Hodgkin's lymphomas in leukemic phase: clinicopathologic correlations. Am J Med 1980;69: 667–674.
280. Macartney JC, Camplejohn RS. DNA flow cytometry of non-Hodgkin's lymphomas. Eur J Cancer 1990;26:635–637.
281. Macartney JC, Camplejohn RS, Morris R, Hollowood K, Clarke D, Timothy A. DNA flow cytometry of follicular non-Hodgkin's lymphoma. J Clin Pathol 1991;44:215–218.
282. Joensuu H, Klemi PJ, Jalkanen S. Biologic progression in non-Hodgkin's lymphoma. A flow cytometric study. Cancer 1990;65:2564–2571.
283. Griffin NR, Howard MR, Quirke P, O'Brien CJ, Child JA, Bird CC. Prognostic indicators in centroblastic-centrocytic lymphoma. J Clin Pathol 1988;41:866–870.
284. Grogan TM, Lippman SM, Spier CM, et al. Independent prognostic significance of a nuclear proliferation antigen in diffuse large cell lymphomas as determined by the monoclonal antibody Ki-67. Blood 1988;71;1157–1160.
285. Bauer KD, Merkel DE, Winter JN, et al. Prognostic implications of ploidy and proliferative activity in diffuse large cell lymphomas. Cancer Res 1986;46:3173–3178.
286. Grierson HL, Wooldridge TN, Purtilo DT, et al. Low proliferative activity is associated with a favorable prognosis in peripheral T-cell lymphoma. Cancer Res 1990;50:4845–4848.
287. Macartney JC, Camplejohn RS, Alder J, Stone MG, Powell G. Prognostic importance of DNA flow cytometry in non-Hodgkin's lymphomas. J Clin Pathol 1986;39:542–546.
288. Durie BG, Young LA, Salmon SE. Human myeloma in vitro colony growth: interrelationships between drug sensitivity, cell kinetics, and patient survival duration. Blood 1983;61;929–934.
289. Boccadoro M, Marmont F, Tribalto M, et al. Early responder myeloma: kinetic studies identify a patient subgroup characterized by very poor prognosis. J Clin Oncol 1989;7:119–125.
290. Kamat D, Laszewski MJ, Kemp JD, et al. The diagnostic utility of immunophenotyping and immunogenotyping in the pathologic evaluation of lymphoid proliferations. Mod Pathol 1990;3:105–112.
291. Grogan TM, Tubbs RR. A double-blind comparative immunotypic study between two institutions phenotyping non-Hodgkin's lymphomas. Am J Clin Pathol 1987;87:478–484.
292. Duque RE, Everett ET, Iturraspe J. Biclonal composite lymphoma. A multiparameter flow cytometric analysis. Arch Pathol Lab Med 1990; 114:176–179.
293. Berliner N, Ault KA, Martin P, Weinberg DS. Detection of clonal excess in lymphoproliferative disease by kappa/lambda analysis: correlation with immunoglobulin gene DNA rearrangement. Blood 1986; 67:80–85.

13

Acute Leukemias

RICARDO E. DUQUE

INTRODUCTION

Acute leukemias are clonal expansions of hematopoietic precursor cells (blasts) that have several common characteristics: (1)

1. Poor responsiveness to regulatory mechanisms.
2. Tendency to have a diminished capacity for normal differentiation.
3. Expansion at the expense of normal elements.

The diminished capacity for normal differentiation and the acquisition of selective advantage for growth result in the unregulated proliferation of the malignant clone. Leukemic cells are probably not "frozen" at a given ontogenic stage. They proliferate in an uncontrolled fashion with a variable maturational impairment. We classify the process as "malignant" due to the presence of diminished normal hematopoiesis and an excess of blasts. Subclassification is based on the degree of resemblance of the blasts to "normal" counterparts.[1] For example, progranulocytic leukemia is characterized by the expansion of cells that resemble (normally transitory) progranulocytes. It is characterized by a composite immunophenotype that is ephemerally expressed on progranulocytes during normal myeloid differentiation. Leukemogenesis might involve the stabilization of normally transitory phenotypes and be paradoxically subservient to normal differentiation programs that have undergone subtle dysregulation (2). Regulatory uncoupling of variable stringency (3) may explain, not the "arrest," by the diminished capacity for normal differentiation and uncontrolled proliferation.

CLASSIFICATION OF LEUKEMIAS

Acute leukemias are currently classified by the French American-British (FAB) criteria (4). This involves separation of lymphoblastic leukemias into three morphologically defined subgroups: L1, L2, and L3 (this last group represents Burkitt's lymphoma in leukemic phase); and non-lymphoid leukemias into seven morphologically and cytochemically defined subgroups: M1 (myeloblastic), M2 (myeloblastic with differentiation), M3 (progranulocytic), M4 (myelomonocytic), M5a (monoblastic), M5b (monocytic), M6 (erythroid), and M7 (megakaryoblastic). Myelodysplastic syndromes, which involve the progressive expansion of ineffective, pluripotential hematopoietic clones, eventually terminate in acute leukemia in slighly fewer than 50% of patients. Their classification is not always straightforward. The FAB group has also established a classification system for myelodysplastic syndromes (5) and, more recently, it has revised criteria to separate more clearly acute myeloid leukemias from myelodysplastic syndromes (6). In response to the growing body of knowledge on immunophenotypic and cytogenetic characteristics of acute leukemias, the Morphologic, Immunologic, and Cytogenetic (MIC) classification has been proposed (7, 8).

In the diagnosis of acute leukemias, the interobserver and intraobserver reproducibility is in the range of 60% to 70% for morphological classification alone, 89% for morphological and cytochemical classification, and 99% when immunophenotypic information is added (9). Some multiinstitutional studies, utilizing FAB criteria, have reported reproducibility in the 70% range (10). In general, morphologically reproducibility is poor because it is subjective. In the classification of non-Hodgkin's lymphomas, for example, certain categories have a consensus diagnosis of as low as 43% (11) (consensus meaning four out of seven expert pathologists). Therefore, this underscores the need to embrace techniques or methods that contribute to minimize subjectivity and increase reproducibility. The growing popularity of the use of flow cytometry as an adjunct to established methods in the diagnosis of hematopoietic neoplasia is a signal that pathologists perceive the above-mentioned problems to be genuine.

IMMUNOLOGIC CLASSIFICATION OF LEUKEMIAS

Ontogeny of Hematopoietic Precursor Cells

In order to classify leukemias immunologically, it is sometimes helpful to refer to what is perceived as "normal" ontogeny and to phenotypically defined stages of cell differentiation. Acute non-T lymphoblastic leukemias, for example, can be separated into six immunophenotypically defined groups (Fig. 13.1). They are distinguished by the expression of cell surface antigens that are either B-cell-restricted

---

[1] Subtle alterations in differentiation programs under the influence of oncogenic forces result in the expansion of cells at a morphologically, cytochemically, and phenotypically defined stage that recapitulates (or is reminiscent of) normal hematopoietic precursors.

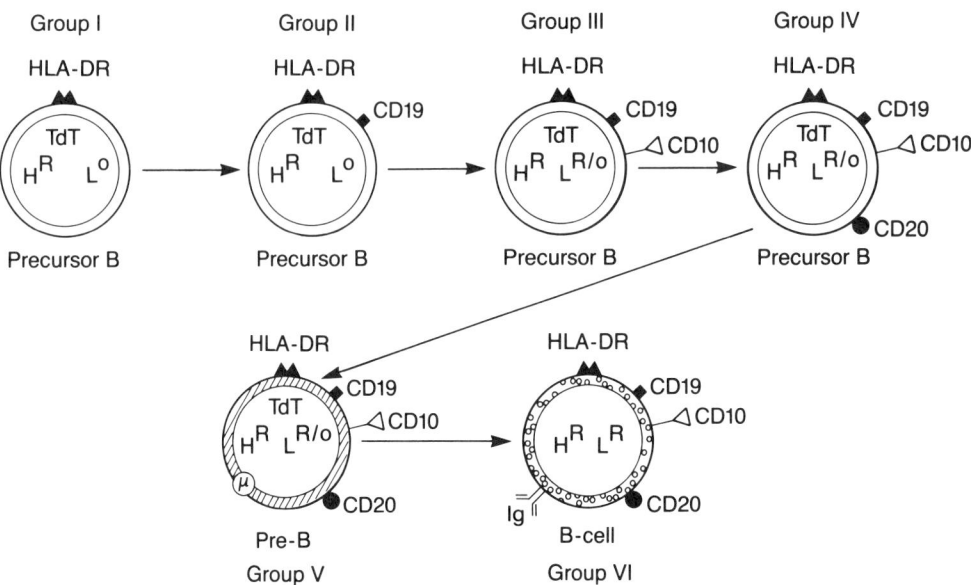

Figure 13.1. B-cell ontogeny and correlations. Conceptually, precursor B-cell leukemias exhibit surface antigens that recapitulate normal bone marrow B-cell precursors. There is a developmentally regulated antigen expression. Acute precursor B-cell leukemias usually exhibit composite immunophenotypes that resemble one of the groups illustrated on the slide. *TdT* is usually present in all precursor B-cell leukemias, with the exception of the mature (surface membrane immunoglobulin positive) B-cell.

(CD19, CD20), or B-cell-associated (CD10 and HLA-DR) (12). The last two groups are recognized by the expression of cytoplasmic μ heavy chain (Group V) or the surface expression of immonoglobin (Group VI). The molecular events that occur during B-cell ontogeny are also alluded to in Figure 13.1. Studies by Korsmeyer, et al. (13) have shown that B-cells exhibit a hierarchy of immunoglobin gene rearrangements wherein heavy chain (μ) gene rearrangement precedes that of light chain genes and κ light chain gene rearrangement precedes that of λ. Recently, Loken et al. have defined four closely related groups of immulogically defined B-cell precursors in normal human bone marrow using multiparameter flow cytometry and a combination of additional monoclonal antibodies (14). The same group has subsequently shown "asynchronous" expression of composite phenotypes in leukemia compared to flow cytometrically defined B-cell development. These data suggest that there is a certain degree of maturation in leukemia and that perhaps deviation from normal cells, rather than the emulation of "caricaturization" of normal cells, determines malignancy (15). Differences between the groups defined by Loken, et al., in terms of the relative proportions of these flow cytometrically defined normal marrow precursors, and the relative incidence of leukemic counterparts are not easily explained. It is not known if leukemic events occur in a completely random fashion. The target cell for leukemagenesis could be a relatively infrequent cell that, because of aberrant or nonfunctional gene rearrangements (3), (which dooms it to maturation failure and death) is undetectable at the level of resolution of two- or three-color analysis by flow cytometry or of single-cell analysis. Oncogenic events would result in the immortalization of an ephemeral cell that would otherwise have died and in the stabilization of its rare composite phenotype, resulting in an acute leukemia expressing an "aberrant" (or asynchronous) cell surface antigen repertoire (3). On the other hand, when analyzed at the single-cell level, normal human tissue is seen to contain cells bearing similar composite phenotypes as some acute lymphoblastic leukemias of either T or non-T lineage; these somewhat rare populations are present in greater relative frequencies in fetal tissue and in regenerating marrow (16). Leukemias then, can be conceived of as expressions of "blocked ontogeny" (17) or "caricatures" of tissue renewal (18). Unusual antigenic expression in leukemias is viewed by others as representing lineage "infidelity" and aberrant gene expression due to misprogramming of differentiation (19).

T-cell ontogeny can be elucidated from the analysis of normal thymic tissue and by the molecular and phenotypic analysis of T-cell malignancies. As illustrated in Figure 13.2, the first expression of T-cell commitment is signaled by the surface expression of CD7 and nuclear TdT. The ensuing composite geno/phenotype is characterized by surface expression of CD5 and CD2, followed by surface coexpression of CD4 and CD8 (thymic phenotype). There are also rearrangement of Tβ and Tτ genes, Tβ gene transcription, T3δ gene transcription, and intracellular T3 accumulation. Finally, Tα transcription occurs and CD3 appears on the surface (20). Figure 13.2 also illustrates the phenotypically defined "stages" of T-cell maturation and their corresponding malignancies. T-cell acute leukemias represent expansions of precursor T-cells. Consequently, TdT is usually positive in addition to T-cell-associated cell surface markers.

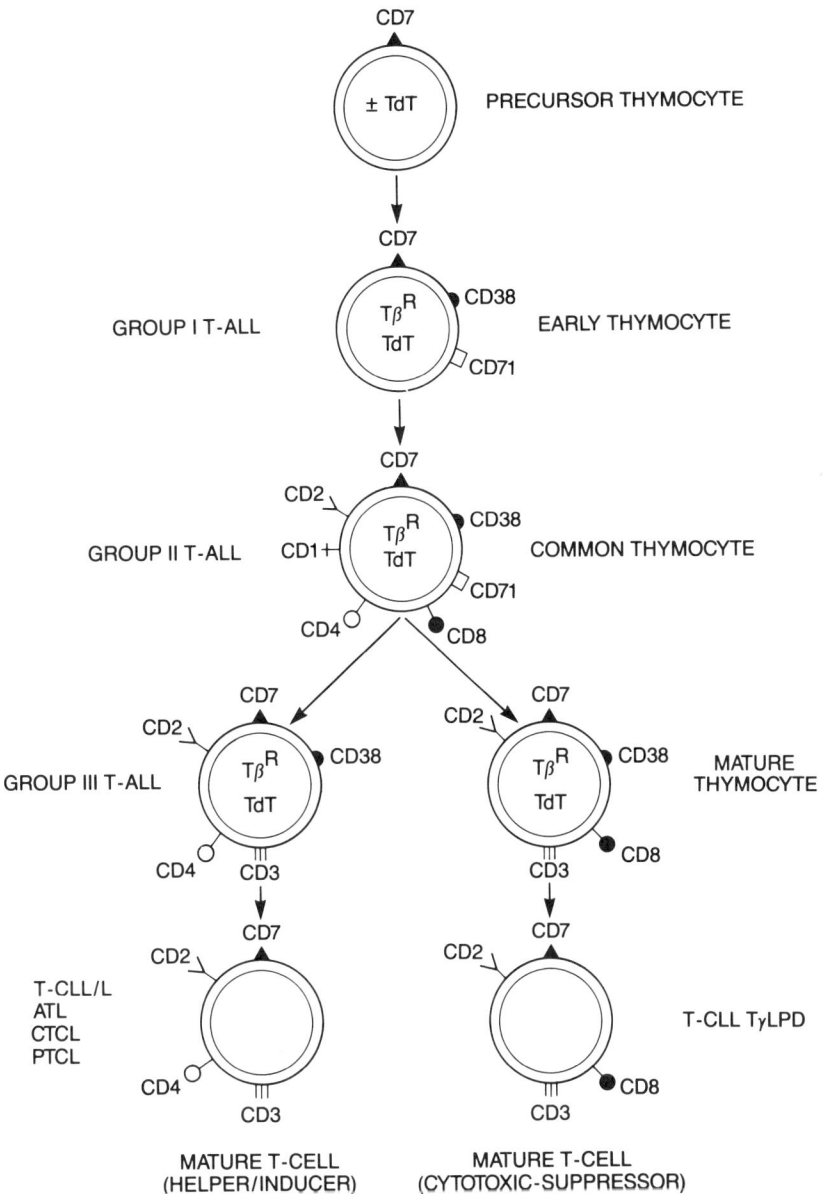

Figure 13.2. T-cell ontogeny and correlations. Similar to B-cells, T-cells exhibit developmentally regulated surface antigen composites that recapitulate T-cell precursors. *CD2, CD5,* and *CD7* are ubiquitous T-cell-associated antigens. In contrast, *CD3, CD4,* and *CD8* are associated with more "mature" cells. *TdT* is almost always positive in acute T-cell leukemias. "Peripheral" or post-thymic malignancies are usually *TdT*-negative.

Myelomonocytic ontogeny involves the almost ubiquitous expression of HLA-DR, CD13, and CD33. As cells commit to myeloid lineage and differentiate to proganulocytes, HLA-DR is no longer expressed. When the cells commit to monocytic lineage, HLA-DR expression is preserved and the more mature cells express CD14 (Fig. 13.3). The monoclonal antibodies CD13 and CD33, for example, will react with virtually all AML's, including some with "undifferentiated" morphology (21). Moreover, the expression of CD33 has been shown to increase in parallel with myeloperoxidase (MPO) activity (22). Surface expression of HLA-DR is virtually ubiquitous in non-T acute leukemias (myeloid, monocytic, and precursor B-cell), with the exception of acute progranulocytic leukemias (FAB M3) where it is usually not expressed (21, 23). This combination of HLA-DR negativity with positive myeloid markers, such as CD13 and/or CD33, in otherwise "undifferentiated" blasts is useful to detect the rare "hypogranular" variant of acute progranulocytic leukemia.

Several monoclonal antibodies that react with platelet-associated antigens, such as CD41 and CD61, can be useful to detect the M7 variant of AML or megakaryoblastic leukemia when morphology is ambiguous. A recent study suggests that this variant may be more prevalent than previously suspected (24). In the detection of erythroleukemias (FAB M6), monoclonal antibody LICR.LON.R10, specific for the major

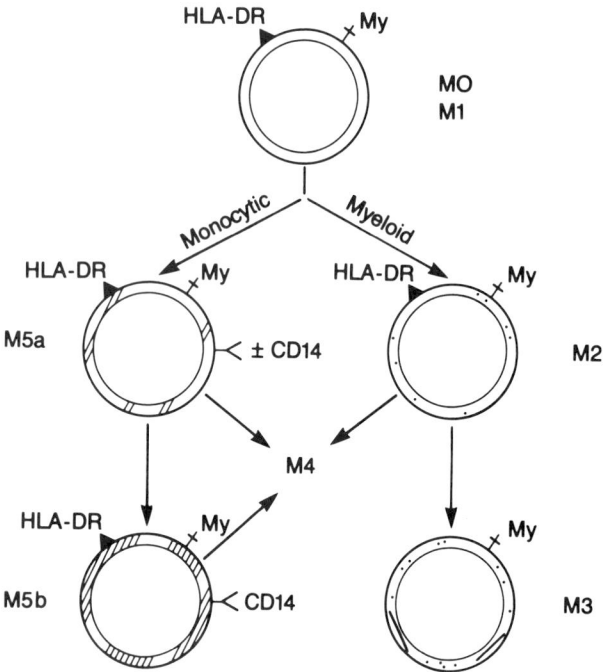

Figure 13.3. Myelomonocytic ontogeny and correlations. HLA-DR is ubiquitous in myelomonocytic cells. The exception is the progranulocyte, which is relatively mature. In contrast, HLA-DR is conserved in the *monocytic* lineage in which the mature cells also express *CD14*. CD13 and CD33 are normally coexpressed on all normal myelomonocytic cells. Frequently, acute myeloid leukemias express one (CD13 or CD33) antigen but not the other.

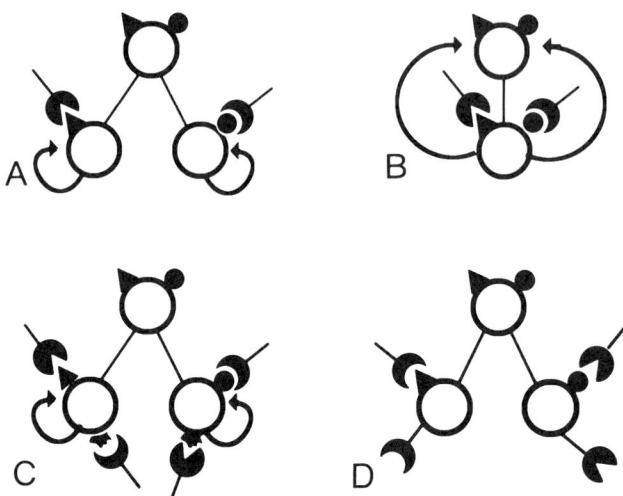

Figure 13.4. Multilineage-associated antigen expression in acute leukemia. Conceptual illustration of possible alternatives in leukemias expressing multilineage-associated immunophenotypes. **A**, Biclonal leukemia. Following an oncogenic event, the progeny of a pluripotential cell undergo lineage commitment. Once committed, each clone has self renewal potential. **B**, Biphenotypic leukemia (lineage promiscuity). The target for leukemogenesis is a multipotential cell that expands without further maturation or unilineage commitment. The progeny, therefore, preserve this multipotentiality and coexpress antigens of different "lineages." **C**, Biphenotypic leukemia (lineage infidelity). Due to genetic misprogramming and "aberrant" antigen expression, leukemic clones unexpectedly express antigens that are aberrant and react with antibodies specific for different lineages. **D**, Nonspecific binding of monoclonal antibodies.

sialoglycoprotein of the erythroid membrane, glycophorin-A, may be helpful (25). In our experience, rare "pure" erythroleukemias (as opposed to acute myeloid leukemias with significant dyserythropoiesis) usually exhibit a myeloid immunophenotype.

Mixed-Lineage Acute Leukemias

A third group of leukemias is being detected and reported with increasing frequency: mixed-lineage, biphenotypic or hybrid leukemias (26–28) (Fig. 13.4). Patients with this group of leukemias tend to have a less favorable prognosis (29). As previously stated, it is a matter of debate whether this group represents aberrant gene expression (19) or so-called "lineage promiscuity" (3). In the former, the expression, on a single blast and its progeny, or antigens though to be of several lineages, is a manifestation of aborted differentiation programs. In the latter, it is thought that there are normal (and infrequent) hematologic presursors that transiently express antigens associated with more than one lineage, indicating multipotentiality and a flexible pattern of gene expression. The immortalization of this cell, due to unknown oncogenic events, results in a leukemic process that has multilineage attributes. In some cases of acute leukemia, two separate clones may be detected and may represent a different category of truly biclonal leukemias (28).

The clinical importance of myeloid antigen expression in acute lymphoblastic leukemia has been demonstrated both in adults (30) and in children (31). In these series, the incidence of myeloid antigen expression was reported to be 33% and approximately 25%, respectively. The coexpression of myeloid antigens had an adverse prognostic impact in both series. Therefore, new therapeutic alternatives must be sought in order to treat this subgroup effectively. Conversely, in a series of myeloid leukemias in children, the coexpression of CD2 and TdT was associated with failed induction chemotherapy (with AML protocols) in 50% of the cases. Each of the patients who failed myeloid induction exhibited a dramatic response to lymphoid-directed therapy (32). More recently, the coexpression of lymphoid-associated markers (CD2 or CD19) in adult acute myeloid leukemias has been associated with a more favorable prognosis following standard AML treatment protocols (33). In summary, it appears from these data that acute lymphoblastic leukemias have a relatively good prognosis, whereas acute myeloid leukemias have a relatively poor prognosis. A third category, i.e., mixed leukemias, have an intermediate prognosis. Morphologic classification, without the benefit of immunophenotypic data, is suboptimal since it does not detect this last group.

Approach to the Diagnosis and Classification of Acute Leukemias

Before attempting to classify an acute leukemic process by flow cytometric immunophenotyping, it is necessary to identify an excess of cells that have the morphology of blasts in an appropriate clinical setting. As mentioned previously, the French-American-British (FAB) classification is widely accepted and establishes morphologic and cytochemical criteria for the diagnosis of acute leukemias or acute leukemic transformation of myelodysplastic or myeloproliferative syndromes (6, 34). It is not clinically useful to perform flow cytometric analysis on myeloproliferative or myelodysplastic syndromes in chronic phase.

The role of immunophenotypic analysis is to provide more reproducible results in terms of the determination of lineage (9, 35, 36). As a result, therapy might be better directed (32, 37). Furthermore, prognosis can be established in some cases. For example, DNA-hyperdiploid acute lymphoblastic leukemias in children are associated with a lesser likelihood of early relapse than their DNA-diploid counterparts (38).

SPECIMEN COLLECTION AND HANDLING

Leukemic blasts are more frequently present in bone marrow and peripheral blood than in other compartments. However, they may be found virtually anywhere. They can be identified by morphologic criteria (39). When immunophenotypic characterization of leukemic blasts is considered, a number of factors may influence which compartment to sample for processing. The relative and absolute number of blasts present in each compartment will play a role in determining the yield of leukemic blasts available for flow cytometric analysis. When the proportion or ratio of leukemic blasts to other normal progenitor cells is low, problems can arise during analysis that may result in ambiguous results. Blasts have light-scatter and immunophenotypic properties that may overlap with those of their normal counterparts (40). Therefore, it is recommended that compartments containing an easily identified excess of blasts be used as the source of material for immunophenotypic analysis. Although peripheral blood may have fewer absolute numbers of blasts, it may be a better source for analysis than bone marrow due to the inherent lack of normal hemopoietic precursors. Peripheral blood and bone marrow aspirate specimens should be collected and transported to the laboratory in sterile, heparinized (14 units/ml Na heparin) vacutainer tubes (Becton Dickinson Vacutainer Systems). Although rapid processing of fresh samples is desirable, these samples may be stored for 24 hr at room temperature (22°C). This will permit processing the following day or shipment to a reference laboratory without compromising the integrity of most immunophenotypic evaluations. Refrigeration, heat, excessive anticoagulant, and extended delays in processing can lead to both ambiguous and erroneous interpretations. Occasionally a "packed" bone marrow from a patient with pancytopenia may result in poor recovery of blasts from either the peripheral blood or a bone marrow aspirate. A bone marrow biopsy may be a more suitable sample since it contains abundant blasts. The bone marrow biopsy is then treated like solid tissues and the blasts are dispersed, as will be described in "Specimen Processing," below. It should be placed on sterile gauze moistened with RPMI 1640 media. The sample should then be transported to the laboratory on wet ice and processed within 4 hr of collection. Cerebrospinal-fluid- (CSF) containing leukemic blasts can be collected in standard tubes and taken to the laboratory at room temperature without anticoagulation. CSF should be stored at 4°C when processing cannot be completed within 2 hr. Fresh-air-dried smears or touch imprints should accompany each specimen at the time of submission.

SPECIMEN PROCESSING

Specimens should be appropriately identified and accessioned. Due to morphologic changes that may occur during transportation, an air-dried slide should be requested from the person obtaining the bone marrow or other source of leukemic cells. Wright-Giemsa staining is mandatory to verify that the sample contains blasts. Samples should be diluted approximately 1:1 with Hank's balanced salt solution (HBSS) (Gibco, Grand Island, NY) and underlayed with an equal volume of a Ficoll-Hypaque solution (Histopaque 1077) (Sigma, St. Louis, MO). Tubes should then be centrifuged for 30 min at 4°C and 400 xg. The mononuclear cells are recovered at the interface and washed a minimum of two times with phosphate-buffered saline (PBS) containing 0.1% NaN_3. Excessive red blood cell contamination is eliminated by incubating the cell pellet for 10 min at room temperature in lysing buffer (8.29 g/L NH_4Cl, 1.0 g/L $KHCO_3$, 37 mg/L EDTA pH 7.2) containing 0.1% NaN_3. Cells are resuspended to a concentration between 1.0–2×10^6/ml in RPMI 1640 media containing 5μg/ml gentamicin, 100 units/ml penicillin, 100μg/ml streptomycin, 25mM HEPES pH 7.2, and 5% v/v newborn calf serum (RPMI 1640/serum). Viability may be determined by the trypan blue dye exclusion test or by propidium iodide staining. Air dried cytospins are made using 1.0–1.5×10^5 cells and stained with Wright-Giemsa and myeloperoxidase.

In the event of a dry tap, touch imprints should be routinely made and morphologic evaluation performed. The biopsy samples are then minced with fine-toothed scissors to release cells from the stroma and particulate material is removed by filtration through a 40 μ mesh. Centrifugation through Ficoll is usually not required due to the small numbers of cells recovered. Excess red blood cells should be lysed and recovered cells should be washed with PBS azide. The final cell pellet should be suspended in RPMI 1640/serum and counted. Viability and cytospin slides should also be prepared. The continued emphasis in preparing Wright-Giemsa cytospin slides is due to the need to monitor the morphologic presence of blasts throughout the processing steps and confirm that the final cell preparation contains a similar constituency of leukemic cells compared to the original sample. Cells can be maintained at 4°C in RPMI 1640/serum for one to two days without significant change in the immunophenotypic profile.

CELL SURFACE STAINING

Cells can be stained indirectly with a primary purified monoclonal antibody (MoAb) or biotin-conjugated MoAb that has been washed and then stained with the secondary (usually polyclonal) antimouse antibody conjugated to fluorochromes, such as fluoroscein isothiocyanate (FITC) or phycoerythrin (PE). The secondary reagent used with biotin is avidin (strepavidin) conjugated to a specific fluorochrome. Alternatively, staining can be accomplished directly with fluorochrome-conjugated MoAb. Advantages and disadvantages can be found in both systems. Advantages to using the indirect system include increased sensitivity, in that several secondary antibodies usually bind to the primary MoAb and amplify the fluorescent signal. Many MoAb are not suitable for the direct conjugation to a particular fluorochrome and, therefore, are restricted for use in indirect staining systems. The disadvantages are increased time, increased washes (which contribute to cell loss during staining), and inability to perform dual-color surface labeling. The direct system allows dual-color surface labeling, speed in processing, and conservation of cells. Directly conjugated MoAb may yield a less intense fluorescent signal. In both systems, 0.5–1.0 $\times 10^6$ cells are incubated with 2–5µg of primary MoAb at 4°C for 15 to 20 min. Cells are incubated in the presence of 0.1% NaN_3 to inhibit modulation and capping of the surface antigens. Cells are washed twice with PBS azide by centrifuging 10 min at 4°C and 400 xg (unit of gravity). Incubation of a secondary antibody, when appropriate, is performed under the same conditions used for primary antibody. The final cell pellet is suspended in a small volume of PBS azide containing 0.01µg/ml propidium iodide. Propidium iodide will pass through the plasma membrane of dead cells, intercalate with nucleic acids, notably DNA, and allow for real-time identification of stained dead cells. Overlap with the fluorescence of phycoerythrin is usually not a problem since the fluorescence intensity of propidium iodide is greater than that of phycoerythrin. Dead and dying cells tend to bind antibodies in a nonspecific fashion, which may introduce problems in the interpretation of results. Cells should also be stained with appropriate negative control antibodies. These are isotype, fluorochrome, and protein concentration matched with the test antibody. Performance characteristic analysis of all test MoAb should be done on appropriate positive reference material.

Monoclonal Antibody Selection

The rationale for constructing "panels" for the diagnosis of leukemias involves the following: a) It is impractical to assign lineage to cells on a purely morphological basis; b)

Table 13.1
Most Commonly Used Antibodies in the Classification of Acute Leukemias

Cluster of Differentiation	Antibody	Specificity
N/A	HLA-DR	B-cells monocytes activ. T cells
CD 19 (gp95)	B4, Leu 12	B-cells
CD 10 (gp 100)	CALLA, J5	Common acute lymphoblastic leukemia antigen
CD 20 (gp 35)	B1, Leu 16	B-cells
CD 2 (gp 50)	T11, Leu 5	T-cells,
CD 5 (gp 67)	T1, Leu1	T-cells, subset B-cells
CD 7 (gp 41)	Leu 9	T-cells
CD 13 (gp 150)	My7	Myeloid cells, Monocytes
CD 14	My4, Mo2, Leu M3	Monocytes
CD33	My9	Myeloid cells, Monocytes

morphological appearance, when supplemented by immunophenotypic data, provides an acceptable degree of reproducibility; c) leukemic cells have composite phenotypes that are suggestive of mono- or multilineage assignment (and thus justify using a battery of antibodies rather than attempting to define leukemia with a single (nonexistent leukemic marker; d) while it is not possible to determine the lineage of a blast on a morphologic basis, it is possible to anticipate the different possibilities and include them in an antibody panel; and e) when analyzing relapsed leukemias, it must be emphasized that one should not anticipate the reexpression of a previously determined phenotypic marker since leukemias can "switch" their lineage (41)). Moreover, in the case of truly biclonal leukemias, one clone may succumb to chemotherapy while the other may prevail, thereby giving the appearance of a lineage "switch."

Table 13.1 is a listing of commonly used monoclonal antibodies for the classification of acute leukemias. As mentioned previously, many other antibodies may be incorporated into a panel and should be used depending on the experience or preference of the person interpreting the results.

Antibody "pairing" is a subject of controversy. When using two- or three-color approaches, many investigators have recommended specific pairs of antibodies that may have more significance than when used alone. This is less than clear. The potential coexpression of antibodies on blastic populations or subpopulations cannot always be anticipated. Therefore, "a priori" pairing will not always be fruitful. Whether one uses multiple cell surface labels or combinations of surface and cytoplasm, surface or cytoplasm and DNA, etc., should depend on specific questions. For example, it may be of interest to determine if a subpopulation of CD13-positive blasts also coexpresses TdT. A leukemic cell suspension with FLS and SSC characteristics that overlap with monocytes and appears to be of ambiguous lineage may present problems that can be resolved by multicolor analysis. One would be interested in defining a CD14−, CD13+, CD19+ subpopulation. These two examples do not imply that one should pair CD13 with TdT or CD13 with

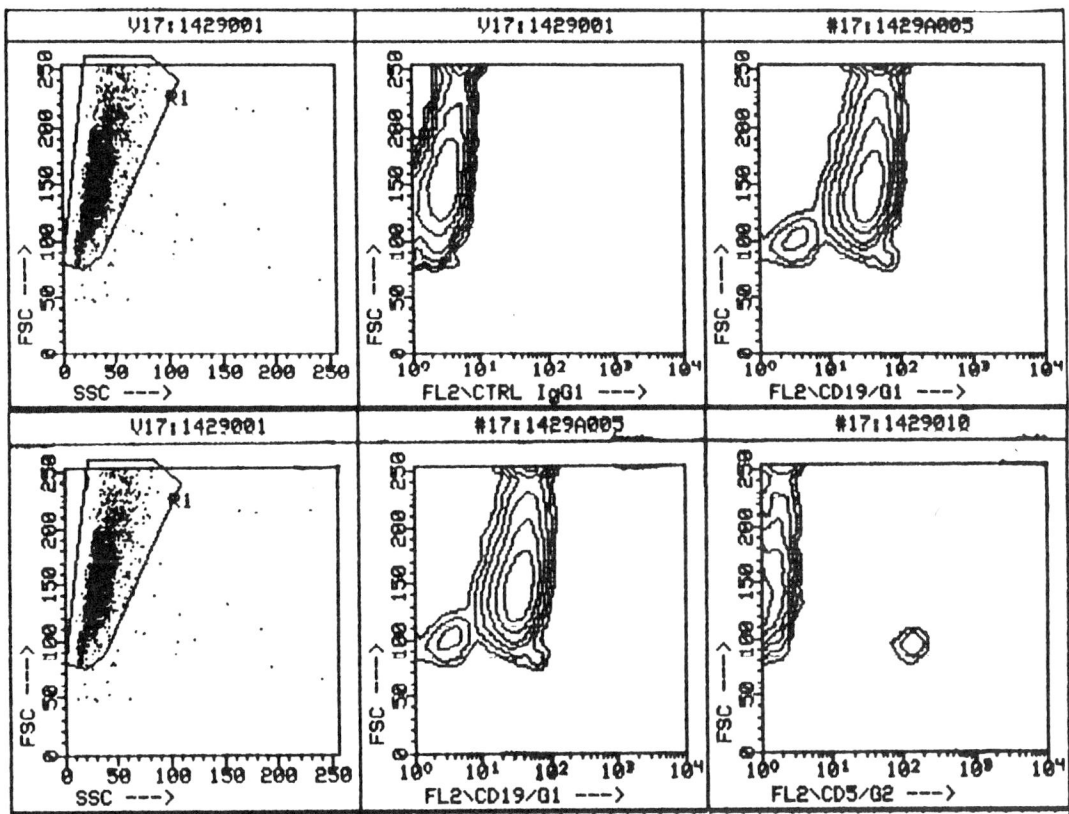

Figure 13.5. The left frame is a dot plot of forward scatter (FSC) vs. side scatter (SSC). The mononuclear cells are gated. In this example, there are virtually no contaminating elements. The middle frame is a contour plot of FSC vs. fluorescence of a fluorochrome- and isotype-matched control (FL2[phycoerythrin]/IgG1). The right frame illustrates the fluorescence of CD19 vs. FSC. The blasts are positive. In contrast, the small cells (normal lymphocytes) are negative except for a minor subpopulation of normal B-cells.

CD19 and CD14. Finally, we favor defining blastic populations by a correlated analysis of FSC and antibody-dependent immunofluorescence; one tube may contain multiple antibodies each of which is displayed as FSC versus fluorescence.

Data Analysis

Mononuclear cells should be gated on contour or dot plots of forward scatter (FSC) vs. side scatter (SSC). The qualitative determination of positivity by comparing dual-parameter contour plots of isotype and light-scatter-matched controls with test cells that are either equal (and therefore negative) or not equal (and therefore positive) is illustrated in Figure 13.5. This provides a simple and easily reproducible method for determining the composite phenotype of leukemic cells. Human peripheral blood or bone marrow samples are composed of heterogeneous populations. Therefore, for any given antibody there will be positive and negative cells. Furthermore, in the case of acute leukemias, the blasts have scatter properties that separate them partially from nonblasts in the preparation (lymphocytes). Lymphocytes are virtually always present since pure bone marrow is almost impossible to obtain. On the other hand, if the analysis is being performed on peripheral blood, the presence of lymphocytes is difficult to exclude. Granulocytes and mature myeloid cells do not present a problem because they have side-scatter properties that are distinct and, furthermore, granulocytes are not usually the object of immunophenotypic analysis. This provides two parameters that will be characteristic of leukemic cells: fluorescence (FL1 or FL2) and forward scatter (FLS). Side scatter is a poor discriminant of mononuclear cells. Consequently, the use of side scatter vs. fluorescence has limited applicability in immunophenotyping. In contrast, as mentioned previously, side scatter (in combination with FLS) is useful to eliminate mature myeloid elements during initial gating. Immunophenotypic analysis should be performed on the FLS- vs. SSC-defined mononuclear fraction. Figure 13.5 is a relatively typical example of an acute lymphoblastic leukemia with L2 morphology. In this preparation, the cell population to be examined contained virtually 100% blasts. Therefore, scatter gating is not critical. In this case the slightly larger cells (blasts) are clearly, (1) positive for CD19 and (2) larger than the negative cells (lymphocytes). (Since monocytes may overlap with leukemic blasts in terms of scatter properties, CD14-positivity in the mononuclear region or gate will determine the monocyte contamination. In some cases, (myelomonocytic variants) the monocytes themselves will be part of the leukemic clone.) The positivity of this FLS- and FL1-defined cluster (CD19-posi-

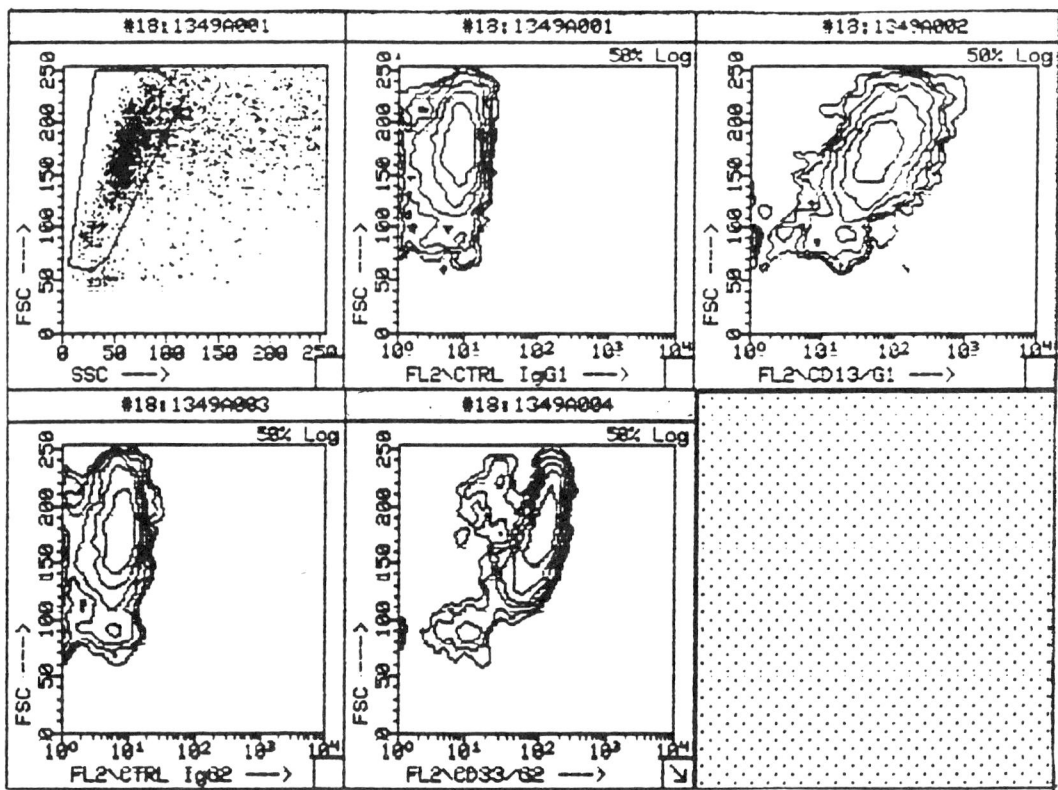

Figure 13.6. The upper left frame is a dot plot of forward scatter (FSC) vs. side scatter (SSC). The gate includes mononuclear cells and excludes neutrophils and debris. The upper middle frame is a contour plot of FSC vs. fluorescence of a fluorochrome- and isotype-matched control (FL2[phycoerythrin]/IgG1). The upper right frame is a contour plot of FSC vs. CD13. The large cells (blasts) are positive. The lower frames are contour plots of fluorochrome and isotype control (FL2[phycoerythrin]/IgG2) (left) and CD33 (middle). The blasts are CD33-positive.

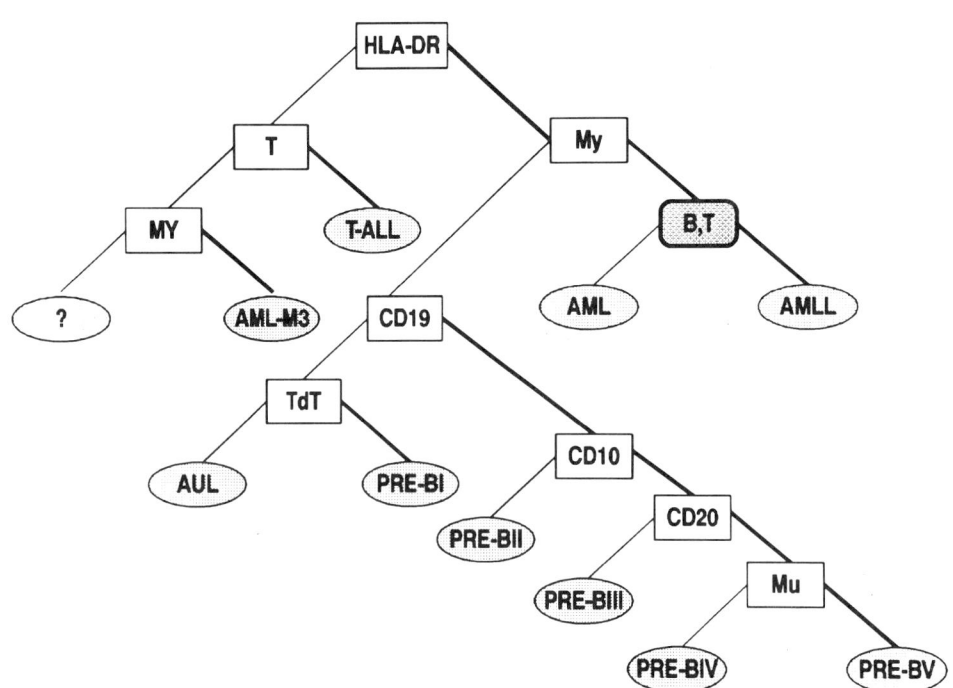

Figure 13.7. Decision-Tree analysis in acute leukemias. The graph is self explanatory. (My = CD13 and/or CD33; T = CD2 and/or CD5 and/or CD7; B = CD19 and/or CD20). Events that are positive result in movement to the right (thick lines). Negative events result in movement to the left (thin lines). A typical case of common acute lymphoblastic leukemia would be HLA-DR-positive (right) with no myeloid-associated antigen expression (left); CD19-positive (right), CD10-positive (right), and CD20-negative (left).

Table 13.2
Conditional Probability of Antigen Coexpression in Childhood Acute Leukemia[a]

	DR	CD19	CD10	CD20	CD13	CD33	CD2	CD5	CD7
DR	100	100	98	100	91	88	0	0	33
CD19	90	100	98	100	73	50	0	0	11
CD10	87	96	100	100	64	38	17	17	22
CD20	19	21	22	100	0	0	0	0	0
CD13	19	17	15	0	100	50	0	0	0
CD33	13	0	7	0	40	100	0	0	22
CD2	6	0	2	0	0	0	100	100	67
CD5	6	0	2	0	0	0	100	100	67
CD7	6	2	2	0	0	25	100	100	100

[a]Conditional probability of antigen coexpression in childhood acute leukemia. The numbers represent the percentage of (columns) that coexpress (rows). For example, 33% of CD7-positive childhood acute leukemias coexpressed HLA-DR, whereas only 6% of HLA-DR-positive leukemias coexpress CD7 n= =59. (University of Florida College of Medicine, Department of Pathology, Division of Hematopathology.)

Table 13.3
Conditional Probability of Antigen Coexpression in Adult Acute Leukemia[a]

	DR	CD19	CD10	CD20	CD13	CD33	CD2	CD5	CD7
DR	100	100	100	100	96	86	75[b]	83[b]	75[b]
CD19	27	100	61	90	14	8	25	11	14
CD10	21	61	100	100	10	7	0	17	0
CD20	10	36	50	100	0	2	0	0	0
CD13	70	36	33	0	100	70	75	67	79
CD33	60	21	24	10	70	100	50	75	71
CD2	3	4	0	0	4	3	100	50	21
CD5	5	4	5	0	6	6	75	100	21
CD7	12	7	0	0	15	14	75	75	100

[a]Conditional probability of antigen coexpression in adult acute leukemia. The numbers represent the percentage of (columns) that coexpress (rows). For example, 90% of CD20-positive leukemias coexpress CD19, whereas only 36% of CD19-positive acute leukemias coexpress CD20. n=114.
[b]In this series, the coexpression of T-cell-associated antigens and HLA-DR only occurred in cases eventually diagnosed as ''mixed-lineage'' acute leukemias. T-cell ALL's (n=2) (University of Florida College of Medicine, Department of Pathology, Division of Hematopathology).

tive blasts) need not be expressed in numerical terms since obviously all the cells that are larger (blasts) are CD19-positive. The numerical designation then, should address the fact that there are a given number of blasts that are positive for CD19 (in this case, >90%). To complement this information, the smaller cells (lymphocytes) react with CD2 and are therefore T-cells. In contrast, the larger cells (blasts) are negative for CD2.

Figure 13.6 is an example of an acute myeloid leukemia that contains numerous nonblastic elements and that requires gating based on light scatter with subsequent analysis of the fluorescence of the mononuclear population. The process of analysis is otherwise similar to that illustrated in Figure 13.5.

Reporting of Data

The composite immunophenotype should be described, i.e., ''HLA-DR(+), CD19(+) blasts present in this sample. These cells constitute approximately 56% of all mononuclear cells examined.''

The alternate method of reporting positively in percentual terms may be confusing. Example:

HLA-DR 62%
CD19 . . . . 75%

It is not clear whether HLR-DR . . 62% means that 62% of the blasts are HLA-DR-positive or whether 62% of the cells are 100% HLA-DR positive. If HLA-DR and CD19 are coexpressed on the same blasts, why is HLA-DR 62% and CD19 75%? The description of qualitative phenomena with numerical designations is not only awkward, but inaccurate.

Interpretation of Data

In an attempt to provide a guideline for the interpretation of immunophenotypic data, we have developed a decision tree (Fig. 13.7) that was used intuitively and was tested retrospectively on approximately 176 randomly selected acute leukemias analyzed in the Flow Cytometry laboratory at the University of Florida College of Medicine, Department of Pathology. The decision tree was formulated in an effort to reproduce the process that led to the diagnosis in each case.

The positivity or negativity of a commonly used panel of monoclonal antibodies was entered in a data base that uses Boolean connectors or operators (Research Information Systems. Encinitas, CA). The columns and rows depicted in Tables 13.2 and 13.3 represent the number of cases (percent of total) that coexpress a pair of antibodies (see figure legends). For example, the most frequently expressed antigen was HLA-DR. The data illustrates the frequency of coexpression of antigens. Thus, for CD19- (across) positive acute leukemias, the coexpression of CD10 (down) is 96% in children and 61% in adults. The coexpression of CD2 is 0% in children and 4% in adults. The coexpression of unanticipated antigens (i.e., myeloid) is 19% (CD13) and 13% (CD33) in children and 36% (CD13) and 21% (CD33) in adults. Therefore, the coexpression of two B-cell-associated/restricted antigens (CD10/CD19) is common (and would be anticipated) while the coexpression of B-cell-associated/restricted antigens and myeloid-associated antigens is less frequent (and less anticipated). The coexpression of B-cell-associated/restricted antigens and T-cell-associated/restricted antigens in acute leukemia is far less common. In contrast, in chronic lymphocytic leukemia (no longer derived from the proliferation of blasts, but of mature cells) the coexpression of B-cell-associated/restricted antigens and CD5, but not other T-cell antigens, is quite common (42). The data illustrated in Tables 13.2 and 13.3 is, therefore, useless unless one is referring to acute leukemias.

When the results were analyzed without regard to age, all cases of HLA-DR (−) acute leukemias (20/176 or 11%) were either T-ALL (eight cases) or AML-M3 (acute progranulocytic leukemia, 12 cases). The pattern of coexpression of all antibodies can be seen in Tables 13.2 and 13.3. This should be used as a guideline only. Invariably, there will be exceptions to any rule. Unusual leukemic processes, such as megakaryoblastic leukemia (M7) are not referred to in this decision tree. In our experience, M7 is usually HLA-DR (−), variably myeloid antigen-positive and almost always positive with anti-platelet antibodies. Similarly, erythroleukemias (M6) are variably positive for myeloid-associated antigens and generally constitute a morphologically defined subgroup.

In conclusion, flow cytometric data provide useful information in the analysis of acute leukemia, not by establishing the diagnosis of acute leukemia but by providing rapid and reproducible data that will aid in selecting in therapeutic regimen. These data must be interpreted by professionals that are familiar with the intricacies and potential pitfalls of flow cytometric immunophenotyping, the clinical and morphological spectrum of leukemias, and the therapeutic implications of their decisions.

REFERENCES

1. Schrier SS. The leukemias and myeloproliferative disorders. In: Rubenstein E, Federman DD, eds. New York: Scientific American Medicine, Scientific American, Inc., 1989.
2. Greaves MF. Differentiation-linked leukemogenesis in lymphocytes. Science 1986;234:697–704.
3. Greaves MF, Chan LC, Furley AJ, Watt SM, Molgaard HV. Lineage promiscuity in hemopoietic differentiation and leukemia. Blood 1986; 67:1–11.
4. Bennett JM, Catovsky D, Daniel MT, et al. Proposals for the classification of the acute leukaemias French-American-British (FAB) co-operative group. Br J Haematol 1976;33:451–458.
5. Bennett JM, Catovsky D, Daniel MT, et al. Proposals for the classification of the myelodysplastic syndromes. Br J Haematol 1982;51:189–199.
6. Bennett JM, Catovsky D, Daniel MT, et al. Proposed revised criteria for the classification of acute myeloid leukemia. A report of the French-American-British Cooperative Group. Ann Intern Med 1985;103:620–625.
7. Morphologic, immunologic, and cytogenetic (MIC) working classification of acute lymphoblastic leukemias. Report of the workshop held in Leuven, Belgium, April 22–23, 1985. First MIC Cooperative Study Group. Cancer Genet Cytogenet 1986;23:189–197.
8. Morphologic, immunologic, and cytogenetic (MIC) working classification of the acute myeloid leukemias. Report of the Workshop held in Leuven, Belgium, September 15–17, 1986. Second MIC Cooperative Study Group. Cancer Genet Cytogenet 1988;30:1–15.
9. Bowman GP, Neame PB, and Soambonsrup P. The contribution of cytochemistry and immunophenotyping to the reproducibility of the FAB classification in acute leukemia. Blood 1986;68:900–905.
10. Head DR, Savage RA, Cerezo L, et al. Reproducibility of the French-American-British Classification of Acute Leukemia: The Southwest Oncology Group Experience. Am J Hematol 1985;18:47–57.
11. National Cancer Institute sponsored study of classification of non-Hodgkin's lymphomas: summary and description of a working formulation for clinical usage. The Non-Hodgkin's Lymphoma Pathologic Classification Project. Cancer 1982;49:2112–2135.
12. Anderson KC, Bates MP, Slaughenhoupt BL, Pinkus GS, Schlossman SF, and Nadler LM. Expression of human B cell-associated antigens on leukemias and lymphomas: a model of human B cell differentiation. Blood 1984;63:1424–1433.
13. Korsmeyer SJ, Hieter PA, Ravetch JV, Poplack DG, Waldmann TA, and Leder, P. Developmental hierarchy of immunoglobin gene rearrangements in human leukemic pre B cells. Proc Natl Acad Sci 1981; 78:7096–7100.
14. Loken MR, Shah VO, Dattilio KL, and Civin CI. Flow cytometric analysis of human bone marrow. II. Normal B lymphocyte development. Blood 1987;70:1316–1324.
15. Hurwitz CA, Loken MR, Graham ML, et al. Asynchronous antigen expression in B lineage acute lymphoblastic leukemia. Blood 1988;72: 299–307.
16. Greaves MF, Hariri G, Newman RA, Sutherland DR, Ritter MA, and Ritz J. Selective expression of the common acute lymphoblastic leukemia (gp 100) antigen on immature lymphoid cells and their malignant counterparts. Blood 1983;61:628–639.
17. Potter VR. Phenotypic diversity in experimental hepatomas: the concept of partially blocked ontogeny. The 10th Walter Hubert Lecture. Br J Cancer 1978;38:1–23.
18. Pierce GB. Neoplasms, Differentiation and Mutations. Am J Pathol 1975;77:103–112.
19. McCulloch EA, Smith LJ and Alder S. Cellular lineages in normal and leukemic hemopoiesis. Prog Clin Biol Res 1983;134:229–244.
20. Littman DR, Newton M, Crommie, et al. Characterization of an expressed CD3-associated Ti gamma-chain reveals C gamma domain polymorphism. Nature 1987;326:85–88.
21. Griffin JD, Mayer RJ, Weinstein HJ, et al. Surface marker analysis of acute myeloblastic leukemia: identification of differentiation-associated phenotypes. Blood 1983;62:557–563.
22. Matutes E, Rodriguez B, Polli N, et al. Characterization of myeloid leukemias with monoclonal antibodies 3C5 and MY9. Hematol Oncol 1985;3:179–186.
23. Volk JR, Kjeldsberg CR, Eyre HJ, and Mary J. T-cell prolymphocytic leukemia. Clinical and immunologic characterization. Cancer 1983;52: 2049–2054.

24. San Miguel JF, Gonzales M, Canizo MC, et al. Leukemias with megakaryoblastic involvement: clinical, hematologic and immunologic characterization. Blood 1988;72:402–407.
25. Greaves MF and Sieff C. Monoclonal antiglycophorin as a probe for erythroleukemias. Blood 1983;61:645–651.
26. Pui CH, Dahl GV, Melvin S, et al. Acute leukaemia with mixed lymphoid and myeloid phenotype. Br J Haematol 1984;56:121–130.
27. Mirro J, Kitchingman GR, Williams DL, Murphy SB, Zipf TF, and Stass SA. Mixed lineage leukemia: the implications for hematopoietic differentiation [Letter]. Blood 1986;68:597–599.
28. Gale RP, Ben Bassat I. Hybrid acute leukaemia. Br J Haematol 1987;65:261–264.
29. Mirro J, Zipf TF, Pui CH, et al. Acute mixed lineage leukemia: clinicopathologic correlations and prognostic significance. Blood 1985;66:1115–1123.
30. Sobol RE, Mick R, Royston I. et al. Clinical importance of myeloid antigen expression in adult acute lymphoblastic leukemia. N Engl J Med 1987;316:1111–1117.
31. Wiersma SR, Ortega J, Sobel E, and Weinberg KI. Clinical importance of myeloid antigen expression in acute lymphoblastic leukemia of childhood. N Engl J Med 1991;324:800–808.
32. Cross AH, Goorha RM, Nuss R et al. Acute myeloid leukemia with T-lymphoid features: a distinct biologic and clinical entity. Blood 1988;72:579–587.
33. Ball ED, David RB, Griffin et al. Prognostic value of lymphocyte surface markers in acute myeloid leukemia. Blood 1991;77:2242–2250.
34. Bennett JM. Classification of the myelodysplastic syndromes. Clin Haematol 1986;15:909–923.
35. Foon KA, Todd RF. 3d. Immunologic classification of leukemia and lymphoma. Blood 1986;68:1–31.
36. Sobol RE, Bloomfield CD, and Royston I. Immunophenotyping in the diagnosis and classification of acute lymphoblastic leukemia. Clin Lab Med 1988;8:151–162.
37. Ortega JA, Nesbit ME Jr, Donaldson MH et al. L-Asparaginase, vincristine, and prednisone for induction of first remission in acute lymphocytic leukemia. Cancer Res 1977;37:535–540.
38. Look AT, Roberson PK, Williams DL, et al. Prognostic importance of blast cell DNA content in childhood acute lymphoblastic leukemia. Blood 1985;65:1079–1086.
39. Bennett JM. The classification of the acute leukemias: cytochemical and morphologic considerations. In: Wiernik PH, Canellos GP, Kyle RA and Schiffer CA eds., Neoplastic Diseases of the Blood. New York: Churchill Livingston, 1985, pp. 201–217.
40. Terstappen LW, and Loken MR. Five-dimensional flow cytometry as a new approach for blood and bone marrow differentials. Cytometry 1988;9:548–556.
41. Stass SA, and Mirro J.Jr. Lineage heterogeneity in acute leukaemia: acute mixed-lineage leukaemia and lineage switch. Clin Haematol 1986;15:811–827.
42. Gale RP, and Foon KA. Chronic lymphocytic leukemia. Recent advances in biology and treatment. Ann Intern Med 1985;103:101–120.

14

Breast Cancer

DAVID W. HEDLEY
Clinical Commentary by **Douglas E. Merkel**

INTRODUCTION: CLINICAL PROBLEMS AND THE SCOPE OF FLOW CYTOMETRY

Clinical Background

Breast cancer remains a major health problem because it is common and because distant metastases, which develop in approximately half of the cases, are rarely curable. There have, nevertheless, been major advances in our understanding of the biology of the disease over the past 20 years that should result in a reduction in overall mortality through the introduction of mammographic screening programs and the use of systemic adjuvant treatment for high-risk disease. Additional benefits to patients' quality of life are the recognition that some primary tumors can be treated by a lesser surgical procedure than mastectomy and by the introduction of steroid hormone receptor assays and medical endocrine therapy to manage hormone sensitive metastatic disease. In addition, cytotoxic chemotherapy is available to palliate the symptoms of advanced disease.

Pending better understanding of causative factors, and the introduction of appropriate public health measures to reduce the incidence of the disease, the main challenges are to improve the detection of early cancers before they disseminate and to develop more effective treatment for metastatic disease. At the present time, both chemotherapy and endocrine therapy can produce major responses in patients with advanced breast cancer. Responses occur more frequently using cytotoxic drugs, but endocrine therapy is better tolerated by patients and capable of producing more durable responses. Unfortunately, irrespective of the treatment type and the magnitude of initial response, drug resistance invariably develops.

For patients with apparently localized primary tumors, it is now accepted that those who subsequently relapse do so because of the presence of occult metastatic disease, which was present at the time of initial diagnosis (1). The use of systemic chemo- or endocrine therapy as an adjunct to primary treatment reduces overall mortality, indicating that it is at least sometimes capable of eradicating metastatic disease when the total tumor cell burden is small. Because of side effects, adjuvant chemotherapy would be best reserved for those patients at significant risk of recurrence, but unfortunately they cannot be identified using current clinicopathological criteria (2).

DNA Content and Proliferation Markers

The large majority of breast cancer studies using flow cytometry have concentrated on identifying patients with early-stage disease and poor prognosis by using single-parameter DNA analysis. Although no strong evidence has emerged to support DNA index as a major independent prognostic variable, a high percentage of cells in S phase appears to indicate an increased probability of recurrence, especially within the first two years following primary treatment. A serious problem is that reliable S-phase estimates are not always obtained, in part due to contaminating normal host cells in the sample. There is an urgent need to improve techniques for measuring tumor cell proliferation, either by more sophisticated computer programs to deconvolute single-parameter DNA histograms, or by the use of additional parameters such as fluorescent-labeled antibodies to cytokeratins, to identify cells of epithelial origin, or to nuclear proliferation markers, such as Ki-67 or PCNA/cyclin.

Compared to cancers that present as palpable breast lumps, little has been reported about the earlier lesions detected by mammography. Here, the problem is often to decide where the abnormality lies on a continuum between ductal hyperplasia and frankly invasive carcinoma and, consequently, whether to proceed to a formal surgical biopsy or simply keep the patient under surveillance. Fine-needle aspiration usually yields material adequate for a cytological diagnosis, with sufficient left over for DNA flow cytometry. This might support a diagnosis of cancer by demonstrating aneuploidy or otherwise give a guide to biological aggression. With the rapid development of population-based mammographic screening programs, and the need to contain the financial costs of a high false-negative surgical biopsy rate, this is an important area for further study. For established early invasive carcinomas, there is a further requirement to identify patients for whom simple excision is adequate, and those for whom the risk of more widespread invasion calls for either mastectomy or breast irradiation.

Treatment Sensitivity

Over the years, attempts have been made to measure estrogen receptors by flow cytometry using fluorescein–tagged estradiol or fluorescent molecules that bind to estrogen receptors, such as coumestrol. More recently, the general availability of monoclonal antibodies to estrogen receptor protein has offered the possibility of developing a flow cytometric immunoassay, although the sister discipline of image cytometry may prove to be the more robust analytical technique in the long run.

In contrast to steroid hormone receptors, there is an emerging interest in functional assays of cytotoxic drug resistance phenotypes using flow cytometry. So far, these primarily attempt to measure p-glycoprotein activity and cellular glutathione content, both of which are thought to affect sensitivity to doxorubicin, the most effective chemotherapeutic agent for treating advanced breast cancer. The excitement here is that pharmacological means of reversing these cell processes are now being developed (3). The assays can be done using fine-needle aspiration samples, and these can be taken sequentially if there is clinical justification. There is thus the possibility not only of identifying the cell processes resulting in clinical treatment failure, but also of monitoring the effects of drug-resistance modulators in individual patients. A future role for flow cytometry in fine-tuning cancer treatment, as opposed to diagnosis or prognosis, is quite promising. The rapid progress being made in understanding drug actions at a molecular level, the ability to design fluorescent probes to measure these specific cell processes, the greater sophistication in flow cytometry instrumentation and computing, and the increasing clinical use of needle-aspiration biopsies to resample tumors during the course of treatment will make this feasible. It is possible therefore that we are only beginning to appreciate the true extent of flow cytometry's potential in clinical breast cancer management.

SAMPLE ACQUISITION

Various sources of tissue have been used for flow cytometry and their suitability depends on the particular assay procedure used. They are briefly outlined here, together with their pros and cons.

Fresh Surgical Biopsies

Because of a need for frozen sections for steroid hormone receptor assays, surgical specimens commonly reach the pathology department unfixed and are the most generally useful material for flow cytometry. Cells can be extracted by scraping the cut surface with a scalpel blade, by mincing, or by enzymatic digestion. For a given volume of tissue, the cell yield varies widely, depending on the amount of fibrous stroma. As assays of viable cell function become more widely used, surgical biopsies will probably be the material of choice when insufficient cells are obtained from a needle aspirate.

Deep-Frozen Tissue

Because of its availability and ease of transport from hospital to laboratory, deep-frozen material surplus to receptor assays is probably the most popular source of tissue for DNA flow cytometry and a large number of studies have been reported over the past 10 years. In our hands, it produced histograms of a quality equivalent to that seen with fresh material, provided that intact tissue was kept frozen until the time of the assay (4). Pulverized material remaining after extraction of the cytosol can also yield DNA histograms and has its advocates (5). Results reported using this approach seem to give a lower overall incidence of DNA aneuploidy than that obtained using intact tissue; a comparative study is needed to ensure that this is not a result of selective tumor cell loss during sample handling.

Paraffin-Embedded Blocks

The use of paraffin-embedded blocks to generate DNA histograms was introduced explicitly to perform large retrospective studies in order to determine the prognostic significance of cellular DNA content, rather than for routine clinical use (6). Ideally, these studies are done using material from patients entered into large prospective clinical trials, where the quality of data management is usually high. Unfortunately, the CV's obtained are poorer than when the sample is run fresh and, almost certainly, the technique underestimates the true incidence of DNA aneuploidy. Determinations of percent-S phase need to be treated with even more caution than determinations obtained from fresh biopsies and, if DNA analysis is seriously considered to be clinically indicated, every effort should be made to obtain fresh tissue, rather than use blocks simply because they are more convenient.

Needle-Aspiration Biopsies

These are either core biopsies, which yield more material and cause more trauma to the patient, or aspirates obtained using a fine (e.g., 23-gauge) needle. The latter approach is more popular in Europe and will usually produce enough cells for a DNA histogram (7–9). Because it is minimally traumatic, it might be possible to obtain sequential samples in order to monitor the short-term effects of treatment. The simplicity and low cost, compared to surgical excision, suggest that serious attention should be given to needle aspirates as the standard source of material; for example, by making flow cytometers more generally capable of handling small samples or by designing biopsy needles specifically for this purpose.

ETHICAL CONSIDERATIONS

Before going on to consider the various clinical applications of flow cytometry, some thought should be given to what we actually intend to do with the information obtained. The real clinical significance of even a simple test like DNA index is not always certain, while the current lack of agreement on interlaboratory standardization and quality assurance is scandalous. Although it may be reasonable to supplement the flow cytometry laboratory's income by providing a clinical

Table 14.1
Prognostic Significance of DNA Flow Cytometry in Breast Cancer

AUTHOR	YEAR	STAGES	No. of cases fresh/ embedded		DNA INDEX(%)			CORRELATIONS WITH DNA INDEX					REMARKS	REFER-ENCE
					=1.0	≠1.0	Multiploid	SIZE	Lymph/node Involvement	ER+	GRADE	AGE		
Beerman	1990	I-III	690	F/E	27	73	12	yes	no	yes	n/a	no	Prognostic significance	11
Dowle	1987	I-II	354	E	40	60	n/a	yes	trend	no	yes	yes	Prognostic significance	12
Dressler	1988	I-II	1331	F	43	57	9	n/a	yes	yes	n/a	no	Includes % S-Phase	5
Ewers	1984	I-IV	638	F	34	66	17	yes	no	n/a	n/a	no		13
Feichter	1988	I-III	300	F	38	62	n/a	trend	yes	yes	yes	trend	Includes % S-Phase	14
Hedley	1987	II	490	E	35	65	11	no	yes	yes	yes	yes	Prognostic significance includes % S-Phase	15
Kallioniemi	1988	I-IV	308	E	36	64	12	yes	yes	yes	yes	yes	Prognostic significance includes % S-Phase	16
Keyhani-Rofaga	1990	NO	165	E	43	57	n/a	n/a	—	n/a	yes	no	Nodes negative only	17
Muss	1989	N0	101	F	45	55	n/a	no	—	no	no	no	Nodes negative only	18
O'Reilly	1990	I-II	140	E	31	69	n/a	no	no	no	yes	n/a	Prognostic significance includes % S-Phase	19
Remvikos	1988	I-IV	206	F	26	74	12	trend	no	n/a	n/a	n/a	used fine-needle sampling	8
Spyratsos	1987		106	F	33	67	4	trend	no	trend	yes	n/a	used fine-needle sampling	9
Taylor	1983	I-III	114	F	21	79	16	no	no	no	n/a	yes		4
Toikkanen	1989	I-IV	351	E	32	68	16	yes	yes	n/a	yes	yes	very long follow up includes S-Phase	20

service, the temptation to overemphasize the value of a test to clinicians who are unfamiliar with the current literature needs to be guarded against. They may go ahead and perform an unjustified mastectomy as a result of your advice. And the paraffin blocks might then find their way into the hands of the patient's lawyers!

SINGLE PARAMETER DNA ANALYSIS: DNA INDEX

A DNA histogram can identify gross chromosomal imbalance, "DNA aneuploidy," and give an indication of tumor cell proliferation kinetics by allowing an approximation of the proportion of cells in S phase. Note that there are no *a priori* grounds for supposing that the two are causally related. Until recently, greater emphasis was placed on the former, partly because it is easier to determine, being "yes or no" phenomenon, and probably also because it was felt intuitively that a clearly abnormal genetic makeup must be bad news. It is possible that more DNA histograms have been generated from breast cancer than from all other malignant diseases combined, and it is beginning to emerge that S phase is, after all, the more promising prognostic determinant. Because they are telling us different things, they will be considered separately, with DNA index considered first for historical reasons.

DNA Index in Early Stage Disease

CORRELATION WITH CYTOGENETIC FINDINGS

Remvikos et al. reported the successful karyotyping of 32 out of 94 previously untreated breast carcinomas and, of these, 25 also had DNA analysis done by flow cytometry (10). Not surprisingly, a correlation was observed between chromosome counts and DNA index, although differences of up to 30% were observed in individual cases. Flow cytometric DNA content tended to overestimate chromosome content. Significantly, some DNA diploid cases showed grossly hypodiploid karyotypes.

CORRELATIONS WITH OTHER PROGNOSTIC FEATURES

Most early studies were done prospectively, using deep-frozen material sent for estrogen receptor assay, and reported correlation with flow cytometry without giving much clinical follow-up. More recent series have tended to use paraffin-embedded blocks, allowing evaluation of DNA index as an independent prognostic feature. Table 14.1 is a summary of the large published series and lists the observed correlations with recognized prognostic features of early breast cancer. The reader is encouraged to study the original papers critically because some of the apparent inconsistencies are due to differences in technique or in definition of terms. It is gener-

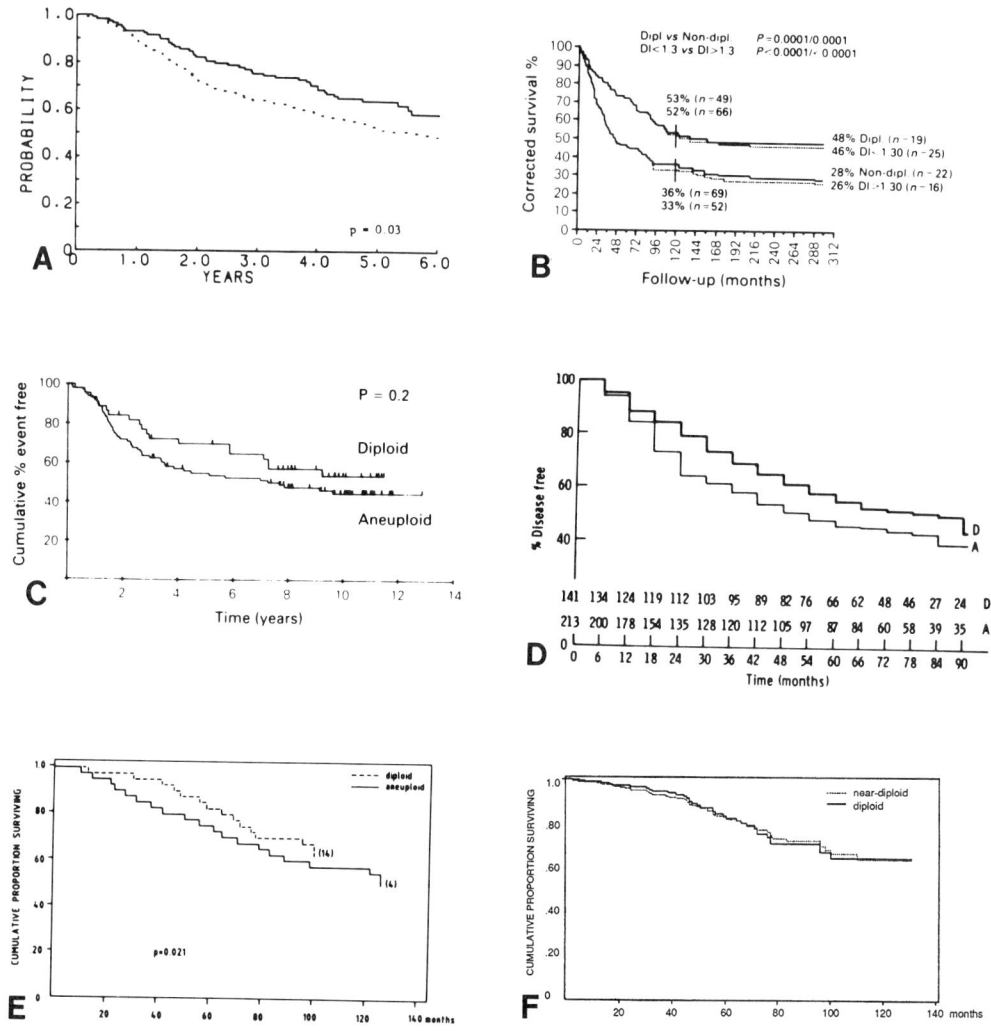

Figure 14.1. Influence of DNA aneuploidy on disease–free survival in operable primary breast cancer. Comparison of results from large series: **A**; Sydney (solid lines: DNA diploid; dotted lines: DNA aneuploid; ref. 15), **B**; Turku (ref. 20), **C**; London (ref. 19), **D**; Nottingham (D: DNA diploid, A: DNA aneuploid; ref. 12) and **E**; Leyden (ref. 11). **F** shows the Leyden patients divided into DNA index ≤ 1.4 compared to > 1.4. In all cases, an abnormal stemline carries a modest increase in the risk of disease recurrence, except that *near-diploid* clones appear to behave like *diploid* tumors.

ally agreed that DNA aneuploidy is found more frequently with estrogen-receptor-negative tumors, although the correlation is weak, while tumor grade appears to be a major correlate. Compared to infiltrating ductal carcinomas, lobular and mucinous types tend to be diploid, while medullary carcinomas are usually aneuploid. The situation with tumor stage is less clear cut. Measuring the exact size of primary breast carcinomas is less easy than might be imagined, which possibly accounts for inconsistent results, but there is a surprising lack of agreement about the extent to which involvement of axillary lymph nodes correlates with DNA index. Even more curious is the failure to reach consensus on whether aneuploidy is more frequent in postmenopausal patients. Although the overall trend is for "good" prognostic features to correlate with DNA index = 1.0, these inconsistencies between the large series will need to be kept in mind when assessing statistical analyses made using Cox models.

PROGNOSTIC SIGNIFICANCE OF DNA INDEX

The first thing to note is that failure to demonstrate an abnormal DNA content does not imply a normal diploid karyotype and that, in general, the higher the resolution achieved, the greater the frequency with which abnormalities are found. As a single variable, however, the results reported from the different centers are surprisingly similar, despite differences in the way authors have interpreted their findings. Figure 14.1 illustrates this encouraging point. Note that whereas the slopes of the survival curves vary depending on how the patients were selected for known prognostic features, without exception they all show a modest but statistically significant improvement in long-term survival for those patients whose primary tumor was DNA diploid. Furthermore, this effect is maintained over prolonged follow-up (20). Because of the trend for this to be associated with other good prognostic features, the effect is less impressive using multivariate anal-

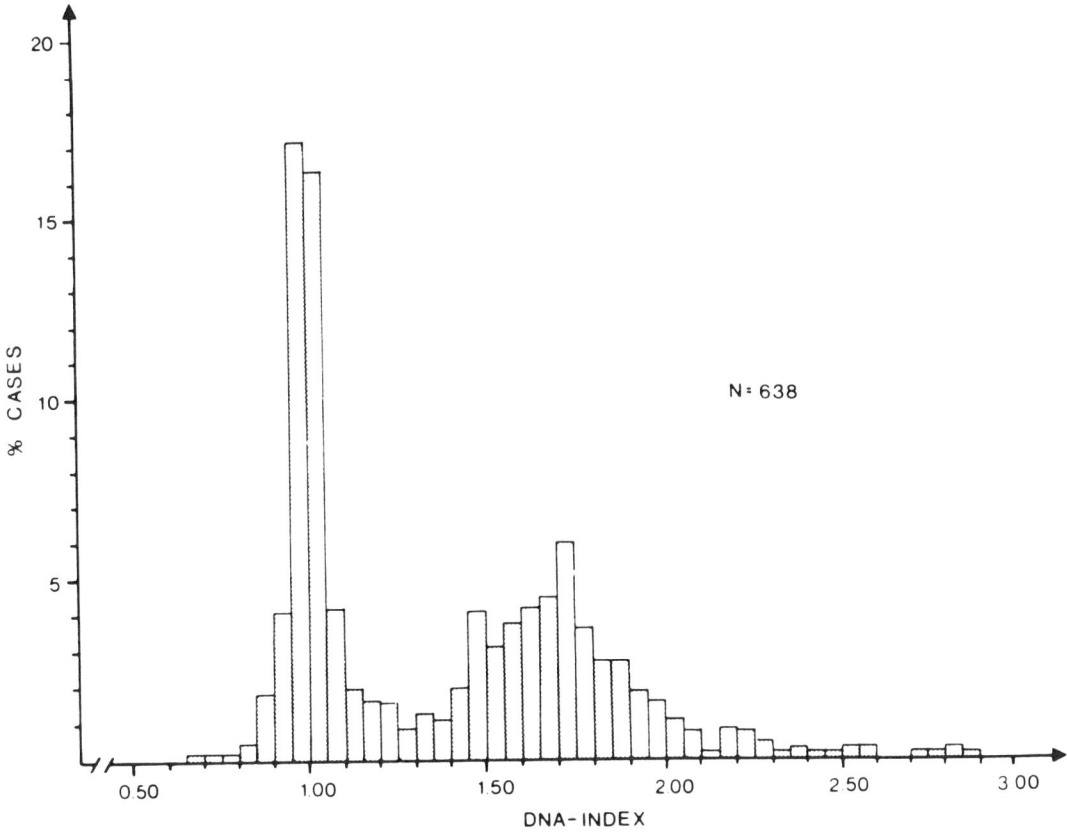

Figure 14.2. Distribution of DNA index, based on 638 cases reported by Ewers et al. (13).

ysis and generally fails to show independent prognostic significance (11, 12, 15, 19, 20). Again, the reader is encouraged to study the original publications critically because the results obtained using Cox models are highly dependent on the extent to which DNA index was found to correlate with the major prognostic variables, especially lymph node status.

There has been considerable interest as to whether some abnormalities of DNA index carry a worse prognosis than others. Intuitively, many of us felt in the early days that tetraploid tumors might be a good prognosis subgroup, while multiple aneuploid peaks, which are seen in about 10% of cases, should be associated with a high risk of relapse. Neither seems to be the case, however. Certainly, there have been no large studies showing that multiploid tumors do particularly badly, although some still maintain that tetraploidy carries a favorable prognosis (16). It should be noted, however, that none of the published frequency distribution histograms of DNA index in breast cancer shows a distinct peak at the 4n level, suggesting that in contrast to, for example, prostate cancer, tetraploidy is not a clearly identifiable subgroup. A consistent finding is that these histograms are bimodal, with DNA indices clustering into either the near-diploid range (i.e., diploid or *DNA index* <1.4) or falling into a roughly normal distribution centered around *DNA index* = 1.7–1.9 (Fig. 14.2). This distribution would be consistent with aneuploidy developing via two distinct mechanisms, one of which involves an initial polyploidization, a possibility that has been discussed by Shackney et al. (21). Interestingly, two papers have reported that the survival of patients with near-diploid tumors is similar to those with a DNA index of 1.0 and better than that seen with DNA indices in the triploid-tetraploid range (DI > 1.4 < 2.0) (11, 20). Note, however, that because these studies used paraffin-embedded material, the rare hypodiploid tumors would have been erroneously assigned a DNA index of > 1.0. It still needs to be determined whether or not these represent a bad prognosis subgroup, as was suggested by Coulson et al. (22). Given the importance of recessive genes in cancer development, this issue requires further study. With this proviso, however, it may be that although we know that we are missing aneuploid peaks in our wide CV "diploid" histograms, this may not be as critical as it appeared to be at one time.

Finally, there is good evidence that hypertetraploidy, i.e., DNA index > 2.1, may be associated with particularly aggressive tumors, especially in premenopausal patients (11, 16). This is based partly on flow cytometry studies, and partly on the results obtained using image cytometry. Only about 10% of samples fall into this group using flow cytometry, but selection of individual cells using image analysis suggests that these grossly abnormal cells often fail to appear as distinct peaks on flow cytometric histograms (23) (Fig. 14.3). Earlier work by Auer, Fallenius and coworkers at the Karolinska Institute in Stockholm had emphasized the bad

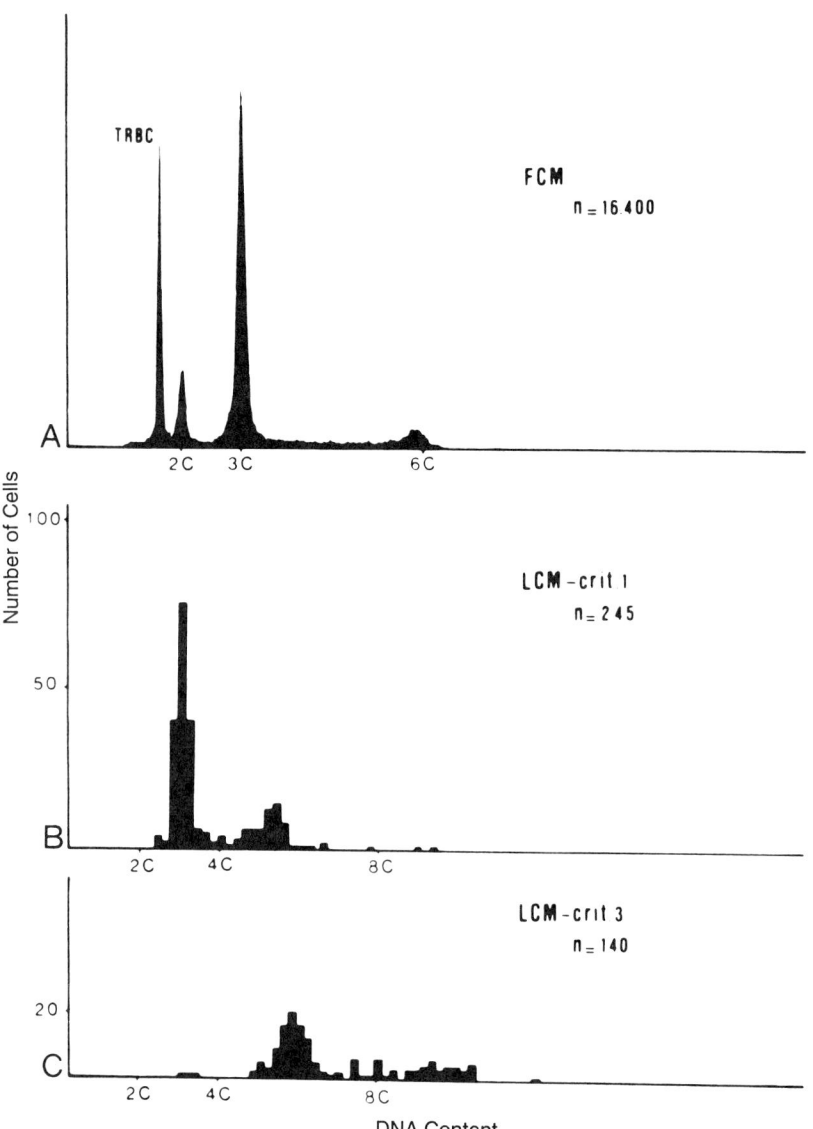

Figure 14.3. Use of image cytometry to demonstrate grossly aneuploid nuclei that were not apparent using flow cytometry. **A,** Flow cytometric DNA histogram showing approximately triploid population, with a G2 + M peak at the 6c position, **B,** DNA distribution of tumor cells measured by image cytometry; note that normal diploid cells at the 2c position were excluded by visual identification. **C,** Selection of tumor cells, based on increased nuclear area or chromatin density, showing individual cells above the 8c level (ref. 23).

prognosis associated with the presence of grossly aneuploid cells identified by microspectrophotometric measurement of DNA content in Feulgen-stained fine-needle aspirates from breast carcinomas (24–26). This would help to explain why none of the published flow cytometry studies has given such a strong correlation with prognosis. As with the apparent difference in survival for near-diploid versus triploid-tetraploid tumors, this effect probably reflects the more powerful influence underlying the different types of mitotic abnormality. In summary, with the probable exception of hypertetraploidy (DI > 2.0), DNA flow cytometry has turned out to be at best only a weak prognostic feature in operable primary breast cancer, with data increasingly hinting that it is an epiphenomenon that reflects underlying mitotic abnormalities rather than a driving force behind tumor progression.

DNA Index in Advanced Breast Cancer

Using image cytometry, Auer has shown that the DNA index of a breast cancer recurrence is usually identical to that seen in the primary tumor, even when relapse occurs many years later (27). Compared to early-stage disease, only a limited number of reports have examined patients with locally advanced or metastatic disease, but these have allowed a possible effect on treatment sensitivity to be investigated. Because survival of patients with recurrent disease is heavily dependent on the response to endocrine therapy, and since expression of estrogen receptors correlates with DNA index, flow cytometric DNA analysis was investigated as a possible predictor of endocrine response in patients with advanced disease. Responses to endocrine therapy were seen in 8/31

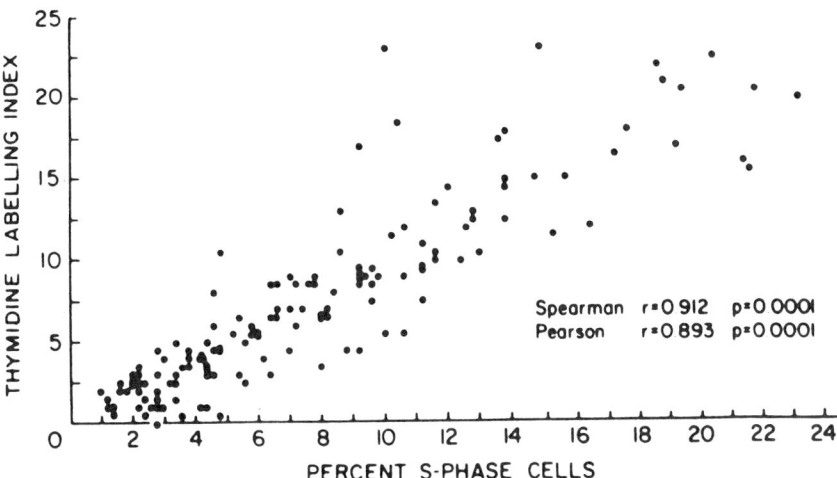

Figure 14.4. Correlation of %S-phase with *thymidine-labeling index*. (Reprinted with permission from McDivitt RW, Stone KR, Craig RB, Meyer JS. A comparison of human breast cancer cell kinetics measured by flow cytometry and thymidine labeling. Lab Invest 1985;52:289–291.)

patients with aneuploid tumors, but in only 1/11 of the diploid group, survival curves for the two groups being superimposable (28). We also failed to observe any effect on response to cytotoxic chemotherapy, regressions being seen in 11/24 diploid and 33/63 aneuploid tumors (unpublished data), a result similar to that recently reported by Masters et al. (29).

DNA Index in Subclinical Disease

The term "subclinical disease" is meant to cover those generally early carcinomas that are found by mammographic screening, but could also be applied to all preinvasive or early-invasive carcinomas. In terms of survival, the prognosis for these lesions is generally excellent and the clinical problem is more one of deciding how extensive a procedure is required in order to achieve local disease control. Previously, a mastectomy was considered standard treatment, but the trend now is toward more conservative surgery. A related problem is the management of patients with abnormalities that fall short of frank malignancy: Are these truly benign, what are the chances of going on to develop cancer, and how should these patients be followed up? The trend for breast carcinomas to be detected at increasingly early stages will probably accelerate with the adoption of population-based mammographic screening programs. There are economic as well as medical considerations to consider when planning management strategies. Compared to the effort that has gone into assessing the prognostic significance of DNA ploidy in more advanced stages, and considering the magnitude of the public health problem, there are disappointingly few studies here; perhaps because clinical flow cytometry laboratories have grown over dependent on paraffin blocks and the leftovers from estrogen receptor assays as a source of material.

Fine-needle aspiration is a virtually nontraumatic office procedure that usually yields sufficient material for DNA cytometry in addition to conventional cytological preparations (8, 9, 30). Because virtually all benign lesions are DNA diploid, the presence of an abnormal peak considerably strengthens a diagnosis of cancer in cytologically borderline cases, although the opposite is not, unfortunately, true (31–33). There is evidence that carcinomas detected by mammographic screening are more likely to be diploid and to have a lower S-phase fraction (34). An important question is whether these characteristics signify a more benign lesion, treatable perhaps by simple excision alone.

PROLIFERATION MARKERS

Estimation of Percentage S-phase Cells: Caveats

Although S phase has been less extensively studied than DNA index, there is little doubt that it is the more powerful prognostic indicator (15, 16, 19, 20, 35–38). The key source literature is a series of publications from Meyer and McDivitt, where the relation between *S-phase* and *thymidine labeling index* (TLI) was demonstrated (39–42) (Fig. 14.4). Although the latter estimation is too laborious to have become established as a standard clinical procedure, an extensive literature has shown it to be a powerful prognostic indicator in both early and advanced breast carcinoma (43–46). Not surprisingly, %S phase was strongly correlated with TLI, although it intended to give an overestimate for low TLI tumors because of the presence of cell debris, because some "S phase" cells are not actually synthesizing DNA, or perhaps because TLI was giving an erroneously low estimate of cell proliferation. It should be noted that these studies used fresh rather than embedded tissue and that S phase was estimated using a rectangular fit. Within the aneuploid group, there was no correlation between actual DNA index and %S-phase. These tumors showed a significantly higher mean TLI than did the diploid, while in the diploid group the overall correlation between S phase and TLI was less good. This is

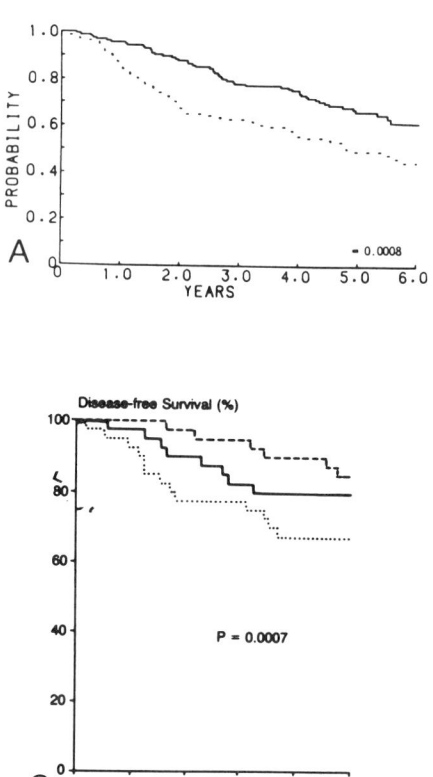
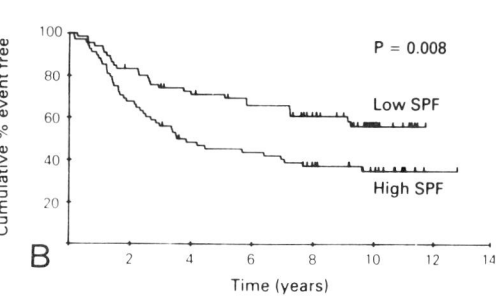
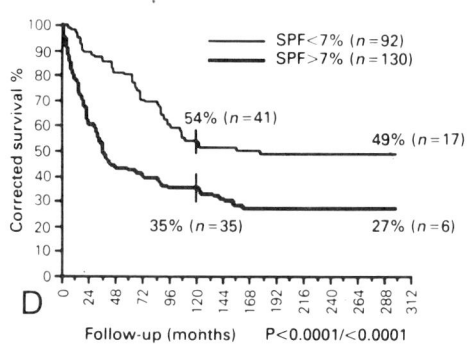

Figure 14.5. Effect of % S-phase on *disease-free survival* in four large series. **A**, Sydney (ref. 15), **B**, London (ref. 19), **C**, Lund (ref. 38), and **D**, Turku (ref. 20). Note the major impact of high % S-phase on *disease-free survival* in the first two years, following which the curves become parallel.

probably because of the variable admixture of diploid host cells, which could not be recognized as such using single-parameter DNA flow cytometry (42). Whereas the frequency distribution histograms of DNA index obtained in all the large series look pretty much the same, reported ranges of %S-phase vary widely due to differences of technique; for this reason individual laboratories need to establish their own cutoff points, rather than use those published in the literature (47–49). The San Antonio group used an iterative procedure to determine which cutoff point for %S phase gave the greatest statistical significance as a predictor of disease survival, although the researchers would need to keep a careful eye open for systematic changes in this over time (50). Thus, although to the uninitiated "percentage of cell in S phase" sounds very objective and scientific, when applied to human breast cancer it is less well validated, and quite probably less reproducible, than is conventional nuclear grading by light microscopy.

Prognostic Significance of %S-Phase

As was the case with TLI, a high %S-phase appears to be a major predictor of poor *disease-free* and overall survival in breast cancer (15, 16, 20, 38). Interestingly, this effect is seen mainly within the first two or three years after diagnosis, the survival curves becoming parallel thereafter (Fig. 14.5). A similar tendency for high proliferation to predict early recurrence can also be seen with TLI. Apart from DNA index, the most important correlate of %S phase is tumor grade, with all studies showing a marked tendency for high-grade tumors to have a high S-phase fraction. Because of this association, S phase tends to lose its impact when grade is included in Cox models, although there is clearly potential for further refinement in techniques for measuring tumor cell proliferation kinetics. An interesting but preliminary study of locally advanced breast cancer suggested that a high %S-phase predicted increased sensitivity to chemotherapy, although it is not certain whether this resulted in improved survival (51). This is a major outstanding question because increased chemo-sensitivity would be an additional justification for using adjuvant therapy to treat high %S-phase nodes negative patients.

Combination of S-Phase and DNA Index

Because these two parameters give fundamentally different information about tumor biology, it seems reasonable to combine them in an attempt to derive a flow cytometric prognostic index. Kallioniemi and coworkers split patients into three groups: diploid, low S phase; hypertetraploid or high S-phase other aneuploid; and the remainder, who consisted of about 50% of the total (16). There were highly significant differences in the survival curves (16). A similar approach has been advocated by the San Antonio group but,

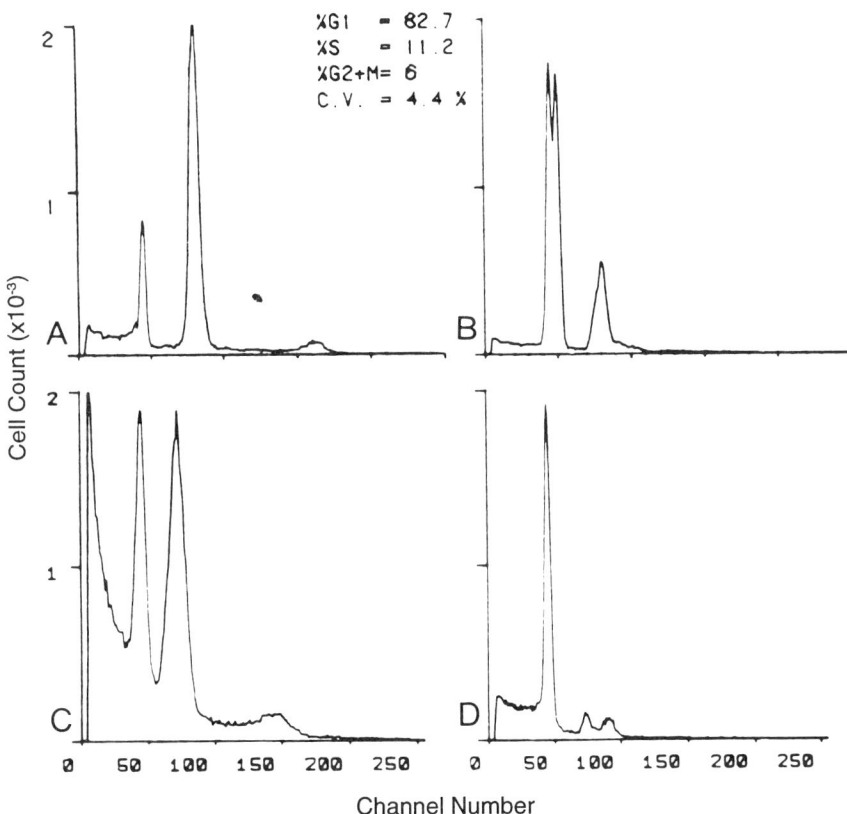

Figure 14.6. Problems with % S-phase estimates using single-parameter DNA flow cytometry to measure nuclei obtained from paraffin blocks. **A**, Satisfactory for analysis; **B**, Multiple aneuploid peaks; **C**, Excessive debris present; **D**, The aneuploid peak is small compared to the diploid population. Note that when the DNA index is 1.0, no appreciation of the proportion of tumor cells compared to infiltrating host cells is given.

for reasons that are not obvious, in their hands S phase did not predict disease-free survival for the DNA aneuploid subgroup. Although S phase could be made statistically significant within the diploid group, this was achieved by a very lopsided split so that out of a total of 345 patients reported, only 15 fell into this category, which would limit its practical clinical value (50).

Factors that can prevent a reliable S-phase estimate being made using a DNA histogram include presence of multiple aneuploid populations, excessive debris, and large preponderance of diploid host cells (15, 47–49) (Fig. 14.6). The most serious problem of all is that although we know that "diploid" DNA histograms are composites of tumor and normal host cells, these cells cannot be distinguished from one another and that S–phase estimates could be saying more about the proportion of infiltrating host cells than about tumor cell proliferation kinetics. At the present time, the flow cytometry community seems to be divided between those who believe that S phase should be offered to clinicians as a routine service and those who interpret the data to mean that we are on the right track investigating proliferation as a marker of cellular malignancy, but maintain that some form of multiparametric analysis will be needed to standardize an assay that is to be reliable enough for clinical decisions to be based on it.

Multiparametric Analysis

This offers the possibility of "cleaning up" DNA histograms to improve S–phase estimates by identifying proliferation markers in addition to S phase or investigating other possible prognostic markers, such as oncogene products. Some of these techniques are dealt with elsewhere, while others are research methods that are unlikely to find a routine clinical application because of their complexity.

The possibility of using forward or orthogonal light scatter to exclude cell debris or gate out subpopulations for DNA analysis has been discussed in meeting abstracts and review articles, but although this sounds pretty neat and well suited to routine clinical practice, definitive studies are required to establish its general applicability. In view of the pleomorphism of human carcinomas, it may prove difficult to isolate "pure" tumor cell populations without then restricting cell cycle analysis to a selected subgroup. A more reliable method for positively identifying the epithelial cell component of a breast carcinoma is probably to counterstain using a fluorescent antibody to cytokeratin. This technique is well described in a recent paper by Zarbo et al., appears to be relatively straightforward, and improves the resolution of DNA index as well as S phase (52). It should prove particularly useful for tumor proliferation when the DNA index is 1.0 and no estimate of the proportion of infiltrating host cells can be obtained from single-parameter DNA analysis.

In addition to markers of cell lineage, proliferation specific antigens have been considered for second parameters to improve tumor cell growth kinetic measurements. Some of these could potentially divide the "G_0/G_1" population into those cells that are in the growth fraction and those that are quiescent or incapable of further division. Examples are Ki-67 and p105 (53–54). Others appear to mark cells that are actively synthesizing DNA and are part of the multi-enzyme complex replitase, which assembles in the nucleus in late G_1. These enzymes are present at low concentrations during other phases, but tend to be more loosely bound and can be removed by washing. Examples that have been measured in combination with DNA include ribonucleotide reductase (56), DNA polymerase α, (57) and DNA polymerase δ-associated protein (PCNA) (58), but detailed clinical evaluation is required. Apart from problems of standardization, a general drawback of this approach could be that, as was the case with single-parameter DNA analysis, these markers do not allow tumor cells to be distinguished from infiltrating host cells. While it should be possible to add a fluorescent anti-keratin antibody as a third parameter, the additional prognostic information obtained would need to be significant before such an approach is adopted for routine use.

An alternative method for studying cell kinetics in detail is to administer the thymidine analog bromodeoxyuridine (BrUdR) to the patient shortly before tumor excision. This compound is incorporated into newly synthesized DNA and can then be identified using monoclonal antibodies (59–61). Potentially, this is a much more powerful and convenient method than measuring TLI. The DNA precursor is given in vivo before the vascular supply is interrupted, and tissue concentrations that are sufficient to saturate the nucleoside transporter are achieved, ensuring that all S–phase cells are labeled. BrUdR uptake can be measured in combination with DNA content. This allows determination of cell cycle parameters such as potential doubling time (T_{pot}), which has been suggested as a possible indicator for employing accelerated radiotherapy fractionation (62) but which was previously accessible only by using pulsed labeled curves and was therefore totally unsuited for clinical use. Hitherto, clinical flow cytometric studies of the cell cycle have confined themselves to S-phase approximations that give, at best, a crude estimate of the true state of affairs even when they are methodologically sound. What really needs answering is the proportion of tumor cells capable of indefinite division and, within this population, the fraction that is actually in the replicative cycle and the fraction that is quiescent (or G_0). The latter could prove to be a major source of resistance to treatment and also presumably account for the relapses that occur many years after ablation of the primary tumor. Considering the potential importance of population kinetics, it is surprising that so little clinical work has been done using BrUdR methods.

TREATMENT SENSITIVITY

In addition to supporting the traditional pathologist's role of determining diagnosis and prognosis, flow cytometry might acquire a more direct role in practical patient management through the development of assays of drug resistance. Both endocrine therapy and chemotherapy are now being understood in terms of their effects on cell function, and this is in accord with an increasing interest among cell biologists for studying viable cell function in flow cytometry rather than confining themselves to structure. It is impossible here to do more than sketch out an exciting and rapidly evolving branch of our discipline.

Estrogen Receptors

About a third of patients with recurrent or metastatic disease show a worthwhile response to treatments that lower circulating estrogens or block estrogen receptors; these responses are limited to tumors that express receptor protein. The standard dextran–coated charcoal (DCC) assay requires more tissue than can be obtained by fine-needle aspiration, gives an average value expressed as fmol/mg protein, and is not completely reliable as a guide to hormone sensitivity. As well as needing less material, a flow cytometric assay might prove a better predictor of response by, for example, identifying receptors in tumor cells when these are only a small minority population in the sample, or distinguishing between low ER tumors, where all cells are weakly positive, and tumors that are essentially ER-negative but have a minor population of strongly positive cells mixed in. Clinical response to endocrine therapy might be expected to differ in these situations.

A spate of flow cytometry methods papers appeared in the early 1980s, using either fluorescent conjugates of estradiol or naturally fluorescent compounds, such as cumestrol, that bind to estrogen receptors (63–65). Although specific fluorescence was detected in ER-positive cell lines, these assays did not enter routine clinical practice. More recently there has been talk of using fluorescent-labeled antibodies to estrogen or progestrone receptor protein, which could presumably allow a greater number of fluorophores to be bound per cell, giving better resolution of weakly positive cells. The technique does not appear to have been evaluated clinically yet, although results using cell lines are encouraging (66, 67). Measurement of estrogen receptors is one area where flow and image cytometry may compete head to head. The latter is now in clinical use and seems to give a better guide to hormone sensitivity than the DCC assay. Unless flow cytometry offers an additional advantage, its adoption as a routine clinical assay must remain in doubt.

Cytotoxic Drug Resistance

This is potentially a very important area for future developments. A common clinical observation is that major responses can occur following the first few courses of chemotherapy, with some tumor deposits appearing to resolve completely. Despite this initially successful treatment, the cancer then regrows, i.e., shows emergence of drug resistance. In part, drug resistance results from the presence of quiescent, clonogenic tumor cells, and some of the earlier flow cytometry studies were aimed at detecting and characterizing such cells. In recent years, however, cellular resistance mechanisms have become much better understood at the molecular level. Basically, to kill a cell, a drug has to accumulate at a sufficient concentration and produce irreversible damage to a target molecule. This molecule is usually DNA, although there is evidence that cell membrane damage may also bring about lethal events. Apart from the antimetabolites and vinca alkaloids, the damage usually results from the generation of reactive-free radicals or oxidizing species. All living organisms are naturally subjected to these stresses, which can be generated by redox cycling xenobiotics, background radioactivity, or simply as a byproduct of oxygen metabolism (68). Cells have evolved protective mechanisms such as membrane transport systems that exclude xenobiotics, free radical scavengers, antioxidants, and repair enzymes. It is now apparent that much of the drug resistance encountered in the clinic results from overexpression of these normal cell processes, and it is perhaps significant that tumors that are derived from epithelial surfaces (i.e., carcinomas), which are subject to bacterial colonization and general wear and tear, tend to show greater resistance to cytotoxic drugs than do the leukemias or lymphomas.

Two drug resistance phenotypes that have been studied using flow cytometry are multidrug resistance mediated by p-glycoprotein and pathways involving glutathione and its related enzymes. A wide range of cytotoxic drugs, including the anthracyclines, epipodophylotoxins and vinca alkaloids, are actively pumped out of cells by a 170 kd transmembrane glycoprotein with ATPase activity, usually referred to as p-glycoprotein (69). Chronic exposure to cytotoxic drugs in vitro results in overexpression of p-glycoprotein, which then confers resistance to these and other drugs. The ubiquitous sulphydryl-containing tripeptide glutathione (GSH) functions as a cytoplasmic-free radical scavenger, binding to electrophiles through its -SH group, either spontaneously or via the action of one of the glutathione-S-transferase isoenzymes (70–73). Additionally, GSH reduces hydrogen peroxide and organic peroxides by undergoing an oxidation/reduction cycle under the actions of GSH peroxidase and reductase. Although attention so far has largely been focused on developing flow cytometric assays of cellular GSH content, the activities of these metabolic pathways may prove to be more relevant to the development of clinical drug resistance (74–76).

With increased understanding of the mechanisms of drug resistance, it has become apparent that some of these processes are amenable to pharmacological blockade, which might then restore sensitivity to chemotherapy (3). Examples currently undergoing clinical trial include agents such as verapamil and cyclosporin A, which block p-glycoprotein (77); buthionine sulphoximine, which depletes cellular glutathione (78); and ethacrinic acid, an inhibitor of glutathione-S-transferase (79). But these processes also protect normal host cells from toxicity. Although drug resistance modulators are an exciting new approach to cancer treatment, they have the potential to increase the severity of side effects. For their rational use there is therefore a need for assays to determine if a particular agent is likely to improve the tumor response. Ideally these assays should be rapid and amenable to fine-needle aspiration biopsy samples so that, if necessary, they could be repeated after administration of the drug resistance modulator but before the cytotoxic agent was given. There is every indication that flow cytometric assays for p-glycoprotein and GSH content could be developed to fulfill this role.

Assays for p-Glycoprotein

These are dealt with in detail elsewhere (see Chs. 28 and 32). There are two fundamentally different flow cytometric approaches, either involving the use of fluorescent antibodies to p-glycoprotein, or examining the influence of p-glycoprotein on cellular uptake of fluorescent compounds. The use of antibodies has received less attention, probably because until recently they were not readily available, and low levels of intracellular antigen can be difficult to measure. Two recent developments are the production of antibodies that recognize extracellular epitopes of p-glycoprotein without requiring prior fixation (80), and the synthesis of polypeptide chains that compete for antigen recognition sites and can therefore be used for specificity controls (81). These refinements should see the introduction of flow cytometry as a method for determining the extent of expression of p-glycoprotein in heterogeneous clinical samples.

The other approach is to examine the inhibitory effect of p-glycoprotein on the uptake of daunomycin or Hoechst 33342, which are fluorescent substrates, by co-incubating a replicate sample with an agent that acts as a competitive inhibitor, such as verapamil or cyclosporin (82–84). This technique was developed by Krishan, who describes it in Ch. 28. Its particular relevance to breast cancer is that this appears to be a tumor where p-glycoprotein is an important cause of acquired drug resistance (85).

Measurement of Cellular Glutathione

Glutathione plays an essential role as a free radical scavenger and in the reduction of peroxides; increased activity of these processes can be associated with resistance to a wide range of agents, including alkylators, anthracyclines and platinum compounds. It is a gross oversimplification to think of GSH as some kind of intracellular blotting paper for soaking up toxic chemical species; its function probably depends more on the activities of glutathione peroxidase or the GSH-S-transferase isoenzymes than on simple GSH content (76, 86), and may well vary at different sites within a cell. Compared to p-glycoprotein, these processes are more complex and less well studied at the clinical level. Evidence using both murine and human breast cancer cells suggests, however, that they will prove to be at least as important, and clinical trials are under way using inhibitors of both GSH synthesis and GSH-S-transferase in conjunction with conventional cytotoxic chemotherapy. The ability to monitor the effects of the agents using rapid flow cytometric assays could have a long-term impact on clinical practice far exceeding that of prognostic markers.

A number of GSH probes have been described (87–91). They work by forming fluorescent adducts through linkage to the -SH group and can be subdivided into those that form this spontaneously and those where the reaction is catalyzed by one of the GSH-S-transferase enzymes. The latter are more specific for GSH because they can be used at low concentrations when binding to other thiols is minimal. Staining is dependent on enzyme activity as well as substrate concentration. In particular, there is concern that monochlorobimane (MCB), currently the most widely used GSH stain, seriously underestimates the results for human cancer. When compared to rodent tumors, human tumors are markedly deficient in the isoenzyme required to conjugate MCB to GSH (92, 93). It is of course possible that by comparing rates of development of fluorescence for different GSH probes, a method for measuring GSH-S-transferase activity could be perfected using flow cytometry.

As already discussed, the relation of GSH biochemistry to clinical drug resistance is likely to prove complex, and clinical studies are needed to determine whether a simple assay of total cell GSH content will show any useful correlation with clinical outcome. For practical purposes, the ability to make rapid single cell measurements using sequential fine-needle aspirates may be more valuable in monitoring response to agents designed to deplate tumor cell GSH than in defining "drug resistance" per se.

COMPARISON OF FLOW AND IMAGE CYTOMETRY IN BREAST CANCER

The two branches of analytical cytology have different strengths and, before concluding this review of applications of flow cytometry to breast cancer, the potential role of image cytometry will be considered. Compared to flow cytometry, data are acquired at a much slower rate; typically only a few hundred cells are examined instead of tens of thousands, and only one or two parameters can be measured at a time. On the other hand, you only **need** a few hundred cells for image analysis, tumor cells can be identified as such by their morphology, and a permanent record is retained. Although classical morphometric features appear to be less useful than expected, these are also readily available and may prove capable of enhancing data obtained through staining with specific probes. Finally, image analysis systems are generally cheaper to acquire and maintain than flow cytometers. In the near future, it is likely that image cytometry will become widely used to quantitate immunocytochemical staining of steroid hormone receptors and, probably, to reveal certain prognostic markers, such as epidermal growth factor receptor and HER2/neu oncogene product (94, 95). The ability to identify individual grossly aneuploid cells, and its apparently major prognostic importance, have been discussed and, in the future, image cytometry may prove the method of choice for quantitating in situ hybridization.

It seems a reasonable prediction that some of the single-parameter analyses currently considered suitable for flow cytometry will prove to be more conveniently measurable by image cytometry. Clinical laboratories have, on the other hand, been slow to realize the full potential of flow cytometry, particularly the possibility of examining cell function in addition to structure and the ability to collect and analyze very complex multiparametric data sets. Preliminary results in the development of functional assays of drug resistance are exciting. They hint that we are only just beginning to explore the full potential of flow cytometry and that the initial trend to regard it as an adjunct to conventional anatomical pathology may have been short-sighted.

REFERENCES

1. Visscher DW, Zarbo RJ, Greenawald KA, Crissman JD. Prognostic significance of morphological parameters and flow cytometric DNA analysis in carcinoma of the breast [Review]. Pathol Annu 1990;1:171–210.
2. Fisher ER, Redmond C, Fisher B, Gordon B. Pathologic findings from the National Surgical Adjuvant Breast and Bowel Projects (NSABP). Cancer 1990;65:212–2127.
3. Coleman CN, Bump EA, Kramer RA. Chemical modifiers of cancer treatment. J Clin Oncol 1988;6:709–733.
4. Taylor IW, Musgrove EA. The influence of age on the DNA ploidy levels of breast tumours. Eur J Cancer Clin Oncol 1983;19(5):623–628.
5. Dressler LG. DNA flow cytometry and prognostic factors in 1331 frozen breast cancer Specimens. Cancer 1988;61:420–427.
6. Hedley DW, Friedlander ML, Taylor IW. Application of DNA flow cytometry to paraffin-embedded archival material for the study of aneuploidy and its clinical significance. Cytometry 1985;6:327–333.
7. Levack PA, Mullen P, Anderson TJ, Miller WR, Forrest APM. DNA analysis of breast tumour fine needle aspirates using flow cytometry. Br J Cancer 1987;56:643–646.
8. Remvikos Y. DNA Flow Cytometry Applied to Fine Needle Sampling of Human Breast Cancer. Cancer 1988;61:1629–1634.
9. Spyratos F, Briffod M, Gentile A, Brunet M, Brault C, Desplaces A. Flow cytometric study of DNA distribution in cytopunctures of benign and malignant breast lesions. Anal Quant Cytol Histol 1987;9:485–494.
10. Remvikos Y, Gerbault-Seurreau M, Vielh P, Zafrani B, Magdelenat H, Dutrillaux B. Relevance of DNA ploidy as a measure of genetic deviation: A comparison of flow cytometry and cytogenetics in 25 cases of human breast cancer. Cytometry 1988;9:612–618.
11. Beerman H, Kluin M, Hermans J, Van De Velde CJH, Cornelisse CJ. Prognostic significance of DNA-ploidy in a series of 690 primary breast cancer patients. Int J Cancer 1990;45:34–39.
12. Dowle CS, Owainati A, Robins A, et al. Prognostic significance of the DNA content of human breast cancer. Br J Surg 1987;74:133–136.
13. Ewers S-B, Langstrom E, Baldetorp B, Killander D. Flow-Cytometric DNA analysis in primary breast carcinomas and clinicopathological correlations. Cytometry 1984;5:408–419.
14. Feichter GE, Mueller A, Kaufmann M, et al. Correlation of DNA flow cytometric results and other prognostic factors in primary breast cancer. Int J Cancer 1988;41:823–828.
15. Hedley DW, Rugg CA, Gelber RD. Association of DNA index and S-Phase fraction with prognosis of node-positive early breast cancer. Cancer Res 1987;47:4729–4735.

16. Kallioniemi O-P, Blanco G, Alavaikko M, Hietanen T, Mattila J, Lauslahti K, Lehtinen M, Koivula T. Improving the prognostic value of DNA flow cytometry in breast cancer by combining DNA index and S-Phase fraction. Cancer 1988;62:2183–2190.
17. Keyhani-Rofagha S, O'Toole RV, Farrar WB, Sickle-Santanello B, DeCenzo J, Young D. Is DNA ploidy an independent prognostic indicator in infiltrative node-negative breast adenocarcinoma. Cancer 1990;65:1577–1582.
18. Muss HB, Kute TE, Case LD, et al. The relation of flow cytometry to clinical and biologic characteristics in women with node negative primary breast cancer. Cancer 1989;64:1894–1900.
19. O'Reilly SM, Camplejohn RS, Barnes DM, et al. DNA index, S-phase fraction, histological grade and prognosis in breast cancer. Br J Cancer 1990;61:671–674.
20. Toikkanen S, Joensuu H, Klemi P. The prognostic significance of nuclear DNA content in invasive breast cancer—a study with long-term follow-up. Br J Cancer 1989;60:693–700.
21. Shackney SE, Smith CA, Miller BW, et al. Model for the genetic evolution of human solid tumors. Cancer Res 1989;49:3344–3354.
22. Coulson PB, Thornthwaite JT, Woolley TW, Sugarbaker EV, Seckinger D. Prognostic indicators including DNA histogram type, receptor content, and staging related to human breast cancer patient survival. Cancer Res 1984;44:4187–4196.
23. Cornelisse CJ, Van Driel-Kulker AM. DNA image cytometry on machine-selected breast cancer cells and a comparison between flow cytometry and scanning cytophotometry. Cytometry 1985;6:471–477.
24. Auer G, Eriksson E, Azavedo E, Caspersson T, Wallgren A. Prognostic significance of nuclear DNA content in mammary adenocarcinomas in humans. Cancer Res 1984;44:394–396.
25. Fallenius AG, Askensten UG, Skoog LK, Auer GU. The reliability of microspectrophotometric and flow cytometric nuclear DNA measurements in adenocarcinomas of the breast. Cytometry 1987;8:260–266.
26. Fallenius AG, Auer GU, Carstensen JM. Prognostic significance of DNA measurements in 409 consecutive breast cancer patients. Cancer 1988;62:331–341.
27. Auer GU, Arrhenius E, Granberg P, Fox CH. Comparison of DNA distributions in primary human breast cancers and their metastases. Eur J Cancer 1980;16:273–277.
28. Stuart-Harris R, Hedley DW, Taylor IW, Levene AL, Smith IE. Tumour ploidy, response and survival in patients receiving endocrine therapy for advanced breast cancer. Br J Cancer 1985;51:573–576.
29. Masters JRW, Camplejohn RS, Millis RR, Rubens RD. Histological grade, elastosis, DNA ploidy and the response to chemotherapy of breast cancer. Br J Cancer 1987;55:455–457.
30. Fallenius AG, Skoog LK, Svane GE, Auer GU. Cytophotometrical and biochemical characterization of nonpalpable, mammographically detected mammary adenocarcinomas. Cytometry 1984;5:426–429.
31. Erhardt K, Auer GU. Mammary carcinoma. Comparison of nuclear DNA content from in situ and infiltrative components. Anal Quant Cytol Histol 1987;9:263–267.
32. Uccelli R, Calugi A, Forte D, Mauro F, Polonio-Balbi P, Vecchione A, Vizzone A, De Vita R. Flow cytometrically determined DNA content of breast carcinoma and benign lesions: correlations with histopathological parameters. Tumori 1986;72:171–177.
33. Palmer JO, McDivitt RW, Stone KR, Rudloff MA, Gonzalez JG. Flow cytometric analysis of breast needle aspirates. Cancer 1988;62:2387–2391.
34. Kallioniemi O-P, Karkkainen A, Auvinen O, Mattila J, Koivula T, Hakama M. DNA Flow cytometric analysis indicates that many breast cancers detected in the first round of mammographic screening have a low malignant potential. Intl J Cancer 1988;42:697–702.
35. Kallioniemi O-P, Hietanen T, Mattila J, Lehtinen M, Lauslahti K, Koivula T. Aneuploid DNA content and high S-Phase fraction of tumour cells are related to poor prognosis in patients with primary breast cancer. Eur J Cancer Clin Oncol 1987;23:277–282.
36. Roos G, Amerlöv C, Emdin S. Retrospective DNA analysis of T3/T4 breast carcinoma using cytophotometry and flow cytometry. A comparative study with prognostic evaluation. Anal Quant Cytol Histol 1988;10:189–194.
37. Klintenberg C, Stal O, Nordenskjold B, Wallgren A, Arvidsson S, Skoog L. Proliferative index, cytosol estrogen receptor and axillary node status as prognostic predictors in human mammary carcinoma. Breast Cancer Res Treat 1986;7:99–106.
38. Sigurdsson H, Baldetorp B, Borg A, et al. Indicators of prognosis in node-negative breast cancer. N Eng J Med 1990;322(15):1045–1053.
39. McDivitt RW, Stone KR, Meyer JS. A Method for dissociation of viable human breast cancer cells that produces flow cytometric kinetic information similar to that obtained by thymidine Labeling. Cancer Res 1984;44:2628–2633.
40. McDivitt RW, Stone KR, Craig RB, Meyer JS. A comparison of human breast cancer cell kinetics measured by flow cytometry and thymidine labeling. Lab Invest 1985;52:287–291.
41. McDivitt RW, Stone KR, Craig RB, Palmer JO, Meyer JS, Bauer WC. A proposed classification of breast cancer based on kinetic information. Cancer 1986;57:269–276.
42. Meyer JS, Coplin MD. Thymidine labeling index, flow cytometric S-Phase measurement, and DNA index in human tumors. Amer J Clin Pathol 1988;89(5):586–596.
43. Tubiana M, Pejovic MH, Koscielny S, Chavaudra N, Malaise E. Growth rate, kinetics of tumor cell proliferation and long-term outcome in human breast cancer. Intl J Cancer 1989;44:17–22.
44. Silvistrini R, Daidone MG, Gaspasini G. Cell Kinetics as a prognostic marker in node-negative breast cancer. Cancer 1985;56:1982–1987.
45. Meyer JS, Lee JY. Relationship of S-phase fraction of breast carcinoma in relapse to duration of remission, estrogen receptor content, therapeutic responsiveness and duration of survival. Cancer Res 1980;40:1890–1896.
46. Meyer JS, Friedman E, McCrate M, Bauer WG. Prediction of early course of breast carcinoma by thymidine labeling. Cancer 1983;51:1879–1886.
47. Kute TE, Gregory B, Galleshaw J, Hopkins M, Buss D, Case D. How reproducible are flow cytometry data from paraffin-embedded blocks? Cytometry 1988;9:494–498.
48. Dressler LG, Seamer L, Owens MA, Clark GM, McGuire WL. Evaluation of a modeling system for S-Phase estimation in breast cancer by flow cytometry. Cancer Res 1987;47:5294–5302.
49. Haag D, Feichter G, Goerttler K, Kaufmann M. Influence of systematic errors on the evaluation of the S-Phase portions from DNA distributions of solid tumors as shown for 328 breast carcinomas. Cytometry 1987;8:377 385.
50. Clark GM, Dressler LG, Owens MA, Pounds G, Oldaker T, McGuire WL. Prediction of relapse or survival in patients with node-negative breast cancer by DNA flow cytometry. N Engl J Med 1989;320:627–633.
51. Remvikos Y, Beuzeboc P, Zajdela A, Voillemot N, Magelenat H, Pouillart P. Correlation of pretreatment proliferative activity of breast cancer with the response to cytotoxic chemotherapy. J Natl Cancer Inst 1989;81:1383–1387.
52. Zarbo RJ, Visscher DW, Crissman JD. Two-Color multiparametric method for flow cytometric DNA analysis of carcinomas using staining for cytokeratin and leukocyte-common antigen. Anal Quant Cytol Histol 1989;11(6):391–402.
53. Van Dierendonck JH, Keijzer R, Van de Velde CJH, Cornelisse CJ. Nuclear distribution of the Ki-67 antigen during the cell cycle: Comparison with growth fraction in human breast cancer cells. Cancer Res 1989;49:2999–3006.
54. Walker RA, Camplejohn RS. Comparison of monoclonal antibody Ki-67 reactivity with grade and DNA flow cytometry of breast carcinomas. Br J Cancer 1988;57:281–283.
55. Clevenger CV, Epstein A, Bauer KD. Modulation of the nuclear antigen p105 as a function of cell-cycle progression. J Cell Physiol 1987;130:336–343.

56. Mann GJ, Musgrove EA, Fox RM, Thelander L. Ribonucleotide reductase M subunit in cellular proliferation, quiescence, and differentiation. Cancer Res 1988;48:5151–5156.
57. Stokke T, Holte H, Erikstein B, Steen HB. Intracellular localization of DNA polymerase. Cytometry 1988;Suppl.2:72.
58. Kurki P, Ogata K, Tan EM. Monoclonal antibodies to proliferating cell nuclear antigen (PCNA)/cyclin as probes for proliferating cells by immunofluorenscence microscopy and flow cytometry. J Immunol Methods 1988;109:49–59.
59. Gratzner HG. Monoclonal antibody to 5-bromo- and 5-iodo- deoxyuridine: A new reagent for detection of DNA replication. Science 1982; 218:474–475.
60. Dolbeare F, Gratzner HG, Pallavicini M, Gray JW. Flow cytometric measurement of total DNA content and incorporated bromodeoxyuridine. Proc Natl Acad Sci USA 1983;80:5573–5577.
61. Riccardi A, Danova M, Dionigi P, Gaetani P, Becrellit, Buttig, Mazzini G, Wilson G. Cell kinetics in leukaemic and solid tumours studied with in vivo bromodeoxyuridine and flow cytometry. Br J Cancer 1989; 59:898–903.
62. McNally NJ. Can cell kinetic parameters predict the response of tumours to radiotherapy? Int J Radiat Biol 1989;56:777–786.
63. Kute TE, Linville C, Barrows G. Cytofluorometric analysis for estrogen receptors using fluorescent estrogen probes. Cytometry 1983;4: 132–140.
64. Van NT, Rober M, Barrows GH, Barlogie B. Estrogen receptor analysis by flow cytometry. Science 1984;224:876–879.
65. Benz C, Wiznitzer I, Lee SH. Flow cytometric analysis of fluorescence-conjugated estradiol (E-BSA-FITC) binding in breast cancer suspensions. Cytometry 1985;6:260–267.
66. Graham II ML, Bunn PA Jr., Jewett PB, Gonzalez-Aller C, Horwitz KB. Simultaneous measurement of progesterone receptors and DNA indices by flow cytometry: Characterization of an assay in breast cancer cell lines. Cancer Res 1989;49:3934–3942.
67. Graham II ML, Dalquist KE, Horwitz KB. Simultaneous measurement of progesterone receptors and DNA indices by flow cytometry: Analysis of breast cancer cell mixtures and genetic instability of the T47D Line. Cancer Res 1989;49:3943–3949.
68. Farber JL, Kyle ME, Coleman JB. Biology of Disease: Mechanisms of cell injury by activated oxygen species. Lab Invest 1990;62(6):670–679.
69. Bradley G, Juranka PF, Ling V. Mechanism of multidrug resistance. Biochim Biophys Acta 1988;948:87–128.
70. Meister A, Anderson ME. Glutathione. Annu Rev Biochem 1983;52: 711–760.
71. Arrick BA, Nathan CF. Glutathione metabolism as a determinant of therapeutic efficacy: A Review. Cancer Res 1984;44:4224–4232.
72. Reed DJ. Regulation of reductive processes by glutathione. Biochem Pharmacol 1986;35:7–13.
73. Russo A, Carmichael JC, Friedman N, et al. The Roles of intracellular glutathione in antineoplastic chemotherapy. Int J Radiat Oncol 1986; 12:1347–1354.
74. Batist G, Tulpule A, Sinha BK, Katki AG, Myers CE, Cowan KH. Overexpression of a novel anionic glutathione transferase in multidrug-resistant human breast cancer cells. J Biol Chem 1986;261(33):15544–15549.
75. Carmichael J, Forrester LM, Lewis AD, Hayes JD, Hayes PC, Wolf CR. Glutathione S-transferase isoenzymes and glutathione peroxidase activity in normal and tumour samples from human lung. Carcinogenesis 1988;9(9):1617–1621.
76. Kramer RA, Zakher J, Kim G. Role of the glutathione redox cycle in acquired and de novo multidrug resistance. Science 1988;241:694–697.
77. Gottesman MM, Pastan I. Clinical trials of agents that reverse multidrug-resistance. J Clin Oncol 1989;7(4):409–411.
78. Mitchell JB, Cook JA, DeGraff W, Glatstein E, Russo A. Keynote address: Glutathione Modulation in Cancer Treatment: Will It Work? Int J Radiat Oncol Biol Phys 1989;16:1289–1295.
79. Tew KD, Bomber AM, Hoffman SJ. Ethacrynic acid and piriprost as enhancers of cytotoxicity in drug resistant and sensitive cell lines. Cancer Res 1988;48:3622–3625.
80. Rothenberg ML, Mickley LA, Cole DE, et al. Expression of the mdr-1/P-170 gene in patients with acute lymphoblastic leukemia. Blood 1989; 74:1388–1395.
81. Georges E, Bradley G, Gariepy J, Ling V. Detection of P-glycoprotein isoforms by gene-specific monoclonal antibodies. Proc Natl Acad Sci USA 1990;87:152–156.
82. Krishan A, Sauerteig A, Gordon K, Swinkin C. Flow cytometric monitoring of cellular anthracycline accumulation in murine leukemic cells. Cancer Res 1986;46:1768–1773.
83. Krishan A. Effect of drug efflux blockers on vital staining of cellular DNA with Hoechst 33342. Cytometry 1987;8:642–645.
84. Nooter K, Sonneveld P, Oostrum R, Herweijer H, Hagenbeck T, Valerio D. Overexpression of the mdr 1 gene blast cells from patients with acute myeloid leukaemia is associated with decreased anthracycline accumulation that can be restored by cyclosporin A. Int J Cancer 1990;45:265–270.
85. Keith WN, Stallard S, Brown R. Expression of mdr1 and gst-pi in human breast tumours: comparison to in vitro chemosensitivity. Br J Cancer 1990;61(5):712–716.
86. Nair S, Singh SV, Samy TSA, Krishan A. Anthracycline resistance in murine leukemic P388 cells: Role of drug efflux and glutathione related enzymes. Biochem Pharmacol 1990;39:723–728.
87. Durand RE, Olive PL. Flow cytometry techniques for studying cellular Thiols. Radiat Res 1983;95:456–470.
88. Treumer J, Valet G. Flow-cytometric determination of glutathione alterations in vital cells by 0-Phthaldialdehyde(OPT) staining. Exp Cell Res 1986;163:518–524.
89. Shrieve DC, Bump EA, Rice GC. Heterogeneity of cellular glutathione among cells derived from a murine fibrosarcoma or a human renal cell carcinoma detected by flow cytometric analysis. J Biol Chem 1988; 263:14107–14114.
90. O'Connor JE, Kimler BF, Morgan MC, Tempas KJ. A flow cytometric assay for intracellular nonprotein thiols using mercury orange. Cytometry 1988;9:529–532.
91. Hedley DW, Hallahan AR, Tripp EH. Flow cytometric measurement of glutathione content of human cancer biopsies. Br J Cancer 1990;61:65–68.
92. Cook JA, Pass HI, Russo A, Iype BA, Mitchell JB. Use of monochlorobimane for glutathione measurements in hamster and human tumor cell lines. Int J Radiat Oncol Biol Phys 1989;16:1321–1324.
93. Simony J, Pujol JL, Radal M, et al. In situ evaluation of growth fraction determined by monoclonal antibody Ki-67 and ploidy in surgically resected non-small cell lung cancers. Cancer Res 1990;50:4382–4387.

CLINICAL COMMENTARY
Douglas E. Merkel

The preceding chapter provides an excellent summary of the scientific rationale for perhaps the largest commercial application of flow cytometric DNA content analysis. Clearly, both aneuploidy and high S–phase fractions predict the earlier relapse of resectable breast cancer. But are these predictions sufficiently robust and independent of other prognostic factors to justify their application to the management of breast cancer patients? If so, then for which patients?

The second question—for which patients can a reliable, objective prognostic marker influence management?—is dependent on the treatment options available to clinicians. The major dilemma faced by medical oncologists managing breast cancer is whether or not to provide systemic adjuvant therapy for patients with pathologic stage-I disease. Those with involved nodes or large (diameter greater than 2 cm) tumors almost always merit some systemic attempt to modify the higher relapse rates predicted for stage-II patients. On the other hand, node–negative patients with tumors smaller than 1 cm have an excellent prognosis and do not require adjuvant systemic therapy. For node–negative patients with tumors between 1.1 and 2.0 cm in diameter who are not candidates for prospective trials, the decision to treat or not must be based on objective prognostic markers such as flow cytometry.

There are other categories of patients for which objective prognostic markers could potentially influence treatment decisions, as treatment options continue to evolve. One example might be postmenopausal, node–positive, estrogen receptor–positive patients for whom either adjuvant hormonal therapy or combined chemohormonal therapy is considered, as in the National Surgical Adjuvant Breast Project Trial B-16. Another example is patients with completely excised ductal carcinoma in situ, for whom the local management options of mastectomy or radiotherapy can be recommended.

If one focuses on papers that compare ploidy to outcome for node–negative patients, the absolute magnitude of the survival advantage enjoyed by patients with diploid tumors is small, and not always detected. This is analogous to the experience with estrogen-receptor status as a prognostic factor in stage-I breast cancer. Just as with estrogen-receptor status, ploidy alone provides inadequate discrimination between good and bad prognoses and cannot be used as the sole determinant of whether or not to prescribe systemic adjuvant therapy.

The magnitude of the prognostic difference described by S-phase fraction for node-negative patients seems considerably greater. Diploid, low S-phase fraction tumors, in particular, form a subset with a distinctly low risk of early relapse. While measurement of S-phase fraction (or other kinetic markers) could reasonably affect the management of stage-I patients, unresolved technical issues currently handicap efforts to apply this finding to individual patients.

The first caveat of reporting and interpreting S-phase fraction measurements is that of interlaboratory variation in results. A particular S-phase fraction value may be classified as "high" in one published series and yet be "low" in the experience of another laboratory. This problem prevented the NIH Consensus Meeting on Early Breast Cancer, held in May 1990, from recommending the use of this marker in clinical practice.

A related issue is the lack of agreement in the dichotomization of a continuous variable, S-phase fraction, into low- and high-risk groups. Usually, this breakpoint is established only by examining the entire data set, and a different proportion of tumors have been assigned to the "low" and "high" S-phase fraction groups within each published series. Not surprisingly for a disease in which only 30% of patients relapse, a distinct majority of stage-I tumors are usually assigned to the "low S-phase fraction" category. Thus, the arbitrary designation of above-median S-phase fractions as "high" by clinical laboratories results in misleading prognostic information from many tumors. Perhaps the only solution is for S-phase fraction to be reported as a percentile result for breast tumors studied within the individual laboratory.

Another important consideration in assessing the clinical importance of S–phase fraction measurement in stage-I breast cancer is that this parameter has seldom been compared to histologic or nuclear grading for independent prognostic significance. Even if an expert histopathologist can obviate the need for S–phase fraction analysis, perhaps the latter objective parameter could substitute for grading in the many situations where grading is not performed, not reliable, or not validated for predictive significance.

To conclude, reliable determination and informed interpretation of tumor S-phase fraction can assist the clinician in the decision of whether or not to employ systemic adjuvant therapy for selected stage-I patients by refining the individual risk estimate. For example, many diploid/low S-phase stage-I patients may be able to safely forego chemotherapy. The relative weight assigned to S-phase fraction and other prognostic indicators in arriving at a global risk estimate must at present be arbitrary, since comprehensive studies large enough to evaluate all possible prognostic variables have not been completed. Such an application also requires the underlying assumption that those patients at greatest risk for relapse derive disproportionate benefit from systemic adjuvant intervention. This crucial assumption is being addressed by the ongoing analysis of tumor specimens from patients enrolled in key prospectively randomized trials.

Perhaps, future refinements in S-phase fraction determination through the use of the gating or two-color strategies described by Dr. Hedley will improve its predictive power. Any such refinement will need to be standardized and validated in multiple laboratories before it can safely be applied in the clinic. Before assays for any particular drug resistance marker will be useful for patient management, the primacy of that mechanism of *in vitro* drug resistance must be established in the much more complex biological and pharmacokinetic environment of the patient.

It is clear that clinical breast cancer is a heterogenous entity and that flow cytometry can help to describe and quantify this variability. The promising results summarized in the preceding chapter should prompt the mandatory inclusion of relevant laboratory assays into the clinical trials that will then expose these heterogenous patients to fundamentally different therapeutic interventions.

15

Gynecological Cancer

MICHAEL L. FRIEDLANDER
Clinical Commentary by **John R. Lurain**

The clinical and prognostic value of cellular DNA content in gynecological cancers has been subjected to detailed study by many groups all over the world. There is now a large body of evidence to indicate that tumor ploidy is an independent and major prognostic variable in ovarian and endometrial cancers and some data to suggest that it may also be of value in some of the other gynecological cancers. These data will be reviewed in detail below, and the limitations as well as the possible place of flow cytometric analysis of DNA content in clinical practice will be discussed.

EPITHELIAL OVARIAN CANCER

Epithelial ovarian cancers vary considerably in their biological behavior but the wide clinical spectrum is only partly reflected in clinico–pathological prognostic variables, such as tumor stage, histological type, histological grade, and volume of residual disease remaining after surgical resection (for review see 1). There is an increased appreciation of the value as well as the limitations of many of these prognostic factors and a growing recognition that tumor ploidy is an independent prognostic variable and correlates closely with patient outcome.

In the first large series, where ploidy of ovarian tumors was determined using static cytometry, it was apparant that tumors could be divided into a near-diploid group and an aneuploid group with a DNA content in a triploid–tetraploid range (2). Patients with tumors in the near–diploid group had both a lower clinical stage and a more favorable outcome than those with DNA-aneuploid tumours (2, 3). The patients studied, however, were heterogeneous with respect to stage, histological type, grade, etc., and it was not possible to draw too many conclusions as to the prognostic value of ploidy in ovarian cancer from these initial observations. With the advent of flow cytometry and the ability to accurately and rapidly determine DNA content in both fresh and, later, paraffin-embedded tissue, more detailed studies became possible. Initial investigations demonstrated a similar bimodal distribution of ploidy in invasive carcinomas, with a clustering of tumors about a diploid mode and a triploid mode (4). The majority of DNA-aneuploid tumors showed evidence of a single aneuploid G_1 DNA peak, but 17% of tumors demonstrated two or more aneuploid populations and were therefore considered to be multiploid. A significant association between tumor stage and ploidy was also evident, with up to 60% of early stage tumors (FIGO Stages 1 and 2) being diploid while the minority of more advanced stage tumors were diploid (4). It was tempting to speculate that the heterogeneity with respect to ploidy among tumors of a similar stage might help to identify patients with different natural histories. However, it was only with the development of a novel technique of obtaining single cell–nuclei suspensions from paraffin-embedded material that it became possible to perform detailed retrospective studies on tumor tissue from patients in whom the outcome was known (5).

The first detailed flow cytometric study was carried out on paraffin-embedded tumor tissue from 128 patients with advanced (FIGO Stages 3 and 4) ovarian cancer who had been entered into a large multicenter Australasian study in which patients had been randomized to receive either combination or sequential chemotherapy with chlorambucil and cisplatin. The details of the flow cytometric study have been reported in full elsewhere (6, 7). Briefly, of the 128 tumors analyzed for DNA content, 27% were classified as diploid and 73% as aneuploid, including 16% that had evidence of more than one aneuploid G1 peak and were considered to be multiploid. Product limit survival analysis demonstrated a highly significant association between tumor ploidy and survival ($P <0.0001$). Patients with DNA-diploid tumors had a median survival of 260 weeks, which was significantly longer than those patients with single aneuploid and multiploid tumors, who had a median survival of 54 weeks. When the effects of histological subtype, grade, clinical stage, volume of residual tumor remaining after initial surgery, and treatment were corrected for by multivariate analysis, only cellular DNA content and stage remained as significant prognostic factors for survival. On further analysis, it was apparant that the relatively good prognosis associated with diploid tumors was limited to those patients with stage three ovarian cancer as patients with stage 4 disease (liver metastases or distant spread beyond the peritoneal cavity) had a poor prognosis irrespective of ploidy (Fig. 15.1).

The finding that tumor ploidy was of prognostic value in patients with advanced ovarian cancer prompted us to study the influence of DNA content on survival of patients with stage 1 and 2 ovarian cancer (8). The relationship of cellular

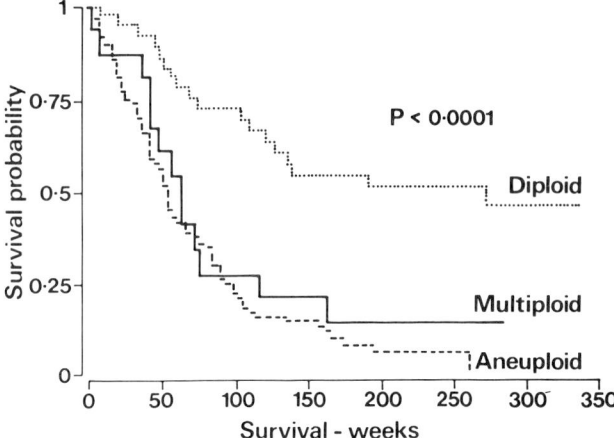

Figure 15.1. Survival by ploidy of patients with FIGO Stages 3 and 4 ovarian cancer. (Reproduced with permission from Friedlander ML: J Clin Oncol 1988;6:282–290.)

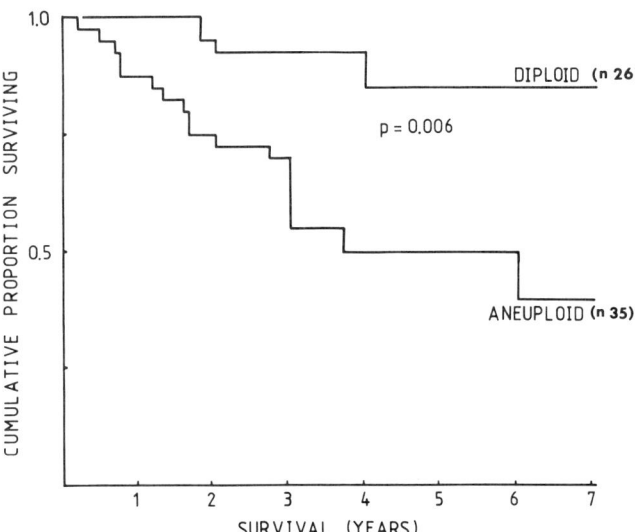

Figure 15.2. Survival by ploidy of patients with FIGO Stages 1 and 2 ovarian cancer.

DNA content to survival was analyzed in 61 patients with early stage epithelial ovarian cancer and 43% of these tumors were classified as diploid and 57% aneuploid, which included 14% of multiploid tumors. There was a significant association between ploidy and grade, with well differentiated tumors being more commonly diploid than aneuploid (P <0.01). There was a significant effect of ploidy on survival and patients with diploid tumors had a significantly longer survival than those with aneuploid tumors (P <0.006) (Fig. 15.2). There was also significant effect of tumor grade on survival, with both well and moderately differentiated tumors having a more favorable prognosis than poorly differentiated tumors. However, in view of the relatively small numbers of patients, a multivariate analysis was not performed to assess the relative importance of stage, histological grade, subtype, and ploidy. These results have been confirmed by a number of other groups and over 1000 patients have been included in these prognostic studies (6–20) (Table 15.1).

The majority of investigators have reported that ploidy is a major predictor of patient outcome and an independent prognostic variable. Interestingly, a number of other investigators (13, 21) have also found that, among patients with advanced disease, those with stage 3 diploid ovarian cancer comprise a biologically distinct sub-group with a relatively good prognosis, while patients with stage 4 disease do uniformly badly, irrespective of tumor DNA content. The reason for this is not known, but it is possible that some stage 3 tumors may have arisen by multifocal tumorigenesis and may not necessarily represent metastatic spread from the ovary. Although this is purely conjecture, it should be noted that there is some evidence to suggest that ovarian tumors that arise by multifocal tumorigenesis do have a relatively good prognosis (22, 23).

It is important to know whether the relative favorable prognosis of patients with FIGO stages 1, 2, and 3 diploid ovarian tumors is in any way related to therapy. This fundamental question has not been fully addressed and still deserves further detailed study. In the multicenter Australasian study, there was a significant prolongation in the time to first treatment failure in patients with aneuploid tumors who received combination therapy initially compared to patients treated with single-agent chlorambucil alone (7). In contrast, there was no significant difference in the time to first or ultimate treatment failure in patients with diploid tumors treated with either sequential or combination therapy. These findings suggest that patients with diploid tumors have a relatively good prognosis, independent of the type of treatment, and support the initial use of a relatively nontoxic treatment in such cases. This could have important therapeutic implications but needs to be confirmed in a larger number of patients. Recently, Murray et al. (18) noted that the five-year survival for patients with diploid stages 1, 2, and 3 ovarian cancer treated with whole abdominal irradiation was 74%, which contrasted with a 22% overall survival of patients with aneuploid tumors also treated with radiotherapy. The results of whole abdominal radiation of patients with aneuploid tumors was poor and suggests that more aggressive therapy is warranted for this population. It is of some interest that ploidy may also help identify patients who have a high risk of recurrence following a negative second look laparotomy and who perhaps would benefit from consolidation therapy (20). There is clearly a need to investigate the relationship between ploidy and response to different treatment modalities in more detail before any conclusions can be drawn. This is an important area deserving of further study as it may permit individualization of treatment according to risk of recurrence and likelihood of response.

Approximately 15% to 20% of ovarian epithelial tumors can be histopathologically classified as being of low potential malignancy (borderline tumors) on the grounds that they have some but not all the features of malignancy (24). These tumors are characterized by a generally indolent course, and

Table 15.1
Summary of Reports on Prognostic Significance of Ploidy in Invasive Ovarian Cancer

Ref.	Patient No.	F.I.G.O. STAGE	% DIP. Diploid	Aneuploid	Median Survival[a] + (MO)	SIG.
7	128	III,IV	27	60	15	$P = 0.001$
8	61	I,II	43	80*	50*	$P = 0.008$
9	37	III,IV	38	NR	12	$P = 0.02$
10	74	III,IV	20	60	24	$P = 0.002$
11	99	III,IV	35	28	12	$P = 0.007$
12	53	I,IV	22	48	18	NS
13	84	II,IV	39	48	19	$P = 0.001$
14	68	III,IV	30	20	12	$P = 0.05$
14	89	I,II	50	NR	NR	$P = 0.0001$
15	153	I-IV	38	NR	18	$P = 0.0002$
16	51	—	48	18	8	$P = 0.0005$
17	90	III/IV	25	26	16	NS
18	40	I-IV	40	NR	18	$P = 0.01$
19	115	I-IV	24	36	23	$P = 0.04$
20	99	I-IV	50	50*	22*	$P = 0.01$

[a]Median Survival estimated from survival curves when data not provided
NR Not Reached
NS Not Significant
* % 5yr survival
REF. = REFERENCE
DIP. = DIPLOID
(MO) = MONTH
SIG. = SIGNIFICANCE

Table 15.2
Summary of Ploidy Studies on Borderline Ovarian Tumors

INVESTI-GATOR (et al.)	REFER-ENCE NO.	PATIENT	STAGE	DIPLOID	ANEU-PLOID
Friedlander	26	53	I-III	47	6
Kuhn	11	8	III	7	1
Erhardt	27	26	I	26	
Fu	28	16	I-III	11	5
Klemi	15	43	—	36	7
Kuhn	29	49	I-III	34	15
Kaern[a]	30	64	I-III	42	22
Padberg	31	70	I-III	27	43

[a]Selected series—DNA analysis on 34 patients who relapsed or died of disease and on 30 matched controls with no recurrence.

the designation of a tumor as a "borderline" or "invasive carcinoma" has important prognostic and therapeutic implications (25). Failure to correctly diagnose the patient with a borderline ovarian tumor could result in the patient receiving inappropriate treatment, in erroneous interpretation of the results of therapy, and in the development of an inconsistent literature. There is growing evidence to suggest that analysis of cellular DNA content may be a useful adjunct to the histopathological diagnosis of borderline tumors (11, 15, 26–31). The majority of borderline tumors are diploid and associated with a good prognosis, while the presence of aneuploidy appears to identify patients with a high risk of recurrence (Table 15.2). This is well demonstrated in a recent study from the Norwegian Radium Institute, where patients with aneuploid borderline tumors had a less than 20% chance of long term survival (30). Similarly, in a recent static cytophotometric study, 30% of patients with aneuploid borderline tumors relapsed or died of disease while none of the patients with diploid tumors had evidence of recurrence (31). It should be emphasized, however, that, as a diploid DNA content can occur in frankly invasive ovarian cancers, the DNA content alone cannot be used as a sole means of differentiating borderline from invasive tumors.

Rather, the finding of aneuploidy in a putative borderline malignant ovarian neoplasm should alert the clinician that the tumor may behave in a more aggressive manner than its histological diagnosis would imply.

ENDOMETRIAL CANCER

A variety of clinical and pathological factors are recognized to be of value in predicting outcome of patients with endometrial cancer. The best established of the prognostic factors include clinical stage, degree of differentiation, and depth of myometrial penetration; these usually form the basis for therapeutic decision making. There is still a need, however, to improve on our ability to predict prognosis in individual patients, and there has been considerable interest in determining whether cellular DNA content is of prognostic value in endometrial cancer. Much of the early work into cellular DNA content in endometrial cancer was performed using the labor intensive technique of static cytometry, in which suitably stained cells were visualized using a microscope based system and the amount of DNA was measured directly by absorption or fluorometric cytometry.

Atkin initially reported a static cytometric study in 186 patients with endometrial cancer and found that patients with near diploid tumors were more likely to have well-differentiated tumors and to have a better prognosis than those with aneuploid tumors (32). The difference in prognosis could not be solely related to the differentiation of tumors as even within the poorly differentiated group patients with near diploid tumors had a significantly better prognosis. A drawback of this study was that few clinical details were provided and no attempt was made to assess the independent prognostic significance of ploidy. A later study, also using static cytometry, reported similar findings and noted that ploidy correlated better with survival than did either clinical stage or histological grade (33). The eight-year survival rates in this study were 79% in patients with diploid tumor and 18% in patients with aneuploid tumors.

More recently, there have been a number of more detailed studies using flow cytometric analysis of DNA content in endometrial cancer. Iverson et al. found tumor ploidy to be of prognostic value in a study of 72 patients with endometrial cancer and, using a Cox multivariate analysis, found tumor ploidy and surgical stage to be the only factors that significantly impacted on prognosis (34). Lindahl et al. reported a study of 110 patients with stage 1 and 2 endometrial cancer who were followed for more than two years (35). The frequency of relapse was significantly higher in patients with nondiploid tumors (28%), and multivariate analysis once again showed that DNA index was the single most important factor in predicting patient outcome. Furthermore, an even better prediction of patient outcome was made possible by combining DNA content, estrogen receptor concentration, and myometrial invasion. Since then, there have been at least five other similar studies reported (36–40). Newbury et al. did a retrospective analysis on 233 patients with endometrial cancer who had a median follow-up of 8.7 years (36). Aneuploidy was evident in 18% of tumors and was associated with adverse histological type, high grade, and depth of invasion (36). 74% of patients with a tumor DNA index >1.5 died of disease while only 13% with a diploid content relapsed and died. Britton et al. reported a study of 256 patients and found that 78% of tumors were diploid, whereas aneuploid and tetraploid patterns accounted for 17% and 5% of the tumors, respectively (37). Only 10% of patients with diploid tumors relapsed, compared by 39% of patients with nondiploid lesions. There was a good correlation between ploidy and outcome with a four-year progression-free survival of 88% for patients with diploid tumors and of 57% for those with nondiploid tumors. When subjected to multivariate analysis, only histological subtype and DNA ploidy maintained significant predictive powers (37).

These data indicate that DNA ploidy is a significant and important prognostic determinant in endometrial cancer and should find clinical application. Combining the ploidy status with more traditional prognostic factors may help to better identify groups of patients who have different risks of recurrence and facilitate more appropriate and rational therapeutic decision making. However, as in the case of epithelial ovarian cancer, more data is needed, particularly with respect to the relationship between tumor ploidy and response to different treatment modalities.

CERVICAL CANCER

There have been a number of studies reported correlating tumor ploidy with survival of patients with cervical cancer. Using static cytometry, Atkin reported a worse prognosis for the near-diploid group in patients with squamous cell carcinoma of the cervix (2). In contrast, Jakobsen et al., have reported a study involving 171 patients with squamous cell carcinoma of the cervix and found that patients with low-ploidy tumors had a significantly better prognosis than those with high-ploidy tumors (44). These findings are quite different to that reported by Atkin using static cytometry and it is unlikely that the difference can be accounted for solely by the different techniques used in determining ploidy and may instead be due to different follow–up times and treatment techniques. Jakobsen has also reported significant correlation between the DNA index and the incidence of pelvic lymph node metastases in patients with stage 1B and 2A tumors, with high-ploidy tumors having a high likelihood of metastasizing (45). Leminen et al. performed a retrospective study in 125 patients with cervical adenocarcinoma and demonstrated that 31% of tumors were aneuploid and that aneuploidy was associated with tumor size, histological grade, clinical stage, and high S-phase fraction. On multivariate analysis, ploidy as well as S-phase fraction were demonstrated to be independent prognostic factors and were valuable in predicting prognosis (46). In a retrospective study of 56 cases of uterine cervical squamous cell carcinomas, Davis et al. determined that almost 70% were aneuploid, but there was no significant correlation between ploidy and outcome (47).

Although there are certainly data to suggest that ploidy analysis may be of value in cervical cancer, it is not possible to draw any firm conclusions at present and more detailed studies are required.

MISCELLANEOUS GYNECOLOGICAL NEOPLASMS

There have been a variety of reports of the flow cytometric analysis of DNA content in other gynecological cancers but, unfortunately, most of these studies have been small and it is difficult to make specific recommendations. The presence of aneuploidy may identify a high-risk group of patients with molar pregnancies (48). In a study of 40 patients who had complete hydatidiform moles, 27 were found to be diploid and 13 aneuploid. Thirty percent of the diploid group required treatment after evacuation, whereas 77% of the aneuploid group required treatment. Further study with larger patient numbers is required before a thorough understanding of the prognostic value of ploidy in molar pregnancy is known. However, DNA flow cytometry may provide a rapid, accurate, and cost effective means for differentiating between complete moles and partial moles, which are nearly always triploid (49), and is already finding clinical application.

Granulosa cell tumors are uncommon ovarian tumors that are generally associated with a good prognosis but, classi-

Table 15.3
Potential Problems and Limitations of Flow Cytometric Analysis of DNA Content in Gynecological Tumors

A. Quality Control and Standardization:
   Sample collection and treatment.
   Sample selection and processing.
   Staining procedures.
   Instrument controls and calibration.
B. Resolution of Technique:
   Coefficient of variation.
   Sensitivity of detection of aneuploid cells.
C. Diagnosis of Diploid tumors:
   Definition of diploid.
   Sample selection—solid tumor or ascites.
D. Analysis of Proliferative Fraction or S phase.
E. Data Interpretation.

cally, patients can relapse many years after diagnosis. There have been a number of studies reported looking at whether analysis of DNA content will help in predicting prognosis in patients with granulosa cell tumor. Klemi et al. reported a study of 23 patients with ovarian granulosa cell tumors where crude survival of patients with aneuploid tumors was more favorable than that of patients with aneuploid tumors (50). However, there are a number of other studies that have been reported that have not shown any relationship between ploidy and outcome (51, 52, 53).

There are also scattered reports on flow cytometric analysis of DNA content in ovarian germ-cell tumors and uterine sarcomas, but further studies are required to determine the role of flow cytometric analysis of DNA content in these tumor types.

LIMITATIONS OF FCM ANALYSIS OF DNA CONTENT

It is reassuring that the majority of studies are in agreement and have demonstrated a close association between tumor ploidy and outcome in patients with epithelial ovarian cancer and endometrial cancer. The uninitiated should, however, be aware of the problems and pitfalls associated with flow cytometric analysis of DNA content and S-phase analysis (54). Apart from differences in quality control and methods of analysis, there are many other confounding problems that can make the interpretation of certain studies very difficult if not impossible. It is beyond the scope of this paper to review the problems in detail, but the potential sources of error are summarized in Table 15.3 and should be taken into account when reading the literature. One problem that comes up frequently relates to the definition of a "diploid" tumor. In the past, the resolution of static cytophotometric estimation of cellular DNA content was such that tumors could only be classified as "near-diploid," and this group almost certainly included tumors with a DNA index (DI) of up to 1.5. A number of investigators using flow cytometry have continued to use a DI of between 1.3–1.5 to define a "near-diploid" group with a more favorable prognosis than the "hyperdiploid" group. We have investigated this and dem-

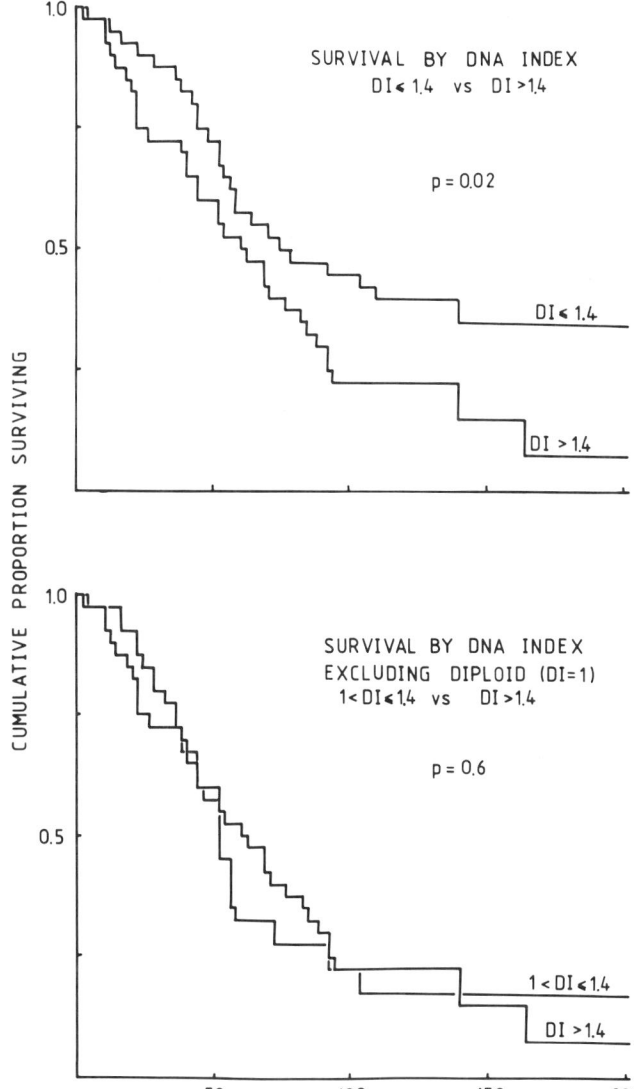

Figure 15.3. Survival of patients with advanced ovarian cancer according to DNA Index (*DI*). The apparent improved prognosis of patients with a *DI* of 1.4 is due solely to the inclusion of diploid tumors (DI = 1) as, following their removal, there is no survival difference between DNA-aneuploid tumors with DI <1.4 or DI> 1.4.

onstrated that patients with "near-diploid" tumors do indeed have a better prognosis, but the statistical significance of this survival difference increases the closer the cutoff DI is to 1 (7). The diploid tumors (DI = 1) were, in fact, the only cohort contributing to the improved survival, and the removal of tumors with a DI of 1 from the analysis revealed that there was no difference in survival of patients with aneuploid tumors that were "near-diploid" or "hyperdiploid" (Fig. 15.3). These findings highlight the importance of the resolution of the technique used when classifying tumors as diploid or aneuploid and may account for some of the conflicting reports in the literature.

The S phase has been reported to have prognostic significance in gynecological cancers, but the results are conflicting and the studies have therefore not been included in this review. Overlapping of normal and neoplastic populations complicates S-phase estimates in diploid tumors and many near-diploid DNA aneuploid tumors. Reports to date indicate that diploid ovarian tumors generally have a lower S-phase fraction (SPF) than aneuploid tumors, but this may reflect comparatively greater contamination of the DNA distribution by normal cells in the diploid tumors relative to the case of DNA aneuploid tumors, where it is sometimes possible to exclude the DNA diploid tumor component (which is often composed of slowly proliferating non–neoplastic cells). In contrast, the estimation of aneuploid S-phase population can be falsely elevated by the presence of diploid doublets and tetraploid normal cells. A further complicating fact is that unlike the estimation of ploidy, which is a stable marker in ovarian and endometrial cancer, there appears to be a significant degree of variability in the SPF in different regions of a tumor. Such findings indicate that, as yet, unaddressed sampling considerations may be critical for an intelligent evaluation of tumor proliferative activity in these biologically complex lesions and that the value of a single SPF estimation may be quite limited. This is reflected in the literature, which is confusing and inconsistent. These findings probably indicate the requirement for better methods for determining proliferative activity before such determinations find clinical application.

These potential problems do not detract from the potential prognostic value of tumor ploidy in gynecological malignancies, particularly ovarian and endometrial cancer, and are mentioned to emphasize the importance of strict quality control and meticulous attention to detail.

CONCLUSIONS

There is now a large body of evidence to support the value of tumor ploidy as an independent prognostic indicator with major significance in patients with ovarian and endometrial cancer. Flow cytometric analysis of Cellular DNA content is now finding clinical application in these tumor types and patients should be stratified in prospective randomized clinical trials on the basis of ploidy. There is still a need for further research in this area, not only in defining the prognostic value of DNA content in gynecological cancer other than endometrial and ovarian cancer, but also in correlating ploidy with response to treatment.

REFERENCES

1. Friedlander ML, Dembo AJ. Prognostic factors in ovarian cancer. Semin Oncol 1991;18;3:205–212.
2. Atkin NB. Modal DNA value and chromosome number in ovarian neoplasia. A clinical and histopathologic assessment. Cancer 1970;27:1064–1073.
3. Atkin NB, Kay R. Prognostic significance of modal DNA value and other factors in 1465 cases. Br J Cancer 1979;10:210–221.
4. Friedlander ML, Taylor IW, Russell P, Musgrove EA, Hedley DH, Tattersall M. Ploidy as a prognostic factor in ovarian cancer. Int J Gynecol Pathol 1983;1:55–62.
5. Hedley DW, Friedlander ML, Taylor IW, Rugg CA, Musgrove EA. Method for analysis of cellular DNA content of paraffin-embedded pathological material using flow cytometry. J Histochem Cytochem 1983;31:1333–1335.
6. Friedlander ML, Hedley DH, Taylor I, et al. Influence of cellular DNA content on survival in advanced ovarian cancer. Cancer Res 1984;44:397–400.
7. Friedlander ML, Hedley DW, Swanson C, Russell P. Prediction of long term survivals by flow cytometric analysis of cellular DNA content in patients with advanced ovarian cancer. J Clin Oncol 1988;6:282–290.
8. Hedley DW, Friedlander ML, Taylor IW. Application of DNA flow cytometry to paraffin-embedded archival material for the study of aneuploidy and its clinical significance. Cytometry 1985;4:327–333.
9. Volm M. Bruggeman A, Gunther M, et al. Prognostic relevance of ploidy, proliferation and resistance predictive tests in ovarian carcinoma. Cancer Res 1985;45:5180–5185.
10. Rodenburg CJ, Cornelisse CJ, Heintz P, et al. Tumor ploidy as a major prognostic factor in advanced ovarian cancer. Cancer 1987;59:317–323.
11. Kuhn W, Kaufmann M, Feichter GE, et al. DNA flow cytometry, clinical and morphological parameters as prognostic factors for advanced malignant and borderline tumors. Gynecol Oncol 1989;33:360–367.
12. Rutgers DH, Wills IS, Schaap AH, Van Lindert ACM. DNA flow cytometry, histological grade, stage, age as prognostic factors in human epithelial ovarian carcinomas. Path Res Pract 1987;182:207–213.
13. Blumenfeld D, Braly PS, Ben-Ezra J, Klevecz RR. Tumor DNA content as a prognostic feature in advanced epithelial ovarian carcinoma. Gynecol Oncol 1987;27:389–398.
14. Kallioniemi OP, Punnonen R, Mattila J, Lehtinen M, Koivula T. Prognostic significance of DNA index, multiploidy and S-phase fraction in ovarian cancer. Cancer 1984;61:334–339.
15. Klemi PJ, Joensuu H, Kiilholma P, Maenpaa J. Clinical significance of abnormal nuclear DNA content in serous ovarian tumors. Cancer 1988;62:2005–2010.
16. Iverson OE. Prognostic value of the flow cytometric DNA index in human ovarian carcinoma. Cancer 1988;61:971–975.
17. Erba E, Ubezio P, Pepe S, et al. Flow cytometric analysis of DNA content in human ovarian cancers. Br J Cancer 1989;60:45–50.
18. Murray K, Hopwood L, Volk D, Wilson JF. Cytoflourometric analysis of the DNA content in ovarian cancer and its relation to patient survival. Cancer 1989;63:2456–2460.
19. Barnabei VM, Miller DS, Bauer KD, Murad TM, Rademaker AW, Lurain JR. Flow cytometric evaluation of epithelial ovarian cancer. Am J Obstet Gynecol 1990;162:1584–1590.
20. Brescia RJ, Barakat RA, Beller U, et al. The prognostic significance of nuclear DNA content in malignant epithelial tumors of the ovary. Cancer 1990;65:141–147.
21. Baak JPA, Schipper NW, Wisse-Brekelmans ECM. The prognostic value of morphometrical features and cellular DNA content in Cisplatin treated late ovarian cancer patients. Br J Cancer 1988;57:503–508.
22. Woodruff JD, Julian CG. Multiple malignancy in the upper genital tract. Am J Obstet Gynecol 1969;103:810–822.
23. Genadry R, Poliakoff S, Rotmensch J, et al. Primary papillary peritoneal neoplasia. Obstet Gynecol 1981;5:730–734.
24. Hart WR. Ovarian epithelial tumors of borderline malignancy (carcinomas of low malignant potential). Human Pathol 1977;8:541–549.
25. Colgan TJ, Norris HJ. Ovarian epithelial tumors of low potential malignancy: A review. Int J Gynecol Oncol 1983;1:367–382.
26. Friedlander ML, Russel P, Taylor IN, Hedley DW, Tattersall MHN. Flow cytometric analysis of cellular DNA content as an adjunct to the diagnosis of ovarian tumors of borderline malignancy. Pathol 1984;16:301–306.
27. Erchardt K, Auer G, Bjorkholm G, et al. Prognostic significance of nuclear DNA content in serous ovarian tumors. Cancer Res 1984;44:2198–2202.

28. Fu Y, Ro J, Regan JW, Hall T, Berek J. Nuclear deoxyribonucleic acid heterogeneity of ovarian borderline malignant serous tumors. Obstet Gynecol 1986;67:478–482.
29. Kuhn W. Kaufmann M, Feichter GE, Schmid H, Hanke J, Rummel HH. Psammoma body content and DNA flow cytometric results as prognostic factors in advanced ovarian cancer. Eur J Gynecol Oncol 1988;9:234–241.
30. Kaern J, Trope C, Kjorstad K, et al. Cellular DNA content as a new prognostic tool in patients with borderline ovarian tumors of the ovary. Gynecol Oncol 1990;38:452–457.
31. Padberg B, Thiedemann C, Arps H, Dietel M. Prognostic significance of nuclear DNA content in ovarian tumors of borderline malignancy. J Cancer Res Clin Oncol 1990;(Suppl)116:520.
32. Atkin NB. Prognostic significance of ploidy level in human tumors. 1. Carcinoma of the uterus. J Natl Cancer Inst 1976;56:909–910.
33. Moberger B, Auer G, Forsslund G, and Moberger G. The prognostic significance in DNA measurements in endometrial carcinoma. Cytometry 1984;5:430–436.
34. Iverson OE. Flow cytometric deoxyribonucleic acid index: A prognostic factor in endometrial carcinoma. Amer J Obstet Gynecol 1986;155:770–776.
35. Lindahl B, Alm P, et al. Prognostic value of flow cytometric DNA measurements in Stage I-II endometrial carcinoma: correlations with steroid receptor concentration, tumor myometrial invasion and degree of differentiation. Anticancer Res 1987;7:791–798.
36. Newbury R, Schuerch C, Godspeed N, Fanning J, Glidewell O, Evans M. DNA content as a prognostic factor in endometrial carcinoma. Obstet Gynecol 1990;76:251–257.
37. Britton LC, Wilson TO, Gaffey TA, et al. Flow cytometric DNA analysis of Stage I endometrial carcinoma. Gynecol Oncol 1989;34:317–322.
38. Quillamor RM, Furlong JW, Hoschner JA, Wynn RM. Relative prognostic significance of DNA flow cytometry and histologic grading in endometrial carcinoma. Gynecol Obstet Invest 1988;26:332–337.
39. Rosenberg P, Wingren S, Simonsen E, Stsal O, Risberg B, Nordenskjsold B. Flow cytometric measurements of DNA index and S-phase on paraffin-embedded early stage endometrial cancer: An important prognostic indicator. Gynecol Oncol 1989;35:50–54.
40. Sorbe B, Risenberg B, Frankendal B. DNA ploidy, morphometry and nuclear grade as prognostic factors in endometrial carcinoma. Gynecol Oncol 1990;38:22–27.
41. Jakobsen A, Kristensen PB, Poulen HK. Flow cytometric classification of biopsy specimens from cervical intra epithelial neoplasia. Cytometry 1983;4:166–170.
42. Fu YS, Regan IW, Richart RM. Definition of precursors. Gynecol Oncol 1980;12:220–231.
43. Bibbo M, Dytch HE, Alenghat E, Bartels PH, Wied GL. DNA ploidy profiles as prognostic indicators in CIN lesions. Am J Clin Pathol 1989;92:261–265.
44. Jakobsen A, Bichel P, Kristensen GB, Nyland M. Prognostic influence of ploidy level and histopathologic differentiation in cervical carcinoma stage IB. Eur J Cancer Clin Oncol 1988;24:969–972.
45. Jakobsen A. Ploidy level and short time prognosis in early cervix cancer. Radiother Oncol 1984;1:271–275.
46. Leminen A, Paavonen J, Vesterinen E, et al. Deoxyribonucleic acid flow cytometric analysis of cervical adenocarcinoma: prognostic significance of deoxyribonucleic acid and ploidy and S-phase fraction. Am J Obstet Gynecol 1990;162:848–853.
47. Davis JR, Aristizabel S, Way DL, Weiner SA, Hicks MJ, Hagaman RM. DNA ploidy, grade and stage in prognosis of uterine cervical cancer. Gynecol Oncol 1989;32:4–7.
48. Martin DA, Sutton GP, Ulbright TM, Sledge GW, Stehman FB, Ehrlich CE. DNA content as a prognostic index in gestational trophoblastic neoplasia. Gynecol Oncol 1989;34:383–388.
49. Large JM, Driscoll SG, Yauner DL, Olivier AP, Mark SD, Weinberg DS. Hydatidiform moles. Application of flow cytometry in diagnosis. Am J Clin Pathol 1988;89:596–600.
50. Klemi PJ, Joensuu H, Salmi T. Prognostic value of flow cytometric DNA content analysis in granulosa cell tumors of the ovary. Cancer 1990;65:1189–1193.
51. Chadha S, Cornelisse CJ, Schaberg A. Flow cytometric DNA ploidy analysis of ovarian granulosa cell tumors. Gynecol Oncol 1990;36:240–245.
52. Hitchcock CL, Norris HJ, Khalifa MA, Wargotz ES. Flow cytometric analysis of granulosa tumors. Cancer 1989;64:2127–2132.
53. Suh KS, Silverberg SG, Rhame JG, Wilkinson DS, Granulosa cell tumor of the ovary. Histopathologic and flow cytometric analysis with clinical correlation. Arch Pathol Lab Med 1990;114:496–501.
54. Dressler LG, Bartow SA. DNA flow cytometry in solid tumors: Practical aspects and clinical applications. Semin Diagnostic Pathol 1989;1:55–82.

CLINICAL COMMENTARY
John R. Lurain

Professor Friedlander and his colleagues initially developed the technique for flow cytometric analysis of paraffin-embedded tissues and were then the first to apply quantitative determination of cellular DNA content and proliferative activity in a detailed retrospective study of patients with ovarian cancer. Many reports now have confirmed their original findings: that there is a strong correlation between DNA ploidy and outcome, and that DNA index is an independent prognostic variable in epithelial ovarian cancer.

At Northwestern University, we have also investigated the relationship of tumor ploidy, DNA index, and S-phase fraction with other known prognostic factors as well as their effect on survival and recurrence for all stages of ovarian cancer (1). S-phase fraction $\geq 18\%$ was the most sensitive predictor of time to recurrence and survival. Multivariate analysis revealed stage, S-phase fraction, residual tumor, and grade to be independently associated with time to recurrence; and stage, age, S-phase fraction, and largest metastasis were factors associated with survival. We believe that the results of flow cytometry can be used in conjunction with other prognostic indicators, especially clinical stage, to tailor treatment protocols based on risk assessment. This may be most useful for two groups of patients: 1) those with early-stage cancers, low DNA index, and low proliferative activity—in order to avoid unnecessary treatment morbidity; and 2) those with especially poor prognosis based on advanced stage and very high proliferative activity (S-phase $\geq 18\%$)—to target for investigational or no therapy.

Flow cytometric analysis of borderline or low malignant-potential tumors of the ovary could be helpful in identifying which of these tumors may behave in a more malignant fashion and, therefore, should be treated with more aggressive postoperative therapy, such as chemotherapy, which is used for true ovarian cancers. Unfortunately, we were unable to confirm a relationship between aneuploidy and recurrence in a study of 50 ovarian tumors of low malignant potential. DNA aneuploidy was demonstrated in only five cases and none of the four recurrences in the series developed in this group (2).

Several studies have now demonstrated that DNA ploidy is a significant determinant in endometrial cancer, independent of other well-established prognostic variables, such as stage, age, tumor grade, and depth of myometrial invasion. These reports have noted that approximately two-thirds of adenocarcinomas of the endometrium are diploid; that well-differentiated tumors tend to be diploid while more anaplastic tumors are commonly aneuploid; that patients with diploid tumors have a better survival than those with aneuploid tumors; and that among patients with poorly differentiated tumors, those with near-diploid DNA index have a better prognosis.

Based on this information, ploidy status may be useful together with traditional risk factors in stratifying endometrial cancer patients for postoperative adjuvant therapy. This could include pelvic radiotherapy for patients with grade-3 tumors and/or deep myometrial invasion, and treatment of positive peritoneal cytology and/or vaginal brachytherapy to prevent local recurrence in low-risk patients. Trials to demonstrate the efficacy of such adjuvant strategies would require a significant data base because of the relatively low recurrence rates in endometrial cancer.

We recently undertook to study lymphatic metastasis of endometrial cancer by flow cytometry (3). Seventy-one percent of the primary tumors were aneuploid. The DNA index of both the primary tumor and the nodal metastasis was a significant predictor of survival, whereas S-phase fraction was not uniformly associated with survival. Consistent ploidy patterns between the primary and metastatic tumors were present only 47% of the time. DNA stem lines in nodal metastases of endometrial carcinoma were an inconsistent reflection of the primary uterine stem lines and provided very little additional useful information.

The reported prognostic significance of DNA content and proliferative activity in malignancies of the uterine cervix has been inconsistent. We determined DNA index and S-phase fraction in a group of 54 women with stage-IB cancer of the cervix treated with radical hysterectomy and pelvic lymphadenectomy as primary therapy (4). DNA ploidy did not correlate with any of the clinical or pathological variables examined. S-phase fraction was associated with tumor grade, mitoses, and patient age. Neither the DNA index nor the S-phase fraction were found to be significantly associated with recurrence or survival in these patients. These results suggest that, although alterations in DNA content or proliferative activity of invasive cancers of the uterine cervix may be indicators of tumor response to radiotherapy (as reported in some studies), they do not reflect the natural history of early stage malignancy.

Flow cytometric analysis of DNA content has been used in an attempt to differentiate partial from complete hydatidiform moles. Partial hydatidiform moles, by definition, are triploid. Molar pregnancies histologically diagnosed as partial moles are most likely complete moles if they have a diploid DNA content. These diploid molar pregnancies require postevacuation chemotherapy more often and for a longer duration than triploid hydatidiform moles (5). Flow cytometry may also be useful in determining which complete hydatidiform moles are at greatest risk for the development of postmolar gestational trophoblastic tumors, as suggested by Martin, et al. (6).

REFERENCES

1. Barnabei VM, Miller DS, Bauer KD, Murad TM, Rademaker AW, Lurain JR. Flow cytometric evaluation of epithelial ovarian cancer. Am J Obstet Gynecol 1990;162:1584–1592.
2. James JS, Miller DS, Eriksen BL, Bauer KD, Murad TM, Lurain JR. Ovarian tumors of low malignant potential: flow cytometric, histologic and clinical correlations. Poster presentation, American College of Obstetricians and Gynecologists, Atlanta, GA, May 22, 1989.
3. Coleman RE, Lurain JR, August CZ, et al. Flow cytometric analysis of clinical stage I endometrial carcinomas with lymph node metastases. Presented at the American College of Obstetricians and Gynecologists District VI Annual Clinical Meeting, Chicago, IL, October 29, 1991.
4. Connor JP, Miller DS, Bauer KD, et al. Flow cytometric evaluation of early invasive cervical cancer. Presented at the American College of Obstetricians and Gynecologists 39th Annual Meeting, New Orleans, LA, May 8, 1991.
5. Lage JM, Berkowitz RS, Rice LW, Goldstein DP, Bernstein MR, Weinberg DS. Flow cytometric analysis of DNA content in partial hydatidiform moles with persistent gestational trophoblastic tumor. Obstet Gynecol 1991;77:111–115.
6. Martin DA, Sutton GP, Ulbright TM, Sledge GW Jr, Stehman FB, Ehrlich CE. DNA content as a prognostic index in gestational trophoblastic neoplasia. Gynecol Oncol 1989;34:383–388.

16

Urological Cancers

T. VINCENT SHANKEY
Clinical Commentary by **James M. Kozlowski**

PROSTATE CANCER

Background

As of 1990, prostate cancer was the most common cancer in males in the U.S., with over 100,000 new cases reported. Prostate cancer is a major cause of cancer-related deaths, with over 30,000 deaths in the U.S. that year. The incidence of prostate cancer increases with age and, with gentrification of the population, it is likely that the clinical incidence of this cancer will increase significantly. In addition, the growing use of screening assays in older males (i.e., serum prostate-specific antigen, or PSA, digital rectal exams, rectal ultrasound) will likely increase the overall detection of prostate cancers.

The natural history of human prostate cancer is poorly understood. The incidence of histologically apparent prostate cancer at autopsy is significantly higher than the incidence of clinical disease in age-matched groups (1, 2), suggesting that in many individuals cancers can progress slowly. In the U.S., the lifetime risk of being clinically diagnosed with prostate cancer is now approximately 9%, and the risk of dying of prostate cancer is less than 3% (3). In contrast, the incidence of histologically apparent prostate cancer at autopsy in the eighth decade is between 40 to 60% (4). Some evidence suggests that the histological prevalence of prostate cancer is similar worldwide. However, there are significant variations in clinical incidence and risk of dying of prostate cancer in different populations (5, 6).

The human prostate is divided into four separate glandular zones plus an anterior fibromuscular zone (Fig. 16.1). The peripheral zone, roughly 70% of the mass of the prostate, comprises the lateral and posterior portions. The central zone is located adjacent to the base of the bladder and comprises about 25% of the mass of the prostate. Anterior to the central zone and located on either side of the prostatic urethra is the transition zone, which forms about 5% of the prostatic mass. A submucosal periurethral gland zone constitutes less than 1% of the prostate (7).

Benign lesions of the prostate (BPH) most frequently arise from the transition zone. Their anatomical location results in a high probability for constriction of the prostatic urethra and for clinical symptoms of voiding difficulties. In contrast, adenocarcinoma of the prostate can arise from the peripheral (roughly 70%), transition (20%), or central (10%) zones (8).

Pathological and Clinical Assessment of Prostate Cancer

A detailed description of pathological grading or clinical staging of prostate cancers is beyond the scope of this chapter. Background information is presented here to provide an outline of the information currently used to predict disease course and to judge the best intervention and treatment. The subsequent review of flow cytometric analysis of prostate cancers will focus on the additional information provided by DNA content measurements for patient management.

Staging provides an assessment of the amount of tumor present in the prostate (tumor volume), penetration through the gland's capsule, presence of tumor cells in the seminal vesicles, and metastatic spread of the primary tumor to other sites (lymph nodes, bone). A number of different staging systems have been proposed and the most commonly used are described in Table 16.1. In the U.S., the system proposed by Whitmore (9) and modified by Jewett (10) is frequently employed, although there is considerable emphasis on the universal use of the TMN (Tumor, Metastices, Nodes) staging system for all cancers. As shown in Table 16.1, the TMN scheme used by the majority of the studies reported by U.S. investigators utilize the system proposed by the American Joint Committee on Cancer (AJCC).

All staging schemes provide a descriptive and qualitative assessment of the extent of disease, with the underlying premise that the greater the extent of disease, the worse the prognosis. They provide a general outline of natural disease course and an indication of where an individual patient is in the general course of the disease. Several important limitations exist in all staging schemes as predictors of disease course for individual patients. Perhaps the most serious problem clinically is that many prostatic tumors are significantly understaged (11). An additional limitation is that, in some patients, low-volume tumors may be more aggressive (life threatening) than high-volume disease. An additional problem of tumor heterogeneity will be discussed below, in both the context of tumor histopathology and in relation to the impact on DNA-content measurements.

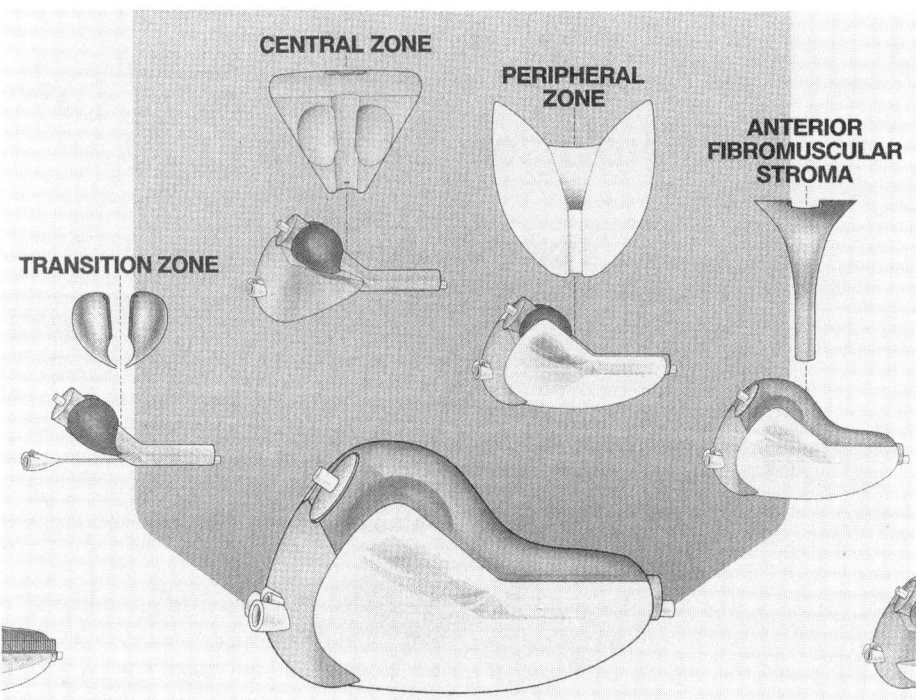

Figure 16.1. Schematic drawing showing the four major areas of the human prostate. The position of the organ shown here is such that the urinary bladder is to the left and the rectum is immediately below the peripheral zone. (Reprinted with permission from from Bruel and Kjaer Instruments, Inc.)

Table 16.1
Criteria for Staging Adenocarcinomas of the Prostate[a]

Clinical Finding	Whitmore/Jewett Classification	AJCC Classification
No Palpable Neoplasm	A	T1
Tumor <3 Microscopic Foci	A1	T1a
Tumor >3 Microscopic Foci	A2	T1b
Palpable Neoplasm	B	T2
Focal, < 1.5 cm, 1 "lobe"	B1	T2a
Diffuse, > 1.5 cm, or > 1 "lobe"	B2	T2b
Local Invasion	C	
Bladder, seminal vesicles, prostate capsule involved; not fixed	C1	T3
Other sites involved; fixed	C2	T4
Metastatic Disease	D	
Only Regional Lymph Nodes Involved	D1	N1-3
Distant Metastatic Sites	D2	M1

[a]Adapted from Murphy WM. Diseases of the urinary bladder, urethra, ureters, and renal pelves. In: Murphy WM, ed. Urological Pathology. Philadelphia: WB Saunders Co. 1989.

Histological grading of prostate cancers provides a well-recognized source of information regarding prognosis, a fact that has been known for more than 50 years (12). A variety of grading systems have been developed for prostatic carcinomas. The most widely used system for tissues, the Gleason's score (13), is based on the degree of glandular differentiation and tumor–stromal patterns. As shown in Figure 16.2, low-Gleason's-score tumors are characterized by well-formed glands, while high-score tumors show loss of glandular structure and infiltrating tumor cells. The overall utility of Gleason's system compared to other tissue-grading schemes has been reported (14) with the recommendation for adoption of Gleason's score as the best reference system for tumor classification.

The initial classification of prostate cancers frequently relies on biopsy specimens. Since many institutions rely on fine-needle-aspiration (FNA) biopsy, it is important to note differences in cytologic-compared to tissue-grading schemes. Since tissue architecture is generally missing from aspiration biopsies, more emphasis is placed on cellular and nuclear characteristics that define the degree of anaplasia. Cytological assessment of aspiration biopsies has been reported to agree with Gleason's scores (15), although the increasing popularity of small-needle tissue-biopsy systems (i.e., Biopty gun), and technical difficulties in aspiration cytology may make this one area where FNA technology is less frequently used.

The disease course of prostate cancers is highly variable. While grade and stage are useful predictors, considerable variation in disease course exists for different individuals presenting with similar disease by these criteria. Overall, individuals with low-stage disease tend to progress at a low rate. The Veterans Administration Cooperative Urological Research Group reported that 6.8% of untreated stage A1 individuals (who received no further therapy) died from their cancers (16). The overall progression rate is dependent on stage, with a low progression rate (2%) for stage A1 and a significantly higher progression rate (33%) for patients with stage A2 disease (17). In low-volume stage A disease (A1), only patients with a high Gleason's score (sum >5)

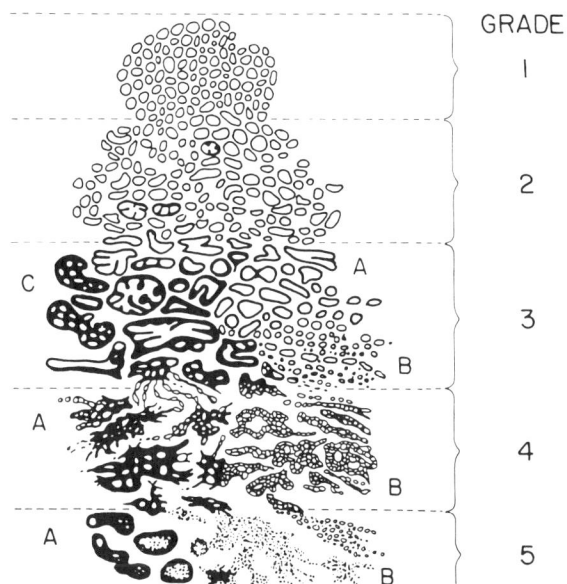

Figure 16.2. Schematic drawing showing the histologic grades of the Gleason's grading scheme based on glandular differentiation and its relationship with the stroma. Grade 1: single glands with a definite, rounded edge. Grade 2: slightly less uniform glands, more loosely grouped, with less sharp but definable edges. Grade 3: more variable glands within same prostate, more variable shape of glands, poorly defined edges; or microglandular tumor with small glands or groups of cells; or rounded masses of papillary and cribriform tumor. Grade 4: poorly defined glands raggedly infiltrating into stroma, fused glands. Grade 5: rounded masses of almost solid cribriform tumor with some central necrosis; or raggedly infiltrating, invading anaplastic carcinoma. (Reprinted from Gleason DF. Classification of prostate carcinomas. Cancer Chemother Rep 1966;50:125–128.)

progressed within four years of diagnosis. This study indicates, however, that some low-stage cancers progress.

Grading schemes, such as the Gleason's score are generally successful in predicting outcome for large groups of patients and are generally better at predicting outcome for individuals with either high or low grade disease. The course for individuals with intermediate grade disease (where the majority of individuals now fall) is quite variable. This is due, in part, to the considerable heterogeneity, with different areas of the prostate containing areas of differing Gleason's score and/or degree of cellular differentiation or anaplasia (18). An additional degree of complexity is provided by the observation that adenocarcinomas arising from different glandular regions of the prostate have different pathological features (8) and may have different biological potentials.

Sample Acquisition and Processing

A detailed overview of the impact of different tissue preparation techniques is presented in Chapter 6. Only brief comments relating to prostate cancers are described here. Early flow cytometry studies were frequently performed on specimens obtained by fine-needle-aspiration biopsies. Although this technique produces a single-cell suspension suitable for flow cytometric analysis, some studies have used nuclei prepared from FNA material using the Vindelov (or a related) procedure (19). No careful comparisons have been published to date demonstrating that intact cells or nuclei from prostate FNA samples are preferred nor that either provide better histograms (less debris, lower C.V.) or more representative tumor material.

Studies of samples isolated from primary prostate tumors that carefully compare the results of different isolation techniques on the recovery of tumor cells or nuclei have not been reported. One report has compared the results of mechanical vs. enzymatic (0.8% collagenase II) disaggregation of fresh prostate cancers (20). The portion of DNA-aneuploid cells was generally higher in mechanically treated suspensions. This report included a number of different primary tumors (bladder, breast, colo-rectal, renal, ovarian, lung, and prostate cancers) and the results did not specify which types of tumors were used for this part of the study.

Technical considerations limited earlier studies to fresh tumor material. Due to the long time course between detection of stage A or B cancer and progression and death, many of these early studies were limited to correlations of DNA ploidy with cytopathological or histopathological (tumor grade or stage) characteristics. With the development of the Hedley technique (21), it was also possible to correlate rapidly DNA content with survival or disease progression by performing retrospective studies. Although many variations to the original technique have been reported, no published study has addressed carefully the effects of modifying the technique specifically for archival prostate cancer specimens.

As a result of a study of interlaboratory variations in DNA-content measurements (C. Hitchcock, manuscript in preparation), we have modified some of the steps in the Hedley technique for paraffin-embedded prostate cancers (22). Following the recommendations of S. Wright (Cytometry Associates) we have compared the results of DNA-content analysis of archival tissues from TURP (transurethral resection of the prostate) specimens prepared by a technique that includes more extensive dewaxing (three incubations for 30–60 min each) and longer rehydration times. As shown in Figure 16.3, these modifications resulted in increased DNA staining intensity, flattened the debris curve, and usually decreased the CV of the G_0/G_1 peak. In some samples, these changes had a significant effect on S-phase calculations and, in most cases, provided a better fit of the cell cycle components to the data points as measured by Chi-square values (in one case from 5.4 to 1.4, using identical model components). To date, we have not compared extensively the effect of these modifications on paraffin-embedded sections of intact prostate specimens and it remains unclear whether these modifications are effective for other tissue samples that contain large amounts of paraffin (i.e., TURP or TURBT—Transurethral Resection of a Prostate or Bladder Tumor—samples).

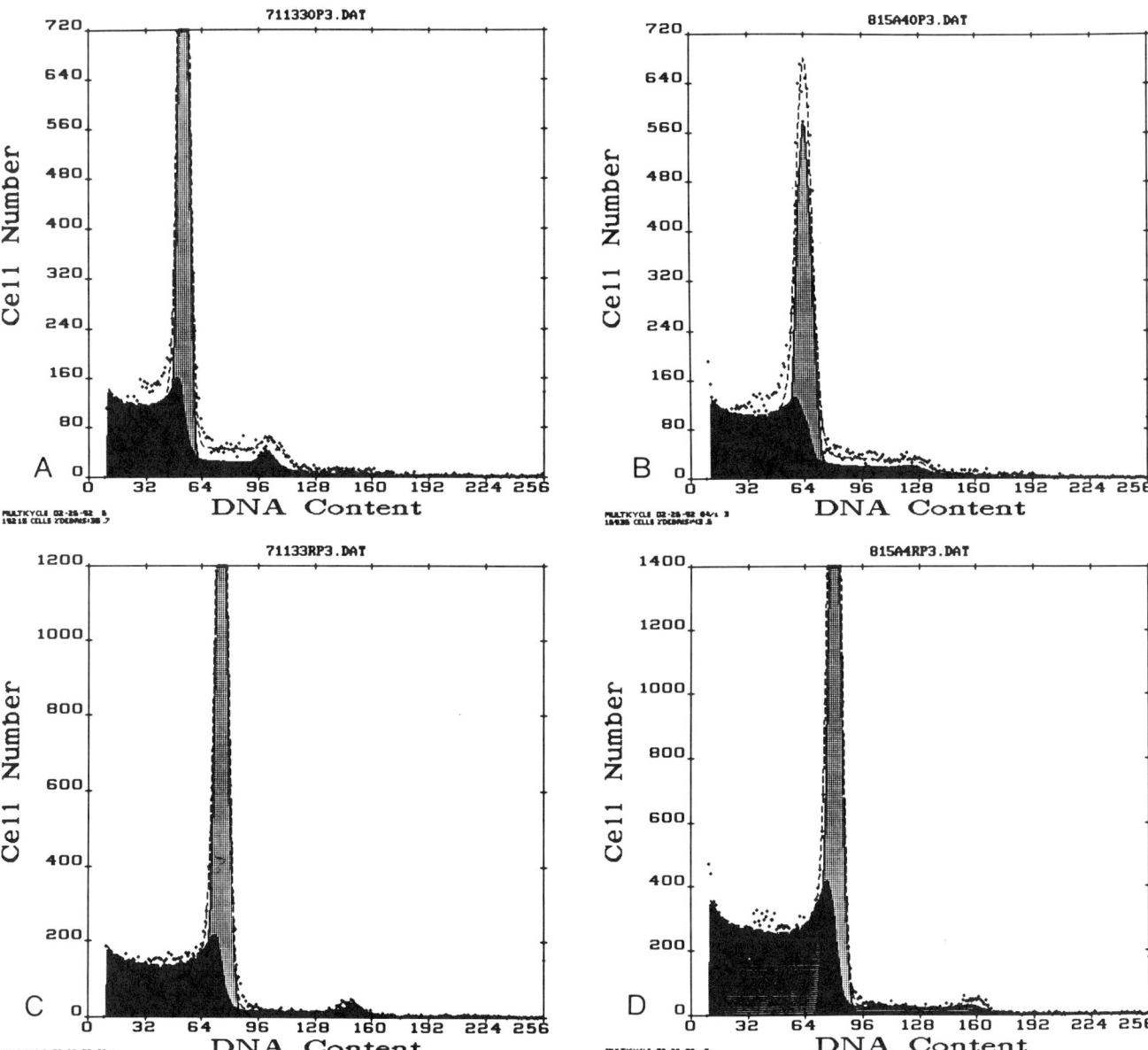

Figure 16.3. Effect of sample preparation on DNA-content analysis of paraffin-embedded prostate carcinoma from stage A2 tumors. Duplicate sets of 50 μ-thick sections were processed by the technique of Hedley et al. (Reprinted with permission from Hedley DW, Friedlander ML, Taylor IW, Rugg CA, Musgrove EA. Method for analysis of cellular DNA content of paraffin-embedded pathological material using flow cytometry. J Histochem Cytochem 1983;31:1333–1335.) (**A** and **B**), or using a revised technique (**C** and **D**). Samples from same tumor tissue were run under identical conditions (Reprinted with permission from Dougherty S, Manion S, Jin J-K, et al. Intra-tumor heterogeneity of stage A2 (T1b) prostatic adenocarcinoma: I. Evaluation of archival tumors by flow cytometry. Cytometry (accepted for publication).).

DNA Content Measurements

DNA INDEX (PLOIDY)

Biopsy Specimens. Many of the earliest studies of the DNA content of prostate cancers involved the use of either aspiration or tissue-biopsy specimens. In addition to the important historical aspects of these studies, their impact will become more critical if prostate screening programs increase the use of multiple-prostate-biopsy sampling.

As shown in Table 16.2, DNA-content analysis of biopsy specimens has been performed for more than 20 years. Most of the earlier studies were reported by European investigators, where fine-needle aspiration of the prostate and other tissues has been widely used for a number of years. These studies have addressed several different issues, including the ability of DNA-content measurements to detect prostate cancers (23, 24) predict local or distant progression (25–29), or predict response to therapy (30, 31). In addition, many studies have included the correlation of DNA content with cytopathological grade or clinical stage (25, 31–35). As

Table 16.2
DNA Content Analysis of Prostatic Carcinoma Aspiration or Biopsy Specimens

Sample	No. of Cases[a]	Significance	Reference
FNA	469	DNA ploidy predicted 5-year survival; ploidy correlated well with cytological grade.	Esposti (25)
Biopsy[b]	76	Analysis of tissue sections; 60% of cases 2C; 28%, 3C; 10%, 4C; and 2%, 6C. Survival of 3C cases significantly lower than 2C or 4C cases. Samples not graded.	Tavares (26)
FNA		Detection of malignancy, used BPH[c] and CaP[d] specimens. 80% of cytologically confirmed CaP detected, 20% graded "suspicious."	Zetterberg (23)
FNA[e]	220	DNA flow cytometry unsuitable due to high false-neg. and false-pos. rates; large no. of insufficient samples.	Sprenger (36)
FNA		DNA ploidy correlated with cytological staging.	Bichel (35)
FNA		DNA ploidy as a predictor of response to estrogen therapy.	Kjaer (30)
FNA[b]	43	DNA ploidy as predictor of disease course in 2 groups of patients selected as responders or nonresponders to hormonal therapy.	Zetterberg (28)
FNA		DNA ploidy correlated well with cytological grading.	Ronstrom (32)
FNA		DNA ploidy correlated well with cytological grading; useful in predicting response to hormonal therapy.	Bocking (31)
FNA	55	DNA ploidy used to detect and classify 35 malignant and 20 BPH specimens; 85% of samples were correctly classified, with an 11% false-neg. and a 4% false-pos. rate.	Seppelt (24)
Biopsy[e]	50	DNA ploidy correlates with histological grade and survival.	Lundberg (27)
FNA[b]	292	Classification scheme based on DNA-content patterns, using prospectively selected groups. DNA ploidy correlated with cytological grade.	Forsslund (33)
FNA[e]	146	Low-grade (1-2), low-stage (T1-2) individuals; 47%, DNA diploid; 46%, DNA tetraploid. DNA ploidy was a significant predictor of 5-year progression.	Adolfsson (34)
FNA[e]	72	Patients followed for 5 years with repeat FNA. 53 individuals with progression (74%); 17 changed from DNA-diploid to DNA-tetraploid and/or aneuploid tumors.	Adolfsson (29)

[a]Actual number of individuals or samples analyzed
[b]Studies performed using image analysis
[c]Benign prostatic hypertrophy
[d]Carcinoma of the prostate
[e]Studies performed using flow cytometry

shown in this table, not all studies have concluded that DNA-content measurements provide useful information for patient management (36). Although the focus of this chapter is on flow cytometric applications, several important image analysis studies are included here.

The results of the majority of these studies have demonstrated that individuals with DNA-diploid tumors have a significantly higher survival rate or a lower rate of progression (local or distant disease) than individuals presenting with DNA-aneuploid tumors. It must be noted that many of these early studies used hormonal therapy (estrogen or castration) in some or all of these patients, frequently without radical prostatectomy (as commonly used in the U.S. for organ-confined disease). While many of these studies report the impact of steroid therapy on DNA-diploid vs. DNA-aneuploid tumors (26), a statistical analysis that would isolate the relative impact of different variables on disease course or survival has not been reported in many cases.

Perhaps the most important clinical issues include the ability to detect prostate cancers as well as the ability to determine disease course or response to therapy of the detected tumor populations. As shown in Table 16.2, the ability of cytometry to "detect" cytologically proven prostate cancer is generally 80% or better (23, 24) as defined using cytologically positive specimens. The goal of many research projects has been to develop automated techniques to identify prostate or other cancers. Perhaps a more intelligent use of the technology is to utilize the trained pathologist to identify tumor cells (particularly from the now more commonly used Biopty gun biopsy) and use the cytometer to make measurements that will predict disease course or therapeutic response. Considering the potential for DNA-content heterogeneity (see below) seen in a significant number of prostate cancers, the use of image cytometry in the analysis of biopsy specimens may become an important tool in the routine pathological workup of incidentally diagnosed prostate cancer.

Radical Prostatectomy Samples. DNA-content analysis has been performed on a large number of radical prostatectomy specimens and on samples obtained by transurethral resections. The majority of studies have used archival paraffin–embedded tissue, which allows retrospective analyses, to determine the predictive value of DNA-content (or other nuclear markers). These studies are summarized in Table 16.3.

Although few studies have specifically focused on BPH, two reports (37, 38) have indicated that a small percentage (7% to 13%) of these samples are DNA aneuploid. The study by Deitch and coworkers (37) reported a high frequency of samples of DI 0.8 to 1.0, and 1.0 to 1.3 (25% overall), although these were not considered DNA-aneuploid by the authors. The use of hyperdiploid fraction (>20% of the events 2 standard deviations above the mean diploid G_0/G_1 peak) in this study as an indicator for aneuploidy is highly questionable, unless appropriate steps are taken to eliminate aggregates (see Chapter 8). Since this study did not report follow up, it remains unclear what potential role DNA aneuploidy may have in predicting the likelihood of subsequent carcinoma of the prostate.

As shown in Table 16.3, in most studies there is a positive correlation for both histological grade and tumor stage with DNA aneuploidy. Overall, DNA-diploid tumors tend to

Table 16.3
Predictive Value of Flow Cytometric DNA-Content Measurements by Stage of Prostate Disease or Cancer

Stage	No. of Cases[a]	Significance	Reference
BPH[b]	177	High-frequency (~25%) of DI 0.8-1.3 samples. Authors considered malignancy (7% of samples) DI outside this range, or >20% HDF[c]. Prognostic value not reported.	Deitch (37)
BPH, A[p]	48	DNA-aneuploid samples in 1/11 stage A1, 9/22 stage A2 and, 2/15 BPH. Trend for DNA-aneuploid tumors to progress (no statistics). Broad CV's.	McIntire (38)
A2[d]	19	DNA-content measurements by flow or image cytometry not predictive of progression.	Mohler (43)
A, B[d]		DNA ploidy of limited value in predicting disease course.	Ritchie (44)
B[d]	261	68% of samples, DNA diploid; 28%, DNA tetraploid; 10%, DNA aneuploid. Correlation for Gleason's score with DNA ploidy. Poor survival (20 yr) for DNA-aneuploid tumors	Montgomery (39)
B, C[d]	88	Combination of DNA aneuploidy or tetraploidy plus seminal vesicle involvement is strong predictor of disease-free interval.	Lee (41)
B, C[d]	73	Subset of patients reported by Lee (41). DNA ploidy not an independent predictor; tumor area and histological grade are better predictors of progression.	Humphrey (45)
C[d]	146	Poor survival (20 yr) for DNA-aneuploid tumors.	Nativ (46)
D1[d]	91	45% of samples, DNA tetraploid; 13% DNA, aneuploid. 10-year survival of DNA-tetraploid or aneuploid tumors significantly worse than DNA-diploid tumors (patients had radical prostatectomy).	Winkler (42)
D1[d]	82	DNA-content analysis of lymph node metastases; presence of DNA-diploid tumor is strong predictor of disease-free survival.	Stephenson (47)
D2[d]	97	21%, DNA tetraploid; 33%, DNA aneuploid—used cRBC as DNA control. Selected patients in good vs. bad outcome groups.	Miller (50)
A, B, C[d]	72	49% of samples, DNA diploid (no differentiation of DNA-tetraploid from DNA-aneuploid tumors). DNA-aneuploid tumors have significantly lower 10-yr survival; combination of Gleason plus DNA ploidy predicts survival.	Fordham (40)

[a]Actual number of individuals or samples analyzed
[b]Fresh Tissue
[c]Hyperdiploid Fraction (excluding inflammatory tail)
[d]Paraffin-embedded tissue-flow cytometry using nuclei isolated by the Hedley technique (ref. 21)

predominate in low-stage disease, while DNA-aneuploid (including DNA-tetraploid) tumors are more common in higher-stage disease. In general, the percentage of DNA-tetraploid tumors (as a percent of all DNA-aneuploid tumors) is greater for low-stage (A, B) tumors, while DNA-aneuploid tumors (DI >1.3 <1.8 and DI >2.2) are more commonly seen in stage D tumors. Overall patients with DNA-tetraploid and other DNA-aneuploid tumors have similar survival in disease progression rates in some studies, although this may depend on tumor stage (see below). The relationship between DNA ploidy and Gleason's score is less clear. In some studies a good correlation was reported (39), whereas in other studies there was marginal (40, 41) or no correlation (42). These results are in contrast to FNA studies, where the majority have reported a good correlation for DNA ploidy with cytological grading (25, 32–34). However, in the case of both Gleason's score and cytological grading, the result is dependent, in part, on the skill of the pathologist. Limitations of both cytometry and pathology indicate either that these criteria cannot be used independently or that flow (or image) cytometry is likely to replace pathological grading or staging in the foreseeable future.

The prognostic significance of DNA ploidy is a clinically important question. Overall, most studies have demonstrated a higher probability for DNA-aneuploid or tetraploid tumors to progress or cause death due to prostate cancer. In one study on stage A cancer, individuals with DNA-aneuploid tumors are more likely to progress (38), although two other studies have reported limited (44) or no predictive value (43) of DNA ploidy. In one study (43), no information was provided on the quality of flow cytometry results (no histograms, no CV's); in addition, the use of cRBC's or fresh human lymphocytes as DNA-content controls reported here for paraffin–embedded samples is likely to reduce the sensitivity and specificity of the DNA-content measurements.

Retrospective studies of stage B cancer have generally shown prognostic significance for DNA ploidy measurements. As shown graphically in Figure 16.4 (39), the probability for disease recurrence in DNA-aneuploid stage B tumors is significantly higher than either DNA-diploid or tetraploid tumors. As shown in this figure, the probability for recurrence and for disease-free survival is similar for DNA-diploid and tetraploid tumors. In this study of 261 individuals, the presence of a DNA-aneuploid (but not a DNA-tetraploid) tumor was a significant prognostic indicator of survival. The study by Lee and coworkers (41) included 29 stage B individuals who were included to test the significance of DNA ploidy and seminal vesicle involvement (stage C) on prognosis (see below). Interestingly, this same group of patients was also used in a subsequent report for multivarate analysis (45) to compare grade, percent tumor area, and DNA ploidy as prognostic variables, with the conclusion that DNA ploidy was not an independent predictive factor.

The results of Lee and coworkers (41) have demonstrated that the combination of DNA ploidy plus seminal vesicle involvement (pathological stage C) is a powerful predictor of disease progression. In this study, DNA-tetraploid (51%) and DNA-aneuploid (7%) tumors were analyzed as a group. As shown in Figure 16.5, the results of this study indicate that DNA–diploid tumors lacking seminal vesicle involvement have an extremely low probability of disease recurrence, while most DNA-aneuploid (including DNA-tetra-

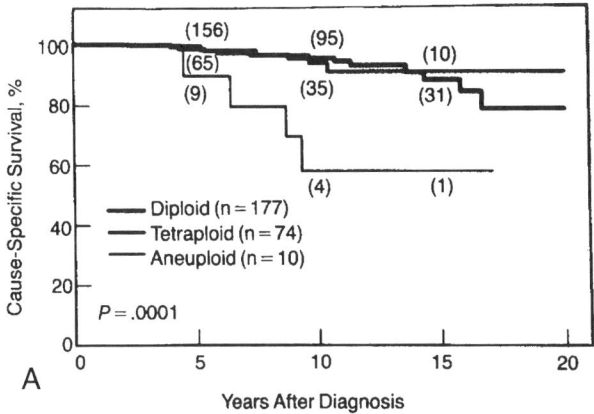

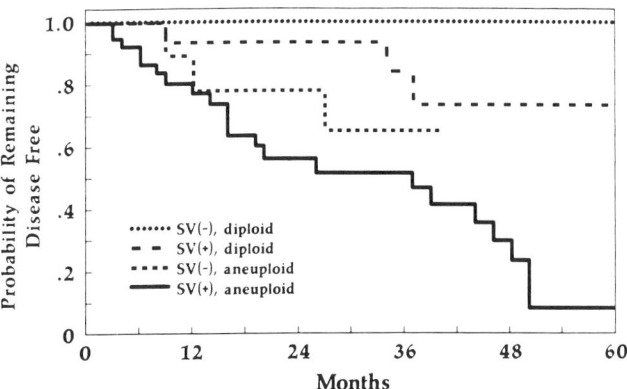

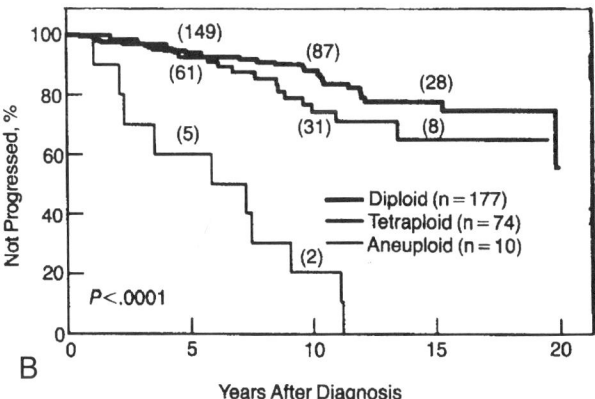

Figure 16.4. Value of DNA ploidy determinations in stage B prostate carcinoma (**A**) in predicting probability of nonprogression; or (**B**) in predicting disease-specific survival for the same stage B individuals (Reprinted with permission from Montgomery BT, Nativ O, Blute ML, et al. Stage B Prostate adenocarcinoma: flow cytometric nuclear DNA ploidy analysis. Arch Surg 1990;125:327–331.). In this study, individuals with DNA-diploid and tetraploid tumors had a similar probability of nonprogression or cancer-related survival, while individuals with other DNA-aneuploid tumors had a significantly higher probability of cancer-related death (P = 0.0001).

Figure 16.5. Impact of DNA ploidy in determining the probability of remaining free of disease using a combination of seminal vesicle involvement plus DNA ploidy in stage C prostate carcinoma. Kaplan-Meier analysis of these same patients demonstrated a significant difference in the probability of remaining disease free for patients with DNA-diploid vs. aneuploid tumors (P = 0.00009). As shown in this figure, the additional use of seminal vesicle involvement allowed further stratification, where DNA-*diploid* individuals with seminal vesicle involvement had a higher probability of remaining disease free than DNA-*aneuploid* individuals without seminal vesicle invasion. (Reprinted with permission from Lee SE, Currin SM, Paulson DF, Walther PJ. Flow cytometric determination of ploidy in prostatic adenocarcinoma: a comparison with seminal vesicle involvement and histopathological grading as a predictor of clinical recurrence. J Urol 1988;140:769–774.).

ploid) tumors with involvement will recur within four years. These results also indicate that DNA-diploid tumors with seminal vesicle involvement have a lower probability for recurrence than DNA-aneuploid tumors without involvement.

Studies of stage D prostate cancers have indicated that DNA ploidy is a significant prognostic indicator for both the analysis of the primary tumor (42) and the analysis of lymph node metastases (47). In the study by Winkler and coworkers (42), DNA-diploid tumors had a significantly lower probability of progression and DNA-tetraploid (45% of samples) and DNA-aneuploid (13%) tumors had a similar, high progression rate (note that in this study all stage D patients received a radical prostatectomy). The disease–related (10-year plus) survival of DNA-diploid tumors was significantly higher than that of individuals with DNA-tetraploid or aneuploid tumors.

Response to Therapy. Predicting the response to therapeutic intervention is an important clinical goal that currently lacks well accepted technology. As indicated in Table 16.2, a number of studies have compared DNA ploidy (biopsy or aspiration samples) with response to hormonal therapy in individuals who did not undergo a radical prostatectomy. In most of these studies, a positive correlation was shown in that DNA-aneuploid tumors generally failed to respond to hormone therapy and cause patients to die of the disease. Unfortunately, the follow-up period in many of these studies was short and, therefore, it is not clear whether a significant number of DNA-diploid individuals did or did not progress later or died of the disease.

At present, roughly 50% of all individuals present with local or distant spread of prostatic cancer. Therapeutic options include hormonal treatment, radiation, and radical prostatectomy. The value of each of these for stage D cancers, particularly radical surgery, has been questioned, and it is not clear if treatment should be initiated at diagnosis or if more conservative management is appropriate. The results of a study of individuals with stage D1 disease by Lieber and coworkers (48), shown in Fig. 16.6, demonstrate the impact of DNA ploidy in predicting the disease–related survival of 126 individuals who underwent radical prostatectomy. As shown here, the 10-year (plus) survival of individuals with DNA-diploid tumors is significantly higher than that of individuals with DNA-aneuploid (including DNA-tetraploid) tumors. The results of this study indicate that individuals with DNA-diploid tumors have a significant response to hormone therapy (~100 vs. 55% 10-year survival), whereas hormonal therapy did not significantly impact survival of individuals with DNA-aneuploid tumors.

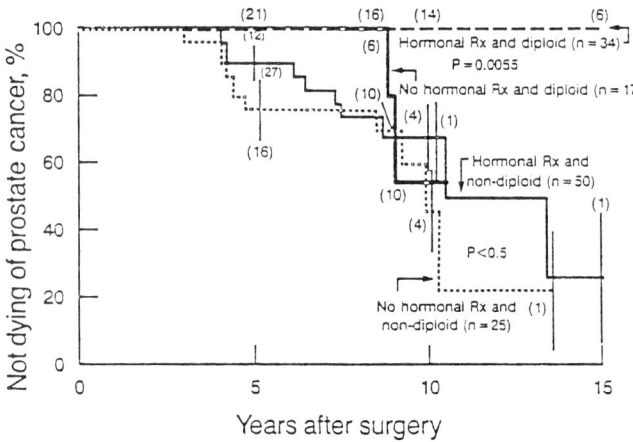

Figure 16.6. The impact of DNA ploidy in predicting response to hormonal therapy (*Rx*) in stage D prostate carcinoma. As shown in these results from 126 patients who received a radical prostatectomy, patients with DNA-*diploid* tumors who received hormonal therapy had a significantly higher survival (P = 0.0055) than DNA-*diploid* patients who received no therapy. For patients with DNA-aneuploid (including tetraploid) tumors, there was no significant impact of hormonal therapy on survival. (Reprinted with permission from Zincke H. The role of pathologic variables and hormonal treatment after radical prostatectomy for stage D1 disease. Oncology 1991;5:129–140.).

Detection of DNA Aneuploid Tumors. As indicated above, the majority of DNA-aneuploid tumors of the prostate are reported to have a DNA-tetraploid or near-tetraploid DNA content (DI ~2). Whereas flow cytometry histograms frequently show an obvious DNA-aneuploid G_0/G_1 population at the 4C position or at a distinct position other than 2C, the detection of a DNA-tetraploid population from samples with high diploid G_2M content (or a large amount of diploid aggregates) is sometimes difficult. An additional point not generally discussed is the criteria used to identify a DNA-aneuploid population with a DNA Index near 1. In some studies of prostate (and other) cancers, a skewed 2C peak, or a peak with a shoulder is interpreted as clear evidence of the presence of a DNA-aneuploid population; this interpretation has been made in reports that do not provide information regarding the CV's of the peaks, or that, perhaps worse, make this interpretation with CV's in the 6% to 10% range (see Chapter 8).

One problem in identifying a DNA-tetraploid population is illustrated in Figure 16.7, which compares the DNA-content histograms for control (area free of histologically apparent tumor from the same level of the same block) and the tumor-containing region from a paraffin-embedded prostate sample. As shown here, the position of the first peak is similar for both control and tumor regions, illustrating the use of normal tissue from the same level of the same block as a diploid-DNA-content control. There is a significantly higher percentage of 4C events in the tumor sample (here 8.9% after correction for debris and aggregates) than in the control. Although the increase in %4C events would make the sample suspicious for the presence of a DNA-tetraploid tumor, the

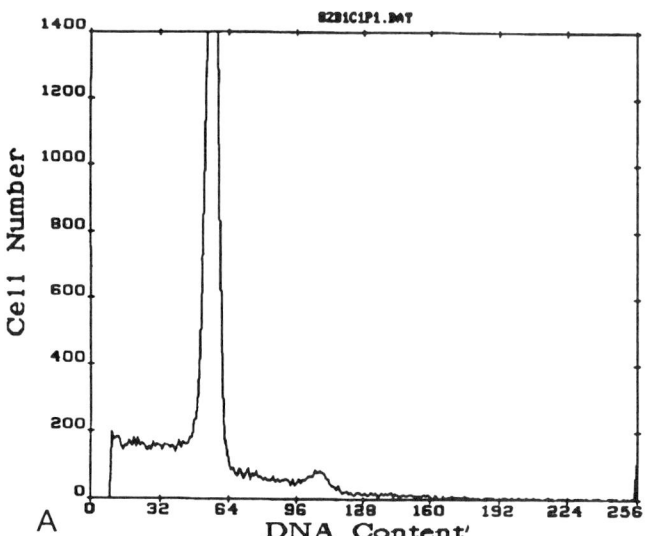

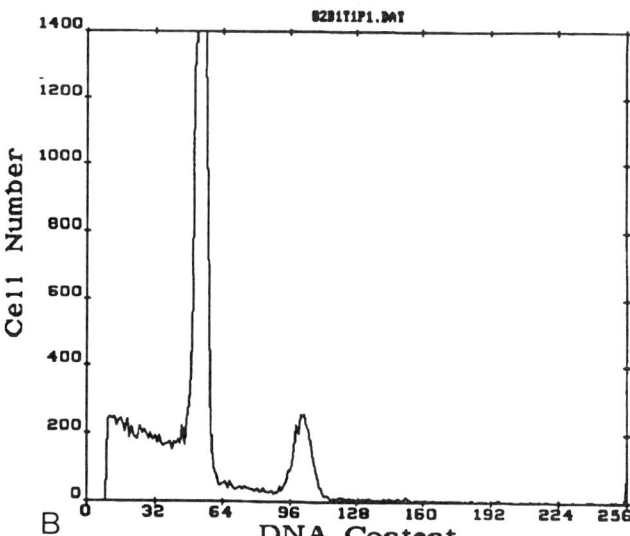

Figure 16.7. Illustration of one problem encountered in identifying the presence of a DNA-tetraploid tumor by DNA-content flow cytometry from archival paraffin-embedded sections of prostate carcinoma. **A** shows the DNA-content histogram of a control area, whereas **B** shows a histogram of a tumor-containing area from the same level of the same block. The tumor area illustrates a problem for histograms showing a high %4C events (here 8.9% after correction for aggregates) with no 8C peak. Subsequent image analysis of the nuclei from this sample demonstrated the presence of a DNA-tetraploid tumor (Reprinted with permission from Dougherty S, Manion S, Jin J-K, et al. Intra-tumor heterogeneity of stage A2 (T1b) prostatic adenocarcinoma: I. Evaluation of archival tumors by flow cytometry. Cytometry (accepted for publication).).

lack of a significant 8C peak (after correction for aggregates) makes this less likely.

Previously reported studies have used statistical methods to define DNA-tetraploid tumors (41, 42, 49), an arbitrary cut-off (40, 50), or have not reported the criteria used. The arbitrary value used for prostate and other tumors is generally from 10% to 15%. The statistical techniques use the analysis of a number of control tissues, calculating the mean %4C events for these controls and then using the mean +3 standard deviations for the upper limit of normal 4C events. The control 4C values reported include 7.87±1.53% (42), and 3.87±0.88% (41). Both of these studies used nuclei from paraffin-embedded material; one study used normal prostate tissue from radical cystoprostatectomies (41), whereas the other used areas of BPH from tumor-containing tissues (42).

Studies in our lab on paraffin-embedded prostate specimens use histologically normal appearing areas to define the %4C events in control tissues. In a study of stage A2 prostate cancers (22), the mean 4C events for control samples (corrected for aggregates using software correction, see Chapter 8) was 2.4±1.77%. Using a statistical upper limit of 8%, the tumor-containing sample shown in Figure 16.7 was interpreted as containing a DNA-tetraploid tumor. In our study of individuals with stage A2 prostate cancer, 24 tumor samples had between 6% and 15% 4C events, after correction for aggregates. Six of these individuals had other tumor blocks with significant DNA-tetraploid populations (>25% of all nuclei analyzed). For the remaining individuals, we analyzed Feulgen stained nuclei (from the same sample used for flow cytometry) using image analysis. For image studies, the analysis included DNA content of all nuclei and an analysis of nuclei with the morphological appearance of tumor (larger size, grainy appearance, prominent nucleoli). By this criteria, all of the samples tested (in a blind fashion intermixed with control samples and samples containing 40% or more 4C events by flow cytometry) with greater than 8% 4C events by flow cytometry were DNA-tetraploid by image analysis. This included the tumor sample shown in Figure 16.7. In addition, one tumor sample with 7% 4C events was determined to contain a DNA-tetraploid tumor by image analysis.

Significance of Tumor DNA-Content Heterogeneity. The presence of multiple foci having different DNA ploidy within a single tumor has been reported in renal cell carcinomas (51), Barrett's esophagus (52), colon carcinomas (53), and mammary adenocarcinomas (54). One published report that demonstrates DNA-content heterogeneity in prostate cancers (55) employed image analysis of tissue sections from radical prostatectomy "whole-mount" specimens. This study mapped 63 separate tumors in prostates from 30 patients, and reported that both DNA-diploid and DNA-aneuploid tumor foci existed in some individuals with stage A2 or B prostate cancer.

Our studies on stage A2 prostate cancer using flow cytometry (22) and image analysis (56) indicate that a significant percent of these individuals have both DNA-diploid and DNA-aneuploid tumors. In the flow cytometry study, different tumor-containing blocks were analyzed until either a DNA-aneuploid population was detected or no more tumor-containing blocks were available for that patient. A total of 154 different blocks were analyzed from 54 individuals. Twenty-six individuals studied had evidence of only DNA-diploid populations (the presence of tumor nuclei was confirmed microscopically and, in some samples, measured by image analysis, only a DNA-diploid tumor was detected). Thirteen individuals demonstrated only DNA-aneuploid tumors, although this may be an underestimation, as not all blocks available for each patient were analyzed. In 14 individuals, DNA-tetraploid tumors were found in one or more blocks and only a DNA-diploid tumor was found in other blocks from the same individual. These results were further confirmed using image analysis of Feulgen-stained tissue sections from 23 of these individuals (56).

The presence of both DNA-diploid and DNA-aneuploid tumor foci in the same prostate could have an important impact on sampling strategies for the analysis of prostate cancers. From our studies of stage A2 prostate samples, the analysis of only one block from each individual would demonstrate the presence of a DNA-tetraploid or aneuploid tumor in roughly 35% of the individuals who were found to have DNA-aneuploid tumors. Considering the prognostic significance of DNA tetraploidy in this group of stage A2 patients (38, 57), these results would suggest that multiple samples should be analyzed in cases where the initial samples are DNA diploid. These results, and those of Greene and coworkers (55) would further suggest that biopsy screening of individuals suspected of having prostate cancer could provide misleading prognostic information in cases where only a DNA-diploid tumor is detected. These individuals could have foci of DNA-tetraploid or aneuploid tumor missed by the biopsy procedure, which could have an impact on the prognosis for the disease.

TUMOR CELL PROLIFERATION

In breast cancers, the impact of S phase as a predictor of disease progression has been extensively investigated and is generally considered a better prognostic marker in premenopausal, node-negative disease than DNA ploidy (see Chapter 14). The role of tumor S phase has not been well studied in prostatic cancers. This may be due, in part, to the fact that the majority of prognostic studies have utilized paraffin-embedded tumor samples. There is a perception that S-phase estimates from paraffin-embedded material are not reliable due to the presence of nuclear debris. While debris and aggregation can have a profound effect on S-phase estimates, mathematical algorithms that can eliminate these factors (from some of the data) are available in a number of software programs (see Chapter 3). These programs can provide reproducible S-phase estimates that are comparable to those from fresh tissues (see Chapter 8).

One report in the literature has addressed the issue of the impact of S-phase measurements on the prognosis of prostatic cancer. In a study of 74 individuals with prostate cancers (stage A, B, C, and D), Kallioniemi and coworkers (58) studied the impact of different debris-subtraction algorithms

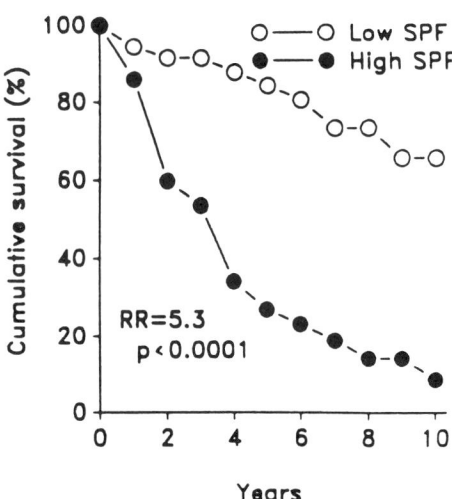

Figure 16.8. Impact of tumor S phase in predicting the relative risk (RR) of death due to prostate cancer on the basis of univariate Cox regression analysis from 74 patients with stage T1 to T4 disease (25 had M1 disease). (Reprinted with permission from Kallioniemi O-P, Visakorpi, Holli K, Heikkinen A, Isola J, Koivula T. Improved prognostic impact of S-phase values from paraffin-embedded breast and prostate carcinomas after correcting for nuclear slicing. Cytometry 1991;12:413–421.).

on S-phase measurements and their subsequent effect on survival statistics. Although debris correction decreased the DNA-diploid S phase (47 individuals) by an average of 36% and decreased the DNA-aneuploid S phase (27 cases) by an average of 24%, S-phase estimate was a significant predictor of survival with or without debris correction. As shown in Figure 16.8, the results of this study demonstrate the predictive value of S phase (shown here corrected for debris) on the survival rate of patients who have prostate cancer. The median value was used to divide all cases into high and low S phase, with the result that these two groups showed a significant difference in survival and relative risk of cancer-related death.

Studies in our lab on S-phase measurements from archival prostate cancer specimens have demonstrated that, in stage A2, DNA-diploid tumors have a lower average S-phase than DNA-tetraploid tumors. The mean S phase of 92 DNA-diploid tumors was $5.8 \pm 2.9\%$, and the mean S phase of 49 DNA-tetraploid tumors was $13.6 \pm 5.1\%$. In this same group of prostate cancer patients, DNA ploidy was a significant predictor of both disease course, survival (57). These results suggest that DNA ploidy and S phase are related and may not be independent predictors of progression in prostate cancers.

Our studies of DNA-tetraploid prostatic cancers have illustrated one important aspect of tumor S-phase determinations. Many of the cell cycle modeling programs use S-fit (59) or related algorithms (see Chapter 3) to calculate from the middle to the upper and lower boundaries of S phase. Since these algorithms begin constructing a polynomial from the 8–12 channels in the center of S phase, any events that perturb the model here can have a profound impact on the final estimate. For DNA-tetraploid tumors, triplet aggregates fall in this location, in the center of the tetraploid-tumor S phase. In our studies of archival prostate tumors, the use of aggregate correction has a significant effect on DNA-tetraploid S-phase values. For the analysis of 49 DNA-tetraploid tumors, software aggregate correction reduced the tumor S phase by an average of nearly 50%. Not all samples show this magnitude of correction. We performed manual counts of aggregates on samples before flow cytometric analysis and compare these results with software aggregation estimates. In general, the manual determination produced a higher aggregation estimate, although it is likely that this represents the bias of visual counting of low frequency (3% to 12%) events.

CORRELATION WITH PSA

The measurement of serum levels of prostate-specific antigen is an important tool in the management of prostate cancer patients. PSA is a 30 kD glycoprotein serine protease that is produced exclusively by epithelial cells in the prostate (60) and is found in (and secreted by) the normal prostate, hyperplasia, and adenocarcinoma of the prostate. Individuals with BPH have serum PSA levels generally above that of control, normal males, whereas most individuals with organ-confined cancers have serum PSA levels above that of most (though not all) cases of BPH. Although the precise role for PSA in random screening for prostate cancer is now controversial, it is clear that it is very useful in monitoring response to therapy and may be useful in predicting disease stage (reviewed in 61).

A study by Stege and coworkers (49) compared the cytosolic levels of PSA and DNA ploidy in FNA samples from 133 patients with different grade and stage prostate tumors. Their studies indicate that the level of PSA is significantly higher in DNA-diploid than in either DNA-tetraploid or aneuploid tumors. Overall, the level of cytosolic PSA decreases with tumors of increasing cytological score. In a study comparing DNA ploidy with serum PSA levels, Nativ and coworkers (62) demonstrated that patients with DNA-tetraploid and DNA-aneuploid tumors had higher serum PSA levels than individuals with DNA-diploid tumors although the range of serum PSA levels in patients with DNA-diploid tumors varied considerably, with 35% of these individuals having PSA levels within the normal range. Unfortunately, this study included stage A through D individuals, and the prognostic value of serum PSA versus DNA ploidy remains uncertain. Together with the above noted study (49), these results indicate that the amount of PSA per cell is lower for DNA-tetraploid and aneuploid than for DNA-diploid tumors and that measurements of DNA ploidy and serum PSA may provide a useful predictor for volume of disease or the likelihood of metastatic lesions.

Future Directions

Important clinical needs for prostate cancer include better markers to predict disease course in individual patients and markers to predict response to different therapies. As indicated above, studies on DNA ploidy and proliferation have provided information that is useful in predicting probable disease course. However, as indicated above, ploidy alone does not provide sufficient information to make predictions for each patient. In addition, the above cited studies are less clear on the impact of DNA-content measurements on stage A and B prostate cancers than on higher stage disease, as ploidy was shown to be a predictor of disease course in some studies, though not in others. Since prostate cancer screening studies may increase the detection of organ-confined disease or of cancers in significantly younger individuals, better markers for predicting disease course in individual patients are clearly needed.

Current models for the initiation and progression of prostate cancers have implicated a number of polypeptide growth factors (reviewed in 63), including fibroblast growth factor (FGF), transforming growth factors (TGF-a and B), and epidermal growth factor (EGF). The pioneering research by Huggins and Hodges (64) demonstrated that growth in the prostate is regulated by androgens, which characteristically control the expression of cell receptors for the above growth factors. In addition, oncogene proteins c-myc and ras have been reported to change the growth or metastatic patterns of model prostate cancer systems (65, 66). Since $p21^{ras}$ protein is uniformly expressed in the glandular epithelium of the normal human prostate (67), the significance of ras transfection experiments on in vivo tumorigenesis is unclear. It is clear, however, that growth factors, including androgens and protooncogene proteins, play important roles.

With this brief background, it would appear that the analysis of proteins that regulate cell proliferation and differentiation and their receptors would provide important targets for research on markers that could predict disease course or response to treatment in prostate cancer. A critical growth regulator of the normal prostate (and of some prostatic tumors) are androgenic steroids. Since some patients with advanced stage disease respond to androgen deprivation therapy at least partially, it is evident that androgens play some role in growth regulation even in metastatic disease. A number of studies on androgen receptor expression in prostate cancers have produced a conflicting picture. Some studies have demonstrated that nuclear-androgen-receptor content correlates with response to therapy and survival (68, 69), whereas others have shown no correlation (70). Part of the discrepancy may be due to the use of polyclonal antibodies and the now documented heterogeneity of prostatic androgen receptors (71). The use of appropriate monoclonal antibodies to physiologically relevant androgen receptors may provide a useful predictor of response to androgen deprivation. Two technical hurdles must be considered, however, before the routine application of flow cytometric measurements of these receptors is possible. First, for applications using archival paraffin-embedded prostate tumor material, it is important to demonstrate the effect of digestion techniques used to isolate nuclei on the expression of nuclear androgen receptors. The second hurdle is the potential impact of tumor heterogeneity on the predictive value of such measurements.

BLADDER CANCERS

Background

The predominant cancer of the bladder in the U.S. is transitional cell cancer (TCC), which make up more than 80% of all cancers of the urinary bladder. Adenocarcinoma of the bladder accounts for approximately 10% and squamous cell cancers add up to approximately 5% of all bladder cancers in the U.S., although squamous cell cancers of the bladder predominate in areas of the world where Schistosomiasis is common. Since transitional epithelium forms the predominant cell type lining the urinary tract from the renal pelvis to the prostatic urethra, transitional cell cancers can arise in the kidney, bladder, prostate, or ureter. TCC of the bladder is the sixth most common cancer in the U.S., with more than 50,000 new cases reported in 1990 and approximately 10,000 deaths attributable to TCC that year. TCC of the renal pelvis and ureter are uncommon, making up less than 1% of all cancers of the urinary tract.

Epidemiological studies have implicated a number of factors in the development of bladder cancer. Exposure to industrial chemicals used in dyestuffs was noted as early as the 1890's as increasing the risk of developing bladder cancer (72). Dietary sweeteners, excessive ingestion of analgesics containing phenacetin, tobacco use, coffee intake, and excessive exposure to automotive exhaust fumes have also been related to increased risk. The apparent long latency period, combined with multiple risk factors, makes a precise definition of relative risk for each of these difficult to predict.

Bladder, and other cancers, are believed to arise as a consequence of initiation (irreversible lesions in DNA) and promotion (continued division or proliferation). One or both steps in the process may be reversible and progression toward neoplasia may require cumulative, added effects of one or more initiating/promoting agents. While many of the environmental agents listed above may be promoting or initiating agents, their role in bladder tumorigenesis is largely unknown. The type of agent influences the type of cancer, as evidenced by the fact that prolonged physical irritation (stones, catheters, or calcified Schistosome eggs) generate squamous cell carcinomas.

A variety of different types of lesions are seen in the human bladder, which may represent different stages of initiation/promotion cycles. Flat lesions (carcinoma in situ) and noninvasive papillary tumors of the transitional epithelium (see below) may represent initiated and promoted cells that have not been stimulated to progress (73). Tumor foci are hypothesized to arise within a region of the epithelium that has undergone initiation/promotion. The concept of a "field

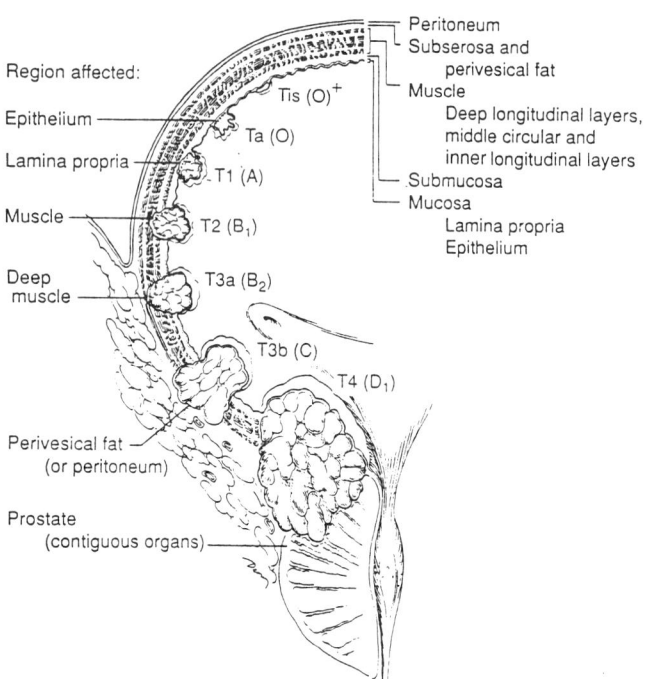

Figure 16.9. Schematic drawing showing the Jewett and Strong (77) staging system, modified by Marshall (78), compared with the UICC staging system for bladder cancers. (Reprinted with permission from P.C. Carroll. Urothelial carcinoma: cancers of the bladder, ureter and renal pelvis. In: Tanagho EA and McAninch JW, eds. Smith's General Urology, 13th ed. E. Norwalk, CT: Appleton & Lange, 1992:343.)

change" for large tissue areas in the bladder (74) may explain why superficial tumors can recur in some individuals after successful removal (or chemotherapy) of the primary tumor. According to this model, tumor foci can arise at multiple sites (or several times) within the field of the transitional epithelium that underwent prior induction/promotion. However, one difficulty with this hypothesis is the apparent clonality of some transitional cell tumors. Karyotypic studies indicate that marker chromosomes are frequently seen in high-grade (75) or invasive (76) transitional cell cancers, and are less frequently demonstrated in superficial tumors. Superficial lesions could have characteristically less evident cytogenetic abnormalities (point mutations, deletions) and could contain multiple subclones of the tumor in different regions of the urothelium. Whether independently arising foci of superficial tumors are genetically identical remains to be determined.

Pathological and Clinical Assessment of Bladder Cancer

Two different staging systems are commonly used for bladder cancers. The first, originally proposed by Jewett and Strong (77) and later modified by Marshall (78), is based on the depth of penetration into the bladder wall and the presence of lymph node or distant metastases. The second system is the TMN staging system as proposed by the Union Internationale Centre le Cancer (UICC). Comparisons of these staging systems are demonstrated in Figure 16.9. Tumors extending into the bladder lumen, and either confined entirely to the bladder epithelium (stage Ta or O) or extending only into the lamina propria (stage T1 or A), are considered to be superficial tumors. Tumors invading into or through muscle (stage T2-3 or B-C) or extending to local nodes (T4Nx) are considered invasive tumors. Carcinoma in situ (Tis), or flat nonpapillary lesions, are generally confined histopathologically to the bladder surface. However, they are recognized as distinct from superficial papillary tumors and frequently progress to invasive disease.

Approximately 70% of all transitional cell cancers of the bladder present as superficial disease (stage Ta, T1, or Tis), and roughly 30% initially present as muscle-invasive disease (stage T2, T3, or T4). Individuals with invasive disease have a significantly poorer prognosis. Superficial disease, however, shows considerable variability in its course. Although 50% or more of the individuals initially presenting with superficial papillary lesions (stage Ta or T1) will have one or more recurrences of their tumor, only 10% to 15% will progress to muscle-invasive cancer (79).

Tumor grade shows a strong correlation with tumor recurrence and progression to invasive disease. Grade 1 tumors show low frequency of progression (10% to 20%), while grade 2 tumors (19% to 37%) and grade 3 tumors (33% to 67%) have increasingly higher frequencies of progression (80). Survival is similarly related to tumor grade, with low-grade tumors showing high overall 10-year survival (>95%) and high-grade tumors showing significantly lower 10-year survival (35%) (81).

Sample Acquisition and Processing

FRESH BLADDER TUMOR SPECIMENS

In a study of the affects of different tissue disaggregation techniques, Smeets and coworkers (76) compared the effects of mechanical scrapping and cutting of transurethral resections of stage Ta through T4 bladder tumors to treatment with collagenase. Their results, from the comparison of 42 tumor specimens, demonstrate that mechanical disaggregation is the better technique. In 19% of these tumor specimens, DNA-aneuploid cells were found in the mechanically dissociated specimen and not in the specimen disaggregated with collagenase. Comparison with other enzymatic digestion techniques (trypsin, pepsin, etc.) have not been reported.

BLADDER WASHINGS AND URINES

As detailed below, exfoliated bladder tumor cells in urine and bladder washings (barbotages) offer a unique opportunity to sample the bladder for the presence of tumor cells. This application is particularly useful in a monitoring program for patients with a previous history of bladder cancer. Two important issues have been addressed in previously reported studies, namely: the appropriate technique(s) to fix exfoliated bladder cells for flow cytometric analysis, and whether urine or bladder washings provided a useful sample for monitoring bladder cancers.

As detailed in Chapter 10, cell fixation conditions have a considerable impact on DNA-content measurements, including CV's, amount of debris, and selective cell loss. The majority of reports for bladder washes or urines have used either alcohol- (or a combination of alcohol plus acetone) or formaldehyde- (formalin) based fixatives or hypotonic cell treatments, which swell cells or release stabilized nuclei. Ratliff and coworkers (82) compared the results of 99 bladder washings fixed either with 70% ethanol or treated with Triton X-100 to lyse cells and release nuclei. Ethanol-fixed samples had a higher false-positive rate (11% vs. 6%) and a lower false-negative rate (29% vs. 36%).

In a comparison of alcohol (plus acetone) fixation to hypotonic cell treatment, Deitch and coworkers (83) demonstrated that the diagnostic accuracy of alcohol-fixed urines and bladder washes was significantly better than paired samples prepared with their hypotonic technique. Fixed cells also demonstrated lower CV's, less debris, and possibly less tendency to show a broadening of the upper side of the G_0/G_1 peak (possibly due to ethanol's fixative effect on neutrophils or monocytes (see Chapter 8)). An additional benefit from ethanol fixation is that cells are suitable for DNA-content analysis after storage in ethanol at 4°C. A multiinstitutional study has demonstrated that although there is some loss of DNA stainability with propidium iodide, ethanol-fixed bladder wash or urine cell samples frequently give similar flow cytometric results when shipped rapidly to other institutions (84). These authors also concluded that the increase in debris and broadening of CV's seen in this study make accurate measurements of DNA index or hyperdiploid fraction unreliable for samples held in ethanol for prolonged periods.

In an earlier study, a low rate of unsatisfactory histograms (3/114 bladder washings) was claimed for the hypotonic cell-swelling technique (85) as compared to values reported in the literature for other fixation techniques. As discussed in that report, a number of other factors influence histogram quality, including the type of specimen (urine vs. washing), presence of interfering cells (RBC's, granulocytes, etc.), and the interval between sample acquisition and fixation or processing. The issue of time interval before fixation is critical, and one study by Konchuba and coworkers (86) demonstrates the negative impact of as little as 3 hrs in delaying fixation.

Techniques that swell or alter the cytoplasmic membrane without fixation (or lyse the cell to release intact nuclei) could restrict multiparameter analysis of DNA content plus cytoplasmic (i.e., cytokeratins) or cell membrane antigens, which are useful in identifying transitional cells and their cancers (see below). In addition, the use of techniques that minimize centrifugation steps improve the quality of the data and significantly reduce cell loss. In our lab, the use of a "no-wash" staining technique for simultaneous flow cytometric analysis of DNA content plus cytokeratins (see Chapter 10) reduced the number of "unacceptable" histograms

(in our study, samples lacking any cytokeratin-positive cells) significantly, compared to cell preparation techniques requiring multiple wash steps.

The issue of whether urine or bladder washings provide useful samples for flow cytometric analysis is not trivial. In most published studies, bladder washings have been used because they are generally perceived to provide more cells and to result in fewer unsatisfactory histograms. Urine would offer a more attractive sample due to the need to catheterize patients to obtain bladder washings. One published study comparing the results of DNA-content analysis on 89 paired urines and bladder-wash samples using a cell-swelling technique (85), demonstrated that the detection rate for 39 cases known to contain transitional cell carcinoma was similar for both types of samples (21 for urines versus 23 for washings). Konchuba and coworkers (86) compared the results of ethanol-fixed paired urine and bladder-wash samples from 77 patients. The rates reported for poor histogram results were significantly lower (3%) for bladder washings than for urines (12%) and the correlation for positive cytologies was higher for bladder washes (12/17) than for urines (4/8).

DNA-Content Measurements of Bladder Tumor Specimens

TRANSITIONAL CELL CANCERS

Fresh Tumor Specimens (TURBT and Cystectomy Specimens). Over the past 20 years, the analysis of specimens from primary tumors of the bladder using both image and flow cytometry has established that changes in DNA-content (DNA aneuploidy) are frequently seen in transitional cell cancers (87, 88). Many reports have focused on the relationship of DNA-content abnormalities with either clinical stage or pathological grade. These have demonstrated that the majority of stage Ta or T1 lesions are diploid, while the majority of T3 or T4 lesions are DNA aneuploid; similarly, low-grade lesions are predominantly DNA diploid while grade 3 lesions are generally DNA aneuploid.

The predictive value of DNA-content measurements are thus limited. Individuals presenting initially with invasive (stage 3-4) lesions almost always have DNA-aneuploid tumors and progress rapidly. The majority of individuals with superficial papillary lesions (generally DNA diploid) will experience recurrence, but not progression of the cancer.

Archival Bladder Tumor Samples. DNA-content measurements of archival bladder tumors have been limited, perhaps due to the results of many earlier studies, which indicate that DNA ploidy alone was a relatively poor predictor of disease course. The earliest reported study of archival paraffin-embedded TCC samples (97) investigated the clinical value of DNA-content analysis. The authors compared the presence of a DNA-aneuploid population to histological grading in a study of 64 samples from patients with grade 1 to 3 lesions. As noted in studies with fresh samples, most grade 1 samples were DNA diploid (18/19), whereas DNA-aneuploid tumors were seen with increasing frequency for

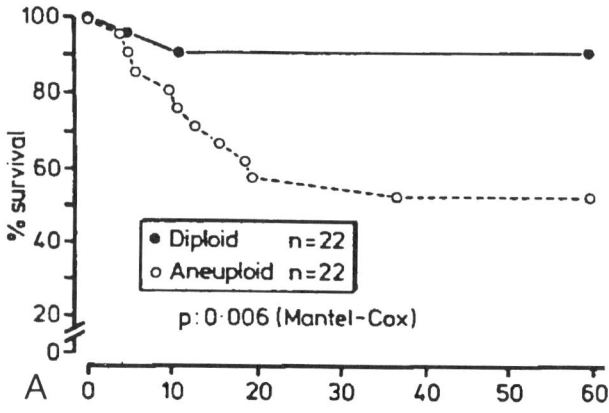

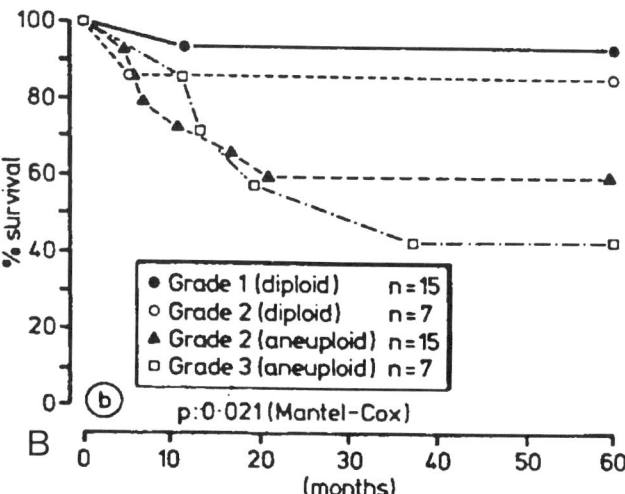

Figure 16.10. Value of DNA ploidy in predicting survival of transitional cell cancers of the bladder. The results shown here from Blomjous, et al. (100) in a retrospective analysis of 44 patients (**A**), show that DNA ploidy was a significant predictor of survival for these individuals. However, as shown in **B**, all patients with grade 1 tumors were DNA *diploid*, while all grade 3 tumors were DNA *aneuploid*. For patients with grade 2 tumors, DNA ploidy was a significant predictor of survival.

diploid tumors progressed. In addition, four individuals considered at risk for progression on the basis of DNA-content histograms did not progress, although the authors argue that the follow-up in some of these cases was too short (<2 years). The use of hyperdiploid fraction (HDF), as in this study, for the analysis of samples from paraffin-embedded material needs careful validation to eliminate or control for the impact of debris and nuclear aggregation (see Chapter 8). Since these controls were not reported, the validity of the HDF discriminator in samples from paraffin-embedded materials is questionable.

A study reported by Blomjous and coworkers (100) demonstrated a predictive value of DNA-content flow cytometry for individuals with DNA-diploid versus DNA-aneuploid tumors. As shown in Figure 16.10, in this study all individuals with grade 1 tumors were DNA diploid and all grade 3 tumors were DNA aneuploid. The only predictive impact was for individuals with grade 2 TCC, where individuals with DNA-diploid grade 2 tumors had significantly longer survival than individuals with DNA-aneuploid tumors at presentation. In this respect, the DNA-diploid tumors acted more like grade 1 tumors and the DNA-aneuploid tumors acted more like grade 3 TCC. Roughly 40% of the individuals in this study had advanced stage (T2 to T4) disease and, although patients with advanced disease received different therapies, no analysis was reported to identify the impact of therapy (or stage) on survival.

Tumor Cell Proliferation. The relationship of histopathologic markers or of disease course to bladder tumor proliferation has been studied by a variety of techniques. Using stathmokinetic techniques and manual counts of mitotic figures, Fulker and coworkers (101) established that the potential doubling time of well-differentiated tumors (22 days) was significantly lower than that of poorly-differentiated tumors (six days). Similar results have been obtained using manual enumeration of proliferating cells following labeling with Ki-67 antibodies (102), which demonstrated a lower percentage of proliferating cells in low-stage tumors (4.3%) compared to invasive lesions (12.3%) and a similar relationship for proliferation in low-grade (4.4% for grades 1 and 2) vs. high-grade (12.2%) tumors.

One of the earliest studies using flow cytometry to analyze cell proliferation of bladder tumor cells and normal bladder urothelium was reported by Tribukait and coworkers in 1979 (103). This study demonstrated a significant difference in S phase for normal mucosa (3.2% to 4.3%) and DNA-diploid tumors (6.4%), with DNA-aneuploid tumors showing the highest average S phase (17%). This study suggested a relationship between tumor grade and S phase, with grade 2 tumors having a lower average S phase than grade 3 tumors. Since all grade 3 tumors in this study were DNA aneuploid and 33% of the grade 2 tumors were DNA aneuploid, it is not clear if the relationship between S phase and grade shown here might not simply reflect the fact that DNA-aneuploid tumors overall have a higher average S phase. An important point raised by these authors was the

grade 2 (4/20) and grade 3 (21/24) individuals. The authors indicated that ploidy did not predict clinical course for any grade, although actuarial survival data were not calculated. Similar results relating DNA ploidy to tumor grade and stage have been reported by Coon and coworkers (98) using archival bladder tumors for flow cytometry.

In a retrospective study of individuals presenting with superficial (Ta) lesions, the presence of "unfavorable" histograms (presence of a DNA-aneuploid population or a hyperdiploid fraction) predicted tumor recurrence in 9/15 individuals (99). Six of the nine individuals with subsequent recurrences had DNA-diploid tumors (and a hyperdiploid fraction <20%) at presentation. In this study, an "unfavorable" histogram correctly predicted progression to invasive disease in eight out of 14 cases; six individuals with DNA-

potential role of tumor proliferation as a biological marker. Due to the lack of patient follow-up, it was impossible to determine if S phase was an independent predictor of disease course.

In a subsequent study of 100 previously untreated patients, Tribukait and coworkers (104) reported the relationship of tumor grade and stage with DNA ploidy and S phase. As in other reported studies, low-grade or stage tumors were predominantly DNA diploid, whereas grade 3 (or invasive-stage) tumors were DNA aneuploid. The relationship of tumor S phase with DNA index (DI) demonstrated the interesting finding that S phase was lower for DI 2 (DNA-tetraploid) tumors, than for either DI 1.5 to 1.7 or DI >2.2 tumors. Although patient follow-up was not reported in this study, these results are of interest because subsequent studies have reported the survival of patients with DNA-tetraploid bladder tumors is significantly higher than the survival of those with DNA-aneuploid tumors with a DI from 1.3 to 1.7 or a DI >2.2.

In two reports, Tribukait and coworkers have reported the results of studies of the relationship between DNA ploidy and S phase for paired samples from carcinoma of the bladder and random mucosal biopsies (104, 105). These studies indicate that for grade 2 or 3 tumors, there is a tendency for histologically normal-appearing bladder mucosa to show DNA aneuploidy, with ploidy values in the mucosal samples approximately twice that of the paired tumor. In contrast, the average S phase for the mucosal samples was less than that of the tumors. These results support the "field effect" concept and indicate that large areas of the urothelium are in a premalignant (or possibly transformed) state in many individuals with high-grade tumors. As the authors indicate, these results suggest that proliferation rates in the histologically normal-appearing urothelium near the tumor may play a critical role in the rate of development (or probability for developing) subsequent tumors following the eradication of the original lesion.

Flow cytometric analysis of bladder tumors using DNA content plus intracellular antigens has demonstrated the potential utility of multiparameter analysis. Using antibodies to cytokeratins in conjunction with DNA-content analysis, Feitz and coworkers (107) demonstrated the improved sensitivity obtained with cytokeratin-antibody staining in detecting DNA-aneuploid bladder tumors, and the role of gating the tumor analysis on cytokeratin-positive cells on tumor S-phase determinations. Using in vitro labeling with BrdUr and dual-parameter DNA content plus anti-BrdUr flow cytometry, Tachibana and coworkers (108) demonstrated a low labeling index (5.1%) for grade 1 tumors, intermediate labeling (8.9%) for grade 2, and a high average index (15.2%) for grade 3 tumors. This study also demonstrated a significantly higher proliferation for invasive (18.7%) versus superficial (8.6%) tumors.

TCC of the Upper Tracts. Two reported flow cytometric studies of TCC of the upper tracts have been paraffin-embedded tumor material and retrospective analysis of the role of DNA ploidy or histopathological criteria as predictors of disease course (109, 110); one study utilized fresh tumor material (111). In a study of 111 patients with TCC of the renal pelvis, Blute and coworkers (109) indicated that DNA ploidy did not provide any additional information in predicting disease course compared with pathological grade and stage for patients with high-stage and/or grade disease. However, individuals with grade 2 DNA-aneuploid tumors had a higher incidence of tumor-related deaths than individuals with DNA-diploid tumors.

In a study of 127 previously untreated individuals with TCC of the ureter and renal pelvis (without any evidence of bladder TCC), Corrado and coworkers (110) used multivariate analysis to test the value of DNA ploidy and histopathological criteria as predictors of disease course. Although the results of this study demonstrate a good overall correlation for DNA ploidy with tumor grade or stage, DNA ploidy was of only marginal significance in predicting outcome. Unlike the results of the study by Blute and coworkers (109), this study failed to demonstrate any significant predictive value for DNA ploidy in individuals presenting with grade 1 or 2 tumors. The authors argued that the differences in the results were due to the inclusion of TCC of the ureters and the shorter follow-up period of their study compared to the previous study (109).

In a prospective study using fresh tumor material from 11 patients with TCC of the renal pelvis or ureter, Oldbring and coworkers (111) showed a positive correlation of DNA aneuploidy and tumor invasiveness. All grade 3 tumors in this study were DNA aneuploid, whereas 50% of the grade 2 tumors were diploid. However, the statistical significance of these data were not reported.

Metastatic Tumors. In a study comparing DNA content of paired archival samples of primary and metastatic bladder cancers, Badalament and coworkers (112) demonstrated that DNA content of primary and metastatic (lymph node) cancers differed in eight of 22 individuals studied. The median survival of patients presenting with DNA-aneuploid tumors (17.5 months) was double that of individuals with DNA-diploid primary bladder cancer. The authors concluded that these data indicate the existence of considerable tumor heterogeneity in TCC of the bladder, caused either by metastases of a tumor other than the primary tumor analyzed by flow cytometry or by genotypic changes at the metastatic site. An alternate possibility is that tumors considered DNA diploid actually contained too few DNA-aneuploid events to be detected, a point made more likely by the shorter average survival of individuals with primary tumors determined to be DNA diploid. Direct proof of the relationship of primary and metastatic tumors by DNA ploidy would perhaps be better performed by DNA-content measurements of tissue sections using image analysis techniques.

Response to Therapy. A number of studies have reported the use of flow cytometry to measure response to chemo- or radiation therapy of bladder tumors. Studies on the response to BCG (bacillus Calmette-Guerin) treatment

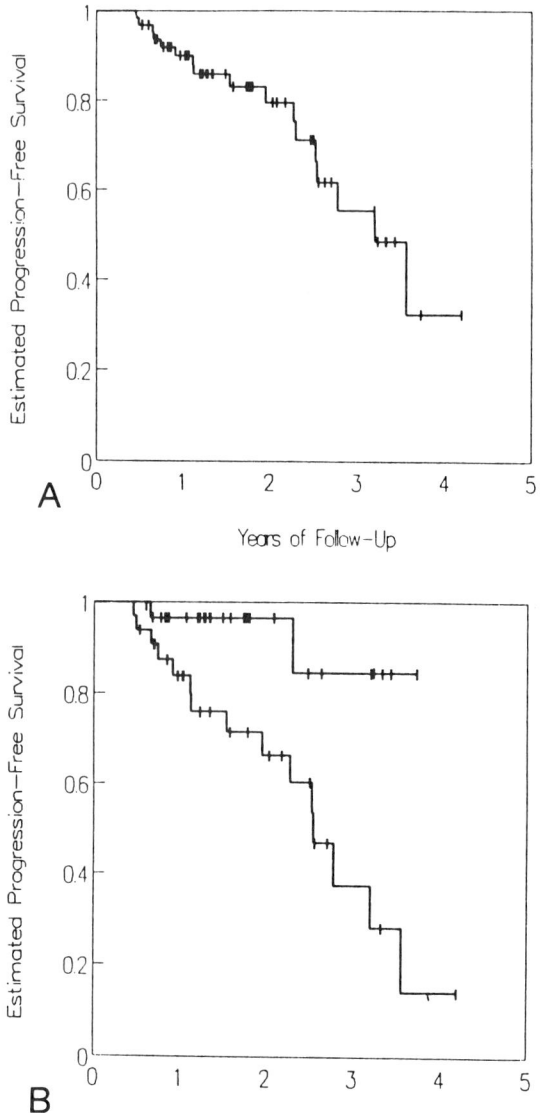

Figure 16.11. Impact of multiparameter (acridine orange) DNA-content flow cytometry on the prediction of response to BCG immunotherapy in 65 patients with bladder TCC. Comparison of the estimated distributions for survival free of progression for the general study population (**A**) with estimates for cases in which the result of flow cytometric analysis at six months was negative or positive (**B**). (Reprinted with permission from Bretton PR, Herr HW, Kimmel M, Fair WR, Whitmore WF, Melamed MR. Flow cytometry as a predictor of response and progression in patients with superficial bladder cancer treated with Bacillus Calmette-Guerin. J Urol 1989;141:1332–1336.).

have generally used a two-parameter analysis (Table 16.6), due in part to the large numbers of neutrophils found in bladder washings for several weeks post-treatment. Using the two-parameter acridine orange staining technique (see details below), the Memorial Sloan Kettering group demonstrated, in patients with low stage bladder cancers, that loss of DNA-aneuploid cells following BCG therapy is a significant predictor of response to therapy (113, 114). Subsequent studies by this group demonstrated that the acridine orange staining technique was more sensitive (though less specific) than conventional cytology in detecting response to BCG therapy (115) and that flow cytometry provided a strong predictor of response (116). As shown in Figure 16.11, individuals who were negative for tumor by flow cytometry at six months post-treatment had a significantly lower probability for recurrent disease than individuals positive for tumor at that time (116).

Using dual-parameter flow cytometry employing simultaneous measurement of DNA content and monoclonal antibodies to cell surface antigens found on urethelial cells, the Memorial Sloan Kettering group demonstrated that the use of these antibodies increased the sensitivity of the flow cytometric assay compared to single-parameter DNA-content measurements. In a study of patients with carcinoma in situ, five out of nine individuals positive for the antigen (Om5) after BCG treatment had tumor recurrences, whereas four individuals who lost cells expressing the antigen had no recurrences up 12 months after therapy (117).

In a study of response to doxorubicin or Mitomycin C treatment, Farsund and coworkers (118) measured exfoliated bladder cells from areas near bladder tumors and from cystoscopically normal-appearing mucosa from two patients with superficial disease. This study is unique because it included detailed studies of multiple areas of the bladder before and during chemotherapy. The results indicate that, in one patient with carcinoma in situ, large areas of the bladder with normal-appearing urothelium potentially exfoliated tumor cells with an identical DNA index as the tumor. Since direct sampling of the normal-appearing mucosa by biopsy was not performed, it is not possible to rule out contamination of the normal mucosal area washings by tumor cells exfoliated from the primary lesion. However, this study does provide an excellent illustration of the use of flow cytometry in monitoring the response of individual patients during chemotherapy.

A preliminary report using flow cytometry to monitor the response to radiation therapy of 28 patients receiving irradiation before cystectomy (119) demonstrated loss of DNA-aneuploid tumors in 12 patients after therapy. All patients had high-stage (T3) and grade tumors (grade 2-3) and all tumors were DNA aneuploid initially. The authors considered the disappearance of DNA-aneuploid tumors to be a potential marker to identify patients who respond to radiation therapy, although significant follow-up periods were not reported. In a study of 61 patients receiving preoperative irradiation and cystectomy, Wijkstrom and coworkers (120) were able to compare the DNA content of pretreatment and cystectomy specimens from 38 individuals. Of the 32 patients with DNA-aneuploid tumors before treatment, 10 showed only DNA-diploid events postirradiation, whereas the other 22 showed no significant change in tumor ploidy. Interestingly, of the six patients with DNA-diploid histograms pretreatment, two had DNA-aneuploid tumors postirradiation. Although the authors suggest that this is evidence of tumor heterogeneity, it is also likely that the use of bladder washings for some samples reduced the sensitivity of the single-parameter technique used here to detect DNA-aneuploid tumors.

Two studies have investigated the role of DNA ploidy measurements in predicting long-term response to radiation therapy of bladder cancers. In a study of 73 individuals with invasive TCC who received radiation (but no cystectomy), all tumors that were initially DNA diploid disappeared, whereas DNA-aneuploid tumors with one cell line disappeared in 55% of the patients and 30% of patients with multiple DNA-aneuploid stem cell lines showed disappearance of tumor (121). Although the authors indicated that patients were followed for five years, radiation response was judged at three months post-treatment, and no long-term actuarial survival information was provided. Although the results of this study are promising, the predictive value of DNA ploidy or S phase for individual patients was not established.

ADENOCARCINOMA AND SQUAMOUS CELL CANCERS OF THE BLADDER

Although primary squamous cell cancer of the bladder accounts for less than 10% of all bladder cancers in the U.S., it is the predominant bladder cancer (or the most common cancer) in some areas of the world. Although tumor grade and stage provide some useful predictive information, accurate assessment of the degree of muscle invasion is often difficult. In a retrospective study of the impact of DNA ploidy in predicting disease course, Winkler and coworkers (122) studied paraffin-embedded tumors from 73 individuals with primary squamous cell bladder cancer. Their results demonstrated no significant correlation of DNA ploidy with tumor grade, a significant correlation of DNA ploidy with tumor stage, and a highly significant correlation of ploidy and overall survival from bladder cancer. An additional finding reported here is the intermediate survival rate for individuals with DNA tetraploid tumors compared to individuals with DNA-diploid or aneuploid tumors.

In a study of 100 cystectomy patients with squamous cell carcinoma, Shaaban and coworkers (123) compared the results of DNA-content analyses of bladder washings with superficial and deep-tumor biopsies. Their results demonstrated a good correlation between tumor grade, but not stage, with DNA ploidy, with biopsy material providing better information than bladder washings in most cases. Although tumor proliferation was significantly higher for invasive compared to superficial tumors, the authors indicate that the S-phase measurement of deep vs. superficial areas of the same tumor sometimes provided different values.

A study on the predictive value of DNA ploidy for disease course in primary squamous cell carcinoma of the male urethra (124) involved a retrospective study of paraffin-embedded samples from 30 individuals. In this study, DNA ploidy was a strong predictor of local or distant metastases. Although DNA ploidy was not a predictor of disease-free survival for low-stage tumors, it was a significant predictor of disease-free survival for high-stage tumors.

Primary adenocarcinoma of the bladder is a relatively rare malignancy in the U.S., accounting for less than 2% of all bladder cancers. Tumors arising at different sites (or in extrophied bladders) are reported to have a different prognosis. Some authors report that tumor grade has a predictive value, with the signet-ring variant being an indicator of poor prognosis. Two reported studies using flow cytometry of archival paraffin-tumor material have measured the prognostic value of DNA ploidy. In a study of 36 patients with primary adenocarcinoma of the bladder, Grignon and coworkers (125) compared tumor stage, histological cell type, and DNA ploidy with disease course. Their results demonstrated that DNA ploidy did not correlate with stage, histological pattern, or disease outcome. The results of a study of archival tumor samples from 38 individuals with primary adenocarcinoma of the bladder by Song and coworkers (126) demonstrated a highly significant correlation for DNA ploidy with disease course. These authors concluded that DNA ploidy was a better indicator of disease course than stage or histological pattern and that DNA ploidy was the most significant predictor of disease course. The reason(s) for the complete discrepancy between the results of these two studies is not clear. Although the staging systems used for both studies was different, the data provided suggest that more individuals with higher-stage tumors were involved in the study by Song and coworkers (126).

Flow Cytometry of Exfoliated Bladder Cancer Cells

SINGLE-PARAMETER DNA CONTENT FLOW CYTOMETRY

Exfoliated bladder cancer cells have been used for urine cytology to detect bladder tumors, as previously noted. The additional sensitivity of flow cytometry, coupled with the ability to rapidly analyze thousands of cells, has made DNA-content flow cytometry an attractive technology to detect bladder cancers. Single-parameter DNA content analysis has been performed on bladder washings since the late 1970's (127) and several early studies addressed the issue of whether bladder washing provided an appropriate sample source (128, 129). Although urines provide a more attractive sample source (they do not require catherization or cystoscopy), the majority of flow cytometry studies have employed bladder washings, or barbotages. Most (though not all) studies have demonstrated that bladder washings provide more cells and a better quality sample (less debris, lower CV's). It is likely that the force of filling and draining the bladder in a washing physically removes greater numbers of tumor cells, possibly due to reduced cell-to-cell adhesion in TCC.

As indicated in Table 16.5, a number of studies have addressed the sensitivity and specificity of DNA-ploidy measurements in the detection of bladder cancers. Overall, these results demonstrate that single-parameter DNA flow cytometry (the majority of these studies using bladder washings) detects 50% or more cytologically or cystoscopically proven

Table 16.5
Single-Parameter DNA-Content Analysis of Exfoliated Bladder Tumor Cells

Author (reference)	No. of Samples or Patients	Definition of "Positive Histogram"	Detection Rate	False-Pos.	Sample Population
Gustafson (94)	229		76% of patients	NR	Ta-1 patients
Ratliff (82)	99	DNA-aneuploid peak or G_0/G_1 <85%, or 4C >15%	71% for EtOH-fixed 64% for TX-100-fixed	11% (EtOH) 6% (TX-100)	Symptomatic patients
deVere White (130)	178	DI > 1.1, or HDF[a] >15%	78% FCM compared to cytology 91% FCM plus cytology	38%	No prior history
Murphy (131)	133	DNA-aneuploid peak or HDF >17%	78% FCM only (95% FCM plus cytology)	17%	105 biopsy pos. + 28 controls
Badalament (134)	228	DNA-aneuploid peak or HDF >16%	49% for single sample 66% for multiple samples	NR	All biopsy pos.
Hadjissotiriou (135)	59	DNA-aneuploid peak or 4C >15%, or HDF >15%	(DNA aneuploidy related to grade and stage)	NR	All active TCC
Cowan (132)	301	DI > 1.05 or < 0.95 or multiple peaks	78% of cytology pos. samples (51 cytology positive cases)	11%	Symptomatic patients
Klein (133)	286	DNA-aneuploid peak or 4C >15%	91% FCM compared with cytoscopy 99% FCM plus cytology	4%	Asymptomatic plus 74 known or prior history
Koss (136)	71	DNA-aneuploid peak or 4C >10%, or HDF >15%	81% (FCM or image pos.) compared to cytology 58% DNA-aneuploid recurred (3 yrs)	36%	Prior history
Hermansen (84)	10/50	DNA-aneuploid peak (Interlab. study by 5 labs. using 10 patient samples)	41/50 overall (82%)	—	All biopsy pos.
Norming (193)	63	DNA-aneuploid peak	97% (Prospective study of predictive value of flow for disease progression)	NR	Gr 3/Tis

[a] Hyperdiploid Fraction

tumors. Important issues must be noted that make these studies difficult to compare. Perhaps the most important is the definition of a positive sample. Some of these studies have compared the results of flow cytometry to cytology (130, 132), or have used cystoscopically proven cancers (133). Cystoscopically proven tumors can have negative cytologies (133), and cytologically positive samples can come from patients whose tumors have not been seen cystoscopically (which is particularly true for upper-tract tumors). If a patient has a prior history of tumor, it is likely that the urologist examining the bladder will increase the level of surveillance. The ability to detect tumors, either cytologically or cystoscopically, is dependent on the skill of the individual pathologist or urologist. Finally, if a flow cytometry sample clearly shows the presence of a DNA-aneuploid TCC when, at the same time, cytology and cystoscopy are negative, and subsequent sampling demonstrates histologically positive TCC, was the earlier flow cytometry result a false-positive?

An additional issue raised by the studies in Table 16.5 is the source of specimens. As indicated, some studies included only known positive specimens (134–136), whereas others included a mixture of known TCC patients plus asymptomatic individuals (133) or utilized a true screening test, including all symptomatic individuals (130). Perhaps the best test of the sensitivity and specificity of a flow cytometric assay to detect tumors would be the inclusion of all appropriately symptomatic individuals (hematuria), with a blind analysis of which patients have a current or previous history of TCC.

A final issue raised by the studies in Table 16.5 is the definition of a "positive histogram." As seen from Table 16.5, different authors use different definitions. Perhaps the most troublesome aspect of many of the studies listed in Table 16.5 is the use of the hyperdiploid fraction to detect TCC. As detailed below, this concept was developed earlier by the Memorial Sloan Kettering group for their two-parameter acridine orange staining technique. Since the single-parameter propidium iodide staining techniques used in the studies listed in Table 16.5 do not eliminate granulocytes, monocytes (see Chapter 8), squamous epithelium, or aggregates (all of which artifactually increase the percent events above the diploid mean channel), the HDF alarm can clearly increase false-positives while it increases the detection of TCC in samples lacking identifiable DNA-aneuploid populations. This point is demonstrated by the low-false-positive rate seen in the dual-parameter technique as reported by the Memorial Sloan Kettering studies (from 2% to 6%, see Table 16.6) compared to false-positive rates as high as 38% for the single-parameter studies shown in Table 16.5.

MULTIPARAMETER FLOW CYTOMETRY

Both urine and bladder wash specimens frequently contain large numbers of red blood cells, granulocytes, squamous epithelial cells, macrophages, and debris, in addition to any exfoliated bladder urothelial cells. With luck (and a good quality specimen) there may be sufficient numbers of bladder tumor cells for detection, provided the tumor has a DNA content that is recognizably distinct from other diploid cells present.

Table 16.6
Multiparameter DNA-Content Analysis of Exfoliated Bladder Tumor Cells

Author	Parameters	No. of Samples or Patients	Detection Rate	False-Pos.	Significance
Collste (138)	DNA/RNA[a]	107	87% overall	6%	1° report 2P technique for clinical sample; noted need for additional marker for dipl. or near-diploid, or rare aneuploid TCC.
Collste (139)	DNA/RNA				Incidence of DNA aneuploidy higher in bladder washes than in biopsies.
Devonec (143)	DNA/RNA	26	NR	NR	Demonstrated utility of FCM + cytology in surveillance.
Devonec (141)	DNA/RNA	110	100% of cytology pos.	14%	DNA aneuploidy increases with stage, good correlation with cytology.
Klein (140)	DNA/RNA	400	81%	2%	Utility of FCM in monitoring disease course or detection in high-risk patients.
Klein (142)	DNA/RNA	>500	88/98%	NR	Detection of Tis with papillary Ca/or Tis as flat lesions. Aneuploid tumors other than 4C are more aggressive.
Klein (146)	DNA/RNA	100	—	2%	Test of false-pos. rate on normal individuals or nonneoplastic diseases of bladder.
Staino-Coico (114)	DNA/RNA	22	(50%)	(28%)	Monitoring BCG-treated patients. 6/12 recurring patients predicted by FCM.
Badalament (115)	DNA/RNA	29	72%	(19%)	103 samples pre- and post-BCG treatment. FCM pos. in 18/25 recurrences. HDF elevated due to granulocytes in 71% of no evidence of disease (NED) patients.
Tetu (147)	DNA/RNA	249	85%	36%	129 patients with prior history of TCC. Some prior treatments make FCM less reliable.
Bretton (116)	DNA/RNA	65	—	NR	Neg FCM at 6 months after BCG treatment is a strong predictor of response; FCM pos. at 6 months is a strong predictor of progression.
Hermansen (149)	DNA/RNA	64	64%	NR	Detection of TCC in 48 women, FCM detection rate lower in women (squamous cells).
Huffman (156)	DNA/Ck[b]	28	86%	NR	1° report of use of cytokeratin antibodies with bladder washes. Ck+ gate removes large % of HDF events.
Orihuella (145)	DNA/RNA + ABH Ag[c]	81	—	NR	DNA ploidy (FCM) plus ABH (slide test) to predict BCG response. Recurrence greatest in aneuploid/ABH neg.
Huffman (117)	DNA/Om5[d]	15	87%	(7%)	Tis patients treated with BCG, 13/15 Om5+ pre-Tx 5/9 Om5+ after Tx had recurrence, 4 Om5- had no tumor at 12 mos. after BCG.
Bretton (144)	DNA/T16[d]	30	95%	30%	Patients with prior TCC. T16 antibody increased sensitivity (with HDF) and decreased specificity.
Wheeless (162)	Slit-scan lesions correctly classified	153	97%	11%	Unique X-Y-Z axis DNA analysis. 66/69 grade 1-2.

[a]Acridine Orange metachromatic staining
[b]Cytokeratins
[c]ABH blood group antigens
[d]Monoclonal antibodies to cell surface antigen

The detection of bladder tumor cells involves, in part, a simple signal-to-noise problem. The pioneering work of the Memorial Sloan Kettering group (137, 138) addressed these problems with the use of a multiparameter flow cytometry technique. Using metachromatic acridine orange (AO) staining of DNA and RNA (DNA stains green and RNA stains red) coupled with pulse-width analysis to gate out granulocytes and squamous epithelial cells, this approach is highly effective in detecting high-grade TCC, carcinoma-in-situ (usually DNA aneuploid), and some papillary TCC lesions (139, 140). In addition, the technique has been utilized to monitor the impact of drug therapy (147), and to monitor surgical patients postcystectomy (148). As outlined in Table 16.6, the Memorial Sloan Kettering group has published extensively the results of its experience with this dual-parameter technique, which it has used to detect bladder tumors (138, 141, 142), monitor disease recurrence (140, 143), monitor response to therapy (114, 115, 144, 145), and determine the impact of squamous epithelial cells on the sensitivity for detecting TCC of the bladder (149).

The AO staining technique is technically rigorous, requires a flow cytometer with pulse-width analysis, and has proven to be difficult to reproduce in some laboratories. However, this dual parameter has a low false-positive rate (146), generally in the 4% to 6% range for this group (excluding studies monitoring BCG-treated patients). The original technique was subsequently modified to include an analysis of the hyperdiploid fraction. This provides a statistical alarm that significantly increases the sensitivity of the AO technique in detecting relatively small numbers of DNA-aneuploid cells or, possibly, an increase in diploid cell proliferation above a normal value. While some studies using a single-parameter propidium iodide stain have relied on the visual inspection of histograms to detect an "inflammatory tail" (133), the subjective nature of this discriminator in single-parameter methods (and the lack of quantitative values) suggests that multiparameter techniques are likely more reliable, sensitive, and reproducible.

Studies of changes in the expression of blood group antigens (150, 151) have suggested that loss of ABH antigen or T- (Thomsen-Freidenreich) antigen expression correlates with high probability of recurrence or progression in superficial tumors or carcinoma in situ. In a prospective study combining acridine-orange-stained cells analyzed by dual-parameter DNA/RNA by flow cytometry with cell surface blood group ABH antigen measurements (using erythrocyte adherence to tissue sections), Orihuela and coworkers (145) analyzed samples from 81 patients with recurrent superficial bladder cancers before and after TUR or BCG treatment. In this study, patients with initially favorable markers (DNA diploid and ABH positive), or who converted to this marker pattern after treatment, had the highest rate of response to therapy. To some extent, these markers were independent of tumor stage or grade. Studies using stimultaneous measurements of DNA and ABH-antigen content by flow cytometry (using FITC-labeled lectins or antibodies to blood group antigens) have been reported (152). This study utilized primary bladder tumor samples and demonstrated some of the technical difficulties of this direct-binding technique, including the need to treat some samples with neuraminidase to obtain positive signals and the apparent reduction in lectin binding in G_2M compared to G_0/G_1 cells.

Antibodies to membrane antigens on bladder cancer cells have been extensively reported in the literature. As indicated in Table 16.6, antibodies developed at the Memorial Sloan Kettering (Om5/T16) have been successfully used with multiparameter flow cytometry to monitor response to BCG. The further development of these antibodies has been reported by Fradet and coworkers (153, 154). Antibodies to some antigens show considerable promise, as they appear to be expressed on low-grade papillary lesions (which are usually DNA diploid) and not on normal transitional cells.

Another multiparameter flow cytometric approach is the use of simultaneous DNA content and cytokeratins, pioneered by Ramaekers and coworkers (155). This technique takes advantage of the epithelial-cell-specific expression of cytokeratin molecules to gate the DNA-content analysis on normal and tumor cells found in exfoliated bladder cell samples. Antibodies to cytokeratins have been used for multiparameter flow cytometric analysis of bladder washings (156) and fresh tissue samples from primary bladder tumors (157–159).

The normal bladder urothelium contains cells expressing cytokeratins #4, #8, and #7 (type I), and 13, 17, 18, and 19 (type II). The pattern of cytokeratin-pair expression at different cell layers of the bladder urothelium is controversial and may be complicated by cross-reactions of some monoclonal antibodies with more than one cytokeratin molecule. Normal basal cells express cytokeratins #13 and #4, whereas intermediate-level cells express cytokeratins #7 and #17 (160). All cell layers of the normal urothelium express cytokeratins #8, #18, and #19, and the superficial cells (umbrella cells) contain only the #8 and #18 pair (161).

Ramaekers and coworkers used a unique antibody to cytokeratin #18 (RGE-53) to demonstrate tumor staining patterns that correlate with bladder tumor stage. Using immunohistochemical techniques, they demonstrated that, in low-grade papillary tumors and normal urothelium only, superficial cells stained with RGE-53; in high-grade, noninvasive tumors, clusters of RGE-53 positive cells were seen scattered throughout the tumor, whereas invasive cells were all RGE-53 positive (158). Using dual-parameter flow cytometry of bladder tumor cells stained with either RGE-53 or a pan-cytokeratin antibody in conjunction with DNA-content measurements, this group reported an increasing ratio of cytokeratin 18 to total cytokeratins as the grade of the tumor increased (157). A similar increase in the %cells staining with RGE-53 antibodies has been reported for increasing grade bladder tumor cells using dual-parameter flow cytometric measurements (159). The authors of this study argued that the expression of cytokeratin 18 measured flow cytometrically is a marker for tumor aggressiveness, as increases in %positive cells coincided with tumor grade and with DNA ploidy. They also argued that mean fluorescence intensity decreased for increasing grade tumors, although the high variability in staining makes the results for different grades or DNA ploidy not statistically significant. Although the results of Ramaekers and coworkers suggest that the percentage of RGE-53-positive cells is dependent on tumor invasiveness, the influence of tumor stage on %positive cells or RGE-53 fluorescence intensity was not reported in this study.

The final technique shown in Table 16.6 is the use of "slit-scan" flow cytometry instrumentation developed by Wheeless and coworkers (162). The instrumentation required for this technique scans particles along three separate axes, and provides spacial information that is not available from normal instrumentation. Although it provides improved sensitivity in detecting TCC from bladder washings, the flow cytometer is sophisticated and expensive (and not available commercially). As pointed out by these authors, the use of some monoclonal antibodies to cell surface markers specific for TCC may provide an equally sensitive and specific technique that can be performed on commercially available instruments (162).

PRACTICAL CONSIDERATIONS FOR BLADDER WASH FLOW CYTOMETRY

A number of factors influence the sensitivity and specificity of DNA-content measurements of exfoliated bladder tumor specimens. Our experience has demonstrated several key points, including: (a) type of sample used, (b) timing of fixation, (c) diploid DNA content controls, and (d) the cell-staining technique used.

Sample Source: Urine vs. Bladder Washing (Barbotage). Previous studies by others have compared the results of single-parameter DNA content flow cytometry on

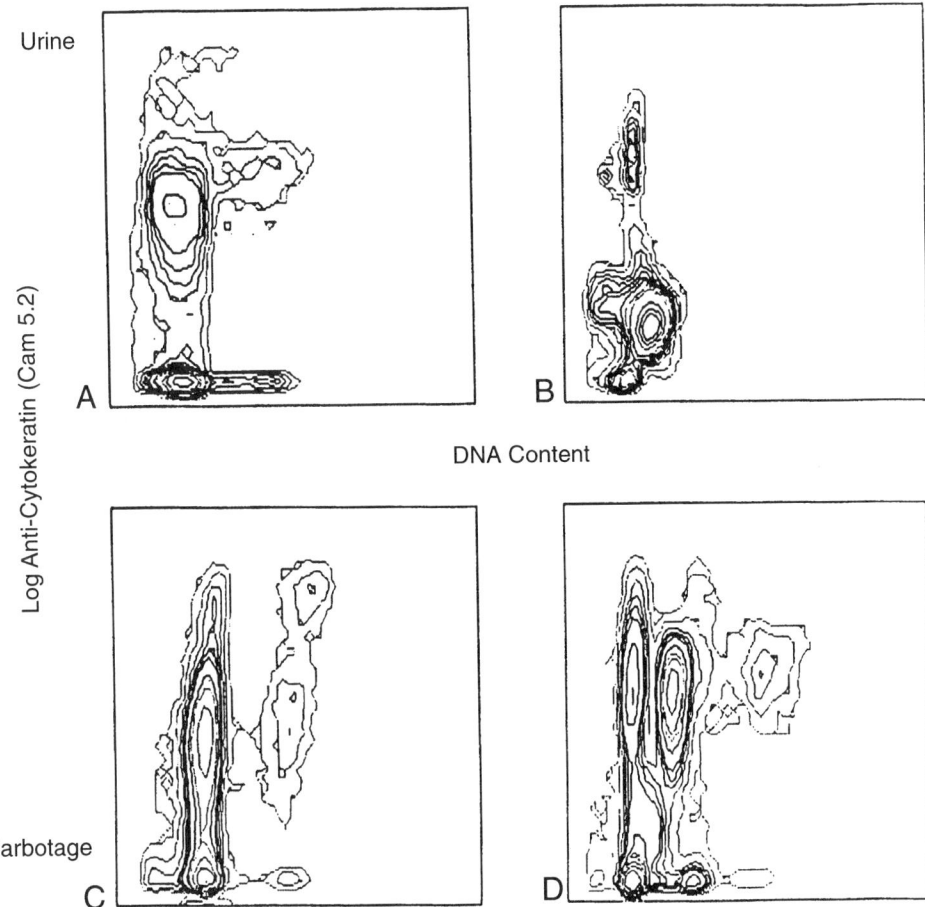

Figure 16.12. Comparison of results of multiparameter *DNA-content* flow cytometric analysis using paired urine (**A** and **B**) and bladder barbotage (**C** and **D**) specimens from two different patients. Samples were stained using antibodies to cytokeratins plus propidium iodide, using a "no-wash" staining technique. Results shown here illustrate the advantages of using barbotage, rather than voided urines for multiparameter analysis. Urines generally showed significantly higher CV's, frequently demonstrated significant numbers of squamous epithelial cells or granulocytes and sometimes failed to demonstrate cytokeratin-positive, DNA-aneuploid populations that were evident in paired barbotages.

paired samples of urine and barbotage specimens (85, 86). These studies have concluded that, although urine is a suitable source of cells for flow cytometry, in most cases, bladder washings provide better quality DNA-content measurements.

We have compared the results of flow cytometric analysis of paired urine and bladder wash samples from more than 45 individuals using DNA-content plus cytokeratin measurements. In most cases, the bladder wash sample demonstrated lower CV's for DNA-content measurements compared with urine taken from the same patient immediately before bladder barbotage. In addition, as shown in Figure 16.12, in two patients a DNA-aneuploid (cytokeratin-positive) cell population was seen in the barbotage, but not in the matched urine. Also shown in this figure is an additional result of this study that, in a number of urines, no cytokeratin-positive cells were seen, whereas cytokeratin-positive cells were evident in the matched barbotage samples. The absence of any cytokeratin-positive cells is indicative of an inadequate sample, since the target cell population (urothelial cells) is not present in the sample.

Cell Fixation. In our experience, a critical way to optimize DNA-content measurements of exfoliated bladder cells is to minimize the time period from when the cells leave the bladder wall to when they are fixed. At the start of our studies, bladder samples were all submitted first to the cytology laboratory, where they were split and fixed, and later sent for flow cytometry. Now all samples are split in the outpatient urology clinic and the portion for flow cytometry is picked up and fixed within four hours of leaving the patient. Better control of fixation and the reduction of time before fixation (coupled with the direct antibody staining technique described below) have significantly improved the quality of flow cytometry results.

Important considerations for clinical applications of flow cytometry are the need for simple, reproducible techniques that minimize or eliminate artifacts. For cell fixation, an additional consideration is the sensitivity of the target antigen to denaturation, masking, or loss by different fixation/permeabilization techniques. Cytokeratin molecules are robust, and different fixation techniques (alcohols, versus aldehydes) do not alter their expression significantly. Although one cytokeratin epitope has been reported lost after fixation and paraffin embedding (epitope of cytokeratin #18 recognized by monoclonal antibody RGE-53), we have not seen a loss of the epitope recognized by this antibody in a number of cytokeratin #18-positive cell lines following fixation in formalin or paraformaldehyde. We have used both alcohol and paraformaldehyde fixation for bladder tumor specimens. Although both techniques give satisfactory results for simultaneous DNA-content plus cytokeratin measurements, we routinely fix clinical specimens in cold alcohol (50% final concentration). Alcohol fixation is a simpler technique because the fixation is not time dependent, alcohol does not need to be removed, and does not require a subsequent permeabilization step.

Diploid-DNA-Content Controls. Clinical measurements of DNA content are dependent on the use of a diploid-DNA-content control. Although there are many arguments about the use of trout, chicken, turkey, etc., red blood cells as internal DNA-content controls, the best positive argument voiced to date is the debater's appetite for fresh trout (what's left after removal of the DNA control cells), chicken, etc.

We add human lymphocytes to all samples, partly as a means to increase the cell yield in samples with low numbers of exfoliated cells and partly for use as a diploid-DNA standard for samples lacking diploid, cytokeratin-positive cells. Lymphocytes are obtained from patients with chronic B-cell leukemia (CLL) proven to be DNA diploid (K. Bauer). Large numbers of cells can be stored frozen as aliquots in liquid nitrogen (in RPMI/15% Fetal Calf Sera/7.5% DMSO). Immediately before use, control lymphocytes are thawed, washed, and added to the bladder wash sample before fixation. While these added lymphocytes are useful diploid-DNA controls for samples lacking diploid cytokeratin-positive cells (seen in few bladder washings), the best DNA control for exfoliated bladder cells is the normal (cytokeratin-positive) urothelial cells.

As a positive control for cytokeratin expression, we routinely use human bladder tumor cell lines that express stable levels of cytokeratins, which are detectable with the monoclonal antibodies used for clinical measurements. These cells are harvested at subconfluence, washed, and fixed after the addition of an equal number of DNA control lymphocytes. The simultaneous analysis of both cytokeratin-positive (bladder tumor cell line) and cytokeratin-negative (lymphocytes) cells provides an excellent control to ensure proper cytokeratin staining. In addition, it provides a useful sample to set up the flow cytometer, including appropriate green and red PMT voltage and color compensation.

Cell-Staining Technique. For the analysis of bladder barbotage samples, we have developed a "no-wash" staining technique. After the removal of the alcohol fixative (one wash with cold Dulbeco's PBS containing 5% calf sera or 2.5% bovine albumin), the reagents are added in a stepwise fashion and, following appropriate incubations, cells are ready for flow cytometric analysis. The principal advantages of this technique are that it greatly cuts down on sample processing time (~ 1.5 hrs compared to 4–5 hrs for indirect antibody techniques) and significantly reduces cell loss during multiple centrifugation steps. The principal disadvantages of the technique are that it requires a direct fluorochrome conjugate of the anticytokeratin antibody, it has significantly higher background than indirect techniques, and it may not be suitable for use on all flow cytometers or with other antibody-fluorochrome conjugates. In some instruments, at the time the flow cytometer is triggered to measure an individual cell, the sensing volume of the instrument will include fluorochrome bound to the cell plus free fluorochrome in and around the cell, or, for some antigen/antibody combinations, the ratio of bound to free fluorochrome may be too low.

Future Directions

A large number of biochemical markers have been extensively investigated as potential signals for bladder tumor recurrence, progression, or response to therapy. These include changes in blood-group antigen expression (151, 163), oncogene proteins (164–167), growth factors and growth factor receptors (168, 169), intermediate filament expression (157, 158), cellular F-actin levels (170), and cell proliferation (101–107) or proliferation-related markers (108). In addition, there are over 100 citations in the literature describing monoclonal antibodies to bladder cancer cells, the most promising of which may be those described by Fradet and coworkers (153, 154)).

The important areas to which flow cytometry will hopefully contribute include the detection of tumor recurrence in previously diagnosed patients, the prediction of which individuals will progress to invasive cancer, and monitoring response to therapy. As indicated above, all three of these areas of application will likely require the use of multiparameter flow cytometric analysis. Promising markers for detection of DNA-diploid tumors are monoclonal antibodies to cell surface antigens specific for transformed urothelium. Used in conjunction with additional markers, such as proliferation-specific markers, growth factor receptors, or Lewis X antigen, multiparameter flow cytometric measurements could provide sensitive and highly specific assessments of the likelihood of progression from superficial to invasive disease. However, what is still needed is a clear understanding of the biological steps in the evolution of the urothelium from premalignant to superficial disease. Considerable interest of late has centered around chromosomal changes in bladder cancers. Recent reports have demonstrated that deletions on the short arm of chromosome 17 are among the most frequent abnormalities in bladder cancer (171). The p53 tumor suppressor gene has been mapped to this area and several recent reports have indicated that alterations in the p53 gene are frequently seen in high-stage (but not low-stage) bladder cancers (172). Although many investigations are now focusing on the expression of p53 in bladder cancer, what is still needed are biochemical (or genetic) markers for events that probably proceed this. With a more detailed understanding of the molecular and cellular events occurring early in tumorigenesis, specific markers for predicting progression and disease course in individual patients should become available.

RENAL CELL CANCERS

Background

Renal cell carcinoma is the predominant cancer of the kidney, accounting for 18,000 new cases of cancer and roughly 9,000 cancer-related deaths in the U.S. annually. This represents roughly 3% of all adult cancers in the U.S. and 85% of all malignant neoplasms involving the kidney. Other names for this tumor include hypernephroma, clear cell carcinoma, and alveolar carcinoma. The other major cancer of the kidney, Wilms tumor (or nephroblastoma) is rare, accounting for roughly 350 new cases per year. It is the most common renal cancer seen in childhood and accounts for roughly 5% of all childhood tumors. Wilms tumor is usually unicentric, although 5% of individuals present with bilateral tumors.

The causes of renal cell carcinoma are unknown. A variety of environmental and occupational hazzards as well as dietary factors, cigarette smoking, hormones, and xenobiological agents excreted by the kidneys have been associated with renal cell carcinoma. Epidemiologic studies have suggested an increased incidence of renal cell carcinoma in leather tanners, petrolium workers, and individuals exposed to cadmium or asbestos. Dietary factors associated with increased risk include caffeine, diuretics, obesity, and excessive ingestion of phenacetin. A rare form of familial renal cell cancer is recognized and the incidence of renal cell carcinoma is significantly increased in individuals with von Hippel-Lindau syndrome and adult polycystic kidney disease.

Wilms tumor can occur in familial and nonfamilial forms. The familial form demonstrates an autosomal dominant inheritance with variable penetrance. The disease arises from a mutation in the developing embryo and cytogenetic analysis indicates that the most frequent abnormality involves the short arm of chromosome 13. The familial form frequently is associated with congenital abnormalities, including cryptorchidism, hypospadias, aniridia, and hemihypertrophy.

Pathological and Clinical Assessment of Renal Cell Cancers

Staging systems commonly used for renal cell cancers include the system proposed by Flocks and Kadesky (173), as modified by Robson and coworkers (174), and the TMN classification system, as modified for renal cell cancers (175). The Robson scheme, commonly used in the U.S., is based on confinement to the kidney parenchyma (stage I), tumor invasion through Gerota's fascia (stage II), involvement of local lymph nodes or renal vasulature (stage III), or tumor invasion of local or distal organs (stage IV). The major deficiency of this staging system is its frequent failure to predict long-term prognosis for all stages. Frequently, patients with stage III disease will have a survival rate that is comparable to patients with stage I or II disease. For the TMN staging system, primary tumor staging (T) is defined by tumor size (T1 or T2), tumor extension into the renal vein or vena cava (T3), or invasion beyond Gerota's fascia (T4). The TMN staging system proposed by the American Joint Committee on Cancer is now similar to that of the UICC, facilitating the comparison of the results of U.S. and European studies.

Renal cell carcinomas arise from tubular epithelial cells of the proximal tubule. Tumors usually arise in the cortex of the kidney and grow to the renal capsule. Different histological cell types have been described, including granular cell, clear cell, and sarcomatoid-appearing (also called spindle cell) types. Clear cell tumors are more common than granular cell tumors, or tumors composed of mixtures of clear, granular, and sarcomatoid cells. Cell type has been related to patient survival, although heterogeneity of cell type in many cancers makes the use of this criteria less useful than tumor staging. Histological grading is less useful than staging for prediction of disease course, in part due to the many different grading schemes that have been proposed.

The overall survival rate for patients presenting with stage T1 disease is 85% to 100% for five-year survival, while individuals with stage T2 or T3a disease have a 60% chance of five-year survival. Patients presenting with metastatic disease have a 5% chance of five-year survival, with a median survival rate of seven months (176). A number of other prognostic factors (tumor size, grade, histological type, patient sex) have been evaluated as predictive markers for disease course, alone or in combination with tumor stage, and generally do not provide additional information compared to stage alone.

DNA Content Measurements

RENAL CELL CARCINOMA

The results of flow cytometry studies of primary and metastatic renal cell carcinoma of the kidney are summarized in Table 16.7. Several important studies that utilized image analysis are included and their importance is noted below.

Cytometric analysis of DNA content of renal cell tumors have generally shown a positive correlation for DNA ploidy and tumor grade (51, 177–180). However, some studies have reported positive correlation for only some grading schemes (180) or have reported no correlation for DNA content and tumor grade (181).

Although most studies have reported a positive correlation for DNA content and tumor stage (178–180, 182), the differences between two stages may not be significant. Studies that include patients with three or four different disease stages frequently show marginal differences for the incidence of DNA aneuploidy between stages 2 and 3, or between stages 3 and 4 tumors (179), although it should be

Table 16.7
Predictive Value of Flow Cytometric DNA-Content Measurements of Renal Cell Carcinoma

Sample	No. of Cases[a]	Significance	Reference
Nuclei[b]	26	210 samples from 26 patients by flow cytometry, 48% diploid only; 9 tumors had both DNA-diploid and aneuploid tumor areas.	Ljungberg (51)
Nuclei[c]	55	Retrospective image study. DNA-diploid tumor in 32/33 10-year survivors; all nonsurvivors DNA aneuploid.	Ljungberg (183)
Cells[b]	55	Prospective study. Malignancy index used to predict patients in good and poor prognosis groups.	Baisch (182)
Nuclei[c]	206	Well-differentiated tumors, DNA aneuploidy predicted subsequent metastatic disease.	Rainwater (178)
FNA[b]	44	FNA compared with analysis of primary tumors; biopsy primary agreed for 18 DNA-diploid, and 15/26 aneuploid tumors.	Ljungberg (186)
Nuclei[b]	48	Stage a significant predictor of survival, DNA ploidy not a significant predictor.	Oosterwijk (179)
Cells[b]	29	No correlation between DNA ploidy, and tumor grade and stage.	Wolman (181)
Nuclei[b]	32	DNA ploidy a significant predictor of survival.	deKernion (176)
Nuclei[c]	103	Positive correlation for DNA ploidy with nuclear grade and mitotic rate; DNA ploidy significant predictor of outcome.	Grignon (180)
Nuclei[b]	59	Prospective study of stage 1 tumors; 1/27 DNA-diploid and 5/32 aneuploid tumors metastasized. Ploidy marginally significant predictor of survival.	Ljungberg (185)
Metastatic Disease			
Cells[b]	43 (11)	DNA ploidy and S phase of 43 primary tumors and 11 metastatic lesions; comparison in 4 patients showed similar DNA ploidy in 3, DNA-diploid primary and DNA-aneuploid metastasis in one.	Chin (187)
Nuclei[b]	23	Image analysis; 6 DNA-diploid tumors had diploid metastases, 1/8 DNA-aneuploid primary with diploid metastases.	Ljungberg (188)
Other Renal Tumors			
Nuclei[c]	56	Wilms' tumors; high survival rate for all patients, DNA-tetraploid tumors slightly lower survival than diploid or other aneuploid.	Rainwater (189)
Nuclei[c]	51	Renal Oncocytomas; 11% grade 1 (do not metasticize) and 24% grade 2 DNA-aneuploid tumors.	Rainwater (192)
Nuclei[c]	22	Squamous cell carcinoma of the collecting system; DNA ploidy correlated with grade and stage, not significant predictor of survival.	Nativ (190)
Nuclei[c]	106	Adenocarcinoma. DNA ploidy showed significant correlation with stage, not grade. Survival advantage for DNA-diploid patients; high incidence of progression in DNA-diploid tumors made ploidy not significant predictor.	Currin (191)

[a]Actual number of individuals or samples analyzed
[b]Fresh tissue
[c]Paraffin-embedded archival tissue

noted that some studies have reported a significant difference between each stage (180). Again, some reports have not found a significant relationship between DNA ploidy and stage (181). As noted in Table 16.7, these studies have used a variety of different sources of tissue and cell or nuclear isolation techniques and it is difficult to attribute different conclusions to strictly methodological issues.

The most important clinical question to be addressed by DNA-content measurements is their ability to predict disease course or outcome. As indicated in Table 16.7, a number of studies have addressed this question, using both prospective and retrospective techniques. Ljungberg and coworkers (183) selected patients from two groups based on 10-year survival, for a study using image analysis of nuclei from paraffin-embedded tumors. All tumors from nonsurviving patients were DNA aneuploid, while 32 out of 33 10-year survivors had DNA-diploid tumors. Baisch and coworkers (182) developed a malignancy index (which includes stage, nuclear grade, histology, and DNA content) to successfully predict the two-year survival rate of 55 stage 1–4 patients who received a nephrectomy. Although DNA content (DNA diploid vs. aneuploid and %SG_2M cells) alone was useful, it had high false-positive and false-negative rates, and its impact in predicting survival in these patients was significantly improved with its incorporation into the malignancy index.

In a study that included 206 individuals with histologically well-differentiated clear cell type cancers, Rainwater and coworkers (178) demonstrated that DNA-aneuploid tumors had a significantly higher probability for the development of metastatic disease, independent of tumor grade or stage. As shown in Figure 16.13, actuarial survival for all stages demonstrated a significant difference for individuals with DNA-diploid versus aneuploid tumors. However, a significant difference was not seen for patients with stage 1 tumors or for pooled survival for stages 2–4 tumors, suggesting that stage contributed significantly to the overall survival prediction. Also shown in this figure is the analysis of individuals who developed metastatic disease postnephrectomy, indicating a survival advantage for individuals with DNA-diploid tumors. In a study of 103 individuals with stages 1–3 tumors who underwent radical nephrectomy, Grignon and coworkers (180) demonstrated a significant correlation for DNA ploidy and overall survival that was independent of stage.

In a prospective study of 32 patients who received a radical nephrectomy, deKernion and coworkers (184) investigated the prognostic significance of DNA ploidy, grade, and

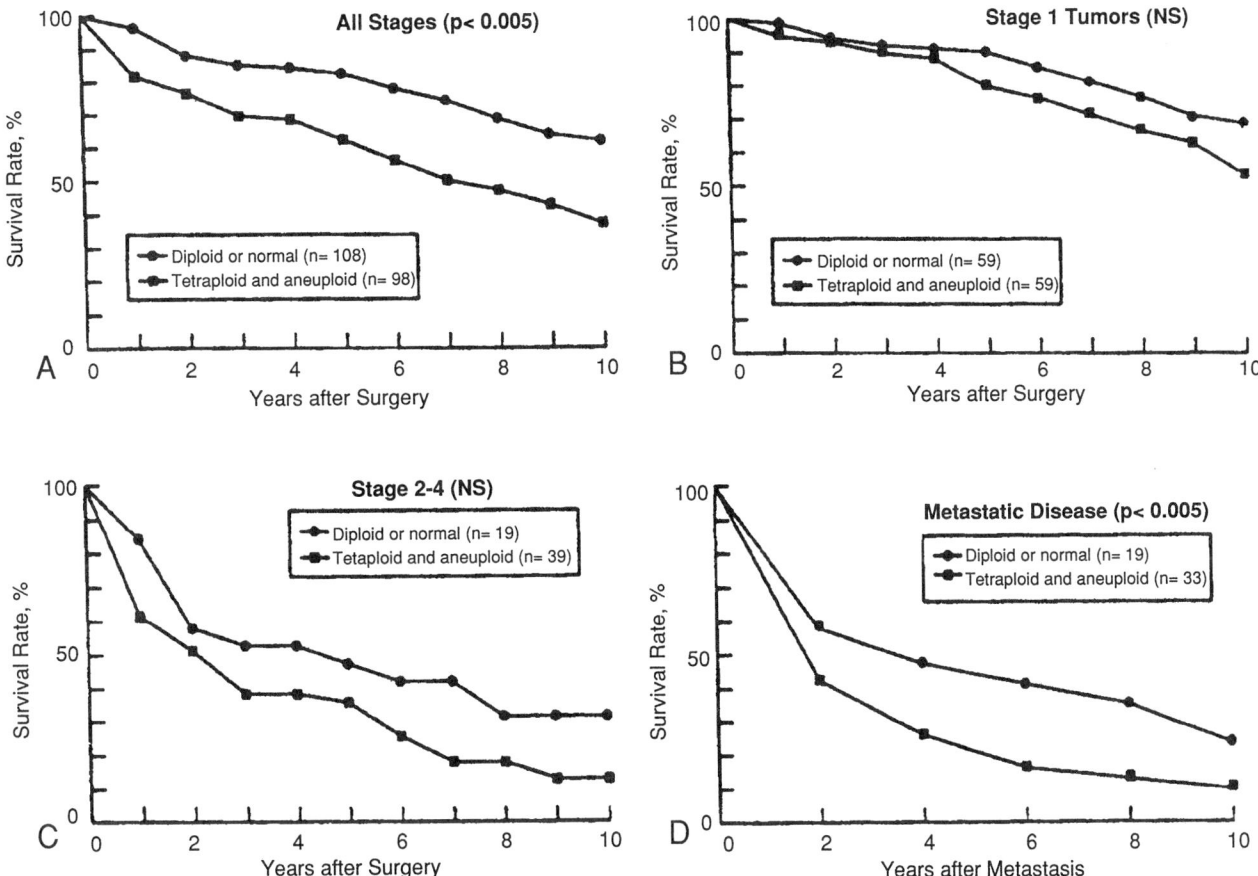

Figure 16.13. Impact of DNA ploidy analysis in renal cell carcinoma. In a retrospective analysis of 206 patients with well-differentiated tumors, Rainwater et al. (178) compared actuarial survival by stage. In (**A**), postoperative survival for all patients demonstrated significant advantage for DNA-diploid, compared with DNA-aneuploid (including tetraploid) individuals. However, DNA ploidy was not a significant predictor for individuals with only stage 1 tumors (**B**), or when comparing postoperative survival for DNA-diploid versus aneuploid stage 2-4 patients (**C**). Ploidy was of only marginal significance in predicting survival for individuals with metastatic disease (**D**).

stage in predicting three-year survival. During this period, 13 out of 19 patients presenting with metastatic disease died and no patients with node-negative (NO) disease died, irrespective of the tumor stage the study included six T3A individuals). In this group of patients, DNA ploidy and grade were significant and independent predictors of death due to disease.

Several studies have shown that DNA ploidy only approaches significance (179, 185) or has no significance (181) in predicting disease course or survival. In a retrospective study using frozen tumor samples from 48 stages 1-4 patients, Oosterwijk and coworkers (179) demonstrated that stage, but not DNA ploidy, was an independent predictor of survival. Although patients with DNA-aneuploid tumors in this study had an increased likelihood of death due to disease, DNA ploidy was not a significant independent predictor for these patients. In a prospective study of 59 individuals with stage 1 disease, Ljungberg and coworkers (185) demonstrated that only one out of 27 individuals with DNA-diploid tumors experienced metastasis within an average follow-up period of 46 months. While there was a significant difference in the survival rates of DNA-diploid and aneuploid individuals, DNA ploidy did not provide a significant predictor of survival.

The reasons for the apparent discrepancies in the role of DNA ploidy in predicting disease course or survival are not clear. As indicated in Table 16.7, most of these studies involved relatively small numbers of patients and several studies involved short follow-up periods. Methods used to prepare and analyze samples are not well documented and few studies provide information on the quality of the data or the criteria used to define aneuploidy. An additional problem is potential tumor heterogeneity. Ljungberg and coworkers (51) reported that a significant percentage of tumors (44% in their study) contain both DNA-diploid and aneuploid areas determined by flow cytometry. While such heterogeneity could provide a basis for some of the discrepancies in the studies discussed above, the results should be interpreted carefully. In one figure, these authors present four areas of the same kidney with 2C tumor, 2C plus 3.7C tumor, and 2C plus 3.8C tumor. In their results section, the authors describe this as an example of a heterogeneous tumor. They interpret the 2C events as normal diploid elements and, thus, imply that the 3.7 and 3.8C peaks are different ploidies. It is more likely that the 3.7 and 3.8C events represent tumor areas with the same DI. Without further evidence, it is difficult to judge the true heterogeneity of these tumors. Further evidence for the heterogeneity of renal cell cancers has been presented by this same group (186). In a comparison of the DNA ploidy of biopsy material with paired surgical samples (and metastatic tumor tissue when available) these authors argued that 20 of 24 primary tumors contained multiple DNA-aneuploid, or aneuploid plus diploid tumor regions (186). Since representative histograms were not presented to evaluate the results directly (and no information is given on the overall quality of the data), it is difficult to determine the impact of potential tumor heterogeneity on the results of these and other studies.

METASTATIC DISEASE

Studies on small numbers of metastatic renal cell cancers have demonstrated that primary and metastatic tumors generally have similar DNA ploidy. In a study of 43 renal cell primary tumors and 11 metastatic tumors, Chin and coworkers (187) were able to compare the results obtained from four patients. In three patients, both primary and metastatic tumors had the same DNA ploidy (one patient with both DNA-diploid primary tumor and metastatic tumor, two patients with similar DI for primary and metastatic tumors and one patient had a DNA-diploid primary tumor and two different DNA-aneuploid metastatic tumors). A study by Ljungberg and coworkers (188) reported on paired samples from four patients using DNA-content flow cytometry or image analysis. This study demonstrated a similar pattern, with six patients showing DNA-diploid tumors in both primary and metastatic sites, seven individuals with similar DNA-aneuploid primary tumors and metastatic tumors, and one individual with a DNA-aneuploid primary tumor and a DNA-diploid metastatic tumor.

OTHER RENAL TUMORS

In a retrospective study of Wilms tumor, Rainwater and coworkers (189) studied the impact of DNA ploidy on the survival of 56 patients. One group of nine patients (diagnosed before the use of chemotherapy) had a low five-year survival rate and, for these individuals, DNA ploidy and survival showed no correlation. In the remaining 47 patients, 13 had DNA-diploid tumors and 34 had DNA-aneuploid (including 23 tetraploid) tumors. In this group of patients, all individuals with DNA-diploid or DNA-aneuploid (other than tetraploid) tumors were alive at five years. These patients demonstrated the unusual pattern whereby individuals with DNA-tetraploid tumors are at increased risk for progression. For individuals with any stage tumor, DNA tetraploidy was not a significant predictor of disease course. However, for individuals with stage 3 or 4 DNA-tetraploid disease, DNA tetraploidy was significant.

In a retrospective study of 28 patients with squamous cell carcinoma of the ureter (four patients), renal pelvis (two patients) or intrarenal collecting system (22 patients), Nativ and coworkers (190) demonstrated that DNA ploidy showed a significant correlation with grade and tumor stage. However, this conclusion could be somewhat biased, as most tumors were high grade and the majority were DNA aneuploid. Although DNA ploidy was not a significant predictor of survival in this study, individuals with high-grade tumors could be separated into fair vs. poor prognosis groups based on DNA ploidy.

The flow cytometric analysis of archival tumor samples from 106 patients with adenocarcinoma of the kidney by Currin and coworkers (191) demonstrated a strong correlation for DNA aneuploidy (including tetraploidy) and stage, and no correlation for ploidy and grade. Although there was a significant difference in the progression of DNA-aneuploid vs. diploid tumors, diploid tumors had no survival advantage when controlled for TMN stage.

Renal oncocytomas are renal parenchymal tumors composed of pure populations of well-differentiated granular cells. In a retrospective study of 44 grade 1 and 21 grade 2 tumors, Rainwater and coworkers (192) demonstrated that nearly 50% of the grade 1 tumors showed either a DNA-aneuploid population (11%), or a high 4C population interpreted by the authors as DNA tetraploid. Since grade 1 tumors rarely invade or metastasize, the authors noted that the high incidence of DNA aneuploidy in this grade tumor was of interest. For grade 2 tumors, 33% were DNA diploid and 43% were DNA tetraploid. The survival rates of individuals with a tumor of either grade were not reported.

FUTURE DIRECTIONS

Approximately one-third of all patients with renal cell carcinoma present with metastatic disease. For these individuals, flow cytometry will likely provide little additional information that is clinically useful. The disease course for these individuals is uniformly progressive to death from the cancer, with an average survival rate of less than one year. Most advanced stage renal cell carcinomas are highly resistant to the chemotherapeutic regimens that are currently available. This resistance likely reflects the normal physiologic mechanisms available to the renal tubular epithelium to pump xenobiological agents (including chemotherapeutic compounds) out of cells, using membrane p-glycoprotein or glutathione (see Chapter 27).

For lower stage and grade renal cell tumors (particularly stage T1 to T3aN0) DNA content measurements likely provide useful information to predict groups of patients at low or high risk for disease progression. At present, the primary treatment for these individuals is radical nephrectomy. Initial studies using biological response modifiers (Interleukin-2) have not provided the hoped for response rate; more recent trials have included interferons or tumor necrosis factor with or without other chemotherapeutic agents. It should be noted, however, that the majority of these studies have utilized individuals with advanced-stage disease. It is possible that stratification of lower stage patients by DNA-content flow cytometry (which is possible in conjunction with other markers, such as drug-resistance determinations) could provide information regarding the potential response of subsets of individuals with low-stage disease who are at significant risk for disease progression.

ACKNOWLEDGMENTS

The author gratefully acknowledges the contributions of Ms. Tina Lubrano, Mrs. Sharon Graham, Dr. Jia-Quin Jin, Dr. Scott Dougherty and Dr. Kevin Gandhi.

This work was supported by research funding from the Departments of Urology and Pathology, Loyola University Medical Center, by the Illinois Division, American Cancer Soc., and by a VA Merit Review Grant.

REFERENCES

1. Edwards CW, Steinthorsson E, Nicholson D. An autopsy study of latent prostatic cancer. Cancer 1953;6:531–554.
2. Scott R, Matchnik D, Nicholson D, et al. Carcinoma of the prostate in elderly men; incidence, growth characteristics, and clinical significance. J Urol 1969;101:602–607.
3. Seidman H, Mushinski M, Geib S, et al. Probabilities of eventually developing or dying of cancer—United States 1985. Cancer 1985;35:35–56.
4. Scardino PT. Early detection of prostate cancer. Urol Clin North Am 1989;16:635–655.
5. Yatani R, Chiggusa I, Akazaki K, et al. Geographic pathology of latent prostatic carcinoma. Int J Cancer 1982;29:611–616.
6. Dhom G. Epidemiologic aspects of latent and clinically manifest carcinoma of the prostate. J Cancer Res Clin Oncol 1983;106:210–218.
7. Whitmore WF. Locoregional prostatic cancer: advances in management. Cancer 1989;65:667–674.
8. McNeal JE. Zonal distribution of prostatic adenocarcinoma: correlation with histologic pattern and direction of spread. Amer. J Surg Pathol 1988;12:897–906.
9. Whitmore WF. Hormone therapy in prostatic cancer. Am J Med 1956;21:697–713.
10. Jewett HJ. The present status of radical prostatectomy for stages A and B prostatic cancer. Urol Clin North Am 1975;24–43.
11. Donohue RE, Miller GJ. Adenocarcinoma of the Prostate: biopsy to whole mount. Urol Clin N Amer 1991;18:449–52.
12. Broders AC. Carcinoma: grading and practical application. Arch Pathol Lab Med 1926;68:376–381.
13. Gleason DF. Classification of prostatic carcinomas. Cancer Chemother Rep 1966;50:125–128.
14. Murphy GP, Whitmore WF. A report on the workshops on the current status of the histologic grading of prostate cancer. Cancer 1979;44:1490–1494.
15. Maksem JA, Resnick MI, Johenning PW. Can a cytological grading system be predictive of Gleason's scores in aspiration biopsy specimens of prostate carcinoma. World J Urol 1987;5:99–102.
16. Byar DB, and the Veterans Administration Cooperative Urological Research Group. Survival of patients with incidentally found microscopic cancer of the prostate; results of a clinical trial of conservative treatment. J Urol 1972;108:908–913.
17. Cantrell BB, deKlerk DP, Eggleston JC, et al. Pathological factors that influence prognosis in stage A prostatic cancer: the influence of extent versus grade. J Urol 1981;125:516–520.
18. Gaeta JF, Asirwatham JE, Miller G, Murphy GP. Histopathologic grading of primary prostate cancer. A new approach to an old problem. J Urol 1980;123:689–693.
19. Adolfsson J, Ronstrom L, Hedlund P-O, Lowhagen T, Carstensen J, Tribukait B. The prognostic value of modal deoxyribonucleic in low grade, low stage unreated prostate cancer. J Urol 1990;144:1404–1407.
20. Frankfurt OS, Slocum HK, Rustum YM, et al. Flow cytometric analysis of DNA aneuploidy in primary and metastatic human solid tumors. Cytometry 1989;5:71–80.
21. Hedley DW, Friedlander ML, Taylor IW, Rugg CA, Musgrove EA. Method for analysis of cellular DNA content of paraffin-embedded pathological material using flow cytometry. J Histochem Cytochem 1983;31:1333–1335.
22. Dougherty S, Manion S, Jin J-K, et al. Intra-tumor heterogeneity of stage A2 (T1b) prostatic adenocarcinoma: I. Evaluation of archival tumors by flow cytometry. Cytometry (accepted for publication).
23. Zetterberg A, Espositi PL. Cytophotometric DNA analysis of aspirated cells from prostatic carcinoma. Acta Cytol 1976:20;46–57.
24. Seppelt Y, Sprenger E, Hedderich J. Investigation of automated DNA diagnosis and grading of prostatic cancer. Anal Quant Cytol, Histol 1986;8:152–157.
25. Espositi PL. Cytological malignancy grading of prostatic carcinoma by transrectal aspiration biopsy. A 5-year followup study of 469 hormone treated patients. Scand J Urol Nephrol 1971;5:199–209.
26. Tavares AS, Costa J, Costa Maia J. Correlation between ploidy and prognosis in prostatic carcinoma. J Urol 109–676.
27. Lundberg S, Carstensen J, Rundquist I. DNA flow cytometry and histophathogical grading of paraffin-embedded prostate biopsy specimens in a survival study. Cancer Res 1982;47:1973–1977.
28. Zetterberg A, Espositi PL. Prognostic significance of nuclear DNA levels in prostatic carcinoma. Scand J Urol Nephrol Suppl 1980;55:53–58.
29. Adolfsson J, Tribukait B. Evaluation of tumor progression by repeated fine needle biopsies in prostate adenocarcinoma: modal deoxyribonucleic acid value and cytological differentiation. J Urol 1990;144:1408–1410.
30. Kjaer TB, Thommesen P, Fredricksen P. DNA content in cells aspirated from carcinoma of the prostate treated with oestrogenic compounds. Urol Res 1979;7:249–251.
31. Bocking A, Aufferman W, Jochman D, et al. DNA grading of malignancy and tumor regression in prostatic carcinoma under hormone therapy. Appl Pathol 1985;3:206–214.
32. Ronstrom L, Tribrikait B. Espositi PL. DNA pattern and cytological findings in fine-needle aspirates of untreated prostatic tumors. A flow cytofluorometric study. Prostate 1981;2:79.
33. Forsslund G, Zetterburg A. Ploidy level determinations in high-grade and low-grade malignant variants of prostatic carcinoma. Cancer Res 1990;50:4281–4285.
34. Adolfsson J, Ronstrom L, Hedlund P-O, Lowhagen T, Carstensen J, Tribukait B. The prognostic value of modal deoxyribonucleic acid in low grade, low stage untreated prostate cancer. J Urol 1990;144:1404–1407.
35. Bichel P, Fredericksen P, Kjaer T, Tommesen P, Vanderlow LL. Flow microfluorometry and transrectal fine-needle biopsy in the classification of human prostatic carcinoma. Cancer 1977;40:1206–1211.
36. Sprenger E, Michaelis WE, Vogt-Schaden M, Olto C. The significance of DNA flow-through fluorescence cytophotometry for the diagnosis of prostate carcinoma. Beitr Path 1976;159:292–293.
37. Deitch AD, Strand MA, deVere White RW. Deoxyribonucleic acid flow cytometry of benign prostatic disease. J Urol 1989;142:759–762.
38. McIntire TL, Murphy WM, Coon JS, et al. The prognostic value of DNA ploidy combined with histologic substaging for incidental carcinoma of the prostate gland. Am J Clin Path 1988;89:370–374.
39. Montgomery BT, Nativ O, Blute ML, et al. Stage B Prostate adenocarcinoma: flow cytometric nuclear DNA ploidy analysis. Arch Surg 1990;125:327–331.
40. Fordham MVP, Burdge AH, Matthews J, Williams G, and Cooke T. Prostatic carcinoma cell DNA content measured by flow cytometry and its relation to clinical outcome. Brit J Surg 1986;73:400–403.
41. Lee SE, Currin SM, Paulson DF, Walther PJ. Flow cytometric determination of ploidy in prostatic adenocarcinoma: a comparison with

seminal vesicle involvement and histopatholgical grading as a predictor of clinical recurrence. J Urol 1988;140:769–774.
42. Winkler HZ, Rainwater LM, Myers RP, et al. Stage D1 prostate adenocarcinoma: significance of nuclear DNA ploidy pattern studied by flow cytometry. Mayo Clin Proc 1988;63:103–112.
43. Mohler JL, Partin AW, Epstein JI, et al. Prediction of prognosis in untreated stage A2 prostatic carcinoma. Cancer 1992;69:511–519.
44. Ritchie AWS, Dorey F, Layfied LJ, Hannah J, Loviekovich H, DeKernion JB. Relationship of DNA content of conventional prognostic factors in clinical localized carcinoma of the prostate. Br J Urol 1988;62:254–260.
45. Humphrey PA, Walther PJ, Currin SM, and Vollmer RT. Histologic grade, DNA ploidy, and intraglandular tumor extent as indicator of tumor progression of clinical stage B prostatic carcinoma. Am J Surg Path 1991;15:1165–1170.
46. Nativ O, Winkler HZ, Raz Y. Stage C prostatic adenocarcinoma: flow cytometric nuclear DNA ploidy analysis. Mayo Clin Proc 1989;64:911–919.
47. Stephenson RA, James BC, Gay H, Fair WR, Whitmore WF, Melamed MR. Flow cytometry of prostate cancer: relationship of DNA content to survival. Cancer Res 1987;47:2504–2509.
48. Zincke H. The role of pathologic variables and hormonal treatment after radical prostatectomy for stage D1 disease. Oncology 1991;5:129–140.
49. Stege R, Lundh B, Tribukait B, Pousette A, Carlstrom K, Hasenson M. Deoxyribonucleic acid ploidy and the direct assay of prostatic acid phosphatase and prostate specific antigen in fine needle aspiration biopsies as diagnostic methods in prostatic carcinoma. J Urol 1990;144:299–302.
50. Miller J, Horsfall DJ, Marshall VR, Rao DM, Leong SY. The prognostic value of deoxgulonucleic and flow cytometric analysis in stage D_2 prostates carcinoma. J Urol 1991;145:1192–1196.
51. Lunberg B, Stirling R, Roos G. DNA content in renal cell carcinoma with reference to tumor heterogeneity. Cancer 1985;56:503–508.
52. Rabinovitch PS, Reid BJ, Haggitt RC, Norwood TH, Rubin CE. Progression to cancer in Barrett's esophagus is associated with genomic instability. Lab Invest 1988;60:65–71.
53. Quirke P, Dyson JED, Dixon MF, Bird CC, Joslin CAF. Heterogeneity of colorectal adenocarcinomas evaluated by flow cytometry and histopathology. Br J Cancer 1985;51:99–106.
54. Askensten UG, von Rosen AK, Nilsson RS, Auer GU. Intratumoral variations in DNA distribution patterns in mammary adenocarcinomas. Cytometry 1988;10:326–333.
55. Greene DR, Taylor SR, Wheeler TM, Scardino PT. DNA ploidy by image analysis of individual foci of prostate cancer: a preliminary report. Cancer Res 1991;51:4084–4089.
56. Jin J-K, Walloch J, Flanigan RC, Herman CJ, Shankey TV. Intratumor heterogeneity of stage A2 (T1b) prostatic adenocarcinoma: II DNA content analysis of tissue sections using image analysis. Cytometry (accepted for publication).
57. Flanigan RC, Dougherty S, Walloch J, Mittleberg B, Waters WB, Manion S, Coggin C, Shankey TV. DNA content and histopathologic factors as predictors of disease course in clinical stage A2 prostatic adenocarcinoma. J Urol (submitted for publication).
58. Kallioniemi O-P, Visakorpi T, Holli K, Heikkinen A, Isola J, Koivula T. Improved prognostic impact of S-phase values from paraffin-embedded breast and prostate carcinomas after correcting for nuclear slicing. Cytometry 1991;12:413–421.
59. Dean PN, Jett JH. Mathematical analysis of DNA distributions derived from flow microfluorometry. J Cell Biol 1974;60:523–527.
60. Wang MC, Papsidero LD, Kuriyama M, Valenzuela LA, Murphy GP Chu TM. Prostate antigen: a new potential marker for prostatic cancer. Prostate 1981;2:89–96.
61. Oesterling JE. Prostate specific antigen: a critical assessment of the most useful tumor marker for adenocarcinoma of the prostate. J Urol 1991;145:907–923.
62. Nativ O, Myers RP, Farrow GM, Therneau TM, Zincke H, Lieber M. Nuclear deoxyribonucleic acid ploidy and serum prostate specific antigen in operable prostatic adenocarcinoma. J Urol 1990;144:303–306.
63. Thompson TC. Growth factors and oncogenes in prostate cancer. Cancer Cells 1990;2:345–354.
64. Huggins C, Hodges CV. Studies on prostate cancer; the effect of castration, of estrogen and of androgen injection on serum phosphatases in metastatic carcinoma of the prostate. Cancer Res 1941;1:293–302.
65. Thompson TC, Southgate J, Kitchner G, Land H. Multistage carcinogenesis induced by ras and myc oncogenes in a reconstituted organ. Cell 1989;56:917–930.
66. Treiger B, Isaacs J. Expression of a v-Ha-ras oncogene in a Dunning rat prostate adenocarcinoma and the development of high metastatic ability. J Urol 1988;140:1580–1586.
67. Chesa PG, Rettig WJ, Melamed MR, Old LJ, Niman HL. Expression of p21ras in normal and malignant human tissues: lack of association with proliferation and malignancy. Proc Natl Acad Sci USA 1989;84:3234–3238.
68. Trachtenberg J, Walsh PC. Correlation of prostatic nuclear androgen receptor content with duration of response and survival following hormonal therapy in advanced prostatic cancer. J Urol 1979;127:466–471.
69. Benson RC, Gorman PA, O'Brien PC, Holicky EL, Veneziale CM. Relationship between androgen receptor binding activity in human prostate cancer and clinical response to endocrine therapy. Cancer 1987;59:1599–1606.
70. Ekman P, Snochowski M, Zetterberg A, Hogberg B, Gustafsson JA. Steroid receptor content in human prostatic carcinoma and response to endocrine therapy. Cancer 1979;44:1173–1181.
71. Chodak GW, Kranc DM, Puy LA, Takeda H, Johnson K, Chang C. Nuclear localization of androgen receptor in heterogeneous samples of normal, hyperplastic and neoplastic prostate. J Urol 1992;147:798–803.
72. Rehn L. Blasengeschwulste bei fuchsin-arbeits. Arch Klin Chirugie. 1896;53:383–392.
73. Droller MJ. Transitional cell cancer: upper tracts and bladder. In: Walsh PC, Gittes RF, Perlmutter AD, Stamey TA, eds. Campbell's urology, 5th ed. Philadelphia: WB Saunders, 1986:1343–1440.
74. Koss LG. Mapping of the urinary bladder: its impact on the concepts of bladder cancer. Human Pathol 1979;10:533–548.
75. Spooner ME, Cooper EH. Chromosomal constitution of transitional cell carcinoma of the urinary bladder. Cancer 1972;29:1401–1412.
76. Smeets AWGB, Pauwels RPE, Beck HLM, Feitz WFJ, Geraedts JPM, Debruyne FMJ, Laarakkers L, Vooijs GP, Ramaekers FCS. Comparison of tissue disaggregation techniques of transitional cell bladder carcinomas for flow cytometry and chromosomal analysis. Cytometry 1987;8:14–19.
77. Jewett JH, Strong GH. Infiltrating carcinoma of the bladder: relationship of the depth of penetration of the bladder wall to incidence of local extension and metastasis. J Urol 1946;55:366–374.
78. Marshall VF. The relationship of the preoperative estimate of the pathologic demonstration of the extent of vesicle neoplasms. J Urol 1952;68:714–723.
79. National Bladder Cancer Collaborative Group A. Superficial bladder cancer: progression and recurrence. J Urol 1983;130:1083–1086.
80. Torti FM. Superficial bladder cancer: the primacy of grade in the development of invasive disease. J Clin Oncol 1987;5:125–130.
81. Jordan AM, Weingarten J, Murphy WM. Transitional cell neoplasms of the urinary bladder: can biologic potential be predicted from tumor grade? Cancer 1987;60:2766–2774.
82. Ratliff JE, Klein FA, White FKH. Flow cytometry of ethanol-fixed versus fresh bladder barbotage specimens. J Urol 1985;133:958–960.
83. Deitch AD, Andreotti VA, Strand MA, Howell L, deVere White RW. Clinically applicable method to preserve urine and bladder washing cells for flow cytometric monitoring of bladder cancer patients. J Urol 1990;143:700–704.

84. Hermansen DK, Melamed MR, Coon JS, Weinstein RS, deVere White R, Deitch AD, Wheeless LL, Reeder JE, Wersto R, Koss LG. Ethanol fixation of bladder irrigation specimens for flow cytometric analysis. A multiinstitutional study from the bladder cancer flow cytometry network. Cancer 1989;63:1780–1789.
85. deVere White RW, Deitch AD, Baker WC, Strand MA. Urine: a suitable sample for deoxyribonucleic acid flow cytometry studies in patients with bladder cancer. J Urol 1988;139:926–928.
86. Konchuba AM, Schellhammer PF, Alexander JP, Wright GL. Flow cytometric study comparing pared bladder washing and voided urine for bladder cancer detection. Urology 1989;33:89–96.
87. Levi PV, Cooper EA, Anderson CK, Path MC, Williams RE. Analysis of DNA content, nuclear size and cell proliferation of transitional cell carcinoma in man. Cancer 1969;23:1074–1085.
88. Fossa SD, Kaalhus O, Scott-Knudsen O. The clinical and histopathological significance of Feulgen DNA-values in transitional cell carcinoma of the human urinary bladder. Europ J Cancer 1977;13:1155–1162.
89. Tribukait B, Esposti PL. Quantitative flow microfluorometric analysis of the DNA in cells from neoplasms of the urinary bladder: correlation of aneuploidy with histological grading and the cytological findings. Urol Res 1978;6:201–205.
90. Jakobsen A, Bichel P, Sell A. Flow cytometric investigation of human bladder carcinoma compared to histological classification. Urol Res 1979;7:109–112.
91. Tribukait B, Gustafson H, Esposti PL. Ploidy and proliferation in human bladder tumors as measured by flow-cytofluorometric DNA-analysis and its relations to histopathology and cytology. Cancer 1979;43:1742–1751.
92. Granberg-Ohman I, Tribukait B, Wijkstro MH, Berlin P, Collste LG. Papillary carcinoma of the urinary-bladder. A study of chromosomal and cytofluorometric DNA analysis. Urol Res 1980;8:87–(93).
93. Gustafson H, Tribukait B, Esposti PL. DNA pattern, histological grade, and multiplicity related to recurrence rate in superficial bladder tumors. Scand J Urol Nephrol 1982;16:135–139.
94. Gustafson H, Tribukait B, Esposti PL. DNA profile and tumour progression in patients with superficial bladder tumours. Urol Res 1982;10:13–18.
95. Wijkstrom H, Gustafson H, Tribukait B. Deoxyribonucleic acid analysis in the evaluation of transitional cell carcinoma before cystectomy. J Urol 1984;132:894–898.
96. Tribukait B. Flow cytometry in assessing the clinical aggressiveness of genito-urinary neoplasias. World J Urol 1987;5:108–122.
97. Murphy WM, Chandler RW, Trafford RM. Flow cytometry of deparaffinized nuclei compared to histological grading for the pathological evaluation of transitional cell carcinomas. J Urol 1986;135:694–697.
98. Coon JS, Schwartz D, Summers JL. et al. Flow cytometer analyzing deparaffinized nuclei in urinary bladder carcinoma. Cancer 1986;57:1594–1599.
99. deVere White RW, Deitch AD, West B, Fitzpatrick JM. The predictive value of flow cytometric information in the clinical management of stage O (Ta) bladder cancer. J Urol 1988;139:279–282.
100. Blomjous CEM, Schipper NW, Baak JPA, van Galen EM, de Voogt HJ, Meyer CJLM. Retrospective study of prognostic importance of DNA flow cytometry of urinary bladder carcinoma. J Clin Pathol 1988;41:21–25.
101. Fulker MJ, Cooper EH, Taraka T. Proliferation and ultrastructure of papillary transitional cell carcinoma of the human bladder. Cancer 1971;27:71–82.
102. Mellon K, Neal DE, Robinson MC, Marsh C, Wright C. Cell cycling in bladder carcinoma determined by monoclonal antibody Ki-67. Br J Urol 1990;66:281–285.
103. Tribukait B, Gustafson H, Esposti P. Ploidy and proliferation in human bladder tumors as measured by flow-cytofluorometric DNA-analysis and its relations to histopathology and cytology. Cancer 1979;43:1742–1751.
104. Tribukait B, Gustafson H, Esposti PL. The significance of ploidy and proliferation in the clinical and biological evaluation of bladder tumours: a study of 100 untreated cases. Brit J Urol 1982;54:130–135.
105. Norming U, Nyman CR, Tribukait B. Comparative flow cytometric deoxyribonucleic acid studies on exophytic tumor and random mucosal biopsies in untreated carcinoma of the bladder. J Urol 1989;142:1442–1447.
106. Norming U, Nyman CR, Tribukait B. Comparative histopathology and deoxyribonucleic acid flow cytometry of random mucosal biopsies in untreated bladder carcinoma. J Urol 1991;145:1164–1168.
107. Feitz WFJ, Beck HLM, Smeets AWGB, Debryune FMJ, Vooijs GP, Herman CJ, Ramaekers FCS. Tissue-specific markers in flow cytometry of urological cancers: cytokeratins in bladder carcinoma. Int J Cancer 1985;36:349–356.
108. Tachibana M, Deguchi N, Jitsukawa S, Baba S, Hata M, Tazaki H. Quantification of cell kinetic characteristics using flow cytometric measurements of deoxyribonucleic acid and bromodeoxyuridine for bladder cancer. J Urol 1991;145:963–967.
109. Blute ML, Tsushima K, Farrow GM, Therneau TM, Lieber MM. Transitional cell carcinoma of the renal pelvis: nuclear deoxyribonucleic acid ploidy studied by flow cytometry. J Urol 1988;140:944–949.
110. Corrado F, Ferri C, Mannini D, Corrado G, Bertoni F, Bachini P, Lelli G, Lieber MM, Song JM. Transitional cell carcinoma of the upper urinary tract: evaluation of prognostic factors by histopathology and flow cytometric analysis. J Urol 1991;145:1159–1163.
111. Oldbring J, Hellsten S, Lindholm K, Mikulowski P, Tribukait B. Flow DNA analysis in the characterization of carcinoma of the renal pelvis and ureter. Cancer 1989;64:2141–2145.
112. Badalament RA, O'Toole RV, Keyhani-Rofagha S, Barkley C, Kenworthy P, Accetta P, Wise H, Perez JF, Drago JR. Flow cytometric analysis of primary and metastatic bladder cancer. J Urol 1990;143:912–916.
113. Klein FA, Herr HW, Whitmore WF, Pinsky CM, Oettgen H, Melamed MR. Automated flow cytometry to monitor intravesical BCG therapy for superficial bladder cancer. Urology 1981;17:310–314.
114. Staino-Coico L, Huffman J, Wolf R, Pinsky CM, Herr HW, Whitmore WF, Oettgen HF, Darzynkiewicz Z, Melamed MR. Monitoring intravesical Bacillus Calmette-Guerin treatment of bladder carcinoma with flow cytometry. J Urol 1985;133:786–788.
115. Badalament RA, Gay H, Whitmore WF, et al. Monitoring intravesical bacillus Calmette-Guerin treatment of superficial bladder carcinoma by serial flow cytometry. Cancer 1986;58:2751–2757.
116. Bretton PR, Herr HW, Kimmel M, Fair, WR, Whitmore WF, Melamed MR. Flow cytometry as a predictor of response and progression in patients with superficial bladder cancer treated with Bacillus Calmette-Guerin. J Urol 1989;141:1332–1336.
117. Huffman JL, Fradet Y, Cordon-Cardo C, Herr HW, Pinsky CM, Oettgen HF, Old LJ, Whitmore WF, Melamed MR. Effect of intravesical Bacillus Calmette-Guerin on detection of a urothelial differentiation antigen in exfoliated cells of carcinoma in situ of the human urinary bladder. Cancer Res 1985;45:5201–5204.
118. Farsund T, Laerum OD, Hostmasrk J, Jordfald G. Local chemotherapeutic effects in bladder cancer demonstrated by selective sampling and flow cytometry. J Urol 1984;131:22–32.
119. Klein FA, Whitmore WF, Wolf RM, Herr HW, Sogani PC, Staino-Coico L, Melamed MR. Presumptive downstaging from preoperative irradiation for bladder cancer as determined by flow cytometry: a preliminary report. Int J Rad Oncol Biol Phys 1983;9:487–494.
120. Wijkstrom H, Gustafson H, Tribukait B. Deoxyribonucleic acid analysis in the evaluation of transitional cell carcinoma before cystectomy. J Urol 1984;132:894–898.
121. Wijkstrom H, Tribukait B. Deoxyribonucleic acid flow cytometry in predicting response to radical radiotherapy of bladder cancer. J Urol 1990;144:646–651.

122. Winkler HZ, Nativ O, Hosaka Y, Farrow GM, Lieber MM. Nuclear deoxyribonucleic acid ploidy in squamous cell bladder cancer. J Urol 1989;141:297–302.
123. Shaaban A, Tribukait B, El-Bedeiwy A-F A, Ghoneim MA. Characterization of squamous cell bladder tumors by flow cytometric deoxyribonucleic acid analysis: a report of 100 cases. J Urol 1990;144:879–883.
124. Winkler HZ, Lieber MM. Primary squamous cell carcinoma of the male urethra: nuclear deoxyribonucleic acid ploidy studied by flow cytometry. J Urol 1988;139:289–303.
125. Grignon DJ, El-Naggar A, Ro JY, Johnson DE, Ayla AG. Deoxyribonucleic acid flow cytometry on primary adenocarcinoma of the bladder: an analysis of 36 cases. J Urol 1989;142:1206–1210.
126. Song J, Farrow GM, Lieber MM. Primary adenocarcinoma of the bladder: favorable prognostic significance of deoxyribonucleic acid diploidy measured by flow cytometry. J Urol 1990;144:1115–1118.
127. Pedersen T, Larsen JK, Krarup T. Characterization of bladder tumours by flow cytometry on bladder washings. Europ Urol 1978;4:351–(55)
128. Pritchett TR, Kanzler BS, Nichols PW, Bakke AC, Hechinger MK, Skinner DG, Parker JW. A simple and practical technique for detecting cancer cells in urine and urinary bladder washings by flow cytometry. Am J Clin Pathol 1985;84:191–196.
129. Chin JL, Huben RP, Nava E, et al. Flow cytometric analysis of DNA content in human bladder tumors and irrigation fluids. Cancer 1985;56:1677–(81).
130. deVere White RW, Olsson CA, Deitch AD. Flow cytometry: role in monitoring transitional cell carcinoma of the bladder. Urology 1986;28:15–(20).
131. Murphy WM, Emerson LD, Chandler RW, Moinuddin SM, Soloway MS. Flow cytometry versus urinary cytology in the evaluation of patients with bladder cancer. J Urol 1986;136:815–819.
132. Cowan DF, Wu B, Young G, Khanna OP. Correlation of histopathological, cytological and flow cytometric findings in neoplastic and nonneoplastic lesions of the bladder. J Urol 1987;138:753–757.
133. Klein FA, White FKH. Flow cytometry deoxyribonucleic acid determinations and cytology of bladder washings: practical experience. J Urol 1988;139:275–278.
134. Badalament RA, Kimmel M, Ajay H, et al. The sensitivity of flow cytometry compared with conventional cytology in the detection of superficial bladder carcinoma. Cancer 1987;59:2078–2085.
135. Hadjissotiriou GG, Green DK, Smith G, et al. Bladder cancer flow cytometry profiles in relation to histological grade and stage. Brit J Urol 1987;60:239–247.
136. Koss LG, Wersto RP, Simmons DA, Deitch D, Herz F, Freed S. Predictive value of DNA measurements in bladder washings. Cancer 1989;64:916–924.
137. Melamed MR, Traganos F, Sharpless T, Darzynkiewicz Z. Urinary cytology automation. Preliminary studies with acridine orange stain and flow-through cytofluorometry. Invest Urol 1976;13:331–338.
138. Collste LG, Darzynkiewicz Z, Traganos F, Sharpless TK, Sogani P, Grabstald, Whitmore WF, Melamed MR. Flow cytometry in bladder cancer detection and evaluation using acridine orange metachromatic mucleic acid staining of irrigation cytology specimens. J Urol 1980;123:478–485.
139. Collste LG, Devonec M, Darzynkiewicz Z, Traganos F, Sharpless TK, Whitmore WF, Melamed MR. Bladder cancer diagnosis by flow cytometry. Correlation between cell samples from biopsy and bladder irrigation fluid. Cancer 1980;45:2389–2394.
140. Klein FA, Herr HW, Sogani PC, Whitmore WF, Melamed MR. Detection and follow-up of carcinoma of the urinary bladder by flow cytometry. Cancer 1982;50:389–395.
141. Devonec M, Darzynkiewicz Z, Kostyra-Claps ML, Collste L, Whitmore WF, Melamed MR. Flow cytometry of low stage bladder tumors: correlation with cytologic and cystoscopic diagnosis. Cancer 1982;49:109–118.
142. Klein FA, Herr, HW, Whitmore WF, Sogani PC, Melamed MR. An evaluation of automated flow cytometry (FCM) in detection of carcinoma in situ of the urinary bladder. Cancer 1982;50:1003–1008.
143. Devonec M, Darzynkiewicz Z, Whitmore WF, Melamed MR. Flow cytometry for follow-up examinations of conservatively treated low stage bladder tumors. J Urol 1981;126:166–170.
144. Bretton PR, Myc A, Cordon-Cardo C, DeAngelis P, Fair WR, Melamed MR. Initial evaluation of a new epithelial antigen (T16) for bivariate flow cytometry of bladder irrigation specimens. Cytometry 1989;10:339–344.
145. Orihuela E, Varadachay S, Herr HW, Melamed MR, Whitmore WF. The practical use of tumor marker determination in bladder washing specimens. Assessing the urothelium of patients with superficial bladder cancer. Cancer 1987;60:1009–1016.
146. Klein FA, Herr HW, Sogani PC, Whitmore WF, Melamed MR. Flow cytometry of normal and nonneoplastic diseases of the bladder: an estimate of the false positive rate. J Urol 1982;127:946–948.
147. Tetu B, Katz RL, Kalter SP, von Eschenback AC, Barlogie B. Acridine-orange flow cytometry of urinary bladder washings for the detection of transitional cell carcinoma of the bladder. The influence of prior local therapy. Cancer 1987;60:1815–1822.
148. Hermansen DK, Badalament RA, Whitmore WF, Fair WR, Melamed MR. Detection of carcinoma in the post-cystectomy urethral remnant by flow cytometric analysis. J Urol 1988;139:304–307.
149. Hermansen DK, Badalament RA, Fair WR, Kimmel M, Whitmore WF, Melamed MR. Detection of bladder carcinoma in females by flow cytometry. Cytometry 1989;10:739–742.
150. Summers JL, Coon JS, Ward RM, Falor WH, Miller AW, Weinstein RS, Prognosis in carcinoma of the urinary bladder based upon tissue blood group ABH and Thomsen-Friedenreich antigen status and karyotype of the initial tumor. Cancer Res 1983;43:934–939.
151. Coon JS, McCall A, Miller AW, Farrow GM, Weinstein RS. Expression of blood-group-related antigens in carcinoma in situ of the urinary bladder. Cancer 1985;56:797–804.
152. Orntoft TF, Petersen SE, Wolf H. Dual-parameter flow cytometry of transitional cell carcinomas. Quantitation of DNA content and binding of carbohydrate ligands in cellular subpopulations. Cancer 1988;61:963–970.
153. Fradet Y, Islam N, Boucher L, Parent-Vaugeois C, Tardif M. Polymorphic expression of a human superficial bladder tumor antigen defined by mouse monoclonal antibodies. Proc Natl Acad Sci USA 1987;84:7227–7231.
154. Fradet Y, Tardif M, Bourget L, Robert J, et al. Clinical cancer progression in urinary bladder tumors evaluated by multiparameter flow cytometry with monoclonal antibodies. Cancer Res 1990;50:432–437.
155. Ramaekers FCS, Beck H, Vooijs GP, Herman CJ. Flow cytometric analysis of mixed cell populations using intermediate filament antibodies. Exp Cell Res 1984;153:249–253.
156. Huffman JL, Garin-Chesa P, Gay H, Whitmore WF, Melamed MR. Flow cytometric identification of human bladder cells using a cytokeratin monoclonal antibody. Ann NY Acad Sci 1986;468:302–315.
157. Feitz WFJ, Beck HLM, Smeets AWGB, Debryune FMJ, Vooijs GP, Herman CJ, Ramaekers FCS. Tissue-specific markers in flow cytometry of urological cancers: cytokeratins in bladder carcinoma. Int J Cancer 1985;36:349–356.
158. Ramaekers FCS, Huysmans A, Moesker O, Schaart G, Herman C, Vooijs P. Cytokeratin expression during neoplastic progression of human transitional cell carcinoma as detected by a monoclonal and a polyclonal antibody. Lab Invest 1985;52:31–38.
159. Hijazi A, Devenoc M, Bouvier R, Revillard J-P. Flow cytometry study of cytokeratin 18 expression according to tumor grade and deoxyribonucleic acid content in human bladder tumors. J Urol 1989;141:522–526.
160. Cooper D, Schermer A, Sun T-T. Classification of human epithelia and their neoplasms using monoclonal antibodies to keratins: strategies, applications, and limitations. Lab Invest 1985;52:243–256.

161. Schaafsma H, Ramaekers FCS, van Muijen GNP, Ooms ECM, Ruiter DJ. Distribution of cytokeratin polypeptides in epithelia of the adult human urinary tract. Histochemistry, 1989;91:151–159.
162. Wheeless LL, Berkan TK, Patten SF, Reeder JE, Robinson RD, Eldidi MM, Hulbert WC, Frank IN. Multidimensional slit-scan detection of bladder cancer. Preliminary clinical results. Cytometry 1986;7: 212–216.
163. Sheinfeld J, Reuter VE, Melamed MR, Fair WR, Morse M, Sogani PC, Herr HW, Whitmore WF, Cordon-Cardo C. Enhanced bladder cancer detection with the Lewis X antigen as a marker of neoplastic transformaiton. J Urol 1990;143:285–288.
164. Malone PR, Visvanathan KV, Ponder BAJ, et al. Oncogenes and bladder cancer. Br J Urol 1895;57:664.
165. Peehl DM, Stamey TA. Oncogenes: a review with relevance to cancers of the urogenital tract. J Urol 1986;135:897.
166. Meyers FJ, Gumerlock PH, Koris SP, et al. Human bladder and colon carcinomas contain activated ras p21. Cancer 1989;63:2177.
167. Viola M, Fromowitz F, Oravez S, Deb S, Schlom J. Ras Oncogene p21 expression is increased in premalignant lesions and high grade bladder carcinoma. J Exp Med 1985;161:1213–1218.
168. Neal DE, Marsh C, Bennett MK. Epidermal-growth factor receptors in human bladder cancer: comparison of invasive and superficial tumours. Lancet 1985;1:366–368.
169. Messing EM, Hanson P, Ulrich P, et al. Epidermal growth factor—interactions with normal and malignant urothelium: in vivo and in situ studies. J Urol 1987;138:1329–1335.
170. Rao JY, Hurst RE, Bales WD, Jones PL, Boss RA, Archer LT, Bell PB, Hemstreet GP. Cellular F-actin levels as a marker for cellular transformation: relationship to cell division and differentiation. Cancer Res 1990;50:2215–2220.
171. Tsai YC, Nichols PW, Hiti AL, et al. Allelic loss of chromosomes 9, 11, and 17 in human bladder cancer. Cancer Res 1990;50:44–47.
172. Sidransky D, Frost P, Von Eschenbach A, et al. Clonal origin of bladder cancer. N Engl J Med 1992;326:737–740.
173. Flocks RH, Kadesky MC. Malignant neoplasms of the kidney: an analysis of 353 patients followed 5 years or more. J Urol 1958;79: 196–201.
174. Robson CJ, Churchill BM, Anderson W. The results of radical nephrectomy for renal cell carcinoma. J Urol 1969;101:297–301.
175. Beahrs OH, Henson DE, Hatter RVP, Myers MH, eds. Manual for Staging Cancer. 3rd ed. Philadelphia, PA: JB Lippincott, 1988; p. 203.
176. deKernion JB, Ramming KP, Smith RB. The natural history of metastatic renal cell carcinoma: a computer analysis. J Urol 1978;120:148–152.
177. Bennington JL, Mayall BH. DNA cytometry on four-micrometer sections of paraffin-embedded human renal adenocarcinomas and adenomas. Cytometry 1983;4:31–39.
178. Rainwater LM, Hosaka Y, Farrow GM, Lieber MM. Well differentiated clear cell carcinoma: significance of nuclear deoxyribonucleic acid patterns studied by flow cytometry. J Urol 1987;137:15–20.
179. Oosterwuk E, Warnaar SO, Zwartenduk J, van der Velde EA, Fleuren GJ, Cornelisse CJ. Relationship between DNA ploidy, antigen expression and survival in renal cell carcinoma. Int J Cancer 1988;42:703–708.
180. Grignon DJ, Ayala AG, El-Naggar A, Wishnow KI, Ro JY, Swanson DA, McLemore D, Giacco GG, Guinee VF. Renal cell carcinoma. A clinicopathologic and DNA flow cytometric analysis of 103 cases. Cancer 1989;64:2133–2140.
181. Wolman SR, Catmuto PM, Golimbu M, Schinella R. Cytogenetic, flow cytometric and ultrastructural studies of twenty-nine nonfamilial human renal carcinomas. Cancer Res 1988;48:2890–2897.
182. Baisch H, Otto U, Kloppel G. Malignancy index based on flow cytometry and histology for renal cell carcinomas and its correlation to prognosis. Cytometry 1986;7:200–204.
183. Ljungberg B, Forsslund G, Stenling R, Zetterberg A. Prognostic significance of the DNA content in renal cell carcinoma. J Urol 1986; 135:422–426.
184. deKernion JB, Mukamel E, Ritchie AWS, Blyth B, Hannah J, Bohman R. Prognostic significance of the DNA content of renal carcinoma. Cancer 1989;64:1669–1673.
185. Ljungberg B, Larsson P, Stenling R, Roos G. Flow cytometric deoxyribonucleic acid analysis in stage I renal cell carcinoma. J Urol 1991; 146:697–699.
186. Ljungberg B, Stenling R, Roos G. Flow cytometric DNA analysis of renal-cell carcinoma. A study of fine needle aspiration biopsies in comparison with multiple surgical samples. Anal Quant Cytol Histol 1987;9:505–508.
187. Chin JL, Pontes E, Frankfurt OS. Flow cytometric deoxyribonucleic acid analysis of primary and metastatic human renal cell carcinoma. J Urol 1985;133:582–585.
188. Ljungberg B, Stenling R, Roos G. Prognostic value of deoxyribonucleic acid content in metastatic renal cell carcinoma. J Urol 1986;136:801–804.
189. Rainwater LM, Hosaka Y, Farrow GM, Kramer SA, Kelalis PP, Lieber MM. Wilms tumors: relationship of nuclear deoxyribonucleic acid ploidy to patient survival. J Urol 1987;138:974–977.
190. Nativ O, Winkler HZ, Reiman HM, Lieber MM. Squamous cell carcinoma of the renal pelvis: nuclear deoxyribonucleic acid ploidy studied by flow cytometry. J Urol 1990;144:23–26.
191. Currin SM, Lee SG, Walther PJ. Flow cytometric assessment of deoxyribonucleic acid content in renal cell adenocarcinoma: does ploidy status enhance prognostic stratification over stage alone? J Urol 1990; 143:458–463.
192. Rainwater LM, Farrow GM, Lieber MM. Flow cytometry of renal oncocytoma: common occurrence of deoxyribonucleic acid polyploidy and aneuploidy. J Urol 1986;135:1167–1171.
193. Norming U, Tribukait B, Gustafson H, Nyman CR, Wang N, Wÿkström H. Deoxyribonucleic acid profile and tumor progression in primary carcinoma in situ of bladder: study of 63 patients with grade 3 lesions. J Urol 1992;147:11–15.

CLINICAL COMMENTARY

James M. Kozlowski

Introduction

Dr. Shankey has constructed a concise, yet impressively comprehensive, summary detailing the current and projected clinical applications of flow cytometry (FCM)/image analysis (IA) for patients with urologic cancers. The following commentary is designed to highlight some of the strengths and weaknesses of these technologies, as viewed from the perspective of a urologic oncologist.

Prostate Cancer

Adenocarcinoma of the prostate is being detected with increasing frequency because of the impact of: (*a*) enhanced patient awareness and (*b*) relatively routine utilization of serum prostate-specific antigen (PSA) determinations and transrectal ultrasonography of prostate (TRUSP) as adjuncts to the digital rectal examination (DRE). Indeed, a serum PSA elevation may be the only abnormal parameter that ultimately prompts systematic, TRUSP-guided quadrant biopsies that then establish the diagnosis of previously unsuspected prostate cancer (stage T1c). In some instances, the latter biopsies will demonstrate prostatic intraepithelial neoplasia (PIN) in the absence of obvious invasive prostate cancer (1).

Multiparameter FCM/IA may oneday play a powerful role in facilitating the accurate biological discrimination of those patients with PIN and/or T1 tumors at greatest risk for subsequent disease progression. The inherent zonal and histopathologic heterogeneity of the human prostate strongly favors the use of (a) biologically representative portions of archival prostate cancer specimens and (b) image analysis that permits the generation of sophisticated information while maintaining the histopathologic "milieu" of the tissue under study. In most instances, fine-needle aspiration (FNA) will not be able to reliably distinguish patients manifesting high-grade PIN from those with early invasive cancers (1).

Assessment of DNA ploidy, percent S-phase activity, as well as other biologically relevant parameters, may also play an important role with respect to the management of patients with stage T2 cancers. For example, there is some evidence to suggest that aneuploid tumors may be less radioresponsive than their diploid counterparts (1). If these perceptions can be validated, such information would be of critical importance to those patients seriously considering this treatment option. Furthermore, it would be highly desirable if these methodologies were applied both retrospectively and prospectively to well-defined populations undergoing radiation monotherapy in order to define more precisely those patients least likely to benefit from such an approach. As noted by Dr. Shankey, over 50% of patients with stage T2b cancers have evidence of microscopic capsular penetration/seminal vesicle involvement not perceived preoperatively. The use of FCM or IA prospectively might assist in the identification of those patients at greatest risk for this development and most likely to benefit from neoadjuvant androgen ablation. Obviously, the use of FCM/IA to help shape treatment paradigms is ultimately dependent upon the accuracy of pretreatment biopsies to reflect the biological "macrocosm" of an admittedly heterogeneous primary tumor. It is essential that comparative analysis be made between the information derived from systematic quadrant biopsies and that provided from careful step-section analysis of radical prostatectomy specimens.

Advanced-stage disease (T3, T4, N+M+) will be clinically apparent in more than 40% of patients at the time of initial presentation. About 60% of patients with untreated stage C disease will exhibit evidence of disease progression at 5 years, with annual progression rates of 10–12% being anticipated. Approximately 85% of patients with stage D1 disease will demonstrate progression within 5 years and those with distant metastases (stage D2) have a median survival of about 30 months with an anticipated 5-year survival rate of approximately 20% (2). Multiparameter FCM and IA may help to further define (a) the need for empiric adjuvant radiation therapy in those patients found to have evidence of microscopic T3 disease following radical prostatectomy; (b) the potential utility of radical prostatectomy in patients with diploid tumors subsequently noted to have pelvic lymph node metastases; and (c) those patients with stage D2 cancers most likely to exhibit prompt relapse following androgen-ablative therapy and requiring early consideration for experimental systemic therapies.

Urothelial Cancers

Approximately 70% of bladder cancers are superficial (stages Tis, Ta, and T1) (3). Although such patients are at risk for the development of multiple tumor recurrences over time (i.e., polychronotropism), only 15–25% of these patients subsequently develop muscle-invasive tumors (4). Conversely, over 60% of patients with muscle-invasive disease (T2-4) do not have a history of antecedent superficial bladder cancer, suggesting that a biological dichotomy exists with respect to these tumor subsets.

These epidemiologic considerations suggest the need for diagnostic tools possessing the following capabilities: (a) the reliable detection of bladder cancer in "at risk" populations; (b) the dissection of superficial tumors permitting the identification of those patients at highest risk for the development of muscle-invasive disease; and (c) monitoring the impact of intravesical therapies designed to control the field change aberration, thus minimizing the frequency/severity of tumor recurrences. Available evidence suggests that cytology and FCM are complementary, rather than mutually exclusive, diagnostic procedures. An analysis of their respective strengths and weaknesses is a necessary prelude to their proper application in the clinical setting.

Bladder washings have become increasingly popular and generally provide a highly cellular sample with superior preservation of morphologic detail. Most, but not all, authorities favor this mechanism of specimen collection despite its inherent disadvantages, which include: (a) It is interventional and mildly uncomfortable. (b) When performed incorrectly, a hypocellular sample has been reported in up to 20% of cases, and (c) The act of catheterization may induce papillary aggregates that could render interpretation (particularly via FCM) more difficult (5, 6). In a study conducted by Murphy and associates, the bladder barbotage technique yielded an average of 30,000 evaluable nuclei per specimen (5). This contrasts with the findings of Cowan and associates, who noted that spontaneously exfoliated samples (i.e., voided urine) yielded an average of 10,000 evaluable nuclei per specimen (7). In addition to bladder washings, voided urine samples may be helpful when tumor is suspected in the proximal/distal urethra.

Virtually all cases of grade 3 transitional cell carcinoma, most squamous cell carcinomas, and adenocarcinoma of the bladder will be detected by cytologic evaluation. The diagnostic yield falls to about 75% for grade 2 tumors and 34% for grade 1 lesions when random analyses are made in a clinic population. The diagnostic yield rises to 98 and 70%, respectively, when urine cytology is used to monitor patients with a documented history of bladder cancer (8). The majority of grade 2–3 tumors are infiltrative, and features of squamous/glandular differentiation will be perceived in over 50% of such cases. Obviously, tumor cells that exfoliate from bladders with carcinoma in situ (CIS) cannot be reliably distinguished from those of invasive neoplasms. It should be emphasized that the latter often present as sessile lesions with a necrotic/ulcerative center covered with a fibrinous exudate. Unless an adjacent field change effect is present, voided urine samples and bladder washings may not contain tumor cells, making it difficult for either cytology or FCM to establish the diagnosis of malignancy.

When an at-risk population is screened, urinary cytodiagnosis has a specificity of 99%, a sensitivity of 85%, a false-positive rate of 11%, and a false-negative rate of 10% (5, 8). Factors contributing to the false-positive rate include (a) the presence of reactive cellular changes associated with urolithiasis, cystitis, and intravesical chemotherapy and (b) the exfoliation of reactive papillary aggregates as a result of previous instrumentation or spontaneous shedding of benign cell clusters from the upper tracts or from a trabeculated bladder (8). Conversely, the false-negative rate is attributable to a lack of spontaneous/induced shedding of most grade 1 and some grade 2 tumors, coupled with the morphologic similarity of some well-differentiated cancers to normal urothelial cells (8).

Since a good correlation has been found between the degrees of ploidy and proliferation in biopsy specimens and exfoliated cell material (9), I infrequently request the submission of fresh tumor material for routine DNA flow cytometry. When additional information is desirable (i.e., grade 2 neoplasms, stage T1 tumors), I prefer that such an analysis be performed following careful scrutiny of the most representative portions of the primary tumor. The use of deparaffinized archival material permits a thorough histologic evaluation of the tumor in question, rather than the empiric submission of "presumably" representative tissue specimens excised from freshly resected tumors.

The ability of FCM to detect bladder cancer may exceed 80% if the population under study consists of patients with a previous history of transitional cell carcinoma. This figure drops to about 65% in patients without a history of urothelial malignancy (5, 6). Although urinary cytology and FCM may detect urothelial neoplasia prior to the development of clinically demonstrable tumors, FCM may be slightly superior in this regard with reports of positive histograms preceding the appearance of visible tumors in over 30% of cases (6). In addition, FCM may be superior to urinary cytology in the assessment of noninvasive superficial bladder cancer and in those patients not having received prior treatment (5). With respect to the latter, the use of intravesical chemotherapy (thiotepa, mitomycin-C, and Adriamycin) may induce a markedly hypocellular sample rendering accurate FCM difficult. Conversely, intravesical immunotherapy with bacillus Calmette-Guerin vaccine (BCG) frequently induces a profound inflammatory reaction that may persist for 6–8 weeks following the conclusion of therapy, negating the accuracy of standard DNA FCM (10). It should be emphasized that the combination of FCM and cytodiagnosis is associated with a detection rate exceeding 90% when "at risk" populations are evaluated (5).

Flow cytometry may not detect the presence of tumor (i.e., false-negative) if (a) the sample size is exceedingly small (less than 5,000–10,000 evaluable nuclei); (b) the tumor is exclusively grade 1 in constitution, thus containing uniformly diploid cells; (c) the tumor is muscle-invasive and not associated with exfoliating urothelium; and (d) immediately following intravesical chemotherapy due to denudation of the bladder lining with resultant hypocellularity (5, 11). Flow cytometry may erroneously establish the diagnosis of cancer (i.e., false-positive) in the presence of: (a) urinary tract infection; (b) urolithiasis; (c) papillary aggregates from a trabeculated bladder/upper tract; (d) samples obtained from ileal conduits/continent urinary reservoirs that contain an abundance of inflammatory debris; and (e) the marked inflammatory response following intravesical BCG therapy. The development of multiparameter FCM (DNA/cytokeratin 18) facilitates exclusion of the "inflammatory tail" characteristic of BCG therapy and permits a more focused analysis of the transformed epithelial population of interest (i.e., cytokeratin 18 positive tumor cells).

Virtually all grade 1 tumors are diploid, and nearly all grade 3 tumors are aneuploid. Bladder tumors of intermediate grade (grade 2) are heterogeneous and may be further stratified on the basis of DNA ploidy, with diploid and aneuploid subsets reported in 40% and 60% of cases, respectively (4, 11). Of note, the 5-year survival of patients in the diploid subgroup approaches 85%. This contrasts with a 60% 5-year survival rate for patients with aneuploid tumors (4). It would appear that FCM has its greatest potential role in the biological dissection of these intermediate grade tumors, as well as the potentially "volatile" T1 cancers.

Sixty-five percent of patients with superficial tumors (stage Ta, T1) have diploid tracings, compared with 15% of patients with advanced stage disease (T2–T4) (4). Virtually all patients with carcinoma in situ have aneuploid tumor systems, with nearly half of these grade 3 tumors possessing multiple aneuploid cell populations (11). With respect to CIS, the primary role of flow cytometry logically involves the detection of early relapse and the more precise assessment of treatment efficacy following the completion of intravesical therapy. For example, a negative bladder wash FCM 6 months following the completion of intravesical BCG therapy is associated with an 85% likelihood of survival free of progression at 30 months (12). The predictive value of FCM is not restricted to patients with CIS. For example, in those patients with stage Ta papillary tumors (grades 1 and 2) having more than one aneuploid tumor incident, development of invasive disease will be noted in more than 65% of these cases, while only one-third of such patients lacking a pattern of repetitive aneuploidy develop muscle-invasive disease (13).

It is a misnomer to discuss bladder cancers as isolated entities. They represent but one component of the "urothelial system," which originates in the region of the collecting ducts and terminates in the proximal urethra. This entire system is at risk to the "field change" aberrations already discussed with respect to bladder cancer. At the present time, flexible and rigid instrumentation exists that permits the endoscopic assessment and sampling of this entire urothelial-lined system. Obviously, such samples are limited in size and contain a modest number of evaluable nuclei. For these reasons, such specimens are routinely submitted for standard histologic and cytologic evaluation. Because of hypocellularity, FCM has played a limited role in the evaluation of these diagnostic samples. Indeed, this setting may be ideal for the use of multiparameter IA.

In conclusion, cytology and FCM should be viewed as complementary diagnostic studies. These diagnostic modalities are best utilized within the context of multiple sequential sampling, which provides the urologic oncologist with a clearer perception of the "biological volatility" of an individual patient's tumor system and permits a more intelligent assessment of this dynamic continuum. As stated previously, FCM is probably best used selectively rather than reflexively in patients with urothelial cancer. In my practice, samples are frequently submitted for flow cytometry (*a*) at the time of initial presentation; (*b*) following a thorough transurethral resection to better assess the residual field change aberration, often obviating the need for "random" bladder biopsies; (*c*) to biologically dissect heterogeneous and potentially volatile neoplasms, such as grade 2 and stage T1 lesions; (*d*) appropriately timed following the completion of intravesical chemotherapy/immunotherapy; (*e*) at the time of surveillance visits in patients at high risk for relapse; and (*f*) as a component of all experimental protocols involving the use of neoadjuvant/adjuvant chemotherapy or attempts at bladder salvage using combined radiation/chemotherapy.

Renal Cell Cancers

FCM and IA have made but a modest impact on our perceptions and management of renal cell cancers. Dr. Shankey has provided a nicely balanced overview of the somewhat conflicting data generated with respect to the prognostic significance of aneuploidy in these tumor systems. Most, but certainly not all, studies suggest that aneuploid renal cell cancers are associated with a greater likelihood of progression and shortened survival. This appears to be true for both localized (stages 1 and 2) and advanced (stages 3 and 4) neoplasms. The finding of frequent aneuploidy in biologically indolent grade 1 oncocytomas would seem to be somewhat incongruous. This observation may be related to the abundant array of abnormal mitochondrial characteristic of these tumors (14) with the exaggerated amounts of mitochondrial DNA accounting for the aberrant FCM findings.

At the present time, there would seem to be very little justification for the preoperative determination of a given renal tumor's ploidy status. Such information would not and should not deter the urologic surgeon from standard management schemes. In my clinical practice, I routinely request determination of DNA content and cell cycle kinetics post-nephrectomy and utilize this information to determine the stringency of follow-up.

With respect to advanced-stage disease, objective sustained remissions have been observed in approximately 15% of patients subjected to interleukin-2 (IL-2)-based regimens. Although modest in scope, these results have stimulated the evaluation of a number of novel immunotherapeutic regimens. Recently, attempts have been made to correlate the results of such immunologically mediated therapy with ploidy status. Unfortunately, there does not appear to be a consistent relationship between the DNA content of a given renal tumor and its likelihood to respond to IL-2-based treatment.

Conclusion

FCM and IA have made the transition from the research laboratory to the clinical arena and have emerged as potentially powerful diagnostic methodologies. Despite their admitted utility, determinations of ploidy status and percent 8-phase activity are merely representative of a wide spectrum of tumor phenotypes which must be assessed in order to provide the urologic oncologist with timely and sophisticated information. It is essential that FCM and IA adapt themselves to the constraints of small sample sizes and to the ever-increasing demands of a multiparameter analysis of tumor subpopulations.

REFERENCES

1. Kozlowski JM, Grayhack JT: Carcinoma of the prostate. In: Adult and Pediatric Urology (Gillenwater, JY, Grayhack JT, Howards SS, and Duckett JW, eds.). Mosby Year Book, Chicago, ed. 2, vol 2, 1991;34:1277–1393.
2. Kozlowski JM, Ellis WJ, Grayhack JT: Advanced Prostate Carcinoma: Early Vs. Late Endocrine Therapy. Urol Clin North Am 1991;18(1):15–24.
3. Ragavan D, Shipley WO, Garnick MB, Russell PJ, Richie JP: Biology and management of bladder cancer. N Engl J Med 1990;322:1129–1138.
4. Blomjous CEM, Schipper NW, Baak JPA, Van Galen EM, De Voogt HJ, Meyer CJLM. Retrospective study of prognostic importance of DNA flow cytometry of urinary bladder carcinoma. J Clin Pathol 1988; 41:21–25.
5. Murphy WM, Emerson LD, Chandler RW, Moinuddin SM, Soloway MS: Flow cytometry versus urinary cytology in the evaluation of patients with bladder cancer. J Urol 1986;136:815–819.
6. De Vere White RW, Meyers FJ, Deitch AD: DNA flow cytometry in transitional carcinoma of the bladder. Prob Urol 1988;2:436–444.
7. Cowan DF, Wu B, Young GP, Khanna OP: Correlation of histopathological, cytological and flow cytometry findings in neoplastic and nonneoplastic lesions of the bladder. J Urol 1987;138:753–757.
8. Dean PJ, Murphy WM: Importance of urinary cytology and future role of flow cytometry. Urology Suppl 1985;26:11–17.
9. Collste LG, Devonec M, Darzynkiewicz Z, Traganos F, Sharpless TK, Whitmore WF Jr, Melamed MR: Bladder cancer diagnosis by flow cytometry. Correlation between cell samples from biopsy and bladder irrigation fluid. Cancer 1980;45:2389.
10. Staiano-Coico L, Huffman J, Wolf R, Pinsky CM, Herr HW, Whitmore WF, Jr, Oettgen HF, Darzynkiewicz Z, Melamed MR. Monitor-

ing intravesicle bacillus Calmette-Guerin treatment of bladder carcinoma with flow cytometry. J Urol 1985;133:786–788.
11. Norming U, Nyman CR, Tribukait B: Comparative flow cytometric deoxyribonucleic acid studies on exophytic tumor and random mucosal biopsies in untreated carcinoma of the bladder. J Urol 1989;142:1442–1447.
12. Bretton PR, Herr HW, Kimmel M, Fair WR, Whitmore WF, Jr, Melamed MR: Flow cytometry as a predictor of response and progression in patients with superficial bladder cancer treated with bacillus Calmette Guerin. J Urol 1989;141:1332–1336.
13. De Vere White RW, Deitch AD, West B, Fitzpatrick JM: The predictive value of flow cytometric information in the clinical management of stage 0 (Ta) bladder cancer. J Urol 1988;139:279–282.
14. Millan JC: Tumors of the kidney. In: Uropathology. Churchill Livingstone, New York, 1989, volume 2, pp 623–701.

17

Colorectal Neoplasia

KENNETH D. BAUER
Clinical Commentary by **Douglas E. Merkel** and **Gershon Y. Locker**

Many features have been shown to have prognostic significance for patients with colorectal cancer. Among these are DNA ploidy and proliferative activity as defined by flow cytometry. A large number of studies have evaluated the importance of DNA ploidy in colorectal cancer. To date, considerable controversy exists as to the significance of this measurement. This chapter will overview the state of this literature, including a discussion of technical features and describe the possible utilization of flow cytometry for more sophisticated biological assessment of colorectal lesions.

BACKGROUND

The disparity between published reports in terms of the prognostic significance of DNA ploidy could stem from a relatively modest prognostic difference due to DNA ploidy. If this were the case, a comparatively large sample size in a study would be required for the data to achieve statistical significance. Table 17.1 summarizes data from 15 previously published reports (1–15) in which the prognostic significance of DNA ploidy was examined in relation to the number of cases studied: No obvious relationship between the number of patients evaluated and the significance of DNA ploidy is apparent. Despite this, the significant proportion of recent reports that indicate little or no prognostic significance of DNA ploidy argues in favor of stratifying patients on the basis of major prognostic features (e.g., stage of disease) when assessing the possible adjunctive value of DNA ploidy.

Another aspect that might be a factor in explaining the controversy is variation in the quality of DNA histograms among the various published series. One variable that serves as a useful quality control parameter for DNA cytometry is the coefficient of variation (CV) about the diploid (and aneuploid) DNA-ploidy peak. Broad CV's can result from instrument misalignment, improper specimen processing, or from DNA staining errors. Alternatively, a single broad peak could reflect the underlying tumor biology—namely the presence of both a DNA-diploid and a near-diploid DNA-aneuploid population with a very modest and unresolvable difference in DNA content. As the CV increases beyond approximately 6%, it becomes increasingly difficult to resolve near-diploid or near-tetraploid populations from their euploid counterparts (Chapters 3 and 8), and the precision of cell cycle fitting decreases (16). Thus, another explanation for the differences in results from study to study could be differences in CV, such that studies including tumors with relatively broad CV values might "miss" near-diploid aneuploid populations that presumably carry a comparatively poor prognosis, and place such patients in a "DNA-diploid" group, thereby incorrectly including patients whose tumors have a comparatively more biologically aggressive behavior in the comparatively favorable DNA-diploid patient subset.

Table 17.2 summarizes CV data from several recently published studies. Based on these data, no obvious relationship is apparent between the prognostic significance of DNA ploidy and the CV data provided in the manuscript. From the standpoint of standardizing DNA flow cytometric analyses, however, it does seem essential that a convention be adopted as to when (i.e., on the basis of CV value) a sample should be rejected as insufficient for DNA ploidy analysis. Our opinion is that the cutoff should be at a value of approximately 5.5% to 6% for tumors in which no second peak is detected.

At least one recent report (2) suggests that near-diploid DNA-aneuploid colonic carcinomas (DNA index <1.2) behave like DNA-diploid tumors. Thus, the more relevant determination may be the identification of a DNA-aneuploid

Table 17.1
Prognostic Significance of DNA Ploidy in Colorectal Cancer in Relation to Risk for Recurrence or Overall Survival

Reference	Number of Cases	Probability
Witzig et al. (1)	694	0.001 (multivariate)
Harlow et al. (2)	69	0.2 (multivariate)
Scott et al. (3)	264	0.003 (multivariate)
Halvorsen, et al. (4)	149	ns[a] (multivariate)
Melamed, et al. (5)	33	ns
Schutte, et al. (6)	279	0.07 (univariate)
Emdin, et al. (7)	37	0.007 (univariate)
Bauer, et al. (8)[c]	97	0.1 (multivariate)
Wolley, et al. (9)[c]	33	nr[b]
Giaretti, et al. (10)	115	0.005 (multivariate)
Wiggers, et al. (11)	350	0.12 (univariate)
Quirke, et al. (12)[d]	125	0.02 (univariate)
Rognum, et al. (13)	100	0.04 (multivariate)
Armitage, et al. (14)	326	ns (multivariate)
Kokal, et al. (15)	77	0.0004 (univariate)

[a] ns = not significant
[b] nr = not reported
[c] colonic cancer only
[d] rectal cancer only

Table 17.2
Coefficient of Variation Data from DNA Flow Cytometric Analysis of Colorectal Cancer

Reference	Specimen Type	n	p	Range	Mean
Harlow et al. (2)	paraffin	69	0.2	2.7–6.6	4.8
Witzig et al. (1)	paraffin	694	0.001	3.5–16.6	7.3
Halvorsen et al. (4)	paraffin	149	ns	3.9–11.1	7.3
Wiggers et al. (11)	paraffin	350	0.12	N.D.[a]	N.D.
Armitage et al. (14)	paraffin	320	ns[b]	N.D.	7.8–8.3%[c]
	fresh			N.D.	8.3–9.7%[c]
Giaretti et al. (10)	paraffin	115	0.005	4–8%	5.6
	fresh			2–6%	2.8
Rognum et al. (13)	fresh	100	0.04[d]	N.D.	N.D.

[a]N.D. not defined
[b]ns not significant
[c]The mean value specified varied as a function of DNA ploidy.
[d]Significance based on comparison of groups with DNA index >1.25 vs DI <= 1.25.

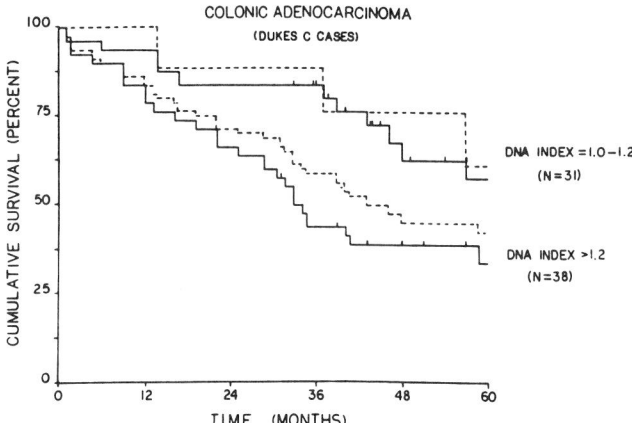

Figure 17.1. Kaplan-Meier survival curves (overall survival) for patients with colonic carcinoma. Astler-Coller Dukes C, having diploid or near-diploid aneuploid tumors with DNA index 1.0–1.2 vs. patients having aneuploid tumors with DNA index >1.2 (solid lines), Mantel-Cox (p = 0.02). Superimposed on survival curves are data from DNA-aneuploid (lower) vs diploid (upper) tumors (dotted lines). (From Harlow SP, Eriksen BL, Poggensee L, et al. Prognostic implications of proliferative activity and DNA ploidy in Astler-Coller Dukes Stage C colonic adenocarcinomas. Cancer Res 51: p2406, 1991, with permission from the publisher.)

population with a DNA index >1.2, rather than a subtle degree of DNA aneuploidy (Fig. 17.1). This approach deemphasizes the importance of a high quality CV. In addition, it could further minimize the false measurement of a "near-diploid DNA-aneuploid population" which, in fact, reflects differences in stainability due to chromatin structure variations in a heterogeneous cell population, rather than a true DNA ploidy abnormality (17, 18). Of note, at least one other recent study (Giaretti et al., 10) provides data suggesting that prognosis is worse for patients with diploid and near-diploid tumors relative to cases with DNA index >1.2.

Another fundamental issue that can profoundly influence the results obtained using DNA flow cytometry is related to tissue sampling. Many reports have documented that colorectal tumors can demonstrate regional differences in DNA ploidy (19–22). Wersto, et al. (23) conclude that DNA aneuploidy may be underestimated in 31% to 40% of tumors when a single sample is measured and no DNA aneuploidy is identified. On this basis, they recommend that at least three separate samples from a specimen should be used to define accurately the DNA ploidy of a colorectal tumor.

Given the fact that diploid nonneoplastic cells can "dilute" a rare neoplastic population in a colorectal cancer specimen, at least one investigation (Jones et al. 24) examined the relationship between the frequency of the observed DNA-aneuploid peak and patient outcome. This study concluded that the prognosis was inversely proportional to the frequency of cells comprising the DNA-aneuploid population in the specimen.

The great majority of studies published to date involving flow cytometric analysis of colorectal cancer include no information regarding the extent of neoplastic cells in the specimen, following routine microscopic evaluation by a pathologist. In a published report that discusses this point, Scott et al. (25) document heterogeneity in DNA ploidy within colorectal cancer specimens. In this study, either biopsy specimens or full thickness specimens were analyzed. Unfortunately, careful histologic scrutiny of the biopsy specimens was not possible. The authors conclude, however, that careful histologic selection of specimens for flow cytometry may reduce the problem of heterogeneous findings within a specimen.

Recent data of Crissman (26) indicate that the largest cellular component of a colorectal tumor is non-neoplastic—on average at least approximately 73% of total cells (Fig. 17.2). Given the fact that these lesions also frequently demonstrate considerable cellular heterogeneity, it appears critical that careful evaluation of the specimen be performed to insure that it contains a sufficient fraction of neoplastic cells prior to the performance of DNA ploidy or cell cycle analysis. Alternatively, in cases in which the contribution of neoplastic cells is very small, microscope-based image analysis (27) or multiparameter analysis coupling DNA measurements with a second marker, to allow for the enrichment of neoplastic cells (28), may be required. Clearly, given the problem of sampling in colorectal cancer, flow cytometric DNA content analysis in the absence of careful morphologic correlation should not be considered appropriate.

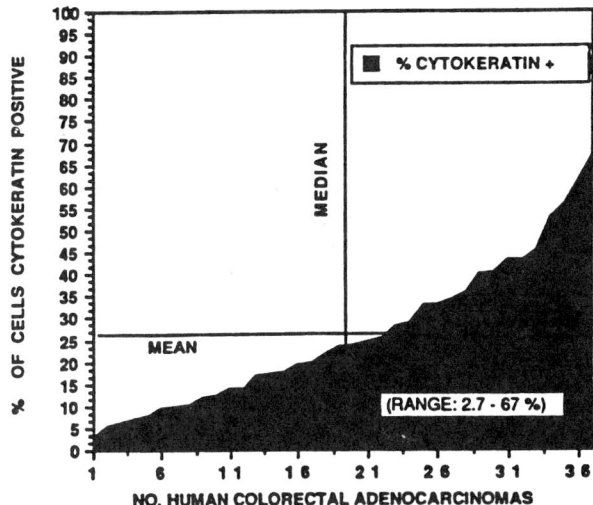

Figure 17.2. Graph demonstrating the distribution of cytokeratin as measured by flow cytometry in 27 colonic adenocarcinomas. The mean proportion of cytokeratin-positive cells, following immunofluorescence staining is 27%. (From Crissman J, Zarbo R, Ma C, Visscher D. Histopathologic parameters and DNA analysis in colorectal adenocarcinomas. Pathol Annual 1989;24(Part 2):122 with permission of the publisher.)

In addition to the issue of noncancer cells influencing the sensitivity of DNA flow cytometry, there is a body of literature at present that suggests that the cancer cells themselves may be heterogeneous with respect to DNA ploidy (i.e., Hiddemann, et al. 19). Further studies are needed to assess the significance of multiple DNA-aneuploid populations in histologically homogeneous tumors as reported by Hiddemann (19) and others.

DNA PLOIDY AND S-PHASE FRACTION IN RELATION TO STAGE AND HISTOLOGIC GRADE

In evaluating the possible adjunctive value of DNA ploidy or tumor proliferative activity in colorectal cancer, it is important first to evaluate possible interrelationships between these parameters and conventional prognostic features. One report of Banner et al. (29) showed a very strong relationship between DNA ploidy and stage of disease. On this basis, these authors hypothesized that the measurement of DNA ploidy on biopsy specimens might provide important information regarding the likelihood of tumor spread prior to definitive surgery. Table 17.3 provides a summary of five studies that examined the relationship between DNA ploidy and stage of disease. In reviewing these data it should be recognized that the staging system varies considerably between institutions (30), such that the different series may reflect staging-system-dependent differences in depth of invasion or metastatic involvement. For most of the cited studies, there appears to be a relationship between the frequency of DNA aneuploidy and stage of disease. However, the results from most studies suggest that the relationship between DNA ploidy and stage of disease, in fact, is rather modest. Nonetheless, these data do suggest that the evaluation of the significance of DNA aneuploidy might be best examined on a stage-controlled basis, given the fundamental importance of depth of invasion and metastatic involvement in predicting the biological aggressiveness of these lesions.

Table 17.4 illustrates the relationship between histologic grade and DNA aneuploidy from three studies. In this case, while there appears to be considerable variation between the different studies, a modest correlation between histologic grade and DNA aneuploidy appears evident.

Far less is known to date regarding the significance of S-phase fraction (SPF) measurements in colorectal cancer relative to that of DNA ploidy. Perhaps the earliest argument for the importance of proliferative activity came from the studies of Grinnell et al. more than 50 years ago (31) who showed a relationship between the frequency of mitoses and survival in colorectal cancer. A more recent study, however, using ^{3}H-thymidine labeling and autoradiographic analysis showed no obvious relationship between SPF and overall or disease-free survival (32). The flow cytometry investigations to date generally argue that proliferative activity may be of value in terms of predicting the biological aggressiveness of the tumor (1, 2, 6, 8, 12). However, very little has been done to assess the relationship between tumor proliferative activity measured by flow cytometry and tumor grade or stage. As early as 1980, however, Temple et al. (33) provided data that suggested a weak relationship between proliferative activity and Dukes stage for colon cancer patients. Another report (34) suggested that this relationship held true for Dukes B vs. Dukes C tumors, preferentially the subpopulation of DNA-aneuploid tumors. This same report (34) suggested that there was no relationship between the histologic grade of the tumor and proliferative activity.

At many points in this chapter, literature promoting the analysis of tumor proliferative activity will be cited. It should be stressed, however, that this measurement is technically more challenging than that of DNA ploidy. Many different algorithms are available for analysis of proliferative activity based on the quantification of cell cycle phase fractions (16). Different algorithms can provide significant differences in the proportions of cells in the various cell cycle phases (16). In addition, the type of DNA fluorochrome used (35), the extent of underlying debris in the tumor specimen (36), and can profoundly influence results.

The sampling issues discussed for DNA ploidy measurements are probably even more critical for the reliable evaluation of SPF. Careful pathologic review to select appropriate tumor regions containing abundant representative neoplastic cells (as discussed above) is crucial for valid measurement of proliferative activity. Finally, it must be recognized that the measurements are relative, rather than absolute. Thus, one must be exceedingly careful in terms of extrapolating results between laboratories. In our opinion, the establishment of a laboratory's own data base, generated on the basis of tumor type and stage of disease, to allow for the statistical evaluation of proliferative activity in one institution offers the best

Table 17.3
DNA Ploidy vs. Stage in Colorectal Cancer

Reference	Percent DNA Aneuploid (aneuploid/total)							
	Dukes A		Dukes B		Dukes C		Dukes D	
Banner et al. (29)	0%	(0/3)	57%	(4/7)	95%	(18/19)	89%	(8/9)
Armitage et al. (14)	52%	(24/46)	58%	(80/139)	63%	(48/76)	66%	(39/59)
Halvorsen, et al. (4)	44%	(8/18)	35%	(25/71)	60%	(21/35)	64%	(16/25)
Scott et al. (3)	29%	(4/14)	49%	(71/144)	57%	(43/75)	55%	(17/31)
Rognum et al. (13)	60%	(12/20)	64%	(23/36)	56%	(14/25)	74%	(14/19)
OVERALL	48%	(48/101)	51%	(203/397)	63%	(144/230)	66%	(94/143)

Table 17.4
DNA Ploidy vs. Histological Grade in Colorectal Cancer

Reference	Percent DNA Aneuploid (aneuploid/total)					
	Well-Diff.[a]		Moderately-Diff.[b]		Poorly-Diff[c]	
Rognum et al. (13)	55%	(6/11)	61%	(42/69)	75%	(15/20)
Halvorsen et al. (4)	53%	(26/53)	44%	(36/81)	50%	(7/14)
Scott et al. (3)	29%	(4/14)	52%	(100/193)	65%	(28/43)
OVERALL	46%	(36/78)	52%	(178/343)	65%	(50/77)

[a]Well-differentiated
[b]Moderately-differentiated
[c]Poorly-differentiated

Table 17.5
Frequency of DNA Aneuploidy in Relation to Size of Colorectal Adenomas

Investigator	Polyp Size		
	<1 cm	1–2 cm	>2 cm
van den Ingh et al. (38)	0% (0/8)	29.7% (11/37)	40% (4/10)
Quirke et al. (39)	0% (0/70)	7.3% (4/55)	16.1% (5/31)
Giaretti et al. (40)	0% (0/8)	25.7% (9/35)	52.4% (11/21)
Murad et al. (41)	25.6% (10/39)	37.5% (9/24)	60% (3/5)
OVERALL	8% (10/125)	21.9% (33/151)	34.3% (23/67)

prospect at present to replicate both the measurement of proliferative activity and its biological significance (2).

DNA ANEUPLOIDY IN THE ABSENCE OF FRANK MALIGNANCY

In reviewing the significance of DNA ploidy in colorectal lesions, it is important to recognize that DNA aneuploidy has been shown to occur in the absence of frank malignancy. A number of investigators, for example, have documented the occurrence of DNA aneuploidy in colorectal adenomas. One parameter that appears to correlate with the biological aggressiveness of these lesions is size, which has been shown to be a useful predictor of carcinomatous change in adenomas (37). Table 17.5 shows the relationship between DNA ploidy and size of colorectal adenomas from four representative reports and indicates an obvious relationship among these features. Another feature of importance in relation to the biological aggressiveness of colorectal adenomas is the degree of dysplasia. A number of studies have examined the relationship between DNA ploidy and the degree of dysplasia in colorectal adenomas (38–46). Goh and Jass (42) indicated an increase in the frequency of DNA aneuploidy of 4% to 36% when comparing adenomas with mild dysplasia vs. adenomas with severe dysplasia. While there have been variations in the conclusions drawn from such investigations, most studies support the notion that DNA aneuploidy is observed more frequently in highly dysplastic lesions. Despite these results, which indirectly support the contention that DNA aneuploidy may correlate with a higher risk for the development of colorectal cancer in these premalignant lesions, very little has been done to date to examine this issue directly through prospective or retrospective longitudinal studies. One study of Scott, et al. (45) examined colonic adenomas in patients who had resection of an invasive colorectal carcinoma subsequent to polypectomy and observed DNA aneuploidy in only 13% of the adenomas. While this argues that DNA ploidy in an adenoma may not predict the likelihood of subsequent development of colorectal cancer, more effort should be focused on this very important issue.

Very little has been done to date to investigate possible DNA content abnormalities in patients with ulcerative colitis. Interestingly, however, Hammarberg et al. (46) showed apparent relationships between the presence of dysplasia and the frequency of DNA aneuploidy in ulcerative colitis patients. In addition, an increased frequency of DNA aneuploidy was observed in this study for patients with a long duration (greater than 20 years) of disease, which is a patient subset at high risk for the development of colorectal cancer. This finding, and more recent studies (47, 48) illustrating that DNA aneuploidy may precede dysplasia, suggest that the presence of DNA aneuploidy could represent an early risk factor for the development of malignancy. Thus, DNA

Table 17.6
DNA Ploidy in Rectal Cancer

Investigator	n	Percent DNA-Aneuploid	Significance
Quirke et al. (12)	125	54% (67/125)	p=0.02 (univariate)
Jass et al. (50)	369	68% (253/369)	p<0.001 (univariate)
			p=0.03 (multivariate)
Scott et al. (51)	121	36% (43/121)	p=0.0024 (univariate)
Fisher et al. (52)	232	53% (123/232)	p=0.29 (multivariate)
			p=0.06, Dukes B + C
Witzig et al. (1)	94	57% (54/94)	p=0.099 (multivariate)

analysis could represent a useful adjunct to conventional histopathology for patients with longstanding ulcerative colitis. More work is needed to verify this interesting observation.

DNA FLOW CYTOMETRY IN RECTAL CANCER

The overwhelming majority of flow-cytometry-based studies published to date evaluating colonic and rectal cancer have pooled data from these anatomical sites. While such studies are valuable and will be referred to occasionally in the following discussion, this chapter will focus on studies reporting results specifically relating to rectal lesions.

A number of reports have investigated the significance of DNA ploidy and proliferative activity in rectal carcinoma. A summary of such studies is presented in Table 17.6. In terms of prognostic significance and using more rigorous multivariate analysis, two of the three studies listed conclude that DNA ploidy is at best a borderline prognostic feature.

DNA tetraploidy is observed at a frequency of approximately 5% to 14% in these lesions. A report of Quirke et al. (12) concludes that tetraploid cases correlate with a biological aggressiveness that is essentially identical to that of DNA-diploid cases. These findings led to the authors' hypothesis that progression in rectal cancer occurs from a DNA-diploid to a DNA-tetraploid to a DNA-aneuploid state. Unfortunately, studies of Jass et al. (50) and Scott et al. (51) suggest that the opposite is true—i.e., the biological aggressiveness of DNA-tetraploid lesions is more consistent with that of DNA-aneuploid tumors than those that are DNA-diploid. One explanation for the possible difference between the studies may be the large variety of definitions that are currently used to classify a DNA histogram as tetraploid (53).

Both Quirke et al. (12) and Witzig et al. (1) conclude that the measurement of both DNA ploidy and proliferative activity (i.e., cell cycle analysis) helps to define prognostically distinct patient subsets: For example, Quirke et al. (12) found that the median proliferative index (%S + G_2M) of patients with diploid DNA ploidy was 24%. When patients with DNA-diploid tumors with higher proliferative activity (S + G_2M) fraction >/25% were evaluated, outcomes essentially identical to those with DNA-aneuploid tumors were noted. Thus, an aggressive subset was defined on the basis of either the presence of a homogeneous DNA-aneuploid population or diploid DNA ploidy and high proliferative activity. Witzig et al. (1) showed that the prognostic significance of a combination of DNA ploidy and proliferative activity based on < or >/20% S + G_2M cells in the DNA-diploid cases is a better predictor of biological aggressiveness than is DNA ploidy alone. These early findings suggest, therefore, that proliferative activity may be a useful feature in rectal cancer; a concept that requires further confirmation.

DNA FLOW CYTOMETRY IN COLONIC CANCER

One of the earliest studies evaluating purely colonic lesions was that of Wolley et al. (9). This study prospectively evaluated the significance of DNA ploidy in 33 patients with Dukes A-D disease and demonstrated that DNA aneuploidy was a powerful adverse prognostic feature. This study from the Montefiore Hospital group was clearly a major catalyst for future investigation of the significance of DNA ploidy in colonic cancer. This study also suggested that DNA aneuploidy was a more frequent occurrence in colonic cancer cases presenting with metastatic disease. A more recent study (54) examining purely colonic cancer patients had similar results. Both of these studies, however, suffer from both the relatively modest number and the heterogeneous nature of the tumors investigated. Given the profound impact of the depth of invasion and metastatic status of the tumor on both prognosis and management, the focus will be on results from other studies to be summarized in which DNA ploidy was examined in the context of particular staging categories.

Stage A and B (Nonmetastatic) Disease

The prognosis for the individual patient with colorectal cancer is strongly influenced by the stage of the tumor (55). For this reason, we will attempt to summarize previous findings in relation to tumor stage. The pioneering work of Dukes (56) involving the staging of rectal cancer provided the foundation for modern staging schemes for colorectal cancer. Though modified from Dukes' original classification, these are often labeled as "Dukes stage." "Dukes" A lesions, confined to the wall of the large bowel (defined as the muscularis propria) without lymph node metastases (57), or alternatively restricted to the mucosa (58) are usually curable by surgical intervention (59). As has been discussed previously (Table 17.4), the incidence of DNA aneuploidy in these lesions appears to be lower than observed in cases with more invasive disease. Rognum et al. (13) demonstrated a marked difference in survival in relation to DNA aneuploidy for 20 stage A colorectal cancer cases. Bauer et al. (8) evaluated 45 colonic cancer patients with stage A and B1 disease. In this latter study, assessment of the prognostic utility of DNA ploidy following multivariate analysis revealed that DNA ploidy was not a significant feature in this subset of colonic cancer patients, although patients with high degrees of DNA aneuploidy (i.e., DNA index >/1.2) had a borderline tendency for more aggressive clinical outcome relative to their counterparts with DNA-diploid or near-diploid tumors. Scott, et al. (3) examined the significance of DNA ploidy in 60 patients with stage A and B1 (tumors invading the muscularis propria) colorectal cancer and also concluded that DNA ploidy did not offer statistically significant prognostic information in this subset. Given the excellent prognosis of these patient subsets, however, the finding of limited if any prognostic predictability of DNA ploidy in both of these studies may reflect the relatively limited (five years) follow-up of patients in both investigations.

In the investigation of Bauer et al. (8), which also included the measurement of proliferative activity, it was concluded, following multivariate analysis, that proliferative activity was a prognostic predictor of more importance than DNA ploidy for patients with stage A and B1 colonic cancer.

Patients with stage B2 disease (extending to the serosa and beyond without evidence of lymph node metastases) generally have a more aggressive clinical course than their counterparts with stage A and B1 disease, despite efforts toward curative resection (60). Given this problem, much effort has recently focused on adjuvant chemotherapy for these patients, as well as for patients with lymph node metastatic disease (61, 62). These studies have yielded promising results in terms of a subset of patients benefitting from adjuvant chemotherapy with the combination of levamisole (an immunomodulator) and 5-fluorouracil, a cell-cycle-specific chemotherapeutic agent.

If DNA ploidy (or proliferative activity) could, in fact, reliably identify a patient subset with particularly aggressive disease, one could conceivably envisage the use of fine-needle aspiration (63) or pretreatment biopsy for flow cytometric analysis to provide assistance in therapeutic decision making. Reaching this lofty goal based on flow cytometry measurements now appears to be some years away.

Witzig et al. (1) examined DNA ploidy in relation to survival for 273 colorectal patients with stage B2 disease and found that DNA ploidy was a statistically significant predictor of disease-free and overall survival. In addition, when combining DNA ploidy and the assessment of proliferative activity (see section on rectal cancer above), a more significant difference in outcome was noted between patients with favorable DNA content tumor features (absence of DNA aneuploidy with %S + G_2M <20%) relative to those with either DNA aneuploidy or DNA diploidy and higher proliferative activity. This study, performed retrospectively on tumors from patients who had been enrolled on one of three prospective adjuvant clinical trials involving the use of adjuvant therapy should be of value in evaluating the possible efficacy of DNA flow cytometry in assisting clincians with adjuvant chemotherapy decisions. An earlier Mayo Clinic investigation of Scott et al. (3) involving 94 patients yielded no significant differences in five-year overall survival for patients with DNA-diploid vs. DNA-aneuploid tumors. A more recent study (64) under the auspices of the Eastern Cooperative Oncology Group (ECOG) further supports the notion that the measurement of proliferative activity is useful in terms of predicting the biological aggressiveness of colorectal tumors. The early results from this study for patients enrolled in randomized clinical trials are promising and should be quite informative when analyses of correlations between therapeutic responsiveness and DNA ploidy and proliferative activity are completed.

Stage C (Lymph Node Metastatic) Disease

When colorectal cancer has metastasized to lymph nodes, the prognosis for patients is usually markedly poorer than that observed in cases in which metastases are not detected. As mentioned previously, DNA aneuploidy is far more common in these lesions. A recent study of Harlow et al. (2) has examined the prognostic utility of DNA flow cytometry in 69 patients with stage C colonic cancer. In this study, DNA ploidy per se did achieve statistical significance, although DNA-aneuploid cases with DNA index >1.2 showed significantly poorer outcome. This study again demonstrated that proliferative activity, based on statistically defined groupings provided a prognostic indicator of more importance than DNA ploidy: Patients whose tumors demonstrated low proliferative activity had a five-year survival of approximately 85%, as compared to a five-year survival of approximately 37% for patients whose tumors demonstrated high proliferative activity. Schutte et al. (6) also show that tumor proliferative activity tends toward more potent prognostication than DNA ploidy in stage C colorectal cancer cases.

The large study of Witzig et al. (1) also examined stage C colorectal cancer cases. DNA ploidy was not a significant predictor among stage C patients in relation to overall survival in this study. However, as described earlier, when cases with DNA-aneuploidy and DNA-diploid cases with a high proliferative activity (%S + G_2M>/20%) were combined, the data reached statistical significance in terms of overall survival in comparison to the more favorable patient subset having DNA-diploid tumors with lower proliferative activity. This study argues that a subset of patients without detectable DNA aneuploidy who have high proliferative activity tend toward biologically aggressive disease.

A recent compilation of data examining both tumor proliferative activity and nodal status in relation to overall survival is shown in Figure 17.3. In looking at the lymph-node-positive cases (stage C), a significantly poorer outcome is noted for patients whose tumors demonstrate moderate- or high-proliferative activity, relative to cases with lower S-phase fraction. Interestingly, the subsets of patients with nonmetastatic lesions (stage A and B) show very similar survival trends relative to those with stage C disease: Stage A and B cases with low and moderate proliferative activity demonstrate relatively favorable five-year survival characteristics similar to stage C cases with low proliferative activity. In addition, stage A and B colonic adenocarcinomas with high proliferative activity demonstrate survival patterns similar to that observed for stage C patients with moderate and high proliferative activity. Such data suggest that nodal metastatic status and tumor proliferative activity are potent predictors of outcome in colonic cancer. Clearly, more stud-

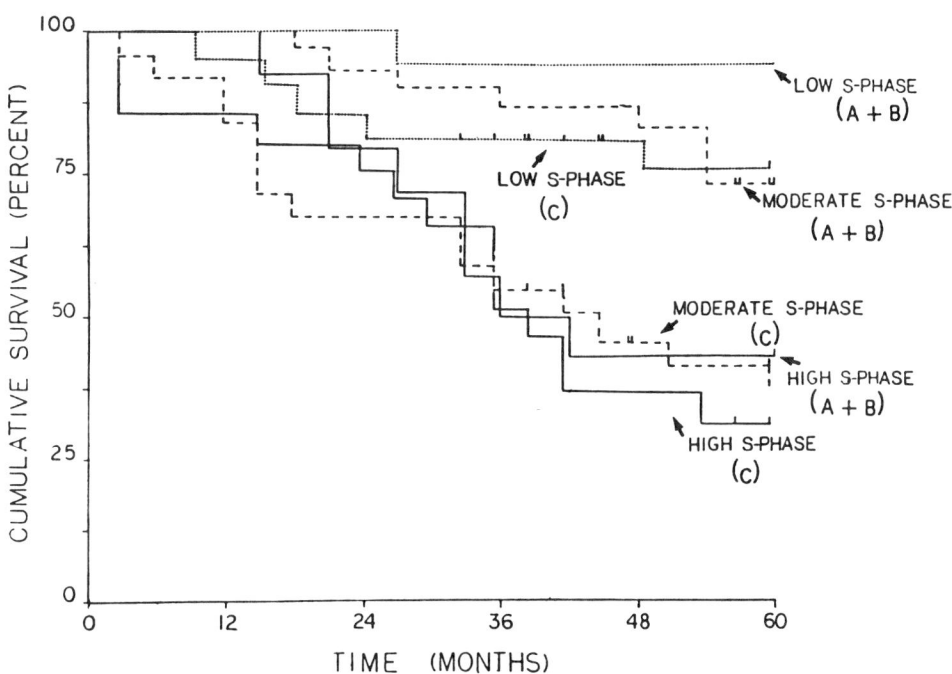

Figure 17.3. Kaplan-Meier survival curves for colonic cancer cases stratified by tumor proliferative activity and nodal metastatic status. (A + B): Stage A + B (nonmetastatic) cases. (C) Stage C, lymph node metastatic cases. Tumor proliferative activity is defined as follows: *Low S-phase fraction:* <mean −0.5 standard error of the mean; *moderate S-phase fraction:* mean +/− 0.5 standard error of the mean; *high S-phase fraction:* > mean + 0.5 standard error of the mean.

ies are needed to confirm the significance of this early finding.

Stage D

The presence of distant metastases, principally to the liver, constitutes stage D colorectal cancer. Prognosis for patients in this stage category is generally very grim. At least four reports have examined the value of DNA ploidy for patients with stage D cancer. Scott et al. (3) suggested that DNA aneuploidy was a significant prognostic feature for patients with stage D and stage C1 disease, but not for the other stages of disease. In contrast, studies of Finan et al. (65), Rognum et al. (50), and Armitage et al. (66) support the notion that DNA ploidy is not a significant prognostic feature for these patients.

One area of emerging interest for the management of patients with metastatic disease involving one lobe of the liver stems from previous reports indicating that a small subset of these patients can be cured by partial hepatectomy (67). This procedure involves considerable morbidity, however, and does not currently appear to benefit a large fraction of these patients. Based on such observations, DNA flow cytometry has been proposed as a possible adjunct in stratifying patients for this procedure based on analysis of tumor DNA ploidy or proliferative activity. Tsushima et al. (68) examined DNA ploidy in this context in 28 patients and found that survival for patients with DNA-aneuploid tumors who had received surgical resection of their hepatic metastases was significantly poorer than that of their DNA-diploid counterparts. A more recent study of Harlow et al. (69) involving 18 patients also indicated that patients with low proliferative activity fared better than their counterparts with higher proliferative activity following surgical resection of hepatic metastases. While these early results are promising, much more needs to be done to identify a possible adjunctive role of flow cytometry for this subset of patients with stage D disease.

Other Lesions of the Gastrointestinal Tract

While earlier studies focused on colorectal tumors, data are rapidly accumulating regarding the possible adjunctive value of DNA flow cytometry for tumors elsewhere in the gastrointestinal tract. One of the more interesting examples is Barrett's esophagus, where a body of literature is accumulating to suggest that DNA aneuploidy and/or elevated proliferative activity may correlate with increased risk for the development of esophageal adenocarcinoma (70, 71). In addition, Robinovitch et al. (72) have shown that Barrett's adenocarcinoma lesions often demonstrate considerable DNA ploidy heterogeneity as do areas of surrounding dysplastic epithelium indicative of genomic instability. Such findings further suggest that the presence of multiple DNA-aneuploid populations in Barrett's esophagus lesions may correlate with a particularly high risk for progression to cancer. Other studies using both flow and image cytometry suggest that the presence of DNA aneuploidy in esophageal squamous cell carcinoma may correlate with tumor stage and histologic grade (73). In addition, the presence of DNA aneuploidy has been shown to correlate with poor prognosis for the patient (74, 75) relative to DNA-diploid esophageal squamous cell carcinomas, and these lesions have been shown to demonstrate intratumor heterogeneity in DNA ploidy (76).

A number of reports have examined the significance of DNA-flow-cytometry-derived features in gastric lesions. Kimura et al. (77) have shown that DNA ploidy is a significant prognostic feature for gastric carcinoma, particularly in advanced (Stage III and IV) lesions. Report of Ohyama et al. (78), who utilized in vitro or in vivo bromodeoxyuridine labeling and multiparameter flow cytometry, documents that low proliferative activity correlates with comparatively favorable prognosis in terms of overall survival for patients with both early- and advanced-stage gastric carcinoma. Tsushima et al. (79) have further shown that DNA aneuploidy correlates with larger tumor size and higher histologic grade in gastric leiomyosarcomas. This same study showed a correlation between DNA ploidy and overall survival for patients with these lesions.

Finally, DNA aneuploidy has been reported to provide significant prognostic information for patients with hepatic and pancreatic tumors. For example, a recent report (80) showed that DNA ploidy status correlated with overall survival for patients with hepatocellular carcinoma, following Cox multivariate analysis. In this study, a particularly unfavorable overall survival was noted for patients with comparatively modest abnormalities in DNA ploidy (i.e., DNA index <1.5) relative to patients with greater degrees of DNA aneuploidy. Similar findings demonstrating strong correlations between DNA ploidy and outcome have been obtained for pancreatic adenocarcinoma cases (81, 82). And, in a study combining DNA flow cytometry and aspiration cytology of solid and papillary epithelial neoplasms of the pancreas (83), both DNA ploidy and SPF were shown to represent useful adjuncts to cytologic evaluation.

Future Applications

Flow cytometry today has been applied to colorectal lesions nearly exclusively for the evaluation of DNA ploidy and tumor proliferative activity. Given the power of this technology for sophisicated multiparameter measurements, the potential for measuring relevant tumor biological features appears nearly limitless. Specific cellular proteins can be reliably quantified using currently available instrumentation, allowing for a broad spectrum of potential measurements of cellular phenotype. One limitation to this approach is that tumor dissociation utilizing general protolytic enzymes can destroy cellular proteins, including the oncoprotein c-*myc* (84). The use of mechanical methods and/or enzymatic dissociation without general proteases should circumvent this problem.

Wirsching et al. (85) have successfully measured carcinoembryonic antigen (CEA) in colorectal cancer specimens, along with cellular DNA, and have shown that high levels of CEA in carcinoma cells can help to separate these cells from normal mucosa as well as inflammatory and stromal elements. This would appear to be a very promising approach to allow for more precise assessment of DNA ploidy and proliferative activity. Bauer et al. (86) have shown the feasibility of simultaneous analysis of cellular DNA content and the proliferation-associated nuclear antigen p105. Analysis of this type may allow for more in-depth evaluation of proliferative activity based on the consideration of the growth fraction, rather than simple the $S-$ or $S + G_2M$ fraction.

Anastasi et al. (87) have utilized a polyclonal nuclear antibody preparation to quantify nuclear prominence from paraffin-embedded Hodgkin's disease specimens. This strategy allowed for the identification of rare DNA-aneuploid populations based on their increased nucleolar prominence. This latter approach, utilizing monoclonal or polyclonal antibodies recognizing proteins associated with specific nuclear domains, may allow for detailed quantification of characteristics defining anaplasia and, as such, provide a less subjective approach for the grading of colorectal cancer. Finally, given the recent efficacy of immunomodulators in the adjuvant treatment of colorectal cancer, the utilization of immunofluorescence approaches to evaluate lymphocyte subpopulations could be a useful early predictor of successful therapeutic responsiveness, particularly when coupled with the simultaneous quantification of a specific mRNA species (88). The above list clearly is only the "tip of the iceberg" in terms of the potential for adjunctive measurements. Thus the future of the flow cytometry as an adjunctive tool in the classification of colorectal cancer appears very promising.

REFERENCES

1. Witzig TE, Loprinzi CL, Gonchoroff NJ, et al. DNA ploidy and cell kinetic measurements as predictors of recurrence and survival in stages B2 and C colorectal adenocarcinoma. Cancer 1991;68:879–888.
2. Harlow SP, Eriksen BL, Poggensee L, et al. Prognostic implications of proliferative activity and DNA aneuploidy in Astler-Coller Dukes stage C colonic adenocarcinomas. Cancer Res 1991;51:2403–2409.
3. Scott N, Wieand HS, Moertel CG, et al. Dukes' stage, tumor site, preoperative plasma CEA level, and patient prognosis related to tumor DNA ploidy pattern. Arch Surg 1987;122:1375–1379.
4. Halvorsen TB, Johannesen E. DNA ploidy, tumor site, and prognosis in colorectal cancer. Scand J Gastroenterol 1990;25:141–148.
5. Melamed MR, Enker WE, Banner P, Janov AJ, Kessler G, Darzynkiewicz Z. Dis. Colon Rectum 1986;19:184–186.
6. Schutte B, Reynders, Wiggers T, et al. Retrospective analysis of the prognostic significance of DNA content and proliferative activity in large bowel carcinoma. Cancer Res 1987;47:5494–5496.
7. Emdin SO, Stenling R, Roos G. Prognostic value of DNA content in colorectal carcinoma. Cancer 1987;60:1282–1287.
8. Bauer KD, Lincoln ST, Vera-Roman JM, et al. Prognostic implications of proliferative activity and DNA aneploidy in colonic adenocarcinomas. Lab Invest 1987;57:329–335.
9. Wolley RC, Schreiber K, Koss LG, Karas M, Sherman A. DNA distribution in human colon carcinoma and its relationship to clinical behavior. Natl Cancer Inst 1982;69:15–22.

10. Giaretti W, Danova M, Geido E, et al. Flow cytometric DNA index in the prognosis of colorectal cancer. Cancer 1991;67:1921–1927.
11. Wiggers T, Arends JW, Schutte B, Volovics L, Bosman FT. A multivariate analysis of pathologic prognostic indicators in large bowel cancer. Cancer 1988;61:386–395.
12. Quirke P, Dixon MF, Clayden AD, et al. Prognostic significance of DNA aneuploidy and cell proliferation in rectal adenocarcinomas. J Pathol 1987;151:285–291.
13. Rognum TO, Lund E, Meling GI, Langmark F. Near diploid large bowel carcinomas have better five-year survival than aneuploid ones. Cancer 1991;68:1077–1081.
14. Armitage NCM, Ballantyne KC, Sheffield JP, Clarke P, Evans DF, Hardcastle JD. A prospective evaluation of the effect of tumor cell DNA content on recurrence in colo-rectal cancer. Cancer 1991;67:2599–2604.
15. Kokal W, Sheibani K, Terz J, Harada R. Tumor DNA content in the prognosis of colorectal carcinoma. JAMA 1986;255:3123–3127.
16. Dean PN. Methods of data analysis in flow cytometry. In: Van Dilla MA, Dean PN, Laerum OD, Melamed MR, eds. Flow Cytometry: Instrumentation and Data Analysis. New York: Academic Press, 1985:195–221.
17. Darzynkiewicz Z, Traganos F, Kapuscinski J, Staiano-Coico L, Melamed MR. Accessibility of DNA in situ to various fluorochromes: relationship to chromatin changes during erythroid differentiation of Friend leukemia cells. Cytometry 1984;5:355–363.
18. Klein FA, White FKH. Flow cytometry deoxyribonucleic acid determinations and cytology of bladder washings: practical experience. J Urol 1988;139:275–278.
19. Hiddeman W. von Bassewitz DB, Kleinemeier H-J, et al. DNA stemline heterogeneity in colorectal cancer. Cancer 1986;58:258–263.
20. Peterson SE, Bichel P, Lorentzen M. Flow cytometric demonstration of tumor-cell subpopulations with different DNA content in human colorectal carcinoma. Eur J Cancer 1978;15:383–386.
21. Peterson SE, Lorentzen M, Bichel P. A mosaic subpopulation structure of human colorectal carcinomas demonstrated by flow cytometry. In: Laerum OD, Lindmo T, Thorud E, eds. Flow Cytometry IV. Bergen: Universitetsforlaget, 1980:412–416.
22. Tribukait B, Hammarberg C, Rubio C. Ploidy and proliferation patterns in colorectal adenocarcinomas related to Duke's classification and to histopathological differentiation. Acta Pathol Microbiol Immunol Scand [A] 1983;91:89–95.
23. Wersto RP, Liblit RL, Deitch D, Koss LG. Variability in DNA measurements in multiple tumor samples of human colonic carcinoma. Cancer 1991;67:106–115.
24. Jones DJ, Moore M, Schofield PF. Refining the prognostic significance of DNA ploidy status in colorectal cancer: a prospective flow cytometry study. Int J Cancer 1988;41:206–210.
25. Scott NA, Grande JP, Weiland LH, Pemberton JH, Beart RW, Lieber MM. Flow cytometric DNA patterns from colorectal cancers—how reproducible are they? Mayo Clin Pro 1987;62:331–337.
26. Crissman JD, Zarbo RJ, Ma CK, Visscher DW. Histopathologic parameters and DNA analysis in colorectal adenocarcinomas. Pathol Annu 1989;24,(Part 2):103–147.
27. Forsslund G, Cedermark B, Ohman U, Erhardt K, Zetterberg A, Auer G. The significance of DNA distribution pattern in rectal carcinoma: a preliminary study. Dis Colon Rectum 1984;27:579–584.
28. Feitz WFJ, Beck HLM, Smeets AWGB, et al. Tissue-specific markers in flow cytometry of urological cancers: cytokeratins in bladder carcinoma. Int J Cancer 1985;36:349–357.
29. Banner BF, De La Vega JET, Roseman DL, Coon JS. Should flow cytometric DNA analysis precede definitive surgery for colon cancer? Ann Surg 1985;202:740–744.
30. Cohen AM, Shank B, Friedman MA. Colorectal cancer. In: DeVita Jr VT, Hellman S, Rosenberg ST, eds. Cancer: Principles and practice of oncology. 3rd ed. Philadelphia: Lippincott, 1989:895–964.
31. Grinnell RS. The grading and prognosis of carcinoma of the colon and rectum. Ann Surg 1939;109:500–533.
32. Meyer JS, Prioleau PG. S-phase fractions of colorectal carcinomas related to pathologic and clinical features. Cancer 1981;48:1221–1228.
33. Temple WJ, Sugarbaker EV, Thornthwaite JT, Hensley GT, Ketcham AS. Correlation of cell cycle analysis with Duke's staging in colon cancer patients. J Surg Res 1980;28:314–318.
34. Tribukait B, Hammarberg C, Rubio C. Ploidy and proliferation patterns in colorectal adenocarcinomas related to Dukes' classification and histopathological differentiation. Acta Pathol Microbiol Immunol Scand [A] 1983;91:89–95.
35. Dean PN, Gray JW, Dolbeare FA. The analysis and interpretation of DNA distributions measured by flow cytometry. Cytometry 1982;3:188–195.
36. Haag D, Feichter G, Goerttler K, Kaufmann M. Influence of systematic errors on the evaluation of S phase portions from DNA distributions of solid tumors as shown for 328 breast carcinomas. Cytometry 1987;8:377–385.
37. Muto T, Bussey HJR, Morson BC. The evolution of cancer of the colon and rectum. Cancer 1975;36:2251–2270.
38. van den Ingh HF, Griffioen G, Cornelisse CJ. Flow cytometric detection of aneuploidy in colorectal adenomas. Cancer Res 1985;45:3392–3397.
39. Quirke P, Fozard JBJ, Dixon MF, Dyson JED, Giles GR, Bird CC. DNA aneuploidy in colorectal adenomas. Br J Cancer 1986;53:477–481.
40. Giaretti W, Sciallero S, Bruno S, Geido E, Aste H, Di Vinci A. DNA flow cytometry of endoscopically examined colorectal adenomas a adenocarcinomas. Cytometry 1988;9:238–244.
41. Murad T, Bauer KD, Scarpelli DG. Histopathologic flow cytometric analysis of adenomatous colonic polyps. Arch Pathol Lab Med 1989;113:1003–1008.
42. Goh HS, Jass JR. DNA content and the adenoma-carcinoma sequence in colorectum. J Clin Pathol 1986;39:387–392.
43. Petrova AS, Subrichina GN, Tschistjakova OV, et al. DNA ploidy and proliferation characteristics of bowel polyps analyzed by flow cytometry compared with cytology and histology. Arch Geschwulstforsch 1986;56:179–191.
44. Hamada S, Itoh R, Fujita S. DNA distribution pattern of the so-called severe dysplasia and small carcinomas of the colon and rectum and its possible significance in the tumor progression. Cancer 1988;61:1555–1562.
45. Scott NA, Weiland LH, Dozois RR, Beart RW, Lieber MM. DNA aneuploidy in solitary colonic adenomas and the future risk of colorectal cancer. Dis Colon Rectum 1988;31:423–426.
46. Hammarberg C, Slezak P, Tribukait B. Early detection of malignancy in ulcerative colitis—a flow cytometric study. Cancer 1984;53:291–295.
47. Loftberg R, Tribukait B, Ost A, Brostrom O, Reichard H. Flow cytometric DNA analysis in longstanding ulcerative colitis: a method of prediction of dysplasia and carcinoma development? Gut 1987;28:1100–1106.
48. Lofberg R, Caspersson T, Tribukait B, Ost A. Comparative DNA analyses in longstanding ulcerative colitis with aneuploidy. Gut 1989;30:1731–1736.
49. Kouri M, Laasonen A, Mecklin J-P, Jarvinen H, Franssila K, Pyrhonen S. Diploid predominance in hereditary nonpolyposis colorectal carcinoma evaluated by flow cytometry. Cancer 1990;65:1825–1829.
50. Jass JR, Mukawa K, Goh HS, Love SB, Capellaro D. Clinical importance of DNA content in rectal cancer measured by flow cytometry. J Clin Pathol 1989;42:254–259.
51. Scott NA, Rainwater LM, Wieand HS, et al. The relative prognostic value of flow cytometric DNA analysis and conventional clinicopathologic criteria in patients with operable rectal carcinoma. Dis Colon Rectum 1987;30:513–520.
52. Fisher E, Siderits RH, Sass R, Fisher B. Value of assessment of ploidy in rectal cancers. Arch Pathol Lab Med 1989;113:525–528.

53. Wersto RP, Liblit RL, Koss LG. Flow cytometric DNA analysis of human solid tumors: a review of the interpretation of DNA histograms. Human Pathol 1991;22:1085–1098.
54. Scivetti P, Danova M, Riccardi A, Fiocca R, Dionigi P, Mazzini G. Prognostic significance of DNA content in large bowel carcinoma: a retrospective flow cytometric study. Cancer Letters 1989;46:213–219.
55. de Leon MP, Sant M, Micheli A, et al. Clinical and pathologic prognostic indicators in colorectal cancer: a population-based study. Cancer 1992;69:626–635.
56. Dukes CE. The classification of cancer of the rectum. J Pathol Bacteriol 1932;35:323–332.
57. Turnbull RB, Kyle K, Watson FR, Spratt J. Cancer of the colon: the influence of the no-touch isolation technique on survival rates. Ann Surg 1967;166:420–425.
58. Astler VB, Coller FA. The prognostic significance of direct extension of carcinoma of the colon and rectum. Ann Surg 1954;139:846–852.
59. Grinnell RS. Lymphatic metastases of carcinoma of the colon and rectum. Ann Surg 1950;131:494–506.
60. Cass AW, Million RR, Pfaff WW. Patterns of recurrence following surgery alone for adenocarcinoma of the colon and rectum. Cancer 1976;37:2861–2865.
61. Moertel CG, Fleming TR, Macdonald JS, et al. Levamisole and fluorouracil for adjuvant therapy of resected colon carcinoma. N Engl J Med 1990;322:352–358.
62. Laurie JA, Moertel CG, Fleming TR, et al. Surgical adjuvant therapy of large-bowel carcinoma: an evaluation of levamisole and the combination of levamisole and fluorouracil. J Clin Oncol 1989;7:1447–1456.
63. Daver A, Bocquillon PG, Page M, et al. Flow cytometric studies of colorectal tumors using fine needle aspiration. Anticancer Res 1987;7:531–534.
64. Benson AB, Bauer KD, Lefkopoulo M, et al. Prognostic implications of proliferative activity and DNA ploidy in colorectal adenocarcinoma: an ECOG study [Abstract]. Presented at Eighty-third Annual Meeting, American Association for Cancer Res. San Diego, CA, May 20–23, 1992.
65. Finan PJ, Quirke P, Dixon MF, Dyson JED, Giles GR, Bird CC. Is DNA aneuploidy a good prognostic indicator in patients with advanced colorectal cancer? Br J Cancer 1986;54:327–330.
66. Armitage NC, Robins RA, Evans DF, Turner DR, Baldwin RW, Hardcastle JD. The influence of tumor cell DNA abnormalities on survival in colorectal cancer. Br J Surg 1985;72:828–830.
67. Adson MA, vanHeerden JA, Adson MH, Wagner JS, Ilstrup DM. Resection of hepatic metastases from colorectal cancer. Arch Surg 1984;119:647–650.
68. Tsushima K, Nagorney DM, Rainwater LM, et al. Prognostic significance of nuclear deoxyribonucleic acid ploidy patterns in resected hepatic metastases from colorectal carcinoma. Surgery 1987;102:635–641.
69. Harlow S, Stryker SJ, Poticha SM, Ujiki ST, Bauer KD. Prognostic significance of flow cytometric DNA analysis of resectable colorectal metastases to the liver [Abstract]. Presented at Meeting of the American Society of Colon and Rectal Surgeons, Toronto, CA, June 13, 1989.
70. Levine DS, Reid BJ, Haggitt RC, et al. Correlation of ultrastructural aberrations with dysplasia and flow cytometric abnormalities in Barrett's epithelium. Gastroenterology 1989;96:355–357.
71. Reid BJ, Haggitt RC, Rubin CE, et al. Barrett's esophagus. Correlation between flow cytometry and histology in detection of patients at risk for adenocarcinoma. Gastroenterology 1987;93:1–11.
72. Rabinovitch PS, Reid BJ, Haggitt RD, et al. Progression to cancer in Barrett's esophagus is associated with genomic instability. Lab Invest 1988;60:65–71.
73. Jin-Ming Y, Li-Hua Y, Guo-Qian, et al. Flow cytometric analysis of DNA content in esophageal carcinoma. Cancer 1989;64:80–82.
74. Matsuura H, Kuwano H, Morita M, et al. Predicting recurrence time of esophageal carcinoma through assessment of histologic factors and DNA ploidy. Cancer 1991;67:1406–1411.
75. Bottger T, Storkel S, Stockle M, et al. DNA image cytometry: a prognostic tool in squamous cell carcinoma of the esophagus? Cancer 1991;67:2290–2294.
76. Sasaki K, Murakami T, Nakamura M, et al. Intratumor heterogeneity in DNA ploidy in esophageal squamous cell carcinomas. Cancer 1991;68:2403–2406.
77. Kimura H, Youemura Y. Flow cytometric analysis of nuclear DNA content in advanced gastric cancer and its relationship with prognosis. Cancer 1991;67:2588–2593.
78. Ohyama S, Yonemura Y, Miyasaki I. Proliferative activity and malignancy in human gastric cancers. Cancer 1992;69:314–321.
79. Tsushima K, Rainwater LM, Goellner JR, van Heerden JA, Lieber MM. Leiomyosarcomas and benign smooth muscle tumors of the stomach: nuclear DNA patterns studied by flow cytometry. Mayo Clin Proc 1987;62:275–280.
80. Fujimoto J, Okamoto E, Yamanaka N, Toyosaka A, Mitsunoba M. Flow cytometric DNA analysis of hepatocellular carcinoma. Cancer 1991;67:939–944.
81. Alanen KA, Joensuu H, Klemi PJ, et al. Clinical significance of nuclear DNA content in pancreatic carcinoma. J Pathol 1990;160:313–320.
82. Eskelinen KA, Lipponen P, Collan Y, et al. Relationship between DNA ploidy and survival in patients with exocrine pancreatic cancer. Pancreas 1991;6:90–95.
83. Wilson MB, Adams DB, Garen PD. Aspiration cytologic, ultrastructural, and DNA cytometric findings of solid and papillary tumors of the pancreas. Cancer 1992;69:2233–2243.
84. Lincoln ST, Bauer KD. Limitations in the measurement of c-*myc* oncoprotein and other nuclear antigens by flow cytometry. Cytometry 1989;10:456–462.
85. Wirsching RP, Lamerz R, Wiebecke B, Demmel N, Liewald F, Valet G. Flow cytometric evaluation of colorectal carcinoma as completion of conventional tumor examination. J Exp Clin Cancer Res 1987;6:117–128.
86. Bauer KD, Clevenger CV, Endow RK, Murad T, Epstein AL, Scarpelli DG. Simultaneous nuclear antigen and DNA content quantitation using paraffin-embedded colonic tissue and multiparameter flow cytometry. Cancer Res 1986;46:2428–2434.
87. Anastasi JA, Bauer KD, Variakojis DV. DNA aneuploidy in Hodgkin's disease. A multiparameter flow cytometric analysis with cytologic correlation. Am J Pathol 1987;128:573–582.
88. Pennline KJ, Pellerito-Bessette F, Umland SP, Siegel MK, Smith SR. Detection of in vivo-induced IL-1 mRNA in murine cells by flow cytometry (FC) and fluorescent in situ hybridization (FISH). Lymphokine Cytokine Res 1992;11:65–71.

CLINICAL COMMENTARY
Douglas E. Merkel and Gershon Y. Locker

This review of the numerous studies that attempt to correlate ploidy or proliferative activity with prognosis of colorectal cancer suggests several areas in which accurate prognostic information could be helpful to the clinician. These are all settings in which interventions of varying toxicity and morbidity—chemotherapy, radiotherapy, total colectomy, or partial hepatectomy—are considered, despite the knowledge that only a portion of treated patients will receive benefit.

Randomized trials completed in the last several years have demonstrated convincingly that the natural history of stage C colon cancer can be improved with adjuvant chemotherapy. Standard therapy with 5-fluorouracil and levamisole is accompanied by significant morbidity and inconvenience, but results in an increase from 47% to 63% (1) in five-year disease-free survival. While treatment is clearly of benefit to many patients, almost half of stage C patients will not relapse following surgical treatment alone. Accurate identification of this subset would permit these "cured" patients to avoid unnecessary treatment. The unexpectedly good survival seen for the stage C low proliferative activity patients in Figure 17.3 (75% at five years) and in stage C/DNA diploid/low proliferative activity subset of the Mayo series (2) (67%) suggests that biological differences can be recognized within patients with lymph-node-positive disease. Clearly, however, the relapse rate in this "good prognosis" subset is too high to safely withhold treatment. Perhaps identification of the low-risk patient can be improved by including other factors, such as the number of involved lymph nodes or tumor grade, in the analysis.

A related problem is patients with stage B_2 colon cancer; as a group, these patients do not obtain a significant benefit when treated with adjuvant chemotherapy. Here, identification of high-risk subsets could permit selective targeting of adjuvant therapy with greater potential for benefit. The recent Mayo series (2) has shown that DNA aneuploid or high-proliferative-activity stage B_2 tumors have a 37% relapse rate at five years which is substantially greater than the 12% seen for tumors that were diploid/low-proliferative-activity and also had not perforated or adhered to adjacent organs. As the patients whose tumors comprised this report were enrolled on trials of adjuvant therapy, it will be of great interest to see whether this high-risk subset has a lower risk of recurrence when randomized to adjuvant treatment.

In rectal carcinoma, both local and distant recurrences are of concern and the current standard of care is to provide postoperative local radiotherapy and systemic chemotherapy to address these two types of recurrences in stage B_2 and C patients (4). While both chemotherapy and radiotherapy decrease the risk of local recurrence, radiotherapy cannot improve distant control and adds substantial morbidity. Local recurrence is not addressed in this chapter, but one series suggests that aneuploidy is associated with a two-fold higher risk of local failure (5). Confirmation of this finding, especially in the context of a randomized trial, could potentially spare some patients the morbidity and expense of pelvic irradiation.

While most patients with colorectal cancer metastatic to liver will have unresectable disease, and most patients who have resectable metastases will experience a recurrence of the disease, approximately 25% of those who undergo resection will survive for five years (6). In this context, an ability to identify potential survivors on the basis of DNA ploidy or proliferative activity, as hinted at by the two small series cited above, could be used to select those patients most likely to profit from partial hepatectomy, thus improving the survival rate for this complex procedure.

The management of patients with longstanding ulcerative colitis is controversial in the absence of any controlled studies of prophylactic colectomy to prevent the development of invasive carcinoma. Severe dysplasia is often used as an indication for surgery and it is enticing to hope that DNA ploidy analysis of biopsy material from patients with longstanding colitis might add to routine histologic assessment. Unfortunately, an association between the finding of DNA aneuploidy in ulcerative colitis and the subsequent development of cancer has not been established by the one group studying this area and any clinical use of DNA ploidy analysis in following such patients must be considered premature.

REFERENCES

1. Moertel C, Fleming T, MacDonald J, Haller D, Lurie J. The intergroup study of 5-FU plus levamisole and levamisole alone as adjuvant therapy for stage C colon cancer. Proc Amer Soc Clin Oncol [Abstract] 1992; 11:161.
2. Witzig TE, Loprinzi CL, Gonchoroff NJ, et al. DNA ploidy and cell kinetic measurements as predictors of recurrence and survival in stage B_2 and C colorectal adenocarcinoma. Cancer 1991;68:879–888.
3. Moertel CG, Loprinzi CL, Witzig TE, et al. The dilemma of stage B_2 colon cancer. Is adjuvant therapy justified? Proc Amer Soc Clin Oncol [Abstracts] 1990;9:108.
4. Gastrointestinal Tumor Study Group. Prolongation of disease. Free interval in surgically treated rectal carcinoma. N Engl J Med 1985;312: 1465–1471.
5. Scott NA, Rainwater LM, Wieland HA, et al. The relative prognostic value of flow cytometric DNA analysis and conventional clinicopathologic criteria in patients with operable rectal carcinoma. Dis Colon Rectum 1987;30:513–520.
6. Hugher K, Simon R, Sorghoubodi S, Sugarbaker P. Patterns of failure after resection for colorectal metastases: a multi-institution study. Surgery 1986;100:278–284.

18

Upper Aerodigestive and Lower Respiratory Tract Tumors

DANIEL W. VISSCHER and JOHN D. CRISSMAN
Clinical Commentary by **Samuel G. Taylor, IV**

BRONCHOGENIC CARCINOMA

Lung cancer is a common disease in Western nations and its incidence is rising, particularly among women (1). In the U.S. alone, an estimated 157,000 new cases were diagnosed in 1990. Tumor stage is the most important prognostic variable in lung cancer. Most oncologists employ the TNM classification as proposed by the American Joint Committee on Cancer (AJC) (2) in which a patient's stage is defined by a combination of a) the gross tumor size and its relationship to normal anatomical structures (T category), b) the presence and location of regional lymph node metastases (N category), and c) the presence of distant organ (i.e., bone, liver, brain) dissemination (M category). Advanced stage at presentation is a major reason for the abysmal survival figures in lung cancer. As shown in Table 18.1, only 16% to 21% of patients in several unselected large series presented with localized (stage I) disease and fully 28% to 41% had distant metastases (stage IV) at initial evaluation.

Several morphological features also impact on lung cancer outcome, most significantly, the histological tumor type, as defined by the World Health Organization Classification (3). Although unusual types are recognized, the vast majority of cases represent a) squamous cell carcinomas, b) adenocarcinomas, c) small cell carcinomas, or d) large cell carcinomas. The relative frequency of each of these in unselected patient series is shown in Table 18.2. Squamous, large-cell, and adenocarcinomas are often referred to collectively as "non-small-cell" carcinoma, since similar primary therapy (i.e., surgical resection) is employed for early stage cases of these types.

Small-cell carcinomas are highly malignant tumors that exhibit a neuroendocrine phenotype. They are usually treated with systemic chemotherapy; first, because they are highly sensitive to *cis*-platinum based regimens and, second, because virtually all cases have distant metastases at presentation (4). Consequently, despite a high rate of chemotherapy-induced remission, the prognosis of small-cell carcinoma is especially dismal, with reported median survivals typically less than one year (4). Less aggressive neuroendocrine lung tumors are well described but less common than small-cell carcinomas. These include carcinoid tumors, which metastasize in 5–10% of cases, and so-called atypical carcinoid tumors (5) (well-differentiated neuroendocrine carcinoma), which have a clinical behavior intermediate between carcinoid and small-cell carcinoma.

Histological factors other than cell type have not frequently been related to prognosis in neoplasms of the lung. In particular, the significance of grade, or differentiation, is rarely evaluated. This reflects partial incorporation of tumor grade into the WHO classification. Most "large-cell" carci-

Table 18.1
Bronchogenic Carcinoma—Stage Distribution at Presentation in Unselected Patient Series

Author (yr)	N	Localized (Pulmonary) Disease	Regional Node Metastases	Distant (Organ) Metastases
Silverberg[a] (1990) (1)	—	21%	30%	39%
Stanford (1976) (61)	3000	20.5%	51%	28.5%
Kemeny (1978) (13)	470	16%	43%	41%

[a]10% patients unknown stage

Table 18.2
Bronchogenic Carcinoma: Distribution by Histologic Type in Unselected Patient Series

Author (N)	Squamous (%)	Adenocarcinoma (%)	Small Cell (%)	Large Cell (%)	Other (%)
Stanford (3000) (61)	1230 (41)	855 (29)	756 (25)	—	147 (5)
El-Torky (4928) (62)	1904 (39)	1134 (23)	821 (17)	1069 (22)	—
Kemeny (470) (13)	230 (49)	142 (30)	54 (12)	—	44 (9)
Vincent (1682) (63)	640 (38)	446 (26)	323 (19)	156 (9)	117 (7)
Totals (10085)	4004 (40)	2577 (26)	1954 (19)	1242 (12)	308 (3)

Table 18.3
Bronchogenic Carcinoma—Survival in Various Series of Surgically-Treated Limited Stage Patients

Author	Stage[a] I (%)	II (%)	III (%)	Follow-up
Mountain (64)	540/795 (68)	81/185 (44)	51/141 (36)	5 yr.
Lipford (6)	62/129 (48)	10/44 (23)	—	5 yr.
Takise (7)	42/54 (78)	—	5/21 (24)	5 yr.
Naruke (65)	348/536 (65)	95/221 (43)	133/718 (18)	5 yr.
Shields (11)	246/422 (58)	46/115 (40)	—	3 yr.
Kataichi (66)	31/65 (48)	6/17 (35)	5/47 (11)	3 yr.
Totals	1269/2001 (63)	237/582 (41)	194/927 (21)	

[a] Stage I: tumor diameter ≤ 5 cm, may involve visceral pleura, no nodal metastases. Stage II: tumor as above with ipsilateral hilar node metastases. Stage III: tumor > 5 cm or extending to chest wall, mediastinum, or within 2.0 cm of carina; tumor of any size with mediastinal, subcarinal or contralateral hilar node metastases.

Table 18.4
Flow Cytometric DNA Analysis of Pulmonary Tumors

Author (yr.)	Tissue Fixation/ Source	DNA Stain	No. of AN/Total (%)	Control	Histology
Tirindelli-Danesi (1987) (22)	Fresh/lobectomy or biopsy	EB + mithramycin	97/101 (96%) 54% multiploid	Normal lung or PBL[a]	All types
Cibas (1989) (18)	Paraffin/lobectomy	DAPI	79/93 (85%) 19% multiploid	None (internal)	Adeno[b]
Zimmerman (1987) (23)	Paraffin/lobectomy or pneumonectomy	PI	45/100 (45%)	None (internal)	Adeno[b] squam[c]
Isobe (1990) (17)	Paraffin/lobectomy	PI	96/125 (77%) 11% multiploid	None (internal)	Adeno[b] squam[c]
Sahin (1990) (26)	Paraffin/lobectomy	PI	85/146 (58%)	None (internal)	Adeno[b] squam[b] large
Volm (1988) (27)	Fresh/lobectomy	PI/DAPI	78/100 (78%)	PBL	Squam
Volm (1985) (21)	Fresh/lobectomy	PI/DAPI	156/187 (83%) 20% multiploid	PBL	Squam[b] large + adeno
Van Bodegom (1989) (16)	Paraffin/lobectomy or pneumonectomy	DAPI	29/52 (56%) 6% multiploid	None (internal)	Squam
Bunn (1983) (20)	Fresh/ bronch wash, bone marrow, effusions, node, liver, lung biopsy	PI	100/113 (88%) 11% multiploid	Human PBL	All types
Carey (1990) (24)	Paraffin/pneumonectomy	PI	19/20 (95%) 45% multiploid	None (internal)	All types
Vindelov (1980) (19)	Fresh/aspiration node metastases	PI	26/29 (90%) 21% multiploid	Mouse lymphocytes	Small cell
Ten Velde (1988) (35)	Paraffin/lobectomy, biopsies	PI	44/67 (65%)	Not specified	Non-small

[a] PBL = peripheral blood lymphocytes.
[b] adeno = adenocarcinoma.
[c] squam = squamous carcinoma.

nomas represent high-grade (i.e., poorly differentiated) squamous or adenocarcinomas, since differentiation is usually apparent with specialized techniques such as electron microscopy. Accordingly, "large-cell" histology is associated with poor patient outcome when compared to other non-small-cell types (6). Conversely, well-differentiated tumors with favorable prognoses are also designated separately. One example is "bronchoalveolar" carcinoma, which is an unusual low-grade form of adenocarcinoma. Nevertheless, tumor grade or differentiation has been reported as a prognostic factor independent of stage in pulmonary tumors of various types (7, 8).

Whether comparably staged squamous carcinomas and adenocarcinomas have similar prognoses is unclear. Some authors have reported significantly better survivals in patients having squamous histology (2, 9–12), however this observation has not been confirmed by others (6, 13, 14, 15). Intratumoral heterogeneity of cellular differentiation is common in lung tumors, which may partially account for disagreements among these studies.

Surgically managed, limited stage, non-small cell carcinomas form the population base of most studies evaluating the prognostic importance of flow cytometric (FCM) DNA content in lung cancer. Thus, for reference purposes, survival data from several representative series of early-stage patients are summarized in Table 18.3.

Results of flow cytometric DNA analysis have been reported for over 1000 lung tumors. Most authors included multiple histologic types (Table 18.4), although two series studied only squamous carcinomas (16, 17) and two others were limited either to adenocarcinoma (18) or small cell carcinoma (19). Small-cell carcinomas are relatively underrepresented in this literature and were reported in significant numbers by only two authors (19, 20) (117 cases, 11% of

Table 18.5
Non-Small-Cell Bronchogenic Carcinoma: Incidence of DNA Aneuploidy by Histologic Type

Author	Squamous (% AN)	Adenocarcinoma (% AN)	Large Cell (% AN)
Cibas (18)	—	79/93 (85%)	—
Zimmerman (23)	22/48 (46%)	16/40 (40%)	—
Isobe (17)	37/58 (63%)	60/68 (88%)	—
Sahin (26)	37/55 (67%)	41/75 (55%)	7/16 (44%)
Volm (21)	80/105 (76%)	51/55 (93%)	25/27 (93%)
Totals	176/266 (66%)	247/331 (74.6%)	32/43 (74.4%)

Table 18.6
Non-Small-Cell Bronchogenic Carcinomas: Distribution and DNA Analysis by T & N Stage

Author	T1 (% AN)[a]	T2 (% AN)	T3 (% AN)	N0 (% AN)	N1 (% AN)	N2 (% AN)
Cibas (18)	50 (86%)	43 (84%)	—	87 (84%)	6 (100%)	—
Tirindelli-Danesi (22)	10 (NA)	45 (NA)	46 (NA)	49 (NA)	23 (NA)	29 (NA)
Zimmerman (23)	NA	NA	NA	74 (39%)	19 (47%)	7 (100%)
Sahin (26)	32 (59%)	94 (63%)	19 (32%)	67 (58%)	37 (57%)	42 (59%)
Volm (squam only) (27)	29 (76%) (T1&2)		76 (78%)	43 (70%)	61 (82%) (N1&2)	
Volm (non-small) (21)	NA	NA	NA	74 (80%)	111 (86%) (N1&2)	
	T1	T2	T3	N0	N1–2	
Total N	92	211	141	394	335	
Incidence Aneuploid (%)	62/82 (76%)	117/166 (70%)	65/95 (68%)	230/345 (67%)	214/283 (76%)	

[a]AN = DNA aneuploid; NA = not available.

total). DNA analysis has most commonly been performed using nuclear suspensions of formalin-fixed, paraffin-embedded lobectomy or pneumonectomy specimens (7/12 series, Table 18.4). Two authors performed combined mechanical and enzymatic dissociation of unfixed tissue samples (21, 22), and mechanical dissociation with Ficoll-Hypaque centrifugation was employed in one study using a broad mixture of specimen types (effusions, washings, small biopsies) (20). Within a given series, uniform dissociation protocols were applied to tumors of different cell types, however no data are provided that relate efficiency of disaggregation (cell yields, coefficients of variation, aneuploid recovery) to histology.

Bronchogenic carcinomas have a high incidence of aneuploidy compared to other visceral malignancies, typically over 70%. However, the reported incidences vary between authors from 45% (23) to 96% (22). Series having greater than 80% aneuploid tumors generally employed multiple-site sampling, numbering up to 10 aliquots per case in one study (24). Higher rates of aneuploidy were also reported by authors who utilized fresh, unfixed tumor cells (20, 21, 22). Heterogeneous case mixture by histologic type may also partially account for differing rates of aneuploidy. As shown in Table 18.5, abnormal DNA content was identified less commonly in squamous carcinomas (176/266, 66%) vs. other types (247/331, 75%). Accordingly, studies with lower aneuploid frequency generally included a higher proportion of squamous tumors (Table 18.4).

Lung tumors also demonstrate significant stemline heterogeneity in comparison to other solid tumors, with multiple DNA aneuploid populations detectable from 6% (16) to 54% (22) (avg. 23%) of cases. Although identification of multiple populations is partly a function of tissue sampling (24), these observations correlate with the high degree of morphologic heterogeneity recognized in pulmonary malignancies, as already noted. None of the publications reviewed, however, has specifically correlated DNA content heterogeneity with light microscopic features.

The relationship between DNA content and tumor stage is summarized in Tables 18.6 and 18.7. When the available data from all series are combined, there is no apparent association between increased tumor size and aneuploidy (T1, 76% aneuploid; T2, 70%; T3, 68% aneuploid). Some studies, however, reported an association between abnormal DNA content and presence of regional lymph node metastases. One series that evaluated the subject in detail (25) reported a statistically meaningful association between aneuploidy and presence of metastatic disease. The overall figures in Table 18.6, however, demonstrate a weak relationship (node negative—67% aneuploid, vs. node positive—76% aneuploid). Detailed information relating ploidy and stage is not available from some publications, and the data at this point are, therefore, inconclusive. However, the overall figures (Table 18.7) reveal a possible relationship between DNA content and clinical stage (I—66% aneuploid vs. II—58% aneuploid vs. III—72% aneuploid vs. IV—92% aneuploid). These findings are similar to other visceral tumor systems, notably adenocarcinomas of the breast and colon, in which marginal correlations between stage parameters and abnormal nodal DNA content have been observed. The mechanism(s) responsible for this finding—evolution of aneuploid clones with metastatic progression vs. increased metastatic potential of aneuploid tumors—remains unresolved.

Table 18.7
Non-Small-Cell Bronchogenic Carcinoma: Distribution and DNA Analysis by Clinical Stage

Author	I (% AN[a])	II (% AN)	III (% AN)	IV (% AN)
Cibas (18)	87 (84%)	6 (100%)	—	—
Tirindelli-Danesi (22)	29 (NA+)	9 (NA)	42 (NA)	21 (NA)
Zimmerman (23)	74 (39%)	19 (47%)	7 (100%)	—
Isobe (17)		54 (70%)	59 (80%)	12 (92%)
Sahin (26)	58 (64%)	30 (57%)	58 (53%)	—
Volm (27)		26 (77%)	79 (77%)	—
Total N (%)	328	64	245	33
Incidence Aneuploid (%)	197/299 (66%)	32/55 (58%)	146/203 (72%)	11/12 (92%)

[a]AN = DNA aneuploid; NA = not available.

Table 18.8
Non-Small-Cell Lung Cancer: Uncorrected Survival by Modal DNA Content

Author	Diploid Range	Aneuploid	Followup
Cibas (18)	10/12 (83%)	47/67 (70%)	6 yr median
Zimmerman (23)	41/45 (91%)	16/29 (55%)	2 yr
Isobe (17)	14/20 (70%)	20/20 (33%)	5 yr
Sahin (26)	23/61 (38%)	19/85 (22%)	5 yr
Volm (27)	14/22 (64%)	21/78 (27%)	5 yr minimum
Volm (21)	17/27 (63%)	49/141 (35%)	2.4 yr
Tirindelli-Danesi (22)	29/33 (88%) [DI = 1–2]	15/32 (47%) [2<DI<1]	12 mo
Totals	148/220 (67%)	187/492 (38%)	

Table 18.9
Non-Small-Cell Lung Carcinoma: Survival by Stage and Modal DNA Content

Author (Type)	Stage I DipR/AN	Stage III DipR/AN	Followup
Van Bodegom[a] (squam) (16)	12/23 (53%) / 14/29 (49%)	—	6 yr
Cibas[b] (adeno) (18)	10/14 (71%) / 47/79 (59%)	—	5 yr
Tirindelli-Danesi[c] (non-small) (22)	16/17 (94%) / 4/7 (57%)	13/15 (87%) / 5/13 (38%)	12 mo
Zimmerman[d] (non-small) (23)	41/45 (91%) / 16/29 (55%)	NA	2 yr
Sahin[e] (squam) (26)	6/7 (83%) / 7/18 (39%)	7/11 (62%) / 3/19 (15%)	6 yr
Volm (squam) (27)	NA	9/16 (56%) / 12/60 (20%)	5 yr
Volm (non-small) (21) min.	NA	11/19 (58%) / 32/102 (31%)	5 yr
Totals	86/106 (81%)/ 88/162 (54%) 174/268 (65%)	40/61 (66%) / 52/194 (27%) 92/255 (36%)	

[a]% AN events prog. sig.
[b]incl. 6 N1 cases.
[c]Group A = DI 1–2, Group B = DI <1 or >2.
[d]N0 group, size not spec.
[e]Outcome in adenocarcinoma not spec., ploidy not sig. in that group.

Eight of the published series, totalling over 900 tumors, evaluated patient outcome in relation to DNA content, most with followup intervals exceeding two years. The uncorrected survival of patients with non–small-cell diploid-range tumors substantially exceeded those with aneuploid tumors—148/220 (67%) diploid range vs. 187/492 (38%) aneuploid (Table 18.8). Furthermore, among both stage-I and stage-III groups, there is an outcome advantage for patients with diploid range tumors—stage-I survival: 81% diploid range vs. 54% aneuploid; stage III-survival: 66% diploid range vs. 27% aneuploid (Table 18.9). Stage-I patient outcome was very similar to the reference population in Table 18.3 (63% reference vs. 65% DNA); however, reported survivals among the stage-III DNA series group was generally better than reference (36% vs. 21%). Possible reasons for this include a short follow-up interval in one series (23) and inclusion of stage-II patients with stage III-survival figures in another (26). Nevertheless, the DNA series population is generally representative of surgically managed, non–small-cell carcinoma patients.

Several aspects of these data deserve specific comment. First, one author reported a correlation between patient outcome and the percent of aneuploid G0/G1 events in the DNA histogram (16). Eighty-three percent (10/12) survived if aneuploid events were less than 10% of total, vs. only 29% (5/17) survival if the proportion of aneuploid events exceeded 10%. In another study, impaired outcome was reported in cases having hypodiploid or hypertetraploid range DNA histograms, but not in those with hyperdiploidy (i.e., DNA index 1–2) (22). The findings of either study have yet to be confirmed and, at present, the standard of histogram interpretation consists of defining populations as "abnormal" (DNA aneuploid) or "normal" (diploid range).

Second, it is not clear that FCM-DNA ploidy has equivalent prognostic significance for all histologic types of non–small-cell lung cancer. Two of the three series (17, 22, 26) that compared squamous carcinoma with adenocarcinoma reported more significant prognostic associations in the former group. Moreover, significant associations between ploidy and outcome were not observed in one series that was limited to adenocarcinomas (18). Finally, few large cell carcinomas have been studied. Therefore, the clinical relevance of DNA content in this group remains undetermined.

Third, a subject that is often not addressed in the FCM literature is whether DNA measurements have prognostic value independent of clinical stage or morphological features. Three of the authors in this review (17, 23, 27) performed multivariate regression analyses (Cox model) and reported that abnormal DNA content predicted for adverse prognosis independently of stage. One author, in fact, found that DNA content was more significant than either T or N stage (23). Tumor grade, however, was not entered into the multivariate analysis of any study. Furthermore, only one study (26) examined critically the relative prognostic value and possible relationship between DNA content and tumor grade. Its authors found that grade and ploidy were not significantly correlated in non–small-cell carcinoma and that DNA content was predictive for survival but grade was not. These results, as well as the relative lack of morphologic–flow cytometric associations in this body of literature, are surprising given the consistent strong relationship between FCM aneuploidy and high tumor grade in other solid tumors as well as the ploidy-grade associations reported in the image analysis literature (28).

Flow cytometric ploidy studies of neuroendocrine lung tumors reflect the spectrum of morphologic and clinical behavior observed in this group as previously described. DNA aneuploidy is reported with high frequency (80–90%) in small cell carcinoma (19, 20), compared to carcinoid (approximately 30%) or atypical carcinoid tumors (approximately 60%) (29, 30, 31). Although DNA content predictably fails to correlate with survival in the limited number of small-cell carcinomas reported (<50 cases to date) (20, 32), DNA-aneuploid tumors may respond more favorably to multiagent chemotherapy than DNA-diploid tumors (33). Jones et al. evaluated a series of well differentiated neuroendocrine tumors (28 carcinoid, 25 atypical carcinoid) and identified a significant relationship between DNA aneuploidy and presence of nodal metastases (29). Patients with diploid-range tumors experienced better five-year survival (diploid range 84% vs. aneuploid 58%); however, DNA content was less predictive than histologic differentiation or nodal status in multivariate analysis.

Finally, synthesis phase fraction (SPF) is rarely reported in this body of literature, although specific reasons are usually not provided. Since most of the studies employed paraffin blocks, we presume baseline elevation due to debris contamination represented a significant obstacle to SPF determinations, particularly for a tumor system in which necrosis is common and often extensive.

Volm et al. correlated SPF values to survival in a clinically heterogeneous series of non–small-cell carcinomas (5-yr. survival low SPF—18/45 (40%) vs. high SPF—16/69 (23%), $p < 0.04$) (34). These authors optimized histogram quality by employing fresh, unfixed tissue samples for FCM analysis, yet reported SPF in only 65% of cases. In a similar group of paraffin-embedded, predominantly advanced stage non-small-cell carcinomas, Ten Velde et al. reported that SPF, but not DNA ploidy, was predictive of short-term (50 week) outcome (high SPF group—15% survival vs. low SPF group—46% survival, $p = .04$, SPF calculated in 73% of cases) (35). None of the series reviewed demonstrated correlation between histologic tumor type and SPF, although the published thymidine labeling index and BrdU literature show higher kinetic indices among small-cell and large-cell carcinomas compared to squamous and adenocarcinomas (36).

Summary

Until early-detection technology is developed and effective systemic anti-tumor therapies become available, the outlook for patients with lung cancer will remain bleak. The prognosis for patients with limited-stage, surgically-treated disease is primarily a function of stage, with a lesser but significant contribution from histopathologic parameters. Flow cytometric DNA analysis has shown that bronchogenic carcinomas have a high frequency of DNA aneuploidy as well as stemline heterogeneity compared to other solid tumors. Although DNA ploidy is only weakly correlated with stage, significant predictive value has been reported for FCM data, particularly squamous carcinomas. These data most likely supplement stage parameters, although their relationship to histologic features remains uncertain. Further evaluation of kinetic data available from DNA histograms is needed and may provide additional useful information, but will probably require modified dissociation protocols and data analysis.

SQUAMOUS NEOPLASIA OF THE UPPER AERODIGESTIVE TRACT

Squamous neoplasia of the head and neck region is a clinically and pathologically heterogeneous disease. The prognosis is highly variable and depends on a number of complex and interrelated factors, most notably stage and morphology. The TNM stage classification is most commonly employed (37), although the details are beyond the scope of this text, since it entails modifications for individual anatomical sites (i.e., oral cavity vs. larynx) as well as incorporation of clinical findings (e.g., vocal cord mobility). Patients with larger tumors, in general, are significantly more likely to have metastases to regional (neck) lymph nodes, a condition that decreases survival by up to 50%. It is also important to recognize that tumors of similar stage in different primary sites of the head and neck region may have different prognoses. Pa-

Table 18.10
Summary of Flow Cytometric Studies in Head and Neck Squamous Carcinomas

Author (yr)	Site(s)	No. of Aneuploid/Total (%)	Tissue Source (Mean CV[a])	Control Population
Johnson (54) (1985)	Various	32/73 (44%)	FFPE[b] (NA[c])	NA
Goldsmith (58) (1986)	Larynx	31/48 (65%)	FFPE (8.1%)	Normal tissue block (same case)
Farrar (50) (1987)	Oral cavity	72/195 (37%)	FFPE (NA)	Normal tissue block (same case)
Kaplan (67) (1988)	Various	19/46 (41%)	FFPE (<10%, half peak)	None (internal)
Kokal (56) (1988)	Various	48/76 (63%)	FFPE (3 samples) (NA)	Chick RBC
Ensley (49) (1989)	Various	181/291 (62%)	Fresh (4%)	Human lymphocytes
Hemmer (48) (1990)	Oral cavity	80/110 (73%)	Fresh (3.4%)	None (internal)
Costello (57) (1990)	Nasopharynx	33/55 (60%)	FFPE (6.2%)	None (internal)
Xiao-li (68) (1990)	Oral cavity	41/70 (59%)	FFPE (NA)	Human lymphocytes (paraffin-embedded)
Rua (69) (1991)	Larynx	89/133 (67%)	FFPE (NA)	None (internal)
Kearsley (55) (1991)	Various	115/172 (67%)	FFPE (4.5%)	None (internal)

[a] CV = coefficient of variation.
[b] FFPE = formalin-fixed, paraffin-embedded.
[c] NA = not available.

tient survival is generally worse for hypopharyngeal or tongue neoplasms. Similarly, within the larynx, glottic tumors tend to be less aggressive than supraglottic tumors.

Outcome in head and neck squamous carcinomas is highly dependent on histopathologic features. Broders' original work comparing tumor differentiation to survival was performed on a series of squamous carcinomas of the lip (38). Traditionally, keratin production by tumor cells, as described by Broders, is most frequently used by pathologists to assess grade in squamous neoplasia, although other features, including vascular invasion, host inflammatory response, pattern of invasion, mitotic index, and nuclear pleomorphism are also important. These have been incorporated into a numerical scheme devised by Jakobsson et al. (39) and modified by Crissman (40). Patients with well-differentiated (low-grade) tumors, according to this system, have five-year survivals up to 88%, compared to 44% for poorly-differentiated (high-grade) groups. In addition, there are a variety of distinctive histologic variants of squamous carcinoma occurring in the upper aerodigestive tract (verrucous carcinoma, spindle cell carcinoma, nasopharyngeal carcinoma), each of which has a characteristic clinical presentation and behavior.

Nuances of stage classification and morphology are critical in this tumor system since therapeutic options are both numerous as well as multidisciplinary (i.e., local surgery/radiation/neck dissection/chemotherapy) and impact the patient outcome. Aerodigestive tract squamous neoplasia, in fact, is a prototypical system in which therapy is tailored to individual patients based on features intrinsic to the tumor. Therefore, objective data relevant to prognosis is highly desirable and potentially useful in clinical decision making.

The most representative flow cytometric studies of head and neck region squamous carcinomas (1269 total cases) are summarized in Table 18.10. Several smaller reports, each less than 50 cases with two utilizing image analysis, account for the remaining 197 tumors studied to date (41–46). Most researchers performed DNA analysis following enzymatic dissociation of formalin–fixed paraffin–embedded tissue, although fresh unfixed samples were used by three authors (42, 47, 48, 49).

There are striking differences in reported DNA content abnormalities between individual series—83% (42) to 37% (50) (avg. 59%). Several factors probably contribute to these discrepancies. First, tissue sources included numerous anatomical sites and primary tumors as well as recurrences or metastatic lesions. Second, the case distribution by stage varied considerably between authors—from all patients without node metastases (50) to all cases having clinically advanced disease (49). Third, technical factors involving tumor dissociation are particularly important as variables in squamous carcinomas. These tumors contain numerous intercellular attachments, making single cell suspensions difficult to achieve. Although most studies employed similar methods, subtle factors such as incubation time and conditions, or enzyme sources and concentrations, may substantially affect cell yield or aneuploid recovery. Sensitivity of technique, moreover, is difficult to compare since G0/G1 peak coefficients of variation are not uniformly reported (Table 18.10). Furthermore, authors employed a wide variety of methods to establish the diploid-range reference population, including mixing of exogenous mononuclear cells, histogram comparison with histologically non-neoplastic tissue from other paraffin blocks of each case, or presumed endogenous stromal-/inflammatory-cell-derived events (Table 18.10). Higher DNA aneuploid frequencies approaching those of image analysis studies (42, 46) are generally reported by authors who utilized fresh tissue. Ensley et al. (51, 52) have carefully evaluated and optimized dissociation methods for squamous carcinomas. The 62% frequency of aneuploidy reported in their large series of cases (291 tumors, mean coefficient of variation 4.2%) (39) is in keeping with other large, recently published, series and probably representative of head and neck region squamous carcinomas.

Many authors provide detailed information comparing the incidence of DNA aneuploidy by tumor stage (Table 18.11). The overall data show an apparent correlation between abnormal DNA content and presence of regional node metastases—node-negative 54% aneuploid vs. node-positive 67% aneuploid. There is a similar trend toward higher incidence of aneuploidy among tumors of larger size—T1

Table 18.11
Relationship Between DNA Aneuploidy and Stage Parameters in Head and Neck Squamous Carcinoma [No. of Aneuploid/Total (%)]

Author (ref)	Node Status		Size			
	Negative	Positive	T1	T2	T3	T4
Goldsmith (58)	22/31 (71%)	9/17 (53%)	8/10 (80%)	4/5 (80%)	10/18 (55%)	9/15 (60%)
Kaplan (67)	8/21 (38%)	11/25 (44%)	7/26 (27%) (T1+T2[a])		12/20 (60%) (T3+T4[b])	
Kokal (56)	19/38 (50%)	32/38 (84%)	17/28 (61%)	21/32 (66%)	13/15 (87%)	—
Ensley (49)	54/78 (69%)	76/118 (64%)	11/17 (65%)	16/31 (52%)	56/84 (67%)	47/64 (73%)
Hemmer (47)	6/17 (35%)	27/30 (90%)	0/5 (0%)	17/25 (68%)	16/17 (94%)	—
Hemmer (48)	—	—	1/13 (8%)	50/65 (77%)	29/32 (90%)	—
Xiao-li (68)	13/28 (46%)	12/16 (75%)	—	—	—	—
Kearsley (55)	60/97 (62%)	54/75 (76%)	17/27 (63%)	28/38 (74%)	23/34 (68%)	
Holm[b] (46)	1/11 (9%)	14/34 (41%)	6/8 (75%)	11/15 (73%)	18/22 (82%)	—
Tytor[b] (45)	14/41 (34%)	8/9 (89%)	3/13 (23%)	19/37 (51%)	—	—
Totals	197/362 (54%)	243/362 (67%)	63/121 (52%)	184/283 (65%)	170/226 (75%)	79/113 (70%)

[a] Image analysis studies.
[b] Excluded from totals.

Table 18.12
Relationship Between DNA Aneuploidy and Differentiation in Head and Neck Squamous Carcinomas [No. of Aneuploid/Total (%)]

Author (ref)	Differentiation		
	Well	Moderate	Poor
Johnson (54)	12/25 (48%)	12/27 (44%)	12/19 (63%)
Kaplan (67)	4/15 (27%)	10/20 (50%)	5/11 (45%)
Kokal (56)	4/11 (36%)	37/53 (70%)	10/12 (83%)
Sakr (53)[a]	—	43/77 (56%)	64/78 (82%)
Feinmesser (41)	8/9 (89%)	16/20 (80%)	1/1 (100%)
Hemmer (48)	8/21 (38%)	59/64 (77%)	23/25 (92%)
Rua (69)	25/49 (51%)	46/61 (75%)	18/23 (78%)
Kearsley (55)	15/30 (50%)	67/102 (66%)	32/40 (80%)
Totals	119/237 (50%)	339/498 (68%)	165/209 (79%)

[a] Nuclear grade, divided into two categories ("low" vs. "high").

63/121 (52%) vs. T2 184/283 (65%) vs. T3 170/226 (75%) vs. T4 79/113 (70%). It is noteworthy that the reported DNA series cases are disproportionately node-positive, large tumors and thus not fully representative of upper aerodigestive squamous neoplasia, which is predominantly node-negative. Nevertheless, the data clearly suggest an association between abnormal DNA content and clinical aggressiveness in this tumor system.

A number of recent studies have compared tumor differentiation (i.e., grade) to flow cytometric data (Table 18.12). This is critical given the proven, unequivocal relationship between morphology and prognosis in this tumor system. Although the findings between individual studies are not uniform, the overall data indicate a relationship between DNA aneuploidy and poor histologic differentiation: well differentiated—50% aneuploid vs. moderately differentiated—68% aneuploid vs. poorly differentiated—79% aneuploid. These findings are expected, since most grade-dependent tumor systems (e.g., transitional, breast, and prostate carcinomas) have shown striking histopathology-flow cytometric correlations. One author associated DNA aneuploidy with angiolymphatic invasion by tumor cells, an observation that has been made in other tumor systems and reflects the higher aneuploid frequency among cases with nodal metastases. DNA aneuploidy is more strongly related to nuclear cytologic features and pattern of invasion than presence of keratinization (53). Although kinetic data are extremely limited in this body of literature, Johnson et al. (54) reported a positive correlation between poor differentiation and high SPF values in cases stratified by DNA content.

Survival data in relation to DNA content is now available in approximately 800 upper aerodigestive tract squamous carcinomas (Table 18.13). Patients with diploid-range tumors have a clear overall outcome advantage—203/343 (59%) vs. 163/449 (36%)—a conclusion supported by authors of most individual studies. However, the data do not yet allow definitive conclusions regarding the clinical utility of DNA analysis.

First, virtually all reported series are heterogeneous with respect to stage, mode of treatment, and primary site. Second, only two authors performed multivariate statistical analyses to correct ploidy results for stage and grade associations (55, 56).

Although a consensus is apparently evolving supporting the predictive value of flow cytometric DNA analysis in head and neck squamous carcinomas, there are individual exceptions. Costello et al. (57) did not identify prognostic significance in a series of nasopharyngeal carcinomas. These tumors, however, are morphologically, clinically, and epide-

Table 18.13
Clinical Outcome by DNA Content in Head and Neck Squamous Carcinoma

Author (ref) (site)	Survival Diploid Range (%)	Survival Aneuploid (%)	% with Node Metastases	Type and Length Follow-up
Farrar (50) (oral cavity)	55/123 (45)	9/73 (13)	0	>5 yr
Kokal (56) (various)	24/25 (96)	14/48 (25)	50	5 yr disease-free
Costello (57) (nasopharynx)	11/22 (48)	14/33 (42)	71	5 yr survival
Holm (46) (various)	5/8 (63)	13/32 (41)	35	4 yr
Goldsmith (58) (larynx)	8/17 (47)	20/23 (87)	35	2 yr disease-free
Tytor (45) (oral cavity)	20/28 (70)	11/22 (50)	18	5 yr
de Braud (70) (various)	18/29 (62)	21/46 (45)	88	3 yr (median)
Xiao-li (68) (oral cavity)	10/15 (66)	2/23 (9)	NA	5 yr disease-free
Rua (69) (larynx)	12/18 (66)	16/35 (45)	0	5 yr survival
Kearsley (55) (various)	40/58 (69)	43/114 (38)	44	5 yr disease-free
Totals	203/343 (59)	153/449 (36)		

miologically distinct from other forms of upper aerodigestive squamous neoplasia.

And, Goldsmith et al. (58) reported a survival advantage in patients with DNA-aneuploid tumors of the larynx—47% diploid range vs. 87% aneuploid. This series, however, has a disproportionately high number of large tumors without recognized node metastases. Although these are considered advanced stage they are known to have a better prognosis. Moreover, since follow-up interval was short (23 mo.) and patient therapy (i.e., radiation) uniform, the results may be better explained by a relationship between abnormal DNA content and radiation-induced tumor regression.

There is evidence from other studies that DNA content may be related to cytotoxic treatment response. Franzen et al. (59) have reported enhanced histologic tumor response to radiotherapy in DNA-aneuploid (vs. diploid range) squamous carcinomas of the oral cavity. In addition, Ensley et al. (49) observed that clinical recurrences following treatment were disproportionately diploid range—54% of recurrences were aneuploid vs. 66% of primaries, although survival data were not presented. Cooke et al. (60) have critically analyzed this issue in a controlled study of advanced stage tumors. They demonstrated a statistically significant increase in median survival among aneuploid-, but not diploid-range, tumors treated with a Cis-platinum based regimen.

Summary

Due to the clinical complexity and unique technical difficulties associated with tissue dissociation in this tumor system, interpretable flow cytometric data for squamous-cell carcinomas of the head and neck region were, until recently, sketchy, evidencing little consensus. Data available on SPF is still limited and a conclusive analysis is not yet possible. Although it appears that there is measurable correlation with patient outcome, the predictive value of DNA content requires more critical appraisal, particularly in relation to tumor stage and other prognostic factors. The possibility of aneuploidy predicting response to adjuvant therapy is an intriguing concept, but at this time the data are preliminary. Larger, more homogeneous, patient series with multivariate analyses of all relevant prognostic data (including ploidy and SPF) will be required to establish the prognostic value of FCM-DNA analysis in squamous carcinomas of the head and neck.

REFERENCES

1. Silberberg E, Boring CC, Squires TS. CA - A Cancer Journal for Clinicians. 1990;40(1):9–26.
2. Mountain CF. The new international staging system for lung cancer. Surg Clin N Am 1987;67:925–935.
3. The World Health Organization histological typing of lung tumours, 2nd ed. Am J Clin Pathol 1982;77:123–136.
4. Kron IL, Harman PK, Mills SE, et al. A reappraisal of limited-stage undifferentiated carcinoma of the lung. J Thorac Cardiovasc Surg 1982; 84:734–737.
5. Mills SE, Walker AN, Cooper PH, Kron IL. Atypical carcinoid tumor of the lung. A clinicopathologic study of 17 cases. Am J Surg Pathol 1982;6:643–654.
6. Lipford EH, III, Eggleston JC, Lillemoe KD, Sears DL, Moore GW, Baker RR. Prognostic factors in surgically resected limited-stage, non-small cell carcinoma of the lung. Am J Surg Pathol 1984;8:357–365.
7. Takise A, Kodama T, Shimosato Y, Watanabe S, Suemasu K. Histopathologic prognostic factors in adenocarcinomas of the peripheral lung less than 2 cm in diameter. Cancer 1988;61:2083–2088.
8. Sorensen JB, Hirsch FR, Olsen J. The prognostic implication of histopathologic subtying of pulmonary adenocarcinoma according to the classification of the World Health Organization. An analysis of 259 consecutive patients with advanced disease. Cancer 1988;62:361–367.
9. Gail MH, Eagan RT, Feld R, et al. Prognostic factors in patients with resected stage I non-small cell lung cancer. A report from the Lung Cancer Study Group. Cancer 1984;54:1802–1813.
10. Shields TW, Humphrey EW, Matthews M, Eastridge CE, Keehn RJ. Pathological stage grouping of patients with resected carcinoma of the lung. J Thorac Cardiovasc Surg 1980;80:400–405.
11. Shields TW. Classification and prognosis of surgically treated patients with bronchial carcinoma: Analysis of VASOG studies. Int J Radiat Oncol Biol Phys 1980;6:1021–1027.
12. Wilkins EW Jr., Scannell JG, Craver JG. Four decades of experience with resections for bronchogenic carcinoma at the Massachusetts General Hospital. J Thorac Cardiovasc Surg 1978;76:364–368.
13. Kemeny MM, Block LR, Braun DW Jr., Martini N. Results of surgical treatment of carcinoma of the lung by stage and cell type. Surg Gynecol Obstet 1978;147:865–871.
14. Greenberg SD, Fraire AE, Kinner BM, Johnson EH. Tumor cell type versus staging in the prognosis of carcinoma of the lung. In: Rosen PP, Fechner RE, eds. Path Ann, Vol. 22, part 2. Norwalk: Appleton and Lange, 1987; pp 387–405.

15. Tosi P, Luzi P, Leoncini L, Miracco C, Gambacorta M, Grossi A. Bronchogenic carcinoma: Survival after surgical treatment according to stage, histologic type and immunomorphologic changes in regional lymph nodes. Cancer 1981;48:2288–2295.
16. Van Bodegom PC, Baak JPA, Stroet-Van Galen C, et al. The percentage of aneuploid cells is significantly correlated with survival in accurately staged patients with stage 1 resected squamous cell lung cancer and long-term follow up. Cancer 1989;63:143–147.
17. Isobe H, Miyamoto H, Shimizu T, et al. Prognostic and therapeutic significance of the flow cytometric nuclear DNA content in non-small cell lung cancer. Cancer 1990;65:1391–1395.
18. Cibas ES, Melamed MR, Zaman MB, Kimmel M. The effect of tumor size and tumor cell DNA content on the survival of patients with stage I adenocarcinoma of the lung. Cancer 1989;63:1552–1556.
19. Vindelov LL, Hansen HH, Christensen IJ, Spang-Thomsen M, et al. Clonal heterogeneity of small-cell anaplastic carcinoma of the lung demonstrated by flow-cytometric DNA analysis. Cancer Res 1980;40:4295–4300.
20. Bunn PA, Jr., Carney DN, Gazdar AF, Whang-Peng J, Matthews MJ. Diagnostic and biological implications of flow cytometric DNA content analysis in lung cancer. Cancer Res 1983;43:5026–5032.
21. Volm M, Drings P, Mattern J, Sonka J, Vogt-Moykopf I, Wayss K. Prognostic significance of DNA patterns and resistance-predictive tests in non-small cell lung carcinoma. Cancer 1985;56:1396–1403.
22. Tirindelli-Danesi D, Teordori L, Mauro F, et al. Prognostic significance of flow cytometry in lung cancer. A 5-year study. Cancer 1987;60:844–851.
23. Zimmerman PV, Bint MH, Hawson GAT, Parsons PG. Ploidy as a prognostic determinant in surgically treated lung cancer. Lancet 1987;530–533.
24. Carey FA, Lamb D, Bird CC. Intratumoral heterogeneity of DNA content in lung cancer. Cancer 1990;65:2266–2269.
25. Volm M, Bak M, Hahn EW, Mattern J, Weber E. DNA and S-phase distribution and incidence of metastasis in human primary lung carcinoma. Cytometry 1988;9:183–188.
26. Sahin AA, Ro JY, El-Naggar AK, et al. Flow cytometric analysis of the DNA content of non-small cell lung cancer. Ploidy as a significant prognostic indicator in squamous cell carcinoma of the lung. Cancer 1990;65:530–537.
27. Volm M, Mattern J, Muller T, Drings P. Flow cytometry of epidermoid lung carcinomas: Relationship of ploidy and cell cycle phases to survival. A five-year follow up study. Anticancer Res 1988;8:105–112.
28. Shimosato Y, Asamura H, Yoshida K, Noguchi M, Nakajima T, Mukai K. Factors possibly affecting the degree of malignancy in adenocarcinoma of the lung. DNA content and bromodeoxyuridine (BrdU) labeling index. Chest 1989;96(Suppl):37S–38S.
29. Jones DJ, Hasleton PS, Moore M. DNA ploidy in bronchopulmonary carcinoid tumours. Thorax 1988;43:195–199.
30. Larsimont D, Kiss R, de Launoit Y, Melamed MR. Characterization of the morphonuclear features and DNA ploidy of typical and atypical carcinoids and small cell carcinomas of the lung. Am J Clin Pathol 1990;94:378–383.
31. Thunnissen FBJM, Van Eijk J, Baak JPA, Schipper NW, Uyterlinde AM, Breederveld RS, Meijer S. Bronchopulmonary carcinoids and regional lymph node metastases. A quantitative pathologic investigation. Am J Pathol 1988;132:119–122.
32. Oud PS, Pahlplatz MMM, Beck JLM, Wiersma-Van Tilburg A, Wagenaar SJ, Vooijs GP. Image and flow DNA cytometry of small cell carcinoma of the lung. Cancer 1989;64:1304–1309.
33. Abe S, Tsuneta Y, Makimura S, Itabashi K, Nagai T, Kawakami Y. Nuclear DNA content as an indicator of chemosensitivity in small-cell carcinoma of the lung. Anal Quant Cytol Histol 1987;9:425–428.
34. Volm M, Hahn EW, Mattern J, Muller T, Vogt-Moykopf I, Weber E. Five-year follow-up study of independent clinical and flow cytometric prognostic factors for the survival of patients with non-small cell lung carcinoma. Cancer Res 1988;48:2923–2928.
35. Ten Velde GPM, Schutte B, Vermeulen A, Volovics A, Reynders MMJ, Blijham GH. Flow cytometric analysis of DNA ploidy level in paraffin-embedded tissue of non-small-cell lung cancer. Eur J Cancer Clin Oncol 1988;24:455–460.
36. Yoshida K, Morinaga S, Shimosato Y, Hayata Y. A cell kinetic study of pulmonary adenocarcinoma by an immunoperoxidase procedure after bromodeoxyuridine labeling. Cancer 1989;64:2284–2291.
37. Sisson GA, Pelzer HJ. Staging system by sites. Problems and refinements. Otolaryngol Clin NA 1985;18:397–402.
38. Broders AC. Carcinoma: grading and practical application. Arch Pathol 1926;2:376–381.
39. Jakobsson PA, Eneroth DM, Killander D, Moberger G, Martensson B. Histologic classification and grading of malignancy in carcinoma of the larynx. Acta Radiol Ther Phys Biol 1973;12:1–8.
40. Crissman JD, Liu WY, Gluckman JL, Cummings G. Prognostic value of histopathologic parameters in squamous cell carcinoma of the oropharynx. Cancer 1984;54:2995–3001.
41. Feinmesser R, Freeman JL, Noyek A. Flow cytometric analysis of DNA content in laryngeal cancer. J Laryngol Otol 1990;104:485–487.
42. Grabel-Pietrusky R, Hornstein OP. Flow cytometric measurement of ploidy and proliferative activity of carcinomas of the oropharyngeal mucosa. Arch Dermatol Res 1982;273:121–128.
43. Gussack GS, Donelly K, Hester R, Dowling E. Flow cytometric DNA analysis of laryngeal carcinomas. Head and Neck Oncology Research Conference, 1987.
44. Sickle-Santanello BJ, Farrar WB, Dobson JL, O'Toole RV, Keyhani-Rofagha S. Flow cytometric analysis of DNA content as a prognostic indicator in squamous cell carcinoma of the tongue. Am J Surg 1986;152:393–395.
45. Tytor M, Franzen G, Olofsson J, Brunk U, Nordenskjold B. DNA content, malignancy grading and prognosis in T1 and T2 oral cavity carcinomas. Br J Cancer 1987;56:647–652.
46. Holm L-E. Cellular DNA amounts of squamous cell carcinomas of the head and neck region in relation to prognosis. Laryngoscope 1982;92:1064–1069.
47. Hemmer J, Schon E, Kriedler J, Haase S. Prognostic implications of DNA ploidy in squamous cell carcinomas of the tongue assessed by flow cytometry. J Cancer Res Clin Oncol 1990;116:83–86.
48. Hemmer J, Kriedler J. Flow cytometric DNA ploidy analysis of squamous cell carcinoma of the oral cavity. Comparison with clinical staging and histologic grading. Cancer 1990;66:317–320.
49. Ensley JF, Maciorowski Z, Hassan M, et al. Cellular DNA content parameters in untreated and recurrent squamous cell cancers of the head and neck. Cytometry 1989;10:334–338.
50. Farrar WB, Artman S, Sickle-Santanello BJ, Decenzo JF, Keyhani-Rofagha S, O'Toole RV. Flow cytometric analysis of DNA content as a prognostic indicator in squamous cell carcinoma of the oral cavity. Head and Neck Oncology Research Conference, 1987.
51. Ensley JF, Maciorowski Z, Pietraszkiewicz H, et al. Solid tumor preparation for flow cytometry using a standard murine model. Cytometry 1987;8:479–487.
52. Ensley JF, Maciorowski Z, Pietraszkiewicz H, et al. Solid tumor preparation for clinical application of flow cytometry. Cytometry 1987;8:488–493.
53. Sakr W, Hussan M, Zarbo RJ, Ensley J, Crissman JD. DNA quantitation and histologic characteristics of squamous cell carcinoma of the upper aerodigestive tract. Arch Pathol Lab Med 1989;113:1009–1014.
54. Johnson TS, Williamson KD, Cramer MM, Peters LJ. Flow cytometric analysis of head and neck carcinoma DNA index and S-phase fraction from paraffin-embedded sections: Comparison with malignancy grading. Cytometry 1985;6:461–470.
55. Kearsley JH, Bryson G, Battistutta D, Collins RJ. Prognostic importance of cellular DNA content in head-and-neck squamous-cell cancers. A comparison of retrospective and prospective series. Int J Cancer 1991;47:31–37.
56. Kokal WA, Gardine RL, Sheibani K, Zak IW, Beatty JD, Riihimaki DU, Wagman LD, Terz J. Tumor DNA content as a prognostic indica-

57. Costello F, Mason BR, Collins RJ, Kearsley JH. A clinical and flow cytometric analysis of patients with nasopharyngeal cancer. Cancer 1990;66:1789–1795.
58. Goldsmith MM, Cresson DS, Postma DS, Askin FB, Pillsbury HC. Significance of ploidy in laryngeal cancer. Am J Surg 1986;152:396–402.
59. Franzen G, Klintenberg C, Risberg B. DNA measurement—An objective predictor of response to irradiation? A review of 24 squamous cell carcinomas of the oral cavity. Br J Cancer 1986;53:643–651.
60. Cooke LD, Cooke TG, Bootz F, Forster G, Helliwell TR, Spiller D, Stell PM. Ploidy as a prognostic indicator in end stage squamous cell carcinoma of the head and neck region treated with cisplatinum. Br J Cancer 1990;61:759–762.
61. Stanford W, Spivey CG Jr., Larsen GL, Alexander JA, Besich WJ. Results of treatment of primary carcinoma of the lung. Analysis of 3,000 cases. J Thorac Cardiovasc Surg 1976;72:441–449.
62. El-Torky M, El-Zeky F, Hall JC. Significant changes in the distribution of histologic types of lung cancer. A review of 4928 cases. Cancer 1990;65:2361–2367.
63. Vincent RG, Pickren JW, Lane WW, et al. The changing histopathology of lung cancer. A review of 1682 cases. Cancer 1977;39:1647–1655.
64. Mountain CF, Lukeman JM, Hammar SP, et al. Lung cancer classification: The relationship of disease extent and cell type to survival in a clinical trials population. J Surg Oncol 1987;35:147–156.
65. Naruke T, Goya T, Tsuchiya R, Suemasu K. Prognosis and survival in resected lung carcinoma based on the new international staging system. J Thorac Cardiovasc Surg 1988;96:440–447.
66. Kitaichi M, Asamoto H, Izumi T, Furuta M. Histological classification of regional lymph nodes in relation to postoperative survival in primary lung cancer. Hum Pathol 1981;12:1000–1005.
67. Kaplan AS, Caldarelli DD, Chacho MS, et al. Retrospective DNA analysis of head and neck squamous cell carcinoma. Arch Otolaryngol Head Neck Surg 1986;112:1159–1162.
68. Xiao-li W, Rui-mei C, Zhi-san W, Wen-lei C, Lian-fu Z. Flow cytometric DNA-ploidy as a prognostic indicator in oral squamous cell carcinoma. Chinese Med J 1990;103:572–575.
69. Rua S, Comino A, Fruttero A, Cera G, Semeria C, Lanzillotta L, Boffetta P. Relationship between histologic features, DNA flow cytometry, and clinical behavior of squamous cell carcinomas of the larynx. Cancer 1991;67:141–149.
70. de Braud F, Ensley JF, Hassan M, Maciorowski Z, Pietraskiewicz H, Sakr W, Kish J, Tapazoglou E, Reed M, Al-Sarraf M. Prospective correlation of clinical outcome in patients with advanced, resectable squamous cell carcinomas of the head and neck (SCCHN) with DNA ploidy from fresh specimens. American Assocation for Cancer Research, [Abstract]. 1989.

CLINICAL COMMENTARY
Samuel G. Taylor, IV

If a test is to have importance in treatment decisions, good treatment alternatives need to exist for which such a test could select. The more difficult a test is to do, the more selective should it be to justify the inconvenience, patient discomfort, expense, and time that such a test might demand. A successful example is the assessment of hormone receptor status in breast cancer. Estrogen receptor status does have an independent prognostic significance in some situations, but, if this were its only advantage, it would be a passing fancy and not a standard-of-care issue. It is not clear at this time whether flow cytometric analysis of tumor specimens will prove to be clinically important or a passing fancy.

Authors Visscher and Crissman have collated data on flow cytometric analysis for bronchogenic carcinoma and upper aerodigestive tract cancers. It is clear that in neither of these situations can the results be considered definitive or offer undisputed indications for performing such an analysis outside of the research laboratory. Nevertheless, one advantage for such an overview is to point out directions and possible leads for more defined research that might lead to clinical relevancy in the future. At this point, it is probably not important to address the impact on accuracy of various methods of specimen preparation and techniques of analysis. Clearly, tissue handling and techniques do impact on the results. But trends from this overview analysis appear consistent and the imprecision of various methodologies does not invalidate the data. What will ultimately be required to demonstrate the practical utility of such information will be the use of flow cytometry in a randomized study of different treatment options. Showing a selective advantage for one form of treatment over another by using flow cytometric parameters would clearly define the value of flow cytometry. The demonstration of some improvement over more traditional techniques in the estimation of prognosis will have dubious clinical value in the long term unless treatment decision-making is altered.

This review emphasizes the high percentage of DNA aneuploidy in lung cancers. In non-small-cell varieties, commonly more than 70% of cases were aneuploid. The highest percentage of aneuploidy occurred in tumors with multiple sites sampled and in those obtained from fresh specimens. The high degree of variability in results obtained from different samples is consistent with the commonly described morphological variability of non-small-cell lung cancers. It is this heterogeneity within tumors that has led to lumping non-small-cell lung cancers as one rather than as multiple clinical entities.

The common association between aneuploidy and a worse outcome is difficult to interpret at present, but certainly is worthy of further definition. Do these observations mean that patients should or should not be selected for surgery or whether they should or should not be selected for chemotherapy on the basis of DNA ploidy? Certainly not at this time. But DNA-ploidy and S-phase fraction should be studied in ongoing clinical trials that are examining the role of chemotherapy, surgery, and radiation. Such studies should use fresh material and obtain multiple samples.

DNA aneuploidy is even more common and therefore even less likely to have predictive value in small-cell lung cancers. When over 85% of tumors are aneuploid, one wonders whether the few diploid ones are a result of sampling or other error. Such a test must have a very high discriminating power to be worthwhile to perform on anyone. Only if the few diploid tumors could be shown to result in substantial benefit to patients due to a different treatment approach than that used for aneuploid tumors, would such testing be likely to be valuable. Clearly, flow cytometry does not have this potential at this time. To state that the high rate of aneuploidy in small-cell lung cancers correlates with the characteristically high chemotherapy responsiveness of these tumors is a glib assumption that we have to carefully avoid. Non-small-cell lung cancers, which also have a high rate of aneuploidy, have a markedly different chemotherapy responsiveness and are considered less likely to respond to chemotherapy than breast cancers for example, which have a lower aneuploidy rate.

If the heterogeneity within each cancer promotes lumping of different types of lung cancer, the heterogeneity between cancers from different regions of the upper aerodigestive tract promotes fragmentation of cancers from this area. As indicated in the review by Visscher and Crissman, prognosis is affected by the degree of differentiation and by the lymphatic draining within a particular site. Because failures are predominately regional, treatment decisions for early cancers are based on the relative toxicities of radiation therapy versus surgery, with both having approximately equal efficacy. For more advanced tumors combined surgery and radiation is often used, with chemotherapy added to increase control of the most advanced lesions.

Chemotherapy is currently being suggested as a possible substitute for surgery in advanced cancers where surgery would cause loss of function. A major question at this time is whether patients can be selected for the option of chemotherapy and radiation therapy over surgery and radiation by parameters that might predict the effectiveness of the nonsurgical approach. There is some indication that flow cytometry may indeed play a role in aiding in such a patient selection. Intriguing work sited by Visscher and Crissman (their references 49, 58–60) suggests that aneuploid tumors may respond best with radiation and chemotherapy. Clearly, such an observation, if confirmed with prospective data, would go far in establishing flow cytometry as clinically valuable tool. One has to caution, however, that tumor differentiation already discriminates somewhat for chemotherapy responsiveness. To be valuable, flow cytometry would have to be shown by multivariate analysis to be more discriminating.

One problem that flow cytometry has to face is that the large, randomized trials that might best address the many issues raised by this overview analysis have been completed only recently or are currently in progress in lung and head and neck cancer. Clinicians need to be vigilant to add such laboratory questions to future studies to optimize sample collection and, then, to begin to define a role for flow cytometry in these malignancies.

19

Pediatric Solid Tumors

DAVID N. SHAPIRO and A. THOMAS LOOK

INTRODUCTION

Solid tumors account for 60% of all pediatric malignancies, with approximately 3700 newly-diagnosed cases per year in the U.S. (1). The spectrum of tumor types that occur in children is much different from that observed in adults. Included among the diverse group of pediatric neoplasms are brain tumors and other tumors of the central nervous system (35%); neuroblastoma (15%); soft tissue sarcomas, including rhabdomyosarcoma (10%); Wilms' tumor (10%); bone tumors, including osteosarcoma (8%); retinoblastoma (5%); and miscellaneous tumors including hepatoblastoma, germ cell tumors, and melanoma (17%) (Table 19.1). Enormous progress has been made in the diagnosis and management of these tumors since the original demonstration of the chemosensitivity of Wilms' tumor to actinomycin-D in 1966; cure rates for most childhood solid tumors have increased by as much as 50% during the last 25 years (2). This is attributable, in large part, to better understanding of prognostically important biologic features, improvements in the precision of clinical staging systems, and the development of more effective treatment, often incorporating a combination of chemotherapy, surgery, and radiotherapy (3).

The ability to cure increasing numbers of children with solid tumors has stimulated efforts to identify at diagnosis those who are likely to fail conventional treatment. Extent of disease prior to therapy, anatomic site of primary disease, and histologic subtype have been used individually and in combination to predict outcome for a variety of solid tumors. However, these criteria are often inadequate to accurately identify, in a prospective fashion, those children who will have a poor response to therapy within a given treatment group. Clearly, tumor cell biology is a major factor contributing to the clinical heterogeneity observed in many pediatric solid tumors. These biologic features are only indirectly reflected by the various clinical presentations found in many childhood neoplasms.

Abnormal cellular DNA content (DNA aneuploidy) has been linked to the rate of cell proliferation, the extent of cellular differentiation, and, ultimately, to prognosis in a variety of pediatric and adult malignant diseases (4–7). The ability to detect this abnormality rapidly, precisely, and reproducibly with flow cytometric techniques offers a quantitative measure of biologic variability that could be used to augment present-day clinical staging systems. Such improvements are being translated into the more judicious selection of therapy, which is especially important in children, for whom the benefits of modern cancer therapy must always be balanced against potential adverse sequelae. It is the purpose of this review to examine the role of ploidy analysis in the diagnosis and management of children with solid tumors as well as to correlate differences in DNA ploidy with the unique clinical and biologic features of these neoplasms.

BRAIN TUMORS

Primary central nervous system (CNS) neoplasms account for almost 20% of all pediatric malignancies and are the most common solid tumors in children. They are exceeded in frequency only by leukemias and lymphomas and have an incidence rate of 2.4 new cases per 100,000 children per year (8). Brain tumors are actually a heterogeneous group of

Table 19.1
Distribution of Solid Tumors for Children Less Than 15 Years of Age

Diagnosis	Percent of Total
Brain and nervous system	35
Glioma	22
Medulloblastoma	8
Ependymona	3
Other	2
Sympathetic nervous system	15
Neuroblastoma	12
Other	3
Kidney	10
Wilms' Tumor	10
Other	<1
Retinoblastoma	5
Bone	8
Osteosarcoma	4
Ewing's sarcoma	3
Chondrosarcoma	<1
Other	<1
Soft tissue sarcomas	10
Rhabdomyosarcoma	5
Fibrosarcoma	1
Others	4
Germ cell tumors	5
Liver	2
Hepatoblastoma	1
Hepatoma	1
Miscellaneous	10
Melanoma	1
Thyroid	2
Other	7

Table 19.2
Histology and DNA Ploidy of Pediatric Gliomas

Tumor Grade	DI = 1.0 (n)	DI >1.00 (n)
Astrocytoma (low-grade)	28	4
Anaplastic Astrocytoma (intermediate-grade)	5	5
Glioblastoma multiforme (high-grade)	1	7

neoplasms whose classification has traditionally been based on the morphologic similarity of the predominant malignant cell type to its presumed normal counterpart. Tumors of the glial series of cells (astrocytomas, oligodendrogliomas, and ependymomas) account for approximately 70% of pediatric brain tumors, while tumors of neuronal progenitors (medulloblastomas) account for 20% of the cases. The remaining tumors are primarily pineal parenchymal tumors (pineoblastomas), meningiomas, and teratomas (9). Certain central nervous system tumors, especially tumors of glial origin, are histologically graded according to the degree of anaplasia of the malignant cells; higher grade tumors generally have a more aggressive clinical course.

Glial Tumors

Although the majority of pediatric central nervous system neoplasms are glial tumors, only about 10% are considered high-grade tumors. Principal prognostic factors have been related to tumor location and histologic grade; patients with lower grade tumors have increased survival times (10). Low-grade astrocytomas have invariably been reported to have a unimodal diploid DNA content and a low mean S-phase labeling index (<4%), as determined by bromodeoxyuridine (BUdR) infusion at the time of craniotomy (11–13). The apparent lack of tumor cell heterogeneity, as reflected by the DNA indexes of low-grade glial neoplasms, has made clinical correlations difficult (Table 19.2). In contrast, high-grade astrocytic tumors, especially glioblastoma multiforme, are almost always DNA aneuploid, may possess multiple tumor stem lines, and have a high mean S-phase labeling index (9.3%) (12, 13). Because of the uniformly aneuploid DNA indexes of high-grade astrocytomas, correlation with prognosis has also usually not been feasible (14–17). However, in a limited study of nine patients with glioblastoma multiforme, decreased survival was noted in those patients whose tumors had greater percentages of cells in the S phase of the cell cycle (18). Few nonastrocytic glial tumors have been evaluated for DNA content by flow cytometry. For cerebral ependymomas, however, the same general pattern has been observed, reflecting a close relationship between tumor cell anaplasia and aneuploidy (19, 20).

Medulloblastoma

Unlike pediatric glial tumors, most medulloblastomas are high-grade malignancies consisting of small, round, undifferentiated cells with hyperchromatic nuclei and abundant mitoses, which arise from the transformation of cells of the

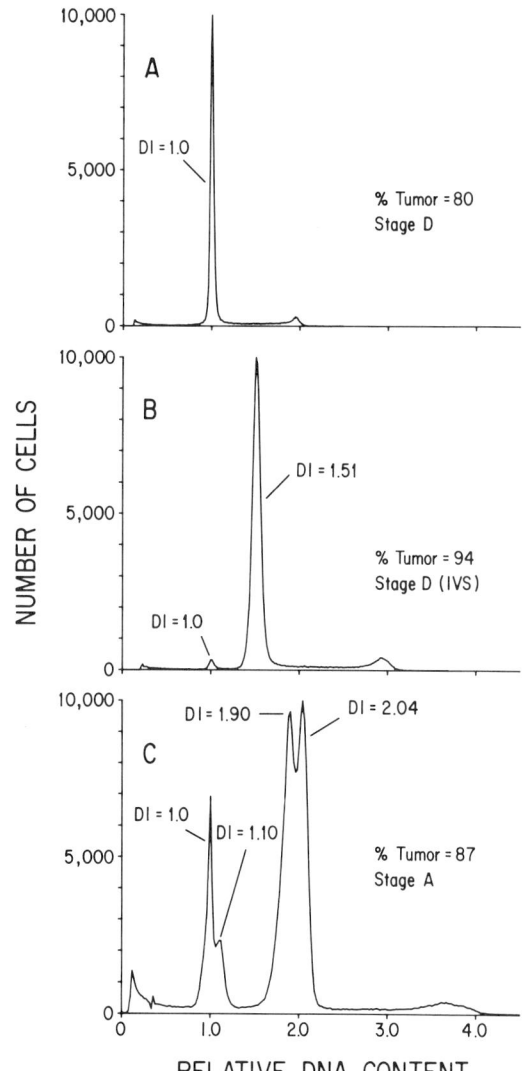

Figure 19.1. Representative DNA histograms demonstrating three discrete categories of neuroblastoma stem–line ploidy. In **A**, the DNA content of the neuroblasts cannot be distinguished from that of residual normal marrow cells (DNA index [DI] = 1.0) in a patient with metastatic tumor. **B** shows results from neuroblasts obtained at biopsy from an infant with Evans' Stage-IV-S disease and hyperdiploid DNA content (DNA index = 1.51). A small peak of diploid G_0/G_1 blood cells is evident, accounting for 2% of the cells analyzed in the histogram and corresponding to the 6% fraction of nontumor cells determined from a differential count on Wright–stained smears. The neuroblasts in **C**, also obtained by tumor biopsy, are those of an infant with localized disease and multiple neuroblastoma stem lines. Peaks of G_0/G_1 cells with DNA indexes of 1.10, 1.90, and 2.04 are evident. (Reproduced with permission from Look AT, et al. N Engl J Med 311:231, 1984.)

external granular layer of the cerebellum. Overall five-year survival rates are between 45% and 60% for children with medulloblastoma, which has allowed meaningful analysis of various prognostic factors, including tumor size, tumor extent, histology, age at diagnosis, and extent of initial resection (21–23).

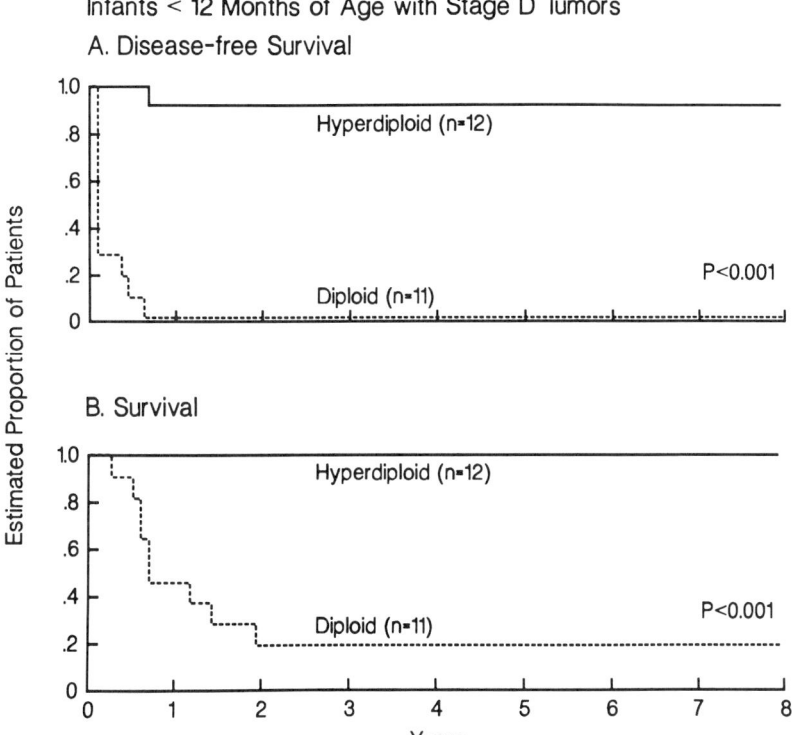

Figure 19.2. Kaplan-Meier analysis of disease-free survival (**A**) and overall survival (**B**) for infants <12 months of age with stage-D (metastatic) tumors who were treated with five courses of cyclophosphamide-doxorubicin chemotherapy, according to the presence of either hyperdiploid (solid line) or diploid (dashed line) tumor stem lines. (Reproduced with permission from Look AT, et al. J Clin Oncol 1991;9: 581.)

Several recent reports have illustrated the prognostic importance of DNA ploidy in medulloblastomas (24–27). These studies, all performed on formalin-fixed, paraffin-embedded tissues, demonstrated that one-half of medulloblastomas have a diploid DNA content while the remainder are DNA aneuploid. Of the aneuploid tumors, virtually all have hyperdiploid DNA content, including approximately 6% with tetraploidy. In a recent trial conducted at St. Jude Children's Research Hospital, patients were treated with a combination of radical resection and radiation therapy; those with hyperdiploid medulloblastomas showed significantly better outcome. Additionally, a correlation was observed between hyperdiploidy and resectability, suggesting that diploid medulloblastomas behave more aggressively. This observation is consistent with the tendency of diploid medulloblastomas to disseminate into the cerebrospinal fluid (25). Diploidy was also noted to correlate with the subsequent development of bone metastases, even in those patients who have achieved good local control (26). Comparison of the G_0/G_1- and S-phase fractions with DNA index revealed no significant correlation, indicating that proliferative rate is not a major biologic factor discriminating these two ploidy groups. Taken together, these results suggest that children with diploid medulloblastomas should be treated with aggressive multimodal therapy, including adjuvant chemotherapy, to prevent recurrence and metastases. By contrast, patients with hyperdiploid tumors appear to do well on current treatment regimens consisting of total resection and adequate radiotherapy.

NEUROBLASTOMA

Neuroblastoma arises from the adrenal medulla and neural crest cells of the sympathetic nervous system, typically presenting as an abdominal mass in young children. It is the most common extracranial solid tumor of childhood and comprises up to 50% of malignancies among infants (28). Children of all ages with localized neuroblastomas are very likely to be cured with contemporary, multimodal therapy, whereas in those with disseminated tumors, the age of the child at diagnosis is a critical factor in predicting outcome (29, 30). Several biologic markers of presumed disease activity, including elevated urinary excretion of catecholamines and cystathionine and elevated serum concentrations of neuron-specific enolase and ferritin, have been advocated as reliable prognostic factors (31–33). Additionally, at least in some series, certain histopathologic features have been associated with decreased survival (34).

DNA ploidy is one of the most strikingly predictive biologic features associated with neuroblasts from infants with unresectable neuroblastoma. The DNA index of tumor cells from these patients has been shown to correlate with clinical outcome as well as with critical tumor-specific molecular ab-

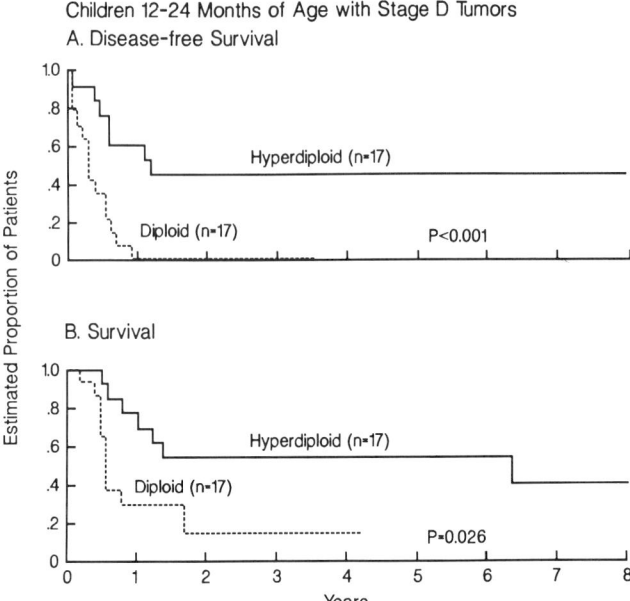

Figure 19.3. Kaplan-Meier analysis of disease-free survival (**A**) and overall *survival* (**B**) for children 12–24 months of age with stage-D tumors who were treated with four-drug combination chemotherapy, according to the presence of either hyperdiploid (solid line) or diploid (dashed line) tumor stem lines. (Reproduced with permission from Look AT, et al. J Clin Oncol 1991;9:581.)

normalities and cytogenetic features. We have demonstrated that the ploidy of neuroblasts from infants with unresectable neuroblastoma, as determined at diagnosis by flow cytometric analysis of cellular DNA content, is closely linked to prognosis (Fig. 19.1) (35, 36). About one-third of the infants with widespread metastatic disease will ultimately fail chemotherapy, and they cannot be identified at diagnosis on the basis of clinical features alone. In this group, each infant (<12 months of age) with hyperdiploid tumor cells had a complete or partial response to five courses of cyclophosphamide and doxorubicin (Fig. 19.2). Over 90% of these very young children with widely disseminated hyperdiploid tumors at diagnosis were in fact cured of their disease. A similar relationship between hyperploidy and prolonged disease-free survival was also evident in children 12-to-24 months of age, with metastatic tumors, who were treated with four-drug regimens (37).

By contrast, none of the infants or children up to 24 months of age with metastatic neuroblastoma and whose tumor cells had a diploid DNA index achieved long-term disease control (Figs. 19.2 and 19.3). The favorable prognostic significance of hyperdiploidy and the dire implications of diploidy in young children with neuroblastoma has subsequently been confirmed by other investigators, using either flow cytometric techniques to determine cellular DNA content or cytogenetic analysis of metaphase chromosomes (38–51). However, despite its clear importance in younger children, tumor cell ploidy does not correlate with disease-free survival in children with unresectable neuroblastoma who

are older than 24 months of age. Older children with disseminated tumors almost invariably fail treatment whether the tumor cells are diploid or hyperdiploid. These results suggest that biologic factors other than DNA content play a role in determining the aggressiveness of neuroblastomas in older children.

The clinical relevance of neuroblast N-*myc* protooncogene copy number was first demonstrated by Brodeur and coworkers (52), who showed that N-*myc* amplification correlates with advanced stage and a high risk of early treatment failure. The association between N-*myc* amplification and rapid disease progression has subsequently been confirmed by others (45–47, 51, 53–55). In infants, N-*myc* gene amplification has been more frequently associated with tumor cell diploidy than with hyperdiploidy (Fig. 19.4). Preliminary results indicate that the combination of a hyperdiploid DNA index and single-copy N-*myc* levels identifies a subset of infants with disseminated tumors who have nearly a 100% complete response rate and long-term disease-free survival. Although only 5% of hyperdiploid tumors in infants have N-*myc* gene amplification, each infant with these findings has had early treatment failure and a fatal outcome (37).

Additionally, comparison of ploidy findings with structural karyotypic alterations has indicated that diploid tumors also have recurrent abnormalities of chromosome structure. Particularly common are deletions of chromosome 1p and double minute chromatin bodies or homogeneously staining regions (36, 37, 49–51, 56–57). Other associations noted with diploid neuroblastomas include a higher mean percentage of cells in S phase as well as a higher frequency of features indicative of an unfavorable histology (34, 38, 46).

These findings have important implications for the clinical management of infants and children with neuroblastoma. An unfavorable combination of advanced disease, diploid DNA content, and increased N-*myc* copy number could be used to identify infants and young children for whom conventional therapy is unlikely to be effective. For this subgroup, one could justify more aggressive therapy. This approach is being investigated in a prospective clinical trial by the Pediatric Oncology Group (one of two large, American pediatric cooperative groups for treating various pediatric solid tumors and leukemias) in which therapy for infants with neuroblastoma is being modified according to the patient's pretreatment DNA index. Patients with hyperdiploid tumors receive conventional therapy with cyclophosphamide and doxorubicin. Those with diploid tumors receive more aggressive therapy with VM-26 and cisplatin, which are agents that induce a response in approximately half of all infants who fail first-line therapy (58). This approach has the advantage of potentially inducing longer remissions in infants with high-risk genetic features (diploidy, N-*myc* amplification) while sparing others the added toxicities of more intensive therapy.

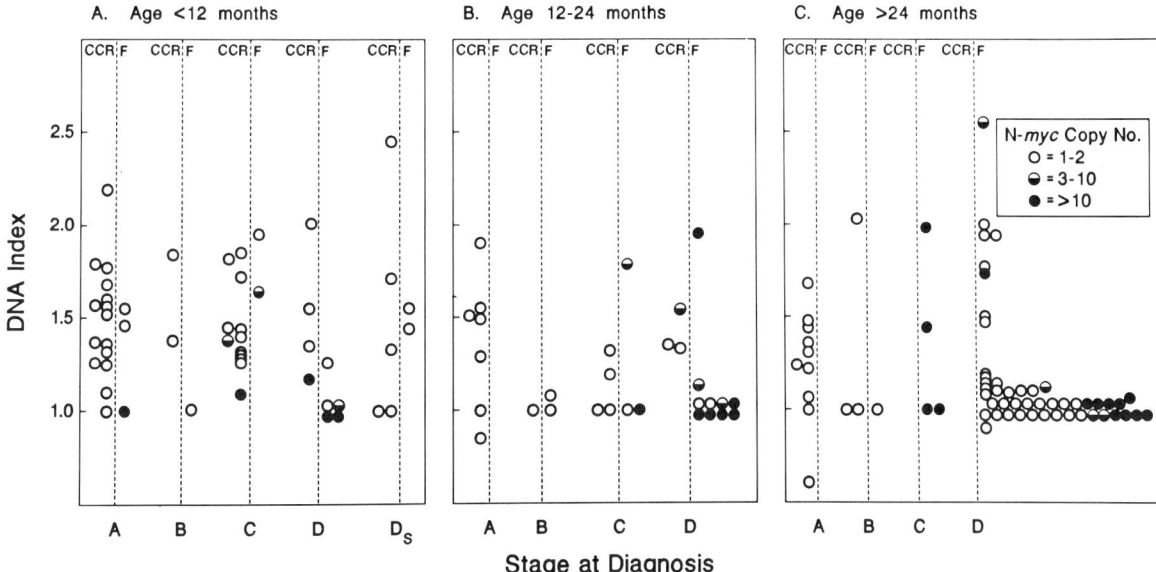

Figure 19.4. Distribution of the DNA indexes of the malignant stem lines in 147 children, according to the stage of the tumor at diagnosis. Values are shown separately for infants <12 months of age (**A**), children 12–24 months (**B**), and children >24 months (**C**). Diploid stem lines (DNA index = 1.0) are shown as two rows of circles for stage-D cases. Cases with single-copy levels of the N-*myc* gene per haploid genome are shown as open circles, those with 3-to-10-fold amplification as half-shaded circles, and those with >10-fold amplification as completely shaded circles. Circles to the left of the dashed line for each stage represent patients in complete remission with a minimum follow-up of 30 months; circles to the right of the dashed line represent patients who have either failed to enter remission or who have developed recurrent disease (*F*). (Reproduced with permission from Look AT, et al. J Clin Oncol 1991;9:581.)

WILMS' TUMOR

Wilms' tumor, the most common abdominal neoplasm in children, accounts for virtually all renal tumors in children less than 15 years of age. With the use of modern combined-modality therapy, nearly 90% of patients with this diagnosis will become long-term survivors (50). However, for those patients with advanced disease or certain unfavorable histologic subtypes, the outlook is generally less optimistic. Approximately 25% of treatment failures and almost 40% of all tumor deaths are accouned for by children with the anaplastic variant of Wilms' tumor, a histopathologic subtype found in less than 10% of all cases (60).

Flow cytometric analysis of the DNA content of Wilms' tumor has been reported to distinguish the anaplastic variant from more favorable histologies (61–66). These studies have shown that nonanaplastic Wilms' tumors usually have DNA indexes less than 1.4 times the DNA content of diploid cells, while anaplastic tumors are invariably hyperdiploid, with DNA indexes ranging from 1.7 to 3.2 times that of diploid cells (Fig. 19.5) (62, 63). The prognostic importance of hyperdiploidy was noted in most cases to correlate with the poor outcome reported for patients with anaplastic histology, although at least one study failed to confirm these observations (Fig. 19.6) (66). Additionally, the effect of disease extent on the prognostic impact of tumor hyperdiploidy has differed between reported studies, with one study reporting finding a significant relationship only for advanced stage patients (63). Other flow cytometric parameters, such as the fraction of S-phase cells, have also been reported to be of value in identifying high-risk patients in one study (64) but not in another (65).

As in neuroblastoma, banded karyotypic analyses of Wilms' tumors have contributed important information beyond that supplied by DNA ploidy and tumor histology (62, 67, 68). Tumors without hyperdiploidy lack major chromosomal rearrangements or translocations. By contrast, complex translocations identified a high-risk group of hyperdiploid patients who had relatively short relapse-free survival intervals compared to those of hyperdiploid patients without translocations (62). It is important to note, however, that the correlation between hyperdiploid tumors and a poor response to treatment is exactly opposite to that observed in neuroblastoma and medulloblastoma. This serves to emphasize that tumor cell ploidy is probably not a primary determinant of drug responsiveness. It appears instead to be a marker of subgroups of tumors with unique cytogenetic and molecular genetic features. It remains to be determined how ploidy relates to critical molecular events leading to malignant transformation in Wilms' tumor, such as the disruption of the newly identified *WT2-1* locus located on chromosome 11p13 (69, 70).

RHABDOMYOSARCOMA

Rhabdomyosarcomas comprise almost 60% of the pediatric soft tissue sarcomas and are thought to arise from striated muscle progenitor cells found throughout the body (71). Combined multiagent chemotherapy, radiotherapy, and sur-

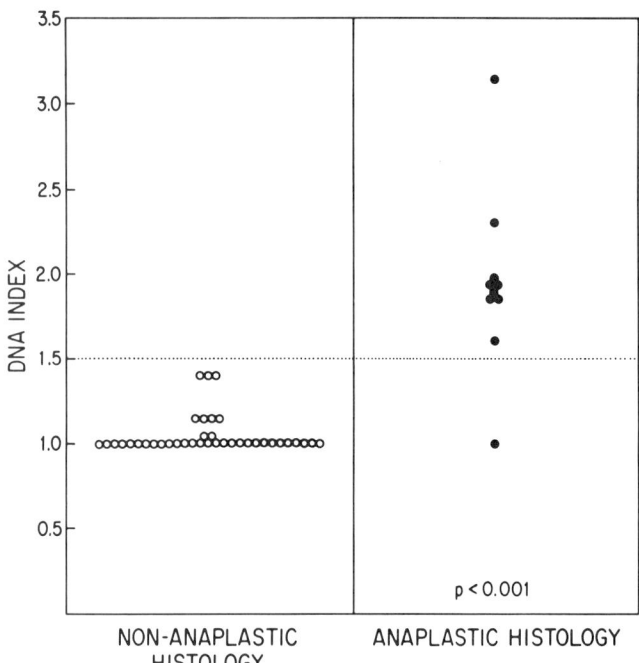

Figure 19.5. Distribution of the DNA indexes of malignant stem lines (with highest ploidy) in 48 patients with Wilms' tumor, according to histologic subtype at diagnosis. A DNA index of 1.5 was chosen as the cutoff for hyperdiploidy because it yielded the best discrimination between the anaplastic and nonanaplastic group. (Reproduced with permission from Douglass EC, et al. J Clin Oncol 1986;4:975.)

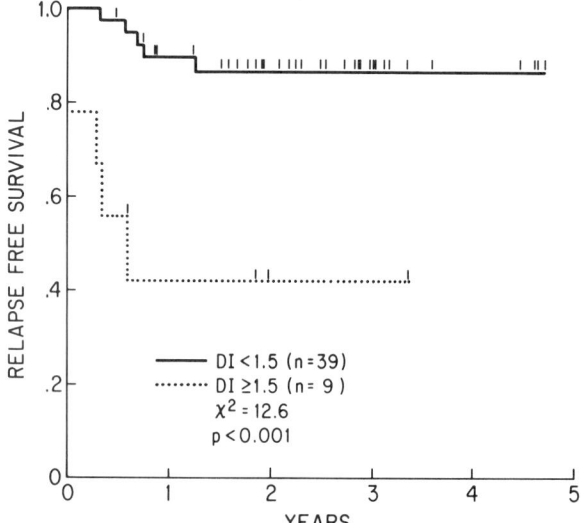

Figure 19.6. Kaplan-Meier analysis of relapse-free survival times for 48 patients with Wilms' tumor according to DNA index at diagnosis. (Reproduced with permission from Douglass EC, et al. J Clin Oncol 1986;4:975.)

gery have substantially improved the clinical outcome in children with rhabdomyosarcoma with cure rates now exceeding 50%. Disease extent, anatomic site of primary disease, and histologic subtype (either embryonal or alveolar) have been used in combination to predict outcome in chil-

dren with rhabdomyosarcoma. Unfortunately, many children with otherwise "favorable" features, whose tumors are indistinguishable by clinical and pathologic characteristics, respond poorly to intensively administered multimodal therapy.

Rhabdomyosarcomas have only recently been investigated for specific ploidy abnormalities as determined by flow cytometry (72–76). Initial reports, examining small numbers of specimens, described only hyperdiploid and near-tetraploid tumor stem lines in rhabdomyosarcoma samples (73, 74). However, in a larger series of patients, we noted that rhabdomyoblast ploidy was divided almost equally between diploid, hyperdiploid, and near-tetraploid tumors (Fig. 19.7) (76). Further, although diploidy was noted to occur in tumors of either embryonal or alveolar histology, hyperdiploidy was exclusively associated with embryonal tumors and near-tetraploidy with alveolar tumors. Although not previously noted, the association of near-tetraploidy with rhabdomyosarcomas of alveolar histology and of hyperdiploid abnormalities with embryonal tumors is consistent with earlier observations (73).

We showed that tumor cell ploidy also has a significant impact on survival in embryonal rhabdomyosarcoma. As in neuroblastoma and medulloblastoma, hyperdiploidy conferred the best prognosis and diploidy the worst (76). Approximately 75% of patients with hyperdiploid rhabdomyosarcomas are long-term survivors, whereas no patients with diploid tumors survived for longer than 18 months (Fig. 19.8). Therapeutic outcomes of patients with near-tetraploid tumors were almost as dismal as the outcomes of patients with diploid tumors. This result is not unexpected, because karyotypic evidence supports the development of tetraploidy by endoreduplication of a primary diploid stem line (77), while intermediate hyperdiploidy may arise through nondisjunction. Diploid tumors, which responded poorly to therapy, also had a lower percentage of cells in S phase compared to hyperdiploid tumors. Most antineoplastic drugs are preferentially cytotoxic to proliferating tumor cells; hence, patients with diploid rhabdomyosarcomas may fail not because of the overgrowth of drug-resistant mutants, but rather due to the relentless expansion of a tumor cell population with a relatively low growth fraction. Similar results, noting an association between lower S-phase percentage and diploidy, have been reported in certain adult solid tumors (4).

As with other pediatric solid tumors, cellular DNA content measurements permit the identification of a subset of patients with rhabdomyosarcoma who are at exceedingly high risk for treatment failure. Patients with tumor cell diploidy respond poorly to current treatment regimens regardless of histology and are obvious candidates for more aggressive or alternative therapy. Conversely, those with hyperdiploid rhabdomyosarcomas appear to respond well to currently available treatment protocols.

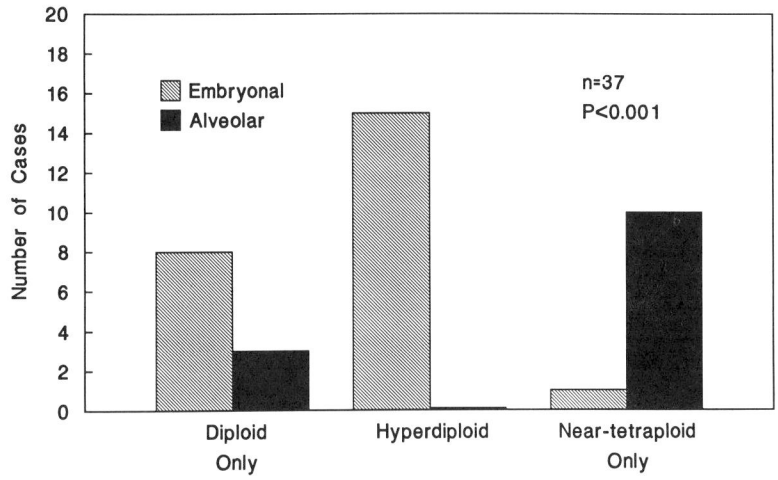

Figure 19.7. Distribution of rhabdomyosarcoma with embryonal or alveolar histology according to ploidy classification. The distribution of diploid tumors between alveolar and embryonal histologic subtypes was not statistically different (P=0.71), whereas hyperdiploidy was associated with embryonal tumors (P<0.001) and near-tetraploidy with alveolar tumors (P<0.001). (Reproduced with permission from Shapiro DN, et al. J Clin Oncol 1991;9:159.)

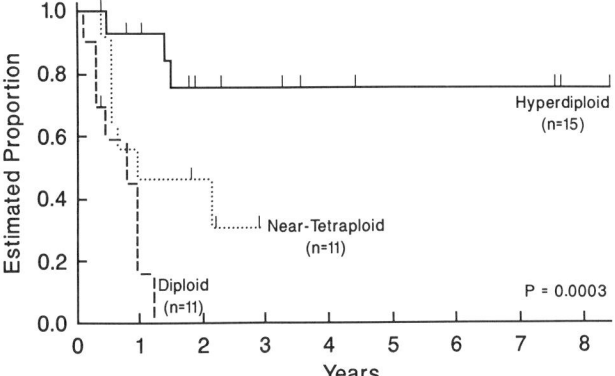

Figure 19.8. Kaplan-Meier analysis of overall survival by ploidy classification at diagnosis for patients with unresectable rhabdomyosarcoma. Patients with hyperdiploid tumors had significantly improved overall survival compared to those with diploid tumors (P<0.0001). (Reproduced with permission from Shapiro DN et al. J Clin Oncol 1991;9:159.)

OSTEOSARCOMA

Osteosarcoma, a spindle-cell tumor that produces malignant osteoid, includes classic high-grade osteosarcoma (approximately 95% of cases) and other infrequent histopathologic subtypes. It is the most common primary bone tumor in children and adolescents. About 90% of cases arise in the long bones of the extremities. Although only 20% of patients have clinically detectable metastatic disease at the time of diagnosis, most patients must have subclinical dissemination because of the high relapse rate for those patients treated with surgical resection alone (78). Several clinical characteristics are thought to be of prognostic significance for patients with osteosarcoma, including extent of disease at diagnosis, histology, and primary site of disease (79). These factors are of limited practical utility, however, because the majority of patients present with similar clinical features. Further, despite the demonstration that multiagent chemotherapy ad-

ministered after definitive surgery improves the relapse-free survival of patients, metastatic disease develops in up to one-half of patients who are usually indistinguishable at diagnosis from those who do not relapse.

Several investigators have reported results of flow cytometric analysis of the DNA content of osteosarcoma cells (80–90). These studies have shown that high-grade osteosarcomas are virtually all aneuploid with up to 50% of samples containing two or more stem lines. Flow cytometric findings of a high degree of aneuploidy have been confirmed by cytogenetic analysis (91). One report demonstrated that paraosteal osteosarcoma tumors, which are histologically well-differentiated and have a relatively favorable prognosis, have diploid DNA indexes that differ markedly from aneuploid high-grade osteosarcomas (86). Additionally, these investigators found that paraosteal tumors had a significantly lower proportion of cells in S phase (8.6%) as compared to high-grade osteosarcomas (18.8%), suggesting another feature to discriminate low-grade from high-grade osteosarcoma.

The determination of the DNA content for patients with high-grade osteosarcoma has also demonstrated the usefulness of tumor cell ploidy for predicting the chemosensitivity of this tumor (92). We have recently shown that near-diploid tumor stem lines coexisted with hyperdiploid lines in 60% of the samples studied (Fig. 19.9). Both the relapse-free and overall survival analysis showed that the presence of a near-diploid tumor stem line was associated with improved outcome, regardless of the presence of additional hyperdiploid stem lines (Fig. 19.10). Pulmonary metastases were noted to develop in only two of the patients with near-diploid lines, in contrast to seven of the 10 with hyperdiploid lines exclusively. Thus, patients with a near-diploid tumor stem line can be expected to respond favorably to currently employed adjuvant therapy, whereas those with only hyperdiploid lines should be considered as candidates for alternative therapy.

In osteosarcoma, as in Wilms' tumor, hyperdiploidy correlates with an adverse clinical outcome. One important dif-

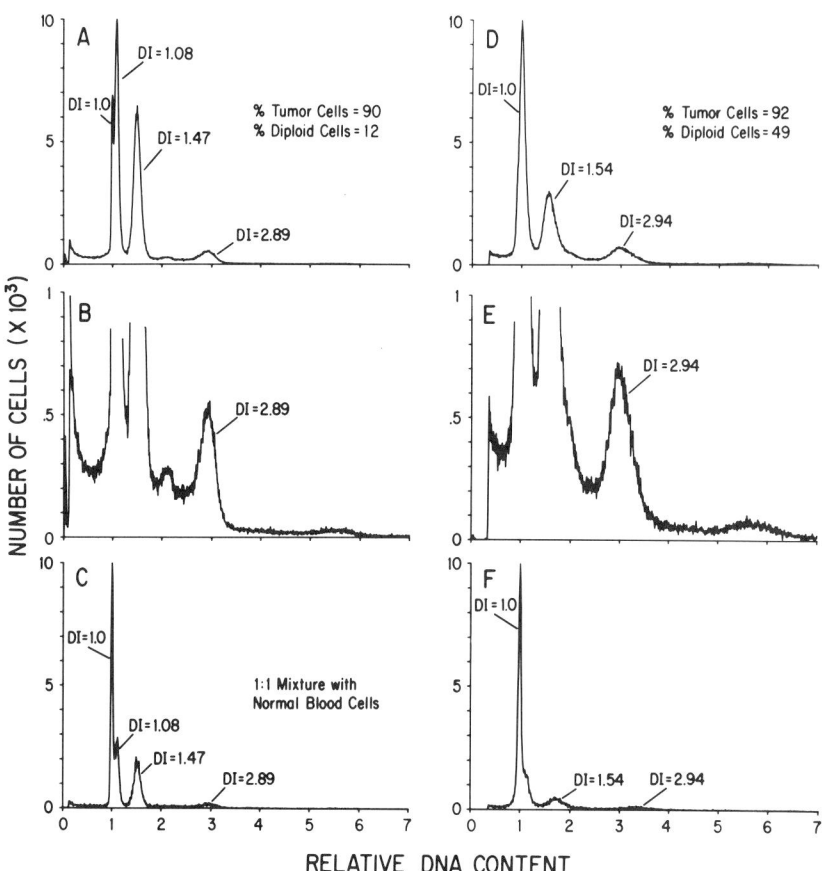

Figure 19.9. Representative DNA histograms demonstrating both diploid and hyperdiploid osteosarcoma stem cell lines. **A** shows three tumor stem lines with DNA indexes of 1.08, 1.47, and 2.89, together with a peak with diploid DNA content (DNA index) of 1.0. The peak of diploid G_0/G_1-phase cells accounts for 12% of the cells analyzed in the histogram and corresponds to the 10% fraction of nontumor cells determined from a differential count on a Wright-stained smear. **B**, An expanded scale plot of the histogram in **A**, shows the S phase and $G_2 + M$–phase cells of the hypertetraploid stem line (DNA index of 2.89). **C** shows the result of flow cytometric analysis of a *1:1* mixture of tumor cells and normal blood cells. The stained normal diploid cells add to the peak corresponding to blood leukocytes in the original sample. **D** through **F** show similar histograms of tumor cells from a patient with osteosarcoma. Forty-nine percent of the cells analyzed in the DNA histogram have a diploid DNA content. Ninety-two percent of the cells were identified as tumor cells, indicating a tumor stem line with DNA content indistinguishable from that of normal *diploid* cells; two hyperdiploid stem lines (DNA indexes of 1.54 and 2.94) were also detected. (Reproduced with permission from Look AT, et al. N Engl J Med 1988;318:1567.)

ference was observed, however. For osteosarcoma, the favorable influence of near-diploidy was apparent even if additional hyperdiploid stem lines were present. Thus, in osteosarcoma, there may be cellular interactions between stem lines of different ploidies affecting tumor growth that have not been observed in other tumors.

OTHER TUMORS

Other relatively uncommon pediatric solid tumors have been examined by flow cytometric analysis for determination of DNA content. Retinoblastoma, a malignant tumor of the embryonic neural retina, was found to generally have aneuploid DNA content by two groups of investigators (93, 94). The proportion of cells in S phase varied widely between individual tumors. It was concluded, in these limited series, that flow cytometric analysis did not improve the classification of retinoblastomas or offer any additional prognostic information superior to that supplied by conventional staging systems.

Ewing's sarcoma, the second most common malignant bone tumor of children and adolescents accounts for about 3% of all childhood solid tumors (1). The cell of origin of Ewing's sarcoma has not been precisely established. However, evidence implicating a cell of neural lineage has been provided by the observation that Ewing's sarcoma and primitive neuroectodermal tumors both share at (11;22) translocation (95, 96). The ploidy of Ewing's sarcoma has been studied in several series analyzing the DNA content of other malignant bone and mesenchymal tumors (73, 84, 88, 97). These tumors were invariably found to be diploid and clinically useful prognostic information could not be ascertained beyond that supplied by established staging schemes.

Finally, rhabdoid tumors are highly malignant neoplasms of unknown cellular origin that usually arise in the kidney, although they occasionally arise in extrarenal sites such as the chest wall and the head and neck. All cases studied by flow cytometry have revealed exclusively diploid cell lines, in agreement with a recent cytogenetic analysis showing only normal-appearing diploid karyotypes in all cases examined (64, 65, 98, 99).

CONCLUSIONS

Flow cytometric analysis of the DNA content of several pediatric solid tumors, including medulloblastoma, neuroblastoma, Wilms' tumor, rhabdomyosarcoma, and osteosarcoma has been shown to be of considerable value for identifying different prognostic subsets of patients. For pediatric solid tumors, both diploidy (neuroblastoma, medulloblastoma,

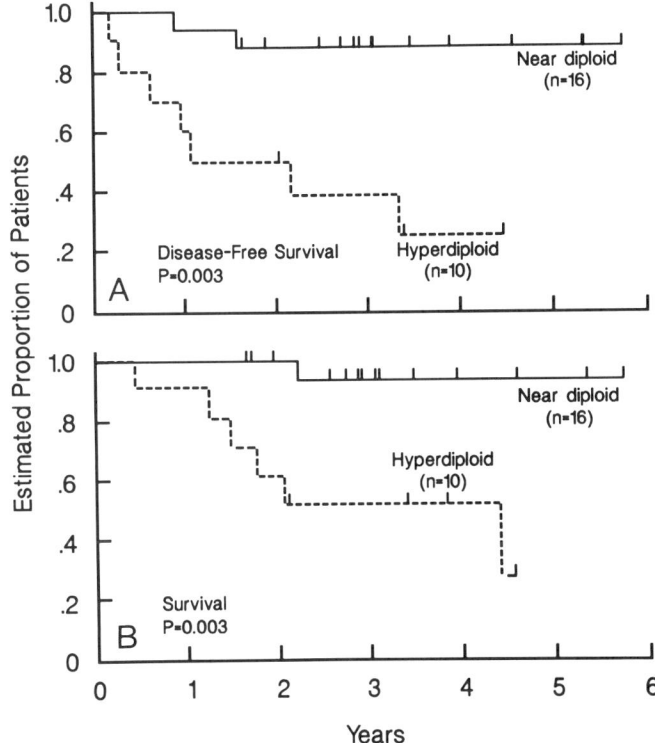

Figure 19.10. Kaplan-Meier analysis of *disease-free survival* (**A**) and overall *survival* (**B**) according to the presence or absence of a near-diploid tumor stem line at diagnosis. The difference between the times to relapse had a high level of significance. (Reproduced with permission from Look AT, et al. N Engl J Med 1988;318:1567.)

Table 19.3
Flow Cytometric Determination of Ploidy in Childhood Tumors

Disease	Ploidy		Reference
	Low-Risk	High-Risk	
Disseminated neuroblastoma of infants	Hyperdiploid	Diploid	35–37
Rhabdomyosarcoma	Hyperdiploid	Diploid, tetraploid	76
Medulloblastoma	Hyperdiploid	Diploid	24–26
Wilms' tumor	Near-diploid	Near-tetraploid	62–66
Osteosarcoma of an extremity	Near-diploid	Near-tetraploid	92

rhabdomyosarcoma) and hyperdiploidy (Wilms' tumor, osteosarcoma) identify a high-risk subset of patients (Table 19.3). Correlations between DNA ploidy and direct measurements of in vitro tumor chemosensitivity or the expression of the multidrug-resistance gene (*mdr*1) product (GP-170) have not been observed, however (100). Thus, tumor cell ploidy is probably not a primary determinant of drug sensitivity. Rather, ploidy appears to be a clinically useful marker of subsets of tumors with unique cytogenetic and molecular genetic features that correlate with the clinical behavior of the tumor. This is especially evident for neuroblastoma, Wilms' tumor, and rhabdomyosarcoma, where unique karyotypic, histopathologic, and molecular genetic characteristics are associated with specific ploidy groups.

There is an obvious need to develop improved therapy for the majority of children with advanced-stage solid tumors. The adverse clinical information conveyed by DNA ploidy values in specific types of pediatric malignancies has definite therapeutic implications. Subsets of patients, indistinguishable by traditional clinicopathologic criteria and at high risk for treatment failure, can be readily identified by flow cytometric determination of DNA content. Patients identified in this fashion are suitable candidates for alternative therapy. Thus, the ultimate goal of clinical staging, the accurate identification of individual patients with high or low probabilities of successful treatment outcome, can be substantially improved by the systematic measurement of tumor cell ploidy and other important biological features of the malignant clone.

ACKNOWLEDGMENTS

Supported in part by the Cancer Center Support (CORE) Grant CA-21765, the Solid Tumor Program Project Grant CA-23099, and The American Lebanese Syrian Associated Charities (ALSAC).

REFERENCES

1. Young JL Jr, Miller RW. Incidence of malignant tumors in U.S. children. J Pediatr 1975;86:254–258.
2. Farber S. Chemotherapy in the treatment of leukemia and Wilms' tumor. JAMA 1966;198:826–836.
3. Balis FM, Holcenberg JS, Poplack DG. In: Pizzo PA, Poplack DG. eds. Principles and Practice of Pediatric Oncology. Philadelphia: J.B. Lippincott, 1989:165–206.
4. Barlogie B, Raber MN, Schumann J, et al. Flow cytometry in clinical cancer research. 1983;43:3982–3997.
5. Friedlander ML, Hedley DW, Taylor IW. Clinical and biological significance of aneuploidy in human tumours. J Clin Pathol 1984;37: 961–974.
6. Merkel DE, Dressler LG, McGuire WL. Flow cytometry, cellular DNA content, and prognosis in human malignancy. J Clin Oncol 1987;5:1690–1703.
7. Merkel DE, McGuire WL. Ploidy, proliferative activity and prognosis. DNA flow cytometry of solid tumors. Cancer 1990;65:1194–1205.
8. Young JL Jr, Ries LG, Silverberg E, Horm JW, Miller RW. Cancer incidence, survival, and mortality for children younger than age 15 years. Cancer 1986;58:598–602.
9. Ertel IJ. Brain tumors in children. Cancer 1980;30:306–321.
10. Duffner PK, Cohen ME, Meyers ME. Survival of children with brain tumors: SEER program 1973–1980. Neurology 1986;36:597–601.
11. Bigner SH, Bjerkvig R, Laerum OD. DNA content and chromosomal composition of malignant human gliomas. Neurol Clin 1985;3:769–784.
12. Hoshino T, Nagashima T, Murovic JA, et al. In situ cell kinetics studies on human neuroectodermal tumors with bromodeoxyuridine labeling. J Neurosurg 1986;64:453–459.
13. Cho KG, Nagashima T, Barnwell S, Hoshino T. Flow cytometric determination of model DNA population in relation to proliferative potential of human intracranial neoplasms. J Neurosurg 1988;69:588–592.
14. Frederiksen P, Reske-Nielsen E, Bichel P. Flow cytometry in tumours of the brain. Acta Neuropathol (Berl) 1978;41:179–183.
15. Hoshino T, Nomura K, Wilson CB, Knebel KD, Gray JW. The distribution of nuclear DNA from human brain-tumor cells. J Neurosurg 1978;49:13–21.

16. Kawamoto K, Herz F, Wolley RC, Hirano A, Kajikawa H, Koss LG. Flow cytometric analysis of the DNA distribution in human brain tumors. Acta Neuropathol (Berl) 1979;46:39–44.
17. Mork SJ, Laerum OD. Modal DNA content of human intracranial neoplasms studied by flow cytometry. J Neurosurg 1980;53:198–204.
18. McKeever PE, Feldenzer JA, McCoy JP, et al. Nuclear parameters as prognostic indicators in glioblastoma patients. J Neuropathol Exp Neurol 1990;49:71–78.
19. Spaar FW, Blech M, Ahyai A. DNA-flow fluorescence—cytometry of ependymomas. Report on ten surgically removed tumours. Acta Neuropathol (Berl) 1986;69:153–160.
20. Horowitz ME, Parham DM, Douglass EC, Kun LE, Houghton JA, Houghton PJ. Development and characterization of human ependymoma xenograft HxBr5. Cancer Res 1987;47:499–504.
21. Allen JC. Childhood brain tumors: Current status of clinical trials in newly diagnosed and recurrent disease. Pediatr Clin North Am 1985;32:633–651.
22. Allen JC, Bloom J, Ertel I, et al. Brain tumors in children: Current cooperative and institutional chemotherapy trials in newly diagnosed and recurrent disease. Semin Oncol 1986;13:110–122.
23. Kopelson G, Linggood RM, Kleinman GM. Medulloblastoma. The identification of prognostic subgroups and implications for multimodality management. Cancer 1983;51:312–319.
24. Tomita T, Yasue M, Engelhard HH, Mclone DG, Gonzalez-Crusse F, Bauer KD. Flow cytometric DNA analysis of medulloblastoma. Prognostic implication of aneuploidy. Cancer 1988;61:744–749.
25. Yasue M, Tomita T, Engelhard H, Gonzalez-Crussi F, Mclone DG, Bauer KD. Prognostic importance of DNA ploidy in medulloblastoma of childhood. J Neurosurg 1989;70:385–391.
26. Tomita T, Das L, Radkowski MA. Bone metastases of medulloblastoma in childhood: Correlation with flow cytometric DNA analysis. J Neurooncol 1990;8:113–120.
27. Kawakami K, Kawamoto K, Oka N, et al. Flow cytometric studies of brain tumors—5: New sensitivity test of antineoplastic agents for brain tumors and its clinical application. No Shinkei Geka 1986;14:627–634.
28. Gale GB, D'Angio GJ, Uri A, Chatten J, Koop CE. Cancer in neonates: The experience at the Children's Hospital of Philadelphia. Pediatrics 1982;70:409–413.
29. Wilson LM, Draper GJ. Neuroblastoma, its natural history and prognosis: A study of 487 cases. BMJ 1974;3:301–307.
30. Nitschke R, Smith EI, Shochat S, et al. Localized neuroblastoma treated by surgery: a Pediatric Oncology Group Study. J Clin Oncol 1988;6:1271–1279.
31. Laug WE, Siegel SE, Shaw KN, Landing B, Baptista J, Gutenstein M. Initial urinary catecholamine metabolite concentrations and prognosis in neuroblastoma. Pediatrics 1978;62:77–83.
32. Zeltzer PM, Marangos PJ, Parma AM, et al. Raised neuron-specific enolase in serum of children with metastatic neuroblastoma. A report from the Children's Cancer Study Group. Lancet 1983;2:361–363.
33. Hann H, Evans A, Siegel S, et al. Prognostic importance of serum ferritin in patients with Stage III and IV neuroblastoma. The CCSG experience. Cancer Res 1985;45:2843–2848.
34. Chatten J, Shimada H, Sather HN, Wong KY, Siegel SE, Hammond GD. Prognostic value of histopathology in advanced neuroblastoma: A report from the Childrens Cancer Study Group. Hum Pathol 1988;19:1187–1198.
35. Look AT, Hayes FA, Nitschke R, McWilliams NB, Green AA. Cellular DNA content as a predictor of response to chemotherapy in infants with unresectable neuroblastoma. N Engl J Med 1984;311:231–235.
36. Look AT. Constellations of genetic abnormalities predict clinical outcome in childhood malignancies. In: Nerk RD. ed. Modern Trends in Leukemia VIII. Berlin: Springer-Verlag 1989:113–120.
37. Look AT, Hayes FA, Shuster JJ, et al. Clinical relevance of tumor cell ploidy and N-*myc* gene amplification in childhood neuroblastoma: A Prediatric Oncology Group Study. J Clin Oncol 1991;9:581–591.
38. Gansler T, Chatten J, Varello M, Bunin GR, Atkinson B. Flow cytometric DNA analysis of neuroblastoma. Correlation with histology and clinical outcome. Cancer 1986;58:2453–2458.
39. Oppedal BR, Storm-Mathisen I, Lie SO, Brandtzaeg P. Prognostic factors in neuroblastoma: Clinical, histopathologic, and immunohistochemical features and DNA ploidy in relation to prognosis. Cancer 1988;62:772–780.
40. Taylor SR, Blatt J, Constantino JP, Roederer M, Murphy RF. Flow cytometric DNA analysis of neuroblastoma and ganglioneuroma. A 10-year retrospective study. Cancer 1988;62:749–754.
41. Suzuki H, Takeuchi K. Flow cytometric DNA analysis of neuroblastoma: correlation with prognostic variables and survival of patients. Jpn J Surg 1988;18:116–118.
42. Abramowsky CR, Taylor SR, Anton AH, Berk AI, Roederer M, Murphy RF. Flow cytometry DNA ploidy analysis and catecholamine secretion profiles in neuroblastoma. Cancer 1989;63:1752–1756.
43. Dominici C, Negroni A, Romeo A, et al. Association of near-diploid DNA content and N-*myc* amplification in neuroblastomas. Clin Exp Metastasis 1989;7:201–211.
44. Brenner DW, Barranco SC, Winslow BH, Shaeffer J. Flow cytometric analysis of DNA content in children with neuroblastoma. J Pediatr Surg 1989;24:204–207.
45. Oppedal BR, Oien O, Jahnsen T, Brandtzaeg P. N-*myc* amplification in neuroblastomas: Histopathological, DNA ploidy, and clinical variables. J Clin Pathol 1989;42:1148–1152.
46. Cohn SL, Rademaker AW, Salwen HR, et al. Analysis of DNA ploidy and proliferative activity in relation to histology and N-*myc* amplification in neuroblastoma. Am J Pathol 1990;136:1043–1052.
47. Taylor SR, Locker J. A comparative analysis of nuclear DNA content and N-*myc* gene amplification in neuroblastoma. Cancer 1990;65:1360–1366.
48. Hayashi Y, Kanda N, Inaba T, et al. Cytogenetic findings and prognosis in neuroblastoma with emphasis on marker chromosome 1. Cancer 1989;63:126–132.
49. Kaneko Y, Kanda N, Maseki N, et al. Different karyotypic patterns in early and advanced stage neuroblastoma. Cancer Res 1987;47:311–318.
50. Hayashi Y, Hanada R, Yamamoto K, Bessho F. Chromosome findings and prognosis in neuroblastoma [letter]. Cancer Genet Cytogenet 1987;29:175–177.
51. Hayashi Y, Inaba T, Hanada R, Yamada M, Nakagome Y, Yamamoto K. Similar chromosomal patterns and lack of N-*myc* gene amplification in localized and IV-S stage neuroblastomas in infants. Med Pediatr Oncol 1989;17:111–115.
52. Brodeur GM, Seeger RC, Schwab M, Varmus HE, Bishop JM. Amplification of N-*myc* in untreated human neuroblastomas correlates with advanced disease stage. Science 1984;224:1121–1124.
53. Seeger RC, Brodeur GM, Sather H, et al. Association of multiple copies of the N-*myc* oncogene with rapid progression of neuroblastomas. N Engl J Med 1985;313:1111–1116.
54. Grady-Leopardi EF, Schwab M, Ablin AR, Rosenau W. Detection of N-*myc* oncogene expression in human neuroblastoma by in situ hybridization and blot analysis: relationship to clinical outcome. Cancer Res 1986;46:3196–3199.
55. Nakagawara A, Ikeda K, Tsuda T, Higashi K. N-*myc* oncogene amplification and prognostic factors on neuroblastoma in children. J Pediatr Surg 1987;22:895–898.
56. Christiansen H, Lampert F. Tumour karyotype discriminates between good and bad prognostic outcome in neuroblastoma. Br J Cancer 1988;57:121–126.
57. Fong CT, Dracopoli NC, White PS, et al. Loss of heterozygosity for the short arm of chromosome 1 in human neuroblastomas: correlation with N-*myc* amplification. Proc Natl Acad Sci USA 1989;86:3753–3757.
58. Hayes FA, Green AA, Casper J, Cornet J, Evans WE. Clinical evaluation of sequentially scheduled cisplatin and VM26 in neuroblastoma: Response and toxicity. Cancer 1981;48:1715–1718.

59. D'Angio GJ, Evans A, Breslow N, et al. The treatment of Wilms' tumor: Results of the Second National Wilms' Tumor Study. Cancer 1981;47:2302–2311.
60. Bonadio JF, Storer B, Norkool P, Farewell VT, Beckwith JB, D'Angio GJ. Anaplastic Wilms' tumor: Clinical and pathologic studies. J Clin Oncol 1985;3:513–520.
61. Tanaka T, Takamatsu T, Sawada T, Kidowaki T, Tozawa M, Kusunoki T. DNA contents in Wilms' tumors. A cytofluorometric study. Cancer 1983;52:1269–1272.
62. Douglass EC, Look AT, Webber B, et al. Hyperdiploidy and chromosomal rearrangements define the anaplastic variant of Wilms' tumor. J Clin Oncol 1986;4:975–981.
63. Rainwater LM, Hosaka Y, Farrow GM, Kramer SA, Kelalis PP, Lieber MM. Wilms' tumors: Relationship of nuclear deoxyribonucleic acid ploidy to patient survival. J Urol 1987;138:974–977.
64. Kumar S, Marsden HB, Cowan RA, Barnes JM. Prognostic relevance of DNA content in childhood renal tumours. Br J Cancer 1989;59:291–295.
65. Schmidt D, Wiedemann B, Keil W, Sprenger E, Harms D. Flow cytometric analysis of nephroblastomas and related neoplasms. Cancer 1986;58:2494–2500.
66. Layfield LJ, Ritchie AW, Ehrlich R. The relationship of deoxyribonucleic acid content to conventional prognostic factors in Wilms' tumor. J Urol 1989;142:1040–1043.
67. Solis V, Pritchard J, Cowell JK. Cytogenetic changes in Wilms' tumors. Cancer Genet Cytogenet 1988;34:223–234.
68. Hohenfellner K, Holl M, Gutjahr P, Zabel B. Cytogenetic findings in Wilms' tumor. Klin Padiatr 1989;201:293–298.
69. Call KM, Glaser T, Ito CY, et al. Isolation and characterization of a zinc finger polypeptide gene at the human chromosome 11 Wilms' tumor locus. Cell 1990;60:509–520.
70. Gessler M, Poustka A, Cavenee W, Neve RL, Orkin SH, Bruns GA. Homozygous deletion in Wilms' tumours of a zinc-finger gene identified by chromosome jumping. Nature 1990;343:774–778.
71. Gaiger AM, Soule EH, Newton WA Jr. Experience of the Intergroup Rhabdomyosarcoma Study. Natl Can Inst 1981;56:19–27.
72. Allsbrook WC Jr, Stead NW, Pantazis CG, Houston JH, Crosby JH. Embryonal rhabdomyosarcoma in ascitic fluid. Immunocytochemical and DNA flow cytometric study. Arch Pathol Lab Med 1986;110:847–849.
73. Molenaar WM, Dam-Meiring A, Kamps WA, Cornelisse CJ. DNA-aneuploidy in rhabdomyosarcomas as compared with other sarcomas of childhood and adolescence. Hum Pathol 1988;19:573–579.
74. Boyle ET Jr, Reiman HM, Kramer SA, Kelalis PP, Rainwater LM, Lieber MM. Embryonal rhabdomyosarcoma of bladder and prostate: nuclear DNA patterns studied by flow cytometry. J Urol 1988;140:1119–1121.
75. Matsuno T, Gebhardt MC, Schiller AL, Rosenberg AE, Mankin HJ. The use of flow cytometry as a diagnostic aid in the management of soft-tissue tumors. J Bone Joint Surg [Am] 1988;70:751–759.
76. Shapiro DN, Parham DM, Douglass ED, et al. Relationship of tumor cell ploidy to histologic subtype and treatment outcome in children and adolescents with unresectable rhabdomyosarcoma. J Clin Oncol 1990;9:159–166.
77. Douglass EC, Valentine M, Etcubanas E, et al. A specific chromosomal abnormality in rhabdomyosarcoma. Cytogenet Cell Genet 1987;45:148–155.
78. Link MP, Gilber F. Osteosarcoma. In: Pizzo PA, Poplack DG, eds. Principles and Practice of Pediatric Oncology. Philadelphia: J.B. Lippincott, 1989;689–711.
79. Simon R. Clinical prognostic factors in osteosarcoma. Cancer Treat Rep 1978;62:193–197.
80. Kreicbergs A, Brostrom LA, Cewrien G, Einhorn S. Cellular DNA content in human osteosarcoma: aspects on diagnosis and prognosis. Cancer 1982;50:2476–2481.
81. Kreicbergs A, Silversward C, Tribukait B. Flow DNA analysis of primary bone tumors. Relationship between cellular DNA content and histopathologic classification. Cancer 1984;53:129–136.
82. Mankin HJ, Gebhardt MC. Advances in the management of bone tumors. Clin Orthop 1985;73–84.
83. Helio H, Karaharju E, Nordling S. Flow cytometric determination of DNA content in malignant and benign bone tumours. Cytometry 1985;6:165–171.
84. Frohn A, Fodisch HJ, Bode U. Fluorescence cytophotometric DNA studies of Ewing and osteosarcomas. Klin Pediatr 1986;198:262–266.
85. Xiang JH, Spanier SS, Benson NA, Braylan RC. Flow cytometric analysis of DNA in bone and soft-tissue tumors using nuclear suspensions. Cancer 1987;59:1951–1958.
86. Hiddemann W, Roessner A, Wormann B, et al. Tumor heterogeneity in osteosarcoma as identified by flow cytometry. Cancer 1987;59:324–328.
87. Bauer HC, Kreicbergs A, Silfversward C, Tribukait B. DNA analysis in the differential diagnosis of osteosarcoma. Cancer 1988;61:1430–1436.
88. Mellin W, Dierschauer W, Hiddemann W, et al. Flow cytometric DNA analysis of bone tumors. Curr Top Pathol 1989;80:115–152.
89. Bauer HC, Kreicbergs A. Feulgen DNA stainability of bone tumors after demineralization. Cytometry 1987;8:590–594.
90. Bauer HC, Kreicbergs A, Silfversward C, Tribukait B. DNA analysis in the differential diagnosis of osteosarcoma. Cancer 1988;61:2532–2540.
91. Biegel JA, Womer RB, Emanuel BS. Complex karyotypes in a series of pediatric osteosarcomas. Cancer Genet Cytogenet 1989;38:89–100.
92. Look AT, Douglass EC, Meyer WH. Clinical importance of near-diploid tumor stem lines in patients with osteosarcoma of an extremity. N Engl J Med 1988;318:1567–1572.
93. Winther J, Ehlers N, Jensen OA, Overgaard J, Prause JU. Predictive value of flow cytometric DNA-analysis on fresh retinoblastoma tissue. Acta Ophthalmol (Copenh) 1988;66:217–219.
94. Helson L, Traganos F, Ellsworth R, Abramson D. Retinoblastoma and cell cycle cytofluorometric analysis. Arch Ophthalmol 1984;102:616–618.
95. Turc-Curel C, Philip T, Berger MP. Chromosomal translocation in Ewing's sarcoma. N Engl J Med 1983;309:492–498.
96. Aurias A, Rimbaut C, Buffe D. Chromosomal translocations in Ewing's sarcoma. N Engl J Med 1983;309:492–497.
97. Bauer HC. DNA cytometry of osteosarcoma. Acta Orthop Scand Suppl 1988;228:1–39.
98. Schmidt D, Leuschner I, Harms D, Sprenger E, Schafer HJ. Malignant rhabdoid tumor. A morphological and flow cytometric study. Pathol Res Pract 1989;184:202–210.
99. Douglass EC, Rowe S, Valentine MC. Malignant rhabdoid tumor: A highly malignant childhood tumor with minimal karyotypic changes. Genes Chromo Cancer 1990;2:210–216.
100. Volm M, Efferth T. Relationship of DNA ploidy to nonresistance of tumors as measured by in vitro test. Cytometry 1990;11:406–410.

20

Soft-Tissue Sarcomas

ENRIQUE BECKMANN
Clinical Commentary by **Rosemary Mazanet**

INTRODUCTION

Tumors of soft tissues manifest themselves morphologically as muscle, fat, fibrous, vascular, and peripheral nerve tissue neoplasms lying outside of the parenchymal organs. Not all tumors of soft tissue are derived from the mesoderm. Soft-tissue neoplasms exclude tumors of lymphoid or myeloid origin but include peripheral nerve tumors that are of ectodermal derivation. This obviously arbitrary grouping of disparate biologic entities is dictated by clinical commonalities at presentation, histologic overlaps, and unavoidable similarities in surgical approaches to their treatment. Tumors derived from mesoderm are called sarcomas and comprise all of the above categories, including those originating in the skeleton and in mesenchymal cells of parenchymal organs.

Most of the discussion in this chapter will center around the soft-tissue tumors. Sarcomas of the pediatric age group are discussed in detail in a separate chapter of this book. Mesenchymal tumors originating in particular viscera or other parenchymal organs are also treated specifically in the respective organ chapters.

The rate of benign to malignant soft-tissue tumors is 100:1 in a general hospital population (1). However, most benign tumors are not biopsied, while most malignant tumors are; therefore, this is probably an underestimate. Soft-tissue sarcomas (STS) represent approximately 1% of all cases of cancer and 2% of cancer deaths in the U.S. (1). Annually, 20 new cases of lung cancer, 15 cases of colon cancer and even six cases of lymphoma are diagnosed for each case of soft-tissue sarcoma (STS) (1).

THE PATHOLOGIST AND THE DIAGNOSIS OF SOFT-TISSUE SARCOMA

The pathologist diagnosing a soft tissue lesion faces several problems:

1. The distinction between neoplastic and non-neoplastic lesions (i.e., granulation tissue or scar tissue vs. vascular tumor or fibrosarcoma).
2. The distinction between benign and malignant neoplasms (i.e., "irritated" lipoma vs. low-grade liposarcoma).
3. The distinction between true sarcomas and pseudosarcomatous soft tissue lesions.
4. The precise histogenetic classification of a soft-tissue tumor into histologic type and subgroup (e.g., there exists a clear overlap between high-grade spindle-cell sarcomas of different histogenesis).
5. The distinction of pseudosarcomatous metastatic carcinoma, sarcomatous variants of carcinomas, lymphoma, and melanoma from true soft-tissue sarcomas.
6. The grading of soft-tissue sarcomas (there are many grading schemes with ill-defined subjective criteria).

All these challenges must be met by the practicing pathologist to convey critical information to the surgeon and the oncologist regarding the likelihood of and time interval to the development of local recurrence or distant metastases, the probability of cure, the outlook with respect to disease-free and overall survival, the relative radiosensitivity and chemosensitivity of the tumor, and information concerning the adequacy of the surgical resection of the tumor.

THE HISTOLOGIC TYPING OF SOFT-TISSUE SARCOMAS

The histologic types and subgroups of soft-tissue neoplasms number at least 86 and 41, respectively (1). Significant progress has been achieved in the standardization of histologic criteria of tumor types and subgroups (1). However, due to the relative rarity of soft-tissue tumors (4500 annually diagnosed in the U.S.), few pathologists have extensive experience with all these categories. Some of the subgroups are rare enough that pathologists may only see them as demonstrations in specialized conferences. There is no doubt that many of these types and subgroups represent definite clinicopathologic entities. However, the usefulness of many of these categories in terms of prognosis and therapy is questionable.

Classifications of STS into "spindle-cell tumors," "small round-cell tumors," and "others" is probably an oversimplification. Nonetheless, this allows some generalizations about their biologic behavior that have therapeutic implications (2). The "spindle-cell tumors" representing the bulk of tumors seen in general practice are those for which most of the information regarding grading and staging is reliably available. The "small round-cell" tumors, on the other hand, are usually rapidly-growing tumors, many occurring predominantly in the pediatric age group and frequently curable with chemotherapy (2). The "other" category is really heterogeneous and includes all those tumors with specific features that permit their definition as individual clinicopathologic entities. While these entities are useful for the proper workup and classification of individual cases, this does not necessarily translate, in common practice, into differences in therapeutic approaches. Recent efforts at approaching relatively homogeneous histologic types as candidates for specific therapeutic approaches may further validate the histologic classification of STS (2). In particular, variables that were not found to be predictive of biological behavior might become independently predictive when analyzed for particular histological types or subgroups rather than, as is the usual approach, together with all other sarcoma histotypes. For example, a study showed that synovial sarcomas with glandular elements and low mitotic indices appear to have better prognosis than those with monophasic histology and high mitotic activity (3). However, as in so many other publications on the subject, no other variables were considered in this publication. Future studies attempting to associate a variable to a clinical end point will have to include nonparametric multivariate analyses to establish independent variables with clinical significance.

The histological typing of STS is frequently difficult and not very reliable. Reproducibility of histological typing was 61% in the French Sarcoma Group among pathologists with expertise in the classification of these tumors (4). In spite of this, histological typing of STS serves practical purposes in directing therapy. Some histological types respond to radiation while others, such as some chondrosarcomas, do not (1). Some, like rhabdomyosarcoma, are exquisitely chemosensitive (1). Some histological types are associated with specific syndromes (neurofibromatoses and malignant schwannoma) or tend to metastasize via the blood or lymphatic streams more commonly than others (epithelioid sarcomas often metastasize to regional lymph nodes) (1). Thus, histological typing provides a starting point for the planning of therapy and may dictate the type of initial and subsequent workup required for the proper management of the individual patient.

THE HISTOLOGICAL GRADING OF SOFT-TISSUE SARCOMAS

Histological grading of STS is a subject even more controversial than histological typing. There are two main reasons for this: (1) There are many different schemes for grading, none of which is universally applied, and (2) the criteria used in all grading schemes are, at best, semiquantitative. As a result, reproducibility of histological grading for STS was 68% among the experts of the French Sarcoma Group and 60% among the members of the Pathology Panel of the Scandinavian Sarcoma Group (4, 5). Because of the currently unsettled nature of grading in STS most practicing pathologists develop their own grading criteria and grading schemes (usually a relatively useless two-tiered low vs. high-grade system), while many do not even grade sarcomas. A case in point is a survey of STS conducted by the American College of Surgeons in the participating hospitals of the cancer programs approved by the Commission on Cancer (6). This survey revealed that between 1977 and 1978, 45% of patient entries had microscopic grade recorded in the pathological diagnosis. Similarly, the second part of this survey between 1983–1984 showed that only 51% had grade recorded. This deficiency is attributable to the lack of consensus among pathologists for criteria for grading STS. Similarly, this survey showed infrequent usage of the American Joint Committee staging system.

There are at least six relatively well-defined histological grading systems (1). Of these, four divide histological types of STS into three grades, one into two, and one into four. Some, but not all, of these systems base grade on one or more variables such as necrosis, mitotic rate, cellularity, degree of differentiation, and degree of pleomorphism. The quantitation of some of these variables, i.e., degree of differentiation, is difficult and subjective at best. Criteria for grade limits for each variable are different in each grading system. Furthermore, these grading systems were developed without regard to biological differences that exist between histological types. In contrast though, certain histological types are too readily accepted as being only high-grade tumors, while only few histological types, such as leiomyosarcomas, are recognized as encompassing the full spectrum of grades. This assignment to "automatic" grades (Fig. 20.1) on the basis of histological type is an inaccurate oversimplification. Cooper and Allen point out in a recent review that while synovial sarcoma has been assigned to an "automatic" high grade and dermatofibrosarcoma protuberans to low grade, examples exist for each of these histotypes where the opposite histological grade has been seen (7). Grading based on biological behavior has been inappropriately confused with histological grading. It is circuitous to reason that—grading criteria being derived, as they have been, from the tumor histotype and the patient's clinical behavior—something additional is being gained by automatically assigning a certain grade to a particular histological type. Prognostic and outcome correlations have been used as justification for the selection of histological grading criteria and for the "automatic" grading of STS based exclusively on histological type. These correlations were reached retrospectively and generally in an era that preceded the modern classification and therapy of STS. Major advances in the surgical, radiotherapeutic, and chemotherapeutic approaches to the treatment of STS have altered dramatically the prognosis of STS, making older conclusions invalid. As with histological type, histological grade variables must be analyzed statistically to establish their prognostic independence. A select few grading variables with prognostic significance, such as the amount of necrosis or the number of mitoses, facilitate the standardization of methods of quantitation and may render histological grading more reliable. Some studies beginning in the late 1980s have analyzed, by multivariate analysis, the collective prognostic significance for all histotypes of STS of necrosis, mitotic rate, type of surgical margins, and size of

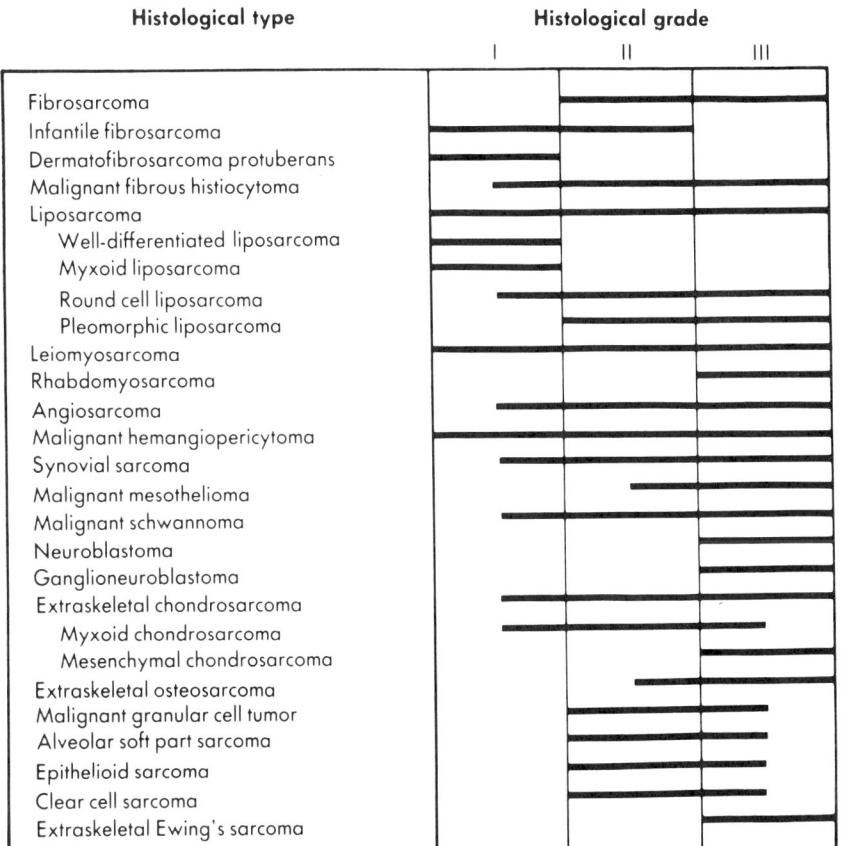

Figure 20.1. Soft-tissue sarcomas. Estimated range of degree of malignancy based on histological type and grade. Grade within the overall range depends on specific histological features, such as cellularity, cellular pleomorphism, mitotic activity, amount of stroma, infiltrative or expansive growth, and necrosis. (Reprinted with permission from Enzinger FM, Weiss SW. Soft tissue tumors. 2nd ed. St. Louis: CV Mosby Company, 1988;13.)

tumor, and have found these variables to be variably powerful independent prognosticators (8–13).

In summary, histological grading of soft-tissue sarcomas is not standardized, the relevant variables are not defined uniformly, the number of grades are not uniform, the reproducibility is poor, and the utilization of grading by practicing pathologists is poor and highly variable. Well-defined standardized criteria applied to individual histological types of tumors in multi-institutional, multinational studies will be necessary, before a useful reliable system emerges. This is required before histological grading can be used optimally in studies where it is compared with other prognostic variables, including DNA flow cytometric measurements.

FLOW CYTOMETRY OF SOFT-TISSUE SARCOMAS

Large studies of flow cytometry of STS are generally retrospective and performed on archival paraffin blocks. There is marked variation in procedures for preparation of nuclei for staining. A single published study has evaluated the efficiency of recovery of cells for analysis (14). The lack of reproducibility and standardization of histological typing and grading make comparisons of data from different studies difficult. The definitions of DNA euploidy and aneuploidy, criteria for distinguishing tetraploidy from diploid G2M peaks, proliferative fraction derivations, and other parameters are not standardized. Instrument bias based on variable flow cell diameters and light beam sizes, types, and shapes, complicate comparisons further. Most studies solely identify ploidy as a relevant factor for prognosis or grading but do not subject data to proper statistical analysis, precluding their reliable identification as true independent variables and preventing assessment of their prognostic power relative to other commonly used variables. Cases for flow cytometric analysis are arbitrarily selected in many series, increasing disproportionately the representation of particular histologic types (i.e., Malignant Fibrous Histiocytomas) and then generalizing conclusions to all sarcomas (15). This diminishes substantially the statistical contributions of rarer histologic types, rendering these generalizations questionable. These and other problems call for cautious interpretation of the available flow cytometry data on STS.

The role of flow cytometry in STS has been assessed in the following areas:

1. The differentiation of benign and malignant soft-tissue tumors by measurement of DNA flow cytometric parameters.
2. The evaluation of DNA flow cytometric parameters as objective and measurable grading criteria.
3. The predictive value of DNA flow cytometric parameters for the development of local recurrences or distant metastases and of metastases-free and overall survival.

THE USE OF DNA FLOW CYTOMETRY FOR THE DISTINCTION OF BENIGN FROM MALIGNANT SOFT-TISSUE LESIONS

As previously stated, a major source of difficulty in the diagnosis of soft-tissue lesions is that morphologically "benign" lesions can metastasize or be locally aggressive, while morphologically "malignant" lesions may be indolent. Also, the distinction of reparative processes from neoplasms is a common source of error. Benign soft-tissue tumors in many series have been found to be diploid (Table 20.1) (14–22). Fibrohistiocytic lesions studied by several authors, such as juvenile xanthogranuloma; nodular fasciitis; fibromatosis, including aggressive cases; juvenile angiofibroma; fibroma; proliferative fasciitis; etc., have all shown diploid DNA content by flow cytometry. Similarly, small numbers of lipomas and atypical lipomas, neurofibromas, neurilemmomas, hemangiomas, soft-tissue leiomyomas, schwannomas, neuromas, ganglioneuromas, myxomas, synovial cysts, pigmented villonodular synovitis, synovial chondromatosis, and lymphangiomas, have also been DNA-diploid by flow cytometry (Table 20.1) (14–22). Benign bone and cartilage tumors have also been uniformly DNA-diploid (14, 23), as have been breast fibroadenomas and giant fibroadenomas (24–26) and benign endometrial neoplasms, such as endolymphatic stromal myosis and endometrial stromal nodule (27, 28). In the instances enumerated above, while the benign tumors were diploid, their malignant counterparts could be either diploid or aneuploid (Table 20.1). Therefore, the finding of a diploid DNA content together with the proper histomorphology is strongly suggestive of a benign diagnosis for all of the lesions enumerated above. More importantly, finding an aneuploid DNA content together with a histologically benign or indeterminate lesion should prompt the pathologist to reevaluate the case. In the case of a diploid lesion with "malignant morphology," the submission of further blocks for DNA flow cytometric evaluation is called for. This is because of the 15%-to-20% incidence of multiple stem lines within soft-tissue tumors that may lead to misclassification of a lesion due to sampling error (29). Moreover, in cases like benign bone tumors (fibrous dysplasia), osteogenic sarcoma, fibroadenomas and cystosarcoma phylloides of the breast, benign cartilaginous tumors (chondroblastoma), and chondrosarcoma, DNA diploidy helps in the distinction between histologic lesions that are morphologically difficult to distinguish from their malignant counterparts (14, 23, 24, 26, 28). Lesions that are difficult to categorize as benign or malignant are frequently classified as indeterminate. For practical purposes, Table 20.1 includes, in the indeterminate category, cases with unique biological properties that are not predictable by morphological criteria. While dermatofibrosarcoma protuberans (DFSP) is generally indolent, some cases are clinically aggressive. Not surprisingly, 28% of DFSP are DNA aneuploid (30). Giant-cell tumors (GCT) of bone are also generally benign, but some are clinically aggressive. Here again 24% of GCT are DNA aneuploid (14). Some lipomas are morphologically atypical and difficult to distinguish from well-differentiated liposarcomas. A single reported case of atypical lipoma has shown diploid DNA content by flow cytometry (19). Leiomyoblastomas are smooth-muscle tumors that show considerable histological pleomorphism, which may lead to their misclassification as leiomyosarcomas. For this reason, 20 reported cases of leiomyoblastomas are shown as "indeterminate" on Table 20.1 (31). Chondroblastomas are generally benign but can be aggressive; they frequently are cellular and simulate low grade chondrosarcomas. Eleven reported cases of chondroblastoma studied by DNA flow cytometry are therefore included in the indeterminate column. Finally, cystosarcoma phylloides is included in the indeterminate category because it can be difficult to distinguish morphologically from some fibroadenomas. Thus, DNA ploidy may be a useful adjunct to histomorphology in establishing a diagnosis in histologically difficult to distinguish cases, such as fibrous dysplasia and osteogenic sarcoma, nodular fasciitis and malignant fibrous histiocytoma, cellular fibroadenoma, cystosarcoma phylloides, and many others.

Benign smooth-muscle tumors of the uterus and gastrointestinal tract do not segregate as well from their malignant counterparts (28, 31). For example, of 41 benign uterine leiomyomata, three were aneuploid but behaved clinically and appeared histologically like typical benign tumors. Similarly, five gastric leiomyomas of 53 and two leiomyoblastomas of 20 were aneuploid in spite of their benign clinical courses and histological appearances. Therefore, DNA ploidy cannot be used as a unique diagnostic criterion to distinguish between benign and malignant smooth-muscle tumors. However, DNA aneuploidy may be used as additional evidence of malignancy in histologically equivocal cases, while DNA diploidy is not as reliable a diagnostic criterion of benignancy. With the exception of smooth-muscle tumors from all locations, all other reported benign lesions seem to show 100% DNA diploid distribution. Indeterminate lesions show a lower proportion of diploid cases and malignant lesions a still lower proportion.

Correlation of particular histomorphologic subtypes of sarcomas with particular DNA ploidy data have been sought. Synovial sarcomas could not be segregated into biphasic or monophasic types based on the DNA ploidy data (32). In contrast, among rhabdomyosarcomas with DNA aneuploid values, the embryonal group showed almost exclusively hyperdiploid values and the alveolar group showed predominantly near-tetraploid aneuploid peaks (33).

In conclusion, cautious use of ploidy data appears to help in the differentiation of benign from malignant lesions, especially in cases where the histology is equivocal. The initial results of particular DNA aneuploid patterns associated with unique histological subsets of rhabdomyosarcoma is promising. In the future, systematic study of DNA patterns in other subsets of STS may permit better standardization of histological subset classification.

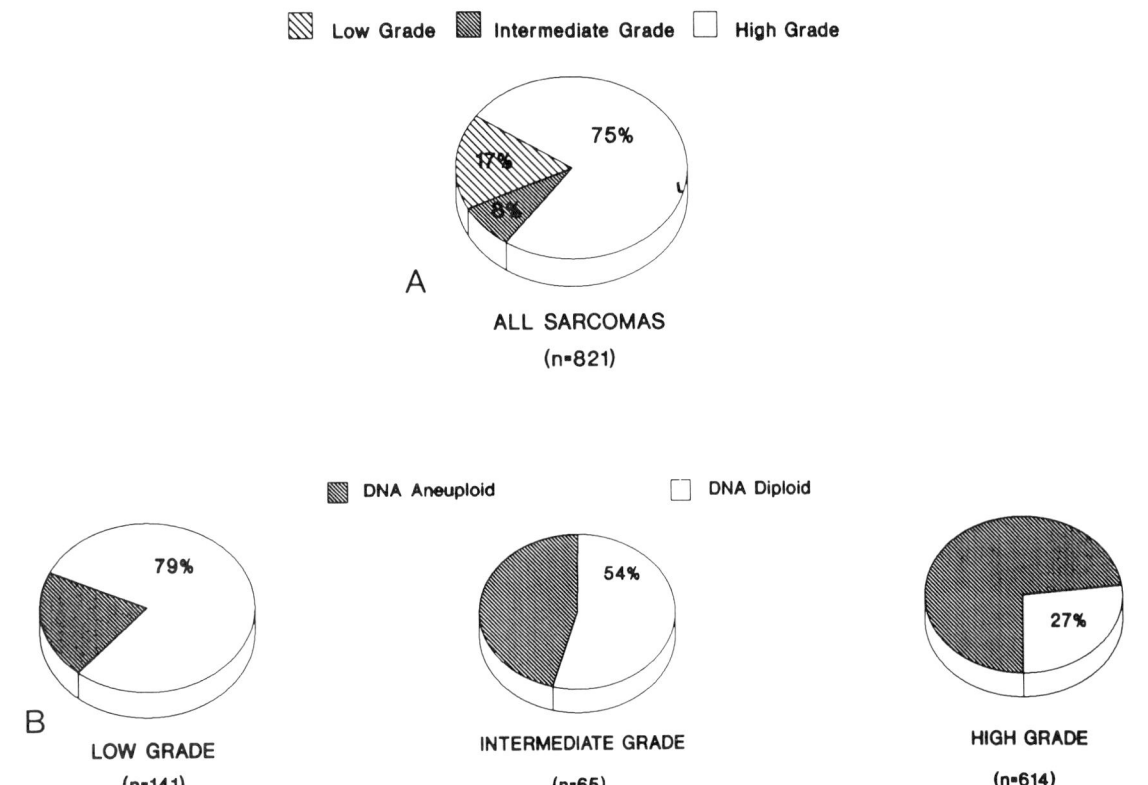

Figure 20.2. **A.** Aggregate proportions of soft-tissue sarcomas by histologic grade. **B.** Correlation of DNA ploidy and histological grade of the aggregate series of soft tissue sarcomas as displayed in **A**.

THE USE OF DNA FLOW CYTOMETRY FOR THE GRADING OF SOFT-TISSUE SARCOMAS

Histological grading of soft-tissue sarcomas presents all of the difficulties enumerated in a preceding section. Nevertheless, grading is generally accepted as being more relevant for management and prognosis within the broad histological categories than specific histological typing (34). Because of this, and in view of the lack of standardized measurable criteria for histological grading, the use of DNA ploidy, proliferative fraction, and other flow cytometrically measurable variables offers opportunities for progress. Even mitotic activity, which is one of the most used and quantifiable criteria in every grading scheme, is a very subjective measurement which is prone to artifactual distortions. Early fixation of tumor material is necessary to accurately measure mitotic counts since delays lead to drops from 13% to 46% of the initial counts, depending on the time interval to fixation. Also altered is the proportion of cells in various phases of the cell cycle, i.e., an increase in cells in the G2 + M phase due to continuation of the cell cycle while cells are awaiting fixation (35).

There is a definite negative correlation between histological grade and percent of DNA diploidy (Table 20.1). This is true for all histological types of sarcomas including STS and others. All reported histological types of sarcomas show some degree of DNA aneuploidy. There is significant variability in the proportion of DNA diploid sarcomas depending on histological type. Aggregate figures of %DNA diploid tumors compiled from most reported cases show 0% for endometrial stromal sarcomas; 15% for gastrointestinal stromal sarcomas; 18% for uterine leiomyosarcomas; 23% for osteogenic sarcomas, soft-tissue leiomyosarcomas, and rhabdomyosarcomas; 27% for malignant fibrous histiocytomas; 37% for neurogenic sarcomas; 40% for chondrosarcomas; 45% for breast stromal sarcomas, including aggressive cystosarcoma phylloides; 49% for liposarcomas; 54% for gastric leiomyosarcomas; 58% for synovial sarcomas; 61% for fibrosarcomas; and 75% for angiosarcomas (14–31). Of the 821 sarcoma cases compiled in Table 20.1, 141(17%) are low grade, 65(8%) are intermediate grade and 615(75%) are high grade (Fig. 20.2). The mean percentage diploid figures for low grade are 79%, for intermediate grade, 54%, and for high grade, 27% (Fig. 20.2). Of most reported sarcomas that were subjected to DNA flow cytometry, three-quarters are high grade, while approximately 200 cases constitute most of the world's collective experience with low and intermediate grade sarcomas. In contrast, total numbers of relatively unselected soft-tissue sarcomas not subjected to flow cytometry published in the last five years as large series, total 949 cases (8–13). These show the following breakdown by grade: Low, 123(13%); intermediate, 292(31%); and high 534(56%). It appears, therefore, that the flow cytometry series show underrepresentation of grade-II tumors and over-

Table 20.1
Soft Tissue Lesions Correlation of Histologic Grade and DNA Ploidy Compilation of Published Results[a]

Type of Lesion	Number DNA Diploid Cases/Total Cases (%)					
	Benign	Indeterminate	Malignancy Grade I	Malignancy Grade II	III-IV	Total Malignant
FIBROHISTIOCYTIC	70/70 (100%)	16/22 (72%)	14/14 (100%)	11/19 (58%)	17/23 (14%)	42/156 (27%)
SYNOVIAL	10/10 (100%)	—	—	—	49/84 (58%)	49/84 (58%)
BONE	41/41 (100%)	54/71[b] (76%)	16/17[c] (94%)	0/1 (0%)	29/174 (17%)	45/192 (23%)
ADIPOSE	20/20 (100%)	1/1[d] (100%)	6/8 (75%)	6/9 (67%)	15/38 (39%)	27/55 (49%)
NEURAL	25/25 (100%)	—	—	—	6/16 (37%)	6/16 (37%)
UTERINE SMOOTH MUSCLE	36/41 (88%)	—	4/8 (50%)	4/20 (20%)	1/21 (5%)	9/49 (18%)
GASTROINTESTINAL SMOOTH MUSCLE	—	9/19 (47%)	—	—	6/39 (15%)	6/39 (15%)
GASTRIC SMOOTH MUSCLE	46/53 (87%)	13/20[e] (65%)	15/21 (71%)	—	9/23 (39%)	24/44 (54%)
SMOOTH MUSCLE	—	—	—	3/6 (50%)	1/11 (9%)	4/17 (23%)
BREAST STROMAL	18/18 (100%)	8/10 (80%)	20/37 (5%)	—	0/7 (0%)	20/44 (45%)
ENDOMETRIAL STROMAL	5/5 (100%)	—	—	—	0/5 (100%)	5/5 (0%)
CARTILAGINOUS	15/15 (100%)	8/11[f] (73%)	23/35 (66%)	—	2/27 (7%)	25/62 (40%)
SKELETAL MUSCLE	—	—	—	—	12/51 (23%)	12/51 (23%)
VASCULAR	13/13 (100%)	—	—	—	3/4 (75%)	3/4 (75%)
FIBROUS	—	—	—	3/3 (100%)	8/15 (53%)	11/18 (61%)
MISCELLANEOUS	—	—	1/1 (100%)	6/7 (86%)	8/23 (35%)	15/31 (48%)

[a] Dermatofibrosarcoma protuberans.
[b] Giant cell tumor of bone.
[c] Parosteal osteogenic sarcoma.
[d] Atypical lipoma.
[e] Leiomyoblastoma.
[f] Chondroblastoma.

The aggregate figures shown in this table are derived from various published series: Rööser B. Prognosis in soft tissue sarcoma. Acta Orthop Scand 1987;58(suppl 225):1–54. Rööser B, Willén H, Gustafson P. Alvegård, Rydholm A. Malignant fibrous histiocytoma of soft tissue: a population-based epidemiologic and prognostic study of 137 patients. Cancer 1991;67:499–505. Xiang J, Spanier SS, Benson NA, Braylan RC. Flow cytometric analysis of DNA in bone and soft-tissue tumors using nuclear suspension. Cancer 1987; 59:1951–1958. Alvegård TA, Berg NO, Baldetorp B, et al. Cellular DNA content and prognosis of high-grade soft tissue sarcoma: the Scandinavian Sarcoma Group experience. J Clin Oncol 1990;8:538–547. Kreicbergs A, Tribukait B, Willems J, Bauer HCF. DNA flow analysis of soft tissue tumors. Cancer 1987;59:128–133. Radio SJ, Wooldridge TN, Linder J. Flow cytometric DNA analysis of malignant fibrous histiocytoma and related fibrohistiocytic tumors. Hum Pathol 1988;19:74–77. Matsuno T, Gebhardt MC, Schiller AL, Rosenberg AE, Mankin HJ. The use of flow cytometry as a diagnostic aid in the management of soft-tissue tumors. J Bone and Joint Surg, June 1988;70-A(suppl 5):751–759. Pettinato G, Manivel JC, De Rosa G, Petrella G, Jaszcz W. Angiomatoid malignant fibrous histiocytoma cytologic, immunohistochemical, ultrastructural, and flow cytometric study of 20 cases. Modern Pathol 1990;3(suppl 4):479.25. El-Naggar A, Barlogie B, Ro J, Ayala A, Batsakis JG. Flow cytometric DNA analysis of soft tissue neoplasms. Lab Invest 1988;58:27A. El-Naggar AK, MacKay B, Sneige N, Batsakis JG. Stromal neoplasms of the breast: a comparative flow cytometric study. J Surg Oncol 1990;44:151–156. El-Naggar AK, Ro JY, McLemore D, Garnsy L. DNA content and proliferative activity of cystosarcoma phyllodes of the breast: potential prognostic significance. Am J Clin Pathol 1990;93:480–485. Murad TM, Hines JR, Beal J, Bauer KD. Histopathological and clinical correlations of cystosarcoma phyllodes. Arch Pathol Lab Med 1988;12:752–756. August CZ, Bauer KD, Lurain J, Murad T. Neoplasms of endometrial stroma: histopathologic and flow cytometric analysis with clinical correlation. Hum Pathol 1989;20:232–237. Dunton CJ, Kelsten ML, Brooks SE, Viglione MJ, Carlson JA, Mikuta JJ. Low-grade stromal sarcoma: DNA flow cytometric analysis and estrogen progesterone receptor data. Gynecol Oncol 1990;37:268–275. El-Naggar AK, Ro JY, McLemore D, Garnsey L, Ordonez N, MacKay B. Gastrointestinal stromal tumors: DNA flow-cytometric study of 58 patients with at least five years of follow-up. Mod Pathol 1989;2(suppl 5):511–515. Eriksen BL, Bauer KD, Caro WA, Roth SI. DNA analysis with cytologic and clinical correlation of dermatofibrosarcoma protuberans. Lab Invest 1988;58:28A. Tsushima K, Rainwater LM, Goellner JR, vanHeerden JA, Lieber MM. Leiomyosarcomas and benign smooth muscle tumors of the stomach: nuclear DNA patterns studied by flow cytometry. Mayo Clin Proc 1987;62:275–280. El-Naggar AK, Ayala AG, Abdul-Karim FW, et al. Synovial sarcoma: a DNA flow cytometric study. Cancer 1990;65:2295–2300. El-Naggar AK, Ro JY, Ayala AG, Hinchey WD, Abdul-Karim F. Angiomatoid malignant fibrous histiocytoma: a DNA flow cytometric analysis of 7 cases. Am J Clin Pathol 1988;90:502A. Tsushima K, Stanhope CR, Gaffey TA, Lieber MM. Uterine leiomyosarcomas and benign smooth muscle tumors: usefulness of nuclear DNA patterns studied by flow cytometry. Mayo Clin Proc 1988;63:248–255. Layfield LJ, Hart J, Neuwirth H, et al. Relation between ploidy and the clinical behavior of phyllodes tumors. Cancer 1989;64:1486–1489. Kreicbergs A, Zetterberg A, Söderberg G. The prognostic significance of nuclear DNA content in chondrosarcoma. Int'l Acad Cytol Anal Quant Cytol, December, 1980;2(suppl 4):272–279. Fernö M, Baldetorp B, Åkerman M. Flow cytometric DNA ploidy analysis of soft tissue sarcomas: a comparative study of preoperative fine needle aspirates and postoperative fresh tissues and archival material. Anal Quant Cytol Histol, August, 1990;12(suppl 4):251–258.

Lesions included are: fibrohistiocytic lesions: dermatofibrosarcoma protuberans, dermatofibroma, juvenile xanthogranuloma, nodular fasciitis, juvenile angiofibroma, fibroma, fibromatosis, proliferative fasciitis, desmoid and malignant fibrous histiocytoma. Synovial lesions: pigmented villonodular synovitis, synovial cysts, synovial chondromatosis and synovial sarcoma. Bone lesions: giant cell tumor, osteoblastoma, aneurysmal bone cyst, fibrous dysplasia and osteogenic sarcoma (including parosteal). Adipose lesions: myxoma, lipoma, atypical lipoma, and liposarcoma. Neural lesions: neuroma, neurilemoma, schwannoma, neurofibroma, ganglioneuroma, and malignant schwannoma. Uterine smooth muscle lesions: Leiomyoma and leiomyosarcoma (includes cellular leiomyoma). Gastrointestinal stromal lesions: No benign lesions only spindle cell sarcomas or indeterminate lesions. Gastric smooth muscle lesions: leiomyoma, leiomyosarcoma and leiomyoblastoma (indeterminate). Breast stromal lesions: fibroadenoma, stromal sarcoma (malignant), and cystosarcoma phylloides. Endometrial stromal lesions: endometrial nodule, endolymphatic stromal myosis, and endometrial stromal sarcoma. Cartilagenous lesions: chondroma, chondroblastoma, and chondrosarcoma. Skeletal muscle lesions: rhabdomyosarcoma. Vascular lesions: hemangioma, lymphangioma, hemangiopericytoma, and angiosarcoma. Fibrous lesions: fibroma and fibrosarcoma. Miscellaneous lesions: mesenchymoma, malignant mesenchymoma, unclassified lesions.

representation of high-grade tumors. In terms of histological types it is also important to point out that 23%, 19%, and 10% of all tumors shown in Table 20.1 are osteogenic sarcomas, malignant fibrous histiocytomas, and synovial sarcomas, respectively. Therefore, less than half of the tumors in this aggregate are accounted for by all other histological types, while some, such as angiosarcomas, represent less than 0.5% of the total. Finally, the great majority of tumors in this aggregate represent soft-tissue tumors from extremities, which are biologically different from their counterparts occurring in the trunk and retroperitoneum, and in the head and neck (1). It is safe to assume, therefore, that conclusions reached from published data are applicable with some degree of certainty to high-grade sarcomas (particularly to malignant fibrous histiocytomas). Conclusions regarding lower grade or rarer histological types of sarcomas should be regarded as preliminary. Future studies must concentrate on defining strict measurable histological criteria that are proven independent variables associated with biological grade for comparison with DNA FCM. Few large soft-tissue multivariate analysis studies have identified such variables as necrosis (measurable extent of necrosis), mitotic index, the presence or absence of vascular invasion, and the type of surgical resection margins (8-13). Multi-institutional, multinational studies using strict criteria for grading, histological type, size, location (head and neck vs. trunk vs. extremities), and type of surgical margins (intra-compartmental vs. extra-compartmental), together with DNA flow cytometry offer the only hope for progress for the future classification of soft-tissue sarcomas into clinically useful categories. The significance of the fact that %DNA diploid figures parallel inversely the histological grade lies primarily in the potential use of DNA content flow cytometric analysis as a tool for standardizing the grading of sarcomas.

An example of the use of DNA flow cytometry to this end is provided by Matsuno et al. (18). Using an algorithm of flow cytometry data, they derived an index of malignancy as follows: 1) Benign lesions: no DNA aneuploidy and percentage of replicating plus dividing cells of less than 11. 2) Low-grade sarcomas (grade 1): no DNA aneuploidy and percentage of replicating plus dividing cells of more than 11 but less than 17. 3) High-grade sarcomas (grade 2 or 3): either DNA aneuploidy or percentage of replicating plus dividing cells of more than 17. Correlation of this algorithm with histological grading (well-, moderately-, and poorly-differentiated sarcomas) was 79%, which is better than the 68% of agreement found among pathologists in the French Sarcoma group and the 60% agreement of the Swedish Sarcoma group. The degree of morphological grading agreement is probably an overestimate since the group evaluating the grades was experienced relative to unselected pathologists. Thus, it would seem that flow cytometry is likely to be a reasonable adjunct to histopathological evaluation when grading STS.

DNA FLOW CYTOMETRIC MEASUREMENTS AND PROGNOSIS IN STS

Studies addressing the prognostic value of DNA flow cytometry in STS can be divided into those that address sarcomas as a group regardless of histology, and those where only a particular histological type is studied. On the basis of the previously discussed data that show marked variability in the proportion of cases showing DNA aneuploidy for each histological type, it is reasonable to be skeptical of studies where sarcomas of disparate histology are grouped together. Nevertheless, due to the rarity of these neoplasms, and because they are approached similarly for clinical management purposes, this study modality persists.

Most of the published studies do not investigate the prognostic value of DNA FCM. Many others only enumerate differences between diploid and aneuploid tumors with respect to the proportion of cases that recur locally, metastasize, or cause demise. These however do not subject their data to appropriate statistical analysis, making it impossible to know whether DNA ploidy is of additional independent value and, if so, what is the relative power of this variable when compared with others.

Another important consideration when dealing with prognosis of STS is the definition of adequate end-points for evaluation. Soft-tissue sarcomas are locally invasive tumors. The rate of post-therapeutic local recurrence depends to a large extent on the adequacy of local treatment, which involves surgical excision with adequate wide margins and, in many instances, radiation therapy (80). Adequate local control leads to reductions of local recurrence rates from 80% to 2%. All of the local recurrences occur approximately within 30 months of initial treatment. Soft-tissue sarcomas also metastasize, most frequently to the lungs. Metastases occur in spite of adequate local control of the tumor. Although local recurrence portends an ominous prognosis, metastases do not develop as a result of local recurrence (8). In all likelihood, a number of micrometastases have already occurred at presentation and only manifest themselves clinically later on (8). Local recurrence in the face of adequate treatment is a manifestation of biological aggressiveness that is related to variables such as tumor size > 10 cm, necrosis, malignancy grade IV, and vascular invasion (8). Local recurrence following inadequate excision is adequately controlled with adequate excision without adverse effect on survival (8). Since local recurrence is due to either inadequate excision or the biology of the tumor, the rate of local recurrence is not a valid end-point for prognostic evaluation. Metastases-free and overall survival are adequate end-points. Few flow cytometry studies have taken this into account. The time of follow-up is also critical for the evaluation of prognostic variables in STS. A minimum follow-up of five years is adequate for high histologic grade tumors. This becomes woefully insufficient when considering low-grade lesions where survival is better measured at 10 or 15 years (7). Extremely rare series carry follow-up to this length (7). To date, no series with flow cytometric data belongs to this category.

PROGNOSTIC VALUE OF DNA FLOW CYTOMETRY IN SOFT-TISSUE SARCOMAS AS A GROUP

In a study by El-Naggar et al., 70 deep STS were divided into low grade (n = 20) and intermediate-high grade (n = 50) (21). None of the 23 patients with diploid tumors died of the disease after a follow-up of 12 to 48 months. In contrast, 28% of the patients with aneuploid tumors died. Local recurrences occurred in 52% of the diploid tumors and 80% of the aneuploid sarcomas. Metastases occurred in 22% of the diploid and 38% of the aneuploid tumors. Among the low-grade diploid tumors, 40% recurred locally and 10% metastasized, while aneuploid low-grade tumors recurred in 70% and metastasized in 20% of the cases. Among the high-grade diploid tumors, 61% recurred locally and 23% metastasized, while aneuploid high-grade tumors recurred locally in 84% and metastasized in 43% of the cases.

Matsuno et al. (18) found no correlation of any of the factors analyzed: %stem line cells, %replicating and dividing cells, %DNA aneuploidy, and mean DNA content, with the rate of local recurrence or distant metastases for a series of 63 sarcomas followed for a mean of 2.75 years. This length of follow-up is too short and Matsuno's cases showed 47.6% metastases when the study was carried out. Both of these factors render conclusions of prognosis unreliable.

Kreicbergs et al., (16) state that in cases where histological grade does not correlate with ploidy, it is possible that ploidy analysis may give additional useful information about the biological behavior of a given tumor. In his series of 81 soft-tissue tumors, 10 high-grade tumors were DNA-diploid, while one low-grade tumor was aneuploid. The data of %S-phase and %G2 + M showed direct correlation with ploidy. DNA-aneuploid tumors had higher G2 + M + S values than diploid tumors, suggesting that these measurements may not be useful. Kreicbergs's study did not evaluate survival, and his conclusions regarding this are speculative.

The Kroese et al., study (22) of 46 STS also concludes with statements regarding prognosis even though prognostic data are not mentioned, save for a single case of an aneuploid juvenile angiofibroma that behaved in an aggressive fashion. This study suggests that STS can be divided into cases that show diploid DNA content, high G1 fractions, and low Ki-67 antigen expression (as measured immunohistochemically), which correlate with low-grade morphology, and those with the opposite characteristics, which correlate with high-grade morphology. Again, as in the series of Kreicbergs (16), the finding of noncompliant cases offers the possibility of refinement of prognostication, particularly for sarcomas of intermediate histological grade. In contrast to the lack of correlation of Ki-67 index and mitotic index in Kroese's series, Ueda et al. (36) found positive correlation (r + 0.428, p<0.02). This discrepancy notwithstanding, low Ki-67 index cases showed a five-year survival of 90% as compared to 30.3% for the high-index cases, supporting Kroese's expectations. The combination of DNA ploidy and Ki-67 measurements by flow cytometry has been carried out successfully on other tumors (22). This combination may prove even more powerful than DNA ploidy or Ki-67 alone for the grading and prognostication of STS.

A study of 240 STS patients from the Swedish Sarcoma group is the only publication addressing the issue of DNA content analysis as an independent prognostic indicator in a large series of histologically unselected tumors of histological grades III and IV (15). The tumor types and risk factors for the 148 tumors included in this study are listed in Table 20.2 (Alvegård, T.A. et al., 1990; reprinted with permission). Risk factors that had been found to be prognostically worse were: histological grade (Broders and Hargrove I through IV), tumor size greater than 10 cm, tumor necrosis longer than 17 mm of the longest cross section after examining all slides, vascular invasion, and male sex. These were compared with DNA ploidy as determined by flow cytometry on propidium iodide-stained nuclei from paraffin-embedded tumors. The criteria used for histogram interpretation were (Fig. 20.3): 1) A broad peak was one where the CV was $3\times$ higher than the CV of TRBC G_0/G_1 peak. 2) Skewedness was assumed to be present if the quotient of the mean and peak channel numbers of the diploid G_0/G_1 peak was less than 0.97 or greater than 1.03. 3) Broad and skewed G_0/G_1 tumor peaks were considered "peri-diploid" because they may contain an aneuploid peak. 4) tetraploid peaks were those where the 4N peak exceeded 20% of the G_0/G_1 and if they were narrow CV < 7.5% and gaussian and if there was an octaploid G2 peak. 5) The tetraploid DI was empirically determined in normal paraffin-embedded samples and ranged from 1.92 to 2.08. 6) Diploid (type 1) and tetraploid (type 5) tumors were considered euploid and types 6 and 7 were considered aneuploid (Fig. 20.3). Types 2, 3, and 4 were assumed to "hide" an aneuploid peak and were therefore also considered aneuploid. A most interesting finding of this study was that while prognosis was better for the euploid (diploid + tetraploid) when compared with the non-diploid group, this difference was not statistically significant. However, by adding into the non-diploid category the tumors with skewed or broad G_0/G_1, this difference became statistically significant, and the best separation was achieved when the group composed of diploid and tetraploid tumors was compared with the group containing aneuploid and peri-diploid tumors. A multivariate analysis was carried out on these 148 STS and extended to 184 cases including also grade-II sarcomas, patients with metastases at diagnoses, patients with head and neck, visceral, and retroperitoneal sarcomas (all of which are known to carry a worse prognosis mostly due to the limitations of the surgical approaches available). DNA aneuploidy was the strongest independent prognostic factor; stronger than malignancy grade IV, tumor size > 10 cm, vascular invasion, and male sex.

Table 20.2
Relationship Between Risk Factors and Tumor Type in 148 Patients With High-Grade Soft Tissue Sarcoma of the Extremities and Trunk

Tumor Type	No. of Patients	Median Age (Range)	Sex		Tumor Size (cm)		Malignancy Grade		Vascular Invasion			Ploidy		5-Year MFS (%)
			Male	Female	≤10	>10	III	IV	0	+	++	Euploid	Aneuploid	
Malignant fibrous histocytoma	70	61 (23–78)	37	33	45	25	23	47	61	5	4	9	61	59
Synovial sarcoma	27	29 (16–70)	16	11	19	8	4	23	18	3	6	16	11	45
Liposarcoma	13	51 (29–68)	9	4	5	8	6	7	13	0	0	9	4	62
Leiomyosarcoma	8	44 (28–63)	2	6	7	1	3	5	6	0	2	0	8	63
Soft tissue sarcoma, unclassified	7	48 (35–66)	3	4	5	2	1	6	5	0	2	1	6	0
Malignant schwannoma	7	58 (21–73)	3	4	3	4	1	6	6	0	1	2	5	43
Malignant mesenchymoma	4	39 (33–58)	2	2	3	1	2	2	3	1	0	1	3	0
Malignant hemangiopericytoma	3	34 (29–39)	1	2	3	0	3	0	0	0	3	2	1	67
Extraskeletal osteosarcoma	3	34 (33–63)	2	1	0	3	0	3	2	0	1	0	3	0
Other sarcoma types*	6	35 (32–63)	4	2	2	4	2	4	3	1	2	4	2	40
Total	148	52 (16–78)	79	69	92	56	45	103	117	10	21	44	104	52

*Spindle cell sarcoma, one patient; epithelioid sarcoma, one patient; clear cell sarcoma, one patient; hemangiosarcoma, one patient; extraskeletal chondrosarcoma, one patient; extraskeletal Ewing's sarcoma, one patient.

Grouping of all of the tumors on the basis of the number of adverse factors present permitted discrimination of a good prognostic group, comprising half of the patients with high grade sarcomas with a five-year metastasis-free survival (MFS) rate of 65%, and a high-risk group, encompassing three to five prognostic indicators (malignancy grade IV, DNA aneuploidy, intratumoral vascular invasion, tumor size > 10 cm, and male sex) with a MFS rate of 35% (Fig. 20.4A). The only patient characteristic required for inclusion in chemotherapeutic trials of soft-tissue sarcomas in most cases is that their tumor be of high histologic grade. On the basis of this selection, it has not been generally possible to show the effectiveness of a variety of chemotherapeutic agents on improvement of overall survival over that achieved with surgery alone or combination surgery and radiation therapy (approximately 60% five-year metastasis-free survival for limb sarcomas). On the basis of the criteria enumerated above it seems reasonable to target the 50% of high-grade sarcoma patients with a 35% five-year MFS for further chemotherapeutic trials. It is possible that advantageous effects of such regimens have been hidden by the lack of effect of chemotherapy on patients with high-histological-grade sarcomas who are otherwise in a better risk category (15).

PROGNOSTIC VALUE OF DNA FLOW CYTOMETRY IN INDIVIDUAL HISTOLOGICAL TYPES OF SARCOMAS

As previously discussed, the value of DNA ploidy analysis for prognostication in STS is dependent, at least in part, on the histological type of sarcoma. In the aggregate data shown in Table 20.1, 58% of the synovial sarcomas were diploid compared with 27% of malignant fibrous histocytomas. The five-year MFS in the cases reported by Alvegård et al. (15) which are included in these figures are: 45% for GIII or GIV synovial sarcoma and 59% for GIII or GIV MFH. The following paragraphs summarize data on DNA flow cytometric measurements in various histological types of sarcomas. Excluded are tumors such as osteogenic sarcoma, Ewing's sarcoma, malignant rhabdoid tumors, rhabdomyosarcomas, and others that have been discussed in the chapter on pediatric tumors. For histotypes where several reported series have reached similar conclusions, only one or a few representative series will be discussed.

Fibrohistiocytic Lesions

Benign and malignant fibrohistiocytic lesions comprise a large proportion of soft-tissue neoplasms (1). The fibromatoses are lesions that appear histologically benign but are locally aggressive. DNA ploidy analysis of fibromatoses has been carried out in a handful of cases, all of which have shown diploid DNA content (18, 22). Thus, it appears that ploidy analysis is noncontributory in the prognostication of fibromatoses. Studies of different fractions of the cell cycle correlated with the degree of aggressiveness have not been carried out.

Dermatofibrosarcoma protuberans (DFSP) is a low-grade, locally-recurring tumor of the skin (17, 30). Few cases of this tumor have been retrospectively studied by DNA flow cytometry. Correlation of DNA aneuploidy has been found with histologic atypia. Aneuploidy has been seen in recurring and non-recurring lesions, although adequate follow-up times required may not have been met (30). Both hypodiploid and hyperdiploid cases have been seen. The proliferative fraction of the aneuploid cases has been higher than that of the diploid cases.

A single case of juvenile angiofibroma that was clinically aggressive and led to the patient's death was found to be DNA aneuploid (22).

Malignant fibrous histiocytoma (MFH) is the commonest soft-tissue sarcoma (1). Most are high-grade sarcomas (1). All general statements made above for sarcomas in general apply to this sarcoma. DNA diploidy is the most powerful

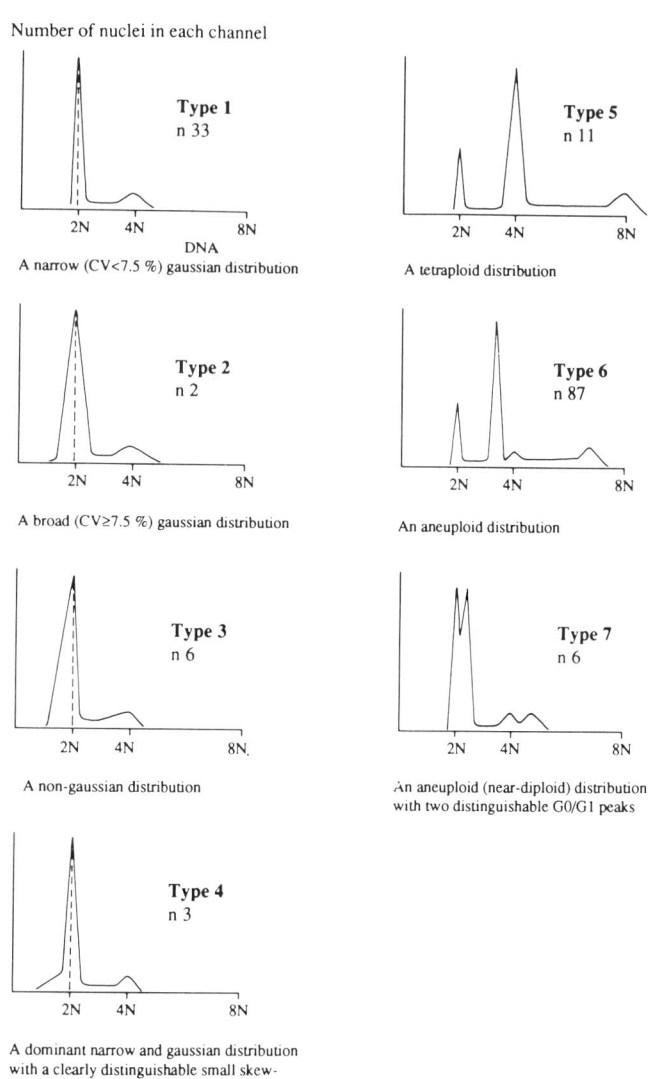

Figure 20.3. Representative examples of different types of DNA histograms; "*n*" designates the number of nuclei in each channel. The nuclear amounts of DNA are shown on the X-axis; "*N*" represents the haploid amounts of DNA. Types 1 and 5 were classified as euploid. Types 2, 3, 4, 6, and 7 were classified as aneuploid. (Reprinted with permission from Alvegård et al.: Cellular DNA content and prognosis of high-grade soft tissue sarcoma: The Scandinavian Sarcoma Group Experience. J Clinical Oncol 1990;8:538–547.)

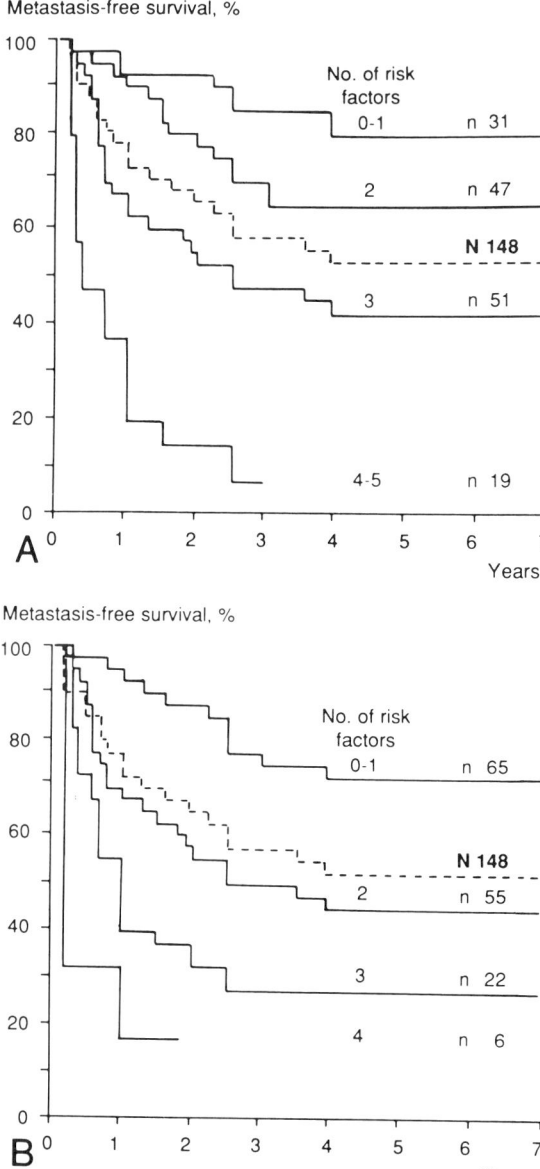

Figure 20.4. Cumulative *MFS* with the five risk factors (Table 20.2), including DNA aneuploidy (**A**), and with four risk factors, excluding DNA aneuploidy (**B**), in 148 patients with malignancy grades III and IV soft-tissue sarcoma of the extremities and trunk. (Reprinted with permission from Alvegård et al. Cellular DNA content and prognosis of high-grade soft tissue sarcoma: The Scandinavian Sarcoma Group Experience. J Clinical Oncol 1990;8:538–547.)

independent prognostic indicator for five-year MFS (15). As with other sarcomas, DNA ploidy correlates with histologic grade (Table 20.1). The proliferative phase is generally larger in MFH than in DFSP, but this does not correlate with ploidy or survival (Table 20.1) (16, 22).

Angiomatoid malignant fibrous histiocytoma (AMFH) is a deep dermal or subcutaneous tumor that affects children and young adults (20). This tumor is regarded as low-grade MFH that, while potentially fatal, rarely metastasizes (1). Of the 14 cases reported by different groups, all tumors, including the recurrent tumors, have been diploid (20, 37). Metastasizing AMFH has not been studied by DNA FCM.

Giant-cell tumors of tendon sheath (GCTTS) are benign lesions that may be rarely confused with MFH. A single study of DNA FCM or GCTTS has been published (38). The great majority of these lesions were diploid. The few aneuploid lesions did not correlate with histologic atypia. None of these cases studied by DNA FCM have recurred or metastasized. While proliferative activity, as measured immunohistochemically by PCNA, correlated with mitotic index, %S-phase was uniformly low and did not correlate.

Synovial Sarcoma

Ploidy status was shown to correlate with patient survival in synovial sarcoma (32). In this study, 46 cases followed for a minimum of five years were selected for analysis on the basis of availability of pretreatment tissue blocks. There was marked heterogeneity in terms of therapy, including excision (not otherwise specified), amputation, preoperative and postoperative radiation or chemotherapy, and combinations thereof. Tumors were classified into monophasic or biphasic, and mitotic counts were obtained. Tumors were not graded histologically any further. The prognostic effect of age, tumor size, mitotic rate, and S phase were tested using a log-rank test. The relationship of variables to survival was tested using the multivariate proportional hazards model of Cox. DNA Ploidy FCM studies showed that 67% of the cases were diploid and 33% we aneuploid. The presence of aneuploidy conferred a relative risk ratio of death of 4.92 as compared to 5.11 for tumor size greater than 5 cm, and 4.7 for age greater than 15 years. The %S-phase was not correlated with survival and mitotic index was only a weak predictor. The proportion of diploid cases found in this study correlates well with other series and with the aggregate figure shown in Table 20.1 (58%). This would suggest that this population is representative of the general group of synovial sarcomas and that conclusions may be cautiously generalized in view of the great variability of treatment modalities that might have affected the outcome in this series.

Leiomyosarcomas

Leiomyosarcomas of soft tissues represent only 7% of STS (1). The aggregate number of those tumors shown in Table 20.1 in which DNA FCM has been done and results have been published is 17. Only the series of Alvegård et al., shows data related to survival (15). Of the eight cases of leiomyosarcoma, all eight were aneuploid and the five-year MFS was 63%. Since all tumors were aneuploid, the effect of ploidy status on survival cannot be evaluated.

Much more common and clinically significant are leiomyosarcomas occurring in the gastrointestinal tract and in the uterus. These tumors, while sharing morphologic and immunohistochemical characteristics, cannot be regarded as the same biologic entity since behavior and therapy differ for each location.

GASTROINTESTINAL STROMAL TUMORS

In a study of 58 gastric, small-intestine, and large-intestine stromal tumors, survival was significantly correlated with DNA content (p<0.001) (29). Furthermore, multivariate regression analysis proved that DNA ploidy is an independent prognostic indicator of clinical outcome in gastrointestinal stromal tumors. There was a correlation of histological typing (indeterminate and malignant) with clinical outcome. DNA ploidy analysis identified further prognostic subgroups within these categories. The median survival for diploid indeterminate tumors was 200 months, compared with 24 months for aneuploid indeterminate tumors. The figures for malignant diploid tumors and malignant aneuploid tumors were 59 and 24 months, respectively.

In a study of 117 gastric smooth-muscle tumors (44 leiomyosarcomas, 53 leiomyomas, and 20 leiomyoblastomas), Tsushima et al. (31) found that aneuploid DNA had a significantly negative impact on survival on patients with leiomyosarcomas of low grade. While a negative impact was also seen in high grade aneuploid tumors, the differences from diploid tumors were not statistically significant.

UTERINE LEIOMYOSARCOMAS

In a retrospective study of 49 uterine leiomyosarcomas, Tsushima et al. (39) found a statistically significant negative influence on survival of DNA aneuploidy. For 28 patients with grade 1 + 2 leiomyosarcomas, the five-year survival rate was 88% for diploid tumors versus 40% for aneuploid tumors. Similarly, 100% of 21 grade 3 + 4 diploid tumors survived five-years, while only 33% survived five-years with aneuploid tumors. Significant differences in survival persisted when diploid and aneuploid tumors were compared between different stages, sizes, and mitotic rates, all of which also had important effects on survival.

OTHER UTERINE SARCOMAS

Endometrial stromal tumors comprise the benign lesions: endolymphatic stromal myosis (ESM) (considered as a low-grade stromal sarcoma), stromal nodule (SN), and frankly-malignant stromal sarcomas (ESS). A single stromal nodule has been found to be diploid (27). ESM and ESS are generally distinguished on the basis of mitotic counts (27). However, this series by August, identified a case of endometrial stromal sarcoma with a high mitotic count but diploid DNA content, no recurrences, and normal survival. Although limited, because of the small number of patients and the short follow-up, this study shows that DNA ploidy can identify as a benign type of stromal lesions that would have been classified as malignant by conventional histology. All cases with ESM or low-grade ESS (LGSS) were diploid, while all high-grade ESS, except for the case described above, were aneuploid. All ESM patients were alive, but one had a recurrence 14 years after diagnosis. This series also shows that high-grade ESS with high %S-phases have a poor prognosis. A Static cytometry study of DNA content showed that two of two cases of LGSS had diploid DNA content (40). In a study by Dunton et al., three cases of LGSS or ESM were analyzed using DNA FCM (28). All three cases showed diploid DNA content but upon rebiopsy of recurrences two cases showed aneuploidy. The %S-phase data were difficult to interpret due to the presence of debris. Conclusions regarding endometrial stromal tumors are impossible at present with respect to prognostication on the basis of DNA FCM data because of the limited number of cases reported.

A group of 11 uterine sarcomas were studied by DNA FCM by K.G. Nelson, et al. (41). Four of these were homologous malignant mixed müllerian tumors (MMMT), four were heterologous MMMTs, two were leiomyosarcomas, and one was a pure heterologous uterine rhabdomyosarcoma. Of these, nine tumors were DNA aneuploid. The two cases with diploid tumors were the rhabdomyosarcoma and a homologous MMMT. Neither one of these tumors had spread at the time of surgery and both patients were alive and free of tumor at one year or more after therapy. This small experience with rare tumors warrants additional larger scale studies to confirm the role of DNA FCM in uterine stromal tumor prognostication.

Stromal Tumors of the Breast

The spectrum of breast stromal tumors includes the benign and giant fibroadenomas, the biologically indeterminate cystosarcoma phyllodes, and the frankly malignant stromal sarcomas. Only one of four studies of these lesions claims that fibroadenomas may be aneuploid (42). This is most likely due to technical problems in this study and is probably incorrect (24). All other studies show clearly that benign fibroadenomas are uniformly diploid. At the other extreme, frankly malignant sarcomas are uniformly aneuploid and are uniformly fatal (24). Cystosarcoma phyllodes can be diploid or aneuploid (24, 26). All studies agree on the fact that diploid cystosarcoma phyllodes pursue a benign clinical course. However, one study (24) claims that this is independent of histological grade while the other (26) shows a correlation of histological grade and DNA ploidy status. Furthermore, a subset of aneuploid cystosarcoma phyllodes cases have a benign clinical course irrespective of histological grade. Only one study analyzed the proliferative index (PI) by FCM (26). The low-grade lesions were diploid and had low PIs and the high-grade lesions all showed high PIs and three of four were aneuploid. In conclusion, DNA ploidy permits distinctions to be made between benign and malignant stromal breast tumors. The cystosarcoma group can be divided into prognostically significant groups on the basis of DNA ploidy and %S-phase measurements. However, in view of the exceptions mentioned above, additional variables will have to be defined to fully segregate cystosarcomas with benign from those with malignant behavior.

Chondrosarcoma

Most of the work on DNA ploidy studies in chondrosarcomas has been done by Kreicbergs's group. Beginning with a study published in 1980, it has been consistently demonstrated and corroborated in subsequent studies that DNA diploid chondrosarcomas have a more favorable clinical course than aneuploid tumors and this is irrespective of location, size, or type of treatment (43). In addition, DNA ploidy was found to be better than histological grading in predicting 10-year survival and development of metastases. The 10-year survival was 77% for diploid tumors and 27% for aneuploid tumors. Of the metastasizing tumors 19% were DNA diploid while 81% were DNA aneuploid. The metastatic rate at 10 years was 18% and 65% for diploid and aneuploid tumors, respectively. Therefore, as with osseous tumors, DNA FCM is valuable for diagnosis, grading, and prognostication of cartilaginous tumors (14).

Alveolar Soft-Part Sarcoma

This tumor, apparently of rhabdomyoblastic origin, is a rare soft part neoplasm with a characteristic light microscopic appearance (44). A single static cytometry, Feulgen-stained DNA content study of a group of 10 of these tumors has been published (44). It showed a correlation between DNA ploidy and histological appearance. Tumors with the usual morphology (n = 8) were diploid while those with pleomorphic features (n = 2) were aneuploid. Also, pleomorphic areas within tumors with usual histology disclosed regional aneuploidy. In terms of prognosis there was no correlation between DNA ploidy and outcome. If this tumor is indeed of rhabdomyoblastic origin, its DNA ploidy characteristics do not match those of rhabdomyosarcomas, which are generally aneuploid and show good correlation between aneuploidy and improved survival (33).

Epithelioid Sarcoma

This is a rare tumor of young adults with a propensity for occurring in the hands and arms, and it is often mistaken for an inflammatory ulcerative process or an ulcerating carcinoma (1). This tumor can have a protracted clinical course, with cases surviving for 25 years before developing recurrence. Multiple recurrences are characteristic. A single study of DNA ploidy by FCM has been published (45). In it, 14 cases of ES followed for at least five years showed 50% aneuploidy. In 12 cases, %S-phase was analyzed. In 11 of these the %S-phase was greater than 5% and eight of these patients died of disease. The one patient with %S-phase of 2% is alive and free of disease. Patients who died of their disease could not be distinguished on the basis of ploidy.

This small series does not permit definitive conclusions regarding the value of DNA FCM in the prognostication of ES. The number of cases is too small and the follow-up time too short. Only multi-institutional, multinational studies can address these issues effectively in such a rare neoplasm.

Clear Cell Sarcomas (CCS)

This rare neoplasm generally affects the feet and ankles of young males. It is usually a deep-seated tumor. It has a protracted course that can go through several local recurrences. It metastasizes frequently to lymph nodes. Its histochemistry, immunohistology, and ultrastructure demonstrate typical melanocytic features (1). A recent study compared 11 CCS to 13 metastatic malignant melanomas (MM) (46). As a group, the CCS showed lower DNA indices than MMs. The percent of aneuploidy was also lower in CCSs than in MMs (56% versus 85%). The mean survival in patients with CCSs and diploid DNA pattern was 68.6 months, versus 8.2 months for those tumors with aneuploid DNA.

In order to draw definitive conclusions regarding the value of DNA FCM in the prognostication of CCSs, this study must be extended.

Kaposi's Sarcoma

There have been three studies of DNA analysis in Kaposi's sarcoma (47, 48, 49). Two of these were performed on both AIDS-associated and classic Kaposi's by Feulgen-stained nuclei and microscopic spectrophotometry (47, 48). The last one was a retrospective flow cytometric analysis of 21 AIDS patients (49). All cases showed diploid DNA values and low %S-phase, suggesting that this is a low-grade neoplasm, or not a neoplasm at all, that behaves in an aggressive fashion due to deficiencies in host defense mechanisms.

Miscellaneous Sarcomas

Malignant schwannomas, liposarcomas, angiosarcomas, and fibrosarcomas have all been studied in small numbers by DNA FCM (Table 20.1). However, none of these groups have been reported as separate histological types. For this reason, and because of the small numbers relative to other more common tumors with which they have been aggregated, it is unwarranted to apply results from other sarcoma studies to these histological types.

CONCLUSIONS

Many sarcomas have been subjected to DNA flow cytometric measurement. It is apparent that DNA ploidy analysis can help in the difficult task of differentiating benign from malignant soft tissue lesions. It is also clear that the degree of aneuploidy increases with the histological grade of sarcomas. It seems that for certain sarcomas, determining the ploidy status provides additional valuable information regarding the biological behavior of a tumor beyond the usual prognostic values, including histological grade. Studies of individual histological types show the most promise in this respect and these must be expanded to increase numbers and enhance detailed analyses. The combination of immunological markers of proliferation and DNA ploidy show promising early results. The rapidly growing area of oncogene expression is finding applications in this field. Significant progress is required to standardize criteria for histological grading, histological typing, DNA nomenclature, preparative procedures, manipulations of data, and statistical analysis. Because of the rarity of these tumors, only cooperative multi-institutional and multinational efforts can achieve the desired results.

Nonsurgical therapy, particularly chemotherapy of soft-tissue sarcomas, is an area of intense controversy. The careful application of flow cytometric measurements to identify patients for protocol randomization may very well resolve some of these issues histotype by histotype.

Editors and reviewers of publications of flow cytometric data should agree on what constitute minimum standards of acceptability in terms of coefficients of variation of peaks, numbers of events counted, numbers of cases studied, etc. Researchers must try to refrain from publishing flow cytometric data on clinical conditions unless they assess whether such data contribute to the diagnosis, management, prognosis, and understanding of these conditions. Most importantly, researchers must demonstrate whether such data offer more information than is already available.

The rapid progress in fine-needle aspiration cytology is fast extending into the area of diagnosis of soft-tissue sarcomas (19, 50). Sarcomas are being successfully diagnosed by this method in a high proportion of cases. DNA flow cytometric and static cytometric measurements in this type of material are likely to grow and become standard practice (50). Needle biopsy is less invasive and results in better local control of disease and survival than incisional or excisional biopsy (8). This practice should be encouraged and diagnosticians must be armed to meet its challenges.

REFERENCES

1. Enzinger FM, Weiss SW. Soft tissue tumors. 2nd ed. St. Louis, Toronto, London: C.V. Mosby, 1988.
2. Brooks JSJ. Soft tissue pathology: conceptual and practical approaches to diagnoses. Am J Clin Pathologists Regional Education Course, 1990.
3. Cagle LA, Mirra JM, Storm FK, Roe DJ, Eilber FR. Histologic features relating to prognosis in synovial sarcoma. Cancer 1987;59:1810–1814.
4. Coindre JM, Trojani M, Contesso G, et al. Reproducibility of a histopathologic grading system for adult soft tissue sarcoma. Cancer 1986; 58:306–309.
5. Alvegård TA, Berg NO. Histopathology peer review of high-grade soft tissue sarcoma: the Scandinavian Sarcoma Group experience. J Clin Oncol 1989;7:1845–1851.
6. Lawrence W, Donegan WL, Natarajan N, Mettlin C, Beart R, Winchester D. Adult soft tissue sarcomas: a pattern of care survey of the American College of Surgeons. Ann Surg, April 1987;205(suppl 4): 349–359.
7. Cooper JE, Allen PW. Low-grade sarcomas. Pathol Annual 1990;Part 2, volume 25:1–18.
8. Rooser B. Prognosis in soft tissue sarcoma. ACTA Orthopaedica Scandinavica 1987;58(suppl 225):1–54.
9. Tsujimoto M, Aozasa K, Ueda T, Morimura Y, Komatsubara Y, Doi T. Multivariate analysis for histologic prognostic factors in soft tissue sarcomas. Cancer 1988;62:994–998.
10. Markhede G, Angervall L, Stener B. A multivariate analysis of the prognosis after surgical treatment of malignant soft-tissue tumors. Cancer 1982;49:1721–1733.
11. Ueda T, Aozasa K, Tsujimoto M, et al. Multivariate analysis for clinical prognostic factors in 163 patients with soft tissue sarcoma. Cancer 1988;62:1444–1450.
12. Mandard AM, Petiot JF, Marnay J, et al. Prognostic factors in soft tissue sarcomas: a multivariate analysis of 109 cases. Cancer 1989;63: 1437–1451.
13. Rööser B, Willén H, Gustafson P, Alvegård TA, Rydholm A. Malignant fibrous histiocytoma of soft tissue: a population-based epidemiologic and prognostic study of 137 patients. Cancer 1991;67:499–505.
14. Xiang J, Spanier SS, Benson NA, Braylan RC. Flow cytometric analysis of DNA in bone and soft-tissue tumors using nuclear suspension. Cancer 1987;59:1951–1958.
15. Alvegård TA, Berg NO, Baldetorp B, et al. Cellular DNA content and prognosis of high-grade soft tissue sarcoma: the Scandinavian Sarcoma Group experience. J Clin Oncol 1990;8:538–547.

16. Kreicbergs A, Tribukait B, Willems J, Bauer HCF. DNA flow analysis of soft tissue tumors. Cancer 1987;59:128–133.
17. Radio SJ, Wooldridge TN, Linder J. Flow cytometric DNA analysis of malignant fibrous histiocytoma and related fibrohistiocytic tumors. Hum Pathol 1988;19:74–77.
18. Matsuno T, Gebhardt MC, Schiller AL, Rosenberg AE, Mankin HJ. The use of flow cytometry as a diagnostic aid in the management of soft-tissue tumors. J Bone and Joint Surg (Am) June 1988;70-A(suppl 5):751–759.
19. Åkerman M, Killander D, Rydholm A, Rööser. Aspiration of musculoskeletal tumors for cytodiagnosis and DNA analysis. Acta Orthop Scand 1987;58:523–528.
20. Pettinato G, Manivel JC, De Rosa G, Petrella G, Jaszcz W. Angiomatoid malignant fibrous histiocytoma: cytologic, immunohistochemical, ultrastructural, and flow cytometric study of 20 cases. Mod Pathol 1990;3(suppl 4):479–487.
21. El-Naggar A, Barlogie B, Ro J, Ayala A, Batsakis JG. Flow cytometric DNA analysis of soft tissue neoplasms. Lab Invest 1988;58:27A.
22. Kroese MCS, Rutgers DH, Wils IS, Van Unnik JAM, Roholl PJM. The relevance of the DNA index and proliferation rate in the grading of benign and malignant soft tissue tumors. Cancer 1990;65:1782–1788.
23. Heliö H, Karaharju E, Nordling S. Flow cytometric determination of DNA content in malignant and benign bone tumours. Cytometry 1985;6:165–171.
24. El-Naggar AK, MacKay B, Sneige N, Batsakis JG. Stromal neoplasms of the breast: a comparative flow cytometric study. J Surg Oncol 1990;44:151–156.
25. El-Naggar AK, Ro JY, McLemore D, Garnsy L. DNA content and proliferative activity of cystosarcoma phyllodes of the breast: potential prognostic significance. Am J Clin Pathol 1990;93:480–485.
26. Murad TM, Hines JR, Beal J, Bauer K. Histopathological and clinical correlations of cystosarcoma phyllodes. Arch Pathol Lab Med 1988;12:752–756.
27. August CZ, Bauer KD, Lurain J, Murad T. Neoplasms of endometrial stroma: histopathologic and flow cytometric analysis with clinical correlation. Hum Pathol 1989;20:232–237.
28. Dunton CJ, Kelsten ML, Brooks SE, Viglione MJ, Carlson JA, Mikuta JJ. Low-grade stromal sarcoma: DNA flow cytometric analysis and estrogen progesterone receptor data. Gynecol Oncol 1990;37:268–275.
29. El-Naggar AK, Ro JY, McLemore D, Garnsey L, Ordonez N, MacKay B. Gastrointestinal stromal tumors: DNA flow-cytometric study of 58 patients with at least five years of followup. Mod Pathol 1989;2(suppl 5):511–515.
30. Eriksen BL, Bauer KD, Caro WA, Roth SI. DNA analysis with cytologic and clinical correlation of dermatofibrosarcoma protuberans. Lab Invest 1988;58:28A.
31. Tsushima K, Rainwater LM, Goellner JR, vanHeerden JA, Lieber MM. Leiomyosarcomas and benign smooth muscle tumors of the stomach: nuclear DNA patterns studied by flow cytometry. May Clin Proc 1987;62:275–280.
32. El-Naggar AK, Ayala AG, Abdul-Karim FW, et al. Synovial sarcoma: a DNA flow cytometric study. Cancer 1990;65:2295–2300.
33. Molenaar WM, Dam-Meiring A, Kamps WA, Cornelisse CJ. DNA-aneuploidy in rhabdomyosarcomas as compared with other sarcomas of childhood and adolescence. Hum Pathol 1988;19:573–579.
34. Costa J, Wesley RA, Glatstein E, Rosenberg SA. The grading of soft tissue sarcomas. Cancer 1984;53:530–541.
35. Donhuijsen K, Schmidt U, Hirche H, vanBeuningen, Budach V. Changes in mitotic rate and cell cycle fractions caused by delayed fixation. Hum Pathol 1990;21:709–714.
36. Ueda T, Aozasa K, Tsujimoto M, et al. Prognostic significance of Ki-67 reactivity in soft tissue sarcomas. Cancer 1989;63:1607–1611.
37. El-Naggar AK, Ro JY, Ayala AG, Hinchey WD, Abdul-Karim F. Angiomatoid malignant fibrous histiocytoma: a DNA flow cytometric analysis of 7 cases. Am J Clin Pathol 1988;90:502A.
38. Han W, Ro J, El-Naggar A, Shin A, Ordonez N, Ayala A. Giant cell tumor of tendon sheath (GCTTS): DNA flow cytometric analysis, proliferating cell nuclear antigen (PCNA) and macrophase (CD-68) immunostaining on 24 cases. Lab Invest, January, 1991;64(suppl 1):4A.
39. Tsushima K, Stanhope CR, Gaffey TA, Lieber MM. Uterine leiomyosarcomas and benign smooth muscle tumors: usefulness of nuclear DNA patterns studied by flow cytometry. Mayo Clin Proc 1988;63:248–255.
40. Goldfarb S, Richart RM, Okagaki T. Nuclear DNA content in endolymphatic stromal myosis. Amer J Obstet Gynecol 1970;106:524.
41. Nelson KG, Haskill JS, Sloan S, et al. Flow cytometric analysis of human uterine sarcomas and cell lines. Cancer Res, June 1987;47:2814–2820.
42. Layfield LJ, Hart J, Neuwirth H, et al. Relation between ploidy and the clinical behavior of phyllodes tumors. Cancer 1989;64:1486–1489.
43. Kreicbergs A, Zetterberg, Söderberg G. The prognostic significance of nuclear DNA content in chondrosarcoma. Int Acad Cytol Anal Quant Cytol Histol December, 1980;2(suppl 4):272–279.
44. Persson S, Willems J-S, Kindblom L-G, Angervall L. Alveolar soft part sarcoma: An Immunohistochemical, cytologic and electron-microscopic study and a quantitative DNA analysis. Virchows Archiv [A] 1988;412:499–513.
45. El-Naggar A, Garcia G. Epithelioid sarcoma (ES): DNA flow cytometric study. Lab Invest, January, 1991;64(suppl 1):4A.
46. El-Naggar A, Ordonez N, Sara A, McLemore D, Batsakis J. Clear cell sarcomas (CCSs) and metastatic soft tissue melanomas (MMs): a comparative flow cytometric analysis. Lab Invest, January, 1991;64(suppl 1):4A.
47. Auerbach H, Brooks JJ. Kaposi's sarcoma: observations and a hypothesis. Lab Invest 1985;52(suppl 1):4A.
48. Sanchez MA, Ames ED, Erhardt K, Auer GU. Analysis of DNA distribution in Kaposi's sarcoma in patients with and without the acquired immune deficiency syndrome. Int'l Acad Cytol Anal Quant Cytol Histol, February, 1988;10(suppl 1):16–20.
49. Fukunaga M, Silverberg SG. Kaposi's sarcoma in patients with acquired immune deficiency syndrome: a flow cytometric DNA analysis of 26 lesions in 21 patients. Cancer 1990;66:758–764.
50. Fernö M, Baldetorp B, Åkerman M. Flow cytometric DNA ploidy analysis of soft tissue sarcomas: a comparative study of preoperative fine needle aspirates and postoperative fresh tissues and archival material. Anal Quant Cytol Histol, August, 1990;12(suppl 4):251–258.

CLINICAL COMMENTARY
Rosemary Mazanet

Malignant soft-tissue sarcomas (STS) are a heterogeneous group of neoplasms that present many diagnostic and therapeutic challenges. The preceding chapter by Dr. Beckmann provides an excellent review of the current problems with the staging system for STS and the rationale and results regarding the application of DNA flow cytometric technology to this tumor. Unfortunately, our progress in developing treatment modalities to prevent local recurrence and metastatic disease has not kept pace with the development of new diagnostic approaches.

The American Joint Committee staging system for soft-tissue sarcomas is largely dependent on grade, which is primarily based on the number of mitoses per 10-high-power microscopic fields. With the exceptions of rhabdomyosarcoma, Kaposi's sarcoma, and mesothelioma, the behavior and treatment of the histological variants of soft tissue sarcoma as a group is generally similar grade-for-grade. The five-year survival rate for soft-tissue sarcomas arising in various anatomical sites is similar when corrected for grade, except for intraabdominal and retroperitoneal tumors, which tend to be large and to have invaded vital organs. In order of importance, the useful pathologic parameters are tumor grade, the extent of the surgical margins (and location of any close margin), the size of the gross lesion in the unfixed pathology specimen, and the histological type. It is true that our knowledge is limited by the small numbers of patients presenting in each of the various subtypes of STS, and Dr. Beckmann correctly states the fact that there is a lack of reproducibility and standardization of histologic subtype classification and STS tumor grading. The correct categorization of histologic subtypes is not impacted by flow cytometry, however, and this area is rapidly advancing due to technical improvements in immunohistochemical staining, electron microscopy, and cytogenetic analysis. The potential value of flow cytometry for STS may be in 1) distinguishing benign from malignant disease by the more accurate evaluation of mitotic activity, which may correlate with metastatic potential and 2) providing reliable prognostic definitions, although it is unclear if this would impact on therapy.

Is the technique reliable and standardized? It is true that the evaluation of mitotic activity is subjective and prone to artifactual distortions, which mandates careful histopathological examination of large tumor specimens. But these problems are not unique to pathological preparation and scrutiny, and will surface in DNA cytometric analysis as well. There is the inherent concern regarding instrument bias. In addition, the definitions of DNA euploidy, aneuploidy, and periploidy, and laboratory criteria for distinguishing tetraploidy from diploidy, are not standard. Not mentioned is the abundant nonneoplastic stromal tissue surrounding STS elements within tumors, which can lead to grave sampling errors. This leads to obvious concern regarding use of fine-needle aspiration for cytometric analysis sampling. Needle biopsy is growing in diagnostic use, with the inherent understanding that an inconclusive result warrants further action. Certainly, a larger biopsy or careful review of an excised specimen would be required to adequately grade a known STS. Similarly, a needle biopsy yielding a diploid DNA histogram could be the result of sampling error.

Is there diagnostic/prognostic value in the test? Dr. Beckmann correctly identifies areas where flow cytometry may be potentially useful in "upgrading" benign lesions to malignant based on aneuploidy. I agree that finding an aneuploid DNA content in a histologically benign or indeterminate lesion should prompt the pathologist to reevaluate the case, but I would not base a therapeutic decision on the DNA histogram alone, given the current data. He also suggests that a diploid lesion with "malignant morphology" should be evaluated further, with other areas sampled for DNA content. This is true, but there is no evidence that a lesion can be determined to be benign based solely on the DNA histogram. The distinction between a local malignant STS and a benign lesion would result in different surgical procedures, a wide excision with a 3–4 cm margin vs. a simple excision, respectively. In any case where there is conflicting evidence or where doubt is present, a wide surgical excision of the tumor should be performed, with subsequent periodic follow-up for evidence of relapse.

Many of the published studies do not investigate the prognostic value of cytometric analysis. This would need to be done in a large multi-institutional setting before any therapeutical medical decision can be based on this test result. DNA analysis may give additional useful information about the biological behavior of a given tumor where ploidy is disparate from mitotic grading. However, for the vast majority of patients, the treatment would not be changed. Nevertheless, the possibility exists that a more reliable estimation of prognosis in individual cases may be useful in determining the intensity of therapy.

Most studies that have looked at the prognostic value of this test have evaluated metastasis-free survival as an end-point. In evaluating this test for the predictive value of developing local recurrences and/or distant metastases, the goal must be clear. Disease-free survival is certainly an important end-point; however, many patients with pulmonary metastases or recurrent local disease can be salvaged surgically. For this reason, disease-free survival may be a less meaningful end-point than overall survival in STS.

Would treatment of patients be changed because of test results? As noted above, the distinction between a malignant and a benign STS may well result in a different surgical procedure and would certainly result in different medical follow-up testing for evidence of early disease recurrence. Unlike breast cancer, however, adjuvant chemotherapy remains unproven for adult soft-tissue sarcomas (with the exception of rhabdomyosarcomas) despite the significant rate of both local recurrence and the development of distant metastatic disease.

Resected low-grade soft tissue sarcomas with 3–4 cm surgical margins are usually cured (80%), and the use of DNA flow cytometry to identify those patients with slightly higher mitotic rates would not lead to changes in management. Of note, patients relapsing with low-grade STS have a higher likelihood of recurring locally than with metastatic disease, in contrast to patients with high-grade lesions. In addition, patients with relapsing low-grade STS respond less well to chemotherapy and radiation than those with higher mitotic rates, arguing for aggressive initial surgery in the low-grade group. As mentioned before, the distinction between benign and potentially malignant STS that would indicate different surgical strategies may be the most important potential use for this diagnostic test.

Future randomized trials of adjuvant chemotherapy in STS should include patients at high risk for metastases (large, high-grade lesions) with a reasonable likelihood of local control. With the low probability of metastatic disease, the risks of adjuvant chemotherapy are not currently warrented in patients with grade-I lesions. To use aneuploidy, irrespective of traditional staging, as a feature to allow enrollment for such studies is not warranted from the data presented. Any such trial should collect information regarding

DNA flow cytometry as well as cytogenetics, molecular biology of retinoblastoma genes, and evidence for drug-resistance mechanisms in eligible patients to determine how these features correlate with survival and disease-free survival.

21

Relative Applicability of Image Analysis and Flow Cytometry in Clinical Medicine

DAVID S. WEINBERG

INTRODUCTION

Image analysis (IA) is a technology that has undergone rapid development since the 1960s, at first mainly stimulated by the needs of the aerospace and defense industries (1). Imaging techniques have found widespread applications in industry, primarily in the field of robotic vision and automated inspection, and there are increasing numbers of medical applications, including radiological imaging and microscopy. In the context of this volume devoted to flow cytometry (FCM), I will attempt to contrast and compare IA and FCM, as many of the current and potential clinical applications are shared by both methods of analysis. I will also try to point out those features of FCM and IA that provide unique advantages for cell analysis and how these methods might be optimally used together.

IA, as defined here, includes automated or semiautomated computer-based methods in which image information is digitized, captured, stored, and subjected to quantitation of image features. This process might also be referred to as "quantitative digital imaging." Other terms commonly employed include "image cytometry," "cytophotometry," "static cytometry," and "microspectrophotometry." This definition of IA excludes standard morphometry, performed with or without aid of computers, as well as systems intended mainly for imaging processing, i.e., altering the image to enhance or suppress visual information. IA, when combined with video microscopy, can measure features of biological interest in cells and tissues, especially when probes and special stains are employed to label specific cellular components. This review is limited to applications using light microscopy (rather than fluorescence) and emphasizes applications in diagnostic pathology. Purely morphometric applications are not discussed, although morphometry has many important potential diagnostic applications (2, 3).

In many respects, the development of clinical applications for IA in diagnostic medicine lags somewhat behind FCM, which is a more "mature" technology. However, there has been a rapid increase in the availability and numbers of applications for IA, and imaging systems will be found with increasing frequency in clinical laboratories, often alongside flow cytometers. The purpose of this review is to familiarize the reader with the basic technology of IA, to compare the features and capabilities of FCM and IA, and to discuss the relative advantages and disadvantages of each technology.

IMAGE ANALYSIS: BASIC TECHNOLOGY

IA systems are usually constructed from the following hardware components, illustrated in Figure 21.1: (a) a device to capture the image, usually a video camera: (b) an image capture board, which samples the continuous (analog) video signal, digitizes the signal, and directs the numerical values to specific memory locations; (c) a computer, which typically houses the image capture board and is used to perform a variety of software-driven operations on the digital image data as well as to provide an interface for the user; (d) a video monitor, which displays the live or stored ("captured") image; (e) storage media, typically magnetic disc or tape (although high-capacity optical media are coming into increasing use); and (f) a user interface, such as a keyboard, mouse, or light pen. Perhaps the most important component of the system is the software, which greatly influences the ease of use of the system and the types of measurements that can be made. Some systems also include extensive software facilities for handling and analyzing data, such as databases and statistics packages.

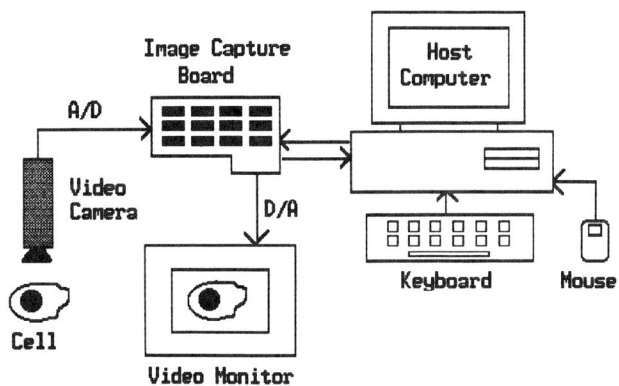

Figure 21.1. Schematic diagram of a typical image analysis configuration. For analysis of cells and tissues, the video camera is coupled to a microscope (not shown).

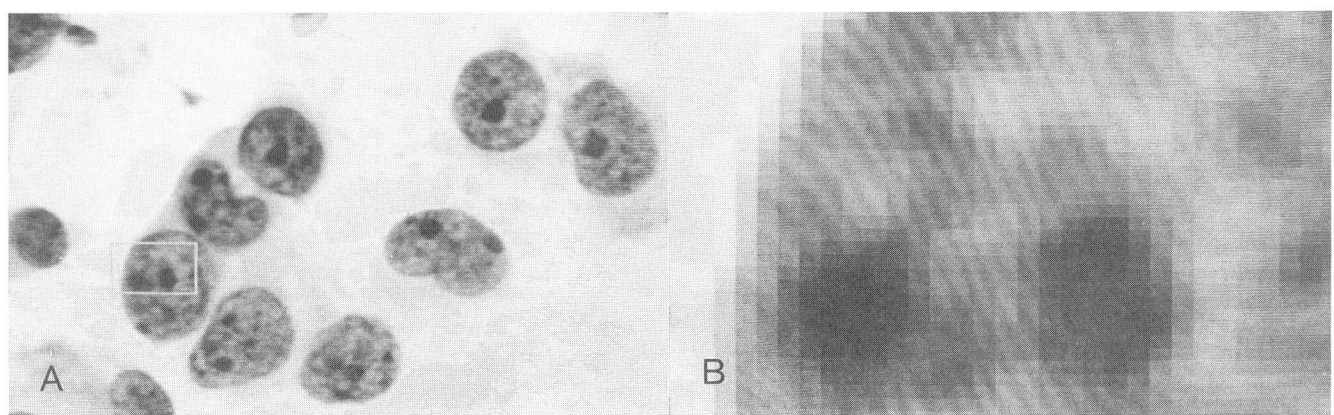

Figure 21.2. Digitization of microscopic image, with pixel storage and display. **A**, A cluster of cells from a case of breast cancer as viewed on the video monitor; the image appears continuous at this level of resolution. The boxed-in area is shown in a magnified view (**B**); it consists of a series of discrete square units (pixels), each displaying a single level of light intensity (grey level). It is the digital values of these grey levels that are stored in computer memory after image capture.

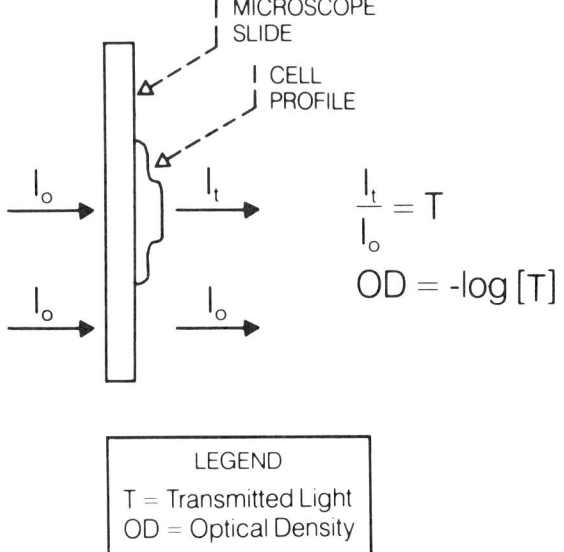

Figure 21.3. Measurement of optical density by image analysis. The relationship between incident light (I_o), transmitted light (I_t), and optical density (OD) is shown for a cell located on a glass slide viewed in cross-section. According to the Beer-Lambert law, the mass of the absorbing substance at any point is linearly proportional to the OD. Therefore, the total optical density of all pixels can be used to measure the quantity of cellular components after appropriate staining, such as nuclear DNA content. (Reprinted with permission from: Bacus JW, Grace LJ. Optical microscope system for standardized cell measurements and analyses. Appl Optics 1987;26:3280–3293.)

Regardless of the specific components used, imaging systems rely on similar principles of operation. The video signal from the camera (or other device) is transferred to an image capture board, which samples and digitizes the analog signal. The digital value of the signal sample is proportional to the amplitude of the video signal, and thus is related to the light intensity of a given portion of the image. Typically, analog-to-digital (A/D) converters are used that distinguish 256 distinct levels of light intensity, or "grey levels," ranging from pure black to pure white. Another important feature of digital conversion is that the continuous image is broken up into discrete picture elements, or "pixels," with each pixel assigned a specific grey level (Fig. 21.2). Each pixel value is stored in a memory location within a frame buffer and, thus, a full frame of the video image is "captured." It is on this stored image data that the host computer can perform a variety of software-driven functions, including distinguishing objects from background (image segmentation) and measuring image features. Typically, images are divided into 512 × 512 horizontal and vertical pixels, although systems with greater pixel resolution are available (at greater cost). The stored image can be displayed in its original form on a video monitor after analog conversion, or the output can be altered to reveal certain features. For example, ranges of grey level can be displayed as different colors ("pseudocolor"), which can be helpful in revealing subtle image features or in coding portions of the image.

Essential for certain types of measurements of biological interest, including nuclear DNA content, is the conversion of light intensity values to optical density (Fig. 21.3). This is because, according to the Beer-Lambert law, the concentration of a subtance at the point being measured is linearly proportional to the optical density. When properly calibrated, optical density measurements at specific wavelengths of light can provide quantitation of staining reactions specific for cellular features. For example, the value of the total optical density of a cell nucleus stained by the Feulgen reaction (which is specific for DNA) provides a measure of the nuclear DNA content. Optical density measurements have many potential applications in quantitative imaging, as will be illustrated later. This function is analogous to fluorescence intensity measurement performed by FCM, in which the fluorescence intensity is linearly related to the amount of dye bound by the cell, and hence to the amount of ligand. Although useful fluorescence measurements can be made by IA systems (4), problems associated with fluorescence decay and preservation of the specimen can make this task difficult. Measurements based instead on absorption (by dyes or other substances) are easily adapted to IA at visual wavelengths, and can overcome many of the technical difficulties associated with fluorescence measurements.

Similar to developments in FCM, the advent of inexpensive yet powerful microcomputers, mass production of electronic components, and market competition have vastly improved the availability of affordable imaging systems. Although it is possible for an individual to construct his/her own FCM system (5), this remains a formidable task. However, rather sophisticated IA systems can be constructed from off-the-shelf components compatible with various computer bus architectures, including PC's. However, the reader is warned that the design of a microscope–based imaging system is fraught with potential technical difficulties, and that a firm knowledge of optics, video technology, IA technology, image processing, and computer programming may be required to produce a useful system (6, 7). Instead, it may be wiser to consider the purchase of an integrated imaging system, as provided by several manufacturers (8–10). It is important that the user evaluate such systems for their ability to accurately perform measurements of interest and that the system be reasonably easy to use (i.e., "user friendly").

The microscope-based IA instrument used in the author's laboratory may serve as an example of such an integrated system. The CAS 200 System (Cell Analysis Systems, Inc., Elmhurst, IL) (8) utilizes two small solid-state video cameras, each with its own optical filter, mounted on a standard light microscope. Similar to use of separate photomultiplier tubes in FCM systems, the use of two cameras sensitive to different wavelengths allows simultaneous two-color analysis. The video signals are transmitted to an IBM "386" computer equipped with an image capture board capable of digitally converting light intensity to 256 possible grey levels. These grey levels are automatically converted to optical density values based on a calibrated "look-up table," and these values are stored in the frame buffer. The central 256 × 256 pixel region of the image is then subjected to analysis and objects are located automatically by image segmentation routines or chosen by the operator. Depending on the software used, a variety of measurements are performed on each object in the field, and each object may be classified by the user. This interactive procedure allows one to compare the features of different classes of objects (e.g., the DNA content of different cell types). These data can be subsequently displayed graphically or analyzed using statistics programs provided with the system software.

Measurements by the CAS 200 System are derived from calibrated conversion of pixel information (8). For example, the area (size) of an object of interest (cell or nucleus) can be calculated from the total number of pixels composing the object. In addition, the optical density can be used to calculate either the concentration or total amount of a given substance, based on the average or total optical density. In this way, nuclear DNA content as well as the concentration of substances detected by immunohistochemical reactions can be quantitated (see later sections). The distribution of optical density, which reflects the "texture" of the object, can be measured in a great number of ways, as will be discussed. It is possible to measure separately the features of both nucleus and cytoplasm within the same cell, a task that is possible only in a more limited way by FCM. Thus, a great number of potentially useful measurements, some direct and others derived, may be made simultaneously on each cell. Up to 200 such features can be routinely measured by some research systems (11), although the usefulness of many of these parameters is uncertain.

Thus, the microscope-based IA system is a powerful analytical tool, capable of multiparametric cellular measurements that are potentially useful in research and diagnosis. In the following sections, I will attempt to further discuss those aspects of cell analysis shared by both flow cytometers and IA systems and to define those capabilities that are unique to each approach.

Flow Cytometry and Image Analysis: Comparison of Technical Features

The technical features of FCM and IA endow each method with certain virtues and inherent limitations. A comparison of some of these features is shown in Table 21.1.

The principal requirement for study of a sample by FCM is the preparation of single-cell suspensions. For certain types of clinical specimens, such as blood, urine, and body fluids, such preparations are easily accomplished. In fact, whole blood can be routinely used for many types of analysis, especially lymphocyte phenotyping. The preparation of single-cell suspensions from solid tissues, however, is more problematic. Lymphoid tissues are easily dispersed by mechanical means, but a variety of more extreme mechanical and/or enzymatic methods are needed for most other tissues (12, 13). The fact that several preparative methods may be needed to handle a variety of tissue types may present difficulties for a clinical laboratory. In addition, there is always concern that the cell population of interest may be lost or greatly reduced during the preparation. Some of these problems may be circumvented by the isolation of cell nuclei from fresh (14, 15) and paraffin–embedded specimens (16). However, the loss of the cell surface and cytoplasm then precludes study of those cell features, and cell markers cannot be used to identify cells of interest. The specimen must also be of adequate size or cellularity to provide sufficient cells for FCM. With IA, on the other hand, one can analyze a great variety of tissue and cell preparations, including touch imprints, cytospins, cytologic smears, and tissue sections. The study of archival material by FCM is largely limited to the measurement of DNA content of tumor cell nuclei isolated by the Hedley technique (17); on the other hand, a vast array of measurements can be performed by IA on paraffin sections of tissue or destained cytologic preparations. It is

Table 21.1
Comparison of Technical Features of Flow Cytometry and Image Analysis

Feature	Flow Cytometry	Image Analysis
Speed	Rapid analysis of large numbers of cells (typically 10,000–20,000)	Relatively slow, fewer cells (typically 100–200)
Measurements	Based on fluorescence and light scatter	Based on transmittance at visual wavelengths
Parameters	Analysis typically based on 4–5 parameters (2 light scatter, 2–3 fluorescence)	Analysis based on large number of direct and derived morphometric and absorbance measurements
Sample preparation	Requires single cell suspensions	Cytocentrifuge preparations, touch imprints, cytologic smears, tissue sections
Morphometry	Usually limited to light-scatter measurements	Large number of morphometric measurements possible, direct and derived
Gating	Based on light scatter or fluorescence	Direct morphologic identification and classification
Morphologic/histologic correlation	None	Measurements correlated with tissue histology or cell morphology
Use of archival material	Limited to DNA analysis (Hedley method) or cryopreserved cells	Paraffin blocks (tissue sections or Hedley preparations); glass slide material
Subset analysis	Based on cell markers and/or light scatter	Based on morphology and/or cell markers
Rare events	Difficult to detect, may be hidden	May select for rare events

even possible to prepare adequate touch imprints from the face of a frozen section block (18), thus sparing valuable research tissue. Of course, in situations in which there are too few cells obtained for FCM, such as cerebrospinal fluids or needle aspirates, IA of a concentrated cytocentrifuge preparation may provide the only means for making cellular measurements. It is possible to envision a routine whereby a tissue specimen is prepared for both IA and FCM to ensure adequate examination. Small specimens and those with low cellularity would be preferentially studied by IA.

One of the principal advantages of FCM is the ability to analyze large numbers of cells at rapid rates, typically 300–1000 cells/second. In this manner, it is possible to collect data on 10^5 to 10^6 cells in a short period of time. The resulting histograms are smoother than those generated by the fewer events collected by IA, and small changes in the histogram can be relied on as statistically significant. Far fewer events are typically recorded using IA, with measurements of 100–200 cells performed in most analyses of single cells. However, in some types of analysis (e.g., nuclear immunostaining reactions, to be discussed), whole microscopic fields that include hundreds of cells may be analyzed at once and many fields can be rapidly measured. With rapid advances in real-time IA and automated cytometry, acquisition of imaging data at flow cytometric rates may be possible in the future, thus diminishing this distinction between flow and static imaging.

The accurate characterization of selective cell types among a mixed-cell population requires some form of interactive gating, and this is accomplished in FCM by the use of light-scatter and fluorescence markers. For example, it is possible to measure the DNA content of epithelial cells admixed with inflammatory and stromal cells by either positive selection using anti-cytokeratin antibodies or by rejecting lymphoid cells stained with lymphocyte–specific antibodies (19). It is routinely possible to identify lymphocytes in whole blood by forward and right-angle light-scatter patterns, and to confirm the identity of the gated population using monoclonal antibodies (20–22). However, it is often difficult to know with absolute certainty the identity of the cell population selected by such gates or whether the cells of interest have been included in the analysis. Tissue architecture, so important in the pathology of solid tumors, is lost upon making single-cell suspensions. Many studies, such as investigation of the relationship between in situ tumor precursor lesions and associated invasive tumor components, absolutely require the preservation of tissue architecture (23–25). For many types of measurements, especially those involving solid tumors, it is far easier and probably more accurate to rely on morphologic gating on tissue sections, which can be used to study specific regions of interest. The study of intact tissues also avoids the problem of cell loss so often encountered in preparing single-cell suspensions from solid tissues. The ability to identify cells of interest by morphology may greatly increase the sensitivity for detection of rare or unusual cell types. IA has been shown in some studies to be more sensitive than FCM for the detection of rare aneuploid cells (26). Such rare events may be "buried" amid the large number of events collected by FCM and may be difficult to distinguish from background noise. Similar to their use in FCM, additional markers, such as cytokeratin, could be used in IA to help identify cells of interest in cytologic preparations. The use of such markers in IA, combined with morphology can allow the positive identification and characterization of several cell types at once in a mixed population. Although there have been attempts to identify morphologic features by FCM, using slit-scan measurement (27), size and cytoplasmic granularity remain the most readily detected morphologic features obtained by FCM methods.

With regard to multiparameter analysis, the most advanced clinical flow cytometers are generally capable of measuring two light-scatter and three fluorescence signals, for a total of five features per cell (28). On the other hand, a vast array of cellular measurements are possible by IA. Many features are measured directly, such as size (area) and optical density. However, many more features can be calculated or derived, including ratio of cytoplasmic area to nuclear area, shape factors, orientation parameters, and texture (8), the number of such features being limited only by the software provided with the system and/or the imagination of the programmer. There are a great many useful and useless measurements among the vast array of cell features that can be detected by IA, and only careful research will reveal which of these are most useful for characterizing cells.

Largely because FCM is a more established clinical laboratory tool than IA, there is greater consensus regarding controls and standards in that field (29). Proper reagent controls for immunofluorescence staining are provided by the manufacturers and are routinely used, as are cellular controls from normal subjects. Acquisition and display of single- and dual-parameter data by FCM differ little among various FCM instruments, and there has been movement toward standardization of data file formats (30). The analysis of flow cytometric histograms usually involves the counting of events in various regions of the histogram as compared to staining controls, although complex curve-fitting routines may be used for certain types of histogram analysis, such as measuring the S-phase content in single-parameter DNA histograms (31). There is still a lack of agreement regarding the optimal methods for performing some clinical tests, which can lead to significant problems. As will be discussed in more detail later, a wide variety of preparative techniques have been used for tumor DNA content analysis, each having its own limitations, and the measurement of S-phase fraction is particularly sensitive to the method of analysis used (32). The use of multicenter studies and check samples has helped to provide standardization in some areas (33, 34), and standards regarding the reporting of DNA histograms have been developed that should help to allow easy comparison of data from different centers (35). However, such multicenter studies have shown striking interlaboratory differences with regard to S-phase measurement (36), demonstrating the need for further efforts in this area.

Controls and standards have not been as well established in imaging as in FCM (37, 38). DNA measurements represent perhaps the most widely applied measurement in clinical IA, and the Feulgen stain has long been regarded as the best dye with which to quantitate nuclear DNA (39). However, just as the fluorescent dyes do not always exhibit stoichiometric binding to DNA, a variety of factors, including the degree of nuclear condensation, can affect Feulgen staining (40). Therefore, the careful selection of normal cellular controls is important. Normal resting lymphocytes, for example, typically show less than diploid DNA content based on Feulgen staining, and should not be used for normal diploid controls for IA (41). For studies using the CAS 200 System, DNA content is calibrated based on the DNA content of normal rat hepatocytes present on a separate area on the same slide as the unknown sample; the additional use of cytologically normal cells in the sample thus allows for the use of both internal and external controls for DNA staining and diploid DNA content. There have been surprisingly few

Table 21.2
Routine Clinical Applications of FCM and IA

Test	FCM	IA
DNA content	+	+
Tumor cell proliferation		
S-phase fraction	+	−
Nuclear antigens	−	+
Phenotyping		
Lymphocyte subsets	+	−
Leukemia	+	−
Lymphoma	+	−
Reticulocyte counts	+	−
Platelet antibodies	+	−
Estrogen/progesterone receptor	−	−
Oncogene expression	−	+
Karyotyping	−	+
Cell classification	−	+/−

studies investigating the precision and reproducibility of DNA measurements by IA, although the work of Taylor and coworkers, using one commercial imaging system, reveals remarkable consistency of DNA content measurements under carefully controlled conditions (42).

Unlike fluorescence staining, which can be directly related to ligand concentration, immunohistochemical reactions are enzymatic and nonlinear. There can be significant daily variation in the quality of immunostaining obtained by even the best laboratories, which makes quantitation of such markers by imaging difficult. It has been possible to standardize immunohistochemical staining reactions, as has been demonstrated with estrogen receptor measurements by IA in breast cancer (43). With known concentrations of the marker of interest external control cell lines can be used for daily calibration, as has been successfully shown for HER-2/neu oncoprotein measurement in breast cancer (44). Further strides in the standardization of imaging hardware, software, computer data file structures, and staining methods are needed to improve the reproducibility of clinically important measurements by IA.

Table 21.2 lists some of the clinical tests that are currently most commonly performed by FCM and IA. Cell-surface-marker analysis is certainly the most frequently performed test in most clinical FCM laboratories, and is best performed using that technology. DNA content and cell-proliferation assays can be performed using either technology, as will be discussed in more detail. Cytoplasmic and nuclear markers may be measured by FCM after permeabilizing the cells (45), although these are more easily assessed in clinical samples using IA. Although there have been reports concerning the use of FCM to quantitate estrogen receptor and oncoproteins in tumor cells (46, 47), such measurements may be accurately made by IA, which allows important morphologic correlation (43, 48). Karyotyping has been adapted to FCM, although the main utility of this technique appears to be for rapid sorting of chromosomes (49). IA has been employed as an aid for karyotyping (50, 51), and there are several commercial instruments dedicated to this task. Fully automated karyotyping by IA may eventually become routine.

The use of IA for multiparameter cell classification is a major potential application of IA that has not yet been fully exploited. I have included this application on the list in Table 21.2 because it is this capability that I feel will provide some of the most important and unique clinical applications of IA. It is the ability to measure large numbers of cellular parameters and to organize these cellular features into automated classification schemes that suits IA to this task. Fully automated cytologic diagnosis and cell classification have long been major goals of IA, particularly with regard to cervical cytology screening. In the mid-1970s, a Cytology Automation Program was established by the National Cancer Institute with the goal of developing automated screening systems for Pap smears (52). Although the goals of this program have not yet been met, there has been a great deal of progress made in this area. Although a clinical imaging instrument designed to perform automated differentials on peripheral blood was marketed (Hematrak), more rapid CBC analyzers based on flow methods and capable of accurate differential counts led to its demise. The ability of an IA system to characterize red blood cell abnormalities has been demonstrated (53), but this technique has not been routinely applied. With regard to the analysis of cells other than peripheral blood, the TICAS-MLD, developed by Bartels, Wied, and associates, is an example of an integrated imaging system capable of performing a variety of automated diagnosis and classification tasks (11, 54). Computer systems can be trained to classify cells by statistical evaluation of multiparameter data from examples of particular cell types (55). A number of studies have demonstrated the ability of such systems to classify cells, primarily in the areas of cervical and urinary tract cytology (56, 57). Neural networks represent a nonalgorithmic method of generating computer learning; such systems are especially adept at tasks involving image recognition and classification (58, 59). Work performed in my laboratory has shown that either statistical or neural network approaches may be successfully applied to automated nuclear grading in breast carcinoma (60). The major stumbling block preventing the wider application of IA for cell classification is the difficulty of distinguishing ''junk'' from real objects by machine vision. Advances in the areas of specimen preparation, scene segmentation algorithms, and artificial intelligence systems should eventually improve the speed and accuracy of automated imaging systems to allow their routine use for cytologic screening and diagnosis. Several commercial systems for automated cervical cytology screening are currently under development (61).

Overall, then, FCM appears to be most useful in the analysis of clinical specimens in which large numbers of cells or nuclei can be easily placed in uniform suspension, and when the analysis of large numbers of events is needed to provide statistically significant data. On the other hand, IA can be used to advantage in situations in which there are too few cells for analysis by FCM, where dispersion of the cells is not easily performed (as for most solid tumors), or where correlations with cell morphology or tissue architecture are essential for accurate analysis. In the following section, I will examine two areas involving clinically important measurements that can be performed by either FCM or IA: DNA content (ploidy), and cell proliferation.

DNA CONTENT

The measurement of nuclear DNA content represents one of the first clinical applications developed for IA. The Feulgen stain is one of the oldest and most well-established methods for quantitative staining of DNA (62) and has been in routine use for more than 30 years. The advent of IA has substantially improved the speed and accuracy of measuring Feulgen staining, and the availability of commercial instruments that can measure DNA content with a high degree of precision (42) has made this method more routinely available. However, FCM has been much more widely applied as a tool for this purpose, partly due to the great speed and precision with which DNA measurements can be made by flow methods and also due to the wider availability of clinical FCM instruments. The clinical laboratory may be faced with a choice concerning whether to use FCM or IA (or both) to measure DNA content, and it is important to understand the advantages and limitations of each method, as previously outlined. It is also important to know whether both methods produce comparable results. The clinical utility of DNA measurement as a prognostic feature for a variety of human neoplasms is discussed by other authors in this volume and will not be addressed here. However, it should be stated that some of the controversy related to contradictory results in the literature may be related to differences in methodology. Understanding the technical limitations of quantitative DNA analysis may prove useful in understanding how such discrepancies may arise.

Several studies have directly compared the results obtained using FCM and IA for DNA analysis of solid tumors in general. Bauer et al. (63) studied 92 solid tumors using the CAS 100 System and FACScan (Becton Dickinson) and found 87% agreement with regard to DNA ploidy. However, nine cases were near-tetraploid by IA alone, and three cases were aneuploid by FCM, but not IA (Table 21.3). In a study in which fresh tumor imprints were compared to Hedley preparations from 31 tumors studied by IA and FCM, respectively, Claud et al. (64) found 87% agreement between the two methods with regard to ploidy description. Some studies have been directed at specific tumor types, including lymphomas (65, 66) and carcinomas of the kidney (67), endometrium (68), lung (69), thyroid (70), prostate (71), bladder (72), and ovary (73). Table 21.3 summarizes the results from several studies examining breast cancer that illustrate most of the features common to such studies comparing FCM and IA. Most studies, regardless of the precise methods used, show a high degree of correlation between IA and FCM with regard to both the description of DNA ploidy sta-

Table 21.3
DNA Ploidy Measurement by Both FCM and IA

Authors	Ref.	No. of Cases	Concordance	Comment
All tumors				
Bauer et al.	(63)	92	87%	9 cases near-tetraploid by IA alone
Claud et al.	(64)	30	90%	
Breast cancer				
Cornelisse et al.	(74)	67	r = 0.96	FCM bettrer for near-diploid cases
Wilbur et al.	(75)	24	65%	IA more sensitive for detecting rare aneuploid cells
Arnelov et al.	(83)	127	78%	IA correlates better with survival than FCM
Cornelisse et al.	(76)	34	r = 0.97	FCM better at detecting near-diploid aneuploidy
Fallenius et al.	(77)	50	98%	tetraploidy missed by FCM
Dawson et al.	(78)	54	91%	some tetraploid cases detected by IA alone

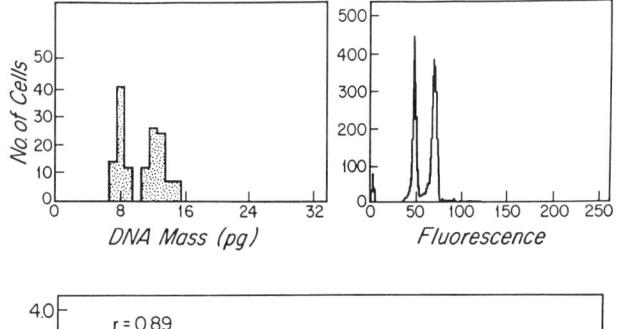
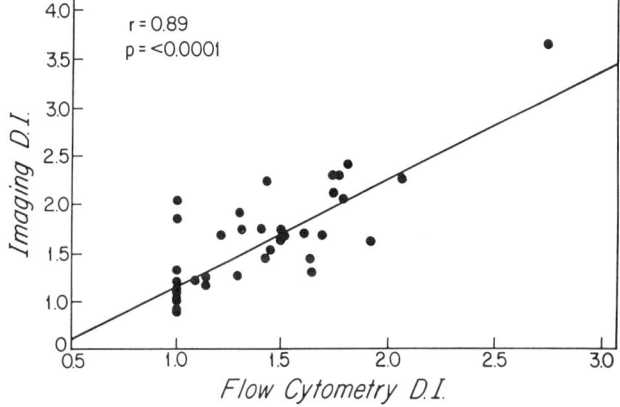

Figure 21.4. Nuclear DNA content measurement by FCM and IA in breast cancer. The top portion of the figure shows DNA histograms obtained by IA (left) and FCM (right). Note the similar appearance of the histograms, each containing diploid and aneuploid (hyperdiploid) G_0/G_1 peaks. The bottom portion of the figure shows the correlation between DI obtained by FCM and IA in 54 cases of breast cancer. Although there is good correlation between the two methods, note that there are two cases that appear near-tetraploid (D.I. = 2.0) by IA, yet are diploid (D.I. = 1.0) by FCM.

tus and DNA index. Several studies indicate the FCM is more sensitive than IA with regard to the detection of near–diploid aneuploid peaks (74–77), probably due to the greater precision of the G_0/G_1 peak obtained by FCM. Some studies note that FCM often fails to detect tetraploid populations seen by IA (77, 78). And some studies indicate that IA may be more sensitive than FCM in the detection of minor populations of aneuploid cells in breast cancer (74–76).

An example of such a comparative study in breast cancer performed in our laboratory is shown in Figure 21.4, in which fresh imprint DNA measured by IA was compared to nuclear preparations from fresh tissue studied by FCM (78). In general, there was good correlation between IA and FCM. Note that in two cases, however, IA revealed tetraploid DNA content in tumors that were diploid by FCM. Subsequent studies have shown that one explanation for this discrepancy is that the tetraploid population may be lost during the preparation of cell suspensions (79). It is also possible that the dilution of tumor cells with benign elements in standard FCM may accentuate the diploid G_0/G_1 peak and obscure the presence of tetraploid cells, which are then interpreted as a G_2/M peak (76). Some breast tumors may include polyploid cells exhibiting multinucleation, in which case nuclear preparations for FCM would not detect tetraploidy.

One of the problems in comparing DNA ploidy by FCM and IA relates to differences in the method of analyzing and classifying DNA histograms. Some investigators make use of the presence of cells having greater than 5c DNA content by IA to detect ''aneuploidy'' (66, 73, 80); the study by Rodenburg et al. (73) shows that the presence of such aneuploid cells in tumors that exhibit otherwise diploid DNA distributions confers a worse prognosis in ovarian cancer. Some studies make use of the histogram classification system developed by Auer (81) for analyzing DNA distributions produced by IA. In this system, histogram types I and III (diploidy, tetraploidy) are associated with a better prognosis in breast cancer than histogram types II and IV (high S phase, aneuploidy) (81). Two studies have shown a relatively poor correlation between IA and FCM with regard to DNA histogram classification according to Auer's method (77, 82). Ultimately, it is the correlation with clinical outcome that is most important, and the method of DNA analysis and histogram interpretation that proves most predictive will find general use. It is interesting that the study of Arnerlov et al. (83) shows that while aneuploidy detected by both FCM and IA is correlated with tumor grade and survival, only results generated by IA appear to be independently predictive of survival.

Table 21.4 summarizes the relative advantages and disadvantages of FCM and IA for DNA ploidy determination. A substantial case can be made for routine application of both techniques for clinical samples. It is clear that aneuploid populations may sometimes be detectable by only one method, as has been discussed above. Despite the greater

Table 21.4
DNA Content by FCM and IA: Advantages and Disadvantages

Advantages	Disadvantages
FCM	
Rapid analysis of large numbers of cells	Requires cell/nuclear suspension; may be difficult for solid tissues
Large numbers of events comprising DNA histogram	Possible cell loss during preparation
Precise G_0/G_1 DNA content measurement, small c.v.; near-diploid peaks resolved	Morphologic correlation not possible; definite cell identification requires markers
Possible to correlate DNA content with other markers	Variable dilution of tumor cells with benign tissue elements
S-phase content can be derived from histogram	Complex DNA histograms may be difficult to model or interpret
Ability to study archival material (paraffin blocks)	Rare events difficult to detect
Several dyes available for staining DNA	Samples are perishable
Data collection may be performed by skilled technician	Requires relatively high degree of technical skill to operate instrument
Standards have evolved for data formats and terminology	Poor reproducibility and lack of standards for S-phase measurement
IA	
Many types of specimen preparation can be used	Relatively slow speed of data collection
Direct morphologic classification and correlation	Relatively few events collected; histograms not as precise as FCM
Multiparameter analysis may include cell markers and morphometric parameters	S-phase content generally not evaluable from DNA histogram (too few events)
Sensitive detection of rare events	Feulgen is the only reliable DNA stain for light microscopy
Archival materials may be studied, including paraffin blocks and cytology slides	Classification and description of DNA histograms not standardized
Staining method is simple, and slides are permanent	Requires professional input for proper field and cell identification

Table 21.5
Methods for Measuring Tumor Cell Proliferation[a]

Mitotic counts (mitotic index)
Tritiated thymidine uptake (thymidine labeling index; TLI)
Flow cytometry
    S-phase estimate
    Bromodeoxyuridine uptake
Immunhistochemistry[b]
    Bromodeoxyuridine uptake
    Ki-67 antibody[c]
    PCNA/cyclin[d]

[a]See text for references.
[b]Labeling may also be measured by FCM.
[c]Frozen sections only.
[d]Frozen and paraffin sections.

precision of the G_0/G_1 peak usually generated by FCM, McFadden et al. (84) have shown that IA can sometimes resolve a broad G_0/G_1 distribution obtained by FCM into separate diploid and aneuploid peaks. The increased sensitivity of IA for minor aneuploid populations is demonstrated in a study by Schneller et al. (26), in which DNA aneuploidy, undetected in routine FCM and random measurement by IA, could be demonstrated by IA after morphologic selection of atypical cells. One method of combining FCM and IA is suggested by Tanke et al. (85), who used FCM to sort cells onto glass slides, where the DNA content could be confirmed using IA. Further studies will be needed to determine the optimal way to apply these techniques for routine clinical analysis.

Although IA may be applied to tissue sections for DNA analysis, there are many technical problems associated with such measurements (40, 86), and the method is probably best applied to studies in which it is essential to correlate DNA content with tissue architecture. Even though some studies have shown a good agreement between FCM and IA with regard to S-phase fraction in tumors (66, 87), the DNA histograms generated by IA usually contain too few events in the S-phase region to be reliable. Proliferative fraction measurement by IA is better performed using alternative methods, as discussed in the next section.

CELL PROLIFERATION

There is growing evidence that, in addition to DNA ploidy, tumor cell proliferation has prognostic significance, particularly in malignant lymphoma (88, 89) and breast cancer (90–92). Some studies indicate that tumor proliferative activity has independent prognostic significance, even if DNA ploidy does not (93). A variety of methods have been used to measure proliferative rate in tumors (see Table 21.5). Mitotic counts are generally regarded as a poor and unreliable measure of proliferation (94). Uptake of radiolabeled thymidine, or thymidine-labeling index (TLI) is a well-established method, and is considered by many to be the "gold standard" for measuring tumor cell kinetics. Although the TLI provides an accurate assessment of S-phase activity and permits histologic correlation, the method is cumbersome and not easily adapted to the clinical laboratory. FCM has been used extensively to determine cell cycle activity, primarily by quantitation of the S-phase portion of single-parameter DNA histograms. This method suffers from a number of technical limitations, however. First, it may be difficult to obtain single-cell suspensions from solid tumors and variable numbers of cells may be lost during the preparative stage. Second, the tumor cells may be greatly diluted by benign stromal or inflammatory cells, which can lead to underestimation of the S-phase fraction. Third, because a single-parameter DNA histogram actually represents a series of overlapping curves (G_0/G_1, S, and G_2/M phases), the complexity

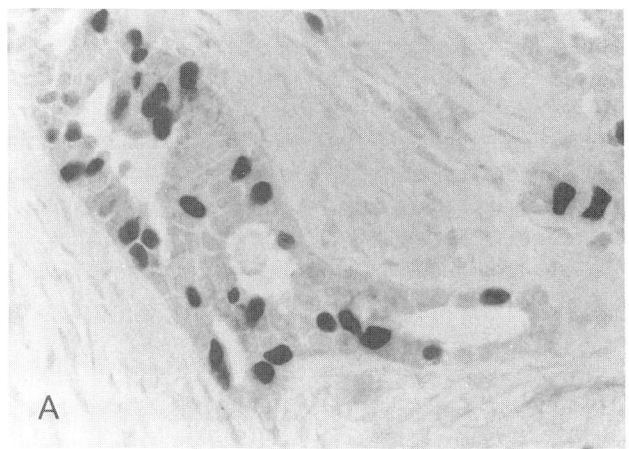

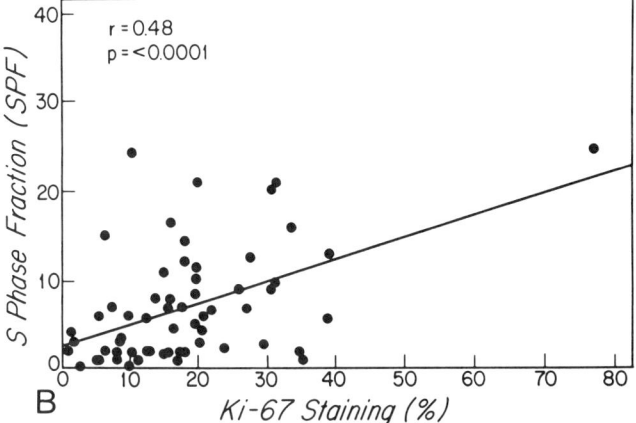

Figure 21.5. Proliferation in breast cancer. **A**, A cryostat section of a case of breast cancer stained with the monoclonal antibody Ki-67. Note the nuclear localization of the staining (dark nuclei) in the cluster of tumor cells, while the surrounding stroma is negative for the proliferation-associated marker. Measurement by IA of the percent tumor nuclear area stained by the antibody was compared to S-phase content measurement by FCM in 54 cases (**B**); there was a significant correlation between the two methods. Discrepancies probably resulted from the fact that IA measurements were performed specifically on tumor cells, whereas the flow cytometric measurements included variable amounts of contaminating stroma and inflammatory cells.

of the histogram may preclude the application of most curve-fitting models used to measure S phase. Multiparameter analysis, particularly the use of monoclonal antibodies to gate on specific cell subsets of interest, can somewhat remedy the problem of tumor cell identification (40). The use of antibodies against incorporated bromodeoxyuridine in combination with DNA-binding dyes (95) may allow more direct measurement of S-phase content, but is not practical for routine clinical use and does not account for contaminating benign cells. The difficulties of performing accurate S-phase determination are illustrated by multicenter studies that show that while the measurement of DNA ploidy is rather reproducible among laboratories, there is very poor reproducibility with regard to S-phase content (36). Such variability makes it unwise for any laboratory to rely on published reference values for reporting S-phase content. The DNA histograms generated by quantitative image includes too few S-phase events to be reliable, and few studies have attempted such measurements (66).

Another problem associated with cell kinetic measurements by FCM is that typically only S-phase (or $S + G_2/M$-phase) activity is assessed. However, a significant proportion of the proliferating cell population may reside in the G_1 phase of the cycle, the duration of which may be extremely variable. There are methods to estimate the G_1-phase fraction by FCM (96), but not by means easily adapted to routine clinical samples. Conceivably, two tumors might have identical S-phase fractions, yet differ greatly in the total fraction of cells in the non-resting state, and may exhibit different growth kinetics and response to cycle-dependent chemotherapeutic agents.

Another approach to measuring proliferative activity in solid tumors is to make use of antibodies against nuclear antigens newly expressed during various phases of the cell cycle. Several such antibodies are commercially available (see Table 21.5 for examples) that may be used for immunohistochemical staining of tissue sections. The antibody most commonly employed is Ki-67, which stains a proliferation-related nuclear antigen in human cells of all lineages (97). This antibody has the useful property of staining nuclei of cells in G_1, S, G_2, and M phases of the cell cycle, but not the nuclei of resting (G_0) phase cells (98). Direct visualization of Ki-67 staining on cryostat secretions of solid tumors provides several advantages over flow cytometric S-phase determination (see Fig. 21.5, top). First, this method avoids all of the problems associated with the preparation of single-cell suspensions. Second, one can make measurements of the tumor cell population specifically by using morphological and histological criteria, excluding irrelevant tissue elements. Third, correlations can be made between histology and tumor proliferation, taking such features as tumor cell heterogeneity and geographic variation in proliferation into account. The nuclear localization of staining is ideally suited for quantitative IA, as will be illustrated.

Several studies have shown that Ki-67 staining correlates with other measures of cell proliferation. Dual-color flow cytometric studies on normal lymphocytes and tumor cell lines have demonstrated the relationship between Ki-67 staining and the progression of cells through the cell cycle (99), and the antibody has been used in flow cytometric studies on human tumors and normal cells to measure proliferative activity (99–101). Some studies have shown an excellent correlation between thymidine-labeling index (TLI) and Ki-67 staining in tumors (102, 103), and the relative ease and reproducibility of Ki-67 staining would provide a definite advantage over TLI. Similarly, Ki–67 staining has been shown to correlate with other in situ markers of cell proliferation, including labeling by bromodeoxyuridine (104), PCNA/cyclin (78, 105) and argyrophilic nucleolar organizer regions (106). PCNA/cyclin is an accessory protein to delta DNA polymerase that appears in the cell nucleus during S phase

(107); staining with antibodies to this protein may prove to be another useful measure of cell cycle activity that may correspond more directly with S-phase activity (108). Studies in our laboratory have shown that although the fraction of cells staining for PCNA/cyclin is always less than with Ki-67 (as expected), there is a good correlation between the expression of these two nuclear markers of cell proliferation (78).

Several studies have demonstrated that the degree of staining with Ki-67 antibody correlates with tumor grade and other prognostic features for a variety of tumor types, including breast carcinoma (109–111), lymphoma (112, 113), meningioma (114), glial and astrocytic brain tumors (115), malignant melanoma (116), and sarcoma (117). In non-Hodgkin's lymphoma, increasing Ki-67 staining is associated with Working Classification grade (112, 113, 118), and some studies have shown that an increasing proliferative fraction measured by Ki-67 staining is associated with a worsened prognosis (119–121). For example, Grogan et al. (122) found that Ki-67 staining is an independent predictor of survival in large-cell lymphomas; a similar result was obtained by Chott et al. (120) in their study of T-cell lymphomas (medium- and large-cell types), in which Ki-67 staining was found to be an important prognostic factor in patients who presented initially with limited disease. In breast cancer, Ki-67 staining has been correlated with nuclear grade and lymph node status (110, 111), both important prognostic features. A preliminary report by Sahin et al. (123) indicates that Ki-67 staining may be an important prognostic feature in node-negative breast cancer. Clearly, more studies need to be performed in order to establish whether Ki-67 staining is an independent prognostic factor in human malignancies.

In most of the studies cited above, Ki-67 staining was measured by estimation or by tedious manual counting. Quantitation of staining for nuclear antigens by IA provides a means of improving the speed, accuracy, and reproducibility of such measurements. The approach that we have employed, using the CAS System, makes use of the fact that the imaging system can distinguish the brown immunoperoxidase reaction product from the green coloration of cell nuclei counterstained with methyl green. Using color filters, the system first creates a "map" of the nuclei in a microscopic field and then superimposes a map of the location of positive immunohistochemical staining. The percent of the total nuclear area stained positive is then rapidly calculated, and the results from sequential fields are accumulated. This approach can be applied to the measurement of any nuclear antigen, and is routinely used to measure hormone receptor expression in tissue sections of breast cancer (124). We have applied this method to measure Ki-67 staining in malignant lymphoma (113) and breast cancer, comparing the results to S-phase estimates provided by single-parameter FCM (Fig. 21.5) Considering the variability of benign stromal elements, discrepancies between IA and FCM measurements in solid tumors that affect only the FCM results are expected.

It is important to realize that the ultimate goal of performing such measurements is to provide prognostic information and that whichever method proves to be most accurate and clinically useful will gain usage. With regard to the measurement of tumor cell proliferation, I feel that immunohistochemical methods combined with IA will prove more accurate and reproducible than FCM because of the advantages provided by direct in situ tumor cell measurement. However, more studies aimed at correlating tumor proliferation measurements with clinical outcome are needed before IA can be more routinely applied.

CONCLUSIONS

New developments in IA will make this technology increasingly attractive and useful for the clinical laboratory. Commercial instruments capable of making reliable and sophisticated measurements are now available, and the price of such integrated imaging systems is highly competitive with FCM. Advances in the fields of image processing and computer programming will eventually overcome the chief limitations on the use of IA, chiefly the relatively slow speed of cell acquisition and the requirement for human supervision of cell selection. The use of high-speed, real-time image processing combined with improved algorithms for scene segmentation make it conceivable that measurements could be performed at rates approaching those attained by FCM. New methods of artificial intelligence might result in machines that can be trained to identify and classify cells more reliably than human observers. Neural networks, in particular, exhibit features that appear highly adaptable to image recognition; such networks have already been successfully trained to perform highly sophisticated tasks involving pattern recognition (125). These technological developments will soon result in imaging instruments capable of routine screening of cervical cytology specimens. With regard to some clinically important measurements, FCM and IA represent complementary technologies, and clinical laboratorians will need to learn how to best take advantage of these methods to optimize their use for specific types of specimens and measurements. Hopefully, this review comparing IA and FCM will provide some guidance in coordinating the use of these powerful diagnostic technologies.

REFERENCES

1. Sheldon K. Probing space by camera. Byte 1987;12:143–148.
2. Marchevsky AM, Gil J, Jeanty H. Computerized interactive morphometry in pathology: Current instrumentation and methods. Hum Pathol 1987;18:320–331.
3. Pesce CM. Defining and interpreting diseases through morphometry. Lab Invest 1987;56:568–575.
4. Arndt-Jovin DJ, Robert-Nicoud M, Kaufman SJ, Jovin TM. Fluorescence digital imaging microscopy in cell biology. Science 1985;230: 247–256.
5. Shapiro HM. Practical Flow Cytometry. 2nd ed. New York: Alan R. Liss, Inc.; 1988:211.
6. Inoue S. Video Microscopy. New York: Plenum Press; 1986.
7. Gonzalez RC, Wintz P. Digital Image Processing. Reading: Addison-Wesley Publishing Company; 1977.
8. Bacus JW, Grace LJ. Optical microscope system for standardized cell measurements and analyses. Appl Optics 1987;26:3280–3293.

9. Kasdan HL, Langford KC, Liberty J, Zachariash M, Deindoerfer FH. High performance pathology workstation using an automated multispectral microscope. Appl Optics 1987;26:3294–3300.
10. Brugal G. Required facilities for image analysis at the microscope in biological and medical applications: The "SAMBA" image processor. In: Burger JS, Ploem K, Goerttler K, eds. Clinical Cytometry and Histometry. San Diego: Academic Press; 1987;3–17.
11. Wied GL, Bartels PH, Bahr GF, Oldfield DG. Taxonomic intracellular system (TICAS) for cell identification. Acta Cytol 1968;12:180–204.
12. Pallavicini MG, Taylor IW, Vindelov LL. Preparation of cell/nuclei suspensions from solid tumors for flow cytometry. In: Melamed MR, Lindmo T, Mendelsohn ML, eds. Flow Cytometry and Sorting. 2nd ed. New York: Wiley-Liss; 1990;187–194.
13. Auer G, Askensten U, Ahrens O. Cytophotometry. Hum Pathol 1989; 20:518–527.
14. Vindelov LL, Christensen IJ, Nissen NI. A detergent-trypsin method for the preparation of nuclei for flow cytometric DNA analysis. Cytometry 1983;3:323–327.
15. Lage J, Driscoll SG, Yavner DL, Olivier A, Mark SD, Weinberg DS. Hydatidiform moles: Application of flow cytometry to diagnosis. Am J Clin Pathol 1988;89:596–600.
16. Hedley DW. Flow cytometry using paraffin-embedded tissue: Five years on. Cytometry 1989;10:229–241.
17. Hedley DW, Friedlander ML, Taylor IW, Rugg CA, Musgrove EA. Method for analysis of cellular DNA content of paraffin-embedded pathological material using flow cytometry. J Histochem Cytochem 1983;31:1333–1335.
18. Suit PF, Bauer TW. DNA quantitation by image cytometry of touch preparations from fresh and frozen tissue. Am J Clin Pathol 1990;94: 49–53.
19. Zarbo RJ, Visscher DW, Crissman JD. Two-color multiparametric method for flow cytometric DNA analysis of carcinomas using staining for cytokeratin and leukocyte-common antigen. Anal Quant Cytol Histol 1989;11:391–402.
20. Lovett EJ, Schnitzer B, Keren DF, Flint A, Hudson JL, McClatchey KD. Application of flow cytometry to diagnostic pathology. Lab Invest 1984;50:115–140.
21. Coon JS, Landay AL, Weinstein RS. Advances in flow cytometry for diagnostic pathology. Lab Invest 1987;57:453–479.
22. Loken MR, Brosnan JM, Bach BA, Ault KA. Establishing optimal lymphocyte gates for immunophenotyping by flow cytometry. Cytometry 1990; 11:453–459.
23. Ibrahim RE, Weinberg DS, Weidner N. Atypical cysts and carcinomas of the kidneys in the phacomatoses. A quantitative DNA study using static and flow cytometry. Cancer 1989;63:148–157.
24. Bibbo M, Dytch HE, Alenghat E, Bartels PH, Weid GL. DNA ploidy profiles as prognostic indicators in CIN lesions. Am J Clin Pathol 1989;92:261–265.
25. Crissman JD, Visscher DW, Kubus MS. Image cytophotometric DNA analysis of atypical hyperplasias and intraductal carcinomas of the breast. Arch Pathol Lab Med 1991;114:1249–1253.
26. Schneller J, Eppich E, Greenebaum E, et al. Flow cytometry and Feulgen cytophotometry in evaluation of effusions. Cancer 1987;59: 1307–1313.
27. Wheeless LL. Slit scanning. In: Melamed MR, Lindmo T, Mendelsohn ML, eds. Flow Cytometry and Sorting. 2nd ed. New York: Wiley-Liss; 1990;109–125.
28. Colvin RB, Preffer FI. New technologies in cell analysis by flow cytometry. Arch Pathol Lab Med 1987;111:628–632.
29. Landay A, Muirhead K, Auer R, Duque R, Green W, Harvath L, Hurley A, Jackson AL, Muirhead K, Nicholson J, Renner P. Clinical Applications of Flow Cytometry: Quality Assurance and Immunophenotyping of Peripheral Blood Lymphocytes (NCCLS Document H42-P, Volume 9, No. 13). Villanova: National Committee for Clinical Laboratory Standards; 1989.
30. Dean PN, Bagwell B, Lindmo T, Murphy RF, Salzman GC. Introduction to flow cytometry data file standard. Cytometry 1990;11:321–322.
31. Dean PN. Methods of data analysis in flow cytometry. In: Van Dilla MA, Dean PN, Laerum OD, Melamed MR, eds. Flow Cytometry: Instrumentation and Data Analysis. London: Academic Press; 1985; 195–221.
32. Sheck LE, Muirhead KA, Horan PK. Evaluation of the S phase distribution of flow cytometric DNA histograms by autoradiography and computer algorithms. Cytometry 1980;1:109–117.
33. Coon JS, Deitch AD, de Vere White RW, et al. Interinstitutional variability in DNA flow cytometric analysis of tumors. The National Cancer Institute's Flow Cytometry Network experience. Cancer 1988;61: 126–130.
34. Coon JS, Deitch AD, White RW, et al. Check samples for laboratory self-assessment in DNA flow cytometry. The National Cancer Institute's Flow Cytometry Network experience. Cancer 1989;63:1592–1599.
35. Hiddemann W, Schumann J, Andreeff M, et al. Convention on nomenclature for DNA cytometry. Cytometry 1984;5:445–446.
36. Wheeless LL. Inter- and intralaboratory variability in DNA flow cytometry of replicate samples [Abstract]. Cytometry 1990;suppl 4: 24.
37. Wied GL, Bartels PH, Bibbo M, Dytch HE. Image analysis in quantitative cytopathology and histopathology. Hum Pathol 1989;20:549–571.
38. Dean P, Mascio L, Ow D, Sudar D, Mullikin J. Proposed standard for image cytometry files. Cytometry 1990;11:561–569.
39. Gill JE, Jotz MM. Further observations on the chemistry of pararosanaline-Feulgen staining. Histochemistry 1976;46:147–160.
40. Herman CJ, McGraw TP, Marder RJ, Bauer KD. Recent progress in clinical quantitative cytology. Arch Pathol Lab Med 1987;111:505–512.
41. Mayall B. Deoxyribonucleic acid cytophotometry of stained human leukocytes. J Histochem Cytochem 1968;17:249-257.
42. Taylor SR, Titus-Ernstoff L, Stitely S. Central values and variation of measured nuclear DNA content in imprints of normal tissues determined by image analysis. Cytometry 1989;10:382–387.
43. Bacus S, Flowers JL, Press MF, Bacus JW, McCarty KS. The evaluation of estrogen receptor in primary breast carcinoma by computer-assisted image analysis. Am J Clin Pathol 1988;90:233–239.
44. Bacus SS, Ruby SG, Weinberg DS, Chin D, Ortiz R, Bacus JW. HER-2/neu oncogene expression and proliferation in breast cancers. Am J Pathol 1990;137:103–111.
45. Clevenger CV, Bauer KD, Epstein AL. A method for simultaneous nuclear immunofluorescence and DNA content quantitation using monoclonal antibodies and flow cytometry. Cytometry 1985;6:208–214.
46. Watson JD. Oncogenes, cancer and analytical cytology. Cytometry 1986;7:400–410.
47. Kelsten ML, Berger MS, Maguire HC, et al. Analysis of c-erbB-2 protein expression in conjunction with DNA content using multiparameter flow cytometry. Cytometry 1990;11:522–532.
48. Czerniak B, Herz F, Wersto RP, et al. Quantitation of oncogene products by computer-assisted image analysis and flow cytometry. J Histochem Cytochem 1990;38:463–466.
49. Gray JW, Dean PN, Fuscoe JC, et al. High-speed chromosome sorting. Science 1987;238:323–329.
50. Huang H. Biomedical image processing. CRC Crit Rev Bioeng 1981; 5:185–271.
51. Castleman K, Wall R. Automatic systems for chromosome identification. In: Caspersson T, Zech L, eds. Chromosome Identification. New York: Academic Press; 1973;77–84.
52. Herman CJ, Bunnag B. Goals of the cytology automation program of the National Cancer Institute. J Histochem Cytochem 1976;24:2–5.
53. Bacus JW. Quantitative red cell morphology. Monogr Clin Cytol 1984;9:1–27.

54. Weid GL, Bibbo M, Bartels PH. Computer analysis of microscopic images: Application in cytopathology. Pathol Annu 1981;16:367–409.
55. Bengtsson E. The meaning of cell features. Anal Quant Cytol Histol 1987;9:212–217.
56. Koss LG, Sherman AB. Image analysis of cells in the sediment of voided urine. Monogr Clin Cytol 1984;9:148–162.
57. Weid GL, Bartels PH, Bibbo M, Dytch HE. Image analysis in quantitative cytopathology and histopathology. Hum Pathol 1989;20:549–571.
58. Wasserman PD. Neural Computing. Theory and Practice. New York: Van Nostrand Reinhold; 1989.
59. Roberts L. Are neural networks like the human brain? Science 1989; 243:481–482.
60. Dawson AE, Austin R, Weinberg DS. Nuclear grading in breast carcinoma by image analysis: Classification by multivariate and neural network analysis. Am J Clin Pathol 1991;95(Suppl):S29–S37.
61. Check WA. Pap smear's future speeding up. CAP Today 1990;4:1.
62. Wied GL. Introduction to Quantitative Chemistry, Vol. 1. New York: Academic Press; 1966.
63. Bauer TW, Tubbs RR, Edinger MG, Suit PF, Gephardt GN, Levin HS. A prospective comparison of DNA quantitation by image and flow cytometry. Am J Clin Pathol 1990;93:322–326.
64. Claud RD, Weinstein RS, Howeedy A, Straus AK, Coon JS. Comparison of image analysis of imprints with flow cytometry for DNA analysis of solid tumors. Mod Pathol 1989;2:463–467.
65. Jones A, Grace J, Hall BE. Comparison of flow cytometry and retrospectively applied static cytometry on lymphoid tissue. Pathology 1990;22:5–9.
66. Felman P, French M, Souchier C, Magaud JP, Gentilhomme O, Bryon PA. Comparison between image and flow DNA cytometry in non-Hodgkin's lymphomas. Pathol Res Pract 1989;185:709–714.
67. Roos G, Stenling R, Ljungberg B. DNA content in renal cell carcinoma. A comparison between flow and static cytometric methods. Scand J Urol Nephrol 1986;20:295–300.
68. Valdes Martin del Campo M, Strang P, Stendahl U, Stenkvist B. DNA determination in endometrial carcinoma by flow and image cytometry. Acta Oncol 1989;28:607–609.
69. Oud PS, Pahlplatz MM, Beck JL, Wiersma-Van Tilburg A, Wagenaar SJ, Vooijs GP. Image and flow DNA cytometry of small cell carcinoma of the lung. Cancer 1989;64:1304–1309.
70. Cusick EL, MacIntosh CA, Krukowski ZH, Ewen SW, Matheson NA. Comparison of flow cytometry with static densitometry in papillary thyroid carcinoma. Br J Surg 1990;77:913–916.
71. Benson MC. Application of flow cytometry and automated image analysis to the study of prostate cancer. NCI Monogr 1988;25–29.
72. Koss L, Wersto RP, Simmons DA, Deitch D, Herz F, Freed SZ. Predictive value of DNA measurements in bladder washings. Comparison of flow cytometry, image cytophotometry, and cytology in patients with a past history of urothelial tumors. Cancer 1989;64:916–924.
73. Rodenburg CJ, Ploem-Zaaijer JJ, Cornelisse CJ, et al. Use of DNA image cytometry in addition to flow cytometry for the study of patients with advanced ovarian cancer. Cancer Res 1987;47:3938–3941.
74. Cornelisse CJ, de Koning HR, Moolenaar AJ, van de Velde CJ, Ploem JS. Image and flow cytometric analysis of DNA content in breast cancer. Relation to estrogen receptor content and lymph node involvement. Anal Quant Cytol 1984;6:9–18.
75. Wilbur DC, Zakowski MF, Kosciol CM, Sojda DF, Pastuszak WT. DNA ploidy in breast lesions. A comparative study using two commercial image analysis systems and flow cytometry. Anal Quant Cytol Histol 1990;12:28–34.
76. Cornelisse CJ, van Driel-Kulker AM. DNA image cytometry on machine-selected breast cancer cells and a comparison between flow cytometry and scanning cytophotometry. Cytometry 1985;6:471–477.
77. Fallenius AG, Askensten UG, Skoog LK, Auer GU. The reliability of microspectrophotometric and flow cytometric nuclear DNA measurements in adenocarcinomas of the breast. Cytometry 1987;8:260–266.
78. Dawson AE, Norton JA, Weinberg DS. Comparative assessment of proliferation and DNA content in breast carcinoma by image analysis and flow cytometry. Am J Pathol 1990;136:1115–1124.
79. Bacus SS, Ruby SG, Chin D, Ortiz R, Weinberg DS. HER-2/neu oncogene overexpression is associated with polyploid DNA content in breast cancers and breast cancer cell lines. Lab Invest 1991;(in press).
80. Watts KC, Husain OA, Campion MJ. Quantitative DNA analysis of low grade cervical intraepithelial neoplasia and human papillomavirus infection by static and flow cytometry. BMJ [Clin Res] 1987;295: 1090–1092.
81. Auer GU, Caspersson TO, Wallgren AS. DNA content and survival in mammary carcinoma. Anal Quant Cytol 1980;2:161–165.
82. Auer GU, Askensten U, Erhardt K, Fallenius A, Zetterberg A. Comparison between slide and flow cytophotometric DNA measurements in breast tumors. Anal Quant Cytol Histol 1987;9:138–146.
83. Arnerlov C, Emdin SO, Roos G, et al. Static and flow cytometric DNA analysis compared to histologic prognostic factors in a cohort of stage T2 breast cancer. Eur J Surg Oncol 1990;16:200–208.
84. McFadden PW, Clowry LJ, Daehnert K, Hause LL, Koethe SM. Image analysis confirmation of DNA aneuploidy in flow cytometric DNA distributions having a wide coefficient of variation of the G0/G1 peak. Am J Clin Pathol 1990;93:637–642.
85. Tanke HJ, van Driel-Kulker AM, Cornelisse CJ, Ploem JS. Combined flow cytometry and image cytometry of the same cytological specimen. J Microsc 1983;130:11–22.
86. Hall TL, Fu YS. Applications of quantitative microscopy in tumor pathology. Lab Invest 1985;53:5–21.
87. Stal O, Klintenberg C, Franzen G, et al. A comparison of static cytofluorometry and flow cytometry for the estimation of ploidy and DNA replication in human breast cancer. Breast Cancer Res Treat 1986;7:15–22.
88. Braylan RC, Diamond LW, Powell ML, Harty-Golder B. Percentage of cells in the S phase of the cell cycle in human lymphoma determined by flow cytometry. Correlation with labeling index and patient survival. Cytometry 1980;1:171–174.
89. Bauer KD, Merkel DE, Winter JN, et al. Prognostic implications of ploidy and proliferative activity in diffuse large cell lymphomas. Cancer Res 1986;46:3173–3178.
90. Clark GM, Dressler LG, Owens MA, Pounds G, Oldaker T, McGuire WL. Prediction of relapse or survival in patients with node-negative breast cancer by DNA flow cytometry. N Engl J Med 1989;320:627–633.
91. Silvestrini R, Daidone MG, Gasparini G. Cell kinetics as a prognostic marker in node-negative breast cancer. Cancer 1985;56:1982–1987.
92. Sigurdsson H, Baldetorp B, Borg A, et al. Indicators of prognosis in node-negative breast cancer. N Engl J Med 1990;322:1045–1053.
93. Visscher DW, Zarbo RJ, Greenawald KA, Crissman JD. Prognostic significance of morphological parameters and flow cytometric DNA analysis in carcinoma of the breast. Pathol Annu 1990;25(Part 1):171–210.
94. Baak JPA. Mitosis counting in tumors [Editorial]. Hum Pathol 1990; 21:683–685.
95. Gray JW, Mayall BH. Monoclonal Antibodies Against Bromodeoxyuridine. New York: Alan R. Liss; 1985.
96. Darzynkiewicz Z, Traganos F, Melamed MR. New cell cycle compartments identified by multiparameter flow cytometry. Cytometry 1980;1:98–108.
97. Gerdes J, Schwab U, Lemke H, Stein H. Production of a mouse monoclonal antibody reactive with a human nuclear antigen associated with cell proliferation. Int J Cancer 1983;31:13–20.
98. Gerdes J, Lemke H, Baisch H, Wacker H-H, Schwab U, Stein H. Cell cycle analysis of a cell proliferation-associated human nuclear antigen defined by the monoclonal antibody Ki-67. J Immunol 1984;133: 1710–1715.
99. Schwarting R, Gerdes J, Niehus J, Jaeschke L, Stein H. Determination of the growth fraction in cell suspensions by flow cytometry using the monoclonal antibody Ki-67. J Immunol Meth 1986;90:65–70.

100. Palutke M, Tabaczka PM, Kukuruga DL, Kantor NL. A method of measuring lymphocyte proliferation in mixed lymphocyte cultures using a nuclear proliferation antigen, Ki-67, and flow cytometry. Am J Clin Pathol 1989;91:417–421.
101. Drach J, Gattringer C, Glassl H, Schwarting R, Stein H, Huber H. Simultaneous flow cytometric analysis of surface markers and nuclear Ki-67 antigen in leukemia and lymphoma. Cytometry 1989;10:743–749.
102. Kamel OW, Franklin WA, Ringus JC, Meyer JS. Thymidine labeling index and Ki-67 growth fraction in lesions of the breast. Am J Pathol 1990;134:107–113.
103. Deshmukh P, Ramsey L, Garewal HS. KI-67 labeling index is a more reliable measure of solid tumor proliferative activity than tritiated thymidine labeling. Am J Clin Pathol 1990;94:192–195.
104. Sasaki K, Matsumura K, Tsuji T, Shinozaki F, Takahashi M. Relationship between labeling indices of Ki-67 and BrdUrd in human malignant tumors. Cancer 1988;62:989–993.
105. Garcia RL, Coltrera MD, Gown AM. Analysis of proliferative grade using anti-PCNA/cyclin monoclonal antibodies in fixed, embedded tissues. Comparison with flow cytometric analysis. Am J Pathol 1989; 134:733–739.
106. Dervan PA, Gilmartin LG, Loftus BM, Carney DN. Breast carcinoma kinetics. Argyrophilic nucleolar organizer region counts correlate with Ki67 scores. Am J Clin Pathol 1989;92:401–407.
107. Kurki P, Ogata K, Tan EM. Monoclonal antibodies to proliferating cell nuclear antigen (PCNA/cyclin) as probes for proliferating cells by immunofluorescence microscopy and flow cytometry. J Immunol Meth 1988;109:49–59.
108. Battersby S, Anderson TJ. Correlation of proliferative activity in breast tissue using PCNA/cyclin (Letter). Hum Pathol 1990;21:781.
109. McGurrin JF, Doria MI, Dawson PJ, Karrison T, Stein H, Franklin WA. Assessment of tumor cell kinetics by immunohistochemistry in carcinoma of breast. Cancer 1987;59:1744–1750.
110. Lelle RJ, Heidenreich W, Stauch G, Gerdes J. The correlation of growth fractions with histologic grading and lymph node status in human mammary carcinoma. Cancer 1987;59:83–88.
111. Bacus SS, Goldschmidt R, Chin D, Moran G, Weinberg DS, Bacus JW. Biologic grading of breast cancer using antibodies to proliferating cells and other markers. Am J Pathol 1989;135:783–792.
112. Gerdes J, Dallenbach F, Lennert K, Lemke H, Stein H. Growth fractions in malignant non-Hodgkin's lymphomas (NHL) as determined in situ with the monoclonal antibody Ki-67. Hematol Oncol 1984;2:365–371.
113. Schwartz BR, Pinkus G, Bacus S, Toder M, Weinberg DS. Cell proliferation in non-Hodgkin's lymphomas. Digital image analysis of Ki-67 staining. Am J Pathol 1989;134:327–336.
114. Roggendorf W, Schuster T, Peiffer J. Proliferative potential of meningiomas determined with the monoclonal antibody Ki-67. Acta Neuropathol (Berl) 1987;73:361–364.
115. Burger PC, Shibata T, Kleihues P. The use of monoclonal antibody Ki-67 in the identification of proliferating cells: Application to surgical neuropathology. Am J Surg Pathol 1986;10:611–617.
116. Kauderwitz P, Braun-Falco O, Ernst M, Landthaler M, Stolz W, Gerdes J. Tumor cell growth fractions in human malignant melanomas and the correlation to histopathologic tumor grading. Am J Pathol 1989;134:1063–1068.
117. Ueda T, Aozasa K, Tsujimoto K. Prognostic significance of Ki-67 reactivity in soft tissue sarcomas. Cancer 1989;63:1607–1611.
118. Weiss LM, Strickler JG, Medeiros LJ, Gerdes J, Stein H, Warnke RA. Proliferative rates of non-Hodgkin's lymphomas as assessed by Ki-67 staining. Hum Pathol 1987;18:1155–1159.
119. Medeiros LJ, Harris NL. Immunohistologic analysis of small lymphocytic infiltrates of the orbit and conjunctiva. Hum Pathol 1990;21:1126–1131.
120. Chott A, Augustin I, Wrba F, Hanak H, Ohlinger W, Radaszkiewicz T. Peripheral T cell lymphomas: A clinicopathologic study of 75 cases. Hum Pathol 1990;21:1117–1125.
121. Gerdes J, Stein H, Pileri S, et al. Prognostic relevance of tumour-cell growth fraction in malignant non-Hodgkin's lymphomas (letter). Lancet 1987;2:448–449.
122. Grogan TM, Lippman SM, Spier CM, et al. Independent prognostic significance of a nuclear proliferation antigen in diffuse large cell lymphomas as determined by the monoclonal antibody Ki-67. Blood 1988;71:1157–1160.
123. Sahin A, Ro J, El-Naggar A, Fitsche H, Blick M, Ayala A. Ki-67 immunostaining in node-negative stage I/II breast carcinoma (NNBC): A significant correlation with prognosis [Abstract]. Lab Invest 1990; 62:88A.
124. Bacus SS, Flowers JL, Press MF, Bacus JW, McCarty KS. The evaluation of estrogen receptor in primary breast carcinoma by computer-assisted image analysis. Am J Clin Pathol 1988;90:233–239.
125. Poggio T, Edelman S. A network that learns to recognize three-dimensional objects. Nature 1990;343:263–266.

SECTION D. APPLICATIONS IN CLINICAL HEMATOLOGY AND IMMUNOLOGY

22

Flow Cytometric Analysis of Red Blood Cells

BRUCE H. DAVIS

INTRODUCTION

The clinical application of flow cytometric (FCM) techniques to the study of red blood cell (RBC) or erythroid disorders has been relatively slow to implement in clinical laboratories. However, there are multiple reasons for this trend to change in the future. RBC analysis by FCM is readily performed and is facilitated by the ease of sample preparation. Issues of cell viability, purification, and clumping rarely present major problems in the study of red cells by flow cytometry. Most RBC assays with clinical application require literally one drop of blood, thus allowing for implementation across the entire range of age from prenatal or fetal blood samples to adult patients. FCM technology has several advantages over the standard clinical laboratory techniques for RBC analysis. Most notable advantages include: (1) improved precision and reproducibility over the subjective manual microscopic techniques currently employed in most clinical immunohematology and hematology laboratories; (2) FCM techniques can be less labor-intensive, translating a more cost-effective clinical assay; (3) FCM techniques can be more quantitative than serologic and manual microscopic techniques with sensitivities rivaling other immunologic techniques, such as ELIZA or RIA; (4) multiparameter FCM techniques allow for better analysis of specific RBC subpopulations or the simultaneous analysis of multiple antigens. Furthermore, the incidence of diseases associated with or due to erythroid abnormalities far excedes those clinical situations necessitating immunophenotypic or DNA content/cell cycle analysis. The diagnostic evaluation and treatment of anemia is a more common medical problem than leukemia or solid tumor neoplastic diseases. Because of the above noted advantages and the fact that many clinical laboratories have already invested in the necessary flow cytometric instrumentation, clinical RBC analysis by FCM will undoubtedly increase in the next decade.

The FCM techniques applied to RBC analysis include standard immunophenotyping, nucleic-acid-specific fluorescence probes to detect cytoplasmic RNA- or DNA-containing structures, labeled ligands for RBC surface receptors, quenching of RBC autofluoresence following enzymatic reactions, and lipophilic fluorescent dyes to probe RBC membrane structure and function (Table 22.1). The clinical applications of these techniques are varied and steadily increasing in number. For the purpose of this review, RBC analysis by FCM can be viewed as having clinical utility in the areas of erythropoietic activity assessment in anemia, immunohematology, or red call antigen evaluation, diagnosis of genetic diseases, and evaluation of acquired clinical conditions affecting RBC structure or function (Table 22.2).

CLINICAL APPLICATIONS IN IMMUNOHEMATOLOGY

Fetal-Maternal Hemorrhage

The escape of fetal cells into the maternal blood circulation, usually through a transplacental route, is now known through flow cytometric studies to occur in every pregnancy (1, 2). Fetal cells can be detected in the maternal circulation in the first trimester of pregnancy, but the volume of such fetal-maternal hemorrhages (FMH) increases as the gestational age increases (3). The volume or frequency of circulating fetal RBC at any given time is usually small, with 98% of women having less than 2.0 ml of whole fetal blood in their peripheral blood (3). The clinical significance of FMH is the immunization of the mother to incompatible fetal blood group antigens, usually of the Rh antigen system,

Table 22.1
Flow Cytometric Techniques Employed for the Study of Red Cell Pathophysiology

Technique	Utility	Reference
Immunofluorescence	Antigen Detection	22,35,51
	Fetal-maternal hemorrhage	5,6,17,23
	Red Cell survival	16,22,34
	Paternity Testing	7,12
	Immune Hemolytic Anemia	4,22,
	Hemoglobinopathy Detection	13
	Genetic Testing	7,22,24,25,30
	Mutagenesis Assay	11,15,28
	Red Cell Cytoskeletal Structure	9
	Loss of PI-linked Proteins	30,47–50
	Erythropoietic activity	55,78
Nucleic Acid Probes	Reticulocyte Analysis	55,70,76
	Reticulocyte Maturity Index	53–55,72,73 76,77,87
	Parasite Infection	18,56–59
	Monitor Leukocyte Filters	31
Fluorescence Quenching	G-6-PD Enzyme levels	14
Lipophilic Probes	Red Cell Survival	19
	Parasite Infection	60
Ligand Binding	Erythropoietin Receptors	21
	Transferrin Receptors	24,33
Biotin Labeling	Red Cell Volume	36–38

Table 22.2
Clinical Applications of Flow Cytometric Red Cell Analysis

Immunohematology and RBC Antigen Detection
  Fetal-Maternal Hemorrhage for Rh Immunization
  Micro-Coombs (DAT) Testing
  Chimera or Mosaicism Analysis
  Rapid, Sensitive Evaluation of Leukocyte Filter Efficacy
  Low Antigen Expression of D^u, Del
  Screen for HLA-negative products to reduce Alloimmunization in multiply-transfused patients (transplantation)
  Paternity Testing
  Red Cell Survial Assay (nonradioactive)
  Red Cell Volume Testing
  Forensic Science

Acquired Red Cell Abnormalities
  Autoimmune Hemolytic Anemia
  Alloimmunization following multiple transfusions
  Parasitic Infection (Malaria, Babesiosis)
  CD35 levels to monitor disease course in AIDS, SLE
  Somatic cell mutagenesis analysis of environmental exposures
  Paroxysmal Nocturnal Hemoglobinuria
  Polycythemia Vera

Genetic Disease Detection
  Abnormal Hemoglobin Detection and Carrier Status
  Glucose-6-Phosphate Dehydrogenase Deficiency
  Fetal cell isolation from maternal blood for subsequent DNA probe/ polymerase chain reaction analysis
  Tn Syndrome
  McLeod Syndrome
  Hereditary Spherocytosis, Ovalocytosis and Elliptocytosis

Erythropoietic Activity
  Reticulocyte enumeration
  Anemia assessment and classification
  Monitor bone marrow transplantation
  Therapeutic monitoring of recombinant erythropoietin
  Monitor chemotherapy and radiation therapy bone marrow toxicity

leading to increased risk of pregnancy complications or fetal death. Current standard medical practice in the U.S. is to administer a standard dose of 300 μg Rh immune globulin (RhIG) to all Rh-negative women following delivery of a Rh-positive infant. This prophylaxis with RhIG will provide protection from immunization of up to 30 ml of fetal blood, corresponding to the detection of ≥0.6% fetal RBC in a background of maternal cells (4). Although risk factors for FMH not protected by the standard RhIG dose have been identified, the majority of women with FMH of 30 ml or more have no associated risk factors (3).

There are a variety of laboratory techniques for the quantitation of FMH (3). Most hospital laboratories use a screening technique to identify positive samples, followed by a quantitative technique, such as the Kleihauer-Betke acid elution technique (5). Yet even in experienced hands, the microscopic acid elution assays are sensitive only to a frequency of 0.5% fetal cells and fraught with errors due to inter-observer subjectivity, a problem circumvented by flow cytometric techniques (6, 7).

Detection of fetal cells in maternal blood by flow cytometric techniques have all employed variations of immunofluorescence techniques with anti-RhD reagents. Elegant high-sensitivity studies with customized "rare-event," high-speed flow cytometry (2, 6) and a two-step procedure using Rh+ cell enrichment with cell sorting (1) have demonstrated a routine detection frequency of 1:100,000 or 0.01%. These methods were necessary to validate the true frequency and degree of FMH, but the technical rigors and equipment specifications of such high-sensitivity techniques make them impractical for routine clinical practice. In clinical practice such high sensitivity is not required, as the practical necessity is to have quantitative precision in the range of 0.5%. Standard indirect immunofluorescence flow cytometry using FITC-labeled anti-IgG reagents has been reported not only to be more accurate than acid elution and serologic techniques in quantitating FMH, but also to demonstrate good precision (CV <15%) at frequencies near 0.1% (7, 8). Since better signal-to-noise fluorescence measurements can be achieved with avidin-FITC and avidin-phycoerythrin immunofluorescence procedures using red cells (2, 9), sensitivities to the level of 0.1% should be achievable for most clinical laboratories equipped with commercial flow cytometric instruments. Given the superior performance of FCM techniques over conventional assays for detecting FMH, it would seem likely that those clinical laboratories performing large numbers of such studies might consider flow cytometric analysis a preferable and cost-effective option.

In Vivo Red Cell Survival and Red Cell Volume Determinations

There are several reasons to determine the lifetime or survival of red cells in the blood circulation. RBC survival studies are indicated to confirm the presence of hemolytic anemia, assist in the validation of the compatibility of transfused cells, determine the clinical significance of reactive alloantibodies defined by in vitro serologic testing, and evaluate new methods for the preservation of RBC transfusion products.

The most commonly utilized method for the clinical determination of RBC survival is to follow the rate of disappearance in ^{51}Cr-labeled red cells following the infusion of 0.5–2.0 ml radioactively-labeled cells (10). The radiolabeled techniques do have disadvantages. These are the real or perceived risks of patient exposure to radioactivity, correction factors required for the decay and elution of the radioactive tag, chemical modification of RBCs during the in vitro labeling step, and the discordance between survival time of small blood volumes compared to infusions of larger volumes (11). Several studies have suggested that flow cytometric techniques could serve as a desirable alternative to the use of radioactive methods (8, 12, 13, 14).

Most FCM red cell survival procedures operate from the principle of quantifying a RBC population with antigenic differences compared to the patient's cells, thereby using the unique RBC surface antigen as a label for the surviving RBC fraction. One approach has been to select donor cells possessing an antigen not found on the recipient's erythrocytes (8, 15). The corollary approach is to select donor cells lacking RBC blood group antigens relative to the recipient (12, 14). The later approach of negative antigen selection has the theoretic advantage of not sensitizing the recipient to foreign antigens. Either approach uses antisera to the relevant antigen in an indirect immunofluorescence technique to quantitate the percentage of donor cells surviving in the recipient's blood circulation over time. FCM techniques for RBC survival studies require the infusions of only 6–10 ml of donor red cells, can be shipped easily to distant reference labs for the FCM analysis, and give comparable results to the standard ^{51}Cr labeling techniques (12, 14). The use of blood group antigenic differences with the common antigens of the ABO, Rh, and MN systems for the marker in FCM red cell survival studies has utility in approximately 98% of the population (14). These flow cytometric methods for RBC survival studies have disadvantages; they have limited utility in the multiply-transfused patient and no role in the study of autologous RBC survival. However, FCM methods for RBC survival studies offer a nonradioactive alternative method that can be applied to the majority of individuals requiring this diagnostic test.

Red cell mass or volume studies, another diagnostic study traditionally requiring the infusion of radioactively labeled red cells, are quantitated by measuring the degree of dilution of labeled cells in subsequent venous blood samples. The test is used to confirm the diagnosis of Polycythemia Vera, where an increased RBC mass is demonstrated. The red cell volume (RCV) measurement has additionally been proposed as an indicator of tranfusion needs in neonates (16, 17). Flow cytometric methods for RCV determinations have been described and validated in man (17, 18) and animals (19). These procedures label autologous red cells with gentle biotinylation, reinfuse the labeled cells, and use streptavidin-FITC to determine the dilution by unlabeled cells in post-infusion blood samples drawn at 10 and 30 minutes. Good correlation between the flow cytometric calculation of RCV and standard ^{51}Cr RBC-labeling techniques have been reported by Cavill et al. (18). The only apparent shortcoming to the use of biotinylated RBC for RCV determinations is the abnormally rapid clearance of these RBCs from the circulation of individuals having recently consumed a meal with eggs or other dietary sources of avidin. The temporary abstinence from eggs seems a small price to pay to avoid radioactive exposure, especially considering that biotin has no known adverse effects (20). The flow cytometric technique for RCV determinations has the additional advantage of utility in the neonatal and pediatric population (17), where the diagnostic use of radioisotopes is either impractical or considered improper.

Utility in Hemolytic Anemia Diagnosis

One of the most common causes of hemolytic anemia in clinical medicine is immunologic lysis or clearance of red cells by opsonized antibodies and complement. Immunologic destruction of red cells can be caused by autoimmune mechanisms, induced by drugs, or occur following the transfusion of incompatible erythrocytes. Standard clinical laboratory serologic agglutination methods, the antiglobulin or Coombs tests, are sufficiently simple and cost-effective to preclude the routine implementation of nonautomated flow cytometric methodologies for DAT testing. Nonetheless, simple immunofluorescence flow cytometric analysis for RBC–associated immunoglobulin has been shown to be superior to serologic antiglobulin quantitation methods, with sensitivity less than 10,000 molecules per red cell (21). The issue of whether any additional clinically useful information is derived from semi-quantitative measurements of low levels of RBC-associated IgG remains unresolved (22). A significant proportion (five % to 10%) of patients with autoimmune hemolytic anemia test negative with serologic methodologies (23). In contrast, increased levels of RBC-associated IgG can be measured in normal blood donors and a subset of patients receiving the anti-hypertensive medication methyldopa, without any laboratory or clinical evidence of hemolytic anemia.

Studies by Van der Meulen et al. (24) demonstrated a clear relationship between the amount of RBC-associated IgG1 and the presence of hemolytic anemia using flow cytometric quantitation of FITC-labeled anti-human IgG binding to red cells. This observation gave birth to the expectation that the increased sensitivity, better quantitation, and lower false-negative rate afforded by flow cytometric techniques over serologic techniques would provide a laboratory parameter with clinical predictive value. Confusing this expectation is recent work by Garratty and Nance (25) demonstrating a correlation between the level of RBC-associated IgG (non-class-specific) and hemolytic anemia in neonates, but not in adult patients, using a similar flow cytometric technique. Quantitation of RBC-associated complement proteins and complement receptor levels by flow cytometry have similarly failed to correlate with clinical observations (24, 26). Undoubtedly, the current availability of flow cytometric technology to clinical immunohematology laboratories will continue to further define the clinical utility of RBC-associated IgG determinations.

A second category of hemolytic disease is secondary to intrinsic red cell defects, either congenital or acquired in origin. The low frequency of these clinical disorders makes current testing for such cellular defects the domain of referral specialty laboratories. Advances in knowledge regarding the protein deficiencies associated with many intrinsic red cell defects manifested as a hemolytic anemia, along with the availability of monclonal antibodies to these proteins, makes flow cytometry ideally suited for rapid diagnostic testing of these diseases.

Hereditary spherocytosis, elliptocytosis, pyropoikilocytosis, and ovalocytosis are genetically transmitted hemolytic diseases with deficiencies or structural defects in the red cell membrane skeleton proteins, such as spectrin, band 4.1, ankyrin, and glycophorin C (27, 28). Flow cytometric immunophenotypic analysis can be easily applied to define the deficient structural protein in patients and genetic carriers with such disorders (29). Paroxysmal nocturnal hemoglobinuria (PNH) is an acquired clonal proliferation of mutated erythroid precursors; it has clinical manifestations of a hemolytic anemia secondary to red cells being abnormally sensitive to the lytic action of complement due to the loss of numerous glycosyl-phosphatidylinositol-linked membrane proteins. Flow cytometeric quantitation of the red-cell-associated Pl-linked proteins, such as acetylcholinesterase, decay-accelerating factor, and homologous restriction factor (CD 59) can readily identify these abnormalities (30, 31). Alternatively, the loss of the neutrophil Pl-linked form of Fc receptor FcRIII (CD16) or other Pl-linked leukocyte surface antigens by FCM can be utilized in the diagnosis of PNH (32–34). Patients with hemolytic anemia due to homozygous or heterozygous deficiency in the enzyme glucose-6-phosphate dehydrogenase, can be identified by a fluorescence-quenching assay by flow cytometric analysis (35). Although all these diseases are relatively infrequent, the FCM techniques offer greater sensitivity and specificity compared to the laborious and time-consuming assays of red cell lysis using acid or osmotic stress that are currently performed by most clinical hematology laboratories.

Applications to Genetic and Paternity Testing

Flow cytometric immunophenotyping methods are ideally suited for the phenotypic characterization of red cell populations. Over 200 red cell antigens have been defined to date (27) with several blood groups, such as Kell (36), being the product of several genes. The increasing availability of monoclonal antibodies and specific antisera to the various RBC antigens allows for the study of antigen mosaicism and chimerism, which can be useful in genetic testing for carrier status of genetic diseases, paternity testing, and analysis of somatic cell mutation in erythroid cells after suspected mutagen exposure.

Flow cytometric techniques have been successfully applied to genetic carrier detection in McLeod syndrome, a rare X-linked disorder affecting the expression of Kell antigens with clinical neuromuscular and hematopoietic manifestations (14, 37). A promising flow cytometric screening technique for hemoglobin S and C has been described by Bigbee et al. (38), although it may be cost effective only in geographic regions that correlate with the racial distribution of hemoglobinopathies. Detection of the RBC chimerisms has utility in the monitoring of bone marrow engraftment, following allogeneic transplantation, by means of a fluorescent microsphere assay (39), which could be readily adapted for flow cytometry. Flow cytometric cell sorting isolation of fetal cells from maternal blood samples, in conjunction with DNA amplification by the polymerase chain reaction technique, has recently been proposed as a non-invasive alternative to amniocentesis or chorionic villus sampling in the prenatal testing for genetic diseases (40). Fetal cells are sorted out from maternal blood in the form of transferrin receptor bearing nucleated red cells, which provide sufficient DNA for molecular DNA probe analysis in pregnancies as early as 15 weeks gestation. The procedure requires only 20 ml of maternal peripheral blood. Perhaps the selection of fetal cells could be further improved with the use of a second fluorescence parameter, such as labeling for the erythropoietin receptor (41), but further studies are required to optimize this approach to routine clinical prenatal genetic testing.

Flow cytometric analysis for the presence and quantity of RBC-associated antigens is ideally suited to the study of zygosity, which has potential for widespread use in cases of disputed parentage and forensic studies. Flow cytometric detection of gene dosage for the red cell antigen groups Rh, Kell, Kidd, Glycophorin A, Glycophorin B (Ss), Gerbich, and Duffy has been demonstrated to have better sensitivity and precision than do conventional manual serologic techniques (9, 42, 43). Diagnostic application of red cell zygosity analysis for paternal and forensic testing is clearly in the developmental stages, but it should complement molecular biologic gene probe techniques in situations where insufficient DNA can be isolated from blood samples.

Another area of potential application of flow cytometric red cell analysis is in the area of somatic cell mutation studies, which have clinical applications in assessment of genetic damage due to environmental insults, such as ionizing radiation, radioactive exposure, or various chemical toxins. Langlois et al. have successfully utilized monoclonal antibodies in flow cytometric studies of Tn antigen and glycophorin A for the study of somatic cell mutation in the Tn syndrome (44, 45) and in victims of a Brazilian radiation accident (46). Development of flow cytometric alternatives to laborious microscopic erythrocyte micronucleus assays for pharmacologic toxicity tests (47) could serve to improve the utility of this mutagenic assay, as well as to reduce the need to sacrifice experimental animals through the use of peripheral blood samples.

Miscellaneous Applications

There are a number of flow cytometric techniques with potential applications to other areas in immunohematology. It was a FCM study that demonstrated the presence of HLA class I antigens in half of blood samples tested (48), placing in doubt the dogmatic belief that leukocyte-depleted blood products offer no risk for alloimmunization in multiply-transfused patients. Thus, this technique could find clinical utility in screening transfusion products for HLA antigens. Quality control of leukocyte-depletion techniques used in transfusion medicine can be performed better and faster with a simple, propidium-iodide-based assay (49). Flow cytometric red cell phenotyping (41) and reticulocyte analysis (50–52) could be combined to provide sensitive indication of erythroid engraftment following autologous or allogeneic bone marrow transplantation, or to determine the therapeutic efficacy of expensive recombinant erythropoietin treatment (53).

Several investigators have described flow cytometric techniques using nucleic acid fluorochromes (54, 55–58) or membrane potential sensitive dyes (59) to quantitate parasitic infections of red cells. Presently these assays appear to be more useful in controlled experimental situations. Unfortunately, these single-color assays lack specificity in separating various *Plasmodium* species and even in distinguishing malaria from other parasitic diseases, such as Babesiosis. However, such assays may prove of clinical utility, perhaps in enumerating the decline in RBC infestation following anti-parasitic therapy in endemic areas. Admittedly, the geographic areas endemic for malaria are not sites with ready accessibility to flow cytometric instrumentation.

Erythroid-specific antigens, such as glycophorin A and transferrin receptor, which occur early in red cell ontogeny (60) are currently utilized in diagnostic hematopathology. Flow cytometric immunotyping of leukemias for erythroid antigens is useful in confirming the diagnosis of acute erythroleukemia (AML M6 subtype) or erythroid blast crisis of chronic myelogenous leukemia. The combination of cell sorting, reticulocyte analysis, and hemoglobin chain analysis has been employed to study the neonatal rate of switching from hemoglobin F to hemoglobin A production (61). This technique may find clinical applicability in large pediatric institutions.

RETICULOCYTE ANALYSIS AND THE RETICULOCYTE MATURITY INDEX

Advantages over Conventional Methodology

Reticulocyte enumeration remains one of the more valuable and inexpensive clinical assays for the evaluation of patients with hematologic disorders resulting in anemia. Determination of the reticulocyte percentage and the absolute reticulocyte count provides an effective indication of the erythropoietic response in an anemic individual. Surprisingly, the methodology to quantify reticulocytes, which can be simply defined as those nonnucleated erythroid cells that are newly released from the hematopoietic tissue and contain higher levels of cellular ribonucleic acid (RNA), has remained a manual technique with poor precision in most clinical laboratories (62, 63). Although the manual microscopic methods of reticulocyte analysis provide reliable information in clinical states of reticulocytosis, confident assessment of ineffective erythropoiesis in the subnormal range is difficult. However, with the recent availability of stable commercial fluorescent reagents for reticulocyte enumeration and discrimination of reticulocyte populations based upon RNA content, and with the availability of flow cytometric instrumentation within clinical laboratories, FCM reticulocyte analysis is rapidly becoming the standard and preferred method for this hematologic laboratory assay.

The advantages of FCM reticulocyte analysis over manual microscopic counting techniques are multiple. FCM techniques provide greater precision and reproducibility compared to the relatively subjective manual technique, which has repeatedly been shown to have inherent inaccuracies (63, 64). Replicate reticulocyte enumeration with flow cytometry is reported consistently with a coefficient of variations of less than 10% (50, 65–76). Additionally, FCM reticulocyte analysis affords the ability to provide a new parameter of erythropoietic activity through the generation of a reticulocyte maturity index (RMI). The more widespread use of flow cytometric technology to reticulocyte enumeration and the standardization of the RMI or RNA index parameter (56–58, 77–79) will clearly make this FCM application in clinical medicine more frequently performed than immunophenotyping and DNA analysis studies combined. The improved precision and ability of FCM reticulocyte analysis to provide clinically relevant information of both increased and inappropriately decreased erythropoeitic activity should result in a renaissance of this hematologic test for anemia assessment.

Methods of Reticulocyte Analysis

Various fluorescent dyes have been reported to be suitable for FCM reticulocyte enumeration (Table 22.3). All methods operate on the principle of identifying the fraction of erythrocytes with higher levels of RNA (Fig. 22.1), although there are relative differences in the quantum efficiency of the fluorescent dyes when bound to RNA (52). Aside from the improved precision and sensitivity these methods offer over manual techniques, the ability for batch analysis provides an avenue for cost savings in the clinical laboratory (56, 58, 75). Immunophenotypic analysis of reticulocytes for the transferrin receptor has also been reported to provide reticulocyte quantification in whole blood samples (80). Enumeration of reticulocytes through the transferrin receptor has the potential shortcomings of identifying only the 20–50% most immature reticulocytes, thereby underestimating the true

Table 22.3
Fluorescent Dyes Utilized for Flow Cytometric Reticulocyte Analysis

Reticulocyte Reagent	Stable RMI	Specificity	Excitation	Emission
Auramine O	Yes[a]	RNA/DNA	435	550
Ethidium Bromide	Yes[b]	RNA/DNA	514	602
Propidium Iodide	Yes[b]	RNA/DNA	535	615
Pyronin Y	Yes[b]	RNA	515	580
Quinolythiacyanine Iodide	Yes	RNA/DNA	509	529
Thiazole Orange	Yes	RNA/DNA	509	533
Acridine Orange	No	RNA/DNA	492	550
$DiOC_1(3)$	No	RNA Membrane potential	482	510
Thioflavin T	No	RNA/DNA	422	487

[a]As utilized in the TOA Sysmex R-1000 automated reticulocyte counter.
[b]May require permeabilization step to allow entry of dye into RBC.

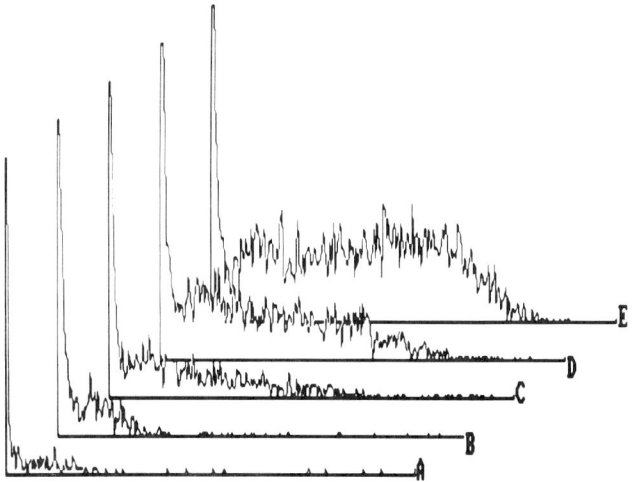

Figure 22.1. Flow cytometric reticulocyte analysis. Fluorescence histograms of thiazole-orange-stained blood samples indicate the degree of heterogeneity of the reticulocyte fluorescence distribution in patient samples. The unstained autofluorescence of an autologous blood sample is used to set the threshold for defining the reticulocytes of greater cytoplasmic RNA content (**A**). Some patients show low erythropoietic activity as indicated by a low frequency of the reticulocytes with a low or normal reticulocyte maturity index (**B, C**). Increased erythropoietic response in anemia is manifested as a reticulocytosis with a higher proportion of highly fluorescent cells or increased RMI (**D, E**).

population size, and of increased cost, due to the expense of monoclonal antibodies. Transferrin-positive reticulocyte enumeration may, however, prove useful as an alternative method to derive a reticulocyte maturity index (53) relative to the calculation of the mean fluorescence intensity or the fraction of highly fluorescent reticulocyte population, and may serve as a sensitive, early indicator of erythropoietic activity.

The optimal reagent for clinical FCM reticulocyte enumeration depends on the experience and type of instrumentation of each clinical laboratory. Thioflavin T has the disadvantages of requiring great attention to the staining times and being unable to generate a reproducible RMI parameter, but its emission and excitation spectral properties are well suited for laboratories equipped with mercury arc based flow cytometers (68, 71). Tanke et al. (77–79) have thoroughly described the feasibility of pyronin Y for FCM reticulocyte enumeration and derived a clinically useful RMI. However, pyronin Y, as well as ethidium bromide and propidium iodide, are impermeable to the cell membrane of "viable" cells and require a fixation or permeabilization step in the staining procedure. Jacobberger et al. (66) have successfully employed the cyanine dye $DiOC_{(1)}3$ for FCM reticulocyte analysis, but this dye has potential shortcomings because of its low quantum yield and fluorescence sensitivity to changes in membrane potential. Acridine orange (52, 70, 80), auramine O (73, 74), and thiazole orange (50–52, 69, 75, 76) appear to have several advantages over other reagents for FCM reticulocyte analysis in laboratories with laser based FCM instruments. Both acridine orange and thiazole orange are vital stains and are readily available from several commercial sources. Specific software programs to simplify the reticulocyte histogram analysis are also available, although there is certainly a need for better software analysis routines to generate absolute reticulocyte counts (Retic % × RBC count) and RMI parameters. Auramine O is the fluorescent reticulocyte reagent used in the fully automated and dedicated Sysmex R-1000 reticulocyte counter, which may be a cost-effective option for laboratories performing a large daily volume of reticulocyte studies. All the fluorescent dyes used for FCM reticulocyte counting share the practical disadvantage of staining the sample lines in the instrument, thus requiring a period of thorough detergent and/or bleach treatment prior to the next flow cytometric application. These various dyes for reticulocyte analysis (Table 22.3) easily allow for the development of two-color assays due to a variety of available spectral properties.

RETICULOCYTE MATURITY INDEX

Reticulocytes undergo a gradual maturation process in the peripheral blood marked by the loss of cellular RNA. The immature reticulocytes, newly released from the bone marrow compartment, are referred to as stress or shift reticulocytes, corresponding to the Heilmeyer stages I and II (81–83). Flow cytometry, by virtue of its ability to derive fluorescence intensity measurements, offers the ability not only to enumerate reticulocytes, but to provide a maturity index based upon the relative fluorescence intensity or RNA content of the reticulocyte population. Since the more immature, "shift" reticulocyte has a higher RNA content, an increase in this population relative to the more mature reticulocytes, corresponding to the Heilmeyer stages III and IV, causes a higher mean fluorescence intensity of the total reticulocyte population and a larger fraction of the highly fluorescent reticulocytes (Fig. 22.1).

Our experience in deriving a reticulocyte maturity index (RMI) using a thiazole orange method of reticulocyte analysis, based upon mean fluorescence intensity measurements of the reticulocyte population, has demonstrated clinical utility in monitoring erythroid engraftment following bone marrow transplantation (50–52, 84). Tanke et al. have similarly proposed a potential utility in RMI measurements that they have termed simply RNA index, and have demonstrated a utility in following the erythroid recovery from radiation and chemotherapy using pyronin Y (78, 79). Unfortunately, the RMI or RNA index derived from pyronin Y requires intra-assay comparison of samples to some standard due to day-to-day differences in the fluorescence intensity of reticulocytes (77). This limitation is an awkward requirement for a busy clinical hematology laboratory. Thiazole orange does not have such a limitation when quality control procedures are employed to insure instrument calibration (50–52).

The automated flow cytometric Sysmex R-1000 (TOA Medical Electronics Co., Kobe, Japan) instrument, which is

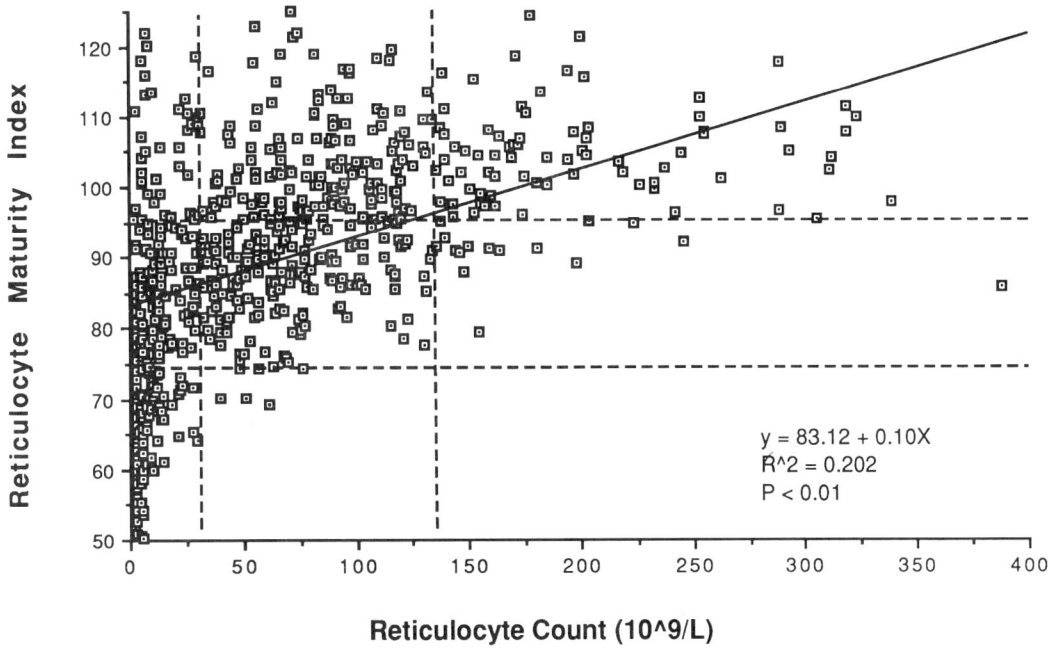

Figure 22.2. Subclassification of anemia by flow cytometric reticulocyte analysis. A relationship between absolute reticulocyte count to a reticulocyte maturity index, based on arbitrary fluorescence units using a thiazole orange technique, is demonstrated for 750 patients. The interrupted lines indicate the reference or normal ranges for reticulocyte counts and RMI values. A subset of truly hypoproductive anemia can be defined by low reticulocyte counts and a low RMI. Patients manifesting an early erythropoietic response to anemia would fall into the region with a normal reticulocyte count, but an elevated RMI.

solely dedicated for reticulocyte analysis, provides a RMI by reporting the percentage of reticulocytes arbitrarily determined as manifesting low (LFR), moderate (MFR), or high (HFR) fluorescence. Kojima et al. (73) reported an increase in the proportion of LFR reticulocytes in patients with suppressed bone marrow function secondary to chemotherapy and an increase in the HFR fraction of samples from patients with hemolytic anemia. The RMI information that can be derived from the R-1000 appears to have two major limitations. Firstly, the precision of the analysis for the MFR and HFR regions is disappointing in samples with low or normal numbers of reticulocytes (74), perhaps due to the analysis of only 30,000 red cells per sample. Secondly, the algorithm for the calculation of the HFR-percent allows only for measuring increased values; the reported reference range of 0.1–3.0% provides no dynamic range for the conditions that have ineffective erythropoiesis and low output of newly synthesized reticulocytes. It is likely that modification in the data analysis software algorithm in the R-1000 could rectify both of these problems.

The clinical utility of RMI determinations, independent of mere reticulocyte enumeration, has been shown to be the earliest indicator of bone marrow engraftment (50–52, 84) and bone marrow erythropoietic responses to cancer therapies (79). Our experience to date has also indicated that the RMI can be of use in the diagnosis of true aplastic anemia and transient drug-induced erythroid hypoplasia. The RMI can identify three patterns of bone marrow engraftment (early, delayed, and failed) following autologous transplantation, while the standard reticulocyte percentage and absolute counts show no significant changes or fluctuations, except those attributable to red cell transfusions (52). It is likely that the RMI parameter will find even broader clinical utility in the physiologic classification of anemic patients and the therapeutic monitoring of pharmacologic treatments of anemia.

When anemic patients are studied for the relationship between the degree of reticulocytosis and the RMI (Fig. 22.2), one can make several observations. First, in patients with increased reticulocytes there is a significant correlation with the RMI values ($p < 0.01$). Specifically, as would be anticipated, a high reticulocyte response in the peripheral blood is associated with a high RMI value, due to a high fraction of newly released RBCs from the bone marrow compartment. Secondly, patients with normal or decreased reticulocytes can be further segregated by the RMI parameter. Particularly in individuals with an abnormally low reticulocyte population, one can further subclassify patients into decreased, normal, or increased RMI values. The clinical significance of this observation is that truly aplastic or hypoproductive anemias can now be segregated from the initial stages of the erythroid response that might be seen in a hemolytic anemia (85), when the numbers of reticulocytes are not elevated but the RMI is elevated. The relationship between the RMI and the true level of erythropoietic response is further demonstrated by the correlation between the RMI and serum erythropoietin levels (Fig. 22.3). The RMI, in contrast to reticulocyte enumeration, RBC count, hematocrit, and hemo-

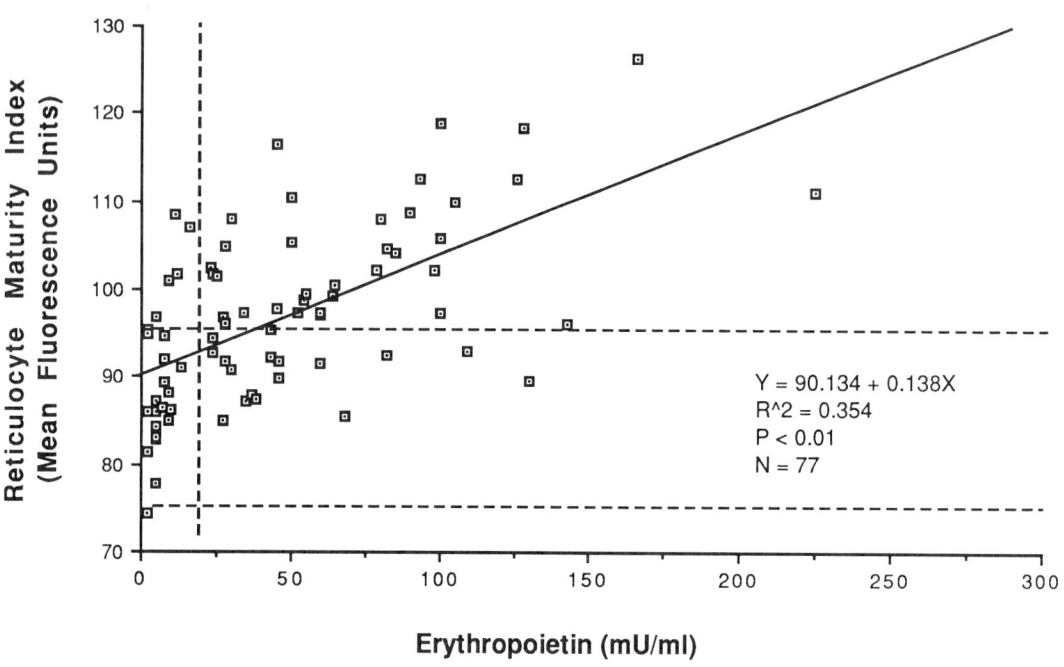

Figure 22.3. Correlation between serum erythropoietic levels and RMI values in anemic patient samples. Randomly selected patient samples show a good correlation between the two indicators of a physiologic response to anemia, not seen between the erythropoietin levels and other standard clinical laboratory RBC or reticulocyte parameters.

Table 22.4
Clinical Utility of the Reticulocyte Maturity Index

Detection of Bone Marrow Engraftment following transplantation
Confirmation of Aplastic Anemia
Improved subclassification of Erythropoeitic Activity in Anemia
    Thalassemia vs. Iron Deficiency
    Anemia of Chronic Disease
    Hypoproductive vs. Early Erythroid Response
Monitor Myelosuppressive Effect of Cancer Chemotherapy and Radiation
    Therapy
Therapeutic Monitoring of Erythropoietin Treatments
Definition of Transfusion Needs in Neonates

globin level, is the only erythroid parameter that correlates with serum erythropoietin levels (86), the growth factor for erythroid production. Clearly, RMI determinations in conjunction with other red-cell parameters (mean cell volume, RDW, mean cell hemoglobin, etc.) could provide the basis for better laboratory diagnosis and classification of anemia.

The RMI has potential diagnostic and therapeutic monitoring applications in numerous clinical conditions, such as detection of drug-related erythroid bone marrow suppression, estimating transfusion requirements for neonates requiring intensive care, and monitoring the therapeutic response either to the suspected deficient substance in nutritional anemias or to recombinant erythropoietin therapy in AIDS or chronic renal failure (Table 22.4). Over 20% of patients with chronic renal failure who are given recombinant erythropoietin either need dose adjustments or fail to elicit an erythropoeitic response to the factor (84). It is likely that reticulocyte analysis utilizing an RMI parameter or two-color analysis for transferrin receptor positivity (53) could be an effective therapeutic monitoring strategy for the clinical use of expensive pharmacologic hematins.

Currently, the RMI parameter, derived either by using arbitrary mean fluorescence intensity measurements on commercial flow cytometric instruments (50–52, 77–79, 86) or frequency value from three arbitrarily defined fluorescent regions on a dedicated reticulocyte counter (70, 71), lacks methods for interinstrument or interinstitution standardization. The clinical transferability of RMI measurements and reference ranges between laboratories is uncertain, given the lack of both calibration materials and an accepted definition of RMI units.

The desire to perform clinical flow cytometric reticulocyte analysis on multiple instruments prompted our laboratory to study the question of the optimal calculated units for the RMI. Clearly the use of arbitrary mean fluorescence units between instruments that may have three or four log amplifiers, use log or linear fluorescence scales, use 256 or 1024 unit scaling of the fluorescent data, and have no available biologic or bead-like standardization reagent makes interinstrument calibration difficult. After examining a number of possible algorithms for RMI calculations, we have instituted a method of reticulocyte enumeration and RMI calculation (87) that gives good correlation between instruments with three- and four-log decade amplifiers (Fig. 22.4). The RMI parameter is now expressed as the fraction of highly fluorescent reticulocytes, defined by those events collected through RBC light-scatter gates having a fluorescence intensity greater than the previously defined 95% interval on the mean fluorescence intensity of reticulocytes of a reference hematologically normal population, relative to the total number of reticulocytes in 50,000 analyzed red cells. The thiazole orange staining technique (50) utilizes an unstained autologous sample to define the cursor placement for definition of reticulocytes from mature erythrocytes. The advantage of expressing the RMI units as the HFR fraction is that this method is easily transfered to all commercial flow cytometers, allows for the use of a variety of reticulocyte dyes, and could be integrated into user-friendly computerized software to allow for the reduction of clerical error in the laboratory. Equally important is the definition of a dynamic normal or reference range with this algorithm for RMI calculation. The HFR fraction reference range is 0.25–0.52; this allows for the identification of patients with subnormal RMI values (Fig. 22.5), in contrast to the Sysmex R-1000 reticulocyte counter in its current configuration. The HFR fractions RMI unit system will need additional validation by other laboratories and the endorsement of laboratory standards organizations, such as the National Committee for Clinical Laboratory Standards, the College of American Pathologists, and the International Committee for Standardization in Haematology. However, the initial experience with this method in anemic patients shows good correlation with the previous method of using arbitrary fluorescence units (Fig. 22.4) and the ability to evaluate and subclassify the erythropoeitic response is retained (Fig. 22.5).

Reticulocyte Data Analysis

The ability of flow cytometric instrumentation to quantitate fluorescence intensity with speed and precision has provided the opportunity both to improve an old manual assay and to define a new laboratory parameter of erythropoiesis, which should result in FCM reticulocyte analysis becoming the preferred method for clinical laboratories. Prior to the final achievement of this goal, standardization of staining methods and FCM instrument operation is essential for accurate, reproducible clinical FCM reticulocyte analysis. Strict temporal control of staining conditions is critical for thioflavin T methods (67, 68, 71). The acridine orange, thiazole orange, and the auramine-O-based Sysmex R-1000 methods of FCM reticulocyte enumeration are less sensitive to the staining conditions (50, 52, 69, 70), but only thiazole orange and the R-1000 give stable RMI determinations (50–52, 73, 74). The current methods of reticulocyte analysis standardization are not fully optimized. Stored rabbit blood (71), preserved human blood products (88), and refrigerated clinical specimens (52) have all been proposed as methods of intralaboratory standardization. A commercial reticulocyte control has recently become available from Streck Laboratories (Omaha, NB) with sufficient stability to serve as a standardizing reagent between laboratories, yet its ability to standardize RMI units is unproven. U.S. organizations concerned with intralaboratory proficiency testing, such as the College of American Pathologists and FAST Systems, Inc. have, at this writing, failed to integrate the mailing of refrigerated blood samples for flow cytometric reticulocyte analysis into their surveys. Once consensus is achieved on standardization of both staining and units of expression for RMI measurements, flow cytometric reticulocyte analysis should become recognized as the reference method for clinical hematology laboratory testing.

Interpretation of flow cytometric histograms for clinical reticulocyte analysis should not be made in a vacuum. Recognition of the clinical conditions that may interfere with results is as important as are standardized, precise flow cytometric analysis methods. It has been observed or predicted that a number of infrequent, but predictable, disease states can interfere with an automated, nonmicroscopic method of reticulocyte analysis (Table 22.5). The importance of parallel microscopic assessment of red cell morphology in such cases can not be stressed enough. Additionally, knowledge of data analysis procedures used to eliminate leukocyte and platelet contamination from the red cell light-scatter analysis region should be understood. For instance, since nucleated cells stained with any of the reticulocyte dyes with specificity for RNA and DNA have a much higher fluorescence intensity than reticulocytes, the use of an upper limit fluorescence intensity gate can be very useful in eliminating leukocyte interference in samples from patients with chronic lymphocytic anemia or high numbers of circulating nucleated red cells. Careful review of histograms by experienced technologists, especially the two-parameter displays of scatter signals vs. fluorescence, can generally detect the infrequent clinical conditions that give erroneous reticulocyte enumeration by flow cytometry.

FUTURE DIRECTIONS

Like virtually all clinical applications of flow cytometry, the future direction of diagnostic red cell testing is one of increasing use of multiparameter analysis. Also anticipated is the development of products that automate and simplify specimen preparation and data analysis. The sheer volume of reticulocyte testing in clinical medicine on an international scale, which involves a significant proportion of biohazardous specimens, will demand the commercial development of automated, autosampling systems for flow-cytometric-based reticulocyte analysis. The instrument development may drift toward the integration of multiparameter flow cytometric technology in laboratory hematology blood cell counters capable of not only reticulocyte analysis, but also CD4 enumerations, leukocyte enzyme measurements, and DNA content/cell cycle analysis. Equally possible is the development of front-end automated sample prep workstations for clinical flow cytometric instrumentation.

Red cell analysis by flow cytometry should continue to find wider usage in the clinical laboratory environment. The methodologies offer more precise, cost-effective, nonradioactive alternatives to many areas of specialized testing in immunohematology and laboratory hematology unrivaled by other techniques. It is likely that flow cytometric instrumentation and molecular genetics testing will work in a symbi-

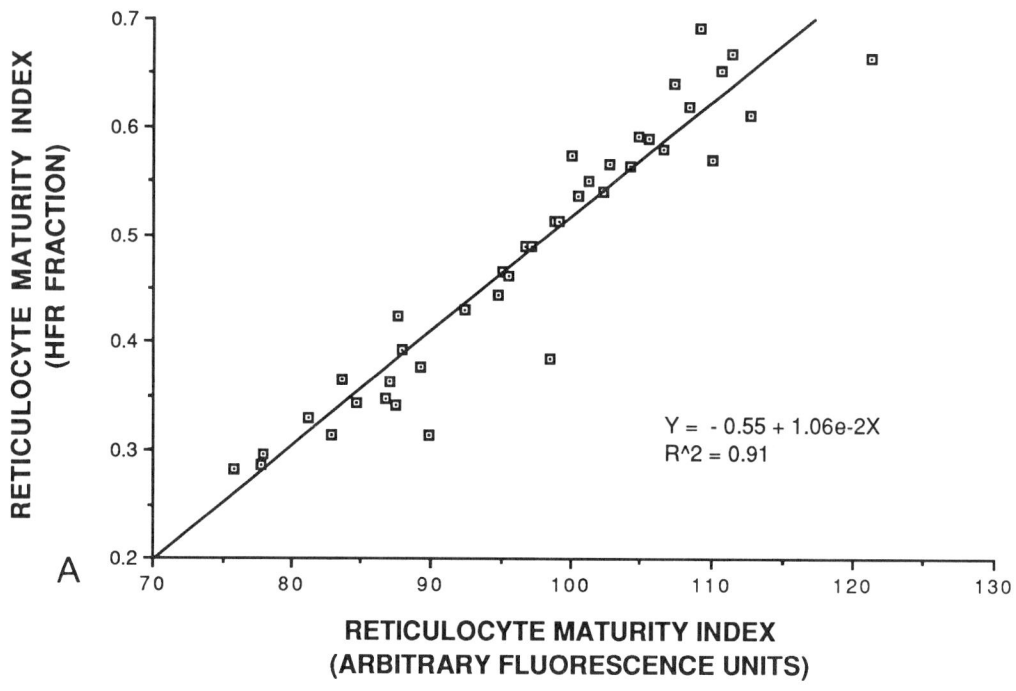

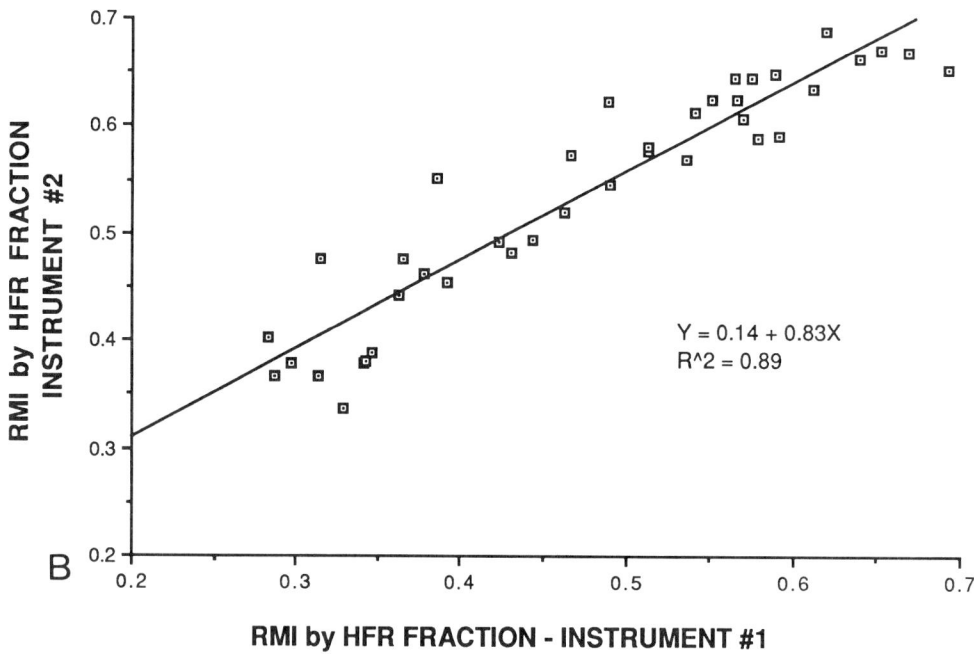

Figure 22.4. Methods of reticulocyte maturity index expression. The highly fluorescent reticulocyte (HFR) fraction, defined as the population of reticulocytes with greater fluorescence intensity than the instrument specific reference range established on a hematologically normal population, correlates well with the previously used method of determining the mean fluorescence intensity of the entire reticulocyte population (**A**). The HFR fraction method of RMI expression gives an equally good correlation between two different flow cytometers (EPICS 541 and Profile II models, Coulter, Inc., Hialeah, FL) on 40 samples analyzed in parallel (**B**).

otic fashion, either through the coordination of cell sorting and PCR (40) or the use of specific nucleic acid sequences, such as fluorescent probes in flow cytometric cell analysis. These developments will need to be paralleled by the availability of better control reagents and methods of analytical standardization beyond mere units of percent-positive or arbitrary fluorescence. Furthermore, the manpower needs for skilled clinical flow cytometric technologists, laboratory scientists, and pathologists knowledgeable in the clinicopathologic correlation of flow cytometric data must be addressed by academic medical centers, professional societies, and the biomedical industry. Flow cytometric erythrocyte

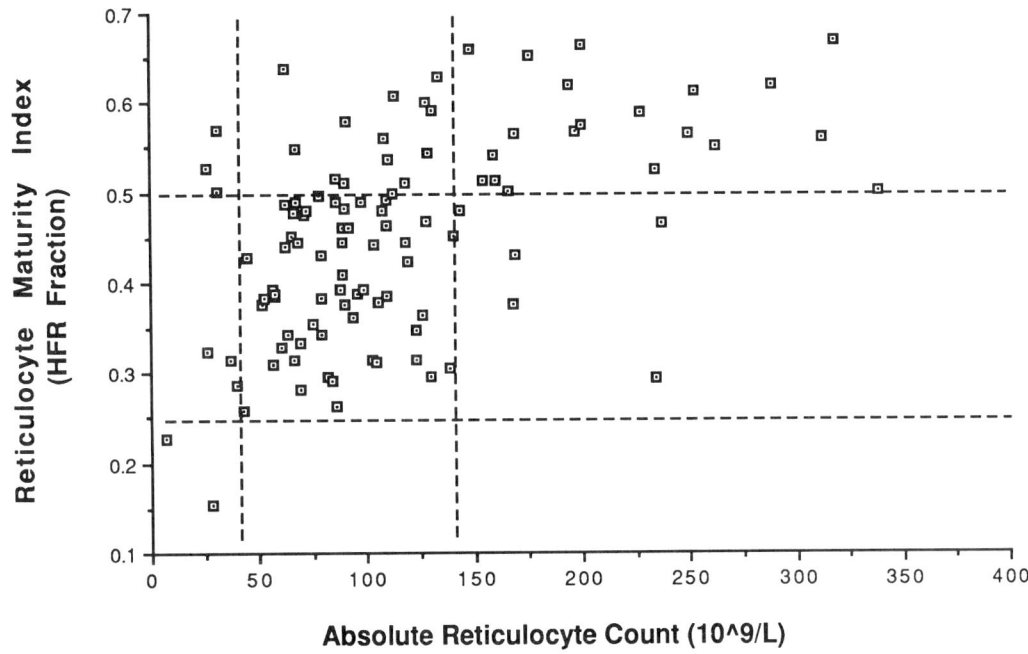

Figure 22.5. Anemia classification by *HFR* fraction *RMI* units and reticulocyte counts. 100 anemic patient samples show a heterogeneity in the *RMI* units expressed as the *HFR* fraction, similar to that seen in Figure 22.3 using a *RMI* expressed in arbitrary fluorescence units, thus retaining the ability to subclassify hypoproliferative and early erythropoietic responses. The interrupted lines define the normal reference range for the reticulocyte count and *RMI*.

Table 22.5
Clinical Conditions with Potential Interference with Flow Cytometric Reticulocyte Analysis

Leukocytosis	Parasitic Infections
Leukemia	Malaria
Leukemoid reaction	Babesiosis
Drugs (autofluorescence)	Thrombocytosis
Nucleated RBCs	Cold Agglutinin Syndromes
Postsplenectomy State	Porphyria (autofluorescence)
Large Platelet Syndromes	Myelodysplastic Syndromes
Bernard-Soulier	
Alport's	Red Cell Inclusions
May-Hegglin Anomaly	Heinz Bodies
Gray Platelet	Howell-Jolly Bodies
"Mediterranean" Macrothrombocytopenia	Pappenheimer Bodies

analysis is undoubtedly poised to be a promising frontier for clinical laboratory testing in this decade.

ACKNOWLEDGMENTS

The creativity and technical assistance of Nancy C. Bigelow, Marc H. Langweiler, Kathy Martin, and Sharon Pickens, the support of the Hitchcock Foundation (Hanover, NH) and of Coulter Diagnostics (Hialeah, FL), and the secretarial assistance of Lisa Dupuis are gratefully acknowledged.

REFERENCES

1. Medearis AL, Hensleigh PA, Parks DR, Herzenberg LA. Detection of fetal erythrocytes in maternal blood postpartum with the fluorescence-activated cell sorter. Am J Obstet Gynecol 1984;148:290–295.
2. Cupp JE, Leary JF, Cernichiari E, Wood JCS, Doherty RA. Rare-Event Analysis Methods for Detection of Fetal Red Blood Cells in maternal Blood. Cytometry 1984;5:138–144.
3. Sebring ES, Polesky HF. Fetomaternal Hemorrhage: Incidence, Risk Factors, Time of Occurence, and Clinical Effects. Transfusion 1990; 30:344–357.
4. Pollack W, Ascarri WO, Kochesky RJ. Studies of Rh prophylaxis relationship between dose of anti-Rh globulin and size of antigen stimulus. Transfusion 1971;11:333–339.
5. Kleihauer E, Braun H, Betke K. Demonstration von fetalem Hemaoglobin in den Erythrocyten eines Blutausstrichs. Klin Wochenschr 1957;35:637–641.
6. Corsetti JP, Cox C, Leary JF, Cox MT, Blumberg N, Doherty RA. Comparison of quantitative acid elution technique and flow cytometry for detecting fetal maternal hemorrhage. Ann Clin Lab Sci 1987;17:197–206.
7. Nance SJ, Nelson JM, Arndt PA, Lam HC, Garratty G. Quantitation of Fetal-Maternal Hemorrhage by Flow Cytometry. Am J Clin Pathol 1989;91:288–292.
8. Nance SJ, Garratty G. Application of Flow Cytometry to Immunohematology. J Immunol Meth 1987;101:127–131.
9. McHugh TM, Reid ME, Stites DP, Chase ES, Casavant CH. Detection of the human erythrocyte surface antigen Gerbich by flow cytometry using human antibodies and phycoerythrin for extreme immunofluorescence sensitivity. Vox Sang 1987;53:231–234.
10. Mollison PL, Engelfriet CP, Contreras M. The Survival of Transfused Red Cells. In: Mollinson, ed. Blood Transfusion in Clinical Medicine, 8th ed. Oxford: Blackwell 1987:807–810.
11. Mollison PL, Johnson CA, Prior DM. Dose dependent destruction of A1 cells by anti-A1. Vox Sang 1978;35:149–152.
12. Issitt PD, Valinsky JE, Marsh WL, DiNapoli J, Gutgsell NS. In vivo red cell destruction by anti-Lu6. Transfusion 1990;30:258–260.
13. Slezak SE, Horan PK. Fluorescent in vivo Tracking of Hematopoietic Cells. Blood 1989;74:2172–2177.
14. Valinsky JE. The Analysis of Red Cells by Flow Cytometry: Applications in Immunohematology. In: Yen A, ed. Flow Cytometry: Ad-

vanced Research and Clinical Applications. CRC Press, Boca Raton, FL, 1989:169–190.

15. Postoway N, Nance S, O'Neill P, Garratty G. Comparison of a Practical Differential Agglutination Procedure to Flow Cytometry in Following the Survival of Transfused Red Cells [Abstract]. Transfusion 1985; 25:453.

16. Jones JG, Holland BM, Hudson IRB, Wardrop CAJ. Total Circulating Red Cells versus Haematocrit as the Primary Descriptor of Oxygen Transport by the Blood. Br J Haematol 1990;76:288–294.

17. Hudson IRB, Cavill IAJ, Cooke A, et al. Biotin Labeling of Red Cells in the Measurement of Red Cell Volume in Preterm Infants. Pediatr Res 1990;28:199–202.

18. Cavill I, Trevett D, Fisher J, Hoy T. The Measurement of the Total Volume of Red Cells in Man: A Non-radioactive Approach Using Biotin. Br J Haematol 1988;70:491–493.

19. Suzuki T, Dale GL. Biotinylated Erythrocytes: in vivo Survival and in vitro Recovery. Blood 1987;70:791–795.

20. Roth KS. Biotin in Clinical Medicine. Am J Clin Nutrition 1981;34: 1967–1974.

21. de Bruin HG, de Leur-Ebeling I, Aaij C. Quantitative determination of the number of FITC-molecules bound per cell in immunofluorescence flow cytometry. Vox Sang 1983;45:373–377.

22. Garratty G. The Significance of IgG on the Red Cell Surface. Transfus Med Rev 1987;1:47–57.

23. Gilliland BC, Baxter E, Evans RS. Red Cell Antibodies in Acquired Immune Hemolytic Anemia with Negative Antiglobulin Serum Tests. N Engl J Med 1971;285:252–256.

24. Van der Meulen FW, De Bruin HG, Goosen PCM, et al. Quantitative aspects of the destruction of red cells sensitized with IgG1 autoantibodies: An application of flow cytofluorometry. Br J Haematol 1980;46: 47–56.

25. Garratty G, Nance SJ. Correlation between in vivo Hemolysis and the Amount of Red Cell-Bound IgG measured by Flow Cytometry. Transfusion 1990;30:617–621.

26. Spycher MO, Spath PJ, The ARC-IVIG Study Group Cologne/Bern. Quantification of Complement Receptor 1 on Erythrocytes: Follow-Up of HIV-1-Infected Patients with AIDS-Related Complex/Walter Reed 5 under Treatment with Intravenous Immunoglobulin. Vox Sang 1990; 59(suppl 1):44–50.

27. Anstee DJ. Blood Group Active Surface Molecules of the Human Red Blood Cell. Vox Sang 1990;58:1–20.

28. Palek J, Lambert S. Genetics of the Red Cell Membrane Skeleton. Semin Hematol 1990;27:290–332.

29. Reid ME, Takakuwa Y, Conboy J, Tchernia G, Mohandas N. Glycophorin C Content of Human Erythrocyte Membrane is Regulated by Protein 4.1. Blood 1990;75(11):2229–2234.

30. Tagushi R, Funahashi Y, Ikezawa H, Nakashima I. Analysis of PI-anchoring Antigens in a Patient of Paroxysmal Nocturnal Hemoglobinuria Reveals Deficiency of 1F5 antigen (CD59), a New Complement-Regulatory Factor. FEBS Letters 1990;261:142–146.

31. Yamashina M, Ueda E, Kinoshita T, et al. Inherited Complete Deficiency of 20-kilodalton Homologous Restriction Factor (CD59) as a Cause of Paroxysmal Nocturnal Hemoglobinuria. N Engl J Med 1990; 323:1184–1189.

32. Huizinga TWJ, van der Schoot CE, Jost C, et al. The PI-linked receptor FcRIII is released on Stimulation of Neutrophils. Nature 1988;333: 667–669.

33. Selvaraj P, Rosse WF, Silber R, Springer TA. The Major Fc Receptor in Blood has a Phosphatidylinositol Anchor and is Deficient in Paroxysmal Nocturnal Hemoglobinuria. Nature 1988;333:565–567.

34. Van der Schoot ED, Huizinga TWJ, Van't Veer-Korthof ET, Wiijmans R, Pinkster J, Von dem Borne AEG. Deficiency of Glycosyl-phosphatidylinositol-Linked Membrane Glycoproteins of Leukocytes in Paroxysmal Nocturnal Hemoglobinuria. Description of a New Diagnostic Cytofluorometric Assay. Blood 1990;76:1853–1859.

35. Van Noorden CJF, Dolbeare F, Aten J. Flow Cytofluorometric Analysis of Enzyme Reactions Based on Quenching of Fluorescence by the Final Reaction Product: Detection of Glucose-6-Phosphate Dehydrogenase Deficiency in Human Erythrocytes. J Histochem Cytochem 1989;37(9):1313–1318.

36. Marsh WL, Redman CM. The Kell Blood Group System: A Review. Transfusion 1990;30:158–167.

37. Valinsky JE, Marsh WL. Identification of Mosaic Blood Types in Female Carriers of X-linked disorders by Flow Cytometry. J Cell Biol 1986;103:519–523.

38. Bigbee WL, Branscomb EW, Weintraub HB, Papayannopoulou TH, Stamatoyannopoulos G. Cell Sorter Immunofluorescence Detection of Human Erythrocytes Labeled in Suspension with Antibodies Specific for Hemoglobin S and C. J Immunol Meth 1981;45:117–127.

39. de Man AJM, Foolen WJG, Van Dijk BA, Kunst VA, de Witte TM. A fluorescent microsphere method for the investigation of erythrocyte chimaerism after allogeneic bone marrow transplantation using antigenic differences. Vox Sang 1988;55:37–41.

40. Bianchi DW, Flint AF, Pizzimenti MF, Knoll JHM, Latt SA. Isolation of Fetal DNA from Nucleated Erythrocytes in Maternal Blood. Proc Natl Acad Sci USA 1990;87:3279–3283.

41. Wognum AW, Lansdrop PM, Humphries RK, Krystal G. Detection and Isolation of the Erythropoietin Receptor using Biotinylated Erythropoietin. Blood 1990;76:697–705.

42. Oien L, Nance S, Arndt P, Garratty G. Determination of zygosity using flow cytometric analysis of red cell antigen strength. Transfusion 1988; 28:541–544.

43. Hasekura H, Ota M, Ito S, Hasegawa Y, Ichinose A, Fukushima H, Ogata H. Flow cytometric studies of the D antigen of various Rh phenotypes with particular reference to D^u and D^{el}. Transfusion 1990;30: 236–238.

44. Langlois RG, Bigbee WL, Jensen RH. Flow cytometric characterization of normal and variant cells with monoclonal antibodies specific for glycophorin. J Immunol 1985;134:4009–4017.

45. Bigbee WL, Langlois RG, Stanker LH, Vanderlaan M, Jensen RH. Flow Cytometric Analysis of Erythrocyte Populations in Tn Syndrome Blood Using Monoclonal Antibodies to Glycophorin A and the Tn Antigen. Cytometry 1990;11:261–271.

46. Langlois RG, Nisbet BA, Bigbee WL, Ridinger DN, Jensen RH. An Improved Flow Cytometric Assay for Somatic Mutations at the Glycophorin A Locus in Humans. Cytometry 1990;11:513–521.

47. Steinheider G, Neth R, Marquardt H. Evaluation of Nongenotoxic and Genotoxic Factors Modulating the Frequency of Micronucleated Erythrocytes in the Peripheral Blood of Mice. Cell Biol Toxicol 1986;2:197–211.

48. Rivera R, Scornik JC. HLA Antigens on Red Cells: Implications for Achieving Low HLA Antigen Content in Blood Transfusions. Transfusion 1989;26:375–379.

49. Bodensteiner DC. A Flow Cytometric Technique to Accurately Measure Post-Filtration White Cell Counts. Transfusion 1989;29:651–653.

50. Davis BH, Bigelow NC. Flow cytometric reticulocyte quantification using thiazole orange provides clinically useful reticulocyte maturity index. Arch Pathol Lab Med 1989;113:684–689.

51. Davis BH, Bigelow NC, Ball ED, Mills M, Cornwell GG. Utility of flow cytometric reticulocyte quantification as a predictor of engraftment in autologous bone marrow transplantation. Am J Hematol 1989; 32:81–87.

52. Davis BH, Bigelow NC. Clinical Flow Cytometric Reticulocyte Analysis. Pathobiology 1990;58:99–106.

53. Ault KA, St. Germain E, Hitchcock S, Melander MP, Gross GG, Hillman RS. Rapid Upregulation of Erythrocyte Transferrin Receptors on Reticulocytes: An Early Indication of Response to Erythropoietin [Abstr]. Blood 1990;76(Suppl 1):130a.

54. Whaun JM, Rittershaus C, Ip SHC. Rapid Identification and Detection of Parasitized Human Red Cells by Automated Flow Cytometry. Cytometry 1983;4:117–122.

55. Bianco AE, Bqattye FL, Brown GV. Plasmodium Faciparum: Rapid Quantitation of Parasitemia in Fixed Malaria Cultures by Flow Cytometry. Exp Parasitol 1986;62:275–279.

56. Franklin RM, Brun R, Grieder A. Microscopic and Flow Cytophotometric analysis of Parasitemia in cultures of Plasmodium faciparum Vitally Stained with Hoechst 33342. Z. Parasitenkd 1986;72:210–214.
57. Hare JD, Buchler DW. Analysis of Plasmodium Faciparum Growth in Culture using Acridine Orange and Flow Cytometry. J Histochem Cytochem 1986;34:215–219.
58. Van Vianen PH, Klayman DL, Lin AJ, Lugt CB, Van Engen AL, Van der Kaay HJ, Mons B. Plasmodium berghei: The Antimalarial Action of Artemisinin and Sodium Artelinate in vivo and in vitro Studied by Flow Cytometry. Exp Parasitol 1990;70:115–123.
59. Jacobberger JW, Horan PK, Hare JD. Analysis of Malaria Parasite Infected Blood by Flow Cytometry. Cytometry 1983;4:228–231.
60. Loken MR, Shah VO, Dttilio KL, Civin CL. Flow Cytometric of Human Marrow. O. Normal Erythroid Development. Blood 1987;69:255–263.
61. Phillips HM, Holland BM, Jones JG, Abdel-Moiz AL, Turner TL, Wardrop CAJ. Definitive Estimate of Percent Hemoglobin F in Neonatal Reticulocytes. Pediatr Res 1988;23:595–597.
62. Gilmer PR, Koepke JA. The reticulocyte. An approach to definition. Am J Clin Pathol 1976;66:262–267.
63. Peebles DA, Hochberg A, Clark TD. Analysis of manual reticulocyte counting. Am J Clin Pathol 1981;76:713–717.
64. Savage RA, Skoog DR, Rabinovitch A. Analytical inaccuracy and imprecision in reticulocyte counting: A preliminary report from the College of American Pathologists' reticulocyte project. Blood Cells 1985;11:97–112.
65. Hackney JR, Cembrowski GS, Prystowsky MB, Kant JA. Automated reticulocyte counting by image analysis and flow cytometry. Lab Med 1989;20:551–555.
66. Jacobberger JW, Horan PK, Hare JD. Flow cytometric analysis of blood cells stained with the cyanine dye $DiOC_1(3)$: reticulocyte quantification. Cytometry 1984;5:589–600.
67. Sage BH, O'Connell JP, Mercolino TJ. A rapid, vital staining procedure for flow cytometric analysis of human reticulocytes. Cytometry 1983;4:222–227.
68. Metzger DK, Charache S. Flow cytometric reticulocyte counting with thioflavin T in a clinical hematology laboratory. Arch Pathol Lab Med 1987;111:540–544.
69. Lee LG, Chen C-H, Chie LA. Thiazole orange: a new dye for reticulocyte analysis. Cytometry 1986;7:508–517.
70. Wearne A, Robin H, Joshua DE, Kronenberg H. Automated enumeration of reticulocytes using acridine orange. Pathology 1985;17:75–77.
71. Corash L, Rheinschmidt M, Lieu S, Meers P, Brew E. Enumeration of reticulocytes using fluoresence-activated flow cytometry. Pathol Immunopathol Res 1988;7:381 394.
72. Vaughan WP, Hall J, Dougherty D, Peebles D. Simultaneous reticulocyte and platelet counting on a clinical flow cytometer. Am J Hematol 1985;18:385–391.
73. Kojima K, Niri M, Setoguchi K, Tsuda I, Tatsumi N. An automated optoelectronic reticulocyte counter. Am J Clin Pathol 1989;92:57–61.
74. Tichelli A, Gratwohl, Driessen A, et al. Evaluation of the Sysmex R-1000. Am J Clin Pathol 1990;93:70–78.
75. Ferguson DJ, Lee S-F, Gordon PA. Evaluation of Reticulocyte Counts by Flow Cytometry in a Routine Laboratory. Am J Hematol 1990;33:13–17.
76. Carter JM, McSweeney PA, Wakem PJ, Nemet AM. Counting Reticulocytes by Flow Cytometry: Use of Thiazole Orange. Clin Lab Haematol 1989;11:267–271.
77. Tanke HJ, Nieuwenhuis IAB, Koper GJM, Slats JCM, Ploem JS. Flow cytometry of human reticulocytes based on RNA fluorescence. Cytometry 1980;1:313–320.
78. Tanke HJ, Rothbarth PH, Vossen JM, Koper GJM, Ploem JS. Flow cytometry of reticulocytes applied to clinical hematology. Blood 1983;61:1091–1097.
79. Tanke HJ, Van Vianen PH, Emiliani FMF, Neuteboom I, de Vogel N, Tates AD, de Bruijin EA, Van Oosterom AT. Changes in erythropoiesis due to radiation or chemotherapy as studies by flow cytometric determination of peripheral blood reticulocytes. Histochemistry 1986;84:544–548.
80. Seligman PA, Allen RH, Kirchanski SJ, Natale PJ. Automated analysis of reticulocytes using fluorescent staining with both acridine orange and an immunofluorescence technique. Am J Hematol 1983;14:57–66.
81. Crouch JY, Kaplow LS. Relationship of reticulocyte age to polychromasia, shift cells, and shift reticulocytes. Arch Pathol Lab Med 1985;109:325–329.
82. Heilmeyer L, Westhaeuser R. Reifungsstadien an ueberlebenden retikulozyten in vitro und ihre bedeutung fuer die schaetzung der taeglichen haemoglobin-produktion in vivo. Z Klin Med 1932;121:361–655.
83. Hilman RS, Finch CA. Erythropoiesis: Normal and Abnormal. Semin Hematol 1972;4:327–438.
84. Ball ED, Mills LE, Cornwell GG, Davis BH, et al. Autologous Bone Marrow Transplantation for Acute Myeloid Leukemia using Monoclonal Antibody-Purged Bone Marrow. Blood 1990;75:1199–1206.
85. Liesveld JL, Rowe JM, Lichtman MA. Variability of the erythropoietic response in autoimmune hemolytic anemia: analysis of 109 cases. Blood 1987;69:820–826.
86. Davis BH, Bigelow NC, Langweiler ML. Utility of the Flow Cytometric Reticulocyte Maturity Index in the Evaluation of Anemia. Amer J Clin Pathol, submitted for publication, 1992.
87. Davis BH, DiCorato M, Bigelow NC, Langweiler ML. Proposal for the Standardization of Flow Cytometric Reticulocyte Maturity Index Measurements in Laboratory Hematology. Cytometry, submitted for publication, 1992.
88. Tsuda I and Tatsumi N. Reticulocytes in Human Preserved Blood as Control Material for Automated Reticulocyte Counters. Am J Clin Pathol 1990;93:109–110.

23

Flow Cytometric Analysis of Platelets

KENNETH A. AULT

BRIEF DESCRIPTION OF PLATELETS

The study of platelets represents a newly emerging arena for flow cytometry in both clinical and research applications. For several reasons that will be explained below, flow cytometry is ideally suited for studying platelets and there are several clinically relevant aspects of platelet pathophysiology and platelet function that can be measured using flow cytometry. In addition, platelets in blood samples frequently create problems for the interpretation of other types of flow cytometric assays. For example, as described below, aggregated platelets frequently contaminate the "lymphocyte gate" used in flow cytometry, and activated platelets can sometimes be found adhering to other cell types in a specimen. Thus, even for those who may not be directly interested in platelets, an understanding of how platelets appear on the flow cytometer may be useful.

This article will attempt to introduce the subject of platelets in a way that will be relevant for flow cytometry in general and will describe several of the clinically useful applications of flow cytometry in the study of platelets. In addition, we shall see that the flow cytometric study of platelets leads very quickly to a discussion of other small particles found in clinical samples and, thus, to some extent, we will discuss other types of "debris" (flow cytometric debris) that may be of clinical importance in the future.

Platelets are annucleate cellular fragments that arise from multinucleated giant cells in the bone marrow known as megakaryocytes. They are released into the circulation by a process of cytoplasmic fragmentation. Although they contain no nucleus they have most of the other normal cellular constituents including mitochondria, ribosomes and mRNA, a cytoskeletal apparatus, and numerous granules that are important in their function. Although they are usually described as discoid and thought of as "plate-like," a more useful mental image for platelets may be of small sponges because platelets have an extensive canalicular system that penetrates their structure. For this reason, although it is bounded by an ordinary plasma membrane, it is sometimes difficult to distinguish the inside from the outside of the platelet. Some of their bizarre immunological behavior, and some of the difficulty in studying them, probably arises from the fact that substances contained within the canalicular structure, although topologically outside the platelet, are not immediately accessible to outside influences and may not be detected by immunological probes.

The array of surface glycoproteins expressed on platelets has been extensively studied. Most of the major surface structures have been assigned known functions and their structure has been determined at the molecular level. Figure 23.1 shows, in cartoon form, the major surface structures of the platelet and indicates their functions. The availability of monoclonal antibodies directed to both functional and structural determinants on these molecules makes possible most of the flow cytometric techniques that we will be discussing.

The most important functional structures inside platelets are their granules. There are several types of granules that, for our purposes, can be divided into three types. First are the α *granules*, which contain a long and growing list of substances. It is interesting and important to note that the contents of α *granules* are essentially the substances that initiate and mediate an inflammatory reaction. They consist of vasoactive, chemotactic, and mitogenic substances that are capable of initiating inflammation and recruiting other cells normally associated with the inflammatory response. In addition, these granules contain substances that are capable of initiating and catalyzing the coagulation cascade. The dense granules, named because of their electron microscopic appearance, contain primarily ADP, which is a major agonist for platelet activation, serotonin, and heparin. Thus the release of dense granule contents is capable of recruiting more platelets to a site of vascular injury in a positive feedback system. Finally, platelets contain lysosomes with the usual array of degradative enzymes. Platelets are not thought of as phagocytic and thus the major role for these lysosomal enzymes is probably after their release from the platelet.

There are essentially three functional responses that platelets are capable of making when they are stimulated. They can release the contents of their granules (release), they can bind to a surface (adhesion), and they can aggregate with other platelets (aggregation). Although we generally think of these three events as being sequential consequences of activation, there is good evidence to suggest that each of these may occur as separate events unrelated to the other two. It may be, however, that in the normal in vivo situation the three events always occur together.

The release reaction is triggered by an influx of calcium and a change in the contractile elements of the cytoskeleton,

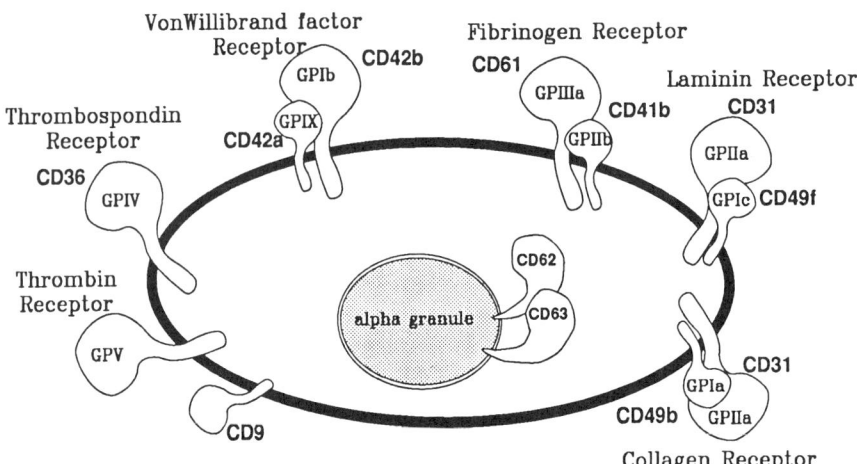

Figure 23.1. Diagram of major platelet surface structures. The major platelet glycoproteins are indicated, with their associated *CD* numbers and their functions, where known. The *GP* designations were originally derived from bands on electrophoretic patterns of platelet protein preparations. The *CD* number indicate the existence of monoclonal antibodies directed against these molecules. The three heterodimers indicated on the right side of the drawing are adhesion molecules of the Integrin family. Not shown on this drawing are a number of other *receptors* for which monoclonal antibodies are not yet widely available. Many of the entities indicated here as single objects are themselves heterodimers—this detail was omitted for clarity. (Adapted from a drawing by AMAC Inc. Westbrook, ME)

which results in the movement of the granules to the plasma membrane, fusion of the granule membrane with the plasma membrane, and release of the granule contents into the extracellular space. Electron microscopy suggests that much of this release occurs into the canalicular system and the released substances must then diffuse to the outside through the canaliculi. One necessary result of this phenomenon is that the membrane of the granule becomes everted and what was the inside of the granule membrane becomes a part of the external membrane of the platelet. This forms the basis for one of the best methods of detecting platelet activation by flow cytometry.

It is generally thought that platelet release is an "all or none" phenomenon involving all of the granules. However there is some evidence for selective degranulation of platelets (1); thus, this deserves further study. Flow cytometry is an excellent tool for studying the correlated expression of markers of release and may help us to understand the release reaction better in the future.

Platelets have an array of adhesion proteins on their surface that rivals that of leukocytes. They are capable of adhering to injured vascular endothelium, to the subendothelial extracellular matrix, to leukocytes, and, of course, to each other. A complete description of the mechanisms of adhesion that have been identified is beyond the scope of this article; however, one can think of most of the adhesion functions of platelets as being mediated by adhesive proteins (integrins) binding to large molecules, such as collagen, fibrinogen, thrombospondin, von Willebrand's factor, etc. (2). In addition to integrins, the platelet expresses at least one of a new class of adhesion molecules variously known as LECAMs or Selectins (3). It is one of these molecules (GMP-140), which is normally sequestered on the inner surface of the α granule membrane and appears on the surface of activated platelets, that has been most extensively used in flow cytometric assays of platelet activation (4).

One particularly interesting aspect of the activation-specific adhesion of platelets is illustrated by the integrin molecule known as gpIIb/IIIa. This is one of the major platelet surface glycoproteins and it is normally in an inactive conformation. When the platelet is activated it undergoes a conformational change that reveals a binding site for fibrinogen and it is this site that mediates a large part of the adhesive function of activated platelets. Monoclonal antibodies that recognize the neo-epitope associated with this conformational change can be used in flow cytometric assays of platelet activation (5).

Thus the platelet, whose major function in vivo is to adhere, has a complex and probably redundant array of adhesion molecules. Since the platelet must circulate in a nonadhesive state and then become sticky only when activated, it has a number of interesting mechanisms for controlling the expression of its adhesive molecules, including sequestering them inside of granules and initiating conformational changes. The surface glycoproteins of the platelet are probably as well studied, and better understood, than those of any other cell type. There is a vast amount of information that the flow cytometrist can use to devise techniques for studying platelet function.

The logical consequence of adhesion is aggregation. However, for platelets, aggregation is not simply a passive process that follows when they become adhesive. It is, rather, a very active and highly regulated process mediated by a powerful positive feedback mechanism. The ADP that is released when platelets are activated is a powerful stimulant for the activation of platelets, thus other platelets are rapidly recruited into a beginning aggregate. In addition, some of the enzymes activated by the coagulation cascade

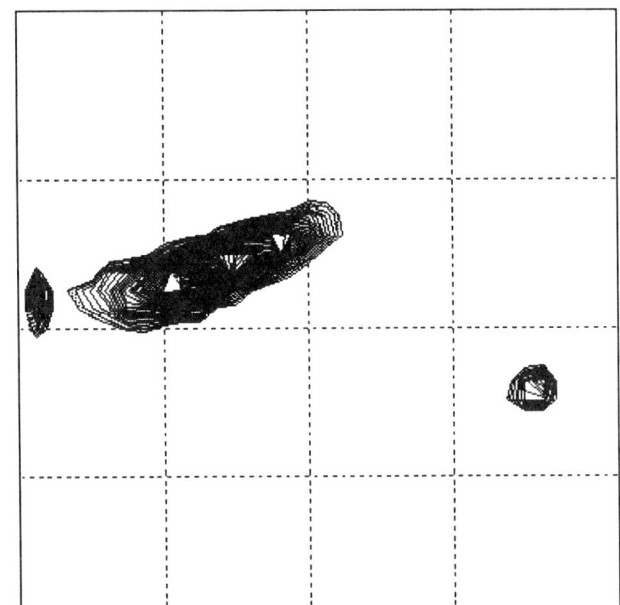

Figure 23.2. The typical "platelet cluster" obtained in whole blood. This figure shows flow cytometric data obtained from fixed whole blood labeled with FITC-antiGPIb. The instrument was triggered on FITC fluorescence so that GPIb-negative particles are not visualized. The typical platelet cluster shows a positive correlation between size (*forward scatter*) and fluorescence. To the left of the platelets are GPIb-positive microparticles that are submicron in size. To the right are some leukocytes that fall above the threshold by virtue of their relatively high levels of autofluorescence. This data was obtained using a FACScan flow cytometer (Becton Dickinson Immunocytometry Systems, Mountain View CA). Both scales are logarithmic (4 decades).

are also powerful platelet agonists (notably thrombin). Thus, once activated, platelets rapidly and specifically aggregate. This aggregation has traditionally formed the predominant method for assaying platelet function as described below.

As one might expect, a positive feedback mechanism must be controlled by an equally powerful inhibitory mechanism. Thus platelet aggregation in vivo is a temporary phenomenon. Platelets once aggregated also deaggregate over a period of about 15 to 30 min under physiological conditions (6). The mechanism for this deaggregation is not well understood. It is thought that the platelet aggregates form a temporary bandage for damaged vessels and that as the platelet aggregate disintegrates it is replaced by the more permanent bond of the coagulation proteins and the inflammatory response.

THE FLOW CYTOMETRIC EVALUATION OF PLATELETS

In blood, platelets are second only to erythrocytes in frequency. Their normal concentration is about 250,000 per mm^3. Thus, they are 20 times less frequent than erythrocytes and about 25 times more frequent than leukocytes. For the purposes of flow cytometry they are, thus, very abundant. One ml of blood contains 2.5×10^8 platelets. In clinical situations, patients frequently become extremely thrombocytopenic with platelet counts on the order of 10,000 per mm^3. In these cases, there are still enough platelets in 1 ml of blood for most flow cytometric analyses. This is in contrast to other methods for studying platelets such as aggregometry and immunochemistry, which may require 10 to 100 ml of blood from a very thrombocytopenic patient to get enough platelets for study.

Normal platelets are very heterogeneous in size when compared to other blood cells. They range from about one to about four μm in diameter. In disease states, it is possible to find platelets that are much larger and, recently, attention has been drawn to the presence of "platelet microparticles" that can range down to less than 0.1μm (7). This four-to-fivefold range in normal diameter results in a 16-to-25-fold range in surface area distribution and more than 100-fold differences in volume. Thus, for almost any surface-associated measurement made on platelets, one can expect, approximately, a one-log-wide distribution and, for volume associated measurements, a two-log-wide distribution. The corollary is that for any flow cytometric assay of platelets there will be a very strong correlation between light scatter and any fluorescence parameter. This constitutes a characteristic flow cytometric signature for platelets (Fig. 23.2).

The relatively small size of platelets as compared with other blood elements means that, for some flow cytometers, they are near the lower limits of resolution. It is of critical importance that anyone studying platelets by flow cytometry be certain that the instrument he is using is capable of clearly resolving 1μm particles. If the platelet population is only partially resolved from instrument noise then some unknown fraction of the data collected may be noise. Techniques described below that allow for identification of platelets using fluorescent as well as light scatter criteria are more reliable than the use of light scatter alone.

The physiological role of platelets is to identify and react with abnormal surfaces in contact with blood. For this reason, they are immediately affected by exposure to any foreign surface. The act of drawing blood, which necessarily results in disruption of vascular integrity and exposure of blood to man made surfaces, causes platelet activation. Any subsequent manipulation of blood, such as mixing, centrifugation, washing of platelets, etc., compounds this problem. Those who have studied platelets for years have developed techniques to minimize these potential artifacts; however, the fact remains that the ability of flow cytometry to identify platelets in a mixture of other cells is one of the strengths of the technology. In general, when platelets are studied using flow cytometry one can minimize the manipulation of the platelets by using impure platelet mixtures and still be quite certain that the properties one is measuring belong to platelets and not to contaminating elements.

The extreme case of using an impure platelet mixture is whole blood. Methods have been devised that allow the flow cytometric measurement of platelet function in whole blood. These techniques have become extremely powerful and have begun to be applied to the study of other reactive cell types. It is fair to say that flow cytometric techniques offer the nearest approximation currently available to an assessment of platelet function in vivo. Nevertheless, one must always question to what extent our results could be affected by any necessary in vitro manipulation of the platelets.

Flow cytometry thus offers a number of advantages for the study of platelets when compared to standard methods. These include: the requirement for relatively small amounts of blood, especially when the patient is thrombocytopenic; the ability to obtain information about the physiological state of platelets with an absolute minimum of manipulation; the ability to positively identify platelets and thus be certain that the measurements are not contaminated by other cell types or by "debris." In addition, flow cytometry has, for the study of platelets, the same major advantages that it has for other applications, namely, the ability to detect functional subpopulations and obtain correlated measurements of several cellular properties simultaneously. These last two advantages, when applied to platelets, have led immediately to fascinating new observations that have never before been possible with any other technique and that had previously not been suspected.

It must be admitted that flow cytometric analysis of platelets also has some disadvantages compared to more conventional methods. The most obvious is the expense, not only of the flow cytometer itself, but of the reagents. A flow cytometer is much more expensive than even the most sophisticated aggregometer, and the flow cytometric methods require the use of fluoresceinated antibodies, usually monoclonals, that are also quite expensive.

Some of the monoclonal reagents, such as markers for platelet activation, are not yet commercially available and must be borrowed from other workers or made in one's own laboratory.

Finally, because platelets are small, even antigens that are expressed in relatively high copy number on the platelet surface result in less fluorescence signal than comparable antigens on leukocytes. For example, a strong signal from a platelet may represent on the order of 10,000 fluorochromes whereas a strong signal from a lymphocyte may be an order of magnitude brighter. Thus sensitivity is an issue. Most investigators feel that the sensitivity of flow cytometric assays is comparable to that of ELISA techniques but is probably less than that of RIA methods. For the applications described below the sensitivity of modern flow cytometers of 300–1000 fluorochromes seems to be perfectly adequate.

PLATELET-ASSOCIATED IMMUNOGLOBULIN

The first application that we will discuss is the measurement of platelet-associated immunoglobulin (PAIg), sometimes referred to as "antiplatelet antibodies." This has a long and somewhat disreputable history as a clinical assay. There have a been a large number of different assays described for PAIg and each one presents a number of difficulties. The flow cytometric assay described here is at least as good as any of the other assays and is considerably better in some ways. The greatest strength of the flow cytometric assay is that it has made clear some of the complications that were not taken into account by previous assays. Thus, we now understand why it has been so difficult to devise a reliable and rational assay in the past. Despite the difficulties in interpretation that remain, most clinicians dealing with thrombocytopenia feel the need for an assay for PAIg and thus, like so many imperfect medical assays, it has a usefulness that goes beyond the technical difficulties and the statistics.

The goal of the measurement of platelet-associated Ig is simply stated. It is desirable to be able to quantitate the amount of Ig on platelets because there are a number of disorders in which the binding of antiplatelet antibodies can be reasonably suspected of resulting in the premature destruction of platelets. Included in the indications for measurement of PAIg are: thrombocytopenia occurring during pregnancy, patients with preexisting autoimmune diseases, lymphomas, HIV infection, and possible drug-induced thrombocytopenia. The resulting immune thrombocytopenia can be a disease of considerable morbidity and frequently is fatal. There is also evidence that, in some cases, the binding of antibodies to platelets may induce platelet dysfunction that leads to bleeding despite normal numbers of platelets. The distinction between immune thrombocytopenia and other forms of thrombocytopenia that may result from decreased platelet production or nonimmune platelet destruction is clinically very important because most of the therapeutic approaches to immune thrombocytopenia are either dangerous (immunosuppression, splenectomy) or expensive (intravenous gamma globulin). There are really only two important diagnostic tests that help in making the diagnosis of immune thrombocytopenia: bone marrow examination and measurement of platelet-associated Ig (PAIg). The bone marrow examination is complicated because it is frequently difficult to determine the rate of platelet production from the appearance of the marrow megakaryocytes. Thus, a good assay for PAIg would be of value in the differential diagnosis of any case of otherwise unexplained thrombocytopenia.

Ideally, such an assay should be sensitive and specific, as should any clinical test, but in addition, it must be useful in the setting of severe thrombocytopenia. Thus, we would like to be able to measure the amount of PAIg in patients whose platelet counts are very low. We would also like to be able to distinguish Ig that is normally found on platelets, bound to Fc receptors, and presumably of no pathological consequence (8–10) from Ig that is bound in the manner of an antiplatelet antibody. Such an antiplatelet antibody would be bound by its Fab region with the Fc region exposed, leading to destruction of platelets by either complement fixation or phagocytosis.

The difficulties that have been encountered in the measurement of PAIg are the following. First, although it is a fairly straightforward matter to measure by immunoassay the total amount of Ig associated with a known number of platelets, and thus determine the average amount of PAIg, it has proven difficult to use such measurements in practice. As noted above, there is a significant level of Ig normally associated with platelets, and this normal level is quite variable in different individuals (11). Thus, one is trying to detect an increased level of PAIg above a variable background. Second, such an assay is more accurate if enough platelets are pooled to result in a fairly high level of total Ig. It is frequently difficult to obtain enough platelets from severely thrombocytopenic patients. Third, if an immunoassay containing, for example, 10^6 platelets is carried out it must be assumed that all of the Ig measured is associated with platelets. There is evidence to suggest that a considerable fraction of the Ig measured may be trapped in interstices between platelets and may not be platelet-associated in vivo (12). In addition, if there are other particles, such as leukocytes or "debris," contaminating the platelet preparation, they may contribute a large amount of the total Ig. Fourth, there is evidence to suggest that the level of PAIg may be a dynamic variable. It may increase with the state of platelet activation (13–15), and antibody once bound to platelets may be capable of "capping" or "shedding" quite rapidly (16, 17). The implication of such observations is that the methods used to isolate the platelets and the conditions under which they are washed and handled may significantly affect the observed levels of PAIg. Finally, the bulk immunoassay approach will totally fail to detect any subpopulations of platelets that may have different levels of PAIg and that may be of clinical significance. Some of our evidence suggests that these latter two possibilities may frequently occur.

The consequence of such difficulties is that there has been a large number of methods developed for the measurement of PAIg and the results of such methods are widely disparate. The amount of normal Ig associated with platelets of controls has been reported to be as low as less than 0.1 fg per platelet to as high as greater than 10 fg per platelet (12). Most investigators have found that platelets from patients with immune thrombocytopenia (ITP) have PAIg elevated from two to 100 times the control levels. The frequency of "positive" results in patients felt by other criteria to have immune thrombocytopenia are reported to be between 70% and 100% (12).

In addition to measuring PAIg, it is also possible to use very similar assays for the measurement of antiplatelet antibody in patients' sera. Such "indirect" assays have proven to be even more variable and difficult to interpret. This appears to be due again to a very wide range of "normal" for such measurements. The range of reported "positive" results in ITP is from 35% to 100%. Most investigators agree that the correlation between direct and indirect tests for PAIg is very poor.

It is of interest that there have been several reports of a correlation between the levels of PAIg in ITP patients and the levels of such seemingly irrelevant substances as albumin (31). Such observations raise the suspicion that, as the platelet count falls, a significant amount of the material that is thought to be platelet-associated may be nonspecifically trapped, or adsorbed, and may not have pathogenic significance.

Given these technical difficulties, what can be said about the clinical information that is available from a measurement of PAIg? As stated above, between 70% to 100% of patients with ITP will have elevated levels of PAIg. There is, in general, an inverse correlation between the levels of PAIg and the platelet count. It is interesting that the slope of this regression increases dramatically as the platelet count drops below 20,000. It is possible that, as the platelet count becomes very low, much of the PAIg that is being measured is not truly platelet-associated.

There is very poor correlation between the results of direct and indirect assays for antiplatelet antibody. In general, when the direct test is elevated, one would expect that there would be enough free antibody in the plasma to result in a positive indirect test. However, this is frequently not the case. Our own experience leads us to suggest that one reason for this is that the variability of the indirect test is so large that there must be a great deal of highly avid antibody in the plasma before it can be reliably detected.

It is generally agreed that the levels of PAIg can fluctuate dramatically with therapy. Frequently, an elevated PAIg will normalize within a few hours to days following therapy with either splenectomy or steroids. In these cases, it is difficult to believe that therapy has so rapidly affected the rate of production of the antibody. It is more likely that dynamic changes, related to the interaction of the antibody with the platelet surface or the interaction of the antibody-coated platelet with the reticuloendothelial system, explain such changes. An additional interesting possibility has been suggested. Ballem, Slichter et al. have presented evidence that antiplatelet antibodies may have profound effects on the rate of platelet production as well as the rate of platelet destruction (32, 33). Thus, to the extent that such antibodies may affect megakaryocyte function, the results of measurement of PAIg may be difficult to relate to the response to therapy.

In essence, the use of the flow cytometer involves much the same approach as outlined above for the immunoassay of PAIg. One must prepare platelets from blood, label them with a fluoresceinated anti-Ig antibody, and then measure the amount of fluorescence associated with the platelets. There are, however, several significant differences. First, when using the flow cytometer, it is not necessary to purify the platelet preparation to nearly the extent required by other methods. This is because the cytometer is capable of directly identifying the platelets even when they are heavily contaminated with erythrocytes and leukocytes. Thus, because the cytometer can reliably identify the platelets, the measure of PAIg is restricted to platelets and contaminating cells do not influence the measurement as they would in an immunoassay. Second, the cytometer is making the measurement of PAIg on individual platelets. If such a measurement is made on 10^5 or even 10^4 platelets the statistical accuracy of the average level of PAIg is extremely good. Thus, it is not necessary to process very much blood in order to obtain enough platelets to analyze in the flow cytometer. One ml of blood from a patient with a platelet count of 10,000 contains 10^6 platelets and should be enough. Third, the measurement is being made with the platelets in suspension. Thus, there is no concern that Ig, which might have been trapped in the interstices between platelets, will influence the measurement. Finally, the flow cytometric method is easily adaptable to the measurement of several different antigens on the same sample. Thus, for example, the measurement of platelet-associated IgG, IgM, and C3 can be accomplished on the same sample with no changes in the basic technique.

Having thus outlined the advantages of the flow cytometric approach, it must be admitted that there are some disadvantages. Although the sensitivity of the fluorescence-based flow cytometric assay appears to be adequate for the measurement of PAIg, it is not, in theory, as sensitive as a radioimmunoassay. Second, the flow cytometric assay has proven to be difficult to quantitate in terms of absolute levels of PAIg. There certainly have been successful methods of calibrating the flow cytometer described, but they are sufficiently awkward that they are not in routine use.

Several workers have described flow cytometric assays for PAIg (34–36). Corash has described a sensitivity of 94% and a specificity of 95% for the diagnosis of immune versus nonimmune thrombocytopenia in a study of 171 patients. Rosenfeld et al. reported similar results in a study of 102 patients and 33 controls. The methods used by these authors included preparation of washed platelets, with three washes in buffer to remove loosely associated Ig. Corash reports that fixation of the washed platelets was possible before they were labeled with fluorescent antibodies. This results in a stable preparation that can be labeled and analyzed at a later time. Our own experience confirms the utility of this approach. In addition, there is some evidence that fixation prior to labeling contributes to a lower background fluorescence on normal platelets. None of these authors report detecting the presence of any subpopulations of platelets during the course of their studies.

We have previously described the method that we use for measurement of PAIg, (37) which is briefly described here. Blood to be analyzed is processed concurrently with a normal control. EDTA anticoagulated blood (a single 7.5 ml lavender top Vacutainer tube) is transferred to a 15 ml conical centrifuge tube. The blood is centrifuged at $200 \times g$ for 20 minutes to sediment erythrocytes. The platelet-rich plasma is transferred to another 15 ml conical tube and centrifuged at $200 \times g$ for 5 min.

The platelet pellet is resuspended in 2 ml of CGS EDTA buffer (13 mM sodium citrate, 10 mM EDTA, and 10 mM dextrose in phosphate buffered saline, pH 7.4). The platelets are again centrifuged at $2000 \times g$ for 5 min. The wash is then repeated using 2 ml of Tyrode's buffer containing 10 mM EDTA and 0.5% (weight/volume) bovine serum albumin. The washed platelets are finally resuspended in 1 ml of the modified Tyrode's buffer and 1 ml of 2% paraformaldehyde is added. The platelets are allowed to fix for at least 2 hr or overnight at 4°C.

After fixation, the platelets are washed twice with Tyrode's buffer, counted, and the concentration is adjusted to 10^7 platelets per ml. Fixed platelets can be stored at 4°C for several days before staining.

To each of four plastic tubes, 100 µl of the fixed platelet suspension is added. The first tube is labeled with FITC-F(ab')$_2$ rabbit anti-human Ig (polyspecific, Cooper Biomedical, Westchester, PA) and biotinylated monoclonal anti-GPIIb/IIIa (AMAC Inc. Westbrook, ME). The second and third tubes receive FITC-F(ab')$_2$ goat antihuman IgG and FITC-F(ab')$_2$ goat antihuman IgM (both from Tago Inc., Burlingame, CA) respectively. The fourth tube receives FITC-F(ab')$_2$ goat antihuman C3 (Accurate Scientific, Westbury, NY). The amount and dilution of each of these reagents is determined by standard titration procedures and the antihuman Ig reagent should not cross react with mouse Ig. The final volume is 200 µl and incubation is carried out for 30 min at RT, followed by two washes with Tyrode's buffer. The first tube then undergoes a second 30-min incubation at RT with PE-avidin (Becton-Dickinson, Mountain View, CA).

Using a FACScan (Becton-Dickinson, Mountain View, CA) data is acquired on 5000 ungated events in list mode. Log amplification is used on forward and side scatter as well as fluorescence. The purpose of the anti-GPIIb/IIIa in the

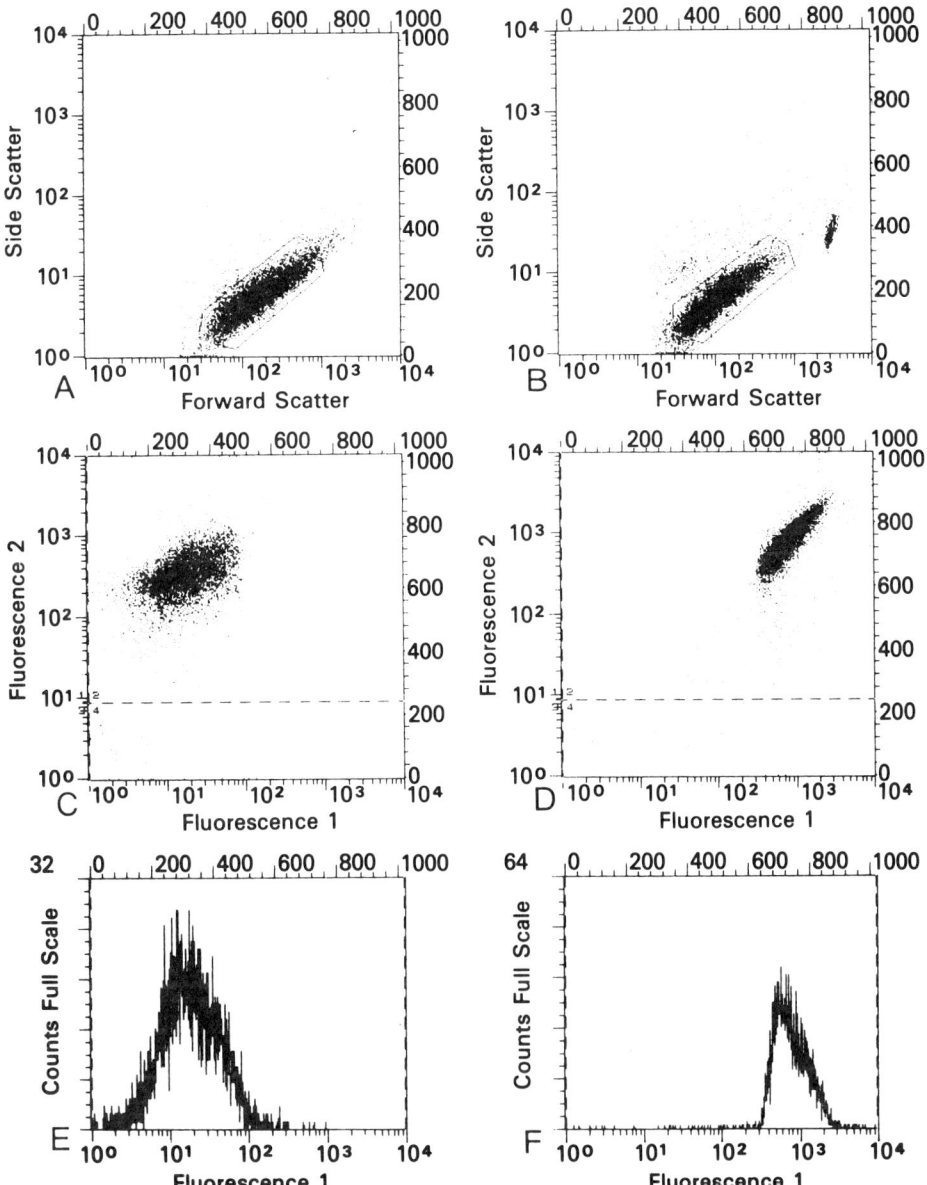

Figure 23.3. Representative data analysis for platelet-associated immunoglobulin. The **A**, **C**, and **D** are from a normal, the **B**, **D**, and **F** are from a patient with a strongly positive antiplatelet antibody. The **A** & **B** show forward light scatter versus side scatter. On **B** a small number of contaminating cells can be seen gated out. **C** and **D** show anti-Ig (*horizontal axis*) versus anti-GPIb (*vertical axis*). In both patients all of the gated events are GPIb positive. The **E** and **F** show the anti-Ig fluorescence histogram. The mean value of this histogram is used to determine the platelet-associated Ig level.

first tube is to determine what proportion of the events gated as platelets on the basis of light scatter are actually platelets (see below). This permits the determination of anti-Ig levels on GPIIb/IIIa-positive platelets and the differentiation of GPIIb/IIIa-negative particles (usually erythrocyte fragments). Representative data from a routine analysis is shown in Figure 23.3. It is necessary to establish a normal range. The upper limit of normal is set at two standard deviations above the mean.

From a technical point of view, there are a number of points that should be discussed. We have discovered that there may be several different populations of particles that appear in patients who have low platelet counts. Some of these are distinct populations of platelets, others appear not to be platelets at all. We do not yet fully understand the significance of these populations, but it is important to be aware that they exist and may profoundly affect the results of tests for PAIg.

First, we have found that in a significant number of patients, and some normals, there appear to be two distinct populations of platelets that bear different levels of PAIg. The most common clinical setting in which we observe this phenomenon is the patient recovering from ITP following treatment with steroids.

There are several possible explanations for this finding, but the correct one is not yet known. It is possible that one of the two populations is platelets that have recently entered the circulation and have bound more or less antibody than the older platelets. Since their size distribution appears to be the same, we cannot be certain which are the new platelets. It is also not clear why there would be such a distinct difference in the binding. It is also possible that one population consists of activated platelets that have expressed more PAIg. At the moment, we have no evidence that would bear on this point, and it is not likely that there would be a significant population of activated platelets circulating. Another possibility is that one population of platelets has altered the amount of PAIg on its surface by some active process akin to capping or shedding, or conceivably by a passive process akin to spherocyte formation by antibody coated erythrocytes. Finally, there is no doubt that we occasionally and sporadically observe a similar phenomenon in normal donors. We have gone to great lengths to create this phenomenon in the laboratory and cannot do so by any manipulation we have tried. Thus, for the moment, we conclude that it is possible for there to be at least two distinct populations of platelets circulating with different levels of PAIg, and that this is most commonly associated with a rapidly rising platelet count during therapy.

Second, we have found that in patients who are severely thrombocytopenic there are frequently particles that appear to be platelets on the flow cytometer, i.e., they have the same light-scatter properties and size-distribution as platelets, but they are not normal platelets. We have called these populations "non–platelet particles" (NPP's) because they appear not to express either GPIb or GPIIb/IIIa. NPP's are most commonly seen when the platelet count is very low (less than 20,000), and are usually not seen in patients who are thrombocytopenic due to marrow failure or chemotherapy. They thus appear to be associated with immune thrombocytopenia; however, the level of NPP's can vary widely in different patients with similar platelet counts.

The most commonly observed type of NPP has very low levels of Ig labeling, less than that of normal platelets. We have shown that many, if not all, of these particles can be labeled with antiglycophorin (Fig. 23.4) and, thus, may be classified as "microspherocytes" or erythrocyte fragments (37) that have been reported in ITP (38, 39). Less commonly seen are NPP's that label very brightly with anti-Ig. These have apparent levels of Ig that are 10^3 higher than normal platelets (Fig. 23.5). We have seen these in patients who have a high probability of immune complexes, and we have postulated that they may be very large complexes.

The presence of NPP's is of considerable importance for the measurement of PAIg because these particles copurify with normal platelets and will thus be included in most assays. If they are of the type that have now levels of Ig they may incorrectly lower the estimate of mean PAIg, and if they are of the high-Ig type they will dramatically elevate the mean PAIg. For this reason, we have developed a two-color assay in which we routinely measure PAIg versus GPIIb/IIIa expression and we consider only the mean Ig levels of GPIIb/IIIa-positive particles.

RETICULATED PLATELETS

The clinical evaluation of a thrombocytopenic patient involves distinguishing between diseases characterized by increased peripheral platelet destruction and those with failure of marrow thrombopoiesis. Usually, this important distinction is inferred from examination of the bone marrow for megakaryocyte number and morphology. Radiolabelled platelet survival studies, by their nature, are not useful clinically and have given inconsistent data in patients with immune thrombocytopenic purpura. It has been suggested that patients with immune thrombocytopenia can have antibody mediated inhibition of platelet production as often as they can have rapid rates of peripheral platelet destruction (40). Furthermore, platelet lifespan studies with homologous platelets cannot be performed in severely thrombocytopenic patients. In this situation, a direct measurement of platelet nucleic acid content could be diagnostically helpful, providing an estimate of megakaryocyte simulation and a measure of the quantity of young platelets released into the circulation.

There is evidence in the literature that the nucleic-acid content of platelets can be used as an estimate of platelet production. Ingram and Coopersmith, in a study of dogs with acute blood loss, reported the release of "reticulated platelets" that they interpreted to be young platelets (41). They employed new methylene blue and a microscopic technique to estimate the number of reticulated platelets. Other workers have suggested that platelets, like reticulocytes, decrease in size, reduce their rate of protein synthesis, and decrease their RNA content as they age (42).

Using the dye known as thiazole orange (43), it is now possible to measure individual platelet nucleic acid content by flow cytometric technique. This dye enters living cells without any pretreatment, displays a very large increase in fluorescence emission on binding to RNA and DNA, and both absorbs and emits at wavelengths similar to fluorescein (509 and 533 nm respectively). It thus lends itself to easy measurement on a standard flow cytometer. Using this technique, excellent correlation between flow cytometric erythrocyte reticulocyte counting and the manual method has been obtained (43). When applied to platelets, this method may provide a clinically useful technique.

We have used thiazole orange obtained from Becton-Dickinson Immunocytometry Systems (Mountain View, CA). The same solution that is commercially available for erythrocyte reticulocyte determination is used. Platelets are prepared as described above for determination of PAIg. Washed, fixed platelets are resuspended at 10^7/ml in Tyrode's buffer containing 10 mM EDTA. 100µl of this suspension is mixed with 900µl of the Thiazole Orange solution and incubated at room temperature for one hr. The samples

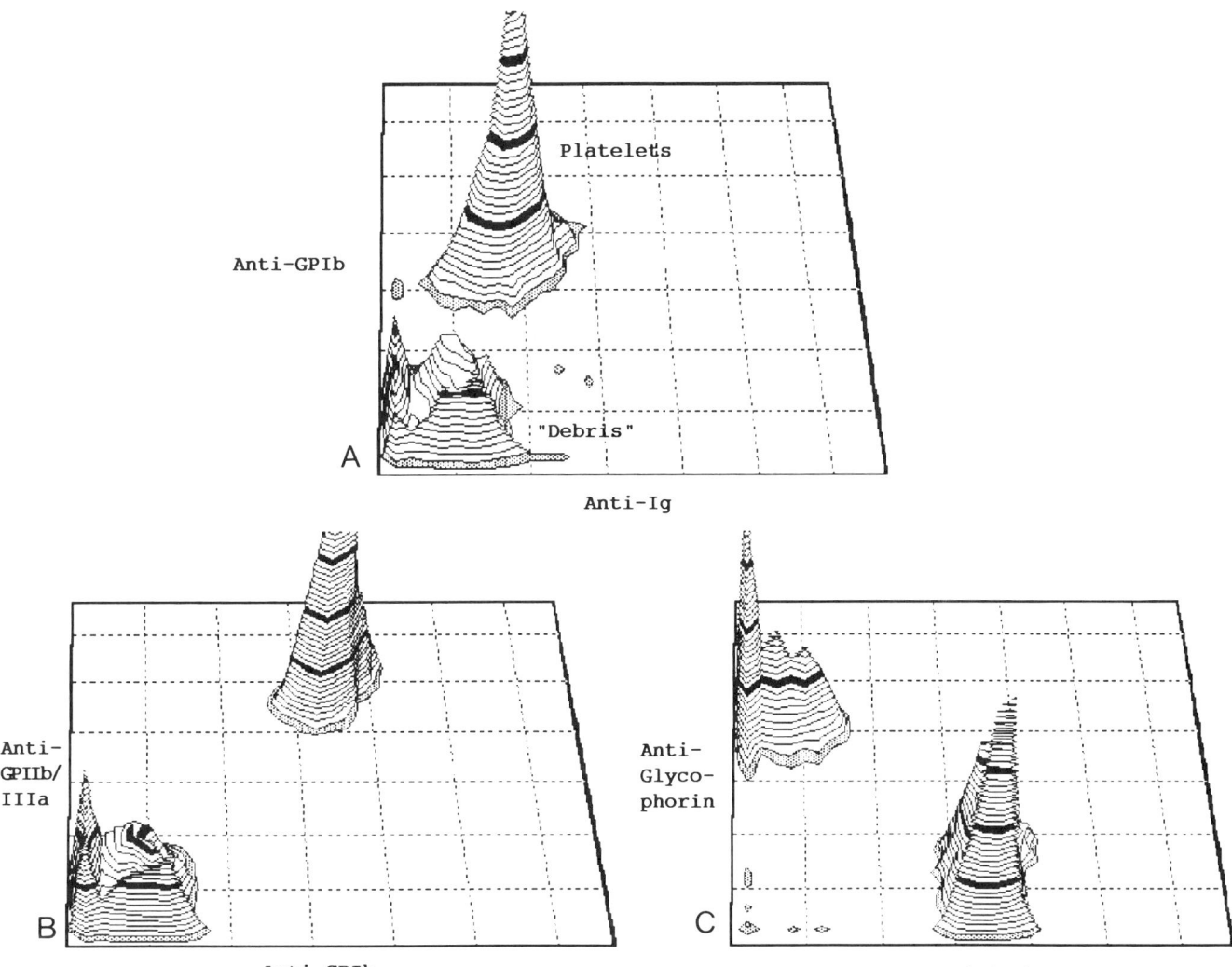

Figure 23.4. Non-platelet particles bearing glycophorin. The **A** shows a typical two-color (anti-Ig versus anti-GPIb) analysis of a patient who had a large number of "non-platelet particles," which do not label with anti-GPIb and have low levels of Ig. The **B** shows the same sample labeled with anti-GPIb and anti-GPIIb/IIIa. The NPP's are negative for both of these platelet-associated markers. The **C** shows labeling with anti-GPIb and anti-glycophorin. The presence of glycophorin on the particles strongly suggests that they are microspherocytes or erythrocyte fragments.

are then analyzed on the flow cytometer. Preliminary experiments have shown that the labeling is complete by one hr and that there is no significant change in fluorescence intensity for at least four hr at room temperature. A normal control sample is prepared at the same time as each patient sample.

Analysis of the samples is carried out using a Becton-Dickinson FACScan flow cytometer. The analysis of thiazole orange labeling is done by measuring both forward light scatter (FSC) and green (540 nm) fluorescence (FL1) using logarithmic amplification. A plot of FSC versus FL1 is generated, showing the expected positive correlation between FSC, which is a measure of particle size, and FL1.

Large platelets containing more nucleic acid are assumed to represent platelet reticulocytes. To determine their frequency, a line with the same slope as the platelet cluster is set to segregate 99% of the platelets of normal controls below the line (Fig. 23.6). The placement of this line is, admittedly, somewhat arbitrary; however, it permits an objective approach to the analysis of patients' samples relative to controls. As shown in Figure 23.6, a thrombocytopenic patient with stimulated thrombopoiesis shows a clear population of platelets that are both larger and contain more nucleic acid. The percent of such platelets is recorded and their absolute number is calculated.

The method we have chosen to determine the "%positive" platelets in patient samples makes use of a normal control in which a marker is set in such a way as to make 99% of the particles "negative." Thus the frequency of "platelet reticulocytes" in normals must by definition be near 1%. This is an arbitrary choice. When a single such marker was used to analyze a series of 20 normals in order to determine the

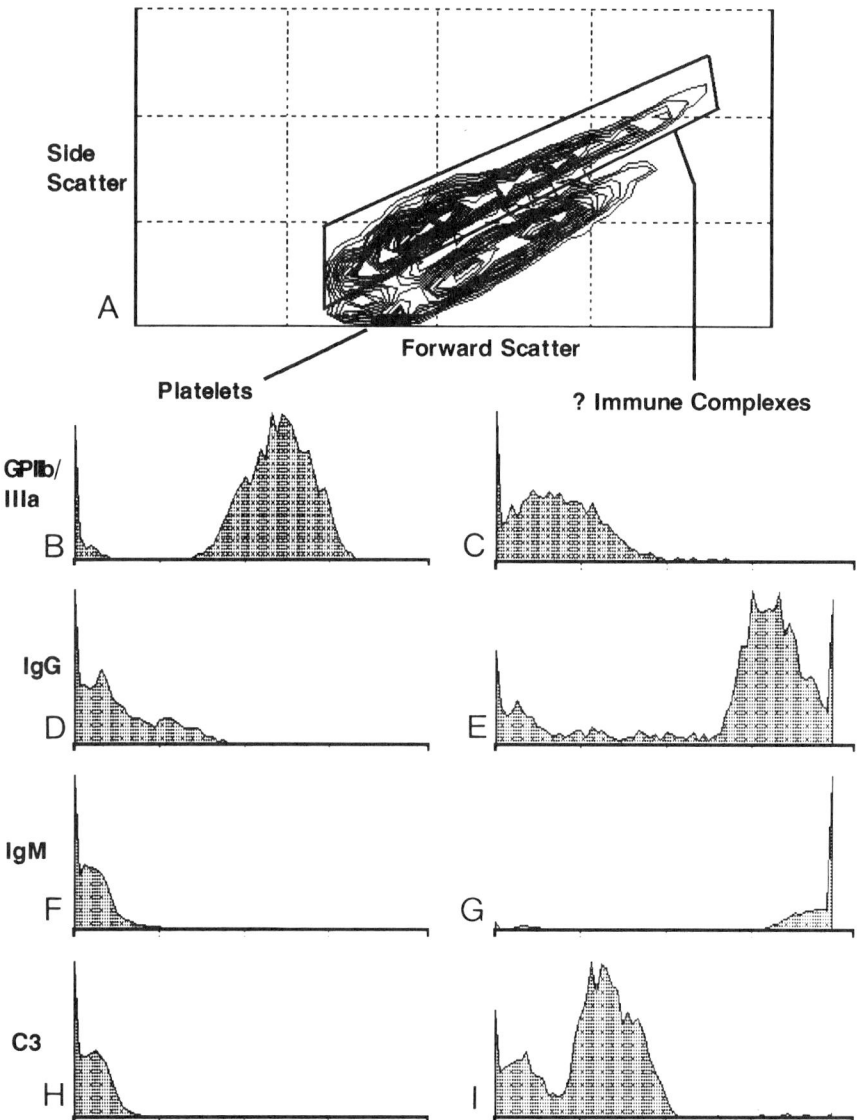

Figure 23.5. Labeling pattern of putative immune complexes. **A** shows the light-scatter pattern of a sample that contains a large amount of "debris" that falls above the usual platelet cluster. **B, D, F** and **H** show the labeling of the platelets, which are positive for GPIIb/IIIa, have low levels of IgG and are negative for IgM and C3. The **C,** **G, E,** and **I** show the labeling of the ? immune complexes that have little or no GPIIb/IIIa but have extraordinarily high levels of IgG, IgM, and C3. Each of the histograms has a four decade logarithmic horizontal scale. Thus the IgM levels of the ? immune complexes is nearly 10^4 times that of the platelets.

variability of the technique, the average percent platelet reticulocytes was 1.0%(+/−1.1 s.d.) and the average absolute number of platelet reticulocytes was 2686/μl³(+/−3070 s.d.).

Figure 23.7 illustrates the results of studies on 400 patients in whom we have measured platelet count, PAIg, and "platelet reticulocytes." It can be clearly seen that the proportion of "platelet reticulocytes" increased in those patients who had platelet counts below about 75,000. The most dramatic increases were seen in those patients who had PAIg levels above the normal range.

These same data were used to calculate the absolute "platelet reticulocyte" level. It is this parameter that should reflect the rate of thrombopoiesis. Elevated levels of absolute "platelet reticulocytes" were observed over a wide range of platelet counts. Two patients with platelet counts greater than 300,000 and negative PAIg had marked elevations of absolute platelet reticulocytes. Both of these patients carry a clinical diagnosis of idiopathic thrombocytosis. Also of note is the frequent occurrence of patients with positive PAIg and severe thrombocytopenia but with normal or low levels of "platelet reticulocytes." It is possible that these patients represent examples of immune thrombocytopenia with a production defect as previously described (40).

Serial studies have been performed in patients who might be expected to have a marked increase in thrombopoiesis. The period of steepest rise in platelet count was accompanied by a marked increase in the level of "platelet reticulocytes."

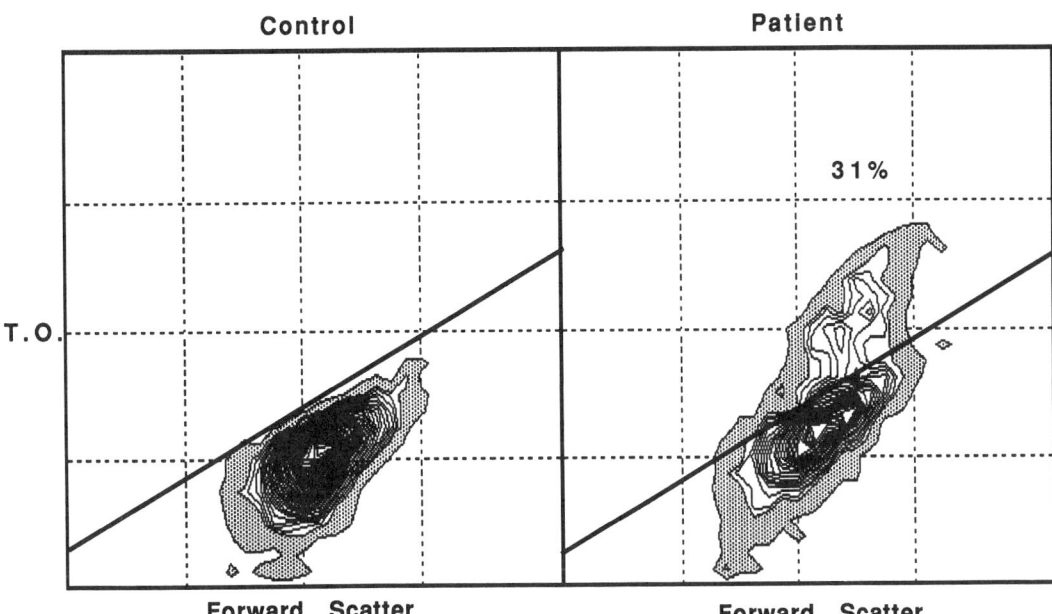

Figure 23.6. Thiazole orange labeling of platelets. The analysis of data for a normal control (**left**) and a patient (**right**) is illustrated. In each panel the horizontal axis is forward light scatter (a measure of particle size), and the vertical axis is thiazole orange fluorescence. Note the positive correlation between size and fluorescence intensity. The sloping line is drawn on the control sample in such a way that it is parallel to the platelet cluster and 99% of the platelets are below the line. The same line is then used to analyze the patient sample. The percentage of platelets falling above the line is recorded. Note that the platelets above the line have approximately the same slope correlation between fluorescence and size as the "normal" platelets, and that they are, on average, larger as judged by light scatter.

This type of behavior is reminiscent of the response described in erythrocyte reticulocytes during response to therapy for pernicious anemia (44).

Our data show that it is possible to identify a population of platelets that have increased nucleic acid content relative to their size. These "platelet reticulocytes" appear to be slightly larger, on average, than the other platelets, as judged by light scatter (see Fig. 23.6). The proportion of these platelets is inversely related to platelet count in patients who are thrombocytopenic. When expressed in absolute numbers, they are increased in clinical settings in which one would expect increased rates of thrombopoiesis.

Based on our results and the data of others (45), we postulate that the "platelet reticulocytes" detected here represent recently released platelets that mature rapidly in the circulation, and that their measurement is an estimate of the rate of thrombopoiesis in the same sense that an erythrocyte reticulocyte count is a measure of erythropoiesis.

A number of questions remain to be answered. From a technical point of view, the counting of these platelets is somewhat arbitrary. The loss of nucleic acid with maturation is, presumably, a continuous process and thus the platelet reticulocytes are not always clearly resolvable from the bulk of platelets. It is likely that the decrease in nucleic acid content is related to the loss of unstable messenger RNA, but this remains to be studied in detail. It remains to be determined exactly how long newly released platelets remain identifiable as "platelet reticulocytes." It is also unknown exactly what nucleic acids are being measured. Thiazole orange will fluoresce in combination with both RNA and DNA.

It is likely that the ability to easily measure the proportion of young platelets will prove to be useful clinically in the differential diagnosis of unexplained thrombocytopenia.

PLATELET ACTIVATION

As described above, the functional responses of platelets are limited to three: release, adhesion, and aggregation. It has recently become clear that it is possible to measure all three in the flow cytometer with all of the advantages of using whole blood techniques and obtaining correlated measurements (6).

Traditionally, platelet release has been measured by testing for the presence of soluble granule contents (serotonin, platelet factor 4, etc.) in the supernatant after platelet stimulation. Aggregation has been measured by preparing "platelet-rich plasma" and measuring the change in light transmission as the platelets aggregate. This "aggregometry" has become well established as a measure of platelet function. Both of these methods suffer from some obvious limitations. They both require manipulation of the platelets prior to making the measurement. They are only applicable to in vitro measurements, and cannot reflect the physiological status of platelets in vivo.

Recently, several groups have described immunological markers of platelet activation that, in combination with flow cytometry, may prove to be of considerable clinical utility. Although beyond the scope of this chapter, there is sufficient

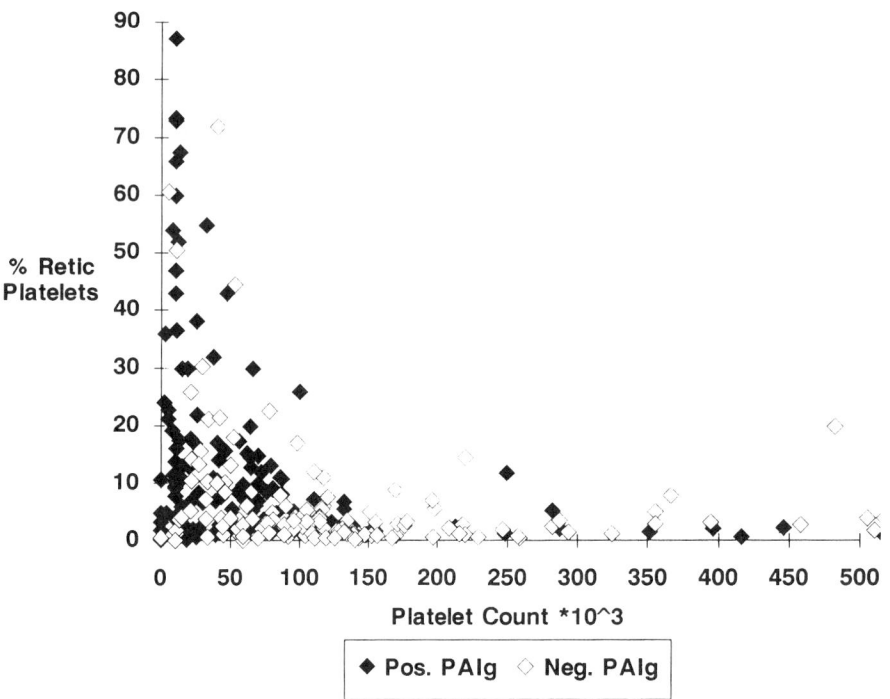

Figure 23.7. The relationship between platelet count and reticulated platelets. The relationship between platelet count and the proportion of "platelet reticulocytes" for 400 patients is shown. Those patients having elevated levels of PAIg are shown by the closed diamonds, those having normal levels of PAIg, by the open diamonds.

Table 23.1
Immunological Markers of Platelet Activation

Antigen		Clone Designation	Ref
α granule membrane proteins	GMP-140 (PADGEM) CD62	S12, W40, KC4	68, 69
Lysosomal proteins	53kD CD63	2.28	54
Fibrinogen receptor	GPIIb/IIIa	PAC1, 7E3	70, 71
Fibrinogen		9F9	72
Thrombospondin		5G11, P10	48, 53
Factor Xa			73

information published to strongly suggest that platelet activation in vivo is important in the pathogenesis of some forms of cardiovascular disease, such as thrombosis and atherosclerosis (46, 47). Platelet activation in vivo has been detected primarily by measurement of the secreted products of activated platelets. These methods are very sensitive but do not allow precise definition of the time or place at which activation is occurring. Thus the availability of immunological markers of platelet activation offers the promise of determining exactly when and where platelets become activated in vivo if the appropriate samples can be obtained.

In addition to the possible clinical applications, the research applications for such markers are too numerous to describe here. The one area of research that our laboratory has emphasized is the use of these new markers to obtain correlated information about the sequence of events occurring during platelet activation and aggregation.

Table 23.1 is a summary of the immunological markers of platelet activation of which we are aware. In general, they fall into three categories.

First are those that are components of the membranes of platelet granules. These have the interesting property that they are not present on the surface of resting platelets but become incorporated into the platelet membrane, in the process of "exocytosis" of the granule contents, when the granule membrane fuses with the platelet surface membrane. These are probably the most spectacular of the activation markers in that they switch from completely negative on resting platelets to strongly positive on platelets that have undergone the release reaction. They have the drawback however that they only appear on platelets that are fully activated in the sense that they have released the contents of their granules. It is quite likely that there are intermediate or more subtle levels of platelet activation that would be missed by these markers.

Second are those epitopes that result from an activation-related change on the platelet surface. The best example of this is PAC1, which binds to a new epitope of GPIIb/IIIa that appears at the time of platelet activation. Such a marker appears to reveal a more subtle, perhaps earlier stage in platelet activation. However, because the antigen is one that is at least potentially present on all platelets, the change from

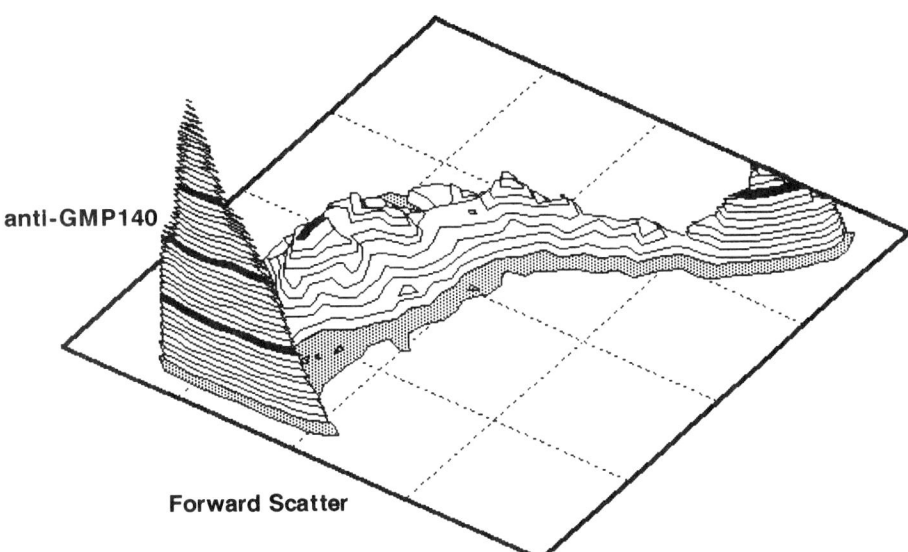

Figure 23.8. Simultaneous measurement of platelet activation and aggregation. This is a flow cytometric analysis of platelets that have been induced to aggregate by addition of 10 μM ADP to heparinized blood. This data was obtained by fixing the blood after 8 min at 37°C. All of the particles in this analysis were GPIb-positive. The horizontal axis is log forward scatter. The *vertical axis* is log fluorescence using phycoerythrin conjugated anti-GMP-140 to detect platelet activation. Within the platelet size range there are both activated and unactivated platelets. In addition, platelet aggregates can be seen extending to the *upper right*. The peak of aggregates in the *upper right* is approximately 10–20 μm in diameter.

resting to activated tends to be less dramatic and, at least in the case of PAC1, the epitope is easily affected by fixation. A number of monoclonal antibodies to GPIIb/IIIa have, to varying degrees, the property of binding more strongly to activated than to resting platelets.

Finally, there are several antigens that appear on the activated platelet because they bind to receptors on the platelet surface. These include fibrinogen itself, thrombospondin, and factor Xa (48, 73). The levels of these substances on activated platelets tend to be low, and the amount can be influenced by factors external to the platelet itself; but the presence or absence of such substances can be of considerable interest because of the functional consequences of their binding.

In addition to these, there are several nonimmunological techniques that have been applied to the flow cytometric study of platelet activation. These include the measurement of Ca^{2+} ion flux using fluorescent dyes that are sensitive to Ca^{2+} concentration (49, 50), and the use of dyes such as mepacrine, which enter platelet dense granules and then are released upon activation (51). Both of these methods, although very powerful, require that the platelets be loaded with the dye prior to activation and, thus, although useful for in vitro research, do not lend themselves to the measurement of in vivo platelet activation.

The protocol we have been using in most of our studies of platelet activation involves immediate fixation of whole blood followed by labeling. In the course of experiments in which platelet agonists were added to the blood prior to fixation we observed that two dramatic changes were taking place in the platelets. First, the platelets became positive for the activation marker (GMP-140 in this case). Second, the size distribution and light scatter of the platelets changed in such a way as to suggest the presence of platelet aggregates. This led us to a series of experiments in which we showed that the flow cytometer was capable of measuring not only the expression of the activation markers but also the phenomenon of aggregation (6). This is illustrated in Figure 23.8. Several interesting observations have come out of these studies. First, platelet aggregation as measured on the flow cytometer has the same general properties as that measured on an aggregometer and we believe that the information is equivalent in terms of clinical usefulness. There are, however, several differences between the aggregation seen in our system and that seen in an aggregometer.

First, the aggregates that we see in whole blood do not become much larger than leukocytes, i.e., about 20 μm, whereas the aggregates that form in platelet-rich plasma in an aggregometer are macroscopic precipitates. This may be due to the lower effective concentration of platelets in whole blood and the presence of so many other particles that decrease the rate of growth of the aggregate.

Second, we find the most of the aggregation that occurs in whole blood is reversible within about 15 min at 37°C, whereas that which occurs in an aggregometer is usually described as irreversible. This difference may be related to the size of the aggregates that form in whole blood, although our experience suggests that a considerable amount of the aggregation seen in an aggregometer is also reversible if one follows the reaction long enough.

Third, we have convincing evidence of the existence of platelet-leukocyte aggregates as well as platelet-platelet aggregates. Preliminary experiments have suggested that activated platelets are most likely to bind to monocytes rather than to erythrocytes, lymphocytes, or granulocytes. This phenomenon may be of importance in view of the recently described structural relationship between several cell-cell adhesion molecules and GMP-140 (52).

Fourth, our data suggests that the phenomena of activation (as measured by expression of GMP-140) and aggregation are quite distinct. In particular, we have observed that platelets that do not express GMP-140 can participate in aggregation. The exact sequence of platelet activation in terms of expression of surface markers and the relation between changes on the surface and aggregation remain to be explored in detail. Other activation markers, such as the activation-dependent epitopes of GPIIb/IIIa are likely to be directly involved in platelet aggregation, whereas perhaps molecules such as GMP-140 are involved in binding of platelets to other types of cells. Flow cytometry will play an important role in this work in the future.

There are a number of areas in which one might expect platelet activation markers to prove to have clinical ability. These can be broadly classified as those applications involving "biocompatibility" issues, platelet storage and transfusion, assessment of cardiovascular disease, and assessment of platelet dysfunction. Although none of these areas has been extensively explored, some data is available.

In the area of biocompatibility, we are referring to the use of platelet activation markers to assess the degree of interaction between platelets and foreign surfaces, such as those encountered in dialysis, cardiopulmonary bypass, vascular protheses, etc. George et al. (53) have reported data using GMP-140 and thrombospondin as markers in patients undergoing cardiopulmonary bypass. They have shown no increase in either GMP-140 or thrombospondin but have noted an increase in platelet membrane microparticles, suggesting that platelet fragmentation, but not release, is taking place. Our own laboratory has shown a significant rise in the proportion of GMP-140-positive platelets during bypass, as have Nieuwenhuis et al. (54). We have also performed "closed loop" bypass experiments in which we observed an initial transient wave of GMP-140-positive platelet aggregates at the initiation of circulation, followed by a significant rise in GMP-140-positive platelets over a period of several hours (55). We have suggested that the results can be best explained by proposing that platelets are initially activated and aggregate in response to the foreign surface, but that many of the platelet aggregates subsequently dissociate, resulting in a mixture of activated and apparently unactivated platelets exiting the bypass. As the activated platelets return to the in vivo circulation, our data suggests, they have a very short survival time and are essentially "filtered" out so that they do not appear in large numbers in the circulation and do not accumulate during bypass. It is entirely possible, as discussed above, that the apparently unactivated platelets that are not expressing GMP-140 are nevertheless abnormal; they may be found to be activated when other markers are used and they may very well be dysfunctional, as has been observed by others (1, 56, 57).

In the area of platelet storage and transfusion, there is abundant evidence that stored platelets undergo progressive activation. This has been shown by examination of shape change, Ca^{2+} concentration (58), release of platelet granule contents (59), and, in our laboratory, by expression of GMP-140 (60). In addition, we have accumulated evidence suggesting that GMP-140-positive platelets have a very short survival in the circulation once transfused. Thus, it seems likely that many of the platelets routinely transfused have become activated during storage and may not contribute very much to improved hemostasis in the recipient. Further studies of the mechanism an kinetics of this process may contribute to more effective storage of platelets in the future.

There is every reason to think that platelets may be involved in several different types of cardiovascular disease and that the ability to assess platelet activation, especially in vivo, may be very useful in these diseases. Platelet deposition may play a role in the primary events of atherosclerosis and obstruction leading to myocardial infarction and stroke (46, 47). Platelets certainly play a role in the reocclusion of vessels after they are opened, either surgically, or after angioplasty. Finally, it is likely that platelets contribute to the process of venous thrombosis as well.

There are several difficulties that one encounters in attempting to use platelet activation to assess the time course or magnitude of each of these cardiovascular events. First, even if every platelet passing through a damaged coronary artery were activated, the activated platelets would be so massively diluted in the systemic circulation that they would be difficult to detect. Most authors put the detection limit of current flow cytometric methods at about 1% to 5% activated platelets (5, 61), and this may be optimistic for measurements made outside of a controlled laboratory situation. Second, there is every reason to believe that activated platelets have a much shorter survival time in the circulation than resting platelets. Thus, the activated platelets probably do not accumulate in the circulation as the disease process continues. For these reasons, with present technology, it seems it will be necessary to look for platelet activation at or near the site of the disease process. For example, in preliminary studies, we have been able to demonstrate high levels of platelet activation in samples drawn directly from a diseased coronary artery, although we were unable to detect any increase above background in samples taken simultaneously from the pulmonary artery.

Despite these difficulties, there may be an important role for measurements of platelet activation in vivo. For example, it might to possible to estimate the rate at which platelets are being activated as they cross a vascular prosthesis, and such a measurement might predict which prostheses are likely to fail. A similar line of reasoning might apply to studying platelet activation across coronary vessels that have been opened by thrombolytic agents or by angioplasty. If, after flow is restored, there is a relatively high level of platelet activation, it might mean that the endothelium is so damaged or that the flow pattern is such that a platelet thrombus is likely to form and lead to reocclusion. Obviously, a great deal more work needs to be done to determine who, when, and where, in vivo studies of platelet activation using markers will be clinically useful.

The final area in which measures of platelet activation may be of clinical utility is in the diagnosis of platelet dysfunction. Use has already been made of low cytometric assays of platelet surface glycoproteins in the diagnosis of genetic platelet disorders (48) and in assessment of platelet modification by plasmin (62). It seems likely that, as we come to understand better the sequence of activation marker displayed on normal platelets, we will be able to use this information to detect subtle abnormalities of platelet function. Of particular importance is the possibility of accurately measuring platelet hyper-function. At the moment there is no suitable, clinically applicable technique for detecting hyper-responsive platelets. There is evidence that platelet hyper-reactivity may be involved in the increased incidence of myocardial infarction in the morning hours (63, 64) and may be an important risk factor for thrombotic disease (47). It has also been shown that platelet hyper-reactivity can accompany the hyperlipidemic state and thus may be involved in the increased incidence of cardiovascular disease in these patients (65, 66). The current flow cytometric methods using fixed blood samples may lend themselves to this problem very nicely.

In addition, it is very likely that powerful new therapeutic interventions for modulating platelet function will become available in the near future. For example, the use of monoclonal antibodies to block functional platelet surface interactions is already being explored (67). The ability to monitor platelet functional status in vivo will be important in the use of these therapies, and flow cytometric approaches once again lend themselves to this type of monitoring.

The recent availability of immunological markers of platelet activation, combined with the unique capability of flow cytometry to identify small subpopulations of platelets and to quantitate the level of markers, now allows us to assess platelet function in an entirely new way. The clinical and research potential of these developments is only beginning to be explored.

REFERENCES

1. Harker LA, Malpass TW, Branson HE, Hessell EA, Slichter SJ. Mechanism of abnormal bleeding in patients undergoing cardiopulmonary bypass: acquired transient platelet dysfunction associated with selective alpha granule release. Blood 1980;56:824–834.
2. Hemler ME. Adhesive protein receptors on hematopoietic cells. Immunol Today 1988;9:109–113.
3. Johnston GI, Cook RG, McEver RP. Cloning of GMP-140, a granule membrane protein of platelets and endothelium: Sequence similarity to proteins involved in cell adhesion and inflammation. Cell 1989;56:1033–1044.
4. Ault KA, Mitchel JG, Rinder HM, Rinder C, Hillman RS. Flow cytometric analysis of plateletes. In: Laerum OD, Bjerknes R, eds. Flow Cytometry in Hematology. Academic Press, London, 1990.
5. Shattil SJ, Cunningham M, Hoxie JA. Detection of activated platelets in whole blood using activation dependent monoclonal antibodies and flow cytometry. Blood 1987;70:307–315.
6. Ault KA, Rinder HM, Mitchell JG, Rinder CS, Lambrew CT, Hillman RS. Correlated measurement of platelet release and aggregation in whole blood. Cytometry 1989;10:448–455.
7. Sims PJ, Faioni EM, Wiedmer T, Shattil SJ. Complement proteins C5b-9 cause release of membrane vesicles from the platelet surface that are enriched in the membrane receptor for coagulation factor Va and express prothrombinase activity. J Biol Chem 1988;263:18205–18212.
8. Karas SP, Rosse WF, Kurlander RJ. Characterization of the Igg-Fc receptor on human platelets. Blood 1982;60:1277–1281.
9. Rosenfeld SI, Looney RJ, Leddy JP, Phipps DC, Abraham GN, Anderson CL. Human platelet Fc receptor for immunoglobulin G. J Clin Invest 1985;76:2317–2322.
10. Steiner M, Luscher EF. Identification of the immunoglobulin G receptor of human platelets. J Biol Chem 1986;261:7230–7236.
11. Blumberg N, Masel D, Stoler M. Disparities in estimates of IgG bound to normal platelets. Blood 1986;67:200–202.
12. Shulman NR, Jordan JV, Jr. Platelet Immunology. In: Colman RW, Hirsh J, Marder VJ, Salzman EW, eds. Hemostasis and Thrombosis. Philadelphia: J. B. Lippincott 1987:497.
13. Fabris F, Casonato A, Randi ML, Luzzatto G, Girolami A. Clinical significance of surface and internal pools of platelet associated immunoglobulins in immune thrombocytopenia. Scand J Haematol 1986;37:215–220.
14. Pfueller SL, David R. Liberation of surface and internal platelet associated IgG during platelet activation. Brit J Haematol 1986;63:785–794.
15. George JN, Saucerman S, Levine SP, Knieriem LK. Immunoglobulin G is a platelet alpha granule secreted protein. J Clin Invest 1985;76:2020–2025.
16. Spycher MO, Nydegger UE. Part of the activating cross linked immunoglobulin G is internalized by human platelets to sites not accessible for enzymatic digestion. Blood 1986;67:12–18.
17. Santoso S, Zimmermann U, Neppert J, Mueller-Eckhardt C. Receptor patching and capping of platelet membranes induced by monoclonal antibodies. Blood 1986;67:343–349.
18. Cines DB, Schreiber AD. Immune thrombocytopenia: Use of a Coomb's antiglobulin test to detect Igg and C3 on platelets. N Engl J Med 1979;300:106.
19. Leporrier M, Dighiero G, Auzemery M, Binet JL. Detection and quantification of platelet bound antibodies with immunoperoxidase. Br J Haematol 1979;42:605.
20. Kernoff LM, Blake KCH, Shackleton D. Influence of the amount of platelet bound IgG on platelet survival and site of sequestration in autoimmune thrombocytopenia. Blood 1980;55:730.
21. Dixon R, Rosse W, Ebbert L. Quantitative determination of antibody in idiopathic thrombocytopenic purpura. N Engl J Med 1975;292:230.
22. Xiaolian P, Bianming W, Di S, Wenning W. Clinical value of detecting platelet associated IgG in patients with idiopathic thrombocytopenic purpura. Acta Acad Med Wuhan 1982;170:1982.
23. Lobuglio AF, Court WS, Vinocur L, et al. Immune thrombocytopenic purpura. N Engl J Med 1983;309:459.
24. Rosse WF, Devine DV, Ware R. Reactions of immunoglobulin G binding ligands with platelets and platelet associated immunoglobulin G. J Clin Invest 1984;73:489.
25. Shaw GM, Axelson J, Maglott JG, LoBuglio AF. Quantification of platelet bound Igg by ^{125}I staphylococcal protein A in immune thrombocytopenic purpura and other thrombocytopenic disorders. Blood 1984;63:154.
26. Janson M, McFarland JG, Aster RH. Quantitative determination of platelet surface alloantigens using a monoclonal probe. Hum Immunol 1986;15:251.
27. Morse BS, Guiliani D, Nussbaum M. Quantitation of platelet associated IgG by radial immunodiffusion. Blood 1981;57:809.
28. Kunicki TJ, Koenig MB, Kristopeit SM, Aster RH. Direction quantitation of platelet bound IgG by electroimmunoassay. Blood 1982;60:54.

29. Hymes K, Shulman S, Karpatkin S. A solid phase radioimmunoassay for bound antiplatelet antibody. Studies on 45 patients with autoimmune platelet disorders. J Lab Clin Med 1979;94:639.
30. Hegde UM, Boes A, Powell DK, Joyner MV. Detection of platelet bound and serum antibodies in autoimmune thrombocytopenia by enzyme linked assay. Vox Sang 1981;41:306.
31. Hotchkiss AJ, Leissinger CA, Smith ME, et al. Evaluation by quantitative acid elution and radioimmunoassay of multiple classes of immunoglobulins and serum albumin associated with platelets in ITP. Blood 1986;67:1126.
32. Ballem PJ, Segal GM, Stratton JR, Gernsheimer T, Adamson JW, Slichter SJ. Mechanisms of thrombocytopenia in chronic autoimmune thrombocytopenic purpura: Evidence of both impaired platelet production and increased platelet clearance. J Clin Inv 1987;80:33–40.
33. Slichter SJ, McFarland JG, Hansen S. Depressed platelet production: a major unrecognized component of autoimmune thrombocytopenia. Blood 1983;62:248.
34. Corash L, Rheinschmidt M. Detection of platelet antibodies with a fluorescence activated flow cytometric technique. In: Rose NR, Friedman H, Fahey JL, eds. Manual of Clinical Laboratory Immunology. American Society for Microbiology, Washington, D.C., 1986.
35. Rosenfeld CS, Nichols G, Bodenstein DC. Flow cytometric measurement of antiplatelet antibodies. Am J Clin Pathol 1987;87:518–522.
36. Lazarchick J, Hall SA. Platelet associated IgG assay using flow cytometric analysis. J Immunol Methods 1986;87:257–265.
37. Ault KA. Flow cytometric measurement of platelet associated immunoglobulin. Pathol Immunopath Res 1988;7:395–408.
38. Gottschall JL, Collins J, Kunicki TJ, Nash R, Aster RH. Effect of hemolysis on apparent values of platelet associated IgG. Am J Clin Pathol 1987;87:218–222.
39. Zucker-Franklin D, Karpatkin S. Red-cell and platelet fragmentation in idiopathic thrombocytopenic purpura. N Engl J Med 1977;297:517–523.
40. Ballem PJ, Segal GM, Stratton JR, Gernsheimer T, Adamson JW, Slichter SJ. Mechanisms of thrombocytopenia in chronic autoimmune thrombocytopenic purpura. Evidence of both impaired platelet production and increased platelet clearance. J Clin Invest 1987;80:33–40.
41. Ingram M, Coopersmith A. Reticulated platelets following acute blood loss. Brit J Haemat 1969;17:225–228.
42. Karpatkin S. Human platelet senescence. Ann Rev Med 1972;23:101–128.
43. Lee LG, Chen CH, Chiu LA. Thiazole Orange, a new dye for reticulocyte analysis. Cytometry 1986;7:508–517.
44. Hillman RS. Characteristics of marrow production and reticulocyte maturation in normal man in response to anemia. J Clin Inv 1969;48:443–453.
45. Kienast J, Schmitz G. Flow cytometric analysis of thiazole orange uptake by platelets: a diagnostic aid in the evaluation of thrombocytopenic disorders. Blood 1990;75:116–121.
46. Fitzgerald D, Roy L, Catella F, Fitzgerald GA. Platelet activation in unstable coronary disease. N Engl J Med 1986;315:983–990.
47. Trip MD, Cats VM, van Capelle FJL, Vreeken J. Platelet hyperreactivity and prognosis in survivors of myocardial infarction. N Engl J Med 1990;322:1549–1554.
48. Marti GE, Magruder L, Schuette WE, Gralnick HR. Flow cytometric analysis of platelet surface antigens. Cytometry 1988;9:448–455.
49. Johnson, PC, Ware JA, Cliveden PB, Smith M, Dvorak AM, Salzman EW. Measurement of ionized calcium in blood platelets with the photoprotein aequorin: Comparision with Quin 2. J Biol Chem 1985;260:2069–2076.
50. Davies TA, Drotts D, Weil GJ, Simons ER. Flow cytometric measurements of cytoplasmic calcium changes in human platelets. Cytometry 1988;9:138–142.
51. Corash L, Mok Y, Rheinschmidt M. Analysis of platelet function using subcellular and surface associated fluorescent probes. In: Taylor LD, Waggoner AS, eds. Applications of Fluorescence in the Biomedical Sciences. New York: Alan R. Liss Inc., 1988;567–584.
52. Marx JL. New family of adhesion proteins discovered. Science 1989;243:1144.
53. George JN, Pickett EB, Saucerman S, McEver RP, Kunicki TJ, Kieffer N, Newman PJ. Platelet surface glycoproteins: Studies on resting and activated platelets and platelet membrane microparticles in normal subjects, and observations in patients during adult respiratory distress syndrome and cardiac surgery. J Clin Invest 1986;78:340–348.
54. Nieuwenhuis HK, van Oosterhout JJG, Rozemuller E, van Iwaarden F, Sixma JJ. Studies with a monoclonal antibody against activated platelets: Evidence that a secreted 53,000 molecular weight lysosome like granule protein is exposed on the surface of activated platelets in the circulation. Blood 1987;70:838–845.
55. Rinder H, Rinder C, Mitchell J, Forest R, Hillman RS, Ault KA. The kinetics of platelet activation and aggregation during cardiopulmonary bypass: a wave of early aggregation is dissociated from a progressive increase in platelet activation. Blood 1988;72:307a.
56. Zilla P, Fasol R, Groscurth P, Klepetko W, Reichenspurner H, Wolner E. Blood platelets in cardiopulmonary bypass operations. Recovery occurs after initial stimulation, rather than continual activation. J Thorac Cardiovasc Surg 1989;379–388.
57. Wenger RK, Lukasiewicz H, Mikuta BS, Niewiarowski S, Edmunds LH. Loss of platelet fibrinogen receptors during clinical cardiopulmonary bypass. J Thorac Cardiovasc Surg 1989;97:235–239.
58. Feinberg H, Sarin MM, Batka EA, Porter CR, Miripol JE, Stewart M. Platelet storage: Changes in cytosolic Ca^{2+}, actin polymerization, and shape. Blood 1988;72:766–769.
59. Murphy S, Gardner FH. Platelet storage at 22°C.: Metabolic morphologic, and functional studies. J Clin Invest 1971;50:370.
60. Rinder H, Murphy M, Ault KA, Mitchell J, Stocks J, Slichter S, Hillman RS. Quantitation of platelet activation during storage: Evidence that activated platelets fail to circulate following transfusion. Blood 1988;72:283a.
61. Johnston GI, Pickett EB, McEver RP, George JN. Heterogeneity of platelet secretion in response to thrombin demonstrated by fluorescence flow cytometry. Blood 69:1401–1403.
62. Adelman B, Michelson AD, Handin RI, Ault KA. Evaluation of platelet glycoprotein Ib by fluorescence flow cytometry. Blood 1985;66:423.
63. Brezinski DA, Toffler GH, Muller JE, Pohjola-Sintonen S, Willich SN, Schafer AI, Czeisler CA, Williams GH. Morning increase in platelet aggregability. Circulation 1988;78:35–40.
64. Trip MD, Cats VM, van Capelle FJL, Vreeken J. Platelet hyperreactivity and prognosis in survivors of myocardial infarction. N Engl J Med 1990;322:1549–1554.
65. Carvalho ACA, Colman RW, Lees RS. Platelet function in hyperlipoproteinemia. N Engl J Med 1974;290:434–438.
66. Shattil SJ, Anaya-Galindo R, Bennett J, Colman RW, Cooper RA. Platelet hypersensitivity induced by cholesterol incorporation. J Clin Invest 1975;55:536–643.
67. Coller BS, Scudder LE, Berger HJ, Iuliucci JD. Inhibition of human platelet function in vivo with a monoclonal antibody. With observations on the newly dead as experimental subjects. Ann Intern Med 1988;109(8):635–638.
68. McEver RP, Martin MN. A monoclonal antibody to a membrane glycoprotein binds only to activated platelets. J Biol Chem 1984;259:9799–9804.
69. Berman CL, Yeo EL, Wencel-Drake JD, Furie BC, Ginsberg MH, Furie, B. A platelet alpha granule membrane protein that is associated with the plasma membrane after activation. J Clin Invest 1986;78:130–137.
70. Shattil SJ, Hoxie JA, Cunningham MC, Brass LF. Changes in the platelet membrane glycoprotein IIb-IIIa complex during platelet activation. J Biol Chem 1985;260:11107–11114.
71. Shattil SJ, Motulsky J, Insel PA, Flaherty L, Brass LF. Expression of fibrinogen receptors during activation and subsequent desensitization of human platelets by epinephrine. Blood 1986;68:1224–1231.

72. Coller BS. A new murine monoclonal antibody reports an activation dependent change in the conformation and/or microenvironment of the platelet glycoprotein IIb/IIIa complex. J Clin Invest 1985;76:101.

73. Tracy PB, Mann KG. Prothrombinase complex assembly on the platelet surface is mediated through the 74,000 dalton component of factor Va. PNAS (USA) 1983;80:2380–2384.

24

Flow Cytometric Analysis of Granulocytes

J. PAUL ROBINSON and WAYNE O. CARTER

INTRODUCTION

Neutrophils are derived from pluripotential stem cells in bone marrow. Both monocytes and granulocytes share a common parental stem cell (GM-CFU–granulocyte-macrophage colony forming unit). Several distinct stages of development are recognized: myeloblast, promyelocyte, myelocyte, metamyelocyte, band cell, and segmented neutrophil. The maturation process is regulated by a number of substances including growth factors produced by circulating monocytes and lymphocytes such as GM-CSF (granulocyte-macrophage colony stimulating factor), M-CSF, G-CSF (1), C3e (2), erythropoietin, and IL-3. Neutrophils are normally stored in the bone marrow for five to seven days, after which mature neutrophils are released into the blood. Factors such as G-CSF and IL-1 play a key role in the release of neutrophils from marrow to the circulating pools (3–5). GM-CSF from T-lymphocytes is implicated in the stimulation of neutrophil progenitors at doses as low as 50-100 U/ml (6). Corticosteroids commonly cause a significant increase in circulating leukocytes, primarily as a result of increased release of neutrophils from bone marrow stores. A secondary effect includes a decreased neutrophil adherence to vascular endothelium, a decreased migration of cells out of the vasculature and a slight prolongation of the neutrophil circulating half-life (7–9).

In blood vessels, two pools of neutrophils are recognized: circulating and marginating. The former circulate throughout the body in the blood stream, while the marginating pool consists of neutrophils attached to endothelial cell surfaces of small capillaries and venules. An approximately equal number of neutrophils occupy each pool in the human. Neutrophils in the circulating pool have a half-life of about 7 hr (10) after which they marginate and emigrate through tissue where they remain functional for one to two days. They are subsequently phagocytosed by macrophages or are disposed of through the mucosal surfaces. In an adult, approximately 1.5×10^9 neutrophils/kg (body weight) are manufactured daily (11). The neutrophil is a cell that is affected by, and can be responsible for, a large number of clinical syndromes. It operates as a primary source of toxic oxygen metabolites as well as a major contributor to the early inflammatory response and is, therefore, a cell of significant importance.

The purpose of this chapter is to examine the role of neutrophils and their functions in the immune system in terms of abnormalities of function and methods for analysis. Several aspects of neutrophil physiology and function will be explored and placed in the context of current experimental techniques. Of primary interest will be the use of flow cytometric techniques now available for single-cell analysis in clinical evaluations. Table 24.1 provides a simple overview of some of the clinical disorders of neutrophils, many of which are discussed in this chapter. In so doing, it is intended to provide some explanation as to the value and utility of each technique. Table 24.2 lists a number of drugs known to affect neutrophil function. These will not be discussed individually, except for the specific neutrophil function defects.

GENERAL OVERVIEW OF FUNCTION

Inflammation has been generally recognized by the classic symptoms of calor (heat), rubor (redness), tumor (swelling), and dolor (pain). It has taken several decades to advance our knowledge beyond a peripheral involvement of the neutrophil in illness and disease. The past decade has seen tremendous advances in the understanding of the physiology and biology of neutrophil function. It was Metchnikoff's description of the phagocytic process as part of a host defense system (12) and the subsequent demonstration of leukocyte chemotaxis that began our present understanding of neutrophil function. The measurement of in vitro chemotaxis by Comandon in 1917 (13) and more recently by Boyden in 1962 (14) provided a means for practical evaluation of neutrophils and other phagocytic cells. Boyden's in vitro techniques quickly spurred several major developments that linked an endogenous chemotactic factor with a major chemotactic factor (15–17). More recently, the linkage between leukocyte recruitment and inflammation has become more apparent with the discovery of a number of glycoproteins (CD11/CD18 complex) whose primary purpose is the adhesion to vessel walls (18). CD11b/CD18 is undoubtedly the same protein that facilitated the earlier observations of C3-coated particle adherence to cell membranes (18, 19). There are many techniques now available for the evaluation of neutrophil function. While this chapter cannot cover in detail all of these methods

Table 24.1
Clinical Disorders of Neutrophil Function

Function	Inherited Disorder	Acquired Disorder
Chemotaxis	Job's syndrome (75, 300–302)	Malnutrition (303)
	SCID (304, 305)	Periodontal Disease (302, 306–309)
	Chediak-Higashi (159–161, 310, 311)	Thermal injury (164, 312–316)
		Diabetes mellitus (317, 318)
	α-mannosidase deficiency	Hodgkin's disease (162)
	Leukocyte adhesion deficiency (149–151)	SLE (319, 320)
	Kartagener's syndrome (73, 75)	Rheumatoid arthritis (166, 167, 321, 322)
	Actin dysfunction (90)	Hepatic cirrhosis (163)
Phagocytosis	Actin dysfunction (323)	Thermal injury (324)
	Tuftsin deficiency (183)	Splenectomy (325)
		Juvenile periodontitis (182)
		Neonate (181)
Microbicidal Killing	Chediak-Higashi (160)	Malnutrition (326, 327)
	MPO deficiency (328)	Thermal injury (329)
	Specific granule deficiency (26, 330–332)	Diabetes mellitus (333)
	CGD (334)	Sepsis
	Actin dysfunction	Hypogammaglobulinemia (335)
		Severe bacterial infections (336)
		Malnutrition (303)
		Hepatic cirrhosis (5539)
		Periodontal disease (30, 337, 338)
		Paraproteinemia (34, 339)
		AIDS (340)
		Splenoctomy (325)
		Rheumatoid arthritis (341)
		Diabetes mellitus (301, 342)
Adherence	Adherence glycoprotein (149–151)	Diabetes mellitus (343)
	Deficiencies (265)	Neonates (288, 289)
	Cystic fibrosis (344)	
Locomotion	Lazy leukocyte syndrome (155, 345)	Cytochalasin B (346)
		Malnutrition (303)
Oxidative Killing	CGD (76, 202, 203, 347)	Neonates (216, 348–350)
	G-6-P Dehydrogenase deficiency (351)	Thermal injury (352)
	Chediak-Higashi (160, 331)	Lassa fever (353)
Granule Functions	MPO deficiency (61–63, 354)	Thermal injury (355)
	Kartagener's syndrome (73–75, 356)	
	Chediak-Higashi (357, 358)	
Membrane Deformability	Chediak-Higashi	Neonates (350)
		Immature neutrophils (359)
Opsonic Defects	Chediak-Higashi (360)	
LTB$_4$ Receptors		Thermal injury (361)
		Rheumatoid arthritis (341, 362)
Fc$_\gamma$III Receptors	Cystic fibrosis (344)	Neonates (363)
		PNH[a] (278, 364–366)

[a]PNH—Paroxysmal nocturnal hemoglobinuria

(listed in Table 24.3), several are discussed in more detail below.

FLOW CYTOMETRIC PROPERTIES OF NEUTROPHILS

Flow cytometry as a tool for evaluation of neutrophils is particularly useful because of the properties neutrophils display on the cytometer. Figure 24.1 shows a typical two-parameter histogram of leukocytes run on a conventional cytometer. The single-parameter projections of the scatter are also shown in this figure. Because of their "granulocytic" properties, neutrophils are easily distinguished from other leukocytes. This property has been one of the easiest methods by which to separate neutrophils for lymphocytes or monocytes. Separation of eosinophils can be further accomplished by observing the light scatter at 90° under polarizing conditions. Many of the studies of neutrophil function evaluated using flow cytometry have taken advantage of the ability to selectively gate the neutrophil population without physical separation of the cells. This is a major advantage over some of the more conventional techniques used in studying neutrophils for several reasons. Firstly, there is less handling of the cells, which reduces activation of various metabolic pathways. Secondly, functional evaluations can be performed faster and in the presence of other cells that can be used as internal controls. Thirdly, properties of other leukocytes can be evaluated as part of a protocol, resulting in a significant time and sample volume reduction. Finally, far fewer cells are usually required when using flow cytometry, so a smaller volume of blood or other tissue is needed from a patient.

Table 24.2
Pharmacologic Alterations of Neutrophil Physiology

Chemical	Chemotaxis	Adherence	Degranulation	Microbicidal Killing	Phagocytosis
Alcohol	(367–369)	(367–369)			
Aminoglycosides				(370)	
Amphotericin B	(73, 371, 372)			(372)	(372, 373)
Aspirin	(367–369)	(367–369)			
Auranofin	(374)			(374, 375)	
Azelastine				(376)	
Chloroamphenicol	(377)				
Clindamycin	(378)				
Colchicine	(368)	(367–369)	(72)		
Cyclophosphamide				(379, 380)	
Dapsone				(381)	
Epirubicin				(382)	
Erythromycin	(378)				
γ-Interferon				(282)	(282)
Gentamycin	(383, 384)				
Ibuprofen	(367–369)	(367–369)		(385)	
Idarubicine				(382)	
Indomethacin	(367)				
Ketoconazole	(372)			(372)	
Naproxin	(367)				
Oxatomide				(386)	
Pentoxifylline	(387)	(387)		(387)	
Phenylbutazone	(367)				
Piroxicam	(367–369)	(367–369)			
Polymixin B				(388)	(389)
Rifampin	(377, 390)				
Steroids	(367)	(367–369)			
Sulphonamides				(356)	
Tetracyclines	(377, 378, 391, 392)			(393)	(394–397)

The specific property of 90° light scatter is thought to be related to refractive properties of the nucleus and cytoplasmic granules and is one of the most useful in the evaluation of neutrophils by flow cytometry. Other methods have been proposed for discrimination of different cell populations by flow cytometry. One in particular has been the use of the metachromatic dye acridine orange (AO), which has been shown to be useful for flow cytometric determinations of differential cell counts, among other things. Acridine orange intercalates into DNA and RNA as well as into the lysosomal granules. This property can be used to accurately discriminate between lymphocytes, monocytes, neutrophils, and eosinophils, although some caution should be used since AO-fluorescence wavelength is altered by changes in pH (20).

Bassoe and coworkers have made a significant contribution to many of the leukocyte quantitation techniques necessary for accurate determination of phagocytosis (21–24). Other methods for establishing a differential cell count using flow cytometry, such as esterase activity, are also useful and rapid. For instance, carboxyfluorescein diacetate (CF-DA), when in the presence of the cell, is rapidly hydrolyzed by cellular esterases to a highly fluorescent molecule, carboxyfluorescein. Figure 24.2 shows the two-parameter representation of leukocytes using 90° light scatter versus fluorescence before and after addition of CF-DA. This useful technique provides another means of identification of neutrophils, thereby also providing some information about the metabolic status of these cells.

PREPARATIVE PROCEDURES FOR EVALUATION OF NEUTROPHIL FUNCTION

Methods for neutrophil preparation vary considerably according to the isolation site and the number of cells required. It is important not to unduly "activate" neutrophils, since their ability to undergo stimulation may provide critically important information. Several preparation methods for functional evaluation have been used, such as Ficoll Hypaque separation using the methods of Boyum (25), percol density gradient separation technique (26), dextran sedimentation of buffy coats (27, 28), erythrocyte lysis using ammonium chloride (20), and many variations of these. One preparation technique that we have found particularly useful for clinical evaluations is a simple technique that we have termed the **overlay** method (29). This method uses only 500 µl of blood, requires little equipment, and is very fast.

Overlay Method: 500 µl of undiluted blood is carefully overlaid onto 1 ml of ficoll in a 2-ml plastic centrifuge bullet. The bullet is left motionless for 20 minutes on the bench at room temperature (Fig. 24.3). The top 250 µl of buffy coat is very carefully removed and washed with PBS.

The resultant leukocyte-rich suspension contains a large number of neutrophils useful for functional analysis. The primary value in using this technique would be the need for a

Table 24.3
Methods for Functional Assessment

Function	Traditional Method	Flow Methods (general reference (398))
Chemotaxis	Boyden chamber (12)	None
	Under agarose (399)	None
H_2O_2 production		DCFH-DA assay (56, 400)
O_2^- production	Cytochrome c reduction	Hydroethidine
		Dihydrorhodamine 123 (401)
Respiratory burst	Chemiluminescence (285, 350, 402–407)	
Bactericidal	S.aureus killing assay (302)	FITC-labeled S. aureus (172, 408, 409)
	AO Uptake (302)	DCF-Texas red (199)
Membrane potential	Spectrofluorometry (105, 106, 108, 110, 410–412)	DiOC6(5) (56, 104, 191, 413–417)
Viability	Trypan blue exclusion	PI exclusion
		Ethidium monoazide
Membrane fluidity	DPH assay (418–420)	
Adhesion glycoproteins	Fluoresceinated receptor (31, 259, 264, 421)	
	^{3}H-fMLP receptors (422–424)	FITC-fMLP (425–427)
Microtubule disruption	Phalacidin	Phalacidin (85)
Membrane structure	NBD phalacidin (428, 429)	
Degranulation	B-glucuronidase (430–432)	AS-B1 (52, 53)
		Lactoferrin (433)
Enzyme activity		LAP assay (434–438)
		Esterase activity (54, 439)
		mCIB
pH measurement		ADB activity (60)
Calcium flux	Quin 2 spectrofluorometry (440)	FITC quenching (171)
	Fura-2 (441, 442)	
	Indo-1 spectro	Indo-1 flow (99, 101, 443–446)
Phagocytosis	Colony counts, S. aureus	FITC-labeled orgs. (172, 176, 447)
	Fluorescence microscopy (302)	3-color flow assay (199)
	Latex phagocytosis (448, 449)	2-color methods (264, 421)
		Latex phagocytosis (450–452)
Pinocytosis		Fluid pinocytosis (187)
Bacterial degradation		AO fluorescence (60)
		DNA measurements (170)

population of minimally-activated neutrophils or for a very rapid technique when a limited cell number is required.

An important consideration when using neutrophils in a flow cytometer is their adhesive properties. Since most functional assays involve lengthy incubations, it is necessary to ensure that neutrophils remain in suspension during the duration of the experiment. Once neutrophils clump, there is little that can be done to separate them without significantly activating them further or damaging them. Neutrophils can be maintained for several hours at 4°C in PBS buffer containing EDTA, glucose, and gelatin (or bovine serum albumin) in the absence of calcium. It is important to understand, however, that this treatment can alter the antigenic expression of adhesion glycoproteins, in particular CD11b. This is discussed in more detail in a later section on these adhesion molecules.

In situations where few cells are available, flow cytometry can be used where other techniques might be less attractive. One such example is the collection of leukocytes from extremely small microenvironments, such as subgingival pockets that exist in periodontal disease. We have used flow cytometry to determine cell function capabilities of neutrophils isolated from a single diseased subgingival pocket (30). In these studies, it was necessary to study neutrophils derived from individual pockets and only a few microliters were available. Clearly the ability to measure functional and phenotypic characteristics by flow cytometry was important in these studies.

POTENTIAL BENEFITS OF USING FLOW CYTOMETRY

An additional benefit of using flow cytometric techniques can be demonstrated in situations where simultaneous measurements of multiple functions are desired. Examples of such combinations include measurements of phagocytosis and bacterial killing, phagocytosis and calcium flux, adhesion glycoproteins and phagocytosis, and many others. A specific study where several neutrophil functions were evaluated concurrently demonstrated that tumor necrosis factor (TNF) caused dose-dependent PMN activations, thereby stimulating phagocytosis, respiratory burst, and C3b expression (31). Each of these functions was evaluated using flow cytometric techniques. Some of these techniques will be discussed later in this chapter. Additionally, there may be situations where functionally distinct subpopulations of neutrophils may exist (32, 33). Few other techniques would be capable of identifying such populations or making simultaneous functional evaluations. A number of fluorescent probes are required to undertake many of these measurements. Most

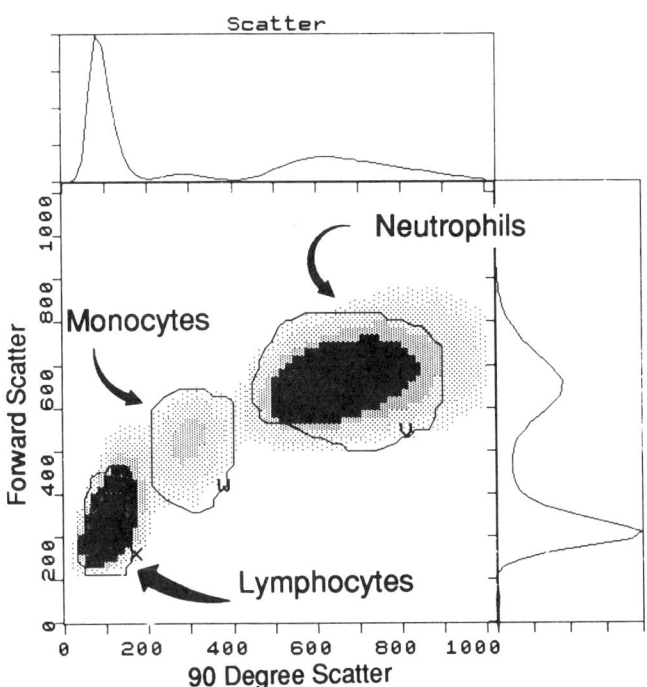

Figure 24.1. Neutrophils can be separated from other leukocytes using dual-parameter light scatter. The abscissa shows the 90° light scatter (90 LS) usually associated with "granularity." The ordinate shows forward-angle light scatter (FALS) normally associated with size. The neutrophils are clearly separated from the monocytes and lymphocytes. The single-parameter "projections" are shown for each parameter.

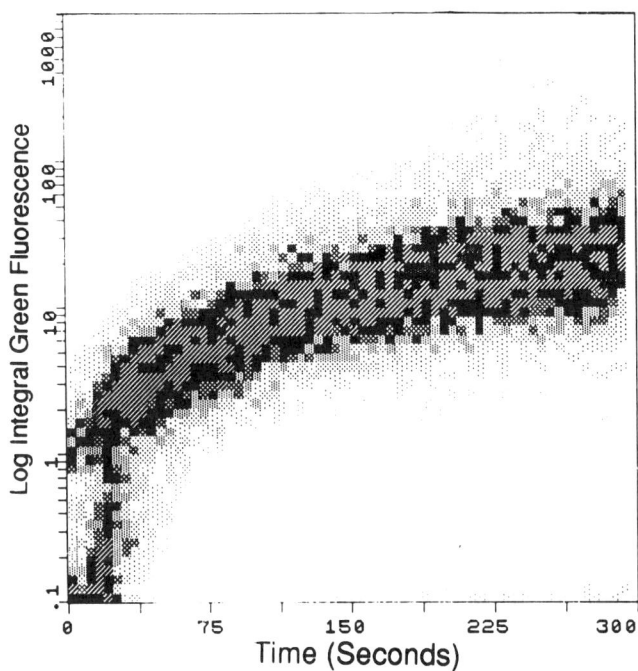

Figure 24.2. CF-DA (1 mM final) was added to a suspension of leukocytes and the increasing fluorescence of a gated neutrophil population was monitored for 5 min (300 seconds). Shown is the change in fluorescence of the cells during that incubation. Viable functional cells immediately hydrolyzed the CF-DA to carboxyfluorescein, which emits a strong green fluorescence when excited by 488 emission (Argon laser).

of the currently available fluorescent probes used for cell function studies are listed in Table 24.4.

EVALUATIONS OF NEUTROPHIL FUNCTION

Neutrophil Mobility

The ability to observe the in vivo mobilization of neutrophils in response to a chemotactic stimulus can only be performed by a test known as the **skin window** assay. An abrasion is made on the skin over which a glass coverslip or, alternatively, a small chamber is placed (34, 35). At several intervals, the coverslip is either removed or the chamber flushed with fresh buffer and leukocyte numbers are evaluated. This is a very difficult test to standardize, particularly in relation to the formation of the skin abrasion, and is considered to be of limited value in evaluating neutrophil function abnormalities, except, perhaps, for severe chemotactic deficiencies. Other in vitro methods are better suited to the determination of chemotactic deficiency and are discussed below.

Granule Development and Function

Neutrophil granules are synthesized at different stages during the maturation period. The release of granule contents from neutrophils is a critical function of the normal neutrophil carrying out its role in the immune response. Primary (azurophilic) granules develop following the myeloblast stage, whereupon they become promyelocytes, which are essentially incompetent neutrophils (36). As the cell further develops into the myelocyte stage, secondary (specific) granules are manufactured, a process that is complete at the metamyelocyte stage, when the immature neutrophil acquires some functional capabilities. In addition to these two well-differentiated granule types, two other types have been proposed and are known as tertiary granules and secretory granules.

The primary granules contain myeloperoxidase (MPO), required for respiratory burst function; and nonoxidative enzymes, including acid hydrolases such as B-glucuronidase, α-mannosidase, and 5'-nucleosidase; lysozyme; neutral proteases, such as cathepsin G and elastase; and cationic proteins (37–40). Further, an increasing number of small peptides, of approximately 30 amino acids in length, known as *defensins*, have been identified that also have important nonoxidative antibactericidal activity in humans (41, 42), rabbits (43) and rats (44).

The secondary granules are formed later in the maturation of the neutrophil and, thus, conditions that result in the release of immature neutrophils may cause a degradation in function related to secondary granules. Contained within these granules are lysozyme, lactoferrin, collagenase, vitamin B_{12} binding protein, cytochrome b, and possibly some of the adhesion glycoproteins (39, 45, 46).

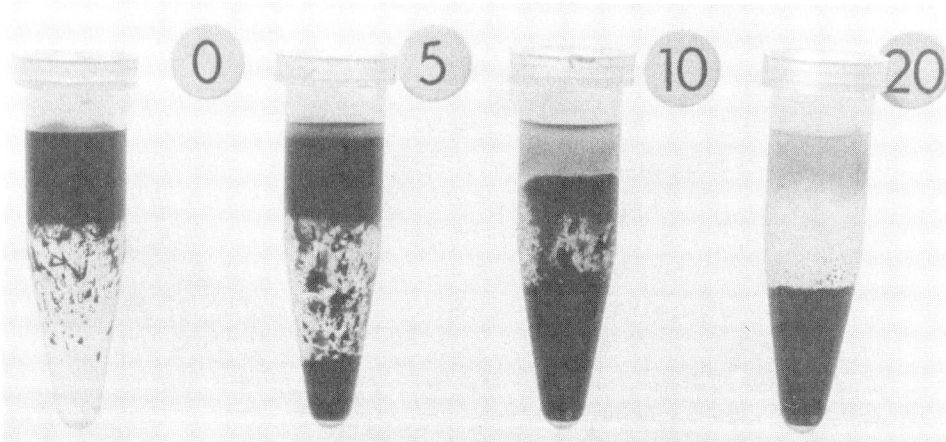

Figure 24.3. A rapid neutrophil isolation technique that we have termed the overlay method is shown. The overlay method is particularly useful in the clinical laboratory as a rapid means of isolating leukocytes for flow cytometric analysis. Heparinized blood (undiluted) is overlayed carefully onto Ficoll-Hypaque and left stationary on the bench at room temperature for 20 min. By carefully removing the top layer of suspension, a red-cell depleted suspension can be achieved. The suspension also contains platelets that can be removed by a gentle centrifugation step. Cells isolated using this technique are the least "activated" of any separation technique.

Tertiary granules contain gelatinase (a metalloproteinase that acts on protein constituents of extracellular matrices (47)), CD11b (MAC-1) receptors, and cytochrome b (47–49). Secretory granules (50, 51) contain gelatinase, although evidence and function for these granules is not well elucidated.

Enzyme Content/Activity

Dolbeare et al. (52, 53) initially demonstrated the presence of phosphatases and glucuronidases by flow cytometry using naphthol derivatives as fluorogenic substrates. Similarly, cellular enzyme activity can be measured by flow cytometry (54, 55) as can esterase activity, using dyes such as dichlorofluorescin diacetate and several others that are listed in Table 24.3 (56–60). The latter assay is more useful as an indicator of the presence of esterase. Because of the variation in the rates of hydrolysis for cellular enzymes, this activity can be used to differentiate cell populations as well as to indicate normal metabolic function. Active metabolism is involved in these hydrolytic reactions, and alterations induced by immunochemical modulators are observable. Some dyes can be very rapidly hydrolyzed to fluorescent compounds directly within a cell. Carboxyfluorescein diacetate is useful in this regard since it is hydrolyzed to the fluorescent carboxyfluorescein (530 nm) very rapidly. The process is complete within a few seconds to a maximum of 5 min in most cells, allowing very rapid evaluation. Several publications attest to the efficacy of the measurement of other enzymes by flow cytometry (54, 55).

Clinical Evaluation

The most common granule deficiency of neutrophils is a myeloperoxidase (MPO) deficiency in which there is a complete or partial deficiency of MPO from the primary granules (61–63). The deficiency is relatively common with an incidence of 5 patients in 10,000 subjects and is characterized by autosomal recessive genetics (63). It has been reported that monocytes from patients with MPO deficiency have increased respiratory burst duration with increased production of superoxide, which may partially compensate for the deficiency (64). Myeloperoxidase deficiency is almost silent clinically except for an increase in susceptibility and severity of *Candida* infections (63). Functionally, neutrophils demonstrate normal chemotaxis, phagocytosis, and degranulation, but a prolonged respiratory burst. The diagnosis is easily made by peroxidase stain of a blood smear. Such deficiencies may become more significant in patients who have another primary condition or receive therapy that may leave them with a reduced immune function.

Congenital specific granule deficiencies have been reported and some neonates have demonstrated deficiencies in specific granule formation (65–71). However, these deficiencies usually result in relatively minor microbial killing abnormalities. Severe recurrent bacterial infections can occur but are the exception.

Microtubule disorders can adversely affect the ability of a neutrophil to degranulate, as demonstrated by colchicine, which interferes with microtubule formation (72). Clinically, patients with microtubule dysfunction, such as Kartagener's syndrome, have recurrent sinus, middle ear, and respiratory infections, but this is more directly related to dysfunctioning cilia and impairment of leukocyte migration and chemotaxis (73–75), as discussed below. Abnormal microtubule metabolism has also been identified in chronic granulomatous disease (76).

Cytoskeleton Function

After activation of neutrophils, extensive movement of the receptor-ligand complexes within the membrane has been demonstrated (77–79), resulting in characteristic shape

Table 24.4
Fluorescence Probes for Neutrophil Studies

Function	Probe	Excitation	Emission	References	Considerations
H_2O_2	DCFH-DA	488	515–575	(56)	Broad emission spectra
O_2^-	HE	488	575–590		
	Bodippy	352			UV Laser required
Calcium	Indo-1	352	420; 525	(453)	UV Laser required
Calcium	fura 2	340	520	(191, 453, 454)	Dual Laser excitation required
Calcium	fura 3				
Calcium	Quin-2	352; 420	525	(440)	UV Laser required
Markers	FITC	488	525		
Markers	PE	488	575		
Markers	Texas Red	610	630		
Markers	APC	532	650		
Esterase	CFDA	488	525	(59)	
Viability	FDA	488	525		
Esterase	Decanoyl Fluor	488	525	(455)	
Esterase	CDF	488	525	(59)	
Esterase	CDF DA	488	525	(59)	
Esterase	CDMDF-DA	488	525	(59)	
Esterase	ADB[2a]	488		(60)	
pH	ADB	488		(60)	
Phosphatases	MFP	488			
GSH	mCIB	352	460–510		UV Laser required
Enzymes	AS-B1			(52, 53)	
Fluidity	DPH	352	420	(418, 419, 456–459)	UV Laser required
Fluidity	TMA-DPH	352	420	(458, 460, 461)	UV Laser required
Fluidity	Pyrenedecanoic acid	360	400/450	(462)	
Actin	NBD-Phallacidin	488	520	(86, 428, 429, 463)	UV Laser required
Actin	PE-Phalloidin	488			
Membrane Potential	$DiOC_5(3)'$	488	505–560		
Membrane Potential	$DiOC_6(3)$	488	505–560	(110)	
Membrane Potential	Rhodamine 123	488		(111)	
Cell Tracking	PKH1	488	525		
Cell Tracking	PKH2	488	575		
pH	AO	488			
Viability[b]	AO	488		(60)	
pH	FDA	488			
pH	DA-dicyanobenz	352	460; 525		UV laser required
Killing	AO	488			Emission is pH dependent
Viability	PI	488	>630		
Viability	EMA	488			
Peptidase	leucyl aminopeptidase	352	525	(438)	UV laser required

[a]ADB: 1,4-diacetoxy-2,3-dicanobenzene
[b]Used to measure bacterial degradation (red DNA)

changes (77). It is likely that this redistribution of the complex occurs within the plane of the plasma membrane utilizing the microfilaments. Evidence for this hypothesis is based partially upon the role of cytoskeletal disruptors, such as cytochalasin B (a fungal metabolite) and chloropromazine. It is thought that cytochalasin B binds to the free ends of the F-actin molecule and thus inhibits fMLP-induced (formyl-methionyl-leucyl-phenylalanine) polymerization of actin in neutrophils (80–82). Evidence of the involvement of regulatory G proteins in the transmembrane signaling after fMLP activation in human neutrophils has recently been demonstrated (83). GP 140, which interacts with the cytoskeleton during activation by wheat germ agglutinin (WGA), has also been implicated as playing a role in neutrophil activation (84). No clear evidence exists as to effective flow cytometric methods for detection of cytoskeletal defects. However, there are well-developed methods for measurement of actin polymerization by flow cytometry (85, 86).

Clinical Evaluation

As discussed above, Kartagener's syndrome, and more specifically immotile cilia syndrome, is an autosomal recessive disorder resulting in a microtubule defect that impairs leukocyte migration and chemotaxis (73, 75). The clinical manifestations of immotile cilia syndrome include recurrent sinusitis, otitis media, and respiratory infections due to the cilia and leukocyte dysfunctions (75). In a separate report, a patient with recurrent bacterial infections and abnormal chemotaxis had excessive neutrophil microtubule assembly (87).

Microfilament disorders inhibit leukocyte locomotion, as demonstrated by in vitro treatment of neutrophils with cytochalasin B, which disrupts actin filaments (88, 89). A clinical report of abnormal actin polymerization involving an infant with impaired chemotaxis has been reported (90).

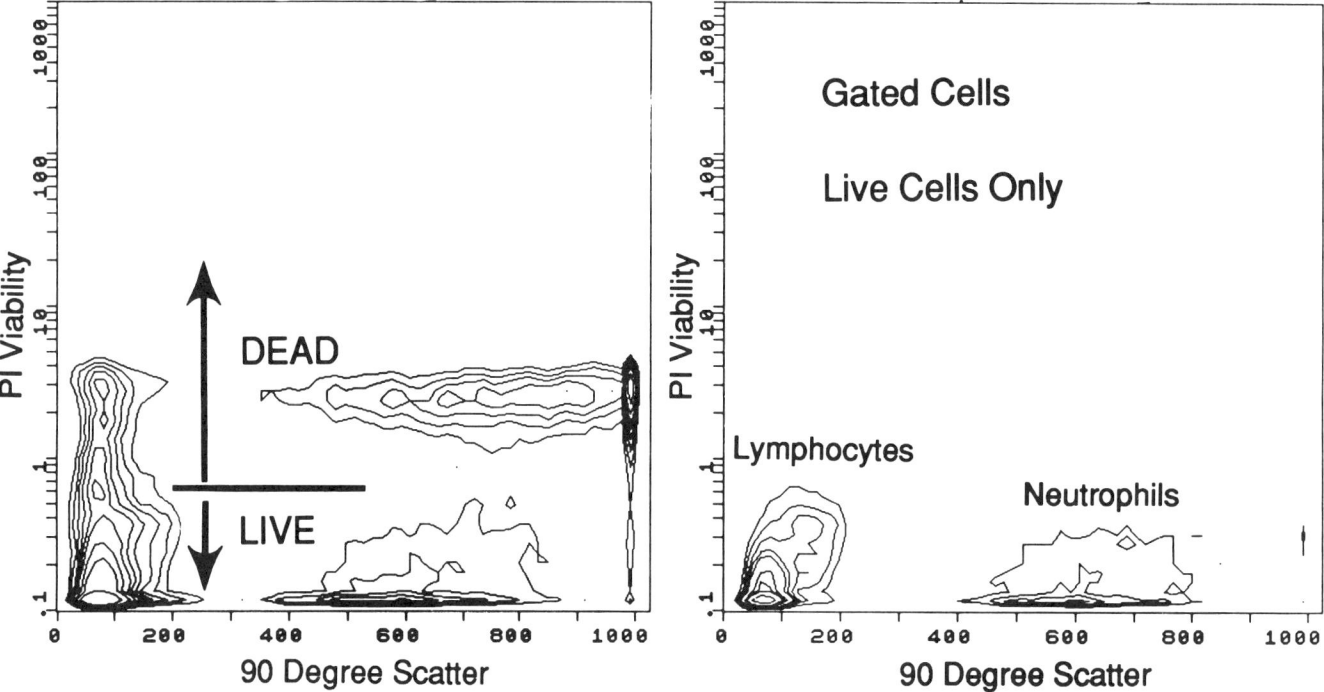

Figure 24.4. Demonstration of the use of *PI* to identify viable and nonviable cells by flow cytometry. *PI* was added to a suspension of cells with a suspected number of dead cells. Using a 2-parameter contour display that shows each population of cells, the lymphocytes and neutrophils can be identified. Cells that take up the *PI* dye are considered nonviable. The histogram on the right shows the viable cells selectively gated from the left histogram.

Membrane Integrity (Viability)

Measurement of cell viability using dyes, such as propidium iodide or fluorescein diacetate, should also be considered functional tests. Failure to evaluate viability in an assay of neutrophil function can lead to erroneous data, since a nonviable population will significantly influence the results. The advantage of flow cytometry is that determinations are made during or immediately after functional measurements. Nonviable cells can be gated out of the analyses (91) by either backgating on the fluorescent population of viable cells or observing scatter alterations of dead cells. An example of this is shown in Figure 24.4, which shows the changing fluorescence of cells that take up the PI stain (dead cells).

Such measurements can be performed rapidly and objectively and have been used successfully in toxicology applications (92) as well as in routine clinical assays. As a simple measure of the effects of xenobiotics on a cell, viability is certainly one of the most straight-forward measurements available using flow cytometry.

METABOLIC FUNCTIONS

Measurement of Cytosolic Free Ca^{2+}

The initial steps of signal transduction following receptor-ligand interaction involve activation of phospholipase c and membrane-bound glycophotidyl inositol. This leads to the release of inositol phosphates and fatty acids, which trigger activation of protein kinase C and subsequent flux of calcium across the plasma membrane. Thus, a direct cellular response can be measured if alterations of calcium concentration can be monitored.

Since calcium plays a critical role in cell function, it is important to be able to determine the extent to which chemical interactions affect the redistribution of this divalent cation. The area is not without controversy, however, since there is not complete agreement on the role of calcium in neutrophil activation during phagocytosis. There are at least two schools of thought on the role of calcium in neutrophil activation. Of importance, is the type and number of different ligand interactions involved. Essentially, when a single ligand-receptor interaction occurs, such as phagocytosis of yeast (via C3b), calcium is not required for respiratory burst (93). However, others have shown a lack of respiratory burst in the absence of calcium (94–96).

Indo-1 is an excellent dye for flow cytometric measurement of free intracellular calcium. This dye has the ability to undergo a fluorescent wavelength emission shift when bound to calcium. Indo-1 is introduced to the cells as an acetoxymethyl ester that undergoes enzymatic hydrolysis in cells to yield free dye. Flow cytometry has proved to be a valuable resource in the evaluation of the role of calcium in neutrophil function. A major spectral change can be measured when indicators of Ca^{2+} penetrate cells and are excited at 357 nm (ultra-violet excitation). Several different Ca^{2+} indicators are now available for use (97–100).

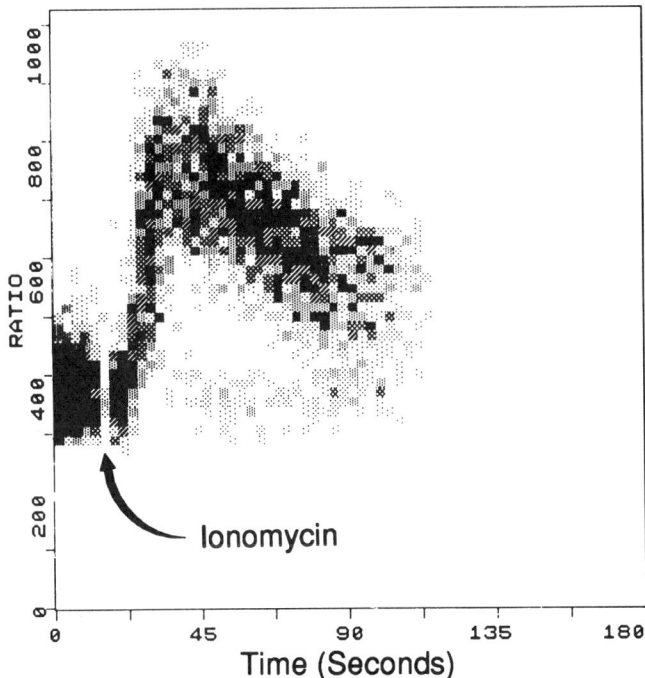

Figure 24.5. A population of Indo-1-loaded cells stimulated by ionomycin to demonstrate the rapid response of viable functional cells. By calibrating the flow cytometer with known concentrations of calcium, accurate measurements of intracellular free calcium can be determined.

Method: Cells are loaded with Indo-1 (final concentration 3 μM) for 15 min at 37°C and then immediately run on the flow cytometer to obtain fluorescence histograms at two emission wavelengths; 395 nm (bound Ca^{2+}) and 525 nm (non-bound calcium). The Ca^{2+} concentration of cells can be determined independently of dye concentration by evaluating the ratios of the two fluorescent emissions. Thus, a high 395/525 nm ratio would indicate bound Ca^{2+}. Ionomycin is used as a positive control for measurement of calcium flux. Ionomycin (3–5 mM) will cause an increase in the BOUND (long wavelength) fluorescence signal (i.e., increase in BOUND [Ca^{2+}] inside the cell).

This measurement is a vary rapid event that can be observed on the flow cytometer in real time given appropriate instrumentation. Accurate determination of intracellular calcium concentration can be made if aliquots of Indo-1 loaded cells are placed in solutions of various known calcium concentrations and treated with an ionophore such as ionomycin. The properties of Indo-1 are well-described in the literature (101). Observations of the real-time alteration in [Ca^{2+}] can be performed using list mode on the flow cytometer. Figure 24.5 shows an example of cells that have been stimulated by ionomycin, demonstrating the rapid alteration in calcium flux as measured by the flow cytometer.

One example of a calcium–requiring activation of human neutrophils is after stimulation by fMLP. In this interaction, fMLP binds to its receptor initiating a G-protein regulated activation of phospholipase C that then hydrolyzes a membrane-bound phospholipid (4,5 bisphosphate) to form both 1,2-diacylglycerol (DAG) and 1,4,5-triphosphate (IP3), which release calcium from intracellular stores primarily in the endoplasmic reticulum (102).

Less-recognized clinical abnormalities include glycogen storage disease type 1b, which is characterized by recurrent bacterial infections. Reduced phagocytic function appears to be related to diminished calcium mobilization and defective calcium stores; it is recognized by decreased elevation of cytosolic calcium to fMLP and decreased mobilization of calcium in response to ionomycin (103).

Membrane Potential

Neutrophils undergoing a receptor-ligand interaction show an increased permeability to ions and a subsequent reduction in transmembrane potential (104–107). A specific inhibitor of chymotrypsin-like enzymes can block the potential change, suggesting that, after the receptor-ligand interaction, a protease is required for the initial reaction (108). By use of the carboxycyanine dye [diO-C_5-(3)], a loss of cell-associated fluorescence is demonstrated, indicative of cell activation and, therefore, a change in resting transmembrane potential (108). Neutrophils that are incapable of responding oxidatively to stimulation do not show this fluorescence shift (109), suggesting a relationship between alterations in transmembrane potential and the generation of oxygen metabolites (108). The dye diffuses into the cells and equilibrates with the external medium. Upon stimulation of the cell, the intracellular dye is displaced by increased uptake of ions, resulting in a reduction of cellular fluorescence (depolarization). With some activating agents (fMLP), the cell will re-equilibrate after a short period and repolarization will be demonstrated by a return to the previous fluorescence intensity.

Using flow cytometry, individual cells can be monitored kinetically, allowing a determination of the rates of polarization. Variations in these rates could be a significant factor in observing alterations in cellular function. This is shown kinetically in Figure 24.6, where a population of neutrophils was stimulated with 100 ng/ml of PMA after being loaded with the carboxycyanine dye diO-C5-(3), as described above. Response is rapid and definitive. Defects are easily observed using this method. However there are a number of problems associated with membrane potential measurements that make this a difficult test to use clinically. Measurements in mitochondrial membrane potential can also be measured directly. This has been demonstrated by Korchak et al. (110) using $DioC_6(3)$ and also by Darzynkiewicz et al. with the mitochondrial-specific dye Rhodamine 123 (111).

Clinical Evaluation

One of the first functional abnormalities to be lost in neutrophils is the normal capabilities of alteration in membrane potential. Our laboratory has observed that neutrophils isolated from blood after 18–24 hours will produce a significant respiratory burst (H_2O_2) or chemiluminescence response but

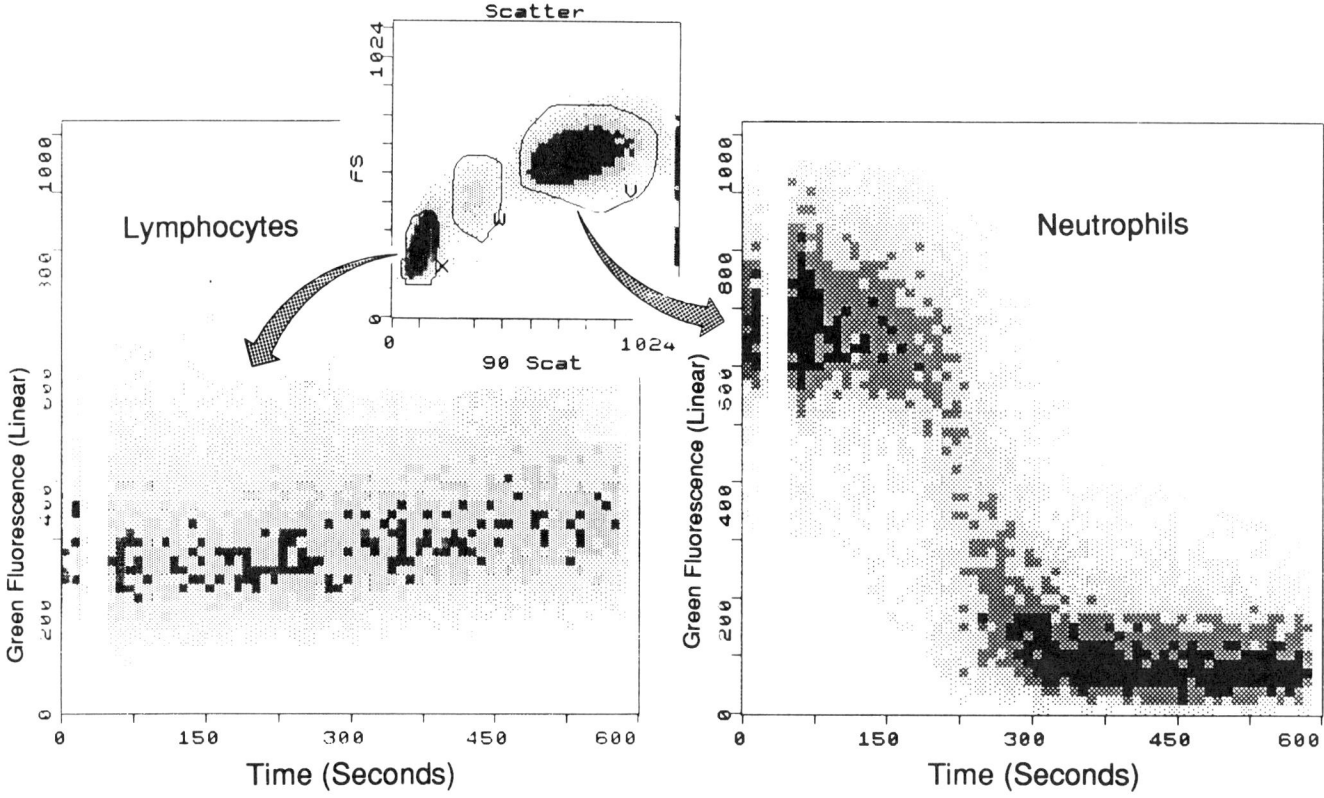

Figure 24.6. A population of neutrophils undergoing a rapid alteration in membrane potential after stimulation by PMA (100 ng/ml). Cells were loaded with DiOC5 (3) for 5 min, placed into the flow cytometer, and stimulated with the PMA. Measurements were begun immediately upon stimulation and maintained for 10 min. The two histograms shown demonstrate the changes in membrane potential of both lymphocytes (left) and neutrophils (right) from the same suspension of leukocytes. Each population was selectively gated and time versus fluorescence was recorded simultaneously. No alteration in the lymphocyte membrane potential of the lymphocyte population to the PMA was recorded. The neutrophil population, however, demonstrated a remarkable and rapid depolarization, consistent with the ionic flux expected with mebrane activation. Neutrophils normally do not repolarize after stimulation with PMA; however, with other activators such as fMLP, repolarization is observed after 4–7 min.

that the membrane potential response is severely deficient (unpublished observations). However, there are no reports of distinct clinical syndromes where abnormal membrane potential has been shown to be of particular importance as an individual phenomenon. Presently no data are available to demonstrate clinical utility.

Chemotaxis

The chemotactic response of the neutrophil is based upon the ability of these cells to determine a gradient of a chemoattractant substance and to direct the movement of the cell toward the source of the attractant. In order to accomplish this, neutrophils have a complex cytoskeletal mechanism involving both microtubules, used to directionally polarize the cell, and microfilaments for cell movement. The process of movement involves a constant attachment and detachment of the neutrophil to a substrate. Chemotactic function is an important neutrophil function and several excellent reviews have been written on the subject (112–116). Chemotaxis, like most neutrophil functions, is not an isolated function, but is part of a complex series of events that occurs during and after activation of the cells. The process of chemotaxis itself cannot presently be measured satisfactorily using flow cytometric technology; however, the ability of a neutrophil population to become activated by common chemotactic agents, such as C5a and formyl methionyl leucyl phenylalanine (fMLP), can be easily evaluated. One useful method is the determination of alterations in fMLP receptors on neutrophils via fluoresceinated fMLP using fluorescein isothiocyanate (FITC). Upon activation of the cells, the appearance of available receptors can be evaluated by measuring the green-associated FITC fluorescence of the neutrophils. GM-CSF has been shown to increase the binding of fMLP to PMN (1).

Neutrophils contain storage pools of fMLP receptors within the specific granules and these receptors are transported to the surface in response to stimulatory signals (117). FMLP binding is saturable with an estimated 50,000 sites/neutrophil and a $K_D = 10^{-14}$ nM (118). The receptor-ligand (radiolabelled) complex has a molecular weight of 55–70 kD (119). The receptor is distinct from C5a (120) or LTB_4 (121). Flow cytometric studies have identified high- and low-affinity binding sites as well as neutrophil subpopulations with varying numbers of receptors (122, 123). Response to fMLP by neutrophils is down-regulated through internalization of

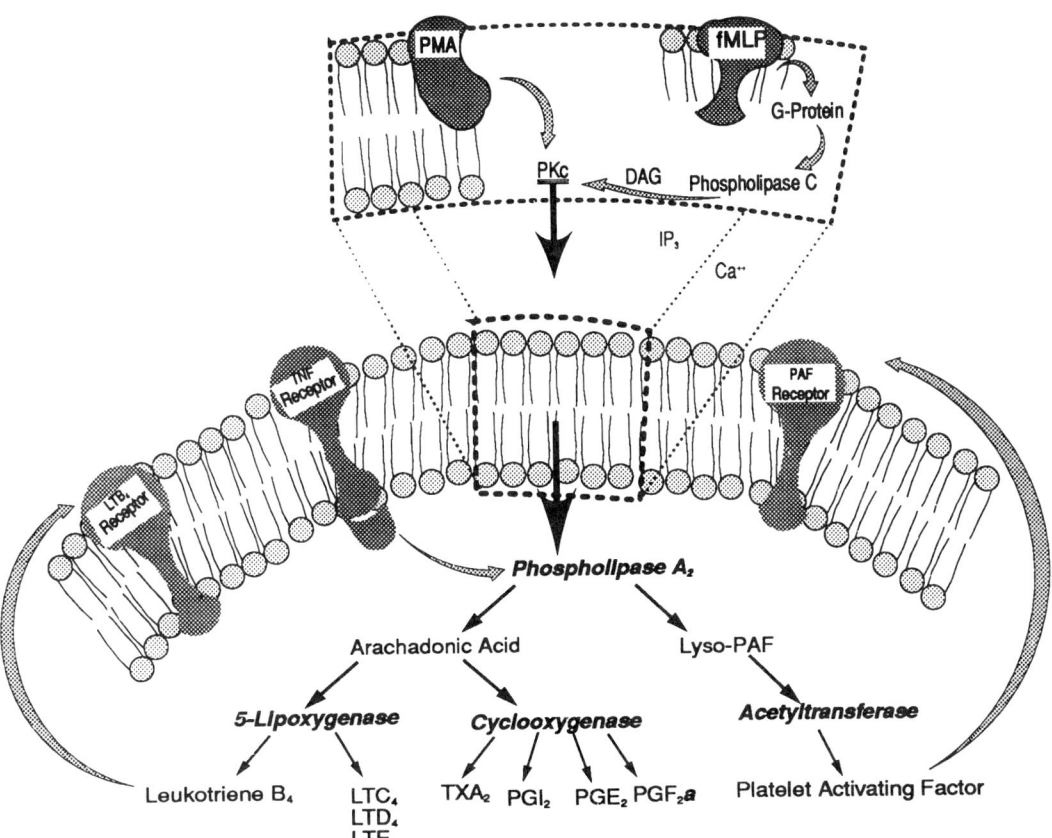

Figure 24.7. A schematic guide for the interaction of some of the neutrophil activation mechanisms and their relationship to cytokines and lipid mediators. Shown are several important receptor-mediated activation mechanisms associated with various inflammatory processes, including endotoxemia or thermal injury.

the receptor-ligand complex; however, recovery from this down-regulation is rapid (20 minutes at 37°C (124, 125). At low concentrations, fMPL will cause the release of vitamin B_{12} binding protein, and B-glucuronidase and lysozyme (47, 126) from secondary granules. However, at higher concentrations, it is known to stimulate the release of most primary and secondary granule contents (127). Since bacterially-derived peptides are remarkably similar to fMLP, the continued presence of fMLP or related molecules could cause a continued stimulation of enzyme release from stimulated neutrophils.

A specific C5a receptor has been identified on human neutrophils with a density of 100,000–300,000 sites/cell (120) and a molecular weight of approximately 44kD (128). Adherent neutrophils undergo exocytosis in the presence of C5a and both specific and azurophilic granules release their contents (129). In vitro, in the absence of cytochalasin B (and adhesion), very high concentrations of C5a are required to stimulate neutrophils (130). Human neutrophils have been shown to be able to activate (131) and inactivate C5a via a specific granule factor (132), thus providing a closed-loop control of responsiveness to this activator.

LTB_4 receptors occur in a low affinity form ($k_D = 3.9 \times 10^{10}$M, ~4400 sites/cell) and in a high affinity form ($K_D = 6 \times 10^{-8}$M, ~270,000 sites/cell) and are quite independent of the previously mentioned receptors (121,133). Neutrophils can be both stimulated by LTB_4 and produce LTB_4 as a product of arachidonic acid metabolism via the lipoxygenase pathway. Several physiological effects have been demonstrated on neutrophils by LTB_4, such as chemotaxis (134), aggregation (135), and degranulation (136). Further, LTB_4 has proved important in the mechanism of neutrophil adhesion to endothelial cells. As noted previously, increased adhesiveness (aggregation) due to LTB_4 has been shown, but these adhesion-promoting properties have also been reversed in the presence of an anti-CD18 monoclonal antibody (137). Figure 24.7 provides a schematic guide for the interaction of some of the neutrophil activation mechanisms and the relationship to cytokines and lipid mediators.

Interleukin 8 (IL-8) is an important chemotactic factor. This factor has numerous nomenclatures including the following: MDNCF–Monocyte-derived neutrophil chemotactic factor) (138); MONAP (139); NAP-1 (140); NAF; and IL-8 (141). The major form of IL-8 is a 72-amino-acid protein, but it has not shown significant homology with other cytokines including, IL-1, TNF, or IFNs (138, 142). IL-8 is capable of stimulating most activation pathways in neutrophils, such as degranulation (139), directional migration, expression of adhesion molecules, activation of the respiratory burst (143), enhanced *Candida* killing (144), and general activation (143, 145). There are specific IL-8 receptors on PMN (146) with about 20,000 high affinity binding sites ($K_d = 8 \times 10^{-10}$). IL-8 also very rapidly regulates its own receptor expression associated with ligand internalization. Further, this down-regulated receptor was shown to be rapidly recycled to the surface of the neutrophil (147).

Products of arachidonic acid metabolism are also well-known chemoattractants. Products of the lipoxygenase pathway, such as 5-HPETE, are converted to 5-HETE and leukotrienes, most of which are known to exert strong chemotactic activity upon neutrophils (148). Regulation of chemotaxis is modulated by the concentration of the chemoattractant. Several chemicals are known to exert enhancement (alcohols, degranulation) or depression (polyene antibiotics) of chemotactic function.

Clinical Evaluation

Defects in neutrophil chemotaxis are usually accompanied by recurrent infections of the skin or respiratory tract. Clinical syndromes with infections usually begin in the infant and are characterized by severe infections caused by organisms normally considered to be of relatively low pathogenicity.

A deficiency in the surface glycoproteins CD11a/CD18 (LFA-1), CD11b/CD18 (CR3), or CD11c/CD18 (p150,95) can lead to chemotactic defects (149). This syndrome, called Leukocyte Adhesion Deficiency (LAD) has been identified by Anderson et al. (149), Arnaout et al. (150), and Ross (151). This is an autosomal recessive disease characterized by recurrent bacterial and fungal infections, impaired pus formation, and poor wound healing (149). The genetic abnormality is caused by defective biosynthesis of the B subunit of the heterodimer complex associated with each glycoprotein adhesion molecule (149, 152–154). Lazy leukocyte syndrome was first identified in children and characterized by gingivitis, stomatitis, and recurrent upper respiratory infections (155). Severe circulating neutropenia was identified, but normal myeloid precursors and mature neutrophils were evident in the bone marrow (155).

Lazy leukocyte syndrome is now recognized to have several variants, including a range in severity of neutropenia, variation in age of onset, and cases with defective phagocytosis and microbial killing. However, all possess a basic functional defect in locomotion (156).

Hyperimmunoglobulin E, or Job's syndrome, is manifested clinically as dermatitis, recurrent staphylococcal skin infections, and elevated serum concentrations of IgE (157). Recurrent staphylococcal pneumonia, otitis externa and media, sinusitis, mucocutaneous candidiasis, and eczema are also seen (157, 158). Neutrophil chemotaxis abnormalities are common, although variable (75).

Chediak-Higashi syndrome is an autosomal recessive disease characterized by severe recurrent pyogenic infections (159). Many cells, including the neutrophils, display abnormal granule formation with giant cytoplasmic lysosomal granules. Neutrophils display abnormal chemotaxis and occasionally decreased numbers of centriole-associated microtubules (75, 160, 161). Other clinical manifestations are granule-related and include partial albinism, associated with a melanocyte dysfunction, and bleeding tendencies due to platelet defects (159).

Several diseases have neutrophil chemotaxis disorders that may be associated with a circulating inhibitor of locomotion or an inhibitor of a serum chemotactic factor. Associated diseases include Hodgkin's disease (162), hepatic cirrhosis (163), thermal injury (164), and severe inflammation (165). Circulating immune complexes in rheumatoid arthritis can act on neutrophils to inhibit chemotaxis (166, 167). Disorders of the microtubule or microfilament system (discussed above) can also inhibit chemotaxis.

Phagocytosis

The phagocytic process can be divided up into a number of clearly defined stages, each of which can fail. These stages are broadly defined as attachment or particle binding and ingestion. Unless phagocytes are able to bind to the microbe, phagocytosis will not take place. By utilizing both opsonized and nonopsonized organisms, both opsonic capacity and phagocytosis can be measured at the same time. Thus, it is important to determine whether abnormal phagocytosis is due to a failure in the opsonization process or to a defect in the ingestion capability of the phagocyte. Since the main cell receptors for phagocytosis are C3b (CR1) and FcR (Fc portion of IgG), it is also possible to evaluate these functional receptors as discussed earlier. An example of a neutrophil phagocytosis of *S. aureus* is shown in Fig. 24.8.[1]

Assays of phagocytosis of bacteria have been developed for flow cytometry by Bassoe et al. and other investigators (20, 21, 168–174). These measurements can be valuable in trauma, such as thermal injury, or in recurrent infections where specific bacteria can be used for assessment of immune function. Immune complexes can also be measured with methods similar to those for bacteria (175).

Several innovative methods for determining phagocytic capacity using flow cytometry have been demonstrated. A major advantage of flow cytometry over other methods is the relatively small number of cells required and the significantly fewer preparative procedures for isolating leukocytes. These are important when using small animals or when evaluating pediatric patients.

The availability of a number of fluorescent probes has increased the number of methods available. One useful method described uses fluorescein heat-killed *Candida albicans* and, after phagocytosis is complete, ethidium bromide (EB) (50 ug/ml) is added. Analysis using ultraviolet (UV) excitation with red and green emissions reveals green internalized organisms, while surface attached, but noninternalized, organisms are red (176). This procedure utilizes the phenomenon of resonance energy transfer between FITC and

---

[1] See Color Plate II between pages 432 and 433.

ethidium bromide. Since EB does not penetrate the cell membrane, only the external organisms are affected by the EB. This provides good discrimination between internal and external organisms. One major advantage of this test is the use of an inexpensive clinical analyzer flow cytometer that does not require expensive lasers and complex optical configurations.

Another interesting use of flow cytometry in evaluating neutrophil function is in the evaluation of phagocytosis of fluorescent-labeled viruses. In this study FITC-labeled *Herpes simplex* viruses (HSV) were phagocytosed by human neutrophils and both internalization and surface binding were determined by flow cytometry. Surface bound virus fluorescence was quenched using a trypan blue quenching procedure (177, 178).

Clinical Evaluation

Abnormal phagocytosis can occur with a variety of clinical disorders. The defect can be associated with the neutrophil itself or with an immunoglobulin or complement defect. Immature neutrophils released from the bone marrow have a defective phagocytosis that may be related to a high negative surface charge (179, 180). Abnormal phagocytosis has also been identified in the neonate and in juvenile periodontitis (181, 182). Tuftsin deficiency is either a familial disorder or is acquired as a consequence of splenectomy; it results in increased susceptibility to infection due to defective neutrophil phagocytosis (183). Tuftsin is a tetrapeptide produced by the spleen that enhances the neutrophil phagocytic ability (184). Clinical findings include respiratory infections such as bronchitis and pneumonia, and enlarged fluctuant lymph nodes (183, 185). Normal actin polymerization and microfilament function are also necessary for phagocytosis and, therefore, actin polymerization defects may interfere with the phagocytic process (90). Complement receptor C3bi deficiency can also result in altered phagocytosis (186).

Pinocytosis

Pinocytosis can also be a useful measure of cell function and several well-defined assays have been developed for flow cytometry (187, 188). fMLP-stimulated pinocytosis studies have demonstrated a linkage between the initial phase of pinocytosis and the characteristic shape changes observed in activated neutrophils. The assay used for these studies uses FITC-dextran and is relatively simple to establish, considering the availability of a flow cytometer. One of the most useful aspects of this assay is the ability to evaluate a large number of concommitant effects such as pH changes, kinetics of the responses, and temperature and ionic concentration effects (187).

Neutrophil Defense Mechanisms

Traditional descriptions of neutrophil defense mechanisms include both oxidative and non-oxidative mechanisms. Rare clinical syndromes may selectively deplete one major component of one or more pathways but, by and large the neutrophil activates many of its defense mechanisms concurrently and often the clinical manifestation of defects in specific components is minimized. One such case would be in MPO deficiency, which is characterized by a reduced bactericidal rate but ultimately has a normal killing capacity of neutrophils.

Oxidative Systems

Oxidative mechanisms require oxygen in significantly larger amounts than resting neutrophils require. A variety of toxic oxygen species is produced both inside and outside the cell. The associated activity is known as the respiratory burst, which results from activation of NADPH oxidases via an electron transfer reaction involving 2 electrons from NADPH through an FAD-flavoprotein utilizing cytochrome b^{245} to oxygen. The superoxide anion produced in this reaction can be converted to H_2O_2 by superoxide dismutase and, in concert with myeloperoxidase (MPO) and a halide (primarily chloride), hypochlorous acid can be produced. Each of these species is capable of damaging ingested bacteria, external microbes, the neutrophil itself, or closely situated tissue.

C-reactive protein has also been proposed to have a modulatory role in neutrophil oxidative burst, inhibiting superoxide release, chemotaxis, degranulation, and phagocytosis of activated neutrophils (189). C-reactive protein has also been demonstrated to bind to the surface of PMA-activated neutrophils (190).

By far the most useful measurement of intracellular H_2O_2 estimation is the dichlorofluorescin diacetate (DCFH-DA) probe technique as proposed by Bass et al. (56). The assay depends upon the incorporation of 2'-7',dichlorofluorescein diacetate (DCFH-DA) into the hydrophobic lipid regions of the cell, where the acetate moieties are cleaved by hydrolytic enzymes to the nonfluorescent molecule 2'-7',dichlorofluorescein (DCFH), which becomes trapped within the cell due to its polarity. Upon cell activation, NADPH oxidase catalyzes the reduction of O_2 to O_2^-, which is further reduced to H_2O_2. The oxidative potential of H_2O_2 and peroxidases are able to oxidize the trapped DCFH to 2'-7',dichlorofluorescein (DCF), which is characteristically fluorescent at 530 nm (the same emission as FITC). Since the green fluorescence produced is proportional to the amount of H_2O_2 generated, it is possible to calibrate this assay to allow the expression of the intracellular production of H_2O_2 in neutrophils in terms of attomoles/cell (56). There are many examples of use of this assay in the literature, including studies on various animals such as rats, mice, and humans (30, 56, 58, 191–199).

A calibration curve can be generated based on data obtained from spectrophotometric and flow cytometric measurements that allows conversion of the fluorescence histograms on the flow cytometer into quantitative estimations of

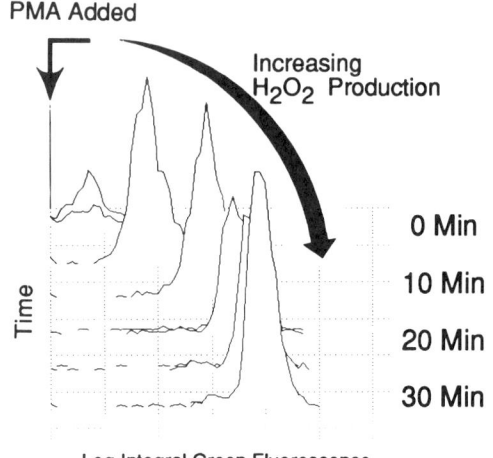

Figure 24.9. A histogram showing the kinetic response of neutrophils stimulated with *PMA* as measured by the H_2O_2 assay (described in the text). As the neutrophils respond to the *PMA*, respiratory burst activation results in production of H_2O_2. Cells had been previously "loaded" with DCFH-DA which becomes intracellularly hydrolysed to DCFH by cellular esterases. Oxidizing conditions (H_2O_2) caused oxidation of the nonfluorescent DCFH to fluorescent DCF, which was measured on the flow cytometer as green fluorescence. Each single-parameter histogram is from measurements taken from a tube sampled seven times over the 30-min reaction period.

H_2O_2 production (56). The assay is a very sensitive measure of a cell's capacity to undergo a respiratory burst in response to a variety of stimuli. While it may be of interest to determine the exact amount of H_2O_2 per cell, it is unnecessary in most clinical situations and an alternative measurement is the relative amount of fluorescence (therefore H_2O_2) produced by the cell before and after stimulation. Figure 24.9 shows the kinetics of the neutrophil response to PMA (100 ng/ml) over a 30-min period. It is not always necessary to measure the kinetics of the entire response. For most clinical evaluations, a beginning and ending measurement at 0 and 30 min is adequate. Figure 24.9 displays the difference in the fluorescence histograms of unstimulated and stimulated cells after a 30-min incubation at 37°C. The fluorescence histograms represent the amount of H_2O_2 (proportional to the amount of green fluorescence) produced by the cell, as described above.

Before the development of the DCF assay for flow cytometry, accurate estimations of intracellular H_2O_2 were very difficult. Several laboratories, including our own, have further developed the assay of Bass et al. for use with microquantities of blood. This is particularly valuable for the evaluation of cell function in experimental models using small animals (mice and rats) and also for pediatric evaluations of cell-function studies. In this respect, several innovative developments have been reported whereby several functions can be determined in μ-quantities of whole blood. Trinkle et al. (199) have used a combination of DCF, phagocytosis, and killing in a few hundred μ-liters of blood. This assay is rapid and comprehensive and would be particularly useful for pediatric patients. The major disadvantage of the procedure however, is the relatively complicated setup required for the flow cytometry since dual-laser excitation is necessary.

Superoxide–Nitroblue tetrazolium (NBT) reduction has been measured flow cytometrically by Blair et al. (200). The method is a variation of a traditional method using a microscope and a glass slide. The presence of oxidative burst enzymes in HL-60 (human leukemia) cells has been demonstrated using NBT using a series of experiments whereby simultaneous measurement of NBT reduction, cell cycle phase, and phagocytosis were made (200, 201). Because of the problems of interpretation of this technique, it is not widely used in the routine clinical laboratory as a flow cytometric assay.

Clinical Evaluation

Chronic granulomatous disease (CGD) is an inherited disorder in which phagocytes have a defective oxidative metabolism and an inability to produce hydrogen peroxide. The clinical picture usually starts with staphylococcal dermatitis and enlarged lymph nodes (202). Pulmonary changes are prominent with bronchopneumonia, hilar lymphadenopathy, and lung abscess formation (202, 203). The inflammatory reactions are usually excessive and often develop into granuloma formation despite appropriate antibiotic treatment. Infections usually begin in infancy or early childhood, however several reports describe the neutrophil abnormality discovered in adults (204–206). The disease is probably a family of diseases with similar clinical manifestations and different defective enzyme systems. Various enzyme deficiencies that have been associated with CGD include cytochrome b_{558} (207–211), glucose-6-phosphate dehydrogenase (212, 213), flavoprotein (214), and a defect in protein or protein phosphorylation p47 or p67 (211, 215). Neonates can also have defective bactericidal activity (216), which may be related to impaired production of the hydroxyl radical from superoxide (67, 216).

Neutrophils isolated from CGD patients fail to produce any fluorescence in the DCF assay. It is important to be able to determine that there is a significant difference between a possible CGD and a normal. Since availability of CGD cells is probably unlikely in most clinical laboratories, the usual procedure is to utilize both stimulated and unstimulated normal cells. The unstimulated normal can serve as a satisfactory control for a possible CGD patient. Thus, by always running an unstimulated normal with a stimulated normal any defect or partial defect should be identified. Several reports demonstrate the detection of possible carriers of CGD using non-flow-cytometry (217) and flow-cytometry-based methods (194, 195). These reports demonstrate the use of the DCF assay whereby neutrophils from heterozygous carriers of CGD produced histograms midway between the unstimulated and stimulated controls (195).

Nonoxidative Bactericidal Mechanisms

Neutrophil cytoplasmic granules contain a variety of different enzymes that do not require oxidative metabolism. These enzymes are released into phagolysosomes and include proteases, such as cathepsin D,E, and G (218), and elastase (219); hydrolytic enzymes, such as phospholipase A_2 (220) and lysozyme (221); bacterial permeability increasing protein (222, 223); lactoferrin (218); and defensins (41, 42).

Cathepsin G, also known as chymotrypsin-like cationic protein, is present in azurophilic granules and has both microbicidal and cytotoxic properties (224, 225). Cathepsin G is a protease that inhibits bacterial oxygen consumption and has microbicidal activity that does not depend on a primary proteolytic attack (226–228).

Elastase is another azurophilic granule component that is important for the degradation of the structural protein elastin (219). Elastase is also capable of degrading bacterial cell wall protein (229) and of potentiating the activity of cathepsin G (230) as well as the lytic activity of lysozyme (229). Elastase is cytotoxic to endothelial cells in culture and may be associated with neutrophil-mediated lung injury in emphysema (231, 232).

Lysozyme is present in both azurophilic and specific granules and acts by hydrolysis of the bacterial cell wall (233). The hydrolytic activity is directed at the *B*-1-4 glycosidic bond between N-acetylglucosamine and *N*-acetylmuramic acid and thus is only effective against selected gram-positive bacteria (228). Lysozyme is capable of killing gram-negative bacteria if the bacteria are first acted on by toxic oxygen products that damage the protective lipid envelope (234).

Bacterial-permeability-increasing protein (BPI) is a cationic protein present in azurophilic granules. BPI contributes to the ability of neutrophils to kill gram-negative bacteria, especially *E. coli* (222, 235, 236). BPI binds to and permeabilizes the bacterial envelope (237). Elsbach and Weiss proposed an initial ionic interaction that eventually activates bacterial phospholipases (235). Neutrophilic phospholipase A_2 in specific granules also increases bacterial envelope permeability and exerts a potent bactericidal effect (220). Another cationic protein with optimal activity at low pH is 37kD cationic protein. It is similar to BPI and is active against several gram-negative bacteria (238).

Lactoferrin is a glycoprotein from specific granules with binding sites for ferric iron (239, 240). Lactoferrin is a member of the iron-binding transferrin family and can also be found in tears, semen, and human milk (241). Lactoferrin exhibits bacteriostatic activity against gram-negative and gram-positive bacteria due to its ability to chelate iron (242, 243). Bactericidal activity of lactoferrin has also been reported against both gram-positive and gram-negative bacteria (244). Lactoferrin deficiency due to specific granule deficiency has been reported in human patients with recurrent infections (223).

Defensins are another group of cationic proteins and are the major protein constituent of azurophilic granules. Defensins have a broad spectrum of activity in vitro against gram-positive and gram-negative bacteria, fungi, and certain enveloped viruses (41, 44, 245–247). Defensins have also demonstrated cytotoxicity of mammalian cells in culture (248).

Despite countless studies and many different probes, a clear consensus about the microenvironment of the phagolysosome is lacking. Studies with fluorescent pH probes indicate the pH rises to 8.0 within a few minutes of phagolysosome formation (249). Within 15 minutes, the pH is neutral and then continues to decrease to 5.5–6.0 within 1–2 hours (249). A lower pH in phagolysosomes of thermal injury patients may contribute to reduced neutrophil intracellular killing due to a lack of initial alkalinization (250).

Due to the complexity of the phagolysosome environment, it is often difficult to separate oxygen-dependent mechanisms from oxygen-independent mechanisms. Considerable synergy also exists between the two mechanisms, which further hinders investigations to differentiate their effects.

Adhesion, Binding to Neutrophil Receptors

Integrins are a superfamily of heterodimeric, transmembrane glycoproteins that act as receptors on leukocyte surfaces (251) promoting several important cellular functions, such as adhesion of phagocytes to surfaces, phagocytosis, and diapedesis. Several related molecules have been identified notably CD11a (LFA-1), CD11b (also called Mo1, OKM-1, CR3, Mac-1), and CD11c (p150,95) (252, 253). The three heterodimers share a common B-unit of 95 kDa which is known as CD18. Each has distinct α-subunits of 177 kDa (LFA-1, CD11a); 165 kDa, (Mac-1, CD11b), and 150 kDa (p150,95, CD11c). The primary role of these molecules is in adhesion-dependent functions and therefore they are found on lymphocytes and monocytes as well as neutrophils. The importance of integrins in normal immune function has been demonstrated in patients with deficiencies in mRNA for CD18 (154). Several reviews of the structure and function of this family of molecules (254, 255) as well as of the related clinical deficiency syndromes (256) have been written.

While the overall function of integrins is to regulate granulocyte diapedesis and migration into inflammatory sites, the mechanism of action is less certain. It is hypothesized that a primary role for the CD11/CD18 complex on phagocytes is the regulation of secretion of toxic oxygen mediators, enzymes, and other secreted products of activation. Thus, if the CD11/CD18 complex is involved in the regulation of these cells, it may be one of the mechanisms that only allows for the secretory response in appropriate circumstances such as adhered or aggregated neutrophils.

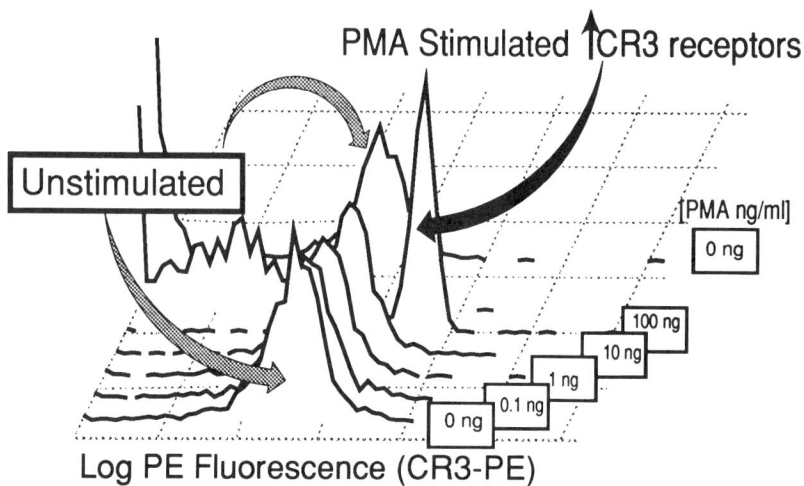

Figure 24.10. Neutrophils stimulated by *PMA* demonstrate an increase in the number of CD11b (*CR3*) receptors as shown in this figure. Shown is a dose response of *PMA* and the resultant increase in the fluorescence from CR3-FITC-labeled antibody bound to the cells.

The CD11/CD18 complex is not essential for neutrophil activation by lipopolysaccharide (LPS) since CD18-deficient patients show normal priming for enhanced release of superoxide anion in response to LPS (257). However, antibodies to CD18 were capable of inducing a defect similar to that described in children with a genetic deficiency of the CD11/CD18 integrins (258). Another report describes profound chemotactic defects in patients suffering from dysmyelopoietic syndromes (DMPS), refractory anemia with excessive blasts (RAEB), and acute nonlymphoblastic leukemia (ANLL). However, in all of the cases reported, no abnormality was detected by flow cytometry for granulocyte integrins (259). Thus, the chemotactic defects observed were unrelated to the integrins in these patients.

Stimulation of neutrophils with PMA directly down-regulated the CD11/CD18 independent mechanism of neutrophil adherence to interleukin-1 (IL-1), tumor necrosis factor (TNF), or LPS pretreated human endothelial cells (260). IL-1 stimulated PMN-endothelial adhesion (261) but IL-1 did not alter PMN degranulation or chemotaxis (262).

CD11b–(CR3; Mac-1) Expression

This receptor is the primary glycoprotein associated with cell adhesion found on neutrophils and regulates the adherence and phagocytosis of particles opsonized with C3bi. Arnaout et al. (263) demonstrated the activation of neutrophils (via increased CD11b expression) by complement fragments generated by new hemodialysis membranes.

CD11b (CR3) has been shown to be critical in the phagocytosis of *S. aureus* and *E. coli*. In one study, *E. coli* enhanced CD11b expression despite not being phagocytosed. This study further demonstrated an initial fall in CD11b expression despite not being phagocytosed. This study further demonstrated an initial fall in CD11b after addition of bacteria to the neutrophils, but a subsequent enhancement of expression after 5 to 10 min (264). In terms of oxidative response, opsonized *E. coli* and *S. aureus* stimulated H_2O_2 production measured by flow cytometry. Nonopsonised *E. coli* did not stimulate H_2O_2 production despite an increase in CD11b. As an example, Figure 24.10 shows the flow cytometric histograms of neutrophil CD11b expression before and after stimulation by PMA (0.1–100 ng/ml). The increased fluorescence indicates increased expression of the CD11b receptor on the neutrophil surface (above normal expression found on unstimulated neutrophils).

Absence of the B-subunits will cause a profound and debilitating deficiency in neutrophil function. An inherited syndrome known as Leukocyte Adhesion Deficiency exists where none of the B-subunits are synthesized (149, 153, 154, 265) (described above). Absence of these adhesion glycoproteins will prevent the chemotaxis and subsequent diapedesis of neutrophils at inflammatory foci. Partial or incomplete B-subunit deficiencies result in less severe functional defects. Several chemotactic stimuli including C5a, LTB_4, and fMLP have been shown to rapidly increase the number of CD11b receptors on neutrophils (266, 267) and, in particular, activation of C5a increases the adhesive characteristic of the neutrophil to the capillary endothelium. Storage pools of CD11b receptors are contained in neutrophils, most likely in granule membranes (50, 268).

Other Functional Neutrophil Receptors

A recent report has described an alteration in the number of CD16-positive neutrophils in HIV-I-infected individuals (269). CD16 ($Fc_\gamma RIII$) is a relatively late differentiation antigen on neutrophils, natural killer (NK) cells, and a subset of T-cells, and is a low-affinity receptor for IgG (270, 271). $Fc_\gamma RIII$ is anchored to the plasma membrane of neutrophils via a phosphatidyl inositol glycan moiety (272, 273) that can be released by chemotactic activation (274). One consistent finding in HIV-infected patients is the frequent observation of neutrophil defects in terms of chemotaxis, phagocytosis, microbicidal killing, and respiratory burst activity (275–277). Other Fc receptors such as Fc$_\gamma$I (CD64) on monocytes and neutrophils and Fc$_\gamma$II (CD32) found on monocytes, neutrophils, eosinophils, basophils, B-cells, and platelets are also important. Some studies have demonstrated that CD32, but not CD64, can trigger activation of the respiratory burst of neutrophils (278–280), whereas others have shown that the glycosylphosphatidylinositol-linked CD16 receptor can also trigger respiratory burst activity (281). Upregulation of CD64 by γ-interferon on neutrophils results in cells that can be activated to trigger an oxidative burst through that receptor (282). Thus, with the possible therapeutic use of γ-interferon for CGD patients, monitoring of upregulation of these receptors on peripheral blood neutrophils may become important. A recent flow cytometry method for quantitation of receptor numbers in peripheral blood neutrophils has been reported (283). Thus, the evaluation of neutrophils, including CD16 receptor expression, may be useful in certain patient groups. An excellent review of Fc receptors has recently been published by van de Winkel and Anderson (284).

Other Neutrophil Dysfunctions

A recent study by Lawton et al. (285) demonstrates the effectiveness of preparations of human immunoglobulin for intravenous use in stimulating oxidative burst and chemiluminescence of isolated human neutrophils. These studies suggest there may be a beneficial therapeutic effect in severe, life-threatening infections. Clear evidence is shown as to the stimulation of oxidative burst but no consistent information can be derived from studies of chemotaxis. The mechanism of action is probably related to the passive immunization with preformed specific antibody. However, there is little doubt that this study demonstrates significant stimulation of neutrophil function directly by the intravenous immunoglobulin. Another recent study has confirmed the above observations of stimulation of oxidative burst in the presence of intravenous immunoglobulin (286). These authors have suggested that inflammatory reactions observed occasionally during infusions in hypogammaglobulinemic patients may well be related to neutrophil activation.

Clinical Prediction by Flow Cytometry

Several studies using flow cytometry have proved of value in the evaluation of trauma. A major study by Valet demonstrates that a multiparametric-multifunctional analysis of neutrophils could accurately predict the course of disease three days in advance of the clinical manifestation of pulmonary or cardiovascular organ failure in 92% of samples from 47 patients (60). Tests used in this study included measurements of phagocytosis and degradation, cell volume, intracellular pH, and esterase activity. A significant value to this study is the use of a small bench type mercury arc-based flow cytometer. Further, studies such as these demonstrate the depth of knowledge that can be gained from a small number of cells when using a multiparametric approach.

Neonatal Infections and Neutrophil Function

An excellent review of the role of complement deficiencies and related neutrophil function abnormalities in neonates has been written by Berger (287). Important considerations in neonates are the combined effects of minor abnormalities of cellular function. Reduced levels of C3, low specific antibody titers, reduced mobility, or reduced numbers of neutrophils can combine to form a critical immune deficiency in these patients. Adhesion glycoproteins are present in significantly reduced numbers on stimulated neonate neutrophils compared to adult neutrophils (288, 289). Several studies have demonstrated an indication for granulocyte transfusions in septic neonatal patients (290–292).

Antibodies against Neutrophils

The normal physiologic removal or destruction of neutrophils occurs constantly; however, accelerated removal can be a result of an antineutrophil antibody. This immune neutropenia might be due to an auto- or alloantibody or may be related to anti-HLA antibodies. In infants, alloantibodies can result from maternal IgG antibodies that cross the placenta, causing severe neonatal neutropenia (293).

Antineutrophil antibody measurements by flow cytometry have become a useful clinical assay (294) because of the relatively small number of neutrophils required and because evaluations can be made of patient sera as well as of antibodies detected on patient neutrophils. The flow cytometric technique is well-suited for children with neutropenia since evaluations can be made using only 1–2 ml of blood. Usually, there are sufficient cells for a simple flow cytometric H_2O_2 screen (using the DCF assay described above) to be carried out on these neutrophils also.

Measurement of Neutrophil Antibodies

One important aspect in performing the antineutrophil assay is in the selection of the normal control neutrophils or neutrophil pool. Because of the possibility of damage to neutrophil surface antigens, the best method for preparation of neutrophils for this assay has proven to be one of the simplest and has already been described above (overlay method). Two washes in buffer remove serum proteins and platelets leaving a leukocyte-rich suspension that is highly suitable for evaluation of antineutrophil antibodies. The various cell populations can easily be light-scatter gated on the flow cytometer. Evaluation of histograms from patient and control sera can determine the presence or absence of antineutrophil antibodies (usually IgG). The histograms shown in Figure 24.11 are indicative of the presence of such antibodies in patient sera. Patient neutrophils can also be evaluated for the presence of surface-bound antibody if sufficient neutrophils can be obtained from the patient.

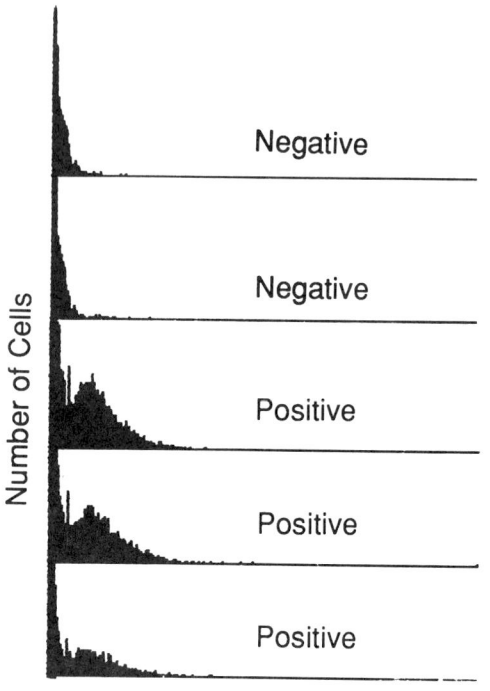

Figure 24.11. The presence of antineutrophil antibodies in patient sera is demonstrated by the increasing fluorescence of gated neutrophils. Control-negative sera demonstrate no fluorescence while the positive control sera are strongly positive.

Antineutrophil Cytoplasmic Antibodies (ANCA)

The presence of antineutrophil cytoplasmic antibodies has been reported in a number of vascular diseases, including uveitis (295) and glomerulonephritis (296, 297). Association of high titers of ANCA is consistent with the diagnosis of Wegener's granulomatosis and several reports have demonstrated positive correlations (295–298). ANCA have also been shown to stimulate oxidative bursts and degranulation in normal neutrophils, effects that were more pronounced after priming with TNF (299). The same study was able to demonstrate the presence of myeloperoxidase on the surface of neutrophils, which was interpreted as indicating that neutrophils have ANCA antigens on their surfaces to interact with ANCA (299). Measurement of ANCA by flow cytometry has been demonstrated (299a) and, in association with immunofluorescence. such measurements may have significant clinical utility.

CONCLUSION

The evaluation of neutrophil function can provide useful information on the capability of immune cells to perform normal operations. This information cannot be achieved by determinations of numbers of neutrophils, or even the presence of neutrophils, at a particular site. Specific defects of neutrophil function can become important, particularly when secondary to other underlying disease. The neutrophil interaction with other cells, particularly endothelial cells, is becoming better defined at the present time. It is clear that the neutrophil is not an isolated, suicidal cell with little effect on other cells, tissues, or organs. It is these relationships that are currently being studied using tools such as specific monoclonal antibodies, a variety of valuable fluorescent dyes, and technologies such as flow cytometry. While many of the studies involving neutrophil function are of a research nature, there are several important evaluations of clinical importance. The neutrophil is the most prolifically produced cell in the human immune system. Future developments in understanding the relationships between neutrophils and other cells will be an important determining factor in the utility of neutrophil function methods for clinical use.

REFERENCES

1. Platzer E, Welte K, Gabrilove JL, et al. Biological activities of a human pluripotent hemopoietic colony-stimulating factor on normal and leukemic cells. J Exp Med 1985;162:178–801.
2. Ghebrehiwet B, Muller-Eberhard HJ. C3e: an acidic fragment of C3 with leukocytosis-inducing activity. J Immunol 1979;123:616–621.
3. Welte K, Platzer E, Lu L, et al. Purification and biochemical characterization of human pluripotent hematopoietic colony-stimulation factor. Proc Natl Acad Sci (USA) 1985;82:1526–1530.
4. Kampschmidt RF, Upchurch HF. Neutrophil release after injections of endotoxin or leukocyte endogenous mediators into rats. J Reticuloendothel Soc 1980;28:191–201.
5. Kampschmidt RF, Upchurch HF. Possible involvement of leukocyte endogenous mediator in granulopoiesis. Proc Soc Exp Biol Med 1977;155:89–93.
6. Welte K. Granulocyte colony-stimulating factor (G-CSF): biochemistry, biology and pathophysiology. Klin Padiatr 1988;200:157–164.
7. Athens JW, Haab OP, Raab SO, et al. Leukokinetic studies. IV. The total blood circulating and marginal granulocyte pools and the granulocyte turnover rate in normal subjects. J Clin Invest 1961;40:989.
8. Boggs DR, Athens JW, Cartwright GE, et al. Leukokinetic studies. IX. Experimental evaluation of a model of granulopoiesis. J Clin Invest 1965;44:643–656.
9. Bishop CR, Athens JW, Boggs DR, et al. Leukokinetic studies. 13. A non-steady-state kinetic evaluation of the mechanism of cortisone-induced granulocytosis. J Clin Invest 1968;47:249–260.
10. Dancey JT, Deubelbeiss KA, Harker LA, et al. Neutrophil kinetics in man. J Clin Invest 1976;58:705–715.
11. Fleidner TM, Cronkite EP, Robertson JS. Granulopoiesis: I. Senescence and random loss of neutrophilic granulocytes in human beings. Blood 1964;24:402.
12. Metchnikoff E. Lectures on the Comparative Pathology of Inflammation. London: Kegan, Paul, Trench, Trubner and Co., 1968, republished by New York: Dover Publications, 1968.
13. Comandon J. Phagocytose in vitro des hématozoaires du Calfat. Comptes rendus hebdomadaires des séances et mémoires de la Société de Biologie 1917;80:314–316.
14. Boyden SV. The chemotactic effect of mixtures of antibody and antigen on polymorphonuclear leukocyte adherence. J Cell Biol 1962;82:347–368.
15. Ward PA. The chemosuppression of chemotaxis. J Exp Med 1966;124:209–226.
16. Fernandez HN, Hugli TE. Primary structural analysis of the polypeptide portion of human C5a anaphylatoxins. I. Evaluation of C3a and C5a leukotaxis in vitro and under stimulated in vivo conditions. J Biol Chem 1978;253:6955–6964.
17. Shin HS, Synderman R, Friedman E, et al. Chemotactic and anaphalytoxic fragment cleaved from the fifth component of guinea pig complement. Science 1968;162:361–363.

18. Springer TA, Teplow DB, Dreyer WJ. Sequence homology of the LFA-1 and Mac-1 leukocyte adhesion glycoproteins and unexpected relation to leukocyte interferon. Nature 1985;314:540–542.
19. Kay AB, Glass EJ, McG Salter D. Leukoattractants enhance complement receptors on human phagocytic cells. Clin Exp Immunol 1979; 38:294–299.
20. Bassoe C-F, Laerum OD, Glette J, et al. Simultaneous measurement of phagocytosis and phagosomal pH by flow cytometry: role of polymorphonuclear neutrophilic leukocyte granules in phagosome acidification. Cytometry 1983;4:254–262.
21. Bassoe C-F, Laerum OD, Solberg CO, et al. Phagocytosis of bacteria by human leukocytes measured by flow cytometry. Proc Soc Exp Biol Med 1983;174:182–186.
22. Bassoe C-F, Laerum OD, Solberg CO, et al. Human peripheral blood phagocyte functions in malignant haematological disorders. Cytometry Suppl [Abstract]. Int Conf Analytical Cytology XI, Hilton Head, SC, 1985.
23. Bassoe C-F, Solberg CO. Phagocytosis of staphylococcus aureus by human leukocytes: quantitation by a flow cytometric and a microbiological method. Acta Pathol Microbiol Immunol Scand [C] 1984;92: 43–50.
24. Bassoe C-F, Solsvik J, Laerum OD. Quantitation of single cell phagocytic capacity by flow cytometry. In: Laerum OD, Lindmo T, Thorud E, eds. Flow Cytometry IV. Oslo: Universitetsforlaget, 1980, pp 170.
25. Boyum A. Isolation of mononuclear cells and granulocytes from human blood. Scand J Clin Lab Invest 1968;21:77–97.
26. Ohno Y, Seligmann BE, Gallin JI. Cytochrome b translocation to human neutrophil plasma membranes and superoxide release. Differential effects of N-formylmethionylleucylphenylalanine, phorbol myristate acetate, and A23187. J Biol Chem 1985;260:2409–2414.
27. Harvath L, Falk W, Leonard EJ. Rapid quantitation of neutrophil chemotaxis: use of a polyvinylpyrrolidone-free polycarbonate membrane in a multiwell assembly. J Immunol Meth 1980;37:39–45.
28. Yancey KB, Lawley TJ, Dersookian M, et al. Analysis of the interaction of human C5a and C5a des Arg with human monocytes and neutrophils: flow cytometric and chemotaxis studies. J Invest Dermatol 1989;92:184–189.
29. Duque RE, Robinson JPaul, Hudson JL, et al. Detection of anti-platelet antibodies in serum by flow cytometry [Abstract]. Proc Soc Anal Cytol 1985;451.
30. Loesche WJ, Robinson JPaul, Flynn M, et al. Reduced oxidative function in gingival crevicular neutrophils in periodontal disease. Infect Immun 1988;56:156–160.
31. Livingston DH, Appel SH, Sonnenfeld G, et al. The effect of tumor necrosis factor-alpha and interferon-gamma on neutrophil function. J Surg Res 1989;46:322–326.
32. Harvath L, Leonard EJ. Two neutrophil subpopulations exist in human blood with different chemotactic activities: separation and chemoattractant binding. Infect Immun 1982;36:443–449.
33. Seligmann B, Malech HL, Melnick DA, et al. An antibody to a subpopulation of neutrophils demonstrates antigenic heterogeneity which correlates with response heterogeneity. Trans Assoc Am Physicians 1984;97:319–324.
34. Ziegler J, Hansen P, Penny R. Leucocyte function in paraproteinaemia. Aust N Z J Med 1975;5:39–43.
35. Zimmerli W, Gallin JI. Monocytes accumulate on Rebuck skin window coverslips but not in skin chamber fluid. A comparative evaluation of two in vivo migration models. J Immunol Meth 1987;96:11–17.
36. Segal EK, Ellegaard J, Borregaard N. Development of the phagocytic and cidal capacity during maturation of myeloid cells. Studies on cells from patients with chronic myelogenous leukaemia. Br J Haematol 1987;67:3–10.
37. Bretz U, Baggiolini M. Biochemical and morphological characterization of azurophil and specific granules of human neutrophilic polymorphonuclear leukocytes. J Cell Biol 1974;63:251–269.
38. Folds JD, Welsh IR, Spitznagel JK. Neutral proteases confined to one class of lysosomes of human polymorphonuclear leukocytes. Proc Soc Exp Biol Med 1972;139:461–463.
39. West BC, Rosenthal AS, Gelb NA, et al. Separation and characterization of human neutrophil granules. Am J Pathol 1974;77:41–66.
40. Dewald B, Rindler Ludwig R, Bretz U, et al. Subcellular localization and heterogeneity of neutral proteases in neutrophilic polymorphonuclear leukocytes. J Exp Med 1975;141:709–723.
41. Ganz T, Selsted ME, Szklarek D, et al. Defensins. Natural peptide antibiotics of human neutrophils. J Clin Invest 1985;76:1427–1435.
42. Selsted ME, Harwig SSL, Ganz T, et al. Primary structures of three human neutrophil defensins. J Clin Invest 1985;76:1436–1439.
43. Selsted ME, Szklarek D, Lehrer RI. Purification and antibacterial activity of antimicrobial peptides of rabbit granulocytes. Infect Immun 1984;45:150–154.
44. Eisenhauer PB, Harwig SL, Szklarek D, et al. Purification and antimicrobial properties of three defensins from rat neutrophils. Infect Immun 1989;57:2021–2027.
45. Kane SP, Peters TJ. Analytical subcellular fractionation of human granulocytes with reference to the localization of vitamin B12-binding proteins. Clin Sci Mol Med 1975;49:171–182.
46. Leffell MS, Spitznagel JK. Association of lactoferrin with lysozyme in granules of human polymorphonuclear leukocytes. Infect Immun 1972;6:761–765.
47. Dewald B, Bretz U, Baggiolini M. Release of gelatinase from a novel secretory compartment of human neutrophils. J Clin Invest 1982;70: 518–550.
48. Petrequin PR, Todd RFI, Devall LJ, et al. Association between gelatinase release and increased plasma membrane expression of MO1 glycoprotein. Blood 1987;69:605–610.
49. Mollinedo F, Schneider DL. Subcellular localization of cytochrome b and ubiquinone in a tertiary granule of resting human neutrophils and evidence for a proton pump ATPase. J Biol Chem 1984;259:7143–7150.
50. Borregaard N, Christensen L, Bejerrum OW, et al. Identification of a highly mobilizable subset of human neutrophil intracellular vesicles that contains tetranectin and latent alkaline phosphatase. J Clin Invest 1990;85:408–416.
51. Borregaard N, Miller LJ, Springer TA. Chemoattractant-regulated mobilization of a novel intracellular compartment in human neutrophils. Science 1987;237:1204–1206.
52. Dolbeare FA, Smith RE. Flow cytoenzymology: Rapid enzyme analysis of single cells. In: Melamed MR, Mullaney PF, Mendelsohn ML, eds. Flow Cytometry and Sorting. New York: Wiley, 1979, p 317.
53. Dolbeare FA, Phares W. Naphthol AS-BI (7-bromo-3-hydroxy-2-naphtho-o-anisidine) phosphatase and naphthol AS-BI b-D-glucuronidase in Chinese hamster ovary cells: biochemical and flow cytometric studies. J Histochem Cytochem 1979;27:120–124.
54. Watson JV. Enzyme kinetic studies in cell populations using fluorogenic substrates and flow cytometric techniques. Cytometry 1980;1:143–151.
55. Treumer J, Valet GK. Flow-cytometric determination of glutathione alterations in vital cells by o-phthaldialdehyde (OPT) staining. Exp Cell Res 1986;163:518–524.
56. Bass DA, Parce JW, DeChatelet LR, et al. Flow cytometric studies of oxidative product formation by neutrophils: a graded response to membrane stimulation. J Immunol 1983;130:1910–1917.
57. Nagel JE, Han K, Coon PJ, et al. Age differences in phagocytosis by polymorphonuclear leukocytes measured by flow cytometry. J Leuk Biol 1986;39:399–407.
58. Robinson JPaul, Bruner LH, Bassoe C-F, et al. Measurement of intracellular fluorescence of human monocytes indicative of oxidative metabolism. J Leuk Biol 1988;43:304–310.
59. Waggoner AS. Fluorescent Probes for Cytometry. In: Melamed MR, Lindmo T, Mendelsohn ML, eds. Cytometry and Sorting. New York, Wiley-Liss, 1990, pp 209–225.

60. Rothe G, Kellermann W, Valet G. Flow cytometric parameters of neutrophil function as early indicators of sepsis- or trauma-related pulmonary or cardiovascular organ failure. J Lab Clin Med 1990;115:52–61.
61. Falloon J, Gallin JI. Neutrophil granules in health and disease. J Allergy Clin Immunol 1986;77:653–662.
62. Nauseef WM, Root RK, Malech HL. Biochemical and immunologic analysis of hereditary myeloperoxidase deficiency. J Clin Invest 1983;71:1297–1307.
63. Parry MF, Root RK, Metcalf JA, et al. Myeloperoxidase deficiency: prevalence and clinical significance. Ann Intern Med 1981;95:293–301.
64. Locksley RM, Wilson CB, Klebanoff SJ. Increased respiratory burst in myeloperoxidase-deficient monocytes. Blood 1983;62:902–909.
65. Gallin JI. Neutrophil specific granule deficiency. Annu Rev Med 1985;36:263–274.
66. Fearon DT. Identification of the membrane glycoprotein that is the C3b receptor of the human erythrocyte, polymorphonuclear leukocyte, B lymphocyte, and monocyte. J Exp Med 1980;152:20–30.
67. Ambruso DR, Altenburger KM, Johnston RB Jr. Defective oxidative metabolism in newborn neutrophils: discrepancy between superoxide anion and hydroxyl radical generation. Pediatrics 1979;64:722–725.
68. Sidiropoulos D, Straume B. The treatment of neonatal isoimmune thrombocytopenia with intravenous immunoglobin (IgG i.v.). Blut 1984;48:383–386.
69. Murphy S, Van Epps D.E. Neutrophil and monocyte function in pediatric patients with recurrent pneumonias: a longitudinal study. Am Rev Respir Dis 1982;126:92–96.
70. Riggs JE, Schochet SS Jr, Gutmann L, et al. Inclusion body myositis and chronic immune thrombocytopenia. Arch Neurol 1984;41:93–95.
71. Lomax KJ, Gallin JI, Rotrosen D, et al. Selective defect in myeloid cell lactoferrin gene expression in neutrophil specific granule deficiency. J Clin Invest 1989;83:514–519.
72. Wright DG, Ungerleider RS, Gallin JI, et al. Pretreatment of filtration leukapheresis donors with colchicine. Blood 1978;52:783–792.
73. Afzelius BA, Ewetz L, Palmblad J, et al. Structure and function of neutrophil leukocytes from patients with the immotile-cilia syndrome. Acta Med Scand 1980;208:145–154.
74. Turner JA, Corkey CW, Lee JY, et al. Clinical expressions of immotile cilia syndrome. Pediatrics 1981;67:805–810.
75. Gallin JI, Wright DG, Malech HL, et al. Disorders of phagocyte chemotaxis. Ann Intern Med 1980;92:520–538.
76. Gallin JI, Buescher ES, Seligmann BE, et al. NIH conference. Recent advances in chronic granulomatous disease. Ann Intern Med 1983;99:657–674.
77. Oliver JM, Berlin RD. Surface and cytoskeletal events regulating leukocyte membrane topography. Semin Hematol 1983;20:282–304.
78. Dykman TR, Hatch JA, Atkinson JP. Polymorphism of the human C3b/C4b receptor. J Exp Med 1984;159:691–703.
79. Jack RM, Ezzel RM, Haartwig J, et al. Differential interaction of the C3b/C4b receptor and MHC class 1 with the cytoskeleton of human neutrophils. J Immunol 1986;137:3996–4003.
80. Brown SS, Spudich A. Mechanism of action of cytochalasin: evidence that it binds to actin filament ends. J Cell Biol 1981;88:487–491.
81. Cooper JA. Effects of cytochalasin and phalloidin on actin. J Cell Biol 1987;105:1473–1478.
82. Wallace PJ, Wersto RP, Packman CH, et al. Chemotactic peptide-induced changes in neutrophil actin conformation. J Cell Biol 1984;99:1060–1065.
83. Särndahl E, Lindroth M, Bengtsson T, et al. Association of ligand-receptor complexes with actin filaments in human neutrophils: a possible regulatory role for a G-protein. J Cell Biol 1989;109:2791–2799.
84. Suchard SJ, Boxer LA. Characterization and cytoskeletal association of a major cell surface glycoprotein, GP 140, in human neutrophils. J Clin Invest 1989;84:484–492.
85. Packman CH, Lichtman MA. Activation of neutrophils: measurement of actin conformational changes by flow cytometry. Blood Cells 1990;16:193–207.
86. Hilmo A, Howard TH. F-actin content of neonate and adult neutrophils. Blood 1987;69:945–949.
87. Gallin JI, Malech HL, Wright DG, et al. Recurrent severe infections in a child with abnormal leukocyte function: possible relationship to increased microtubule assembly. Blood 1978;51:919–933.
88. Hartwig JH, Stossel TP. Cytochalasin B and the structure of actin gels. J Mol Biol 1979;134:539–553.
89. Maruyama K, Hartwig JH, Stossel TP. Cytochalasin B and the structure of actin gels. II. Further evidence for the splitting of F-actin by cytochalasin B. Biochim Biophys Acta 1980;626:494–500.
90. Boxer LA, Hedley Whyte ET, Stossel TP. Neutrophil actin dysfunction and abnormal neutrophil behavior. N Engl J Med 1974;291:1093–1099.
91. Bjerknes R. Flow cytometric assay for combined measurement of phagocytosis and intracellular killing of Candida albicans. J Immunol Meth 1984;72:229–241.
92. Aeschbacher M, Reinhardt CA, Zbinden G, et al. A rapid cytotoxicity assay using dye retention and exclusion in two-parameter flow cytometry. Int Conf Prac Tox 1986;24:467.
93. Della Bianca V, Grzeskowiak M, Rossi F. Studies on molecular regulation of phagocytosis and activation of the NADPH oxidase in neutrophils: IgG- and C3b-mediated ingestion and associated respiratory burst independent of phospholipid turnover and Ca^{2+} transients. J Immunol 1990;144:1411–1417.
94. Rossi F. The O2-forming oxidase of phagocytes: nature, mechanisms of activation and function. Biochim Biophys Acta 1986;853:65.
95. Lambeth JD. Activation of the respiratory burst oxidase in neutrophils: on the role of membrane-derived second messengers, Ca++, and protein kinase C. J Bioenerg Biomembr 1988;20:709–733.
96. Grzeskowiak M, Della-Bianca V, Cassatella MA, et al. Complete dissociation between the activation of phosphoinositide turnover and of NADPH oxidase by formyl-methionyl-leucyl-phenylalanine in human neutrophils depleted of Ca2+ and primed by X subthreshold doses of phorbol 12,myristate 13,acetate. Biochem Biophys Res Commun 1986;135:785–794.
97. Grynkiewicz, G, Poenie M, Tsien RY. A new generation of calcium indicators with greatly improved fluorescence properties. J Biol Chem 1985;260:3440–3450.
98. Tsien RY, Pozzan T, Rink TJ. Calcium homeostasis in intact lymphocytes: cytoplasmic free calcium monitored with a new intracellularly trapped fluorescent indicator. J Cell Biol 1982;94:325–334.
99. Rabinovitch PS, June CH, Grossmann A, et al. Heterogeneity among T cells in intracellular free calcium responses after mitogen stimulation with PHA or anti-CD3. Simultaneous use of indo-1 and immunofluorescence with flow cytometry. J Immunol 1986;137:952–961.
100. Vandenberghe PA, Ceuppens JL. Flow cytometric measurement of cytoplasmic free calcium in human peripheral blood T lymphocytes with fluo-3, a new fluorescent calcium indicator. J Immunol Meth 1990;127:197–205.
101. Tsien RY. New tetracarboxylate chelators for fluorescence measurement and photochemical manipulation of cytosolic free calcium concentrations. Soc Gen Physiol Ser 1986;40:327–345.
102. Streb H, Irvine RF, Berridge MJ, et al. Release of Ca2+ from a nonmitochondrial intracellular store in pancreatic acinar cells by inositol-1,4,5-triphosphate. Nature 1983;306:67–69.
103. Kilpatrick K, Garty BZ, Lundquist KF, et al. Impaired metabolic function and signaling defects in phagocytic cells in glycogen storage disease type 1b. J Clin Invest 1990;86:196–202.
104. Seligmann BE, Chused TM, Gallin JI. Human neutrophil heterogeneity identified using flow microfluorometry to monitor membrane potential. J Clin Invest 1981;68:1125–1131.
105. Seligmann BE, Gallin JI. Comparison of indirect probes of membrane potential utilized in studies of human neutrophils. J Cell Physiol 1983;115:105–115.

106. Seligmann BE, Gallin JI. Use of lipophilic probes of membrane potential to assess human neutrophil activation. Abnormality in chronic granulomatous disease. J Clin Invest 1980;66:493–503.
107. Seligmann BE, Gallin JI. Neutrophil activation studied using two indirect probes of membrane potential which respond by different fluorescence mechanisms. Adv Exp Med Biol 1982;141:335–349.
108. Duque RE, Phan SH, Sulavik MC, et al. Inhibition by Tosyl-L-phenylalanyl chloromethyl ketone of membrane potential changes in rat neutrophils. J Biol Chem 1983;258:8123–8128.
109. Seligmann BE, Gallin EK, Martin DL, et al. Interaction of chemotactic factors with human polymorphonuclear leukocytes: studies using a membrane potential-sensitive cyanine dye. J Membr Biol 1980;52: 257–272.
110. Korchak HM, Rich AM, Wilkenfeld C, et al. A carbocyanine dye, DiOC6(3), acts as a mitochondrial probe in human neutrophils. Biochem Biophys Res Commun 1982;8:1495–1501.
111. Darzynkiewicz Z, Staiano-Coico L, Melamed MR. Increased mitochondrial uptake of rhodamine 123 during lymphocyte stimulation [Abstract]. Proc Natl Acad Sci USA 1981;77:6696.
112. Gallin J, Snyderman R. Leukocyte chemotaxis. Fed Proc 1983;42: 2851–2862.
113. Hayashi H. A review on the natural mediators of inflammatory leucotaxis. Acta Pathol Jpn 1982;322:271–284.
114. Sharma JN, Mohsin SS. The role of chemical mediators in the pathogenesis of inflammation with emphasis on the kinin system. Exp Pathol 1990;38:73–96.
115. Rotrosen D, Gallin JI. Disorders of phagocyte function. Annu Rev Immunol 1987;5:127–150.
116. Snyderman R, Pike MC. Transductional mechanisms of chemoattractant receptors on leukocytes. Contemp Top Immunobiol 1984;14:1–28.
117. Jesaitis AJ, Naemura JR, Painter RG, et al. Intracellular localization of N-formyl chemotactic receptor and Mg2+ dependent ATPase in human granulocytes. Biochim Biophys Acta 1982;719:556–568.
118. Williams LT, Snyderman R, Pike MC, et al. Specific receptor sites for chemotactic peptides on human polymorphonuclear leukocytes. Proc Natl Acad Sci USA 1977;74:1204–1208.
119. Niedel J, Davis J, Cuatrecasas P. Covalent affinity labelling of the formyl peptide chemotactic receptor. J Biol Chem 1980;255:7063–7066.
120. Chenoweth DE, Hugli TE. Demonstration of specific C5a receptor on intact polymorphonuclear leukocytes. Proc Natl Acad Sci USA 1978; 75:3943–3947.
121. Goldman DW, Goetzl EJ. Specific binding of leukotriene B4 to receptors on human polymorphonuclear leukocytes. J Immunol 1982;129: 1600–1604.
122. Mackin WM, Huang CK, Becker EL. The formylpeptide chemotactic receptor on rabbit peritoneal neutrophils. I. Evidence for two binding sites with different affinities. J Immunol 1982;129:1608–1611.
123. Seligmann BE, Chused TM, Gallin JI. Differential binding of chemoattractant peptide to subpopulations of human neutrophils. J Immunol 1984;133:2641–2646.
124. Sullivan SJ, Zigmond SH. Chemotactic peptide receptor modulation in polymorphonuclear leukocytes. J Cell Biol 1980;85:703–711.
125. Niedel JE, Kahane I, Cuatrecasas P. Receptor mediated internalization of fluorescent chemotactic peptide by human neutrophils. Science 1979;205:1412–1414.
126. Smith RJ, Iden SS, Bowman BJ. Activation of the human neutrophil secretory process with 5(S),12(R)-dihydroxy-6,14-cis-8,10-trans-eicosatetraenoic acid. Inflammation 1984;8:365–384.
127. Becker EL, Sigman M, Oliver JM. Superoxide production induced in rabbit polymorphonuclear leukocytes by synthetic chemotactic peptides and A23187. Am J Pathol 1979;95:81–97.
128. Huey R, Hugli TE. Characterization of a C5a receptor on human polymorphonuclear leukocytes (PMN). J Immunol 1985;135:2063–2068.
129. Becker EL, Showell HJ, Henson PM, et al. The ability of chemotactic factors to induce lysosomal enzyme release. I. The characteristics of the release, the importance of surfaces and the relation of enzyme release to chemotactic responsiveness. J Immunol 1974;112:2047–2054.
130. Bentwood BJ, Henson PM. C5a-induced degranulation from human neutrophils in the absence of cytochalasin B. Fed Proc 1980;39:798.
131. Wright DG, Gallin JI. Modulation of the inflammatory response by products released from human polymorphonuclear leukocytes during phagocytosis: generation and inactivation of the chemotactic factor C5a. Inflammation 1975;13:23–39.
132. Wright DG, Gallin JI. A functional differentiation of human neutrophil granules: generation of C5a by a specific (secondary) granule product and inactivation of C5a by a azurophil (primary) granule product. J Immunol 1977;119:1068–1076.
133. Kreisle RA, Parker CW. Specific binding of leukotriene B4 to a receptor on human polymorphonuclear leukocytes. J Exp Med 1983;157: 628–641.
134. Goetzl EJ, Pickett WC. The human polymorphonuclear leukocyte chemotactic activity of complex hydroxy-eicosatetraenoic acids. (HETES). J Immunol 1980;125:1789–1791.
135. Ford-Hutchinson AW, Bray MA, Doig MU, et al. Leukotriene B4, a potent chemokinetic and aggregating substance released from polymorphonuclear leukocytes. Nature 1980;286:264–265.
136. Bokoch GM, Reed PW. Effect of various lipoxygenase metabolites of arachidonic acid on degranulation of polymorphonuclear leukocytes. J Biol Chem 1981;256:5317–5320.
137. Wallis WJ, Hickstein DD, Schwartz BR, et al. Monoclonal antibody-defined functional epitopes on the adhesion promoting glycoprotein complex (CDw18) of human neutrophils. Blood 1986;67:1007–1013.
138. Yoshimura T, Matsushima K, Tanaka S, et al. Purification of a human monocyte-derived neutrophil chemotactic factor that has peptide sequence similarity to other host defense cytokines. Proc Natl Acad Sci USA 1987;84:9233–9237.
139. Schroeder JM, Mrowietz U, Morita E, et al. Purification and partial characterization of a human monocyte derived, neutrophil-activation peptide that lacks interleukin 1 activity. J Immunol 1987;139:3474–3483.
140. Larsen CG, Anderson AO, Appella E, et al. The neutrophil-activating protein (NAP-1) is also chemotactic for T lymphocytes. Science 1989; 243:1464–1466.
141. Westwick J, Li SW, Camp RD. Novel neutrophil-stimulation peptides. Immunol Today 1989;10:146–147.
142. Lindley I, Aschauer H, Seifert JM, et al. Synthesis and expression in Escherichia coli of the gene encoding monocyte-derived neutrophil-activation factor: biological equivalence between natural and recombinant neutrophil-activation factor. Proc Natl Acad Sci USA 1988;85: 9199–9203.
143. Peveri P, Walz A, Dewald B, et al. A novel neutrophil-activating factor produced by human mononuclear phagocytes. J Exp Med 1988; 167:1547–1559.
144. Djeu JY, Matsushima K, Oppenheim JJ, et al. Functional activation of human neutrophils by recombinant monocyte-derived neutrophil chemotactic factor/IL-8. J Immunol 1990;144:2205–2210.
145. Thelen M, Peveri P, Kernen P, et al. Mechanism of neutrophil activation by NAF, a novel monocyte-derived peptide agonist. Fed Am Soc Exp Biol 1988;2:2702–2706.
146. Samanta AK, Oppenheim JJ, Matsushima K. Identification and characterization of specific receptors for monocyte-derived neutrophil chemotactic factor (MDNCF) on human neutrophils. J Exp Med 1989; 169:1185–1189.
147. Samanta AK, Oppenheim JJ, Matsushima K. Interleukin 8 (monocyte-derived neutrophil chemotactic factor) dynamically regulates its own receptor expression on human neutrophils. J Biol Chem 1990; 265:183–189.
148. Snyderman R, Goetzl EJ. Molecular and cellular mechanisms of leukocyte chemotaxis. Science 1981;213:830–837.
149. Anderson DC, Springer TA. Leukocyte adhesion deficiency: an inherited defect in the Mac-1, LFA-1 and p150,95 glycoproteins. Annu Rev Med 1987;38:175–194.

150. Arnaout MA, Spits H, Terhorst C, et al. Deficiency of a leukocyte surface glycoprotein (LFA-1) in two patients with Mo1 deficiency: effects of cell activation on Mo1/LFA-1 surface expression in normal and deficient leukocytes. J Clin Invest 1984;74:1291–1300.
151. Ross GD. Clinical and laboratory features of patients with an inherited deficiency of neutrophil membrane complement receptor type 3 (CR3) and the related membrane antigens LFA-1 and p150,95. J Clin Immunol 1986;6:107–113.
152. Birembaut P, Legrand YJ, Bariety J, et al. Histochemical and ultrastructural characterization of subendothelial glycoprotein microfibrils interacting with platelets. J Histochem Cytochem 1982;30:75–80.
153. Elsbach P: Degradation of microorganisms by phagocytic cells. Rev Infect Dis 1980;2:106–128.
154. Kishimoto TK, Hollander N, Roberts TM, et al. Heterogeneous mutations in the beta subunit common to the LFA-1, Mac-1, and p150,95 glycoproteins cause leukocyte adhesion deficiency. Cell 1987;50:193–202.
155. Miller ME, Oski FA, Harris MB. Lazy-leukocyte syndrome. A new disorder of neutrophil function. Lancet 1971;1:665–669.
156. Foroozanfar N, Lutterloch MJ, Thomas C, et al. Persistent mandibular infection in three patients with lazy and incompetent phagocyte syndromes. J Maxillofac Surg 1983;11:124–127.
157. Buckley RH, Sampson HA. The hyperimmunoglobulinemia E syndrome. In Franklin EC, ed. Clinical Immunology Update. New York: Elsevier, 1981.
158. Donabedian H, Gallin JI. The hyperimmunoglobulin E recurrent-infection (Job's) syndrome. A review of the NIH experience and the literature. Medicine 1983;62:195–208.
159. Blume RS, Wolff SM. The Chediak-Higashi syndrome: studies in four patients and a review of the literature. Medicine 1972;51:247–280.
160. Root RK, Rosenthal AS, Balestra DJ. Abnormal bactericidal, metabolic, and lysosomal functions of Chediak-Higashi syndrome leukocytes. J Clin Invest 1972;51:649–665.
161. Gallin JI, Klimerman JA, Padgett GA, et al. Defective mononuclear leukocyte chemotaxis in the Chediak-Higashi syndrome of humans, mink, and cattle. Blood 1975;45:863–870.
162. Ward PA, Berenberg JL. Defective regulation of inflammatory mediators in Hodgkin's disease. Supernormal levels of chemotactic-factor inactivator. N Engl J Med 1974;290:76–80.
163. DeMeo AN, Anderson BR. Defective chemotaxis associated with a serum inhibitor in cirrhotic patients. N Engl J Med 1972;286:735–740.
164. Davis JM, Dineen P, Gallin JI. Neutrophil degranulation and abnormal chemotaxis after thermal injury. J Immunol 1980;124:1467–1471.
165. Ginsburg I, Quie PG. Modulation of human polymorphonuclear leukocyte chemotaxis by leukocyte extracts, bacterial products, inflammatory exudates, and polyelectrolytes. Inflammation 1980;4:301–311.
166. Attia WM, Clark HW, Brown TM, et al. Inhibition of polymorphonuclear leukocyte migration by sera of patients with rheumatoid arthritis. Ann Allergy 1982;48:21–24.
167. Ito S, Mikawa H, Shinomiya K, et al. Suppressive effect of IgA soluble immune complexes on neutrophil chemotaxis. Clin Exp Immunol 1979;37:436–440.
168. Szejda P, Parce JW, Seeds MS, et al. Flow cytometric quantitation of oxidative product formation by polymorphonuclear leukocytes during phagocytosis. J Immunol 1984;133:3303–3307.
169. Tulp A, Barnhoorn MG. A separation chamber to sort cells and cell organelles by weak physical forces. V. A sector-shaped chamber and its application to the separation of peripheral blood cells. J Immunol Meth 1984;69:281–295.
170. Bassoe C-F, Bjerknes R. Phagocytosis by human leukocytes, phagosomal pH and degradation of seven species of bacteria measured by flow cytometry. J Med Microbiol 1985;19:115–125.
171. Bjerknes R, Bassoe C-F. Phagocyte C3-mediated attachment and internalization: flow cytometric studies using a fluorescence quenching technique. Blut 1984;49:315–323.
172. Bassoe C-F. Processing of staphylococcus aureus and zymosan particles by human leukocytes measured by flow cytometry. Cytometry 1984;5:86–91.
173. Bjerknes R, Bassoe C-F. Human leukocyte phagocytosis of zymosan particles measured by flow cytometry. Acta Patol Microbiol Immunol Scand [C] 1983;91:341–348.
174. Bjerknes R, Bassoe C-F, Sjursen H, et al. Flow cytometry for the study of phagocyte functions. Rev Infect Dis 1989;11:16–33.
175. Terstappen LWMM, De Grooth BG, Nolten GM, et al. Flow cytometric determination of circulating immune complexes with the indirect granulocyte phagocytosis test. Cytometry 1985;6:316–320.
176. Fattorossi A, Nisini R, Pizzolo JG, et al. New, simple flow cytometry technique to discriminate between internalized and membrane-bound particles in phagocytosis. Cytometry 1989;10:320–325.
177. Ma D, Chapman GJV, Chen S, et al. Flow cytometry with crystal violet to detect intracytoplasmic fluorescence in viable human lymphocytes. Demonstration of antibody entering living cells. J Immunol Meth 1987;104:195–200.
178. Bjerknes R, Bassoe C-F. Phagocyte C3-mediated attachment and internalization: Flow cytometric studies using a fluorescence quenching technique. Blut 1984;49:315–323.
179. Lightman MA. Rheology of leukocytes, leukocyte suspensions, and blood in leukemia. Possible relationship to clinical manifestations. J Clin Invest 1973;52:350–358.
180. Giordano GF, Lichtman MA. The role of sulfhydryl groups in human neutrophil adhesion, movement and particle ingestion. J Cell Physiol 1973;82:387–395.
181. Miller ME. Phagocyte function in the neonate: selected aspects. Pediatrics 1979;64:709–712.
182. Suzuki JB, Collison BC, Falkler WA Jr., et al. Immunologic profile of juvenile periodontitis. II. Neutrophil chemotaxis, phagocytosis and spore germination. J Periodontol 1984;55:461–467.
183. Najjar VA. Defective phagocytosis due to deficiencies involving the tetrapeptide tuftsin. J Pediatr 1975;87:1121–1124.
184. Fridkin M, Najjar VA. Tuftsin: its chemistry, biology, and clinical potential. Crit Rev Biochem Mol Biol 1989;24:1–40.
185. Najjar VA. The clinical and physiological aspects of tuftsin deficiency syndromes exhibiting defective phagocytosis. Klin Wochenschr 1979;57:751–756.
186. Buescher ES, Gaither T, Nath J, et al. Abnormal adherence-related functions of neutrophils, monocytes, and Epstein-Barr virus-transformed B cells in a patient with C3bi receptor deficiency. Blood 1985;65:1382–1390.
187. Davis BH, McCabe E, Langweiler M. Characterization of f-Met-Leu-Phe-stimulated fluid pinocytosis in human polymorphonuclear leukocytes by flow cytometry. Cytometry 1986;7:251–262.
188. Romano EL, Pascual CJ, Suarez G. Quantitation of antiplatelet antibodies by radioimmunoassay in sera of patients with immune thrombocytopenia. Haematologica (Pavia) 1981;66:597–604.
189. Shephard EG, Anderson R, Beer SM, et al. Neutrophil lysosomal degradation of human CRP: CRP-derived peptides modulate neutrophil function. Clin Exp Immunol 1988;73:139.
190. Shephard EG, Beer SM, Anderson R, et al. Generation of biologically active C-reactive protein peptides by a neutral protease on the membrane of phorbol myristate acetate-stimulated neutrophils. J Immunol 1989;143:2974–2981.
191. Stelzer GT, Robinson JPaul. Flow cytometric evaluation of leukocyte function. Diag Clin Immunol 1988;5:223–231.
192. Duque RE, Phan SH, Hudson JL, et al. Functional defects in phagocytic cells following thermal injury. Application of flow cytometric analysis. Am J Path 1985;118:116–127.
193. Bass DA, Olbrantz P, Szejda P, et al. Subpopulations of neutrophils with increased oxidative product formation in blood of patients with infection. J Immunol 1986;136:860–866.
194. Hassan NF, Campbell DE, Douglas SD. Flow cytometric analysis of oxidase activity of neutrophils from chronic granulomatous disease patients. Adv Exp Med Biol 1988;239:73–78.

195. Hassan NF, Campbell DE, Douglas SD. Phorbol myristate acetate induced oxidation of 2′,7′-dichlorofluorescin by neutrophils from patients with chronic granulomatous disease. J Leuk Biol 1988;43:317–322.
196. Patel AK, Hallett MB, Campbell AK. Threshold responses in production of reactive oxygen metabolites in individual neutrophils detected by flow cytometry and microfluorimetry. Biochem J 1987;248:173–180.
197. Wolber RA, Duque RE, Robinson JPaul, et al. Oxidative product formation in irradiated neutrophils. A flow cytometric analysis. Transfusion 1987;27:167–170.
198. Lepoivre M, Roche AC, Tenu JP, et al. Identification of two macrophage populations by flow cytometry monitoring of oxidative burst and phagocytic functions. Biol Cell 1986;57:143–146.
199. Trinkle LS, Wellhausen SR, McLeish KR. A simultaneous flow cytometric measurement of neutrophil phagocytosis and oxidative burst in whole blood. Diagn Clin Immunol 1987;5:62–68.
200. Blair OC, Carbone R, Sartorelli AC. Differentiation of HL-60 promyelocytic leukemia cells monitored by flow cytometric measurement of nitro blue tetrazolium (NBT) reduction. Cytometry 1985;6:54–61.
201. Blair OC, Carbone R, Sartorelli AC. Differentiation of HL-60 promyelocytic leukemia cells: simultaneous determination of phagocytic activity and cell cycle distribution by flow cytometry. Cytometry 1986;7:171–177.
202. Johnston RB, Newman SL. Chronic granulomatous disease. Pediatr Clin North Am 1977;24:365–376.
203. Caldicott WJ, Baehner RL. Chronic granulomatous disease of childhood. Am J Roentgenol Radium Ther Nucl Med 1968;103:133–139.
204. Balfour HH Jr., Shehan JJ, Speicher CE, et al. Chronic granulomatous disease of childhood in a 23-year-old man. JAMA 1971;217:960–961.
205. Chusid MJ, Parrillo JE, Fauci AS. Chronic granulomatous disease. Diagnosis in a 27-year-old man with Mycobacterium fortuitum. JAMA 1975;233:1295–1296.
206. Dilworth JA, Mandell GL. Adults with chronic granulomatous disease of "childhood." Am J Med 1977;63:233–243.
207. Segal AW, Jones OT. The subcellular distribution and some properties of the cytochrome b component of the microbicidal oxidase system of human neutrophils. Biochem J 1979;182:181–188.
208. Segal AW, Cross AR, Garcia RC, et al. Absence of cytochrome b-245 in chronic granulomatous disease. A multicenter European evaluation of its incidence and relevance. N Engl J Med 1983;308:245–251.
209. Segal AW, Jones OT. Novel cytochrome b system in phagocytic vacuoles of human granulocytes. Nature 1978;276:515–517.
210. Gabig TG, Lefker BA. Molecular heterogeneity in chronic granulomatous disease: a human model of defective phagocyte superoxide production. Free Radic Biol Med 1985;1:65–69.
211. Heyworth PG, Curnutte JT, Nauseef WM, et al. Neutrophil nicotinamide adenine dinucleotide phosphate oxidase assembly. Translocation of p47-phox and p67-phox requires interaction between p47-phox and cytochrome b558. J Clin Invest 1991;87:352–356.
212. Gabig TG. Leukocyte abnormalities. Med Clin North Am 1980;64:647–666.
213. Gray GR, Stamatoyannopoulos G, Naiman SC, et al. Neutrophil dysfunction, chronic granulomatous disease, and non-spherocytic haemolytic anaemia caused by complete deficiency of glucose-6-phosphate dehydrogenase. Lancet 1973;2:530–534.
214. Gabig TG, Lefker BA. Deficient flavoprotein component of the NADPH-dependent O2-.-generating oxidase in the neutrophils from three male patients with chronic granulomatous disease. J Clin Invest 1984;73:701–705.
215. Kleinberg ME, Malech HL, Rotrosen D. The phagocyte 47-kilodalton cytosolic oxidase protein is an early reactant in activation of the respiratory burst. J Biol Chem 1990;265:15577–15583.
216. Mills EL, Thompson T, Bjorksten B, et al. The chemiluminescence response and bactericidal activity of polymorphonuclear neutrophils from newborns and their mothers. Pediatrics 1979;63:429–434.
217. Kragballe K, Borregaard N, Brandrup F, et al. Relation of monocyte and neutrophil oxidative metabolism to skin and oral lesions in carriers of chronic granulomatous disease. Clin Exp Immunol 1981;43:390–398.
218. Root RK, Cohen MS. The microbicidal mechanisms of human neutrophils and eosinophils. Rev Infect Dis 1981;3:565–598.
219. Ohlsson K, Olsson I, Spitznagel K. Localization of chymotrypsin-like cationic protein, collagenase and elastase in azurophil granules of human neutrophilic polymorphonuclear leukocytes. Hoppe Seylers Z Physiol Chem 1977;358:361–366.
220. Franson R, Weiss J, Martin L, et al. Phospholipase A activity associated with membranes of human polymorphonuclear leucocytes. Biochem J 1977;167:839–841.
221. Neeman N, Lahav M, Ginsburg I. The effect of leukocyte hydrolases on bacteria. II. The synergistic action of lysozyme and extracts of PMN, macrophages, lymphocytes, and platelets in bacteriolysis. Proc Soc Exp Biol Med 1974;146:1137–1145.
222. Weiss J, Olsson I. Cellular and subcellular localization of the bactericidal/permeability-increasing protein of neutrophils. Blood 1987;69:652–659.
223. Breton Gorius J, Mason DY, Buriot D, et al. Lactoferrin deficiency as a consequence of a lack of specific granules in neutrophils from a patient with recurrent infections. Detection by immunoperoxidase staining for lactoferrin and cytochemical electron microscopy. Am J Pathol 1980;99:413–428.
224. Odeberg H, Olsson I. Antibacterial activity of cationic proteins from human granulocytes. J Clin Invest 1975;56:1118–1124.
225. Clark RA, Olsson I, Klebanoff SJ. Cytotoxicity for tumor cells of cationic proteins from human neutrophil granules. J Cell Biol 1976;70:719–723.
226. Odeberg H, Olsson I. Mechanisms for the microbicidal activity of cationic proteins of human granulocytes. Infect Immun 1976;14:1269–1275.
227. Shafer WM, Onunka VC, Martin LE. Antigonococcal activity of human neutrophil cathepsin G. Infect Immun 1986;54:184–188.
228. Strominger JL, Tipper DJ. Structure of bacterial walls: the lysozyme substrate. In Osserman (ed): Lysozyme. New York, Academic Press, 1974, pp 169-184.
229. Thorne KJ, Oliver RC, Barrett AJ. Lysis and killing of bacteria by lysosomal proteinases. Infect Immun 1976;14:555–563.
230. Odeberg H, Olsson I. Microbicidal mechanisms of human granulocytes: synergistic effects of granulocyte elastase and myeloperoxidase or chymotrypsin-like cationic protein. Infect Immun 1976;14:1276–1283.
231. Smedly LA, Tonnesen MG, Sandhaus RA, et al. Neutrophil-mediated injury to endothelial cells. Enhancement by endotoxin and essential role of neutrophil elastase. J Clin Invest 1986;77:1233–1243.
232. Gruber DF, Laws AB, O'Halloran KP. Biochemical and physiological alterations in canine neutrophils separated by lysis or Percoll gradient isolation technologies. Immunopharmacol Immunotoxicol 1990;12:93–104.
233. Spitznagel JK, Dalldorf FG, Leffell MS, et al. Character of azurophil and specific granules purified from human polymorphonuclear leukocytes. Lab Invest 1974;30:774–785.
234. Miller TE. Killing and lysis of gram-negative bacteria through the synergistic effect of hydrogen peroxide, ascorbic acid, and lysozyme. J Bacteriol 1969;98:949–955.
235. Elsbach P, Weiss J. Oxygen-dependent and oxygen-independent mechanisms of microbicidal activity of neutrophils. Immunol Lett 1985;11:159–163.
236. Weiss J, Kao L, Victor M, et al. Oxygen-independent intracellular and oxygen-dependent extracellular killing of Escherichia coli S15 by human polymorphonuclear leukocytes. J Clin Invest 1985;76:206–212.
237. Weiss J, Elsbach P, Olsson I, et al. Purification and characterization of a potent bactericidal and membrane active protein from the granules

of human polymorphonuclear leukocytes. J Biol Chem 1978;253: 2664–2672.
238. Shafer WM, Martin LE, Spitznagel JK. Late intraphagosomal hydrogen ion concentration favors the in vitro antimicrobial capacity of a 37-kilodalton cationic granule protein of human neutrophil granulocytes. Infect Immun 1986;53:651–655.
239. Aisen P, Listowsky I. Iron transport and storage proteins. Annu Rev Biochem 1980;49:357–393.
240. Moguilevsky N, Retegui LA, Masson PL. Comparison of human lactoferrins from milk and neutrophilic leucocytes. Relative molecular mass, isoelectric point, iron-binding properties and uptake by the liver. Biochem J 1985;229:353–359.
241. Masson PL, Heremans JF, Dive C. An iron binding protein common to many external secretions. Clin Chim Acta 1966;14:735.
242. Bishop JG, Schanbacher FL, Ferguson LC, et al. In vitro growth inhibition of mastitis-causing coliform bacteria by bovine apo-lactoferrin and reversal of inhibition by citrate and high concentrations of apo-lactoferrin. Infect Immun 1976;14:911–918.
243. Oram JD, Reiter B. Inhibition of bacteria by lactoferrin and other iron-chelating agents. Biochim Biophys Acta 1968;170:351–365.
244. Arnold RR, Russell JE, Champion WJ, et al. Bactericidal activity of human lactoferrin: differentation from the stasis of iron deprivation. Infect Immun 1982;35:792–799.
245. Lehrer RI, Daher K, Ganz T, et al. Direct inactivation of viruses by MCP-1 and MCP-2, natural peptide antibiotics from rabbit leukocytes. J Virol 1985;54:467–472.
246. Miyasaki KT, Bodeau AL, Ganz T, et al. In vitro sensitivity or oral, gram-negative, facultative bacteria to the bactericidal activity of human neutrophil defensins. Infect Immun 1990;58:3934–3940.
247. Selsted ME, Szklarek D, Ganz T, et al. Activity of rabbit leukocyte peptides against Candida albicans. Infect Immun 1985;49:202–206.
248. Lichtenstein AK, Ganz T, Nguyen TM, et al. Mechanism of target cytolysis by peptide defensins. Target cell metabolic activities, possibly involving endocytosis, are crucial for expression of cytotoxicity. J Immunol 1988;140:2686–2694.
249. Cech P, Lehrer RI. Phagolysosomal pH of human neutrophils. Blood 1984;63:88–95.
250. Bjerknes R, Vindenes H. Neutrophil dysfunction after thermal injury alteration of phagolysosomal acidification in patients with large burns. Burns 1989;15:77–81.
251. Hynes RO: Integrins: a family of cell surface receptors. Cell 1987;48: 549–554.
252. Sanchez-Madrid N, Nagy J, Robbins E, et al. A human leukocyte differentiation antigen family with distinct alpha-subunit and common beta-subunits: the lymphocyte function associated antigen (LFA-1), the C3bi complement receptor (OKM-1/Mac-1) and the p150,95 molecule. J Exp Med 1983;158:1785–1803.
253. Springer TA, Anderson DC. The importance of the Mac-1, LFA-1 glycoprotein family in monocyte and granulocyte adherence, chemotaxis and migration to inflammatory sites: insight from an experiment of nature. In Ciba Foundation Symposium. London: Pitman, 1986, pp 102–106.
254. Arnaout MA. Structure and function of the leukocyte adhesion molecules CD11/CD18. Blood 1990;75:1037–1050.
255. Rosen H, Law SK. The leukocyte cell surface receptor(s) for the iC3b product of complement. Curr Top Microbiol Immunol 1990;153:99–122.
256. Styrt B. History and implications of the neutrophil glycoprotein deficiencies. Am J Hematol 1989;31:288–297.
257. Wright SD, Detmers PA, Aida Y, et al. CD18-deficient cells respond to lipopolysaccharide in vitro. J Immunol 1990;144:2566-2571.
258. Nathan C, Srimal S, Farber C, et al. Cytokine-induced respiratory burst of human neutrophils: dependence on extracellular matrix proteins and CD11/CD18 integrins. J Cell Biol 1989;109:1341–1349.
259. Ricevuti G, Mazzone A, Notario A. Definition of CD 11a, b, c, and CD 18 glycoproteins on chemotactically deficient granulocyte membranes in patients affected by myeloid disorders. Acta Haemotal (Basel) 1989;81:126–130.
260. Dobrina A, Carlos TM, Schwartz BR, et al. Phorbol ester causes down-regulation of CD11/CD18-independent neutrophil adherence to endothelium. Immunolgy 1990;69:429–434.
261. Pober JS, Bevilacqua MP, Mendrick DL, et al. Two distinct monokines, interleukin 1 and tumor necrosis factor, each independently induce biosynthesis and transient expression of the same antigen on the surface of cultured human vascular endothelial cells. J Immunol 1986;136:1680–1687.
262. Georgilis K, Schaefer C, Dinarello CA, et al. Human recombinant interleukin 1 beta has no effect on intracellular calcium or on functional responses of human neutrophils. J Immunol 1987;138:3403-3407.
263. Arnaout MA, Hakim RM, Todd RF III, et al. Increased expression of an adhesion-promoting surface glycoprotein in the granulocytopenia of hemodialysis. N Eng J Med 1985;312:457–462.
264. Gordon DL, Rice JL, McDonald PJ. Regulation of human neutrophil type 3 complement receptor (iC3b receptor) expression during phagocytosis of Staphylococcus aureus and Escherichia coli. Immunology 1989;67:460–465.
265. Anderson DC, Schmalstieg FC, Finegold MJ, et al. The severe and moderate phenotypes of heritable Mac-1, LFA-1 deficiency: their quantitative definition and relation to leukocyte dysfunction and clinical features. J Infect Dis 1985;152:668–689.
266. Miller LJ, Bainton DF, Borregaard N, et al. Stimulated mobilization of monocyte Mac-1 and p150,95 adhesion proteins from an intracellular vesicular compartment to the cell surface. J Clin Invest 1987;80: 535–544.
267. Berger M, Wetzler EM, Wallis RS. Tumor necrosis factor is the major monocyte product that increases complement receptor expression on mature human neutrophils. Blood 1988;71:151–158.
268. Bainton DF, Miller LJ, Kishimoto TK, et al. Leukocyte adhesion receptors are stored in peroxidase negative granules of human neutrophils. J Exp Med 1987;166:1641–1653.
269. Boros P, Gardos E, Bekesi GJ, et al. Change in expression of $Fc_{gamma}RIII$ (CD16) on neutrophils from human immunodeficiency virus-infected individuals. Clin Immunol Immunopathol 1990;54:281–289.
270. Fleit HB, Wright SD, Unkeless JC. Human neutrophil Fcgamma receptor distribution and structure. Proc Natl Acad Sci USA 1982;79: 3275–3279.
271. Lanier LL, Kipps TJ, Phillips JH. Functional Properties of a unique subset of cytotoxic CD3+ T lymphocytes that express Fc receptors for IgG (CD16/Leu-11 antigen). J Exp Med 1985;162:2089–2107.
272. Koss LG, Czerniak B, Herz F, et al. Flow cytometric measurements of DNA and other cell components in human tumors: a critical appraisal. Hum Pathol 1989;20:528–548.
273. Selvaraj P, Rosse WF, Silber R, et al. The major Fc receptor in blood has a phosphatidylinositol anchor and is deficient in paroxysmal nocturnal haemoglobinuria. Nature (London) 1988;333:565–567.
274. Huizinga TW, van der Schoot CE, Jost C, et al. The PI-linked receptor FcRIII is released on stimulation of neutrophils. Nature 1988;333: 667–669.
275. Nielsen H, Kharazami A, Faber V. Blood monocyte and neutrophil functions in the acquired immune deficiency syndrome. Scand J Immunol 1986;24:291–296.
276. Kinne TJ, Gupta S. Antibody-dependent cellular cytotoxicity by polymorphonuclear leucocytes in patients with AIDS and AIDS-related complex. J Clin Lab Immunol 1989;30:153–156.
277. Ellis M, Gupta S, Galant S, et al. Impaired neutrophil function in patients with AIDS or AIDS-related complex: a comprehensive evaluation. J Infect Dis 1988;158:1268–1276.
278. Huizinga TWJ, van Kemenade F, Koenderman L, et al. The 40-kDa Fcgamma receptor (FcRII) on human neutrophils is essential for the IgG-induced respiratory burst and IgG-induced phagocytosis. J Immunol 1989;142:2365–2369.

279. Huizinga TWJ, Dolman KM, van der Linden NJM, et al. Phosphatidylinositol-linked FcRIII mediates exocytosis of neutrophil granule proteins, but does not mediate initiation of the respiratory burst. J Immunol 1990;144:1432–1437.
280. Tosi MF, Berger M. Functional differences between the 40 kDa and 50 to 70 kDa IgG Fc receptors on human neutrophils revealed by elastase treatment and antireceptor antibodies. J Immunol 1988;141(6):2097–2103.
281. Crockett Torabi E, Fantone JC. Soluble and insoluble immune complexes activate human neutrophil NADPH oxidase by distinct Fc gamma receptor-specific mechanisms. J Immunol 1990;145:3026–3032.
282. Akerley WL III, Guyre PM, Davis BH. Neutrophil activation through high-affinity Fc gamma receptor using a monomeric antibody with unique properties. Blood 1991;77:607–615.
283. Davis BH. Measurement of receptor numbers by flow cytometry [Abstract]. International Society for Analytical Cytology, Bergen, Norway, August 25–30, 1991.
284. van de Winkel JG, Anderson CL. Biology of human immunoglobulin G Fc receptors. J Leukoc Biol 1991;49:511–524.
285. Lawton JWM, Robinson JPaul, Till GO. The effects of intravenous immunoglobulin on the in vitro function of human neutrophils. Immunopharmacology 1989;18:97–105.
286. Maródi L, Kalmár A, Karmazsin L. Stimulation of the respiratory burst and promotion of bacterial killing in human granulocytes by intravenous immunoglobulin preparations. Clin Exp Immunol 1990;79:164–169.
287. Berger M. Complement deficiency and neutrophil dysfunction as risk factors for bacterial infection in newborns and the role of granulocyte transfusion in therapy. Rev Infect Dis 1990;12(Suppl 4):S401–S409.
288. Charon JA, Metzger Z, Hoffeld JT, et al. An in vitro study of neutrophils obtained from the normal gingival sulcus. J Periodont Res 1982;17:614–625.
289. Anderson DC, Becker-Freeman KL, Heerdt B, et al. Abnormal stimulated adherence of neonatal granulocytes; impaired induction of surface MAC-1 by chemotactic factors or secretagogues. Blood 1987;70:740–750.
290. Christensen RD, Rothstein G, Anstall HB, et al. Granulocyte transfusions in neonates with bacterial infection, neutropenia, and depletion of mature marrow neutrophils. Pediatrics 1982;70:1–6.
291. Laurenti F, Ferro R, Isacchi G, et al. Polymorphonuclear leukocyte transfusion for the treatment of sepsis in the newborn infant. J Pediatr 1981;98:118–123.
292. Wheeler JG, Chauvenet AR, Johnson CA, et al. Buffy coat transfusions in neonates with sepsis and neutrophil storage pool depletion. Pediatrics 1987;79:422–425.
293. Clay ME, Kline WE. Detection of granulocyte antigens and antibodies: current perspectives and approaches. In Arlington VA, ed. Current Concepts in Transfusion Therapy. American Association of Blood Banks, 1985, pp 183–265.
294. Robinson JPaul, Duque RE, Boxer LA, et al. Measurement of antineutrophil antibodies (NAB) by flow cytometry: simultaneous detection of antibodies against monocytes and lymphocytes. Diag Clin Immunol 1987;5:163–170.
295. Young DW. The antineutrophil antibody in uveitis. Br J Ophthalmol 1991;75:208–211.
296. Mustonen J, Soppi E, Pasternack A, et al. Clinical significance of autoantibodies against neutrophil cytoplasmic components in patients with renal disease. Am J Nephrol 1990;10:482–488.
297. Brouwer E, Cohen Tervaert JW, Horst G, et al. Predominance of IgG1 and IgG4 subclasses of anti-neutrophil cytoplasmic autoantibodies (ANCA) in patients with Wegener's granulomatosis and clinically related disorders. Clin Exp Immunol 1991;83:379–386.
298. Venning MC, Quinn A, Broomhead V, et al. Antibodies directed against neutrophils (C-ANCA and P-ANCA) are of distinct diagnostic value in systemic vasculitis. Q J Med 1990;77:1287–1296.
299. Falk RJ, Terrell RS, Charles LA, et al. Anti-neutrophil cytoplasmic autoantibodies induce neutrophils to degranulate and produce oxygen radicals in vitro. Proc Natl Acad Sci USA 1990;87:4115–4119.
299a. Peter JB, Wormsley SB, Dawkins RL. Anti Neutrophil Cytoplasm Antibody (ANCA) in systemic vasculitis: clinical utility of quantitation by flow cytometry. Urology 1988;38(suppl 1):99.
300. Gaither TA, Gallin JI, Lida K, et al. Deficiency in C3b receptors on neutrophils of patients with chronic granulomatous disease and hyperimmunoglobulin-E recurrent infection (Job's) syndrome. Inflammation 1984;8:429–444.
301. White CJ, Gallin JI. Phagocyte defects. Clin Immunol Immunopathol 1986;40:50–61.
302. Bellinati-Pires R, Melki SE, Colletto GMDD, et al. Evaluation of a fluorochrome assay for assessing the bactericidal activity of neutrophils in human phagocyte dysfunctions. J Immunol Meth 1989;119:189–196.
303. Schopfer K, Douglas SD. Neutrophil functions in children with kwashiorkor. J Lab Clin Med 1976;88:450–461.
304. Lizard G, Chardonnet Y, Chignol MC, et al. Evaluation of mitochondrial content and activity with nonyl-acridine orange and rhodamine 123: flow cytometric analysis and comparison with quantitative morphometry. Cytotechnology 1990;3:179–188.
305. Leung-Tack J, Tavera C, Gensac MC, et al. Modulation of phagocytosis-associated respiratory burst by human cystatin C: role of the N-terminal tetrapeptide Lys-Pro-Pro-Arg. Exp Cell Res 1990;188:16–22.
306. Clark RA, Page RC, Wilde G. Defective neutrophil chemotaxis in juvenile periodontitis. Infect Immun 1977;18:694–700.
307. Miller DR, Lamster IB, Chasens AI. Role of the polymorphonuclear leukocyte in periodontal health and disease. J Clin Periodontol 1984;11:1–15.
308. Wilson ME, Zambon JJ, Suzuki JB, et al. Generalized juvenile periodontitis, defective neutrophil chemotaxis and Bacteroides gingivalis in a 13-year-old female. A case report. J Periodontol 1985;56:457–463.
309. Smith QT, Hinrichs JE, Melnyk RS. Gingival crevicular fluid myeloperoxidase at periodontitis sites. J Periodont Res 1986;21:45–55.
310. Clark RA, Kimball HR. Defective granulocyte chemotaxis in the Chediak-Higashi syndrome. J Clin Invest 1971;50:2645–2652.
311. Goldstein IM. Neutrophil degranulation. Contemp Top Immunobiol 1984;14:189–219.
312. Altman LC, Furukawa CT, Klebanoff SJ. Depressed mononuclear leukocyte chemotaxis in thermally injured patients. J Immunol 1977;119:199.
313. Deitch EA. The relationship between thermal injury and neutrophil membrane functions as measured by chemotaxis, adherence and spreading. Burns 1984;10:264–270.
314. Grogan JB. Suppressed in vitro chemotaxis of burn neutrophils. J Trauma 1976;16:985–988.
315. Ninnemann JL, Ozkan AN, Sullivan JJ. Hemolysis and suppression of neutrophil chemotaxis by a low molecular weight component of human burn patient sera. Immunol Lett 1985;10:63–69.
316. Warden GD, Mason AD, Pruitt BA Jr. Suppression of leukocyte chemotaxis by chemotherapeutic agents used in the management of thermal injuries. Ann Surg 1975;181:363–369.
317. Iacono VJ, Singh S, Golub LM, et al. In vivo assay of crevicular leukocyte migration. Its development and potential applications. J Periodontol 1985;56:56–62.
318. McMullen JA, van Dyke TE, Horoszewicz HU, et al. Neutrophil chemotaxis in individuals with advanced periodontal disease and a genetic predisposition to diabetes mellitus. J Periodont Res 1981;52(4):167–173.
319. Giorgi JV, Cheng H-L, Margolick JB, et al. Quality control in the flow cytometric measurement of T-lymphocyte subsets: The Multicenter AIDS Cohort Study experience. Clin Immunol Immunopathol 1990;55:173–186.
320. Parker JW, Adelsberg B, Azen SP, et al. Leukocyte immunophenotyping by flow cytometry in a multisite study: standardization, qual-

ity control, and normal values in the transfusion safety study. Clin Immunol Immunopathol 1990;55:187–220.
321. Elmgreen J, Hansen TM. Subnormal sensitivity of neutrophils to complement split product C5a in rheumatoid arthritis: relation to complement catabolism and disease extent. Ann Rheum Dis 1985;44:514–518.
322. Weisman SJ, Berkow RL, Plautz G, et al. Glycoprotein-180 deficiency: genetics and abnormal neutrophil activation. Blood 1985;65:696–704.
323. Teshima T, Shibuya T, Harada M, et al. Granulocyte-macrophage colony-stimulating factor suppresses induction of neutrophil alkaline phosphatase synthesis by granulocyte colony-stimulating factor. Exp Hematol 1990;18:316–321.
324. Grogan JB. Altered neutrophil phagocytic function in burn patients. J Trauma 1976;16:734–738.
325. Wysocki H, Wierusz-Wysocka B, Karon H, et al. Polymorphonuclear neutrophils function in splenectomized patients. Clin Exp Immunol 1989;75:392–395.
326. Toniolo C, Crisma M, Moretto V, et al. $N\alpha$-formylated and tert-butyloxycarbonylated Phe-(Leu-Phe)$_n$ and (Leu-Phe)$_n$ peptides as agonists and antagonists of the chemotactic formylpeptide receptor of the rabbit peritoneal neutrophil. Biochim Biophys Acta Gen Subj 1990;1034:67–72.
327. Slocombe RF, Watson GL, Killingsworth CR. Effect of deferoxamine pretreatment on acute pneumonic pasteurellosis and neutrophil oxidative metabolism in calves. Can J Vet Res 1990;54:227–231.
328. Stendahl O, Coble BI, Cahlgren C, et al. Myeloperoxidase modulates the phagocytic activity of polymorphonuclear neutrophil leukocytes. Studies with cells from a myeloperoxidase-deficient patient. J Clin Invest 1984;73:366–373.
329. Fahey JL: Doing it right: Measuring T cell subsets by flow cytometry. Clin Immunol Immunopathol 1990;55:171–172.
330. Raphael GD, Davis JL, Fox PC, et al. Glandular secretion of lactoferrin in a patient with neutrophil lactoferrin deficiency. J Allergy Clin Immunol 1989;84:914–919.
331. Ganz T, Metcalf JA, Gallin JI, et al. Microbicidal/cytotoxic proteins of neutrophils are deficient in two disorders: Chediak-Higashi syndrome and "specific" granule deficiency. J Clin Invest 1988;82:552–556.
332. Gallin JI. Neutrophil specific granules: a fuse that ignites the inflammatory response. Clin Res 1984;32:320–328.
333. Fletcher J, Haynes AP, Crouch SM. Acquired abnormalities of polymorphonuclear neutrophil function. Blood Rev 1990;4:103–110.
334. Cohen MS, Leong PA, Simpson DM. Phagocytic cells in periodontal defense. Periodontal status of patients with chronic granulomatous disease of childhood. J Periodontol 1985;56:611–617.
335. Steerman RL, Snyderman R, Leikin SL, et al. Intrinsic defect of the polymorphonuclear leukocyte resulting in impaired chemotaxis and phagocytosis. Clin Exp Immunol 1971;9:939–946.
336. McCall CE, Caves J, Cooper R, et al. Functional characteristics of human toxic neutrophils. J Infect Dis 1971;124:68–75.
337. Cianciola LJ, Genco RJ, Patters MR, et al. Defective polymorphonuclear leukocyte function in a human periodontal disease. Nature (London) 1977;265:445–447.
338. Yoneda M, Maeda K, Aona M. Suppression of bactericidal activity of human polymorphonuclear leukocytes by Bacteroides gingivalis. Infect Immun 1990;58:406–411.
339. Robinson JPaul, Penny R. Chemiluminescence response in normal human phagocytes. II Effect of paraproteins. J Clin Lab Immunol 1982;7:219–221.
340. Murphy PM, Lane HC, Fauci AS, et al. Impairment of neutrophil bactericidal capacity in patients with AIDS. J Infect Dis 1988;158:627–630.
341. Poubelle PE, Bourgoin S, McColl SR, et al. Altered formation of leukotriene B4 in vitro by synovial fluid neutrophils in rheumatoid arthritis. J Rheumatol 1989;16:280–284.
342. Wierusz Wysocka B. Disturbances of neutrophil granulocyte function in diabetes. Part II. Mechanisms responsible for impaired neutrophil granulocyte functions. Mater Med Pol 1988;20:255–257.
343. Kelly MK, Brown JM, Thong YH. Neutrophil and monocyte adherence in diabetes mellitus, alcoholic cirrhosis, uraemia and elderly patients. Int Arch Allergy Appl Immunol 1985;78:132–138.
344. Berger M, Sorensen RU, Tosi MF, et al. Complement receptor expression on neutrophils at an inflammatory site, the Pseudomonas-infected lung in cystis fibrosis. J Clin Invest 1989;84:1302–1313.
345. Ng DS, Blass KG. A rapid, sensitive method for accurate determination of the lecithin/sphingomyelin ratio by thin-layer chromatography and reflectance spectrofluorometry. J Chromatogr 1979;163:37–46.
346. Hartwig JH, Stossel TP. Cytochalasin B dissolves actin gels by breaking actin filaments. J Cell Biol 1979;79:M11741.
347. Tauber AI, Borregaard N, Simons E, et al. Chronic granulomatous disease: a syndrome of phagocyte oxidase deficiencies. Medicine (Baltimore) 1983;62:286–309.
348. Ambruso DR, Altenburger KM, Johnson RB Jr. Effective oxidative metabolism in newborn neutrophils: discrepancy between superoxide anion and hydroxyl radical generation. Pediatrics 1979;64:722.
349. Ambruso DR. Decreased hydroxyl radical generation and lactoferrin content in cord blood neutrophils. Pediatr Res 1982;16:198A.
350. Abramson JS, Mills EL, Sawyer MK, et al. Recurrent infections and delayed separation of the umbilical cord in an infant with abnormal phagocytic cell locomotion and oxidative response during particle phagocytosis. J Pediatr 1981;99:887–894.
351. Cooper MR, McCall CE, DeChatelet LR. Stimulation of leukocyte hexose monophosphate shunt activity by ascorbic acid. Infect Immun 1971;3:851–853.
352. Burns WH, Barbour GM, Sandford GR. Molecular cloning and mapping of rat cytomegalovirus DNA. Virology 1988;166:140–148.
353. Roberts PJ, Cummins D, Bainton AL, et al. Plasma from patients with severe Lassa fever profoundly modulates f-met-leu-phe induced superoxide generation in neutrophils. Br J Haematol 1989;73:152–157.
354. Ross DW, Kaplow LS. Myeloperoxidase deficiency. Increased sensitivity for immunocytochemical compared to cytochemical detection of enzyme. Arch Pathol Lab Med 1985;109:1005–1006.
355. Alexander JW. Serum and leukocyte lysosomal enzymes. Derangements following severe thermal injury. Arch Surg 1967;95:482–490.
356. Malech HL, Gallin JI. Current concepts: immunology. Neutrophils in human diseases. N Engl J Med 1987;317:687–694.
357. Haak RA, Ingraham LM, Baehner RL, et al. Membrane fluidity in human and mouse Chediak-Higashi leukocytes. J Clin Invest 1979;64:138–144.
358. Takeuchi KH, Swank RT. Inhibitors of elastase and cathepsin G in Chédiak-Higashi (beige) neutrophils. J Biol Chem 1989;264:7431–7436.
359. Liang D-C, Ma S-W, Lin-Chu M, et al. Granulocyte/macrophage colony-forming units from cord blood of premature and full-term neonates: its role in ontogeny of human hemopoiesis. Pediatr Res 1988;24:701–702.
360. Cairo MS, Vandeven C, Toy C, et al. Fluorescent cytometric analysis of polymorphonuclear leukocytes in Chediak-Higashi Syndrome: diminished C3bi receptor expression (OKM1) with normal granular cell density. Pediatr Res 1988;24:673–676.
361. Brom J, Köller M, Schönfeld W, et al. Decreased expression of leukotriene B$_4$ receptor sites on polymorphonuclear granulocytes of severely burned patients. Prostaglandins Leukot Essent Fatty Acids 1988;34:153–159.
362. Belch JJF, O'Dowd A, Ansell D, et al. Leukotriene B$_4$ production by peripheral blood neutrophils in rheumatoid arthritis. Scand J Rheumatol 1989;18:213–219.
363. Masuda K, Kinoshita Y, Kobayashi Y. Heterogeneity of Fc receptor expression in chemotaxis and adherence of neonatal neutrophils. Pediatr Res 1989;25:6–10.

364. Huizanga TWJ, Kerst M, Nuyens JH, et al. Binding characteristics of dimeric IgG subclass complexes to human neutrophils. J Immunol 1989;142:2359–2364.
365. Ueda E, Kinoshita T, Nojima J, et al. Different membrane anchors of FcgammaRIII (CD16) on K/NK-lymphocytes and neutrophils: protein- vs lipid-anchor. J Immunol 1989;143:1274–1277.
366. Huizanga TW, de Haas M, Kleijer M, et al. Soluble Fc gamma receptor III in human plasma originates from release by neutrophils. J Clin Invest 1990;86:416–423.
367. Gallin JI. Abnormal phagocyte chemotaxis: pathophysiology, clinical manifestations and management of patients. Rev Infect Dis 1981;3: 1196.
368. Malech HL, Root RK, Gallin JI. Structural analysis of human neutrophil migration: centriole, microtubule and microfilament orientation and function during chemotaxis. J Cell Biol 1977;75:666.
369. Weissmann G. From Auden to arichidonate: a tribute and a hypothesis. Cell Immunol 1983;82:117.
370. Ferrari FA, Pagani A, Marconi M, et al. Inhibition of candidacidal activity of human neutrophil leukocytes by aminoglycoside antibiotics. Antimicrob Agents Chemother 1980;17:87–88.
371. Bjorksten B, Ray C, Quie PG. Inhibition of human neutrophil chemotaxis and chemiluminescence by amphotericin B. Infect Immun 1976; 14:315–317.
372. Roilides E, Walsh TJ, Rubin M, et al. Effects of antifungal agents on the function of human neutrophils in vitro. Antimicrob Agents Chemother 1990;34:196–201.
373. Chan CK, Balish E. Inhibition of granulocyte phagocytosis of Candida albicans by amphotericin B. Can J Microbiol 1978;243:363–364.
374. Hafstrom I, Seligmann BE, Friedman MM, et al. Auranofin affects early events in human polymorphonuclear neutrophil activation by receptor-mediated stimuli. J Immunol 1984;132:2007–2014.
375. Herlin T, Fogh K, Christiansen NO, et al. Effect of auranofin on eicosanoids and protein kinase C in human neutrophils. Agents Actions 1989;28:121–129.
376. Taniguchi K, Urakami M, Takanaka K. Effects of various drugs on superoxide generation, arachidonic acid release and phospholipase A2 in polymorphonuclear leukocytes. Jpn J Pharmacol 1988;46:275–284.
377. Forsgren A, Schmeling D. Effect of antibiotics on chemotaxis of human leukocytes Antimicrob Agents Chemother 1977;11:580–584.
378. Esterly NB, Furey NL, Flanagan LE. The effect of antimicrobial agents on leukocyte chemotaxis. J Invest Dermatol 1978;70:51–55.
379. Etzioni A, Obedeanu N, Benderly A, et al. Saethre-Chotzen syndrome associated with defective neutrophil chemotaxis. Acta Paediatr Scand 1990;79:375–379.
380. Katayama T, Ohtsuka T, Wakamura K, et al. Effect of glutaraldehyde on NADPH oxidase system of guinea pig polymorphonuclear leukocytes. Arch Biochem Biophys 1990;278:431–436.
381. Stendahl O, Molin L, Dahlgren C. The inhibition of polymorphonuclear leukocyte cytotoxicity by dapsone: A possible mechanism in the treatment of dermatitis herpetiformis. J Clin Invest 1978;62:214.
382. Cairo MS, Toy C, Sender L, et al. Effect of idarubicin and epirubicin on in vitro polymorphonuclear function. J Leuk Biol 1990;47:224–233.
383. Goodhart GL. Effect of aminoglycosides on the chemotactic response of human polymorphonuclear leukocytes. Antimicrob Agents Chemother 1977;12:540–542.
384. Khan AJ, Evans HE, Glass L, et al. Abnormal neutrophil chemotaxis and random migration induced by aminoglycoside antibiotics. J Lab Clin Med 1979;93:295–300.
385. Altman RD. Neutrophil activation: an alternative to prostaglandin inhibition as the mechanism of action for NSAIDs. Semin Arthritis Rheum 1990;19:1–5.
386. Taniguchi K, Masuda Y, Takanaka K. Action sites of antiallergic drugs on human neutrophils. Jpn J Pharmacol 1990;52:101–108.
387. Hand WL, Butera ML, King-Thompson NL, et al. Pentoxifylline modulation of plasma membrane functions in human polymorphonuclear leukocytes. Infect Immun 1989;57:3520–3526.
388. Bannatyne RM, Harnett NM, Lee KY, et al. Inhibition of the biologic effects of endotoxin on neutrophils by polymyxin B sulfate. J Infect Dis 1977;136:469–474.
389. Downey RJ, Pisano JC. Some effects of antimicrobial compounds on phagocytosis in vitro. J Reticuloendothel Soc 1965;2:75–88.
390. Gray GD, Knight KA, Talley CA. Rifampin has paradoxical effects on leukotaxis. Fed Proc 1980;39:878.
391. Belsheim J, Gnarpe H, Persson S. Tetracyclines and host defense mechanisms: interference with leukocyte chemotaxis. Scand J Infect Dis 1979;11:141–145.
392. Gnarpe H, Belsheim J. Tetracyclines and host defense mechanisms. Doxycycline polymethaphosphate sodium complex (DMSC) compared to doxycycline with regard to influence on some host defense mechanisms. Microbios 1978;22:45–49.
393. Hamada M, Nishio I, Baba A, et al. Enhanced DNA synthesis of cultured vascular smooth muscle cells from spontaneously hypertensive rats. Difference of response to growth factor, intracellular free calcium concentration and DNA synthesizing cell cycle. Atherosclerosis 1990;81:191–198.
394. Forsgren A, Schmeling D, Quie PG. Effect of tetracycline on the phagocytic function of human leukocytes. J Infect Dis 1974;130:412–418.
395. Gnarpe H, Leslie D. Tetracyclines and host defense mechanisms. Doxycycline interference with phagocytosis of Escherichia coli. Microbios 1974;10A:127–138.
396. Bokoch GM, Quilliam LA. Guanine nucleotide binding properties of rap1 purified from human neutrophils. Biochem J 1990;267:407–411.
397. Teahan CG, Totty N, Casimir CM, et al. Purification of the 47 kDa phosphoprotein associated with the NADPH oxidase of human neutrophils. Biochem J 1990;267:485–489.
398. Pasini FL, Capecchi PL, Pasqui AL, et al. Adenosine blocks calcium entry in activated neutrophils and binds to flunarizine-sensitive calcium channels. Immunopharmacol Immunotoxicol 1990;12:77–91.
399. Nelson RD, Herron MJ. Agarose method for human neutrophil chemotaxis. Methods Enzymol 1988;162:50–58.
400. Deitch EA, Gelder F, McDonald JC. Prognostic significance of abnormal neutrophil chemotaxis after thermal injury. J Trauma 1982;22: 199–204.
401. Rothe G, Oser A, Valet GK. Dihydrorhodamine 123: a new flow cytometric indicator for respiratory burst activity in neutrophil granulocytes. Naturwissenschaften 1988;75:354–355.
402. Abramson JS, Mills EL, Giebink GS, et al. Depression of monocyte and polymorphonuclear leukocyte oxidative metabolism and bactericidal capacity by influenza A virus. Infect Immun 1982;35:350–355.
403. Robinson JPaul, Wakefield D, Breit SN, et al. Chemiluminescent response to pathogenic organisms: normal human polymorphonuclear leukocytes. Infect Immun 1984;43:744–752.
404. Robinson JPaul, Wakefield D, Graham DM, et al. The chemiluminescent response of normal human leukocytes to Chlamydia trachomatis. Diag Immunol 1985;3:119–125.
405. Gale R, Bertouch JV, Bradley J, et al. Direct activation of neutrophil chemiluminescence by the rheumatoid sera and synovial fluid. Ann Rheum Dis 1983;42:158–162.
406. Palmblad J, Gyllenhammar H, Lindgren JA, et al. Effects of leukotrienes and F-Met-Leu-Phe on oxidative metabolism of neutrophils and eosinophils. J Immunol 1984;132:3041–3045.
407. DeChatelet LR, Long GD, Shirley PS, et al. Mechanism of the luminol-dependent chemiluminescence of human neutrophils. J Immunol 1982;129:1589–1593.
408. Bassoe C-F. Flow cytometric studies on phagocyte function in bacterial infections. Acta Pathol Microbiol Immunol Scand [C] 1984;92: 167–171.
409. Bassoe C-F, Bjerknes R. The effect of serum opsonins on the phagocytosis of Staphylococcus aureus and zymosan particles, measured by

flow cytometry. Acta Pathol Microbiol Immunol Scand [C] 1984;92: 51–58.
410. Cohen HJ, Newburger PE, Chovaniec ME, et al. Opsonized zymosan-stimulated granulocytes–activation and activity of the superoxide-generating system and membrane potential changes. Blood 1981;58:975–982.
411. Whitin JC, Clark RA, Simons ER, et al. Effects of the myeloperoxidase system on fluorescent probes of granulocyte membrane potential. J Biol Chem 1981;256:8904–8906.
412. Tatham PE, Delves PJ, Shen L, et al. Chemotactic factor-induced membrane potential changes in rabbit neutrophils monitored by the fluorescent dye 3,3'-dipropylthia–dicarbocyanine iodide. Biochim Biophys Acta 1980;602:285–298.
413. Fletcher MP, Seligmann BE. Monitoring human neutrophil granule secretion by flow cytometry: secretion and membrane potential changes assessed by light scatter and a fluorescent probe of membrane potential. J Leuk Biol 1985;37:431–447.
414. Nerl C, Valet GK, Schendel DJ, et al. Early transmembrane potential changes of lymphocytes in mixed lymphocyte cultures as detected by flow-cytometry. Naturwissenschaften 1982;69:292–294.
415. Shaprio HM. Flow cytometric probes of early events in cell activation. Cytometry 1981;1:301–312.
416. Witkowski J, Michlem HS. Decreased membrane potential of T-lymphocytes in ageing mice: flow cytometric studies with a carbocyanine dye. Immunol 1985;56:307–313.
417. Fletcher MP, Halpern GM. Effects of low concentrations of arachidonic acid derived mediators on the membrane potential and respiratory burst responses of human neutrophils as assessed by flow cytometry. Fundam Clin Pharmacol 1990;4:65–77.
418. Robinson JPaul, Phan SH, Fantone JC. Neutrophil membrane fluidity alteration with activation [Abstract]. Fed Proc 1986;45:(abstract 901): 309.
419. Rolland JM, Dimitropoulos K, Bishop A, et al. Fluorescence polarization assay by flow cytometry. J Immunol Meth 1985;76:1–10.
420. Fox MH, Delohery TM. Membrane fluidity measured by fluorescence polarization using an EPICS V cell sorter. Cytometry 1987;8:20–25.
421. Vandenbroucke-Grauls CMJE, Thijssen HMWM, Tetteroo PAT, et al. Increased expression of leucocyte adherence-related glycoproteins by polymorphonuclear leucocytes during phagocytosis of staphylococci on an endothelial surface. Scand J Immunol 1989;30:91–98.
422. Alteri E, Leonard EJ. N-formylmethionyl-leucyl-[3H]phenylalanine binding, superoxide release, and chemotactic responses of human blood monocytes that repopulate the circulation during leukapheresis. Blood 1983;62:918–923.
423. De Togni P, Della-Bianca V, Bellavite P, et al. Studies on stimulus-response coupling in human neutrophils. II. Relationships between the effects of changes of external ionic composition on the properties of N-formylmethyl leucyl phenylalanine receptors and on the respiratory and secretory response. Biochim Biophys Acta 1983;755:506–513.
424. Tennenberg SD, Zemlan FP, Solomkin JS. Characterization of N-formyl-methionyl-leucyl-phenylalanine receptors on human neutrophils: effects of isolation and temperature on receptor expression and functional activity. J Immunol 1988;141:3937–3944.
425. Sklar LA, Finney DA. Analysis of ligand-receptor interactions with the fluorescence activated cell sorter. Cytometry 1982;3:161–165.
426. Leonard EJ, Noer K, Skeel A. Analysis of human monocyte chemoattractant binding by flow cytometry. J Leuk Biol 1985;38:403–413.
427. Psychoyos S, Smith CW. Enhancement of N-formyl-methionyl-leucyl-phenylalanine (fMLP) binding to isolated human neutrophils. Agents Actions 1989;26:372–377.
428. Howard TH, Wang D, Berkow RL. Lipopolysaccharide modulates chemotactic peptide-induced actin polymerization in neutrophils. J Leuk Biol 1990;47:13–24.
429. Wang D, Berry K, Howard TH. Kinetic analysis of chemotactic peptide-induced actin polymerization in neutrophils. Cell Motil Cytoskeleton 1990;16:80–87.
430. Della-Bianca V, Bellavite P, De Togni P, et al. Studies on stimulus-response coupling in human neutrophils. I. Role of monovalent cations in the respiratory and secretory response to N-formylmethionyl leucyl phenylalanine. Biochim Biophys Acta 1983;755:497–505.
431. Fishman WH, Springer B, Brunetti R. Application of an improved glucuronidase assay method of the study of human blood β-glucuronidase. J Biol Chem 1948;173:449–456.
432. Lucisano YM, Mantovani B. Lysosomal enzyme release from polymorphonuclear leukocytes induced by immune complexes of IgM and of IgG. J Immunol 1984;132:2015–2020.
433. Butler TW, Heck LW, Huster WJ, et al. Assessment of total immunoreactive lactoferrin in hematopoietic cells using flow cytometry. J Immunol Meth 1988;108:159–170.
434. Haskill S, Becker S, Johnson T, et al. Simultaneous three color and electronic cell volume analysis with a single UV excitation source. Cytometry 1983;3:359–366.
435. Becker S, Halme J, Haskill S. Heterogeneity of human peritoneal macrophages: cytochemical and flow cytometric studies. J Reticuloendothel Soc 1983;33:127–138.
436. Haskill S, Becker S. Flow cytometric analysis of macrophage heterogeneity and differentiation: utilization of electronic cell volume and fluorescent substrates corresponding to common macrophage markers. J Reticuloendothel Soc 1982;32:273–285.
437. Gorvel JP, Mishal Z, Liegey F, et al. Conformational change of rabbit aminopeptidase N into enterocyte plasma membrane domains analyzed by flow cytometry fluorescence energy transfer. J Cell Biol 1989;108:2193–2200.
438. Mahoney KH, Miller BE, Heppner GH. FACS quantitation of leucine aminopeptidase and acid phosphatase on tumor-associated macrophages from metastatic and nonmetastatic mouse mammary tumors. J Leuk Biol 1985;38:573–585.
439. Dive C, Workman P, Watson JV. Inhibition of intracellular esterases by antitumour chloroethyl-nitrosoureas. Measurement by flow cytometry and correlation with molecular carbamoylation activity. Biochem Pharmacol 1988;37:3987–3993.
440. Elferink JGR. Deierkauf M. Suppressive action of cobalt on exocytosis and respiratory burst in neutrophils. Am J Physiol 1989;257: C859–C864.
441. Ozaki Y, Kume S. Functional responses of aequorin-loaded human neutrophils. Comparison with fura-2-loaded cells. Biochim Biophys Acta 1988;972:113–119.
442. Kuroki M, Takeshige K, Minakami S. ATP-induced calcium mobilization in human neutrophils. Biochim Biophys Acta 1989;1012:103–106.
443. Ransom JT, DiGiusto DL, Cambier JC. Single cell analysis of calcium mobilization in anti-immunoglobulin-stimulated B lymphocytes. J Immunol 1986;136:45–57.
444. Griffioen AW, Rijkers GT, Keij J, et al. Measurement of cytoplasmic calcium in lymphocytes using flow cytometry. J Immunol Meth 1989; 120:23–27.
445. Jennings LK, Dockter ME, Wall CD, et al. Calcium mobilization in human platelets using indo-1 and flow cytometry. Blood 1989;74: 2674–2680.
446. Lee KH, White KL, Robinson JPaul, et al. Tumor necrosis factor activates G protein-mediated calcium redistribution from bound to free form in 30A5 preadipocytes. Mol Endocrinol 1990;4:1671–1678.
447. Cantinieaux B, Hariga C, Courtoy P, et al. Staphylococcus aureus phagocytosis. A new cytofluorometric method using FITC and paraformaldehyde. J Immunol Meth 1989;121:203–208.
448. Korn ED, Weisman RA. Phagocytosis of latex beads by Acanthamoeba. II. Electron microscopic study of the initial events. J Cell Biol 1967;34:219–227.
449. Leffell MS, Spitznagel JK. Fate of human lactoferrin and myeloperoxidase in phagocytizing human neutrophils: effects of immunoglobulin G subclasses and immune complexes coated on latex beads. Infect Immun 1975;12:813–820.

COLOR PLATE II

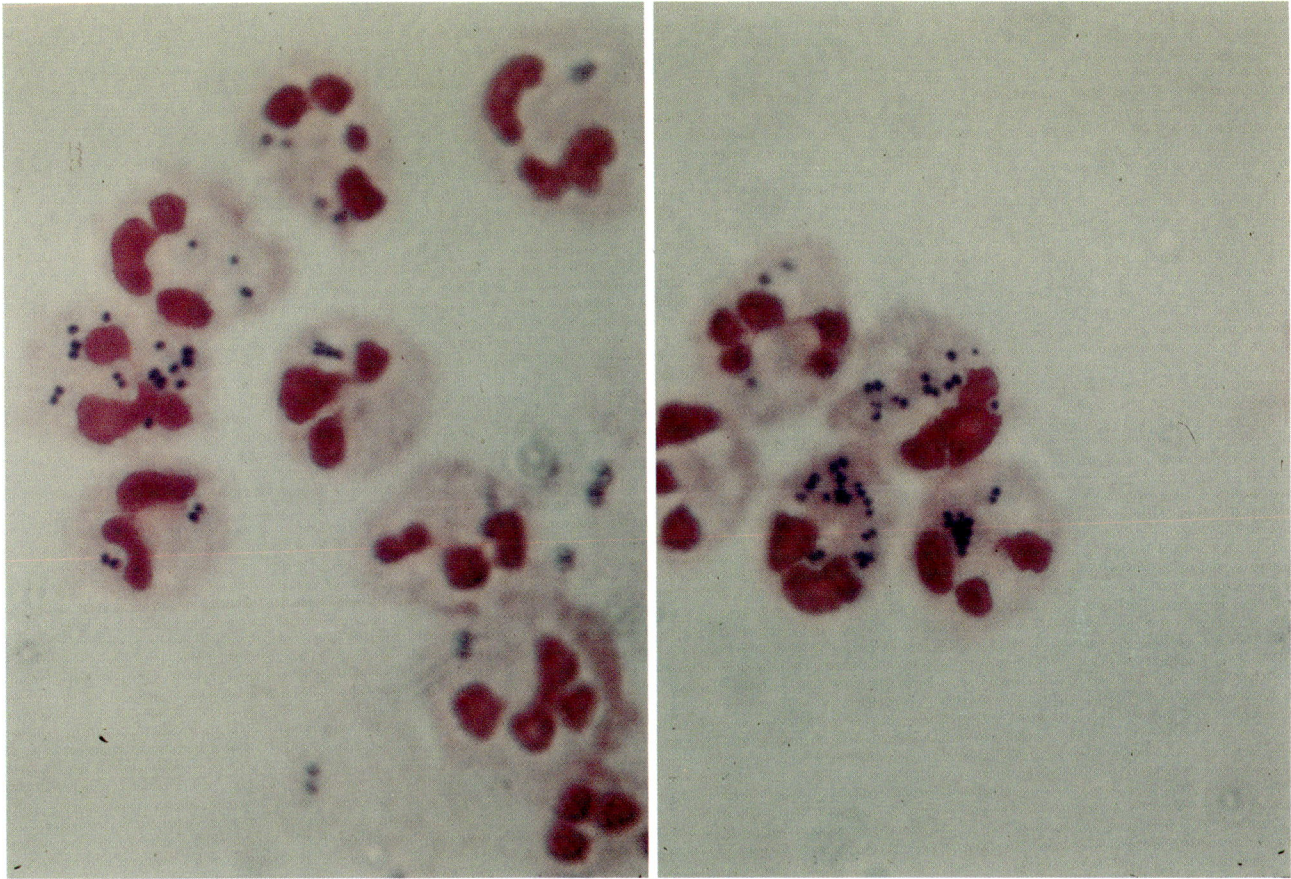

Figure 24.8. Neutrophils containing *S. aureus* (stained dark blue) after a 30-min incubation. Present are several neutrophils containing the organisms as well as some free organisms (Wright stain).

450. Dunn PA, Tyrer HW. Quantitation of neutrophil phagocytosis, using fluorescent latex beads. Correlation of microscopy and flow cytometry. J Lab Clin Med 1981;98:374–381.
451. Caldwell KD, Cheng ZQ, Hradecky P, et al. Separation of human and animal cells by steric field-flow fractionation. Cell Biophys 1984;6:233–251.
452. Christensen J, Leslie RG. Quantitative measurement of Fc receptor activity on human peripheral blood monocytes and the monocyte-like cell line, U937, by laser flow cytometry. J Immunol Methods 1990;132:211–219.
453. Lopez M, Olive D, Mannoni P. Analysis of cytosolic ionized calcium variation in polymorphonuclear leukocytes using flow cytometry and Indo-1 AM. Cytometry 1989;10:165–173.
454. Naccache PH, McColl SR, Caon AC, et al. Arachidonic acid-induced mobilization of calcium in human neutrophils: evidence for a multicomponent mechanism of action. Br J Pharmacol 1989;97:461–468.
455. Uchida T, Hosaka S, Miura K. Direct measurement of phagolysosomal esterase activity. Biochem Biophys Res Commun 1985;127:584–589.
456. Stocker R, Winterhalter KH, Richter C. Increased fluorescence polarization of 1,6-Diphenyl-1,3,5-Hexatriene in the phorbol myristate acetate stimulated plasma membrane of human neutrophils. FEBS Lett 1982;144:199–203.
457. Collard JG, De Wildt A. Localization of the lipid probe 1,6-diphenyl-1,3,5 hexatriene (DPH) in intact cells by fluorescence microscopy. Exp Cell Res 1978;116:447–450.
458. Collins JM, Grogan WM. Comparison between flow cytometry and fluorometry for the kinetic measurement of membrane fluidity parameters. Cytometry 1989;10:44–49.
459. Böck G, Huber LA, Wick G, et al. Use of a FACS III for fluorescence depolarization with DPH. J Histochem Cytochem 1989;37:1653–1658.
460. Duportail G, Weinreb A. Photochemical changes of fluorescent probes in membranes and their effect on the observed fluorescence anisotropy values. Biochim Biophys Acta 1983;736:171–177.
461. Haugland RP. Handbook of Fluorescent Probes and Research Chemicals. Eugene, Or: Molecular Probes, Inc., 1985.
462. Masuda M, Komiyama Y, Murakami T, et al. Decrease of polymorphonuclear leukocyte membrane fluidity in uremic patients on hemodialysis. Nephron 1990;54:36–41.
463. Kolotila MP, Diamond RD. Stimulation of neutrophil actin polymerization and degranulation by opsonized and unopsonized Candida albicons Hyphae and Zymosan. Infect Immun 1988;56(8):2016–2022.

25

Role of Flow Cytometric Evaluation of Congenital and Acquired Immunodeficiencies

ALAN L. LANDAY and JOHN L. SULLIVAN

INTRODUCTION

The clinical immunology laboratory is often called upon for input in measuring components of the immune system in patients with immunodeficiency disease. Surface-marker assays performed by flow cytometry are useful in diagnosing some immunodeficiencies but are not needed for others (1, 2).

While a thorough workup of patients with immunodeficiency diseases can be expensive, it is important to establish the diagnosis, determine the prognosis, and direct the treatment. For instance, patients with common variable immunodeficiency disease may require gammaglobulin therapy or be helped by cimetidine therapy. In other cases, such as IgA-deficiency, it is important to be aware of the possibility of an anaphylactic response to blood transfusion. As the genetics of diseases such as adenosine-deaminase-deficiency are understood and antenatal screening tests are introduced, patients will benefit from both genetic counseling and gene therapy.

Other conditions that are not due to a primary deficit in the immune system can lead to recurrent infections in an individual. Genetic diseases, such as cystic fibrosis, impair mucosal immunity and set up a cycle of recurrent infections similar to that seen in primary immunodeficiency disease. Other conditions, including diabetes or disorders caused by chemo- or radiotherapy, can damage their respiratory or gastrointestinal tracts, which sets them up for infectious complications.

Although the lymphoid immune system is important in host defense, other hematopoietic elements also play an important role. Macrophages, granulocytes, and mast cells are necessary for host defense against intracellular microbial agents, pyogenic bacteria, and some parasites. Patients with chronic granulomatous disease can have defects in respiratory burst and ineffective killing by phagocytes (3). Complement components can play an important role in host defense by generating chemotactic factors and opsonizing bacteria for phagocytosis.

Several factors need to be considered in laboratory evaluation of the immune compromised host. Age, sex, family history, and initial presentations are useful guides for determining which workup is most helpful. Family history is important because the genetics of many conditions are understood. The initial evaluation of the immunodeficient patient should include screening tests for humoral and cellular immunity that are available in most hospital labs. Serum protein electrophoresis is useful in screening for total γ-globulin and identifying monoclonal gammopathies that can be associated with immunodeficiency disorders (associated with a predisposition to infection). The most common deficiencies involve humoral immunity, so quantitation of serum IgG, IgA, and IgM should be performed. Normal serum immunoglobulins do not rule out a humoral immunodeficiency because subtle deficiencies, such as deficient IgG responses to polysaccharide antigens and IgG-subclass deficiencies, may be clinically important.

Specific antibody reactivity can be assessed by examining ABO blood group isohemagglutinins. Most individuals from Western countries will also have received diphtheria, pertussis, and tetanus vaccines. One can give a booster dose of these antigens and measure an effective humoral response (4, 5). Other active immunizations recommended by the World Health Organization for evaluating humoral responses include bacteriophage φX174, Hemophilus influenza polysaccharide, Neisseria meningitis polysaccharide, and monomeric flagellin. Measurement of the total hemolytic comple-

Table 25.1
Phenotyping of Cells in Humoral Immunodeficiency Disease[a]

Disease	T-Cells	B-Cells
X-linked Hypogammaglobulinemia	Normal CD4+-cells are CD45RA (naive) CD4/CD8 normal	Few mature Cells (CD20+CD21−) Pre-B Cells present
Transient Hypogammaglobulinemia of infancy	Usually normal May have CD4 ↓	Normal
Common variable Immunodeficiency	CD4(WNL)[b] CD8 ↑ CD4 ↓ CD8(WNL) CD4(WNL) CD8(WNL)	Normal
IgA deficiency	Normal number CD3+	Normal number Most IgA+ are IgM+, IgD+

[a]Immunophenotyping studies are not useful in studying selective IgG-deficiency, hyper-IgM immunodeficiency, or Jobs Syndrome (hyper-IgE disease).
[b]WNL, within normal limits.

Table 25.2
Phenotyping of Cells in Primarily T- or Combined T- and B-Cell Deficiencies

Disease	T-Cells	B-Cells
Severe combined immunodeficiency	3 T-cell phenotypes (1) $CD3^-CD4^-CD8^-CD2^-$ (2) $CD3^-CD4^-CD8^-CD38^+CD71^+$ (3) $CD3^+CD4^+CD8^+CD38^+$	Increased Number $CD19^+/CD20^+$
Adenosine deaminase deficiency	Absent	Absent
Bare lymphocyte syndrome	Normal number $CD3^+CD2^+$; $CD4^+$ are CD45Ra (naive) $CD8^+$ are $CD28^+$ (cytotoxic) HLA class I and/or class II absent	Normal number High density IgM/IgD HLA Class I and/or class II absent
DiGeorge's syndrome	↓ CD3, ↓ CD8, ±CD4 ↓$CD3^+$ ↓$CD8^+$, ±$CD4^+$	Normal Number
Ataxia-telangiectasia	↑$CD4^+CD8^+$ ↑$CD3^+$/TCRγδ$^+$	Normal Number
Wiskott-Aldrich syndrome	↓$CD3^+CD43^+$	Normal Number

Table 25.3
Suggested Monoclonal Antibody for Evaluating Primary Immunodeficiency Disease

Primary Panel
1. FITC IgG$_1$/PE IgG$_2$ - Isotype control
2. CD45/CD14 - Gating antibodies
3. CD3/CD4 - T-Helper/inducer
4. CD3/CD8 - T-Suppressor cytotoxic
5. CD3/CD16/56 - NK-cells
6. CD10/CD19 - Mature and immature B-cells

Extended Panel
1. CD3/TCRαB - Congenital TCR defect
2. CD3/TCRγδ - Congenital TCR defect
3. CD3/CD43 - Sialophorin defect in Wiskott-Aldrich syndrome
4. CD4/CD45RA - Naive T-cells in X-linked agammaglobulinemia
5. CD4/CD8 - Immature T-cells in SCID CD4/CD38 CD4/CD71
6. CD20/sIg or K/λ - Selected Ig Deficiency
7. CD20/HLA-DR, HLA-ABC - Bare lymphocyte syndrome
8. CD45/CD18 - Leukocyte adhesion deficiency

ment can easily rule out deficiencies of the complement system.

The simplest screening tests for cell–mediated immunity involve a complete blood count (CBC) and skin testing. The CBC reveals lymphocytopenia, granulocytopenia, and leukemias that may appear during immune deficient states. Skin testing is a quick and reliable method for screening cellular immunity and can be done as an inexpensive alternative to in vitro proliferative assays. The exception would be in the screening of young children, where mitogen testing is useful. In adults and older children, skin testing with antigens to mycobacteria (PPD), candida albicans, trichophyton, streptococcal, tetanus, and diphtheria are employed (Mérieux CMI Multitest). A positive reaction is indicated by the size of induration 48 to 72 hr after intracutaneous inoculation. The site should also be examined at 24 hr for a possible arthus reaction due to the formation of antibody–antigen complexes.

Following screening, more extensive analyses using functional studies and flow cytometric immunophenotyping can be performed. Much of what is understood about the ontogeny of immune responses in man has been made possible by the use of well-defined monoclonal antibodies. In each disease discussed below, the current understanding of the pathogenesis of the condition, including immunophenotyping studies, is reviewed (Tables 25.1 and 25.2).

HUMORAL IMMUNODEFICIENCY DISEASE

Brutons X-Linked Agammaglobulinemia (XLA)

In 1952, Colonel Ogden Bruton began the modern era of immunodeficiency disease analysis. Dr. Bruton examined an eight-year-old boy who had had several bouts of pneumonia and performed a serum protein electrophoresis which revealed the child to have absent gammaglobulin. (Actually, it was very low but with the zonal electrophoresis techniques of 1952, it appeared absent) (6). Using gammaglobulin replacement therapy, Colonel Bruton introduced the first therapy of a primary immunodeficiency disorder.

Patients suffering from XLA begin experiencing difficulties around the age of six months, at which time the nadir of serum immunoglobulin is reached. Children with this disease do not synthesize immunoglobulin and when maternal IgG has been catabolized they become sensitive to infection by pyogenic bacteria.

The disease is X-linked recessive and occurs in 1 in 100,000 live births. By looking at glucose-6-phosphate dehydrogenase activity in B-cells of women heterozygous for X-linked disease, Conley et al. found B-cell precursors that expressed the defective allele and were not able to mature to become peripheral B-cells. Female carriers can be detected by using recombinant DNA probes to analyze X-chromosome inactivation (7). Women who are carriers will show random X inactivation in T-cells but preferential activity of the single normal X in B-cells.

Routine laboratory tests include immunoglobulin quantitation. The IgG level will be less that 2g/liter and IgA and

IgM will be undetectable. Surface marker analysis is useful in diagnosing this condition. B-lymphocytes are markedly decreased or absent in blood and bone marrow. Plasma cells are virtually absent from lymph nodes or bone marrow. Pre-B-cells are present in bone marrow (8). Those few B-cells present express CD20 but usually not CD21 (9), a marker expressed on 80% of mature B-cells. The CD38 marker (immature/activated B-cells) is also expressed, indicating an overall arrest early in B-cell development. T-cell number and function are normal (10–13).

Transient Hypogammaglobulinemia of Infancy (THI)

THI is a relatively common disorder (14, 15). Although up to 20% of infants may be delayed in the development of their immunoglobulin synthesizing capabilities, it is highly unlikely for them to develop problems. Children with THI may have markedly decreased antibody in the serum, a condition that persists beyond six months of age but usually recovers after one-to-two years of age.

Laboratory features of THI that distinguish it from XLA include a normal IgM level, good antibody responses to immunogens such as tetanus and diphtheria, and a normal number of circulating B-cells. T-cell subset analysis in some patients with THI has revealed a decreased number of CD4-lymphocytes, suggesting that THI may result from a delayed CD4 T-inducer function (15).

Common Variable Immunodeficiency (CVID)

CVID is a heterogeneous form of hypogammaglobulinemia that usually begins in early adult life with symptoms of recurrent pyogenic infections (16). This disorder should not be confused with the well-characterized acquired immunodeficiency syndrome (AIDS) caused by infection with the human immunodeficiency virus (HIV). In CVID, upper respiratory tract infections—leading to permanent respiratory tract damage—and bowel manifestations, such as giardiasis—leading to malabsorption—are frequent presenting features (17).

The most useful laboratory tests are immunoglobulin quantitation for IgG, IgA, and IgM. When hypogammaglobulinemia is found in a young adult, CVID is the most likely diagnosis. Hypogammaglobulinemia in an older individual is a key feature heralding B-cell lymphoproliferative disease and may be associated with cancer chemotherapy.

Surface-marker studies are useful for pinpointing a specific diagnosis and following certain forms of therapy. CVID patients have normal numbers of circulating B-cells, as identified by CD20 or surface immunoglobulin. Some studies have suggested that B-cells express an activated phenotype, indicating a maturation defect following the activation phase (12). This may explain nodular hyperplasia in the gut, which is seen in some patients with CVID. If, once activated, B-cells cannot mature to plasma cells, they might be destroyed or trapped at a postactivation stage, which could lead to mucosal lymphoid nodules.

T-cell studies have indicated a heterogeneity in CVID (18, 19). Some patients have decreased CD4-cells with normal CD8-cells, while others have normal numbers of CD4-cells and increased numbers of CD8-cells. Patients with excessive CD8-suppressor-cell function can be successfully treated with cimetidine (20, 21). When treatment is successful, CD8-cell number goes down and immunoglobulin levels increase. The defect in antibody production in CVID is probably due to heterogeneous causes (deficient helper activity, increased suppressor activity, intrinsic B-cell unresponsiveness, autoantibody to T- or B-cells) all of which result in abnormal terminal differentiation of B-cells.

IgA Deficiency

A selective deficiency of IgA is the most common immunodeficiency disease, with an incidence of about 1 in 700. Patients with selective IgA deficiency are usually clinically asymptomatic. Recent studies have suggested that patients with concomitant IgG_2 or IgG_3 subclass deficiency may be the ones most likely to develop problems with pulmonary infections (22, 23). In some instances, there is an association of celiac disease (gluten sensitive enteropathy) and autoimmune disease with IgA deficiency (24).

IgA levels are below .05 g/liter while the IgG and IgM levels are normal (IgG subclass may be associated with some cases, but may go undetected on routine immunoglobulin studies). Patients have normal T- and B-cell phenotypes. Conley et al. have shown, by means of two-color immunofluorescence, that more than 80% of the IgA+ B-cells coexpress surface IgM and IgD, suggesting a maturational arrest in B-cells (25). Others have suggested that a defective helper T-cell function or excessive suppressor function may be responsible (26). Further studies are needed to better define these mechanisms.

Other Selective Humoral Immunodeficiencies

A wide variety of rare selective immunodeficiencies have been described. IgG_2- and IgG_3-deficiencies are found in association with IgA-deficiency and may be associated with pyogenic sinopulmonary infections and poor antibody responses to bacterial capsular polysaccharide antigens (27). Due to population heterogeneity and low levels of IgG_2 and IgG_4, it is difficult to detect these deficiencies before age 10. Another uncommon antibody deficiency involves specific antibody deficiency in the face of near-normal immunoglobulin concentrations. These patients suffer from impaired response to bacterial polysaccharides not associated with IgG-subclass-deficiency (28). Surface markers have not been shown to be useful in evaluating these patients.

Hyper IgM immunodeficiency can be X-linked or acquired in patients who have an interrupted maturation of B-cells (29). IgM levels are elevated while IgA and IgG are at low levels or absent. Surface-marker analysis shows B-cells that express IgM but lack IgG or IgA (30). This deficiency may be due to a lack of "Switch T-cells" that mediate the isotype switch from IgM to IgG or IgA (31, 32). Monoclonal antibodies that might define this T-cell population are not available.

Hyperimmunoglobulin E syndrome (Job's syndrome) is a rare condition in which dermatitis is usually accompanied by recurrent sinopulmonary infections (33). The disorder occurs in early childhood and patients are described as having coarse facial features. They suffer from staphylococcal abscesses at multiple sites including skin, mucosa, and internal organs. Laboratory findings show elevated IgE, leukocyte chemotactic defects, mild eosinophilia, and poor IgG memory response. T-cell markers, including CD4, CD8, and CD2 are normally expressed and are not helpful in the diagnosis of this disease.

In some patients with B-cell deficiencies there is a deficiency in ecto-5'-nucleotidase activity, an enzyme found on the plasma membrane of T- and B-cell subpopulations. Ecto-5'-nucleotidase appears to be a maturational marker, since adult T- and B-cells have a higher activity than thymocytes or cord blood cells. Two-color immunofluorescence studies show that patients with hypogammaglobulinemia or agammaglobulinemia have decreased levels of ecto-5'-nucleotidase on their B-cells. Because of this reduction, B-cells are thought to be blocked in maturation.

PRIMARILY T- OR COMBINED T- AND B-CELL DEFICIENCIES

Severe Combined Immunodeficiency (SCID)

In this disease, infants succumb quickly to a variety of bacterial, fungal, and viral infections unless they receive a bone marrow transplant. The condition can be identified as an X-linked autosomal recessive trait or as a sporadic form of it (34–36). A classic presentation includes "failure to thrive" which can be seen as early as three months of age.

The primary defect is in the T-cell lineage, resulting in few circulating T-cells (34). The few T-cells that are present respond poorly to mitogens or allogeneic cells. Laboratory studies reveal serum IgG, IgA, and IgM to be absent or low and there is a marked lymphocytopenia (less than 1,000 lymphocytes/mm³). B-cells are frequently present in increased numbers, as determined by the expression of surface immunoglobulin and CD19/CD20. Despite the presence of B-cells, antibody responses to most antigens are markedly deficient (37, 38). The condition in which normal serum immunoglobulins accompany severe T-cell-deficiency is called Nezelof's Syndrome. Although this syndrome is less severe than other forms of SCID, these patients are highly susceptible to fatal infections (13).

Several studies have divided patients with SCID into three major groups based on their T-lymphocyte profiles (10). One group of patients had no cells that expressed antigens characteristic of mature T-cells (i.e., CD3, CD4, CD8, or CD2 antigens), and few cells, if any, expressed other markers for immature activated T-cells. Another group had no mature T-cells, but had cells bearing markers for immaturity (CD38) as well as the transferrin receptor (CD71) and thus had a phenotype consistent with stage I thymocytes. The third group had a large percentage of cells coexpressing certain T-cell antigens (CD3, CD4, CD8) as well as CD38. This phenotype is consistent with stage-three cortical thymocytes. Some investigators have found evidence of engraftment of maternal T-cells, which complicates the interpretation of T-cell phenotype. The difference in the functional capabilities of cells in these three groups may be related to the stage of maturation of the lymphocytes. More recently, two patients with low expression of the T-cell receptor CD3 complex have been described. Their cells were unresponsive to mitogenic doses of CD3 but responded to CD2. These cases demonstrate how newly available monoclonal reagents and flow cytometry can define immunodeficiency disease (39).

The B-cell population in patients with SCID is variable as well (40). Some patients have no mature B-cells, while others have normal or elevated proportions of circulating B-cells. Regardless of the number of B-cells detected by surface immunoglobulin labeling or with monoclonal antibodies specific for B-cells (CD19), CD20, or CD21) these cells are poor producers of antibody in vivo or in vitro. This probably reflects the absence of helper T-cells and an intrinsic defect of B-cell function. Few studies have looked at subsets of T- or B-cells from this group of patients. A recent study by Gougheon et al. (41) on B-cell lymphocyte subsets from SCID patients has shown the absence of activation antigens on B-lymphocytes for these patients as compared to activation antigens on B-lymphocytes from age-matched infants. The SCID patients' B-cells show decreased expression of the 4F2 and transferrin receptor (CD71) antigens that correlate with a functionally resting state on the SCID B-cells. Analysis of the expression of other B-cell membrane antigens of the SCID B-cells reveal that the vast majority are SIgM$^+$/SIgD$^+$ and express the phenotypic characteristics of mature B-cells, such as CD19, CD20, CD21, CD22 and CD24. Furthermore, SCID B-cells express CD45RO, CD29, and the adhesion molecule LFA1 with the same frequency and intensity as that seen in age-matched infants. However, a phenotypic abnormality is found on the peripheral B-cells from all patients with tested SCID, since they continue to express the p120 antigen recognized by the AL1 monoclonal antibody. This antigen is expressed exclusively on bone marrow and immature B-cells (TDT+ pre B- and IgM$^+$/IgD$^-$-cells) but never on peripheral B-cells. The persistence of AL1 expression on peripheral SCID B-cells suggests a defect in B-cell development. This is supported by studies of Conley et al., who have demonstrated nonrandom X inactivation in B-cells of obligate carrier females of X-linked SCID (42).

Another group of patients with SCID consists of those with a deficiency of the purine salvage pathway enzyme, adenosine deaminase (43). Patients with adenosine deaminase deficiency have only a few lymphocytes with either a T- or B-cell phenotype. Cells that express immature T- and B-cell markers (CD1 or CD38) are also deficient.

Other antigens sometimes found on SCID lymphocytes are those associated with natural-killer- (NK) cells (CD16) (40, 44). These have been reported to be normal, elevated, or decreased in patients with SCID. Likewise, NK-function in these patients also varies and does not necessarily correspond to NK-phenotype or frequency.

The treatment of choice for SCID is the transplantation of bone marrow from either an HLA identical donor or a nonidentical donor whose marrow is purged to remove T-cells. Seventeen transplants reported by Buckley and co-workers showed that immunologic responses improved or became normal in 12 out of 15 surviving patients (45). The percentage of lymphocytes displaying T-cell antigens increased markedly, usually three to four months after transplantation, and this was followed by an increase in the function of these cells four to seven months after transplantation. In two patients in whom NK-cells (CD16+) were followed, an increase in CD16+ cells preceded the increase in T-cells. Because SCID often has a genetic basis, flow cytometry becomes useful as a prenatal diagnostic aid in families with one child with SCID. In one study of six at-risk families, fetal blood was obtained by fetoscopy and analyzed for T- and B-cells. Three fetuses were determined to have SCID by lymphocyte phenotyping coupled with adenosine deaminase determinations. The other three children were confirmed to be normal. Because lymphocyte phenotyping can be done on very small volumes of blood (.3 to 1 ml) it is well suited for making such diagnoses during at-risk pregnancies (46, 47).

One rare form of immunodeficiency is called bare lymphocyte syndrome. These patients often have normal numbers of T- and B-cells but their cells lack class I or II antigens and, on occasion, both class I and class II expression is deficient. Although T-cell numbers were normal in one patient studied by Clement et al. (48), most of the T-cells were CD8+ T-cells and only a few CD4+ T-cells were present. The CD4 cells expressed a phenotype characteristic of immature cells CD4+/CD45RA+ whereas CD8 cells expressed antigens typically found on cytotoxic precursor cells CD8+/CD28+. Cells with a suppressor phenotype CD8+/CD28 were virtually absent. Also, the B-cells of these patients had a high density of surface IgM and IgD, thus resembling B-cells from cord blood. Clement et al. (48) also found the absence of B-cells reactive with the CD21 monoclonal antibody in one patient. The relevance of these findings to disease pathogenesis is not clearly understood.

DiGeorge's Syndrome with Congenital Thymic Aplasia

In the classic presentation of DiGeorge's syndrome, cellular immunity is markedly decreased and there is deficient humoral immunity as well. The diagnosis may be suspected at birth from the child's characteristic facies: low set ears, hypertelorism, micrognathia, and an antimongoloid slant of the eyes. The finding of hypocalcemia with tetany also raises the suspicion of DiGeorge's syndrome. The chest X-ray will show an absence of the usually prominent thymic shadow found in the anterior mediastinum.

The pathogenesis of DiGeorge's syndrome is relatively well understood. During embryonic development, the third and fourth pharyngeal pouches form the thymus gland. In DiGeorge's patients, there is interference with this process during the 12th week of fetal development. The cause of this abnormality is unclear, but vascular anomalies have been suggested. The thymus in an affected baby is either absent or poorly developed at birth. Although earlier studies suggested that the thymus was totally absent, more recent studies suggest that most patients have some degree of T-cell function and all have some relatively mature T-cells (positive for CD3) in the peripheral blood. The severe form of the disease, when the thymus is completely absent, is treated by thymic transplantation or by implants of fetal thymic epithelium. However, in the majority of patients with thymic tissue present, T-cell maturation does occur spontaneously. Laboratory findings include a decreased number of circulating T-cells with decreased CD3-lymphocytes, a greater relative decrease in CD8+-lymphocytes than CD4+-cells, and a normal number of B-lymphocytes. Surface-marker assays can be helpful. The helper suppressor ratio, which is usually around 2:0 in immunologically normal individuals, is elevated in DiGeorge's syndrome patients. Patients with DiGeorge's syndrome also have normal natural-killer-cells (44, 49, 50).

Ataxia-Telangiectasia

Ataxia-telangiectasia is a rare condition known to be transmitted by a single autosomal recessive gene (13). As the name suggests, the clinical presentation is manifested by progressive cerebellar ataxia noted early in childhood and in telangiectasia, usually on the bulbar conjunctiva. This condition shows that 80% of the individuals have a marked decrease in serum IgA, IgG2, and IgE (51, 52), poor cellular immune responses, and increased α-fetoprotein levels. As with CVID- and IgA-deficiency, autoantibodies are commonly found in the serum of these patients. This immunodeficiency is associated with recurrent respiratory infections.

The inheritance pattern of ataxia-telangiectasia is autosomal recessive. DNA from these individuals is sensitive to damage by ionizing radiation. Studies in SCID mice show problems with DNA repair and recombinase deficiency that may be similar to defects that occur in ataxia-telangiectasia. While some workers have found a slightly reduced percentage of CD3-cells, this is not of sufficient magnitude to be useful in diagnosis. More recently, it has been reported that two-thirds of patients have an increased population of α/B⁻ γ/δ+ CD3+ T-cells (53). Clinical findings and α-fetoprotein levels are the most useful diagnostic tools in these patients. Surface markers are useful in determining the phenotype of T-cell leukemias that may occur in these individuals. The chromosomal breakage that occurs in these patients makes them particularly vulnerable to neoplastic transformation. Indeed, several cases of T-cell leukemia of CD8-cells have been reported (54).

Wiskott-Aldrich Syndrome

This condition presents with a triad of eczematoid dermatitis, thrombocytopenia, and recurrent opportunistic infections (55). The disease appears early in childhood, with infections providing the most life-threatening features. The disease is an X-linked recessive abnormality. IgA and IgE levels are increased in serum; the IgG level is normal. IgM levels are decreased and isohemagglutinins are usually absent. Despite having normal levels of IgG2, these patients have poor antibody response to carbohydrates. This points out the fallacy of equating antibody specificities with any particular isotype (56). All serum antibodies, however, are metabolized at a greater rate than normal. After developing relatively normal portions of T-cells during infancy, these children have a gradual decline in T-cell numbers, both in peripheral blood and in lymphoid organs. Eventually, they become lymphocytopenic. As would be expected, T-cell function declines. There is no change, however, in the CD4:CD8 ratio, and it may be useful to follow the total CD3-cells to help predict clinical problems with infections. Absence of CD43 (sialophorin) antigen, a surface glycoprotein that is involved in T-cell activation and proliferation, has been reported on lymphocytes and platelets and may provide some explanation for immunological deficiency in Wiskott patients (57). Splenectomy is suggested for controlling thrombocytopenia and bone marrow transplantation can be curative (58).

Complete Absence of Natural-Killer-Cells

There is a single case described of an adolescent who was found to have complete absence of natural-killer-cells, as demonstrated by immunophenotyping with monoclonal antibodies directed against CD16 and CD56 antigens. This young girl had complete absence of functional NK-cells and exhibited susceptibility to human herpesvirus infections, suffering near fatal varicella zoster and cytomegalovirus infections (59).

Leukocyte-Adhesion Deficiency (CD11/CD18)

This is a phagocytic cell defect characterized by the failure of cells to express the molecules of the CD11a, CD11b, and CD11c/CD18 family. Clinical manifestations of this disease include recurrent infections and compromised inflammatory responses. Granulocytes, monocytes, and lymphocytes do not express the β chain of the leukocyte adhesion molecule (CD18) and will also fail to express the α chains CD11a, CD11b, and CD11c. There is variability in the expression of these molecules ranging from absent to low-level expression. The amount of antigen expression does correlate with the clinical severity of the disease. One should, therefore, utilize antibodies against the CD11a, CD11b, and CD11c as well as the CD18 antigens in order to characterize these types of patients. In this disease, as compared to the other immunodeficiencies, the major cell involved is the neutrophil and therefore flow cytometric analysis should focus on that particular cell type (60).

Use of Immunophenotyping in Clinical Evaluation of Patients with Primary Immunodeficiencies

Immunophenotyping of lymphocytes is becoming an important component of the evaluation of immunodeficiency disorders (Table 25.3). In patients with hypogammaglobulinemia, for instance, B-cells may be absent. Conversely, if B-cells are present, they may not be functional. Their presence is important because helper T-cells are necessary for B-cell responses to most antigens as well as for T-effector cell function. In SCID, these cells are absent. This is manifested by the inability of the patient cells to respond to foreign antigens by either immunoglobulin production or cell-mediated immunity. Therefore, lymphocyte immunophenotyping is important in the diagnosis of SCID. It is primarily of academic interest whether the T-cells are representative of stage one or stage three thymocytes since neither cell type is capable of the full range of immune functions. Perhaps the most interesting application for immunophenotyping in SCID is for assessing the emergence of functional lymphocytes after bone marrow transplant, since phenotypically immature T-cells precede the appearance of T-cell function. Lastly, immunophenotyping can be potentially valuable in the prenatal diagnosis of SCID in conjunction with enzyme determinations.

In T-cell deficiency, immunophenotyping may or may not be useful for diagnostic studies. The most important feature is the function of the lymphocytes since cells from immunodeficient patients may not be functional. Immunophenotyping may be less important for diagnosis or therapy. However, T-cell quantitation is an important component for proper interpretation of functional data. In addition, studies of subsets of CD4 or CD8 cells that are responsible for specific functions may be helpful to define the basis for a demonstrated functional defect.

APPLICATIONS OF FLOW CYTOMETRY IN MONITORING HIV DISEASE

Immune Subset Alterations in HIV Infection in Adults

The individual course of disease progression between HIV infection and development of AIDS is highly variable. Flow cytometric evaluation using monoclonal antibodies against cell surface differentiation antigens provides a powerful tool to assess the extent of immunologic damage and to predict survival in individual patients (Table 25.4). A classification scheme based on nonrandom development of immunologic abnormalities assessed by flow cytometry is useful in staging both symptomatic and asymptomatic HIV-infected individuals (61). In addition, flow-cytometric information can be used both to determine whether treatment with antiviral drugs is appropriate or whether it has been effective in ar-

Table 25.4
Immunophenotypic Profiles in HIV Disease

	Heterosexual Control		Asymptomatic HIV-seropositive homosexual		Late AIDS	
	Percent	Abs. no.	Percent	Abs. no.	Percent	Abs. no.
Total leukocytes	—	8000	—	5800	—	3200
Total lymphocytes	20	1600	30	1740	25	800
Total T-cells	73	1168	70	1218	72	576
Total B-cells	13	208	10	174	8	64
Total NK-cells	15	240	12	209	11	88
$CD4^+$ (T H/I)	46	736	30	522	2	16
$CD8^+$ (T S/C)	27	432	41	713	70	560
CD4/CD8 ratio	1.72		0.73		0.03	
			Proportion of CD8-expressing			
HLA-DR	10		24		18	
CD38	20		32		81	
CD57	19		35		42	

Table 25.5
Suggested Monoclonal Antibody (MAb) Combinations for Routine Immunophenotyping

Monoclonal Antibody Combination	Purpose
1. IgG/IgG2[a]	Isotype control
2. CD45/CD14[b]	Percent lymphocytes in the analysis region
3. CD3/CD4	T-helper/inducer subset
4. CD3/CD8	T-suppressor/cytotoxic subset
5. CD3/CD16 + CD56	Total T-cells/total NK-cells
6. CD3/CD19	Total B-cells

[a]For each combination, the first MAb is labeled with fluorescein isothiocyanate and the second with phycoerythrin.
[b]This MAb combination can be run first to determine the appropriate region on the forward-scatter vs. right-angle scatter display for the lymphocyte population. All other samples from the same patient must be run using this region. IgG, immunoglobulin G.

resting immunologic deterioration. As described here, $CD4^+$-cell level and the proportion of $CD8^+$-cells that are $CD38^+$ (marker of activation and immaturity) are the two most informative lymphocyte subset parameters for predicting survival in HIV infection (62–64). Results from the Multicenter AIDS Cohort Study support monitoring of the $CD4^+$-lymphocyte count at three- to six-month intervals during the HIV-seropositive period (64). For routine monitoring of HIV infected individuals, the panel of antibodies described in Table 25.5 would be sufficient. Additional combinations of antibodies may be included ($CD8^+$, $CD38^+$) if they are of interest to the laboratory. Three-color flow cytometry using newer dyes, such as Percp or ECD, may be necessary for some antibody combinations because of overlap of subsets. For example, CD8 and CD38 antigens are expressed on T-cells and NK-cells and in order to accurately evaluate the $CD8^+$ and $CD38^+$ T-cells, one would require three-color flow cytometry. Both CD3 and CD8 monoclonals conjugated to Percp are commercially available and allow three-color cytometry on most flow cytometers.

$CD4^+$ CELLS AND SUBSETS

The measurement of the CD4-cell number has become an important part of the management and follow-up of the patient with HIV infection. In addition to the use of the CD4 count as a prognostic indicator, the CD4 count is also found to be useful as an entry criteria in clinical trials and as a marker to evaluate the efficacy of antiretroviral drugs. Although patients infected with HIV-1 have been documented as having an increased incidence of serious infections that depend on humoral immunity, the AIDS defining diagnoses are most often those associated with deficiencies in cellular mediated immune responses. The decline of $CD4^+$ T-cells correlates with the degree of cellular immunodeficiency and results in the inability to suppress endogenous and infectious agents, such as *Pneumocystis carinii* (PCP), *toxoplasma gondii*, and the herpesviruses, and to combat exogenous infections with agents such as *Candida albicans, cryptococcus neoformans* and *mycobacterium avium intracellulare*. The absolute CD4-cell count is the product of the peripheral blood white count, the percentage of lymphocytes determined on an automated differential, and the fraction of $CD4^+$-lymphocytes determined usually by flow cytometric analysis. Recent publications have shown that the $CD4^+$-cell percentages are slightly more predictive for AIDS survival than are the $CD4^+$-cell absolute counts (65). A CD4 count of $1000/mm^3$ in HIV infected individuals is not associated with clinically significant immunosuppression. Patients have a progressive decline in CD4-cell counts following infection. When the CD4 count falls below 200–500 per mm^3, many patients remain asymptomatic while others begin to experience generalized lymphadenopathy, oral thrush, weight loss, and diarrhea. A herpes zoster infection may demonstrate a defect in cellular immune responses and may be the first clue to prompt the evaluation of an otherwise healthy appearing individual for infection with HIV-1. Over the next several years, patients have a further decline to less than 200 CD4 cells/mm^3. This decline correlates with an increased risk of opportunistic infection with pathogens such as PCP, *toxoplasma gondii*, cytomegalovirus, and *cryptococcus neoformans* (Fig. 25.1). It is at this level of immunosuppression that an AIDS-defining illness occurs. Eighty-five percent of HIV-1-infected patients with a CD4 count below

Table 25.6
Functional Role for Cell-Surface Antigens (Adhesion molecules, enzymes, growth factor receptors)

Cluster Differentiation	Ligand	Cellular Distribution
CD2	CD58 receptor	T-cell, NK subset
CD3	T-cell antigen receptor complex	T-cells
CD4	Class II MHC/HIV receptor	T-cell subsets, monocytes
CD8	Class I MHC	T and NK subset
CD10	Neutral endopeptidase	Pre-B cells, PMN
CD11a	LFA-1	Leukocytes
CD11b	CR3, cell adhesion molecule	Mono/myeloid, NK, T-cell subset
CD11c	CR4, cell adhesion molecule	Mono/myeloid, NK, T-cell subset
CD13	Amino endopeptidase	Monocyte, PMN
CD15	X hapten	PMN, activated T subset
CD16	IgG receptor III	NK, PMN, T subset
CD18	Cell adhesion β subunit	Leukocytes
CD21	Complement receptor-2	B-cells
CD23	Fc ε receptor	B-cells
CD25	IL-2 receptor	T and B subsets, activated T-cells
CD26	Dipeptidylpeptidase IV	T subset
CD32	IgG receptor II	Monocyte, PMN, B cells
CD35	CR1	B-cells, monocyte, PMN, RBC
CD40	Homology NGF receptor	B-cells
CD41a	gp IIb/IIIa	Platelets
CD43	Leucosialin	T-cells, PMN
CD44	Pgpl homing receptor	Leukocytes
CD45	Leukocyte common antigen	Leukocytes T and B
CD54	ICAM	Activated BT, macrophages
CD56	N-CAM	NK, T subset
CD57	Myelin-associated glycoprotein	NK, T subset
CD58	LFA 3	Leukocytes
CD64	IgG receptor I	Monocytes, macrophages
CD71	Transferrin receptor	Proliferating cells

that of 200 cells/mm^3 will develop AIDS within three years. The mean time course from the initial HIV infection to development of AIDS is approximately eight to 10 years.

Following the demonstration of the efficacy of zidovudine (azidothymidine/AZT; retrovir) in the treatment of patients with AIDS and AIDs-related complex, the AIDS clinical trial group protocol 019 showed AZT to be more efficacious in selective asymptomatic patients with HIV-1 disease, slowing down the rate of progression to either AIDS or advanced AIDS-related complex (66, 67). This trial demonstrated a benefit in patients with CD4 counts of 200–400 CD4 cells/mm^3. In patients with less than 200 CD4-cells/mm^3, the major cause of death is pneumonia due to PCP. In patients with this degree of depletion of CD4-cells, prophylaxis with nebulized pentamidine or with trimethoprim/sulfametroxasole reduces the rate of infection with PCP.

In the clinical management of the HIV$^+$ patients, if the CD4 count is greater than 600 cells/mm^3, patients should be seen in follow-up in six months with a repeat CD4-cell count. If the count is 500–600 cells/mm^3, patients should be followed-up every three months with a repeat CD4-cell count. If the count falls below 500 cells/mm^3, the test should be repeated at least one week later to confirm the results. This repeat is important since there are a number of factors that can produce variable CD4 counts and one needs to confirm the value. If on repeat, the CD4 count is confirmed less than 500, therapy should be initiated with AZT and patients should be followed for symptoms and drug toxicity (68, 69).

For patients with CD4 counts of less than 500 but greater than 200, the CD4 counts should be repeated every six months. Once the CD4 count falls below 300, the count should be repeated every three months. If the count is less than 200, and confirmed on repeat, patients should receive prophylaxis for PCP. Current recommendations are that even in patients whose count is below 200 there may be additional benefit for CD4-cell counts because of the institution of multidrug regimens to deal with the development of resistant strains of HIV. Besides following-up these patients for therapy, the CD4 levels are highly prognostic for predicting survival, especially during their end-stage disease, where death occurs within one year in about 80% of untreated people whose CD4 level has fallen to 10% CD4$^+$ cells/mm^3.

Enumeration of functional CD4$^+$-cell subsets with Leu8, CD29 (4B4 helper/inducer) or CD45RA (Leu18, 2H4, suppressor/inducer) antibodies offers no additional information compared with CD4$^+$-cell measurement alone for evaluating HIV disease progression (62). Although several authors originally suggested that there is preferential early loss of one or another of the CD4$^+$ subsets during a certain stage of HIV disease (70–72), it is now generally accepted that various functional subpopulations of CD4$^+$-cells defined by these antibodies are lost concurrently.

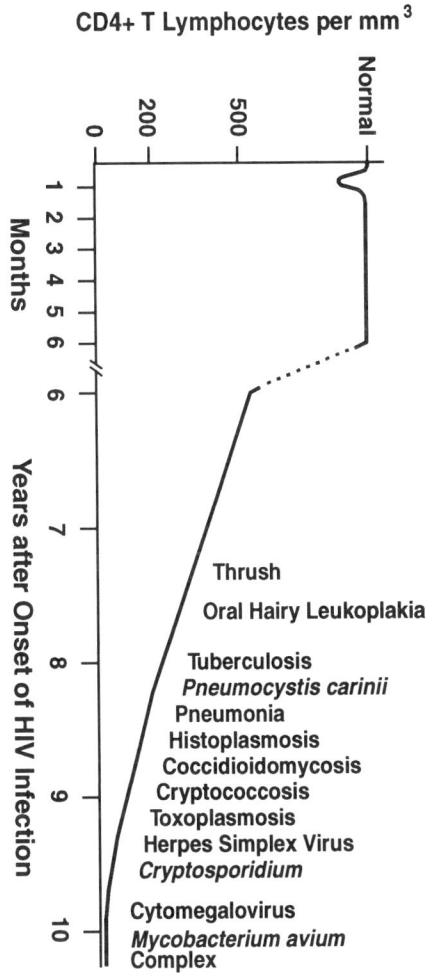

Figure 25.1. Correlation of CD counts and development of opportunistic infections.

Other dual-color flow cytometry studies initiated on CD4+ subsets reveal that AIDS patients have increased numbers of activated CD4+-cells that express CD38 (Leu17) or HLA-DR. These activated cells are not observed in seropositive men without AIDS and may result from or reflect HIV replication in frank AIDS. The observation of activated CD4+-cells was made on AIDS patients who were not on antiretroviral therapy, and studies are now under way to determine whether the proportion of activated CD4+ cells is diminished in individuals who are on antiretroviral therapy. In other studies, interleukin-2 receptor (CD25) expression, if present, was below detectable levels, although some investigators have shown a reduction in CD25 expression on cells from HIV-positive individuals (73).

CD8+ CELLS AND SUBSETS

Throughout the course of HIV disease, the total T-cell levels remain fairly constant in spite of the fall in the CD4+-cell number. This is due to the concomitant CD8+ lymphocytosis. A role for homeostatic control of the total T-cell number irrespective of the CD4+ or CD8+ phenotype has been suggested (65). About 20% of the CD8+-cells in HIV-seronegative people are CD38+. There is a progressive increase in the proportion of CD38+- (activated and immature) CD8+-cells throughout the course of disease, beginning as soon as there is evidence of HIV infection (seroconversion). In acute Epstein-Barr virus and cytomegalovirus infection, the CD38+-CD8+-cell number also rises quickly during active infection, then diminishes within months one or two. The decrease occurs at about the time that the herpesvirus infections are cleared by the host. In HIV infection, the CD38+-CD8+-lymphocytosis never abates, and the high CD38+ CD8+ level throughout HIV disease may reflect persistent immune stimulation by HIV. At least some of the CD38+-cells may be anti-HIV T-cytotoxic cells. A high level of CD38+ CD8+ cells, for example > 50%, is an extremely poor prognostic sign in HIV-infected people.

Increases in both HLA-DR+ and HLA-DR− subsets occur in HIV-infected individuals. HLA-DR+-CD8+-cells are extremely rare in HIV seronegatives (3–8%) and HLA-DR+ CD8+ cells represent an average of 20% to 30% of the CD8+-cells. In one study, the HLA-DR+ CD8+ proportion was higher in seropositive men than in individuals with AIDS, so it is possible that at least some HLA-DR+-CD8+-cells represent a protective component of the specific anti-HIV immune responses. However, the identification of a high HLA-DR+ CD8+ level as a poor prognostic sign by Stites et al. suggests the alternative hypothesis: That the HLA-DR+ CD8+ cells may accelerate the CD4+-cell decline or serve as a marker of HIV activity (74).

TOTAL T-CELLS

The total T-cell level remains relatively stable throughout the period of HIV infection until around the time that AIDS develops. Total leukocyte count, lymphocyte percentage, and absolute lymphocyte number are also relatively stable throughout the pre-AIDS period, and are comparable to levels in healthy normal controls. Late in HIV disease, the leukocyte and lymphocyte numbers begin to fall, and lymphopenia and panleukopenia herald the final stages of HIV disease progression. The CD3 percentage remains stable throughout HIV disease and averages approximately 75 ± 10%. In very ill patients who become lymphopenic and leukopenic, the CD3 percentage may be abnormally low and values of 30% to 50% are not uncommon. In such patients, most lymphocytes (i.e., up to 80%) are CD8-positive and almost none are CD4 positive.

GAMMA/DELTA T-CELL RECEPTOR BEARING T-CELLS

The T-cell receptor for antigen is comprised of a two-chain molecule that has specificity for antigen and is expressed together with CD3 on the T-cell membrane. The two-chain antigen-specific molecule or heterodimer can be one of two types. Ninety to 95% of T-cells in normal adult peripheral blood express the α-β heterodimer. The rest express the γ-δ heterodimer. In normal healthy adults, most γ-δ-expressing T-cells (CD3$^+$) do not express CD4 or CD8. This population is thus sometimes referred to as the CD4$^-$CD8$^-$ T-cells. In HIV infection, the γ-δ T-cells, including those that are CD4$^-$CD8$^-$, are elevated (75, 76). In routine immunophenotyping, these cells may cause the CD3 percentage to be significantly higher than the sum of the CD4 plus CD8 percentages.

NK-LYMPHOCYTES

NK-cells have been shown to lyse HIV-infected target cells in vitro. In addition, they are the effector cells for antibody-dependent cellular cytotoxicity and play a role in the first-line defense against a variety of pathogens. The number of NK-cells measured by CD16 (Leu 11) monoclonal antibody does not decrease during HIV disease despite a dramatic decrease in NK function during the later stages, i.e., AIDS. It is generally believed that this decrease in function may reflect a failure of CD4-cell IL-2 production in vivo because NK functional activity is restored in vitro by the addition of IL-2. With respect to cell surface phenotype, recent reports suggest that NK-cells in HIV-infected people gradually lose CD16 and CD56 coexpression (77–79). This work is being extended now in several laboratories and dual-color or three-color enumeration of NK-cell phenotype is likely to prove prognostically useful as an adjunct to T-cell assessments currently in place.

B-LYMPHOCYTES

Average B-cell percentages and absolute numbers are decreased in the circulation of HIV-infected people compared to normal, healthy controls, but these values are seldom outside the normal range. In AIDS, there are more immature B-cells and a few express the common acute lymphocytic leukemia antigen (CD10) (80). There are also activated B-cells that spontaneously secrete immunoglobulin. Although B-cell enumeration is important to complete the immunophenotyping panel, prognostic significance has not been associated with specific B-cell alterations at various stages of HIV disease.

IMMUNE SUBSET ALTERATIONS IN HIV INFECTION IN CHILDREN

AIDS in children was first reported in 1983. Since that time, much has been learned about the diagnosis and treatment of this disorder in infants and young children, though new challenges continue in the area of laboratory medicine. Today, it is clear that the flow cytometry laboratory plays a central role in the diagnosis and monitoring of newborn children infected with HIV. The opportunity to extensively evaluate a child for an immunologic defect using minimal blood volume is possible due to advances in instrumentation and reagents. However, it is important to understand that values in children may be quite different from those in adults. These differences may be reflected in the overall distribution of phenotypic markers present on cells from newborns and young children. Alternatively, the markers may be present although the absolute number of phenotypically distinct cells will be quite different compared to adults. With HIV disease progression, many children show a similar pattern of alteration in lymphocyte phenotype as seen in adults with HIV infection. The initial decline in CD4 percentage and number is paralleled with an increase in percentage and number of CD8-cells, resulting in a fall of the CD4:CD8 ratio. In a subset of HIV-infected neonates, disease progression is very rapid and death usually occurs within a few months of age. In these patients, there is no CD4 lymphopenia. With the differences in normal ranges for absolute lymphocyte counts in adults and children, recommendations have been established for the institution of therapy in HIV-positive infants. As is highlighted above, children normally have a higher lymphocyte count. A count of less than 1500 CD4 cells/mm^3 in HIV-antibody positive infants younger than 12 months of age, or of less than 1000 CD4 cells/mm^3 in an infant 12 to 24 months of age, indicates a need to initiate PCP prophylaxis (81, 82, 83). In addition, infants with HIV infection may also have an increased number of CD19- and CD20-B-cells and these subsets should be followed. While HIV infection results in a low number of circulating CD4-cells, the rate of decline throughout the course of HIV disease may be different and should be monitored regularly.

Similar changes in T-cell subsets have been recognized in infants by increases in CD8/CD38, CD8/CD57, and CD8/DR in HIV-infected children. Little is currently understood about these changes with respect to pathogenesis and prognosis. Studies are under way to try and determine their role in HIV disease.

CHOICE OF IMMUNOPHENOTYPING PANEL

In clinical medicine, a complete immunophenotypic profile includes enumeration of several lymphocyte subsets. Many of the monoclonal antibodies used for lymphocyte subset enumeration react with important functional structures on the cell surface, e.g., anti-CD3 antibodies react with the T-cell receptor for antigen (Table 25.6). Those antibodies with similar reactivity, i.e., those that react with the same cell-surface molecule, are clustered for naming purposes and have been given the designations CD1 through CD78. The CD in these names stands for cluster differentiation. Specific immunophenotyping information reported on patients should include both the cluster designation examined and the particular monoclonal antibody utilized, e.g., CD4 [OKT4A], CD3 [T3], or CD8 [Leu 2a].

A typical two-color immunofluorescence staining panel for evaluating HIV-infected individuals is provided in Table 25.5. Fluorescein and phycoerythrin are the two fluorochromes recommended and all antibodies are directly labeled. Two-color as opposed to single-color immunofluorescence analysis is strongly recommended. Less blood is required (a factor especially important for pediatric speci-

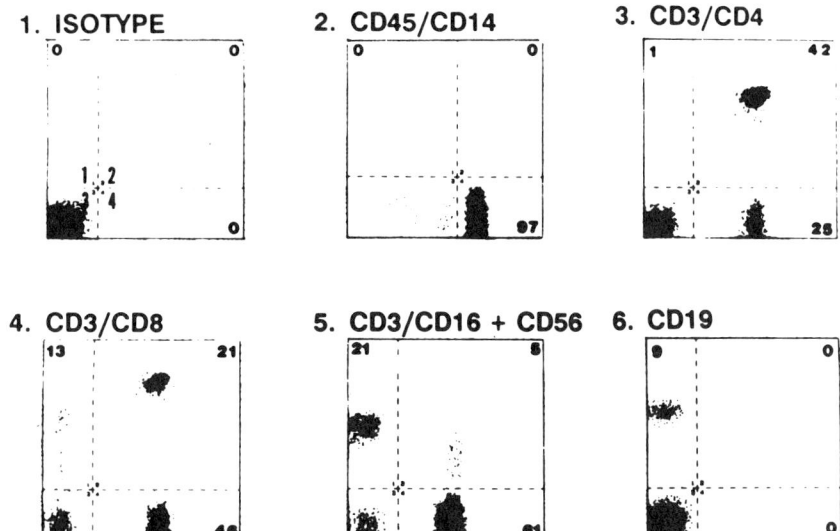

Figure 25.2. Dot plots of dual–color immunofluorescence staining. The sample number and antibodies correspond to those listed in Table 25.2 and the cursors (dotted horizontal and perpendicular lines) are set according to the isotype control, except for sample 2, where the perpendicular cursor is set to distinguish CD45^{bright} (lymphocytes) and CD45^{dim} (granulocytes). The numbers on the histogram give the percentage of cells shown in **1** (PE-stained only), **2** (dual–stained), and **4** (FITC only). The unstained cells are in **3**. Increasing FITC-staining is toward the right along the X-axis and increasing PE-staining is toward the top of each histogram along the Y-axis.

mens) and there are fewer tubes to handle. Furthermore, far more information can be obtained for a lower cost. Using two-color staining, it is possible to determine whether CD4$^+$- (helper/inducer) and CD8$^+$- (suppressor/cytotoxic) cells are T-cells. Monocytes are CD4$^+$ CD3$^-$ and NK-cells are CD8$^+$ CD3$^-$. Therefore, they would be excluded from analysis when using a two-color technique based on CD3 positivity. Finally, T-cells and NK-cells can be enumerated simultaneously to make sure that the T-cells that are CD56$^+$ are not included in the NK-cell count as well as in the T-cell count (Fig. 25.2).

The monoclonal antibodies (MAbs) in the first two samples of Table 25.5 are controls. The isotype control establishes whether non-specific binding of IgG$_1$ or IgG$_2$ occurs with each patient's cells, and indicates where to set the integration cursors (boundaries that are set to determine the percentage of positively stained cells). These isotype controls are chosen such that they match the isotypes of the specific antibodies used in the recommended panel. If there is non-specific binding with the isotype control or with other reagents, adding 50 μl of neat normal human sera to the sample tube just prior to adding the MAbs may reduce the non-specific binding.

Sample two (Table 25.5) should be run first for each patient in order to determine where the lymphocytes fall on the forward-scatter × right-angle scatter histogram. The CD45 reagent identifies all the leukocytes while CD14 stains the monocytes. The lymphocytes are differentiated from neutrophils based on their density of CD45 (CD45^{dim} neutrophils CD45^{bright} lymphocytes). The MAbs in this sample allow the percentage of lymphocytes (CD45^{bright}CD14$^-$) in the gated light-scatter analysis region to be determined. This presumes that the flow cytometry operator has set the lymphocyte analysis region to include all the lymphocytes. The percentage of events in the selected lymphocyte region that are CD45^{bright}CD14$^-$ should be between 90% and 100% on all specimens and blood should be restained if the value is not at least 85%. The percentage of lymphocytes obtained from leukocyte samples should be used to correct the values from the other MAb samples.

QUALITY CONTROL FOR LYMPHOCYTE IMMUNOPHENOTYPING

There are several checks on immunophenotyping results that experienced cytometrists use to determine whether flow results are accurate. First, the "lymphosum" or sum of [T + B + NK] should approach 100% after each of these values is corrected for the "percentage of lymphocytes in the lymphocyte region," as determined from the CD45/CD14 tube (sample two of the panel in Table 25.5). This serves as a check on the overall accuracy of the immunophenotyping results. For example, if the measured values are T = 70%, B = 10%, and NK = 12%, CD45^{bright}CD14$^-$ = 92%; then the corrected values are T = 76%, B = 11%, NK = 13% and lymphosum = 100%. This result is reasonable. If the lymphosum is 85% or 120%, however, the cytometrist should check for clerical errors, cursor setting errors, light-scatter distortions, or staining errors. Additionally, this variability may be due to some underlying biological process. Either all or select samples from the panel can be restained and rerun to validate or rectify the results. A second useful check is whether the sum of CD4$^+$ + CD8$^+$ approaches the CD3$^+$-cells within ± 10% of the CD3$^+$ value. In some people, this will not be the case for one of two reasons. First, the NK-cells express CD8$^+$, and if a person has many

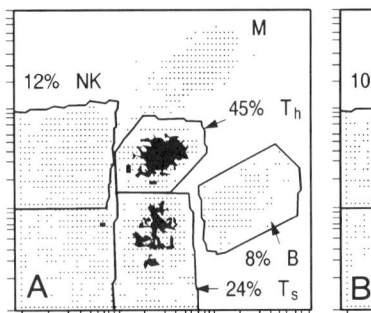
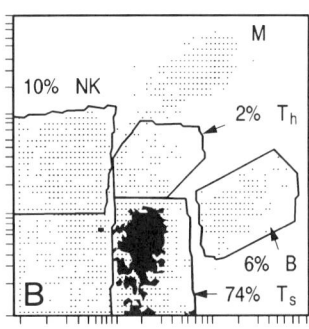

Figure 25.3. Cocktail staining: (**A**) normal, healthy individual and (**B**) AIDS patient (HIV–infected). Leukocyte–enriched preparations were stained using monoclonal antibody cocktail (19). The unlabeled cells in the bottom left-hand corner of the HIV-infected patients' histogram represent contaminated neutrophils or unlysed red cells. M, monocytes. $T_h = CD3^+ CD4^+$ $T_s = CD3^+ CD8^+ = CD19^+ CD20^+$; NK = $CD16^+ CD3^-$; M = $CD14^+$.

$CD8^+$-NK-cells, their $CD4^+ + CD8^+$ values may be greater than the $CD3^+$ value by > 10%. Second, the presence of the recently described subset of $CD3^+$ 4^- 8^- $\gamma = \delta$ cells will tend to lower the $CD4^+ + CD8^+$ value, making it less than the CD3 value by > 10%. Keeping these two exceptions in mind, checking whether the $CD4^+ + CD8^+$ values and $CD3^+$ values are approximately equal in number can prevent many gross errors. Another check is to see that the $CD3^+$ values in tubes three, four, five, and six are identical.

Flow-cytometric evaluation of the immune system is an extremely valuable technology for prognosis and for assessing the efficacy of therapy in HIV disease. Research is under way to improve clinical flow cytometry in order to simplify quality control and guarantee accuracy and reproducibility of results. At present, the best way to be certain that a CD4 value is accurate is to obtain a complete immunophenotypic profile (Table 25.5) on the individual and use the checks for accuracy of the data suggested above. Recently, a MAb cocktail method has been proposed that allows one to monitor the major lymphocyte subsets in a single sample. This method optimizes accuracy of $CD4^+$-lymphocyte assessment and provides simultaneous enumeration of four other lymphocyte subsets ($CD3^+$ $CD8^+$, $CD3^+$ $CD4^-$ $CD8^-$, $CD19^+$ $CD20^+$, and $CD16^+$ $CD56^+$ $CD8^-$) (Fig. 25.3). This method has the potential for providing accurate $CD4^+$ cell counts on small volumes of blood (100–200 µl), which would be valuable in monitoring pediatric patients with HIV infection. This method requires careful titration of the reagents as well as further development of specialized analytical software to streamline data processing before it will be ready for routine clinical application (84).

Many of the issues described above have been discussed by a national panel of experts. A consensus document (H42-P) for immunophenotypic analysis by flow cytometry has been published by the National Committee for Clinical Laboratory Standards and is available from this organization for use in the laboratory (85).

CONCLUSIONS FOR PHENOTYPING HIV DISEASE

Flow cytometric enumeration of lymphocyte subsets is an extremely valuable technology to aid in understanding the biology of HIV disease and in assessing the immune system of individuals during disease progression (86) (Table 25.6). With respect to the biology of HIV disease, three- and four-color immunofluorescence promises to add further to our knowledge. Cell sorting experiments to test the function of phenotypic subsets is under way. With respect to patient care, additional studies and analysis of existing data, especially in conjunction with soluble markers of immune activation, including neopterin, β_2-microglobulin, and p24, are being conducted in order to improve the utility of CD4 T-lymphocyte measurements for staging HIV disease, selecting participants for trials, and monitoring therapy. Newer flow cytometry systems promise to simplify quality control and guarantee increased accuracy and reproducibility of results. Finally, the widespread need for more economical means of obtaining T-lymphocyte subset values, including CD4 levels, should propel the development and maturation of clinical flow cytometric immunophenotyping.

REFERENCES

1. Eibl M, Griscelli C, Seligmann M, et al. Primary immunodeficiency diseases: report of a WHO sponsored meeting. Immunodefic Rev 1989; 1:173–205.
2. Nicholson JKA. Use of flow cytometry in the evaluation and diagnosis of primary and secondary immunodeficiency diseases. Arch Pathol Lab Med 1989;113:598–605.
3. Curnette JT. Classification of chronic granulomatous disease. Hematol Oncol Clin N America 1988;2:241–252.
4. World Health Organization Scientific Group: Immunodeficiency. Clin Immunol Immunopathol 1979;13:297–359.
5. World Health Organization Scientific Group: Primary immunodeficiency diseases: report of a World Health Organization scientific group. Clin Immunol Immunopathol 1986;40:166–196.
6. Bruton OC. Agammaglobulinemia. Pediatrics 1952;9:722–727.
7. Conley ME, Brown P, Pickard AR, et al. Expression of the gene defect in X-linked agammaglobulinemia. N Engl J Med 1986;315:564–567.
8. Pearl ER, Vogler LB, Okos AJ, et al. B lymphocyte precursors in human bone marrow: an analysis of normal individuals and patients with antibody-deficiency states. J Immunol 1978;120:1169–1175.
9. Conley ME. B cells in patients with X-linked agammaglobulinemia. J Immunol 1985;134:3070–3074.
10. Reinherz EL, Cooper MD, Schlossman SF. Abnormalities of T cell maturation and regulation in human beings with immunodeficiency disorders. J Clin Invest 1981;68:699–705.
11. Buckley RH, Gard S, Schiff R, et al. T cells and T-cell subsets in a large population of patients with primary immunodeficiency. Birth Defects 1983;19:187–191.
12. Tedder TF, Crain MJ, Kubagawa H, et al. Evaluation of lymphocyte differentiation in primary and secondary immunodeficiency diseases. J Immunol 1985;135:1786–1791.
13. Rosen FS, Cooper MD, Wedgwood RJP. The primary immunodeficiencies. N Engl J Med 1984;311:235–242.
14. McGeady SJ. Transient hypogammaglobulinemia of infancy: need to reconsider name and definition. J Pediatr 1987;110:47–50.
15. Siegel RL, Issekutz T, Schwaber JF, et al. Deficiency of T helper cells in transient hypogammaglobulinemia of infancy. N Engl J Med 1981; 305:1307–1313.

16. Geha RS, Schneeberger E, Merler E, et al. Heterogeneity of "acquired" or common variable agammaglobulinemia. N Engl J Med 1974;291:1–6.
17. Dawson J, Hadgson HJF, Pepys MB, et al. Immunodeficiency, malabsorption and secretory diarrhea, a new syndrome. Am J Med 1979;67:540–546.
18. Reinherz EL, Rubinstein AJ, Geha RS, et al. Abnormalities of immunoregulatory T cells in disorders of immune function. N Engl J Med 1979;301:1018–1022.
19. Reinherz EL, Geha R, Wohl EM, et al. Immunodeficiency associated with loss of T4+ inducer T-cell function. N Engl J Med 1981;304:811–816.
20. White WB, Ballow M. Modulation of suppressor cell activity by cimetidine in patients with common variable hypogammaglobulinemia. N Engl J Med 1985;312:198–202.
21. Sahasrabudhe DM, McCune CS, O'Donnell RW, Henshaw EC. Inhibition of suppressor T lymphocytes (Ts) by cimetidine. J Immunol 1987;138:2760–2763.
22. Bjorkander J, Bakem B, Oxelius V, et al. Impaired lung function in patients with IgA deficiency and low levels of IgG2 or IgG3. N Engl J Med 1985;313:720–724.
23. Lane P, MacLennan I. Impaired lung function in patients with IgA deficiency and low levels of IgG2 or IgG3. N Engl J Med 1986;314:924–925.
24. Ammann AJ, Hong R. Selective IgA deficiency: presentation of 30 cases and a review of the literature. Medicine 1971;50:223–236.
25. Conley ME, Cooper MD. Immature IgA B cells in IgA deficient patients. N Engl J Med 1981;305:495–497.
26. Atwater JS, Tomasi TB Jr. Suppressor cells and IgA deficiency. Clin Immunol Immunopathol 1978;9:379–384.
27. Umetsu DT, Ambrosino DM, Quinti I, et al. Recurrent sinopulmonary infection and impaired antibody response to bacterial capsular polysaccharide antigen in children with selective IgG subclass deficiency. N Engl J Med 1985;313:1247–1251.
28. Ambrosino DM, Siber GR, Chilmonczyk BA, Jernberg JB, Finberg RW. An immunodeficiency characterized by impaired antibody responses to polysaccharides. N Engl J Med 1987;316:790–793.
29. Brahmi Z, Lazarus KH, Hodes ME, et al. Immunologic studies of three family members with the immunodeficiency with hyper-IgM syndrome. J Clin Immunol 1983;3:127–134.
30. Schwaber JF, Lazarus H, Rosen FS. IgM-restricted production of immunoglobulin by lymphoid cell lines from patients with immunodeficiency with hyper IgM (Dysgammaglobulinemia). Clin Immunol Immunopathol 1981;19:91–97.
31. Levitt D, Haber P, Rich K, et al. Hyper IgM immunodeficiency a primary dysfunction of B lymphocyte isotype switching. J Clin Invest 1983;72:1650–1657.
32. Mayer L, Kwan SP, Thompson C, et al. Evidence for a defect in "switch" T cells in patients with immunodeficiency and hyperimmunoglobulinemia M. N Engl J Med 1986;314:409–413.
33. Hutto JO, Bryan CS, Green FL, et al. Cryptococcosis of the colon resembling Crohn's disease in a patient with the hyperimmunoglobulinemia E-recurrent infection (Job's) syndrome. Gastroenterology 1988;94:808–815.
34. Gupta S, Good RA. Markers of human lymphocyte subpopulations in primary immunodeficiency and lymphoproliferative disorders. Semin Hematol 1980;17:1–29.
35. Spirer Z, Zakuth V, Tzechoval E, et al. Lack of proliferation to alloantigen in a sibling of two infants with severe combined immunodeficiency (SCID). Clin Immunol Immunopathol 1982;24:286–291.
36. Gelfand EW, Dosch HM. Diagnosis and classification of severe combined immunodeficiency disease. Birth Defects 1983;19:65–72.
37. Businco L, Pandolfi F, Rossi P, et al. Selective defect of a T helper subpopulation in severe combined immunodeficiency. J Clin Immunol 1981;1:125–130.
38. Neudorf S, Kersey J, Filipovich A. Lymphoid progenitor cells in severe combined immunodeficiency. J Clin Immunol 1985;5:26–30.
39. Alarcon B, Terhorst C, Arniaz-Villena A, Perez-Aciego P, Reguerio JR. Congenital T-cell receptor immunodeficiencies in man. Immunodefic Rev 1990;2:1–16.
40. Peter HH, Friedrich W, Dopfer R, et al. NK cell function in severe combined immunodeficiency (SCID): evidence of a common T and NK cell defect in some but not all SCID patients. J Immunol 1983;131:2332–2339.
41. Gougheon M-L, Drean G, LeDeist F, et al. Human severe combined immunodeficiency disease: phenotypic and functional characteristics of peripheral B lymphocytes. J Immunol 1990;145:2873–2879.
42. Conley MH, Lavole A, Briggs C, Brown P, Guerra C, Puck JM. Non random X chromosome inactivation in B cells from carriers of X chromosome-linked severe combined immunodeficiency. Proc Natl Acad Sci USA 1988;85:3090.
43. Parkman R, Gelfand EW, Rosen FS, et al. Severe combined immunodeficiency and adenosine deaminase deficiency. N Engl J Med 1975;292:714–719.
44. Sirianni MC, Buscinco L, Seminara R, Aiuti F. Severe combined immunodeficiencies, primary T-cell defects and DiGeorge syndrome in humans: characterization by monoclonal antibodies and natural killer cell activity. Clin Immunol Immunopathol 1983;28:361–370.
45. Buckley RH, Schiff SE, Sampson HA, et al. Development of immunity in human severe primary T cell deficiency following haploidentical bone marrow stem cell transplantation. J Immunol 1986;136:2398–2407.
46. Durandy A, Griscelli C, Dumez Y, et al. Antenatal diagnosis of severe combined immunodeficiency from fetal cord blood. Lancet 1982;1:852–853.
47. Levinsky RJ, Linch DC, Beverly CL, Rodeck C. Prenatal exclusion of severe combined immunodeficiency. Arch Dis Child 1982;57:958–960.
48. Clement LT, Plaeger-Marshall S, Haas A, Saxon A, Martin AM. Bare lymphocyte syndrome: consequences of absent class II major histocompatibility antigen expression for B lymphocyte differentiation and function. J Clin Invest 1988;81:669–675.
49. Buscinco L, Rubaltelli FF, Paganelli R, et al. Results in two infants with the DiGeorge syndrome-effects on long-term TP. Clin Immunol Immunopathol 1986;39:222–230.
50. Durandy A, LeDeist F, Fischer A, Griscelli C. Impaired T8 lymphocyte-mediated suppressive activity in patients with partial DiGeorge syndrome. J Clin Immunol 1986;6:265–270.
51. McFarlin DD, Strober W, Waldman TA. Ataxia telangiectasia. Medicine (Baltimore) 1972;51:281–307.
52. Trompeter RS, Layward L, Hayward AR. Primary and secondary abnormalities of T cell subpopulations. Clin Exp Immunol 1978;34:388–392.
53. Carbonari M, Cherchi M, Paganelli R, et al. Relative increase of T cells expressing the gamma/delta rather than the alpha/beta receptor in ataxia-telangiectasia. N Engl J Med 1990;322:73–76.
54. Butterworth SV, Taylor AMR. A comparison of fresh and cultured T lymphocytes from patients with ataxia telangiectasia using T-cell subset markers and chromosome translocations. Int J Cancer 1987;39:678–681.
55. Lum LG, Tubergen DG, Corash L, et al. Splenectomy in the management of the thrombocytopenia of the Wiskott-Aldrich syndrome. N Engl J Med 1980;301:892–896.
56. Nahm MN, Blaese RM, Crain MJ, Briles DE. Patients with Wiskott-Aldrich syndrome have normal IgG2 levels. J Immunol 1986;137:3484–3487.
57. Remold-O'Donnell E, Rosen FS. Sialophorin (CD43) and the Wiskott-Aldrich Syndrome. Immunodefic Rev 1990;2:151–174.
58. Parkman R, Rappeport J, Geha R, et al. Complete correction of the Wiskott-Aldrich syndrome by allogeneic bone-marrow transplantation. N Engl J Med 1978;298:921–927.
59. Biron CA, Byron KS, Sullivan JL. Severe herpesvirus infections in an adolescent without natural killer cells. N Engl J of Med 1989;320:1731–1735.

60. Dana N, Todd RF, Pitt J, Springer TA, Arnaout MA. Deficiency of a surface membrane glycoprotein (Mol) in man. J Clin Invest 1984;73: 153–159.
61. Zolla-Pazner S, DesJarlais DC, Friedlan SR, et al. Non-random development of immunologic abnormalities after infection with human immunodeficiency virus: implications for immunologic classification of the disease. Proc Natl Acad Sci USA 1987;84:5405–5408.
62. Giorgi JV, Nishanian PG, Schmid I, Hultin LE, Cheng H, Detels R. Selective alterations in immunoregulatory lymphocyte subsets in early HIV (human T-lymphotropic virus type II/lymphadenopathy-associated virus) infection. J Clin Immunol 1987;7:140–150.
63. Giorgi JV, Fahey JL, Smith DC, et al. Early effects of HIV on CD4 lymphocytes *in vivo*. J Immunol 1987;138:3725–3730.
64. Giorgi JV, Detels R. T-cell subset alterations in HIV-infected homosexual men: NIAID multicenter AIDS cohort study. Clin Immunol Immunopathol 1989;52:10–18.
65. Taylor JMG, Fahey JL, Detels R, Giorgi JV. CD4 percentage, CD4 number and CD4:CD8 ratio in HIV infection: which to choose and how to use. J Acquir Immune Defic Syndr 1989;2:114–124.
66. Fischl MA, Richmen DD, Grieco MH, et al. The efficacy of azidothymidine (AZT) in the treatment of patients with AIDS and AIDS-related complex. N Engl J Med 1987;317:185–191.
67. Volberding PA, Lagakos SW, Koch MA, et al. Zidovudine in asymptomatic human immunodeficiency virus infection: a controlled trial in persons with fewer than 500 CD4-positive cells per cubic millimeter. N Engl J Med 1990;322:941–949.
68. State-of-the-art conference on azidothymidine therapy for early HIV infection. Am J Med 1990;69:334–335.
69. Machado SG, Gail MH, Ellenberg SS. On the use of laboratory markers as surrogates for clinical endpoints in the evaluation of treatment of HIV infection. J Acquir Immune Defic Syndr 1990;3:1065–1073.
70. Fletcher MA, Azen SP, Adelsberg B, et al. Immunophenotyping in a multicenter study: the transfusion safety study experience. Clin Immunol Immunopathol 1989;52:38–47.
71. Nicholson JKA, McDougal JS, Spira TJ, Croas GD, Jones BM, Reinherz EL. Immunoregulatory subsets of the T helper and T suppressor cell populations in homosexual men with chronic unexplained lymphadenopathy. J Clin Invest 1984;73:191–201.
72. Nicholson JKA, McDougal JS, Jaffe HW, et al. Exposure to human T-lymphotropic virus type III/lymphadenopathy-associated virus and immunologic abnormalities in asymptomatic homosexual men. Ann Intern Med 1985;103:37–42.
73. Bach BA, Campbell D, Robertson MJ, Ullery S, Borowitz MJ. T-cell activation markers (HLA-DR, CD38, CD25, CD69) are expressed discordantly in HIV infected individuals. Blood 1989;74:339a.
74. Sites DP, Casavant CH, McHugh TM, et al. Flow cytometric analysis of lymphocyte phenotypes in AIDS using monoclonal antibodies and simultaneous dual immunofluorescence. Clin Immunol Immunopathol 1986;38:161–177.
75. Margolick JB, Carey V, Munoz A, et al. Development of antibodies to HIV-1 is associated with an increase in circulating $CD3^+$ $CD4^-$ $CD8^-$ lymphocytes. Clin Immunol Immunopathol 1989;51:348–361.
76. Margolick JD, Scott ER, Odaka N, Saah AJ. Flow cytometric analysis of γ T cells and natural killer cells in HIV-1 infection. Clin Immunol Immunopathol 1991;58:126–138.
77. Vuillier F, Bianco NE, Montagnier L, Dighiero G. Selective depletion of low density $CD8^+$, $CD16^+$ lymphocytes during HIV infection. AIDS Res Hum Retroviruses 1988;4:121–129.
78. Ojo-Amaize E, Nishanian PG, Heitjan DF, et al. Serum and effector-cell antibody-dependent cellular cytotoxicity (ADCC) activity remains high during human immunodeficiency virus (HIV) disease progression. J Clin Immunol 1989;9:454–461.
79. Robinson WE, Mitchell WM, Chambers WH, Schuftman SS, Montefiori DC, Oeltmann TW. Natural killer cell infection and inactivation *in vitro* by the human immunodeficiency virus. Hum Pathol 1988;19:535–540.
80. Martinez-Maza O, Crabb E, Mitsuyasu RT, Fahey JL, Giorgi JV. Infection with the human immunodeficiency virus (HIV) is associated with an *in vivo* increase in B lymphocyte activation and immaturity. J Immunol 1987;138:3720–3724.
81. Denny TN, Niven P, Skuza C, et al. Age-related changes in lymphocyte phenotypes in healthy children. Pediatr Res 1990;27:155.
82. Connor E, Bagarazzi M, McSherry G, et al. Clinical and laboratory correlates of pneumocystis carinii pneumonia in children infected with HIV. JAMA 1991;265(13):1693–1697.
83. Guidelines for prophylaxis against pneumocystis carinii pneumonia for children infected with human immunodeficiency virus. MMWR 40: RR2-March 15, 1991.
84. Liu CM, Muirhead KA, George SP, Landay AL. Flow cytometric monitoring of human immunodeficiency virus infected patients: simultaneous enumeration of five lymphocyte subsets. Am J Clin Pathol 1989;92:721–728.
85. National Committee for Clinical Standards. Clinical applications of flow cytometry quality assurance and immunophenotyping of peripheral blood lymphocytes. NCCLS Publication H42-P. Villanova, PA: NCCLS.
86. Landay AL, Ohlsson-Wilhelm B, Giorgi JV. Application of flow cytometry for the study of HIV infection. AIDS 1990;4:479–497.

26

Role of Flow Cytometry in Clinical Transplantation

JUAN C. SCORNIK

In the early 1980s, the first reports describing the application of flow cytometry for pretransplant crossmatching and transplant monitoring were published (1, 2). These have remained the two major areas of clinical application of flow cytometry in transplantation and each has expanded considerably as we enter the 1990s. Transplant monitoring is one more application of cell phenotyping where the detection and quantification of cell lineages in different tissues and body fluids can give information on the status of the immune response to the graft. The flow cytometry crossmatch is a unique application to transplantation and, essentially, represents the detection and quantification of alloantibodies against the graft.

MEASUREMENT OF ALLOANTIBODIES

The importance of antibodies against leukocytes from the transplant donor was documented more than two decades ago. Most transplants performed in the presence of these antibodies were rejected within minutes or hours after transplantation (3). Since then, the pretransplant crossmatch has been routinely used to detect antibodies in the serum of the recipient reacting with lymphocytes from the donor. The pretransplant crossmatch is universally performed with a highly standardized complement-dependent cytotoxicity technique (4). The routine application of this technique has been highly successful in preventing hyperacute rejections, which now occur only sporadically. The use of flow cytometry to measure alloantibodies is appealing because of its potential for being more sensitive, quantitative, objective, and independent of variations in visual scoring of cell death. It would detect antibodies whether or not they bound complement and it would generate a permanent graphical record of the results that could be used for discussions with clinicians and surgeons. For these reasons, a growing number of laboratories have begun to use this technique and experience on its clinical significance is advancing at a fast pace. As this process develops further, there is little doubt that the number of laboratories using flow cytometry and the number of clinical situations that would benefit from its use will expand.

Components of the Test

The test is performed with antibodies from the patient and cells from the transplant donor or other cells bearing the appropriate antigens. Details of the technique have been described elsewhere (5).

SOURCE OF ANTIBODIES

Serum or plasma are almost universally used as a source of antibodies. Despite the potential variability of using serum or plasma from different patients treated with different medications, very few factors have been identified that may interfere with the test. One is the presence of murine antibodies in patients treated for rejection with OKT3. These antibodies bind to T-cells and may be detected by fluorescein-conjugated anti-human IgG antibodies that cross react with mouse antibodies. Anti–idiotypic antibodies produced by the patient may inhibit the binding of alloantibodies (6). Inhibition can also be produced by soluble donor HLA molecules, presumably produced by the allograft (7). Another source of antibodies are cell eluates produced to study the specificity of the antibodies (8) or eluates from tissue homogenates to recover antibodies that may have been responsible for allograft rejection (9).

SOURCE OF CELLS

Peripheral blood lymphocytes, lymph node cells, or spleen cells are most frequently used. Lymphocytes are isolated from peripheral blood by ficoll hypaque centrifugation. In contrast to lymphocyte phenotyping, where monoclonal antibodies can be added to whole blood, a rather homogeneous population of mononuclear cells is preferred for detection of alloantibodies. This is because erythrocytes, and probably granulocytes, express variable amounts of HLA antigens (10) and would introduce a source of variability in the test. Cells can be used fresh or cryopreserved, although there are no published studies to establish that frozen cells and fresh lymphocytes give comparable results.

TYPES OF TARGET CELLS

T-lymphocytes are appropriate target cells to detect anti-HLA antibodies. T-cells express high amounts of HLA constitutively (11), so that variations due to the effect of interferon or other lymphokines are not as significant as in other cells (12). In addition, there is minimal nonspecific binding of IgG to T-lymphocytes, with a concomitantly high signal-to-noise ratio. B-lymphocytes are more complex target cells. They express more HLA Class I molecules than T-

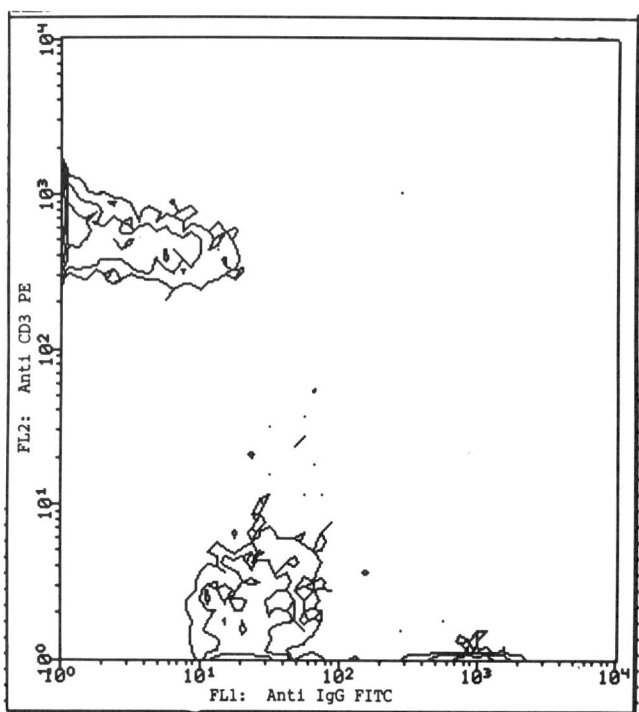

Figure 26.1. Contour map of peripheral blood mononuclear cells stained sequentially with normal human serum, fluorescein-conjugated anti-human IgG (*FL1*) and phycoerythrin-conjugated anti-CD3 (*FL2*). The T-cell cluster shows minimal fluorescence due to IgG whereas the non-T-cell cluster shows an heterogeneous pattern of cells with higher fluorescence intensity.

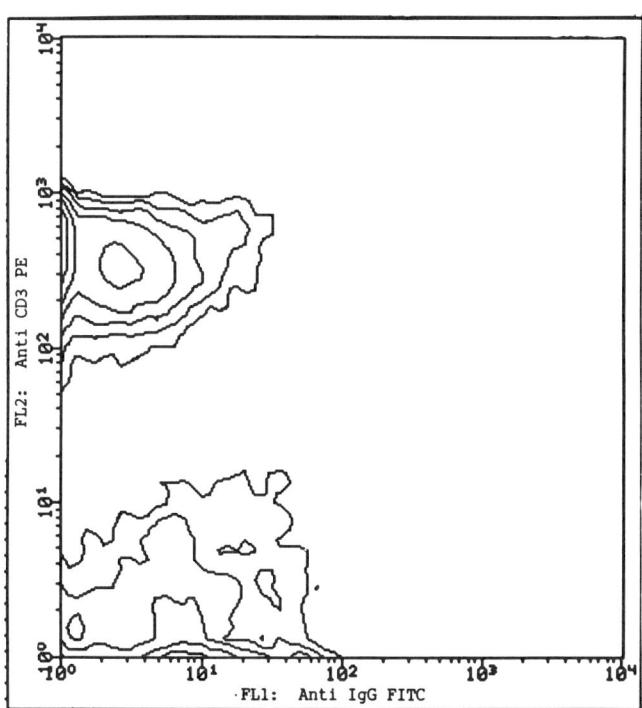

Figure 26.2. Spleen cells stained as described in Figure 26.1. The T- and non-T cell clusters show little fluorescence due to IgG.

lymphocytes (11) and antibodies reacting preferentially with B-cells may still have Class I specificity. In addition, B-cells express Class II HLA molecules, so that alloantibodies reacting with B-cells and not with T-cells may also have Class II specificity. Furthermore, B-cells have an active Fc receptor that binds normal IgG under regular test conditions (1). Such binding can be more prominent in the presence of antigen-antibody complexes or aggregated IgG. Finally, B-cells also bear surface immunoglobulin, which could be detected if an anti–whole Ig reagent is used as a second antibody. Therefore, in contrast to T-lymphocytes, detection of IgG binding to B-lymphocytes may or may not indicate the presence of an anti-B-cell antibody. In regard to NK-cells, they represent the most significant source of IgG receptor-mediated binding and potential interference with the test. The binding of normal IgG to NK-cells is considerable and accounts for most of the nonspecific fluorescence observed in non-T-cells (13). NK-cell interference is observed often with peripheral blood lymphocytes (Fig. 26.1) and less with lymph node or spleen lymphocytes (Fig. 26.2). In summary, T-cells are the best target cells for detection of alloantibodies and they are best identified in two-color studies by using any of the T-cell-specific monoclonal antibodies conjugated to phycoerythrin. Anti-B-cell antibodies are also best detected in two-color studies by using B-cell-specific monoclonal antibodies such as anti-CD20. However, whereas the presence of anti-T-cell alloantibodies almost always indicates anti-HLA Class I specificity, positive reactions with B-cells could be due to antibodies directed to Class I antigens, Class II antigens, or immune complexes.

Alloantibodies against monocytes that cross-react with endothelial cells have been associated with early graft rejection (14). Flow cytometry, however, does not provide a reliable way to measure anti-monocyte antibodies because of the high Fc receptor activity that these cells display. Similar limitations apply to the use of polymorphonuclear leukocytes, which also capture significant amounts of IgG from normal serum. Platelets have also been used as target cells (15, 16). Platelets express 80,000 to 120,000 HLA molecules per cell, which is comparable to the number of HLA antigens expressed by lymphocytes (17). However, it appears that HLA expression on platelets is more variable from individual to individual than HLA on lymphocytes (17). In addition, the expression of individual HLA specificities on platelets can be subject to significant variation from person to person (18). Thus, platelets can be used to detect anti-HLA antibodies but, as target cells, they are not as reliable as T-lymphocytes. Platelet specific alloantibodies may cause positive reactions with platelets but not with lymphocytes. However, care should be exercised in the interpretation of the results, as platelets also display IgG Fc receptor activity and can bind immune complexes (19). The clinical significance of the platelet crossmatch relates to patients who become refractory to platelet transfusions. However, in the majority of cases, refractoriness is caused by anti-HLA antibodies (20), so that detecting antibodies to T-lymphocytes by flow cytometry may be predictive of refractoriness to platelet transfusions from that donor. Further studies are necessary to test this possibility. Finally, Epstein-Barr virus transformed lymphoblastoid cell lines represent a valuable source of target cells that are likely to have a wider application in clinical transplantation. A large panel of HLA homozygous cell lines has been developed over the years (21) and are now available through the reagent bank of the American Society of Histocompatibility and Immunogenetics. Most cell lines are extremely well characterized in terms of the Class I and Class II HLA antigens they express and are invaluable reagents to study the fine specificity of individual sera. Even though these cell lines display variable Fc receptor activity, the binding of normal IgG is usually very low, allowing one to conduct tests under very favorable signal-to-noise conditions.

The Flow Cytometry Crossmatch

The clinical significance of the flow cytometry crossmatch is determined by the specificity and isotype of the antibodies detected, the sensitivity of the test, and the correlation between test results and transplantation outcome.

SPECIFICITY

By standard complement-dependent cytotoxicity, most antibodies that react with donor T-lymphocytes are of HLA Class I specificity (22). Characterization of antibodies detected by flow cytometry, either by panel studies (23) or by inhibition by purified HLA Class I molecules (24), have also established specificity against HLA Class I. There have been sporadic reports in the literature describing antibodies recognizing antigens other than HLA by standard cytotoxicity (25), and that has also been the case for flow cytometry (26). The nature of these alloantigens present on T-cells is not known. As mentioned before, anti-HLA Class I antibodies are unquestionably relevant for transplantation outcome and the T-cell flow cytometry crossmatch has the potential of detecting such antibodies with a high degree of sensitivity.

SENSITIVITY

Flow cytometry is a more sensitive test than the standard complement-dependent cytotoxicity technique (1, 24, 27) and cytotoxicity enhanced by antiglobulin (1, 23). However, when comparing dilutions of different sera, there are variations from patient to patient. Thus, in some cases, flow cytometry has been found to be more sensitive than standard cytotoxicity only by a few double dilutions and, in other cases, by as much as 500 times (1). Some sera show reactivity by cytotoxicity but not by flow cytometry (23). This may be due to IgM antibodies, which can induce complement lysis when only a few molecules bind to the target cell. When the predominant antibody is IgG, flow cytometry is more sensitive because IgG is less efficient in binding complement. It has been estimated that about 2000 IgG molecules per cell are necessary to trigger complement binding and induce lysis (28). However, some IgG anti-HLA antibodies do not bind complement at all, a fact that could be due to different IgG subclass composition or to different binding properties of antibodies cross-reacting with related epitopes (29). Flow cytometry, therefore, can detect the same antibodies that produce complement-dependent cytotoxicity with higher sensitivity and, in addition, antibodies that do not bind complement.

ANTIBODY CLASS

Most studies addressing the clinical significance of the pretransplant crossmatch have been performed with patients undergoing renal transplantation. The typical patients with high levels of preformed antibodies had had prior pregnancies or transplants and, in addition, had received blood transfusions (27). Since these patients were challenged with multiple antigenic stimulations, the resulting antibody population is made up almost exclusively of IgG antibodies. On the other hand, autoantibodies, which have been shown in several studies not to be detrimental to the graft, are consistently IgM (30). However, IgM autoantibodies do not react equally with lymphocytes from different donors and sometimes they react less or do not react at all with the patient's own lymphocytes. In this situation, it may not be possible to distinguish IgM autoantibodies from alloantibodies. IgM alloantibodies can be damaging to the graft, as demonstrated by the rapid rejection that occurs with most ABO incompatible grafts (31). How frequent is the presence of IgM anti-HLA antibodies in renal transplant candidates and what is the effect of such antibodies on the graft is not known. At present, flow cytometry is designed to detect IgG alloantibodies, which are the ones known to be the most prevalent and to have the capacity to damage the graft. However, it may be prudent to also look systematically for the presence of IgM alloantibodies at least until their clinical significance is resolved.

CLINICAL SIGNIFICANCE

IgG antibodies against donor lymphocytes that are detected by flow cytometry but not by standard cytotoxicity can damage the graft through several possible mechanisms. Lymphocytes (32), monocytes or macrophages (33), and polymorphonuclear leukocytes (34) are all capable of inducing antibody-dependent cell-mediated cytotoxicity in the presence of very low concentrations of IgG antibodies. Also, the very presence of anti-donor antibodies at the time of transplantation is an indication of a pre-immunized state of the recipient against donor antigens, which can lead to restimulation of antibody production to levels sufficient to bind complement.

Significance in Kidney Transplantation. A number of studies have evaluated the correlation between a positive flow cytometry crossmatch and the incidence of graft loss or rejection episodes. Table 26.1 summarizes three representative studies that are useful to discuss the clinical relevance of

Table 26.1
Significance of the Flow Cytometry (FCM) Crossmatch in Kidney Transplant Survival

Type of Patients	Number of Patients	Graft Survival at (months)	% Graft Survival		Immunosuppression	Reference
			FC-Positive	FC-Negative		
PRA−	103	12	48	70[d]	Aza[b]	35
PRA+[a]	45	12	36	76[d]	Aza	35
Primary, PRA−	146	1	92	93	Cy-A[c]	36
Primary, PRA+	50	1	70	93[d]	Cy-A	36
Regrafts	35	1	44	79[d]	Cy-A	36
Primary	89	12	83	78	Cy-A	37
Regrafts	47	12	48	83[d]	Cy-A	37

[a]PRA+: Panel-reactive antibody reacting with >10% of panel cells
[b]Aza: Azathioprine
[c]Cy-A: Cyclosporine
[d]Difference between the FC-positive and negative groups is statistically significant

the flow cytometry crossmatch. The first study reported that graft survival was less in patients who had a positive flow cytometry crossmatch. This effect was seen in patients who received pretransplant transfusions and were either sensitized (> 10% panel reactivity) or not sensitized. These patients were immunosuppressed with azathioprine and prednisone (35). In the other two studies, patients were immunosuppressed with cyclosporine and prednisone with or without the addition of azathioprine. The main difference between the first and the subsequent studies is that in the cylosporine era, when there is improved survival of first cadaveric transplants, there is no difference between the flow cytometry positive and flow cytometry negative groups in unsensitized patients. In contrast, in sensitized patients or patients receiving a second or subsequent transplant, a positive flow cytometry crossmatch is associated with poorer graft survival (36, 37). Comparable results have been obtained in most other studies (38–40). Despite technical differences in the performance of the flow cytometry crossmatch from one study to another, such as one-color vs. two-color techniques or consideration of T-cell antibodies only as opposed to T-cell plus B-cell antibodies, it is of interest that no association has been reported between a positive flow cytometry crossmatch and hyperacute rejections. However, there is also agreement that most graft losses occur early after transplantation In some reports, especially those with a smaller number of patients, there has not been any difference in graft survival between flow cytometry positive and negative patients. However the flow cytometry positive group had clearly a higher incidence of rejection episodes (41, 42). The correlation between a positive flow cytometry crossmatch and graft loss appears to be higher for kidneys obtained from older female or nontrauma donors (43).

A point for discussion is why the clinical significance of a positive flow cytometry crossmatch in sensitized patients is different from nonsensitized patients. One possible explanation is that nonsensitized patients are a low risk group for both presence of antibodies and graft survival and therefore the probability that a positive test is a false-positive is higher than in high risk groups (44). However, the differences can also be explained by the profound immunobiological differences between sensitized and nonsensitized patients. Sensitization often occurs in patients who had prior transplants or pregnancies and received multiple transfusions. Serum antibodies can persist for long periods of time but they also can, in the absence of further transfusions, decline to very low levels (45). In this case, antibodies could be detectable only by flow cytometry but they would be the expression of minimal baseline antibody production by greatly expanded clones. In contrast, nonsensitized patients may respond with low level antibodies to one or a few blood transfusions. This response is usually temporary (46) and is the expression of minimal clonal expansion. In both cases, the standard cytotoxicity crossmatch would be negative and the flow cytometry crossmatch positive. However, in the case of sensitized patients, the antibodies would be a marker of a significant preexistent T- and B-cell clonal expansion whereas, in nonsensitized patients, it would be a marker of a minimal effect produced by prior exposure to alloantigens. Still, it is clear that even in sensitized patients, a positive flow cytometry crossmatch is compatible with successful transplantation (47).

In summary, the presence of a positive flow cytometry crossmatch is a relative rather than absolute contraindication for transplantation. From the clinical standpoint, a positive flow cytometry crossmatch in a sensitized patient is a factor that definitely reduces the probability of success. The decision of whether to perform the transplant or not will also be influenced by other factors such as HLA-A, B, DR compatibility, positive or negative B-cell crossmatch, or probability for that particular patient to find another crossmatch-compatible donor within a reasonable period of time. For nonsensitized patients, the incidence of a flow cytometry positive crossmatch is very low and does not correlate with graft outcome, so that this test is not indicated under these circumstances.

Heart and Liver Transplants. Very little is known about the value of flow cytometry crossmatch in heart and liver transplantation. Patients awaiting heart or liver transplants are different from kidney transplant candidates in that the incidence of sensitization is significantly lower. In consequence, the incidence of positive flow cytometry crossmatches is expected to be very low. For patients who do have preformed cytotoxic antibodies against donor antigens, the effect on heart transplants appears to be immunologically similar to that of kidney transplants (48). However, the effect of antibodies on liver transplants is not always detrimental; high antibody levels appear to be required to cause detectable damage (49, 50).

Bone Marrow Transplants. Cytotoxic antibodies against donor antigens are associated with lack of engraftment in bone marrow transplants (51). Little is known about the significance of antibodies detected only by flow cytometry. One patient evaluated in our laboratory had an antibody directed against non-HLA antigens present on cells from an HLA-identical sibling. The antibody was detected by flow cytometry but not by standard cytotoxicity. Repeated attempts at transplantation were followed by lack of engraftment (rejection) (26). Since rejection is infrequent when donors are HLA-identical siblings, it is likely that the antibodies played a pathogenic role in this case. With the increasing use of unrelated bone marrow transplant donors, anti-donor antibodies will probably be found more frequently. Since bone marrow cells are given intravenously, there is immediate and full contact between antibodies and target cells so that it is conceivable that detrimental effects will occur even at low antibody concentrations.

ANTIBODIES PRODUCED AFTER TRANSPLANTATION

In contrast to the pretransplant crossmatch, where alloantibodies are almost exclusively IgG, antigraft antibodies produced after transplantation can be IgM or IgG. Such antibodies detected by flow cytometry were found in about 40% of the patients undergoing rejection episodes (52). Interestingly, IgG antibodies were always associated with rejection, whereas IgM antibodies were also produced by patients who had stable renal function (52). Although the presence of IgG antibodies was a very specific indicator for rejection, many rejection episodes, nevertheless, occurred in the absence of such antibodies. This lack of sensitivity for the diagnosis of rejection significantly limits the clinical usefulness of the test.

The B-Cell Flow Cytometry Crossmatch

A number of reports have addressed the significance of antibodies directed to B-cells as detected by flow cytometry. The results have been contradictory and inconclusive. In most reports, "B-cells" were in fact non-T-cells (1, 37, 40, 41) and the true nature of the target cells remained undefined. In addition, in none of the studies was the possibility that positive reactions could be due to the presence of immune complexes or aggregated IgG addressed. The interpretation of a flow cytometry B-cell-positive crossmatch is further complicated by the fact that the presence of low levels of cytotoxic anti-B-cell antibodies may not be associated with decreased graft survival (30). Thus, while the T-cell flow cytometry crossmatch complements the standard cytotoxicity crossmatch and provides further information, the B-cell flow cytometry crossmatch is less useful than its standard cytotoxicity counterpart and provides little help in deciding whether a transplant should or should not be done.

Other Applications

Flow cytometry is a valuable technique to quantitate anti-HLA antibodies. For a given serum specimen, a relationship can be established between different dilutions of such a specimen and the corresponding mean fluorescence intensity with a given target cell. Changes in antibody concentration over time can in this way be accurately measured (45). This information has been useful as an indirect measurement of antibody synthesis activity in a given patient (45). This approach has also been used to determine whether treatment with immunosuppressive agents is capable of producing a gradual decrease in antibody concentrations (53).

Flow cytometry can also be used as a more sensitive technique to measure panel antibody reactivity. For example, patients who show little or no reactivity in conventional panel studies may have low–level antibodies reacting with one or more panel cells. When present, these panel reactive antibodies detected by flow cytometry in multiparous patients have been associated with a high risk of producing broadly reactive lymphocytotoxic antibodies after transfusions (27). At the same time, patients who do not have such antibodies are less likely to be sensitized by blood transfusions (54), so they can be safely included in pretransplant transfusion protocols. More recently, flow cytometry panel reactivity has been studied as a risk factor for graft survival. Preliminary results suggest an association between this test and renal transplant outcome (55).

Technical Considerations

Calibration of the instrument can be accomplished with single- or multiple-fluorescence beads. The use of multiple-fluorescence beads with different fluorescein content allows the construction of a standard curve so that channel displacements can actually be expressed as the number of fluorescein molecules measured (56). Serum from one broadly sensitized patient (1) or pooled sera from several such patients (36) are used as positive controls, whereas normal serum known not to have alloantibodies is used as a negative control.

DEFINITION OF A POSITIVE REACTION

Results of individual patients are expressed in relationship to the normal serum control. The mean channel number is the most used parameter of fluorescence intensity but the peak (mode) or median can also be used. The relationship between patient and control can be expressed as the difference between the two (channel displacement) or the ratio between the two. Since most current instruments measure fluorescence intensity in log scale, the experimental/control ratio should reflect more accurately than channel displacement the relationship between patient and control, especially when the mean channel fluorescence intensity of the control varies from run to run. However, in practice, it has not been demonstrated that one way of expressing the results is more advantageous than the other. The Kolmogorov-Smirnov method, which compares differences between two specimens along all their fluorescence peaks (57), can be a very convenient and accurate way to express results (27). However, this method can only establish if a given specimen is positive or negative and does not provide quantitative information, except when the experimental and control fluorescence peaks overlap to a significant degree. The cutoff between a positive and negative reaction must be established in each laboratory. One normal serum control should be chosen as the baseline against which the channel displacement or ratio of 20 other normal sera should be measured. The cutoff point should be the mean value of the 20 determinations plus two or three SD.

CHANGES OF LYMPHOCYTE SUBSETS AFTER TRANSPLANTATION

The purpose of evaluating lymphocyte subsets in transplant recipients is to have the possibility to detect rejection during its early stages and to differentiate rejection from other conditions that compromise graft function. To date, these objectives have not been fully accomplished, but as more markers are being identified and the functional significance of different subsets is being defined, this area represents an exciting and promising field for future development.

PERIPHERAL BLOOD

The most consistent change in T-cell subsets after organ transplantation is a decrease of CD4-lymphocytes and an increase of CD8-lymphocytes (2, 58, 59). In patients who are immunosuppressed with either azathioprine or cyclosporine and who have functionally stable grafts, the decrease in the CD4 subset occurs mainly in a subgroup bearing the 4B4 antigen, which identifies cells with helper/inducer function (59). The remaining CD4-cells represent actually increased proportions of two particular subgroups. One is functionally characterized as suppressor/inducer and coexpresses the CD45RA marker (59). The second group consists of CD4 lymphocytes expressing the Leu 7 or CD57 antigen (60). These cells are large granular lymphocytes but do not have NK activity nor do they respond to allogeneic cells or mitogens (60). At present, the function of this cell subset remains obscure. However, the expansion of lymphocytes bearing the CD4 and CD57 markers occurs only in patients who are immunosuppressed with azathioprine but not with cyclosporine (60). Within the CD8 subset, the increase is due mainly to CD45-positive cells that display suppressor/effector function (59). As mentioned before, these changes were seen in stable renal allograft recipients and were not associated with rejection episodes (59).

In terms of correlation with graft rejection, early studies suggested that acute rejection occurs rarely in patients displaying the low CD4/CD8 ratios typical of the post transplant period (2). However, this observation was not confirmed in subsequent reports (61). Later, it was suggested that when rejection did occur in patients with low CD4/CD8 ratios, the response to anti-rejection therapy was poor compared with rejecting patients having higher CD4/CD8 ratios (62). Another marker apparently associated with acute rejections is HLA-DR on CD8-lymphocytes (63). DR expression on T-lymphocytes is usually very low and it was increased during rejection in CD8-lymphocytes but not in CD4-lymphocytes. HLA-DR expression was restored to normal levels after reversal of the rejection episode (63). Another marker of T-cell activation, the IL2 receptor, was not found to be increased during rejection (59), although it was not studied separately in CD4- and CD8-lymphocytes. Further studies on these activation markers are necessary to establish their clinical usefulness. It is pertinent to mention that decreased CD4/CD8 ratios and high HLA-DR expression on CD8-cells is also seen during CMV infection, which stresses the nonspecific nature of these changes (63, 64).

In terms of monitoring therapy with murine monoclonal antibodies, anti-CD3 is the only one widely used at the present time, mainly as a treatment of acute rejection. A significant reduction of CD3-positive lymphocytes occurs during therapy (65). Monitoring for concentrations of CD3-lymphocytes may be useful in cases of retreatment with anti-CD3. As some patients develop anti-mouse antibodies, the effectiveness of the treatment is less and it may require more prolonged treatment periods (65). Thus, measurement of CD3-positive lymphocytes is helpful in determining the length of treatment in these patients.

CHANGES IN THE GRAFT

In contrast to peripheral blood, where changes in T-cell subpopulations are inconsistently associated with acute rejection, a number of reports have described immunohistological manifestations of acute rejection. As compared with nonrejecting kidneys, the total number of infiltrating leukocytes is increased and may consist of a predominantly CD4-positive population (66), a CD8-positive population (67), or a preponderant monocyte infiltration (67, 68). Markers of lymphocyte activation such as HLA-DR or IL2 receptor have been consistently associated with acute rejection, including the expression of HLA-DR in tubular cells (69). Although flow cytometry has not been extensively used for these purposes, its application to the analysis of biopsy material has been documented (70). Whereas the number of monoclonal antibodies that can be used depends on the numbers of lymphocytes obtained, a yield of about 10^5 lymphocytes has been sufficient to study a number of useful markers by flow cytometry (70). Acute rejection was associated with higher numbers of total lymphocytes retrieved; increased proportion of CD8-lymphocytes and CD16 lymphocytes; increased proportion of CD8- and CD16-lymphocytes expressing HLA-DR; and a decrease of CD8/CD11-lymphocytes (suppressor effectors) and CD4/CD45-positive lymphocytes (suppressor inducers) (70). Combined analysis of these variables provided a stronger correlation with acute rejection. Recently, the recognition of macrophages in the graft expressing CD14 has been shown to be a sensitive and specific marker of acute rejection (71). Thus, the application of flow cytometry to the study of cells obtained directly from the graft and the availability of an increasing number of reagents with improved diagnostic value will probably increase the usefulness of this methodology in the clinical setting.

Changes in the cell composition of the urine sediment have also been described in acute transplant rejection (72). While very little is known about the use of flow cytometry in this area, it is an obvious application for future development.

CHANGES AFTER BONE MARROW TRANSPLANTATION

After bone marrow ablation and reconstitution with allogeneic bone marrow cells, there is a period of one-to-three months in which the absolute number and proportion of T-cells is depressed (73). This period may be longer in patients receiving T-cell depleted bone marrow transplants (74). It is during this time that life-threatening bacterial, viral, and fungal opportunistic infections occur. Acute graft-vs.-host disease is also often seen in the immediate post transplant period. Thus, lymphocyte phenotyping could offer the opportunity to evaluate the normal repopulation process and determine if it is proceeding at a normal pace or not. There is a significant elevation of NK-cells during the first three months after transplantation. These cells bear the CD16 and CD57 markers and display NK cytotoxic activity against susceptible targets (73). The proportion of NK-cells can be 50% or even higher, in contrast to normal individuals in which it is usually less than 10% (75). A subpopulation of NK-cells expressing low amounts of CD8 is also increased in the post-bone-marrow-transplant period (73). This is important to take into account because it could lead to overestimations of the proportion of CD8-positive T-cytotoxic/suppressor lymphocytes in single-color determinations. During the recovery period typical helper T-lymphocytes (CD3/CD4) and cytotoxic/suppressor T-lymphocytes (CD3/CD8) appear in approximately equal numbers after transplantation (73, 74). Helper T-lymphocytes continue their recovery process slowly thereafter until eventually they reach the normal proportion of about double the number of CD8-lymphocytes. Both CD4 and CD8 T-lymphocytes express elevated but variable amount of HLA-DR (73, 76). However, another marker of T-cell activation, the IL2 receptor, is not expressed (76). B-cells, as identified by the presence of CD19, CD20, or surface immunoglobulin, appear to recover faster than T-cells, so that their proportion is increased from normal during the early and late post-transplant periods (73). A significant proportion of these B-cells also expresses the CD5 marker, which was originally thought to be a pan T antigen (73). This CD5 positive B-cell phenotype is characteristic of chronic lymphocytic leukemias. However, the B-cells appearing after transplantation have been shown to be polyclonal in nature and not neoplastic (73). A population of CD3-lymphocytes not bearing neither CD4 nor CD8 markers and presumably associated with the expression of the γ/δ T-cell receptor has also been found in increased proportion during the early post-transplant period (74).

In summary, the recovery period after bone marrow transplantation is characterized by the presence of high numbers of NK-cells and the presence of cells with unusual phenotypes, such as NK-cells with dim expression of CD8- and, CD5-positive B-cells, and CD4- and CD8-negative T-cells. The qualitative and quantitative changes, however, are highly variable from one patient to another, as is the clinical course and the incidence of complications during this period. Therefore, it has been difficult to correlate these changes with clinical parameters such as graft-vs-host disease or infections.

Future Projections

In trying to project what the future will be for the clinical applications of flow cytometry in transplantation, one has to anticipate where transplantation will be, for example, five-to-10 years from now. A key factor is immunosuppression, and breakthroughs in immunosuppressive therapy are likely to happen within the next few years. New drugs, lymphokines, lymphokine receptors, monoclonal antibodies, and donor antigens in different forms, are all areas of active investigation aiming to achieve a higher degree of specific tolerance for grafts. Flow cytometry may become indispensable not only to measure anti-donor antibodies but also to measure the inhibition of known antibodies by anti-idiotypic antibodies produced by the patients. Administration of monoclonal antibodies or lymphokines may produce selective changes in lymphocyte subpopulations that may require close monitoring with flow cytometry; and techniques to grow and expand cells infiltrating the grafts may become more widely used, with their phenotypic characterization being part of the required analysis. Flow cytometry is now an established technique both in the clinical laboratory in general and in the particular area of transplantation. Future developments will create new fields of clinical application that will require responsive clinical laboratories. It will be up to those working in clinical flow cytometry to be fully prepared to implement changes and adopt new techniques to meet the demands of a changing clinical environment.

REFERENCES

1. Garovoy MR, Rheinschvidt M, Bigos M, et al. Flow cytometry analysis: A high technology crossmatch technique facilitating transplantation. Transplant Proc 1983;15:1939–1943.
2. Cosimi AB, Colvin RB, Burton RC, et al. Use of monoclonal antibodies to T cell subsets for immunologic monitoring and treatment in recipients of renal allografts. N Engl J Med 1981;305:308–314.

3. Kissmayer-Nielsen F, Olsen F, Petersen V, Fjeldborg O. Hyperacute rejection of kidney allografts associated with preexisting humoral antibodies against donor cells. Lancet 1966;2:662–665.
4. Amos DB. Cytotoxicity testing. In: Ray JG Jr, Hare DB, Pedersen PD, Mullaly DI, eds. NIAID manual of tissue typing techniques. Washington, D.C., 1976; DHEW publication No. (NIH) 77–545:25–28.
5. Scornik JC. Flow cytometry crossmatch. In: AA Zachary, GA Teresi, eds. ASHI Laboratory Manual, 2nd edition. American Society for Histocompatibility and Immunogenetics, 1990:325–331.
6. Reed E, Hardy M, Benvenisty A, et al. Effect of antiidiotypic antibodies to HLA on graft survival in renal allograft recipients. N Engl J Med 1987;316:1450–1455.
7. Davies H, Pollard SG, Calne RY. Soluble HLA antigens in the circulation of liver graft recipients. Transplantation 1989;524–527.
8. Fuller AA, Trevithick JE, Rodey GE, Parham P, Fuller TC. Topographic map of the HLA-A2 CREG epitopes using human alloantibody probes. Hum Immunol 1990;28:284–305.
9. Soulillou JP, Monzon-Cambon A, Dubois C, et al. Immunological studies of eluates of 83 rejected kidneys. Transplantation 1981;32:368–374.
10. Everett ET, Kao KJ, Scornik JC. Class I HLA molecules of human erythrocytes. Quantification and transfusion effects. Transplantation 1987;44:123–129.
11. Trucco M, Petris S, Garotta G, Ceppellini R. Quantitative analysis of cell surface HLA structures by means of monoclonal antibodies. Hum Immunol 1980;3:233–243.
12. Halloran PF, Urmson J, Van Der Meide P, Antewied D. Regulation of MHC expression *in vivo* II. IFN inducers and recombinant IFN α modulate MHC antigen expression in mouse tissues. J Immunol 1989;142:4241–4247.
13. Bray RA, Lebeck LK, Gebel HM. The flow cytometric crossmatch. Dual-color analysis of T cell and B cell reactivities. Transplantation 1989;48:834–840.
14. Paul LC, Baldwin WM, van Es LA. Vascular endothelial alloantigens in renal transplantation. Transplantation 1985;40:117–123.
15. Cook DJ, Scornik JC. Purified HLA antigens to probe human alloantibody specificity. Hum Immunol 1985;14:234–244.
16. Wang GX, Terashita GY, Terasaki PI. Platelet crossmatching for kidney transplants by flow cytometry. Transplantation 1989;48:959–961.
17. Kao KJ, Cook DJ, Scornik JC. Quantitative analysis of platelet surface HLA by W6/32 anti-HLA monoclonal antibody. Blood 1986;68:627–632.
18. Kao KJ, Scornik JC, McQueen C. Evaluation of individual specificities of Class I HLA on platelets by a newly developed monoclonal antibody. Hum Immunol 1990;27:285–297.
19. Moore A, Nachman RL. Platelet Fc receptor. Increased expression in myeloproliferative disorders. J Clin Invest 1981;67:1064–1071.
20. Dahlke MB, Weiss KL. Platelet transfusion from donors mismatched for crossreactive HLA antigens. Transfusion 1984;24:299–302.
21. Yang SY, Milford E, Hämmerling U, Dupont B. Description of the reference panel of B-lymphoblastoid cell lines for factors of the HLA system. In: B Dupont, ed. Immunobiology of HLA. New York: Springer-Verlag, 1989:11–19.
22. Konoeda Y, Terasaki PI, Wakisaka A, Park MS, Mickey MR. Public determinants of HLA indicated by pregnancy antibodies. Transplantation 1986;41:253–259.
23. Rodey G, Bollig B, Oldfather J, et al. Extra reactivities detected in flow-cytometry-positive, CDC-negative crossmatches are definable HLA specificities. Transplant Proc 1987;19:778–779.
24. Cook CJ, Scornik JC. Serum antibody reactivity of broadly sensitized patients with HLA-matched peripheral blood T lymphocytes. Transplantation 1986;41:447–453.
25. Opelz G, Territo MC, Gale RP, Sparkes R. Sensitization against non-HLA antigens following bone marrow graft rejection. Tissue Antigens 1977;9:209–219.
26. Scornik JC, Elfenbein G, Graham-Pole J, et al. Role of anti-donor antibodies in bone marrow transplant rejection: Evaluation by flow cytometry and effect of plasma exchanges. Transplant Proc 1989;21:2974–2975.
27. Scornik JC, Ireland JE, Howard RJ, Pfaff WW. Assessment of the risk for broad sensitization by blood transfusions. Transplantation 1984;37:249–253.
28. Schreiber AD, Frank MM. Role of antibody and complement in the immune clearance and destruction of erythrocytes. II. Molecular nature of IgG and IgM complement-fixing sites and effects of their interactions with serum. J Clin Invest 1972;51:583–590.
29. Lublin DM, Grumet FC. Mechanisms of the CYNAP phenomenon. Evidence in the Bw49/Bw50 model for epitopes with different spacial orientation of antibody. Hum Immunol 1982;4:137–145.
30. Brown WE. Laboratory and clinical management of the highly sensitized organ transplant recipient. Hum Immunol 1989;26:245–260.
31. Cook DJ, Graver B, Terasaki PI. ABO incompatibility in cadaver donor kidney allografts. Transplant Proc 1987;19:4549–4552.
32. Van Boxel JA, Paul WE, Frank MM, Green I. Antibody-dependent lymphoid cell-mediated cytotoxicity. Role of lymphocytes bearing a receptor for complement. J Immunol 1973;110:1027–1032.
33. Scornik JC, Cosenza H. Antibody-dependent cell-mediated cytotoxicity. III. Two functionally different effector cells. J Immunol 1974;113:1527–1532.
34. Gale RP, Zighelboim J. Modulation of polymorphonuclear leukocyte-mediated antibody-dependent cellular cytotoxicity. J Immunol 1974;113:1793–1797.
35. Stabile C, Bernhardt JP, Colombe BW, et al. Study of pre-sensitization by flow cytometry in cadaveric kidney recipients. Proc EDTA 1985;22:622–626.
36. Cook DJ, Terasaki PI, Iwaki Y, Terashita GY, Lau M. An approach to reducing early kidney transplant failure by flow cytometry crossmatching. Clinical Transplant 1987;1:253–256.
37. Kerman RH, van Buren CT, Lewis R, et al. Improved graft survival for flow cytometry and antihuman globulin crossmatch-negative retransplant recipients. Transplantation 1990;49:52–56.
38. Thistlethwaite JR, Buckingham M, Stuart JK, et al. T cell immunofluorescence flow cytometry cross-match results in cadaver donor renal transplantation. Transplant Proc 1987;19:722–724.
39. Chapman JR, Deierhoi MH, Carter NP, Ting A, Morris PJ. Analysis of flow cytometry and cytotoxicity crossmatches in renal transplantation. Transplant Proc 1985;17:2480–2481.
40. Mahoney RJ, Ault KA, Given SR, et al. The flow cytometric crossmatch and early renal transplant loss. Transplantation 1990;49:527–535.
41. Lazda VA, Pollack R, Mozes MF, Jonasson O. The relationship between flow cytometer crossmatch results and subsequent rejection episodes in cadaver renal allograft recipients. Transplantation 1988;45:562–565.
42. Talbot D, Givan AL, Shenton BK, et al. The prospective value of the preoperative flow cytometric crossmatch assay in renal transplantation. Transplantation 1990;49:809–810.
43. Cook DJ, Terasaki PI, Iwaki Y, et al. Donor factors that influence flow cytometry crossmatching. Transplant Proc 1988; 20(Suppl 1):81–83.
44. Meyer KB, Pauker SG. Screening for HIV. Can we afford the false positive rate? N Engl J Med 1987;317:238–242.
45. Scornik JC, Salomon DR, Howard RJ, Pfaff WW. Evaluation of antibody synthesis in broadly sensitized patients. Transplantation 1988;45:95–100.
46. Norman DJ, Barry JM, Wetzsteon PJ. Successful cadaveric kidney transplantation in patients highly sensitized by blood transfusions. Transplantation 1985;39:253–255.
47. Raftery MJ, Malik STA, Tidman N. Successful renal transplantation despite a positive fluorescence-activated cell sorter crossmatch following plasma exchange of donor-specific antibodies. Transplantation 1986;41:131–134.
48. Singh G, Thompson M, Griffith B, et al. Histocompatibility in cardiac transplantation with reference to immunopathology of positive serologic crossmatch. Clin Immunol Immunopathol 1983;28:56–66.

49. Fung JJ, Makowka L, Griffin M, et al. Successful sequential liver-kidney transplantation in patients with performed lymphocytotoxic antibodies. Clin Transplant 1987;1:187–194.
50. Starzl TE, Demetris AJ, Todo S, et al. Evidence for hyperacute rejection of human liver grafts: The case of the canary kidneys. Clin Transplant 1989;3:37–45.
51. Anasetti C, Amos D, Beatty PG, et al. Effect of HLA compatibility on engraftment of bone marrow transplants in patients with leukemia or lymphoma. N Engl J Med 1989;320:197–204.
52. Scornik JC, Salomon DR, Lim PB, Howard RJ, Pfaff WW. Post-transplant antidonor antibodies and graft rejection. Evaluation by two-color flow cytometry. Transplantation 1989;47:287–290.
53. Scornik JC, Salomon DR, Fennell RS, Howard RJ, Pfaff WW. Evaluation of immunosuppression as a treatment for pre-sensitization in renal transplant candidates. Clin Transplant [In Press].
54. Scornik JC, Ireland JE, Salomon DR, et al. Pretransplant blood transfusion in patients with previous pregnancies. Transplantation 1987;43:449–450.
55. Cicciarelli J. Flow cytometry PRA. In: Paul Terasaki, ed. Visuals of the Clinical Histocompatibility Workshop. Kahuku, Hawaii: One Lambda, Inc., publishers, 1990, pp. 20.
56. Cook DJ, Rhodes CL. Standardization of the flow cytometry crossmatch. In: Paul Terasaki, ed. Visuals of the Clinical Histocompatibility Workshop. Kahuku, Hawaii: One Lambda, Inc., publishers, 1990, pp. 21–23.
57. Young IT. Proof without prejudice: Use of the Kolmogorov-Smirnov test for the analysis of histograms from flow systems and other sources. J Histochem Cytochem 1977;25:935–939.
58. Chatenoud L, Chkoff N, Kreis H, Bach JF. Interest and limitations of monoclonal anti-T cell antibodies for the follow-up of renal transplant patients. Transplantation 1983;36:45–50.
59. Ramos EL, Turka LA, Leggat JE, Wood IG, Milford EL, Carpenter CB. Decrease in phenotypically defined T helper inducer cells (T4+4B4+) and increase in T suppressor effector cells (T8+4B4+) in stable renal allograft recipients. Transplantation 1989;47:465–471.
60. Lengendre CM, Forbes RDC, Loertscher R, Guttman RD. CD4+/Leu 7+ large granular lymphocytes in long-term renal allograft recipients. Transplantation 1989;47:964–971.
61. Morris PJ, Carter NP, Cullen PR, Thompson JF, Wood RFM. Role of T cell subset monitoring in renal allograft recipients. N Engl J Med 1982;306:1110–1111.
62. van Es A, Tanke HJ, Baldwin WM, Oljans PJ, Ploem JS, van Es LA. Ratios of T-lymphocyte subpopulations predict survival of cadaveric renal allografts in adult patients on low-dose corticosteroid therapy. Clin Exp Immunol 1983;52:13–20.
63. van Es A, Baldwin WM, Oljans PJ, Tanke JH, Ploem JS, van Es LA. Expression of HLA-DR on T lymphocytes following renal transplantation and association with graft rejection episodes and cytomegalovirus infection. Transplantation 1984;37:65–69.
64. Mordacchini M, Viarnello A, Calconi G, et al. T-lymphocyte subsets in the early and late phases of cytomegalovirus infeciton in renal transplant recipients. Clin Transplant 1990;4:9–13.
65. First MR, Schroeder TJ, Hurtubise PE, et al. Successful retreatment of allograft rejection with OKT3. Transplantation 1989;47:88-91.
66. Hall BM, Bishop GA, Farnsworth A, et al. Identification of the cellular subpopulations infiltrating rejecting cadaver renal allografts. Transplantation 1984;37:564–570.
67. Hancock WW, Thomson NM, Atkins RC. Comparison of interstitial cellular infiltrate identified with monoclonal antibodies in renal biopsies of rejecting human allografts. Transplantation 1983;35:458–463.
68. Waltzer WC, Miller F, Arnold A, Anaise D, Rappaport FT. Immunohistologic analysis of human renal allograft disfunction. Transplantation 1987;43:100–105.
69. Seron D, Alexopoulos E, Raftery MJ, Hartley RB, Cameron JS. Diagnosis of rejection in renal allograft biopsies using the presence of activated and proliferating cells. Transplantation 1989;47:811–816.
70. Tötterman TH, Hanås E, Gergström R, Larsson E, Tufveson G. Immunologic diagnosis of kidney rejection using FACS analysis of graft-infiltrating functional and activated T and NK cell subsets. Transplantation 1989;47:817-823.
71. Bogman MJ, Dooper IMM, van de Winkel JGJ, et al. Diagnosis of renal allograft rejection by macrophage immunostaining with a CD14 monoclonal antibody, WT14. Lancet 1989;2:235–238.
72. Segasothy M, Birch DF, Fairley KF, Kincaid-Smith P. Urine cytologic profile in renal allograft recipients determined by monoclonal antibodies. Transplantation 1989;47:482–487.
73. Ault KA, Antiu JH, Ginesburg D, et al. Phenotype of recovering lymphoid cell populations after marrow transplantation. J Exp Med 1985;161:1483–1502.
74. Gratama JW, Fibbe WE, Visser JWM, et al. CD3+, 4+ and/or 8+ T cells and CD3+, 4−, 8− T cells repopulate at different rates after allogenic bone marrow transplantation. Bone Marrow Transplant 1989;4:291–296.
75. Lanier LL, Loken MR. Human lymphocyte subpopulations identified by using three color immunofluorescence and flow cytometry analysis. J Immunol 1984;132:151–156.
76. Atkinson K. T cell subpopulations defined by monoclonal antibodies after HLA-identical sibling marrow transplantation. II. Activated and functional subsets of helper-inducer and cytotoxic-suppressor subpopulations defined by two-color fluorescence flow cytometry. Bone Marrow Transplant 1986;1:121–132.

SECTION E. EMERGING CLINICAL AND RESEARCH APPLICATIONS

27

Flow Cytometric Monitoring of Cellular Resistance to Cancer Chemotherapy

AWTAR KRISHAN and ANTONIETA SAUERTEIG

INTRODUCTION

Drug resistance continues to be a major mechanism for the survival of cells that otherwise would succumb to the toxic effects of potent drugs used in cancer chemotherapy. Over the course of cellular evolution, several protective mechanisms have developed to either prevent the toxic drug effects or to repair the damage caused by such agents to vital cellular constituents and mechanisms.

EXTRACELLULAR PARAMETERS IN DRUG RESISTANCE

At the level of the whole organism, drug resistance, especially that of malignant tumor cells, may be influenced by several extracellular factors or mechanisms that may involve bioavailability of the drug at the target site (influenced by circulation, tumor necrosis), drug excretion and metabolism (into nonactive metabolites), and anatomic sanctuaries (e.g., blood–brain barrier). Discussion of these and other extracellular factors and mechanisms that may influence cellular drug effects are beyond the scope of this review, which will instead focus on the role of flow cytometry in the detection of markers and mechanisms of cellular drug resistance. The reader is referred to several excellent reviews and books that have dealt with various mechanisms of drug resistance (1–3). Thus, for the purpose of this review, it is assumed that a drug is available in its potent form (unmetabolized or after production of an active metabolite) to exert its toxic effects on the target cell.

CELLULAR FACTORS IN DRUG RESISTANCE

It may be important to realize that drug effects on a cell may be cytostatic, which may result in loss of growth and proliferation, or cytotoxic, which may cause cell death. The cytostatic effects of a drug may be exercised by effects on clonogenicity, in which case a cell may live but cease to proliferate, or may involve differentiation into a more mature terminal cell that, after its normal progression and function, will die. The reader is referred to an excellent article dealing with the subject of drug-induced differentiation as a means for achieving cytostasis (4). It is important to point out that a toxic drug may kill a target cell by either interacting with cellular constituents essential for its survival (e.g., specific proteins or target enzymes) or clonogenicity (DNA), or by the generation of products (e.g., free radicals) that may inactivate or physically damage important cellular constituents (e.g., microtubules) or functions needed for survival. A different mechanism may involve interaction at the cell membrane level, which may lead to the generation of signals for self destruction (5–7) or other metabolic activities that may alter various essential cellular processes. There is strong evidence to suggest that some of the phenomena involving cell surface and drug interactions may play an imporant role in cellular drug toxicity and resistance (8–11).

Having excluded the extracellular factors that affect drug toxicity and resistance, we have drug transport (influx, retention, and efflux) as the major cellular process that determines cellular response to a cytotoxic drug.

Drug Transport; Drug Influx

For a cytotoxic drug to interact with target cellular constituents, it must enter the cell by one of the mechanisms involving passive diffusion, facilitated (carrier mediated), or energy-dependent active transport. There is evidence to suggest that tumor-cell barriers to drug influx may either be dictated by the plasma membrane composition or the presence of cell-surface resident molecules. Thus, it has been shown that in certain drug-resistant cells incubation with a detergent (e.g., Tween-80) or Amphotericin-B (a polyene antibiotic) can increase cellular drug uptake (12, 13).

As shown in Figure 27.1, laser flow cytometry provides a powerful means for monitoring cellular transport and retention of fluorescent drugs, such as anthracyclines (14–16). This method can also be used as a rapid tool for monitoring changes in drug retention caused by a secondary agent or to study the effect of other variables, such as temperature, pH, cell-surface charge, and cell-membrane composition. In an earlier study, we combined cytotoxicity studies and laser flow cytometry (for quantitation of anthracycline fluorescence) to monitor the effect of Amphotericine-B on cellular retention of doxorubicin in drug-resistant cells (17). Amphotericin-B has been shown to react with sterols in the plasma membrane and cause formation of 4-10Å pores (18). Several earlier studies suggested that Amphotericine-B can enhance

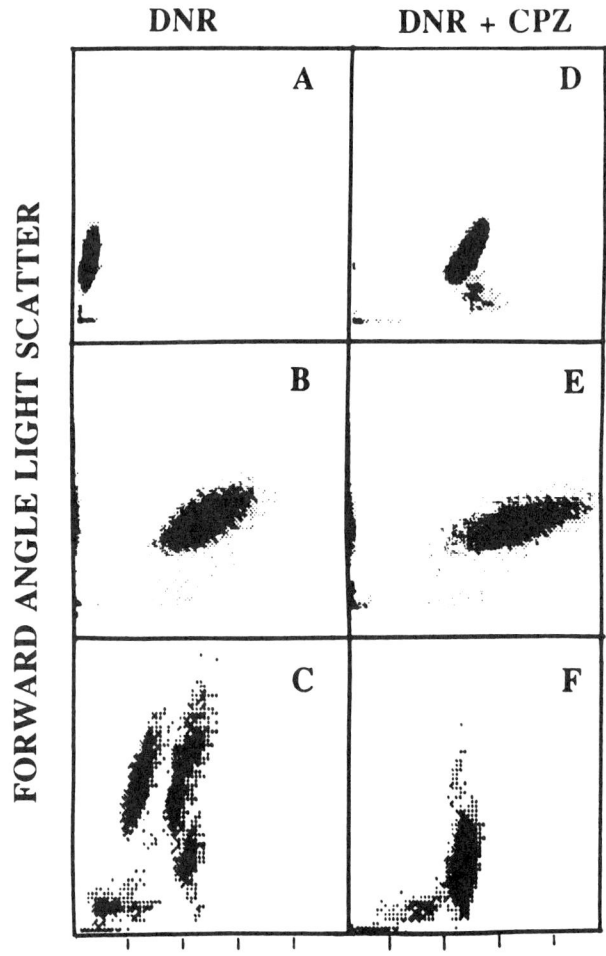

Figure 27.1. Two-parameter dot plots of cells incubated with daunorubicin (*DNR*) alone or in combination with an efflux blocker (chlorpromazine, *CPZ*). In drug-resistant P388/R-84 cells (**A**), coincubation with *CPZ* (**D**) significantly enhances drug retention. A similar but not as pronounced effect is seen in human cells (**B**) incubated with the efflux blocker (**E**). In cells retrieved from a human leukemic blood sample (**C**), three major subpopulations can be recognized on the basis of drug fluorescence. Coincubation with *CPZ* leads to enhancement of drug fluorescence. Forward-angle light scatter is also reduced due to pronounced effects of *CPZ* on cell membrane. DNR and CPZ concentration used was 3.5 and 25 μM respectively and cells were incubated at 37°C for 60 min.

cytotoxicity to a variety of drugs and increase adriamycin uptake by 20% in HeLa cells (19–21). Our flow cytometric data clearly showed that pre- but not co- or postincubation with Amphotericin-B had a pronounced effect on cellular retention of fluorescent doxorubicin in tumor cells and cells from several normal tissues. In bone marrow cells the effect of Amphotericin-B on cellular doxorubicin retention was more heterogeneous, with some populations being more sensitive than others.

It is conceivable that cellular interaction may modulate transport and retention of a drug and this in turn may lead to chemoresistance. That this is in fact a reality in heterogeneous cell populations has been documented in the elegant work of Heppner, Yamashima, and their colleagues (22, 23), and more recently by Teicher et al. (24), who have compared in vitro to in vivo sensitivity of tumor cells. That this phenomena may be related to cellular drug transport is indicated by the work of Durand et al. (25), who have shown that, in spheroids, cellular interaction may determine cellular drug retention. Additional support for this hypothesis was recently documented by Pelletier et al. (26), who reported that, with increased confluence of two colon cancer cells in monolayer cultures, cellular uptake of drugs known to enter by passive diffusion as well as chemosensitivity was reduced.

EFFLUX

Recent studies from several laboratories have indicated that rapid efflux of natural products used as drugs in cancer chemotherapy (e.g., anthracyclines, vinca-alkaloids) may result in reduced cellular retention and chemosensitivity (27–29). In vitro studies on multidrug resistant (MDR) cell lines have identified a family of MDR genes (30–33) and a putative 170 Kd P-glycoprotein gene product (34–39), which is believed to act as an efflux pump. As mentioned earlier, laser flow cytometry provides a powerful tool for monitoring of cellular retention of fluorescent drugs and for identification of heterogeneity in drug retention. This method can also be used for monitoring the effect of drugs that may block efflux and enhance cellular drug retention and thus chemosensitivity. The following will summarize several of our flow cytometric studies on anthracycline retention and efflux in doxorubicin resistant murine leukemic P388/R-84 cells as well as human leukemic and solid tumor cells. In one of our initial flow cytometric studies, we reported that phenothiazine chlorpromazine causes a marked enhancement of cellular doxorubicin retention and the effect was cell cycle phase related, being more pronounced in S phase and in proliferating cells than in cells in other phases of the cell cycle or in a nonproliferating status (40). In a subsequent paper (41), we compared cellular retention, efflux, and cytotoxicity of three closely related anthracyclines, daunorubicin, AD-32, and THP-AdR. Laser flow cytometry played a major role in this study and complemented data obtained by soft agar clonogenic and fluorometric assays.

HETEROGENEITY OF DOXORUBICIN RETENTION AND DIFFERENTIAL RESPONSE TO EFFLUX BLOCKERS

Unlike tissue culture cell lines, which show more or less homogeneous cellular retention of anthracyclines, human tumor cells retrieved from either leukemic blood, ascites, pleural fluid, or disaggregated from solid tumors often show extensive heterogeneity in both drug retention and in response to drugs known to block drug efflux and enhance retention. Several of our key observations using laser flow cytometric analysis of doxorubicin and daunorubicin fluorescence were summarized in a paper published in 1987

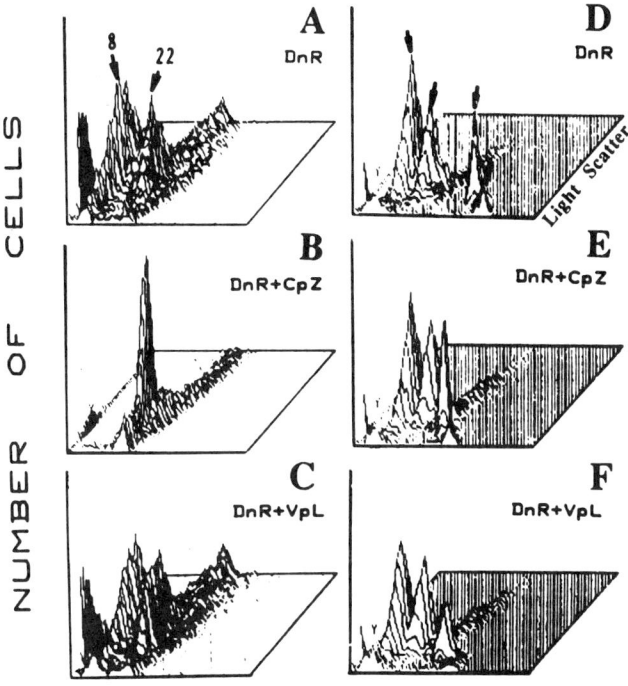

DRUG FLUORESCENCE

Figure 27.2. DNR retention in cells retrieved from gradient separation of two human pleural fluid samples from patients with lung malignancies. Note in both cases (**A, D**), several populations (*arrows*) on the basis of drug retention can be recognized. Coincubation of cells in **A** with *CPZ* (**B**) but not verapamil (*VPL*) (**C**) resulted in enhancement of drug retention and emergence of a single fluorescent population. In samples **E**, and **F**, the two efflux blockers did not alter the heterogeneity of drug retention as compared to the sample incubated with *DNR* alone. (Modified, with permission, after Figures 27.5 and 27.7. From Krishan A, Sauerteig A, Gordon K, Swinkin C. Flow cytometric monitoring of cellular anthracycline accumulation in murine leukemic cells. Cancer Res 1986;46:1768–1773.)

(42). Data from this study showed that on the basis of daunorubicin fluorescence, multiple populations can be identified in human tumors (Figure 27.2). Some of the populations with low drug retention were sensitive to efflux blockers, such as verapamil or phenothiazines, and would enhance their retention, whereas others were not affected, indicating that either their efflux pump was not sensitive to the particular efflux blocker or that their low drug retention was due to factors other than efflux. The value of laser flow cytometry for rapid monitoring of cellular drug retention and heterogeneity and for identification and possible use of an appropriate efflux blocker was also stressed in this publication. A particularly useful role of laser flow cytometry in clinical oncology may be that changes in retention of a fluorescent drug in a heterogeneous tumor cell population may be useful for monitoring of tumor progression and emergence of resistant cells.

The use of laser flow cytometry for monitoring of cellular transport (influx, retention, efflux) has been valuable in our understanding and correlation of this paramater with cytotoxicity of anthracyclines and rhodamines. Laser flow cytometry has similarly been used in monitoring cellular content of dihydrofolate reductase (an enzyme involved in cytotoxicity of methotrexate) and multiple drug resistance (43). Similarly, data has been published by Teicher et al. (44) using a Platinum-Rhodamine complex as fluorescent probes for monitoring cellular transport and retention. It is hoped that with the recent availability of easy to use, high efficiency flow cytometers, similar studies will be extended to other cancer chemotherapeutic (and other therapeutic) drugs that either have fluorescent analogs or can be easily tagged with a fluorescent molecule.

pH and Drug Transport

Although intracellular pH does not fluctuate over a wide range, the extracellular environment around a cell can significantly change from alkaline to acidic or vice versa. Cellular drug transport can in turn be significantly affected by the pH of the surrounding extracellular fluids. For example, in any protocol using intravesical irrigation of the bladder with a chemotherapeutic agent, one has to bear in mind the highly acidic nature of the urine and the effect this may have on cellular transport and cytotoxicity of a drug. As shown in Figure 27.3, flow cytometry offers a unique tool for monitoring the effect of *pH* on cellular transport and fluorescence of a drug. In these time versus cellular daunorubicin fluorescence two-parameter scattergrams, incubation of cells at the alkaline *pH* (Fig. 27.3C) increased cellular drug fluorescence approximately 20-fold as compared to cells incubated at the acidic *pH* (Fig. 27.3A). It should be noted that *pH* had no effect on excitation or emission characteristics of the drug itself and the differences observed must be related to the effect of *pH* on cellular transport of daunorubicin. For further in-depth discussion of pH, cellular drug transport, cytotoxicity and flow cytometry, the reader is referred to a recent publication of Alabaster et al. (45).

Cell Membrane and Transport

Besides extracellular pH, cell-membrane potential and cell-surface charge may alter cellular transport and in turn modulate cytotoxicity. Several studies have suggested that changes in composition of lipid moieties in the cell membrane can alter cellular resistance to doxorubicin (8–10, 46). In an earlier study, we have used a free-flow electrophoresis system to sort cells on the basis of their cell-surface charge and then study their drug retention characteristics with flow cytometry (47). A positive correlation between cell-surface charge and drug retention could be seen in P388 cells even though neuraminidase treatment (to remove sialic acid and decrease electrophoretic mobility) had no effect on doxorubicin retention. In a series of papers emanating from the laboratory of Dr. Lampidis and his colleagues, correlation between chemical charge of a drug and its cellular accumulation and cytotoxicity has been demonstrated (48). These studies show that in drug resistant Friend leukemia cells as

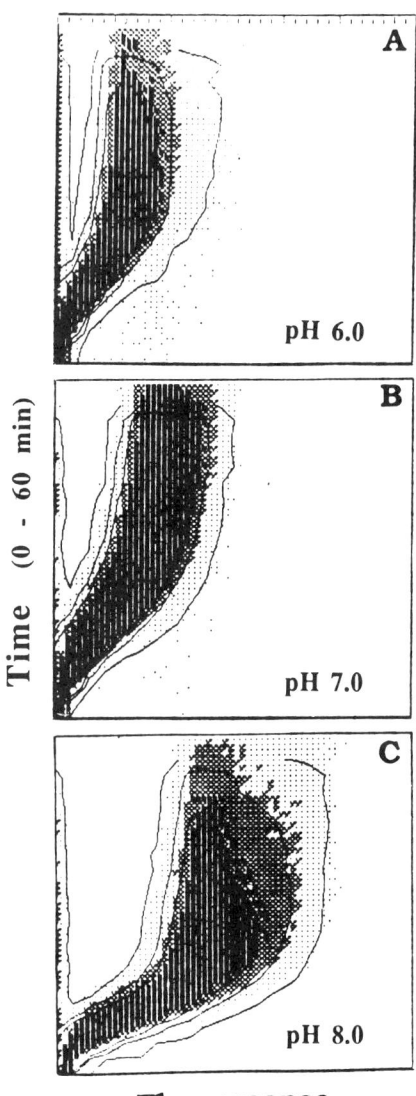

Figure 27.3. Two-parameter scattergrams of drug fluorescence (linear) versus time (0–60 min) of P388 cells incubated with daunorubicin 3.5 µM at 37°C. Cells incubated at alkaline pH of 8.0 (**C**) had significantly higher drug fluorescence than those incubated at neutral (**B**) or acidic pH of 6.0 (**A**). (Reprinted with permission from Ablex Publishing Corporation.)

well as in several normal cell types (cardiac fibroblasts, epithelial cells), lower membrane potential correlates with reduced retention of positively charged drugs and resistance to cytotoxicity. It is important to note that several reagents and fluorochromes are commercially available for monitoring cell-membrane potential and surface charge by laser flow cytometry. Correlation of these parameters with other factors involved in cellular drug transport and resistance may provide information of therapeutic value.

MDR Phenotype and P-glycoprotein

As mentioned above, it has become increasingly evident that reduced cellular drug retention may be due to the presence of an energy-dependent efflux pump. Elegant work from several laboratories, especially those of Dr. Ling in Toronto, Canada, and investigators at the National Institutes of Health (Fojo, Gottesman, Pastan) and other institutions (Houseman, Biedler, Roninson, Borst), has identified the genes responsible for MDR, transfected cells with the MDR gene to confer resistance both in vitro and in vivo, and identified the sequence and chromosomal location of the gene as well as monitored gene expression in various cells, tissues, and organs. The MDR genes, messenger RNA, and the putative gene product P-glycoprotein have been studied in depth. The reader is referred to some excellent review articles on the general phenomena of MDR, its molecular biology and its biological significance (49–53). Although flow cytometry may be a valuable tool for in situ studies on the MDR gene, most of the published flow cytometric work has focussed on the detection of the P-glycoprotein in drug resistant cells. It should be emphasized that the MDR gene and its putative gene product, P-glycoprotein, is not specifically a unique feature of tumor cells but is also present in normal cells, especially those of secretory nature, or in endothelial cells, where it may control intracellular transport of important biological molecules. Overexpression of P-glycoprotein can accompany several biochemical stimuli and may serve as an important means of controlling cellular transport of hormones and natural products as well as secretory functions (54, 55). The P-glycoprotein molecule is plasma membrane resident with a small part of the molecule sticking out of the cell surface. Several immunohistochemical studies have also identified cytoplasmic location of the antigen reacting with the P-glycoprotein-specific antibodies.

ANTIBODIES AGAINST P-GLYCOPROTEIN; C219 AND RELATED MONOCLONALS

Chinese hamster cells made resistant in vitro to colchicine by Ling and Thompson (12) have been a predominant focus of studies on MDR as well as for generation of mono- and polyclonal antibodies. In one of the earliest studies, Kartner et al. (56) produced several P-glycoprotein-specific monoclonal antibodies by immunizing mice with SDS solubilized plasma membranes of multidrug resistant Chinese hamster ovary (CHO) and human cell lines. Out of eight stable hybridoma clones, three monoclonal antibodies specific for antigen in CHO cells were isolated while five other monoclonals recognized antigens common to CHO and human drug-resistant cells. These eight MOB's could recognize three specially distinct epitopes, and four monoclonals (C11, C26, C36, and C219) belonging to Group I could recognize Chinese hamster, Syrian hamster, mouse, and human multidrug resistant cells. In Western blots, reactivity of C219 with 170 K, P-glycoprotein correlated with the level of resistance. Results from indirect immunofluorescence staining of drug-resistant cells with C219 and a fluorescent second antibody confirmed the above results. These authors also demonstrated the use of C219 for detection of drug-resistant cells by laser flow cytometry. Single-parameter histograms in Figure 27.3 of this communication clearly show that in both col-

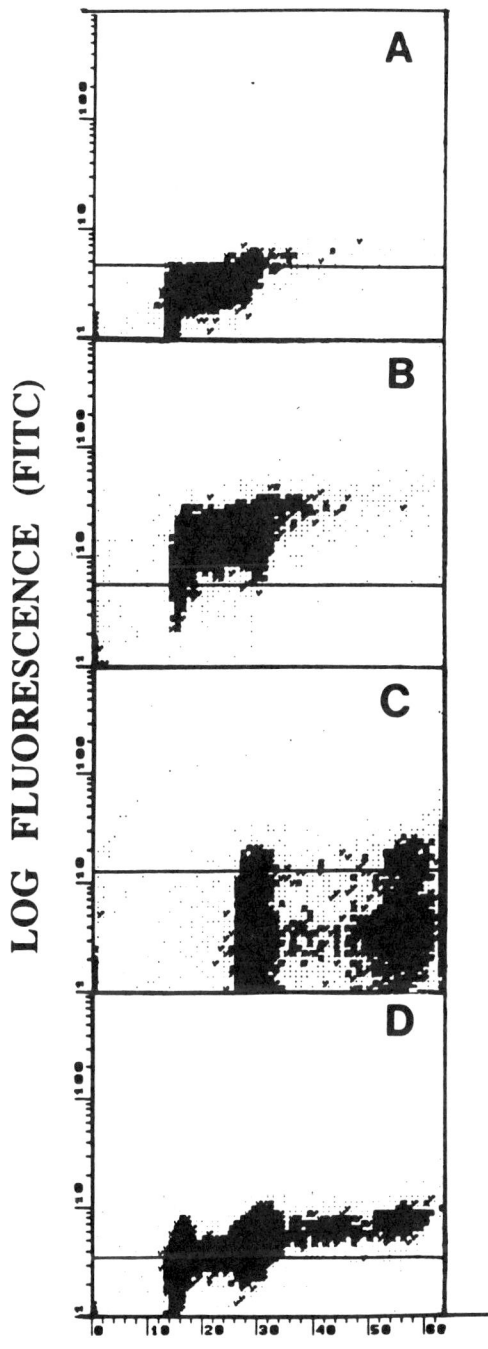

LINEAR FLUORESCENCE (PI)

Figure 27.4. Dot plots (log FITC versus linear propidium-iodide fluorescence) of Chinese hamster cell lines AuxB$_1$ (**A**), ChrC$_5$ (drug resistant; (**B**), a human lung (**C**), and human renal cell carcinoma lines (**D**). Cells were fixed, reacted with P-glycoprotein antibody (C219) obtained from Centocor, Inc., and a goat antimouse FITC-labeled antibody. Note high expression of P-glycoprotein in the drug-resistant ChrC$_5$ (**B**) and the renal tumor cells (**D**).

chicine-resistant, Chinese hamster multidrug-resistant cells, and vinblastine-resistant human lymphoid (CCRF-CEM) cells, fluorescence of C219-second antibody FITC correlated with the level of resistance. Recently the C219 monoclonal that was most widely reactive with the MDR cells from various species has become commercially available through Centocor, Inc. (Malvern, PA). Kits are available for use of the antibody for staining of cells by anti-peroxidase, avidin/Biotin, or other indirect methods, as well as for use of the FITC conjugated antibody for flow cytometric work. In our laboratory, we have extensively used this monoclonal (C219) for staining of cells by both the immunoperoxidase and the indirect second FITC-labeled antibody method. We have obtained excellent results in both cytospins and smears stained by the immunoperoxidase method, as well as by use of the FITC-labeled second antibody procedure. Figure 27.4 shows some of the recent data obtained in our laboratory on double staining of cells for DNA content and P-glycoprotein expression. Although our results with the C219 antibody used by the indirect (second antibody) method have been reasonably consistent, we have not succeeded in getting the FITC-labeled antibody to work in our system. It is possible further methodological work may be needed to get the direct antibody to work.

JSB-1 and MDR (Ab-1) are two other commercially available P-glycoprotein specific antibodies. JSB-1 was produced by Scheper et al. (57) by immunizing mice with Chinese hamster (ChrC$_5$) cells and is commercially available from BIO/CAN America, Inc. (Portland, ME). MDR (Ab-1) is an affinity-purified polyclonal antibody raised in rabbits against the peptide in P-glycoprotein's C-terminal cytoplasmic domain and sold by Oncogene Science, Inc. (Manhasset, NY). Besides these commercially available P-glycoprotein-specific antibodies, several other P-glycoprotein-specific antibodies have been described in the literature (58–62). Some of these antibodies, such as MRK 16 and 20, are especially valuable as they react with the cell-membrane resident part of the P-glycoprotein molecule and thus can stain live cells.

It is interesting to note that very few published reports have presented technically acceptable dual-parameter histograms of P-glycoprotein-positive cells. In most cases, data presented is in the shape of single-parameter histograms and, in some recent publications, the quality of histograms is not good enough to discriminate between noise and signal. An exception is the use of C219 (with FITC-labeled second antibody) and propidium iodide (for DNA content) reported by Epstein et al. (63). Use of multiple antibodies to detect different parts of the P-glycoprotein epitope has been recently recommended as a means for getting enhanced detection of cells with low-level expression (64). We have, in a recent study, compared the three commercially available antibodies for their use in flow cytometric detection of P-glycoprotein-positive cells (65). Our results show that whereas C219 and JSB-1 are excellent for detection of P-glycoprotein-positive ChrC$_5$ cells, the cross-reactivity of JSB-1 with drug sensitive AUXB$_1$ and P388 cells was also detected. The MDR (Ab-1) antibody that was raised in rabbits against the peptide gave reasonably good results comparable to that of C219.

Xenobiotic Detoxification

Besides drug influx and MDR-related drug efflux, several other cellular mechanisms may circumvent the cytotoxic effects of a drug (66, 67). Cellular detoxification of xenobiotics and free radicals by glutathione (GSH) and related enzymes (glutathione S-transferase, glutathione peroxidases) play a major role in the protection of cells from a variety of physical (e.g., radiation) and chemical agents (68–72). Elevated levels of GSH and related enzymes have been reported in several drug-resistant cells. Similarly, depletion of cellular GSH content by buthionine sulfoxamine (BSO) has been shown to enhance chemosensitivity (70). Several laboratory studies suggest that modulation of GSH and related cellular enzymes may play an important role in the resistance of cells to chemotherapeutic agents.

In one of our recent studies (67), we have shown that in doxorubicin-resistant P388/R-84 cells, blocking of efflux by trifluoperazine can reduce cellular resistance from 150- to 60-fold. Prior depletion of cellular GSH content (by incubation with BSO for 24 hr) in combination with efflux blocking can reduce resistance from 150- to eight-fold, thus showing that efflux blocking and xenobiotic detoxification can play a major role in overcoming cellular resistance to doxorubicin. Several recent studies have also shown that changes in cellular GSH content can be easily monitored by laser flow cytometry using monochlorobimane (mBCL) as a GSH-specific fluorochrome (see also Chapter 31). The method, as thoroughly investigated in two papers (73, 74), is relatively easy and involves incubation of cells with 20 μM mBCL at room temperature for 5 min. UV excitation (351-364 nm) from an argon laser is used for collection of GSH-mBCL-conjugate fluorescence emitted between 465-505 nm. In most cases, this simple method gives excellent results and allows for rapid monitoring of GSH content and its heterogeneity in tumor cells. As the staining is done in live cells, this method can be used in a multiparametric setting to analyze heterogeneity as well as correlate GSH content with presence or absence of other phenotypic markers. It should be noted that even though this is a simple and reliable method for monitoring of cellular GSH content, several factors may influence measurement of GSH content by this method. mBCL itself is nonfluorescent and formation of the fluorescent product GSH-mBCL-conjugate is catalyzed by glutathione-S-transferase (GST). Thus, formation of the product will depend on cellular GST activity, which is known to be variable in content and form (three isoforms of GST π, α, μ). Thus, cells with high GST content or activity may catalyze more of the conjugate formation and, thus, discrepancy between flow cytometric determination using mBCL and other biochemical methods for GSH determination (e.g., Tietze's enzymatic method, (75) may become evident. Another artifact may be seen in cells incubated with BSO for GSH depletion. In a recent study (76), we have shown that although the enzymatic method shows that a 24-hr incubation with 100 μM of BSO causes a 95% reduction of GSH content, the flow cytometric mBCL method does not detect this significant reduction. It is presumed that mBCL in GSH-depleted cells may react with other sulphahydryl groups and thus give erroneous results. Figure 27.5, taken from this report, shows the potential of laser flow cytometry for simultaneous monitoring of drug efflux and GSH content of cells (76).

In a recent study, Poot et al. (77) have used two new halogenated dyes (5-chloromethylfluorescein-diacetate, CMFDA and 5-chloromethyleosin-diacetate, CMEDA) as probes for cellular thiol and GSH content. These reagents, unlike mBCL, are excited by 488 or 514 nm argon laser lines. By using Hoechst 33342 as a DNA stain, these authors monitored cellular thiol content in relation to cell cycle phase. Data in this report suggests that these dyes may measure the total level of free intracellular thiols rather than GSH alone. However, it is clear that these reagents will be valuable for monitoring of cellular thiol content in a multiparametric setting.

Proliferation and Cell Cycle

Besides drug transport, membrane charge, and cellular detoxification mechanisms, there are several other cellular factors that determine cellular response to a drug. It is well known that most of our cancer chemotherapeutic agents have proliferation-dependent or cell cycle phase-related cytotoxic effects. Thus, some of the well-known agents will kill only cycling cells or cells in a certain phase of the cell cycle. Flow cytometry provides an excellent tool for the monitoring of cell cycle phase distribution (by estimation of DNA content) of a heterogeneous population. Several vital dyes (e.g., Hoechst 33342) are available, although some of these may be effluxed by the MDR-related P-glycoprotein pump. In a recent study (78), we have shown that MDR cells may efflux the DNA binding dye, Hoechst 33342, and that the use of an efflux blocker will lead to the generation of excellent DNA distribution histograms from cells that are difficult to stain by the vital (dye alone) method. Use of DNA binding fluorochromes in parallel with P-glycoprotein-specific monoclonals is an accepted method and can provide important information, especially if one is dealing with a heterogeneous population of cells containing both normal (diploid) and aneuploid tumor cells.

The availability of anti-BrdUrd monoclonals (79) provides one with means for correlating cellular proliferation, as determined by quantitation of DNA synthesis, with other parameters related to resistance. The literature on the role of cell cycle phase and cellular proliferation in chemosensitivity covers one of the oldest areas in flow cytometry and includes several excellent earlier articles. Suffice it to say that the combination of analytical methods available for monitoring of cell cycle phase, proliferation status with the newer reagents (proliferation-related antibodies, KI-67), and other markers of resistance (P-glycoprotein, GSH) should make it

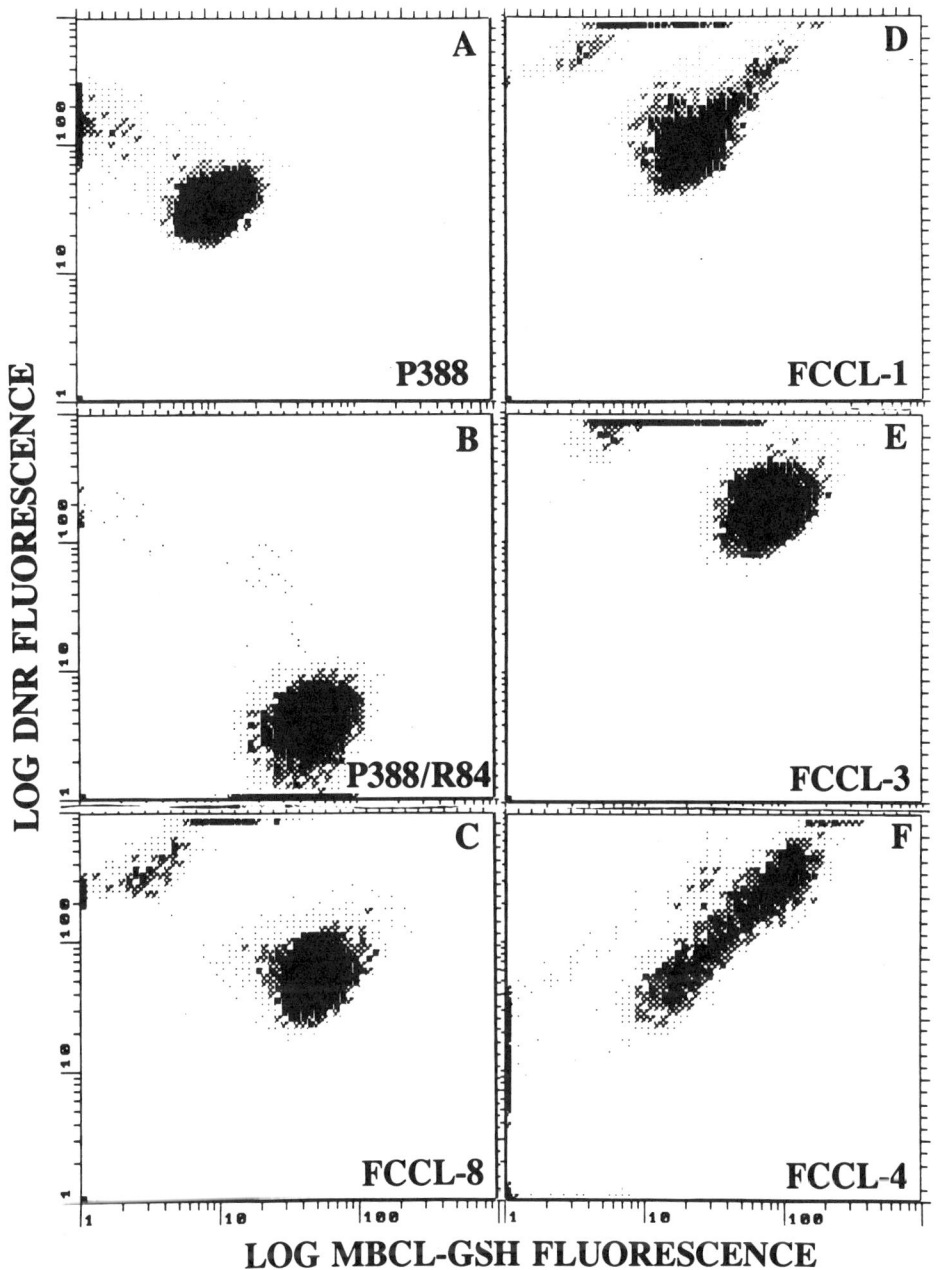

Figure 27.5. Two-parameter dot plots of DNR and GSH-mBCL fluorescence show low retention of DNR with high GSH-mBCL fluorescence in *P388/R-84* cells (**B**). In contrast, all human tumor cell lines examined had higher GSH-mBCL fluorescence and DNR retention than that of the *P388* cells (**A**). Some of the human tumor cell lines (**E** and **F**) show extensive heterogeneity in DNR retention as well as in GSH-mBCL fluorescence. (Reproduced with permission from Nair S, Singh S, Krishan A. Flow cytometric monitoring of glutathione content and anthracycline retention in tumor cells. Cytometry 1991;12:336–342.)

worthwhile to undertake correlative studies on drug resistance in a multiparametric setting.

FUTURE DIRECTIONS

In conclusion, this brief review shows that flow cytometry can be a valuable tool in the study of cellular resistance to chemotherapeutic agents, especially those used in management of cancer patients. One can study drug influx, retention and efflux, heterogeneity, sensitivity to selected efflux blockers, P-glycoprotein expression, glutathione content, alterations in cell membrane and changes in membrane potential and surface charges, which may affect drug retention, and, thus, chemosensitivity.

At present, direct quantitation of intracellular drug content, which correlates to a large degree with cytotoxicity, can be undertaken only for fluorescent drugs (anthracyclines), fluorescent analogs (methotrexate), or drug-fluorescent conjugates (Platinum-Rhodamine). With the availability of fluorescent analogs or drugs tagged with fluorescent labels, flow cytometry may be a valuable tool in monitoring cellular transport and a means for modulating drug retention. For this purpose, we have to seek the help of synthetic chemists (or companies that market reagents) for production and testing of labeled drugs or their fluorescent analogs.

In the detection of MDR-related P-glycoprotein, some of the commercially available monoclonals (e.g., C219) are valuable tools for flow cytometric detection of this important molecule. It is hoped that wider (commercial) availability of reagents that recognize other parts of the P-glycoprotein epitope, especially those resident on the outer cell surface (e.g., MRK 16,20), will make it feasible to routinely use flow cytometry for detection of P-glycoprotein in combination with other phenotypic markers.

In the important area of cellular detoxification, there is a need for commercial availability of reagents (e.g., fluorescent enzyme substrates) that can help us to quantitate the intracellular content of enzymes involved in this important protective mechanism. Similarly, there is a need for monoclonals that may recognize and help us quantitate the expression of three major isoforms of glutathione S-transferase. It is hoped that, within the next few years, reagents for flow cytometric determination of drug-resistance-related markers will become widely available and flow cytometry will find wider use in the management of cancer patients.

REFERENCES

1. Curt GA, Clendeninn NJ, Chabner BA. Drug resistance in cancer. Cancer Treat Rep 1984;68:87–99.
2. Fox BW, Fox M. (Eds) Antitumor drug resistance, volume 72. In: Handbook of Experimental Pharmacology. Berlin: Springer-Verlag, 1984.
3. Kessel D, ed. Resistance to antineoplastic drugs. 1st ed. Boca Raton, FL: CRC Press, Inc., 1989.
4. Sartorelli AC, Ishiguro K, King CL, Morin MJ, Reiss M. Mechanisms involved in the induction of malignant cell differentiation. Adv Enzyme Regul 1986;25:507–529.
5. Kerr JF, Wyllie AH, Currie AR. Apoptosis: a biological phenomenon with wide-ranging implications in tissue kinetics. Br J Cancer 1972;26:239–257.
6. Krishan A, Frei E III. Morphological basis for the cytolytic effect of vinblastine and vincristine on cultured human leukemic lymphoblasts. Cancer Res 1975;35:497–501.
7. Wyllie AH, Kerr JF, Currie AR. Cell death: the significance of apoptosis. Int Rev Cytol 1980;68:251–306.
8. Ganapathi R, Krishan A. Effect of cholesterol content of liposomes on the encapsulation, efflux, and toxicity of adriamycin. Biochem Pharmacol 1984;33:698–700.
9. Kessel D. Membrane alterations associated with progressive adriamycin resistance. Biochem Pharmacol 1982;31:2691–2693.
10. Ramu A, Glaubiger D, Magrath IT, Joshi A. Plasma membrane lipid structural order in doxorubicin-sensitive and resistant P388 cells. Cancer Res 1983;43:5533–5537.
11. Tritton TR, Yee G. The anticancer agent adriamycin can be actively cytotoxic without entering cells. Science 1982;217:248–250.
12. Ling V, Thompson LH. Reduced permeability in CHO cells as a mechanism of resistance to colchicine. J Cell Physiol 1974;84:103–116.
13. Presant CA, Carr D. Amphotericin B (Fungizone) enhancement of nitrogen mustard uptake by human tumor cells. Biochem Biophys Res Commun 1980;93:1067–1073.
14. Krishan A, Ganapathi R. Laser flow cytometric studies on intracellular fluorescence of anthracyclines. Cancer Res 1980;40:3895–3900.
15. Durand RE. Flow cytometry studies of intracellular adriamycin in multicell spheroids in vitro. Cancer Res 1981;41:3495–3498.
16. Tapiero H, Fourcade A, Vaigot P, Farhi JJ. Comparative uptake of adriamycin and daunorubicin in sensitive and resistant friend leukemia cells measured by flow cytometry. Cytometry 1982;2:298–302.
17. Sutton DD, Arnow PM, Lampen JO. Effects of high concentrations of Nystatin upon glycolysis and cellular permeability in yeast. Proc Soc Exp Biol Med 1961;108:170–175.
18. Krishan A, Sauerteig A, Gordon K. Effect of amphotericin B on adriamycin transport in P388 cells. Cancer Res 1985;45:4097–4102.
19. Medoff G, Valeriote F, Dieckman J. Potentiation of anticancer agents by amphotericin B. Natl Cancer Inst 1981;67:131–135.
20. Presant CA, Valeriote F, Proffitt R, Metter G. Amphotericin B-Interactions with nitrosoureas and other antineoplastic agents. In: Prestayko AW, Crooke ST, Baler LH, Carter SK, Schein PS, eds. Nitrosoureas: Current status and new developments. New York: Academic Press Inc., 1981:343–360.
21. Valeriote F, Medoff G, Dieckman J. Potentiation of cytotoxicity of anticancer agents by several different polyene antibiotics. Natl Cancer Inst 1984;72:435–439.
22. Heppner GH, Miller BE. Therapeutic implications of tumor heterogeneity. Semin Oncol 1989;16:91–105.
23. Yamashina K, Miller BE, Heppner GH. Macrophage-mediated induction of drug-resistant variants in a mouse mammary tumor cell line. Cancer Res 1986;46:2396–2401.
24. Teicher BA, Herman TS, Holden SA, et al. Tumor resistance to alkylating agents conferred by mechanisms operative only in vivo. Science 1990;247:1457–1461.
25. Durand RE, Vanderbyl SL. Tumor resistance to therapy: a genetic or kinetic problem? Cancer Commun 1989;1:277–283.
26. Pelletier H, Millot JM, Chauffert B, Manfait M, Genne P, Martin F. Mechanisms of resistance of confluent human and rat colon cancer cells to anthracyclines: Alteration of drug passive diffusion. Cancer Res 1990;50:6626–6631.
27. Danø K. Active outward transport of daunomycin in resistant Ehrlich ascites tumor cells. Biochim Biophys Acta 1973;323:466–483.
28. Skovsgaard T. Mechanisms of resistance to daunorubicin in Ehrlich ascites tumor cells. Cancer Res 1978;38:1785–1791.
29. Inaba M, Kobayashi H, Sakurai Y, Johnson RK. Active efflux of daunorubicin and adriamycin in sensitive and resistant sublines of P388 leukemia. Cancer Res 1979;39:2200–2203.
30. Van der Bliek A, Van der Velde-Koerts T, Ling V, Borst P. Overexpression and amplification of five genes in a multidrug resistant Chinese hamster ovary cell line. Mol Cell Biol 1986;6:1671–1678.
31. Fojo AT, Ueda K, Slamon DJ, Poplack DG, Gottesman MM, Pastan I. Expression of a multidrug-resistance gene in human tumors and tissues. Proc Natl Acad Sci USA 1987;84:265–269.
32. Goldstein LJ, Galski H, Fojo A, et al. Expression of a multidrug resistance gene in human cancers. Natl Cancer Inst 1989;81:116–124.
33. Chen C, Clark D, Ueda K, Pastan I, Gottesman MM, Roninson I. Genomic organization of the human multidrug resistance (MDR1) gene and origin of P-glycoproteins. J Biol Chem 1990;265:506–514.
34. Beck WT, Mueller TJ, Tanzer LR. Altered surface membrane glycoproteins in vinca-alkaloid resistant human leukemic lymphoblasts. Cancer Res 1979;39:2070–2076.
35. Cornwell MM, Tsuruo T, Gottesman MM, Pastan I. ATP-binding properties of P-glycoprotein from multidrug-resistant KB cells. FASEB J 1987;1:51–54.
36. Gerlach JH, Bell DR, Karakousis C, et al. P-glycoprotein in human sarcoma: evidence for multidrug resistance. J Clin Oncol 1987;5:1452–1460.
37. Juliano RL, Ling V. A surface glycoprotein modulating drug permeability in Chinese hamster ovary cell mutants. Biochim Biophys Acta 1976;455:152–162.
38. Kartner N, Shales M, Riordan JR, Ling V. Daunorubicin-resistant chinese hamster ovary cells expressing multidrug resistance and a cell-surface P-glycoprotein. Cancer Res 1983;43:4413–4419.
39. Riordan JR, Deuchars K, Kartner N, Alon N, Trent J. Amplification of P-glycoprotein genes in multidrug-resistant mammalian cell lines. Nature 1985;316:817–819.
40. Krishan A, Sauerteig A, Wellham L. Flow cytometric studies on modulation of cellular adriamycin retention by phenothiazines. Cancer Res 1985;45:1046–1051.

41. Krishan A, Sauerteig A, Gordon K, Swinkin C. Flow cytometric monitoring of cellular anthracycline accumulation in murine leukemic cells. Cancer Res 1986;46:1768–1773.
42. Krishan A, Sridhar KS, Davila E, Vogel C, Sternheim W. Patterns of anthracycline retention modulation in human tumor cells. Cytometry 1987;8:306–314.
43. Assaraf YG, Seamer LC, Schimke RT. Characterization by flow cytometry of fluorescein-methotrexate transport in Chinese hamster ovary cells. Cytometry 1989;10:50–55.
44. Teicher BA, Holden SA, Jacobs JL, Abrams MJ, Jones AG. Intracellular distribution of a platinum-rhodamine 123 complex in cis-Platinum sensitive and resistant human squamous carcinoma cell lines. Biochem Pharmacol 1986;35:3365–3369.
45. Alabaster O, Woods T, Ortiz-Sanchez V, Jahangeer S. Influence of microenvironmental pH on adriamycin resistance. Cancer Res 1989;49:5638–5643.
46. Guffy MM, North JA, Burns CP. Effect of cellular fatty acid alteration on adriamycin sensitivity in cultured L1210 murine leukemia cells. Cancer Res 1984;44:1863–1866.
47. Nair S, Horton A, Leif RC, Krishan A. Electrophoretic mobility studies on doxorubicin-resistant and sensitive murine P388 leukemic cells. Cytometry 1988;9:232–237.
48. Lampidis TJ, Savaraj N, Valet GK, Trevorrow K, Fourcade A, Tapiero H. Relationship to chemical charge of anti-cancer agents to increased accumulation and cytotoxicity in cardiac and tumor cells: Relevance to multi-drug resistance. Anticancer Drugs In: Tapiero H, Robert J, and Lampidis TJ eds., Colloque INSERM John Libbey Eurotex Ltd. 1989;191:29–38.
49. Beck WT. The cell biology of multiple drug resistance. Biochem Pharmacol 1987;36:2879–2887.
50. Chabner BA, Gottesman MM. Meeting highlights: William Guy Forbeck Foundation think tank on "multidrug resistance in cancer chemotherapy." J Natl Cancer Inst 1988;80:391–394.
51. Moscow JA, Cowan KH. Multidrug resistance. J Natl Cancer Inst 1988;80:14–20.
52. Kartner N, Ling V. Multidrug resistance in cancer. Scientific American 1989;260:44–51.
53. Kaye SB. The multidrug resistance phenotype. Br J Cancer 1989;58:691–694.
54. Cordon-Cardo C, O'Brien JP, Casals D, et al. Multidrug-resistance gene (P-glycoprotein) is expressed by endothelial cells at blood-brain barrier sites. Proc Natl Acad Sci USA 1989;86:695–698.
55. Arceci RJ, Baas F, Raponi R, Horwitz SB, Housman D, Croop JM. Multidrug resistance gene expression is controlled by steroid hormones in the secretory epithelium of the uterus. Mol Reprod Dev 1990;25:101–109.
56. Kartner N, Evernden-Porelle D, Bradley G, Ling V. Detection of P-glycoprotein in multidrug resistant cell lines by monoclonal antibodies. Nature 1985;16:820–823.
57. Scheper RJ, Bulte JW, Brakkee JG, et al. Monoclonal antibody JSB-1 detects a highly conserved epitode on the P-glycoprotein associated with multi-drug-resistance. Int J Cancer 1988;42:389–394.
58. Danks MK, Metzger DW, Ashmum RA, Beck WT. Monoclonal antibodies to glycoproteins of vinca alkaloid-resistant human leukemic cells. Cancer Res 1985;45:3220–3224.
59. Hamada H, Tsuruo T. Functional role for the 170- to 180-kDa glycoprotein specific to drug-resistant tumor cells as revealed by monoclonal antibodies. Proc Natl Acad Sci USA 1986;83:7785–7789.
60. Marquardt D, McCrone S, Center MS. Mechanisms of multidrug resistance in HL60 cells: detection of resistance-associated proteins with antibodies against synthetic peptides that correspond to deduced sequence of P-Glycoprotein. Cancer Res 1990;50:1426–1430.
61. Tanaka S, Currier SJ, Bruggemann EP, Ueda K, et al. Use of recombinant P-glycoprotein fragments to produce antibodies to the multidrug transporter. Biochem Biophys Res Commun 1990;166:180–186.
62. Zeheb R, Beittenmiller HF, Horwitz SB. Use of antibodies to probe membrane glycoproteins associated with drug-resistant J774.2 cells. Biochem Biophys Res Commun 1987;143:732–739.
63. Epstein J, Barlogie B. Tumor resistance to chemotherapy associated with expression of the multidrug resistance phenotype. Cancer Bull 1989;41:41–44.
64. Grogan T, Dalton W, Rybski J, et al. Optimization of immunocytochemical P-glycoprotein assessment in multidrug-resistant plasma cell myeloma using three antibodies. Lab Invest 1990;63:815–824.
65. Krishan A, Sauerteig A, Stein J. A comparison of three commercially available antibodies for flow cytometric monitoring of P-glycoprotein expression in tumor cells. Cytometry. 1991;12:731–742.
66. Deffie AN, Alam T, Seneviratne C, et al. Multifactorial resistance to adriamycin: relationship to DNA repair, glutathione transferase activity, drug efflux, and P-glycoprotein in cloned cell lines of adriamycin-sensitive and resistant P388 leukemia. Cancer Res 1988;48:3595–3602.
67. Nair S, Singh SV, Samy TSA, Krishan A. Anthracycline resistance in murine leukemic P388 cells: Role of drug efflux and glutathione related enzymes. Biochem Pharmacol 1990;39:723–728.
68. Meister A, Anderson ME. Glutathione. Annu Rev Biochem 1983;52:711–760.
69. Bradley AA, Nathan CF. Glutathione metabolism as a determinant of therapeutic efficacy: a review. Cancer Res 1984;44:4224–4232.
70. Hamilton TC, Winker MA, Louie KG, et al. Augmentation of adriamycin, melphalan, and cisplatin cytotoxicity in drug-resistant and sensitive human ovarian carcinoma cell lines by buthionine sulfoximine mediated glutathione depletion. Biochem Pharmacol 1985;34:2583–2586.
71. Kramer RA, Zakher J, Kim G. Role of glutathione redox cycle in acquired and de novo multidrug resistance. Science 1988;241:694–697.
72. Russo A, Mitchell JB. Potentiation and protection of doxorubicin cytotoxicity by cellular glutathione modulation. Cancer Treat Rep 1985;69:1293–1296.
73. Rice GC, Bump EA, Shrieve DC, Lee W, Kovacs M. Quantitative analysis of cellular glutathione by flow cytometry utilizing monochlorobimane: some applications to radiation and drug resistance in vitro and in vivo. Cancer Res 1986;46:6105–6110.
74. Shrieve DC, Bump EA, Rice GC. Heterogeneity of cellular glutathione among cells derived from a murine fibrosarcoma or a human renal cell carcinoma detected by flow cytometric analysis. J Biol Chem 1988;263:14107–14114.
75. Teitze F. Enzymatic method for quantitative determination of nanogram amounts of total and oxidized glutathione: application to mammalian blood and other tissues. Anal Biochem 1969;27:502–522.
76. Nair S, Singh S, Krishan A. Flow cytometric monitoring of glutathione content and anthracycline retention in tumor cells. Cytometry 1991;12:336–342.
77. Poot M, Kavanagh TJ, Kang HC, Haugland RP, Rabinovitch PS. Flow cytometric analysis of cell cycle-dependent changes in cell thiol level by combining a new laser dye with Hoechst 33342. Cytometry 1991;12:184–187.
78. Krishan A. Effect of drug efflux blockers on vital staining of cellular DNA with Hoechst 33342. Cytometry 1987;8:642–645.
79. Gratzner HG. Monoclonal antibody to 5-bromo- and 5-iododeoxyuridine: a new reagent for detection of DNA replication. Science 1982;107:474–475.
80. Gray JW. Monoclonal antibodies against bromodeoxyuridine. Cytometry 1985;6:499–674.

28

Cell-Associated Receptor Quantitation

ROBERT A. HOFFMAN, DIETHER J. RECKTENWALD and ROBERT F. VOGT, JR.

INTRODUCTION

Flow cytometry is routinely used to classify cells into discrete types, such as "positive" or "negative" for a surface antigen. The further classification, on the basis of amount of fluorescence or quantity of antigen, is less frequently used and, then, only the relative number of cells classified as "dim" or "bright" is generally reported. Historically, these terms originated from visual analysis by fluorescence microscopy, a technique that did not permit objective quantitative measurements. The infrequent use of the quantitative abilities of the flow cytometer may also be based on the difficulties in aligning, calibrating, and controlling the earlier generations of cytometers as well as on their rather primitive data analysis capabilities. The more recent models of flow cytometers have a much improved stability, are easier to align and control, and make use of advances in computers. Therefore, instrument-related factors no longer present major obstacles to quantitative fluorescence measurements. Instead, the primary impediments lie in the standardization of the fluorescence measurements and in the characterization of the binding chemistry between surface molecules and their labeled ligands (1, 2). This chapter will address these issues, examining the theoretical considerations and practical aspects of routinely obtaining quantitative results.

Although the chapter will not review the applications of quantitative surface protein measurements, it is clear that many biological functions are controlled by surface receptors for critically important intercellular signals. Quantitative flow cytometric measurements for surface fluorescence staining have already been applied to the study of cell differentiation (3), neoplastic transformation, and leukemia (4, 5, 6), to phenotypic variation (7), cell growth, and cycling (8, 9) and to other aspects of cellular physiology and pathology (10).

A variety of clinical determinations in immunohematology use fluorescence measurements (11), and quantitative flow cytometric crossmatches for transplanted tissues appear to provide better predictive value for long-term graft acceptance (12). Other clinical applications include the measurement of antiplatelet antibodies in immune thrombocytopenia (see Chapter 23) and of antineutrophil antibodies (see Chapter 24). There is undoubtedly more useful information in the quantitation of surface proteins that would come to light if this parameter were routinely studied. In addition, regular fluorescence calibration provides an augmented level of quality assurance and instrument evaluation that becomes increasingly important with the expanding clinical applications of immunophenotyping by flow cytometry (2, 13, 14).

Quantitative measurements may be either relative or absolute. For example, the relative amount of immunofluorescence for a particular antibody provides information about the amount of antigen on the cell surface. Even if the system is not calibrated in absolute terms, it is still possible to determine how much antigen a particular sample has relative to another. Within the same sample one can also determine the relative amount of antigen, as demonstrated in Figure 28.1. As shown here, the fluorescence histogram of CD8-stained lymphocytes overlayed with a control histogram of unstained lymphocytes demonstrates CD8 staining with three regions, corresponding to unstained, dimly-stained, and brightly-stained cells. The dimly-stained population includes primarily natural killer cells, while the brightly-stained population includes supressor and cytotoxic T-cells. This simple example illustrates the advantage of utilizing the antigen quantitation inherent in immunofluorescence analysis rather than classifying the cells only as positive or negative.

Relative quantitation of surface molecules is often useful, but quantitation in absolute numbers (i.e., number of molecules per cell) is intellectually more satisfying and allows comparison between different methodologies, staining reagents, instrumentation, and cell types. This is much more difficult, however, since reference materials and methods do not usually exist. It is not always practical to determine the number of surface molecules to a high degree of accuracy. For this reason, two levels of accuracy, that we will call estimates and quantitative measurements will be considered. Some practical problems in quantitating surface ligands will be discussed.

INSTRUMENT CALIBRATION

For fluorescence measurements, instrument calibration requires running a sample of known fluorescence characteristics. In theory, the sample could be a solution of fluorochrome, but most flow cytometers are not configured to measure the steady background fluorescence from a stream of dye. The practical means to calibrate flow cytometric fluorescence is to use fluorochrome-labeled particles with fluo-

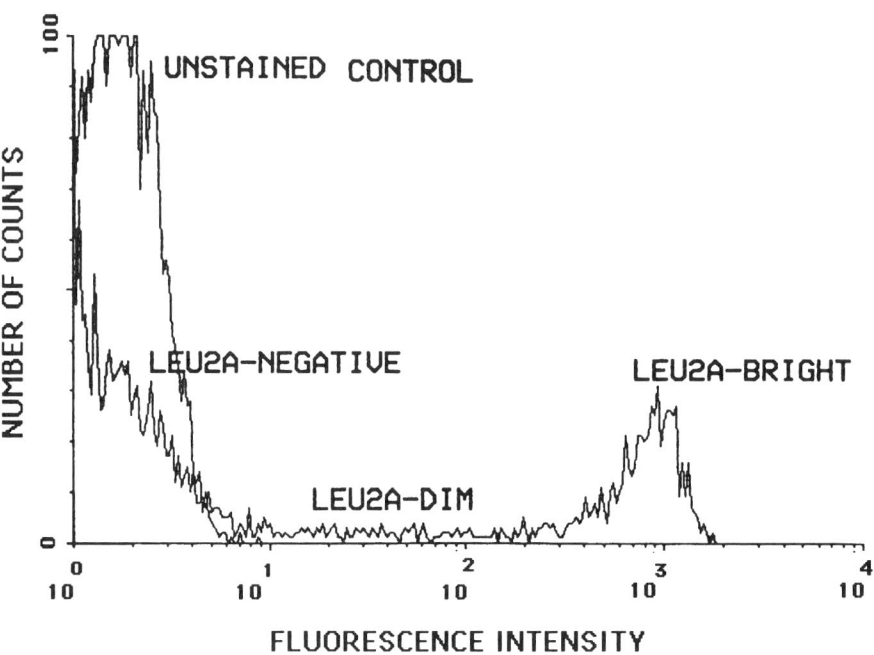

Figure 28.1. Fluorescence histograms of CD8-stained lymphocytes and an unstained control sample. The CD8-stained sample shows a bright positive population as well as dimly-stained and unstained cells.

rescence spectra identical to those of the labeled analyte. The particle may be biological (15) or plastic beads (16, 17); some advantage may be gained by simultaneously analyzing a mixture of the two (2, 14). In any case, the fluorescence value of the particles should be expressed in units of fluorochrome molecules in solution rather than be the actual number of bound molecules. Since the fluorescence of bound molecules is often quenched to a varying degree, the unit of fluorochrome molecules in solution is a more meaningful measure of the amount of fluorescence; it is often abbreviated as MESF (molecules of equivalent soluble fluorochrome).

While the assignment of MESF values to particulate standards is theoretically straightforward compared to fluorochrome solutions, technical difficulties can make it problematic, especially at low concentration of fluorochrome (2). The different chemical nature of various fluorochromes can add further complications. Since MESF calibration has been done mostly with fluorescein isothiocyanate (FITC), these conjugates are currently the most reliable reagents for quantitative fluorescence measurements.

An alternative is to calibrate in terms of ligand molecules, although this is more difficult and requires separate calibration for each ligand of interest. The most accurate approach is to double-label the ligand with radioactive and fluorescent labels (18, 19, 20). The bound ligand is measured both radiometrically and fluorometrically. A second approach is to use beads containing known numbers of a reactive group to quantitate the amount of bound fluorescent ligand. Beads with known numbers of antimouse binding sites are commercially available (16, 17). This approach should be used with caution since the antimouse beads bind both active and inactive antibody. Only active antibody will react with antigen. Thus one must either assume that both active and inactive antibody have the same amount of fluorescence or be able to determine the relative amount of fluorescence from both active and inactive antibody.

An alternative is to use a biological standard with a known number of binding sites (5, 21). For example, one could store frozen aliquots of cells that could be thawed as needed to calibrate a new fluorescent reagent. The frozen cells could be analyzed at one point in time to determine the number of one or more antigens. One has to assume or demonstrate that the antigens remain stable on the frozen cells with the passage of time. Erythrocytes might be used for such a purpose. For less accurate results, one could use a cultured cell line or even a common cell type with relatively stable antigen. The common lymphocyte cell antigens CD3, CD4, and CD45 are fairly constant in normal individuals (15, 22). The number of CD4 and CD45 antigens per cell are particularly uniform from one individual to another (15, 22).

Whatever means of calibration are used, it is important that the experimenter understand the limitations of the method so that the appropriate uncertainties can be attached to any result using the calibration method. If the calibration involves equilibrium binding, additional considerations apply, as discussed below.

ESTIMATES OF LIGAND BINDING

Any determination of ligand binding has inherent errors and uncertainties. When an effort is made to understand and quantitate the uncertainty, we will call the determination a quantitative measurement. When the results are taken at face value, without critical error analysis, we will call the determination an estimate. Estimates are very useful and may provide all the information one needs. For example, if one is using indirect immunofluorescence staining with primary mouse monoclonal antibody, it is reasonable to assume that a second, polyclonal antimouse IgG antibody will react in nearly the same way with all mouse IgG of the same subclass. With this assumption, there are two methods for estimating antibody binding sites.

Estimates Using Fluorescent Secondary Antibody

The first method relies on using a fluorescent conjugate of antimouse IgG. One first uses the conjugate to indirectly stain a reference cell population with a known number of primary antibody binding sites. The fluorescence intensity is then related to a known number of primary binding sites. The same antimouse conjugate can then be used to estimate the number of binding sites of various other mouse monoclonal antibodies (5, 21, 23). It may not be important even to know the number of primary binding sites, but, rather, the number relative to a reference binding site or cell type is important. The expression of antigenic polymorphism can be studied in this way (24, 25). It must be remembered that fluorescence detection electronics in some flow cytometers is sensitive to fluorescence distribution (see Chapter 5), and that similar numbers of fluorescence on different size cells can give different fluorescence measurements.

Estimates Using Fluorescent Primary Antibody

A second way to apply quantitative antimouse IgG binding is to prepare reference beads containing calibrated numbers of active antimouse IgG molecules. A commercial product (Simply Cellular Microbeads, Flow Cytometry Standards Corp.) is available. The antimouse IgG beads can be reacted with a fluorescence-conjugated mouse antibody. Then the fluorescence from the reacted beads can be related to the number of anti-mouse binding sites on the bead. From this measurement, the fluorescence due to a particular number of the fluorescence-conjugated mouse monoclonal antibodies can be determined. Potential error due to inactive antibody has been discussed above.

THERMODYNAMICS AND KINETICS OF EQUILIBRIUM BINDING

All of the quantitative binding studies discussed in this chapter are based on reversible reactions. All of those reactions can be assigned to one of two types: namely ligand binding to cell receptors or antibody binding to cell-associated antigens. In the most general sense, cell-associated antigens can be viewed as receptors for antibodies. Therefore, the following discussion of equilibrium binding is restricted to the discussion of ligand binding to a receptor. An excellent experimental example for the subsequent discussion can be found in Sklar and Finney (26).

A Simple 1:1 Association-Dissociation Model

In a first approximation, the reaction of a ligand with a receptor (or an antibody with an antigen) can be expressed as

$$R + L = RL \qquad (28.1)$$

Three constants are commonly used to describe the kinetic and thermodynamic (equilibrium) properties of such a system: the association rate constant k_a, the dissociation rate constant k_d, and the complex dissociation equilibrium constant K_D. For cell surface receptor studies, the number of receptors per cell is also important and will be discussed below.

The kinetic and equilibrium behavior of such a simple system can be easily described by the equations:

$$\frac{d[RL]}{dt} = k_a \cdot [R][L] - k_d \cdot [RL] \qquad (28.2)$$

$$\frac{d[R]}{dt} = \frac{d[L]}{dt} = -k_a[R][L] + K_d[RL] \qquad (28.3)$$

Where [R], [L], and [RL] represent the equilibrium concentrations of receptor, ligand, and receptor-ligand complex.

Figure 28.2 shows the approach to equilibrium for the reaction. The curves follow exponential functions and the reaction is more than 99% complete after seven halftimes. When equilibrium is reached, then

$$k_a \cdot [R][L] = k_d[RL] \qquad (28.4)$$

or

$$\frac{[R][L]}{[RL]} = \frac{k_d}{k_a} = K_D \qquad (28.5)$$

$$[RL] = \frac{[R]_0 \cdot [L]}{K_D + [L]} \qquad (28.6)$$

That is, the ratio of dissociation constant over association constant is the equilibrium-dissociation constant. For the determination of binding constants, [RL] is often referred to as bound ligand b and [L] is referred to as free ligand f. With these new notations, equation (28.6) can be rearranged into two forms suitable for linear regression analysis. (t is the total receptor concentration $[R]_0$)

$$b = \frac{t \cdot f}{K_D + f} \qquad (28.7)$$

$$b/f = \frac{t}{K_D} - \frac{1}{K_D} \cdot b \qquad (28.8)$$

or

$$1/b = \frac{K_D}{t} \cdot \frac{1}{f} + \frac{1}{t} \qquad (28.9)$$

Equation (28.8) is known as the Scatchard equation. A plot of b/f vs. b results in a straight line if homogeneous binding applies, with a slope of $-1/K_D$ and an ordinate intercept of t/K_D. Another way to obtain t is to read the

Figure 28.2. Approach of the reaction R + L = RL to equilibrium ($[R]_0 = 0.7$, $[L]_0 = 1.0$, $k_a = 100$, $k_d = 0.001$).

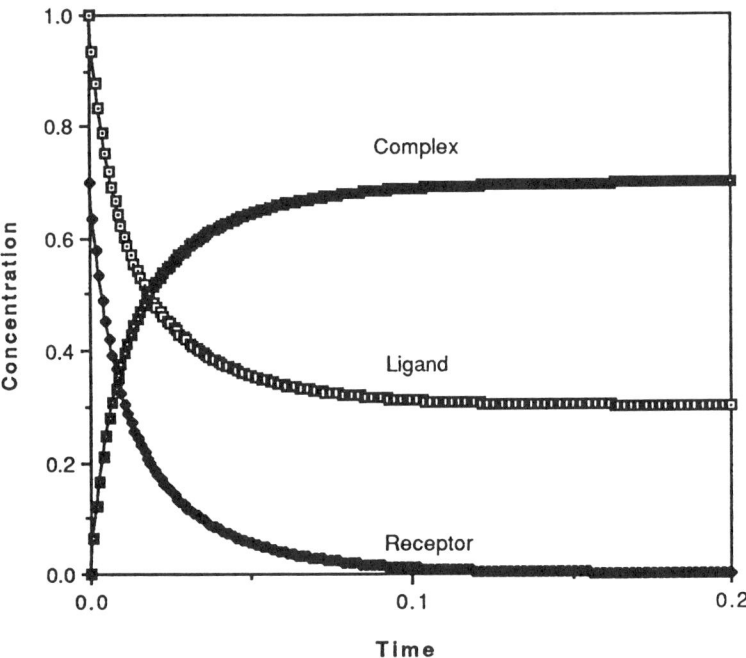

graph at infinite concentration of free ligand. There, b/f = 0 or b = t, i.e., t can be determined from the abscissa intercept of the line.

The transformation of equation (28.6) into (28.9) is known as Langmuir plot or as a Michaelis Menten plot. A graph of 1/b vs. 1/f (for ideal binding behavior) is linear with an ordinate intercept of 1/t and a slope of K_D/t.

Figure 28.3a shows a calculated 1:1 binding curve, with t = 10.0 and K_D = 2.0, and a relative error in b of 5% (1 standard deviation). At a ligand concentration equaling K_D, a 50% saturation is achieved; at 2.5 times K_D, the saturation is below 75%. Even without the additional effects discussed below, even for a simple 1:1 model, it is hard to estimate complete binding from saturation curves directly. To obtain better than 99% binding, the free ligand concentration has to be greater than 100 times K_D. The data in Figure 28.3a is graphed in Figure 28.3b as a Scatchard plot and in Figure 28.3c, as a Langmuir plot.

In this simple model—a more adequate model for cell surface binding will be discussed below—a cell potentially has many receptors and receptor-ligand complexes, but not free ligand, and the total cell fluorescence obtained in the reaction with a fluorescent ligand is proportional to the concentration of the complex RL. If the total number of receptors on a cell or a change in number of receptors is of interest, a measurement of the total cell fluorescence could provide the information directly only if all of the receptors are occupied by ligand and the fluorescence per ligand is known. However, equation (28.1) shows, that the reaction is time-(rate) dependent. In a mathematical sense, it never reaches equilibrium, even though, for experimental purposes, many reactions are fast enough to reach equilibrium within experimental error in seconds. The ramifications of an equilibrium reaction expressed in equation (28.4) are more significant from the point of view of receptor quantitation. For the simple mechanism under discussion here, a very high concentration of ligand could be used to bring the system into full saturation within experimental error. As mentioned above, a concentration at 100 times K_D is needed for 99% saturation, whereas only 7 "halftimes" bring a reaction to >99% completion. Generally, the concentration effect on the estimation of cell receptor numbers is much more pronounced than the kinetic effect.

In real systems, most ligands show low-affinity ("nonspecific") binding to many structures in addition to high-affinity ("specific") binding. For this reason, and because of autofluorescence limitations discussed in section IV below, a measurement at 100 · K_D rarely yields reliable information about the number of receptors. In equations 28.1 to 28.4 concentrations are expressed as free ligand [L], receptor [R] and complex [RL] concentrations. A change of any of the concentrations during the measurement will cause a readjustment of the system to a new equilibrium state. Therefore, cells must not be washed for cell receptor quantitation studies unless the dissociation rate for RL is much longer than the duration of the experiment after the wash step is performed. Because of the fluorescence from free ligand, the instrument also has to be capable of reliable discrimination between free (DC signal in a flow cytometer) and bound (AC signal) fluorescence. This discrimination only works if the detector/amplifier system is not at saturation. For the enumeration of subpopulation fractions (percentages) cell washing is adequate as long as dissociation does not affect the separation of the populations of interest.

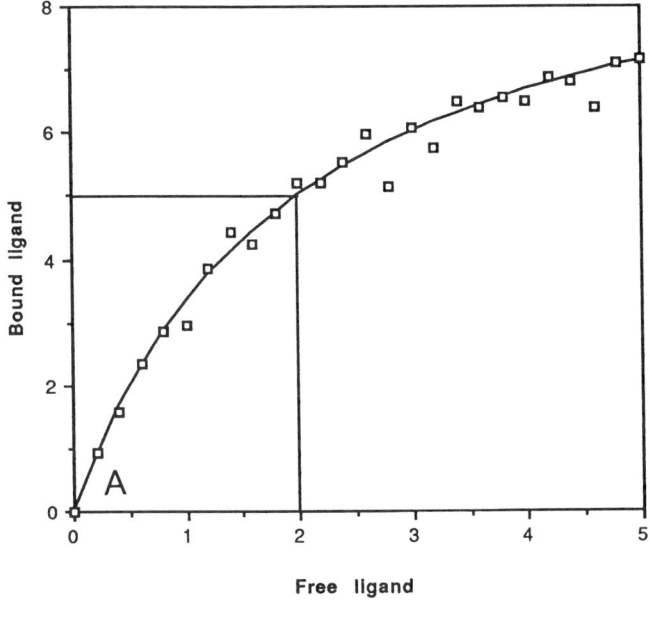
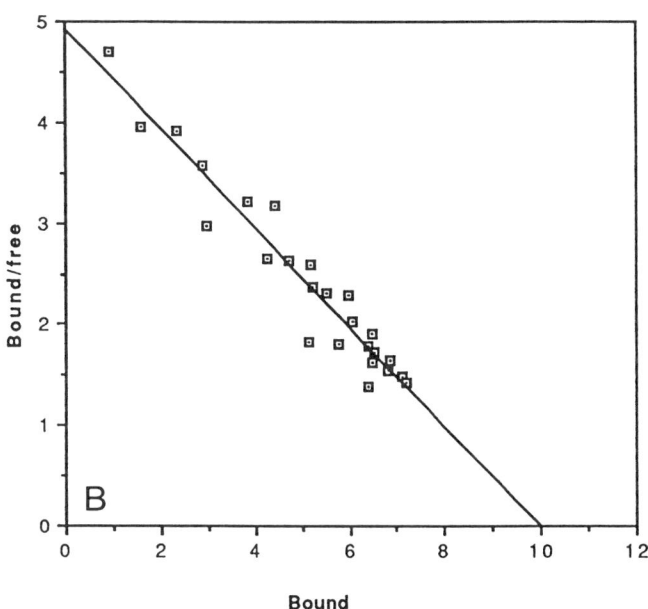
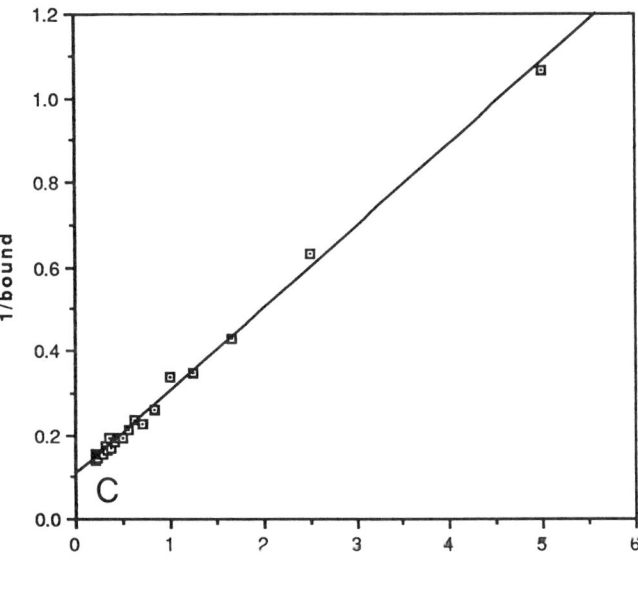

Figure 28.3. Calculated binding behavior for a 1:1 model with 10 units of receptor sites and a dissociation constant of 2 units with a 5% Gaussian experimental noise. **A**. Saturation curve with lines showing 50% binding at a *free ligand* concentration of 2 units. **B**. Scatchard plot of the data in **A** with an ordinate intercept of 4.9 ± 0.2, a slope of −0.49 ± 0.03 and a r² of 0.96. The abscissa intercept is 10, as expected. **C**. Langmuir plot of the data in **A** with an ordinate intercept of 0.105 ± 0.004, a slope of 0.196 ± 0.003, and a r² of 0.998, resulting in a t of 9.5 ± 0.4 and a K_D of 1.9 ± 0.1.

LANGMUIR BINDING ISOTHERM

The simple 1:1 model described above oversimplifies the reaction of a ligand with receptors on a cell surface. A more adequate model analyzes the binding of ligands to many identical, distinguishable, independent sites, without interaction between bound ligands. If $[R]_{tot}$ is the total receptor concentration, $[R]_{bound}$ is the concentration of receptors with ligand, [L] is the ligand concentration, and Ka is a binding constant, then

$$\frac{[R]_{bound}/[R]_{tot}}{1 - \frac{[R]_{bound}}{[R]_{tot}}} = Ka[L] \qquad (28.10)$$

(27). This equation is also known as the Langmuir binding equation, which was derived for the absorption of gas molecules onto surfaces.

Equation 28.10 can be reformulated as:

$$[R]_{bound} = \frac{Ka\,[R]_{total} \cdot [L]}{1 + Ka\,[L]} \qquad (28.11)$$

Ka is an association constant. The corresponding dissociation constant Kd is 1/Ka. With this change, equation 28.12 becomes

$$[R]_{bound} = \frac{[R]_{total} \cdot [L]}{K_D + [L]} \qquad (28.12)$$

Table 28.1
Effect of Fluorophor/Protein (F/P)[a]

Ratios on the fluorescent quantum yield of fluorescein protein conjugates		
F/P	Rel. fluor. intensity	Intensity/(F/P)
6.9	3.6	0.52
12	5.8	0.48
18	4.5	0.25

[a]Goat-antimouse IgG fluorescein conjugates prepared from the same antibody under the same reaction condition were used to stain lymphocytes that had been prereacted with anti-CD4 antibody. Fluorescence was measured by flow cytometry. The staining intensity obtained with a directly fluoresceinated CD4 antibody preparation was set to 1.

Equation 28.12 describes the same binding behavior as equation 28.6 for the simple 1:1 case and binding constants can be determined as described.

Experimental Considerations

Cell surface receptor binding studies are performed experimentally by reacting a number of cells (n) with $[R]_{tot}/n$ receptors with the concentration $[L]_{total}$ of a fluorescent ligand. In the case of flow cytometric measurements, the resulting fluorescence of the cells is measured. It is reasonable to assume that the measured fluorescence is:

$$F = C \cdot [R]_{bound}/n.$$

where C is a constant. The concentration [L] of the free ligand is not directly available experimentally in most cases and cannot be calculated easily from $[L]_{total}$ and F, since C is not known. Sometimes, experimental conditions can be selected, i.e., very few cells in a large volume of ligand solution, where the free ligand concentration change due to receptor binding is negligible. Then, a binding constant K_D and the value of $C \cdot [R]_{total}$ can be obtained through Scatchard or Michaelis Menten analysis, provided the above models apply.

DETERMINATION OF C

C can be described as

$$C = f \cdot (F/P) \cdot (Q.Y.)$$

f is an instrument sensitivity factor,
F/P is the number of fluorophors per ligand (protein),
Q.Y. is the fluorescent quantum yield of the fluorophor.

When the assumption can be made that $(F/P) \cdot (Q.Y.)$ is stable and constant (for example, when a fluorophor is exposed to different chemical environments) C can be obtained from the measurement of fluorescent polymer particles with a known fluorophor content, as described for the calibration of flow cytometers above. However, the most commonly used fluorophor, fluoroscein, shows a significant dependence of its fluorescence yield on its chemical environment. Table 28.1 shows the relative fluorescence efficiencies of three immunoglobulin conjugates as an example.

OTHER BINDING MODELS

Many more binding mechanisms than the ones described above apply to the reaction of ligands with cell surface receptors. Bivalent binding of antibodies to cell surface antigens is one of many examples. A detailed mathematical treatment of additional models can be found in (27).

MODEL INDEPENDENT BINDING DATA ANALYSIS

As indicated above, binding studies make assumptions about the binding mechanism for the experimental design. Recently, an approach has been proposed that allows the determination of free and bound ligand for cell surface receptor studies independent of the binding mechanism (28). The method, called "isoparametric titration," is based on the assumption that the binding of a fluorescent ligand to a cell can be quantitated in relative units by flow cytometry. By measuring cell fluorescence (relative bound concentration) as a function of cell concentration and total ligand concentration, linear graphs of total ligand concentration vs. concentration of cells can be obtained. The ordinate intercept of such graphs yields the free ligand concentration at a given saturation level (constant cell fluorescence). By expressing all concentrations, including the cell concentration in molar units (1 Mole of cells is $6.023 \cdot 10^{23}$ cells), the slope of the lines is the average number of ligands per cell at a given saturation level. The slopes and intercepts at different cell fluorescence intensities can be used to determine binding constants by model dependent analysis. Figure 28.4 explains the technique step by step.

As one can see from Figure 28.4A, the isoparametric titration method depends on the accuracy of the data. Some of the curves fall close together and should not intersect. In addition, the method requires a series of several interpolations from derived curves to obtain the final answer, which could cause considerable error propagation. The method has been carefully evaluated only in one well-characterized monovalent receptor-ligand system (epidermal growth factor), where the labeled ligand was purified to a uniform F/P ratio. The application of isoparametric titration in calibrating results obtained by immunophenotyping remains, therefore largely untested, although preliminary results have shown the feasibility of the method (29). The method should, in theory, provide the best possible binding data whenever the required experimental precision can be obtained. Even though it is independent of any particular binding model, it is still dependent on a reaction in equilibrium; therefore, at least 10 halftimes should pass after mixing the components and before the measurements are performed.

PROBLEMS AND PITFALLS

Nonspecific Fluorescence

As discussed earlier, many labeled ligands show low-affinity, high-capacity binding to a variety of cellular molecules; this has no biological significance. This nonspecific binding (NSB) must be accounted for in determining the contribution

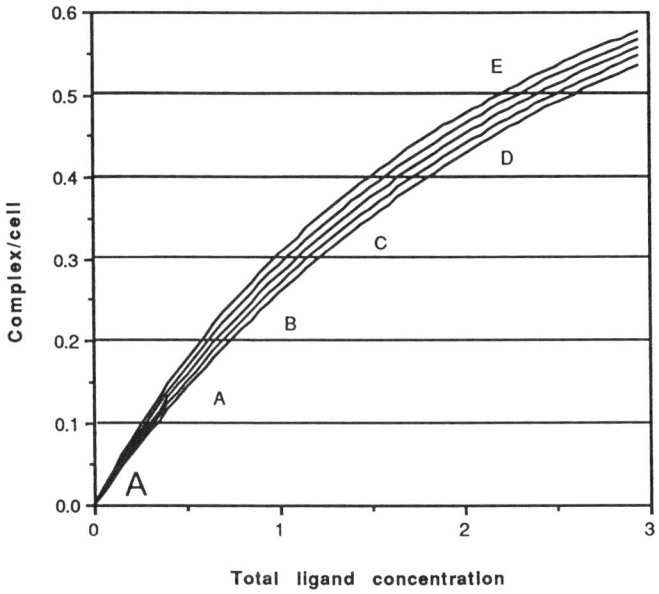
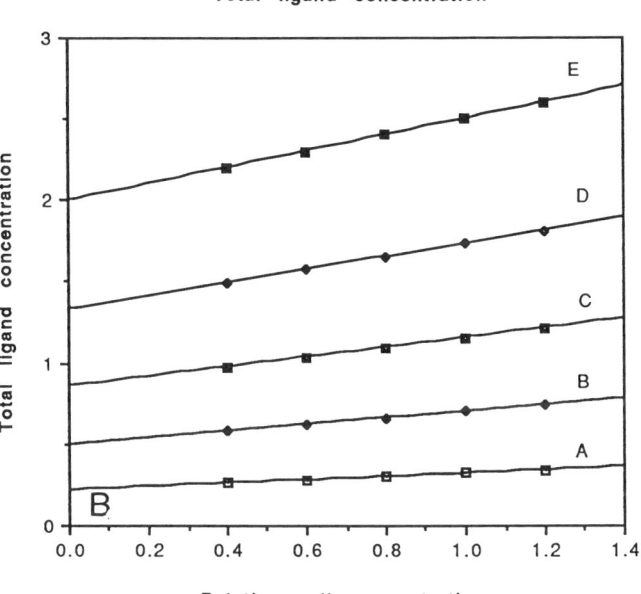
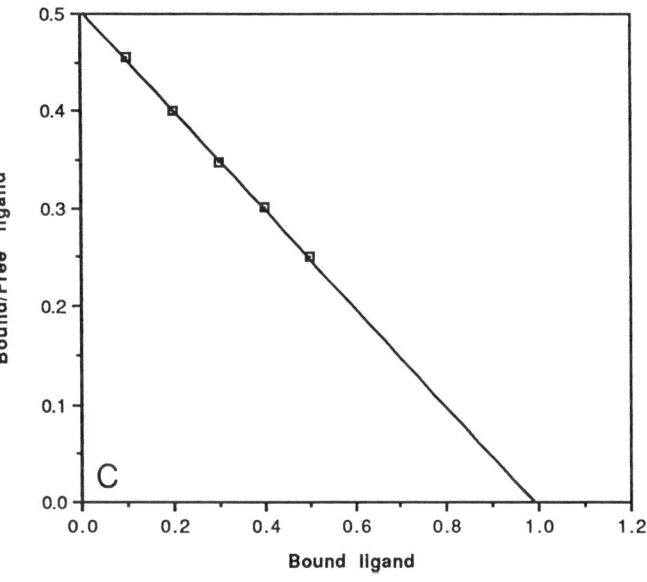

Figure 28.4. Isoparametric binding analysis according to Chatelier RC et al. (28). The lines in **A** were calculated assuming a 1:1 model and a K_D of 2.0. Measure the cell fluorescence as a function of *total ligand concentration* at several cell concentrations by flow cytometry. Correct for autofluorescence and nonspecific binding as described by Chatelier RC. (28). Use the graph of corrected cell fluorescence obtained by flow cytometry vs. total ligand concentration (**A**) to obtain total ligand and cell concentrations at different fluorescence levels (intercepts of lines *A, B, C, D, E* with the curves in **A**). Plot these results as total ligand concentration versus cell concentration to obtain lines *A, B, C, D, E* in **B**. (Intercepts and slopes: A 0.22, 0.10; B 0.50, 0.20; C 0.86, 0.30; D 1.33, 0.40; E 2.00, 0.50). Use the slope and intercept of lines *A, B, C, D* in **B** and use the results for the Scatchard analysis shown in **C** (Intercept = 0.503, slope = −0.51).

of specific (ligand-receptor) binding in the total binding observed. An unfortunate practice has developed in immunophenotyping of determining a "general" NSB by measuring the binding of an "irrelevant"-labeled antibody (i.e., one that is not directed against any cell surface protein). In fact, this approach cannot give a proper measure of NSB for any particular ligand-receptor system. The correct approach, long used in biochemical and immunochemical assays, is to compete against specific binding of the labeled ligand with an excess (usually 100-fold) of unlabeled ligand. While competition assays may not be required for routine measurements in every system, they should certainly be used in any careful evaluation of methods for surface protein measurements.

Fluorescence measurements are further complicated by the inherent fluorescence of certain cell molecules, particularly the flavinoids. This autofluorescence contributes to total fluorescence and, again, must be accounted for in determining fluorescence due to specific binding. The problem can become especially severe when low levels of receptor are being measured on activated cells that show considerable autofluorescence. Controls that contain an excess of unlabeled ligand should account for both nonspecific binding and autofluorescence, since the residual fluorescence measured will be a combination of the two.

High-Density Receptors

When receptors are present in very high density, it is possible to have self quenching of the fluorochrome molecules due to close proximity of adjacent fluorescent ligands. This effect does not necessarily require high numbers of ligands per cell, but only that they be in high density with mean distances between ligands of 100 Å or less (30). If staining is done under conditions that patch or cap the ligand, one should be alert to the possibility of reduced fluorescence (31).

Low-Density Receptors

If the ligand under study is present in too low a number per cell, it may not be possible to measure accurately the ligand fluorescence. In general, the limiting factor is cell autofluorescence. Since autofluorescence is usually more intense when blue excitation is used, it may be possible to obtain higher sensitivity by using higher wavelength excitation—i.e., 515 nm instead of 488 nm for FITC—or by using a fluorochrome, such as phycoerythrin (PE) or allophycocyanin (APC), which can be excited with green and red light, respectively (32, 33). In some cases, when instrument sensitivity rather than cell autofluorescence may be the limiting factor, one may be able to adjust the laser power or select filters for the best signal-to-noise ratio. It should be kept in mind that the best signal-to-noise performance may not be at the highest laser power, since fluorochromes often saturate at lower light intensities than autofluorescence. Low ligand number is particularly limiting when only a fraction of the cells express the receptor. If a subpopulation of cells positive for the receptor is not clearly resolved, then one must resort to models and assumptions about the relative contribution from the negative and positive cells in the fluorescence distribution (23).

Instrument Calibration

A method for calibrating (in absolute terms) or controlling (in terms relative to a stable sample) the flow cytometer is essential for quantitative measurements. The minimum requirement for being able to compare data from one day to the next is a stable sample, such as a bead or fixed biological sample, that can be used as a reference for each experiment within a day's run and for experiments performed on different days or even on different instruments. The more closely the stable sample resembles the real sample, the better it will control the instrument for the experiment being performed. For the quantitative measurements considered in this chapter, it is not particularly important that the stable sample be calibrated in terms of fluorescence intensity, since fluorescence intensity is not simply related to ligand binding. It is necessary, however, to be able to insure that the instrument is stable at least during the course of an experiment or that the change from an initial point be measured so that the data can be corrected for instrument changes.

CONCLUSIONS

The expansion of clinical and research applications for immunophenotyping by fluorescence flow cytometry has set the stage for the next phase in the evolution of this technology: the quantitative measurement of fluorescence signals and the binding events they signify. While the cytometers themselves are capable of very sensitive and precise fluorescence measurements, difficulties with the fluorochromes and the chemistry of ligand-receptor interactions remain impediments to a comprehensive and uniform system of standardized measurements. As these issues are addressed through continuing research and development, those who use flow cytometry should be aware of the unfolding potential as well as the pitfalls associated with quantitative fluorescence measurements.

REFERENCES

1. Shapiro HM. Quantitative immunofluorescence measurements and standards: practical approaches. Clin Immuno Newsletter 1991;11(4): 49–54.
2. Vogt RF, Cross GD, Phillips DL, Henderson LO. Interlaboratory study of cellular fluorescence intensity measurements with fluorescein-labeled microbead standards. Cytometry 1991;12:525–536.
3. Caldwell CW, Patterson WP. Relationship between T200 antigen expression and stages of B-cell differentiation in resurgent hyperplasia of bone marrow. Blood 1987;70:1165–1172.
4. Anderson KC, Bates MP, Slaughenhoupt BL, Pinkus GS, Schlossman SF, Nadler LM. Expression of human B-cell-associated antigens on leukemias and lymphomas: a model of human B-cell differentiation. Blood 1984;63:1422–1433.
5. Poncelet P, Lavabre-Bertrand T, Carayon P. Quantitative phenotypes of B chronic lymphocytic leukemia B cells established with monoclonal antibodies from the B cell protocol. In: Reinherz EL, Haynes BF, Nadler LM, Bernstein ID, eds. Leukocyte Typing II, vol. 2. New York: Springer-Verlag, 1986:229–343.
6. Duperray C, Klein B, Durie BGM, et al. Phenotypic analysis of human myeloma cell lines. Blood 1989;73:566–572.
7. van Bockstaele D, Berneman Z, Muylle L, et al. Flow cytometric analysis of erythrocytic D antigen density profile. Vox Sang 1986;51:40–46.
8. Matsui Y, Shapiro HM, Sheehy MJ, et al. Differential expression of T cell differentiation antigens and major histocompatibility antigens on activated T cells during the cell cycle. Eur J Immunol 1986;16:248–251.
9. Rudolph N, Ohlsson-Wilhelm B, Leary J, and Rowley P. Single-cell analysis of the relationship among transferrin receptors, proliferation, and cell cycle phase in K562 cells. Cytometry 1985;6:151–158.
10. Bohn B. Flow cytometry: a novel approach for the quantitative analysis of receptor-ligand interactions on surfaces of living cells. Mol Cell Endocrinol 1980;20:1–15.
11. Nance TN. Applications of flow cytometry in blood transfusion science. In: More, SB ed. *Progress in Immunohematology*. Arlington, VA: American Association of Blood Banks, 1988;1–29.
12. Bray RA, Lebeck LK, Gebel HM. The flow cytometric crossmatch. Dual-color analysis of T cell and B cell reactivities. Transplantation 1989;48:834–840.
13. Caldwell C, Maggi J, Henry L and Taylor H. Fluorescence intensity as a quality control parameter in clinical flow cytometry. Am J Clin Pathol 1987;88:447–456.
14. Vogt RF, Cross GD, Phillips DL and Henderson LO. A model system evaluating fluorescein-labeled microbeads as internal standards to calibrate fluorescence intensity on flow cytometers. Cytometry 1989;10: 294–302.
15. Brown MC, Hoffman RA, and Kirchanski SJ. Controls for flow cytometers in hematology and cellular immunology. Ann NY Acad Sci 1986;468:93–103.
16. Schwartz A. *Monograph: fluorescent microbead standards*. Flow Cytometry Standards Corp., Research Triangle Park, North Carolina. 1988.
17. Vogt RF, Marti GE and Schwartz A. Quantitative calibration of fluorescence intensity for clinical and research applications of leukocyte immunophenotyping by flow cytometry. In: Tyrer HW, ed, Critical Is-

sues in Biotechnology and Bioengineering, vol 1. Norwood, New Jersey: Ablex Publishing Company, (In press).

18. Loken MR and Herzenberg LA. Analysis of cell populations with a fluorescence activated cell sorter. Ann NY Acad Sci 1975;254:163–171.

19. de Bruin HG, de Leur-Ebeling I, Aaij C. Quantitative determination of the number of FITC-molecules bound per cell in immunofluorescence flow cytometry. Vox Sang. 1983;45:373–377.

20. Van Wauwe JP, De Mey JR and Goossens JG. OKT3: a monoclonal anti-human T lymphocyte antibody with potent mitogenic properties. J Immunol 1980;124:2708–2713.

21. Poncelet P and Carayon P. Cytofluorometric quantification of cell-surface antigens by indirect immunofluorescence using monoclonal antibodies. J Immunol Methods 1985;85:65–74.

22. Poncelet P, Bikque A, Lavabre T, et al. Quantitative expression of human lymphocytes membrane antigens: definition of normal densities measured in immuno-cytometry with the QIFI assay. Cytometry Suppl 1991;5:82–83.

23. Dux R, Kindler-Rohrborn A, Lennartz K, and Rajewsky MF. Calibration of fluorescence intensities to quantify antibody binding surface determinants of cell subpopulations by flow cytometry. Cytometry 1991; 12:422–428.

24. Fuller TC, Trevithick JE, Fuller AA, et al. Antigenic polymorphism of the T4 differentiation antigen expressed on human T helper/inducer lymphocytes. Hum Immunol 1984;9:89–102.

25. Kornprobst M, Rouger PH, Goosens D, et al. Determination of Rh(D) genotype: use of human monoclonal antibodies in flow cytometry. J Clin Pathol 1986;39:1039–1042.

26. Sklar LA and Finney DA. Analysis of ligand-receptor interactions with the fluorescence activated cell sorter. Cytometry 1982;3:161–165.

27. Connors KA. Binding Constants. New York: John Wiley & Sons, 1987.

28. Chatelier RC, Ashcroft RG, Lloyd CJ, et al. Binding of fluoresceinated epidermal growth factor to A431 cell subpopulations studied using a model-independent analysis of flow cytometric fluorescence data. EMBO J 1986;5:1181–1186.

29. Vogt RF, Ethridge SF and Henderson LO. Calibrated binding capacities of antibody-coated microbeads for labeled antibodies against the CD4 surface marker. Cytometry 1990;suppl 4:75.

30. Stryer L. Energy transfer: a spectroscopic ruler. Ann Rev Biochem 1987;47:819–846.

31. Cuchens MA and Buttke TM. A kinetic study of membrane immunoglobulin caping by flow cytometry. Cytometry 1984;5:601–609.

32. Loken MR, Keif, and Kelley KA. Comparison of helium-neon and dye lasers for the excitation of allophycocyanin. Cytometry 1987;8:96–100.

33. Shapiro H. Practical Flow Cytometry. Second Edition. New York: Alan R. Liss, 1990:156–157.

29

Detection of Minimal Residual Disease by Flow Cytometry

DANIEL H. RYAN

INTRODUCTION

The flow cytometer seems especially suited for the detection of rare events. It is capable of analyzing multiple parameters on hundreds of thousands of individual cells in a few minutes and requires relatively straightforward sample preparation. There are a number of potentially important clinical applications, such as the detection of bacteria in whole blood (1), identification of HIV-infected lymphocytes (2), analysis of fetal red cells or leukocytes in maternal blood (3), and detection of minimal residual disease in malignant neoplasms. In spite of this, there is as yet no routine clinical application of rare event analysis by flow cytometry. The major limitation in the past has been not the instrumentation, but the biological specificity of markers used to detect rare events. In recent years, however, there has been an increased understanding of differentiation events in leukemia that may now make it possible to develop a flow cytometry assay capable of providing useful clinical information based on detection of minimal residual disease. The discussion of minimal residual disease in this chapter is applicable to any flow cytometry based rare cell assay.

CLINICAL OVERVIEW OF MINIMAL RESIDUAL DISEASE

Why Monitor Residual Disease?

The detection of small numbers of residual tumor cells in patients who appear to be in complete remission has long been a "holy grail" of medical research. Many types of malignancies are responsive to current anticancer chemotherapeutic drugs, often resulting in a complete clinical remission. However, in some patients, residual tumor cells that are too few in number to be detectable by standard techniques may proliferate and later cause relapse. The consequence of this is that cancer chemotherapy must be continued in all patients for a period of time that is adequate to eliminate residual clonogenic tumor cells in as many patients as possible. In patients who have achieved complete remission, the physician is in effect "flying blind" in that he or she is treating a disease for which there are no clinical signs or symptoms by which to judge therapeutic effect. Given the same therapeutic regimen, some patients are cured, some relapse, and some suffer significant complications of therapy. Clearly, a test that would differentiate between patients who need more therapy from those who are already cured would offer the possibility of improved outcome.

EARLY DETECTION OF RELAPSE

Acute leukemia is a useful model to frame the questions that need to be addressed. At diagnosis, a patient with acute leukemia may be carrying a tumor burden of 10^{12}–10^{13} cells. Induction chemotherapy typically results in a rapid drop in the number of leukemic cells, which become morphologically undetectable when their number falls below 5% of the total bone marrow nucleated cells. This represents roughly 10^{11} cells. Below this level, the leukemic tumor burden can no longer be measured, and the patient continues to be treated according to protocol. If, at some point, the leukemic stem cells become drug resistant or are still present after discontinuation of chemotherapy, they may proliferate but are not detectable until once again 10^{11} abnormal cells are present. The question is: Would it make a difference if the relapsing tumor cells could be detected at a tumor burden of 10^8 cells, rather than 10^{11} cells? This hypothesis cannot be tested until a reliable method exists to detect such a small number of malignant cells.

EARLY DETECTION CELLS

In a similar fashion, if it were possible to measure the smallest number of tumor cells that could be responsible for clinical relapse, it would be possible to identify patients who were already cured of their disease and would not require further chemotherapy. In childhood acute lymphoblastic leukemia (ALL), this might reduce the duration of maintenance chemotherapy for some children. It remains to be seen whether this is a practical goal. Recent evidence suggests that neoplastic cells persist in childhood ALL throughout the first 18 months of the standard two years of maintenance chemotherapy (4).

EARLY ASSESSMENT OF CHEMOTHERAPEUTIC EFFECT

The rate at which tumor cells decrease during induction chemotherapy, as measured by bone marrow tumor burden at day 14 of induction therapy, has been correlated with ultimate clinical outcome in childhood acute lymphoblastic leukemia in the CCSG 160 series. Detection of residual leukemic cells by more sensitive methods may allow more accurate estimation of effectiveness of tumor-cell killing by induction chemotherapy as a prognostic variable.

EARLY DETECTION OF OCCULT METASTATIC DISEASE

Adjuvant chemotherapy is effective in reducing relapses in some patients with node-negative breast carcinoma (5), suggesting the presence of micrometastases in these patients at the time of treatment. Detection of such micrometastases may identify patients who would benefit from adjuvant therapy and spare others the risks of treatment. It is an open question, however, whether the presence of micrometastases in the bone marrow is an accurate reflection of micrometastases elsewhere.

ASSESSMENT OF TUMOR CONTAMINATION OF AUTOTRANSPLANTED MARROW

A potential problem with autologous bone marrow transplantation (BMT) in diseases such as neuroblastoma or acute leukemia is that residual tumor cells at undetectable levels may be returned to the patient. It is not clear that this represents an actual clinical problem. A retrospective comparison of data from multiple transplant centers suggests that "purging" protocols to remove residual tumor may be beneficial (6), but a controlled trial testing the effect of purging on relapse has not been done. Insofar as purging may represent a useful clinical maneuver, it would be a significant step forward to be able to monitor the success of the depletion procedures in individual pateint marrow samples.

What Sensitivity is Required for Clinical Usefulness?

Clearly, no currently conceivable in vitro method is capable of detecting a single tumor cell in a patient. Even if the contents of four bone marrow aspirates (about 3×10^8 cells) could be subjected to a method capable of detecting a single tumor cell in the entire sample, this would still represent at least 10,000 tumor cells in the entire bone marrow. However, these considerations do not necessarily defeat the purpose of residual leukemia detection for a number of reasons.

Only a small percentage of the total neoplastic population may represent stem cells (i.e., cells capable of self-renewal), which are the only cells relevant to patient survival. This percentage is not known for any human tumor, but may be in the range of 1 cell in 1,000–10,000, judging by in vitro colony assays. Therefore, detection of 10^8 total cells may represent only 10^4 stem cells.

Drug resistance is a major problem in the treatment of residual malignant disease. There is evidence that stable drug-resistant malignant cells arise spontaneously by mutation with a certain frequency, independently of the presence of a selecting agent (7). Therefore, assuming a fixed mutation rate, the number of drug-resistant mutant clones in a tumor is related to the number of cell divisions that have occurred, and therefore to the total tumor burden. This is the theoretical basis for the Goldie-Coldman model (8) of mutation to drug resistance in malignant tumors. In this model, the probability of emergence of multiple drug-resistant clones rises as tumor burden increases, particularly when the rate of tumor cell death or differentiation is substantial (9). If one assumes that 0.01% of the total leukemic cells are stem cells with an independent mutation rate for resistance to two different chemotherapeutic agents, then the likelihood of there being no doubly resistant stem cell (i.e., "curability") drops from 97.5% to 4% as the tumor burden rises from 10^8 to 10^9 cells (10). Thus, in this model, it makes little difference whether the relapse is treated at a tumor burden of 10^{12} vs 10^{11} cells, but it does make a difference whether treatment is begun at 10^{11} vs 10^8 cells.

In chronic myelogenous leukemia (CML) patients treated with bone marrow transplantation, residual disease is almost universally present in the first year after BMT. However, it is now becoming apparent that these cells, identified by PCR or even conventional cytogenetics, gradually disappear *without further treatment* in patients who go on to sustained remission (11). Therefore, measurements of residual disease that are quantitative in nature may be very useful in assessing whether the remaining tumor cells are proliferating or gradually disappearing for whatever reason.

The conclusion, at present, still must be that this question can only be answered by clinical therapeutic trials, and the answer is very likely to be different for each neoplasm.

Clinical Considerations Affecting Detection of Minimal Residual Disease

PHENOTYPIC SHIFT

One of the most important factors potentially limiting detection of relapsing malignancy is phenotypic changes from diagnosis to relapse. This may be a greater problem when multiple markers are needed to provide specificity for tumor cells as opposed to normal cells. In acute leukemia, phenotypic changes from lymphoid to myeloid phenotype occur, but very infrequently (4% of cases in a recent study (12)), while shifts in individual markers, such as from CD10-positive to CD10-negative, are somewhat more common (about 15% (13)). Some observed shifts may represent a growth advantage of one subclone over another, as has been observed in acute myeloid leukemia (AML) (14). Detailed analysis using multiple markers suggests that 40% of AML cases contain two phenotypic subpopulations, and 35% show three or more subpopulations. At a molecular level, alterations in gene rearrangement patterns in ALL were observed at relapse in seven of 11 patients, even though immunophenotype was unchanged (15). In four of these patients, at least one identical rearranged allele was found. These results suggest that the loss of a specific rearranged sequence, as detected by polymerase chain reaction (PCR), may occur at relapse in ALL, but that phenotypic markers may prove to be more stable.

ANATOMIC VARIABILITY IN LOCALIZATION OF RESIDUAL DISEASE

Variability in leukemic infiltration measured in bone marrow from different anatomical sites has been observed in patients with relapsing ALL (16) and has been documented to occur in the transplantable Brown Norway rat myelocytic leukemia (BNML) model. Martens has shown that the smaller the tumor burden, the larger the variability in numbers of tumor cells measured in different bones (17). Prior to chemotherapy of animals injected with BNML cells, the average ratio of highest to lowest frequency in 16 specimens from five different bones was about four. After suboptimal chemotherapy, minimal residual disease was detected by flow cytometry using a leukemia-associated marker at day 30, but the variation from sample to sample was large, averaging 7,000-fold. At day 31, larger numbers of residual cells were present, and the variability was reduced to 140-fold (17). This may be related to statistical variations in early repopulation of bones with small numbers of residual leukemia cells. A possible way to solve this problem is to study the peripheral blood, rather than the bone marrow. Circulating leukemic cells may better reflect total marrow tumor burden in the same way that the blood reticulocyte count is a better way of estimating total bone marrow erythroid production than is a bone marrow biopsy.

GROWTH RATE OF RESIDUAL TUMOR CELLS

The hypothesis upon which methods for early detection of tumor relapse is based is that a therapeutic advantage is derived from treating a smaller number of tumor cells. The clinical usefulness of residual tumor detection depends on the rate of progression from the minimal detectable tumor burden to a tumor burden at which this therapeutic advantage is lost. Recent evidence based on detection of T-ALL cells in the blood and bone marrow (18) suggests that a three-log increase in these cells occurs over an average of four months, allowing sufficient time for early detection of relapse by frequent monitoring of peripheral blood.

METHODS FOR DETECTION OF MINIMAL RESIDUAL DISEASE

No currently available assay fulfills all of the desirable criteria listed in Table 29.1 for a minimal residual disease assay, but we are much closer to these goals now than even two years ago. Some of these goals may be more clinically important than others. For instance, the time course over which residual tumor cells proliferate and become resistant to therapy is not known, but may be much longer than the turnaround of even cumbersome assays, which would be at most one to two weeks. On the other hand, an objective endpoint based on the use of defined reagents is likely to be critical for any test that must be performed routinely in large numbers of laboratories. It is becoming more evident that it will probably be clinically necessary to quantitate the level of residual disease, rather than simply providing a yes or no answer.

Table 29.1
The Ideal Test for Minimal Residual Disease

Technical
• Ability to detect rare populations
—low analytical and biological background
—enough positive events counted to achieve reliable result
• Not excessively labor intensive
• Objective endpoint
• Reasonably rapid turnaround time
• Potential for automation
• Quantitative
• Uses only reagents characterized at a molecular level

Clinical
• Predicts clinical outcome
• Forms a basis for therapeutic decisions in individual patients

As shown in Table 29.2, each method to detect minimal residual disease has technical strengths and limitations, which have been recently reviewed (10). The summary in Table 29.2 reflects only a best guess of the potential of each technique based on its performance to date and will be undoubtedly (hopefully!) proven to be an underestimate of the utility of one or more of these techniques. Note that the biological endpoint is not the same for each type of assay. For instance, immunofluorescence and flow cytometry have as their endpoint the detection of viable cells; cytogenetics, the division of cells in vitro; PCR, DNA alone; clonogenic assays, that cells divide at least five times in culture. The presence of nonviable or nonproliferating cells detected by PCR might not be as clinically relevant as the presence of colony-forming cells, for instance. It is possible that more than one test in combination may be required to provide the information necessary to make clinical decisions.

MORPHOLOGY

The analysis of cell shape (morphology), whether by light microscope, electron microscope, computer aided morphometry, or light scatter, lacks the requisite specificity to distinguish rare malignant cells from immature or reactive normal cells. Conventional morphology remains subject to variation in reagent quality, even though the major dyes that produce the Romanowsky effect have been identified, since the breakdown products of these dyes affect the staining quality.

SECRETED PROTEINS

Detection of secreted proteins by malignant cells is, in general, also severely limited by lack of specificity, although detection of recurrent cancer is possible if the tumor burden is moderately high (19).

ANEUPLOID DNA CONTENT

Technical limitations significantly limit the detection of rare aneuploid tumor cells. The detection of a rare aneuploid

Table 29.2
Some Technical Characteristics of Methods for Detection of Minimal Residual Disease

Technique	Cell Equivalents Typically Analyzed	Technical Limit of Sensitivity	Labor Intensity	Potential for Automation	Quantitation	Objectivity of Endpoint	Characterized Reagents	Specificity	Type of Event Identified
Morphology	100–1,000	0.1%	low	low	yes	low	no	very low	viable cells
DNA ploidy analysis	10,000–50,000	0.5%	low	high	yes	high	yes	high	intact cells
Cytogenetics	20–100	1%	high	low	yes	high	yes	very high	cells dividing in vitro
FISH	100–1,000	1%	high	mod. low	yes	mod. high	yes	high	intact cells
Southern blot	1,000,000	2%	mod. high	mod. high	semi-quant	high	yes	high	DNA fragment
PCR	1,000,000	<0.0001%	mod. high	high	no	high	yes	extremely high	RNA, DNA fragment
Flow cytometry	50,000–1,000,000	0.1–0.01%	low	high	yes	high	yes	variable	viable cells
Immunocytochemistry	10,000–100,000	0.1–0.001%	high	mod. low	semi-quant	moderate	yes	mod. high	viable cells
Colony assays	200,000	0.0005%	high	low	yes	mod. low	variable	low	clonogenic cells

stemline is limited by debris (when DI < 1.0), normal cells in S phase (1.0 < DI < 2.0), or clumps and debris (DI>2.0). Detection of an aneuploid stemline at a < 1% level is very difficult.

DNA TRANSLOCATIONS OR REARRANGEMENTS

Conventional Cytogenetics. The conventional method for detection of translocations is visualization of metaphase chromosomes using banding techniques. This method is labor-intensive, both in terms of analysis time and technical skill required, and is not suitable for screening thousands of cells. Another limitation is that mitosis is required to produce a karyotype, and leukemia cells frequently demonstrate less proliferative activity in vitro than do normal cell types, thereby obscuring the quantitative relationship between normal and abnormal cells in the sample. Fusion of malignant cells with cycling cells may address this problem, but such premature chromatin condensation assays (20) are unlikely to be technically suitable for detection of rare cells. The effect of culture conditions on number and type of tumor cells that grow out will remain unpredictable until more is known about the growth factor requirements for optimal proliferation of malignant cells.

Chromosome Analysis in Flow. Metaphase chromosomes can be directly identified by two-color flow cytometry using pairs of dyes, such as chromomycin A3 (specific for GC base pairs) and Hoechst 33258 (specific for AT base pairs), to improve resolution of the larger human chromosomes (21). Translocations that alter the DNA content or average base pair composition of chromosomes, such as t(9; 22), the Philadelphia chromosome, can be detected by this method. Unfortunately, the abnormal chromosome is usually not separable enough in DNA content from the normal chromosomes or debris to allow detection of rare abnormal cells (22).

Fluorescence In-Situ Hybridization (FISH). Individual chromosomes can be labeled in interphase cells using DNA probes for chromosome-specific repetitive sequences (23). This is a potentially important advance, since it eliminates the need for induction of mitosis in the tumor cells. In a study of patients post-BMT (24), detection of residual host cells by FISH with X or Y chromosome probes was strongly correlated with graft failure or leukemic relapse. Conventional cytogenetics failed to detect host cells in over 75% of the cases in which FISH was positive. This technique was also capable of detecting rare leukemic cells two months before relapse in a patient with AML using a chromosome 7 probe. This specific application is limited to those malignancies with loss or gain of a whole chromosome, and may be fairly time-consuming if one is screening tens of thousands of cells rather than hundreds, as in this study. It is also limited by the occurrence of aneuploidy in 0.1–1% of normal cells (25). Using two different fluorochromes to label probes for two chromosomes, it is possible to identify chromosome translocation by colocalization of the two colors. For rare cell detection, this technique is currently limited by the frequency with which the two separate normal chromosomes will overlie each other in the visualized nucleus, which occurs in about 1% of the cells (26).

Southern Blot DNA Hybridization. If a known gene lies within 5–10 kb of a translocation breakpoint, a population of cells bearing the translocation can be identified by Southern blot of DNA digested with restriction endonuclease and probed for the marker gene. Clinical examples include *bcr/abl* in chronic myelogenous leukemia (CML) and *bcl-2* in follicular lymphoma. The same diagnostic principle is used to detect clonal rearrangements of the immunoglobulin (Ig) and T-cell antigen receptor (TcR) genes in lymphoid neoplasms (27). Since the detection system involves analysis of total DNA from a clinical sample, sensitivity depends on the ability to detect a particular band among a diffuse background of numerous bands arising from normal clones. The detection limit is about 5%, too low to permit the detection of residual lymphoma in peripheral blood prior to relapse (28), although pre-enrichment steps may increase this to 0.1% (29). An additional consideration is the limited quantitative accuracy of the technique, especially if combined with pre-enrichment steps.

Polymerase Chain Reaction. The most sensitive assay for detection of gene translocation or rearrangement is polymerase chain reaction (PCR), which has a sensitivity of up to 0.001%. This technique requires knowledge of the DNA (or cDNA) sequence near the breakpoint site in order to construct probes for DNA amplification ("primers") as well as a probe to detect the amplified DNA. Translocations such as *bcr/abl* and *bcl-2* are suitable for PCR, since the breakpoint site is limited to one or two very circumscribed regions in nearly all patients or occurs within an intron, allowing cDNA amplification by PCR. Application of PCR technology to Ig or TcR gene rearrangements requires sequencing of the rearrangement site in the original tumor from each patient, since each rearrangement will be slightly different, depending on the particular variable (V), diversity (D), and joining (J) region used in the neoplastic clone, as well as on the insertion of additional nucleotides in N regions. It should be noted that PCR can itself be used to generate this sequence. PCR has the significant advantages of having molecularly characterized reagents, a fairly unequivocal endpoint, significant potential for automation, minimal sample requirement, and low background. A major disadvantage is the difficulty in quantifying the results. A simple, reliable method of calibrating the amplification achieved by a given PCR run on a given sample would be a major advance in the clinical application of this technique. When PCR is run for more than 30 cycles, as is required for detection of rare cells, the final amount of amplified DNA sequence is not proportional to the starting amount. Methods are being developed for relative quantitation of messenger RNAs (30) (competitive PCR), but they remain somewhat technically cumbersome.

It has been amply demonstrated that PCR provides more sensitivity than any known method for detection of a rare cell population. Sensitivities of 0.001–0.0001% have been documented in mixing experiments with cells positive for *bcr/abl* (31), *bcl-2* (32), and clonal Ig or TcR rearrangements (33). Two-stage PCR assays, which use a second set of primers to further amplify DNA sequences, demonstrate sensitivities on the higher end of this range, but may be more subject to false-postive results. The extraordinary sensitivity of PCR makes it susceptible to false-positive results due to the contamination of samples with trace amounts of DNA used for positive controls. Several technical precautions have been suggested to limit the occurence of false-positive PCR results (34). Even with implementation of these recommendations, it may be difficult to avoid occasional false-positive results. Another approach to deal with this problem is to split all samples into two aliquots, one of which is held in reserve to be analyzed independently in case the other is positive. PCR detects specific DNA sequences, and hence may be positive when only DNA fragments, and not intact cells, are present. This may be of clinical significance in the early phases of chemotherapy, when dead cells and debris are present.

CELL MARKERS

Flow Cytometry. Identification of rare cells (0.01–0.001%) is technically feasible with instruments already in use in thousands of clinical laboratories, and monoclonal antibodies can be readily standardized to insure reagent uniformity. A major advantage of flow cytometry is the ability to easily perform multiparameter analysis using antibodies labeled with different fluorochromes. The major limiting factor in the application of flow cytometry to minimal residual disease has been the lack of specificity of cell surface neoplasia-associated markers. Another significant limitation, particularly in bone marrow, is the nonspecific binding of antibody to various types of mature cells.

Immunofluorescence Microscopy. Detection of cell markers by immunofluorescence microscopy (or immunocytochemical techniques) offers the advantage that morphological parameters can be used, for instance, in recognizing false-positivity on cell aggregates, debris, and mature myeloid cells. Thousands of cells can be screened by a skilled operator using a good microscope with a 45× oil power objective. However, these techniques are not as objective and require significantly more operator time than an equivalent flow cytometric measurement. Automation of the screening procedure may enhance the practicality of this technique (35). Multiparameter analysis is possible with these methods, but is best interpretable when an intracellular marker is combined with a surface marker.

Intracytoplasmic and intranuclear markers are very useful in identifying certain residual leukemic populations, but present the additional difficulty of increased background staining and nonspecific binding in comparison with surface staining of viable cells. For the important intranuclear lymphoid marker terminal deoxynucleotidyl transferase (TdT), manual immunofluorescence microscopy has the advantage that the characteristic speckled structure of the nuclear staining can be evaluated in order to reduce the number of false-positive events. Potentially, however, the speed and objectivity of flow cytometry makes it a worthwhile goal to adapt these assays to flow analysis.

CLONOGENIC ASSAYS

All of the assays for minimal residual disease discussed so far will detect any intact cell, or even any isolated DNA fragment, as positive if the distinguishing tumor marker is present. Only clonogenic assays can distinguish cells that are capable of proliferation. The greatest difficulty is to distinguish the tumor cells from the more numerous normal progenitor cells, which may grow or aggregate under the culture conditions. This is a significant problem in leukemia since there is substantial overlap between the growth requirements of leukemic and normal hematopoietic progenitors (36). Specificity for residual malignant disease requires that some additional characteristics of the colony be determined. Detection of residual leukemic colonies using cytogenetics (37), surface markers (38), or DNA sequences identified by

PCR (39) is technically feasible but impractical for routine clinical diagnosis at present. A vexing problem with clonogenic assays of residual disease has been the fastidious growth requirements of many abnormal cells. Clonogenic assays for certain malignancies require accessory cells or conditioned medium to support the growth of the plated tumor cells (40). Until such reagents can be characterized at the molecular level, these assays will be difficult to bring into clinical practice. Heterogeneity of cells in a given type of malignancy with regard to growth factor requirements has been clearly shown for multiple myeloma (41), AML (42), and ALL (43), making it difficult to predict which growth factor to use in which patient. It should be recognized that the cells that can be driven into cell cycle in an in vitro environment may not represent true tumor stem cells, but only cells capable of responding to the special microenvironment created in the assay system.

MARKERS OF MINIMAL RESIDUAL DISEASE

The Elusive Goal of Tumor Specificity

The detection of residual tumor cells requires, first of all, a characteristic by which the neoplastic cells can be distinguished from nearly all resting and regenerating normal cells in the tissue being studied. It is clear from Table 29.2 that there are several techniques that are technically capable of detecting a rare population of 0.1–0.001% positive cells. However, routine clinical monitoring of minimal residual disease has not been prevented as much by technical limitations as by the lack of truly tumor-specific markers. The hope of identifying single tumor-specific antigens with monoclonal antibodies has been unrealized, as "tumor-specific" markers turned out to be differentiation- or proliferation-related proteins of general biochemical importance, as in the case of the neutral endopeptidase, CD10 (44). Combinations of markers to identify an aberrant phenotype show more clinical potential.

This effort has been most productive in the leukemias, since the normal pattern of differentiation is best understood in the hematopoietic system. There are three ways in which an atypical phenotype may be relatively specific for an abnormal cell: 1) The markers expressed are lineage-appropriate, but are not expressed in the proper combination (asynchronous differentiation); 2) one or more markers from two or more different lineages are expressed (mixed lineage differentiation); and 3) the phenotype is found in some normal cells, but not in those in the tissue being examined (abnormal tissue localization). There are several important biological problems raised by the occurence of mixed lineage and asynchronous differentiation in leukemia (45). Use is being made of all three possibilities in devising assays for minimal residual leukemia.

ASYNCHRONOUS DIFFERENTIATION

It has been clinically useful to classify neoplasms based on morphologic and phenotypic resemblance to some stage of normal differentiation. Recently, detailed phenotypic "maps" of normal erythroid (46), B-lymphocyte (47, 48), and myeloid (49) differentiation in the bone marrow have been constructed based on two- and three-color flow cytometric analysis using monoclonal antibodies to cell surface proteins that are classified according to the CD nomenclature (50).

When leukemic cells are compared to their putative normal counterparts based on these detailed descriptions, differences in patterns of marker expression are observed. Studies of human B-cell precursors suggest that as normal lineages are described with greater detail and subtlety, differences between the phenotype of neoplastic cells and their normal counterparts become more evident. In normal B-lymphocyte differentiation, there is an ordered progression of display of B-cell-associated markers, with CD10, CD34, and TdT decreasing with differentiation as CD20, cytoplasmic Ig μ chain (cyto μ), and CD45 increase. In contrast, the expression of CD20, TdT (51), and CD34 (52) by the neoplastic counterparts of these B-cell precursors (B-cell precursor ALL) is random with respect to each other and to CD10 or cyto μ. For example, the phenotype TdT+/cyto μ+ is found in <0.5% of normal marrow CD10+ B-cell precursors, but is present in 38% of patients with CD10+ B-cell precursor acute lymphoblastic leukemia. Similar observations have been made in myeloid leukemia when detailed phenotypic analysis is performed (53). Although the TdT+/cyto μ+ phenotype is not sufficiently specific to allow detection of rare ALL cells in bone marrow, some of the asynchronous AML phenotypes, in particular CD34+/CD15+, may be sufficiently atypical to permit detection of 0.01% residual cells (14).

Neoplastic cells are often viewed as "frozen" or blocked at a particular stage of differentiation. Another perspective suggested by recent literature is to consider cells as having multiple "differentiation programs." Such programs proceed in synchrony in normal cells, but may be blocked at different stages in neoplastic cells, leading to phenotypically asynchronous differentiation (45).

MIXED LINEAGE DIFFERENTIATION

Mixed lineage differentiation has been clearly documented in about 15% of B-precursor ALL cases that coexpress myeloid markers CD11b, CD13, or CD33 (54) with the lymphoid marker TdT. Multiple myeloma B-cells frequently express several myeloid associated markers, such as CD13, CD15, and CD14 (55). Conversely, expression of the lymphoid-associated marker TdT is fairly common (15–20%) among patients with AML (56). If small subpopulations of TdT-positive cells are sought in AML patients, 75% of cases can be shown to possess at least 0.1% TdT+/myeloid antigen+ cells (57).

The hypothesis of "lineage promiscuity" proposed by Greaves et al. (58) suggests that early progenitor cells express multiple markers thought to be lineage specific, losing them gradually as lineage committment takes place. However, the evidence thus far suggests that early progenitors express fewer lineage-associated markers such as CD33 or CD13, than do later progenitors (CFU-GM). Multiparameter flow cytometry has shown little evidence of expression of markers found on mature myeloid lineage cells (CD14, CD11b, CD16, CD15 [59, 60]) by CD34+ cells, which have been shown to contain virtually all cells capable of reconstituting hematopoisesis (61). Bone marrow TdT+ lymphoid precursor cells do not appear to express the myeloid-associated markers CD13, CD33, CD14, or CD15 (57, 62). The data so far tend to support the alternate hypothesis that a leukemic cell coexpressing myeloid- and lymphoid-associated markers represents aberrant differentiation as opposed to expansion of a very rare normal cell type. However it should be recognized that this hypothesis is very difficult to prove, due in large part to inherent difficulties in performing multiparameter analysis on very rare cell populations.

Whether leukemic cells expressing mixed lineage differentiation represent expansions of very rare normal cells or aberrant differentiation, they are sufficiently distinctive in some patients to allow the detection of residual leukemic cells using CD13 or CD33 in combination with TdT. These assays are currently being used to evaluate the occurence of residual disease in acute leukemia patients cells coexpressing TdT and a myeloid marker (about 15% of acute leukemia), with initial promising results (63). Recent evidence that ALL colony forming cells are CD33+ even in those patients in whom the bulk leukemic cells are CD33− (64) suggests that assays for residual TdT+/CD33+ cells may be applicable even to cases not obviously showing mixed lineage differentiation.

MARKERS OF ABNORMAL TISSUE LOCALIZATION

In spite of the dearth of markers absolutely specific for malignancy, there are several circumstances in which tumor associated markers are found on normal cells in some tissues, but not in the tissue being analyzed. This approach has had some success in identifying small numbers of neoplastic cells in bone marrow or blood. The thymocyte phenotype of T-cell ALL (TdT plus a T-cell marker) is characteristic of most thymocytes, but exceedingly rare in normal blood or bone marrow. CD2+/TdT+ or CD7+/TdT+ cells can be detected in small numbers in bone marrow (<0.3%), and may represent prothymocytes (65), but cytoplasmic CD3 (cCD3) and CD5, which appear later than CD2 and CD7 in T-cell development, are expressed by less than 0.01% of blood or bone marrow cells (65). Preliminary data from van Dongen's group using an immunofluorescence microscopy assay indicate that over 90% of T-ALL patients who do relapse show detectable abnormal cells in both blood and bone marrow an average of 15 weeks prior to clinical relapse (18). All 15 patients without detectable residual cells remain in remission more than four years from diagnosis. Results from Campana et al. (63) with a similar assay are also encouraging: All 19 patients with detectable CD5+/TdT+ (T-ALL) or CD13+/TdT+ (mixed lineage leukemia) cells in the bone marrow relapsed after an average of 14.5 weeks, while 70% of the 23 patients without detectable residual cells remain in clinical remission (median observation time of seven months since last assay). Two of the patients not detected by the assay suffered isolated CNS relapse.

By first staining with CD5 antibody followed by anti-TdT after paraformaldehyde fixation, rare CD5+/TdT+ cells can be identified in bone marrow by flow cytometry (66). Using this assay, the upper limit of CD5+/TdT+ cells in bone marrow is 0.13% in resting marrow and 0.04% in regenerating marrow, as opposed to <0.01% using the same markers in a manual immunofluorescence assay. The reason for this discrepancy is unclear, and may be related to the ability to visually identify and discount debris or aggregates with manual microscopy or to increased sensitivity of the flow cytometry assay to weakly staining cells.

Identification of metastatic epithelial tumor in bone marrow has been studied extensively using epithelial-associated markers. Redding et al. (67) used epithelial membrane antigen (EMA) in an immunohistochemistry technique to identify metastatic breast carcinoma in the bone marrow and observed that 26% of node-negative patients with apparently localized disease showed metastatic disease using this technique. Later follow-up shows that detection of marrow micrometastases has a modest predictive value for later bony relapse, but many patients without detectable tumor in the bone marrow do relapse as well (68). EMA and another epithelial marker, HMFG, may not be the ideal markers for future investigations since it has become evident that they react with occasional plasma cells and other bone marrow elements, to a level of 5–10% of cells in some patients (69). Several studies have suggested that cytokeratin is a much more specific marker in bone marrow (70), detectable on <0.001% of normal bone marrow cells, but present on most breast, colon, and small-cell lung tumors.

A flow cytometry assay capable of detecting small numbers of metastatic neuroblastoma cells has been described by Frantz et al. (71). Using phycoerythrin-conjugated antibodies to three myeloid markers and CD45, over 70% of the normal bone marrow cells, including nearly all of the cells with high 90° light scatter, fluoresce red and can be gated out, reducing the potential for background binding with FITC-conjugated antibody to the neuroblastoma-associated marker GD2. The number of cells in normal marrow that are negative for the myeloid markers but are GD2+ in the intensity range of neuroblastoma is in the range of 0.005%. Using this assay, residual disease has been identified during clinical remission in a patient four weeks prior to relapse. The major problem with flow cytometry assays for detection of residual metastatic solid tumor is the propensity of the tumor cells to clump and hence be lost during washing steps or not included in the light scatter gates during flow cytometry analysis. Clumping of tumor cells is obvious even in routine bone marrow smears from patients with metastatic disease and is increased by centrifugation steps during staining of the cells.

MARKERS OF CLONALITY

Since neoplastic proliferations are generally monoclonal, and normal cell populations polyclonal, identification of a substantial monoclonal population in a tissue is suggestive of the presence of a neoplasia (although not proof of malignancy!).

Neoplastic B-cells generally show a restricted intensity of expression of surface Ig compared with the characteristically broad range of intensity of normal surface Ig bearing B-cells (72). Careful mathematical analysis of fluorescence intensity distributions of B-cell populations stained for κ and λ light chain can detect the presence of a 2% to 5% population of clonal cells. Combination of this technique with specific B-cell markers to reduce background staining by non-B-cells has been reported to increase this sensitivity to better than 1% (73).

As discussed above, clonal gene rearrangements of the Ig or TcR gene can be used to detect neoplastic lymphoid cells. A potential biological limit to sensitivity is the presence of normal clones with the same rearrangement as the leukemic cells. PCR studies have shown, however, that normal cells with a rearrangement identical to a given leukemic clone are not found at a detection limit of 0.0001–0.001%. PCR assays fail to detect any cells bearing the t(14;18) *bcl-2* rearrangement in normal tissues. Therefore, assays based on sensitive detection of clonally rearranged DNA remain very specific for a given neoplastic clone. Unfortunately, information about specific cells is lost if total cellular DNA is examined. A method for reliably performing in situ PCR would be especially useful in this regard (73a).

FUTURE DIRECTIONS IN FLOW CYTOMETRIC DETECTION OF RESIDUAL DISEASE

At present, no flow cytometry assay has been shown to be capable of predicting relapse by identifying residual malignant cells in blood or bone marrow of patients in morphological remission. What has contributed to the difficulty in establishing this much needed clinical assay?

Methodological Considerations

NONSPECIFIC BINDING OF ANTIBODIES

Nonspecific binding of antibody reagents is a problem particularly in bone marrow, where significant numbers of myeloid and monocytic cells bearing Fc receptors are present. Three types of human Fc receptors for IgG have been identified (74). All of these receptors will avidly bind aggregated murine or human IgG, but only FcRγ I binds significant amounts of monomeric immunoglobulin. Therefore, it is important to reduce the amount of aggregated Ig in the reagents used for these types of assays. Figure 29.1A shows the effect of spinning a commercial murine IgG1 FITC-conjugated control antibody at 14,000 g for 15 min (to remove aggregated Ig) on the background binding to peripheral blood mononuclear cells with low 90° light scatter. In order to more clearly portray the proportion of cells above a given fluorescence intensity, this figure shows the *cumulative* histogram of fluorescence intensity on the X-axis versus the total number of cells above this level of fluorescence on the Y-axis. Note that the shape of the cumulative histogram at lower levels of fluorescence is not greatly affected; it is only the number of relatively rare background events that is significantly reduced by spinning the antibody. Centrifugation of antibodies is thus very useful in assays used to detect rare residual cells, even though it is not generally necessary for routine clinical studies. The type of cells and antibody reagents used also affect the degree of nonspecific binding observed. IgG Fc receptor type I binds irrelevant monomeric murine *IgG2a* or IgG3, but not *IgG1* or IgG2b antibodies. Therefore the type of antibody used will affect the degree of nonspecific binding observed in a heterogeneous population of cells, as shown in Figure 29.1C. Monoclonal antibodies generally show much lower nonspecific binding than polyclonal antibodies, including F(ab) fragments.

In hematopoietic cells, the background fluorescence intensity observed with PE-conjugated control antibody is similar to that observed with FITC-conjugated controls using the appropriate filtration for each fluorochrome, as shown in Figure 29.1C. There is an advantage to using PE in cells with significant autofluorescence since natural autofluorescent compounds are poorly excited at longer wavelengths. However, hematopoietic cells, and in particular lymphocytes, appear to show very little autofluorescence.

In order to determine whether the difference between the number of positive events in the stained sample as compared with the control is significant, it is necessary to know the variability associated with these determinations. Figure 29.2A shows the cumulative histograms of four replicate samples stained with the same conjugated control IgG1 preparation. The mean ± two standard deviations (95% confidence limits) is shown in Figure 29.2B. The variability in this figure can be compared with the 95% confidence limits associated with the Poisson statistical variability (Fig. 29.2C) in enumerating rare events. Note that Poisson statistics determine a large proportion of the actual experimental variability when few positive events are acquired.

Figure 29.2D shows the mean experimental cumulative histogram compared with a cumulative histogram derived from an idealized normal distribution with the same mean fluoresence intensity. Note that the background level of positive cells varies considerably with the cutoff value for fluorescence intensity, but the relationship between these two variables is not simple. There appears to be an inflection point in the curve at about channel 110. Below this point, background decreases dramatically with even slight increases in fluorescence cutoff value; above this point, the reduction in background with increasing fluorescence intensity is more modest. The experimental histogram is strikingly similar to a normal distribution below this point, but displays

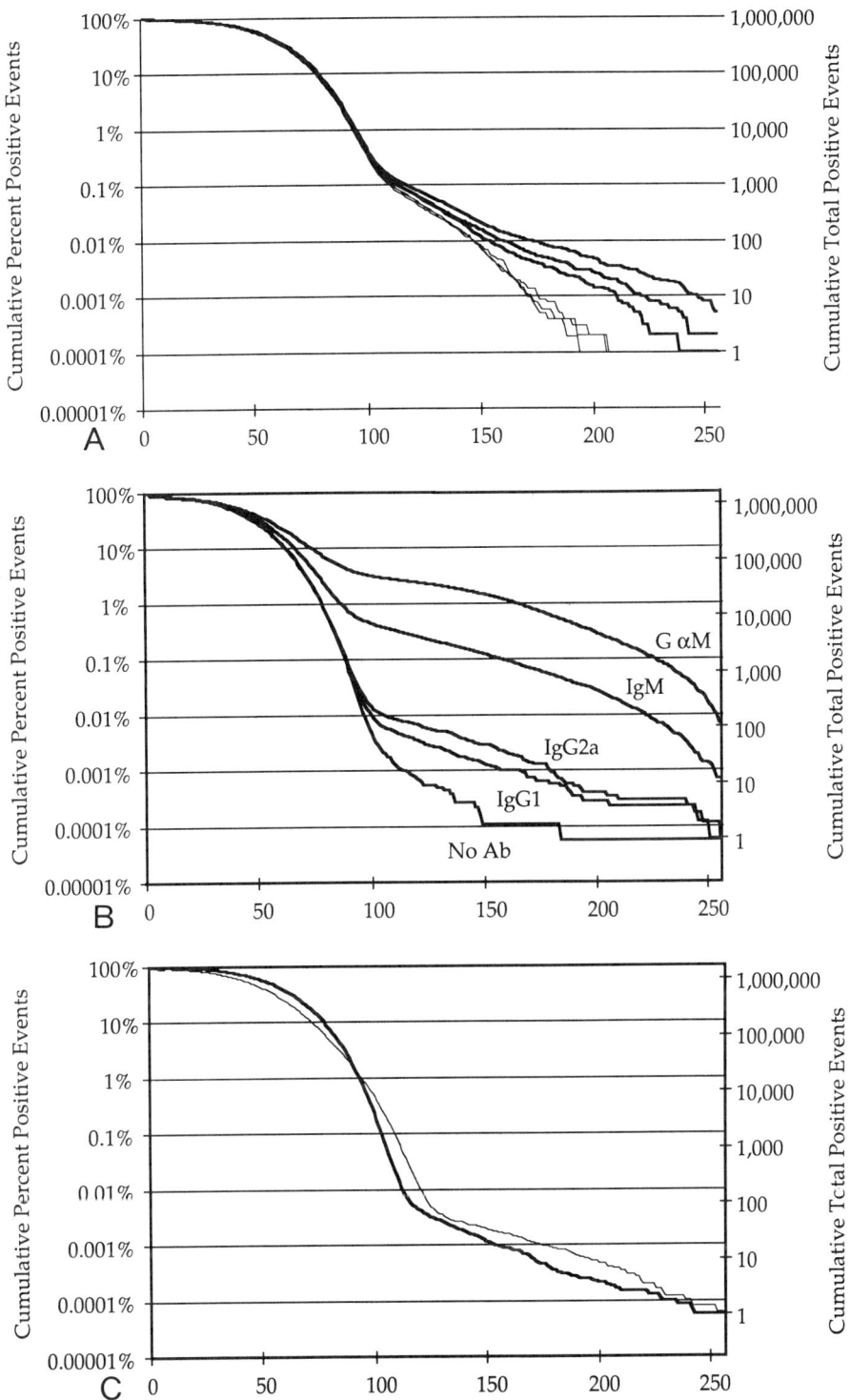

Figure 29.1. Cumulative histograms of fluorescence intensity (X-axis) vs. percentage of cells with fluorescence intensity above this level (Y-axis-left). On the Y-axis at right is the absolute number of cells with fluorescence intensity above the level indicated on the X-axis. This representation differs from conventional flow cytometry histograms in that the quantity on the Y-axis is not cells per channel, but all cells *above* a given channel. Mononuclear cells were separated by Ficoll/hypaque from the peripheral blood of a normal subject and stained for analysis on an EPICS C flow cytometer. Approximately 1,800,000 total cells were analyzed per condition. **A,** Triplicate aliquots of cells were stained with FITC-conjugated mouse IgG1 control antibody (Coulter Immunology) that was spun at 14,000 g for 15 min (light lines) or unspun (bold lines). **B,** Cells were stained in duplicate with either no antibody, FITC-conjugated irrelevant mouse *IgG1*, FITC-conjugated irrelevant mouse *IgG2a*, FITC-conjugated irrelevant mouse *IgM* (all from Coulter Immunology), or FITC-conjugated polyclonal goat anti-mouse IgG (Tago). In all cases, antibodies were spun at 14,000 g before use. **C,** Cells were stained with both FITC-conjugated irrelevant mouse IgG1 and PE-conjugated irrelevant mouse IgG1 (Coulter Immunology) with standard filtration for FITC and PE. The cumulative histograms shown are the mean of triplicate determinations of the green fluorescence (bold lines) or red fluorescence signals (light lines).

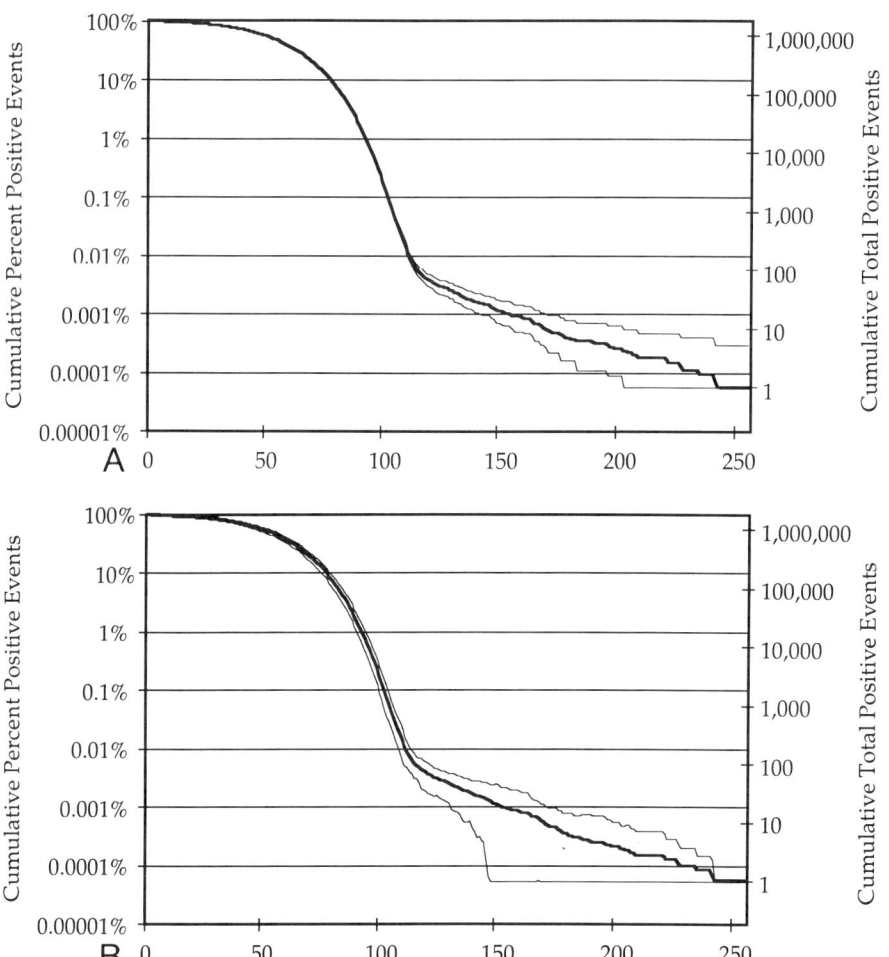

Figure 29.2. Cumulative histograms of cells prepared and stained as described for Figure 29.1 (all antibodies spun before use). **A,** Four replicate samples separately stained with FITC-conjugated irrelevant mouse IgG1. Note the increasing variability at higher fluorescence channels with smaller numbers of positive events counted. **B,** Mean values (bold) from histograms in **A** for each channel and additional histograms representing two standard deviations above and below the mean of cumulative values for each channel. The standard deviations were derived from the actual data in **A**. **C,** Mean values for each channel (bold) and additional histograms representing two standard deviations above and below the mean of cumulative values for each channel-based only on the minimum statistical variation in these values expected by the Poisson distribution of rare events. Note the similarity to the actual variation shown in **B**. **D,** The mean histogram shown in **B** and **C** superimposed on a cumulative histogram derived from an idealized Gaussian distribution with a similar mean and variance. Note that the two histograms are similar up to channel 110, where the experimental histogram shows an inflection in the curve. The range of fluorescence intensity of several commonly used hematopoietic markers is shown on the X-axis for comparison.

a significantly greater number of positive events above this fluorescence intensity in the range of expression of some commonly used lineage-associated markers, as shown on the X-axis of Figure 29.2D.

This suggests that careful attention needs to continue to be directed toward the characteristics and preparation of monoclonal antibodies used for detection of rare events, based on already existing knowledge. The shape of the cumulative histogram suggests that the rare, very fluorescent events observed when large numbers of events are analyzed may not simply represent the "tail of the curve," but distinct populations. The number of cells in this part of the control fluorescence histogram may vary significantly from person to person, as shown in Figure 29.3, even when the same light-scatter gating strategy is used.

Physical depletion of monocytes and myeloid cells in the sample prior to staining and flow cytometry analysis could potentially reduce the number of nonspecifically binding cells, but at a cost of increased manipulation of the cells, which is likely to promote cell clumping. Multiparameter flow cyometry offers the possibility of gating out cells responsible for nonspecific binding with one color and identifying the malignant cells using one or two additional colors (71, 75, 76). This technique reduces, but does not eliminate, the variable number of nonspecific positive events observed from sample to sample.

Enhancement of signal intensity would improve detection of rare events, but only if the nonspecific binding is kept constant or reduced. Use of fluorescent beads tagged with anti-Ig antibody provides a 200-fold increase in signal inten-

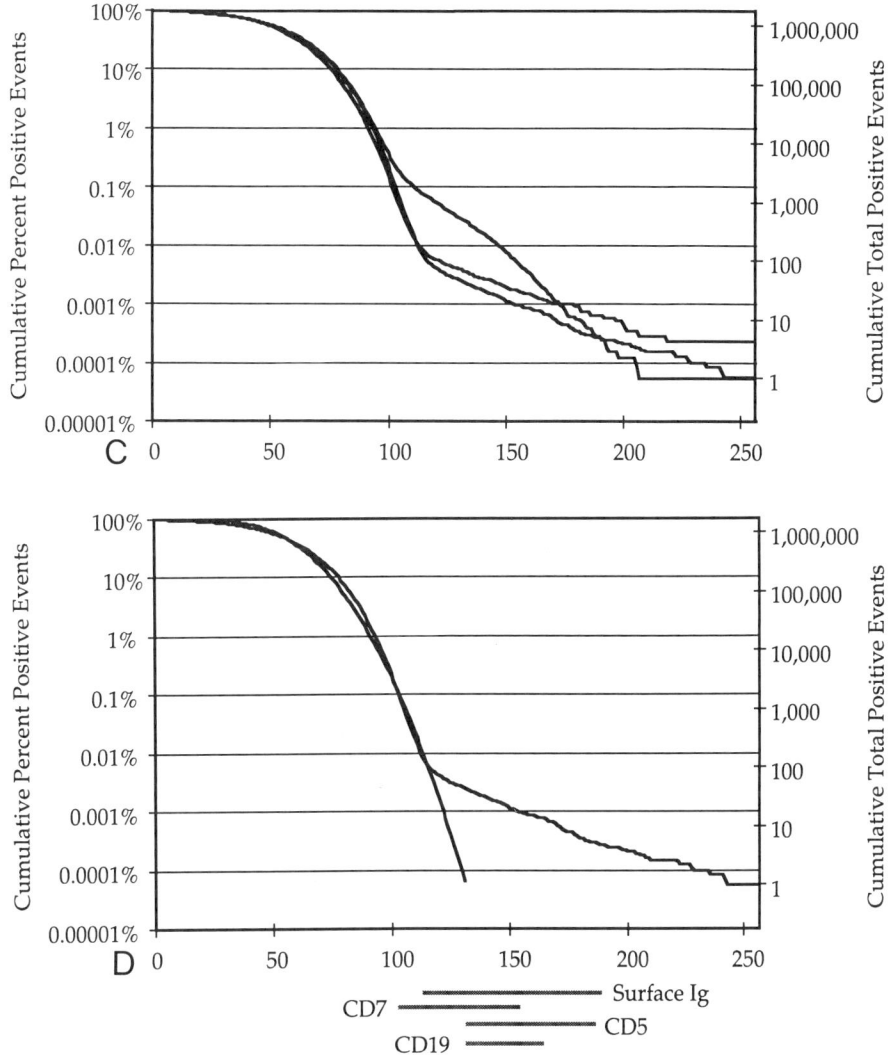

Figure 29.2—continued

sity over indirect immunofluorescence (3) and actually reduces noise in staining of red cells, but it results in higher noise when staining leukocytes, which can adhere to or phagocytize the beads. Lansdorp et al. have recently described preliminary results with an amplification system using a biotinylated probe labeled with at least two cycles of streptavidin phycoerythrin (PE) followed by biotinylated anti-PE (77). Using erythropoietin as a probe for the erythropoietin receptor, and CD34 or glycophorin as the FITC-conjugated second parameter, they were able to clearly identify erythropoietin receptors on CD34-/glycophorin+ maturing erythroid cells, a significant accomplishment, given the small number of erythropoietin receptors (<500/cell) on these cells.

This may be an appropriate time to focus on the technical limitations preventing us from detecting residual cells at very low concentrations in peripheral blood. In this context, it may be instructive to study the nature of those particles or cells that are responsible for the background nonspecific binding observed in flow cytometry assays using instruments capable of high-speed analysis and sorting (78). These events may represent cellular aggregates, cellular debris, aggregated antibody bound to cells or debris, or a rare cell expressing Fc receptors. The broad fluorescence intensity distribution of events with high nonspecific binding of antibody suggests that debris or aggregates may be responsible. Knowledge of the nature of nonspecifically binding events may allow development of strategies (choice of scatter gates, inhibition of aggregation during staining, gating with specific antibodies using multicolor staining, etc.) to reduce nonspecific binding and, equally important, to reduce the variability in nonspecific binding that occurs from patient to patient.

Those aspects of flow cytometric analysis that now seem routine for standard clinical assays need to be reevaluated in the context of assays for minimal residual disease. Aggregation of antibody reagents and choice of Ig isotype have been mentioned above. Minor variations in light-scatter gating, which may not affect standard flow cytometry results, may significantly increase nonspecific binding if debris or cell aggregates are included. Fixation of stained cells with paraformaldehyde is standard for routine flow cytometry, but the

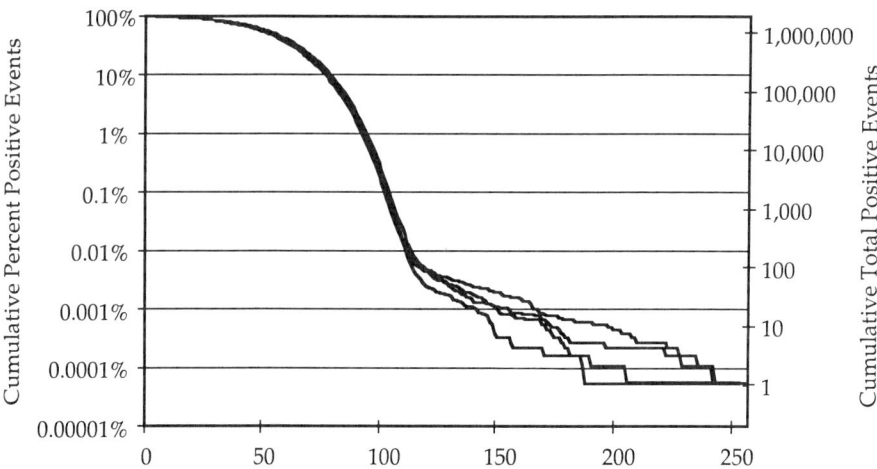

Figure 29.3. Cumulative histograms of cells prepared and stained as described for Figure 29.1 (all antibodies spun before use). Mean cumulative histograms of three normal subjects whose cells were stained with FITC conjugated irrelevant mouse IgG1. Note the variability above fluorescent channel 110. Gating logic was based on a bitmap of forward and 90° light scatter and was similar for all three samples.

effect of fixation on background positive events needs to be carefully evaluated for each flow cytometry rare cell assay.

CELL CLUMPING

Improved techniques to inhibit cellular aggregation will be especially helpful in the diagnosis of metastatic solid tumor (to prevent loss of the true positive events) and hematologic malignancies (to prevent aggregation of normal cells causing nonspecific binding). Recent advances in understanding the mechanisms of cell adhesion suggest possible avenues to explore in inhibiting these undesirable adhesion events. Adhesion of cells is not nonspecific, but is based on receptor-ligand interactions of adhesive proteins on the cell surface. Adhesion of malignant epithelial cells to extracellular matrix is in part mediated by integrins (79), particularly VLA-5 and $\alpha_v\beta_4$. There are additional adhesion proteins capable of mediating binding of malignant cells to each other or to normal hematopoietic cells. The peptide sequences responsible for adhesion recognition are known for some of these receptor-ligand pairs, and the adhesive interaction can be blocked by synthetic peptides. For instance, binding of $\alpha_v\beta_3$ (vitronectin receptor) to vitronectin is blocked by RGDS (79), binding of $\alpha_2\beta_1$ (VLA-2) to collagen is blocked by KDGEAC, binding of $\alpha_4\beta_1$ (VLA-4) to fibronectin is blocked by LDV-containing peptides (80), and binding of ELAM-1 (endothelial protein) to neutrophils and colon tumor cells is blocked by soluble sialyl-Lewis[x] (81). It may be possible to inhibit aggregation in vitro by the use of blocking reagents such as these.

Another possibility is to eliminate centrifugation steps entirely by simply lysing red cells and staining the nucleated cells with a carefully calibrated concentration of antibody, as used in commercially available methods for whole blood quantitation of CD4+ cells ("Q-prep"; Coulter Electronics). This would require fairly strong expression of the marker by the tumor cells and special steps to reduce nonspecific binding to cell debris.

INTRACELLULAR MARKERS

Given the success of immunofluorescence microscopy based assays using a surface marker plus TdT, it is of considerable importance to devise methods of improved detection of intracellular markers. Permeabilization techniques induce significant alteration of the cell surface, which promotes clumping during the centrifugation steps. A careful balance needs to be maintained between obtaining adequate permeabilization to allow the antibody to enter but not allow the antigen to leak out. Nonspecific binding of antibody is higher in these assays since the antibody has access to intracellular compartments instead of just the cell surface. Furthermore, light scatter properties of normal cells may be altered, making it more difficult to use light scatter to gate out myeloid or monocytic cells typically responsible for high nonspecific binding (66). Using cDNA clones derived from hybridomas, it is now possible to produce monoclonal antibodies in *E. coli* (82). Using this technique, it is possible to delete regions unnecessary for antigen binding and construct an antibody consisting only of a single V_H region that binds to specific antigen with an affinity in the 20 nM range. A directly conjugated antibody of this type might require less aggressive cell permeabilization, which would tend to reduce cell aggregation as well as minimize loss or denaturation of the antigen itself. The same approach could be used to modify immunoglobulins to decrease FcR-mediated binding without the potentially damaging effects of enzyme treatment.

NUMBER OF POSITIVE EVENTS REQUIRED

Given the uncertain origin of the background events encountered in rare cell analysis and the variability associated with enumeration of small numbers of events, it is somewhat risky to base clinical conclusions on the observation of a small number of positive events in a test sample as compared with the control. This is also a concern with immunofluorescence microscopy, where there is the additional variability of the human decision-making process that is required to categorize an event as "positive." This same type of problem has been faced in the past by markers of commercial hematology platelet counting instruments, since platelets represent a much rarer cell than red cells in whole blood. The advance that permitted routine clinical application was to accept as valid platelets only those populations with the anticipated mathematical distribution of volumes (log-normal) expected for platelets.

The goal for a reliable **clinical** assay should be a method that has a low enough background and high enough number of counted events to show the positive events at the lower limit of sensitivity as a distinct population, separated from the negative cells by a definite "valley," rising to a peak value, and then tapering back to baseline. How the number of positive events affects the appearance of the positive population can be demonstrated by repeatedly analyzing the same sample and counting progressively fewer total numbers of cells (Fig. 29.4). A cell population centered around channel 165 is clearly seen when 500 positive events are acquired (Fig. 29.4B), but is less evident when only 80 (Fig. 29.4H) or 29 (Fig. 29.4J) positive events are counted. The number of positive events required depends on: 1) statistical variability of replicate measurements of positive and control samples, 2) fluorescence intensity of the positive population, 3) background nonspecific binding, and 4) the variability of fluorescence intensity of the positive population. The fluorescence intensity distribution of the positive population is an important variable that deserves consideration. The range of fluorescence intensity that includes 75% of normal CD3+ lymphocytes, such as were used in Figure 29.4, is 35 log channels (i.e., from channel 155 to 190). If the cells showed more variability in fluorescence intensity, they would correspondingly be more difficult to recognize visually as a distinct population. We have measured the range of fluorescence intensity of 75% of positive cells from 47 samples stained with hematopoietic differentiation markers in 18 patients with CLL, ALL, AML, or lymphoma. The mean range (in log channels of fluorescence) is 53 ± 5 (S.E.M.). The corresponding value for normal blood cells (lymphocytes, neutrophils, monocytes) stained with surface markers CD3, CD7, CD19, CD16, CD13, CD33, CD14, and surface Ig, taken from our own data and that of others (83) is 37 ± 4 (S.E.M.). The greater variability of antigen expression by malignant cells may make it necessary to count more positive events in the case of a neoplastic population than is demonstrated in Figure 29.4.

These data suggest that acquisition of more than 100 positive events may be required to generate a recognizable positive population distribution that can unequivocally be distinguished from merely an elongation of the "tail" of the population of nonspecifically binding cells. Even though an appropriate control antibody is used, each monoclonal antibody has a unique sequence and may bind in an unpredictable fashion to nonspecific events (debris, DNS, dead cells, granulocytes, etc.) or may specifically bind to aggregates containing cell fragments. Therefore, it is a desirable goal to collect enough positive events to show that the histogram of the positive population has an expected distribution (usually unimodal and roughly normally distributed), based on the marker and cell type studied. The same logic applies to multiparameter analysis, particularly when analyzing double-positive cells in two-color histograms. "Double-positive" cells are frequently found scattered in the upper right quadrant of such two-color histograms when large numbers of cells are counted in a complex tissue such as bone marrow. Due to the frequency with which such scattered events are observed for any marker combination (representing aggregates of cells or cells plus debris, dead cells, Fc receptor bearing cells, etc.), it is necessary to be careful in interpreting such data, especially when the double-positive cells do not form a recognizable population on the two-paremeter histogram. This is an issue particularly when rare populations are too few in number to form a distinctive population in the two-parameter histogram. In such cases, the flow cytometry data must be confirmed by other techniques, such as analysis of clonally expanded cells (84).

For clinical applications requiring detection limits of 0.01% positive events, it would be necessary to count at least 1,000,000 total events, which generally requires 5 to 10 min on a commercial flow cytometer. Detection of 0.001% positive events in 10,000,000 or more total cells may require adaptations of existing instrumentation for the purpose. An outboard multiparameter hardware/software system for high-speed counting (100,000 cells/sec), logic-gating, and count-rate error checking that can be attached to commercial flow cytometers has been described by Leary et al. (78). Practical limitations on sample volume may make it impractical to routinely sample more than this number of cells from the peripheral blood of patients with malignancy who are often leukopenic, but bone marrow aspirations may yield larger numbers of cells.

Biological Considerations

A major diagnostic limitation of tumor markers is that unless these markers are selected on the basis of fundamental knowledge of normal and neoplastic differentiation, the rationale for their use is purely empirical and likely to erode as additional clinical data are gathered. Therefore, it is not surprising that the best tumor-associated markers currently available are frequently based on fundamental observations of normal cell differentiation and carcinogenesis. The small number of genuinely useful markers of this nature reflects our inadequate understanding of the basic nature of the malignant transformation of cells.

We now have several markers that appear to fulfill the major requirements for detection of minimal residual disease, among them several abnormal bone marrow pheno-

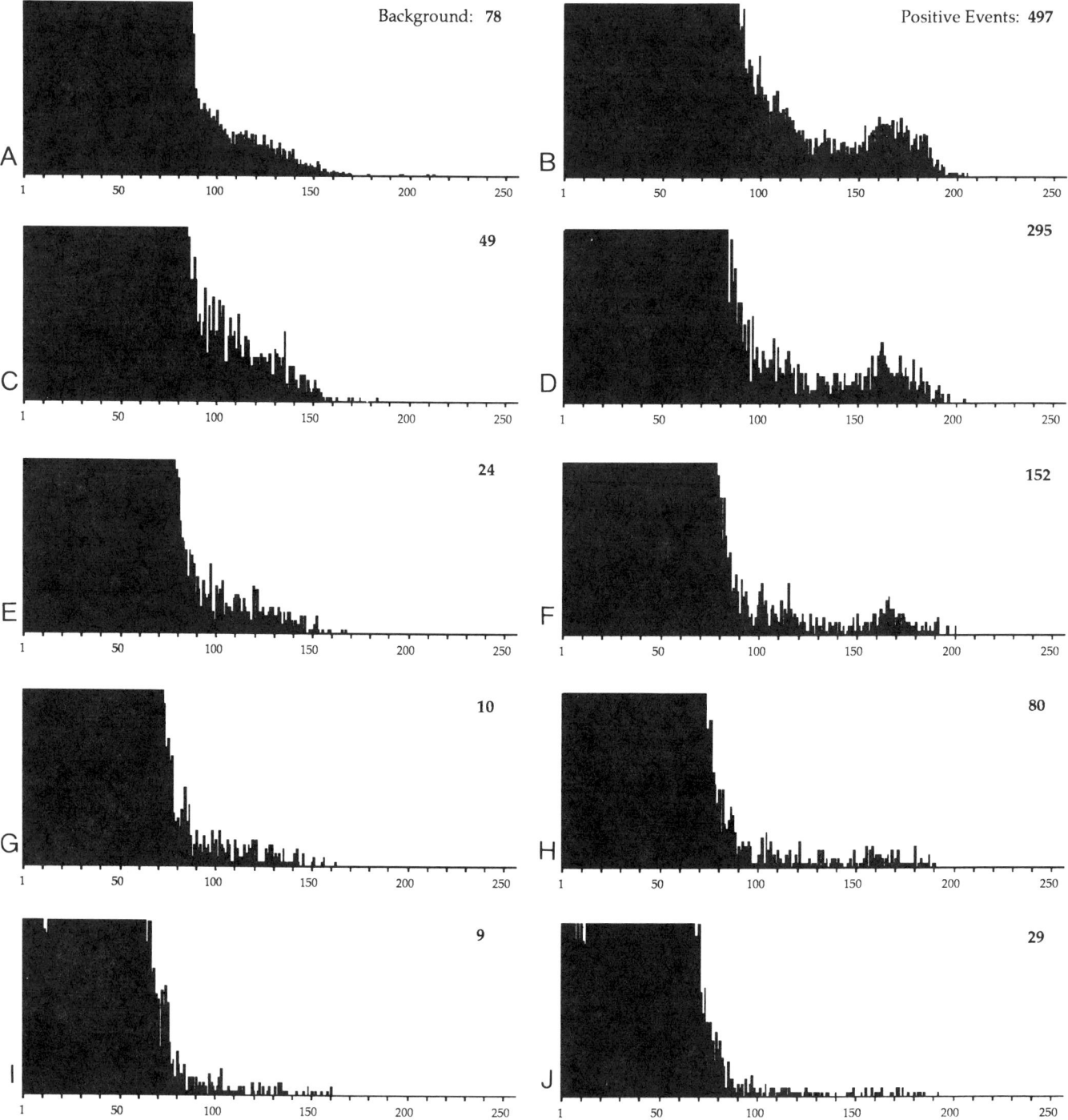

Figure 29.4. Standard flow cytometry histograms (not cumulative histograms) of fluorescence intensity vs. cells per channel for mononuclear cells prepared as described in Figure 29.1. Cells were stained with FITC-conjugated irrelevant mouse IgG2a and analyzed alone or mixed with a separate aliquot of mononuclear cells stained with FITC-conjugated anti-CD3 (Coulter Immunology) to give a final concentration of 0.5% CD3+ cells. Varying numbers of total cells, based on forward vs. 90° scatter lymphocyte gates were counted: 100,000 (a, b); 54,000 (c, d); 27,000 (e, f); 14,000 (g, h); 7,400 (i, j). **A, C, E, G, I** show mononuclear cells stained with FITC-conjugated irrelevant mouse IgG2a alone, while panels (**B, D, F, H, J**) show a mixture of mononuclear cells stained with FITC-conjugated IgG2a and mononuclear cells stained with FITC-conjugated anti-CD3 to yield a final concentration of 0.5% CD3+ cells. The total number of positive events (> channel 150) for each histogram is shown in the upper right corner.

Table 29.3
Examples of Assays for Minimal Residual Disease

Cell Marker
• T-ALL (cCD3 or CD5 vs. TdT)
• B-cell precursor ALL (CD13 or CD33 vs. TdT; ~15% of cases)
• AML (CD13 or CD33 vs. TdT; ~15% of cases)

PCR
• T-ALL (TcR δ)
• B-cell precursor ALL (Ig or TcR δ)
• CML (*bcr/abl*)
• Follicular lymphoma (*bcl-2*)

types (cCD3+/TdT+, CD5+/TdT+, CD13+/TdT+) and several translocations or rearrangements that are amenable to PCR at the DNA or RNA level (*bcr/abl, bcl-2*, retinoic acid receptor αIg, and TcR gene rearrangements). Since only a few malignancies are characterized by predictable translocations of genetic material suitable for PCR analysis, development of cell marker based residual detection assays remains important.

How should these assays be used in clinical decision making? More specifically, does flow cytometry have a role detection of minimal residual disease, given that the only published assay systems shown to detect residual disease in patients are immunofluorescence microscopy and PCR?

T-CELL ALL AS A MODEL FOR RESIDUAL DISEASE ASSAYS

The sensitivity of PCR and immunofluorescence microscopy methods in detecting residual T-ALL may be comparable (85). PCR has the advantage of simplicity (once the tumor-specific gene rearrangement has been cloned), but is not quantitative. Given the fact that residual tumor cells persist for 18 months to two years in virtually all patients with ALL or CML during chemotherapy or after BMT (4), the necessity for accurate quantitation is evident. Immunofluorescence microscopy offers the ability to quantitate residual tumor, but the assay involves the screening of tens of thousands of cells by a trained observer and the endpoint is not completely objective. Flow cytometry offers the advantages of accurate quantitation and an objective endpoint, and is less demanding of technical time.

Table 29.3 shows several residual disease assays that are under active investigation to determine their prognostic utility. T-ALL is currently the only malignancy in which both cell marker- and PCR-based residual disease assay is available for the majority of patients with the disease. There are two approaches that could greatly aid in bringing a residual disease assay into the clinical laboratory: 1) development of a flow cytometry based assay that provides the sensitivity and specificity of the immunofluorescence microscopy assay for minimal residual disease, and 2) a clinical trial of the comparative utility of cell marker assay (i.e. cCD3/TdT or CD5/TdT) vs PCR (δ TcR rearrangement) in predicting relapse in patients with T-ALL. A flow cytometry assay may be more suitable for most clinical laboratories, particularly if it can be shown that it provides the same clinical information, or perhaps additional quantitative information, than PCR.

It is possible that future applications of rare event analysis may use flow cytometry and PCR in tandem: PCR to detect residual disease, and flow to quantitate its presence, with each method confirming the results of the other and minimizing the risk of false-positives. It is certainly not unreasonable for the physician to have at least two laboratory tests at his or her disposal before making important decisions regarding chemotherapy for leukemia.

DETECTION OF MINIMAL RESIDUAL DISEASE IN PERIPHERAL BLOOD

Peripheral blood analysis has important technical advantages over bone marrow: 1) easily obtainable serial samples, 2) lower background nonspecific binding, and 3) less cellular heterogeneity, and hence more reproducible scatter gating. A successful flow cytometric application of intracellular staining techniques to the detection of minimal residual disease may serve as a model system for applications in solid tumors, such as the detection of cytokeratin positive cells in blood. Circulating tumor cells may express fewer adhesion receptors and be less subject to clumping. Furthermore, they may be more indicative of the metastatic potential of the tumor. The major potential problem with peripheral blood analysis is that the number of tumor cells may be well below any reasonable limit of sensitivity. This may depend greatly on the type of malignancy being analyzed. For instance, we had little success in finding strongly CD10+ cells in the peripheral blood prior to relapse in children with CD10+ B-cell precursor ALL even with an assay capable of detecting 0.001–0.01% positive cells (10). In one patient, less than 0.001% strongly CD10-positive cells were identifiable in the peripheral blood at a time when the bone marrow contained 14% lymphoblasts. In contrast, a notable feature of T-ALL is its propensity to seed peripheral tissues (86), reflected in the surprisingly equal sensitivity of peripheral blood and bone marrow assays in the detection of residual T-ALL (87). An advantage of peripheral blood analysis is that rare normal precursors found in peripheral blood may be phenotypically distinct from those in bone marrow and, thus, more easily distinguishable from leukemic cells. For instance, normal TdT+ lymphoid precursors in peripheral blood are CD10−/CD19−/CD9− (88), a phenotype distinctly different from both normal bone marrow lymphoid precursors and their neoplastic counterpart, ALL.

MULTIPARAMETER DETECTION OF ABERRANT PHENOTYPES

Multiparameter analysis of neoplastic cells has attracted continued interest because of the promise of increased specificity. All of the current applications of cell marker analysis to residual disease detection rely on at least two markers. The preliminary results of Wörmann et al. (14) suggest that nearly all cases of AML show either asynchronous or mixed lineage differentiation when studied with a panel of markers in a three-color immunofluorescence assay using FITC, PE and duochrome labels. Aberrant phenotypes were found on 60 or 61 patients, including mixed lineage differentiation (CD2 T-cell marker expression) and asynchronous differentiation (CD34+/CD15+, CD13+/CD33−, CD34+ neutrophils and monocytes, and CD16 expression on immature cells with low 90° scatter). Residual aberrant phenotypes were observed in >0.5% of marrow cells from 28 of 35 patients in clinical remission. The clinical significance of these residual cells is not yet determined. The sensitivity of detection will depend on the particular marker combination used. For instance the CD34+/CD15+ phenotype is quite atypical, as discussed earlier, but other phenotypes may not be as distinctive.

Aberrant populations in three-, four- and five-dimensional space that cannot be appreciated by conventional analysis can be identified by sophisticated data analysis techniques, such as principal component and biplot analysis. Multivariate analysis of three-parameter data is utilized in newer automated clinical leukocyte differential instruments, such as the STKS (Coulter Electronics). Rare populations (0.01%) of human lymphocytes labeled by Y chromosome fluorescence in-situ hybridization can be detected by more complex application of this technique to six flow cytometry parameters (78), identifying combinations of parameters that maximize the distance between the rare abnormal cells and nonspecifically binding normal cells when projected onto two-dimensional histograms. This specific application may permit low-risk fetal detection of genetic diseases by the sorting of fetal cells in maternal blood for subsequent PCR analysis of specific genes. A potential problem with multiparameter analysis is that one may only be looking at subsets of the leukemic cells, thereby reducing the number of positive events. Principal component analysis minimizes this problem by identifying abnormal tendencies in a minor population as a whole, and may be a means of "flagging" a sample as abnormal. Furthermore, in AML, residual cells may not be all that rare after initial chemotherapy, and multivariate analysis may be suited for answering important clinical questions such as whether a patient should be given another round of chemotherapy.

Several additional problems are associated with this multiparameter approach to residual leukemia detection. The first of these is that the more specific and detailed the description of the leukemia cell, the fewer patients any particular description will fit. An approach to this problem is to characterize cells with multiple combinations of antibodies, thus increasing the likelihood of finding at least one aberrant phenotype expressed in each patient. An associated risk is that phenotypic shifts, which occur at relapse in a significant proportion of patients with ALL, will be more commonly observed the finer and more detailed the leukemic phenotypic description becomes. If the combination of markers used for detection has a functional significance for cell survival or proliferation, it is possible that phenotypic shifts will be less frequently observed.

THE BIOLOGY OF RESIDUAL CELLS

It is likely that the mere detection of residual cells will not be sufficient to provide accurate prognostic information. The next step will be to define characteristics of residual cells with respect to growth and drug sensitivity. Flow cytometry can be an endpoint for drug sensitivity assays, as shown by Asselin et al. (89), and could potentially be an endpoint for in vitro assays of response to specific growth factors. The sensitivity of immunologically based assays for metastatic tumor cells in bone marrow may be increased by allowing the cells to grow in culture before testing (90). Progeny of leukemic colonies in AML may be identifiable by aberrant marker expression (91, 38) (CD20+/CD15+, CD34+/CD65+) and could be tested in liquid culture for response to chemotherapeutic agents or growth factors. These research applications may allow treatment to be based on the characteristics of the residual cells, and not just the cells obtainable at diagnosis, which may have been significantly changed during the course of disease and treatment.

REFERENCES

1. Mansour J, Robson J, Arndt C, Schulte H. Detection of *Escherichia coli* in blood using flow cytometry. Cytometry 1985;6:186.
2. Cory J, Ohlsson-Wilhelm B, Brock E, Sheaffer N, Steck M, Eyster M, Rapp F. Detection of human immunodeficiency virus infected lymphoid cells at low frequency by flow cytometry. J Immunol Meth 1987;105:71.
3. Cupp JE, Leary JF, Cernichiari E, Wood JCS, Doherty RA. Rare-event analysis methods for detection of fetal red blood cells in maternal blood. Cytometry 1984;5:138–144.
4. Yamada M, Wasserman R, Lange B, Reichard BA, Womer RB, Rovera G. Minimal residual disease in childhood B-lineage lymphoblastic leukemia. N Engl J Med 1990;323:448–455.
5. Bonadonna G, Brusamolino E, Balagussa P, et al. Combination chemotherapy as an adjuvant treatment in operable breast cancer. New Engl J Med 1976;294:405.
6. Gorin NC, Aegerter P, Auvert B, et al. Autologous bone marrow transplantation for acute myelocytic leukemia in first remission: a European survey of the role of marrow purging. Blood 1990;75:1606–1614.
7. Goldie J, Coldman A. Genetic instability in the development of drug resistance. Semin Oncol 1985;12:222–230.
8. Goldie JH, Coldman AJ. A mathematical model for relating the drug sensitivity of tumors to their spontaneous mutation rate. Cancer Treat Rep 1979;63:1727–1733.
9. Goldie JH, Coldman AJ. Quantitative model for multiple levels of drug resistance in clinical tumors. Cancer Treat Rep 1983;67:923–931.
10. Ryan D, Van Dongen JJM. Detection of minimal residual disease in acute leukemia by monoclonal antibodies. In: Foon, K. and Bennett, J, eds. Immunologic Approaches to the Classification and Management of Lymphomas and Leukemias. Boston: Kluwer Academic Publishers, 1988.
11. Huges TP, Goldman JM. Biological importance of residual leukaemic cells after BMT for CML: does the polymerase chain reaction help? Bone Marrow Transplant 1990;5:3–6.
12. Pui C, Raimondi S, Behm F, et al. Shifts in blast cell phenotype and karyotype at relapse of childhood lymphoblastic leukemia. Blood 1986;68:1306–1310.

13. Greaves M, Paxton A, Janossy G, Pain C, Johnson S, Lister T. Acute lymphoblastic leukemia associated antigen. III. Alterations in expression during treatment and relapse. Leuk Res 1980;4:1–14.
14. Wörmann B, Könemann S, Safford M, et al. Immunophenotypic heterogeneity of leukemic cells in AML at diagnosis and after chemotherapy. Blood 1990;76:337a.
15. Raghavachar A, Thiel E, Bartram CR. Analyses of phenotype and genotype in acute lymphoblastic leukemias at first presentation and relapse. Blood 1987;70:1079–1083.
16. Hann I, Jones P, Evans D. Discrepancy of bone marrow aspirations in acute lymphoblastic leukemia in relapse. (Letter) Lancet 1977;i:1215–1216.
17. Martens ACM, Schultz FW, Hagenbeek A. Nonhomogeneous distribution of leukemia in the bone marrow during minimal residual disease. Blood 1987;70:1073–1078.
18. van Dongen JJM. Human T cell differentiation: Basic aspects and their clinical applications [Thesis]. Rotterdam, the Netherlands: Erasmus University, 1990:466–468.
19. Cohn S, Lincoln S, Rosen S. Present status of serum tumor markers in diagnosis, prognosis, and evaluation of therapy. Cancer Invest 1986; 4:305–327.
20. Hittelman WN, Agbor P, Petkovic I, et al. Detection of leukemic clone maturation in vivo by premature chromatin condensation. Blood 1988;72:1950–1960.
21. Langlois R, Yu L-C, Gray J, et al. Quantitative Karyotyping of Human Chromosomes by Dual Beam Flow Cytometry. Science 1983; 220:620.
22. Arkesteijn GJA, Martens ACM, Hagenbeek A. Bivariate flow karyotyping in human Philadelphia-positive chronic myelocytic leukemia. Blood 1988;72:282–286.
23. Pinkel D, Straukme T, Gray J. Cytogenetic analysis using quantitative, high sensitivity, fluorescence in situ hybridization. Proc Natl Acad Sci USA 1986;83:2934–2938.
24. Homge M, Pereira J, Kernan NA et al. Detection of residual/recurrent host cells after allogeneic BMT using fluorescence-in-situ-hybridization (FISH). Blood 1990;76:545a.
25. Eastmond D, Pinkel D. Detection of aneuploidy and aneuploidy-inducing agents in human lymphocytes using fluorescence in situ hybridization with chromosome-specific DNA probes. Mutat Res 1990; 234:303–318.
26. Gray JW, Kuo WL, Liang J, et al. Analytical approaches to detection and characterization of disease-linked chromosome aberrations. Bone Marrow Transplant 1990;14–19.
27. Griesser H, Tkachuk D, Reis MD, Mak TW. Review: Gene rearrangements and translocations in lymphoproliferative diseases. Blood 1989; 1402–1415.
28. Horning SJ, Galili N, Cleary M, Sklar J. Detection of non-Hodgkin's lymphoma in the peripheral blood by analysis of antigen receptor gene rearrangements: Results of a prospective study. Blood 1990;75:1139–1144.
29. Bregni M, Siena S, Neri A, et al. Minimal residual disease in acute lymphoblastic leukemia detected by immune selection and gene rearrangement analysis. J Clin Oncol 1989;7:338–343.
30. Gilliland G, Perrin S, Blanchard K, Bun HF. Analysis of cytokine mRNA and DNA: Detection and quantitation by competitive polymerase chain reaction. PNAS 1990;87:2725–2729.
31. Kawasaki ES, Clark SS, Coyne MY, et al. Diagnosis of chronic myeloid and acute lymphocytic leukemias by detection of leukemia-specific mRNA sequences amplified in vitro. Proc Natl Acad Sci USA 1988;85:5698.
32. Lee MS, Chang KS, Cabanillas F, Freireich EJ, Trujillo JM, Stass SA. Detection of minimal residual cells carrying the t(14;18) by DNA sequence amplification. Science 1987;237:175–178.
33. Hansen-Hagge TE, Yokota S, Bartram CR. Detection of minimal residual disease in acute lymphoblastic leukemia by in vitro amplification of rearranged T-cell receptor δ chain sequences. Blood 1989;74: 1762–1767.
34. Kwok S, Higuchi R. Avoiding false positives with PCR. Nature 1989; 339:237–238.
35. Lee BR, Haseman DB, Reynolds CP. A digital image microscopy system for rare-event detection using fluorescent probes. Cytometry 1989;10:256–262.
36. Begley CG, Metcalf D, Nicola NA. Binding characteristics and proliferative action of purified granulocyte colony-stimulating factor (G-CSF) on normal and leukemic human promyelocytes. Exp Hematol 1988;16:71.
37. Estrov Z, Grunberger T, Dube I, Wang Y, Freedman M. Detection of residual acute lymphoblastic leukemia cells in cultures of bone marrow obtained during remission. New Engl J Med 1986;315:538.
38. Gerhartz HH, Schmetzer H. Detection of minimal residual disease in acute myeloid leukemia. Leukemia 1990;4:508–516.
39. Turhan AG, Eaves CJ, Eaves AC, Humphries RK. Quantitation of normal and leukemic progenitors by polymerase chain reaction (PCR) detection of bcr/abl transcripts in individual hemopoietic colonies. Blood 1990;76:247a.
40. Estrov Z, Freedman MH. Growth requirements for human acute lymphoblastic leukemia cells: Refinement of a clonogenic assay. Cancer Res 1988;48:5901–5907.
41. Zhang XG, Klein B, Bataille R. Interlukin-6 is a potent myeloma-cell growth factor in patients with aggressive multiple myeloma. Blood 1989;74:11–13.
42. Park LS, Waldron PE, Friend D, et al. Interlukin-3, GM-CSF, and G-CSF receptor expression on cell lines and primary leukemia cells: Receptor heterogeneity and relationship to growth factor responsiveness. Blood 1989;74:56–65.
43. Eder M, Ottmann OG, Hansen-Hagge TE, et al. Effects of recombinant human IL-7 blast cell proliferation in acute lymphoblastic leukemia. Leukemia 1990;4:533–540.
44. LeBien T, McCormak R. The common acute lymphoblastic leukemia antigen (CD10)—emancipation from a functional enigma. Blood 1989;73:625.
45. Ryan DH. Phenotypic heterogeneity in acute leukemia. Clin Chim Acta 1991;206:9–23.
46. Loken MR. Shah VO, Dattilio KL, Civin CI. Flow cytometric analysis of human bone marrow. I. Normal erythroid development. Blood 1987;69:255.
47. Ryan D, Kossover S, Mitchell S, Frantz C, Hennessy L, Cohen H. Subpopulations of common acute lymphoblastic leukemia antigen-positive lymphoid cells in normal bone marrow identified by hematopoietic differentiation antigens. Blood 1986;68:417.
48. Loken MR, Shah VO, Dattilio KL, Civin CI. Flow cytometric analysis of human bone marrow. II. Normal B lymphocyte development. Blood 1987;70:1316–1324.
49. Civin C. Human monomyeloid cell membrane antigens. Exp Hematol 1990;18:461.
50. Knapp W, Rieber P, Dorken B, et al. Towards a better definition of human leucocyte surface molecules. Immunology 1989;10:253.
51. Ryan DH, Chapple CW, Kossover SA, Sandberg AA, Cohen HJ. Phenotypic similarities and differences between CALLA-positive acute lymphoblastic leukemia cells and normal CALLA-positive B cell precursors. Blood 1987;70:814.
52. Hurwitz CA, Loken MR, Graham ML, et al. Asynchronous antigen expression in B lineage acute lymphoblastic leukemia. Blood 1988;72: 299–307.
53. Wormann B, Safford M, Konemann S, et al. Identification of clonal subpopulations in ANLL by multidimensional flow cytometry. Blood 1989;74:22a.
54. Bradstock KF, Kirk J, Grimsley PG, Kabral A, Hughes WG. Unusual immunophenotypes in acute leukemias: Incidence and clinical correlations. Brit J Haematol 1989;72:512–518.
55. Grogan T, Durie B, Vela E, et al. Myelomonocytic antigen positive multiple myeloma. Blood 1989;73:763.

56. Bradstock KF, Kirk J, Grimsley PG, Kabral A, Hughes WG. Unusual immunophenotypes in acute leukemias: Incidence and clinical correlations. Brit J Haematol 1989;72:512–518.
57. Adriaansen HJ, van Dongen JJM, Kappers-Klunne MC, et al. Terminal deoxynucleotidyl transferase positive subpopulations occur in the majority of ANLL: Implications for the detection of minimal disease. Leukemia 1990;4:404–410.
58. Greaves MF, Chan LC, Furley AJW, Watt SM, Molgaard HV. Lineage promiscuity in hemopoietic differentiation and leukemia. Blood 1986;67:1–11.
59. Nagler A, Greenberg PL, Lanier LL, Phillips JH. The effects of recombinant interleukin 2-activated natural killer cells on autologous peripheral blood hematopoietic progenitors. J Exp Med 1988;168:47–54.
60. Civin CI, Banquerigo ML, Strauss LC, Loken MR. Antigenic analysis of hematopoiesis. VI. Flow cytometric characterization of My-10-positive progenitor cells in normal human bone marrow. Exp Hematol 1987;15:10–17.
61. Berenson RJ, Andrews RG, Bensinger WI, et al. Antigen CD34+ marrow cells engraft lethally irradiated baboons. J Clin Invest 1988;81:951–955.
62. Adriaansen HJ, Hooijkaas H, Kappers-Klunne MC, Hählen K, van't Veer MB, van Dongen JJM. Double marker analysis for terminal deoxynucleotidyl transferase and myeloid antigens in acute nonlymphocytic leukemia patients and healthy subjects. Haematol Blood Transf 1990;33:41–49.
63. Campana D, Coustan-Smith E, Janossy G. The immunologic detection of minimal residual disease in acute leukemia. Blood 1990;76:163.
64. Hudson AM, Makrynikola V, Kabral A, Bradstock KF. Immunophenotypic analysis of clonogenic cells in acute lymphoblastic leukemia using an in vitro colony assay. Blood 1989;74:2112–2120.
65. Van Dongen J, Hooijkaas H, Comans-Bitter M, et al. Human bone marrow cells positive for terminal deoxynucleotidyl transferase (TdT), HLA-DR, and a T cell marker may represent prothymocytes. J Immunol 1985;135:3144–3150.
66. Gore SD, Kastan MB, Goodman SN, Civin CI. Detection of minimal residual T cell acute lymphoblastic leukemia by flow cytometry. J Immunol Meth 1990;132:275–286.
67. Redding W, Coombes R, Monaghan P, et al. Detection of micrometastases in patients with primary breast cancer. Lancet 1983;ii:1271–1273.
68. Mansi JL, Berger U, Easton D, et al. Micrometastases in bone marrow in patients with primary breast cancer: evaluation as an early predictor of bone metastases. Brit Med J 1987;295:1093–1096.
69. Thor A, Viglione MJ, Ohuchi N, et al. Comparison of monoclonal antibodies for the detection of occult breast carcinoma metastases in bone marrow. Breast Cancer Res Treat 1988;11:133–145.
70. Schlimok G, Funke I, Holzmann B, et al. Micrometastatic cancer cells in bone marrow: In vitro detection with anticytokeratin and in vivo labeling with anti-17-1A monoclonal antibodies. Proc Natl Acad Sci USA 1987;84:8672–8676.
71. Frantz C, Ryan D, Cheung NV, Duerst RE, Wilbur DC. Sensitive detection of rare metastic human neuroblastoma cells in bone marrow by two-color immunofluorescence and cell sorting. Adv Neuroblastoma Res 1988;2:249–262.
72. Ault K. Detection of small numbers of monoclonal B lymphocytes in the blood of patients with lymphoma. New Engl J Med 1979;300:1401–1405.
73. Letwin BW, Wallace PK, Muirhead KA, Hensler GL, Kashatus WH, Horan PK. An improved clonal excess assay using flow cytometry and B-cell gating. Blood 1990;75:1178–1185.
73a. Haase AT, Retyel EF, Staskus KA. Amplification and detection of lentival DNA inside cells. DNAS 1990;87:4971–4975.
74. Anderson C, Looney J. Human Leukocyte IgG Fc Receptors. Immunol Today 1986;7:264–266.
75. Ryan D, Mitchell S, Hennessy L, Bauer K, Horan P, Cohen H. Improved detection of rare CALLA-positive cells in peripheral blood using multiparameter flow cytometry. J Immunol Methods 1984;74:115.
76. Kristensen JS, Ellegaard J, Hokland P. A two-color flow cytometry assay for detection of hairy cells using monoclonal antibodies. Blood 1987;70:1063–1068.
77. Landsdorp PM, Wognum AW, Smith C, Krystal G. Direct analysis of growth factor receptor expression on hemopoietic cells by stepwise amplified immunofluorescence staining and multiparameter flow cytometry. Blood 1990;76:151a.
78. Leary JF, Ellis SP, McLaughlin SR, et al. High-resolution separation of rare-cell types. In: Kampala D, Todd P. eds. ACS symposium service vol 464. Washington: American Chemical Society. 1991:26–40.
79. Albelda SM, Buck CA. Integrins and other cell adhesion molecules. FASEB J 1990;4:2868–2880.
80. Garcia-Pardo A, Wayner EA, Carter WG, Ferreira OC. Human B lymphocytes define an alternative mechanism of adhesion to fibronectin. J Immunol 1990;144:3361–3366.
81. Phillips ML, Nudelman E, Gaeta FCA, et al. ELAM-1 mediates cell adhesion by recognition of a carbohydrate ligand, sialyl-Le^x. Science 1990;250:1130–1132.
82. Ward ES, Gussow D, Griffiths AD, Jones PT, Winter G. Binding activities of a repertoire of single immunoglobulin variable domains secreted from *Escherichia coli*. Nature 1989;341:544–546.
83. Terstappen LWMM, Hollander Z, Meiners H, Loken MR. Quantitative comparison of myeloid antigens on five lineages of mature peripheral blood cells. J Leuk Biol 1990;48:138–148.
84. Uckun F, Muraguchi A, Ledbetter J, et al. Biphenotypic leukemic lymphocyte precursors in CD2+CD19+ acute lymphoblastic leukemia and their putative normal counterparts in human fetal hematopoietic tissues. Blood 1989;73:1000–1015.
85. Campana D, Yokota S, Coustan-Smith E, et al. The detection of residual acute lymphoblastic leukemic cells with immunologic methods and polymerase chain reaction: A comparative study. Leukemia 1990;4:609–614.
86. Bloomfield C, Gajl-Peczalska K. The clinical relevance of lymphocyte surface markers in leukemia and lymphoma. Curr Top Hematol 1980;3:175–240.
87. Van Dongen JJM, Hooijkaas H, Adriaansen H, Hahlen K, Van Zanen G. Detection of minimal residual acute lymphoblastic leukemia by immunological marker analysis: Possibilities and limitations. In: Hagenbeek A, Lowenberg B, eds. Minimal Residual Disease in Acute Leukemia 1986. Dordrecht, Martinus Nijhoff Publishers, 1986, pp 113–133.
88. Graham Smith R, Kitchens RL. Phenotypic heterogeneity of TdT+ cells in the blood and bone marrow: Implications for surveillance of residual leukemia. Blood 1989;74:312–319.
89. Asselin B, Ryan D, Frantz C, et al. In vitro and in vivo killing of acute lymphoblastic leukemia cells by L-asparaginase. Cancer Research 1989;49:4363–4368.
90. Joshi SS, Novak DJ, Messbarger L, et al. Levels of detection of tumor cells in human bone marrow with or without prior culture. Bone Marrow Transplant 1990;6:179–183.
91. Delwel R, van Gurp R, Bot F, Touw I, Löwenberg B. Phenotyping of acute myelocytic leukemia (AML) progenitors: An approach for tracking minimal numbers of AML cells among normal bone marrow. Leukemia 1988;2:814–819.

30

Flow Cytometric Assays of Cell-Mediated Cytotoxicity

THOMAS M. ELLIS

INTRODUCTION

The in vitro assessment of cell–mediated cytotoxicity (CMC) represents a fundamental tool for assessing the effector arm of the cellular immune response and, consequently, has been utilized extensively in basic and clinical research. Most of these assessments are designed to address issues related to the quantitation of CMC mediated by whole populations of heterogeneous cell types; that is, are cytolytic cells present and, if so, what are their relative levels of cytolytic activity? In more specialized applications, there exists a further need to assess or quantitate the CMC reaction at defined stages of the cytolytic process or to evaluate the relative abilities of effector cell subsets to mediate cytotoxicity. These latter problems require the application of single-cell cytotoxicity techniques and/or cell selection methods prior to assay.

The capability of flow cytometry to sensitively evaluate multiple parameters of cells and to select signal-defined subsets of cells from heterogeneous populations makes it well suited for detailed studies of the effector cell:target cell interaction, assessed simultaneously at the single-cell and population levels. Results from a number of laboratories attest to the utility of flow cytometry to reproducibly and sensitively quantify the CMC reaction in a variety of experimental systems (1–3). Thus, flow cytometry offers the ability to characterize and quantitate the ability of lymphocytes to bind to and kill targets as well as the ability to study the biochemical changes occurring in effector cells during this interaction. However, despite the potential of flow cytometry to studies of CMC and the fact that such techniques are no longer considered new, it is noteworthy that flow-based cytotoxic assays have not received general acceptance in the immunologic or cell-biologic literature and that articles dealing with this subject are restricted almost entirely to methods-oriented publications. To understand the reasons for the limited acceptance of flow-based cytotoxicity assays among immunologists and to subsequently define the role of flow cytometry in assessing CMC, it is critical to understand the basic cytometric techniques for assessing CMC, their advantages and disadvantages, and how cytometric CMC techniques compare to other technologies currently available for such assessments.

CYTOLYTIC EFFECTOR CELLS

Cytolytic T-Lymphocytes (CTL)

Cytolytic T-lymphocytes (CTL) mediate a form of CMC that is characterized as antigen-specific, major-histocompatibility complex (MHC)-restricted cytolysis and are important in the host response to viruses, tumors, and transplanted tissues. Although CTL recognition of specific foreign antigens is mediated by the T-cell antigen receptor (TCR), recognition and binding of CTL to targets also requires the participation of accessory cell adhesion molecules on the CTL, including CD8, CD4, and CD11a (LFA-1) (4). CTL are generated from cytolytically inactive precursors following stimulation by specific antigen bearing cells in the presence of factors generated by helper T-cells and other accessory cells. Inasmuch as the frequency of CTL specific for a particular antigen is normally quite low, in vitro manipulations are normally required to increase the numbers of specific CTL to levels of detection. Although CTL can exhibit either a CD4 or CD8 phenotype, a majority of CTL activity usually observed in most systems is mediated by CD8+-cells (5).

A minor subset of CD3+CD56+ T-cells also exhibits a form of non-antigen-specific cytotoxicity similar to that observed for NK-cells; these cells are referred to as non-MHC-restricted cytolytic T-cells (6).

Natural-Killer- (NK) Cells

NK cells comprise a class of non-T-, non-B-lymphocytes capable of mediating CMC without prior activation against a limited range of tumor or virally-infected target cells. Target cell recognition occurs in an immunologically nonspecific fashion and via receptor(s) that have not been fully identified. NK-cells are comprised of CD56+ large granular lymphocytes that express neither the T-cell receptor nor surface immunoglobulin. Although NK-cells normally kill only a very limited range of susceptible target cells, exposure to appropriate cytokines induces a more potent form of cytotoxicity against most tumor cell types known as lymphocyte-activated killer (LAK) cytotoxicity (7, 8).

MECHANISMS OF CELL-MEDIATED CYTOTOXICITY

The CMC reaction comprises a complex sequence of events that ultimately result in the lysis of the target cell. Despite

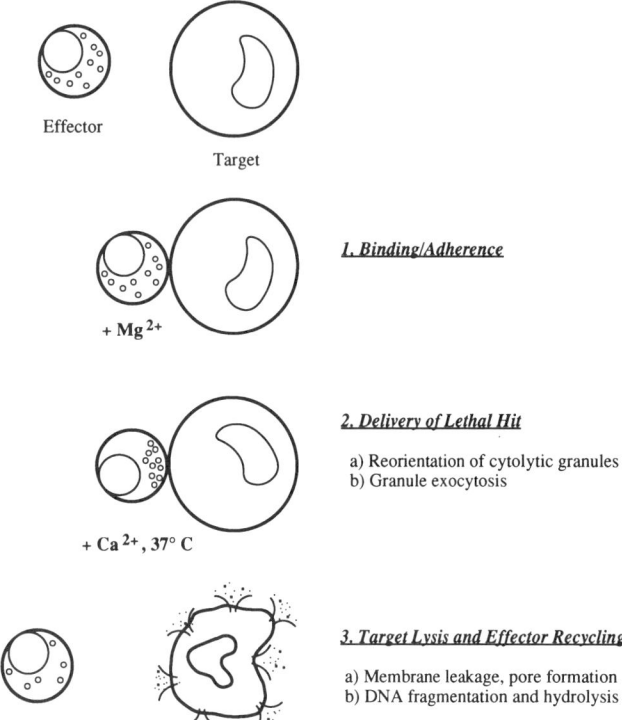

Figure 30.1. Diagram representing the distinct stages of cell-mediated cytolysis.

considerable differences among the cell types capable of mediating CMC and the manner in which they recognize tumor cells, the same basic elements of the CMC mechanism are employed by all cytotoxic cells (4, 9).

The process can be divided into three discrete stages, which include: a) recognition and binding of the cytolytic cell to the target cell, b) delivery of the "lethal hit" to the target cell, and c) target cell lysis and recycling of the effector cell to other targets (Fig. 30.1). Initial target cell binding is a reversible step requiring productive interactions between multiple receptors on the effector cell and membrane molecules on the target cell. In addition to the T-cell receptor (TCR), CD8 (10), CD4 (11), and LFA-1 (12) molecules are required for productive CTL interactions with target cells. Although the target recognition structure(s) on NK-cells has not been identified, LFA-1 molecules participate in target cell binding (13). These interactions require magnesium, can occur at temperatures below those required for lysis, and are enhanced in vitro through close effector-target cell contact provided by centrifugation of effector and target cells together (14, 15).

The exact mechanisms underlying induction of the lytic mechanism and cytolysis of the target cell are not completely understood, although cytolysis may involve multiple independent mechanisms. Most current evidence favors the predominant role for an exocytic mechanism of cytotoxicity in which specialized effector cell cytoplasmic granules are released to the boundary between the effector and target cell membrane. A 60 kd granule constituent called perforin mediates the formation of pores within the target cell membrane, leading to a loss of membrane integrity and leakage of macromolecules (4). Other granule constituents induce target cell DNA fragmentation and hydrolysis (4, 16, 17). Non-granule mediators, such as tumor necrosis factor (TNF), may play accessory roles in the lytic event. Although a perforin-mediated lytic event is currently favored as the predominant event in target cell cytolysis, there continues to exist controversy as to the exact contribution of this and other molecules to overall cytotoxicity (4, 17).

It has been well documented that the triggering of cellular lytic machinery involves events and surface molecules that are distinct from those involved in binding (4, 9). Triggering for lysis is both temperature- and calcium-dependent and involves a morphological reorganization of structures within the cytolytic cell, leading to the delivery of molecules that are lethal to the target cell (18). Once the lethal proteins have been released from the cytolytic cell, the cytolytic cell may detach and "recycle" to other targets even before target cell death is complete. However, for many types of cytolytic cell populations, a minority of cells appear capable of target cell recognition but fail to mediate the lethal hit stage of cytolysis. During target cell lysis, morphologic alterations in the target cell are accompanied by Ca^{2+} influx and followed by chromatin condensation (18). Later, DNA fragmentation occurs and, ultimately, release from the cell of cytosolic macromolecules. It is this latter event which is measured using target cell ^{51}Cr release assays (4, 17).

Logically, most assays of cytolytic cell function rely on quantitation of the terminal stage of this process, i.e., cell death. Most assays rely on detecting the accompanying loss of cell membrane integrity to establish cell death, although such approaches suffer from "false-positives" due to leakage of markers from viable cells. Furthermore, the loss of target cell membrane integrity represents a late event in the course of the lytic reaction and, therefore, may not represent the optimal parameter for "real-time" assessments of cytotoxicity (4). Additionally, more detailed investigation of the mechanism underlying the acquisition or expression of cytolytic functions requires techniques capable of dissecting the individual components of the CMC reaction. The ideal assay of CMC would allow quantitative assessments of all stages of the CMC reaction.

METHODS FOR ASSESSING CELL-MEDIATED CYTOTOXICITY AND EFFECTOR/TARGET CELL INTERACTIONS

Most assessments of CMC simply require measurement of CMC at the population level and, therefore, commonly involve the use of radionuclide release assays. A second, less frequently employed method involves the measurement of the CMC reaction at the single cell level. Although single cell cytotoxicity assays are technically demanding, they allow the quantitation of all stages of the CMC reaction.

RADIONUCLIDE RELEASE ASSAYS

Radionuclide release assays represent the most commonly performed approach for assessing CMC. These assays involve incubating the cytolytic effector cell populations together with target cells that are loaded with a radionuclide (i.e., ^{51}Cr) and, subsequently, assessing the amount of ^{51}Cr released into the medium as an index of target cell damage. The ^{51}Cr release assay allows sensitive assessments of cytotoxicity while utilizing few effector and target cells (in the range of 2,500–10,000 target cells) and can easily accommodate a large number of different sample types in a single experiment. Furthermore, a significant practical advantage is that the collection of raw data from ^{51}Cr release assays is easily automated by using a large capacity gamma counter, thereby minimizing the technical time required for the performance of such assays.

The major limitation of this assay is that, in its usual form, it assesses only the final outcome of the cytolytic reaction mediated by whole populations of cells and does not easily allow for the dissection of the events involved in the CMC reaction. Additionally, it requires the use of radionuclides, which present significant storage, safety, and disposal problems for laboratories. The assay can be further complicated by high spontaneous release of ^{51}Cr from cells or poor loading of some cell types, particularly fresh tumor cells. These problems can severely limit the ability of this assay to sensitively and reproducibly quantitate cytotoxicity.

SINGLE-CELL CYTOTOXICITY ASSAYS

Single-cell cytotoxicity assays offer a powerful approach for analyzing and quantifying CMC at each phase of the cytolytic reaction (19). The technique involves the generation of effector/target conjugates by centrifugation of the effector and target cells together and their incubation in a pellet for a brief period to allow the establishment of stable effector/target conjugates. This incubation is carried out at temperatures (i.e., 20°C) that do not permit delivery of the lethal hit. The pellets are carefully resuspended to maintain intact conjugates and are immobilized on microscope slides in agarose. The slides are incubated at 37°C, stained with a vital dye, and are evaluated microscopically for the determination of: a) percentages of effector cells bound to targets, b) percentages of nonviable target cells, and c) percentages of bound, but not killed, target cells.

The utility of the single cell cytotoxicity assay is that it allows analysis and quantitation of the effector/target cell interaction at each stage of the cytolytic reaction. Thus, it allows for the determination of the percentages of target binding cells, cytolytically active cells, kinetics of target cytolysis, affinity of target cell binding, multiplicity of target binding by effectors, and numerous other parameters that can be directly determined using simple modifications in the study design. The major drawback to current single cell cytotoxicity methods is that they are technically difficult and laborious and suffer from the requisite variability associated with visual assessments involving a limited number of cells.

FLOW CYTOMETRIC ASSESSMENTS OF CELL-MEDIATED CYTOTOXICITY

Flow cytometry offers the technical capability to automate single-cell cytotoxicity assays while providing simultaneous multiparameter assessments of effector/target cell interactions at the population level. Furthermore, flow cytometry allows for the quantitative dissection of the CMC reaction in real time, thereby enabling assessments of the physiologic events that accompany target cell binding and cytolysis. Finally, flow cytometry uniquely offers a means to sort cytolytic cells based on functional parameters, such as the ability to deliver a lethal hit or binding-induced increases in cytosolic Ca^{2+} levels.

Various flow cytometric methods have been developed for characterizing and quantifying the effector-target cell interaction. These range from simple assessments of total target cell lysis by heterogeneous populations of effector cells to multiparameter analysis of the interactions between individual effector and target cells. The basic element of a flow-based assay is the ability to discriminate effector cells from both viable and nonviable target cells. Most flow cytometric approaches to CMC analysis favor the use of one or more fluorochromes, either alone or in conjunction with cell volume measurements, to provide the higher resolution required for discrimination of single effector cells, single target cells, and effector/target conjugates. Effector/target conjugates can be resolved easily by using two-parameter analysis on single-laser instruments and exploiting either the larger size of most targets commonly used in such assays or labeling the target cell with a second fluorochrome. Thus, cytometric data describing the interactions between effector and target cells can be acquired using as few as two parameters in those cases where effector and target cells can be distinguished by cell volumes (Figure 30.2). In addition to allowing for the discrimination between comparably sized effector and target cells, this approach allows for the detection of effector-target conjugates based on the simultaneous expression of fluorescent signal and larger particle size. Some investigators have also exploited the loss of such fluorescent dyes from nonviable target cells as an index of CMC, although uptake of propidium iodide (PI) is more specific and less susceptible to the problem of leakage. By combining multiple fluorochromes to discriminate effector from target cells and to identify nonviable cells, one can determine the relative numbers of individual effector and target cells, effector cells conjugated to viable and nonviable target cells, and the percentage of dead target cells (Figure 30.3). Furthermore, one can sort functionally active effector cells based on their ability to bind and kill target cells (20–22). Table 30.1 depicts the enrichment of cytolytically active NK-cells from mixed populations by sorting effector-target conjugates containing dead target cells defined by PI uptake.

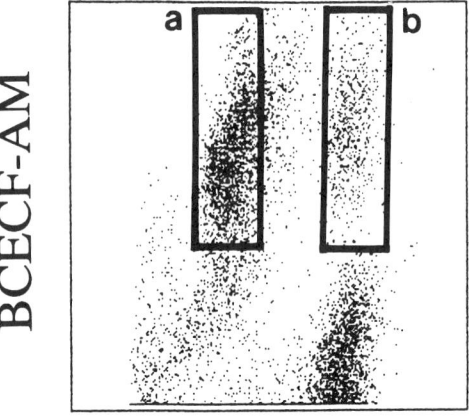

Figure 30.2. Flow cytometric analysis of conjugate formation between NK cells and K562 target cells. BCECF-AM (1mM)-loaded lymphocytes were centrifuged with an equal number of K562 cells, gently resuspended, and analyzed on a FACS Analyzer using logarithmic amplification. Conjugates are identifiable in *Panel B* based on coexpression of the BCECF and increased cell volume signals. Single lymphocytes appear in *Panel A*.

Table 30.1
Fluorescent Probes Used For Cytometric Assessments of Cell-Mediated Cytotoxicity

Probe	Excitation Max. (nm)	Emission Max. (nm)	Comments
FITC	494	520	Significant leakage; fluorescence pH-dependent
FDA (AM)	494	520	Significant leakage; more stable than FITC
c'FDA (AM)	494	520	Leakage; more stable than FDA
BCECF (AM)	494	520	Leakage; more stable than c'FDA
Hydroethidine (dihydroethidium)	482	616	Significant leakage
Tetramethyl rhodamine	541	572	Significant leakage
PKH-1	488	525	Membrane bound; stable
F-18	494	520	Membrane bound; stable
Propidium iodide (PI)	493	630	DNA-binding vital stain; stable
Hoechst 33342	343	483	DNA-binding; leakage occurs with incubation

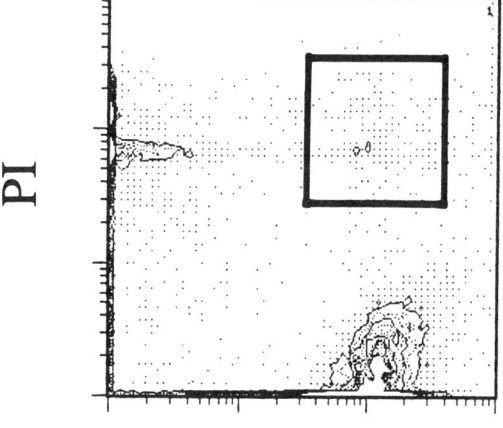

Figure 30.3. Detection of nonviable target cells in effector-target cell conjugates. Conjugates were established using BCECF-AM loaded lymphocytes and K562 target cells and incubated at 37°C for 2 hr. *PI* (propidium iodide) was added to the suspension and cells were analyzed on a FACS Analyzer. Signals corresponding to conjugates between effector cells and nonviable target cells are identifiable within the box.

Investigators have successfully applied single-parameter flow-based analysis for the quantitation of CMC by whole populations and have suggested the utility of using cytometric CMC assays in place of ^{51}Cr release assays. McGinnes et al. loaded K562 cells with carboxyfluorescein diacetate (c'FDA) and then measured the relative loss of this fluorochrome from these cells upon incubation with NK cell-containing populations (3). Although this cytometric approach offers a simple nonradioactive alternative to the ^{51}Cr release assay, the endpoints of c'FDA loss, which constitute "cell death," can be somewhat arbitrary and confounded by leakiness of c'FDA from viable cells. Vitale et al. measured NK-cell cytolytic activity against K562 target cells simply by differences in orthogonal light scatter between effector cells, viable K562 cells, and killed K562 cells (23). K562 cells were easily distinguishable from effector cells based on increased orthogonal scatter, and the orthogonal scatter was observed to increase when K562 cells were lysed by NK cells. Although the results obtained by this method were highly correlated with uptake of PI, it was not confirmed that the PI and orthogonal scatter peaks coincided directly. In consideration of the known morphologic alterations that occur during events leading up to lysis and the nature of cellular characteristics that influence orthogonal scatter, caution must be exercised in extrapolating these results to other target cell types. Even simple substitution of PI uptake-based assays for ^{51}Cr release has demonstrated the utility of flow cytometry in quantifying CMC at the single cell level (24). The exploitation of the larger target cell volume facilitates the use of a vital dye, such as PI, for the detection of killed targets. Zarcone et al. used PI to detect cell death in target cells that were distinguishable from effectors based on cell volume (25). Competitive inhibition of target cell lysis by unrelated target cells could also be evaluated using this technique, provided that the primary target cells were distinguishable from their competitor by volume signals. In general, this approach showed a high correlation with results obtained using standard ^{51}Cr release assays, although ^{51}Cr release appeared to be more sensitive at lower E:T ratios.

More sophisticated approaches, utilizing additional fluorochromes and optical parameters, allow highly detailed analysis and quantitation of the CMC reaction. Storkus et al. analyzed and sorted conjugates formed between FDA-labeled NK effector cells and Hoechst 33342-stained tumor cells using dual-laser flow cytometry and were able to easily distinguish and quantitate NK-cells bound to a variety of targets, many of which were not distinguishable from NK-cells by cell volume alone (20). Conjugates remained intact during cell sorting, indicating their ability to withstand the shearing forces of cytometric analysis and sorting. Furthermore, the use of the DNA fluorochrome Hoechst 33342 allowed the investigators to determine whether target cells in different cell cycle phases exhibited differential sensitivity to binding and killing by NK cells. A similar approach was used by Lebow et al. to quantitate and sort conjugates between murine cytotoxic T-cells and target cells, in which effector cells were labeled with Hoechst 33342 and targets loaded with FITC (21). This approach allowed these investigators to quantitate and sort populations containing single and multiple effector-bound conjugates. Although conjugates were disrupted during the sorting procedure unless they were fixed prior to sorting with paraformaldehyde, this would not present a problem for experiments that do not require intact conjugates following sorting. Alternate approaches exploited the higher resolution offered by two-color systems using single-laser flow cytometers. Cavarec et al. studied conjugate formation between c'FDA-labelled, lymphokine-activated killer cells and hydroethidine-(dihydroethidium) stained lymphoblastoid targets (26). However, significant dye leakage occurred during incubations lasting longer than 15 min and prohibited incubation of cells for periods longer than one hour. The combined use of a single laser and fluorescein-based dyes plus rhodamine also offers a similar multicolor approach. (1).

Although these studies demonstrate the utility of flow cytometric analysis for measurements of CMC, the real potential of flow cytometry to studies of CMC resides in its ability to assess novel questions regarding the biology of CMC using real-time physiologic events during the effector/target interaction. Flow cytometric analysis has been utilized successfully to assess temporal changes in Ca^{2+} flux in NK-cells during binding to NK-sensitive and NK-resistant target cells (22). This approach provided the first direct demonstration that Ca^{2+} flux was associated with the delivery of the lethal hit to sensitive targets and did not occur when effector cells bound nonsensitive target cells. Sorting of conjugates based on levels of NK-cell Ca^{2+} flux revealed that the elevation of Ca^{2+} correlated with NK cells that bound more strongly to target cells. These studies demonstrated that the ability of a target cell to induce NK-cell Ca^{2+} flux correlated with its sensitivity to lysis and that such flux does not result solely from effectors binding to target cells. Flow cytometric analysis also allowed real-time comparisons of the kinetics of effector/target conjugate formation between various populations of cytolytic cells and targets, and provided evidence for the existence of mechanisms that prevent the binding of multiple effectors to a single target cell (2).

A significant problem encountered in many flow-based cytotoxicity assays is the relatively poor stability of dyes used to distinguish effector and target cells during incubation times (2–4 hr) required for complete expression of target cell binding and cytotoxicity. Initial experiments successfully detected CMC mediated by effector cells labeled with FITC (fluorescein isothiocyanate), although leakiness of the cells to this molecule and its known spectral sensitivity to intracellular pH limited its utility in this assay (2). Subsequent approaches utilized low fluorescence fluorescein-based esters that readily crossed cell membranes and were hydrolyzed intracellularly into hydrophilic fluorescent products that were membrane impermeable (3). Effector cells loaded with the more stable carboxy-derivative of FDA, c'FDA, and BCECF-AM (2'7'-bis (2-carboxymethyl)-5-(and 6) carboxyfluorescein acetoxymethylester) have subsequently been used successfully for the assessment of CMC, using either loss of fluorochrome from the cell or PI uptake as an index of cell death (3, 27). Shi et al. added c'FDA at the end of the cytotoxicity reaction and noted that target cell death assessed by the loss of c'FDA was routinely higher than that determined by PI uptake (27). Lebow et al. observed that measurements of conjugation between Hoechst 33342-labeled effector cells were not easily resolvable after 60 min of incubation, due to leakage of Hoechst 33342 from the target cell (21). Similar leakage of other fluorochromes hampers flow cytometry-based assessment of CMC. Although c'FDA and BCECF-AM appear to be less leaky than FDA or FITC, they nevertheless undergo measurable leakage from cells that can affect signal resolution during longer incubation periods. The development of new fluorochromes may overcome this problem and facilitate the application of flow cytometry in CMC experiments requiring longer incubation times. Slezak and Horan recently developed the dye PKH-1, which stably binds the cytoplasmic membrane, excites at 488 nm, and emits at 525 nm (28). PKH-1 used in conjunction with PI allowed the enumeration of NK target-cell conjugates and the detection of PI-stained nonviable targets. Stable expression of PKH-1 allowed long-term cytotoxicity assays (> 18 h). A similar approach utilized the fluorescein-based membrane probe octadecylamine-fluorescein isothiocyanate (F-18) in conjunction with PI to stably label target cells for discrimination of effector-target conjugates (29). This technique allowed for the quantitation of cytolytic activity in whole blood, although the presence of antibodies in serum may have a significant effect on the effector-target interaction. Although it has been pointed out that the incorporation of these dyes into the cell membrane does not affect target cell binding or cytolysis, it is not yet clear whether this is universally true for all effector cell systems.

Other problems concern the definition of "cytolysis." Although PI uptake is considered to be an acceptable index of cell death, the use of PI to detect killed targets may provide lower estimates of total target cell lysis, due to the DNA fragmentation that occurs during cytolysis. The simple measurement of fluorochrome loss as an index of cytotoxicity is hampered by the high levels of spontaneous loss of these molecules from viable cells and often poses a greater problem than the spontaneous release of radionuclides.

Interestingly, despite the possible disruptive influence of shear forces on effector-target conjugate quantitation, significant shearing does not appear to pose a threat to accurate quantitation or collection, based on comparisons between flow cytometry and visual assessments of conjugate formation. Shearing or other disruptive influences are of greater concern when attempting to sort intact conjugates; they may also disrupt the weaker effector/target interactions. Although shearing does not appear to present a problem in most of the systems studied to date, a majority of laboratories using flow cytometry to analyze CMC have utilized effector-target systems exhibiting relatively strong intercellular binding (i.e., NK-cell, cytolytic T-cell lines). Less is known about the effects of flow cytometry analysis on the stability of weaker effector/target cell interactions.

COMPARISONS OF CONVENTIONAL AND FLOW CYTOMETRIC ASSAYS FOR CMC

Despite the flexibility and multiparametric nature of cytometry-based assays, these techniques are better suited for nonroutine, highly specialized dissection of the cytolytic reaction at the single cell level and, therefore, offer a significant improvement over manual single cell assays performed using agarose-immobilized cells. It has been argued by many authors that, compared to ^{51}Cr release assays, cytometric assays for cytotoxicity are a) easier to perform and more reproducible, b) avoid the use of radionuclides, c) require less time for labeling, harvesting, and analysis of samples, d) allow assessments of larger sample sizes, e) offer easier data analysis, and f) are less expensive. However, in practical terms, cytometric assays have not been accepted by investigators for the routine evaluation of cytotoxicity due to a number of attributes that make cytometric determinations less attractive for routine assessments of total cellular cytotoxicity.

The major disadvantage of flow cytometric techniques for assessing cytolytic activity in bulk effector populations is that these techniques would be quite cumbersome when performed at the scale on which experimental assessments of CMC are normally carried out. In normal circumstances, assay conditions are usually performed in replicate, using titrations of effector cells to achieve defined effector-target cell ratios. It is thus not unusual for such assays to comprise over 100 individual samples, which represents a sizable number for individual analysis by flow cytometry. Despite the fact that one might require 1 min to analyze each sample cytometrically and 3–5 min per sample for radionuclide counting, the latter analyses are easily performed automatically, after hours, by automated radionuclide counters. On the other hand, cytometric analysis requires hands-on manipulations by a highly skilled operator.

Although it has also been suggested that flow cytometry offers a less expensive alternative to ^{51}Cr release assays, careful consideration of the real costs to most laboratories of both techniques suggest otherwise. Clearly, the cost of purchasing, maintaining, and operating a flow cytometer must be considered, and these costs alone make it prohibitively expensive for routine analyses. Even when using a centralized core facility for cytometric analyses, the investigator is responsible for users' fees and must schedule around instrument availability. Even if a laboratory has access to a dedicated cytometer available for these assessments, the cost of the considerable technical time required for flow cytometric sample processing entails a significant cost for the performance of these assays. For larger experiments, the length of time before some samples are analyzed may alter the conjugate or cytotoxic frequencies. Any cost advantage of cytometric techniques over radionuclide methods offered by less expensive reagents can be easily offset by these considerations.

Although problems with radionuclide handling and disposal will continue to be an important consideration in the choice of assays, it is notable that many flow cytometry-based cytolytic assays, while not involving radionuclides, nevertheless utilize toxic and carcinogenic agents that present significant handling and disposal problems in their own right, although these issues have been largely overshadowed by radionuclide issues.

Despite the fact that flow cytometry does not provide a practical routine substitute for ^{51}Cr release assays for the assessment of cytotoxic activity at the population level, it does represent a practical and powerful improvement over visual single cell cytotoxicity assays by virtue of its ability to resolve multiple parameters of the lytic event in real time. Furthermore, flow cytometry offers the ability to objectively and rapidly assess CMC simultaneously at the single cell and population levels.

SUMMARY

The development of flow cytometric techniques to assess CMC reactions offers a novel approach for the study of cytolytic cell populations, albeit one which has not been widely accepted by the immunology research community. Instead, cytometry-based CMC assays have been largely relegated to reports in technique-oriented publications. Similar situations are frequently encountered in other applications of flow cytometry due to the fact that, while novel cytometric techniques might be glitzy and fashionable, they may offer few real advantages over existing techniques. The lack of acceptance of these techniques by immunologists reflects the fact that flow cytometric assays of cytotoxicity do not represent reasonable routine alternatives to the ^{51}Cr release assays. Instead, these assays represent a complement to the ^{51}Cr release assay by providing a highly specialized approach best suited either for addressing specific issues relating to cytotoxicity at the single cell level, or for assays in which the

^{51}Cr release assay is less well suited (i.e., poor labeling). In this regard flow cytometric assays provide a powerful alternative to current methods of evaluating CMC at the single cell level and dissecting the components of the lytic process. However, for the assessment of cytolytic activity in bulk effector cell populations, the ^{51}Cr release assay continues to be the technique of choice due to its simplicity and convenience.

ACKNOWLEDGMENTS

Supported by Grant CA48069 from the National Cancer Institute.

REFERENCES

1. Berke G. Enumeration of lymphocyte-target cell conjugates by cytofluorimetry. Eur J Immunol 1985;15:337-340.
2. Perez P, Bluestone JA, Stephany DA, Segal DM. Quantitative measurements of the specificity and kinetics of conjugate formation between cloned cytotoxic T lymphocytes and splenic target cells by dual parameter flow cytometry. J Immunol 1985;134:478-485.
3. McGinnes K, Chapman G, Marks R, Penny R. A fluorescence NK assay using flow cytometry. J Immunol Meth 1986;86:7-15.
4. Henkart PA. Mechanism of lymphocyte-mediated cytotoxicity. Annu Rev Immunol 1985;3:31-58.
5. Townsend A, Bodmer H. Antigen recognition by class-I restricted T lymphocytes. Annu Rev Immunol 1989;7:601-624.
6. Lanier LL, My Le A, Civin CI, Loken MR, Phillips JH. The relationship of CD16 (Leu-11) and Leu 19 (NKH-1) antigen expression on human peripheral blood NK cells and cytotoxic T lymphocytes. J Immunol 1986;136:4480.
7. McMannis JD, Fisher RI, Creekmore SP, Braun DP, Harris JE, Ellis TM. In vivo effects of recombinant IL-2, I. Isolation of circulating Leu 19+ lymphokine-activated killer effector cells from cancer patients receiving recombinant IL-2. J Immunol 1988;140:1335-1340.
8. Phillips JH, Lanier LL. Dissection of the lymphokine-activated killer phenomenon. Relative contributions of peripheral blood natural killer cells and T lymphocytes to cytolysis. J Exp Med 1986;164:1193.
9. Herberman RB, Reynolds CW, Ortaldo JR. Mechanism of cytotoxicity by natural killer (NK) cells. Annu Rev Immunol 1986;4:651-680.
10. MacDonald HR, Glasebrook AL, Cerrotini JC. Clonal heterogeneity in the functional requirement for Lyt-2/3 molecules on cytolytic T lymphocytes: analysis by antibody blocking and selective trypsinization. J Exp Med 1982;156:1711.
11. Biddison WE, Rao PE, Talle MA, Goldstein G, Shaw S. Possible involvement of the T4 molecule in T cell recognition of class II HLA antigens: evidence from studies of CTL-target cell binding. J Exp Med 1984;159:783.
12. Martz E, Heagy W, Gromkowski SH. The mechanism of CTL-mediated killing: monoclonal antibody analysis of the roles of killer and target cell membrane proteins. Immunol Rev 1983;72:73.
13. Krensky AM, Sanchez-Madrid F, Robbins E, Nagy J, Springer TA, Burakoff SJ. The functional significance, distribution and structure of LFA-1, LFA-2, and LFA-3: cell surface antigens associated with CTL-target interaction. J Immunol 1983;131:611-619.
14. Martz E. Mechanism of specific tumor-cell lysis by alloimmune T-lymphocytes: resolution and characterization of discrete steps in the cellular interaction. Contemp Top Immunobiol 1977;7:301-361.
15. Goldstein PA, Smith ET. The lethal hit stage of mouse T and non-T cell mediated cytolysis: differences in cation requirements and characterization of an analytical cation pulse method. Eur J Immunol 1976;6:31-37.
16. Tschopp J, Jongeneel CV. Cytotoxic T lymphocyte mediated cytolysis. Biochemistry 1988;27:2641-2646.
17. Tschopp J, Nabholz M. Perforin-mediated target cell lysis by cytolytic T lymphocytes. Annu Rev Immunol 1990;8:279-302.
18. Kupfer A, Singer SJ. Cell biology of cytotoxic and helper cell functions: immunofluorescence microscopic studies of single cells and cell couples. Annu Rev Immunol 1989;7:309-337.
19. Bonavida B, Bradley TP, Grimm EA. Frequency determination of killer cells by a single cell cytotoxicity assay. Methods Enzymol 1983; 93:270.
20. Storkus WJ, Balber AE, Dawson JR. Quantitation and sorting of vitally stained natural killer cell-target cell conjugates by dual beam flow cytometry. Cytometry 1986;7:163-170.
21. Lebow LT, Stewart CC, Perelson AS, Bonavida B. Analysis of lymphocyte-target cell conjugates by flow cytometry. I. Discrimination between killer and non-killer lymphocytes bound to targets and sorting of conjugates containing one or multiple lymphocytes. Nat Immun Cell Growth Reg 1986;5:221-237.
22. Edwards BS, Nolla HA, Hoffman RR. Relationship between target cell recognition and temporal fluctuations in intracellular Ca^{2+} of human NK cells. J Immunol 1989;143:1058-1065.
23. Vitale M, Neri LM, Comani S, et al. Natural killer function in flow cytometry. II. Evaluation of NK lytic activity by means of target cell morphological changes detected by right angle scatter. J Immunol Meth 1989;121:115-120.
24. Landay AL, Zarcone D, Grossi CE, Bauer K. Relationship between target cell cycle and susceptibility to natural killer lysis. Cancer Res 1987;47:2767-2770.
25. Zarcone D, Tilden AB, Cloud G, Friedman HM, Landay A, Grossi CE. Flow cytometry evaluation of cell-mediated cytotoxicity. J Immunol Meth 1986;94:247-255.
26. Cavarec L, Quillet-Mary A, Fradelizi D, Conjeaud H. An improved double fluorescence flow cytometry method for the quantification of killer cell/target cell conjugate formation. J Immunol Methods 1990; 130:251-261.
27. Shi T-X, Tong MJ, Bohman R. The application of flow cytometry in the study of natural killer cell cytotoxicity. Clin Immunol Immunopathol 1987;45:356-365.
28. Slezak SE, Horan PK. Cell-mediated cytotoxicity. A highly sensitive and informative flow cytometric assay. J Immunol Methods 1989;117: 205-214.
29. Radosevic K, Garritsen HSP, Van Graft M, De Grooth B, Greve J. A simple and sensitive flow cytometric assay for the determination of the cytotoxic activity of human natural killer cells. J Immunol Methods 1990;135:81-89.

31

Measurements of Cell Physiology: Ionized Calcium, pH, and Glutathione

PETER S. RABINOVITCH, CARL H. JUNE, and TERRANCE J. KAVANAGH

The flow cytometer can be used to measure a variety of functional parameters that are of increasing interest to cell biologists. The recent development of a number of new fluorescent probes now permits the measurement of various intracellular free ion concentrations in single living cells. Among these ions are calcium, magnesium, sodium, potassium, and hydrogen (pH). In addition, intracellular glutathione, crucial for maintaining the physiological redox state, can be easily and accurately measured by flow cytometry. Most previously available techniques to measure cellular activation parameters determined the mean value for a mixed population of cells. The flow cytometer has the unique capacity to permit the measurement of physiologic parameters in large numbers of single cells; it allows correlation with other parameters, such as immunophenotype and cell cycle; and, finally, it reveals heterogeneity within the cell population, sometimes even in cells that were previously thought to be homogeneous. In this chapter, flow cytometric techniques to measure intracellular calcium concentration, pH, and glutathione as well as their applications are described.

INTRACELLULAR IONIZED CALCIUM

Introduction

Eukaryotic cells in their resting state maintain an internal calcium ion concentration that is far below that of the extracellular environment. Ionized calcium has an important role as a mediator of transmembrane signal transduction, and elevations in intracellular ionized calcium concentration ($[Ca^{2+}]_i$) regulate diverse cellular processes. Measurement of $[Ca^{2+}]_i$ in living cells is thus of considerable interest to investigators over a broad area of immunology and cell biology.

Calcium influx is thought to be initiated by membrane depolarization, which opens voltage-gated channels, or by the binding of ligands to receptor-operated channels. The latter case is more commonly encountered by investigators using flow cytometry, as these mechanisms predominate in cells that are not electrically excitable; in this case, the binding of agonist to its specific membrane receptor activates enzymatic processes that result in the activation of phospholipase C. In many cases, this process requires the intervening activation of a guanine nucleotide binding (G) protein. Phospholipase C causes the hydrolysis of a membrane phospholipid, phosphatidylinositol 4,5-bisphosphate (PIP2), which yields a water soluble product, inositol 1,4,5-trisphosphate (IP3), and a lipid, 1,2-diacylglycerol (DAG). IP3 then causes the release of calcium from intracellular stores while DAG, in conjunction with calcium ions, activates protein kinase C. Thus, a single agonist can result in the production of at least two "second messengers," making this pathway a unique, bifurcating system. Calcium is therefore a "third messenger" that controls numerous cellular processes, activating a broad variety of enzyme systems, both as a cofactor and in conjunction with the calcium-binding protein calmodulin. While it appears clear that the initial elevation of ionized calcium is due to the release of intracellular calcium stores, little is known about regulation of the influx of calcium from extracellular sources that is necessary to sustain the response.

Indicators of Intracellular Ionized Calcium Concentration

Until 1982, it was not possible to measure $[Ca^{2+}]_i$ in small intact cells and attempts to measure cytosolic free calcium were restricted mostly to large invertebrate cells where the use of microelectrodes was possible. Bioluminescent indicators, such as aequorin, a calcium-sensitive photoprotein, are well suited for certain applications (1). Their greatest limitation is the necessity for loading into cells by microinjection or other forms of plasma membrane disruption. $[Ca^{2+}]_i$ was first measured in diverse populations of cells with the development of quin2 (2). The indicator was easily loaded into small intact cells using a chemical technique developed by Tsien (3). Cells are incubated in the presence of the acetoxymethyl ester of quin2. This uncharged form is cell permeant and diffuses freely into the cytoplasm, where it serves as a substrate for esterases. Hydrolysis releases the tetraanionic form of the dye that is trapped inside the cell. Unfortunately, quin2 has several disadvantages that limit its application to flow cytometry (4). Its relatively low extinction coefficient and quantum yield have made detection of the dye at low concentrations difficult; at higher concentrations, quin2 itself

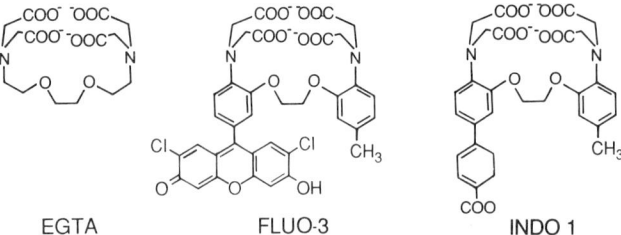

Figure 31.1. Structure of the [Ca^{2+}]$_i$ probes *indo-1* and *fluo-3* as well as, for reference, the structure of the calcium chelator *EGTA*. Note the identical calcium-binding domain in all three compounds.

buffers the [Ca^{2+}]$_i$. Subsequently, Grynkiewicz et al. (5) described a new family of highly fluorescent calcium chelators that overcome most of the above limitations. One of these days, indo-1 ([1-[2amino-5-[6-carboxylindol-2-yl]-phenoxy]-2-[2'-amino-5'-methlyphenoxy] ethane N,N,N'N'-tetraacetic acid]) (Fig. 31.1), has spectral properties that make it especially useful for analysis with flow cytometry. In particular, indo-1 exhibits large changes in fluorescent emission wavelength upon calcium binding (Fig. 31.2). As described below, use of the ratio of intensities of fluorescence at two wavelengths allows calculation of [Ca^{2+}]$_i$ independent of variability in cellular size or intracellular dye concentration. The only significant drawback to the use of indo-1 is the requirement for ultraviolet (UV) excitation.

A practical alternative to indo-1 became available upon the description by Minta et al. (6) of a fluorescein-based, calcium-sensitive probe, fluo-3 (Fig. 31.1). This dye exhibits an increase in fluorescence intensity with increasing [Ca^{2+}]$_i$ (Fig. 31.3). Fluo-3 is less sensitive than quin2 and indo-1 at detecting small changes in [Ca^{2+}]$_i$, in part, because the Kd is higher (400 nM); conversely, this may be an advantage if the experimental situation involves the distinciton between stimuli, all of which produce large (above 400 nM) [Ca^{2+}]$_i$ responses. An additional advantage with fluo-3 is that it can be used with other probes such as caged calcium chelators that may themselves require UV excitation (7). Furthermore, the use of fluo-3 permits, for the first time, the simultaneous use of other UV-excitable probes for flow cytometry, such as those used for cell cycle analysis or measurement of intracellular glutathione, allowing correlation of these parameters with calcium responses. The primary disadvantage of fluo-3 is that it does not have fluorescence properties that allow ratiometric determinations. Therefore, calibration on a flow cytometer is more complicated because the signal is proportional to cell size and dye concentration as well as [Ca^{2+}]$_i$. Thus, the ability to measure responses in subsets of cells is more limited because the broad distribution of fluorescence intensities of unstimulated cells often results in an overlapping distribution of the values from stimulated and unstimulated cells. This problem can be minimized by loading cells in the presence of pluronic F-127 (see below), which minimizes the cell-to-cell variation in loading with fluo-3 (8), or by the simultaneous use of a second dye that serves as an indicator of the magnitude of dye loading in an individual cell. Rijkers et al. (9) have suggested using the ratio of fluo-3 to SNARF-1 (SemiNaphthoRhodaFluor) fluorescence for this purpose (see subsequent section on pH).

Preparation of Cells for Indo-1 or Fluo-3 Analysis

Uptake and retention of indo-1 and fluo-3 is facilitated by the use of their penta-acetoxymethyl esters, using the scheme described above. Approximately 20% of the total dye is trapped in this manner during typical loadings. After loading, the extracellular dye should be diluted 10- to 100-fold before flow cytometric analysis (10). A typical protocol for cell preparation might be:

1. Incubate $0.5 - 20.0 \times 10^6$ cells/ml in medium with 1–3 μM fluo-3 or indo-1 (acetoxymethyl esters) at 37°C for 30 min. For cells that are difficult to load, or that result in heterogeneous loading, 0.02% (final vol/vol) pluronic detergent F-127 (Molecular Probes) may be added to the above before incubation, with subsequent incubation at 30°C (11, 12). An additional option for the use of fluo-3 is to retard transport of the deesterified dye out of the cell by adding 4 mM (final concentration) probenecid during loading. In initial experiments, we suggest using lymphocytes, as they are easily and reliably loaded. Later, when the technique is validated on the flow cytometer, other cell types can be utilized.

2. For experiments with simultaneous immunofluorescence, incubate aliquots of the above with saturating concentrations of R-phycoerythrin (PE)-conjugated antibody (with fluo-3) or fluorescein isocyanate (FITC)- and/or PE-conjugated antibody (with indo-1) at room temperature for 30 min. Antibodies should be azide-free in order not to poison cellular metabolism.

3. Centrifuge the above for 6 min at 180g at room temperature. Gently resuspend (do not vortex) in medium at the desired cell concentration (usually $\sim 10^6$/ml) at room temperature. The medium used for resuspension and analysis can be dictated primarily by the metabolic requirements of the cells, subject only to reasonable pH buffering (preferably not solely bicarbonate, as it will become alkaline) and the requirement that mM concentrations of calcium be present. If analysis of Ca^{2+} released from intracellular stores (independent of influx of extracellular Ca^{2+}) is desired, addition of 5 mM EGTA (ethylene glycol bis (β-aminoethyl ether)-N,N,N',N'-tetraacetic acid) to the cell suspension will reduce Ca^{2+} to ~ 20 nM, thus abolishing the usual cross-membrane gradient.

4. Hold cells at room temperature ($\sim 22°C$) until ready for analysis. Leakage and compartmentalization of both dyes is accelerated at higher temperatures. Warm an aliquot of cells to 37°C for at least 5 min before analysis. Analyze at 37°C. For a 10-min analysis, use at least 0.5 ml (usually containing $\sim 10^6$ cells), so that cell transit is rapid from the warm sample chamber to the flow cell (this is usually through unheated small caliber tubing). [Ca^{2+}]$_i$ responses are highly temperature-dependent and reduced or even absent calcium responses will be seen at lower temperatures.

5. After obtaining a baseline measurement for a sample, kinetic analyses are typically performed by quickly ceasing flow, removing the sample container, adding agonist, and rapidly restarting flow (more sophisticated and rapid sample injection schemes are possible, but for the great majority of applications, the above will suffice).

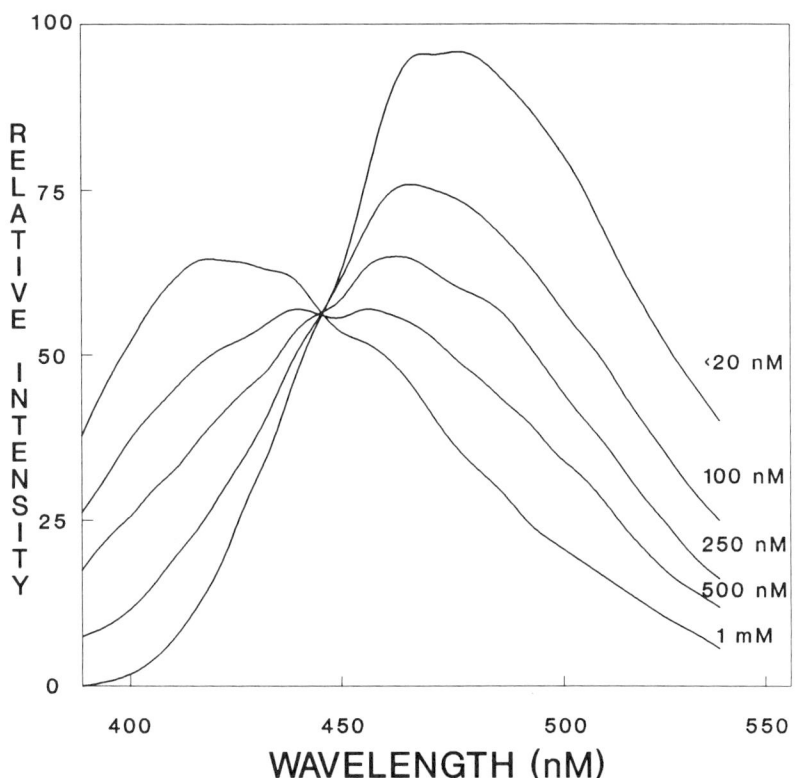

Figure 31.2. Emission spectra for indo-1 as a function of ionized calcium concentration. Fluorescence excited at 356 nm was measured in a spectrofluorimeter. At wavelengths below 400 nm, emission in the absence of calcium drops to a few percent of that seen in the presence of calcium. About 500 nm, there is more than a fourfold converse difference. The [Ca^{2+}] of EGTA buffer solutions used is shown at the right.

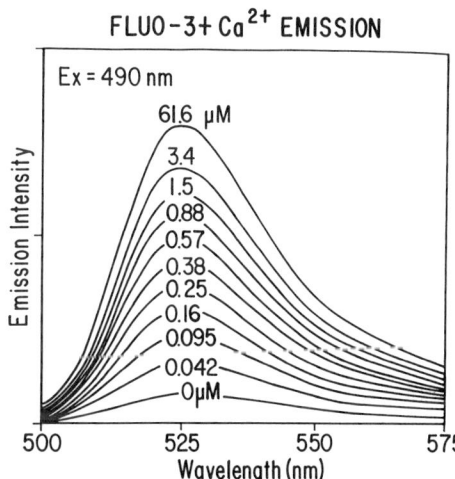

Figure 31.3. Emission spectra for fluo-3 as a function of ionized calcium concentration. The concentration of Ca^{2+} in the buffer solutions is indicated for each individual spectrum. (Courtesy of R. Haugland, Molecular Probes, Eugene, OR).

6. Calibrate using one of the approaches discussed subsequently. If using ionomycin, it is necessary to be scrupulous in removing residual ionophore before analyzing the next sample. Use washes with bleach and/or dimethyl sulfoxide (DMSO) to facilitate the removal of traces of ionomycin.

One incidental benefit of the above loading strategy is that this procedure, like the more familiar use of fluorescein diacetate or carboxyfluorescein diacetate, allows one to distinguish between live and dead cells. The latter will not retain the hydrophilic impermeant dye and can be excluded by gating during subsequent analysis. The lower limit of useful intracellular loading concentrations of both indo-1 and fluo-3 is determined by the sensitivity of fluorescence detection of the flow cytometer, and the upper limit is determined by avoidance of buffering of [Ca^{2+}]$_i$ by the presence of the calcium chelating dye itself. In practice, one should use the least amoung of dye that is necessary to reliably quantitate the fluorescence signal. In the case of indo-1, the dye has excellent fluorescence characteristics (30-fold greater quantum yield than quin2; (5) and useful ranges of indo-1 loading are much lower than the millimolar amounts required with quin2. In using the newer dye fluo-3, one should begin with the same lower concentration that is found to be suitable with indo-1 and increase loading only if detection against background autofluorescence becomes limiting (autofluorescence in the green spectrum is greater than in the blue or violet spectrum used by indo-1). For human peripheral blood T-cells, we have found adequate detection of both dyes at or above 1 μM (for indo-1 intracellular concentrations achieved with this loading are ~5 μM). Buffering of [Ca^{2+}]$_i$ in human T-cells can be observed as a slight delay in the rise in [Ca^{2+}]$_i$ and a retarded rate of return of [Ca^{2+}]$_i$ to baseline values. In the case of indo-1, this was seen when loading concentrations above 3 μM (22 μM intracellular concentration) were used. A reduction in peak [Ca^{2+}]$_i$ occurred at even higher indol-1 concentrations (10). Chused et al. (13) have observed slightly greater sensitivity of murine B-cells to indo-1 buffering, recommending a loading concentration of no greater than 1 μM. In side by side comparisons, we have found that calcium transients in B-cells are much more sensitive to the effects of buffering by indo-1 than are T-cells. For human platelets, a 2 μM indo-1 loading concentration has been reported (14). Rates of loading of the dye esters can be expected to vary between cell types, perhaps as a consequence of variations in intracellular esterase activity. In peripheral human blood, more rapid rates of loading are seen in platelets and monocytes than in lymphocytes. Even within one cell type, donor- or treatment-specific factors may affect loading; for example, lower rates of indo-1 loading are seen in splenocytes from aged mice than from young mice (15).

Indo-1 has been found to be remarkably nontoxic to cells subsequent to loading. Analysis of the proliferative capacity of human T-lymphocytes (10) loaded with indo-1 has shown no adverse effects on the ability of cells to enter and complete three rounds of the cell cycle. Similar results have been obtained with murine B-lymphocytes (13). This is especially pertinent to the viable sorting of indo-1 loaded cells based on $[Ca^{2+}]_i$, as described subsequently.

Flow Cytometric Analysis Technique

The instrumental setup for fluo-3 analysis is simply that used for routine FITC analysis, i.e., 488 nm excitation and detection of emission in a band centered near 525 nm. Considerations for indo-1 analysis are slightly more complex. The absorption maximum of indo-1 is between 330 and 350 nm, depending upon the presence of calcium (5); this is well suited to excitation at either 351–356 nm from an argon-ion laser, or 337–356 nm from a krypton-ion laser, and (although spectrally slightly less optimal) almost the same performance can be obtained by excitation at 325 nm using a helium-cadmium Laser. Laser power requirements depend upon the choice of emission filters and the optical efficiency of the instrument; however, under optimal optical conditions (i.e., using quartz flow cells) as little as 5–10 mW excitation appears satisfactory, within the range available from helium-cadmium lasers. In this regard, the stability of the intensity of the excitation source is less important in this application because of the use of the ratio of fluorescence emissions.

An increase in $[Ca^{2+}]_i$ is detected with indo-1 as an increase in the ratio of fluorescence intensity from a lower to a higher emission wavelength. The optimal strategy is to select bandpass filters so that the collection of light near the isosbestic point is minimized and the collection of fluorescence that exhibits the largest variation in calcium-sensitive emission is maximized. The choice of filters used to select these wavelengths is dictated by the spectral characteristics of the shift in indo-1 emission upon binding to calcium (Fig. 31.2). The original spectral curves published for indo-1 (5) did not depict the large amounts of indo-1 emission in the blue-green and green wavelengths; in practice, we find that there is more light available in the blue region than in the violet region, although the dynamic range of the calcium-sensitive changes in the violet region exceeds that of the blue region (Fig. 31.2). The choice of filter combinations can lead to substantial differences in the magnitude of observed changes in the violet-blue ratio, perhaps best summarized by the value R_{max}/R, the range of change in indo-1 ratio observed from resting intracellular calcium (R) to saturated calcium (R_{max}). Filters centered at 530 nm (green) and 395 nm (violet) result in a R_{max}/R ratio of 9.0 (individual instruments may vary somewhat). Replacing the 395-nm emission filter with one centered on the peak "violet" emission of the calcium-bound indo-1 dye (405 nm) reduces R_{max}/R to 7.9. Instead, replacing the 530-nm filter with one centered on the peak "blue" emission of calcium-free indo-1 dye (485 nm) reduces R_{max}/R to 6.6. Thus, a greater dynamic range of the indo-1 ratio is observed when light nearer to the isosbestic (450 nm) is avoided. Note that if a 525- to 530-nm bandpass filter is used for the longer indo-1 emission wavelength, then this same filter and photomultiplier tube can be used for (temporally displaced and gated) FITC emission.

Calibration of Fluorescence to $[Ca^{2+}]_i$

Prior to the development of indo-1, $[Ca^{2+}]_i$ determination with quin2 fluorescence was sensitive to cell size and intracellular dye concentration as well as $[Ca^{2+}]_i$. This made necessary the calibration of each loaded sample (in fact, strictly speaking of each individual assay) by determination of the fluorescence intensity of the dye at known Ca^{2+} concentrations, or at least at zero and saturating Ca^{2+}. This is still the case for analyses with fluo-3. In contrast, with indo-1, use of the Ca^{2+}-dependent shift in dye emission wavelength allows the ratio of fluorescence intensities of the dye at the two wavelengths to be used to calculate $[Ca^{2+}]_i$:

$$[Ca^{2+}]_i = Kd \cdot \frac{(R - R_{min})}{(R_{max} - R)} \cdot \frac{S_{f2}}{S_{b2}} \quad [Eq. 31.1]$$

where Kd is the effective dissociation constant (250 nM at 37°C, pH 7.05), R, R_{min}, and R_{max} are the fluorescence intensity ratios of violet-blue fluorescence at the experimental, zero, and saturating $[Ca^{2+}]_i$, respectively; and S_{f2}/S_{b2} is the ratio of the blue fluorescence intensity of the calcium-free and calcium-bound dye, respectively (5). Note that the Kd varies dramatically as a function of temperature, pH, and ionic strength. The term S_{f2}/S_{b2} is a constant that depends in largest part on the filters used for indo-1 analysis. Determination of R_{max}, R_{min} and S_{f2}/S_{B2} for calibration of indo ratios allows the direct determination of $[Ca^{2+}]_i$ using the above equation. Determintion of R_{max} is easily performed by addition of the calcium ionophore ionomycin to a cell sample (1–3 μg/ml final). Unfortunately, simply adding EGTA to cells treated with ionomycin and metabolic poisons does not reduce the fluorescence ratio of cells to R_{min}. The details of a strategy that makes use of a spectrofluorimeter in addition to the flow cytometer has been previously described (10). Because the ratiometric analysis is independent of cell size and total intracellular dye concentration as well as of instrumental variation in efficiency of excitation or emission detection, it is not necessary to measure the fluorescence of the dye in the calcium-free and calcium-saturated states for each individual assay. In principle, it is sufficient to calibrate the instrument once, after which only R is measured for each subsequent analysis. In practice, the minimum quality control should include determination of the value of R_{max}/R (cells treated with ionomycin/resting cells) for each experiment; day-to-day optical variations in the flow cytometer are usually minimal (with the same filter set) and, thus, a narrow range of R_{max}/R values should be obtained.

As an alternative to the combined use of a flow cytometer with a spectrofluorimeter, Parks et al. (16) have proposed that, with minor modification, the flow cytometer may be used as a spectrofluorimeter. In essence, the fluorescence of a steady stream of dye is measured by the photomultiplier and the photomultiplier voltage is analyzed as in a standard spectrofluorimeter, or, even beter, the voltage is electronically converted to a pulse for processing by the flow cytometer. An alternative approach to create such a pulse is to rapidly strobe the laser, such that the beam is illuminated for only µsec intervals. Preliminary experience with this technique suggests that it is possible to determine all of the constants necessary for calibration directly on the flow cytometer using EGTA-depleted and calcium-saturated indo-1 solutions.

An alternative approach to calibration can be based upon a regression curve that relates R to treated cells suspended in a series of precisely prepared calcium buffers. Chused et al. (13) have suggested that ionomycin plus a cocktail of metabolic poisons be used to collapse the calcium gradient to zero ($[Ca^{2+}]_i = [Ca^{2+}]_o$). Thus, this technique allows one to estimate $[Ca^{2+}]_i$ without the need to determine R_{min}, S_{f2} or S_{b2}, although it is subject to limitations in the precision with which one can prepare a series of calcium buffers that yield known and reproducible free calcium concentrations. Accuracy of prediction of ionized Ca^{2+} concentration in buffer solutions depends on a variety of interacting factors, so that care must be exercised in formulating Ca^{2+} standards. The ionized calcium concentration in an EGTA buffer system depends on the magnesium concentration; other metals such as aluminum, iron, and lanthanum also bind avidly to EGTA (17). In addition, the dissociation constant of Ca^{2+}-EGTA is a function of pH, temperature, and ionic strength (review 1, 19). For example, in an EGTA buffer (total EGTA 2 mM, total Ca^{2+} 1 mM, ionic strength 0.1 at 37°C), changing the pH from 7.4 to 7.0 can result in the ionized calcium increasing by more than 200 nM, a change that is approximately twice the magnitude of that found in resting cells and that is easily measured on a flow cytometer. Finally, it is important to prepare the buffers using the "pH metric technique" (19), in part because of the varying purity of commercially available EGTA (20). Detailed methods and a computer program that will determine $[Ca^{2+}]$ as a function of pH, total calcium and magnesium concentrations, temperature, and ionic strength are available from one of this chapter's authors (C. June). A practical alternative for obtaining buffers is made possible by the recent commercial availability of carefully prepared calcium buffer solutions (Molecular Probes, Eugene, OR).

Display and Analysis of Results

Analyses with fluo-3 require the observation of changing fluorescence intensities, while the analysis of indo-1 (and of fluo-3, if normalized for cellular loading differences by using the ratio to fluorescence of a second dye) requires determination of a ratio of two fluorescence intensities. Fortunately, commercial flow cytometers all have some provision for a direct calculation of the value of the fluorescence ratio, either by analog circuitry or by digital computation. If cellular indo-1 loading is extremely heterogeneous, it may be desirable to work with a logarithmic conversion of "violet" and "blue" emission intensities in order to observe a broader range of cellular fluorescence. In this case, the instrument must permit the logarithm of the ratio to be calculated by **subtraction** of the log "blue" from the log "violet" signals (10).

Plotted as a histogram of the ratio values, quiescent cell populations stained with indo-1 show narrow distributions of ration, even when cellular loading with indo-1 is very heterogeneous, and coefficients of variation (CVs) of less than 10% are common (Figs. 31.4A and 31.4I). For simple fluo-3 analyses, variation in dye loading will result in wider CVs (Figs. 31.4B and 31.4J). In either case, the effects of perturbation of $[Ca^{2+}]_i$ by agonists can be noted by changes in the histogram profiles, observed by storing histograms sequentially over time, with subsequent analysis of data.

A more informative and elegant display is obtained by a bivariate plot of indo-1 ratio or fluo-3 fluorescence intensity vs. time. The bivariate data can be displayed as "dot plots" on which the indo-1 ratio or fluo-3 intensity of each cell (proportional to $[Ca^{2+}]_i$) is plotted on the Y-axis vs. time on the X-axis (Figures 4A and 4B). Alternatively, the data can be presented as "isometric plots" in which the X-axis represents time, the Y-axis is indo-1 ratio or fluo-3 intensity, and the Z-axis is the number of cells (Figs. 31.4K and 31.4L). In these bivariate plots, kinetic changes in $[Ca^{2+}]_i$ are seen with much greater resolution, limited only by the number of channels on the time axis, the interval of time between each channel, and the rate of cell analysis. Changes in the fraction of cells responding, the mean magnitude of the response, and the heterogeneity of the responding populaiton are best observed with these displays. For example, it can be seen in Figures 31.4A and 31.4K that for $CD4^+$ T-cells at 1 min after the addition of antibody to CD3, the response is highly heterogeneous, with some cells not yet responding and other cells achieving very high levels of $[Ca^{2+}]_i$. Figures 31.4B and 31.4L show that when the analysis of fluo-3 and indo-1 in the same cells are compared, the presence of the broader CV of the measurement with fluo-3 results in a diminished ability to resolve the heterogeneity of the response.

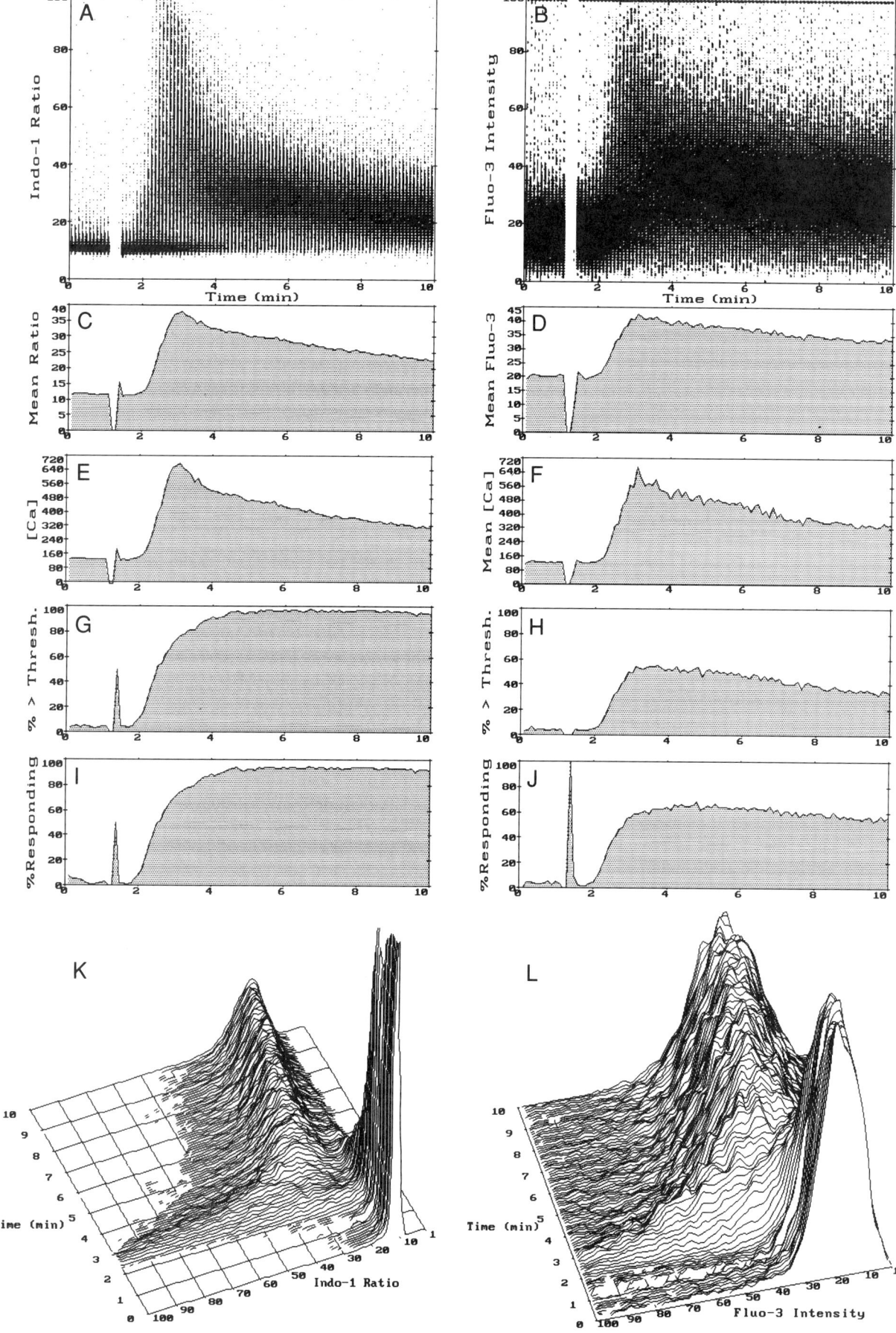

Figure 31.4. A-L

For many purposes, it is simpler and preferable to examine and compare a univariate parameter describing the cellular response vs. time. There are several such parameters that can be derived from the bivariate data for this purpose. Calculation of the mean Y-axis value for each X-axis time interval allows presentation of the data as mean ratio or fluorescence intensity vs. time (10), as shown in Figures 31.4C and 31.4D. Calibration of the ratio or intensity to $[Ca^{2+}]_i$ allows data presentation in the same manner as traditionally displayed by spectrofluorimetric analysis, i.e., mean $[Ca^{2+}]_i$ vs. time (Figs. 31.4E and 31.4F). While this presentation yields much of the information of interest in an easily displayed format, data relating to the heterogeneity of the $[Ca^{2+}]_i$ response are lost. Some of this information can be displayed by a calculation of the "proportion of responding cells." If a threshold value of the resting ratio distribution is chosen, i.e., one at which only 5% of control cells are above the threshold value, the proportion of cells responding by ratio elevations above this threshold vs. time yields a presentation that is informative of the heterogeneity of the response (Figs. 31.4G and 31.4H). A more sophisticated measure of the proportion of responding cells can be derived by the application of histogram subtraction techniques as illustrated in Figures 31.4I and 31.4J: The measurement of unperturbed cells in the initial period of observation can be used as the "control" fluorescence or ratio distribution; at each subsequent time interval, the cumulative subtraction technique (21) has been used to subtract the control distribution from the observed distribution. Comparison of Figure 31.4H with Figure 31.4J shows that the latter technique is less sensitive to the shape and width of the control distribution; this difference is more dramatic in Figure 31.5, described below.

Simultaneous Analysis of $[Ca^{2+}]_i$ and Other Fluorescence Parameters

The most common application for use of additional fluorochromes with $[Ca^{2+}]_i$ analysis will be the determination of cellular immunophenotype simultaneously with the $[Ca^{2+}]_i$ assay, allowing alterations in $[Ca^{2+}]_i$ to be examined in, and correlated with, specific immunophenotypic subsets. FITC- and PE-conjugated antibodies can be used with indo-1, and PE and allophycocyanine (APC) (or other red-excited dyes) can be used with fluo-3. Only the fluo-3/PE combination can be performed with a single-laser flow cytometer, the other combinations require a dual-laser flow cytometer.

Numerous examples of the analysis of $[Ca^{2+}]_i$ in immunophenotypically defined subsets have been described (review, 22). Several illustrative examples will demonstrate the utility of this approach. Figure 31.5 shows the simultaneous analysis of CD4+ and CD8+ subsets of PBL, performed using two-color immunofluorescence simultaneously with indo-1. This figure is also chosen in order to illustrate the importance of the interaction of receptors on the cell surface (in this case by cross-linking) and the relative contributions of internal and external calcium stores.

Addition of an anti-CD3 antibody alone (3 µg/ml, UCHT-1, Dako Corp) in this experiment had only a small effect upon $[Ca^{2+}]_i$ in either CD4+ or CD8+ subsets (see 2- to 4-min interval of Fig. 31.5). This antibody differs from that of Figure 31.4, in that the latter produces a greater $[Ca^{2+}]_i$ response in the absence of further cross-linking. When saturating amounts of goat antimouse antibody is added (at time 4 min), an intracellular calcium response is seen that is greater in CD4+ cells than in CD8+ cells. As the extracellular medium used in this experiment contained little Ca^{2+}, the response from 4 to 7.5 min represents mobilization from intracellular stores. This response is characteristically of short duration (compare Figs. 31.5C and 31.5D to Figures 31.4E and 31.4F), which is the strongest evidence that the extended phase of elevated $[Ca^{2+}]_i$ is maintained by influx of extracellular Ca^{2+}. To illustrate that calcium channels have, in fact, been opened by the cross-linked stimulus in Figure 31.5, Ca^{2+} was added to the extracellular medium at time 7.5 min; this is accompanied by an immediate elevation of $[Ca^{2+}]_i$, with a sustained response. Again, the magnitude of

Figure 31.4. Comparison of indo-1 and fluo-3 measurement of $[Ca^{2+}]_i$ and methods of displaying and analyzing calcium signaling in single cells as a function of time. Human PBL were loaded simultaneously with 3 µg/ml indo-1 and 3 µg/ml fluo-3, stained with PE-CD4 mAb, and then stimulated with anti-CD3 antibody (10 µg/ml G19-4, a gift from Dr. Jeff Ledbetter, Oncogen, Seattle, WA) at approximately 1 min after the start of analysis (note gap in data acquisition). Violet and green indo-1 emission (UV-excited) and green fluo-3 and orange PE emission (temporally delayed 488-nm excited) were collected simultaneously for each cell. Only results gated from PE-CD4+ cells are displayed. In A and B, the results are displayed as a "dot plot" in which $[Ca^{2+}]_i$ is plotted for each cell analyzed on a 100 × 100 pixel grid, where the X-axis is time and the Y-axis is indo-1 ratio or fluo-3 fluorescence intensity. The number of cells per pixel is displayed by darkness that ranges over 12 levels. At the bottom, K and L, isometric plots of the same experiment in A are shown; sequential histograms are plotted in which the X-axis represents time, the Y-axis $[Ca^{2+}]_i$ or fluo-3 intensity, and the Z-axis number of cells. The mean indo-1 ratio vs. time is shown in C and fluo-3 intensity vs. time is shown in D. The data were converted to calcium concentration vs. time by calibration, using measured constants in equation 1 for indo-1 (E) and by calibration with buffer solutions for fluo-3 (F), and the values were plotted vs. time. Note that whereas the shapes of the curves for mean indo-1 ration (C) and fluo-3 intensity (D) are different (particularly in an apparently slower return to baseline after 4 min in (D), once converted to $[Ca^{2+}]_i$, the two measurements are essentially identical. The apparent discrepancy in C vs. D is related to the different values of Km for indo-1 and fluo-3. The percent cells responding with $[Ca^{2+}]_i$ elevated beyond two standard deviations above the mean of the cells before antibody stimulation ("percent cells above threshold") are plotted for indo-1 (G) and fluo-3 (H). The percent responding cells calculated by cumulative curve subtraction (see text) are plotted in I and J for indo-1 and fluo-3, respectively. Data analysis in this and subsequent figures was performed with a software program written by one of the authors (P.S. Rabinovitch) ("MultiTime," Phoenix Flow Systems, San Diego, CA).

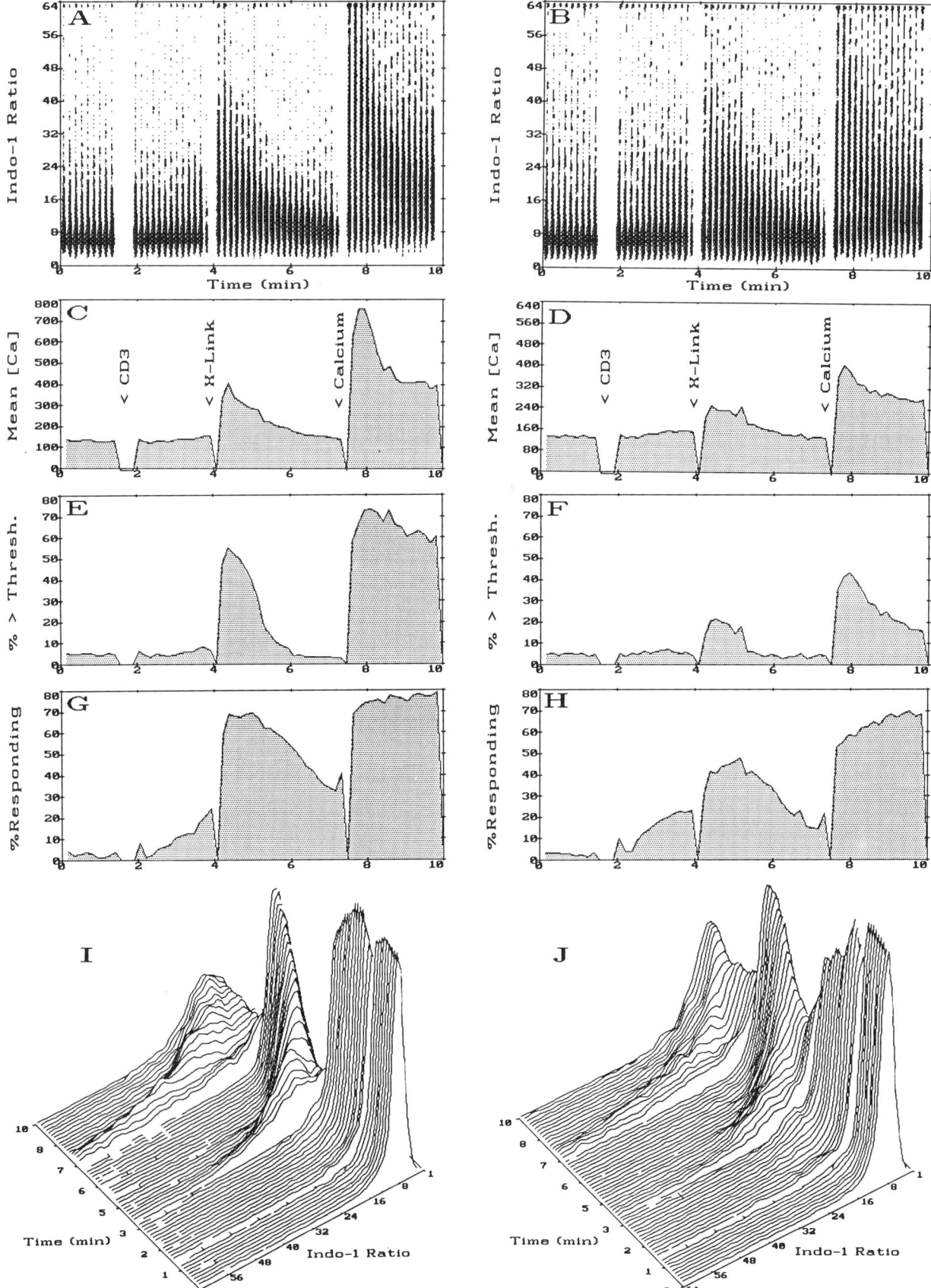

Figure 31.5. A-J

this final response is greater in CD4+ cells than in CD8+ cells.

Comparison of the percent of cells above a threshold (Figs 31.5E and 31.5F) with the proportion of responding cells calculated by curve subtraction (Figs. 31.5G and 31.5H) shows a substantial difference between the two measures: The latter calculation detects more responding cells than the former. This difference is greater in Figure 31.5 than in Figure 31.4, due in largest part to the presence in Figure 31.5 of a greater fraction of cells with elevated $[Ca^{2+}]_i$ in the "resting" control population; in such a case, a threshold set to exclude all but 5% of the "resting" population may be required to be at a relatively high $[Ca^{2+}]_i$ value, whereas the method of curve subtraction avoids this problem (21).

As a second example, Figure 31.6 shows the analysis of murine splenocytes. While approximately 50% of murine splenocytes are B-cells, only approximately 10% are T-cells (CD5-positive), as seen from the bivariate analysis of FITC-CD20 vs. PE-CD5 (Fig. 31.6A). When a T-cell stimulus (antibody to the T-cell receptor, CD3) is applied to the splenocytes, only a very small response is visible in the total population. In contrast, when the indo-1 analysis is gated upon the CD5$^+$ CD20$^-$ cells, then the response of the minority subpopulation of T-cells can be seen to be vigorous, and, as expected, the CD5$^-$CD20$^+$ cells show no response (Fig. 31.6B). Conversely, when a B-cell stimulus (anti-IgM) is applied, the response of the immunophenotypically identified B-cells shows twice the magnitude of that seen in the ungated total population of splenocytes, while the T-cell subset, as expected, shows no response (Fig. 31.6C).

Using other probes excited by visible light, it is possible to analyze additional physiologic responses in cells simultaneously with $[Ca^{2+}]_i$. The simultaneous analysis of membrane potential and $[Ca^{2+}]_i$ has been accomplished by several groups (23, 24), and simultaneous analysis of $[Ca^{2+}]_i$ and pH_i will be illustrated in a subsequent section of this chapter.

Sorting on the Basis of $[Ca^{2+}]_i$ Responses

The ability of the flow cytometric analysis with indo-1 to observe small proportions of cells with different $[Ca^{2+}]_i$ responses than the majority of cells suggests that the flow cytometer may be useful to identify and sort variants in the population for their subsequent biochemical analysis or growth. Results of artificial mixing experiments with Jurkat (T-cell) and K562 (myeloid cell) leukemia lines indicate that subpopulations of cells with variant $[Ca^{2+}]_i$ comprising <1% of total cells could be accurately identified (10). Goldsmith and Weiss (25, 26) have reported the use of sorting on the basis of indo-1 fluorescence to identify mutant Jurkat cells that fail to mobilize $[Ca^{2+}]_i$ in response to CD3 stimulus, in spite of the expression of structurally normal CD3/Ti complexes. These experiments suggest that sorting on the basis of indo-1 fluorescence can be an important tool for the selection and identification of genetic variants in the biochemical pathways leading to Ca^{2+} mobilization and cell growth and differentiation.

Critical Aspects in the Analysis of $[Ca^{2+}]_i$ by Flow Cytometry

COMPLICATIONS IN LOADING AND INTRACELLULAR ENVIRONMENT

Potential problems to be aware of in the use of $[Ca^{2+}]_i$ dye indicators include compartmentalization, leakage or secretion of dye, quenching by heavy metals, and incomplete deesterification of dye ester. The flow cytometric analysis of $[Ca^{2+}]_i$ using indicator dyes is predicated on achieving uniform distribution of the dye within the cytoplasm. In several cell types, the related dye fura-2 has been reported to be compartmentalized within organelles (27, 28). In bovine aortic endothelial cells, fura-2 has been reported to be localized to mitochondria; however, under those conditions, indo-1 remained diffusely cytoplasmic (29). Thus, it is possible that there will be fewer problems with compartmentalization of indo-1 than with fura-2. Some cell types, such as neutrophils and monocytes, and some cell lines (as opposed to primary cells) appear to be more susceptible to compartmentalization. In addition, compartmentalization is enhanced by prolonged incubation of cells at 37°C. In general, it is advisable to examine the cellular distribution of indo-1 or fluo-3 microscopically and, in each new application, to confirm the expected behavior of the dye. This is done by determining the ratio of R_{max} to R as a control for each experiment, as described below. In addition, one should store indo-1 loaded cells at room temperature after loading and use the cells promptly after loading. Since heavy metals may quench the

Figure 31.5. Analysis of $[Ca^{2+}]_i$ in human PBL stained with indo-1 and gated to show PE-positive CD4 cells (**A, C, E, G, I**) and PE-Texas-red-tandem-conjugate (ECD, Coulter Corp.) positive CD8 cells (**B, D, F, H, J**). Cells were suspended in medium with 5 mM EGTA, which reduced available extracellular Ca^{2+} to approximately 20 nM (10). At 1 min after the start of analysis, 3 μg/ml anti-CD3 mAb UCHT-1 was added (Dako Corp; this antibody by itself produces only a small $[Ca^{2+}]_i$ response), and at 4 min, 20 μg/ml goat antimouse antibody was added. The return to baseline after a brisk response is characteristic of mobilization of Ca^{2+} from intracellular stores, without influx of extracellular Ca^{2+} (see text). At 7 min, 10 mM $CaCl_2$ was added to the medium. **A** and **B** display "dot plots" of indo-1 ratio vs. time; **C** and **D** show calibrated $[Ca^{2+}]_i$ vs. time; **E** and **F** show the percent of cells above a threshold vs. time (the threshold ratio is set such that only 5% of "resting" cells are above that ratio, see text); **G** and **H** show the percent responding cells vs. time; and **I** and **J** show isometric displays of the indo-1 ratio vs. time.

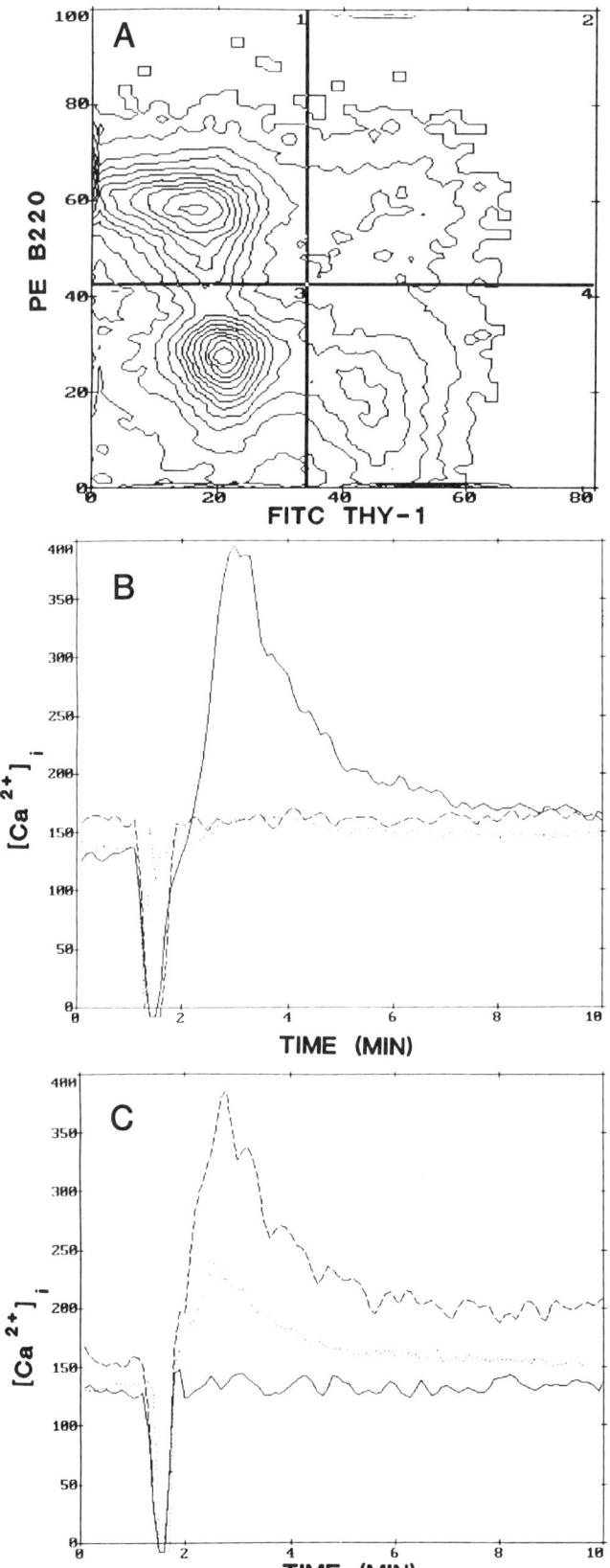

indicator dye fluorescence, the use of the membrane permeant heavy metal chelator diethylenetriaminepentaacetic acid (TPEN) may be advisable for cell lines that contain increased amounts of heavy metals (30). In situations in which loaded dye is not retained intracellularly for a sufficient length of time, the dye is often being actively secreted and the use of probenecid, a blocker of organic anion transport, may be beneficial (27).

If, for a particular cell type loaded with indo-1, the magnitude of change between R and R_{max} is in good agreement with the values predicted from spectral curves of indo-1 in a cell-free buffer, then it would be unlikely that the dye is in a compartment inaccessible to cytoplasmic Ca^{2+}, in a form unresponsive to $[Ca^{2+}]_i$ (i.e., still esterified, see below), or in a cytoplasmic environment in which the spectral properties of the dye have been altered.

It has been suggested that both fura-2 and indo-1 (and by extension, presumably fluo-3) may be incompletely deesterified within some cell types (31, 32). Since the fluorescence of the ester has little spectral dependence upon changes in Ca^{2+}, the presence of this dye form could lead to false estimates of $[Ca^{2+}]_i$. Again, results of calibration experiments are helpful in excluding this possibility. Further, it has been proposed that since indo-1 fluorescence, but not that of the indo-1 ester, is quenched in the presence of mM concentrations of Mn^{2+}, then, in the presence of ionomycin, MN^{2+} can be used as a further test of complete hydrolysis of the indo-1 ester within cells (31).

Because of its more recent introduction, there is relatively little information available at this time regarding the relative importance of each of the above concerns when using fluo-3. To illustrate, however, the possibility that each dye may be subject to different intracellular processing and environments, Figure 31.7 shows results obtained with human peripheral blood lymphocytes (PBL) loaded simultaneously with indo-1 and fluo-3. Using UV and 488-nm excitation, each dye was independently but simultaneously analyzed in the same cells. After exposure to ionomycin, indo-1 showed the expected response associated with calcium influx; however, after an initial increase in fluorescence, fluo-3 intensity declined at the same time that the indo-1 ratio remained stably maximal. The reasons for the different behaviors of the two dyes is not yet clear, although it almost certainly is related to one or more of the points discussed above. This experiment demonstrates graphically the need

Figure 31.6. Analysis of splenic B- and T-cells. B6 mouse spleen cells were stained with PE-thy-1 (T-cells) and FITC B220 (B-cells). **A** shows the bivariate plot of FITC vs. PE. Fifteen percent of splenocytes are T-cells (*quadrant 4*) and 44% are B-cells (*quadrant 1*). **B** shows the result of stimulation with anti-CD3 mAb 2C11 (25 μg/ml). The ungated total cell response is small (*dotted line*), whereas the response gated upon the T-cell region of **A** is approximately sevenfold larger (*solid line*), and B-cells (*dashed line*) show no response. **C** shows stimulation with goat antimouse Ig (25 μg/ml): the response gated to show only B-cells (*dashed line*) is twice as large as that of all cells (*dotted line*) and T-cells (*solid line*) show no response.

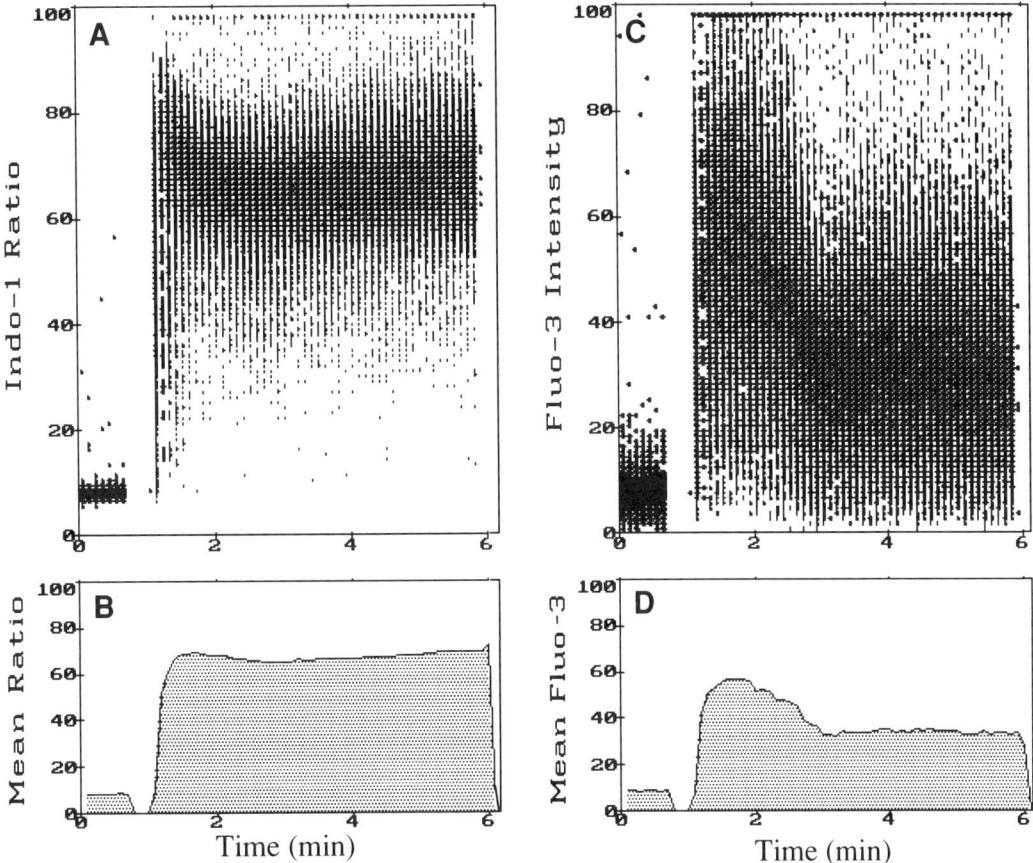

Figure 31.7. Ionomycin treatment of CD4$^+$ peripheral blood T-cells simultaneously loaded with indo-1 and fluo-3. The indo-1 ratio is shown in **A** and **B**, and the fluo-3 intensity is shown in **C** and **D**. Note the partial return to baseline seen in fluo-3 intensity and not seen in the indo-1 ratio. This observation was variable over a series of experiments (data not shown).

for care in assessing correspondence to [Ca^{2+}]$_i$ by indicator dyes.

UNSTABLE OR BROAD BASELINE DISTRIBUTIONS

Under typical conditions, the baseline [Ca^{2+}]$_i$ distribution should show little (<3%) variation from sample to sample and should, with indo-1, be narrow and, most often, follow a normal (Gaussian) distribution. Some cell lines may have asymmetric distributions and altered mean values of "resting" [Ca^{2+}]$_i$, which can often be ascribed to a subpopulation of cells with elevated [Ca^{2+}]$_i$. This may result from the impaired viability of some cells, "spontaneous" activation of some subset of cells, or presumably, biological heterogeneity within the cell population (for example, cells within certain phases of the cell cycle). Sometimes, the baseline will start at a normal level and then rise with time. This may be due to the failure to completely remove from the sample lines an agonist from a previous experiment. The most common problem has been residual calcium ionophore; this can be efficiently removed by first washing the sample lines with DMSO and then scavenging residual ionophore by washing with a buffer containing 2% bovine serum albumin.

POOR CELLULAR RESPONSE

Failure of the cells to show proper response to various treatments may be due either to difficulty with the cells or with the instrument. To differentiate between these, the cells should be stimulated with the calcium ionophore ionomycin and the magnitude of R$_{max}$ to R should be determined. If the indo-1 ratio of R$_{max}$/R increases by the amount described previously, or if the fluo-3 fluorescence increases by approximately five-fold, then the instrument is functioning properly. If the increase is less than expected, then one should obtain an independent preparation of cells, such as murine thymocytes or human PBL. If these cells load properly and also respond poorly, then the instrument alignment should be checked. With indo-1, not uncommonly, the violet or blue signals may not be properly focused, or there might be interference from a second laser. This problem can be pinpointed by analyzing separately the blue and violet signals after ionophore treatment; the violet signal should increase approximately threefold and the blue signal should decrease approximatley twofold (Fig. 31.2).

If the instrument is functioning properly, then the problem may be in the cells. The cells must be loaded with sufficient indo-1 or fluo-3 to be detectable well above autofluorescence. If there is any question, this should be checked independently with fluorescence microscopy, especially if one has become familiar with the expected fluorescence of loaded cells examined in this manner. If the cells are too dim or excessively bright or if the dye is compartmentalized, the ability to detect calcium signals will be impaired. For unknown reasons, the calcium signaling of B-cells, and not T-cells, is particularly sensitive to overloading with indo-1 (10, 13). The cells must be suspended in media that contains calcium; occasionally, responses will appear blunted because of the inadvertent resuspension of cells in a medium than contains no added calcium.

EFFECT OF ANTIBODY LABELING

In the simultaneous analysis of $[Ca^{2+}]_i$ and immunofluorescence, consider that the use of the antibody probe can itself alter the cellular $[Ca^{2+}]_i$. It is becoming increasingly clear that the binding of monoclonal antibodies (mAbs) to cell surface proteins can alter $[Ca^{2+}]_i$, even when these proteins are not previously recognized as part of a signal transducing pathway (10, 33–38). For example, antibody binding to CD4 will reduce CD3-mediated $[Ca^{2+}]_i$ signals; if the anti-CD4 mAb is cross-linked to the CD3 complex, as with a goat–antimouse mAb, the CD3 signals are augmented (39, 40). Antibody binding to CD8 has similar effects (unpublished data).

As a consequence of these concerns, a reciprocal staining strategy should be used whenever possible so that the cellular subpopulation of interest is unlabeled while undesired cell subsets are identified by mAb staining. The CD4$^+$ subset in PBL may be identified, for example, by staining with a combination of CD8, CD20, and CD11 mAbs (10), and the CD5$^+$ subset can be identified by staining with CD16, CD20, and HLA-DR mAbs (35). Finally, it is important to reiterate that, when staining cells with mAbs for functional studies, antibodies must be azide-free for metabolic processes to be uninhibited. Commercial antibody preparations may thus require dialysis before use.

INTRINSIC LIMITATIONS OF FLOW CYTOMETRY

Flow cytometry is unable to detect heterogeneity of cellular calcium concentrations within a single cell, and there are reports from assays using digital video microscopy that, in some situations, calcium transients may be present only within compartmentalized sublocations (41, 42). It has been reported that the photoprotein aequorin may in some circumstances help to detect changes in cytosolic calcium not reported by indo-1 (43), although use of the two indicators is complementary because aequorin cannot measure Ca^{2+} in single cells (44). In addition, there is evidence that calcium elevations occurring after cellular stimulation may be oscillatory rather than sustained (45, 46), thus raising the possibility that some cellular processes controlled by calcium may be frequency-modulated as well as amplitude-modulated. Since flow cytometry cannot measure the calcium concentration inside a single cell as a function of time, it is not possible to distinguish whether there is a subpopulation of cells that is responding with a sustained response or, alternatively, whether there are two populations of cells, one that has elevated calcium concentration and one that has basal levels. In spite of these limitations, determination of $[Ca^{2+}]_i$ in large numbers of single cells using flow cytometry with indo-1 offers great practical advantages and allows measurements of a kind that is not possible to achieve by means of the alternative techniques that are currently available.

Applications of the Flow Cytometric Analysis of $[Ca^{2+}]_i$

The flow cytometric assay of cellular calcium concentration has already been applied to a wide variety of cells, providing interesting and sometimes unexpected results. Examples of the initial applications of the technique are presented in recent reviews (47, 48). One of the first observations that was made readily quantifiable by the flow cytometric analysis was that there is great heterogeneity in the response of lymphocytes to mitogenic agonists. Using simultaneous immunofluorescence, some of this heterogeneity can be shown to be related to immunophenotypic subsets; for instance, CD4$^+$ cells show more vigorous responses to phytohemagglutinin, concanavalin A, and mAb to CD3 (see Fig. 31.5) than do CD8$^+$ cells (10, 49). Considerable use of this approach has been made in the demonstration of differences between $[Ca^{2+}]_i$ activation requirements of different cell subsets and subset specificities of activation pathways. The effects of antibody binding to cell surface molecules, sometimes a complication in labeling experiments (see above), has been extensively employed to analyze signaling mechanisms, and augmented, or even new relationships have been probed by cross-linking antibodies on the cell surface (39, 50). Flow cytometric measurements with indo-1 have been performed to date with all nucleated blood cell types. Applications of fluo-3 have been reported with most types, with reports appearing at a rapidly increasing pace. The combination of sensitivity, reliability, and ability to analyze large numbers of cells within cell subsets has made the flow cytometric assay of $[Ca^{2+}]_i$ the preferred technique for a broad spectrum of research applications.

There are many potentially exciting clinical applications of the flow cytometric assay of cellular calcium concentration (47). One topical example is the study of the effects of the human immunodeficiency virus (HIV) infection. In in vitro studies, the HIV-1 retrovirus was found to impair signal

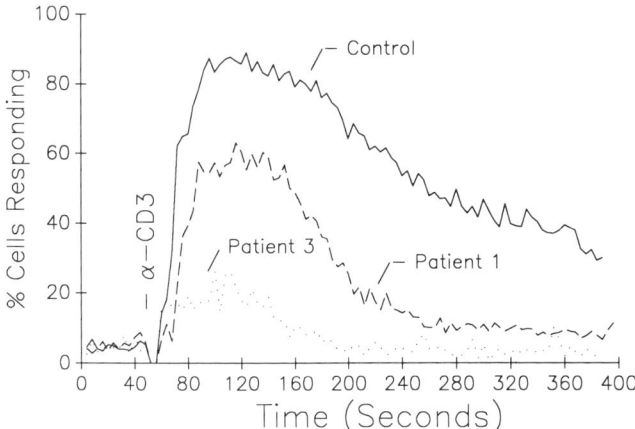

Figure 31.8. PBL from asymptomatic HIV-infected patients or an uninfected control donor were stained with FITC-CD8 and PE-CD5. The cells were loaded with indo-1, stimulated with anti-CD3 antibody 38.1 (1 μg/ml), and changes in calcium were monitored in the CD4+ (CD5+CD8−) cells. The time course of the percent of responding cells is shown.

transduction in CD4 cells, the major target of HIV-1 (51). Recent studies suggest that CD4 T-cells from HIV-infected patients also have impaired responses to T-cell receptor stimulation (G.P. Linette et al., unpublished). Due to the necessity of analyzing live, retrovirus-infected cells on the flow cytometer, a closed fluidics system is used to prevent the generation of aerosols. Lymphocytes from patients with early-stage HIV infection (Walter Reed stage II-III) have been isolated, loaded with indo-1, and the CD4 subset has been analyzed by gating on the CD5$^{bright+}$ CD8$^-$ subset of cells. More than 80% of HIV-infected patients have anti-CD3-induced calcium responses that are below that of age- and sex-matched control donors. There is substantial patient-to-patient variability, with some patients having minimally diminished responses and others displaying severely reduced signaling (Fig. 31.8). Thus, both normal CD4 T-cells after HIV infection in vitro, and the CD4 T-cells from patients with early stage HIV infection have impaired signal transduction. It is likely that the mechanism of impaired signaling may differ in vitro vs. in vivo. During in vitro infections, more than 50% of the CD4 cells can be shown to be infected. In contrast, <1% of CD4 cells from patients with early-stage HIV infection can be demonstrated to be infected, so that an indirect mechanism must exist for the virus to impair signal transduction pathways in uninfected cells. In any case, the existence of this abnormality may prove to be a sensitive test of the immune system function in HIV infections. Current studies in several laboratories are exploring whether or not this application may provide prognostic information and whether the assay may be useful for the management of this chronic illness. Study of HIV infection does illustrate the utility of flow cytometry in the clinical setting in which only one immunological subset of cells, or a portion of cells, in that subset may show an altered response. Demonstration of such heterogeneity in [Ca^{2+}]$_i$ signals would have been impossible to discern in conventional assays carried out in a fluorimeter where only the mean calcium response is recorded.

INTRACELLULAR pH

Introduction

The pH$_i$ of mammalian cells is ~7.2, and pH$_i$ appears to be closely regulated (the coefficient of variation measured by flow cytometry is ~5%). In mammalian cells, pH$_i$ is regulated by at least three mechanisms, including Na^+_o/H^+_i exchange, sodium-dependent Cl^-_o/HCO$_{(3)}^-{}_i$, and HCO$_{(3)}^-{}_o$/Cl^-_i exchange. Acid extrusion is primarily accomplished by the Na$^+$/H$^+$ antiport and by sodium-dependent Cl^-_o/HCO$_{(3)}^-{}_i$, while HCO$_{(3)}^-{}_o$/Cl^-_i exchange has the major role for base extrusion. All of these mechanisms appear to be stimulated by a variety of growth factors and by phorbol esters, presumably through the activation of protein kinase C. It was initially proposed that an alkaline shift in pH$_i$ itself might be a signal to regulate cell activation. A number of reports purporting to show a role for pH$_i$ in cellular activation are derived from measurements made in bicarbonate-free medium. Thus, nonphysiologic (HCO$_{(3)}$-free) buffers were used so that only the Na$^+$/H$^+$ antiporter could operate and, thus, alkalinization was the dominant response. Recent reports, performed in physiologic medium, show that cellular activation is, in fact, accompanied by acidification in many cases and that the net effect (acidification or alkalinization) is a function of the medium in which the experiment is conducted. Thus studies of pH$_i$ are important physiologic tools for the elucidation of various ion exchange mechanisms but cannot be taken as evidence per se of cellular activation (52).

Indicators of Intracellular pH

Until recently, the most commonly used probes were modifications of fluorescein—fluorescein diacetate being the first-generation pH probe—followed by carboxyfluorescein diacetate (COFDA). Both of these dyes are limited by relatively poor retention inside loaded cells. An improved fluorescein probe is 2′,7′-bis-carboxyethyl-5(6)-carboxyfluorescein (BCECF). As with indo-1 and fluo-3, BCECF is loaded into cells using the acetoxymethyl ester; after hydrolysis, it has a negative charge of −4 or −5 and, therefore, leaks more slowly than COFDA. The pKa of BCECF, 6.98, is near the pH$_i$ of resting cells, and there is a pH-dependent shift in the excitation wavelength, making it possible to use the ratio of fluorescence signals to correct for differences in loading and cell size. In addition, BCECF fluorescence excited at 450 nm is pH-independent, while fluorescence at 500 nm is pH-dependent, allowing ratiometric fluorescence

Table 31.1
Fluorochromes for Ratiometric Determination of pH Using Flow Cytometry

Probe +	pKa	Fluorescence (nm) Excitation	Emission	References
Bis-carboxyethyl-carboxyfluorescein Acetoxymethyl ester (BCECF AM)	6.98	ratio 439/490 488	535 ratio 520/620	Paradiso et al. (130) Musgrove et al. (131)
Diacetoxy-dicyanobenzene (ADB); yields Dicyanhydroquinone (DCH) after de-esterification	8.0	~350	ratio 425/540	Cook and Fox (132) Musgrove et al. (131)
Carboxy SNARF-1 acetoxymethyl acetate (SNARF-1)	7.50	514 or 530	ratio 575/670	Whitaker et al. (55)

emission analysis (Table 31.1). In both cases, the magnitude of pH-dependent ratio shifts is relatively modest (for example, when the pH is raised from 6.5 to 7.5, the ratio of fluorescence intensities after excitation of FDA at 436 nm and 495 nm increases by only 1.45 to 1.55 times (53)); this is **much** less of a shift than is seen with SNARF-1 (see below) and, thus, these dyes are largely supplanted by the latter.

Subsequently, several UV-excited pH probes were developed. The most useful of these is 1,4-diacetoxy-2,3-dicyanobenzene (ADB). The cell-permeant ADB is hydrolyzed and trapped intracellularly to yield 2,3-dicyanhydroquinone (DCH) (54). DCH fluorescence is pH-dependent, while the esterified forms exhibit pH-independent fluorescence; therefore, care must exercised to ensure complete deesterification of the ADB. Ratiometric determinations of pH are possible using a single excitation source by measuring fluorescence emission at 429 nm and 477 nm (Table 31.1). As with measurement of $[Ca^{2+}]_i$, many laboratories will find the requirement for UV excitation to be the primary limitation for use of this probe.

Recently, the most useful probe for pH_i measurement has been introduced; a compound termed SNARF-1 (SemiNaphthoRhodaFluor) (55). This is the first pH probe to have fluorescence characteristics that provide truly useful and sensitive ratiometric properties. SNARF-1 appears to be the most promising reagent for use in flow cytometry, as it has convenient excitation spectra and exhibits large changes in pH-dependent fluorescence (Fig. 31.9). The acid form of SNARF-1 has absorption maxima at 518 and 548 nm, while the base form excites maximally at 574 nm. The emission of SNARF-1 in acid is maximal at 587 nm and the basic form emits maximally at 636 nm; there is an isosbestic point at 610 nm. In practice, either the 530-nm line of a krypton-ion laser or the 514-nm line of an argon-ion laser are optimal for excitation, while emission should be collected at both 575 nm and 640 nm and the ratio of these signals calculated. Excitation of SNARF-1 at 488 nm is only slightly less optimal; because it has the advantage of permitting simultaneous analysis of FITC probes (i.e., immunofluorescence), this will probably be the most commonly used excitation wavelength.

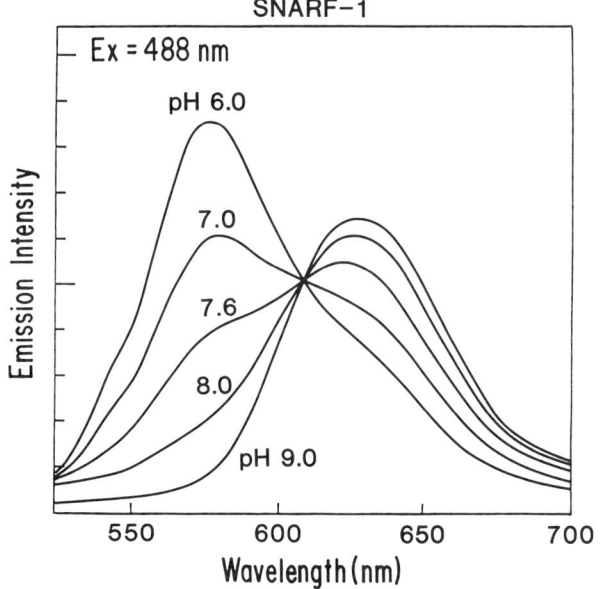

Figure 31.9. Fluorescence emission spectra of SNARF-1 as a function of pH. Excitation was at 488 nm. (Courtesy of R. Haughland, Molecular Probes, Eugene, OR).

Preparation of Cells for SNARF-1 Analysis

As with indo-1, the optimal conditions for loading must be empirically determined for each cell type. Cells must be loaded only to the point where the fluorescence signal is above autofluorescence and sufficient for detection by the flow cytometer. A typical loading protocol is:

1. Incubate cells ($\leq 2 \times 10^7$/ml) in buffered (for example, HEPES (4-(2-hydroxyethyl)-1-piperazine-ethanesulfonic acid), see above for rationale for using bicarbonate-free buffer or not) medium with 2–6 μM carboxy SNARF-1 (diacetate) at 30°C to 37°C for 30 min.
2. Centrifuge and resuspend in fresh medium at the desired cell concentration (usually ~10^6/ml). Store cells at 22°C and protect from light until analysis. The cell pellet will appear faintly pink.
3. Warm aliquot of SNARF-1 loaded cells to 37°C for 5 to 10 min before analysis. The medium and sheath fluid should be buffered saline. The rate of cell analysis can vary; commonly, cells are analyzed at 200 to 300 cells/sec.

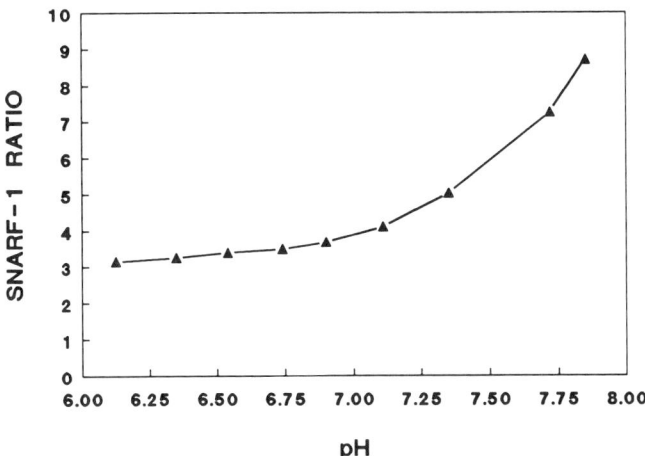

Figure 31.10. Calibration of the SNARF-1 ratio of fluorescence (670/575 nm) to pH. Jurkat cells were loaded with SNARF-1 and resuspended at 37°C in buffers containing 135 nM K+ with varying pH in the presence of nigericin. Excitation was at 514 nm.

Flow Cytometric Analysis of SNARF-1-Loaded Cells

As described above, SNARF-1 emission should be collected using orange and red bandpass filters centered at 570 nm (orange) and 640 to 670 nm (red). The ratio of red/orange fluorescence intensity is proportional to increasing pH (Fig. 31.10). See the corresponding section above on indo-1 analyses for further discussion of ratiometric analysis.

SNARF-1 is a vital dye and, therefore, dead cells are efficiently excluded from analysis by electronically gating on cells that have fluorescence. If FITC fluorescence is measured simultaneously, as for determination of cell subsets, suitable gating on this parameter would also be used. Using single-beam 488-nm illumination for analysis of FITC fluorescence requires compensation for overlap into the green FITC bandpass from the leftward tail of the SNARF-1 emission spectrum. This is analogous to the routine color compensation used when analyzing PE with FITC, except that even greater crossover compensation may be required when the SNARF-1 fluorescence is bright. Alternatively, cells can be stained with amino-methycoumarin acetic acid (AMCA)-conjugated antibodies and fluorescence analyzed using UV illumination, thereby avoiding the need for compensation.

The effective Kd of SNARF-1 for protons is 7.5, which is near the pH$_i$ of resting cells (pH$_i$ = ~7.2). The pH$_i$ of stimulated cells may decrease or increase, depending on cell type, stimulus, and medium composition (52). As seen in Fig. 31.9, described below, SNARF-1 has a greater sensitivity in reporting alkaline shifts than acidification.

Calibration and Data Analysis

For each experiment, it is necessary to construct a calibration curve so that one may determine the fluorescence channel number as a function of pH$_i$. Cells are suspended in a series of high potassium buffers of different pH (with otherwise identical ionic composition). The cells should be loaded with the pH probe and treated with the proton ionophore nigericin (1 μg/ml), added in order to equalize pH$_i$ and to buffer pH (56). Nigericin produces the equilibrium $[K^+_i]/[K^+_o]$ = $[H^+_i]/[H^+_o]$. Thus, in the presence of nigericin, pH$_i$ can be calculated with the knowledge of $[K^+_i]$, $[K^+_o]$ and the buffer pH. Equilibrate tube at 37°C for 15 min, and measure SNARF-1 fluorescence ratio at 37°C. Note that nigericin is toxic to human lymphocytes at >5 μg/ml, manifested by loss of cellular fluorescence.

A calibration curve can be constructed by plotting values of pH$_i$ vs. the SNARF-1 fluorescence ratio (Fig. 31.10). The ratio changes ~3-fold between pH 6.0 and pH 8.0. Most of the ratio shift occurs between pH 7 and pH 8, as the slope of the curve is much less between pH 6 and pH 7.

Simultaneous Analysis of pH and Other Fluorescence Parameters

As mentioned above, the fluorescence emission properties of SNARF-1 allow simultaneous excitation of FITC probes using the same 488-nm laser. Analysis of pH$_i$ in immunophenotypically-defined cell subsets using FITC-conjugated mAb is thus very straightforward. Similarly, single-laser $[Ca^{2+}]_i$ measurements with fluo-3 simultaneously with pH$_i$ are easily performed. When a second UV laser is available, $[Ca^{2+}]_i$ measurement with indo-1 is preferred. Fig. 31.11 illustrates the analysis of $[Ca^{2+}]_i$ and pH$_i$ in splenic T-cells. FITC-CD4 mAb was used to gate the analysis specifically upon this subset of murine splenocytes (not shown), and the ratios of UV-excited green and violet indo-1 emission and 488-nm-excited red and orange SNARF-1 emission were each calculated. The elevation in $[Ca^{2+}]_i$ reached its peak within 5 min after antibody to the T-cell receptor (CD3) was added and, subsequently, slowly declined (Fig. 31.11A). Intracellular pH$_i$ reached its peak more slowly, 10 min after addition (a 0.2 unit alkalinization), and thereafter also slowly declined toward baseline (Fig. 31.10B).

INTRACELLULAR GLUTATHIONE
Introduction

Glutathione (glutamylcysteinylglycine, GSH) is an important antioxidant tripeptide thiol that is involved in the scavenging of toxic oxygen products (57–59). In addition, GSH is involved in a number of other important reactions in the cell, including conjugation of xenobiotics, amino acid transport, and deoxyribonucleotide synthesis (57). GSH is also important for the maintenance of cellular thiol redox status, and its redox state has been proposed to be a major determinant of the functioning of a number of enzymes (60–66) and of the integrity of cytoskeletal proteins (67). The biochemical pathways of GSH synthesis, renewal, and utilization are outlined in Fig. 31.12.

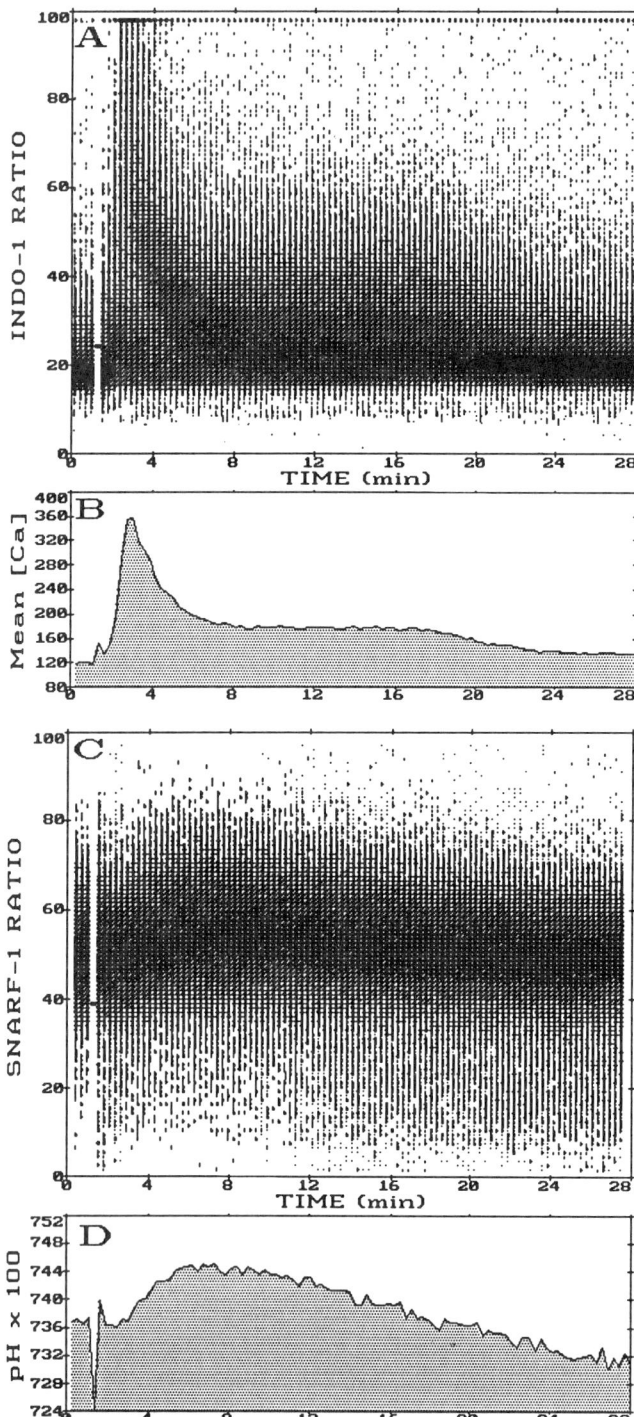

Figure 31.11. Simultaneous analysis of murine splenocytes loaded with indo-1 and carboxy SNARF-1. Murine splenocytes were loaded with indo-1 and SNARF-1 and stained with FITC-conjugated anti-CD4 mAb. Fluorescence was excited by the UV lines of a krypton-ion laser and the 488-nm line from an argon-ion laser. The ratio of indo-1 fluorescence (395 nm/525 nm) and of SNARF-1 fluorescence (665 nm/575 nm) is displayed vs. time. The analysis was gated on the FITC-CD4 fluorescence at 525 nm. Anti-CD3 antibody 2C11 was added during the gap in the analysis. The calcium and pH values for individual cells are plotted as dot plots in the upper panels, and the mean response vs. time is plotted in the lower panels. $[Ca^{2+}]_i$ increased from 0.13 μM to 0.36 μM and pH_i increased 0.2 units.

These roles for GSH have important consequences upon cell physiology and mitogenic activation. In general, augmenting cellular GSH with cysteine delivery agents such as N-acetycysteine, oxathiayolidine carboxylic acid, or glutathione esters has a positive effect on cell growth, whereas depletion of GSH inhibits many cellular functions (57). GSH is known to influence cell growth and replication at various levels including G_1/S-phase transition (68, 69) and very early events in mitogen-induced cell activation (70). We will discuss later in this chapter some evidence that GSH affects very early steps of signal transduction. Other aspects of cell growth, such as protein synthesis, are also influenced by GSH levels, since γ-glutamyl transferase-dependent amino acid transport utilizes GSH (71). Interestingly, although activation steps that are important for cell replication of lymphocytes are GSH-dependent, those activation steps that are necessary for acquisition of differentiated functions are not necessarily GSH-dependent (72), although some workers have reported that GSH content can influence these functions as well (73, 74).

GSH has also been shown to affect the responsiveness of cells to various cytokines. For example, intracellular GSH has been shown to enhance the action of IL-2 in lymphocytes (72, 73, 75, 76), whereas extracellular GSH has been shown to decrease the effects of granulocyte/macrophage colony stimulating factor (GM-CSF) and platelet-derived growth factor (PDGF) on granulopoiesis and fibroblast proliferation, respectively (77, 78). For GM-CSF, this effect is thought to be mediated through direct competition by GSH for the GM-CSF receptor. In fact, GSSG, the oxidized form of GSH was found to have the capacity to stimulate the GM-CSF receptor directly (78). The mechanisms responsible for the inhibitory effects of GSH on PDGF-stimulated fibroblast growth are less well understood.

The activity of redox-responsive oncogene products and DNA-binding factors responsible for gene regulation may also be affected by the GSH status of the cell (79-85). For instance, the interaction of c-*fos* and c-*jun* products are known to be redox-sensitive (79). Another example is that of NFkB. Roederer and colleagues (83) have shown that this nuclear binding factor can be inhibited from binding to regulatory DNA binding domains by n-acetylcysteine, a drug that increases the GSH level of cells.

Finally, GSH has recently received attention because of its role in a number of pathological disease states, including tumor cell resistance to chemotherapeutic agents (98, Chapters 14 and 27, this volume), idiopathic pulmonary fibrosis (86, 87), and acquired immune deficiency syndrome (83, 88-91).

Fluorescent Indicators of Intracellular GSH

Several years ago Durand and Olive (92) published a review of fluorescent indicators for thiols, including GSH, that might prove useful for flow cytometric purposes. The prob-

OVERVIEW OF GLUTATHIONE METABOLISM

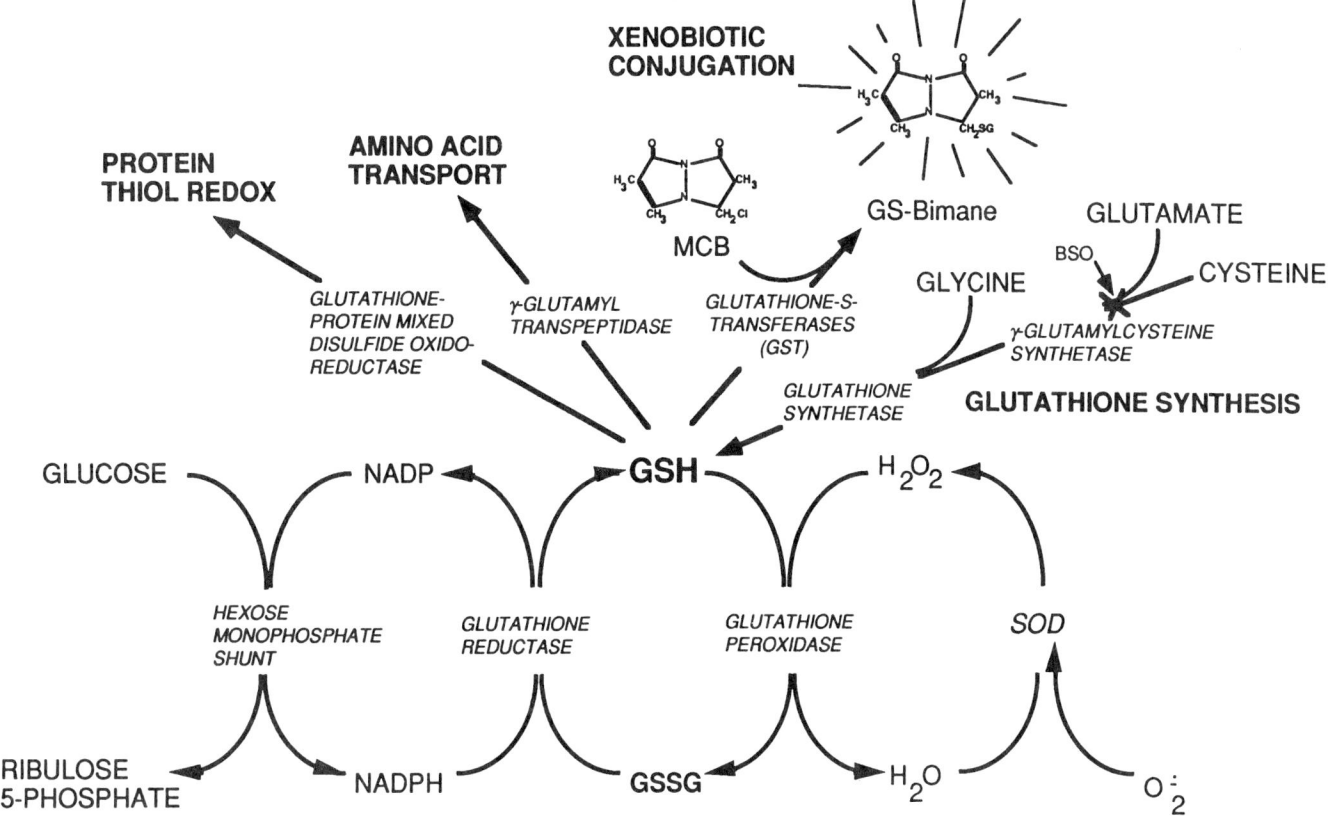

Figure 31.12. Overview of glutathione (*GSH*) metabolism. *GSH* is synthesized in a two-step process (upper right hand portion of figure): cysteine is metabolized to g-glutamylcysteine (g-CC) by g-GC synthetase and then to *GSH* by *GSH* synthetase. In the glutathione redox cycle (lower portion of figure) glutathione peroxidase uses *GSH* to reduce hydrogen peroxide and other peroxides and, in the process, *GSH* becomes oxidized (*GSSG*). Reduction of *GSSG* to *GSH* by glutathione reductase requires *NADPH*, supplied primarily by the hexose monophosphate shunt (also called the pentose phosphate shunt). *GSH* can also serve as a cosubstrate for glutathione transferase-mediated conjugation of xenobiotics (shown here is the conjugation of monochlorobimane (*MCB*), to the fluorescent glutathione-bimane conjugate, top of figure). *GSH* also has important roles in amino acid transport and in the maintenance of protein thiol redox status (upper left of figure).

lem with all of the dyes reviewed was one of specificity for GSH. Durand and Olive (92) concluded that a promising reagent at the time for flow cytometric determination of GSH was monobromobimane (MBB). Poot et al. (93) used this reagent in their investigation of human diploid fibroblast GSH metabolism, and employed n-ethylmaleimide-treated "blanks" in order to control for nonspecific protein thiol binding.

Another reagent that has since been introduced for flow cytometric evaluation of GSH is o-phthaldialdehyde (OPT) (94). OPT is supposedly more specific for GSH than MBB because, in addition to reacting with the cysteine thiol group of GSH, it also reacts with the amino terminus of GSH, forming a highly fluorescent cyclic structure. However, this dye has the same disadvantage as MBB of potentially binding to other nonprotein thiols in the cells such as cysteine and γ-glutamylcysteine. Nonetheless, OPT still has the advantage that, when bound to protein thiols, it has slightly different spectral characteristics than when bound to GSH (higher blue fluorescence and less green fluorescence), allowing one to simultaneously measure the sulfhydryl redox status of protein and of low molecular weight thiols (94).

A major advance in the measurement of GSH by flow cytometry was made by Rice et al. (95), who used mono**chloro**bimane (MCB) to detect changes in GSH status of individual normal and tumor cells. Many papers have since been published applying this technique to a number of interesting problems in biology and medicine (70, 96–102). This dye owes its specificity for GSH to the fact that it is conjugated to GSH by glutathione-S-transferases (GSTs) (see below) and has relatively low nonenzymatic reactivity toward GSH and other thiols (103).

Another reagent, chloromethylfluorescein diacetate (CMFDA) has recently been described for measuring intracellular GSH (104). Once inside the cell, this dye is deesterified to a hydrophilic cell impermeant product chloromethylfluorescein (CMF) by nonspecific esterases. CMF is then presumably metabolized in a manner similar to MCB, yielding a GSH-MF fluorescent conjugate. The advantage provided by this dye is that excitation and emission characteristics are similar to fluorescein once conjugated to GSH, allowing one to use this dye simultaneously with UV excitable dyes—for instance, with Hoechst 33342 for simultaneous GSH and DNA determinations (see below). The principal disadvantage of this dye is that it is apparently less specific for GSH than MCB (104).

Performing and Calibrating GSH Analysis

FACTORS THAT AFFECT THE UTILITY OF MCB FOR GSH MEASUREMENTS

We and others have found that the conditions required to adequately stain cells for GSH content with MCB depend upon several factors, including the concentration of dye used, staining time, and temperature. Cook et al. (96) have recently reported that 50 μM MCB are sufficient to label most of the GSH in rodent cells; however, it appears that the Km toward MCB for human GSTs is much higher, suggesting that an appropriate concentration may be closer to 1 mM MCB. Moreover, there are different isozymes of GST in different cell types; classes described include α, μ, and π forms, grouped according to their mobility on isoelectric focusing gels. The GST isozymes are homo- and heterodimes of peptides coded from multiple genes (105). Heterogeneity in the expression of the different isozymes of GST in different cell types, coupled with the fact that different GST isozyme classes show different reactivity toward MCB (GST-μ > GST-α > GST-π) (96, 101), means that one must be very careful to substantiate flow cytometric data with other conventional biochemically based assays of GSH (see below). This caveat is especially relevant when examining human tumor cells, many of which express primarily GST-π, the isoform with the lowest reactivity toward MCB. Caution is also appropriate in the use of MCB for measurement of GSH in human lymphocytes, since there is differential expression of a GST-π form (GST-λ), depending upon culture conditions and growth state (106), and since GST-μ expression has been shown to be polymorphic in PBL from humans, as indicated by reactivity toward trans-stilbene oxide (TSO) and benzo-a-pyrene-4,5-oxide (BPO) (106–110). In fact, if GST-μ shows the same polymorphic distribution in the human population toward MCB as it does towards TSO and BPO, then MCB/FCM would be a convenient assay for the polymorphic expression of this isozyme, the expression of which has been associated with a decreased risk for certain cancers (108–111).

Figure 31.13A shows a comparison of human T-cells stained with MCB at two different concentrations. It is obvious that the fluorescence of these cells is much higher when one stains with 1 mM MCB. We have also found, in contrast to Cook et al. (96), that the rate of GSH conjugation in human cells is temperature–dependent. As seen in Figure 31.13A, a significantly faster rate of fluorescence development is achieved when cells are incubated at 37°C rather than at room temperature, and the relative difference is greater with 60 μM MCB than with 1000 μM MCB. This result is consistent with the observations of Seidegård and colleagues (108) who showed that the maximum activity toward TSO occurs at 40°C in peripheral blood mononuclear cells. In the studies published by Cook et al. (96) and Ublacker et al. (101), cells were reacted with MCB at room temperature (although it is mentioned in a footnote to the Cook paper that they did not see a temperature dependence in the cell types they examined). Both of these groups have addressed the limitations of the MCB/FCM method when measuring GSH in human cells (especially tumor cells). The specificity of the various GST isozymes toward MCB as a function of temperature needs to be more fully explored.

Figure 31.13B shows that rodent cells (Rat-1 fibroblasts) exhibit more rapid and higher MCB staining with lower concentrations than human lymphocytes, which is consistent with Cook et al. (96). Note, however, that appreciable differences between 60 μM and 1000 μM MCB still remain, and that the temperature used for incubation has a smaller, but still reproducible, effect than is seen with human T-cells. Thus, for both rodent cells (Rat-1 fibroblasts) and human cells (PBL), the concentration of dye used can substantially affect the amount of fluorescence seen on the cell sorter. The choice of a dye concentration depends upon the cell type being used, the Km for MCB for the GST isozymes of those cells, and the proportion of GSH one wishes to derivatize within the cell. Even with these higher concentrations of MCB, not all of the GSH has been derivatized. The basis for this incomplete derivatization has been attributed to the inhibition of GST by the GSH-bimane conjugate (96).

Nonetheless, we have found that lower concentrations of MCB (60 μM) may be used with human lymphocytes to reflect qualitative differences in GSH level. If one wishes only to determine if there is a correspondence of some parameter with relative GSH content, and not the exact amount of GSH present, these lower concentrations of MCB may suffice (see calibration section below).

PRACTICAL CONSIDERATIONS FOR STAINING WITH MCB

Stock solutions of the dye are prepared either in ethanol or DMSO, such that at the final working concentration the total

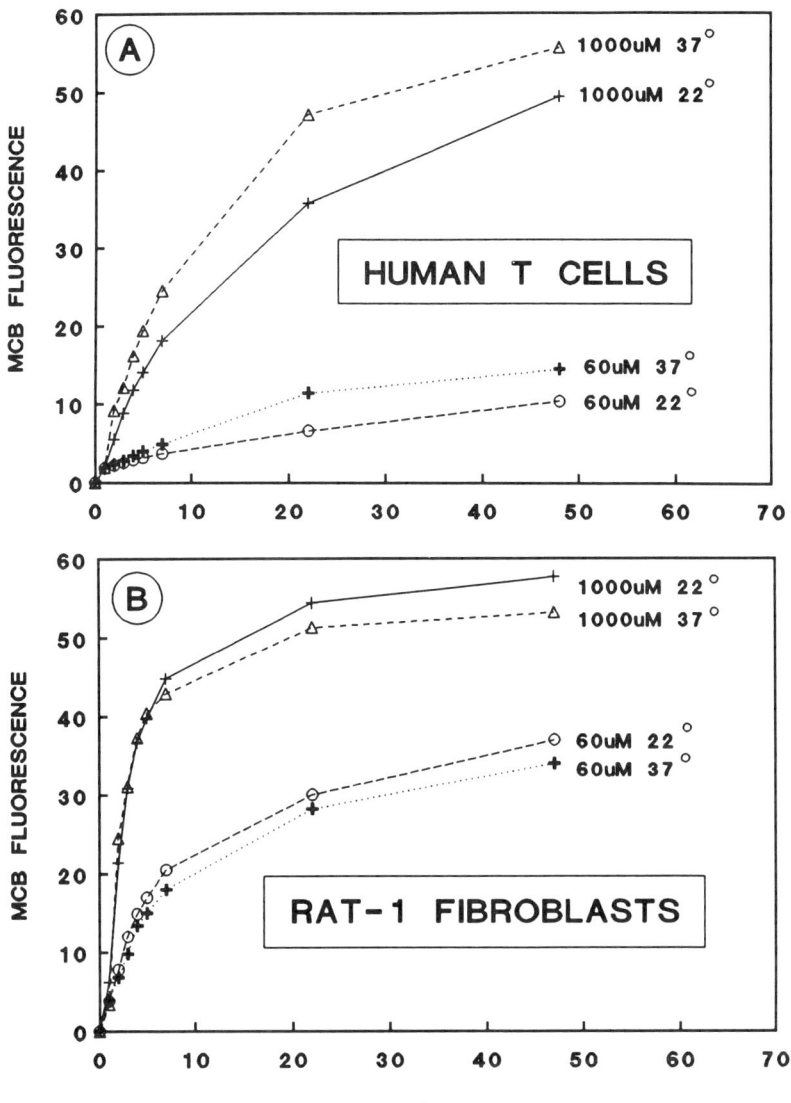

Figure 31.13. Influence of temperature and concentration on the kinetics of fluorescence development in MCB-stained *human T-cells* and *rat 1 fibroblasts*. *Human T-cells* from normal donors were isolated after density-gradient centrifugation. Cells were then left at room temperature or prewarmed in a 37°C water bath and placed in a fluorescence-activated cell sorter (Epics Elite, Coulter Corp., Hialeah, FL) with excitation at 328 nm and emission detected at 410–480 nm. After determining the baseline autofluorescence, the cells were quickly removed from the machine and an aliquot of these cells was transferred to another tube containing MCB that had been held at the appropriate temperature. This tube was quickly placed back into the machine and fluorescence development monitored over time. **A** shows MCB-stained *human T-cells,* **B** shows *rat-1 fibroblasts*. Both cell types developed fluorescence at a faster rate when stained with MCB at 1000 µM than when stained at 60 µM. *Human T-cells* showed a greater temperature dependence for this reaction than did *rat-1 fibroblasts*.

volume of the ethanol or DMSO does not exceed 1% of the cell suspension. These reagents can be stored at 4°C in the dark for up to one month. Cells are incubated with the desired concentration of MCB in medium, for the length of time desired, at 22°C or 37°C. After staining, cells are kept on ice and protected from light until analysis is performed on the flow cytometer.

MCB is nonfluorescent, but when conjugated to GSH by GST it is converted to the fluorescent GSH-bimane product. This reaction occurs in most cells, but since it does require GST, rates of formation of fluorescent product are variable depending upon the activities and specificities of the various isoenzymes of GST among cell types. However, there may also be pools of GSH unavailable for conjugation by this reagent in intact cells, such as within mitochondria.

CMFDA

As with MCB, CMFDA concentrations are expected to be cell-type-specific. We have previously reported experiments with Rat-1 fibroblasts using this reagent. Cells are stained in the dark with a 50 µM solution (final) for 10 min (or longer if nonspecific esterase activity is low) at 37°C. Stock solu-

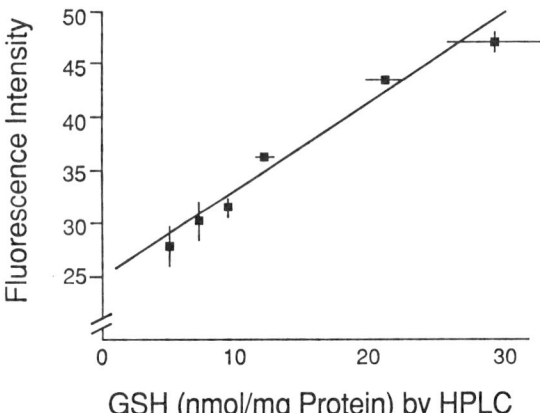

Figure 31.14. Correlation between GSH content measured by HPLC and MCB/flow cytometry. Chinese hamster V79 cells were treated with 0, 1, 3, 6, 9, or 12 mM BSO for 12 hr. Cells were then processed for flow cytometry (45 mM MCB for 10 min at 37°C) or for HPLC determination of total reduced GSH content. (Reprinted with permission from Kavanagh TJ, Martin GM, Livesey TC, Rabinovitch PS. Direct evidence of intercellular sharing of glutathione via metabolic cooperation. J Cell Physiol 1988;137:353–359.)

tions are made up with 5 mM in DMSO. If the cells are to be stained in medium, care should be taken to minimize or exclude serum in the medium, since serum contains variable esterase activity that will convert the reagent extracellularly to a cell impermeant form.

Calibration and Data Analysis

CALIBRATION BY HIGH-PERFORMANCE LIQUID CHROMATOGRAPHY (HPLC) AND TIETZE ASSAY

Since MCB fluorescence is dependent upon GST activity, and since CMFDA stains thiols and is not necessarily specific for GSH (although under most circumstances GSH is the major low-molecular-weight-thiol in cells), it is important to calibrate the level of fluorescence measured by flow cytometry with a more specific biochemical measure of GSH. We have used HPLC analysis and various doses of the GSH-depleting agent buthionine sulfoximine (BSO; an inhibitor of γ-glutamylcysteine synthetase) to ascertain the correlation between MCB fluorescence and GSH content in matched samples of Chinese hamster V79 cells (112). For this cell type, there is a very good correlation between mean cellular fluorescence by flow cytometry and GSH content by HPLC (Fig. 31.14).

In order to determine MCB specificity for GSH, one can simply run stained cell lysates over an HPLC column with fluorescence detection and compare with known low-molecular-weight-thiol standards that have been derivatized with MBB (70). When this is done, we find that >99% of the MCB fluorescence is associated with the GSH peak in protein-free cytosolic extracts of normal human PBL.

Another method of calibration is the Tietze recycling assay (113) or a modified version designed to handle multiple samples in a microtiter plate (114).

Since many flow cytometric applications of the use of MCB will be concerned with relative differences in GSH content among different cell populations, a highly relevant calibration procedure is to examine cells with different GSH content by both FCM and HPLC or biochemical assay. MCB is easily used to discriminate between the degrees of GSH depletion after treatments that oxidize or deplete intracellular GSH. Figures 31.15A and 31.15B show a series of histograms of human CD4+ and CD8+ lymphocytes pretreated with various doses of 1-chloro-2,4-dinitrobenzene (CDNB; a GSH depleter). A dose-dependent reduction in MCB fluorescence (after correction by subtraction of autofluorescence) is seen with increasing concentrations of CDNB in both T-cell subsets. The measurement of the relative decline of MCB fluorescence as a function of CDNB concentration used for MCB depletion reveals that the lower concentration of MCB (60 μM) shows a greater relative depletion with increasing doses of CDNB than does staining with 1 mM MCB (Fig. 31.15C), and a closer correspondence with GSH concentration, as determined by HPLC (Fig. 31.15E). This suggests that a greater proportion of the MCB staining at 1 mM is attributable to nonspecific staining of protein thiols and that, in spite of the lower proportion of total GSH stained with 60 μM MCB (Fig. 31.13A), this staining concentration may yield a more linear relationship to alterations in GSH content in this cell type. Figures 31.15D and 31.15F show 25% to 40% CDNB-resistant staining with MBB (higher with higher MBB concentration), a result that is consistent with the staining of other non-GSH thiols by MBB. For each staining protocol used in Figure 31.15, there was little difference between CD4+ PBL, CD8+ PBL, and CD4−CD8− PBL (primarily B-cells). Data, such as that in Figure 31.15, suggest that, while absolute and intercell-type comparisons of GSH using MCB may be more difficult, measurement of **relative** differences may be much more reliable in human lymphocytes. This result is in agreement with the study of rodent cells by Ublacker et al. (101), but is in contrast with the results they obtained with human tumor and monkey cells. The latter showed disparity between flow cytometric GSH quantitation and biochemical assay. As described previously, this may be related to the predominance of the GSH-π isoform in these tumor cells, which has the highest Km for MCB. Thus, while MCB fluorescence can, in some cell types, yield excellent correlation with GSH, this may not be the case in other cell types, and reliance upon the flow cytometric determinations should be made only after careful study of the relationship of MCB fluorescence to independent assays of GSH content.

GST ACTIVITY

As discussed above, heterogeneity among cell types exists with respect to the total GST activity, the isozymes expressed, and their Km toward MCB. It is therefore important to address the question of GST activity toward MCB in a particular cell type in order to insure that one is using the

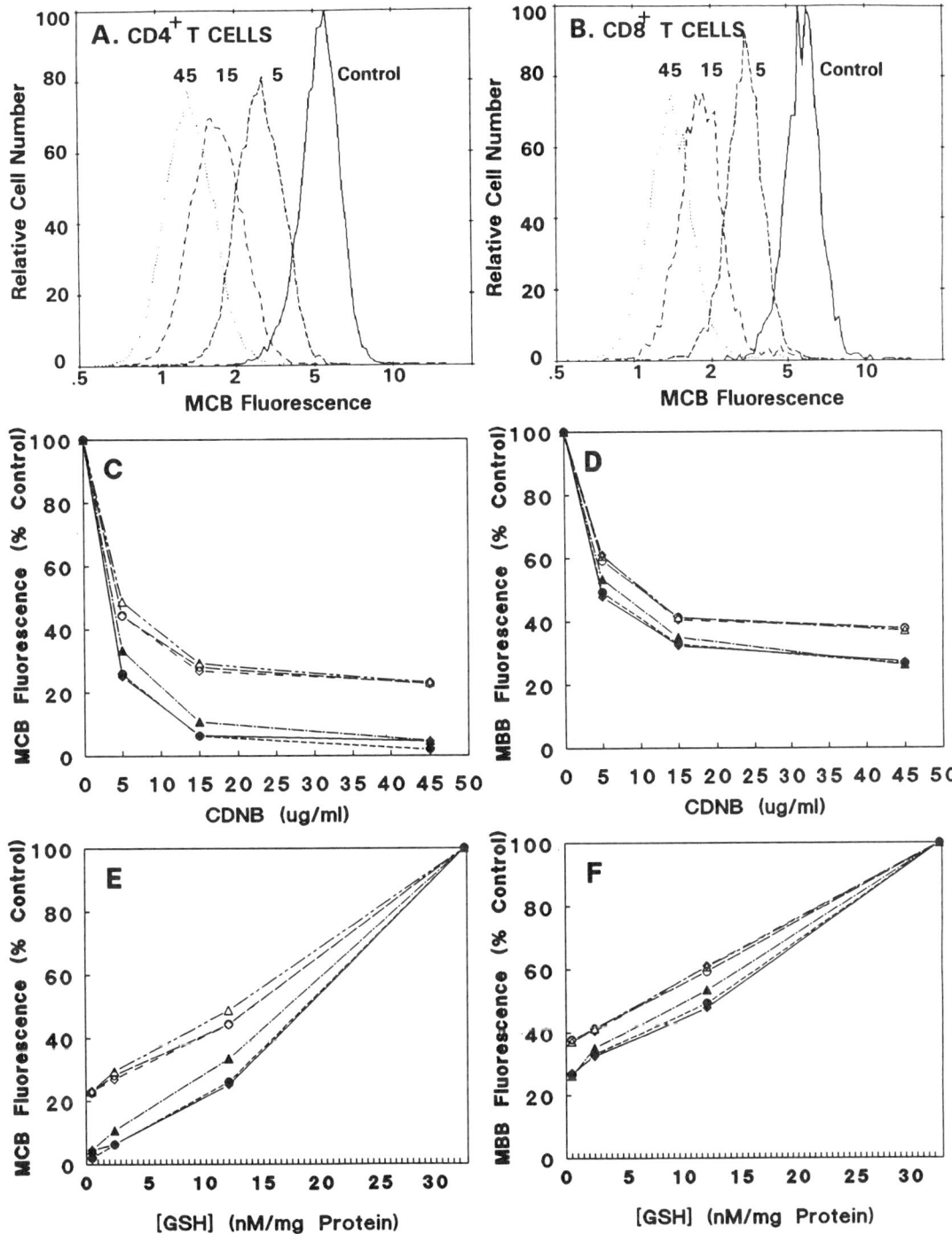

Figure 31.15. Effect of treatment with 1-chloro-2,4-dinitrobenzene (CDNB) on the GSH content of human CD4+ and CD8+ PBL as determined with MCB/flow cytometry and HPLC. A and B show MCB (60 μM) fluorescence histograms of CD4+ and CD8+ cells pretreated with CDNB at 0, 5, 15, and 45 mg/ml for 15 min at 37°C. C shows the effect of using two different concentrations of MCB (60μM, closed symbols, and 1 mM, open symbols) on the measurement of CDNB-induced decline in fluorescence as a percentage of the fluorescence of the nontreated controls. D shows a similar analysis for MBB (60 μM and 1 mM). There is a similar proportionate decline in fluorescence for CD4+ cells (*diamonds*), CD8+ cells (*circles*) and CD4−,CD8− cells (*triangles*), regardless of which concentration of MCB or MBB was used. E compares the level of GSH-bimane conjugate (as determined by HPLC in PBL stained with 1 mM MBB) with the mean relative fluorescence of CD4+ cells, CD8+ cells, and CD4−,CD8− cells stained with either 60 μM or 1 mM MCB. F shows a similar comparison for 60 μM ro 1mM MBB. Autofluorescence of unstained cells has been subtracted from all fluorescence intensities plotted in panels C–F.

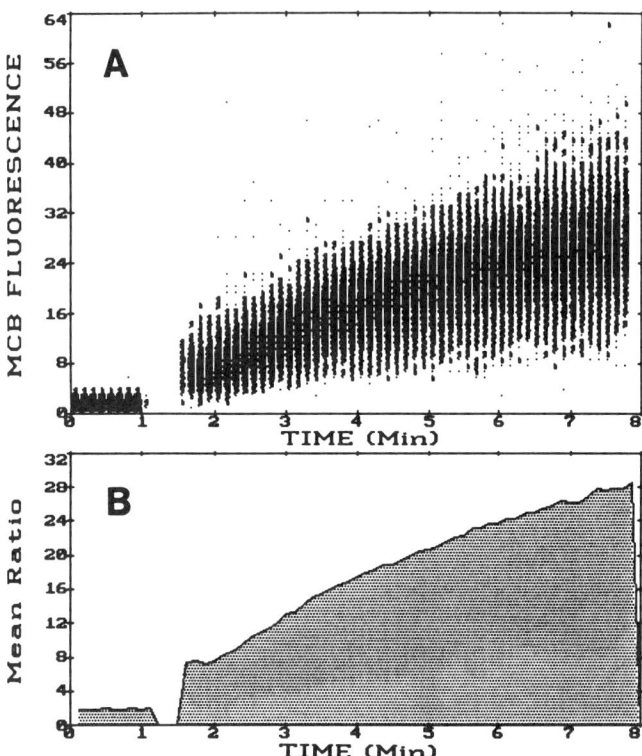

Figure 31.16. MCB fluorescence cytogram (**A**) showing the kinetics of MCB conjugation in human CD4+ lymphocytes by flow cytometry. CD4+ cells isolated on density gradients from peripheral blood were resuspended in RPMI medium at a concentration of 10^6 cells/ml, and warmed to 37°C. After establishing a baseline (auto)fluorescence for 1 min, a 0.5-ml aliquot of the cells was then transferred to a prewarmed (37°C) tube containing MCB, such that the final concentration was 1 mM. This tube was quickly placed back in the cell sorter and the analysis was continued for a total of 8 min. One not only can determine the mean GST activity from such data, but also an indication of intercellular heterogeneity in the GST activity. The mean MCB fluorescence vs. time is shown in **B**.

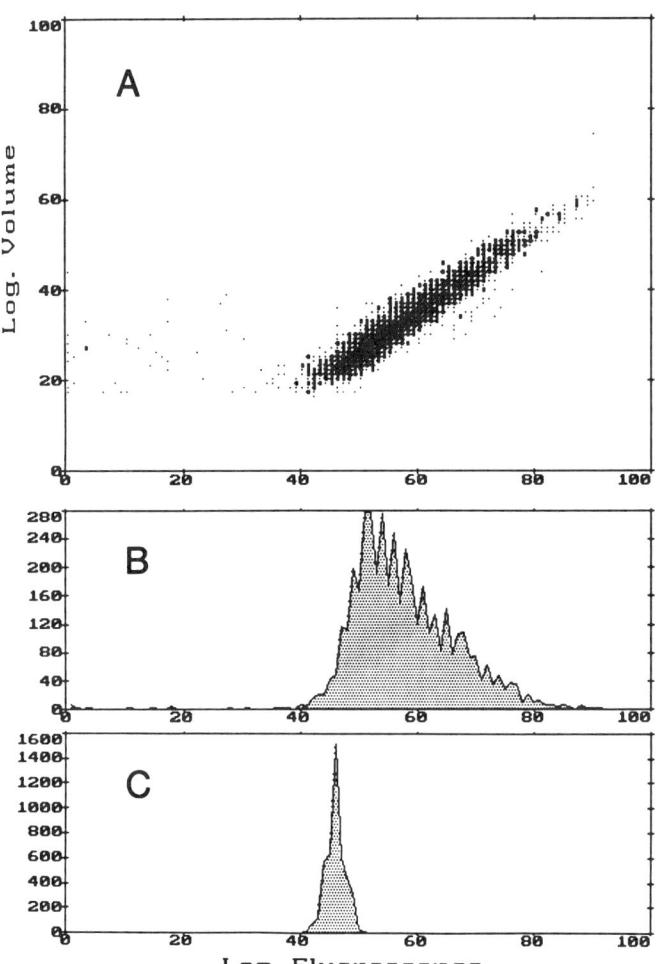

Figure 31.17. Relationship between cell size and GSH content as determined by MCB/flow cytometry. **A** shows a bivariate cytogram (log electronic volume vs. log fluorescence) of rat-1 fibroblasts stained with MCB. There is a direct relationship between cell size and fluorescence intensity, indicating that the concentration of GSH in these cells is tightly controlled, independent of cell size. **B** and **C** show histograms of fluorescence data not corrected for volume and corrected for volume, respectively. The logarithmic axes correspond to a three-decade range.

appropriate MCB staining concentration, incubation time, and temperature. Watson et al. (115) and Cook et al. (96) have described the use of MCB to assess GST activity in cells. An example of this type of analysis is presented in Figure 31.16. Cells in suspension are held at room temperature or prewarmed to 37°C, and a baseline autofluorescence is established for 30 sec to 1 min. Analysis is briefly interrupted and a predetermined aliquot of these cells is added to a room temperature or prewarmed tube containing MCB. The suspension is then quickly analyzed for accumulation of fluorescence over time (usually 10 min) at the preferred temperature. Initial rates of fluorescence accumulation are indicative of GST activity.

DATA ANALYSIS

Since GSH **concentration** but not absolute **content,** is regulated in the cell, it is important to compensate for heterogeneity in cell volume when assessing cellular GSH by flow cytometry. We have found that simultaneous measurement of GSH-dependent fluorescence and electronic cell volume yields data that can easily be adjusted for cell volume bias in GSH content (Fig. 31.17A). By acquiring data on log scale for both volume and MCB fluorescence, one can more easily distinguish between control and GSH-depleted cell populations. Compensation for volume by analyzing the ratio of log fluorescence to log volume, allows one to appreciate diffrences that would otherwise be complicated by volume effects (112) (Figures 31.17B and 31.17C). MCB analysis on instruments that rely on sizing of the cells with forward-angle light scatter can be made either by gating on a specific cell size to minimize volume effects, or by displaying bivariate histograms of the ratio of log fluorescence to log forward scatter.

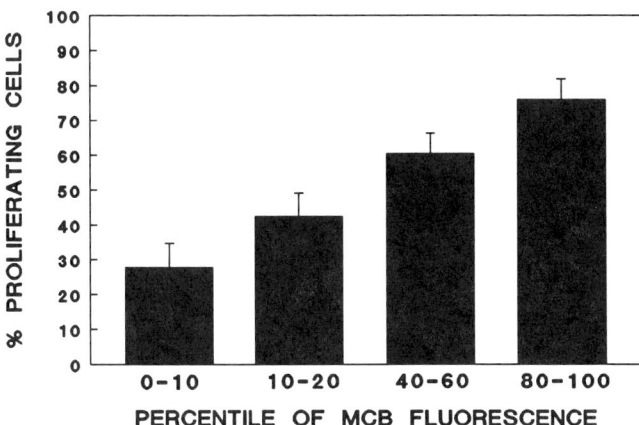

Figure 31.18. Proliferation of human CD4+ lymphocytes sorted on MCB fluorescence intensity. Human lymphocytes isolated from peripheral blood by density gradient centrifugation were stained with MCB (60 μM for 15 min at 37°C) and PE anti-CD4 mAb. CD4+ cells were then sorted according to their MCB fluorescence for the lowest 10%, 10th to 20th percentile, 40th to 60th percentile, and the highest 20% of the fluorescence histogram. Cells were then plated into 96-well plates precoated with anti-CD3 mAb in BrdU-containing medium. After three days, cells were harvested and assessed for proliferation by the BrdU/Hoechst method (129). There is a direct correlation between the MCB fluorescence intensity of the unstimulated cells and their subsequent ability to proliferate. (Data kindly provided by Dr. A. Grossman.)

Role of GSH in Cell Activation

A large number of studies have shown that GSH is important for cell activation and replication. Most of these studies relied on the disruption of normal GSH homeostasis with drugs that either deplete or enhance GSH content in order to show a GSH dependence for cell growth or function (69, 116–124).

An alternative approach to the study of lymphocyte activation is to sort on the basis of GSH content and then to assess growth in the sorted cells after they are allowed to recover to their pretreatment levels of GSH (70). There is a direct correlation between initial GSH content in unstimulated CD4+ human lymphocytes and the capacity for subsequent cell proliferation (Fig. 31.18). In an attempt to ascertain the GSH-sensitive steps in cell proliferation, we stained cells sorted on their GSH content with acridine orange in order to assess their RNA and DNA content simultaneously; this allows one to distinguish activated G_1 cells from those that remain quiescent (G_0) (122). Figure 31.19 shows that cells with low GSH remain in G_0 after mitogen stimulation, suggesting that there is an early step in cell activation that is blocked in low GSH cells.

The effect of GSH content on very early steps in transmembrane signal transduction can be directly evaluated by using indo-1 to assess mitogen-induced changes in intracellular calcium. Figure 31.20 shows the effects of pretreating CD4+ cells with n-ethylmaleimide (NEM) on their ability to respond to stimulation with anti-CD3 mAb. As is apparent

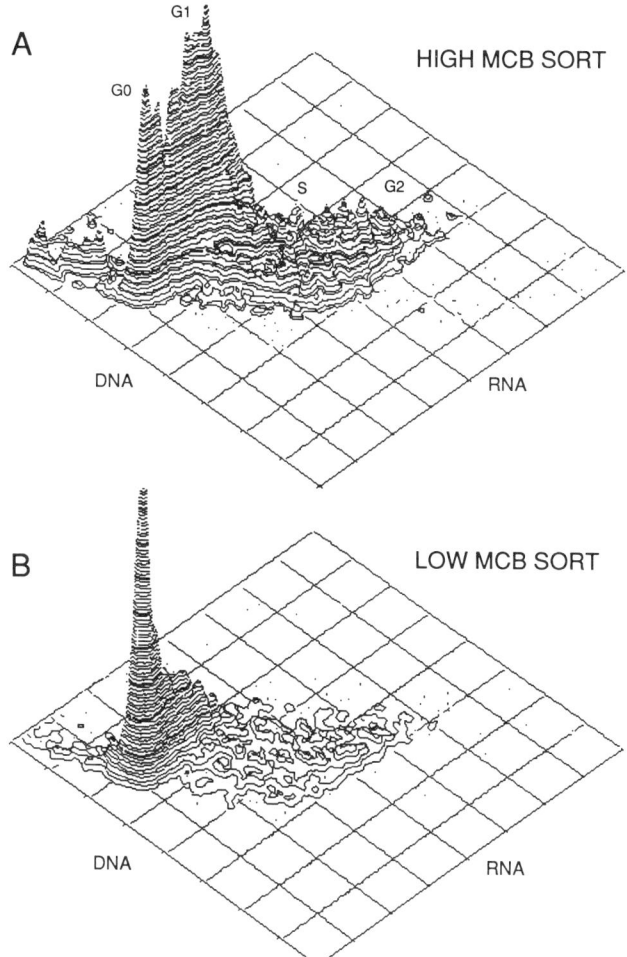

Figure 31.19. Effect of GSH status on the ability of CD4+ cells to become activated by anti-CD3 mAb, as determined by RNA synthesis. Cells were stained with 60 μM MCB and sorted on GSH content (upper and lower 30% of fluorescence histogram distribution), allowed to recover for 24 hr, and stimulated with solid phase anti-CD3 mAb. After 48 hr, cells were fixed in ethanol, stained with acridine orange, and analyzed by flow cytometry for RNA and DNA content. Cells with high GSH (high MCB sort) show the ability to exit G_0 (A), as demonstrated by elevated RNA content in G_1, S, and G_2 cells, whereas cells sorted for low GSH content have nearly all remained in G_0 (B). (Reprinted with permission from Kavanagh TJ, Grossman A, Jaecks EP, et al. Proliferative capacity of human peripheral blood lymphocytes sorted on the basis of glutathione content. J Cell Physiol 1990;145: 472–480.)

from the figure, this reagent, which effectively depletes GSH (Fig. 31.20B), as well as some protein thiols (126), dramatically reduced calcium signaling in CD4+ cells. In contrast to the results with T-cells, B-cells pretreated with NEM do not show similar reductions in calcium signaling (data not shown), indicating that these two classes of lymphocytes have very different thiol-redox requirements for their activation. To directly assess the relationship of GSH to transmembrane signaling, we have sorted CD4+ cells on the basis of their GSH content and subsequently evaluated mitogen-induced changes in indo-1 ratio. As shown in Figure 31.21, cells with high GSH show more vigorous $[Ca^{2+}]_i$

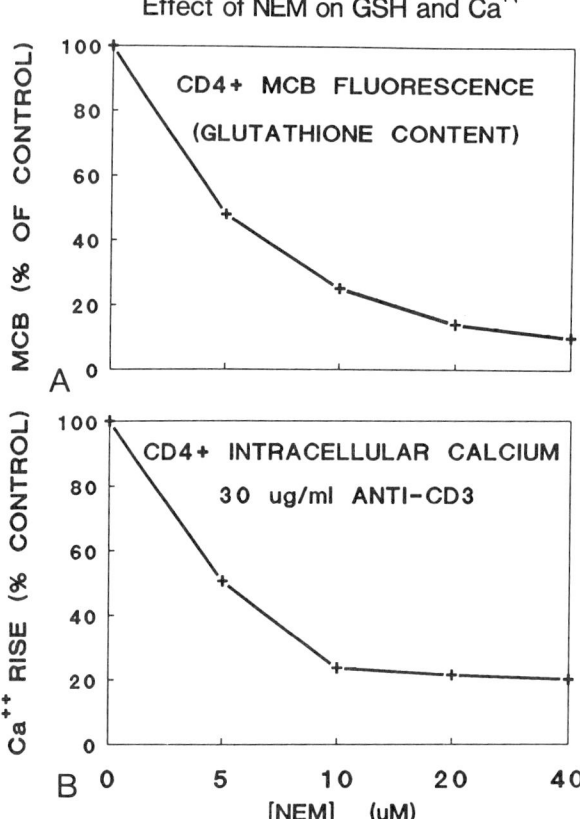

Figure 31.20. Effect of depletion of intracellular thiols upon intracellular calcium signaling. Indo-1-loaded cells were exposed to various concentrations of NEM, which progressively deplete GSH (**A**). The cells were stimulated with 30 μg/ml anti-CD3 mAb and $[Ca^{2+}]_i$ was examined. There is a progressive inhibition of the rise in $[Ca^{2+}]_i$ as a function of increasing NEM concentration (**B**).

responses than cells with low GSH. These results suggest the possibility that certain components of the transmembrane signaling apparatus in at least some cell types are sensitive to thiol oxidation. It is known, for instance, that p56*lck* contains a zinc finger motif and this should be sensitive to oxidation. Freed et al. (126) have assessed the influence of membrane thiol groups in lymphocyte proliferation, IL-2 production, and receptor expression, and have reported similar findings. Surface thiol groups are also important for the activation of neutrophils and monocytes (127).

GSH Variations During the Cell Cycle: Simultaneous Evaluation of DNA and GSH

The ability of CMFDA to be used as a thiol stain excited with visible light has made it possible to simultaneously evaluate GSH levels and DNA content in viable proliferating populations of cells (104). Figure 31.22 shows a bivariate cytogram of Rat-1 fibroblasts stained for GSH with CMFDA and DNA with Hoechst 33342. As one can appreciate from the figure, there is a twofold increase in fluorescence in G_2 cells; this increase appears consistent with their increase in cell volume. The ability to evaluate GSH and DNA simultaneously will surely facilitate the identification of drug-resistant tumor cells from biopsy material, allowing one to ascertain relationships between aneuploidy and changes in GSH content and/or GST activity.

Sorting on the Basis of GSH: Isolation of GSH Variants

Since MCB is a viable stain, one can sort cells on the basis of their GSH content. Lee and Siemann (128) have used MCB and cell sorting to isolate human ovarian tumor cells with increased GSH content that were also resistant to adriamycin. However, the cells reverted to normal GSH after growth in culture over time. Using repeated rounds of MCB staining and sorting, a number of Rat-1 fibroblast variants that have stable alterations in their basal GSH content have been isolated (Fig. 31.23). These cells have not yet been characterized as to the basis of their altered GSH level. However, likely possibilities include mutations in the genes coding one of the two enzymes responsible for GSH synthesis (γ-glutamylcysteine synthetase; glutathione synthetase), differential expression of GST isozymes, or deficiencies in cysteine transport. High GSH variants may have mutations

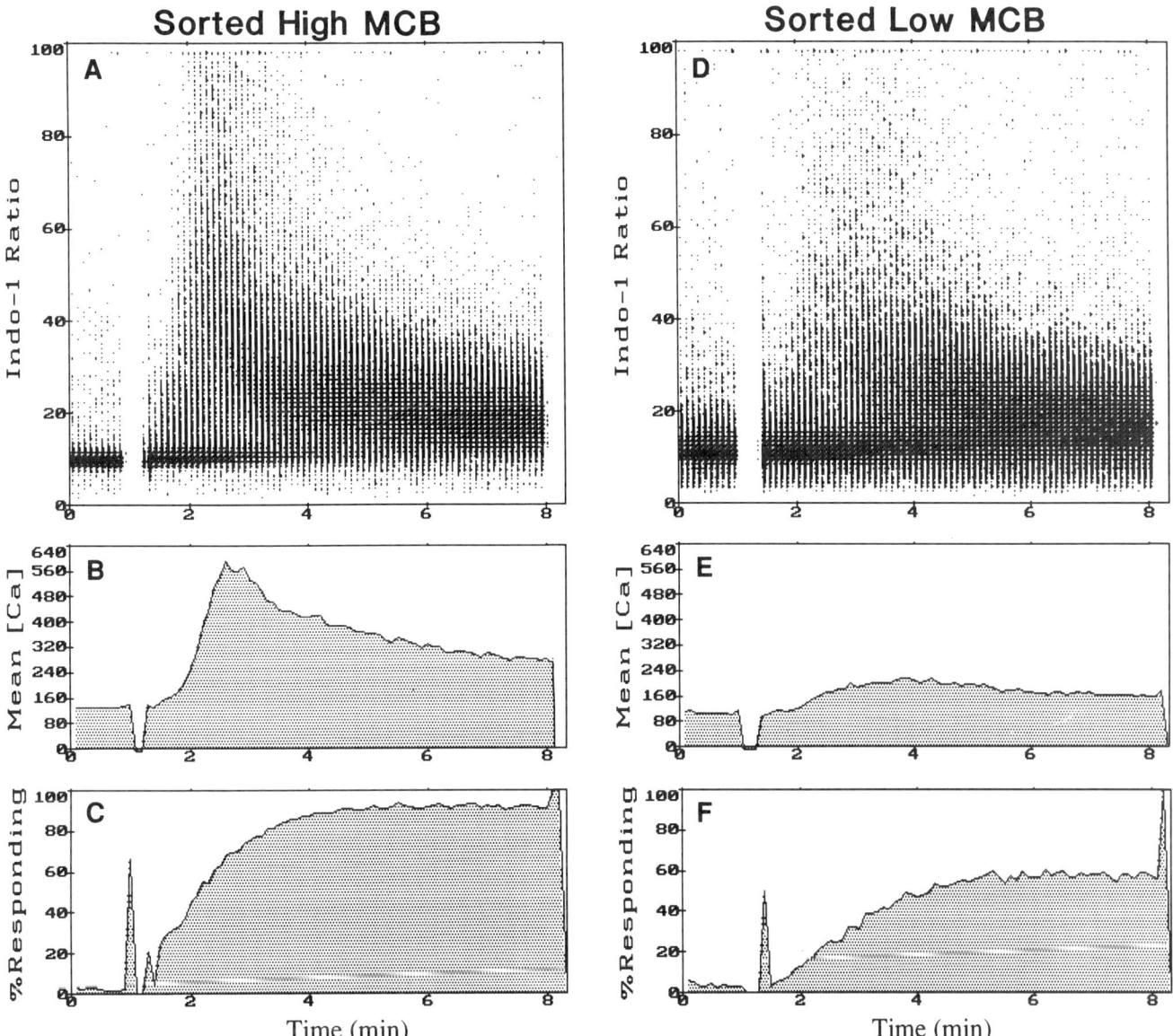

Figure 31.21. Effect of intracellular GSH content on anti-CD3 mAb-stimulated transmembrane signal transduction in CD4+ human T-lymphocytes. Human PBL were stained with 60 mM MCB and PE-anti-CD4 mAb. CD4+ cells were sorted as the upper and lower 30% of the MCB fluorescence distribution, allowed to recover for 24 hr, and then loaded with either indo-1 AM or restained with MCB. Cells restained with MCB retained their originally sorted MCB fluorescence differences (not shown). Cells loaded with indo-1 were analyzed for $[Ca^{2+}]_i$ content following stimulus with anti-CD3 mAb (as described in Fig. 31.4). **A** and **D** show the kinetics of $[Ca^{2+}]_i$ transients in cells sorted on high and low GSH content, respectively. There is a diminished response in cells with low GSH content. **B** and **E** show the mean $[Ca^{2+}]_i$ calculated from the data presented in **A** and **D**, respectively. There is a higher mean and peak calcium response in the high GSH-sorted cells than in the low GSH-sorted cells. **C** and **F** show the percentage of responding cells calculated from the data presented in **A** and **D**, respectively. There is an appreciably higher percentage of cells that are able to respond in the high GSH-sorted cells than in the low GSH-sorted cells.

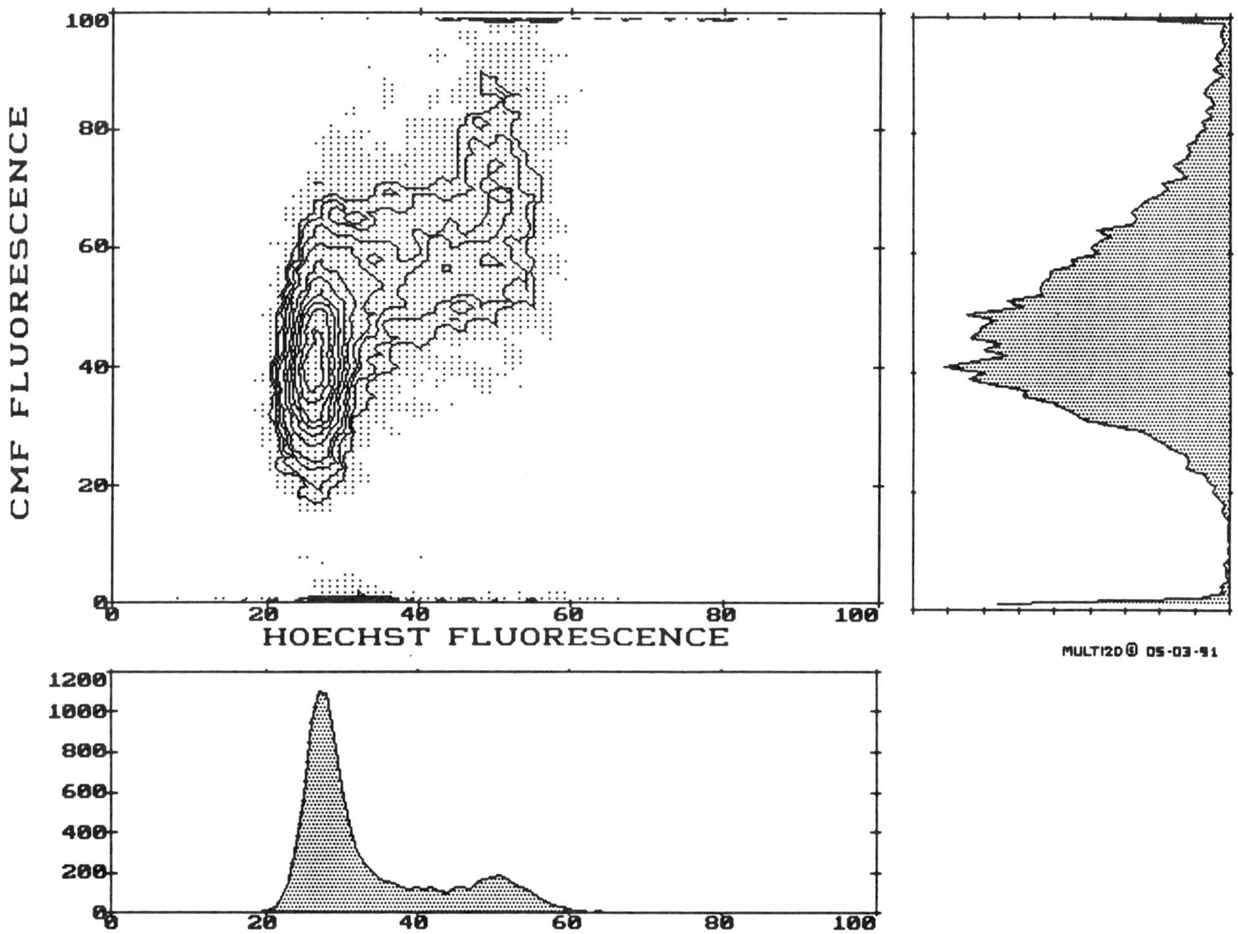

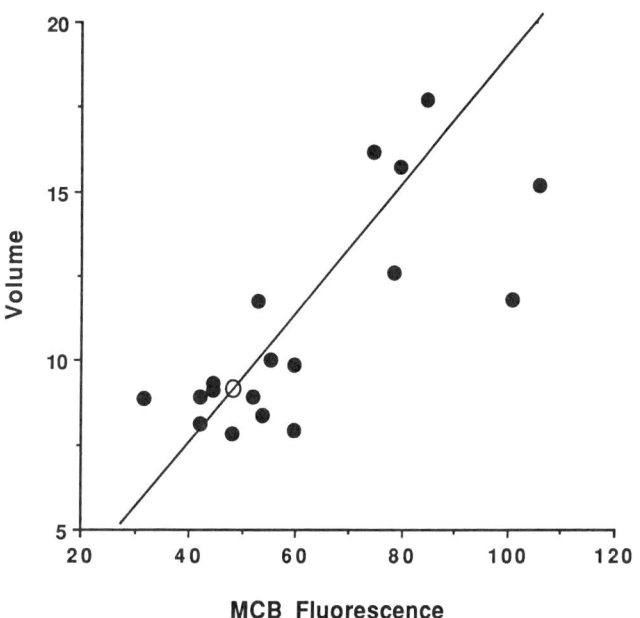

Figure 31.22. Simultaneous measurement of DNA and thiol content in rat-1 fibroblasts. Live cells were stained with Hoechst 33342 (10 mM) for 30 min at 37°C. During the last 10 min of Hoechst staining, the cells were stained with CMFDA (10 mg/ml). Cells were then analyzed with a dual-laser cell sorter. DNA-specific fluorescence was measured with UV excitation and blue emission. The temporarily delayed signal from CMF fluorescence was measured with 488-nm excitation and green emission. Cell aggregates were gated from the analysis by displaying Hoechst-42 fluorescence vs. peak area (not shown).

Figure 31.23. MCB fluorescence and electronic cell volume of rat-1 fibroblast GSH variants isolated after several rounds of MCB staining and viable sorting. Rat-1 fibroblasts were stained with MCB (45 μM for 10 min at 37°C), placed in a cell sorter, and sorted on the highest and lowest 5% of the fluorescence histogram. Cells were then allowed to recover and grow to confluence, at which time the process was repeated until separate populations of cells appeared. The cells were then individually sorted into wells of a 96-well tissue culture plate and grown from single cell clones. The figure represents mean volume and fluorescence of these variants after analysis on a FACS Analyzer. The nonselected parental cells are shown as an open symbol. Note that clones lying above the line from the origin through the parental cells have lower intracellular MCB *concentrations* than the parental cells (although many such cells had larger volumes), while clones below this line have an increased intracellular MCB concentration.

in the regulation of these genes resulting in increased amounts of product, such that there is an abnormally high level of GSH synthesis or conjugation.

ACKNOWLEDGMENTS

We thank J.A. Ledbetter and A. Grossmann for many valuable comments, D.L. Eaton for providing HPLC analysis, and J.C. Jinneman and R.C. Lee for their much-valued technical assistance. This work ws supported in part by the Naval Medical Research and Development Command, Research Task No. M0095.003-1007, and by the National Institutes of Health Grant AG01751. The opinions and assertions expressed herein are those of the authors and are not to be construed as official or reflecting the views of the Navy Department or the naval service at large.

REFERENCES

1. Blinks JR, Wier WG, Hess P, Prendergast FG. Measurement of Ca^{2+} concentrations in living cells. Prog Biophys Mol Biol 1982;40:1–114.
2. Tsien RY, Pozzan T, Rink TJ. Calcium homeostasis in intact lymphocytes: cytoplasmic free calcium monitored with a new, intracellularly trapped fluorescent indicator. J Cell Biol 1982;94:325–334.
3. Tsien RY. A non-disruptive technique for loading calcium buffers and indicators into cells. Nature 1981;290:527–528.
4. Ransom JT, DiGiusto DL, Cambier JC. Single cell analysis of calcium mobilization in anti-immunoglobulin-stimulated B lymphocytes. J Immunol 1986;136:54–57.
5. Grynkiewicz G, Poenie M, Tsien RY. A new generation of $Ca2+$ indicators with greatly improved fluorescence properties. J Biol Chem 1985;260:3440–3450.
6. Minta A, Kao JPY, Tsien RY. Fluorescent indicators for cytosolic calcium based on rhodamine and fluorescein chromophores. J Biol Chem 1989;264:8171–8178.
7. Tsien RY. Fluorescent indicators of ion concentrations. Methods Cell Biol 1989;30:127–156.
8. Vandenberghe PA, Ceuppens JL. Flow cytometric measurement of cytoplasmic free calcium in human peripheral blood T-lymphocytes with fluo-3, a new fluorescent calcium indicator. J Immunol Methods 1990;127:197–205.
9. Rijkers GT, Justement LB, Griffioen AW, Cambier JC. Improved method for measuring intracellular Ca^{++} with fluo-3. Cytometry 1990;11:923–927.
10. Rabinovitch PS, June CH, Grossmann A, Ledbetter JA. Heterogeneity among T cells in intracellular free calcium responses after mitogen stimulation with PHA or anti-CD3. Simultaneous use of indo-1 and immunofluorescence with flow cytometry. J Immunol 1986;137:952–961.
11. Cohen LB, Salzberg BM, Davila HV, et al. Changes in axon fluorescence during activity: Molecular probes of membrane potential. J Membr Biol 1974;19:1–36.
12. Poenie M, Alderton J, Steinhardt R, Tsien R. Calcium rises abruptly and briefly throughout the cell at the onset of anaphase. Science 1986;233:886–889.
13. Chused TM, Wilson HA, Greenblatt D, et al. Flow cytometric analysis of murine splenic B lymphocyte cytosolic free calcium response to anti-IgM and anti-IgD. Cytometry 1987;8:396–404.
14. Davies TA, Drotts D, Weil GJ, Simons ER. Flow cytometric measurements of cytoplasmic calcium change in human platelets. Cytometry 1988;9:138–142.
15. Miller RA, Jacobson B, Weil G, Simons ER. Diminished calcium influx in lectin-stimulated T cells from old mice. J Cell Physiol 1987;132:337–42.
16. Parks DR, Nozaki T, Dunne JF, Peterson LL. Flow cytometer adaptation for quantitation of immunofluorescent reagents and for calibration of dyes for measuring cellular Ca^{2+} and pH. Cytometry 1987;(Suppl 1):104.
17. Martell AE, Smith RM. Critical Stability Constants. Volume 1: Amino Acids. New York: Plenum Press, 1968;269–272.
18. Harafuji H, Ogawa, Y. Re-examination of the apparent binding constant of ethylene glycol bis(beta-aminoethyl ether)-N,N,N',N'-tetraacetic acid with calcium around neutral pH. J Biochem 1980;87:1305–1312.
19. Moisescu DG, Pusch H. A pH-metric method for the determination of the relative concentration of calcium to EGTA. Pflugers Archiv 1975;355:R122.
20. Miller DJ, Smith GL. EGTA purity and the buffering of calcium ions in physiological solutions. Am J Physiol 1984;246:C160-C166.
21. Overton, WR. Modified histogram subtraction technique for analysis of flow cytometric data. Cytometry 1988;9:619–626.
22. Rabinovitch PS, June CH. Intracellular ionized calcium, membrane potential, and pH. In: Olmerod MG, ed. Flow Cytometry: A Practical Approach. Oxford: Oxford Press, 1990;161–185.
23. Lazzari KG, Proto PJ, Simons ER. Simultaneous measurement of stimulus-induced changes in cytoplasmic Ca^{2+} and in membrane potential of human neutrophils. J Biol Chem 1986;261:9710–9713.
24. Ishida Y, Chused TM. Heterogeneity of lymphocyte calcium metabolism is caused by a T cell-specific calcium-sensitive potassium channel and sensitivity of the calcium ATPase pump to membrane potential. J Exp Med 1988;168:839–852.
25. Goldsmith MA, Weiss A. Isolation and characterization of a T-lymphocyte somatic mutant with altered signal transduction by the antigen receptor. Proc Natl Acad Sci USA 1987;84:6879–6883.
26. Goldsmith MA, Weiss A. Early signal transduction by the antigen receptor without commitment to T cell activation. Science 1988;240:1029–1031.
27. De Virgilio F, Steinberg TH, Swanson JA, Silverstein SC. Fura-2 secretion and sequestration in macrophages. A blocker of organic anion transport reveals that these processes occur via a membrane transport system for organic anions. J Immunol 1988;140:915–920.
28. Malgawli A, Milani D, Meldolesi J, Pozzan T. Fura-2 measurement of cytosolic free Ca^{2+} in monolayers and suspensions of various types of animal cells. J Cell Biol 1987;105:2145–2155.
29. Steinberg SF, Bilezikian JP, Al-Awqati Q. Fura-2 fluorescence is localized to mitochondria in endothelial cells. Am J Physiol 1987;253(Pt 1):C744–C747.
30. Arslan P, DeVirgilio F, Beltrane M, Tsien RY, Dozzan T. Cytosolic Ca^{2+} homeostasis in Ehrlich and Yoshida carcinomas. J Biol Chem 1985;260:2719–2725.
31. Luckhoff A. Measuring cytosolic free calcium concentration in endothelial cells with indo 1: The pitfall of using the ratio of two fluorescence intensities recorded at different wavelengths. Cell Calcium 1986;7:233–248.
32. Scanlon M, Williams DA, Fay FS. A $Ca2+$-insensitive form of fura-2 associated with polymorphonuclear leukocytes. Assessment and accurate $Ca2+$ measurement. J Biol Chem 1987;262:6308–6312.
33. Anasetti C, Martin PJ, June CH, et al. Induction of calcium flux and enhancement of cytolytic activity in natural killer cells by cross-linking of the sheep erythrocyte binding protein (CD2) and the Fc-receptor (CD16). J Immunol 1987;139:1772–1779.
34. Geppert TD, Wacholtz MC, Davis LS, Lipsky PE. Activation of human T4 cells by cross-linking class I MHC molecules. J Immunol 1988;140:2155–2164.
35. June CH, Rabinovitch PS, Ledbetter JA. CD5 antibodies increase intracellular ionized calcium concentration in T cells. J Immunol 1987;138:2782–2792.
36. Pezzutto A, Dörken B, Rabinovitch PS, Ledbetter JA, Moldenhauer G, Clark EA. CD19 monoclonal antibody HD37 inhibits anti-immunoglobulin-induced B cell activation and proliferation. J Immunol 1987;138:2793–2799.
37. Pezzutto A, Rabinovitch PS, Dorken B, Moldenhauer G, Clark EA. Role of the CD22 human B cell antigen in B cell triggering by anti-immunoglobulin. J Immunol 1988;188:1791–1795.

38. Wilson HA, Greenblatt D, Taylor CW, et al. The B lymphocyte calcium response to anti-Ig is diminished by membrane immunoglobulin cross-linkage to the Fc gamma receptor. J Immunol 1987;138:1712–1718.
39. Ledbetter JA, June CH, Grosmaire LS, Rabinovitch PS. Crosslinking of surface antigens causes mobilization of intracellular ionized calcium in T lymphocytes. Proc Natl Acad Sci USA 1987;84:1384–1388.
40. Ledbetter JA, June CH, Rabinovitch PS, Grossmann A, Tsu TT, Imboden JB. Signal transduction through CD4 receptors: stimulatory vs. inhibitory activity is regulated by CD4 proximity to the CD3/T cell receptor. Eur J Immunol 1988;18:525–532.
41. Poenie M, Tsien RY, Schmitt-Verhulst AM. Sequential activation and lethal hit measured by $[Ca^{2+}]_i$ in individual cytolytic T cells and targets. EMBO J 1987;6:2223–2232.
42. Williams DA, Becker PL, Fay FS. Regional changes in calcium underlying contraction of single smooth muscle cells. Science 1987;235:1644–1648.
43. Ware JA, Smith M, Salzman EW. Synergism of platelet-aggregating agents. Role of elevation of cytoplasmic calcium. J Clin Invest 1987;80:267–271.
44. Cobbold PH, Rink TJ. Fluorescence and bioluminescence measurement of cytoplasmic free calcium. Biochem J 1987;248:313–328.
45. Ambler SK, Poenie M, Tsien RY, Taylor P. Agonist-stimulated oscillations and cycling of intracellular free calcium in individual cultured muscle cells. J Biol Chem 1988;263:1952–1959.
46. Wilson HA, Greenblatt D, Poenie M, Finkelman FD, Tsien RY. Crosslinkage of B lymphocyte surface immunoglobulin by anti-Ig or antigen induces prolonged oscillation of intracellular ionized calcium. J Exp Med 1987;166:601–606.
47. June CH, Rabinovitch PS. Flow cytometric measurement of cellular ionized calcium concentration. Pathol Immunopathol Res 1988;7:409–432.
48. Rabinovitch PS, June CH. Measurement of intracellular free calcium and membrane potential. In: Melamed MR, Lindmo T, Mendelsohn ML, eds. Flow cytometry and cell sorting. Second Edition. New York: Wiley-Liss, 1990; 651–668.
49. Grossmann A, Rabinovitch PS. Flow cytometry with indo-1 reveals variation in intracellular free calcium within T-cell subsets and between donors after mitogen stimulation. In: Burger G, Ploem JS, Goerttler K, eds. Clinical cytometry and histometry. London: Academic Press, 1987;192–194.
50. Ledbetter JA, Rabinovitch PS, Hellström I, Hellström KE, Grosmaire LS, June CH. Role of CD2 crosslinking in cytoplasmic calcium responses and T cell activation. Eur J Immunol 1988;18:1601–1608.
51. Linette GP, Hartzman RJ, Ledbetter JA, June CH. HIV-1 infected T cells exhibit a selective transmembrane signalling defect through the CD3/antigen receptor pathway. Science 1988;241:573–576.
52. Thomas, RC. Bicarbonate and pHi response Nature 1989;337:601.
53. de Grooth BG, van Dam M, Swart NC, Willemsen A, Greve J. Multiple wavelength illumination in flow cytometry using a single arc lamp and a dispersing element. Cytometry 1987;8:445–452.
54. Valet G, Raffael A, Moroder L, Wünsch E, Ruhenstroth-Bauer G. Fast intracellular pH determination in single cells by flow-cytometry. Naturwissenschaften 1981;68:265–266.
55. Whitaker JE, Haugland RP, Prendergast FG. Seminaphtho-fluoresceins and -rhodafluors: dual fluorescence pH indicators. Biophys J 1988;53:197a.
56. Thomas JA, Buchsbaum RN, Zimniak A, Racker E. Intracellular pH measurements in Ehrlich ascites tumor cells utilizing spectroscopic probes generated in situ. Biochemistry 1979;18:2210–2218.
57. Meister A. Glutathione metabolism and its selective modification. J Biol Chem 1988;263:17205–17208.
58. Meister A, Anderson ME. Glutathione. Ann Rev Biochem 1983;52:711–760.
59. Reed DJ. Glutathione: Toxicological implications. Annu Rev Pharmacol Toxicol 1990;30:603–631.
60. Brodie AE, Reed DJ. Cellular recovery of glyceraldehyde-3-phosphate dehydrogenase activity and thiol status after exposure to hydroperoxides. Arch Biochem Biophys 1990;276:212–218.
61. Joshi TG, Schirch V. The role of a critical sulfhydryl group in the mechanism of serine hydroxymethyltransferase. Ann NY Acad Sci 1990;585:339–345.
62. Nalecz KA, Müller M, Zambrowicz EB, Wojtczak L, Azzi A. Significance and redox state of SH groups in pyruvate carrier isolated from bovine heart mitochondria. Biochim Biophys Acta 1990;1016:272–279.
63. Niroomand F, Rössle R, Mülsch A, Böhme E. Under anaerobic conditions, soluble guanylate cyclase is specifically stimulated by glutathione. Biochem Biophys Res Commun 1989;161:75–80.
64. Sun Y, Oberley LW. The inhibition of catalase by glutathione. Free Radic Biol Med 1989;7:595–602.
65. Suzuki M, Capparelli AW, Jo OD, Yanagawa N. Thiol redox and phosphate transport in renal brush-border membrane. Biochim Biophys Acta 1990;1021:85–90.
66. Zeigler DM. Role of reversible oxidation-reduction of enzyme thiols-disulfides in metabolic regulation. Ann Rev Biochem 1985;54:305–329.
67. Leung MF, Chou IN. Relationship between 1-chloro-2,4-dinitrobenzene-induced cytoskeletal perturbations and cellular glutathione. Cell Biol Toxicol 1989;5:51–66.
68. Messina JP, Lawrence DA. Cell cycle progression of glutathione-depleted human peripheral blood mononuclear cells is inhibited at S phase. J Immunol 1989;143:1974–1981.
69. Suthanthiran M, Anderson ME, Sharma VK, Meister A. Glutathione regulates activation-dependent DNA synthesis in highly purified normal human T lymphocytes stimulated via the CD2 and CD3 antigens. Proc Natl Acad Sci USA 1990;87:3342–3347.
70. Kavanagh TJ, Grossmann A, Jaecks EP, et al. Proliferative capacity of human peripheral blood lymphocytes sorted on the basis of glutathione content. J Cell Physiol 1990;145:472–480.
71. Meister A. 5-Oxoprolinuria (pyroglutamic aciduria) and other disorders of the γ-glutamyl cycle. In: Wyngaarden JB, Stanbury JB, Fredrickson DS, Goldstein JL, Brown MS, eds. Metabolic basis of inherited disease. New York: McGraw Hill, 1983;348–359.
72. Smyth MJ. Glutathione modulates activation-dependent proliferation of human peripheral blood lymphocyte populations without regulating their activated function. J Immunol 1991;146:1921–1927.
73. Liang SM, Liang CM, Hargrove ME, Ting CC. Regulation by glutathione of the effect of lymphokines on differentiation of primary activated lymphocytes J Immunol 1991;146:1909–1913.
74. Dröge W, Pottmeyer GC, Schmidt H, Nick S. Glutathione augments the activation of cytotoxic T lymphocytes in vivo. Immunobiology 1986;172:151–156.
75. Gmünder H, Roth S, Eck HP, Gallas H, Mihm S, Dröge W. Interleukin-2 mRNA expression, lymphokine production and DNA synthesis in glutathione-depleted T cells. Cell Immunol 1990;130:520–528.
76. Liang CM, Lee N, Cattell D, Liang SM. Glutathione regulates interleukin-2 activity on cytotoxic T-cells. J Biol Chem 1989;264;13519–13523.
77. Cantin A, Larivee P, Begin R. Extracellular glutathione suppresses human lung fibroblast proliferation. Am J Respir Cell Mol Biol 1990;2:79–85.
78. Fetsch J, Maurer HR. Glutathione: An in vitro granulopoiesis inhibitor at nanomolar concentration, isolated from calf spleen. Exp Hematol 1990;18:322–325.
79. Abate C, Patel L, Rauscher FJ III, Curran T. Redox regulation of fos and jun DNA binding activity in vitro. Science 1990;249:1157–1161.
80. Hentze MW, Rouault TA, Harford JB, Klausner RD. Oxidation-reduction and the molecular mechanism of a regulatory RNA-protein interaction. Science 1989;244:357–359.

81. Keyse SM, Applegate LA, Tromvoukis Y, Tyrrell RM. Oxidant stress leads to transcriptional activation of the human heme oxygenase gene in cultured skin fibroblasts. Mol Cell Biol 1990;10:4967–4969.
82. Najita L, Sarnow P. Oxidation-reduction sensitive interaction of a cellular 50-kDa protein with an RNA hairpin in the 5' noncoding region of the poliovirus genome. Proc Natl Acad Sci USA 1990;87:5846–5850.
83. Roederer M, Staal F, Raju PA, Ela SW, Herzenberg LA, Herzenberg LA. Cytokine-stimulated human immunodeficiency virus replication is inhibited by N-acetyl-L-cysteine. Proc Natl Acad Sci USA 1990;87:4884–4888.
84. Rokutan K, Thomas JA, Sies H. Specific S-thiolation of a 30-kDa cytosolic protein from rat liver under oxidative stress. Eur J Biochem 1989;179:233–239.
85. Storz G, Tartaglia LA, Ames BN. Transcriptional regulation of oxidative stress-inducible genes: Direct activation by oxidation. Science 1990;248:189–194.
86. Cantin AM, Hubbard RC, Crystal RG. Glutathione deficiency in the epithelial lining fluid of the lower respiratory tract in idiopathic pulmonary fibrosis. Am Rev Respir Dis 1989;139:370–372.
87. Cantin AM, North SL, Hubbard RC, Crystal RG. Normal alveolar epithelial lining fluid contains high levels of glutathione. J Appl Physiol 1987;63:152–157.
88. Buhl R, Jaffe HA, Holroyd KJ, et al. Systemic glutathione deficiency in symptom-free HIV-seropositive individuals. Lancet 1989;2:1294–1298.
89. Buhl R, Jaffe HA, Holroyd KJ, et al. Glutathione deficiency and HIV [Letter]. Lancet 1990;335:546.
90. Eck HP, Gmünder H, Hartmann M, Petzoldt D, Daniel V, Dröge W. Low concentrations of acid-soluble thiol (cysteine) in the blood plasma of HIV-1-infected patients. Biol Chem Hoppe Seyler 1989;370:101–108.
91. Kalebic T, Kinter A, Poli G, Anderson ME, Meister A, Fauci AS. Suppression of human immunodeficiency virus expression in chronically infected monocytic cells by glutathione, glutathione ester, and N-acetylcysteine. Proc Natl Acad Sci USA 1991;88:986–990.
92. Durand RE, Olive PE. Flow cytometry techniques for studying cellular thiols. Radiat Res 1983;95:456–470.
93. Poot M, Verkerk A, Koster JF, Jongkind JF. De novo synthesis of glutathione in human fibroblasts during in vitro ageing and in some metabolic diseases as measured by a flow cytometric method. Biochim Biophys Acta 1986;883:580–584.
94. Treumer J, Valet G. Flow-cytometric determination of glutathione alterations in vital cells by o-phthaldialdehyde (OPT) staining. Exp Cell Res 1986;163:518–524.
95. Rice GC, Bump EA, Shrieve DC, Lee W, Kovacs M. Quantitative analysis of cellular glutathione in Chinese hamster ovary cells by flow cytometry utilizing monochlorobimane: Some applications to radiation and drug resistance in vitro and in vivo. Cancer Res 1986;46:1–6.
96. Cook JA, Iype SN, Mitchell JB. Differential specificity of monochlorobimane for isozymes of human and rodent glutathione s-transferases. Cancer Res 1991;51:1606–1612.
97. Cook JA, Pass HI, Russo A, Iype S, Mitchell JB. Use of monochlorobimane for glutathione measurements in hamster and human tumor cell lines. Int J Radiat Oncol Biol Phys 1989;16:1321–1324.
98. Hedley DW, Hallahan AR, Tripp EH. Flow cytometric measurement of glutathione content of human cancer biopsies. Br J Cancer 1990;61:65–68.
99. Kavanagh TJ, Martin GM, Livesey JC, Rabinovitch PS. Direct evidence of intercellular sharing of glutathione via metabolic cooperation. J Cell Physiol 1988;137:353–359.
100. Shrieve DC, Bump EA, Rice GC. Heterogeneity of cellular glutathione among cells derived from a murine fibrosarcoma or a human renal cell carcinoma detected by flow cytometric analysis. J Biol Chem 1988;263:14107–14114.
101. Ublacker GA, Johnson JA, Siegel FL, Mulcahey RT. Influence of glutathione S-transferases on cellular glutathione determination by flow cytometry using monochlorobimane. Cancer Res 1991;51:1783–1788.
102. Woronicz JD, Rice GC. Simple modification of a commercial flow cytometer to triple laser excitation. Simultaneous five-color fluorescence detection. J Immunol Methods 1989;120:291–296.
103. Radkowsky AE, Kosower EM. Bimanes. 17. (Haloalkyl)-1,5-diazabicyclo[33.0] octadienediones (halo-9,10-dioxa-bimanes): Reactivity toward the tripeptide thiol, glutathione. J Am Chem Soc 1986;108:4527–4531.
104. Poot M, Kavanagh TJ, Kang HC, Haugland RP, Rabinovitch PS. Flow cytometric analysis of cell cycle-dependent changes in cell thiol level by combining a new laser dye with Hoechst 33342. Cytometry 1991;12:184–187.
105. Picket CB, Lu AYH. Glutathione-S-transferases: Gene structure regulation and biological function. Annu Rev Biochem 1989;58:743–764.
106. Jones SM, Brooks BA, Langley SC, Idle JR, Hirom PC. Glutathione transferase activities of cultured human lymphocytes. Carcinogenesis 1988;9:395–398.
107. Jones SM, Idle JR, Hirom PC. Differential expression of glutathione transferases by native and cultured human lymphocytes. Biochem Pharmacol 1988;37:4586–4590.
108. Seidegård J, De Pierre JW, Birberg W, Pilotti A, Pero RW. Characterization of soluble glutathione transferase activity in resting mononuclear leukocytes from human blood. Biochem Pharmacol 1984;33:3053–3058.
109. Seidegård J, Pero RW, Markowitz MM, Roush G, Miller DG, Beattie EJ. Isoenzyme(s) of glutathione transferase (class Mu) as a marker for the susceptibility to lung cancer: A follow up study. Carcinogenesis 1990;11:33–36.
110. Wiencke JK, Kelsey KT, Lamela RA, Toscano WJ. Genetic polymorphisms in carcinogen metabolism predict substrate-induced cytogenetic damage in humans. Prog Clin Biol Res 1990;340B:137–147.
111. Seidegård J, Pero RW, Stille B. Identification of the trans-stilbene oxide-active glutathione transferase in human mononuclear leukocytes and in liver as GST1. Biochem Genet 1989;27:253–261.
112. Kavanagh TJ, Martin GM, El-Fouly MH, Trosko JE, Chang C-C, Rabinovitch PS. Flow cytometry and scrape-loading/dye transfer as a rapid quantitative measure of intercellular communication in vitro. Cancer Res 1987;47:6046–6051.
113. Tietze F. Enzymatic method for quantitative determination of nanogram amounts of total and oxidized glutathione. Application to mammalian blood and other tissues. Anal Biochem 1969;27:502–522.
114. Baker MA, Cerniglia GJ, Zaman A. Microtiter plate assay for the measurement of glutathione and glutathione disulfide in large numbers of biological sample. Anal Biochem 1990;190:360–365.
115. Watson JV, Dive C, Workman P. Measurement of dynamic cellular events. In: Ormerod MG, ed. Flow cytometry: a practical approach. Oxford: IRL Press, 1990;241–264.
116. Fidelus RK. The generation of oxygen radicals: a positive signal for lymphocyte activation. Cell Immunol 1988;113:175–182.
117. Fidelus RK, Ginouves P, Lawrence D, Tsan MF. Modulation of intracellular glutathione concentrations alters lymphocyte activation and proliferation. Exp Cell Res 1987;170:269–275.
118. Fidelus RK, Tsan MF. Enhancement of intracellular glutathione promotes lymphocyte activation by mitogen. Cell Immunol 1986;97:155–163.
119. Fischman CM, Udey MC, Kurtz M, Wedner HJ. Inhibition of lectin-induced lymphocyte activation by 2-cyclohexene-1-one: Decreased intracellular glutathione inhibits an early event in the activation sequence. J Immunol 1981;127:2257–2262.
120. Hamilos DL, Wedner HJ. The role of glutathione in lymphocyte activation. I. Comparison of inhibitory effects of buthionine sulfoximine and 2-cyclohexene-1-one by nuclear size transformation. J Immunol 1985;135:2740–2747.
121. Hamilos DL, Zelarney P, Mascali JJ. Lymphocyte proliferation in glutathione-depleted lymphocytes: Direct relationship between gluta-

thione availability and the proliferative response. Immunopharmacology 1989;18:223–235.
122. Lacombe P, Kraus L, Fay M, Pocidalo J-J. Glutathione status during the mitogenic response of rat splenocytes. Effects of oxygen concentration: FO2 21% versus FO2 7%. Biochimie 1986;68:555–563.
123. Lacombe P, Kraus L, Fay M, Pocidalo J-J. Glutathione status of rat thymocytes and splenocytes during the early events of their ConA proliferative responses. Biochimie 1987;69:37–44.
124. Shaw JP, Chou I. Elevation of intracellular glutathione content associated with mitogenic stimulation of quiescent fibroblasts. J Cell Physiol 1986;129:193–198.
125. Traganos F, Darzynkiewicz Z, Sharpless T, Melamed MR. Simultaneous staining of ribonucleic and deoxyribonucleic acids in unfixed cells using acridine orange in a flow cytofluorometric system. J Histochem Cytochem 1977;25:46–52.
126. Freed BM, Lempert N, Lawrence DA. The inhibitory effects of N-ethylmaleimide, colchicine and cytochalasins on human T cell functions. J Immunopharmacol 1989;11:459–465.
127. Pettit CM, Hall ND. Surface thiol group involvement in neutrophil and monocyte activation. Biochem Soc Trans 1990;18:305–306.
128. Lee F, Siemann DW. Isolation by flow cytometry of a human ovarian tumor cell subpopulation exhibiting a high glutathione content phenotype and increased resistance to adriamycin. Int J Radiat Oncol Biol Phys 1989;16:1315–1319.
129. Rabinovitch PS, Kubbies M, Chen YC, Schindler D, Hoehn H. BrdU-Hoechst flow cytometry: A unique tool for quantitative cell cycle analysis. Exp Cell Res 1988;174:309–318.
130. Paradiso AM, Tsien RY, Machen TE. Digital image processing of intracellular pH in gastric oxyntic and chief cells. Nature 1987;325:447–450.
131. Musgrove E, Rugg C, Hedley D. Flow cytometric measurement of cytoplasmic pH: A critical evaluation of available fluorochromes. Cytometry 1986;7:347–355.
132. Cook JA, Fox MH. Intracellular pH measurements using flow cytometry with 1,4-diacetoxy-2,3-dicyanobenzene. Cytometry 1988;9:441–447.

32

Microsphere-Based Fluorescence Immunoassays Using Flow Cytometry Instrumentation

THOMAS M. McHUGH and MACK J. FULWYLER

INTRODUCTION

The laboratory measurement of a soluble analyte is often performed utilizing antibody-based assays (immunologic reactions). Early immunoassays relied on the visible precipitation of immune complexes in an agar-based gel. The use of solid phases, such as erythrocytes (hemagglutination), on which the immunologic reaction occurs increased the sensitivity and allowed quantitation of the analyte. The addition of a radioactive label (radioimmunoassay, RIA) allowed greater sensitivity and more accurate quantitation of the analyte. The radioactive tag has largely been replaced by an enzyme that catalyzes the visible color change of a substrate, permitting the safe and economical use of enzyme immunoassays (EIA). Various solid supports have been used in immunoassays, including test tubes, wells of microtiter plates, erythrocytes, paper discs, and microspheres.

Immunoassays are broadly categorized according to their assay format, including: (a) the detection of antibody or antigen; (b) heterogeneous or homogeneous systems; and (c) solid-phase or liquid-phase assays. The detection of antibody indicates that the assay is designed to detect a specific antibody in the test sample generally by using its corresponding antigen to capture the antibody. The antibody then must be detected and quantified. The detection of antigen typically relies on the capture of the antigen from the test sample using an antibody.

A heterogeneous immunoassay involves the sequential addition of reagents, incubation (to capture the analyte), removal of excess sample by washing, and subsequent addition of an indicator reagent (labeled with radioactivity, enzyme, or fluorescence). The signal results from multiple steps, and the greater the signal the higher the concentration of the analyte in the test sample. In a homogeneous assay, the test sample and the indicator are incubated simultaneously and analysis is performed without separate steps. Often, homogeneous assays are competitive, meaning that a known amount of labeled (radioactive, enzyme, or fluorescence) analyte is added to the test sample and competes with the unlabeled analyte in the test sample. In the competitive assays, the greater the signal, the lower the concentration of the analyte in the test sample.

Solid-phase immunoassays employ a physical support on which the reaction can occur, whereas liquid-phase assays occur in solution without the need for a solid support to anchor the reactants. Of utmost concern in the development or selection of an immunoassay is the signal-to-noise ratio (S:N). The analytic sensitivity of immunoassays can be defined by the lowest concentration of analyte producing a signal significantly and reproducibly seen over background. The S:N is affected by all phases of the assay, including reagent purity and quality, characteristics of the test sample (neat serum, lipemic samples, clear fluid), and the type of indicator used. Experimentation is necessary to determine the optimal indicator, test sample dilution, reagent grade, etc. The higher the S:N the more confidence there is that the result represents a true positive.

Two recent developments have significantly altered the field of immunology. The first was the development of monoclonal antibodies that provided essentially an unlimited supply of an immunologic reagent with defined specificity. The second was the development of flow cytometry, allowing rapid and sensitive analysis of cell populations, often by using monoclonal antibodies. While flow cytometry has primarily been applied to the analysis of biological samples, this instrument has also demonstrated a capacity to perform immunoassays for soluble analyte, such as those previously done by RIA or EIA. While this ability to measure soluble analyte was described early in the history of flow cytometry, it is only in the last few years that the techniques have been investigated in an organized fashion.

HISTORY OF FLOW MICROSPHERE IMMUNO-ASSAY (FMIA)

The original idea for using microspheres of multiple sizes as solid-phase accumulators to collect and concentrate soluble analyte for subsequent detection by flow cytometry was patented by Fulwyler, in 1976, and described by Horan and Wheeless (1). Horan et al. (2) first demonstrated this technology in detecting rheumatoid factor using IgG-coated 19.5 μm polystyrene microspheres and fluorescein isothiocyanate (FITC)-conjugated anti-human IgM. The samples were analyzed with a Coulter TPS-1 flow cytometer and the assay was able to detect levels of rheumatoid factor below those detected with an agglutination immunoassay procedure.

Since this early work, various investigators have explored the utility of using microspheres in flow cytometric analysis of soluble analytes. Microspheres are used in many other configurations, such as agglutination procedures and batch EIA. This review will focus on the use of microspheres in flow cytometric assays, particularly for applications having potential clinical utility.

DESCRIPTION OF THE TECHNIQUE

Nonfluorescent microspheres in the range of 1–15 μm in diameter are used as the solid phase for the detection of soluble analytes. Numerous variations to the basic (immunology) methodology have been applied to flow microsphere immunoassays. In the simplest form, a capture reagent (usually antigen or antibody) is coated onto the microsphere surface and a test sample is added. Detection of analyte in the sample can either be in a homogeneous or heterogeneous format. The indicator is generally a fluorescence-conjugated soluble reagent, although an application has been described using submicron fluorescent microspheres coated with the appropriate detection reagent. For the detection of specific antibody, the corresponding antigen is added to the microspheres followed by a fluorescent antispecies immunoglobulin reagent. Antigen capture assays usually start with an antibody-coated microsphere followed by a test sample and then a fluorescently labeled anti-antigen reagent.

The analysis of microspheres after reaction with the test sample and the measurement of fluorescence signal reagent is performed using flow cytometry. The microsphere population is detected by either forward-angle light scatter or electronic volume, which can easily detect microspheres in the 1–15 μm diameter range. The fluorescence signal of the microspheres is then displayed on a single-parameter histogram and the mean or mode of the fluorescent peak is used as the measure of signal intensity. Since the flow cytometer can accurately distinguish particles based upon size, it is possible to use different size microspheres simultaneously for discrete analyses of the different populations. Generally, a difference of 1 μm in diameter allows identification of different microsphere populations. This allows the user to coat different size microspheres, mix them, and perform simultaneous yet discrete assays. In the multiple microsphere assay, the number of different analytes to be measured must be determined. Then, the appropriate microsphere populations are selected. Once the different size populations are coated with their specific capture reagent the microspheres are mixed and incubated with test sample. A fluorescent reagent(s) is added to identify the presence of the analyte in question. The sample is analyzed on the flow cytometer using size (forward-angle light scatter or electronic volume) and often right-angle light scatter to separate the different populations. Each population is selected by ''gating'' and the fluorescence for that population is displayed. Using this technique, it is possi-

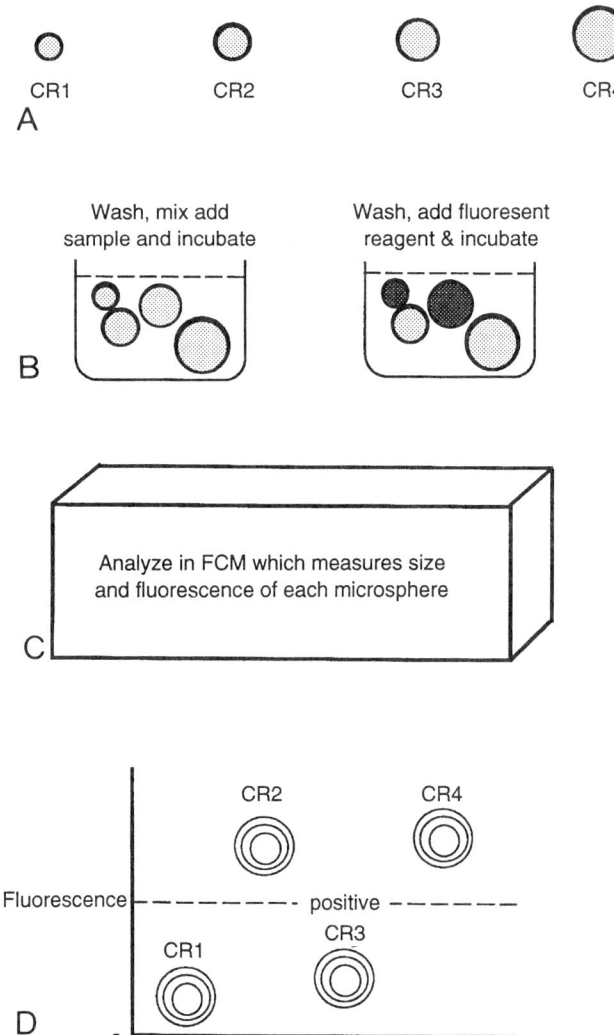

Figure 32.1. A representation of FMIA showing four different sizes of microspheres (**A**) each coated with a different capture reagent (*CR 1–4*); the microspheres are mixed and incubated with a test sample and a fluorescent reagent (**B**). The microspheres are analyzed for size and fluorescence in a flow cytometer (**C**) and the size vs. fluorescence is displayed (**D**) for the determination of positive/negative and for quantitation of the positive signal.

ble to analyze simultaneously for multiple analytes in one assay tube. The magnitude of the fluorescent signal indicates the concentration of each analyte present in the sample. The method offers increased sensitivity due to the use of fluorescence and to the improved optical and spectrophotometric capabilities of flow cytometers.

As shown in Figure 32.1A each of four size classes of microspheres of 3, 5, 10, and 15 μm in diameter is coated with a specific capture reagent. Although the capture reagent is commonly antibody or antigen, in principle, the capture reagent can be any material with sufficient binding affinity for the fluid-phase (dissolved) analyte. Once coated, the several sizes of microspheres are mixed and an aliquot of the test sample is added to the microsphere mixture as shown in

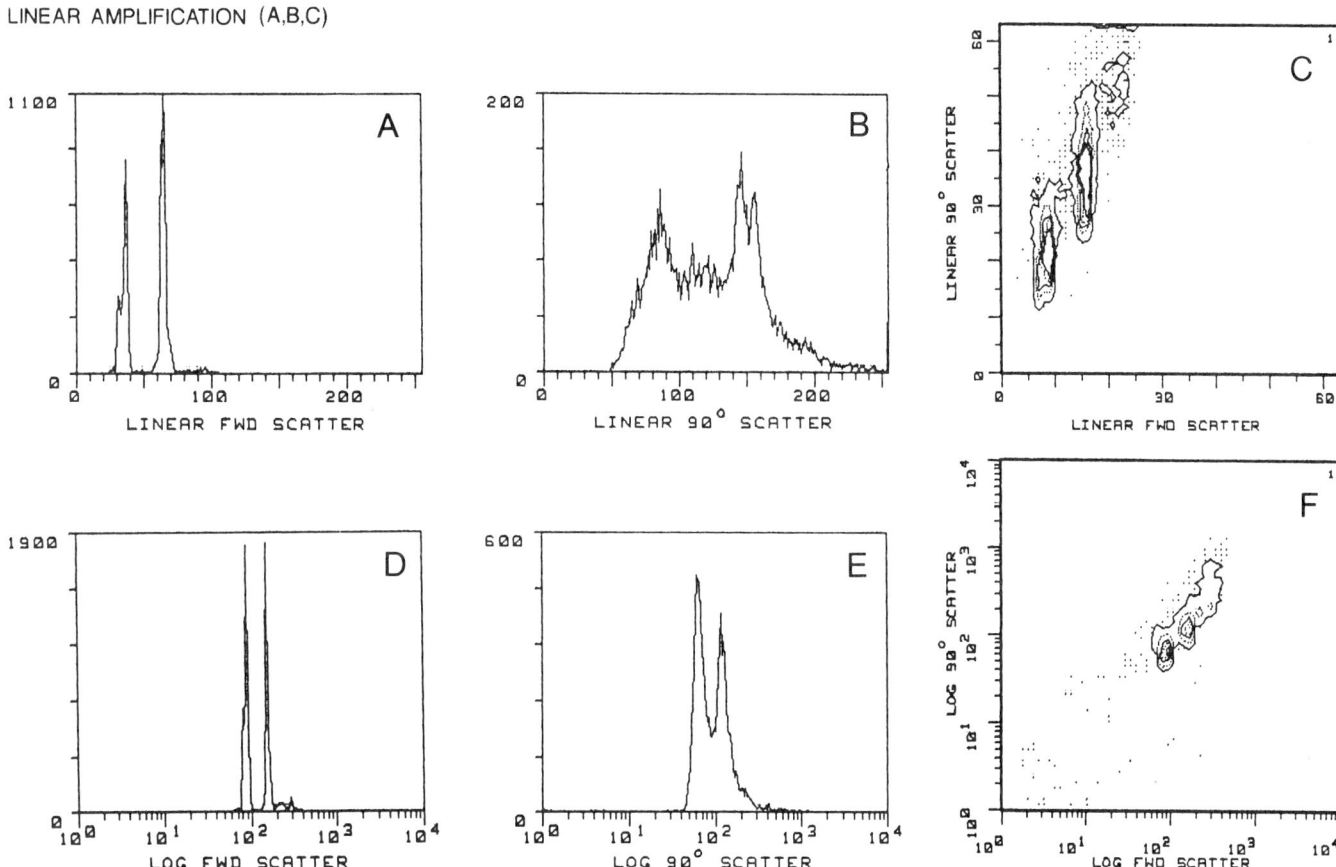

Figure 32.2. Demonstration of the ability to separate by flow cytometry two microsphere populations of 5- and 7-μm using linear (A, B, C) and logarithmic (D, E, F) amplified forward and right-angle light scatter.

Figure 32.1B. Following an incubation step, the microspheres are washed and resuspended in a fluorescent second-step reagent that will bind to the analytes captured by the appropriate microspheres. The microspheres are again washed and analyzed using flow cytometry (Figure 32.1C). The microspheres are analyzed individually as they flow through a light beam produced by a laser or other light source. The generated optical signals reveal the size of each microsphere and the fluorescence (if any) of bound second-step reagent. The data obtained resembles that shown in Figure 32.1D, which represents two-dimensional distribution histograms. The axes represent fluorescence intensity (Y-axis) and light-scatter or volume (X-axis) signals, both plotted on a logarithmic scale. The axis coming out of the page represents the number of events. Isocount contour lines are shown delineating the peaks of the distributions. The microspheres that have picked up their appropriate analyte become fluorescent (populations 2 and 4) and are shifted up the fluorescent axis, compared to unlabeled microsphere populations 1 and 3 in this illustration. Negatively stained microspheres remain near the fluorescence base line while positively stained microspheres increase in fluorescence in proportion to analyte concentration. When properly standardized against a known concentration of analyte, the position of the microsphere population on the fluorescence axis is directly related to the concentration of analyte in the sample.

Detection of Specific Antibody Using Capture-Reagent-Coated Microspheres

The study by Horan et al. (2) was the first to document the ability to detect specific antibody using microspheres coated with antigen. In this application, purified IgG from patients with multiple myeloma was coated onto the surface of 19.5 μm microspheres. The microspheres were incubated with test serum thought to contain rheumatoid factor (IgM anti-human IgG). After incubation, FITC anti-human IgM was added and the microspheres were analyzed by flow cytometry using a Coulter TPS-1. These authors were able to compare the FMIA to a standard agglutination method for the detection of rheumatoid factor. The FMIA had a 60-fold increase in sensitivity as compared to agglutination. In this study, the reactivity of rheumatoid factor to different IgG subclasses was evaluated using only a single size of microsphere. Microspheres were coated with IgG1, IgG2, or IgG3

Table 32.1
Comparison of Anti-CMV and Anti-HSV Antibody Levels by FMIA, LA, and IFA

	CMV					HSV			
			FMIA results					FMIA results	
No. of Samples	Titer by Latex Agglutination	$\bar{X}^a$ MFC	MFC range[b]	% of Samples Above Cutoff[c]	No. of Samples	Titer by Immuno Fluorescence Assay	$\bar{X}$ MFC	MFC Range	% of Samples Above Cutoff
31	Negative	4	1–18	16	20	<1:10	6	1–20	35
9	Und. – 1:4	23	10–38	100	27	1:10–1:20	90	31–163	100
15	1:8–1:16	84	15–140	100	15	1:40–1:80	189	149–823	100
10	1:32–1:64	304	103–502	100	10	1:160–1:320	888	516–1510	100
5	1:128–1:256	617	514–704	100	1	>1:640	3230	—	—
5	1:512–1:1024	3003	1870–4535	100	—	—	—	—	—

[a]: X = mean MFC value (MFC = mean fluorescence channel)
[b]: MFC range = lowest and highest MFC
[c]: % of samples above cutoff = >7.1 is positive for anti-CMV and >4.2 positive for anti-HSV
(From: McHugh TM, Miner RC, Logan LH, Stites DP. Simultaneous detection of antibodies to cytomegalovirus and herpes simplex virus by using flow cytometry and a microsphere-based fluorescence immunoassay. J Clin Microbiol 1988;26:1957–1961.)

(IgG4 was not used) and serum was analyzed for reactivity. The results demonstrated that rheumatoid factor bound well to the three IgG subclasses evaluated. The potential for coating microspheres of different sizes for simultaneous analysis was discussed but was not demonstrated.

The study by McHugh et al. (3) was the first to show the simultaneous reaction of multiple analytes using more than one size microsphere. This study used 5- and 7-μm microspheres, which are shown using forward- (linear amplification) and right-angle (logarithmic amplification) light scatter in Figure 32.2. The 5-μm and 7-μm polystyrene microspheres were coated with purified herpes simplex virus (HSV) and cytomegalovirus (CMV) antigens respectively. The microspheres were mixed and incubated with diluted patient serum followed by biotinylated antihuman IgG and streptavidin-phycoerythrin (PE). The microspheres were analyzed with a Becton Dickinson FACS Analyzer. The FMIA was compared to the standard methods of latex agglutination (LA) for CMV antibody, and slide indirect immunofluorescence (IFA) for HSV antibody. The FMIA was able to detect specific antibody at concentrations eight- to 16-fold less than detectable with the standard methods. The FMIA detected 35% of presumed HSV-negative patients as antibody-positive and 16% of the presumed CMV-negative patients as positive (see Table 32.1).

Fulwyler et al. (4) described the simultaneous detection of specific antibodies to four different HIV proteins. An expanded study was later published by Scillian et al. (5). They showed excellent correlation between the FMIA for HIV antibodies in serum from patients known to be HIV infected. They also examined 35 homosexual men who seroconverted to HIV as demonstrated by EIA and western blot analysis. The FMIA used four microspheres of 5, 7, 10, and 15-μm in diameter, each coated separately with purified recombinant-DNA-produced HIV proteins (p31, gp120, p24, and gp41, respectively). The coated and mixed microspheres were incubated with dilutions of serum and followed by FITC-conjugated antihuman IgG or IgM. These were analyzed using a Becton Dickinson FACScan flow cytometer. In these 35 individuals, 9 (26%) had specific anti-HIV antibody detectable by FMIA earlier than it was detectable by either EIA or western blot. FMIA also had a significantly lower false-positive rate than either EIA or western blot. This study underscored the increased sensitivity and specificity of FMIA (specificity due to the use of recombinant-DNA-produced proteins) over the standard methods. The relevance of these "early" seroconverters to diagnosing HIV infection is as yet unclear. Further studies comparing specific antibody, HIV culture, and use of the polymerase chain reaction are underway. The FMIA also demonstrated the ability to quantify specific antibody and the potential utility of this measurement. Patients were evaluated for levels of anti-p24 by FMIA and the levels appeared to correlate with the rate of progression to AIDS. This finding has prompted a larger study of the role of anti-p24 in HIV infection and an evaluation of the FMIA in monitoring patients infected with HIV.

Wilson et al. (6, 7) used 3–10-μm polstrene and 1–3-μm polyacrylamide microspheres coated with either human light chains or cell membrane fragments. The coated microspheres were incubated with mouse monoclonal antibodies, followed by FITC-conjugated anti-mouse IgG, and analyzed on a Becton Dickinson FACS 440. These studies showed the potential for using microspheres to screen antibodies for reactivity to (a) desired antigen(s) as well as the capability for long-term storage of the microspheres with passively coated antigen. The coated microspheres were stored and used as stable reagents to screen hybridoma supernatants for reactivity to known cell surface antigens.

In a direct comparison of FMIA to EIA, McHugh et al. (8) described the detection and quantitation of anti-*Candida* antibodies in patient serum. In this study, three different sizes of polystyrene microspheres (5-, 7- and, 9.5-μm) were used. The microspheres were coated with three distinct antigen preparations from *Candida albicans*. The coated and mixed microspheres were incubated with patient serum, followed by a FITC antihuman IgG, and analyzed on a Becton Dickinson FACS Analyzer. The same three antigen preparations were used to coat wells of 96-well microtiter plates. The same dilutions of patient serum were incubated in the coated wells, followed by a peroxidase-conjugated antihuman IgG and then o-phenylenediamine peroxidase substrate. The absorbance was read at 490 nm. The FMIA was able to provide complete discrimination of antibody levels between the healthy controls and the infected patients. This study documented the increased ability to semi-quantify antibody levels with FMIA compared to EIA. The discrete fluorescent signal and the increased signal-to-noise ratio of the FMIA, as compared to the EIA, were the major factors for the improved performance of the FMIA. In this study, the three distinct antigen preparations performed equally well in the FMIA.

Lim et al. (9) showed the sensitive detection of human antimouse antibodies (HAMA) using FMIA. Polystyrene microspheres of 9.5 μm were coated with mouse monoclonal antibody (OKT3, anti-CD3). The coated microspheres were incubated with patient serum, followed by FITC-labeled antihuman IgG, and analyzed with a Becton Dickinson FACScan. Patients had been treated in vivo with OKT3 for prevention of graft rejection following solid organ transplantation. All patient sera had been previously tested by EIA. The FMIA correlated well with the EIA in patients with high levels of HAMA. However, a number of patients identified as lacking HAMA by EIA were positive by FMIA. The increase in detection rate of HAMA-positive samples is probably due to the increased sensitivity of FMIA and preliminary results suggest this to be the case. However, further studies are underway to determine the utility of these low-level antibody results and the potential for false-positive reactions.

Detection of Soluble Analyte by Antibody or Other Capture-Reagent-Coated Microspheres

The first demonstration of capturing soluble analyte with antibody-coated microspheres for flow cytometric analysis was by Lisi et al. (10). In their study, 1–5-μm or 30–40-μm polyacrylamide, or 40–50-μm dextran microspheres were coated with antihuman IgG. Human IgG in either buffer or serum was added, followed by FITC antihuman IgG. Analysis was performed using an Ortho Spectrum III. This was done as a homogeneous assay without a wash step (to remove the unbound fluorescent reagent). The assay sensitivity was calculated as 1–10 ng human IgG/ml, which was comparable or slightly less sensitive than an RIA performed in parallel. The lack of increase in sensitivity of this FMIA was probably due to the microsphere and the capture antibody. Subsequent work by other investigators has shown that both polyacrylamide and dextran microspheres are less desirable than polystyrene for maximum sensitivity. This assay did however perform well in a one-step no-wash format. The flow cytometer is generally set to analyze a fluorescent signal only when it is associated with a size pulse. This means that any remaining fluorescent tag that is not bound to the microsphere does not influence the result. Additionally, since the microspheres are drawn into the sensing region using sheath fluid to produce a laminar flow, much of the unbound fluorescent reagent may dissociate into the sheath.

Saunders et al. (11) demonstrated the sensitive detection of soluble horseradish peroxidase (HRP) using FMIA. In their application, 10-μm polystyrene microspheres, (or in a noncompetitive assay, soluble HRP captured by nonfluorescent 10-μm microspheres) were "sandwiched" with a fluorescent 0.1-μm anti-HRP-coated microsphere. This assay for soluble HRP was compared to a standard RIA for HRP and showed a similar sensitivity of 10^{-14} M. This FMIA was also performed as a one-step no-wash procedure. Since the fluorescence tag was smaller than the size detection limit set on the flow cytometer, the nonbound fluorescent 0.1-μm microspheres did not influence the assay signal.

Two studies have been published showing the capture of immune complexes using human C1q-coated microspheres. In the first study by McHugh et al. (12) 5-μm polystyrene microspheres were coated with the purified human C1q component of complement. The coated microspheres were incubated with patient serum, washed, and incubated with FITC anti-human IgG. The microspheres were analyzed on a Becton Dickinson FACS Analyzer. The FMIA was compared to a standard immune complex assay using human C1q-coated nitrocellulose discs with a fluorometer method (FIAX, International Diagnostic Technology, Sanata Clara, CA) employing FITC antihuman IgG. The FMIA was able to detect immune complexes in the patient sera tested and showed a threefold increase in sensitivity over the FIAX method. A second study (13) used this immune complex method to detect viral antigens in immune complexes in the serum of patients infected with the human immunodeficiency virus (HIV). Human C1q-coated 5-μm microspheres were incubated with patient serum, then with a pool of mouse monoclonal antibodies to HIV core and envelope proteins, and biotinylated antimouse IgG. The fluorescent tag was streptavidin-PE and samples were analyzed with a Becton Dickinson FACS Analyzer. The excitation light was filtered to provide 546-nm light to maximally excite PE, resulting in an assay with increased sensitivity. The presence of HIV-containing immune complexes was documented in most infected patients, including those identified as lacking circulating HIV antigen by conventional HIV p24 antigen EIA.

Lindmo et al. (14) used two microspheres in an antigen capture assay with the novel addition of using capture antibodies of different afinities directed toward the same antigen. Microspheres of 7- and 10-μm were coated with anticarcinoembryonic antigen (CEA) antibodies; the 7-μm microspheres were coated with an antibody of high affinity (association constant of 3.2×10^{10}) for CEA; and the 10-μm microspheres were coated with a lower affinity anti-CEA antibody (association constant of 3.3×10^9). The microspheres were mixed and incubated with purified CEA, a third biotinylated monoclonal anti-CEA antibody (directed toward a different CEA epitope), and streptavidin-PE. The assay was a one-step, no-wash procedure. By gating on the separate populations, they demonstrated an increased dynamic range over standard assays using the combination of two antibodies with different affinities. They estimated that

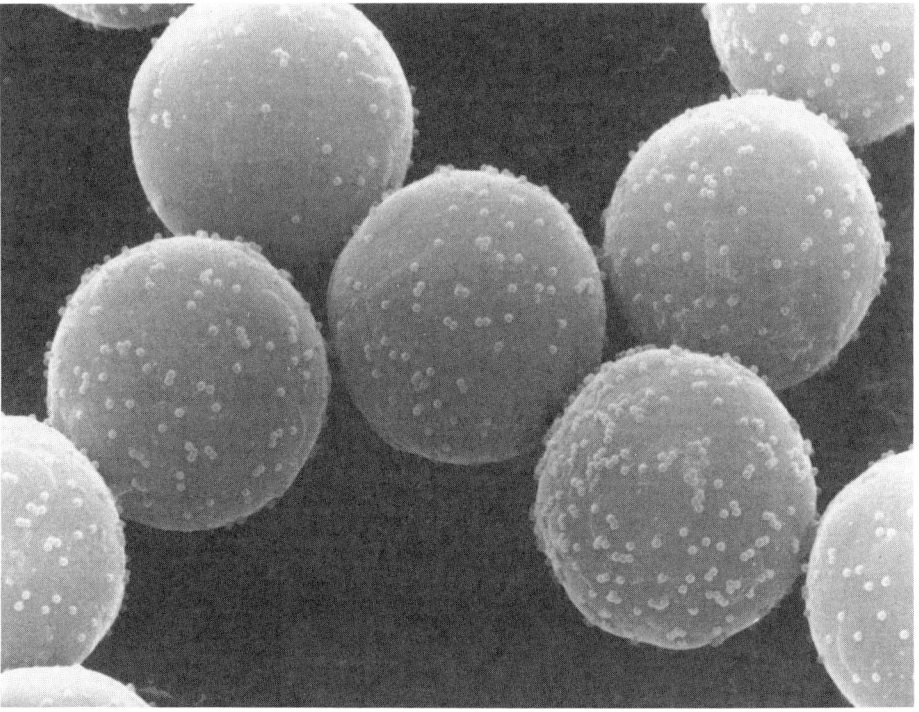

Figure 32.3. A scanning electron micrograph of 10-μm microspheres coated with antihorseradish peroxidase (HRP) binding HRP-coated 0.25-μm microspheres. (From Saunders GC, Jett JH, Martin JC. Amplified flow-cytometric separation-free fluorescence immunoassays. Clin Chem 1985;31:2020–2023.)

the dynamic range of their FMIA using one antibody-coated microsphere would provide a 100-fold range of detection, whereas the inclusion of the second microsphere population with a second antibody of different affinity extends the dynamic range 1000-fold. In samples with high concentrations of CEA, the result is read using the low-affinity antibody-coated microspheres (10 μm), while samples with low concentrations of CEA are analyzed using the high-affinity-coated microspheres (7 μm).

Other Applications of Flow Cytometric Microsphere Assays

Saunders et al. (15) have described the detection of ligand binding to DNA-coated 10-μm microspheres. In their study, DNA-coated microspheres were incubated with a known concentration of mithramycin, a DNA binding antibiotic that fluoresces. Increasing concentrations of the antibiotic actinomycin-D was added and competed with mithramycin for binding sites on the DNA. The binding of actinomycin-D on the microspheres resulted in lower fluorescent signals due to lower amounts of bound mithramcyin.

Sensitivity of FMIA

Published reports indicate that the sensitivity of FMIA is considerably greater than that obtained by other techniques. However, a careful evaluation of the sensitivity of FMIA versus batch microsphere based EIA has not been performed. While most studies suggest that FMIA is more sensitive than the standard EIA method, there is no clear evidence to conclusively prove this. In the first comparison of FMIA to standard methods, Horan et al. (2) demonstrated the increased sensitivity of a flow cytometric immunoassay for the detection of rheumatoid factor over that obtained with conventional agglutination procedure. This study showed the ability to detect rheumatoid factor in samples scored as negative by the agglutination assay. The S/N ratio with the FMIA was 60, with a large dynamic range. In a later study of the detection of human IgG with antihuman IgG-coated microspheres, Lisi et al. (10) concluded that their FMIA assay was less sensitive than a standard RIA. This lack of sensitivity was possibly due to the system evaluated, including the microsphere composition (polyacrylamide or dextran versus polystyrene) and the coupling method used to attach antibody to the microspheres. All later studies confirm the hypothesis that a flow cytometer should be capable of detecting analyte with fluorescence at lower concentrations than methods such as RIA. Saunders et al. (11) however, were the first to document that an antigen capture immunoassay using microspheres and flow cytometry could be more sensitive than other methods. In this study, the detection of soluble HRP was similar to that obtained with radioimmunoassay (10^{-14}M).

McHugh et al. (3) showed an increased sensitivity in detecting anti-CMV and HSV-specific antibody by FMIA, compared to slide IFA and LA. In this study, FMIA was able to detect antibody at eight- to 16-fold lower concentrations than the slide IFA and LA (see Table 32.2). In a subsequent study (8), FMIA was shown to be more sensitive than a standard EIA for the detection of specific antibody to *Candida albicans*. Scillian et al. (5) showed the

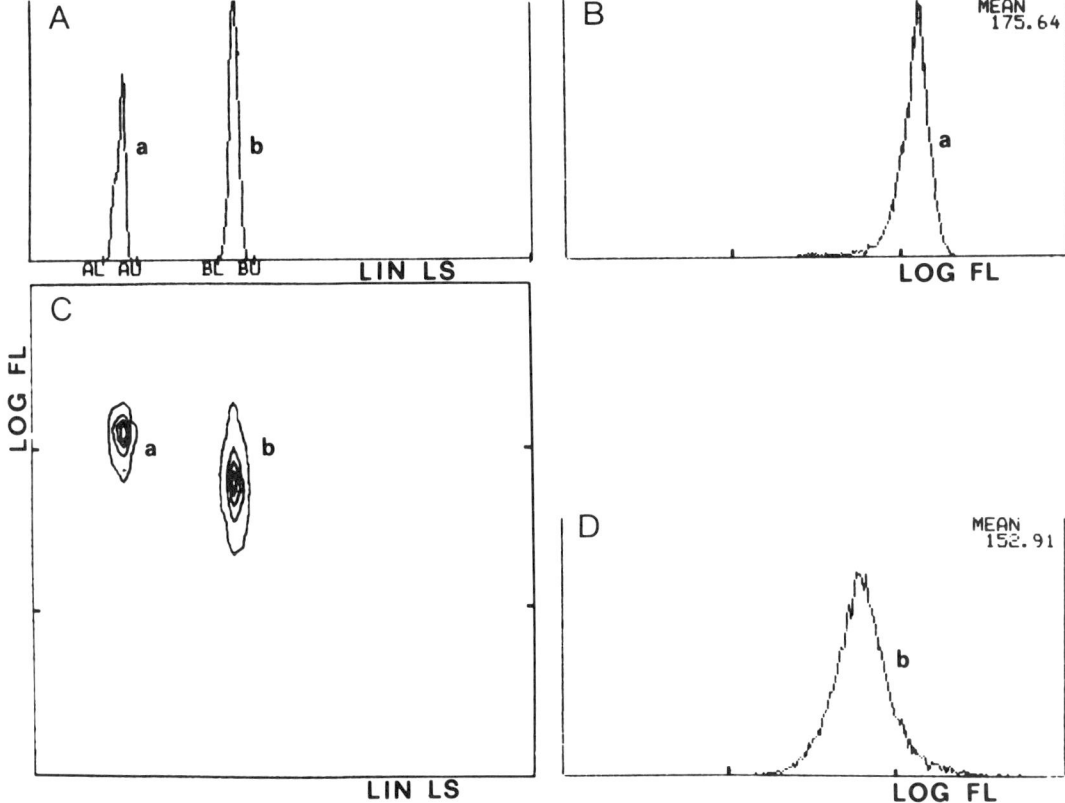

Figure 32.4. FMIA for carcinoembryonic antigen (CEA) using anti-CEA-coated microspheres in the presence of 50 μg CEA per liter; **A** shows the separation of the 7- and 10-μm microspheres by linear forward-angle light scatter; **C** shows the correlated forward-angle light scatter versus logarithmic fluorescence with high-affinity anti-CEA coated microspheres (7 μm, population a) and low-affinity anti-CEA-coated microspheres (10 μm, population b); **B** and **D** are the single-parameter phycoerythrin signals for the 7-μm and the 10-μm microspheres, respectively. (From Lindmo T, Bormer O, Ugelstad J, Nustad K. Immunometric assay by flow cytometry using mixtures of two particle types of different affinity. J Immunol Meth 1990;126:183–189.)

Table 32.2
Dilution Study of a CMV and HSV Seropositive Sample Assayed by FMIA, LA, and IFA

Dilution of the Positive Sample[a]	CMV Results		HSV Results	
	FMIA MFC[b]	Latex Agglutination	FMIA MFC	Immuno Fluorescence Assay
1:40	487	+[c]	613	+
1:80	269	+	290	+
1:160	190	+	174	+
1:320	94	+	88	+
1:640	65	+/−[c]	63	−[c]
1:1280	41	−	51	−
1:2560	20	−	21	−
1:5120	9	−	8	−
1:10240	5	−	3	−
1:20480	2	−	2	−

[a]: positive sample was diluted into a serum sample negative for antibodies to CMV and HSV
[b]: MFC using a final serum dilution of 1:20 with >7.1 positive for anti-CMV and >4.2 positive for anti-HSV
[c]: + = positive, − = negative, +/− = weak positive
(From McHugh TM, Miner RC, Logan LH, Stites DP. Simultaneous detection of antibodies to cytomyalovirus and herpes simplex virus by using flow cytometry and a microsphere-based fluorescence immunoassay. J Clin Microbiol 1988;26:1957–1961.)

sensitive detection of anti-HIV antibodies as compared to EIA and WB and demonstrated the relative increase in sensitivity of FMIA by documenting the detection of specific antiviral antibody in samples that were negative by other techniques. Lim et al. (9) showed the relative increase in sensitivity of FMIA for detecting human antimouse IgG, where samples negative by EIA were clearly positive by FMIA.

Figure 32.5 shows a comparison of EIA to FMIA in detecting low levels of specific antibody using antigen- (human IgG) coated microspheres. This plot shows the increased signal-to-noise ratio that can be obtained with FMIA. In these experiments, the ability to detect low levels of specific antibody was significantly greater with the FMIA (minimum detectable level of 0.075 ng/ml) as compared to EIA (minimum detectable level of 1 ng/ml). Along with the ability to detect lower levels of analyte with FMIA, the S/N is much higher with FMIA, resulting in greater confidence in the result. Using microspheres coupled with specific antibody to capture soluble analyte, the sensitivity of the FMIA has been compared to EIA. Figure 32.5 also shows the curves for signal/noise versus analyte concentration comparing the FMIA with EIA. The FMIA gave a signal that is three times more than background noise at a concentration of 3 pg of analyte/ml. This is a 20-fold increase in analytic sensitivity of the FMIA over that seen with EIA.

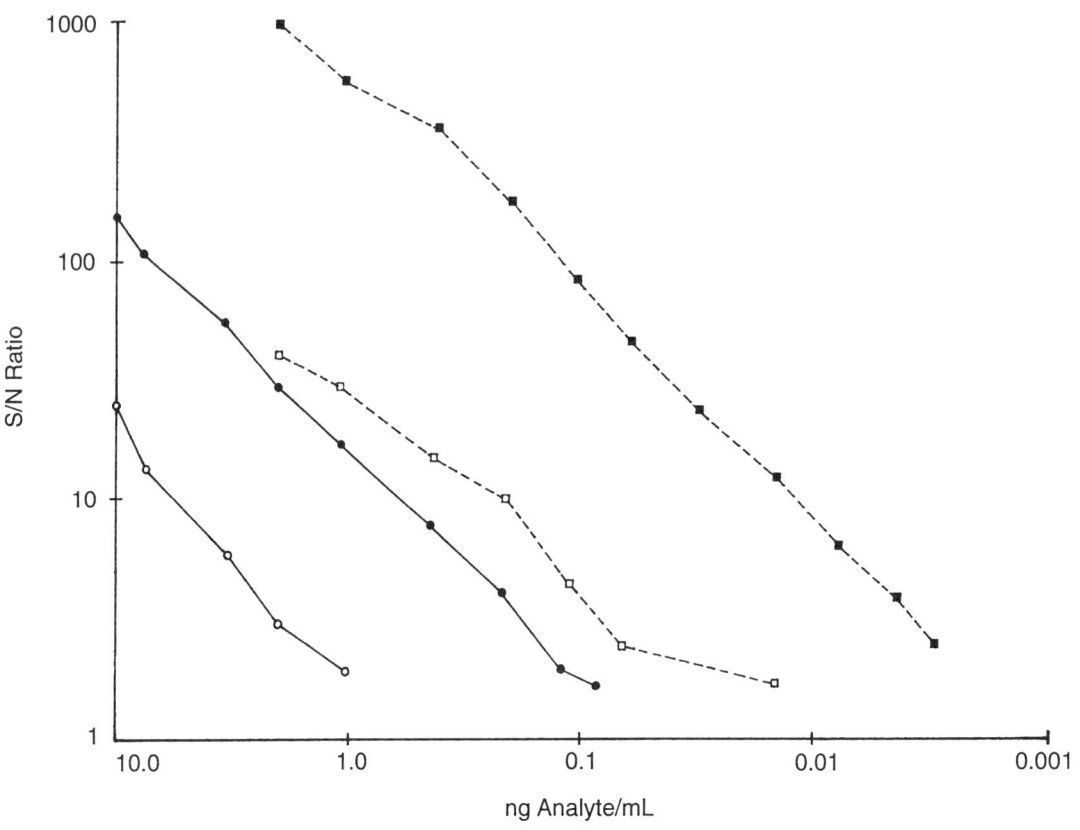

Figure 32.5. Shows the signal-to-noise ratio (S/N) in the lower range for the detection of antibody or antigen by FMIA as compared to a standard EIA. Detection of antibody by FMIA (●———●), antibody by EIA (○———○), antigen by FMIA (■- - - - -■) and antigen by EIA (□- - - - -□).

Limitations of FMIA

While FMIA has advantages over current techniques, including the ability to rapidly and simultaneously detect multiple analytes and flexibility, there are important drawbacks to this method. The most immediate of these is the high cost and complexity of the flow cytometry instrumentation used to perform these assays.

For laboratories equipped with a flow cytometer, the use of FMIA may be attractive. Recent models of air-cooled laser-based flow cytometers are significantly less complex and less expensive than previous instrumentation. However, current instrumentation still requires an individual with knowledge of the theory and practice of flow cytometry to allow the accurate investigation of new techniques. The data analysis programs of commercial cytometers are directed toward cell analysis and consequently are more complex than that needed for FMIA. Smaller, more streamlined data analysis packages would be helpful and such systems are now being developed commercially.

The speed at which FMIA can be performed is also a limitation. At present, using EIA, it is possible to scan a 96-well plate in <20 seconds resulting in 96 discrete analyses. Using contemporary flow cytometers data from 96 FMIA samples takes at least 30 min to acquire. Automated delivery of the microspheres to the flow cytometer would be an improvement and would release the operator for other duties.

Sample processing is time consuming, especially if centrifugal washing steps are used. In addition, a significant loss of microspheres in centrifugation-based washing steps occurs. As was shown by Jolley et al. (16), it is possible to use vacuum filtration to wash microspheres with little or no loss. However, filtration becomes more difficult when assaying a sample with a high protein content such as serum. Efforts are underway to design a vacuum device for the automated washing and quantitative recovery of microspheres and preliminary results are encouraging.

The use of iron-containing microspheres, which would allow for their removal with a magnetic field, is currently being evaluated. Preliminary results show encouraging progress both in the easy handling of the microspheres and in the sensitivity of the assays for analyte detection.

Only recently have coated microsphere reagents become commercially available for FMIA. At present, a variety of analyte systems are available (Advanced Biosystems, Redwood City, CA). Investigators may experiment with attaching their particular capture reagent to the microsphere surface. Numerous papers have been published describing the attachment of capture reagents to polystyrene and other solid surfaces. There are numerous suppliers of microspheres of different sizes and surface chemistry (Bangs Laboratories,

Carmel, IN; Polysciences, Warrington, PA; Seradyn, Indianapolis, IN). An article on the methods involved in these microsphere immunoassays has been published (17) and other methods have been described for protein attachment to microspheres (18–22).

The first decision to be made is to select the type and size of microsphere. Size is a simple parameter to select, as any microsphere which can be detected by light-scatter or electronic volume signal will work. The choice of attaching the capture reagent to the microsphere surface by absorption or by covalent binding is an area to be explored by the investigator. Fluorescent-labeled second antibody reagents are easily obtained and numerous manufacturers now supply "flow-cytometry-grade" reagents that can be used with satisfactory results.

Dynamic Range

The feasible dynamic range of FMIA has not been extensively explored. In our experience, using a single-capture reagent, we have observed a dynamic range of at least 100-fold. An ingenious means of extending the dynamic range of the system has been explored by Lindmo et al. (14), as discussed above.

Simultaneous Detection of Multiple Analytes

We have discussed the selection of several sizes of microspheres that are distinguishable by light scatter or by their volume signal. By careful selection of microsphere diameter, eight or more sizes could be selected, permitting the simultaneous analysis of eight different analytes. The number of distinguishable microspheres can be greatly increased by adding a fluorescent dye to the polymer of the microsphere. For example, by adding a set amount of a phycoerythrin red fluorochrome to a second class of eight microsphere sizes, 16 distinguishable classes are now available. If a third group of eight microsphere classes is created by adding twice the amount of phycoerythrin dye to the microspheres, they become distinguishable from the undyed and from the lightly dyed microspheres. The scheme is straight-forward and the number distinguishable classes of microspheres becomes very large.

Insensitivity to Microsphere Carryover and Microsphere Aggregation

Many analytic methods are sensitive to the carryover of processed sample from one reaction vessel to another. By its nature, FMIA is relatively insensitive to such contamination because of its "digital" nature. It is, of course, sensitive to the admixture of an unprocessed sample into a second unprocessed sample. However, the contamination of one batch of processed microspheres by carry-over of processed microspheres from another well will have little effect. The positivity of a sample is indicated by the mean fluorescence channel of a distribution histogram representing thousands of microspheres. Carry-over of 2% of one processed sample into the next well can represent about 200 microspheres, the admixture of which would not affect the mean fluorescence channel determination.

FMIA is relatively forgiving of background aggregation of microspheres. Aggregates of microspheres can certainly occur, although in our work we have not found this to be a problem. Microsphere aggregates present prior to capture-reagent coating can usually be dispersed by vortexing the sample. Aggregated microspheres give a light-scatter or fluorescence signal that is the summation of the individual microspheres. By careful selection of microsphere size and color (if a dye is added to the microspheres), one can arrange the microspheres so that potential aggregates give signals located away from the position of single microspheres, permitting software discrimination.

Potential Clinical Applications of FMIA

Descriptions of studies already published provide a sense of the state of utility of FMIA. Currently, this technology is primarily a research tool that is being applied in basic or clinical research studies. Application in the clinical laboratory for diagnostic work needs further validation. In the meantime, the technique lends itself well to clinical research studies where a sensitive assay technique is required. The simultaneous use of both immunological-based assays and nonimmunological-based ligand-binding assays, as demonstrated by Saunders et al. (15) suggests that a number of clinical applications will be forthcoming. The clinical applications for this technology will be decided after extensive clinical trials of the basic techniques that are now just being explored.

REFERENCES

1. Horan PK, Wheeless LL. Quantitative single cell analysis and sorting. Science 1977;198:149–157.
2. Horan PK, Schenck EA, Abraham GN, Kloszewski MD. Fluid phase particle fluorescence analysis: rheumatoid factor specificity evaluated by laser flow cytophotometry. In: Nakamure RM, Dito WR, Tucker ES, eds. Immunoassays in the Clinical Laboratory. New York: Alan R Liss, 1979:187.
3. McHugh TM, Miner RC, Logan LH, Stites DP. Simultaneous detection of antibodies to cytomegalovirus and herpes simplex virus by using flow cytometry and a microsphere-based fluorescence immunoassay. J Clin Microbiol 1988;26:1957–1961.
4. Fulwyler MJ, McHugh TM, Schwadron R, et al. Immunoreactive bead (IRB) assay for the quantitative and simultaneous flow cytometric detection of multiple soluble analytes. Cytometry Suppl 1988;9:19.
5. Scillian JJ, McHugh TM, Busch MP, et al. Early detection of antibodies against rDNA-produced HIV proteins with a flow cytometric assay. Blood 1989;73:2041–2048.
6. Wilson MR, Witherspoon JS. A new microsphere-based immunofluorescence assay using flow cytometry. J Immunol Methods 1988;107:225–230.
7. Wilson MR, Mulligan SP, Raison RL. A new microsphere-based immunofluorescence assay for antibodies to membrane-associated antigens. J Immunol Methods 1988;107:231–237.
8. McHugh TM, Wang YJ, Chong HO, Blackwood LL, Stites DP. Development of a microsphere-based fluorescent immunoassay and its comparison to an enzyme immunoassay for the detection of antibodies to

three antigen preparations from *Candida albicans*. J Immunol Methods 1989;116:213–219.
9. Lim VL, Gumbert M, Garovoy MR. A flow cytometric method for the detection of the development of antibody to Orthoclone OKT3. J Immnol Methods 1989;121:197–201.
10. Lisi PJ, Huang CW, Hoffman RA, Teipel JW. A fluorescence immunoassay for soluble antigens employing flow cytometric detection. Clin Chim Acta 1982;120:171–179.
11. Saunders GC, Jett JH, Martin JC. Amplified flow-cytometric separation-free fluorescence immunoassays. Clin Chem 1985;31:2020–2023.
12. McHugh TM, Stites DP, Casavant CH, Fulwyler MJ. Flow cytometric detection and quantitation of immune complexes using human C1q-coated microspheres. J Immunol Methods 1986;95:57–61.
13. McHugh TM, Stites DP, Busch MP, Krowka JF, Stricker RB, Hollander H. Relation of circulating levels of human immunodeficiency virus (HIV) antigen, antibody to p24, and HIV-containing immune complexes in HIV-infected patients. J Infect Dis 1988;158:1088–1091.
14. Lindmo T, Bormer O, Ugelstad J, Nustad K. Immunometric assay by flow cytometry using mixtures of two particle types of different affinity. J Immunol Methods 1990;126:183–189.
15. Saunders GC, Martin JC, Jett JH, Perkins A. Flow cytometric competitive binding assay for determination of actinomycin-D concentrations. Cytometry 1990;11:311–313.
16. Jolley ME, Wang C-HJ, Ekenberg SJ, Zuelke MS, Kelso, DM. Particle concentration fluorescence immunoassay (PCFIA): a new, rapid immunoassay technique with high sensitivity. J Immunol Methods 1984;67:21–35.
17. Fulwyler MJ, McHugh TM. Flow microsphere immunoassay for the quantitative and simultaneous detection of multiple soluble analytes. In: Crissman HA, Darzyneckiewicz Z, eds. Methods in Cell Biology. New York: Academic Press, 1990;33:613–629.
18. Cantarero LA, Butler JE, Osborne JW. The absorptive characteristics of proteins for polystyrene and their significance in solid-phase immunoassays. Anal Biochem 1980;105:375–382.
19. Colvin M, Smolka A, Chang M, Rembaum A. The covalent binding of enzymes and immunoglobulins to hydrophilic microspheres. In: Rembaum A, Tokes ZA, eds. Microspheres: medical and biological applications. Boca Raton, FL: CRC Press 1988;1–13.
20. Nilsson K, Mosback K. Immobilization of enzymes and affinity ligands to various hydroxyl group carrying supports using highly reactive sulfonyl chlorides. Biochem Biophys Res Comm 1981;102:449–457.
21. Nustad K, Danielson H. Reith A, et al. Monodisperse polymer particles in immunoassays and cell separation. In: Rembaum A, Tokes ZA, eds. Microspheres: medical and biological applications. Boca Raton, FL: CRC Press, 1988:53–75.
22. Rembaum A, Yen SPS, Cheong E, et al. Functional polymeric microspheres based on 2-hydroxyethyl methacrylate for immunochemical studies. Macromolecules 1976;9:328–336.

33

Flow Cytometry in Skin Disorders

LISA STAIANO-COICO

The epidermis is a self-renewing, stratified squamous epithelium whose primary function is to protect the body against invasion by microorganisms and to aid in the maintenance of fluid-electrolyte homeostasis (1). The epidermis is comprised predominantly of keratinocytes, which express keratin as their intermediate filament (Fig. 33.1). A small population of cells, known as melanocytes, also exists within the epidermis. These cells are dendritic in appearance, reside in the basal layer of the epidermis, and are uniquely characterized by their ability to produce the pigment melanin in specialized organelles known as melanosomes (Fig. 33.2). Finally, a third population of dendritic cells, known as Langerhans cells, can also be found in the epidermis. They are located in the suprabasal layers and can be identified on the basis of HLA-DR or CD-1 expression (Fig. 33.3). Langerhans cells are the major immunocompetent cells of the epidermis (2). In addition, at any given time, mononuclear (including resident γδ T-cells) and polymorphonuclear leukocytes may also be seen in the epidermis (3, 4).

A wide variety of immunologic and nonimmunologic diseases can exhibit cutaneous manifestations (3). The present chapter will highlight some benign and neoplastic diseases of the epidermal keratinocyte and the epidermal melanocyte, both of which have been extensively studied using flow cytometric techniques.

DISEASES OF THE EPIDERMAL KERATINOCYTE

As described above, the epidermis is comprised primarily of keratinocytes. Keratinocytes are a diverse population of cells

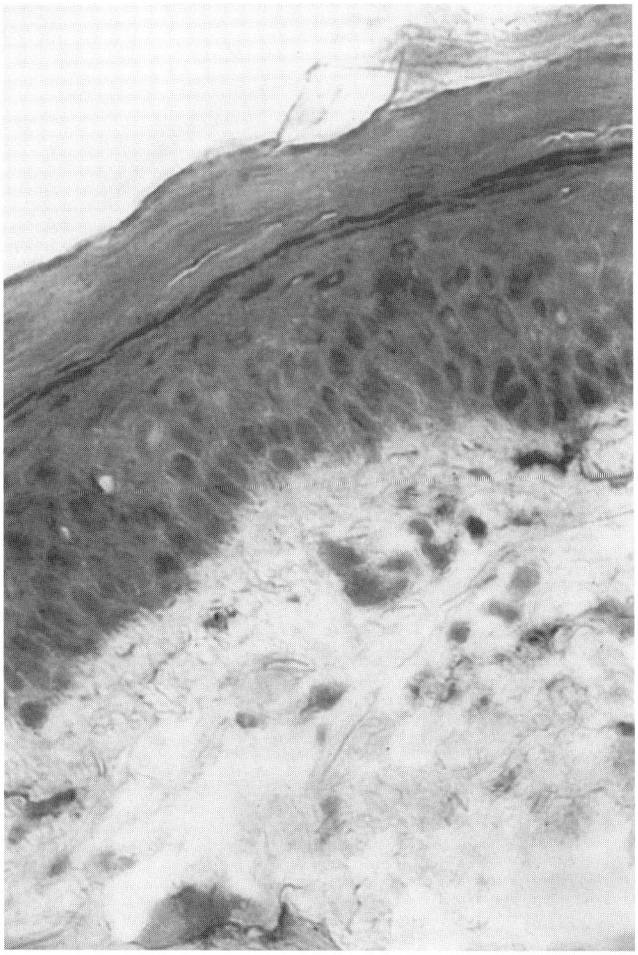

Figure 33.1. Reactivity of normal skin with anti-keratin monoclonal antibody AE-3. Immunoperoxidase staining clearly identifies the epidermal keratinocytes, which show intense cytoplasmic reactivity with AE-3 (×400).

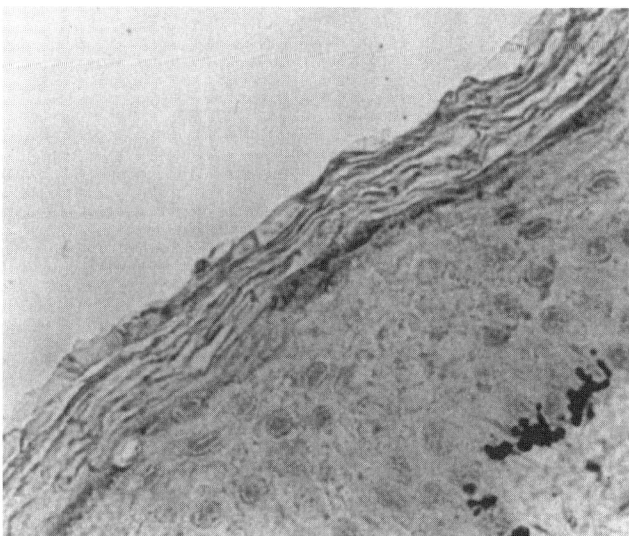

Figure 33.2. Reactivity of normal skin with monoclonal antibody TA99. TA99 is specific for a 75 kD glycoprotein found in mature melanosomes. Immunoperoxidase staining identifies melanocytes that reside in the basal layer of the epidermis (×600).

Figure 33.3. Reactivity of normal skin with monoclonal antibody anti-CD-1. Anti-CD-1 reactivity marks the Langerhans cells, which reside in the suprabasal layers of the epidermis (×600).

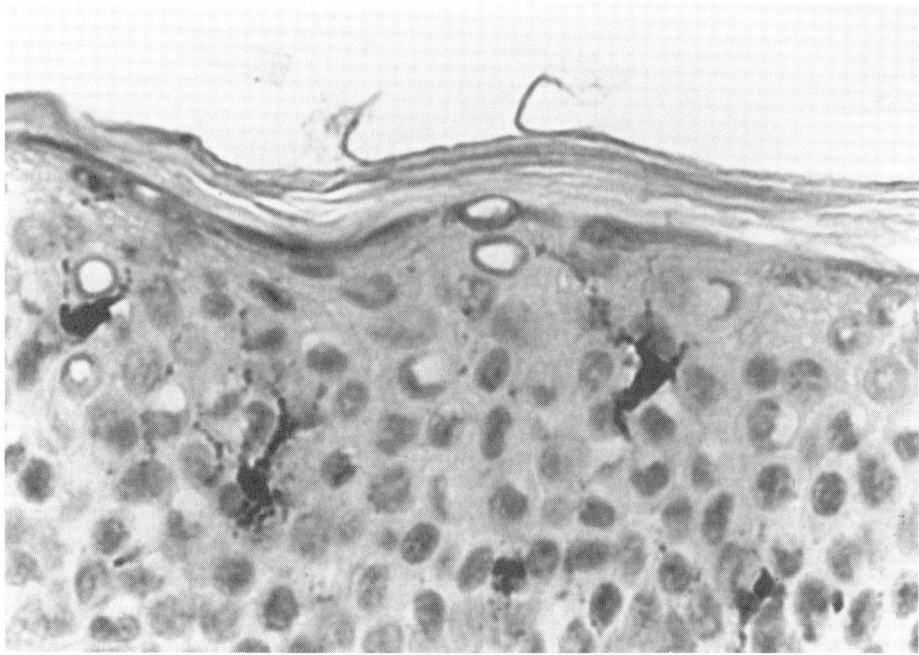

that differ in their proliferative capacity, biochemical profiles, and patterns of gene expression, depending upon their level of maturation and location within the epidermis (1).

The innermost, or basal, layer contains the proliferative, stemcell compartment of the epidermis. There is substantial evidence to support the view that basal keratinocytes are heterogeneous with respect to their morphology, cell kinetics, and function (5–7). Multiparameter FCM techniques have been particularly useful in identifying the presence of slowly- and rapidly-dividing basal layer keratinocytes (8–14). The heterogeneity of the epidermal basal cell has recently been reviewed by Clausen and Potten (6). A unique feature of the basal keratinocyte is its expression of the low-molecular-weight keratin pair K5 (58 kD) and K14 (50 kD). As a basal keratinocyte commits to terminal differentiation, it withdraws from the cell cycle, leaves the basal layer, and begins its migration toward the surface of the skin.

The first several cell layers superficial to the basal layer are known as spinous layers. The spinous cell is postmitotic, and characterized by a shift in keratin synthesis from the K5/K14 pair to the higher-molecular-weight K1 (67 kD)/k10 (58.6 kD) pair.

As the keratinocyte continues its maturation, it begins to synthesize involucrin, a precursor protein that is eventually cross-linked beneath the plasma membrane during the final stages of differentiation. Once the keratinocyte reaches the granular layer, keratin synthesis stops. It begins to synthesize fillagrin, which is involved in the bundling of keratins to form macrofibrils, and synthesizes loricrin, another precursor protein of the terminally differentiated, cornified envelope (CE) (15).

Finally, as the keratinocyte undergoes the final steps in its differentiation pathway, calcium influx occurs, activating epidermal transglutaminase, which, in turn, catalyzes the crosslinking of proteins on the inner surface of the plasma membrane. Lytic enzymes are released that destroy most of the cells' internal constituents, resulting in the final evolution of the flattened enucleated CE, which exists as a mosaic among other CEs in a matrix of intercellular lipids and is referred to as the stratum corneum. CEs within the stratum corneum eventually desquamate from the skin's surface to be replaced by other migrating and terminally differentiating keratinocytes.

Disruption of the delicate balance between self-renewal and differentiation can result in benign and/or neoplastic disease within the epidermis. Such disease can be associated with alterations in: a) cell proliferation; b) the patterns of keratin expression, including the appearance or disappearance of particular keratin species; c) the number of basal-like cell layers; d) the morphologic appearance of the cells; and e) the inflammatory cell infiltrates.

Over the past 20 years, FCM has been employed in the diagnosis and prognosis of keratinocyte disease (9, 11, 16, 17). Initial clinical studies relied heavily on single-parameter measurement of DNA ploidy and proliferative activity of normal and abnormal keratinocytes. More recently however, investigators have taken advantage of some of the differentiation-related changes that occur during disease progression, and have developed multiparameter FCM approaches to study keratinocyte pathogenesis in detail. Presented below are some of the more common clinical applications of single- and multiparameter FCM in the diagnosis and prognosis of benign, premalignant, and malignant keratinocyte disorders.

Figure 33.4. Histologic appearance of psoriatic skin. Hemotoxylin and eosin staining of a cross-section of psoriatic skin reveals keratinocyte hyperplasia, parakeratosis and dense inflammatory cell infiltrate of both the epidermis and the dermis.

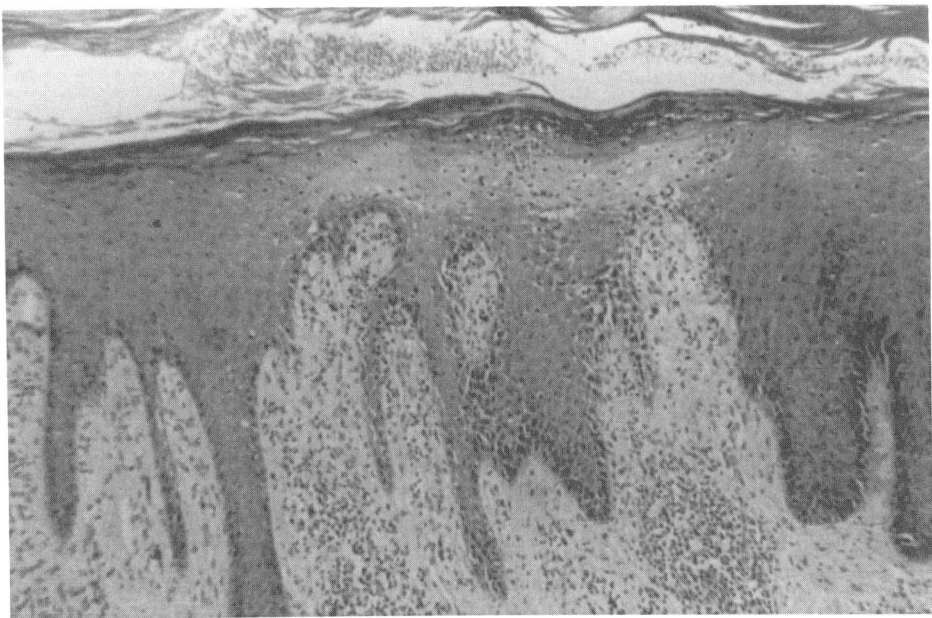

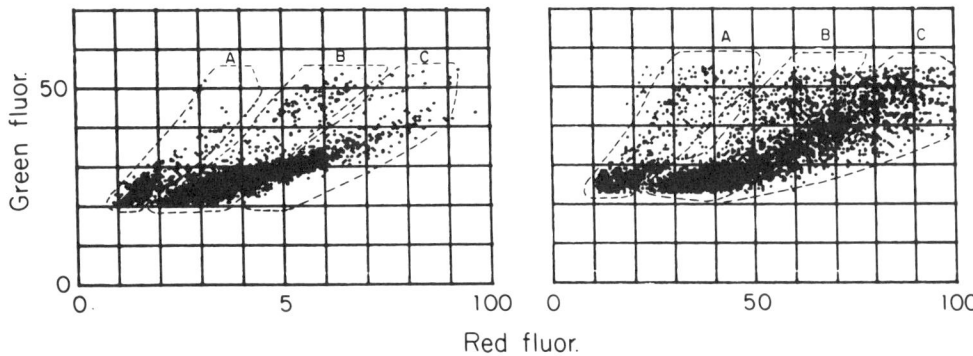

Figure 33.5. Dual-parameter RNA (red fluor.)/DNA (green fluor) scattergrams of keratinocytes from nonlesional (left) and psoriatic (right) epidermis. Three subpopulations of cells ("A", "B", and "C") are evident in both cases, based upon RNA content. Psoriatic skin shows a great increase in the number of proliferating cells. In addition, "B" keratinocytes from psoriatic lesions (right) show elevated RNA content compared to nonlesional skin (left). "C" compartment cells are difficult to assess due to considerable nonspecific staining with acridine orange (The cytogram depicts the total cell population excluding doublets and dead cells).

FCM MEASUREMENT OF NORMAL AND BENIGN KERATINOCYTE DISORDERS

Psoriasis is a common, chronic, hyperproliferative skin disorder that presents clinically as raised, scaly, erythematous plaques. The epidermis is characterized by hyperplasia and parakeratosis and may contain an extensive inflammatory cell infiltrate composed of both poly- and mononuclear cells (Fig. 33.4) (18–21).

Measurement of DNA content in normal epidermis has revealed uniformly diploid DNA content with a low proliferative fraction ($\leq 10\%$). Significantly higher numbers of cycling cells are observed in keratinocytes derived from involved psoriatic epidermis compared to epidermis from healthy donors (22–26). Interestingly, normal-appearing (nonlesional) skin from psoriatic patients also shows elevated numbers of cycling cells, although not as high as observed for active psoriatic plaques (26). The number of cycling cells decreased as the plaques resolved.

The addition of total cellular RNA content [via staining with acridine orange (AO)] as a second FCM parameter further resolves three subpopulations of nucleated cells in normal and psoriatic epidermis that differ in their RNA content (Fig. 33.5). The first subpopulation, "A," is comprised of small cells with low RNA content and long cell generation times (Fig. 33.5, left) (13). Purification of basal keratinocytes by immunochemical methods results in keratinocyte suspensions that are greatly enriched in "A" cells. Moreover, these small "A" cells are more highly clonogenic than the larger epidermal keratinocytes, suggesting that the "stem cell" of the epidermis resides within the "A" cell compartment. A second subpopulation of keratinocytes (designated as "B") can also be identified. This subpopulation is composed of larger more rapidly dividing cells that have significantly greater amounts of cellular RNA. Finally, a third subpopulation "*C*" is comprised of even larger, more superficial squamous cells that are nondividing and exhibit yellowish cytoplasmic fluorescence due to the presence of keratohyalin granules (13).

Analysis of cell cycle distributions after staining with AO revealed that psoriatic keratinocytes had the highest number of cycling cells (25%) compared to normal keratinocytes; nonlesional skin from the same patients had slightly lower numbers of cycling cells (21%), while keratinocytes from healthy subjects had the lowest numbers of proliferative cells (10%). Examination of psoriasis patients two to three weeks following the commencement of antipsoriatic therapy, showed a slight decrease in proliferative fraction in responsive patients; proliferative activity alone, however, was not a particularly sensitive marker of early treatment response (26).

In contrast to DNA content, RNA content appeared to be a more useful clinical parameter for monitoring early treatment responses in psoriasis. An initial comparison prior to therapy revealed that the mean RNA content of the "*A*" keratinocyte subpopulation was similar in normal, uninvolved, and psoriatic skin (Fig. 33.5). A significant increase in the mean RNA content of "*B*" keratinocytes in lesional psoriatic skin, compared with uninvolved skin or skin from control subjects, was apparent (Fig. 33.5, right) (26). Following therapy, the cellular RNA content of "*B*" keratinocytes from responsive patients decreased to quasinormal levels. By contrast, nonresponsive patients did not show a decrease in cellular RNA content after therapy, suggesting that RNA content may have clinical value as a prognostic indicator of treatment response in psoriasis (26).

As described above, the stratum corneum of psoriatic skin is parakeratotic (i.e., contains nucleated keratinocytes) compared to normal skin (Fig. 33.4). Bauer et al. exploited this dissimilarity using an FCM technique designed to identify the proportion of nucleated cells in the stratum corneum of psoriatic plaques and to correlate their disappearance with response to antipsoriatic therapy (27). Briefly, superficial corneocytes were obtained by scraping the surface of either active psoriatic plaques or normal skin. The cells were then stained with propidium iodide (PI). Enucleated corneocytes showed a relatively low-intensity diffuse staining pattern, while nucleated corneocytes showed a strong peak of intense red fluorescence corresponding to the intact nucleus. The ratio of area/peak (A/P) (an estimate of cellular PI distribution), when plotted as a function of the integrated red fluorescence, identified two discrete populations of corneocytes (Fig. 33.6). *Window A* contained enucleated corneocytes that exhibited high A/P ratios and relatively low red fluorescence; *window B* contained nucleated corneocytes with lower A/P ratios and relatively high red fluorescence. The percentage of nucleated corneocytes decreased as a function of successful antipsoriatic therapy, suggesting that this method might also be of potential clinical use in monitoring treatment responses in psoriasis.

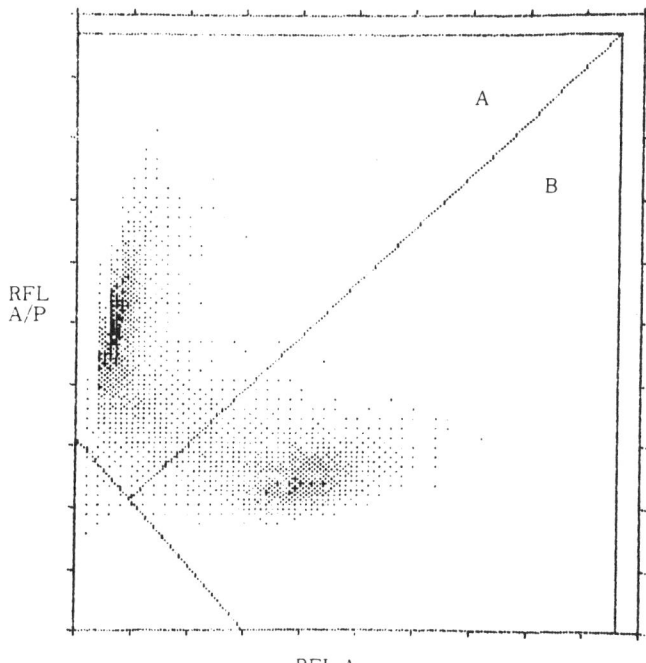

Figure 33.6. Bivariate histogram of propidium-iodide-stained corneocytes from psoriatic skin. The integrated area of red fluorescence (RFL A) is depicted o the *x*-axis and the ratio of area/peak of red fluorescence (RFL A A/P) is depicted on the *y*-axis. Two subpopulations of corneocytes are identified. *Window A* contains enucleated corneocytes; *window B* contains nucleated corneocytes.

An extensive series of polyclonal and monoclonal antibodies have been shown to bind to various determinants within the epidermis. Table 33.1 lists a few representative antibodies known to react with basal and suprabasal epidermal cells. It is not intended to serve as a complete list, but to highlight the ability of antibodies to distinguish basal from suprabasal keratinocytes and keratinocytes from nonkeratinocytes. Antibodies have been employed in FCM assays to identify epidermal subpopulations in normal and psoriatic skin.

Monoclonal antibodies EL-2 and 4F2 recognize the same cell surface glycoprotein (although not necessarily the same epitope) on basal layer keratinocytes (28, 29). In normal epidermis, approximately 30% of the cells are EL-2$^+$ (29), while 20% are 4F2$^+$ (28). A significant increase in the number of EL2$^+$ epidermal cells is observed in both nonlesional (35%) and lesional (54%) psoriatic skin. This corresponds with a similar increase in 4F2$^+$ basal cells (50%) in active psoriasis. These results likely reflect an expanded basal cell population in psoriasis. In a series of psoriasis patients, the number of EL-2$^+$ cells fell to near normal levels (39%) as patients responded to antipsoriatic therapy, suggesting that EL2 (and possibly 4F2) may also be useful markers of clinical response in psoriasis.

As briefly outlined above, in normal epidermis, basal keratinocytes express keratins K5 and K14, while suprabasal cells express predominantly keratins K1 and K10. The study of isolated keratins from psoriatic skin by two-dimensional

Table 33.1
Monoclonal Antibodies Reactive with Epidermis

Name	Specificity	Localization in Normal Epidermis	Localization in Psoriasis
4F2	Surface glycoprotein 40kD/80kD	Basal layer	Basal layer
EL-2	same as above (different epitope?)	Basal layer	Basal layer
PKSE60	Keratin 10	Suprabasal layers	Suprabasal layers
PAb601	K5/K14?	Basal layer	Lowest 3 layers
$K_s8.12$	K13/K16	Absent	Suprabasal layers
MVI	Vimentin	Non-keratinocytes	Non-keratinocytes
HLA-DR	HLA-DR	Langerhans cells	Keratinocytes/Langerhans cells/lymphocytes
α-CD-1	CD-1	Langerhans cells	Langerhans cells
α-TGFα	TGFα	Strongly reactive in basal layer	Increased reactivity
α-Involucrin	Involucrin	Granular layer	Granular & spinous layer

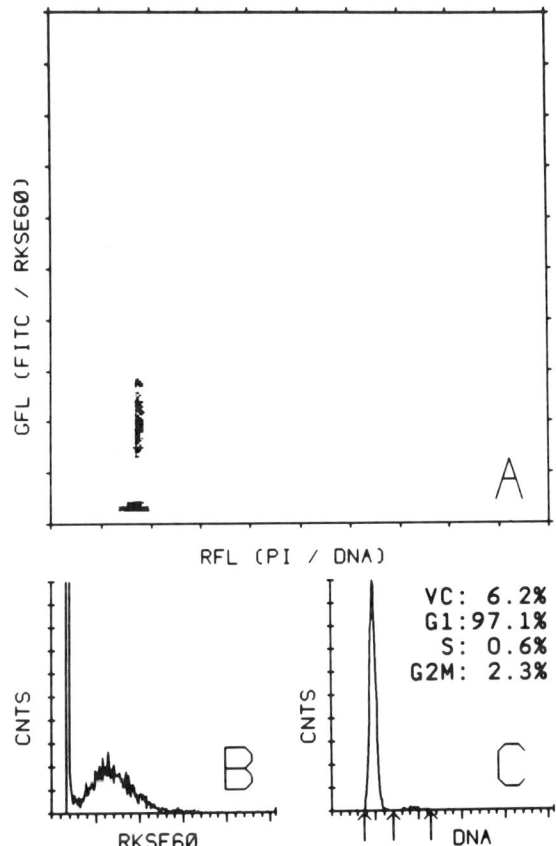

Figure 33.7. Dual-parameter histogram of keratin K10 expression (green fluorescence) and DNA content (red fluorescence) in normal epidermis (A). K10+ keratinocytes can be identified by increased green fluorescence (B). Note that K10-reactive keratinocytes exhibit quiescent-like DNA distributions, with the majority of cells residing in the G_1 phase of the cell cycle (C). This is in keeping with K10's suprabasal reactivity in the epidermis.

mis. A number of investigators have taken advantage of the differential expression of keratins in psoriatic and normal epidermis to examine more closely keratin expression and cell proliferation in normal skin and psoriatic plaques (34–36). In one such study, Van Erp et al. (37), utilized anti-keratin antibodies reactive with basal keratinocytes (Anti-K5/K14), suprabasal keratinocytes from normal epidermis (Anti-K10), and suprabasal keratinocytes from psoriatic epidermis (Anti-K16). Finally anti-vimentin was used to identify nonkeratinocytes (i.e., melanocytes, Langerhans cells, and inflammatory cells) in the epidermis of normal and psoriatic skin.

Figure 33.7 shows a typical bivariate cytogram of K10 expression and DNA content in normal epidermis. Anti-K10 reactivity was primarily associated with noncycling G_1 phase cells. Conversely, expression of K5/K14, which is located in the germinative epidermal basal layer, was more closely associated with proliferating cells. Again, an increase in basal cell number was observed in psoriatic compared to normal epidermis. The expression of K10 was decreased in psoriatic keratinocytes. Anti-K16-reactivity, which was usually absent in normal skin, was elevated in psoriatic skin. The percentages of K16+ cells, however, varied widely among psoriatic lesions (37). The number of nonkeratinocytes within the epidermal cell suspension was reduced in lesional psoriatic skin, presumably due, at least in part, to the relative increase in keratinocyte number in psoriatic epidermis. This study showed the potential role of anti-keratin and anti-vimentin antibodies in assessing changes that occur during the onset of psoriasis.

Table 33.2 summarizes some of the salient features of normal, non-lesional and lesional psoriatic keratinocytes that have been quantitated by flow cytometry. It is apparent that the normal-appearing (nonlesional) keratinocytes from psoriasis patients more closely resemble psoriatic keratinocytes than normal keratinocytes. Multiparameter FCM technology will continue to play an important role in the examination of the pathogenesis of psoriasis.

Psoriasis is also associated with the influx of inflammatory cells into both the epidermis and the dermis. Activated IL-2-receptor-positive T-cells are found in psoriatic plaques. Gamma interferon and the γ-interferon-induced inflammatory protein IP-10 are also observed (38). A number of stud-

(2-D) gel electrophoresis showed a decrease in K1 in conjunction with the specific appearance of two hyperproliferation-associated keratins, K6 and K16 (30–34). 2-D gel electrophoresis can provide extremely quantitative data on the amounts and types of keratins expressed in psoriatic epidermis; it cannot provide information on whether the changes in keratin expression are occurring generally or are more confined to particular subsets of keratinocytes within the epider-

Table 33.2
FCM Parameters that Distinguish Normal Keratinocytes from Psoriatic Keratinocytes

Parameter Measured	Expression in Lesional and Non-Lesional Psoriatic Skin Compared to Normal Keratinocytes	
	Non-Lesional	Lesional
Proliferative fraction	Elevated	Elevated
RNA content	Normal	Elevated
Percent nucleated corneocytes	N.D.[a]	Elevated
Number of basal cells	Elevated	Elevated
Number of K16-expressing keratinocytes	N.D.	Elevated
Number of HLA-DR-expressing keratinocytes	Absent	Present

[a]: Not determined.

ies suggest that immune activation may be important to the pathogenesis of psoriasis (38–41). HLA-DR molecules are critical in various immunologic responses (42). While normal keratinocytes do not express HLA-DR, HLA-DR expression by keratinocytes has been noted in a number of dermatoses (40, 41, 43, 44) and can be induced in vitro by treatment with γ-interferon (45). A recent study has shown that psoriatic keratinocytes express HLA-DR on their cell surface and that such expression can be reversed by successful antipsoriatic therapy (42). The role of HLA-DR-expressing psoriatic keratinocytes is yet to be defined, but it further supports the concept that the immune system contributes to the pathogenesis of psoriasis.

X-linked ichthyosis is an inherited deficiency of steroid sulfatase (STS) that is clinically manifest in dry, excessively scaly skin (46, 47). Single-beam and, more recently, dual-beam flow cytometry were able to detect X-chromosomal deletions in the DNA of affected individuals (48, 49). Moreover, this technique was used in combination with a series of cDNA probes to identify an Xp22 microdeletion, which is associated with both ichthyosis and ocular albinism (49). Such studies show the great potential of using multiparameter FCM as part of a strategy to locate and identify the loci of disease-associated genetic defects.

PLOIDY IN PREMALIGNANT AND MALIGNANT KERATINOCYTE DISEASE

The demonstrated presence of a nondiploid DNA stemline by FCM has been correlated with poor prognosis in a number of tumor systems, including colorectal, breast, and bladder cancers (reviews, 17, 50). Studies on these and other types of tumors estimate that at least 70% of tumors exhibit aneuploidy (50).

Initial studies were, therefore, aimed at ploidy determination in premalignant and malignant dermatoses. These include Bowen's disease, actinic keratosis, keratocanthoma (KA), basal cell carcinoma (BCC), and squamous cell carcinoma (SCC). Most laboratories have identified aneuploidy within all of these groups (51–53). The significance of aneuploidy in premalignant lesions is not clear. Long-term follow up studies will be necessary to establish whether premalig-

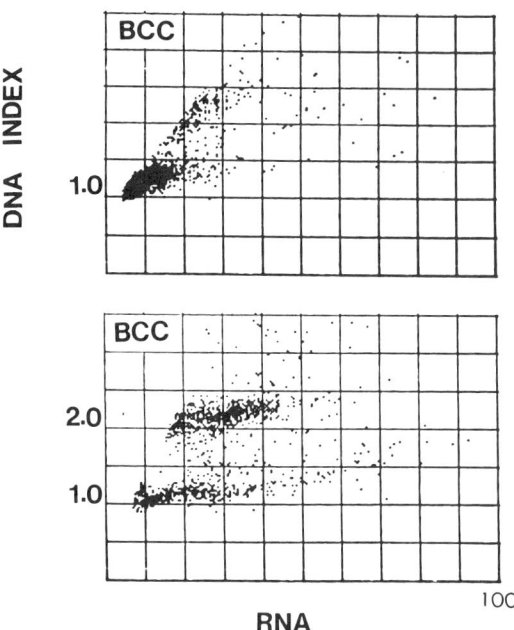

Figure 33.8. Dual-parameter RNA/DNA cytogram of two basal cell carcinomas (BCC). Note the increase in the proportion of low-RNA containing cells (i.e., the lack of differentiation) compared with normal (see Fig. 33.5, top). The more marked tetraploidy of the tumor in the lower panel has been observed in association with more aggressive behavior.

nant lesions that are aneuploid are at higher risk for disease progression than diploid lesions.

Basal cell carcinomas are generally slow-growing tumors that can ecome locally invasive, but rarely metastasize (54). The majority of BCCs examined have been diploid with relatively low proliferative fractions (16). A low prevalence of tetraploidy/aneuploidy has been also been noted. The keratinocyte subpopulations of 35 BCCs of the face were examined after staining with AO and compared to those of normal skin (Fig. 33.8). Thirty of 35 specimens were diploid, with similar proliferative indices as normal skin. However, instead of being composed of three cell subpopulations based on RNA content, BCCs were composed primarily of low RNA "A" cells (Fig. 33.8, top). Parallel measurements on the same specimens also showed a dramatic increase in basal cell numbers based upon EL-2 reactivity. In the five larger, more aggressive BCCs, keratinocytes showed differences not only in RNA content, but also in DNA content, exhibiting tetraploid stemlines (DNA index of 2.0; Fig. 33.8, bottom).

One area of particular interest to dermatologists is the examination of keratoacanthomas, which are tumors composed of well-differentiated squamous epithelium, with a central keratin mass opening on the skin's surface, that resolve spontaneously. KAs are generally considered to be benign. Even when using specific histologic criteria, however, it is sometimes difficult to distinguish KA from well-differentiated squamous cell carcinoma (WDSCC) (55). Indeed, there have been a number of reports in which patients initially diagnosed as having KA subsequently died of metastatic SCC (56, 57). It was thought that DNA ploidy analysis might be a useful parameter to distinguish KA from WDSCC. Comparative studies of DNA content in KA and WDSCC have shown similar prevalence of aneuploidy and proliferative fraction (53). Attempts to correlate the presence of aneuploidy in KA with distinctive features of the tumor (i.e., higher degree of atypia etc.) have also been unsuccessful suggesting that single-parameter DNA content analysis will not be clinically useful in this regard. It is possible that correlated measurement of differentiation-specific markers and DNA content may prove more useful in identifying WDSCC from those subsets of tumors initially classified as KA.

FCM MEASUREMENT OF DNA PLOIDY IN NORMAL AND MALIGNANT MELANOCYTES

As described above, the epidermal melanocyte resides within the basal layer of the epidermis (Fig. 33.2). It is derived from cells originating in the neural crest during embryonic development. Since melanocytes produce the pigment melanin, early changes in the proliferative behavior are often recognized as an enlarged area of pigmentation within the skin.

Studies to date support the theory that the transformation of normal melanocytes to metastatic melanoma cells occurs in a stepwise fashion, involving at least five stages. These progressive stages involve development of: a) an acquired nevus; b) a dysplastic nevus; c) a primary melanoma with radial growth; d) a primary melanoma with vertical growth and, finally, e) a metastatic melanoma (58, 59). Congenital lesions, particularly large (>20 cm) congenital hairy pigmented nevi (CHPN), also show a relatively high lifetime risk for melanoma development compared to the general population (60, 61).

A number of morphologic and histologic criteria have been correlated with disease recurrence and poor prognosis among individuals with melanoma. These include: mitotic activity, nuclear morphology, nucleolar size, thickness of the lesion (Breslow thickness), and depth of invasion (Clarks level) (58, 59, 62, 63). As in other tumor systems with documented premalignant lesions, it is difficult to identify those subsets of patients who will progress to melanoma from those who will not. It is equally difficult in early thin primary melanomas to discriminate individuals at higher risk for metastatic disease. It is particlarly in the premalignant lesions (i.e., dysplastic nevi and giant CHPN) and in early stage I melanomas that FCM measurement of DNA ploidy has been explored as an adjunctive test for disease diagnosis and prognosis (64–71).

Many of the studies in melanoma (both prospective and retrospective) have utilized single-parameter DNA measurements to compare the ploidy levels of normal and neoplastic melanocytes. Stenziger et al. measured DNA content of benign acquired nevi and congenital melanocytic nevi (61). None of the acquired nevi were aneuploid, while approximately 10% of the congenital nevi were aneuploid. Further examination revealed that 75% of the aneuploidy occurred in relatively large (>20 cm in diameter) lesions. An important point arising from the study of large nevi is the variability in ploidy that can occur in multiple samples from the same lesion. This highlights the need to be cautious when making clinical interpretations based upon DNA measurement of a single site taken from a relatively large lesion. Other studies also have confirmed the presence of aneuploidy in CHPN and dysplastic nevi. Interestingly, some investigators have observed a low prevalence of aneuploidy in acquired nevi. The significance of aneuploidy in frankly benign nevi is not clear; it may be artifactual due to poor specimen preparation or, alternatively, it may indicate that FCM can detect a minor subset of benign nevi with very early premalignant changes not apparent by any morphologic criteria. Long-term studies on such patients would be necessary to determine the ultimate clinical significance of these findings.

Preparation of single-cell suspensions from excised nevi or melanomas usually results in a mixture of neoplastic and non-neoplastic cells. Melanoma cells may constitute a relatively small population of cells in comparison to the background of normal melanocytes, epidermal cells, stromal tissue, and inflammatory cells. In such instances, single-parameter FCM measurement may not be sensitive enough to detect occult tumor cell populations. Multiparameter FCM approaches have been successfully used to increase the sensitivity of aneuploidy detection in solid tumors (72–74). A recent study by Kamino and Ratech utilized such an approach, which combined DNA-content measurement with S100 protein expression in stage I melanomas (71). S100 is an extremely acidic calcium-binding cytoplasmic protein. Expression of S100 is confined predominantly to melanocytes, Langerhans, cells and Schwann cells. Detection of aneuploidy increased from 10.8% in single-parameter DNA analysis to 32.4% when S100 expression was added as a second parameter. In addition, S100 was heterogeneously expressed in certain early Stage I melanomas. The clinical significance of the observed heterogeneity is not presently known.

While DNA ploidy is a notable parameter, the importance of proliferative fractions in diploid specimens cannot be ignored (50). To date, relatively few studies have examined the proliferative fractions on large numbers of melanocytic lesions. These studies should be undertaken to discern whether quantitation of proliferative activity will increase the sensitivity for detecting high-risk early lesions.

The majority of studies suggest that DNA ploidy analysis may be a useful adjunctive tool for the identification and classification of individuals with premalignant nevi or stage I melanomas who are at higher risk for disease progression and/or recurrence. It is not clear, however, whether routine clinical measurement of single-parameter DNA content in melanoma will impart any added advantage over the already well-established morphologic and histologic criteria for melanoma assessment.

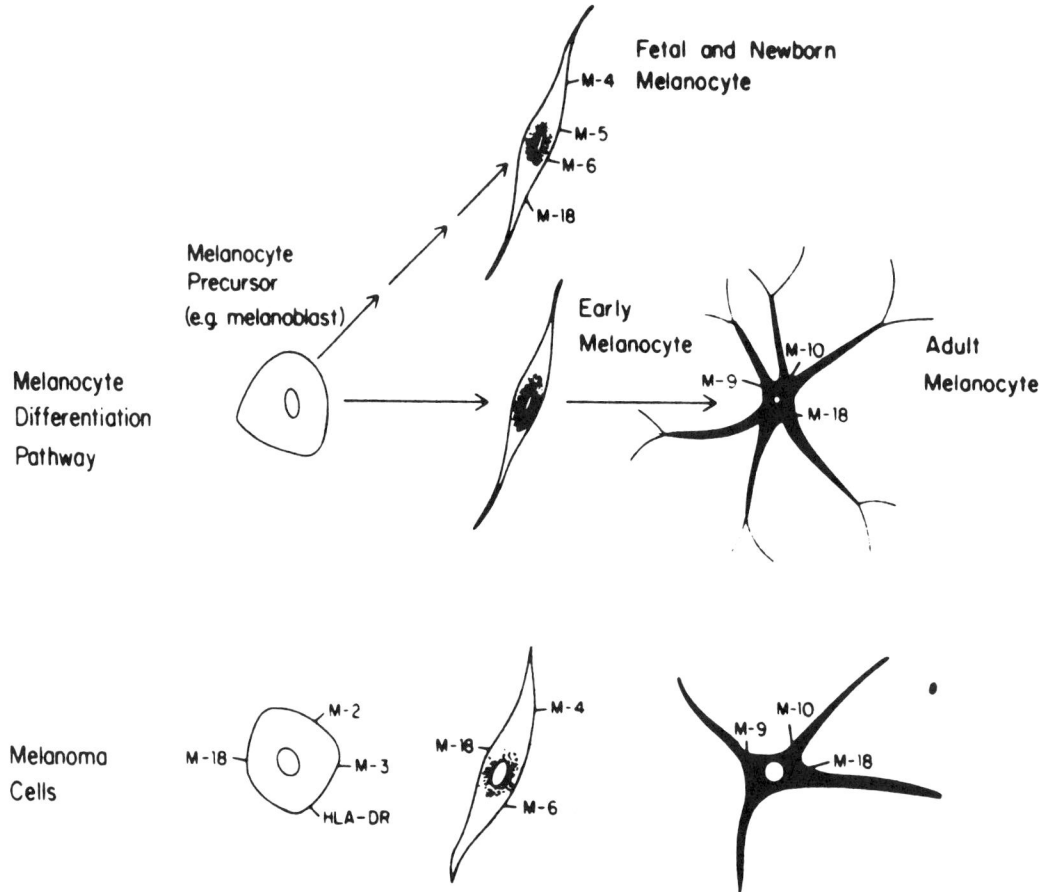

Figure 33.9. Proposed pathway of melanocyte differentiation. Antigens M4, M5 and M6 are expressed at an early-to-intermediate stage of melanocyte differentiation. Antigens M9 and M10 are expressed on mature melanocytes. Melanoma cells follow a similar differentiation pathway.

FCM ASSESSMENT OF DIFFERENTIATION-ASSOCIATED AND MELANOMA-ASSOCIATED ANTIGENS

A major thrust of ongoing research in melanomas is the identification and staging of melanocytes and melanoma cells on the basis of specific antigenic expression. A large series of antibodies have been generated that detect both differentiation-associated and melanoma-associated antigens (75–90).

Houghton et al. have described a series of antigens that defined the differentiated phenotype of normal melanocytes (85, 89). The phases of differentiation were defined based upon antigenic reactivity with normal melanocytes and cultured melanoma cell lines. The rudimentary differentiation pathway is shown in Figure 33.9. Markers of "early" melanocyte (termed melanoblast) differentiation were those antigens expressed on melanomas, but not on fetal, newborn, or adult normal melanocytes. "Intermediate" markers were expressed predominantly on fetal and newborn melanocytes, while "late" markers were expressed on all three types of melanocytes. Melanomas classified as "early" were generally epitheloid in nature and relatively nonpigmented. Those classified as "intermediate" resembled the bipolar phenotype of fetal and newborn melanocytes; melanoma cells expressing "late" markers resembled the adult, polydendritic melanocytes with intense pigmentation.

Classification schemes have also been developed to identify the major stages in melanoma development (83, 89, 90). A number of antigens can distinguish normal melanocytes from melanocytic nevi, radially and vertically growing primary melanomas, and metastatic melanomas. These antigens have been reviewed in detail elsewhere and, therefore, only a few of the antigens will be discussed below (89). The appearance or increased expression of Epidermal Growth Factor (EGF), Nerve Growth Factor (NGF), melanotransferrin (p97) and melanoma-specific chondroitin sulfate proteoglycan (mCSP) correlate with progessive stages of melanoma development. Likewise, expression of the ganglioside GD2 is highly specific for a small subset of melanomas, while ganglioside GD3 is overexpressed on metastatic melanoma cells, but weakly expressed on nevi.

Not only do specific antigens appear during disease progression, but antigens may also decrease or disappear during malignant transformation. Notable among these antigens are the decreased expression of Class I HLA molecules in advanced disease and the loss of adenosine deaminase binding protein in malignant melanocytes (77, 82, 85).

FCM APPLICATIONS FOR MONITORING PATIENTS UNDERGOING IMMUNOTHERAPY

As described above, a number of melanoma-associated antigens have been identified (89, 90). Melanoma-specific antibodies have been identified in patient serum (91). In vitro, patient lymphocytes have also been shown to be reactive with soluble melanoma antigens (92). These observations, combined with the occurrence of heavy lymphocytic infiltrates in melanomas, would all suggest that immunotherapy might be an effective treatment modality for metastatic melanoma (93, 94). Both cell-mediated and antibody-mediated approaches have been explored for treatment of patients with metastatic melanoma (93–95).

One such approach has been adoptive immunotherapy via generation of melanoma-specific lymphocyte-activated killer (LAK) cells (93, 94). Initial studies utilizing combination IL-2/LAK therapy are encouraging, although prediction of treatment response is difficult. Investigators are presently examining whether HlLA-DR expression on the melanomas is correlated with treatment response.

An alternate approach has utilized target-directed antibodies for the treatment of melanoma (review, 95). This strategy is based on the specific localization of antibodies in melanoma cells, thereby increasing antibody-dependent cell lysis (ADCC) or complement-dependent cell lysis of the tumor cells (95). Target-directed antibodies can also be conjugated to toxins, such as ricin, which would aid in tumor destruction. In order for immunotherapy to be effective, the antibodies would have to be directed against antigens that are relatively specific for the melanoma (i.e., not expressed in high levels in normal tissue) and not subject to down-modulation or significant antigenic heterogeneity. Histochemical examination of tissue sections and of melanoma cell lines derived from single tumors suggests that melanoma cells within an individual tumor may be heterogenous with respect to antigenic expression (75, 77, 82, 84, 87, 98–101). Such antigenic heterogeneity has tremendous implication for the ultimate utility of antibody-mediated target therapy in melanoma. Rare populations of melanoma cells that express low levels of a particular antigen may escape the cell-killing effects of a target-directed antibody. A tumor expressing relatively high levels of a particular antigen might be expected to respond significantly better to an antibody directed against that antigen than a tumor in which antigenic expression is low or heterogeneous.

One of the major strengths of flow cytometry lies in its ability to detect subpopulational heterogeneity within a morphologically homogeneous tumor. Quantitation of melanoma-associated antigens before and following immunotherapy is an area in which FCM may be of significant clinical use.

Four major cell surface melanoma-related antigens have been used as targets for immunotherapy. These include two gangliosides (GD2 and GD3) and two glycoproteins (p97 and mCSP). Antibodies directed against all four antigens (GD2, GD3, p97, and mCSP) have been utilized in FCM assays to quantitate expression on normal and transformed melanocytes (review, 89). Antigenic heterogeneity has been demonstrated for all four antigens on melanomas. One example is the heterogeneity of mCSP expression, which has been described in vitro and in vivo employing monoclonal antibody 9.2.27 (80, 96, 97, 99, 100). Using cell sorting techniques, Lindmo et al. isolated stable cell lines that expressed high or low levels of mCSP from a heterogeneously expressing parental cell line (99). They also showed that mCSP expression was not cell cycle phase related (101). The intratumor localization of monoclonal antibody 9.2.27 in vivo has been quantitated by FCM and immunohistochemistry (100). FCM revealed that the percentage of tumor cells that bound 9.2.27 was dose-dependent in vivo, but that the saturation of binding sites varied among tumors and patients. In a complementary fashion, immunoperoxidase staining of histologic sections from the same tumors provided information on the distribution of mCSP and 9.2.27 reactivity within the tumor. Not only can antigens vary on melanoma cells, but monoclonal antibodies directed against particular antigens can recognize different epitopes of the same molecule (80). This can further lead to variations in stainability as well as effectiveness in binding to the target antigen in melanoma (review, 95). Recent studies would suggest that "cocktails" of antibodies might prove more effective in immunotherapy than any single antibody (82). Since screening of antigenic expression is accomplished rapidly using FCM, the development of antibody panels that can be screened by single- or multiparameter FCM on individual tumors might lead to new and better strategies for immunotherapy in melanoma.

REFERENCES

1. Fuchs E. Epidermal differentiation: The bare essentials. J Cell Biol 1990;111:2808–2814.
2. Stingl G, Tamaki K, Katz SI. Origin and function of epidermal Langerhans cells. Immunol Rev 1980;53:149.
3. Braverman IM. Skin Signs of Systemic Disease. 2nd ed. Philadelphia: Saunders, 1981.
4. Brenner MB, Strominger JL, Kranget MS. The γδ T cell receptor. Adv Immunol 1988;43:133–192.
5. Lavker RM, Sun TT. Heterogeneity in epidermal basal keratinocytes: Morphological and functional correlations. Science 1982;215:1239–1241.
6. Potten CS, Wichmann HE, Loeffler M, Dobek K, Major D. Evidence for discrete cell kinetic subpopulations in mouse epidermis based upon mathematical analysis. Cell Tissue Kinet 1982;15:302–329.
7. Clausen OPF and Potten CS. Heterogeneity of keratinocytes in the epidermal basal cell layer. J Cutan Pathol 1990;17:129–143.
8. Clausen OPF, Thorud E, Aarnaes E. Evidence of rapid and slow progression of cells through G_2 phase in mouse epidermis. Labelled and unlabelled cells in S phase after administration of tritiated thymidine. Virchows Arch [B] 1980;34:1–11.
9. Thorud E and Volden G. Flow cytometry (FCM) of human epidermal cells. A preparation method for epidermal cells and demonstration of

circadian variations in the proportion of S-phase cells. Arch Dermatol Res 1980;269:137–145.
10. Clausen OPF, Thorud E, Aarnaes E. Evidence of rapid and slow progression of cells through G_2, phase in mouse epidermis: a comparison between phase durations measured by different methods. Cell Tissue Kinet 1981;14:227–240.
11. Clausen OPF. Flow cytometry of keratinocytes. J Cutan Pathol 1983; 10:33–51.
12. Kimmel M, Darzynkiewicz Z, Staiano-Coico L. Stathmokinetic analysis of human epidermal cells in vitro. Cell Tissue Kinet 1986;19:289–304.
13. Staiano-Coico L, Higgins PJ, Darzynkiewicz Z, et al. Human keratinocyte culture. Identification and staging of epidermal cell subpopulations. J Clin Invest 1986;77:396–404.
14. Clausen OPF, Kirkhus B, Schjolberg AR. Cell cycle progression kinetics of regenerating mouse epidermal cells: An in vivo study combining DNA flow cytometry, cell sorting, and 3H-Thd autoradiography. J Invest Dermatol 1990;86:402–405.
15. Mehrel T, Hohl D, Rothnagel JA, et al. Identification of a major keratinocyte cell envelope protein, loricrin. Cell 1990;61:1103–1112.
16. Frenz G. Flow cytometry DNA analysis of normal, premalignant and malignant human epidermal tissues. Stockholm: L. Klinken Dekan. Almquist and Wiksell Periodical Co., 1986, pp. 1–44.
17. Melamed MR, Staiano-Coico L. Flow cytometry in clinical cytology specimens. In: Melamed MR, Lindmo T, eds. Flow Cytometry and Cell Sorting. 2nd ed. New York: Wiley-Liss, Inc., 1990, pp. 755–772.
18. Farber EM, van Scott EJ. Epidermis: disorders of cell kinetics and differentiation. In: Fitzpatrick TB, Eisen AZ, Wolff K, Freedberg IM, and Austen KF, eds. Dermatology in General Medicine, 2nd ed. New York: McGraw-Hill, 1979, pp. 233–247.
19. Cormane RH. Immunopathology of psoriasis. Arch Dermatol Res 1984;276:45.
20. Bernard BA, Robinson SM, Vanderele S, Mansbridge JN and Darmon M. Abnormal maturation pathway of keratinocytes in psoriatic skin. Br J Dermatol 1985;112:647–653.
21. Weinstein GD, McCullough JL and Ross PA. Cell kinetic basis for pathophysiology of psoriasis. J Invest Dermatol 1985;85:579–583.
22. Bauer FW and De Grood RM. Impulse cytophotometry in psoriasis. Br J Dermatol 1975;93:225–227.
23. Bauer FW and De Grood RM. Improved technique for epidermal cell cycle analysis. Br J Dermatol 1976;95:565.
24. Bauer FW, Crombag NHCMN, De Grood RM and De Jongh GJ. Flow cytometry as a tool for the study of cell kinetics in epidermis. Investigations of normal epidermis. Br J Dermatol 1980;102:629–639.
25. Larsen JK, Frentz G, Moller U and Christensen IJ. A method for flow cytometric cell cycle analysis of normal and psoriatic human epidermis based on a detergent/citric acid technique for suspension of nuclei. Virchows Arch [B] 1985;48:247–259.
26. Staiano-Coico L, Gottlieb AB, Barazani L, Carter DM. RNA, DNA and cell surface characteristics of lesional and nonlesional psoriatic skin. J Invest Dermatol 1987;88:646–651.
27. Bauer FW, Boezeman JBM, DeGrood RM, Koopman RJJ. Flow cytometric investigations of corneocytes from psoriatic scales and the effects of ingram therapy. J Dermatol 1986;13:175–178.
28. Patterson JAK, Eisinger M, Haynes BF, Berger CL, Edelson RL. Monoclonal antibody 4F2 reactive with basal layer keratinocytes: studies in the normal and a hyperproliferative state. J Invest Dermatol 1984;83:210–213.
29. Gottlieb AB, Posnett DN, Crow MK, Horikoshi T, Mayer L and Carter DM. Purification and in vitro growth of human epidermal basal keratinocytes using a monoclonal antibody. J Invest Dermatol 1985; 85:299–303.
30. Skerrow D and Hunter I. Protein modifications during the keratinization of normal and psoriatic human epidermis. Biochim Biophys Acta 1978;537:474–484.
31. Baden HP, McGilvray N, Cheng CK, Lee LD, Kubilus J. The keratin polypeptides of psoriatic epidermis. J Invest Dermatol 1978;70:294–297.
32. Matoltsy AG, Matoltsy MN, Cliffel PJ. Characterization of keratin polypeptides in normal and psoriatic horny cells. J Invest Dermatol 1983;80:185–188.
33. McGuire J and Lightfoot OM. Two keratins MW 50,000 and 56,000 are synthesized by psoriatic epidermis. Br J Dermatol 1984;111:27–37.
34. Weiss RA, Eichner R and Sun T-T. Monoclonal antibody analysis of keratin expression in epidermal diseases: a 48- and 56-kdalton keratin as molecular markers for hyperproliferative keratinocytes. J Cell Biol 1984;98:1397–1406.
35. Bauer FW, Boezeman JBM, Engelen LV, DeGrood RM and Ramaekers CS. Monoclonal antibodies for epidermal population analysis. J Invest Dermatol 1986;87:72–75.
36. DeMare S, van Erp PEJ, van de Kerkhof PCM. Epidermal hyperproliferation assessed by the monoclonal antibody $k_s8.12$ on frozen sections. J Invest Dermatol 1989;92:130–131.
37. van Erp PEJ, Rijzewijk JJ, Boezeman JBM, et al. Flow cytometric analysis of epidermal subpopulations from normal and psoriatic skin using monoclonal antibodies against intermediate filaments. Amer J Pathol 1989;135:865–870.
38. Gottlieb AB, Luster AD, Posnett DN, Carter DM. Detection of a γ-Interferon-induced protein IP-10 in psoriatic plaques. J Exp Med 1988;168:941–948.
39. Krueger GG, Jederberg WW, Ogden BE, Reese DL. Inflammatory and immune cell function in psoriasis. II. Monocyte function lymphokine production. J Invest Dermatol 1978;71:195.
40. Lampert IA. Expression of HLA-DR (Ia-like) antigen on epidermal keratinocytes in human dermatoses. Clin Exp Immunol 1984;57:93.
41. Volc-Platzer B, Majdic O, Knapp W, et al. Evidence of HLA-DR antigen biosynthesis by human keratinocytes in disease. J Exp Med 1984;159:1784.
42. Giles RC, Capra JD. Structure, function and genetics of human class II molecules. Adv Immunol 1985;37:1.
43. Gottlieb AB, Lifshitz B, Fu SM, Staiano-Coico L, Wang CY and Carter DM. Expression of HLA-DR molecules by keratinocytes, and presence of Langerhans cells in the dermal infiltrate of active psoriatic plaques. J Exp Med 1986;164:1013–1028.
44. Aiba S, Tagami H. HLA-DR antigen expression on the keratinocyte surface in dermatoses characterized by lymphocyte exocytosis (e.g. pityriasis rosea). Br J Dermatol 1984;111:285.
45. Basham TY, Nickoloff BJ, Merigan TC, Morhenn VB. Recombinant γ-Interferon induces HLA-DR expression on cultured human keratinocytes. J Invest Dermatol 1984;83:88.
46. Frost P, van Scott EJ. Ichthyosiform dermatoses. Arch Dermatol 1966;94:113–126.
47. Kaloustian VM Der, Kurban AK. Genetic diseases of the skin. Berlin, Heidelberg, New York: Springer, 1979:26–34.
48. Cooke A, Gillard EF, Yates JRW, et al. X chromosome deletion detectable by flow cytometry in some patients with steroid sulphatase deficiency (x-linked ichthyosis). Hum Genet 1988;7:49–52.
49. Schnur RE, Trask BJ, van den Engh G, et al. An Xp22 microdeletion associated with ocular albinism and ichthyosis: Approximation of breakpoints and estimation of deletion size by using cloned DNA probes and flow cytometry. Am J Hum Genet 1989;45:706–720.
50. Raber MN, Barlogie B. DNA flow cytometry of human solid tumors. In: Melamed MR, Lindmo T and Mendelsohn ML, eds. Flow cytometry and cell sorting. 2nd ed. New York: Wiley-Liss, 1990, pp. 745–754.
51. Newton JA, Camplejohn RS and McGibbon DH. A flow cytometric study of the significance of DNA aneuploidy in cutaneous lesions. Br J Dermatol 1987;117:169–174.
52. Newton JA, Camplejohn RS and McGibbon DH. Aneuploidy in Bowen's disease. Br J Dermatol 1986;114:691–694.

53. Randall MB, Geisinger R, Kute TE, Buss DH and Prichard RW. DNA content and proliferative index in cutaneous squamous cell carcinoma and keratoacanthoma. Amer J Clin Pathol 1990;93:259–262.
54. Pollack SV, Goslen JB, Sherertz EF and Jegasothy BV. The biology of basal cell carcinoma: A review. J Am Acad Dermatol 1982;7:569–577.
55. Kern WH and McCray MK. The histopathologic differentiation of keratoacanthoma and squamous cell carcinoma of the skin. J Cutan Pathol 1980;7:318–325.
56. Jackson IT. Diagnostic problem of keratoacanthoma. Lancet 1969;1:490–492.
57. Schnur PL and Bozzo P. Metastasizing keratoacanthomas. Plast Reconstr Surg 1978;62:258–262.
58. Clark WH, From L, Bernadina EA, Mihm MC. The histogenesis and biologic behavior of primary human malignant melanomas of the skin. Cancer Res 1969;29:705–727.
59. Clark WH Jr., Elder DE, Gurerry D IV, Epstein MN, Greene MH, Van Horn M. A study of tumor progression: The precursor lesion of superficial spreading and nodular melanoma. Hum Pathol 1984;15:1147–1165.
60. Aper JC. Congenital nevi. The controversy rages on. Arch Dermatol 1985;121:734.
61. Stenziger W, Suter L, Schumann J. DNA aneuploidy in congenital melanocyte nevi: suggestive evidence for premalignant changes. J Invest Dermatol 1984;82:569–572.
62. Breslow A. Thickness, cross-sectional areas and depth of invasion in the prognosis of cutaneous melanoma. Ann Surg 1970;1972:902–908.
63. Brocker E-B, Suter L, Brüggen J, Ruiter DJ, Macher E and Sorg C. Phenotypic dynamics of tumor progression in human malignant melanoma. Int J Cancer 1985;36:29–35.
64. Sondergaard K, Larson JK, Moller U, Christensen IJ and Hou-Jensen K. DNA ploidy-characteristics of human malignant melanoma analyzed by flow cytometry and compared with histology and clinical course. Virchows Archiv [B] 1983;42:43–52.
65. Frankfurt OS, Slocum HK, Rustum YM, et al. Flow cytometric analysis of DNA aneuploidy in primary and metastatic human solid tumors. Cytometry 1984;5:71–80.
66. Buchner T, Hiddemann W, Wormann B, et al. Differential pattern of DNA-aneuploidy in human malignancies. Pathol Res Pract 1985;179:310–317.
67. Lindholm C, Hofer P, Jonsson H and Tribukait B. Flow DNA-cytometric findings of paraffin-embedded primary cutaneous melanomas related to prognosis. Virchows Archiv [B] 58:147–151.
68. Von Roenn JH, Kheir SM, Wolter JM and Coon JS. Significance of DNA abnormalities in primary malignant melanoma and nevi, a retrospective flow cytometric study. Cancer Res 1986;46:3192–3195.
69. Kheir SM, Bines SD, Vonroenn JH, Soong S-J, Urist, MM and Coon JS. Prognostic significance of DNA aneuploidy in stage I cutaneous melanoma. Ann Surg 1988;207:455–461.
70. Newton JA, Camplejohn RS and McGibbon DH. The flow cytometry of melanocytic skin lesions. Br J Cancer 1988;58:606–609.
71. Kamino H, Ratech H. Improved detection of aneuploidy in malignant melanoma using multiparameter flow cytometry for S100 protein and DNA content. J Invest Dermatol 1989;93:392–396.
72. Ramaekers FCS, Puts JJG, Moesker O, et al. Antibodies to intermediate filament proteins in the immunohistochemical identification of human tumors: An overview. Histochem J 1983;15:691–713.
73. Ramaekers FCS, Huysmans A, Moesker O, Schaart G, Vooijs GP, Herman CJ. Cytokeratin expression during neoplastic progression of human transitional cell carcinomas as detected by a monoclonal and polyclonal antibody. Lab Invest 1985;52:31–38.
74. Garin Chesa P, Gay H, Whitmore WF, Jr., Melamed MR. Flow cytometric identification of human bladder cells using a cytokeratin monoclonal antibody. NY Acad Sci 1986;468:302–315.
75. Albino AP, Lloyd KO, Houghton AN, Oettgen HF and Old LJ. Heterogeneity in surface antigen and glycoprotein expression of cell lines derived from different melanoma metastases of the same patient. J Exp Med 1981;154:1764–1778.
76. Brown JP, Woodbury RG, Hart CE, Hellström I and Hellström KE. Quantitative analysis of melanoma-associated antigen p97 in normal and neoplastic tissues. Proc Natl Acad Sci USA 1981;78:539–543.
77. Burchiel SW, Martin JC, Imai K, Ferrone S and Warner NL. Heterogeneity of HLA-A,B, Ia-like, and melanoma-associated antigen expression by human melanoma cell lines analyzed with monoclonal antibodies and flow cytometry. Cancer Res 1982;42:4110–4115.
78. Atkinson B, Ernst CS, Ghrist BFD, et al. Identification of melanoma-associated antigens using fixed tissue screening of antibodies. Cancer Res 1984;44:2577–2581.
79. Cheresh DA, Reisfeld RA and Varki AP. O-acetylation of disaloganglioside GD3 by human melanoma cells creates a unique antigenic determinant. Science 1984;225:844–846.
80. Harper JR, Bumol TF and Reisfeld RA. Characterization of monoclonal antibody 155.8 and partial characterization of its proteoglycan antigen on human melanoma cells. J Immunol 1984;132:2096–2104.
81. Ross AH, Grob P, Bothwell M, et al. Characterization of nerve growth factor receptor in neural crest tumors using monoclonal antibodies. Proc Natl Acad Sci USA 1984;81:6681–6685.
82. Natali PG, Bigotti A, Cavaliere R, et al. Heterogeneous expression of melanoma-associated antigens and HLA antigens by primary and multiple metastatic lesions removed from patients with melanoma. Cancer Res 1985;45:2882–2889.
83. Holzman B, Bröcker EB, Lehmann JM, et al. Tumor progression in human malignant melanoma: five stages defined by their antigenic phenotypes. Int J Cancer 1987;39:466–471.
84. Houghton AN, Real FX, Davis LJ, Cordon-Cardo C and Old LJ. Phenotypic heterogeneity of melanoma. Relation to the differentiation program of melanoma cells. J Exp Med 1987;164:812–829.
85. Houghton AN, Albino AP, Cordon-Cardo C, Davis LJ and Eisinger M. Cell surface antigens of human melanocytes and melanoma. Expression of adenosine deaminase binding protein is extinguished with melanocyte transformation. J Exp Med 1988;167:197–212.
86. Thomson TM, Real FX, Murakami S, Cordon-Cardo C, Old LJ and Houghton AN. Differentiation antigens of melanocytes and melanoma: Analysis of melanosome and cell surface markers of human pigmented cells with monoclonal antibodies. J Invest Dermatol 1988;90:459–466.
87. Berd D, Herlyn M, Koprowski H and Mastrangelo MJ. Flow cytometric determination of the frequency and heterogeneity of expression of human melanoma-associated antigens. Cancer Res 1989;49:6840–6844
88. Anichini A, Mortarini R, Berti E and Parmiani G. Multiple VLA antigens on a subset of melanoma clones. Hum Immunol 1990;28:119–122.
89. Houghton AN, Herlyn M, Ferrone S. Melanoma antigens. In: Balch C, Houghton AN, and Sober AS, eds. *Cutaneous Melanomas*. 2nd Edition. Philadelphia: JB Lippincott, 1992 pp. 130–143.
90. Holzmann B, Bröcker EB, Lehmann JM, Ruiter DJ, Macher E, Sorg C. Phenotypic dynamics of tumor progression in human malignant melanoma. Int J Cancer 1985;36:29–35.
91. Morton DL, Malmgren RA, Homes EC, Ketcham AS. Demonstration of antibodies against human malignant melanoma by immunofluorescence. Surgery 1968;64:233–239.
92. Jehn VW, Nathanson L, Schwartz RS. In vitro lymphocyte stimulation by a soluble antigen from malignant melanoma. N Engl J Med 1970;283:329–333.
93. Rosenberg SA, Lotze MT, Muul LM, et al. Observations on the systemic administration of autologous lymphokine-activated killer cells and recombinant interleukin-2 to patients with metastatic cancer. N Engl J Med 1985;313:1485–1492.
94. Dutcher JP, Creekmore S, Weiss GR, et al. A phase II study of interleukin-2 and lymphokine-activated killer cells in patients with metastatic malignant melanoma. J Clin Oncol 1989;7:477–485.

95. Houghton AN, Chapman PB, Bajorin D. Antibodies in cancer therapy: Clinical applications. In: Rosenberg SA, Helman S, Devita V, eds. Principles and Practice of Biologic Therapy of Cancer. Philadelphia: J.B. Lippincott (In press), 1992.
96. Berd D, Maguire HC Jr, and Mastrangelo MJ. Induction of cell-mediated immunity to autologous melanoma cells and regression of metastases after treatment with a melanoma cell vaccine preceded by cyclophosphamide. Cancer Res 1986;46:2572–2577.
97. Schroff RW, Woodhouse CS, Foon KA, et al. Intratumor localization of monoclonal antibody in patients with melanoma treated with antibody to a 250,000-dalton melanoma-associated antigen. J Natl Cancer Inst 1985;74:299–306.
98. Thurin J, Thurin M, Kimoto Y, et al. Monoclonal antibody-defined correlations in melanoma between levels of G_{D2} and G_{D3} antigens and antibody-mediated cytoxicity. Cancer Res 1987;47:1229–1233.
99. Lindmo T, Davies C, Fodstad O and Morgan AC. Stable quantitative differences of antigen expression in human melanoma cells isolated in flow cytometric cell sorting. Int J Cancer 1984;34:507–512.
100. Morgan AC, Woodhouse C, Bartholemew R and Schroff R. Human melanoma-associated antigens: Analysis of antigenic heterogeneity by molecular, serologic and flow-cytometric approaches. Mol Immunol 1986;23:193–200.
101. Lindmo T, Davies C, Rofstad EK, Fodstad O and Sundan A. Antigen expression in human melanoma cells in relation to growth conditions and cell-cycle distribution. Int J Cancer 1984;33:167–171.
102. Leong SPL, Bolen JL, Chee DO, et al. Cell-cycle-dependent expression of human melanoma membrane antigen analyzed by flow cytometry. Cancer 1985;55:1276–1283.

34

Molecular Biology and Flow Cytometry I: Analytical Considerations

SETH P. HARLOW, EARL A. TIMM, JR., DOROTHY E. LEWIS, and CARLETON C. STEWART

INTRODUCTION

During the past decade, tremendous strides have been made in the fields of molecular biology and cytogenetics. Knowledge gained in these fields has greatly enhanced our understanding of the nature of the malignant process as well as of a number of other genetically based diseases. The focus of this chapter will be to briefly explain how flow cytometric techniques can be used in conjunction with molecular biological techniques in order to gain even more information than either can achieve alone. We will concentrate, in part, on the study of the cellular genes known to be important in the malignant process, the protooncogenes and tumor suppressor genes. These methods, however, can be easily extrapolated to the study of other genes as well.

The discovery of the transforming genes of retroviruses, termed "oncogenes," ushered in a new age in the way researchers and clinicians alike viewed and studied the malignant process. When Stehelin, in 1976 (1), demonstrated that a gene very similar to a viral oncogene was present in normal nonmalignant cells, the notion of cellular protooncogenes was begun. This group of genes (protooncogenes) are thought to be intimately involved with the cells' proliferative processes and it is thought that activation of these genes by a variety of mechanisms may lead to the development of the malignant phenotype. Since the discovery of protooncogenes, which act in a dominant fashion, (i.e., activation at a single allele can lead to a malignant change in a cell) another category of tumor-related genes has been identified. These are the tumor suppressor genes or recessive oncogenes. As their name suggests, these gene products tend to inhibit or suppress the malignant phenotype of tumor cells. The effect of tumor suppressor genes on the development or progression of malignancy is seen only after their inactivation by deletion or mutation. Because there are two functioning or wild type alleles present in the diploid genome, inactivation of both allelic sites would be necessary to see the effects on a cell, hence the term "recessive oncogene." The number of genes identified that fall into these categories grows every year. Presently, over 40 protooncogenes have been identified and, currently, at least four to nine genes are thought to act as tumor suppressor genes (2–3).

The ability to rapidly identify and quantify the activation or inactivation of these protooncogenes and tumor suppressor genes in a rapid cell by cell basis would be a powerful tool from both a research and a clinical perspective. Not only could important information regarding the basic mechanisms of the malignant process be garnered, but this information could also be used in the clinical arena, with levels being correlated to patient prognosis, and be of potential benefit in choosing appropriate therapeutic options. In fact, there have been several small clinical studies performed that correlated protooncogene activation with patient prognosis, demonstrating the potential clinical importance of these measurements (4–7).

MOLECULAR BIOLOGY OF ONCOGENES AND TUMOR SUPPRESSOR GENES

The following is a brief description of the mechanisms of protooncogene activation and suppressor gene inactivation that are known to take place in human malignancies. Refer to Figure 34.1 for a conceptualization of each process. Certainly, other mechanisms may be involved but, as yet, have not been identified as being important in human tumors.

Activation of Protooncogenes

GENE AMPLIFICATION

There are normally two copies of each gene within each cell. A gene is considered amplified when the absolute number of that gene is increased above two copies in a given cell. These areas of gene amplification may be located either on the same chromosome as the parent gene (heterogeneous staining regions on cytogenetic analysis) or as separate entities termed double minute chromosomes. These changes produce a concurrent increase in the DNA content of the cell. Flow cytometry has been extensively used to detect this increased DNA content.

GENE REARRANGEMENTS, TRANSLOCATIONS

This phenomenon occurs when there is a change in the location of a gene either on the same chromosome or to another chromosome. Gene rearrangements are normal phenomena required for the production of unique cellular products.

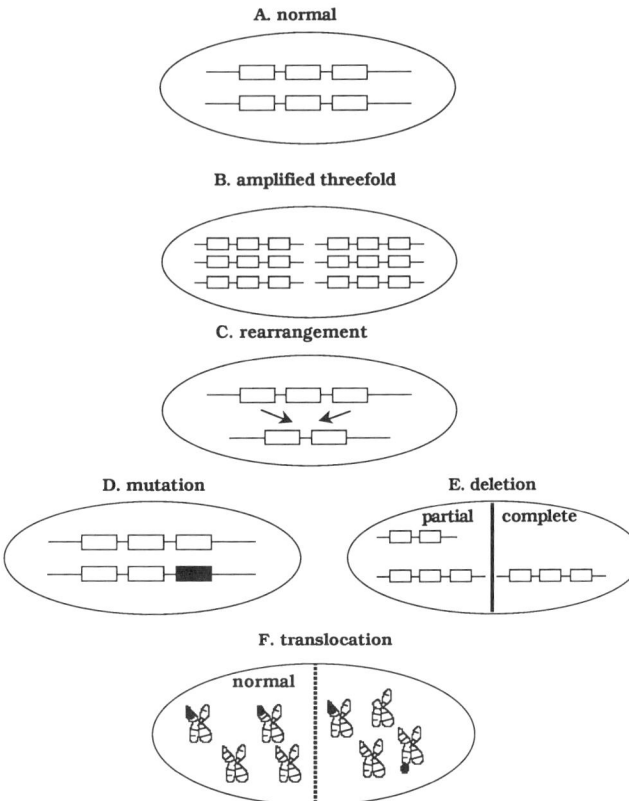

Figure 34.1. Genetic Changes Associated with Cancer. The normal genetic complement (**A**) consists of two alleles on two separate chromosomes. The actual genes, represented by the rectangles, are usually segmented with DNA sequences (*lines*) between the segments' introns. In cancer, additional copies of the genes can be made (**B**) e.g., threefold amplification would result in six copies. (**C**) Rearrangements occur when segments of a gene are moved to a different location. (**D**) Mutations result in altered DNA sequences in one of the segments. A partial loss of a gene (**E**) results in a partial or complete loss of one or more of the segments. There may also be a complete loss of one copy of the gene. Normally, genes are located on two separate chromosomes (**F**). Entire chromosomal segments can translocate to an entirely different chromosome. The gene's function, located at the breakpoint, can be altered as a result of the translocation.

These rearrangements occur within the gene segments themselves and are highly regulated. The most studied rearrangements include immunoglobulin genes (8) and T-cell receptor genes (9). When the rearrangement occurs to an inappropriate location in the genome, (i.e., if it is under the control of a strong promoter region from another locus or if it loses a negative control from its own site) abnormalities of gene expression can result. In addition, novel gene products can be created by the fusion of different genes. Any of these abnormal changes may play a role in the malignant process.

MUTATION

Activating mutations lead to gene protein products that have increased activity when compared to those of the wild type genes. Confirmational changes caused by a mutation could lead to abnormal regulation, an extended half life, or an overall increased activity of the protein product.

Inactivation of Tumor Suppressor Gene

GENE DELETIONS

They occur when the absolute number of a specific gene is decreased in a given cell, resulting in that cell's decreased DNA content. Deletions may involve an entire chromosome or a single gene or portion of a gene.

GENE MUTATIONS

Point mutations may lead to gene inactivation if the protein product is inactive or significantly less active than the wild type protein.

Molecular techniques are available to assess these mechanisms of activation or inactivation, with varying degrees of usefulness and reliability. Flow cytometric methods can be utilized in concert with these molecular techniques to give a tremendous amount of information about specific gene function from the DNA level to the mRNA to the protein expression level. We will briefly describe some of the more important molecular techniques that can be used in tandem with flow cytometry, but we will focus most of our attention on one particular technique, the polymerase chain reaction (PCR), which, we feel, has the greatest potential for use with flow cytometry.

MOLECULAR TECHNIQUES IN ANALYSIS OF GENOMIC DNA AND RNA

Analysis of DNA

The important parameters to analyze in protooncogene studies at the DNA level include the presence or absence of gene amplifications, gene mutations, or gene rearrangements. These kinds of studies have resulted in the creation of molecular diagnostics. The most frequently used methods to determine these changes include the following procedures:

SOUTHERN HYBRIDIZATION (10)

This technique has been the standard by which other techniques have been compared in the study of gene amplification and gene rearrangements. Briefly, the technique employs the digestion of genomic DNA by restriction enzymes. These enzymes break DNA at specific sequence sites leading to distinct fragments. This step is followed by agarose gel electrophoresis, separating fragments according to molecular weight. The material on the gel is transferred to a solid support, either nitrocellulose filters or nylon membranes, by capillary action. The DNA is dried and immobilized on its solid support and hybridization with a gene-specific probe is then performed. Quantitation can be achieved by scintillation counting or densitometry of auto radiograms if a radiolabeled probe is used, or by densitometric techniques if chemiluminescent probes are employed. Determination of gene rearrangements can be made by alterations in the electrophoretic migration pattern of the DNA fragments.

DOT AND SLOT HYBRIDIZATION (11)

These techniques are simplified methods for quantitating gene copy numbers. Basically, a serial dilution of genomic or digested DNA is "spotted" onto a solid support (i.e., nitrocellulose paper) then dried and immobilized. Hybridization with a gene-specific probe is then performed and quantitated by appropriate methods. Quantitation, in a relative fashion, can be made by comparison to hybridization of a reporter gene probe or by comparison to DNA extracted from cells with a known copy number.

The advantages of these techniques over Southern analysis are that they are easier to perform and take less time. In addition, the likelihood of errors in quantitation due to differences in transfer of DNA from the agarose gel to the solid support, as is needed with Southern analysis, is removed. Disadvantages of this technique are that no information concerning gene rearrangements can be gleaned and that the specificity of probe hybridization is not as easily verified as with Southern analysis.

The detection of gene mutations can theoretically be evaluated with either of these techniques. For this to be reliable, one needs mutation-specific probes and high stringency conditions. Gene mutations also can be determined by DNA sequencing techniques, which are the most accurate but are more involved techniques, requiring specialized instrumentation.

Analysis of RNA

Any of the above alterations in DNA will lead to changes in the cell only if the alterations are transcribed. Thus, the measurement of mRNA is one important proof that the genomic change is effectively altering the cell's behavior. The repertoire of mRNA's within the cell at any given moment provides information on what genes are being expressed and the level of their expression. We have termed this the "molecular phenotype" of the cell, in contrast to the thousands of functional proteins on the cell's membrane, identified by labeled antibodies to them, which have been termed the "immunophenotype." Techniques used for mRNA quantitation are given below.

NORTHERN HYBRIDIZATION (12)

In Northern analysis, RNA is extracted from cells and denatured with glyoxal and dimethylsulfoxide (DMSO) for agarose gel electrophoresis, which separates RNA according to size. The RNA is then transferred to a solid support (nitrocellulose or nylon membranes), similar to Southern analysis, before drying, immobilization, and hybridization to a specific probe. Quantitation can then be performed in a similar fashion to Southern analysis.

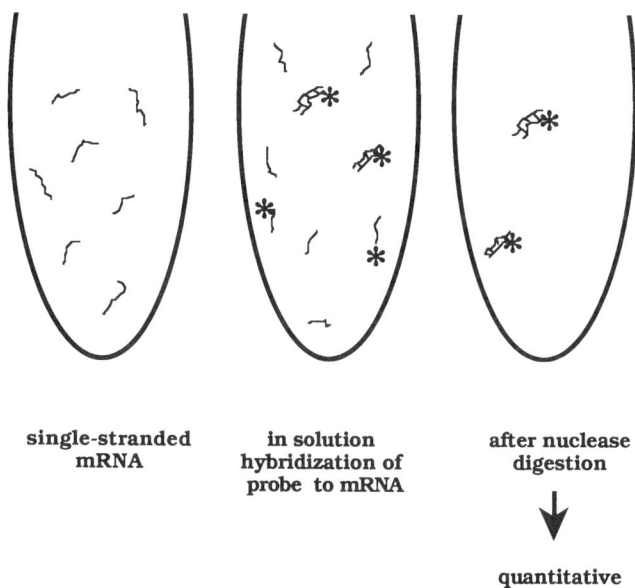

Figure 34.2. Nuclease Protection Assay. An appropriate probe is hybridized to single-stranded mRNA. A nuclease that digests single- but not double-stranded oligonucleotides is added. After sufficient time, the remaining material is quantitated with an appropriate method.

DOT AND SLOT HYBRIDIZATION TECHNIQUES

These methods are similar to the procedure described for DNA analysis except that extracted RNA is used. Similar advantages and disadvantages are found with the RNA techniques as with the DNA techniques. In addition, mutations of specific mRNA can also be identified by using mutation-specific probes and high stringency conditions.

NUCLEASE PROTECTION ASSAYS (13)

These techniques employ hybridization of a specific labeled probe to target RNA or cDNA molecules in solution (Fig. 34.2) The hybridization of the probe yields a double-stranded molecule (DNA:RNA, DNA:cDNA, or RNA:RNA) in the region of the target sequence but any nontarget molecules of RNA or cDNA remain single-stranded. If an excess of probe is used, all of the RNA or cDNA that is complementary to that probe will be hybridized. If an appropriate nuclease, such as RNAse or nuclease S1, is used, all single-stranded molecules (including unhybridized probe) are digested leaving only the double-stranded hybrids. This provides a very sensitive means for the quantitation of gene expression.

In addition to the techniques described above, in-situ hybridization to determine gene copy number and gene expression has been used for these determinations as well. A more complete discussion of these techniques will follow later in this chapter.

Polymerase Chain Reaction

When evaluating the utility of the previously described methods for use on a relatively large scale, as in the analysis

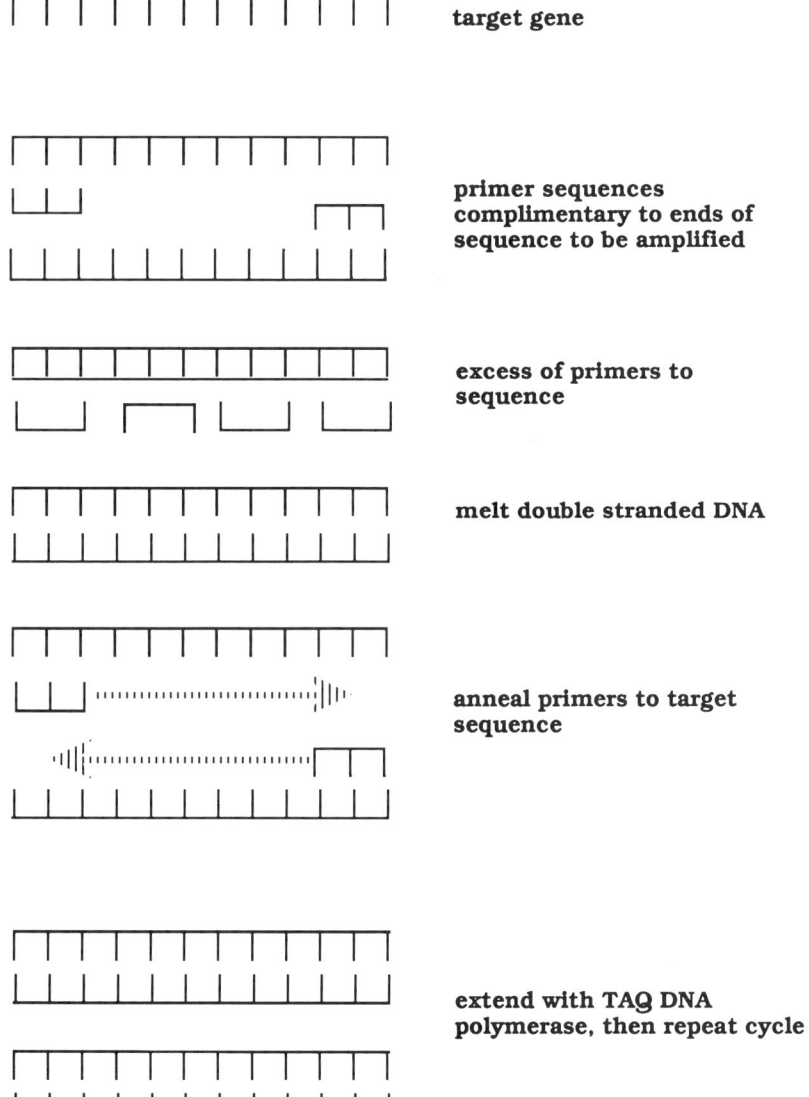

Figure 34.3. The Polymerase Chain Reaction. The sequence of the *target gene* is first determined. This data is generally available from the literature or from computer banks such as "Gene Bank." Oligonucleotide Primers of 15- to 30-mer lengths that span the region to be amplified must be made. *DNA* or *mRNA* is then isolated from cells. If mRNA is obtained, a cDNA template is prepared using the sense oligonucleotide primer, reverse transcriptase, and a cocktail of *dNTP*. (mRNA has the complementary sequence of the gene, which is the sense sequence). After this step, the PCR is the same as for extracted *DNA*. The *DNA* or cDNA templates are then added to a solution containing an excess of primers, *dNTP*'s, and taq polymerase and placed in the thermocycler. The temperature is increased to 94°C to melt the double-stranded *DNA*. The temperature is then reduced to 55°C to allow hybridization of the primers to the template strands. The mixture is then heated to 72°C, allowing the taq polymerase to extend the primer and making new copies of each strand between the primers. The cycles are repeated to geometrically amplify the previously produced copies. Under appropriate conditions, there is a doubling of target molecules with every cycle. Therefore, after 20 cycles, if one was starting with just two target molecules, a total of 2^{20} or 1,048,576 molecules would be generated.

of clinical tumor specimens, a major problem is that large amounts of starting material (DNA, mRNA) are required for each technique (with the exception of in-situ hybridization). While this may not pose a particular problem for studies of gene function utilizing cell lines, where a large number of cells can be obtained, it could pose a significant obstacle to analysis of clinical specimens. Frequently, the amount of tumor tissue available from clinical specimens is limited by tumor size and the need of tissue for histopathology or hormone receptor analysis. The requirement for large amounts of starting material can be circumvented by the use of the polymerase chain reaction.

The polymerase chain reaction was first described by Mullis et al. in 1985 (14). The technique allows the exponential amplification of specific DNA sequences to levels up to several million-fold over starting levels (Fig. 34.3). It requires oligonucleotide primers that are specific for a certain gene sequence, the appropriate buffer, ion, and nucleotide

COMBINING PCR AND FLOW CYTOMETRY

Current flow cytometric methods can accurately determine gene expression at the protein level with the use of fluorochrome-labeled monoclonal antibodies. Antibodies to cell membrane components or internal products also can be used to identify specific cell types. In addition, DNA-content measurements can be made and cells can be sorted by ploidy or phase of the cell cycle. Because PCR requires DNA or RNA from only a small number of cells, specific populations can be sorted for determinations of gene frequency and/or gene expression at the mRNA level. If one is faced with a heterogeneous cell population, such as is seen in clinical tumor specimens, where there is always a variable percentage of non-neoplastic inflammatory and stromal cells, one can sort the neoplastic cells from a heterogeneous cell suspension for molecular phenotyping. It is then possible to make intelligent assessments of gene studies without the confounding effect of inappropriate cell contamination.

General Methodology

The following will be a brief description of the methods used to prepare cells for PCR techniques, the methods for performing quantitative PCR, and the methods for determining gene mutations with PCR. A schematic is shown in Figure 34.4.

CELL PREPARATION

DNA and RNA can be obtained from fixed as well as unfixed cells for use in PCR. Because high-molecular-weight DNA is not needed for most PCR applications, DNA extracted from paraffin-embedded material has been used for this purpose (16). Standard deparaffinization of the specimen must be performed prior to the release of the DNA from nuclei. However, due to the lability of mRNA molecules, reliable quantitation of RNA from paraffin-embedded material has not yet been possible. Fresh fixed cells, however, have been used successfully for mRNA quantitation as well as DNA quantitation procedures. Fixation with formaldehyde or ethanol has been found to be compatible with DNA-quantitating PCR procedures. Fixation of cells in 50% to 70% ethanol has been found to be optimal for mRNA-quantitating PCR procedures (17).

DNA AND RNA AMPLIFICATION

After cell sorting and collection in nuclease-free tubes, the cells are processed to release their DNA and RNA for the PCR reaction. Currently, we employ the methods described by Higuchi (18) for the lysis of cells to obtain DNA and RNA for PCR reaction. These techniques use nonionic detergents (NP40 and Tween 20) for cell lysis and proteinase K for protein digestion to release cellular DNA (note, only NP40 can be used for mRNA release as Tween 20 will inhibit the reverse transcription reaction). Following this processing, the cell lysates can then be used for reverse tran-

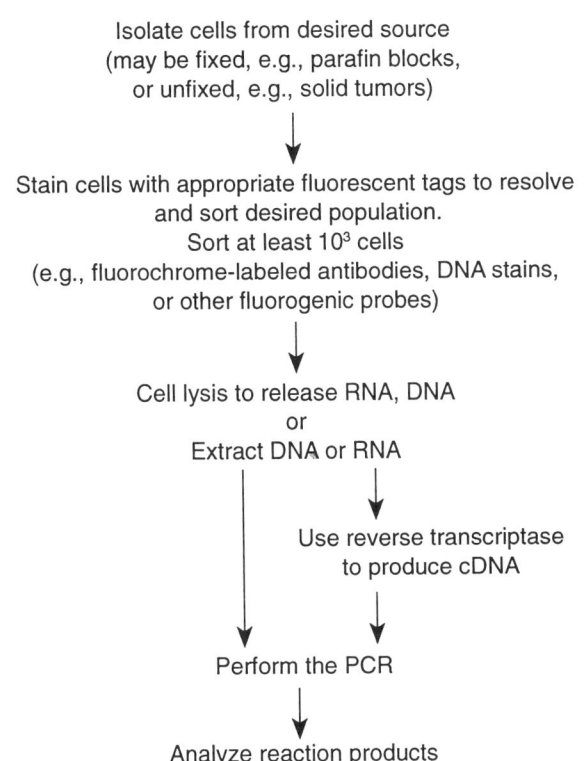

Figure 34.4. Measuring Genomic or message sequences by combining Flow Cytometry and PCR

concentrations, for the reaction to proceed. The enzyme Thermus Aquaticus DNA polymerase (taq DNA polymerase) allows automation of the procedure. This enzyme is stable at the high temperatures required for denaturation of double-stranded DNA molecules (94–95°C). The procedure can now be simply performed in a machine known as a "thermal cycler." This machine rapidly adjusts temperature of specimens placed in it to allow the sequential steps of DNA denaturation, DNA annealing to oligonucleotide primers, and polymerase extension to take place.

By carefully selecting oligonucleotide primers to span known intron sequences, one can differentiate genomic DNA products from mRNA-derived products by product length (15). With appropriate controls, PCR can be used in a quantitative fashion for the measurement of gene frequency (amplification or deletion), gene rearrangement, and gene expression at the mRNA level (if a reverse transcription of the mRNA is performed prior to PCR). In addition, by using appropriate modifications of the standard PCR protocol, a number of other assays, including detection and quantitation of gene mutations, can also be accomplished. This technique's wide range of quantitative and qualitative abilities, combined with its need for only small amounts of starting material (cellular DNA, RNA), clearly makes it the method of the future for not only clinical but also basic research applications of molecular phenotyping.

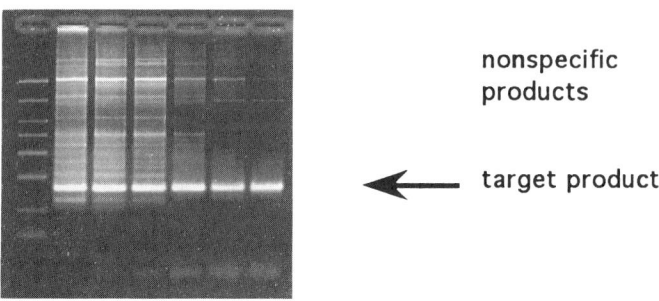

LANE	SAMPLE (#cells)
1	MW Marker
2	50,000 cells
3	10,000 cells
4	1,000 cells
5	100 cells
6	10 cells
7	1 cell

PCR of Gastrin Gene From HL60 Cells

Figure 34.5. Sensitivity of *PCR* To Detect mRNA. Note that the nonspecific products are seen more frequently the greater the number of cells used.

scription (mRNA studies) or go directly to PCR for gene amplification studies.

In Figure 34.5, the sensitivity of the *PCR* reaction is illustrated in the detection of genomic Gastrin gene sequences in cultured HL60 cells. *PCR* product bands can be seen with dilutions containing from 1 to 50,000 cells. In general, best results are achieved with 100 to 1000 cells, a number that can easily be sorted in less than 1 hr. Since cells can be fixed prior to processing by flow cytometry, problems with viability and the effects of sorting them are avoided.

ANALYSIS OF PCR PRODUCTS

Proper analysis of PCR reaction products requires determinations of product length (in nucleotides) and quantity. Standard methods to determine these are gel electrophoresis with quantitation of product bands by scintillation counting, optical densitometry, or serial dilutions.

Recently, high-performance liquid chromatography (HPLC) has been adapted to the analysis of oligonucleotides, giving accurate measurements of both product size and quantity. This method uses a nonporous anion exchange column for separating DNA fragments and quantitates the products based on optical density. The procedure is both rapid and accurate with results reported to be reproducible to within ± 10%, a result that is superior to gel electrophoresis (19).

Strategies for Quantitative PCR

Because of the exponential amplification of DNA target sequences by the polymerase chain reaction, small variations in conditions between tubes at the start of a reaction can lead to large discrepancies in results by the end of the reaction. For accurate quantitation of the PCR reaction, an internal standard in each tube should be employed to gauge results. Two methods have been described to achieve these ends.

DIFFERENTIAL PCR (20)

In this method, PCR primers for two different gene sequences are added to each reaction, creating two PCR products that are similar in length but can be distinguished by hybridization with gene specific probes or by restriction endonuclease digestion. This technique has been used to quantitate gene amplification where a nonamplified control gene sequence located on the same chromosome but at a distance from the studied gene is also subjected to PCR. A recent study using this method reported an incidence of *c-myc* amplification in ovarian carcinoma specimens as being 17% (21).

COMPETITIVE PCR (22)

We feel that this method offers the most accurate and sensitive way of utilizing PCR as a quantitative tool. It is reported to accurately quantitate less than 1 pg of target from 1 ng of total mRNA, and it can detect twofold differences in target concentrations. This method requires the addition of known amounts of a competitive template that uses the same primers as the target gene sequence. As shown in Figure 34.6, one makes a competitive template that has the same nucleotide sequence as the target molecule, with the exception of a point mutation, creating a novel restriction site not found in the target sequence. Quantitation of each product fragment is made after restriction endonuclease digestion cuts the competitive product into two shorter fragments. By running multiple tubes with decreasing quantities of competitive template, one can determine the quantity of target sequence in a sample by extrapolating where the quantity of target product equals that of the competitor product. At that point, the initial quantity of target sequence would be equal to that of the competitor sequence added. If one knows the number of cells used at the start of the reaction, one can then determine the average number of target molecules per cell to an accuracy of approximately 0.1 a-moles. Presynthesized competitors have recently been made that can accommodate a number of different primer pairs, so that the same competitive template can be used

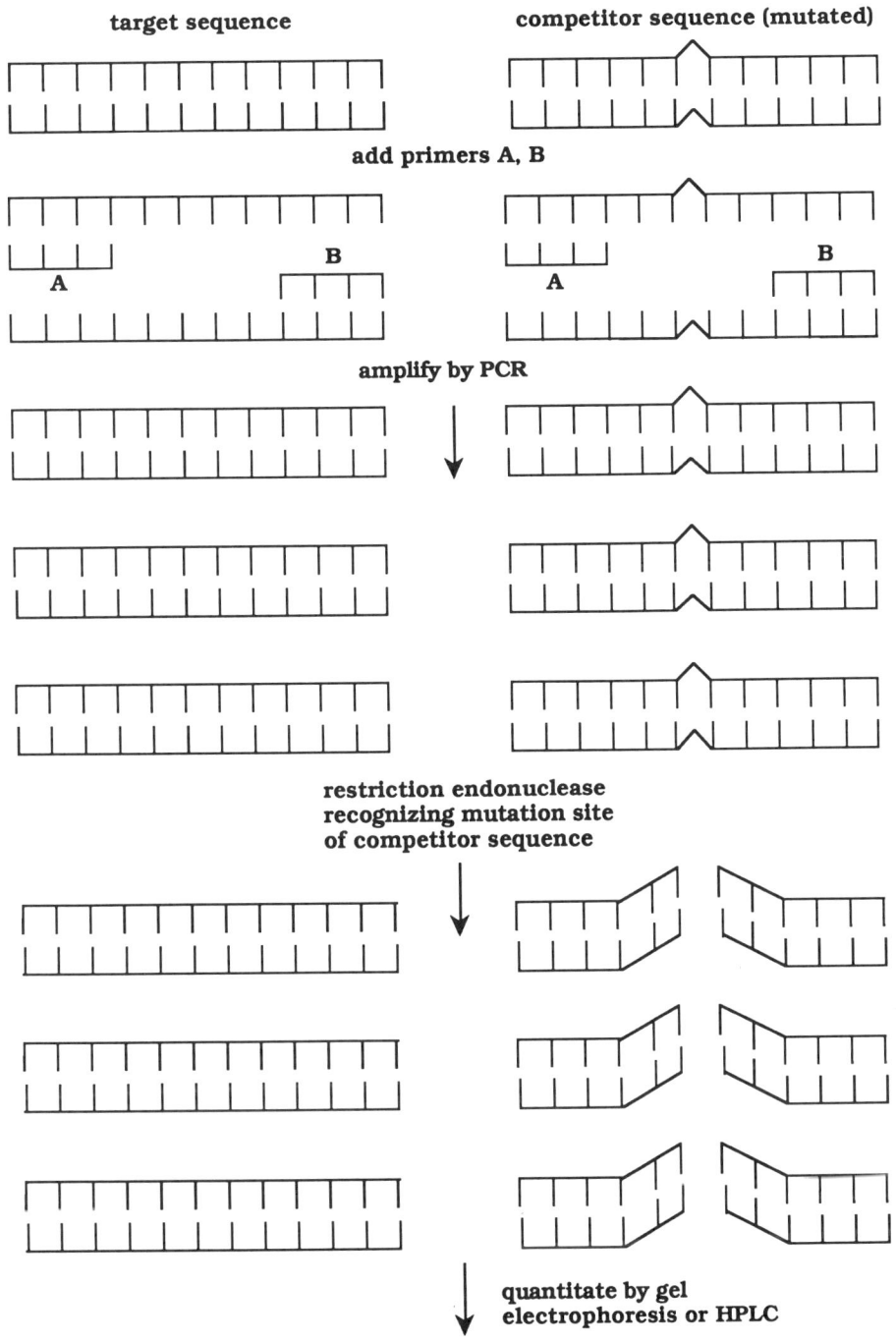

Figure 34.6. Schematic representation of the technique of competitive PCR for quantitation of genomic DNA. For quantitation of mRNA species, a single-stranded cRNA with the restriction site mutation is employed as the competitive template and added before the reverse transcription step.

to study multiple genes (23). The theoretical advantages of this method are that the affinity of primer annealing to the target and competitor templates are identical and the rate of polymerase extension is nearly equal, as both products are of the same length and similar base content (differing only in one or two bases). This method can be used for quantitation from DNA templates or mRNA templates if the appropriate cRNA competitor is made.

Strategies for Mutation Detection by PCR

Multiple strategies for detection of gene mutations by PCR techniques have been devised. The applicability of any one strategy to a given laboratory depends on that laboratory's accessibility to the resources needed for that strategy to be performed. A brief overview of these strategies will follow.

DIRECT SEQUENCING OF PCR PRODUCTS (24, 25)

PCR methodology produces relatively large quantities of oligonucleotide product for sequencing purposes. If one devises PCR primers that incorporate regions where activating mutations are known to occur, one can then simply sequence the PCR product and compare it to the known nucleotide sequence for that region. Point mutations can be accurately detected and the specific base substitutions can be identified.

A potential disadvantage of this method is that if a mutation is found in only a relatively small proportion of the cells analyzed, it may be "diluted" by the overabundant normal wild type genes. Exponential amplification can lead to greater disparities, and rare mutations could be missed. Flow cytometric sorting of cell subpopulations could reduce or eliminate this problem. In addition there is a definite rate of misincorporation of nucleotides by the taq DNA polymerase enzyme, (average of 1 in 10,000 nucleotides per cycle) which, if it occurs in an early PCR cycle, can lead to false-positive results.

HYBRIDIZATION OF THE PCR PRODUCT USING MUTATION-SPECIFIC PROBES (26)

PCR can be used to generate larger quantities of product for analysis by hybridization techniques as described earlier. The probe selected must be highly specific for the mutated region.

CREATION OF ALTERED OLIGONUCLEOTIDE PRIMERS FOR MUTATION DETECTION

Two different techniques have been described. One technique is to alter an oligonucleotide primer at its 3' end, such that it forms a restriction site when it lies immediately upstream of a specific point mutation (27). Therefore, if a mutated species is present in the sample solution, the restriction enzyme digests the product and a specific band is detectable by gel electrophoresis.

A second technique is to create multiple oligonucleotide primers, each specific for a known point mutation (i.e., ras gene mutation at codons 12, 13, 61) at its 3' end (28). If PCR conditions of annealing temperature, Mg++ ion concentration, and dNTP concentration are made sufficiently stringent, only a mutated target will be amplified. This method can also be modified such that mutations in any gene of a family of genes can be detected simultaneously. An example would be the three genes of the Ras family (K-ras, H-ras, N-ras). This is done by substituting inosine residues (complementary to all nucleotide bases) at all locations in the primer sequence where there is no homology between different members of the gene family (28). In this way, more rapid screening of samples for these mutations can be performed.

DENATURING GRADIENT GEL ELECTROPHORESIS (29)

This method makes use of the slight changes in DNA duplex melting temperatures incurred by base substitutions. When properly performed, it can detect mutations at a single base pair. However, it does not identify the location of the mutation nor the base that is mutated. It is useful, however, for large-scale screening of genes to detect mutations that may be responsible for disease.

Strategies for Detection of Translocations by PCR

At present, detection of gene rearrangements or chromosome translocations by PCR is possible only if the translocation is known to occur at a specific site. (For example, the translocation seen in patients with Chronic Myelogenous Leukemia {t(9,22) translocation} presents as a novel mRNA fusion product called *bcr-abl*.) By designing PCR primers specific for the DNA sequence that gives this product, one lying in each of the two genes, one can identify this translocation very accurately by PCR techniques (30). Another translocation (14;18) seen in follicular and some nonfollicular lymphomas can also be detected by similar methods using PCR techniques (31). In many known translocations, the break point may cover a region anywhere among several hundred nucleotides. Identification of break point sites could potentially be made by using PCR primers spanning large lengths of DNA, and then sequencing to find the break point. It is likely that cleaver strategies will be devised that will simplify and replace this time-consuming and unreliable approach.

In summary, a wealth of information can be obtained by combining flow cytometry to sort specific cell populations from heterogeneous cell suspensions with presently available molecular biology techniques. Using specific primers in combination with PCR techniques, it is possible to obtain the "molecular phenotype" of the sorted cells. Combining these techniques will make possible the accurate detection of oncogene activation or tumor suppressor gene inactivation from small clinical specimens, which was not previously possible. The clinician will then have the ability to use this information for assessing patient prognosis, predicting patient response to therapy, or for a host of other clinically relevant questions that the study of these genes has recently answered.

FLUORESCENCE IN SITU HYBRIDIZATION IN SUSPENSION (FISHES) AND DETECTION BY FLOW CYTOMETRY

The development of various in situ hybridization techniques during the past few years has provided important tools for the analysis of DNA and RNA within fixed cells attached to microscope slides. Many detection systems use a biotin-avidin method. A biotinylated nucleotide is incorporated into the probe and, after performing the in situ hybridization, the hybridized probe is detected by fluorescent-labeled avidin. Originally, the preparation consisted mainly of tissues or cells fixed to prepared slides, as in the work of Singer and Ward (32). Many other laboratories now apply such methods in their work. While the actual microscopic visualization can yield determination of positive and negative cells within specimens, their quantitation is tedious. In some recent work, in situ hybridization using cells in suspension (33), employing the biotin-avidin system, has been used to detect the hybridized probe within the cell. Combining this technique with flow cytometry could result in a rapid means for molecular phenotyping on a cell-by-cell basis.

General Methodology for FISHES

PREPARATION OF THE CELLS

The most promising results have been from Bauman and Bentvelzen (34), Bayer and Bauman (35), and Timm and Stewart (36). One of the most immediate problems confronted during the procedure is the loss of cells from the suspension as they adhere to the tube surface. Bauman et al. (34) address this problem by reducing the volumes used in the washes. Another method, developed by Timm and Stewart (36), uses minimal amounts of nuclease-free bovine serum albumin (BSA) in all washes. It was found that 500 μg/ml BSA is just enough to prevent cell adherence to the tube but does not interfere with the in situ hybridization. The latter method provides for a high recovery of the cells after each wash.

The choice of fixatives for fixation of the cells and permeabilization of the cell membrane appears to be paraformaldehyde (32, 34, 37, 38, 39). We have found that ultrapure formaldehyde (36) works as well as paraformaldehyde. These fixatives open large enough holes in the cell membrane for small nucleic acid probes to enter and, at the same time, the mRNA stays fixed inside the cell. After fixation, the cells are stored in 70% ethanol, which not only slows the degradation of the mRNA but also enlarges openings in the cell membrane. The integrity of the cell's internal structures is also maintained, so that the localization of the probe within the cell can be viewed microscopically. The cells can be stored for days or weeks in the ethanol (37), but gradual degradation of the mRNA will occur.

Different labeling methods are used to tag the nucleic acid probes so that they can be detected on the flow cytometer. The most popular method is the biotin-avidin system. Avidin has a high affinity for biotin and, when avidin is coupled with fluorescein (FL-avidin) or some other fluorochrome, a biotinylated probe can be either detected microscopically or analyzed on a flow cytometer.

In one method, the biotin is coupled to dUTP (deoxyuridine triphosphate) and this deoxynucleotide is incorporated into the probe. The biotinated dUTP is available from many vendors. Bauman et al. (34), Singer et al. (32), Bresser et al. (37), as well as many others have used this method of detection for in situ hybridization. The biotin-avidin system does provide good sensitivity, but some problems arise with this method. Cells have endogenous biotin to which the fluoresceinated avidin binds. This generalized binding raises the noise level when the probe is being detected with the fluoresceinated avidin. Since the avidin–biotin link has a high affinity, it is difficult to wash out the nonspecific binding. Although Bresser and Evinger-Hodges (37) claim to have detected as low as 10 copies of mRNA, the consensus among other researchers seems to be a much higher number (500-1000 copies) for the minimal level of detection.

In the biotin-avidin system, there is a need to block endogenous biotin within the cells, which requires extra steps or chemicals. Bauman et al. (34) and Bayer et al. (35) appear to use the reagent Tween-20 (polyoxyethylene–sorbitan monolaurate), which is also used in ELISA (enzyme-linked immunoabsorbent assays) to reduce background found in that particular biotin-avidin procedure.

Recently, a new method for nucleic acid probe labeling and detection has been introduced by the Boehringer-Mannheim Company (Indianapolis, Indiana 46250). The product is digoxigenin (Dig), which is steroid derived from the digitalis plant. This molecule is coupled to dUTP and can be detected using fluorescent-labeled antibodies directed against the Dig molecules. Since digoxigenin is not endogenous to mammalian cells, there should be a lower nonspecific background due to nonspecific binding by the antibody. As shown in Figure 34.7, antibodies bind nonspecifically to dead cell components so the advantages of Dig over Biotin are partially negated. When the biotin-avidin and Dig procedures were compared, it was found that the biotin-avidin system consistently produced too much background, even after extensive washes, to allow for the detection of low copy numbers of mRNA. The digoxigenin system provided a higher signal-to-noise ratio and the method was easier to apply (unpublished results).

PREPARATION OF NUCLEIC ACID PROBES

The probes used for in situ hybridization include both single- and double-stranded DNA. Double-stranded DNA (dsDNA) provides stable probes, but these probes present problems of complimentary strand competition and the possible hybridization by the complimentary strand to other targets in the cell. This latter problem can be checked by testing with single-stranded complimentary probes. Single-stranded DNA (ssDNA) probes provide a stable alternative to single-stranded RNA (ssRNA) probes. Bauman et al. (34) and Bayer et al. (35) have obtained good results using single-stranded RNA (ssRNA) probes. The work of Taneja and Singer (39) on double-stranded DNA probes versus single-stranded DNA probes provides valuable insight on the comparisons of dsDNA and ssDNA probes used for in situ hybridization. Their paper states that there was little difference in the results between the two types of probes, and Timm et al. (36, unpublished data) have found similar results. In dealing with mid-to-high copy number of mRNA targets (>500 copies), there seems to be relatively little difference between the type of probe used. This may change when hybridizing to mRNAs having low copy number (<500 cop-

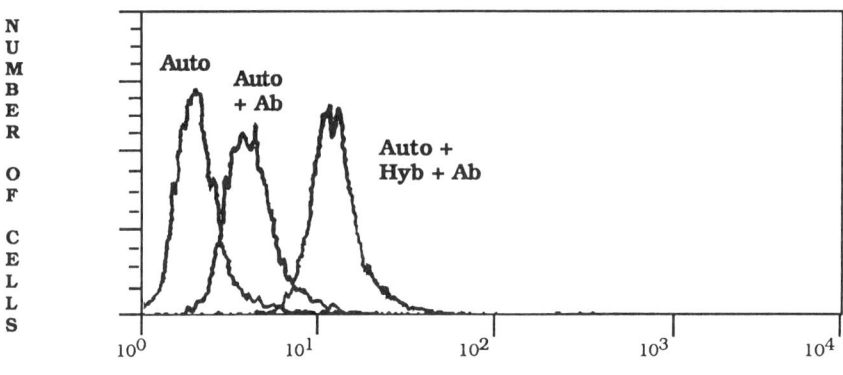

Figure 34.7. Flow Cytometric Analysis of FISHES. L-929 cells were fixed and analyzed without further treatment (*auto*), treated with fluoresceinated sheep anti digoxigenin (*Auto + Ab*) or hybridized with a digoxigenin-labeled α actin probe followed by antibody (*Auto + Hyb + Ab*). Cells were analyzed using a FACScan.

ies), where highly-specific single-stranded probes would be preferred.

The use of a single-stranded probe of 50-200 bases is optimal but it limits the amount of signal because the number of Dig (or Biotin) molecules that can be incorporated in the probe is dependent on the number of thymidine residues. Timm and Stewart (36) have employed the polymerase chain reaction to construct double-stranded probes or the asymmetric PCR to produce single-stranded probes. The label, which is either biotin-16-dUTP or Dig-11-dUTP, is added during the PCR. The ratio of the labeled dUTP to its analog, deoxythymidine triphosphate (dTTP), is around one part dUTP to two parts dTTP. While asymmetric PCR does produce a small quantity of double-stranded component, purification of the single-stranded component does not appear to be necessary because the probe can be added at a temperature such that the double-stranded portion will not melt or hybridize to a target. The labeled single-stranded probe provides the readout.

In a variation of the PCR, one large sequence of double-stranded DNA (e.g., 1000 bases) can be made to the desired nucleotide sequence. This longer probe, with its concomitant increase in label, can provide a much higher signal. After synthesis, the probe is digested into smaller pieces using a restriction enzyme to facilitate its entry into cells before being used as a probe (Timm, unpublished). The disadvantage of this approach, however, is that rehybridization of the probe to itself reduces its sensitivity and that targets of similar sequence homology to those desired may hybridize to some of the fragments, leading to misinterpretation. This approach is similar to the preparation of long RNA probes, employed by both Bauman et al. (34) and Bayer et al. (35). They construct long ssRNA probes transcribed from cloned sequences and these probes are degraded into smaller pieces using limited alkaline hydrolysis. These probes are efficient due to the lack of a competing complementary strand, although the area of the target hybridized cannot be easily determined. Also, RNA probes are sensitive to possible RNase digestion, so extra care must be taken to preserve these probes during hybridization.

There are several other methods for labeling nucleic acid probes for use in in situ hybridization. The classical method, used for double-stranded DNA, is nick translation. A series of nicks are made in the double-stranded DNA by one enzyme and a second enzyme is used to incorporate labeled nucleotides starting at the nick (40). In another method, used by Bresser et al. (37), a photobiotin reagent is used to label double- or single-stranded nucleic acid probes. In the photobiotin method, a biotin molecule with a linker arm is chemically attached to thymidine residues in the strand and bound by subjecting the mixture to ultraviolet light. More recently, Taneja et al. (38, 39) have used oligonucleotide probes that are directed at specific sites on the target mRNA. These oligonucleotide probes are about 50 bases in length and they are labeled with biotin-16-dUTP by means of a terminal transferase that adds the nucleotides to the 3' end of the DNA sequence. A similar methodology could be employed to label the probes with Dig-11-dUTP. It is also possible to incorporate thymidine analogs with an amino-linked molecule during oligonucleotide synthesis on a DNA synthesizer (Applied Biosystems, Foster City, CA 94404). The probe can then be directly labeled with an appropriate fluorochrome.

IN SITU HYBRIDIZATION TECHNIQUES

Proof of specificity is generally lacking in many publications and may result in misinformation. Several controls need to be performed with internally consistent results before valid conclusions can be drawn. This is largely accomplished by the utilization of antisense and sense probes (34, 35, 36) in systems where specific messages are or are not induced. The antisense probe is the complement of the target sequence in the hybridization and the sense strand is the sequence equivalent to the target area of the strand. As shown in Figure 34.8, if there are no nucleic acid sequences in the cell complementary to the sense probe, then the signal from the sense

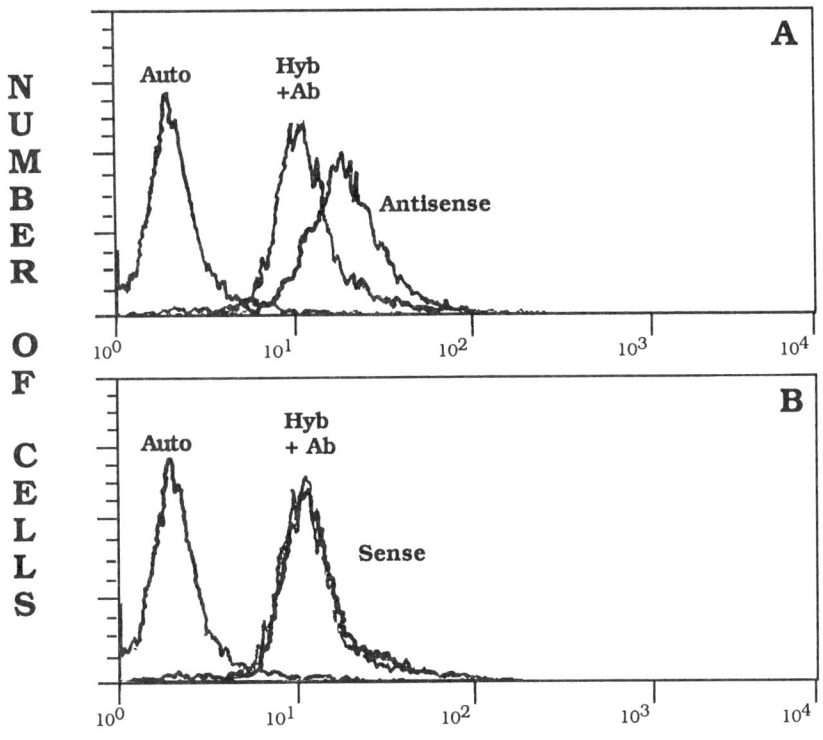

Figure 34.8. Comparison of Antisense and Sense Probe Binding. Hybridization of Dig-labeled *sense* or *antisense* probes were performed using fixed L-929 cells. The *antisense* probe (**A**) produced a good signal-to-noise ratio, allowing for detection of positive cells, but the *sense* probe (**B**) did not yield a significant signal over the noise.

probe should fall within the background noise region. The antisense probe should bind to the target mRNA and survive the stringency washes that are designed to remove nonspecific binding of the probes.

The in situ hybridization and all the washes are done in microcentrifuge tubes although Timm et al. (36) used 12 × 75 polystyrene tubes for the stringency washes in order to use larger volumes (up to 3.0 ml). The hybridization reaction is often done in a 2X to 6X SSC (standard saline citrate) buffer where a 1X SSC is approximately 0.15 M sodium chloride. The range of temperatures for in situ hybridization is from 37°C to 50°C. While most hybridizations have been performed at lower stringencies and hybridization temperatures to insure that the probe will bind, a lower SSC concentration and higher temperature will mean a higher stringency in the hybridization, thereby insuring that only the target molecule will be bound by the probes. This strategy can improve the signal-to-noise ratio of the system and the reliability of the results. The rationale for using less stringent conditions appears to be that more target molecules will be hybridized by the probes. If this approach is used, high stringency washes can be used to remove mismatches.

The length of time for hybridization varies between experimenters, but the times must be long enough for the probes to enter the cell and diffuse to the target. Bresser et al. (37) and Taneja et al. (39) have stated that, when short probes (50-200 bases) are used, longer incubations allow for more nonspecific binding. They state that the optimal time is about 30 min, after which the rate of specific hybridization of the probe levels off. Other results suggest that longer incubation times improve signal provided nonspecific binding is reduced by high stringency washes.

The rate of reaction also depends upon the probe concentration, but these amounts will vary because of differences in the amount of expected target and cell concentration in the reaction. Probes can be added at a concentration of about 10 ng per 10^6 cells, as done by Timm et al. (36), so that the concentration is greater than the expected total number of target molecules, in order to insure optimal hybridization. These calculations are rarely made and the amount of probe added for the in situ hybridization is determined on a case-by-case basis. According to Bresser et al. (37), the amount of probe added for in situ hybridizations ranges on the average from 0.06–4.0 μg/ml of probe added, with the reaction volume varying between 10–20 μl. Our studies using probes prepared by PCR have revealed that 1–25 μg/ml work best for our conditions; these concentrations have been generally employed by investigators using cells attached to microscope slides. High-stringency washes appear to be the best method for reducing nonspecific binding.

Another problem that occurs with this method or any other in situ hybridization is one of choosing the proper probe and targeting the probe. Bresser et al. (37) have determined that the optimal size probe for in situ hybridization is between 50 and 200 bases. According to Taneja et al. (39), probes that are too small will enter the nucleus where the nucleic acid concentration is very high, resulting in high nonspecific binding. Large probes, greater than 200 bases, have more difficulty entering through the fixed cell membrane (37).

FLOW CYTOMETRY DETECTION

Quantitation of the hybridized probes can be accomplished by running the cell suspensions through a flow cytometer. Bauman's group employs a two-laser Rijswijk Experimental Light Activated Cell Sorter (RELACS III Coherent, Palo Alto, CA). Timm and Stewart (36) use a FACSCAN flow cytometer (Becton Dickinson, Mountain View, CA 94039). Both devices produce similar results in reference to light scatter and fluorescence intensity. The important parameters for measuring successful hybridization of the probe are the autofluorescence of the fixed samples before the hybridization; fluorescence of the sample after the hybridization when no probe is added; background caused by the fluorescent second-step labeling agent itself; and the signal due to the hybridization of the probe.

Bauman et al. (34), and Timm and Stewart (36) have observed that the hybridization procedure alone, minus probes, causes an increase in the fluorescence of the cells above their autofluorescence. (Fig 34.7) This background increases further when the fluorescent label, either fluorescein-avidin or fluorescein-anti Dig, is added. It is important to hybridize to the target mRNA using a probe and a probe detection system that will elevate the signal above this noise. While both the biotin/avidin and Dig/antibody systems provide good sensitivity and a good signal-to-noise ratio when microscopy is used, the high backgrounds due to the generalized background fluorescence are not yet acceptable for detection of low copy number by flow cytometry. When viewing hybridized probes in cells attached to microscope slides the reaction product appears as bright points of fluorescence. The bright points of varying size appear over a dimmer contrasting fluorescent background throughout the cell. When a cell passes a laser beam, the fluorescence from all sources is integrated by the photomultiplier tube and all spatial information of the bright points on dim backgrounds is lost. The integrated fluorescences of the dim background may far exceed the integrated fluorescence of the bright points, thereby rendering them unresolvable by flow cytometry. The large RNA probes (digested) as used by Bauman et al. (34) or the digoxigenin-labeled DNA probes used by Timm and Stewart (36) have accomplished this task for cells that express (>2000 copies) of message. Both groups have elevated the signal from the antisense probes above the background noise. Bayer and Bauman (35) have resolved cells with β-globin mRNA transcripts in various hemopoietic cells where the copy number of the mRNA transcripts range from an estimated 2,000–20,000 per cell, depending on the cell type studied. Timm and Stewart (36) have resolved γ-actin mRNA transcripts in murine L929 fibroblasts (Fig. 34.8). Both of the above mentioned mRNA transcripts are found in relative abundance in their respective cell types. The real problems will occur when trying to detect mRNA transcripts found in low-copy numbers (<500 copies per cell) by using the flow cytometer, where the signal from the background may overshadow the probe signal. It is this problem that needs to be solved before flow cytometry will be useful in providing reliable molecular phenotyping.

FUTURE APPROACH: FLUORESCENCE IN SITU PCR (FLIP) BY FLOW CYTOMETRY

Unless a primary system can be developed that eliminates the nonspecific binding problem introduced by second reagents or unless second reagents can be found that do not bind to dead cell components, it is not likely that the signal-to-noise ratio problems will be improved in the future using this strategy. We have been working on a new approach to this problem by specifically amplifying the signal at a greater rate than the noise, using the in situ polymerase chain reaction. Before such a system can be adapted, there are several problems that must be dealt with. Since amplification must occur in each cell acting as the reaction vessel, the target sequences must be available for primer hybridization and the cell must remain intact through the temperature extremes required. The cell must also be permeabilized so that the enzymes and substrates can enter the cell, but it must not be as permeable as to allow the products to escape from the cell. The specific sequences desired need to be amplified, but not the undesired sequences. Finally, problems associated with the interference of product concentration on the PCR efficiency within the small volume of the cell must be dealt with.

Haase et al. (41) claim to have performed in situ PCR to detect lentiviral DNA. Our own experience indicates that after 10–15 cycles of in situ PCR, amplification no longer occurs, as the products rehybridize to themselves more readily than the primers, thereby completely terminating product production. Unfortunately, amplification of undesired sequences can still take place, thereby increasing noise and misinterpretation of data.

While we have successfully performed in situ PCR utilizing the methods described above for fixation, permeabilization, and hybridization of primers to the cellular targets, we have not improved the signal-to-noise ratio. The procedure was done by incorporating biotin-16-dUTP during the in situ PCR using fixed cells in suspension. This problem is illustrated in Figure 34.9, which shows an increase of over two orders of magnitude in both the signal and the noise. Thermocycling itself causes a threefold increase in background. When cells are completely processed in the absence of taq DNA polymerax, there is a greater than 20-fold increase in fluorescence above the baseline. This fluorescence

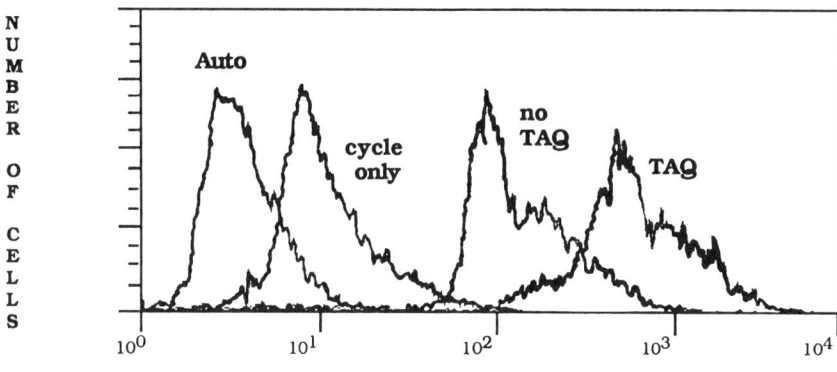

Figure 34.9. Fos expression in macrophage measured by fluorescence in situ PCR (FLIP). Mouse bone marrow was cultured for six days in CSF1 to produce a subconfluent monolayer of macrophages. Following the procedure of Willman et al., (49) who demonstrated maximal fos expression 30 min after addition of CSF1 to macrophage cultures, we removed CSF1 from the medium and read it 24 hr later. Cells were harvested 30 min later and fixed. One sample was not treated (auto); in a second sample, the cells underwent 15 complete cycles on the thermocycler (cycle only), autofluorescence increased by a factor of about 3. In a third and fourth sample, selected primers, reverse transcriptase, dNTP, and Biotin-UTP were added and a cDNA produced. In one of these tubes, no taq was added while to a second tube taq was added. The two tubes underwent 15 complete cycles. After washing and resuspension in PBS, the cells in both tubes were incubated with Fl-avidin, washed, and analyzed using the FACScan.

is due to increased autofluorescence due to the thermocycling, nonspecific FL-avidin fluorescence, and the specific binding of FL-avidin to biotin-16-dUTP trapped nonspecifically in the cell. A further increase by one order of magnitude is seen when taq is present. Our results suggest that a coamplification of nonspecific and specific sequences occurs, causing the signal from the specific sequences to become lost in the total signal. While we view these results as encouraging, further experimentation with the development of new strategies will be required before reliable fluorescence in situ PCR (FLIP) can be applied to molecular phenotyping using flow cytometry.

Applications of FISHES and FLIP

One future application of the FISHES and FLIP techniques will be the measurement of oncogene expression on a cell-by-cell basis utilizing the flow cytometer. When the sensitivity of the techniques becomes such that mRNAs expressed at a very low copy number (>100 copies per cells) can be easily detected, it will be possible to use the procedures to detect aberrations in a cell population from a normal level of gene expression using the flow cytometer.

A powerful future application of the FISHES or FLIP techniques would be the colabeling of surface or internal antigens to detect changes in levels of gene products simultaneously with changes in mRNA expression on a cell-by-cell basis. Timm (unpublished data) has found that the level of surface labeling using the F4/80 antibody against murine bone marrow macrophages remains stable even after the 30 cycles of temperature changes commonly used in the polymerase chain reaciton. Therefore, it is possible to detect oncogene mRNA transcript levels and oncogene product levels in the same cell by analyzing the cells on a flow cytometer.

These techniques can be used as an initial indicator of changes from the norm. If warranted by the data, more conventional techniques to detect gene expression would follow. It is likely that further research will lead to the ability to simultaneously determine the immunophenotype and molecular phenotype of normal and abnormal cells on a cell-by-cell basis using flow cytometry.

FLOW CYTOMETRIC USE OF β-GALACTOSIDASE AS A REPORTER GENE

The well-characterized *Escherichia coli lacZ* gene has been used in mammalian and insect cells as a marker to detect the expression of introduced genetic elements such as promoter/enhancer sequences. This system is ideal for use as a reporter because it is stable, the expression of enzymatic activity is not harmful to cells, and there are a variety of methods to detect expression, including immunocytochemical and chromogenic enzymatic assays. The *lacZ* gene encodes the β-galactosidase (β-gal) enzyme, which is a glycoside hydrolase. Because the first 27 amino acids of the protein can be replaced without affecting enzyme activity, exogenous gene sequences can be added to a truncated β-gal gene product and the transfection frequency can be monitored.

In 1988, Nolan et al. introduced the use of a fluorescent substrate for β-gal that can be used on the flow cytometer (42). This has resulted in a significant technical advance because, although other substrates for the enzyme are useful in tissue sections and in single-cell suspensions, the other substrates all result in cellular death. In contrast, Nolan's method allows for living cells to be selected using flow cytometry after uptake and cleavage of the fluorescent substrate. The flow cytometer can also give quantitative information about expression levels in individual cells and, thus,

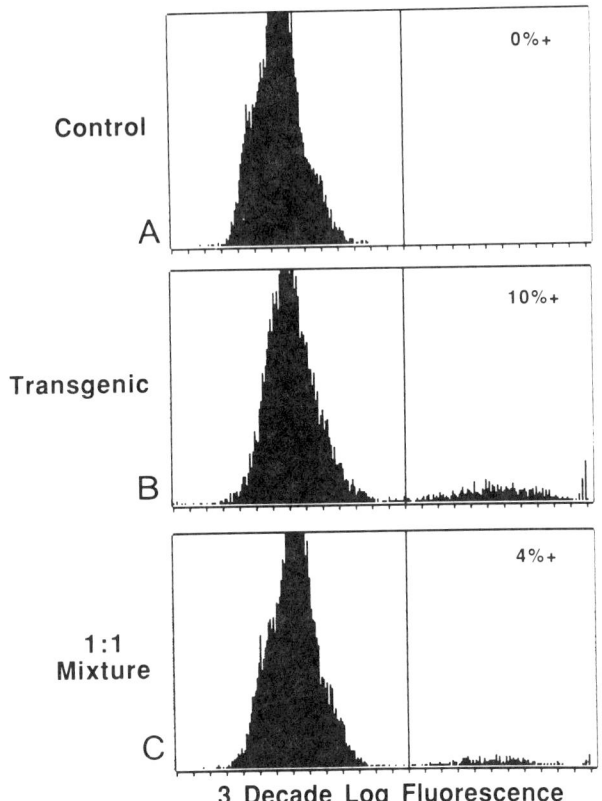

Figure 34.10. Flow cytometric detection of β-gal expression in control (**A**) or transgenic (**B**) liver cells. Note that 10% of the cells are positive in **B**. A 1:1 mixture (**C**) shows that nonspecific uptake did not occur.

reveal heterogeneous expression patterns with complex cellular mixtures.

General Methodology

The fluorescent substrate for β-gal, fluorescein di-β-D-galactoside (FDG), is loaded into viable cells (2.0 μM FDG) using a brief hypotonic shock at 37°C for 1 min. Immediately after the entrance of the substrate, the cells are cooled to 4°C in isotonic medium to help prevent leakage of free fluorescein. It was determined that heating the cells to 15°C allows the fluorescein to leak out; hence, it is important to do the procedure using iced medium and to measure fluorescence using a flow cytometer with ice cold sheath fluid (42, 43). Because this protocol was developed for sorting, it was important to prevent leakage because of the time it takes to do the sort. It may be possible to avoid the stressful hypotonic shock if analysis is all that is required. The samples could then be processed immediately after uptake of substrate. In that case, 20 μM FGD is required for loading of the cells. As can be seen in Figure 34.10, if the β-gal fluorescence is monitored as a function of time, leakage of fluorescein and substrate uptake into previously negative cells can be monitored and enzymatic rate can also be recorded.

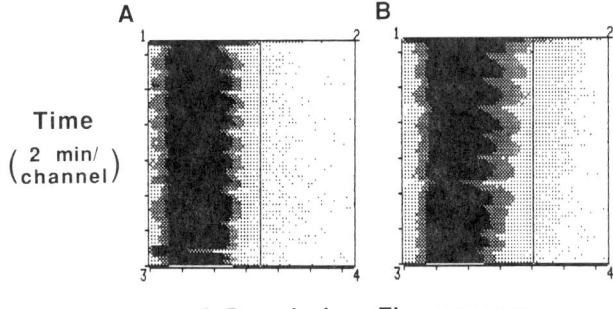

Figure 34.11. Time (2 min/channel, y axis) as a parameter used to monitor leakage and uptake of free fluorescein during the β-gal assay. Log green fluorescence is shown on the x axis. The cursor was originally placed to detect the rare positive cells in the cellular mixture. As can be seen in **A**, there is no change in the fluorescence intensity of the negative cells as a function of time. In contrast, the fluorescence of the negative cells does increase with time in another experiment, shown in **B**. This indicates that leakage of fluorescein and uptake by negative cells was a problem in the **B** experiment.

METHODS FOR INTRODUCTION OF LACZ

Several pUC19 based vectors that express β-gal under the control of different viral proteins have been produced (44). These include vectors with the Herpes simplex virus thymidine kinase HSV TK promoter, the human cytomegalovirus immediate early promoter (enhancer CMV IEP), the SV40 early promoter (origin), and the adenovirus 2 major late promoter AD2 MLP. The vectors typically contain an RNA splice signal, a polyadenylation signal, and a unique Not 1 restriction site from which β-gal can be excised for reinsertion of alternative DNA coding sequences.

Another approach, which is used to stably incorporate β-gal into cells, is to use *lacZ* encoding retroviruses. If the retrovirus is transcriptionally defective, the regulatory abilities of endogenous transcription control sequences can be explored. Thus, using a pre-B cell line 70Z/3, Kerr et al. showed that *lacZ* was expressed in a pre–B cell line only when the pre-B cell line was treated with LPS to cause differentiation (45).

CURRENT APPLICATIONS FOR THE β-GAL ASSAY ON FLOW CYTOMETERS

The development of Drosophila has been studied using the β-gal reporter system linked to the P-element reporter to identify genomic elements that can regulate transcription at a distance. β-gal expression was examined in developing Drosophila tissue with many specific cell types and tissue-specific patterns of expression (46). This technique has now been applied to minced tissue specimens in single-cell suspensions. Such a system allows detection and subsequent selection of marked cells at different developmental stages to determine the effect of environment on cell fate. Such a study is made possible by the existence of over 10,000 different β-gal expression Drosophila mutants (47).

An assay for detection of infectious HIV was recently described by Roederer et al. (48). A fusion gene was constructed that contained the HIV LTR as the promoter for the β-gal coding sequence. This construct was activated 20- 40-fold by the presence of the transactivating protein, tat. The presence of HIV in clinical samples could then be monitored and quantitated by the expression of HIV-LTR driven β-gal activity after interaction with patient-derived tat protein.

An example of the usage of the β-gal detection system in a transgenic mouse liver is shown in Figure 34.10. The al anti-trypsin promoter was fused to β-gal coding sequences and tansgenic animals were produced. As can be seen, about 10% of the cells in the liver of the transgenic mouse expressed β-gal. In addition, mixtures between dissociated normal and transgenic liver cells showed no exchange of the fluorescein to the normal cells. A method to monitor leakage of fluorescein and subsequent uptake by nonfluorescent cells as a function of time during the experiment also was devised as shown in Figure 34.11. Time was used as a parameter and was plotted versus log green fluorescence as shown in Panel A. No leakage of fluorescein occurred over a 2-hr period; however, in another experiment, shown in B, there was significant increase in the background fluorescence of the originally negative cells as a function of time.

Future Applications

The analysis of DNA transcription rates, the expression of various transgenes, and the isolation of control elements for gene expression are all future applications for the fluorescence β-gal detection system. For gene therapy protocols, the presence of any enzymatic fluorescence marker inside the cells could facilitate the detection and study of rare cells containing the new genetic information.

REFERENCES

1. Stehelin D, Varmus HE, Bishop JM. Purification of DNA complementary to nucleotide sequences required for neoplastic transformation of fibroblasts by avian sarcoma viruses. J Mol Biol 1976;101:349.
2. Stewart CC. Flow cytometric analysis of oncogene expression in human neoplasias. Arch Pathol Lab Med 1989;113:634–640.
3. Spandidos DA, Anderson M. Oncogenes and onco-suppressor genes: their involvement in cancer. J of Pathol. 1989;157:1–10.
4. Seeger RC, Brodeur GM, Sather H, et al. Association of multiple copies of n-myc oncogene with rapid progression of neuroblastomas, N Engl J Med, 1985;313:1111–1116.
5. Slamon DJ, Clark GM, Wong SG, et al. Human breast cancer: correlation of relapse and survival with amplification. Science 1987;235:177–182.
6. Parks HC, Lillicrop K, Howell A, Craig R. K. C-Erb B2 mRNA expression in human breast tumors: comparison with C-erb B2 DNA amplification and correlation with prognosis. Br J Cancer Jan. 6, 1990; 61(1) 39–45.
7. Varley JM, Swallow JE, Brammer WJ, et al. Alterations to either C-erbB2 (neu) or C-Myc proto-oncogenes in breast carcinomas correlated with poor short term prognosis. Oncogene 1987;1:423.
8. Hozumi N, Tonegawa S. Evidence of somatic rearrangements of immunoglobulin genes coding for variable and constant regions. Proc Nat Acad Sci USA 1976;73:3628–3632.
9. Kronenberg M, Siu G, Hood LE, Shastri N. The molecular genetics of the T-cell antigen receptor and T-cell antigen recognition. Ann Rev Immunol 1986;4:529–592.
10. Southern EM. Detection of specific sequences among DNA fragments separated by gel electrophoresis. J Mol Biol 1975;98:503–517.
11. Kaftos FC, Jones CW. Fcfstratiadis A. Determination of nucleic acid sequence hemologies and relative concentrations by a dot hybridization procedure. Nucleic Acids Res 1971;F:1541–1552.
12. Alwine JC, Kemp DJ, Stark GR. Method for detection of specific RNA's in agarose gels by transfer to diazobenzyloxymethyl-paper and hybridization with DNA probes. Proc Nat Acad Sci USA 1977;74: 5350–5354.
13. Johnson MA, McCrae MA. A rapid and sensitive solution hybridization assay for the quantitative determination of specific viral RNA sequences. J Virol Methods 1988;22 (2–3)247–254.
14. Mullis KB, Faloona F, Scharf SJ, Saiki RK, Horn GT, Erlich, HA. Specific enzyme amplification of DNA in vitro: The polymerase chain reaction. Cold Spring Harbor Symp. Quant Biol 1986;51:263–273.
15. Kawasaki ES, Clark SS, Coyne MY, et al. Diagnosis of chronic myeloid and acute lymphocytic leukemias by detection of leukemia-specific RNA sequences amplified in vitro. Proc Nat Acad Sci USA 1988;85: 5698–5702.
16. Imprain CC, Saiki RK, Erlich HA, Templit RL. Analysis of DNA extracted from formation fixed, paraffin embedded tisues by enzymatic amplification and hybridization with sequence specific oligonucleotides. Biochem Biophys Res Comm 1987;142:710–716.
17. Khochbin S, Grunwald D, Pabion M, Lawrence JJ. Recovery of RNA from flow sorted fixed cells. Cytometry 1990;11:869–874.
18. Higuchi R. In: Erlich HA, ed. PCR technology-principles and applications for DNA amplification. New York: Stockton Press. 1989;31–38.
19. Katz ED, Wong M W. Rapid analysis and purification of polymerase chain reaction products by high-performance liquid chromatography. Biotechniques 1990;8:546–555.
20. Frye RA, Benz CC, Liu, E. Detection of amplified oncogenes by differential polymerase chain reaction. Oncogene 1989;4,1153–1157.
21. Schrieber G, Dubeau L. C-Myc proto-oncogene amplification detected by polymerase chain reaction in archival human ovarian carcinomas. Am J Path 1990;(137)3:653–658.
22. Gilliland G, Perrin S, Bunn HF. Competitive PCR for quantitation of mRNA In: Innis MA, Gelfand DH, Sninsky JJ, White TJ, eds. PCR Protocols: a guide to methods and applications. San Diego, CA: Academic Press Inc., 1990;60–69.
23. Wong AM, Mark DF. Quantitative PCR Protocols: A guide to methods and applications In: Innis MA, Gelfand DH, Sninsky JJ, White JJ eds. PCR Protocols: a guide to methods and applications. San Diego: Academic Press Inc., 1990;70–75.
24. Sanger F, Coulson AR. A rapid method for determining sequences in DNA primed synthesis with DNA polymerase. J Mol Biol 1975;94: 441–448.
25. Maxam AM, Gilbert W. A new method for sequencing DNA. Proc Nat Acad Sci USA 1977;74:560–564.
26. Verlaan-deVries M, Bogaard ME, Vanden Elst H, Ban Boom JH, Vander Els AJ, Bes JL. A dot blot procedure for mutated ras oncogenes using synthetic oligodeoxynucleotides. Gene 1986;50:313–320.
27. Haliassos A, Chomel JC, Groundjouan S, Kruh J, Kitzis A. Detection of minority point mutations by modified PCR techniques: A new approach for a sensitive diagnosis of tumor progression markers. Nucleic Acids Res 1989;(17)20:8093–8099.
28. Ehlen T, Bubeau L. Detection of ras point mutations by polymerase chain reaction using mutation specific, inosine containing oligonucleotide primers. Biochemical and Biophysical Research Communications 1989;(160)2:441–447.
29. Myers RM, Sheffield VC, Cox DR. Mutation detection by PCR, ge-clamps, and denaturing gradient gel electrophoresis. In: Erlich HA, ed. PCR technology-principles and applications for DNA amplifications. New York: Stockton Press, 1989;71–88.

30. Kawaski ES, Clark SS, Coyne MY, et al. Diagnosis of chronic myeloid and acute lymphocytic leukemias by detection of leukemia specific mRNA sequences amplified in vitro. Proc Nat Acad Sci USA 1988;85:5698–5702.
31. Lee MS, Chang R, Cabanillas E J, et al. Detection of minimal residual cells carrying the T (14:18) by DNA sequence amplification. Science 1987;237:175–178.
32. Singer RH, Ward DC. Actin gene expression visualized in chicken muscle tissue by using in situ hybridization with a biotinylated nucleotide analog. Proc Nat Acad Sci USA 1982;79:7331–7335.
33. Bauman JGJ, van Dekken H. Flow cytometry of fluorescent in situ hybridization to detect specific RNA and DNA sequences. Acta Histochem Suppl (Jena) 1989;37:65–69.
34. Bauman JGJ, Bentvelzen P. Flow cytometric detection of ribosomal RNA in suspended cells by fluorescent in situ hybridization. Cytometry 1988;9:517–520.
35. Bayer JA, Bauman JGJ. Flow cytometric detection of b-globin mRNA in murine haemopoietic tissues using fluorescent in situ hybridization. Cytometry 1990;11:132–143.
36. Timm EA Jr, Stewart CC. Fluorescent in situ hybridization in suspension (FISHES) using digoxigenin labeled probes and flow cytometry. Biotechniques 1992;12:362–365.
37. Bresser J, Evinger-Hodges MJ. Comparison and optimization of in situ hybridization procedures yielding rapid sensitive mRNA detections. Gen Anal Tech 1987;4:89–104.
38. Taneja K, Singer RH. Detection and localization of actin isoforms in chicken muscles cells by in situ hybridization using biotinated oligonucleotide probes. J Cell Biochem 1990;44:241–252.
39. Taneja K, Singer RH. Use of oligodeoxynucleotide probes for quantitative in situ hybridization to actin mRNA. Anal Biochem 1987;166:389–398.
40. Sambrook J, Fritsch EF, Mantiatis T. Molecular cloning: A laboratory manual, 2nd ed. Cold Spring Harbor Laboratory Press, 1989.
41. Haase AT, Retzel EF, Staskus KA. Amplification and detection of lentiviral DNA inside cells. Proc Natl Acad Sci USA 1990;87:4971–4975.
42. Nolan GP, Fiering S, Nicholas JF, Herzenberg LA. Fluorescence-activated cell analysis and sorting of viable mammalian cells based on β-D-galactosidase activity after transduction of *Escherichia coli lacZ*. Proc Natl Acad Sci USA 1988;85:2603–2607.
43. MacGregor GR, Nolan GP, Fiering S, Roederer M, Herzenberg LA. Use of *E. coli* lacZ (β-Galactosidase) as a reporter gene. In: Murry EJ, Walker JM, eds. Methods in Molecular Biology: Gene expression *in vivo*, Vol. 7. Clifton, NJ: Humana Press, Inc., 1989;1–19.
44. MacGregor GR, Caskey CT. Construction of plasmids that express *E. coli* β-galactosidase in mammalian cells. Nucleic Acids Res 1989;17:2365.
45. Kerr WG, Nolan GP, Serafini AT, Herzenberg LA. Transcriptionally defective retroviruses containing *lacZ* for the in situ detection of endogenous genes and developmentally regulated chromatin. In: Cold Spring Harbor Symposia on Quantitative Biology, Vol. LIV. New York: Cold Spring Harbor Laboratory Press, 1989;767–776.
46. O'Kane CJ, Gehring WH. Detection in situ of genomic regulatory elements in Drosophila. Proc Natl Acad Sci USA 1987;84:9123–9127.
47. Krasnow M, Nolan GP. Whole animal cell sorting to study drosophila development [Abstract]: Cytometry 1990;(suppl 4):13.
48. Roederer MM, Fiering S, Nolan G, Herzenberg L. The Development of an assay for infectious HIV: use of FACS-GAL to measure Gene Activity in Individual cells [Abstract]. Cytometry 1990; (supplement 4) p 40.
49. Willman CL, Stewart CC, Griffith JK, Stewart SJ, Tomasi TB. Differential expression and regulation of the c-src and c-fgr protooncogenes in myelomonocytic cells. Proc Natl Acad Sci USA 1987;84:4480–4484.

35

Molecular Biology and Flow Cytometry II: Clinical Potential

STEPHEN C. PEIPER and ALVIN W. MARTIN

INTRODUCTION

The technology of flow cytometric analysis of leukocytes stained with monoclonal antibodies to epitopes expressed on the plasma membrane and of tumor cell nuclei stained with DNA intercalating dyes has evolved to the point of common utilization as a front-line study for the characterization of human malignancies. Athough initially operationally defined as differentiation markers, several of the molecules recognized by immunodiagnostic reagents have been shown to function as cell surface receptor structures operative in cytoadhesion to other cells, viruses, and extracellular matrix. The current repertoire of immunologic reagents and DNA-intercalating dyes was established prior to the explosion in the application of molecular biology to characterize the genetic changes responsible for the transformed phenotype. Insight into structural genetic abnormalities associated with specific hematopoietic malignancies and retroviral oncogenes has provided a guide for the identification of genes—and of the products they encode—that are involved in the pathogenesis of malignant transformation. Furthermore, molecular biological approaches have been applied to elucidate mechanisms responsible for metastatic potential and drug resistance of human malignancies Two novel classes of proteins have emerged as regulators of critical pathways that control cellular growth: products of cellular protooncogenes, which promote proliferation (review, 1), and products of tumor suppressor genes, which provide a layer of regulation for proliferation (review, 2). Protooncogenes may encode polypeptides that function as secreted growth factors, plasma membrane receptors for growth factors, cytoplasmic intermediaries in signal transduction, and nuclear factors that control transcription. Tumor suppressor genes have been shown to encode products active in regulating signal transduction in the cytoplasm; transcriptional activity of selected genes; and, potentially, as "receptor phosphatases" expressed on the plasma membrane. Typically, transformation results from activating mutations in cellular oncogenes associated with the acquisition of function or from the inactivation of a suppressor gene resulting in the loss of function. Thus, the action of oncogenes is dominant and that of tumor suppressor genes is recessive.

However, none of the monoclonal antibodies to cell surface differentiation antigens utilized in routine immunodiagnostics recognize a cell surface receptor involved in the modulation of cell growth, with the possible exception of CD45 antibodies, which bind to a member of the "receptor phosphatase" family. A full repertoire of immunological reagents for the products encoded by protooncogenes and tumor suppressor genes is not yet available, in part, because of the rapidity of development in this field and their limited immunogenicity due to the high degree of conservation between species. Thus, none of the candidate products encoded by protooncogenes and tumor suppressor genes have been fully characterized for their direct application in the clinical arena. However, several have emerged as potentially important for the characterization of tumors and will be discussed further in the context of utility for flow cytometric analysis in the future. It is predicted that the typing reagents of the future will be directed to the products of protooncogenes, which will be employed to assign lineage as well as to gain insight into prognostic features.

CELL SURFACE RECEPTORS

Growth factor receptor structures on the plasma membrane fall into generic categories, two of the best studied being those with tyrosine-specific kinase activity, which are members of the immunoglobulin gene superfamily; and those wih primary sequence motifs characteristic of a novel cytokine receptor family, which do not have homology to the immunoglobulin gene superfamily and lack kinase activity. The topologic features of these receptors are schematically depicted in Figure 35.1. Genes encoding several members of the tyrosine kinase family have been transduced in oncogenic retroviruses, i.e., v-erb-B, v-fms, and v-kit. Molecular detective work has demonstrated that the receptor for epidermal growth factor (EGF-R) is the normal cognate of v-erb-B (3), the receptor for macrophage colony stimulating factor (CSF-1-R) is v-fms (4), and the receptor for stem cell factor (c-kit) is v-kit (5–7). Monoclonal antibodies that recognize epitopes specific for each of these receptors have been used to characterize normal and malignant cells. Flow cytometric analysis has been preferentially utilized in the study of hematopoietic neoplasms, primarily because of logistical considerations.

Macrophage-colony-stimulating factor (CSF-1) stimulates the formation of the mononuclear phagocyte system from primitive hemopoietic progenitors and activates the physiologic functions of these cells, which include mono-

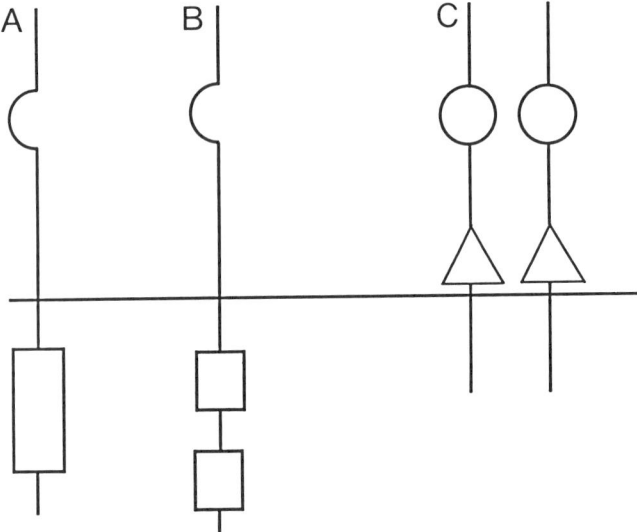

Figure 35.1. Schematic representation of cell surface receptor structures for growth factors. **A**, tyrosine kinase family with continuous cytoplasmic kinase domain, ie EGF-R and *neu*; **B**, tyrosine kinase family with interrupted kinase domain, i.e., CSF-1-R and c-*kit*; **C**, cytokine receptor family, including receptors for IL-2, IL-3, IL-4, IL-6, IL-7, Epo, GM-CSF, and G-CSF, composed of multiple subunits. Key to schematic: immunoglobulin domain, open circle; tyrosine kinase domain, rectangle; fibrinectin domain, closed circle; tryptophan-serine-X-tryptophan-serine motif, triangle.

cytes, macrophages, and osteoclasts (review, 8). It also acts on decidual cells, trophoblasts, and microglial cells. An inactivating mutation in the gene encoding CSF-1, which results in complete absence of the factor, is the molecular defect in murine congenital osteopetrosis due to a lack of osteoclasts (9, 10). Soluble CSF-1 is produced from both secreted and membrane-bound forms by fibroblasts, epithelial cells of the gravid uterus, thymic epithelium, keratinocytes, osteoblasts, astrocytes, and mesothelial cells. The production of CSF-1 by endometrial cells during pregnancy and the expression of the CSF-1-R by decidual cells and trophoblasts has raised speculation regarding the role of this factor in placental development (11).

The receptor for CSF-1 is a transmembrane glycoprotein that has an extracellular moiety that binds CSF-1 and a cytoplasmic domain with tyrosine kinase activity. Structurally, it is closely related to the receptor for platelet-derived growth factor and c-*kit* (12). Monoclonal antibodies to the CSF-1-R have been used to study its expression in normal and leukemic hematopoietic cells (13). Based on the ability of CSF-1 to stimulate bone marrow cells to form macrophage colonies, it is presumed that primitive hematopoietic progenitors express CSF-1-R, albeit in low numbers. Blood and bone marrow leukocytes that bind monoclonal antibodies to the CSF-1-R have forward-angle and orthogonal light scatter properties characteristeric of monocytes. Cells isolated by fluorescence-activated cell sorting had cytomorphology typical of monocytes and expressed α-napthyl-butyrate esterase activity. Flow cytometry analysis of acute leukemias revealed that 30% of acute myelogenous leukemias (AML) in children and 15% in adults bound the anti-CSF-1-R monoclonal antibodies, but blasts from patients with acute lymphoblastic leukemia (ALL) lacked reactivity. Although 62% of childhood AMLs with monocytic differentiation expressed the CSF-1-R, five of the 15 positive pediatric and three of four positive adult cases had granulocytic differentiation without evidence of a monoblastic component. In contrast to the findings in normal leukocytes, there was poor correlation between the expression of CD14, a cell surface receptor for complexes of lipopolysaccharide (LPS) and LPS-binding protein found on monocytes and macrophages, and CSF-1-R in AML. Among 17 cases of CD14-negative childhood AML, seven expressed CSF-1-R. In adults, all four adult patients with CSF-1-R-positive leukemic cells lacked reactivity with MY4, a CD14 antibody, and all CD14-positive cases lacked expression of CSF-1-R. When analyzed for the presence of mRNA transcripts encoding CSF-1-R, the incidence of expression in AML is higher than observed by flow cytometric analysis, but also without correlation to objective evidence of monocytic differentiation (14). A subset of AML that expresses CSF-1-R mRNA also expresses mRNA transcripts encoding CSF-1 (14). Since coexpression of these two genes in mouse fibroblasts confers the transformed phenotype (15), such leukemic cells may have an autocrine mechanism of growth, in which the CSF-1 produced by the cell interacts with its receptor, expressed by the same cell, and is autostimulatory.

The viral cognate of the CSF-1-R gene is activated by point mutations in the segment encoding the extracellular region and a deletion resulting in the truncation of the cytoplasmic domain (16, 17). Furthermore, virally-induced AML in mice may be associated with a viral integration in the CSF-1-R gene that increases its expression (18). No such rearrangements have been detected in the analogous regions of this gene in cases of CSF-1-R-positive human AML (13). There is disagreement about whether activating mutations in the CSF-1-R gene occur in myeloid malignancies in humans. While one group failed to demonstrate activating point mutations in AML (19), another study found that six of 30 patients with chronic myelomonocytic leukemia and seven of 24 patients with AML with monocytic differentiation (French-American-British group M4: 5/22, M5: 2/2) had genetic evidence of an activating mutation (20).

Although CSF-1 and its receptor have not been clearly implicated in the pathogenesis of AML, the expression of CSF-1-R appears to discriminate between AML and ALL. Flow cytometric analysis with monoclonal antibodies to this receptor may help to identify cases of AML having the potential for an autostimulatory growth mechanism by CSF-1 production.

A second receptor-ligand system of relevance to normal and leukemic hematopoietic cells consists of c-*kit* and its ligand, designated stem cell factor. The v-*kit* gene encodes the oncoprotein of the HZ4-feline sarcoma virus (21). Its cellular homologue, c-*kit*, encodes a product with the predicted topologic features of a transmembrane receptor with tyrosine-activity (22). The locus encoding c-*kit* was found to be deleted in mice with the white spotting locus (W), associated with defects in the development of hematopoietic cells, germ cells, and melanocytes of neural crest origin (23). The mice are anemic and have impairments in the development of hematopoietic progenitors. A similar phenotype is evident in mice with the steel (Sl) locus, which encodes the ligand for c-*kit* (24).

A monoclonal antibody developed by immunization with leukemic cells from a patient with AML (French-American-British group M1 type) (25) has been shown to bind an epitope expressed in the extracellular domain of the 150-Kd glycoprotein encoded by c-*kit* and to block its interaction with stem cell factor (26). This antibody, designated YB5.B8 has novel clinical and biological features. Flow cytometric analysis of normal bone marrow with YB5.B8 reveals that a small subpopulation of bone marrow mononuclear cells, constituting up to 3%, express c-*kit* (27). Hematopoietic progenitors that give rise to granulocyte, macrophage, granulocyes can be isolated by selecting for binding to YB5.B8. Differentiated hematopoietic precursors, which constitute the majority of bone marrow cells, lack expression of c-*kit*, as predicted by the low levels of c-*kit* mRNA transcripts detected in normal marrow (26). The relevance of this receptor-ligand system in early events of hematopoiesis is evidenced by the finding that coincubation of this monoclonal antibody in hematopoietic progenitor assays resulted in partial inhibition of the ability of crude hematopoietin preparations to promote colony formation (27). Further studies with recombinant colony stimulating factors show that YB5.B8 inhibits the action of granulocyte-macrophage colony-stimulating factor (GM-CSF) on hematopoietic stem cells, but does not alter the ability of granulocyte-stimulating factor (G-CSF) or interleukin (IL)-3 to induce colony formation (28).

RNA transcripts encoding the c-*kit* receptor protein were demonstrable in 13 of 19 (68%) of patients with AML (26). Rearrangements of the c-*kit* gene have not been detected. However, flow cytometric analysis of c-*kit* expression by indirect immunofluorescence indicated that only 22 of 71 (31%) patients bound YB5.B8 (29). There was a strong association between c-*kit* expression and an unfavorable clinical prognosis, as evidenced by the lower rate of induction of complete remission (four of 16 in the c-*kit*-positive group; 28 of 34 in the c-*kit*-negative group) and the shorter interval of survival. This factor appeared to be independent of other features known to be associated with a poor prognosis in AML, such as age at diagnosis, blood leukocyte count, and the presence of a preleukemic state. While the numbers of patients studied are small, this preliminary observation raises the possibility that c-*kit* and stem cell factor may play a role in the pathogenesis or evolution of the outcome of AML.

Another growth factor receptor of the tyrosine-kinase family that has been implicated as a determinant of biological behavior in human tumors is the epidermal growth factor receptor, the cellular counterpart of the *erb*-B oncogene. This cell surface molecule expressed by nonhematopoietic cells can also be analyzed by flow cytometry, since numerous monoclonal antibodies to epitopes in the extracellular domain are available. It should be noted that some of these monoclonal antibodies demonstrate reactivity with blood group A oligosaccharide and are not suitable for tumor typing (30). The major drawback to this approach is the difficulty in preparing cell suspensions from solid tumors. In spite of this logistical obstacle, preliminary clinical data regarding the expression of this receptor will be briefly discussed.

The EGF-R functions as the plasma membrane receptor structure for both epidermal growth factor and transforming growth factor (review, 31). The latter molecule is frequently secreted by malignant tumors and can confer the transformed phenotype to cells that express EGF-R (32). The transforming potential of the gene encoding this receptor may be activated by amplification; increased expression has been reported in glioblastoma multiforme (33) and mammary carcinoma (34). Amplification of the EGF-R gene is often accompanied by mutations that result in the expression of receptor molecules that are structurally and functionally altered (35). It is of note that high-grade gliomas may express high levels of transforming growth factor α, an activating ligand of the EGF-R. The clinical significance of overexpression of EGF-R, with or without coexpression of transforming growth factor α is difficult to assess in glioblastoma multiforme, since this is typically a rapidly progressive disease.

The potential prognostic importance of EGF-R expression has evolved to a clearer focus in mammary carcinoma. Initial studies have been performed using ligand-binding analysis of solubilized membrane fractions by the method of Scatchard and immunocytochemical analysis of frozen sections of tumors. Mammary carcinomas positive for EGF-R had other factors associated with a poor prognosis (34). Expression of EGF-R was positively correlated with increasing size, the presence of metastases to axillary lymph nodes, and anaplasia, as measured by nuclear grade and inversely correlated with the presence of estrogen receptors (36). More importantly, patients with EGF-R-positive mammary carcinomas had a shorter relapse-free interval and overall survival than those with EGF-R-negative tumors that achieved stastical significance. Multivariate analysis showed that EGF-R status was statically significant and second only to nodal status in prognostic power. Among patients lacking axillary metastases, the prognostic significance of EGF-R expression was maintained, indicating that this may be an important parameter in determining which of these patients require further therapy after resection.

This is in contrast to the finding with the *neu* (also designated HER-2 and c-*erb*-B-2) oncogene. This gene, which also encodes a transmembrane receptor with a cytoplasmic tyrosine-kinase domain, is closely related to the EGF-R at the level of primary structure (37). A putative ligand for this receptor has recently been identified (38), but has not yet been well characterized. Initial studies indicated that patients with amplified (>3) copies of the *neu* gene determined by Southern blotting have a high incidence of recurrence and a shorter survival (39). This finding could not be uniformly reproduced and subsequent immunocytochemical analysis of expression of the *neu* gene product showed that expression was correlated with a poor prognosis only in patients with metastases to regional lymph nodes and was not predictive of clinical outcome in node-negative patients (40). However, the *neu* gene has been implicated in the pathogenesis of mammary carcinoma in elegant animal studies. Transgenic mice carrying a *neu* gene activated by a point mutation in the transmembrane spanning domain, under the transcriptional control of elements that direct expression in mammary tissues, had a high incidence of mammary carcinomas (41). Thus, it is conceivable that activation of the *neu* gene could play a role in the pathogenesis of selected cases of mammary carcinoma. Expression of the *neu* gene product has been analyzed by immunocytochemical analysis of frozen and paraffin-embedded tissues and flow cytometric analysis (personal observation) using commercially available monoclonal antibodies. A significant correlation between cell surface *neu* expression, but not cytoplasmic staining, detected by immunocytochemistry, and an unfavorable clinical course has been demonstrated (42). This suggests that flow cytometry may be an effective approach to measure *neu* oncoprotein expression on the plasma membrane. Thus, there is an evolving rationale for the analysis of cell surface growth factor receptors in cell suspensions from mammary carcinomas in parallel with flow cytometric analysis of DNA content.

A second family of growth factor receptors, designated the cytokine-receptor family (43), that transduce the signaling events of several hemopoietins has become of increasing interest in normal and leukemic/lymphomatous hematopoietic cells. This family includes receptors for G-CSF (44), GM-CSF (45), IL-2 (46), IL-3 (47), IL-4 (48), IL-6 (49), IL-7 (50), and erythropoietin (51), as well as prolactin and growth hormone (52). The features shared by these receptors, as schematically represented in Figure 35.1, include the presence of characteristically spaced cysteine residues in the amino terminal region, presumed to be involved in ligand binding (which contribute to a shared secondary conformation); a sequence motif encoding tryptophan-serine-X-tryptophan-serine (X is any amino acid) near the transmembrane spanning region; a fibrinonectin type III module, also associated with cytoadhesion molecules and presumed to be involved in protein-protein interactions; and the requirement for more than one polypeptide chain to form a biologically active receptor complex. Receptors of this group lack both homology to the immunoglobulin gene superfamily and a tyrosine-kinase domain. The study of these receptors has been impeded by the low level of expression by normal and malignant cells (53). They have been identified and the genes molecularly cloned using recombinant cytokines as ligands in many instances. However, the binding proteins encoded by the cloned cDNAs are insufficient to encode a biologically active receptor. In general, the cDNAs encode cell surface glycoproteins that bind ligand at low affinity, fail to undergo ligand-induced internalization, and do not mediate a signal to stimulate growth. The complementary component that reconstitutes an active receptor complex has been identified in the case of IL-2 (46) and IL-6 (54) and, with relative certainty, for GM-CSF (55). The mechanism for ligand-induced signal transduction for these receptors is undetermined. Preliminary evidence suggests that the cytoplasmic portion of the ligand-binding proteins are not critical for signaling (54) and that the second subunit is the transducer of messages to cytoplasmic circuitry (56).

There are several mechanisms by which members of the cytokine-receptor-gene family have been implicated in hematologic malignancies. A recent report described a novel murine retrovirus that causes a distinctive myeloproliferative disorder. Infection with myeloproliferative leukemia virus (MPLV) leads to the development of an acute leukemia in adult mice, which is accompanied by a progressive polycythemia (57). When analyzed in hematopoietic progenitor assays, erythroid differentiation occured without the addition of erythropoietin (Epo) and was not abrogated by the addition of neutralizing antibodies to Epo, indicating that the mechanism responsible for leukemic transformation and polycythemia was not autocrine production of Epo. Moreover, the content of erythroid progenitors was also increased. Analysis of the MPLV genome revealed that a truncated gene encoding a member of the cytokine-receptor family was transduced into the viral envelope gene (58). This indicates that mutations in cytokine-receptor genes may encode products capable of immortalizing hematopoietic progenitors, rendering them independent of CSF activity for proliferation and allowing them to escape normal mechanisms that regulate proliferation. Indeed, this notion has been corroborated by the finding that point mutations can activate the Epo receptor to confer immortalization of cells without stimulation by ligand (59).

The potential involvement of the CSF-cytokine receptor system in the pathogenesis of myeloproliferative disorders is further evidenced by the finding that hematopoietic stem cells (HSC) from patients with two clonal disorders of multipotential progenitors show an exaggerated response to CSFs. Polycythemia vera (PV) is a clonal disease of HSC that has hyperplasia of predominantly erythroid, but also granulocytic and megakaryocytic lineages. Several studies have shown that erythroid precursors in PV are approximately 10-fold hypersensitive to (60–62), or independent of (63, 64), Epo. Subsequent analyses demonstrate that erythroid precursors express normal numbers of low-affinity Epo receptors, but are deficient of high-affinity receptors (65). HSC in PV are also 100-fold hypersensitive to the effect of IL-3 (66), a hematopoietin that stimulates proliferation of early progenitors and trilineage hematopoiesis, but analyses of number and affinity of IL-3 receptors have not yet been reported.

Juvenile chronic myelogenous leukemia (JCML) is characterized by differentiation abnormalities in the three myeloid lineages: neutrophilia, expression of fetal hemoglobin, and thrombocytopenia. This variant lacks the t(9;22) translocation, but is a clonal, trilineage expansion in which progenitors of granulocytes and macrophages may proliferate spontaneously in vitro. This spontaneous proliferation is dependent on the presence of adherent cells and is mediated by GM-CSF (67). Since JCML monocytes do not produce levels of GM-CSF (68), the finding that hematopoietic stem cells in JCML are 10-fold hypersensitive to GM-CSF (69) is plausible. Similarly, a variety of experimental strategies have been employed to demonstrate that leukemic myeloblasts are stimulated to proliferate by CSFs (review, 70), expanding the clonogenic compartment.

The molecular defect in hematopoietic proliferative disorders that show hypersensitivity to CSFs could be localized to the receptor structure, cytosolic signaling circuitry, and nuclear targets that control transcription and entry into the cell cyle. To date, no immunological reagents are available for members of the cytokine receptor family, with the exception of the IL-2 receptor. However a novel system for the analysis of a repertoire of cytokine receptors, designated Fluorokine assay, is commercially available. It exploits the availability of recombinant CSFs, which are modified with biotin and detected with fluoresceinated avidin or directly conjugated with phycoerythrin. Although many of the CSF receptors are expressed in extremely low levels on normal and leukemic hematopoietic cells, as determined by radioligand-binding studies, the level of fluorochrome amplification achieved can be detected by the current generation of flow cytometers. It is important to verify the reactivity to the fluorokine reagents by competitive inhibition with unlabeled cytokine and multiparameter analysis of leukocyte subpopulations discriminated by light-scatter properties to confirm the pattern of reactivity with normal cells.

Since several studies have demonstrated that GM-CSF promotes the growth of clonogenic cells among leukemic myeloblasts, we used the fluorokine assay to detect GM-CSF receptors in cases of acute myelogenous leukemia. Receptor structures were demonstrated on leukemic myeloblasts in 82% (nine of 11) of cases of AML by multiparameter analysis (Martin AW and Peiper SC, unpublished observations). Representative histograms are shown in Figures 35.2 and 35.3. The identity of the leukemic population was confirmed by analyzing forward-angle and orthogonal light scatter properties to subdivide the events, which were analyzed by electronic bit-mapped gating for fluorescence analysis. Leukemic cells were positively identified among cells in the selected light-scatter gates by coexpression of CD33. Immunofluorescence staining with monoclonal antibodies may artifactually inhibit the fluorokine assay, so it is critical to perform fluorokine binding prior to incubations with immunologic reagents. Quantitative analysis of cytokine-receptor expression may help to dissect the mechanism for CSF hypersensitivity in myeloproliferative disorders by determining whether there is hyperexpression of CSF receptors, which is, potentially, the result of increased transcription due to a genetic rearrangement, translocation, or amplification.

The CD45 molecule, also known as the leukocyte common antigen, is the prototype of a novel, rapidly expanding class of transmembrane glycoproteins expressed on the plasma membrane that have cytoplasmic domains with phosphatase activity (71). It is speculated that the extracellular domain, which is antigenically distinguishable in different leukocyte subsets, can bind to specific ligands, which results in the activation of the phosphatase activity of the cytoplasmic tail. CD45 molecules dephosphorylate, thus activating the tyrosine kinase activity of, *lck* (72), a cytoplasmic peptide involved in signal transduction during T-lymphocyte activation by antigen. T-lymphocytes that lack CD45 are unable to undergo activation by antigen. Upon reconstitution of CD45 activity, the ability to proliferate in response to antigenic stimulation is restored (73). It is postulated that basal activities of receptor phosphatases could suppress the action of tyrosine-kinases, thus antagonizing proliferation and activation, the generic function of a tumor suppressor gene.

Nucleic acid probes derived from the CD45 gene have been used to molecularly clone genes encoding other members of this human gene family. One new member of the family has been localized to a region of human chromosome 3 that is frequently deleted in pulmonary and renal cell carcinomas (74). Analysis of the gene encoding this receptor phosphatase indicates that it is frequently deleted in carcinomas of the lung and in cell lines derived from renal cell carcinoma. If the transformed phenotype is reversed in carcinoma cell lines following genetic reconstitution of the receptor phosphatase activity, there would be presumptive evidence that this is a tumor suppressor gene. Such a finding could engender a high level of interest in the analysis of cell surface receptor phosphatase molecule expression by tumor cells.

CYTOSOLIC SIGNAL TRANSDUCTION

Multiple cytoplasmic constituents have been implicated in the transduction of mitogenic signals from the plasma membrane to the nucleus. The most extensively studied, the polypeptides encoded by the *ras* family of genes, were the first products of activated protooncogenes to be identified in human tumors by a murine fibroblast transforming assay (75, 76). Polypeptides encoded by *ras* genes are localized to the internal aspect of the plasma membrane and bind and cleave guanosine triphosphate (GTP) (77). Oncogenic forms of *ras* genes contain point mutations (78–80) that encode proteins capable of binding GTP, the so-called GTPase activity but have an impaired ability to cleave GTP (81). Thus, they are constitutively activated by GTP binding and are unable to return to an inactive state. Activating mutations in *ras* genes have been consistently found in a variety of types of human tumors, including carcinomas of the colon, lung, pancreas, thyroid, and genitourinary tract; in melanomas;

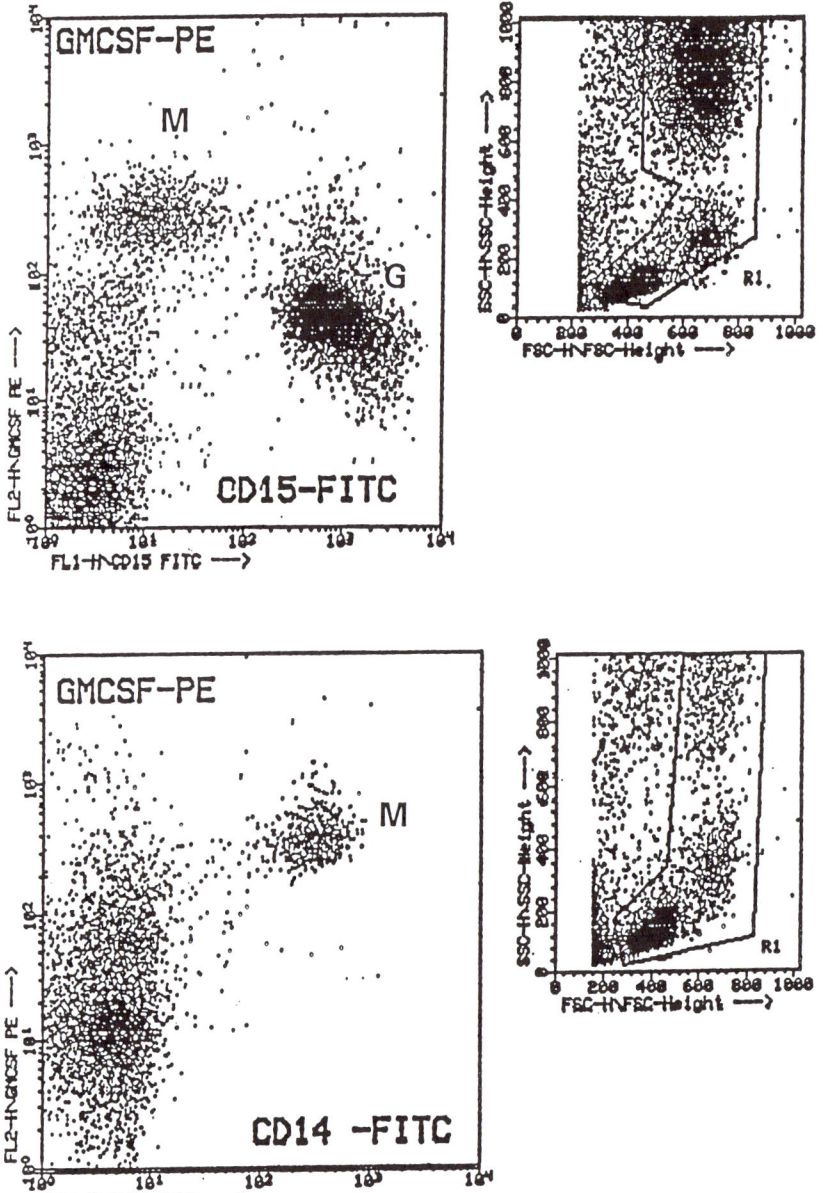

Figure 35.2. Characterization of normal leukocytes by simultaneous fluorokine and immunofluorescence analysis. Flow cytometric analysis of cells sequentially stained with phycoerythrin-labeled (PE) GM-CSF and a fluorescein-labeled CD15 monoclonal antibody (*upper panel*) reveals distinct populations that include granulocytes (G), which have intermediate intensity fluorescence for GM-CSF and bright fluorescence for CD15. Monocytes (M) have dim expression of CD15, but high levels of GM-CSF-R. Dual analysis with the GM-CSF fluorokine and a CD14 monoclonal antibody (*lower panel*) confirms that monocytes (M) express high levels of the GM-CSF-R. (Material kindly provided by Dr. Dan Collins, R & D Systems.)

and in acute lymphoblastic and myeloblastic leukemias (review, 1). High levels of mRNA transcripts encoding *ras* genes have also been demonstrated in leukemias and lymphomas (82). An interrelationship between point mutations and high levels of mRNA transcripts has not been firmly established. Immunologic reagents have been prepared to normal and oncogenic *ras* oncoproteins. Several groups have employed these reagents to measure *ras* oncoprotein levels in hematopoietic malignancies by flow cytometry (83, 84). In contrast to immunoblotting analysis of protein and Northern analysis for mRNA levels, which examine extracts of heterogeneous tumor populations, analytic strategies that employ flow cytometry permit the analysis of individual, detergent-permeabilized cells that can be confirmed to be tumor cells by multiparameter analysis. Moreover, concomitant DNA analysis can allow for the quantitation of *ras* oncoproteins in aneuploid stemlines and at varying stages of the cell cycle. Preliminary evidence suggests that, in plasma cell myeloma, the expression of high levels of *ras* oncoproteins is more frequently found in cases with an aneu-

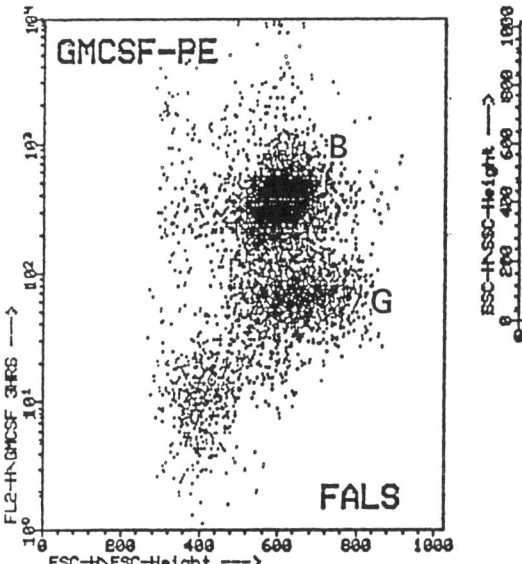
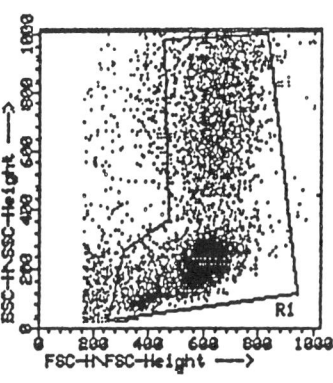

Figure 35.3. Characterization of leukemic myeloblasts by fluorokine assay. Representation of histogram data as GM-CSF-binding versus size discriminates three subpopulations: myeloblasts (B) with intense activity, differentiated granulocytes (G) with intermediate levels of ligand binding, and residual lymphoid cells that lack activity. (Material kindly provided by Dr. Dan Collins, R & D Systems.)

ploid stemline than in those with a diploid DNA content. This high-level expression was found to be associated with a poor prognosis for survival in the small group of patients studied (85).

A GTPase-activating protein (GAP), which binds to the ras oncoprotein-GTP complex, is the only molecule identified to date that may be necessary for the transforming activity of ras proteins. The cloning of the neurofibromatosis type I gene revealed that it encodes a GAP protein (86) that functions as a recessive oncogene. Subsequently, by translocation in the Philadelphia chromosome (87), the bcr gene—on chromosome 22, which is juxtaposed to the abl gene, located on chromosome 9—may also be a GAP protein (88). It is predicted that the characterization of GAP proteins in normal and malignant cells, perhaps by flow cytometry, will be an area of increasing interest.

NUCLEAR FACTORS

Recent evidence has implicated the involvement of protooncogene-encoded nuclear polypeptides in the regulation of transcription (review, 89). The cellular cognates of several retroviral oncogenes that induce lymphohematopoietic malignancies in infected animals (i.e., c-rel, c-myb, c-myc, c-ets) are expressed by normal and leukemic hematopoietic progenitors (90–93). In addition, a group of cellular protooncogenes encoding polypeptides with amino acid sequences predictive of DNA-binding proteins have been activated in acute leukemias by juxtaposition to the T-cell receptor gene through a translocation (94), the formation of a chimeric transcriptional regulatory protein through a translocation (95, 96), and by the insertional activation by retroviral integration (97). The expression of nuclear protooncogene products may be induced by factors that activate resting cells. Their requirement for differentiation and maturation is evidenced by the inhibition of hematopoietic colony formation when c-myb expression is molecularly abrogated by antisense c-myb oligonucleotides (98), and by the inhibition of pharmacologically-induced terminal differentiation in HL-60 cells (99) and phytohemagglutinin-induced lymphocyte blastogenesis (100) by antisense c-myc oligonucleotides.

The vast majority of experiments to determine the expression of nuclear polypeptides encoded by protooncogenes have been performed using oligonucleotide probes for Northern blot or in situ hybridization analysis of mRNA transcripts. The permeabilization techniques employed to quantitate the expression of the proliferating cell nuclear antigen (PCNA) by flow cytometric immunofluorescence analysis (101) have empowered the quantitation of several other nuclear polypeptides. Kastan and coworkers (102) have used monoclonal antibodies to demonstrate by dual-parameter immunofluorescence analysis that CD34-positive hematopoietic progenitors express the products of c-myb, c-myc, and c-fos. Lymphocytes and monocytes expressed lower levels of the polypeptides encoded by c-myc and c-myb than did progenitors, whereas granulocytes did not contain detectable levels. This supports the hypothesis that the levels of oncoproteins expressed by hematopoietic cells are related to their program of differentiation and maturation. While monoclonal antibodies to nuclear oncoproteins may not prove to be useful typing reagents for assessing hematopoietic lineage, they may provide valuable insight into the capability of acute leukemias to differentiate in response to recombinant cytokines and CSFs and the entry of chronic leukemias into accelerated phases. Cyclins, first described in clams and subsequently in yeast prior to their characterization in mammalian cells, are nuclear proteins that control progression

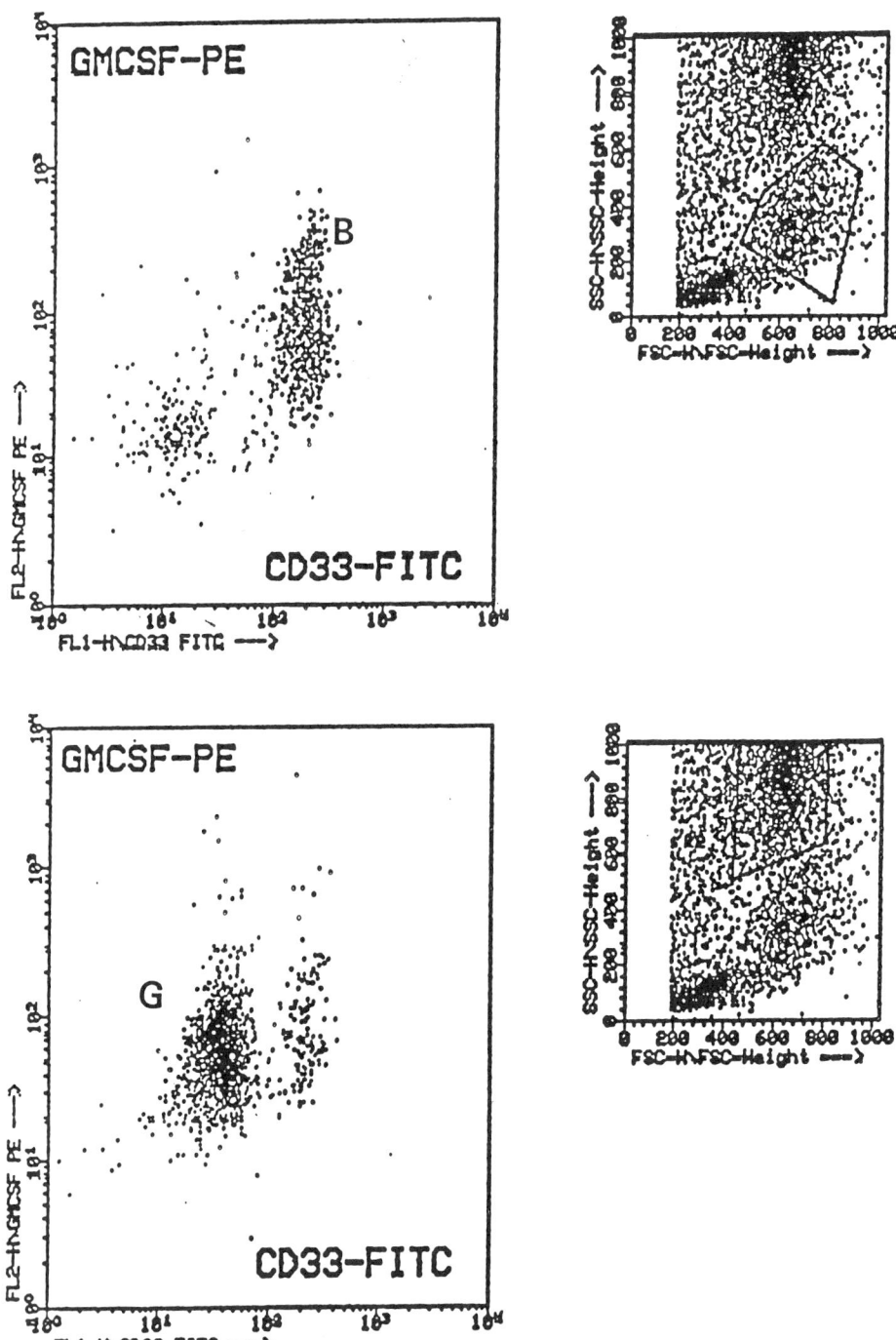

Figure 35.4. Dual-parameter analysis of leukemic myeloblasts. Bone marrow cells from a patient with AML were stained sequentially with GM-CSF-PE and CD33-FITC. Subpopulations were resolved by electronic bit-mapped gating of forward versus orthogonal light-scatter properties for fluorescence analysis. Display of cells with light-scatter properties characteristic of leukemic myeloblasts (B) have high level expression of GM-CSF-R and CD33, as shown in the *upper panel*. Those with light-scatter properties of differentiated granulocytes (G), as shown in the *lower panel*, show dim expression of GM-CSF-R and CD33. (Material kindly provided by Dr. Dan Collins, R & D Systems.)

through the cell cycle (reviews, 103, 104). The unique feature that characterizes cyclins is precipitous degradation synchronized with the phases of the cell cycle. Several categories of cyclins can be recognized, those associated with the G_1, S, and GM phases of the cell cycle. The effects of cyclins are mediated through their association with a protein kinase.

Recently, genes encoding cyclins have been associated with human malignancies. The PRAD1 gene, which is clonally rearranged by fusion with the parathyroid hormone gene in parathyroid adenomas, has homology to G_1 cyclin genes. It is overexpressed in parathyroid adenomas (105) and amplified in 15% to 20% of mammary and squamous cell carcinomas. PRAD1 is located in the q13 region of the long arm of chromosome 11, which is the site of the t(11;14) translocation that is frequently seen in chronic lymphocytic leukemia. Preliminary evidence indicates that PRAD1 is near the breakpoint on chromosome 11q, designated *bcl*-1, which is juxtaposed to the immunoglobulin heavy chain gene locus on chromosome 14. PRAD1 mRNA transcripts are expressed in high levels in the B-cell leukemias that have the translocation. Thus, the overexpression of a cyclin may result in the disruption of the cell cycle in an indolent lymphoproliferative disorder characterized by increased production of cells that typically have an extremely small proportion of cells in the S and G_2/M phases of the cell cycle.

The gene encoding p53 is a tumor suppressor gene that is currently recognized as the most frequent site of mutations in human cancers (review, 106). The p53 polypeptide is a negative regulator of the cell cycle. Inactivation of p53 by mutation or interaction with oncoproteins encoded by DNA tumor viruses can result in malignant transformation. Reduction to homozygosity with abnormalities in both p53 alleles frequently occurs in colonic carcinomas; in cancers of the lung, brain, liver, and breast; and in chronic myelogenous leukemia during blast crisis. All members of six families with the Li-Fraumeni syndrome—a rare familial condition associated with a high rate of malignancy of various organs—who have been studied, carry point mutations in the p53 gene on one of their chromosomes (107). Thus, inheritance of a mutation in the p53 gene may predispose a person to develop malignancy. The point mutations that have been detected are located in highly conserved regions of the p53 gene and those occuring in specific tumors are tightly clustered. Mutant p53 molecules have increased stability and are therefore present in high levels in transformed cells. The abnormal forms are not transported to the nucleus and accumulate in the cytoplasm. These characteristics should allow the use of available monoclonal antibodies to p53 to identify tumors with p53 mutations based on high-level, (cytoplasmic) expression. This could be accomplished by flow cytometric analysis on differentially permeabilized and solubilized cells.

SUMMARY

Molecular genetics has become the vanguard technology for the characterization of human tumors. The advances have been rapid and have provided valuable insights into the mechanisms of malignant transformation. Unlike phenotypic analysis for the description of cellular differentiation, many of the genetic changes are specific for malignancy and more strongly associated with clinical outcome.

However, the future of flow cytometry is not dim. It provides the single most powerful tool for the analysis of individual cells. It is dependent however, upon the repertoire of available reagents. Since monoclonal antibodies are the reagents best suited for flow cytometric analysis, it will be critical to utilize novel antibodies directed against the key molecules that regulate cellular proliferation and differentiation because multiple parameters can be measured simultaneously! For the most part these molecules have been recognized as the products of oncogenes and tumor suppressor genes through progress in molecular biology. By updating the repertoire of reagents that recognize critical cellular targets, flow cytometric studies will continue to play a central role in tumor characterization that will complement molecular genetic analyses.

ACKNOWLEDGMENTS

The authors wish to thank Dr. Samuel R. Wellhausen and LouAnn Eskildsen for their advice and assistance. We thank Judy Hollkamp for her help in preparing the manuscript.

REFERENCES

1. Bishop JM. Molecular themes in oncogenesis. Cell 1991;64:235–248.
2. Marshall CJ. Tumor suppressor genes. Cell 1991;64:313–326.
3. Downward J, Yarden Y, Mayes E, et al. Close similarity of epidermal growth factor receptor and v-erb-B oncogene protein sequences. Nature 1982;307:521–527.
4. Sherr CJ, Rettenmeir CW, Sacca R, et al. The c-fms proto-oncogene product is related to the receptor for the mononuclear phagocyte growth factor, CSF-1. Cell 1985;41:665–676.
5. Huang E, Nocka K, Beier DR, et al. The hematopoietic growth factor KL is encoded by the Sl locus and is the ligand of the c-kit receptor, the gene product of the W locus. Cell 1990;63:225–233.
6. Williams DE, Eisenman J, Baird A, et al. Identification of a ligand for the c-kit proto-oncogene. Cell 1990;63:167–174.
7. Zsebo M, Williams DA, Geissler EN, et al. Stem cell factor is encoded at the Sl locus of the mouse and is the ligand for the c-kit tyrosine kinase receptor. Cell 1990;63:213–224.
8. Stanley ER. Role of colony stimulating factor-1 in monocytopoiesis and placental development. In: Mahowald AE, ed. Genetics of pattern formation and growth control. New York: Wiley Liss 1990;165–180.
9. Wiktor-Jedrzejczak W, Bartocci A, Ferrante AW Jr, et al. Total absence of colony-stimulating factor 1 in the macrophage-deficient osteopetrotic (op/op) mouse. Proc Natl Acad Sci USA 1990;87:4828–4832.
10. Yoshida H, Hayashi S-I, Kunisada T, et al. The murine mutation osteopetrosis is in the coding region of the macrophage colony stimulating factor gene. Nature 1990;345:442–444.
11. Pollard JW, Bartocci A, Arceci R, Orlofsky A, Ladner MB, Stanley ER. Apparent role of the macrophage growth factor, CSF-1, in placental development. Nature 1987;330:484–486.
12. Yarden Y, Escobedo JA, Kuang WJ, et al. Structure of the receptor for platelet-derived growth factor helps define a family of closely related growth factor receptors. Nature 1986;323:226–232.
13. Ashmun RA, Look AT, Roberts WM, et al. Monoclonal antibodies to the human CSF-1 receptor (c-fms proto-oncogene product) detect epitopes on normal mononuclear phagocytes and on human myeloid leukemic blast cells. Blood 1989;73:827–837.
14. Rambaldi A, Wakamiya N, Vallenga E, et al. Expression of the macrophage colony stimulating factor and c-fms genes in human acute myeloblastic leukemia cells. J Clin Invest 1988;81:1030–1035.
15. Roussel MF, Dull TJ, Rettenmier CW, Ralph P, Ullrich A, Sherr CJ. Transforming potential of the c-fms proto-oncogene (CSF-1 receptor). Nature 1987;325:549–555.
16. Roussel MF, Downing JR, Rettenmier CW, Sherr CJ. A point mutation in the extracellular domain of the human CSF-1 receptor (c-fms

proto-oncogene product) activates its transforming potential. Cell 1988;55:979–988.
17. Woolford J, McAuliffe A, Rohrschneider LR. Activation of the feline c-fms proto-oncogene: Multiple alterations are required to generate a fully transformed phenotype. Cell 1988;55:965–977.
18. Gisselbrecht S, Fichelson S, Sola B, et al. Frequent c-fms activation by proviral insertion in mouse myeloblastic leukaemias. Nature 1987;329:259–261.
19. Sherr CJ. Colony-stimulating factor-1 receptor. Blood 1990;75:1–12.
20. Ridge SA, Worwood M, Oscier D, Jacobs A, Padua RA. FMS mutations in myelodysplastic, leukemic and normal subjects. Proc Natl Acad Sci USA 1990;87:1377–1380.
21. Besmer P, Murphy JE, George PC, et al. A new acute transforming feline retrovirus and relationship of its oncogene v-kit with the protein kinase gene family. Nature 1986;320:415.
22. Qui F, Ray P, Brown K, et al. Primary structure of c-kit. Relationship with the CSF-1/PDGF receptor kinase family-oncogene activation of v-kit involves deletion of extracellular domain and C terminus. EMBO J 1988;7:1003–1010.
23. Chabot B, Stephenson DA, Chapman VM, Besmer P, Bernstein A. The proto-oncogene c-kit encoding a transmembrane tyrosine kinase receptor maps to the mouse W locus. Nature 1988;335:88-89.
24. Flanagan JG, Leder P. The kit ligand: a cell surface molecule altered in steel mutant fibroblasts. Cell 1990;63:185–194.
25. Gadd SJ, Ashman LK. A murine monoclonal antibody specific for a cell-surface antigen expressed by a subgroup of human myeloid leukaemias. Leukemia Res 1985;9:1329–1333.
26. Lerner NB, Nocka KH, Cole SR, et al. Monoclonal antibody YB5.B8 identifies the human c-kit protein product. Blood 1991;77:1876–1883.
27. Cambareri, Ashman LK, Cole SR, Lyons AB. A monoclonal antibody to a human mast cell/myeloid leukaemia-specific antigen binds to normal haemopoietic progenitor cells and inhibits colony formation in vitro. Leukemia Res 1988;12:929–939.
28. Ashman LK, Cambareri, Eglinton JM. A monoclonal antibody that inhibits the action of GM-CSF on normal but not leukaemic progenitors. Leukemia Res 1990;14:637–644.
29. Ashman LK, Roberts MM, Gadd SJ, Cooper SJ, Juttner CA. Expression of a 150-kD cell surface antigen identified by monoclonal antibody YB5.B8 is associated with poor prognosis in acute non-lymphoblastic leukaemia. Leukemia Res 1988;12:923–928.
30. Jones NR, Rossi ML, Gregorion M, Hughes JT. Epidermal growth factor receptor expression in 72 meningiomas. Cancer 1990;66:152–155.
31. Gill GN, Bertics PJ, Stanton JB. Epidermal growth factor and its receptor. Mol Cell Endocrinol 1987;51:169–186.
32. Goustin AS, Leof EB, Shipley GD, Moses HL. Growth factors and cancer. Cancer Res 1986;46:1015–1029.
33. Wong AJ, Bigner SH, Bigner DD, Kinzler KW, Hamilton SR, Vogelstein B. Increased expression of the epidermal growth factor receptor gene in malignant gliomas is invariably associated with gene amplification. Proc Natl Acad Sci USA 1987;84:6899–6903.
34. Sainsbury JR, Farndon JR, Needham GK, Malcolm AJ, Harris AL. Epidermal-growth-factor receptor status as predictor of early recurrence of and death from breast cancer. Lancet 1987;1398–1402.
35. Yamazaki J, Fukai Y, Ueyama Y, et al. Amplification of the structurally and functionally altered epidermal growth factor receptor gene (c-erbB) in human brain tumors. Mol Cell Biol 1988;3:1816–1820.
36. Delarue JC, Friedman S, Mouriesse H, May-Levin F, Sancho-Garnier, Contesso G. Epidermal growth factor receptor in human breast cancers: correlation with estrogen and progesterone receptors. Breast Cancer Res and Treat 1988;11:173–178.
37. Bargmann CI, Hung M-C, Weinberg RA. The neu oncogene encodes an epidermal growth factor receptor-related protein. Nature 1986;319:226–230.
38. Lupu R, Colomer R, Zugmaier G, et al. Direct interaction of a ligand for the erbB2 oncogene product with the EGF receptor and p185-erbB2. Science 1990;249:1552-1555.
39. Slamon DJ, Clark GM, Wong SG, Levin WJ, Ullrich A, McGuire WL. Human breast cancer: correlation of relapse and survival with amplification of the HER-2/neu oncogene. Science 1987;235:177–182.
40. Slamon DJ, Godolphin W, Jones LA, et al. Studies of the HER-2/neu proto-oncogene in human breast and ovarian cancer. Science 1989;244:707–713.
41. Muller WJ, Sinn E, Pattengale PK, Wallace R, Leder P. Single-step induction of mammary adenocarcinoma in transgenic mice bearing the activated c-neu oncogene. Cell 1988;54:105–115.
42. De Potter CR, Beghin C, Makar AP, Vandekerckhove D, Roels HJ. The neu-oncogene protein as a predictive factor for haematogenous metastases in breast cancer patients. Int J Cancer 1990;45:55–58.
43. D'Andrea AD, Fasman GD, Lodish HF. Erythropoietin receptor and interleukin-2 receptor beta chain: a new receptor family. Cell 1989;58:1023–1024.
44. Fukunaga R, Ishizaka-Ikeda E, Seto Y, Nagata S. Expression cloning of a receptor for murine granulocyte colony-stimulating factor. Cell 1990;61:341–350.
45. Gearing DP, King JA, Gough NM, Nicola NA. Expression cloning of a receptor for human granulocyte-macrophage colony-stimulating factor. EMBO J 1989;8:3667–3676.
46. Hatakeyama M, Tsudo M, Minamoto S, et al. Interleukin-2 receptor beta chain gene: generation of the three receptor forms by cloned human alpha and beta chain cDNAs. Science 1989;244:551–556.
47. Itoh N, Yonehara S, Schreurs J, et al. Cloning of an interleukin-3 receptor gene: a member of a distinct receptor gene family. Science 1990;247:324–327.
48. Mosley B, Beckmann MP, March CJ, et al. The murine interleukin-4 receptor: molecular cloning and characterization of secreted and membrane bound forms. Cell 1989;59:335–348.
49. Yamasaki K, Taga T, Hirata Y, et al. Cloning and expression of the human interleukin-6 (BSF-2/IFNB2) receptor. Science 1988;241:825–828.
50. Goodwin RG, Friend D, Ziegler SF, et al. Cloning of the human and murine interleukin-7 receptors: demonstration of a soluble form and homology to a new receptor superfamily. Cell 1990;60:941–951.
51. D'Andrea AD, Lodish HF, Wong GG. Expression cloning of the murine erythropoietin receptor. Cell 1989;57:277–285.
52. Bazan JF. A novel family of growth factor receptors. Biochem Biophys Res Commun 1989;164:788–796.
53. Urdal D, Price V, Sassenfeld HM, Cosman D, Gillis S, Park LS. Molecular characterization of colony stimulating factors and their receptors: human interleukin-3. In: Orlic D, ed. Molecular and Cellular Controls of Hematopoiesis. New York, Ann NY Acad Sci 1989:167–176.
54. Taga T, Hibi M, Hirata Y, et al. Interleukin-6 triggers the association of its receptor with a possible signal transducer, gp130. Cell 1898;58:573–581.
55. Kitamura T, Hayashida K, Sakamaki K, Yokota T, Arai K-i, Miyajima A. Reconstitution of functional receptors for human granulocyte/macrophage colony-stimulating factor (GM-CSF): evidence that the protein encoded by the AIC2B cDNA is a subunit of the murine GM-CSF receptor. Proc Natl Acad Sci USA 1991;88:5082–5086.
56. Hibi M, Murakami M, Saito M, Hirano T, Taga T, Kishimoto T. Molecular cloning and expression of an IL-6 signal transducer, gp130. Cell 1990;63:1149–1157.
57. Wendling F, Penciolelli JF, Charon M, Tambourin P. Factor-independent erythropoietic progenitor cells in leukemia induced by the myeloproliferative leukemia virus. Blood 1989;73:1161–1167.
58. Souyri M, Vigon I, Penciolelli J-F, Heard J-M, Tambourin P, Wendling F. A putative truncated cytokine receptor gene transduced by the myeloproliferative leukemia virus immortalizes hematopoietic progenitors. Cell 1990;63:1137–1147.
59. Yoshimura A, Longmore G, Lodish H. Point mutation in the exoplasmic domain of the erythropoietin receptor resulting in hormone-independent activation and tumorigenicity. Nature 1990;348:647–649.

60. Zanjani ED, Lutton JD, Hoffman R, Wasserman LR. Erythroid colony formation by polycythemia vera marrow in vitro. Dependence on erythropoietin. J Clin Invest 1977;59:841–844.
61. Golde DW, Bersch N, Cline MJ. Polycythemia vera. Hormonal modulation of erythropoiesis in vitro. Blood 1977;49:399–405.
62. Casadevall N, Vainchenker W, Lacombe C, et al. Erythroid progenitors in polycythemia vera. Demonstration of their hypersensitivity to erythropoietin using serum-free cultures. Blood 1982;59:447–451.
63. Eaves AC, Krystal G, Cashman JD, Eaves CJ. Polycythemia vera: in vitro analysis of regulatory defects. In: Zanjani ED, Tarassoli M, Ascensao TL, eds. Regulation of Erythropoiesis. New York: PMA Publishing Corp, 1988:523–535.
64. Eridani S, Dudley JM, Sawyer BM, Pearson TC. Erythropoietic colonies in a serum-free system: results in primary proliferative polycythemia and thrombocythemia. Br J Haematol 1987;67:387–391.
65. Means RT Jr, Krantz SB, Sawyer ST, Gilbert HS. Erythropoietin receptors in polycythemia vera. J Clin Invest 1989;84:1340–1344.
66. Dai CH, Krantz SB, Means RT Jr, Horn ST, Gilbert HS. Polycythemia vera blood burst-forming units-erythroid are hypersensitive to interleukin-3. J Clin Invest 1991;87:391–396.
67. Gualtieri RJ, Emanuel PD, Zuckerman KS, et al. Granulocyte-macrophage colony-stimulating factor is an endogenous regulator of cell proliferation in juvenile chronic myelogenous leukemia. Blood 1989;74:1360–1367.
68. Emanuel PD, Bates LJ, Zhu S-W, Castleberry RP, Gualtieri RJ, Zuckerman KS. GM-CSF dysregulation in juvenile chronic myelogenous leukemia. Blood 1990;76:267a.
69. Emanuel PD, Bates LJ, Castleberry RP, Gualtieri RJ, Zuckerman KS. Selective hypersensitivity to granulocyte-macrophage colony stimulating factor by juvenile chronic myeloid leukemia hematopoietic progenitors. Blood 1991;77:925–929.
70. Oster W, Mertelsmann R, Herrmann F. Role of colony-stimulating factors in the biology of acute myelogenous leukemia. Int J Cell Cloning 1989;7:13–29.
71. Tonks NK, Charbonneau H, Diltz CD, Fischer EH, Walsch KA. Demonstrations that the leukocyte common antigen CD45 is a protein Tyrosine Phosphate Biochemistry 1988;27:8695–8701.
72. Mustelin T, Coggeshall KM, Altman A. Rapid activation of the T-cell tyrosine protein kinase pp56^{lck} by the CD45 phosphotyrosine phosphatase. Proc Natl Acad Sci USA 1989;86:6302–6306.
73. Pingel JT, Thomas ML. Evidence that the leukocyte-common antigen is required for antigen-induced T lymphocyte proliferation. Cell 1989;58:1055–1065.
74. LaForgia S, Morse B, Levy J, et al. Receptor protein-tyrosine phosphatase gamma is a candidate tumor suppressor gene at human chromosome region 3p21. Proc Natl Acad Sci USA 1991;88:5036–5040.
75. Krontiris TG, Cooper GM. Transforming activity of human tumor DNAs. Proc Natl Acad Sci USA 1981;78:1181–1184.
76. Shih C, Padhy LC, Murray M, Weinberg RA. Transforming genes of carcinomas and neuroblastomas introduced into mouse fibroblasts. Nature 1981;290:261–264.
77. Scolnick EM, Papageorge AG, Shih TY. Guanine-nucleotide binding activity as an assay for src protein of rat-derived murine sarcoma viruses. Proc Natl Acad Sci USA 1979;76:5355–5359.
78. Reddy EP, Reynolds RK, Santos E, Barbacid M. A point mutation is responsible for the acquisition of transforming properties by the T24 human bladder carcinoma oncogene. Nature 1982;300:149–152.
79. Tabin CJ, Bradley SM, Bargmann CI, et al. Mechanism of activation of a human oncogene. Nature 1982;300:143–149.
80. Taparowsky E, Suard Y, Fasano O, Shimizu K, Goldfarb M, Wigler M. Activation of the T24 bladder carcinoma transforming gene is linked to a single amino acid change. Nature 1982;300:762–765.
81. Gibbs JB, Sigal IS, Poe M, Scolnick EM. Intrinsic GTPase activity distinguishes normal and oncogenic ras p21 molecules. Proc Natl Acad Sci USA 1984;81:5704–5708.
82. Shen WPV, Aldrich TH, Venta-Perez G, Franza BR Jr, Furth ME. Expression of normal and mutant ras proteins in human acute leukemia. Oncogene 1987;1:157–165.
83. Andreeff M, Slater DE, Bressler J, Furth ME. Cellular ras oncogene expression and cell cycle measured by flow cytometry in hematopoietic cell lines. Blood 1986;67:676–681.
84. Takeda T, Krause JR, Carey JL, McCoy JP Jr. Detection of the ras p21 gene product in human leukemias by flow cytometry. J Clin Lab Anal 1989;3:108–115.
85. Tsuchiya H, Epstein J, Selvanayagam P, et al. Correlated flow cytometric analysis of H-ras p21 and nuclear DNA in mutiple myeloma. Blood 1988;72:796–800.
86. Xu G, O'Connell P, Viskochil D, et al. The neurofibromatosis type 1 gene encodes a protein related to GAP. Cell 1990;62:599–608.
87. Heisterkamp N, Stephenson JR, Groffen J, et al. Localization of the c-abl oncogene adjacent to a translocation breakpoint in chronic myelocytic leukaemia. Nature 1983;306:239-242.
88. Diekmann D, Brill S, Garrett MD, et al. Bcr encodes a GTPase-activating protein for p21^{rac}. Nature 1991;351:400–402.
89. Lewin B. Oncogenic conversion by regulatory changes in transcription factors. Cell 1991;64:303–312.
90. Slamon DJ, deKiernon JB, Verma IM, Cline MJ. Expression of cellular oncogenes in human malignancies. Science 1984;224:256–262.
91. Thompson CB, Challoner PB, Nieman PE, Groudine M. Expression of the c-myb proto-oncogene during cellular proliferation. Nature 1986;319:374–380.
92. Emilia G, Donelli A, Ferrari S, et al. Cellular levels of mRNA from c-myc, c-myb, and c-fes onc-genes in normal myeloid and erythroid precursors of human bone marrow: an in situ hybridization study. Br J Haematol 1986;62:287–292.
93. Lee J, Mehta K, Blick MB, Gutterman JU, Lopez-Berenstein G. Expression of c-fos, c-myb, and c-myc in human monocytes: correlation with monocytic differentiation. Blood 1987;69:1542–1545.
94. Chen Q, Yang CY-C, Tsan JT, et al. Coding sequences of the tal-1 gene are disrupted by chromosome translocation in human T cell leukemia. J Exp Med 1990;172:1403–1408.
95. Kamps MP, Murre C, Sun X, Baltimore D. A new homeobox gene contributes to the DNA binding domain of the t(1;19) translocation protein in pre-B ALL. Cell 1990;60:547–555.
96. Nourse J, Mellentin JD, Galili N, et al. Chromosomal translocation t(1;19) results in synthesis of a homeobox fusion mRNA that codes for a potential chimeric transcription factor. Cell 1990;60:535–545.
97. Morisha K, Parker DS, Mucenski ML, Jenkins NA, Copeland NG, Ihle JN. Retroviral activation of a novel gene encoding a zinc finger protein in IL-3 dependent myeloid leukemia cell lines. Cell 1988;54:831–840.
98. Gewirtz AM, Calabretta B. A c-myb antisense oligodeoxynucleotide inhibits normal human hematopoiesis in vitro. Science 1988;242:1303–1306.
99. Holt JT, Redner RL, Nienhuis AW. An oligomer complementary to c-myc mRNA inhibits proliferation of HL-60 promyelocytic cells and induces differentiation. Mol Cell Biol 1988;8:963–973.
100. Heikkila R, Schwab G, Wickstrom E, et al. A c-myc antisense oligodeoxynucleotide inhibits entry into S phase but not progress from G0 to G1. Nature 1987;328:445–449.
101. Clevenger CV, Bauer KD, Epstein AL. A method for simultaneous nuclear immunofluorescence and DNA content quantitation using monoclonal antibodies and flow cytometry. Cytometry 1985;6:208–214.
102. Kastan MB, Stone KD, Civin CI. Nuclear oncoprotein expression as a function of lineage, differentiation stage, and proliferative status of normal human hematopoietic cells. Blood 1989;74:1517–1524.
103. Murray AW, Kirschner MW. Dominoes and clocks: the union of two views of the cell cycle. Science 1989;246:614–621.
104. North G. Cell cycle: Starting and stopping. Nature 1991;351:604–605.

105. Motokura T, Bloom LT, Kim HG, et al. A novel cyclin encoded by a bcl1-linked candidate oncogene. Nature 1991;350:512–515.
106. Levine AJ, Momand J, Finlay CA. The p53 tumor suppressor gene. Nature 1991;351:453–456.
107. Malkin D, Li FP, Strong LC, et al. Germ line p53 mutations in a familial syndrome of breast cancer, sarcomas, and other neoplasms. Science 1990;250:1233–1238.

36

Expert Systems for Cytometry Data Analysis

GARY C. SALZMAN and PETER H. BARTELS

INTRODUCTION

Data analysis for flow cytometry continues to be a highly interactive time-consuming task requiring significant skill on the part of the investigator. As flow cytometry has moved into the clinic, where large numbers of patient samples are being analyzed, the data analysis task has become a bottleneck. Typical commercial flow cytometers can collect data much faster than it can be processed and fully utilized by the clinician. Methods to automate the data analysis task are essential if flow cytometry is to become a tool for routine use in the clinical laboratory. Expert systems technology may provide just such a tool.

An expert system is a computer program that uses specific knowledge about a task and procedures that emulate human reasoning to solve a problem usually addressed only by a human expert. Knowledge is here defined as the information about a problem that enables a human expert to solve it. This knowledge is frequently in the form of heuristics or "rules of thumb" and is mostly symbolic rather than numeric. Expert systems are distinguished from conventional computer programs by their ability to manipulate symbolic knowledge.

Expert systems began as a subfield of artificial intelligence research. Between the early 1970s and mid 1980s, expert system development tools were large complex programs written in a list processing language, LISP. These programs were expensive, ran only on large, costly LISP language workstations (1), and required a knowledge engineer to work with a domain expert to develop an expert system. Now there are many personal computer-based expert system development tools that can be used by the domain experts to develop their own expert systems. In the early days of expert systems development, expectations were high and the technology was oversold. Now expert systems have entered the mainstream of computer science tools. They should be considered along with other software tools for a variety of data analysis problems.

In this chapter we outline the major features of expert systems, including knowledge representation issues and ways to deal with uncertainty. We include an appendix on cluster analysis, which is used in partitioning flow cytometry list mode data. We discuss several examples, including attempts at automatically analyzing flow immunophenotyping data and a system for histopathology diagnostic expert systems. There are many tutorial articles (2–4) and books (5–8) on expert systems.

WHY EXPERT SYSTEMS?

Expertise in flow cytometry data analysis in the clinical laboratory usually reside in only a few members of the clinical laboratory team. These individuals are not always available for consultation when they are needed. These human experts also retire or leave the laboratory and their expertise, often developed over many years, is lost. An expert system serving as a data analysis assistant enables the knowledge of the human experts to be preserved indefinitely. It also enables the transfer of the expertise developed at a major clinical laboratory to a new clinical laboratory. An expert system may be able to provide results that are more consistent than those provided by different human experts. The expert system can provide these results at a low cost.

Tasks for Expert Systems

Expert systems are suitable for jobs in which a large amount of specialized knowledge is needed about a narrow domain, e.g., flow immunophenotyping. They are also suitable for tasks that require complex decision sequences and those in which the data and knowledge are uncertain. Expert systems are appropriate for tasks in which a brute force search through all possible outcomes would be extremely time-consuming. Tasks involving common-sense and reasoning, however, are inappropriate for expert systems. Common-sense knowledge is global rather than domain-specific and is developed by humans over many years. At present, no expert system could hope to capture enough of this knowledge to behave with common sense. Other inappropriate tasks for expert systems are those that can be carried out by a conventional computer program using well-defined procedures and those tasks for which no human expertise exists.

Expectations of Expert Systems

An expert system should perform at or near the level of a human expert. It should degrade only gradually, as the limit of its knowledge is reached. It should be able to address difficult problems and use complex rules. It should be able to use and manipulate symbolically represented knowledge such as "the fluorescence from CD4 is bright." An expert system should be able to examine its own reasoning and be able to give the user an explanation of reasoning used to reach conclusions. Theses expectations are achieved in a number of expert systems (9) and expert system development tools (10).

Expert Systems and Conventional Programs

A conventional program, such as one in FORTRAN or C, manipulates data usually in the form of numbers using rules in the form of precisely defined equations. Algorithms are used to iterate through the data to arrive at a fixed set of solutions that are presented with certainty. A conventional program can be characterized as Data + Algorithm = Program.

An expert system manipulates symbolically-represented knowledge using rules based on heuristics, or "rules of thumb." An inference process arrives at a best solution among a number of possible solutions. An expert system can reach conclusions even when information is missing and provides results that may be uncertain. An expert system can be characterized as Knowledge + Inference = Expert System.

FLOW CYTOMETRY: DATA TYPES AND STRUCTURE

Flow cytometric data analysis involves a considerable diversity of data types. There are the flow cytometric variables—typically four to eight for each data point. These are numeric, vector type data. Then, there are numeric data derived from the measured variables: data descriptive of profiles, such as a ploidy distribution or descriptive statistics of multivariate distributions formed by subpopulations of cells. There are symbolic data: names of cell types, names for clusters of data points, expressions specifying the brightness of fluorescence markers, and profiles of results from immunophenotyping. In clinical cytometry there are symbolic data providing clinical symptoms, assessment of diagnostic clues, results from clinical laboratory tests, and a full set of patient anamnestic data. Beyond that, numeric/statistical and symbolic data may be considered that provide information on diagnostic categories, prognosis, and response of patients in a given treatment regimen to therapy. The analysis of flow cytometric data thus requires expertise in several distinct, although related domains.

Multidimensional Data

For the assessment of the directly recorded flow cytometric data, the task essentially comes down to an examination of a very large set of p-dimensional vectors, with p representing the number of sensed variables, i.e., the external dimensionality (11). Such an assessment is directed as a series of well-defined questions, such as: is this a homogeneous set of data or are there indications that the data fall into two or more distinct subsets? How are the data points distributed in the feature space for a single population or for each of any detected subpopulations? What is the proportional distribution of data points for each of the subpopulations, and can one reliably classify each data point into a given subpopulation? What is the correlation structure for the different observed variables in each of the subpopulations? Other typical questions are: do the data reflect a particular pattern of subpopulations, and what are the biologic and immunobiologic implications of presence or absence of given subpopulations?

Subpopulations

The concept of a subpopulation implies that, among the cells represented by the measured data vectors, there are subsets whose cells are more similar to each other than they are to cells falling into other subsets. However, since the similarity in feature values and feature correlations is expressed in a p-dimensional space, detection of these subpopulations requires visualization of a value distribution in a p-dimensional space. This can be greatly aided by computer graphic displays or multivariate clustering algorithms. Both of these processes can be guided very effectively by an expert system.

Data Reduction

The principal challenge here is one of data reduction without loss of diagnostic information and with elucidation of the structure of the recorded data. There is the option to attain a data reduction by considering only the intrinsic dimensionality, e.g., by using projections onto principal components. This can be very effective, although one has to be aware that the principal components transformation for an entire dataset is a heuristic, and not an optimum representation for most of the subpopulations. There is the option to view a bivariate display of data points as an "image." Then image processing methods may be applied to detect and delineate clusters. This may bring very substantial savings in computational effort in subsequent iterative clustering procedures by providing good starting estimates for number of clusters, cluster centroids, and dispersion. Another option is to provide guidance for an exhaustive examination of the extension of clusters and membership of each data point in p-space by providing computer graphic displays, or by guiding the user in the correct application of statistical subsampling schemes that make algorithmic procedures feasible.

Bringing Data Types Together

Once data reduction has led to a more compact representation of the data, for instance by the descriptive statistics of clusters in feature space, expertise is required in the choice of appropriate methods for the analysis, classification, and interpretation of the derived data. This involves multivariate algorithmic procedures, where the choice of parameters controlling the processing has a significant impact on the outcome. A comprehensive insight into how the algorithms work and what the underlying assumptions are is required here and may be provided by an expert consulting system. For the numeric data discussed so far (immunophenotyping data sets), expert systems may provide guidance in methodology of data reduction, data description, and data analysis by multivariate algorithmic procedures. In clinical flow cytometry, these numeric data are nearly always related to other data provided by the clinical laboratory, the attending physician, and patient records. Most of these data are of a categorical type, i.e., symbolic, descriptive, or conceptual. Analysis here is carried out by the inference processor of a suitable expert system. The final result is often the determination of a most likely diagnosis or course of treatment for the patient, given the facts and combined evidence. The expert system may be designed to provide its guidance in an interactive mode, where the user becomes involved both in supplying information and in receiving instructions during the interpretation process. The expert system may also be designed to process all recorded information from a given cytometric run and all pertinent information in a knowledge base in a fully automatic mode and to present the user with the results, together with a documented reasoning sequence that explains the final recommended course of action or diagnostic decision.

The efficacy of an expert system substituting for an experienced flow cytometry data analyst depends critically on two aspects of expert system design, knowledge representation and uncertainty management.

REPRESENTING KNOWLEDGE

Flow cytometry data analysis involves a great diversity of problems. The knowledge representation scheme must allow a sufficiently rich, logically adequate, and efficient manipulation of the data. If the examination of the data can be conducted by satisfying a series of independent conditions, a rule-based approach to knowledge representation may be a good choice. If the assessment of the data is based on evaluating declarative descriptions rather than conditional relationships, then a relational knowledge representation, such as an associative net, would serve better. If a particularly information-rich representation is required, a frame-based knowledge representation would be chosen. It is not unusual today to have expert systems employing different knowledge representation schemes and having a number of different *inference processors*. Many expert systems employ a knowledge representation scheme structured as some form of a network, with chunks of knowledge arranged in "nodes" connected to other nodes by links. Thus, the diversity of analytic procedures encountered in flow cytometry may be accommodated best by a variety, and even an integration, of different knowledge representation schemes.

Humans represent and solve problems in terms of concepts. "CD2-negative" and "side scatter high" are concepts in the mind of an immunologist. These concepts are also symbols that can be manipulated by an expert system.

Symbolic reasoning is the manipulation of a set of symbols to solve a problem. These symbols can be combined into patterns that are matched against other patterns to arrive at conclusions. For the expert system, these symbols are merely bit patterns in the computer. The expert system has no deep knowledge of the concepts represented by the symbols. Symbolic reasoning requires that the knowledge be represented in a specific form. Knowledge representation methods include semantic networks, rule-based systems, frame-based systems, object-oriented programming, and predicate logic systems. We discuss several of these methods below.

Knowledge is information about the world that enables an expert to make decisions. For expert systems, specific knowledge about a narrow domain of expertise is more important than general intelligence. Shallow knowledge is information about a specific situation. It is relatively easy to acquire. It can be used to build expert systems where the reasoning is obvious. Deep knowledge about a problem is needed to solve difficult problems. Model-based reasoning is an example of the use of deep knowledge in an expert system (12). Here, a cartoon model is used to represent the various states in a diagnostic problem. Good explanation systems benefit from deep, structured knowledge about a **problem domain**.

Knowledge can be divided into five components (5). These are 1) naming, the use of proper nouns to label entities; 2) describing, the use of adjectives to label attributes of the entities; 3) organizing, the classification of entities into tree structures; 4) relating, the establishment of links between entities; and 5) constraining, the placing of conditions and limits on the attributes of the entities. These components will be used in the discussion of frames as a knowledge representation scheme below.

A human expert has a great deal of knowledge in a narrow domain and uses heuristics based on experience and facts relevant to a specific situation to reason about a problem and arrive at a conclusion. In the expert system model of the human expert, the computer knowledge base contains the general knowledge about a domain in the form of data structures and rules; a computer fact base or working memory contains specific facts about the problem at hand; and a computer inference engine emulates human reasoning.

Knowledge bases for medical diagnostic expert systems are developed incrementally because the medical knowledge is rarely complete (7). The knowledge base is modular so that there are no complex interactions among the chunks of knowledge. The disease knowledge is structured so that similar aspects of different diseases can be accessed efficiently. The control knowledge, which describes the procedures for doing a diagnosis, is separated from the disease knowledge; this step helps with knowledge base maintenance and with explanations of the expert system's reasoning.

Semantic Networks

Semantic networks (13) represent knowledge by creating connections among the entities in a domain. The entities are represented by nodes in a network. The nodes are connected by arcs that are labeled to indicate the relationship between the nodes. "Nephrotic syndrome IS-A-TYPE-OF clinical state" can be represented by the two nodes "Nephrotic syndrome" and "clinical state" connected by an arc labeled "IS-A-TYPE-OF." Numerous other syndromes can be connected to the clinical state node by similar arcs. The network forms a large database of multiply interconnected nodes. Reasoning with semantic nets consists of moving along the arcs in search of specific patterns of nodes and arcs. Semantic networks are not used in most current expert systems because the networks quickly become too large to search effectively and a number of concepts cannot be easily expressed in these networks. The knowledge representation schemes described below provide the needed expressiveness.

Rule-Based Knowledge Representation

Rules are a natural way to represent knowledge. A rule has two parts, the premise and the conclusion. A rule is represented as IF (premise) THEN (conclusion). The premise has one or more patterns containing collections of symbols. An example of a premise containing a single pattern is (Associated with monoclonal antibody panel B1 is some cluster with FSC low.) This pattern can be translated into six symbols: (panel B1 cluster ?clus FSC low). One of the symbols is a variable (?clus). This pattern matches a symbolic fact, such as (panel B1 cluster clus3 FSC low), which represents the concept that the forward-scatter component of cell cluster 3 has low intensity, i.e., the histogram of cell number versus forward scatter intensity has a major peak at a low-scatter intensity. Premises can contain multiple patterns, which are logically ANDed or ORed together. Here is a typical rule:

```
IF
    (panel B1 cluster ?clus FSC low)
AND
    (panel B1 cluster ?clus SSC low)
AND
    (panel B1 cluster ?clus HLA-DR high)
THEN
    (assert (disease is ALL with CF 0.7))
```

The English translation of this rule is, "IF associated with monoclonal antibody panel B1 is some cluster for which FSC is low AND associated with monoclonal antibody panel B1 is the same cluster with SSC low AND associated with monoclonal antibody panel B1 is the same cluster with HLA-DR high THEN assert into the fact base the pattern that the disease is ALL with a certainty factor of 0.7."

The premise in the rule above contains three patterns that are ANDed together. If the fact base contains three patterns for the same cluster that match the three patterns in the premise, then the rule "fires" and the fact pattern (disease is ALL with CF 0.7) is asserted into the fact base. CF is a certainty factor, which indicates the degree of confidence (uncertainty) in the fact. Uncertainty is addressed again later in this chapter.

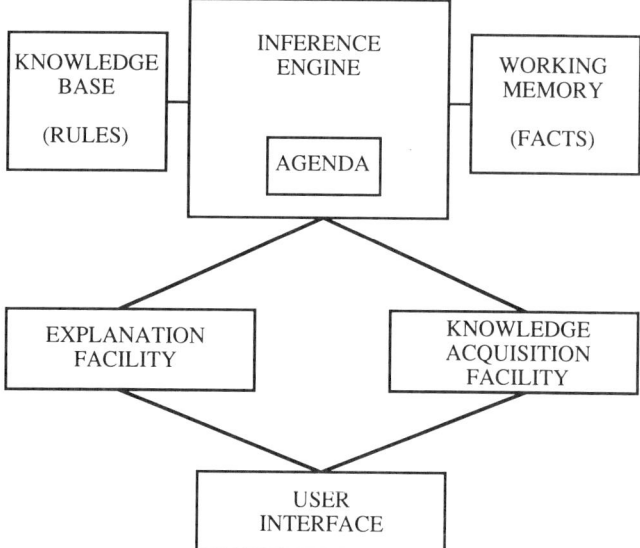

Figure 36.1. The *inference engine* matches facts from the *working memory* with patterns in the rules from the *knowledge base*. If the patterns match, a rule is "fired" adding new facts to the working memory. The agenda contains a prioritized list of rules whose patterns are satisfied by facts in the working memory. The *explanation facility* explains the reasoning of the system to a user. The *knowledge acquisition facility* is an automatic way for the user to enter knowledge into the system. The intelligence in the system is contained in the knowledge base rules. (Adapted from Giorratano J, Riley G. Expert systems. Principles and programming. Boston: PWS-KENT Pub. Co., 1989.)

Figure 36.1 shows a block diagram of a rule-based expert system. The knowledge base contains the general knowledge about the problem domain as a fixed set of rules. The working memory has a list that contains the specific fact patterns about the problem. This list changes as the expert system moves toward a solution to the problem. The inference engine performs the symbolic reasoning task. It contains the methodology for pattern matching in the knowledge base and working memory. It also contains an internal set of rules that control its reasoning strategy. The inference engine requests information from the user if needed, resolves conflicts between rules, presents the user with the final results, and provides an explanation of its reasoning. The inference engine matches fact patterns from the working memory with patterns in the rule premises. When a rule premise is satisfied by pattern matches from the working memory, the rule "fires" and the patterns in the conclusion part of the rule are asserted into the working memory, i.e., they become new facts in the fact base in the working memory.

In the inference process, facts match patterns in a rule premise and produce new facts. These new facts match a premise in another rule causing it to fire. This is called rule chaining. It is a method for traversing a tree-like knowledge structure. There are two chaining strategies, forward chaining and backward chaining. Forward chaining is a data-driven, bottom-up method. A fact is observed and matched against a rule. The rule fires and selects the next fact to be tested. The process continues until a goal state is reached. In flow immunophenotyping, the data are the cluster centroids for each of the antibody panels. The goal state is a recommended disease diagnosis. Forward chaining is a good strategy when the number of possible outcomes (goals) is small. Backward chaining is just the reverse of forward chaining. It begins with a hypothesis (rule conclusion) that a certain outcome is given. The backward chaining mechanism then examines the patterns in the premise of the rule and searches for facts in working memory that match the patterns in the premise. If some patterns remain unmatched, the backward chaining inference engine looks at other rules in the knowledge base to find conclusion patterns that would match the premise patterns in the first rule. Inferencing proceeds backward through the knowledge-base tree until facts are found to support the chain of rules satisfying the hypothesis. Backward chaining is a goal-driven method and is good to use when the number of possible outcomes is large.

MYCIN (6, 7) was one of the first medical expert systems developed. It provided interactive medical consultation on diagnosis and therapy for bacterial infections in blood. Its knowledge base contained 450 rules. MYCIN used a backward chaining inference engine and provided an explanation facility that was invoked when a user responded with "why?" to a response from the expert system. Each rule contained a single chunk of knowledge. Certainty factors (discussed below) associated with the rules were used to deal with uncertain knowledge such as "if premise A is true, there is evidence to **suggest** that conclusion B is true." MYCIN also used metarules, which are control rules for other rules, to guide the expert system through the search tree to confirm a hypothesis.

Because the domain-specific knowledge was separated from the inference engine, it was possible to remove this knowledge from the MYCIN expert system to create EMYCIN, which stands for essential MYCIN (14). EMYCIN is used to acquire a knowledge base from a system designer and then to interpret the knowledge base to give advice to clients. EMYCIN has been used to develop an expert system for bleeding disorders (15) and one for pulmonary function analysis (16).

Immunophenotyping by flow cytometry is the technique in which fluorescently-labeled monoclonal antibodies are used to identify a variety of neoplasias and hematologic disorders (17–19). An EMYCIN-based expert system has been developed for assisting in leukemia diagnosis using cell surface immunophenotyping (20). Investigators were particularly interested in studying knowledge-acquisition methods. Their goal was to develop an expert system that emulated a skilled clinician in interpreting laboratory immunophenotyping reports. They achieved 73% agreement with the clinician. They made recordings of the clinician as he talked his way through the analysis of patient data and then developed EMYCIN rules to analyze the data. They concluded that this approach was unsatisfactory for a variety of reasons, including the observation that the transcripts "gave no guidance about the higher-level structure of the task." They also noted that the transcript set they used was not representative of the range of leukemias encountered in practice.

Preliminary reports of another approach to automating flow immunophenotyping have been presented (21, 22). Mathematical cluster analysis (Appendix A) is used to analyze list mode flow cytometry data to identify the groups of cells responding to panels of fluorescently-labeled monoclonal antibodies. The cluster centroids are then translated into symbolic facts and placed in working memory. For example,

(panel CD20 cluster clus1 FSC low)
(panel CD20 cluster clus1 SSC low)
(Panel CD20 cluster clus1 FL1 high)
(panel CD20 cluster clus1 FL2 negative)

are four fact patterns that represent one of the clusters for a monoclonal antibody panel consisting of fluorescently-labeled anti-CD20 and of propidium iodide (PI), which is used to label dead cells. The cluster-labeled clus1 has low forward scatter (FSC), low side scatter (SSC), high fluorescence channel 1 (CD20), and very low fluorescence channel 2 (PI), indicating that the cells in cluster 1 are viable and are brightly labeled by CD20. The knowledge base consists of a series of rules using the CLIPS expert system development tool (8). CLIPS (C Language Integrated Production System) was developed by the U.S. National Aeronatuics and Space Administration for use in building expert systems. A CLIPS rule that matches the above set of facts is

(defrule Mature-B "Check for mature B cells"
  (panel ?panel-label cluster ?cluster-label FSC low)
  (panel ?panel-label cluster ?cluster-label SSC low)
  (panel ?panel-label cluster ?cluster-label FL1 high)
  (panel ?panel-label cluster ?cluster-label FL2 low)
=>
  (assert (panel ?panel-label cluster ?cluster-label represents mature B cells))).

The phrase "defrule" stands for define rule. The name of the rule is Mature-B and is followed by a text comment stating the use of the rule. The English translation of the first pattern is, "a monoclonal antibody panel with some panel label has a cluster with some cluster label for which FSC is low." The panel labels must be the same for each of the patterns and the cluster labels must be the same for each of the patterns for the rule to fire. When the rule fires, the conclusion of the rule is activated. The conclusion is to assert into the fact base the pattern, "the panel with the panel label matched by the premise has a cluster with cluster label matched by the premise that represents mature B-cells."

The rule format is

IF (pattern) AND (pattern) AND (pattern) ... {premise}
THEN (=>)
   (assert new fact into working memory). {conclusion}

The question mark terms in each pattern are variables. ?panel-label in the rule premise patterns matches with CD20 in the fact patterns above, and ?cluster-label matches with clus 1, which is a label for cluster number 1. The four patterns on the premise side of the rule match the four fact patterns in the working memory, so the rule fires and asserts the fact (panel CD20 cluster clus1 represents mature B-cells) into working memory alongside the existing patterns that support the new fact.

The preliminary version of this expert system used a tree structure with multiple antibody panels at the nodes. The leaves of the tree were diagnostic categories such as acute lymphoblastic leukemia (ALL). The decision boundaries on the rule premises that determined the channel ranges representing "low," "medium," and "high" could be tuned to obtain correct results on a small training set of known cases. The system achieved less than 50% correct results on a larger test set of cases. It failed because of the rigid decision boundaries at the nodes of the tree. This problem may be correctable by using the concept of uncertainty in rules. This is discussed as a separate topic below.

An unconventional rule-based expert system for flow cytometry data has been developed over the past six years for assisting in the automated diagnosis of several types of solid tumors (23–25). The DIAGNOS1 system uses multifactorial analysis to evaluate list mode flow cytometry data from a test set of known normal and abnormal samples. The evaluation identifies a set of factors providing the most significant discrimination between the normal and abnormal samples. An unknown sample can be classified from three-parameter list mode data in 2 min on an IBM PC/AT clone. To begin, the CALC (calculation) module reads the list mode data, creates bivariate histograms, and automatically establishes gates for counting different types of cells. A tumor-specific procedure, DBPHPI, derives 50 parameters for each sample from gates on four bivariate histograms. It writes the data into a database. A procedure called LEARN screens the database to find the five parameters with the most significant differences in means between normal and abnormal samples. It writes these data into a second database. LEARN then applies multifactorial analysis to these data, generating 26 additional parameters. LEARN then rescreens this set of parameters. The aneuploidy parameter and the three parameters with the greatest difference between normal and abnormal samples are chosen as diagnostic indicators for screening unknown samples. DIAGNOS1 correctly classified 91% of malignant cervical specimens and 88% of normal specimens (24). DIAGNOS1 correctly classified 84% of lung cancer specimens (25). The error rates are sufficiently high for this program not to be used alone for diagnosis. It may, however, serve well as an intelligent assistant for automating the analysis of clinical laboratory data.

DIAGNOS1 is, perhaps, properly classified as a conventional program rather than as an expert system. Part of the permanent knowledge base and inference engine are mixed together in the FORTRAN language procedures CALC, DBPHPI, and LEARN. Part of the permanent knowledge base is in a database created the first time the learn procedure runs. The working memory is also a database filled by CALC for each specimen. Whatever its program classification, DIAGNOS1 appears to be a useful and versatile rule-based system.

Predicate Logic as Knowledge Representation

Predicate logic grew out of work in formal logic (5, 13). A predicate is a verb linking two arguments. A predicate and its arguments form a clause, which is similar to a pattern in a rule-based expert system. For example, (clus3 has-value-of-FSC low) is a clause with predicate, "has-value-of-FSC," and arguments, "clus3" and "low." Prolog (26) is an example of a language that uses predicate logic clauses to build expert systems.

Prolog has been used to develop a hematopathology expert system for flow cytometry (27). The expert system uses four separate knowledge bases of clauses. The differentiates knowledge base contains the three of information about the differentiation pathways for pleuripotent stem cells. The expresses knowledge base indicates which antigenic determinants are expressed on a cell's surface. The marker knowledge base links the cluster differentiation group nomenclature to the names of cell-lineage-specific markers. The malignancy knowledge base contains clauses linking cell types to disease states. Given a set of positive and negative markers for a blast cell population, the expert system provides a differential diagnosis. If the user gives the expert system a diagnosis, it will list the markers characteristic of that diagnosis.

Frame-Based Knowledge Representation

Frames (5, 28) were developed as a way to combine declarative knowledge, such as the fact patterns in the working memory, and procedural knowledge, such as a function that computes numerical values. Frames are packaged knowledge representing stereotyped situations. Frames exist in a hierarchy, such that child frames inherit attributes from parent frames. Frames include the components of knowledge mentioned earlier: naming, describing, organizing, relating, and constraining. A frame structure is shown below:

```
Name of frame               (naming)
Name of parent frame        (organizing)
slot 1 name, slot 1 value   (describing and relating)
    if-needed predicate     (constraining)
    if-added predicate      (constraining)
slot 2 name, slot 2 value   (describing and relating)
    if-needed predicate     (constraining)
    if-added predicate      (constraining)
```

Each slot has a unique name that is local to the frame. A slot may have one or more values. The value may be missing when the slot is first defined. The optional "if-needed" predicate must be proved before the slot value can be accessed. This predicate can be used to invoke a function or procedure to calculate the values needed when the slot is accessed. The optional "if-added" predicate must be proved before information can be placed in the value position of the slot. This predicate can call a function that checks the validity of the value to be placed in the slot. These two predicates act to maintain database integrity by serving as filters for storage and retrieval.

The permanent knowledge about a disease can be represented by frames. The descending hierarchy stem cell, lymphoid stem cell, Null-ALL cell, and Common-ALL cell can be stored as a series of frames (29). The slots are cluster differentiation (CD) groups and other sites recognized by monoclonal antibodies. The values are the responses to the antibodies for the specific disease state. " + + " is a strong response, " + " is a weaker response, and " − " is a negative response.

```
frame: Lymphoid stem cell
parent:     stem cell
slot:   Tdt         value: + +
slot:   HLA-DR      value: + +
slot:   CD34        value: + +

frame: Null-ALL
parent:     lymphoid stem cell
slot:   CD19        value: + +
slot:   CD24        value: + +

frame: Common-ALL
parent:     Null-ALL
slot:   CD34        value: +
slot:   CD10        value: +
slot:   CD20        value: +
```

The Null-ALL frame inherits all the slots and values from its parent, the lymphoid stem cell frame. Only the new slots, CD19 and CD24, are shown in the Null-ALL frame. If the value of the CD34 slot in Null-ALL is needed, the expert system searches up the hierarchy until it finds the appropriate value in the Lymphoid stem cell frame. The CD34 slot is shown in the Common-ALL frame because the value is changed from that inherited from the Lymphoid stem cell frame through the Null-ALL frame. The Common-ALL frame has a total of seven slots, five of them inherited.

Specific knowledge about the immunophenotyping of a patient can be represented in frames. For example, here is a frame for patient immunophenotyping with a cocktail of fluoresceinated anti-CD20 (FL1) and PI (propidium iodide) (FL2):

```
frame:      CD20
parent:     patient 12345
slot:   FSC mean    value:
slot:   SSC mean    value:
slot:   FL1 mean    value:
slot:   FL2 mean    value:

instance:   cluster 1
parent:     CD20
slot:   FSC mean    value: 95.
slot:   SSC mean    value: 83.
slot:   FL1 mean    value: 164.
        if-added:   check-limits
        if-needed:  check-brightness
slot:   FL2 mean    value: 43.
        if-added:   check-limits
        if-needed:  check-viability

instance:   cluster 2
parent:     CD20
slot:   FSC mean    value: 99.
slot:   SSC mean    value: 75.
slot:   FL1 mean    value: 90.
        if-added:   check-limits
        if-needed:  check-brightness
slot:   FL2 mean    value: 223.
        if-added:   check-limits
        if needed:  check-viability
```

The CD20 frame has four slots for the four flow cytometric parameters: forward scatter (FSC), side scatter (SSC), fluorescence 1 (FL1) (CD20), and fluorescence 2 (FL2) (PI). No values are in the slots. These slots are placeholders so that the slots will be inherited by the instances, which are frames at the bottom of the hierarchy that cannot have children. The slot values in the instances are the result of cluster analysis on list mode flow cytometric data. Cluster 1 has low values for FSC and SSC, identifying this as a lymphocyte cluster. Cluster 1 also has a low value for FL2, which means that the cells in this cluster excluded PI and are, therefore, viable. Cluster 1 has a high value for CD20 and would be designated as having a " + + " response. Cluster 2 is also lymphocytes, but it has a high value of FL2 (PI) and so represents dead cells. The check-viability predicate is a function call that tests the value of FL2 mean (PI) and returns a decision, "viable" or "dead," so that the cluster can be used in the analysis or ignored, respectively. The check-limits predicate is a function that verifies that the value entered into the frame slot is between channels 0 and 255. The check-brightness is a function that returns " − " if the value is less than channel 100, " + " if the value is between 101 and 150, and " + + " if the value is greater than 150.

Rules and frames can interact to provide an easily-maintained expert system. Attached predicates, such as "check-limits," are rules attached to slots in frames. They serve to maintain database integrity when reading or writing data. Rules driven by the inference engine can act on knowledge stored in a frame structure. For example, a rule could interrogate the frame system to determine if there exists a **viable** cell cluster that has a CD20 fluorescence value (FL1) of " + + ." Parsaye and Chisnell (5) give a detailed discussion of knowledge representation with frames and the use of rules with frames.

Frames and predicate logic can be mapped to relational databases to take advantage of these ready-made tools (5). A relational database (RDB) consists of tables called relations. The table name and attributes or fields (columns) in the table are contained in a list called a schema. The table is filled

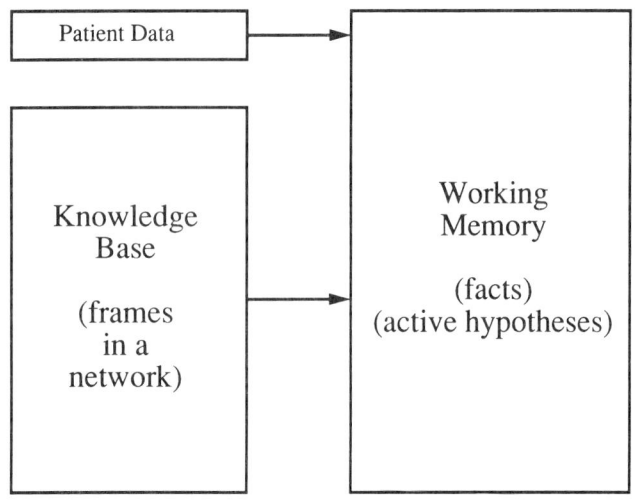

Figure 36.2. The supervisory program in PIP places the *patient data* into *working memory*, which initially contains a list of facts. The program then tries to match the facts in the working memory with the facts stored in frame slots in the *knowledge base*. Frames with the best matches are moved into the working memory to become active diagnostic hypotheses. (Reprinted with permission from Parker SG, Gorry GA, Kassirer JP, Schwartz WB. Towards the simulation of clinical cognition: taking a present illness by computer. In: Clancey WJ, Shortliffe EH, eds. Readings in medical artificial intelligence, the first decade. Reading, MA: Addison-Wesley, 1984;131–159.)

with records (rows), which have values for each field. The collection of slots in a frame is called a frame schema. A slot is equivalent to an RDB attribute and the value of a slot is the same as the value of an RDB attribute. A clause schema in predicate logic contains a predicate name and list of arguments. It is similar to an RDB schema. An argument is identical to an attribute and a predicate logic value is the same as an RDB value. A predicate logic fact, a frame instance, and an RDB record contain the same information. Attached predicates can be used to insure relational database integrity. A frame-based expert system can be used with existing relational database files by creating a frame that is the parent of an RDB file. Then, the relational database can be used to store the frame part of the knowledge base. It can be accessed by rules and operations on frames.

PIP (Present Illness of Patient) is a program that demonstrates the use of frame-based knowledge representation (30). The goal of PIP is to emulate the clinician gathering information and making a diagnosis of a patient with edema. Figure 36.2 shows a block diagram of the processes used by the program to reach a diagnosis. The supervisory program collects information about the problem from the patient (*patient data*) and stores this information in short-term *working memory* as a series of facts. Permanent knowledge in the form of a collection of frames is stored in a *knowledge base*. The frames are interconnected in a semantic network. After the patient data are stored in working memory, the supervisory program tries to find matches between the facts in working memory and the values of slots in the frames in the knowledge base. When matches are found, hypotheses are generated by copying the relevant frames from the knowledge base into working memory. These hypotheses are evaluated by the supervisory program and either accepted or rejected. The most likely final hypotheses are presented in rank order at program completion. Part of one frame in the knowledge base is shown below:

NAME:	Nephrotic syndrome
IS-A-TYPE-OF:	Clinical State
FINDING:	Low serum albumin concentration
FINDING:	>5g/24 hr proteinuria
MUST-NOT-HAVE:	Proteinuria absent
IS-SUFFICIENT:	Both massive pedal edema and >5g/24 hr proteinuria

MAJOR-SCORING:
    Serum albumin concentrations
        low: 1.0
        high: −1.0
    Proteinuria
        >5g/24 hr: 1.0
        heavy: 0.5
        either absent or light: −1.0
MAY-BE-CAUSED-BY:
    Acute glomerulonephritis
    Nephrotic drugs
MAY-BE-COMPLICATED-BY:
    hypovolemia
    cellulitis
MAY-BE-CAUSE-OF:
    sodium retention
DIFFERENTIAL DIAGNOSIS:
    if pulmonary emboli present, consider: renal vein thrombosis.

This frame differs in detail from the prototype frame described earlier but has many of the same features. NAME is the frame name, and IS-A-TYPE-OF is a semantic network verb linking the frame nephrotic syndrome to its parent clinical state. FINDINGs are description slots containing values that describe the clinical state. MUST-NOT-HAVE is a constraint slot that causes this frame to be rejected as a hypothesis if the slot value is true (proteinuria absent). IS-SUFFICIENT is a constraint slot that causes the frame to be accepted as a hypothesis if the value of the slot is true. MAJOR-SCORING is a set of constraint slots that assign scores to the frame so that it can be ranked among the accepted slots. The MAY-BE-... set of slots are links to other frames in the network of frames in the knowledge base. The DIFFERENTIAL DIAGNOSIS slot in this frame causes the frame named renal vein thrombosis to be moved into the working memory if pulmonary emboli are present. The PIP expert system program contains 70 frames associated with 20 diseases and numerous clinical and physiological states. A typical frame in PIP has five to 10 findings, three or four exclusionary rules, 10–20 scoring parameters, and five to 10 links to other frames.

Some of the frames in the knowledge base have daemons, which are small, independent computer programs that scan the working memory looking for facts to match their frame's slots and slot values. Frames can be in one of four states. A dormant frame exists only in the knowledge base and its daemons are inactive. The daemons in semiactive frames have access to the facts in the working memory. Active frames have been copied into the working memory. Accepted frames are part of hypotheses in the working memory that have become facts. When a frame daemon finds a match with facts in the working memory, the entire frame is activated and becomes a hypothesis in the working memory. The values in the scoring slots determine the quality of the fit of a hypothesis to the patient data.

When a frame becomes active and moves to the working memory, the frames in the knowledge base that are linked to the active frame through MAY-BE-... slots become semiactive. Their daemons can then scan the working memory looking for facts to match their slot values. If matches are found, these semiactive frames become active and cause dormant frames linked to them to become semiactive. Hypothesis generation can thus expand rationally through the semantic network of frames.

PIP can access and use knowledge when it is needed. Its goal-directed approach to hypothesis generation means that only pertinent information is gathered and that all the branches in a search tree need not be specified at the start. The knowledge base is a database of clusters of related facts (frames) organized into a semantic network.

MANAGING UNCERTAINTY IN EXPERT SYSTEMS

A similar diversity of methods is required for the problem of uncertainty management in expert systems. Uncertainties arising from randomness are best handled by probabilistic procedures; uncertainties due to vagueness are best managed by procedures from possibility theory and fuzzy set methodology. In practice, uncertainty management in expert systems is still very much in the research and development stage and it presents a variety of difficult and by no means resolved problems (31–35).

For some aspects of flow cytometric data analysis well-established, probabilistic methodology exists, is appropriate, and is computationally feasible. This applies, for example, to the algorithmic processing of the multivariate numerical data and for the derived, statistical/descriptive numerical data. The large sample sizes common in flow cytometry allow excellent estimates of descriptive statistics, such as cluster centroids and cluster covariance structure. The large sample sizes also allow sensitive rare event detection in these data. The data are of limited dimensionality, especially if one restricts the analysis to the intrinsic dimensionality, which is typically no higher than three to four. The dependence structure of these numerical data is defined by each subpopulation's variance/covariance matrix or, more commonly by some nonparametric but numerical estimate of the probability density in p-space.

Data analysis guidance provided by a consultant expert system really does not require uncertainty management, as it involves an almost wholly deterministic procedural sequence. Implementation in the form of an expert system is more a question of convenience in development. The flow cytometric data, though, merely serve as a subset of all the information considered in arriving at an interpretation of a patient's condition and diagnostic classification. Thus, one must distinguish the flow cytometric data's dimensionality from the dimensionality and dependence structure of the information offered by the symbolic data during the inference process. The reasoning sequence may involve a substantial number of propositions, rules, declarative statements, and symbolic variables, such as clinical symptoms and patient anamnestic data. A large patient data base offers a sufficient number of entries so that estimates of probabilities could be based on frequency counts. In this case, there exists a solid basis for an assessment of uncertainty based on probabilistic measures. It is not uncommon, though, that probabilities are expressed not as based on frequency counts, but as "subjective" probabilities, which are based on a personal judgment. They may very well, in some instances, have a good basis in some expert's long-term personal experience and ability to express it as a probability. Nevertheless, it is not possible to delimit such estimates by confidence limits. Consequently, the propagation of uncertainty through a series of propositions—each assessed in terms of a subjective probability—may lead to the limited reliability of the final conclusion. However, even if the data base provided objective, frequency-based probabilities for each proposition in the reasoning sequence, one faces a feasibility problem. The different propositions in a knowledge base are, in practice, not independent, and neither are the associated probabilities. For even a limited number of propositions, the updating and computation of conditional probabilities simply becomes not feasible. Yet, the dependence structure of the elements in a reasoning sequence should not be neglected. If it is, it may lead to seriously flawed uncertainty assessments for the final result. This is the problem inherent in non-probabilistic approaches to uncertainty assessment, that propositional independence is assumed, but practically never given in real-world data. A simple example may illustrate the problem. Let it be assumed that a certain diagnostic alternative is suggested by the outcome observed for a given diagnostic clue. The next proposition may strongly reinforce that conclusion. The resulting, updated cumulative certainty may, in fact, lead to the conclusion that this diagnostic alternative is the most likely one suggested by the evidence. However, if one considered that the outcome from the second proposition is possibly fully correlated to the outcome of the first, the second proposition has not really added any new information. The updated, greatly increased certainty for the diagnosis may be entirely unwarranted.

Since diagnostic reasoning almost always involves highly-dependent propositions, the sensitivity of an automated inference procedure to this situation, and the reliability of certainty assessment for the final diagnostic recommendation, must be very carefully considered. In practical terms, this means the utilization of some form of probabilistic methodology. To address these difficulties, probabilistic methods for uncertainty management in inference networks were developed that did not require the computation of the full joint-probability distribution of all propositions (36–38).

In Bayesian belief networks, only the conditional dependencies of those variables known to be causally and functionally dependent are considered. This makes the computational load feasible. Moreover, since a Bayesian belief network continuously adjusts its probabilities by propagating the effects of new, incoming information throughout the net, even inference nets with uncertainties originally estimated as subjective probabilities have, in practice, been found to provide a stable and probabilistically supported final belief in a proposition.

It is fortunate that the representation of information in a Bayesian belief network is so compatible with the organizational structure of rule-based and associative net representations of knowledge. In a Bayesian belief network, each proposition is represented as a node. The node's entity may assume different values, i.e., a given cluster of data points may have been recorded as having either no fluorescence, low fluorescence, or very bright fluorescence. The nodes store the prior probability vector for these alternatives. It relates to a descendent node via a link with which a conditional probability matrix is associated. In this matrix, the conditional probabilities are stored for observing a certain alternative outcome for the entity at the second node, given that a certain alternative outcome in the first node is found. This is when the prior probabilities for the second node are assigned. Let it be assumed now that an observation is made concerning the alteratives at the second node. The external information is entered in the form of a likelihood vector at the second node. With the prior probability vector for the second node and this likelihood vector, the belief vector for the second node can be updated from the initialization state. Furthermore, the updating at the second node is now propagated, via the link matrix, to the likelihood vector at the first node, and its belief vector is updated. Usually, the first node has as a descendent node not only the second, but another, third, node. Since the belief vector in the first node has been updated, the initial prior probability vector at the third node should be updated via the matrix linking the first and third node. In a general network, belief updating thus involves two processes: the top-down updating of prior probabilities from parent node to descendent nodes, and the bottom-up dating of likelihood vectors and beliefs from descendent nodes to parent nodes as new information comes in. Provisions have to be made for the node initiating the update cycle not to be addressed in the subsequent updating of prior probabilities, as this would create an endless loop.

At this time, uncertainty management in expert systems is largely handled on a heuristic basis. Practical experience in several fields of application has clearly confirmed the need for proper consideration of the propositional dependence structure; however, the lack of proper consideration of

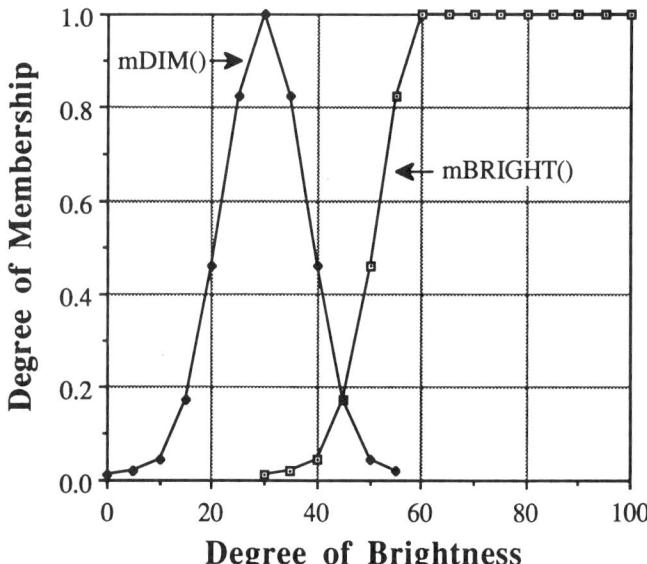

Figure 36.3. Truth values as a function of brightness for two fuzzy set membership functions. Given a value for degree of brightness the membership function *mDIM* returns a truth value stating the degree of membership in the fuzzy set DIM. For example, the fluorescence mean of a cluster of cells has a degree of brightness of 50. Its degree of membership (truth value) in the fuzzy set of DIM clusters is 0.05 and its degree of membership (truth value) in the set of BRIGHT clusters is 0.45. A cluster with a degree of brightness of 80 would have a degree of membership of 1.0 in the fuzzy set of BRIGHT clusters and a degree of membership of 0.0 in the fuzzy set of DIM clusters.

this dependence in many heuristic uncertainty management schemes has not been reported to have led to serious impairment of performance. For clinical diagnosis and selection of treatment data, however, this issue will require very careful consideration.

Uncertainty is an important concept in the development of expert systems because both facts and rules are uncertain to a degree. The degree of certainty is given a numerical score or certainty value. A theory is needed of how to assign certainty values to premises in a rule, how to combine the uncertainties in the premises to obtain a certainty value for a rule, and how to combine the certainty values for rules to determine the certainty value for the conclusions of a set of rules. Sources of uncertainty include: a) lack of precision in measurements, b) inconsistency or incompleteness of data, and c) lack of precision in the concepts that are used in the expert system. Measurement precision is handled with standard statistical methods and is not discussed here. Inconstency and incompleteness of data are addressed using the theory of certainty factors as in the MYCIN program (6, 7). Impreciseness in concepts is a common problem with expert systems and is attacked by using fuzzy logic (39, 40). This section presents fuzzy logic and certainty theory and discusses several examples to show how these ideas are used in expert systems.

In conventional two-valued logic and set theory, a fact can be true or false and an object can be or not be a member

Table 36.1
Excerpt from the Knowledge Base of the Colonic Lesion Diagnostic Expert System[a]

	Normal	Adenoma	Adenocarcinoma
Regularity of gland structure			
Regular	0.40	0.35	−0.45
Moderately irreg.	0.20	0.50	0.20
Severely irreg.	−0.40	−0.10	0.40
Crowding of cells in gland			
None	0.40	−0.40	0.10
Moderate	−0.20	0.40	0.40
Severe	−0.50	0.50	0.40

[a] The three columns are diagnostic outcomes. Two diagnostic clues or rules are shown. For each clue, a clue value or certainty factor is given. Here, the clue values have a range from −1.0 (exclusion of the diagnosis for which this clue value is observed) to 1.0 (absolute certainty for the diagnosis). A clue value of 0.0 indicates that this clue value gives no information useful for diagnosis. (Reprinted with permission from Weber JE, Bartels PH, Griswold W, Kuhn W, Paplanus SH, Graham AR. Colonic lesion expert system. Performance evaluation. Anal Quant Cytol Histol 1988;10:150–159.)

of a set. This law of the excluded middle requires that a proposition be either true or false. Fuzzy logic and fuzzy sets embrace a more general concept of a multivalued possibility theory that includes two-valued logic as a subset (39). Fuzzy logic uses a truth value in the range [0-1] to estimate the likelihood that a fact or rule is true. Similarly, fuzzy set theory assigns a membership value or degree of membership to show the likelihood that a given object is a member of an identified fuzzy set. Figure 36.3 shows graphs of two fuzzy set membership functions. Given a value for degree of brightness, the membership function *mDIM* returns a degree of membership in the fuzzy set DIM. For example, the fluorescence mean of a cluster of cells has a degree of brightness of 50. Its degree of membership in the fuzzy set of DIM clusters is 0.05 and its degree of membership in the fuzzy set of BRIGHT clusters is 0.45. Fuzzy set degrees of membership values are not required to sum to 1.0. A cluster with a degree of brightness of 80 would have a degree of membership of 1.0 in the fuzzy set of BRIGHT clusters and a degree of membership of 0.0 in the fuzzy set of DIM clusters. A monoclonal antibody (mAb) fluorescence degree of brightness could be used to create two fact patterns for matching against rules in a knowledge base: (mAb is bright 0.45) and (mAb is dim 0.05).

A complete algebra exists for manipulating fuzzy sets. mF(x) and mG(x) are membership functions for the fuzzy sets F and G. x is an element of a set X that contains the fuzzy sets F and G. Membership in the complement of F is given by m(complement of F) = 1 − mF. The fuzzy set F is a subset of the fuzzy set G if and only if mF(x) is less than mG(x) for every x. The fuzzy set H represents the union of fuzzy sets F and G if mH(x) is the maximum of mF(x) and mG(x) for every x. H is the intersection of F and G if mH(x) is the minimum of mF(x) and mG(x) for every x.

In common with fuzzy logic, certainty theory assigns a degree of belief to a fact or rule using certainty factors or confidence factors (CF) (5). Certainty factors can be placed on a scale from 0 to 100, with 0 being absolute falsehood and 100 being absolute truth. Certainty theory, combined with fuzzy set theory, provides a mechanism to assign certainty factors to patterns in rules connected by ANDs and ORs, to assign certainty factors to rules, and to determine certainty factors when rules are combined. To combine patterns F and G connected by AND we use the definition of intersection from fuzzy set theory: CF(F AND G) = minimum(CF(F), CF(G)). To combine patterns F and G connected by OR we use the definition of union from fuzzy set theory: CF(F OR G) = maximum (CF(F), CF(G)). Consider the following rule:

IF
    (CD20 cluster ?clus FSC low CF 60)
AND
    (CD20 cluster ?clus SSC low CF 70)
AND
    (CD20 cluster ?clus FL1 bright CF 65)
THEN
    (CD20 cluster ?clus is mature B cells).

This is a rule that might be appropriate for flow immunophenotyping. For the CD20 monoclonal antibody panel, it says that if some cluster (?clus) has low forward scatter AND this same cluster has low side scatter AND this same cluster has bright fluorescence for the CD20 antibody THEN tis cluster represents mature B-cells. Each of the patterns in the premise has been assigned a certainty factor indicating the degree of belief in the truth of this pattern. To find the degree of belief in the overall premise consisting of these three patterns, one applies the rule: CF(pattern 1 AND pattern 2 AND pattern 3) = minimum (60,70,65) = 60.

Certainty factors are also assigned to rules. If the above rule has a CF of 80, there is only an 80% certainty that the rule is valid. The CF for the conclusion of the rule is calculated with the formula CF(conclusion) = CF(premise) * CF(rule) / 100. The certainty factor for the conclusion of the above rule is CF(CD20 cluster is mature B-cells) = 60 * 80 / 100 = 48. This conclusion pattern becomes a fact in working memory: (CD20 cluster clus1 is mature B-cells CF48), where clus1 is the particular cluster that matched the rule premises.

A hypothesis may be supported by the conclusions of several rules. To assign a degree of belief in the hypothesis we need to calculate a certainty factor for the hypothesis. Given the certainty factors for the conclusions of two rules, CF1 and CF2, the combined certainty factor supporting some hypothesis is given by CF1 + CF2 − CF1 * CF2 / 100. More than two rules can be combined two at a time using this formula.

A colonic lesion diagnostic expert system has been developed recently that makes significant use of certainty factors (41, 42). The knowledge base for the expert system is stored in a spreadsheet table. Table 36.1 shows an excerpt from the knowledge base of the colonic lesion diagnostic expert system. The three-column headings are diagnostic outcomes. Two diagnostic clues or rules are shown. For each clue, a clue value or certainty factor is given. Here, the clue values have a range from -1.0 (exclusion of the diagnosis for which this clue value is observed) to 1.0 (absolute certainty for the diagnosis). A value of 0.0 indicates that this clue gives no information useful for diagnosis. The clue values are based on opinions of clinical pathologists experienced in examining colon tissue sections. A clue and its value are based on the belief that the clue applies to the diagnostic outcome. For example, if the gland shape is regular, the colon is diagnosed as normal with a certainty factor of 0.40, a moderately positive value. If the gland shape is regular, the colon can also be diagnosed as adenocarcinoma with a certainty factor of -0.45, an unlikely possibility.

The expert system starts by selecting the clue that provides the best discrimination among the diagnostic outcomes. The clue values are used to update the certainty factors for the diagnostic outcomes. The clue providing the next best discrimination is then selected and the diagnostic outcome certainty factors are updated again using the rules above for combining certainty factors. The process repeats until the certainty factor for one of the diagnostic outcomes rises above some threshold near 1.0 and becomes the recommended diagnosis. This diagnostic expert system is part of an image-understanding system for histopathology (11).

SUMMARY AND CONCLUSION

Multivariate flow cytometry data analysis continues to be a highly interactive and time-consuming task. Expert systems are computer programs that use domain-specific knowledge to solve problems usually only solved by a human expert. These tools offer the exciting possibility to automate some parts of flow cytometry data analysis. A well-designed expert system will be able to do the job faster and, perhaps, more consistently than a human expert. These systems may be particularly useful in the clinical environment, where the same protocols may be used on a daily basis. Expert systems may enable nonexpert analysts in the clinical laboratory to carry out multivariate flow cytometry data analysis at the level of a human expert.

This chapter stresses knowledge representation because the ability of an expert system to competently solve a problem in a domain depends critically on the domain knowledge being correctly represented in the computer. Rule-based schemes are the oldest and most commonly used ways to represent knowledge. Powerful inferencing schemes for pattern matching enable rule-based expert systems to solve problems quickly. Frames enable the representation of more complex knowledge by the creation of hierarchically organized active knowledge bases. They are active because each slot in a frame can have attached procedures that are invoked when the slot is read or written by the inference engine in the expert system. Medical diagnostic knowledge is uncertain to a degree. This uncertainty can be dealt with in expert systems through the formalism of certainty factors or fuzzy sets.

Expert systems technology has been around now for about 20 years. It has moved beyond the early days of extraordinary expectations and less than extraordinary delivery into the mainstream as another software development tool. Many high-quality expert systems are used routinely in areas ranging from credit card approvals, to computer assembly, to adhesive formulation. Expert systems are likely to make their way into the clinical laboratory over the next few years.

ACKNOWLEDGMENTS

Part of this work was performed under the auspices of the U.S. Department of Energy and was supported by the following U.S. National Institutes of Health Grants: 1R01 GM26857 (GCS), 1R01 CA54518 (GCS), The National Flow Cytometry and Sorting Research Resource [RR01315] (GCS), and 1P01 CA38548 (PHB).

Appendix
Cluster Analysis of List Mode Flow Cytometric Data

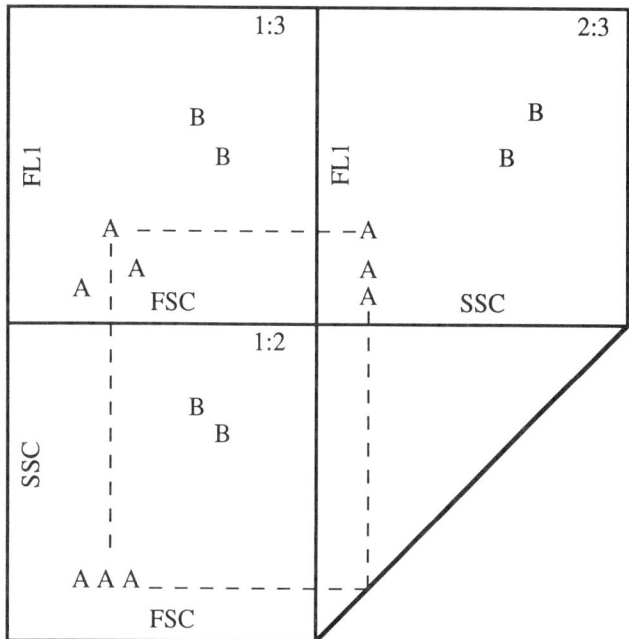

Figure 36.4. All the bivariate dot plot histograms for three-parameter data. Each bivariate plot has its origin in the lower left-hand corner. The plots are arranged so that parameter 1 *(FSC)* is the X-axis for all the boxes in column 1. The bivariate plots are arranged so that one can identify the same data point in each of the plots. Any point *(letters A and B)* can be located in any of the boxes by following the *dashed line*. The 45° line acts as a mirror.

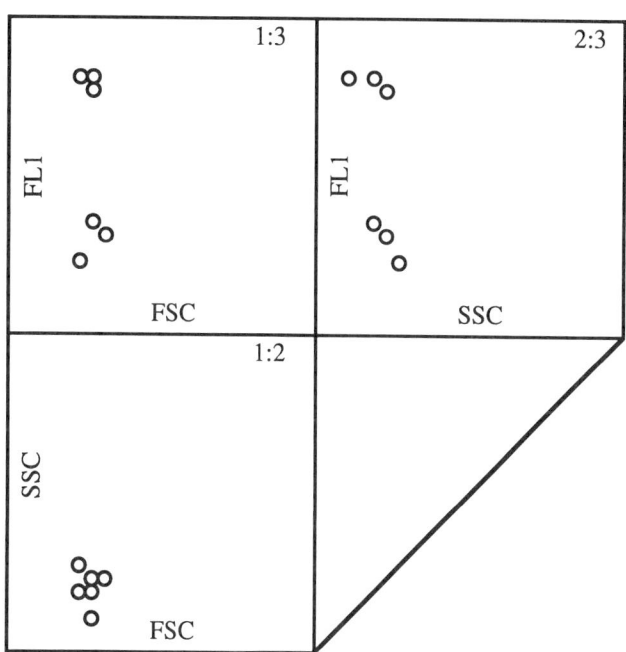

Figure 36.5. Three bivariate dot plots showing six data points split into two clusters. The clusters overlap in the lower bivariate plot (parameter 1 v. 2) but are separated in the upper plots (1:3 and 2:3).

Multivariate flow cytometry data are normally examined using a series of bivariate dot plots that provide two-dimensional views of the data. Figure 36.4 shows three bivariate plots for three-parameter data arranged so that common data points can be located easily. Figures 36.5 and 36.6 demonstrate the clustering of data by hand using the brushplot technique (43). Figure 36.5 shows three bivariate plots containing six data points that separate into two obvious clusters shown in the upper plots. In Figure 36.6, a polygon has been drawn around one of the clusters of three points. The points inside the polygon have become solid circles. It is now clear which points in the lower bivariate plot belong to which of the clusters. This technique is available commercially (Becton Dickinson, San Jose, CA, USA). This method for "by-hand" clustering works well when there are distinct clusters in at least one of the bivariate views of the data. This approach to clustering is error-prone when there are many data points and the clusters overlap in most of the bivariate plots.

Cluster analysis is a mathematical technique for automatically partitioning multivariate data. It is used in flow cytometry for separating subpopulations present in multivariate list mode data (29, 30, 44–52). The data may be partitioned into disjoint subsets by algorithms such as K-means, or it may be arranged into a hierarchy of increasing distances by numerous hierarchical clustering algorithms (53–55). Hierarchical methods are not discussed here.

The K-means algorithm divides the multivariate data into K clusters with different centroids. The number of clusters is specified by the user. Figure 36.7 illustrates the algorithm for two-parameter data. In **I**, nine data points are assigned at random to one of two clusters, *A* and *B*. The mean of each cluster is then computed and marked by the large boldface letters *A* and *B*. The two centroids nearly overlap, as one would expect for randomly assigned points. In **II**, point 2 has been reassigned from cluster *A* to cluster *B* and the means have been recomputed. The cluster means are separating. In **III**, point *4* has been reassigned to cluster *B* and point *5* has been reassigned to cluster *A* and the means have been recalculated. In panel **IV**, all the points are assigned to their proper clusters and the cluster means are now within their own clusters. In the K-means algorithm, each point is reassigned from its original cluster to each of the other clusters in turn. This iterative process continues until only a few points are changing clusters on each iteration.

How do we know when a point is in the correct cluster? This question is answered at each iteration by computing the within-cluster sum of squares, which is the sum of the squared distances from each point in a cluster to its cluster

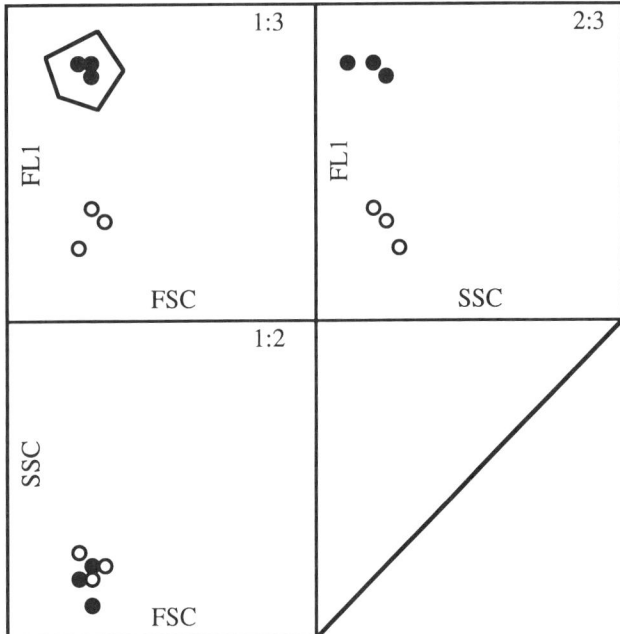

Figure 36.6. The same three bivariate dot plots as in Fig. 36.5. A polygon has been drawn around the points in one of the clusters and those points have become solid circles in each of the dot plots. The points in the two overlapping clusters in the lower dot plot can now be distinguished. This "by-hand" clustering becomes much more complicated and error-prone under realistic conditions, when there are many more data points and the clusters overlap in most of the dot plots.

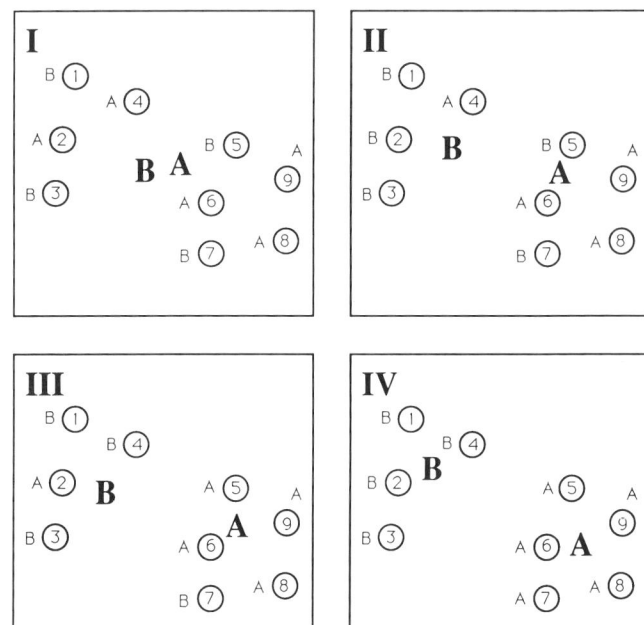

Figure 36.7. Illustration of the K-means clustering algorithm for two-parameter data. In I, nine data points have been assigned at random to two clusters, A and B. The cluster means are shown as large boldface letters. In II, point 2 has been reassigned to cluster B and the means have been recomputed. In III, point 4 has been reassigned to cluster B and point 5 has been reassigned to cluster A. In IV, all the points have been reassigned to their correct clusters. The means are now within their clusters.

mean. The objective function is the sum of all the within-cluster sum of squares. The within-cluster sum of squares is illustrated in Figure 36.8 for panels **I** and **IV** from Figure 36.7. The *dashed lines* are the distances from each member of cluster A to its mean and the *solid lines* are the distances from each member point of cluster B to its mean. We seek to minimize the objective function at each iteration through the data. If a point is moved from cluster *1* to cluster *2* and the objective function increases, that point is moved from cluster *2* to cluster *3*. If the new objective function is less than what it was when the point was in cluster *1*, the point is left in cluster *3* and another point is moved.

What is the correct number of clusters? This question is addressed by several authors (56, 57). One statistic is illustrated in Figure 36.9. *WSS* is the sum of the within-cluster sum of squared distances (w) and X is the grand mean, i.e., the mean of the cluster means. *BSS* is the between-cluster sum of squared distances (b). The *statistic* is essentially the ratio of *BSS* to *WSS*. One proceeds by computing this statistic as the number of clusters increase from two to some maximum number chosen by the user. The statistic increases monotonically until it reaches the correct number of clusters and then drops slightly before again increasing monotonically. The local peak marks the correct number of clusters. The statistic is not very robust. It is sensitive to the number of data points and to the amount of overlap between the clusters.

The K-means algorithm using the standard Euclidean distance metric performs best when the clusters are "spherical"

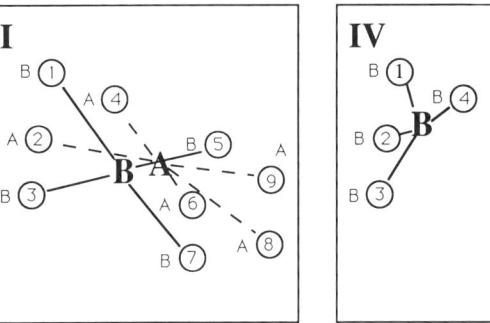

Figure 36.8. Illustration of the within-cluster sum of squares calculation. **I** shows the distances from the cluster means to their respective points. The within-cluster sum of squares is the sum of the squared distances from each point in a cluster to its cluster mean. The objective function is the sum of the within-cluster sum of squares for all the clusters. The objective function is smaller for **IV** than it is for **I**.

with a Gaussian distribution of points. This algorithm will incorrectly split two adjacent cigar-shaped clusters across the middle of each cluster. This problem can be corrected by using the Mahalanobis distance, which takes into account the covariance structure of the data (44).

Starting the clustering by randomly assigning points to clusters is inefficient. In a new approach to this problem, a principal components transformation is used to reduce the dimensionality of the dataset to the first two principal

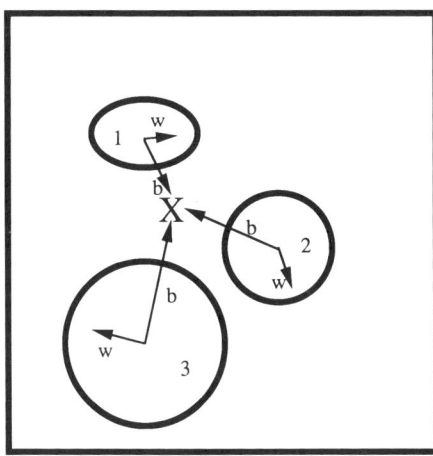

$$\text{Statistic} = \frac{\text{BSS}/(\text{Nclus} - 1)}{\text{WSS}/(\text{Nsamples} - \text{Nclus})}$$

Figure 36.9. Illustration of one cluster stopping rule (48). Three clusters are shown. *WSS* is the within-cluster sum of squared distances (*w*). *BSS* is the between-cluster sum of squared distances (*b*). *X* is the grand mean of the clusters. *Nclus* is the number of clusters and *Nsamples* is the total number of data points. A local maximum in the statistic determines the correct number of clusters.

components (51). The points on this new bivariate plot are processed with a hexagonal segmentation routine and two-dimensional Gaussian smoothing (58). Peaks above a manually-set threshold are detected and used to start cluster formation using the density method. The technique has been applied to flow immunophenotyping data. Another approach used the first two principal components as a bivariate "image" (52). This image is processed using an expert system guided scene segmentation algorithm to find the centroids of the clusters in the data. The clusters are assumed to form convex elliptically-shaped objects in the scene. The method was tested successfully with a number of generated datasets.

Artificial neural networks have been used successfully to cluster aquatic flow cytometry data (59). Two approaches were evaluated. The self-organizing feature map functions as an unsupervised learning algorithm and clusters the input data into a specified number of classes. The backpropagation method requires a learning set to partition the dataset into a fixed number of clusters. After training, it rapidly classifies unknown data.

Clustering methods have been available for many years. These approaches to multivariate data analysis must be tested carefully and used correctly to avoid serious misinterpretations of data. They offer an automated approach to consistently classifying multivariate data. Clustering may be particularly useful for flow immunophenotyping in the clinical laboratory.

REFERENCES

1. Salzman GC, Krall RB, Marinuzzi JG. Knowledge-engineering software. A demonstration of a high-end tool. Anal Quant Cytol Histol 1988;10:219–224.
2. Bartels PH, Weber JE. Expert systems in histopathology. I. Introduction and Overview. Anal Quant Cytol Histol 1989a;11:1–7.
3. Bartels PH, Hiessel H. Expert systems in histopathology. II. Knowledge representation and rule-based systems. Anal Quant Cytol Histol 1989b;11:147–153.
4. Bartels PH, Thompson D. Expert systems in histopathology. II. Representation of knowledge as "structured objects." Anal Quant Cytol Histol 1989c;11:367–374.
5. Parsaye K, Chignell M. Expert systems for experts. New York: John Wiley and Sons, 1988.
6. Buchanan BG, Shortliffe EH, eds. Rule-based expert systems. Reading, MA: Addison-Wesley, 1985.
7. Clancey WJ, Shortliffe EH. Introduction: Medical Artificial Intelligence Programs. In: Clancey WJ, Shortliffe EH, eds. Readings in medical artificial intelligence, the first decade. Reading, MA: Addison-Wesley, 1984:1–17.
8. Giarratano J, Riley G. Expert systems. Principles and programming. Boston: PWS-KENT Pub Co, 1989.
9. Waterman DA. A guide to expert systems. Reading, MA: Addison-Wesley, 1985.
10. Harmon P, Maus R, Morrissey W. Expert system tools and applications. New York: John Wiley and Sons, 1988.
11. Patrick EA. Fundamentals of pattern recognition. Englewood Cliffs, NJ: Prentice Hall, 1972;411–477.
12. Bartels PH, Bibbo M, Graham A, Paplanus S, Shoemaker RL, Thompson D. Image understanding system for histopathology. Anal Cell Pathol 1989d;1:195–214.
13. Shadbolt N. Knowledge representation in man and machine. In: Forsyth R, ed. Expert Systems. Principles and case studies. 2nd ed. London: Chapman and Hall Computing, 1989:142–172.
14. van Melle W, Shortliffe EH, Buchanan BG. EMYCIN: A knowledge engineer's tool for constructing rule-based expert systems. In: Buchanan BG, Shortliffe EH, eds. Rule-based expert systems. The MYCIN experiments of the Stanford Heuristic Programming Project. Reading, MA: Addison-Wesley, 1985;302–313.
15. Bennett JS, Engelmore RS. Experience using MYCIN. In: Buchanan BG, Shortliffe EH, eds. Rule-based expert systems. The MYCIN experiments of the Stanford Heuristic Programming Project. Reading, MA: Addison-Wesley, 1985;314–330.
16. Aikins JS, Kunz JC, Shortliffe EH, Fallat RJ. PUFF: An expert system for interpretation of pulmonary function data. In: Clancey WJ, Shortliffe EH, eds. Readings in Medical Artificial Intelligence. The First Decade. Reading, MA: Addison-Wesley, 1984.
17. Loken MR, Shah VO, Hollander Z, Civin CI. Flow cytometric analysis of normal B lymphoid development. Pathol Immunopathol Res 1988;7:357–370.
18. Melamed MR, Lindmo T, Mendelsohn ML, eds. Flow cytometry and sorting. 2nd ed. New York: Wiley-Liss, 1990.
19. Stewart CC. Multiparameter flow cytometry for clinical applications. In: Salzman GC, ed. New technologies in cytometry. Proc SPIE 1989;1063;150–160.
20. Fox J, Myers CD, Greaves MF, Pegram S. Knowledge acquisition for expert systems: experience in leukaemia diagnosis. Methods Inf Med 1985;24:65–72.
21. Salzman GC, Stewart CC, Duque RE. Expert systems for flow cytometry data analysis: a preliminary report. In: Salzman GC, ed. New technologies in cytometry and molecular biology. Proc SPIE 1990a;1206:98–105.
22. Salzman GC, Duque RE, Braylan RC, Stewart CC. A CLIPS expert system for clinical flow cytometry data analysis. In: Proceedings of the 1st CLIPS user's group conference, Houston, Texas. 13–15 August 1990b. (in press).

23. Valet G, Russmann L, Wirsching R. Automated flow-cytometric identification of colo-rectal tumor cells by simultaneous DNA, CEA-antibody and cell volume measurements. J Clin Chem Clin Biochem 1984; 22:935–942.
24. Valet G, Warnecke HH, Kahle H. Automated diagnosis of malignant and other abnormal cells by flow-cytometry using the DIAGNOS1 program system. In: Burger G, Ploem JS, Goerttler K eds. Clinical cytometry and histometry. London: Academic Press, 1987:58–65.
25. Liewald F, Demmel N, Wirsching R, Kahle H, Valet G. Intracellular pH, esterase activity, and DNA measurements of human lung carcinomas by flow cytometry. Cytometry 1990;11:341–348.
26. Clocksin WF, Mellish CS. Programming in PROLOG. 2nd ed. New York: Springer-Verlag, 1984.
27. Bagwell CB. New horizons: expert systems for flow cytometry. Cytometry Suppl 1988;3:89–93.
28. Minsky M. A framework for representing knowledge. In: Winston P, ed. The psychology of computer vision. New York: McGraw-Hill, 1975.
29. Andreeff M. Flow cytometry of leukemia. In: Melamed MR, Lindmo T, Mendelsohn ML, eds. Flow cytometry and sorting. 2nd ed. New York: Wiley-Liss, 1990, 697–724.
30. Pauker SG, Gorry GA, Kassirer JP, Schwartz WB. Towards the simulation of clinical cognition: taking a present illness by computer. In: Clancey WJ, Shortliffe EH, eds. Readings in medical artificial intelligence, the first decade. Reading, MA: Addison-Wesley, 1984:131–159.
31. Shafer G. Probability judgment in artificial intelligence and expert systems. Stat Sci 1987;2:3–16.
32. Lindley DV. The probability approach to the treatment of uncertainty in artificial intelligence and expert systems. Stat Sci 1987;2:17–24.
33. Zadeh LA. The role of fuzzy logic in the management of uncertainty in expert systems. Fuzzy Sets Systems 1983;11:199–228.
34. Dubois D, Prade H. Possibility theory: An approach to computerized processing of uncertainty. New York: Plenum Press, 1988.
35. de Mantaras L. Approximate reasoning models. Chichester, U.K.: Ellis Horwood Ltd., 1990.
36. Duda RO, Hart PE, Nilsson NJ: Subjective Bayesian methods for rule-based inference systems. Proc AFIPS Nat Computer Conference 1976; 47:1075–1082.
37. Nilsson NJ. Probabilistic logic. Artif Intelligence J 1986;28:71–88.
38. Pearl J. Probabilistic reasoning in intelligent systems: methods of plausible inference. San Mateo, CA: Morgan Kaufman Publ, 1988.
39. Zadeh LA. Fuzzy sets as a basis for a theory of possibility. Fuzzy Sets Systems 1978;1:3–28.
40. Kaufmann A, Gupta MM. Fuzzy mathematical models in engineering and management science. Amsterdam: Elsevier Science Publishers, 1988.
41. Weber JE, Bartels PH, Griswold W, Kuhn W, Paplanus SH, Graham AR. Colonic lesion expert system. Performance evaluation. Anal Quant Cytol Histol 1988;10:150–159.
42. Weber JE, Bartels PH. Colonic lesion expert system. Evaluation of sensitivity. Anal Quant Cytol Histol 1989;11:249–254.
43. Becker RA, Cleveland WS, Wilks AR. Dynamic graphics for data analysis. Stat Sci 1987;2:355–395.
44. Bartels PH, Olson GB, Jeter WS, Wied GL. Evaluation of unsupervised learning algorithms in computer analysis of lymphocytes. Acta Cytol 1974;18:376–388.
45. Salzman GC, Crowell JM, Goad CA, et al. A flow-system multiangle light scattering instrument for cell characterization. Clin Chem 1975; 21:1297–1304.
46. Salzman GC, Crowell JM, Hansen KM, Ingram M, Mullaney PF. Gynecologic specimen analysis by multiangle light scattering in a flow system. J Histochem Cytochem 1976;24:308–314.
47. Crowell JM, Hiebert RD, Salzman GC, Price BJ, Cram LS, Mullaney PF. A light-scattering system for high-speed cell analysis. IEEE Trans Biomed Eng 1978;BME-25:519–526.
48. Genter FC, Salzman GC. A statistical approach to the classification of biological cells from their diffraction patterns. J Histochem Cytochem 1979;27:268–272.
49. Murphy RF. Automated identification of subpopulations in flow cytometric list mode data using cluster analysis. Cytometry 1985;6:302–309.
50. Conrad MP. A rapid, nonparametric clustering scheme for flow cytometry data. Pattern Recog 1987;20:229–235.
51. Kosugi Y, Sato R, Genka S, Shitara N, Takakura K. An interactive multivariate analysis of FCM data. Cytometry 1988;9:405–408.
52. Weber JE, Bartels PH. Statistical identification of subpopulations for flow cytometric data. In: Salzman GC, ed. New technologies in cytometry and molecular biology. Proc SPIE 1990;1206:19–30.
53. Spath H. Cluster dissection and analysis. Theory, FORTRAN programs and examples. New York: Halsted Press, John Wiley and Sons, 1985.
54. Everitt B. Cluster analysis. 2nd ed. New York: Halsted Press, John Wiley and Sons, 1986.
55. Hartigan JA. Clustering algorithms. New York: John Wiley and Sons, 1975.
56. Ratkowsky DA. A stopping rule and clustering method of wide applicability. Botany Gazette 1984;145:518–523.
57. Milligan GW, Cooper MC. An examination of procedures for determining the number of clusters in a data set. Psychometrika 1985;50:159–179.
58. Kosugi Y, Ikebe J, Shitara N, Takakura K. Graphical presentation of multidimensional flow histogram using hexagonal segmentation. Cytometry 1986;7:291–294.
59. Frankel DS, Olson RJ, Frankel SL, Chisholm SW. Use of a neural net computer system for analysis of flow cytometric data of phytoplankton populations. Cytometry 1989;10:540–550.

GLOSSARY

Algorithms. Precisely specified procedures for iterating through data to arrive at a fixed set of solutions with certainty. The procedures are frequently in the form of mathematical equations.

Artificial intelligence. A branch of computer science focused on developing computer programs that appear to a human observer to be intelligent.

Backward chaining. Goal-driven method of rule chaining in which one begins with a final goal and searches for rules whose conclusions match premises in the rule at the goal. This is a good strategy when the number of possible outcomes is large.

Bayesian belief network. A semantic network in which the nodes are connected by conditional probabilities that guide the inference process.

Certainty factors. A scheme for assigning a degree of belief to a fact or rule.

Cluster analysis. Algorithm for grouping together similar data points in a multivariate dataset.

Common-sense knowledge. The global knowledge of the everyday world possessed by human beings as a result of their experience.

Expert system. A computer program that uses specific knowledge about a task and procedures that emulate human reasoning to solve a problem usually addressed only by a human expert.

Domain expert. An expert in the area in which an expert system is to be developed. A clinical pathologist who uses flow cytometry would be an appropriate domain expert for the development of an expert system for flow immunophenotyping.

Forward chaining. Data-driven method of rule chaining in which facts match premises of rules causing rules to fire and thus place more facts in the fact base. The new facts match other rule premises causing them to fire and eventually reach a final conclusion. This is a good strategy when the number of possible outcomes is small.

Frames. Knowledge representation scheme in which fact patterns and procedural knowledge are combined in a hierarchy of parent and child frames. The permanent knowledge about a disease can be represented by frames.

Fuzzy logic. Extension of a two-valued logic system that allows entities to belong partially to two adjacent sets. Fuzzy logic enables numeric values to be assigned to concepts such as bright and dim.

Heuristics. "Rules of thumb," developed through experience with a domain, that are used by experts in solving complex problems.

Inference engine. Part of an expert system program that does the pattern matching in support of the inference process.

Inference process. Symbol manipulation using rules that arrive at a best solution among a number of possible solutions.

Knowledge. Information about a problem that enables a human expert to solve it. The components of knowledge for expert system development are naming, describing, organizing, relating, and constraining.

Knowledge acquisition. That part of the expert system building process in which the rules are obtained from the domain expert typically through interviews.

Knowledge base. Part of an expert system program in which the rules are stored.

Knowledge engineer. A computer scientist knowledgeable about expert systems who works with domain experts to develop expert systems.

Knowledge representation. Methods for representing knowledge about a domain. These methods include semantic networks, rules, frames, and predicate logic.

LISP. A list processing programming language that uses symbol manipulation rather than numeric processing to carry out a programming task.

Multivariate or multidimensional data. Data that contain more than one piece of information about an object. List mode data collected with a flow cytometer that measures light scattering as well as fluorescence are an example of multivariate data.

Predicate logic. Formal logic equivalent of rules for expert system development. Predicate logic is used by the Prolog expert system development language.

Principal components analysis. A procedure for reducing the dimensionality of multivariate data by projecting the data onto a new set of variates that are linear combinations of the original variates.

Rule chaining. Two rules are connected by rule chaining when conclusions in a first rule produce facts that match premises in a second rule causing that rule to fire. The two types of rule chaining are forward and backward.

Rule firing. When fact patterns match all the premises of a rule, the rule conclusions are added to the fact data base. A rule is said to have fired when this occurs.

Rules. Elements in a knowledge representation scheme in which the knowledge about a domain is captured as a set of rules containing premises and conclusions.

Semantic network. Knowledge representation scheme in which knowledge is mapped onto a network of interconnected nodes and arcs. The nodes represent entities that are described typically by nouns and the arcs represent connections between the entities that are described typically by verbs.

Slot. Part of a frame that contains a name, a value, and one or more predicates.

Uncertainty management. Procedures for combining the outcomes of rules, each of which are certain only to a degree. Estimates of uncertainty are attached to both rules and facts.

Working memory. Storage for facts in an expert system program. Facts are added or removed from the working memory as rules in the knowledge base fire.

Appendix

Workshop-Defined Cluster Groups for Monoclonal Antibodies to Human Leukocytes

Cluster	Specificity Group	MW of Antigen	Periph. Blood Normal Range (% of Lymphs)	Monoclonal Antibodies
CD1a	(Langerhans' cells)	49/12K		Leu6, T6, OKT6, BL6, NA1/34, VIT6
1b	Common thymocyte	45K	0%	4A76, NUT2, WM-25
1c		43K		L161, M241, 7C6, PHM3
CD2	E-rosette receptor LFA-3 (CD58) receptor T, NK-cells	45–50K	83 ± 5%	Leu5b, T11, 35.1, 6F10-3, 9.6, OKT11, 9-2, MT910
CD2R	Activated T-cells			D66, T11.3, VIT13
CD3	T-cell Ag receptor associated	22–28K	75 ± 7%	Leu4, 41F, UCHT1, X35, OKT3, T3, CRIS-7
CD4	T-helper/inducer cells, monocytes	59K	45 ±10%	Leu3a, T4, 91.D6, BL4, OKT4(4A)
CD5	Mature thymocytes, T-cells, some sIg$^+$ CLL, some T-ALL	67K	72 ± 7%	Leu1, OKT1, HH9, BL1a, L17F12, T101, T1, DK23 6-2, 10.2, UCHT2
CD6	Thymocytes, mature T-cells, some B-cells	100K	0%	T12, 12.1, T411
CD7	T-cells, NK cells, most T-ALL (Fc$_u$R?)	40K	75%	Leu-9, CL1.3, G3-7 4H9, 3A1, T55, DK24
CD8	T cytotoxic/suppressor cells Class I MHC receptor α chain β chain	32K	28 ± 9%	OKT8, T-8, Leu2a, UCHT4, M236, T811, B9-11, OKT5 T8/2T8, 5H7
CD9	Lymphopoetic bone marrow progenitor cells	24K	0%	BA2, ALB6, FMC56
CD10	Common acute lymphoblastic leukemia Ag (CALLA)	100K	0%	anti-CALLA, ALB1, J5, BA-3, VILA1
CD11a	Leukocytes α-chain LFA-1 (β2 family)	180K	30 ± 5	2F12, CRIS-3, 25.3.1, VEP13
11b	NK subset, monocytes (C3bi recept)	150K		OKM1, Leu15, Mo1, MAC1, 5A4.C5

Cluster	Specificity Group	MW of Antigen	Periph. Blood Normal Range (% of Lymphs)	Monoclonal Antibodies
11c	α-chain Leu CAMb (β2 family) NK subset, monocytes, granulocytes α-chain MAC 1 (β2 family)	170K		KiM1, L29, BL-4H4
CDw12	Monocytes, PMN, platelets	90–120K	0%	M67
CD13	Monocytes, granulcyts some CFU-C (marrow) (Aminopeptidase N)	150K	0%	MY7, 44H, TUK1, MOU28, MCS-2
CD14	Monocytes (>90%), some granulocytes (Folicular dendritic reticular cells)	55K	0%	LeuM3, MY4, Mo2, VIM13, MoP15, UCHM1, Bear-2
CD15	Monocytes (80–100%) (gpLNPIII) mature granulocytes (>95%) (Reed-Sternberg cells)		0%	LeuM1, VIM-D5, 80H5, My1, MCS1
CD16	IgG Fc receptor on NK cells and neutrophils Fc$_{gamma}$ receptor (FcRIII)	50–60K	15 ± 7%	Leu11, L23, HUNK2, VEP13, 3G8, BW209, 80H3
CDw17	PMN, monocytes, platelets (Lactosylceramide)		0%	T5A7, 6035, GO35, Huly-m13
CD18	Leukocytes (integrin β2 chain LEU CAM complexes w LFA α-chain: CD11a,b,c)	95K	(100%)	BL5, M232, 11H6, CLB54, MHM23
CD19	B-cells	95K	10 ± 5%	Leu12, B4, HD37, AB-1
CD20	B-cells (ion channel?)	32–37K	10 ± 5%	Leu16, B1, 1F5
CD21	mature B-cells (CR_2)/(C3dR) EBV receptor	140K	5–15%	LeuCR$_2$, HB5, BL13, B2, CRII, BL-10
CD22	B-cells (95% blood B-cells, mantle & germinal center B-cells in nodes)	135K	10 ± 5%	Leu14, B3, HD39, To15, S-HCL1, BL-9
CD23	Fc$_E$ (low-affinity) recept. (FceRII) B-cell subset, activated eosinophils	45–50K		BLAST-2, B6, MHM
CD24	Immature leukocytes, some B-cells, granulocytes	45, 55, 65K		BA-1, VIBE3, ALB9
CD25	Interleukin-2 recept. β chain (activated T-cells and B-cells (Hairy cell leukemias)	55–60K	0%	anti-IL 2 recept., TAC, 7G7/B6, B1.49.9, IL-2R1, 2A3 ACT1, 33.B3.1
CD26	activated T-cells (Dipeptidylpeptidase IV)	120K	0%	134-2C2, Ta1, TS145
CD27	T-cell subset, plasma cells	55K		OKT18a, VIT14, S152

APPENDIX

Cluster	Specificity Group	MW of Antigen	Periph. Blood Normal Range (% of Lymphs)	Monoclonal Antibodies
CD28	T-cell subset	44K		9.3, Kolt-2
CD29	Regulatory subpopulations of CD4 and CD8 cells, platelets gpIIa (integrin β1-chain; VLA1-6)	135K		4B4, K20, A-1A5
CD30	Activated T- and B-cells (Reed-Sternberg cells)	120K		Ki-1, BerH4, HSR4
CD31	PMN, monocytes, platelets (gpIIa)	130–140K	0%	SG134, TM3, HEC-75, ES12F11
CDw32	Fc$_{gamma}$ RII (PMN, monocytes, platelets, B-cells)	40K	0%	2E1, C1KM5, IV3, 41H16, IV.3, KB61
CD33	Early myeloid progenitors (myelogenous leukemia)	67K	0%	MY9, LeuM6, L4F3, H153
CD34	Early myeloid progenitors (myeloid and lymphoblastic leukemia)	105–120K	0%	12-8, B1-3C5, HPCA-1, My10, ICH-3
CD35	B-cells, PMN, monocytes (CR$_1$) C3b receptor	220K		TO5, CBO4, J3D3
CD36	Monocytes, plts (gpIV)	90K	0%	5F1, C1Meg1, ESIVC7
CD37	B-cells	40–45K		G28-1, HD28, HH1, G28-1, BL14
CD38	Plasma cells, activated T-cells, Lymphocyte progenitors	45K	0%	OKT10, Leu-17, T16, HB7
CD39	B-cells, macrophages	70–100K		G28-2, AC2
CD40	B-cell subset (some carcinomas) (homology to NGF receptor)	50K		G28-5, MA6
CD41a	Platelets gpIIb/IIIa (integrin β3 family) (Fibrinogen receptor)		0%	J15, PL273, PBM 6.4
CD41b	Platelets gpIIb (integrin β3 family)	125+25K		PLT-1
CD42a	Platelets (gpIX) (von Willebrand's factor receptor)	23K	0%	FMC25, BL-H6, GR-P
CD42b	Platelets (gpIb)	25/135K	0%	PHN89, AN51, GN287
	Platelet CD42a/42b (gpIbIX complex)			HLe-1
CD43	T-cells, granulocytes, monocytes, brain (leukosialin)	95K		G10-2, OTH 71C5, G19-1, MEM-59

Cluster	Specificity Group	MW of Antigen	Periph. Blood Normal Range (% of Lymphs)	Monoclonal Antibodies
CD44	T-cells, pre-B, brain, granulocytes, RBC	80–95K		1-173, GRHL1, BRIC35, F10-44-2, 33-3B3
CD45	Pan-leukocyte (leucocyte common antigen)	200K	(100%)	T29/33, T-200/LCA, 2311, ROS 220, KC56
CD45RA	T-subsets, B-cells, granulocytes, monocytes (restricted T200)	220K		73.5, G1-15, F8-11-13
CD45RB	T-subsets, B-cells, granulocytes, monocytes (restricted T200)	220K		PTD/26
CD45RO	T-subsets (some CD4 and CD8 cells), NK cells, and some B-cells	220K	77 ± 15%	Leu-18, 2H4, 3AC5, HB10, HB11, UCHL1
CD46	Leukocytes (membrane cofactor protein)	56/66K		HULYM5, 122-2, J48
CD47	broad pan-leukocyte (N-linked glycan)	47–52K		BRIC 126, CIKM1, BRIC 125
CD48	Leukocytes (Platelet-linked)	41K		WM68, J4-57, LO-MN25
CDw49a	Platelets (α1-chain VLA-1)	200K		
CDw49b	Platelets gpIa, cultured T-cells (α2-chain VLA-2)	165K		Gi14, CLB-thromb/4
CDw49c	Platelets (α3-chain VLA-3)	135+25K		
CDw49d	Thymocytes, T-cells, B-cells, monocytes, Langerhans' cells (epidermis) (α4-chain VLA-4)	150K		
CDw49e	Platelets gpIc (α5-chain VLA-5)	140K		
CDw49f	Platelets gpIc, (T-cells) (α6-chain VLA-6)	140K		GoH3
CDw50	Leukocytes	108/148K		101-1D2, 140-1
CD51	Platelets (α-chain Vitronectin receptor) (integrin β3 and β4 family)			NKI-M7, NKI-M9, 13C2, 23C6
CDw52	Leukocytes	21–28K		097, YTH66.9, YTH34.5
CD53	Leukocytes	32–40K		HI29, HI36, MEM-53, HD77
CD54	Broad specificity (ICAM-1)			7F7, WEHI-CAMI, RR1/1

Cluster	Specificity Group	MW of Antigen	Periph. Blood Normal Range (% of Lymphs)	Monoclonal Antibodies
CD55	Broad specif. (Decay Accelerating Factor, DAF)			BRIC 110, BRIC 128, 143-30, F2B-7.2
CD56	NK (pan-NK Ag), T-subset	135/220K		Leu 19, NKH-1, FP2-11.14, L185
CD57	NK, T-cell subsets B-cell subsets, (brain)	110K	20 ± 7%	Leu 7, L183, L186
CD58	Leukocytes, epithelium (LFA-3)	40–65K		G26, BRIC 5, TS2/9
CD59	Broad specificity	18–20K		YTH53.1, MEM-43
CDw60	T-cell subset (NeuAc-NeuAc-Gal-)			M-T32, M-T21, M-T41, UM4D4
CD61	Platelets gpIIIa (integrin β3 chain) (Vitronectin Receptors)	105K		Y2/51, VL-PL2, BL-E6, CLB-thromb/1
CD62	Platelet activation (GMP-140) (PADEGM LEC-CAM)	140K		CLB-thromb/6, RUU-SP1.18.1
CD63	Platelets activation, monocytes, some T-, B-cells, granulocytes. (α-granules)	53K		RUU-SP2.28, CLB-gran/12
CD64	Monocytes (FcRI)	75K		MAb32.2, MAb22
CDw65	Granulocytes, monocytes (Ceramide-dodecasaccharide 4c)			VIM2, VIM8, HE10, CF4
CD66	Granulocytes	180–200K		CLB-gran/10, YTH71.3
CD67	Granulocytes	100K		B13.9, G10F5, JML-H16
CD68	Macrophages	110K		Ki-M6, Ki-M7, EBM11, Y2/131, Y-1/82A
CD69	Activated B- and T-cells	28/32K		MLR3, L78, FN50, BL-Ac/p26
CDw70	Activated B- and T-cells (Reed-Sternberg cells)			Ki-24, HNE 51, HNC 142
CD71	Proliferating cells, macrophages (Transferrin receptor)			OKT-9, 138-18, 120-2A3, MEM-75, VIP-1, Nu-TfR2, BK19.9, B3/25
CD72	Pan B-cell (B-cell activation marker?)	39/43K	10± 5%	BU-40, BU-41, S-HCL2, J3-109
CD73	B-cell subset (70% CD19+) T-cell subset (20% CD3+) (ecto-5'-nucleotidase)	69K	20%	1E9.28.1, AD2, 7G2.2.11

Cluster	Specificity Group	MW of Antigen	Periph. Blood Normal Range (% of Lymphs)	Monoclonal Antibodies
CD74	B-cells, monocytes (HLA II assoc. invar. chain)	33/35/41K		LN2, BU-43, BU-45
CDw75	mature B-cells (germinal center) T-cell subset	53K		LN1, HH2, EBU-141
CD76	mature B-cells T-cell subset	67/85K		HD66, CRIS-4
CD77	B-cells (Gb3) (Globotriaosylceramide)			38.13(BLA), 424/4A11, 424/3D9
CDw78	B-cells, monocytes			Anti Ba, 1588, LO-panB-a

Other Specificities, not clustered

Regulatory subpopulations of CD4+ and CD8+ cells; some B-cells, neutrophils, and monocytes		80K	68 ± 7%	Leu-8, TQ1

HLA Antigens (class II):

HLA-DR: B-cells, monocytes, activated T-cells		28–34K	11 ± 4%	anti-HLA-DR, I2, ILR2
HLA-DQ: B-cells, monocytes		27, 32K	10 ± 5%	Leu-10
HLA-DP: B-calls, monocytes		25, 35K	10 ± 5%	anti-HLA-DP

Platelet and Endothelial Cell Receptor Complexes

CD 36 (gpIIIbouIV)	Thrombosponden Receptor
CD61 + CD 41b (gpIIIa + gpIIb/IIb)	Vitronectin (Fibrinogin) Receptor
CD49f (gpIc)	Fibronectin Receptor
CD31 + CD49f (gpIIa + gpIc)	Laminin Receptor
CD49b (gpIIa + gpIa)	Collagen Receptor
CD42a + CD42b (gpIX + gpIb/gpIb)	Von Willebrand's Factor Receptor
CD54 (I-CAM)	Receptor for LFA-1 (CD11a)

List compiled by: T. Vincent Shankey and Thomas M. Ellis
Loyola Univ. Med. Cntr.
1/91

Index

Page numbers followed by t or f indicate tables or figures, respectively.

A

7AAD. See 7-Amino-actinomycin D
Absorption filters, 79
Acetic acid fixation, of dissociated tumor cells, from fresh specimens, 96
Acid hydrolase(s), 409
Acquired immunodeficiency syndrome. See also HIV disease
  role of glutathione in, 520
Acridine orange, 24
  in cytochemical detection of BrdUrd incorporation, 31
  in flow cytometric analysis of DNA denaturation, 27, 27f
  in flow cytometric analysis of neutrophils, 407
  in flow cytometric analysis of reticulocytes, 377t, 378
  staining, 205
    of bladder cancer specimens, 285–286, 289–290
    in lymphoma, 205
    of psoriatic keratinocytes, 547–548
  use of, 119
Actin
  dysfunction, 406t
  polymerization, flow cytometric analysis of, 411
Actinic keratosis, aneuploidy in, 550
Actinomycin D, 23
Activation antigen(s), detection of, 68
Activation antigen 4F2
  expression, in severe combined immunodeficiency, 438
  in immunophenotyping of lymphoma, 221
    prognostic significance of, 225
  in immunophenotyping of plasma-cell tumors, 224
Acute lymphoblastic leukemia, CSF-1-R gene expression in, 574
Acute lymphocytic leukemia
  B-precursor, markers of, 484–485
  childhood, neoplastic cell persistence in, 479
  lymphoid tissue involvement, 221
  myeloid antigen expression in, clinical significance of, 238
  relapse
    anatomic variability in, 481
    and phenotypic shift, 480
  T-cell
    antigen expression in, 244
    as model for residual disease assays, 493
    thymocyte phenotype of, 485
Acute myeloid leukemia
  antigen expression in, clinical significance of, 238
  c-kit gene expression in, 575
  CSF-1-R gene expression in, 574
  immunophenotyping of, 237, 238f
  M3, antigen expression in, 244

markers of, 484–485
multiparameter flow cytometry in, 493–494
M7 variant
  antigen expression in, 244
  detection of, 237
phenotypic subpopulations in, 480
ADB. See 1,4-Diacetoxy-2,3-dicyanobenzene
ADCC. See Antibody-dependent cell-mediated cytotoxicity
Adenocarcinoma
  of bladder, 281
    diagnosis of, 302–303
    flow cytometric analysis of, 187
  pulmonary, 319–320
    flow cytometric DNA analysis of, 320t, 320–321
    ploidy, prognostic significance of, 321t, 321–323
Adenoma(s)
  colonic, chromosomal changes in, 5t
  colorectal, aneuploidy in
    relation to aggressiveness, 310
    relation to dysplasia, 310
    relation to size, 310, 310t
  pleomorphic, chromosomal changes in, 5t
Adenoma-carcinoma sequence, in colorectal cancer, 7, 310
Adenosine deaminase deficiency, 435, 438–439
  phenotyping of cells in, 436t
Adherent cells, preparation of, for cell surface immunofluorescence, 148, 152
Adhesion glycoproteins, 405, 408
Adoptive immunotherapy, for melanoma, 553
Adult T-cell lymphomas/leukemias, 222
AE-3 (monoclonal antibody), 545f
Aequorin, 505, 516
Aerodigestive tract squamous neoplasia, 323–326
Affinity, definition of, 143
Aggregation. See also Cell clumping
  and background nonspecific binding observed, 489
  blocking reagents, 490
  effects on DNA content and cell cycle analysis, 133f, 133–138
Aggregation modeling, 53, 123
  based on probability of aggregate formation, 135f, 135–136, 136f
  effect on S-phase measurements, 138f
  practical use of, 138–139
AIL. See Angioimmunoblastic lymphadenopathy
Alcohol(s), 157–158
  effects of
    on cellular parameters, 161, 283
    on neutrophil function, 407t
  as permeabilizing agent, 158
Algorithms, 586
  definition of, 600

ALL. See Acute lymphocytic leukemia; Axial light loss
Allelotype, definition of, 7
Alloantibodies
  detection of, in transplantation, 449
  measurement of, 449–454
    components of test for, 449–451
    positive reaction, definition of, 453–454
    source of antibodies for, 449
    source of cells for, 449
    target cells for, 449–451
    technical considerations, 453–454
  platelet-specific, 450
Allophycocyanin, 64
  fluorescent emission characteristics of, 112t, 146f
  in immunofluorescence applications, 146
  spectral properties of, 112t, 146f
Allowable miss rate, throughput determination from, 91–92
Alveolar soft-part sarcoma
  histologic grading of, 345f
  prognostic value of DNA flow cytometry in, 354
AMCA. See Amino-methyl-coumarin-acetic acid
7-Amino-actinomycin D, use of, 119
Aminoglycosides, effects on neutrophil function, 407t
Amino-methyl-coumarin-acetic acid, 115, 167, 168t
AML. See Acute myeloid leukemia
Ammonium citrate dextrose, blood collection in, 148
Amorphous gating, 87
Amphotericin B, effects of
  on cellular retention of doxorubicin in drug-resistant cells, 459–460
  on neutrophil function, 407t
Amplifier(s), performance, assessment of, 182–184, 183f
Analog-to-digital conversion, 86–87, 123–124, 124f
  quality control, 178
ANCA. See Antineutrophil cytoplasmic antibodies
Androgen(s), in prostate cancer, 281
Anemia. See also Hemolytic anemia
  assessment of, by flow cytometric analysis of red blood cells, 373
  classification of, by HFR fraction RMI units and reticulocyte counts, 380–381, 383f
  subclassification of, by flow cytometric reticulocyte analysis, 379f, 379
Aneuploid tumor(s), detection of, 168
Aneuploidy. See also DNA-aneuploid cells; Near-diploid aneuploid populations; Near-tetraploid aneuploid populations; Ploidy
  and absence of aneuploidy, mixed regions of, 122

609

Aneuploidy—(*continued*)
  in breast cancer, prognostic significance of, 250, 250f, 251–252, 252f
  of bronchogenic carcinoma, incidence of, by histologic type, 321, 321t
  in colorectal adenomas
    relation to aggressiveness, 310
    relation to size, 310, 310t
  detection of
    flow cytometric, 118
      versus image analysis, 362
      number of tissue samples required for, 122–123
  development of, genomic instability and, 8
  diagnosis of, 124–125
  and DNA histogram, 44
  evolution of, 8
  in lymphoid neoplasms, 215–216
  in lymphoma, prognostic significance of, 225
  in malignant disease, 331
  in minor population of total cell population, 128–129
  in myeloma, 224
  oncogene activation and, 8–9
  in ovarian cancer, 263–265, 269
  prognostic significance of, 10
  of prostate cancer, 275–277
    detection of, 278f, 278–279
  pseudo, 120f, 121, 211
  in transitional cell cancer of bladder, 283–284, 284f
  tumor suppressor gene inactivation and, 9–10
Angioimmunoblastic lymphadenopathy, 224
Angiomatoid malignant fibrous histiocytoma, prognostic value of DNA flow cytometry in, 351
Angiosarcoma(s)
  histologic grading of, 345f
  percent of DNA diploidy, 347
  prognostic value of DNA flow cytometry in, 355
Anthracyclines
  cellular transport and retention of
    in drug-resistant cells, flow cytometric monitoring of, 460
    flow cytometric monitoring of, 459, 460f
  cytotoxicity, flow cytometric monitoring of, 461, 465
Antibody(ies). *See also* Alloantibodies; Autoantibodies; Crossmatch, flow cytometric; Monoclonal antibody(ies); Polyclonal antibodies
  centrifugation before use, effects of, 144, 486–488, 487f–488f
  deficiency, 437
  in diagnosis and classification of lymphomas, 204–205, 305t
  divalent, 143
  fluorescence, dependence on degree of labeling, 112–113
  for immunohistochemical staining, 367
  to intracellular antigens, 162–165
  labeled at different dye/protein ratios, brightness of, 111–112, 112f
  multivalent, 143
  against neutrophils, 421
    measurement of, 421, 422f
  nonspecific binding of
    in detection of minimal residual disease, 486–490
    to Fc receptors, 144–145, 486–490
  panel of, for diagnosis and classification of lymphomas, 204–205, 305t
  to proliferation-associated cellular constituents, 28–29
  target-directed, treatment of melanoma with, 553
  valency of, 143
Antibody-dependent cell-mediated cytotoxicity, 67, 451
Antibody reagents, concentration, 144
Antifading medium, 146
Antigen(s), expression, aberrant, in T-cell lymphoma, 214f, 215, 220f
Antigen–antibody complexes, structure of, 143
Antigen–antibody interactions, 143
  affinity of, 143
  avidity of, 143
Antigen-presenting cell(s), 65, 67
Anti-idiotypic antibodies, in transplant recipients, 449
Antineutrophil cytoplasmic antibodies, 422
Antiplatelet antibodies. *See* Immunoglobulin(s), platelet-associated
Antitumor drug(s)
  mechanism of action of, 31, 34
  nucleolar changes caused by, 26–27
APC. *See* Allophycocyanin
Arc lamp(s), 73–75
  useful lifetime of, 74–75
Argon-ion laser(s), 63–64, 76, 115
  excitation wavelengths available from, 75, 75t
  in multicolor detection, 114
Artificial intelligence, definition of, 600
Aspirin, effects on neutrophil function, 407t
Association-dissociation model, for ligand binding to receptor, 471–472, 472f–473f
Association rate constant (k_a), 471
Astrocytoma(s)
  DNA content and cell cycle analysis, aggregation modeling in, 134f, 135–138, 136f–137f
  fractional allelic loss, as prognostic factor, 8
  Ki-67 staining, correlation with tumor grade and other prognostic features, 368
  pediatric
    incidence of, 332
    ploidy of, 332t
Ataxia-telangiectasia, 439–440
  phenotyping of cells in, 436t
Auramine O, for flow cytometric reticulocyte analysis, 377t, 378
Auranofin, effects on neutrophil function, 407t
Autoantibodies
  in ataxia-telangiectasia, 439
  in transplant recipient, 451
Autofluorescence, 167, 472, 475, 486
  control for, in cell surface immunofluorescence, 147–148
Autolysis, DNA staining with, 121
Autoradiography, 13
Average histogram comparison, of DNA histograms, 58–59, 60f
Avidity, definition of, 143
Axial light loss, 155
Azelastine, effects on neutrophil function, 407t
Azidothymidine. *See* Zidovudine
AZT. *See* Zidovudine

B

Bacillus Calmette-Guerin immunotherapy, in bladder cancer, flow cytometric findings related to, 285–286, 286f, 303
Backgating, in whole blood cellular immunofluorescence, 149–150
Backward chaining, 588–589
  definition of, 600
Bacterial-permeability-increasing protein, 419
Baisch (box) method, of DNA data analysis, 45
Bandpass filters, 79
Barbotage. *See* Bladder cancer, bladder washing/urine samples
Bare lymphocyte syndrome, 439
  phenotyping of cells in, 436t
Barrett's esophagus
  DNA-content heterogeneity, 122, 279
  DNA flow cytometry, 313–314
  DNA index (ploidy), prognostic significance of, 129
Basal cell carcinoma
  aneuploidy in, 550
  ploidy in, 550
  RNA/DNA cytogram of, 550, 550f
Baseline restoration, 84
Basophil(s), development of, 64
Bayesian belief network(s), 594
  definition of, 600
BCECF. *See* 2′,7′-Bis-carboxyethyl-5(6)-carboxyfluorescein
B cell(s), 64–65. *See also* Lymphoma(s), B-cell
  antigenic heterogeneity of, 66–67
  in bare lymphocyte syndrome, 439
  cell-surface-differentiation antigens, 66f
  cell surface immunofluorescence, 145–146
  clonality. *See also* Clonal B-cell expansion
    markers of, 486
  in common variable immunodeficiency, 437
  deficiency, 438
  development, 65
  in DiGeorge's syndrome, 439
  flow cytometric crossmatch, 453
  in HIV disease, 444
  in IgA deficiency, 437
  immunophenotyping
    analysis of data, 153, 153f
    in clinical evaluation of primary immunodeficiency, 436t, 440
  membrane antigens, 65
  monoclonal. *See also* Clonal B-cell expansion
    detection of, 205, 207–210
  ontogeny of, 236, 236f
  phenotyping of
    in combined T- and B-cell deficiencies, 436t
    in humoral immunodeficiency, 435t
    in severe combined immunodeficiency, 438–439
  subsets, 67
    changes in, after bone marrow transplantation, 455
  as target cells for detection of alloantibodies, 449–450
  in X-linked agammaglobulinemia, 436–437
bcl-2
  expression, in non-Hodgkin's lymphoma, prognostic significance of, 225
  translocation, detection of, 482–483
bcr. *See* Breakpoint cluster region
bcr/abl, translocation, detection of, 482–483
bcr gene, 579
Becton-Dickinson Paint-a-Gate, 87–88
Beer-Lambert law, 360f, 360
Between-cluster sum of squares, 598, 598f–599f
B5 fixative, 158
  effects on DNA histograms from archival material, 102, 103f

Biological quality control, 190–195
Biological standards, in quantitative
 calibration of fluorescence intensity, 470
Biopsy, of solid tumors, specimen
 procurement and handling, 93
Biotin-avidin immunofluorescence systems,
 145, 165, 240
 in fluorescent in situ hybridization, 564–
  565
Biotin-labeled antibodies, 165
 for cell surface staining, in leukemia, 240
2′,7′-Bis-carboxyethyl-5(6)-
  carboxyfluorescein, 517
 fluorescence emission characteristics of,
  517–518, 518t
 loading, 517–518
Bitmap gates, 149, 149f
Bladder cancer, 281–293
 biological volatility of, 303–304
 bladder washing/urine samples, 290–291,
  302
  processing, 282–283
 blood group antigen expression in, 290
 carcinoma in situ, 281–282
 clinical assessment of, 282, 302–303
 cytokeratin expression in, 290–291, 291f,
  303
 detection, flow cytometric analysis for, 303
  multi-parameter, 288–290, 289t, 303
  single-parameter, 287–288, 288t
 diagnosis of, 302
 DNA content analysis, 283–287, 303
 DNA distributions, from fresh versus
  archival material, 101t
 epidemiology of, 281, 302
 exfoliated cells
  flow cytometric analysis of, 287–290,
   302–303
   cell fixation for, 291–292
   cell staining technique for, 292
   diploid-DNA-content controls, 292
   multi-parameter, 288–290, 289t
   practical considerations for, 290–292
   single-parameter, 287–288, 288t
  processing, 282–283
 field change aberrations in, 281–282, 304
 flow cytometric analysis of
  cytokeratin 18 as potential marker for,
   168, 303
  future of, 292–293
  multiparameter analysis, 285
  using antibodies to cytokeratins, 168,
   303
 fresh specimens, processing, 282
 invasive, 282
 markers for, 292
 membrane antigen expression in, 290
 metastatic, DNA content analysis of, 285
 monoclonal antibodies to, 292
 pathological assessment of, 282
 ploidy
  description, flow cytometry versus image
   analysis for, 364
  and response to radiation therapy, 287
  possible initiation/promotion cycles, 281–
   282
 prognosis for, 282
 ras gene mutations in, 169
 recurrence, 282
  detection of, 292–293
 response to therapy, flow cytometric
  findings related to, 285–287, 286f
 sample
  acquisition, 282–283
  processing, 282–283
 S-phase determinations, 284–285

staging, 282, 282f
tumorigenesis, 281
urinary cytodiagnosis, 302–303
Blood. See also Peripheral blood; Whole
 blood cellular immunofluorescence
 neoplastic cells in, detection of, 485–486
Bodipy, 114
Bone marrow
 erythropoietic responses, assessment of,
  flow cytometric reticulocyte analysis
  in, 379–380, 380f
 neoplastic cells in, detection of, 485–486
Bone marrow aspirate(s)
 contamination, by peripheral blood, 204
 in leukemia, collection and handling, 240
Bone marrow biopsy
 cells obtained from, 204
 in leukemia, 240
Bone marrow transplant
 autologous, tumor contamination of,
  assessment of, 480
 engraftment
  detection of, 376
  reticulocyte maturity index in, 379
 flow cytometric crossmatch for, clinical
  significance of, 453
 lymphocyte subsets in, changes in, after
  transplantation, 455
 in severe combined immunodeficiency, 439
Bone tumor(s). See also specific tumor
 DNA ploidy, 346, 348t
 pediatric, incidence of, 331, 331t
Bouin's fixative, 158
 effects on DNA histograms from archival
  material, 102, 103f
Bowen's disease, aneuploidy in, 550
BPE. See B-phycoerythrin
B-phycoerythrin, absorption spectrum of, 114
BPI. See Bacterial-permeability-increasing
 protein
Brain tumor(s). See also specific tumor
 pediatric, 331–333
  incidence of, 331, 331t
BrdUrd
 antibodies, 31
  in correlation of cellular proliferation
   with drug resistance, 464
  and DNA content analysis, in flow
   cytometric analysis of bladder
   cancer, 285
 incorporation
  in DNA replication, 24, 31–32
   detection of, 31–32
  in kinetic analysis of breast cancer cell
   growth, 256
  in measurement of S-phase fraction, 367
Breakpoint cluster region, 5
Breast cancer, 247–261
 allelic imbalance in, 7, 7f
 allelotypes of, 7
 aneuploidy in, prognostic significance of,
  250, 250f, 251–252, 252f
 versus benign lesions, discrimination of,
  253
 carcinogenesis, accumulation of genetic
  events in, 7
 cell growth kinetics
  BrdUrd incorporation as marker of, 256
  enzyme markers of, 256
  proliferation-specific antigens as markers
   of, 256
 cell proliferation in, prognostic significance
  of, 366, 367f
 chemotherapy for, 247
 clinical background, 247
 cytokeratin expression in, 168

deep-frozen tissue, 248
detected by mammography, management
 of, 247
dissociation, 95
DNA content, 247
DNA content analysis, flow cytometry
 versus image analysis for, 364–365,
 365f
DNA content and cell cycle analysis,
 aggregation modeling in, 134f, 135–
 138, 137f–138f
DNA content heterogeneity, 122
DNA distributions, from fresh versus
 archival material, 101t
DNA flow cytometry in, prognostic
 significance of, 249t, 252
DNA index in, 249–253
 in advanced disease, 252–253
 combination with S-phase, 254–255
 in early disease, 249–252
  correlation with cytogenetic findings,
   249
  correlation with other prognostic
   features, 249t, 249–250
  distribution of, 251, 251f
  prognostic significance of, 250f, 250–
   252
 in subclinical disease, 253
drug resistance in, 247
 flow cytometric assays of, 256–257
 monitoring, 248
endocrine therapy for, 247
 response to, 252–253
estrogen receptor status, 261
 determination of, 247–248
 flow cytometric assays of, 256
 image cytometry of, 256
fine needle aspiration biopsy, 247–248, 253
flow cytometric analysis in
 clinical utility of, 261
 ethical considerations, 248–249
 versus image cytometry, 258
 potential applications of, 247–248
fractional allelic imbalance in, 9, 9f
HER-2/c-erB/neu oncogene amplification
 in, 6
Her-2/neu (Erb-b) oncogene in, 169
neu gene expression in, 576
hypertetraploidy in, prognostic significance
 of, 251
intracellular glutathione in, measurement
 of, 257–258
Ki-67 staining
 in cell growth kinetics measurements,
  256
 correlation with tumor grade and other
  prognostic features, 368
metastases, 247
 in bone marrow, detection of, 485
needle-aspiration biopsy specimens, 248
node-negative
 prognostic factors in, and management,
  261
 S-phase estimates in, prognostic value
  of, 128, 128t
paraffin-embedded blocks, 248
percentage S-phase in
 clinical utility of, 261
 correlation with other prognostic
  indicators, 261
 correlation with thymidine-labeling
  index, 253f, 253–254
 estimation of
  caveats, 253–254
  problems encountered in, 255, 255f
 prognostic significance of, 254f, 254

Breast Cancer—(*continued*)
  ploidy status, 249–250, 261
    and oncogene activation, 9
  prognostic factors in, effects on treatment, 260–261
  proliferation markers, 247, 253–256
    multiparametric analysis, 255–256
    nuclear antigen p105 as, 256
  sample acquisition, 248
  S-phase fraction in, 249, 261
  steroid hormone receptor assays, 247–248
    specimen for, 248
  subclinical disease
    definition of, 253
    management of, 253
  surgical biopsy specimens, 248
  treatment of, 247, 260–261
  treatment sensitivity, 247–248, 256–258
Breast lesions, FNA samples, flow cytometric analysis of, 94
Breast stromal sarcoma(s)
  percent of DNA diploidy, 347
  prognostic value of DNA flow cytometry in, 354
Broadened polynomial model component, 50–51, 51f
Broadened rectangle model component, 49–50, 50f
Broadened trapezoid model component, 50, 50f
5-Bromodeoxyuridine. *See* BrdUrd
Bronchoalveolar carcinoma
  prognosis for, 320
  surgically treated, limited stage, survival with, 320, 320f
Bronchogenic carcinoma, 319–323. *See also* Lung cancer
  aneuploidy
    incidence of, by histologic type, 321, 321t
    prognostic significance of, 328
  distribution by histologic type, 319t
  flow cytometric analysis of, clinical and research utility of, 328
  stage distribution at presentation, 319t
Brown Norway rat myelocytic leukemia model, relapse, anatomic variability in, 481
Bruton's X-linked agammaglobulinemia. *See* X-linked agammaglobulinemia
BSS, 598–599, 599f
Burkitt's lymphoma, flow cytometric findings in, 217t, 221–222, 222f
n-Butyrate, inhibition of histone deacetylase, 15

C
c-abl gene, 5
Calcium
  cytosolic free, in neutrophils, measurement of, 412–413
  intracellular ionized (Ca^{2+}), 505–517
    detection of, 68
    flow cytometric analysis of
      applications of, 516–517, 517f
      calibration of fluorescence, 508–509
      complications in loading of indicators, 513–515, 515f
      critical aspects in, 513–516
      display and analysis of results, 509–511, 510f–513f
      effect of antibody labeling, 516
      effects of intracellular environment, 513–515, 515f
      limitations of, 516
      poor cellular response in, 515–516
      simultaneous analysis with other fluorescence parameters, 511–513, 512f–513f
      simultaneously with pH_i, 519, 520f
      sorting on the basis of, 513
      technique, 508
      unstable or broad baseline distributions in, 515
    indicators of, 505–506
    measurement of, 86
  in neutrophil activation, 412
  as third messenger, 505
Calcium buffers, to collapse calcium gradient to zero, 509
Calibration, commercially available materials for, 180t–181t
Calibrite particles, 189
Calmodulin, 505
cAMP-dependent H1 kinase. *See* Kinase A
Cancer, 3. *See also* Breast cancer; Colon cancer; Colorectal cancer; Endometrial cancer; Ovarian cancer; Prostate cancer
  chromosomal changes in, 5t
  as genetic disease of somatic cells, 3
Cancer chemotherapy
  cellular resistance to. *See also* Drug resistance
    flow cytometric monitoring of, 459–466
    future of, 465–466
  effects of, early assessment of, 479–480
Candida, antibodies against, flow microsphere immunoassay for, 539–540
Carboxyfluorescein diacetate, 517
  in identification of neutrophils, 407, 409f
Carcinoembryonic antigen
  capture of, with flow microsphere immunoassay, 539–540, 541f
  in colorectal cancer, 314
Carcinoid tumor(s), of lung, 319
  atypical, 319
  DNA ploidy, 323
Cardiovascular disease, platelets in, 400
C5a receptor, in neutrophils, 415
Carnoy's fixative, effects on DNA histograms from archival material, 102, 103f
Cartilage tumor(s), DNA ploidy, 346, 348t
Cascade Blue, 115, 167, 168t
  spectral properties of, 112t
Cathepsin G, 409, 419
C-banding, 4
CBC. *See* Complete blood count
C3bi deficiency, 417
CD1, in immunophenotyping of lymphoma, 221
CD1a, 603
CD1b, 603
CD1c, 603
CD2, 67, 603
  cellular distribution of, 444t
  in immunophenotyping of leukemia, 236–237, 237f, 238, 240t, 242f, 243t, 243–244
  in immunophenotyping of lymphoma, 205, 205t, 208f, 214f
  ligand, 444t
CD3, 65, 67, 152, 184–185, 603
  cellular distribution of, 444t
  expression
    after bone marrow transplantation, 455
    in ataxia-telangiectasia, 439
    on cytolytic T cells, 497
    in DiGeorge's syndrome, 439
    in HIV disease, 443–444
    by lymphocytes in peripheral blood, changes in, after transplantation, 454
    in severe combined immunodeficiency, 438
  in immunophenotyping of leukemia, 236–237, 237f
  in immunophenotyping of lymphoma, 205, 205t, 208f–209f, 214f, 221
  ligand, 444t
CD3ε, 65–66
CD4, 67, 152, 184–185, 603
  cellular distribution of, 444t
  expression
    after bone marrow transplantation, 455
    in bare lymphocyte syndrome, 439
    in common variable immunodeficiency, 437
    on cytolytic T cells, 497–498
    in DiGeorge's syndrome, 439
    in HIV disease, 441, 443–444
    by lymphocytes in graft, changes in, after transplantation, 454–455
    by lymphocytes in peripheral blood, changes in, after transplantation, 454
    in severe combined immunodeficiency, 438
    in transient hypogammaglobulinemia of infancy, 437
  fluorochrome used to detect, effect on flow cytometric results, 68, 68f
  in immunophenotyping of leukemia, 236–237, 237f
  in immunophenotyping of lymphoma, 205, 205t, 208f, 214f, 221
  ligand, 444t
  staining, for identification of lymphocytes, 192
CD5, 603
  expression, after bone marrow transplantation, 455
  in immunophenotyping of leukemia, 211f, 236–237, 237f, 240t, 242f, 243t, 244
  in immunophenotyping of lymphoma, 205, 205t, 208f, 214f, 219–220
CD6, 603
CD7, 603
  in immunophenotyping of leukemia, 236–237, 237f, 240t, 242f, 243t
  in immunophenotyping of lymphoma, 205, 205t, 208f, 214f, 221
CD8, 67, 150, 184, 603
  cellular distribution of, 444t
  expression
    after bone marrow transplantation, 455
    in bare lymphocyte syndrome, 439
    in common variable immunodeficiency, 437
    on cytolytic T cells, 497–498
    in cytomegalovirus infection, 443
    in DiGeorge's syndrome, 439
    in Epstein-Barr virus infection, 443
    in HIV disease, 441, 443–444
    by lymphocytes in graft, changes in, after transplantation, 454–455
    by lymphocytes in peripheral blood, changes in, after transplantation, 454
    in severe combined immunodeficiency, 438
  in immunophenotyping of leukemia, 236–237, 237f
  in immunophenotyping of lymphoma, 205, 205t, 208f, 214f, 221
  ligand, 444t
CD9, 603
CD10, 603
  cellular distribution of, 444t

expression
  on cytolytic T cells, 498
  in HIV disease, 444
  in neoplasia, 493
  in immunophenotyping of leukemia, 236, 240t, 242f, 243t, 244
  in immunophenotyping of lymphoma, 205t, 209f, 220, 222f
  in immunophenotyping of plasma-cell tumors, 224
ligand, 444t
CD11, expression, by lymphocytes in graft, changes in, after transplantation, 455
CD11a, 67, 603
  cellular distribution of, 444t
  expression
    on cytolytic T cells, 497
    in leukocyte adhesion deficiency, 440
    in immune function, 419
  ligand, 444t
CD11b, 603
  cellular distribution of, 444t
  expression
    in leukemia, 484
    in leukocyte adhesion deficiency, 440
    in neutrophil activation, 420, 420f
    on neutrophils, 405, 408
    in immune function, 419
  ligand, 444t
  receptors, on neutrophils, 420, 420f
CD11c, 604
  cellular distribution of, 444t
  expression, in leukocyte adhesion deficiency, 440
  in hairy cell leukemia, 223, 223f
  in immune function, 419
  ligand, 444t
CD13, 604
  cellular distribution of, 444t
  expression
    in leukemia, 484–485
    in neoplasia, 494
  in immunophenotyping of leukemia, 237, 238f, 240t, 240–241, 242f, 243t, 244
  ligand, 444t
CD14, 149, 149f, 604
  expression
    in acute leukemia, 574
    in leukemia, 484
    by lymphocytes in graft, changes in, after transplantation, 455
  in immunophenotyping of leukemia, 237, 238f, 240t, 240–241
  in immunophenotyping of lymphoma, 205, 205t
  staining, for identification of lymphocytes, 190–192, 191f
CD15, 604
  cellular distribution of, 444t
  expression
    in leukemia, 484
    in neoplasia, 494
  ligand, 444t
CD16, 66, 152, 421, 604
  cellular distribution of, 444t
  expression
    after bone marrow transplantation, 455
    in HIV disease, 444
    by lymphocytes in graft, changes in, after transplantation, 454–455
    on neutrophils, 420
    in severe combined immunodeficiency, 439
  ligand, 444t
CD18, 604
  cellular distribution of, 444t

expression
  in leukocyte adhesion deficiency, 440
  on neutrophils, 405, 408
  in immune function, 419–420
ligand, 444t
CD19, 65, 152, 604
  expression
    after bone marrow transplantation, 455
    in HIV disease, 444
    in severe combined immunodeficiency, 438
  in immunophenotyping of leukemia, 210, 211f, 236, 238, 240t, 240–241, 242f, 243t, 243–244
  in immunophenotyping of lymphoma, 205, 205t, 205, 205t, 208f–209f, 219
CD20, 145–146, 604
  expression
    after bone marrow transplantation, 455
    in HIV disease, 444
    in leukemia, 484
    in severe combined immunodeficiency, 438
    in X-linked agammaglobulinemia, 437
  in hairy cell leukemia, 223, 223f
  in immunophenotyping of leukemia, 210, 211f, 219, 236, 240t, 242f, 243t
  in immunophenotyping of lymphoma, 205, 205t, 208f–209f, 210, 213f, 219
CD21, 604
  cellular distribution of, 444t
  expression, in severe combined immunodeficiency, 438
  ligand, 444t
CD22, 604
  expression, in severe combined immunodeficiency, 438
CD23, 604
  cellular distribution of, 444t
  ligand, 444t
CD24, 604
  expression, in severe combined immunodeficiency, 438
  in immunophenotyping of lymphoma, 222
CD25, 67, 604
  cellular distribution of, 444t
  expression, in HIV disease, 443
  fluorochrome used to detect, effect on flow cytometric results, 68, 68f
  in hairy cell leukemia, 223, 223f
  in immunophenotyping of lymphoma, 205, 205t, 208f, 219
  ligand, 444t
CD26, 604
  cellular distribution of, 444t
  ligand, 444t
CD27, 604
CD28, 605
  expression, in bare lymphocyte syndrome, 439
CD29, 67, 605
  expression, in severe combined immunodeficiency, 438
CD30, 605
CD31, 605
  + CD49f, 608
CD32, 421
  cellular distribution of, 444t
  ligand, 444t
CD33, 605
  expression
    in leukemia, 484–485
    in neoplasia, 494
  in immunophenotyping of leukemia, 237, 238f, 240t, 242f, 243t, 244
CD34, 64, 605

expression
  in leukemia, 484
  in neoplasia, 494
CD35, 605
  cellular distribution of, 444t
  ligand, 444t
CD36, 605, 608
CD37, 605
CD38, 150, 605
  expression
    in cytomegalovirus infection, 443
    in Epstein-Barr virus infection, 443
    in HIV disease, 441, 443–444
    in severe combined immunodeficiency, 438
    in X-linked agammaglobulinemia, 437
  in immunophenotyping of plasma-cell tumors, 224
CD39, 605
CD40, 605
  cellular distribution of, 444t
  ligand, 444t
CD41a, 605
  cellular distribution of, 444t
  ligand, 444t
CD41b, 605
CD42a, 605
  + CD42b, 608
CD42b, 605
CD43, 605
  cellular distribution of, 444t
  expression, in Wiskott-Aldrich syndrome, 440
  ligand, 444t
CD44, 67, 606
  cellular distribution of, 444t
  ligand, 444t
CD45, 67, 149, 149f, 606
  in antigen-induced T-cell activation, 577
  cellular distribution of, 444t
  expression, by lymphocytes in graft, changes in, after transplantation, 455
  in immunophenotyping of lymphoma, 205, 205t, 213
  ligand, 444t
  staining, for identification of lymphocytes, 190–192, 191f
CD46, 606
CD47, 606
CD48, 606
CD49b, 608
CD49f, 608
CD51, 606
CD53, 606
CD54, 606, 608
  cellular distribution of, 444t
  ligand, 444t
CD55, 607
CD56, 66, 152, 607
  cellular distribution of, 444t
  expression
    on cytolytic T cells, 497
    in HIV disease, 444
    in immunophenotyping of plasma-cell tumors, 224
  ligand, 444t
CD57, 67, 607
  cellular distribution of, 444t
  expression
    after bone marrow transplantation, 455
    in HIV disease, 444
    by lymphocytes in peripheral blood, changes in, after transplantation, 454
  ligand, 444t
CD58, 67, 607

CD58—(continued)
  cellular distribution of, 444t
  ligand, 444t
CD59, 607
CD61, 607
  + CD41b, 608
CD62, 607
CD63, 607
CD64, 421, 607
  cellular distribution of, 444t
  ligand, 444t
CD66, 607
CD67, 607
CD68, 607
CD69, 607
CD71, 607
  cellular distribution of, 444t
  expression, in severe combined immunodeficiency, 438
  in immunophenotyping of lymphoma, 205, 205t, 208f, 219, 221
  ligand, 444t
CD72, 607
CD73, 607
CD74, 608
CD76, 608
CD77, 608
CD antigen(s), 64, 143, 444, 444t
  on platelets, 388f
  in quantitative calibration of fluorescence intensity, 470
CDw12 antigen, 604
CDw17 antigen, 604
CDw32 antigen, 605
CDw49a antigen, 606
CDw49b antigen, 606
CDw49c antigen, 606
CDw49d antigen, 606
CDw49e antigen, 606
CDw49f antigen, 606
CDw50 antigen, 606
CDw52 antigen, 606
CDw60 antigen, 607
CDw65 antigen, 607
CDw70 antigen, 607
CDw75 antigen, 608
CDw78 antigen, 608
CD11/CD18 complex, in immune function, 419–420
CD4/CD8 ratio, changes in, after transplantation, 454
CD4 counts
  absolute, 154–155, 441
  in HIV disease, 441–443
CD45RA antigen, 606
  expression, by lymphocytes in peripheral blood, changes in, after transplantation, 454
CD2R antigen, 603
CD45RB antigen, 606
CD45RO antigen, 606
  expression, in severe combined immunodeficiency, 438
CDRs. See Complementarity-determining region(s)
Cell activation
  measurement of, 68
  role of glutathione in, 520
  thiol-redox requirements for, 527f, 527–528, 528f–529f
Cell adhesion, blocking reagents, 490
Cell adhesion molecules, 497
Cell age, in cell cycle, 43, 43f
Cell-associated receptor(s), quantitation, 469–476
Cell clumping, 147, 161

effects on detection of rare events, 490
  of tumor cells, 485–486
Cell cycle, 13–34. See also Yeast, cell cycle
  cell synchronization in, 30–31
  cellular DNA content as marker of cell position in, 19f, 22–23
  and cellular response to chemotherapeutic agents, 464–465
  chromatin changes during, 14–21
  flow cytometric analysis of, 22–32
    multiparametric, 23–29
      advantages of, 23–24
      applications of, 23–24
  flow cytometric components, 117f, 117–118
  intermitotic times in, distribution of, 32
  markers, in classification of tumors, 34
  metabolic features of cells in, 24–25
  number of cells versus cell age in, 43, 43f
  parameters, estimation of, effects of debris, 132–133
  phases, 13. See also G_0 phase; G_2 phase; G_1 phase; Mitosis; M phase; S phase
    DNA content during, 42–43, 43f
    duration of, estimation of, based on DNA content frequency histograms, 30, 30f
  probabilistic and deterministic compartments, 32
  progression, and cell size, 21–22
  subcompartments, 32
  univariate DNA content analysis in, 22–23
    kinetic information in, 29–30
    limitations of, 22–23
Cell cycle analysis
  aggregation modeling in, 133–138
  confidence estimation in, 139
  data, clinical interpretation of, 139–140
  of DNA content histogram, 125–126
  practical challenges and problems in, 126–139
  results, variability in, 126
Cell cycle fractions, in lymphoma, 211–212, 214f
Cell cycle kinetics, in lymphoma, 216
Cell enucleation, 95
  techniques, evaluation of, 96
Cell fixation, 144
Cell generation time, 13
  heterogeneity, in cell populations, 32
Cell growth, 13
  coordination of, 21–22
  kinetics measurements, p105 antigen in, in breast cancer, 256
  in minimal residual disease, 481, 494
  in tumor classification, 34
  unbalanced, detection of, 25
Cell kinetics
  analysis of, 29–32
  exponential-like components of, 32–33
  intercellular variability of, 32
  markers predictive of, 23–24
  stochastic elements in, 32–34
  studies of, tumor dissociation for, 96
  in tumor classification, 34
Cell-mediated cytotoxicity. See also Antibody-dependent cell-mediated cytotoxicity
  assessment of, 498–499
  cytolytic effector cells, 497
  effector/target cell interactions, assessment of, 498–499
  flow cytometric assessment of, 497–502
    advantages of, 497
    versus conventional assays, 502
    fluorescent probes for, 499–501, 500t
  mechanisms of, 497–498

radionuclide release assays of, 499, 502–503
  single-cell assays of, 499
  stages of, 498, 498f
Cell-mediated immunity, 65
  screening tests for, 436
Cell physiology, measurements of, 505–531
Cell population(s)
  metabolic and kinetic heterogeneity of, 33–34
  properties of, 32–34
  snapshot (static analysis) of, 23–24, 29
  synchronized, progression of, measured sequentially, 24
Cell proliferation. See also Tumor cell proliferation
  in breast cancer, prognostic significance of, 366, 367f
  and cellular response to chemotherapeutic agents, 464–465
  cellular RNA metabolism and, 21–22
  correlation of
    with drug resistance, BrdUrd antibodies in, 464
    with Ki-67 staining, 367, 367f
  in lymphoma, prognostic significance of, 366
  markers of, 21, 34
  measurement of, 366–368
  proteins associated with, detection of, 28–29
Cell reproduction, 13. See also Cell cycle
Cell size
  and cell cycle progression, 21–22
  measurement of, 24
Cell sorting, 89
  on basis of intracellular glutathione content, 528–531, 530f
  flow cytometric, by analysis of variant intracellular ionized calcium (Ca^{2+}) measurement, 513
  for phenotyping in HIV disease, 446
Cell staining, 144
Cell-surface antigen(s), functional role for, 442t
Cell surface immunofluorescence, 143–155
  for absolute and differential counts, 154–155
  analysis of data, 153–154, 153f–155f
  analysis of routine samples, 148
  background fluorescence, reduction of, 148
  cell preparations for, 144, 148
  compensation/subtraction in, 150–151
  controls for, 147–148
  detection, by flow cytometry versus microscopy, 146–147
  direct versus indirect immunofluorescence, 147
  future of, 154–155
  nonspecific binding caused by Fc interactions, 144–145
  number of cells to analyze, 147
  problems encountered in, 147–148, 152–153, 153t
  with polyclonal antibodies, 145
  quantitation, 154, 469–476
    clinical applications, 469
    instrument calibration for, 469–470, 476
    relative versus absolute, 469, 470f
  reagent concentration for, 144
  staining and fixation methods, 144
  steric hindrance in, 145–146
  troubleshooting, 147t, 147–148
  whole blood, 148–150
Cell surface receptor(s), 573–577
  binding studies

experimental considerations, 474
high-density receptors in, 475–476
Cell surface receptor(s)—*(continued)*
    isoparametric titration, 474, 475f
    low-density receptors in, 476
    nonspecific fluorescence in, 474–475
    quantitation, 469–476
Cell suspension(s). *See also* Single-cell suspensions
    from lymphomas, 204
    preparation of, 118–119, 144
Cellular detoxification, of xenobiotics, 464, 466
Cellular DNA
    fluorescence intensity of, and actual content, 23
    as marker of cell position in cell cycle, 19f, 22–23
    univariate content analysis, 22–23
        kinetic information in, 29–30
Cellular RNA
    cell cycle–related changes in, 24–25
    cell (tissue) type-specific variation in, 24–25
    metabolism, and cell proliferation, 21–22
    in psoriasis, 547f, 547–548
    in tumor cells, as prognostic marker, 21–22
Cell viability
    assessment of, 412, 412f
    methods for, 408t
    determination of, 144, 147, 192
Cell washing, effect on cell receptor quantitation, 472
Central nervous system tumor(s), pediatric, 331–333
    incidence of, 331, 331t
Cerebrospinal fluid, in leukemia, collection and handling, 240
Certainty factors, 595–596
    definition of, 600
Certainty theory, 595–596
Cervical cancer, ploidy in, 266, 270
CF-DA. *See* Carboxyfluorescein diacetate
c-fos, immunofluorescence measurement of, using flow cytometry, 169t, 170, 172
Chamber-based flow cell design, 73
Chediak-Higashi syndrome, 406t, 416
Chelation, dissociation of solid tumors, 94
Chemical dissociation, of solid tumors, 94
Chemotaxis. *See* Neutrophil(s), chemotaxis
Chemotherapeutic agents
    fluorescent analogs, 461, 465
    proliferation-dependent or cell cycle phase–related, 464–465
    tagged with fluorescent molecules, 461, 465
Chemotherapy. *See* Cancer chemotherapy; Drug resistance; *specific drug*
Chinese hamster ovary cells
    BrdUrd incorporation in, 31–32, 33f
    histone phosphorylation in, 18
    multidrug resistance, monoclonal antibodies specific for antigen in, 462
Chi-square analysis, 46–47, 48f, 139. *See also* Reduced chi-square
Chloroamphenicol, effects on neutrophil function, 407t
5-Chloromethyleosin-diacetate, as probe for cellular thiol or GSH content, 464
Chloromethylfluorescein diacetate
    in flow cytometric determination of intracellular glutathione, 464, 522–524
    and DNA content analysis, 528, 530f
    as probe for cellular thiol, 464

Chlorpromazine, 411
    effects on cellular retention of daunorubicin, in drug-resistant cells, 459–460, 460f
CHO cells. *See* Chinese hamster ovary cells
Chondroblastoma, DNA ploidy, 346, 348t
Chondrosarcoma
    DNA ploidy, 346, 348t
    extraskeletal, histologic grading of, 345f
    mesenchymal, histologic grading of, 345f
    myxoid, histologic grading of, 345f
    percent of DNA diploidy, 347
    prognostic value of DNA flow cytometry in, 354
Chromatin
    changes
        during cell cycle, 14–21
        history of studies of, 14
    flow cytometric analysis of, 26
    maturation, 15
    of noncycling cells, 19–20
    poly(ADP)ribosylation, 15–16
    structure of, nascent versus mature, 15
Chromium-51 release assay, of cell-mediated cytotoxicity, 499, 502–503
Chromomycin, 119, 482
Chromosomal abnormalities, 3. *See also* Aneuploidy
Chromosomal loci, associated with cancer, 6, 6t
Chromosomal translocation, 4f, 5
Chromosome(s)
    changes, in carcinogenesis, 4
    gains, 4
    genetic analysis at level of, 3–5
    losses, 4
    rearrangements, 4. *See also* Gene(s), deletions
Chromosome analysis, in flow, for detection of minimal residual disease, 482t, 482
Chromosome banding, 3–4
Chromosome cycle, 13. *See also* Cell cycle
Chronic granulomatous disease, 406t, 418, 435
    therapy for, 421
Chronic lymphocytic leukemia
    B-cell
        immunophenotyping, 210, 211f
        S-phase cells in, prognostic significance of, 226
    flow cytometric findings in, 219
    T-cell, flow cytometric findings in, 219
Chronic lymphoproliferative disorder(s), flow cytometric findings in, 218–219
Chronic myelogenous leukemia
    Philadelphia chromosome in, 3, 5
    residual disease in, measurement of, 480
    translocations in, detection of, 482
Chymotrypsin, used to dissociate tumors, 95
Citrate ion, as chelating agent, 94
Citrate-propidium-iodide, as enucleation-DNA-staining solution, 95
c-*kit*, 574–575
Clear cell sarcoma
    histologic grading of, 345f
    prognostic value of DNA flow cytometry in, 354–355
Clindamycin, effects on neutrophil function, 407t
CLIPS, 589
Clonal B-cell expansion
    detection of, 58, 207–210, 213–215
    κ and λ distributions in, 207–210, 209f–210f
Clonal evolution, 3
Clonal excess assay(s), of B cells, 209–210
Clonogenic assays

for detection of minimal residual disease, 482t, 483–484
tumor dissociation for, 96
Clumping. *See* Cell clumping
Cluster analysis, 589
    definition of, 600
    of list mode data, 597–599
Cluster of differentiation. *See under* CD
Cluster groups, workshop-defined, for monoclonal antibodies to human leukocytes, 603–608. *See also* clusters of differentiation (CD).
CMC. *See* Cell-mediated cytotoxicity
CMEDA. *See* 5-Chloromethyleosin-diacetate
CMFDA. *See* Chloromethylfluorescein diacetate
CML. *See* Chronic myelogenous leukemia
c-myb antigen, immunofluorescence measurement of, using flow cytometry, 169t, 170, 172
c-myc antigen
    immunofluorescence measurement of, using flow cytometry, 169t, 170, 172
    localization
        fixation artifacts with, 160
        fixative techniques for, 160
c-myc gene
    expression
        independent verification of, 165
        in non-Hodgkin's lymphoma, prognostic significance of, 225
        in prostate cancer, 281
    overexpression, in ovarian cancer, 9
Coagulant fixative(s), 157, 157t
Coefficient of variation, 118
    calculation of, 89, 125
    of DNA-aneuploid populations, 129
    from DNA flow cytometric analysis of colorectal cancer, 307, 308t
    of peaks, dissimilarity, 129
    for tuning flow cytometer, 121
Colchicine, effects of, on neutrophil function, 407t
Collagenase, used to dissociate tumors, 95
Collection angle, of forward-angle light scatter sensor, 78
Colon cancer. *See also* Colorectal cancer
    aneuploidy, prognostic significance of, 311–313
    dissociation, 95–96, 98f, 100f
    DNA-content heterogeneity, 122, 279
    DNA distributions, from fresh versus archival material, 101t
    DNA flow cytometry in, 311–313
    hepatic metastases
        to single liver lobe, management of, 313
        surgical resection of, 313
    proliferative activity, 311–313, 313f
    ras gene mutations in, 169
    stage A and B (nonmetastatic), DNA flow cytometry in, 311–312, 313f
    stage B_2, adjuvant therapy for, 317
    stage C (lymph node metastatic)
        adjuvant therapy for, 316–317
        DNA flow cytometry in, 312–313, 313f
    stage D, DNA flow cytometry in, 313
Colonic lesion(s), diagnosis of, expert system for, 595–596
Colony-stimulating factors, 64
    hypersensitivity to, 576–577
Color compensation
    effect of filter selection and signal amplification on, 185–187, 186f
    monitoring, 189
        fluorochrome-labeled particles for, 180t–181t, 189

Color compensation—(*continued*)
   optimization of, 184–189
   in two-color analysis
      effect of under- and overcompensation on, 187f, 188, 188f
      instrument set-up and monitoring for, 185–187, 186f
Colorectal adenoma(s), aneuploidy in
   relation to aggressiveness, 310
   relation to dysplasia, 310
   relation to size, 310, 310t
Colorectal cancer
   adenoma-carcinoma sequence in, 7, 310
   carcinoembryonic antigen in, 314
   cellular heterogeneity, 308
   dissociated cells, 314
   fixation, 99
   DNA flow cytometric analysis, 307–317
      coefficient of variation in, 307, 308t
      effects of noncancer cells on, 308–309
      future of, 314
      morphologic correlation with, 308
      and tissue sampling, 308
   DNA Index, and survival, 307–308, 308f
   hepatic metastases
      DNA ploidy and proliferative activity, prognostic significance of, 317
      surgical treatment of, 317
   loss of heterozygosity in, 9
   near-diploid DNA-aneuploid, biologic behavior of, 307–308
   neoplastic versus non-neoplastic cell fractions, 308–309, 309f
   ploidy
      evolution pathway of, 8
      versus histologic grade, 309–310, 310t
      prognostic significance of, 307t, 307–308, 308f
      versus stage, 309–310, 310t
   S-phase fraction, evaluation of, 309–310
   therapeutic responsiveness, evaluation of, 314
Color filter(s), 181
Combined T- and B-cell deficiencies, 438–440
   phenotyping of cells in, 436t
Common-sense knowledge, 585
   definition of, 600
Common variable immunodeficiency, 435, 437
   phenotyping of cells in, 435t
Compensation, in multicolor detection, 113–114
Complement, in host defense, 435
Complementarity-determining region(s), 143
Complete blood count, 436
Congenital hairy pigmented nevi, 551
Congenital thymic aplasia, 439
Constant region(s), 65
Convolution, in DNA model(s)/modeling, 47–49
Cornified envelope, 546
Coulter Elite flow cytometer, 115
Coumarin-phalloidin, spectral properties of, 112t
Covalent labeling reagents, properties of, 111–112
C219 (P-glycoprotein-specific antibody), in flow cytometric detection of drug-resistant cells, 462–463
CPZ. *See* Chlorpromazine
CR3. *See* CD11b antigen
C-reactive protein, in neutrophil respiratory burst, 417
C region. *See* Constant region(s)
Cross-linking fixative(s), 158

Crossmatch
   flow cytometric
      antibody class in, 451
      B-cell, 453
      clinical significance of, 451–453
      sensitivity of, 451
      specificity of, 451
      in transplantation, 449
   pretransplant, standardized complement-dependent cytotoxicity technique, 449
CSF-1. *See* Macrophage colony-stimulating factor
Cumulative subtraction, for immunofluorescence analysis, 57f, 57–58
Curve-fitting, 125–126
Cutaneous T-cell lymphoma(s), flow cytometric findings in, 222
CV. *See* Coefficient of variation
CV_{GI}, 22
CVID. *See* Common variable immunodeficiency
Cy-3 (to Cy-7), 168t
Cyanine3.18, brightness of antibodies labeled with
   dependence on degree of labeling, 113
   at different dye/protein ratios, 111–112, 112f
Cyanine5.12, monomers and dimers, absorption spectra of, 113f
Cyanine5.18, brightness of antibodies labeled with
   dependence on degree of labeling, 113
   at different dye/protein ratios, 111–112, 112f
Cyanine dye(s), 155
Cyanine dye derivative(s), 111, 112t. *See also specific derivative*
   spectral properties of, 112t
Cyanine3-reactive dye, 114
Cyanine5-reactive dye, 115
CyChrome, 115
Cyclin(s), 17, 17f, 579–581
   class A, 17f
   class B, 17f, 18
   G_1, 17, 17f, 18, 21
   G_2, 17f, 18, 21
   proliferating cell nuclear antigen as, 28
Cyclophosphamide, effects on neutrophil function, 407t
Cyclosporin A, in blockade of drug resistance processes, 257
Cystic fibrosis, 406t
Cystosarcoma phylloides
   DNA ploidy, 346, 348t
   percent of DNA diploidy, 347
   prognostic value of DNA flow cytometry in, 354
Cytochalasin B, 411
Cytochemistry. *See* Cell surface immunofluorescence; Intracellular antigen(s), immunofluorescence measurement of
Cytochrome b, 409–410
Cytogenetic changes, in cancer, 4, 5t
Cytogenetics, conventional, for detection of minimal residual disease, 482t, 482
Cytokeratin(s)
   antibodies to
      in breast cancer analysis, 255–256
      and DNA content analysis, in flow cytometric analysis of bladder cancer, 285
      in tumor analysis, 168
   and DNA content, simultaneous measurement of, in bladder cancer, 290–291, 291f

immunofluorescence measurement of, using flow cytometry, 168, 169t
monoclonal antibody, available as direct conjugate, 166t, 166, 166f
as tumor marker in bone marrow, 485
Cytokeratin 18
   immunofluorescence measurement of, using flow cytometry, 168, 169t
   as potential marker for bladder tumor, 168
Cytokine(s), 64–65
   cell responsiveness to, role of glutathione in, 520
Cytokine-receptor family, 576–577
Cytokinesis, 13
Cytomegalovirus
   detection of, using capture-reagent-coated microspheres, 538, 538t, 540, 541t
   infection, CD8 and CD38 expression in, 443
Cytophotometry, 359
Cytoplasmic cycle, 13. *See also* Cell cycle
Cytoplasmic Ig μ chain, expression, in leukemia, 484
Cytosolic signal transduction, 577–579
Cytostatic drugs, 459
   tumor sensitivity to, predictors of, 34
Cytotoxic drugs, 459

D
DABCO, 146
Daemons, 592–593
Dansyl-NH-CH3, spectral properties of, 112t
DAPI, 23
Dapsone, effects on neutrophil function, 407t
Data
   multidimensional, 586
   numeric, 586
   symbolic, 586
Data acquisition system, 86–87
Data analysis. *See also* DNA model(s)/modeling; Expert systems
   theoretical aspects of, 41–60
Data reduction, 586
Data types, flow cytometric, 586–587
Daunorubicin
   cellular retention of, in drug-resistant cells, 257, 459–460, 460f
   cellular transport and fluorescence of, and pH, 461, 462f
DCC gene, 7
DCFH-DA. *See* Dichlorofluorescein diacetate
DCH. *See* 2,3-Dicyanohydroquinone
Dead-cell gating, 192–193. *See also* Cell viability
   for analysis of lymphoma, 204
Debris, 387
   in DNA content analysis, 123, 130, 130f
   effects on estimation of cell cycle parameters, 132–133
   from fixed paraffin-embedded materials, 101, 104
   in flow cytometric assays, and background nonspecific binding observed, 130f, 130, 489
   from paraffin-block sections, 101, 104
Defensin(s), 409, 419
Deleted in colorectal cancer gene. *See* DCC gene
Denaturing gradient gel electrophoresis, for mutation detection, 564
Dendritic cells, in epidermis, 545
Dense granules, platelet, 387
Dependency, of DNA model components, 45–46
Dermatofibrosarcoma protuberans

DNA ploidy, 346, 348t
  histologic grading of, 345f
  prognostic value of DNA flow cytometry in, 351
Detection windows, 113
Detergent lysis, of tumor cells, 95
DFSP. See Dermatofibrosarcoma protuberans
1,4-Diacetoxy-2,3-dicyanobenzene, 518t, 518
Diacylglycerol, 505
DIAGNOS1, 590
Dichlorofluorescein assay, of neutrophil H_2O_2 production, 418
Dichlorofluorescein diacetate, in assessment of neutrophil H_2O_2 production, 417–418, 418f
Dichroic-interference filters, 79
2,3-Dicyanohydroquinone, 518t, 518
Differential counts, use of flow cytometer for, 154–155
Differentiation
  asynchronous, in neoplasia, 484
  drug-induced, for cytostasis, 459
  mixed lineage, in neoplasia, 484–485
DiGeorge's syndrome, 439
  pathogenesis of, 439
  phenotyping of cells in, 436t
Digitonin, 158
Digoxigenin, in fluorescent in situ hybridization, 565–568, 566f
Dihydrofolate reductase
  cellular content of, and multidrug resistance, 461
  detection of, 28–29
$DiOC_1(3)$, for flow cytometric reticulocyte analysis, 377t, 378
Diode laser(s), 75–76
Diploidy, 4, 118
Direct immunofluorescence, 147
  advantages and disadvantages of, 165–166
Dissociation equilibrium constant (K_D), 471
Dissociation rate constant (k_d), 471
DNA. See also Cellular DNA
  aberrations, in cancer cells, 3
  content, versus cell age, 42–43, 43f
  genomic
    analysis of, molecular techniques in, 558–559
    quantitation of, competitive polymerase chain reaction for, 562–563, 563f
  measurement of, inaccuracies in, 22
  melting curves, 27
  sensitivity to denaturation, 27–28
  subunit of, monoclonal antibody, available as direct conjugate, 166t, 167
  synthesis, relative rates of, 55–56
DNA analysis
  of colorectal neoplasia, 307–317
  of FNA samples, 94
  in lymphoma, 206, 210–212, 215–216, 225–226
  of nuclear suspensions, 95
  performance surveys, 198
  quality control, 194–195
  of testicular aspirates, 94
DNA-aneuploid cells
  characterization, 211
  defining, 99–100
  loss of
    after fixation, 99
    during tumor dissociation, 96
DNA aneuploidy. See Aneuploidy
DNA-binding factors, redox-responsive, and glutathione, 520
DNA-coated microspheres, ligand binding to, detection of, 540

DNA content
  analysis
    background debris curve, 130, 130f
    in bladder cancer, 283–287
    effect of fixation protocols on, 160–161
    effects of cell or nuclear aggregation, 133f, 133–138
    in endometrial cancer, 265–266
    flow cytometric
      versus image analysis, advantages and disadvantages, 366t
      running the sample, 121–125
    by image analysis, 364–365, 365t
    number of cells for, 121–122
    number of tissue samples required for, 122–123
    in ovarian cancer, 263–265
      limitations of, 267f, 267–268
    practical considerations for, 117–140
    preparation of cell suspensions for
      from fresh tissue, 118–119
      from paraffin-embedded tissue, 119
    in prostate cancer, 274f, 274–281
    in renal cell carcinoma, 293–296
    simultaneously with intracellular glutathione, 528, 530f
    skewed peaks on, 121
    standards for, 124–125
    tissue sampling for, 122–123
    in unfixed samples, 122
  aneuploid, 118
  diploid, 118
  euploid, 118
  haploid, 118
  and intracellular (or cell surface) antigens, simultaneous measurement of, 168
    in bladder cancer, 290–291, 291f
    in lymphoma, 206
  standards for, 124–125
  tetraploid, 118
DNA content histogram
  analysis
    effect of debris, 123
    range of data in, 123
  of aneuploid tumor, 118
  background debris curve, 130, 130f
  cell cycle analysis of, 125–126
  digitized, linearity, and number of channels, 123–124, 124f
  effect of fixation protocols on, 160–161, 161f
  effects of nuclear slicing on, 131f, 131–132
  Gaussian broadening of, 117, 117f, 125
  interpretation of, 117
  nonbiological results on, 129
  from paraffin-embedded materials, 130f, 131–132
DNA data analysis, 41–56. See also DNA model(s)/modeling
  future of, 56
  inside-out, 45, 47f
  outside-in, 45, 45f
  simple methods, 45
    accuracy of, 45, 46f
DNA denaturation
  acid-induced, in tumor characterization, 28
  in situ sensitivity to, 27–28
DNA-diploid cells, in archival specimens, identification of, control specimens for, 194–195
DNA-diploid reference cells, 99–101
DNA histogram(s)
  and aneuploidy, 44
  average histogram comparison, 58–59, 60f
  classification of, flow cytometry versus image analysis for, 365

comparison methods, 58–59
  for doublet discrimination, 85, 86f
  D-value comparison, 58, 59f
  from fixed, paraffin-embedded versus unfixed samples, 101–102
  in high-grade soft-tissue sarcoma, interpretation of, 350, 352f
  ideal, 43f, 43–44
  interpretation of, 126
  modeling process applied to, 45–49
  nonlinear least-squares analysis and models, 44t, 44–45
  number of cells versus cell age relation, 43, 43f
  from paraffin-embedded materials, 41, 100–101, 101t
  and signal broadening, 44, 44f
  theory, 42–44
DNA Index, 118, 184, 211
  in breast cancer, 249–253
  in colorectal cancer, and survival, 307–308, 308f
  determination, composite control materials for, 194, 195f
  2.0 peak, 129
  prognostic significance of, in ovarian cancer, 267f, 267–269
  of prostate cancer
    biopsy specimens, 274–275, 275t
    radical prostatectomy samples, 275–277, 276t
DNA ligase, recognition points in chromatin, 15
DNA model(s)/modeling, 45–49
  components, 49–54. See also Gaussian function
    aggregates, 52–53, 53f
    broadened polynomial, 50–51, 51f
    broadened rectangle, 49–50, 50f
    broadened trapezoid, 50, 50f
    broadening, 48
    dependency, 45–46
    exponential, 51, 51f
    hardware aggregation compensation, 54
    histogram-dependent, 51, 53
    multi-cut, 51–52, 52f, 130
    single-cut, 51–52, 52f
    software aggregation compensation, 53–54
  construction of, 54
    conservative approach, 55
    and number of counts, 55
    and overlapping distributions, 55
    use of published models for, 55
  continuous aggregate theory, 53
  convolution, 47–49
  definition of, 45
  discrete aggregate theory, 52–53
  doublets in, 52–53, 84–85
  gradient search method, 46
  iterative process for, 48–49
  limitations of, 60
  linearization, 46–47
  mathematical analysis, 53
  quadruplets, 52–53
  rules of the road, 54–55
  standardization of, 56
  triplets, 52–53
  two- or more parameter, 56
DNA ploidy. See Ploidy
DNA polymerase, detection of, 28–29
DNA probes, 4–5
DNA replication
  in cell cycle, 13
  enzymes of, 21
  detection of, 28–29

DNA replication—(*continued*)
  measurement of, in multiparameter flow cytometry, 24
DNAse
  to prevent cell clumping, 147
  used to dissociate tumors, 95
DNA staining
  with cell enucleation, combination of, 95
  of cells, 119–121
    artifacts, 119–121
    with autolysis, 121
    differences in, 119–120
    DNA/dye ratio in, 120–121
    dye accessibility and binding in, 119–120
    dyes and dye classes for, 119
    nonstoichiometric, 120f, 120–121
    sample preservation, 121
  in lymphomas, 205–206
DNA topoisomerase II, detection of, 28–29
DNR. *See* Daunorubicin
Domain expert, definition of, 601
Dot hybridization
  for DNA, 559
  for RNA, 559
Double minute chromosomes, 4, 557
Double-positive cells, analysis of, 491
Doublet(s)
  in DNA analysis, 84–85
  in DNA model(s)/modeling, 52–53
    effects of, on DNA content analysis, 133
  orientation and fluorescence profile, 133f, 134
  types of, 85
Doublet discrimination, 84–85
  graphical approaches to, 85
  by peak vs. integral pulse plot, 85, 86f
  pulse-shape, 134f, 134–135
  by pulse width vs. integral pulse amplitude, 85, 86f
Doxorubicin
  cellular resistance to, cell membrane and, 461–462
  cellular retention of
    differential response to efflux blockers, 460–461, 461f
    in drug-resistant cells, effects of amphotericin B on, 459–460
    heterogeneity of, 460–461, 461f
  therapy, in bladder cancer, flow cytometric findings related to, 286
Droplets, for sorting, 89
Drosophila, studied using β-gal reporter system, 570
Drug efflux, cellular, 460
  energy-dependent pump for, 462
  and GSH content of cells, simultaneous monitoring of, 464, 465f
Drug influx, cellular
  of cytotoxic drugs, 459–461
  tumor-cell barriers to, 459
Drug resistance
  in breast cancer, 247
    flow cytometric assays of, 256–257
    monitoring, 248
  in cancer chemotherapy
    and chemotherapeutic effectiveness, 480
    flow cytometric monitoring of, 459–466
    role of glutathione in, 257, 520
  cellular parameters in, 459–465
  extracellular parameters in, 459
  mechanisms of, 257
  pharmacologic blockade of, 257
Drug retention, cellular
  differential response to efflux blockers, 460–461, 461f
  in drug-resistant cells, 257, 459–460, 460f

effects of cellular interactions on, 460
flow cytometric monitoring of, 459, 460f
Drug transport, cellular
  cell membrane and, 461–462
  of cytotoxic drugs, 459–461
  effects of cellular interactions on, 460
  flow cytometric monitoring of, 460–461, 461f
  and pH, 461
Duochrome, 114
D-value comparison, of DNA histograms, 58, 59f
Dye(s)
  for cell surface immunofluorescence, 155
  in DNA analysis, 194
  DNA-binding, 119
  DNA-specific, 117
  exclusion or retention, in flow cytometric definition of dead cells, 192
  fluorescent, for flow cytometric reticulocyte analysis, 377t, 378
  for measurement of intracellular antigens, 167, 168t
  vital, 464
Dye laser(s), 75
Dysgerminoma, chromosomal changes in, 5t

E
ECD, use of, 441
Ecto-5'-nucleotidase, deficiency, 438
EDTA. *See* Ethylenediaminetetraacetic acid; Ethylene-diaminoacetate
EGTA. *See* Ethyleneglycol (2-aminoethylether)-N-N'-tetraacetic acid
Elastase(s), 409, 419
  cleavage of immunoglobulin, 145
  used to dissociate tumors, 95
Electronic fluorescence spectral compensation, 84
Electronic signal processing, quality control, 178
Electron microscopy, fixation for, validation of technique, 164–165
Elliptical gating, 87
Elliptocytosis, 375–376
EL-2 (monoclonal antibody), reaction with epidermis, in psoriasis, 548, 549t
EM. *See* Electron microscopy
EMA. *See* Ethidium monoazide
Embryonal tumor(s), chromosomal changes in, 5t
EMYCIN, 589
Endolymphatic stromal myosis
  DNA ploidy in, 346, 348t
  prognostic value of DNA flow cytometry in, 353–354
Endometrial cancer
  DNA content in, 265–266
  flow cytometric analysis, using antibodies to cytokeratins, 168
  lymphatic metastasis, flow cytometric analysis of, 270
  ploidy in, 265–266, 268, 270, 346, 348t
    description, flow cytometry versus image analysis for, 364
  prognostic factors in, 265
Endometrial stromal nodule
  DNA ploidy, 346, 348t
  prognostic value of DNA flow cytometry in, 353–354
Endometrial stromal sarcoma(s)
  percent of DNA diploidy, 347
  prognostic value of DNA flow cytometry in, 353–354
Endometrial stromal tumor(s), prognostic value of DNA flow cytometry in, 353–354

Enzymatic digestion
  of paraffin-embedded materials, 119
  of rehydrated sections from fixed, paraffin-embedded materials, 104
Enzymatic dissociation, of solid tumors, 94–95
  evaluation of, 95–96
  from rehydrated sections from fixed, paraffin-embedded materials, 104–105
Enzyme(s)
  with fluorogenic substrate, 115
  used to dissociate tumors, 94–95, 104–105
Enzyme immunoassay, 535
  versus flow microsphere immunoassay, 540–541, 541t, 542f
Eosinophil(s)
  development of, 64
  flow cytometric properties of, 406–407
Ependymoma
  cerebral, pediatric, 332
  pediatric, incidence of, 332
EPICS flow cytometer, optical system in, 78
EPICS PRISM, 87–88
Epidermal growth factor, receptor, 573
  expression, in human tumors, 575
Epidermis
  basal layer, 546
  cells within, 545, 545f–546f
  DNA content, measurement of, 547
  monoclonal antibodies reactive with, 548, 549t
Epirubicin, effects of, on neutrophil function, 407t
Epithelial membrane antigen, in detection of metastatic epithelial tumor, 485
Epithelioid sarcoma
  histologic grading of, 345f
  prognostic value of DNA flow cytometry in, 354
Equilibrium binding, thermodynamics and kinetics of, 471–474
Erythrocyte(s). *See* Red blood cell(s)
Erythroid cell(s), progenitors for, 64
Erythroleukemia
  antigen expression in, 244
  detection of, 237–238
Erythromycin, effects on neutrophil function, 407t
Erythropoietin receptor(s), identification of, flow cytometric, 489
Esterase(s), flow cytometric analysis of, 410
EtBr. *See* Ethidium bromide
Ethanol
  effects of
    on cellular parameters, 283
    on DNA histograms from archival material, 102, 103f
  fixation, 157
    for bladder-wash or urine cell samples, 283
    of dissociated tumor cells, from fresh specimens, 96, 99, 101f
Ethidium, fluorescence, after cell growth in presence of BrdUrd, 31, 32f
Ethidium bromide, 119
  for flow cytometric reticulocyte analysis, 377t, 378
Ethidium monoazide, in cell viability determination, 144, 192
Ethylenediaminetetraacetic acid, blood collection in, 148
Ethylene-diaminoacetate, as chelating agent, 94
Ethyleneglycol (2-aminoethylether)-N-N'-tetraacetic acid
  as chelating agent, 94

structure of, 506f
Euploidy, 118
Event discrimination, 86
Ewing's sarcoma, 338
  chromosomal changes in, 5t
  extraskeletal, histologic grading of, 345f
Excitation light beam, 71, 72f
  horizontally shortened, 77
  narrowed, effect on sensitivity, 77, 77f
  shape of, 76–77
  vertical dimension of, 77
  widening, 76–77, 77f
Excitation light source, 73–77. See also Arc lamp(s); Laser(s)
Expert systems
  and conventional programs, 586
  for cytometry data analysis, 585–601
  definition of, 585, 601
  expectations of, 585–586
  knowledge representation for, 587–593
  tasks for, 585
  uncertainty management in, 593–596
Explanation facility, 588f
Exponential model component, 51, 51f

F
FAB criteria, for classification of leukemias, 235
FAI. See Fractional allelic imbalance
FAL. See Fractional allelic loss
4F2 antigen. See Activation antigen 4F2
FCM. See Flow cytometry
Fc receptors
  expression, on neutrophils, 420–421
  nonspecific binding of antibody to, 144–145, 486–490
FCS file format. See Flow cytometry standard file format
Fetal–maternal hemorrhage, flow cytometric analysis of red blood cells in, 373–374
α-fetoprotein, in ataxia-telangiectasia, 439
FEU. See Fluorescent equivalent units
Feulgen stain, 363–364
Fibroadenoma(s)
  breast
    DNA ploidy, 346, 348t
    prognostic value of DNA flow cytometry in, 354
  DNA ploidy, 346, 348t
  giant
    DNA ploidy, 346, 348t
    prognostic value of DNA flow cytometry in, 354
Fibrohistiocytic lesions, prognostic value of DNA flow cytometry in, 351–352
Fibroma, DNA ploidy, 346, 348t
Fibromatosis
  DNA ploidy, 346, 348t
  prognostic value of DNA flow cytometry in, 351
Fibrosarcoma(s)
  histologic grading of, 345f
  infantile, histologic grading of, 345f
  percent of DNA diploidy, 347
  prognostic value of DNA flow cytometry in, 355
Fillagrin, biosynthesis of, 546
Filter(s), 71, 72f, 78–79, 181
  absorption, 79
  bandpass, 79
  color, 181
  dichroic-interference, 79
  interference, 79, 181
  longpass, 79
  monitoring, 181
  notch, 79

selection, 181
  and color compensation, 185–187, 186f
  shortpass, 79
Fine-needle aspiration
  in breast cancer, 247–248, 253
  samples, storage of, 99
  of soft-tissue sarcomas, 355
  of solid tumors, 93–94
FISH. See Fluorescence in situ hybridization
FISHES. See Fluorescent in situ hybridization, en suspension
Fission yeast. See Saccharomyces pombe
FITC. See Fluorescein-isothiocyanate
Fixation. See also Cell fixation
  of dissociated tumor cells
    from fresh specimens, 96–99
    from sections of fixed, paraffin-embedded tissue, 102, 103f
  effects of
    on antigen detection, 158–159
    on background nonspecific binding observed, 489–490
Fixation/permeabilization
  for intracellular antigen quantitation, 157t, 157–161
  agents for, 157, 157t, 158
  protocols for, and DNA content analysis, 160–161
FLIP. See Fluorescent in situ polymerase chain reaction
FLM curves, 13
Flow aneuploidy, 215. See also Aneuploidy and DNA aneuploidy cells.
Flow cell, 72f, 72–73
  capillary, 73
  chambered, 73
  closed flow design, 73
  hydrodynamic focusing within, 71–72, 85
  lens coupling to, 73
  optical sensitivity of, 73
  orientation of cells within, 73
  quartz, 73
  sample delivery rate, 71–72
  sense-in-air design, 73
  sensing portion of, 73
Flow cytometer(s), 71–92. See also Flow cell
  accuracy, 90
  beam dump, 77–78
  beam-shaping optics, 76–77
  calibration, 121, 197, 469–470, 476
  cell sorters, 89
  clinical applications, 76
  components of, 71
  data acquisition system, 86–87
    analog-to-digital converter, 86–87
    event discriminator, 86
    time in, 87
  data storage, 87
  electronic pulse integrators used in, 82
  excitation light beam, 71, 72f
    horizontally shortened, 77
    narrowed, 77, 77f
    shape of, 76–77
    vertical dimension of, 77
    widening, 76–77, 77f
  excitation light source, 73–77. See also Arc lamp(s); Laser(s)
  filters, 71, 72f, 78–79, 181
  flow system, 71–73
  fluidics system, 71–72
  fluorescence and side-scatter sensors, 78–80
  forward-angle light scatter sensor, 77–78
  future developments in, 76
  gating capability, 87–88
  lens, 73
  light collection optics, 72f, 73, 78

collimating radiometer design, 78
  imaging radiometer design, 78
  list mode data format, 88
  multilaser, 114–115
  operation of, 71
  performance, objective measurements of, 89–92
  performance assessment, 178
  performance monitoring
    record keeping, 189–190
    strategy, 189–190
    using constant instrument settings, 179f, 179–181
    using constant intensity settings, 181, 182f
  photodetector, 71, 72f, 77–78
    fluorescence and side-scatter, 79–80
  precision, 76, 89–90
  pulse amplification, 83
    linear, 83
    logarithmic, 83
  pulse processing, 83–86
    baseline restoration, 84
    doublet discrimination, 84–85
    electronic fluorescence spectral compensation, 84
  pulse signals generated by
    integral pulse, 80–81, 81f, 82
    peak pulses, 80–82, 81f
    types of, 80–82
  pulse width, 91
  quality control, 177–190
  ratio circuitry, 86
  resolution, 197
  running the sample on, for DNA content analysis, 121–125
  sample core stream, 71–72
    diameter, 71
  sample delivery systems
    pressure-driven, 71
    syringe-based, 71
  sample suspension, 71
  sensitivity, 76, 90–91, 91f, 197
  sheath fluid, 71–72
  signal processing
    electronics for, 80–86
    quality control, 177–178
    types of, 178, 178f
  skewed peaks, 121
  specimen analysis conditions, optimization, 195, 196f
  as spectrofluorimeter, in measurement of intracellular ionized calcium, 509
  standardization, 197
  throughput, 76, 91–92
    determination, from allowable miss rate, 91–92
  throughput rate, and doublets, 85
  tuning, criteria for, 121
Flow cytometry. See also Multiparameter flow cytometry
  advantages of, 362, 364
  in cell-cycle research, 13–14
  cell kinetic measurements by, 367
  controls for, 363
  data structure, 586–587
  data types, 586–587
  history of, 13
  in HIV disease, 440–445
  immunoassay technique utilizing, 535–543
  measurement of panel antibody reactivity, 453
  measurement process, quality control, 177–178
  and microscopy, differences in detection by, 146–147

Flow cytometry—(continued)
  plugs in, 147–148
  and polymerase chain reaction, tandem use of, in detection of residual disease, 493
  problems encountered in, 147–148
  standards for, 363
  three-color, 114
    in HIV disease, 441
  two-color analysis
    effect of under- and overcompensation on, 187f, 188
    instrument set-up and monitoring for, 185–187, 186f
Flow cytometry standard file format, 88, 89f
Flow microsphere immunoassay
  for *Candida* antibodies, 539–540
  clinical applications of, 543
  for cytomegalovirus, 538, 538t, 540, 541t
  detection of soluble analyte, 539–540
  detection of specific antibody using capture-reagent-coated microspheres, 537–539
  dynamic range of, 543
  versus enzyme immunoassay, 540–541, 541t, 542f
  for herpes simplex virus, 538, 538t, 540, 541t
  history of, 535–536
  for HIV proteins, 538
  for human antimouse antibodies, 539
  limitations of, 542–543
  microsphere aggregation and, 543
  microsphere carryover and, 543
  microspheres for, 536f, 536–537, 542–543
  sensitivity of, 540–541
  simultaneous detection of multiple analytes, 543
  technique for, 536–537
Fluo-3
  emission spectra, upon calcium binding, 506, 507f
  loading
    complications of, 514–515, 515f
    poor cellular response with, 515–516
    preparation of cells for, 506–508
    ranges of, 507
  in measurement of intracellular ionized calcium
    calibration of fluorescence, 508–509
    display and analysis of results, 509–511, 510f–511f
  structure of, 506f
Fluorescein, 111
  bound to IgG at different dye/protein ratios, average quantum yields of, 112f
  dichlorotrizinyl-, spectral properties of, 112t
  for evaluation of HIV disease, 444
  fluorescence, quenching, 113
  iodoacetamido-, spectral properties of, 112t
  maleimido-, spectral properties of, 112t
Fluorescein diacetate, 517
  in cell viability determination, 144
Fluorescein-isothiocyanate, 63–64, 143, 167, 168t
  antibodies conjugated to, 166t, 167
    at different dye/protein ratios, brightness of, 111–112, 112f
  effect on flow cytometric results, 68, 68f
  fluorescence, increased, compensation/subtraction for, 150, 151f
  fluorescent emission characteristics of, 112t, 146f

  in immunofluorescence applications, 146
  and phycoerythrin
    spectral overlap between emission spectra of, 84, 84f
    in two-color analysis
      effect of under- and overcompensation on, 187f, 188
      instrument set-up and monitoring for, 185–187, 186f
    in quantitative calibration of fluorescence intensity, 470
  spectral properties of, 112t, 146f
Fluorescein-isothiocyanate-CD5, in two-color analysis, effect of under- and overcompensation on, 188f, 188
Fluorescein-isothiocyanate-PE combination, 150
Fluorescence in situ hybridization
  cell preparation for, 565
  for detection of minimal residual disease, 482t, 482
  en suspension, 564–569
  and flow cytometric detection, 568
  methodology, 565–568
  mRNA measurement, 314, 567–568
  nucleic acid probes
    choosing, 567–568
    preparation of, 565–566
Fluorescence intensity. *See also* Intensity scale(s)
  versus percentage of cells with fluorescence intensity above this level, cumulative histograms of, 486, 487f
  quantitative calibration for, 469–470, 476
  relative, conversion to biologically or clinically meaningful units, 184
  stability, 184
  stepwise amplification of, 488–489
Fluorescence microscopy, photobleaching in, 146
Fluorescence quenching, 112–113, 146
  with high-density receptors, 475–476
Fluorescence sensor, 78–80
Fluorescent drug(s), cellular transport and retention of, flow cytometric monitoring of, 459, 460f
Fluorescent dye(s). *See also* Dye(s)
  for flow cytometric reticulocyte analysis, 377t, 378
Fluorescent equivalent units, 90, 91f
Fluorescent in situ polymerase chain reaction, 568–569, 569f
Fluorescent reagent(s), 111–115. *See also specific reagent*
  aggregates, on antibody surface, 113
  applications of, 111
  for covalent labeling, properties of, 111–112
  future of, 115
  low-molecular-weight, 111, 112t, 114
  for multi-color immunofluorescence, 113–115
  three-color, 115
Fluorochrome(s), 63–64
  DNA-specific, 23
  effect on flow cytometric results, 68, 68f
  fluorescent emission characteristics of, 146f
  suitable for quantitation of intracellular antigens, 167, 168t
  tandem, 64
Fluorokine assay, for detection of GM-CSF receptors, in acute myeloid leukemia, 577, 578f–579f
Fluorophore(s)
  average Φ of, 111–112, 112f
  blue-emitting, 115

  bound to IgG at different dye/protein ratios, average quantum yields of, 112f
  fluorescence properties of, 111
  hydrophobicity, modification of, 113
  as physiological indicators, 111
  red-emitting, 115
Fluorophor/protein (F/P), fluorescence efficiencies of, 474
FMC7, in immunophenotyping of leukemia, 219, 223
FMIA. *See* Flow microsphere immunoassay
fMLP. *See* Formyl-methionyl-leucyl-phenylalanine
4F2 (monoclonal antibody), reaction with epidermis, in psoriasis, 548, 549t. *See also* Activation antigen
FNA. *See* Fine-needle aspiration
Follicular lymphoma. *See* Lymphoma(s), follicular
Formaldehyde, fixation with, 158
Formalin, 158
  fixation, of dissociated tumor cells, from fresh specimens, 96
Formic acid, 158
Formyl-methionyl-leucyl-phenylalanine
  activation of neutrophils, 411
    calcium requirement for, 413
  binding, in neutrophils, 414–415, 415f
  receptors, in neutrophils, 414, 415f
Forward-angle light scatter sensor, 77–78
  collection angle of, 78
  detectors, 77–78
  optics, 77–78
Forward-angle light scatter signal, monitoring instrument precision with, 90
Forward chaining, 588–589
  definition of, 601
Forward scatter, 207
  and immunofluorescence, correlation, in lymphoma, 207, 208f
  versus log side scatter, of whole blood cell preparation, 148–150, 149f
  versus side scatter, in leukemia, 241f, 241–243, 242f
Fractional allelic imbalance, in breast cancer, 9, 9f
Fractional allelic loss, 7–8
Frames, 590–593
  accepted, 593
  active, 593
  definition of, 601
  dormant, 593
  semiactive, 593
French-American-British criteria. *See* FAB criteria
FS sensor. *See* Forward-angle light scatter sensor
Full-peak method, for calculation of coefficient of variation, 89
Full-width-at-half-maximum, 90
Fura-2
  compartmentalization, 513
  loading, complications of, 513–515, 515f
Fuzzy logic, 594–595
  definition of, 601
Fuzzy set membership functions, 594f, 595
FWHM. *See* Full-width-at-half-maximum

G

β-galactosidase
  fluorescent substrate for, 569–570
  as reporter gene, flow cytometric use of, 569–571, 570f
Ganglioneuroblastoma, histologic grading of, 345f

Ganglioneuroma(s), DNA ploidy, 346, 348t
Ganglioside GD2
  as marker for metastatic neuroblastoma, 485
  in melanoma, 552–553
Ganglioside GD3, in melanoma, 552–553
GAP. *See* GTPase-activating protein
Gas laser(s), 75
Gastric carcinoma, ploidy, prognostic significance of, 314
Gastric lesions, DNA flow cytometry of, 314
Gastrointestinal stromal sarcoma(s), percent of DNA diploidy, 347
Gastrointestinal tract tumor(s)
  DNA flow cytometry for, 313–314
  smooth-muscle, 346
  stromal, prognostic value of DNA flow cytometry in, 353
Gating, 87–88
  amorphous, 87
  of cells responsible for nonspecific binding, in multiparameter flow cytometry, 488
  dead-cell, 192–193
  for analysis of lymphoma, 204
  elliptical, 87
  for flow cytometry versus image analysis, 362
  light-scatter
    effects on background nonspecific binding observed, 489–490
    of lymphocytes, 190
  live, in analysis of lymphoma samples, 206
  morphologic, with solid tumors, 362
  rectilinear, 87
Gaussian beam profile, 76
Gaussian function, 41–42, 48–49, 49f, 54
  intuitive derivation of, 41–42, 42f
Gelatinase, 410
Gene(s)
  amplification, 557
    in oncogene overexpression, 5
  deletions, 558, 558f
  genetic analysis at level of, 5–7
  mutations, 558, 558f
    detection of, 559
  rearrangements, 557–558, 558f
  translocation, 557–558, 558f
    detection of, by polymerase chain reaction, 564
Genetic analysis
  at chromosomal level, 3–5
  at gene level, 5–7
Genetic testing, flow cytometric techniques in, 376
Genomic instability
  in clonal evolution, 8
  and development of aneuploidy, 8
Gentamycin, effects on neutrophil function, 407t
Germ cell tumor(s)
  ovarian, flow cytometric analysis of, 267
  pediatric, incidence of, 331, 331t
GFAP, immunofluorescence measurement of, using flow cytometry, 169t
G_2/G_1 ratio(s), 118
  aberrant, 129
Giant-cell tumors of bone, DNA ploidy, 346
Giant-cell tumors of tendon sheath, prognostic value of DNA flow cytometry in, 351
Giemsa trypsin (G)-banding, 4
Gleason's score, 272, 273f
Glial tumor(s), pediatric
  grading of, 332
  histology of, 332t
  ploidy of, 332t

prognostic factors for, 332
Glioblastoma multiforme, pediatric, 332
  ploidy of, 332t
Glucose-6-phosphate dehydrogenase deficiency, 376, 406t
β-glucuronidase, 409
Glucuronidase(s), flow cytometric analysis of, 410
Glutaraldehyde, 158
  fixation, of dissociated tumor cells, from fresh specimens, 96
Glutathione
  biological functions of, 257, 519
  biosynthesis of, 519, 521f
  in drug-resistant cells, 257, 464, 520
  effects on cell functions, 520
  intracellular, 519–531
    correlation between HPLC measurement and MCB/flow cytometry, 524, 524f
    flow cytometric analysis of
      calibration, 524, 525f
      correlation with biochemical assay, 524
      correlation with HPLC measurements, 524, 524f–525f
      data analysis, 526, 526f
      technique, 522–524
    flow cytometric monitoring of, 464
    fluorescent indicators of, 520–522
    measurement of, 257–258
    relationship to cell size, 526, 526f
    sorting on basis of, 528–531, 530f
    variations during cell cycle, 528, 530f
  metabolism of, 519, 521f
    in clinical drug resistance, 257
  redox cycle, 519, 521f
  role in cell activation, 520, 527f, 527–528
Glutathione peroxidase, in drug-resistant cells, 464
Glutathione reductase, 521f
Glutathione S-transferase, 257–258
  activity toward MCB, 522, 524
    in different cell types, 522, 524–526, 526f
  in drug-resistant cells, 464
  isoforms of, 464, 466, 522, 524
Glycogen storage disease type 1b, 413
Glycophorin A, in diagnostic hematology, 377
Glycoprotein p97, in melanoma, 552–553
GM-CFU. *See* Granulocyte-macrophage colony-forming unit
GM-CSF. *See* Granulocyte-macrophage colony-stimulating factor
G_2-M cyclin(s), 17f, 18
GMP-140, expression of, in platelet activation, 399–400
Goodness of fit, between model and observed data, 46
G_0 phase, DNA content during, 42–43, 43f
G_1 phase, 13
  DNA content during, 42–43, 43f, 117f, 117–118, 129–130
G_2 phase, 13
  DNA content during, 42–43, 43f, 117f, 117–118, 129–130
gpIIb/IIIa, 388
  as marker of platelet activation, 398t, 398–399
G-proteins, 5
G_{10} cells, 24
Graft-versus-host disease, after bone marrow transplantation, 455
α Granules, platelet, 387–388
Granulocyte(s)
  cell-surface-differentiation antigens, 66f

development of, 64
flow cytometric analysis of, 405–422
in host defense, 435
stem cell origin of, 405
Granulocyte-macrophage colony-forming unit, 405
Granulocyte-macrophage colony-stimulating factor, 405
  cell responsiveness to, role of glutathione in, 520
Granulosa cell tumors, ovarian, flow cytometric analysis of, 266–267
Growth-associated kinase. *See* Kinase G
Growth factor(s). *See also* Epidermal growth factor; Platelet-derived growth factor
  cell surface receptors, 573, 574f
  in prostate cancer, 281
GTPase-activating protein, 579
Gynecological cancer, flow cytometric analysis of, 263–270. *See also* Cervical cancer; Endometrial cancer; Ovarian cancer
  limitations of, 267–268

H
Hairy cell leukemia, flow cytometric findings in, 223, 223f
Half-peak method, for calculation of coefficient of variation, 89
Haploidy, 118
HCL. *See* Hairy cell leukemia
Heart transplantation, flow cytometric crossmatch for, clinical significance of, 452–453
HeCd laser(s). *See* Helium-cadmium laser(s)
HeLa cells
  high mobility group protein content of, 20
  histone phosphorylation in, 18
Helium-cadmium laser(s), 75, 115
  excitation wavelengths available from, 75, 75t
  output wavelengths, 75–76
Helium-neon laser(s), 75
  excitation wavelengths available from, 75, 75t
Hemangioma(s), DNA ploidy, 346, 348t
Hemangiopericytoma(s), risk factors in, 351t
Hematological malignancy, pathogenesis of, 576
Hematologic malignancy. *See also specific malignancy*
  cytogenetic analysis of, 3
  DNA-content changes in, 3
Hematopoietic cells, oncoproteins expressed by, 579
Hematopoietic elements, in host defense, 435
Hematopoietic precursor cells, ontogeny of, 235–238
Hematopoietic stem cell(s), 64
Hemoglobin F, neonatal switching from, detection of, 377
Hemolytic anemia, 374
  diagnosis of, flow cytometric techniques for, 375–376
  etiology of, 375
HeNe laser(s). *See* Helium-neon laser(s)
Heparin, blood collection in, 148
Hepatic cirrhosis, neutrophil chemotaxis in, 416
Hepatoblastoma, pediatric, incidence of, 331, 331t
Hepatocellular carcinoma, ploidy, prognostic significance of, 314
HER-2/c-erB/neu oncogene
  amplification of, in carcinoma, 6
  in breast cancer, 169

HER-2/c-erB/neu oncogene—(continued)
  immunofluorescence measurement of, using flow cytometry, 168–169, 169t
  overexpression, and DNA content, 9
  oncoprotein, measurement of, in breast cancer, 363
Hereditary spherocytosis, 375–376
Herpes simplex virus, detection of, using capture-reagent-coated microspheres, 538, 538t, 540, 541t
Heuristics, 585–586
  definition of, 601
High Mobility Group proteins, 16, 20–21
  HMG I family of, 20
  types of, 20
High-performance liquid chromatography, of oligonucleotides, 562
Histogram(s), 178
  one-parameter, 88
  two-parameter, 88
    elliptical gating in, 87
Histogram data, 88. See also DNA histogram(s)
Histone(s)
  acetylation, 27
    during cell cycle, 14–15
  dephosphorylation, 19
  methylation, 15
  modification, in noncycling cells, 19–20
  phosphorylation, 16–19, 27
    sites of, 18–19
  poly(ADP)ribosylation, 15–16
  synthesis, 21
    in S phase, 15
  ubiquitination, 16
Histone H1
  phosphorylation, 16
  poly(ADP)ribosylation, 16
  subtypes of, 18–19
Histone H1°, 19–20
Histone H2A, phosphorylation, during cell cycle, 19
Histone H3, phosphorylation, during cell cycle, 19
Histone H5, in avian erythrocytes, 20
Histone H1 kinase. See Kinase G
Histone kinases, 16–19
HIV, detection of, assay, using reporter gene, 570–571
HIV-associated p24 antigen, immunofluorescence measurement of, using flow cytometry, 168–170, 169t
HIV disease
  B cells in, 444
  $CD4^+$ cells and subsets in, 441–443
  $CD8^+$ cells and subsets in, 443
  clinical management of, 442
  flow cytometric analysis of cellular calcium concentration in, 516–517, 517f
  flow cytometry in, 440–445
  γ/δ-receptor-bearing T cells in, 443–444
  immune subset alterations in
    in adults, 440–444
    in children, 444
  immunophenotypic profiles in, 440, 441t
  immunophenotyping in
    choice of panel for, 444–445
    monoclonal antibody combinations for, 441t, 441, 444–445
  natural killer cells in, 444
  neutrophil defects in, 420–421
  phenotyping in, conclusions for, 446
  platelet-associated immunoglobulins in, flow cytometric analysis of, 390
  survival prediction in
    and CD4 counts, 442

lymphocyte subset parameters for, 441
  total T cells in, 443
HIV proteins, detection of, using capture-reagent-coated microspheres, 538
HK-I (H). See Kinase A
HLA
  antibodies against, flow cytometric quantitation of, 453
  expression
    on lymphocytes, 450
    on platelets, 450
HLA class I molecules
  antibodies against, in transplant recipients, detection of, 449–450
  screening of transfusion products for, 376
HLA class II molecules, 608
HLA-DR
  expression
    after bone marrow transplantation, 455
    in HIV disease, 443
    by keratinocytes, 550
    by lymphocytes in graft, changes in, after transplantation, 454–455
    by lymphocytes in peripheral blood, changes in, after transplantation, 454
  in immunophenotyping of leukemia, 236–237, 238f, 243t, 244
  in immunophenotyping of lymphoma, 205, 205t, 208f, 221
HMG proteins. See High Mobility Group proteins
Hodgkin's disease
  aneuploidy in, 314
  flow cytometric findings in, 217t, 217–218
  neutrophil chemotaxis in, 416
Hoechst 33258
  DNA specificity, 119, 482
  fluorescence, after cell growth in presence of BrdUrd, 31, 32f
Hoechst 33342, 24, 119, 464
Homogeneously stained regions, 4
Horseradish peroxidase, detection of, with flow microsphere immunoassay, 539, 540f
Host defense. See also Immunodeficiency
  normal, 435
H-ras, 5
HSR. See Homogeneously stained regions
Human antimouse antibodies, flow microsphere immunoassay for, 539, 541
Human T-cell lymphotropic virus, T-cell neoplasia with, 222
Humoral immunity, 65
  evaluation of, 435
Humoral immunodeficiency, 436–438
  phenotyping of cells in, 435t
Hyaluronidase, used to dissociate tumors, 95
Hybridoma technology, and development of flow cytometry, 63
Hydatidiform mole(s). See Molar pregnancy
Hyper IgM immunodeficiency, 437–438
Hyperimmunoglobulin E, 416, 438
Hypertetraploidy, in breast cancer, prognostic significance of, 251
Hypogammaglobulinemia. See also Common variable immunodeficiency; Severe combined immunodeficiency; Transient hypogammaglobulinemia of infancy; X-linked agammaglobulinemia
  lymphocyte immunophenotyping in, 440
Hypomethylation, 8

I

IA. See Image analysis
Ibuprofen, effects on neutrophil function, 407t

Idarubicin, effects on neutrophil function, 407t
IF. See Immunofluorescence
Ig. See Immunoglobulin(s)
IgA
  deficiency, 435, 437
    phenotyping of cells in, 435t
  diffusion coefficient of, 162
IgG
  antimouse
    fluorescent conjugate of, in estimation of ligand binding, 471
    reference beads containing calibrated numbers of, 471
  detection of, with flow microsphere immunoassay, 539–540
  diffusion coefficient of, 162
  in transplant recipient, 451
IgG_2 deficiency, 437
IgG_3 deficiency, 437
IgM
  diffusion coefficient of, 162
  nonspecific intracellular staining, 163
Image analysis, 359–368
  advantages of, 364
  analog-to-digital conversion, 360
  automated, 364
  basic technology, 359–364
  CAS 200 System, 361
  cellular measurements possible with, 362–363
  clinical applications of, 359, 363t, 363–364
  controls and standards for, 363
  definition of, 359
  DNA content analysis, 363–365, 365t
  versus flow cytometry, comparison of technical features, 361–364, 362t
  fluorescence measurements by, 360–361
  future of, 368
  grey levels, 306f, 360
  hardware components for, 359, 359f
  microscope-based system, 361
  for multiparameter cell classification, 364
  optical density measurement by, 360f, 360
  pixels, 306f, 360
  software, 359
Image cytometry, 359
  in breast cancer, 248, 252f, 252–253, 258
Immotile cilia syndrome, 411
Immune complexes, capture of, with flow microsphere immunoassay, 539
Immune system
  cell lineages of, 64–66
  complexity of, 63–64
  dynamics of, 68
Immune thrombocytopenia
  diagnosis of, 391
  platelet-associated immunoglobulins in, flow cytometric analysis of, 390
  treatment of, 390–391
Immunoassay(s), 535. See also Enzyme immunoassay; Flow microsphere immunoassay; Microsphere-based fluorescence immunoassays
  classification of, 535
  competitive, 535
  for detection of antibody or antigen, 535
  heterogeneous, 535
  homogeneous, 535
  signal-to-noise ratio in, 535
  solid-phase, 535
Immunoblotting, confirmation of flow cytometric antigen measurements with, 165
Immunocompromised patient. See also HIV disease

aggressive B-cell lymphomas in, 222
  laboratory evaluation of, 435
Immunocytochemistry, of proliferation-associated cellular constituents, 28–29
  limitations of, 29
Immunodeficiency. See also HIV disease
  flow cytometric evaluation of, 435–446
Immunofluorescence. See also Cell surface immunofluorescence; Indirect immunofluorescence
  analysis regions, selection of, and variability of results, 193
  data, reporting methods, 193–194
  four-color, for phenotyping in HIV disease, 446
  multi-color, 184
    development of, 63–64
    fluorescent labels for, 113–115
  percent-positive results, 193–194
  in lymphoma, 206–207
  quality control, 190–194
  quantitative, 154
  three-color, for phenotyping in HIV disease, 446
  two-color, in evaluation of HIV disease, 445f
    staining panel for, 444–445
Immunofluorescence analysis, 56–58
  cumulative subtraction for, 57f, 57–58
  normalized subtraction for, 56f, 56–57
Immunofluorescence microscopy, residual disease assay, 482t, 483
  in T-cell ALL, 493
Immunoglobulin(s). See also IgA; IgG; IgM
  in ataxia-telangiectasia, 439
  μ chains, immunofluorescence measurement of, using flow cytometry, 169t, 170
  in common variable immunodeficiency, 437
  genes, 65
    clonal rearrangement, in detection of neoplastic lymphoid cells, 482, 486
  heavy chains, 65
    antibodies against, in immunophenotyping of lymphoma, 205
    monoclonal antibody against, available as direct conjugate, 166t
  intravenous, stimulation of neutrophils by, 421
  κ and λ distributions, in analysis of clonal B-cell expansion, 207–210, 209f–210f
  light chains, 65
    antibodies against, in immunophenotyping of lymphoma, 205
    in follicular lymphoma, 219
    monoclonal antibody against, available as direct conjugate, 166t
    in myeloma, 224
  nonspecific intracellular staining, 163, 192
  platelet-associated
    flow cytometric analysis of, 390–394
      advantages and disadvantages of, 392
      data analysis, 393f, 393
      indications for, 390
      non-platelet particles in, 394, 395f–396f
      protocols for, 392–393
      specimen preparation and handling for, 392
    immunoassay for, 391–392
    measurement of
      clinical significance of, 391–392
      direct versus indirect techniques, 391

methods for, 391
problems encountered in, 391
secretion, 65
serum
  in humoral immunodeficiency, 435
  in T-cell deficiency, 438
  in Wiskott-Aldrich syndrome, 440
  in X-linked agammaglobulinemia, 436–437
Immunohematology, flow cytometric analysis of red blood cells in, 373–377
Immunophenotype, 559
Immunophenotyping
  of acute myeloid leukemia, 237, 238f
  of acute progranulocytic leukemia, 237
  of B-cell chronic lymphocytic leukemia, 210, 211f
  of B cells
    analysis of data, 153, 153f
    in clinical evaluation of primary immunodeficiency, 436t, 440
  of benign lymphoid hyperplasia, 216–217, 217t
  in clinical evaluation of primary immunodeficiency, 436t, 440
  flow
    automated, 589
    principal components analysis in, 599
  of FNA samples, 94
  in HIV disease
    choice of panel for, 444–445
    monoclonal antibody combinations for, 441t, 441, 444–445
  of leukemia, 211f, 236–237, 237f, 238, 240t, 242f, 243t, 243–244
  of lymphocytes
    analysis of data, 153–154, 153f–155f
    choice of reagents for, 151t, 151–152
    in clinical evaluation of primary immunodeficiency, 436t, 440
    guidelines for, 152, 152t
    in HIV disease, quality control for, 445–446
    in hypogammaglobulinemia, 440
    internal consistency checks for, 152, 152t
    monoclonal antibodies for, 151t, 151–152
    in severe combined immunodeficiency, 440
  of lymphomas, antibodies used for, 205, 205t, 208f–209f, 214, 219–221, 225
  performance surveys, 198
  of plasma cell tumors, 224
  of red blood cells, for genetic or paternity testing, 376
  of reticulocytes, through transferrin receptor, 377–378
  simultaneously with intracellular ionized calcium (Ca^{2+}) assay, 511–513, 512f–513f
  in T-cell deficiencies, 436t, 440
  of T cells, in clinical evaluation of primary immunodeficiency, 436t, 440
Immunostaining, 363
  of lymphoma, 204–205
Immunotherapy
  for melanoma, flow cytometric monitoring during, 553
  for renal cell carcinoma, 304
Indirect immunofluorescence, 147
  antigen saturation for, 162–163
  versus direct, for quantitation of intracellular antigens, 165–167
  limitations of, 165
  reagent titration for, 144
Indo-1, 506
  as Ca^{2+} indicator, 412–413, 413f
  calibration of fluorescence, 508–509

display and analysis of results, 509–511, 510f–513f
emission spectra, upon calcium binding, 506, 507f
loading
  complications of, 513–515, 515f
  poor cellular response with, 515–516
  preparation of cells for, 506–508
  ranges of, 507
  structure of, 506f
Indomethacin, effects on neutrophil function, 407t
Infection(s). See also Opportunistic infection(s)
  after bone marrow transplantation, 455
  neonatal, and neutrophil function, 421
  predisposition to, with immunodeficiency, 435
Inference engine, 588f, 588
  definition of, 601
Inference process, 588f, 588
  definition of, 601
Inference processors, 587
Inflammation, 405
  neutrophil chemotaxis in, 416
In situ hybridization, 4–5, 559. See Fluorescence in situ hybridization; Fluorescent in situ hybridization techniques, 566–568
Inositol 1,4,5-triphosphate, 505
Inside-out DNA analysis, 45, 47f
Instrumentation, 64, 68–69, 71–92. See also Flow cytometer(s)
Instrument performance monitoring. See Flow cytometer(s), performance monitoring
Integral pulse(s), 80–81, 81f, 82
Integrin(s), 388, 388f, 419
  in cell adhesion, 490
  functions of, 419–420
Intensity scale(s)
  calibration of, 184
  constant, for troubleshooting sample preparation effects, 184, 185f
  standardization of, 184
Interference filter(s), 79, 181
γ-Interferon
  in chronic granulomatous disease, 421
  effects of, on neutrophil function, 407t
Interleukin 2
  cell responsiveness to, role of glutathione in, 520
  expression, in HIV disease, 444
  receptor, expression
    after bone marrow transplantation, 455
    by lymphocytes in graft, changes in, after transplantation, 454–455
  response of CD16- NK cells to, 67
Interleukin 8
  effects on neutrophils, 415–416
  nomenclature for, 415
Intermediate filament proteins, immunofluorescence measurement of, using flow cytometry, 167–168, 169t
Intermodel error estimation, 139
Interphase cytogenetics, 5
Intracellular antigen(s)
  antibodies to, 162–165
    antigen saturation, 162–163
    cell and antibody controls, 162–165
    isotype controls, 163–164
    negative controls, 164f, 164
    positive controls, 163f, 164
    specificity of, 165
  expression
    independent verification of, 165

Intracellular antigen(s)—(*continued*)
  validation of, 164–165
  immunofluorescence measurement of, 157–172
    indirect versus direct techniques, 165–167
  quantitation
    applications of, 167–172
    dyes for, 167, 168t
    fixation/permeabilization for, 157t, 157–161
      paraformaldehyde/permeabilization protocols for, 161–162
    by flow cytometry, 157
    fluorochromes suitable for, 167, 168t
Intramodel error estimation, 139
Intranuclear antigen(s). *See also* c-fos, c-myb, c-myc, Ki-67, p53, p105, PCNA, SV40 T, TdT antigens
  effect of fixative protocols on, 159, 159f
ISH. *See* In situ hybridization
Isoparametric titration, 474, 475f
Isotype controls, 163–164, 193, 193f
Iterative process, for DNA model(s)/modeling, 48–49

J
Job's syndrome, 406t, 416, 438
Joining region(s), 65
J region. *See* Joining region(s)
JSB-1 (P-glycoprotein-specific antibody), in flow cytometric detection of drug-resistant cells, 463
Juvenile angiofibroma
  DNA ploidy, 346, 348t
  prognostic value of DNA flow cytometry in, 351
Juvenile chronic myelogenous leukemia, 577
Juvenile xanthogranuloma, DNA ploidy, 346, 348t

K
Kaposi's sarcoma, prognostic value of DNA flow cytometry in, 355
Kartagener's syndrome, 406t, 410
Karyorrhexis, 121
Karyolysis, 121
Karyotyping, by image analysis, 363–364
Keratin(s), biosynthesis of, 546, 548–549, 549f
  in psoriasis, 548–549
Keratinocyte(s)
  epidermal, 545, 545f
    basal layer, 546
    benign disorders of, flow cytometric analysis of, 547–550
    differentiation, 546
    diseases of, 545–551
      flow cytometric analysis of, 546–551
    maturation of, 546
  malignant disorders of, ploidy in, 550–551
  premalignant disorders of, ploidy in, 550–551
  psoriatic
    flow cytometric analysis of, 547f, 547–549
    versus normal, flow cytometric discrimination of, 549, 550t
Keratoacanthoma
  aneuploidy in, 550
  versus well-differentiated squamous cell carcinoma, 550–551
Ketoconazole, effects on neutrophil function, 407t

Ki-1 antibody, in immunophenotyping of lymphomas, 221
Ki-67 antibody, 28
  applications in clinical studies, 170
  available as direct conjugate, 166t, 167
  immunofluorescence measurement of, using flow cytometry, 169t, 170
  staining
    in breast cancer, in cell growth kinetics measurements, 256
    correlation with other measures of cell proliferation, 367, 367f
    correlation with tumor grade and other prognostic features, 368
    measurement of, 368
Kinase(s). *See also* Histone kinases
Kinase A, 16
Kinase G, 16–17, 17f
K-means algorithm, 597–599, 598f
Knowledge, definition of, 585, 601
Knowledge acquisition, 588f, 589
  definition of, 601
Knowledge base
  definition of, 601
  maintenance of, 587
Knowledge engineer, definition of, 601
Knowledge representation
  definition of, 601
  for expert systems, 587–593
  frame-based, 587, 590–593
  predicate logic as, 590
  rule-based, 588–590
Komogorov-Smirnov D-values, 58
Komogorov-Smirnov method, 454
K-ras, 5
  in colorectal carcinogenesis, 7
Krypton laser(s), 115
  excitation wavelengths available from, 75, 75t
KS D-value. *See* Komogorov-Smirnov D-values

L
Lactoferrin, 419
lacZ gene, 569
  incorporation into cells, methods for, 570
λ, 46–47
Langerhans cells, 545, 546f
Langmuir binding equation, 473
Langmuir binding isotherm, 473–474
Langmuir plot, 472, 473f
Large granular lymphoproliferative disorder, flow cytometric findings in, 224
Laryngeal carcinoma, aneuploidy, prognostic significance of, 326
Laser(s). *See also* Argon-ion laser(s); Helium-cadmium laser(s); Helium-neon laser(s); Krypton laser(s)
  air-cooled, 73, 75, 92, 115
  beam profile, 75
  diode, 75–76
  dye, 75
  as excitation light source, 73–77
  for excitation of blue-fluorescing dyes, 115
  excitation wavelengths available from, 75, 75t
  frequency-doubled neodymium-YAG, 75
  gas, 75
  low-power, 75–76, 92
  modes of, 75
  power of, 75
  solid-state, 75
  water-cooled, 75–76, 115
Lazy leukocyte syndrome, 406t, 416
LDS-751
  in cell viability determination, 144

exclusion or retention, in flow cytometric definition of dead cells, 192
LEC-CAM, 388
Leiomyoblastoma(s)
  DNA ploidy, 346, 348t
  gastric, DNA ploidy, 346
  gastrointestinal, prognostic value of DNA flow cytometry in, 353
Leiomyoma(s)
  gastric, DNA ploidy, 346
  gastrointestinal, prognostic value of DNA flow cytometry in, 353
  soft-tissue, DNA ploidy, 346, 348t
Leiomyosarcoma(s)
  gastric, percent of DNA diploidy, 347
  gastrointestinal, prognostic value of DNA flow cytometry in, 353
  histologic grading of, 345f
  prognostic value of DNA flow cytometry in, 353
  risk factors in, 351t
  soft-tissue, percent of DNA diploidy, 347
  uterine
    percent of DNA diploidy, 347
    prognostic value of DNA flow cytometry in, 353
Lennert's lymphoma, 221
Leu 8 antigen, 67
Leukemia(s). *See also* Acute lymphocytic leukemia; Acute myeloid leukemia; Chronic lymphocytic leukemia; Chronic myelogenous leukemia; Erythroleukemia
  aberrant phenotypes in, multiparameter detection of, 493–494
  acute, 235–244
    characteristics of, 235
    classification of, 239
    conditional probability of antigen coexpression
      in adults, 243t, 243–244
      in children, 243t, 243–244
    decision-tree analysis in, 242f
    diagnosis of, 239
    flow cytometric analysis of
      cell surface staining for, 240t, 240–241
      specimen collection and handling for, 239
      specimen processing for, 239–240
    lymphoblastic, CSF-1-R gene expression in, 574
    mixed-lineage, 238, 238f
    phenotypic shift in, 480
    relapse, early detection of, 479
  automated diagnosis of, 589–590
  biphenotypic, 238, 238f
  cells expressing mixed lineage differentiation in, 484–485
  chronic, flow cytometric findings in, 219
  classification of, 235
    immunologic, 235–244
  diagnosis of, interobserver and intraobserver reproducibility, 235
  erythroid, 235
  flow cytometric analysis of, data
    analysis of, 241f, 241–243, 242f
    interpretation of, 242f, 243–244
    reporting of, 243
  hybrid, 238, 238f
  lineage promiscuity in, 238, 238f, 484–485
  megakaryoblastic, 235
    antigen expression in, 244
    detection of, 237
  minimal residual disease, markers of, 484–486
  molecular biology of, 579–581
  monoblastic, 235

monocytic, 235
multiparameter analysis in, 493–494
myeloblastic, 235
myeloblastic with differentiation, 235
myelomonocytic, 235
pediatric, incidence of, 331
precursor B-cell, 235–236, 236f
progranulocytic, 235
   acute, immunophenotyping of, 237
prolymphocytic, flow cytometric findings in, 219
ras gene mutations in, 169
T-cell
   adult, 222
   in ataxia-telangiectasia, 439–440
Leukemogenesis, 235
Leukocyte(s), 64
   activation, detection of, 68
   monoclonal antibodies to, workshop-defined cluster groups for, 603–608
   progenitors for, 64
   recruitment, and inflammation, 405
Leukocyte adhesion deficiency, 406t, 416, 440
Leukocyte-depletion techniques, quality control in, 376
Leukocyte differentiation antigens. *See also under* CD
   lineage infidelity of, 66
   monoclonal antibodies against, 64
LFA-1. *See also* CD11a
   expression, on cytolytic T cells, 498
LHR. *See* Lymphocyte homing receptors
Li-Fraumeni syndrome, 581
Ligand binding. *See also* Equilibrium binding
   estimation of, 470–471
      using fluorescent primary antibody, 471
      using fluorescent secondary antibody, 471
   nonspecific, 474–475
Light scatter analysis, in characterization of lymphoid neoplasms, 207
Light-scatter gating
   effects on background nonspecific binding observed, 489–490
   of lymphocytes, 190
Lineage promiscuity, in leukemia, 238, 238f, 484–485
Linear amplifier system, 83
Lipoma(s)
   atypical, DNA ploidy, 346, 348t
   chromosomal changes in, 5t
   DNA ploidy, 346, 348t
Liposarcoma(s)
   chromosomal changes in, 5t
   histologic grading of, 345f
   percent of DNA diploidy, 347
   prognostic value of DNA flow cytometry in, 355
   risk factors in, 351t
LISP, 585
   definition of, 601
Lissamine rhodamine sulfonamide, spectral properties of, 112t
List mode data, 88, 178
   cluster analysis of, 597–599
Live gating. *See also* Dead-cell gating
   in analysis of lymphoma samples, 206
Liver transplantation, flow cytometric crossmatch for, clinical significance of, 452–453
Logarithmic amplifier system, 83
LOH. *See* Loss of heterozygosity
Longpass filters, 79
Loricin, biosynthesis of, 546

Loss of heterozygosity, 6–7. *See also* Allelotype
   in colorectal cancer, 9
Lower respiratory tract tumor(s), 319–329. *See also* Lung cancer
LTB_4 receptors, in neutrophils, 415, 415f
Lung cancer. *See also* Bronchogenic carcinoma
   dissociation, 95
   DNA content heterogeneity, 122
   epidemiology of, 319
   flow cytometric analysis of, clinical and research utility of, 328
   flow cytometric DNA analysis of, 320t, 320–321
      dissociation protocols for, 321
   incidence of, 319
   large-cell, 319–320
      ploidy, prognostic significance of, 321t, 321–323
   non-small-cell, 319, 319t
      distribution and DNA analysis
         by clinical stage, 321, 322t
         by T&N stage, 321, 321t
      flow cytometric DNA analysis of, 320t, 320–321
      heterogeneity of, 328
      incidence of DNA aneuploidy, 328
      by histologic type, 321, 321t
      survival
         by modal DNA content, 322t, 322
         by stage and modal DNA content, 322t, 322
   ploidy
      description, flow cytometry versus image analysis for, 364
      prognostic significance of, 322–323
   prognosis for
      and histologic tumor type, 319
      morphologic factors in, 319
      and tumor stage, 319
   small-cell, 319, 319t
      DNA ploidy, 323, 328–329
      flow cytometric DNA analysis of, 320t, 320–321
      neuroendocrine phenotype, 319
      prognosis for, 319
      treatment of, 319
   stemline heterogeneity of, 321
   synthesis phase fraction (SPF)
      determination of, 323
      prognostic significance of, 323
   TNM classification of, 319
Lung tumors, FNA samples, flow cytometric analysis of, 94
Lymphadenopathy. *See also* Angioimmunoblastic lymphadenopathy
   flow cytometric findings in, 216, 217t
Lymphangioma(s), DNA ploidy, 346, 348t
Lymphoblastoid cell lines, Epstein-Barr virus–transformed, as target cells for detection of alloantibodies, 451
Lymphocyte(s)
   antigenic phenotype, and function, 67–68
   cytotoxicity induced by, in transplant recipient, 451
   flow cytometric properties of, 406–407
   identification of
      CD45 antibody staining for, 190–192, 191f
      immunofluorescence for, 190, 191f
      light scatter for, 190, 191f
   immunophenotyping
      analysis of data, 153–154, 153f–155f
      choice of reagents for, 151t, 151–152

in clinical evaluation of primary immunodeficiency, 436t, 440
      guidelines for, 152, 152t
      in HIV disease, quality control for, 445–446
      in hypogammaglobulinemia, 440
      internal consistency checks for, 152, 152t
      monoclonal antibodies for, 151t, 151–152, 446, 446f
      in severe combined immunodeficiency, 440
   isolation of, 148
   light scatter, gating, 190
   lineages of, 64–66
   mitogenic stimulation of, RNA and DNA content changes with, 24, 25f
   phenotyping, after bone marrow transplantation, 455
   preparations of, 144
   response to mitogenic agonists, heterogeneity in, 516, 527
   subsets, 63, 66–67
      changes in, after transplantation, 454–455
      monoclonal antibodies for identification of, 151t, 151–152
Lymphocyte-activated killer cytotoxicity, 497
Lymphocyte homing receptors, expression, in lymphoma, prognostic significance of, 225
Lymphoid hyperplasia(s)
   benign
      flow cytometric findings in, 216–217, 217t
      immunophenotyping, 216–217, 217t
   in homosexual men, ploidy, 215–216
Lymphoid immune system, 435
Lymphoid neoplasm(s). *See also* Leukemia(s); Lymphoma(s)
   aneuploidy in, 215–216
   characterization of, 227
   light scatter analysis in, 207
   DNA rearrangements in, detection of, 482
   flow cytometric analysis of
      findings in, 217t, 217–227
      practical considerations, 226
      sample preparation, 203–204
      sensitivity of, 227
Lymphoma(s). *See also* Burkitt's lymphoma; Hodgkin's disease; Non-Hodgkin's lymphoma
   aggressive
      cell cycle kinetics in, 216
      flow cytometric findings in, 220–222
      prognosis for, 224
   aneuploidy in, 215–216
      prognostic significance of, 225
   antigen expression in, prognostic significance of, 224–225
   B-cell
      aggressive, in immunocompromised patient, 222
      detection of, 205, 205t
      prognosis for, 224
   cell cycle fractions in, 211–212, 214f
   cell cycle kinetics in, 216
   cell proliferation in, prognostic significance of, 366
   cell suspensions prepared from, 204
   cellular antigens, analysis of, 206–210, 212–215
   circulating neoplastic cells in, prognostic significance of, 225
   classification of, 203, 216
      antibodies used in, 204–205, 305t
   clonal expansion in, 213–215

Lymphoma(s)—(continued)
    analysis of, 207–210
    cutaneous T-cell, flow cytometric findings in, 222
    detection of, 213–215
        clinical applications, 215
    diagnosis of, 203, 216
        antibodies used in, 204–205, 305t
    diffuse
        large cell, B-cell type, flow cytometric findings in, 217t, 220
        mixed cell, flow cytometric findings in, 220
        mixed or large cell, T-cell type, flow cytometric findings in, 217t
        small B-cell, antigen expression in, prognostic significance of, 225
        small cleaved, flow cytometric findings in, 220
    DNA analysis, 206, 210–212, 215–216, 225–226
    DNA content
        correlated multiparameter analysis of, 210, 212f
        prognostic significance of, 225–226
        significance of, 215–216
    DNA distributions, from fresh versus archival material, 101t
    DNA staining in, 205–206
    flow cytometric analysis of, 203–227
        advantages of, 227
        correlations with histology, 216–227
        indications for, 226
        practical considerations, 226
        sensitivity of, 227
    flow cytometric findings in, prognostic significance of, 224–226
    follicular
        antigen expression in, prognostic significance of, 225
        circulating neoplastic cells in, 225
        flow cytometric findings in, 217t, 219–220
        large-cell, κ and λ distributions in, 210, 210f
        low-grade, κ and λ distributions in, 209f
    grade, and ploidy, 216
    high-grade, ploidy in, 216
    immunoblastic, flow cytometric findings in, 221
    immunophenotyping, antibodies used for, 205, 205t
    immunostaining, 204–205
    intermediate differentiation, 220
    intracellular antigens, staining, 205
    Ki-67 staining, correlation with tumor grade and other prognostic features, 368
    large B-cell, aneuploid, flow cytometric analysis of, 213f
    large-cell, flow cytometric findings in, 221
    lineage identification, 213
    low-grade
        cell cycle kinetics in, 216
        ploidy in, 216
    lymphoblastic, flow cytometric findings in, 217t, 220, 220f, 221
    lymphocytic
        κ and λ distributions in, 209
        small, flow cytometric findings in, 219
    mantle zone, 220
    multiple staining, 206
    pediatric, incidence of, 331
    platelet-associated immunoglobulins in, flow cytometric analysis of, 390
    ploidy
        characterization, 211, 213f
        description, flow cytometry versus image analysis for, 364
        significance of, 215–216
    proliferative fraction in, prognostic significance of, 225–226
    ras gene mutations in, 169
    RNA index in, prognostic significance of, 226
    sample
        analysis of, 206
        preparation of, 203–204
        staining, 204–206
        storage of, 204
        transportation of, 204
    small, noncleaved cell. See also Burkitt's lymphoma
        non-Burkitt, 221–222
    small lymphocytic, flow cytometric findings in, 217t
    S-phase cells in, prognostic significance of, 226
    stage differentiation, 213
    T-cell
        aberrant expression in, 215
        adult, 222
        almost clonotypic reagent for, 210
        detection of, 205, 205t
        flow cytometric findings in, 221
        immunological recognition of, 215
        lymphoblastic
            flow cytometric analysis of, 215, 220f
            flow cytometric findings in, 220f, 220
        peripheral
            flow cytometric analysis of, 214f, 215
            flow cytometric findings in, 221
        prognosis for, 224–225
Lymphoproliferative disorder(s). See also Leukemia(s); Lymphoma(s)
    flow cytometric analysis
        future of, 226–227
        sample preparation, 203–204
Lymphosum, 445
Lysolecithin, 158
Lysosome(s), platelet, 387
Lysozyme, 409, 419

M
mAb. See Monoclonal antibody(ies)
Mac-1. See CD11b antigen
Macrophage(s)
    cytotoxicity induced by, in transplant recipient, 451
    in host defense, 435
Macrophage colony-stimulating factor, receptor, 573–574
Mahalanobis distance, 599
Major histocompatibility complex, 65
Malignancy, phenotypic shift in, 480
Malignant fibrous histiocytoma
    histologic grading of, 345f
    percent of DNA diploidy, 347
    prognostic value of DNA flow cytometry in, 351–352
    risk factors in, 351t
Malignant granular cell tumor, histologic grading of, 345f
Malignant hemangiopericytoma, histologic grading of, 345f
Malignant histiocytosis, 221
Malignant mixed müllerian tumor(s), prognostic value of DNA flow cytometry in, 354
Malignant transformation, 573
    of melanocytes, antigen expression during, 552–553

Mammary adenocarcinoma, DNA-content heterogeneity, 279
α-mannosidase, 409
    deficiency, 406t
Marquardt compromise method, of DNA data analysis, 46–47
Marquardt method, of DNA data analysis, 44t, 44–45
Mast cell(s), in host defense, 435
Maturation-Promoting Factor, 17f, 18
MBB. See Monobromobimane
MCB. See Monochlorobimane
McLeod syndrome, carrier status, testing for, 376
mCSP. See Melanoma-specific chondroitin sulfate proteoglycan
MDR (Ab-1) (P-glycoprotein-specific antibody), in flow cytometric detection of drug-resistant cells, 463
MDR gene. See Multidrug resistance gene
Mechanical dissociation, of solid tumors, 94
    evaluation of, 95–96
Medulloblastoma, pediatric, 332–333
    incidence of, 332
    ploidy, prognostic significance of, 333, 338–339, 339t
Megakaryocytes, 387
Melanocyte(s), 545
    antigenic expression, flow cytometric assessment of, 552–553
    differentiation, pathway of, 552, 552f
    malignant, flow cytometric measurement of ploidy in, 551
    malignant transformation, antigen expression during, 552–553
    normal, flow cytometric measurement of ploidy in, 551
Melanoma
    chromosomal changes in, 5t
    development of, 551
        stages of, 552
        antigen expression during, 552–553
    dissociation, 95
    immunotherapy for, flow cytometric monitoring during, 553
    Ki-67 staining, correlation with tumor grade and other prognostic features, 368
    pediatric, incidence of, 331, 331t
    prognosis for, morphologic and histologic correlates with, 551
    recurrence, morphologic and histologic correlates with, 551
    treatment of, with target-directed antibodies, 553
Melanoma-associated antigens
    expression of, flow cytometric assessment of, 552–553
    as targets for immunotherapy, 553
Melanoma-specific chondroitin sulfate proteoglycan, 552–553
Melanosome(s), 545, 545f
Membrane potential, flow cytometric analysis of, simultaneously with intracellular ionized calcium (Ca^{2+}) measurement, 513
Meningioma(s)
    chromosomal changes in, 5t
    Ki-67 staining, correlation with tumor grade and other prognostic features, 368
    pediatric, incidence of, 332
Mercury arc lamp(s)
    as excitation light source, 73–74
    spectral intensity distribution for, 74, 74f
Merkel cell carcinoma, S-phase fraction, clinical significance of, 140, 140t

Mesenchymoma(s), risk factors in, 351t
MESF (molecules of equivalent soluble fluorochrome), 470
Mesothelioma, chromosomal changes in, 5t
Messenger RNA, quantitation, 559
Metachronous tumor(s), 8
Metarules, 589
Metastasis, 3
Metastatic lesions, FNA samples, flow cytometric analysis of, 94
Methanol, fixation with, 157
Methotrexate, cytotoxicity, flow cytometric monitoring of, 461, 465
MHC. See Major histocompatibility complex
MIC classification. See Morphologic, Immunologic, and Cytogenetic classification
Michaelis Menten plot, 472
Microbeads, fluorescent
  in estimation of ligand binding, 471
  as standards for quantitative calibration of fluorescence intensity, 470
Microscope image, digitization of, 360f
Microscopy. See also Electron microscopy; Immunofluorescence microscopy
  and flow cytometry, differences in detection by, 146–147
Microspectrophotometry, 359
Microsphere(s), uses of, 179
Microsphere-based fluorescence immunoassays, 535–543
Microspherocytes, 394
Minimal residual disease
  cell biology in, 494
  cell growth in, 481, 494
  detection of
    cell clumping in, 490
    cell markers in, 482t, 483, 490–493
    by chromosome analysis in flow, 482
    clinical considerations in, 480–481
    by clonigenic assay, 482t, 483–484
    conventional cytogenetics for, 482t, 482
    by DNA ploidy analysis, 481–482, 482t
    by DNA translocations or rearrangements, 482t, 482–483
    by fluorescence in situ hybridization, 482
    by immunofluorescence microscopy, 482t, 483
    methods for, 481–484, 482t
    by morphology, 481, 482t
    number of positive events required, 490–491, 492f
    in peripheral blood, 493
    by polymerase chain reaction, 482t, 483
    by protein secretion, 481, 482t
    sensitivity required for clinical usefulness, 480
    by Southern blot DNA hybridization, 482
  drug sensitivity in, 494
  flow cytometric detection of, 479–494, 482t, 483
    biological considerations, 491–494
    future of, 486–494
    methodological considerations, 486–491
    nonspecific binding of antibodies in, 486–490
  ideal test for, 481, 481t
  localization
    anatomic variability in, 481
    markers of, 485–486
  markers of, 484–486
  monitoring, rationale for, 479–480
Mithramycin, 119
Mitomycin C therapy, in bladder cancer, flow cytometric findings related to, 286

Mitosis, 13
  abnormal, 8
  cell entrance to, 17f, 18
  growth threshold for, 21
Mitotic nondisjunction, 8
MMMT. See Malignant mixed müllerian tumor(s)
Mo1. See CD11b antigen
Modal chromosome number, 4
Molar pregnancy, flow cytometric analysis in, 266, 270
Molecular biology, and flow cytometry, 573–581
  analytical aspects, 557–571
Molecular genetics, 3, 581
Molecular phenotype, 559, 561
Monobromobimane, as GSH-specific fluorochrome, 521
Monochlorobimane
  as GSH-specific fluorochrome, 258, 464, 521–522
    cell sorting on basis of, 528–531, 530f
    factors affecting utility of, 522, 523f
    staining with, practical considerations for, 522–523
Monoclonal antibody(ies), 63
  antigen–antibody complexes formed by, 143
  to bladder cancer, 292
  to B-cells, 608
  for cell surface staining, in leukemia, 240t, 240–241
  cocktail
    for immunophenotyping lymphocyte subsets, 446, 446f
    for immunotherapy of melanoma, 553
  combinations, for immunophenotyping in HIV disease, 441t, 441, 444–445
  consisting of single V_H region, construction of, 490
  cross-reactions with unrelated antigens, 165
  in detection of acute leukemias, 237–238
  for detection of rare events, preparation of, 486–488, 487f–488f
  direct conjugates to intracellular antigens, 165–166, 166t
  for evaluation of primary immunodeficiency, 436t, 436
  to HLA antigens (class II), 608
  to human leukocyte differentiation antigens, 64
  to human leukocytes, workshop-defined cluster groups for, 603–608
  for identification of lymphocyte subsets, 151t, 151–152
  to monocytes, 608
  to neutrophils, 608
  to nuclear oncoproteins, 169t, 579
  P-glycoprotein-specific, 462–463
  to platelet and endothelial cell receptor complexes, 608
  reactive with epidermis, 548, 549t
  to regulatory subpopulations of $CD4^+$ and $CD8^+$ cells, 608
  specificity of, 165
Monoclonal gammopathy, 435
Monocyte(s)
  alloantibodies against, in transplant recipients, 450
  cytotoxicity induced by, in transplant recipient, 451
  flow cytometric properties of, 406–407
  isolation of, 148
  stem cell origin of, 405
Monocyte/macrophage(s)

cell-surface-differentiation antigens, 66f
development of, 64
Mononuclear cells, isolation of, 148
Morphologic, Immunologic, and Cytogenetic classification, of leukemias, 235
MPF. See Maturation-Promoting Factor
M phase, 13
  DNA content during, 42–43, 43f
MPO deficiency, 406t
Multialkali photocathode, 79
Multi-cut model component, 51–52, 52f, 130
Multidimensional data, 586
  definition of, 601
Multidrug resistance, in tumor, and chemotherapeutic effectiveness, 480
Multidrug resistance gene, 460, 462
Multiparameter flow cytometry, 23–29, 184, 362
  analysis of cell cycle, 23–29
  in bladder cancer, 285
  combined with cell kinetics measurements, 24
  development of, 63–64
  and gating of cells responsible for nonspecific binding, 488
  in leukemia, 493–494
  of neoplastic cells, 493–494
  of proliferation markers in breast cancer, 255–256
  and stepwise amplified immunofluorescence staining, 488–489
Multiple myeloma, ras gene expression in, 169
Multiplets, types of, 85
Multiploidy, 122
  detection of, number of tissue samples required for, 122–123
Multivariate data, 593
  cluster analysis of, 597–599
  definition of, 601
Mutation(s), 558
  detection of, by polymerase chain reaction, 563–564
MYCIN, 589
Mycosis fungoides, 222
Myeloblasts, leukemic, dual-parameter analysis of, 580f
Myelodysplastic syndrome(s), 235
Myeloid cell(s), progenitors for, 64
Myeloma. See also Multiple myeloma
  antigenic expression in, 224
  proliferative fraction in, prognostic significance of, 226
Myelomonocytic ontogeny, 237, 238f
Myeloperoxidase, 409
  immunofluorescence measurement of, using flow cytometry, 169t
Myeloperoxidase deficiency, 410, 417
Myeloproliferative disorders, pathogenesis of, 576
Myeloproliferative leukemia virus, 576
Myxoma(s), DNA ploidy, 346, 348t

N
NA. See Numerical aperture
Naproxen, effects on neutrophil function, 407t
Nasopharyngeal carcinoma, flow cytometric analysis of, prognostic significance of, 325–326
Natural killer cell(s), 64–65
  antigenic heterogeneity of, 66–67
  in cell-mediated cytotoxicity, 497
  cell-surface-differentiation antigens, 66f
  changes in, after bone marrow transplantation, 455

Natural killer cell(s)—(continued)
  complete absence of, 440
  functions of, 65–66
    correlation with antigenic phenotype, 67
  in HIV disease, 444
  interference with alloantibody detection, in transplant recipients, 450
  in severe combined immunodeficiency, 439
  subsets of, 67
  target recognition structures on, 498
NBD-amine, spectral properties of, 112t
NBF. See Neutral-buffered formalin
N-CAM. See Neural cell adhesion molecule
Near-diploid aneuploid peaks, detection of, flow cytometry versus image analysis for, 365
Near-diploid aneuploid populations, 211, 250
  clinical interpretation of, 139–140
  detection of, 126–127, 127f–128f
Near-diploid cells, 4
Near-tetraploid aneuploid populations, 129
  clinical interpretation of, 139–140
Needle biopsy. See also Fine-needle aspiration
  in breast cancer, 248
  of soft-tissue sarcomas, 355, 357
Negative control(s), 164f, 164, 193–194
Nephroblastoma. See Wilms tumor
neu gene, 575–576. See also HER-2/c-erB/neu oncogene
Neural cell adhesion molecule, 66
Neural networks, 368, 599
Neurilemmoma(s), DNA ploidy, 346, 348t
Neuroblastoma
  chromosomal changes in, 5t
  histologic grading of, 345f
  metastases, detection of, 485–486
  N-myc oncogene amplification in, 5–6
  pediatric, 333–334
    biologic markers of, 333
    diploidy, prognostic significance of, 334, 334f
    hyperdiploidy, prognostic significance of, 334, 334f
    incidence of, 331, 331t, 333
    karyotype with, 334
    management of, 334
    metastatic, 334
    N-myc copy number, prognostic significance of, 334, 335f
    ploidy, 332f–333f, 334
      prognostic significance of, 333–334, 338–339, 339t
    prognostic factors for, 333
Neurofibroma(s), DNA ploidy, 346, 348t
Neurogenic sarcoma(s), percent of DNA diploidy, 347
Neuroma(s), DNA ploidy, 346, 348t
Neutral-buffered formalin, effects on DNA histograms from archival material, 102, 103f
Neutral protease, 409
Neutrophil(s)
  activation, mechanisms of, 414–415, 415f
  adhesion, 419–421
  adhesion glycoproteins, 405, 408
    assessment of, methods for, 408t
  adhesive properties of, 405, 408, 415, 415f
  antibodies against, 421
    measurement of, 421, 422f
  bacterial degradation, assessment of, methods for, 408t
  bactericidal activity, assessment of, 419
    methods for, 408t
  calcium flux, measurement of, methods for, 408t

C5a receptor, 415
chemoattractants, 415–416
chemotaxis, 405
  assessment of, 414–416
    methods for, 408t
  defects, 420
    clinical evaluation of, 416
  in chronic granulomatous disease, 418
circulating, 405
clinical disorders of, 405, 406t
cytoskeletal function, assessment of, 410–411
cytosolic free calcium in, measurement of, 412–413
degranulation, assessment of, methods for, 408t
development of, 64, 405
enzyme content/activity, assessment of, 410
  methods for, 408t
Fc receptor expression on, 420–421
flow cytometric analysis of, in evaluation of trauma, 421
flow cytometric properties of, 406–407
functions of, 405
  flow cytometric analysis of, 408t
    advantages of, 408–409
    fluorescent probes for, 408–409, 411t
    preparative techniques for, 407–408
  methods for, 408t
  overview of, 405–406
granules
  clinical evaluation of, 411
  development, assessment of, 409–410
  function of, assessment of, 409–410
H_2O_2 production, assessment of, 417–418, 418f
  methods for, 408t
isolation, overlay method for, 407–408, 410f
in leukocyte adhesion deficiency, 440
90°; light scatter, 407, 409f
LTB_4 receptors, 415, 415f
marginating pool of, 405
maturation of, 405
membrane fluidity, assessment of, methods for, 408t
membrane integrity, assessment of, 412
membrane potential
  abnormal, 413–414
  assessment of, 413, 414f
    methods for, 408t
membrane structure, assessment of, methods for, 408t
metabolic functions, assessment of, 412–414
microtubule abnormalities, 410–411
  assessment of, methods for, 408t
mobility, assessment of, 409
in neonatal infections, 421
nonoxidative bactericidal mechanisms, assessment of, 419
O_2 production, assessment of, methods for, 408t
phagocytosis
  abnormal, 417
  assessment of, 416–417
    methods for, 408t
  clinical evaluation of, 417
  determination of, 407
  flow cytometric analysis of
    fluorescent probes for, 416–417
    utility of, 416
pH measurement, methods for, 408t
physiology, drugs affecting, 405, 407t
pinocytosis, assessment of, 417
  methods for, 408t

release, 405
respiratory burst, 417
  assessment of, methods for, 408t
stem cell origin of, 405
stimulation of, by intravenous immunoglobulin, 421
storage, 405
subsets, identification of, 408
viability, assessment of, 412, 412f
  methods for, 408t
Neutrophil defense mechanisms, assessment of, 417–419
Neutrophil receptors, binding to, 419–421
Nezelof's syndrome, 438
Nitroblue tetrazolium, reduction, in assay of neutrophil oxidative system, 418
NK cells. See Natural killer cell(s)
N-myc copy number, in pediatric neuroblastoma, prognostic significance of, 334, 335f
N-myc oncogene, amplification, 5–6
Nodular fasciitis, DNA ploidy, 346, 348t
Non-convergence, 47
Non-Hodgkin's lymphoma
  aggressive, 218
  classification of, 218, 235
  clonal B-cell expansion in, detection of, 213–214
  definition of, 218
  flow cytometric findings in, 217t, 218–222
  indolent. See Non-Hodgkin's lymphoma, low-grade
  κ and λ distributions in, 207–209, 213
  Ki-67 staining, correlation with tumor grade and other prognostic features, 368
  low-grade, flow cytometric findings in, 218–220
  oncogene expression in, prognostic significance of, 225
  pathogenesis of, 218
  RNA index in, prognostic significance of, 226
Nonlinear least-squares analysis and models, 44t, 44–46, 139
Nonparametric analysis, 58
Non-platelet particles, 394, 395f–396f
Nonspecific binding, 474–475
  of antibodies
    in detection of minimal residual disease, 486–490
    to Fc receptors, 144–145, 486–490
  and debris, 130f, 130, 489
Normalized subtraction, for immunofluorescence analysis, 56f, 56–57
Northern hybridization, 559
Notch filters, 79
No wash technique, 166–167
  for bladder cancer samples, 283
NP-40, 158, 167
NPP. See Non-platelet particles
N-ras, 5
Nuclear constituents, analysis of, 26
Nuclear factors, in regulation of transcription, 579–581
Nuclear protein(s). See also Histone(s)
  changes during cell cycle, 26, 26f
  immunofluorescence measurement of, using flow cytometry, 169t, 170–172
  isolation method and, 20
  nonhistone, 20
Nuclear RNA, changes during cell cycle, 26, 27f
Nuclear size, 26
Nuclear slicing, effects on DNA content histogram, 131f, 131–132

INDEX

Nuclear suspensions
  DNA analysis of, 95
  preparation of, 118–119
Nuclease protection assay, 559, 559f
Nucleolar antigen(s), monoclonal antibody to, available as direct conjugate, 166t, 167
Nucleolar segregation, 26–27
Nucleosome, structure of, during cell cycle, 14–15
Nucleus
  isolated, flow cytometric analysis of, 26
  protein content of
    changes during cell cycle, 26, 26f
    isolation method and, 20
Numerical aperture, 78

O

Occult metastatic disease, early detection of, 480
OKM-1. See CD11b antigen
OKT3, transplant recipients treated with, antibodies in, 449
Oligodendroglioma, pediatric, incidence of, 332
Oligonucleotide primers, for mutation detection, 564
Omni Fix, effects of, on DNA histograms from archival material, 102, 103f
Oncocytoma(s), renal, DNA content analysis, 296, 304
Oncogene(s), 4, 28, 557. See also c-abl gene; c-myc gene; Protooncogenes; ras gene(s)
  activation, and aneuploidy, 8–9
  antibodies to, 163
  as cell cycle marker in tumor classification, 34
  cytoplasmic, immunofluorescence measurement of, using flow cytometry, 168–170, 169t
  expression
    measurement of, 569
    in non-Hodgkin's lymphoma, prognostic significance of, 225
    in prostate cancer, 281
  molecular biology of, 557–558
  products, redox-responsive, and glutathione, 520
  recessive, 557
o-phthaldialdehyde, in flow cytometric determination of intracellular glutathione, 521–522
Opportunistic infection(s), risk of, in HIV disease, and CD4 counts, 441–442, 443f
OPT. See o-phthaldialdehyde
Optical alignment
  monitoring, 179f, 179–181, 182f
  optimization of, 178–179
Optical processing, 178
Optical system, quality control, 178–181
Optical to electronic conversion, quality control, 178
Organ transplantation. See Transplantation
Ornithine decarboxylase, immunofluorescence measurement of, using flow cytometry, 169t
Osteogenic sarcoma(s)
  DNA ploidy, 346, 348t
  percent of DNA diploidy, 347
Osteosarcoma
  dissociation, 99f
  extraskeletal
    histologic grading of, 345f
    risk factors in, 351t
  pediatric
    histopathologic subtypes of, 337
  incidence of, 331, 331t, 337
  ploidy, prognostic significance of, 337–338, 338f, 338–339, 339f, 339t
  prognostic factors for, 337
Outside-in DNA analysis, 45, 45f
Ovalocytosis, 375–376
Ovarian cancer
  chromosomal changes in, 5t
  c-myc overexpression in, 9
  dissociation, 95
  epithelial
    borderline, 264–265
    DNA content analysis in, 265, 265t, 269
    DNA content, and survival, 263–264, 264f, 269
    DNA content analysis in, limitations of, 267f, 267–268
    flow cytometric analysis of, 263–265
    ploidy, prognostic significance of, 263–264, 264f, 265t, 268–269
    S-phase fraction in, 268–269
  fractional allelic loss, as prognostic factor, 8
  HER-2/c-erB/neu oncogene amplification in, 6
  ploidy, description, flow cytometry versus image analysis for, 364
  p53 overexpression in, and aneuploidy, 9
Overlay method, for neutrophil isolation, 407–408, 410f
Oxatomide, effects on neutrophil function, 407t

P

p150, 95. See CD11c antigen
PAIg. See Immunoglobulin(s), platelet-associated
Pancreatic carcinoma
  ploidy, prognostic significance of, 314
  proliferative activity, prognostic significance of, 314
p24 antigen, HIV-associated. See HIV-associated p24 antigen
p34 antigen, 28
p53 antigen
  immunofluorescence measurement of, using flow cytometry, 169t, 170
  measurement of, simultaneously with DNA content measurement, 171f, 172
p105 antigen, 28, 29f
  analysis of, simultaneously with DNA content analysis, in colorectal cancer, 314
  in cell growth kinetics measurements, in breast cancer, 256
  effect of fixative protocols on, 159, 159f
  expression, independent verification of, 165
  immunofluorescence measurement of, using flow cytometry, 169t
p120 antigen, expression, in severe combined immunodeficiency, 438
Papain, used to dissociate tumors, 95
Paraffin-embedded materials
  antigen expression in, independent verification of, 165
  in breast cancer, 248
  deparaffinization and rehydration of, 104
  DNA histograms from, 41, 100–101, 101t, 130f, 131–132
  DNA staining in, 206
  fixed
    debris from, 101, 104
    tumor cell dissociation from, 100–105
      clinical applications of, 100
      handling, 102–104
      histological examination, 122
      nuclei recovered from, measurement of proteins in, 165
      preparation of cell suspensions from, 119
      rehydrated sections, enzymatic digestion of, 104
      sample storage, 105
      sectioning, 104
Paraformaldehyde, 144, 158
  fixation, of dissociated tumor cells, from fresh specimens, 96
Paraformaldehyde/permeabilization protocol(s), for quantitation of intracellular antigens, 161–162
Parametric analysis, 41
Parasitic infections, in red blood cells, quantitation, 376–377
Paroxysmal nocturnal hemoglobinuria, 375–376
Passive resistor-capacitor networks, 82
Paternity testing, flow cytometric techniques in, 376
PCA-1 antibody, in immunophenotyping of plasma-cell tumors, 224
PCA-2 antibody, in immunophenotyping of plasma-cell tumors, 224
PC-1 antibody, in immunophenotyping of plasma-cell tumors, 224
p34^{cdc2} kinase
  as cell cycle marker in tumor classification, 34
  in phosphorylation of HMG I, 21
  in regulation of cell cycle, 16–17, 17f, 18
PCNA. See Proliferating cell nuclear antigen
PE. See Phycoerythrin
Peak pulse(s), 80–82, 81f
  height (amplitude), 80–82, 81f
Pentoxyfylline, effects on neutrophil function, 407t
Pepsin
  cleavage of immunoglobulin, 145
  dissociation and enucleation of sections of rehydrated tissue from fixed, paraffin-embedded materials, 104, 104f–105f
  used to dissociate tumors, 95
Percent-positive cells, 68, 193–194, 206
PerCP. See Peridinin-chlorophyll-a-protein
Perforin, 498
Peridinin-chlorophyll-a-protein, 64, 115
  use of, 441
Peripheral blood
  analysis of cell surface molecules in, 148
  detection of minimal residual disease in, 493
  in leukemia, collection and handling, 240
  lymphocyte subsets in, changes in, after transplantation, 454
Permeabilizing agent(s), 158
P-glycoprotein, 460
  antibodies against, 462–463
  in drug-resistant cells, 257, 462
  expression, in lymphoma, prognostic significance of, 225
  flow cytometric detection of, 257, 465–466
pH, intracellular
  and cellular activation, 517
  flow cytometric analysis of, 517–519. See also SNARF-1
    calibration for, 519
    data analysis, 519
    simultaneously with other fluorescence parameters, 519, 520f
  indicators of, 517–518
  ratiometric determination of, 517–518, 518t, 519

pH, intracellular—(*continued*)
  regulation of, 517
Phagocytosis. *See* Neutrophil(s), phagocytosis
Phagolysosome(s), 419
Ph¹ chromosome. *See* Philadelphia chromosome
Phenotypic shift, in malignancy, 480
Phenylbutazone, effects on neutrophil function, 407t
Φ, 111
Philadelphia chromosome, 3, 4f, 5
Phosphatase(s)
  activities, in cell cycle phases, 19
  flow cytometric analysis of, 410
Phosphatidylinositol 4,5-bisphosphate, 505
Phospholipase C, 505
Phosphotyrosine, monoclonal antibody, available as direct conjugate, 166t
Photobleaching, 146
Photodetector(s), 71, 72f, 77–78
  fluorescence and side-scatter, 79–80
  integral pulses, 80–81, 81f, 82
  outputs, 80
  peak pulses, 80–82, 81f
  solid-state, 80
Photodiode, 77–78
Photomultiplier tube, 79, 80f
  fluorescence-detection, monitoring instrument precision with, 90
  gain of, 79, 80f
  spectral response range, 79, 80f
  spectral sensitivity of, 79–80
Phycobiloprotein(s), 64, 114
Phycoerythrin, 64, 167, 168t
  effect on flow cytometric results, 68, 68f
  for evaluation of HIV disease, 444
  and fluorescein-isothiocyanate, spectral overlap between emission spectra of, 84, 84f
  in immunofluorescence applications, 146
Phycoerythrin-CD20, in two-color analysis, effect of under- and overcompensation on, 188f, 188
Phycoerythrin-R, spectral properties of, 112t
Phycoerythrin-RITC tandem conjugate, 168t
Phycoerythrin-TR tandem conjugate, 150–151
Physarum polycephalum, histone phosphorylation in, 18
Physiological indicator probes, 111
PI. *See* Propidium iodide
Pigmented villonodular synovitis, DNA ploidy, 346, 348t
Pineoblastoma, pediatric, incidence of, 332
PIP (Present Illness of Patient), 592f, 592–593
Piroxicam, effects on neutrophil function, 407t
Plasma-cell tumors, flow cytometric findings in, 223–224
Platelet-associated immunoglobulins. *See* Immunoglobulin(s), platelet-associated
Platelet cluster, 389, 389f
Platelet count, and reticulated platelets, relationship between, 396, 398f
Platelet-derived growth factor, cell responsiveness to, role of glutathione in, 520
Platelet–leukocyte aggregates, 399–400
Platelet microparticles, 389
Platelet reticulocytes, 394–397
Platelets. *See also* Immunoglobulin(s), platelet-associated
  activation, 387–388, 397–401
    and aggregation, simultaneous measurement of, 399f, 399
    detection of, in vivo, 398, 400–401
    flow cytometric analysis of, protocol for, 399
    immunological markers of, 397–399, 398t
      clinical significance of, 400–401
  adhesion, 387–388, 397
  aggregation, 387–389, 397
  aggregometry, 397
  biocompatibility studies with, 400
  canalicular system, 387–388
  in cardiovascular disease, 400
  constituents of, 387
  crossmatch, 450
  deaggregation, 389
  degranulation, 387–388
  dysfunction, diagnosis of, 401
  flow cytometric analysis of, 387–401
    advantages of, 390
    artifacts in, 390
    cost of, 390
    resolution required for, 389–390
    sensitivity required for, 390
    specimen handling for, 390
    in whole blood, 389f, 390
  functions of, 387, 397
    therapeutic modulation of, 401
  granules in, 387–388
    membrane proteins, 398, 398t
  HLA expression on, 450
  hyperresponsive, detection of, 401
  lifespan studies, 394
  normal concentration of, 389
  nucleic acid content, measurement of, 394–395, 397f
  ontogeny of, 387
  release reaction, 387–388, 397
  reticulated, 394–397
  size of, 389
  storage, 400
  structure of, 387
  subsets
    detection of, 390
    with different levels of platelet-associated immunoglobulins, 393–394
  surface glycoproteins on, 387–388, 388f
  as target cells for detection of alloantibodies, 450
  transfusion, 400
    refractoriness to, 450–451
Platinum-rhodamine, cytotoxicity, flow cytometric monitoring of, 461, 465
PLL. *See* Leukemia(s), prolymphocytic
Ploidy. *See also* Aneuploidy
  allele losses and, 10
  in cervical cancer, 266, 270
  characterization, in lymphoma, 211, 213f
  clinical interpretation of, 139–140
  in colonic cancer, 311–313
  in colorectal cancer, prognostic significance of, 307t, 307–308, 308f
  definition of, 4
  determination, flow cytometry versus image analysis for, 364–366, 365t, 365f
  in endometrial cancer, 265–266, 268, 270
  of gastric carcinoma, prognostic significance of, 314
  of hepatocellular carcinoma, prognostic significance of, 314
  in malignant and premalignant disorders of keratinocytes, 550–551
  in melanocytes, flow cytometric measurement of, 551
  in molar pregnancy, 266, 270
  in ovarian cancer, 263–265, 268–269
  of pancreatic carcinoma, prognostic significance of, 314
  of pediatric solid tumors, clinical significance of, 331, 338–339
  of prostate cancer
    biopsy specimens, 274–275, 275t
    correlation with prostate-specific antigen, 280
    prognostic significance of, 275t, 275–276, 276t, 277f
    radical prostatectomy samples, 275–277, 276t
  in rectal cancer, 311, 311t
  of tumor, relation between molecular-genetic changes and, 8
Pluripotential stem cells, 405
PMT. *See* Photomultiplier tube
Pneumocystis carinii pneumonia, in HIV disease, 441–442
Poly(ADP)ribosylation, 15
  of histone, 15–16
Polyclonal antibodies
  problems with, in cell surface immunofluorescence, 145
  specificity of, 165
Polyclonal sera, antigen–antibody complexes formed by, 143
Polycythemia vera, 576
  test for, 375
Polymerase chain reaction, 559–561, 560f. *See also* Fluorescent in situ polymerase chain reaction
  cell preparation for, 561
  competitive, 562–563, 563f
  detection of translocations by, 564
  differential, 562
  DNA and RNA amplification for, 561–562
  and flow cytometry, combination of, 561–564
  mutation detection by, 563–564
  quantitative, 562–563
  reaction products
    analysis of, 562
    direct sequencing of, 564
    hybridization of, using mutation-specific probes, 564
  residual disease assay, 482t, 483
    in T-cell ALL, 493
  sensitivity of, 562, 562f
Polymixin B, effects on neutrophil function, 407t
Polymorphonuclear leukocytes, as target cells for detection of alloantibodies, 450
Population ratios, determination of, 111
Positive control(s), 163f, 164, 193–194
PRAD1 gene, 580–581
p21 ras oncoprotein, in myeloma, 224
Predicate logic
  definition of, 601
  as knowledge representation, 590
Principal components analysis, 599
  definition of, 601
Problem domain, 587
Prognostic factor(s)
  aneuploidy as, 10
  cell cycle markers as, 34
  cellular RNA in tumor cells as, 21–22
  fractional allelic loss as, 7–8
Prolactin, intranuclear, effect of fixative protocols on, 159, 160f
Proliferating cell nuclear antigen, 28
  cell-cycle-related expression of, 170–172
  immunofluorescence measurement of, using flow cytometry, 169t, 170
  localization
    fixation artifacts with, 160

fixative techniques for, 160
monoclonal antibody to, available as direct conjugate, 166t, 167
PCNA/cyclin, 367–368
quantitation, 579
Proliferation. *See* Cell proliferation
Proliferation-associated antigens, antibodies to, 28–29
Proliferation-specific antigens, as markers of cell growth kinetics, in breast cancer, 256
Proliferative fasciitis, DNA ploidy, 346, 348t
Proliferative fraction, in lymphoma, prognostic significance of, 225–226
Prolog, 590
Pronase
  dissociation and enucleation of sections of rehydrated tissue from fixed, paraffin-embedded materials, 105
  used to dissociate tumors, 95
Propidium iodide, 113, 114f, 119, 167
  in demonstration of cell viability, 412, 412f
  for DNA staining, 205
    in lymphoma, 205–206
  for flow cytometric reticulocyte analysis, 377t, 378
  staining, of corneocytes, from psoriatic skin, 548, 548f
Prostate, anatomy of, 271, 272f
Prostate cancer
  androgen receptors in, flow cytometric analysis of, 281
  aneuploidy, 275–277
    detection of, 278f, 278–279
  aspiration biopsy, cytologic assessment of, 272
  clinical assessment of, 272–273, 301–302
  diagnosis of, 301–302
  DNA content analysis, 274f, 274–281, 302
  DNA content and cell cycle analysis, aggregation modeling with, 138f, 138–139
  DNA-content heterogeneity, significance of, 279
  DNA Index (ploidy), 302
    of biopsy specimens, 274–275, 275t
    correlation with prostate-specific antigen, 280
    description, flow cytometry versus image analysis for, 364
    plus seminal vesicle involvement, prognostic significance of, 276–277, 277f
    prognostic significance of, 129, 275t, 275–276, 276t, 277f
    of radical prostatectomy samples, 275–277, 276t
  epidemiology of, 271
  flow cytometric analysis of, future directions for, 281, 302
  Gleason's score for, 272, 273f
  grading, 272–273
  incidence of, 271
  mortality of, 271
  natural history of, 271–272, 302
  oncogene expression in, 281
  outcome, and grading, 272–273
  paraffin-embedded material
    detection of aneuploidy in, 278f, 278–279
    processing of, 273, 274f
  pathological assessment of, 271–272
  prostate-specific antigen determination in, 280, 301–302
  ras gene mutations in, 169
  sample
    acquisition of, 273
    processing of, 273
  S-phase measurements, 279–280, 280f, 302
  staging, 271, 272t
  therapeutic response, possible predictive factors, 275, 277, 278f, 302
  tumor cell proliferation, 279–280
  tumorigenesis, 281
Prostate-specific antigen
  determination, in prostate cancer, 301–302
  and DNA ploidy, in prostate cancer, 280
Protein(s), synthesis, in cell growth, 21
Protein A, antibody binding, 145
Proteinase K, dissociation and enucleation of sections of rehydrated tissue from fixed, paraffin-embedded materials, 105
Protein G, antibody binding, 145
Protein kinase C, 505
Protooncogenes, 5, 557. *See also* Oncogene(s)
  activation of, 10, 557–558
  immunofluorescence measurement of, using flow cytometry, 169t, 170–172
  nuclear polypeptides encoded by, 579
  products encoded by, 573
Pseudo-aneuploidy, 120f, 121, 211
Psoriasis, 547f, 547–549
  inflammatory response in, 549
  pathogenesis of, 549–550
p53 tumor suppressor gene, 581
  expression of, 172
    in bladder cancer, 292–293
    in oncogenesis, 6–7
  overexpression, in ovarian cancer, 9
  point mutations of, in breast cancer, 7
Pulmonary fibrosis, idiopathic, role of glutathione in, 520
Pulse width
  maximum throughput determination from, 91
  vs. integral pulse amplitude, doublet discrimination by, 85, 86f
Pyronin Y, 24
  for flow cytometric reticulocyte analysis, 377t, 378
Pyropoikilocytosis, 375–376

Q
Q-banding, 4
QC. *See* Quality control
Q prep, 148
Quality control, 177–198
  biological, 190–195
    data-collection-analysis related, 190
    patient-related variables, 190
    reagent-related variables, 190
    specimen-related variables, 190
  commercially available materials for, 180t–181t
  DNA analysis, 194–195
  immunofluorescence, 190–194
  for immunophenotyping lymphocytes in HIV disease, 445–446
  instrument, 177–190
  issues, for flow cytometry, 177, 177t
  in leukocyte-depletion techniques, 376
  materials, for troubleshooting, 196, 197t
  optical system, 178–181
  record keeping, 189
  signal processing, 181–189
  strategy, 189–190
  successful, keys to, 195–198
Quantitative digital imaging, 359
Quiescent cells
  DNA and RNA distributions in, 24, 25f
  G_1. *See* G_{1Q} cells

nuclear protein content, 26
size of, 24
Quin2, 505–506
Quinolythiacyanine iodide, for flow cytometric reticulocyte analysis, 377t

R
Radiation accident(s), somatic cell mutation studies in, 376
Radiation therapy
  in bladder cancer, flow cytometric findings related to, 286–287
  for rectal cancer, 317
Radioimmunoassay, 535
Radiometer design
  collimating, 78
  imaging, 78
Radionuclide release assays, of cell-mediated cytotoxicity, 499
Rare event analysis, by flow cytometry, 479. *See also* Fetal–maternal hemorrhage; HIV disease
ras gene(s)
  activation of, 5
  expression
    in human malignancy, 577–579
    in prostate cancer, 281
  mutant, dosage of, 9
  mutation, in genetic evolution of tumor cell population, 9
  polypeptides encoded by, 577
ras protein
  immunofluorescence measurement of, using flow cytometry, 168–169, 169t
  recombinant, absorption of anti-ras antibodies with, 163–164
Rat sarcoma genes. *See* ras gene(s)
Rb1 gene, 6
  expression of, in breast cancer, 7
  mutation, in retinoblastoma tumorigenesis, 6
RC networks. *See* Resistor-capacitor networks
Receptor(s)
  high-density, fluorescence reduction with, 475–476
  ligand binding to, 471–474
    association-dissociation model, 471–472, 472f–473f
  low-density, fluorescence distribution with, 476
Receptor protein tyrosine phosphatase, 577
Recessive oncogene, 557
Recombinant erythropoietin therapy, response to, assessment of, 380
Rectal cancer
  chemotherapy for, 317
  DNA flow cytometry in, 311, 311t
  radiation therapy for, 317
  recurrence, 317
Rectilinear gating, 87
Red 613, 114, 168t
Red blood cell(s)
  abnormalities, image analysis system for characterization of, 364
  antigens, flow cytometric analysis for, 376
  biotinylated, 375
  flow cytometric analysis of, 373–383
    advantages of, 373
    clinical utility of, 373, 374t
    future of, 381–383
    techniques, 373, 373t
  immunologic destruction of, 375
  immunophenotyping, for genetic or paternity testing, 376
  intrinsic defects, 375–376

Red blood cell(s)—(*continued*)
  lysis, 148
  mass, flow cytometric analysis of, 375
  mutagenic assay, 376
  parasitic infections in, quantitation, 376–377
  survival in vivo, flow cytometric analysis of, 374–375
  transfusions, 374–375
  volume studies, flow cytometric, 375
Reduced chi-square, 54
Reed-Sternberg cell(s), 217–218
Relapse, early detection of, 479
Relational databases, 591–592
Renal cancer, flow cytometric analysis, using antibodies to cytokeratins, 168
Renal cell carcinoma, 293–297
  aneuploidy, prognostic significance of, 294–296, 304
  causes of, 293
  clinical assessment of, 293
  DNA content
    heterogeneity, 279
    and stage, 293–294
  DNA content analysis, 293–296
  epidemiology of, 293
  flow cytometric analysis of, 304
    future of, 296–297
  fractional allelic loss, as prognostic factor, 8
  immunotherapy, 304
  incidence of, 293
  metastatic, flow cytometric analysis of, 296
  pathological assessment of, 293
  ploidy
    description, flow cytometry versus image analysis for, 364
    and grade, 293
    and stage, 293–294
  prognosis for, 293
  staging, 293
  survival, 293
    and DNA ploidy, 294t, 294–296, 295f
    predictors of, 294t, 294–296, 295f
  tumorigenesis, 293
Renal transplantation
  flow cytometric crossmatch for, 451
    clinical significance of, 451–452, 452f
  outcome, and flow cytometric panel antibody reactivity, 453
Residual malignant disease. *See also* Minimal residual disease
  detection of, 479–480
Resistor-capacitor networks, passive, 82
Restriction enzymes, 558
Reticulocyte(s), 377. *See also* Platelet reticulocytes
  immunophenotyping, through transferrin receptor, 377–378
Reticulocyte analysis, flow cytometric
  advantages over conventional methodology, 377
  automated, 378–379
  clinical conditions that interfere with, 381, 383t
  data analysis, 381
  fluorescent dyes for, 377t, 377–378
  methods of, 377–378
  standardization, 381
Reticulocyte maturity index, 377–381
  calculation and expression, methods of, 380–381, 382f–383f
  clinical utility of, 379f, 379–380, 380f, 380t
  diagnostic and therapeutic applications of, 380
  HFR fraction RMI units for, 380–381, 383f

Retinoblastoma, 338
  chromosomal changes in, 5t
  hereditary, tumorigenesis, 6
  pediatric, incidence of, 331, 331t
  sporadic, tumorigenesis, 6
  tumorigenesis, Rb gene mutation in, 6
Retrovir. *See* Zidovudine
Reverse (R)-banding, 4
RGE-53, and DNA content, simultaneous measurement of, in bladder cancer, 290
Rhabdoid tumor(s), 338
Rhabdomyosarcoma
  alveolar, chromosomal changes in, 5t
  histologic grading of, 345f
  histomorphologic subtypes of, and DNA ploidy, 346
  pediatric, 335–336
    incidence of, 331, 331t, 335
    ploidy, 336, 337f
      prognostic significance of, 336, 337f, 338–339, 339t
    therapeutic response with, 336
  percent of DNA diploidy, 347
Rheumatoid arthritis, neutrophil chemotaxis in, 416
Rh immune globulin, 374
Rhodamine, 111
  cytotoxicity, flow cytometric monitoring of, 461
Rhodamine derivative(s), 167, 168t. *See also* TRITC; XRITC
Rhodamine dyes, aggregates, on antibody surface, 113
Ribonucleotide reductase
  detection of, 28–29
  immunofluorescence measurement of, using flow cytometry, 168–170, 169t
Ribosomal RNA, as predictive marker of malignancy, 21–22
Richter's syndrome, 219
Rifampin, effects on neutrophil function, 407t
RMI. *See* Reticulocyte maturity index
RNA. *See also* Cellular RNA; Ribosomal RNA
  genomic, analysis of, molecular techniques in, 559
RNA index
  of aggressive lymphomas, 221
  in non-Hodgkin's lymphoma, prognostic significance of, 226
RPE. *See* R-phycoerythrin
R-phycoerythrin
  absorption spectrum of, 114
  fluorescent emission characteristics of, 146f
Rule-based knowledge representation, 588–590
Rule chaining, 588
  definition of, 601
Rule firing, 588, 588f
  definition of, 601
Rules, 588
  definition of, 601
Running track paradigm, for number of cells versus cell age relation, 43, 43f

S
S100, expression, in melanocytes, 551
Saccharomyces cerevisiae
  cdc2 gene, product of, 16, 17f
  cdc25 gene, phosphatase encoded by, 17f
  G_1-specific cyclins in, 16–17, 17f
  nim1 gene, product of, 17f
  suc1 gene, product of, 17f
  wee1 gene, product of, 17f
Saccharomyces pombe

$p34^{cdc2}$ kinase, 16–17, 17f
  protein phosphatases, in regulation of cell cycle, 19
  sds22+, 19
Saponin, 158
Sarcoma(s). *See also* Alveolar soft part sarcoma; Clear cell sarcoma; Epithelioid sarcoma; Soft-tissue sarcoma(s); Synovial sarcoma
  chromosomal changes in, 5t
  dissociation, 95–96, 99f
  histomorphologic subtypes of, and DNA ploidy, 346
  Ki-67 staining, correlation with tumor grade and other prognostic features, 368
Scatchard equation, 471
Scatchard plot, 472, 473f
Schwannoma(s)
  DNA ploidy, 346, 348t
  prognostic value of DNA flow cytometry in, 355
  risk factors in, 351t
SCID. *See* Severe combined immunodeficiency
Secretory granules, neutrophil, 409–410
Selectins, 388
Selective immunodeficiency, 437–438
Semantic network, 587–588
  definition of, 601
Sense-in-air flow cell design, 73
Severe combined immunodeficiency, 406t, 438–439
  lymphocyte immunophenotyping in, 440
  phenotyping of cells in, 436t
  prenatal diagnosis of, 439
  treatment of, 439
Sezary syndrome, 222
SFit method, of DNA data analysis, 45
Shackney model, for evolution of aneuploid stemlines, 8–9, 9f
Shortpass filters, 79
Side scatter, 207
Side-scatter sensor, 78–80
Signal broadening, and DNA histogram, 44, 44f
Signal processing
  linear versus logarithmic, 182–184, 183f
  quality control, 181–189
Simply Cellular Microbeads, 471
Single-cell cytotoxicity assays, 499
Single-cell suspensions, preparation of, 361
  from excised nevi or melanomas, 551
Single-cut model component, 51–52, 52f
Singlet(s), orientation and fluorescence profile, 133f, 134
Skin disorders. *See also* Melanoma
  flow cytometry in, 545–553
Skin testing, 436
Skin window test, 409
Slime mold. *See Physarum polycephalum*
Slit-scan flow cytometry, 290
Slot(s), 590
  definition of, 601
Slot hybridization
  for DNA, 559
  for RNA, 559
Small round-cell tumors, biologic behavior of, 343
Smooth-muscle tumor(s)
  benign versus malignant, 346
  gastric, prognostic value of DNA flow cytometry in, 353
SNARF-1, 506
  fluorescence emission spectra for, 518, 518f

as probe for pH, measurement, 518
  cell preparation for, 518
  flow cytometric analysis of loaded cells, 519, 519f
Soft-tissue sarcoma(s), 343–358
  adjuvant chemotherapy for, 357–358
  aneuploidy, prognostic significance of, 350–351
  benign versus malignant, 357
    relative incidence of, 343
  diagnosis of, 343, 355
  flow cytometric analysis of, 345–355, 357
    diagnostic/prognostic value of, 357
    future of, 355
    reliability, 357
    standardization, 357
  grading, 357
    DNA flow cytometry for, 347–349
  high-grade
    DNA histograms in, interpretation of, 350, 352f
    risk factors in, 350, 351t
  histologic grade, and percent of DNA diploidy, 347
  histologic grading of, 344–345, 345f
    prognostic significance of, 344–345
  histologic subtypes of, 357
  histologic typing of, 343–344
    and biologic behavior, 344
    and therapy, 344
  index of malignancy, using flow cytometry data, 349
  local recurrences, 349
  metastases, 349
  metastasis-free survival, 351, 352f, 357
  mitotic activity, 347
  multiple stem lines in, 346
  needle biopsy of, 355, 357
  pediatric, incidence of, 331, 331t
  ploidy, 345
  prognosis for, DNA flow cytometric measurements and, 349–350
  prognostic indicators for, 351
  prognostic value of DNA flow cytometry in, 350–351
    by individual tumor type, 351–355
  survival rate for, 356–357
  therapy for, 355–358
Soft-tissue tumor(s)
  benign versus malignant, DNA flow cytometry for distinction of, 346
  diploid lesion with malignant morphology, 346, 357
  histologic grade, and DNA ploidy, correlation of, 346, 348t
  subgroups of, 343
Solid-state detectors, 80
Solid-state laser(s), 75
Solid tumors. *See also specific tumor*
  automated diagnosis of, 590
  biopsy, 93
  chromosomal changes in, 4, 5t
  chromosomal loci involved in, 6t, 7
  cytogenetic analysis of, 3
    problems in, 4
  dissociation
    chemical, 94
    criteria for, 93, 93t
    enzymatic, 94–95
      evaluation of, 95–96
    in fresh-tissue samples, 94–100
    mechanical, 94
      evaluation of, 95–96
    protocols, evaluation of, 95–96
    from sections of fixed, paraffin-embedded tissue, 100–105

fixation, 102, 103f
  initial tissue handling, 102
  tissue sampling for, 102
  technical considerations, 93–105
  yield of subpopulations of tumor cell, 95–96
DNA content analysis, flow cytometry versus image analysis for, 364, 365t
DNA-content changes in, 3
DNA histograms, outside-in analysis method, 45
fine-needle aspiration of, 93–94
genetic evolution of, model for, 8–9, 9f, 10
interphase cytogenetics, 5
morphologic gating with, 362
pediatric, 331–339
  biologic features of, 331
  diagnosis of, 331
  heterogeneity of, 331
  incidence of, 331t, 331
  outcome predictors, 331
  ploidy, clinical significance of, 331, 338–339
  relative incidence of, 331t
  spectrum of tumor types, 331t, 331
  therapy for, 339
proliferative activity in, 367
sampling, 93–94
tumorigenesis, accumulation of genetic events in, 7–8
Soluble analyte, detection of, with flow microsphere immunoassay, 539–540
Somatic cell mutation studies, 376
Somatic cells
  cancer as genetic disease of, 3
  genomic stability of, 8
Somatic mutation theory, 3
Southern blot DNA hybridization, 6, 6f, 558–559
  for detection of minimal residual disease, 482t, 482
Specific granule deficiency, 406t, 410
Spectrofluorimetry, in measurement of intracellular ionized calcium, 508–509
Spermiogenesis, histone acetylation in, 15
S phase, 13–15
  calculation, in mixed DNA-diploid and aneuploid populations, 128
  cell commitment to enter, size threshold for, 21
  cells in, in lymphoma, prognostic significance of, 226
  DNA content during, 42–43, 43f, 117f, 118
  estimation, 49
    interlaboratory variations with, 363
    in mixed DNA-diploid and aneuploid populations, 128
    prognostic value of, in node-negative breast cancer, 128, 128t
    simple analysis strategies for, 45
  histone acetylation in, 14–15
  histone synthesis in, 15
  in lymphoma, prognostic significance of, 226
  measurements
    in prostate cancer, 279–280, 280f
    in transitional cell cancer of bladder, 284–285
  overlapping, in mixed DNA-diploid and aneuploid populations, 128
  relative rate of DNA synthesis across, 55–56
S-phase fraction
  in breast cancer, 249, 261
    combination with DNA index, 254–255
    prognostic significance of, 254f, 254

in colorectal cancer, evaluation of, 309–310
  determination, interlaboratory differences in, 261
  estimation, image analysis for, 366
  measurement of, BrdUrd in, 367
  prognostic categories based on, 140, 140t
  prognostic significance of, in gynecological cancer, 267–269
Spindle-cell tumor(s), biologic behavior of, 343
Splenocyte(s), intracellular ionized calcium (Ca^{2+}), flow cytometric analysis of, tandem immunophenotyping with, 513, 514f
Squamous cell carcinoma
  aneuploidy in, 550
  of bladder, 281
    diagnosis of, 302–303
    flow cytometric analysis of, 187
  cervical, 266
  dissociation, 95
  esophageal, DNA flow cytometry, 314
  of head and neck
    aneuploidy
      incidence of, by tumor stage, 324–325, 325t
      and tumor differentiation, 325, 325t
    dissociated cells, ethanol fixation of, 101f
    dissociation, 96, 97f
    DNA content
      abnormal, and biologic behavior of tumor, 324–325
      analysis of, 324–325
      and survival, 325, 326t
      and therapeutic responsiveness, 326
    DNA distributions, from fresh versus archival material, 101t
    flow cytometric analysis, 323–326
      prognostic significance of, 325–326
      summary of, 324t, 324
      using antibodies to cytokeratins, 168
    grading, 324
    heterogeneity of, 323
    outcome, and histopathologic features, 324
    prognostic factors for, 323–324
    TNM staging of, 323
    treatment of, 329
  pulmonary, 319–320
    flow cytometric DNA analysis of, 320t, 320–321
    ploidy, prognostic significance of, 321t, 321–323
  renal, DNA content analysis, 296
  of ureter, DNA content analysis, 296
  well-differentiated, versus keratoacanthoma, 550–551
Staining
  in DNA analysis, 194
  of leukemic cells, 240
  of lymphoma, 204–206
Standard file format, for flow cytometry data, 88, 89f
Standardization
  commercially available materials for, 180t–181t
  of DNA model(s)/modeling, 56
  of flow cytometers, 197
  of flow cytometric analysis of soft-tissue sarcomas, 357
  of flow cytometric reticulocyte analysis, 381
  of intensity scale(s), 184
START point, in yeast cell cycle, 17–18

Stathmokinesis, 24
  combined with flow cytometry, 31
Static cytometry, 359
Statin, 29
Stemline(s), 4
  DNA-aneuploid, 8
Steric hindrance, in cell surface immunofluorescence, 145–146
Steroids, effects on neutrophil function, 407t
Steroid sulfatase, deficiency, 550
Storage, of dissociated tumor cells, 99
Streptavidin, 145
Streptavidin-PE/TR tandem conjugate, 150
Subpopulations, 586
  in multivariate list mode data, 597
Sulphonamides, effects on neutrophil function, 407t
Suppressor gene(s). *See* Tumor suppressor gene(s)
Suspension cells, preparations of, 144
SV40 T antigen
  immunofluorescence measurement of, using flow cytometry, 169t
  multiparameter quantitation of, paraformaldehyde/permeabilization protocols for, 161
Symbolic reasoning, 587
Synchronous tumor(s), 8
Synovial chondromatosis, DNA ploidy, 346, 348t
Synovial cyst(s), DNA ploidy, 346, 348t
Synovial sarcoma
  chromosomal changes in, 5t
  histologic grading of, 345f
  histomorphologic subtypes of, and DNA ploidy, 346, 347f
  percent of DNA diploidy, 347
  prognostic value of DNA flow cytometry in, 353
  risk factors in, 351t
Sysmex R-1000 reticulocyte counter, 378–379, 381

T
TA99 (monoclonal antibody), 545f
TBP. *See* Tetraphenylboron
TCC. *See* Transitional cell cancer
T cell(s), 64–65. *See also* Combined T- and B-cell deficiencies; Lymphoma(s), T-cell
  activation, 65
    detection of, 68
    antigenic heterogeneity of, 66–67
  in bare lymphocyte syndrome, 439
  CD4
    activation, role of glutathione in, 527f, 527–528, 528f–529f
    flow cytometric analysis of calcium concentration in, in HIV disease, 516–517, 517f
  cell-surface-differentiation antigens, 66f
  in common variable immunodeficiency, 437
  cytolytic
    in cell-mediated cytotoxicity, 497
    non-MHC-restricted, 497
    target recognition structures on, 498
  cytotoxic/suppressor, 67
  deficiencies, 438–440
    immunophenotyping in, 436t, 440
  development, 65, 65f
  in DiGeorge's syndrome, 439
  γ/δ-receptor-bearing
    in HIV disease, 443–444
    in transplant recipient, 455
  helper/inducer, 67
  in IgA deficiency, 437
  immunophenotyping, in clinical evaluation of primary immunodeficiency, 436t, 440
  intranuclear prolactin, localization and quantitiation of, 159
  memory, 67
  naive, 67
  ontogeny of, 236–237, 237f
  phenotyping of
    in combined T- and B-cell deficiencies, 436t
    in humoral immunodeficiency, 435t
  proliferation, after indo-1 loading, 508
  in psoriasis, 549
  in severe combined immunodeficiency, 438–439
  subsets, 66–67
    changes in, after bone marrow transplantation, 455
    in graft, changes in, after transplantation, 454–455
    in peripheral blood, changes in, after transplantation, 454
  switch, deficiency, 437–438
  as target cells for detection of alloantibodies, 449–450
  in transient hypogammaglobulinemia of infancy, 437
  in Wiskott-Aldrich syndrome, 440
T-cell antigen receptor, 497–498
  gene, 65
  clonal rearrangement, in detection of neoplastic lymphoid cells, 482, 486
3T3 cells, DNA and RNA distributions, in quiescence versus exponential growth, 24, 25f
TcR. *See* T-cell antigen receptor
TdT. *See* Terminal deoxynucleotidyl transferase
Teratoma, pediatric, incidence of, 332
Terminal deoxynucleotidyl transferase expression, in leukemia, 484
  immunofluorescence measurement of, using flow cytometry, 169t, 172
  in immunophenotyping of leukemia, 236, 236f–237f, 238
  in immunophenotyping of lymphoma, 221
  monoclonal antibody, available as direct conjugate, 166t
Testicular aspirates, DNA analysis of, 94
Testicular seminoma, chromosomal changes in, 5t
Tetracyclines, effects on neutrophil function, 407t
Tetraphenylboron, as chelating agent, 94
Tetraploidization, 8, 10
Tetraploidy, 118. *See also* Hypertetraploidy
Texas Red, 64, 113, 115
  antibodies labeled with at different dye/protein ratios, brightness of, 111–112, 112f
  bound to IgG at different dye/protein ratios, average quantum yields of, 112f
  covalent binding to PE, 114
  fluorescent emission characteristics of, 146f
  in immunofluorescence applications, 146
  spectral properties of, 112t
T-γ lymphoproliferative disorder, flow cytometric findings in, 224
Thermal injury, neutrophil chemotaxis in, 416
THI. *See* Transient hypogammaglobulinemia of infancy
Thiazole orange
  for flow cytometric reticulocyte analysis, 377t, 378, 378f, 381
  in measurement of platelet nucleic acid content, 394–395, 397f
Thioflavin T, for flow cytometric reticulocyte analysis, 377t, 378
Thrombocytopenia, platelet-associated immunoglobulins in, flow cytometric analysis of, 390
Thrombopoiesis, rate of, measurement of, 394–397
Thymidine kinase, detection of, 28–29
Thymidine-labeling index, 366
  correlation with percentage S-phase, in breast cancer, 253f, 253–254
  and Ki-67 staining, in tumors, 367
Thymidylate synthase, immunofluorescence measurement of, using flow cytometry, 169t, 170
Thymocyte(s), phenotype, of T-cell ALL, 485
Thyroid cancer, ploidy, description, flow cytometry versus image analysis for, 364
Thyroid tumors, FNA samples, flow cytometric analysis of, 94
TICAS-MLD, 364
Tietze recycling assay, for calibration of flow cytometric analysis of intracellular glutathione, 524
Time
  as parameter in kinetic experiments, 87
  as quality control parameter, 87, 195, 196f
Time-lapse cinematography, in cell-cycle research, 13
Tissue
  paraffin-embedded, gross evaluation and histological examination, 122
  unfixed samples, gross evaluation and histological examination, 122
Tissue autolysis, 102
Tissue dissociation, 118–119
TLI. *See* Thymidine-labeling index
Tn syndrome, somatic cell mutation studies in, 376
TR. *See* Texas Red
Transferrin receptor. *See also* CD71 antigen
  in diagnostic hematology, 377–378
Transgenic mouse liver, use of β-gal reporter system in, 570f, 571
Transient hypogammaglobulinemia of infancy, 437
  phenotyping of cells in, 435t
Transitional cell cancer
  of bladder, 281
    clonality, 282
    diagnosis of, 302–303
    DNA content analysis, 283–287
      for archival specimens, 283–284
      for fresh specimens (TURBT and cystectomy specimens), 283
      prognostic significance of, 284, 284f
    tumor cell proliferation, 284–285
    tumorigenesis, 281–282
  of upper urinary tract, flow cytometric analysis of, 285
Transplantation
  antibodies produced after, flow cytometric detection of, 453
  changes of lymphocyte subsets after detection of, 454–455
    in graft, 454–455
    in peripheral blood, 454
  role of flow cytometry in, 449–455. *See also* Alloantibodies; Crossmatch, flow cytometric
  future of, 455

Tri-Color, 115
Triplet(s), orientation and fluorescence profile, 133f, 134
Triploidy, 4
TRITC, 167, 168t
TRITC-amines, spectral properties of, 112t
Triton X-100, 158, 167
Trypan blue exclusion, in cell viability determination, 144
Trypsin
  dissociation and enucleation of sections of rehydrated tissue from fixed, paraffin-embedded materials, 104
  used to dissociate tumors, 94–95
Tuftsin, 417
  deficiency, 406t, 417
Tumor(s). *See also* Solid tumors
  classification of, cell cycle markers in, 34
  diagnosis and classification of, 26
  dissociation, technical considerations, 93–105
  DNA content heterogeneity, 122
  grading, 34
  histology, 34
  ploidy, relation between molecular-genetic changes and, 8
Tumor cell(s)
  detection of, 479
  dissociated
    fixation, 96–99
    from fresh samples, storage of, 99
    handling, 96–99
  enucleation, 95
    techniques, evaluation of, 96
Tumor cell populations
  clonal origin, 4
  cytogenetic classification of, 4
  modal chromosome number, 4
  stemline, 4
Tumor cell proliferation
  measurement of, 366–368
    immunohistochemical methods combined with image analysis for, 368
    methods for, 366t
    prognostic significance of, 368
  prognostic significance of, 366
  in prostate cancer, 279–280
  in transitional cell cancer of bladder, 284–285
Tumorigenesis
  genetic factors in, 3–8
  multistage model for, 10
Tumor markers, 491–493
Tumor necrosis factor, 498

Tumor progression, 3
Tumor suppressor gene(s), 6–7, 557
  chromosomal loci of, 6, 6t
  in colorectal carcinogenesis, 7
  inactivation, 10, 558
    and aneuploidy, 9
  products encoded by, 573
Tween-20, 158, 167
Tyrosine kinase family, genes, 573

U
Ubiquitin, 16
Ubiquitination, of histones, 16
Ulcerative colitis
  DNA content heterogeneity, 122
  DNA ploidy analysis of, 317
  relation between dysplasia and DNA aneuploidy in, 310–311, 317
Ultralite, 167, 168t
Ultraviolet excitation, 74
Uncertainty management
  definition of, 601
  in expert systems, 593–596
    heuristic, 594
    probabilistic methods for, 593–594
Upper aerodigestive tract tumor(s), 319–329. *See also* Squamous cell carcinoma, of head and neck
  flow cytometric analysis of, clinical and research utility of, 328
  treatment of, 329
Urological cancer(s), 271–304. *See also* Bladder cancer; Prostate cancer; Renal cell carcinoma; Urothelial cancer(s)
Urothelial cancer(s). *See also* Bladder cancer
  epidemiology of, 302
  field change aberrations in, 281–282, 304
Uterine sarcoma(s), prognostic value of DNA flow cytometry in, 353–354
Uterus, smooth-muscle tumors, 346

V
Variable region(s), 65
Verapamil, in blockade of drug resistance processes, 257
v-*erb*-B, 573
v-*fms*, 573
Vimentin, immunofluorescence measurement of, using flow cytometry, 169t
Vita Blue, 168t
v-*kit*, 573–574
V region. *See* Variable region(s)

W
Waldenström's macroglobulinemia, 219
Western blot analysis, confirmation of flow cytometric antigen measurements with, 165
Whole blood cellular immunofluorescence, 148–150
Wilms' tumor
  anaplastic variant, 335
  chromosomal changes in, 5t
  DNA content analysis, 296
  epidemiology of, 293
  familial, 293
  incidence of, 293, 331t, 331, 335
  karyotypic analyses of, 335
  ploidy, prognostic significance of, 335, 336f, 338–339, 339t
Wiskott-Aldrich syndrome, 440
  phenotyping of cells in, 436t
Within-cluster sum of squares, 598, 598f–599f
Working memory, 588f, 588
  definition of, 601
Wright-Giemsa staining, of leukemic blasts, 240
WSS, 598–599, 599f

X
Xenobiotics, cellular detoxification of, 464, 466
Xenon arc lamp(s)
  as excitation light source, 74
  spectral intensity distribution for, 74, 74f
XLA. *See* X-linked agammaglobulinemia
X-linked agammaglobulinemia, 436–437
X-linked hypogammaglobulinemia, phenotyping of cells in, 435t
X-linked ichthyosis, 550
X-rhodamine, bound to IgG at different dye/protein ratios, average quantum yields of, 112f
X-RITC, 115, 167, 168t
  antibodies labeled with at different dye/protein ratios, brightness of, 111–112, 112f
X-RITC-amines, spectral properties of, 112t
X-RITC-antibody, 113

Y
YB5.B8 antibody, 575
Yeast, cell cycle, START point, 17–18

Z
Zidovudine, therapy, in HIV disease, 442

ID	Author	Title	Journal	Yr	Vol:Pages
5473		Possible involvement of	Soc.Exp.Biol.Me	77	155:89-93
142	Abramson,J.S.	Depression of monocyte a	Infect.Immun.	82	35:350-355
194	Abramson,J.S.	Recurrent infections and	J.Pediatr.	81	99:887-894
2653	Aeschbacher,M	A rapid cytotoxicity ass	Int.Conf.Prac.T	86	24:467-
5521	Afzelius,B.A.	Structure and function o	Acta.Med.Scand.	80	208:145-145
5705	Afzelius,B.A.	Structure and function o	Acta Med.Scand.	80	208:145-154
1382	Alexander,J.W	Serum and leukocyte lyso	Arch.Surg.	67	95:482-490
779	Alteri,E.	N-formylmethionyl-leucyl	Blood	83	62:918-923
5522	Altman,L.C.	Depressed mononuclear le	J.Immunol.	77	119:199-199
5345	Altman,R.D.	Neutrophil activation: a	Semin.Arthritis	90	19:1-5
5527	Ambruso,D.R.	Decreased hydroxyl radic	Pediatr.Res.	82	16:198A-198A
5526	Ambruso,D.R.	Effective oxidative meta	Pediatrics	79	64:722-722
5766	Ambruso,D.R.	Defective oxidative meta	Pediatrics.	79	64:722-725
5717	Anderson,D.C.	Abnormal stimulated adhe	Blood	87	70:740-750
1404	Anderson,D.C.	The severe and moderate	J.Infect.Dis.	85	152:668-689
5482	Anderson,D.C.	Leukocyte adhesion defic	Ann.Rev.Med.	87	38:175-194
5723	Arnaout,M.A.	Structure and function o	Blood	90	75:1037-1050
2044	Arnaout,M.A.	Increased expression of	N.Eng.J.Med.	85	312:457-462
2047	Arnaout,M.A.	Deficiency of a leukocyt	J.Clin.Invest.	84	74:1291-1300
5690	Athens,J.W.	Leukokinetic studies. IV	J.Clin.Invest.	61	40:989-989
5483	Bainton,D.F.	Leukocyte adhesion recep	J.Exp.Med.	87	166:1641-1653
5756	Balfour,H.H.,	Chronic granulomatous di	JAMA.	71	217:960-961
5523	Bannatyne,R.M	Inhibition of the biolog	J.Infect.Dis.	77	136:469-474
70	Bass,D.A.	Flow cytometric studies	J.Immunol.	83	130:1910-1917
544	Bassoe,C-F.	Processing of staphyloco	Cytometry	84	5:86-91
25	Bassoe,C-F.	Flow cytometric studies	Acta Pathol.Mic	84	92:167-171
48	Bassoe,C-F.	The effect of serum opso	Acta Pathol.Mic	84	92:51-58
522	Bassoe,C-F.	Phagocytosis by human le	J.Med.Microbiol	85	19:115-125
51	Bassoe,C-F.	Simultaneous measurement	Cytometry	83	4:254-262
55	Bassoe,C-F.	Phagocytosis of bacteria	Proc.Soc.Exp.Bi	83	174:182-186
5710	Bassoe,C-F.	Human Peripheral Blood P	Cytometry Suppl	85	:-
3416	Bassoe,C-F.	Phagocytosis of Staphylo	Acta.Pathol.Mic	84	92:43-50
5711	Bassoe,C-F.	Quantitation of single c	Flow Cytometry	80	Universitetsfor
5677	Becker,E.L.	The ability of chenmotac	J.Immunol.	74	112:2047-2054
5678	Becker,E.L.	Superoxide production in	Am.J.Pathol.	79	95:81-97
562	Becker,S.	Heterogeneity of human p	J.Reticuloendot	83	33:127-138
3556	Bellinati-Pir	Evaluation of a fluoroch	J.Immunol.Meth.	89	119:189-196
5524	Belsheim,J.	Tetracyclines and host d	Scand.J.Infect.	79	11:141-145
5680	Bentwood,B.J.	C5a-induced degranulatio	Fed.Proc.	80	39:798-
5331	Berger,M.	Complement deficiency an	Rev.Infect.Dis.	90	12 Suppl 4:S40
5640	Berger,M.	Tumor necrosis factor is	Blood	88	71:151-158
5692	Bishop,C.R.	Leukokinetic studies. 13	J.Clin.Invest.	68	47:249-260
2906	Bjerkirens,R.	Phagocyte C3-mediated at	Blut.	84	49:315-323
536	Bjerknes,R.	Flow cytometric assay fo	J.Immunol.Meth.	84	72:229-241
549	Bjerknes,R.	Human leukocyte phagocyt	Acta Pathol.Mic	83	91:341-348
531	Bjerknes,R.	Phagocyte C3-mediated at	Blut.	84	49:315-323
2231	Bjerknes,R.	Flow cytometry for the s	Rev.Infect.Dis.	89	11:16-33
5525	Bjorksten,B.	Inhibition of human neut	Infect.Immun.	76	14:315-317
768	Blair,O.C.	Differentiation of HL-60	Cytometry	85	6:54-61
1831	Blair,O.C.	Differentiation of HL-60	Cytometry	86	7:171-177
5691	Boggs,D.R.	Leukokinetic studies. IX	J.Clin.Invest.	65	44:643-656
5666	Bokoch,G.M.	Effect of various lipoxy	J.Biol.Chem.	81	256:5317-5320
4883	Boros,P.	Change in expression of	Clin.Immunol.Im	90	54:281-289
5486	Borregaard,N.	Identification of a high	J.Clin.Invest.	90	85:408-416
5480	Borregaard,N.	Chemoattractant-regulate	Science	87	237:1204-1206
5709	Boxer,L.A.	Neutrophil actin dysfunc	N.Engl.J.Med.	74	291:1093-1099
5657	Boyden,S.V.	The chemotactic effect o	J.Cell Biol.	62	82:347-368
1255	Boyum,A.	Isolation of mononuclear	Scand.J.Clin.La	68	21:97-77
5695	Bretz,U.	Biochemical and morpholo	J.Cell Biol.	74	63:251-269
2199	Brom,J.	Decreased expression of	Prostaglandins	88	34:153-159
5621	Brown,S.S.	Mechanism of action of c	J.Cell Biol.	81	88:487-491
5741	Buescher,E.S.	Abnormal adherence-relat	Blood.	85	65:1382-1390
2792	Burns,W.H.	Molecular cloning and ma	Virology.	88	166:140-148

ID	Author	Title	Journal	Year Vol:Pages
2826	Butler,T.W.	Assessment of total immu	J.Immunol.Meth.	88 108:159-170
1934	Cairo,M.S.	Fluorescent cytometric a	Pediatr.Res.	88 24:673-676
5750	Caldicott,W.J	Chronic granulomatous di	Am J Roentgenol	68 103:133-139
3719	Cantinieaux,B	Staphylococcus aureus ph	J.Immunol.Meth.	89 121:203-208
5529	Chan,C.K.	Inhibition of granulocyt	Can.J.Microbiol	78 243:363-364
1489	Charon,J.A.	An in vitro study of neu	J.Perio.Res.	82 17:614-625
5671	Chenoweth,D.E	Demonstration of specifi	Proc.Natl.Acad.	78 75:3943-3947
5714	Christensen,R	Granulocyte transfusions	Pediatrics	82 70:1-6
5757	Chusid,M.J.	Chronic granulomatous di	JAMA.	75 233:1295-1296
1496	Cianciola,L.J	Defective polymorphonucl	Nature (London)	77 265:445-447
5532	Clark,R.A.	Defective granulocyte ch	J.Clin.Invest.	71 50:2645-2652
1500	Clark,R.A.	Defective neutrophil che	Infect.Immun.	77 18:694-700
5619	Clay,M.E.	Detection of granulocyte	Current Concept	85American Associ
739	Cohen,H.J.	Opsonized zymosan-stimul	Blood	81 58:975-982
1512	Cohen,M.S.	Phagocytic cells in peri	J.Periodontol.	85 56:611-617
5656	Comandon,J.	Phagocytose in vitro des	Compres rendus	17 80:314-316
5622	Cooper,J.A.	Effects of cytochalasin	J.Cell Biol.	87 105:1473-1478
5533	Cooper,M.R.	Stimulation of leukocyte	Infect.Immun.	71 3:851-853
5509	Dancey,J.T.	Neutrophil kinetics in m	J.Clin.Invest.	76 58:705-715
2046	Darzynkiewicz	Increased mitochondrial	Proc.Natl.Acad.	81 77:6696-
1762	Davis,B.H.	Characterization of f-Me	Cytometry	86 7:251-262
1285	Davis,J.M.	Neutrophil degranulation	J.Immunol.	80 124:1467-1471
742	De Togni,P.	Studies on stimulus-resp	Biochim.Biophys	83 755:506-513
776	DeChatelet,L.	Mechanism of the luminol	J.Immunol.	82 129:1589-1593
701	Deitch,E.A.	The relationship between	Burns	84 10:264-270
1283	Deitch,E.A.	Prognostic significance	J.Trauma	82 22:199-204
4811	Della Bianca,	Studies on molecular reg	J.Immunol.	90 144:1411-1417
740	Della-Bianca,	Studies on stimulus-resp	Biochim.Biophys	83 755:497-505
5464	Dewald,B.	Release of gelatinase fr	J.Clin.Invest.	82 70:518-550
5701	Dewald,B.	Subcellular localization	J.Exp.Med.	75 141:709-723
5758	Dilworth,J.A.	Adults with chronic gran	Am J Med	77 63:233-243
1943	Dive,C.	Inhibition of intracellu	Biochem.Pharmac	88 37:3987-3993
5643	Djeu,J.Y.	Functional activation of	J.Immunol.	90 144:2205-2210
5078	Dobrina,A.	Phorbol ester causes dow	Immunology	90 69:429-434
800	Dolbeare,F.A.	Naphtol AS-BI (7-bromo-3	J.Histochem.Cyt	79 27:120-124
1839	Dolbeare,F.A.	Flow cytoenzymology: Rap	Flow Cytometry	79Wiley
5531	Downey,R.J.	Some effects of antimicr	J.Reticuloendot	65 2:75-88
738	Duque,R.E.	Inhibition by Tosy-L-phe	J.Biol.Chem.	83 258:8123-8128
2233	Duque,R.E.	Detection of anti-platel	Proc.Soc.Anal.C	85 Abstract 451:-
5631	Dykman,T.R.	Polymorphism of the huma	J.Exp.Med.	84 159:691-703
3564	Eisenhauer,P.	Purification and antimic	Infect.Immun.	89 57:2021-2027
4614	Elferink,J.G.	Suppressive action of co	Am.J.Physiol.	89 257:C859-C864
2842	Ellis,M.	Impaired neutrophil func	J.Infect.Dis.	88 158:1268-1276
1271	Elmgreen,J.	Subnormal sensitivity of	Ann.Rheum.Dis.	85 44:514-518
5534	Esterly,N.B.	The effect of antimicrob	J.Invest.Dermat	78 70:51-55
1360	Falloon,J.	Neutrophil granules in h	J.Allergy Clin.	86 77:653-662
2583	Fattorossi,A.	New, simple flow cytomet	Cytometry	89 10:320-325
5704	Fearon,D.T.	Identification of the me	J.Exp.Med.	80 152:20-30
5661	Fernandez,H.N	Primary structural analy	J.Biol.Chem.	78 253:6955-6964
5535	Ferrari,F.A.	Inhibition of candidacid	Antimicrob.Agen	80 17:87-88
1446	Fishman,W.H.	Application of an improv	J.Biol.Chem.	48 173:449-456
5693	Fleidner,T.M.	Granulopoiesis: I. Senes	Blood	64 24:402-402
5719	Fleit,H.B.	Human neutrophil Fcgamma	Proc.Natl.Acad.	82 79:3275-3279
5329	Fletcher,J.	Acquired abnormalities o	Blood.Rev.	90 4:103-110
5052	Fletcher,M.P.	Effects of low concentra	Fundam.Clin.Pha	90 4:65-77
14	Fletcher,M.P.	Monitoring human neutrop	J.Leuk.Biol.	85 37:431-447
5696	Folds,J.D.	Neutral proteases confin	Proc.Soc.Exp.Bi	72 139:461-463
5667	Ford-Hutchins	Leukotriene B4, a potent	Nature	80 286:264-265
5537	Forsgren,A.	Effect of antibiotis on	Antimicrob.Agen	77 11:580-584
5538	Forsgren,A.	Effect of tetracycline o	J.Infect.Dis	74 130:412-418
1605	Fox,M.H.	Membrane fluidity measur	Cytometry	87 8:20-25
5747	Fridkin,M.	Tuftsin: its chemistry,	Crit.Rev.Bioche	89 24:1-40
5760	Gabig,T.G.	Leukocyte abnormalities	Med Clin.North.	80 64:647-666
5754	Gabig,T.G.	Deficient flavoprotein c	J Clin.Invest.	84 73:701-705

ID	Author	Title	Journal	Yr	Vol:Pages
5762	Gabig,T.G.	Molecular heterogeneity	J Free.Radic.Bi	85	1:65-69
5590	Gaither,T.A.	Deficiency in C3b recept	Inflammation	84	8:429-444
719	Gale,R.	Direct activation of neu	Ann.Rheum.Dis.	83	42:158-162
807	Gallin,J.	Leukocyte chemotaxis	Fed.Proc.	83	42:2851-2862
5648	Gallin,J.I.	Abnormal phagocyte chemo	Rev.Infect.Dis.	81	3:1196-
5591	Gallin,J.I.	Neutrophil specific gran	Clin.Res.	84	32:320-328
817	Gallin,J.I.	Neutrophil specific gran	Annu.Rev.Med.	85	36:263-274
964	Gallin,J.I.	NIH conference. Recent	Ann.Intern.Med.	83	99:657-674
5557	Gallin,J.I.	Defective mononuclear le	Blood	75	45:863-870
5558	Gallin,J.I.	Recurrent severe infecti	Blood	78	51:919-933
5706	Gallin,J.I.	Disorders of phagocyte c	Ann.Intern.Med.	80	92:520-538
5579	Ganz,T.	Microbicidal/cytotoxic p	J.Clin.Invest.	88	82:552-556
5470	Ganz,T.	Defensins. Natural pepti	J.Clin.Invest.	85	76:1427-1435
5637	Georgilis,K.	Human recombinant interl	J.Immunol.	87	138:3403-3407
5467	Ghebrehiwet,B	C3e: an acidic fragment	J.Immunol.	79	123:616-621
5743	Giordano,G.F.	The role of sulfhydryl g	J Cell.Physiol.	73	82:387-395
5542	Gnarpe,H.	Tetracyclines and host d	Microbios.	78	22:45-49
5543	Gnarpe,H.	Tetracyclines and host d	Microbios.	74	10A:127-138
5668	Goetzl,E.J.	The human polymorphonucl	J.Immunol.	80	125:1789-1791
5672	Goldman,D.W.	Specific binding of leuk	J.Immunol.	82	129:1600-1604
5544	Goodhart,G.L.	Effect of aminoglycoside	Antimicrob.Agen	77	12:540-542
3708	Gordon,D.L.	Regulation of human neut	Immunology	89	67:460-465
3605	Gorvel,J.P.	Conformational change of	J.Cell Biol.	89	108:2193-2200
5545	Gray,G.D.	Rifampin has paradoxical	Fed.Proc.	80	39:878-878
5761	Gray,G.R.	Neutrophil dysfunction,	Lancet.	73	2:530-534
3604	Griffioen,A.W	Measurement of cytoplasm	J.Immunol.Meth.	89	120:23-27
1376	Grogan,J.B.	Suppressed in vitro chem	J.Trauma	76	16:985-988
1377	Grogan,J.B.	Altered neutrophil phago	J.Trauma	76	16:734-738
1457	Grynkiewicz,G	A new generation of Calc	J.Biol.Chem.	85	260:3440-3450
205	Haak,R.A.	Membrane fluidity in hum	J.Clin.Invest.	79	64:138-144
961	Hafstrom,I.	Auranofin affects early	J.Immunol.	84	132:2007-2014
5708	Hartwig,J.H.	Cytochalasin B and the s	J.Mol.Biol.	79	134:539-553
5649	Hartwig,J.H.	Cytochalasin B dissolves	J.Cell Biol.	79	79:M11741-
984	Harvath,L.	Two neutrophil subpopula	Infect.Immun.	82	36:-
563	Haskill,S.	Flow cytometric analysis	J.Reticuloendot	82	32:273-285
560	Haskill,S.	Simultaneous three color	Cytometry	83	3:359-366
1279	Hayashi,H.	A review on the natural	Acta Pathol.Jpn	82	322:271-284
3706	Herlin,T.	Effect of auranofin on e	Agents Actions	89	28:121-129
5763	Heyworth,P.G.	Neutrophil nicotinamide	J Clin.Invest.	91	87:352-356
1851	Hilmo,A.	F-actin content of neona	Blood	87	69:945-949
913	Huey,R.	Characterization of a C5	J.Immunol.	85	135:2063-2068
5606	Huizinga,T.W.	The PI-linked receptor F	Nature	88	333:667-669
5638	Hynes,R.O.	Integrins: a family of c	Cell.	87	48:549-554
1576	Iacono,V.J.	In vivo assay of crevicu	J.Periodontol.	85	56:56-62
5632	Jack,R.M.	Differential interaction	J.Immunol.	86	137:3996-4003
4654	Jennings,L.K.	Calcium mobilization in	Blood	89	74:2674-2680
5508	Jesaitis,A.J.	Intracellular localizati	Biochim.Biophys	82	719:556-568
5749	Johnston,R.B.	Chronic granulomatous di	Pediatr.Clin.No	77	24:365-376
5472	Kampschmidt,R	Neutrophil release after	J.Reticuloendot	80	28:191-201
5697	Kane,S.P.	Analytical subcellular f	Clin.Sci.Mol.Me	75	49:171-182
5663	Kay,A.B.	Leukoattractants enh	Clin.Exp.Immuno	79	38:294-299
1683	Kelly,M.K.	Neutrophil and monocyte	Int.Arch.Allerg	85	78:132-138
5547	Khan,A.J.	Abnormal neutrophil chem	J.Lab.Clin.Med.	79	93:295-300
5724	Kinne,T.J.	Antibody-dependent cellu	J.Clin.Lab.Immu	89	30:153-156
5639	Kishimoto,T.K	Heterogeneous mutations	Cell.	87	50:193-202
5764	Kleinberg,M.E	The phagocyte 47-kilodal	J Biol Chem.	90	265:15577-1558
743	Korchak,H.M.	A carbocyanine dye, DiOC	Biochem.Biophys	82	8:1495-1501
3602	Koss,L.G.	Flow cytometric measurem	Hum.Pathol.	89	20:528-548
5673	Kreisle,R.A.	Specific binding of leuk	J.Exp.Med.	83	157:628-641
3593	Kuroki,M.	ATP-induced calcium mobi	Biochim.Biophys	89	1012:103-106
5720	Lanier,L.L.	Functional Properties of	J.Exp.Med.	85	162:2089-2107
2522	Larsen,C.G.	The neutrophil-activatin	Science	89	243:1464-1466
5715	Laurenti,F.	Polymorphonuclear leukoc	J.Pediatr.	81	98:118-123
4789	Lawton,J.W.M.	The effects of intraveno	Immunopharmacol	89	18:97-105

4782	Lee,K.H.	Tumor necrosis factor ac		90	
5698	Leffell,M.S.	Association of lactoferr	Infect.Immun.	72	6:761-765
1043	Leonard,E.J.	Analysis of human monocy	J.Leuk.Biol.	85	38:403-413
5744	Lightman,M.A.	Rheology of leukocytes,	J Clin.Invest.	73	52:350-358
5627	Lindley,I.	Synthesis and expression	Proc.Natl.Acad.	88	85:9199-9203
2805	Livingston,D.	The effect of tumor necr	J.Surg.Res.	89	46:322-326
1165	Locksley,R.M.	Increased respiratory bu	Blood	83	62:902-909
1842	Loesche,W.J.	Reduced oxidative functi	Infect.Immun.	88	56:156-160
1537	Lucisano,Y.M.	Lysosomal enzyme release	J.Immunol.	84	132:2015-2020
5644	Ma,D.	Flow Cytometry with crys	J.Immunol.Meth.	87	104:195-200
5674	Mackin,W.M.	The formalpeptide chemot	J.Immunol.	82	129:1608-1611
4704	Mahonyey,K.H.	FACS quantitation of leu	J.Leuk.Biol.	85	38:573-585
5582	Malech,H.L.	Current concepts: immuno	N.Engl.J.Med.	87	317:687-694
5650	Malech,H.L.	Structural analysis of h	J.Cell Biol.	77	75:666-
4806	Maródi,L.	Stimulation of the respi	Clin.Exp.Immuno	90	79:164-169
5707	Maruyama,K.	Cytochalasin B and the s	Biochim.Biophys	80	626:494-500
5550	McCall,C.E.	Functional characteristi	J.Infec.Dis.	71	124:68-75
2347	McMullen,J.A.	Neutrophil Chemotaxis in	J.Periodontal	81	52(4):167-173
5655	Metchnikoff,E	Lectures on the Comparat		68	Kegan, Paul, Tr
1516	Miller,D.R.	Role of the polymorphonu	J.Clin.Periodon	84	11:1-15
5500	Miller,L.J.	Stimulated mobilization	J.Clin.Invest.	87	80:535-544
5745	Miller,M.E.	Phagocyte function in th	Pediatrics.	79	64:709-712
5759	Mills,E.L.	The chemiluminescence re	Pediatrics.	79	63:429-434
5466	Mollinedo,F.	Subcellular localization	J.Biol.Chem.	84	259:7143-7150
5578	Murphy,P.M.	Impairment of neutrophil	J.Infect.Dis.	88	158:627-630
1258	Nagel,J.E.	Age differences in phago	J.Leukocyte.Bio	86	39:399-407
5746	Najjar,V.A.	Defective phagocytosis d	J Pediatr.	75	87:1121-1124
5742	Najjar,V.A.	The clinical and physiol	Klin.Wochenschr	79	57:751-756
3744	Nathan,C.	Cytokine-induced respira	J.Cell Biol.	89	109:1341-1349
5703	Nauseef,W.M.	Biochemical and immunolo	J.Clin.Invest.	83	71:1297-1307
2007	Nelson,R.D.	Agarose method for human	Methods Enzymol	88	162:50-58
737	Nerl,C.	Early transmembrane pote	Naturwissen.	82	69:292-294
5675	Niedel,J.	Covalent affinity labell	J.Biol.Chem.	80	255:7063-7066
5478	Niedel,J.E.	Receptor mediated intern	Science	79	205:1412-1414
5722	Nielsen,H.	Blood monocyte and neutr	Scand.J.Immunol	86	24:291-296
1409	Ninnemann,J.L	Hemolysis and suppressio	Immunol.Lett.	85	10:63-69
5589	Ohno,Y.	Cytochrome b translocati	J.Biol.Chem.	85	260:2409-2414
5630	OLiver,J.M.	Surface and cytoskeletal	Semin.Hematol.	83	20:282-304
2019	Ozaki,Y.	Functional responses of	Biochim.Biophys	88	972:113-119
4828	Packman,C.H.	Activation of neutrophil	Blood Cells	90	16:193-207
730	Palmblad,J.	Effects of leukotrienes	J.Immunol.	84	132:3041-3045
5765	Parry,M.F.	Myeloperoxidase deficien	Ann.Intern.Med	81	95:293-301
86	Parry,M.F.	Myeloperoxidase deficien	Ann.Intern.Med.	81	95:293-301
5465	Petrequin,P.R	Association between gela	Blood	87	69:605-610
5628	Peveri,P.	A novel neutrophil-activ	J.Exp.Med.	88	167:1547-1559
5633	Platzer,E.	Biological activities of	J.Exp.Med.	85	162:1788-1801
5468	Platzer,E.	Biological activities of	J.Exp.Med.	85	162:178-801
5635	Pober,J.S.	Two distinct monokines,	J.Immunol.	86	136:1680-1687
2050	Prince,H.E.	Early activation marker	Diag.Immunol.	86	4:306-311
3476	Psychoyos,S.	Enhancement of N-formyl-	Agents Actions	89	26:372-377
1616	Rabinovitch,P	Heterogeneity among T ce	J.Immunol.	86	137:952-961
2296	Ransom,J.T.	Single cell analysis of	J.Immunol.	86	136:45-57
5574	Raphael,G.D.	Glandular secretion of l	J.Allergy Clin.	89	84:914-919
2802	Ricevuti,G.	Definition of CD 11a, b,	Acta.Haematol.(	89	81:126-130
1383	Robinson,J.Pa	Measurement of intracell	J.Leuk.Biol.	88	43:304-3100
1384	Robinson,J.Pa	Measurement of anti-neut	Diag.Clin.Immun	87	5:163-170
1246	Robinson,J.Pa	Chemiluminescence respon	J.Clin.Lab.Immu	82	7:219-221
1840	Robinson,J.Pa	Neutrophil membrane flui	Fed.Proc.	86	45: (abstract
542	Robinson,J.Pa	Chemiluminescent respons	Infect.Immun.	84	43:744-752
1254	Robinson,J.Pa	The chemiluminescent res	Diag.Immunol.	85	3:119-125
1806	Rolland,J.M.	Fluorescence polarizatio	J.Immunol.Meth.	85	76:1-10
273	Romano,E.L.	Quantitation of antiplat	Haematologica (	81	66:597-604
5565	Root,R.K.	Abnormal bactericidal, m	J.Clin.Invest.	72	51:649-665
1267	Ross,D.W.	Myeloperoxidase deficien	Arch.Pathol.Lab	85	109:1005-1006

ID	Author	Title	Journal	Year	Citation
2048	Ross,G.D.	Clinical and laboratory	J.Clin.Immunol.	86	6:107-113
4857	Rothe,G.	Flow cytometric paramete	J.Lab.Clin.Med.	90	115:52-61
2809	Rothe,G.	Dihydrorhodamine 123: a	Naturwissenscha	88	75:354-355
5583	Rotrosen,D.	Disorders of phagocyte f	Annu.Rev.Immuno	87	5:127-150
3472	Samanta,A.K.	Identification and chara	J.Exp.Med.	89	169:1185-1189
4692	Samanta,A.K.	Interleukin 8 (monocyte-	J.Biol.Chem.	90	265:183-189
5481	Sanchez-Madri	A human leukocyte differ	J.Exp.Med.	83	158:1785-1803
4661	Särndahl,E.	Association of ligand-re	J.Cell Biol.	89	109:2791-2799
5555	Schopfer,K.	Neutrophil functions in	J.Lab.Clin.Med.	76	88:450-461
5624	Schroeder,J.M	Purification and partial	J.Immunol.	87	139:3474-3483
5752	Segal,A.W.	Absence of cytochrome b-	N.Engl.J Med	83	308:245-251
5753	Segal,A.W.	Novel cytochrome b syste	Nature	78	276:515-517
5751	Segal,A.W.	The subcellular distribu	Biochem.J	79	182:181-188
5479	Segal,E.K.	Development of the phago	Br.J.Haematol.	87	67:3-10
5588	Seligmann,B.	An antibody to a subpopu	Trans.Assoc.Am.	84	97:319-324
84	Seligmann,B.E	Human neutrophil heterog	J.Clin.Invest.	81	68:1125-1131
31	Seligmann,B.E	Differential binding of	J.Immunol.	84	133:2641-2646
986	Seligmann,B.E	Interaction of chemotact	J.Membr.Biol.	80	52:257-272
752	Seligmann,B.E	Use of lipophilic probes	J.Clin.Invest.	80	66:493-503
771	Seligmann,B.E	Neutrophil activation st	Adv.Exp.Med.Bio	82	141:335-349
736	Seligmann,B.E	Comparison of indirect p	J.Cell Physiol.	83	115:105-115
5646	Selsted,M.E.	Primary Structures of th	J.Clin.Invest.	85	76:1436-1439
5647	Selsted,M.E.	Purification and antibac	Infect.Immun.	84	45:150-154
5721	Selvaraj,P.	The major Fc receptor in	Nature (London)	88	333:565-567
759	Shapiro,H.M.	Flow cytometric probes o	Cytometry	81	1:301-312
5336	Sharma,J.N.	The role of chemical med	Exp.Pathol.	90	38:73-96
5645	Shephard,E.G.	Neutrophil lysosomal deg	Clin.Exp.Immuno	88	73:139-
4555	Shephard,E.G.	Generation of biological	J.Immunol.	89	143:2974-2981
5664	Shin,H.S.	Chemotactic and anaphaly	Science	68	162:361-363
75	Sklar,L.A.	Analysis of ligand-recep	Cytometry	82	3:161-165
1572	Smith,Q.T.	Gingival crevicular flui	J.Periodont.Res	86	21:45-55
5681	Smith,R.J.	Activation of the human	Inflammation	84	8:365-384
5475	Snyderman,R.	Molecular and cellular m	Science	81	213:830-837
178	Snyderman,R.	Transductional mechanism	Contemp.Top.Imm	84	14:1-28
5484	Springer,T.A.	The importance of the Ma	Ciba Foundation	86	Pitman
5694	Springer,T.A.	Sequence homology of the	Nature	85	314:540-542
5556	Steerman,R.L.	Intrinsic defect of the	Clin.Exp.Immuno	71	9:939-946
1843	Stelzer,G.T.	Flow cytometric evaluati	Diag.Clin.Immun	88	5:223-231
2337	Stendahl,O.	Myeloperoxidase Modulate	J.Clin.Invest.	84	73:366-373
5651	Stendahl,O.	The inhibition of polymo	J.Clin.Invest.	78	62:214-
5620	Streb,H.	Release of Ca^{2+} from a n	Nature	83	306:67-69
3679	Suchard,S.J.	Characterization and cyt	J.Clin.Invest.	89	84:484-492
5477	Sullivan,S.J.	Chemotactic peptide rece	J.Cell Biol.	80	85:703-711
5748	Suzuki,J.B.	Immunologic profile of j	J Periodontol.	84	55:461-467
29	Szejda,P.	Flow cytometric quantita	J.Immunol.	84	133:3303-3307
4858	Taniguchi,K.	Action sites of antialle	Jpn.J.Pharmacol	90	52:101-108
981	Tatham,P.E.	Chemotactic factor-induc	Biochim.Biophys	80	602:285-298
5566	Tauber,A.I.	Chronic granulomatous di	Medicine.(Balti	83	62:286-309
2003	Tennenberg,S.	Characterization of N-fo	J.Immunol.	88	141:3937-3944
513	Terstappen,L.	Flow cytometric determin	Cytometry	85	6:316-320
5629	Thelen,M.	Mechanism of neutrophil	Fed.Am.Soc.Exp.	88	2:2702-2706
1728	Treumer,J.	Flow-cytometric determin	Exp.Cell Res.	86	163:518-524
2844	Trinkle,L.S.	A simultaneous flow cyto	Diagn.Clin.Immu	87	5:62-68
1617	Tsien,R.Y.	New tetracarboxylate che	Soc.Gen.Physiol	86	40:327-345
1458	Tsien,R.Y.	Calcium homeostasis in i	J.Cell Biol.	82	94:325-334
45	Tulp,A.	A separation chamber to	J.Immunol.Meth.	84	69:281-295
5567	Turner,J.A.	Clinical expressions of	Pediatrics	81	67:805-810
5057	Vandenberghe,	Flow cytometric measurem	J.Immunol.Meth.	90	127:197-205
3665	Vandenbroucke	Increased expression of	Scand.J.Immunol	89	30:91-98
5018	Waggoner,A.S.	Fluorescent Probes for C	Cytometry and S	90	Wiley-Liss
33	Wallace,P.J.	Chemotactic peptide-indu	J.Cell Biol.	84	99:1060-1065
5670	Wallis,W.J.	Monoclonal antibody-defi	Blood	86	67:1007-1013
5658	Ward,P.A.	The chemosuppression of	J.Exp.Med.	66	124:209-226
1417	Warden,G.D.	Suppression of leukocyte	Ann.Surg.	75	181:363-369

1836	Watson,J.V.	Enzyme kinetic studies i	Cytometry	80	1:143-151
5652	Weissmann,G.	From Auden to arichidona	Cell.Immunol.	83	82:117-
5396	Welte,K.	[Granulocyte colony-stim	Klin.Padiatr.	88	200:157-164
5469	Welte,K.	Purification and biochem	Proc.Natl.Acad.	85	82:1526-1530
5700	West,B.C.	Separation and character	Am.J.Pathol.	74	77:41-66
5626	Westwick,J.	Novel neutrophil-stimula	Immunology Toda	89	10:146-147
5716	Wheeler,J.G.	Buffy coat transfusions	Pediatrics	87	79:422-425
5586	White,C.J.	Phagocyte defects	Clin.Immunol.Im	86	40:50-61
758	Whitin,J.C.	Effects of the myelopero	J.Biol.Chem.	81	256:8904-8906
5676	Williams,L.T.	Specific receptor sites	Proc.Natl.Acad.	77	74:1204-1208
1566	Wilson,M.E.	Generalized juvenile per	J.Periodontol.	85	56:457-463
1208	Witkowski,J.	Decreased membrane poten	Immunol.	85	56:307-13
5682	Wright,D.G.	Modulation of the inflam	Inflammation	75	13:23-39
5683	Wright,D.G.	A functional differentia	J.Immunol.	77	119:1068-1076
5570	Wright,D.G.	Pretreatment of filtrati	Blood	78	52:783-792
5642	WRIGHT,S.D.	CD18-deficient cells res	J.Immunol.	90	144:2566-2571
5623	Yoshimura,T.	Purification of a human	Proc.Natl.Acad.	87	84:9233-9237
786	Ziegler,J.	Leucocyte function in pa	Aust.N.Z.J.Med.	75	5:39-43
5584	Zimmerli,W.	Monocytes accumulate on	J.Immunol.Meth.	87	96:11-17